Genetics

A Conceptual Approach

sixth edition

Benjamin A. Pierce

Southwestern University

w.h.
freeman

Macmillan Learning

New York

Vice President, STEM: Ben Roberts
Executive Editor: Lauren Schultz
Development Editor: Maria Lokshin
Executive Marketing Manager: Will Moore
Marketing Assistant: Cate McCaffery
Director of Content: Clairissa Simmons
Content Development Manager, Biology: Amber Jonker
Lead Content Developer, Genetics: Cassandra Korsvik
Senior Media and Supplements Editor: Amy Thorne
Assistant Editor: Shannon Moloney
Director, Content Management Enhancement: Tracey Kuehn
Managing Editor: Lisa Kinne
Project Management: J. Carey Publishing Service
Manuscript Editor: Norma Sims Roche
Director of Design, Content Management: Diana Blume
Interior and Cover Design: Blake Logan
Illustrations: Dragonfly Media Group
Illustration Coordinator: Janice Donnola
Photo Editor: Christine Buese
Photo Researcher: Richard Fox
Senior Production Supervisor: Paul Rohloff
Composition: codeMantra
Printing and Binding: LSC Communications
Cover and Title Page Illustration: Echo Medical Media/PDB data entry 5F9R

Library of Congress Control Number: 2016955732

ISBN-13: 978-1-319-05096-2
ISBN-10: 1-319-05096-4

Printed in the United States of America

First printing

W. H. Freeman and Company
One New York Plaza
Suite 4500
New York, NY 10004-1562

www.macmillanlearning.com

To my parents, Rush and Amanda Pierce;
my children, Sarah Pierce Dumas and Michael Pierce;
and my genetic partner, friend, and soul mate
for 36 years, Marlene Tyrrell

Contents in Brief

1   Introduction to Genetics   1
2   Chromosomes and Cellular Reproduction   17
3   Basic Principles of Heredity   47
4   Sex Determination and Sex-Linked Characteristics   81
5   Extensions and Modifications of Basic Principles   109
6   Pedigree Analysis, Applications, and Genetic Testing   145
7   Linkage, Recombination, and Eukaryotic Gene Mapping   173
8   Chromosome Variation   217
9   Bacterial and Viral Genetic Systems   251
10   DNA: The Chemical Nature of the Gene   287
11   Chromosome Structure and Organelle DNA   311
12   DNA Replication and Recombination   339
13   Transcription   373
14   RNA Molecules and RNA Processing   399
15   The Genetic Code and Translation   429
16   Control of Gene Expression in Bacteria   461
17   Control of Gene Expression in Eukaryotes   491
18   Gene Mutations and DNA Repair   515
19   Molecular Genetic Analysis and Biotechnology   559
20   Genomics and Proteomics   605
21   Epigenetics   641
22   Developmental Genetics and Immunogenetics   663
23   Cancer Genetics   691
24   Quantitative Genetics   715
25   Population Genetics   749
26   Evolutionary Genetics   779
     Reference Guide to Model Genetic Organisms   A1
     Working with Fractions: A Review   B1
     Glossary   C1
     Answers to Selected Problems   D1
     Index   E1

Contents

Letter from the Author   xvii
Preface   xviii

Chapter 1 Introduction to Genetics   1

Albinism in the Hopis   1

1.1 Genetics Is Important to Us Individually, to Society, and to the Study of Biology   2

The Role of Genetics in Biology   4
Genetic Diversity and Evolution   4
DNA in the Biosphere   5
Divisions of Genetics   5
Model Genetic Organisms   6

1.2 Humans Have Been Using Genetic Techniques for Thousands of Years   7

The Early Use and Understanding of Heredity   7
The Rise of the Science of Genetics   9
The Cutting Edge of Genetics   11

1.3 A Few Fundamental Concepts Are Important for the Start of Our Journey into Genetics   12

Chapter 2 Chromosomes and Cellular Reproduction   17

The Blind Men's Riddle   17

2.1 Prokaryotic and Eukaryotic Cells Differ in a Number of Genetic Characteristics   18

2.2 Cell Reproduction Requires the Copying of the Genetic Material, Separation of the Copies, and Cell Division   20

Prokaryotic Cell Reproduction by Binary Fission   20
Eukaryotic Cell Reproduction   20
The Cell Cycle and Mitosis   23
Genetic Consequences of the Cell Cycle   26

CONNECTING CONCEPTS Counting Chromosomes and DNA Molecules   27

2.3 Sexual Reproduction Produces Genetic Variation Through the Process of Meiosis   28

Meiosis   28
Sources of Genetic Variation in Meiosis   30

CONNECTING CONCEPTS Mitosis and Meiosis Compared   34

The Separation of Sister Chromatids and Homologous Chromosomes   34
Meiosis in the Life Cycles of Animals and Plants   35

Chapter 3 Basic Principles of Heredity   47

The Genetics of Blond Hair in the South Pacific   47

3.1 Gregor Mendel Discovered the Basic Principles of Heredity   49

Mendel's Success   49
Genetic Terminology   50

3.2 Monohybrid Crosses Reveal the Principle of Segregation and the Concept of Dominance   51

What Monohybrid Crosses Reveal   52

CONNECTING CONCEPTS Relating Genetic Crosses to Meiosis   54

The Molecular Nature of Alleles   54
Predicting the Outcomes of Genetic Crosses   56
The Testcross   60
Genetic Symbols   60

CONNECTING CONCEPTS Ratios in Simple Crosses   61

3.3 Dihybrid Crosses Reveal the Principle of Independent Assortment   61

Dihybrid Crosses   61
The Principle of Independent Assortment   62
Relating the Principle of Independent Assortment to Meiosis   63
Applying Probability and the Branch Diagram to Dihybrid Crosses   64
The Dihybrid Testcross   65

3.4 Observed Ratios of Progeny May Deviate from Expected Ratios by Chance 67

The Chi-Square Goodness-of-Fit Test 67

Chapter 4 Sex Determination and Sex-Linked Characteristics 81

The Sex of a Dragon 81

4.1 Sex Is Determined by a Number of Different Mechanisms 83

Chromosomal Sex-Determining Systems 83
Genic Sex Determination 85
Environmental Sex Determination 85
Sex Determination in *Drosophila melanogaster* 87
Sex Determination in Humans 87

4.2 Sex-Linked Characteristics Are Determined by Genes on the Sex Chromosomes 89

X-Linked White Eyes in *Drosophila* 89
Nondisjunction and the Chromosome Theory of Inheritance 90
X-Linked Color Blindness in Humans 92
Symbols for X-Linked Genes 94
Z-Linked Characteristics 94
Y-Linked Characteristics 94

CONNECTING CONCEPTS Recognizing Sex-Linked Inheritance 97

4.3 Dosage Compensation Equalizes the Amount of Protein Produced by X-Linked and Autosomal Genes in Some Animals 97

The Lyon Hypothesis 98
Mechanism of Random X Inactivation 99

Chapter 5 Extensions and Modifications of Basic Principles 109

The Odd Genetics of Left-Handed Snails 109

5.1 Additional Factors at a Single Locus Can Affect the Results of Genetic Crosses 110

Types of Dominance 110
Penetrance and Expressivity 113
Lethal Alleles 114
Multiple Alleles 114

5.2 Gene Interaction Takes Place When Genes at Multiple Loci Determine a Single Phenotype 116

Gene Interaction That Produces Novel Phenotypes 117
Gene Interaction with Epistasis 118

CONNECTING CONCEPTS Interpreting Phenotypic Ratios Produced by Gene Interaction 122

Complementation: Determining Whether Mutations Are at the Same Locus or at Different Loci 124
The Complex Genetics of Coat Color in Dogs 124

5.3 Sex Influences the Inheritance and Expression of Genes in a Variety of Ways 126

Sex-Influenced and Sex-Limited Characteristics 126
Cytoplasmic Inheritance 128
Genetic Maternal Effect 130
Genomic Imprinting 131

5.4 Anticipation Is the Stronger or Earlier Expression of Traits in Succeeding Generations 133

5.5 The Expression of a Genotype May Be Influenced by Environmental Effects 133

Environmental Effects on the Phenotype 134
The Inheritance of Continuous Characteristics 134

Chapter 6 Pedigree Analysis, Applications, and Genetic Testing 145

The Mystery of the Missing Fingerprints 145

6.1 The Study of Genetics in Humans Is Constrained by Special Features of Human Biology and Culture 146

6.2 Geneticists Often Use Pedigrees To Study the Inheritance of Characteristics in Humans 147

Symbols Used in Pedigrees 147
Analysis of Pedigrees 148
Autosomal Recessive Traits 148
Autosomal Dominant Traits 149

X-Linked Recessive Traits 150
X-Linked Dominant Traits 151
Y-Linked Traits 152
Genetic Mosaicism 154

6.3 Studying Twins and Adoptions Can Help Us Assess the Importance of Genes and Environment 154

Types of Twins 154
Concordance in Twins 155
A Twin Study of Asthma 156
Adoption Studies 156

6.4 Genetic Counseling and Genetic Testing Provide Information to Those Concerned about Genetic Diseases and Traits 157

Genetic Counseling 157
Genetic Testing 158
Interpreting Genetic Tests 163
Direct-to-Consumer Genetic Testing 163
Genetic Discrimination and Privacy 164

Chapter 7 Linkage, Recombination, and Eukaryotic Gene Mapping 173

Linked Genes and Bald Heads 173

7.1 Linked Genes Do Not Assort Independently 174

7.2 Linked Genes Segregate Together While Crossing Over Produces Recombination Between Them 176

Notation for Crosses with Linkage 176
Complete Linkage Compared with Independent Assortment 177
Crossing Over Between Linked Genes 178
Calculating Recombination Frequency 180
Coupling and Repulsion 181

CONNECTING CONCEPTS Relating Independent Assortment, Linkage, and Crossing Over 182

Evidence for the Physical Basis of Recombination 183
Predicting the Outcomes of Crosses with Linked Genes 184
Testing for Independent Assortment 185
Gene Mapping with Recombination Frequencies 187
Constructing a Genetic Map with a Two-Point Testcross 188

7.3 A Three-Point Testcross Can Be Used to Map Three Linked Genes 189

Constructing a Genetic Map with a Three-Point Testcross 190

CONNECTING CONCEPTS Stepping Through the Three-Point Cross 195

Effects of Multiple Crossovers 197
Mapping Human Genes 198
Mapping with Molecular Markers 199
Locating Genes with Genome-Wide Association Studies 199

7.4 Physical-Mapping Methods Are Used to Determine the Physical Positions of Genes on Particular Chromosomes 200

Somatic-Cell Hybridization 201
Deletion Mapping 201
Physical Chromosome Mapping Through Molecular Analysis 203

7.5 Recombination Rates Exhibit Extensive Variation 203

Chapter 8 Chromosome Variation 217

Building a Better Banana 217

8.1 Chromosome Mutations Include Rearrangements, Aneuploidy, and Polyploidy 218

Chromosome Morphology 218
Types of Chromosome Mutations 219

8.2 Chromosome Rearrangements Alter Chromosome Structure 220

Duplications 220
Deletions 222
Inversions 224
Translocations 227
Fragile Sites 229
Copy-Number Variations 229

8.3 Aneuploidy Is an Increase or Decrease in the Number of Individual Chromosomes 230

Types of Aneuploidy 230
Effects of Aneuploidy 230
Aneuploidy in Humans 232
Uniparental Disomy 235
Genetic Mosaicism 236

8.4 Polyploidy Is the Presence of More Than Two Sets of Chromosomes 236

Autopolyploidy 237
Allopolyploidy 238
The Significance of Polyploidy 240

Chapter 9   Bacterial and Viral Genetic Systems   251

The Genetics of Medieval Leprosy 251

9.1 Bacteria and Viruses Have Important Roles in Human Society and the World Ecosystem 252

Life in a Bacterial World 252
Bacterial Diversity 253

9.2 Genetic Analysis of Bacteria Requires Special Methods 253

Techniques for the Study of Bacteria 254
The Bacterial Genome 255
Plasmids 255

9.3 Bacteria Exchange Genes Through Conjugation, Transformation, and Transduction 257

Conjugation 258
Natural Gene Transfer and Antibiotic Resistance 264
Transformation in Bacteria 265
Bacterial Genome Sequences 266
Horizontal Gene Transfer 266
Bacterial Defense Mechanisms 267

9.4 Viruses Are Simple Replicating Systems Amenable to Genetic Analysis 267

Techniques for the Study of Bacteriophages 268
Transduction: Using Phages To Map Bacterial Genes 269
CONNECTING CONCEPTS Three Methods for Mapping Bacterial Genes 271
Gene Mapping in Phages 272
Plant and Animal Viruses 273
Human Immunodeficiency Virus and AIDS 274
Influenza 276
Rhinoviruses 276

Chapter 10   DNA: The Chemical Nature of the Gene   287

Arctic Treks and Ancient DNA 287

10.1 Genetic Material Possesses Several Key Characteristics 288

10.2 All Genetic Information Is Encoded in the Structure of DNA or RNA 288

Early Studies of DNA 289
DNA As the Source of Genetic Information 290
Watson and Crick's Discovery of the Three-Dimensional Structure of DNA 293
RNA As Genetic Material 295

10.3 DNA Consists of Two Complementary and Antiparallel Nucleotide Strands That Form a Double Helix 296

The Primary Structure of DNA 296
Secondary Structures of DNA 299
CONNECTING CONCEPTS Genetic Implications of DNA Structure 301

10.4 Special Structures Can Form in DNA and RNA 302

Chapter 11   Chromosome Structure and Organelle DNA   311

Telomeres and Childhood Adversity 311

11.1 Large Amounts of DNA Are Packed into a Cell 312

Supercoiling 313
The Bacterial Chromosome 314
Eukaryotic Chromosomes 314
Changes in Chromatin Structure 317

11.2 Eukaryotic Chromosomes Possess Centromeres and Telomeres 319

Centromere Structure 319
Telomere Structure 320

11.3 Eukaryotic DNA Contains Several Classes of Sequence Variation 321

The Denaturation and Renaturation of DNA 321
Types of DNA Sequences in Eukaryotes 321
Organization of Genetic Information in Eukaryotes 322

11.4 Organelle DNA Has Unique Characteristics 322

Mitochondrion and Chloroplast Structure 322
The Endosymbiotic Theory 323
Uniparental Inheritance of Organelle-Encoded Traits 324
The Mitochondrial Genome 326
The Evolution of Mitochondrial DNA 328
Damage to Mitochondrial DNA Associated with Aging 329
Mitochondrial Replacement Therapy 329
The Chloroplast Genome 330
Movement of Genetic Information Between Nuclear, Mitochondrial, and Chloroplast Genomes 331

Chapter 12   DNA Replication and Recombination   339

Topoisomerase, Replication, and Cancer 339

12.1 Genetic Information Must Be Accurately Copied Every Time a Cell Divides 340

12.2 All DNA Replication Takes Place in a Semiconservative Manner 340

Meselson and Stahl's Experiment 341
Modes of Replication 343
Requirements of Replication 346
Direction of Replication 346
CONNECTING CONCEPTS The Direction of Synthesis in Different Modes of Replication 348

12.3 Bacterial Replication Requires a Large Number of Enzymes and Proteins 348

Initiation 348
Unwinding 348
Elongation 350
Termination 353
The Fidelity of DNA Replication 353
CONNECTING CONCEPTS The Basic Rules of Replication 354

12.4 Eukaryotic DNA Replication Is Similar to Bacterial Replication but Differs in Several Aspects 354

Eukaryotic Origins of Replication 354
DNA Synthesis and the Cell Cycle 354

The Licensing of DNA Replication 355
Unwinding 355
Eukaryotic DNA Polymerases 355
Nucleosome Assembly 356
The Location of Replication Within the Nucleus 357
Replication at the Ends of Chromosomes 357
Replication in Archaea 360

12.5 Recombination Takes Place Through the Alignment, Breakage, and Repair of DNA Strands 360

Models of Recombination 361
Enzymes Required for Recombination 362
Gene Conversion 363

Chapter 13   Transcription   373

Death Cap Poisoning 373

13.1 RNA, Consisting of a Single Strand of Ribonucleotides, Participates in a Variety of Cellular Functions 374

An Early RNA World 374
The Structure of RNA 374
Classes of RNA 375

13.2 Transcription Is the Synthesis of an RNA Molecule from a DNA Template 376

The Template 377
The Substrate for Transcription 379
The Transcription Apparatus 379

13.3 Bacterial Transcription Consists of Initiation, Elongation, and Termination 381

Initiation 381
Elongation 383
Termination 384
CONNECTING CONCEPTS The Basic Rules of Transcription 386

13.4 Eukaryotic Transcription Is Similar to Bacterial Transcription but Has Some Important Differences 386

Transcription and Nucleosome Structure 386
Promoters 386
Initiation 387

Elongation 389
Termination 389

13.5 Transcription in Archaea Is More Similar to Transcription in Eukaryotes Than to Transcription in Bacteria 390

Chapter 14   RNA Molecules and RNA Processing   399

A Royal Disease 399

14.1 Many Genes Have Complex Structures 400

Gene Organization 400
Introns 402
The Concept of the Gene Revisited 403

14.2 Messenger RNAs, which Encode Proteins, Are Modified after Transcription in Eukaryotes 403

The Structure of Messenger RNA 404
Pre-mRNA Processing 405
RNA Splicing 407
Alternative Processing Pathways 409
RNA Editing 412

CONNECTING CONCEPTS Eukaryotic Gene Structure and Pre-mRNA Processing 413

14.3 Transfer RNAs, which Attach to Amino Acids, Are Modified after Transcription in Bacterial and Eukaryotic Cells 414

The Structure of Transfer RNA 414
Transfer RNA Gene Structure and Processing 415

14.4 Ribosomal RNA, a Component of the Ribosome, Is Also Processed after Transcription 416

The Structure of the Ribosome 416
Ribosomal RNA Gene Structure and Processing 417

14.5 Small RNA Molecules Participate in a Variety of Functions 418

RNA Interference 418
Small Interfering RNAs and MicroRNAs 419
Piwi-Interacting RNAs 420
CRISPR RNA 420

14.6 Long Noncoding RNAs Regulate Gene Expression 421

Chapter 15   The Genetic Code and Translation   429

A Child Without a Spleen 429

15.1 Many Genes Encode Proteins 430

The One Gene, One Enzyme Hypothesis 430
The Structure and Function of Proteins 433

15.2 The Genetic Code Determines How the Nucleotide Sequence Specifies the Amino Acid Sequence of a Protein 435

Breaking the Genetic Code 436
The Degeneracy of the Code 438
The Reading Frame and Initiation Codons 439
Termination Codons 440
The Universality of the Code 440

CONNECTING CONCEPTS Characteristics of the Genetic Code 440

15.3 Amino Acids Are Assembled into a Protein Through Translation 441

The Binding of Amino Acids to Transfer RNAs 441
The Initiation of Translation 442
Elongation 445
Termination 446

CONNECTING CONCEPTS A Comparison of Bacterial and Eukaryotic Translation 448

15.4 Additional Properties of RNA and Ribosomes Affect Protein Synthesis 449

The Three-Dimensional Structure of the Ribosome 449
Polyribosomes 449
Messenger RNA Surveillance 449
Folding and Posttranslational Modifications of Proteins 452
Translation and Antibiotics 452

Chapter 16   Control of Gene Expression in Bacteria   461

Operons and the Noisy Cell 461

16.1 The Regulation of Gene Expression Is Critical for All Organisms 462

Genes and Regulatory Elements 463
Levels of Gene Regulation 463
DNA-Binding Proteins 464

16.2 Operons Control Transcription in Bacterial Cells 465

Operon Structure 465
Negative and Positive Control: Inducible and Repressible Operons 466
The *lac* Operon of *E. coli* 468
lac Mutations 471
Positive Control and Catabolite Repression 475
The *trp* Operon of *E. coli* 476

16.3 Some Operons Regulate Transcription Through Attenuation, the Premature Termination of Transcription 477

Attenuation in the *trp* Operon of *E. coli* 477
Why Does Attenuation Take Place in the *trp* Operon? 481

16.4 Other Sequences Control the Expression of Some Bacterial Genes 481

Bacterial Enhancers 482
Antisense RNA 482
Riboswitches 482
RNA-Mediated Repression Through Ribozymes 483

Chapter 17   Control of Gene Expression in Eukaryotes   491

Genetic Differences That Make Us Human 491

17.1 Eukaryotic Cells and Bacteria Share Many Features of Gene Regulation but Differ in Several Important Ways 492

17.2 Changes in Chromatin Structure Affect the Expression of Eukaryotic Genes 493

DNase I Hypersensitivity 493
Chromatin Remodeling 493
Histone Modification 494
DNA Methylation 497

17.3 The Initiation of Transcription Is Regulated by Transcription Factors and Transcriptional Regulator Proteins 497

Transcriptional Activators and Coactivators 497
Transcriptional Repressors 499

Enhancers and Insulators 500
Regulation of Transcriptional Stalling and Elongation 500
Coordinated Gene Regulation 501

17.4 Some Eukaryotic Genes Are Regulated by RNA Processing and Degradation 502

Gene Regulation Through RNA Splicing 502
The Degradation of RNA 503

17.5 RNA Interference Is an Important Mechanism of Gene Regulation 504

Small Interfering RNAs and MicroRNAs 505
Mechanisms of Gene Regulation by RNA Interference 505
The Control of Development by RNA Interference 506
RNA Crosstalk 506

17.6 The Expression of Some Genes Is Regulated by Processes That Affect Translation or by Modifications of Proteins 507

CONNECTING CONCEPTS A Comparison of Bacterial and Eukaryotic Gene Control 508

Chapter 18   Gene Mutations and DNA Repair   515

Lou Gehrig and Expanding Nucleotide Repeats 515

18.1 Mutations Are Inherited Alterations in the DNA Sequence 516

The Importance of Mutations 516
Categories of Mutations 517
Types of Gene Mutations 518
Functional Effects of Mutations 520
Suppressor Mutations 521
Mutation Rates 525

18.2 Mutations May Be Caused by a Number of Different Factors 526

Spontaneous Replication Errors 527
Spontaneous Chemical Changes 528
Chemically Induced Mutations 528
Radiation 532

18.3 Mutations Are the Focus of Intense Study by Geneticists 532

Detecting Mutagens with the Ames Test 532
Effects of Radiation Exposure in Humans 533

Contents

18.4 Transposable Elements Can Cause Mutations 534

General Characteristics of Transposable Elements 534
The Process of Transposition 535
The Mutagenic Effects of Transposition 536
Transposable Elements in Bacteria 538
Transposable Elements in Eukaryotes 539

CONNECTING CONCEPTS Types of Transposable Elements 543

Transposable Elements in Genome Evolution 543

18.5 A Number of Pathways Can Repair DNA 544

Mismatch Repair 544
Direct Repair 545
Base-Excision Repair 545
Nucleotide-Excision Repair 546

CONNECTING CONCEPTS The Basic Pathway of DNA Repair 547

Repair of Double-Strand Breaks 547
Translesion DNA Polymerases 548
Genetic Diseases and Faulty DNA Repair 548

Chapter 19 Molecular Genetic Analysis and Biotechnology 559

Editing the Genome with CRISPR-Cas9 559

19.1 Genetics Has Been Transformed by the Development of Molecular Techniques 560

Key Innovations in Molecular Genetics 561
Working at the Molecular Level 561

19.2 Molecular Techniques Are Used to Cut and Visualize DNA Sequences 562

Recombinant DNA Technology 562
Restriction Enzymes 562
Engineered Nucleases 564
CRISPR-Cas Genome Editing 564
Separating and Viewing DNA Fragments 567
Locating DNA Fragments with Probes 568

19.3 Specific DNA Fragments Can Be Amplified 568

The Polymerase Chain Reaction 569
Gene Cloning 571

19.4 Molecular Techniques Can Be Used to Find Genes of Interest 576

DNA Libraries 576
In Situ Hybridization 578
Positional Cloning 579

19.5 DNA Sequences Can Be Determined and Analyzed 582

Dideoxy Sequencing 582
Next-Generation Sequencing Technologies 584
DNA Fingerprinting 586

19.6 Molecular Techniques Are Increasingly Used to Analyze Gene Function 589

Forward and Reverse Genetics 589
Creating Random Mutations 589
Targeted Mutagenesis 589
Transgenic Animals 590
Knockout Mice 591
Silencing Genes with RNAi 593
Using RNAi to Treat Human Disease 594

19.7 Biotechnology Harnesses the Power of Molecular Genetics 595

Pharmaceutical Products 595
Specialized Bacteria 595
Agricultural Products 595
Genetic Testing 596
Gene Therapy 596

Chapter 20 Genomics and Proteomics 605

Building a Chromosome for Class 605

20.1 Structural Genomics Determines the DNA Sequences and Organization of Entire Genomes 606

Genetic Maps 607
Physical Maps 608
Sequencing an Entire Genome 608
The Human Genome Project 610
What Exactly Is the Human Genome? 613
Single-Nucleotide Polymorphisms 613
Copy-Number Variations 615

Bioinformatics 615
Metagenomics 616
Synthetic Biology 617

20.2 Functional Genomics Determines the Functions of Genes by Using Genomic Approaches 617

Predicting Function from Sequence 618
Gene Expression 619
Gene Expression and Reporter
   Sequences 622
Genome-Wide Mutagenesis 623

20.3 Comparative Genomics Studies How Genomes Evolve 624

Prokaryotic Genomes 624
Eukaryotic Genomes 625
The Human Genome 628

20.4 Proteomics Analyzes the Complete Set of Proteins Found in a Cell 629

Determination of Cellular Proteins 629
Affinity Capture 631
Protein Microarrays 631
Structural Proteomics 632

Chapter 21   Epigenetics   641

Epigenetics and the Dutch Hunger Winter 641

21.1 What Is Epigenetics? 642

21.2 Several Molecular Processes Lead to Epigenetic Changes 643

DNA Methylation 643
Histone Modifications 645
Epigenetic Effects of RNA Molecules 647

21.3 Epigenetic Processes Produce a Diverse Set of Effects 647

Paramutation 647
Behavioral Epigenetics 650
Epigenetic Effects of Environmental
   Chemicals 651
Epigenetic Effects on Metabolism 652
Epigenetic Effects in Monozygotic
   Twins 652
X Inactivation 652
Epigenetic Changes Associated
   with Cell Differentiation 654
Genomic Imprinting 655

21.4 The Epigenome 656

Chapter 22   Developmental Genetics and Immunogenetics   663

The Origin of Spineless Sticklebacks 663

22.1 Development Takes Place Through Cell Determination 664

Cloning Experiments on Plants 665
Cloning Experiments on Animals 665

22.2 Pattern Formation in *Drosophila* Serves as a Model for the Genetic Control of Development 666

The Development of the Fruit Fly 666
Egg-Polarity Genes 667
Segmentation Genes 670
Homeotic Genes in *Drosophila* 671
Homeobox Genes in Other Organisms 672

CONNECTING CONCEPTS The Control of
Development 673

Epigenetic Changes in Development 674

22.3 Genes Control the Development of Flowers in Plants 674

Flower Anatomy 674
Genetic Control of Flower Development 674

CONNECTING CONCEPTS Comparison of Development
in *Drosophila* and Flowers 676

22.4 Programmed Cell Death Is an Integral Part of Development 676

22.5 The Study of Development Reveals Patterns and Processes of Evolution 678

22.6 The Development of Immunity Occurs Through Genetic Rearrangement 679

The Organization of the Immune System 680
Immunoglobulin Structure 682
The Generation of Antibody Diversity 682
T-Cell-Receptor Diversity 684
Major Histocompatibility Complex Genes 684
Genes and Organ Transplants 685

Chapter 23   Cancer Genetics   691

Palladin and the Spread of Cancer 691

23.1 Cancer Is a Group of Diseases Characterized by Cell Proliferation 692

Tumor Formation 693
Cancer As a Genetic Disease 693

The Role of Environmental Factors in Cancer 695

23.2 Mutations in Several Types of Genes Contribute to Cancer 696

Oncogenes and Tumor-Suppressor Genes 696
Genes That Control the Cell Cycle 698
DNA-Repair Genes 702
Genes That Regulate Telomerase 702
Genes That Promote Vascularization and the Spread of Tumors 703
MicroRNAs and Cancer 703
Cancer Genome Projects 704

23.3 Epigenetic Changes Are Often Associated with Cancer 705

23.4 Colorectal Cancer Arises Through the Sequential Mutation of a Number of Genes 705

23.5 Changes in Chromosome Number and Structure Are Often Associated with Cancer 706

23.6 Viruses Are Associated with Some Cancers 708

Retroviruses and Cancer 708
Human Papillomavirus and Cervical Cancer 709

Chapter 24 Quantitative Genetics 715

Corn Oil and Quantitative Genetics 715

24.1 Quantitative Characteristics Are Influenced by Alleles at Multiple Loci 716

The Relation Between Genotype and Phenotype 716
Types of Quantitative Characteristics 718
Polygenic Inheritance 719
Kernel Color in Wheat 719
Determining Gene Number for a Polygenic Characteristic 721

24.2 Statistical Methods Are Required for Analyzing Quantitative Characteristics 721

Distributions 721
Samples and Populations 722

The Mean 723
The Variance and Standard Deviation 723
Correlation 724
Regression 725
Applying Statistics to the Study of a Polygenic Characteristic 727

24.3 Heritability Is Used to Estimate the Proportion of Variation in a Trait That Is Genetic 728

Phenotypic Variance 728
Types of Heritability 730
Calculating Heritability 730
The Limitations of Heritability 732
Locating Genes That Affect Quantitative Characteristics 734

24.4 Genetically Variable Traits Change in Response to Selection 736

Predicting the Response to Selection 737
Limits to the Response to Selection 738
Correlated Responses to Selection 739

Chapter 25 Population Genetics 749

The Wolves of Isle Royale 749

25.1 Genotypic and Allelic Frequencies Are Used To Describe the Gene Pool of a Population 751

Mathematical Models for Understanding Genetic Variation 751
Calculating Genotypic Frequencies 751
Calculating Allelic Frequencies 752

25.2 The Hardy–Weinberg Law Describes the Effect of Reproduction on Genotypic and Allelic Frequencies 753

Genotypic Frequencies at Hardy–Weinberg Equilibrium 754
Closer Examination of the Hardy–Weinberg Law 755
Implications of the Hardy–Weinberg Law 755
Extensions of the Hardy–Weinberg Law 756
Testing for Hardy–Weinberg Proportions 756
Estimating Allelic Frequencies with the Hardy–Weinberg Law 757

25.3 Nonrandom Mating Affects the Genotypic Frequencies of a Population 758

25.4 Several Evolutionary Forces Can Change Allelic Frequencies 760

Mutation 760
Migration 762
Genetic Drift 763
Natural Selection 766

CONNECTING CONCEPTS The General Effects of Forces That Change Allelic Frequencies 771

Chapter 26 Evolutionary Genetics 779

Taster Genes in Spitting Apes 779

26.1 Evolution Occurs Through Genetic Change within Populations 780

Biological Evolution 780
Evolution as a Two-Step Process 781
Evolution in Bighorn Sheep 781

26.2 Many Natural Populations Contain High Levels of Genetic Variation 782

Molecular Variation 782

26.3 New Species Arise Through the Evolution of Reproductive Isolation 784

The Biological Species Concept 784
Reproductive Isolating Mechanisms 784
Modes of Speciation 786
Genetic Differentiation Associated with Speciation 791

26.4 The Evolutionary History of a Group of Organisms Can Be Reconstructed by Studying Changes in Homologous Characteristics 791

The Alignment of Homologous Sequences 793
The Construction of Phylogenetic Trees 793

26.5 Patterns of Evolution Are Revealed by Molecular Changes 794

Rates of Molecular Evolution 794
The Molecular Clock 796
Evolution Through Changes in Gene Regulation 797
Genome Evolution 798

Reference Guide to Model Genetic Organisms A1

The Fruit Fly *Drosophila melanogaster* A2
The Bacterium *Escherichia coli* A4
The Nematode Worm *Caenorhabditis elegans* A6
The Plant *Arabidopsis thaliana* A8
The Mouse *Mus musculus* A10
The Yeast *Saccharomyces cerevisiae* A12

Working with Fractions: A Review B1

Glossary C1

Answers to Selected Problems D1

Index E1

Letter from the Author

[Marlene Tyrrell]

I still remember the excitement I felt when I was in your place, taking my first genetics course. I was intrigued by the principles of heredity, which allow one to predict what offspring will look like even before they are born. I was fascinated to learn that these principles have their foundation in the chemistry of an elegant molecule called DNA. And I was captivated to find that genetics underlies evolution, the process responsible for life's endless diversity and beauty. These elements of genetics still impress and excite me today. One of the great things about teaching genetics is the chance to convey that excitement to students.

This book has been written in many different places: in my office at Southwestern University, on the back porch of my home overlooking the hills of central Texas, in airports and hotel rooms around the country. Regardless of location, whenever I write, I try to imagine that I'm sitting with a small group of students, having a conversation about genetics. My goal as the author of *Genetics: A Conceptual Approach* is to have that conversation with you. I want to become a trusted guide on your journey through introductory genetics. In this book, I've tried to share some of what I've learned in my years of teaching genetics. I provide advice and encouragement at places where students often have difficulty, and I tell stories of the people, places, and experiments of genetics—past and present—to keep the subject relevant, interesting, and alive. My goal is to help you learn the necessary details, concepts, and problem-solving skills while encouraging you to see the elegance and beauty of the larger landscape.

At Southwestern University, my office door is always open, and my students often drop by to share their own approaches to learning, things that they have read about genetics, and their experiences, concerns, and triumphs. I learn as much from my students as they learn from me, and I would love to learn from you—by email (pierceb@southwestern.edu), by telephone (512-863-1974), or in person (Southwestern University, Georgetown, Texas).

Ben Pierce

PROFESSOR OF BIOLOGY AND
HOLDER OF THE LILLIAN NELSON PRATT CHAIR
SOUTHWESTERN UNIVERSITY

Preface

The main goals of *Genetics: A Conceptual Approach* have always been to help students uncover and make connections between the major concepts of genetics. Throughout the five preceding editions of this book, its accessible writing style, simple and instructive illustrations, and useful pedagogical features have helped students develop a fuller understanding of genetics.

Hallmark Features

CONCEPTS

Epistasis is the masking of the expression of one gene by another gene at a different locus. The epistatic gene does the masking; the hypostatic gene is masked. Epistatic alleles can be dominant or recessive.

✔ CONCEPT CHECK 7

A number of all-white cats are crossed, and they produce the following types of progeny: $^{12}/_{16}$ all-white, $^3/_{16}$ black, and $^1/_{16}$ gray. What is the genotype of the black progeny?

a. *Aa*   c. *A_B_*
b. *Aa Bb*   d. *A_b*

■ **Key Concepts and Connections** Throughout the book, I've included features to help students focus on the major concepts of each topic.

  ■ *Concepts boxes* throughout each chapter summarize the key points of the preceding section. *Concept Checks* allow students to quickly assess their understanding of the material they've just read. Concept Checks are in multiple-choice or short-answer format, and their answers are given at the end of each chapter.

  ■ *Connecting Concepts* sections compare and contrast processes or integrate ideas across sections and chapters to help students see how different genetics topics relate to one another. All major concepts in each chapter are listed in the *Concepts Summary* at the end of the chapter.

■ **Accessibility** The conversational writing style of this book has always been a favorite feature for both students and instructors. In addition to carefully walking students through each major concept of genetics, I invite them into the topic with an **introductory story**. These stories include relevant examples of diseases or other biological phenomena to give students a sample of what they'll be learning in a chapter. More than a third of the introductory stories in this edition are new.

■ **Clear, Simple Illustration Program** The attractive and instructive figures have proved to be an effective learning tool for students throughout the past five editions and continue to be a signature feature of the new edition. Each figure has been carefully rendered to highlight main points and to step the reader through experiments and processes. Most figures include text that walks students through the graphical presentation. Illustrations of experiments reinforce the scientific method by first proposing a hypothesis, then pointing out the methods and results, and ending with a conclusion that reinforces concepts explained in the text.

---

42. *Nicotiana glutinosa* ($2n = 24$) and *N. tabacum* ($2n = 48$) are two closely related plants that can be intercrossed, but the F_1 hybrid plants that result are usually sterile. In 1925, Roy Clausen and Thomas Goodspeed crossed *N. glutinosa* and *N. tabacum* and obtained one fertile F_1 plant (R. E. Clausen and T. H. Goodspeed. 1925 *Genetics* 10:278–284). They were able to self-pollinate the flowers of this plant to produce an F_2 generation. Surprisingly, the F_2 plants were fully fertile and produced viable seeds. When Clausen and Goodspeed examined the chromosomes of the F_2 plants, they observed 36 pairs of chromosomes in metaphase I and 36 individual chromosomes in metaphase II. Explain the origin of the F_2 plants obtained by Clausen and Goodspeed and the numbers of chromosomes observed.

■ **Emphasis on Problem Solving** One of the things that I've learned in my 36 years of teaching is that students learn genetics best through problem solving. Working through an example, equation, or experiment helps students see concepts in action and reinforces the ideas explained in the text. In the book, I help students develop problem-solving skills in a number of ways. **Worked Problems** walk students through each step of a difficult concept. **Problem Links** spread throughout each chapter point to end-of-chapter problems that students can work to test their understanding of the material they have just read, all with answers in the back of the book so that students can check their results. I provide a wide range of end-of-chapter problems, organized by chapter section and split into Comprehension Questions, Application Questions and Problems, and Challenge Questions. Some of these questions, marked by a data analysis icon, draw on examples from published, and cited, research articles.

New to the Sixth Edition

NEW SaplingPlus for *Genetics: A Conceptual Approach* The sixth edition is now fully supported in SaplingPlus. This comprehensive and robust online teaching and learning platform incorporates online homework with the e-Book, all instructor and student resources, and powerful gradebook functionality. Students benefit from just-in-time hints and feedback specific to their misconceptions to develop their problem-solving skills, while instructors benefit from automatically graded homework and robust gradebook diagnostics.

NEW Active learning components One of my main goals for this new edition is to provide better resources for active learning. In this edition, I have added Think-Pair-Share questions, which require students to work, and learn, in groups. These questions not only focus on the genetics topics covered in the chapter, but also tie them to genetics in medicine, agriculture, and other aspects of human society. An online instructor guide provides resources for instructors leading the in-class discussion.

- *Chapter Opening Think-Pair-Share Questions* get students to discuss the chapter opening story itself and to connect it with what they know about genetics.
- *End-of-Chapter Think-Pair-Share Questions* provide more challenging problem solving for students to work on in groups and encourage them to discuss the bigger-picture aspects of the material they learned in the chapter. They also allow students to connect the material they have learned to broader genetics topics.

THINK-PAIR-SHARE

- Most cells are unable to copy the ends of chromosomes, and therefore chromosomes shorten with each cell division. This limits the number of times a cell can divide. In germ cells and stem cells, however, an enzyme called telomerase lengthens the telomeres and prevents chromosome shortening. Thus, these cells are not limited in the number of times they can divide. All cells have the gene for telomerase, but most somatic cells don't express it, and they produce no telomerase. Why don't somatic cells express telomerase and have unlimited division?

- The introduction to this chapter discussed recent research showing that children who experience early childhood stresses have shorter telomeres. How might this information be used in a practical sense?

New and Reorganized Content

The sixth edition addresses recent discoveries in genetics corresponding to our ever-changing understanding of inheritance, the molecular nature of genetic information, epigenetics, and genetic evolution. This edition also focuses on updating the new research techniques that have become available to geneticists in the past few years. For example, I have expanded coverage of CRISPR-Cas systems and reorganized the chapter on molecular genetic analysis.

New and updated content includes

- New section on DNA in the biosphere (Chapter 1)
- New sections on genetic mosaicism and pharmacogenetic testing (Chapter 6)
- Expanded discussion of aneuploidy in humans (Chapter 8)
- New section on the importance of bacterial and viral genetics; new section on bacterial defense mechanisms; new section on rhinoviruses (Chapter 9)
- Updated discussion of chromatin structure; new section on mitochondrial replacement therapy (Chapter 11)
- Updated discussion of licensing of DNA replication; updated discussion of the end-replication problem for telomeres (Chapter 12)
- Expanded discussion of Piwi-interacting RNAs; revised section on CRISPR RNA; expanded discussion of long noncoding RNAs (Chapter 14)
- Expanded discussion of enhancers and insulators; expanded discussion of gene regulation through RNA splicing; new section on RNA crosstalk; expanded discussion of translational control of gene expression (Chapter 17)
- Significant reorganization to focus on methods currently in use; significant updates on new technologies; new section on CRISPR-Cas genome editing; expanded section on engineered nucleases (Chapter 19)
- Updated methods in genomics; new sections "What Exactly Is the Human Genome?" and "RNA Sequencing" (Chapter 20)
- Updates for cancer statistics; expanded discussion of telomerase in human cancers; expanded discussion of genetics of tumor metastases (Chapter 23)
- A new section on population variation; an expanded discussion on the reproductive isolation of apple maggot flies (Chapter 26)

NEW Introductory Stories Each chapter begins with a brief **introductory story** that illustrates the relevance of a genetic concept that students will learn in the chapter. These stories—a favorite feature of past editions—give students a glimpse of what's going on in the field of genetics today and help to draw the reader into the chapter. Among new introductory story topics are "The Sex of a Dragon," "The Genetics of Medieval Leprosy," "Editing the Genome with CRISPR-Cas9," "Building a Chromosome for Class," and "The Wolves of Isle Royale." End-of-chapter problems specifically address concepts discussed in many of the introductory stories, both old and new.

Media and Supplements

For this edition, we have thoroughly revised and refreshed the extensive set of online learning tools for *Genetics: A Conceptual Approach*. All of the new media resources for this edition will be available in our new SaplingPlus system.

SaplingPlus is a comprehensive and robust online teaching and learning platform that also incorporates all instructor resources and gradebook functionality.

Student Resources in SaplingPlus for *Genetics: A Conceptual Approach*

SaplingPlus provides students with media resources designed to enhance their understanding of genetic principles and improve their problem-solving ability.

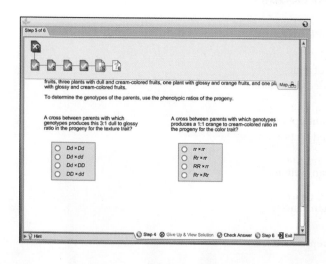

■ **Detailed Feedback for Students** Homework questions include hints, wrong-answer feedback targeted to students' misconceptions, and fully worked out solutions to reinforce concepts and to build problem-solving skills.

■ **The e-Book** The e-Book contains the full contents of the text as well as embedded links to important media resources (listed following).

■ **Updated and New Problem-Solving Videos** offer students valuable help by reviewing basic problem-solving strategies. The problem-solving videos demonstrate an instructor working through problems that students find difficult in a step-by-step manner.

■ **New Online Tutorials** identify where students have difficulty with a problem and route them through a series of steps in order to reach the correct answer. Hints and feedback at every step guide students along the way, as if they were working the problem with an instructor. Complete solutions are also included.

- **Updated and New Animations/Simulations** help students understand key processes in genetics by outlining them in a step-by-step manner. All of the animations and simulations include assessment questions to help students evaluate whether they understood the concept or technique they viewed.

- **Comprehensively Revised Assessment** All media resources have undergone extensive rewriting, reviewing, and accuracy checking.

- **Online Reading Quizzes,** covering the key concepts in each chapter, allow instructors to assess student preparedness before class and to identify challenging areas.

- **New Online Homework.**
  SaplingPlus offers robust, high-level homework questions with hints and wrong-answer feedback targeted to students' misconceptions as well as detailed worked-out solutions to reinforce concepts. Online Homework includes select end-of-chapter Application Problems from the text, converted into a variety of auto-graded formats. It also includes a variety of Sapling Genetics questions, curated for alignment with the text. These questions can also be used for quizzing or student practice. The questions are tagged by difficulty level.

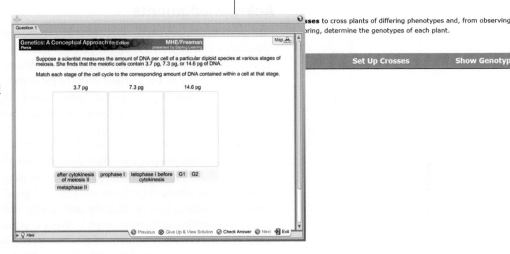

- The printable **Test Bank** contains at least 50 multiple-choice and short-answer questions per chapter. The Test Bank questions are also available in a downloadable Diploma format.

- **New In-Class Activities** contribute to active learning of some of the more challenging topics in genetics. Ten activities (15–45 minutes in length) allow students to work in groups to apply what they have learned to problems ranging from gene mapping to statistical analysis to interpreting phylogenetic trees. Each activity includes clicker questions and multiple-choice assessment questions.

- *Nature Genetics* **Articles with Assessment** engage students with primary research and encourage critical thinking. Specifically selected for both alignment with text coverage and exploration of identified difficult topics, the *Nature Genetics* articles include assessment questions that can be automatically graded. Some of the open-ended (non-multiple-choice) questions are also suitable for use in flipped classrooms and active learning discussions either in class or online.

Instructor Resources in SaplingPlus for *Genetics: A Conceptual Approach*

- **Updated Clicker Questions** allow instructors to integrate active learning into the classroom and to assess students' understanding of key concepts during lectures. Available in PowerPoint format, numerous questions are based on the Concept Check questions featured in the textbook.

- **Updated Lecture PowerPoint Files** have been developed to minimize preparation time for new users of the book. These files offer suggested lectures, including key illustrations and summaries, that instructors can adapt to their teaching styles.

- **Layered PowerPoint Slides** deconstruct key concepts, sequences, and processes from the textbook illustrations, allowing instructors to present complex ideas step by step.

- **Textbook Illustrations and Tables** are offered as high-resolution JPEG files. Each image has been fully optimized to increase type sizes and adjust color saturation. These images have been tested in a large lecture hall to ensure maximum clarity and visibility. Images are presented in both labeled and unlabeled formats.

- The **Solutions and Problem-Solving Manual** (written by Jung Choi and Mark McCallum) contains complete answers and worked-out solutions to all questions and problems in the textbook. The Solutions Manual is also available in print (ISBN: 1-319-08870-8).

Acknowledgments

I am indebted to many people for help with this and previous editions of *Genetics: A Conceptual Approach.* I learned much from my genetics teachers: Ray Canham, who first exposed me to genetics and instilled in me a life-long love for the subject; and Jeff Mitton, who taught me the art of genetic research. I've learned from the thousands of genetics students who have filled my classes over the past 36 years, first at Connecticut College, then at Baylor University, and now at Southwestern University. Their intelligence, enthusiasm, curiosity, and humor have been a source of motivation and pleasure throughout my professional life. I have also learned from students worldwide who have used earlier editions of this book and kindly shared with me— through emails and phone calls—their thoughts about the book and how it could be improved.

I am grateful for the wonderful colleagues who surround me daily at Southwestern University and whose friendship, advice, and good humor sustain my work. The small classes, close interaction of students and faculty, and integration of teaching and research have made working at Southwestern University personally and professionally rewarding. I thank Edward Burger, President of Southwestern University and Alisa Gaunder, Dean of the Faculty, for sustaining this supportive academic environment and for their continued friendship and collegiality.

Writing a modern science textbook requires a team effort, and I have been blessed with an outstanding team at W. H. Freeman and Macmillan Learning. Managing Director Susan Winslow has been a champion of the book for a number of years; I value her support, strategic vision, and commitment to education. Lauren Schultz, Executive Editor, has been a great project leader. She has been a continual source of encouragement, support, and creative ideas, as well as a good friend and colleague. Working daily with Development Editor Maria Lokshin has been a wonderful experience. Maria's hard work, passion for excellence, superior knowledge of genetics, great organizational skills, and good humor made crafting this edition rewarding and fun, in spite of a demanding schedule. I am also grateful to Lisa Samols, Director of Development, for shepherding the development of this edition and for great insight at key points.

Norma Sims Roche was an outstanding manuscript editor, making numerous suggestions that kept the text accurate and consistent and that also greatly improved its readability. Project Editor Jennifer Carey expertly managed the production of this sixth edition. Her dedication to excellence in all phases of the production process has been a major factor in making the book a success. I thank Dragonfly Media Group for creating and revising the book's illustrations and Janice Donnola for coordinating the illustration program. Quade Paul (Echo Medical Media) designed the cover image (from a concept by Emiko Paul). Thanks to Paul Rohloff at W. H. Freeman and Sofia Buono at codeMantra for coordinating the composition and manufacturing phases of production. Blake Logan developed the book's design. I thank Christine Buese and Richard Fox for photo research. Amy Thorne, Cassandra Korsvik, Amber Jonker, Clairissa Simmons, Amanda Nietzel, Elaine Palucki, and Emiko Paul developed the excellent media and supplements that accompany the book. I am grateful to Jung Choi and Mark McCallum for writing solutions to new end-of-chapter problems. Robert Fowler, Marcie Moehnke, Ellen France, Amy McMillan, Daniel Williams, Douglas Thrower, Victor Fet, and Usha Viveganananthan developed and reviewed assessment questions.

As always, I am grateful to the Macmillan Learning sales representatives, regional managers, and STEM specialists, who introduce my book to genetic instructors throughout the world. I have greatly enjoyed working with this sales staff; their expertise, hard work, and good service are responsible for the success of Macmillan books.

A number of colleagues served as reviewers of this book, kindly lending me their technical expertise and teaching experience. Their assistance is gratefully acknowledged. Any remaining errors are entirely my own.

Marlene Tyrrell—my spouse and best friend for 36 years—our children and their spouses—Sarah, Matt, Michael, and Amber—and now my F_2 progeny—Ellie, Beckett, and Caroline—provide love, support, and inspiration for everything I do.

My gratitude goes to the reviewers of this new edition of *Genetics: A Conceptual Approach*:

Kirk Anders
Gonzaga University

Catalina Arango
Saint Joseph's University

Glenn Barnett
Central College

Paul W. Bates
University of Minnesota Duluth

Christine Beatty
Benedictine University

Aimee Bernard
University of Colorado Denver

Jim Bonacum
University of Illinois Springfield

Gregory Booton
Ohio State University

Indrani Bose
Western Carolina University

Aaron Cassill
University of Texas at San Antonio

Brian Chadwick
Florida State University

Helen Chamberlin
Ohio State University

Bruce Chase
University of Nebraska at Omaha

Craig Coleman
Brigham Young University

Claire Cronmiller
University of Virginia

Claudette Davis
George Mason University

Wu-Min Deng
Florida State University

Eric Domyan
Utah Valley University

Teresa Donze-Reiner
West Chester University

Robert Dotson
Tulane University

Erastus Dudley
Huntingdon College

Iain Duffy
Saint Leo University

Richard Duhrkopf
Baylor University

Edward Eivers
California State University, Los Angeles

Cheryld L. Emmons
Alfred University

Victor Fet
Marshall University

Christy Fillman
University of Colorado

Robert G. Fowler
San Jose State University

Jeffrey French
North Greenville University

Joseph Gar
West Kentucky Community & Technical College

Julie Torruellas Garcia
Nova Southeastern University

J. Yvette Gardner
Clayton State University

Michael Gilchrist
University of Tennessee

William Gilliland
DePaul University

James Godde
Monmouth College

Elliott S. Goldstein
Arizona State University

Cynthia van Golen
Delaware State University

Steven W. Gorsich
Central Michigan University

Christine Gray
St. Mary's University

John Gray
University of Toledo

Eli Greenbaum
University of Texas at El Paso

Briana Gross
University of Minnesota Duluth

Danielle Hamill
Ohio Wesleyan University

Janet Hart
Massachusetts College of Pharmacy and Health Sciences

Ken Hillers
California Polytechnic State University, San Luis Obispo

Debra Hinson
Dallas Baptist University

Carina Endres Howell
Lock Haven University

Colin Hughes
Florida Atlantic University

Jeba Inbarasu
Metropolitan Community College, South Omaha Campus

Diana Ivankovic
Anderson University

David Kass
Eastern Michigan University

Cathy Silver Key
North Carolina Central University

Brian Kreiser
University of Southern Mississippi

Tim Kroft
Auburn University at Montgomery

Judith Leatherman
University of Northern Colorado

Melanie Lee-Brown
Guilford College

James Lodolce
Loyola University Chicago

Joshua Loomis
East Stroudsburg University

Michelle Mabry
Davis & Elkins College

Cindy S. Malone
California State University, Northridge

Endre Mathe
University of Debrecen, Hungary/Vasile Goldis University of Arad, Romania

Karen McGinnis
Florida State University

Thomas McGuire
Penn State Abington

Amy McMillan
SUNY Buffalo State

Melissa Merrill
North Carolina State University

Marcie Moehnke
Baylor University

Srinidi Mohan
University of New England

Jessica L. Moore
Western Carolina University

Jeanelle Morgan
University of North Georgia

Shahid Mukhtar
University of Alabama at Birmingham

Mary Murnik
Ferris State University

Todd Nickle
Mount Royal University

Margaret Olney
Saint Martin's University

Kavita Oommen
Georgia State University

Jessica Pamment
DePaul University

Tatiana Tatum Parker
Saint Xavier University

Ann Paterson
Williams Baptist College

Alexandra Peister
Morehouse College

Uwe Pott
University of Wisconsin-Green Bay

Eugenia Ribeiro-Hurley
Fordham University

Michael Robinson
Miami University

Helena Schmidtmayerova
Florida International University

Wendy Shuttleworth
Lewis-Clark State College

Walter Sotero
University of Central Florida

Agnes Southgate
College of Charleston

Ernest C. Steele, Jr.
Morgan State University

Tara Stoulig
Southeastern Louisiana University

Ann Sturtevant
Oakland University

Sandra Thompson-Jaeger
Christian Brothers University

Douglas Thrower
University of California at Santa Barbara

Kathleen Toedt
Housatonic Community College

Harald Vaessin
Ohio State University

Willem Vermerris
University of Florida

Meenakshi Vijayaraghavan
Tulane University

Usha Vivegananthan
Mohawk College

Yunqiu Wang
University of Miami

Daniel Williams
Winston-Salem State University

Jennifer Wolff
Carleton College

Kathleen Wood
University of Mary Hardin-Baylor

Chuan Xiao
University of Texas at El Paso

Eric Yager
Albany College of Pharmacy and Health Sciences

Malcolm Zellars
Georgia State University

Introduction to Genetics

A Hopi pueblo on Black Mesa. Albinism, a genetic condition, arises with high frequency among the Hopi people and occupies a special place in the Hopi culture. [Ansel Adams/National Park Archives at College Park, MD.]

Albinism in the Hopis

Rising a thousand feet above the desert floor, Black Mesa dominates the horizon of the Enchanted Desert and provides a familiar landmark for travelers passing through northeastern Arizona. Black Mesa is not only a prominent geological feature, but also, more significantly, the ancestral home of the Hopi Native Americans. Fingers of the mesa reach out into the desert, and alongside or on top of each finger is a Hopi village. Most of the villages are quite small, having only a few dozen inhabitants, but they are incredibly old. One village, Oraibi, has existed on Black Mesa since A.D. 1150 and is the oldest continuously occupied settlement in North America.

In 1900, Aleš Hrdlička, an anthropologist and physician working for the American Museum of Natural History, visited the Hopi villages of Black Mesa and reported a startling discovery. Among the Hopis were 11 white people—not Caucasians, but white Hopi Native Americans. These Hopis had a genetic condition known as albinism (**Figure 1.1**).

Albinism is caused by a defect in one of the enzymes required to produce melanin, the pigment that darkens our skin, hair, and eyes. People with albinism either don't produce melanin or produce only small amounts of it and, consequently, have white hair, light skin, and no pigment in the irises of their eyes. Melanin normally protects the DNA of skin cells from the damaging effects of ultraviolet radiation in sunlight, and melanin's presence in the developing eye is essential for proper eyesight.

The genetic basis of albinism was first described by the English physician Archibald Garrod, who recognized in 1908 that the condition was inherited as an autosomal recessive trait, meaning that a person must receive two copies of an albino mutation—one from each parent—to have albinism. In recent years, the molecular nature of the mutations that lead to albinism has been elucidated. Albinism in humans is caused by a mutation in any one of several different genes that control the synthesis and storage of melanin; many different types of mutations can occur at each gene, any one of which may lead to albinism. The form of albinism found among the Hopis is most likely oculocutaneous albinism (albinism affecting the eyes and skin) type 2, caused by a defect in the *OCA2* gene on chromosome 15.

The Hopis are not unique in having people with albinism among the members of their tribe. Albinism is found in almost all human ethnic groups and is described in ancient writings: it has probably been present since humankind's beginnings. What is unique about the Hopis is the high frequency of albinism in their population. In most human groups, albinism is rare, present in only about 1 in 20,000 individuals. In the villages on Black Mesa, it reaches a frequency of 1 in 200, a hundred times higher than in most other populations.

Why is albinism so frequent among the Hopis? The answer to this question is not completely known, but geneticists who have studied albinism in the Hopis speculate that

1.1 Albinism among the Hopi Native Americans. The Hopi girl in the center of this photograph, taken around 1900, displays albinism. [© The Field Museum, #CSA118. Charles Carpenter.]

the high frequency of the albino mutation is related to the special place that albinism occupied in the Hopi culture. For much of their history, the Hopis considered members of their tribe with albinism to be important and special. People with albinism were considered attractive, clean, and intelligent. Having a number of people with albinism in one's village was considered a good sign, a symbol that the people of the village contained particularly pure Hopi blood. Members of the tribe with albinism performed in Hopi ceremonies and held positions of leadership, often as chiefs, healers, and religious leaders.

Hopis with albinism were also given special treatment in everyday activities. The Hopis have farmed small garden plots at the foot of Black Mesa for centuries. Every day throughout the growing season, the men of the tribe trekked to the base of Black Mesa and spent much of the day in the bright southwestern sunlight tending their corn and vegetables. With little or no melanin pigment in their skin, people with albinism are extremely susceptible to sunburn and have an increased incidence of skin cancer when exposed to the sun. Furthermore, many don't see well in bright sunlight. Therefore, the male Hopis with albinism were excused from farming and allowed to remain behind in the village with the women of the tribe, performing other duties.

Throughout the growing season, the men with albinism were the only male members of the tribe in the village with the women during the day, and thus they enjoyed a mating advantage, which helped to pass on their albino genes. In addition, the special considerations given to Hopis with albinism allowed them to avoid the detrimental effects of albinism—increased skin cancer and poor eyesight. The small size of the Hopi tribe probably also played a role by allowing chance to increase the frequency of the albino mutation. Regardless of the factors that led to the high frequency of albinism, the Hopis clearly respected and valued the members of their tribe who possessed this particular trait. Unfortunately, people with genetic conditions in many societies are often subject to discrimination and prejudice. ▶ **TRY PROBLEMS 1 AND 26**

THINK-PAIR-SHARE

- Albinism occupied a special place in the Hopi culture; individuals who possessed this trait were valued by members of the tribe. What are some examples of genetic traits that, in contrast, sometimes result in discrimination and prejudice?

- Albinism in humans can be caused by mutations in any one of several different genes. This situation, in which the same phenotype may result from variation in several different genes, is referred to as genetic heterogeneity. Is genetic heterogeneity common? Are most genetic traits in humans the result of variation in a single gene, or are there many genetic traits that result from variation in several genes, as albinism does?

Genetics is one of the most rapidly advancing fields of science, with important new discoveries reported every month. Look at almost any major news source, and chances are that you will see articles related to genetics: on the sequencing of new genomes, such as those of the king cobra, minke whale, and loblolly pine; on the discovery of genes that affect major diseases, including multiple sclerosis, depression, and cancer; on analyses of DNA from long-extinct organisms such as a 700,000-year-old Pleistocene horse; or on the identification of genes that affect skin pigmentation, height, and learning ability in humans. Even among advertisements, you are likely to see ads for genetic testing to determine a person's ancestry or the pedigree of your dog. These new findings and applications of genetics often have significant economic and ethical implications, making the study of genetics relevant, timely, and interesting.

This chapter introduces you to genetics and reviews some concepts that you may have encountered in your general biology course. We begin by considering the importance of genetics to each of us, to society, and to students of biology. We then turn to the history of genetics and how the field as a whole developed. The final part of the chapter presents some fundamental terms and principles of genetics that are used throughout the book.

1.1 Genetics Is Important to Us Individually, to Society, and to the Study of Biology

Albinism among the Hopis illustrates the important role that genes play in our lives. This one genetic alteration, among the 20,000 genes that humans possess, completely changes the life of a Hopi who possesses it. It alters his or her occupation, role in

Hopi society, and relations with other members of the tribe. We all possess genes that influence our lives in significant ways. Genes affect our height, weight, hair color, and skin pigmentation. They affect our susceptibility to many diseases and disorders (**Figure 1.2**) and even contribute to our intelligence and personality. Genes are fundamental to who and what we are.

Although the science of genetics is relatively new compared with sciences such as astronomy and chemistry, people have understood the hereditary nature of traits and practiced genetics for thousands of years. The rise of agriculture began when people started to apply genetic principles to the domestication of plants and animals. Today, the major crops and animals used in agriculture are quite different from their wild progenitors, having undergone extensive genetic alteration that increased their yields and provided many desirable traits, such as disease and pest resistance, special nutritional qualities, and characteristics that facilitate harvest. The Green Revolution, which expanded food production throughout the world in the 1950s and 1960s, relied heavily on the application of genetic methods and principles (**Figure 1.3**). Today, genetically engineered corn, soybeans, and other crops constitute a significant proportion of all the food produced worldwide.

1.3 In the Green Revolution, genetic techniques were used to develop new high-yielding strains of crops. (Left) Norman Borlaug, a leader in the development of new varieties of wheat that led to the Green Revolution. Borlaug was awarded the Nobel Peace Prize in 1970. (Right) Modern, high-yielding rice plant (left) and traditional rice plant (right). [Left: Bettmann/CORBIS. Right: IRRI.]

(a)   **(b)**

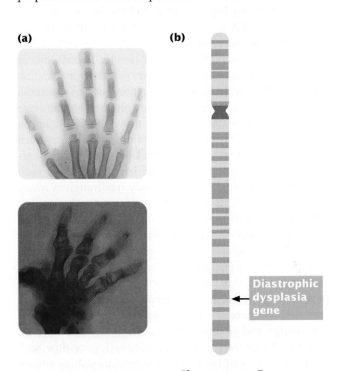

Chromosome 5

1.2 Genes influence susceptibility to many diseases and disorders. (a) An X-ray of the hand of a person suffering from diastrophic dysplasia (bottom), a hereditary growth disorder that results in curved bones, short limbs, and hand deformities, compared with an X-ray of a normal hand (top). (b) Diastrophic dysplasia is due to a defect in the *SLC26A2* gene on chromosome 5. [Part a: (top) Biophoto Associates/Science Source; (bottom) Reprinted from *Cell*, 78(6) Johanna Hästbacka, et al., The diastrophic dysplasia gene encodes a novel sulfate transporter: Positional cloning by fine-structure linkage disequilibrium mapping, pp. 1073–1087, ©1994 with permission from Elsevier. Permission conveyed through Copyright Clearance Center, Inc. Elsevier. Courtesy of Prof. Eric Lander, Whitehead Institute, MIT.]

The pharmaceutical industry is another area in which genetics plays an important role. Numerous drugs and food additives are synthesized by fungi and bacteria that have been genetically manipulated to make them efficient producers of these substances. The biotechnology industry employs molecular genetic techniques to develop and mass-produce substances of commercial value. Antimalarial drugs, growth hormone, insulin, clotting factor, antiviral drugs, enzymes, antibiotics, vaccines, and many other compounds are now produced commercially by genetically engineered bacteria and other organisms (**Figure 1.4**).

1.4 The biotechnology industry uses molecular genetic methods to produce substances of economic value. [REUTERS/Jerry Lampen.]

Genetics has also been used to generate bacterial strains that remove minerals from ore, break down toxic chemicals, and help produce biofuels.

Genetics also plays a critical role in medicine. Physicians recognize that many diseases and disorders have a hereditary component, including rare genetic disorders such as sickle-cell anemia and Huntington disease as well as many common diseases such as asthma, diabetes, and hypertension. Advances in genetics have resulted in important insights into the nature of diseases such as cancer and in the development of diagnostic tests, including tests that identify disease-causing mutations as well as pathogens. Genomic data are helping to usher in the era of personalized medicine. Rapid, low-cost sequencing methods now allow us to obtain a person's complete genome sequence, which provides important information about that person's susceptibilities to diseases and likely responses to particular treatments. And gene therapy—the direct alteration of genes to treat human diseases—has now been administered to thousands of patients, although its use is still experimental and limited.

THINK-PAIR-SHARE Question 1

The Role of Genetics in Biology

Although an understanding of genetics is important to all people, it is critical to the student of biology. Genetics provides one of biology's unifying principles: all organisms use genetic systems that have a number of features in common. Genetics also undergirds the study of many other biological disciplines. Evolution, for example, is genetic change taking place over time, so the study of evolution requires an understanding of genetics. Developmental biology relies heavily on genetics: tissues and organs develop through the regulated expression of genes (**Figure 1.5**). Even such fields as taxonomy, ecology, and animal behavior are making increasing use of genetic methods. The study of almost any field of biology or medicine is incomplete without a thorough understanding of genes and genetic methods.

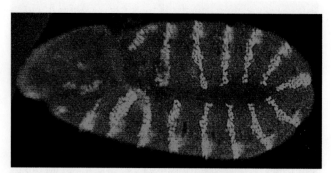

1.5 The key to development lies in the regulation of gene expression. This early fruit-fly embryo illustrates the localized expression of the *engrailed* gene, which helps determine the development of body segments in the adult fly. [Steven Paddock.]

Genetic Diversity and Evolution

Life on Earth exists in a tremendous array of forms and features, occupying almost every conceivable environment. Life is also characterized by adaptation: many organisms are exquisitely suited to the environment in which they are found. The history of life is a chronicle of new forms of life emerging, old forms disappearing, and existing forms changing.

Despite their tremendous diversity, living organisms have an important feature in common: all use similar genetic systems. The complete set of genetic instructions for any organism is its **genome**, and all genomes are encoded in nucleic acids—either DNA or RNA. The coding system for genomic information is also common to all life: genetic instructions are in the same format and, with rare exceptions, the code words are identical. Likewise, the process by which genetic information is copied and decoded is remarkably similar for all forms of life. These common features suggest that all life on Earth evolved from the same primordial ancestor, which arose between 3.5 billion and 4 billion years ago. Biologist Richard Dawkins describes life as a river of DNA that runs through time, connecting all organisms past and present.

That all organisms have similar genetic systems means that the study of one organism's genes reveals principles that apply to other organisms. Investigations of how bacterial DNA is replicated (copied), for example, provide information that applies to the replication of human DNA. It also means that genes can often function in foreign cells, which makes genetic engineering possible. Unfortunately, the similarity of genetic systems is also the basis for diseases such as AIDS (acquired immune deficiency syndrome), in which viral genes are able to function—sometimes with alarming efficiency—in human cells.

Life's diversity and adaptations are products of evolution, which is simply genetic change over time. Evolution is a two-step process: first, inherited differences arise randomly, and then the proportion of individuals with particular differences increases or decreases. Genetic variation is therefore the foundation of all evolutionary change and is ultimately the basis of all life as we know it. Techniques of molecular genetics are now routinely used to decipher evolutionary relationships among organisms; for example, a recent analysis of DNA from Neanderthal fossils has yielded new information concerning the relationship between Neanderthals and modern humans, demonstrating that Neanderthals and the ancestors of modern humans interbred some 30,000 to 40,000 years ago. Genetics and the study of genetic variation are critical to understanding the past, present, and future of life. **> TRY PROBLEM 17**

THINK-PAIR-SHARE Question 2

CONCEPTS

Our genes affect many of our physical features as well as our susceptibility to many diseases and disorders. Genetics contributes to advances in agriculture, pharmaceuticals, and medicine and is fundamental to modern biology. All organisms use similar genetic systems, and genetic variation is the foundation of the diversity of all life.

✔ CONCEPT CHECK 1

What are some of the implications of all organisms having similar genetic systems?

a. That all life forms are genetically related
b. That research findings on one organism's gene function can often be applied to other organisms
c. That genes from one organism can often function in another organism
d. All of the above

DNA in the Biosphere

Each DNA molecule is very small, but because all cells contain genetic information, there is a tremendous amount of DNA in the world. Scientists estimate that the total amount of DNA in the biosphere is 5.3×10^{31} megabase pairs (millions of base pairs), altogether weighing some 50 billion tons. Storing the information content of the world's DNA would require 10^{21} computers, each with the average storage capacity of the world's four most powerful supercomputers.

Scientists are now cataloging and measuring the world's biodiversity through analysis of DNA. For example, researchers aboard the ship *Tara* surveyed the world's oceans for organisms by isolating DNA from seawater during a three-and-a-half-year voyage. They collected 35,000 seawater samples and extracted DNA from each. The DNA was then sequenced and analyzed, revealing the presence of 150,000 genetically distinct types of eukaryotes. Most of these eukaryotes were newly discovered single-celled organisms. The researchers also detected 5000 viruses, only 39 of which were previously known to science. (See the Suggested Readings for this chapter in your Sapling Plus for the reference to this study and many others mentioned in the book.)

Divisions of Genetics

The study of genetics consists of three major subdisciplines: transmission genetics, molecular genetics, and population genetics (**Figure 1.6**). Also known as classical genetics, **transmission genetics** encompasses the basic principles of heredity and how traits are passed from one generation to the next. This subdiscipline addresses the relation between chromosomes and heredity, the arrangement of genes on chromosomes, and gene mapping. Here, the focus is on the individual organism—how an individual inherits its genetic makeup and how it passes its genes to the next generation.

Molecular genetics concerns the chemical nature of the gene itself: how genetic information is encoded, replicated,

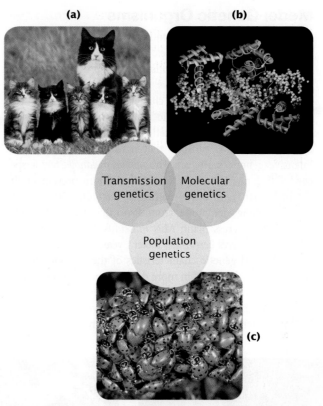

1.6 Genetics can be divided into three major subdisciplines. [Top left: Juniors Bildarchiv/Alamy. Top right: Martin McCarthy/Getty Images. Bottom: Stuart Wilson/Science Source.]

and expressed. It includes the cellular processes of replication, transcription, and translation (by which genetic information is transferred from one molecule to another) and gene regulation (the processes that control the expression of genetic information). The focus in molecular genetics is the gene and its structure, organization, and function.

Population genetics explores the genetic composition of populations (groups of individuals of the same species) and how that composition changes geographically and with the passage of time. Because evolution is genetic change, population genetics is fundamentally the study of evolution. The focus of population genetics is the group of genes found in a population.

Division of the study of genetics into these three subdisciplines is convenient and traditional, but we should recognize not only that they overlap, but also that each one can be further divided into a number of more specialized fields, such as chromosomal genetics, biochemical genetics, quantitative genetics, and so forth. Alternatively, genetic studies can be subdivided by organism (fruit fly, corn, or bacterial genetics), and each of these organisms may be studied at the levels of transmission, molecular, and population genetics. Modern genetics is an extremely broad field, encompassing many interrelated subdisciplines and specializations. **▶ TRY PROBLEM 18**

Model Genetic Organisms

Through the years, genetic studies have been conducted on thousands of different species, including almost all major groups of bacteria, fungi, protists, plants, and animals. Nevertheless, a few species have emerged as **model genetic organisms**—organisms with characteristics that make them particularly useful for genetic analysis and about which a tremendous amount of genetic information has accumulated. Six model organisms that have been the subject of intensive genetic study are *Drosophila melanogaster*, a species of fruit fly; *Escherichia coli*, a bacterium present in the gut of humans and other mammals; *Caenorhabditis elegans*, a soil-dwelling nematode (roundworm); *Arabidopsis thaliana*, the thale-cress plant; *Mus musculus*, the house mouse; and *Saccharomyces cerevisiae*, baker's yeast (**Figure 1.7**). The life cycles and genetic characteristics of these model genetic organisms are described in more detail in the Reference Guide to Model Genetic Organisms located at the end of this book (pp. A1–A13). This Reference Guide will be a useful resource as you encounter these organisms throughout the book.

At first glance, these lowly and sometimes unappreciated creatures might seem unlikely candidates for model genetic organisms. However, all possess life cycles and traits that make them particularly suitable for genetic study, including a short generation time, large but manageable numbers of progeny, adaptability to a laboratory environment, and the ability to be housed and propagated inexpensively. Other species that are frequently the subjects of genetic research and considered genetic models include *Neurospora crassa* (bread mold), *Zea mays* (corn), *Danio rerio* (zebrafish), and *Xenopus laevis* (clawed frog). Although not generally considered a model genetic organism, *Homo sapiens* has also been subjected to intensive genetic scrutiny; special techniques for the genetic analysis of humans are discussed in Chapter 6.

The value of model genetic organisms is illustrated by the use of zebrafish to identify genes that affect skin pigmentation in humans. For many years, geneticists recognized that differences in pigmentation among human ethnic groups were genetic (**Figure 1.8a**), but the

(a)

Drosophila melanogaster
Fruit fly (pp. A2–A3)

(b)

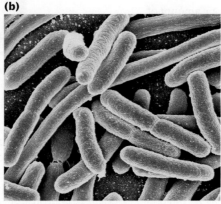

Escherichia coli
Bacterium (pp. A4–A5)

(c)

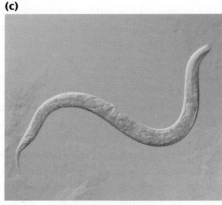

Caenorhabditis elegans
Nematode (pp. A6–A7)

(d)

Arabidopsis thaliana
Thale-cress plant (pp. A8–A9)

(e)

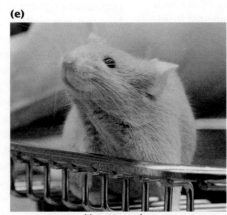

Mus musculus
House mouse (pp. A10–A11)

(f)

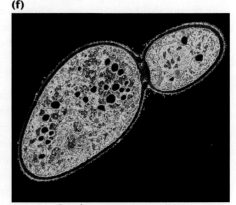

Saccharomyces cerevisiae
Baker's yeast (pp. A12–A13)

1.7 Model genetic organisms are species with features that make them useful for genetic analysis. [Part a: © Alfred Schauhuber/ImageBROKER/Alamy Stock Photo. Part b: Pasieka/Science Source. Part c: Sinclair Stammers/Science Source. Part d: Peggy Greb/ARS/USDA. Part e: AP Photo/Joel Page. Part f: Biophoto Associates/Science Source.]

(a)

(b)

Normal zebrafish   *Golden* mutant

1.8 The zebrafish, a model genetic organism, has been instrumental in helping to identify genes encoding pigmentation differences among humans. (a) Human ethnic groups differ in degree of skin pigmentation. (b) The zebrafish *golden* mutation is caused by a gene that controls the amount of melanin in melanosomes. [Part a: (left) Barbara Penoyar/Getty Images; (center) Amos Morgan/Getty Images; (right) Stockbyte/Getty Images. Part b: Keith Cheng/Jake Gittlen, Cancer Research Foundation, Penn State College of Medicine.]

genes causing these differences were largely unknown. The zebrafish has become an important model in genetic studies because it is a small vertebrate that produces many offspring and is easy to rear in the laboratory. The mutant zebrafish called *golden* has light pigmentation due to the presence of fewer, smaller, and less dense pigment-containing structures called melanosomes in its cells (**Figure 1.8b**).

Keith Cheng and his colleagues hypothesized that light skin in humans might result from a mutation that is similar to the *golden* mutation in zebrafish. Taking advantage of the ease with which zebrafish can be manipulated in the laboratory, they isolated and sequenced the gene responsible for the *golden* mutation and found that it encodes a protein that takes part in calcium uptake by melanosomes. They then searched a database of all known human genes and found a similar gene called *SLC24A5*, which encodes a protein that has the same function in human cells. When they examined human populations, they found that light-skinned Europeans often possess one form of this gene, whereas darker-skinned Africans, East Asians, and Native Americans usually possess a different form. Many other genes also affect pigmentation in humans, as illustrated by the mutations in the *OCA2* gene that produce albinism among the Hopis (discussed in the introduction to this chapter). Nevertheless, *SLC24A5* appears to be responsible for 24% to 38% of the differences in pigmentation between Africans and Europeans.

This example illustrates the power of model organisms in genetic research. However, we should not forget that all organisms possess unique characteristics and the genetics of model organisms do not always accurately reflect the genetic systems of other organisms.

CONCEPTS

The three major divisions of genetics are transmission genetics, molecular genetics, and population genetics. Transmission genetics examines the principles of heredity; molecular genetics deals with the gene and the cellular processes by which genetic information is transferred and expressed; population genetics concerns the genetic composition of groups of organisms and how that composition changes geographically and over time. Model genetic organisms are species that have received special emphasis in genetic research; they have characteristics that make them useful for genetic analysis.

✔ CONCEPT CHECK 2

Would the horse make a good model genetic organism? Why or why not?

1.2 Humans Have Been Using Genetic Techniques for Thousands of Years

Although the science of genetics is young—almost entirely a product of the past 100 years or so—people have been using genetic principles for thousands of years.

The Early Use and Understanding of Heredity

The first evidence that people understood and applied the principles of heredity in earlier times is found in the domestication of plants and animals, which began between approximately 10,000 and 12,000 years ago; early farming villages appeared in the Middle East between 11,000 and 11,500 years ago. The first domesticated organisms included wheat, peas, lentils, barley, dogs, goats, and sheep (**Figure 1.9a**). By 4000 years ago, genetic techniques of selective breeding were already in use in the Middle East. The Assyrians and Babylonians developed several hundred varieties of date palms that differed in fruit size, color, taste, and time of ripening (**Figure 1.9b**). Other crops and domesticated animals were developed by cultures in Asia, Africa, and the Americas in the same period.

Ancient writings demonstrate that early humans were also aware of their own heredity. Hindu sacred writings dating to 2000 years ago suggested that many traits are inherited from the father and that differences between siblings are produced by the mother. The Talmud, the Jewish book of religious laws based on oral traditions dating back thousands of years, presents an uncannily accurate understanding of the inheritance of hemophilia. It directs that, if a woman bears two sons who die of bleeding after circumcision, any additional sons that she bears should not be circumcised; nor should the sons of her sisters be circumcised. This advice accurately corresponds to the X-linked pattern of inheritance of hemophilia (discussed further in Chapter 6).

Some early concepts of heredity were incorrect, but reflect human interest in heredity and our attempts to explain the inheritance of traits. The ancient Greeks gave careful consideration to human reproduction and heredity. Greek philosophers developed the concept of **pangenesis**, in which specific particles, later called gemmules, carry information from various parts of the body to the reproductive organs, from which they are passed to the embryo at the moment of conception (**Figure 1.10**). Although incorrect, the concept of pangenesis was highly influential and persisted until the late 1800s.

The concept of pangenesis led the ancient Greeks to propose the notion of the **inheritance of acquired characteristics**, according to which traits acquired in a person's lifetime become incorporated into that person's hereditary information and are passed on to offspring; for example, they proposed that people who developed musical ability through diligent study would produce children who were innately endowed with musical ability. Jean-Baptiste Lamarck (1744–1829) was a proponent of this idea and incorporated it into his theory of biological change. The notion of the inheritance of acquired characteristics is also no longer accepted, but it remained popular through the twentieth century.

Although the ancient Romans contributed little to an understanding of human heredity, they successfully developed a number of techniques for animal and plant breeding; their techniques were based on trial and error rather than any general concept of heredity. Little new information was added to the understanding of genetics in the next 1000 years.

Additional developments in our understanding of heredity occurred during the seventeenth century. Dutch eyeglass makers began to put together simple microscopes in the late 1500s, enabling Robert Hooke (1635–1703) to discover cells in 1665. Microscopes provided naturalists with new and exciting vistas on life. Perhaps it was excessive enthusiasm for this new world of the very small that gave rise to the idea of **preformationism**: that inside the egg or sperm there exists a fully formed miniature adult, a *homunculus*, which simply enlarges in the course of development (**Figure 1.11**). Preformationism meant that all traits were inherited from only one parent—from the father if the homunculus was in the sperm or from the mother if it was in the egg. Although many observations suggested that offspring possess a mixture of traits from both parents, preformationism remained a popular concept throughout much of the seventeenth and eighteenth centuries.

Another early notion of heredity was **blending inheritance**, which proposed that the traits of offspring are a blend, or mixture, of parental traits. This idea suggested that

(a)

(b)

1.9 Ancient peoples practiced genetic techniques in agriculture. (a) Modern wheat, with larger and more numerous seeds that do not scatter before harvest, was produced by interbreeding at least three different wild species. (b) Assyrian bas-relief sculpture showing artificial pollination of date palms at the time of King Assurnasirpalli II, who reigned from 883 to 859 B.C. [Part a: Scott Bauer/ARS/USDA. Part b: Lower register: Image copyright © The Metropolitan Museum of Art. Image source: Art Resource, NY.]

(a) Pangenesis concept

1 According to the pangenesis concept, genetic information from different parts of the body...

2 ...travels to the reproductive organs...

3 ...where it is transferred to the gametes.

Sperm

Zygote

Egg

(b) Germ-plasm theory

1 According to the germ-plasm theory, germ-line tissue in the reproductive organs...

2 ...contains a complete set of genetic information...

3 ...that is transferred directly to the gametes.

Sperm

Zygote

Egg

1.10 Pangenesis, an early concept of inheritance, compared with the modern germ-plasm theory.

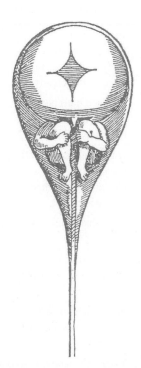

1.11 Preformationists in the seventeenth and eighteenth centuries believed that sperm or eggs contained a fully formed human (the homunculus). Shown here is a drawing of a homunculus inside a sperm. [Science Source.]

the genetic material itself blends, much as blue and yellow pigments blend to make green paint; it also suggested that after having been blended, genetic differences could not be separated in future generations, just as green paint cannot be separated into blue and yellow pigments. Some traits do *appear* to exhibit blending inheritance; however, we realize today that individual genes do not blend.

The Rise of the Science of Genetics

In 1676, Nehemiah Grew (1641–1712) reported that plants reproduce sexually. With this information, a number of botanists began to experiment with crossing plants and creating hybrids, including Gregor Mendel (1822–1884; **Figure 1.12**), who went on to discover the basic principles of heredity. Mendel's conclusions, which were not widely known in the scientific community until 35 years after their publication, laid the foundation for our modern understanding of heredity, and he is generally recognized today as the father of genetics.

Developments in cytology (the study of cells) in the 1800s had a strong influence on genetics. Robert Brown (1773–1858) described the cell nucleus in 1833. Building on the work of others, Matthias Jacob Schleiden (1804–1881) and Theodor Schwann (1810–1882) proposed the concept that came to be known as the **cell theory** in 1839. According to this theory, all life is composed of cells, cells arise only from preexisting cells, and the cell is the fundamental unit of structure and function in living organisms. Biologists interested in heredity began to examine cells to see what took place in the course of cell reproduction. Walther Flemming (1843–1905) observed the division of chromosomes in 1879 and published a superb description of mitosis. By 1885, biologists generally recognized that the cell nucleus contains the hereditary information.

Charles Darwin (1809–1882), one of the most influential biologists of the nineteenth century, put forth the theory of evolution through natural selection and published his ideas in *On the Origin of Species* in 1859. Darwin recognized that heredity was fundamental to evolution, and he conducted extensive genetic crosses with pigeons and other organisms. He never understood the nature of inheritance, however, and this lack of understanding was a major omission in his theory of evolution.

In the last half of the nineteenth century, cytologists demonstrated that the nucleus had a role in fertilization. Near the close of that century, August Weismann (1834–1914) finally laid to rest the notion of the inheritance of acquired characteristics. He cut off the tails of mice for 22 consecutive generations and showed that the tail length in descendants remained stubbornly long. Weismann proposed the **germ-plasm theory**, which holds that the cells in the reproductive organs carry a complete set of genetic information that is passed to the egg and sperm (see Figure 1.10b).

The year 1900 was a watershed in the history of genetics. Gregor Mendel's pivotal 1866 publication on experiments with pea plants, which revealed the principles of heredity, was rediscovered, as considered in more detail in Chapter 3. Once the significance of his conclusions was recognized, other biologists immediately began to conduct similar genetic studies on mice, chickens, and other organisms. The results of these investigations showed that many traits indeed follow Mendel's rules. Some of the early concepts of heredity are summarized in **Table 1.1**.

In 1902, after the acceptance of Mendel's theory of heredity, Walter Sutton (1877–1916) proposed that genes, the units of inheritance, are located on chromosomes. Thomas Hunt Morgan (1866–1945) discovered the first mutant fruit fly in 1910 and used fruit flies to unravel many details of transmission genetics. Ronald A. Fisher (1890–1962), John B. S. Haldane (1892–1964), and Sewall Wright (1889–1988) laid the foundation for population genetics in the 1930s by integrating Mendelian genetics and evolutionary theory.

Geneticists began to use bacteria and viruses in the 1940s; the rapid reproduction and simple genetic systems of these organisms allowed detailed study of the organization and structure of their genes. At about this same time, evidence accumulated that DNA was the repository of genetic information. James Watson (b. 1928) and Francis Crick

1.12 Gregor Mendel is the father of modern genetics. Mendel first discovered the principles of heredity by crossing different varieties of pea plants and analyzing the transmission of traits in subsequent generations. [Hulton Archive/Getty Images.]

TABLE 1.1	Early concepts of heredity	
Concept	**Proposed**	**Correct or Incorrect**
Pangenesis	Genetic information travels from different parts of the body to reproductive organs.	Incorrect
Inheritance of acquired characteristics	Acquired traits become incorporated into hereditary information.	Incorrect
Preformationism	Miniature organism resides in sex cells, and all traits are inherited from one parent.	Incorrect
Blending inheritance	Genes blend and mix.	Incorrect
Germ-plasm theory	All cells contain a complete set of genetic information.	Correct
Cell theory	All life is composed of cells, and cells arise only from cells.	Correct
Mendelian inheritance	Traits are inherited in accord with defined principles.	Correct

(1916–2004), along with Maurice Wilkins (1916–2004) and Rosalind Franklin (1920–1958), described the three-dimensional structure of DNA in 1953, ushering in the era of molecular genetics (see Chapter 10).

By 1966, the chemical structure of DNA and the system by which it determines the amino acid sequence of proteins had been worked out. Advances in molecular genetics led to the first recombinant DNA experiments in 1973, which provided techniques for combining genetic material from different sources and touched off another revolution in genetic research. Walter Gilbert (b. 1932) and Frederick Sanger (1918–2013) developed methods for sequencing DNA in 1977. The polymerase chain reaction (PCR), a technique for quickly amplifying tiny amounts of DNA, was developed by Kary Mullis (b. 1944) and others in 1983. PCR subsequently become one of the most widely used tools in molecular biology.

In 1990, gene therapy was used for the first time to treat human genetic disease in the United States, and the Human Genome Project was launched. By 1995, the first complete DNA sequence of a free-living organism—the bacterium *Haemophilus influenzae*—had been determined, and the first complete sequence of a eukaryotic organism (yeast) was reported a year later. A rough draft of the human genome sequence was reported in 2000 (see Chapter 20), and the sequence was essentially completed in 2003, bringing in another new era in genetics (**Figure 1.13**). ▶ TRY PROBLEMS 23 AND 25

The Cutting Edge of Genetics

With exciting advances being made every year, genetics remains at the forefront of biological research. New, rapid

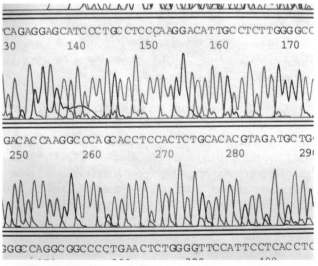

1.13 The human genome was completely sequenced for the first time in 2003. This chromatograph represents the DNA sequences from a small portion of one human gene. [© Science Museum/Science & Society Picture Library—All rights reserved.]

methods for sequencing DNA are being used to sequence the genomes of numerous species, from scorpions to sheep to trout. Recently, whole-genome sequences were obtained for more than 2600 Icelanders, providing a detailed view of the genetic diversity of a nation. Analysis of DNA from ancient bones has demonstrated that several different species of humans roamed Earth as recently as 30,000 years ago. Powerful modern genetic techniques are being used to identify genes that influence agriculturally important characteristics such as size in cattle, domestication in chickens, speed in racehorses, and leaf shape in corn. DNA analysis is now routinely used to identify and convict criminals or prove the innocence of suspects.

The power of the new methods being used to identify and analyze genes is illustrated by genetic studies of myocardial infarction (heart attack) in humans. Physicians have long recognized that heart attacks run in families, but finding specific genes that contribute to an increased risk has, until recently, been difficult. An international team of geneticists examined the DNA of 26,000 people in 10 countries for single-nucleotide differences in their DNA (called single-nucleotide polymorphisms, or SNPs) that might be associated with an increased risk of heart attack. This study and others identified several genes that affect the risk of coronary artery disease and early heart attacks. These findings may make it possible to identify people who are predisposed to heart attacks and allow early intervention that might prevent an attack from happening. Analyses of SNPs are helping to locate genes that affect all types of traits, from eye color and height to glaucoma and cancer.

Information about sequence differences among organisms is also a source of new insights about evolution. For example, scientists recently sequenced the whole genomes of multiple gorillas, including individuals of the eastern and western species. The study revealed that eastern and western populations began diverging some 150,000 years ago and stopped exchanging genes some 20,000 years ago. Eastern gorillas appear to have experienced a long-term population decline and currently have very low levels of genetic variation, which threatens their long-term survival as a species.

In recent years, scientists have discovered that alterations to DNA and chromosome structure that do not involve the base sequence of the DNA play an important role in gene expression. These alterations, called epigenetic changes, affect our appearance, behavior, and health and are currently the focus of intensive research (see Chapter 21). Other studies demonstrate that RNA is a key player in many aspects of gene function. The discovery in the late 1990s of tiny RNA molecules called small interfering RNAs and microRNAs led to the recognition that these molecules play central roles in gene expression and development. A powerful new method called CRISPR/Cas9 uses another group of small RNAs to precisely edit specific DNA sequences in living cells. This new system is now being widely used in both research and biotechnology.

Geneticists were recently able to design and synthesize, from scratch, an entirely artificial chromosome in yeast cells. The cells containing this chromosome grew just as well as those with a natural chromosome. In the field of proteomics, computer programs are being developed to model the structure and function of proteins using DNA sequence information. All of this information provides us with a better understanding of numerous biological processes and evolutionary relationships. The flood of new genetic information requires the continuous development of sophisticated computer programs to store, retrieve, compare, and analyze genetic data. That need has given rise to the field of bioinformatics, which merges molecular biology and computer science.

As sequencing becomes more affordable, the focus of DNA-sequencing efforts is shifting from the genomes of different species to individual differences within species. In the not-too-distant future, each person will probably possess a copy of his or her entire genome sequence, which can be used to help assess the risk of acquiring various diseases and to tailor their treatment should they arise. The use of genetics in agriculture—in both traditional breeding and genetic engineering—continues to improve the productivity of domesticated crops and animals, helping to feed the world population. This ever-widening scope of genetics raises significant ethical, social, and economic issues.

This brief overview of the history of genetics, from the first domestication of crops to present-day whole-genome sequencing, is not intended to be comprehensive; rather, it is designed to provide a sense of the accelerating pace of advances in genetics. In the chapters to come, we will learn more about the experiments and the scientists who helped shape the discipline of genetics.

THINK-PAIR-SHARE Question 3

CONCEPTS

Humans first applied genetic methods to the domestication of plants and animals between 10,000 and 12,000 years ago. Developments in plant hybridization and cytology in the eighteenth and nineteenth centuries laid the foundation for the field of genetics today. After Mendel's work was rediscovered in 1900, the science of genetics developed rapidly, and today it is one of the most active areas of science.

✔ **CONCEPT CHECK 3**

How did developments in cytology in the nineteenth century contribute to our modern understanding of genetics?

1.3 A Few Fundamental Concepts Are Important for the Start of Our Journey into Genetics

Undoubtedly, you learned some genetic principles in other biology classes. Let's take a few moments to review some fundamental genetic concepts.

CELLS ARE OF TWO BASIC TYPES: EUKARYOTIC AND PROKARYOTIC Structurally, cells consist of two basic types, although evolutionarily, the story is more complex (see Chapter 2). Prokaryotic cells lack a nuclear membrane and do not generally possess membrane-bounded organelles, whereas eukaryotic cells are more complex, possessing a nucleus and membrane-bounded organelles such as chloroplasts and mitochondria.

THE GENE IS THE FUNDAMENTAL UNIT OF HEREDITY The precise way in which a gene is defined often varies depending on the biological context. At the simplest level, we can think of a gene as a unit of information that encodes a genetic characteristic. We will expand this definition as we learn more about what genes are and how they function.

GENES COME IN MULTIPLE FORMS CALLED ALLELES A gene that specifies a characteristic may exist in several forms, called alleles. For example, a gene for coat color in cats may exist as an allele that encodes black fur or as an allele that encodes orange fur.

GENES CONFER PHENOTYPES One of the most important concepts in genetics is the distinction between traits and genes. Traits are not inherited directly. Rather, genes are inherited, and genes, along with environmental factors, determine the expression of traits. The genetic information that an individual organism possesses is its genotype; the trait is its phenotype. For example, the albinism seen in some Hopis is a phenotype, and the information in *OCA2* genes that causes albinism is a genotype.

GENETIC INFORMATION IS CARRIED IN DNA AND RNA Genetic information is encoded in the molecular structure of nucleic acids, which come in two types: deoxyribonucleic acid (DNA) and ribonucleic acid (RNA). Nucleic acids are polymers consisting of repeating units called nucleotides; each nucleotide consists of a sugar, a phosphate, and a nitrogenous base. The nitrogenous bases in DNA are of four types: adenine (A), cytosine (C), guanine (G), and thymine (T). The sequence of these bases encodes genetic information. DNA consists of two complementary nucleotide strands. Most organisms carry their genetic information in DNA, but a few viruses carry it in RNA. The four nitrogenous bases of RNA are adenine, cytosine, guanine, and uracil (U).

THINK-PAIR-SHARE Question 4

GENES ARE LOCATED ON CHROMOSOMES The vehicles of genetic information within a cell are chromosomes. (**Figure 1.14**), which consist of DNA and associated proteins. The cells of each species have a characteristic number of chromosomes; for example, bacterial cells normally possess a single chromosome, human cells possess 46, and pigeon cells possess 80. Each chromosome carries a large number of genes.

CHROMOSOMES SEPARATE THROUGH THE PROCESSES OF MITOSIS AND MEIOSIS The processes of mitosis and meiosis ensure that a complete set of an organism's chromosomes exists in each cell that results from cell division. Mitosis is the separation of chromosomes in the division of somatic (nonsex) cells. Meiosis is the pairing and separation of chromosomes in the division of sex cells to produce gametes (reproductive cells).

GENETIC INFORMATION IS TRANSFERRED FROM DNA TO RNA TO PROTEIN Many genes encode characteristics by specifying the structure of proteins. Genetic information is first transcribed from DNA into RNA, and then RNA is translated into the amino acid sequence of a protein.

MUTATIONS ARE PERMANENT CHANGES IN GENETIC INFORMATION THAT CAN BE PASSED FROM CELL TO CELL OR FROM PARENT TO OFFSPRING Gene mutations affect the genetic information of a single gene; chromosome mutations alter the number or the structure of chromosomes and therefore usually affect many genes.

MANY TRAITS ARE AFFECTED BY MULTIPLE FACTORS Many traits are affected by multiple genes that interact in complex ways with one another and with environmental factors. Human height, for example, is affected by many genes as well as by environmental factors such as nutrition.

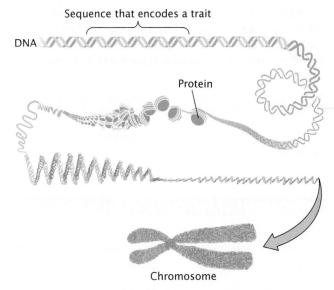

1.14 Genes are carried on chromosomes, which consist of highly compacted DNA and proteins.

EVOLUTION IS GENETIC CHANGE Evolution can be viewed as a two-step process: first, genetic variation arises, and second, some genetic variants increase in frequency, whereas other variants decrease in frequency.

▶ TRY PROBLEM 25

CONCEPTS SUMMARY

- Genetics is central to the life of every person: it influences a person's physical features, personality, intelligence, and susceptibility to numerous diseases.

- Genetics plays important roles in agriculture, the pharmaceutical industry, and medicine. It is central to the study of biology.

- All organisms use similar genetic systems. Genetic variation is the foundation of evolution and is critical to understanding all life.

- A tremendous amount of DNA exists in the biosphere. Scientists use environmental DNA to study and analyze biodiversity.

- The study of genetics can be broadly divided into transmission genetics, molecular genetics, and population genetics.

- Model genetic organisms are species about which much genetic information exists because they have characteristics that make them particularly amenable to genetic analysis.

- The use of genetics by humans began with the domestication of plants and animals.

- Ancient Greeks developed the concepts of pangenesis and the inheritance of acquired characteristics, both of which were later disproved. Ancient Romans developed practical measures for the breeding of plants and animals.

- Preformationism suggested that a person inherits all of his or her traits from one parent. Blending inheritance proposed that offspring possess a mixture of the parental traits. These ideas were later shown to be incorrect.

- By studying the offspring of crosses between varieties of peas, Gregor Mendel discovered the principles of heredity. Developments in cytology in the nineteenth century led to the understanding that the cell nucleus is the site of heredity.

- In 1900, Mendel's principles of heredity were rediscovered. Population genetics was established in the early 1930s, followed by bacterial and viral genetics. The structure of DNA was discovered in 1953, stimulating the rise of molecular genetics.

- The first human whole-genome sequence was completed in 2003.

- There are two basic types of cells: prokaryotic and eukaryotic.

- The set of alleles that determines a trait is termed the genotype; the trait that they produce is the phenotype.

- Genes are located on chromosomes, which are made up of nucleic acids and proteins and are partitioned into daughter cells through the process of mitosis or meiosis.

- Genetic information is expressed through the transfer of information from DNA to RNA to proteins.

- Evolution requires genetic change in populations.

IMPORTANT TERMS

genome (p. 4)
transmission genetics (p. 5)
molecular genetics (p. 5)
population genetics (p. 5)
model genetic organism (p. 6)
pangenesis (p. 8)
inheritance of acquired
   characteristics (p. 8)
preformationism (p. 8)
blending inheritance (p. 8)
cell theory (p. 10)
germ-plasm theory (p. 10)

ANSWERS TO CONCEPT CHECKS

1. d

2. No, because horses are expensive to house, feed, and propagate, they have too few progeny, and their generation time is too long.

3. Developments in cytology in the 1800s led to the identification of parts of the cell, including the cell nucleus and chromosomes. The cell theory focused the biologists' attention on the cell, eventually leading to the conclusion that the nucleus contains the hereditary information.

COMPREHENSION QUESTIONS

Answers to questions and problems preceded by an asterisk can be found at the end of the book.

Section 1.1

*1. How did Hopi culture contribute to the high incidence of albinism among members of the Hopi tribe?

2. Outline some of the ways in which genetics is important to all of us.

3. Give at least three examples of the role of genetics in society today.

4. Briefly explain why genetics is crucial to modern biology.

5. List the three traditional subdisciplines of genetics and summarize what each covers.

6. What are some characteristics of model genetic organisms that make them useful for genetic studies?

Section 1.2

7. When and where did agriculture first arise? What role did genetics play in the development of the first domesticated plants and animals?

8. Outline the concept of pangenesis and explain how it differs from the present-day germ-plasm theory.

9. What does the concept of the inheritance of acquired characteristics propose and how is it related to the concept of pangenesis?

10. What is preformationism? What did it have to say about how traits are inherited?

11. Define blending inheritance and contrast it with preformationism.

12. How did developments in botany in the seventeenth and eighteenth centuries contribute to the rise of modern genetics?

13. List some advances in genetics made in the twentieth century.

14. Briefly explain the contribution that each of the following people made to the study of genetics.

  a. Matthias Schleiden and Theodor Schwann

  b. August Weismann

  c. Gregor Mendel

  d. James Watson and Francis Crick

  e. Kary Mullis

Section 1.3

15. What are the two basic cell types (from a structural perspective) and how do they differ?

16. Summarize the relations between genes, DNA, and chromosomes.

APPLICATION QUESTIONS AND PROBLEMS

Section 1.1

*17. How are genetics and evolution related?

*18. For each of the following genetic topics, indicate whether it focuses on transmission genetics, molecular genetics, or population genetics.

   a. Analysis of pedigrees to determine the probability of someone inheriting a trait

   b. Study of people on a small island to determine why a genetic form of asthma is prevalent on the island

   c. Effect of nonrandom mating on the distribution of genotypes among a group of animals

   d. Examination of the nucleotide sequences found at the ends of chromosomes

   e. Mechanisms that ensure a high degree of accuracy in DNA replication

   f. Study of how the inheritance of traits encoded by genes on sex chromosomes (sex-linked traits) differs from the inheritance of traits encoded by genes on nonsex chromosomes (autosomal traits)

19. How does the picture in **Figure 1.6a** illustrate transmission genetics?

20. Describe some of the ways in which your own genetic makeup affects you as a person. Be as specific as you can.

21. Describe at least one trait that appears to run in your family (appears in multiple members of the family). Does this trait run in your family because it is an inherited trait or because it is caused by environmental factors that are common to family members? How might you distinguish between these possibilities?

Section 1.2

*22. Genetics is said to be both a very old science and a very young science. Explain what is meant by this statement.

*23. Match each of the descriptions (*a* through *d*) with the correct theory or concept listed below.

   Preformationism

   Pangenesis

   Germ-plasm theory

   Inheritance of acquired characteristics

   a. Each reproductive cell contains a complete set of genetic information.

   b. All traits are inherited from one parent.

   c. Genetic information may be altered by the use of a characteristic.

   d. Cells of different tissues contain different genetic information.

24. Briefly explain why each of the following theories is incorrect:

   a. Pangenesis

   b. Preformationism

   c. Blending inheritance

   d. Inheritance of acquired characteristics

Section 1.3

*25. Compare and contrast the following terms:

   a. Eukaryotic and prokaryotic cells

   b. Gene and allele

   c. Genotype and phenotype

   d. DNA and RNA

   e. DNA and chromosome

CHALLENGE QUESTIONS

Introduction

*26. The type of albinism that arises with high frequency among Hopi Native Americans (discussed in the introduction to this chapter) is most likely oculocutaneous albinism type 2, due to a defect in the *OCA2* gene on chromosome 15. Do some research on the Internet to determine how the phenotype of this type of albinism differs from the phenotypes of other forms of albinism in humans and to identify the mutated genes that result in those phenotypes. Hint: Visit the Online Mendelian Inheritance in Man Website (http://www.ncbi.nlm.nih.gov/omim/) and search the database for albinism.

Section 1.1

27. We now know a great deal about the genetics of humans, and humans are the focus of many genetic studies. What are some of the reasons humans have been the focus of intensive genetic study?

Section 1.3

*28. Suppose that life exists elsewhere in the universe. All life must contain some type of genetic information, but alien genomes might not consist of nucleic acids and have the same features as those found in the genomes of life on Earth. What might be the common features of all genomes, no matter where they exist?

29. Choose one of the ethical or social issues in parts *a* through *e* and give your opinion on the issue. For background information, you might read one of the articles on ethics marked with an asterisk in the Suggested Readings section for Chapter 1 in your Sapling Plus.

a. Should a person's genetic makeup be used in determining his or her eligibility for life insurance?

b. Should biotechnology companies be able to patent newly sequenced genes?

c. Should gene therapy be used in people?

d. Should genetic testing be made available for inherited disorders for which there is no treatment or cure?

30. A 45-year-old woman undergoes genetic testing and discovers that she is at high risk for developing colon cancer and Alzheimer disease. Because her children have 50% of her genes, they may also be at an increased risk for these diseases. Does she have a moral or legal obligation to tell her children and other close relatives about the results of her genetic testing?

31. Suppose that you could undergo genetic testing at age 18 for susceptibility to a genetic disease that would not appear until middle age and has no available treatment.

a. What would be some of the possible reasons for having such a genetic test and some of the possible reasons for not having the test?

b. Would you personally want to be tested? Explain your reasoning.

THINK-PAIR-SHARE QUESTIONS

Think-Pair-Share questions are designed to be worked in collaboration with other students. First THINK about the question, then PAIR up with one or more other students, and finally SHARE your answers and work together to arrive at a solution.

Section 1.1

1. Bob says that he is healthy and has no genetic diseases such as hemophilia or Down syndrome. Therefore, he says, genetics plays little role in his life. Do you think Bob is correct in his conclusion? Why or why not?

2. Are mutations good or bad? Explain your answer.

Section 1.2

3. Do you support or oppose the development of genetically engineered foods (genetically modified organisms, or GMOs)? Find someone who takes the opposite position and discuss this question with them. Think about the economic and environmental benefits, health risks, ecological effects, and social impact of their use. List some reasons for and against genetically engineering the foods we eat.

Section 1.3

4. Why do you think all organisms use nucleic acids for encoding genetic information? Why not use proteins or carbohydrates? What advantages might DNA have as the source of genetic information?

Chromosomes and Cellular Reproduction

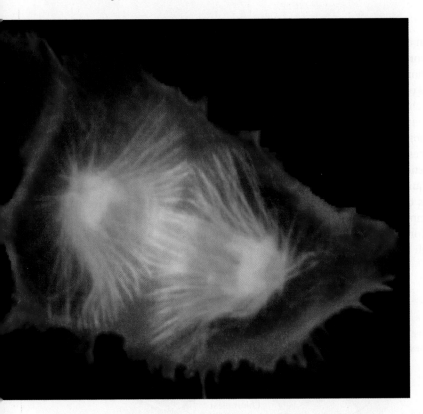

A rat kangaroo kidney cell undergoing mitosis, the process through which each new cell receives a complete copy of the genetic material. Chromosomes are shown in blue. [Courtesy of Julie Canman and Ted Salman.]

The Blind Men's Riddle

In a well-known riddle, two blind men by chance enter a department store at the same time, go to the same counter, and both order five pairs of socks, each pair a different color. The salesclerk is so befuddled by this strange coincidence that he places all ten pairs (two black pairs, two blue pairs, two gray pairs, two brown pairs, and two green pairs) into a single shopping bag, gives the bag of socks to one blind man, and gives an empty bag to the other. The two blind men happen to meet on the street outside, where they discover that one of their bags contains all ten pairs of socks. How do the blind men, without seeing and without any outside help, sort out the socks so that each man goes home with exactly five pairs of different colored socks? Can you come up with a solution to the riddle?

By an interesting coincidence, cells face the same challenge as that of the blind men. Most organisms possess two sets of genetic information, one set inherited from each parent. Before cell division, the DNA in each chromosome replicates; after replication, there are two copies—called sister chromatids—of each chromosome. At the end of cell division, it is critical that each of the two new cells receives a complete copy of the genetic material, just as each blind man needs to go home with a complete set of socks.

The solution to the riddle is simple. Socks are sold as pairs; the two socks of a pair are typically connected by a thread. As a pair is removed from the bag, the men each grasp a different sock of the pair and pull in opposite directions. When the socks are pulled tight, one of the men can take a pocket knife and cut the thread connecting the pair. Each man then deposits his single sock in his own bag. At the end of the process, each man's bag will contain exactly two black socks, two blue socks, two gray socks, two brown socks, and two green socks.*

Remarkably, cells employ a similar solution for separating their chromosomes into new daughter cells. As we learn in this chapter, the replicated chromosomes line up at the center of a cell undergoing division, and like the socks in the riddle, the sister chromatids of each chromosome are pulled in opposite directions. Like the thread connecting two socks of a pair, a molecule called cohesin holds the sister chromatids together until it is severed by a molecular knife called separase. The two resulting chromosomes separate, and the cell divides. This process ensures that a complete set of chromosomes is deposited in each cell.

In this analogy, the blind men and cells differ in one critical regard: if the blind men make a mistake, one man ends up with an extra sock and the other is a sock short, but no great

* This analogy is adapted from K. Nasmyth, *Annual Review of Genetics* 35:673–745, 2001.

harm results. The same cannot be said for human cells. Errors in chromosome separation, producing cells with too many or too few chromosomes, are frequently catastrophic, leading to cancer, miscarriage, or—in some cases—a child with severe disabilities.

THINK-PAIR-SHARE
- In the blind men's riddle, two blind men must sort out ten pairs of socks so that each man gets exactly five pairs of different colored socks. In the analogy, is it important that the men are blind? In a cell, what does the blindness represent?

This chapter explores the process of cell reproduction and explains how a complete set of genetic information is transmitted to new cells. In prokaryotic cells, reproduction is relatively simple because prokaryotic cells tend to possess only one chromosome per cell. In eukaryotic cells, multiple chromosomes must be copied and distributed to each of the new cells, making cell reproduction more complex. Cell division in eukaryotes takes place through mitosis or meiosis, processes that serve as the foundation for much of eukaryotic genetics.

Grasping the processes of mitosis and meiosis requires more than simply memorizing the sequences of events that take place in each stage, although these events are important. The key is to understand how genetic information is apportioned in the course of cell reproduction through the dynamic interplay of DNA synthesis, chromosome movement, and cell division. These processes also underlie the transmission of genetic information from parent to progeny and are the basis of the similarities and differences between parents and their offspring.

2.1 Prokaryotic and Eukaryotic Cells Differ in a Number of Genetic Characteristics

Biologists have traditionally classified all living organisms into two major groups, the prokaryotes and the eukaryotes (**Figure 2.1**). A **prokaryote** is a unicellular organism with a relatively simple cell structure. A **eukaryote** has a

Prokaryote

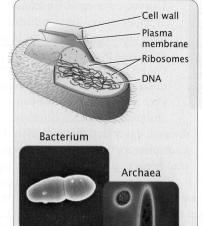

Eukaryote

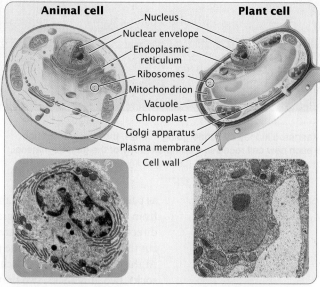

	Prokaryotic cells	**Eukaryotic cells**
Nucleus	Absent	Present
Cell diameter	Relatively small, from 1 to 10 μm	Relatively large, from 10 to 100 μm
Genome	Usually one circular DNA molecule	Multiple linear DNA molecules
DNA	Not complexed with histones in bacteria; some histones in archaea	Complexed with histones
Amount of DNA	Relatively small	Relatively large
Membrane-bounded organelles	Absent	Present

2.1 Prokaryotic and eukaryotic cells differ in structure. [Photographs (left to right): Dr. Gary D. Gaugler/Newscom; Dr. Kari Lounatmaa/Science Source; Dr. Gopal Murti/Phototake; Biophoto Associates/Science Source.]

compartmentalized cell structure with components bounded by intracellular membranes; eukaryotes may be either unicellular or multicellular.

Research indicates, however, that the division of life is not so simple. Although all prokaryotes are similar in cell structure, they include at least two fundamentally distinct types: the **bacteria** (also called eubacteria or "true bacteria") and the **archaea** ("ancient bacteria"). An examination of equivalent DNA sequences reveals that bacteria and archaea are as distantly related to each other as they are to the eukaryotes. Thus, from an evolutionary perspective, there are three major groups of organisms: bacteria, archaea, and eukaryotes. Although bacteria and archaea are similar in cell structure, some genetic processes in archaea (such as transcription) are more similar to those in eukaryotes, and the archaea may be closer evolutionarily to eukaryotes than to bacteria. Some researchers propose that eukaryotes arose within the archaea, and thus that there are two primary domains of life: the bacteria and another domain that includes the archaea and the eukaryotes. The evolutionary relationships among bacteria, archaea, and eukaryotes remain uncertain, however, and are the focus of current research. In this book, the prokaryotic–eukaryotic distinction will be made frequently, but important bacterial–archaeal differences will also be noted.

From the perspective of genetics, a major difference between prokaryotic and eukaryotic cells is that a eukaryotic cell has a nuclear envelope, which surrounds the genetic material to form a **nucleus** and separates the DNA from the other cellular contents. In prokaryotic cells, the genetic material is in close contact with other components of the cell—a property that has important consequences for the way in which gene expression is controlled (**Figure 2.2**).

Another fundamental difference between prokaryotes and eukaryotes lies in the packaging of their DNA. In eukaryotes, DNA associates closely with a special class of proteins, the **histones**, to form tightly packed chromosomes (**Figure 2.3**). This complex of DNA and histone proteins, called **chromatin**, is the stuff of eukaryotic chromosomes. Histone proteins help regulate the accessibility of DNA to enzymes and other proteins that copy and read the DNA,

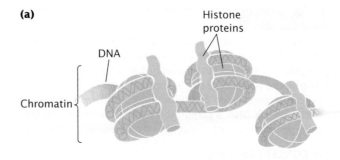

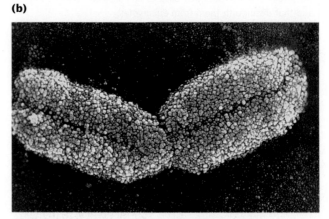

2.3 Eukaryotic chromosomes consist of DNA and histone proteins. (a) DNA wraps around the histone proteins to form chromatin, the material that makes up chromosomes. (b) A eukaryotic chromosome. [Part b: Biophoto Associates/Science Source.]

and they also enable the DNA to fit into the nucleus. Eukaryotic DNA must separate from the histones before the genetic information in the DNA can be read. Archaea also have some histone proteins that complex with DNA. The histones found in archaea are related to but distinct from those found in eukaryotes, and the structure of archaeal chromatin is different from that found in eukaryotes. Bacteria do not possess histones, and their DNA does not exist in the highly ordered, tightly packed arrangement found in eukaryotic cells. The copying and reading of DNA are therefore simpler processes in bacteria.

The genes of prokaryotic cells are generally located on a single circular molecule of double-stranded DNA—the chromosome of a prokaryotic cell. In eukaryotic cells, genes are located on multiple, usually linear DNA molecules (multiple chromosomes). Eukaryotic cells therefore require mechanisms to ensure that a copy of each chromosome is faithfully transmitted to each new cell. However, this generalization—a single circular chromosome in prokaryotes and multiple linear chromosomes in eukaryotes—is not always true. A few bacteria have more than one chromosome, and important bacterial and archaeal genes are frequently found on other DNA molecules called *plasmids* (see Chapter 9). Furthermore, in some eukaryotes, a few genes are located on circular DNA molecules found in certain organelles, such as mitochondria and chloroplasts (see Chapter 11).

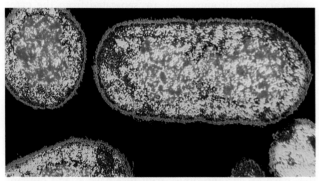

2.2 Prokaryotic DNA (shown in red) is neither surrounded by a nuclear membrane nor complexed with histone proteins.
[A. Barry Dowsett/Science Source.]

Organisms are classified as prokaryotes or eukaryotes, and the prokaryotes consist of archaea and bacteria. A prokaryote is a unicellular organism that lacks a nucleus, and its genome is usually a single chromosome. Eukaryotes may be either unicellular or multicellular, their cells possess a nucleus, their DNA is complexed with histone proteins, and their genomes consist of multiple chromosomes.

✔ CONCEPT CHECK 1

List several characteristics that bacteria and archaea have in common and that distinguish them from eukaryotes.

Viruses are neither prokaryotic nor eukaryotic because they do not possess the structure of a cell. Viruses are actually simple structures composed of an outer protein coat surrounding nucleic acid (either DNA or RNA; **Figure 2.4**). All known viruses can reproduce only within host cells, and their evolutionary relationship to cellular organisms is uncertain. Their simplicity and small genomes make viruses

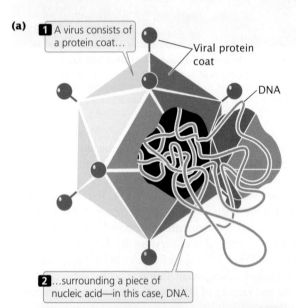

(a)

1 A virus consists of a protein coat…

Viral protein coat

DNA

2 …surrounding a piece of nucleic acid—in this case, DNA.

(b)

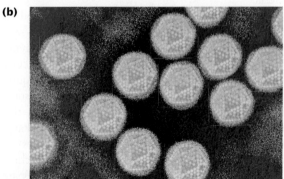

2.4 A virus is a simple replicative structure consisting of protein and nucleic acid. (a) Structure of a virus. (b) A micrograph of adenoviruses. [Part b: BSIP/Science Source.]

useful for studying some molecular processes and for some types of genetic analyses.

2.2 Cell Reproduction Requires the Copying of the Genetic Material, Separation of the Copies, and Cell Division

For any cell to reproduce successfully, three fundamental events must take place: (1) its genetic information must be copied, (2) the copies must be separated from each other, and (3) the cell must divide. All cellular reproduction includes these three events, but the processes that lead to these events differ in prokaryotic and eukaryotic cells because of their structural differences.

Prokaryotic Cell Reproduction by Binary Fission

When a prokaryotic cell reproduces, its circular chromosome replicates, and the cell divides in a process called *binary fission* (**Figure 2.5**). Replication usually begins at a specific place on the circular chromosome, called the origin of replication. In a process that is not well understood, the origins of the two newly replicated chromosomes move away from each other and toward opposite ends of the cell. In at least some prokaryotes, proteins bind near the origins and anchor the new chromosomes to the plasma membrane at opposite ends of the cell. Finally, a new cell wall forms between the two chromosomes, producing two cells, each with an identical copy of the chromosome. Under optimal conditions, some bacterial cells divide every 20 minutes. At this rate, a single bacterial cell could produce a billion descendants in a mere 10 hours.

Eukaryotic Cell Reproduction

Like prokaryotic cell reproduction, eukaryotic cell reproduction requires the processes of DNA replication, copy separation, and division of the cytoplasm. However, the presence of multiple DNA molecules requires a more complex mechanism to ensure that exactly one copy of each DNA molecule ends up in each of the new cells.

Eukaryotic chromosomes are separated from the cytoplasm by the nuclear envelope. The nucleus has a highly organized internal scaffolding, called the nuclear matrix, that consists of a network of protein fibers. The nuclear matrix maintains precise spatial relations among the components of the nucleus and takes part in DNA replication, the expression of genes, and the modification of gene products before they leave the nucleus.

EUKARYOTIC CHROMOSOMES Each eukaryotic species has a characteristic number of chromosomes per cell: potatoes have 48 chromosomes, fruit flies have 8, and humans

(a)

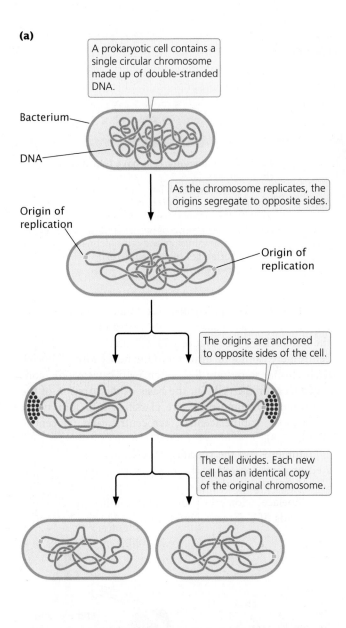

A prokaryotic cell contains a single circular chromosome made up of double-stranded DNA.

Bacterium

DNA

Origin of replication

As the chromosome replicates, the origins segregate to opposite sides.

Origin of replication

The origins are anchored to opposite sides of the cell.

The cell divides. Each new cell has an identical copy of the original chromosome.

(b)

2.5 Prokaryotic cells reproduce by binary fission. (a) The process of binary fission. (b) A bacterial cell undergoing binary fission. [Part b: Lee D. Simon/Science Source.]

have 46. There appears to be no special relation between the complexity of an organism and its number of chromosomes per cell.

In most eukaryotic cells, there are two sets of chromosomes. The presence of two sets is a consequence of sexual reproduction: one set is inherited from the male parent and the other from the female parent. Each chromosome in one set has a corresponding chromosome in the other set; together, the two chromosomes constitute a **homologous pair** (**Figure 2.6**). Human cells, for example, have 46 chromosomes, constituting 23 homologous pairs.

The two chromosomes of a homologous pair are usually alike in structure and size, and each carries genetic information for the same set of hereditary characteristics (the sex chromosomes are an exception and will be discussed in Chapter 4). For example, if a gene on a particular chromosome encodes a characteristic such as hair color, another copy of the gene (each copy is called an *allele*) at the same position on that chromosome's homolog *also* encodes hair color. However, these two alleles need not be identical: one might encode brown hair and the other might encode blond hair.

Cells that carry two sets of genetic information are **diploid**. In general, the ploidy of the cell indicates how many sets of genetic information it possesses. The reproductive cells of eukaryotes (such as eggs, sperm, and spores), and even the nonreproductive cells of some eukaryotic organisms, contain a single set of chromosomes and are **haploid**. The cells of some other eukaryotes contain more than two sets of genetic information and are therefore called **polyploid** (see Chapter 8).

CONCEPTS

Cells reproduce by copying their genetic information, separating the copies, and then dividing. Because eukaryotic cells possess multiple chromosomes, mechanisms exist to ensure that each new cell receives one copy of each chromosome. Most eukaryotic cells are diploid, and their two chromosome sets can be arranged in homologous pairs. Haploid cells contain a single set of chromosomes.

✔ **CONCEPT CHECK 2**

Diploid cells have
a. two chromosomes.
b. two sets of chromosomes.
c. one set of chromosomes.
d. two pairs of homologous chromosomes.

CHROMOSOME STRUCTURE The chromosomes of eukaryotic cells are larger and more complex than those found in prokaryotes, but each unreplicated chromosome nevertheless consists of a single molecule of DNA. Although linear, the DNA molecules in eukaryotic chromosomes are highly folded and condensed; if stretched out, some human chromosomes would be several centimeters long—thousands of

(a)

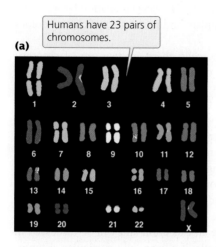

Humans have 23 pairs of chromosomes.

1 2 3 4 5
6 7 8 9 10 11 12
13 14 15 16 17 18
19 20 21 22 X

(b)

A *diploid* organism has two sets of chromosomes organized as *homologous* pairs.

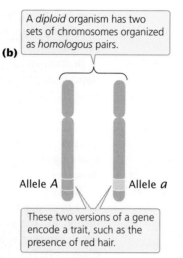

Allele *A*   Allele *a*

These two versions of a gene encode a trait, such as the presence of red hair.

2.6 Diploid eukaryotic cells have two sets of chromosomes. (a) A set of chromosomes from a female human cell. Each pair of chromosomes has been hybridized to a uniquely colored probe, giving it a distinct color. (b) The chromosomes are present in homologous pairs. Each pair consists of two chromosomes that are alike in size and structure and carry information for the same characteristics. [Part a: Courtesy of Dr. Thomas Ried and Dr. Evelin Schrock.]

times as long as the span of a typical nucleus. To package such a tremendous length of DNA into the small volume of the nucleus, each DNA molecule is coiled around histone proteins and tightly packed, forming a rod-shaped chromosome. Most of the time, the chromosomes are thin and difficult to observe, but before cell division, they condense further into thick, readily observed structures; it is at this stage that chromosomes are usually studied.

A functional chromosome has three essential elements: a centromere, a pair of telomeres, and origins of replication. The **centromere** appears as a constricted region on the chromosome (**Figure 2.7**). It serves as the attachment point for spindle microtubules—the filaments responsible for moving chromosomes in cell division. Before cell division,

a multiprotein complex called the kinetochore assembles on the centromere; later, spindle microtubules attach to the kinetochore (see Figure 2.7). Chromosomes lacking a centromere cannot be drawn into the newly formed nuclei; such chromosomes are lost, often with catastrophic consequences for the cell. On the basis of the location of the centromere, chromosomes are classified into four types: metacentric, submetacentric, acrocentric, and telocentric (**Figure 2.8**).

Telomeres are the specific DNA sequences and associated proteins located at the tips of whole linear chromosomes (see Figure 2.7). Just as plastic tips protect the ends of a shoelace, telomeres protect and stabilize the chromosome ends. If a chromosome breaks, producing new ends,

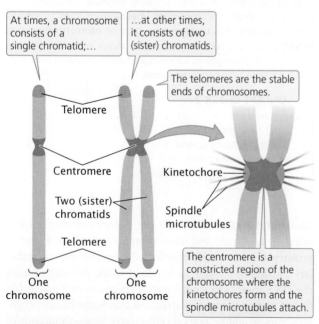

At times, a chromosome consists of a single chromatid;…

…at other times, it consists of two (sister) chromatids.

The telomeres are the stable ends of chromosomes.

Telomere

Centromere   Kinetochore

Two (sister) chromatids   Spindle microtubules

Telomere

One chromosome   One chromosome

The centromere is a constricted region of the chromosome where the kinetochores form and the spindle microtubules attach.

2.7 Each eukaryotic chromosome has a centromere and telomeres.

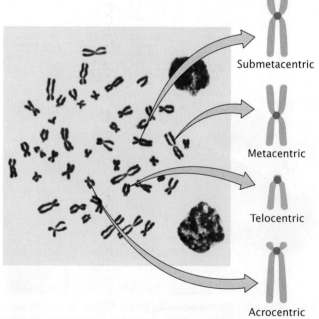

Submetacentric

Metacentric

Telocentric

Acrocentric

2.8 Eukaryotic chromosomes can be divided into in four major types based on the position of the centromere. [Micrograph by Don W. Fawcett/Science Source.]

the chromosome is degraded at the newly broken ends. Telomeres provide chromosome stability. Research shows that telomeres also participate in limiting cell division and may play important roles in aging and cancer (discussed in Chapter 12).

Origins of replication are the sites where DNA synthesis begins; unlike centromeres and telomeres, they are not easily observed by microscopy. Their structure and function will be discussed in more detail in Chapter 12.

In preparation for cell division, each chromosome replicates, making a copy of itself, as already mentioned. These two initially identical copies, called **sister chromatids**, are held together at the centromere (see Figure 2.7). Each sister chromatid consists of a single molecule of DNA.

THINK-PAIR-SHARE Question 1

CONCEPTS

Functional chromosomes contain a centromere, telomeres, and origins of replication. The kinetochore is the multi-protein point of attachment for the spindle microtubules. Telomeres are the stabilizing ends of a chromosome. Origins of replication are sites where DNA synthesis begins. Sister chromatids are copies of a chromosome held together at the centromere.

✔ CONCEPT CHECK 3

What would be the result if a chromosome did not have a kinetochore?

The Cell Cycle and Mitosis

The **cell cycle** is the series of stages through which a cell passes from one division to the next (**Figure 2.9**). It is through the cell cycle that the genetic instructions for all characteristics are accurately passed from parent to daughter cells.

The cell cycle takes place in cells that are actively dividing. A new cycle begins after a cell has divided and produced two new cells. Each new cell grows, develops, and carries out the functions specific to its cell type. At the end of the cycle, the cell divides to produce two cells, which can then undergo additional cell cycles. Progression through the cell cycle is regulated at key transition points called **checkpoints**, which allow or prohibit the cell's progression to the next stage. Checkpoints ensure that all cellular components are present and in good working order, and checkpoints are necessary to prevent cells with damaged or missing chromosomes from proliferating. Defects in checkpoints can lead to unregulated cell growth, as is seen in some cancers. The molecular basis of these checkpoints will be discussed in Chapter 23.

The cell cycle consists of two major phases. The first is **interphase**, the period between cell divisions, in which the cell grows, develops, and functions. In interphase, critical events necessary for cell division also take place. The second major phase is the **M (mitotic) phase**, the period of active cell division. The M phase includes **mitosis**, the process of nuclear division, and **cytokinesis**, or cytoplasmic division. Let's take a closer look at the details of interphase and the M phase.

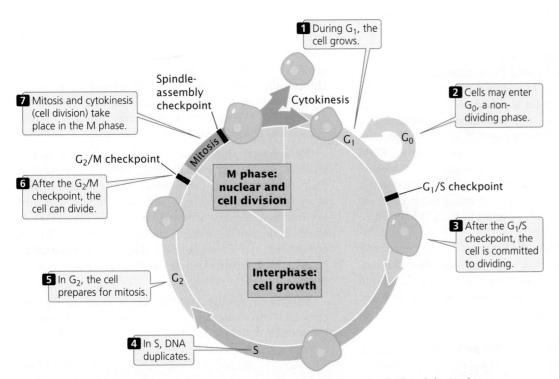

2.9 The cell cycle consists of interphase (divided into three stages: G₁, S, and G₂) and the M phase.

INTERPHASE Interphase is the extended period of growth and development between cell divisions. Although little activity can be observed with a light microscope, the cell is quite busy: DNA is being synthesized, RNA and proteins are being produced, and hundreds of biochemical reactions necessary for cellular functions are taking place. In addition to growth and development, interphase includes several checkpoints.

By convention, interphase is divided into three stages: G_1, S, and G_2 (see Figure 2.9). Interphase begins with G_1 (for gap 1). In G_1, the cell grows and proteins necessary for cell division are synthesized; this stage typically lasts several hours. Near the end of G_1, a critical checkpoint, termed the G_1/S checkpoint, holds the cell in G_1 until the cell has all of the enzymes and proteins necessary for the replication of DNA. After this checkpoint has been passed, the cell is committed to divide.

Before reaching the G_1/S checkpoint, cells may exit the active cell cycle in response to regulatory signals and pass into a nondividing phase called G_0, a stable state during which cells usually maintain a constant size. They can remain in G_0 for an extended time, even indefinitely, or they can reenter G_1 and the active cell cycle. Many cells never enter G_0; rather, they cycle continuously. On the other hand, many cells spend most of their life span in G_0.

After G_1, the cell enters the S phase (for DNA synthesis), in which each chromosome is duplicated. Although the cell is committed to divide after the G_1/S checkpoint has been passed, DNA synthesis must take place before the cell can proceed to mitosis. If DNA synthesis is blocked (by drugs or by a mutation), the cell will not normally be able to undergo mitosis. Before the S phase, each chromosome is unreplicated; after the S phase, each chromosome is composed of two sister chromatids (see Figure 2.7).

After the S phase, the cell enters G_2 (gap 2). In this stage, several additional biochemical events necessary for cell division take place. The G_2/M checkpoint is reached near the end of G_2. This checkpoint is passed only if the cell's DNA is completely replicated and undamaged. Unreplicated or damaged DNA can inhibit the activation of some proteins that are necessary for mitosis to take place. After the G_2/M checkpoint has been passed, the cell is ready to divide and enters the M phase. Although the length of interphase varies from cell type to cell type, a typical dividing mammalian cell spends about 10 hours in G_1, 9 hours in S, and 4 hours in G_2 (see Figure 2.9).

Throughout interphase, the chromosomes are in a relaxed, but by no means uncoiled, state, and individual chromosomes cannot be seen with a microscope. This condition changes dramatically when interphase draws to a close and the cell enters the M phase.

THE M PHASE The M phase is the part of the cell cycle in which the copies of the cell's chromosomes (sister chromatids) separate and the cell undergoes division. The separation of sister chromatids in the M phase is a critical process that results in a complete set of genetic information for each of the resulting cells. Biologists usually divide the M phase into six stages: the five stages of mitosis (prophase, prometaphase, metaphase, anaphase, and telophase), illustrated in **Figure 2.10**, and cytokinesis. It's important to keep in mind that the M phase is a continuous process and that its separation into these six stages is somewhat arbitrary.

Prophase. As a cell enters prophase, the chromosomes condense, becoming more compact and visible under a light microscope. A group of proteins called **condensins** bind to the chromosomes and bring about condensation. Because the chromosomes were duplicated in the preceding S phase, each chromosome consists of two sister chromatids attached at the centromere. The mitotic spindle, an organized array of microtubules that move the chromosomes in mitosis, forms. In animal cells, the spindle grows out from a pair of centrosomes that migrate to the opposite sides of the cell. Within each centrosome is a special organelle, the centriole, which is also composed of microtubules. Some plant cells do not have centrosomes or centrioles, but they do have mitotic spindles.

Prometaphase. Disintegration of the nuclear membrane marks the start of prometaphase. Spindle microtubules, which until now have been outside the nucleus, enter the nuclear region. The spindle microtubules are composed of subunits of a protein called tubulin (**Figure 2.11**). During prometaphase, tubulin molecules are added to and removed from the microtubules, causing them to undergo repeated cycles of growth and shrinkage. When the end of a microtubule encounters a kinetochore, the microtubule becomes stabilized. Eventually, each chromosome becomes attached to microtubules from opposite spindle poles: for each chromosome, a microtubule from one of the centrosomes anchors to the kinetochore of *one* of the sister chromatids; a microtubule from the opposite centrosome then attaches to the *other* sister chromatid, anchoring the chromosome to both of the centrosomes. This arrangement is known as chromosome bi-orientation.

Metaphase. During metaphase, the chromosomes become arranged in a single plane, called the metaphase plate, between the two centrosomes. The centrosomes, now at opposite ends of the cell, with microtubules radiating outward from each one and meeting in the middle of the cell, center at the spindle poles. A spindle-assembly checkpoint ensures that each chromosome is aligned on the metaphase plate and attached to spindle microtubules from opposite poles.

Tension is generated at the kinetochore as the two conjoined chromatids are pulled in opposite directions by the spindle microtubules. This tension is required for the cell to pass through the spindle-assembly checkpoint. If one chromatid is attached to a microtubule but the other is not, no tension is generated, and the cell is unable to progress to the next stage of mitosis. The spindle-assembly checkpoint is

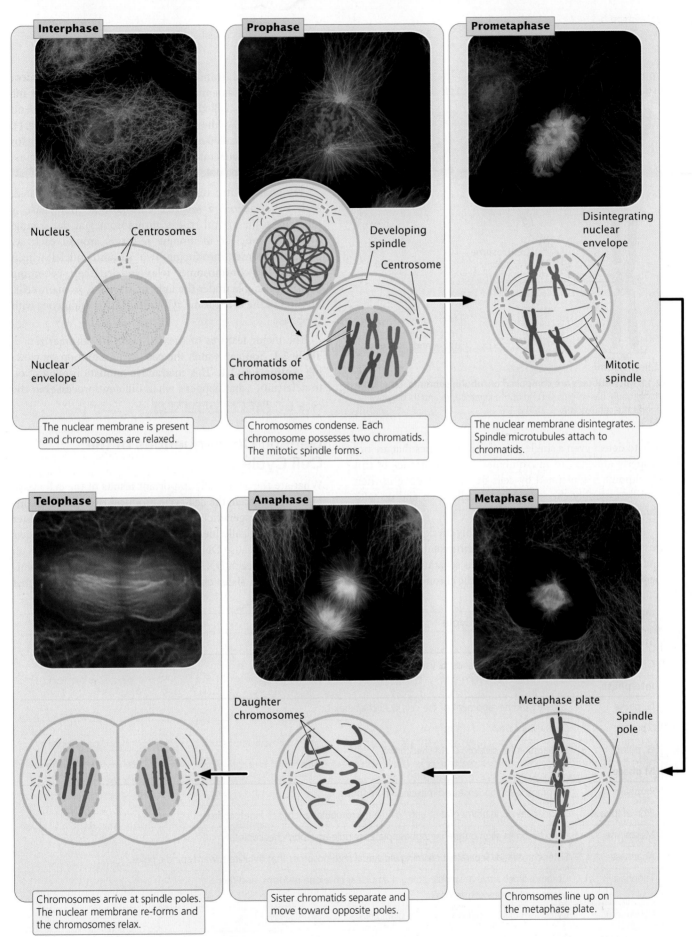

Interphase

Nucleus
Centrosomes
Nuclear envelope

The nuclear membrane is present and chromosomes are relaxed.

Prophase

Developing spindle
Centrosome
Chromatids of a chromosome

Chromosomes condense. Each chromosome possesses two chromatids. The mitotic spindle forms.

Prometaphase

Disintegrating nuclear envelope
Mitotic spindle

The nuclear membrane disintegrates. Spindle microtubules attach to chromatids.

Telophase

Chromosomes arrive at spindle poles. The nuclear membrane re-forms and the chromosomes relax.

Anaphase

Daughter chromosomes

Sister chromatids separate and move toward opposite poles.

Metaphase

Metaphase plate
Spindle pole

Chromsomes line up on the metaphase plate.

2.10 Mitosis is divided into six stages. [Photographs (top row, bottom row center, bottom row right): Dr. Torsten Wittman/Science Source; (bottom row left): Jennifer Waters/Science Source.]

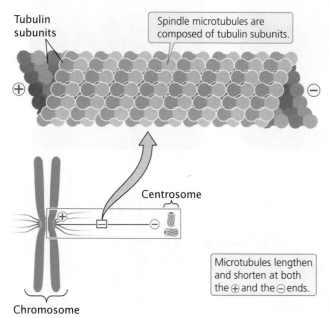

Tubulin subunits

Spindle microtubules are composed of tubulin subunits.

Centrosome

Microtubules lengthen and shorten at both the ⊕ and the ⊖ ends.

Chromosome

2.11 Microtubules are composed of tubulin subunits. Each microtubule has its plus (+) end at the kinetochore and its negative (−) end at the centrosome.

able to detect even a single pair of chromosomes that are not properly attached to microtubules. The importance of this checkpoint is illustrated by cells that are defective in their spindle-assembly checkpoint; these cells often end up with abnormal numbers of chromosomes.

Anaphase. After the spindle-assembly checkpoint has been passed, the connection between sister chromatids breaks down and the sister chromatids separate. This separation marks the beginning of anaphase, during which the chromosomes move toward opposite spindle poles. Chromosome movement is due to the disassembly of tubulin molecules at both the kinetochore end (called the + end) and the spindle pole end (called the − end) of the spindle microtubule (see Figure 2.11). Special proteins called molecular motors disassemble tubulin molecules and generate forces that pull the chromosome toward the spindle pole.

Telophase. After the sister chromatids have separated, each is considered a separate chromosome. Telophase is marked by the arrival of the chromosomes at the spindle poles. The nuclear membrane re-forms around each set of chromosomes, producing two separate nuclei within the cell. The chromosomes relax and lengthen, becoming indistinguishable under the light microscope. In many cells, division of the cytoplasm (cytokinesis) is simultaneous with telophase.

The major features of the cell cycle are summarized in **Table 2.1**. You can watch the cell cycle in motion by viewing **Animation 2.1**. This interactive animation allows you to determine what happens when different processes in the cycle fail. **TRY PROBLEM 24**

Genetic Consequences of the Cell Cycle

What are the genetically important results of the cell cycle? From a single cell, the cell cycle produces two cells that contain the same genetic instructions. The resulting daughter cells are genetically identical with each other and with their parent cell because DNA synthesis in the S phase creates an exact copy of each DNA molecule, giving rise to two genetically identical sister chromatids. Mitosis then ensures that

TABLE 2.1	Features of the cell cycle
Stage	**Major Features**
G_0 phase	Stable, nondividing period of variable length.
Interphase	
G_1 phase	Growth and development of the cell; G_1/S checkpoint.
S phase	Synthesis of DNA.
G_2 phase	Preparation for division; G_2/M checkpoint.
M phase	
Prophase	Chromosomes condense and mitotic spindle forms.
Prometaphase	Nuclear envelope disintegrates, and spindle microtubules anchor to kinetochores.
Metaphase	Chromosomes align on the metaphase plate; spindle-assembly checkpoint.
Anaphase	Sister chromatids separate, becoming individual chromosomes that migrate toward spindle poles.
Telophase	Chromosomes arrive at spindle poles, the nuclear envelope re-forms, and the condensed chromosomes relax.
Cytokinesis	Cytoplasm divides; cell wall forms in plant cells.

one of the two sister chromatids from each replicated chromosome passes into each new cell.

Another genetically important result of the cell cycle is that each of the cells produced contains a full complement of chromosomes: there is no net reduction or increase in chromosome number. Each cell also contains approximately half the cytoplasm and organelle content of the original parent cell, but no precise mechanism analogous to mitosis ensures that organelles are evenly divided. Consequently, not all cells resulting from the cell cycle are identical in their cytoplasmic content.

THINK-PAIR-SHARE Question 2

CONCEPTS

The active cell cycle phases are interphase and the M phase. Interphase consists of G_1, S, and G_2. In G_1, the cell grows and prepares for cell division; in the S phase, DNA synthesis takes place; in G_2, other biochemical events necessary for cell division take place. Some cells enter a quiescent phase called G_0. The M phase includes mitosis and cytokinesis and is divided into prophase, prometaphase, metaphase, anaphase, and telophase. The cell cycle produces two genetically identical cells, each of which possesses a full complement of chromosomes.

✔ CONCEPT CHECK 4

Which is the correct order of stages in the cell cycle?

a. G_1, S, G_2, prophase, metaphase, anaphase
b. S, G_1, G_2, prophase, metaphase, anaphase
c. Prophase, S, G_1, G_2, metaphase, anaphase
d. S, G_1, G_2, anaphase, prophase, metaphase

CONNECTING CONCEPTS

Counting Chromosomes and DNA Molecules

The relations among chromosomes, chromatids, and DNA molecules frequently cause confusion. At certain times, chromosomes are unreplicated; at other times, each has undergone replication and possesses two chromatids (see Figure 2.7). Chromosomes sometimes consist of a single DNA molecule; at other times, they consist of two DNA molecules. How can we keep track of the number of these structures in the cell cycle?

There are two simple rules for counting chromosomes and DNA molecules: (1) to determine the number of chromosomes, count the number of functional centromeres; (2) to determine the number of DNA molecules, first determine if sister chromatids are present. If sister chromatids are present, the chromosome has replicated, and the number of DNA molecules is twice the number of chromosomes. If sister chromatids are not present, the chromosome has not replicated, and the number of DNA molecules is the same as the number of chromosomes.

Let's examine a hypothetical cell as it passes through the cell cycle (**Figure 2.12**). At the beginning of G_1, this diploid cell has two complete sets of chromosomes, for a total of four chromosomes. Each chromosome is unreplicated and consists of a single molecule of DNA, so there are four DNA molecules in the cell during G_1. In the S phase, each DNA molecule is copied. The two resulting DNA molecules combine with histones and other proteins to form sister chromatids. Although the amount of DNA doubles in the S phase, the number of chromosomes remains the same because the sister chromatids are tethered together and share a single functional centromere. At the end of the S phase, this cell still contains four chromosomes, each with two sister chromatids; so $4 \times 2 = 8$ DNA molecules are present.

	G_1	S	G_2	Prophase and prometaphase	Metaphase	Anaphase	Telophase and cytokinesis
Number of chromosomes per cell	4	4	4	4	4	8	4
Number of DNA molecules per cell	4	4 → 8	8	8	8	8	4

2.12 The number of chromosomes and the number of DNA molecules change in the course of the cell cycle. The number of chromosomes per cell equals the number of functional centromeres. The number of DNA molecules per cell equals the number of chromosomes when the chromosomes are unreplicated (no sister chromatids are present) and twice the number of chromosomes when sister chromatids *are* present.

Through prophase, prometaphase, and metaphase, the cell has four chromosomes and eight DNA molecules. At anaphase, however, the sister chromatids separate. Each now has its own functional centromere, so each is considered a separate chromosome. Until cytokinesis, the cell contains eight unreplicated chromosomes; thus, there are still eight DNA molecules present. After cytokinesis, the eight chromosomes (and eight DNA molecules) are distributed equally between two daughter cells, so each daughter cell contains four chromosomes and four DNA molecules—the number present at the beginning of the cell cycle.

In summary, the number of chromosomes increases only in anaphase, when the two chromatids of a chromosome separate and become distinct chromosomes. The number of chromosomes decreases only through cytokinesis. The number of DNA molecules increases only in the S phase and decreases only through cytokinesis.

▶ TRY PROBLEM 28

THINK-PAIR-SHARE Question 3

2.3 Sexual Reproduction Produces Genetic Variation Through the Process of Meiosis

If all reproduction were accomplished through mitosis, life would be quite dull because mitosis produces only genetically identical progeny. With only mitosis, you, your children, your parents, your brothers and sisters, your cousins, and many people whom you don't even know would be clones—copies of one another. Only the occasional mutation would introduce any genetic variation. All organisms reproduced in this way for the first 2 billion years of Earth's existence (and some organisms still do today). Then, about 1.5 billion to 2 billion years ago, something remarkable evolved: cells that produce genetically variable offspring through sexual reproduction.

The evolution of sexual reproduction is among the most significant events in the history of life. As will be discussed in Chapters 24–25, the pace of evolution depends on the amount of genetic variation present. By shuffling the genetic information from two parents, sexual reproduction greatly increases the amount of genetic variation and allows for accelerated evolution. Most of the tremendous diversity of life on Earth is a direct result of sexual reproduction.

Sexual reproduction consists of two processes. The first is **meiosis**, which leads to gametes in which the number of chromosomes is reduced by half. The second process is **fertilization**, in which two haploid gametes fuse and restore the number of chromosomes to its original diploid value.

Meiosis

The words *mitosis* and *meiosis* are sometimes confused. They sound a bit alike and may appear superficially similar. But don't be deceived. The outcomes of mitosis and meiosis are radically different, and several unique events that have important genetic consequences take place only in meiosis.

How does meiosis differ from mitosis? Mitosis consists of a single nuclear division and is usually accompanied by a single cell division. Meiosis, on the other hand, consists of two divisions. After mitosis, the chromosome number in the newly formed cells is the same as that in the original cell, whereas meiosis causes the chromosome number in the newly formed cells to be reduced by half. Finally, mitosis produces genetically identical cells, whereas meiosis produces genetically variable cells. Let's see how these differences arise.

Like mitosis, meiosis is preceded by an interphase that includes G_1, S, and G_2 stages. Meiosis consists of two distinct processes: meiosis I and meiosis II, each of which includes a cell division. The first division, which comes at the end of meiosis I, is termed the reduction division because the number of chromosomes per cell is reduced by half (**Figure 2.13**). The second division, which comes at the end of meiosis II, is sometimes termed the equational division. The events of meiosis II are similar to those of mitosis. However, meiosis II differs from mitosis in that chromosome number has already been halved in meiosis I, so the cell does not begin with the same number of chromosomes as it does in mitosis (see Figure 2.13).

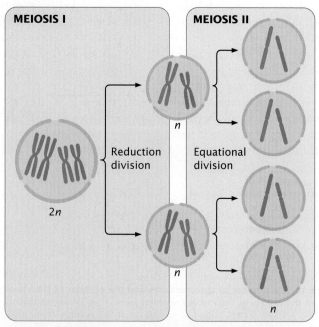

2.13 Meiosis includes two cell divisions. In this illustration, the original cell is diploid ($2n$) with four chromosomes. After two meiotic divisions, each resulting cell is haploid ($1n$) with two chromosomes.

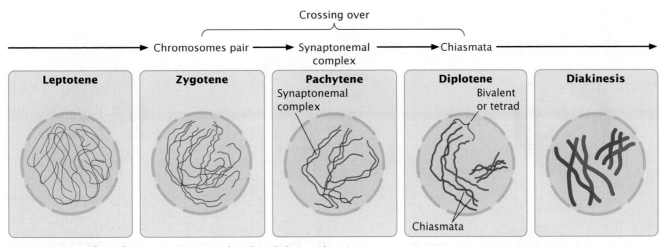

2.14 Substages of prophase I. Crossing over takes place during prophase I.

MEIOSIS I During interphase, the chromosomes are relaxed and indistinguishable as diffuse chromatin. **Prophase I** is a lengthy stage, divided into five substages (**Figure 2.14**). In leptotene, the chromosomes condense and become visible. In zygotene, the chromosomes continue to condense; homologous chromosomes pair up and begin **synapsis**, a very close pairing association. Each homologous pair of synapsed chromosomes, called a **bivalent** or **tetrad**, consists of four chromatids. In pachytene, the chromosomes become shorter and thicker, and a three-part synaptonemal complex develops between homologous chromosomes. The function of the synaptonemal complex is unclear, but the chromosomes of many cells deficient in this complex do not separate properly.

Crossing over, in which homologous chromosomes exchange genetic information, takes place in prophase I. Crossing over generates genetic variation (as we will see shortly) and is essential for the proper alignment and separation of homologous chromosomes. Each location where two chromosomes cross is called a *chiasma* (plural, *chiasmata*). In diplotene, the centromeres of the paired chromosomes move apart, but the two homologs remain attached at each chiasma. Near the end of prophase I, the nuclear membrane breaks down and the spindle forms, setting the stage for metaphase I. The stages of meiosis are outlined in **Figure 2.15**.

Metaphase I is initiated when homologous pairs of chromosomes align along the metaphase plate (see Figure 2.15). A microtubule from one spindle pole attaches to one chromosome of a homologous pair, and a microtubule from the other pole attaches to the other member of the pair. **Anaphase I** is marked by the separation of homologous chromosomes. The two chromosomes of a homologous pair are pulled toward opposite poles. Although the homologous chromosomes separate, the sister chromatids remain attached and travel together. In **telophase I**, the chromosomes arrive at the spindle poles and the cytoplasm divides.

MEIOSIS II The period between meiosis I and meiosis II is **interkinesis**, in which the nuclear membrane re-forms around the chromosomes clustered at each pole, the spindle breaks down, and the chromosomes relax. The cells then pass through **prophase II**, in which the events of interkinesis are reversed: the chromosomes recondense, the spindle re-forms, and the nuclear envelope once again breaks down. In interkinesis in some types of cells, the chromosomes remain condensed and the spindle does not break down. These cells move directly from cytokinesis into **metaphase II**, which is similar to metaphase of mitosis: the replicated chromosomes line up on the metaphase plate, with the sister chromatids facing opposite poles.

In **anaphase II**, the sister chromatids separate and the chromatids are pulled to opposite poles. Each chromatid is now a distinct chromosome. In **telophase II**, the chromosomes arrive at the spindle poles, a nuclear envelope re-forms around the chromosomes, and the cytoplasm divides. The chromosomes relax and are no longer visible.

CONCEPTS

Meiosis consists of two distinct processes: meiosis I and meiosis II. Meiosis I includes the reduction division, in which homologous chromosomes separate and chromosome number is reduced by half. In meiosis II (the equational division), sister chromatids separate.

✔ **CONCEPT CHECK 5**

Which of the following events takes place in metaphase I?
a. Crossing over occurs.
b. Chromosomes condense.
c. Homologous pairs of chromosomes line up on the metaphase plate.
d. Individual chromosomes line up on the metaphase plate.

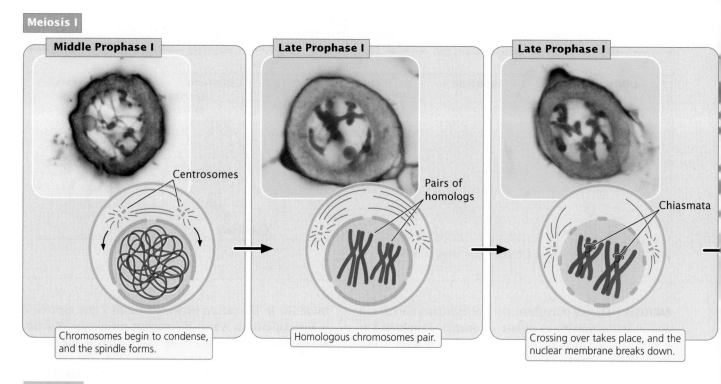

2.15 Meiosis is divided into stages. [University of Wisconsin Plant Teaching Collection, photographs by Michael Clayton.]

The major events of meiosis are summarized in **Table 2.2**. To examine the details of meiosis and the consequences of its failure, take a look at **Animation 2.2**.

(A)

Sources of Genetic Variation in Meiosis

What are the overall consequences of meiosis? First, meiosis comprises two divisions, so each original cell produces four cells (although there are exceptions to this generalization, as, for example, in many female animals; see Figure 2.20b). Second, chromosome number is reduced by half, so cells produced by meiosis are haploid. Third, cells produced by meiosis are genetically different from one another and from the parent cell. These genetic differences result from two processes that are unique to meiosis: crossing over and the random separation of homologous chromosomes.

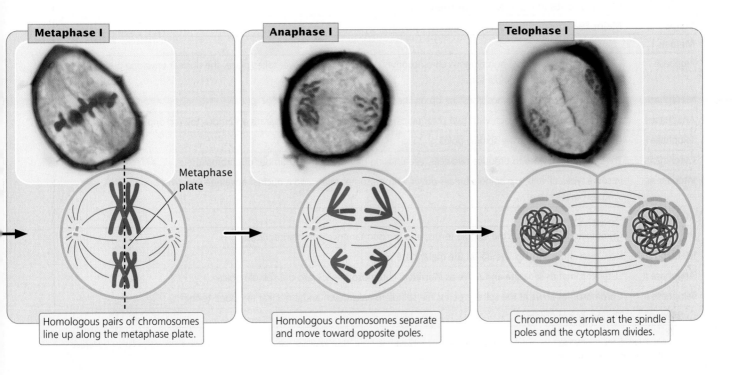

Metaphase I

Metaphase plate

Homologous pairs of chromosomes line up along the metaphase plate.

Anaphase I

Homologous chromosomes separate and move toward opposite poles.

Telophase I

Chromosomes arrive at the spindle poles and the cytoplasm divides.

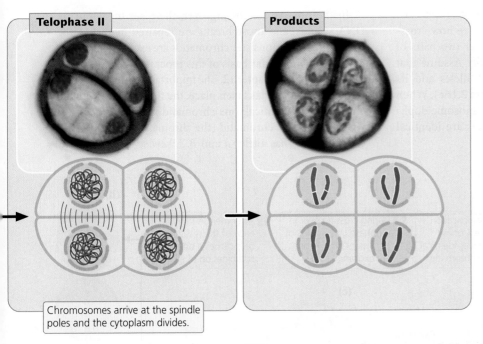

Telophase II

Chromosomes arrive at the spindle poles and the cytoplasm divides.

Products

CROSSING OVER Crossing over, which takes place in prophase I, refers to the exchange of genetic material between nonsister chromatids (chromatids from different homologous chromosomes). Evidence from yeast suggests that crossing over is initiated in zygotene, before the synaptonemal complex develops, and is not completed until near the end of prophase I (see Figure 2.14). In other organisms, crossing over is initiated after the formation of the synaptonemal complex; in still others, there is no synaptonemal complex.

After crossing over has taken place, the sister chromatids are no longer identical. Crossing over is the basis for

TABLE 2.2	Major events in each stage of meiosis
Stage	**Major Features**
Meiosis I	
Prophase I	Chromosomes condense, homologous chromosomes synapse, crossing over takes place, the nuclear envelope breaks down, and the mitotic spindle forms.
Metaphase I	Homologous pairs of chromosomes line up on the metaphase plate.
Anaphase I	The two chromosomes (each with two chromatids) of a homologous pair separate and move toward opposite poles.
Telophase I	Chromosomes arrive at the spindle poles.
Cytokinesis	The cytoplasm divides to produce two cells, each having half the original number of chromosomes.
Interkinesis	In some types of cells, the spindle breaks down, chromosomes relax, and a nuclear envelope re-forms, but no DNA synthesis takes place.
Meiosis II	
Prophase II*	Chromosomes condense, the spindle forms, and the nuclear envelope disintegrates.
Metaphase II	Individual chromosomes line up on the metaphase plate.
Anaphase II	Sister chromatids separate and move as individual chromosomes toward the spindle poles.
Telophase II	Chromosomes arrive at the spindle poles; the spindle breaks down and a nuclear envelope re-forms.
Cytokinesis	The cytoplasm divides.

*Only in cells in which the spindle has broken down, chromosomes have relaxed, and the nuclear envelope has re-formed in telophase I. Other types of cells proceed directly to metaphase II after cytokinesis.

intrachromosomal **recombination**, the creation of new combinations of alleles on a chromatid. To see how crossing over produces genetic variation, consider two pairs of alleles, which we will abbreviate *Aa* and *Bb*. Assume that one chromosome possesses the *A* and *B* alleles and its homolog possesses the *a* and *b* alleles (**Figure 2.16a**). When DNA is replicated in the S phase, each chromosome duplicates, and so the resulting sister chromatids are identical (**Figure 2.16b**).

In the process of crossing over, there are breaks in the DNA strands, and those breaks are repaired in such a way that segments of nonsister chromatids are exchanged (**Figure 2.16c**). The molecular basis of this process will be described in more detail in Chapter 12. The important thing here is that, after crossing over has taken place, the two sister chromatids are no longer identical: one chromatid has alleles *A* and *B*, whereas its sister chromatid (the chromatid that underwent crossing over) has alleles *a* and *B*. Likewise, one chromatid

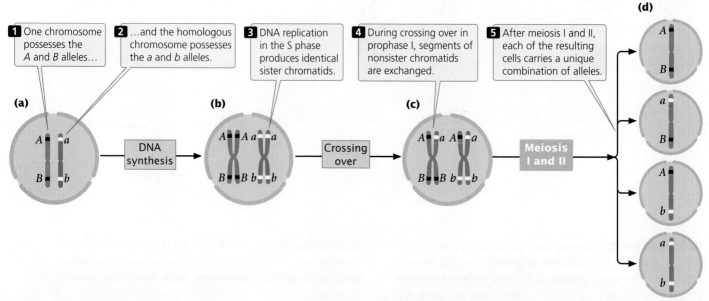

2.16 Crossing over produces genetic variation.

of the other chromosome has alleles *a* and *b*, and the other chromatid has alleles *A* and *b*. Each of the four chromatids now carries a unique combination of alleles: *A B*, *a B*, *A b*, and *a b*. Eventually, the two homologous chromosomes separate, each going into a different cell. In meiosis II, the two chromatids of each chromosome separate, and thus each of the four cells resulting from meiosis carries a different combination of alleles (**Figure 2.16d**). You can see how crossing over affects genetic variation by viewing **Animation 2.3**.

RANDOM SEPARATION OF HOMOLOGOUS CHROMO-SOMES The second process of meiosis that contributes to genetic variation is the random distribution of chromosomes in anaphase I after their random alignment in metaphase I. To illustrate this process, consider a cell with three pairs of chromosomes, I, II, and III (**Figure 2.17a**). One chromosome

of each pair is maternal in origin (I_m, II_m, and III_m); the other is paternal in origin (I_p, II_p, and III_p). The chromosome pairs line up in the center of the cell in metaphase I, and in anaphase I the chromosomes of each homologous pair separate. Note that each cell gets one chromosome from each of the three pairs (I, II, and III).

How each pair of homologs aligns and separates is random and independent of how other pairs align and separate (**Figure 2.17b**). By chance, all the maternal chromosomes might migrate to one side and all the paternal chromosomes to the other. After division, one cell would contain chromosomes I_m, II_m, and III_m, and the other, I_p, II_p, and III_p. Alternatively, the I_m, II_m, and III_p chromosomes might move to one side and the I_p, II_p, and III_m chromosomes to the other. These different migration patterns would produce different combinations of chromosomes in the resulting cells (**Figure 2.17c**).

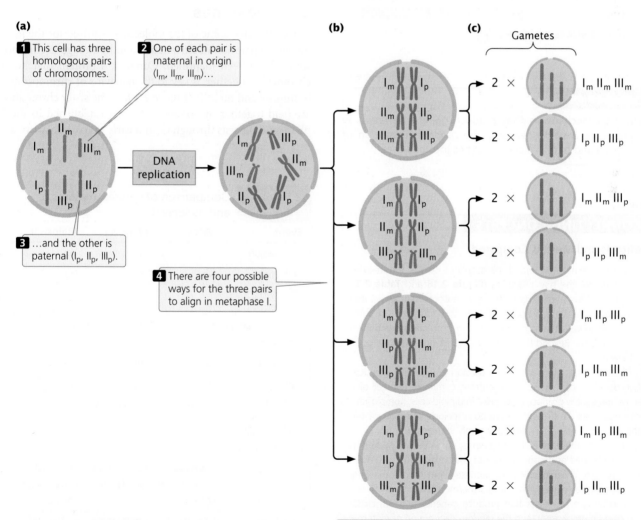

(a)

1 This cell has three homologous pairs of chromosomes.

2 One of each pair is maternal in origin (I_m, II_m, III_m)...

3 ...and the other is paternal (I_p, II_p, III_p).

DNA replication

4 There are four possible ways for the three pairs to align in metaphase I.

(b)   **(c)**   Gametes

2 × I_m II_m III_m

2 × I_p II_p III_p

2 × I_m II_m III_p

2 × I_p II_p III_m

2 × I_m II_p III_p

2 × I_p II_m III_m

2 × I_m II_p III_m

2 × I_p II_m III_p

2.17 The random distribution of chromosomes in meiosis produces genetic variation. In this example, the cell possesses three homologous pairs of chromosomes.

Conclusion: Eight different combinations of chromosomes in the gametes are possible, depending on how the chromosomes align and separate in meiosis I and II.

There are four ways in which the chromosomes in a diploid cell with three homologous pairs can migrate, producing a total of eight different combinations of chromosomes in the gametes. In general, the number of possible combinations is 2^n, where n equals the number of homologous pairs. As the number of chromosome pairs increases, the number of combinations quickly becomes very large. In humans, who have 23 pairs of chromosomes, 2^{23}, or 8,388,608, different combinations of chromosomes are made possible by the random separation of homologous chromosomes. You can explore the random distribution of chromosomes by viewing **Animation 2.3**. The genetic consequences of this process, termed independent assortment, will be explored in more detail in Chapter 3.

In summary, crossing over shuffles alleles on the *same* chromosome into new combinations, whereas the random distribution of maternal and paternal chromosomes shuffles alleles on *different* chromosomes into new combinations. Together, these two processes are capable of producing tremendous amounts of genetic variation among the cells resulting from meiosis. **▶ TRY PROBLEMS 35 AND 36**

THINK-PAIR-SHARE Question 4

CONCEPTS

The two mechanisms that produce genetic variation in meiosis are crossing over and the random separation of maternal and paternal chromosomes into gametes.

CONNECTING CONCEPTS

Mitosis and Meiosis Compared

Now that we have examined the details of mitosis and meiosis, let's compare the two processes (**Figure 2.18** and **Table 2.3**). In both mitosis and meiosis, the chromosomes condense and become visible; both processes include the movement of chromosomes toward the spindle poles, and both are accompanied by cell division. Beyond these similarities, the processes are quite different.

Mitosis results in a single cell division and usually produces two daughter cells. Meiosis, in contrast, comprises two cell divisions and usually produces four cells. In diploid cells, homologous chromosomes are present before both meiosis and mitosis, but the pairing of homologs takes place only in meiosis.

Another difference is that in meiosis, chromosome number is reduced by half as a consequence of the separation of homologous pairs of chromosomes in anaphase I, but no chromosome reduction takes place in mitosis. Furthermore, meiosis is characterized by two processes that produce genetic variation: crossing over (in prophase I), and the random distribution of maternal and paternal chromosomes (in anaphase I). There are normally no equivalent processes in mitosis.

Mitosis and meiosis I also differ in the behavior of chromosomes in metaphase and anaphase and their outcomes. In metaphase I of meiosis, *homologous pairs* of chromosomes line up on the metaphase plate, whereas *individual chromosomes* line up on the metaphase plate in metaphase of mitosis (and in metaphase II of meiosis). In anaphase I of meiosis, *paired chromosomes* separate and migrate toward opposite spindle poles, each chromosome possessing two chromatids attached at the centromere. In contrast, in anaphase of mitosis (and in anaphase II of meiosis), *sister chromatids* separate, and each chromosome that moves toward a spindle pole is unreplicated. Therefore, in contrast with both mitosis and meiosis II, meiosis I produces nonidentical daughter cells. **▶ TRY PROBLEMS 30 AND 31**

THINK-PAIR-SHARE Questions 5 and 6

The Separation of Sister Chromatids and Homologous Chromosomes

In recent years, some of the molecules required for the joining and separation of chromatids and homologous chromosomes have been identified. **Cohesin**, a protein that holds chromatids together, is key to the behavior of chromosomes in mitosis and meiosis (**Figure 2.19a**). The sister chromatids are held together by cohesin, which is established in the S phase and persists through G_2 and early mitosis. In anaphase

TABLE 2.3	Comparison of mitosis, meiosis I, and meiosis II		
Event	**Mitosis**	**Meiosis I**	**Meiosis II**
Cell division	Yes	Yes	Yes
Reduction in chromosome number	No	Yes	No
Genetic variation produced	No	Yes	No
Crossing over	No	Yes	No
Random distribution of maternal and paternal chromosomes	No	Yes	No
Metaphase	Individual chromosomes line up	Homologous pairs line up	Individual chromosomes line up
Anaphase	Chromatids separate	Homologous chromosomes separate	Chromatids separate

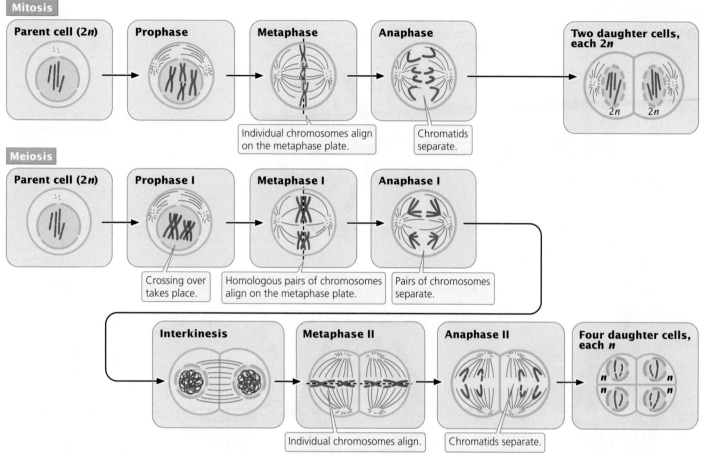

2.18 Mitosis and meiosis compared.

of mitosis, cohesin along the entire length of the chromosome is broken down by an enzyme called separase, allowing the sister chromatids to separate.

As we have seen, mitosis and meiosis differ fundamentally in the behavior of chromosomes in anaphase (see Figure 2.18). Why do homologs separate in anaphase I of meiosis, whereas chromatids separate in anaphase of mitosis and anaphase II of meiosis? It is important to note that the forms of cohesin used in mitosis and meiosis differ. At the beginning of meiosis, the meiosis-specific cohesin is found along the entire length of a chromosome's arms (**Figure 2.19b**). The cohesin also acts on the chromosome arms of homologs at the chiasmata, the crossover points between homologous chromosomes, tethering two homologs together at their ends.

In anaphase I, cohesin along the chromosome arms is broken, allowing the two homologs to separate. However, cohesin at the centromere is protected by a protein called shugoshin, which means "guardian spirit" in Japanese. Because of the protective action of shugoshin, the centromeric cohesin remains intact and prevents the separation of the two sister chromatids during anaphase I. Shugoshin is subsequently degraded. At the end of metaphase II, the centromeric cohesin—no longer protected by

shugoshin—breaks down, allowing the sister chromatids to separate in anaphase II, just as they do in mitosis (see Figure 2.19b). **▶ TRY PROBLEM 32**

THINK-PAIR-SHARE Question 7

CONCEPTS

Cohesin holds sister chromatids together during the early part of mitosis. In anaphase, cohesin breaks down, allowing sister chromatids to separate. In meiosis, cohesin is protected at the centromeres during anaphase I, so homologous chromosomes, but not sister chromatids, separate in meiosis I. The breakdown of centromeric cohesin allows sister chromatids to separate in anaphase II of meiosis.

✔ CONCEPT CHECK 6

How does shugoshin affect sister chromatids in meiosis I and meiosis II?

Meiosis in the Life Cycles of Animals and Plants

The overall result of meiosis is four haploid cells that are genetically variable. Let's see where meiosis fits into the life cycles of a multicellular animal and a multicellular plant.

(a) Mitosis

1 Sister kinetochores orient toward different poles,...

2 ...and cohesin keeps sister chromatids together.

3 The breakdown of cohesin allows sister chromatids to separate.

Spindle fibers

Cohesin

Chromatid

Breakdown of cohesin

Metaphase

Anaphase

(b) Meiosis

4 Cohesin along chromosome arms holds homologs together at chiasmata.

5 Cohesin along chromosome arms breaks down, allowing homologs to separate,...

6 ...but cohesin at the centromere is protected by shugoshin.

7 Shugoshin is degraded. Cohesin at the centromeres breaks down, allowing chromatids to separate.

Chiasma

Shugoshin

Breakdown of cohesin along arms

Breakdown of centromeric cohesin

Metaphase I

Anaphase I

Anaphase II

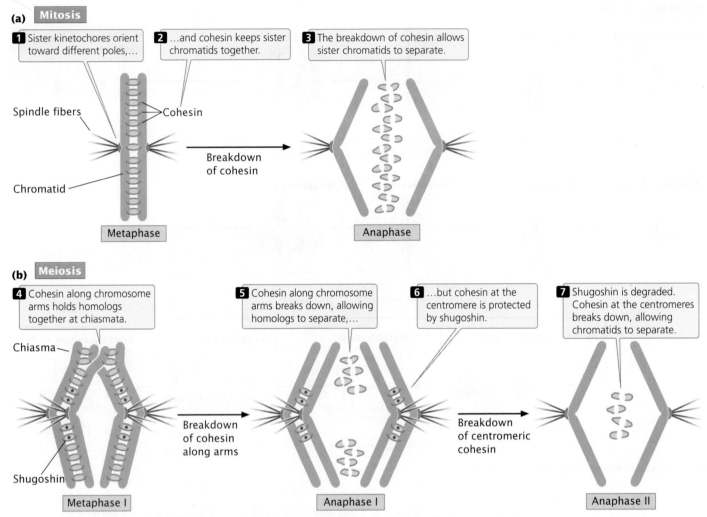

2.19 Cohesin controls the separation of chromatids and chromosomes in mitosis and meiosis.

MEIOSIS IN ANIMALS The production of gametes in a male animal, a process called **spermatogenesis**, takes place in the testes. There, diploid primordial germ cells divide mitotically to produce diploid cells called **spermatogonia** (**Figure 2.20a**). Each spermatogonium can undergo repeated rounds of mitosis, giving rise to numerous additional spermatogonia. Alternatively, a spermatogonium can initiate meiosis and enter prophase I. Now called a **primary spermatocyte**, the cell is still diploid because the homologous chromosomes have not yet separated. Each primary spermatocyte completes meiosis I, giving rise to two haploid **secondary spermatocytes** that then undergo meiosis II, with each producing two haploid **spermatids**. Thus, each primary spermatocyte produces a total of four haploid spermatids, which mature and develop into sperm.

The production of gametes in a female animal, a process called **oogenesis**, begins much as spermatogenesis does. Within the ovaries, diploid primordial germ cells divide mitotically to produce **oogonia** (**Figure 2.20b**). Like spermatogonia, oogonia can undergo repeated rounds of mitosis or they can enter meiosis. When they enter prophase I, these still-diploid cells are called **primary oocytes**. Each primary oocyte completes meiosis I and divides.

At this point, oogenesis begins to differ from spermatogenesis. In oogenesis, cytokinesis is unequal: most of the cytoplasm is allocated to one of the two haploid cells, the **secondary oocyte**. The smaller cell, which contains half of the chromosomes but only a small part of the cytoplasm, is called the **first polar body**; it may or may not divide further. The secondary oocyte completes meiosis II, and, again,

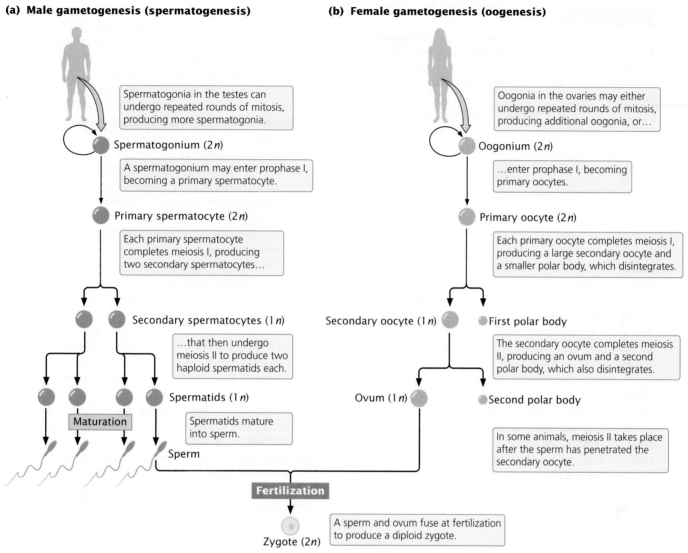

2.20 Gamete formation in animals.

cytokinesis is unequal—most of the cytoplasm passes into one of the cells. The larger cell, which acquires most of the cytoplasm, is the **ovum**, the mature female gamete. The smaller cell is the **second polar body**. Only the ovum is capable of being fertilized, and the polar bodies usually disintegrate. Oogenesis, then, produces a single mature gamete from each primary oocyte.

In mammals, oogenesis differs from spermatogenesis in another way. The formation of sperm takes place continuously in a male throughout his adult reproductive life. The formation of female gametes, however, is often a discontinuous process and may take place over a period of years. Oogenesis begins before birth; at this time, oogonia initiate meiosis and give rise to primary oocytes. Meiosis

is then arrested, stalled in prophase I. Thus, a female is born with primary oocytes arrested in prophase I. In humans, this period of suspended animation may last 30 or 40 years, until rising hormone levels stimulate one or more of the primary oocytes to recommence meiosis. The first division of meiosis is completed, and a secondary oocyte is ovulated from the ovary (a process called ovulation). In humans and many other species, the second division of meiosis is then delayed until contact with the sperm. When a sperm penetrates the outer layer of the secondary oocyte, the second meiotic division takes place, the second polar body is extruded from the ovum, and the nuclei of the sperm and newly formed ovum fuse, giving rise to the zygote.

MEIOSIS IN PLANTS Most multicellular plants and algae have a complex life cycle that includes two distinct structures (generations): a multicellular diploid sporophyte and a multicellular haploid gametophyte. These two generations alternate; the sporophyte produces haploid spores through meiosis, and the gametophyte produces haploid gametes through mitosis (**Figure 2.21**). This type of life cycle is sometimes called alternation of generations. In this cycle, the immediate products of meiosis are called spores, not gametes; the spores undergo one or more mitotic divisions to produce gametes. Although the terms used to describe the plant life cycle are somewhat different from those commonly used for animals (and from some of those employed so far in this chapter), the processes in plants and animals are basically the same: in both, meiosis leads to a reduction in chromosome number, producing haploid cells.

In flowering plants, the sporophyte is the obvious, vegetative part of the plant; the gametophyte consists of only a few haploid cells within the sporophyte. The flower, which is part of the sporophyte, contains the reproductive structures. In some species, both male and female reproductive structures are found in the same flower; in other species, they exist in different flowers. In either case, the male part of the flower, the stamen, contains diploid reproductive cells called **microsporocytes**, each of which undergoes meiosis to produce four haploid **microspores** (**Figure 2.22a**). Each microspore divides mitotically, producing an immature pollen grain consisting of two haploid nuclei. One of these nuclei, called the tube nucleus, directs the growth of a pollen tube. The other, termed the generative nucleus, divides mitotically to produce two sperm cells. The pollen grain, with its two haploid nuclei, is the male gametophyte.

The female part of the flower, the ovary, contains diploid cells called **megasporocytes**, each of which undergoes meiosis to produce four haploid **megaspores** (**Figure 2.22b**), only one of which survives. The nucleus of the surviving megaspore divides mitotically three times, producing a total of eight haploid nuclei that make up the female gametophyte, otherwise known as the embryo sac. Division of the cytoplasm then produces separate cells, one of which becomes the egg.

When the plant flowers, the stamens open and release pollen grains. Pollen lands on a flower's stigma—a sticky platform that sits on top of a long stalk called the style. At the base of the style is the ovary. If a pollen grain germinates, it grows a tube down the style into the ovary. The two sperm cells pass down this tube and enter the embryo sac (**Figure 2.22c**). One of the sperm cells fertilizes the egg cell, producing a diploid

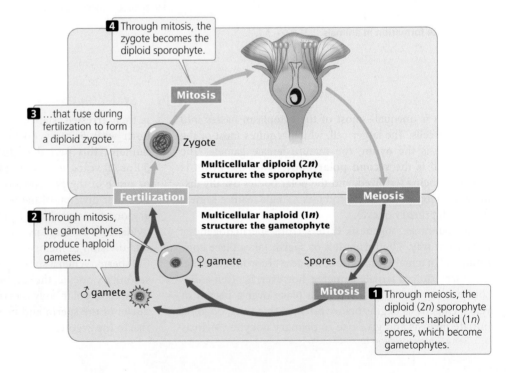

2.21 Plants alternate between diploid and haploid life stages.
(Female, ♀; male, ♂).

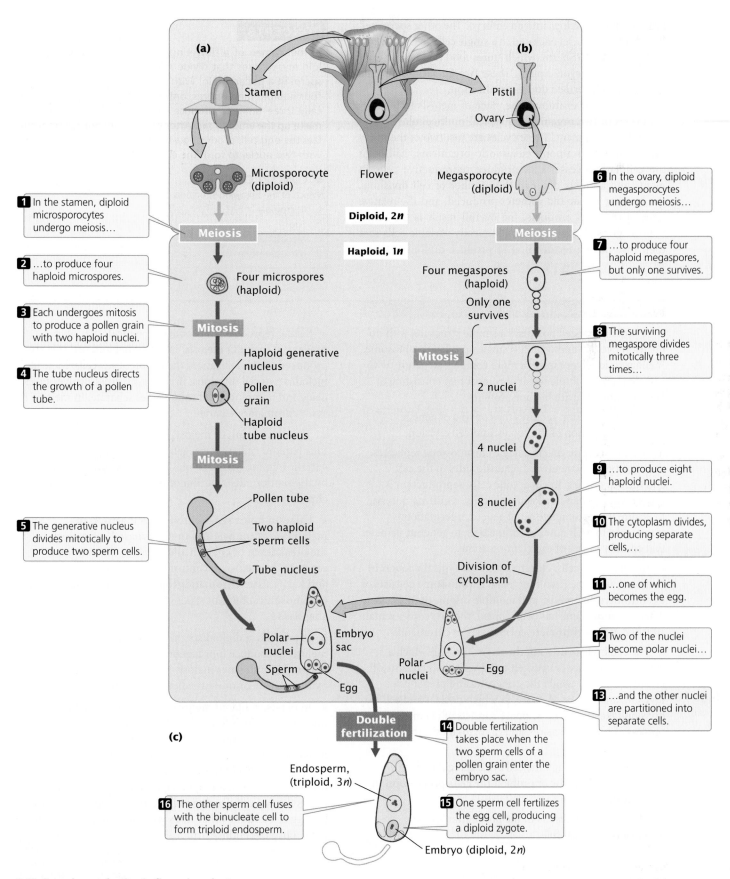

(a)

Stamen

Flower

(b)

Pistil

Ovary

Microsporocyte (diploid)

Megasporocyte (diploid)

Diploid, 2*n*

1 In the stamen, diploid microsporocytes undergo meiosis…

Meiosis

6 In the ovary, diploid megasporocytes undergo meiosis…

Meiosis

Haploid, 1*n*

2 …to produce four haploid microspores.

Four microspores (haploid)

Four megaspores (haploid)

Only one survives

7 …to produce four haploid megaspores, but only one survives.

3 Each undergoes mitosis to produce a pollen grain with two haploid nuclei.

Mitosis

Mitosis

8 The surviving megaspore divides mitotically three times…

4 The tube nucleus directs the growth of a pollen tube.

Haploid generative nucleus

Pollen grain

Haploid tube nucleus

2 nuclei

4 nuclei

Mitosis

8 nuclei

9 …to produce eight haploid nuclei.

5 The generative nucleus divides mitotically to produce two sperm cells.

Pollen tube

Two haploid sperm cells

Tube nucleus

Division of cytoplasm

10 The cytoplasm divides, producing separate cells,…

11 …one of which becomes the egg.

Polar nuclei

Embryo sac

Sperm

Egg

Polar nuclei

Egg

12 Two of the nuclei become polar nuclei…

13 …and the other nuclei are partitioned into separate cells.

(c)

Double fertilization

14 Double fertilization takes place when the two sperm cells of a pollen grain enter the embryo sac.

Endosperm, (triploid, 3*n*)

16 The other sperm cell fuses with the binucleate cell to form triploid endosperm.

15 One sperm cell fertilizes the egg cell, producing a diploid zygote.

Embryo (diploid, 2*n*)

2.22 Sexual reproduction in flowering plants.

zygote, which develops into an embryo. The other sperm cell fuses with two nuclei enclosed in a single cell, giving rise to a 3*n* (triploid) endosperm, which stores food that will be used later by the embryonic plant. These two fertilization events are collectively termed double fertilization.

We have now examined the place of meiosis in the sexual cycles of two organisms, a typical multicellular animal and a flowering plant. These cycles are just two of the many variations found among eukaryotic organisms. Although the cellular processes that produce reproductive cells in plants and animals differ in the number of cell divisions, the number of haploid gametes produced, and the relative sizes of the final products, the overall result is the same: meiosis gives rise to haploid, genetically variable cells that then fuse during fertilization to produce diploid progeny.

> TRY PROBLEMS 38 AND 40

CONCEPTS

In the stamen of a flowering plant, meiosis produces haploid microspores that divide mitotically to produce haploid sperm in a pollen grain. Within the ovary, meiosis produces four haploid megaspores, only one of which divides mitotically three times to produce the eight haploid nuclei that make up the embryo sac. After pollination, one sperm fertilizes the egg cell, producing a diploid zygote; the other fuses with two nuclei to form the endosperm.

✔ **CONCEPT CHECK 8**

Which of the following structures is diploid?
a. Microspore
b. Egg
c. Megaspore
d. Microsporocyte

CONCEPTS SUMMARY

■ A prokaryotic cell possesses a simple structure, with no nuclear envelope and usually a single circular chromosome. A eukaryotic cell possesses a more complex structure, with a nucleus and multiple linear chromosomes consisting of DNA complexed with histone proteins.

■ Cell reproduction requires the copying of genetic material, separation of the copies, and cell division.

■ In a prokaryotic cell, the single chromosome replicates, the two copies move toward opposite sides of the cell, and the cell divides. In eukaryotic cells, reproduction is more complex than in prokaryotic cells, requiring mitosis, to ensure that a complete set of genetic information is transferred to each new cell, or meiosis, to generate gametes that contain a copy of each chromosome.

■ In eukaryotic cells, chromosomes are typically found in homologous pairs. Each functional chromosome consists of a centromere, telomeres, and multiple origins of replication. After a chromosome has been copied, the two copies remain attached at the centromere, forming sister chromatids.

■ The cell cycle consists of the stages through which a eukaryotic cell passes between cell divisions. It consists of (1) interphase, in which the cell grows and prepares for division, and (2) the M phase, in which nuclear and cell division take place. The M phase consists of (1) mitosis, the process of nuclear division, and (2) cytokinesis, the division of the cytoplasm.

■ Progression through the cell cycle is regulated at checkpoints that allow or prohibit the cell's progression to the next stage.

■ Mitosis usually results in the production of two genetically identical cells.

■ Sexual reproduction produces genetically variable progeny. It includes meiosis, in which haploid sex cells are produced, and fertilization, the fusion of sex cells. Meiosis includes two cell divisions. In meiosis I, crossing over takes place and homologous chromosomes separate. In meiosis II, sister chromatids separate.

■ The usual result of meiosis is the production of four haploid cells that are genetically variable. Genetic variation in meiosis is produced by crossing over and by the random distribution of maternal and paternal chromosomes.

■ Cohesin holds sister chromatids together. In metaphase of mitosis and in metaphase II of meiosis, the breakdown of cohesin allows sister chromatids to separate. In meiosis I, centromeric cohesin remains intact and keeps sister chromatids together so that homologous chromosomes, but not sister chromatids, separate in anaphase I.

■ In animals, a diploid spermatogonium undergoes meiosis to produce four haploid sperm cells. A diploid oogonium undergoes meiosis to produce one large haploid ovum and one or more smaller polar bodies.

■ In plants, a diploid microsporocyte in the stamen undergoes meiosis to produce four pollen grains, each with two haploid sperm cells. In the ovary, a diploid megasporocyte undergoes meiosis to produce four haploid megaspores. One of these megaspores divides mitotically three times to produce eight haploid nuclei, one of which forms the egg cell. During pollination, one sperm fertilizes the egg cell and the other fuses with two haploid nuclei to form a triploid endosperm.

IMPORTANT TERMS

prokaryote (p. 18)	cell cycle (p. 23)	bivalent (p. 29)	secondary spermatocyte
eukaryote (p. 18)	checkpoint (p. 23)	tetrad (p. 29)	(p. 36)
bacteria (p. 19)	interphase (p. 23)	crossing over (p. 29)	spermatid (p. 36)
archaea (p. 19)	M (mitotic) phase	metaphase I (p. 29)	oogenesis (p. 36)
nucleus (p. 19)	(p. 23)	anaphase I (p. 29)	primary oocyte (p. 36)
histone (p. 19)	mitosis (p. 23)	telophase I (p. 29)	secondary oocyte
chromatin (p. 19)	cytokinesis (p. 23)	interkinesis (p. 29)	(p. 36)
virus (p. 20)	prophase (p. 24)	prophase II (p. 29)	first polar body (p. 36)
origin of replication	condensin (p. 24)	metaphase II (p. 29)	ovum (p. 37)
(p. 20)	prometaphase (p. 24)	anaphase II (p. 29)	second polar body
homologous pair (p. 21)	metaphase (p. 24)	telophase II (p. 29)	(p. 37)
diploid (p. 21)	anaphase (p. 26)	recombination (p. 32)	oogonium (p. 38)
haploid (p. 21)	telophase (p. 26)	cohesin (p. 34)	microsporocyte (p. 38)
polyploid (p. 21)	meiosis (p. 28)	spermatogenesis (p. 36)	microspore (p. 38)
centromere (p. 22)	fertilization (p. 28)	spermatogonium (p. 36)	megasporocyte (p. 38)
telomere (p. 22)	prophase I (p. 29)	primary spermatocyte	megaspore (p. 38)
sister chromatid (p. 23)	synapsis (p. 29)	(p. 36)	

ANSWERS TO CONCEPT CHECKS

1. Bacteria and archaea are prokaryotes. They differ from eukaryotes in possessing no nucleus, a genome that usually consists of a single circular chromosome, and a smaller amount of DNA.

2. b

3. The kinetochore is the point at which spindle microtubules attach to the chromosome during cell division. If the kinetochore were missing, spindle microtubules would not attach to the chromosome, the chromosome would not be drawn into a newly formed nucleus, and the resulting daughter cells would be missing a chromosome.

4. a

5. c

6. During anaphase I, shugoshin protects cohesin at the centromeres from the action of separase, so centromeric cohesin remains intact, and the sister chromatids remain together. Subsequently, shugoshin breaks down, so centromeric cohesin is cleaved by separase in anaphase II, and the sister chromatids separate.

7. d

8. d

WORKED PROBLEMS

Problem 1

A student examines a thin section of an onion-root tip and records the number of cells that are in each stage of the cell cycle. She observes 94 cells in interphase, 14 cells in prophase, 3 cells in prometaphase, 3 cells in metaphase, 5 cells in anaphase, and 1 cell in telophase. If the complete cell cycle in an onion-root tip requires 22 hours, what is the average duration of each stage in the cycle? Assume that all cells are in the active cell cycle (not G_0).

›› Solution Strategy

What information is required in your answer to the problem?

The average duration of each stage of the cell cycle.

What information is provided to solve the problem?

- The numbers of cells in different stages of the cell cycle.
- A complete cell cycle requires 22 hours.

For help with this problem, review:

The Cell Cycle and Mitosis, in Section 2.2.

›› Solution Steps

This problem is solved in two steps. First, we calculate the proportions of cells in each stage of the cell cycle, which correspond to the amount of time that an average cell spends in each stage. For example, if cells spend 90% of their time in interphase, then at any given moment, 90% of the cells will be in interphase. The second step is to convert the proportions into lengths of time, which is done by multiplying the proportions by the total time of the cell cycle (22 hours).

The proportion of cells at each stage is equal to the number of cells found in that stage divided by the total number of cells examined:

> **Hint:** The total of all the proportions should equal 1.0.

Interphase	$^{94}/_{120}$	$= 0.783$
Prophase	$^{14}/_{120}$	$= 0.117$
Prometaphase	$^{3}/_{120}$	$= 0.025$
Metaphase	$^{3}/_{120}$	$= 0.025$
Anaphase	$^{5}/_{120}$	$= 0.042$
Telophase	$^{1}/_{120}$	$= 0.008$

To determine the average duration of each stage, multiply the proportion of cells in each stage by the time required for the entire cell cycle:

Interphase	0.783×22 hours $= 17.23$ hours
Prophase	0.117×22 hours $= 2.57$ hours
Prometaphase	0.025×22 hours $= 0.55$ hour
Metaphase	0.025×22 hours $= 0.55$ hour
Anaphase	0.042×22 hours $= 0.92$ hour
Telophase	0.008×22 hours $= 0.18$ hour

> **Hint:** The total time all stages should eq 22 hours.

Problem 2

A cell in G_1 of interphase has 8 chromosomes. How many chromosomes and how many DNA molecules will be found per cell as this cell progresses through the following stages: G_2, metaphase of mitosis, anaphase of mitosis, after cytokinesis in mitosis, metaphase I of meiosis, metaphase II of meiosis, and after cytokinesis of meiosis II?

≫ Solution Strategy

What information is required in your answer to the problem?

The number of chromosomes and the number of DNA molecules present per cell at different stages of the cell cycle and meiosis.

What information is provided to solve the problem?

The number of chromosomes in the cell in G_1.

For help with this problem, review:

Connecting Concepts: Counting Chromosomes and DNA Molecules, in Section 2.2.

≫ Solution Steps

> **Hint:** These two rules are important for answering the question.

Remember the rules about counting chromosomes and DNA molecules: (1) to determine the number of chromosomes, count the functional centromeres; (2) to determine the number of DNA molecules, determine whether sister chromatids exist. If sister chromatids are present, the number of DNA molecules is two times the number of chromosomes. If the chromosomes are unreplicated (don't contain sister chromatids), the number of DNA molecules equals the number of chromosomes. Think carefully about when and how the numbers of chromosomes and DNA molecules change in the course of mitosis and meiosis.

> **Recall:** Chromosome number increases only when chromatids separate. The number of DNA molecules increases only in the S phase.

The number of DNA molecules increases only in the S phase, when DNA replicates; the number of DNA molecules decreases only when the cell divides. Chromosome number increases only when sister chromatids separate in anaphase of mitosis and in

anaphase II of meiosis (homologous chromosomes, not chromatids, separate in anaphase I of meiosis). Like the number of DNA molecules, chromosome number is reduced only by cell division.

Let's now apply these principles to the problem. A cell in G_1 has 8 chromosomes, and sister chromatids are not present, so 8 DNA molecules are present per cell in G_1. DNA replicates in the S phase, and now each chromosome consists of two chromatids, so in G_2, $2 \times 8 = 16$ DNA molecules are present per cell. However, the two copies of each DNA molecule remain attached at the centromere, so there are still only 8 chromosomes present. As the cell passes through prophase and metaphase of the cell cycle, the number of chromosomes and the number of DNA molecules remain the same, so at metaphase there are 16 DNA molecules and 8 chromosomes. In anaphase, the sister chromatids separate and each becomes an independent chromosome; at this point, the number of chromosomes increases from 8 to 16. This increase is temporary, lasting only until the cell divides in telophase or subsequent to it. The number of DNA molecules remains at 16 in anaphase. The number of DNA molecules and chromosomes per cell is reduced by cytokinesis after telophase because the 16 chromosomes and DNA molecules are now distributed between two cells. Therefore, after cytokinesis, each cell has 8 DNA molecules and 8 chromosomes, the same numbers that were present at the beginning of the cell cycle.

Now, let's trace the numbers of DNA molecules and chromosomes through meiosis. At G_1, there are 8 chromosomes and 8 DNA molecules. The number of DNA molecules increases to 16 in the S phase, but the number of chromosomes remains at 8 (each chromosome has two chromatids).

The cell therefore enters metaphase I having 16 DNA molecules and 8 chromosomes. In anaphase I of meiosis, homologous chromosomes separate, but the number of chromosomes remains at 8. After cytokinesis, the original 8 chromosomes are distributed between two cells, so the number of chromosomes per cell falls to 4 (each with two chromatids). The original 16 DNA molecules are also distributed between two cells, so the number of DNA molecules per cell is 8. There is no DNA synthesis in interkinesis, and each cell still maintains 4 chromosomes and 8 DNA molecules through metaphase II. In anaphase II, the two sister chromatids of each chromosome separate, temporarily raising the number of chromosomes per cell to 8, whereas the number of DNA molecules per cell remains at 8. After cytokinesis, the chromosomes and DNA molecules are again distributed between two cells, providing 4 chromosomes and 4 DNA molecules per cell. These results are summarized in the following table:

Stage	Number of chromosomes per cell	Number of DNA molecules per cell
G_1	8	8
G_2	8	16
Metaphase of mitosis	8	16
Anaphase of mitosis	16	16
After cytokinesis of mitosis	8	8
Metaphase I of meiosis	8	16
Metaphase II of meiosis	4	8
After cytokinesis of meiosis II	4	4

COMPREHENSION QUESTIONS

Section 2.1

1. What are some genetic differences between prokaryotic and eukaryotic cells?

2. Why are viruses often used in the study of genetics?

Section 2.2

3. List three fundamental events that must take place in cell reproduction.

4. Outline the process by which prokaryotic cells reproduce.

5. Name three essential structural elements of a functional eukaryotic chromosome and describe their functions.

6. Sketch and identify four different types of chromosomes based on the position of the centromere.

7. List the stages of interphase and the major events that take place in each stage.

8. What are checkpoints? List some of the important checkpoints in the cell cycle.

9. List the stages of mitosis and the major events that take place in each stage.

10. Briefly describe how the chromosomes move toward the spindle poles during anaphase.

11. What are the genetically important results of the cell cycle and mitosis?

12. Why are the two cells produced by the cell cycle genetically identical?

Section 2.3

13. What are the stages of meiosis and what major events take place in each stage?

14. What are the major results of meiosis?

15. What two processes unique to meiosis are responsible for genetic variation? At what point in meiosis do these processes take place?

16. How does anaphase I of meiosis differ from anaphase of mitosis?

17. Briefly explain why sister chromatids remain together in anaphase I but separate in anaphase II of meiosis.

18. Outline the processes of spermatogenesis and oogenesis in animals.

19. Outline the processes of male gamete formation and female gamete formation in plants.

APPLICATION QUESTIONS AND PROBLEMS

Introduction

*20. Answer the following questions about the blind men's riddle, presented in the introduction to this chapter.

a. What do the two socks of a pair represent in the cell cycle?

b. In the riddle, each blind man buys his own pairs of socks, but the clerk places all the pairs in one bag. Thus, there are two pairs of socks of each color in the bag (two black pairs, two blue pairs, two gray pairs, etc.). What do the two pairs (four socks in all) of each color represent?

c. What is the thread that connects the two socks of a pair?

d. What is the molecular knife that cuts the thread holding the two socks of a pair together?

e. What in the riddle performs the same function as spindle microtubules?

f. What would happen if one man failed to grasp his sock of a particular pair? How does that outcome relate to events in the cell cycle?

Section 2.1

21. A cell has a circular chromosome and no nuclear membrane. Its DNA is complexed with some histone proteins. Does this cell belong to a bacterium, an archaean, or a eukaryote? Explain your reasoning.

Section 2.2

22. A certain species has three pairs of chromosomes: an acrocentric pair, a metacentric pair, and a submetacentric pair. Draw a cell of this species as it would appear in metaphase of mitosis.

23. Examine **Figure 2.6a**. What type of chromosome (metacentric, submetacentric, acrocentric, or telocentric) is chromosome 1? What about chromosome 4?

*24. A biologist examines a series of cells and counts 160 cells in interphase, 20 cells in prophase, 6 cells in prometaphase, 2 cells in metaphase, 7 cells in anaphase, and 5 cells in telophase. If the complete cell cycle requires 24 hours, what is the average duration of the M phase in these cells? Of metaphase?

25. In what stage of mitosis is the cell illustrated in the **chapter opening figure** (p. 17)?

Section 2.3

26. A certain species has three pairs of chromosomes: one acrocentric pair and two metacentric pairs. Draw a cell of this species as it would appear in the following stages of meiosis:

a. Metaphase I

b. Anaphase I

c. Metaphase II

d. Anaphase II

27. Construct a table similar to that in **Figure 2.12** for the different stages of meiosis, giving the number of chromosomes per cell and the number of DNA molecules per cell for a cell that begins with 4 chromosomes (two homologous pairs) in G_1. Include the following stages in your table: G_1, S, G_2, prophase I, metaphase I, anaphase I, telophase I (after cytokinesis), prophase II, metaphase II, anaphase II, and telophase II (after cytokinesis).

*28. A cell in G_1 of interphase has 12 chromosomes ($2n = 12$). How many chromosomes and DNA molecules will be found per cell when this original cell progresses to the following stages?

a. G_2 of interphase

b. Metaphase I of meiosis

c. Prophase of mitosis

d. Anaphase I of meiosis

e. Anaphase II of meiosis

f. Prophase II of meiosis

g. After cytokinesis following mitosis

h. After cytokinesis following meiosis II

29. How are the events that take place in spermatogenesis and oogenesis similar? How are they different?

*30. All of the following cells, shown in various stages of mitosis and meiosis, come from the same rare species of plant.

a. What is the diploid number of chromosomes in this plant?

b. Give the names of each stage of mitosis or meiosis shown.

c. Give the number of chromosomes and number of DNA molecules per cell present at each stage.

*31. The amount of DNA per cell of a particular species is measured in cells found at various stages of meiosis, and the following amounts are obtained:

Amount of DNA per cell in picograms (pg)

_____3.7 pg   _____7.3 pg   _____14.6 pg

Match the amounts of DNA above with the corresponding stages of meiosis (*a* through *f*, on the adjoining page). You may use more than one stage for each amount of DNA.

Stage of meiosis

a. G_1

b. Prophase I

c. G_2

d. Following telophase II and cytokinesis

e. Anaphase I

f. Metaphase II

*32. How would each of the following events affect the outcome of mitosis or meiosis?

a. Mitotic cohesin fails to form early in mitosis.

b. Shugoshin is absent during meiosis.

c. Shugoshin does not break down after anaphase I of meiosis.

d. Separase is defective.

*33. A cell in prophase II of meiosis contains 12 chromosomes. How many chromosomes would be present in a cell from the same organism if it were in prophase of mitosis? Prophase I of meiosis?

34. A cell has 8 chromosomes in G_1 of interphase. Draw a picture of this cell with its chromosomes at the following stages. Indicate how many DNA molecules are present at each stage.

a. Metaphase of mitosis

b. Anaphase of mitosis

c. Anaphase II of meiosis

d. Diplotene of meiosis I

*35. The fruit fly *Drosophila melanogaster* (left) has four pairs of chromosomes, whereas the house fly *Musca domestica* (right) has six pairs of chromosomes. In which species would you expect to see more genetic variation among the progeny of a cross? Explain your answer.

[© Graphic Science/Alamy Stock Photo.]

[© Debug/iStock Photo.]

*36. A cell has two pairs of submetacentric chromosomes, which we will call chromosomes I_a, I_b, II_a, and II_b (chromosomes I_a and I_b are homologs, and chromosomes II_a and II_b are homologs). Allele *M* is located on the long arm of chromosome I_a, and allele *m* is located at the same position on chromosome I_b. Allele *P* is located on the short arm of chromosome I_a, and allele *p* is located at the same position on chromosome I_b. Allele *R* is located on chromosome II_a and allele *r* is located at the same position on chromosome II_b.

a. Draw these chromosomes, identifying genes *M*, *m*, *P*, *p*, *R*, and *r*, as they might appear in metaphase I of meiosis. Assume that there is no crossing over.

b. Taking into consideration the random separation of chromosomes in anaphase I, draw the chromosomes (with genes identified) present in all possible types of gametes that might result from this cell's undergoing meiosis. Assume that there is no crossing over.

37. A horse has 64 chromosomes and a donkey has 62 chromosomes. A cross between a female horse and a male donkey produces a mule, which is usually sterile. How many chromosomes does a mule have? Can you think of any reasons for the fact that most mules are sterile?

*38. Normal somatic cells of horses have 64 chromosomes ($2n = 64$). How many chromosomes and DNA molecules will be present in the following types of horse cells?

Cell type	Number of chromosomes	Number of DNA molecules
a. Spermatogonium	_____	_____
b. First polar body	_____	_____
c. Primary oocyte	_____	_____
d. Secondary spermatocyte	_____	_____

[© Dozornaya/iStock Photo.]

39. Indicate whether each of the following cells is haploid or diploid.

Cell type	Haploid or diploid?
Microspore	_____
Primary spermatocyte	_____
Microsporocyte	_____
First polar body	_____
Oogonium	_____
Spermatid	_____
Megaspore	_____
Ovum	_____
Secondary oocyte	_____
Spermatogonium	_____

*40. A primary oocyte divides to give rise to a secondary oocyte and a first polar body. The secondary oocyte then divides to give rise to an ovum and a second polar body.

   a. Is the genetic information found in the first polar body identical with that found in the secondary oocyte? Explain your answer.

   b. Is the genetic information found in the second polar body identical with that in the ovum? Explain your answer.

CHALLENGE QUESTIONS

Section 2.3

41. From 80% to 90% of the most common human chromosome abnormalities arise because the chromosomes fail to divide properly in oogenesis. Can you think of a reason why failure of chromosome division might be more common in female gametogenesis than in male gametogenesis?

42. On average, what proportion of the genome in the following pairs of humans would be exactly the same if no crossing over took place? (For the purposes of this question only, we will ignore the special case of the X and Y sex chromosomes and assume that all genes are located on nonsex chromosomes.)

   a. Father and child

   b. Mother and child

   c. Two full siblings (offspring that have the same two biological parents)

   d. Half siblings (offspring that have only one biological parent in common)

   e. Uncle and niece

   f. Grandparent and grandchild

*43. Female bees are diploid, and male bees are haploid. The haploid males produce sperm and can successfully mate with diploid females. Fertilized eggs develop into females and unfertilized eggs develop into males. How do you think the process of sperm production in male bees differs from sperm production in other animals?

 THINK-PAIR-SHARE QUESTIONS

Section 2.2

1. A chromosome consists of two sister chromatids. Does the genetic information on the two sister chromatids come from only one parent or from both parents? Explain your reasoning.

2. Are homologous pairs of chromosomes present in mitosis? Explain your reasoning.

3. A cell has 8 chromosomes in metaphase II of meiosis. How many chromosomes and DNA molecules will be present per cell in this same organism at the following stages?

   a. Prophase of mitosis

   b. Metaphase I of meiosis

   c. Anaphase of mitosis

   d. Anaphase II of meiosis

   e. Anaphase I of meiosis

   f. After cytokinesis that follows mitosis

   g. After cytokinesis that follows meiosis II

Section 2.3

4. What is the difference between sister chromatids and homologous chromosomes?

5. List as many similarities and differences between mitosis and meiosis as you can. Which differences do you think are most important, and why?

6. Describe how and where each of the following terms applies to mitosis, meiosis, or both: (1) replication; (2) pairing; (3) separation.

7. Do you know of any genetic diseases or disorders that result from errors in mitosis or meiosis? How do errors in mitosis or meiosis bring about these diseases?

Basic Principles of Heredity

Blond hair occurs in 5%–10% of dark-skinned Solomon Islanders.
Research demonstrates that blond hair in this group is a recessive trait and has a different genetic basis from blond hair in Europeans. [© Anthony Asael/Danita Delimont/Alamy Stock Photo.]

The Genetics of Blond Hair in the South Pacific

A thousand miles northeast of Australia lies an ancient chain of volcanic and coral islands known as the Solomons (**Figure 3.1**). The Solomon Islands were first inhabited some 30,000 years ago, when Neanderthals still roamed northern Europe. Today, the people of the Solomons are culturally diverse, but consist largely of Melanesians, a group that also inhabits other South Pacific islands. Most people from the Solomon Islands have dark skin. Remarkably, 5%–10% also have strikingly blond hair; in fact, people of the Solomon Islands have the highest frequency of blond hair outside of Europe.

How did the Solomon Islanders get their blond hair? A number of hypotheses have been proposed over the years. Some suggested that the blond islanders had naturally dark hair that was bleached by the sun and salt water. Others proposed that the blond hair color was caused by diet. Still others suggested that it was the result of genes for blond hair left by early European explorers.

The mystery of the blond Solomon Islanders was solved in 2012 by geneticists Eimear Kenny and Sean Myles and their colleagues. Their research demonstrated that blond hair on the islands is, in fact, caused by a gene, but not one left by Europeans—blond hair in Solomon Islanders and in Europeans has completely separate evolutionary origins.

To search for the origin of blond hair among the people of the Solomon Islands, the geneticists collected saliva and hair samples from over 1200 people on the islands, from which they then extracted DNA. In a type of analysis known as a genome-wide association study, they looked for statistical associations between the presence of blond hair and thousands of genetic variants scattered across the genome. Right away, they detected a strong correlation between the presence of blond hair and a particular genetic variant located on the short arm of chromosome 9. This region of chromosome 9 contains the tyrosinase-related protein 1 gene (*TYRP1*), which encodes an enzyme known to play a role in the production of melanin and to affect pigmentation in mice. The researchers found a single base difference between the DNA of islanders with blond hair and that of islanders with dark hair: the blonds had a thymine (T) base instead of cytosine (C) in their *TYRP1* gene.

Further research showed that blond hair in Solomon Islanders is a recessive trait, meaning that blonds carry two copies of the blond version of the gene (*TT*)—one inherited from each parent. Dark hair is dominant: dark-haired islanders carry either

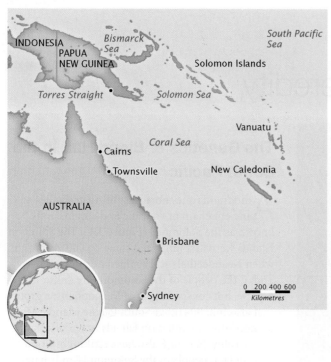

3.1 Map of the Solomon Islands.

one (*CT*) or two (*CC*) copies of the dark-hair version of the gene. Thus, many dark-haired islanders are heterozygous, carrying a hidden copy of the blond gene that can be passed on to their offspring. A DNA analysis of 900 Solomon Islanders demonstrated that over 40% of dark-haired islanders carry a blond gene. Interestingly, the C to T mutation in the *TYRP1* gene that causes blond hair in Solomon Islanders is rare outside of the South Pacific, suggesting that the mutation arose independently within the Melanesian population. There is no evidence that the gene was inherited from Europeans.

THINK-PAIR-SHARE Question 1

The genetics of blond hair in Solomon Islanders differs from that in Europeans in other ways as well. In Europeans, variations in at least eight different genes have been associated with blond hair. In 2015, researchers examined one of these genes (called *KITLG*) and found that the mutation causing blond hair occurred not in the gene itself, but in a region of DNA that affects the expression of the *KITLG* gene. The *KITLG* gene produces a protein that is involved in a number of functions, including melanocyte development and melanin synthesis.

THINK-PAIR-SHARE
- Why is knowing the genetic basis of a trait such as blond hair important? Why would scientists go to the trouble to investigate the genetic basis of blond hair in Solomon Islanders?

- If a blond-haired person from northern Europe mated with a blond Solomon Islander, what proportion of their offspring would be expected to have blond hair? Explain your reasoning.

This chapter is about the principles of heredity: how genes—such as the one for blond hair in Solomon Islanders—are passed from generation to generation and how factors such as dominance influence their inheritance. The principles of heredity were first put forth by Gregor Mendel, so we begin this chapter by examining Mendel's scientific achievements. We then turn to simple genetic crosses in which a single characteristic is examined. We consider some techniques for predicting the outcome of genetic crosses, and then turn to crosses in which two or more characteristics are examined. We see how the principles applied to simple genetic crosses and the ratios of offspring they produce can serve as the key to understanding more complicated crosses. The chapter ends with a discussion of statistical tests for analyzing crosses.

Throughout this chapter, a number of concepts are interwoven: Mendel's principles of segregation and independent assortment; probability; and the behavior of chromosomes. These concepts might at first appear to be unrelated, but they are actually different views of the same phenomenon because the genes that undergo segregation and independent assortment are located on chromosomes. In this chapter, we examine these different views and clarify how they are related.

3.1 Gregor Mendel Discovered the Basic Principles of Heredity

It was in the early 1900s that the principles of heredity first became widely known among biologists. Surprisingly, these principles had been discovered some 44 years earlier by an Augustinian priest named Gregor Johann Mendel (1822–1884) (**Figure 3.2**).

Mendel was born in what is now part of the Czech Republic. Although his parents were simple farmers with little money, he received a sound education and was admitted to the Augustinian monastery in Brno in September 1843. After graduating from seminary, Mendel became an ordained priest and was appointed to a teaching position in a local school. He excelled at teaching, and the abbot of the monastery recommended him for further study at the University of Vienna, which he attended from 1851 to 1853. There, Mendel enrolled in the newly opened Physics Institute and took courses in mathematics, chemistry, entomology, paleontology, botany, and plant physiology. It was probably there that Mendel acquired knowledge of the scientific method that he later applied so successfully to his genetic experiments.

After two years of study in Vienna, Mendel returned to Brno, where he taught school and began his experimental work with pea plants. He conducted breeding experiments from 1856 to 1863 and presented his results publicly at meetings of the Brno Natural Science Society in 1865. Mendel's paper based on these lectures was published in 1866. However, in spite of widespread interest in heredity, the effect of his research on the scientific community was minimal. At the time, no one seemed to have noticed that Mendel had discovered the basic principles of inheritance.

In 1868, Mendel was elected abbot of his monastery, and increasing administrative duties brought an end to his teaching and, eventually, to his genetic experiments. He died at the age of 61 on January 6, 1884, unrecognized for his contribution to genetics.

The significance of Mendel's discovery was not recognized until 1900, when three botanists—Hugo de Vries, Erich von Tschermak-Seysenegg, and Carl Correns—began independently conducting similar experiments with plants and arrived at conclusions similar to those of Mendel. Coming across Mendel's paper, they interpreted their results in accord with his principles and drew attention to his pioneering work.

Mendel's Success

Mendel's approach to the study of heredity was effective for several reasons. Foremost was his choice of experimental subject, the pea plant *Pisum sativum* (**Figure 3.3**), which offered clear advantages for genetic investigation. The plant is easy to cultivate, and Mendel had the monastery garden and greenhouse at his disposal. Compared with some other plants, peas grow relatively rapidly, completing an entire generation in a single growing season. By today's standards, one generation per year seems frightfully slow—fruit flies complete a generation in 2 weeks and bacteria in 20 minutes—but Mendel was under no pressure to publish quickly and was able to follow the inheritance of individual characteristics for several generations. Had he chosen to work on an organism with a longer generation time—horses, for example—he might never have discovered the basis of inheritance. Pea plants also produce many offspring—their seeds—which allowed Mendel to detect meaningful mathematical ratios in the traits he observed in the progeny. The numerous varieties of peas that were available to Mendel were also crucial to his success because they differed in various traits and were genetically pure. Mendel was therefore able to begin with plants of variable, known genetic makeup.

Much of Mendel's success can be attributed to the seven characteristics of pea plants that he chose for study (see Figure 3.3). He avoided characteristics that display a range of variation; instead, he focused his attention on those that exist in two easily differentiated forms, such as white versus gray seed coats, round versus wrinkled seeds, and inflated versus constricted pods.

THINK-PAIR-SHARE Question 2

Finally, Mendel was successful because he adopted an experimental approach and interpreted his results by using mathematics. Unlike many earlier investigators who simply described the *results* of crosses, Mendel formulated *hypotheses* based on his initial observations and then conducted additional crosses to test his hypotheses. He kept careful

3.2 Gregor Johann Mendel discovered the principles of heredity by experimenting with peas. [James King-Holmes/Science Source.]

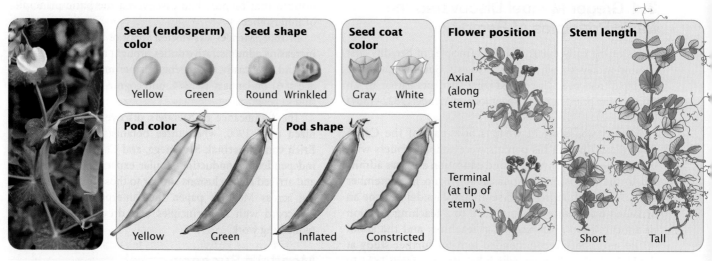

3.3 Mendel used the pea plant *Pisum sativum* in his studies of heredity. He examined seven characteristics that appeared in the seeds and in plants grown from the seeds. [Photograph by Charles Stirling/Alamy.]

records of the numbers of progeny possessing each trait and computed ratios of the different traits. He was adept at seeing patterns in detail and was patient and thorough, conducting his experiments for 10 years before attempting to write up his results. ❯ **TRY PROBLEM 13**

CONCEPTS

Gregor Mendel put forth the basic principles of inheritance, publishing his findings in 1866. Much of Mendel's success can be attributed to the seven characteristics of pea plants that he studied and his experimental approach.

✔ CONCEPT CHECK 1

Which of the following factors did not contribute to Mendel's success in his study of heredity?

a. His use of the pea plant
b. His study of plant chromosomes
c. His adoption of an experimental approach
d. His use of mathematics

Genetic Terminology

Before we examine Mendel's crosses and the conclusions that he drew from them, a review of some terms commonly used in genetics will be helpful (**Table 3.1**). The term *gene* is a word that Mendel never knew. It was not coined until 1909, when Danish geneticist Wilhelm Johannsen first used it. The definition of *gene* varies with the context of its use, so its definition will change as we explore different aspects of heredity. For our present use in the context of genetic crosses, we define a **gene** as an inherited factor that determines a characteristic.

Genes frequently come in different versions called **alleles** (**Figure 3.4**). In Mendel's crosses, seed shape was determined by a gene that exists as two different alleles: one allele encodes

round seeds and the other encodes wrinkled seeds. All alleles for any particular gene will be found at a specific place on a chromosome called the **locus** for that gene. (The plural of locus is *loci*; it's bad form in genetics—and incorrect—to speak of "locuses.") Thus, there is a specific place—a locus—on a chromosome in pea plants where the shape of seeds is determined. This locus may be occupied by an allele for round seeds or by one for wrinkled seeds. We will use the term *allele* when referring to a specific version of a gene; we will use the term *gene* to refer more generally to any allele at a locus.

TABLE 3.1	Summary of important genetic terms
Term	**Definition**
Gene	An inherited factor (encoded in the DNA) that helps determine a characteristic
Allele	One of two or more alternative forms of a gene
Locus	Specific place on a chromosome occupied by an allele
Genotype	Set of alleles possessed by an individual organism
Heterozygote	An individual organism possessing two different alleles at a locus
Homozygote	An individual organism possessing two of the same alleles at a locus
Phenotype or trait	The appearance or manifestation of a characteristic
Characteristic or character	An attribute or feature possessed by an organism

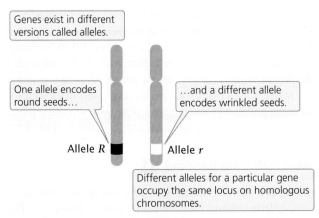

Genes exist in different versions called alleles.

One allele encodes round seeds...

...and a different allele encodes wrinkled seeds.

Allele *R*   Allele *r*

Different alleles for a particular gene occupy the same locus on homologous chromosomes.

3.4 At each locus, a diploid organism possesses two alleles located on different homologous chromosomes. The alleles identified here refer to traits studied by Mendel.

The **genotype** is the set of alleles that an individual organism possesses. A diploid organism with a genotype consisting of two identical alleles is **homozygous** at that locus. One that has a genotype consisting of two different alleles is **heterozygous** at the locus.

A **phenotype** is the manifestation or appearance of a characteristic. This term can refer to any type of characteristic—physical, physiological, biochemical, or behavioral. Thus, the condition of having round seeds is a phenotype, a body weight of 50 kilograms (50 kg) is a phenotype, and having sickle-cell anemia is a phenotype. In this book, the term *characteristic* or *character* refers to a general feature such as eye color; the term *trait* or *phenotype* refers to specific manifestations of that feature, such as blue or brown eyes.

A given phenotype arises from a genotype that develops within a particular environment. The genotype determines the potential for development; it sets certain limits, or boundaries, on that development. How the phenotype develops within those limits is determined by the effects of other genes and of environmental factors, and the balance between these effects varies from characteristic to characteristic. For some characteristics, differences between phenotypes are determined largely by differences in genotype. In Mendel's peas, for example, the genotype, not the environment, largely determined the shape of the seeds. For other characteristics, environmental differences are more important. The height reached by an oak tree at maturity is a phenotype that is strongly influenced by environmental factors, such as the availability of water, sunlight, and nutrients. Nevertheless, the tree's genotype imposes some limits on its height: an oak tree will never grow to be 300 meters (almost 1000 feet) tall, no matter how much sunlight, water, and fertilizer are provided. Thus, even the height of an oak tree is determined to some degree by genes. For many characteristics, both genes and environment are important in determining phenotypic differences.

An obvious but important point is that only the alleles of the genotype are inherited. Although the phenotype is determined, at least to some extent, by the genotype, organisms do not transmit their phenotypes to the next generation. The distinction between genotype and phenotype is one of the most important principles of modern genetics. The next section describes Mendel's careful observation of phenotypes through several generations of breeding experiments. These experiments allowed him to deduce not only the genotypes of individual pea plants, but also the rules governing their inheritance.

CONCEPTS

Each phenotype results from a genotype developing within a specific environment. The alleles of the genotype, not the phenotype, are inherited.

✔ **CONCEPT CHECK 2**

What is the difference between a locus and an allele? What is the difference between genotype and phenotype?

3.2 Monohybrid Crosses Reveal the Principle of Segregation and the Concept of Dominance

Mendel started with 34 varieties of peas and spent 2 years selecting those varieties that he would use in his experiments. He verified that each variety was pure-breeding (homozygous for each of the traits that he chose to study) by growing the plants for two generations and confirming that all offspring were the same as their parents. He then carried out a number of crosses between the different varieties. Although peas are normally self-fertilizing (each plant mates with itself), Mendel conducted crosses between different plants by opening the buds before the anthers (male sex organs) were fully developed, removing the anthers, and then dusting the stigma (female sex organ) with pollen from a different plant's anthers (**Figure 3.5**).

Mendel began by studying **monohybrid crosses**—crosses between parents that differed in a single characteristic. In one experiment, Mendel crossed a pea plant that was pure-breeding (homozygous) for round seeds with one that was pure-breeding for wrinkled seeds (see Figure 3.5). This first generation of a cross is called the **P (parental) generation**.

After crossing the two varieties in the P generation, Mendel observed the offspring that resulted from the cross. The seed shape phenotype develops as soon as the seed matures because seed traits are determined by the newly formed embryo within the seed. For characteristics associated with the plant itself, such as stem length, the phenotype doesn't develop until the plant grows from the seed; for these characteristics, Mendel had to wait until the following spring, plant the seeds, and then observe the phenotypes of the plants that germinated.

The offspring of the parents in the P generation are the **F₁ (first filial) generation**. When Mendel examined the

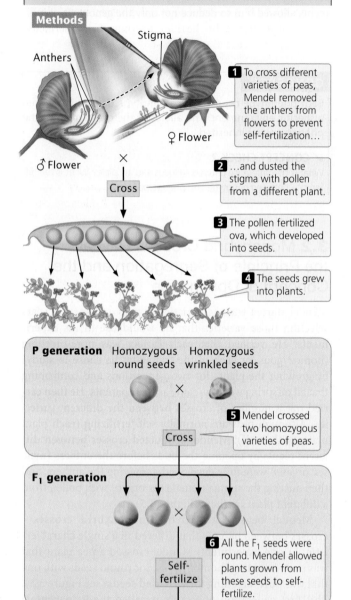

Experiment

Question: When peas with two different traits—round and wrinkled seeds—are crossed, will their progeny exhibit one of those traits, both of those traits, or an intermediate trait?

Methods

Stigma

Anthers

♀ Flower

♂ Flower

×

Cross

1 To cross different varieties of peas, Mendel removed the anthers from flowers to prevent self-fertilization…

2 …and dusted the stigma with pollen from a different plant.

3 The pollen fertilized ova, which developed into seeds.

4 The seeds grew into plants.

P generation Homozygous round seeds Homozygous wrinkled seeds

×

Cross

5 Mendel crossed two homozygous varieties of peas.

F₁ generation

×

Self-fertilize

6 All the F₁ seeds were round. Mendel allowed plants grown from these seeds to self-fertilize.

Results

F₂ generation

Fraction of progeny seeds

5474 round seeds   ¾ round

1850 wrinkled seeds   ¼ wrinkled

7 ¾ of F₂ seeds were round and ¼ were wrinkled, a 3 : 1 ratio.

Conclusion: The traits of the parent plants do not blend. Although F₁ plants display the phenotype of one parent, both traits are passed to F₂ progeny in a 3 : 1 ratio.

3.5 Mendel conducted monohybrid crosses.

F₁ generation of this cross, he found that they expressed only one of the phenotypes present in the parental generation: all the F₁ seeds were round. Mendel carried out 60 such crosses and always obtained this result. Furthermore, he conducted **reciprocal crosses**: in one cross, pollen (the male gamete) was taken from a plant with round seeds, and in its reciprocal cross, pollen was taken from a plant with wrinkled seeds. Reciprocal crosses gave the same result: all the F₁ seeds were round.

THINK-PAIR-SHARE Question 3

Mendel wasn't content with examining only the seeds arising from these monohybrid crosses, however. The following spring, he planted the F₁ seeds, cultivated the plants that germinated from them, and allowed the plants to self-fertilize, producing a second generation—the **F₂ (second filial) generation**. Both of the traits from the P generation emerged in the F₂ generation; Mendel counted 5474 round seeds and 1850 wrinkled seeds in the F₂ (see Figure 3.5). He noticed that the numbers of the round and wrinkled seeds constituted approximately a 3 to 1 ratio; that is, about $^3/_4$ of the F₂ seeds were round and $^1/_4$ were wrinkled. Mendel conducted monohybrid crosses for all seven of the characteristics that he studied in pea plants, and in all of the crosses he obtained the same result: all of the F₁ resembled only one of the two parents, but both parental traits emerged in the F₂ in an approximate ratio of 3 : 1.

What Monohybrid Crosses Reveal

Mendel drew several conclusions from the results of his monohybrid crosses. First, he reasoned that, although the F₁ plants display the phenotype of only one parent, they must inherit genetic factors from both parents because they transmit both parental phenotypes to the F₂ generation. The presence of both round and wrinkled seeds in the F₂ plants could be explained only if the F₁ plants possessed both round and wrinkled genetic factors that they had inherited from the P generation. He concluded that each plant must therefore possess two genetic factors encoding a characteristic.

The genetic factors (now called alleles) that Mendel discovered are, by convention, designated with letters: the allele for round seeds is usually represented by R and the allele for wrinkled seeds by r. The plants in the P generation of Mendel's cross possessed two identical alleles: RR in the round-seeded parent and rr in the wrinkled-seeded parent (**Figure 3.6a**).

The second conclusion that Mendel drew from his monohybrid crosses was that the two alleles in each plant separate when gametes are formed, and one allele goes into each gamete. When two gametes (one from each parent) fuse to produce a zygote, the allele from the male parent unites with the allele from the female parent to produce the genotype of the offspring. Thus, Mendel's F₁ plants inherited an R

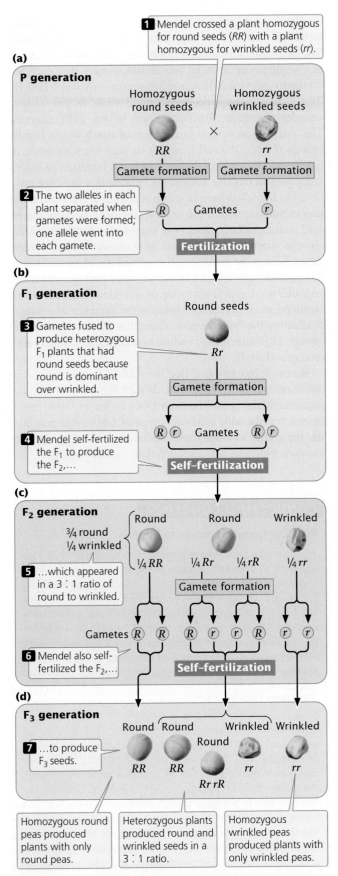

(a)

P generation

1 Mendel crossed a plant homozygous for round seeds (*RR*) with a plant homozygous for wrinkled seeds (*rr*).

Homozygous round seeds   Homozygous wrinkled seeds

RR   ×   *rr*

Gamete formation   Gamete formation

2 The two alleles in each plant separated when gametes were formed; one allele went into each gamete.

R   Gametes   r

Fertilization

(b)

F₁ generation

Round seeds

3 Gametes fused to produce heterozygous F₁ plants that had round seeds because round is dominant over wrinkled.

Rr

Gamete formation

4 Mendel self-fertilized the F₁ to produce the F₂,…

R r   Gametes   R r

Self-fertilization

(c)

F₂ generation

¾ round
¼ wrinkled

Round   Round   Wrinkled

¼ *RR*   ¼ *Rr*   ¼ *rR*   ¼ *rr*

5 …which appeared in a 3 : 1 ratio of round to wrinkled.

Gamete formation

Gametes   R R   R r   r R   r r

6 Mendel also self-fertilized the F₂,…

Self-fertilization

(d)

F₃ generation

Round Round   Wrinkled Wrinkled
Round

7 …to produce F₃ seeds.

RR   *RR*   *rr*   *rr*
Rr rR

Homozygous round peas produced plants with only round peas.

Heterozygous plants produced round and wrinkled seeds in a 3 : 1 ratio.

Homozygous wrinkled peas produced plants with only wrinkled peas.

3.6 Mendel's monohybrid crosses revealed the principle of segregation and the concept of dominance.

allele from the round-seeded plant and an *r* allele from the wrinkled-seeded plant (**Figure 3.6b**). However, only the trait encoded by the round allele (*R*) was *observed* in the F₁: all the F₁ progeny had round seeds. Those traits that appeared unchanged in the F₁ heterozygous offspring Mendel called **dominant**, and those traits that disappeared in the F₁ heterozygous offspring he called **recessive**. In plants, alleles for dominant traits are often symbolized with uppercase letters (e.g., *R*), while alleles for recessive traits are often symbolized with lowercase letters (e.g., *r*). When dominant and recessive alleles are present together, the recessive allele is masked, or suppressed. The concept of dominance was the third important conclusion that Mendel derived from his monohybrid crosses.

Mendel's fourth conclusion was that the two alleles of an individual plant separate with equal probability into the gametes. When plants of the F₁ (with genotype *Rr*) produced gametes, half of the gametes received the *R* allele for round seeds and half received the *r* allele for wrinkled seeds. The gametes then paired randomly to produce the following genotypes in equal proportions among the F₂: *RR*, *Rr*, *rR*, *rr* (**Figure 3.6c**). Because round (*R*) is dominant over wrinkled (*r*), there were three round-seeded progeny (*RR*, *Rr*, *rR*) for every wrinkled-seeded progeny (*rr*) in the F₂. This 3 : 1 ratio of round-seeded to wrinkled-seeded progeny that Mendel observed in the F₂ could be obtained only if the two alleles of a genotype separated into the gametes with equal probability.

The conclusions that Mendel drew about inheritance from his monohybrid crosses have been further developed and formalized into the principle of segregation and the concept of dominance. The **principle of segregation** (Mendel's first law; see **Table 3.2**) states that each individual diploid organism possesses two alleles for any particular characteristic,

TABLE 3.2	Comparison of the principles of segregation and independent assortment		
Principle	**Observation**	**Stage of Meiosis***
Segregation (Mendel's first law)	1. Each individual organism possesses two alleles encoding a trait.	Before meiosis
	2. Alleles separate when gametes are formed.	Anaphase I
	3. Alleles separate in equal proportions.	Anaphase I
Independent assortment (Mendel's second law)	Alleles at different loci separate independently.	Anaphase I

*Assumes that no crossing over occurs. If crossing over takes place, then segregation and independent assortment may also occur in anaphase II of meiosis.

one inherited from the maternal parent and one from the paternal parent. These two alleles segregate (separate) when gametes are formed, and one allele goes into each gamete. Furthermore, the two alleles segregate into gametes in equal proportions. The **concept of dominance** states that when two different alleles are present in a genotype, only the trait encoded by one of them—the dominant allele—is observed in the phenotype.

Mendel confirmed these principles by allowing his F_2 plants to self-fertilize and produce an F_3 generation. He found that the plants grown from the wrinkled seeds—those displaying the recessive trait (rr)—produced an F_3 in which all plants produced wrinkled seeds. Because his wrinkled-seeded plants were homozygous for wrinkled alleles (rr), only wrinkled alleles could be passed on to their progeny (**Figure 3.6d**).

The plants grown from round seeds—the dominant trait—fell into two types (see Figure 3.6c). Following self-fertilization, about $^2/_3$ of these plants produced both round-seeded and wrinkled-seeded progeny in the F_3 generation. These plants were heterozygous (Rr), so they produced $^1/_4$ RR (round), $^1/_2$ Rr (round), and $^1/_4$ rr (wrinkled) progeny, giving a 3 : 1 ratio of round to wrinkled in the F_3. About $^1/_3$ of the plants grown from round seeds were of the second type; they produced only the round-seeded trait in the F_3. These plants were homozygous for the round allele (RR) and could thus produce only round-seeded offspring in the F_3 generation. Mendel planted the seeds obtained in the F_3 and carried these plants through three more rounds of self-fertilization. In each generation, $^2/_3$ of the round-seeded plants produced round and wrinkled offspring, whereas $^1/_3$ produced only round offspring. These results are entirely consistent with the principle of segregation.

CONCEPTS

The principle of segregation states that each individual organism possesses two alleles that encode a characteristic. These alleles segregate when gametes are formed, and one allele goes into each gamete. The concept of dominance states that when the two alleles of a genotype are different, only the trait encoded by dominant allele is observed.

✔ CONCEPT CHECK 3

How did Mendel know that each of his pea plants carried two alleles encoding a characteristic?

The Molecular Nature of Alleles

Let's take a moment to consider in more detail exactly what an allele is and how it determines a phenotype. Although Mendel had no information about the physical nature of the genetic factors in his crosses, modern geneticists have now determined the molecular basis of those factors and how they encode a trait such as wrinkled peas.

Alleles, such as the R and r alleles that encode round and wrinkled peas, usually represent specific DNA sequences. The locus that determines whether a pea is round or wrinkled is a sequence of DNA on pea chromosome 5 that encodes a protein called starch-branching enzyme isoform I (SBEI). The R allele, which produces round seeds in pea plants, encodes a normal, functional form of the SBEI enzyme. This enzyme converts a linear form of starch into a highly branched form. The r allele, which encodes wrinkled seeds, is a different DNA sequence that contains a mutation or error; it encodes an inactive form of the enzyme that does not produce the branched form of starch and leads to the accumulation of sucrose within the rr seed (the pea). Because the rr seed contains a large amount of sucrose, the developing seed absorbs water and swells. Later, as the seed matures, it loses water. Because rr seeds absorb more water and expand more during development, they lose more water during maturation and afterward appear shriveled or wrinkled. The r allele for wrinkled seeds is recessive because the presence of a single R allele in the heterozygote allows the plant to synthesize enough SBEI enzyme to produce branched starch and, therefore, round seeds.

Research has revealed that the r allele contains an extra 800 base pairs of DNA that disrupt the normal coding sequence of the gene. The extra DNA appears to have come from a transposable element, a type of DNA sequence that has the ability to move from one location in the genome to another, which we will discuss further in Chapter 18.

CONNECTING CONCEPTS

Relating Genetic Crosses to Meiosis

We have now seen how the results of monohybrid crosses are explained by Mendel's principle of segregation. Many students find that they enjoy working genetic crosses but are frustrated by the abstract nature of the symbols. Perhaps you feel the same at this point. You may be asking, "What do these symbols really represent? What does the genotype RR mean in regard to the biology of the organism?" The answers to these questions lie in relating the abstract symbols of crosses to the structure and behavior of chromosomes, the repositories of genetic information (see Chapter 2).

In 1900, when Mendel's work was rediscovered and biologists began to apply his principles of heredity, the relation between genes and chromosomes was still unclear. The theory that genes are located on chromosomes (the **chromosome theory of heredity**) was developed in the early 1900s by Walter Sutton, then a graduate student at Columbia University. Through the careful study of meiosis in insects, Sutton documented the fact that each homologous pair of chromosomes consists of one maternal chromosome and one paternal chromosome. Showing that these pairs segregate independently into gametes in meiosis, he concluded that this process is the biological basis for Mendel's principles of heredity. German cytologist and embryologist Theodor Boveri came to similar conclusions at about the same time.

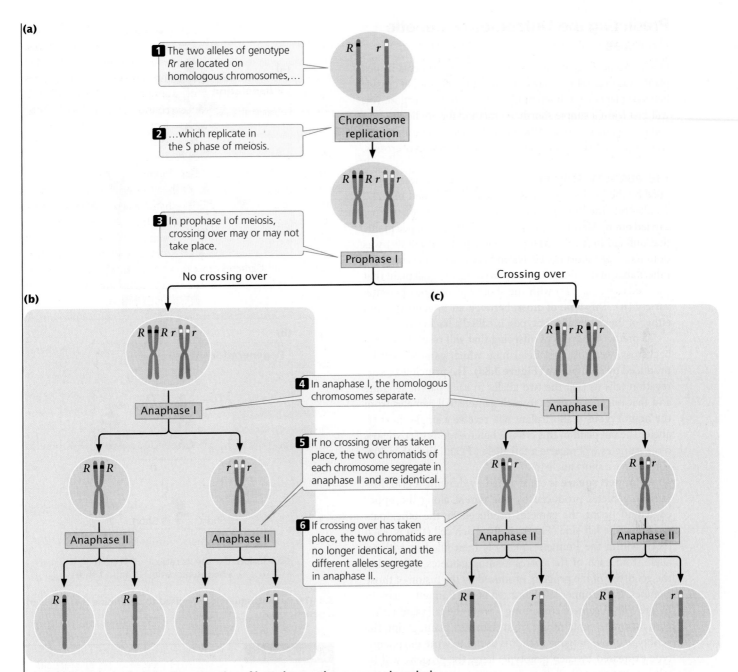

(a)

1 The two alleles of genotype *Rr* are located on homologous chromosomes,…

Chromosome replication

2 …which replicate in the S phase of meiosis.

3 In prophase I of meiosis, crossing over may or may not take place.

Prophase I

No crossing over   Crossing over

(b)   **(c)**

Anaphase I   Anaphase I

4 In anaphase I, the homologous chromosomes separate.

5 If no crossing over has taken place, the two chromatids of each chromosome segregate in anaphase II and are identical.

Anaphase II   Anaphase II   Anaphase II   Anaphase II

6 If crossing over has taken place, the two chromatids are no longer identical, and the different alleles segregate in anaphase II.

3.7 Segregation results from the separation of homologous chromosomes in meiosis.

The symbols used in genetic crosses, such as *R* and *r*, are just shorthand notations for particular sequences of DNA in the chromosomes that encode particular phenotypes. The two alleles of a genotype are found on different but homologous chromosomes. One chromosome of each homologous pair is inherited from the mother and the other is inherited from the father. In the S phase of meiotic interphase, each chromosome replicates, producing two copies of each allele, one on each chromatid (**Figure 3.7a**). The homologous chromosomes segregate in anaphase I, thereby separating the two different alleles (**Figure 3.7b and c**). This chromosome segregation is the basis of the principle of segregation. In anaphase II of meiosis, the two chromatids of each replicated chromosome separate, so each gamete resulting from meiosis carries only a single allele at each locus, as Mendel's principle of segregation predicts. If crossing over has taken place in prophase I of meiosis, then the two chromatids of each replicated chromosome are no longer identical, and the segregation of different alleles takes place at anaphase I and anaphase II (see Figure 3.7c).

Mendel knew nothing about chromosomes; he formulated his principles of heredity entirely on the basis of the results of the crosses that he carried out. Nevertheless, we should not forget that these principles work because they are based on the behavior of actual chromosomes in meiosis. ▶ **TRY PROBLEM 30**

Predicting the Outcomes of Genetic Crosses

One of Mendel's goals in conducting his experiments in pea plants was to develop a way to predict the outcomes of crosses between plants with different phenotypes. In this section, you will first learn a simple shorthand method for predicting outcomes of genetic crosses (the Punnett square), and then you will learn how to use probability to predict the results of crosses.

THE PUNNETT SQUARE The Punnett square was developed by the English geneticist Reginald C. Punnett in 1917. To illustrate the Punnett square, let's examine another cross carried out by Mendel. By crossing two varieties of pea plants that differed in height, Mendel established that tall (T) was dominant over short (t). He tested his theory concerning the inheritance of dominant traits by crossing an F_1 tall plant that was heterozygous (Tt) with the short homozygous parental variety (tt). This type of cross, between an F_1 genotype and either of the parental genotypes, is called a **backcross**.

To predict the types of offspring that will result from this backcross, we must first determine which gametes will be produced by each parent (**Figure 3.8a**). The principle of segregation tells us that the two alleles in each parent separate and that one allele passes to each gamete. All gametes from the homozygous tt short plant will receive a single short (t) allele. The tall plant in this cross is heterozygous (Tt), so 50% of its gametes will receive a tall allele (T) and the other 50% will receive a short allele (t).

A **Punnett square** is constructed by drawing a grid, listing the gametes produced by one parent along the upper edge, and listing the gametes produced by the other parent down the left side (**Figure 3.8b**). Each cell (that is, each block within the Punnett square) is then filled in with an allele from each of the corresponding gametes, generating the genotype of the progeny produced by the fusion of those gametes. In the upper left-hand cell of the Punnett square in Figure 3.8b, a gamete containing T from the tall plant unites with a gamete containing t from the short plant, giving the genotype of the progeny (Tt). It is useful to write the phenotype expressed by each genotype; here, the progeny will be tall because the tall allele is dominant over the short allele. This process is repeated for all the cells in the Punnett square.

By simply counting, we can determine the types of progeny produced and their ratios. In Figure 3.8b, two cells contain tall (Tt) progeny and two cells contain short (tt) progeny, so the genotypic ratio expected for this cross is 2 Tt to 2 tt (a 1 : 1 ratio). Another way to express this result is to say that we expect $\frac{1}{2}$ of the progeny to have genotype Tt (and the tall phenotype) and $\frac{1}{2}$ of the progeny to have genotype tt (and the short phenotype).

In this cross, the genotypic ratio and the phenotypic ratio are the same, but this outcome need not be the case for all crosses. Try completing a Punnett square for the cross in which the F_1 round-seeded plants in Figure 3.6b undergo self-fertilization (you should obtain a phenotypic ratio of 3 round to 1 wrinkled and a genotypic ratio of 1 RR to 2 Rr to 1 rr).

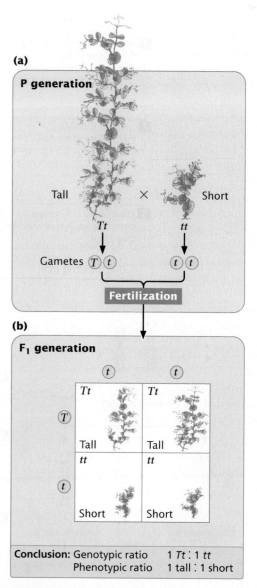

(a)

P generation

Tall   ×   Short

Tt tt

Gametes T t t t

Fertilization

(b)

F_1 generation

	t	t
T	Tt Tall	Tt Tall
t	tt Short	tt Short

Conclusion: Genotypic ratio   1 Tt : 1 tt
Phenotypic ratio   1 tall : 1 short

3.8 The Punnett square can be used to determine the results of a genetic cross.

CONCEPTS

The Punnett square is a shorthand method of predicting the genotypic and phenotypic ratios of progeny from a genetic cross.

✔ **CONCEPT CHECK 4**

If the F_1 plant depicted in Figure 3.6b is backcrossed to the parent with round seeds, what proportion of the progeny will have wrinkled seeds? (Use a Punnett square.)

a. $\frac{3}{4}$ c. $\frac{1}{4}$

b. $\frac{1}{2}$ d. 0

PROBABILITY AS A TOOL OF GENETICS Another method for determining the outcome of a genetic cross is to use the rules of probability, as Mendel did with his crosses. **Probability** expresses the likelihood of the occurrence of a particular event. It is the number of times that a particular event takes place, divided by the number of all possible outcomes. For example, a deck of 52 cards contains only one king of hearts. The probability of drawing one card from the deck at random and obtaining the king of hearts is $^1/_{52}$ because there is only one card that is the king of hearts (one event) and there are 52 cards that can be drawn from the deck (52 possible outcomes). The probability of drawing a card and obtaining an ace is $^4/_{52}$ because there are four cards that are aces (four events) and 52 cards (possible outcomes). Probability can be expressed either as a fraction ($^4/_{52}$ in this case) or as a decimal number (0.077 in this case).

The probability of a particular event may be determined by knowing something about *how* or *how often* the event takes place. We know, for example, that the probability of rolling a six-sided die and getting a four is $^1/_6$ because the die has six sides and any one side is equally likely to end up on top. So, in this case, understanding the nature of the event—the shape of the thrown die—allows us to determine its probability. In other cases, we determine the probability of an event by making a large number of observations. When a weather forecaster says that there is a 40% chance of rain on a particular day, this probability was obtained by observing a large number of days with similar atmospheric conditions and finding that it rains on 40% of those days. In this case, the probability has been determined empirically (by observation).

THE MULTIPLICATION RULE Two rules of probability are useful for predicting the ratios of offspring from genetic crosses. The first is the **multiplication rule**, which states that the probability of two or more independent events taking place together is calculated by multiplying their independent probabilities.

To illustrate the use of the multiplication rule, let's again consider the roll of a die. The probability of rolling one die and obtaining a four is $^1/_6$. To calculate the probability of rolling a die twice and obtaining two fours, we can apply the multiplication rule. The probability of obtaining a four on the first roll is $^1/_6$ and the probability of obtaining a four on the second roll is $^1/_6$, so the probability of rolling a four on both is $^1/_6 \times ^1/_6 = ^1/_{36}$ (**Figure 3.9a**). The key indicator for applying the multiplication rule is the word *and*; in the example just considered, we wanted to know the probability of obtaining a four on the first roll *and* a four on the second roll. (It may have been some time since you worked with fractions. If you are rusty and need a review, turn to the appendix at the end of the book entitled Working with Fractions [pp. B1–B3].)

For the multiplication rule to be valid, the events whose joint probability is being calculated must be independent—the outcome of one event must not influence the outcome of the other. For example, the number that comes up on one roll of the die has no influence on the number that comes up on the next roll, so these events are independent. However, if we wanted to know the probability of being hit on the head with a hammer and going to the hospital on the same day, we could not simply apply the multiplication rule and multiply the two probabilities together because the two events are not independent—being hit on the head with a hammer certainly influences the probability of going to the hospital.

THE ADDITION RULE The second rule of probability frequently used in genetics is the **addition rule**, which states that the probability of any of two or more mutually exclusive events is calculated by adding the probabilities of the events. Let's look at this rule in concrete terms. To obtain the

(a) The multiplication rule

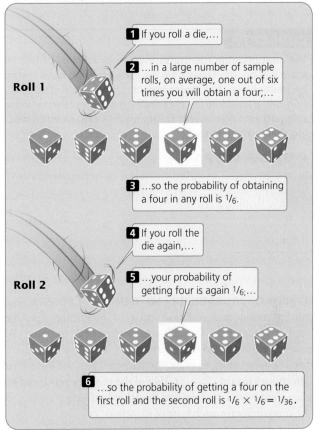

1 If you roll a die,…

2 …in a large number of sample rolls, on average, one out of six times you will obtain a four;…

Roll 1

3 …so the probability of obtaining a four in any roll is $^1/_6$.

4 If you roll the die again,…

Roll 2

5 …your probability of getting four is again $^1/_6$;…

6 …so the probability of getting a four on the first roll and the second roll is $^1/_6 \times ^1/_6 = ^1/_{36}$.

(b) The addition rule

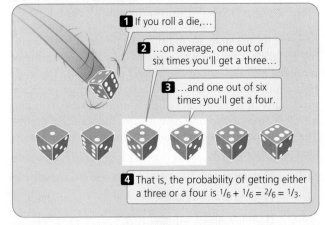

1 If you roll a die,…

2 …on average, one out of six times you'll get a three…

3 …and one out of six times you'll get a four.

4 That is, the probability of getting either a three or a four is $^1/_6 + ^1/_6 = ^2/_6 = ^1/_3$.

3.9 The multiplication and addition rules can be used to determine the probability of combinations of events.

probability of throwing a die once and rolling *either* a three *or* a four, we would use the addition rule, adding the probability of obtaining a three ($\frac{1}{6}$) to the probability of obtaining a four (again, $\frac{1}{6}$), or $\frac{1}{6} + \frac{1}{6} = \frac{2}{6} = \frac{1}{3}$ (**Figure 3.9b**). The key indicators for applying the addition rule are the words *either* and *or*.

For the addition rule to be valid, the events whose probability is being calculated must be mutually exclusive, meaning that one event excludes the possibility of the other. For example, you cannot throw a single die just once and obtain both a three and a four because only one side of the die can be on top. These events are mutually exclusive.

CONCEPTS

The multiplication rule states that the probability of two or more independent events taking place together is calculated by multiplying their independent probabilities. The addition rule states that the probability that any of two or more mutually exclusive events taking place is calculated by adding their probabilities.

✔ CONCEPT CHECK 5

If the probability of being blood-type A is $\frac{1}{8}$ and the probability of being blood-type O is $\frac{1}{2}$, what is the probability of being either blood-type A or blood-type O?

a. $\frac{5}{8}$ c. $\frac{1}{10}$
b. $\frac{1}{2}$ d. $\frac{1}{16}$

APPLYING PROBABILITY TO GENETIC CROSSES The multiplication and addition rules of probability can be used in place of the Punnett square to predict the ratios of progeny expected from a genetic cross. Let's first consider a cross between two pea plants heterozygous for the locus that determines height, $Tt \times Tt$. Half of the gametes produced by each plant have a T allele, and the other half have a t allele, so the probability for each type of gamete is $\frac{1}{2}$.

The gametes from the two parents can combine in four different ways to produce offspring. Using the multiplication rule, we can determine the probability of each possible combination. To calculate the probability of obtaining TT progeny, for example, we multiply the probability of receiving a T allele from the first parent ($\frac{1}{2}$) by the probability of receiving a T allele from the second parent ($\frac{1}{2}$). The multiplication rule should be used here because we need the probability of receiving a T allele from the first parent *and* a T allele from the second parent—two independent events. The four types of progeny from this cross and their associated probabilities are

TT	(T gamete and T gamete)	$\frac{1}{2} \times \frac{1}{2} = \frac{1}{4}$	tall
Tt	(T gamete and t gamete)	$\frac{1}{2} \times \frac{1}{2} = \frac{1}{4}$	tall
tT	(t gamete and T gamete)	$\frac{1}{2} \times \frac{1}{2} = \frac{1}{4}$	tall
tt	(t gamete and t gamete)	$\frac{1}{2} \times \frac{1}{2} = \frac{1}{4}$	short

Notice that there are two ways for heterozygous progeny to be produced: a heterozygote can either receive a T allele from the first parent and a t allele from the second or receive a t allele from the first parent and a T allele from the second.

After determining the probabilities of obtaining each progeny genotype, we can use the addition rule to determine the overall phenotypic ratios. Because of dominance, a tall plant can have genotype TT, Tt, or tT, so, using the addition rule, we find the probability of tall progeny to be $\frac{1}{4} + \frac{1}{4} + \frac{1}{4} = \frac{3}{4}$. Because only one genotype (tt) encodes the short phenotype, the probability of short progeny is simply $\frac{1}{4}$.

Two methods have now been introduced to work genetic crosses: the Punnett square and the probability method. At this point, you may be asking, "Why bother with probability rules and calculations? The Punnett square is easier to understand and just as quick." This is true for simple monohybrid crosses. For tackling more complex crosses that assess genes at two or more loci, however, the probability method is both clearer and quicker than the Punnett square.

CONDITIONAL PROBABILITY Thus far, we have used probability to predict the chances of producing certain types of progeny given only the genotypes of the parents. Sometimes, however, we have additional information that modifies, or *conditions*, the probability, a situation termed **conditional probability**. For example, assume that we cross two heterozygous pea plants ($Tt \times Tt$) and obtain a tall progeny plant. What is the probability that this tall plant is heterozygous (Tt)? You might assume that the probability would be $\frac{1}{2}$, the probability of obtaining heterozygous progeny in a cross between two heterozygotes. In this case, however, we have some additional information—the phenotype of the progeny plant—which modifies that probability. When two heterozygous individuals are crossed, we expect $\frac{1}{4}$ TT, $\frac{1}{2}$ Tt, and $\frac{1}{4}$ tt progeny. We know that the plant in question is tall, so we can eliminate the possibility that it has genotype tt. Tall progeny must be either genotype TT or genotype Tt, and in a cross between two heterozygotes, these genotypes occur in a 1 : 2 ratio. Therefore, the probability that a tall progeny plant is heterozygous (Tt) is two out of three, or $\frac{2}{3}$.

THE BINOMIAL EXPANSION AND PROBABILITY When probability is used, it is important to recognize that there may be several different ways in which a set of events can occur. Consider two parents who are both heterozygous for albinism, a recessive condition in humans that causes reduced pigmentation in the skin, hair, and eyes (**Figure 3.10**; see also the introduction to Chapter 1). When two parents heterozygous for albinism mate ($Aa \times Aa$), the probability of their having a child with albinism (aa) is $\frac{1}{4}$, and the probability of their having a child with normal pigmentation (AA or Aa) is $\frac{3}{4}$. Suppose we want to know the probability of this couple having three children with albinism. In this case, there is

3.10 Albinism in humans is usually inherited as a recessive trait. [Richard Hutchings/Science Source.]

only one way in which this can happen: their first child has albinism *and* their second child has albinism *and* their third child has albinism. Here, we simply apply the multiplication rule: $\frac{1}{4} \times \frac{1}{4} \times \frac{1}{4} = \frac{1}{64}$.

Suppose we now ask what the probability is of this couple having three children, one with albinism and two with normal pigmentation. This situation is more complicated. The first child might have albinism, whereas the second and third might be unaffected; the probability of this sequence of events is $\frac{1}{4} \times \frac{3}{4} \times \frac{3}{4} = \frac{9}{64}$. Alternatively, the first and third children might have normal pigmentation, whereas the second might have albinism; the probability of this sequence is $\frac{3}{4} \times \frac{1}{4} \times \frac{3}{4} = \frac{9}{64}$. Finally, the first two children might have normal pigmentation and the third albinism; the probability of this sequence is $\frac{3}{4} \times \frac{3}{4} \times \frac{1}{4} = \frac{9}{64}$. Because *either* the first sequence *or* the second sequence *or* the third sequence produces one child with albinism and two with normal pigmentation, we apply the addition rule and add the probabilities: $\frac{9}{64} + \frac{9}{64} + \frac{9}{64} = \frac{27}{64}$.

If we want to know the probability of this couple having five children, two with albinism and three with normal pigmentation, figuring out *all* the different combinations of children and their probabilities becomes more difficult. This task is made easier if we apply the binomial expansion.

The binomial takes the form $(p + q)^n$, where p equals the probability of one event, q equals the probability of the alternative event, and n equals the number of times the event occurs. For figuring the probability of two out of five children with albinism,

p = the probability of a child having albinism ($\frac{1}{4}$)

q = the probability of a child having normal pigmentation ($\frac{3}{4}$)

The binomial for this situation is $(p + q)^5$ because there are five children in the family ($n = 5$). The expansion of this binomial is

$$(p + q)^5 = p^5 + 5p^4q + 10p^3q^2 + 10p^2q^3 + 5pq^4 + q^5$$

Each of the terms in the expansion provides the probability of one particular combination of traits in the children. The first term in the expansion (p^5) equals the probability of having five children with albinism because p is the probability of albinism. The second term ($5p^4q$) equals the probability of having four children with albinism and one with normal pigmentation, the third term ($10p^3q^2$) equals the probability of having three children with albinism and two with normal pigmentation, and so forth.

To obtain the probability of any combination of events, we insert the values of p and q. Thus, the probability of having two out of five children with albinism is

$$10p^2q^3 = 10\,(\tfrac{1}{4})^2(\tfrac{3}{4})^3 = {}^{270}/_{1024} = 0.26$$

We can easily figure out the probability of any desired combination of albinism and pigmentation among five children by using the other terms in the expansion.

How did we expand the binomial in this example? In general, the expansion of any binomial $(p + q)^n$ consists of a series of $n + 1$ terms. In the preceding example, $n = 5$, so there are $5 + 1 = 6$ terms: p^5, $5p^4q$, $10p^3q^2$, $10p^2q^3$, $5pq^4$, and q^5. To write out the terms, first figure out their exponents. The exponent of p in the first term is always the power to which the binomial is raised, or n. In our example, $n = 5$, so our first term is p^5. The exponent of p decreases by one in each successive term, so the exponent of p is 4 in the second term (p^4), 3 in the third term (p^3), and so forth. The exponent of q is 0 (no q) in the first term and increases by one in each successive term, increasing from 0 to 5 in our example.

Next, we determine the coefficient of each term. The coefficient of the first term is always 1, so in our example, the first term is $1p^5$, or just p^5. The coefficient of the second term is always the same as the power to which the binomial is raised; in our example, this coefficient is 5, and the term is $5p^4q$. For the coefficient of the third term, we look back at the preceding term, multiply the coefficient of the preceding term (5 in our example) by the exponent of p in that term (4), and then divide by the number of that term (second term, or 2). So the coefficient of the third term in our example is $(5 \times 4)/2 = {}^{20}/_2 = 10$, and the term is $10p^3q^2$. We follow this procedure for each successive term. The coefficients for the terms in the binomial expansion can also be determined from Pascal's triangle (**Table 3.3**). The exponents and coefficients for each term in the first five binomial expansions are given in **Table 3.4**.

Another way to determine the probability of any particular combination of events is to use the following formula:

$$P = \frac{n!}{s!t!} p^s q^t$$

where P equals the overall probability of event X with probability p occurring s times and alternative event Y with

TABLE 3.3	Pascal's triangle

The numbers in each row represent the coefficients of each term in the binomial expansion $(p + q)^n$.

n	Coefficients
	1
1	1 1
2	1 2 1
3	1 3 3 1
4	1 4 6 4 1
5	1 5 10 10 5 1
6	1 6 15 20 15 6 1

Note: Each number in the triangle, except for the first, equal to the sum of the two numbers directly above it.

TABLE 3.4	Coefficients and terms for the binomial expansion $(p + q)^n$ for $n = 1$ through 5

n	Binomial Expansion
1	$a + b$
2	$a^2 + 2ab + b^2$
3	$a^3 + 3a^2b + 3ab^2 + b^3$
4	$a^4 + 4a^3b + 6a^2b^2 + 4ab^3 + b^4$
5	$a^5 + 5a^4b + 10a^3b^2 + 10a^2b^3 + 5ab^4 + b^5$

probability q occurring t times. For our albinism example, event X would be the occurrence of a child with albinism ($^1/_4$) and event Y would be the occurrence of a child with normal pigmentation ($^3/_4$); s would equal the number of children with albinism (2) and t would equal the number of children with normal pigmentation (3). The ! symbol stands for factorial, and it means the product of all the integers from n to 1. In this example, $n = 5$; so $n! = 5 \times 4 \times 3 \times 2 \times 1$. Applying this formula to obtain the probability of two out of five children having albinism, we obtain

$$P = \frac{5!}{2!3!}(^1/_4)^2(^3/_4)^3 = \frac{5 \times 4 \times 3 \times 2 \times 1}{2 \times 1 \times 3 \times 2 \times 1}(^1/_4)^2(^3/_4)^3 = 0.26$$

This value is the same as that obtained with the binomial expansion. **> TRY PROBLEMS 25, 26, AND 27**

THINK-PAIR-SHARE Question 4

The Testcross

A useful tool for analyzing genetic crosses is the **testcross**, in which one individual of unknown genotype is crossed with another individual with a homozygous recessive genotype for the trait in question. Figure 3.8 illustrates a testcross (in this case, it is also a backcross). A testcross tests, or reveals, the genotype of the first individual.

Suppose you were given a tall pea plant with no information about its parents. Because tallness is a dominant trait in peas, your plant could be either homozygous (TT) or heterozygous (Tt) for the dominant allele, but you would not know which. You could determine its genotype by performing a testcross. If the plant were homozygous (TT), a testcross would produce all tall progeny ($TT \times tt \rightarrow$ all Tt); if the plant were heterozygous (Tt), half of the progeny would be tall and half would be short ($Tt \times tt \rightarrow {}^1/_2\ Tt$ and $^1/_2\ tt$). When a testcross is performed, any recessive allele in the unknown genotype will be expressed in the progeny because it will be paired with a recessive allele from the homozygous recessive parent. **> TRY PROBLEMS 18 AND 21**

CONCEPTS

The binomial expansion can be used to determine the probability of a particular set of events. A testcross is a cross between an individual with an unknown genotype and one with a homozygous recessive genotype. The outcome of the testcross can reveal the unknown genotype.

Genetic Symbols

As we have seen, genetic crosses are usually depicted with symbols that designate the different alleles. The symbols used for alleles are usually determined by the community of geneticists who work on a particular organism, and therefore there is no universal system for designating symbols. In plants, lowercase letters are often used to designate recessive alleles and uppercase letters to designate dominant alleles. Two or three letters may be used for a single allele: the recessive allele for heart-shaped leaves in cucumbers is designated *hl*, and the recessive allele for abnormal sperm-head shape in mice is designated *azh*.

In animals, the most common allele for a characteristic—called the **wild type** because it is the allele usually found in the wild—is often symbolized by one or more letters and a plus sign ($+$). The letter or letters chosen are usually based on a mutant (less common) phenotype. For example, the recessive allele that encodes yellow eyes in the Oriental fruit fly is represented by *ye*, whereas the allele for wild-type eye color is represented by ye^+. At times, the letters for the wild-type allele are dropped and the allele is represented simply by a plus sign.

Superscripts and subscripts are sometimes added to distinguish between genes: Lfr_1 and Lfr_2 represent dominant mutant alleles at different loci that produce lacerate leaf margins in opium poppies; El^R represents an allele in goats that restricts the length of the ears.

A slash may be used to distinguish the two alleles present in an individual genotype. For example, the genotype of a goat that is heterozygous for restricted ears might be written El^+/El^R, or simply $+/El^R$. If genotypes at more than one locus are presented together, a space separates the genotypes. For example, a goat heterozygous for a pair of alleles that produces restricted ears and heterozygous for another pair of alleles that produces goiter can be designated El^+/El^R G/g. Sometimes it is useful to designate the possibility of several genotypes. An underline in a genotype, such as $A\_$, indicates that any allele is possible. In this case, $A\_$ might include both AA and Aa genotypes.

CONNECTING CONCEPTS

Ratios in Simple Crosses

Now that we have had some experience with genetic crosses, let's review the ratios that appear in the progeny of simple crosses, in which a single locus is under consideration and one of the alleles is dominant over the other. Understanding these ratios and the parental genotypes that produce them will enable you to work simple genetic crosses quickly, without resorting to the Punnett square. Later in this chapter, we use these ratios to work more complicated crosses that include several loci.

There are only three phenotypic ratios to understand (**Table 3.5**). The first is the 3 : 1 ratio, which arises in a simple genetic cross when both of the parents are heterozygous for a dominant trait ($Aa \times Aa$). The second phenotypic ratio is the 1 : 1 ratio, which results from the mating of a heterozygous parent and a homozygous recessive parent. The homozygous parent in this cross must carry two recessive alleles ($Aa \times aa$) because a cross between a homozygous dominant parent and a heterozygous parent ($AA \times Aa$) produces offspring displaying only the dominant trait.

TABLE 3.5	Phenotypic ratios for simple genetic crosses (crosses for a single locus) with dominance	
Phenotypic Ratio	**Genotypes of Parents**	**Genotypes of Progeny**
3 : 1	$Aa \times Aa$	$^3/_4 A\_ : ^1/_4 aa$
1 : 1	$Aa \times aa$	$^1/_2 Aa : ^1/_2 aa$
Uniform progeny	$AA \times AA$	All AA
	$aa \times aa$	All aa
	$AA \times aa$	All Aa
	$AA \times Aa$	All $A\_$

The third phenotypic ratio is not really a ratio: all the offspring have the same phenotype (uniform progeny). Several combinations of parental genotypes can produce this outcome (see Table 3.5). A cross between any two homozygous parents—either between two parents of the same homozygous genotype ($AA \times AA$ or $aa \times aa$) or between two parents with different homozygous genotypes ($AA \times aa$)—produces progeny all having the same phenotype. Progeny of a single phenotype can also result from a cross between a homozygous dominant parent and a heterozygote ($AA \times Aa$).

If we are interested in the ratios of genotypes instead of phenotypes, there are still only three outcomes to remember (**Table 3.6**): the 1 : 2 : 1 ratio, produced by a cross between two heterozygotes; the 1 : 1 ratio, produced by a cross between a heterozygote and a homozygote; and uniform progeny, produced by a cross between two homozygotes. These simple phenotypic and genotypic ratios, and the parental genotypes that produce them, provide the key to understanding crosses not only for a single locus but also, as you will see in the next section, for multiple loci.

TABLE 3.6	Genotypic ratios for simple genetic crosses (crosses for a single locus)	
Genotypic Ratio	**Genotypes of Parents**	**Genotypes of Progeny**
1 : 2 : 1	$Aa \times Aa$	$^1/_4 AA : ^1/_2 Aa : ^1/_4 aa$
1 : 1	$Aa \times aa$	$^1/_2 Aa : ^1/_2 aa$
	$Aa \times AA$	$^1/_2 Aa : ^1/_2 AA$
Uniform progeny	$AA \times AA$	All AA
	$aa \times aa$	All aa
	$AA \times aa$	All Aa

3.3 Dihybrid Crosses Reveal the Principle of Independent Assortment

We now extend Mendel's principle of segregation to some more complex crosses that include alleles at multiple loci. Understanding the nature of these crosses will require an additional principle: the principle of independent assortment.

Dihybrid Crosses

In addition to his work on monohybrid crosses, Mendel crossed varieties of peas that differed in *two* characteristics—that is, he performed **dihybrid crosses**. For example, he crossed one homozygous variety that had seeds that were round and yellow with another homozygous variety that had seeds that were wrinkled and green. The seeds of all the F_1 progeny were round and yellow. He then allowed the F_1 to self-fertilize and obtained the following progeny in the F_2: 315 round, yellow seeds; 101 wrinkled, yellow seeds; 108 round, green seeds; and 32 wrinkled, green seeds. Mendel recognized that these traits appeared in a ratio of approximately 9 : 3 : 3 : 1; that is, $^9/_{16}$ of the progeny were round and yellow,

$^3/_{16}$ were wrinkled and yellow, $^3/_{16}$ were round and green, and $^1/_{16}$ were wrinkled and green.

The Principle of Independent Assortment

Mendel carried out a number of dihybrid crosses for pairs of characteristics and always obtained a 9 : 3 : 3 : 1 ratio in the F_2. This ratio makes perfect sense in regard to the principle of segregation and the concept of dominance if we add a third principle, which Mendel recognized in his dihybrid crosses: the **principle of independent assortment** (Mendel's second law). This principle states that alleles at different loci separate independently of one another (see Table 3.2).

A common mistake is to think that the principle of segregation and the principle of independent assortment refer to two different processes. The principle of independent assortment is really an extension of the principle of segregation. The principle of segregation states that the two alleles at a locus separate when gametes are formed; the principle of independent assortment states that, when these two alleles separate, their separation is independent of the separation of alleles at *other* loci.

Let's see how the principle of independent assortment explains the results that Mendel obtained in the dihybrid cross described above. Each pea plant possesses two alleles encoding each characteristic, so the parent plants must have had genotypes *RR YY* and *rr yy* (**Figure 3.11a**). The principle of segregation tells us that the alleles at each locus separate, and that one allele for each locus passes into each gamete. The gametes produced by the round, yellow parent therefore contained alleles *RY*, whereas the gametes produced by the wrinkled, green parent contained alleles *ry*. These two types of gametes united to produce the F_1, all with genotype *Rr Yy*. Because round is dominant over wrinkled and yellow is dominant over green, the phenotype of the F_1 was round and yellow.

When Mendel allowed the F_1 plants to self-fertilize to produce the F_2, the alleles at each locus separated, and one of those alleles passed into each gamete. This event is where the principle of independent assortment becomes important. The two pairs of alleles can separate in two ways: (1) *R* separates with *Y*, and *r* separates with *y*, to produce gametes *RY* and *ry*, or (2) *R* separates with *y*, and *r* separates with *Y*, to produce gametes *Ry* and *rY*. The principle of independent assortment tells us that the alleles at each locus separate independently; thus, both kinds of separation take place with equal frequency, and all four types of gametes (*RY*, *ry*, *Ry*, and *rY*) are produced in equal proportions (**Figure 3.11b**). When these four types of gametes are combined to produce the F_2 generation, the progeny consist of $^9/_{16}$ round and yellow, $^3/_{16}$ wrinkled and yellow, $^3/_{16}$ round and green, and $^1/_{16}$ wrinkled and green, resulting in a 9 : 3 : 3 : 1 phenotypic ratio (**Figure 3.11c**).

THINK-PAIR-SHARE Question 5

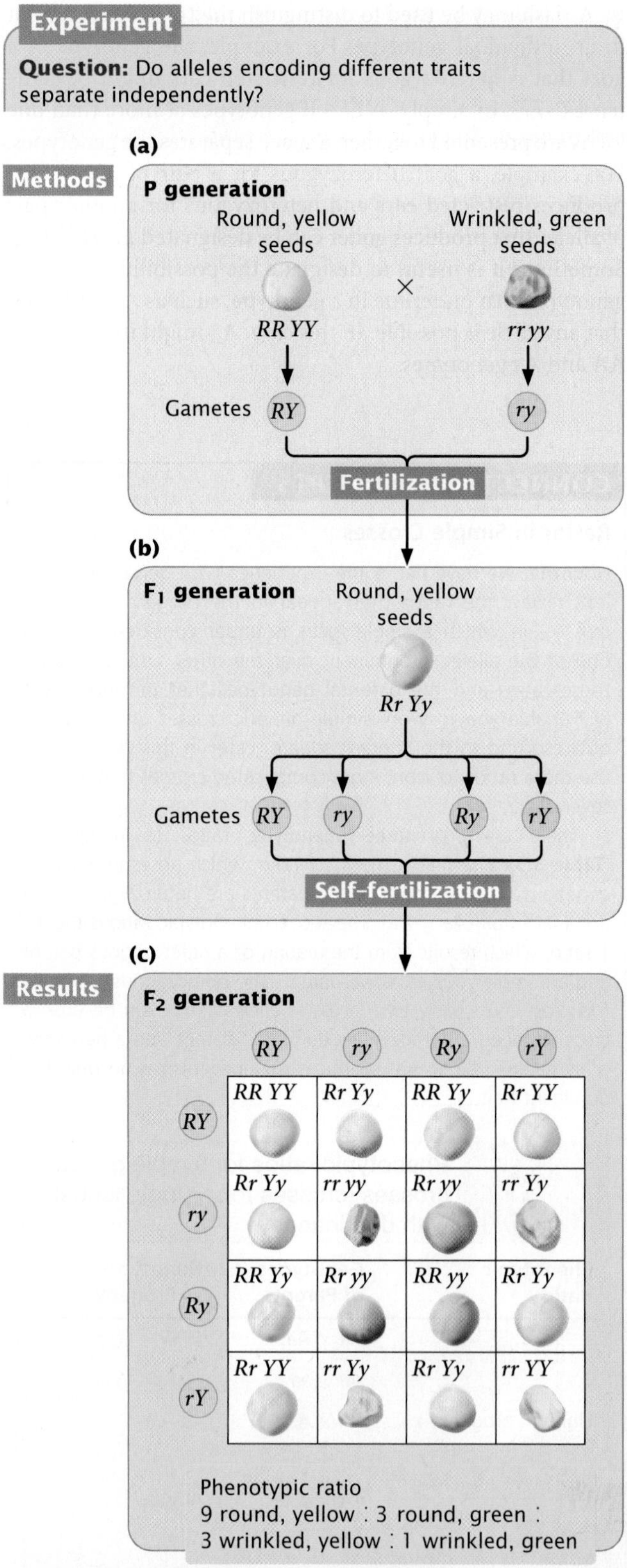

Question: Do alleles encoding different traits separate independently?

(a)

Methods

P generation

Round, yellow seeds   ×   Wrinkled, green seeds

RR YY   *rr yy*

Gametes   *RY*   *ry*

Fertilization

(b)

F_1 generation   Round, yellow seeds

Rr Yy

Gametes   *RY*   *ry*   *Ry*   *rY*

Self-fertilization

(c)

Results

F_2 generation

Phenotypic ratio
9 round, yellow : 3 round, green : 3 wrinkled, yellow : 1 wrinkled, green

Conclusion: The allele encoding color separated independently of the allele encoding seed shape, producing a 9 : 3 : 3 : 1 ratio in the F_2 progeny.

3.11 Mendel's dihybrid crosses revealed the principle of independent assortment.

Relating the Principle of Independent Assortment to Meiosis

An important qualification of the principle of independent assortment is that it applies to characteristics encoded by loci located on different chromosomes. Like the principle of segregation, it is based wholly on the behavior of chromosomes in meiosis. Each pair of homologous chromosomes separates independently of all other pairs in anaphase I of meiosis (**Figure 3.12**), so genes located on different pairs of homologs will assort independently. Genes that happen to be located on the same chromosome will travel together during anaphase I of meiosis and will arrive at the same destination—within the same gamete (unless crossing over takes place). So genes located on the same chromosome do not assort independently (unless they are located sufficiently far apart that crossing over takes place in every meiotic division, a situation that will be discussed fully in Chapter 7).

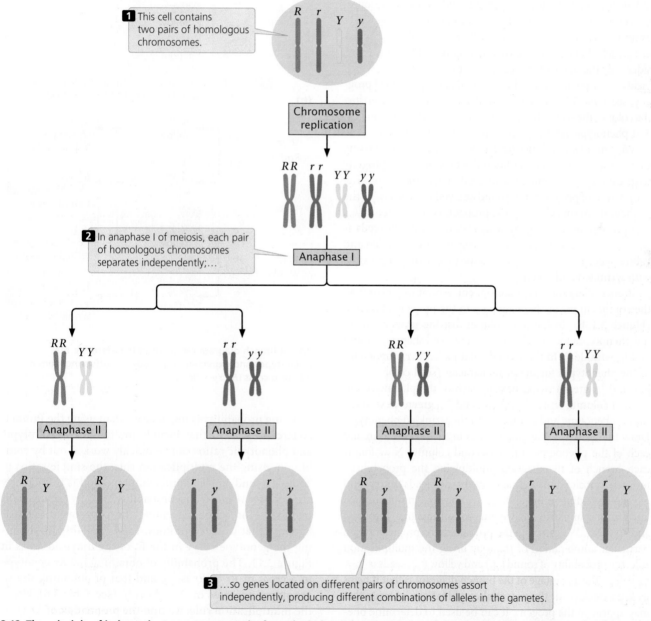

1 This cell contains two pairs of homologous chromosomes.

Chromosome replication

2 In anaphase I of meiosis, each pair of homologous chromosomes separates independently;…

Anaphase I

Anaphase II   Anaphase II   Anaphase II   Anaphase II

3 …so genes located on different pairs of chromosomes assort independently, producing different combinations of alleles in the gametes.

3.12 The principle of independent assortment results from the independent separation of chromosomes in anaphase I of meiosis.

Applying Probability and the Branch Diagram to Dihybrid Crosses

When the genes at two loci separate independently, a dihybrid cross can be understood as two monohybrid crosses. Let's examine Mendel's dihybrid cross ($Rr\,Yy \times Rr\,Yy$) by considering each characteristic separately (**Figure 3.13a**). If we consider only the shape of the seeds, the cross was $Rr \times Rr$, which yields a 3 : 1 phenotypic ratio ($\frac{3}{4}$ round and $\frac{1}{4}$ wrinkled progeny; see Table 3.5). Next, we consider the other characteristic, the color of the seeds. The cross was $Yy \times Yy$, which produces a 3 : 1 phenotypic ratio ($\frac{3}{4}$ yellow and $\frac{1}{4}$ green progeny).

We can now combine these monohybrid ratios by using the multiplication rule to obtain the proportion of progeny with different combinations of seed shape and color. The proportion of progeny with round and yellow seeds is $\frac{3}{4}$ (the probability of round) $\times \frac{3}{4}$ (the probability of yellow) $= \frac{9}{16}$. The proportion of progeny with round and green seeds is $\frac{3}{4} \times \frac{1}{4} = \frac{3}{16}$; the proportion of progeny with wrinkled and yellow seeds is $\frac{1}{4} \times \frac{3}{4} = \frac{3}{16}$; and the proportion of progeny with wrinkled and green seeds is $\frac{1}{4} \times \frac{1}{4} = \frac{1}{16}$.

Branch diagrams are a convenient way of organizing all the combinations of characteristics in the progeny of a cross (**Figure 3.13b**). In the first column, list the proportions of the phenotypes for one character (here, seed shape: $\frac{3}{4}$ round and $\frac{1}{4}$ wrinkled). In the second column, list the proportions of the phenotypes for the next character (seed color: $\frac{3}{4}$ yellow and $\frac{1}{4}$ green) twice, next to each of the phenotypes in the first column: write "$\frac{3}{4}$ yellow" and "$\frac{1}{4}$ green" next to the round phenotype and again next to the wrinkled phenotype. Draw lines between the phenotypes in the first column and each of the phenotypes in the second column. Now follow each branch of the diagram, multiplying the probabilities for each trait along that branch. One branch leads from round to yellow, yielding round and yellow progeny. Another branch leads from round to green, yielding round and green progeny, and so forth. We calculate the probability of progeny with a particular combination of traits by using the multiplication rule: the probability of round ($\frac{3}{4}$) and yellow ($\frac{3}{4}$) seeds is $\frac{3}{4} \times \frac{3}{4} = \frac{9}{16}$. The advantage of the branch diagram is that it helps us keep track of all the potential combinations of traits that may appear in the progeny. It can be used to determine phenotypic or genotypic ratios for any number of characteristics.

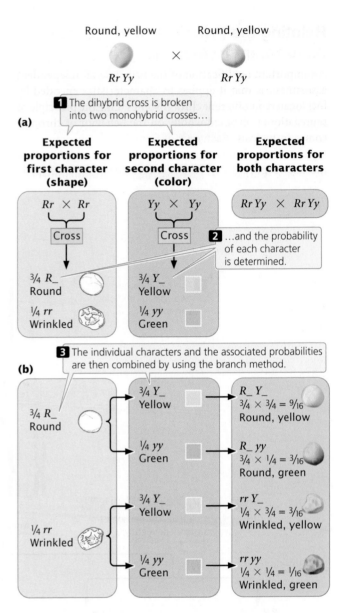

3.13 A branch diagram can be used to determine the phenotypes and expected proportions of offspring from a dihybrid cross ($Rr\,Yy \times Rr\,Yy$).

Using probability is much faster than using the Punnett square for crosses that include multiple loci. Genotypic and phenotypic ratios can be quickly worked out by combining, using the multiplication rule, the simple ratios in Tables 3.5 and 3.6. The probability method is particularly efficient if we need the probability of only a *particular* phenotype or genotype among the progeny of a cross. Suppose that we need to know the probability of obtaining the genotype $Rr\,yy$ in the F_2 of the dihybrid cross in Figure 3.11. The probability of obtaining the Rr genotype in a cross of $Rr \times Rr$ is $\frac{1}{2}$, and that of obtaining the yy genotype in a cross of $Yy \times Yy$ is $\frac{1}{4}$ (see Table 3.6). Using the multiplication rule, we find the probability of $Rr\,yy$ to be $\frac{1}{2} \times \frac{1}{4} = \frac{1}{8}$.

To illustrate the advantage of the probability method, consider the cross *Aa Bb cc Dd Ee* × *Aa Bb Cc dd Ee*. Suppose that we want to know the probability of obtaining offspring with the genotype *aa bb cc dd ee*. If we use a Punnett square to determine this probability, we might be working on the solution for months. However, we can quickly figure the probability of obtaining this one genotype by breaking this cross into a series of single-locus crosses:

Progeny cross	Genotype	Probability
Aa × *Aa*	*aa*	$1/4$
Bb × *Bb*	*bb*	$1/4$
cc × *Cc*	*cc*	$1/2$
Dd × *dd*	*dd*	$1/2$
Ee × *Ee*	*ee*	$1/4$

The probability of an offspring from this cross having genotype *aa bb cc dd ee* is now easily obtained by using the multiplication rule: $1/4 \times 1/4 \times 1/2 \times 1/2 \times 1/4 = 1/256$. This calculation assumes that the genes at these five loci all assort independently.

Now that you've had some experience working genetic crosses, explore Mendel's principles of heredity by setting up some of your own crosses in **Animation 3.1**.

Ⓐ

CONCEPTS

A cross including several characteristics can be worked by breaking it down into single-locus crosses and using the multiplication rule to determine the proportions of combinations of characteristics (provided that the genes assort independently).

The Dihybrid Testcross

Let's practice using the branch diagram by determining the types and proportions of phenotypes in a dihybrid testcross between the round and yellow F_1 pea plants (*Rr Yy*) obtained by Mendel in his dihybrid cross and the wrinkled and green pea plants (*rr yy*), as depicted in **Figure 3.14**. First, break the cross down into a series of single-locus crosses. The cross *Rr* × *rr* yields $1/2$ round (*Rr*) progeny and $1/2$ wrinkled (*rr*) progeny. The cross *Yy* × *yy* yields $1/2$ yellow (*Yy*) progeny and $1/2$ green (*yy*) progeny. Using the multiplication rule, we find the proportion of round and yellow progeny to be $1/2$ (the probability of round) × $1/2$ (the probability of yellow) = $1/4$. Four combinations of traits appear in the offspring in the following proportions: $1/4$ *Rr Yy*, round yellow; $1/4$ *Rr yy*, round green; $1/4$ *rr Yy*, wrinkled yellow; and $1/4$ *rr yy*, wrinkled green.

THINK-PAIR-SHARE Question 6

3.14 A branch diagram can be used to determine the phenotypes and expected proportions of offspring from a dihybrid testcross (*Rr Yy* × *rr yy*).

WORKED PROBLEM

The principles of segregation and independent assortment are important not only because they explain how heredity works, but also because they provide the means for predicting the outcome of genetic crosses. This predictive power has made genetics a powerful tool in agriculture and other fields, and the ability to apply the principles of heredity is an important skill for all students of genetics. Practice with genetic problems is essential for mastering the basic principles of heredity; no amount of reading and memorization can substitute for the experience gained by deriving solutions to specific problems in genetics.

You may find genetics problems difficult if you are unsure of where to begin or how to organize a solution to the problem. In genetics, every problem is different, so no common series of steps can be applied to all genetics problems. Logic and common sense must be used to analyze a problem and arrive at a solution. Nevertheless, certain steps can facilitate the process, and solving the following problem will serve to illustrate those steps.

In mice, black coat color (*B*) is dominant over brown (*b*), and a solid pattern (*S*) is dominant over a white-spotted pattern (*s*). Color and spotting are controlled by genes that assort independently. A homozygous black, spotted mouse is crossed with a homozygous brown, solid mouse. All the F₁ mice are black and solid. A testcross is then carried out by mating the F₁ mice with brown, spotted mice.

a. Give the genotypes of the parents and the F₁ mice.

b. Give the genotypes and phenotypes, along with their expected ratios, for the progeny expected from the testcross.

Solution Strategy

What information is required in your answer to the problem?

First, determine what question or questions the problem is asking. Is it asking for genotypes, or genotypic ratios, or phenotypic ratios? This problem asks you to provide the *genotypes* of the parents and the F₁, the *expected genotypes* and *phenotypes* of the progeny of the testcross, and their *expected proportions*.

What information is provided to solve the problem?

Next, determine what information is provided that will be necessary to solve the problem. This problem gives important information about the dominance relations of the traits involved and the genes that encode them:

- Black is dominant over brown.
- Solid is dominant over white-spotted.
- The genes for the two characteristics assort independently.
- Symbols for the different alleles: *B* for black, *b* for brown, *S* for solid, and *s* for spotted.

It is often helpful to write down the symbols at the beginning of the solution:

> *B*—black   *S*—solid
> *b*—brown   *s*—white-spotted

Next, write out the crosses given in the problem:

| P | Homozygous black, spotted | × | Homozygous brown, solid |

F₁   Black, solid

Testcross   Black, solid   ×   Brown, spotted

For help with this problem, review:

If you need help solving the problem, review those sections of the chapter that cover the relevant information. For this problem, review Sections 3.2 and 3.3.

Solution Steps

STEP 1 Write down any genetic information that can be determined from the phenotypes alone.

From their phenotypes and the statement that they are homozygous, you know that the P-generation mice must be *BB ss* and *bb SS*. The F₁ mice are black and solid, both of which are dominant traits, so the F₁ mice must possess at least one black allele (*B*) and one solid allele (*S*). At this point, you may not be certain about the other alleles, so you can represent the genotype of the F₁ as *B_ S_*, where _ means that any allele is possible. The brown, spotted mice used in the testcross must be *bb ss* because both brown and spotted are recessive traits that will be expressed only if two recessive alleles are present. Record these genotypes on the crosses that you wrote out:

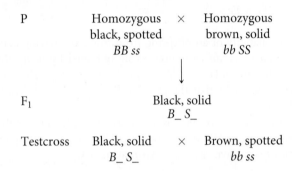

STEP 2 Break the problem down into smaller parts.

First, determine the genotype of the F₁. After this genotype has been determined, you can predict the results of the testcross and determine the genotypes and phenotypes of the progeny of the testcross.

Second, because this cross includes two independently assorting loci, it can be conveniently broken down into two single-locus crosses: one for coat color and the other for spotting.

Third, you can use a branch diagram to determine the proportion of progeny of the testcross with different combinations of the two traits.

STEP 3 Work the different parts of the problem.

Start by determining the genotype of the F₁ progeny. Mendel's first law indicates that the two alleles at a locus separate, one going into each gamete. Thus, the gametes produced by the black, spotted parent contain *B s* and the gametes produced by the brown, solid parent contain *b S*, which combine to produce F₁ progeny with the genotype *Bb Ss*:

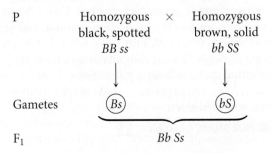

Use the F_1 genotype to work the testcross ($Bb\ Ss \times bb\ ss$), breaking it into two single-locus crosses. First, consider the cross for coat color: $Bb \times bb$. Any cross between a heterozygote and a homozygous recessive genotype produces a 1 : 1 phenotypic ratio of progeny (see Table 3.5):

$$Bb \times bb$$
$$\downarrow$$
$$\tfrac{1}{2}\ Bb\ \text{black}$$
$$\tfrac{1}{2}\ bb\ \text{brown}$$

Next, consider the cross for spotting: $Ss \times ss$. This cross is also between a heterozygote and a homozygous recessive genotype and produces $\tfrac{1}{2}$ solid (Ss) and $\tfrac{1}{2}$ spotted (ss) progeny (see Table 3.5).

$$Ss \times ss$$
$$\downarrow$$
$$\tfrac{1}{2}\ Ss\ \text{solid}$$
$$\tfrac{1}{2}\ ss\ \text{spotted}$$

Finally, determine the proportions of progeny with combinations of these characters by using a branch diagram.

$\tfrac{1}{2}\ Bb$ black
- $\tfrac{1}{2}\ Ss$ solid ⟶ $Bb\ Ss$ black, solid $\tfrac{1}{2} \times \tfrac{1}{2} = \tfrac{1}{4}$
- $\tfrac{1}{2}\ ss$ spotted ⟶ $Bb\ ss$ black, spotted $\tfrac{1}{2} \times \tfrac{1}{2} = \tfrac{1}{4}$

$\tfrac{1}{2}\ bb$ brown
- $\tfrac{1}{2}\ Ss$ solid ⟶ $bb\ Ss$ brown, solid $\tfrac{1}{2} \times \tfrac{1}{2} = \tfrac{1}{4}$
- $\tfrac{1}{2}\ ss$ spotted ⟶ $bb\ ss$ brown, spotted $\tfrac{1}{2} \times \tfrac{1}{2} = \tfrac{1}{4}$

STEP 4 **Check all work.**

As a last step, reread the problem, checking to see if your answers are consistent with the information provided. You have used the genotypes $BB\ ss$ and $bb\ SS$ in the P generation. Do these genotypes encode the phenotypes given in the problem? Are the F_1 progeny phenotypes consistent with the genotypes that you assigned? The answers are consistent with the information.

> ❯❯ Now that we have stepped through a genetics problem together, try your hand at **Problem 33** at the end of this chapter.

3.4 Observed Ratios of Progeny May Deviate from Expected Ratios by Chance

When two individual organisms of known genotype are crossed, we expect certain ratios of genotypes and phenotypes among the progeny; these expected ratios are based on the Mendelian principles of segregation, independent assortment, and dominance. The ratios of genotypes and phenotypes *actually* observed among the progeny, however, may deviate from these expectations.

For example, in German cockroaches, brown body color (Y) is dominant over yellow body color (y). If we cross a brown, heterozygous cockroach (Yy) with a yellow cockroach (yy), we expect a 1 : 1 ratio of brown (Yy) and yellow (yy) progeny. Among 40 progeny, we therefore expect to see 20 brown and 20 yellow offspring. However, the observed numbers might deviate from these expected values; we might in fact see 22 brown and 18 yellow progeny.

Chance plays a critical role in genetic crosses, just as it does in flipping a coin. When you flip a coin, you expect a 1 : 1 ratio—$\tfrac{1}{2}$ heads and $\tfrac{1}{2}$ tails. If you flipped a coin 1000 times, the proportion of heads and tails obtained would probably be very close to that expected 1 : 1 ratio. However, if you flipped the coin 10 times, the ratio of heads to tails might be quite different from 1 : 1. You could easily get 6 heads and 4 tails, or 3 heads and 7 tails, just by chance. You might even get 10 heads and 0 tails. The same thing happens in genetic crosses. We may expect 20 brown and 20 yellow cockroaches, but 22 brown and 18 yellow progeny *could* arise as a result of chance.

The Chi-Square Goodness-of-Fit Test

If you expected a 1 : 1 ratio of brown and yellow cockroaches but the cross produced 22 brown and 18 yellow cockroaches, you probably wouldn't be too surprised, even though it wasn't a perfect 1 : 1 ratio. In this case, it seems reasonable to assume that chance produced the deviation between the expected and the observed results. But if you observed 25 brown and 15 yellow cockroaches, would you still assume that this result represents a 1 : 1 ratio? Something other than chance might have caused this deviation. Perhaps the inheritance of this characteristic is more complicated than was assumed, or perhaps some of the yellow progeny died before they were counted. Clearly, we need some means of evaluating how likely it is that chance is responsible for a deviation between the observed and the expected numbers.

To evaluate the role of chance in producing deviations between observed and expected values, a statistical test called the **chi-square goodness-of-fit test** is used. This test provides information about how well observed values fit expected values. Before we learn how to use this test, however, it is important to understand what it does and does not indicate about a genetic cross. The chi-square test cannot tell us whether a genetic cross has been correctly carried out, whether the results are correct, or whether we have chosen the correct genetic explanation for the results. What it does indicate is the *probability* that the difference between the observed and the expected values is due to chance. In other words, it indicates the likelihood that chance alone could produce the deviation between the expected and the observed values.

If we expected 20 brown and 20 yellow progeny from a genetic cross, the chi-square test gives us the probability that we might observe 25 brown and 15 yellow progeny simply owing to chance deviations from the expected 20 : 20 ratio. This hypothesis—that chance alone is responsible for a deviation between observed and expected values—is sometimes called the *null hypothesis*. Statistical tests such as the chi-square test cannot prove that the null hypothesis is correct, but they can help us decide whether we should reject it. When the probability calculated with the chi-square test is high, we assume that chance alone produced the deviation, and we do not reject the null hypothesis. When the probability is low, we assume that some factor other than chance—some significant factor—produced the deviation; for example, the mortality rate of the yellow cockroaches might be higher than that of the brown cockroaches. When the probability that chance produced the deviation is low, we reject the null hypothesis.

To use the chi-square goodness-of-fit test, we first determine the expected results. The chi-square test must always be applied to *numbers* of progeny, not to proportions or percentages. Let's consider a locus for coat color in domestic cats, for which black color (*B*) is dominant over gray (*b*). If we crossed two heterozygous black cats (*Bb* × *Bb*), we would expect a 3 : 1 ratio of black and gray kittens. Imagine that a series of such crosses yields a total of 50 kittens—30 black and 20 gray. These numbers are our *observed* values. We can obtain the *expected* numbers by multiplying the expected proportions by the total number of observed progeny. In this case, the expected number of black kittens is $\frac{3}{4} \times 50 = 37.5$ and the expected number of gray kittens is $\frac{1}{4} \times 50 = 12.5$.

The chi-square (χ^2) value is calculated by using the following formula:

$$\chi^2 = \Sigma \frac{(\text{observed} - \text{expected})^2}{\text{expected}}$$

where Σ means the sum. We calculate the sum of all the squared differences between observed and expected values divided by the expected values. To calculate the chi-square value for our black and gray kittens, we first subtract the number of *expected* black kittens from the number of *observed* black kittens ($30 - 37.5 = -7.5$) and square this value: $-7.5^2 = 56.25$. We then divide this result by the expected number of black kittens, $56.25/37.5 = 1.5$. We repeat the calculations on the number of expected gray kittens: $(20 - 12.5)^2/12.5 = 4.5$. To obtain the overall chi-square value, we sum the (observed – expected)^2/expected values: $1.5 + 4.5 = 6.0$.

The next step is to determine the probability associated with this calculated chi-square value, which is the probability that the deviation between the observed and the expected results is due to chance. This step requires us to compare our calculated chi-square value (6.0) with theoretical values in a chi-square table that have the same degrees of freedom. The degrees of freedom represent the number of ways in which the expected classes are free to vary. For a chi-square goodness-of-fit test, the degrees of freedom are equal to $n - 1$, in which n is the number of different expected phenotypes. Here, we lose one degree of freedom because the total number of expected progeny must equal the total number of observed progeny. In our example, there are two expected phenotypes (black and gray), so $n = 2$, and the degree of freedom equals $2 - 1 = 1$.

TABLE 3.7	Critical values of the χ^2 distribution								
	P								
df	0.995	0.975	0.9	0.5	0.1	0.05*	0.025	0.01	0.005
1	0.000	0.000	0.016	0.455	2.706	3.841	5.024	6.635	7.879
2	0.010	0.051	0.211	1.386	4.605	5.991	7.378	9.210	10.597
3	0.072	0.216	0.584	2.366	6.251	7.815	9.348	11.345	12.838
4	0.207	0.484	1.064	3.357	7.779	9.488	11.143	13.277	14.860
5	0.412	0.831	1.610	4.351	9.236	11.070	12.832	15.086	16.750
6	0.676	1.237	2.204	5.348	10.645	12.592	14.449	16.812	18.548
7	0.989	1.690	2.833	6.346	12.017	14.067	16.013	18.475	20.278
8	1.344	2.180	3.490	7.344	13.362	15.507	17.535	20.090	21.955
9	1.735	2.700	4.168	8.343	14.684	16.919	19.023	21.666	23.589
10	2.156	3.247	4.865	9.342	15.987	18.307	20.483	23.209	25.188
11	2.603	3.816	5.578	10.341	17.275	19.675	21.920	24.725	26.757
12	3.074	4.404	6.304	11.340	18.549	21.026	23.337	26.217	28.300
13	3.565	5.009	7.042	12.340	19.812	22.362	24.736	27.688	29.819
14	4.075	5.629	7.790	13.339	21.064	23.685	26.119	29.141	31.319
15	4.601	6.262	8.547	14.339	22.307	24.996	27.488	30.578	32.801

P, probability; *df*, degrees of freedom.
*Most scientists assume that, when *P* < 0.05, a significant difference exists between the observed and the expected values in a chi-square test.

Now that we have our calculated chi-square value and have figured out the associated degrees of freedom, we are ready to obtain the probability from a chi-square table (**Table 3.7**). The degrees of freedom are given in the left-hand column of the table and the probabilities are given at the top; within the body of the table are chi-square values associated with these probabilities. First, we find the row for the appropriate degrees of freedom; for our example with 1 degree of freedom, it is the first row of the table. Then, we find our calculated chi-square value (6.0) among the theoretical values in this row. The theoretical chi-square values increase, and the probabilities decrease, from left to right. Our chi-square value of 6.0 falls between the value of 5.024, associated with a probability of 0.025, and the value of 6.635, associated with a probability of 0.01. Thus, the probability associated with our chi-square value is less than 0.025 and greater than 0.01. So there is less than a 2.5% probability that the deviation that we observed between the expected and the observed numbers of black and gray kittens could be due to chance.

Most scientists use the 0.05 probability level as their cutoff value: if the probability of chance being responsible for the deviation between observed and expected values is greater than or equal to 0.05, they accept that chance may be responsible for the deviation. When the probability is less than 0.05, scientists assume that chance is not responsible and that a significant difference from the expected values exists. The expression *significant difference* means that a factor other than chance is responsible for the deviation between the observed and expected values. In regard to the kittens, perhaps one of the genotypes had a greater mortality rate before the progeny were counted, or perhaps other genetic factors skewed the observed ratios.

In choosing 0.05 as the cutoff value, scientists have agreed to assume that chance is responsible for deviations between observed and expected values unless there is strong evidence to the contrary. Bear in mind that even if we obtain a probability of, say, 0.01, there is still a 1% probability that the deviation between the observed and the expected values is due to nothing more than chance. Calculation of the chi-square value is illustrated in **Figure 3.15**. ▶ **TRY PROBLEM 38**

THINK-PAIR-SHARE Question 7 👥

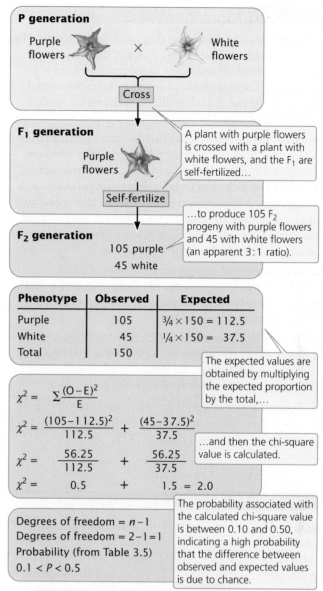

Phenotype	Observed	Expected
Purple	105	$3/4 \times 150 = 112.5$
White	45	$1/4 \times 150 = 37.5$
Total	150	

The expected values are obtained by multiplying the expected proportion by the total,…

$$\chi^2 = \sum \frac{(O-E)^2}{E}$$

$$\chi^2 = \frac{(105-112.5)^2}{112.5} + \frac{(45-37.5)^2}{37.5}$$

$$\chi^2 = \frac{56.25}{112.5} + \frac{56.25}{37.5}$$

$$\chi^2 = 0.5 + 1.5 = 2.0$$

…and then the chi-square value is calculated.

Degrees of freedom = $n-1$
Degrees of freedom = $2-1=1$
Probability (from Table 3.5)
$0.1 < P < 0.5$

The probability associated with the calculated chi-square value is between 0.10 and 0.50, indicating a high probability that the difference between observed and expected values is due to chance.

Conclusion: No significant difference between observed and expected values.

3.15 A chi-square goodness-of-fit test is used to determine the probability that the difference between observed and expected values is due to chance.

CONCEPTS

Differences between observed and expected ratios among the progeny of a cross can arise by chance. The chi-square goodness-of-fit test can be used to evaluate whether deviations between observed and expected numbers are likely to be due to chance or to some other significant factor.

✔ CONCEPT CHECK 7

A chi-square test comparing observed and expected numbers of progeny is carried out, and the probability associated with the calculated chi-square value is 0.72. What does this probability represent?

a. Probability that the correct results were obtained
b. Probability of obtaining the observed numbers
c. Probability that the difference between observed and expected numbers is significant
d. Probability that the difference between observed and expected numbers could be due to chance

CONCEPTS SUMMARY

■ Gregor Mendel discovered the principles of heredity. His success can be attributed to his choice of the pea plant as an experimental organism, the use of characteristics with a few easily distinguishable phenotypes, his experimental approach, the use of mathematics to interpret his results, and careful attention to detail.

■ Genes are inherited factors that determine a characteristic. Alternative forms of a gene are called alleles. The alleles are located at a specific place, called a locus, on a chromosome, and the set of alleles that an individual organism possesses is its genotype. A phenotype is the manifestation or appearance of a characteristic and may refer to a physical, physiological, biochemical, or behavioral characteristic. Only the genotype—not the phenotype—is inherited.

■ The principle of segregation states that a diploid individual organism possesses two alleles encoding a trait and that these two alleles separate in equal proportions when gametes are formed.

■ The concept of dominance indicates that when two different alleles are present in a heterozygote, only the trait encoded by one of them, the dominant allele, is observed in the phenotype. The other allele is said to be recessive.

■ The two alleles of a genotype are located on homologous chromosomes. The separation of homologous chromosomes in anaphase I of meiosis brings about the segregation of alleles.

■ Probability is the likelihood that a particular event will occur. The multiplication rule states that the probability of two or more independent events occurring together is calculated by multiplying the probabilities of the independent events. The addition rule states that the probability of any of two or more mutually exclusive events occurring is calculated by adding the probabilities of the events.

■ The binomial expansion can be used to determine the probability of a particular combination of events.

■ A testcross, which reveals the genotype (homozygote or heterozygote) of an individual organism that has a dominant trait, consists of crossing that individual with one that has the homozygous recessive genotype.

■ The principle of independent assortment states that genes encoding different characteristics separate independently when gametes are formed. Independent assortment is based on the random separation of homologous pairs of chromosomes in anaphase I of meiosis; it takes place when genes encoding different characteristics are located on different pairs of chromosomes.

■ Observed ratios of progeny from a genetic cross may deviate from the expected ratios owing to chance. The chi-square goodness-of-fit test can be used to determine the probability that a difference between observed and expected numbers could be due to chance.

IMPORTANT TERMS

gene (p. 50)
allele (p. 50)
locus (p. 50)
genotype (p. 51)
homozygous (p. 51)
heterozygous (p. 51)
phenotype (p. 51)
monohybrid cross (p. 51)
P (parental) generation (p. 51)

F_1 (first filial) generation (p. 51)
reciprocal cross (p. 52)
F_2 (second filial) generation (p. 52)
dominant (p. 53)
recessive (p. 53)
principle of segregation (Mendel's first law) (p. 53)

concept of dominance (p. 54)
chromosome theory of heredity (p. 54)
backcross (p. 56)
Punnett square (p. 56)
probability (p. 56)
multiplication rule (p. 57)
addition rule (p. 57)

conditional probability (p. 58)
testcross (p. 60)
wild type (p. 60)
dihybrid cross (p. 61)
principle of independent assortment (Mendel's second law) (p. 62)
chi-square goodness-of-fit test (p. 67)

ANSWERS TO CONCEPT CHECKS

1. b

2. A locus is a place on a chromosome where genetic information encoding a characteristic is located. An allele is a version of a gene that encodes a specific trait. A genotype is the set of alleles possessed by an individual organism, and a phenotype is the manifestation or appearance of a characteristic.

3. The traits encoded by both alleles appeared in the F_2 progeny.

4. d

5. a

6. Both the principle of segregation and the principle of independent assortment refer to the separation of alleles in anaphase I of meiosis. The principle of segregation says that these alleles separate, and the principle of independent assortment says that they separate independently of alleles at other loci.

7. d

WORKED PROBLEMS

Problem 1

Short hair (*S*) in rabbits is dominant over long hair (*s*). The following crosses are carried out, producing the progeny shown. Give all possible genotypes of the parents in each cross.

	Parents	Progeny
a.	short × short	4 short and 2 long
b.	short × short	8 short
c.	short × long	12 short
d.	short × long	3 short and 1 long
e.	long × long	2 long

›› Solution Strategy

What information is required in your answer to the problem?

All possible genotypes of the parents in each cross.

What information is provided to solve the problem?

- Short hair is dominant over long hair.
- Phenotypes of the parents in each cross.
- Phenotypes and number of progeny of each cross.

For help with this problem, review:
Connecting Concepts: Ratios in Simple Crosses, in Section 3.2.

›› Solution Steps

For this problem, it is useful to first gather as much information about the genotypes of the parents as possible on the basis of their phenotypes. We can then look at the types of progeny produced to provide the missing information.

a. short × short   4 short and 2 long

Because short hair is dominant over long hair, a rabbit with short hair could be either *SS* or *Ss*. The two long-haired offspring must be homozygous (*ss*) because long hair is recessive and will appear in the phenotype only when two alleles for long hair are present. Because each parent contributes one of the two alleles found in the progeny, each parent must be carrying the *s* allele and must therefore be *Ss*.

b. short × short   8 short

The short-haired parents could be *SS* or *Ss*. All eight of the offspring are short haired (*S_*), so at least one of the parents is likely to be homozygous (*SS*); if both parents

Note: The problem asks for all possible genotypes of the parents.

Note: When both parents are heterozygotes (*Ss* × *Ss*) we expect to see a 3 : 1 ratio in progeny, but just by chance the progeny exhibited a 4 : 2 ratio.

were heterozygous, we would expect $^1/_4$ of the progeny to be long haired (*ss*), but we do not observe any long-haired progeny. The other parent could be homozygous (*SS*) or heterozygous (*Ss*); as long as one parent is homozygous, all the offspring will be short haired. It is theoretically possible, although unlikely, that both parents are heterozygous (*Ss* × *Ss*). If both were heterozygous, we would expect two of the eight progeny to be long haired. Although no long-haired progeny are observed, it is possible that just by chance no long-haired rabbits would be produced among the eight progeny of the cross.

c. short × long   12 short

The short-haired parent could be *SS* or *Ss*. The long-haired parent must be *ss*. If the short-haired parent were heterozygous (*Ss*), half of the offspring would be expected to be long haired, but we don't see any long-haired progeny. Therefore, this parent is most likely homozygous (*SS*). It is theoretically possible, although unlikely, that the parent is heterozygous and just by chance no long-haired progeny were produced.

d. short × long   3 short and 1 long

On the basis of its phenotype, the short-haired parent could be homozygous (*SS*) or heterozygous (*Ss*), but the presence of one long-haired offspring tells us that the short-haired parent must be heterozygous (*Ss*). The long-haired parent must be homozygous (*ss*).

e. long × long   2 long

Because long hair is recessive, both parents must be homozygous for a long-hair allele (*ss*).

Problem 2

In cats, black coat color is dominant over gray. A female black cat whose mother is gray mates with a gray male. If this female has a litter of six kittens, what is the probability that three will be black and three will be gray?

» Solution Strategy

What information is required in your answer to the problem?

The probability that in a litter of six kittens, three will be black and three will be gray.

What information is provided to solve the problem?

- Black is dominant over gray.
- The mother of the litter is black and her mother is gray.
- The father of the litter is gray.

For help with this problem, review:

The Binomial Expansion and Probability, in Section 3.2.

» Solution Steps

Hint: We can determine the female parent's genotype from her phenotype and her mother's phenotype.

Because black (*G*) is dominant over gray (*g*), a black cat may be homozygous (*GG*) or heterozygous (*Gg*). The black female in this problem must be heterozygous (*Gg*) because her mother is gray (*gg*) and she must have inherited one of her mother's alleles. The gray male must be homozygous (*gg*) because gray is recessive. Thus, the cross is

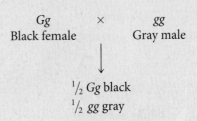

$$Gg \qquad \times \qquad gg$$
Black female   Gray male

$$^1/_2 \; Gg \text{ black}$$
$$^1/_2 \; gg \text{ gray}$$

Recall: The binomial expansion can be used to determine the probability of different combinations of traits in the progeny of a cross.

We can use the binomial expansion to determine the probability of obtaining three black and three gray kittens in a litter of six. Let *p* equal the probability of a kitten being black and *q* equal the probability of a kitten being gray. The binomial is $(p + q)^6$, expansion of which is

Hint: See pp. 59–60 for an explanation of how to expand the binomial.

$$(p + q)^6 = p^6 + 6p^5q + 15p^4q^2 + 20p^3q^3 + 15p^2q^4 + 6p^1q^5 + q^6$$

The probability of obtaining three black and three gray kittens in a litter of six is provided by the term $20p^3q^3$. The probabilities of *p* and *q* are both $^1/_2$, so the overall probability is $20(^1/_2)^3(^1/_2)^3 = \,^{20}/_{64} = \,^5/_{16}$.

Problem 3

In corn, purple kernels are dominant over yellow kernels, and full kernels are dominant over shrunken kernels. A corn plant that has purple and full kernels is crossed with a plant that has yellow and shrunken kernels, and the following progeny are obtained:

purple, full	112
purple, shrunken	103
yellow, full	91
yellow, shrunken	94

What are the most likely genotypes of the parents and progeny? Test your genetic hypothesis with a chi-square test.

» Solution Strategy

What information is required in your answer to the problem?

a. The genotypes of parents and progeny.

b. A chi-square test comparing the observed and expected results and the interpretation of the chi-square test.

What information is provided to solve the problem?

- Purple kernels are dominant over yellow kernels, and full kernels are dominant over shrunken kernels.

- The phenotypes of the parents.
- The phenotypes and numbers of progeny of the cross.

For help with this problem, review:

Sections 3.3 and 3.4.

≫ Solution Steps

The best way to begin this problem is by breaking the cross down into simple crosses for a single characteristic (seed color or seed shape):

P   purple × yellow   full × shrunken

F_1 112 + 103 = 215 purple   112 + 91 = 203 full

        91 + 94 = 185 yellow   103 + 94 = 197 shrunken

> **Hint:** A good strategy in a cross involving multiple characteristics is to analyze the results for each characteristic separately.

In this cross, purple × yellow produces approximately $\frac{1}{2}$ purple and $\frac{1}{2}$ yellow (a 1 : 1 ratio). A 1 : 1 ratio is usually the result of a cross between a heterozygote and a homozygote. Because purple is dominant, the purple parent must be heterozygous (Pp) and the yellow parent must be homozygous (pp). The purple progeny produced by this cross will be heterozygous (Pp), and the yellow progeny must be homozygous (pp).

> **Recall:** The multiplication rule states that the probability of two or more independent events occurring together is calculated by multiplying their independent probabilities.

Now let's examine the other character. Full × shrunken produces $\frac{1}{2}$ full and $\frac{1}{2}$ shrunken, or a 1 : 1 ratio, so these progeny phenotypes are also produced by a cross between a heterozygote (Ff) and a homozygote (ff); the full-kernel progeny will be heterozygous (Ff) and the shrunken-kernel progeny will be homozygous (ff).

Now combine the two crosses and use the multiplication rule to obtain the overall genotypes and the proportions of each genotype:

P   Purple, full   ×   Yellow, shrunken
        $Pp\,Ff$ $pp\,ff$

F_1 $Pp\,Ff = \frac{1}{2}$ purple × $\frac{1}{2}$ full = $\frac{1}{4}$ purple, full

        $Pp\,ff = \frac{1}{2}$ purple × $\frac{1}{2}$ shrunken = $\frac{1}{4}$ purple, shrunken

        $pp\,Ff = \frac{1}{2}$ yellow × $\frac{1}{2}$ full = $\frac{1}{4}$ yellow, full

        $pp\,ff = \frac{1}{2}$ yellow × $\frac{1}{2}$ shrunken = $\frac{1}{4}$ yellow, shrunken

Our genetic calculations predict that from this cross, we should see $\frac{1}{4}$ purple, full-kernel progeny; $\frac{1}{4}$ purple, shrunken-kernel progeny; $\frac{1}{4}$ yellow, full-kernel progeny; and $\frac{1}{4}$ yellow, shrunken-kernel progeny. A total of 400 progeny were produced, so $\frac{1}{4} \times 400 = 100$ of each phenotype are expected. Therefore, the observed numbers do not fit the expected numbers exactly.

Could the difference between what we observe and what we expected be due to chance? If the probability is high that chance alone is responsible for the difference between observed and expected values, we will assume that the progeny have been produced in the 1 : 1 : 1 : 1 ratio predicted for the cross. If the probability that the difference between observed and expected values is due to chance is low, we will assume that the progeny really are not in the predicted ratio and that some other, *significant* factor must be responsible for the deviation from our expectations.

The observed and expected numbers are

Phenotype	Observed	Expected
purple, full	112	$\frac{1}{4} \times 400 = 100$
purple, shrunken	103	$\frac{1}{4} \times 400 = 100$
yellow, full	91	$\frac{1}{4} \times 400 = 100$
yellow, shrunken	94	$\frac{1}{4} \times 400 = 100$

> **Hint:** See Figure 3.15 for help on how to carry out a chi-square test.

To determine the probability that the difference between observed and expected numbers is due to chance, we calculate a chi-square value using the formula $\chi^2 = \Sigma[(\text{observed} - \text{expected})^2/\text{expected}]$:

$$\chi^2 = \frac{(112 - 100)^2}{100} + \frac{(103 - 100)^2}{100}$$

$$+ \frac{(91 - 100)^2}{100} + \frac{(94 - 100)^2}{100}$$

$$= \frac{12^2}{100} + \frac{3^2}{100} + \frac{9^2}{100} + \frac{6^2}{100}$$

$$= \frac{144}{100} + \frac{9}{100} + \frac{81}{100} + \frac{36}{100}$$

$$= 1.44 + 0.09 + 0.81 + 0.36 = 2.70$$

The next step is to determine the probability associated with this calculated chi-square value, which is the probability that the deviation between the observed and the expected results is due to chance. This step requires us to compare our calculated chi-square value (2.70) with theoretical values in a chi-square table that have the same degrees of freedom. The degrees of freedom for a chi-square goodness-of-fit test are $n - 1$, where n equals the number of expected phenotypic classes. In this case, there are four expected phenotypic classes, so the degrees of freedom equal 4 − 1 = 3. We must now look up the chi-square value in a chi-square table (see Table 3.7). We select the row corresponding to 3 degrees of freedom and look along this row to find our calculated chi-square value. The calculated chi-square value of 2.7 lies between 2.366 (a probability of 0.5) and 6.251 (a probability of 0.1). The probability (P) associated with the calculated chi-square value is therefore 0.5 > P > 0.1. This P is the probability that the difference between what we observed and what we expected is due to chance, which in this case is relatively high, so chance is probably responsible for the deviation. We can conclude that the progeny *do* appear in the 1 : 1 : 1 : 1 ratio predicted by our genetic explanation.

COMPREHENSION QUESTIONS

Section 3.1

1. Why was Mendel's approach to the study of heredity so successful?

2. What is the difference between genotype and phenotype?

Section 3.2

3. What is the principle of segregation? Why is it important?

4. How are Mendel's principles different from the concept of blending inheritance discussed in Chapter 1?

5. What is the concept of dominance?

6. What are the addition and multiplication rules of probability and when should they be used?

7. Give the genotypic ratios that may appear among the progeny of simple crosses and the genotypes of the parents that may give rise to each ratio.

8. What is the chromosome theory of heredity? Why was it important?

Section 3.3

9. What is the principle of independent assortment? How is it related to the principle of segregation?

10. In which phases of mitosis and meiosis are the principles of segregation and independent assortment at work?

Section 3.4

11. How is the chi-square goodness-of-fit test used to analyze genetic crosses? What does the probability associated with a chi-square value indicate about the results of a cross?

APPLICATION QUESTIONS AND PROBLEMS

Introduction

12. If blond hair in the Solomon Islanders had originated from early European explorers, what would you predict the researchers would have found when they conducted their genetic study of the islanders?

Section 3.1

13. What characteristics of an organism would make it suitable for studies of the principles of inheritance? Can you name several organisms that have these characteristics?

Section 3.2

*14. In cucumbers, orange fruit color (R) is dominant over cream fruit color (r). A cucumber plant homozygous for orange fruit is crossed with a plant homozygous for cream fruit. The F_1 are intercrossed to produce the F_2.

a. Give the genotypes and phenotypes of the parents, the F_1, and the F_2.

b. Give the genotypes and phenotypes of the offspring of a backcross between the F_1 and the orange-fruited parent.

c. Give the genotypes and phenotypes of a backcross between the F_1 and the cream-fruited parent.

15. **Figure 1.1** (p. 2) shows three girls, one of whom has albinism. Could the three girls shown in the photograph be sisters? Why or why not?

16. J. W. McKay crossed a stock melon plant that produced tan seeds with a plant that produced red seeds and obtained the following results (J. W. McKay. 1936. *Journal of Heredity* 27:110–112).

Cross	F_1	F_2
tan ♀ × red ♂	13 tan seeds	93 tan, 24 red seeds

a. Explain the inheritance of tan and red seeds in this plant.

b. Assign symbols for the alleles in this cross and give genotypes for all the individual plants.

*17. White (w) coat color in guinea pigs is recessive to black (W). In 1909, W. E. Castle and J. C. Phillips transplanted an ovary from a black guinea pig into a white female whose ovaries had been removed. They then mated this white female with a white male. All the offspring from the mating were black (W. E. Castle and J. C. Phillips. 1909. *Science* 30:312–313).

[Wegner/ARCO/Nature Picture Library; Nigel Cattlin/Alamy.]

a. Explain the results of this cross.

b. Give the genotype of the offspring of this cross.

c. What, if anything, does this experiment indicate about the validity of the pangenesis and the germ-plasm theories discussed in Chapter 1?

*18. In cats, blood-type A results from an allele (I^A) that is dominant over an allele (i^B) that produces blood-type B. There is no O blood type. The blood types of male and female cats that were mated and the blood types of their kittens follow. Give the most likely genotypes for the parents of each litter.

	Male parent	Female parent	Kittens
a.	A	B	4 with type A, 3 with type B
b.	B	B	6 with type B
c.	B	A	8 with type A
d.	A	A	7 with type A, 2 with type B
e.	A	A	10 with type A
f.	A	B	4 with type A, 1 with type B

19. **Figure 3.8** shows the results of a cross between a tall pea plant and a short pea plant.

 a. What phenotypes and proportions will be produced if a tall F_1 plant is backcrossed to the short parent?

 b. What phenotypes and proportions will be produced if a tall F_1 plant is backcrossed to the tall parent?

20. Joe has a white cat named Sam. When Joe crosses Sam with a black cat, he obtains $\frac{1}{2}$ white kittens and $\frac{1}{2}$ black kittens. When the black kittens are interbred, all the kittens that they produce are black. On the basis of these results, would you conclude that white or black coat color in cats is a recessive trait? Explain your reasoning.

*21. In sheep, lustrous fleece results from an allele (L) that is dominant over an allele (l) for normal fleece. A ewe (adult female) with lustrous fleece is mated with a ram (adult male) with normal fleece. The ewe then gives birth to a single lamb with normal fleece. From this single offspring, is it possible to determine the genotypes of the two parents? If so, what are their genotypes? If not, why not?

[Jeffrey van Daele/FeaturePics.]

*22. Alkaptonuria is a metabolic disorder in which affected people produce black urine. Alkaptonuria results from an allele (a) that is recessive to the allele for normal metabolism (A). Sally has normal metabolism, but her brother has alkaptonuria. Sally's father has alkaptonuria, and her mother has normal metabolism.

 a. Give the genotypes of Sally, her mother, her father, and her brother.

 b. If Sally's parents have another child, what is the probability that this child will have alkaptonuria?

 c. If Sally marries a man with alkaptonuria, what is the probability that their first child will have alkaptonuria?

23. Suppose that you are raising Mongolian gerbils. You notice that some of your gerbils have white spots, whereas others have solid coats. What type of crosses could you carry out to determine whether white spots are due to a recessive or a dominant allele?

24. Hairlessness in American rat terriers is recessive to the presence of hair. Suppose that you have a rat terrier with hair. How can you determine whether this dog is homozygous or heterozygous for the hairy trait?

*25. What is the probability of rolling one six-sided die and obtaining the following numbers?

 a. 2

 b. 1 or 2

 c. An even number

 d. Any number but a 6

*26. What is the probability of rolling two six-sided dice and obtaining the following numbers?

 a. 2 and 3

 b. 6 and 6

 c. At least one 6

 d. Two of the same number (two 1s, or two 2s, or two 3s, etc.)

 e. An even number on both dice

 f. An even number on at least one die

*27. In a family of seven children, what is the probability of obtaining the following numbers of boys and girls?

 a. All boys

 b. All children of the same sex

 c. Six girls and one boy

 d. Four boys and three girls

 e. Four girls and three boys

28. Phenylketonuria (PKU) is a disease that results from a recessive gene. Suppose that two unaffected parents produce a child with PKU.

 a. What is the probability that a sperm from the father will contain the PKU allele?

 b. What is the probability that an egg from the mother will contain the PKU allele?

 c. What is the probability that their next child will have PKU?

 d. What is the probability that their next child will be heterozygous for the PKU gene?

*29. In German cockroaches, curved wings (cv) are recessive to normal wings (cv^+). A homozygous cockroach having normal wings is crossed with a homozygous cockroach having curved wings. The F_1 are intercrossed to produce the F_2. Assume that the pair of chromosomes containing the locus for wing shape is metacentric. Draw this pair of chromosomes as it would appear in the parents, the F$_1$, and each class of F$_2$ progeny at metaphase I of meiosis. Assume that no crossing over takes place. At each stage, label a location for the alleles for wing shape (*cv* and *cv*$^+$) on the chromosomes.

*30. In guinea pigs, the allele for black fur (*B*) is dominant over the allele for brown (*b*) fur. A black guinea pig is crossed with a brown guinea pig, producing five F$_1$ black guinea pigs and six F$_1$ brown guinea pigs.

a. How many copies of the black allele (*B*) will be present in each cell of an F$_1$ black guinea pig at the following stages: G$_1$, G$_2$, metaphase of mitosis, metaphase I of meiosis, metaphase II of meiosis, and after the second cytokinesis following meiosis? Assume that no crossing over takes place.

b. How many copies of the brown allele (*b*) will be present in each cell of an F$_1$ brown guinea pig at the same stages as those listed in part *a*? Assume that no crossing over takes place.

Section 3.3

31. In watermelons, bitter fruit (*B*) is dominant over sweet fruit (*b*), and yellow spots (*S*) are dominant over no spots (*s*). The genes for these two characteristics assort independently. A homozygous plant that has bitter fruit and yellow spots is crossed with a homozygous plant that has sweet fruit and no spots. The F$_1$ are intercrossed to produce the F$_2$.

a. What are the phenotypic ratios in the F$_2$?

b. If an F$_1$ plant is backcrossed with the bitter, yellow-spotted parent, what phenotypes and proportions are expected in the offspring?

c. If an F$_1$ plant is backcrossed with the sweet, unspotted parent, what phenotypes and proportions are expected in the offspring?

32. **Figure 3.11** shows the results of a dihybrid cross involving seed shape and seed color.

a. What proportion of the round and yellow F$_2$ progeny from this cross is homozygous at both loci?

b. What proportion of the round and yellow F$_2$ progeny from this cross is homozygous at least at one locus?

*33. In cats, curled ears result from an allele (*Cu*) that is dominant over an allele (*cu*) for normal ears. Black color results from an independently assorting allele (*G*) that is dominant over an allele for gray (*g*).
[Jean-Michel Labat/Science Source.]
A gray cat homozygous for curled ears is mated with a homozygous black cat with normal ears. All the F$_1$ cats are black and have curled ears.

a. If two of the F$_1$ cats mate, what phenotypes and proportions are expected in the F$_2$?

b. An F$_1$ cat mates with a stray cat that is gray and possesses normal ears. What phenotypes and proportions of progeny are expected from this cross?

*34. The following two genotypes are crossed: *Aa Bb Cc dd Ee* × *Aa bb Cc Dd Ee*. What will the proportion of the following genotypes be among the progeny of this cross?

a. *Aa Bb Cc Dd Ee*

b. *Aa bb Cc dd ee*

c. *aa bb cc dd ee*

d. *AA BB CC DD EE*

35. In mice, an allele for apricot eyes (*a*) is recessive to an allele for brown eyes (*a*$^+$). At an independently assorting locus, an allele for tan coat color (*t*) is recessive to an allele for black coat color (*t*$^+$). A mouse that is homozygous for brown eyes and black coat color is crossed with a mouse having apricot eyes and a tan coat. The resulting F$_1$ are intercrossed to produce the F$_2$. In a litter of eight F$_2$ mice, what is the probability that two will have apricot eyes and tan coats?

36. In cucumbers, dull fruit (*D*) is dominant over glossy fruit (*d*), orange fruit (*R*) is dominant over cream fruit (*r*), and bitter cotyledons (*B*) are dominant over non-bitter cotyledons (*b*). The three characters are encoded by genes located on different pairs of chromosomes. A plant homozygous for dull, orange fruit and bitter cotyledons is crossed with a plant that has glossy, cream fruit and non-bitter cotyledons. The F$_1$ are intercrossed to produce the F$_2$.

a. Give the phenotypes and their expected proportions in the F$_2$.

b. An F$_1$ plant is crossed with a plant that has glossy, cream fruit and non-bitter cotyledons. Give the phenotypes and expected proportions among the progeny of this cross.

*37. Alleles *A* and *a* are located on a pair of metacentric chromosomes. Alleles *B* and *b* are located on a pair of acrocentric chromosomes. A cross is made between individuals having the following genotypes: *Aa Bb* × *aa bb*.

a. Draw the chromosomes as they would appear in each type of gamete produced by these individuals.

b. For each type of progeny resulting from this cross, draw the chromosomes as they would appear in a cell at G$_1$, G$_2$, and metaphase of mitosis.

Section 3.4

*38. J. A. Moore investigated the inheritance of spotting patterns in leopard frogs (J. A. Moore. 1943. *Journal of Heredity* 34:3–7). The pipiens phenotype had the normal spots that give leopard frogs their name. In contrast, the burnsi phenotype lacked spots on its back. Moore carried out the following crosses, producing the progeny indicated.

Parent phenotypes	Progeny phenotypes
burnsi × burnsi	39 burnsi, 6 pipiens
burnsi × pipiens	23 burnsi, 33 pipiens
burnsi × pipiens	196 burnsi, 210 pipiens

a. On the basis of these results, is the burnsi phenotype most likely inherited as a dominant trait or as a recessive trait?

b. Give the most likely genotypes of the parent in each cross (use B for the burnsi allele and B^+ for pipiens allele).

c. Use a chi-square test to evaluate the fit of the observed numbers of progeny to the number expected on the basis of your proposed genotypes.

39. In the 1800s, a man with dwarfism who lived in Utah produced a large number of descendants: 22 children, 49 grandchildren, and 250 great-grandchildren (see the illustration of a family pedigree to the right), many of whom also exhibited dwarfism (F. F. Stephens. 1943. *Journal of Heredity* 34:229–235). The type of dwarfism found in this family is called Schmid-type metaphyseal chondrodysplasia, although it was originally thought to be achondroplastic dwarfism. Among the families of this kindred, dwarfism appeared only in members who had one parent with dwarfism. When one parent exhibited dwarfism, the following numbers of children were produced.

Family in which one parent had dwarfism	Children with normal stature	Children with dwarfism
A	15	7
B	4	6
C	1	6
D	6	2
E	2	2
F	8	4
G	4	4
H	2	1
I	0	1
J	3	1
K	2	3
L	2	1
M	2	0
N	1	0
O	0	2
Total	52	40

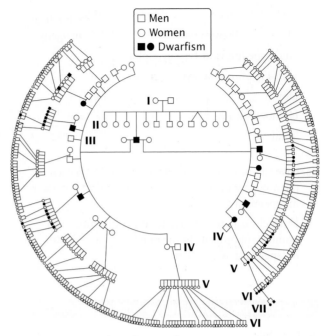

Legend: □ Men; ○ Women; ■● Dwarfism

[Data from *The Journal of Heredity* 34:232.]

a. With the assumption that Schmid-type metaphyseal chondrodysplasia is rare, is this type of dwarfism inherited as a dominant or recessive trait? Explain your reasoning.

b. On the basis of your answer for part *a*, what is the expected ratio of children with dwarfism to children with normal stature in the families given in the table? Use a chi-square test to determine if the total number of children with each phenotype in these families (52 with normal stature, 40 with dwarfism) is significantly different from the number expected.

c. Use chi-square tests to determine if the numbers of children with each phenotype in family C (1 with normal stature, 6 with dwarfism) and in family D (6 with normal stature, 2 with dwarfism) are significantly different from the numbers expected on the basis of your proposed mode of inheritance. How would you explain these deviations from the overall ratio expected?

40. Pink-eye and albino are two recessive traits found in the deer mouse *Peromyscus maniculatus*. In pink-eyed mice, the eye is devoid of color and appears pink because of the blood vessels within it. Albino mice are completely lacking color, both in their fur and in their eyes. F. H. Clark crossed pink-eyed mice with albino mice; the resulting F_1 had normal coloration in their fur and eyes. He then crossed these F_1 mice with mice that were pink eyed and albino and obtained the following progeny. It is hard to distinguish between mice that are albino and mice that are both pink eyed and albino, so he combined these two phenotypes (F. H. Clark. 1936. *Journal of Heredity* 27:259–260).

Phenotype	Number of progeny
wild-type fur, wild-type eye color	12
wild-type fur, pink-eye	62
albino ⎫	78
albino, pink-eye ⎭	
Total	152

a. Give the expected numbers of progeny with each phenotype if the genes for pink-eye and albino assort independently.

b. Use a chi-square test to determine if the observed numbers of progeny fit the number expected with independent assortment.

*41. In the California poppy, an allele for yellow flowers (*C*) is dominant over an allele for white flowers (*c*). At an independently assorting locus, an allele for entire petals (*F*) is dominant over an allele for fringed petals (*f*). A plant that is homozygous for yellow and entire petals is crossed with a plant that has white and fringed petals. A resulting F$_1$ plant is then crossed with a plant that has white and fringed petals, and the following progeny are produced: 54 yellow and entire; 58 yellow and fringed; 53 white and entire; and 10 white and fringed.

a. Use a chi-square test to compare the observed numbers of progeny having each phenotype with those expected for the cross.

b. What conclusion can you draw from the results of the chi-square test?

c. Suggest an explanation for the results.

CHALLENGE QUESTIONS

Section 3.2

42. Dwarfism is a recessive trait in Hereford cattle. A rancher in western Texas discovers that several of the calves in his herd are dwarfs, and he wants to eliminate this undesirable trait from the herd as rapidly as possible. Suppose that the rancher hires you as a genetic consultant to advise him on how to breed the dwarfism trait out of the herd. What crosses would you advise the rancher to conduct to ensure that the allele causing dwarfism is eliminated from the herd?

*43. A geneticist discovers an obese mouse in his laboratory colony. He breeds this obese mouse with a normal mouse. All the F$_1$ mice from this cross are normal in size. When he interbreeds two F$_1$ mice, eight of the F$_2$ mice are normal in size and two are obese. The geneticist then intercrosses two of his obese mice, and he finds that all the progeny from this cross are obese. These results lead the geneticist to conclude that obesity in mice results from a recessive allele.

A second geneticist at a different university also discovers an obese mouse in her laboratory colony. She carries out the same crosses as the first geneticist and obtains the same results. She also concludes that obesity in mice results from a recessive allele.

One day the two geneticists meet at a genetics conference, learn of each other's experiments, and decide to exchange mice. They both find that, when they cross two obese mice from the different laboratories, all the offspring are normal; however, when they cross two obese mice from the same laboratory, all the offspring are obese. Explain their results.

44. Albinism is a recessive trait in humans (see the introduction to Chapter 1). A geneticist studies a series of families in which both parents have normal pigmentation and at least one child has albinism.

The geneticist reasons that both parents in these families must be heterozygotes and that albinism should appear in $\frac{1}{4}$ of their children. To his surprise, the geneticist finds that the frequency of albinism among the children of these families is significantly greater than $\frac{1}{4}$. Can you think of an explanation for the higher-than-expected frequency of albinism among these families?

45. Two distinct phenotypes are found in the salamander *Plethodon cinereus*: a red form and a black form. Some biologists have speculated that the red phenotype is due to an allele that is dominant over an allele for black. Unfortunately, these salamanders will not mate in captivity, so the hypothesis that red is dominant over black has never been tested.

One day, a genetics student is hiking through the forest and finds 30 female salamanders, some red and some black, laying eggs. The student places each female with her eggs (about 20 to 30 eggs per female) in a separate plastic bag and takes them back to the lab. There, the student successfully raises the eggs until they hatch. After the eggs have hatched, the student records the phenotypes of the juvenile salamanders, along with the phenotypes

[George Grall/Getty Images.]

of their mothers. Thus, the student has the phenotypes for 30 females and their progeny, but no information is available about the phenotypes of the fathers.

Explain how the student can determine whether red is dominant over black with this information on the phenotypes of the females and their offspring.

THINK-PAIR-SHARE QUESTIONS

Introduction

1. About 40% of Solomon Islanders carry a gene for blond hair, and yet only 5%–10% of these people actually have blond hair. Why is the proportion of people with blond hair only 5%–10% when so many people carry genes for blond hair?

2. Why was Mendel's success dependent on his studying characteristics that exhibit only two easily distinguished phenotypes, such as white versus gray seed coats and round versus wrinkled seeds? Would he have been less successful if he had instead studied traits such as seed weight or leaf length, which vary much more in their phenotypes? Explain your answer.

Section 3.2

3. Geneticists often carry out reciprocal crosses when they are studying the inheritance of traits. Why do geneticists use reciprocal crosses?

4. Red hair in humans is inherited as a recessive trait. Bill and Sarah both have black hair. They marry and have four children, three of whom have red hair. Bill says it isn't genetically possible for two black-haired people to have $3/4$ red-haired children, and he accuses Sarah of infidelity. Sarah says Bill is a homozygous dominant idiot and knows nothing about genetics. Who is correct and why?

Section 3.3

5. Are Mendel's principles of segregation and independent assortment even relevant today in the age of genomics, when it is possible to sequence an organism's entire genome and determine all of its genetic information? Why is it important to study these principles, and how can they be used?

6. In cats, short hair (L) is dominant over long hair (l) and stripes (A) are dominant over solid color (a). A cat with genotype $Ll\ Aa$ mates with a cat that is $Ll\ aa$, and they produce a litter of six kittens. What is the probability that four of the six kittens will have both long hair and stripes?

Section 3.4

7. In corn, purple kernels (P) are dominant over yellow kernels (p) and starchy kernels (Su) are dominant over sugary kernels (su). A corn plant grown from a purple and starchy kernel is crossed with a plant grown from a yellow and sugary kernel, and the following progeny (kernels) are produced:

Phenotype	Number
purple, starchy	150
purple, sugary	142
yellow, starchy	161
yellow, sugary	115

Formulate a hypothesis about the genotypes of the parents and offspring in this cross. Perform a chi-square goodness-of-fit test comparing the observed progeny with the numbers expected based on your genetic hypothesis. What conclusion can you draw based on the results of your chi-square test? Can you suggest an explanation for the observed results?

Sex Determination and Sex-Linked Characteristics

In European traditions, dragons are often portrayed as large lizards.
This statue adorns the Dragon Bridge in Ljubljana, Slovenia. [© Matej Kastelic/
Alamy Stock Photo.]

The Sex of a Dragon

Dragons are the stuff of kids' tales and nightmares—giant fire-breathing reptiles that destroy everything in their path. These mythical creatures have a long history, appearing in folk traditions of both Europe and Asia. Almost always reptilian, dragons were originally depicted as snakelike, but since the Middle Ages they have been increasingly drawn with legs, often imagined as large lizards. In European traditions, dragons are usually evil, but in Asian myths, dragons are a symbol of benevolent power and sometimes of fertility.

Real dragons are found in the deserts of central Australia. Known as central bearded dragons (*Pogona vitticeps*), these lizards grow up to 2 feet in length (**Figure 4.1**). Like some of their mythical relatives, they possess numerous spines that run from head to tail. Their color varies widely, from brown to gray, red, yellow, orange, or even white. Bearded dragons derive their name from a pouch that projects from the underside of the chin, often appearing dark in color like a man's beard. When threatened, they open their mouths and hiss, imitating their mythological namesake (but, disappointingly, without the fire). But it is through their sex that bearded dragons have become best known to geneticists.

As we learn in this chapter, organisms employ an amazing variety of mechanisms to determine sex—whether an individual is male or female. In some, such as humans, sex is determined by chromosomes: females possess two identical sex chromosomes (X chromosomes), and males possess a single X and a different chromosome, called Y. In others, such as birds, butterflies, and some reptiles, sex is still chromosomal, but it is the male that possesses two identical sex chromosomes (called Z chromosomes) and the female that has two different sex chromosomes (called Z and W). There are even some animals for which sex is not determined by sex chromosomes at all, but is primarily a function of the environment: in turtles and alligators, sex is influenced by the temperature at which the eggs are incubated, with males produced at one temperature and females at another.

For many years, scientists thought that sex determination could be genetic or environmental, but not both. In 2015, however, scientists reported the results of field and laboratory studies demonstrating unequivocally that sex in bearded dragons is determined by sex chromosomes *and* the temperature at which their eggs are incubated.

4.1 A central bearded dragon (*Pogona vitticeps*). [© Juniors Bildarchiv GmbH/Alamy Stock Photo.]

A female bearded dragon lays a clutch of 11 to 30 eggs in a nest in the sand. The female then leaves the nest, and the eggs develop unattended, warmed by the desert sun. Eventually, baby dragons emerge from the eggs.

Bearded dragons have sex chromosomes: males are normally ZZ and females are ZW. Scientists discovered, however, that over 20% of wild-caught females examined in one year had ZZ sex chromosomes, which are expected only in males. By conducting laboratory experiments, the scientists discovered that when they incubated eggs with ZZ embryos at temperatures below 32°C, the offspring were all males (as expected), but as they increased the incubation temperature, more and more ZZ embryos developed as females. At 36°C, almost all the ZZ individuals were females. They concluded that ZZ females found in the wild probably hatched from warm nests. Matings between ZZ males and ZZ females produced viable and fertile offspring, all ZZ; for unknown reasons, ZZ females themselves lay twice as many eggs as normal ZW females. In contrast to ZZ lizards, which can be male or female depending on incubation temperature, ZW lizards are always female. One concern of the researchers was how these lizards might respond to future climate change. If the environment warms significantly, only females may be produced, leading to the extinction of the species.

 THINK-PAIR-SHARE

- Climate warming could lead to all-female populations of bearded dragons, resulting in extinction of the species. What are some other potential effects of climate warming on natural populations of organisms?

- High incubation temperatures reverse the sex of ZZ bearded dragons, causing them to develop as females instead of males. Imagine that temperature also caused sex reversal in ZW individuals, so that high temperatures caused ZW individuals to develop as males instead of females (this doesn't actually happen in bearded dragons). What would be the result of a mating between a normal ZW female and a sex-reversed ZW male?

In Chapter 3, we studied Mendel's principles of segregation and independent assortment and saw how these concepts explain much about the nature of inheritance. After Mendel's principles were rediscovered in 1900, biologists began to conduct genetic studies on a wide array of different organisms. As they applied Mendel's principles more widely, they observed exceptions that made it necessary to devise extensions of those basic principles. In this chapter, we explore one of those extensions: the inheritance of characteristics encoded by genes located on the sex chromosomes, which often differ between males and females (**Figure 4.2**). These characteristics, and the genes that produce them, are referred to as sex linked. To understand the inheritance of sex-linked characteristics, we must first know how sex is determined—why some members of a species are male and others are female. Thus, the first part of this chapter focuses on sex determination. The second part examines how characteristics encoded by genes on the sex chromosomes are inherited and some of the consequences of this inheritance. In Chapter 5, we will explore some additional ways in which sex and inheritance interact.

As we consider sex determination and sex-linked characteristics, it will be helpful to keep two important principles in mind. First, there are several different mechanisms of sex determination, and ultimately, a species' mechanism of sex

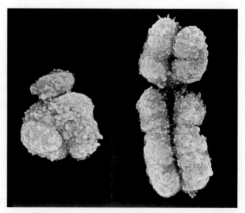

4.2 The human male sex chromosome (Y, at the left) differs from the human female sex chromosome (X, at the right) in size and shape. [Biophoto Associates/Science Source.]

determination controls its inheritance of sex-linked characteristics. Second, like other pairs of chromosomes, the X and Y sex chromosomes pair in the course of meiosis and segregate, but throughout most of their length, they are not homologous (their gene sequences do not encode the same characteristics): most genes on the X chromosome are different from genes on the Y chromosome. Consequently, males and females do not possess the same numbers of alleles at sex-linked loci. This difference in the number of sex-linked alleles produces distinct patterns of inheritance in males and females.

4.1 Sex Is Determined by a Number of Different Mechanisms

Sexual reproduction is the formation of offspring that are genetically distinct from their parents; most often, two parents contribute genes to their offspring, and those genes are assorted into new combinations through meiosis. Among most eukaryotes, sexual reproduction consists of two processes that lead to an alternation of haploid and diploid cells: meiosis produces haploid gametes (spores in plants), and fertilization produces diploid zygotes (**Figure 4.3**).

The term **sex** refers to sexual phenotype. Most organisms have two sexual phenotypes: male and female. The fundamental difference between males and females is gamete size: males produce small gametes, and females produce relatively large gametes (**Figure 4.4**).

The mechanism by which sex is established is termed **sex determination**. We define the sex of an individual organism in reference to its phenotype. Sometimes an individual organism has chromosomes or genes that are normally associated with one sex but an anatomy corresponding to the opposite sex. For instance, the cells of human females normally have two X chromosomes, and the cells of males have one X chromosome and one Y chromosome. A few rare individuals have male anatomy although their cells each contain two X chromosomes. Even though these people are

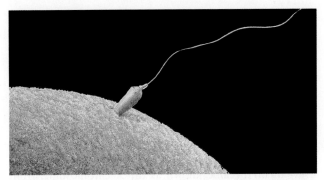

4.4 Male and female gametes differ in size. In this photograph, a human sperm (with flagellum) penetrates a human egg. [Francis Leroy, Biocosmos/Science Source.]

genetically female, we refer to them as male because their sexual phenotype is male. (As we see later in the chapter, these XX males usually have a small piece of the Y chromosome that is attached to another chromosome.)

Gender is not the same as sex. Biological sex refers to the anatomical and physiological phenotype of an individual. Gender is a category assigned by the individual or others based on behavior and cultural practices. One's gender need not coincide with one's biological sex.

CONCEPTS

In sexual reproduction, two parents contribute genes to produce an offspring that is genetically distinct from both parents. In most eukaryotes, sexual reproduction consists of meiosis, which produces haploid gametes (or spores), and fertilization, which produces a diploid zygote.

✔ CONCEPT CHECK 1

What process causes the genetic variation seen in offspring produced by sexual reproduction?

As discussed in the introduction to this chapter, there are many ways in which sex differences arise. In some species, both sexes are present in the same organism, a condition termed **hermaphroditism**; organisms that bear both male and female reproductive structures are said to be **monoecious** (meaning "one house"). Species in which an individual organism has either male or female reproductive structures are said to be **dioecious** ("two houses"). Humans are dioecious. Among dioecious species, sex may be determined chromosomally, genetically, or environmentally.

Chromosomal Sex-Determining Systems

The chromosome theory of heredity (see Chapter 3) states that genes are located on chromosomes, which serve as

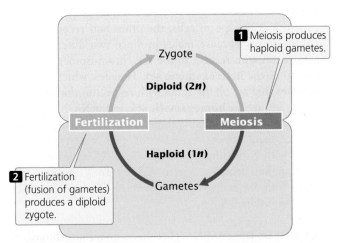

4.3 In most eukaryotic organisms, sexual reproduction consists of an alternation of haploid (1*n*) and diploid (2*n*) cells.

Zygote

Diploid (2*n*)

Fertilization   **Meiosis**

Haploid (1*n*)

Gametes

1 Meiosis produces haploid gametes.

2 Fertilization (fusion of gametes) produces a diploid zygote.

vehicles for the segregation of genes in meiosis. Definitive proof of this theory was provided by the discovery that the sex of certain insects is determined by the presence or absence of particular chromosomes.

In 1891, Hermann Henking noticed a peculiar structure in the nuclei of cells from male insects. Understanding neither its function nor its relation to sex, he called this structure the X body. Later, Clarence E. McClung studied the X body in grasshoppers and recognized that it was a chromosome. McClung called it the accessory chromosome, but it eventually became known as the X chromosome, from Henking's original designation. McClung observed that the cells of female grasshoppers had one more chromosome than the cells of male grasshoppers, and he concluded that accessory chromosomes played a role in sex determination. In 1905, Nettie Stevens and Edmund Wilson demonstrated that in grasshoppers and other insects, the cells of females have two X chromosomes, whereas the cells of males have a single X. In some insects, they counted the same number of chromosomes in the cells of males and females but saw that one chromosome pair was different: two X chromosomes were found in female cells, whereas a single X chromosome plus a smaller chromosome, which they called Y, was found in male cells.

Stevens and Wilson also showed that the X and Y chromosomes separate into different cells in sperm formation: half of the sperm receive an X chromosome and the other half receive a Y. All eggs produced by the female in meiosis receive one X chromosome. A sperm containing a Y chromosome unites with an X-bearing egg to produce an XY male, whereas a sperm containing an X chromosome unites with an X-bearing egg to produce an XX female. This distribution of X and Y chromosomes in sperm accounts for the 1:1 sex ratio observed in most dioecious organisms (**Figure 4.5**). Because sex is inherited like other genetically determined characteristics, Stevens and Wilson's discovery that sex is associated with the inheritance of a particular chromosome also demonstrated that genes are located on chromosomes.

As Stevens and Wilson found for insects, sex in many organisms is determined by a pair of chromosomes, the **sex chromosomes**, which differ between males and females. The nonsex chromosomes, which are the same for males and females, are called **autosomes**. We think of sex in organisms with sex chromosomes as being determined by the presence of the sex chromosomes, but in fact, the individual genes located on the sex chromosomes, in conjunction with genes on the autosomes, are usually responsible for the sexual phenotypes.

XX-XO SEX DETERMINATION The mechanism of sex determination in the grasshoppers studied by McClung, called the XX-XO system, is one of the simplest mechanisms of chromosomal sex determination. In this system, females have two X chromosomes (XX) and males possess a single X chromosome

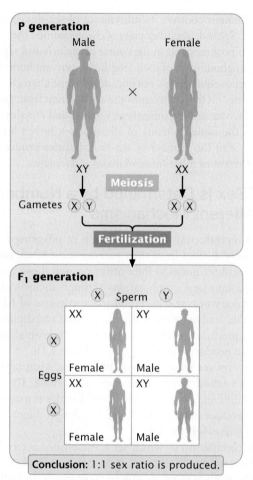

4.5 Inheritance of sex in organisms with X and Y chromosomes results in equal numbers of male and female offspring.

(XO). There is no O chromosome—the letter O signifies the absence of a sex chromosome.

In meiosis in females, the two X chromosomes pair and then separate, and one X chromosome enters each haploid egg. In males, the single X chromosome segregates in meiosis to half the sperm cells; the other half receive no sex chromosome. Because males produce two different types of gametes with respect to the sex chromosomes, they are said to be the **heterogametic sex**. Females, which produce gametes that are all the same with respect to the sex chromosomes, are the **homogametic sex**. In the XX-XO system, the sex of an individual is therefore determined by which type of male gamete fertilizes the egg. X-bearing sperm unite with X-bearing eggs to produce XX zygotes, which develop into females. Sperm lacking an X chromosome unite with X-bearing eggs to produce XO zygotes, which develop into males.

XX-XY SEX DETERMINATION In many species, the cells of males and females have the same number of chromosomes, but the cells of females have two X chromosomes (XX) and

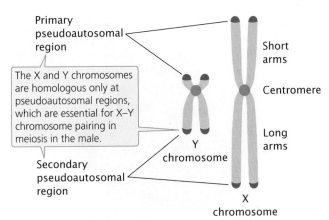

Primary pseudoautosomal region

Short arms

Centromere

The X and Y chromosomes are homologous only at pseudoautosomal regions, which are essential for X–Y chromosome pairing in meiosis in the male.

Long arms

Y chromosome

Secondary pseudoautosomal region

X chromosome

4.6 The X and Y chromosomes in humans differ in size and genetic content. They are homologous only in the pseudoautosomal regions.

the cells of males have a single X chromosome and a smaller sex chromosome, the Y chromosome (XY). In humans and many other organisms, the Y chromosome is acrocentric (**Figure 4.6**), not Y-shaped, as is often assumed. In this sex-determining system, the male is the heterogametic sex—half of his gametes have an X chromosome and half have a Y chromosome. The female is the homogametic sex—all her eggs contain an X chromosome. Many organisms, including some plants, insects, and reptiles and all mammals (including humans), have the XX-XY sex-determining system. Other organisms have variations of the XX-XY system; for example, the duck-billed platypus has an interesting system in which females have five pairs of X chromosomes and males have five pairs of X and Y chromosomes.

THINK-PAIR-SHARE Question 1

Although the X and Y chromosomes are not generally homologous, they do pair and segregate into different cells in meiosis. They can pair because these chromosomes are homologous in small regions called the **pseudoautosomal regions** (see Figure 4.6), in which they carry the same genes. In humans, there are pseudoautosomal regions at both tips of the X and Y chromosomes.

ZZ-ZW SEX DETERMINATION In the ZZ-ZW sex-determining system, found in the bearded dragons discussed in the chapter introduction, the female is heterogametic and the male is homogametic. To prevent confusion with the XX-XY system, the sex chromosomes in this system are called Z and W, but the chromosomes do not resemble Zs and Ws. Females in this system are ZW; after meiosis, half of the eggs have a Z chromosome and the other half have a W chromosome. Males are ZZ; all sperm contain a single Z chromosome. The ZZ-ZW system is found in birds, some reptiles, butterflies, some amphibians, and some fishes.

THINK-PAIR-SHARE Question 2

Genic Sex Determination

In some organisms, sex is genetically determined, but there are no obvious differences in the chromosomes of males and females: there are no sex chromosomes. These organisms have **genic sex determination**: genotypes at one or more loci determine the sex of an individual. Scientists have observed genic sex determination in some plants, fungi, protozoans, and fishes.

It is important to understand that even in chromosomal sex-determining systems, sex is actually determined by individual genes. In mammals, for example, a gene (*SRY*, discussed later in this chapter) located on the Y chromosome determines the male phenotype. In both genic sex determination and chromosomal sex determination, sex is controlled by individual genes; the difference is that with chromosomal sex determination, the sex chromosomes also look different in males and females.

Environmental Sex Determination

In a number of organisms, sex is determined fully or in part by environmental factors. A fascinating example of environmental sex determination is seen in the marine mollusk *Crepidula fornicata*, also known as the common slipper limpet (**Figure 4.7**). Slipper limpets live in stacks, one on top of another. Each limpet begins life as a swimming larva. The first larva to settle on a solid, unoccupied substrate develops into a female limpet. It then produces chemicals that attract other larvae, which settle on top of it. These larvae develop into males, which then serve as mates for the limpet below. After a period of time, the males on top develop into females and, in turn, attract additional larvae that settle on top of the stack, develop into males, and serve as mates for the limpets under them. Limpets can form stacks of a dozen or more animals; the uppermost animals are always

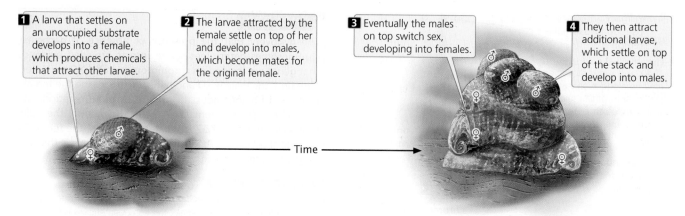

4.7 In *Crepidula fornicata*, **the common slipper limpet, sex is determined by an environmental factor—the limpet's position in the stack.**

male. This type of sexual development is called **sequential hermaphroditism**; each individual animal can be both male and female, although not at the same time. In *Crepidula fornicata*, sex is determined environmentally by the limpet's position in the stack.

Environmental factors are also important in determining sex in some reptiles; the sexual phenotype of many turtles, crocodiles, alligators, and a few birds is affected by temperature during embryonic development. In turtles, for example, warm incubation temperatures produce more females, whereas cool temperatures produce males. In alligators, the reverse is true. In a few species, sex chromosomes determine whether individuals are male or female, but environmental factors can override this chromosomal sex determination. As discussed in the introduction to this chapter, bearded dragons are normally male when ZZ and female when ZW, but when the eggs are incubated at high temperatures, ZZ individuals develop into phenotypic females. Some of the different types of sex determination are summarized in **Table 4.1**.

Now that we have surveyed some of the different ways in which sex can be determined, we will examine one mechanism (the XX-XY system) in detail. Sex determination is XX-XY in both fruit flies and humans, but as we will see, the ways in which the X and Y chromosomes determine sex in these two organisms are quite different.

▶ **TRY PROBLEMS 4 AND 21**

CONCEPTS

In genic sex determination, sex is determined by genes at one or more loci, but there are no obvious differences in the chromosomes of males and females. In environmental sex determination, sex is determined fully or in part by environmental factors.

✔ CONCEPT CHECK 3

How do chromosomal, genic, and environmental sex-determining systems differ?

TABLE 4.1	Some common sex-determining systems			
System	**Mechanism**		**Heterogametic Sex**	**Organisms**
XX-XO	Females XX	Males X	Male	Some grasshoppers and other insects
XX-XY	Females XX	Males XY	Male	Many insects, fishes, amphibians, reptiles; mammals, including humans
ZZ-ZW	Females ZW	Males ZZ	Female	Butterflies, birds; some reptiles and amphibians
Genic sex determination	No distinct sex chromosomes Sex determined by genes on undifferentiated chromosomes		Varies	Some plants, fungi, protozoans, and fishes
Environmental sex determination	Sex determined by environmental factors		None	Some invertebrates, turtles, alligators

Sex Determination in *Drosophila melanogaster*

The fruit fly *Drosophila melanogaster* has eight chromosomes: three pairs of autosomes and one pair of sex chromosomes. Usually, females have two X chromosomes, and males have an X chromosome and a Y chromosome. In the 1920s, Calvin Bridges proposed that sex in *Drosophila* was determined not by the number of X and Y chromosomes, but rather by the balance of female-determining genes on the X chromosome and male-determining genes on the autosomes. He suggested that a fly's sex is determined by the so-called X : A ratio: the number of X chromosomes divided by the number of haploid sets of autosomes. Normal flies possess two haploid sets of autosomes and either two X chromosomes (females) or one X chromosome and a Y chromosome (males). Bridges proposed that an X : A ratio of 1.0 produces a female fly and that an X : A ratio of 0.5 produces a male fly. He also suggested that an X : A ratio between 1.0 and 0.5 produces an intersex fly, with a mixture of male and female characteristics. An X : A ratio of less than 0.5 or greater than 1.0 produces developmentally abnormal flies called metamales and metafemales, respectively. When Bridges and others examined flies with different numbers of sex chromosomes and autosomes, the X : A ratio appeared to correctly predict the phenotypic sex of the flies (**Table 4.2**).

Although the X : A ratio correctly *predicts* the sexual phenotype, recent research suggests that the *mechanism* of sex determination is not a balance between X-linked genes and autosomal genes, as Bridges proposed. Researchers have located a number of genes on the X chromosome that affect sexual phenotype, but few autosomal sex-determining genes (required for the X : A ratio hypothesis) have been identified. New evidence suggests that genes on the X chromosome are the primary sex determinant. The influence of the number of sets of autosomes on sex is indirect, affecting the timing of developmental events and therefore how long sex-determining genes on the X chromosome are active. For example, XX flies with three autosomal sets (XX, AAA) have an X : A ratio of 0.67 and develop an intersex phenotype. In these flies, the presence of three autosomal sets causes a critical developmental stage to shorten, not allowing female factors encoded on the X chromosomes enough time to accumulate, with the result that the flies end up with an intersex phenotype. Thus, the number of autosomal sets of chromosomes influences sex determination in *Drosophila*, but not through the action of autosomal genes, as envisioned by Bridges.

CONCEPTS

Although the sexual phenotype of a fruit fly is predicted by the X : A ratio, sex is actually determined by genes on the X chromosome.

Sex Determination in Humans

Humans, like *Drosophila*, have XX-XY sex determination, but in humans, maleness is primarily determined by the presence of a particular gene (*SRY*) on the Y chromosome. The phenotypes that result from abnormal numbers of sex chromosomes, which arise when the sex chromosomes do not segregate properly in meiosis or mitosis, illustrate the importance of the Y chromosome in human sex determination.

TURNER SYNDROME People who have **Turner syndrome** are female and often have underdeveloped secondary sex characteristics. This syndrome is seen in 1 of 3000 female births. Affected women are frequently short and have a low hairline, a relatively broad chest, and folds of skin on the neck. Their intelligence is usually normal. Most women who have Turner syndrome are sterile. In 1959, Charles Ford used new techniques to study human chromosomes and discovered that cells from a 14-year-old girl with Turner syndrome had only a single X chromosome (**Figure 4.8**); this chromosome

TABLE 4.2	Chromosome complements and sexual phenotypes in *Drosophila*		
Sex-Chromosome Complement	Haploid Sets of Autosomes	X : A Ratio	Sexual Phenotype
XX	AA	1.0	Female
XY	AA	0.5	Male
XO	AA	0.5	Male
XXY	AA	1.0	Female
XXX	AA	1.5	Metafemale
XXXY	AA	1.5	Metafemale
XX	AAA	0.67	Intersex
XO	AAA	0.33	Metamale
XXXX	AAA	1.3	Metafemale

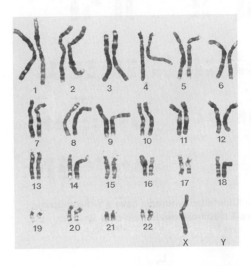

4.8 People with Turner syndrome have a single X chromosome in their cells. [Dept. of Clinical Cytogenetics, Addenbrookes Hospital/Science Photo Library/Science Source.]

complement is usually referred to as XO. Many people with Turner syndrome have some cells that are XX and other cells that are XO, a situation referred to as mosaicism (see p. 154).

There are no known cases in which a person is missing both X chromosomes, an indication that at least one X chromosome is necessary for human development. Presumably, embryos missing both Xs spontaneously abort in the early stages of development.

KLINEFELTER SYNDROME People who have **Klinefelter syndrome**, which has a frequency of about 1 in 1000 male births, have cells with one or more Y chromosomes and multiple X chromosomes. The cells of most males with this condition are XXY (**Figure 4.9**), but the cells of a few males with Klinefelter syndrome are XXXY, XXXXY, or XXYY. Men with this condition frequently have small testes and reduced facial and pubic hair. They are often taller than normal and sterile; most have normal intelligence.

POLY-X FEMALES In about 1 in 1000 female births, the infant's cells possess three X chromosomes, a condition often referred to as **triple-X syndrome**. These individuals have no distinctive features other than a tendency to be tall and thin. Although a few are sterile, many menstruate regularly and are fertile. The incidence of intellectual disability among triple-X females is slightly greater than that in the general population, but most XXX females have normal intelligence. Much rarer are females whose cells contain four or five X chromosomes. These females usually have normal female anatomy but are intellectually disabled and have a number of physical problems. The severity of intellectual disability increases as the number of X chromosomes increases beyond three.

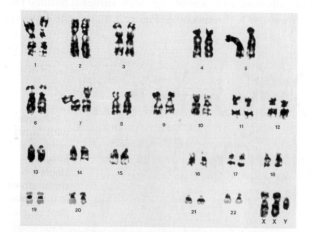

4.9 People with Klinefelter syndrome have a Y chromosome and two or more X chromosomes in their cells. [Biophoto Associates/Science Source.]

XYY MALES Males with an extra Y chromosome (XYY) occur with a frequency of about 1 in 1000 male births. These individuals have no distinctive physical characteristics other than a tendency to be several inches taller than the average of XY males. Their IQ is usually within the normal range; some studies suggest that learning difficulties may be more common than in XY males.

THE ROLE OF SEX CHROMOSOMES The phenotypes associated with sex-chromosome anomalies allow us to make several inferences about the role of sex chromosomes in human sex determination:

1. The X chromosome contains genetic information essential for both sexes; at least one copy of an X chromosome is required for human development.

2. The male-determining gene is located on the Y chromosome. A single copy of this chromosome, even in the presence of several X chromosomes, usually produces a male phenotype.

3. The absence of the Y chromosome usually results in a female phenotype.

4. Genes affecting fertility are located on the X and Y chromosomes. A female usually needs at least two copies of the X chromosome to be fertile.

5. Additional copies of the X chromosome may upset normal development in both males and females, producing physical problems and intellectual disabilities that increase as the number of extra X chromosomes increases.

THE MALE-DETERMINING GENE IN HUMANS The Y chromosome in humans, and in all other mammals, is of paramount importance in producing a male phenotype. However, scientists have discovered a few rare XX males whose cells apparently lack a Y chromosome. For many years, these males presented an enigma: How could a male phenotype exist without a Y chromosome? Close examination eventually revealed a small part of the Y chromosome attached to another chromosome, usually the X. This finding indicates that it is not the entire Y chromosome that determines maleness in humans; rather, it is a gene on the Y chromosome.

Early in development, all humans possess undifferentiated gonads and both male and female reproductive ducts. Then, about 6 weeks after fertilization, a gene on the Y chromosome becomes active. This gene causes the neutral gonads to develop into testes, which begin to secrete two hormones: testosterone and Mullerian-inhibiting substance. Testosterone induces the development of male characteristics, and Mullerian-inhibiting substance causes the degeneration of the female reproductive ducts. In the absence of this male-determining gene, the neutral gonads become ovaries, and female features develop.

The male-determining gene in humans, called the **sex-determining region Y (*SRY*) gene**, was discovered in 1990 (**Figure 4.10**). This gene is found in XX males and is missing from rare XY females; it is also found on the Y chromosome of other mammals. Definitive proof that *SRY* is the male-determining gene came when scientists placed a copy of this gene in XX mice by means of genetic engineering. The XX mice that received this gene, although sterile, developed into anatomical males.

The *SRY* gene encodes a protein called a transcription factor (see Chapter 13) that binds to DNA and stimulates the transcription of other genes that promote the differentiation of the testes. Although *SRY* is the primary determinant of maleness in humans, other genes (some X linked, others Y linked, and still others autosomal) also have roles in fertility and the development of sexual phenotypes.

THINK-PAIR-SHARE Question 3

CONCEPTS

The presence of the *SRY* gene on the Y chromosome causes a human embryo to develop as a male. In the absence of this gene, a human embryo develops as a female.

✔ CONCEPT CHECK 4

What is the phenotype of a person who has XXXY sex chromosomes?

a. Klinefelter syndrome
b. Turner syndrome
c. Poly-X female

ANDROGEN-INSENSITIVITY SYNDROME Although the *SRY* gene is the primary determinant of sex in human embryos, several other genes influence sexual development, as illustrated by women with androgen-insensitivity syndrome. These individuals have female external sexual characteristics. Indeed, most are unaware of their condition until they reach puberty and fail to menstruate. Examination by a gynecologist reveals that the vagina ends blindly and that the uterus, oviducts, and ovaries are absent. Inside the abdominal cavity, a pair of testes produce levels of testosterone normally seen in males. The cells of a woman with androgen-insensitivity syndrome contain an X and a Y chromosome.

How can a person be female in appearance when her cells contain a Y chromosome and she has testes that produce testosterone? The answer lies in the complex relation between genes and sex in humans. In a human embryo with a Y chromosome, the *SRY* gene causes the gonads to develop into testes, which produce testosterone. Normally, testosterone stimulates embryonic tissues to develop male characteristics. But for testosterone to have its effects, it must bind to an androgen receptor. This receptor is defective in females with androgen-insensitivity syndrome; consequently, their cells do not respond to testosterone, and female characteristics develop. The gene for the androgen receptor is located on the X chromosome, so people with this condition always inherit it from their mothers.

Androgen-insensitivity syndrome illustrates several points about the influence of genes on a person's sex. First, this condition demonstrates that human sexual development is a complex process, influenced not only by the *SRY* gene on the Y chromosome, but also by other genes found elsewhere. Second, it shows that most people carry genes for both male and female characteristics, as illustrated by the fact that those with androgen-insensitivity syndrome have the capacity to develop female characteristics even though they have male chromosomes. Indeed, the genes for most male and female secondary sex characteristics are present not on the sex chromosomes, but on autosomes. The key to maleness and femaleness lies not in the genes, but in the control of their expression. **▶ TRY PROBLEM 17**

4.2 Sex-Linked Characteristics Are Determined by Genes on the Sex Chromosomes

In Chapter 3, we learned several basic principles of heredity that Mendel discovered in his experiments with pea plants. A major extension of these Mendelian principles is the pattern of inheritance exhibited by **sex-linked characteristics**: characteristics determined by genes located on the sex chromosomes. Genes on the X chromosome determine **X-linked characteristics**; those on the Y chromosome determine **Y-linked characteristics**. Because the Y chromosome of many organisms contains little genetic information, most sex-linked characteristics are X linked. Males and females differ in their sex chromosomes, so the pattern of inheritance for sex-linked characteristics differs from that exhibited by genes located on autosomes.

X-Linked White Eyes in *Drosophila*

The first person to provide an explanation for sex-linked inheritance was the American biologist Thomas Hunt Morgan

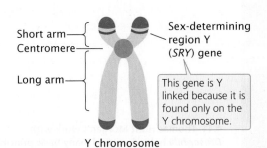

Short arm—
Centromere—
Long arm—

Sex-determining region Y (*SRY*) gene

This gene is Y linked because it is found only on the Y chromosome.

Y chromosome

4.10 The *SRY* gene, located on the Y chromosome, causes the development of male characteristics.

(**Figure 4.11**). Morgan began his career as an embryologist, but the discovery of Mendel's principles inspired him to begin conducting genetic experiments, initially on mice and rats. In 1909, Morgan switched his research to *Drosophila melanogaster*; a year later, he discovered among the flies of his laboratory colony a single male that possessed white eyes, in stark contrast to the red eyes of normal fruit flies. This fly had a tremendous effect on Morgan's career as a biologist and on the future of genetics.

To investigate the inheritance of the white-eye trait in fruit flies, Morgan systematically carried out a series of genetic crosses. First, he crossed pure-breeding red-eyed females with his white-eyed male, producing F_1 progeny that had red eyes (**Figure 4.12a**). Morgan's results from this initial cross were consistent with Mendel's principles: a cross between a homozygous dominant individual and a homozygous recessive individual produced heterozygous offspring exhibiting the dominant trait. These results suggested that white eyes are a simple recessive trait. When Morgan crossed the F_1 flies with each other, however, he found that all the female F_2 flies possessed red eyes, but half the male F_2 flies had red eyes and the other half had white eyes. This finding was clearly not the expected result for a simple recessive trait, which should appear in one-fourth of both male and female F_2 offspring.

To explain this unexpected result, Morgan proposed that the locus affecting eye color is on the X chromosome (i.e., that eye color is X linked). He also recognized that eye-color alleles are present on the X chromosome only; no homologous allele is present on the Y chromosome. Because the cells of females possess two X chromosomes, females can be homozygous or heterozygous for the eye-color alleles. The cells of males, on the other hand, possess only a single X chromosome and can carry only a single eye-color allele. Males, therefore, cannot be homozygous or heterozygous, but are said to be **hemizygous** for X-linked loci.

To verify his hypothesis that the white-eye trait is X linked, Morgan conducted additional crosses. He predicted that a cross between a white-eyed female and a red-eyed male would produce all red-eyed females and all white-eyed

males (**Figure 4.12b**). When Morgan performed this cross, the results were exactly as predicted. Note that this cross is the reciprocal of the original cross and that the two reciprocal crosses produced different results in the F_1 and F_2 generations. Morgan also crossed the F_1 heterozygous females with their white-eyed father, the red-eyed F_2 females with white-eyed males, and white-eyed females with white-eyed males. In all these crosses, the results were consistent with Morgan's conclusion that the white-eye trait is an X-linked characteristic. You can view the results of Morgan's crosses in **Animation 4.1**.

Nondisjunction and the Chromosome Theory of Inheritance

When Morgan crossed his original white-eyed male with homozygous red-eyed females, all 1237 of the progeny had red eyes, except for 3 white-eyed males. Morgan attributed these white-eyed F_1 males to the occurrence of further random mutations. However, flies with these unexpected phenotypes continued to appear in his crosses. Although uncommon, they appeared far too often to be due to spontaneous mutation. Calvin Bridges, who was one of Morgan's students, set out to investigate the genetic basis of these exceptions.

Bridges found that the exceptions arose only in certain strains of white-eyed flies. When he crossed a white-eyed female from one of those strains with a red-eyed male, about 5% of the male offspring had red eyes and about 5% of the female offspring had white eyes. In this cross, the expected result is that every male fly should inherit its mother's X chromosome and should have the genotype X^wY and white eyes. Furthermore, every female fly should inherit a dominant red-eye allele on its father's X chromosome, along with a white-eye allele on its mother's X chromosome; thus, all the female progeny should be X^+X^w and have red eyes (see the F_1 progeny in Figure 4.12b). The continual appearance of red-eyed males and white-eyed females in this cross was therefore unexpected.

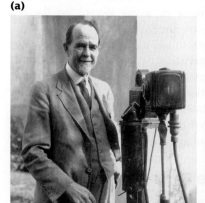

4.11 Thomas Hunt Morgan's work with *Drosophila* helped unravel many basic principles of genetics, including X-linked inheritance. (a) Morgan. (b) The Fly Room, where Morgan and his students conducted genetic research. [Part a: AP/Wide World Photos. Part b: American Philosophical Society.]

Experiment

Question: Are white eyes in fruit flies inherited as an autosomal recessive trait?

Methods Perform reciprocal crosses.

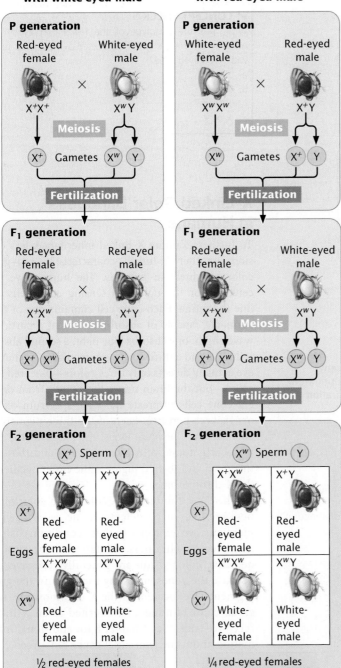

(a) Red-eyed female crossed with white-eyed male

(b) White-eyed female crossed with red-eyed male

Results

½ red-eyed females
¼ red-eyed males
¼ white-eyed males

¼ red-eyed females
¼ white-eyed females
¼ red-eyed males
¼ white-eyed males

Conclusion: No. The results of reciprocal crosses are consistent with X-linked inheritance.

4.12 Morgan's crosses demonstrated the X-linked inheritance of white eyes in fruit flies. (a) Original and F_1 crosses. (b) Reciprocal crosses.

BRIDGES'S EXPLANATION To explain the appearance of red-eyed males and white-eyed females in his cross, Bridges hypothesized that the exceptional white-eyed females of this strain actually possessed two X chromosomes and a Y chromosome (X^wX^wY). In *Drosophila*, flies with XXY sex chromosomes normally develop as females, in spite of possessing a Y chromosome (see Table 4.2). About 90% of the time, the two X chromosomes of the X^wX^wY females separate from each other in anaphase I of meiosis, with an X and a Y chromosome entering one gamete and a single X entering another gamete (**Figure 4.13**). When these gametes are fertilized by sperm from a normal red-eyed male, white-eyed males and red-eyed females are produced. About 10% of the time, the two X chromosomes in the females fail to separate in anaphase I of meiosis, a phenomenon known as **nondisjunction**. When nondisjunction of the Xs occurs, half of the eggs receive two copies of the X chromosome and the other half receive only a Y chromosome (see Figure 4.13). When these eggs are fertilized by sperm from a normal red-eyed male, four combinations of sex chromosomes are produced. An egg with two X chromosomes that is fertilized by an X-bearing sperm produces an $X^+X^wX^w$ zygote, which usually dies. When an egg carrying two X chromosomes is fertilized by a Y-bearing sperm, the resulting zygote is X^wX^wY, which develops into a white-eyed female. An egg with only a Y chromosome that is fertilized by an X-bearing sperm produces an X^+Y zygote, which develops into a normal red-eyed male. If the egg with only a Y chromosome is fertilized by a Y-bearing sperm, the resulting zygote has two Y chromosomes and no X chromosome, and dies. Nondisjunction of the X chromosomes among X^wX^wY white-eyed females therefore produces a few white-eyed females and red-eyed males, which is exactly what Bridges found in his crosses. The results of Bridges's crosses are further explained in **Animation 4.1**. (A)

CONFIRMATION OF BRIDGES'S HYPOTHESIS Bridges's hypothesis predicted that the white-eyed females from his crosses would possess two X chromosomes and one Y chromosome and that the red-eyed males would possess a single X chromosome. To verify his hypothesis, Bridges examined the chromosomes of his flies and found precisely what he had predicted. The significance of Bridges's study is not that it explained the appearance of an occasional odd fly in his culture, but that he was able to link the inheritance of a specific gene (*w*) to the presence of a specific chromosome (X). This association between genotype and chromosomes provided unequivocal evidence that sex-linked genes are located on the X chromosome and confirmed the chromosome theory of inheritance. ▶ **TRY PROBLEM 22**

Question: In a cross between a white-eyed female and a red-eyed male, why are a few white-eyed females and red-eyed males produced?

Hypothesis: White-eyed females and red-eyed males in F_1 result from nondisjunction in an XXY female.

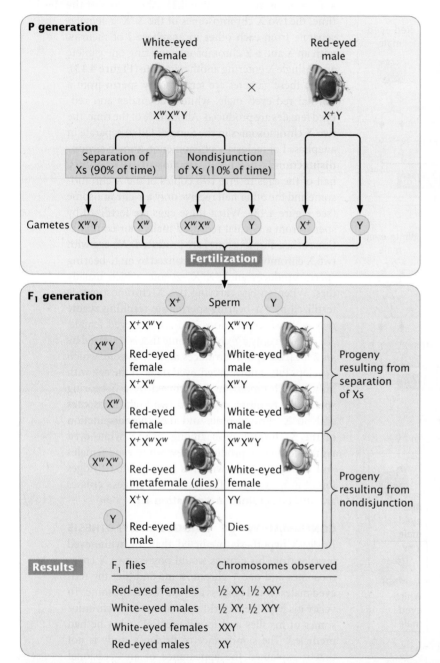

P generation

White-eyed female   ×   Red-eyed male

$X^w X^w Y$ $X^+ Y$

Separation of Xs (90% of time)   Nondisjunction of Xs (10% of time)

Gametes: $X^w Y$ X^w $X^w X^w$ Y X^+ Y

Fertilization

F_1 generation

Sperm: X^+ Y

	X^+	Y	
$X^w Y$	$X^+ X^w Y$ Red-eyed female	$X^w Y Y$ White-eyed male	Progeny resulting from separation of Xs
X^w	$X^+ X^w$ Red-eyed female	$X^w Y$ White-eyed male	
$X^w X^w$	$X^+ X^w X^w$ Red-eyed metafemale (dies)	$X^w X^w Y$ White-eyed female	Progeny resulting from nondisjunction
Y	$X^+ Y$ Red-eyed male	YY Dies	

F_1 flies	Chromosomes observed
Red-eyed females	½ XX, ½ XXY
White-eyed males	½ XY, ½ XYY
White-eyed females	XXY
Red-eyed males	XY

Conclusion: The white-eyed females and red-eyed males in the F_1 result from nondisjunction of the X chromosomes in an XXY female.

4.13 Bridges's crosses proved that the gene for white eyes is located on the X chromosome.

By showing that the appearance of rare phenotypes is associated with the inheritance of particular chromosomes, Bridges proved that sex-linked genes are located on the X chromosome and that the chromosome theory of inheritance is correct.

✔ CONCEPT CHECK 5

What was the genotype of the few F_1 red-eyed males obtained by Bridges when he crossed a white-eyed female with a red-eyed male?

a. X^+
b. $X^w X^+ Y$
c. $X^+ Y$
d. $X^+ X^+ Y$

X-Linked Color Blindness in Humans

To further examine X-linked inheritance, let's consider another X-linked characteristic: red–green color blindness in humans. The human eye perceives color through light-sensing cone cells that line the retina. Each cone cell contains one of three pigments capable of absorbing light of a particular wavelength: one absorbs blue light, a second absorbs red light, and a third absorbs green light. The human eye actually detects only three colors—blue, red, and green—but the brain mixes the signals from different cone cells to create the wide spectrum of colors that we perceive. Each of the three pigments is encoded by a separate locus; the locus for the blue pigment is found on chromosome 7, and those for the green and the red pigments lie close together on the X chromosome.

The most common types of human color blindness are caused by defects of the red and green pigments; we will refer to these conditions as red–green color blindness. Mutations that produce defective color vision are generally recessive, and because the genes encoding the red and the green pigments are located on the X chromosome, red–green color blindness is inherited as an X-linked recessive trait. (See Chapter 8 for additional information about the genes that cause red–green color blindness.)

We will use the symbol X^c to represent an allele for red–green color blindness and the symbol X^+ to represent an allele for normal color vision. Females possess two X chromosomes, so there are three possible genotypes among females: $X^+ X^+$ and $X^+ X^c$, which produce normal color vision, and X^{c}X^c, which produces color blindness. Males have only a single X chromosome and two possible genotypes: X$^+$Y, which produces normal color vision, and X^{c}Y, which produces color blindness.

Let's consider what happens when a woman homozygous for normal color vision mates with a color-blind man (**Figure 4.14a**). All the gametes produced by the woman contain an allele for normal color vision. Half of the man's gametes receive the X chromosome, with the color-blindness allele, and the other half receive the Y chromosome, which carries no alleles affecting color vision. When an X^c-bearing sperm unites with the X$^+$-bearing egg, a heterozygous female with normal color vision (X$^+$X^c) is produced. When a Y-bearing sperm unites with the X$^+$-bearing egg, a hemizygous male with normal color vision (X$^+$Y) is produced.

Now consider the reciprocal cross between a color-blind woman and a man with normal color vision (**Figure 4.14b**). The woman produces only X^c-bearing gametes. The man produces some gametes that contain the X chromosome and others that contain the Y chromosome. Males inherit the X chromosome from their mothers, so, because both of the mother's X chromosomes bear the X^c allele, all the male offspring of this cross will be color blind. In contrast, females inherit an X chromosome from each parent; thus, all the female offspring of this cross will be heterozygous with normal color vision. A female is color blind only when she inherits color-blindness alleles from both parents, whereas a male need only inherit a color-blindness allele from his mother to be color blind; for this reason, color blindness and most other rare X-linked recessive traits are more common in males than in females.

In these crosses for color blindness, notice that an affected woman passes the X-linked recessive trait to her sons, but not to her daughters, whereas an affected man passes the trait to his grandsons through his daughters, but never to his sons. X-linked recessive characteristics may therefore appear to alternate between the sexes, appearing in females in one generation and in males in the next generation.

Recall that the X and Y chromosomes pair in meiosis because they are homologous at the small pseudoautosomal regions. Genes in these regions of the X and Y chromosomes are homologous, just like those on autosomes, and they exhibit autosomal patterns of inheritance, rather than the sex-linked inheritance seen for most genes on the X and Y chromosomes.

THINK-PAIR-SHARE Questions 4 and 5

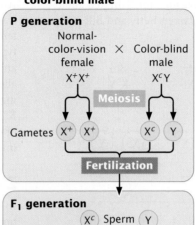
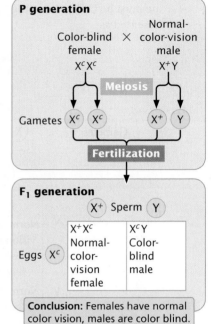

(a) Normal female and color-blind male

(b) Reciprocal cross

4.14 Red–green color blindness is inherited as an X-linked recessive trait in humans.

WORKED PROBLEM

Now that we understand the pattern of X-linked inheritance, let's apply our knowledge to answer a specific question.

Betty has normal color vision, but her mother is color blind. Bill is color blind. If Bill and Betty marry and have a child together, what is the probability that the child will be color blind?

Solution Strategy

What information is required in your answer to the problem?
The probability that Bill and Betty's child will be color blind.

What information is provided to solve the problem?
The phenotypes of Betty, Betty's mother, and Bill.

Solution Steps

Because color blindness is an X-linked recessive trait, Betty's color-blind mother must be homozygous for the color-blindness allele (X^{c}X^c). Females inherit one X chromosome from each of their parents, so Betty must have inherited a color-blindness allele from her mother. Because Betty has normal color vision, she must have inherited an allele for normal color vision (X$^+$) from her father; thus, Betty is heterozygous (X^+X^c). Bill is color blind. Because males are hemizygous for X-linked alleles, he must be (X^cY). A mating between Betty and Bill is represented as

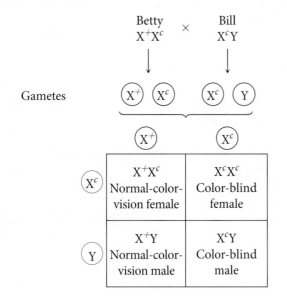

Thus, the overall probability that their child will be color blind is $^1/_2$.

THINK-PAIR-SHARE Question 6

>> Get some additional practice with X-linked inheritance by working **Problem 24** at the end of this chapter.

CONCEPTS

Characteristics determined by genes on the sex chromosomes are called sex-linked characteristics. In organisms with XX-XY sex determination, diploid females have two alleles at each X-linked locus, whereas diploid males possess a single allele at each X-linked locus. Females inherit X-linked alleles from both parents, but males inherit a single X-linked allele from their mothers.

✔ CONCEPT CHECK 6

Hemophilia (reduced blood clotting) is an X-linked recessive disease in humans. A woman with hemophilia mates with a man who exhibits normal blood clotting. What is the probability that their child will have hemophilia?

Symbols for X-Linked Genes

There are several different ways to record genotypes for X-linked characteristics. Sometimes these genotypes are written in the same way as those for autosomal characteristics. In this case, hemizygous males are simply given a single allele: for example, the genotype of a female *Drosophila* with white eyes

is *ww*, and the genotype of a white-eyed hemizygous male is *w*. Another method is to include the Y chromosome, designating it with a diagonal slash (/). With this method, the white-eyed female's genotype is still *ww*, and the white-eyed male's genotype is *w/*. Perhaps the most useful method is to record the X and Y chromosomes in the genotype, designating the X-linked alleles with superscripts, as is done in this chapter. With this method, a white-eyed female is X^wX^w and a white-eyed male is X^wY. The use of Xs and Ys in the genotype has the advantage of reminding us that the genes are X linked and that the male must always have a single allele, inherited from the mother.

Z-Linked Characteristics

In organisms with ZZ-ZW sex determination, males are the homogametic sex (ZZ) and carry two sex-linked (usually referred to as Z-linked) alleles; thus, males may be homozygous or heterozygous. Females are the heterogametic sex (ZW) and possess only a single Z-linked allele. The inheritance of Z-linked characteristics is the same as that of X-linked characteristics, except that the pattern of inheritance in males and females is reversed.

One example of a Z-linked characteristic is the cameo phenotype in Indian blue peafowl (*Pavo cristatus*). In these birds, the wild-type plumage is a glossy metallic blue. The female peafowl is ZW and the male is ZZ. Cameo plumage, which produces brown feathers, results from a Z-linked allele (Z^{ca}) that is recessive to the wild-type blue allele (Z^{Ca+}). If a blue female ($Z^{Ca+}W$) is crossed with a cameo male ($Z^{ca}Z^{ca}$), all the F_1 females are cameo ($Z^{ca}W$) and all the F_1 males are blue ($Z^{Ca+}Z^{ca}$), as shown in **Figure 4.15**. When the F_1 are interbred, $^1/_4$ of the F_2 are blue males ($Z^{Ca+}Z^{ca}$), $^1/_4$ are blue females ($Z^{Ca+}W$), $^1/_4$ are cameo males ($Z^{ca}Z^{ca}$), and $^1/_4$ are cameo females ($Z^{ca}W$). The reciprocal cross of a cameo female with a homozygous blue male produces an F_1 generation in which all offspring are blue and an F_2 generation consisting of $^1/_2$ blue males ($Z^{Ca+}Z^{Ca+}$ and $Z^{Ca+}Z^{ca}$), $^1/_4$ blue females ($Z^{Ca+}W$), and $^1/_4$ cameo females ($Z^{ca}W$).

In organisms with ZZ-ZW sex determination, the female always inherits her W chromosome from her mother, and she inherits her Z chromosome, along with any Z-linked alleles, from her father. In this system, the male inherits Z chromosomes, along with any Z-linked alleles, from both his mother and his father. This pattern of inheritance is the reverse of that of X-linked alleles in organisms with XX-XY sex determination. **> TRY PROBLEM 33**

Y-Linked Characteristics

Y-linked traits—also called holandric traits—exhibit a distinct pattern of inheritance. These traits are present only in males, because only males possess a Y chromosome, and are always inherited from the father. Furthermore, all male offspring of a male with a Y-linked trait will display this trait because every male inherits his Y chromosome from his father. **> TRY PROBLEM 45**

gene that determines maleness, such as the *SRY* gene found in humans today (**Figure 4.16**). This step took place in mammals about 250 million years ago. Any individual organism with a copy of the chromosome containing this gene then became male. Additional mutations occurred on the proto-Y chromosome affecting traits that are beneficial only in males, such as the bright coloration male birds use to attract females and the antlers a male elk uses in competition with other males. The genes that encode these types of traits are advantageous only if they are present in males. Natural selection favors the suppression of crossing over for most of the length of the X and Y chromosomes because this suppression prevents genes that encode male traits from appearing in females. Crossing over can still take place between the two X chromosomes in females, but there is little crossing over between the X and the Y chromosomes, except for the small pseudoautosomal regions in which the X and the Y chromosomes continue to pair in meiosis, as stated earlier.

For reasons that are beyond the scope of this discussion, the lack of crossing over led to (and continues to lead to) an accumulation of mutations and a loss of genetic material from the Y chromosome (see Figure 4.16). Over millions of years, the Y chromosome slowly degenerated, losing DNA and genes until it became greatly reduced in size and contained little genetic information. This degeneration produced the Y chromosome found in males today. Indeed, the Y chromosomes of humans and many other organisms are small and contain little genetic information; therefore, few characteristics exhibit Y-linked inheritance. Some researchers have predicted that the human Y chromosome will continue to lose genetic information in the future and will completely disappear from the species in about 10 million years, a disheartening prospect for those of us with a Y chromosome (and perhaps some of those with two Xs). However, new research suggests that decay of the human Y chromosome has come to a halt and that no genes have

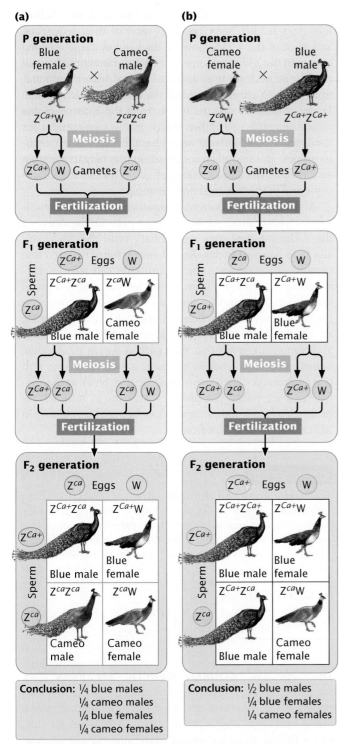

4.15 The cameo phenotype in Indian blue peafowl is inherited as a Z-linked recessive trait. (a) Blue female crossed with cameo male. (b) Reciprocal cross of cameo female with homozygous blue male.

EVOLUTION OF THE Y CHROMOSOME

Research on sex chromosomes has led to the conclusion that the X and Y chromosomes in many organisms evolved from a pair of autosomes. The first step in this evolutionary process took place when one member of a pair of autosomes acquired a

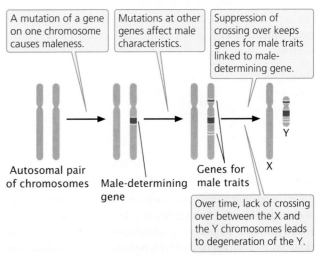

4.16 Evolution of the Y chromosome.

been lost in the last 6 million years. The genes that remain on the Y chromosome appear to be remarkably stable. Internal recombination within the Y chromosome (described in the next section) may have aided in slowing down or preventing the complete decay of the human Y chromosome.

Among animals, sex chromosomes have independently evolved from autosomes many times. This phenomenon has even been observed in mammals: the sex chromosomes of monotremes (egg-laying mammals such as the duck-billed platypus) evolved from different autosomes than did the sex chromosomes of placental and marsupial mammals.

THINK-PAIR-SHARE Question 7

CHARACTERISTICS OF THE HUMAN Y CHROMOSOME

The genetic sequence of most of the human Y chromosome was determined as part of the Human Genome Project (see Chapter 20). This work revealed that about two-thirds of the Y chromosome consists of short DNA sequences that are repeated many times and contain no active genes. The other third consists of just a few genes. Only about 350 genes have been identified on the human Y chromosome, compared with thousands on most chromosomes, and only about half of those identified encode proteins. Some of the protein-encoding genes found on the Y chromosome have homologous genes on the X chromosome. The function of most Y-linked genes is poorly understood; many appear to influence male sexual development and fertility. Others play a role in gene regulation and protein stability. Some are expressed throughout the body, but many are expressed predominately or exclusively in the testes. Although the Y chromosome has relatively few genes, research in *Drosophila* suggests that it carries genetic elements that affect the expression of numerous genes on autosomes and X chromosomes.

A surprising feature revealed by sequencing is the presence of eight massive palindromic sequences on the Y chromosome. A palindrome is a sentence or word, such as "rotator," that reads the same backward and forward. A palindromic sequence in DNA is a sequence that reads the same on both strands of the double helix, creating two nearly identical copies stretching out from a central point, such as

Arm 1

5′–TGGGAG...CTCCCA–3′
3′–ACCCTC...GAGGGT–5′

Arm 2

Thus, a palindromic sequence in DNA appears twice, very much like the two copies of a DNA sequence that are found on two homologous chromosomes. And just as it does with homologous chromosomes, recombination can take place between the two palindromic sequences on the

Y chromosome. As already mentioned, the X and the Y chromosomes are not homologous at most of their sequences, and most of the Y chromosome does not undergo crossing over with the X chromosome. This lack of interchromosomal recombination leads to an accumulation of deleterious mutations on the Y chromosome and the loss of genetic material. Evidence suggests that the two arms of the Y chromosome recombine with each other, which partly compensates for the absence of recombination between the X and the Y chromosomes. This internal recombination may help to maintain some sequences and functions of genes on the Y chromosome and prevent its total degeneration.

Although the palindromic sequences afford opportunities for recombination, which helps prevent the decay of the Y chromosome over evolutionary time, they occasionally have harmful effects. Recent research has revealed that recombination between the palindromes can lead to rearrangements of the Y chromosome that cause anomalies of sexual development. In some cases, recombination between the palindromes leads to deletion of the *SRY* gene, producing an XY female. In other cases, recombination deletes other Y-chromosome genes that take part in sperm production. Sometimes, recombination produces a Y chromosome with two centromeres; such a chromosome may break as the centromeres are pulled in opposite directions in mitosis. The broken Y chromosomes may be lost in mitosis, resulting in XO cells and Turner syndrome.

CONCEPTS

Y-linked traits exhibit a distinct pattern of inheritance: they are present only in males, and all male offspring of a male with a Y-linked trait inherit the trait. Palindromic sequences within the Y chromosome can undergo internal recombination, but such recombination may lead to chromosome anomalies.

✔ CONCEPT CHECK 7

What unusual feature of the Y chromosome allows some recombination among the genes found on it?

THE USE OF Y-LINKED GENETIC MARKERS DNA sequences in the Y chromosome undergo mutation with the passage of time and thus vary among individual males. These mutations create variations in DNA sequence that, like Y-linked traits, are passed from father to son and can therefore be used as genetic markers to study male ancestry. Although the markers themselves do not encode any physical traits, they can be detected with the use of molecular methods. Mutations can readily accumulate in the Y chromosome because so much of it is nonfunctional. Many of these mutations are unique; they arise only once and are passed down through the generations. Individual males possessing the same set of mutations are therefore assumed to be related, and the distribution of these genetic markers on Y chromosomes provides clues about the genetic relationships of present-day people.

Y-linked genetic markers have been used to study the offspring of Thomas Jefferson, principal author of the Declaration of Independence and third president of the United States. In 1802, a political enemy accused Jefferson of fathering a child by his slave Sally Hemings, but the evidence was circumstantial. Hemings, who worked in the Jefferson household and accompanied Jefferson on a trip to Paris, had five children. Jefferson was accused of fathering the first child, but rumors about the paternity of the other children circulated as well. Descendants of Hemings's children maintained that they were part of the Jefferson line, but some Jefferson descendants refused to recognize their claim.

To resolve this long-standing controversy, geneticists examined markers from the Y chromosomes of male-line descendants of Hemings's first son (Thomas Woodson), her last son (Eston Hemings), and a paternal uncle of Thomas Jefferson with whom Jefferson had Y chromosomes in common (descendants of Jefferson's uncle were used because Jefferson himself had no verified male descendants). Geneticists determined that Jefferson possessed a rare and distinctive set of genetic markers on his Y chromosome. The same markers were also found on the Y chromosomes of the male-line descendants of Eston Hemings. The probability of such a match arising by chance is less than 1%. The markers were not found on the Y chromosomes of the descendants of Thomas Woodson. Together with the circumstantial historical evidence, these matching markers suggest that Jefferson was the father of Eston Hemings, but not Thomas Woodson.

Y-chromosome sequences have also been used extensively to examine past patterns of male migration and the genetic relationships among different human populations. Female lineages can be traced through sequences on mitochondrial DNA, which are inherited from the mother (see Chapter 11).

CONNECTING CONCEPTS

Recognizing Sex-Linked Inheritance

What features should we look for to identify a trait as sex linked? A common misconception is that any genetic characteristic in which the phenotypes of males and females differ must be sex linked. In fact, the expression of many *autosomal* characteristics differs between males and females. The genes that encode these characteristics are the same in both sexes, but their expression is influenced by sex hormones. The different sex hormones of males and females cause the same genes to generate different phenotypes in males and females.

Another misconception is that any characteristic that is found more frequently in one sex than in the other is sex linked. A number of autosomal traits are expressed more commonly in one sex than in the other. These traits are said to be sex-influenced. Some autosomal traits are expressed in only one sex; these traits are said to be sex-limited. Both sex-influenced and sex-limited characteristics will be considered in more detail in Chapter 5.

Several features of sex-linked characteristics make them easy to recognize. Y-linked traits are found only in males, but observing a trait only in males does not guarantee that it is Y linked because some autosomal characteristics are expressed only in males. Y-linked traits are unique, however, in that all the male offspring of an affected male express the father's phenotype. Furthermore, a Y-linked trait can be inherited only from the father's side of the family. Thus, a Y-linked trait can be inherited only from the paternal grandfather (the father's father), never from the maternal grandfather (the mother's father).

X-linked characteristics also exhibit a distinctive pattern of inheritance. X linkage is a possible explanation when reciprocal crosses give different results. If a characteristic is X-linked, a cross between an affected male and an unaffected female will not give the same results as a cross between an affected female and an unaffected male. For almost all autosomal characteristics, reciprocal crosses give the same result. We should not conclude, however, that when the reciprocal crosses give different results, the characteristic is necessarily X linked. Other sex-associated forms of inheritance, considered in Chapter 5, also produce different results in reciprocal crosses. The key to recognizing X-linked inheritance is to remember that a male always inherits his X chromosome from his mother, not from his father. Thus, an X-linked characteristic is not passed directly from father to son; if a male clearly inherits a trait from his father—and his mother is not heterozygous—it cannot be X linked.

4.3 Dosage Compensation Equalizes the Amount of Protein Produced by X-Linked and Autosomal Genes in Some Animals

In species with XX-XY sex determination, differences between males and females in their number of X chromosomes present a special problem in development. In females, there are two copies of the X chromosome and two copies of each autosome, so genes on the X chromosomes and on autosomes are "in balance." In males, however, there is only a single X chromosome, while there are two copies of every autosome. Because the amount of a protein produced is often a function of the number of gene copies encoding that protein, males are likely to produce smaller amounts of a protein encoded by X-linked genes than of a protein encoded by autosomal genes. This difference can be detrimental because protein concentration often plays a critical role in development.

Some animals have overcome this problem by evolving mechanisms to equalize the amounts of protein produced by the single X and by two autosomes in the heterogametic sex. These mechanisms are referred to as **dosage compensation**. In fruit flies, dosage compensation is achieved by a doubling of the activity of the genes on the X chromosome of males, but not of females. In placental mammals, the expression of dosage-sensitive genes on the X chromosomes of both males and females has increased, coupled with inactivation of one of the X chromosomes in females, so that expression of X-linked and autosomal genes is balanced in both males and females.

For unknown reasons, the presence of sex chromosomes does not always produce problems of gene dosage, and dosage compensation of X-linked genes is not universal. A number of animals do not exhibit obvious mechanisms of dosage compensation; these animals include butterflies and moths, birds, some fishes, and even the duck-billed platypus. As we see in the next section, even in placental mammals, a number of genes escape dosage compensation.

The Lyon Hypothesis

In 1949, Murray Barr observed condensed, darkly staining bodies in the nuclei of cells from female cats (**Figure 4.17**); these structures became known as **Barr bodies**. Mary Lyon proposed, in 1961, that the Barr body was an inactive X chromosome; her hypothesis (now generally accepted for placental mammals) has become known as the **Lyon hypothesis**. She suggested that, within each female cell, one of the two X chromosomes is inactivated. Which X chromosome is inactivated is random; if a cell contains more than two X chromosomes, all but one of them are inactivated. The number of Barr bodies present in human cells with different complements of sex chromosomes is shown in **Table 4.3**. Lyon's hypothesis led to important insights into the process of development, the expression of X-linked traits, and X-linked genetic diseases.

One of the outstanding geneticists of the twentieth century, Mary Lyon (**Figure 4.18**) conducted significant research in mouse genetics, including important work on mutagenesis, chromosome inversions, and the *t* complex, a genetic element on mouse chromosome 17 that causes some chromosomes to be preferentially transmitted during meiosis. Lyon also helped develop many techniques that are used today in mouse genetics, helping to make the mouse an important model genetic organism.

4.18 Mary Lyon. [© Miron Latyszewski, Courtesy MRC Harwell/Mary Lyon.]

As a result of X inactivation, female placental mammals are functionally hemizygous at the cellular level for X-linked genes. In females that are heterozygous at an X-linked locus, approximately 50% of the cells express one allele and 50% express the other allele; thus, in heterozygous females, proteins encoded by both alleles are produced, but not within the same cell. This functional hemizygosity means that the cells in an individual female are not identical with respect to the expression of the genes on the X chromosome; females are mosaics for the expression of X-linked genes.

Random X inactivation takes place early in development—in humans, it occurs within the first few weeks of development.

(a)

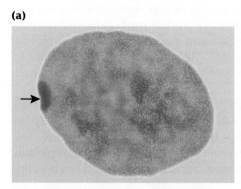

(b)

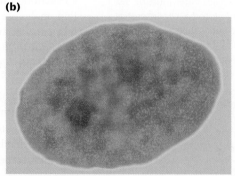

4.17 A Barr body is an inactivated X chromosome. (a) Female cell with a Barr body (indicated by arrow). (b) Male cell without a Barr body. [Chris Bjornberg/Science Source.]

TABLE 4.3	Number of Barr bodies in human cells with different complements of sex chromosomes	
Sex Chromosomes	**Syndrome**	**Number of Barr Bodies**
XX	None	1
XY	None	0
XO	Turner	0
XXY	Klinefelter	1
XXYY	Klinefelter	1
XXXY	Klinefelter	2
XXXXY	Klinefelter	3
XXX	Triple-X	2
XXXX	Poly-X female	3
XXXXX	Poly-X female	4

After an X chromosome has become inactivated in a cell, it remains inactive in that cell and in all somatic cells that descend from that cell. Thus, neighboring cells tend to have the same X chromosome inactivated, producing a patchy pattern (mosaic) for the expression of an X-linked characteristic in heterozygous females.

This patchy distribution of gene expression can be seen in tortoiseshell and calico cats (**Figure 4.19**). Although many genes contribute to coat color and pattern in domestic cats, a single X-linked locus determines the presence of orange color. There are two possible alleles at this locus: X^+, which produces non-orange (usually black) fur, and X^o, which produces orange fur. Males are hemizygous and thus may be black (X^+Y) or orange (X^oY), but not black and orange. (Rare tortoiseshell males can arise from the presence of two X chromosomes, X^+X^oY.) Females may be black (X^+X^+), orange (X^oX^o), or tortoiseshell (X^+X^o), with the tortoiseshell pattern arising from a patchy mixture of black and orange fur. Each orange patch is a clone of cells derived from an original cell in which the black allele was inactivated, and each black patch is a clone of cells derived from an original cell in which the orange allele was inactivated.

The Lyon hypothesis suggests that the presence of variable numbers of X chromosomes should not affect the phenotype in mammals because any X chromosomes in excess of one should be inactivated. However, people with Turner syndrome (XO) differ from XX females, and those with Klinefelter syndrome (XXY) differ from XY males. How do the phenotypes of these conditions arise in the face of dosage compensation?

The phenotypes associated with these conditions probably arise because some X-linked genes escape inactivation. Indeed, the nature of X inactivation is more complex than originally envisioned. Studies of individual genes now reveal that only about 75% of X-linked human genes are permanently inactivated. About 15% completely escape X inactivation, meaning that these genes produce twice as much protein in females as they do in males. The remaining 10% are inactivated in some females but not in others. The reason for this variation among females is not known. Furthermore, recent research indicates that X inactivation does not actually equalize dosage of many X-linked and autosomal genes in humans and mice. **> TRY PROBLEM 44**

Mechanism of Random X Inactivation

Random inactivation of X chromosomes requires two steps. In the first step, the cell somehow assesses, or counts, how many X chromosomes are present. In the second step, one X chromosome is selected to become the active X chromosome and all others are inactivated.

Although many details of X-chromosome inactivation remain unknown, several genes and sequences that participate in the process have been identified. Foremost among them is a gene called *Xist* (for *X-i*nactive *s*pecific *t*ranscript). On the X chromosomes destined to be inactivated, the *Xist* gene is active, producing a 17,000-nucleotide-long RNA molecule that coats the X chromosome and inactivates the genes on it by recruiting protein complexes that alter chromatin structure. On the X chromosome destined to be active, other genes repress the activity of *Xist* so that the *Xist* RNA never coats the X chromosome, and genes on this chromosome remain active.

4.19 The patchy distribution of color on tortoiseshell cats results from the random inactivation of one X chromosome in females. [Robert Adrian Hillman/Shutterstock.]

CONCEPTS

In placental mammals, all but one X chromosome are inactivated in each cell; which of the X chromosomes is inactivated is random and varies from cell to cell.

✔ CONCEPT CHECK 8

How many Barr bodies does a male with XXXYY chromosomes have in each of his cells?

CONCEPTS SUMMARY

- Sexual reproduction is the production of offspring that are genetically distinct from their parents. Most organisms have two sexual phenotypes—males and females. Males produce small gametes; females produce large gametes.

- The mechanism by which sex is specified is termed sex determination. Sex may be determined by differences in specific chromosomes, genotypes, or environment.

- The sex chromosomes of males and females differ in number and appearance. The homogametic sex produces gametes that are all identical with regard to sex chromosomes; the heterogametic sex produces gametes that differ in their sex-chromosome composition.

- In the XX-XO system of sex determination, females possess two X chromosomes, whereas males possess a single X chromosome. In the XX-XY system, females possess two X chromosomes, whereas males possess a single X chromosome and a single Y chromosome. In the ZZ-ZW system, males possess two Z chromosomes, whereas females possess a Z chromosome and a W chromosome.

- Some organisms have genic sex determination, in which genotypes at one or more loci determine the sex of an individual organism. Still others have environmental sex determination.

- In *Drosophila melanogaster*, sex can be predicted by the X : A ratio but is primarily determined by genes on the X chromosome.

- In humans, sex is ultimately determined by the presence or absence of the *SRY* gene located on the Y chromosome.

- Sex-linked characteristics are determined by genes on the sex chromosomes. X-linked characteristics are encoded by genes on the X chromosome, and Y-linked characteristics are encoded by genes on the Y chromosome.

- A female inherits X-linked alleles from both parents; a male inherits X-linked alleles from his female parent only.

- Y-linked characteristics are found only in males and are passed from a father to all of his sons.

- The sex chromosomes evolved from autosomes. Crossing over between the X and the Y chromosomes has been suppressed, but palindromic sequences within the Y chromosome allow for internal recombination on the Y chromosome. This internal recombination sometimes leads to chromosome rearrangements that can adversely affect sexual development.

- In placental mammals, one of the two X chromosomes in females is normally inactivated. Which X chromosome is inactivated is random and varies from cell to cell. Some X-linked genes escape X inactivation, and other X-linked genes may be inactivated in some females but not in others. X inactivation is controlled by the *Xist* gene.

IMPORTANT TERMS

sex (p. 83)
sex determination (p. 83)
hermaphroditism (p. 83)
monoecious (p. 83)
dioecious (p. 83)
sex chromosome (p. 84)
autosome (p. 84)
heterogametic sex (p. 84)
homogametic sex (p. 84)
pseudoautosomal region (p. 85)
genic sex determination (p. 85)
sequential hermaphroditism (p. 86)
Turner syndrome (p. 87)
Klinefelter syndrome (p. 88)
triple-X syndrome (p. 88)
sex-determining region Y (*SRY*) gene (p. 89)
sex-linked characteristic (p. 89)
X-linked characteristic (p. 89)
Y-linked characteristic (p. 89)
hemizygosity (p. 98)
nondisjunction (p. 91)
dosage compensation (p. 97)
Barr body (p. 98)
Lyon hypothesis (p. 98)

ANSWERS TO CONCEPT CHECKS

1. Meiosis

2. b

3. In chromosomal sex determination, males and females have chromosomes that are distinguishable. In genic sex determination, sex is determined by genes, but the chromosomes of males and females are indistinguishable. In environmental sex determination, sex is determined fully or in part by environmental effects.

4. a

5. c

6. All of their male offspring will have hemophilia, and none of their female offspring will have hemophilia, so the overall probability of hemophilia in their offspring is $\frac{1}{2}$.

7. Eight large palindromes allow crossing over within the Y chromosome.

8. Two Barr bodies.

WORKED PROBLEMS

Problem 1

A fruit fly has XXXYY sex chromosomes; all the autosomal chromosomes are normal. What sexual phenotype does this fly have?

>> Solution Strategy

What information is required in your answer to the problem?
The sexual phenotype of a fly with sex chromosomes XXXYY.

What information is provided to solve the problem?
- The fly has sex chromosomes XXXYY.
- All autosomal chromosomes are normal.

For help with this problem, review:
Sex Determination in *Drosophila melanogaster* in Section 4.1.

>> Solution Steps

Sex in fruit flies is predicted by the X : A ratio—the ratio of the number of X chromosomes to the number of haploid autosomal sets. An X : A ratio of 1.0 produces a female fly; an X : A ratio of 0.5 produces a male. If the X : A ratio is greater than 1.0, the fly is a metafemale; if it is less than 0.5, the fly is a metamale; if the X : A ratio is between 1.0 and 0.5, the fly is intersex.

This fly has three X chromosomes and normal autosomes, so the X : A ratio in this case is $^3/_2$, or 1.5. Thus, this fly is a metafemale.

> **Recall:** *Drosophila melanogaster* normally has two sets of autosomes.

Problem 2

In *Drosophila melanogaster*, forked bristles are caused by an allele (X^f) that is X linked and recessive to an allele for normal bristles (X^+). Brown eyes are caused by an allele (b) that is autosomal and recessive to an allele for red eyes (b^+). A female fly that is homozygous for normal bristles and red eyes mates with a male fly that has forked bristles and brown eyes. The F_1 are intercrossed to produce the F_2. What will be the phenotypes and proportions of the F_2 flies from this cross?

>> Solution Strategy

What information is required in your answer to the problem?
Phenotypes and proportions of the F_2 flies.

What information is provided to solve the problem?
- Forked bristles are X-linked recessive.
- Brown eyes are autosomal recessive.
- Phenotypes of the parents of the cross.
- The F_1 are intercrossed to produce the F_2.

For help with this problem, review:
X-linked Color Blindness in Humans in Section 4.2.
Section 3.3 in Chapter 3.

>> Solution Steps

This problem is best worked by breaking the cross down into two separate crosses, one for the X-linked genes that determine the type of bristles and one for the autosomal genes that determine eye color.

Let's begin with the autosomal characteristics. A female fly that is homozygous for red eyes (b^+b^+) is crossed with a male with brown eyes. Because brown eyes are recessive, the male fly must be homozygous for the brown-eye allele (bb). All the offspring of this cross will be heterozygous (b^+b) and will have red eyes:

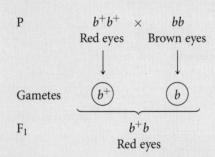

The F_1 are then intercrossed to produce the F_2. Whenever two individual organisms heterozygous for an autosomal recessive characteristic are crossed, $^3/_4$ of the offspring will have the dominant trait and $^1/_4$ will have the recessive trait; thus, $^3/_4$ of the F_2 flies will have red eyes and $^1/_4$ will have brown eyes:

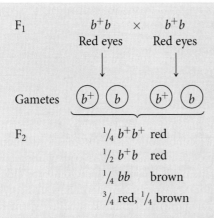

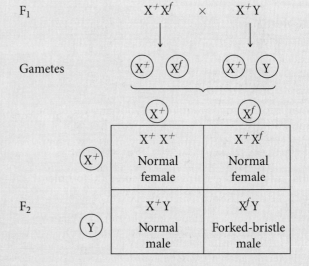

$^1/_2$ normal female, $^1/_4$ normal male, $^1/_4$ forked-bristle male

Next, we work out the results for the X-linked characteristic. A female that is homozygous for normal bristles (X^+X^+) is crossed with a male that has forked bristles (X^fY). The female F_1 from this cross are heterozygous (X^+X^f), receiving an X chromosome with a normal-bristle allele (X^+) from their mother and an X chromosome with a forked-bristle allele (X^f) from their father. The male F_1 are hemizygous (X^+Y), receiving an X chromosome with a normal-bristle allele (X^+) from their mother and a Y chromosome from their father:

Recall: Females have two X-linked alleles, but males have only a single X-linked allele.

To obtain the phenotypic ratio in the F_2, we now combine these two crosses by using the multiplication rule and a branch diagram:

Recall: The multiplication rule states that the probability of two independent events occurring together is the multiplication of their independent probabilities.

P X^+X^+ × X^fY
           Normal   Forked
           bristles   bristles
                ↓   ↓
Gametes X^+ X^f Y

F_1 $^1/_2\ X^+X^f$ normal bristles
          $^1/_2\ X^+Y$ normal bristles

When these F_1 are intercrossed, $^1/_2$ of the F_2 will be normal-bristle females, $^1/_4$ will be normal-bristle males, and $^1/_4$ will be forked-bristle males:

Eye color	Bristle and sex	F_2 phenotype	Probability
red ($^3/_4$)	normal female ($^1/_2$)	red normal female	$^3/_4 × ^1/_2 = ^3/_8$ $= ^6/_{16}$
	normal male ($^1/_4$)	red normal male	$^3/_4 × ^1/_4 = ^3/_{16}$
	forked-bristle male ($^1/_4$)	red forked-bristle male	$^3/_4 × ^1/_4 = ^3/_{16}$
brown ($^1/_4$)	normal female ($^1/_2$)	brown normal female	$^1/_4 × ^1/_2 = ^1/_8$ $= ^2/_{16}$
	normal male ($^1/_4$)	brown normal male	$^1/_4 × ^1/_4 = ^1/_{16}$
	forked-bristle male ($^1/_4$)	brown forked-bristle male	$^1/_4 × ^1/_4 = ^1/_{16}$

Hint: The branch diagram is a convenient way of keeping up with all the different combinations of traits.

COMPREHENSION QUESTIONS

Section 4.1

1. What is considered to be the fundamental difference between males and females of most organisms?

2. How do monoecious organisms differ from dioecious organisms?

3. Describe the XX-XO system of sex determination. In this system, which is the heterogametic sex and which is the homogametic sex?

*4. How does sex determination in the XX-XY system differ from sex determination in the ZZ-ZW system?

5. What is the pseudoautosomal region? How does the inheritance of traits encoded by genes in this region differ from the inheritance of other Y-linked characteristics?

6. What is meant by genic sex determination?

7. How does sex determination in *Drosophila* differ from sex determination in humans?

8. Give the typical sex-chromosome complement found in the cells of people with Turner syndrome, with Klinefelter syndrome, and with androgen-insensitivity syndrome. What is the sex-chromosome complement of triple-X females?

Section 4.2

9. What characteristics are exhibited by an X-linked trait?

10. Explain how Bridges's study of nondisjunction in *Drosophila* helped prove the chromosome theory of inheritance.

11. What characteristics are exhibited by a Y-linked trait?

Section 4.3

12. Explain why tortoiseshell cats are almost always female and why they have a patchy distribution of orange and black fur.

13. What is a Barr body? How is it related to the Lyon hypothesis?

APPLICATION QUESTIONS AND PROBLEMS

Introduction

14. As discussed in the introduction to this chapter, sex in bearded dragons is determined by both sex chromosomes and temperature. What, if any, would be the effect of unusually cool temperatures during the breeding season on the sex ratio of bearded lizards? What would be the effect of unusually warm temperatures on the sex ratio?

Section 4.1

*15. What is the sexual phenotype of fruit flies having the following chromosomes?

	Sex chromosomes	Autosomal chromosomes
a.	XX	all normal
b.	XY	all normal
c.	XO	all normal
d.	XXY	all normal
e.	XYY	all normal
f.	XXYY	all normal
g.	XXX	all normal
h.	XX	four haploid sets
i.	XXX	four haploid sets
j.	XXX	three haploid sets
k.	X	three haploid sets
l.	XY	three haploid sets
m.	XX	three haploid sets

16. If nondisjunction of the sex chromosomes takes place in meiosis I in the male in **Figure 4.5**, what sexual phenotypes and proportions of offspring will be produced?

*17. For each of the following chromosome complements, what is the phenotypic sex of a person who has

a. XY with the *SRY* gene deleted?

b. XX with a copy of the *SRY* gene on an autosomal chromosome?

c. XO with a copy of the *SRY* gene on an autosomal chromosome?

d. XXY with the *SRY* gene deleted?

e. XXYY with one copy of the *SRY* gene deleted?

18. A normal female *Drosophila* produces abnormal eggs that contain all (a complete diploid set) of her chromosomes. She mates with a normal male *Drosophila* that produces normal sperm. What will the sex of the progeny from this cross be?

19. In certain salamanders, the sex of a genetic female can be altered, changing her into a functional male; these salamanders are called sex-reversed males. When a sex-reversed male is mated with a normal female, approximately $2/3$ of the offspring are female and $1/3$ are male. How is sex determined in these salamanders? Explain the results of this cross.

20. In some mites, males pass genes to their grandsons, but they never pass genes to their sons. Explain.

*21. In organisms with the ZZ-ZW sex-determining system, from which of the following possibilities can a female inherit her Z chromosome?

	Yes	No
Her mother's mother	_____	_____
Her mother's father	_____	_____
Her father's mother	_____	_____
Her father's father	_____	_____

Section 4.2

*22. When Bridges crossed white-eyed females with red-eyed males, he obtained a few red-eyed males and white-eyed females (see **Figure 4.13**). What types of offspring would be produced if these red-eyed males and white-eyed females were crossed with each other?

*23. Joe has classic hemophilia, an X-linked recessive disease. Could Joe have inherited the gene for this disease from the following people?

	Yes	No
a. His mother's mother	_____	_____
b. His mother's father	_____	_____
c. His father's mother	_____	_____
d. His father's father	_____	_____

*24. In *Drosophila*, yellow body color is due to an X-linked gene that is recessive to the gene for gray body color.

[Courtesy Dr. Masa-Toshi Yamamoto, Drosophila Genetic Resource Center, Kyoto Institute of Technology.]

a. A homozygous gray female is crossed with a yellow male. The F_1 are intercrossed to produce the F_2. Give the genotypes and phenotypes, along with the expected proportions, of the F_1 and F_2 progeny.

b. A yellow female is crossed with a gray male. The F_1 are intercrossed to produce the F_2. Give the genotypes and phenotypes, along with the expected proportions, of the F_1 and F_2 progeny.

c. A yellow female is crossed with a gray male. The F_1 females are backcrossed with gray males. Give the genotypes and phenotypes, along with the expected proportions, of the F_2 progeny.

d. If the F_2 flies in part *b* mate randomly, what are the expected phenotypes and proportions of flies in the F_3?

25. Coat color in cats is determined by genes at several different loci. At one locus on the X chromosome, one allele (X^+) encodes black fur; another allele (X^o) encodes orange fur. Females can be black (X^+X^+), orange (X^oX^o), or a mixture of orange and black called tortoiseshell (X^+X^o). Males are either black (X^+Y) or orange (X^oY). Bill has a female tortoiseshell cat named Patches. One night Patches escapes from Bill's house, spends the night

out, and mates with a stray male. Patches later gives birth to the following kittens: one orange male, one black male, two tortoiseshell females, and one orange female. Give the genotypes of Patches, her kittens, and the stray male with which Patches mated.

*26. Red–green color blindness in humans is due to an X-linked recessive gene. Both John and Cathy have normal color vision. After 10 years of marriage to John, Cathy gave birth to a color-blind daughter. John filed for divorce, claiming that he is not the father of the child. Is John justified in his claim of nonpaternity? Explain why. If Cathy had given birth to a color-blind son, would John be justified in claiming nonpaternity?

27. Red–green color blindness in humans is due to an X-linked recessive gene. A woman whose father is color blind possesses one eye with normal color vision and one eye with color blindness.

a. Propose an explanation for this woman's vision pattern. Assume that no new mutations have spontaneously arisen.

b. Would it be possible for a man to have one eye with normal color vision and one eye with color blindness?

*28. Bob has XXY chromosomes (Klinefelter syndrome) and is color blind. His mother and father have normal color vision, but his maternal grandfather is color blind. Assume that Bob's chromosome abnormality arose from nondisjunction in meiosis. In which parent and in which meiotic division did nondisjunction take place? Assume no crossing over has taken place. Explain your answer.

29. Xg is an antigen found on red blood cells. This antigen is caused by an X-linked allele (X^a) that is dominant over an allele for the absence of the antigen (X^-). The inheritance of these X-linked alleles was studied in children with chromosome abnormalities to determine where nondisjunction of the sex chromosomes took place. For each type of mating in parts *a* through *d*, indicate whether nondisjunction took place in the mother or in the father and, if possible, whether it took place in meiosis I or meiosis II (assume no crossing over).

a. $X^aY \times X^-X^- \rightarrow X^a$ (Turner syndrome)

b. $X^aY \times X^aX^- \rightarrow X^-$ (Turner syndrome)

c. $X^aY \times X^-X^- \rightarrow X^aX^-Y$ (Klinefelter syndrome)

d. $X^aY \times X^aX^- \rightarrow X^-X^-Y$ (Klinefelter syndrome)

30. The Talmud, an ancient book of Jewish civil and religious laws, states that, if a woman bears two sons who die of bleeding after circumcision (removal of the foreskin from the penis), any additional sons that she has should not be circumcised. (The bleeding is most likely due to the X-linked disorder hemophilia.) Furthermore, the Talmud states that the sons of her sisters must not be circumcised, whereas the sons of her

brothers should be. Is this religious law consistent with sound genetic principles? Explain your answer.

31. Craniofrontonasal syndrome (CFNS) is a birth defect in which premature fusion of the cranial sutures leads to abnormal head shape, widely spaced eyes, nasal clefts, and various other skeletal abnormalities. George Feldman and his colleagues looked at several families in which offspring had CFNS and recorded the results shown in the following table (G. J. Feldman. 1997. *Human Molecular Genetics* 6:1937–1941).

| Family number | Parents | | Offspring | | | |
| | Father | Mother | Normal | | CFNS | |
			Male	Female	Male	Female
1	normal	CFNS	1	0	2	1
5	normal	CFNS	0	2	1	2
6	normal	CFNS	0	0	1	2
8	normal	CFNS	1	1	1	0
10a	CFNS	normal	3	0	0	2
10b	normal	CFNS	1	1	2	0
12	CFNS	normal	0	0	0	1
13a	normal	CFNS	0	1	2	1
13b	CFNS	normal	0	0	0	2
7b	CFNS	normal	0	0	0	2

a. On the basis of these results, what is the most likely mode of inheritance for CFNS?

b. Give the most likely genotypes of the parents in family 1 and in family 10a.

32. Miniature wings (X^m) in *Drosophila* result from an X-linked allele that is recessive to the allele for long wings (X^+). Give the genotypes of the parents in each of the following crosses.

Male parent	Female parent	Male offspring	Female offspring
a. long	long	231 long, 250 miniature	560 long
b. miniature	long	610 long	632 long
c. miniature	long	410 long, 417 miniature	412 long, 415 miniature
d. long	miniature	753 miniature	761 long
e. long	long	625 long	630 long

*33. In chickens, congenital baldness is due to a Z-linked recessive gene. A bald rooster is mated with a normal hen. The F_1 from this cross are interbred to produce the F_2. Give the genotypes and phenotypes, along with their expected proportions, among the F_1 and F_2 progeny.

34. If the blue F_1 females in **Figure 4.15b** are backcrossed to the blue males in the P generation, what phenotypes and proportions of offspring will be produced?

35. Red–green color blindness is an X-linked recessive trait in humans. Polydactyly (extra fingers and toes) is an autosomal dominant trait. Martha has normal fingers and toes and normal color vision. Her mother is normal in all respects, but her father is color blind and polydactylous. Bill is color blind and polydactylous. His mother has normal color vision and normal fingers and toes. If Bill and Martha marry, what phenotypes and proportions of children can they produce?

36. A *Drosophila* mutation called *singed* (*s*) causes the bristles to be bent and misshapen. A mutation called *purple* (*p*) causes the fly's eyes to be purple in color instead of the normal red. Flies homozygous for *singed* and *purple* were crossed with flies that were homozygous for normal bristles and red eyes. The F_1 were intercrossed to produce the F_2, and the following results were obtained.

Cross 1

P	male, singed bristles, purple eyes × female, normal bristles, red eyes
F_1	420 female, normal bristles, red eyes
	426 male, normal bristles, red eyes
F_2	337 female, normal bristles, red eyes
	113 female, normal bristles, purple eyes
	168 male, normal bristles, red eyes
	170 male, singed bristles, red eyes
	56 male, normal bristles, purple eyes
	58 male, singed bristles, purple eyes

Cross 2

P	female, singed bristles, purple eyes × male, normal bristles, red eyes
F_1	504 female, normal bristles, red eyes
	498 male, singed bristles, red eyes
F_2	227 female, normal bristles, red eyes
	223 female, singed bristles, red eyes
	225 male, normal bristles, red eyes
	225 male, singed bristles, red eyes
	78 female, normal bristles, purple eyes
	76 female, singed bristles, purple eyes
	74 male, normal bristles, purple eyes
	72 male, singed bristles, purple eyes

a. What are the modes of inheritance of *singed* and *purple*? Explain your reasoning.

b. Give genotypes for the parents and offspring in the P, F_1, and F_2 generations of cross 1 and cross 2.

37. The following two genotypes are crossed: $Aa\ Bb\ Cc\ X^+X^r \times Aa\ BB\ cc\ X^+Y$, where a, b, and c represent alleles of autosomal genes and X^+ and X^r represent X-linked alleles in an organism with XX-XY sex determination. What is the probability of obtaining genotype $aa\ Bb\ Cc\ X^+X^+$ in the progeny?

*38. Miniature wings in *Drosophila* result from an X-linked allele (X^m) that is recessive to the allele for long wings (X^+). Sepia eyes are produced by an autosomal allele (s) that is recessive to an allele for red eyes (s^+).

   a. A female fly that has miniature wings and sepia eyes is crossed with a male that has normal wings and is homozygous for red eyes. The F_1 flies are intercrossed to produce the F_2. Give the phenotypes, as well as their expected proportions, of the F_1 and F_2 flies.

   b. A female fly that is homozygous for normal wings and has sepia eyes is crossed with a male that has miniature wings and is homozygous for red eyes. The F_1 flies are intercrossed to produce the F_2. Give the phenotypes, as well as their expected proportions, of the F_1 and F_2 flies.

39. Suppose that a recessive gene that produces a short tail in mice is located in the pseudoautosomal region. A short-tailed male mouse is mated with a female mouse that is homozygous for a normal tail. The F_1 mice from this cross are intercrossed to produce the F_2. Give the phenotypes, as well as their proportions, of the F_1 and F_2 mice.

*40. A color-blind woman and a man with normal color vision have three sons and six daughters. All the sons are color blind. Five of the daughters have normal color vision, but one of them is color blind. The color-blind daughter is 16 years old, is short for her age, and has not undergone puberty. Explain how this girl inherited her color blindness.

Section 4.3

*41. How many Barr bodies would you expect to see in a human cell containing the following chromosomes?

   a. XX   d. XXY   g. XYY
   b. XY   e. XXYY   h. XXX
   c. XO   f. XXXY   i. XXXX

42. A woman with normal chromosomes mates with a man who also has normal chromosomes.

   a. Suppose that, in the course of oogenesis, the woman's sex chromosomes undergo nondisjunction in meiosis I; the man's chromosomes separate normally. Give all possible combinations of sex chromosomes that this couple's children might inherit and the number of Barr bodies that you would expect to see in each of the cells of each child.

   b. What chromosome combinations and numbers of Barr bodies would you expect to see if the chromosomes separate normally in oogenesis, but nondisjunction of the sex chromosomes takes place in meiosis I of spermatogenesis?

43. What is the most likely sex and genotype of the cat shown in **Figure 4.19**?

*44. Anhidrotic ectodermal dysplasia is an X-linked recessive disorder in humans characterized by small teeth, no sweat glands, and sparse body hair. This trait is usually seen in men, but women who are heterozygous carriers of the trait often have irregular patches of skin with few or no sweat glands (see the illustration below).

   a. Explain why women who are heterozygous carriers of a recessive gene for anhidrotic ectodermal dysplasia have irregular patches of skin lacking sweat glands.

   b. Why does the distribution of the patches of skin lacking sweat glands differ among the females depicted in the illustration, even between the identical twins?

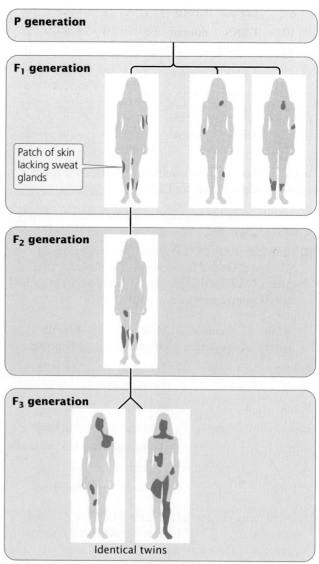

P generation

F₁ generation

Patch of skin lacking sweat glands

F₂ generation

F₃ generation

Identical twins

[After A. P. Mange and E. J. Mange, *Genetics: Human Aspects*, 2nd ed. (Sinauer Associates, 1990), p. 133.]

CHALLENGE QUESTIONS

Section 4.2

*45. A geneticist discovers a male mouse with greatly enlarged testes in his laboratory colony. He suspects that this trait results from a new mutation that is either Y linked or autosomal dominant. How could he determine whether the trait is autosomal dominant or Y linked?

Section 4.3

46. Human females who are heterozygous for an X-linked recessive allele sometimes exhibit mild expression of the trait. However, such mild expression of X-linked traits in females who are heterozygous for X-linked alleles is not seen in *Drosophila*. What might cause this difference in the expression of X-linked genes between human females and female *Drosophila*? (Hint: In *Drosophila*, dosage compensation is accomplished by doubling the activity of genes on the X chromosome of males.)

47. Identical twins (also called monozygotic twins) are derived from a single egg fertilized by a single sperm, creating a zygote that later divides into two (see Chapter 6). Because identical twins originate from a single zygote, they are genetically identical.

Caroline Loat and her colleagues examined nine measures of social, behavioral, and cognitive ability in 1000 pairs of identical male twins and 1000 pairs of identical female twins (C. S. Loat et al. 2004. *Twin Research* 7:54–61). They found that, for three of the measures (prosocial behavior, peer problems, and verbal ability), the two male twins of a pair tended to be more alike in their scores than were two female twins of a pair. Propose a possible explanation for this observation. What might this observation indicate about the location of genes that influence prosocial behavior, peer problems, and verbal ability?

48. Occasionally, a mouse X chromosome is broken into two pieces and each piece becomes attached to a different autosome. In this event, the genes on only one of the two pieces undergo X inactivation. What does this observation indicate about the mechanism of X-chromosome inactivation?

THINK-PAIR-SHARE QUESTIONS

Section 4.1

1. The duck-billed platypus has a unique mechanism of sex determination: females have five pairs of X chromosomes ($X_1X_1X_2X_2X_3X_3X_4X_4X_5X_5$) and males have five pairs of X and Y chromosomes ($X_1Y_1X_2Y_2X_3Y_3X_4Y_4X_5Y_5$). Do you think each of the X and Y chromosome pairs in males assorts independently of other X and Y pairs during meiosis? Why or why not?

2. Most organisms with XX-XY sex determination have pseudoautosomal regions, portions of the X and Y chromosomes that are homologous. Would you predict that organisms with ZZ-ZW sex determination have pseudoautosomal regions of homology between Z and W chromosomes? Explain your answer.

3. Both men and women produce testosterone, but concentrations of testosterone in the blood are generally higher in men than in women. However, the testosterone levels of some XX females fall within the range of testosterone levels of XY men. This overlap has created controversy within women's sports. Testosterone is known to increase muscle mass and enhance some types of athletic performance, so some people have suggested that women with naturally high testosterone levels have an unfair competitive advantage. In 2011, the International Association of Athletics Federations (IAAF) adopted a policy that limits levels of testosterone in female athletes, saying that female athletes must not have a blood testosterone concentration greater than 10 nanomoles per liter (nmol/L), a level typically seen in men. Some elite female athletes have natural testosterone levels above this limit and have challenged the policy. Do you think that it is fair for XX females with naturally high testosterone (levels typically found in XY males) to compete in women's sports? Do they have an unfair advantage in competition with other women? What about male athletes with naturally high levels of testosterone? Do they have an unfair advantage over other males? In general, what role does genetics play in athletic competition—do some individuals have genes that give them an unfair advantage in competition?

Section 4.2

4. How is the inheritance of X-linked traits different from the inheritance of autosomal traits? How is the inheritance of X-linked and autosomal traits similar? List as many differences and similarities as you can.

5. On average, what proportion of X-linked genes in the first individual are the same (inherited from a common ancestor) as those in the second individual?
   a. A male and his mother
   b. A female and her mother
   c. A male and his father
   d. A female and her father
   e. A male and his brother
   f. A female and her sister
   g. A male and his sister
   h. A female and her brother
6. Red–green color blindness is an X-linked recessive trait. Susan has normal color vision, but her father is color blind. Susan marries Bob, who has normal color vision.
   a. What is the probability that Susan and Bob will have a color-blind son?
   b. Susan and Bob have a daughter named Betty, who has normal color vision. If Betty marries a man with normal color vision, and they have a son, what is the probability that the son will be color blind?
7. As pointed out in the section Evolution of the Y Chromosome, some researchers have predicted that the human Y chromosome will continue to lose genetic information in the future and will completely disappear from the species in about 10 million years. What would happen if the Y chromosome disappeared from the human species?

SaplingPlus Self-study tools that will help you practice what you've learned and reinforce this chapter's concepts are available online. Go to www.macmillanlearning.com/PierceGenetics6e.

Extensions and Modifications of Basic Principles

The direction of shell coiling in *Lymnaea* snails is determined by genetic maternal effect. Shown here are two *Lymnaea stagnalis*: a snail with a left-handed (sinistral) shell on the left and a snail with a right-handed (dextral) shell on the right. [Courtesy of Dr. Reiko Kuroda.]

The Odd Genetics of Left-Handed Snails

At the start of the twentieth century, Mendel's work on inheritance in pea plants became widely known (see Chapter 3), and a number of biologists set out to verify his conclusions by conducting crosses with other organisms. Biologists quickly confirmed that Mendel's principles applied not just to peas, but also to corn, beans, chickens, mice, guinea pigs, humans, and many other organisms. At the same time, biologists began to discover exceptions—traits whose inheritance was more complex than that of the simple dominant and recessive traits that Mendel had observed. One of these exceptions was the spiral of a snail's shell.

The direction of coiling in snail shells is called chirality. Most snail shells spiral downward in a clockwise or right-handed direction. These shells are said to be dextral. A few snails have shells that coil in the opposite direction, spiraling downward in a counterclockwise or left-handed direction. These shells are said to be sinistral. Most snail species have shells that are all dextral or all sinistral; only in a few rare instances do both dextral and sinistral shells coexist in the same species.

In the 1920s and 1930s, Arthur Boycott, of the University of London, investigated the genetics of shell coiling in *Lymnaea peregra*, a common pond snail in Britain. In this species, most snails are dextral, but a few sinistral snails occur in some populations. Boycott learned from amateur naturalists of a pond near Leeds, England, where an abnormally high number of sinistral snails could be found. He obtained four sinistral snails from this location and began to investigate the genetics of shell chirality.

Boycott's research was complicated by the fact that these snails are hermaphroditic, meaning that a snail can self-fertilize, or *self* (mate with itself). If a suitable partner is available, the snails are also capable of *outcrossing*—mating with another individual. Boycott found that if he isolated a newly hatched snail and reared it alone, it would eventually produce offspring, so he knew that it had selfed. But when he placed two snails together and one produced offspring, he had no way of knowing whether it had mated with itself or with the other snail. Boycott's research required rearing large numbers of snails in isolation and in pairs, raising their offspring, and determining the direction of shell coiling for each offspring. To facilitate the work, he enlisted the aid of several amateur scientists. One of his helpers was Captain C. Diver, a friend who worked as an assistant for the British Parliament. Since Parliament met for only part of the year, Diver had time on his hands and eagerly enlisted to assist with the research. Together, Boycott, Diver, and other assistants carried out numerous breeding experiments, selfing and crossing snails and raising the progeny in jam jars. They eventually raised more than 6000 broods and determined the direction of coiling in a million snails.

Initially, their results were puzzling—shell coiling did not appear to conform to Mendel's principles of heredity. Eventually, they realized that dextral was dominant over sinistral, but with a peculiar twist: the phenotype of a snail was determined *not* by its own genotype, but by the genotype of its mother. This phenomenon—determination of the phenotype by the genotype of the mother—is called genetic maternal effect. Genetic maternal effect often arises because the maternal parent produces a substance, encoded by her own genotype, that is deposited in the cytoplasm of the egg and which influences early development of the offspring.

The substance that determines the direction of shell coiling in snails has never been isolated. However, in 2009, Reiko Kuroda and her colleagues demonstrated that the direction of coiling in *Lymnaea* snails is determined by the orientation of cells when the embryo is at an early developmental stage—specifically, the eight-cell stage. By gently pushing on the cells of eight-cell embryos, they were able to induce offspring whose mother's genotype was dextral to develop as sinistral snails; similarly, they induced the offspring of mothers whose genotype was sinistral to develop as dextral snails by pushing on the cells in the opposite direction.

THINK-PAIR-SHARE

- The introduction to this chapter discusses the genetic basis of chirality in snails and the research of Arthur Boycott, whose work established the mode of inheritance for this trait. In the course of his research, Boycott enlisted the aid of several amateur scientists—men who were not trained as scientists and had other jobs. The research would have been impossible without the aid of these individuals. In the past, amateur scientists such as these often made important contributions to science, but the practice is less frequent today. Discuss some possible reasons for the decline in contributions to research by amateur scientists. Are there any areas where amateur scientists still actively contribute?

- Genetic maternal effect is often seen in mammals. For example, research shows that the maternal genotype influences adult body size in mice. Why might these types of genetic effects be more common in mammals than in other organisms such as fishes, amphibians, or reptiles?

Boycott's research on the direction of coiling in snails demonstrated that not all characteristics are inherited as simple dominant and recessive traits like the shapes and colors of peas that Mendel described. This demonstration doesn't mean that Mendel was wrong; rather, it indicates that Mendel's principles are not, by themselves, sufficient to explain the inheritance of all genetic characteristics. Our modern understanding of genetics has been greatly enriched by the discovery of a number of modifications and extensions of Mendel's basic principles, which are the focus of this chapter.

5.1 Additional Factors at a Single Locus Can Affect the Results of Genetic Crosses

In Chapter 3, we learned that the principle of segregation and the principle of independent assortment enable us to predict the outcomes of genetic crosses. Here, we examine several additional factors acting at individual loci that can alter the phenotypic ratios predicted by Mendel's principles.

Types of Dominance

One of Mendel's important contributions to the study of heredity is the concept of dominance—the idea that although an individual organism possesses two different alleles for a characteristic, the trait encoded by only one of the alleles is observed in the phenotype. With dominance, the heterozygote possesses the same phenotype as one of the homozygotes.

Mendel observed dominance in all the traits he chose to study extensively, but he was aware that not all characteristics exhibit dominance. He conducted some crosses to look at the length of time that pea plants take to flower. For example, when he crossed two homozygous varieties that differed in their flowering time by an average of 20 days, the length of time taken by the F_1 plants to flower was intermediate between those of the two parents. When the heterozygote has a phenotype intermediate between the phenotypes of the two homozygotes, the trait is said to display *incomplete dominance*.

COMPLETE AND INCOMPLETE DOMINANCE Dominance can be understood in regard to how the phenotype of the heterozygote relates to the phenotypes of the two homozygotes. In the example presented in the upper panel of **Figure 5.1**, flower color potentially ranges from red to white. One homozygous genotype, A^1A^1, produces red pigment, resulting in red flowers; another, A^2A^2, produces no pigment, resulting in white flowers. Where the heterozygote falls in the range of phenotypes determines the type of dominance. If the heterozygote (A^1A^2) produces the same amount of pigment as the A^1A^1 homozygote, resulting in red flowers, then the A^1 allele displays **complete dominance** over the A^2 allele; that is, red is dominant over white. If, on the other hand, the heterozygote produces no pigment, resulting in flowers with the same color as the A^2A^2 homozygote (white), then the A^2 allele is completely dominant, and white is dominant over red.

When the phenotype of the heterozygote falls in between the phenotypes of the two homozygotes, dominance is incomplete. With **incomplete dominance**, the heterozygote need not be exactly intermediate between the two homozygotes (see the lower panel of Figure 5.1); it might be a slightly lighter shade of red or a slightly pink shade of white. As long as the heterozygote's phenotype can be differentiated from

those of the two homozygotes and falls between them, dominance is incomplete.

Incomplete dominance is also exhibited in the fruit color of eggplant. When a homozygous plant that produces purple fruit (PP) is crossed with a homozygous plant that produces white fruit (pp), all the heterozygous F_1 (Pp) plants produce violet fruit (**Figure 5.2a**). When the F_1 are crossed with each other, $^1/_4$ of the F_2 are purple (PP), $^1/_2$ are violet (Pp), and $^1/_4$ are white (pp), as shown in **Figure 5.2b**. Note that this 1 : 2 : 1 ratio is different from the 3 : 1 ratio that we would observe if eggplant fruit color exhibited complete dominance. Another example of incomplete dominance is feather

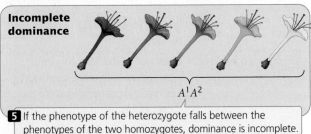

5.1 The type of dominance exhibited by a trait depends on how the phenotype of the heterozygote relates to the phenotypes of the homozygotes.

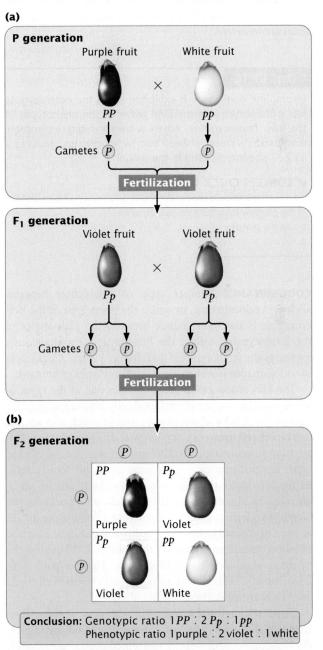

5.2 Fruit color in eggplant is inherited as an incompletely dominant trait.

color in chickens. A cross between a homozygous black chicken and a homozygous white chicken produces F_1 chickens that are gray. If these gray F_1 chickens are intercrossed, they produce F_2 birds in a ratio of 1 black : 2 gray : 1 white.

We should now add the 1 : 2 : 1 ratio to the phenotypic ratios for simple crosses presented in Chapter 3 (see Table 3.5). A 1 : 2 : 1 phenotypic ratio arises in the progeny of a cross between two parents that are heterozygous for a character that exhibits incomplete dominance ($Aa \times Aa$). The genotypic ratio among these progeny is also 1 : 2 : 1. When a trait displays incomplete dominance, the genotypic ratios and phenotypic ratios of the offspring are the *same* because each genotype has its own phenotype. The important thing to remember about dominance is that it affects the way that genes are *expressed* (the phenotype), but not the way that genes are *inherited*.

CONCEPTS

Incomplete dominance is exhibited when the heterozygote has a phenotype intermediate between the phenotypes of the two homozygotes. When a trait exhibits incomplete dominance, a cross between two heterozygotes produces a 1 : 2 : 1 phenotypic ratio in the progeny.

✔ CONCEPT CHECK 1

If an F_1 eggplant in Figure 5.2 is used in a testcross, what proportion of the progeny from this cross will be white?

a. All the progeny   c. $\frac{1}{4}$

b. $\frac{1}{2}$ d. 0

CODOMINANCE Another type of interaction between alleles is **codominance**, in which the phenotype of the heterozygote is not intermediate between the phenotypes of the homozygotes; rather, the heterozygote simultaneously expresses the phenotypes of both homozygotes. An example of codominance is seen in the MN blood types of humans.

The MN blood-group locus encodes one of the types of antigens on the surface of red blood cells. Unlike foreign antigens of the ABO and Rh blood groups (which also encode red-blood-cell antigens), foreign MN antigens do not elicit a strong immunological reaction; therefore, the MN blood types are not routinely considered in blood transfusions. At the MN locus, there are two alleles: the L^M allele, which encodes the M antigen; and the L^N allele, which encodes the N antigen. Homozygotes with genotype $L^M L^M$ express the M antigen on the surface of their red blood cells and have the M blood type. Homozygotes with genotype $L^N L^N$ express the N antigen and have the N blood type. Heterozygotes with genotype $L^M L^N$ exhibit codominance and express both the M and the N antigens; they have blood-type MN.

Some students might ask why the pink flowers illustrated in the lower panel of Figure 5.1 exhibit incomplete dominance—that is, why is this outcome not an example of codominance? The flowers would exhibit codominance

only if the heterozygote produced both red and white pigments, which then combined to produce a pink phenotype. However, in our example, the heterozygote produces only red pigment. The pink phenotype comes about because the amount of pigment produced by the heterozygote is less than the amount produced by the $A^1 A^1$ homozygote. So in that case, the alleles clearly exhibit incomplete dominance, not codominance. The differences between complete dominance, incomplete dominance, and codominance are summarized in **Table 5.1**. ▶ **TRY PROBLEM 13**

LEVEL OF PHENOTYPE OBSERVED MAY AFFECT DOMINANCE Many phenotypes can be observed at several different levels, including the anatomical level, the physiological level, and the molecular level. The type of dominance exhibited by a characteristic depends on the level at which the phenotype is examined. This dependency is seen in cystic fibrosis, a common genetic disorder in Caucasians that is usually considered to be a recessive disease. People who have cystic fibrosis produce large quantities of thick, sticky mucus, which plugs up the airways of the lungs and clogs the ducts leading from the pancreas to the intestine, causing frequent respiratory infections and digestive problems. Even with medical treatment, people with cystic fibrosis suffer chronic, life-threatening medical problems.

The gene responsible for cystic fibrosis resides on the long arm of chromosome 7. It encodes a protein termed *cystic fibrosis transmembrane conductance regulator* (CFTR), which acts as a gated channel in the cell membrane and regulates the movement of chloride ions into and out of the cell. People with cystic fibrosis have a mutated, dysfunctional form of CFTR that causes the channel to stay closed, so chloride ions build up in the cell. This buildup causes the formation of thick mucus and produces the symptoms of the disease.

Most people have two copies of the normal allele for CFTR and produce only functional CFTR protein. Those

TABLE 5.1	Differences between complete dominance, incomplete dominance, and codominance
Type of Dominance	**Definition**
Complete dominance	Phenotype of the heterozygote is the same as the phenotype of one of the homozygotes.
Incomplete dominance	Phenotype of the heterozygote is intermediate (falls within the range) between the phenotypes of the two homozygotes.
Codominance	Phenotype of the heterozygote includes the phenotypes of both homozygotes.

with cystic fibrosis possess two copies of the mutated CFTR allele and produce only the defective CFTR protein. Heterozygotes, who have one normal and one defective CFTR allele, produce both functional and defective CFTR protein. Thus, at the molecular level, the alleles for normal and defective CFTR are codominant because both alleles are expressed in the heterozygote. However, because one functional allele produces enough functional CFTR protein to allow normal chloride ion transport, heterozygotes exhibit no adverse effects, and the mutated CFTR allele appears to be recessive at the physiological level. The type of dominance expressed by an allele, as illustrated in this example, is thus a function of the phenotypic effect of the allele that we observe.

THINK-PAIR-SHARE Question 1

CHARACTERISTICS OF DOMINANCE Several important characteristics of dominance should be emphasized. First, dominance is a result of interactions between genes at the same locus (allelic genes); in other words, dominance is *allelic* interaction. Second, dominance does not alter the way in which the genes are inherited; it influences only the way in which they are expressed as a phenotype. The allelic interaction that characterizes dominance is therefore interaction between the *products* of the genes. Finally, the type of dominance exhibited frequently depends on the level at which the phenotype is examined. As seen for cystic fibrosis, an allele may exhibit codominance at one level and be recessive at another level.

CONCEPTS

Dominance entails interactions between genes at the same locus (allelic genes) and is an aspect of the phenotype; dominance does not affect the way in which genes are inherited. The type of dominance exhibited by a characteristic frequently depends on the level at which the phenotype is examined.

✔ **CONCEPT CHECK 2**

How do complete dominance, incomplete dominance, and codominance differ?

Penetrance and Expressivity

In the genetic crosses presented thus far, we have considered only the interactions of alleles and have assumed that every individual organism having a particular genotype expresses the expected phenotype. We have assumed, for example, that in peas, the genotype *Rr* always produces round seeds and that the genotype *rr* always produces wrinkled seeds. For some characteristics, however, such an assumption is incorrect: the genotype does not always produce the expected phenotype, a phenomenon termed **incomplete penetrance**.

Incomplete penetrance is seen in human polydactyly, the condition of having extra fingers or toes (**Figure 5.3**). There are several different forms of human polydactyly, but the trait is usually caused by a dominant allele. Occasionally, people possess the allele for polydactyly (as evidenced by the fact that their children inherit the polydactyly), but nevertheless have a normal number of fingers and toes. In these cases, the gene for polydactyly is not fully penetrant. **Penetrance** is defined as the percentage of individual organisms having a particular genotype that express the expected phenotype. For example, if we examined 42 people having an allele for polydactyly and found that only 38 of them were polydactylous, the penetrance would be $^{38}/_{42} = 0.90$ (90%).

A related concept is that of **expressivity**, the degree to which a trait is expressed. In addition to incomplete penetrance, polydactyly exhibits variable expressivity. Some polydactylous people possess extra fingers or toes that are fully functional, whereas others possess only a small tag of extra skin.

Incomplete penetrance and variable expressivity are due to the effects of other genes and environmental factors that can alter or completely suppress the effect of a particular gene. For example, a gene may encode an enzyme that produces a particular phenotype only within a limited temperature range. At higher or lower temperatures, the enzyme does not function, so the phenotype is not expressed; the allele encoding such an enzyme is therefore penetrant only within a particular temperature range (see also Environmental Effects on the Phenotype in Section 5.5). Many characters exhibit incomplete penetrance and variable expressivity; thus, the mere presence of a gene does not guarantee its expression.

⟩ TRY PROBLEM 15

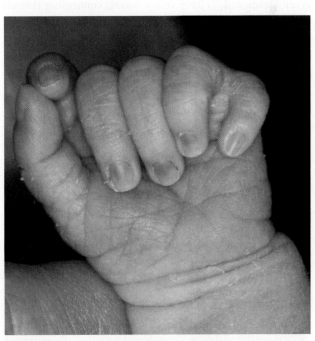

5.3 Human polydactyly (extra digits) exhibits incomplete penetrance and variable expressivity. [SPL/Science Source.]

Penetrance is the percentage of individuals having a particular genotype that express the associated phenotype. Expressivity is the degree to which a trait is expressed. Incomplete penetrance and variable expressivity result from the influence of other genes and environmental factors on the phenotype.

✔ **CONCEPT CHECK 3**

How does incomplete dominance differ from incomplete penetrance?

a. Incomplete dominance refers to alleles at the same locus; incomplete penetrance refers to alleles at different loci.

b. Incomplete dominance ranges from 0% to 50%; incomplete penetrance ranges from 51% to 99%.

c. In incomplete dominance, the heterozygote is intermediate between the homozygotes; in incomplete penetrance, heterozygotes express phenotypes of both homozygotes.

d. In incomplete dominance, the heterozygote is intermediate between the homozygotes; in incomplete penetrance, some individuals do not express the expected phenotype.

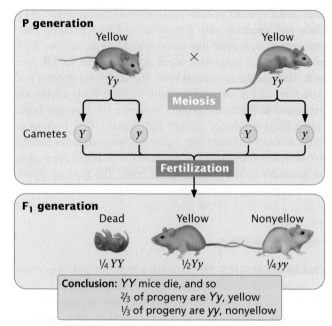

5.4 The 2 : 1 ratio produced by a cross between two yellow mice results from a lethal allele.

Lethal Alleles

A **lethal allele** causes death at an early stage of development—often before birth—so that some genotypes do not appear among the progeny. An example of a lethal allele, originally described by Erwin Baur in 1907, is found in snapdragons. The *aurea* strain in these plants has yellow leaves. When two plants with yellow leaves are crossed, $^2/_3$ of the progeny have yellow leaves and $^1/_3$ have green leaves. When green is crossed with green, all the progeny have green leaves; when yellow is crossed with green, however, $^1/_2$ of the progeny have green leaves and $^1/_2$ have yellow leaves, confirming that all yellow-leaved snapdragons are heterozygous.

Another example of a lethal allele is one that determines yellow coat color in mice. A cross between two yellow heterozygous mice produces an initial genotypic ratio of $^1/_4$ *YY*, $^1/_2$ *Yy*, and $^1/_4$ *yy*, but the homozygous *YY* mice die early in development and do not appear among the progeny, resulting in a 2 : 1 ratio of *Yy* (yellow) to *yy* (nonyellow) in the offspring (**Figure 5.4**). A 2 : 1 ratio is almost always produced by a recessive lethal allele, so observing this ratio among the progeny of a cross between individuals with the same phenotype is a strong clue that one of the alleles is lethal. In this example, as in that of yellow leaves in snapdragons, the lethal allele (*Y*) is recessive because it causes death only in homozygotes. Unlike its effect on *survival*, the effect of the yellow allele on *color* is dominant; in both mice and snapdragons, a single copy of the allele in heterozygotes produces a yellow color. These examples illustrate the point made earlier (pp. 112–113) that the type of dominance depends on the aspect of the phenotype examined.

Many lethal alleles in nature are recessive, but lethal alleles can also be dominant; in this case, homozygotes and heterozygotes for the allele die. Truly dominant lethal alleles cannot be transmitted unless they are expressed after the onset of reproduction. ▶ TRY PROBLEM 17

A lethal allele causes death at an early developmental stage, so that one or more genotypes are missing from the progeny of a cross. Lethal alleles therefore modify the ratio of progeny resulting from a cross.

✔ **CONCEPT CHECK 4**

A cross between two green corn plants yields $^2/_3$ progeny that are green and $^1/_3$ progeny that are yellow. What is the genotype of the green progeny?

a. *WW*
b. *Ww*
c. *ww*
d. *W_* (*WW* and *Ww*)

Multiple Alleles

Most of the genetic systems that we have examined so far consist of two alleles. In Mendel's peas, for instance, one allele encoded round seeds and another encoded wrinkled seeds; in cats, one allele produced a black coat and another produced a gray coat. For some loci, however, more than two alleles are present within a group of organisms—the locus has **multiple alleles** (which may also be referred to as an *allelic series*). Although there may be more than two alleles present within a *group* of organisms, the genotype of each individual diploid organism still consists of only two alleles. The inheritance of characteristics encoded by multiple alleles is no different from the inheritance of characteristics encoded by two alleles, except that a greater variety of genotypes and phenotypes are possible.

PLUMAGE PATTERNS IN DUCKS An example of multiple alleles is found at a locus that determines the plumage pattern of mallard ducks. One allele, M, produces the wild-type *mallard* pattern. A second allele, M^R, produces a different pattern called *restricted*, and a third allele, m^d, produces a pattern termed *dusky*. In this allelic series, restricted is dominant over mallard and dusky, and mallard is dominant over dusky: $M^R > M > m^d$. The six genotypes possible with these three alleles, and their resulting phenotypes, are

Genotype	Phenotype
$M^R M^R$	restricted
$M^R M$	restricted
$M^R m^d$	restricted
MM	mallard
Mm^d	mallard
$m^d m^d$	dusky

In general, the number of genotypes possible will be $[n(n + 1)/2]$, where n equals the number of different alleles at a locus. Working crosses with multiple alleles is no different from working crosses with two alleles; Mendel's principle of segregation still holds, as shown in the cross between a restricted duck and a mallard duck (**Figure 5.5**).

> **TRY PROBLEM 19**

THE ABO BLOOD GROUP IN HUMANS Another multiple-allele system is found at the locus for the ABO blood group, which determines your ABO blood type. This locus, like the MN locus, encodes antigens on the surface of red blood cells. The three common alleles for the ABO blood-group locus are I^A, which encodes the A antigen; I^B, which encodes the B antigen; and i, which encodes no antigen (O). We can represent the dominance relations among the ABO alleles as follows: $I^A > i$, $I^B > i$, $I^A = I^B$. The I^A and the I^B alleles are dominant over i and are codominant with each other; the AB phenotype is due to the presence of an I^A allele and an I^B allele, which results in the production of A and B antigens on red blood cells. A person with genotype ii produces neither antigen and has blood-type O. The six common genotypes at this locus and their phenotypes are shown in **Figure 5.6a**.

The body produces antibodies against any foreign antigens (see Figure 5.6a). For instance, a person with blood-type A produces anti-B antibodies because the B antigen is foreign to that person. A person with blood-type B produces anti-A antibodies, and a person with blood-type AB produces neither anti-A nor anti-B antibodies because neither A nor B antigen is foreign to that person. A person with blood-type O possesses no A or B antigens; consequently, that person produces both anti-A antibodies and anti-B antibodies. The presence of antibodies against foreign ABO antigens means that successful blood transfusions are possible only between people with certain compatible blood types (**Figure 5.6b**).

The inheritance of alleles at the ABO locus is illustrated by a paternity suit against the movie actor Charlie Chaplin.

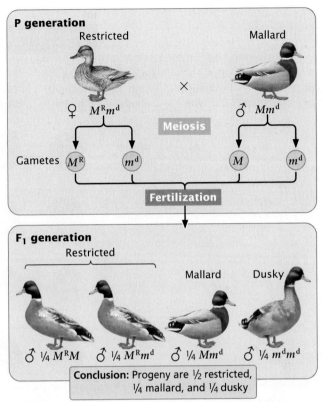

5.5 Mendel's principle of segregation applies to crosses with multiple alleles. In this example, three alleles determine the type of plumage in mallard ducks: M^R (restricted) > M (mallard) > m^d (dusky). Note that the phenotypes of male and female mallard ducks differ. The restricted pattern of a female duck is shown for the female parent. All the progeny in the F_1 have male plumage patterns.

In 1941, Chaplin met a young actress named Joan Barry, with whom he had an affair. The affair ended in February 1942, but 20 months later, Barry gave birth to a baby girl and claimed that Chaplin was the father. Barry then sued for child support. At this time, blood typing had just come into widespread use, and Chaplin's attorneys had Chaplin, Barry, and the child blood typed. Barry had blood-type A, her child had blood-type B, and Chaplin had blood-type O. Could Chaplin have been the father of Barry's child?

Your answer should be no. Joan Barry had blood-type A, which can be produced by either genotype $I^A I^A$ or genotype $I^A i$. Her baby possessed blood-type B, which can be produced by either genotype $I^B I^B$ or genotype $I^B i$. The baby could not have inherited the I^B allele from Barry (Barry could not carry an I^B allele if she were blood-type A); therefore, the baby must have inherited the i allele from her. Barry must have had genotype $I^A i$, and the baby must have had genotype $I^B i$. Because the baby girl inherited her i allele from Barry, she must have inherited the I^B allele from her father. Having blood-type O, produced only by genotype ii, Chaplin could not have been the father of Barry's child. While blood types can be used to exclude the possibility of paternity (as in this case), they cannot prove that a particular person is the parent of a child because many different people have the same blood type.

(a)

(b)

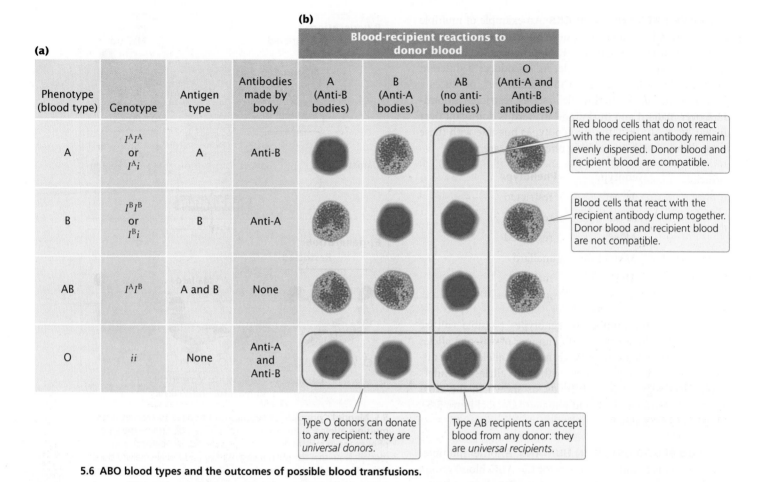

Phenotype (blood type)	Genotype	Antigen type	Antibodies made by body	Blood-recipient reactions to donor blood			
				A (Anti-B bodies)	B (Anti-A bodies)	AB (no anti-bodies)	O (Anti-A and Anti-B antibodies)
A	$I^A I^A$ or $I^A i$	A	Anti-B				
B	$I^B I^B$ or $I^B i$	B	Anti-A				
AB	$I^A I^B$	A and B	None				
O	ii	None	Anti-A and Anti-B				

Red blood cells that do not react with the recipient antibody remain evenly dispersed. Donor blood and recipient blood are compatible.

Blood cells that react with the recipient antibody clump together. Donor blood and recipient blood are not compatible.

Type O donors can donate to any recipient: they are *universal donors*.

Type AB recipients can accept blood from any donor: they are *universal recipients*.

5.6 ABO blood types and the outcomes of possible blood transfusions.

In the course of the trial to settle the paternity suit against Chaplin, three pathologists testified that it was genetically impossible for Chaplin to have fathered the child. Nevertheless, the jury ruled that Chaplin was the father and ordered him to pay child support and Barry's legal expenses.

> **TRY PROBLEM 24**

COMPOUND HETEROZYGOTES Different alleles often give rise to the same phenotype. For example, cystic fibrosis, as we saw earlier in this section, arises from defects in alleles at the *CFTR* locus, which encodes a protein that controls the movement of chloride ions into and out of the cell. Over a thousand different alleles at the *CFTR* locus that can cause cystic fibrosis have been discovered worldwide. Because cystic fibrosis is an autosomal recessive condition, one must normally inherit two defective *CFTR* alleles to have cystic fibrosis. In some people with cystic fibrosis, these two defective alleles are identical, meaning that the person is homozygous. Other people with cystic fibrosis are heterozygous, possessing two different defective alleles. An individual who carries two different alleles at a locus that result in a recessive phenotype is referred to as a **compound heterozygote**.

CONCEPTS

More than two alleles (multiple alleles) may be present at a locus within a group of organisms, although each individual diploid organism still has only two alleles at that locus. A compound heterozygote possesses two different alleles that result in a recessive phenotype.

✔ **CONCEPT CHECK 5**

How many genotypes are possible at a locus with five alleles?

a. 30   c. 15
b. 27   d. 5

5.2 Gene Interaction Takes Place When Genes at Multiple Loci Determine a Single Phenotype

In the dihybrid crosses that we examined in Chapter 3, each locus had an independent effect on the phenotype. When Mendel crossed a homozygous pea plant that produced round and yellow seeds (*RR YY*) with a homozygous plant that produced wrinkled and green seeds (*rr yy*), and then allowed the F_1 to self-fertilize, he obtained F_2 progeny in the following proportions:

$^9/_{16}$ $R\_\ Y\_$	round, yellow
$^3/_{16}$ $R\_\ yy$	round, green
$^3/_{16}$ $rr\ Y\_$	wrinkled, yellow
$^1/_{16}$ $rr\ yy$	wrinkled, green

In this example, the two loci showed two kinds of independence. First, the genes at each locus were independent in their *assortment* in meiosis, which produced the 9 : 3 : 3 : 1 ratio of phenotypes in the progeny, in accord with Mendel's principle of independent assortment. Second, the genes were independent in their *phenotypic expression*: the *R* and *r* alleles affected only the shape of the seed and had no influence on the color of the seed; the *Y* and *y* alleles affected only color and had no influence on the shape of the seed.

Frequently, genes exhibit independent assortment but do not act independently in their phenotypic expression; instead, the effects of genes at one locus depend on the presence of genes at other loci. This type of interaction between the effects of genes at different loci (genes that are not allelic) is termed **gene interaction**. With gene interaction, it is the products of genes at different loci that interact to produce new phenotypes that are not predictable from the single-locus effects alone. In our consideration of gene interaction, we will focus primarily on interactions between the effects of genes at two loci, although interactions among genes at three, four, or more loci are common.

CONCEPTS

In gene interaction, genes at different loci contribute to the determination of a single phenotypic characteristic.

✔ CONCEPT CHECK 6

How does gene interaction differ from dominance?

Gene Interaction That Produces Novel Phenotypes

Let's begin by examining gene interaction in which genes at two loci interact to produce a single characteristic. Fruit color in the pepper *Capsicum annuum* is determined in this way. Certain types of peppers produce fruits in one of four colors: red, peach, orange (sometimes called yellow), or cream (white). If a homozygous plant with red peppers is crossed with a homozygous plant with cream peppers, all the F_1 plants have red peppers (**Figure 5.7a**). When the F_1 are crossed with each other, the F_2 show a ratio of 9 red : 3 : peach : 3 orange : 1 cream (**Figure 5.7b**). This dihybrid ratio (see Chapter 3) is produced by a cross between two plants that are both heterozygous for two loci ($Y^+y\ C^+c \times Y^+y\ C^+c$). In this example, the *Y* locus and the *C* locus interact to produce a single phenotype—the color of the pepper:

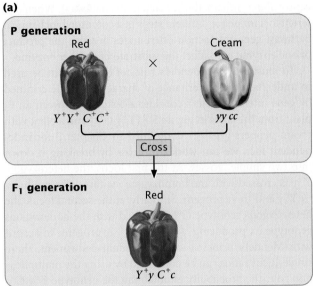

(a)

P generation

Red × Cream

$Y^+Y^+\ C^+C^+$ $yy\ cc$

Cross

F_1 generation

Red

$Y^+y\ C^+c$

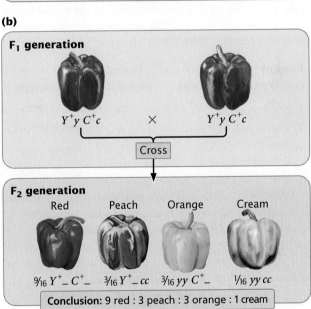

(b)

F_1 generation

$Y^+y\ C^+c$ × $Y^+y\ C^+c$

Cross

F_2 generation

Red   Peach   Orange   Cream

$^9/_{16}\ Y^+\_\ C^+\_$ $^3/_{16}\ Y^+\_\ cc$ $^3/_{16}\ yy\ C^+\_$ $^1/_{16}\ yy\ cc$

Conclusion: 9 red : 3 peach : 3 orange : 1 cream

5.7 Interaction between genes at two loci determines a single characteristic, fruit color, in the pepper *Capsicum annuum*.

Genotype	Phenotype
$Y^+\_\ C^+\_$	red
$Y^+\_\ cc$	peach
$yy\ C^+\_$	orange
$yy\ cc$	cream

Color in peppers results from the relative amounts of red and yellow carotenoids, compounds that are synthesized in a complex biochemical pathway. The *C* locus encodes one enzyme (phytoene synthase, an early step in the pathway), and the *Y* locus encodes a different enzyme (capsanthin capsorubin synthase, the last step in the pathway). When different loci influence different steps in a common biochemical pathway, gene interaction often arises because the product of one enzyme may affect the substrate of another enzyme.

To illustrate how Mendel's rules of heredity can be used to understand the inheritance of characteristics determined by gene interaction, let's consider a testcross between an F_1 plant from the cross in Figure 5.7 ($Y^+y\ C^+c$) and a plant with cream peppers ($yy\ cc$). As outlined in Chapter 3 for independent loci, we can work this cross by breaking it down into two simple crosses. At the first locus, the heterozygote Y^+y is crossed with the homozygote yy; this cross produces $^1\!/_2\ Y^+y$ and $^1\!/_2\ yy$ progeny. Similarly, at the second locus, the heterozygous genotype C^+c is crossed with the homozygous genotype cc, producing $^1\!/_2\ C^+c$ and $^1\!/_2\ cc$ progeny. In accord with Mendel's principle of independent assortment, these single-locus ratios can be combined by using the multiplication rule: the probability of obtaining the genotype $Y^+y\ C^+c$ is the probability of Y^+y ($^1\!/_2$) multiplied by the probability of C^+c ($^1\!/_2$), or $^1\!/_2 \times ^1\!/_2 = ^1\!/_4$. The probabilities of the progeny genotypes resulting from the testcross are

Progeny genotype	Probability at each locus			Overall probability	Phenotype
$Y^+y\ C^+c$	$^1\!/_2 \times ^1\!/_2$	=		$^1\!/_4$	red peppers
$Y^+y\ cc$	$^1\!/_2 \times ^1\!/_2$	=		$^1\!/_4$	peach peppers
$yy\ C^+c$	$^1\!/_2 \times ^1\!/_2$	=		$^1\!/_4$	orange peppers
$yy\ cc$	$^1\!/_2 \times ^1\!/_2$	=		$^1\!/_4$	cream peppers

When you work problems involving gene interaction, it is especially important to determine the probabilities of single-locus genotypes and to multiply the probabilities of *genotypes*, not phenotypes, because the phenotypes cannot be determined without considering the effects of the genotypes at all the contributing loci. ▶ TRY PROBLEM 25

Gene Interaction with Epistasis

Sometimes the effect of gene interaction is that one gene masks (hides) the effect of another gene at a different locus, a phenomenon known as **epistasis**. In the examples of gene interaction that we have already examined, genes at different loci interacted to determine a single phenotype, but one gene did not *mask* the effect of a gene at another locus, meaning that there was no epistasis. Epistasis is similar to dominance, except that dominance entails the masking of genes at the *same* locus (allelic genes). In epistasis, the gene that does the masking is called an **epistatic gene**; the gene whose effect is masked is a **hypostatic gene**. Epistatic genes may be recessive or dominant in their effects.

RECESSIVE EPISTASIS Recessive epistasis is seen in the genes that determine coat color in Labrador retrievers. These dogs may be black, brown (frequently called chocolate), or yellow; their different coat colors are determined by interactions between genes at two loci (although a number of other loci also help to determine coat color; see pp. 124-125). One locus determines the type of pigment produced by the skin cells: a dominant allele B encodes black pigment, whereas a recessive allele b encodes brown pigment. Alleles at a second locus affect the *deposition* of the pigment in the shaft of the hair; dominant allele E allows dark pigment (black or brown) to be deposited, whereas recessive allele e prevents the deposition of dark pigment, causing the hair to be yellow. The presence of genotype ee at the second locus therefore masks the expression of the black and brown alleles at the first locus. The genotypes that determine coat color and their phenotypes are

Genotype	Phenotype
$B\_\ E\_$	black
$bb\ E\_$	brown
$B\_\ ee$	yellow
$bb\ ee$	yellow

If we cross a black Labrador that is homozygous for the dominant alleles ($BB\ EE$) with a yellow Labrador that is homozygous for the recessive alleles ($bb\ ee$) and then intercross the F_1, we obtain progeny in the F_2 in a 9 : 3 : 4 ratio:

P $BB\ EE$ × $bb\ ee$
       Black   Yellow

                ↓

F_1 $Bb\ Ee$
             Black

                ↓ Intercross

F_2 $^9\!/_{16}\ B\_\ E\_$ black
        $^3\!/_{16}\ bb\ E\_$ brown
        $^3\!/_{16}\ B\_\ ee$ yellow $\Big\}\ ^4\!/_{16}$ yellow
        $^1\!/_{16}\ bb\ ee$ yellow

Notice that yellow dogs can carry alleles for either black or brown pigment, but these alleles are not expressed in their coat color.

In this example of gene interaction, allele e is epistatic to B and b because e masks the expression of the alleles for black and brown pigments, and alleles B and b are hypostatic to e. In this case, e is a recessive epistatic allele because two copies of e must be present to mask the expression of the black and brown pigments. ▶ TRY PROBLEM 29

Another recessive epistatic gene determines a rare blood type called the Bombay phenotype; this gene masks the expression of alleles at the ABO locus. As mentioned earlier in the chapter, the alleles at the ABO locus encode antigens on the red blood cells; the antigens consist of short chains of carbohydrates embedded in the outer membranes of red blood cells. The difference between the A and the B antigens is a function of chemical differences in the terminal sugar of the chain. The I^A and I^B alleles actually encode different enzymes, which add sugars designated A or B to the ends of the carbohydrate chains (**Figure 5.8**). The common substrate on which these enzymes act is a molecule called compound H. Apparently, either the enzyme encoded by the *i* allele adds no sugar to compound H or no functional enzyme is specified.

In most people, a dominant allele (*H*) at the H locus encodes an enzyme that makes compound H, but people with the Bombay phenotype are homozygous for a recessive mutation (*h*) that encodes a defective enzyme. The defective enzyme is incapable of making compound H, and because compound H is not produced, no ABO antigens are synthesized. Thus, the expression of the alleles at the ABO locus depends on the genotype at the H locus.

Genotype	Compound H present	ABO phenotype
$H\_\,I^A I^A$, $H\_\,I^A i$	Yes	A
$H\_\,I^B I^B$, $H\_\,I^B i$	Yes	B
$H\_\,I^A I^B$	Yes	AB
$H\_\,ii$	Yes	O
$hh\,I^A I^A$, $hh\,I^A i$, $hh\,I^B I^B$, $hh\,I^B i$, $hh\,I^A I^B$, and $hh\,ii$	No	O

In this example, the alleles at the ABO locus are hypostatic to the recessive *h* allele.

The Bombay phenotype provides us with a good opportunity for considering how epistasis can arise when genes affect a series of steps in a biochemical pathway. The ABO antigens are produced in a multistep biochemical pathway (see Figure 5.8) that depends on enzymes that make compound H and on other enzymes that convert compound H into the A or B antigen. Note that blood-type O may arise in one of two ways: (1) from failure to add a terminal sugar to compound H (genotype $H\_\,ii$) or (2) from failure to produce compound H (genotype hh __). Many cases of epistasis arise in similar ways. A gene (such as *h*) that has an effect on an early step in a biochemical pathway will be epistatic to genes (such as I^A and I^B) that affect subsequent steps because the effects of the genes in a later step depend on the product of the earlier reaction.

DOMINANT EPISTASIS In *recessive* epistasis, the presence of two recessive alleles (the homozygous genotype) inhibits the expression of an allele at a different locus. In *dominant* epistasis, however, only a single copy of an allele is required to inhibit the expression of an allele at a different locus.

Dominant epistasis is seen in the interaction of two loci that determine fruit color in summer squash, which is commonly found in one of three colors: yellow, white, or green.

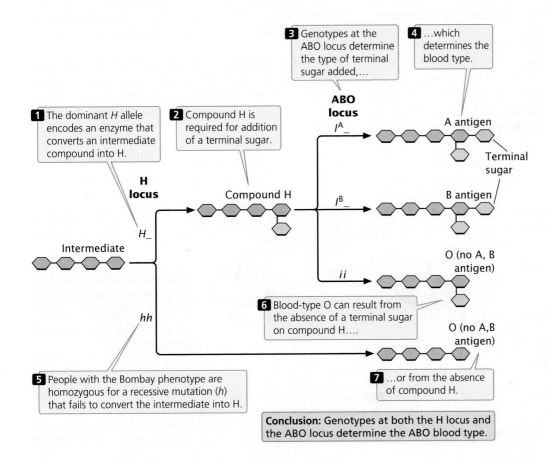

3 Genotypes at the ABO locus determine the type of terminal sugar added,...

4 ...which determines the blood type.

1 The dominant *H* allele encodes an enzyme that converts an intermediate compound into H.

2 Compound H is required for addition of a terminal sugar.

H locus

ABO locus

$I^A\_$

A antigen

Terminal sugar

Compound H

$I^B\_$

B antigen

Intermediate

$H\_$

ii

O (no A, B antigen)

6 Blood-type O can result from the absence of a terminal sugar on compound H....

hh

O (no A,B antigen)

5 People with the Bombay phenotype are homozygous for a recessive mutation (*h*) that fails to convert the intermediate into H.

7 ...or from the absence of compound H.

Conclusion: Genotypes at both the H locus and the ABO locus determine the ABO blood type.

5.8 Expression of the ABO antigens depends on alleles at the H locus. The H locus encodes a precursor to the antigens, called compound H. Alleles at the ABO locus determine which types of terminal sugars are added to compound H.

When a homozygous plant that produces white squash is crossed with a homozygous plant that produces green squash and the F_1 plants are crossed with each other, the following results are obtained:

P   Plants with   ×   Plants with
white squash   green squash

↓

F_1 Plants with
white squash

↓ Intercross

F_2 $^{12}/_{16}$ plants with white squash

$^{3}/_{16}$ plants with yellow squash

$^{1}/_{16}$ plants with green squash

How can gene interaction explain these results?

In the F_2, $^{12}/_{16}$, or $^{3}/_{4}$, of the plants produce white squash and $^{3}/_{16} + ^{1}/_{16} = ^{4}/_{16} = ^{1}/_{4}$ of the plants produce colored squash. This outcome is the familiar 3 : 1 ratio produced by a cross between two heterozygotes, which suggests that a dominant allele at one locus inhibits the production of pigment, resulting in white progeny. If we use the symbol W to represent the dominant allele that inhibits pigment production, then genotype $W\_$ inhibits pigment production and produces white squash, whereas ww allows pigment production and results in colored squash.

Among those ww F_2 plants with pigmented fruit, we observe $^{3}/_{16}$ yellow and $^{1}/_{16}$ green (a 3 : 1 ratio). This observation suggests that a second locus determines the type of pigment produced in the squash, with yellow ($Y\_$) dominant over green (yy). This locus is expressed only in ww plants, which lack the dominant inhibitory allele W. We can assign

the genotype ww $Y\_$ to plants that produce yellow squash and the genotype ww yy to plants that produce green squash. The genotypes and their associated phenotypes are

$W\_ Y\_$	white squash
$W\_ yy$	white squash
$ww Y\_$	yellow squash
$ww yy$	green squash

Allele W is epistatic to Y and y: it masks the expression of these pigment-producing genes. Allele W is a dominant epistatic allele because, in contrast with e in Labrador retriever coat color and with h in the Bombay phenotype, a single copy of the allele is sufficient to inhibit pigment production.

Yellow pigment in the squash is most likely produced in a two-step biochemical pathway (**Figure 5.9**). A colorless (white) compound (designated A in Figure 5.9) is converted by enzyme I into green compound B, which is then converted into compound C by enzyme II. Compound C is the yellow pigment in the fruit. Plants with the genotype ww produce enzyme I and may be green or yellow, depending on whether enzyme II is present. When allele Y is present at a second locus, enzyme II is produced and compound B is converted into compound C, producing a yellow fruit. When two copies of allele y, which does not encode a functional form of enzyme II, are present, the fruit remains green. The presence of W at the first locus inhibits the conversion of compound A into compound B; plants with genotype $W\_$ do not make compound B, and their fruit remains white, regardless of which alleles are present at the second locus.

DUPLICATE RECESSIVE EPISTASIS Finally, let's consider duplicate recessive epistasis, in which two recessive alleles at either of two different loci are capable of suppressing a phenotype. This type of epistasis is illustrated by albinism in snails.

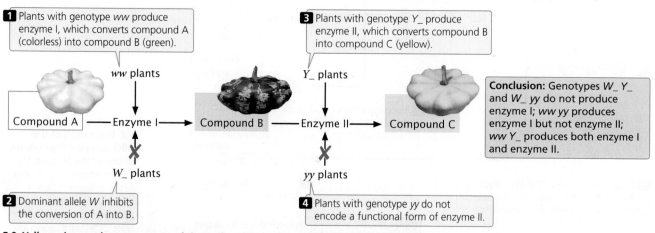

1 Plants with genotype ww produce enzyme I, which converts compound A (colorless) into compound B (green).

3 Plants with genotype $Y\_$ produce enzyme II, which converts compound B into compound C (yellow).

ww plants

$Y\_$ plants

Compound A —— Enzyme I —→ Compound B —— Enzyme II —→ Compound C

$W\_$ plants

yy plants

2 Dominant allele W inhibits the conversion of A into B.

4 Plants with genotype yy do not encode a functional form of enzyme II.

Conclusion: Genotypes $W\_ Y\_$ and $W\_ yy$ do not produce enzyme I; $ww yy$ produces enzyme I but not enzyme II; $ww Y\_$ produces both enzyme I and enzyme II.

5.9 Yellow pigment in summer squash is produced in a two-step biochemical pathway.

Albinism, the absence of pigment, is a common genetic trait in many plants and animals. Pigment is almost always produced through a multistep biochemical pathway; thus, albinism may entail gene interaction. Robert T. Dillon and Amy R. Wethington found that albinism in the common freshwater snail *Physa heterostropha* can result from the presence of two recessive alleles at either of two different loci. They collected inseminated snails from a natural population and placed them in cups of water, where the snails laid eggs. Some of the eggs hatched into albino snails. When two of these albino snails were crossed, all of the F_1 were pigmented. When the F_1 were intercrossed, the F_2 consisted of $^9/_{16}$ pigmented snails and $^7/_{16}$ albino snails. How did this 9 : 7 ratio arise?

The 9 : 7 ratio seen in the F_2 snails can be understood as a modification of the 9 : 3 : 3 : 1 ratio obtained when two individuals heterozygous for two loci are crossed. The 9 : 7 ratio arises when dominant alleles at both loci ($A\_ B\_$) produce pigmented snails and any other genotype produces albino snails:

P	*aa BB*	×	*AA bb*
	Albino		Albino

$$\downarrow$$

F_1 *Aa Bb*
Pigmented

$$\downarrow \text{Intercross}$$

F_2 $^9/_{16}$ *A_ B_* pigmented

$^3/_{16}$ *aa B_* albino ⎫
$^3/_{16}$ *A_bb* albino ⎬ $^7/_{16}$ albino
$^1/_{16}$ *aa bb* albino ⎭

The 9 : 7 ratio in these snails is probably the result of a two-step pathway of pigment production (**Figure 5.10**). Pigment (compound C) is produced only when compound A is converted into compound B by enzyme I and compound B is converted into compound C by enzyme II. At least one dominant allele *A* at the first locus is required to produce enzyme I; similarly, at least one dominant allele *B* at the second locus is required to produce enzyme II. Albinism arises from the absence of compound C, which may happen in one of three ways. First, two recessive alleles at the first locus (genotype *aa B_*) may prevent the production of enzyme I, so that compound B is never produced. Second, two recessive alleles at the second locus (genotype *A_ bb*) may prevent the production of enzyme II; in this case, compound B is never converted into compound C. Third, two recessive alleles may be present at both loci (*aa bb*), causing the absence of both enzyme I and enzyme II. In this example of gene interaction, *a* is epistatic to *B*, and *b* is epistatic to *A*; *both* are recessive epistatic alleles because the presence of two copies of either allele *a* or allele *b* is necessary to suppress pigment production. This example differs from the suppression of coat color in Labrador retrievers in that recessive alleles at either of two loci are capable of suppressing pigment production in the snails, whereas recessive alleles at a single locus suppress pigment expression in the dogs.

CONCEPTS

Epistasis is the masking of the expression of one gene by another gene at a different locus. The epistatic gene does the masking; the hypostatic gene is masked. Epistatic alleles can be dominant or recessive.

✔ CONCEPT CHECK 7

A number of all-white cats are crossed, and they produce the following types of progeny: $^{12}/_{16}$ all-white, $^3/_{16}$ black, and $^1/_{16}$ gray. What is the genotype of the black progeny?

a. *Aa*   c. *A_B_*
b. *Aa Bb*   d. *A_b*

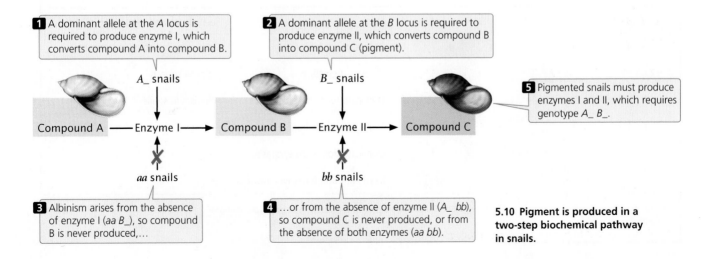

1 A dominant allele at the *A* locus is required to produce enzyme I, which converts compound A into compound B.

2 A dominant allele at the *B* locus is required to produce enzyme II, which converts compound B into compound C (pigment).

5 Pigmented snails must produce enzymes I and II, which requires genotype *A_ B_*.

*A_* snails   *B_* snails

Compound A —— Enzyme I —— Compound B —— Enzyme II—— Compound C

✗ *aa* snails   ✗ *bb* snails

3 Albinism arises from the absence of enzyme I (*aa B_*), so compound B is never produced,...

4 ...or from the absence of enzyme II (*A_ bb*), so compound C is never produced, or from the absence of both enzymes (*aa bb*).

5.10 Pigment is produced in a two-step biochemical pathway in snails.

CONNECTING CONCEPTS

Interpreting Phenotypic Ratios Produced by Gene Interaction

A number of modified phenotypic ratios that result from gene interaction are shown in **Table 5.2**. Each of these examples represents a modification of the basic 9 : 3 : 3 : 1 dihybrid ratio. In interpreting the genetic basis of these modified ratios, we should keep several points in mind. First, the inheritance of the genes producing these characteristics is no different from the inheritance of genes encoding simple genetic characters. Mendel's principles of segregation and independent assortment still apply; each individual organism possesses two alleles at each locus, which separate in meiosis, and genes at the different loci assort independently. The only difference is in how the *products* of the genotypes interact to produce the phenotype. Thus, we cannot consider the expression of genes at each locus separately; instead, we must take into consideration how the genes at different loci interact.

A second point is that, in the examples that we have considered, the phenotypic proportions were always in sixteenths because, in all the crosses, pairs of alleles segregated at two independently assorting loci. The probability of inheriting one of the two alleles at a locus is $1/2$. Because there are two loci, each with two alleles, the probability of inheriting any particular combination of genes is $(1/2)^4 = 1/16$. For a trihybrid cross, the progeny proportions should be in sixty-fourths, because $(1/2)^6 = 1/64$. In general, the progeny proportions should be in fractions of $(1/2)^{2n}$, where n equals the number of loci with two alleles segregating in the cross.

Crosses rarely produce exactly 16 progeny; therefore, modifications of the dihybrid ratio are not always obvious. Modified dihybrid ratios are more easily seen if the number of individuals of each phenotype is expressed in sixteenths:

$$\frac{x}{16} = \frac{\text{number of progeny with a phenotype}}{\text{total number of progeny}}$$

where $x/16$ equals the proportion of progeny with a particular phenotype. If we solve for x (the proportion of the particular phenotype in sixteenths), we have

$$x = \frac{\text{number of progeny with a phenotype} \times 16}{\text{total number of progeny}}$$

For example, suppose that we cross two homozygotes, interbreed the F_1, and obtain 63 red, 21 brown, and 28 white F_2 individuals. Using the preceding formula, we find the phenotypic ratio in the F_2 to be red = $(63 \times 16)/112 = 9$; brown = $(21 \times 16)/112 = 3$; and white = $(28 \times 16)/112 = 4$. The phenotypic ratio is 9 : 3 : 4.

A final point to consider is how to assign genotypes to the phenotypes in modified ratios that result from gene interaction. Don't try to *memorize* the genotypes associated with all the modified ratios in Table 5.2. Instead, practice relating modified ratios to known ratios, such as the 9 : 3 : 3 : 1 dihybrid ratio. Suppose that we obtain $15/16$ green progeny and $1/16$ white progeny in a cross between two plants. If we compare this 15 : 1 ratio with the standard 9 : 3 : 3 : 1 dihybrid ratio, we see that $9/16 + 3/16 + 3/16$ equals $15/16$. All the genotypes associated with these proportions in the dihybrid cross (*A_ B_*, *A_ bb*, and *aa B_*) must give the same phenotype, the green progeny. Genotype *aa bb* makes up $1/16$ of the progeny in a dihybrid cross—the white progeny in this cross.

In assigning genotypes to phenotypes in modified ratios, students sometimes become confused about which letters to assign to which phenotype. Suppose that we obtain the following phenotypic ratio: $9/16$ black : $3/16$ brown : $4/16$ white. Which genotype do we assign to the brown progeny, *A_ bb* or *aa B_*? Either answer is correct because the letters are just arbitrary symbols for the genetic information. The important thing to realize about this ratio is that the brown phenotype arises when two recessive alleles are present at one locus.

TABLE 5.2	Modified dihybrid phenotypic ratios due to gene interaction					
	Genotype					
Ratio*	**A_ B_**	**A_ bb**	**aa B_**	**aa bb**	**Type of Interaction**	**Example Discussed in Chapter**
9 : 3 : 3 : 1	9	3	3	1	None	Seed shape and seed color in peas
9 : 3 : 4	9	3	4		Recessive epistasis	Coat color in Labrador retrievers
12 : 3 : 1	12		3	1	Dominant epistasis	Color in squash
9 : 7	9	7			Duplicate recessive epistasis	Albinism in snails
9 : 6 : 1	9	6		1	Duplicate interaction	—
15 : 1	15			1	Duplicate dominant epistasis	—
13 : 3	13		3		Dominant and recessive epistasis	—

*Each ratio is produced by a dihybrid cross (*Aa Bb* × *Aa Bb*). Shaded bars represent combinations of genotypes that give the same phenotype.

WORKED PROBLEM

A homozygous strain of yellow corn is crossed with a homozygous strain of purple corn. The F_1 are intercrossed, producing an ear of corn with 119 purple kernels and 89 yellow kernels (the progeny). What is the genotype of the yellow kernels?

Solution Strategy

What information is required in your answer to the problem?

The genotype of the yellow kernels.

What information is provided to solve the problem?

- A homozygous yellow corn plant is crossed with a homozygous purple corn plant.

- The numbers of purple and yellow progeny produced by the cross.

Solution Steps

We should first consider whether the cross between yellow and purple strains might be a monohybrid cross for a simple dominant trait, which would produce a 3 : 1 ratio in the F_2 ($Aa \times Aa \rightarrow \frac{3}{4} A\_$ and $\frac{1}{4} aa$). Under this hypothesis, we would expect 156 purple progeny and 52 yellow progeny:

Phenotype	Genotype	Observed number	Expected number
purple	$A\_$	119	$\frac{3}{4} \times 208 = 156$
yellow	Aa	89	$\frac{1}{4} \times 208 = 52$
Total		208	

We see that the expected numbers do not closely fit the observed numbers. If we performed a chi-square test (see Chapter 3), we would obtain a calculated chi-square value of 35.08, which has a probability much less than 0.05, indicating that it is extremely unlikely that, when we expect a 3 : 1 ratio, we would obtain 119 purple progeny and 89 yellow progeny just by chance. Therefore, we can reject the hypothesis that these results were produced by a monohybrid cross.

Another possible hypothesis is that the observed F_2 progeny are in a 1 : 1 ratio. However, we learned in Chapter 3 that a 1 : 1 ratio is produced by a cross between a heterozygote and a homozygote ($Aa \times aa$), and in this cross, both original parental strains were homozygous. Furthermore, a chi-square test comparing the observed numbers with an expected 1 : 1 ratio yields a calculated chi-square value of 4.32, which has a probability of less than 0.05.

Next, we should look to see if the results can be explained by a dihybrid cross ($Aa\ Bb \times Aa\ Bb$). A dihybrid cross results in phenotypic proportions that are in sixteenths. We can apply the formula given earlier in the chapter to determine the number of sixteenths for each phenotype:

$$x = \frac{\text{number of progeny with a phenotype} \times 16}{\text{total number of progeny}}$$

$$x_{\text{(purple)}} = \frac{119 \times 16}{208} = 9.15$$

$$x_{\text{(yellow)}} = \frac{89 \times 16}{208} = 6.85$$

Thus, purple and yellow appear in an approximate ratio of 9 : 7.

We can test this hypothesis with a chi-square test:

Phenotype	Genotype	Observed number	Expected number
purple	?	119	$\frac{9}{16} \times 208 = 117$
yellow	?	89	$\frac{7}{16} \times 208 = 91$
Total		208	

$$\chi^2 = \sum \frac{(\text{observed} - \text{expected})^2}{\text{expected}}$$

$$= \frac{(119 - 117)^2}{117} + \frac{(89 - 91)^2}{91}$$

$$= 0.034 + 0.044 = 0.078$$

Degree of freedom $= n - 1 = 2 - 1 = 1$

$$P > 0.05$$

The probability associated with the chi-square value is greater than 0.05, indicating that there is a good fit between the observed results and a 9 : 7 ratio.

We now need to determine how a dihybrid cross can produce a 9 : 7 ratio and what genotypes correspond to the two phenotypes. A dihybrid cross without epistasis produces a 9 : 3 : 3 : 1 ratio:

$$Aa\ Bb \times Aa\ Bb$$

$$\downarrow$$

$$A\_ B\_\ \frac{9}{16}$$
$$A\_ bb\ \frac{3}{16}$$
$$aa\ B\_\ \frac{3}{16}$$
$$aa\ bb\ \frac{1}{16}$$

Because $\frac{9}{16}$ of the progeny from the corn cross are purple, purple must be produced by genotypes $A\_ B\_$; in other words, individual kernels that have at least one dominant allele at the first locus and at least one dominant allele at the second locus are purple. The proportions of all the other genotypes ($A\_ bb$, $aa\ B\_$, and $aa\ bb$) sum to $\frac{7}{16}$, which is the proportion of the progeny in the corn cross that are yellow, so any individual kernel that does not have a dominant allele at both the first and the second locus is yellow.

>> Now test your understanding of epistasis by working **Problem 26** at the end of the chapter.

Complementation: Determining Whether Mutations Are at the Same Locus or at Different Loci

How do we know whether different mutations that affect a characteristic occur at the same locus (are allelic) or at different loci? In fruit flies, for example, *white* is an X-linked recessive mutation that produces white eyes instead of the red eyes found in wild-type flies; *apricot* is an X-linked recessive mutation that produces light orange eyes. Do the *white* and *apricot* mutations occur at the same locus or at different loci? We can use the complementation test to answer this question.

To carry out a **complementation test** on recessive mutations (*a* and *b*), parents that are homozygous for different mutations are crossed, producing offspring that are heterozygous. If the mutations are allelic (occur at the same locus), then the heterozygous offspring have only mutant alleles (*a b*) and exhibit a mutant phenotype:

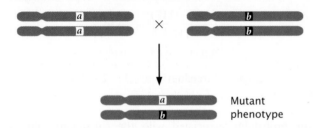

If, on the other hand, the mutations occur at different loci, each of the homozygous parents possesses wild-type genes at the other locus (*aa b⁺b⁺* and *a⁺a⁺ bb*); so the heterozygous offspring inherit a mutant allele and a wild-type allele at each locus. In this case, the presence of a wild-type allele complements the mutation at each locus, and the heterozygous offspring have the wild-type phenotype:

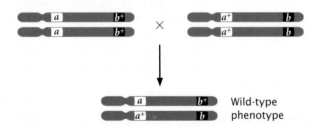

Complementation has taken place if an individual organism possessing two recessive mutations has a wild-type phenotype, indicating that the mutations are at nonallelic genes. There is a lack of complementation when two recessive mutations occur at the same locus, producing a mutant phenotype.

When the complementation test is applied to *white* and *apricot* mutations, all the heterozygous offspring have light-colored eyes, demonstrating that white eyes and apricot eyes are produced by mutations that occur at the same locus and are allelic.

THINK-PAIR-SHARE Question 2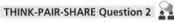

The Complex Genetics of Coat Color in Dogs

The genetics of coat color in dogs is an excellent example of the complex interactions between genes that may take part in the determination of a phenotype. Domestic dogs come in an amazing variety of shapes, sizes, and colors. For thousands of years, people have been breeding dogs for particular traits, producing the large number of types that we see today. Each breed of dog carries a selection of alleles from the ancestral dog gene pool; these alleles define the features of a particular breed. The genome of the domestic dog was completely sequenced in 2004, greatly facilitating the study of canine genetics.

Here, we consider four loci (in the list that follows) that are important in producing many of the noticeable differences in color and pattern among breeds of dogs. In interpreting the genetic basis of differences in the coat colors of dogs, consider how the expression of a particular gene is modified by the effects of other genes. Keep in mind that additional loci not listed here can modify the colors produced by these four loci and that not all geneticists agree on the genetics of color variation in some breeds.

1. **Agouti (*A*) locus.** This locus has five common alleles that determine the depth and distribution of color in a dog's coat:

   A^s Solid black pigment.

   a^w Agouti, or wolflike gray. Hairs encoded by this allele have a salt-and-pepper appearance, produced by a band of yellow pigment on a black hair.

   a^y Yellow. The black pigment is markedly reduced, so the entire hair is yellow.

   a^s Saddle markings (dark color on the back, with extensive tan markings on the head and legs).

   a^t Bicolor (dark color over most of the body, with tan markings on the feet and eyebrows).

Alleles A^s and a^y are generally dominant over the other alleles, but the dominance relations are complex and not yet completely understood.

2. **Black (*B*) locus.** This locus determines whether black pigment can be formed. The actual color of a dog's coat depends on the effects of genes at other loci (such as the *A* and *E* loci). Two alleles are common:

B   Allows black pigment to be produced.

b   Black pigment cannot be produced; pigmented dogs can be chocolate, liver, tan, or red.

Allele *B* is dominant over allele *b*.

3. **Extension (*E*) locus.** Four alleles at this locus determine where the genotype at the *A* locus is expressed. For example, if a dog has the *A*ˢ allele (solid black) at the *A* locus, then black pigment will either be extended throughout the coat or be restricted to some areas, depending on the alleles present at the *E* locus. Areas where the *A* locus is not expressed may appear yellow, red, or tan, depending on the presence of particular alleles at other loci. When *A*ˢ is present at the *A* locus, the four alleles at the *E* locus have the following effects:

*E*ᵐ   Black mask with a tan coat.

E   The *A* locus is expressed throughout (solid black).

*e*ᵇʳ   Brindle, in which black and yellow are in layers to give a tiger-striped appearance.

e   No black in the coat, but the nose and eyes may be black.

The dominance relations among these alleles are poorly known.

4. **Spotting (*S*) locus.** Alleles at this locus determine whether white spots will be present. There are four common alleles:

S   No spots.

*s*ⁱ   Irish spotting: numerous white spots.

*s*ᵖ   Piebald spotting: various amounts of white.

*s*ʷ   Extreme white piebald: almost all white.

Allele *S* is completely dominant over alleles *s*ⁱ, *s*ᵖ, and *s*ʷ; alleles *s*ⁱ and *s*ᵖ are dominant over allele *s*ʷ (*S* > *s*ⁱ, *s*ᵖ > *s*ʷ). The relation between *s*ⁱ and *s*ᵖ is poorly defined; indeed, they may not be separate alleles. Genes at other poorly known loci also modify spotting patterns.

To illustrate how genes at these loci interact in determining a dog's coat color, let's consider a few examples.

LABRADOR RETRIEVER Labrador retrievers (**Figure 5.11a**) may be black, brown, or yellow. Most are homozygous *A*ˢ*A*ˢ *SS*; thus, they vary only at the *B* and *E* loci. The *A*ˢ allele allows dark pigment to be expressed; whether a dog is black depends on which genes are present at the *B* and *E* loci. As discussed earlier in the chapter, all black Labradors must carry at least one *B* allele and one *E* allele (*B_ E_*). Brown dogs are homozygous *bb* and have at least one *E* allele (*bb E_*). Yellow dogs are a result of the presence of *ee* (*B_ ee* or *bb ee*). Labrador retrievers are homozygous for the *S* allele, which produces a solid color; the few white spots that appear in some dogs of this breed are due to other modifying genes.

BEAGLE Most beagles (**Figure 5.11b**) are homozygous *a*ˢ*a*ˢ *BB* *s*ᵖ*s*ᵖ, although other alleles at these loci are occasionally present. The *a*ˢ allele produces the saddle markings—dark back and sides, with tan head and legs—that are characteristic of the breed. Allele *B* allows black to be produced, but its distribution is limited by the *a*ˢ allele. Most beagles are *E_*, but the genotype *ee* does occasionally arise, leading to a few all-tan beagles. White spotting in beagles is due to the *s*ᵖ allele.

DALMATIAN Dalmatians (**Figure 5.11c**) have an interesting genetic makeup. Most are homozygous *A*ˢ*A*ˢ *EE* *s*ʷ*s*ʷ, so they vary only at the *B* locus. Notice that these dogs possess genotype *A*ˢ*A*ˢ *EE*, which allows for a solid coat that would be black, if genotype *B_* were present, or brown (called liver), if genotype *bb* were present. However, the presence of the *s*ʷ allele produces a white coat, masking the expression of the solid color. The dog's color appears only in the pigmented spots, which are due to the presence of an allele at yet another locus that allows the color to penetrate in a limited number of spots.

Table 5.3 gives the common genotypes of other breeds of dogs. ▶ **TRY PROBLEM 33**

(a)    (b)    (c)

5.11 Coat color in dogs is determined by interactions between genes at a number of loci. (a) Most Labrador retrievers are genotype *A*ˢ*A*ˢ *SS*, varying only at the *B* and *E* loci. (b) Most beagles are genotype *a*ˢ*a*ˢ *BB* *s*ᵖ*s*ᵖ. (c) Dalmatians are genotype *A*ˢ*A*ˢ *EE* *s*ʷ*s*ʷ, varying at the *B* locus, which makes the dogs black (*B_*) or brown (*bb*). [Part a: imagebroker/Alamy. Part b: RFcompany/age fotostock. Part c: Stockbyte/Getty Images.]

TABLE 5.3	Common genotypes in different breeds of dogs	
Breed	**Usual Homozygous Genotypes***	**Other Alleles Present Within the Breed**
Basset hound	*BB EE*	a^y, a^t S, s^p, s^i
Beagle	$a^s a^s$ *BB* $s^p s^p$	*E, e*
English bulldog	*BB*	A^s, a^y, a^t E^m, E, e^{br} S, s^i, s^p, s^w
Chihuahua		A^s, a^y, a^s, a^t *B, b* E^m, E, e^{br}, e S, s^i, s^p, s^w
Collie	*BB EE*	a^y, a^t $s^i, s^{\,w}$
Dalmatian	$A^s A^s$ *EE* $s^w s^w$	*B, b*
Doberman	$a^t a^t$ *EE SS*	*B, b*
German shepherd	*BB SS*	a^y, a, a^s, a^t E^m, E, e
Golden retriever	$A^s A^s$ *BB SS*	*E, e*
Greyhound	*BB*	A^s, a^y E, e^{br}, e S, s^p, s^w, s^i
Irish setter	*BB ee SS*	A^s, a^t
Labrador retriever	$A^s A^s$ *SS*	*B, b* *E, e*
Poodle	*SS*	A^s, a^t *B, b* *E, e*
Rottweiler	$a^t a^t$ *BB EE SS*	
St. Bernard	$a^y a^y$ *BB*	E^m, E s^i, s^p, s^w

*Most dogs in the breed are homozygous for these genes; a few individual dogs may possess other alleles at these loci.
Source: Data from M. B. Willis, *Genetics of the Dog* (London: Witherby, 1989).

5.3 Sex Influences the Inheritance and Expression of Genes in a Variety of Ways

In Chapter 4, we considered characteristics encoded by genes located on the sex chromosomes (sex-linked traits) and how their inheritance differs from the inheritance of characteristics encoded by autosomal genes. X-linked traits, for example, are passed from father to daughter but never from father to son, and Y-linked traits are passed from father to all sons. Here, we examine additional influences of sex, including the effect of the sex of an individual organism on the expression of autosomal genes, on characteristics determined by genes located in the cytoplasm, and on characteristics for which the genotype of only the maternal parent determines the phenotype of the offspring. Finally, we look at situations in which the expression of autosomal genes is affected by the sex of the parent from whom the genes are inherited.

Sex-Influenced and Sex-Limited Characteristics

Sex-influenced characteristics are determined by autosomal genes and are inherited according to Mendel's principles, but they are expressed differently in males and females. In this case, a particular trait is more readily expressed in one sex; in other words, the trait has higher penetrance in one of the sexes.

For example, the presence of a beard on some goats is determined by an allele at an autosomal locus (B^b) that is dominant in males and recessive in females. In males, a single beard allele is required for the expression of this trait: both the homozygote ($B^b B^b$) and the heterozygote ($B^b B^+$) have beards, whereas the $B^+ B^+$ male is beardless. In contrast, females require two beard alleles in order for this trait to be expressed: the homozygote $B^b B^b$ has a beard, whereas the heterozygote ($B^b B^+$) and the other homozygote ($B^+ B^+$) are beardless.

Genotype	Males	Females
$B^+ B^+$	beardless	beardless
$B^+ B^b$	bearded	beardless
$B^b B^b$	bearded	bearded

The key to understanding the expression of the beard allele is to look at the heterozygote. In males (for which the presence of a beard is dominant), the heterozygous genotype produces a beard, but in females (for which the absence of a beard is dominant), the heterozygous genotype produces a goat without a beard.

Figure 5.12a illustrates a cross between a beardless male ($B^+ B^+$) and a bearded female ($B^b B^b$). The alleles separate

(a)

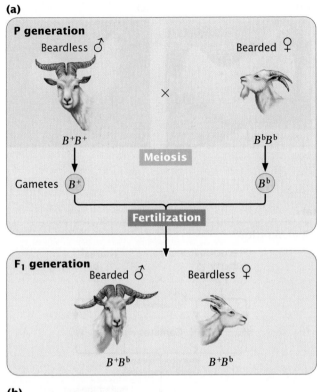

(b)

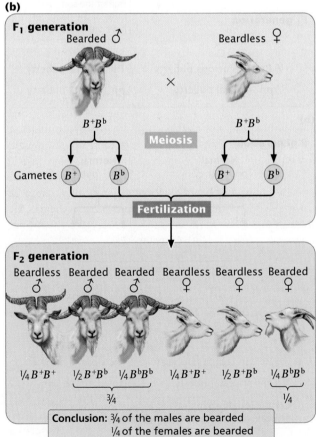

Conclusion: ¾ of the males are bearded
¼ of the females are bearded

5.12 Genes that encode sex-influenced characteristics are inherited according to Mendel's principles but are expressed differently in males and females.

into gametes according to Mendel's principle of segregation, and all the F_1 are heterozygous (B^+B^b). Because the trait is dominant in males and recessive in females, all the F_1 males will be bearded and all the F_1 females will be beardless. When the F_1 are crossed with each other, $^1/_4$ of the F_2 progeny are B^bB^b, $^1/_2$ are B^bB^+, and $^1/_4$ are B^+B^+ (**Figure 5.12b**). Because male heterozygotes are bearded, $^3/_4$ of the males in the F_2 possess beards; because female heterozygotes are beardless, only $^1/_4$ of the females in the F_2 are bearded.

A **sex-limited characteristic** is encoded by autosomal genes that are expressed in only one sex; the trait has zero penetrance in the other sex. In domestic chickens, for example, some males display a plumage pattern called cock feathering (**Figure 5.13a**). Other males and all females display a pattern called hen feathering (**Figure 5.13b and c**). Cock feathering is an autosomal recessive trait that is limited to males. Because the trait is autosomal, the genotypes of males and females are the same, but the phenotypes produced by these genotypes differ between males and females:

Genotype	Male phenotype	Female phenotype
HH	hen feathering	hen feathering
Hh	hen feathering	hen feathering
hh	cock feathering	hen feathering

An example of a sex-limited characteristic in humans is male-limited precocious puberty. There are several types of precocious puberty in humans, most of which are not genetic. Male-limited precocious puberty, however, results from an autosomal dominant allele (*P*) that is expressed only in males; females with this allele are normal in phenotype. Males with precocious puberty undergo puberty at an early age, usually before the age of 4. At this time, the penis enlarges, the voice deepens, and pubic hair develops. There is no impairment of sexual function; affected males are fully fertile. Most are short as adults because the long bones stop growing after puberty.

Because the trait is rare, affected males are usually heterozygous (*Pp*). A male with precocious puberty who mates with a woman who has no family history of this condition will transmit the allele for precocious puberty to half of their children (**Figure 5.14a**), but it will be expressed only in the sons. If one of the heterozygous daughters (*Pp*) mates with a male who exhibits normal puberty (*pp*), half of their sons will exhibit precocious puberty (**Figure 5.14b**). Thus, a sex-limited characteristic can be inherited from either parent, although the trait appears in only one sex. **▶ TRY PROBLEM 35**

THINK-PAIR-SHARE Questions 3, 4, and 5

5.13 A sex-limited characteristic is encoded by autosomal genes that are expressed in only one sex. An example is cock feathering in chickens, an autosomal recessive trait that is limited to males. (a) Cock-feathered male. (b) Hen-feathered female. (c) Hen-feathered male. [Part a: Moment Open/Getty Images. Part b: Guy J. Sagi/Shutterstock. Part c: James Marshall/Corbis.]

(a)

(b)

(c)

CONCEPTS

Sex-influenced characteristics are encoded by autosomal genes that are more readily expressed in one sex. Sex-limited characteristics are encoded by autosomal genes whose expression is limited to one sex.

✔ CONCEPT CHECK 9

How do sex-influenced and sex-limited characteristics differ from sex-linked characteristics?

Cytoplasmic Inheritance

Mendel's principles of segregation and independent assortment are based on the assumption that genes are located on chromosomes in the nucleus of the cell. For most genetic characteristics, this assumption is valid, and Mendel's principles allow us to predict the types of offspring that will be produced in a genetic cross. Not all the genetic material of a cell is found in the nucleus, however. Some characteristics are encoded by genes located in the cytoplasm, and these characteristics exhibit **cytoplasmic inheritance**.

A few organelles, notably chloroplasts and mitochondria, contain DNA. The human mitochondrial genome contains about 16,569 nucleotides of DNA, encoding 37 genes. Compared with the nuclear genome, which contains some 3 billion nucleotides encoding some 20,000 genes, the mitochondrial genome is very small; nevertheless, mitochondrial and chloroplast genes encode some important characteristics. The molecular characteristics of this extranuclear DNA are discussed in Chapter 11; here, we focus on *patterns* of cytoplasmic inheritance.

Cytoplasmic inheritance differs from the inheritance of characteristics encoded by nuclear genes in several important respects. A zygote inherits nuclear genes from both parents, but typically, all its cytoplasmic organelles, and thus all its cytoplasmic genes, come from only one of the gametes, usually the egg. A sperm from the male parent generally contributes only a set of nuclear genes. Thus, most cytoplasmically inherited traits are present in both males and females and

(a)

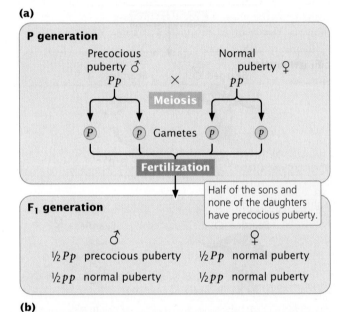

P generation

Precocious puberty ♂ × Normal puberty ♀
Pp pp

Meiosis

P p Gametes p p

Fertilization

F₁ generation

> Half of the sons and none of the daughters have precocious puberty.

♂

½ Pp precocious puberty

½ pp normal puberty

♀

½ Pp normal puberty

½ pp normal puberty

(b)

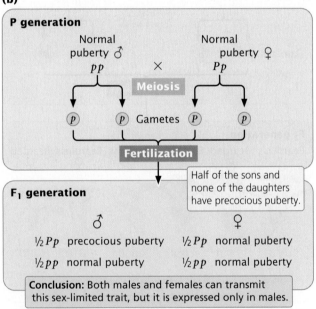

P generation

Normal puberty ♂ × Normal puberty ♀
pp Pp

Meiosis

p p Gametes P p

Fertilization

F₁ generation

> Half of the sons and none of the daughters have precocious puberty.

♂

½ Pp precocious puberty

½ pp normal puberty

♀

½ Pp normal puberty

½ pp normal puberty

Conclusion: Both males and females can transmit this sex-limited trait, but it is expressed only in males.

5.14 Sex-limited characteristics are inherited according to Mendel's principles. Precocious puberty is an autosomal dominant trait that is limited to males.

are passed from mother to offspring, never from father to offspring. Reciprocal crosses, therefore, give different results when cytoplasmic genes encode a trait. In a few organisms, however, cytoplasmic genes are inherited from the male parent only or from both parents.

Cytoplasmically inherited characteristics frequently exhibit extensive phenotypic variation because no mechanism analogous to mitosis or meiosis ensures that cytoplasmic genes are evenly distributed during cell division. Thus, different cells and different individual offspring will contain various proportions of cytoplasmic genes.

Consider mitochondrial genes. Most cells contain thousands of mitochondria, and each mitochondrion contains from 2 to 10 copies of the mitochondrial DNA (mtDNA). Suppose that half of the mitochondria in a cell contain a normal wild-type copy of mtDNA and the other half contain a mutated copy (**Figure 5.15**). During cell division, the mitochondria segregate into the progeny cells

at random. Just by chance, one cell may receive mostly mutated mtDNA and another cell may receive mostly wild-type mtDNA. Therefore, different progeny from the same mother, and even different cells within an individual offspring, may vary in their phenotypes. Traits encoded by chloroplast DNA (cpDNA) are similarly variable. The characteristics exhibited by cytoplasmically inherited traits are summarized in **Table 5.4**.

VARIEGATION IN FOUR-O'CLOCKS In 1909, cytoplasmic inheritance was recognized by Carl Correns as an exception to Mendel's principles. Correns, one of the biologists who rediscovered Mendel's work, studied the inheritance of leaf variegation in the four-o'clock plant, *Mirabilis jalapa*. Correns found that some of the leaves and stems of one variety of four-o'clock were variegated, displaying a mixture of green and white splotches. He also noted that some branches of the variegated strain had all-green leaves; other branches had all-white leaves. Each branch produced flowers, so Correns was able to cross flowers from variegated, green, and white branches in all combinations (**Figure 5.16**). The seeds from flowers on green branches always gave rise to green progeny, no matter whether the pollen was from a green, white, or variegated branch. Similarly, flowers on white branches always produced white progeny. Flowers on the variegated branches gave rise to green, white, and variegated progeny in no particular ratio.

Correns's crosses demonstrated the cytoplasmic inheritance of variegation in four-o'clocks. The phenotypes of the offspring were determined entirely by the maternal parent, never by the paternal parent (the source of the pollen). Furthermore, the production of all three phenotypes by flowers on variegated branches is consistent with cytoplasmic inheritance. The white color in these plants is caused by a defective gene in cpDNA, which results in a failure to produce the green pigment chlorophyll. Cells from green branches contain normal chloroplasts only, cells from white branches contain abnormal chloroplasts only, and cells from variegated branches contain a mixture of normal and abnormal chloroplasts. In the flowers from variegated branches, the random segregation of chloroplasts in the course of oogenesis produces some egg cells with only normal cpDNA,

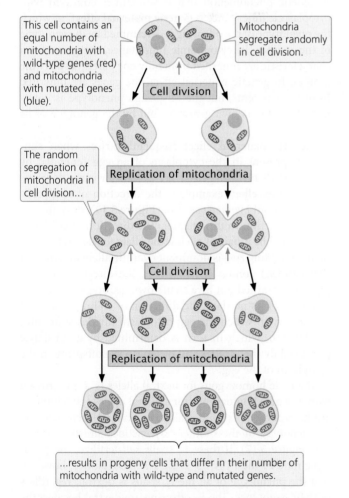

This cell contains an equal number of mitochondria with wild-type genes (red) and mitochondria with mutated genes (blue).

Mitochondria segregate randomly in cell division.

Cell division

Replication of mitochondria

The random segregation of mitochondria in cell division...

Cell division

Replication of mitochondria

...results in progeny cells that differ in their number of mitochondria with wild-type and mutated genes.

5.15 Cytoplasmically inherited characteristics frequently exhibit extensive phenotypic variation because different cells and different individual offspring may contain different proportions of cytoplasmic genes.

TABLE 5.4	Characteristics of cytoplasmically inherited traits
1.	Present in males and females.
2.	Usually inherited from one parent, typically the maternal parent.
3.	Reciprocal crosses give different results.
4.	Exhibit extensive phenotypic variation, even within a single family.

Experiment

Question: How is stem and leaf color inherited in the four-o'clock plant?

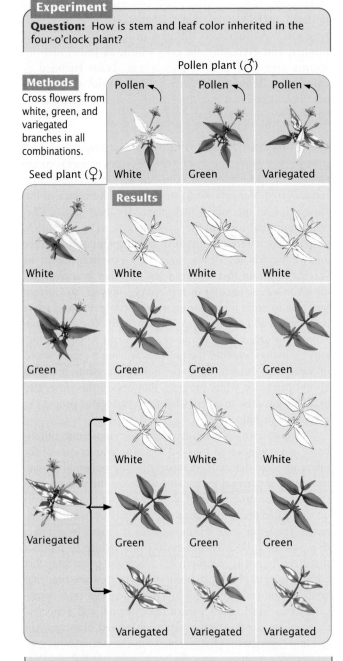

Pollen plant (♂)

Methods

Cross flowers from white, green, and variegated branches in all combinations.

Seed plant (♀)

	Pollen — White	Pollen — Green	Pollen — Variegated
Results			
White	White	White	White
Green	Green	Green	Green
Variegated	White / Green / Variegated	White / Green / Variegated	White / Green / Variegated

Conclusion: The phenotype of the progeny is determined by the phenotype of the branch from which the seed originated, not from the branch on which the pollen originated. Stem and leaf color exhibit cytoplasmic inheritance.

5.16 Crosses for stem and leaf color in four-o'clocks illustrate cytoplasmic inheritance.

which develop into green progeny; other egg cells with only abnormal cpDNA, which develop into white progeny; and still other egg cells with a mixture of normal and abnormal cpDNA, which develop into variegated progeny.

MITOCHONDRIAL DISEASES A number of human diseases (mostly rare ones) that exhibit cytoplasmic inheritance have been identified. These disorders arise from mutations in mtDNA, most of which occur in genes encoding components of the electron-transport chain, which generates most of the ATP (adenosine triphosphate) in aerobic cellular respiration. One such disease is Leber hereditary optic neuropathy (LHON). People who have this disorder experience rapid loss of vision in both eyes, which results from the death of cells in the optic nerve. This loss of vision typically occurs in early adulthood (usually between the ages of 20 and 24), but it can occur any time after adolescence. There is much clinical variation in the severity of the disease, even within the same family. Leber hereditary optic neuropathy exhibits cytoplasmic inheritance: the trait is passed from a mother to all her children, sons and daughters alike.

Genetic Maternal Effect

A genetic phenomenon that is sometimes confused with cytoplasmic inheritance is **genetic maternal effect**, in which the phenotype of the offspring is determined by the genotype of the mother. In cytoplasmic inheritance, the genes for a characteristic are inherited from only one parent, usually the mother. In genetic maternal effect, the genes are inherited from both parents, but the offspring's phenotype is determined not by its own genotype, but by the genotype of its mother.

Genetic maternal effect frequently arises when substances present in the cytoplasm of an egg (encoded by the mother's nuclear genes) are pivotal in early development. An excellent example is the direction of shell coiling in the snail *Lymnaea peregra* (**Figure 5.17**), described in the introduction to this chapter. The direction of coiling is determined by a pair of alleles; the allele for dextral (right-handed) coiling (s^+) is dominant over the allele for sinistral (left-handed) coiling (s). However, the direction of coiling is determined not by a snail's own genotype, but by the genotype of its mother. The direction of coiling is affected by the way in which the egg cytoplasm divides soon after fertilization, which in turn is determined by a substance produced by the mother and passed to the offspring in the cytoplasm of the egg.

If a male homozygous for dextral alleles (s^+s^+) is crossed with a female homozygous for sinistral alleles (ss), all the F_1 are heterozygous (s^+s), but all have sinistral shells because the genotype of the mother (ss) encodes sinistral coiling (see Figure 5.17). If these F_1 snails self-fertilize, the genotypic ratio of the F_2 is 1 s^+s^+ : 2 s^+s : 1 ss. Notice, however, that the phenotype of all the F_2 snails is dextral, regardless of their genotypes. The F_2 offspring are dextral because the genotype of their mother (s^+s), which encodes a dextral shell, determines their phenotype. With genetic maternal effect, the phenotype of the progeny is not necessarily the

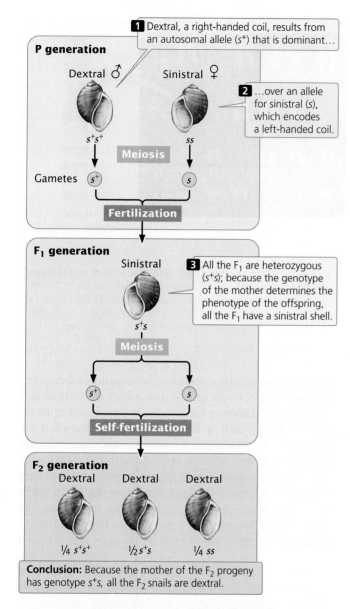

1 Dextral, a right-handed coil, results from an autosomal allele (s^+) that is dominant...

P generation

Dextral ♂   Sinistral ♀

2 ...over an allele for sinistral (s), which encodes a left-handed coil.

s^+s^+ ss

Meiosis

Gametes s^+ s

Fertilization

F$_1$ generation

Sinistral

3 All the F$_1$ are heterozygous (s^+s); because the genotype of the mother determines the phenotype of the offspring, all the F$_1$ have a sinistral shell.

s^+s

Meiosis

s^+ s

Self-fertilization

F$_2$ generation

Dextral   Dextral   Dextral

¼ s^+s^+ ½ s^+s ¼ ss

Conclusion: Because the mother of the F$_2$ progeny has genotype s^+s, all the F$_2$ snails are dextral.

5.17 In genetic maternal effect, the genotype of the maternal parent determines the phenotype of the offspring. Chirality (the direction of shell coiling) in the snail *Lymnaea peregra* is a characteristic that exhibits genetic maternal effect.

same as the phenotype of the mother because the progeny's phenotype is determined by the mother's *genotype*, not her phenotype. Neither the male parent's nor the offspring's own genotype has any role in the offspring's phenotype. However, a male does influence the phenotype of the F$_2$ generation: by contributing to the genotypes of his daughters, he affects the phenotypes of their offspring. Genes that exhibit genetic maternal effect are therefore transmitted through males to future generations. In contrast, genes that exhibit cytoplasmic inheritance are transmitted through only one of the sexes (usually the female). ▸ **TRY PROBLEM 39**

CONCEPTS

Characteristics exhibiting cytoplasmic inheritance are encoded by genes in the cytoplasm and are usually inherited from one parent, most commonly the mother. In genetic maternal effect, the genotype of the mother determines the phenotype of the offspring.

✔ CONCEPT CHECK 10

How might you determine whether a particular trait is due to cytoplasmic inheritance or to genetic maternal effect?

Genomic Imprinting

A basic tenet of Mendelian genetics is that the parental origin of a gene does not affect its expression and, therefore, reciprocal crosses give identical results. We have seen, however, that there are some genetic characteristics—those encoded by X-linked genes and cytoplasmic genes—for which reciprocal crosses do not give the same results. In these cases, males and females do not contribute the same genetic material to the offspring. Males and females do contribute the same number of autosomal genes to their offspring, and paternal and maternal autosomal genes have long been assumed to have equal effects. However, the expression of some autosomal genes is significantly affected by their parental origin. This differential expression of genetic material depending on whether it is inherited from the male or female parent is called **genomic imprinting**. Although the number of imprinted genes is unknown, recent research suggests that more genes may be imprinted than previously suspected (see pp. 654–655 in Chapter 21).

A gene that exhibits genomic imprinting in both mice and humans is *Igf2*, which encodes a protein called insulin-like growth factor 2 (Igf2). Offspring inherit one *Igf2* allele from their mother and one from their father. The paternal copy of *Igf2* is actively expressed in the fetus and placenta, but the maternal copy is completely silent (**Figure 5.18**). Both male and female offspring possess *Igf2* genes; the key to whether the gene is expressed is the sex of the parent transmitting the gene. In the present example, the gene is expressed only when it is transmitted by a male parent. In some way that is not completely understood, the paternal *Igf2* allele (but not the maternal allele) promotes placental and fetal growth; when the paternal copy of *Igf2* is deleted in mice, a small placenta and low-birth-weight offspring result. In other genomically imprinted traits, only the copy of the gene transmitted by the female parent is expressed.

Genomic imprinting has been implicated in several human disorders, including Prader–Willi and Angelman syndromes. Children with Prader–Willi syndrome have small hands and feet, short stature, poor sexual development, and intellectual disability. These children are small at birth and nurse poorly, but as toddlers they develop voracious appetites and frequently become obese. Many people with

(a)

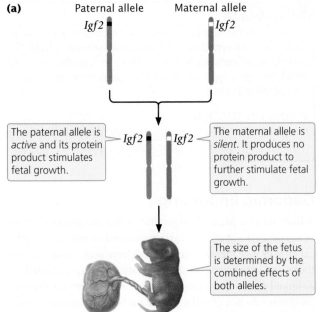

Paternal allele · Maternal allele

Igf2 · *Igf2*

The paternal allele is *active* and its protein product stimulates fetal growth.

Igf2 · *Igf2*

The maternal allele is *silent*. It produces no protein product to further stimulate fetal growth.

The size of the fetus is determined by the combined effects of both alleles.

(b)

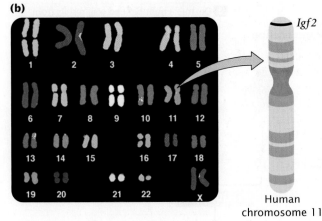

Igf2

Human chromosome 11

5.18 Genomic imprinting of the *Igf2* gene in mice and humans affects fetal growth. (a) The paternal *Igf2* allele is active in the fetus and placenta, whereas the maternal allele is silent. (b) The human *Igf2* locus is on the short arm of chromosome 11; the locus in mice is on chromosome 7 [Courtesy of Dr. Thomas Ried and Dr. Evelin Schrock.]

Prader–Willi syndrome are missing a small region on the long arm of chromosome 15. The deletion of this region is always inherited from the *father*. Thus, children with Prader–Willi syndrome lack a paternal copy of genes on the long arm of chromosome 15.

The deletion of this same region of chromosome 15 can also be inherited from the *mother*, but this inheritance results in a completely different set of symptoms, producing Angelman syndrome. Children with Angelman syndrome exhibit frequent laughter, uncontrolled muscle movement, a large mouth, and unusual seizures. They are missing a maternal copy of genes on the long arm of chromosome 15. For normal development to take place, copies of this region of chromosome 15 from both male and female parents are apparently required.

Many imprinted genes in mammals are associated with fetal growth. Imprinting has also been reported in plants, with differential expression of paternal and maternal genes in the endosperm, which, like the placenta in mammals, provides nutrients for the growth of the embryo. The mechanism of imprinting is still under investigation, but the methylation of DNA—the addition of methyl (CH_3) groups to DNA nucleotides (see Chapters 10 and 17)—is essential to the process. In mammals, methylation is erased in the germ cells each generation and then reestablished in the course of gamete formation, with sperm and eggs undergoing different levels of methylation, which then cause the differential expression of male and female alleles in the offspring. Some of the different ways in which sex interacts with heredity are summarized in **Table 5.5**.

EPIGENETICS Genomic imprinting is just one form of a phenomenon known as **epigenetics**. Most traits are encoded by genetic information that resides in the sequence of the nucleotide bases of DNA—the genetic code, which will be discussed in Chapter 15. However, some traits are determined by alterations to DNA—such as DNA methylation—that affect the way in which DNA sequences are expressed. These changes are often stable and heritable in the sense that they are passed from one cell to another. For example, in genomic imprinting, whether a gene passes through the egg or the sperm determines how much methylation of the DNA takes place. The pattern of methylation on a gene is copied when the DNA is replicated and therefore remains on the gene as it is passed

TABLE 5.5	Influences of sex on heredity
Genetic Phenomenon	**Phenotype Determined by**
Sex-linked characteristic	Genes located on the sex chromosomes
Sex-influenced characteristic	Autosomal genes that are more readily expressed in one sex
Sex-limited characteristic	Autosomal genes whose expression is limited to one sex
Genetic maternal effect	Nuclear genotype of the maternal parent
Cytoplasmic inheritance	Cytoplasmic genes, which are usually inherited from only one parent
Genomic imprinting	Genes whose expression is affected by the sex of the transmitting parent

from cell to cell through mitosis. However, the methylation may be modified or removed when the DNA passes through a gamete, so a gene that is methylated in sperm may be unmethylated when it is eventually passed down to a daughter's egg. Ultimately, the amount of methylation determines whether the gene is expressed in the offspring.

These types of reversible changes to DNA that influence the expression of traits are termed epigenetic marks. The inactivation of one of the X chromosomes in female mammals (discussed in Chapter 4) is another type of epigenetic change. We will consider epigenetic changes in more detail in Chapter 21.

CONCEPTS

In genomic imprinting, the expression of a gene is influenced by the sex of the parent transmitting the gene to the offspring. Epigenetic marks are reversible changes in DNA that do not alter the base sequence but may affect how a gene is expressed.

✔ **CONCEPT CHECK 11**

What type of epigenetic mark is responsible for genomic imprinting?

5.4 Anticipation Is the Stronger or Earlier Expression of Traits in Succeeding Generations

Another genetic phenomenon that is not explained by Mendel's principles is **anticipation**, in which a genetic trait becomes more strongly expressed, or is expressed at an earlier age, as it is passed from generation to generation. In the early 1900s, several physicians observed that many patients with moderate to severe myotonic dystrophy—an autosomal dominant muscle disorder—had ancestors who were only mildly affected by the disease. These observations led to the concept of anticipation. The concept quickly fell

out of favor with geneticists, however, because there was no obvious mechanism to explain it; traditional genetics held that genes are passed unaltered from parents to offspring. Geneticists tended to attribute anticipation to observational bias.

Research has now reestablished anticipation as a legitimate genetic phenomenon. The mutation that causes myotonic dystrophy consists of an unstable region of DNA that can increase in size as the gene is passed from generation to generation. The age of onset and the severity of the disease are correlated with the size of the unstable region; an increase in the size of the region over generations produces anticipation. The phenomenon has now been implicated in a number of genetic diseases. We will examine these interesting types of mutations in more detail in Chapter 18.

CONCEPTS

Anticipation is the stronger or earlier expression of a genetic trait in succeeding generations. It is caused by an unstable region of DNA that increases in size from generation to generation.

5.5 The Expression of a Genotype May Be Influenced by Environmental Effects

In Chapter 3, we learned that each phenotype is the result of a genotype that develops within a particular environment; each genotype may produce several different phenotypes, depending on the environmental conditions in which development takes place. For example, a fruit fly that is homozygous for the *vestigial* mutation (*vgvg*) develops reduced wings when raised at a temperature below 29°C, but the same genotype develops much longer wings when raised at 31°C (**Figure 5.19**).

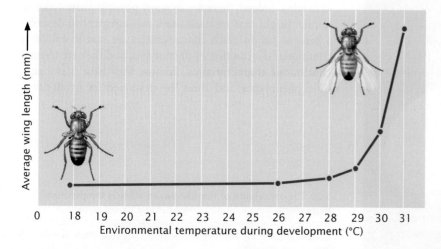

5.19 The expression of the *vestigial mutation* in *Drosophila* is temperature dependent. When reared at temperatures below 29°C, flies homozygous for *vestigial* have greatly reduced wings; at temperatures above 31°C, the flies develop normal wings. [Data from M. H. Harnly, *Journal of Experimental Zoology* 56:363–379, 1936.]

Average wing length (mm) →

Environmental temperature during development (°C)

0   18   19   20   21   22   23   24   25   26   27   28   29   30   31

For most of the characteristics we have discussed so far, the effect of the environment on the phenotype has been slight. Mendel's peas with genotype *yy*, for example, developed green seeds regardless of the environment in which they were raised. Similarly, people with genotype $I^A I^A$ have the A antigen on their red blood cells regardless of their diet, socioeconomic status, or family environment. For other phenotypes, however, environmental effects play a more important role.

Environmental Effects on the Phenotype

The phenotypic expression of some genotypes depends critically on the presence of a specific environment. For example, the *himalayan* allele in rabbits produces dark fur at the extremities of the body—on the nose, ears, and feet (**Figure 5.20**). The dark pigment develops, however, only when a rabbit is reared at a temperature of 25°C or lower; if a Himalayan rabbit is reared at 30°C, no dark patches develop. The expression of the *himalayan* allele is thus temperature dependent; an enzyme necessary for the production of dark pigment is inactivated at higher temperatures. The pigment is restricted to the nose, feet, and ears of a Himalayan rabbit because the animal's core body temperature is normally above 25°C and the enzyme is functional only in the cells of the relatively cool extremities. The *himalayan* allele is an example of a **temperature-sensitive allele,** an allele whose product is functional only at certain temperatures. Similarly, vestigial wings in *Drosophila melanogaster* are caused by a temperature-dependent mutation (see Figure 5.19).

Environmental factors also play an important role in the expression of a number of human genetic diseases. Phenylketonuria (PKU) is due to an autosomal recessive allele that causes intellectual disability. The disorder arises from a defect in an enzyme that normally metabolizes the amino acid phenylalanine. When this enzyme is defective, phenylalanine is not metabolized, and its buildup causes neurological

damage in children. A simple environmental change, putting an affected child on a low-phenylalanine diet, prevents the development of intellectual disability.

THINK-PAIR-SHARE Question 6

These examples illustrate the point that genes and their products do not act in isolation; rather, they frequently interact with environmental factors. Occasionally, environmental factors alone can produce a phenotype that is the same as the phenotype produced by a genotype; such a phenotype is called a **phenocopy**. In fruit flies, for example, the autosomal recessive mutation *eyeless* produces greatly reduced eyes. The eyeless phenotype can also be produced by exposing the larvae of normal flies to sodium metaborate.

CONCEPTS

The expression of many genes is modified by the environment. A phenocopy is a phenotype produced by environmental effects that mimics the phenotype produced by a genotype.

✔ **CONCEPT CHECK 12**

How can you determine whether a phenotype such as reduced eyes in fruit flies is due to a recessive mutation or is a phenocopy?

The Inheritance of Continuous Characteristics

So far, we've dealt primarily with characteristics that have only a few distinct phenotypes. In Mendel's peas, for example, the seeds were either smooth or wrinkled, yellow or green; the coats of dogs were black, brown, or yellow; blood types were A, B, AB, or O. Such characteristics, which have a few easily distinguished phenotypes, are called **discontinuous characteristics**.

Not all characteristics exhibit discontinuous phenotypes, however. Human height is an example; people do not come in just a few distinct heights, but rather display a wide range of heights. Indeed, there are so many possible phenotypes of human height that we must use a measurement to describe a person's height. Such characteristics are said to exhibit a continuous distribution of phenotypes, and they are termed **continuous characteristics**. Because they have many possible phenotypes and must be described in quantitative

5.20 The expression of the himalayan allele is temperature dependent. This rabbit was reared below 25°C. Its pigment is restricted to the extremities where the body temperature falls below 25°C and the enzyme that produces pigment is functional. [Petra Wegner/Alamy.]

terms, continuous characteristics are also called **quantitative characteristics**.

Continuous characteristics frequently arise because genes at many loci interact to produce the phenotypes. When a single locus with two alleles encodes a characteristic, there are three genotypes possible: *AA*, *Aa*, and *aa*. With two loci, each with two alleles, there are $3^2 = 9$ genotypes possible. The number of genotypes encoding a characteristic is 3^n, where *n* equals the number of loci, each with two alleles, that influence the characteristic. For example, when a characteristic is determined by eight loci, each with two alleles, there are $3^8 = 6561$ different genotypes possible for this characteristic. If each genotype produces a different phenotype, many phenotypes will be possible. The slight differences between the phenotypes will be indistinguishable, and the characteristic will appear to be continuous. Characteristics encoded by genes at many loci are called **polygenic characteristics**.

The converse of polygeny is **pleiotropy**, in which one gene affects multiple characteristics. Many genes exhibit pleiotropy. Phenylketonuria, mentioned earlier, results from a recessive allele; people homozygous for this allele, if untreated, exhibit intellectual disability, blue eyes, and light skin color. The lethal allele that causes yellow coat color in mice is also pleiotropic. In addition to its lethality and its effect on coat color, the gene causes a diabetes-like condition, obesity, and increased propensity to develop tumors.

The phenotypes of many continuous characteristics are also influenced by environmental factors. In these cases, each genotype is capable of producing a range of phenotypes, and the particular phenotype that results depends on both the genotype and the environmental conditions in which the genotype develops. For example, there may be only three genotypes at a single locus, but because each genotype produces a range of phenotypes associated with different environments, the phenotype of the characteristic will exhibit a continuous distribution. Many continuous characteristics are both polygenic and influenced by environmental factors; such characteristics are called **multifactorial characteristics** because many factors help determine the phenotype. Human height is an example of a multifactorial characteristic; variation in height is determined by a number of genes as well as environmental factors such as nutrition.

The inheritance of continuous characteristics may appear to be complex, but the alleles at each locus follow Mendel's principles and are inherited in the same way as alleles encoding simple, discontinuous characteristics. However, because many genes participate, because environmental factors influence the phenotype, and because the phenotypes do not sort out into a few distinct types, we cannot observe the distinct ratios that have allowed us to interpret the genetic basis of discontinuous characteristics. To analyze continuous characteristics, we must employ special statistical tools, which will be discussed in Chapter 24.

▸ TRY PROBLEM 45

CONCEPTS

Discontinuous characteristics exhibit a few distinct phenotypes; continuous characteristics exhibit a range of phenotypes. A continuous characteristic is frequently produced when genes at many loci and environmental factors combine to determine a phenotype.

✔ CONCEPT CHECK 13

What is the difference between polygeny and pleiotropy?

CONCEPTS SUMMARY

- Dominance is an interaction between genes at the same locus (allelic genes) and can be understood in regard to how the phenotype of the heterozygote relates to the phenotypes of the homozygotes.

- Dominance is complete when a heterozygote has the same phenotype as one homozygote. It is incomplete when the heterozygote has a phenotype intermediate between those of two parental homozygotes. In codominance, the heterozygote exhibits traits of both parental homozygotes.

- The type of dominance does not affect the inheritance of an allele; it does affect the phenotypic expression of the allele. The classification of dominance depends on the level at which the phenotype is examined.

- Penetrance is the percentage of individuals having a particular genotype that exhibit the expected phenotype. Expressivity is the degree to which a character is expressed.

- Lethal alleles cause the death of an individual possessing them at an early stage of development and may alter the phenotypic ratios resulting from a cross.

- The existence of multiple alleles refers to the presence of more than two alleles at a locus within a group of individuals. The presence of multiple alleles increases the number of genotypes and phenotypes that are possible.

- Gene interaction refers to interaction between genes at different loci to produce a single phenotype. An epistatic gene at one locus suppresses, or masks, the expression of hypostatic genes at other loci. Gene interaction frequently produces phenotypic ratios that are modifications of dihybrid ratios.

- Sex-influenced characteristics are encoded by autosomal genes that are expressed more readily in one sex. Sex-limited characteristics are encoded by autosomal genes that are expressed in only one sex.

■ In cytoplasmic inheritance, the genes that encode a characteristic are found in the organelles and are usually inherited from a single (typically maternal) parent. In genetic maternal effect, an offspring inherits genes from both parents, but the nuclear genes of the mother determine the offspring's phenotype.

■ Genomic imprinting refers to differential expression of autosomal genes depending on the sex of the parent transmitting the genes. Epigenetic effects such as genomic imprinting are caused by alterations to DNA—such as DNA methylation—that do not affect the DNA base sequence.

■ Anticipation refers to the stronger expression of a genetic trait, or its expression at an earlier age, in succeeding generations.

■ Phenotypes are often modified by environmental effects. A phenocopy is a phenotype produced by an environmental effect that mimics a phenotype produced by a genotype.

■ Continuous characteristics are those that exhibit a wide range of phenotypes; they are frequently produced by the combined effects of many genes and environmental effects.

IMPORTANT TERMS

complete dominance (p. 111)
incomplete dominance (p. 111)
codominance (p. 112)
incomplete penetrance (p. 113)
penetrance (p. 113)
expressivity (p. 113)
lethal allele (p. 114)
multiple alleles (p. 114)

compound heterozygote (p. 116)
gene interaction (p. 117)
epistasis (p. 118)
epistatic gene (p. 118)
hypostatic gene (p. 118)
complementation test (p. 124)
complementation (p. 124)
sex-influenced characteristic (p. 126)

sex-limited characteristic (p. 127)
cytoplasmic inheritance (p. 128)
genetic maternal effect (p. 130)
genomic imprinting (p. 131)
epigenetics (p. 132)
anticipation (p. 133)
temperature-sensitive allele (p. 134)
phenocopy (p. 134)

discontinuous characteristic (p. 134)
continuous characteristic (p. 134)
quantitative characteristic (p. 135)
polygenic characteristic (p. 135)
pleiotropy (p. 135)
multifactorial characteristic (p. 135)

ANSWERS TO CONCEPT CHECKS

1. b

2. With complete dominance, the heterozygote expresses the same phenotype as one of the homozygotes. With incomplete dominance, the heterozygote has a phenotype that is intermediate between those of the two homozygotes. With codominance, the heterozygote has a phenotype that simultaneously expresses the phenotypes of both homozygotes.

3. d

4. b

5. c

6. Gene interaction is interaction between genes at different loci. Dominance is interaction between alleles at a single locus.

7. d

8. Cross a bulldog homozygous for *brindle* with a Chihuahua homozygous for *brindle*. If the two *brindle* genes are allelic, all the offspring will be brindle: $bb \times bb \rightarrow$ all bb (brindle). If, on the other hand, brindle in the two breeds is due to recessive genes at different loci, then none of the offspring will be brindle: $a^+a^+ bb \times aa\, b^+b^+ \rightarrow a^+a\, b^+b$.

9. Both sex-influenced and sex-limited characteristics are encoded by autosomal genes whose expression is affected by the sex of the individual organism possessing the genes. Sex-linked characteristics are encoded by genes on the sex chromosomes.

10. Cytoplasmically inherited traits are encoded by genes in the cytoplasm, which is usually inherited only from the female parent. Therefore, a trait due to cytoplasmic inheritance will always be passed through females. Traits due to genetic maternal effect are encoded by autosomal genes and can therefore be passed through males, although any individual organism's trait is determined by the genotype of the maternal parent.

11. Methylation of DNA.

12. Cross two eyeless flies and cross an eyeless fly with a wild-type fly. Raise the offspring of both crosses in the same environment. If the trait is due to a recessive mutation, all the offspring of the two eyeless flies should be eyeless, whereas at least some of the offspring of the eyeless and wild-type flies should be wild type. If the trait is a phenocopy, there should be no differences in the progeny of the two crosses.

13. Polygeny refers to the influence of multiple genes on the expression of a single characteristic. Pleiotropy refers to the effect of a single gene on the expression of multiple characteristics.

WORKED PROBLEMS

Problem 1

A geneticist crosses two yellow mice with straight hair and obtains the following progeny:

$\frac{1}{2}$ yellow, straight

$\frac{1}{6}$ yellow, fuzzy

$\frac{1}{4}$ gray, straight

$\frac{1}{12}$ gray, fuzzy

a. Provide a genetic explanation for the results and assign genotypes to the parents and progeny of this cross.

b. What additional crosses might be carried out to determine if your explanation is correct?

›› Solution Strategy

What information is required in your answer to the problem?

a. A genetic explanation for the inheritance of color and hair type in the mice. Genotypes of the parents.

b. Examples of other crosses that might be carried out to determine if the explanation given in your answer to part *a* is correct.

What information is provided to solve the problem?

- Phenotypes of the parents.
- Phenotypes and proportions of different types of progeny.

For help with this problem, review:
Lethal Alleles in Section 5.1 and ratios for simple genetic crosses (Table 3.5).

›› Solution Steps

Hint: Examine the progeny ratios for each trait separately.

a. This cross concerns two separate characteristics—color and hair type. First, let's look at the inheritance of color. Two yellow mice are crossed, producing $\frac{1}{2} + \frac{1}{6} = \frac{3}{6} + \frac{1}{6} = \frac{4}{6} = \frac{2}{3}$ yellow mice and $\frac{1}{4} + \frac{1}{12} = \frac{3}{12} + \frac{1}{12} = \frac{4}{12} = \frac{1}{3}$ gray mice. We learned in this chapter that a 2 : 1 ratio is often produced when a recessive lethal allele is present:

Recall: A 2 : 1 ratio is usually produced by a lethal allele.

$$Yy \times Yy$$

$$\downarrow$$

YY $\frac{1}{4}$ die

Yy $\frac{1}{2}$ yellow, becomes $\frac{2}{3}$

yy $\frac{1}{4}$ gray, becomes $\frac{1}{3}$

Now, let's examine the inheritance of the hair type. Two mice with straight hair are crossed, producing $\frac{1}{2} + \frac{1}{4} = \frac{2}{4} + \frac{1}{4} = \frac{3}{4}$ mice with straight hair and $\frac{1}{6} + \frac{1}{12} = \frac{2}{12} + \frac{1}{12} = \frac{3}{12} = \frac{1}{4}$ mice with fuzzy hair. We learned in Chapter 3 that a 3 : 1 ratio is usually produced by a cross between two individuals heterozygous for a simple dominant allele:

$$Ss \times Ss$$

$$\downarrow$$

SS $\frac{1}{4}$ straight

Ss $\frac{1}{2}$ straight $\Big\}$ $\frac{3}{4}$ straight

ss $\frac{1}{4}$ fuzzy

We can now combine both loci and assign genotypes to all the individual mice in the cross:

$$Yy\,Ss \quad \times \quad Yy\,Ss$$

Yellow, straight   Yellow, straight

$$\downarrow$$

Phenotype	Genotype	Probability at each locus	Combined probability
yellow, straight	$Yy\ S\_$	$\frac{2}{3} \times \frac{3}{4}$	$= \frac{6}{12} = \frac{1}{2}$
yellow, fuzzy	$Yy\ ss$	$\frac{2}{3} \times \frac{1}{4}$	$= \frac{2}{12} = \frac{1}{6}$
gray, straight	$yy\ S\_$	$\frac{1}{3} \times \frac{3}{4}$	$= \frac{3}{12} = \frac{1}{4}$
gray, fuzzy	$yy\ ss$	$\frac{1}{3} \times \frac{1}{4}$	$= \frac{1}{12}$

b. We could carry out a number of different crosses to test our hypothesis that yellow is a recessive lethal and straight is dominant over fuzzy. For example, a cross between any two yellow mice should always produce $\frac{2}{3}$ yellow and $\frac{1}{3}$ gray offspring, and a cross between two gray mice should produce only gray offspring. A cross between two fuzzy mice should produce only fuzzy offspring.

Problem 2

In some sheep, horns are produced by an autosomal allele that is dominant in males and recessive in females. A horned female is crossed with a hornless male. One of the resulting F_1 females is crossed with a hornless male. What proportion of the male and female progeny from this cross will have horns?

›› Solution Strategy

What information is required in your answer to the problem?

Proportions of male and female progeny that have horns.

What information is provided to solve the problem?

- The presence of horns is due to an autosomal gene that is dominant in male and recessive in females.

- A horned female is crossed with a hornless male. A resulting F_1 female is crossed with a hornless male to produce progeny.

For help with this problem, review:

Sex-Influenced and Sex-Limited Characteristics in Section 5.3.

›› Solution Steps

> **Hint:** Write out the genotypes and the associated phenotype for each sex.

The presence of horns in these sheep is an example of a sex-influenced characteristic. Because the phenotypes associated with the genotypes differ between the two sexes, let's begin this problem by writing out the genotypes and phenotypes for each sex. We will let H represent the allele that encodes horns and H^+ represent the allele that encodes hornless. In males, the allele for horns is dominant over the hornless allele, which means that males homozygous (HH) and heterozygous (H^+H) for this gene are horned. Only males homozygous for the recessive hornless allele (H^+H^+) are hornless. In females, the allele for horns is recessive, which means that only females homozygous for this allele (HH) are horned; females heterozygous (H^+H) and homozygous (H^+H^+) for the hornless allele are hornless. The following table summarizes the genotypes and their associated phenotypes:

> **Recall:** When a trait is dominant, both the homozygote and the heterozygote express the trait in their phenotypes.

Genotype	Male phenotype	Female phenotype
HH	horned	horned
HH^+	horned	hornless
H^+H^+	hornless	hornless

In the problem, a horned female is crossed with a hornless male. From the preceding table, we see that a horned female must be homozygous for the allele for horns (HH), and that a hornless male must be homozygous for the hornless allele (H^+H^+), so all the F_1 will be heterozygous; the F_1 males will be horned and the F_1 females will be hornless, as shown in the following diagram:

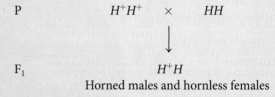

P $\qquad H^+H^+ \qquad \times \qquad HH$

$F_1 \qquad\qquad H^+H$

Horned males and hornless females

A heterozygous hornless F_1 female (H^+H) is then crossed with a hornless male (H^+H^+):

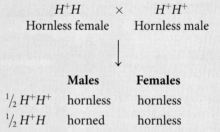

$$H^+H \qquad \times \qquad H^+H^+$$

Hornless female $\qquad$ Hornless male

	Males	**Females**
$\frac{1}{2}\,H^+H^+$	hornless	hornless
$\frac{1}{2}\,H^+H$	horned	hornless

Therefore, $\frac{1}{2}$ of the male progeny will be horned, but none of the female progeny will be horned.

COMPREHENSION QUESTIONS

Section 5.1

1. What are some important characteristics of dominance?

2. What is incomplete penetrance and what causes it?

Section 5.2

3. What is gene interaction? What is the difference between an epistatic gene and a hypostatic gene?

4. What is a recessive epistatic gene?

5. What is a complementation test and what is it used for?

Section 5.3

6. What characteristics are exhibited by a cytoplasmically inherited trait?

7. What is genomic imprinting?

8. What is the difference between genetic maternal effect and genomic imprinting?

9. What is the difference between a sex-influenced gene and a gene that exhibits genomic imprinting?

Section 5.4

10. What characteristics do you expect to see in a trait that exhibits anticipation?

Section 5.5

11. What are continuous characteristics and how do they arise?

APPLICATION QUESTIONS AND PROBLEMS

Sections 5.1 through 5.4

12. Match each of the following terms with its correct definition (parts *a* through *i*).

_____ Phenocopy

_____ Pleiotropy

_____ Polygenic trait

_____ Penetrance

_____ Sex-limited trait

_____ Genetic maternal effect

_____ Genomic imprinting

_____ Sex-influenced trait

_____ Anticipation

a. The percentage of individuals with a particular genotype that express the expected phenotype.

b. A trait determined by an autosomal gene that is more easily expressed in one sex.

c. A trait determined by an autosomal gene that is expressed in only one sex.

d. A trait that is determined by an environmental effect and has the same phenotype as a genetically determined trait.

e. A trait determined by genes at many loci.

f. The expression of a trait is affected by the sex of the parent that transmits the gene to the offspring.

g. The trait appears earlier or is more severe in succeeding generations.

h. A gene affects more than one phenotype.

i. The genotype of the maternal parent influences the phenotype of the offspring.

Section 5.1

*13. Palomino horses have a golden yellow coat, chestnut horses have a brown coat, and cremello horses have a coat that is almost white. A series of crosses between the three different types of horses produced the following offspring:

Cross	Offspring
palomino × palomino	13 palomino, 6 chestnut, 5 cremello
chestnut × chestnut	16 chestnut
cremello × cremello	13 cremello
palomino × chestnut	8 palomino, 9 chestnut
palomino × cremello	11 palomino, 11 cremello
chestnut × cremello	23 palomino

a. Explain the inheritance of the palomino, chestnut, and cremello phenotypes in horses.

b. Assign symbols for the alleles that determine these phenotypes, and list the genotypes of all parents and offspring given in the preceding table.

(a)

(b)

(c)

Coat color: (a) palomino; (b) chestnut; (c) cremello. [Part a: Keith J. Smith/Alamy. Part b: © jirijura/iStock Photo.com. Part c: Olga_i/Shutterstock.]

14. The L^M and L^N alleles at the MN blood-group locus exhibit codominance. Give the expected genotypes and phenotypes and their ratios in progeny resulting from the following crosses.

a. $L^M L^M \times L^M L^N$

b. $L^N L^N \times L^N L^N$

c. $L^M L^N \times L^M L^N$

d. $L^M L^N \times L^N L^N$

e. $L^M L^M \times L^N L^N$

*15. Assume that long ear lobes in humans are an autosomal dominant trait that exhibits 30% penetrance. A person who is heterozygous for long ear lobes mates with a person who is homozygous for normal ear lobes. What is the probability that their first child will have long ear lobes?

16. Club foot is one of the most common congenital skeletal abnormalities, with a worldwide incidence of about 1 in 1000 births. Both genetic and nongenetic factors are thought to be responsible for club foot. C. A. Gurnett et al. (2008. *American Journal of Human*

Genetics 83:616–622) identified a family in which club foot was inherited as an autosomal dominant trait with incomplete penetrance. They discovered a mutation in the *PITXI* gene that caused club foot in this family. Through DNA testing, they determined that 11 people in the family carried the *PITXI* mutation, but only 8 of these people had club foot. What is the penetrance of the *PITXI* mutation in this family?

*17. When a Chinese hamster with white spots is crossed with another hamster that has no spots, approximately $\frac{1}{2}$ of the offspring have white spots and $\frac{1}{2}$ have no spots. When two hamsters with white spots are crossed, $\frac{2}{3}$ of the offspring possess white spots and $\frac{1}{3}$ have no spots.

  a. What is the genetic basis of white spotting in Chinese hamsters?

  b. How might you go about producing Chinese hamsters that breed true for white spotting?

18. In the early 1900s, Lucien Cuénot studied the genetic basis of yellow coat color in mice (discussed on p. 114). He carried out a number of crosses between two yellow mice and obtained what he thought was a 3 : 1 ratio of yellow to gray mice in the progeny. The following table gives Cuénot's actual results, along with the results of a much larger series of crosses carried out by Castle and Little (W. E. Castle and C. C. Little. 1910. *Science* 32:868–870).

Progeny resulting from crosses of yellow × yellow mice

Investigators	Yellow progeny	Non-yellow progeny	Total progeny
Cuénot	263	100	363
Castle and Little	800	435	1235
Both combined	1063	535	1598

  a. Using a chi-square test, determine whether Cuénot's results are significantly different from the 3 : 1 ratio that he thought he observed. Are they different from a 2 : 1 ratio?

  b. Determine whether Castle and Little's results are significantly different from a 3 : 1 ratio. Are they different from a 2 : 1 ratio?

  c. Combine the results of Castle and Little and Cuénot and determine whether they are significantly different from a 3 : 1 ratio and a 2 : 1 ratio.

  d. Offer an explanation for the different ratios that Cuénot and Castle and Little obtained.

*19. In the pearl-millet plant, color is determined by three alleles at a single locus: Rp^1 (red), Rp^2 (purple), and rp (green). Red is dominant over purple and green, and purple is dominant over green ($Rp^1 > Rp^2 > rp$). Give the expected phenotypes and ratios of offspring produced by the following crosses.

  a. $Rp^1/Rp^2 \times Rp^1/rp$

  b. $Rp^1/rp \times Rp^2/rp$

  c. $Rp^1/Rp^2 \times Rp^1/Rp^2$

  d. $Rp^2/rp \times rp/rp$

  e. $rp/rp \times Rp^1/Rp^2$

20. If there are five alleles at a locus, how many genotypes can there be at this locus? How many different kinds of homozygotes can there be? How many genotypes and homozygotes can there be with eight alleles at a locus?

21. Turkeys have black, bronze, or black-bronze plumage. Examine the results of the following crosses:

Parents	Offspring
Cross 1: black and bronze	all black
Cross 2: black and black	$\frac{3}{4}$ black, $\frac{1}{4}$ bronze
Cross 3: black-bronze and black-bronze	all black-bronze
Cross 4: black and bronze	$\frac{1}{2}$ black, $\frac{1}{4}$ bronze, $\frac{1}{4}$ black-bronze
Cross 5: bronze and black-bronze	$\frac{1}{2}$ bronze, $\frac{1}{2}$ black-bronze
Cross 6: bronze and bronze	$\frac{3}{4}$ bronze, $\frac{1}{4}$ black-bronze

Do you think these differences in plumage arise from incomplete dominance between two alleles at a single locus? If yes, support your conclusion by assigning symbols to each allele and providing genotypes for all turkeys in the crosses. If your answer is no, provide an alternative explanation and assign genotypes to all turkeys in the crosses.

22. In rabbits, an allelic series helps to determine coat color: C (full color), c^{ch} (chinchilla, gray color), c^h (Himalayan, white with black extremities), and c (albino, all-white). The C allele is dominant over all others, c^{ch} is dominant over c^h and c, c^h is dominant over c, and c is recessive to all the other alleles. This dominance hierarchy can be summarized as $C > c^{ch} > c^h > c$. The rabbits in the following list are crossed and produce the progeny shown. Give the genotypes of the parents for each cross:

Phenotypes of parents	Phenotypes of offspring
a. full color × albino	$\frac{1}{2}$ full color, $\frac{1}{2}$ albino
b. Himalayan × albino	$\frac{1}{2}$ Himalayan, $\frac{1}{2}$ albino
c. full color × albino	$\frac{1}{2}$ full color, $\frac{1}{2}$ chinchilla
d. full color × Himalayan	$\frac{1}{2}$ full color, $\frac{1}{4}$ Himalayan, $\frac{1}{4}$ albino
e. full color × full color	$\frac{3}{4}$ full color, $\frac{1}{4}$ albino

23. In this chapter, we considered Joan Barry's paternity suit against Charlie Chaplin and how, on the basis of blood types, Chaplin could not have been the father of her child.

a. What blood types are possible for the father of Barry's child?

b. If Chaplin had possessed one of these blood types, would that prove that he fathered Barry's child?

*24. A woman has blood-type A M. She has a child with blood-type AB MN. Which of the following blood types could *not* be that of the child's father? Explain your reasoning.

George	O	N
Tom	AB	MN
Bill	B	MN
Claude	A	N
Henry	AB	M

Section 5.2

25. In chickens, comb shape is determined by alleles at two loci (R, r and P, p). A walnut comb is produced when at least one dominant allele R is present at one locus and at least one dominant allele P is present at a second locus (genotype $R\_ P\_$). A rose comb is produced when at least one dominant allele is present at the first locus and two recessive alleles are present at the second locus (genotype $R\_ pp$). A pea comb is produced when two recessive alleles are present at the first locus and at least one dominant allele is present at the second (genotype $rr P\_$). If two recessive alleles are present at the first and at the second locus ($rr pp$), a single comb is produced. Progeny with what types of combs and in what proportions will result from the following crosses?

a. $RR\ PP \times rr\ pp$

b. $Rr\ Pp \times rr\ pp$

c. $Rr\ Pp \times Rr\ Pp$

d. $Rr\ pp \times Rr\ pp$

e. $Rr\ pp \times rr\ Pp$

f. $Rr\ pp \times rr\ pp$

(a) **(b)**

(c) **(d)**

Comb shape: (a) walnut; (b) rose; (c) pea; (d) single. [Parts a and d: Robert Dowling/CORBIS. Part b: Robert Maier/Animals Animals. Part c: Daphne Godfrey Trust/Animals Animals.]

*26. Tatuo Aida investigated the genetic basis of color variation in the medaka (*Aplocheilus latipes*), a small fish found in Japan (T. Aida. 1921. *Genetics* 6:554–573). Aida found that genes at two loci (B, b and R, r) determine the color of the fish: fish with a dominant allele at both loci ($B\_ R\_$) are brown, fish with a dominant allele at the B locus only ($B\_ rr$) are blue, fish with a dominant allele at the R locus only ($bb R\_$) are red, and fish with recessive alleles at both loci ($bb rr$) are white. Aida crossed a homozygous brown fish with a homozygous white fish. He then backcrossed the F_1 with the homozygous white parent and obtained 228 brown fish, 230 blue fish, 237 red fish, and 222 white fish.

a. Give the genotypes of the backcross progeny.

b. Use a chi-square test to compare the observed numbers of backcross progeny with the number expected. What conclusion can you make from your chi-square results?

c. What results would you expect for a cross between a homozygous red fish and a white fish?

d. What results would you expect if you crossed a homozygous red fish with a homozygous blue fish and then backcrossed the F_1 with a homozygous red parental fish?

27. A variety of opium poppy (*Papaver somniferum* L.) with lacerate leaves was crossed with a variety that has normal leaves. All the F_1 had lacerate leaves. Two F_1 plants were interbred to produce the F_2. Of the F_2, 249 had lacerate leaves and 16 had normal leaves. Give genotypes for all the plants in the P, F_1, and F_2 generations. Explain how lacerate leaves are determined in the opium poppy.

28. E. W. Lindstrom crossed two corn plants with green seedlings and obtained the following progeny: 3583 green seedlings, 853 virescent-white seedlings, and 260 yellow seedlings (E. W. Lindstrom. 1921. *Genetics* 6:91–110).

a. Give the genotypes for the green, virescent-white, and yellow progeny.

b. Explain how color is determined in these seedlings.

c. Is there epistasis among the genes that determine color in the corn seedlings? If so, which gene is epistatic and which is hypostatic?

*29. A dog breeder liked yellow and brown Labrador retrievers. In an attempt to produce yellow and brown puppies, he mated a yellow Labrador male and a brown Labrador female. Unfortunately, all the puppies produced in this cross were black. (See pp. 124–125 for a discussion of the genetic basis of coat color in Labrador retrievers.)

a. Explain this result.

b. How might the breeder go about producing yellow and brown Labradors?

(a)

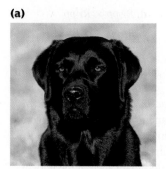

(b)

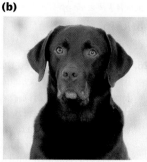

(c)

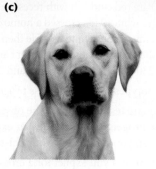

Coat color in Labrador retrievers: (a) black; (b) brown; (c) yellow. [Parts a and b: Juniors Bildarchiv/Alamy. Part c: c. byatt-norman/Shutterstock.]

30. When a yellow female Labrador retriever was mated with a brown male, half of the puppies were brown and half were yellow. The same female, when mated with a different brown male, produced only brown offspring. Explain these results.

*31. A summer-squash plant that produces disc-shaped fruit is crossed with a summer-squash plant that produces long fruit. All the F_1 have disc-shaped fruit. When the F_1 are intercrossed, F_2 progeny are produced in the following ratio: $^9/_{16}$ disc-shaped fruit : $^6/_{16}$ spherical fruit : $^1/_{16}$ long fruit. Give the genotypes of the F_2 progeny.

32. Some sweet-pea plants have purple flowers and others have white flowers. A homozygous variety of sweet pea that has purple flowers is crossed with a homozygous variety that has white flowers. All the F_1 have purple flowers. When these F_1 self-fertilize, the F_2 appear in a ratio of $^9/_{16}$ purple to $^7/_{16}$ white.

a. Give genotypes for the purple and white flowers in these crosses.

b. Draw a hypothetical biochemical pathway to explain the production of purple and white flowers in sweet peas.

*33. Refer to pages 124–125 for a discussion of how coat color and pattern are determined in dogs.

a. Why are Irish setters not black in color?

b. Can a poodle crossed with any other breed produce spotted puppies? Why or why not?

c. If a St. Bernard is crossed with a Doberman, what will be the coat color of the offspring: solid, yellow, saddle, or bicolor?

d. If a Rottweiler is crossed with a Labrador retriever, what will be the coat color of the offspring: solid, yellow, saddle, or bicolor?

Section 5.3

34. Male-limited precocious puberty results from a rare, sex-limited autosomal allele (*P*) that is dominant over the allele for normal puberty (*p*) and is expressed only in males. Bill underwent precocious puberty, but his brother Jack and his sister Beth underwent puberty at the usual time, between the ages of 10 and 14. Although Bill's mother and father underwent normal puberty, two of his maternal uncles (his mother's brothers) underwent precocious puberty. All of Bill's grandparents underwent normal puberty. Give the most likely genotypes for all the relatives mentioned in this family.

*35. In some goats, the presence of horns is produced by an autosomal gene that is dominant in males and recessive in females. A horned female is crossed with a hornless male. The F_1 offspring are intercrossed to produce the F_2. What proportion of the F_2 females will have horns?

36. In goats, a beard is produced by an autosomal allele that is dominant in males and recessive in females. We'll use the symbol B^b for the beard allele and B^+ for the beardless allele. Another independently assorting autosomal allele that produces a black coat (*W*) is dominant over the allele for white coat (*w*). Give the phenotypes and their expected proportions for the following crosses.

a. B^+B^b *Ww* male × B^+B^b *Ww* female

b. B^+B^b *Ww* male × B^+B^b *ww* female

c. B^+B^+ *Ww* male × B^bB^b *Ww* female

d. B^+B^b *Ww* male × B^bB^b *ww* female

37. Cock feathering in chickens is an autosomal recessive trait that is limited to males. List all possible genotypes for the chicken shown in

a. Figure 5.13a

b. Figure 5.13b

c. Figure 5.13c

38. J. K. Breitenbecher (1921. *Genetics* 6:65–86) investigated the genetic basis of color variation in the four-spotted cowpea weevil (*Bruchus quadrimaculatus*). The weevils were red, black, white, or tan. Breitenbecher found that four alleles (*R*, R^b, R^w, and *r*) at a single locus determine color. The alleles exhibit a dominance hierarchy, with red (*R*) dominant over all other alleles, black (R^b) dominant over white (R^w) and tan (*r*), white dominant over tan, and tan recessive to all others ($R > R^b > R^w > r$). The following genotypes encode each of the colors:

RR, RR^b, RR^w, Rr	red
R^bR^b, R^bR^w, R^br	black
R^wR^w, R^wr	white
rr	tan

Color variation in this species is limited to females: males carry color genes but are always tan regardless of their genotype. For each of the following crosses carried out by Breitenbecher, give all possible genotypes of the parents.

Parents	Progeny
a. tan ♀ × tan ♂	78 red ♀, 70 white ♀, 184 tan ♂
b. black ♀ × tan ♂	151 red ♀, 49 black ♀, 61 tan ♀, 249 tan ♂
c. white ♀ × tan ♂	32 red ♀, 31 tan ♂
d. black ♀ × tan ♂	3586 black ♀, 1282 tan ♀, 4791 tan ♂
e. white ♀ × tan ♂	594 white ♀, 189 tan ♀, 862 tan ♂
f. black ♀ × tan ♂	88 black ♀, 88 tan ♀, 186 tan ♂
g. tan ♀ × tan ♂	47 white ♀, 51 tan ♀, 100 tan ♂
h. red ♀ × tan ♂	1932 red ♀, 592 tan ♀, 2587 tan ♀
i. white ♀ × tan ♂	13 red ♀, 6 white ♀, 5 tan ♀, 19 tan ♂
j. red ♀ × tan ♂	190 red ♀, 196 black ♀, 311 tan ♂
k. black ♀ × tan ♂	1412 black ♀, 502 white ♀, 1766 tan ♂

*39. The direction of shell coiling in the snail *Lymnaea peregra* (discussed in the introduction to this chapter) results from genetic maternal effect. An autosomal allele for a right-handed, or dextral, shell (s^+) is dominant over the allele for a left-handed, or sinistral, shell (s). A pet snail called Martha is sinistral and reproduces only as a female (the snails are hermaphroditic). Indicate which of the following statements are true and which are false. Explain your reasoning in each case.

a. Martha's genotype *must* be ss.

b. Martha's genotype *cannot* be s^+s^+.

c. All the offspring produced by Martha *must* be sinistral.

d. At least some of the offspring produced by Martha *must* be sinistral.

e. Martha's mother *must* have been sinistral.

f. All of Martha's brothers *must* be sinistral.

40. If the F_2 dextral snails with genotype s^+s in **Figure 5.17** undergo self-fertilization, what phenotypes and proportions are expected to occur in the progeny?

41. Hypospadias, a birth defect in human males in which the urethra opens on the shaft instead of at the tip of the penis, results from an autosomal dominant gene in some families. Females who carry the gene show no effects. Is this birth defect an example of an X-linked trait, a Y-linked trait, a sex-limited trait, a sex-influenced trait, or genetic maternal effect? Explain your answer.

42. In unicorns, two autosomal loci interact to determine the type of tail. One locus controls whether a tail is present at all; the allele for a tail (T) is dominant over the allele for tailless (t). If a unicorn has a tail, then alleles at a second locus determine whether the tail is curly or straight. Farmer Baldridge has two unicorns with curly tails: when he crosses them, $1/2$ of the progeny have curly tails, $1/4$ have straight tails, and $1/4$ do not have a tail. Give the genotypes of the parents and progeny in Farmer Baldridge's cross. Explain how he obtained the 2 : 1 : 1 phenotypic ratio in his cross.

43. In 1983, a sheep farmer in Oklahoma noticed in his flock a ram that possessed increased muscle mass in his hindquarters. Many of the offspring of this ram possessed the same trait, which became known as the callipyge phenotype (*callipyge* is Greek for "beautiful buttocks"). The mutation that caused the callipyge phenotype was eventually mapped to a position on the sheep chromosome 18.

When the male callipyge offspring of the original mutant ram were crossed with normal females, they produced the following progeny: $1/4$ male callipyge, $1/4$ female callipyge, $1/4$ male normal, and $1/4$ female normal. When the female callipyge offspring of the original mutant ram were crossed with normal males, all of the offspring were normal. Analysis of the chromosomes of these offspring of callipyge females showed that half of them received a chromosome 18 with the allele encoding callipyge from their mother. Propose an explanation for the inheritance of the allele for callipyge. How might you test your explanation?

Section 5.5

44. Which of the following statements describes an example of a phenocopy? Explain your reasoning.

a. Phenylketonuria results from a recessive mutation that causes light skin as well as intellectual disability.

b. Human height is influenced by genes at many different loci.

c. Dwarf plants and mottled leaves in tomatoes are caused by separate genes that are linked.

d. Vestigial wings in *Drosophila* are produced by a recessive mutation. This trait is also produced by high temperature during development.

e. Intelligence in humans is influenced by both genetic and environmental factors.

*45. Long ears in some dogs are an autosomal dominant trait. Two dogs mate and produce a litter in which 75% of the puppies have long ears. Of the dogs with long ears in this litter, $1/3$ are known to be phenocopies. What are the most likely genotypes of the two parents of this litter?

46. The fly with vestigial wings shown in the lower-left corner of **Figure 5.19** is crossed with the fly with normal wings shown in the upper-right corner of the figure. If the progeny are reared at 31°C, what percentage will have vestigial wings?

CHALLENGE QUESTIONS

Section 5.1

47. Pigeons have long been the subject of genetic studies. Indeed, Charles Darwin bred pigeons in the hope of unraveling the principles of heredity but was unsuccessful. A series of genetic investigations in the early 1900s worked out the hereditary basis of color variation in these birds. W. R. Horlancher was interested in the genetic basis of kiteness, a color pattern that consists of a mixture of red and black stippling of the feathers. He carried out the following crosses to investigate the genetic relation of kiteness to black and red feather color (W. R. Horlancher. 1930. *Genetics* 15:312–346).

Cross	Offspring
kitey × kitey	16 kitey, 5 black, 3 red
kitey × black	6 kitey, 7 black
red × kitey	18 red, 9 kitey, 6 black

a. On the basis of these results, propose a hypothesis to explain the inheritance of kitey, black, and red feather color in pigeons. (Hint: Assume that two loci are involved and some type of epistasis occurs.)

b. For each of the preceding crosses, test your hypothesis by using a chi-square test.

Section 5.3

48. Suppose that you are tending a mouse colony at a genetic research institute, and one day you discover a mouse with twisted ears. You breed this mouse with twisted ears and find that the trait is inherited. Both male and female mice may have twisted ears, but when you cross a twisted-eared male with a normal-eared female, you obtain results that differ from those obtained when you cross a twisted-eared female with a normal-eared male: the reciprocal crosses give different results. Describe how you would determine whether this trait results from a sex-linked gene, a sex-influenced gene, genetic maternal effect, a cytoplasmically inherited gene, or genomic imprinting. What crosses would you conduct, and what results would be expected with these different types of inheritance?

THINK-PAIR-SHARE QUESTIONS

Section 5.1

1. Over a thousand different alleles at the *CFTR* locus have been discovered that can cause cystic fibrosis. What difficulties might the presence of so many different alleles at this locus create for the diagnosis and treatment of cystic fibrosis?

Section 5.2

2. Could you carry out a complementation test on two dominant mutations? Why or why not?

Section 5.3

3. In goats, a beard is produced by an autosomal allele that is dominant in males and recessive in females. Is it possible to cross two bearded goats and obtain a beardless male offspring? Why or why not? What about a bearded female offspring?

4. In chickens, cock feathering is an autosomal recessive trait that is limited to males. Is it possible to cross a cock-feathered male with a hen-feathered female and get male progeny that are cock-feathered? Explain your reasoning.

5. Suppose you observed a new mutant phenotype, notched ears, that appears only in male mice. How might you go about determining whether notched ears is a Y-linked trait or a sex-limited trait? What crosses would you carry out to distinguish between these two modes of inheritance?

6. Phenylketonuria (PKU) is an autosomal recessive disease that results from a defect in an enzyme that normally metabolizes the amino acid phenylalanine; when this enzyme is defective, high levels of phenylalanine cause brain damage. In the past, most children with PKU became intellectually disabled. Fortunately, intellectual disability can be prevented in these children by carefully controlling the amount of phenylalanine in the diet. The diet is usually applied during childhood, when brain development is taking place. As a result of this treatment, many people with PKU now reach reproductive age. Children born to women with PKU (who are no longer on a phenylalanine-restricted diet) frequently have low birth weight, developmental abnormalities, and intellectual disabilities. However, children of men with PKU do not have these problems.

a. Provide an explanation for these observations.

b. What type of genetic effect is this? Explain your reasoning.

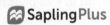 **SaplingPlus** Self-study tools that will help you practice what you've learned and reinforce this chapter's concepts are available online. Go to www.macmillanlearning.com/PierceGenetics6e.

Pedigree Analysis, Applications, and Genetic Testing

Fingerprints are unique to each person. A few people have a condition known as adermatoglyphia (ADG), in which fingerprints are completely absent; this condition is inherited as an autosomal dominant trait. Shown here is a human fingerprint superimposed on DNA sequence information. [Phanie/Science Source.]

The Mystery of the Missing Fingerprints

In 2007, a 29-year-old Swiss woman attempted to enter the United States. Although her appearance matched the photograph on her passport, she failed a fingerprint check, not because her fingerprints matched those of a terrorist, but rather because she had no fingerprints at all—her fingers were completely devoid of any prints. She was delayed for several hours as puzzled immigration officials tried to decide what to do about a person with no fingerprints.

Fingerprints are among our most unique and permanent traits. No two individuals, not even identical twins, share the same fingerprints. Fingerprints are technically termed *epidermal ridges* or *dermatoglyphic patterns*; these patterns are found on our fingers, toes, palms, and the soles of our feet. Epidermal ridges appear long before birth—they are fully formed by the 17th week of pregnancy—and are permanent for life. Research shows that fingerprint patterns are clearly influenced by heredity, but random factors also play a role. One of the first scientists to study fingerprints was Francis Galton, a cousin of Charles Darwin. In the late 1800s, Galton established that no two individuals have the same fingerprints and showed that fingerprints of relatives are more similar than those of unrelated people.

The complete absence of fingerprints, as exhibited by the Swiss woman at the airport, is an extremely rare condition known as adermatoglyphia (ADG). Dubbed "immigration delay disease" because of the hassle it creates when people with the condition attempt to cross borders, ADG has been documented in only a few people from four families around the world. In ADG, fingerprints are absent at birth and never develop. Otherwise, the disorder produces no harmful effects.

In 2011, geneticists in Israel and Switzerland solved the mystery of missing fingerprints in people with ADG. Janna Nousbeck and her colleagues examined the condition in a large Swiss family in which some members had normal fingerprints and other members were missing them entirely (**Figure 6.1**). In this family, ADG exhibits

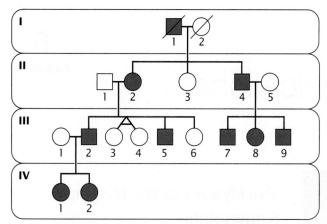

6.1 Pedigree of Swiss family with adermatoglyphia (absence of fingerprints). Squares represent males; circles represent females. Colored squares and circles are people with adermatoglyphia.

the hallmark characteristics of an autosomal dominant trait. The researchers took blood samples from family members who lacked fingerprints and from those with normal fingerprints. They extracted DNA from the blood and genotyped the family members for 6000 single-nucleotide polymorphisms (SNPs), which are DNA sequences that vary in a single nucleotide. By comparing the SNPs of family members with and without fingerprints, they were able to determine that the gene for ADG was located in a specific interval on the long arm of chromosome 4. One of the genes in this region is *SMARCAD1*, which encodes a short form of a protein found exclusively in the skin. Sequencing of the gene revealed that family members with ADG possessed a mutation not found in those with fingerprints. The mutation causes abnormal splicing in RNA transcribed from the gene, with the result that the RNA is less stable. How the decreased stability of this RNA leads to ADG is not known, but scientists hope that the identification of this gene will lead to a better understanding of how fingerprints develop.

THINK-PAIR-SHARE
- If identical twins share 100% of the same genes, why do they have different fingerprints?

- What characteristics of the pedigree in Figure 6.1 suggest that adermatoglyphia in this family is inherited as an autosomal dominant trait?

The absence of fingerprints is just one of a large number of human traits and diseases that are currently the focus of intensive genetic research. In this chapter, we consider human genetic characteristics and examine three important techniques used by geneticists to investigate these characteristics: pedigrees, twin studies, and adoption studies. At the end of the chapter, we see how the information garnered with these techniques can be used in genetic counseling and prenatal diagnosis.

Keep in mind as you read this chapter that many important characteristics are influenced by both genes and environment, and that separating these factors in humans is always difficult. Studies of twins and adopted people are designed to distinguish the effects of genes and environment, but such studies are based on assumptions that may be difficult to meet for some human characteristics, particularly behavioral ones. Therefore, it's always prudent to interpret the results of such studies with caution.

6.1 The Study of Genetics in Humans Is Constrained by Special Features of Human Biology and Culture

Humans are both the best and the worst of all organisms for genetic study. On the one hand, we know more about human anatomy, physiology, and biochemistry than about that of

most other organisms, so many well-characterized traits are available for study. Families often keep detailed records about their members extending back many generations. Furthermore, a number of important human diseases have a genetic component, so the incentive for understanding human inheritance is tremendous. On the other hand, the study of human genetic characteristics presents some major challenges.

First, controlled matings are not possible. With other organisms, geneticists carry out specific crosses to test their hypotheses about inheritance. We have seen, for example, that the testcross provides a convenient way to determine whether an individual with a dominant trait is homozygous or heterozygous. Unfortunately (for the geneticist at least), matings between humans are usually determined by romance, family expectations, or—occasionally—accident rather than by the requirements of science.

Another challenge is that humans have a long generation time. Human reproductive age is not normally reached until 10 to 14 years after birth, and most people do not reproduce until they are 18 years of age or older; thus, generation time in humans is usually about 20 years. This long generation time means that even if geneticists could control human crosses, they would have to wait an average of 40 years just to observe the F_2 progeny. In contrast, generation time in *Drosophila* is 2 weeks; in bacteria, it's a mere 20 minutes.

Finally, human family size is generally small. Observation of even the simple genetic ratios that we learned in Chapter 3 would require a substantial number of progeny in each family. When parents produce only 2 children, the detection of a 3 : 1 ratio is impossible. Even an extremely large family of 10 to 15 children would not permit the recognition of a dihybrid 9 : 3 : 3 : 1 ratio.

Although these special constraints make genetic studies of humans more complex, understanding human heredity is tremendously important. Therefore, geneticists have been forced to develop techniques that are uniquely suited to human biology and culture. **▶ TRY PROBLEM 18**

CONCEPTS

The study of human inheritance is constrained by the inability to control genetic crosses, a long generation time, and a small number of offspring.

6.2 Geneticists Often Use Pedigrees To Study the Inheritance of Characteristics in Humans

An important technique used by geneticists to study human inheritance is the analysis of pedigrees. A **pedigree** is a pictorial representation of a family history, essentially a family tree that outlines the inheritance of one or more characteristics. When a particular characteristic or disease is observed in a person, a geneticist often studies the family of that affected person by drawing a pedigree.

Symbols Used in Pedigrees

The symbols commonly used in pedigrees are summarized in **Figure 6.2**. Males in a pedigree are represented by squares, females by circles. A horizontal line drawn between two symbols representing a man and a woman indicates a mating; children are connected to their parents by vertical lines extending downward from the parents. The pedigree shown in **Figure 6.3a** illustrates a family with Waardenburg syndrome, an autosomal dominant type of deafness that may be accompanied by fair skin, a white forelock, and visual problems (**Figure 6.3b**). People who exhibit the trait of interest are represented by filled circles and squares; in the pedigree of Figure 6.3a, the filled symbols represent members of the family who have Waardenburg syndrome. Unaffected members are represented by open circles and squares. The person from whom the pedigree is initiated is called the **proband** and is usually designated by the letter P and an arrow (IV-2 in Figure 6.3a).

Let's look closely at Figure 6.3 and consider some additional features of a pedigree. Each generation in a pedigree is identified by a roman numeral; within each generation, family members are assigned arabic numerals, and children in each family are listed in birth order from left to right. Person

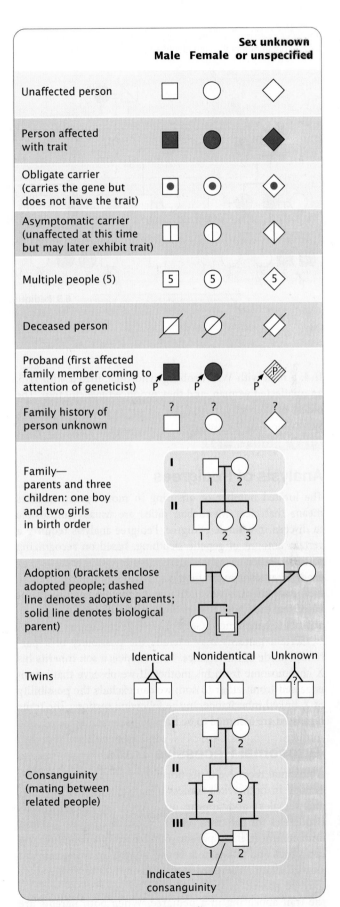

6.2 Standard symbols are used in pedigrees.

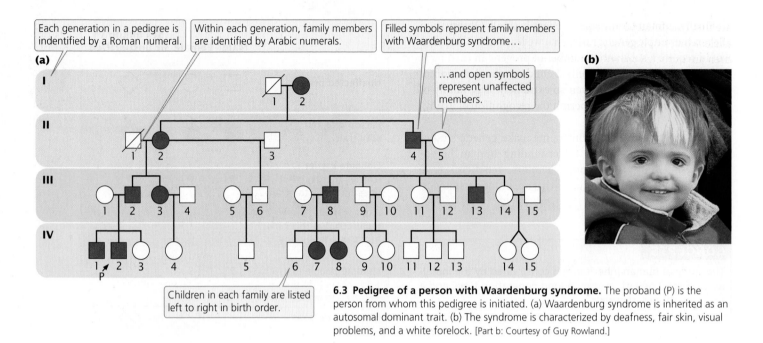

(a)

Each generation in a pedigree is indentified by a Roman numeral.

Within each generation, family members are identified by Arabic numerals.

Filled symbols represent family members with Waardenburg syndrome...

...and open symbols represent unaffected members.

Children in each family are listed left to right in birth order.

6.3 Pedigree of a person with Waardenburg syndrome. The proband (P) is the person from whom this pedigree is initiated. (a) Waardenburg syndrome is inherited as an autosomal dominant trait. (b) The syndrome is characterized by deafness, fair skin, visual problems, and a white forelock. [Part b: Courtesy of Guy Rowland.]

(b)

II-4, a man with Waardenburg syndrome, mated with II-5, an unaffected woman, and they produced five children. The oldest of their children is III-8, a male with Waardenburg syndrome, and the youngest is III-14, an unaffected female.

> **TRY PROBLEM 19a**

Analysis of Pedigrees

The limited number of offspring in most human families means that clear Mendelian ratios are usually impossible to discern in a single pedigree. Pedigree analysis requires a certain amount of genetic sleuthing, based on recognizing patterns associated with different modes of inheritance. For example, autosomal dominant traits should appear with equal frequency in both sexes and should not skip generations, provided that the trait is fully penetrant (see p. 113 in Chapter 5) and not sex-influenced (see pp. 126–128 in Chapter 5).

Certain patterns may exclude the possibility of a particular mode of inheritance. For instance, a son inherits his X chromosome from his mother. If we observe that a trait is passed from father to son, we can exclude the possibility of X-linked inheritance. In the following sections, the traits discussed are assumed to be fully penetrant and rare.

Autosomal Recessive Traits

Autosomal recessive traits normally appear with equal frequency in both sexes (unless penetrance differs in males and females) and appear only when a person inherits two alleles for the trait, one from each parent. If the trait is uncommon, most parents of affected offspring are heterozygous and unaffected; consequently, the trait seems to skip generations (**Figure 6.4**). Frequently, a recessive allele may be passed on for a number of generations without the trait appearing in a pedigree. When both parents are

heterozygous, approximately one-fourth of the offspring are expected to express the trait, but this ratio will not be obvious unless the family is large. In the rare event that both parents are affected by an autosomal recessive trait, all the offspring will be affected.

When a recessive trait is rare, most people outside affected families are homozygous for the normal allele. Thus, when an affected person mates with someone outside the family ($aa \times AA$), usually none of the children display the trait, although all will be carriers (i.e., heterozygous). A recessive trait is more likely to appear in a pedigree when two people within the same family mate because there is a greater chance of both parents carrying the same recessive allele. Mating between closely related people is called **consanguinity**. In the pedigree shown in Figure 6.4, individuals III-3 and III-4

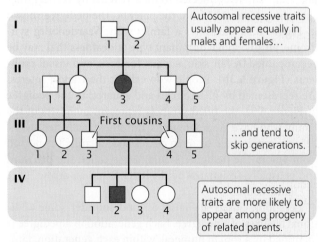

Autosomal recessive traits usually appear equally in males and females...

First cousins

...and tend to skip generations.

Autosomal recessive traits are more likely to appear among progeny of related parents.

6.4 Autosomal recessive traits normally appear with equal frequency in both sexes and often skip generations.

are first cousins, and both are heterozygous for the recessive allele; when two heterozygotes mate, one-fourth of their children are expected to have the recessive trait.

CONCEPTS

Autosomal recessive traits appear with equal frequency in males and females. Affected children are commonly born to unaffected parents who are carriers of the gene for the trait, and the trait tends to skip generations. Recessive traits appear more frequently among the offspring of consanguineous matings.

✔ CONCEPT CHECK 1

Autosomal recessive traits often appear in pedigrees in which there have been consanguineous matings because these traits

a. tend to skip generations.

b. appear only when both parents carry a copy of the gene for the trait, which is more likely when the parents are related.

c. usually arise in children born to parents who are unaffected.

d. appear equally in males and females.

A number of human metabolic diseases are inherited as autosomal recessive traits. One of them is Tay–Sachs disease. Children with Tay–Sachs disease appear healthy at birth but become listless and weak at about 6 months of age. Gradually, their physical and neurological conditions worsen, leading to blindness, deafness, and, eventually, death at 2 to 3 years of age. The disease results from the accumulation of a lipid called G_{M2} ganglioside in the brain. A normal component of brain cells, G_{M2} ganglioside is usually broken down by an enzyme called hexosaminidase A, but children with Tay–Sachs disease lack this enzyme. Excessive G_{M2} ganglioside accumulates in the brain, causing swelling and, ultimately, neurological symptoms. Heterozygotes have only one normal copy of the allele encoding hexosaminidase A and produce only about half the normal amount of the enzyme. However, this amount is enough to ensure that G_{M2} ganglioside is broken down normally, and heterozygotes are usually healthy.

Autosomal Dominant Traits

Autosomal dominant traits appear in both sexes with equal frequency, and both sexes are capable of transmitting these traits to their offspring. Every person with a dominant trait must have inherited the allele from at least one parent; autosomal dominant traits therefore do not skip generations (**Figure 6.5**). Exceptions to this rule arise when people acquire the trait as a result of a new mutation or when the trait has incomplete penetrance.

If an autosomal dominant allele is rare, most people displaying the trait are heterozygous. When one parent is affected and heterozygous and the other parent is unaffected, approximately half of the offspring will be affected. If both

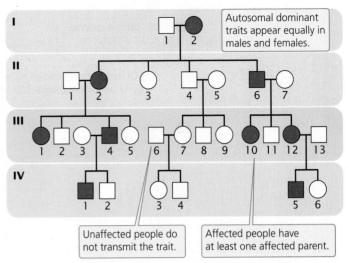

6.5 **Autosomal dominant traits normally appear with equal frequency in both sexes and do not skip generations.**

parents have the trait and are heterozygous, approximately three-fourths of the children will be affected. Unaffected people do not transmit the trait to their descendants, provided that the trait is fully penetrant. In Figure 6.5, we see that none of the descendants of II-4 (who is unaffected) have the trait.

THINK-PAIR-SHARE Question 1

CONCEPTS

Autosomal dominant traits appear in both sexes with equal frequency. An affected person has an affected parent (unless the person carries new mutations), and the trait does not skip generations. Unaffected people do not transmit the trait.

✔ CONCEPT CHECK 2

When might you see an autosomal dominant trait skip generations?

One condition that is usually considered to be an autosomal dominant trait is familial hypercholesterolemia, an inherited disease in which blood cholesterol is greatly elevated owing to a defect in cholesterol transport. Cholesterol is normally transported throughout the body in small soluble particles called lipoproteins (**Figure 6.6**). A principal lipoprotein in the transport of cholesterol is low-density lipoprotein (LDL). When an LDL molecule reaches a cell, it attaches to an LDL receptor, which then moves the LDL through the cell membrane into the cytoplasm, where it is broken down and its cholesterol is released for use by the cell.

Familial hypercholesterolemia occurs when there is a defect in the gene that normally encodes the LDL receptor. The disease is usually considered an autosomal dominant disorder because heterozygotes are deficient in LDL receptors

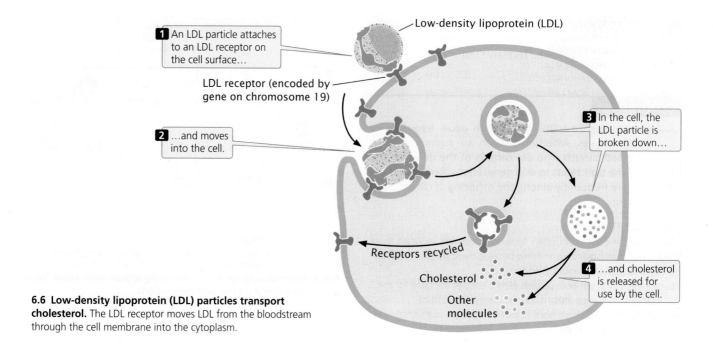

6.6 **Low-density lipoprotein (LDL) particles transport cholesterol.** The LDL receptor moves LDL from the bloodstream through the cell membrane into the cytoplasm.

and have elevated blood levels of cholesterol, which lead to increased risk of coronary artery disease. People who are heterozygous for familial hypercholesterolemia have blood LDL levels that are twice normal levels and usually have heart attacks by the age of 35.

Very rarely, a person inherits two defective alleles for the LDL receptor. Such individuals don't make any functional LDL receptors; their blood cholesterol levels are more than six times normal levels, and they may suffer a heart attack as early as age 2 and almost inevitably by age 20. Because homozygotes are more severely affected than heterozygotes, familial hypercholesterolemia is said to be incompletely dominant. However, homozygotes are rarely seen, and the common heterozygous form of the disease appears as a simple dominant trait in most pedigrees.

X-Linked Recessive Traits

X-linked recessive traits have a distinctive pattern of inheritance (**Figure 6.7**). First, these traits appear more frequently in males than in females because males need inherit only a single copy of the allele to display the trait, whereas females must inherit two copies of the allele, one from each parent, to be affected. Second, because a male inherits his X chromosome from his mother, affected males are usually born to unaffected mothers who carry an allele for the trait. Because the trait is passed from unaffected female to affected male to unaffected female, it tends to skip generations (see Figure 6.7). When a woman is heterozygous, approximately half of her sons will be affected and half of her daughters will be unaffected carriers. For example, we know that females I-2, II-2, and III-7 in Figure 6.7 are carriers because they transmit the trait to approximately half of their sons. Individuals like these

females, whose heterozygous genotype can be definitively determined from the pedigree, are called *obligate carriers*.

A third important characteristic of X-linked recessive traits is that they are not passed from father to son because a son inherits his father's Y chromosome, not his X. In Figure 6.7, there is no case in which both a father and his son are affected. All daughters of an affected man, however, will be carriers (if their mother is homozygous for the normal allele). When a woman displays an X-linked recessive trait, she must be homozygous for the trait, and all of her sons will also display the trait.

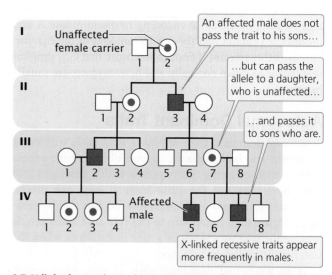

6.7 **X-linked recessive traits appear more often in males than in females and are not passed from father to son.** Obligate carriers are indicated in this pedigree to help illustrate the X-linked recessive mode of inheritance.

An example of an X-linked recessive trait in humans is hemophilia A, also called classic hemophilia. Hemophilia results from the absence of a protein necessary for blood to clot. The complex process of blood clotting consists of a cascade of reactions that includes more than 13 different clotting factors. For this reason, there are several types of clotting disorders, each due to a glitch in a different step of the clotting pathway.

Hemophilia A results from abnormal or missing factor VIII, one of the proteins in the clotting cascade. The gene for factor VIII is located on the tip of the long arm of the X chromosome, so hemophilia A is an X-linked recessive disorder. People with hemophilia A bleed excessively; even small cuts and bruises can be life threatening. Spontaneous bleeding occurs in joints such as elbows, knees, and ankles, producing pain, swelling, and erosion of the bone. Fortunately, bleeding in people with hemophilia A can be controlled by administering concentrated doses of factor VIII. The inheritance of hemophilia A is illustrated by the family of Queen Victoria of England (**Figure 6.8**).

X-Linked Dominant Traits

X-linked dominant traits appear in both males and females, although they often appear more frequently in females than in males. Each person with an X-linked dominant trait must have an affected parent (unless the person possesses a new mutation or the trait has incomplete penetrance). X-linked dominant

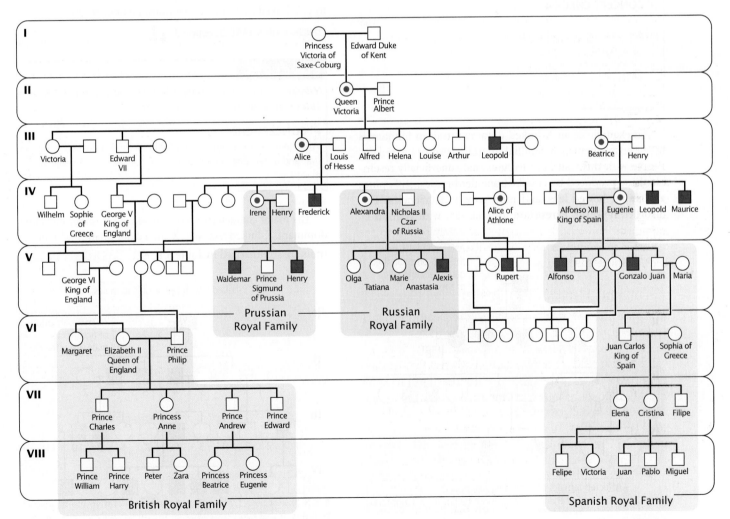

6.8 Classic hemophilia in the royal families of Europe. This blood-clotting disorder is inherited as an X-linked recessive trait.

traits do not skip generations (**Figure 6.9**); affected men pass the trait to all their daughters and none of their sons, as is seen in the children of I-1 in Figure 6.9. In contrast, affected women (if heterozygous) pass the trait to about half of their sons and about half of their daughters, as seen in the children of III-6 in the pedigree. As with X-linked recessive traits, a male inherits an X-linked dominant trait only from his mother; the trait is not passed from father to son. This fact is what distinguishes X-linked dominant inheritance from autosomal dominant inheritance, in which a male can inherit the trait from his father. A female, on the other hand, inherits an X chromosome from both her mother and her father, so females can receive an X-linked dominant trait from either parent.

THINK-PAIR-SHARE Question 2

CONCEPTS

X-linked dominant traits affect both males and females. Affected males must have affected mothers (unless the males possess a new mutation), and they pass the trait to all their daughters.

✔ CONCEPT CHECK 4

A male affected with an X-linked dominant trait will have what proportion of offspring affected with the trait?

a. $\frac{1}{2}$ sons and $\frac{1}{2}$ daughters
b. All sons and no daughters
c. All daughters and no sons
d. $\frac{3}{4}$ daughters and $\frac{1}{4}$ sons

An example of an X-linked dominant trait in humans is hypophosphatemia, or familial vitamin-D-resistant rickets. People with this trait have features that superficially resemble those produced by rickets: bone deformities, stiff spines and joints, bowed legs, and mild growth deficiencies. This disorder, however, is resistant to treatment with vitamin D,

which normally cures rickets. X-linked hypophosphatemia results from the defective transport of phosphate, especially in cells of the kidneys. People with this disorder excrete large amounts of phosphate in their urine, resulting in low levels of phosphate in the blood and reduced deposition of minerals in the bone. The disorder is treated with high doses of calcitriol (a hormonally active form of vitamin D) and phosphate. As is common with X-linked dominant traits, males with hypophosphatemia are often more severely affected than females.

Y-Linked Traits

Y-linked traits exhibit a specific, easily recognized pattern of inheritance. Only males are affected, and the trait is passed from father to son. If a man is affected, all his male offspring should also be affected, as is the case for I-1, II-4, II-6, III-6, and III-10 of the pedigree in **Figure 6.10**. Y-linked traits do not skip generations. As mentioned in Chapter 4, little genetic information is found on the human Y chromosome. Maleness (encoded by the *SRY* gene) is one of the few traits in humans that has been shown to be Y linked. Because each male has a single Y chromosome, there is only one copy of each Y-linked allele: therefore, Y-linked traits are neither dominant nor recessive.

THINK-PAIR-SHARE Question 3

CONCEPTS

Y-linked traits appear only in males and are passed from a father to all his sons.

✔ CONCEPT CHECK 5

What features of a pedigree would distinguish between a Y-linked trait and a trait that is rare, autosomal dominant, and sex-limited to males?

The major characteristics of autosomal recessive, autosomal dominant, X-linked recessive, X-linked dominant, and Y-linked traits are summarized in **Table 6.1**. **▶ TRY PROBLEM 22**

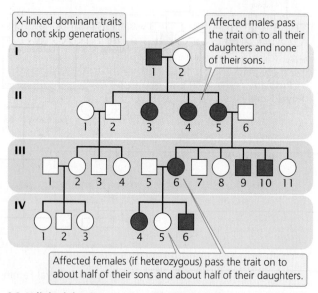

6.9 X-linked dominant traits affect both males and females. An affected male must have an affected mother.

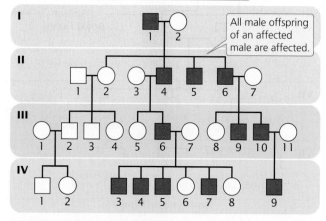

6.10 Y-linked traits appear only in males and are passed from a father to all his sons.

TABLE 6.1	Pedigree characteristics of autosomal recessive, autosomal dominant, X-linked recessive, X-linked dominant, and Y-linked traits

Autosomal Recessive Trait

1. Usually appears in both sexes with equal frequency.

2. Tends to skip generations.

3. Affected offspring are usually born to unaffected parents.

4. When both parents are heterozygous, approximately one-fourth of the offspring will be affected.

5. Appears more frequently among the children of consanguineous marriages.

Autosomal Dominant Trait

1. Usually appears in both sexes with equal frequency.

2. Both sexes transmit the trait to their offspring.

3. Does not skip generations.

4. Affected offspring must have an affected parent unless they possess a new mutation.

5. When one parent is affected (heterozygous) and the other parent is unaffected, approximately half of the offspring will be affected.

6. Unaffected parents do not transmit the trait.

X-Linked Recessive Trait

1. Usually more males than females are affected.

2. Affected sons are usually born to unaffected mothers; thus, the trait skips generations.

3. Approximately half of a carrier (heterozygous) mother's sons are affected.

4. Never passed from father to son.

5. All daughters of affected fathers are carriers.

X-Linked Dominant Trait

1. Both males and females are usually affected; often, more females than males are affected.

2. Does not skip generations. Affected sons must have an affected mother; affected daughters must have either an affected mother or an affected father.

3. Affected fathers pass the trait to all their daughters.

4. Affected mothers (if heterozygous) pass the trait to half of their sons and half of their daughters.

Y-Linked Trait

1. Only males are affected.

2. Passed from father to all sons.

3. Does not skip generations.

WORKED PROBLEM

The following pedigree represents the inheritance of a rare disorder in an extended family. What is the most likely mode of inheritance for this disease? (Assume that the trait is fully penetrant.)

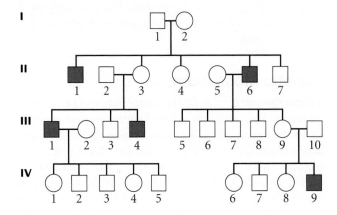

Solution Strategy

What information is required in your answer to the problem?

The most likely mode of inheritance for the trait shown in the pedigree.

What information is provided to solve the problem?

- The pedigree, which includes information about the sex and family relationships of affected individuals.

- The trait is rare.

Solution Steps

To answer this question, we should consider each mode of inheritance and determine which, if any, we can eliminate. The trait appears only in males, so autosomal dominant and autosomal recessive modes of inheritance are unlikely because traits with these modes appear equally in males and females. Additionally, autosomal dominance can be eliminated because some affected individuals do not have an affected parent.

The trait is observed only among males in this pedigree, which might suggest Y-linked inheritance. However, affected men should pass a Y-linked trait to all their sons, which is not the case here; II-6 is an affected man who has four unaffected male offspring. Therefore, we can eliminate Y-linked inheritance.

X-linked dominance can be eliminated because affected men should pass an X-linked dominant trait on to all their female offspring, and II-6 has an unaffected daughter (III-9).

X-linked recessive traits often appear more commonly in males, and affected males are usually born to unaffected female carriers; the pedigree shows this pattern of inheritance. If the disorder is an X-linked trait, about half the sons of a heterozygous carrier mother should be affected. II-3 and III-9 are suspected carriers, and about half of their male children (three of five) are affected. Another important characteristic of an X-linked recessive trait is that it is not passed from father to son. We observe no father-to-son transmission in this pedigree. X-linked recessive is therefore the most likely mode of inheritance.

>> For additional practice, try to determine the mode of inheritance for the pedigrees in **Problem 24** at the end of the chapter.

Genetic Mosaicism

A factor that can complicate the interpretation of pedigrees is the presence of mosaicism. A long-held assumption in genetics was that every cell in the body contains the same genetic information. Yet from time to time, geneticists discovered exceptions to this rule. In 1976, researchers published the curious case of a woman who had two different blood types. Most of her red blood cells were type O, but a few cells were type A. In 1998, a 52-year-old woman needed a kidney transplant. When her three children were genetically tested to see if any of them might be a potential donor, two of the children had genotypes that could not have come from their mother, even though she had given birth to them. Further testing revealed that the mother actually had two different types of cells in her body: the genotype of her blood cells was different from that of her skin and some organs. These individuals, in which different cells of the body have different genetic constitutions, are referred to as **genetic mosaics** or chimeras.

Genetic mosaicism can arise from a number of different processes. Often, somatic mutations or errors in the separation of chromosomes during early development lead to genetically distinct groups of cells within the body. There can also be exchanges of cells and DNA between twins or between mother and child during pregnancy. Mosaicism was once thought to be rare, but the increasing use of DNA sequencing has revealed that it may be common. For example, in 2012, researchers sequenced skin cells and discovered that 30% of the cells contained genetic variants called copy-number variations (CNVs, see Chapter 8). Another study examined CNVs in diverse tissues from six people and also found major differences in CNVs, even within a single person.

6.3 Studying Twins and Adoptions Can Help Us Assess the Importance of Genes and Environment

Twins and adoptions provide natural experiments for separating the effects of genes and environmental factors on differences in traits. These two techniques have been widely used in genetic studies.

Types of Twins

Twins are of two types. **Dizygotic** (nonidentical) **twins** arise when two separate eggs are fertilized by two different sperm, producing genetically distinct zygotes. **Monozygotic** (identical) **twins** result when a single egg, fertilized by a single sperm, splits early in development into two separate embryos.

Because monozygotic twins result from a single egg and sperm (a single, "mono," zygote), they're genetically identical (except for epigenetic differences—see Chapter 21—or rare somatic mutations); thus, identical twins share 100% of their genes (**Figure 6.11a**). Dizygotic twins (**Figure 6.11b**), on the other hand, share, on average, only 50% of their genes, which is the same percentage that any pair of siblings shares. Like other siblings, dizygotic twins may be of the same sex or of different sexes. The only difference between dizygotic twins and other siblings is that dizygotic twins are the same age and shared the same uterine environment. Dizygotic twinning often runs in families, and the tendency to produce dizygotic twins is influenced by both heredity and environmental factors. There appears to be little genetic tendency for producing monozygotic twins. The frequency of dizygotic twinning in the United States has doubled since 1980, largely because of fertility treatments, which often stimulate women to ovulate multiple eggs.

THINK-PAIR-SHARE Question 4

(a)

(b)

6.11 Two types of twins. Monozygotic twins (a) are identical; dizygotic twins (b) are nonidentical. [Part a: f4foto/Alamy. Part b: Courtesy of Randi Rossignol.]

TABLE 6.2	Concordance of monozygotic and dizygotic twins for several traits	
	Concordance (%)	
Trait	**Monozygotic**	**Dizygotic**
1. Heart attack (males)	39	26
2. Heart attack (females)	44	14
3. Bronchial asthma	47	24
4. Cancer (all sites)	12	15
5. Epilepsy	59	19
6. Death from acute infection	7.9	8.8
7. Rheumatoid arthritis	32	6
8. Multiple sclerosis	28	5

Sources: (1 and 2) B. Havald and M. Hauge, U.S. Public Health Service Publication 1103 (1963), pp. 61–67; (3, 4, 5, and 6) B. Havald and M. Hauge, *Genetics and the Epidemiology of Chronic Diseases* (U.S. Department of Health, Education, and Welfare, 1965); (7) J. S. Lawrence, *Annals of Rheumatic Diseases* 26:357–379, 1970; (8) G. C. Ebers et al., *American Journal of Human Genetics* 36:495, 1984.

Concordance in Twins

Comparisons of dizygotic and monozygotic twin pairs can be used to assess the importance of genetic and environmental factors in producing differences in a characteristic. Such assessments are often made by calculating the concordance for a trait. If both members of a twin pair have a trait, the twins are said to be *concordant*; if only one member of the pair has the trait, the twins are said to be *discordant*. **Concordance** is the percentage of twin pairs that are concordant for a trait. Because identical twins share 100% of their genes and dizygotic twins share, on average, only 50%, genetically influenced traits should exhibit higher concordance in monozygotic twins. For instance, when one member of a monozygotic twin pair has epilepsy (**Table 6.2**), the other twin of the pair has epilepsy about 59% of the time; thus, the monozygotic concordance for epilepsy is 59%. When a dizygotic twin has epilepsy, however, the other twin has epilepsy only 19% of the time (19% dizygotic concordance). The higher concordance in monozygotic twins suggests that genes influence epilepsy, a finding supported by the results of other family studies of this disease. In contrast, the concordance of death from acute infection is similar in monozygotic and dizygotic twin pairs, suggesting that death from infection has little inherited tendency. Concordances for several additional human traits and diseases are listed in Table 6.2.

Although the hallmark of a genetic influence on a particular trait is higher concordance in monozygotic twins than in dizygotic twins, high concordance in monozygotic twins by itself does not signal a genetic influence. Monozygotic twins usually share the same environment—they are raised in the same home, have the same friends, attend the same school—so high concordance may be due to their common genes or to their common environment. If the high concordance is due to environmental factors, then dizygotic twins, who also share the same environment, should have a concordance just as high as that of monozygotic twins. When genes influence the trait, however, monozygotic twin pairs should exhibit higher concordance than dizygotic twin pairs because monozygotic twins share a greater percentage of their genes. It is important to note that any discordance between monozygotic twins is usually due to environmental factors because monozygotic twins are genetically identical. The concordance of monozygotic twins for epilepsy, for example, is considerably less than 100% (see Table 6.2), suggesting that in addition to genetic influences, environmental factors affect variation in this trait.

The use of twins in genetic research rests on the important assumption that when concordance for monozygotic twins is greater than that for dizygotic twins, it is because monozygotic twins are more similar in their genes and not because they have experienced a more similar environment. The degree of environmental similarity between monozygotic twins and dizygotic twins is assumed to be the same. This assumption may not always be correct, particularly for human behaviors. Because they look alike, identical twins may be treated more similarly by parents, teachers, and peers than are nonidentical twins. Evidence of this similar treatment is seen in the past tendency of parents to dress identical twins alike. In spite of this potential complication, twin studies have played a pivotal role in the study of human genetics.

⟩ TRY PROBLEM 30

THINK-PAIR-SHARE Question 5

A Twin Study of Asthma

To illustrate the use of twins in genetic research, let's consider a study of asthma. Asthma is characterized by constriction of the airways and the secretion of mucus into the air passages, causing coughing, labored breathing, and wheezing (**Figure 6.12**). Severe cases can be life threatening. Asthma is a major health problem in industrialized countries and appears to be on the rise. The incidence of childhood asthma varies widely around the globe; some of the highest rates (from 21% to 27%) are found in Australia, the United Kingdom, Sweden, and Brazil.

A number of environmental stimuli are known to precipitate asthma attacks, including dust, pollen, air pollution, respiratory infections, exercise, cold air, and emotional stress. Allergies frequently accompany asthma, suggesting that asthma is a disorder of the immune system, but the precise relation between immune function and asthma is poorly understood. Numerous studies have shown that genetic factors are important in asthma.

A genetic study of childhood asthma was conducted as a part of the Twins Early Development Study in England, an ongoing research project that studies more than 15,000 twin pairs born in the United Kingdom between 1994 and 1996. These twins were assessed for language, cognitive development, behavioral problems, and academic achievement at ages 7 and 9, and the genetic and environmental contributions to a number of their traits were examined. In the asthma study, researchers looked at a sample of 4910 twin pairs at age 4. Parents of the twins were asked whether either of their twins had been prescribed medication to control asthma; those children receiving asthma medication were considered

6.12 Twin studies show that asthma, characterized by constriction of the airways, is caused by a combination of genetic and environmental factors. Inhalers are often used to deliver asthma medication to the lungs. [Stockbyte/Getty Images.]

to have asthma. The concordance for the monozygotic twins (65% among 1658 twin pairs) was significantly higher than that for the dizygotic twins (37% among 3252 twin pairs), and the researchers concluded that, among the 4-year-olds included in the study, asthma was strongly influenced by genetic factors. The fact that even monozygotic twins were discordant 35% of the time indicates that environmental factors also play a role in asthma.

> **CONCEPTS**
>
> Higher concordance for monozygotic twins than for dizygotic twins indicates that genetic factors play a role in determining differences in a trait. Less than 100% concordance for monozygotic twins indicates that environmental factors play a significant role.
>
> ✔ **CONCEPT CHECK 7**
>
> A trait exhibits 100% concordance for both monozygotic and dizygotic twins. What conclusion can you draw about the role of genetic factors in determining differences in the trait?
>
> a. Genetic factors are extremely important.
> b. Genetic factors are somewhat important.
> c. Genetic factors are unimportant.
> d. Both genetic and environmental factors are important.

Adoption Studies

Another technique used by geneticists to analyze human inheritance is the study of families involved in adoptions. This approach is one of the most powerful for distinguishing the effects of genes and environment on characteristics.

For a variety of reasons, many children are separated from their biological parents soon after birth and adopted by adults with whom they have no genetic relationship. These adoptees share no more genes with their adoptive parents, on average, than any two randomly chosen individuals share; however, they do share an environment with their adoptive parents. In contrast, the adoptees share 50% of their genes with each of their biological parents, but do not share the same environment with them. If adoptees and their adoptive parents show similarities in a characteristic, these similarities can be attributed to environmental factors. If, on the other hand, adoptees and their biological parents show similarities, these similarities are probably due to genetic factors. Comparisons of adoptees with their adoptive parents and with their biological parents can therefore help to define the roles of genetic and environmental factors in the determination of human variation. For example, adoption studies were instrumental in showing that schizophrenia has a genetic basis. Adoption studies have also shown that obesity is at least partly influenced by genetics (**Figure 6.13**).

Adoption studies assume that the environments of biological and adoptive families are independent (i.e., not more alike than would be expected by chance). This assumption may not always be correct because adoption agencies

Question: Is body mass index (BMI) influenced by genetic factors?

Methods

Compare the body mass indexes of adopted children with those of their adoptive and biological parents.

Results

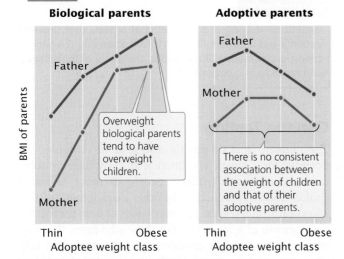

Conclusion: Genetic factors influence body mass index.

6.13 Adoption studies demonstrate that obesity is influenced by genetics. In these studies, obesity was measured by body mass index (BMI). [Redrawn with the permission of the *New England Journal of Medicine* 314:195, 1986.]

carefully choose adoptive parents and may select a family that resembles the adoptee's biological family. Thus, some of the similarity between adoptees and their biological parents may be due to these similar environments and not to shared genetic factors. In addition, adoptees and their biological mothers share the same environment during prenatal development. ❯ **TRY PROBLEM 33**

CONCEPTS

Similarities between adoptees and their genetically unrelated adoptive parents indicate that environmental factors affect a particular characteristic; similarities between adoptees and their biological parents indicate that genetic factors influence the characteristic.

✔ **CONCEPT CHECK 8**

What assumptions underlie the use of adoption studies in genetics?

a. Adoptees have no contact with their biological parents after birth.

b. The adoptive parents and biological parents are not related.

c. The environments of biological and adoptive parents are independent.

d. All of the above.

6.4 Genetic Counseling and Genetic Testing Provide Information to Those Concerned about Genetic Diseases and Traits

Our knowledge of human genetic diseases and disorders has expanded rapidly in recent years. The Online Mendelian Inheritance in Man Web site now lists more than 23,000 human genetic diseases, disorders, genes, and traits that have a simple genetic basis. Research has provided a great deal of information about the inheritance, chromosomal location, biochemical basis, and symptoms of many of these genetic traits, diseases, and disorders. This information is often useful to people who have a genetic condition.

Genetic Counseling

Genetic counseling is a field that provides information to patients with genetic disorders and others who are concerned about hereditary conditions. It is an educational process that helps patients and family members deal with many aspects of having a genetic condition: its diagnosis, symptoms and treatment, and mode of inheritance. Genetic counseling also helps patients and their families cope with the psychological and physical stress that may be associated with the disorder. Clearly, all of these considerations cannot be handled by a single health professional; most genetic counseling is done by a team that may include counselors, physicians, medical geneticists, and laboratory personnel. **Table 6.3** lists some common reasons for seeking genetic counseling.

TABLE 6.3	Common reasons for seeking genetic counseling
1.	A person knows of a genetic disease in the family.
2.	A couple has given birth to a child with a genetic disease, birth defect, or chromosome abnormality.
3.	A couple has a child who is intellectually disabled or has a close relative who is intellectually disabled.
4.	An older woman becomes pregnant or wants to become pregnant. There is disagreement about the age at which a prospective mother who has no other risk factors should seek genetic counseling; many experts suggest that it should be age 35 or older.
5.	Husband and wife are closely related (e.g., first cousins).
6.	A couple experiences difficulties achieving a successful pregnancy.
7.	A pregnant woman is concerned about exposure to an environmental substance (drug, chemical, or virus) that causes birth defects.
8.	A couple needs assistance in interpreting the results of a prenatal or other test.
9.	Both prospective parents are known carriers for a recessive genetic disease or both belong to an ethnic group with a high frequency of a genetic disease.

Genetic counseling usually begins with a diagnosis of a condition that may have a genetic basis. On the bases of a physical examination, biochemical tests, DNA testing, chromosome analysis, family history, and other information, a physician determines the cause of the condition. An accurate diagnosis is critical because appropriate treatment and the probability of passing the condition on depend on the diagnosis. For example, there are a number of different types of dwarfism, caused by chromosome abnormalities, single-gene mutations, hormonal imbalances, and environmental factors. People who have dwarfism resulting from an autosomal dominant gene have a 50% chance of passing the condition on to their children, whereas people who have dwarfism caused by a rare recessive gene have a low likelihood of doing so.

When the nature of the condition is known, a genetic counselor meets with the patient and members of the patient's family and explains the diagnosis. A family pedigree may be constructed, and the probability of transmitting the condition to future generations can be calculated for different family members. The counselor helps the family interpret the genetic risks and explains various available reproductive options, including prenatal diagnosis, artificial insemination, and in vitro fertilization. Family members often have questions about genetic testing that may be available to help determine whether they carry a particular genetic mutation. The counselor helps them decide whether genetic testing is appropriate for them and which tests to apply. After the test results are in, the genetic counselor usually helps family members interpret the results.

A family's decision about future pregnancies frequently depends on the magnitude of the genetic risk, the severity and effects of the condition, the importance they place on having children, and their religious and cultural views. Traditionally, genetic counselors have been trained to apply *nondirected* counseling, which means that they provide information and facilitate discussion, but do not bring their own opinions and values into the discussion. The goal of nondirected counseling is for the family to reach its own decision on the basis of the best available information.

Because of the growing number of genetic tests available and the complexity of assessing genetic risk, there is now some movement away from completely nondirected counseling. The goal is still to provide the family with information about all options and to reach the decision that is best for them, but that goal may sometimes require the counselor to recommend certain options, much as a physician recommends the most appropriate medical treatments for his or her patient.

Who does genetic counseling? In the United States, over 7000 health professionals are currently certified in genetics by the American Board of Medical Genetics or the American Board of Genetic Counseling. About half of them are specifically trained in genetic counseling, usually through a special two-year master's degree program that provides education in both genetics and counseling. Most of the remainder are physicians and scientists certified in medical or clinical genetics. Because of the shortage of genetic counselors and medical geneticists, information about genetic testing and genetic risk is often conveyed by primary care physicians, nurses, and social workers. **▶ TRY PROBLEM 10**

CONCEPTS

Genetic counseling is an educational process that provides patients and their families with information about a genetic condition, its medical implications, the mode of inheritance, and reproductive options.

Genetic Testing

The ultimate goal of genetic testing is to recognize the potential for a genetic condition at an early stage. In some cases, genetic testing allows people to make informed choices about reproduction. In other cases, genetic testing allows early intervention that may lessen or even prevent the development of a condition. For those who know that they are at risk for a genetic condition, genetic testing may help alleviate anxiety associated with the uncertainty of their situation. Genetic testing includes prenatal testing and postnatal testing.

Prenatal genetic tests are those that are conducted before birth and now include procedures for diagnosing several hundred genetic diseases and disorders (**Table 6.4**). The major purpose of prenatal tests is to provide families with the information that they need to make choices during pregnancies and, in some cases, to prepare for the birth of a child with a genetic condition. Several approaches to prenatal diagnosis are described in the following sections.

ULTRASONOGRAPHY Some genetic conditions can be detected through direct visualization of the fetus. This is most commonly done by **ultrasonography**—usually referred to as *ultrasound*. In this technique, high-frequency sound is beamed into the uterus; when the sound waves encounter dense tissue, they bounce back and are transformed into a picture (**Figure 6.14**). In this way, the size and position of the fetus can be determined, and conditions such as neural-tube defects (defects in the development of the spinal column and the skull) and skeletal abnormalities can be detected. Ultrasound is a standard procedure performed during pregnancy to estimate the age of the fetus, determine its sex, and check for the presence of developmental disorders or other problems.

AMNIOCENTESIS Traditional prenatal testing requires fetal tissue, which can be obtained in several ways. The most widely used method is **amniocentesis**, a procedure for obtaining a sample of amniotic fluid from the uterus of

TABLE 6.4	Examples of genetic diseases and disorders that can be detected prenatally and the techniques used in their detection
Disorder	**Method of Detection**
Chromosome abnormalities	Examination of a karyotype from cells obtained by amniocentesis or chorionic villus sampling. Some forms can be detected by DNA analysis of maternal blood.
Cleft lip and palate	Ultrasound
Cystic fibrosis	DNA analysis of cells obtained by amniocentesis or chorionic villus sampling
Dwarfism	Ultrasound or X-ray; some forms can be detected by DNA analysis of cells obtained by amniocentesis or chorionic villus sampling
Hemophilia	Fetal blood sampling* or DNA analysis of cells obtained by amniocentesis or chorionic villus sampling
Lesch–Nyhan syndrome	Biochemical tests on cells obtained by amniocentesis or chorionic villus sampling
Neural-tube defects	Initial screening with maternal blood test, followed by biochemical tests on amniotic fluid obtained by amniocentesis or by the detection of birth defects with the use of ultrasound
Osteogenesis imperfecta	Ultrasound or X-ray (brittle bones)
Phenylketonuria	DNA analysis of cells obtained by amniocentesis or chorionic villus sampling
Sickle-cell anemia	Fetal blood sampling* or DNA analysis of cells obtained by amniocentesis or chorionic villus sampling
Tay–Sachs disease	Biochemical tests on cells obtained by amniocentesis or chorionic villus sampling

*A sample of fetal blood is obtained by inserting a needle into the umbilical cord.

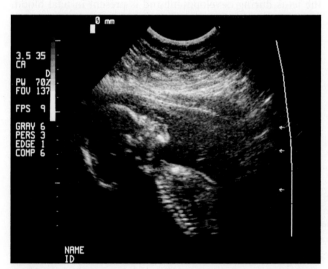

6.14 Ultrasonography can be used to detect some genetic disorders in a fetus and to locate the fetus during amniocentesis and chorionic villus sampling. [PhotoDisc/Media Bakery.]

a pregnant woman (**Figure 6.15**). Amniotic fluid—the substance that fills the amniotic sac and surrounds the developing fetus—contains fetal cells that can be cultured and used for genetic testing.

Amniocentesis is routinely performed as an outpatient procedure either with or without the use of a local anesthetic (see Figure 6.15). Genetic tests are then performed on the cultured cells. Complications with amniocentesis (mostly miscarriage) are uncommon, arising in only about 1 in 400 procedures.

CHORIONIC VILLUS SAMPLING A major disadvantage of amniocentesis is that it generally cannot be performed until about the 15th to 18th week of a pregnancy (although some obstetricians successfully perform amniocentesis earlier). The cells obtained by amniocentesis must then be cultured before genetic tests can be performed, which requires yet more time. For these reasons, genetic information about the fetus may not be available until the 17th or 18th week of pregnancy. By this stage, if abortion is chosen, it carries a risk of complications and is often stressful for the parents. **Chorionic villus sampling** (CVS) can be performed earlier (between the 10th and 12th weeks of pregnancy) and collects a larger amount of fetal tissue, which eliminates the necessity of culturing the cells.

In CVS, a catheter—a soft plastic tube—is placed in contact with the chorion, the outer layer of the placenta (**Figure 6.16**). Suction is then applied, and a small piece of the chorion is removed. Although the chorion is composed of fetal cells, it is a part of the placenta, which is expelled from the uterus after birth; the tissue that is removed is not actually from the fetus. This tissue contains millions of actively dividing cells that can be used directly in many genetic tests. Chorionic villus sampling has a somewhat higher risk of complications than amniocentesis; the results of several studies suggest that this procedure may increase the incidence of limb defects in the fetus when performed earlier than the 10th week of pregnancy.

Fetal cells obtained by amniocentesis or by CVS can be used to prepare a **karyotype**, which is a picture of a complete set of metaphase chromosomes. Karyotypes can be studied

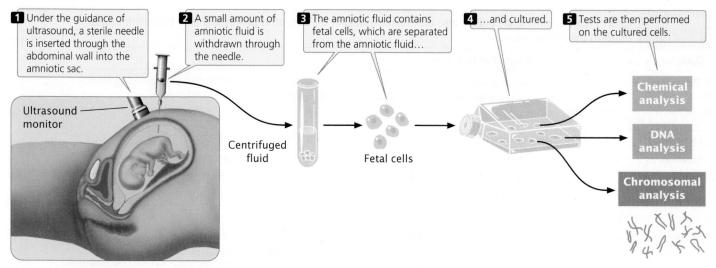

1 Under the guidance of ultrasound, a sterile needle is inserted through the abdominal wall into the amniotic sac.

2 A small amount of amniotic fluid is withdrawn through the needle.

3 The amniotic fluid contains fetal cells, which are separated from the amniotic fluid...

4 ...and cultured.

5 Tests are then performed on the cultured cells.

Ultrasound monitor

Centrifuged fluid

Fetal cells

Chemical analysis

DNA analysis

Chromosomal analysis

6.15 Amniocentesis is a procedure for obtaining fetal cells for genetic testing.

for chromosome abnormalities (see Chapter 8). Biochemical analyses can be conducted on fetal cells to determine the presence of particular metabolic products of genes. To detect genetic diseases for which the DNA sequence of the causative gene has been determined, the DNA sequence (DNA testing; see Chapter 19) can be examined for defective alleles.

MATERNAL BLOOD SCREENING TESTS Increased risk of some genetic conditions can be detected by examining levels of certain substances in the blood of the mother (referred to as a **maternal blood screening test**). However, these tests do not determine the presence of a genetic problem; rather, they simply indicate that the fetus is at increased risk of a problem and hence are referred to as *screening* tests. When

increased risk is detected, follow-up tests (additional blood screening tests, ultrasound, amniocentesis, or all three) are usually conducted.

One substance examined in maternal screening tests is α-fetoprotein, a protein that is normally produced by the fetus during development and is present in fetal blood, amniotic fluid, and the mother's blood during pregnancy. The level of α-fetoprotein is significantly higher than normal when the fetus has a neural-tube defect or one of several other disorders. Some chromosome abnormalities produce lower-than-normal levels of α-fetoprotein. Measuring the amount of α-fetoprotein in the mother's blood gives an indication of these conditions.

The American College of Obstetricians and Gynecologists recommends that physicians offer all pregnant women

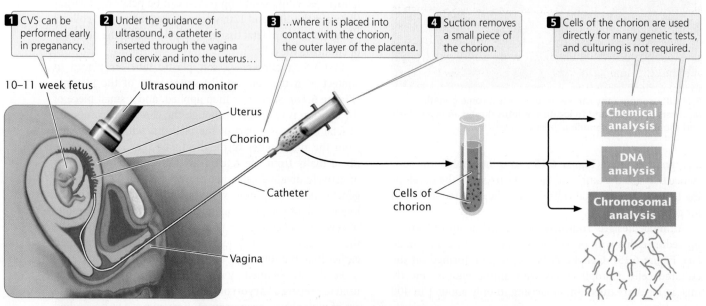

1 CVS can be performed early in preganancy.

2 Under the guidance of ultrasound, a catheter is inserted through the vagina and cervix and into the uterus...

3 ...where it is placed into contact with the chorion, the outer layer of the placenta.

4 Suction removes a small piece of the chorion.

5 Cells of the chorion are used directly for many genetic tests, and culturing is not required.

10–11 week fetus

Ultrasound monitor

Uterus

Chorion

Catheter

Vagina

Cells of chorion

Chemical analysis

DNA analysis

Chromosomal analysis

6.16 Chorionic villus sampling (CVS) is another procedure for obtaining fetal cells for genetic testing.

maternal blood screening tests. One typical test, carried out between the 11th and 13th weeks of pregnancy, measures human chorionic gonadotropin (hCG, a pregnancy hormone) and a substance called pregnancy-associated plasma protein A (PAPP-A). When the fetus has certain chromosome abnormalities, the level of PAPP-A tends to be low and the level of hCG tends to be high. The risk of a chromosome abnormality is calculated on the basis of the levels of hCG and PAPP-A in the mother's blood along with the results of ultrasound. Another test, referred to as the quad screen, measures the levels of four substances: α-fetoprotein, hCG, estriol, and inhibin. The risk of chromosome abnormalities and certain other birth defects is calculated on the basis of the combined levels of the four substances plus the mother's age, weight, ethnic background, and number of fetuses. The quad screen successfully detects Down syndrome (due to three copies of chromosome 21) 81% of the time.

NONINVASIVE PRENATAL GENETIC DIAGNOSIS Prenatal tests that use only maternal blood are highly desirable because they are noninvasive and pose no risk to the fetus. In addition to maternal blood screening tests, which measure chemical substances produced by the fetus or placenta, these noninvasive tests include procedures collectively called **noninvasive prenatal genetic diagnosis**, which directly examine fetal DNA found in maternal blood. These tests can be performed as early as the 10th week of pregnancy.

During pregnancy, a few fetal cells are released into the mother's circulatory system, where they mix with and circulate with her blood. Recent advances have made it possible to separate these fetal cells from maternal blood cells (a procedure called **fetal cell sorting**) with the use of lasers and automated cell-sorting machines. The fetal cells obtained can be cultured for chromosome analysis or used as a source of fetal DNA for molecular testing (see Chapter 19). Maternal blood also contains free-floating fragments of fetal DNA, which is released when fetal cells break down. This fetal DNA can be sequenced and tested for mutations. Tests are also available to determine the number of copies of genetic variants, and thus the number of chromosomes carried by the fetus, so that chromosome abnormalities such as Down syndrome can be detected. Noninvasive prenatal genetic diagnosis is now being used to determine the blood type of the fetus, to detect Down syndrome and other chromosomal disorders, and to identify mutations that cause genetic diseases such as cystic fibrosis and thalassemia (a blood disorder).

Prenatal genetic diagnosis based on DNA in the mother's blood occasionally reveals information about the mother's genetic makeup or health. For example, sometimes DNA analysis detects the presence of a Y chromosome, but ultrasound reveals a female fetus. In a few rare instances, this finding is due to a previously undiagnosed sex chromosome anomaly of the mother (although there are other potential explanations as well). Occasionally, testing can reveal the presence of undiagnosed cancer in the mother. Whether and how this type of incidental genetic information should be conveyed to the mother, who did not request genetic testing on herself, remains the subject of discussion among genetic counselors.

Noninvasive prenatal genetic diagnosis creates the potential to use a single blood sample from the mother to test for hundreds of genetic diseases and even for ordinary traits in the fetus. Indeed, researchers have demonstrated that they can sequence the entire fetal genome from a maternal blood sample. This possibility raises a number of social and ethical questions about the use of such information in reproductive decisions.

PREIMPLANTATION GENETIC DIAGNOSIS Prenatal genetic tests provide today's prospective parents with increasing amounts of information about the health of their future children. New reproductive technologies provide couples with options for using this information. One of these technologies is in vitro fertilization. In this procedure, hormones are used to induce ovulation. The ovulated eggs are surgically removed from the surface of the ovary, placed in a laboratory dish, and fertilized with sperm. The resulting embryo is then implanted in the uterus. Thousands of babies resulting from in vitro fertilization have now been born.

Genetic testing can be combined with in vitro fertilization to allow the implantation of embryos that are free of a specific genetic defect. Called **preimplantation genetic diagnosis** (PGD), this technique enables people who carry a genetic defect to avoid producing a child with the disorder. For example, if a woman is a carrier of an X-linked recessive disease, approximately half of her sons are expected to have the disease. Through in vitro fertilization and PGD, an embryo without the disorder can be selected for implantation in her uterus.

The procedure begins with the production of several single-celled embryos through in vitro fertilization. The embryos are allowed to divide several times until they reach the 8- or 16-cell stage. At this point, one cell is removed from each embryo and tested for the genetic abnormality. Removing a single cell at this early stage does not harm the embryo. Once the embryos that are free of the disorder have been identified, a healthy embryo is selected and implanted in the woman's uterus.

Preimplantation genetic diagnosis requires the ability to conduct genetic testing on a single cell. Such testing is possible with the use of the polymerase chain reaction, through which minute quantities of DNA can be amplified (replicated) quickly (see Chapter 19). Once the cell's DNA has been amplified, the DNA sequence is examined. Preimplantation genetic diagnosis has resulted in the birth of thousands of healthy children. Its use raises a number of ethical concerns because it can be used as a means of selecting for or against genetic traits that have nothing to do with medical concerns. For example, it can potentially be used to select for a child with genes for a certain eye color or genes for increased height.

NEWBORN SCREENING Testing for genetic disorders in newborn infants is called **newborn screening**. All states in the United States and many other countries require by law that newborn infants be tested for some genetic diseases and conditions. In 2006, the American College of Medical Genetics recommended mandatory screening for 29 conditions (**Table 6.5**), and many states have now adopted this list for newborn testing. These genetic conditions were chosen because early identification can lead to effective treatment. For example, as mentioned in Chapter 5, phenylketonuria is an autosomal recessive disease that, if not treated at an early age, can result in intellectual disability. But early intervention, through the administration of a modified diet, prevents this outcome.

TABLE 6.5	Genetic conditions recommended for mandatory screening by the American College of Medical Genetics
Medium-chain acyl-CoA dehydrogenase deficiency	
Congenital hypothyroidism	
Phenylketonuria	
Biotinidase deficiency	
Sickle-cell anemia (Hb SS disease)	
Congenital adrenal hyperplasia (21-hydroxylase deficiency)	
Isovaleric acidemia	
Very long chain acyl-CoA dehydrogenase deficiency	
Maple syrup (urine) disease	
Galactosemia	
Hb S/β-thalassemia	
Hb S/C disease	
Long-chain L-3-hydroxyacyl-CoA dehydrogenase deficiency	
Glutaric acidemia type I	
3-Hydroxy-3-methyl glutaric aciduria	
Trifunctional protein deficiency	
Multiple carboxylase deficiency	
Methylmalonic acidemia (mutase deficiency)	
Homocystinuria (due to cystathionine β synthase deficiency)	
3-Methylcrotonyl-CoA carboxylase deficiency	
Hearing loss	
Methylmalonic acidemia (Cbl A,B)	
Propionic acidemia	
Carnitine uptake defect	
β-Ketothiolase deficiency	
Citrullinemia	
Argininosuccinic acidemia	
Tyrosinemia type I	
Cystic fibrosis	

PRESYMPTOMATIC GENETIC TESTING In addition to testing for genetic diseases in fetuses and newborns, testing of healthy adults for genes that might predispose them to a genetic condition in the future is possible. This type of testing is known as **presymptomatic genetic testing**. For example, presymptomatic testing is available for members of families in which an autosomal dominant form of breast cancer occurs. In this case, early identification of the disease-causing allele allows for closer surveillance and the early detection of tumors. Presymptomatic testing is also available for some genetic diseases for which no treatment is available, such as Huntington disease, an autosomal dominant disease that leads to slow physical and mental deterioration in middle age.

HETEROZYGOTE SCREENING Another form of genetic testing in adults is **heterozygote screening**. In this type of screening, members of a population are tested to identify heterozygous carriers of recessive disease-causing alleles—people who are healthy but have the potential to produce children with a particular disease.

Testing for Tay–Sachs disease is a successful example of heterozygote screening. In the general population of North America, the frequency of Tay–Sachs disease is only about 1 person in 360,000. Among Ashkenazi Jews (descendants of Jewish people who settled in eastern and central Europe), the frequency is 100 times as great. A simple blood test is used to identify Ashkenazi Jews who carry the allele for Tay–Sachs disease. If a man and woman are both heterozygotes, approximately one-fourth of their children are expected to have Tay–Sachs disease. Couples identified as heterozygous carriers may use that information in deciding whether to have children. A prenatal test is also available for determining if the fetus of an at-risk couple will have Tay–Sachs disease. Screening programs have led to a significant decline in the number of children of Ashkenazi ancestry born with Tay–Sachs disease (now fewer than 10 children per year in the United States).

PHARMACOGENETIC TESTING Genetic testing is becoming increasingly important in guiding drug treatment. In some cases, a drug's effectiveness is influenced by a patient's genotype; in other cases, individuals with certain genotypes may be more likely to suffer from adverse drug reactions.

Warfarin provides a good example of how genetic testing can be used in conjunction with drug treatment. Warfarin is the most common anticoagulant (blood thinner) used worldwide. It is often administered to people with certain heart problems and after surgery to prevent the formation of clots in the blood vessels, which can obstruct blood flow and lead to tissue death. An effective dose of warfarin is critical: too little, and blood clots are not prevented; too much, and internal bleeding results. Unfortunately, people vary greatly in their responses to warfarin, and some of this variation is due to genes. For example, *CYP2C9* is a gene that encodes an enzyme that metabolizes warfarin. Over 30 different alleles occur at this locus. People who are homozygous for the *CYP2C9*1* allele metabolize warfarin normally, but individuals who are homozygous or heterozygous for the *CYP2CP9*2* or *CYP2CP9*3* alleles metabolize warfarin at a much lower rate and therefore require a lower dose. If given the usual dose of warfarin, these people are at greater risk of bleeding. Genetic variation at *CYP2CP* and another locus called *VKORC1* accounts for up to 30% of the variation in response to warfarin dose and risk of bleeding. Some hospitals are screening patients for variation at these genes to help determine the proper warfarin dose to administer. A study conducted by Vanderbilt University in Nashville, Tennessee found that over 90% of all patients carried one or more genetic variants that affected their response to five common drugs, including warfarin. **▶ TRY PROBLEM 12**

THINK-PAIR-SHARE Questions 6 and 7

CONCEPTS

Genetic testing is used to screen newborns for genetic diseases, detect individuals who are heterozygous for recessive diseases, detect disease-causing alleles in those who have not yet developed symptoms of a disease, and detect defective alleles in unborn babies. Preimplantation genetic diagnosis combined with in vitro fertilization allows for the selection of embryos that are free from specific genetic diseases. Genetic testing is also being used to guide individual drug treatment.

✔ CONCEPT CHECK 9

How does preimplantation genetic diagnosis differ from prenatal genetic testing?

Interpreting Genetic Tests

Today, more than a thousand genetic tests are clinically available, and several hundred more are available through research studies. Future research will greatly increase the number and complexity of genetic tests. Many of these tests will be for complex multifactorial diseases that are influenced by both genetics and environment, such as coronary artery disease, diabetes, asthma, some types of cancer, and depression.

Interpreting the results of genetic tests is often complicated by several factors. First, some genetic diseases can be caused by many different mutations. For example, more than a thousand different mutations at a single locus can cause cystic fibrosis, an autosomal recessive disease in which chloride ion transport is defective (see Chapter 5). Genetic tests typically screen for only the most common mutations; uncommon and rare mutations are not detected. Therefore, a negative result does not mean that a genetic defect is absent; it indicates only that the person being tested does not have one of the common mutations. The DNA of an affected person can be examined to determine

the nature of the mutation, and other family members can then be screened for the same mutation, but this option is not possible if affected family members are unavailable or unwilling to be tested.

A second problem lies in interpreting the results of genetic tests. For a classic genetic disease such as Tay–Sachs disease, the inheritance of two copies of the gene virtually ensures that a person will have the disease. However, this is not the case for many genetic diseases for which penetrance is incomplete and environmental factors play a role. For these conditions, carrying a disease-predisposing mutation only elevates a person's risk of acquiring the disease. The risk associated with a particular mutation is a statistical estimate, based on the average effect of the mutation on many people. In this case, the calculated risk may provide little useful information to a specific person. It is also important to keep in mind that for many genetic traits and disorders, no genetic test exists.

CONCEPTS

Interpreting genetic tests is complicated by the presence of multiple causative mutations, incomplete penetrance, and the influence of environmental factors.

Direct-to-Consumer Genetic Testing

Genome sequencing and new molecular technologies now make it possible to determine a person's risk of a large number of diseases and disorders. **Direct-to-consumer genetic tests** attempt to make this information available to any consumer, without involving a health care provider. Several companies sold such tests in the United States between 2007 and 2013, until their sale was restricted by the U.S. Food and Drug Administration (FDA) in 2013. These tests are able to screen for a large array of genetic conditions in adults and children—everything from single-gene disorders such as cystic fibrosis to multifactorial conditions such as obesity, cardiovascular disease, potential for athletic performance, and predisposition to nicotine addiction.

Most direct-to-consumer genetic tests were advertised and ordered through the Internet. Geneticists, public health officials, and consumer advocates raised a number of concerns about direct-to-consumer genetic testing, including concerns that some tests are offered without appropriate information and genetic counseling and that consumers are often not equipped to interpret the results. A few direct-to-consumer genetic tests for individual genetic disorders have now been approved by the FDA for use in the United States; more extensive direct-to-consumer genetic testing is available in some other countries. Direct-to-consumer tests are also available for paternity testing and for determining ancestry. **▶ TRY PROBLEM 15**

THINK-PAIR-SHARE Question 8

Genetic Discrimination and Privacy

With the development of so many new genetic tests, concerns have been raised about the privacy of genetic information and the potential for genetic discrimination. Research shows that many people at risk for genetic diseases avoid genetic testing because they fear that the results would make it difficult for them to obtain insurance or that the information might adversely affect their employability. Some of those who do seek genetic testing pay for it themselves and use aliases to prevent the results from becoming part of their health records. Fears about genetic discrimination have been reinforced by past practices. In the 1970s, some people who were identified as carriers of sickle-cell anemia (an autosomal recessive disorder that is common among African Americans) faced employment discrimination and had difficulty obtaining health insurance, in spite of the fact that carriers are healthy.

In response to these concerns, the U.S. Congress passed the **Genetic Information Nondiscrimination Act** (GINA) in 2008. This law prohibits health insurers from using genetic information to make decisions about health-insurance coverage and rates. It also prevents employers from using genetic information in employment decisions and prohibits health insurers and employers from asking or requiring a person to take a genetic test. Results of genetic testing receive some degree of protection by other federal regulations that cover the uses and disclosure of individual health information. However, GINA covers health insurance and employment only; it does not apply to life, disability, and long-term care insurance. **▶ TRY PROBLEM 16**

CONCEPTS

The growing number of genetic tests and their increasing complexity has raised several concerns, including concerns about genetic discrimination and the privacy of test results.

CONCEPTS SUMMARY

■ Constraints on the genetic study of human traits include the inability to conduct controlled crosses, long generation time, small family size, and the difficulty of separating genetic and environmental influences. Pedigrees are often used to study the inheritance of traits in humans.

■ Autosomal recessive traits typically appear with equal frequency in both sexes and tend to skip generations. When both parents are heterozygous for a particular autosomal recessive trait, approximately one-fourth of their offspring will have the trait. Recessive traits are more likely to appear in families with consanguinity (mating between closely related people).

■ Autosomal dominant traits usually appear equally in both sexes and do not skip generations. When one parent is affected and heterozygous for an autosomal dominant trait, approximately half of the offspring will have the trait. Unaffected people do not normally transmit an autosomal dominant trait to their offspring.

■ X-linked recessive traits appear more frequently in males than in females. When a woman is a heterozygous carrier for an X-linked recessive trait and a man is unaffected, approximately half of their sons will have the trait and half of their daughters will be unaffected carriers. X-linked traits are not passed from father to son.

■ X-linked dominant traits appear in both males and females but more frequently in females. They do not skip generations. Affected men pass an X-linked dominant trait to all their daughters but none of their sons. Heterozygous women pass the trait to half of their sons and half of their daughters.

■ Y-linked traits appear only in males and are passed from a father to all his sons.

■ A trait's higher concordance in monozygotic than in dizygotic twins indicates a genetic influence on the trait; less than 100% concordance in monozygotic twins indicates environmental influences on the trait.

■ Similarities between adopted children and their biological parents indicate the importance of genetic factors in the expression of a trait; similarities between adopted children and their genetically unrelated adoptive parents indicate the influence of environmental factors.

■ Genetic counseling provides information and support to people concerned about hereditary conditions in their families.

■ Genetic testing includes prenatal diagnosis, screening for disease-causing alleles in newborns, the detection of people heterozygous for recessive alleles, presymptomatic testing for the presence of a disease-causing allele in at-risk people, and testing to evaluate the effectiveness and safety of drug treatment.

■ The interpretation of genetic tests may be complicated by the presence of numerous causative mutations, incomplete penetrance, and the influence of environmental factors.

■ The availability of direct-to-consumer genetic tests has raised concerns about the adequacy of the information provided and the absence of genetic counseling.

■ Genetic testing has raised concerns about genetic discrimination and the privacy of test results. The Genetic Information Nondiscrimination Act prohibits the use of genetic information in decisions on health insurability and employment.

IMPORTANT TERMS

pedigree (p. 147)
proband (p. 147)
consanguinity (p. 148)
genetic mosaic (p. 154)
dizygotic twins (p. 154)
monozygotic
    twins (p. 154)
concordance (p. 155)

genetic counseling (p. 157)
ultrasonography (p. 158)
amniocentesis (p. 158)
chorionic villus sampling
    (CVS) (p. 159)
karyotype (p. 159)
maternal blood
    screening test (p. 160)

noninvasive prenatal genetic
    diagnosis (p. 161)
fetal cell sorting (p. 161)
preimplantation genetic
    diagnosis (PGD) (p. 161)
newborn screening (p. 162)
presymptomatic genetic
    testing (p. 162)

heterozygote
    screening (p. 162)
direct-to-consumer genetic
    test (p. 163)
Genetic Information
    Nondiscrimination
    Act (GINA) (p. 164)

ANSWERS TO CONCEPT CHECKS

1. b

2. It might skip generations when a new mutation arises or the trait has incomplete penetrance.

3. If it is X-linked recessive, the trait will not be passed from father to son.

4. c

5. If the trait were Y linked, an affected male would pass it on to all his sons, whereas if the trait were autosomal and sex-limited, affected heterozygous males would pass it on to only half of their sons on average.

6. d

7. c

8. d

9. Preimplantation genetic diagnosis determines the presence of disease-causing genes in an embryo at an early stage, before it is implanted in the uterus and initiates pregnancy. Prenatal genetic testing determines the presence of disease-causing genes or chromosomes in a developing fetus.

WORKED PROBLEMS

Problem 1

Joanna has "short fingers" (brachydactyly). She has two older brothers who are identical twins; both have short fingers. Joanna's two younger sisters have normal fingers. Joanna's mother has normal fingers, and her father has short fingers. Joanna's paternal grandmother (her father's mother) has short fingers; her paternal grandfather (her father's father), who is now deceased, had normal fingers. Both of Joanna's maternal grandparents (her mother's parents) have normal fingers. Joanna marries Tom, who has normal fingers; they adopt a son named Bill, who has normal fingers. Bill's biological parents both have normal fingers. After adopting Bill, Joanna and Tom produce two children: an older daughter with short fingers and a younger son with normal fingers.

a. Using standard symbols and labels, draw a pedigree illustrating the inheritance of short fingers in Joanna's family.

b. What is the most likely mode of inheritance for short fingers in this family?

c. If Joanna and Tom have another biological child, what is the probability (based on your answer to part *b*) that this child will have short fingers?

>> Solution Strategy

What information is required in your answer to the problem?

a. A pedigree to represent the family, drawn with correct symbols and labeling.

b. The most likely mode of inheritance for short fingers.

c. The probability that Joanna and Tom's next child will have short fingers.

What information is provided to solve the problem?

The phenotypes of Joanna and Tom and their family members.

For help with this problem, review:

The information on pedigrees in Section 6.2.

>> Solution Steps

Hint: See Figure 6.2 for a review of symbols used in a pedigree.

a. In the pedigree for the family, use filled circles (females) and filled squares (males) to represent family members with the trait of interest (short fingers). Connect Joanna's identical twin brothers to the line above by drawing diagonal lines that have a horizontal line between them. Enclose Bill, the adopted child of Joanna and Tom, in brackets; connect him to his biological parents by drawing a diagonal line and to his adoptive parents by a dashed line.

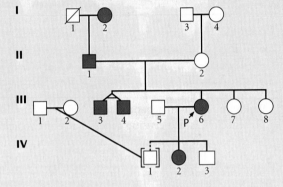

See Table 6.1 for a review of the characteristics of different modes of inheritance.

b. The most likely mode of inheritance for short fingers in this family is autosomal dominant. The trait appears equally in males and females and does not skip generations. When one parent has the trait, it appears in approximately half of that parent's sons and daughters, although the number of children in the families is small. We can eliminate Y-linked inheritance because the trait is found in females as well as males. If short fingers were X-linked recessive, females with the trait would be expected to pass the trait to all their sons, but Joanna (III-6), who has short fingers, produced a son with normal fingers. For X-linked dominant traits, affected men should pass the trait to all their daughters; because male II-1 has short fingers and produced two daughters without short fingers (III-7 and III-8), we know that the trait cannot be X-linked dominant. The trait is unlikely to be autosomal recessive because it does not skip generations and because approximately half the children of affected parents have the trait.

c. If having short fingers is an autosomal dominant trait, Tom must be homozygous (*bb*) because he has normal fingers. Joanna must be heterozygous (*Bb*) because she and Tom have produced both short-fingered and normal-fingered offspring. In a cross between a heterozygote and homozygote, half the progeny are expected to be heterozygous and the other half homozygous ($Bb \times bb \rightarrow \frac{1}{2}\, Bb, \frac{1}{2}\, bb$), so the probability that Joanna's and Tom's next biological child will have short fingers is $\frac{1}{2}$.

Problem 2

Concordance for a series of traits was measured in monozygotic twins and dizygotic twins; the results are shown in the following table. For each trait, indicate whether the concordances suggest genetic influences, environmental influences, or both. Explain your reasoning.

	Concordance (%)	
Characteristic	**Monozygotic**	**Dizygotic**
a. ABO blood type	100	65
b. Diabetes	85	36
c. Coffee drinking	80	80
d. Smoking	75	42
e. Schizophrenia	53	16

>> Solution Strategy

What information is required in your answer to the problem?

For each trait, whether it is influenced by genetic factors, environmental factors, or both.

What information is provided to solve the problem?

Concordance for each trait in monozygotic and dizygotic twins.

For help with this problem, review:

Concordance in Twins in Section 6.3.

>> Solution Steps

a. The concordance for ABO blood type in monozygotic twins is 100%. This high concordance in monozygotic twins does not, by itself, indicate a genetic basis for the trait. Because concordance for ABO blood type is substantially lower in dizygotic twins, we would be safe in concluding that genes play a role in determining differences in ABO blood types.

b. The concordance for diabetes is substantially higher in monozygotic twins than in dizygotic twins; therefore, we can conclude that genetic factors play some role in susceptibility to diabetes. The fact that monozygotic twins show a concordance less than 100% suggests that environmental factors also play a role.

c. Both monozygotic and dizygotic twins exhibit the same high concordance for coffee drinking, so we can conclude that there is little genetic influence on coffee drinking. The fact that monozygotic twins show a concordance less than 100% suggests that environmental factors play a role.

d. The concordance for smoking is lower in dizygotic twins than in monozygotic twins, so genetic factors appear to influence the tendency to smoke. The fact that monozygotic twins show a concordance less than 100% suggests that environmental factors also play a role.

e. Monozygotic twins exhibit substantially higher concordance for schizophrenia than do dizygotic twins, so we can conclude that genetic factors influence this psychiatric disorder. Because the concordance of monozygotic twins is substantially less than 100%, we can conclude that environmental factors play a role in the disorder as well.

COMPREHENSION QUESTIONS

Section 6.1

1. What three factors complicate the task of studying the inheritance of human characteristics?

Section 6.2

2. Who is the proband in a pedigree? Is the proband always found in the last generation of the pedigree? Why or why not?

3. For each of the following modes of inheritance, describe the features that will be exhibited in a pedigree in which the trait is present: autosomal recessive, autosomal dominant, X-linked recessive, X-linked dominant, and Y-linked inheritance.

4. How does the pedigree of an autosomal recessive trait differ from the pedigree of an X-linked recessive trait?

5. Other than the fact that a Y-linked trait appears only in males, how does the pedigree of a Y-linked trait differ from the pedigree of an autosomal dominant trait?

Section 6.3

6. What are the two types of twins and how do they arise?

7. Explain how a comparison of concordance in monozygotic and dizygotic twins can be used to determine the extent to which the expression of a trait is influenced by genes or by environmental factors.

8. How are adoption studies used to separate the effects of genes and environment in the study of human characteristics?

Section 6.4

9. What is genetic counseling?

*10. Give at least four different reasons for seeking genetic counseling.

11. Briefly define newborn screening, heterozygote screening, presymptomatic genetic testing, and prenatal diagnosis.

*12. Compare the advantages and disadvantages of amniocentesis versus chorionic villus sampling for prenatal diagnosis.

13. What is preimplantation genetic diagnosis?

14. How does heterozygote screening differ from presymptomatic genetic testing?

*15. What are direct-to-consumer genetic tests? What are some of the concerns about these tests?

*16. What activities does the Genetic Information Nondiscrimination Act prohibit?

17. How might genetic testing lead to genetic discrimination?

APPLICATION QUESTIONS AND PROBLEMS

Section 6.1

*18. If humans have characteristics that make them unsuitable for genetic analysis, such as long generation time, small family size, and uncontrolled crosses, why do geneticists study humans? Give several reasons why humans have been the focus of so much genetic study.

Section 6.2

*19. Joe is color blind. Both his mother and his father have normal vision, but his mother's father (Joe's maternal grandfather) is color blind. All Joe's other grandparents have normal color vision. Joe has three sisters—Patty, Betsy, and Lora—all with normal color vision. Joe's oldest sister, Patty, is married to a man with normal color vision; they have two children, a 9-year-old color-blind boy and a 4-year-old girl with normal color vision.

a. Using standard symbols and labels, draw a pedigree of Joe's family.

b. What is the most likely mode of inheritance for color blindness in Joe's family?

c. If Joe marries a woman who has no family history of color blindness, what is the probability that their first child will be a color-blind boy?

d. If Joe marries a woman who is a carrier of the color-blind allele, what is the probability that their first child will be a color-blind boy?

e. If Patty and her husband have another child, what is the probability that the child will be a color-blind boy?

20. Consider the pedigree shown in **Figure 6.3**.

a. If individual IV-7 married a person who was unaffected with Waardenburg syndrome, what is the probability that their first child would have Waardenburg syndrome? Explain your reasoning.

b. If individuals IV-4 and IV-5 mated and produced a child, what is the probability that the child would have Waardenburg syndrome? Explain your reasoning.

21. Many studies have suggested a strong genetic predisposition to migraine headaches, but the mode of inheritance is not clear. L. Russo and colleagues examined migraine headaches in several families, two of which are shown below (L. Russo et al. 2005. *American Journal of Human Genetics* 76:327–333). What is the most likely mode of inheritance for migraine headaches in these families? Explain your reasoning.

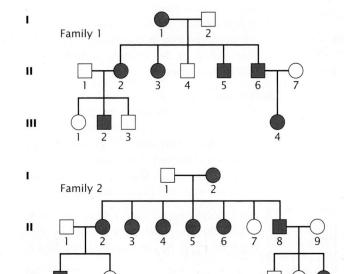

*22. Dent disease is a rare disorder of the kidney in which reabsorption of filtered solutes is impaired and there is progressive renal failure. R. R. Hoopes and colleagues studied mutations associated with Dent disease in the following family (R. R. Hoopes et al. 2005. *American Journal of Human Genetics* 76:260–267):

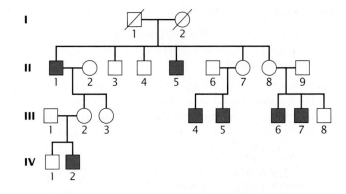

a. On the basis of this pedigree, what is the most likely mode of inheritance for the disease? Explain your reasoning.

b. Based your answer to part *a*, give the most likely genotypes for all family members in the pedigree.

23. A man with a specific unusual genetic trait marries an unaffected woman and they have four children. Pedigrees of this family are shown in parts *a* through *e*, but the presence or absence of the trait in the children is not indicated. For each type of inheritance, indicate how many children of each sex are expected to express the trait by filling in the appropriate circles and squares. Assume that the trait is rare and fully penetrant.

a. Autosomal recessive trait

b. Autosomal dominant trait

c. X-linked recessive trait

d. X-linked dominant trait

e. Y-linked trait

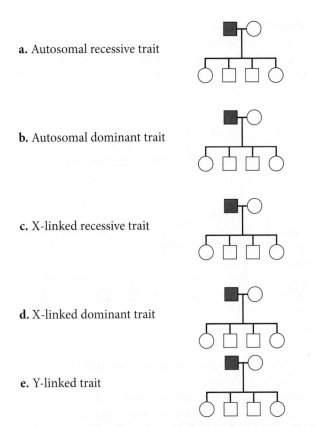

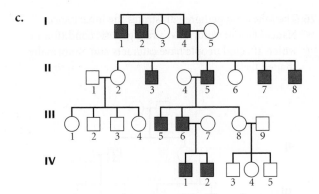

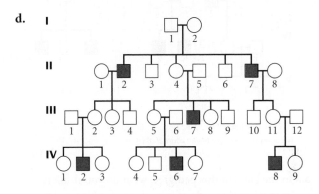

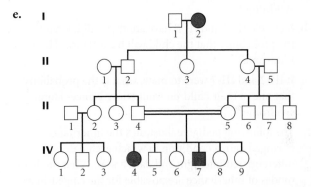

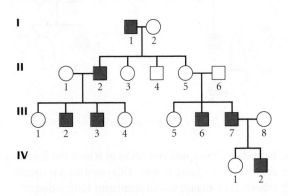

*24. For each of the following pedigrees, give the most likely mode of inheritance, assuming that the trait is rare. Carefully explain your reasoning.

a.

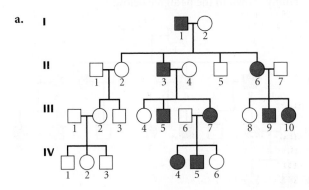

b.

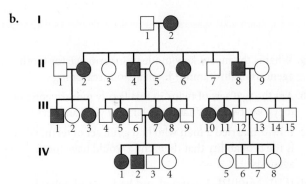

25. The trait represented in the following pedigree is expressed only in the males of the family. Is the trait Y linked? Why or why not? If you believe that the trait is not Y linked, propose an alternative explanation for its inheritance.

26. The following pedigree illustrates the inheritance of Nance–Horan syndrome, a rare genetic condition in which affected people have cataracts and abnormally shaped teeth.

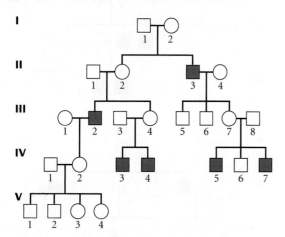

[Source: D. Stambolian, R. A. Lewis, K. Buetow, A. Bond, and R. Nussbaum. 1990. *American Journal of Human Genetics* 47:15.]

a. On the basis of this pedigree, what do you think is the most likely mode of inheritance for Nance–Horan syndrome?

b. If couple III-7 and III-8 have another child, what is the probability that the child will have Nance–Horan syndrome?

c. If III-2 and III-7 were to mate, what is the probability that one of their children would have Nance–Horan syndrome?

27. The following pedigree illustrates the inheritance of ringed hair, a condition in which each hair is differentiated into light and dark zones. What mode or modes of inheritance are possible for the ringed-hair trait in this family?

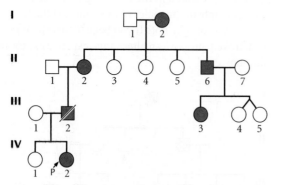

[Source: L. M. Ashley and R. S. Jacques. 1950. *Journal of Heredity* 41:83.]

28. Ectrodactyly is a rare condition in which the fingers are absent and the hand is split. This condition is usually inherited as an autosomal dominant trait. Ademar

Freire-Maia reported the appearance of ectrodactyly in a family in São Paulo, Brazil, whose pedigree is shown here. Is this pedigree consistent with autosomal dominant inheritance? If not, what mode of inheritance is most likely? Explain your reasoning.

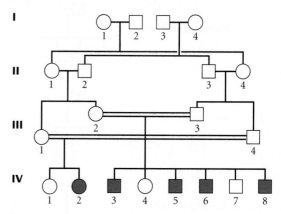

[Source: A. Freire-Maia. 1971. *Journal of Heredity* 62:53.]

29. The complete absence of one or more teeth (tooth agenesis) is a common trait in humans—indeed, more than 20% of humans lack one or more of their third molars. However, more severe tooth agenesis, defined as the absence of six or more teeth, is less common and is frequently an inherited condition. L. Lammi and colleagues examined tooth agenesis in the Finnish family shown in the pedigree below.

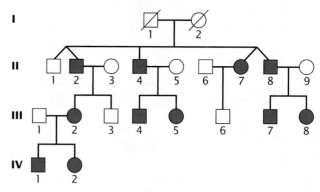

[Source: L. Lammi. 2004. *American Journal of Human Genetics* 74:1043–1050.]

a. What is the most likely mode of inheritance for tooth agenesis in this family? Explain your reasoning.

b. Are the two sets of twins in this family monozygotic or dizygotic twins? What is the basis of your answer?

c. If IV-2 married a man who had a full set of teeth, what is the probability that their child would have tooth agenesis?

d. If III-2 and III-7 married and had a child, what is the probability that their child would have tooth agenesis?

Section 6.3

*30. A geneticist studies a series of characteristics in pairs of monozygotic and dizygotic twins, obtaining the concordances listed below. For each characteristic, indicate whether these concordances suggest genetic influences, environmental influences, or both. Explain your reasoning.

Characteristic	Concordance (%)	
	Monozygotic	**Dizygotic**
Migraine headaches	60	30
Eye color	100	40
Measles	90	90
Club foot	30	10
High blood pressure	70	40
Handedness	70	70
Tuberculosis	5	5

31. On the basis of the concordances shown in **Table 6.2**, is variation in rheumatoid arthritis influenced by genetic factors, environmental factors, or both? Explain your reasoning.

32. M. T. Tsuang and colleagues studied drug dependence in male twin pairs (M. T. Tsuang et al. 1996. *American Journal of Medical Genetics* 67:473–477). They found that 82 out of 313 monozygotic twin pairs were concordant for abuse of one or more illicit drugs, whereas 40 out of 243 dizygotic twin pairs were concordant for the same trait. Calculate the concordances for drug abuse in these monozygotic and dizygotic twins. On the basis of these data, what conclusion can you draw concerning the roles of genetic and environmental factors in drug abuse?

*33. In a study of schizophrenia (a mental disorder including disorganization of thought and withdrawal from reality), researchers looked at the prevalence of the disorder in the biological and adoptive parents of people who were adopted as children; they found the following results:

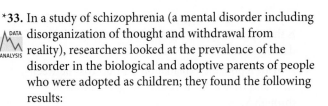

	Prevalence of schizophrenia	
Adoptees	**Biological parents**	**Adoptive parents**
With schizophrenia	12	2
Without schizophrenia	6	4

[Source: S. S. Kety et al., 1978, in *The Nature of Schizophrenia: New Approaches to Research and Treatment*, L. C. Wynne, R. L. Cromwell, and S. Matthysse, Eds. New York: Wiley, 1978, pp. 25–37.]

What can you conclude from these results concerning the role of genetics in schizophrenia? Explain your reasoning.

34. Which conclusions are supported by **Figure 6.13**?

a. Adoptive fathers of obese children have a higher BMI than adoptive fathers of thin children.

b. Adoptive mothers of thin children have a lower BMI than adoptive mothers of obese children.

c. Biological fathers of obese children have a higher BMI than adoptive fathers of thin children.

d. Both a and b.

e. Both a and c.

Section 6.4

35. What, if any, ethical issues might arise from the widespread use of noninvasive prenatal genetic diagnosis, which can be carried out much earlier than amniocentesis or chorionic villus sampling?

CHALLENGE QUESTIONS

Section 6.1

36. Many genetic studies, particularly those of recessive traits, have focused on small, isolated human populations, such as those on islands. Suggest one or more advantages that isolated populations might have for the study of recessive traits.

Section 6.2

37. Draw a pedigree that represents an autosomal dominant trait, sex-limited to males, and that excludes the possibility that the trait is Y linked.

38. A. C. Stevenson and E. A. Cheeseman studied deafness in a family in Northern Ireland and recorded the following pedigree (A. C. Stevenson and E. A. Cheeseman. 1956. *Annals of Human Genetics* 20:177–231).

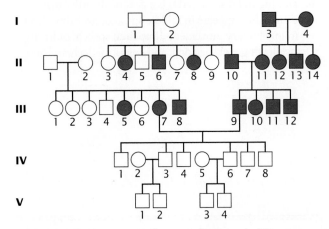

[Source: A. C. Stevenson and E. A. Cheeseman. 1956. *Annals of Human Genetics* 20:177–231.]

a. If you consider only generations I through III, what is the most likely mode of inheritance for this type of deafness?

b. Provide a possible explanation for the cross between III-7 and III-9 and the results for generations IV through V.

Section 6.3

39. Dizygotic twinning often runs in families, and its frequency varies among ethnic groups, whereas monozygotic twinning rarely runs in families, and its frequency is quite constant among ethnic groups. These observations have been interpreted as evidence of a genetic basis for variation in dizygotic twinning but little genetic basis for variation in monozygotic twinning. Can you suggest a possible reason for these differences in the genetic tendencies toward dizygotic and monozygotic twinning?

 THINK-PAIR-SHARE QUESTIONS

Section 6.2

1. How would the presence of incomplete penetrance affect the interpretation of pedigrees? Do you think it would more strongly affect the interpretation of autosomal recessive or autosomal dominant traits? Explain your reasoning.

2. Explain why X-linked recessive traits are often more common in males but X-linked dominant traits are often more common in females.

3. Why are there so few Y-linked traits in humans? (Hint: See Chapter 4.)

Section 6.3

4. The frequency of dizygotic twinning is higher in older women. Can you think of some possible evolutionary reasons for the occurrence of more dizygotic twins in older women?

5. Trait X displays 98% concordance in both monozygotic and dizygotic twins. Do these concordances suggest that trait X is influenced by (a) genetic factors; (b) environmental factors; or (c) both genetic and environmental factors? Explain your reasoning.

6. Huntington disease (HD) is a genetic disorder that usually first appears in middle age and, over time, leads to involuntary movements, impaired speech, difficulty swallowing and speaking, and cognitive decline. There is no cure for HD, and affected individuals eventually die, usually after 10 to 15 years of progressively worsening symptoms. Huntington disease is inherited as an autosomal dominant disorder. Presymptomatic genetic testing for HD is available for people with a family history of the disease.

Individuals who have a parent with HD have a 50% chance of inheriting the HD gene and eventually having HD. Some of these individuals want to know if they will eventually get the disease, and they undergo presymptomatic genetic testing. Others do not want to know, given that there is no cure or effective treatment for the disease. If you were at risk for HD, would you want to undergo genetic testing? What are some reasons for and against having a genetic test for HD?

7. Many genetic counselors will not provide presymptomatic genetic testing for Huntington disease to people below the age of 18. Why are there concerns about offering this and many other genetic tests to minors? What types of presymptomatic genetic testing might be appropriate for minors?

8. Do you think direct-to-consumer tests that provide information about the risk of developing medical conditions (such as breast cancer or Alzheimer disease) should be made available to the public? List some arguments for and against providing this service.

Linkage, Recombination, and Eukaryotic Gene Mapping

Pattern baldness is a hereditary trait. Recent research demonstrated that a gene for pattern baldness is linked to genetic markers located on the X chromosome, leading to the discovery that pattern baldness is influenced by variation in the androgen-receptor gene. [Jose Luis Pelaez Inc/ Age Fotostock.]

Linked Genes and Bald Heads

For many, baldness is the curse of manhood. Twenty-five percent of men begin balding by age 30 and almost half are bald to some degree by age 50. In the United States, baldness affects more than 40 million men, and hundreds of millions of dollars are spent each year on hair-loss treatment. Baldness is not just a matter of vanity: it is associated with some medically significant conditions, including heart disease, high blood pressure, and prostate cancer.

Baldness can arise for a number of different reasons, including illness, injury, drugs, and heredity. The most common type of baldness seen in men is pattern baldness—technically known as androgenic alopecia—in which hair is lost prematurely from the front and top of the head. More than 95% of hair loss in men is pattern baldness. Although pattern baldness is also seen in women, it is usually expressed weakly as mild thinning of the hair. The trait is stimulated by male sex hormones (androgens), as evidenced by the observation that males castrated at an early age rarely become bald (though this is not recommended as a preventive treatment).

A strong hereditary influence on pattern baldness has long been recognized, but its exact mode of inheritance has been controversial. An early study suggested that it was autosomal dominant in males and recessive in females, and was thus an example of a sex-influenced trait (see Chapter 5). Other evidence, and common folklore, suggested that a man inherits baldness from his mother's side of the family—that the trait exhibits X-linked inheritance.

In 2005, geneticist Axel Hillmer and his colleagues set out to locate the gene that causes pattern baldness. While they suspected that the gene might be located on the X chromosome, they had no idea where on the X chromosome it might reside. To identify the location of the gene, they conducted a linkage analysis study, in which they looked for an association between the inheritance of pattern baldness and the inheritance of genetic variants known to be located on the X chromosome. The genetic variants used in the study were single-nucleotide polymorphisms (SNPs, pronounced "snips"), which are positions in the genome at which individuals in a population differ in a single nucleotide. The geneticists studied the inheritance of pattern baldness and SNPs in 95 families in which at least two brothers developed pattern baldness at an early age.

Hillmer and his colleagues found that pattern baldness and SNPs from the X chromosome were not inherited independently, as predicted by Mendel's principle of independent assortment. Instead, they tended to be inherited together, which occurs when genes are physically linked on the same chromosome and segregate together in meiosis.

As we will learn in this chapter, linkage between genes is broken down over time by a process called recombination, or crossing over, and the frequency of recombination between genes is usually related to the distance between them. In 1911, Thomas Hunt Morgan and his student Alfred Sturtevant demonstrated, in fruit flies, that the locations of genes can be mapped by determining the rates of recombination between them. By using this mapping method in the families with pattern baldness, Hillmer and his colleagues demonstrated that the gene for pattern baldness is closely linked to SNPs located on the short arm of the X chromosome. This region includes the androgen-receptor gene, which encodes a protein that binds male sex hormones. Given the clear involvement of male hormones in the development of pattern baldness, the androgen-receptor gene seemed a likely candidate for causing pattern baldness. Further analysis revealed that certain alleles of the androgen-receptor gene are closely associated with the inheritance of pattern baldness, and that variation in the androgen-receptor gene is almost certainly responsible for much of the differences in pattern baldness seen in the families examined. Additional studies conducted in 2008 found that genes on chromosomes 3 and 20 also appear to contribute to the expression of pattern baldness. **> TRY PROBLEM 13**

> **THINK-PAIR-SHARE**
>
> - Common folklore says that if a young man wants to know whether he will become bald, he should look at his mother's father. Based on the information provided in the introduction to this chapter, is this folklore scientifically accurate? Why or why not?

This chapter explores the inheritance of genes located on the same chromosome. These linked genes do not strictly obey Mendel's principle of independent assortment; rather, they tend to be inherited together. This tendency requires a new approach to understanding their inheritance and predicting the types of offspring that will be produced by a cross. A critical piece of information necessary for making these predictions is the arrangement of the genes on the chromosomes; thus, it will be necessary to think about the relation between genes and chromosomes. A key to understanding the inheritance of linked genes is to make the conceptual connection between the genotypes in a cross and the behavior of chromosomes in meiosis.

We begin our exploration of linkage by comparing the inheritance of two linked genes with the inheritance of two genes that assort independently. We then examine how recombination breaks up linked genes. Next, we use our knowledge of linkage and recombination for predicting the results of genetic crosses in which genes are linked as well as for mapping genes. Later in the chapter, we focus on physical methods of determining the chromosomal locations of genes. The final section examines variation in rates of recombination.

7.1 Linked Genes Do Not Assort Independently

Chapter 3 introduced Mendel's principles of segregation and independent assortment. Let's take a moment to review these two important concepts. The principle of segregation states

that each diploid organism possesses two alleles at a locus that separate in meiosis, and that one allele passes into each gamete. The principle of independent assortment provides additional information about the process of segregation: it tells us that the two alleles at a locus separate independently of alleles at other loci.

The independent separation of alleles results in *recombination*, the sorting of alleles into new combinations. Consider a cross between individuals that are homozygous for two different pairs of alleles: $AA\ BB \times aa\ bb$. The first parent, $AA\ BB$, produces gametes with the alleles $A\ B$, and the second parent, $aa\ bb$, produces gametes with the alleles $a\ b$, resulting in F_1 progeny with genotype $Aa\ Bb$ (**Figure 7.1**). Recombination means that when one of the F_1 progeny reproduces, the combination of alleles in its gametes may differ from the combinations in the gametes of its parents. In other words, the F_1 may produce gametes with new allele combinations $A\ b$ or $a\ B$ in addition to parental gametes $A\ B$ or $a\ b$.

Mendel derived his principles of segregation and independent assortment by observing the progeny of genetic crosses, but he had no idea what biological processes produced these phenomena. In 1903, Walter Sutton proposed a biological basis for Mendel's principles, called the chromosome theory of heredity, which holds that genes are found on chromosomes (see Chapter 3). Let's restate Mendel's two principles in relation to the chromosome theory of heredity. The principle of segregation states that a diploid organism possesses two alleles for a characteristic, each of which is located at the same position, or locus, on one of the two

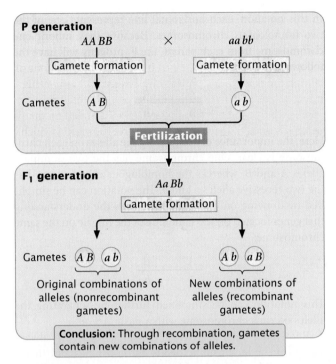

7.1 Recombination is the sorting of alleles into new combinations.

homologous chromosomes. These chromosomes segregate in meiosis, and each gamete receives one homolog. The principle of independent assortment states that, in meiosis, each pair of homologous chromosomes assorts independently of other homologous pairs. With this new perspective, it is easy to see that, because the number of chromosomes in most organisms is limited, there are certain to be more genes than chromosomes, so some genes must be present on the same chromosome and should not assort independently. Genes located close together on the same chromosome are called **linked genes** and belong to the same **linkage group**. Linked genes travel together in meiosis, eventually arriving at the same destination (the same gamete), and are not expected to assort independently. All of the characteristics examined by Mendel in peas did display independent assortment, and the first genetic characteristics studied in other organisms also seemed to assort independently. How could genes be carried on a limited number of chromosomes and yet assort independently?

This apparent inconsistency between the principle of independent assortment and the chromosome theory of heredity soon disappeared as biologists began finding genetic characteristics that did not assort independently. One of the first cases was reported in sweet peas by William Bateson, Edith Rebecca Saunders, and Reginald C. Punnett in 1905. They crossed a homozygous strain of peas that had purple flowers and long pollen grains with a homozygous strain that had red flowers and round pollen grains. All the F₁ had purple flowers and long pollen grains, indicating that purple was dominant over red and long was dominant over

round. When they intercrossed the F₁, however, the resulting F₂ progeny did not appear in the 9 : 3 : 3 : 1 ratio expected with independent assortment (**Figure 7.2**). An excess of F₂ plants had purple flowers and long pollen or red flowers and round pollen (the parental phenotypes). Although Bateson, Saunders, and Punnett were unable to explain these results, we now know that the two loci that they examined lie close together on the same chromosome and therefore do not assort independently.

THINK-PAIR-SHARE Question 1

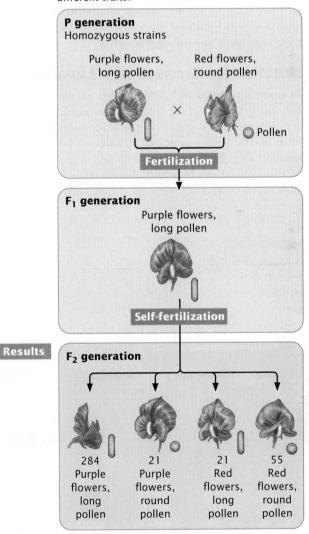

Conclusion: F₂ progeny do not appear in the 9 : 3 : 3 : 1 ratio expected with independent assortment.

7.2 Nonindependent assortment of flower color and pollen shape in sweet peas.

7.2 Linked Genes Segregate Together While Crossing Over Produces Recombination Between Them

Genes that are close together on the same chromosome usually segregate as a unit and are therefore inherited together. However, genes occasionally switch from one homologous chromosome to the other through the process of crossing over (see Chapter 2), as illustrated in **Figure 7.3**. Crossing over results in recombination: it breaks up the associations of genes that are close together on the same chromosome. Linkage and crossing over can be seen as processes that have opposite effects: linkage keeps particular genes together, and crossing over mixes them up, producing new combinations of genes. In Chapter 5, we considered a number of exceptions and extensions to Mendel's principles of heredity. The concept of linked genes adds a further complication to interpretations of the results of genetic crosses. However, with an understanding of how linkage affects heredity, we can analyze crosses for linked genes and successfully predict the types of progeny that will be produced.

Notation for Crosses with Linkage

In analyzing crosses with linked genes, we must know not only the genotypes of the individuals crossed, but also the arrangement of the genes on the chromosomes. To keep track of this arrangement, we introduce a new system of notation for presenting crosses with linked genes. Consider a cross between an individual homozygous for dominant alleles at two linked loci and another individual homozygous for recessive alleles at those loci ($AA\ BB \times aa\ bb$). For linked genes, it's necessary to write out the specific alleles as they are arranged on each of the homologous chromosomes:

$$\frac{A \qquad B}{A \qquad B} \times \frac{a \qquad b}{a \qquad b}$$

In this notation, each horizontal line represents one of the two homologous chromosomes. Because they inherit one chromosome from each parent, the F_1 progeny will have the following genotype:

$$\frac{A \qquad B}{a \qquad b}$$

Here, the importance of designating the alleles on each chromosome is clear. One chromosome has the two dominant alleles A and B, whereas the homologous chromosome has the two recessive alleles a and b. The notation can be simplified by drawing only a single line, with the understanding that genes located on the same side of the line lie on the same chromosome:

$$\frac{A \qquad B}{a \qquad b}$$

This notation can be simplified further by separating the alleles on each chromosome with a slash: AB/ab.

Remember that the two alleles at a locus are always located on different homologous chromosomes and therefore must lie on opposite sides of the line. Consequently, we would *never* write the genotypes as

$$\frac{A \qquad a}{B \qquad b}$$

because the alleles A and a can *never* be on the same chromosome. It is also important to keep the genes in the same order on both sides of the line; thus, we would *never* write

$$\frac{A \qquad B}{b \qquad a}$$

because it would imply that alleles A and b are allelic (at the same locus).

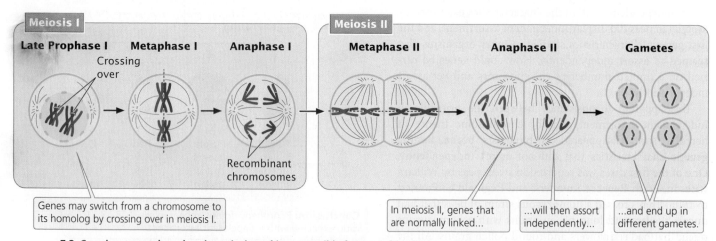

Meiosis I

Late Prophase I   **Metaphase I**   **Anaphase I**

Crossing over

Recombinant chromosomes

Genes may switch from a chromosome to its homolog by crossing over in meiosis I.

Meiosis II

Metaphase II   **Anaphase II**   **Gametes**

In meiosis II, genes that are normally linked...

...will then assort independently...

...and end up in different gametes.

7.3 Crossing over takes place in meiosis and is responsible for recombination.

Complete Linkage Compared with Independent Assortment

In this section, we first consider what happens to genes that exhibit complete linkage, meaning that they are located very close together on the same chromosome and do not exhibit crossing over. Genes are rarely completely linked, but by assuming that no crossing over takes place, we can see the effect of linkage more clearly. We then consider what happens when genes assort independently. Finally, we consider the results obtained if the genes are linked but exhibit some crossing over.

A testcross reveals the effects of linkage. For example, if a heterozygous individual is test-crossed with a homozygous recessive individual ($Aa\ Bb \times aa\ bb$), the alleles that are present in the gametes contributed by the heterozygous parent will be expressed in the phenotype of the offspring because the homozygous parent cannot contribute dominant alleles that might mask them. Consequently, traits that appear in the progeny reveal which alleles were transmitted by the heterozygous parent.

Consider a pair of linked genes in tomato plants. One of the genes affects the type of leaf: an allele for mottled leaves (m) is recessive to an allele that produces normal leaves (M). Nearby on the same chromosome, another gene determines the height of the plant: an allele for dwarf (d) is recessive to an allele for tall (D). Testing for linkage can be done with a testcross, which requires a plant that is heterozygous for both characteristics. A geneticist might produce this heterozygous plant by crossing a variety of tomato that is homozygous for normal leaves and tall height with a variety that is homozygous for mottled leaves and dwarf height:

$$\text{P} \qquad \frac{M \quad D}{M \quad D} \times \frac{m \quad d}{m \quad d}$$

$$\downarrow$$

$$\text{F}_1 \qquad \frac{M \quad D}{m \quad d}$$

The geneticist would then use this F_1 heterozygote in a testcross, crossing it with a plant that is homozygous for mottled leaves and dwarf height:

$$\frac{M \quad D}{m \quad d} \times \frac{m \quad d}{m \quad d}$$

The results of this testcross are diagrammed in **Figure 7.4a**. The heterozygote produces two types of gametes: some with the $\underline{M \quad D}$ chromosome and others with the $\underline{m \quad d}$ chromosome. Because no crossing over takes place, these gametes are the only types produced by the heterozygote. Notice that the gametes produced by the heterozygote contain only combinations of alleles that were present in its parents (the P generation):

either the allele for normal leaves together with the allele for tall height (M and D) or the allele for mottled leaves together with the allele for dwarf height (m and d). Gametes that contain only original combinations of alleles that were present in the parents are **nonrecombinant gametes**, or *parental gametes*.

The homozygous parent in the testcross produces only one type of gamete, which contains chromosome $\underline{m \quad d}$. When its gametes pair with the two types of gametes generated by the heterozygous parent (see Figure 7.4a), two types of progeny result: half have normal leaves and are tall:

$$\frac{M \quad D}{m \quad d}$$

and half have mottled leaves and are dwarf:

$$\frac{m \quad d}{m \quad d}$$

These progeny display the original combinations of traits present in the P generation and are **nonrecombinant progeny**, or *parental progeny*. No new combinations of the two traits, such as normal leaves with dwarf height or mottled leaves with tall height, appear in the offspring because the genes affecting the two traits are completely linked and are inherited together. New combinations of traits could arise only if the physical connection between M and D or between m and d were broken.

These results are distinctly different from the results that are expected when genes assort independently (**Figure 7.4b**). If the M and D loci assorted independently, the heterozygous plant ($Mm\ Dd$) would produce four types of gametes: two nonrecombinant gametes containing the original combinations of alleles ($M\ D$ and $m\ d$) and two gametes containing new combinations of alleles ($M\ d$ and $m\ D$). Gametes with new combinations of alleles are called **recombinant gametes**. With independent assortment, nonrecombinant and recombinant gametes are produced in equal proportions. These four types of gametes join with the single type of gamete produced by the homozygous parent of the testcross to produce four kinds of progeny in equal proportions (see Figure 7.4b). The progeny with new combinations of traits formed from recombinant gametes are termed **recombinant progeny**.

CONCEPTS

A testcross in which one of the individuals is heterozygous for two completely linked genes yields two types of progeny, each type displaying one of the original combinations of traits present in the P generation. In contrast, independent assortment produces four types of progeny in a 1 : 1 : 1 : 1 ratio—two types of recombinant progeny and two types of nonrecombinant progeny in equal proportions.

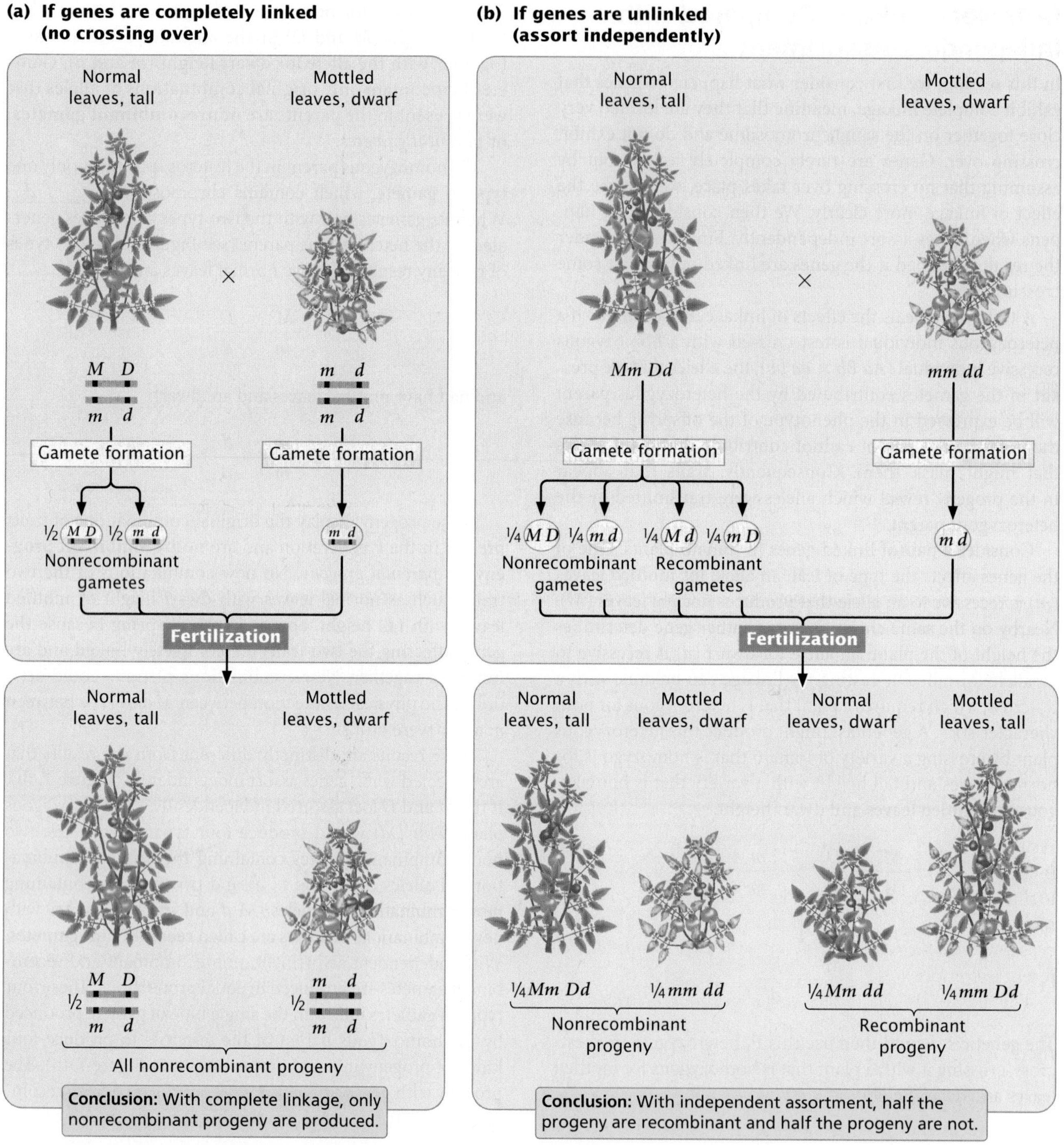

**(a) If genes are completely linked
(no crossing over)**

**(b) If genes are unlinked
(assort independently)**

7.4 A testcross reveals the effects of linkage. Results of a testcross for two loci in tomatoes that determine leaf type and plant height.

Crossing Over Between Linked Genes

Usually, there is some crossing over between genes that lie on the same chromosome, which produces new combinations of traits. Genes that exhibit crossing over are said to be incompletely linked. Let's see how incomplete linkage affects the results of a cross.

THEORY The effect of crossing over on the inheritance of two linked genes is shown in **Figure 7.5**. Crossing over, which takes place in prophase I of meiosis, is the exchange of genetic material between nonsister chromatids (see Figures 2.16 and 2.18). After a single crossover has taken place, the two chromatids that did not participate in

(a) No crossing over

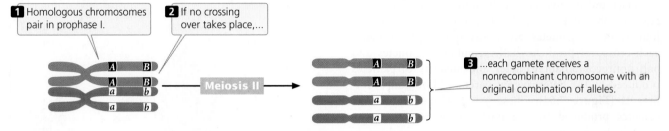

1 Homologous chromosomes pair in prophase I.

2 If no crossing over takes place,...

Meiosis II

3 ...each gamete receives a nonrecombinant chromosome with an original combination of alleles.

(b) Crossing over

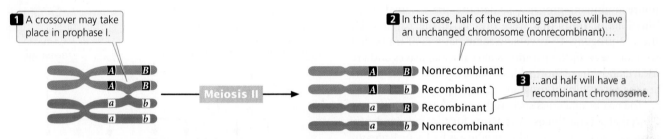

1 A crossover may take place in prophase I.

2 In this case, half of the resulting gametes will have an unchanged chromosome (nonrecombinant)...

Meiosis II

Nonrecombinant
Recombinant
Recombinant
Nonrecombinant

3 ...and half will have a recombinant chromosome.

7.5 A single crossover produces half nonrecombinant gametes and half recombinant gametes.

crossing over are unchanged; gametes that receive these chromatids are nonrecombinants. The other two chromatids, which did participate in crossing over, now contain new combinations of alleles; gametes that receive these chromatids are recombinants. For each meiosis in which a single crossover takes place, two nonrecombinant gametes and two recombinant gametes are produced. This result is the same as that produced by independent assortment (see Figure 7.4b), so if crossing over between two loci takes place in every meiosis, it is impossible to determine whether the genes are on the same chromosome and crossing over took place or whether the genes are on different chromosomes.

For closely linked genes, however, crossing over does not take place in every meiosis. In meioses in which there is no crossing over, only nonrecombinant gametes are produced. In meioses in which there is a single crossover, half the gametes are recombinants and half are nonrecombinants (because a single crossover affects only two of the four chromatids). Because each crossover leads to half recombinant gametes and half nonrecombinant gametes, the total percentage of recombinant gametes is always half the percentage of meioses in which crossing over takes place. Even if crossing over between two genes takes place in every meiosis, only 50% of the resulting gametes are recombinants. Thus, the frequency of recombinant gametes is always half the frequency of crossing over, and the maximum proportion of recombinant gametes is 50%.

CONCEPTS

Linkage between genes causes them to be inherited together and reduces recombination; crossing over breaks up the associations of such genes. In a testcross for two linked genes, each crossover produces two recombinant gametes and two nonrecombinants. The frequency of recombinant gametes is half the frequency of crossing over, and the maximum frequency of recombinant gametes is 50%.

✔ CONCEPT CHECK 1

For single crossovers, the frequency of recombinant gametes is half the frequency of crossing over because

a. a testcross between a homozygote and a heterozygote produces $1/2$ heterozygous and $1/2$ homozygous progeny.
b. the frequency of recombination is always 50%.
c. each crossover takes place between only two of the four chromatids of a homologous pair.
d. crossovers take place in about 50% of meioses.

APPLICATION Let's apply what we have learned about linkage and recombination to a cross between tomato plants that differ in the genes that encode leaf type and plant height. Assume now that these genes are linked and that some crossing over takes place between them. Suppose a geneticist carried out the testcross described earlier:

$$\frac{M \quad D}{m \quad d} \times \frac{m \quad d}{m \quad d}$$

When crossing over takes place between the genes for leaf type and height, two of the four gametes produced are recombinants. When there is no crossing over, all four resulting gametes are nonrecombinants. Because each crossover produces half recombinant gametes and half nonrecombinant gametes, the majority of gametes produced by the heterozygous parent will be nonrecombinants (**Figure 7.6a**). These gametes then unite with gametes produced by the homozygous recessive parent, which contain only the recessive alleles, resulting in mostly nonrecombinant progeny and a few recombinant progeny (**Figure 7.6b**). In this testcross, we see that 55 of the progeny have normal leaves and are tall and that 53 have mottled leaves and are dwarf. These plants are the nonrecombinant progeny, containing the original combinations of traits that were present in the parents. Of the 123 progeny, 15 have new combinations of traits that were not seen in the parents: 8 have normal leaves and are dwarf, and 7 have mottled leaves and are tall. These plants are the recombinant progeny.

The results of a cross such as the one illustrated in Figure 7.6 reveal several things. A testcross for two independently assorting genes is expected to produce a 1 : 1 : 1 : 1 phenotypic ratio in the progeny. The progeny of this cross clearly do not exhibit such a ratio, so we might suspect that the genes are not assorting independently. When linked genes undergo some crossing over, the result is mostly nonrecombinant progeny and a few recombinant progeny. This result is what we observe among the progeny of the testcross illustrated in Figure 7.6, so we conclude that the two genes show evidence of linkage with some crossing over.

Calculating Recombination Frequency

The percentage of recombinant progeny produced in a cross is called the **recombination frequency** (or *rate of recombination*), which is calculated as follows:

$$\text{recombination frequency} = \frac{\text{number of recombinant progeny}}{\text{total number of progeny}} \times 100\%$$

In the testcross shown in Figure 7.6, 15 progeny exhibit new combinations of traits, so the recombination frequency is

$$\frac{8 + 7}{55 + 53 + 8 + 7} \times 100\% = \frac{15}{123} \times 100\% = 12.2\%$$

Thus, 12.2% of the progeny exhibit new combinations of traits resulting from crossing over. The recombination frequency can also be expressed as a decimal fraction (0.122).

> TRY PROBLEM 15

THINK-PAIR-SHARE Question 2

(a)

Normal leaves, tall ✕ Mottled leaves, dwarf

Meioses with and without crossing over together result in less than 50% recombination on average.

M D / m d   m d / m d

Gamete formation   Gamete formation

No crossing over   Crossing over

M D m d   M D m d   M d m D m d

Nonrecombinant gametes (100%)   Nonrecombinant gametes (50%)   Recombinant gametes (50%)

Fertilization

(b)

Normal leaves, tall   Mottled leaves, dwarf   Normal leaves, dwarf   Mottled leaves, tall

M D / m d   m d / m d   M d / m d   m D / m d

55   53   Progeny number   8   7

Nonrecombinant progeny   Recombinant progeny

Conclusion: With linked genes and some crossing over, nonrecombinant progeny predominate.

7.6 Crossing over between linked genes produces nonrecombinant and recombinant offspring. In this testcross, genes are linked, and there is some crossing over.

Coupling and Repulsion

In crosses for linked genes, the arrangement of alleles on the homologous chromosomes is critical in determining the outcome of the cross. For example, consider the inheritance of two genes in the Australian blowfly, *Lucilia cuprina*. In this species, one locus determines the color of the thorax: a purple thorax (p) is recessive to the normal green thorax (p^+). A second locus determines the color of the puparium: a black puparium (b) is recessive to the normal brown puparium (b^+). The loci for thorax color and puparium color are located close together on the same chromosome. Suppose that we test-cross a fly that is heterozygous at both loci with a fly that is homozygous recessive at both. Because these genes are linked, there are two possible arrangements on the chromosomes of the heterozygous fly. The dominant alleles for green thorax (p^+) and brown puparium (b^+) might reside on one chromosome of the homologous pair, and the recessive alleles for purple thorax (p) and black puparium (b) might reside on the other homologous chromosome:

$$\frac{p^+ \quad b^+}{p \quad b}$$

This arrangement, in which wild-type alleles are found on one chromosome and mutant alleles are found on the other chromosome, is referred to as the **coupling**, or **cis**, **configuration**. Alternatively, one chromosome might carry the alleles for green thorax (p^+) and black puparium (b), and the other chromosome might carry the alleles for purple thorax (p) and brown puparium (b^+):

$$\frac{p^+ \quad b}{p \quad b^+}$$

This arrangement, in which each chromosome contains one wild-type and one mutant allele, is called the **repulsion**, or **trans**, **configuration**. Whether the alleles in the heterozygous parent are in coupling or repulsion determines which phenotypes will be most common among the progeny of a testcross.

When the alleles of the heterozygous parent are in the coupling configuration, the most numerous progeny types are those with a green thorax and brown puparium and those with a purple thorax and black puparium (**Figure 7.7a**). However, when the alleles are in repulsion, the most numerous progeny types are those with a green thorax and black

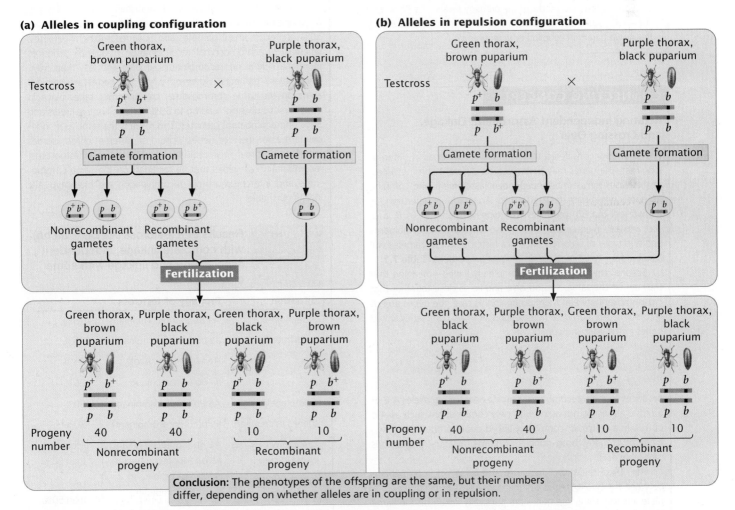

(a) Alleles in coupling configuration

(b) Alleles in repulsion configuration

Conclusion: The phenotypes of the offspring are the same, but their numbers differ, depending on whether alleles are in coupling or in repulsion.

7.7 The arrangement (coupling or repulsion) of linked genes on a chromosome affects the results of a testcross.
Linked loci in the Australian blowfly (*Lucilia cuprina*) determine the color of the thorax and that of the puparium.

puparium and those with a purple thorax and brown puparium (**Figure 7.7b**). Notice that the genotypes of the parents in Figure 7.7a and b are the same ($p^+p\ b^+b \times pp\ bb$) and that the dramatic difference in the phenotypic ratios of the progeny in the two crosses results entirely from the configuration—coupling or repulsion—of the chromosomes. Knowledge of the arrangement of the alleles on the chromosomes is essential to accurately predicting the outcome of crosses in which genes are linked.

THINK-PAIR-SHARE Question 3

CONCEPTS

The arrangement of linked alleles on the chromosomes is critical for determining the outcome of a cross. When two wild-type alleles are on one homologous chromosome and two mutant alleles are on the other, the alleles are in the coupling configuration; when each chromosome contains one wild-type allele and one mutant allele, the alleles are in repulsion.

✔ CONCEPT CHECK 2

The following testcross produces the progeny shown: $Aa\ Bb \times aa\ bb$ → 10 $Aa\ Bb$, 40 $Aa\ bb$, 40 $aa\ Bb$, 10 $aa\ bb$. Were the A and B alleles in the $Aa\ Bb$ parent in coupling or in repulsion?

CONNECTING CONCEPTS

Relating Independent Assortment, Linkage, and Crossing Over

We have now considered three situations for genes at different loci. First, the genes may be located on different chromosomes; in this case, they exhibit independent assortment and combine randomly when gametes are formed. An individual heterozygous at two loci ($Aa\ Bb$) produces four types of gametes ($A\ B$, $a\ b$, $A\ b$, and $a\ B$) in equal proportions: two types of nonrecombinants and two types of recombinants. In a testcross, these gametes will result in four types of progeny in equal proportions (**Table 7.1**).

Second, the genes may be completely linked—meaning that they are on the same chromosome and lie so close together that crossing over between them is rare. In this case, the genes do not recombine. An individual heterozygous for two completely linked genes in the coupling configuration,

$$\frac{A \qquad B}{a \qquad b}$$

produces only nonrecombinant gametes containing alleles $A\ B$ or $a\ b$; the alleles do not assort into new combinations such as $A\ b$ or $a\ B$. In a testcross, completely linked genes produce only two types of progeny, both nonrecombinants, in equal proportions (see Table 7.1).

The third situation, incomplete linkage, is intermediate between the two extremes of independent assortment and complete linkage. Here, the genes are physically linked on the same chromosome, which prevents independent assortment. However,

occasional crossovers break up the linkage and allow the genes to recombine. With incomplete linkage, an individual heterozygous at two loci produces four types of gametes—two types of recombinants and two types of nonrecombinants—but the nonrecombinants are produced more frequently than the recombinants because crossing over between these loci does not take place in every meiosis. In the testcross, these gametes result in four types of progeny, with the nonrecombinants more frequent than the recombinants (see Table 7.1).

Earlier in the chapter, the term *recombination* was defined as the sorting of alleles into new combinations. We've now considered two types of recombination that differ in their mechanisms. Interchromosomal recombination takes place between genes located on *different* chromosomes. It arises from independent assortment—the random segregation of chromosomes in anaphase I of meiosis—and is the kind of recombination that Mendel discovered while studying dihybrid crosses. A second type of recombination, intrachromosomal recombination, takes place between genes located on the *same* chromosome. This recombination arises from crossing over—the exchange of genetic material in prophase I of meiosis. Both recombination mechanisms produce new allele combinations in the gametes, so they cannot be distinguished by examining the types of gametes produced. Nevertheless, they can often be distinguished by the *frequencies* of types of gametes: interchromosomal recombination produces 50% nonrecombinant gametes and 50% recombinant gametes, whereas intrachromosomal recombination frequently produces more than 50% nonrecombinant gametes and less than 50% recombinant gametes. However, when the genes are very far apart on the same chromosome, crossing over takes place in every meiotic division, leading to 50% recombinant gametes and 50% nonrecombinant gametes. This result is the same as in independent assortment of genes located on different chromosomes (interchromosomal recombination). Thus, intrachromosomal recombination of genes that lie far apart on the same chromosome and interchromosomal recombination are phenotypically indistinguishable.

TABLE 7.1	Results of a testcross ($Aa\ Bb \times aa\ bb$) with complete linkage, independent assortment, and linkage with some crossing over	
Situation	**Progeny of Testcross**	
Independent assortment	$Aa\ Bb$ (nonrecombinant)	25%
	$aa\ bb$ (nonrecombinant)	25%
	$Aa\ bb$ (recombinant)	25%
	$aa\ Bb$ (recombinant)	25%
Complete linkage (genes in coupling)	$Aa\ Bb$ (nonrecombinant)	50%
	$aa\ bb$ (nonrecombinant)	50%
Linkage with some crossing over (genes in coupling)	$Aa\ Bb$ (nonrecombinant) $\Big\}$ $aa\ bb$ (nonrecombinant)	More than 50%
	$Aa\ bb$ (recombinant) $\Big\}$ $aa\ Bb$ (recombinant)	Less than 50%

Evidence for the Physical Basis of Recombination

Walter Sutton's chromosome theory of inheritance, which stated that genes are physically located on chromosomes (see Chapter 3), was supported by Nettie Stevens and Edmund Wilson's discovery that sex was associated with a specific chromosome in insects and by Calvin Bridges's demonstration that nondisjunction of X chromosomes was related to the inheritance of eye color in *Drosophila* (see Chapter 4). Further evidence for the chromosome theory of heredity came in 1931, when Harriet Creighton and Barbara McClintock (**Figure 7.8**) obtained evidence that intrachromosomal recombination was the result of physical exchange between chromosomes. Creighton and McClintock discovered a strain of corn that had an abnormal chromosome 9, containing a densely staining knob at one end and a small piece of another chromosome attached to the other end. This aberrant chromosome allowed them to visually distinguish the two members of a homologous pair.

Creighton and McClintock focused on the inheritance of two traits in corn determined by genes on chromosome 9. At one locus, a dominant allele (*C*) produced colored kernels, whereas a recessive allele (*c*) produced colorless kernels. At a second, linked locus, a dominant allele (*Wx*) produced starchy kernels, whereas a recessive allele (*wx*) produced waxy kernels. They obtained a plant that was heterozygous at both loci in repulsion, with the alleles for colored and waxy on the aberrant chromosome and the alleles for colorless and starchy on the normal chromosome:

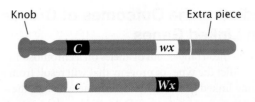

They then crossed this heterozygous plant with one that was homozygous for colorless and heterozygous for waxy (with both chromosomes normal):

$$\frac{C}{c} \quad \frac{wx}{Wx} \times \frac{c}{c} \quad \frac{Wx}{wx}$$

This cross produces different combinations of traits in the progeny, but the only way that colorless and waxy progeny can arise is through crossing over in the doubly heterozygous parent:

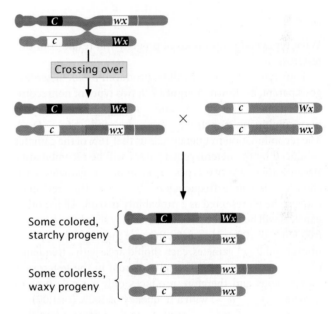

Note: Not all progeny genotypes are shown.

Notice that, if crossing over entails physical exchange between the chromosomes, then the colorless, waxy progeny resulting from recombination should have a chromosome with an extra piece, but not a knob. Furthermore, some of the colored, starchy progeny should have a chromosome with a knob, but not the extra piece. This outcome is precisely what Creighton and McClintock observed, confirming the chromosomal theory of inheritance. Curt Stern provided a similar demonstration by using chromosomal markers in *Drosophila* at about the same time. We will examine the molecular basis of recombination in more detail in Chapter 12.

7.8 Barbara McClintock (left) and Harriet Creighton (right) provided evidence that genes are located on chromosomes. [Karl Maramorosch/Courtesy of Cold Spring Harbor Laboratory Archives.]

Predicting the Outcomes of Crosses with Linked Genes

Knowing the arrangement of alleles on a chromosome allows us to predict the types of progeny that will result from a cross entailing linked genes and to determine which of these types will be the most numerous. Determining the *proportions* of the types of offspring requires an additional piece of information: the recombination frequency. The recombination frequency provides us with information about how often the alleles in the gametes appear in new combinations, allowing us to predict the proportions of offspring phenotypes that will result from a specific cross that entails linked genes.

In cucumbers, smooth fruit (t) is recessive to warty fruit (T) and glossy fruit (d) is recessive to dull fruit (D). Geneticists have determined that these two genes exhibit a recombination frequency of 16%. Suppose that we cross a plant that is homozygous for warty and dull fruit with a plant that is homozygous for smooth and glossy fruit, and then carry out a testcross using the F_1:

$$\frac{T \quad D}{t \quad d} \times \frac{t \quad d}{t \quad d}$$

What types and proportions of progeny will result from this testcross?

Four types of gametes will be produced by the heterozygous parent, as shown in **Figure 7.9**: two types of nonrecombinant gametes ($\underline{T \quad D}$ and $\underline{t \quad d}$) and two types of recombinant gametes ($\underline{T \quad d}$ and $\underline{t \quad D}$). The recombination frequency tells us that 16% of the gametes produced by the heterozygous parent will be recombinants. Because there are two types of recombinant gametes, each should arise with a frequency of $^{16}/_2 = 8\%$. This frequency can also be represented as a probability of 0.08. All the other gametes will be nonrecombinants, so they should arise with a frequency of $100 - 16 = 84\%$. Because there are two types of nonrecombinant gametes, each should arise with a frequency of $^{84}/_2 = 42\%$ (or 0.42). The other parent in the testcross is homozygous and therefore produces only a single type of gamete ($\underline{t \quad d}$) with a frequency of 100% (or 1.00).

Four types of progeny result from the testcross (see Figure 7.9). The expected proportion of each progeny type can be determined by using the multiplication rule (see Chapter 3), multiplying together the probability of each gamete. For example, testcross progeny with warty and dull fruit

$$\frac{T \quad D}{t \quad d}$$

appear with a frequency of 0.42 (the probability of inheriting a gamete with chromosome $\underline{T \quad D}$ from the heterozygous parent) $\times$ 1.00 (the probability of inheriting a gamete with chromosome $\underline{t \quad d}$ from the recessive parent) = 0.42. The proportions of the other types of F_2 progeny can be calculated in a similar manner (see Figure 7.9). This method can be used for predicting the outcome of any cross with linked genes for which the recombination frequency is known.

THINK-PAIR-SHARE Question 4

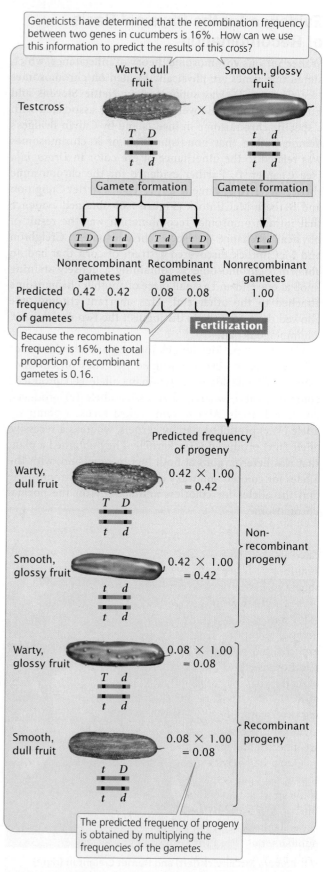

7.9 The recombination frequency allows us to predict the proportions of offspring expected for a cross entailing linked genes.

Testing for Independent Assortment

In some crosses, it is obvious that the genes are linked because there are clearly more nonrecombinant progeny than recombinant progeny. In other crosses, the difference between independent assortment and linkage isn't as obvious. For example, suppose we did a testcross for two pairs of alleles, such as *Aa Bb* × *aa bb*, and observed the following numbers of progeny: 54 *Aa Bb*, 56 *aa bb*, 42 *Aa bb*, and 48 *aa Bb*. Is this outcome the 1 : 1 : 1 : 1 ratio we would expect if *A* and *B* assorted independently? Not exactly, but it's pretty close. Perhaps these genes assorted independently and chance produced the slight deviations between the observed numbers and the expected 1 : 1 : 1 : 1 ratio. Alternatively, the genes might be linked, but considerable crossing over might be taking place between them, and so the number of nonrecombinants is only slightly greater than the number of recombinants. How do we distinguish between the role of chance and the role of linkage in producing deviations from the results expected with independent assortment?

We encountered a similar problem in crosses in which genes were unlinked: the problem of distinguishing between deviations due to chance and those due to other factors. We addressed this problem (in Chapter 3) with the chi-square goodness-of-fit test, which helps us evaluate the likelihood that chance alone is responsible for deviations between the numbers of progeny that we observe and the numbers that we expect according to the principles of inheritance. Here, we are interested in a different question: Is the inheritance of alleles at one locus independent of the inheritance of alleles at a second locus? If the answer to this question is yes, then the genes are assorting independently; if the answer is no, then the genes are probably linked.

A possible way to test for independent assortment is to calculate the expected probability of each progeny type, assuming independent assortment, and then use the chi-square goodness-of-fit test to evaluate whether the observed numbers deviate significantly from the expected numbers. With independent assortment, we expect $\frac{1}{4}$ of each phenotype: $\frac{1}{4}$ *Aa Bb*, $\frac{1}{4}$ *aa bb*, $\frac{1}{4}$ *Aa bb*, and $\frac{1}{4}$ *aa Bb*. This expected probability of each genotype is based on the multiplication rule (see Chapter 3). For example, if the probability of *Aa* is $\frac{1}{2}$ and the probability of *Bb* is $\frac{1}{2}$, then the probability of *Aa Bb* is $\frac{1}{2} \times \frac{1}{2} = \frac{1}{4}$. In this calculation, we are making two assumptions: (1) that the probability of each single-locus genotype is $\frac{1}{2}$, and (2) that genotypes at the two loci are inherited independently ($\frac{1}{2} \times \frac{1}{2} = \frac{1}{4}$).

One problem with this approach is that a significant chi-square value can result from a violation of either assumption. If the genes are linked, then the inheritance of genotypes at the two loci is not independent (assumption 2), and we will get a significant deviation between observed and expected numbers. But we can also get a significant deviation if the probability of each single-locus genotype is not $\frac{1}{2}$ (assumption 1), even when the genotypes are assorting independently. We may obtain a significant deviation, for example, if individuals with one genotype have a lower probability

of surviving or if the penetrance of a genotype is not 100%. We could test both assumptions by conducting a series of chi-square tests, first testing the inheritance of genotypes at each locus separately (assumption 1) and then testing for independent assortment (assumption 2). However, a faster method is to test for independence in genotypes with a *chi-square test of independence*.

THE CHI-SQUARE TEST OF INDEPENDENCE The chi-square test of independence allows us to evaluate whether the segregation of alleles at one locus is independent of the segregation of alleles at another locus without making any assumption about the probability of single-locus genotypes. In the chi-square goodness-of-fit test, the expected value is based on a theoretical relation: for example, the expected ratio in a genetic cross. For the chi-square test of independence, the expected value is based strictly on the observed values, along with the assumption that they are independent.

To illustrate this analysis, let's examine the results of a cross between German cockroaches, in which yellow body (*y*) is recessive to brown body (y^+) and curved wings (*cv*) are recessive to straight wings (cv^+). A testcross ($y^+y\ cv^+cv \times yy\ cvcv$) produces the progeny shown in **Figure 7.10a**. If the segregation of alleles at each locus is independent, then the proportions of progeny with y^+y and *yy* genotypes should be the same for cockroaches with genotype cv^+cv and for cockroaches with genotype *cvcv*. The converse is also true: the proportions of progeny with cv^+cv and *cvcv* genotypes should be the same for cockroaches with genotype y^+y and for cockroaches with genotype *yy*.

To determine whether the proportions of progeny with genotypes at the two loci are independent, we first construct a table of the observed numbers of progeny, somewhat like a Punnett square, except that we put the genotypes that result from the segregation of alleles at one locus along the top and the genotypes that result from the segregation of alleles at the other locus along the side (**Figure 7.10b**). Next, we compute the total for each row, the total for each column, and the grand total (the sum of all row totals or the sum of all column totals, which should be the same). These totals will be used to compute the expected values for the chi-square test of independence.

Our next step is to compute the expected numbers for each combination of genotypes (each cell in the table) under the assumption that the segregation of alleles at the *y* locus is independent of the segregation of alleles at the *cv* locus. If the segregation of alleles at each locus is independent, the expected number for each cell can be computed with the following formula:

$$\text{expected number} = \frac{\text{row total} \times \text{column total}}{\text{grand total}}$$

For the cell of the table corresponding to genotype $y^+y\ cv^+cv$ (the upper-left cell of the table in Figure 7.10b), the expected number is

(a)

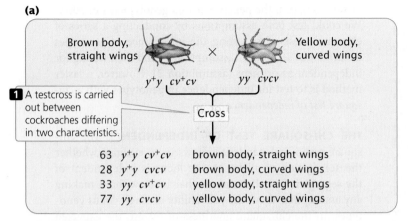

Brown body, straight wings × Yellow body, curved wings

$y^+y \ cv^+cv$ $yy \ cvcv$

1 A testcross is carried out between cockroaches differing in two characteristics.

Cross

63 $y^+y \ cv^+cv$ brown body, straight wings
28 $y^+y \ cvcv$ brown body, curved wings
33 $yy \ cv^+cv$ yellow body, straight wings
77 $yy \ cvcv$ yellow body, curved wings

(b) Contingency table

2 To test for independent assortment of alleles encoding the two traits, a table is constructed...

3 ...with genotypes for one locus along the top...

Segregation of y^+ and y

4 ...and genotypes for the other locus along the left side.

5 Numbers of each genotype are placed in the table cells, and the row totals, column totals, and grand total are computed.

	y^+y	yy	Row totals
cv^+cv	63	33	96
$cv \ cv$	28	77	105
Column totals	91	110	201
			Grand total

Segregation of cv^+ and cv (left side label)

(c)

Genotype	Number observed	Number expected $\left(\dfrac{\text{row total} \times \text{column total}}{\text{grand total}}\right)$
$y^+y \ cv^+cv$	63	$\dfrac{96 \times 91}{201} = 43.46$
$y^+y \ cvcv$	28	$\dfrac{105 \times 91}{201} = 47.54$
$yy \ cv^+cv$	33	$\dfrac{96 \times 110}{201} = 52.46$
$yy \ cvcv$	77	$\dfrac{105 \times 110}{201} = 57.46$

6 The expected numbers of progeny, assuming independent assortment, are calculated.

(d)

7 A chi-square value is calculated.

$$\chi^2 = \sum \frac{(\text{observed} - \text{expected})^2}{\text{expected}}$$

$$= \frac{(63 - 43.46)^2}{43.46} + \frac{(28 - 47.54)^2}{47.54} + \frac{(33 - 52.54)^2}{52.54} + \frac{(77 - 57.46)^2}{57.46}$$

$$= 8.79 + 8.03 + 7.27 + 6.64$$

$$= 30.73$$

(e)

$$\text{df} = (\text{number of rows} - 1) \times (\text{number of columns} - 1)$$
$$\text{df} = (2 - 1) \times (2 - 1) = 1 \times 1 = 1$$
$$P < 0.005$$

8 The probability is less than 0.005, indicating that the *difference* between numbers of observed and expected progeny is probably not due to chance.

Conclusion: The genes for body color and type of wing are not assorting independently and must be linked.

$$\frac{96 \ (\text{row total}) \times 91 (\text{column total})}{201 \ (\text{grand total})} = \frac{8736}{201} = 43.46$$

The expected numbers for each cell calculated by this method are given in **Figure 7.10c**.

We now calculate a chi-square value by using the same formula that we used for the chi-square goodness-of-fit test in Chapter 3:

$$x^2 = \sum \frac{(\text{observed} - \text{expected})^2}{(\text{expected})}$$

Recall that $\sum$ means "sum" and that we are adding together the (observed − expected)^2/expected values for the four types of progeny. The observed and expected numbers of cockroaches from the testcross give us a calculated chi-square value of 30.73 (**Figure 7.10d**).

To determine the probability associated with this chi-square value, we need the degrees of freedom. Recall from Chapter 3 that the degrees of freedom are the number of ways in which the observed classes are free to vary from the expected values. In general, for the chi-square test of independence, the degrees of freedom equal the number of rows in the table minus 1, multiplied by the number of columns in the table minus 1 (**Figure 7.10e**), or

$$\text{df} = (\text{number of rows} - 1) \times (\text{number of columns} - 1)$$

In our example, there are two rows and two columns, and so the degrees of freedom are

$$\text{df} = (2 - 1) \times (2 - 1) = 1 \times 1 = 1$$

Therefore, our calculated chi-square value is 30.73, with 1 degree of freedom. We can use Table 3.7 to find the associated probability. Looking at Table 3.7, we find that our calculated chi-square value is larger than the largest chi-square value given for 1 degree of freedom, which has a probability of 0.005. Thus, our calculated chi-square value has a probability of less than 0.005. This very small probability indicates that the genotypes are not in the proportions that we would expect if independent assortment were taking place. Our conclusion, then, is that these genes are not assorting independently and must be linked. As in the case of the chi-square goodness-of-fit test, geneticists generally consider that any chi-square value for the test of independence with a probability of less than 0.05 means that the observed values are significantly different from the expected values and is therefore evidence that the genes are not assorting independently. **> TRY PROBLEM 16**

7.10 A chi-square test of independence can be used to determine if genes at two loci are assorting independently.

Gene Mapping with Recombination Frequencies

Thomas Hunt Morgan and his students developed the idea that physical distances between genes on a chromosome are related to their rates of recombination. They hypothesized that crossover events take place more or less at random up and down the chromosome and that two genes that lie far apart are more likely to undergo a crossover than are two genes that lie close together. They proposed that recombination frequencies could provide a convenient way to determine the order of genes along a chromosome and would give estimates of the relative distances between the genes. Chromosome maps calculated by using the genetic phenomenon of recombination are called **genetic maps**. In contrast, chromosome maps calculated by using physical distances along the chromosome (often expressed as numbers of base pairs) are called **physical maps**.

Distances on genetic maps are measured in **map units** (abbreviated m.u.); one map unit equals a 1% recombination rate. Map units are also called **centiMorgans** (cM), in honor of Thomas Hunt Morgan. Genetic distances measured with recombination rates are approximately additive: if the distance from gene A to gene B is 5 m.u., the distance from gene B to gene C is 10 m.u., and the distance from gene A to gene C is 15 m.u., then gene B must be located between genes A and C. On the basis of the map distances just given, we can draw a simple genetic map for genes A, B, and C, as shown here:

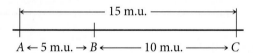

We could just as plausibly draw this map with C on the left and A on the right:

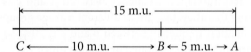

Both maps are correct and equivalent because, with information about the relative positions of only three genes, the most that we can determine is which gene lies in the middle. If we obtained distances to an additional gene, then we could position A and C relative to that gene. An additional gene D, examined through genetic crosses, might yield the following recombination frequencies:

Gene pair	Recombination frequency (%)
A and D	8
B and D	13
C and D	23

Notice that C and D exhibit the highest rate of recombination; therefore, C and D must be farthest apart, with genes A and B between them. Using the recombination frequencies and remembering that 1 m.u. = 1% recombination, we can now add D to our map:

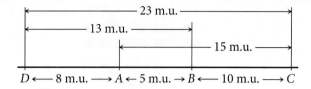

Thus, by doing a series of crosses between pairs of genes, we can construct genetic maps showing the linkage arrangements of a number of genes.

Two points about constructing chromosome maps from recombination frequencies should be emphasized. First, recall that we cannot distinguish between genes on different chromosomes and genes located far apart on the same chromosome. If genes exhibit 50% recombination, the most that can be said about them is that they belong to different linkage groups, either on different chromosomes or far apart on the same chromosome.

The second point is that a testcross for two genes that are far apart on the same chromosome tends to underestimate the true physical distance between them because the cross does not reveal double crossovers that might take place between the two genes (**Figure 7.11**). A double crossover arises when two separate crossover events take place between two loci. (For now, we will consider only double crossovers that take place between two of the four chromatids of a homologous pair—two-strand double crossovers. Double crossovers that take place among three and four chromatids will be considered later, in the section Effects of Multiple Crossovers.) Whereas a single crossover between two genes produces combinations of alleles that were not present on the original parental chromosomes, a second crossover between the same two genes reverses the effects of the first, thus restoring the original parental combination of alleles (see Figure 7.11). We therefore cannot distinguish between the progeny produced by two-strand double crossovers and the progeny produced when there is no crossing over at all. As we see in the next section, however, we can detect double crossovers if we examine a third gene that lies between the two crossovers. Because double crossovers between two genes go undetected, map distances will be underestimated whenever double crossovers take place. Double crossovers are more frequent between genes that are far apart; therefore, genetic maps based on short distances are usually more accurate than those based on longer distances.

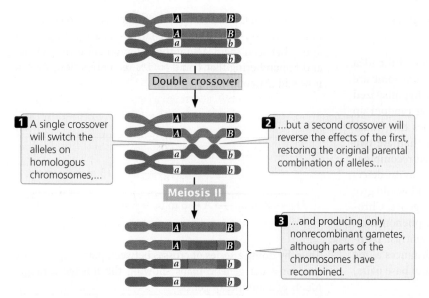

1 A single crossover will switch the alleles on homologous chromosomes,...

2 ...but a second crossover will reverse the effects of the first, restoring the original parental combination of alleles...

3 ...and producing only nonrecombinant gametes, although parts of the chromosomes have recombined.

7.11 A two-strand double crossover between two linked genes produces only nonrecombinant gametes.

CONCEPTS

Recombination frequencies can be used to determine the order of the genes on a chromosome and calculate the approximate distances from one gene to another. In genetic maps, 1% recombination equals 1 map unit, or 1 centiMorgan. Double crossovers between two genes go undetected, so map distances between distant genes tend to underestimate genetic distances.

✔ CONCEPT CHECK 3

How does a genetic map differ from a physical map?

Constructing a Genetic Map with a Two-Point Testcross

Genetic maps can be constructed by conducting a series of testcrosses. In each testcross, one of the parents is heterozygous for a different pair of genes, and recombination frequencies are calculated for each pair of genes. A testcross between two genes is called a **two-point testcross**, or simply a two-point cross. Suppose that we carried out a series of two-point crosses for four genes, *a*, *b*, *c*, and *d*, and obtained the following recombination frequencies:

Gene loci in testcross	Recombination frequency (%)
a and *b*	50
a and *c*	50
a and *d*	50
b and *c*	20
b and *d*	10
c and *d*	28

We can begin constructing a genetic map for these genes by considering the recombination frequency for each pair of

genes. The recombination frequency between *a* and *b* is 50%, which is the recombination frequency expected with independent assortment. Therefore, genes *a* and *b* may either be on different chromosomes or be very far apart on the same chromosome; we will place them in different linkage groups with the understanding that they may or may not be on the same chromosome:

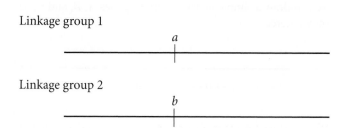

Linkage group 1

a

Linkage group 2

b

The recombination frequency between *a* and *c* is 50%, indicating that they, too, are in different linkage groups. The recombination frequency between *b* and *c* is 20%, so these genes are linked and separated by 20 map units:

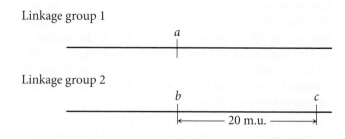

Linkage group 1

a

Linkage group 2

b   *c*

|←——— 20 m.u. ———→|

The recombination frequency between *a* and *d* is 50%, indicating that these genes belong to different linkage groups, whereas genes *b* and *d* are linked, with a recombination frequency of 10%. To decide whether gene *d* is 10 m.u.

to the left or to the right of gene *b*, we must consult the *c*-to-*d* distance. If gene *d* is 10 m.u. to the left of gene *b*, then the distance between *d* and *c* should be approximately the sum of the distance between *b* and *c* and between *c* and *d*: 20 m.u. + 10 m.u. = 30 m.u. If, on the other hand, gene *d* lies to the right of gene *b*, then the distance between gene *d* and gene *c* will be much shorter, approximately 20 m.u. − 10 m.u. = 10 m.u. Again, the summed distances will be only approximate because any double crossovers between the two genes will be missed and the map distance will be underestimated.

By examining the recombination frequency between *c* and *d*, we can distinguish between these two possibilities. The recombination frequency between *c* and *d* is 28%, so gene *d* must lie to the left of gene *b*. Notice that the sum of the recombination frequency between *d* and *b* (10%) and between *b* and *c* (20%) is greater than the recombination frequency between *d* and *c* (28%). As already discussed, this discrepancy arises because double crossovers between the two outer genes go undetected, causing an underestimation of the true map distance. The genetic map of these genes is now complete:

Linkage group 1

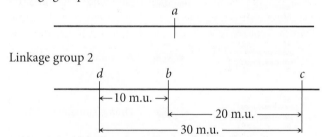

Linkage group 2

> TRY PROBLEM 27

7.3 A Three-Point Testcross Can Be Used to Map Three Linked Genes

While genetic maps can be constructed from a series of testcrosses for pairs of genes, this approach is not particularly efficient because numerous two-point crosses must be carried out to establish the order of the genes and because double crossovers will be missed. A more efficient mapping technique is a testcross for three genes—a **three-point testcross**, or three-point cross. With a three-point cross, the order of the three genes can be established in a single set of progeny, and some double crossovers can usually be detected, providing more accurate map distances.

Consider what happens when crossing over takes place among three hypothetical linked genes. **Figure 7.12** illustrates a pair of homologous chromosomes of an individual that is heterozygous at three loci (*Aa Bb Cc*). Notice that the genes are in the coupling configuration: all the dominant alleles are on one chromosome (*A   B   C*) and all the recessive alleles are on the other chromosome (*a   b   c*). Three types of crossover events can take place between these three genes: two types of single crossovers (see Figure 7.12a and b) and a double crossover (see Figure 7.12c). In each type of crossover, two of the resulting chromosomes are recombinants and two are nonrecombinants.

Notice that in the recombinant chromosomes resulting from the double crossover, the outer two alleles are the same as in the nonrecombinants, but the middle allele is different. This result provides us with an important clue about the order of the genes. In progeny that result from a double crossover, only the middle allele should differ from the alleles present in the nonrecombinant progeny.

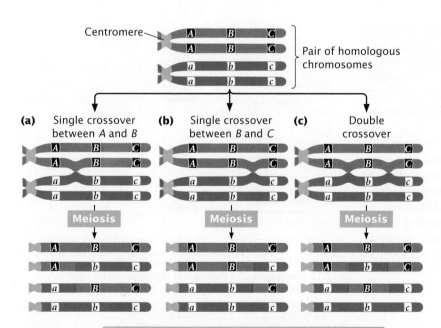

Conclusion: Recombinant chromosomes resulting from the double crossover have only the middle gene altered.

7.12 Three types of crossovers can take place among three linked loci.

Constructing a Genetic Map with a Three-Point Testcross

To examine gene mapping with a three-point testcross, let's consider three recessive mutations in the fruit fly *Drosophila melanogaster*. In this species, scarlet eyes (*st*) are recessive to wild-type red eyes (*st*⁺), ebony body color (*e*) is recessive to wild-type gray body color (*e*⁺), and spineless (*ss*)—that is, the presence of small bristles—is recessive to wild-type normal bristles (*ss*⁺). The loci encoding these three characteristics are linked and located on chromosome 3.

We will refer to these three loci as *st*, *e*, and *ss*, but keep in mind that either the recessive alleles (*st*, *e*, and *ss*) or the dominant alleles (*st*⁺, *e*⁺, and *ss*⁺) may be present at each locus. So when we say that there are 10 m.u. between *st* and *ss*, we mean that there are 10 m.u. between the loci at which mutations *st* and *ss* occur; we could just as easily say that there are 10 m.u. between *st*⁺ and *ss*⁺.

To map these genes, we need to determine their order on the chromosome and the genetic distances between them. First, we must set up a three-point testcross: a cross between a fly heterozygous at all three loci and a fly homozygous for recessive alleles at all three loci. To produce flies heterozygous for all three loci, we might cross a stock of flies that are homozygous for wild-type alleles at all three loci with flies that are homozygous for recessive alleles at all three loci:

$$ P \qquad \frac{st^+ \quad e^+ \quad ss^+}{st^+ \quad e^+ \quad ss^+} \times \frac{st \quad e \quad ss}{st \quad e \quad ss} $$

$$ \downarrow $$

$$ F_1 \qquad \frac{st^+ \quad e^+ \quad ss^+}{st \quad e \quad ss} $$

The order of the genes has been arbitrarily assigned because at this point, we do not know which one is the middle gene. Additionally, the alleles in the F_1 heterozygotes are in coupling configuration (because all the wild-type dominant alleles were inherited from one parent and all the recessive mutations from the other parent), although the testcross can also be done with alleles in repulsion.

In the three-point testcross, we cross the F_1 heterozygotes with flies that are homozygous for all three recessive mutations. In many organisms, it makes no difference whether the heterozygous parent in the testcross is male or female (provided that the genes are autosomal), but in *Drosophila*, no crossing over takes place in males. Because crossing over in the heterozygous parent is essential for determining recombination frequencies, the heterozygous flies in our testcross must be female. So we mate female F_1 flies that are heterozygous for all three traits with male flies that are homozygous for all the recessive traits:

$$ \frac{st^+ \quad e^+ \quad ss^+}{st \quad e \quad ss} \text{ Female} \times \frac{st \quad e \quad ss}{st \quad e \quad ss} \text{ Male} $$

The progeny of this cross are listed in **Figure 7.13**. For each locus, two classes of progeny are produced: progeny that are heterozygous, displaying the dominant trait, and progeny that are homozygous, displaying the recessive trait. With two classes of progeny possible for each of the three loci, there will be $2^3 = 8$ classes of phenotypes possible in the progeny.

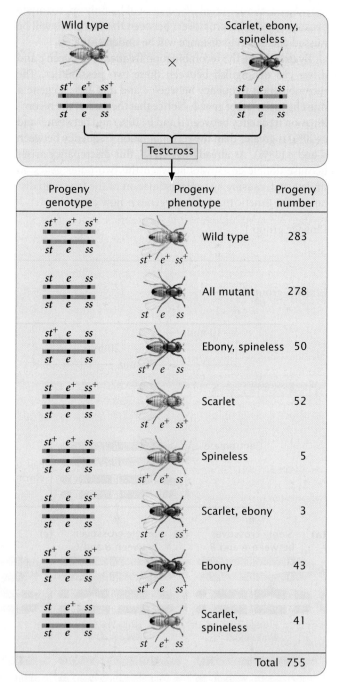

7.13 The results of a three-point testcross can be used to map linked genes. In this three-point testcross of *Drosophila melanogaster*, the recessive mutations leading to scarlet eyes (*st*), ebony body color (*e*), and spineless bristles (*ss*) are at three linked loci. The order of the loci has been assumed arbitrarily. Each phenotypic class includes both male and female flies; the sex of the pictured flies is random.

In this example, all eight phenotypic classes are present, but in some three-point crosses, one or more of the phenotypes may be missing if the number of progeny is limited. Nevertheless, the absence of a particular class can provide important information about which combination of traits is least frequent and, ultimately, about the order of the genes, as we will see.

To map the genes, we need information about where and how often crossing over has taken place. In the homozygous recessive parent, the two alleles at each locus are the same, and so crossing over will have no effect on the types of gametes produced: with or without crossing over, all gametes from this parent have a chromosome with three recessive alleles (st e ss). In contrast, the heterozygous parent has different alleles on its two chromosomes, and so crossing over can be detected. The information that we need for mapping, therefore, comes entirely from the gametes produced by the heterozygous parent. Because chromosomes contributed by the homozygous parent carry only recessive alleles, whatever alleles are present on the chromosome contributed by the heterozygous parent will be expressed in the progeny.

As a shortcut, we often do not write out the complete genotypes of the testcross progeny, listing instead only the alleles expressed in the phenotype, which are the alleles inherited from the heterozygous parent. This convention is used in the discussion that follows.

CONCEPTS

To map genes, information about the location and number of crossovers in the gametes that produced the progeny of a cross is needed. An efficient way to obtain this information is to use a three-point testcross, in which an individual that is heterozygous at three linked loci is crossed with an individual that is homozygous recessive at the same three loci.

✔ CONCEPT CHECK 4

Write the genotypes of all recombinant and nonrecombinant progeny expected from the following three-point cross:

$$\frac{m^+ \qquad p^+ \qquad s^+}{m \qquad p \qquad s} \times \frac{m \qquad p \qquad s}{m \qquad p \qquad s}$$

DETERMINING THE GENE ORDER The first task in mapping the genes is to determine their order on the chromosome. In Figure 7.13, we arbitrarily listed the loci in the order st, e, ss, but we had no way of knowing which of the three loci was between the other two. We can now identify the middle locus by examining the double-crossover progeny.

First, determine which progeny are the nonrecombinants: they will be the two most numerous classes of progeny (even if crossing over takes place in every meiosis, the nonrecombinants will constitute at least 50% of the progeny). Among the progeny of the testcross in Figure 7.13, the most numerous are those with all three dominant traits

(st^+ e^+ ss^+) and those with all three recessive traits (st e ss).

Next, identify the double-crossover progeny. These progeny should always have the two least numerous phenotypes because the probability of a double crossover is always less than the probability of a single crossover. The least common progeny among those listed in Figure 7.13 are progeny with spineless bristles (st^+ e^+ ss) and progeny with scarlet eyes and ebony body (st e ss^+), so they are the double-crossover progeny.

Three orders of genes on the chromosome are possible: the eye-color locus could be in the middle (e st ss), the body-color locus could be in the middle (st e ss), or the bristle locus could be in the middle (st ss e). To determine which gene is in the middle, we can draw the chromosomes of the heterozygous parent with all three possible gene orders and then see if a double crossover produces the combination of genes observed in the double-crossover progeny. The three possible gene orders and the types of progeny produced by their double crossovers are

The only gene order that produces chromosomes with the set of alleles observed in the least numerous progeny or double crossovers (st^+ e^+ ss and st e ss^+ in Figure 7.13) is the one in which the ss locus for bristles lies in the middle (gene order 3). Therefore, this order (st ss e) must represent the correct sequence of genes on the chromosome.

With a little practice, we can quickly determine which locus is in the middle without writing out all the gene orders. The phenotypes of the progeny are expressions of the alleles inherited from the heterozygous parent. Recall that when we looked at the results of double crossovers (see Figure 7.12), only the alleles at the middle locus differed from those in the nonrecombinants. If we compare the nonrecombinant progeny with the double-crossover progeny, they should differ only in alleles at the middle locus (**Table 7.2**).

Let's compare the alleles in the double-crossover progeny st^+ e^+ ss with those in the nonrecombinant progeny st^+ e^+ ss^+. We see that both have an allele for red eyes (st^+) and both have an allele for gray

TABLE 7.2	Steps in determining gene order in a three-point cross

1. Identify the nonrecombinant progeny (two most numerous phenotypes).

2. Identify the double-crossover progeny (two least numerous phenotypes).

3. Compare the phenotypes of double-crossover progeny with the phenotypes of nonrecombinant progeny. They should be alike in two characteristics and differ in one.

4. The characteristic that differs between the double crossover and the nonrecombinant progeny is encoded by the middle gene.

body (e^+), but the nonrecombinants have an allele for normal bristles (ss^+), whereas the double crossovers have an allele for spineless bristles (ss). Because the bristle locus is the only one that differs between these two types of progeny, it must lie in the middle. We would obtain the same results if we compared the other class of double-crossover progeny (st e ss^+) with the other class of nonrecombinant progeny (st e ss). Again, the only locus that differs is the one for bristles. Don't forget that the nonrecombinants and the double crossovers should differ at only one locus; if they differ at two loci, the wrong classes of progeny are being compared. **Animation 7.1** illustrates how to determine the order of three linked genes.

CONCEPTS

To determine the middle locus in a three-point cross, compare the double-crossover progeny with the nonrecombinant progeny. The double crossovers will be the two least common classes of phenotypes; the nonrecombinants will be the two most common classes of phenotypes. The double-crossover progeny should have the same alleles as the nonrecombinant types at two loci and different alleles at the locus in the middle.

✔ CONCEPT CHECK 5

A three-point testcross is carried out between three linked genes. The resulting nonrecombinant progeny are $s^+ r^+ c^+$ and $s r c$ and the double-crossover progeny are $s r c^+$ and $s^+ r^+ c$. Which is the middle locus?

DETERMINING THE LOCATIONS OF CROSSOVERS When we know the correct order of the loci on the chromosome, we can rewrite the phenotypes of the testcross progeny in Figure 7.13 with the alleles in the correct order so that we can determine where crossovers have taken place (**Figure 7.14**).

Among the eight classes of progeny, we have already identified two classes as nonrecombinants (st^+ ss^+ e^+ and st ss e) and two classes as double crossovers

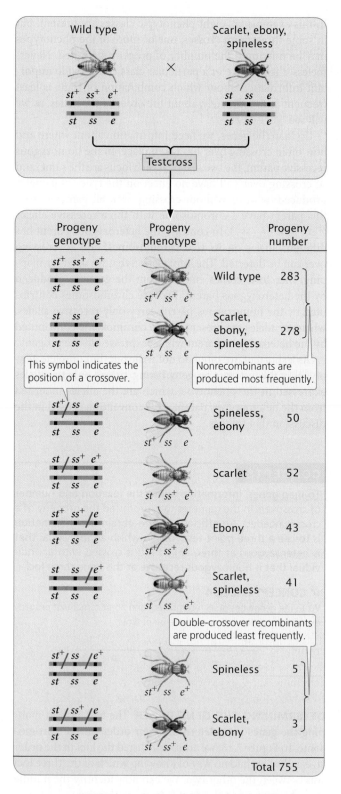

7.14 Writing the results of a three-point testcross with the loci in the correct order allows the locations of crossovers to be determined. These results are from the testcross illustrated in Figure 7.13, with the loci shown in the correct order. The location of a crossover is indicated by a slash (/). Each phenotypic class includes both male and female flies; the sex of the pictured flies is random.

(st^+ ss e^+ and st ss^+ e). The other four classes contain progeny that resulted from a chromosome that underwent a single crossover: two underwent single crossovers between *st* and *ss*, and two underwent single crossovers between *ss* and *e*.

To determine where the crossovers took place in these single-crossover progeny, we can compare the alleles found in these progeny with those found in the nonrecombinants, just as we did for the double crossovers. For example, consider the progeny with chromosome st^+ ss e . The first allele (st^+) came from the nonrecombinant chromosome st^+ ss^+ e^+ , and the other two alleles (*ss* and *e*) must have come from the other nonrecombinant chromosome st ss e through crossing over:

$$
\begin{array}{ccc}
st^+ & ss^+ & e^+ \\
\hline
st & ss & e
\end{array}
\;\to\;
\begin{array}{ccc}
st^+ & ss^+ & e^+ \\
\times \\
st & ss & e
\end{array}
\;\to\;
\begin{array}{ccc}
st^+ & ss & e \\
\hline
st & ss^+ & e^+
\end{array}
$$

This same crossover also produces the st ss^+ e^+ progeny.

This method can also be used to determine the location of crossing over in the other two types of single-crossover progeny. Crossing over between *ss* and *e* produces st^+ ss^+ e and st ss e^+ chromosomes:

$$
\begin{array}{ccc}
st^+ & ss^+ & e^+ \\
\hline
st & ss & e
\end{array}
\;\to\;
\begin{array}{ccc}
st^+ & ss^+ & e^+ \\
& \times \\
st & ss & e
\end{array}
\;\to\;
\begin{array}{ccc}
st^+ & ss^+ & e \\
\hline
st & ss & e^+
\end{array}
$$

We now know the locations of all the crossovers, which are marked with red slashes in Figure 7.14.

CALCULATING THE RECOMBINATION FREQUENCIES
Next, we can determine the map distances, which are based on the frequencies of recombination. We can calculate recombination frequency by adding up all of the recombinant progeny, dividing this number by the total number of progeny from the cross, and multiplying the number obtained by 100%. To determine the map distances accurately, we must include all crossovers (both single and double) that take place between two genes.

Recombinant progeny that possess a chromosome that underwent crossing over between the eye-color locus (*st*) and the bristle locus (*ss*) include the single crossovers (st^+ / ss e and st / ss^+ e^+) and the two double crossovers (st^+ / ss / e^+ and st / ss^+ / e) (see Figure 7.14). There are a total of 755 progeny, so the recombination frequency between *ss* and *st* is:

st − *ss* recombination frequency =

$$
\frac{50 + 52 + 5 + 3}{755} \times 100\% = 14.6\%
$$

The distance between the *st* and *ss* loci can be expressed as 14.6 m.u.

The map distance between the bristle locus (*ss*) and the body-color locus (*e*) is determined in the same manner. The recombinant progeny that possess a crossover between *ss* and *e* are the single crossovers st^+ ss^+ / e and st ss / e^+ and the double crossovers st^+ / ss / e^+ and st / ss^+ / e . The recombination frequency is:

ss − *e* recombination frequency =

$$
\frac{43 + 41 + 5 + 3}{755} \times 100\% = 12.2\%
$$

Thus, the map distance between *ss* and *e* is 12.2 m.u.

Finally, calculate the map distance between the outer two loci, *st* and *e*. This map distance can be obtained by summing the map distances between *st* and *ss* and between *ss* and *e* (14.6 m.u. + 12.2 m.u. = 26.8 m.u.). We can now use the map distances to draw a map of the three genes on the chromosome:

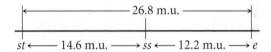

A genetic map of *D. melanogaster* is illustrated in **Figure 7.15**.

INTERFERENCE AND THE COEFFICIENT OF COINCIDENCE Map distances give us information not only about the distances that separate genes, but also about the proportions of recombinant and nonrecombinant gametes that will be produced in a cross. For example, knowing that genes *st* and *ss* on chromosome 3 of *D. melanogaster* are separated by a distance of 14.6 m.u. tells us that 14.6% of the gametes produced by a fly heterozygous at these two loci will be recombinants. Similarly, 12.2% of the gametes from a fly heterozygous for *ss* and *e* will be recombinants.

Theoretically, we should be able to calculate the proportion of double-recombinant gametes by using the multiplication rule (see Chapter 3). Applying this rule, we should find that the proportion (probability) of gametes with double crossovers between *st* and *e* is equal to the probability of recombination between *st* and *ss* multiplied by the probability of recombination between *ss* and *e*, or 0.146 × 0.122 = 0.0178. Multiplying this probability by the total number of progeny gives us the *expected* number of double-crossover progeny from the cross: 0.0178 × 755 = 13.4. But only 8 double crossovers—considerably fewer than the 13 expected—were observed in the progeny of the cross (see Figure 7.14).

This phenomenon is common in eukaryotic organisms. The calculation assumes that each crossover event is independent and that the occurrence of one crossover does not influence the occurrence of another. But crossovers are frequently *not* independent events: the occurrence of one

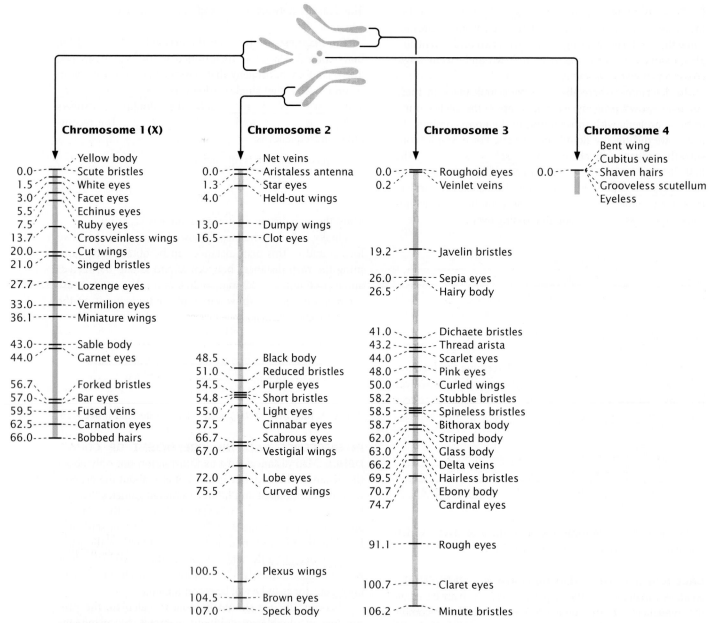

7.15 *Drosophila melanogaster* has four linkage groups corresponding to its four pairs of chromosomes. These genes were mapped using recombination frequencies. Distances between genes within a linkage group are in map units. Note that the small chromosome 4 never undergoes recombination.

crossover tends to inhibit additional crossovers in the same region of the chromosome, so double crossovers are less frequent than expected.

The degree to which one crossover interferes with additional crossovers in the same region is termed the **interference**. To calculate the interference, we first determine the **coefficient of coincidence**, which is the ratio of observed double crossovers to expected double crossovers:

coefficient of coincidence =

$$\frac{\text{number of observed double crossovers}}{\text{number of expected double crossovers}}$$

For the loci that we mapped on chromosome 3 of *D. melanogaster* (see Figure 7.14), we find that

coefficient of coincidence =

$$\frac{5 + 3}{0.146 \times 0.122 \times 755} = \frac{8}{13.4} = 0.6$$

which indicates that we are actually observing only 60% of the double crossovers that we expected on the basis of the single-crossover frequencies. The interference is calculated as

interference = 1 − coefficient of coincidence

So the interference for our three-point cross is:

$$\text{interference} = 1 - 0.6 = 0.4$$

This interference value tells us that 40% of the double-crossover progeny expected will not be observed because of interference. When interference is complete and no double-crossover progeny are observed, the coefficient of coincidence is 0 and the interference is 1.

Sometimes a crossover *increases* the probability of another crossover taking place nearby and we see *more* double-crossover progeny than expected. In this case, the coefficient of coincidence is greater than 1 and the interference is negative.

Most eukaryotic organisms exhibit interference, which causes crossovers to be more widely spaced than would be expected on a random basis. Interference was first observed in crosses of *Drosophila* in the early 1900s, yet despite years of study, the mechanism by which interference occurs is still not well understood. One proposed model of interference suggests that crossovers occur when stress builds up along the chromosome. According to this model, a crossover releases stress for some distance along the chromosome. Because a crossover relieves the stress that causes crossovers, additional crossovers are less likely to occur in the same area. Recent research shows that an enzyme called topoisomerase II is required for crossover interference. Topoisomerase enzymes function to relieve stress along the chromosome, such as torsional stress that occurs during the unwinding of DNA (see Chapters 11 and 12). The fact that a topoisomerase enzyme is required for crossover interference supports the stress release model of crossover interference. ▸ **TRY PROBLEM 29**

THINK-PAIR-SHARE Questions 5 and 6

CONCEPTS

The coefficient of coincidence equals the number of double crossovers observed divided by the number of double crossovers expected on the basis of single-crossover frequencies. The interference equals 1 minus the coefficient of coincidence; it indicates the degree to which one crossover interferes with additional crossovers.

✔ CONCEPT CHECK 6

In analyzing the results of a three-point testcross, a student determines that the interference is −0.23. What does this negative interference value indicate?

a. Fewer double crossovers took place than expected on the basis of single-crossover frequencies.

b. More double crossovers took place than expected on the basis of single-crossover frequencies.

c. Fewer single crossovers took place than expected.

d. A crossover in one region interferes with additional crossovers in the same region.

CONNECTING CONCEPTS

Stepping Through the Three-Point Cross

We have now examined the three-point cross in considerable detail, and we have seen how the information derived from it can be used to map a series of three linked genes. Let's briefly review the steps required to map genes using a three-point cross.

1. **Write out the phenotypes and numbers of progeny produced by the three-point cross.** The progeny phenotypes will be easier to interpret if you use allelic symbols for the traits (such as st^+ e^+ ss).

2. **Write out the genotypes of the original parents used to produce the triply heterozygous F_1 individual** in the testcross and, if known, the arrangement (coupling or repulsion) of the alleles on their chromosomes.

3. **Determine which phenotypic classes among the progeny of the testcross are the nonrecombinants and which are the double crossovers.** The nonrecombinants will be the two most common phenotypes; the double crossovers will be the two least common phenotypes.

4. **Determine which locus lies in the middle.** Compare the alleles present in the double crossovers with those present in the nonrecombinants; each class of double crossovers should be like one of the nonrecombinants at two loci and should differ at one locus. The locus that differs is the middle one.

5. **Rewrite the phenotypes with the genes in correct order.**

6. **Determine where crossovers must have taken place to give rise to the recombinant progeny phenotypes.** To do so, compare each phenotype with the phenotype of the nonrecombinant progeny.

7. **Determine the recombination frequencies.** Add the numbers of the progeny that possess a chromosome with a single crossover between a pair of loci. Add the double crossovers to this number. Divide this sum by the total number of progeny from the cross, and multiply by 100%; the result is the recombination frequency between the loci, which is the same as the map distance.

8. **Draw a map of the three loci.** Indicate which locus lies in the middle, and indicate the distances between them.

9. **Determine the coefficient of coincidence and the interference.** The coefficient of coincidence is the number of observed double-crossover progeny divided by the number of expected double-crossover progeny. The expected number can be obtained by multiplying the product of the two single-recombination probabilities by the total number of progeny from the cross.

WORKED PROBLEM

In *D. melanogaster*, cherub wings (*ch*), black body (*b*), and cinnabar eyes (*cn*) result from recessive alleles that are all located on chromosome 2. A homozygous wild-type fly was mated with a cherub, black, cinnabar fly, and the resulting F_1 females were test-crossed with cherub, black, cinnabar males. The following progeny were produced by the testcross:

ch	b^+	cn	105
ch^+	b^+	cn^+	750
ch^+	b	cn	40
ch^+	b^+	cn	4
ch	b	cn	753
ch	b^+	cn^+	41
ch^+	b	cn^+	102
ch	b	cn^+	5
	Total		1800

a. Determine the linear order of the genes on the chromosome (which gene is in the middle?).

b. Calculate the map distances between the three loci.

c. Determine the coefficient of coincidence and the interference for these three loci.

Solution Strategy

What information is required in your answer to the problem?

The order of the genes on the chromosome, the map distances among the genes, the coefficient of coincidence, and the interference.

What information is provided to solve the problem?

- A homozygous wild-type fly was mated with a cherub, black, cinnabar fly, and the resulting F_1 females were test-crossed with cherub, black, cinnabar males.

- The numbers of the different types of flies appearing among the progeny of the testcross.

Solution Steps

a. We can represent the crosses in this problem as follows:

$$P \quad \frac{ch^+ \quad b^+ \quad cn^+}{ch^+ \quad b^+ \quad cn^+} \times \frac{ch \quad b \quad cn}{ch \quad b \quad cn}$$

$$\downarrow$$

$$F_1 \quad \frac{ch^+ \quad b^+ \quad cn^+}{ch \quad b \quad cn}$$

$$\text{Testcross} \quad \frac{ch^+ \quad b^+ \quad cn^+}{ch \quad b \quad cn} \times \frac{ch \quad b \quad cn}{ch \quad b \quad cn}$$

Note that at this point we do not know the order of the genes; we have arbitrarily put b in the middle.

The next step is to determine which of the testcross progeny are nonrecombinants and which are double crossovers. The nonrecombinants should be the most frequent phenotype, so they must be the progeny with phenotypes

encoded by ch^+ b^+ cn^+ and ch b cn. These genotypes are consistent with the genotypes of the parents, given earlier. The double crossovers are the least frequent phenotypes and are encoded by ch^+ b^+ cn and ch b cn^+.

We can determine the gene order by comparing the alleles present in the double crossovers with those present in the nonrecombinants. The double-crossover progeny should be like one of the nonrecombinants at two loci and unlike it at one locus; the allele that differs should be in the middle. Compare the double-crossover progeny ch b cn^+ with the nonrecombinant ch b cn. Both have cherub wings (ch) and black body (b), but the double-crossover progeny have wild-type eyes (cn^+), whereas the nonrecombinants have cinnabar eyes (cn). The locus that determines cinnabar eyes must be in the middle.

b. To calculate the recombination frequencies among the genes, we first write the phenotypes of the progeny, with the genes encoding them, in the correct order. We have already identified the nonrecombinant and double-crossover progeny, so the other four progeny types must have resulted from single crossovers. To determine *where* single crossovers took place, we compare the alleles found in the single-crossover progeny with those in the nonrecombinants. Crossing over must have taken place where the alleles switch from those found in one nonrecombinant to those found in the other nonrecombinant. The locations of the crossovers are indicated with a slash:

ch		cn	/ b^+	105	single crossover
ch^+		cn^+	b^+	750	nonrecombinant
ch^+ /	cn		b	40	single crossover
ch^+ /	cn	/ b^+		4	double crossover
ch		cn	b	753	nonrecombinant
ch /	cn^+		b^+	41	single crossover
ch^+		cn^+ /	b	102	single crossover
ch /	cn^+ /	b		5	double crossover
Total				1800	

Next, we determine the recombination frequencies and draw a genetic map:

ch–cn recombination frequency $=$
$$\frac{40 + 4 + 41 + 5}{1800} \times 100\% = 5\%$$

cn–b recombination frequency $=$
$$\frac{105 + 4 + 102 + 5}{1800} \times 100\% = 12\%$$

ch–b map distance $= 5\% + 12\% = 17\%$

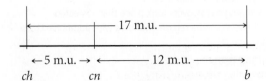

c. The coefficient of coincidence is the number of observed double crossovers divided by the number of expected double crossovers. The number of expected double crossovers is obtained by multiplying the probability of a crossover between *ch* and *cn* (0.05) by the probability of a crossover between *cn* and *b* (0.12), then multiplying the product by the total number of progeny from the cross (1800):

$$\text{coefficient of coincidence} = \frac{4 + 5}{0.05 \times 0.12 \times 1800} = 0.83$$

Finally, the interference is equal to 1 − the coefficient of coincidence:

$$\text{interference} = 1 - 0.83 = 0.17$$

>> To increase your skill with three-point crosses, try working **Problem 30** at the end of this chapter.

Effects of Multiple Crossovers

So far, we have examined the effects of double crossovers taking place between only two of the four chromatids (strands) of a homologous pair (two-strand double crossovers). Double crossovers that include three and even four of the chromatids of a homologous pair may also take place (**Figure 7.16**). As we have seen, if we examine only the alleles at loci on either side of both crossover events, two-strand double crossovers result in no new combinations of alleles, and no recombinant gametes are produced (see Figure 7.16). Three-strand double crossovers result in two of the four gametes being recombinant, and four-strand double crossovers result in all four gametes being recombinant (see Figure 7.16). Thus, two-strand double crossovers produce 0% recombination, three-strand double crossovers produce 50% recombination, and four-strand double crossovers produce 100% recombination. The overall result is that all types of double crossovers, taken together, produce an average of 50% recombinant gametes.

As we have also seen, two-strand double crossovers cause alleles on either side of the crossovers to remain the same and produce no recombinant progeny. Three-strand and four-strand crossovers produce recombinant progeny, but these progeny are the same types as those produced by single crossovers. Consequently, some multiple crossovers go undetected when the progeny of a genetic cross are observed. Therefore, map distances based on recombination rates underestimate the true physical distances between genes because some multiple crossovers are not detected among the progeny of a cross. When genes are very close together, multiple crossovers are unlikely, and genetic distances based on recombination rates accurately correspond to the physical distances on the chromosome. But as the distance between genes increases, more multiple crossovers are likely, and the discrepancy between genetic distances (based on recombination rates) and physical distances increases. To correct for this discrepancy, geneticists have developed mathematical **mapping functions**, which relate recombination frequencies to actual physical distances between genes

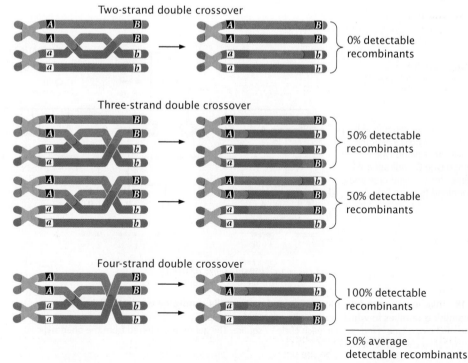

7.16 Effects of two-, three-, and four-strand double crossovers on recombination between two genes.

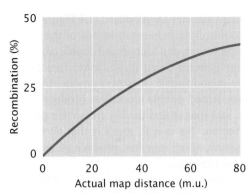

7.17 Recombination rates underestimate the true physical distance between genes at higher map distances.

(**Figure 7.17**). Most of these functions are based on the Poisson distribution, which predicts the probability of multiple rare events. With the use of such mapping functions, map distances based on recombination rates can be more accurately estimated.

Mapping Human Genes

Efforts to map human genes are hampered by the inability to perform specific crosses and by the small number of progeny in most human families. Geneticists are often restricted to analyses of pedigrees, which may be incomplete and may provide limited information. Nevertheless, a large number of human traits have been successfully mapped with the use of pedigree data to analyze linkage. Because the number of progeny from any one mating is usually small, data from several families and pedigrees are usually combined to test for independent assortment. The methods used in these types of analyses are complex, but

an example will illustrate how linkage can be detected from pedigree data.

One of the first documented demonstrations of linkage in humans was between the locus for nail–patella syndrome and the locus that determines the ABO blood types. Nail–patella syndrome is an autosomal dominant disorder characterized by abnormal fingernails and absent or rudimentary kneecaps. The ABO blood types are determined by an autosomal locus with multiple alleles (see Chapter 5). Linkage between the genes encoding these two traits was established in families in which both traits segregate. A pedigree for part of one such family is illustrated in **Figure 7.18**.

Nail–patella syndrome is rare, so we can assume that people who have this trait are heterozygous (Nn); unaffected people are homozygous (nn). The ABO genotypes of individuals can be inferred from their phenotypes and the types of offspring they produce. Female I-2 in Figure 7.18, for example, has blood-type B, which has two possible genotypes: $I^B I^B$ or $I^B i$ (see Figure 5.6). Because some of her offspring are blood-type O (genotype ii) and must have therefore inherited an i allele from each parent, female I-2 must have genotype $I^B i$. Similarly, the presence of blood-type O offspring in generation II indicates that male I-1, with blood-type A, must also carry an i allele and therefore has genotype $I^A i$. The parents of this family are

$$I^A i\ Nn \times I^B i\ nn$$

In generation II, we can see that the genes for nail–patella syndrome and the blood types do not appear to assort independently. All children in generation II with nail–patella syndrome have either blood-type B or blood-type O; all those with blood-type A have normal nails and kneecaps. If the genes encoding nail–patella syndrome and the ABO blood types

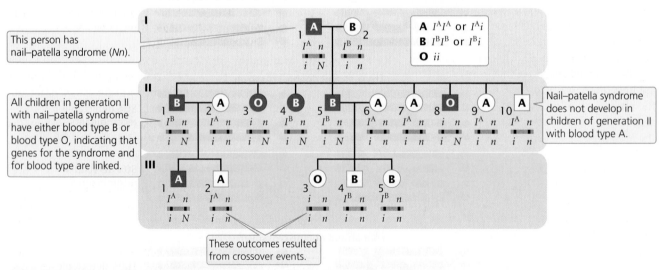

7.18 Linkage between ABO blood type and nail–patella syndrome was established by examining families in which both traits segregate. The pedigree shown here is for one such family. The ABO blood type is indicated in each circle or square. The genotype, inferred from the phenotype, is given below each circle or square.

assorted independently, we would expect that some children in generation II would have blood-type A and nail–patella syndrome, inheriting both the I^A and N alleles from their father. The observed outcome indicates that the arrangements of the alleles on the chromosomes of the crossed parents are

$$\frac{I^A \quad n}{i \quad N} \times \frac{I^B \quad n}{i \quad n}$$

The pedigree indicates that there is no recombination among the offspring (generation II) of these parents, but there are two instances of recombination among the off-spring in generation III. Individuals II-1 and II-2 have the following genotypes:

$$\frac{I^B \quad n}{i \quad N} \times \frac{I^A \quad n}{i \quad n}$$

Their child III-2 has blood-type A and does not have nail–patella syndrome, so he must have genotype

$$\frac{I^A \quad n}{i \quad n}$$

and must have inherited both the i and the n alleles from his father. These alleles are on different chromosomes in the father, so crossing over must have taken place. Crossing over must also have taken place to produce child III-3.

In the pedigree in Figure 7.18, thirteen children are off-spring of matings in which the genes encoding nail–patella syndrome and ABO blood types segregate; two of the thir-teen children are recombinant progeny. On this basis, we might assume that the loci for nail–patella syndrome and ABO blood types are linked, with a recombination fre-quency of $^2/_{13} = 0.154$. However, it is possible that the genes *are* assorting independently, and that the small number of children just makes it seem as though the genes are linked.

To determine the probability that genes are actually linked, geneticists often calculate **lod** (logarithm of odds) **scores**. To obtain a lod score, we calculate (1) the probability of obtain-ing the observed results under the assumption that the genes are linked with a specified degree of recombination and (2) the probability of obtaining the observed results under the assumption of independent assortment. We then determine the ratio of these two probabilities, and the logarithm of this ratio is the lod score. Suppose that the probability of obtain-ing a particular set of observations under the assumption of linkage and a certain recombination frequency is 0.1, and that the probability of obtaining the same observations under the assumption of independent assortment is 0.0001. The ratio of these two probabilities is $^{0.1}/_{0.0001} = 1000$, the logarithm of which (the lod score) is 3. Thus, linkage with the specified recombination frequency is 1000 times as likely as independent assortment to produce what was observed. A lod score of 3 or higher is usually considered convincing evidence for linkage. **▶ TRY PROBLEM 36**

Mapping with Molecular Markers

For many years, gene mapping was limited in most organ-isms by the availability of **genetic markers**: variable genes with easily observable phenotypes whose inheritance could be studied. Traditional genetic markers include genes that encode easily observable characteristics such as flower color, seed shape, blood types, or biochemical differences. The pau-city of these types of characteristics in many organisms lim-ited mapping efforts.

In the 1980s, new molecular techniques made it possible to examine variations in DNA itself, providing an almost unlim-ited number of genetic markers. The earliest of these molecular markers consisted of restriction fragment length polymor-phisms (RFLPs), which are variations in DNA sequence that can be detected by cutting the DNA with restriction enzymes. Later, methods were developed for detecting variable numbers of short DNA sequences repeated in tandem, called micro-satellites. Now DNA sequencing allows the direct detection of individual variations in the DNA nucleotides. All of these methods have expanded the availability of genetic markers and greatly facilitated the creation of genetic maps.

Gene mapping with molecular markers is done in essen-tially the same manner as mapping performed with traditional phenotypic markers: the cosegregation of two or more mark-ers is studied, and map distances are based on the rates of recombination between markers. These methods and their use in mapping are presented in more detail in Chapters 19 and 20.

Locating Genes with Genome-Wide Association Studies

The traditional approach to mapping genes, which we have learned in this chapter, is to examine progeny phenotypes in genetic crosses or among individuals in a pedigree, look-ing for associations between the inheritance of a particular phenotype and the inheritance of alleles at other loci. This type of gene mapping is called **linkage analysis** because it is based on the detection of physical linkage between genes, as measured by the rate of recombination, in the progeny of a cross. Linkage analysis has been a powerful tool in the genetic analysis of many different types of organisms, includ-ing fruit flies, corn, mice, and humans.

An alternative approach to mapping genes is to conduct **genome-wide association studies**, looking for nonrandom associations between the presence of a trait and alleles at many different loci scattered across the genome. Unlike link-age analysis, this approach does not trace the inheritance of genetic markers and traits in a genetic cross or family. Rather, it looks for associations between traits and particular suites of alleles in a *population*.

Imagine that we are interested in finding genes that con-tribute to bipolar disorder, a psychiatric illness characterized by severe depression and mania. When a mutation that pre-disposes a person to bipolar disorder first arises in a popula-tion, it will occur on a particular chromosome and will be associated with a specific set of alleles on that chromosome. In the example illustrated in **Figure 7.19**, the D^- mutation first arises on a chromosome that has alleles A^2, B^2, and C^4, and therefore the D^- mutation is initially linked to the A^2, B^2, and C^4 alleles. A specific set of linked alleles such as this is called a **haplotype**, and the nonrandom association between alleles in a haplotype is called **linkage disequilibrium**. Because of the physical linkage between the bipolar mutation and the other alleles of the haplotype, bipolar disorder and the haplotype will tend to be inherited together. Crossing over, however, breaks up the association between the alleles of the haplotype (see Figure 7.19), reducing the linkage disequilibrium between them. How long the linkage disequilibrium persists over evolutionary time depends on the frequency of recombination between alleles at different loci. When the loci are far apart, linkage disequilibrium breaks down quickly; when the loci are close together, crossing over is less common and linkage disequilibrium persists longer. The important point here is that linkage disequilibrium provides information about the distance between genes. A strong association between a trait such as bipolar disorder and a set of linked genetic markers indicates that one or more genes contributing to bipolar disorder are likely to be near the genetic markers.

In recent years, geneticists have mapped millions of **single-nucleotide polymorphisms** (SNPs), which are positions in the genome where individuals vary in a single nucleotide base (see Chapter 20). Recall that human SNPs were used in a linkage analysis that located the gene responsible for pattern baldness, as discussed in the introduction to this chapter. It is now possible to quickly and inexpensively genotype people for hundreds of thousands or millions of SNPs. This genotyping has provided the genetic markers needed for conducting genome-wide association studies in which SNP haplotypes of people who have a particular disease, such as bipolar disorder, are compared with the haplotypes of healthy people. Nonrandom associations between SNPs and the disease suggest that one or more genes that contribute to the disease are closely linked to the SNPs. Genome-wide association studies do not usually locate specific genes: rather, they associate the inheritance of a trait or disease with a specific chromosomal region. After such an association has been established, geneticists can examine the chromosomal region for genes that might be responsible for the trait. Genome-wide association studies have been instrumental in the discovery of genes or chromosomal regions that affect a number of genetic diseases and important human traits, including bipolar disorder, height, skin pigmentation, eye color, body weight, coronary artery disease, blood-lipid concentrations, diabetes, heart attacks, bone density, and glaucoma, among others. Genome-wide association studies are also being used to locate and study genes that affect traits in numerous other organisms.

THINK-PAIR-SHARE Question 7

CONCEPTS

The development of molecular techniques for examining variation in DNA sequences has provided a large number of genetic markers that can be used to create genetic maps and study linkage relations. Genome-wide association studies examine the nonrandom association of genetic markers and phenotypes to locate genes that contribute to the expression of traits.

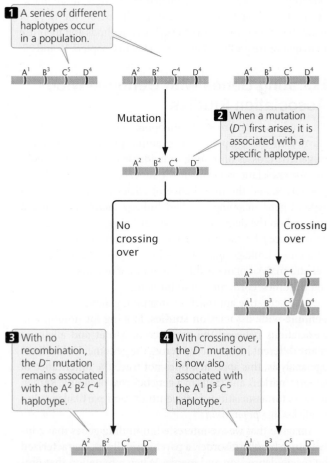

7.19 Genome-wide association studies are based on the nonrandom association of a mutation (D^-) that produces a trait with closely linked genes that constitute a haplotype.

7.4 Physical-Mapping Methods Are Used to Determine the Physical Positions of Genes on Particular Chromosomes

Genetic maps reveal the relative positions of genes on a chromosome, but they do not provide information that allows us to place groups of linked genes on particular chromosomes. Furthermore, the units of a genetic map do not always precisely correspond to physical distances on the chromosome because a number of factors other than physical distances between genes (such as the type and sex of the organism) can influence recombination. Because of these limitations, physical-mapping methods that do not rely on recombination frequencies have been developed.

Somatic-Cell Hybridization

One method used for positioning genes on chromosomes is **somatic-cell hybridization**, which requires the fusion of different types of cells. Most mature somatic (nonsex) cells can undergo only a limited number of divisions and therefore cannot be grown continuously. However, cells that have been altered by viruses or derived from tumors that have lost the normal constraints on cell division will divide indefinitely; this type of cell can be cultured in the laboratory to produce a **cell line**.

Cells from two different cell lines can be fused by treating them with polyethylene glycol or other agents that alter their plasma membranes. The resulting cell possesses two nuclei and is called a **heterokaryon**. The two nuclei of a heterokaryon eventually also fuse, generating a hybrid cell that contains chromosomes from both cell lines. If human and mouse cells are mixed in the presence of polyethylene glycol, their fusion results in human–mouse somatic-cell hybrids (**Figure 7.20**). The hybrid cells tend to lose chromosomes as they divide, and for reasons that are not understood, chromosomes from one of the species are lost preferentially. In human–mouse somatic-cell hybrids, the human chromosomes tend to be lost, whereas the mouse chromosomes are retained. Eventually, the chromosome number stabilizes when all but a few of the human chromosomes have been lost. Chromosome loss is random and differs among cell lines. The presence of these "extra" human chromosomes in the mouse genome makes it possible to assign human genes to specific chromosomes.

Mapping genes by using somatic-cell hybridization requires a panel of different hybrid cell lines. Each cell line is examined microscopically and the human chromosomes that it contains are identified. The cell lines of the panel are chosen so that they differ in the human chromosomes that they have retained. For example, one hybrid cell line might possess human chromosomes 2, 4, 7, and 8, whereas another might possess chromosomes 4, 19, and 20. Each cell line in the panel is examined for evidence of a particular human gene. The human gene can be detected by looking either for the gene itself (as discussed in Chapter 19) or for the protein that it produces. Correlation of the presence of the gene with the presence of specific human chromosomes often allows the gene to be assigned to the correct chromosome. For example, if a gene is detected in both of the aforementioned cell lines, the gene must be on chromosome 4 because it is the only human chromosome common to both cell lines (**Figure 7.21**). Somatic-cell hybridization is used less often today because DNA sequencing and mapping with

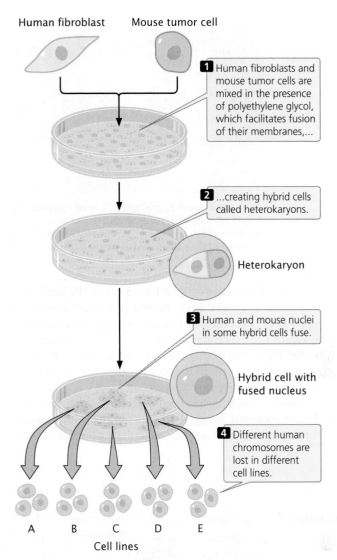

1 Human fibroblasts and mouse tumor cells are mixed in the presence of polyethylene glycol, which facilitates fusion of their membranes,...

2 ...creating hybrid cells called heterokaryons.

Heterokaryon

3 Human and mouse nuclei in some hybrid cells fuse.

Hybrid cell with fused nucleus

4 Different human chromosomes are lost in different cell lines.

A B C D E
Cell lines

7.20 Somatic-cell hybridization can be used to determine which chromosome contains a gene of interest.

molecular markers can often establish the chromosomal locations of linkage groups.

Deletion Mapping

Another method for determining the chromosomal location of a gene is **deletion mapping**. Special staining methods have been developed that reveal characteristic banding patterns on chromosomes (see Chapter 8). The absence of one or more of the bands that are normally seen on a chromosome reveals the presence of a chromosomal *deletion* (loss of a chromosome segment). We can assign genes to regions of chromosomes by studying the association between a phenotype or product encoded by a gene and particular chromosomal deletions.

In deletion mapping, an individual that is homozygous for a recessive mutation in the gene of interest is crossed with an

Human chromosomes present

Cell line	Gene product present	1	2	3	4	5	6	7	8	9	10	11	12	13	14	15	16	17	18	19	20	21	22	X
A	+		+		+			+	+															
B	+	+	+		+				+	+	+	+	+	+										
C	–															+		+		+				+
D	+		+		+		+	+	+															
E	–												+								+			
F	+				+															+	+			

7.21 Somatic-cell hybridization is used to assign a gene to a particular human chromosome. A panel of six hybrid cell lines, each containing a different subset of human chromosomes, is examined for the presence of the gene product (such as an enzyme). Four of the cell lines (A, B, D, and F) express the gene product. The only chromosome common to all four of these cell lines is chromosome 4, indicating that the gene is located on this chromosome.

individual that is heterozygous for a deletion (**Figure 7.22**). If the gene of interest is in the region of the chromosome represented by the deletion (the red part of the chromosomes in Figure 7.22), then approximately half of the progeny will display the mutant phenotype (see Figure 7.22a). If the gene is not within the deleted region, then all of the progeny will be wild type (see Figure 7.22b).

Deletion mapping has been used in the past to reveal the chromosomal locations of a number of human genes. For example, Duchenne muscular dystrophy is a disease that causes progressive weakening and degeneration of the muscles (see introduction to Chapter 19). From its X-linked pattern of inheritance, the mutant allele causing this disorder was known to be on the X chromosome, but its precise location was uncertain. Examination of a number of patients with the disease who also possessed small deletions allowed researchers to position the gene on a small segment of the short arm of the X chromosome. Like somatic-cell hybridization, deletion mapping is used less often today because of the widespread availability of DNA sequencing. ▶ **TRY PROBLEM 39**

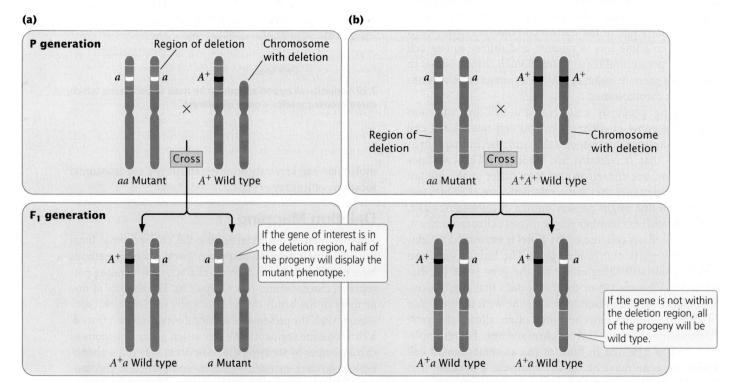

7.22 Deletion mapping can be used to determine the chromosomal location of a gene. An individual homozygous for a recessive mutation in the gene of interest (*aa*) is crossed with an individual heterozygous for a deletion.

7.23 In situ hybridization is another technique for determining the chromosomal location of a gene. The red and green fluorescence spots are produced by probes that are specific to different DNA sequences on chromosome 15. [Wessex Reg. Genetics Centre, Wellcome Images.]

Physical Chromosome Mapping Through Molecular Analysis

So far, we have explored methods that determine the chromosomal location of a gene indirectly. Researchers now have the information and technology to see where a gene actually lies. Described in more detail in Chapter 19, fluorescence in situ hybridization (FISH) is a method for determining the chromosomal location of a particular gene through molecular analysis. This method requires the creation of a probe for the gene, which is a single-stranded DNA complement to the gene of interest. The probe fluoresces under high-frequency light so that it can be visualized. The probe binds to the DNA sequence of the gene of interest on the chromosome. The presence of fluorescence from the bound probe reveals the location of the gene on a particular chromosome (**Figure 7.23**).

In addition to allowing us to see where a gene is located on a chromosome, modern laboratory techniques now allow researchers to identify the precise location of a gene at the nucleotide level. For example, with DNA sequencing (described fully in Chapter 19), physical distances between genes can be determined in numbers of base pairs.

7.5 Recombination Rates Exhibit Extensive Variation

In recent years, geneticists have studied variation in rates of recombination and have found that they vary widely—among species, among and along chromosomes of a single species, and even between males and females of the same species. For example, about twice as much recombination takes place in humans as in mice and rats. Within the human genome, recombination rates vary among chromosomes, with chromosomes 21 and 22 having the highest rates and chromosomes 2 and 4 having the lowest rates. Researchers have also detected differences between male and female humans in rates of recombination: the autosomal chromosomes of females undergo about 50% more recombination than do the autosomal chromosomes of males.

Geneticists have found numerous recombination *hotspots*, where recombination is at least 10 times as high as the average elsewhere in the genome. The human genome contains an estimated 25,000 to 50,000 such recombination hotspots, and approximately 60% of all crossovers take place in them. For humans, recombination hotspots tend to be found near, but not within, active genes. Recombination hotspots have been detected in the genomes of other organisms as well. Other chromosomal regions, such as those near centromeres, often display reduced rates of recombination.

■ Linked genes do not assort independently. In a testcross for two completely linked genes (with no crossing over), only nonrecombinant progeny are produced. When two genes assort independently, recombinant progeny and nonrecombinant progeny are produced in equal proportions. When two genes are linked with some crossing over between them, more nonrecombinant progeny than recombinant progeny are produced.

■ Recombination frequency is calculated by summing the number of recombinant progeny, dividing by the total number of progeny produced in the cross, and multiplying by 100%. The frequency of recombinant gametes is half the frequency of crossing over, and the maximum frequency of recombinant gametes is 50%.

■ Coupling and repulsion refer to the arrangement of alleles on a chromosome. Whether alleles are in coupling or in repulsion determines which combination of phenotypes will be most frequent in the progeny of a testcross.

■ Interchromosomal recombination takes place among genes located on different chromosomes through the random segregation of chromosomes in meiosis. Intrachromosomal recombination takes place among genes located on the same chromosome through crossing over.

- A chi-square test of independence can be used to determine whether genes are linked.

- Recombination rates can be used to determine the relative order of genes and distances between them on a chromosome. One percent recombination equals one map unit. Maps based on recombination rates are called genetic maps; maps based on physical distances are called physical maps.

- Genetic maps can be constructed by examining recombination rates from a series of two-point crosses or by examining the progeny of a three-point testcross.

- Interference is the degree to which one crossover interferes with additional crossovers in the same area. The interference equals 1 minus the coefficient of coincidence, which is the observed number of double crossovers divided by the expected number of double crossovers.

- Some multiple crossovers go undetected; thus, genetic maps based on recombination rates underestimate the true physical distances between genes.

- Human genes can be mapped by examining the cosegregation of traits in pedigrees.

- A lod score is the logarithm of the ratio of the probability of obtaining the observed progeny with the assumption of linkage to the probability of obtaining the observed progeny with the assumption of independent assortment. A lod score of 3 or higher is usually considered evidence for linkage.

- Molecular techniques that allow the detection of differences in DNA sequence have greatly facilitated gene mapping.

- Genome-wide association studies locate genes that affect particular traits by examining the nonrandom association of a trait with sets of genetic markers from across the genome.

- Physical mapping methods are used to determine the chromosomal locations of genes.

- Nucleotide sequencing is another method of physically mapping genes.

- Rates of recombination vary widely, differing among species, among and along chromosomes within a single species, and even between males and females of the same species.

IMPORTANT TERMS

linked genes (p. 175)
linkage group (p. 175)
nonrecombinant (parental) gamete (p. 177)
nonrecombinant (parental) progeny (p. 177)
recombinant gamete (p. 177)
recombinant progeny (p. 177)
recombination frequency (p. 180)

coupling (cis) configuration (p. 181)
repulsion (trans) configuration (p. 181)
genetic map (p. 187)
physical map (p. 187)
map unit (m.u.) (p. 187)
centiMorgan (cM) (p. 187)
two-point testcross (p. 188)

three-point testcross (p. 189)
interference (p. 194)
coefficient of coincidence (p. 194)
mapping function (p. 197)
lod (logarithm of odds) score (p. 199)
genetic marker (p. 199)
linkage analysis (p. 199)
genome-wide association study (p. 199)

haplotype (p. 200)
linkage disequilibrium (p. 200)
single-nucleotide polymorphism (SNP) (p. 200)
somatic-cell hybridization (p. 201)
cell line (p. 201)
heterokaryon (p. 201)
deletion mapping (p. 201)

ANSWERS TO CONCEPT CHECKS

1. c

2. Repulsion

3. Genetic maps are based on rates of recombination; physical maps are based on physical distances.

4.

$m^+ p^+ s^+$	$m^+ p s$	$m p^+ s^+$	$m^+ p^+ s$	$m p s^+$	$m^+ p s^+$	$m p^+ s$	$m p s$
$m\ p\ s$	$m\ p s$	$m\ p\ s$	$m\ p\ s$	$m p s$	$m\ p s$	$m\ p\ s$	$m p s$

5. The *c* locus

6. b

WORKED PROBLEMS

Problem 1

In guinea pigs, white coat (*w*) is recessive to black coat (*W*) and wavy hair (*v*) is recessive to straight hair (*V*). A breeder crosses a guinea pig that is homozygous for white coat and wavy hair with a guinea pig that is black with straight hair. The F$_1$ are then crossed with guinea pigs having white coats and wavy hair in a series of testcrosses. The following progeny are produced from these testcrosses:

black, straight	30	
black, wavy	10	
white, straight	12	
white, wavy	31	
Total	83	

a. Are the genes that determine coat color and hair type assorting independently? Carry out chi-square tests to test your hypothesis.

b. If the genes are not assorting independently, what is the recombination frequency between them?

>> Solution Strategy

What information is required in your answer to the problem?

a. Whether the genes are assorting independently, along with a chi-square value, degrees of freedom, and P value to evaluate your hypothesis.

b. If the genes are not assorting independently, the recombination frequency between them.

What information is provided to solve the problem?

- White coat is recessive to black coat and wavy hair is recessive to straight hair.
- The numbers of different types of progeny from a series of testcrosses.

For help with this problem, review:

Crossing Over Between Linked Genes, Calculating Recombination Frequency, and Testing for Independent Assortment in Section 7.2.

>> Solution Steps

a. Assuming independent assortment, outline the crosses conducted by the breeder:

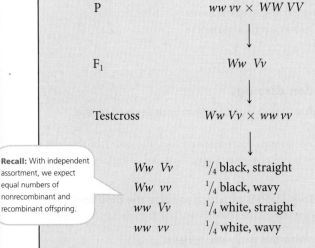

P $ww\ vv \times WW\ VV$

F$_1$ $Ww\ Vv$

Testcross $Ww\ Vv \times ww\ vv$

$Ww\ Vv$	$^1/_4$ black, straight
$Ww\ vv$	$^1/_4$ black, wavy
$ww\ Vv$	$^1/_4$ white, straight
$ww\ vv$	$^1/_4$ white, wavy

Recall: With independent assortment, we expect equal numbers of nonrecombinant and recombinant offspring.

Because a total of 83 progeny were produced in the testcrosses, we expect $^1/_4 \times 83 = 20.75$ of each. The observed numbers of progeny from the testcross (30, 10, 12, 31) do not appear to fit the expected numbers (20.75, 20.75, 20.75, 20.75) well, so independent assortment may not have taken place.

To test the hypothesis, carry out a chi-square test of independence. Construct a table with the genotypes of the first locus along the top and the genotypes of the second locus along the side. Compute the totals for the rows and columns and the grand total.

Hint: See Figure 7.10 for a review of how to carry out a chi-square test of independence.

	Ww	ww	Row totals
Vv	30	12	42
vv	10	31	41
Column totals	40	43	83 ⟵ Grand total

The expected value for each cell of the table is calculated with the following formula:

$$\text{expected number} = \frac{\text{row total} \times \text{column total}}{\text{grand total}}$$

Using this formula, we find the expected values (given in parentheses) to be

	Ww	ww	Row totals
Vv	30 (20.24)	12 (21.76)	42
vv	10 (19.76)	31 (21.24)	41
Column totals	40	43	83 ⟵ Grand total

Using these observed and expected numbers, we find the calculated chi-square value to be

$$\chi^2 = \sum \frac{(\text{observed} - \text{expected})^2}{\text{expected}}$$

$$= \frac{(30 - 20.24)^2}{20.24} + \frac{(10 - 19.76)^2}{19.76} +$$

$$\frac{(12 - 21.76)^2}{21.76} + \frac{(31 - 21.24)^2}{21.24}$$

$$= 4.71 + 4.82 + 4.38 + 4.48 = 18.39$$

The degrees of freedom for the chi-square test of independence are df = (number of rows − 1) × (number of columns − 1). There are two rows and two columns, so the degrees of freedom are

$$df = (2 - 1) \times (2 - 1) = 1 \times 1 = 1$$

> **Recall:** The probability associated with the chi-square value is the probability that the difference between observed and expected values is due to chance.

In Table 3.5, the probability associated with a chi-square value of 18.39 and 1 degree of freedom is less than 0.005, indicating that chance is very unlikely to be responsible for the differences between the observed numbers and the numbers expected with independent assortment. The genes for coat color and hair type have therefore not assorted independently.

b. To determine the recombination frequencies, identify the recombinant progeny. Using the notation for linked genes, write the crosses:

P $\quad \dfrac{w \quad v}{w \quad v} \quad \times \quad \dfrac{W \quad V}{W \quad V}$

$\downarrow$

F$_1$ $\qquad\qquad \dfrac{W \quad V}{w \quad v}$

Testcross $\quad \dfrac{W \quad V}{w \quad v} \quad \times \quad \dfrac{w \quad v}{w \quad v}$

$\downarrow$

$\dfrac{W \quad V}{w \quad v}$ 30 black, straight (nonrecombinant progeny)

$\dfrac{w \quad v}{w \quad v}$ 31 white, wavy (nonrecombinant progeny)

$\dfrac{W \quad v}{w \quad v}$ 10 black, wavy (recombinant progeny)

$\dfrac{w \quad V}{w \quad v}$ 12 white, straight (recombinant progeny)

The recombination frequency is

recombination frequency =

$$\frac{10 + 12}{30 + 31 + 10 + 12} \times 100\% = \frac{22}{83} \times 100\% = 26.5$$

> **Recall:** The recombination frequency = $\dfrac{\text{number of recombinant progeny}}{\text{total number progeny}} \times 100\%$.

Problem 2

A series of two-point crosses were carried out among seven loci (*a*, *b*, *c*, *d*, *e*, *f*, and *g*), producing the following recombination frequencies. Using these recombination frequencies, map the seven loci, showing their linkage groups, the order of the loci in each linkage group, and the distances between the loci of each group:

Loci	Recombination frequency (%)	Loci	Recombination frequency (%)
a and *b*	10	*c* and *d*	50
a and *c*	50	*c* and *e*	8
a and *d*	14	*c* and *f*	50
a and *e*	50	*c* and *g*	12
a and *f*	50	*d* and *e*	50
a and *g*	50	*d* and *f*	50
b and *c*	50	*d* and *g*	50
b and *d*	4	*e* and *f*	50
b and *e*	50	*e* and *g*	18
b and *f*	50	*f* and *g*	50
b and *g*	50		

≫ Solution Strategy

What information is required in your answer to the problem?

The linkage groups for the seven loci, the order of the loci within each linkage group, and the map distances between the loci.

What information is provided to solve the problem?

Recombination frequencies for each pair of loci.

For help with this problem, review:

Constructing a Genetic Map with a Two-Point Testcross in Section 7.2.

≫ Solution Steps

To work this problem, remember that 1% recombination equals 1 map unit. The recombination frequency between *a* and *b* is 10%, so these two loci are in the same linkage group, approximately 10 m.u. apart.

> **Hint:** A recombination frequency of 50% means that genes at the two loci are assorting independently (located in different linkage groups).

Linkage group 1

$$a \quad\quad b$$
$$\leftarrow 10 \text{ m.u.} \rightarrow$$

The recombination frequency between *a* and *c* is 50%, so *c* must lie in a second linkage group.

Linkage group 1

$$a \quad\quad b$$
$$\leftarrow 10 \text{ m.u.} \rightarrow$$

Linkage group 2

$$c$$

The recombination frequency between *a* and *d* is 14%, so *d* is located in linkage group 1. Is locus *d* 14 m.u. to the right or to the left of locus *a*? If *d* is 14 m.u. to the left of *a*, then the *b*-to-*d* distance should be 10 m.u. + 14 m.u. = 24 m.u. On the other hand, if *d* is to the right of *a*, then the distance between *b* and *d* should be 14 m.u. − 10 m.u. = 4 m.u. The *b*–*d* recombination frequency is 4%, so *d* is 14 m.u. to the right of *a*. The updated map is

> **Hint:** To determine whether locus *d* is to the right or the left of locus *a*, look at the *b* to *d* distance.

Linkage group 1
$$a \longleftarrow 14 \text{ m.u.} \longrightarrow d$$
$$b$$
$$\longleftarrow 10 \text{ m.u.} \longrightarrow \leftarrow 4 \text{ m.u.} \rightarrow$$

Linkage group 2

$$c$$

The recombination frequencies between each of loci *a*, *b*, and *d* and locus *e* are all 50%, so *e* is not in linkage group 1 with *a*, *b*, and *d*. The recombination frequency between *e* and *c* is 8%, so *e* is in linkage group 2:

Linkage group 1
$$a \longleftarrow 14 \text{ m.u.} \longrightarrow d$$
$$b$$
$$\longleftarrow 10 \text{ m.u.} \longrightarrow \leftarrow 4 \text{ m.u.} \rightarrow$$

Linkage group 2
$$c \quad\quad\quad e$$
$$\longleftarrow 8 \text{ m.u.} \longrightarrow$$

There is 50% recombination between *f* and all the other genes, so *f* must belong to a third linkage group:

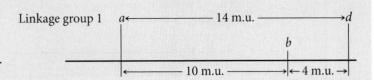

Linkage group 1
$$a \longleftarrow 14 \text{ m.u.} \longrightarrow d$$
$$b$$
$$\longleftarrow 10 \text{ m.u.} \longrightarrow \leftarrow 4 \text{ m.u.} \rightarrow$$

Linkage group 2
$$c \quad\quad\quad e$$
$$\longleftarrow 8 \text{ m.u.} \longrightarrow$$

Linkage group 3
$$f$$

Finally, position locus *g* with respect to the other genes. The recombination frequencies between *g* and loci *a*, *b*, and *d* are all 50%, so *g* is not in linkage group 1. The recombination frequency between *g* and *c* is 12%, so *g* is a part of linkage group 2. To determine whether *g* is 12 m.u. to the right or left of *c*, consult the *g*–*e* recombination frequency. Because this recombination frequency is 18%, *g* must lie to the left of *c*:

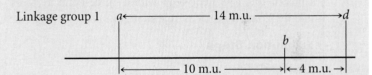

Linkage group 1
$$a \longleftarrow 14 \text{ m.u.} \longrightarrow d$$
$$b$$
$$\longleftarrow 10 \text{ m.u.} \longrightarrow \leftarrow 4 \text{ m.u.} \rightarrow$$

Linkage group 2
$$g \longleftarrow 18 \text{ m.u.} \longrightarrow e$$
$$c$$
$$\longleftarrow 12 \text{ m.u.} \longrightarrow \leftarrow 8 \text{ m.u.} \rightarrow$$

Linkage group 3
$$f$$

Note that the *g*-to-*e* distance (18 m.u.) is shorter than the sum of the *g*-to-*c* (12 m.u.) and *c*-to-*e* distances (8 m.u.) because of undetectable double crossovers between *g* and *e*.

> **Recall:** Because some double crossovers may go undetected, the distance between two distant genes (such as *g* and *e*) may be less than the sum of shorter distances (such as *g* to *c* and *c* to *e*).

Problem 3

Ebony body color (*e*), rough eyes (*ro*), and brevis bristles (*bv*) are three recessive mutations that occur in fruit flies. The loci for these mutations have been mapped and are separated by the following map distances:

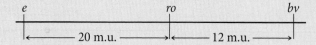

The interference between these genes is 0.4.

A fly with ebony body, rough eyes, and brevis bristles is crossed with a fly that is homozygous for the wild-type traits. The resulting F_1 females are test-crossed with males that have ebony body, rough eyes, and brevis bristles; 1800 progeny are produced. Give the expected numbers of phenotypes in the progeny of the testcross.

≫ Solution Strategy

What information is required in your answer to the problem?

The expected numbers of different progeny phenotypes produced by the testcross.

What information is provided to solve the problem?

- The map distances among the three loci.
- The interference among the loci.
- A testcross is carried out and 1800 progeny are produced.

For help with this problem, review:

Constructing a Genetic Map with a Three-Point Testcross and the Worked Problem in Section 7.3.

≫ Solution Steps

The crosses are

P
$$\frac{e^+ \quad ro^+ \quad bv^+}{e^+ \quad ro^+ \quad bv^+} \times \frac{e \quad ro \quad bv}{e \quad ro \quad bv}$$

F_1
$$\frac{e^+ \quad ro^+ \quad bv^+}{e \quad ro \quad bv}$$

Testcross
$$\frac{e^+ \quad ro^+ \quad bv^+}{e \quad ro \quad bv} \times \frac{e \quad ro \quad bv}{e \quad ro \quad bv}$$

Hint: The order of the genes is provided by the genetic map.

In this case, we know that *ro* is the middle locus because the genes have been mapped. Eight classes of progeny will be produced by this cross:

e^+	ro^+	bv^+	Nonrecombinant
e	ro	bv	Nonrecombinant
e^+ / ro	bv	single crossover between *e* and *ro*
e / ro^+	bv^+	single crossover between *e* and *ro*
e^+	ro^+ / bv	single crossover between *ro* and *bv*
e	ro / bv^+	single crossover between *ro* and *bv*
e^+ / ro / bv^+	double crossover
e / ro^+ / bv	double crossover

To determine the numbers of each type, use the map distances, starting with the double crossovers. The expected number of double crossovers is equal to the product of the single-crossover probabilities:

$$\text{expected number of double crossovers} = 0.20 \times 0.12 \times 1800$$
$$= 43.2$$

Recall: The presence of interference means that not all expected double crossovers will be observed.

However, there is some interference, so the observed number of double crossovers will be less than the expected. The interference is 1 − coefficient of coincidence, so the coefficient of coincidence is

$$\text{coefficient of coincidence} = 1 - \text{interference}$$

The interference is given as 0.4, so the coefficient of coincidence equals $1 - 0.4 = 0.6$. Recall that the coefficient of coincidence is

$$\text{coefficient of coincidence} =$$
$$\frac{\text{number of observed double crossovers}}{\text{number of expected double crossovers}}$$

Rearranging this equation, we obtain

$$\frac{\text{number of observed}}{\text{double crossovers}} = \frac{\text{coefficient of}}{\text{coincidence}} \times \frac{\text{number of expected}}{\text{double crossovers}}$$

$$= 0.6 \times 43.2 = 26$$

A total of 26 double crossovers should be observed. Because there are two classes of double crossovers (e^+ / ro / bv^+ and e / ro^+ / bv), we expect to observe 13 of each class.

Next, we determine the number of single-crossover progeny. The genetic map indicates that the distance between e and ro is 20 m.u., so 360 progeny (20% of 1800) are expected to have resulted from recombination between these two loci. Some of them will be single-crossover progeny and some will be double-crossover progeny. We have already determined that the number of double-crossover progeny is 26, so the number of progeny resulting from a single crossover between e and ro is 360 − 26 = 334, which will be divided

equally between the two single-crossover phenotypes (e / ro^+ / bv^+ and e^+ / ro bv).

The distance between ro and bv is 12 m.u., so the number of progeny resulting from recombination between these two genes is 0.12 × 1800 = 216. Again, some of these recombinants will be single-crossover progeny and some will be double-crossover progeny. To determine the number of

progeny resulting from a single crossover, subtract the double crossovers: 216 − 26 = 190. These single-crossover progeny will be divided between the two single-crossover phenotypes (e^+ ro^+ / bv and e ro / bv^+), so there will be $190/2$ = 95 of each of these phenotypes. The remaining progeny will be nonrecombinants, and they can be obtained by subtraction: 1800 − 26 − 334 − 190 = 1250; there are two nonrecombinants (e^+ ro^+ bv^+ and e ro bv), so there will be $1250/2$ = 625 of each. The numbers of the various phenotypes are listed here:

e^+	ro^+	bv^+	625	nonrecombinant
e	ro	bv	625	nonrecombinant
e^+ / ro	bv	167	single crossover between e and ro
e / ro^+	bv^+	167	single crossover between e and ro
e^+	ro^+ / bv	95	single crossover between ro and bv
e	ro / bv^+	95	single crossover between ro and bv
e^+ / ro / bv^+	13	double crossover
e / ro^+ / bv	13	double crossover
Total	1800	

COMPREHENSION QUESTIONS

Section 7.1

1. What does the term *recombination* mean? What are two causes of recombination?

Section 7.2

2. In a testcross for two genes, what types of gametes are produced with (a) complete linkage, (b) independent assortment, and (c) incomplete linkage?

3. What effect does crossing over have on linkage?

4. Why is the frequency of recombinant gametes always half the frequency of crossing over?

5. What is the difference between genes in coupling configuration and genes in repulsion? How does the arrangement of linked genes (whether they are in coupling or repulsion) affect the results of a genetic cross?

6. How would you test to see if two genes are linked?

7. What is the relationship between recombination frequency and a centiMorgan?

8. Why do calculated recombination frequencies between pairs of loci that are located far apart underestimate the true genetic distances between loci?

Section 7.3

9. Explain how to determine, using the numbers of progeny from a three-point cross, which of three linked loci is the middle locus.

10. What does the interference tell us about the effect of one crossover on another?

11. What is a lod score and how is it calculated?

Section 7.4

12. List some of the methods for physically mapping genes and explain how they are used to position genes on chromosomes.

APPLICATION QUESTIONS AND PROBLEMS

Introduction

*13. The introduction to this chapter described the search for genes that determine pattern baldness in humans. In 1916, Dorothy Osborn suggested that pattern baldness is a sex-influenced trait (see Chapter 5) that is dominant in males and recessive in females. More research suggested that pattern baldness is an X-linked recessive trait. Would you expect to see independent assortment between genetic markers on the X chromosome and pattern baldness if (a) pattern baldness is sex-influenced and (b) if pattern baldness is X-linked recessive? Explain your answer.

Section 7.2

14. In the snail *Cepaea nemoralis*, an autosomal allele causing a banded shell (B^B) is recessive to the allele for an unbanded shell (B^O). Genes at a different locus determine the background color of the shell; here, yellow (C^Y) is recessive to brown (C^{Bw}). A banded, yellow snail is crossed with a homozygous brown, unbanded snail. The F_1 are then crossed with banded, yellow snails (a testcross).

   a. What will the results of the testcross be if the loci that control banding and color are linked with no crossing over?

   b. What will the results of the testcross be if the loci assort independently?

   c. What will the results of the testcross be if the loci are linked and 20 m.u. apart?

   [Picture Press/Getty Images.]

*15. In silkmoths (*Bombyx mori*), red eyes (*re*) and white-banded wings (*wb*) are encoded by two mutant alleles that are recessive to those that produce wild-type traits (*re*⁺ and *wb*⁺); these two genes are on the same chromosome. A moth homozygous for red eyes and white-banded wings is crossed with a moth homozygous for the wild-type traits. The F_1 have wild-type eyes and wild-type wings. The F_1 are crossed with moths that have red eyes and white-banded wings in a testcross. The progeny of this testcross are

wild-type eyes, wild-type wings	418
red eyes, wild-type wings	19
wild-type eyes, white-banded wings	16
red eyes, white-banded wings	426

   a. What phenotypic proportions would be expected if the genes for red eyes and for white-banded wings were located on different chromosomes?

   b. What is the rate of recombination between the gene for red eyes and the gene for white-banded wings?

*16. A geneticist discovers a new mutation in *Drosophila melanogaster* that causes the flies to shake and quiver. She calls this mutation *quiver* (*qu*) and determines that it is due to an autosomal recessive gene. She wants to determine whether the gene encoding the quiver phenotype is linked to the recessive gene encoding vestigial (reduced) wings (*vg*). She crosses a fly homozygous for quiver and vestigial traits with a fly homozygous for the wild-type traits and then uses the resulting F_1 females in a testcross. She obtains the following flies from this testcross:

vg⁺	*qu*⁺	230
vg	*qu*	224
vg	*qu*⁺	97
vg⁺	*qu*	99
Total		650

   Are the genes that cause vestigial wings and the quiver phenotype linked? Do a chi-square test of independence to determine whether the genes have assorted independently.

*17. In cucumbers, heart-shaped leaves (*hl*) are recessive to normal leaves (*Hl*) and having numerous fruit spines (*ns*) is recessive to having few fruit spines (*Ns*). The genes for leaf shape and for number of spines are located on the same chromosome; findings from mapping experiments indicate that they are 32.6 m.u. apart. A cucumber plant having heart-shaped leaves and numerous spines is crossed with a plant that is homozygous for normal leaves and few spines. The F_1 are crossed with plants that have heart-shaped leaves and numerous spines. What phenotypes and phenotypic proportions are expected in the progeny of this cross?

18. In tomatoes, tall (*D*) is dominant over dwarf (*d*) and smooth fruit (*P*) is dominant over pubescent fruit (*p*), which is covered with fine hairs. A farmer has two tall and smooth tomato plants, which we will call plant A and plant B. The farmer crosses plants A and B with the same dwarf and pubescent plant and obtains the following numbers of progeny:

	Progeny of	
	Plant A	**Plant B**
Dd Pp	122	2
Dd pp	6	82
dd Pp	4	82
dd pp	124	4

   a. What are the genotypes of plant A and plant B?

   b. Are the loci that determine the height of the plant and pubescence linked? If so, what is the rate of recombination between them?

   c. Explain why different proportions of progeny are produced when plant A and plant B are crossed with the same dwarf pubescent plant.

19. Alleles *A* and *a* reside at a locus on the same chromosome as a locus with alleles *B* and *b*. *Aa Bb* is crossed with *aa bb* and the following progeny are produced:

Aa Bb	5
Aa bb	45
aa Bb	45
aa bb	5

What conclusion can be drawn about the arrangement of the genes on the chromosome in the *Aa Bb* parent?

20. Daniel McDonald and Nancy Peer determined that eyespot (a clear spot in the center of the eye) in flour beetles is caused by an X-linked allele (*es*) that is recessive to the allele for the absence of eyespot (*es⁺*). They conducted a series of crosses to determine the distance between the gene for eyespot and a dominant X-linked gene for striped (*St*), which causes white stripes on females and acts as a recessive lethal (is lethal when homozygous in females or hemizygous in males). The following cross was carried out (D. J. McDonald and N. J. Peer. 1961. *Journal of Heredity* 52:261–264).

$$\female \ \frac{es^+ \quad St}{es \quad St^+} \times \frac{es \quad St^+}{Y} \ \male$$

$$\downarrow$$

$\dfrac{es^+ \quad St}{es \quad St^+}$	1630
$\dfrac{es \quad St^+}{es \quad St^+}$	1665
$\dfrac{es \quad St}{es \quad St^+}$	935
$\dfrac{es^+ \quad St^+}{es \quad St^+}$	1005
$\dfrac{es \quad St^+}{Y}$	1661
$\dfrac{es^+ \quad St^+}{Y}$	1024

a. Which progeny are the recombinants and which progeny are the nonrecombinants?

b. Calculate the recombination frequency between *es* and *St*.

c. Are some potential genotypes missing among the progeny of the cross? If so, which ones, and why?

*21. Recombination frequencies between three loci in corn are shown in the following table:

Loci	Recombination frequency (%)
R and *W₂*	17
R and *L₂*	35
W₂ and *L₂*	18

What is the order of the genes on the chromosome?

22. In tomatoes, dwarf (*d*) is recessive to tall (*D*) and opaque (light-green) leaves (*op*) are recessive to green leaves (*Op*). The loci that determine height and leaf color are linked and separated by a distance of 7 m.u. For each of the following crosses, determine the phenotypes and proportions of progeny produced.

a. $\dfrac{D \quad Op}{d \quad op} \times \dfrac{d \quad op}{d \quad op}$

b. $\dfrac{D \quad op}{d \quad Op} \times \dfrac{d \quad op}{d \quad op}$

c. $\dfrac{D \quad Op}{d \quad op} \times \dfrac{D \quad Op}{d \quad op}$

d. $\dfrac{D \quad op}{d \quad Op} \times \dfrac{D \quad op}{d \quad Op}$

23. In German cockroaches, bulging eyes (*bu*) are recessive to normal eyes (*bu⁺*) and curved wings (*cv*) are recessive to straight wings (*cv⁺*). These two traits are encoded by autosomal genes that are linked. A cockroach has genotype *bu⁺bu cv⁺cv*, and the genes are in repulsion. Which of the following sets of genes will be found in the most common gametes produced by this cockroach?

a. *bu⁺ cv⁺*

b. *bu cv*

c. *bu⁺bu*

d. *cv⁺cv*

e. *bu cv⁺*

Explain your answer.

24. In *Drosophila melanogaster*, ebony body (*e*) and rough eyes (*ro*) are encoded by autosomal recessive genes found on chromosome 3; they are separated by 20 m.u. The gene that encodes forked bristles (*f*) is X-linked recessive and assorts independently of *e* and *ro*. Give the phenotypes of progeny and their expected proportions when a female of each of the following genotypes is test-crossed with a male.

a. $\dfrac{e^+ \qquad ro^+}{e \qquad ro} \quad \dfrac{f^+}{f}$

b. $\dfrac{e^+ \qquad ro}{e \qquad ro^+} \quad \dfrac{f^+}{f}$

25. Honeybees have haplodiploid sex determination: queens (females) are diploid, developing from fertilized eggs, whereas drones (males) are haploid, developing from unfertilized eggs. Otto Mackensen studied linkage relations among eight mutations in honeybees (O. Mackensen. 1958. *Journal of Heredity* 49:99–102). The following table gives the results of two of Mackensen's crosses including three recessive mutations: *cd* (cordovan body color), *h* (hairless), and *ch* (chartreuse eye color).

Queen genotype	Phenotypes of drone (male) progeny
$\dfrac{cd \quad h^+}{cd^+ \quad h}$	294 cordovan, 236 hairless, 262 cordovan and hairless, 289 wild type
$\dfrac{h \quad ch^+}{h^+ \quad ch}$	3131 hairless, 3064 chartreuse, 96 chartreuse and hairless, 132 wild type

a. Only the genotype of the queen is given. Why is the genotype of the male parent not needed for mapping these genes? Would the genotype of the male parent be required if we examined female progeny instead of male progeny?

b. Determine the nonrecombinant and recombinant progeny for each cross and calculate the map distances between *cd*, *h*, and *ch*. Draw a linkage map illustrating the linkage arrangements among these three genes.

26. Perform a chi-square test of independence on the data provided in **Figure 7.2** to determine if the genes for flower color and pollen shape in sweet peas are assorting independently. Give the chi-square value, degrees of freedom, and associated probability. What conclusion would you draw about the independent assortment of these genes?

*27. A series of two-point crosses were carried out among seven loci (*a*, *b*, *c*, *d*, *e*, *f*, and *g*), producing the following recombination frequencies. Map the seven loci, showing their linkage groups, the order of the loci in each linkage group, and the distances between the loci of each group.

Loci	Recombination frequency (%)	Loci	Recombination frequency (%)
a and *b*	50	*c* and *d*	50
a and *c*	50	*c* and *e*	26
a and *d*	12	*c* and *f*	50
a and *e*	50	*c* and *g*	50
a and *f*	50	*d* and *e*	50
a and *g*	4	*d* and *f*	50
b and *c*	10	*d* and *g*	8
b and *d*	50	*e* and *f*	50
b and *e*	18	*e* and *g*	50
b and *f*	50	*f* and *g*	50
b and *g*	50		

28. R. W. Allard and W. M. Clement determined recombination rates for a series of genes in lima beans (R. W. Allard and W. M. Clement. 1959. *Journal of Heredity* 50:63–67). The following table lists paired recombination frequencies for eight of the loci (*D*, *Wl*, *R*, *S*, *L₁*, *Ms*, *C*, and *G*) that they mapped. On the basis of these data, draw a series of genetic maps for the different linkage groups of the genes, indicating the distances between the genes. Keep in mind that these frequencies are estimates of the true recombination frequencies and that some error is associated with each estimate. An asterisk beside a

recombination frequency indicates that the recombination frequency is significantly different from 50%.

Recombination frequencies (%) among seven loci in lima beans

	Wl	*R*	*S*	*L₁*	*Ms*	*C*	*G*
D	2.1*	39.3*	52.4	48.1	53.1	51.4	49.8
Wl		38.0*	47.3	47.7	48.8	50.3	50.4
R			51.9	52.7	54.6	49.3	52.6
S				26.9*	54.9	52.0	48.0
L₁					48.2	45.3	50.4
Ms						14.7*	43.1
C							52.0

*Significantly different from 50%.

Section 7.3

*29. Raymond Popp studied linkage among genes for pink eye (*p*), shaker-1 (*sh-1*, which causes circling behavior, head tossing, and deafness), and hemoglobin (*Hb*) in mice (R. A. Popp. 1962. *Journal of Heredity* 53:73–80). He performed a series of testcrosses, in which mice heterozygous for pink eye, shaker-1, and hemoglobin 1 and 2 were crossed with mice that were homozygous for pink eye, shaker-1, and hemoglobin 2.

$$\frac{P \; Sh\text{-}1 \; Hb^1}{p \; sh\text{-}1 \; Hb^2} \times \frac{p \; sh\text{-}1 \; Hb^2}{p \; sh\text{-}1 \; Hb^2}$$

The following progeny were produced:

Progeny genotype	Number
$\dfrac{p \; sh\text{-}1 \; Hb^2}{p \; sh\text{-}1 \; Hb^2}$	274
$\dfrac{P \; Sh\text{-}1 \; Hb^1}{p \; sh\text{-}1 \; Hb^2}$	320
$\dfrac{P \; Sh\text{-}1 \; Hb^2}{p \; sh\text{-}1 \; Hb^2}$	57
$\dfrac{p \; Sh\text{-}1 \; Hb^1}{p \; sh\text{-}1 \; Hb^2}$	45
$\dfrac{P \; Sh\text{-}1 \; Hb^2}{p \; sh\text{-}1 \; Hb^2}$	6
$\dfrac{p \; sh\text{-}1 \; Hb^1}{p \; sh\text{-}1 \; Hb^2}$	5
$\dfrac{p \; sh\text{-}1 \; Hb^2}{p \; sh\text{-}1 \; Hb^2}$	0
$\dfrac{p \; sh\text{-}1 \; Hb^1}{p \; sh\text{-}1 \; Hb^2}$	1
Total	708

a. Determine the order of these genes on the chromosome.

b. Calculate the map distances between the genes.

c. Determine the coefficient of coincidence and the interference among these genes.

30. Waxy endosperm (*wx*), shrunken endosperm (*sh*), and yellow seedlings (*v*) are encoded by three recessive genes in corn that are linked on chromosome 5. A corn plant homozygous for all three recessive alleles is crossed with a plant homozygous for all the dominant alleles. The resulting F_1 are then crossed with a plant homozygous for the recessive alleles in a three-point testcross. The progeny of the testcross are

wx	*sh*	*V*	87
Wx	*Sh*	*v*	94
Wx	*Sh*	*V*	3,479
wx	*sh*	*v*	3,478
Wx	*sh*	*V*	1,515
wx	*Sh*	*v*	1,531
wx	*Sh*	*V*	292
Wx	*sh*	*v*	280
Total			10,756

a. Determine the order of these genes on the chromosome.

b. Calculate the map distances between the genes.

c. Determine the coefficient of coincidence and the interference among these genes.

31. Priscilla Lane and Margaret Green studied the linkage relations of three genes affecting coat color in mice: mahogany (*mg*), agouti (*a*), and ragged (*Rg*). They carried out a series of three-point crosses, mating mice that were heterozygous at all three loci with mice that were homozygous for the recessive alleles at these loci (P. W. Lane and M. C. Green. 1960. *Journal of Heredity* 51:228–230). The following table lists the progeny of the testcrosses:

a	*Rg*	+	1
+	+	*mg*	1
a	+	+	15
+	*Rg*	*mg*	9
+	+	+	16
a	*Rg*	*mg*	36
a	+	*mg*	76
+	*Rg*	+	69
Total			213

Note: + represents a wild-type allele.

a. Determine the order of the loci that encode mahogany, agouti, and ragged on the chromosome, the map distances between them, and the coefficient of coincidence and the interference among these genes.

b. Draw a picture of the two chromosomes in the triply heterozygous mice used in the testcrosses, indicating which of the alleles are present on each chromosome.

32. Fine spines (*s*), smooth fruit (*tu*), and uniform fruit color (*u*) are three recessive traits in cucumbers, the genes for which are linked on the same chromosome. A cucumber plant heterozygous for all three traits is used in a testcross, and the following progeny are produced by this testcross:

S	*U*	*Tu*	2
s	*u*	*Tu*	70
S	*u*	*Tu*	21
s	*u*	*tu*	4
S	*U*	*tu*	82
s	*U*	*tu*	21
s	*U*	*Tu*	13
S	*u*	*tu*	17
Total			230

a. Determine the order of these genes on the chromosome.

b. Calculate the map distances between the genes.

c. Determine the coefficient of coincidence and the interference among these genes.

d. List the genes found on each chromosome in the parents used in the testcross.

33. In *Drosophila melanogaster*, black body (*b*) is recessive to gray body (b^+), purple eyes (*pr*) are recessive to red eyes (pr^+), and vestigial wings (*vg*) are recessive to normal wings (vg^+). The loci encoding these traits are linked, with the following map distances:

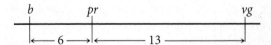

The interference among these genes is 0.5. A fly with a black body, purple eyes, and vestigial wings is crossed with a fly homozygous for a gray body, red eyes, and normal wings. The female progeny are then crossed with males that have a black body, purple eyes, and vestigial wings. If 1000 progeny are produced by this testcross, what will be the phenotypes and proportions of the progeny?

34. *Sepia eyes, spineless bristles,* and *striped body* are three recessive mutations in *Drosophila* found on chromosome 3. A genetics student crosses a fly homozygous for the alleles encoding sepia eyes, spineless bristles, and striped body with a fly homozygous for the wild-type alleles—encoding red eyes, normal bristles, and solid body. The female progeny are then test-crossed with males that have sepia eyes, spineless bristles, and striped body. Assume that the interference between these genes is 0.2 and that 400 progeny flies are produced by the testcross. Based on the map distances provided in **Figure 7.15**, predict the phenotypes and proportions of the progeny resulting from the test cross.

35. Shown below are eight DNA sequences from different individuals.

	Nucleotide Position															
	1				5					10					15	
Sequence 1. . . .	T	C	T	G	G	A	T	C	A	T	C	A	C	A	T	. . .
Sequence 2. . . .	A	C	A	G	C	A	T	C	A	T	T	A	C	G	T	. . .
Sequence 3. . . .	T	C	A	G	G	A	T	C	A	T	T	A	C	T	A	. . .
Sequence 4. . . .	T	C	A	G	G	A	T	C	A	T	T	A	C	A	T	. . .
Sequence 5. . . .	A	C	A	G	C	A	T	C	A	T	T	A	C	G	T	. . .
Sequence 6. . . .	T	C	T	G	G	A	T	C	A	T	C	A	C	A	T	. . .
Sequence 7. . . .	T	C	A	G	G	A	T	C	A	T	T	A	C	A	T	. . .
Sequence 8. . . .	A	C	A	G	C	A	T	C	A	T	T	A	C	G	T	. . .

a. Give the nucleotide positions of all single-nucleotide polymorphisms (SNPs; nucleotide positions where individuals vary in which base is present) in these sequences.

b. How many different haplotypes (sets of linked variants) are found in these eight sequences?

c. Give the haplotype of each sequence by listing the specific bases at each variable position in that particular haplotype. (Hint: See **Figure 20.8.**)

***36.** A group of geneticists are interested in identifying genes that may play a role in susceptibility to asthma. They study the inheritance of genetic markers in a series of families that have two or more asthmatic children. They find an association between the presence or absence of asthma and a genetic marker on the short arm of chromosome 20 and calculate a lod score of 2 for this association. What does this lod score indicate about genes that may influence asthma?

Section 7.4

37. A panel of cell lines was created by human–mouse somatic-cell hybridization. Each line was examined for the presence of human chromosomes and for the production of an enzyme. The following results were obtained:

Cell line	Enzyme	Human chromosomes 1	2	3	4	5	6	7	8	9	10	17	22
A	−	+	−	−	−	+	−	−	−	−	−	+	−
B	+	+	+	−	−	−	−	−	+	−	−	+	+
C	−	+	−	−	−	+	−	−	−	−	−	−	+
D	−	−	−	−	+	−	−	−	−	−	−	−	−
E	+	+	−	−	−	−	−	−	+	−	+	+	−

On the basis of these results, which chromosome has the gene that encodes the enzyme?

***38.** A panel of cell lines was created by human–mouse somatic-cell hybridization. Each cell line was examined for the presence of human chromosomes and for the production of three enzymes. The following results were obtained:

Cell line	Enzyme 1	2	3	4	Human chromosomes 8	9	12	15	16	17	22	X
A	+	−	+	−	−	+	−	+	+	−	−	+
B	+	−	−	−	−	+	−	−	+	+	−	−
C	−	+	+	+	−	−	−	−	−	+	−	+
D	−	+	+	+	+	−	−	−	+	−	−	+

On the basis of these results, give the chromosomal locations of the genes encoding enzyme 1, enzyme 2, and enzyme 3.

***39.** The locations of six deletions have been mapped to a *Drosophila* chromosome, as shown in the following deletion map. Recessive mutations *a*, *b*, *c*, *d*, *e*, and *f* are known to be located in the same region as the deletions, but the order of the mutations on the chromosome is not known.

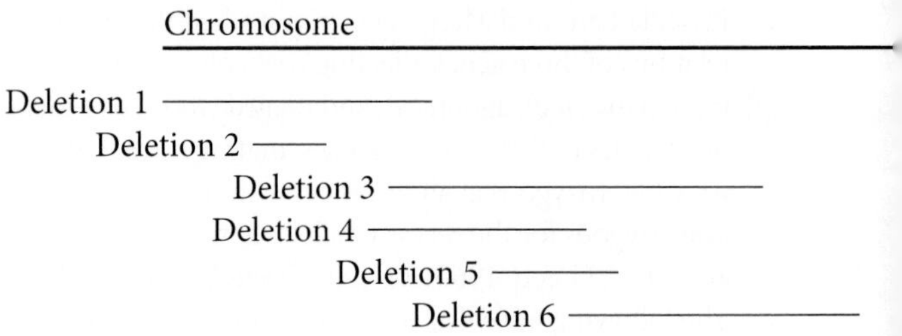

When flies homozygous for the recessive mutations are crossed with flies heterozygous for the deletions, the following results are obtained, in which "m" represents a mutant phenotype and a plus sign (+) represents the wild type. On the basis of these data, determine the relative order of the seven mutant genes on the chromosome.

Deletion	Mutations *a*	*b*	*c*	*d*	*e*	*f*
1	m	+	m	+	+	m
2	m	+	+	+	+	+
3	+	m	m	m	m	+
4	+	+	m	m	m	+
5	+	+	+	m	m	+
6	+	m	+	m	+	+

Section 7.5

40. Transferrin is a blood protein that is encoded by the transferrin locus (*Trf*). In house mice, the two alleles at this locus (*Trf*^{a} and *Trf*^{b}) are codominant and encode three types of transferrin:

Genotype	Phenotype
Trf^a/*Trf*^{a}	Trf-a
Trf^a/*Trf*^{b}	Trf-ab
Trf^b/*Trf*^{b}	Trf-b

The dilution locus, found on the same chromosome, determines whether the color of a mouse is diluted or full; an allele for dilution (*d*) is recessive to an allele for full color (*d*$^+$):

Genotype	Phenotype
d$^+$*d*$^+$	*d*$^+$ (full color)
d$^+$*d*	*d*$^+$ (full color)
dd	*d* (dilution)

Donald Shreffler conducted a series of crosses to determine the map distance between the transferrin locus and the dilution locus (D. C. Shreffler. 1963. *Journal of Heredity* 54:127–129). The following table presents a series of crosses carried out by Shreffler and the progeny resulting from these crosses.

a. Calculate the recombination frequency between the *Trf* and the *d* loci by using the pooled data from all the crosses.

b. Which crosses represent recombination in male gamete formation and which crosses represent recombination in female gamete formation?

c. On the basis of your answer to part *b*, calculate the frequency of recombination among male parents and female parents separately.

d. Are the rates of recombination in males and females the same? If not, what might produce the difference?

A mouse with the dilution trait. [L. Montoliu, W. S. Oetting, D. C. Bennett. Color Genes. 3/2010. European Society for Pigment Cell Research. (http://www.espcr.org/micemut) 03/2013.]

Cross	♂		♀	d^+ Trf-ab	d^+ Trf-b	d Trf-ab	d Trf-b	Total
1	$\frac{d^+\ Trf^a}{d\ Trf^b}$	×	$\frac{d\ Trf^b}{d\ Trf^b}$	32	3	6	21	62
2	$\frac{d\ Trf^b}{d\ Trf^b}$	×	$\frac{d^+\ Trf^a}{d\ Trf^b}$	16	0	2	20	38
3	$\frac{d^+\ Trf^a}{d\ Trf^b}$	×	$\frac{d\ Trf^b}{d\ Trf^b}$	35	9	4	30	78
4	$\frac{d\ Trf^b}{d\ Trf^b}$	×	$\frac{d^+\ Trf^a}{d\ Trf^b}$	21	3	2	19	45
5	$\frac{d^+\ Trf^b}{d\ Trf^a}$	×	$\frac{d\ Trf^b}{d\ Trf^b}$	8	29	22	5	64
6	$\frac{d\ Trf^b}{d\ Trf^b}$	×	$\frac{d^+\ Trf^b}{d\ Trf^a}$	4	14	11	0	29

Column header above: **Progeny phenotypes**

THINK-PAIR-SHARE QUESTIONS

Section 7.1

1. Species A has $2n = 10$ chromosomes. Species B has $2n = 40$ chromosomes. On average, will two randomly selected genes from species A be more likely, less likely, or equally likely to assort independently than two randomly selected genes from species B? Explain your reasoning.

Section 7.2

2. The recombination frequency between genes *a* and *b* is 20%. What is the frequency of crossing over between genes *a* and *b*? Explain your reasoning.

3. Is it possible to have a recombination frequency of 80% between two genes? Is it possible for two genes to be separated by 80 map units? Why or why not?

4. The rate of crossing over between two linked genes (*r* and *w*) is 0.44. The following cross is carried out: *Rr Ww* × *rr ww*. What proportion of the gametes of the *Rr Ww* parent were recombinant gametes? Explain your reasoning and state any assumptions you made.

Section 7.3

5. Why is interference important? Why do we calculate it in a three-point cross? Why don't we calculate interference in a two-point cross?

6. Three recessive mutations in *Drosophila melanogaster*, *roughoid* (*ru*, small rough eyes), *javelin* (*jv*, cylindrical bristles), and *sepia eyes* (*se*, dark brown eyes) are linked. A three-point cross was carried out and the following progeny obtained:

jv^+	ru^+	se^+	37
jv^+	ru^+	se	2
jv^+	ru	se	14
jv^+	ru	se^+	146
jv	ru	se^+	2
jv	ru	se	35
jv	ru^+	se	154
jv	ru^+	se^+	10

a. Determine the order of the genes on the chromosome.

b. Determine which progeny contain single crossovers and which contain double crossovers and indicate where among the genes the crossovers occurred.

c. Calculate the map distances among the genes.

d. Calculate the coefficient of coincidence and interference among the genes.

7. Compare and contrast linkage analysis and genome-wide association studies. How are they similar? How are they different?

Chromosome Variation

Many varieties of bananas have multiple sets of chromosomes.
[Frankie Angel/Alamy.]

Building a Better Banana

Bananas and plantains (collectively referred to as bananas) are the world's most popular fruit. In many developing countries, they are a critically important source of food, providing starch and calories for hundreds of millions of people. In industrialized countries, more bananas are consumed than any other fruit; for example, Americans consume as many pounds of bananas as apples and oranges combined. Over 100 million tons of bananas are produced annually worldwide.

There is no concrete biological distinction between bananas and plantains, but the term *banana* generally refers to the sweeter forms that are eaten uncooked, while the term *plantain* is applied to bananas that are peeled when unripe and cooked before eating. Cultivated bananas differ from their wild relatives by being seedless, which makes them more edible but hinders their reproduction. Farmers propagate bananas vegetatively, by cutting off parts of existing plants and coaxing them to grow into new plants. Because bananas are propagated vegetatively, many cultivated banana plants are genetically identical.

From a genetic standpoint, bananas are interesting because many varieties have multiple sets of chromosomes. Most eukaryotic organisms in nature are diploid ($2n$), with two sets of chromosomes. Others, such as fungi, are haploid (n), with a single set of chromosomes. Cultivated bananas are often polyploid, with more than two sets of chromosomes ($3n$, $4n$, or higher). Most strains of cultivated bananas were created by crossing plants within and between two diploid species: *Musa acuminata* (genome = AA) and *Musa balbisiana* (genome = BB). Many cultivated bananas are triploid, with three sets of chromosomes, consisting of AAA, AAB, or ABB, and some bananas even have four sets of chromosomes (tetraploid), consisting of AAAA, AAAB, AABB, or ABBB.

In spite of their worldwide importance as a food, modern cultivated bananas are in trouble. The strain most often sold in grocery stores—the Cavendish—is threatened by disease and pests; in recent years, a soil fungus has devastated crops in Asia. The Cavendish's predecessor, called Gros Michel (Big Mike), was the banana of choice until disease wiped it out in the 1950s and 1960s. Because vegetative propagation produces genetically identical plants, cultivated bananas are particularly vulnerable to attack by pathogens and pests.

To help develop a better banana—more disease and pest resistant, as well as more nutritious—geneticists launched an international effort to sequence the genome of the banana, producing a draft sequence in 2012. This research demonstrated that the banana genome consists of over 500 million base pairs (bp) of DNA and contains 36,500 protein-encoding genes. Using this genome sequence, scientists have already identified several genes that play a role in resistance to fungal diseases and are exploring ways to better breed and genetically engineer bananas.

THINK-PAIR-SHARE
- What are some possible advantages to producing cultivated bananas that are polyploid? What might be some disadvantages?

- What is a genetically modified food? Are the currently consumed, polyploid bananas genetically modified? Explain your reasoning.

Most species have a characteristic number of chromosomes, each with a distinct size and structure, and all the tissues of an organism (except for gametes) generally have the same set of chromosomes. Nevertheless, variations in chromosome number—such as the extra sets of chromosomes seen in bananas—do periodically arise. Variations may also arise in chromosome structure: individual chromosomes may lose or gain parts, and the order of genes within a chromosome may be altered. These variations in the number and structure of chromosomes are termed **chromosome mutations**, and they frequently play an important role in agriculture and evolution.

We begin this chapter by reviewing some of the basic concepts of chromosome structure. We then consider the different types of chromosome mutations and their features, phenotypic effects, and influences on evolution.

8.1 Chromosome Mutations Include Rearrangements, Aneuploidy, and Polyploidy

Before we consider the different types of chromosome mutations, their effects, and how they arise, it may be helpful to review and expand on the basics of chromosome structure that we learned in Chapter 2.

Chromosome Morphology

Each functional eukaryotic chromosome has a centromere, to which spindle microtubules attach, and two telomeres, which stabilize the chromosome ends (see Figure 2.7). Chromosomes are classified into four basic types:

1. **Metacentric.** The centromere is located approximately in the middle, so the chromosome has two arms of equal length.

2. **Submetacentric.** The centromere is displaced toward one end, creating a long arm and a short arm. (On human chromosomes, the short arm is designated by the letter p and the long arm by the letter q.)

3. **Acrocentric.** The centromere is near one end, producing a long arm and a knob, or satellite, at the other end.

4. **Telocentric.** The centromere is at or very near the end of the chromosome (see Figure 2.8).

The complete set of chromosomes possessed by an organism is called its *karyotype*. An organism's karyotype is usually presented as a picture of metaphase chromosomes lined up in descending order of their size (**Figure 8.1**). Karyotypes are prepared from actively dividing cells, such as white blood cells, bone-marrow cells, or cells from meristematic tissues of plants. After treatment with a chemical (such as colchicine) that prevents them from entering anaphase, the cells are chemically preserved. They are then burst open to release the chromosomes onto a microscope slide, and the chromosomes are stained and photographed. The photograph is then enlarged, and the individual chromosomes are cut out and arranged in a karyotype. For human chromosomes, karyotypes are routinely prepared by automated machines, which scan a slide using a video camera attached to a microscope, looking for a chromosome spread (a group of chromosomes that are well separated). When a spread has been

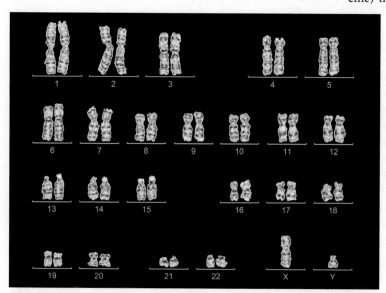

8.1 A human karyotype consists of 46 chromosomes. A karyotype for a male is shown here; a karyotype for a female would have two X chromosomes. [© Pr Philippe Vago © ISM/Phototake.]

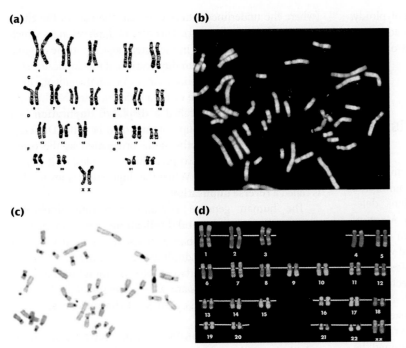

(a)

(b)

(c)

(d)

8.2 Chromosome banding is revealed by special staining techniques. (a) G banding. (b) Q banding. (c) C banding. (d) R banding. [Part a: Leonard Lessin/Science Source. Parts b and c: University of Washington Pathology Department. http://pathology.washington.edu. Part d: © Dr. Ram Verma/Phototake.]

located, the camera takes a picture of the chromosomes, the image is digitized, and the chromosomes are sorted and arranged electronically by a computer.

Preparation and staining techniques help to distinguish among chromosomes of similar size and shape. For instance, special preparation and staining of chromosomes with a dye called Giemsa reveals G bands, which distinguish areas of DNA that are rich in adenine–thymine (A–T) base pairs (**Figure 8.2a**; see also Chapter 10). Q bands (**Figure 8.2b**) are revealed by staining chromosomes with quinacrine mustard and viewing the chromosomes under ultraviolet light; variation in the brightness of Q bands results from differences in the relative numbers of cytosine–guanine (C–G) and adenine–thymine base pairs. Other techniques reveal C bands (**Figure 8.2c**), which are regions of DNA occupied by centromeric heterochromatin, and R bands (**Figure 8.2d**), which are rich in cytosine–guanine base pairs.

Types of Chromosome Mutations

Chromosome mutations can be grouped into three basic categories: chromosome rearrangements, aneuploidy, and polyploidy (**Figure 8.3**). Chromosome rearrangements alter the *structure* of chromosomes: for example, a piece of a chromosome may be duplicated, deleted, or inverted. In aneuploidy, the *number* of chromosomes is altered: one or more

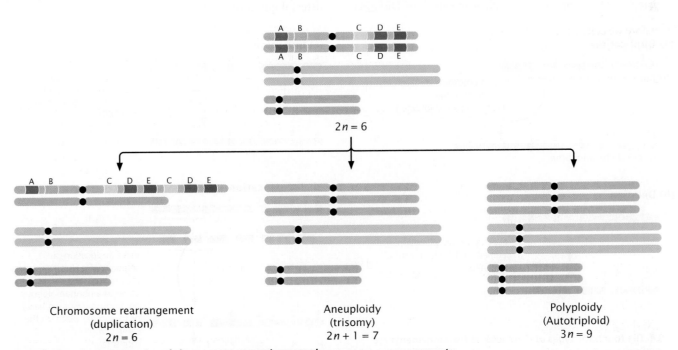

Chromosome rearrangement (duplication)
$2n = 6$

Aneuploidy (trisomy)
$2n + 1 = 7$

Polyploidy (Autotriploid)
$3n = 9$

8.3 The three basic categories of chromosome mutations are chromosome rearrangements, aneuploidy, and polyploidy. Duplications, trisomy, and autotriploidy, respectively, are examples of these categories of mutations.

individual chromosomes are added or deleted. In polyploidy, one or more complete *sets* of chromosomes are added. A polyploid is any organism that has more than two sets of chromosomes ($3n$, $4n$, $5n$, or more).

THINK-PAIR-SHARE Question 1

8.2 Chromosome Rearrangements Alter Chromosome Structure

Chromosome rearrangements are chromosome mutations that change the structures of individual chromosomes. The four basic types of rearrangements are duplications, deletions, inversions, and translocations (**Figure 8.4**). Many of these chromosome rearrangements originate when double-stranded breaks occur in the DNA molecule found within a chromosome. Double-stranded breaks in DNA often cause cell death, so organisms have evolved elaborate mechanisms to repair breaks by reconnecting the broken ends of DNA (see pp. 547–548 in Chapter 18). If the two broken ends are rejoined correctly, the original chromosome is restored, and no chromosome rearrangement results. Sometimes the wrong ends are connected, however, leading to a chromosome rearrangement. Chromosome rearrangements can also arise through errors in crossing over or when crossing over occurs between repeated DNA sequences.

Duplications

A **chromosome duplication** is a mutation in which part of the chromosome has been doubled (see Figure 8.4a). Consider a chromosome with segments AB•CDEFG, in which • represents the centromere. A duplication might include the EF segments, giving rise to a chromosome with segments AB•CDEF<u>EF</u>G (where the underlined letters represent the part of the chromosome that has changed). This type of duplication, in which the duplicated segment is immediately adjacent to the original segment, is called a **tandem duplication**. If the duplicated segment is located some distance from the original segment, either on the same chromosome or on a different one, the chromosome rearrangement is called a **displaced duplication**. An example of a displaced duplication would be AB•CDEFG<u>EF</u>. A duplication can have the same orientation as the original sequence, as in the two preceding examples, or it can be inverted: AB•CDEF<u>FE</u>G. When the duplication is inverted, it is called a **reverse duplication**.

The human genome contains numerous duplicated sequences called **segmental duplications**, which are defined as duplications greater than a thousand base pairs (bp) in length. Most segmental duplications are *intrachromosomal* duplications (that is, the two copies are found on the same chromosome), but others are *interchromosomal* duplications (the two copies are found on different chromosomes).

EFFECTS OF CHROMOSOME DUPLICATIONS An individual that is homozygous for a duplication carries that duplication on both homologous chromosomes, and an individual that is heterozygous for a duplication has one normal chromosome and one chromosome with the duplication. In heterozygotes (**Figure 8.5a**), problems in chromosome pairing arise at prophase I of meiosis because the two chromosomes are not homologous throughout their length. The pairing and synapsis of homologous regions require that one or both chromosomes loop and twist so that these regions are able to line up (**Figure 8.5b**). The appearance of this characteristic loop structure in meiosis is one way to detect duplications.

(a) Duplication

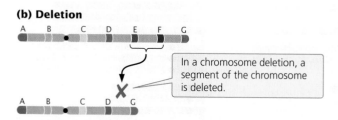

(b) Deletion

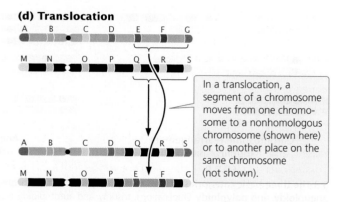

8.4 The four basic types of chromosome rearrangements are duplications, deletions, inversions, and translocations.

(c) Inversion

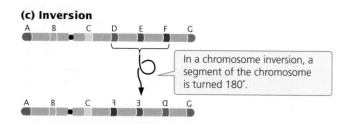

(d) Translocation

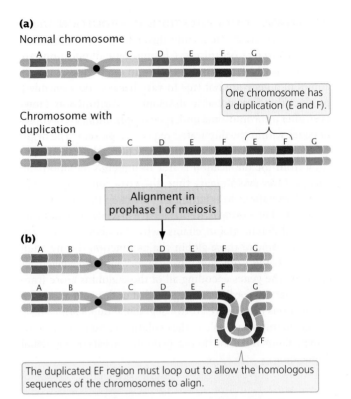

(a)
Normal chromosome

Chromosome with duplication

One chromosome has a duplication (E and F).

Alignment in prophase I of meiosis

(b)

The duplicated EF region must loop out to allow the homologous sequences of the chromosomes to align.

8.5 In an individual heterozygous for a duplication, the duplicated region loops out during pairing in prophase I.

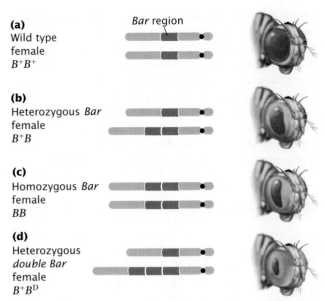

(a)
Wild type female
B^+B^+

Bar region

(b)
Heterozygous *Bar* female
B^+B

(c)
Homozygous *Bar* female
BB

(d)
Heterozygous *double Bar* female
B^+B^D

8.6 The Bar phenotype in *Drosophila melanogaster* results from an X-linked duplication. (a) Wild-type fruit flies have normal-sized eyes. (b) Flies that are heterozygous and (c) homozygous for the *Bar* mutation have smaller, bar-shaped eyes. (d) Flies with *double Bar* have three copies of the duplication and much smaller bar-shaped eyes.

Duplications may have major effects on the phenotype. Among fruit flies, for example, a fly with the *Bar* mutation has a reduced number of facets in the eye, making the eye smaller and bar shaped instead of oval (**Figure 8.6**). The *Bar* mutation results from a small duplication on the X chromosome that is inherited as an incompletely dominant, X-linked trait: heterozygous female flies have somewhat smaller eyes (the number of facets is reduced; see Figure 8.6b), whereas in homozygous female and hemizygous male flies, the number of facets is greatly reduced (see Figure 8.6c). Occasionally, a fly carries three copies of the *Bar* duplication on its X chromosome; in flies with this mutation, termed *double Bar*, the number of facets is extremely reduced (Figure 8.6d).

Duplications and deletions often arise from **unequal crossing over**, in which chromosomes misalign during crossing over. Unequal crossing over is frequently the cause of red–green color blindness in humans. Perception of color is affected by red and green opsin genes, which are found on the X chromosome and are 98% identical in their DNA sequence. Most people with normal color vision have one red opsin gene and one green opsin gene (although some people have more than one copy of each). Occasionally, two paired X chromosomes in a female do not align properly in prophase I, and unequal crossing over takes place. The unequal crossing over produces one chromosome with an extra opsin gene and one chromosome that is missing an opsin gene (**Figure 8.7**; see also Figure 18.13). When a male inherits the chromosome that is missing one of the opsin genes, red–green color blindness results.

UNBALANCED GENE DOSAGE How does chromosome duplication alter the phenotype? After all, gene sequences are not altered by duplications, and no genetic information is missing; the only change is the presence of additional copies of normal sequences. The answer to this question is not well understood, but the effects are most likely due to imbalances in the amounts of gene products (abnormal gene dosage). The amount of a particular protein synthesized by a cell is often

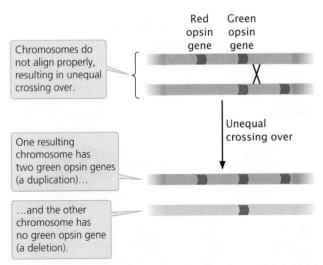

Red opsin gene   Green opsin gene

Chromosomes do not align properly, resulting in unequal crossing over.

Unequal crossing over

One resulting chromosome has two green opsin genes (a duplication)…

…and the other chromosome has no green opsin gene (a deletion).

8.7 Unequal crossing over produces duplications and deletions.

directly related to the number of copies of its corresponding gene: an individual organism with three functional copies of a gene often produces 1.5 times as much of the protein encoded by that gene as an individual with two copies. Because developmental processes require the interaction of many proteins, they often depend critically on proper gene dosage. If the amount of one protein increases while the amounts of others remain constant, problems can result (**Figure 8.8**). Duplications can have severe consequences when the precise balance of gene products is critical to cell function (**Table 8.1**).

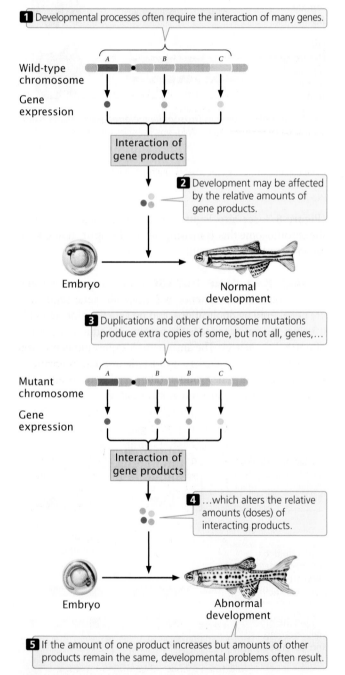

1 Developmental processes often require the interaction of many genes.

Wild-type chromosome

Gene expression

Interaction of gene products

2 Development may be affected by the relative amounts of gene products.

Embryo

Normal development

3 Duplications and other chromosome mutations produce extra copies of some, but not all, genes,...

Mutant chromosome

Gene expression

Interaction of gene products

4 ...which alters the relative amounts (doses) of interacting products.

Embryo

Abnormal development

5 If the amount of one product increases but amounts of other products remain the same, developmental problems often result.

8.8 Unbalanced gene dosage leads to developmental abnormalities.

IMPORTANCE OF DUPLICATIONS IN EVOLUTION Duplications have arisen frequently throughout the evolution of many eukaryotic organisms. Chromosome duplications are one way in which new genes evolve. In many cases, existing copies of a gene are not free to vary because they encode a product that is essential to development or function. However, after a chromosome undergoes duplication, extra copies of genes within the duplicated region are present. The original copy can provide the essential function, whereas an extra copy from the duplication is free to undergo mutation and change. Over evolutionary time, the extra copy may acquire enough mutations to assume a new function that benefits the organism. For example, humans have a series of genes that encode different globin chains, which function as oxygen carriers. Some of these globin chains function during adult stages and others function during embryonic and fetal development. The genes encoding all of these globins arose from an original primordial globin gene that underwent a series of duplications (see Figure 26.16). Segmental duplications have played an important role in the evolution of human chromosomes; about 4% of the human genome consists of segmental duplications.

CONCEPTS

A chromosome duplication is a mutation that doubles part of a chromosome. In individuals heterozygous for a chromosome duplication, the duplicated region of the chromosome loops out when homologous chromosomes pair in prophase I of meiosis. Duplications often have major effects on the phenotype, possibly by altering gene dosage. Segmental duplications are common within the human genome.

✔ CONCEPT CHECK 1

Chromosome duplications often result in abnormal phenotypes because

a. developmental processes depend on the relative amounts of proteins encoded by different genes.

b. extra copies of the genes within the duplicated region do not pair in meiosis.

c. the chromosome is more likely to break when it loops in meiosis.

d. extra DNA must be replicated, which slows down cell division.

Deletions

A second type of chromosome rearrangement is a **chromosome deletion**: the loss of a chromosome segment (see Figure 8.4b). A chromosome with segments AB•CDEFG that undergoes a deletion of segment EF would generate the mutated chromosome AB•CDG.

A large deletion can be easily detected because the chromosome is noticeably shortened. In individuals heterozygous for deletions, the normal chromosome must loop out

TABLE 8.1	Effects of some human chromosome rearrangements		
Type of Rearrangement	**Chromosome**	**Disorder**	**Symptoms**
Duplication	4, short arm	—	Small head, short neck, low hairline, reduced growth, and intellectual disability
Duplication	4, long arm	—	Small head, sloping forehead, hand abnormalities
Duplication	7, long arm	—	Delayed development, asymmetry of the head, fuzzy scalp, small nose, low-set ears
Duplication	9, short arm	—	Characteristic facial features, variable intellectual disability, high and broad forehead, hand abnormalities
Deletion	5, short arm	*Cri-du-chat* syndrome	Small head, distinctive cry, widely spaced eyes, round face, intellectual disability
Deletion	4, short arm	Wolf–Hirschhorn syndrome	Small head with high forehead, wide nose, cleft lip and palate, severe intellectual disability
Deletion	4, long arm	—	Small head, mild to moderate intellectual disability, cleft lip and palate, hand and foot abnormalities
Deletion	7, long arm	Williams–Beuren syndrome	Distinctive facial features, heart defects, cognitive impairment
Deletion	15, long arm	Prader–Willi syndrome	Feeding difficulty at early age but becoming obese after 1 year of age, mild to moderate intellectual disability
Deletion	18, short arm	—	Round face, large and low-set ears, mild to moderate intellectual disability
Deletion	18, long arm	—	Distinctive mouth shape, small hands, small head, intellectual disability

during the pairing of homologs in prophase I of meiosis (**Figure 8.9**) to allow the homologous regions of the two chromosomes to align and undergo synapsis. This looping out generates a structure that looks very much like that seen in individuals heterozygous for duplications.

EFFECTS OF DELETIONS The phenotypic consequences of a deletion depend on which genes are located in the deleted region. If the deletion includes the centromere, then the chromosome will not segregate in meiosis or mitosis and will usually be lost. Many deletions are lethal in the homozygous state because all copies of any essential genes located in the deleted region are missing.

Even individuals heterozygous for a deletion may have multiple defects for three reasons. First, the heterozygous condition may produce imbalances in the amounts of gene products similar to those produced by extra gene copies. Second, normally recessive mutations on the homologous chromosome lacking the deletion may be expressed when the wild-type allele has

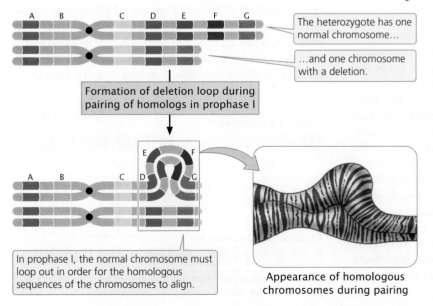

The heterozygote has one normal chromosome…

…and one chromosome with a deletion.

Formation of deletion loop during pairing of homologs in prophase I

In prophase I, the normal chromosome must loop out in order for the homologous sequences of the chromosomes to align.

Appearance of homologous chromosomes during pairing

8.9 In an individual heterozygous for a deletion, the normal chromosome loops out during chromosome pairing in prophase I.

8.10 The Notch phenotype is produced by a chromosome deletion that includes the *Notch* gene. (Left) Normal wing venation. (Right) Wing venation produced by a *Notch* mutation. [Spyros Artavanis-Tsakonas, Kenji Matsuno, and Mark E. Fortini.]

CONCEPTS

A chromosome deletion is a mutation in which a part of a chromosome is lost. In individuals heterozygous for a deletion, the normal chromosome loops out during prophase I of meiosis. Deletions cause recessive genes on the homologous chromosome to be expressed and may cause imbalances in gene products.

✔ **CONCEPT CHECK 2**

What is pseudodominance and how is it produced by a chromosome deletion?

been deleted (and is no longer present to mask the recessive allele's expression). The expression of a normally recessive mutation is referred to as **pseudodominance**, and it is an indication that one of the homologous chromosomes has a deletion.

Third, some genes must be present in two copies for normal function. When a single copy of a gene is not sufficient to produce a wild-type phenotype, that gene is said to be **haploinsufficient**. A series of X-linked wing mutations in *Drosophila* are known as *Notch* mutations. These mutations often result from chromosome deletions. *Notch* deletions behave in a dominant manner: when heterozygous for a *Notch* deletion, a fly has wings that are notched at the tips and along the edges (**Figure 8.10**). The *Notch* gene is therefore haploinsufficient. Females that are homozygous for a *Notch* deletion (or males that are hemizygous) die early in embryonic development. The *Notch* gene, which is deleted in *Notch* mutants, encodes a receptor that normally transmits signals received from outside the cell to the cell's interior and is important in fly development. The deletion acts as a recessive lethal because the loss of all copies of the *Notch* gene prevents normal development.

CHROMOSOME DELETIONS IN HUMANS In humans, a deletion on the short arm of chromosome 5 is responsible for *cri-du-chat* syndrome. The name (French for "cry of the cat") derives from the peculiar, catlike cry of infants with this syndrome. A child who is heterozygous for this deletion has a small head, widely spaced eyes, a round face, and is intellectually disabled. Deletion of part of the short arm of chromosome 4 results in another human disorder, Wolf–Hirschhorn syndrome, which is characterized by seizures, severe intellectual disability, and delayed growth. A deletion of a tiny segment of chromosome 7 causes haploinsufficiency of the gene encoding elastin and a few other genes and leads to a condition known as Williams–Beuren syndrome, which is characterized by distinctive facial features, heart defects, high blood pressure, and cognitive impairments. The effects of some deletions in human chromosomes are summarized in Table 8.1.

Inversions

A third type of chromosome rearrangement is a **chromosome inversion**, in which a chromosome segment is inverted—turned 180 degrees (see Figure 8.4c). If a chromosome originally had segments AB•CDEFG, then chromosome AB•CFEDG represents an inversion that includes segments DEF. For an inversion to take place, the chromosome must break in two places. Inversions that do not include the centromere, such as AB•CFEDG, are termed **paracentric inversions** (*para* meaning "next to"), whereas inversions that include the centromere, such as ADC•BEFG, are termed **pericentric inversions** (*peri* meaning "around").

Inversion heterozygotes are common in many organisms, including a number of plants, some species of *Drosophila*, mosquitoes, and grasshoppers. Inversions may have played an important role in human evolution: G-banding patterns reveal that several human chromosomes differ from those of chimpanzees by only a pericentric inversion (**Figure 8.11**).

EFFECTS OF INVERSIONS Individual organisms with inversions have neither lost nor gained any genetic material; only the DNA sequence has been altered. Nevertheless, these mutations often have pronounced phenotypic effects. An inversion may break a gene into two parts, with one part moving to a new location and destroying the function of that gene. Even when the chromosome breaks lie between genes, phenotypic effects may arise from the inverted gene order.

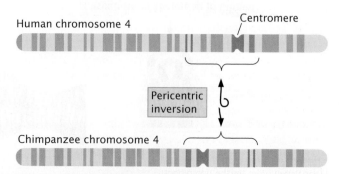

8.11 Chromosome 4 in humans and in chimpanzees differs by a pericentric inversion. [Courtesy of Dr. Christine Harrison.]

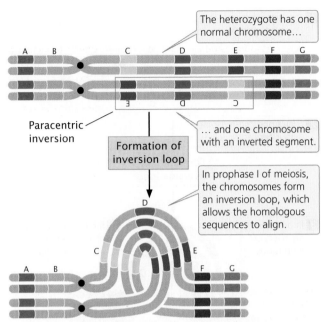

8.12 In an individual heterozygous for a paracentric inversion, the chromosomes form an inversion loop during pairing in prophase I.

Many genes are regulated in a position-dependent manner; if their positions are altered by an inversion, their expression may be altered, an outcome referred to as a **position effect**. For example, when an inversion moves a wild-type allele (that normally encodes red eyes) at the *white* locus in *Drosophila* to a chromosomal region that contains highly condensed and inactive chromatin, the wild-type allele is not expressed in some cells, resulting in a eye consisting of red and white spots.

INVERSIONS IN MEIOSIS When an individual is homozygous for a particular inversion, no special problems arise in meiosis, and the two homologous chromosomes can pair and separate normally. However, when an individual is heterozygous for an inversion, the gene order of the two homologs differs, and the homologous sequences can align and pair only if the two chromosomes form an inversion loop (**Figure 8.12**).

Individuals heterozygous for inversions also exhibit reduced recombination among genes located in the inverted region. The frequency of crossing over within the inversion is not actually diminished, but when crossing over does take place, the outcome is abnormal gametes that do not give rise to viable offspring, and thus no recombinant progeny are observed. Let's see why this happens.

Figure 8.13 illustrates the results of crossing over within a paracentric inversion. Suppose that an individual is heterozygous for an inversion (see Figure 8.13a), with one wild-type, nonmutated chromosome (AB•CDEFG) and one inverted chromosome (AB•EDCFG). In prophase I of meiosis, an inversion loop forms, allowing the homologous sequences to pair

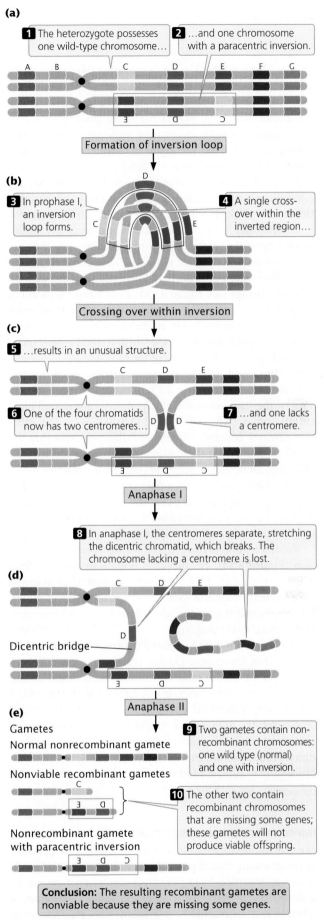

8.13 In a heterozygous individual, a single crossover within a paracentric inversion leads to abnormal gametes.

up (see Figure 8.13b). If a single crossover takes place in the inverted region (between segments C and D in Figure 8.13), an unusual structure results (see Figure 8.13c). The two outer chromatids, which did not participate in crossing over, contain original, nonrecombinant gene sequences. The two inner chromatids, which did participate in crossing over, are highly abnormal: each has two copies of some genes and no copies of others. Furthermore, one of the four chromatids now has two centromeres, and is therefore referred to as a **dicentric chromatid**; the other lacks a centromere and is an **acentric chromatid**.

In anaphase I of meiosis, the centromeres are pulled toward opposite poles, and the two homologous chromosomes separate. This action stretches the dicentric chromatid across the center of the nucleus, forming a structure called a **dicentric bridge** (see Figure 8.13d). Eventually, the dicentric bridge breaks as the two centromeres are pulled farther apart. Spindle microtubules do not attach to the acentric fragment, so that fragment does not segregate to a spindle pole and is usually lost when the nucleus re-forms.

In the second division of meiosis, the sister chromatids separate, and four gametes are produced (see Figure 8.13e). Two of the gametes contain the original, nonrecombinant chromosomes (AB•CDEFG and AB•EDCFG). The other two gametes contain recombinant chromosomes that are missing some genes; these gametes will not produce viable offspring. Thus, no recombinant progeny result when crossing over takes place within a paracentric inversion. The key is to recognize that crossing over still takes place, but when it does so, the resulting recombinant gametes are not viable, so no recombinant progeny are observed.

Recombination is also reduced within a pericentric inversion (**Figure 8.14**). No dicentric bridges or acentric fragments are produced, but the recombinant chromosomes have too many copies of some genes and no copies of others, so gametes that receive the recombinant chromosomes cannot produce viable progeny.

Figures 8.13 and 8.14 illustrate the results of single crossovers within inversions. Double crossovers in which both crossovers are on the same two strands (two-strand double crossovers) result in functional recombinant chromosomes (to see why functional gametes are produced by double crossovers, try drawing the results of a two-strand double crossover). Thus, even though the overall rate of recombination is reduced within an inversion, some viable recombinant progeny may still be produced through two-strand double crossovers. **▶ TRY PROBLEM 27**

IMPORTANCE OF INVERSIONS IN EVOLUTION Inversions can play important evolutionary roles by suppressing recombination among a set of genes. As we have seen, crossing over within an inversion in an individual that is heterozygous for a pericentric or paracentric inversion leads to unbalanced gametes and no recombinant progeny. This suppression of recombination allows alleles adapted to a specific environment to remain together, unshuffled by recombination.

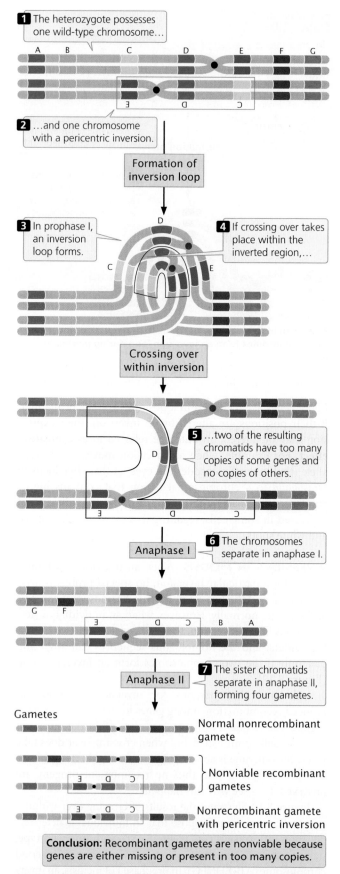

1 The heterozygote possesses one wild-type chromosome...

2 ...and one chromosome with a pericentric inversion.

Formation of inversion loop

3 In prophase I, an inversion loop forms.

4 If crossing over takes place within the inverted region,...

Crossing over within inversion

5 ...two of the resulting chromatids have too many copies of some genes and no copies of others.

Anaphase I

6 The chromosomes separate in anaphase I.

Anaphase II

7 The sister chromatids separate in anaphase II, forming four gametes.

Gametes

Normal nonrecombinant gamete

Nonviable recombinant gametes

Nonrecombinant gamete with pericentric inversion

Conclusion: Recombinant gametes are nonviable because genes are either missing or present in too many copies.

8.14 In a heterozygous individual, a single crossover within a pericentric inversion leads to abnormal gametes.

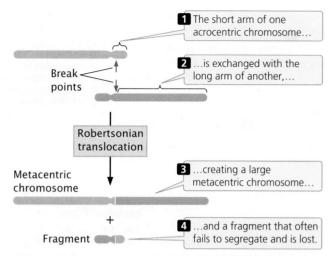

1 The short arm of one acrocentric chromosome…

2 …is exchanged with the long arm of another,…

Break points

Robertsonian translocation

3 …creating a large metacentric chromosome…

Metacentric chromosome

+

4 …and a fragment that often fails to segregate and is lost.

Fragment

8.15 In a Robertsonian translocation, the short arm of one acrocentric chromosome is exchanged with the long arm of another.

Translocations

A **translocation** entails the movement of genetic material between nonhomologous chromosomes (see Figure 8.4d) or within the same chromosome. Translocation should not be confused with crossing over, in which there is an exchange of genetic material between *homologous* chromosomes.

In a **nonreciprocal translocation**, genetic material moves from one chromosome to another without any reciprocal exchange. Consider the following two nonhomologous chromosomes: AB•CDEFG and MN•OPQRS. If chromosome segment EF moves from the first chromosome to the second without any transfer of segments from the second chromosome to the first, a nonreciprocal translocation has taken place, producing chromosomes AB•CDG and MN•OP<u>EF</u>QRS. More commonly, there is a two-way exchange of segments between the chromosomes, resulting in a **reciprocal translocation**. A reciprocal translocation between chromosomes AB•CDEFG and MN•OPQRS might give rise to chromosomes AB•CD<u>QRS</u> and MN•OP<u>EFG</u>.

EFFECTS OF TRANSLOCATIONS Translocations can affect a phenotype in several ways. First, they can physically link genes that were formerly located on different chromosomes. These new linkage relations may affect gene expression (a position effect), as genes translocated to new locations may come under the control of different regulatory sequences or other genes that affect their expression.

Second, the chromosome breaks that bring about translocations may take place within a gene and disrupt its function. Molecular geneticists have used these types of effects to map human genes. Neurofibromatosis, a genetic disease characterized by numerous fibrous tumors of the skin and nervous tissue, results from an autosomal dominant mutation. Linkage studies first placed the locus that, when mutated, causes neurofibromatosis on chromosome 17, but its precise location was unknown. Geneticists later narrowed down the location

when they identified two patients with neurofibromatosis who possessed a translocation affecting chromosome 17. They assumed that these patients had developed neurofibromatosis because one of the chromosome breaks that occurred in the translocation disrupted a particular gene. Sequencing of DNA from the regions around the breaks eventually led to the identification of the gene responsible for neurofibromatosis.

Deletions frequently accompany translocations. In a **Robertsonian translocation**, for example, the long arms of two acrocentric chromosomes become joined to a common centromere through a translocation, generating a metacentric chromosome with two long arms and another chromosome with two very short arms (**Figure 8.15**). The smaller chromosome is often lost because very small chromosomes do not have enough mass to segregate properly during mitosis and meiosis. The result is an overall reduction in chromosome number. As we will see, Robertsonian translocations are the cause of some cases of Down syndrome, a chromosome disorder discussed later in this chapter.

TRANSLOCATIONS IN MEIOSIS The effect of a translocation on chromosome segregation in meiosis depends on the nature of the translocation. Let's consider what happens in an individual heterozygous for a reciprocal translocation. Suppose that the original chromosomes are AB•CDEFG and M•NOPQRST (designated N_1 and N_2 respectively, for normal chromosomes 1 and 2) and that a reciprocal translocation takes place, producing chromosomes AB•CD<u>QRST</u> and M•NOP<u>EFG</u> (designated T_1 and T_2, respectively, for translocated chromosomes 1 and 2). An individual heterozygous for this translocation would possess one normal copy and one translocated copy of each chromosome (**Figure 8.16a**). Each of these chromosomes contains segments that are homologous to segments of two other chromosomes. Thus, when the homologous segments pair in prophase I of meiosis, crosslike configurations consisting of all four chromosomes form (**Figure 8.16b**).

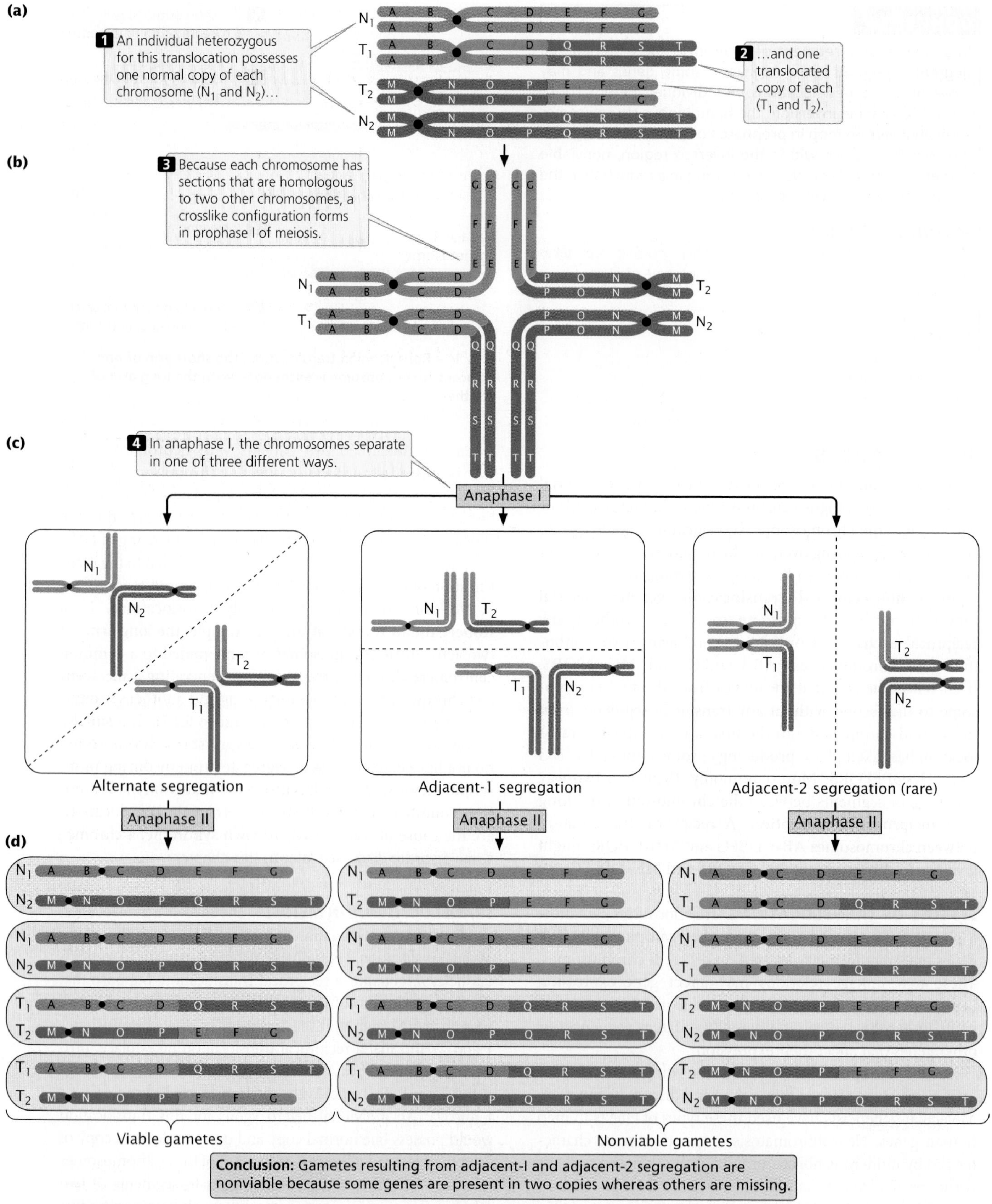

(a)

1 An individual heterozygous for this translocation possesses one normal copy of each chromosome (N_1 and N_2)...

2 ...and one translocated copy of each (T_1 and T_2).

(b)

3 Because each chromosome has sections that are homologous to two other chromosomes, a crosslike configuration forms in prophase I of meiosis.

(c)

4 In anaphase I, the chromosomes separate in one of three different ways.

Anaphase I

Alternate segregation

Adjacent-1 segregation

Adjacent-2 segregation (rare)

Anaphase II

Anaphase II

Anaphase II

(d)

Viable gametes

Nonviable gametes

Conclusion: Gametes resulting from adjacent-1 and adjacent-2 segregation are nonviable because some genes are present in two copies whereas others are missing.

8.16 In an individual heterozygous for a reciprocal translocation, crosslike structures form in homologous pairing.

Notice that N_1 and T_1 have homologous centromeres (in both chromosomes, the centromere is between segments B and C); similarly, N_2 and T_2 have homologous centromeres (between segments M and N). Normally, homologous centromeres separate and move toward opposite poles in anaphase I of meiosis. With a reciprocal translocation, the chromosomes may segregate in three different ways. In **alternate segregation** (**Figure 8.16c**), N_1 and N_2 move toward one pole and T_1 and T_2 move toward the opposite pole. In **adjacent-1 segregation**, N_1 and T_2 move toward one pole and T_1 and N_2 move toward the other pole. In both alternate and adjacent-1 segregation, homologous centromeres segregate to opposite poles. **Adjacent-2 segregation**, in which N_1 and T_1 move toward one pole and T_2 and N_2 move toward the other, is rare because the two homologous chromosomes usually separate in meiosis.

The products of the three segregation patterns are illustrated in **Figure 8.16d**. As you can see, each of the gametes produced by alternate segregation possesses one complete set of the chromosome segments. These gametes are therefore functional and can produce viable progeny. In contrast, gametes produced by adjacent-1 and adjacent-2 segregation are not viable because some chromosome segments are present in two copies, whereas others are missing. Because adjacent-2 segregation is rare, most gametes are produced by alternate or adjacent-1 segregation. Therefore, approximately half of the gametes from an individual heterozygous for a reciprocal translocation are expected to be functional.

▶ TRY PROBLEM 28

THINK-PAIR-SHARE Question 2

CONCEPTS

In translocations, parts of chromosomes move to other non-homologous chromosomes or to other regions of the same chromosome. Translocations can affect the phenotype by causing genes to move to new locations, where they come under the influence of new regulatory sequences, or by breaking genes and disrupting their function.

✔ CONCEPT CHECK 4

What is the outcome of a Robertsonian translocation?

a. Two acrocentric chromosomes
b. One large metacentric chromosome and one very small chromosome with two very short arms
c. One large metacentric and one large acrocentric chromosome
d. Two large metacentric chromosomes

Fragile Sites

Some chromosomes contain **fragile sites** (**Figure 8.17**), which are sites that develop constrictions or gaps when the cells are grown in culture and that are prone to breakage under certain conditions. More than 100 fragile sites have been identified on human chromosomes.

Human fragile sites fall into two groups. *Common fragile sites* are present in all humans and are a normal feature of chromosomes. Common fragile sites are often the locations of chromosome breakage and rearrangements in cancer cells, leading to chromosome deletions, translocations, and other chromosome rearrangements. *Rare fragile sites* are found in few people and exhibit Mendelian inheritance. Rare fragile sites are often associated with genetic disorders, such as intellectual disability. Most of them consist of expanding nucleotide repeats, in which the number of repeats of a set of nucleotides is increased (see Chapter 18).

One of the most intensively studied rare fragile sites is located on the human X chromosome and is associated with **fragile-X syndrome**, a disorder that includes intellectual disability. Fragile-X syndrome, which exhibits X-linked inheritance and arises with a frequency of about 1 in 5000 male births, has been shown to result from an increase in the number of repeats of a CGG trinucleotide.

Molecular studies of fragile sites have shown that many of these sites are more than 100,000 bp in length and include one or more genes. Fragile sites are often replicated late in S phase. At these sites, the enzymes that replicate DNA may stall while unwinding of the DNA continues (see Chapter 12), leading to long stretches of DNA that are unwound and vulnerable to breakage. In spite of recent advances in our understanding of fragile sites, their nature is not completely understood.

Copy-Number Variations

Chromosome rearrangements have traditionally been detected by examining chromosomes with a microscope. Visual examination identifies chromosome rearrangements on the basis of changes in the overall size of a chromosome, alteration of banding patterns revealed by chromosome staining, or the behavior of chromosomes in meiosis. Microscopy, however, can detect only large chromosome rearrangements, typically those that are at least 5 million base pairs in length.

With the completion of the Human Genome Project (see Chapter 20), detailed information about DNA sequences found on individual human chromosomes became available. Using this information, geneticists can now examine the number of copies of specific DNA sequences present in a cell and detect duplications, deletions, and

8.17 Fragile sites are chromosomal regions susceptible to breakage under certain conditions. Shown here is a fragile site on the human X chromosome.
[Courtesy of Dr. Christine Harrison.]

other chromosome rearrangements that cannot be observed with microscopy alone. This work has been greatly facilitated by the availability of microarrays (see Chapter 20), which allow the simultaneous detection of hundreds of thousands of specific DNA sequences across the genome. Because these methods measure the number of copies of particular DNA sequences, the variations that they detect are called **copy-number variations** (CNVs). Copy-number variations include duplications and deletions that range in length from thousands of base pairs to several million base pairs. Many of these variants encompass at least one gene and may encompass several genes. Recently developed techniques for rapidly sequencing the genome now make it possible to detect not only copy-number variations, but also other chromosome rearrangements such as inversions and translocations. Chromosome rearrangements and copy-number variations are collectively referred to as **structural variants**.

Recent studies of copy-number variations have revealed that submicroscopic chromosome duplications and deletions are quite common: research suggests that each person may possess as many as a thousand copy-number variations. Many of these variations probably have no observable phenotypic effects, but some have now been implicated in a number of diseases and disorders. For example, Janine Wagenstaller and her colleagues studied copy-number variation in 67 children with unexplained intellectual disability and found that 11 (16%) of them had duplications or deletions. Copy-number variations have also been associated with osteoporosis, autism, schizophrenia, and a number of other diseases and disorders. ▶ **TRY PROBLEM 20**

> **CONCEPTS**
>
> Variations in the number of copies of particular DNA sequences (copy-number variations) are surprisingly common in the human genome.

8.3 Aneuploidy Is an Increase or Decrease in the Number of Individual Chromosomes

In addition to chromosome rearrangements, chromosome mutations include changes in the number of chromosomes. Variations in chromosome number can be classified into two basic types: **aneuploidy**, which is a change in the number of individual chromosomes, and **polyploidy**, which is an increase in the number of chromosome sets.

Aneuploidy can arise in several ways. First, a chromosome may be lost in the course of mitosis or meiosis if, for example, its centromere is deleted. Loss of the centromere prevents the spindle microtubules from attaching, so the chromosome fails to move to the spindle pole and does not become incorporated into a nucleus after cell division. Second, the small chromosome generated by a Robertsonian translocation may be lost in mitosis or meiosis. Third, aneuploidy may arise through nondisjunction, the failure of homologous chromosomes or sister chromatids to separate in meiosis or mitosis (see p. 91 in Chapter 4). Nondisjunction leads to some gametes or cells that contain an extra chromosome and other gametes or cells that are missing a chromosome (**Figure 8.18**). ▶ **TRY PROBLEM 29**

Types of Aneuploidy

We consider four types of common aneuploid conditions in diploid organisms: nullisomy, monosomy, trisomy, and tetrasomy.

1. **Nullisomy** is the loss of both members of a homologous pair of chromosomes. It is represented as $2n - 2$, where n refers to the haploid number of chromosomes. Thus, among humans, who normally possess $2n = 46$ chromosomes, a nullisomic zygote has 44 chromosomes.

2. **Monosomy** is the loss of a single chromosome, represented as $2n - 1$. A human monosomic zygote has 45 chromosomes.

3. **Trisomy** is the gain of a single chromosome, represented as $2n + 1$. A human trisomic zygote has 47 chromosomes. The gain of a chromosome means that there are three homologous copies of one chromosome. Most cases of Down syndrome, discussed later in this section, result from trisomy of chromosome 21.

4. **Tetrasomy** is the gain of two homologous chromosomes, represented as $2n + 2$. A human tetrasomic zygote has 48 chromosomes. Tetrasomy is not the gain of *any* two extra chromosomes, but rather the gain of two homologous chromosomes, so that there are four homologous copies of a particular chromosome.

More than one aneuploid mutation may occur in the same individual. An individual that has an extra copy of each of two different (nonhomologous) chromosomes is referred to as being double trisomic; this condition is represented as $2n + 1 + 1$. Similarly, a double monosomic individual has two fewer nonhomologous chromosomes than normal ($2n - 1 - 1$), and a double tetrasomic individual has two extra pairs of homologous chromosomes ($2n + 2 + 2$).

THINK-PAIR-SHARE Question 3 🗪

Effects of Aneuploidy

One of the first aneuploids to be recognized was a fruit fly with a single X chromosome and no Y chromosome, discovered by Calvin Bridges in 1913 (see p. 87 in Chapter 4). Another early study of aneuploidy focused on mutants in the Jimson weed (*Datura stramonium*). A. Francis Blakeslee began breeding this plant in 1913, and he observed that

(a) Nondisjunction in meiosis I

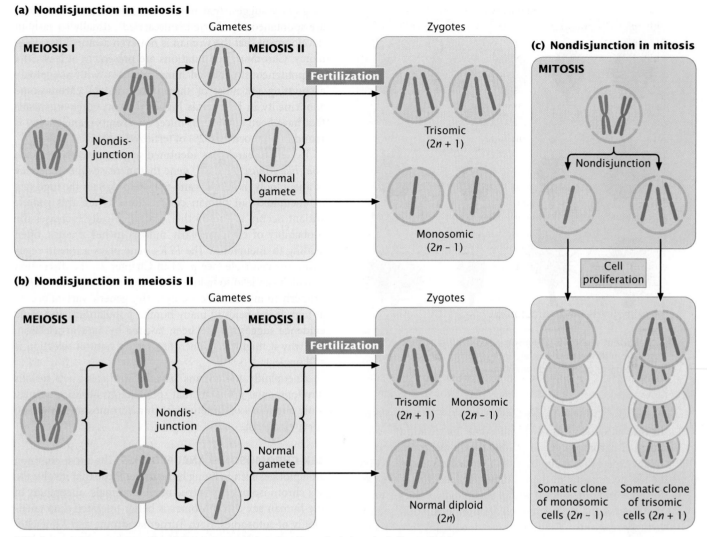

(b) Nondisjunction in meiosis II

(c) Nondisjunction in mitosis

8.18 Aneuploids can be produced through nondisjunction in meiosis I, meiosis II, or mitosis.
The gametes that result from meioses with nondisjunction may combine with a gamete (with blue chromosome) that results from normal meiosis to produce aneuploid zygotes.

crosses with several Jimson-weed mutants produced unusual ratios of progeny. For example, the *globe* mutation (which produces a globe-shaped seedcase) was dominant, but was inherited primarily from the female parent. When *globe* mutants self-fertilized, only 25% of the progeny had the globe phenotype. If the *globe* mutation were strictly dominant, Blakeslee should have seen 75% of the progeny with the trait (see Chapter 3), so the 25% that he observed was unusual. Blakeslee isolated 12 different mutants (**Figure 8.19**) that exhibited peculiar patterns of inheritance. Eventually, John Belling demonstrated that these 12 mutants were in fact trisomics. *Datura stramonium* has 12 pairs of chromosomes ($2n = 24$), and each of the 12 mutants was trisomic for a different chromosome pair. The aneuploid nature of the mutants explained the unusual ratios that Blakeslee had observed in the progeny. Many of the extra chromosomes in the trisomics were lost in meiosis, so fewer than 50% of the

gametes carried the extra chromosome, and the proportion of trisomics in the progeny was low. Furthermore, the pollen containing an extra chromosome was not as successful in fertilization, and trisomic zygotes were less viable.

Aneuploidy usually alters the phenotype drastically. In most animals and many plants, aneuploidies are lethal. Because aneuploidy affects the number of gene copies, but not their nucleotide sequences, the effects of aneuploidy are most likely due to abnormal gene dosage. Aneuploidy alters the dosage for some, but not all, genes, disrupting the relative concentrations of gene products and often interfering with normal development.

A major exception to the relation between gene number and gene dosage pertains to genes on the mammalian X chromosome. In mammals, X-chromosome inactivation ensures that males (who have a single X chromosome) and females (who have two X chromosomes) receive the same functional

Seed cases

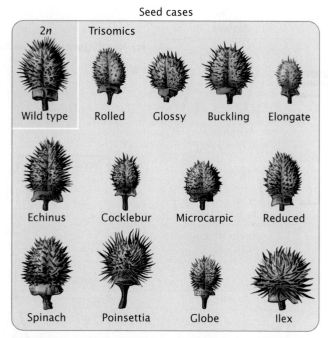

8.19 Mutant seedcases in Jimson weed (*Datura stramonium*) result from different trisomies. Each seedcase phenotype results from trisomy of a different chromosome.

dosage for X-linked genes (see pp. 98–99 in Chapter 4 for further discussion of X-chromosome inactivation). Additional X chromosomes in mammals are inactivated, so we might expect that aneuploidy of the sex chromosomes would be less detrimental in these animals. Indeed, this is the case for mice and humans, for whom aneuploidy of the sex chromosomes is the most common form of aneuploidy seen in living organisms. Y-chromosome aneuploidy is probably relatively common because there is so little information in the Y chromosome.

THINK-PAIR-SHARE Question 4

CONCEPTS

Aneuploidy, the loss or gain of one or more individual chromosomes, may arise from the loss of a chromosome subsequent to translocation or from nondisjunction in meiosis or mitosis. It disrupts gene dosage and often has severe phenotypic effects.

✔ CONCEPT CHECK 5

A diploid organism has $2n = 36$ chromosomes. How many chromosomes will be found in a trisomic member of this species?

Aneuploidy in Humans

For unknown reasons, a high percentage of all human embryos that are conceived possess chromosome abnormalities. Findings from studies of women who are attempting

pregnancy suggest that more than 30% of all conceptions are spontaneously aborted (miscarried), usually so early in development that the woman is not even aware of her pregnancy. Chromosome mutations are present in at least 50% of spontaneously aborted human fetuses, with aneuploidy accounting for most of them. This rate of chromosome abnormality in humans is higher than in other organisms that have been studied; in mice, for example, aneuploidy is found in no more than 2% of fertilized eggs.

Recent research has identified a genetic variant at the *Polo-like Kinase 4* (*PLK4*) gene that increases the frequency of aneuploidy in humans and may help explain the high rate of aneuploidy in human conceptions. When this genetic variant occurs in the mother, it dramatically increases the probability of errors in early mitosis in her zygotes, often leading to aneuploidy. The *PLK4* gene plays a role in regulating the centriole (see p. 24 in Chapter 2), the disruption of which can lead to failure of the chromosomes to separate properly in mitosis. Interestingly, this genetic variant occurs at high frequencies in many human populations, and some evidence suggests it has been favored by natural selection, but why it might have been favored by natural selection is still unclear.

Aneuploidy in humans usually produces such serious developmental problems that spontaneous abortion results. Only about 2% of all fetuses with a chromosome mutation survive to birth.

SEX-CHROMOSOME ANEUPLOIDIES The most common aneuploidies seen in living humans are those that involve the sex chromosomes. As is true of all mammals, aneuploidy of the human sex chromosomes is better tolerated than aneuploidy of autosomes. Both Turner syndrome and Klinefelter syndrome (see Figures 4.8 and 4.9) result from aneuploidy of the sex chromosomes.

AUTOSOMAL ANEUPLOIDIES Autosomal aneuploidies resulting in live births are less common than sex-chromosome aneuploidies in humans, probably because there is no mechanism of dosage compensation for autosomes. Most embryos with autosomal aneuploidies are spontaneously aborted, though occasionally fetuses with aneuploidies of some of the small autosomes, such as chromosome 21, complete development. Because these chromosomes are small and carry relatively few genes, the presence of extra copies is less detrimental than it is for larger chromosomes.

DOWN SYNDROME In 1866, John Langdon Down, physician and medical superintendent of the Earlswood Asylum in Surrey, England, noticed a remarkable resemblance among a number of his intellectually disabled patients: all of them possessed a broad, flat face, a small nose, and oval-shaped eyes. Their features were so similar, in fact, that he felt that they might easily be mistaken for children from the same family. Down did not understand the cause of their

(a)

(b)

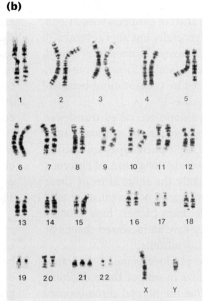

8.20 Down syndrome (a) is caused by trisomy of chromosome 21 (b). [Part a: George Doyle/Getty Images. Part b: L. Wilatt, East Anglian Regional Genetics Service/Science Photo Library/Science Source.]

intellectual disability, but his original description faithfully records the physical characteristics of people with this genetic form of intellectual disability. In his honor, the disorder is today known as Down syndrome.

Down syndrome, also known as **trisomy 21**, is the most common autosomal aneuploidy in humans (**Figure 8.20a**). The incidence of Down syndrome in the United States is similar to that worldwide—about 1 in 700 human births—although the incidence increases among children born to older mothers. Approximately 92% of those who have Down syndrome have three full copies of chromosome 21 (and therefore a total of 47 chromosomes), a condition termed **primary Down syndrome** (**Figure 8.20b**). Primary Down syndrome usually arises from spontaneous nondisjunction in egg formation: about 75% of the nondisjunction events that cause Down syndrome are maternal in origin, most arising in meiosis I. Most children with Down syndrome are born to unaffected parents, and the failure of the chromosomes to divide has little hereditary tendency. A couple who has conceived one child with primary Down syndrome has only a slightly higher risk of conceiving a second child with Down syndrome (compared with other couples of similar age who have not had any children with Down syndrome). Similarly, the couple's relatives are not more likely to have a child with primary Down syndrome.

THINK-PAIR-SHARE Question 5 👥

About 4% of people with Down syndrome are not trisomic for a complete chromosome 21. Instead, they have 46 chromosomes, but an extra copy of part of chromosome 21 is attached to another chromosome through a translocation. This condition is termed **familial Down syndrome** because

it has a tendency to run in families. The phenotypic characteristics of familial Down syndrome are the same as those of primary Down syndrome.

Familial Down syndrome arises in offspring whose parents are carriers of chromosomes that have undergone a Robertsonian translocation, most commonly between chromosome 21 and chromosome 14: the long arm of 21 and the short arm of 14 exchange places (**Figure 8.21**). This exchange produces one chromosome that includes the long arms of

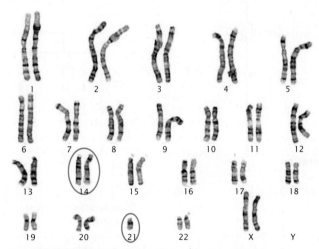

8.21 The translocation of part of chromosome 21 to another chromosome results in familial Down syndrome. Here, the long arm of chromosome 21 is attached to chromosome 14. This karyotype is from a translocation carrier, who is phenotypically normal but is at increased risk for producing children with Down syndrome. [© Centre for Genetics Education for and on behalf of the Crown in right of the State of New South Wales.]

chromosomes 14 and 21 and another, very small chromosome that consists of the short arms of chromosomes 21 and 14. The small chromosome is generally lost after several cell divisions. Although exchange between chromosomes 21 and 14 is the most common cause of familial Down syndrome, the condition can also be caused by translocations between 21 and other chromosomes, such as 15.

People with this type of translocation, called **translocation carriers**, do not have Down syndrome. Although they possess only 45 chromosomes, their phenotypes are normal because they have two copies of the long arms of chromosomes 14 and 21, and apparently the short arms of these chromosomes (which are lost) carry no essential genetic information. Although translocation carriers have a completely normal phenotype, they have an increased chance of producing children with Down syndrome.

When a translocation carrier produces gametes, the translocation chromosome segregates in one of three different ways. First, it may separate from the normal chromosomes 14 and 21 in anaphase I of meiosis (**Figure 8.22a**). In this type of segregation, half of the gametes produced will have the translocation chromosome and no other copies of chromosomes 21 and 14; the fusion of such a gamete with a normal gamete will give rise to a translocation carrier. The other half of the gametes produced by this first type of segregation will be normal, each with a single copy of chromosomes 21 and 14, and will result in normal offspring.

Alternatively, the translocation chromosome may separate from chromosome 14 and pass into the same cell with the normal chromosome 21 (**Figure 8.22b**). This type of segregation produces abnormal gametes only; half will have two functional copies of chromosome 21 (one normal and one attached to chromosome 14) and the other half will lack chromosome 21. If a gamete with the two functional copies of chromosome 21 fuses with a normal gamete carrying a single copy of chromosome 21, the resulting zygote will have familial Down syndrome. If a gamete lacking chromosome 21 fuses with a normal gamete, the resulting zygote will have monosomy 21 and will be spontaneously aborted.

In the third type of segregation, the translocation chromosome and the normal copy of chromosome 14 segregate together (**Figure 8.22c**). This pattern is presumably rare because the two centromeres are both derived from chromosome 14 and usually separate from each other. All the gametes produced by this process are abnormal: half result in monosomy 14 and the other half result in trisomy 14. All are spontaneously aborted.

Thus, only three of the six types of gametes that can be produced by a translocation carrier will result in the birth of a baby, and theoretically, these gametes should arise with equal frequency. One-third of the offspring of a translocation carrier should be translocation carriers like their parent, one-third should have familial Down syndrome, and

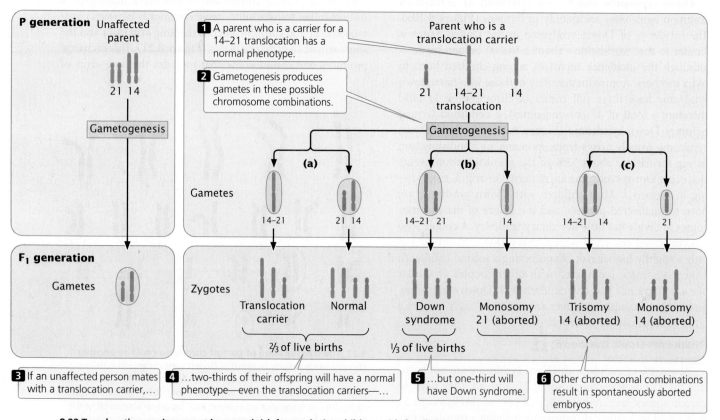

8.22 Translocation carriers are at increased risk for producing children with familial Down syndrome.

one-third should be normal. In reality, however, fewer than one-third of the children born to translocation carriers have Down syndrome, which suggests that some of the embryos with Down syndrome are spontaneously aborted.

▶ **TRY PROBLEM 32**

OTHER HUMAN TRISOMIES Few autosomal aneuploidies in humans besides trisomy 21 result in live births. **Trisomy 18**, also known as **Edward syndrome**, arises with a frequency of approximately 1 in 8000 live births. Babies with Edward syndrome have severe intellectual disability, low-set ears, a short neck, deformed feet, clenched fingers, heart problems, and other disabilities. Few live for more than a year after birth. **Trisomy 13** has a frequency of about 1 in 15,000 live births and produces features that are collectively known as **Patau syndrome**. Characteristics of this condition include severe intellectual disability, a small head, sloping forehead, small eyes, cleft lip and palate, extra fingers and toes, and numerous other problems. About half of children with trisomy 13 die within the first month of life, and 95% die by the age of 3. Rarer still is **trisomy 8**, which arises with a frequency ranging from about 1 in 25,000 to 1 in 50,000 live births. This aneuploidy is characterized by intellectual disability, contracted fingers and toes, low-set malformed ears, and a prominent forehead. Most individuals who have this condition are mosaics, having some cells with three copies of chromosome 8 and other cells with the usual two copies (see the discussion on genetic mosaicism below). Complete trisomy 8, in which all cells in the body have three copies of chromosome 8, usually results in spontaneous abortion.

ANEUPLOIDY AND MATERNAL AGE Most cases of Down syndrome and other types of aneuploidy in humans arise from maternal nondisjunction, and the frequency of aneuploidy increases with maternal age (**Figure 8.23**). Why maternal age is associated with nondisjunction is not known for certain. Female mammals are born with primary oocytes suspended in the diplotene substage of prophase I of meiosis. Just before ovulation, meiosis resumes and the first division is completed, producing a secondary oocyte. At this point, meiosis is suspended again and remains so until the secondary oocyte is penetrated by a sperm. The second meiotic division takes place immediately before the nuclei of egg and sperm unite to form a zygote.

Primary oocytes may remain suspended in diplotene for many years before ovulation takes place and meiosis recommences. Components of the spindle and other structures required for chromosome segregation may break down during the long arrest of meiosis, leading to more aneuploidy in children born to older mothers. According to this theory, no age effect is seen in males because sperm are produced continuously after puberty with no long suspension of the meiotic divisions.

ANEUPLOIDY AND CANCER Many tumor cells have extra chromosomes or missing chromosomes, or both; some types of tumors are consistently associated with specific chromosome mutations, including aneuploidy and chromosome rearrangements. The role of chromosome mutations in cancer will be explored in Chapter 23.

CONCEPTS

In humans, sex-chromosome aneuploidies are more common than autosomal aneuploidies. X-chromosome inactivation prevents problems of gene dosage for X-linked genes. Down syndrome results from three functional copies of chromosome 21, either through trisomy (primary Down syndrome) or a Robertsonian translocation (familial Down syndrome).

✔ **CONCEPT CHECK 6**
Briefly explain why, in humans and other mammals, sex-chromosome aneuploidies are more common than autosomal aneuploidies.

Uniparental Disomy

Normally, the two chromosomes of a homologous pair are inherited from two different parents—one from the father and one from the mother. The development of molecular techniques that facilitate the identification of specific DNA sequences (see Chapter 19) has made the determination of the parental origins of chromosomes possible. Surprisingly,

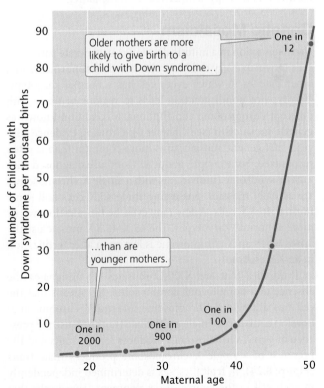

8.23 The incidence of primary Down syndrome and other aneuploidies increases with maternal age.

these techniques have revealed that sometimes both chromosomes are inherited from the same parent, a condition termed **uniparental disomy**.

Many cases of uniparental disomy probably originate as trisomies. Although most autosomal trisomies are lethal, a trisomic embryo can survive if one of the three chromosomes is lost early in development. If, just by chance, the two remaining chromosomes are both from the same parent, uniparental disomy results.

Uniparental disomy violates the rule that children affected with a recessive disorder appear only in families in which both parents are carriers. For example, cystic fibrosis is an autosomal recessive disease; typically, both parents of an affected child are heterozygous for a cystic fibrosis mutation on chromosome 7. However, for a small proportion of people with cystic fibrosis, only one of the parents is heterozygous for a cystic fibrosis mutation. In these cases, cystic fibrosis is due to uniparental disomy: the person with cystic fibrosis has inherited from the heterozygous parent two copies of the chromosome 7 that carries the defective cystic fibrosis allele and no copy of the normal allele from the other parent.

Uniparental disomy has also been observed in Prader–Willi syndrome, a rare condition that arises when a paternal copy of a gene on chromosome 15 is missing. Although most cases of Prader–Willi syndrome result from a chromosome deletion that removes the paternal copy of the gene (see pp. 131–132 in Chapter 5), 20% to 30% of the cases arise when both copies of chromosome 15 are inherited from the mother and no copy is inherited from the father.

Genetic Mosaicism

Nondisjunction in a mitotic division may generate patches of cells in which every cell has a chromosome abnormality and other patches in which every cell has a normal karyotype. This type of nondisjunction leads to regions of tissue with different chromosome constitutions, a condition known as **genetic mosaicism** (see Chapter 6). Growing evidence suggests that genetic mosaicism is more common than is often recognized. For example, about 50% of those diagnosed with Turner syndrome (individuals with a single X chromosome) are actually mosaics, possessing some 45,X cells and some normal 46,XX cells. A few may even be mosaics for two or more abnormal karyotypes. The 45,X/46,XX mosaic usually arises when an X chromosome is lost soon after fertilization in an XX embryo.

Fruit flies that are XX/XO mosaics (O designates the absence of a homologous chromosome; XO means that the cell has a single X chromosome and no Y chromosome) develop a mixture of male and female traits because the presence of two X chromosomes produces female traits and the presence of a single X chromosome produces male traits (**Figure 8.24**). In fruit flies, sex is determined independently in each cell in the course of development. Those cells that are XX express female traits; those that are XO express male traits. Such sexual mosaics are called **gynandromorphs**.

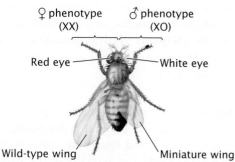

♀ phenotype (XX)   ♂ phenotype (XO)

Red eye   White eye

Wild-type wing   Miniature wing

8.24 Genetic mosaicism for the sex chromosomes produces a gynandromorph. This XX/XO gynandromorph fruit fly carries one wild-type X chromosome and one X chromosome with recessive alleles for white eyes and miniature wings. The left side of the fly has a normal female phenotype because the cells are XX and the recessive alleles on one X chromosome are masked by the presence of wild-type alleles on the other. The right side of the fly has a male phenotype with white eyes and miniature wing because the cells are missing the wild-type X chromosome (are XO), allowing the white and miniature alleles to be expressed.

Normally, X-linked recessive genes are masked in heterozygous females, but in XX/XO mosaics, any X-linked recessive genes present in the cells with a single X chromosome will be expressed.

THINK-PAIR-SHARE Question 6

CONCEPTS

In uniparental disomy, an individual organism has two copies of a chromosome from one parent and no copy from the other. Uniparental disomy may arise when a trisomic embryo loses one of the triplicate chromosomes early in development. In genetic mosaicism, different cells within the same individual organism have different chromosome constitutions.

8.4 Polyploidy Is the Presence of More Than Two Sets of Chromosomes

As discussed in the introduction to this chapter, some organisms (such as bananas) possess more than two sets of chromosomes—they are polyploid. Polyploids include *triploids* ($3n$), *tetraploids* ($4n$), *pentaploids* ($5n$), and organisms with even higher numbers of chromosome sets.

Polyploidy is common in plants and is a major mechanism by which new plant species have evolved. Approximately 40% of all flowering-plant species and 70% to 80% of grasses are polyploids. A number of agriculturally important plants are polyploid, including wheat, oats, cotton, potatoes, and sugarcane. Polyploidy is less common in animals but is found in some invertebrates, fishes, salamanders, frogs, and lizards. No naturally occurring, viable polyploids are known in birds or mammals.

We consider two major types of polyploidy: **autopoly-ploidy**, in which all chromosome sets are from a single species, and **allopolyploidy**, in which chromosome sets are from two or more species.

Autopolyploidy

Autopolyploidy is caused by accidents of mitosis or meiosis that produce extra sets of chromosomes, all derived from a single species. Nondisjunction of all chromosomes in mitosis in an early $2n$ embryo, for example, doubles the chromosome number and produces an autotetraploid ($4n$), as depicted in **Figure 8.25a**. An autotriploid ($3n$) may arise when nondisjunction in meiosis produces a diploid gamete that then fuses with a normal haploid gamete to produce a triploid zygote (**Figure 8.25b**). Alternatively, triploids may arise from a cross between an autotetraploid that produces $2n$ gametes and a diploid that produces $1n$ gametes. Nondisjunction can be artificially induced by colchicine, a chemical that disrupts spindle formation. Colchicine is often used to induce polyploidy in agriculturally and ornamentally important plants.

Because all the chromosome sets in autopolyploids are from the same species, they are homologous and attempt to align in prophase I of meiosis, which usually results in sterility. Consider meiosis in an autotriploid (**Figure 8.26**).

In meiosis in a diploid cell, two homologs pair and align, but in autotriploids, three homologs are present. One of the three homologs may fail to align with the other two, and this unaligned chromosome will segregate randomly (see Figure 8.26a). Which gamete gets the extra chromosome will be determined by chance and will differ for each homologous trio of chromosomes. The resulting gametes will have two copies of some chromosomes and one copy of others. Even if all three chromosomes align, two chromosomes must segregate to one gamete and one chromosome to the other (see Figure 8.26b). Occasionally, the presence of a third chromosome interferes with normal alignment, and all three chromosomes move to the same gamete (see Figure 8.26c).

No matter how the three homologous chromosomes align, their random segregation will create **unbalanced gametes**, with various numbers of chromosomes. A gamete produced by meiosis in such an autotriploid might receive, say, two copies of chromosome 1, one copy of chromosome 2, three copies of chromosome 3, and no copies of chromosome 4. When the unbalanced gamete fuses with a normal gamete (or with another unbalanced gamete), the resulting zygote has different numbers of the four types of chromosomes. This difference in number creates unbalanced gene dosage in the zygote, which is often lethal. For this reason, triploids do not usually produce viable offspring.

(a) Autopolyploidy through mitosis

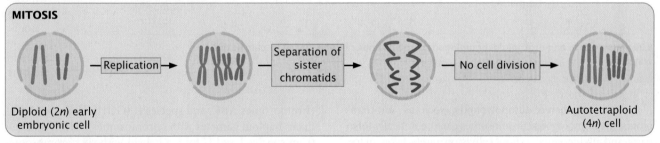

(b) Autopolyploidy through meiosis

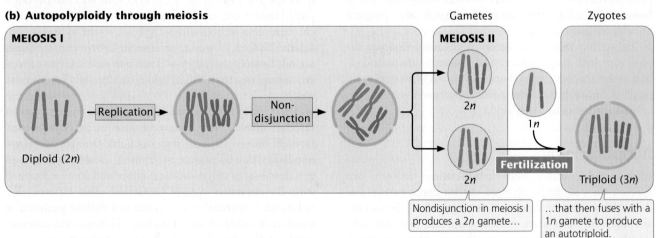

8.25 Autopolyploidy can arise through nondisjunction in mitosis or meiosis.

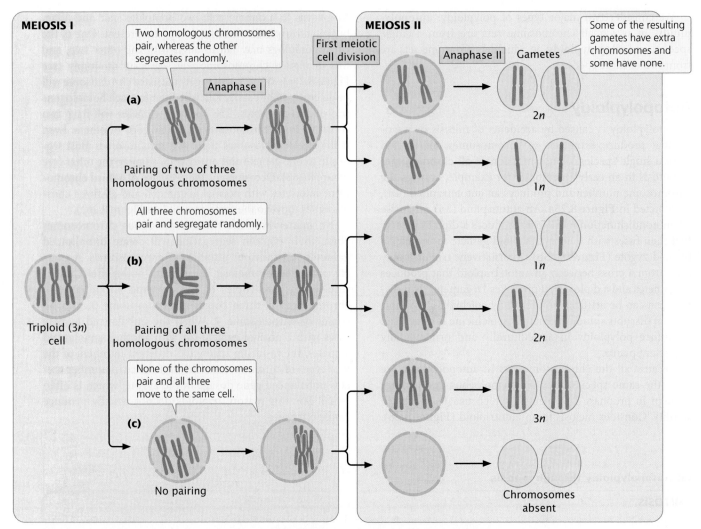

8.26 In meiosis in an autotriploid, homologous chromosomes can pair, or fail to pair, in three ways. This example illustrates the pairing and segregation of a single homologous set of chromosomes.

In even-numbered autopolyploids, such as autotetraploids, the homologous chromosomes can theoretically form pairs and divide equally. Equal division rarely takes place, however, so these types of autotetraploids also produce unbalanced gametes.

The sterility that usually accompanies autopolyploidy has been exploited in agriculture. As discussed in the introduction to this chapter, triploid bananas ($3n = 33$) are sterile and seedless. Similarly, seedless triploid watermelons have been created and are now widely sold.

Allopolyploidy

Allopolyploidy arises from hybridization between two species; the resulting polyploid carries chromosome sets derived from two or more species. **Figure 8.27** shows how allopolyploidy can arise from two species that are sufficiently related that hybridization takes place between them. Species I (AABBCC, $2n = 6$) produces haploid gametes with chromosomes ABC, and species II (GGHHII, $2n = 6$) produces haploid gametes with chromosomes GHI. If gametes from species I and II fuse, a hybrid with six chromosomes (ABCGHI) is created. The hybrid has the same number of chromosomes as both diploid species, so the hybrid is considered diploid. However, because the hybrid chromosomes are not homologous, they will not pair and segregate properly in meiosis; thus, this hybrid is functionally haploid and sterile.

The sterile hybrid is unable to produce viable gametes through meiosis, but it may be able to perpetuate itself through mitosis (asexual reproduction). On rare occasions, nondisjunction takes place in a mitotic division, which leads to a doubling of chromosome number and an allotetraploid with chromosomes AABBCCGGHHII. This type of allopolyploid, consisting of two combined diploid genomes, is sometimes called an **amphidiploid**. Although the chromosome number has doubled compared with what was present in each of the parental species, the amphidiploid is

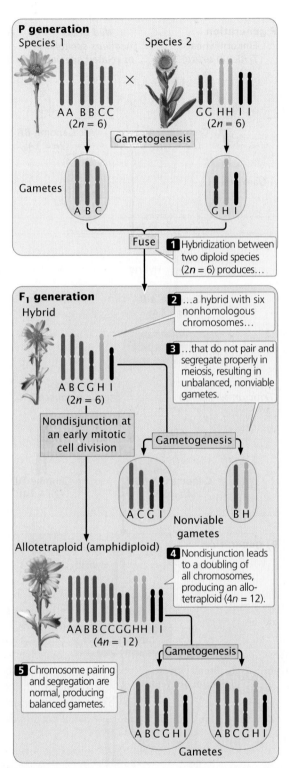

P generation

Species 1 × Species 2

A A   B B   C C
(2n = 6)

G G   H H   I I
(2n = 6)

Gametogenesis

Gametes

A B C   G H I

Fuse

1 Hybridization between two diploid species (2n = 6) produces...

F₁ generation

Hybrid

A B C G H I
(2n = 6)

2 ...a hybrid with six nonhomologous chromosomes...

3 ...that do not pair and segregate properly in meiosis, resulting in unbalanced, nonviable gametes.

Nondisjunction at an early mitotic cell division

Gametogenesis

A C G I   B H

Nonviable gametes

Allotetraploid (amphidiploid)

4 Nondisjunction leads to a doubling of all chromosomes, producing an allotetraploid (4n = 12).

A A B B C C G G H H I I
(4n = 12)

5 Chromosome pairing and segregation are normal, producing balanced gametes.

Gametogenesis

A B C G H I   A B C G H I

Gametes

8.27 Most allopolyploids arise from hybridization between two species followed by chromosome doubling.

functionally diploid: every chromosome has one and only one homologous partner, which is exactly what is required for proper segregation in meiosis. The amphidiploid can now undergo normal meiosis to produce balanced gametes with six chromosomes each.

George Karpechenko created polyploids experimentally in the 1920s. Cabbages (*Brassica oleracea*, 2n = 18) and radishes (*Raphanus sativa*, 2n = 18) are agriculturally important plants now, as they were then, but only the leaves of the cabbage and the roots of the radish are normally consumed. Karpechenko wanted to produce a plant that had cabbage leaves and radish roots so that no part of the plant would go to waste. Because both cabbages and radishes possess 18 chromosomes, Karpechenko was able to cross them successfully, producing a hybrid with 2n = 18, but, unfortunately, the hybrid was sterile. After several crosses, Karpechenko noticed that one of his hybrid plants produced a few seeds. When planted, these seeds grew into plants that were viable and fertile. Analysis of their chromosomes revealed that the plants were allotetraploids, with 2n = 36 chromosomes. To Karpechencko's great disappointment, however, the new plants possessed the roots of a cabbage and the leaves of a radish.

WORKED PROBLEM

Species I has 2n = 14 and species II has 2n = 20. Give all possible chromosome numbers that may be found in the following individuals:

a. An autotriploid of species I.

b. An autotetraploid of species II.

c. An allotriploid formed from species I and species II.

d. An allotetraploid formed from species I and species II.

Solution Strategy

What information is required in your answer to the problem?

All possible chromosome numbers for individuals with the type of polyploidy indicated.

What information is provided to solve the problem?

- Species I has 2n = 14 and species II has 2n = 20.
- The type of polyploidy the individual possesses.

Solution Steps

The haploid number of chromosomes (n) for species I is 7, and that for species II is 10.

a. A triploid individual is 3n. A common mistake is to assume that 3n means three times as many chromosomes as in a normal individual, but remember that normal individuals are 2n. Because n for species I is 7 and all the chromosome sets of an autopolyploid are from the same species, 3n = 3 × 7 = 21.

b. An autotetraploid is 4n with all chromosome sets from the same species. The n for species II is 10, so 4n = 4 × 10 = 40.

c. A triploid is 3n. By definition, an allopolyploid must have chromosome sets from two different species. An allotriploid could have $1n$ from species I and $2n$ from species II or $(1 \times 7) + (2 \times 10) = 27$. Alternatively, it could have $2n$ from species I and $1n$ from species II, or $(2 \times 7) + (1 \times 10) = 24$. Thus, the number of chromosomes in an allotriploid could be 24 or 27.

d. A tetraploid is $4n$. By definition, an allotetraploid must have chromosome sets from at least two different species. An allotetraploid could have $3n$ from species I and $1n$ from species II or $(3 \times 7) + (1 \times 10) = 31$; or $2n$ from species I and $2n$ from species II or $(2 \times 7) + (2 \times 10) = 34$; or $1n$ from species I and $3n$ from species II or $(1 \times 7) + (3 \times 10) = 37$. Thus, the number of chromosomes could be 31, 34, or 37.

>> For additional practice, try **Problem 38** at the end of this chapter.

The Significance of Polyploidy

In many organisms, cell volume is correlated with nuclear volume, which in turn is determined by genome size. Thus, the increase in chromosome number in polyploidy is often associated with an increase in cell size, and many polyploids are physically larger than diploids. Breeders have used this effect to produce plants with larger leaves, flowers, fruits, and seeds. The hexaploid ($6n = 42$) genome of wheat probably contains chromosomes derived from three different wild species (**Figure 8.28**). As a result, the seeds of modern wheat are larger than those of its ancestors. Many other cultivated plants are also polyploid (**Table 8.2**).

Polyploidy is less common in animals than in plants for several reasons. As discussed, allopolyploids require hybridization between different species, which happens less frequently in animals than in plants. Animal behavior often prevents interbreeding among species, and the complexity of animal development causes most interspecific hybrids to be nonviable. Many of the polyploid animals that do arise are in groups that reproduce through parthenogenesis (a type of reproduction in which the animal develops from an unfertilized egg). Thus, asexual reproduction may facilitate the development of polyploids, perhaps because the perpetuation of hybrid individuals through asexual reproduction provides greater opportunities for nondisjunction than does sexual reproduction. Only a few human polyploid babies have been reported, and most died within a few days of birth. Polyploidy—usually triploidy—is seen in about 10% of all spontaneously aborted human fetuses.

Although complete polyploidy, in which all cells are polyploid, is lethal in humans, significant numbers of autopolyploid cells do occur in some human tissues. For example, some liver cells and heart cells are polyploid. Although the reason for polyploidy in these cells is unknown, the large cell size that results from polyploidy might be advantageous under certain circumstances. Polyploid cells are also found

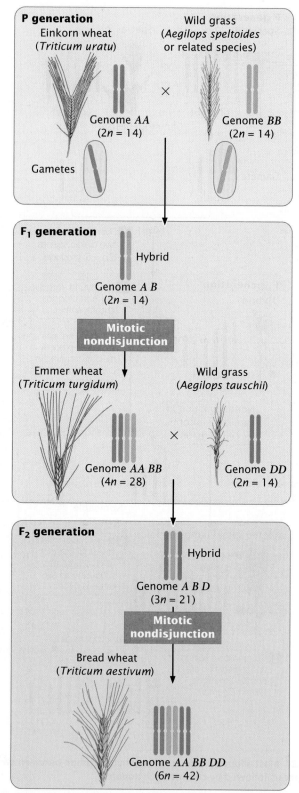

8.28 Modern bread wheat, *Triticum aestivum*, is a hexaploid with genes derived from three different species. Two diploid species, *T. uratu* ($n = 14$) and probably *Aegilops speltoides* or a related species ($n = 14$), originally crossed to produce a diploid hybrid ($2n = 14$) that underwent chromosome doubling to create *T. turgidum* ($4n = 28$). A cross between *T. turgidum* and *A. tauschii* ($2n = 14$) produced a triploid hybrid ($3n = 21$) that then underwent chromosome doubling to eventually produce *T. aestivum*, which is a hexaploid ($6n = 42$).

Plant	Type of Polyploidy	Chromosome Sets	Chromosome Number
Potato	Autopolyploid	$4n$	48
Banana	Autopolyploid	$3n$	33
Peanut	Autopolyploid	$4n$	40
Sweet potato	Autopolyploid	$6n$	90
Tobacco	Allopolyploid	$4n$	48
Cotton	Allopolyploid	$4n$	52
Wheat	Allopolyploid	$6n$	42
Oats	Allopolyploid	$6n$	42
Sugarcane	Allopolyploid	$8n$	80
Strawberry	Allopolyploid	$8n$	56

TABLE 8.2   Examples of polyploid crop plants

Source: After F. C. Elliot, *Plant Breeding and Cytogenetics* (New York: McGraw-Hill, 1958).

in many cancers. One study suggested that extra sets of chromosomes occur in 37% of all cancers.

THINK-PAIR-SHARE Question 7

IMPORTANCE OF POLYPLOIDY IN EVOLUTION Polyploidy, particularly allopolyploidy, often gives rise to new species and has been particularly important in the evolution of flowering plants. Occasional genome doubling through polyploidy has been a major contributor to evolutionary success in several groups. For example, *Saccharomyces cerevisiae* (yeast) is a tetraploid, having undergone whole-genome duplication about 100 million years ago. The vertebrate genome has duplicated twice, once in the common ancestor of jawed vertebrates and again in the ancestor of fishes. Certain groups of vertebrates, such as some frogs and some fishes, have undergone additional duplications. Cereal plants have undergone several genome-duplication events.

The types of chromosome mutations we have discussed in this chapter are summarized in **Table 8.3**.

CONCEPTS

Polyploidy is the presence of extra chromosome sets. Autopolyploids possess extra chromosome sets from the same species; allopolyploids possess extra chromosome sets from two or more species. Problems in chromosome pairing and segregation often lead to sterility in autopolyploids, but many allopolyploids are fertile.

✔ CONCEPT CHECK 7

Species A has $2n = 16$ and species B has $2n = 14$. How many chromosomes would be found in an allotriploid of these two species?

a. 21 or 24   c. 22 or 23
b. 42 or 48   d. 45

TABLE 8.3   Different types of chromosome mutations

Chromosome Mutation	Definition
Chromosome rearrangement	Change in chromosome structure
Chromosome duplication	Duplication of a chromosome segment
Chromosome deletion	Deletion of a chromosome segment
Inversion	Chromosome segment inverted 180 degrees
Paracentric inversion	Inversion that does not include the centromere in the inverted region
Pericentric inversion	Inversion that includes the centromere in the inverted region
Translocation	Movement of a chromosome segment to a nonhomologous chromosome or to another region of the same chromosome
Nonreciprocal translocation	Movement of a chromosome segment to a nonhomologous chromosome or to another region of the same chromosome without reciprocal exchange
Reciprocal translocation	Exchange between segments of nonhomologous chromosomes or between regions of the same chromosome
Aneuploidy	Change in number of individual chromosomes
Nullisomy	Loss of both members of a homologous pair
Monosomy	Loss of one member of a homologous pair
Trisomy	Gain of one chromosome, resulting in three homologous chromosomes
Tetrasomy	Gain of two homologous chromosomes, resulting in four homologous chromosomes
Polyploidy	Addition of entire chromosome sets
Autopolyploidy	Polyploidy in which extra chromosome sets are derived from the same species
Allopolyploidy	Polyploidy in which extra chromosome sets are derived from two or more species

CONCEPTS SUMMARY

- The three basic categories of chromosome mutations are (1) chromosome rearrangements, which are changes in the structures of chromosomes; (2) aneuploidy, which is an increase or decrease in chromosome number; and (3) polyploidy, which is the presence of extra chromosome sets.

- Chromosome rearrangements include duplications, deletions, inversions, and translocations.

- In individuals heterozygous for a duplication, the duplicated region forms a loop when homologous chromosomes pair in meiosis. Duplications often have pronounced effects on the phenotype owing to unbalanced gene dosage. Segmental duplications are common in the human genome.

- In individuals heterozygous for a deletion, one of the chromosomes loops out during pairing in meiosis. Deletions may cause recessive alleles to be expressed.

- Pericentric inversions include the centromere; paracentric inversions do not. In individuals heterozygous for an inversion, the homologous chromosomes form inversion loops in meiosis, and reduced recombination is observed within the inverted region.

- In translocation heterozygotes, the chromosomes form crosslike structures in meiosis, and the segregation of chromosomes produces unbalanced gametes.

- Fragile sites are constrictions or gaps that appear at particular regions on the chromosomes of cells grown in culture and are prone to breakage under certain conditions.

- Copy-number variations (CNVs) are differences in the number of copies of DNA sequences and include duplications and deletions. These variants are common in the human genome; some are associated with diseases and disorders.

- Nullisomy is the loss of two homologous chromosomes; monosomy is the loss of a single chromosome; trisomy is the addition of a single chromosome; and tetrasomy is the addition of two homologous chromosomes.

- Aneuploidy usually causes drastic phenotypic effects because it leads to unbalanced gene dosage.

- Primary Down syndrome is caused by the presence of three full copies of chromosome 21. Familial Down syndrome is caused by the presence of two normal copies of chromosome 21 and a third copy that is attached to another chromosome through a translocation.

- Uniparental disomy is the presence of two copies of a chromosome from one parent and no copy from the other. Genetic mosaicism is caused by nondisjunction in an early mitotic division that leads to different chromosome constitutions in different cells of a single individual.

- All the chromosomes in an autopolyploid derive from one species; chromosomes in an allopolyploid come from two or more species.

IMPORTANT TERMS

chromosome mutation (p. 218)
metacentric (p. 218)
submetacentric (p. 218)
acrocentric (p. 218)
telocentric (p. 218)
chromosome rearrangement (p. 220)
chromosome duplication (p. 220)
tandem duplication (p. 220)
displaced duplication (p. 220)
reverse duplication (p. 220)
segmental duplication (p. 220)
unequal crossing over (p. 221)
chromosome deletion (p. 222)

pseudodominance (p. 224)
haploinsufficiency (p. 224)
chromosome inversion (p. 224)
paracentric inversion (p. 224)
pericentric inversion (p. 224)
position effect (p. 225)
dicentric chromatid (p. 226)
acentric chromatid (p. 226)
dicentric bridge (p. 226)
translocation (p. 227)
nonreciprocal translocation (p. 227)
reciprocal translocation (p. 227)
Robertsonian translocation (p. 227)

alternate segregation (p. 229)
adjacent-1 segregation (p. 229)
adjacent-2 segregation (p. 229)
fragile site (p. 229)
fragile-X syndrome (p. 229)
copy-number variations (CNVs) (p. 230)
structural variants (p. 230)
aneuploidy (p. 230)
polyploidy (p. 230)
nullisomy (p. 230)
monosomy (p. 230)
trisomy (p. 230)
tetrasomy (p. 230)
Down syndrome (trisomy 21) (p. 233)

primary Down syndrome (p. 233)
familial Down syndrome (p. 233)
translocation carrier (p. 234)
Edward syndrome (trisomy 18) (p. 235)
Patau syndrome (trisomy 13) (p. 235)
trisomy 8 (p. 235)
uniparental disomy (p. 236)
genetic mosaicism (p. 236)
gynandromorph (p. 236)
autopolyploidy (p. 237)
allopolyploidy (p. 237)
unbalanced gametes (p. 237)
amphidiploidy (p. 238)

ANSWERS TO CONCEPT CHECKS

1. a

2. Pseudodominance is the expression of a normally recessive mutation. It is produced when the dominant wild-type allele in a heterozygous individual is absent due to a deletion on one chromosome.

3. c

4. b

5. 37

6. Dosage compensation prevents the expression of additional copies of X-linked genes in mammals, and there is little information in the Y chromosome, so extra copies of the X and Y chromosomes do not have major effects on development. In contrast, there is no mechanism of dosage compensation for autosomes, and so extra copies of autosomal genes are expressed, upsetting developmental processes and causing the spontaneous abortion of aneuploid embryos.

7. c

WORKED PROBLEMS

Problem 1

A chromosome has the following segments, where • represents the centromere:

$$A B C D E • F G$$

What types of chromosome mutations are required to change this chromosome into each of the following chromosomes? (In some cases, more than one chromosome mutation may be required.)

a. A B E • F G

b. A E D C B • F G

c. A B A B C D E • F G

d. A F • E D C B G

e. A B C D E E D C • F G

» Solution Strategy

What information is required in your answer to the problem?

The types of chromosome mutations that would lead to the chromosome shown.

What information is provided to solve the problem?

- The original segments found on the chromosome.
- The segments that are present after the mutations.

For help with this problem, review:

Section 8.2.

» Solution Steps

a. The mutated chromosome (A B E • F G) is missing segment CD, so this mutation is a deletion.

b. The mutated chromosome (A E D C B • F G) has one and only one copy of each original segment, but segment

B C D E has been inverted 180 degrees. Because the centromere has not changed its location and is not in the inverted region, this chromosome mutation is a paracentric inversion.

c. The mutated chromosome (A B A B C D E • F G) is longer than normal, and we see that segment AB has been duplicated. This mutation is a tandem duplication.

d. The mutated chromosome (A F • E D C B G) is normal in length, but the order of the segments and the location of the centromere have changed; this mutation is therefore a pericentric inversion of region (B C D E • F).

e. The mutated chromosome (A B C D E E D C • F G) contains a duplication (C D E) that is also inverted, so this chromosome has undergone a duplication and a paracentric inversion.

Problem 2

Species I is diploid ($2n = 4$) with chromosomes AABB; related species II is diploid ($2n = 6$) with chromosomes MMNNOO. Give the chromosomes that would be found in individuals with the following chromosome mutations:

a. Autotriploidy in species I.

b. Allotetraploidy including species I and II.

c. Monosomy in species I.

d. Trisomy in species II for chromosome M.

e. Tetrasomy in species I for chromosome A.

f. Allotriploidy including species I and II.

g. Nullisomy in species II for chromosome N.

›› Solution Strategy

What information is required in your answer to the problem?

The chromosomes found in individuals with each type of chromosome mutation.

What information is provided to solve the problem?

- Species I is diploid with $2n = 4$.
- Species I has chromosomes AABB.
- Species II is diploid with $2n = 6$.
- Species II has chromosomes MMNNOO.

For help with this problem, review:
Sections 8.3 and 8.4.

›› Solution Steps

> **Hint:** First determine the haploid genome complement for each species. For species I, $n = 2$ with chromosomes AB, and for species II, $n = 3$ with chromosomes MNO.

a. An autotriploid is $3n$, with all the chromosomes coming from a single species, so an autotriploid of species I would have chromosomes AAABBB ($3n = 6$).

b. An allotetraploid is $4n$, with the chromosomes coming from more than one species. An allotetraploid could consist of $2n$ from species I and $2n$ from species II, giving the allotetraploid ($4n = 2 + 2 + 3 + 3 = 10$)

chromosomes AABBMMNNOO. An allotetraploid could also possess $3n$ from species I and $1n$ from species II ($4n = 2 + 2 + 2 + 3 = 9$; AAABBBMNO) or $1n$ from species I and $3n$ from species II ($4n = 2 + 3 + 3 + 3 = 11$; ABMMMNNNOOO).

c. A monosomic individual is missing a single chromosome, so monosomy in species I would result in $2n - 1 = 4 - 1 = 3$. The monosomy might include either of the two chromosome pairs, giving chromosomes ABB or AAB.

d. Trisomy requires an extra chromosome, so trisomy in species II for chromosome M would result in $2n + 1 = 6 + 1 = 7$ (MMMNNOO).

e. A tetrasomic individual has two extra homologous chromosomes, so tetrasomy in species I for chromosome A would result in $2n + 2 = 4 + 2 = 6$ (AAAABB).

f. An allotriploid is $3n$ with the chromosomes coming from two different species, so an allotriploid could be $3n = 2 + 2 + 3 = 7$ (AABBMNO) or $3n = 2 + 3 + 3 = 8$ (ABMMNNOO).

g. A nullisomic individual is missing both chromosomes of a homologous pair, so a nullisomy in species II for chromosome N would result in $2n - 2 = 6 - 2 = 4$ (MMOO).

COMPREHENSION QUESTIONS

Section 8.1

1. List the three basic categories of chromosome mutations and define each one.

Section 8.2

2. Why do extra copies of genes sometimes cause drastic phenotypic effects?

3. Draw a pair of chromosomes as they would appear during synapsis in prophase I of meiosis in an individual heterozygous for a chromosome duplication.

4. What is haploinsufficiency?

5. What is the difference between a paracentric and a pericentric inversion?

6. How can inversions in which no genetic information is lost or gained cause phenotypic effects?

7. Explain, with the aid of a drawing, how a dicentric bridge is produced when crossing over takes place in an individual heterozygous for a paracentric inversion.

8. Explain why recombination is suppressed in individuals heterozygous for paracentric and pericentric inversions.

9. How do translocations in which no genetic information is lost or gained produce phenotypic effects?

10. Sketch the chromosome pairing and the different segregation patterns that can arise in an individual heterozygous for a reciprocal translocation.

11. What is a Robertsonian translocation?

Section 8.3

12. List four major types of aneuploidy.

13. What is the difference between primary Down syndrome and familial Down syndrome? How does each type arise?

14. What is uniparental disomy and how does it arise?

15. What is genetic mosaicism and how does it arise?

Section 8.4

16. What is the difference between autopolyploidy and allopolyploidy? How does each arise?

17. Explain why autopolyploids are usually sterile, whereas allopolyploids are often fertile.

APPLICATION QUESTIONS AND PROBLEMS

Section 8.1

18. Examine the karyotypes shown in **Figure 8.1** and **Figure 8.2**. Are the individuals from whom these karyotypes were made males or females?

*19. Which types of chromosome mutations

 a. increase the amount of genetic material in a particular chromosome?

 b. increase the amount of genetic material in all chromosomes?

 c. decrease the amount of genetic material in a particular chromosome?

 d. change the position of DNA sequences in a single chromosome without changing the amount of genetic material?

 e. move DNA from one chromosome to a nonhomologous chromosome?

Section 8.2

*20. A chromosome has the following segments, where • represents the centromere:

$$A B • C D E F G$$

What types of chromosome mutations are required to change this chromosome into each of the following chromosomes? (In some cases, more than one chromosome mutation may be required.)

 a. A B A B • C D E F G   **f.** A B • E D C F G

 b. A B • C D E A B F G   **g.** C • B A D E F G

 c. A B • C F E D G   **h.** A B • C F E D F E D G

 d. A • C D E F G   **i.** A B • C D E F C D F E G

 e. A B • C D E

21. A chromosome initially has the following segments:

$$A B • C D E F G$$

Draw the chromosome, identifying its segments, that would result from each of the following mutations.

 a. Tandem duplication of DEF

 b. Displaced duplication of DEF

 c. Deletion of FG

 d. Paracentric inversion that includes DEFG

 e. Pericentric inversion of BCDE

22. The following diagram represents two nonhomologous chromosomes:

$$A B • C D E F G$$
$$R S • T U V W X$$

What type of chromosome mutation would produce each of the following groups of chromosomes?

 a. A B • C D   **c.** A B • T U V F G
    R S • T U V W X E F G   R S • C D E W X

 b. A U V B • C D E F G   **d.** A B • C W G
    R S • T W X   R S • T U V D E F X

*23. The *Notch* mutation is a deletion on the X chromosome of *Drosophila melanogaster*. Female flies heterozygous for *Notch* have an indentation on the margins of their wings; *Notch* is lethal in the homozygous and hemizygous conditions. The *Notch* deletion covers the region of the X chromosome that contains the locus for white eyes, an X-linked recessive trait. Give the phenotypes and proportions of progeny produced in the following crosses.

 a. A red-eyed Notch female is mated with a white-eyed male.

 b. A white-eyed Notch female is mated with a red-eyed male.

 c. A white-eyed Notch female is mated with a white-eyed male.

24. The green-nose fly normally has six chromosomes: two metacentric and four acrocentric. A geneticist examines the chromosomes of an odd-looking green-nose fly and discovers that it has only five chromosomes; three of them are metacentric and two are acrocentric. Explain how this change in chromosome number might have taken place.

*25. A wild-type chromosome has the following segments:

$$A B C • D E F G H I$$

Researchers have found individuals that are heterozygous for each of the following chromosome mutations. For each mutation, sketch how the wild-type and mutated chromosomes would pair in prophase I of meiosis, showing all chromosome strands.

 a. A B C • D E F D E F G H I   **c.** A B C • D G F E H I

 b. A B C • D H I   **d.** A B E D • C F G H I

26. For the chromosomes shown in **Figure 8.12**, draw the chromatids that would result from a two-strand double crossover: one crossover between C and D and the other crossover between D and E.

*27. As discussed in this chapter, crossing over within a pericentric inversion produces chromosomes that have extra copies of some genes and no copies of other genes. The fertilization of gametes containing chromosomes with duplications or deletions often result in children with syndromes characterized by developmental delay, intellectual disability, abnormal development of organ systems, and early death. Maarit Jaarola and colleagues examined individual sperm cells of a male who was heterozygous for a pericentric inversion on chromosome 8 and determined that crossing over took place within the pericentric inversion in 26% of the meiotic divisions (M. Jaarola, R. H. Martin, and T. Ashley. 1998. *American Journal of Human Genetics* 63:218–224).

Assume that you are a genetic counselor and that a couple seeks counseling from you. Both the man and the woman are phenotypically normal, but the woman is heterozygous for a pericentric inversion on chromosome 8. The man is karyotypically normal. What is the probability that this couple will produce a child with a debilitating syndrome as the result of crossing over within the pericentric inversion?

*28. An individual heterozygous for a reciprocal translocation possesses the following chromosomes:

$$\text{A B} \bullet \text{C D E F G}$$

$$\text{A B} \bullet \text{C D V W X}$$

$$\text{R S} \bullet \text{T U E F G}$$

$$\text{R S} \bullet \text{T U V W X}$$

a. Draw the pairing arrangement of these chromosomes in prophase I of meiosis.

b. Diagram the alternate, adjacent-1, and adjacent-2 segregation patterns in anaphase I of meiosis.

c. Give the products that result from alternate, adjacent-1, and adjacent-2 segregation.

Section 8.3

*29. Red–green color blindness is a human X-linked recessive disorder. A young man with a 47,XXY karyotype (Klinefelter syndrome) is color blind. His 46,XY brother is also color blind. Both parents have normal color vision. Where did the nondisjunction that gave rise to the young man with Klinefelter syndrome take place? Assume that no crossing over took place in prophase I of meiosis.

30. Junctional epidermolysis bullosa (JEB) is a severe skin disorder that results in blisters over the entire body. The disorder is caused by autosomal recessive mutations at any one of three loci that help to encode laminin 5, a major component in the dermal–epidermal basement membrane. Leena Pulkkinen and colleagues described a male newborn who was born with JEB and died at 2 months of age (L. Pulkkinen et al. 1997. *American Journal of Human Genetics* 61:611–619); the child had healthy, unrelated parents. Chromosome analysis revealed that the infant had 46 normal-appearing chromosomes. Analysis of DNA showed that his mother was heterozygous for a JEB-causing allele at the *LAMB3* locus, which is on chromosome 1. The father had two normal alleles at this locus. DNA fingerprinting demonstrated that the male assumed to be the father had, in fact, conceived the child.

a. Assuming that no new mutations occurred in this family, explain the presence of an autosomal recessive disease in the child when the mother is heterozygous and the father is homozygous normal.

b. How might you go about proving your explanation? Assume that a number of genetic markers are available for each chromosome.

31. Some people with Turner syndrome are 45,X/46,XY mosaics. Explain how this mosaicism could arise.

*32. Bill and Betty have had two children with Down syndrome. Bill's brother has Down syndrome and his sister has two children with Down syndrome. On the basis of these observations, indicate which of the following statements are most likely correct and which are most likely incorrect. Explain your reasoning.

a. Bill has 47 chromosomes.

b. Betty has 47 chromosomes.

c. Bill and Betty's children each have 47 chromosomes.

d. Bill's sister has 45 chromosomes.

e. Bill has 46 chromosomes.

f. Betty has 45 chromosomes.

g. Bill's brother has 45 chromosomes.

33. In mammals, sex-chromosome aneuploids are more common than autosomal aneuploids, but, in fishes, sex-chromosome aneuploids and autosomal aneuploids are found with equal frequency. Offer a possible explanation for these differences between mammals and fishes. (Hint: Think about why sex-chromosome aneuploids are more common than autosomal aneuploids in mammals.)

*34. A young couple is planning to have children. Knowing that there have been a substantial number of stillbirths, miscarriages, and fertility problems on the husband's side of the family, they see a genetic counselor. A chromosome analysis reveals that, whereas the woman has a normal karyotype, the man possesses only 45 chromosomes and is a carrier of a Robertsonian translocation between chromosomes 22 and 13.

a. List all the different types of gametes that might be produced by the man.

b. What types of zygotes will develop when each of gametes produced by the man fuses with a normal gamete produced by the woman?

c. If trisomies and monosomies entailing chromosomes 13 and 22 are lethal, approximately what proportion of the surviving offspring are expected to be carriers of the translocation?

35. Using breeding techniques, Andrei Dyban and V. S. Baranov (*Cytogenetics of Mammalian Embryonic Development*. Oxford: Oxford University Press, Clarendon Press; New York: Oxford University Press, 1987) created mice that were trisomic for each of

the different mouse chromosomes. They found that only mice with trisomy 19 developed. Mice that were trisomic for all other chromosomes died in the course of development. For some of these trisomics, the researchers plotted the length of development (number of days after conception before the embryo died) as a function of the size of the mouse chromosome that was present in three copies (see the adjoining graph). Summarize their findings and provide a possible explanation for the results.

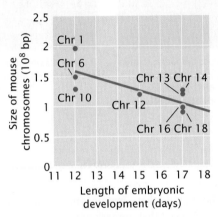

[E. Torres, B. R. Williams, and A. Amon. 2008. *Genetics* 179:737–746, Fig. 2B.]

Section 8.4

36. Species I has $2n = 16$ chromosomes. How many chromosomes will be found per cell in each of the following mutants in this species?

a. Monosomic

b. Autotriploid

c. Autotetraploid

d. Trisomic

e. Double monosomic

f. Nullisomic

g. Autopentaploid

h. Tetrasomic

37. Species I is diploid ($2n = 8$) with chromosomes AABBCCDD; related species II is diploid ($2n = 8$) with chromosomes MMNNOOPP. What types of chromosome mutations do individuals with the following sets of chromosomes have?

a. AAABBCCDD

b. MMNNOOOOPP

c. AABBCDD

d. AAABBBCCCDDD

e. AAABBCCDDD

f. AABBDD

g. AABBCCDDMMNNOOPP

h. AABBCCDDMNOP

*38. Species I has $2n = 8$ chromosomes and species II has $2n = 14$ chromosomes. What would the expected chromosome numbers be in individuals with the following chromosome mutations? Give all possible answers.

a. Allotriploidy including species I and II

b. Autotetraploidy in species II

c. Trisomy in species I

d. Monosomy in species II

e. Tetrasomy in species I

f. Allotetraploidy including species I and II

39. Suppose that species I in **Figure 8.27** had $2n = 10$ and species II in that figure had $2n = 12$. How many chromosomes would be present in the allotetraploid at the bottom of the figure?

40. Consider a diploid cell that has $2n = 4$ chromosomes: one pair of metacentric chromosomes and one pair of acrocentric chromosomes. Suppose that this cell undergoes nondisjunction, giving rise to an autotriploid cell ($3n$). The triploid cell then undergoes meiosis. Draw the different types of gametes that could result from meiosis in the triploid cell, showing the chromosomes present in each type. To distinguish between the different metacentric and acrocentric chromosomes, use a different color to draw each metacentric chromosome; similarly, use a different color to draw each acrocentric chromosome. (Hint: See **Figure 8.27**.)

41. Assume that the autotriploid cell in **Figure 8.26** has $3n = 30$ chromosomes. For each of the gametes produced by this cell, give the chromosome number of the zygote that would result if the gamete fused with a normal haploid gamete.

42. *Nicotiana glutinosa* ($2n = 24$) and *N. tabacum* ($2n = 48$) are two closely related plants that can be intercrossed, but the F_1 hybrid plants that result are usually sterile. In 1925, Roy Clausen and Thomas Goodspeed crossed *N. glutinosa* and *N. tabacum* and obtained one fertile F_1 plant (R. E. Clausen and T. H. Goodspeed. 1925 *Genetics* 10:278–284). They were able to self-pollinate the flowers of this plant to produce an F_2 generation. Surprisingly, the F_2 plants were fully fertile and produced viable seeds. When Clausen and Goodspeed examined the chromosomes of the F_2 plants, they observed 36 pairs of chromosomes in metaphase I and 36 individual chromosomes in metaphase II. Explain the origin of the F_2 plants obtained by Clausen and Goodspeed and the numbers of chromosomes observed.

43. What would be the chromosome number of progeny resulting from the following crosses in wheat (see **Figure 8.28**)? What type of polyploid (allotriploid, allotetraploid, etc.) would result from each cross?

a. Einkorn wheat and emmer wheat

b. Bread wheat and emmer wheat

c. Einkorn wheat and bread wheat

44. Karl and Hally Sax crossed *Aegilops cylindrica* ($2n = 28$), a wild grass found in the Mediterranean region, with *Triticum vulgare* ($2n = 42$), a type of wheat (K. Sax and H. J. Sax. 1924. *Genetics* 9:454–464). The resulting F_1 plants from this cross had 35 chromosomes. Examination of metaphase I in the F_1 plants revealed the presence of 7 pairs of chromosomes (bivalents) and 21 unpaired chromosomes (univalents).

a. If the unpaired chromosomes segregate randomly, what possible chromosome numbers will appear in the gametes of the F_1 plants?

b. What does the appearance of the bivalents in the F_1 hybrids suggest about the origin of *Triticum vulgare* wheat?

Aegilops cylindrica, jointed goatgrass. [Sam Brinker, MNR-NHIC, 2008/Canadian Food Inspection Agency.]

Triticum vulgare, wheat. [Michael Hieber/123RF.com.]

CHALLENGE QUESTIONS

Section 8.3

45. Red–green color blindness is a human X-linked recessive disorder. Jill has normal color vision, but her father is color blind. Jill marries Tom, who also has normal color vision. Jill and Tom have a daughter who has Turner syndrome and is color blind.

a. How did the daughter inherit color blindness?

b. Did the daughter inherit her X chromosome from Jill or from Tom?

46. Progeny of triploid tomato plants often contain parts of an extra chromosome, in addition to the normal complement of 24 chromosomes (J. W. Lesley and M. M. Lesley. 1929. *Genetics* 14:321–336). Mutants with a part of an extra chromosome are referred to as secondaries. James and Margaret Lesley observed that secondaries arise from triploid ($3n$), trisomic ($3n + 1$), and double trisomic ($3n + 1 + 1$) parents, but never from diploids ($2n$). Give one or more possible reasons that secondaries arise from parents that have unpaired chromosomes but not from parents that are normal diploids.

47. Mules result from a cross between a horse ($2n = 64$) and a donkey ($2n = 62$), have 63 chromosomes, and are almost always sterile. However, in the summer of 1985, a female mule named Krause who was pastured with a male donkey gave birth to a male foal (O. A. Ryder et al. 1985. *Journal of Heredity* 76:379–381). Blood tests established that the foal, appropriately named Blue Moon, was the offspring of Krause and that Krause was indeed a mule. Both Blue Moon and Krause were fathered by the same donkey (see the accompanying pedigree). The foal, like his mother, had 63 chromosomes—half of them horse chromosomes and the other half donkey chromosomes. Analyses of genetic markers showed that, remarkably, Blue Moon seemed to have inherited a complete set of horse chromosomes from his mother, instead of the random mixture of horse and donkey chromosomes that would be expected with normal meiosis. Thus, Blue Moon and Krause were not only mother and son, but also brother and sister.

a. With the use of a diagram, show how, if Blue Moon inherited only horse chromosomes from his mother, Krause and Blue Moon are mother and son as well as sister and brother.

b. Although rare, additional cases of fertile mules giving birth to offspring have been reported. In these cases, when a female mule mates with a male horse, the offspring is horselike in appearance, but when a female mule mates with a male donkey, the offspring is mulelike in appearance. Is this observation consistent with the idea that the offspring of fertile female mules inherit only a set of horse chromosomes from their mule mothers? Explain your reasoning.

c. Can you suggest a possible mechanism for how fertile female mules might pass on a complete set of horse chromosomes to their offspring?

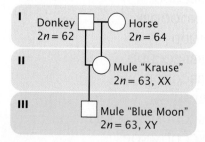

Section 8.4

48. Humans and many other complex organisms are diploid, possessing two sets of genes, one inherited from the mother and one from the father. However, a number of eukaryotic organisms spend most of their life cycles in a haploid state. Many of these eukaryotes, such as *Neurospora* and yeast, still undergo meiosis and sexual reproduction, but most of the cells that make up the organism are haploid.

Considering that haploid organisms are fully capable of sexual reproduction and generating genetic variation, why are most complex eukaryotes diploid? In other words, what might be the evolutionary advantage of existing in a diploid state instead of a haploid state? And why might a few organisms, such as *Neurospora* and yeast, exist as haploids?

THINK-PAIR-SHARE QUESTIONS

Section 8.1

1. Why do species usually have a characteristic number of chromosomes? Why don't we see many species in which chromosome number varies within the species, with some individuals, say, having $2n = 20$ and others having $2n = 24$?

Section 8.2

2. An individual is heterozygous for a reciprocal translocation, with the following chromosomes:

$$A \cdot B C D E F$$

$$A \cdot B C V W X$$

$$R S T \cdot U D E F$$

$$R S T \cdot U V W X$$

a. Draw a picture of these chromosomes pairing in prophase I of meiosis.

b. Draw the products of alternate, adjacent-1, and adjacent-2 segregations.

c. Explain why the fertility of this individual is likely to be less than the fertility of an individual without a translocation.

Section 8.3

3. Monozygotic (identical) twins arise when a single egg, fertilized by a single sperm, divides early in development, giving rise to two genetically identical embryos. A recent study examined a pair of monozygotic twins in which one of the twins had trisomy 21 and the other twin had two normal copies of chromosome 21. Explain how two identical twins could differ in their number of chromosomes.

4. The human X chromosome is about the same size as human chromosomes 4 and 5. Trisomy of the X chromosome occurs about once in every thousand human female births, and yet trisomy 4 and trisomy 5 are almost never seen among living humans. Why do these differences in the frequency of trisomic individuals occur, in spite of similar chromosome size?

5. Geneticists have been exploring ways to suppress the expression of the extra chromosome 21 in individuals with Down syndrome in hopes of preventing the medical problems and intellectual disability of individuals with trisomy 21. One approach involves modifying a gene that is already present in human cells and using it to suppress the expression of the extra copy of chromosome 21. What approach, do you think, are they taking, and what may be some of the challenges for using it in patients?

6. In Chapter 2, we learned that the two chromosomes of a homologous pair carry information for the same characteristics but are not identical, having different alleles. Would you ever expect the two chromosomes of a homologous pair to be identical? If so, how might that occur? If not, why not?

Section 8.4

7. Trisomics often have more developmental problems than triploids. Can you suggest a reason why?

 **SaplingPlus**   Self-study tools that will help you practice what you've learned and reinforce this chapter's concepts are available online. Go to www.macmillanlearning.com/PierceGenetics6e.

Bacterial and Viral Genetic Systems

Woman with leprosy, a disease caused by the bacterium *Mycobacterium leprae*. The study of *M. leprae* DNA isolated from ancient skeletons of medieval Europeans with leprosy has provided information about the evolution of this bacterium. [Reuters/Rupak de Chowdhuri (India).]

The Genetics of Medieval Leprosy

Leprosy, one of the most feared diseases of history, was well known in ancient times, and people with leprosy were frequently ostracized from society. Although leprosy is successfully treated today with antibiotics, it remains a major public health problem: from 2 million to 3 million people worldwide are disabled by leprosy, and over 200,000 new cases are reported each year. In its severest form, leprosy causes paralysis, blindness, and disfigurement. Leprosy is caused by the bacterium *Mycobacterium leprae*, which infects cells of the nervous system—although human genes do play a role in susceptibility to this disease.

In 2013, geneticists isolated DNA of *M. leprae* from five skeletons of medieval Europeans who exhibited signs of leprosy. Their comparisons of the gene sequences of these ancient bacteria with those of modern strains provided insight into the evolution of this organism. Scientists had previously determined the genome sequence of *M. leprae* and found that it contains 3,268,203 base pairs of DNA, 1 million base pairs fewer than the genomes of other mycobacteria. In most bacterial genomes, the vast majority of the DNA encodes proteins—little DNA lies between the protein-encoding genes. In contrast, only 50% of the DNA of *M. leprae* encodes proteins. *M. leprae* also has 2300 fewer genes than its close relative *M. tuberculosis*. An incredible 27% of *M. leprae*'s genome consists of nonfunctional copies of genes (called pseudogenes) that have been inactivated by mutations. Its reduced DNA content, fewer functional genes, and large number of pseudogenes suggest that, evolutionarily, the genome of *M. leprae* has undergone massive decay over time, losing DNA and acquiring mutations that have inactivated many of its genes. Although the reasons for this decay are not known, it helps account for *M. leprae*'s long generation time—14 days in humans, an incredibly long replication time for a bacterium—and the inability of scientists to culture the bacteria in the laboratory.

Leprosy was common in Europe until the Middle Ages, when it disappeared from the population. Why did it disappear from Europe, despite remaining common in many other parts of the world? To address this question, geneticists extracted DNA from the bones and teeth of five medieval skeletons (dating from the eleventh to the fourteenth centuries) exhumed from cemeteries in Denmark, Sweden, and the United Kingdom. They separated out the DNA of *M. leprae* and determined whole-genome sequences for the bacteria. They then compared the DNA sequences of these medieval strains of *M. leprae* with those of modern strains from India, Thailand, the United States, Brazil, and other locations.

This analysis revealed that the ancient strains of *M. leprae* were remarkably similar to modern strains. Three of the ancient strains were most closely related to modern strains from Iran and Turkey, suggesting a Middle Eastern–European connection to the disease. Some of the ancient strains were closely related to modern *M. leprae* currently found in the United States, suggesting that these North American bacteria originated in Europe.

The close similarity between ancient European strains and a modern virulent strain from North America suggests that leprosy did not disappear from Europe because it lost its virulence. More likely, improved social conditions, changes in the immunity of Europeans, or the presence of other infectious diseases brought about the demise of leprosy in Europe. This study of leprosy illustrates the importance of genetic studies of bacteria to human health and shows how modern tools of evolutionary and molecular genetics are being applied to our understanding of bacterial biology.

> **THINK-PAIR-SHARE**
>
> - In the ancient world, leprosy was greatly feared, and people with the disease were often ostracized from society. Why was leprosy so feared? Are there modern diseases that evoke similar fears and for which infected people are ostracized? If so, give some examples. What characteristics of a disease produce this response on the part of society? Are any of these modern diseases caused by bacteria or viruses?
>
> - Genetic analysis of *Mycobacterium leprae*, the bacterium that causes leprosy, reveals that its genome has undergone decay over time, losing DNA and acquiring mutations that make some of its genes nonfunctional. What might be some potential reasons for this evolutionary decay of its genome?

In this chapter, we examine some of the genetic properties of bacteria and viruses, and the mechanisms by which they exchange and recombine their genes.

9.1 Bacteria and Viruses Have Important Roles in Human Society and the World Ecosystem

The genetic systems of bacteria and viruses are studied because these organisms play critically important roles in human society and the world ecosystem. Viruses infect all living organisms and are, by far, the most common biological element on Earth. Bacteria in the oceans produce 50% of the oxygen in the air and remove roughly 50% of the carbon dioxide. Bacteria and viruses also play critical roles in agriculture, serving as economically important pathogens of crops and domesticated animals, but also providing nitrogen, phosphorus, and other essential nutrients to our food plants. Bacteria are found naturally in the mouth, in the gut, and on the skin, where they are essential to human function and ecology. These complex communities of bacteria protect us from disease and are essential for digestion and other physiological functions.

Bacteria and viruses have immense medical significance because they cause many human diseases. Infectious diseases are one of the leading causes of death worldwide, although many of these diseases can be controlled through antibiotics and vaccines. Bacteria have also been harnessed to produce a number of economically important substances, including drugs, hormones, food additives, and other chemicals and compounds. And viruses are now being used in gene therapies.

Bacteria and viruses are also important in the study of genetics because they possess a number of characteristics that make them suitable for genetic analysis (**Table 9.1**). Their simple genetic systems have a number of features in common with the genetic systems of humans and other more complex organisms, so information gleaned from the genetic study of bacteria and viruses often provides important insight into genetic principles that are applicable to many other organisms.

Life in a Bacterial World

Humans like to think that we rule the world, but we are clearly in a minor position compared with bacteria. Bacteria first evolved some 3.5 billion years ago, 2 billion years

TABLE 9.1	Advantages of using bacteria and viruses for genetic studies
1. Reproduction is rapid.	
2. Many progeny are produced.	
3. The haploid genome allows all mutations to be expressed directly.	
4. Asexual reproduction simplifies the isolation of genetically pure strains.	
5. Growth in the laboratory is easy and requires little space.	
6. Genomes are small.	
7. Techniques are available for isolating and manipulating their genes.	
8. They have medical importance.	
9. They can be genetically engineered to produce substances of commercial value.	

before the first eukaryotes appeared (or perhaps even earlier, as suggested by some evidence). Today, bacteria are found in every conceivable environment, including boiling springs, highly saline lakes, and beneath more than 2 miles of ice in Antarctica. They are found at the top of Mt. Everest and at the bottoms of the deepest oceans. They are also present on and in *us*—in astounding numbers! Within the average human gut, there are approximately 10 trillion bacteria, ten times the total number of cells in the entire human body. No one knows how many bacteria populate the world, but an analysis conducted by scientists in 1998 estimated that the total number of living bacteria on Earth exceeded 5 million trillion trillion (5×10^{30}).

Bacteria are not only numerically vast, but also constitute the majority of life's diversity. The total number of *described* species of bacteria is less than 10,000, compared with about 1.4 million plants, animals, fungi, and single-celled eukaryotes. But the number of described species of bacteria falls far short of the true microbial diversity.

Species of bacteria are typically described only after they have been cultivated and studied in the laboratory. Because only a few species are amenable to laboratory culture, it was impossible for many years to identify and study most bacteria. Then, in the 1970s, molecular techniques for analyzing DNA became available and opened up a whole new vista on microbial diversity. These techniques revealed several important facts about bacteria. First, many of the relations among bacteria that microbiologists had worked out on the basis of their physical and biochemical traits turned out to be incorrect. Bacteria once thought to be related were in fact genetically quite different. Second, molecular analysis showed that members of one group of microbes—now called the archaea—were as different from other bacteria as they are from eukaryotes. Third, molecular analysis revealed that the number of different types of bacteria is astounding.

In 2007, Luiz Roesch and his colleagues set out to determine exactly how many types of bacteria exist in a gram of soil. They obtained soil samples from four locations: Brazil, Florida, Illinois, and Canada. From these soil samples, they extracted and purified bacterial DNA. Using this DNA, they determined the sequences of a gene present in all bacteria, the 16S rRNA gene. Each bacterial species has a unique 16S rRNA gene sequence, so the researchers could determine how many species of bacteria existed in each soil sample by counting the different DNA sequences.

Roesch's results were amazing. The number of different bacterial species in each gram of soil ranged from 26,140 for samples from Brazil to 53,533 for Canadian samples. Many unusual bacteria were detected that appeared dissimilar to any previously described groups of bacteria. Another interesting finding was that soil from agricultural fields harbored considerably fewer species than did soil from forests.

This study and others demonstrate that bacterial diversity far exceeds that of multicellular organisms, and undoubtedly, numerous groups of bacteria have yet to be discovered. Like it or not, we truly live in a bacterial world. **>> TRY PROBLEM 17**

THINK-PAIR-SHARE Question 1

Bacterial Diversity

Prokaryotes, as we saw in Chapter 2, are unicellular organisms that lack nuclear membranes and membrane-bounded organelles. For many years, biologists considered all prokaryotes to be related, but genome sequence information now provides convincing evidence that prokaryotes are divided into at least two distinct groups: the archaea and the eubacteria. The archaea are a group of diverse prokaryotes that are frequently found in extreme environments, such as hot springs and the bottoms of oceans. The eubacteria (usually referred to simply as *bacteria*, as we do in this book) include most of the familiar bacterial species. Although superficially similar in their cell structure, bacteria and archaea are distinct in their genetic makeup, and the differences between them are as great as those between bacteria and eukaryotes. In fact, the archaea are *more* similar to eukaryotes than to bacteria in a number of molecular features and genetic processes.

Bacteria are extremely diverse and come in a variety of shapes and sizes. Some are rod-shaped, whereas others are spherical or helical. Most are much smaller than eukaryotic cells, but at least one species isolated from the guts of fish is almost 1 mm long and can be seen with the naked eye. Some bacteria are photosynthetic. Others produce stalks and spores, superficially resembling fungi.

Bacteria have long been considered simple organisms that lack much of the cellular complexity of eukaryotes. However, recent evidence points to a number of similarities and parallels in bacterial and eukaryotic structure. For example, a bacterial protein termed FtsZ, which plays an integral part in bacterial cell division, is structurally similar to eukaryotic tubulin proteins, which are subunits of microtubules and help to segregate chromosomes in mitosis and meiosis (see Chapter 2). Like eukaryotes, bacteria have proteins that help to condense DNA. Other bacterial proteins function much as cytoskeletal proteins do in eukaryotes, helping to give bacterial cells shape and structure. And, although bacteria don't undergo mitosis and meiosis, replication of the bacterial chromosome precedes binary fission, and there are bacterial processes that ensure that one copy of the chromosome is allocated to each daughter cell.

9.2 Genetic Analysis of Bacteria Requires Special Methods

Heredity in bacteria is fundamentally similar to heredity in more complex organisms. However, the bacterial haploid genome and the small size of bacteria (which makes observation of their phenotypes difficult) necessitate different approaches and methods.

(a)

(b)

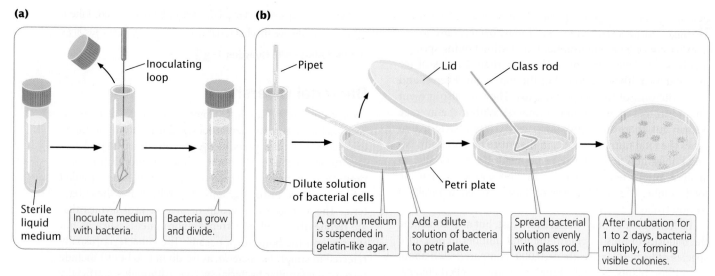

Inoculating loop

Pipet

Lid

Glass rod

Sterile liquid medium

Inoculate medium with bacteria.

Bacteria grow and divide.

Dilute solution of bacterial cells

Petri plate

A growth medium is suspended in gelatin-like agar.

Add a dilute solution of bacteria to petri plate.

Spread bacterial solution evenly with glass rod.

After incubation for 1 to 2 days, bacteria multiply, forming visible colonies.

9.1 Growing bacteria in the laboratory. Bacteria can be grown (a) in liquid medium (broth) or (b) on solid medium.

Techniques for the Study of Bacteria

Microbiologists have defined the nutritional needs of a number of bacteria and developed culture media for growing them in the laboratory. These culture media typically contain a carbon source, essential elements such as nitrogen and phosphorus, certain vitamins, and other required ions and nutrients. Wild-type, or **prototrophic**, bacteria can use these simple ingredients to synthesize all the compounds that they need for growth and reproduction. A medium that contains only the nutrients required by prototrophic bacteria is termed a **minimal medium**.

Mutant strains called **auxotrophs** lack one or more enzymes necessary for synthesizing essential compounds and will grow only on medium supplemented with those essential molecules. For example, auxotrophic strains that are unable to synthesize the amino acid leucine will not grow on minimal medium, but *will* grow on medium to which leucine has been added. A **complete medium** contains all the substances, such as the amino acid leucine, required by bacteria for growth and reproduction.

Cultures of bacteria are often grown in test tubes that contain sterile liquid medium, or *broth* (**Figure 9.1a**). A few bacteria are added to a broth tube, in which they grow and divide until all the nutrients are used up or—more

commonly—until the concentration of their waste products becomes toxic to them. Bacteria can also be grown on agar plates (**Figure 9.1b**), in which melted agar is suspended in growth medium and poured into the bottom half of a petri plate. The agar solidifies when cooled and provides a solid, gelatin-like base for bacterial growth. In a process called *plating*, a dilute solution of bacteria is spread over the surface of the agar. As each bacterium grows and divides, it gives rise to a visible clump of genetically identical cells (a **colony**). Genetically pure strains can be isolated by collecting bacteria from a single colony and transferring them to a new broth tube or agar plate. The chief advantage of this method is that it allows one to isolate and count bacteria, which individually are too small to see without a microscope.

Microbiologists often study phenotypes that affect the appearance of the colony (**Figure 9.2**) or can be detected by simple chemical tests. Auxotrophs are commonly studied phenotypes. Suppose we want to detect auxotrophs that cannot synthesize leucine (*leu*⁻ mutants). We first spread bacteria on a petri plate containing medium that includes leucine; both prototrophs that have the *leu*⁺ allele and auxotrophs that have the *leu*⁻ allele will grow on it (**Figure 9.3**). Next, using a technique called *replica plating*, we transfer a few cells from each of the colonies on the original plate to two new plates:

(a)

(b)

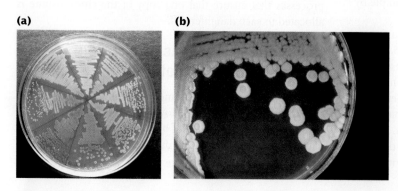

9.2 Bacterial colonies have a variety of phenotypes.
(a) *Serratia marcescens* with color variation. (b) *Bacillus cereus*; colony shape varies among different strains of this species.
[Part a: Courtesy Dr. Robert Shanks, Department of Ophthalmology, University of Pittsburgh. Part b: Biophoto Associates/Science Source.]

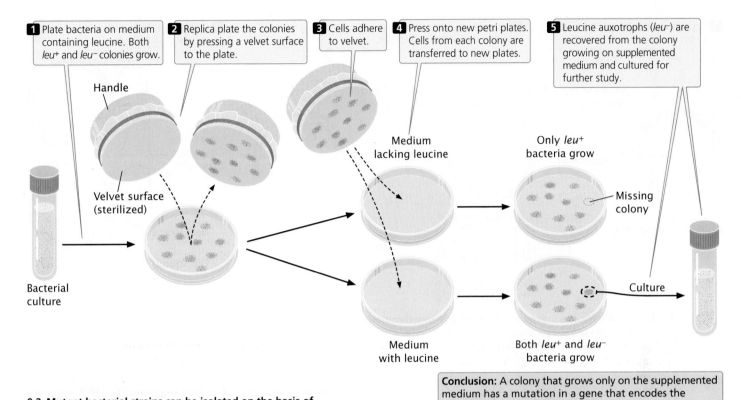

1 Plate bacteria on medium containing leucine. Both *leu+* and *leu−* colonies grow.

2 Replica plate the colonies by pressing a velvet surface to the plate.

3 Cells adhere to velvet.

4 Press onto new petri plates. Cells from each colony are transferred to new plates.

5 Leucine auxotrophs (*leu−*) are recovered from the colony growing on supplemented medium and cultured for further study.

Handle

Velvet surface (sterilized)

Bacterial culture

Medium lacking leucine

Only *leu+* bacteria grow

Missing colony

Medium with leucine

Both *leu+* and *leu−* bacteria grow

Culture

9.3 Mutant bacterial strains can be isolated on the basis of their nutritional requirements.

Conclusion: A colony that grows only on the supplemented medium has a mutation in a gene that encodes the synthesis of an essential nutrient.

one plate contains medium to which leucine has been added; the other plate contains medium lacking leucine. A medium that lacks an essential nutrient, such as the medium lacking leucine, is called a *selective* medium. The *leu+* bacteria will grow on both media, but the *leu−* mutants will grow only on the selective medium supplemented by leucine because they cannot synthesize their own leucine. Any colony that grows on medium that contains leucine, but not on medium that lacks leucine, consists of *leu−* bacteria. The auxotrophs that grow only on the supplemented medium can then be cultured for further study. Auxotrophic mutants are often used to study the results of genetic crosses and other genetic manipulations.

The Bacterial Genome

Most bacterial genomes that have been studied consist of a circular chromosome that contains a single double-stranded DNA molecule several million base pairs (bp) in length (**Figure 9.4**). For example, the *E. coli* genome has approximately 4.6 million base pairs of DNA. However, some bacteria contain multiple chromosomes. For example, *Vibrio cholerae*, which causes cholera, has two circular chromosomes, and *Rhizobium meliloti* has three chromosomes. There are even a few bacteria that have linear chromosomes. Most bacterial chromosomes consist largely of sequences that encode proteins. For example, more than 90% of the DNA in *E. coli* encodes proteins. In contrast, only about 1% of human DNA encodes proteins.

THINK-PAIR-SHARE Question 2

Plasmids

In addition to having a chromosome, many bacteria possess **plasmids**: small, usually circular DNA molecules that are distinct from the bacterial chromosome. Some plasmids are present in many copies per cell, whereas others are present in only one or two copies. In general, plasmids carry genes that are not essential to bacterial function but may

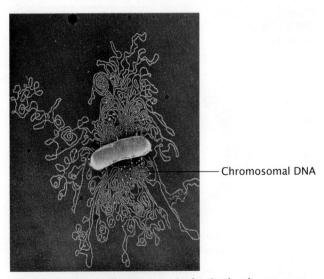

9.4 Most bacterial cells possess a single, circular chromosome. The chromosome shown here is emerging from a ruptured bacterial cell. [Dr. Gopal Murti/Science Source.]

Chromosomal DNA

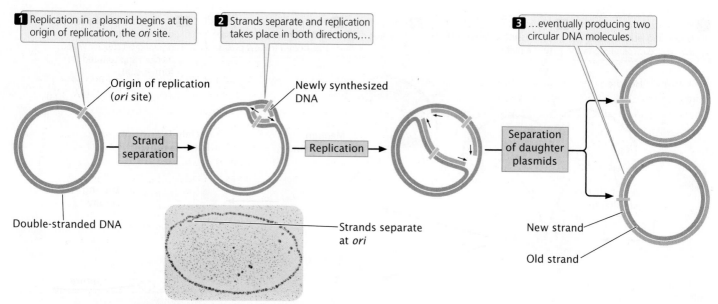

9.5 A plasmid replicates independently of its bacterial chromosome. Replication begins at the origin of replication (*ori*) and continues around the circle. In this diagram, replication is taking place in both directions; in some plasmids, replication is in one direction only. [Photograph: Biology Pics/Science Source.]

play important roles in the life cycle and growth of their bacterial hosts. There are many different types of plasmids; *E. coli* alone is estimated to have more than 270 different types of naturally occurring plasmids. Some plasmids promote mating between bacteria; others contain genes that kill other bacteria. Of importance to human health, plasmids are responsible for the spread of antibiotic resistance among bacteria. Plasmids are also used extensively in genetic engineering (see Chapter 19).

Most plasmids are circular and several thousand base pairs in length, although plasmids as large as several hundred thousand base pairs have also been found. Each plasmid possesses an origin of replication, a specific DNA sequence where DNA replication is initiated (see Chapter 2). The origin of replication allows a plasmid to replicate independently of the bacterial chromosome (**Figure 9.5**). **Episomes** are plasmids that are capable of replicating freely and are able to integrate into bacterial chromosomes. The **F (fertility) factor** of *E. coli* (**Figure 9.6**) is an episome that controls mating and gene exchange between *E. coli* cells, a process we will discuss shortly.

CONCEPTS

Bacteria can be studied in the laboratory by growing them on liquid or solid media. A typical bacterial genome consists of a single circular chromosome that contains several million base pairs. Some bacterial genes may be present on plasmids, which are small, circular DNA molecules that replicate independently of the bacterial chromosome.

✔ CONCEPT CHECK 1

Which of the following statements is true of plasmids?

a. They are composed of RNA.
b. They normally exist outside of bacterial cells.
c. They possess only a single strand of DNA.
d. They contain an origin of replication.

These genes regulate plasmid transfer to other cells.

These sequences regulate insertion into the bacterial chromosome.

These genes control plasmid replication.

F factor

9.6 The F factor, a circular episome of *E. coli*, contains a number of genes that regulate its transfer into a bacterial cell, replication, and insertion into the bacterial chromosome. Replication is initiated at *ori*. Insertion sequences (see Chapter 18) *IS3* and *IS2* control insertion into the bacterial chromosome and excision from it.

9.3 Bacteria Exchange Genes Through Conjugation, Transformation, and Transduction

Bacteria exchange genetic material by three different mechanisms, all entailing some type of DNA transfer and recombination between the transferred DNA and the bacterial chromosome.

1. **Conjugation** takes place when genetic material passes directly from one bacterium to another (**Figure 9.7a**). In conjugation, two bacteria lie close together and a connection forms between them. A plasmid or a part of the bacterial chromosome passes from one cell (the donor) to the other (the recipient). After conjugation, crossing over may take place between homologous sequences in the transferred DNA and the chromosome of the recipient cell. In conjugation, DNA is transferred only from donor to recipient, with no reciprocal exchange of genetic material.

2. **Transformation** takes place when a bacterium takes up DNA from the medium in which it is growing (**Figure 9.7b**). After transformation, recombination may

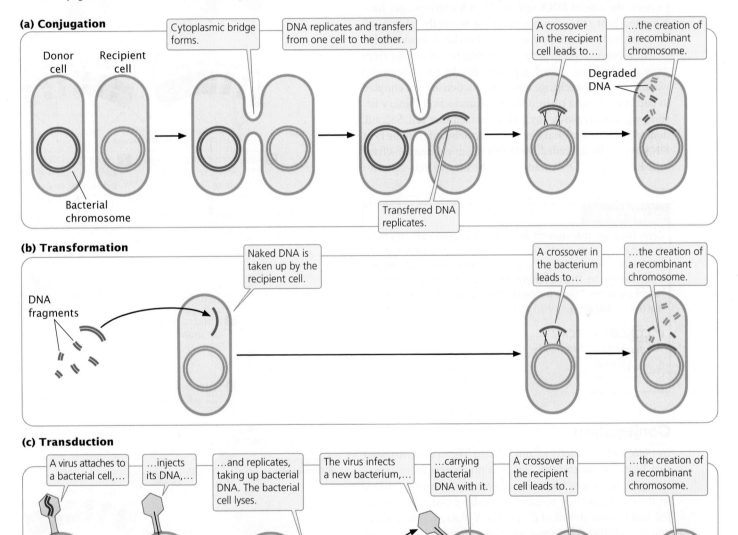

(a) Conjugation

Donor cell   Recipient cell

Cytoplasmic bridge forms.

DNA replicates and transfers from one cell to the other.

A crossover in the recipient cell leads to…

…the creation of a recombinant chromosome.

Degraded DNA

Bacterial chromosome

Transferred DNA replicates.

(b) Transformation

DNA fragments

Naked DNA is taken up by the recipient cell.

A crossover in the bacterium leads to…

…the creation of a recombinant chromosome.

(c) Transduction

A virus attaches to a bacterial cell,…

…injects its DNA,…

…and replicates, taking up bacterial DNA. The bacterial cell lyses.

The virus infects a new bacterium,…

…carrying bacterial DNA with it.

A crossover in the recipient cell leads to…

…the creation of a recombinant chromosome.

9.7 Conjugation, transformation, and transduction are the three processes of gene transfer in bacteria. For the transferred DNA to be stably inherited, all three processes require the transferred DNA to undergo recombination with the bacterial chromosome.

take place between the introduced genes and those of the bacterial chromosome.

3. **Transduction** takes place when bacterial viruses (bacteriophages) carry DNA from one bacterium to another (**Figure 9.7c**). Inside the bacterium, the newly introduced DNA may undergo recombination with the bacterial chromosome.

Not all bacterial species exhibit all three types of DNA transfer. Conjugation takes place more frequently in some species than in others. Transformation takes place to a limited extent in many species of bacteria, but laboratory techniques can increase the rate of DNA uptake. Most bacteriophages have a limited host range, so transduction normally takes place between bacteria of the same or closely related species only.

These processes of genetic exchange in bacteria differ from diploid eukaryotic sexual reproduction in two important ways. First, DNA exchange and reproduction are not coupled in bacteria; bacteria often undergo reproduction (binary fission) without receiving any DNA from another cell. Second, donated genetic material that is not recombined into the host DNA is usually degraded, so the recipient cell remains haploid. Each type of DNA transfer can be used to map genes.

CONCEPTS

DNA may be transferred between bacterial cells through conjugation, transformation, or transduction. Each type of DNA transfer consists of a one-way movement of genetic information to the recipient cell, sometimes followed by recombination. These processes are not connected to reproduction in bacteria.

✔ **CONCEPT CHECK 2**

Which process of DNA transfer in bacteria requires a virus?

a. Conjugation
b. Transduction
c. Transformation
d. All of the above

Conjugation

In 1946, Joshua Lederberg and Edward Tatum demonstrated that bacteria can transfer and recombine genetic information, paving the way for the use of bacteria in genetic studies. In the course of their research, Lederberg and Tatum studied auxotrophic strains of *E. coli*. The Y10 strain required the amino acids threonine (and was thus genotypically *thr⁻*) and leucine (*leu⁻*) and the vitamin thiamine (*thi⁻*) for growth, but did not require the vitamin biotin (*bio⁺*) or the amino acids phenylalanine (*phe⁺*) and cysteine (*cys⁺*); the genotype of this strain can be written as *thr⁻ leu⁻ thi⁻ bio⁺ phe⁺ cys⁺*. The Y24 strain had the opposite set of alleles: it required biotin, phenylalanine, and cysteine in its medium, but it did not require threonine, leucine, or thiamine; its genotype was *thr⁺ leu⁺ thi⁺ bio⁻ phe⁻ cys⁻*. In one experiment, Lederberg and Tatum mixed Y10 and Y24 bacteria together and plated them on minimal medium (**Figure 9.8**). Each strain was also plated separately on minimal medium.

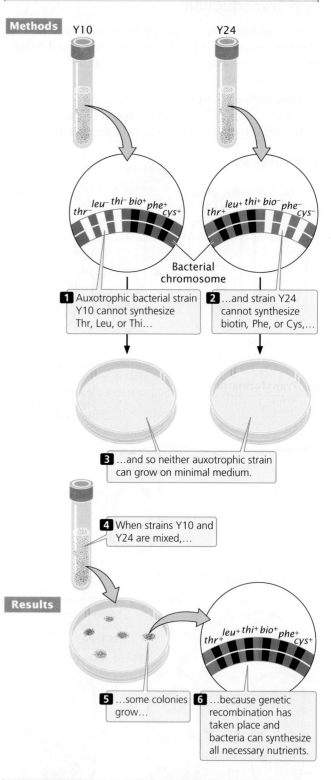

Experiment

Question: Do bacteria exchange genetic information?

Methods

Y10   Y24

thr⁻ leu⁻ thi⁻ bio⁺ phe⁺ cys⁺   *thr⁺ leu⁺ thi⁺ bio⁻ phe⁻ cys⁻*

Bacterial chromosome

1 Auxotrophic bacterial strain Y10 cannot synthesize Thr, Leu, or Thi…

2 …and strain Y24 cannot synthesize biotin, Phe, or Cys,…

3 …and so neither auxotrophic strain can grow on minimal medium.

4 When strains Y10 and Y24 are mixed,…

Results

5 …some colonies grow…

6 …because genetic recombination has taken place and bacteria can synthesize all necessary nutrients.

thr⁺ leu⁺ thi⁺ bio⁺ phe⁺ cys⁺

Conclusion: Yes, genetic exchange and recombination took place between the two mutant strains.

9.8 Lederberg and Tatum's experiment demonstrated that bacteria undergo genetic exchange.

Alone, neither Y10 nor Y24 grew on minimal medium: each strain required nutrients that were absent. Strain Y10 was unable to grow because it required threonine, leucine, and thiamine, which were absent in the minimal medium; strain Y24 was unable to grow because it required biotin, phenylalanine, and cysteine, which were also absent from the minimal medium. When Lederberg and Tatum mixed the two strains, however, a few colonies did grow on the minimal medium. These prototrophic bacteria must have had genotype thr^+ leu^+ thi^+ bio^+ phe^+ cys^+. Where had they come from?

If mutations were responsible for the prototrophic colonies, then some colonies should also have grown on the plates containing Y10 or Y24 alone, but no bacteria grew on those plates. Multiple simultaneous mutations ($thr^- \rightarrow thr^+$, $leu^- \rightarrow leu^+$, and $thi^- \rightarrow thi^+$ in strain Y10 or $bio^- \rightarrow bio^+$, $phe^- \rightarrow phe^+$, and $cys^- \rightarrow cys^+$ in strain Y24) would have been required for either strain to become prototrophic by mutation, which was very improbable. Lederberg and Tatum concluded that some type of genetic transfer and recombination had taken place:

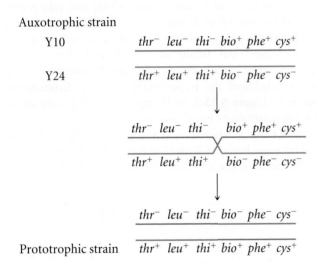

What they did not know was *how* it had taken place.

To study this problem, Bernard Davis constructed a U-shaped tube (**Figure 9.9**) that was divided into two compartments by a filter with fine pores. This filter allowed liquid medium to pass from one side of the tube to the other, but the pores of the filter were too small to allow the passage of bacteria. Two auxotrophic strains of bacteria were placed on opposite sides of the filter, and suction was applied alternately to the ends of the U-tube, causing the medium to flow back and forth between the two compartments. Despite hours of incubation in the U-tube, bacteria plated on minimal medium did not grow; there had been no genetic exchange between the strains. The exchange of bacterial genes clearly

required direct contact, or conjugation, between the bacterial cells.

F⁺ AND F⁻ CELLS In many bacterial species, conjugation depends on a plasmid or DNA element that is present in the donor cell and absent in the recipient cell. In *E. coli*, this plasmid is known as the fertility (F) factor. Cells that contain the F factor are referred to as F⁺, and cells lacking the F factor are F⁻.

The F factor is an episome that contains an origin of replication and a number of genes required for conjugation (see Figure 9.6). For example, some of the genes in this plasmid encode sex **pili** (singular, pilus), slender extensions of the cell membrane. A cell containing the F factor produces sex pili, one of which makes contact with a receptor on an F⁻ cell (**Figure 9.10**) and pulls the two cells together. DNA is then transferred from the F⁺ cell to the F⁻ cell. Conjugation can take place only between a cell that possesses the F factor and a cell that lacks the F factor.

Experiment

Question: How did the genetic exchange seen in Lederberg and Tatum's experiment take place?

Methods

Auxotrophic strain A   Auxotrophic strain B

← Airflow →

Strain A   Strain B

Two auxotrophic strains were separated by a filter that allowed mixing of medium but not bacteria.

No prototrophic bacteria were produced

Results

| Minimal medium | Minimal medium | Minimal medium | Minimal medium |

No growth   No growth   No growth   No growth

Conclusion: Genetic exchange requires direct contact between bacterial cells.

9.9 Davis's U-tube experiment.

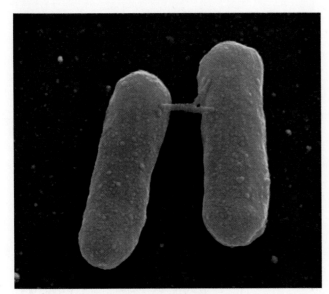

9.10 Sex pili connect F⁺ and F⁻ cells during bacterial conjugation. [Eye of Science/Science Source.]

In most cases, the only genes transferred during conjugation between an F⁺ and F⁻ cell are those on the F factor (**Figure 9.11a and b**). Transfer is initiated when one of the DNA strands on the F factor is nicked at an origin of transfer (*oriT*). One end of the nicked DNA separates from the circular F plasmid and passes into the recipient cell (**Figure 9.11c**). Replication takes place on the F factor, proceeding around the circular plasmid in the F⁺ cell and replacing the transferred strand. Because the F factor in the F⁺ donor cell is always nicked at the *oriT* site, this site always enters the recipient cell first, followed by the rest of the plasmid. Thus, the transfer of genetic material has a defined direction. Inside the recipient cell, the single strand replicates, producing a circular, double-stranded copy of the F plasmid (**Figure 9.11d and e**). If the entire F factor is transferred to the recipient F⁻ cell, that cell becomes an F⁺ cell.

Hfr CELLS Conjugation transfers genetic material in the F plasmid from F⁺ to F⁻ cells, but it does not account for the transfer of chromosomal genes observed by Lederberg and Tatum. In Hfr (high-frequency recombination) bacterial strains, the F factor is integrated into the bacterial chromosome (**Figure 9.12**). Hfr cells behave like F⁺ cells, forming sex pili and undergoing conjugation with F⁻ cells.

In conjugation between Hfr and F⁻ cells (**Figure 9.13a**), the integrated F factor is nicked, and the end of the nicked strand moves into the F⁻ cell (**Figure 9.13b**), just as it does in conjugation between F⁺ and F⁻ cells. But because, in an Hfr cell, the F factor has been integrated into the bacterial chromosome, the chromosome follows the F factor into the recipient cell. How much of the bacterial chromosome is transferred depends on the length of time that the two cells remain in conjugation.

Inside the recipient cell, the donor DNA strand replicates (**Figure 9.13c**), and crossing over between it and the original chromosome of the F⁻ cell (**Figure 9.13d**) may take place. This chromosomal gene transfer between Hfr and F⁻ cells explains how the recombinant prototrophic cells observed by Lederberg and Tatum were produced. After crossing over has taken place in the recipient cell, the donated strand is degraded and the recombinant recipient chromosome remains (**Figure 9.13e**), to be replicated and passed on to later generations by binary fission.

In a mating between an Hfr and an F⁻ cell, the F⁻ cell almost never becomes F⁺ or Hfr because the F factor is nicked in the middle at the initiation of strand transfer, which places

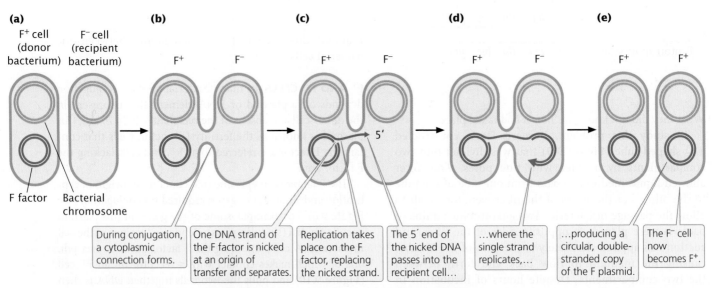

(a)
F⁺ cell (donor bacterium)   F⁻ cell (recipient bacterium)

F factor   Bacterial chromosome

(b) F⁺   F⁻

(c) F⁺   F⁻
5′

(d) F⁺   F⁻

(e) F⁺   F⁺

During conjugation, a cytoplasmic connection forms.

One DNA strand of the F factor is nicked at an origin of transfer and separates.

Replication takes place on the F factor, replacing the nicked strand.

The 5′ end of the nicked DNA passes into the recipient cell...

...where the single strand replicates,...

...producing a circular, double-stranded copy of the F plasmid.

The F⁻ cell now becomes F⁺.

9.11 The F factor is transferred during conjugation between an F⁺ and an F⁻ cell.

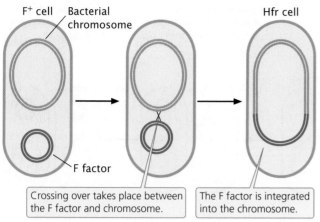

Crossing over takes place between the F factor and chromosome.

The F factor is integrated into the chromosome.

9.12 The F factor is integrated into the bacterial chromosome in an Hfr cell.

part of the F factor at the beginning and part at the end of the strand that is transferred. To become F$^+$ or Hfr, the recipient cell must receive the entire F factor, which requires that the entire donor chromosome be transferred. This event happens rarely because most conjugating cells break apart before the entire chromosome has been transferred.

The F plasmid in an F$^+$ cell integrates into the bacterial chromosome, causing the F$^+$ cell to become Hfr, at a frequency of only about 1 in 10,000. This low frequency accounts for the low rate of recombination observed by Lederberg and Tatum in their F$^+$ cells. The F factor is excised from the bacterial chromosome at a similarly low rate, causing a few Hfr cells to become F$^+$.

F′ CELLS When an F factor is excised from the bacterial chromosome, a small amount of the bacterial chromosome may be removed with it, and these chromosomal genes will then be carried with the F plasmid (**Figure 9.14**). Cells containing an

F plasmid with some bacterial genes are called F prime (F′) cells. For example, if an F factor integrates into a chromosome at a position adjacent to the *lac* genes (genes that enable a cell to metabolize the sugar lactose), the F factor may pick up *lac* genes when it is excised, becoming F′*lac*. F′ cells can conjugate with F$^-$ cells because F′ cells possess the F plasmid, with all the genetic information necessary for conjugation and DNA transfer. Characteristics of different mating types of *E. coli* (cells with different types of F) are summarized in **Table 9.2**.

During conjugation between an F′ cell and an F$^-$ cell, the F plasmid is transferred to the F$^-$ cell, which means that any genes on the F plasmid, including those from the bacterial chromosome, may be transferred to the F$^-$ recipient cell (see Figure 9.14). This process produces partial diploids, or merozygotes, which are cells with two copies of some genes, one on the bacterial chromosome and one on the newly introduced F plasmid. The outcomes of conjugation between different mating types of *E. coli* are summarized in **Table 9.3**.

CONCEPTS

Conjugation in *E. coli* is controlled by an episome called the F factor. Cells containing the F factor (F$^+$ cells) are donors of DNA; cells lacking the F factor (F$^-$ cells) are recipients. In Hfr cells, the F factor is integrated into the bacterial chromosome; these cells donate DNA to F$^-$ cells at a high frequency. F′ cells contain a copy of the F plasmid with some bacterial genes.

✔ **CONCEPT CHECK 3**

Conjugation between an F$^+$ and an F$^-$ cell usually results in

a. two F$^+$ cells.
b. two F$^-$ cells.
c. an F$^+$ and an F$^-$ cell.
d. an Hfr cell and an F$^+$ cell.

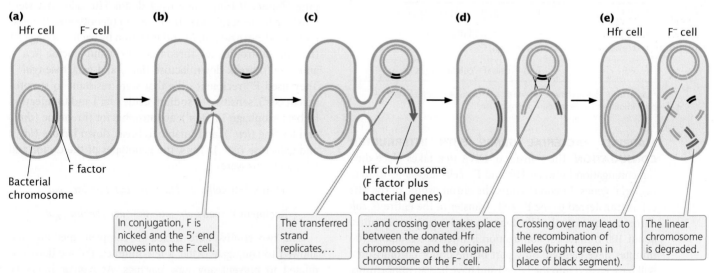

In conjugation, F is nicked and the 5′ end moves into the F$^-$ cell.

The transferred strand replicates,...

...and crossing over takes place between the donated Hfr chromosome and the original chromosome of the F$^-$ cell.

Crossing over may lead to the recombination of alleles (bright green in place of black segment).

The linear chromosome is degraded.

Hfr chromosome (F factor plus bacterial genes)

9.13 Bacterial genes may be transferred from an Hfr cell to an F$^-$ cell in conjugation.

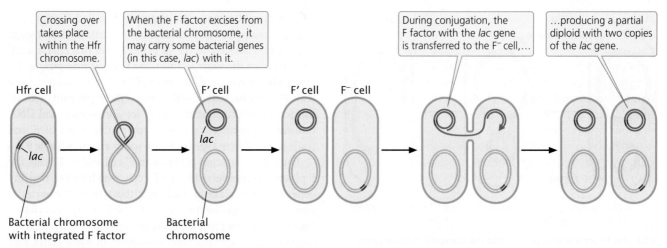

9.14 An Hfr cell may be converted into an F′ cell when the F factor is excised from the bacterial chromosome and carries bacterial genes with it. Conjugation between an F′ cell and an F⁻ cell produces a partial diploid.

TABLE 9.2	Characteristics of *E. coli* cells with different types of F factor	
Type	**F Factor Characteristics**	**Role in Conjugation**
F⁺	Present as separate circular plasmid	Donor
F⁻	Absent	Recipient
Hfr	Present, integrated into bacterial chromosome	High-frequency donor
F′	Present as separate circular plasmid, carrying some bacterial genes	Donor

TABLE 9.3	Results of conjugation between cells with different F factors
Conjugating Cells	**Cell Types Present after Conjugation**
F⁺ × F⁻	Two F⁺ cells (F⁻ cell becomes F⁺)
Hfr × F⁻	One Hfr cell and one F⁻ cell (no change)*
F′ × F⁻	Two F′ cells (F⁻ cell becomes F′)

*Rarely, the F⁻ cell becomes F⁺ in an Hfr × F⁻ conjugation if the entire chromosome is transferred during conjugation.

MAPPING BACTERIAL GENES WITH INTERRUPTED CONJUGATION The transfer of DNA that takes place during conjugation between Hfr and F⁻ cells allows us to map bacterial genes. In conjugation, the chromosome of the Hfr cell is transferred to the F⁻ cell. Transfer of the entire *E. coli* chromosome from the Hfr donor to the F⁻ recipient requires about 100 minutes; if conjugation is interrupted before 100 minutes have elapsed, only part of the donor chromosome will have passed into the F⁻ cell and have had an opportunity to recombine with the recipient chromosome.

Chromosome transfer always begins within the integrated F factor and proceeds in a defined direction, so genes are transferred according to their sequence on the chromosome. The times required for individual genes to be transferred indicate their relative positions on the chromosome. In most genetic maps, distances are expressed as recombination frequencies; however, in bacterial gene maps constructed with interrupted conjugation, the basic unit of distance is a minute. View **Animation 9.1** to see how genes are mapped using interrupted conjugation.

WORKED PROBLEM

To illustrate the method of mapping genes with interrupted conjugation, let's look at a cross analyzed by François Jacob and Elie Wollman, who developed this method of gene mapping (**Figure 9.15a**). They used donor Hfr cells that were sensitive to the antibiotic streptomycin (genotype *str*ˢ), resistant to sodium azide (*azi*ʳ) and infection by bacteriophage T1 (*ton*ʳ), prototrophic for threonine (*thr*⁺) and leucine (*leu*⁺), and able to break down lactose (*lac*⁺) and galactose (*gal*⁺). They used F⁻ recipient cells that were resistant to streptomycin (*str*ʳ), sensitive to sodium azide (*azi*ˢ) and to infection by bacteriophage T1 (*ton*ˢ), auxotrophic for threonine (*thr*⁻) and leucine (*leu*⁻), and unable to break down lactose (*lac*⁻) and galactose (*gal*⁻). Thus, the genotypes of the donor and recipient cells were

Donor Hfr cells: *str*ˢ *leu*⁺ *thr*⁺ *azi*ʳ *ton*ʳ *lac*⁺ *gal*⁺

Recipient F⁻ cells: *str*ʳ *leu*⁻ *thr*⁻ *azi*ˢ *ton*ˢ *lac*⁻ *gal*⁻

The two strains were mixed in complete medium and allowed to conjugate. After a few minutes, the medium was diluted to prevent any new pairings. At regular intervals, a sample of cells was removed and agitated vigorously in a kitchen blender to halt all conjugation and DNA transfer. The cells from each sample were plated on a selective medium that contained streptomycin and lacked leucine and threonine. The Hfr donor cells were streptomycin sensitive (str^s) and would not grow on this medium. The F⁻ recipient cells were auxotrophic for leucine and threonine, and they also failed to grow on this medium. Only recipient cells that had undergone conjugation and received at least the leu^+ and thr^+ genes from the Hfr donors could grow on this medium. All of these str^r leu^+ thr^+ cells were then tested for the presence of other genes that might have been transferred from the donor Hfr strain.

Because Jacob and Wollman used streptomycin to kill all the donor cells, they were not able to examine the transfer of the str^s gene. All the cells that grew on the selective medium were leu^+ thr^+, so we know that those genes were transferred. In **Figure 9.15b**, the percentages of str^r leu^+ thr^+ cells receiving specific alleles (azi^r, ton^r, lac^+, and gal^+) from the Hfr donor cells are plotted against the duration of conjugation. What is the order in which the genes are transferred, and what are the distances among them?

Solution Strategy

What information is required in your answer to the problem?

The order of the genes on the bacterial chromosome and the distances between them.

What information is provided to solve the problem?

- The donor cells were str^s leu^+ thr^+ azi^r ton^r lac^+ gal^+ and the recipient cells were str^r leu^- thr^- azi^s ton^s lac^- gal^-.

- The percentages of recipient cells with different traits that appeared at various times after the start of conjugation (see Figure 9.15b).

Solution Steps

The first donor gene to appear in the recipient cells (at about 9 minutes) was azi^r. Gene ton^r appeared next (after about 10 minutes), followed by lac^+ (at about 18 minutes) and by gal^+ (after 25 minutes). These transfer times indicate the order of gene transfer and the relative distances among the genes.

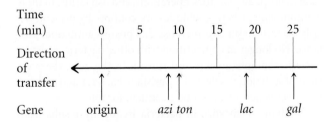

Notice that the frequency of gene transfer from donor to recipient cells decreased with distance from the origin of transfer. For example, about 90% of the recipients received the azi^r allele, but only about 30% received the gal^+ allele. The lower percentage for gal^+ is due to the fact that some

Question: How can interrupted conjugation be used to map bacterial genes?

Methods

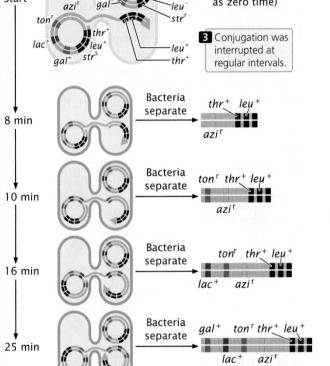

1 An Hfr cell with genotype str^s thr^+ leu^+ azi^r ton^r lac^+ gal^+...

2 ...was mated with an F⁻ cell with genotype str^r thr^- leu^- azi^s ton^s lac^- gal^-.

Genes transferred: leu^+ and thr^+ (first selected genes, defined as zero time)

3 Conjugation was interrupted at regular intervals.

Results

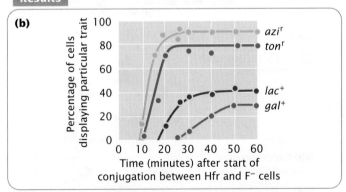

Conclusion: The transfer times indicate the order and relative distances between genes and can be used to construct a genetic map.

9.15 Jacob and Wollman used interrupted conjugation to map bacterial genes.

conjugating cells spontaneously broke apart before they were disrupted by the blender. The probability of spontaneous disruption increases with time, so fewer cells had an opportunity to receive genes that were transferred later.

> » For additional practice mapping bacterial genes with interrupted conjugation, try **Problem 22** at the end of the chapter.

DIRECTIONAL TRANSFER AND MAPPING Different Hfr strains of a given species of bacteria have the F factor integrated into the bacterial chromosome at different sites and in different orientations. Gene transfer always begins within the F factor, and the orientation and position of the F factor determine the direction and starting point of gene transfer. **Figure 9.16a** shows that, in strain Hfr1, the F factor is

integrated between *leu* and *azi*; the orientation of the F factor at this site dictates that gene transfer will proceed in a counterclockwise direction around the circular chromosome. Genes from this strain will be transferred in the following order:

$$\leftarrow leu–thr–thi–his–gal–lac–pro–azi$$

In strain Hfr5, the F factor is integrated between the *thi* and the *his* genes (**Figure 9.16b**) and in the opposite orientation. Here, gene transfer will proceed in a clockwise direction:

$$\leftarrow thi–thr–leu–azi–pro–lac–gal–his$$

Although the starting point and direction of transfer may differ between two strains, the relative distance in minutes between any two genes is constant.

THINK-PAIR-SHARE Question 3

CONCEPTS

Conjugation can be used to map bacterial genes by mixing Hfr and F⁻ cells of different genotypes and interrupting conjugation at regular intervals. The amounts of time required for individual genes to be transferred from the Hfr to the F⁻ cells indicate the relative positions of the genes on the bacterial chromosome.

✔ CONCEPT CHECK 4

Interrupted conjugation was used to map three genes in *E. coli*. The donor genes first appeared in the recipient cells at the following times: *gal*, 10 minutes; *his*, 8 minutes; *pro*, 15 minutes. Which gene is in the middle?

(a)
Hfr1

1 Transfer always begins within F, and the orientation of F determines the direction of transfer.

2 In Hfr1, F is integrated between the *leu* gene and the *azi* gene;....

thr
thi   leu
F factor
his   azi
Chromosome
gal   pro
lac

3 ...so the genes are transferred beginning with *leu*.

leu thr thi his gal lac pro azi
Genetic map

(b)
Hfr5

4 In Hfr5, F is integrated between *thi* and *his*.

thr
thi   leu
F factor
his   azi
Chromosome
gal   pro
lac

5 F has the opposite orientation in this chromosome; so the genes are transferred beginning with *thi*.

thi thr leu azi pro lac gal his
Genetic map

9.16 The orientation of the F factor in an Hfr strain determines the direction of gene transfer. Arrowheads indicate the origin and direction of transfer.

Natural Gene Transfer and Antibiotic Resistance

Antibiotics are substances that kill bacteria. Their development and widespread use has greatly reduced the threat of infectious disease and saved countless lives. But many pathogenic bacteria have developed resistance to antibiotics, particularly in environments where antibiotics are routinely used, such as hospitals, livestock operations, and fish farms. In these environments, where antibiotics are continually present, the only bacteria to survive are those that possess antibiotic resistance. No longer in competition with other bacteria, resistant bacteria multiply quickly and spread. In this way, the presence of antibiotics selects for resistant bacteria and reduces the effectiveness of antibiotic treatment for infections.

Antibiotic resistance in bacteria frequently results from the action of genes located on *R plasmids*, small circular plasmids that can be transferred by conjugation. Some drug-resistant R plasmids convey resistance to several antibiotics simultaneously. Ironic but plausible sources of some of the resistance genes found in R plasmids are the microbes that produce antibiotics in the first place.

THINK-PAIR-SHARE Question 4

Transformation in Bacteria

A second way in which DNA can be transferred between bacteria is through transformation (see Figure 9.7b). Transformation played an important role in the initial identification of DNA as the genetic material, as we will see in Chapter 10.

Transformation requires both the uptake of DNA from the surrounding medium and its incorporation into a bacterial chromosome or a plasmid. It may occur naturally when dead bacteria break down and release DNA fragments into the environment. In soil and marine environments, transformation may be an important route of genetic exchange for some bacteria. Transformation is also an important technique for transferring genes to bacteria in the laboratory.

MECHANISM OF TRANSFORMATION Cells that can take up DNA through their cell membranes are said to be **competent**. Some species of bacteria take up DNA more easily than others: competence is influenced by growth stage, the concentration of available DNA in the environment, and other environmental factors. The DNA that a competent cell takes up need not be bacterial: virtually any type of DNA (bacterial or otherwise) can be taken up by competent cells under the appropriate conditions.

As a DNA fragment enters the cell in the course of transformation (**Figure 9.17**), one of the strands is broken up, whereas the other strand moves across the membrane and may pair with a homologous region and become integrated into the bacterial chromosome. Its integration into the recipient chromosome requires two crossover events, after which the remaining single-stranded DNA is degraded by bacterial enzymes. In some bacteria, double-stranded DNA moves across the cell membrane and is integrated into the bacterial chromosome. Cells that receive genetic material through transformation are called **transformants**.

Bacterial geneticists have developed techniques for increasing the frequency of transformation in the laboratory in order to introduce particular DNA fragments or whole plasmids into cells. They have also developed strains of bacteria that are more competent than wild-type cells.

Treatment with calcium chloride, heat shock, or an electrical field makes bacterial membranes more porous and permeable to DNA. The efficiency of transformation can also be increased by using high concentrations of DNA. These techniques enable researchers to transform bacteria such as *E. coli*, which are not naturally competent.

GENE MAPPING WITH TRANSFORMATION Transformation, like conjugation, is used to map bacterial genes, especially in those species that do not undergo conjugation or transduction. Transformation mapping requires two strains of bacteria that differ in several genetic traits; for example, the recipient strain might be $a^- \, b^- \, c^-$ (auxotrophic for three nutrients), and the donor strain might be $a^+ \, b^+ \, c^+$ (prototrophic for the same three nutrients) (**Figure 9.18**). DNA from the donor strain is isolated, purified, and fragmented. The recipient strain is treated to increase its competence, and DNA from the donor strain is added to the medium. Fragments of the donor DNA enter the recipient cells and undergo recombination with homologous DNA sequences on the bacterial chromosome.

We can map bacterial genes by observing the rate at which two or more genes are transferred to the recipient chromosome together, or **cotransformed**. When the donor DNA is fragmented before transformation, genes that are physically closer together on the bacterial chromosome are more likely to be present on the same DNA fragment and transferred together, as shown for genes a^+ and b^+ in Figure 9.18. Genes that are far apart are unlikely to be present on the same DNA fragment and are rarely cotransformed. Inside the cell, DNA becomes incorporated into the recipient chromosome through recombination. If two genes are close together on the same fragment, any two crossovers are likely to take place on either side of the two genes, allowing both to become part of the recipient chromosome. If the two genes are far apart, there may be one crossover between them, allowing one gene, but not the other, to recombine with the recipient chromosome. Thus, two genes are more likely to be incorporated into the recipient chromosome together when they are close together on the donor chromosome, and genes located far apart are rarely cotransformed. Therefore, the frequency

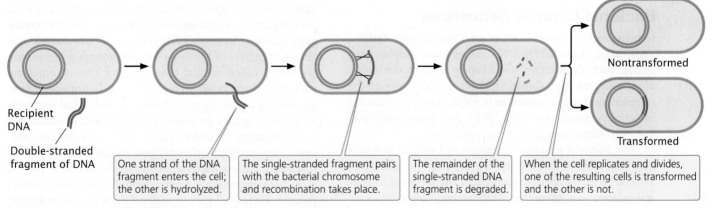

Recipient DNA

Double-stranded fragment of DNA

| One strand of the DNA fragment enters the cell; the other is hydrolyzed. | The single-stranded fragment pairs with the bacterial chromosome and recombination takes place. | The remainder of the single-stranded DNA fragment is degraded. | When the cell replicates and divides, one of the resulting cells is transformed and the other is not. |

Nontransformed

Transformed

9.17 Genes can be transferred between bacteria through transformation.

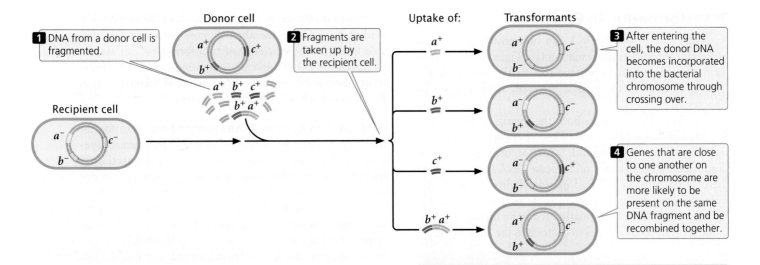

9.18 Transformation can be used to map bacterial genes.

of cotransformation can be used to map bacterial genes. If genes *a* and *b* as well as genes *b* and *c* are frequently cotransformed, but genes *a* and *c* are rarely cotransformed, then gene *b* must be between *a* and *c*—the gene order is *a*, *b*, *c*. ▶ TRY PROBLEM 24

CONCEPTS

Genes can be mapped in bacteria by taking advantage of transformation—the ability of bacteria to take up DNA from the environment and incorporate it into their chromosomes through crossing over. The relative rates at which pairs of genes are cotransformed indicates the distance between them: the higher the rate of cotransformation, the closer the genes are on the bacterial chromosome.

✔ CONCEPT CHECK 5

A bacterial strain with genotype *his⁻ leu⁻ thr⁻* is transformed with DNA from a strain that is *his⁺ leu⁺ thr⁺*. A few *leu⁺ thr⁺* cells and a few *his⁺ thr⁺* cells are found, but no *his⁺ leu⁺* cells are observed. Which genes are farthest apart?

Bacterial Genome Sequences

Genetic maps serve as the foundation for the more detailed information provided by DNA sequencing, such as gene content and organization (see Chapter 19 for a discussion of gene sequencing). Geneticists have now determined the complete nucleotide sequences of more than two thousand bacterial genomes (see Table 20.1), and many additional microbial sequencing projects are under way.

Most bacterial genomes contain from 1 million to 4 million base pairs of DNA, but a few are much smaller (e.g., 580,000 bp in *Mycoplasma genitalium*) and some are considerably larger (e.g., more than 7 million bp in *Mesorhizobium loti*). The small size of bacterial genomes (relative to those

found in multicellular eukaryotes, which often have billions of base pairs of DNA) is thought to be an adaptation for rapid cell division, because the rate of cell division is limited by the time required to replicate the DNA. On the other hand, the lack of mobility in most bacteria requires metabolic and environmental flexibility, so genome size and content are likely to reflect a balance between the opposing evolutionary forces of gene loss to maintain rapid reproduction and gene acquisition to ensure flexibility.

The functions of a substantial proportion of genes in all bacteria have not been determined. Certain genes, particularly those with related functions, tend to reside next to one another, but these clusters are in very different locations in different species, suggesting that bacterial genomes are constantly being reshuffled. Comparisons of the gene sequences of pathogenic and benign bacteria are helping to identify genes implicated in disease and may suggest new targets for antibiotics and other antimicrobial agents.

Horizontal Gene Transfer

The availability of bacterial genome sequences has provided evidence that many bacteria have acquired genetic information from other species of bacteria—and sometimes even from eukaryotic organisms—in a process called **horizontal gene transfer**. In most eukaryotes, genes are passed only among members of the same species through reproduction (a process called vertical transmission); that is, genes are passed from one generation to the next. In horizontal gene transfer, genes can be passed between individual members of different species by nonreproductive mechanisms, such as conjugation, transformation, and transduction. Evidence suggests that horizontal gene transfer has taken place repeatedly among bacteria. For example, as much as 17% of *E. coli*'s genome has been acquired from other bacteria through horizontal gene transfer. Of medical significance, some pathogenic bacteria have acquired the genes necessary for infection, whereas others have acquired genes that confer resistance to antibiotics.

Because of the widespread occurrence of horizontal gene transfer, many bacterial chromosomes are a mixture of genes inherited through vertical transmission and genes acquired through horizontal gene transfer. This observation has caused some biologists to question whether the species concept is appropriate for bacteria. A *species* is often defined as a group of organisms that are reproductively isolated from other groups, have a set of genes in common, and evolve together (see Chapter 26). Because of horizontal gene transfer, the genes of one bacterial species are not isolated from the genes of other species, making the traditional species concept difficult to apply. Horizontal gene transfer also muddies the determination of the ancestral relationships among bacteria. The reconstruction of ancestral relationships is usually based on genetic similarities and differences: organisms that are genetically similar are assumed to have descended from a recent common ancestor, whereas organisms that are genetically distinct are assumed to be more distantly related. Through horizontal gene transfer, however, even distantly related bacteria may share some genes and thus appear to have descended from a recent common ancestor. The nature of species and how to classify bacteria are currently controversial topics within the field of microbiology.

Bacterial Defense Mechanisms

Bacteria and archaea are found in almost every conceivable environment, often thriving where other organisms perish. These organisms face significant challenges, however, the most serious of which are viruses. Indeed, scientists have estimated that there are ten times as many viruses as bacteria and archaea, so viruses are a constant threat. Not surprisingly, bacteria and archaea have evolved a number of defense mechanisms to prevent the entry and reproduction of viruses and other invading DNA.

To invade a cell, a virus must first attach to the cell wall or membrane and inject its DNA or RNA into the cell. Many bacteria and archaea have evolved mechanisms to prevent viral attachment and injection of the viral DNA or RNA into the cell. Some bacteria turn off the expression of receptors to which viruses attach; other bacteria secrete polysaccharide coats that limit the access of viruses to those receptors. Bacteria can also produce proteins on the cell membrane that prevent viruses from injecting their nucleic acids into the cell. In other cases, a bacterium that has been infected self-destructs, so that the virus within it is unable to replicate and spread to other bacteria. Other defense mechanisms block virus replication.

Another type of bacterial defense consists of restriction-modification systems. Bacteria produce enzymes called restriction endonucleases that recognize and cleave double-stranded DNA at specific nucleotide sequences (see Chapter 19). These enzymes cut up viral DNA, destroying the viral genome before it can replicate. Bacteria protect their own DNA from cleavage by modifying the recognition sequences on their DNA, usually by the addition of methyl groups. Restriction enzymes have been extensively used by geneticists in genetic engineering (see Chapter 19).

Another recently discovered defense mechanism in bacteria and archaea is CRISPR-Cas systems (see p. 420 in Chapter 14). These systems are analogous to the immune systems of vertebrates in that they recognize and remember DNA from specific pathogens. After DNA from a virus, plasmid, or other element has invaded a bacterial cell, bacterial proteins cut up the foreign DNA and incorporate short pieces of it into the bacterial chromosome. The foreign pieces are inserted as spacers into sequences called clustered regularly interspaced short palindromic repeats (CRISPRs). The CRISPR sequences are later transcribed and processed into small RNA molecules called CRISPR RNA (crRNA), which form complexes with proteins and sometimes with other RNA molecules. These complexes then recognize and cleave foreign DNA with sequences that are complementary to the crRNAs, destroying the foreign DNA. Because the bacterial cell carries the inserted pieces of foreign DNA in its own chromosome, it can "remember" previously encountered pathogens and quickly destroy their DNA.

CRISPR-Cas systems are diverse, with a number of different types that vary in the molecules involved and the details of their mechanisms. They are widespread in bacteria and archaea: about 85% of all archaea and close to 50% of all bacteria use CRISPR-Cas systems as defenses against pathogens. Geneticists are currently using these systems as powerful tools for genetic engineering (see Chapter 19).

THINK-PAIR-SHARE Question 5

9.4 Viruses Are Simple Replicating Systems Amenable to Genetic Analysis

All organisms—plants, animals, fungi, and bacteria—are infected by viruses. A **virus** is a simple replicating structure made up of nucleic acid surrounded by a protein coat (see Figure 2.4). Viruses come in a great variety of shapes and sizes (**Figure 9.19**). Enveloped viruses have an outer lipid envelope that is derived from the host's cell membrane. Some viruses have DNA as their genetic material, whereas others have RNA; the nucleic acid may be double stranded or single stranded, linear or circular.

Viruses that infect bacteria (bacteriophages, or phages for short) have played a central role in genetic research since the late 1940s. They are ideal for many types of genetic research because they have small and easily manageable genomes, reproduce rapidly, and produce large numbers of progeny. Bacteriophages have two alternative life cycles: the *lytic* cycle and the *lysogenic* cycle. In the lytic cycle, a phage attaches to a receptor on the bacterial cell wall and injects its DNA into the cell (**Figure 9.20**). Inside the host cell, the host DNA is degraded, and phage DNA is replicated, transcribed, and translated, producing more phage DNA and phage proteins. New phage particles are assembled from these components.

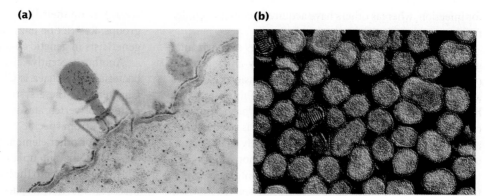

9.19 Viruses have a variety of structures and sizes.
(a) Bacteriophage T4 (bright orange).
(b) Influenza A virus (green structures).
[Part a: Biozentrum, University of Basel/ Science Source. Part b: Eye of Science/ Science Source.]

The phages then produce an enzyme that breaks open the host cell, releasing the new phages. **Virulent phages** reproduce strictly through the lytic cycle and always kill their host cells.

Temperate phages can undergo either the lytic or the lysogenic cycle. The lysogenic cycle begins like the lytic cycle (see Figure 9.20), but inside the cell, the phage DNA integrates into the bacterial chromosome, where it remains as an inactive **prophage**. The prophage is replicated along with the bacterial DNA and is passed on when the bacterium divides. Certain stimuli, such as ultraviolet light and some chemicals, can cause the prophage to dissociate from the bacterial chromosome and enter the lytic cycle, producing new phage particles and lysing the cell.

THINK-PAIR-SHARE Question 6 👥

Techniques for the Study of Bacteriophages

Viruses reproduce only within host cells, so bacteriophages must be cultured in bacterial cells. Phages can be grown in large liquid cultures of bacteria to generate large numbers of offspring, but to study the characteristics of individual phages, we must isolate them on petri plates. To do so, we can mix phages and bacteria together and plate them on solid medium. A high concentration of bacteria is used so that the colonies will grow into one another and produce a continuous layer of bacteria, or *lawn*, on the agar. As the bacteria grow, an individual phage infects a single bacterial cell and goes through its lytic cycle. Many new phages are released from the lysed cell and infect additional cells, and the cycle is

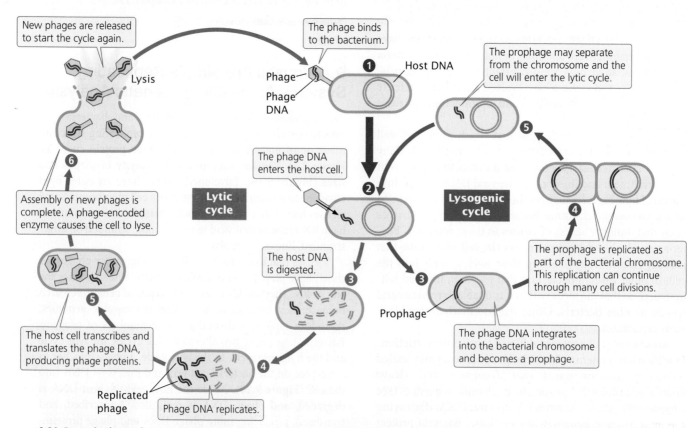

9.20 Bacteriophages have two alternative life cycles: lytic and lysogenic.

repeated. Because the bacteria are growing on solid medium, the diffusion of the phages is restricted, so only nearby cells are infected. After several rounds of phage reproduction, a clear patch of lysed cells, called a **plaque**, appears on the plate (**Figure 9.21**). Each plaque represents a single phage that multiplied and lysed many cells. Plating of a known volume of a dilute solution of phages on a bacterial lawn and counting the number of plaques that appear is a way to determine the original concentration of phages in the solution.

CONCEPTS

Viral genomes may be DNA or RNA, circular or linear, and double or single stranded. Bacteriophages are used in many types of genetic research.

✔ CONCEPT CHECK 6

In which bacteriophage life cycle does the phage DNA become incorporated into the bacterial chromosome?

a. Lytic

b. Lysogenic

c. Both lytic and lysogenic

d. Neither lytic nor lysogenic

Transduction: Using Phages To Map Bacterial Genes

In our discussion of bacterial genetics in Section 9.3, three mechanisms of gene transfer were identified: conjugation, transformation, and transduction (see Figure 9.7). Let's take a closer look at transduction, in which genes are transferred between bacteria by viruses. In **generalized transduction**, any gene may be transferred. In **specialized transduction**, only a few genes are transferred.

GENERALIZED TRANSDUCTION Joshua Lederberg and Norton Zinder discovered generalized transduction in 1952 while trying to produce recombination in the bacterium *Salmonella typhimurium* by conjugation. They mixed a strain of *S. typhimurium* that was $phe^+ trp^+ tyr^+ met^- his^-$ with a strain that was $phe^- trp^- tyr^- met^+ his^+$ (**Figure 9.22**) and

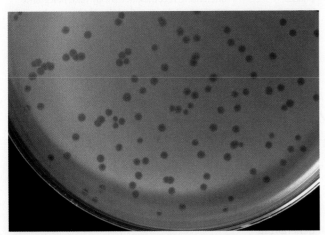

9.21 Plaques are clear patches of lysed cells on a lawn of bacteria. [© Carolina Biological Supply Company/Phototake.]

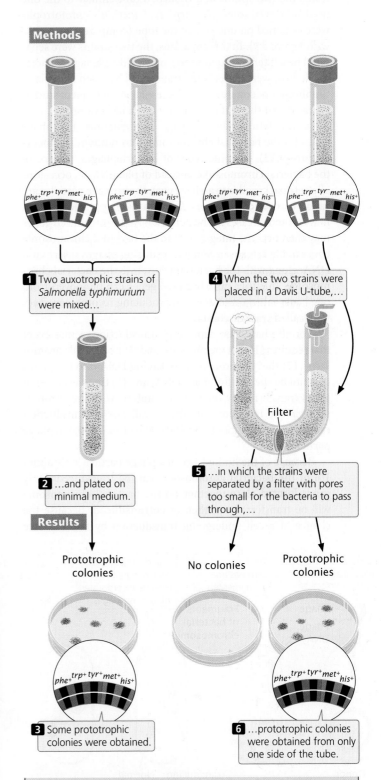

Experiment

Question: Does genetic exchange between bacteria always require cell-to-cell contact?

Methods

$phe^+ trp^+ tyr^+ met^- his^-$ $phe^- trp^- tyr^- met^+ his^+$ $phe^+ trp^+ tyr^+ met^- his^-$ $phe^- trp^- tyr^- met^+ his^+$

1 Two auxotrophic strains of *Salmonella typhimurium* were mixed...

4 When the two strains were placed in a Davis U-tube,...

Filter

2 ...and plated on minimal medium.

5 ...in which the strains were separated by a filter with pores too small for the bacteria to pass through,...

Results

Prototrophic colonies

No colonies

Prototrophic colonies

$phe^+ trp^+ tyr^+ met^+ his^+$

$phe^+ trp^+ tyr^+ met^+ his^+$

3 Some prototrophic colonies were obtained.

6 ...prototrophic colonies were obtained from only one side of the tube.

Conclusion: Genetic exchange did not take place through conjugation. A phage was later shown to be the agent of transfer.

9.22 The Lederberg and Zinder experiment.

plated the mixture on minimal medium. A few prototrophic recombinants (*phe⁺ trp⁺ tyr⁺ met⁺ his⁺*) appeared, suggesting that conjugation had taken place. However, when they tested the two strains in a U-shaped tube similar to the one used by Davis, some *phe⁺ trp⁺ tyr⁺ met⁺ his⁺* prototrophs were obtained on one side of the tube (compare Figure 9.22 with Figure 9.9). In this apparatus, the two strains were separated by a filter with pores too small for the passage of bacteria, so how were genes being transferred between bacteria in the absence of conjugation? The results of subsequent studies revealed that the agent of transfer was a bacteriophage.

In the lytic cycle of phage reproduction, the phage degrades the bacterial chromosome into random fragments (**Figure 9.23**). In some types of bacteriophages, a piece of the bacterial chromosome, instead of phage DNA, occasionally gets packaged into a phage coat; these phage particles are called **transducing phages**. If a transducing phage infects a new cell and releases the bacterial DNA, the introduced genes may then become integrated into the bacterial chromosome by a double crossover. Some transducing phages insert viral DNA, along with the bacterial DNA, into the bacterial chromosome. In either case, bacterial genes can be moved from one bacterial strain to another, producing recombinant bacteria called **transductants**.

Not all phages are capable of transduction, a rare event that requires (1) that the phage degrade the bacterial chromosome, (2) that the process of packaging DNA into the phage coat not be specific for phage DNA, and (3) that the bacterial genes transferred by the virus recombine with the chromosome in the recipient cell. The overall rate of transduction ranges from only about 1 in 100,000 to 1 in 1,000,000 phage particles.

Because of the limited size of a phage particle, only about 1% of the bacterial chromosome can be transduced. Only genes located close together on the bacterial chromosome will be transferred together, or **cotransduced**. Because the chance of a cell undergoing transduction by two separate

phages is exceedingly small, we can assume that any cotransduced genes are located close together on the bacterial chromosome. Thus, rates of cotransduction, like rates of cotransformation, give us an indication of the physical distances between genes on a bacterial chromosome.

To map genes by using transduction, two bacterial strains with different alleles at several loci are used. The donor strain is infected with phages (**Figure 9.24**), which reproduce within the cells. When the phages have lysed the donor cells, a suspension of the lysate, containing progeny phages, is mixed with the recipient strain, and the mixture is then plated on several different kinds of media to determine the phenotypes of the transducing progeny phages.

> **TRY PROBLEM 31**

CONCEPTS

In transduction, bacterial genes are packaged into a phage coat, transferred to another bacterium by the virus, and incorporated into the recipient bacterial chromosome by crossing over. Bacterial genes can be mapped with the use of generalized transduction.

✔ CONCEPT CHECK 7

In gene mapping using generalized transduction, bacterial genes that are cotransduced are

a. far apart on the bacterial chromosome.
b. on different bacterial chromosomes.
c. close together on the bacterial chromosome.
d. on a plasmid.

SPECIALIZED TRANSDUCTION Like generalized transduction, specialized transduction involves gene transfer from one bacterium to another by a phage, but here, only genes near particular sites on the bacterial chromosome are transferred. This process requires temperate bacteriophages

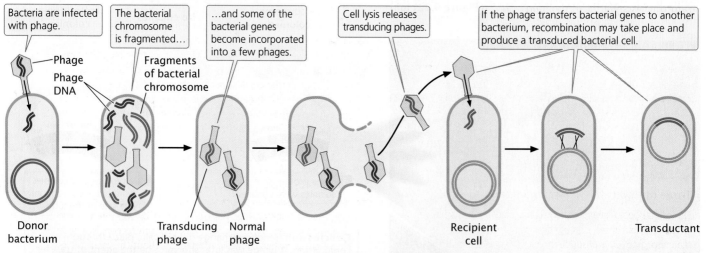

| Bacteria are infected with phage. | The bacterial chromosome is fragmented… | …and some of the bacterial genes become incorporated into a few phages. | Cell lysis releases transducing phages. | If the phage transfers bacterial genes to another bacterium, recombination may take place and produce a transduced bacterial cell. |

Phage
Phage DNA
Fragments of bacterial chromosome

Donor bacterium · Transducing phage · Normal phage · Recipient cell · Transductant

9.23 Genes can be transferred from one bacterium to another through generalized transduction.

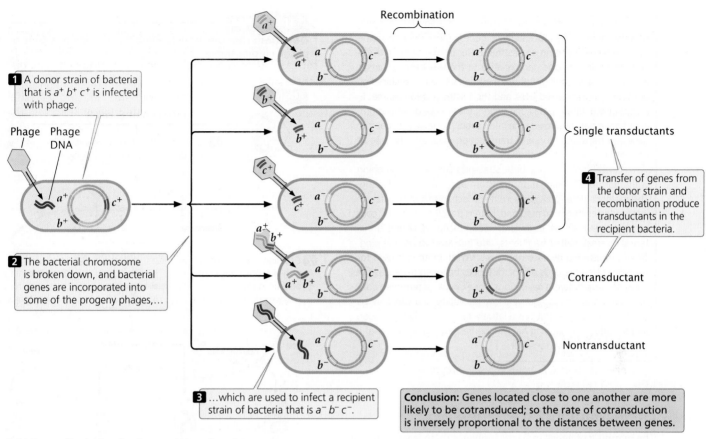

9.24 Generalized transduction can be used to map genes.

that use specific prophage integration sites on the bacterial chromosome. When the lysogenic cycle ends, the prophage may be imperfectly excised from the bacterial chromosome, carrying with it a small part of the bacterial DNA adjacent to the integration site. A phage carrying this DNA then injects it into another bacterial cell in the next round of infection. This process resembles conjugation in F′ cells, in which the F plasmid carries genes from one bacterium into another (see Figure 9.14). Specialized transduction requires phages that use specific integration sites; many phages integrate randomly and exhibit only generalized transduction.

CONCEPTS

Specialized transduction transfers only those bacterial genes located near the site of prophage integration.

CONNECTING CONCEPTS

Three Methods for Mapping Bacterial Genes

Three methods of mapping bacterial genes have now been outlined: (1) interrupted conjugation mapping; (2) transformation mapping; and (3) transduction mapping. These methods have important similarities and differences.

Mapping with interrupted conjugation is based on the time required for genes to be transferred from one bacterium to another by means of cell-to-cell contact. The key to this technique is that the bacterial chromosome itself is transferred, so the order of genes and the time required for their transfer provide information about the positions of the genes on the chromosome. In contrast with other mapping methods, the distance between genes is measured not in recombination frequencies, but in units of time required for genes to be transferred. Here, the basic unit of distance is a minute.

In transformation mapping, DNA from a donor strain is isolated, broken up, and mixed with a recipient strain of a different genotype. Some DNA fragments pass into the recipient cells, where they may recombine with the bacterial chromosome. The unit of transfer here is a random fragment of the chromosome. Loci that are close together on the donor chromosome tend to be on the same DNA fragment, so rates of cotransformation provide information about the relative positions of genes on the chromosome.

Transduction mapping also relies on the transfer of genes between bacteria that differ in two or more traits, but here, the vehicle of gene transfer is a bacteriophage. In a number of respects, transduction mapping is similar to transformation mapping. Small fragments of DNA are carried by the phage from donor to recipient bacteria, and rates of cotransduction, like rates of cotransformation, provide information about the relative distances between the genes.

All of these methods use a common strategy for mapping bacterial genes. The movement of genes from donor to recipient is detected by using strains that differ in two or more traits, and the transfer of one gene relative to the transfer of others is examined. Additionally, all three methods rely on recombination between the transferred DNA and the bacterial chromosome. In mapping with interrupted conjugation, the relative order and timing of gene transfer provide the information necessary to map the genes; in transformation and transduction mapping, the rate of cotransfer provides this information.

In conclusion, the same basic strategies are used for mapping with interrupted conjugation, transformation, and transduction. The methods differ principally in their mechanisms of transfer: in interrupted conjugation mapping, DNA is transferred though contact between bacteria; in transformation mapping, DNA is transferred as small naked fragments; and in transduction mapping, DNA is transferred by bacteriophages. Which method is used for mapping often depends on the distances between genes, as the methods have different resolutions. For example, interrupted conjugation mapping has relatively low resolution and can only be used to map genes that are relatively far apart. Transformation and transduction mapping have higher resolutions and are useful when genes are close together.

Gene Mapping in Phages

The mapping of genes in bacteriophages themselves depends on homologous recombination between phage chromosomes and therefore requires the application of the same principles that are applied to mapping genes in eukaryotic organisms (see Chapter 7). Crosses are made between viruses that differ in two or more genes, and recombinant progeny phages are identified and counted. The proportion of recombinant progeny is then used to estimate the distances between the genes and to determine their linear order on the chromosome.

In 1949, Alfred Hershey and Raquel Rotman examined rates of recombination in the bacteriophage T2, which has single-stranded DNA. They studied recombination between genes in two strains that differed in plaque appearance and host range (the bacterial strains that the phages could infect). One strain was able to infect and lyse type B *E. coli* cells, but not type B/2 cells (making this strain of phage wild type with a normal host range, or h^+), and produced abnormal plaques that were large with distinct borders (r^-). The other strain was able to infect and lyse *both* B *and* B/2 cells (mutant host range, h^-) and produced wild-type plaques that were small with fuzzy borders (r^+).

Hershey and Rotman crossed the $h^+ r^-$ and $h^- r^+$ strains of T2 by infecting type B *E. coli* cells with a mixture of the two strains. They used a high concentration of phages so that most cells could be simultaneously infected by both strains (**Figure 9.25**). Within the bacterial cells, homologous recombination occasionally took place between the chromosomes of the different bacteriophage strains, producing $h^+ r^+$ and $h^- r^-$ chromosomes, which were then packaged

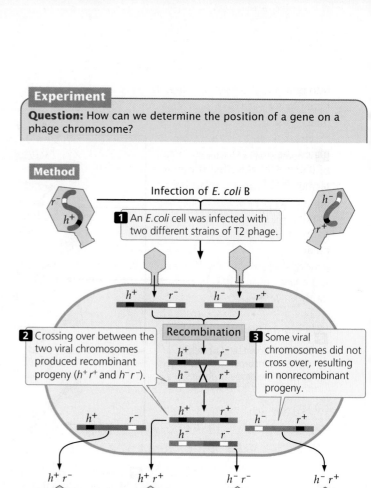

Experiment

Question: How can we determine the position of a gene on a phage chromosome?

Method

Infection of *E. coli* B

1 An *E.coli* cell was infected with two different strains of T2 phage.

2 Crossing over between the two viral chromosomes produced recombinant progeny ($h^+ r^+$ and $h^- r^-$).

Recombination

3 Some viral chromosomes did not cross over, resulting in nonrecombinant progeny.

$h^+ r^-$	$h^+ r^+$	$h^- r^-$	$h^- r^+$
Nonrecombinant phage produces cloudy, large plaques	Recombinant phage produces cloudy, small plaques	Recombinant phage produces clear, large plaques	Nonrecombinant phage produces clear, small plaques

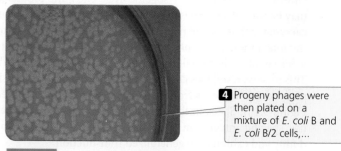

4 Progeny phages were then plated on a mixture of *E. coli* B and *E. coli* B/2 cells,...

Results

Genotype	Plaques	Designation	
$h^- r^+$	42	Parental progeny	
$h^+ r^-$	34	76%	
$h^+ r^+$	12	Recombinant progeny	
$h^- r^-$	12	24%	

5 ...which allowed all four genotypes of progeny to be identified.

6 The percentage of recombinant progeny allowed the h^- and r^- mutations to be mapped.

$$RF = \frac{\text{recombinant plaques}}{\text{total plaques}} = \frac{(h^+ r^+) + (h^- r^-)}{\text{total plaques}}$$

Conclusion: The recombination frequency indicates that the distance between *h* and *r* genes is 24%.

9.25 Hershey and Rotman developed a technique for mapping viral genes. [Dr. Steven R. Spilatro.]

into new phage particles. When the cells lysed, the recombinant phages were released, along with nonrecombinant $h^+ r^-$ phages and $h^- r^+$ phages.

Hershey and Rotman diluted the progeny phages and plated them on a bacterial lawn that consisted of a *mixture* of B and B/2 cells. Phages carrying the h^+ allele (which conferred the ability to infect only B cells) produced a cloudy plaque because the B/2 cells were not lysed. Phages carrying the h^- allele produced a clear plaque because all the cells within the plaque were lysed. The r^+ phages produced small plaques, whereas the r^- phages produced large plaques. The genotypes of the progeny phages could therefore be determined by the appearance of the plaques (see Figure 9.25 and **Table 9.4**).

In this type of phage cross, the recombination frequency (*RF*) between the two genes can be calculated by using the following formula:

$$RF = \frac{\text{recombinant plaques}}{\text{total plaques}}$$

In Hershey and Rotman's cross, the recombinant plaques were $h^+ r^+$ and $h^- r^-$; so the recombination frequency was

$$RF = \frac{(h^+r^+) + (h^-r^-)}{\text{total plaques}}$$

Recombination frequencies can be used to determine the distances between genes and their order on the phage chromosome, just as they are used to map genes in eukaryotes.

> **TRY PROBLEMS 30 AND 34**

CONCEPTS

To map phage genes, bacterial cells are infected with viruses that differ in two or more traits. Recombinant plaques are counted, and rates of recombination are used to determine the linear order of the genes on the chromosome and the distances between them.

Plant and Animal Viruses

Thus far, we have primarily considered viruses that infect bacteria. Viruses also infect plants and animals, and some are important pathogens in these organisms. What we have learned about bacteriophages has important implications for the study of viruses that infect these more complex organisms.

Viral genomes may be encoded in either DNA or RNA and may be double stranded or single stranded. Double-stranded DNA viruses include adenoviruses, which are responsible for gastroenteritis and some types of pneumonia; herpesviruses, which causes genital herpes and cold sores; papillomaviruses, which are associated with some cases of cervical cancer (see Chapter 23); and numerous viruses

TABLE 9.4	Progeny phages produced from $h^- r^+ \times h^+ r^-$	
Phenotype	**Genotype**
Clear and small	$h^- r^+$
Cloudy and large	$h^+ r^-$
Cloudy and small	$h^+ r^+$
Clear and large	$h^- r^-$

that infect other vertebrate and invertebrate animals. Single-stranded DNA viruses include parvovirus (parvo), which is highly infectious and sometimes lethal in dogs.

Almost all viruses that infect plants have RNA genomes. RNA is also the genetic material of some human viruses, including those that cause colds, influenza, polio, and AIDS. The medical and economic importance of RNA viruses has encouraged their study.

RNA viruses capable of integrating into the genomes of their hosts, much as temperate phages insert themselves into bacterial chromosomes, are called **retroviruses** (**Figure 9.26a**). Because the retroviral genome is RNA, whereas that of the host is DNA, a retrovirus must produce **reverse transcriptase**, an enzyme that can synthesize complementary DNA (cDNA) from either an RNA or a DNA template. A retrovirus uses reverse transcriptase to copy its RNA genome into a single-stranded DNA molecule. The reverse transcriptase enzyme—or sometimes the host DNA polymerase—then copies this single-stranded DNA, creating a double-stranded DNA molecule that is integrated into a host chromosome. A viral genome incorporated into a host chromosome is called a **provirus**. The provirus is replicated by host enzymes when the host chromosome is duplicated (**Figure 9.26b**).

When conditions are appropriate, the provirus undergoes transcription to produce numerous copies of the original viral RNA genome. This RNA encodes viral proteins and serves as genomic RNA for new viral particles. As these new viruses escape the cell, they collect patches of the cell membrane to use as their envelopes.

All known retroviral genomes have three genes in common: *gag*, *pol*, and *env*, each encoding a precursor protein that is cleaved into two or more functional proteins. The *gag* gene encodes proteins that make up the viral protein coat. The *pol* gene encodes reverse transcriptase and an enzyme called **integrase**, which inserts the viral DNA into a host chromosome. The *env* gene encodes the glycoproteins that appear on the surface of the viral envelope.

Some retroviruses contain **oncogenes** (see Chapter 23) that may stimulate cell division and cause the formation of tumors. The first retrovirus to be isolated, the Rous sarcoma virus, was originally recognized by its ability to produce connective-tissue tumors (sarcomas) in chickens.

THINK-PAIR-SHARE Question 7

(a)

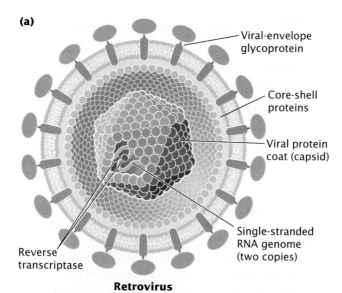

Viral-envelope glycoprotein

Core-shell proteins

Viral protein coat (capsid)

Single-stranded RNA genome (two copies)

Reverse transcriptase

Retrovirus

9.26 A retrovirus uses reverse transcription to incorporate its RNA into the host DNA. (a) Structure of a typical retrovirus. Two copies of the single-stranded RNA genome and the reverse transcriptase enzyme are shown enclosed within a protein capsid. The capsid is surrounded by a viral envelope that is studded with viral glycoproteins. (b) The retrovirus life cycle.

Human Immunodeficiency Virus and AIDS

An example of a retrovirus is human immunodeficiency virus (HIV), which causes acquired immune deficiency syndrome (AIDS). AIDS was first recognized in 1982, when a number of homosexual males in the United States began to exhibit symptoms of a new immune-system-deficiency disease. In that year, Robert Gallo proposed that AIDS was caused by a retrovirus; eventually, research by Gallo, Luc Montagnier, Françoise Barré-Sinoussi, Jay Levy, and others identified HIV as the causative agent of AIDS. Between 1983 and 1984, as the AIDS epidemic became widespread, the HIV retrovirus was isolated from people with the disease. AIDS is now known to be caused by two different immuno-deficiency viruses, HIV-1 and HIV-2, which together have infected more than 71 million people worldwide since the beginning of the epidemic. Of those infected, 34 million have died. Most cases of AIDS are caused by HIV-1, which now has a global distribution; HIV-2 is found primarily in western Africa.

HIV illustrates the importance of genetic recombination in viral evolution. Studies of the DNA sequences of HIV and other retroviruses reveal that HIV-1 is closely related to the simian immunodeficiency virus found in chimpanzees (SIV_{cpz}). Many wild chimpanzees in Africa are infected with SIV_{cpz}, although it doesn't cause AIDS-like symptoms in these animals. SIV_{cpz} is itself a hybrid that resulted from recombination between a retrovirus found in the red-capped mangabey (a monkey) and a retrovirus found in the greater

(b)

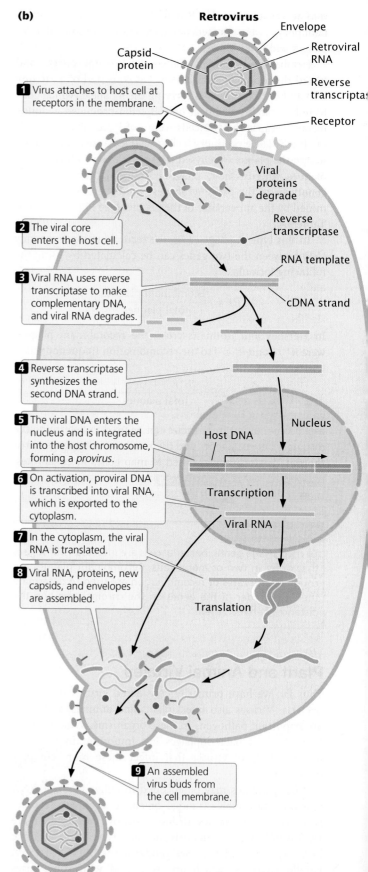

Retrovirus

Envelope

Capsid protein

Retroviral RNA

Reverse transcripta[se]

1 Virus attaches to host cell at receptors in the membrane.

Receptor

Viral proteins degrade

2 The viral core enters the host cell.

Reverse transcriptase

3 Viral RNA uses reverse transcriptase to make complementary DNA, and viral RNA degrades.

RNA template

cDNA strand

4 Reverse transcriptase synthesizes the second DNA strand.

5 The viral DNA enters the nucleus and is integrated into the host chromosome, forming a *provirus*.

Nucleus

Host DNA

6 On activation, proviral DNA is transcribed into viral RNA, which is exported to the cytoplasm.

Transcription

Viral RNA

7 In the cytoplasm, the viral RNA is translated.

8 Viral RNA, proteins, new capsids, and envelopes are assembled.

Translation

9 An assembled virus buds from the cell membrane.

spot-nosed monkey (**Figure 9.27**). Apparently, one or more chimpanzees became infected with both viruses; recombination between the viruses produced SIV_{cpz}, which was then transmitted to humans through contact with infected chimpanzees. In humans, SIV_{cpz} underwent significant evolution to become HIV-1, which then spread throughout the world to produce the AIDS epidemic. Several independent transfers of SIV_{cpz} to humans gave rise to different groups (strains) of HIV-1. In the case of HIV-1 groups O and P, SIV_{cpz} first jumped to gorillas, then passed from gorillas to humans. The M group of HIV-1 has spread throughout the world and is responsible for most cases of HIV worldwide. The O group has infected about 100,000 people in west central Africa. Groups N and P have infected only a small number of people in Africa. HIV-2 evolved from a different retrovirus, SIV_{sm}, found in sooty mangabeys.

HIV is transmitted between humans by sexual contact and through any type of blood-to-blood contact, such as that caused by the sharing of dirty needles by drug users. HIV can also be transmitted between mother and child during pregnancy or in breast milk. Until screening tests were developed to identify HIV-infected blood, transfusions and clotting factors used to treat hemophilia were sources of infection as well.

HIV principally attacks a class of blood cells called helper T lymphocytes, or simply helper T cells (**Figure 9.28**). HIV enters a helper T cell, undergoes reverse transcription, and integrates into the chromosome. The virus reproduces rapidly, destroying the T cell as new virus particles escape from the cell. Because helper T cells are central to immune function, people with AIDS have a diminished immune response; most deaths among AIDS patients are caused by secondary infections that develop because they have lost the ability to fight off pathogens.

The HIV genome is 9749 nucleotides long and carries *gag*, *pol*, *env*, and six other genes that regulate the life cycle of the virus. HIV's reverse transcriptase is very error prone, giving the virus a high mutation rate and allowing it to evolve rapidly, even within a single host. This rapid evolution makes the development of an effective vaccine against HIV particularly difficult. Genetic variation within the human population also affects the virus. To date, more than 10 loci in humans that affect HIV infection and the progression of AIDS have been identified.

CONCEPTS

A retrovirus is an RNA virus that integrates into a host chromosome by using reverse transcription to make a DNA copy of its RNA genome. Human immunodeficiency virus, the causative agent of AIDS, is a retrovirus. It evolved from related retroviruses found in other primates.

✔ CONCEPT CHECK 8

What enzyme is used by a retrovirus to make a DNA copy of its genome?

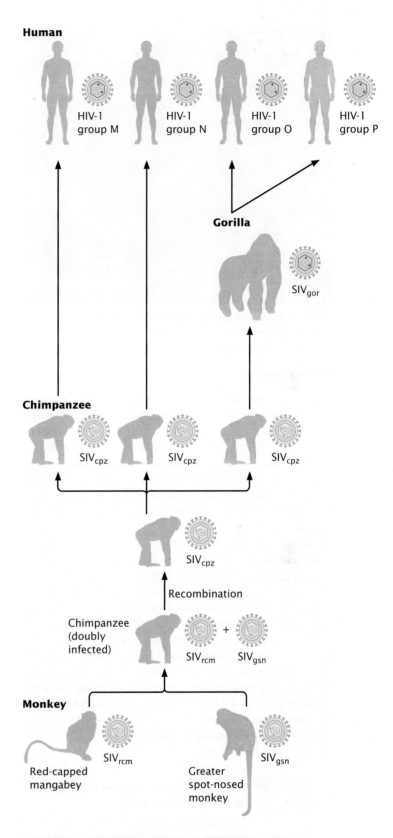

9.27 HIV-1 evolved from a similar virus (SIV_{cpz}) found in chimpanzees and was transmitted to humans. SIV_{cpz} arose from recombination between retroviruses from red-capped mangabeys and from greater spot-nosed monkeys.

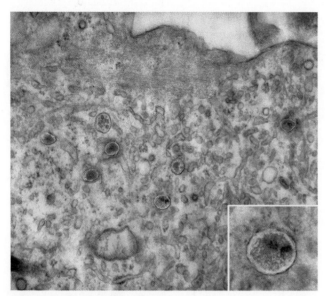

9.28 HIV principally attacks helper T lymphocytes. This electron micrograph shows a T cell (green) infected with HIV (orange). [Thomas Deerinck, NCMIR/Science Source.]

Influenza

Influenza demonstrates how rapidly changes in a pathogen can arise through recombination of its genetic material. Influenza, commonly called flu, is a respiratory disease caused by influenza viruses. In the United States, from 5% to 20% of the population is infected with influenza annually, and though most cases are mild, an estimated 36,000 people die from influenza-related causes each year. At certain times, particularly when new strains of influenza virus enter the human population, there are worldwide epidemics (called pandemics); for example, in 1918, the Spanish flu virus killed an estimated 20 million to 100 million people worldwide.

Influenza viruses are RNA viruses that infect birds and mammals. The three main types are influenza A, influenza B, and influenza C. Most cases of the common flu are caused by influenza A and B. Influenza A is divided into subtypes on the basis of two proteins, hemagglutinin (HA) and neuraminidase (NA), found on the surface of the virus. The HA and NA proteins affect the ability of the virus to enter host cells and the host organism's immune response to infection. There are 16 types of HA and 9 types of NA, which can exist in a virus in different combinations. For example, common subtypes of influenza A circulating in humans today are H1N1 and H3N2 (**Table 9.5**), along with several types of influenza B.

TABLE 9.5	Strains of influenza virus responsible for major flu pandemics	
Year	Influenza Pandemic	Strain
1918	Spanish flu	H1N1
1957	Asian flu	H2N2
1968	Hong Kong flu	H3N2
2009	Swine flu	H1N1

Although the influenza virus is an RNA virus, it is not a retrovirus: its genome is not copied into DNA and incorporated into a host chromosome. The influenza viral genome consists of seven or eight pieces of RNA enclosed in a viral envelope. Each piece of RNA encodes one or two of the virus's proteins. The virus enters a host cell by attaching to specific receptors on the cell membrane. After the viral particle has entered the cell, the viral RNA is released, copied, and translated into viral proteins. Viral RNA molecules and viral proteins are then assembled into new viral particles, which exit the cell and infect additional cells.

One of the dangers of the influenza virus is that it evolves rapidly, so that new strains appear frequently. The virus evolves in two ways. First, the enzyme that copies the viral RNA is especially prone to making mistakes, so new mutations are continually introduced into the genome of each viral strain. This type of continual change is called **antigenic drift**. Second, major changes in the viral genome occasionally take place through **antigenic shift**, in which genetic material from different strains is combined in a process called reassortment. Reassortment may take place when a host is simultaneously infected with two different strains. When the RNAs of both strains are replicated within the cell, RNA segments from two different strains may be incorporated into the same viral particle, creating a new strain. For example, in 2002, reassortment occurred between the H1N1 and H3N2 subtypes, creating a new H1N2 strain that contained the HA from H1N1 and the NA from H3N2. New strains produced by antigenic shift are responsible for most pandemics because no one has immunity to the radically different virus that is produced.

Birds harbor the most different strains of influenza A, but humans are not easily infected with bird influenza. New strains in humans are thought to arise most often from viruses that reassort in pigs, which can be infected by viruses from both humans and birds. In 2009, a new strain of H1N1 influenza (called swine flu) emerged in Mexico and quickly spread throughout the world. This virus arose from a series of reassortment events that combined gene sequences from human, bird, and pig influenza viruses (**Figure 9.29**). Farming practices that raise pigs and birds in close proximity may facilitate reassortment among avian, swine, and human strains of influenza.

Rhinoviruses

Rhinoviruses are very small RNA viruses that infect epithelial cells of the respiratory tract and are responsible for respiratory infections, including the common cold. Although the symptoms of rhinovirus infections are usually mild, they constitute a serious public health burden: colds force many people to miss work, make millions of people miserable, and occasionally lead to medical complications such as pneumonia or exacerbation of asthma and cystic fibrosis. Respiratory infections caused by rhinoviruses remain resistant to treatment, even in the era of modern medicine.

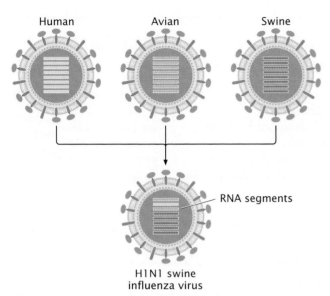

Human   Avian   Swine

RNA segments

H1N1 swine
influenza virus

9.29 New strains of influenza virus are created by reassortment of genetic material from different strains. A new H1N1 virus (swine flu) that appeared in 2009 contained genetic material from avian, swine, and human viruses.

Rhinoviruses have a genome of single-stranded RNA, consisting of approximately 7200 nucleotides, that encodes only 11 proteins. Rhinoviruses have considerable genetic diversity and evolve rapidly. These viruses have a high mutation rate, caused by the inability of their RNA polymerase enzymes (which copy the RNA genome of the virus) to detect and correct mistakes made during the copying process. In addition, multiple viral strains may simultaneously infect the same cell and recombine, leading to the emergence of new rhinovirus strains.

Rhinoviruses that cause colds in humans are highly specialized and are adapted to the human ecosystem. Curiously, they evolved from enteroviruses—viruses that inhabit the gut and can cause acute intestinal infections. Research suggests that rhinoviruses evolved long ago and are adapted to replicate and spread between human cells efficiently without killing their host, leading to relatively mild effects on the humans they infect. Genetic analysis demonstrates that there are three species of human rhinoviruses, called rhinovirus A, rhinovirus B, and rhinovirus C. Within each species, there are numerous variants, called serotypes. Recombination within and between species constantly generates new types of rhinoviruses.

CONCEPTS

Influenza is caused by RNA viruses. New strains of influenza viruses appear through antigenic shift, in which new viral genomes are created through the reassortment of RNA molecules of different strains. Rhinoviruses are small RNA viruses that often cause respiratory infections.

CONCEPTS SUMMARY

■ Bacteria and viruses are well suited to genetic studies: they are small, have a small haploid genome, undergo rapid reproduction, and produce large numbers of progeny through asexual reproduction.

■ The bacterial genome normally consists of a single, circular molecule of double-stranded DNA. Plasmids are small pieces of bacterial DNA that can replicate independently of the bacterial chromosome.

■ DNA can be transferred between bacteria by conjugation, transformation, or transduction.

■ Conjugation is the transfer of genetic material from one bacterial cell to another by direct contact between the two cells. Conjugation is controlled by an episome called the F factor. The time it takes for individual genes to be transferred during conjugation provides information about the order of the genes, and the distances between them, on the bacterial chromosome.

■ Bacteria take up DNA from the environment through the process of transformation. Frequencies of cotransformation provide information about the physical distances between chromosomal genes.

■ Complete genome sequences of many bacterial species have been determined. This sequence information indicates that horizontal gene transfer—the movement of DNA between species—is common in bacteria.

■ Viruses are replicating structures with DNA or RNA genomes that may be double stranded or single stranded and linear or circular.

■ Bacterial genes may be incorporated into phage coats and transferred to other bacteria by phages in a process called transduction. Rates of cotransduction can be used to map bacterial genes.

■ Phage genes can be mapped by infecting bacterial cells with two different phage strains and counting the recombinant plaques produced by the progeny phages.

■ A number of viruses have RNA genomes. Retroviruses encode reverse transcriptase, an enzyme used to make a DNA copy of the viral genome, which then integrates into the host genome as a provirus. HIV is a retrovirus that is the causative agent for AIDS.

■ Influenza is caused by RNA viruses. These viruses evolve both through small changes that take place by mutation (antigenic drift) and through major changes that take place via the reassortment of genetic material from different strains within a host (antigenic shift). Rhinoviruses are small RNA viruses that often cause respiratory infections.

IMPORTANT TERMS

prototrophic (p. 254)	transformation (p. 257)	virulent phage (p. 268)	transductant (p. 270)
minimal medium (p. 254)	transduction (p. 258)	temperate phage (p. 268)	cotransduction (p. 270)
auxotrophic (p. 254)	pilus (p. 259)	prophage (p. 268)	retrovirus (p. 273)
complete medium (p. 254)	competence (p. 265)	plaque (p. 269)	reverse transcriptase (p. 273)
colony (p. 254)	transformant (p. 265)	generalized	provirus (p. 273)
plasmid (p. 255)	cotransformation (p. 265)	transduction (p. 269)	integrase (p. 273)
episome (p. 256)	horizontal gene	specialized	oncogene (p. 273)
F (fertility) factor (p. 256)	transfer (p. 266)	transduction (p. 269)	antigenic drift (p. 276)
conjugation (p. 257)	virus (p. 267)	transducing phage (p. 270)	antigenic shift (p. 276)

ANSWERS TO CONCEPT CHECKS

1. d

2. b

3. a

4. *gal*

5. *his* and *leu*

6. b

7. c

8. Reverse transcriptase

WORKED PROBLEMS

Problem 1

DNA from a strain of bacteria with genotype $a^+ b^+ c^+ d^+ e^+$ was isolated and used to transform a strain of bacteria that was $a^- b^- c^- d^- e^-$. The transformants were tested for the presence of donated genes. The following genes were cotransformed:

a^+ and d^+ b^+ and e^+ c^+ and d^+ c^+ and e^+

What is the order of genes *a*, *b*, *c*, *d*, and *e* on the bacterial chromosome?

» Solution Strategy

What information is required in your answer to the problem?

The order of genes *a*, *b*, *c*, *d*, and *e* on the bacterial chromosome.

What information is provided to solve the problem?

- The donor cells were $a^+ b^+ c^+ d^+ e^+$ and the recipient cells were $a^- b^- c^- d^- e^-$.

- The combinations of genes that were cotransformed.

For help with this problem, review:
Transformation in Bacteria in Section 9.3.

» Solution Steps

> **Recall:** The rate at which genes are cotransformed is inversely proportional to the distance between them: genes that are close together are frequently cotransformed, whereas genes that are far apart are rarely cotransformed.

In this transformation experiment, gene c^+ is cotransformed with both gene e^+ and gene d^+, but genes e^+ and d^+ are not cotransformed; therefore, the *c* locus must be between the *d* and *e* loci:

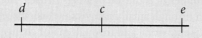

Gene e^+ is also cotransformed with gene b^+, so the *e* and *b* loci must be located close together. Locus *b* could be on either side of locus *e*. To determine whether locus *b* is on the same side of *e* as locus *c*, we look to see whether genes b^+ and c^+ are cotransformed. They are not; so locus *b* must be on the opposite side of *e* from *c*:

Gene a^+ is cotransformed with gene d^+, so the *a* and *d* loci must be located close together. If locus *a* were located on the same side of *d* as locus *c*, then genes a^+ and c^+ would be cotransformed. Because these genes display no cotransformation, locus *a* must be on the opposite side of locus *d*:

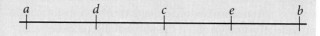

Problem 2

Consider three genes in *E. coli*: *thr*⁺, *ara*⁺, and *leu*⁺ (which give the cell the ability to synthesize threonine, arabinose, and leucine, respectively). All three of these genes are close together on the *E. coli* chromosome. Phages are grown in a *thr*⁺ *ara*⁺ *leu*⁺ strain of bacteria (the donor strain). The phage lysate is collected and used to infect a strain of bacteria that is *thr*⁻ *ara*⁻ *leu*⁻. The recipient bacteria are then tested on selective medium lacking leucine. Bacteria that grow and form colonies on this medium (*leu*⁺ transductants) are then replica-plated on medium lacking threonine and on medium lacking arabinose to see which are *thr*⁺ and which are *ara*⁺.

Another group of the recipient bacteria are tested on medium lacking threonine. Bacteria that grow and form colonies on this medium (*thr*⁺ transductants) are then replica-plated on medium lacking leucine and onto medium lacking arabinose to see which are *ara*⁺ and which are *leu*⁺. Results from these experiments are as follows:

Selected gene	Cells with cotransduced genes (%)
leu⁺	3 *thr*⁺
	76 *ara*⁺
thr⁺	3 *leu*⁺
	0 *ara*⁺

How are the loci arranged on the chromosome?

≫ Solution Strategy

What information is required in your answer to the problem?

The order of genes *thr*, *leu*, and *ara* on the bacterial chromosome.

What information is provided to solve the problem?

- The genes are located close together on the *E. coli* chromosome.
- The donor strain is *thr*⁺ *ara*⁺ *leu*⁺ and the recipient strain is *thr*⁻ *ara*⁻ *leu*⁻.
- The percentage of cells with cotransduced genes.

For help with this problem, review:

Transduction: Using Phages to Map Bacterial Genes in Section 9.4.

≫ Solution Steps

Notice that, when we select for *leu*⁺ (the top half of the table), most of the selected cells are also *ara*⁺. This finding indicates that the *leu* and *ara* genes are usually cotransduced and are therefore located close together. In contrast, *thr*⁺ is only rarely cotransduced with *leu*⁺, indicating that *leu* and *thr* are much farther apart. On the basis of these observations, we know that *leu* and *ara* are closer together than are *leu* and *thr*, but we don't yet know the order of the three genes—whether *thr* is on the same side of *ara* as *leu* or on the opposite side:

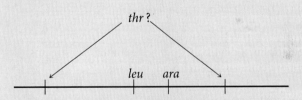

Hint: Genes located close together are more likely to be cotransduced than are genes located far apart.

Notice that, although the cotransduction frequency for *thr* and *leu* is 3%, no *thr*⁺ *ara*⁺ cotransductants are observed. This finding indicates that *thr* is closer to *leu* than to *ara*, and therefore *thr* must be on the opposite side of *leu* from *ara*:

Hint: We can determine the position of *thr* with respect to the other two genes by looking at the cotransduction frequencies when *thr*⁺ is selected.

COMPREHENSION QUESTIONS

Section 9.1

1. What are some advantages of using bacteria and viruses for genetic studies?

Section 9.2

2. Explain how auxotrophic bacteria are isolated.

3. What is the difference between complete medium and minimal medium? How are complete media and

minimal media to which one or more nutrients have been added (selective media) used to isolate auxotrophic mutants of bacteria?

Section 9.3

4. Briefly explain the differences between F⁺, F⁻, Hfr, and F′ cells.

5. What types of matings are possible between F⁺, F⁻, Hfr, and F′ cells? What outcomes do these matings produce? What is the role of the F factor in conjugation?

6. Explain how interrupted conjugation, transformation, and transduction can be used to map bacterial genes. How are these methods similar and how are they different?

7. What is horizontal gene transfer and how might it take place?

Section 9.4

8. List some of the characteristics that make bacteriophages ideal organisms for many types of genetic studies.

9. What types of genomes do viruses have?

10. Briefly describe the differences between the lytic cycle of virulent phages and the lysogenic cycle of temperate phages.

11. Briefly explain how genes in phages are mapped.

12. How does specialized transduction differ from generalized transduction?

13. Briefly describe the genetic structure of a typical retrovirus.

14. Explain how a retrovirus, which has an RNA genome, is able to integrate its genetic material into that of a host having a DNA genome.

15. What are the evolutionary origins of HIV-1 and HIV-2?

16. Most humans are not easily infected by avian influenza. How, then, do DNA sequences from avian influenza become incorporated into human influenza?

APPLICATION QUESTIONS AND PROBLEMS

Section 9.1

*17. Suppose you want to compare the species of bacteria that exist in a polluted stream with the species that exist in an unpolluted stream. Traditionally, bacteria have been identified by growing them in the laboratory and comparing their physical and biochemical properties. You recognize that you will be unable to culture most of the bacteria that reside in the streams. How might you go about identifying the species in the two streams without culturing them in the laboratory?

Section 9.3

18. John Smith is a pig farmer. For the past five years, Smith has been adding vitamins and low doses of antibiotics to his pig food; he says that these supplements enhance the growth of the pigs. Within the past year, however, several of his pigs died from infections of common bacteria, which failed to respond to large doses of antibiotics. Can you explain the increased rate of mortality due to infection in Smith's pigs? What advice might you offer Smith to prevent this problem in the future?

19. Rarely, the conjugation of Hfr and F⁻ cells produces two Hfr cells. Explain how this event takes place.

20. In **Figure 9.8**, what do the red and blue parts of the DNA labeled by balloon 6 represent?

21. Austin Taylor and Edward Adelberg isolated some new strains of Hfr cells that they then used to map several genes in *E. coli* by using interrupted conjugation

(A. L. Taylor and E. A. Adelberg. 1960. *Genetics* 45:1233–1243). In one experiment, they mixed cells of Hfr strain AB-312, which were *xyl⁺ mtl⁺ mal⁺ met⁺* and sensitive to phage T6, with F⁻ strain AB-531, which was *xyl⁻ mtl⁻ mal⁻ met⁻* and resistant to phage T6. The cells were allowed to undergo conjugation. At regular intervals, the researchers removed a sample of cells and interrupted conjugation by killing the Hfr cells with phage T6. The F⁻ cells, which were resistant to phage T6, survived and were then tested for the presence of genes transferred from the Hfr strain. The results of this experiment are shown in the accompanying graph. On the basis of these data, give the order of the *xyl, mtl, mal*, and *met* genes on the bacterial chromosome and indicate the minimum distances between them.

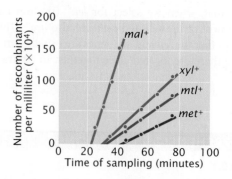

*22. Three different Hfr donor strains are mixed with separate samples of an F⁻ strain, and the following mapping data are provided from studies of interrupted conjugation:

		Appearance of genes in F⁻ cells				
Hfr1:	Genes	b^+	d^+	c^+	f^+	g^+
	Time*	3	5	16	27	59
Hfr2:	Genes	e^+	f^+	c^+	d^+	b^+
	Time	6	24	35	46	48
Hfr3:	Genes	d^+	c^+	f^+	e^+	g^+
	Time	4	15	26	44	58

*Minutes after start of conjugation

Construct a genetic map for these genes, indicating their order on the bacterial chromosome and the distances between them.

23. In **Figure 9.16**, which gene on the bacterial chromosome will be transferred last in strain Hfr5?

24. DNA from a strain of *Bacillus subtilis* with the genotype $trp^+ tyr^+$ was used to transform a recipient strain with the genotype $trp^- tyr^-$. The following numbers of transformed cells were recovered:

Genotype	Number of transformed cells
$trp^+ tyr^-$	154
$trp^- tyr^+$	312
$trp^+ tyr^+$	354

What do these results suggest about the linkage of the *trp* and *tyr* genes?

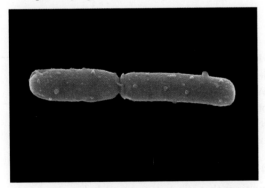

B. subtilis. [Oxford Scientific/Getty Images.]

***25.** DNA from a strain of *Bacillus subtilis* with genotype $a^+ b^+ c^+ d^+ e^+$ is used to transform a strain with genotype $a^- b^- c^- d^- e^-$. Pairs of genes are checked for cotransformation, and the following results are obtained:

Pair of genes	Cotransformation	Pair of genes	Cotransformation
a^+ and b^+	No	b^+ and d^+	No
a^+ and c^+	No	b^+ and e^+	Yes
a^+ and d^+	Yes	c^+ and d^+	No
a^+ and e^+	Yes	c^+ and e^+	Yes
b^+ and c^+	Yes	d^+ and e^+	No

On the basis of these results, what is the order of the genes on the bacterial chromosome?

26. DNA from a bacterial strain that is $his^+ leu^+ lac^+$ is used to transform a strain that is $his^- leu^- lac^-$. The following percentages of cells are transformed:

Donor strain	Recipient strain	Genotype of transformed cells	Percentage of all cells
$his^+ leu^+ lac^+$	$his^- leu^- lac^-$	$his^+ leu^+ lac^+$	0.02
		$his^+ leu^+ lac^-$	0.00
		$his^+ leu^- lac^+$	2.00
		$his^+ leu^- lac^-$	4.00
		$his^- leu^+ lac^+$	0.10
		$his^- leu^- lac^+$	3.00
		$his^- leu^+ lac^-$	1.50

a. What conclusions can you draw about the order of these three genes on the chromosome?

b. Which two genes are closest?

27. Rollin Hotchkiss and Julius Marmur studied transformation in the bacterium *Streptococcus pneumoniae* (R. D. Hotchkiss and J. Marmur. 1954. *Proceedings of the National Academy of Sciences of the United States of America* 40:55–60). They examined four mutations in this bacterium: penicillin resistance (*P*), streptomycin resistance (*S*), sulfanilamide resistance (*F*), and the ability to use mannitol (*M*). They extracted DNA from strains of bacteria with different combinations of different mutations and used this DNA to transform wild-type bacterial cells ($P^+ S^+ F^+ M^+$). The results from one of their transformation experiments are shown here.

Donor DNA	Recipient DNA	Transformants	Percentage of all cells
$M S F$	$M^+ S^+ F^+$	$M^+ S F^+$	4.0
		$M^+ S^+ F$	4.0
		$M S^+ F^+$	2.6
		$M S F^+$	0.41
		$M^+ S F$	0.22
		$M S^+ F$	0.0058
		$M S F$	0.0071

a. Hotchkiss and Marmur noted that the percentage of cotransformation was higher than would be expected on a random basis. For example, the results show that 2.6% of the cells were transformed into M and 4% were transformed into S. If the M and S traits were inherited independently, the expected probability of cotransformation of M and S (M S) would be $0.026 \times 0.04 = 0.001$, or 0.1%. However, they observed 0.41% M S cotransformants, four times more than they expected. What accounts for the relatively high frequency of cotransformation of the traits they observed?

b. On the basis of the results, what conclusion can you draw about the order of the M, S, and F genes on the bacterial chromosome?

c. Why is the rate of cotransformation for all three genes (M S F) almost the same as the rate of cotransformation for M F alone?

28. In the course of a study on the effects of the mechanical shearing of DNA, Eugene Nester, A. T. Ganesan, and Joshua Lederberg studied the transfer, by transformation, of sheared DNA from a wild-type strain of *Bacillus subtilis* (his_2^+ aro_3^+ tyr_2^+ aro_1^+ tyr_1^+ aro_2^+) to strains of bacteria carrying a series of mutations (E. W. Nester, A. T. Ganesan, and J. Lederberg. 1963. *Proceedings of the National Academy of Sciences of the United States of America* 49:61–68). They reported the following rates of cotransformation between his_2^+ and the other genes (expressed as cotransfer rate).

Genes	Rate of cotransfer
his_2^+ and aro_3^+	0.015
his_2^+ and tyr_2^+	0.10
his_2^+ and aro_1^+	0.12
his_2^+ and tyr_1^+	0.23
his_2^+ and aro_2^+	0.05

On the basis of these data, which gene is farthest from his_2^+? Which gene is closest?

29. C. Anagnostopoulos and I. P. Crawford isolated and studied a series of mutations that affected several steps in the biosynthetic pathway leading to tryptophan in the bacterium *Bacillus subtilis* (C. Anagnostopoulos and I. P. Crawford. 1961. *Proceedings of the National Academy of Sciences of the United States of America* 47:378–390). Seven of the strains that they used in their study are listed here, along with the mutation found in each strain.

Strain	Mutation
T3	T^-
168	I^-
168PT	I^-
TI	I^-
TII	I^-
T8	A^-
H25	H^-

To map the genes for tryptophan synthesis, they carried out a series of transformation experiments on strains having different mutations and determined the percentage of recombinants among the transformed bacteria. Their results were as follows:

Recipient	Donor	Percentage of recombinants
T3	168PT	12.7
T3	T11	11.8
T3	T8	43.5
T3	H25	28.6
168	H25	44.9
TII	H25	41.4
TI	H25	31.3
T8	H25	67.4
H25	T3	19.0
H25	TII	26.3
H25	TI	13.4
H25	T8	45.0

On the basis of these two-point crosses, determine the order of the genes and the distances between them. Where more than one cross was completed for a pair of genes, average the recombination rates from the different crosses. Draw a map of the genes on the chromosome.

Section 9.4

*30. Two mutations that affect plaque morphology in phages (a^- and b^-) have been isolated. Phages carrying both mutations (a^- b^-) are mixed with wild-type phages (a^+ b^+) and added to a culture of bacterial cells. Once the phages have infected and lysed the bacteria, samples of the phage lysate are collected and cultured on plated bacteria. The following numbers of plaques are observed:

Plaque phenotype	Number
$a^+ b^+$	2043
$a^+ b^-$	320
$a^- b^+$	357
$a^- b^-$	2134

What is the frequency of recombination between the a and b genes?

*31. T. Miyake and M. Demerec examined proline-requiring mutations in the bacterium *Salmonella typhimurium* (T. Miyake and M. Demerec. 1960. *Genetics* 45:755–762). On the basis of complementation testing, they found four proline auxotrophs: *proA*, *proB*, *proC*, and *proD*. To determine whether *proA*, *proB*, *proC*, and *proD* loci were located close together on the bacterial chromosome, they conducted a transduction experiment. Bacterial strains that were *proC*$^+$ and had mutations at *proA*, *proB*, or *proD* were used as donors. The donors were infected with bacteriophages, and progeny phages were allowed to infect recipient bacteria with genotype *proC*$^-$ *proA*$^+$ *proB*$^+$ *proD*$^+$. The recipient bacteria were then plated on a selective medium that allowed only *proC*$^+$ bacteria to grow. After this, the *proC*$^+$ transductants were plated on selective media to reveal their genotypes at the other three *pro* loci. The following results were obtained:

Donor genotype	Transductant genotype	Number
proC$^+$ *proA*$^-$ *proB*$^+$ *proD*$^+$	*proC*$^+$ *proA*$^+$ *proB*$^+$ *proD*$^+$	2765
	proC$^+$ *proA*$^-$ *proB*$^+$ *proD*$^+$	3
proC$^+$ *proA*$^+$ *proB*$^-$ *proD*$^+$	*proC*$^+$ *proA*$^+$ *proB*$^+$ *proD*$^+$	1838
	proC$^+$ *proA*$^+$ *proB*$^-$ *proD*$^+$	2
proC$^+$ *proA*$^+$ *proB*$^+$ *proD*$^-$	*proC*$^+$ *proA*$^+$ *proB*$^+$ *proD*$^+$	1166
	proC$^+$ *proA*$^+$ *proB*$^+$ *proD*$^-$	0

a. Why are there no *proC*$^-$ genotypes among the transductants?

b. Which genotypes represent single transductants and which represent cotransductants?

c. Is there evidence that *proA*, *proB*, and *proD* are located close to *proC*? Explain your answer.

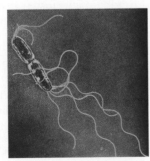

S. typhimurium. [Kwangshin Kim/Science Source.]

*32. A geneticist isolates two mutations in a bacteriophage. One mutation causes clear plaques (c), and the other produces minute plaques (m). Previous mapping experiments have established that the genes responsible for these two mutations are 8 m.u. apart. The geneticist mixes phages with genotype $c^+ m^+$ and genotype $c^- m^-$ and uses the mixture to infect bacterial cells. She collects the progeny phages and cultures a sample of them on plated bacteria. A total of 1000 plaques are observed. What numbers of the different types of plaques ($c^+ m^+$, $c^- m^-$, $c^+ m^-$, $c^- m^+$) should she expect to see?

33. The geneticist carries out the same experiment described in Problem 32, but this time she mixes phages with genotypes $c^+ m^-$ and $c^- m^+$. What results are expected from *this* cross?

*34. A geneticist isolates two bacteriophage r mutants (r_{13} and r_2) that cause rapid lysis. He carries out the following crosses and counts the number of plaques listed here:

Genotype of parental phage	Progeny	Number of plaques
$h^+ r_{13}^- \times h^- r_{13}^+$	$h^+ r_{13}^+$	1
	$h^- r_{13}^+$	104
	$h^+ r_{13}^-$	110
	$h^- r_{13}^-$	2
Total		$\overline{216}$
$h^+ r_2^- \times h^- r_2^+$	$h^+ r_2^+$	6
	$h^- r_2^+$	86
	$h^+ r_2^-$	81
	$h^- r_2^-$	7
Total		$\overline{180}$

a. Calculate the recombination frequencies between r_2 and h and between r_{13} and h.

b. Draw all possible linkage maps for these three genes.

*35. *E. coli* cells are simultaneously infected with two strains of phage λ. One strain has a mutant host range, is temperature sensitive, and produces clear plaques (genotype h st c); another strain carries the wild-type alleles (genotype h^+ st^+ c^+). Progeny phages are collected from the lysed cells and are plated on bacteria. The following numbers of different progeny phages are obtained:

Progeny phage genotype	Number of plaques
$h^+\,c^+\,st^+$	321
$h\,c\,st$	338
$h^+\,c\,st$	26
$h\,c^+\,st^+$	30
$h^+\,c\,st^+$	106
$h\,c^+\,st$	110
$h^+\,c^+\,st$	5
$h\,c\,st^+$	6

a. Determine the order of the three genes on the phage chromosome.

b. Determine the map distances between the genes.

c. Determine the coefficient of coincidence and the interference (see pp. 193–195 in Chapter 7).

36. A donor strain of bacteria with alleles $a^+\,b^+\,c^+$ is infected with phages to map the donor chromosome using generalized transduction. The phage lysate from the bacterial cells is collected and used to infect a second strain of bacteria that are $a^-\,b^-\,c^-$. Bacteria with the a^+ allele are selected, and the percentage of cells with cotransduced b^+ and c^+ alleles are recorded.

Donor	Recipient	Selected allele	Cells with cotransduced allele (%)
$a^+\,b^+\,c^+$	$a^-\,b^-\,c^-$	a^+	25 b^+
		a^+	3 c^+

Is gene b or gene c closer to gene a? Explain your reasoning.

37. A donor strain of bacteria with genotype $leu^+\,gal^-\,pro^+$ is infected with phages. The phage lysate from the bacterial cells is collected and used to infect a second strain of bacteria that are $leu^-\,gal^+\,pro^-$. The second strain is selected for leu^+, and the following cotransduction data are obtained:

Donor	Recipient	Selected allele	Cells with cotransduced allele (%)
$leu^+\,gal^-\,pro^+$	$leu^-\,gal^+\,pro^-$	leu^+	47 pro^+
		leu^+	26 gal^-

Which genes are closest, *leu* and *gal* or *leu* and *pro*?

38. Viruses from which organism contributed the most RNA to the H1N1 influenza virus shown at the bottom of **Figure 9.29**?

CHALLENGE QUESTION

Section 9.3

39. A group of genetics students mix two auxotrophic strains of bacteria: one is $leu^+\,trp^+\,his^-\,met^-$ and the other is $leu^-\,trp^-\,his^+\,met^+$. After mixing the two strains, they plate the bacteria on minimal medium and observe a few prototrophic colonies ($leu^+\,trp^+\,his^+\,met^+$). They assume that some gene transfer has taken place between the two strains. How can they determine whether the transfer of genes is due to conjugation, transduction, or transformation?

THINK-PAIR-SHARE QUESTIONS

Section 9.1

1. Luiz Roesch and colleagues conducted a study to determine how many types of bacteria exist in a gram of soil at different locations around the world. They found that soil from agricultural fields harbored considerably fewer species of bacteria than did soil from forests (see Life in a Bacterial World). Propose some possible reasons that agricultural fields might have less bacterial diversity than forests.

Section 9.2

2. One advantage of using bacteria and viruses for genetic study is the fact that they have haploid genomes. Explain why a haploid genome facilitates genetic analysis.

Section 9.3

3. A series of Hfr strains that have genotype $m^+\,n^+\,o^+\,p^+\,q^+\,r^+$ are mixed with an F$^-$ strain that has genotype $m^-\,n^-\,o^-\,p^-\,q^-\,r^-$. Conjugation is interrupted at regular intervals and the order of the appearance of genes from the Hfr strain is determined in the recipient cells. The order of gene transfer for each Hfr strain is

Hfr5	$m^+\,q^+\,p^+\,n^+\,r^+\,o^+$
Hfr4	$n^+\,r^+\,o^+\,m^+\,q^+\,p^+$
Hfr1	$o^+\,m^+\,q^+\,p^+\,n^+\,r^+$
Hfr9	$q^+\,m^+\,o^+\,r^+\,n^+\,p^+$

What is the order of genes on the circular bacterial chromosome? For each Hfr strain, give the location of the F factor in the chromosome and its polarity.

4. Antibiotic resistance genes are often found on R plasmids (see Natural Gene Transfer and Antibiotic Resistance). A likely source of the R plasmids is bacteria that produce the antibiotic. Why would some bacteria produce antibiotics (chemicals that kill bacteria) and why would they carry R plasmids?

5. Bacteria have evolved numerous mechanisms to prevent the invasion of foreign viral DNA (see Bacterial Defense Mechanisms). Yet clearly some bacteria have evolved competence, the ability to take up foreign DNA from the environment. Why do bacteria take up naked DNA from the environment and yet exclude DNA from viruses?

Section 9.4

6. Researchers have recently discovered giant viruses that are 1 μm in length, the same size as some bacterial cells. The genomes of these viruses contain over 2 million base pairs of DNA, which is more DNA than is found in many bacterial genomes, and their genomes contain hundreds—in some cases, thousands—of genes. Given these observations and what you know about viruses, should viruses be considered living or nonliving? Give arguments for and against considering viruses as living organisms.

7. RNA viruses often undergo rapid evolution. What aspects of their biology contribute to their high rate of evolution? What are some consequences of their rapid evolution?

DNA: The Chemical Nature of the Gene

Greenland, one of Earth's most extreme environments, was originally settled by the Saqqaq people. The genome of a 4000-year-old Saqqaq male was sequenced in 2010. The remarkable stability of DNA makes analysis of genomes from ancient remains possible. [Alex Hibbert/age fotostock.]

Arctic Treks and Ancient DNA

Greenland is the world's largest island, consisting of over 830,000 square miles (2,200,000 square kilometers), but the vast majority of the land is permanently buried under hundreds of feet of ice. It is one of Earth's most extreme environments. Temperatures along the coast rise a few degrees above freezing during summer days, but then drop to far below zero during much of the winter. With limited daylight (the Sun moves above the horizon for only a few hours on winter days), extreme cold, and winds reaching hurricane force, Greenland has a dangerously inhospitable environment.

Yet in spite of these severe conditions, Arctic peoples have continuously occupied Greenland for almost 5000 years. The earliest inhabitants were the Saqqaq people, who occupied small settlements on Greenland's coast from around 4800 to 2500 years ago. The Saqqaq lived in small tents and hunted marine mammals and seabirds. The origin of the Saqqaq people had long been a mystery. Did they descend from Native Americans who migrated from Asia into the New World and later moved to Greenland? Or did they descend from the same group that gave rise to the Inuit people, who currently inhabit the New World Arctic? Or did they perhaps originate from yet another group that migrated independently from Asia to Greenland after the ancestors of both the Inuit and Native Americans entered the New World?

The mystery of Saqqaq origin was solved in 2010, when geneticists determined the entire DNA sequence of a 4000-year-old Saqqaq male—nicknamed Inuk—whose remains were recovered from an archaeological site on the western coast of Greenland. Scientists extracted DNA from four hair tufts found in the permafrost. Despite the great age of the samples, they were able to successfully determine Inuk's entire genome sequence, consisting of over 3 billion base pairs of DNA.

By comparing Inuk's DNA with sequences from known populations, the geneticists were able to demonstrate that the Saqqaq are most closely related to the Chukchi, a present-day group of indigenous people from Russia. This finding indicates that the Saqqaq originated from hunters who trekked from Siberia eastward across Alaska and Canada to Greenland, arriving in the New World independently of the groups that gave rise to Native Americans and the Inuit. Further analysis of Inuk's DNA revealed that he was dark-skinned and brown-eyed, had blood type A+, and was probably going bald.

DNA, with its double-stranded spiral, is among the most elegant of all biological molecules, but the double helix is not just a beautiful structure; it also gives DNA incredible stability and permanence, as evidenced by the condition of Inuk's 4000-year-old

DNA. In an even more remarkable feat, geneticists extracted DNA from the femur of a 45,000-year-old human (*Homo sapiens*) and sequenced his entire genome. And in 2013 researchers sequenced the genome of a long-extinct ancestral horse from DNA extracted from a 700,000-year-old bone fragment recovered from the permafrost of Yukon, Canada.

THINK-PAIR-SHARE

■ What do you think might be some of the problems associated with isolating and sequencing DNA from ancient samples, such as that of the 4000-year-old Saqqaq man from Greenland?

This chapter focuses on how DNA was identified as the source of genetic information and on how it encodes the genetic instructions for all life. We begin by considering the basic requirements for the genetic material and the history of the study of DNA—how its relation to genes was uncovered and its structure determined. The history of DNA illustrates several important points about the nature of scientific research. As with so many important scientific advances, the structure of DNA and its role as the genetic material were not discovered by any single person, but were gradually revealed over a period of almost 100 years, thanks to the work of many investigators. Our understanding of the relation between DNA and genes was enormously enhanced in 1953, when James Watson and Francis Crick, analyzing data provided by Rosalind Franklin and Maurice Wilkins, proposed a three-dimensional structure for DNA that brilliantly illuminated its role in genetics.

After reviewing the discoveries that led to our current understanding of DNA, we examine DNA structure. While the structure of DNA is important in its own right, the key genetic concept is the relation between the structure and the function of DNA, or how its structure allows it to serve as the genetic material.

10.1 Genetic Material Possesses Several Key Characteristics

Life is characterized by tremendous diversity, but the coding instructions for all living organisms are written in the same genetic language—that of nucleic acids. Surprisingly, the idea that genes are made of nucleic acids was not widely accepted until after 1950. This skepticism was due in part to a lack of knowledge about the structure of deoxyribonucleic acid (DNA). Until the structure of DNA was understood, no one knew how DNA could store and transmit genetic information.

Even before nucleic acids were identified as the genetic material, biologists recognized that, whatever the nature of the genetic material, it must possess four important characteristics:

1. **The genetic material must contain complex information.**
   First and foremost, the genetic material must be capable of storing large amounts of information—instructions for the traits and functions of an organism.

2. **The genetic material must replicate faithfully.**
   Every organism begins life as a single cell. To produce a complex multicellular organism like yourself, this single cell must undergo billions of cell divisions. At each cell division, the genetic instructions must be accurately transmitted to descendant cells. And when organisms reproduce and pass genes to their progeny, the genetic instructions must be copied with fidelity.

3. **The genetic material must encode the phenotype.**
   The genetic material (the genotype) must have the capacity to be expressed as a phenotype—to code for traits. The product of a gene is often a protein or an RNA molecule, so there must be a mechanism for genetic instructions in the DNA to be copied into RNAs and proteins.

4. **The genetic material must have the capacity to vary.**
   Genetic information must have the ability to vary because different species, and even individual members of the same species, differ in their genetic makeup.

THINK-PAIR-SHARE Question 1

CONCEPTS

The genetic material must carry large amounts of information, replicate faithfully, express its coding instructions as phenotypes, and have the capacity to vary.

✔ CONCEPT CHECK 1

Why was the discovery of the structure of DNA so important for understanding genetics?

10.2 All Genetic Information Is Encoded in the Structure of DNA or RNA

Although our understanding of how DNA encodes genetic information is relatively recent, the study of DNA structure stretches back more than 100 years (**Figure 10.1**).

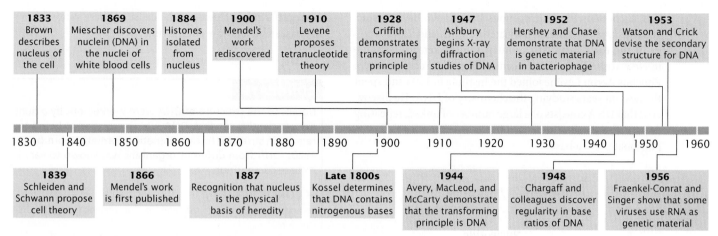

| **1833** Brown describes nucleus of the cell | **1869** Miescher discovers nuclein (DNA) in the nuclei of white blood cells | **1884** Histones isolated from nucleus | **1900** Mendel's work rediscovered | **1910** Levene proposes tetranucleotide theory | **1928** Griffith demonstrates transforming principle | **1947** Ashbury begins X-ray diffraction studies of DNA | **1952** Hershey and Chase demonstrate that DNA is genetic material in bacteriophage | **1953** Watson and Crick devise the secondary structure for DNA |

1830   1840   1850   1860   1870   1880   1890   1900   1910   1920   1930   1940   1950   1960

| **1839** Schleiden and Schwann propose cell theory | **1866** Mendel's work is first published | **1887** Recognition that nucleus is the physical basis of heredity | **Late 1800s** Kossel determines that DNA contains nitrogenous bases | **1944** Avery, MacLeod, and McCarty demonstrate that the transforming principle is DNA | **1948** Chargaff and colleagues discover regularity in base ratios of DNA | **1956** Fraenkel-Conrat and Singer show that some viruses use RNA as genetic material |

10.1 Many people have contributed to our understanding of the structure of DNA.

Early Studies of DNA

In 1868, Johann Friedrich Miescher (**Figure 10.2**) graduated from medical school in Switzerland. Influenced by an uncle who believed that the key to understanding disease lay in the chemistry of tissues, Miescher traveled to Tübingen, Germany, to study under Ernst Felix Hoppe-Seyler, an early leader in the emerging field of biochemistry. Under Hoppe-Seyler's direction, Miescher turned his attention to the chemistry of pus, a substance of clear medical importance. Pus contains white blood cells, which have large nuclei, and Miescher developed a method for isolating these nuclei. The minute amounts of nuclear material that he obtained were insufficient for a thorough chemical analysis, but he did establish that the nuclear material contained a novel substance that was slightly acidic and high in phosphorus. This material, which we now know must have consisted of DNA and protein, Miescher called *nuclein*. The substance was later renamed *nucleic acid* by one of his students.

By 1887, several researchers had independently concluded that the physical basis of heredity lies in the nucleus. Chromatin was shown to consist of nucleic acid and proteins, but which of these substances was actually the genetic information was not clear. In the late 1800s, Albrecht Kossel carried out further work on the chemistry of DNA and determined

10.2 Johann Friedrich Miescher performed the first chemical analysis of DNA. (a) Portrait of Miescher. (b) Miescher's laboratory in Tübingen, Germany. [Part a: SPL/Science Source. Part b: Courtesy of the University of Tübingen Library Image Database, Tübingen, Federal Republic of Germany.]

that it contains four nitrogenous bases: adenine, cytosine, guanine, and thymine (abbreviated A, C, G, and T).

In the early twentieth century, the Rockefeller Institute in New York City became a center for nucleic acid research. Phoebus Aaron Levene joined the Institute in 1905 and spent the next 40 years studying the chemistry of DNA. He discovered that DNA consists of a large number of linked, repeating units, called **nucleotides**; each nucleotide contains a sugar, a phosphate, and a base.

Nucleotide

Levene incorrectly proposed that DNA consists of a series of four-nucleotide units, each containing all four bases—adenine, guanine, cytosine, and thymine—in a fixed sequence. This concept, known as the tetranucleotide hypothesis, implied that the structure of DNA was not variable enough to make it the genetic material. The tetranucleotide hypothesis contributed to the idea that protein is the genetic material because the structure of protein, with its 20 different amino acids, could be highly variable.

As additional studies of the chemistry of DNA were completed in the 1940s and 1950s, the notion of DNA as a simple, invariant molecule began to change. Erwin Chargaff and his colleagues carefully measured the amounts of the four bases in DNA from a variety of organisms, and they found that DNA from different organisms varies greatly in base composition. This finding disproved the tetranucleotide hypothesis. They discovered that, within each species, there is some regularity in the ratios of the bases: the amount of adenine is always equal to the amount of thymine (A = T), and the amount of guanine is always equal to the amount of cytosine (G = C) (**Table 10.1**). These findings became known as

Chargaff's rules. However, the cause of these ratios among the bases was unknown at the time.

CONCEPTS

Details of the structure of DNA were worked out by a number of scientists. At first, DNA was interpreted as being too regular in structure to carry genetic information, but by the 1940s, DNA from different organisms was shown to vary in its base composition.

✔ CONCEPT CHECK 2

Levene made which contribution to our understanding of DNA structure?

a. He determined that the nucleus contains DNA.
b. He determined that DNA contains four nitrogenous bases.
c. He determined that DNA consists of nucleotides.
d. He determined that the nitrogenous bases of DNA are present in regular ratios.

DNA As the Source of Genetic Information

While chemists were working out the structure of DNA, biologists were attempting to identify the carrier of genetic information. Mendel identified the basic rules of heredity in 1866, but he had no idea about the physical nature of hereditary information. By the early 1900s, biologists had concluded that genes resided on chromosomes, which were known to contain both DNA and protein. Two sets of experiments, one conducted on bacteria and the other on viruses, provided pivotal evidence that DNA, rather than protein, was the genetic material.

THE DISCOVERY OF THE TRANSFORMING PRINCIPLE
An initial step in identifying DNA as the source of genetic information came with the discovery of a phenomenon called *transformation* (described in Section 9.3). This phenomenon was first observed in 1928 by Fred Griffith, an English physician whose special interest was the bacterium that causes pneumonia: *Streptococcus pneumoniae*. Griffith had succeeded in isolating several different strains of *S. pneumoniae* (type I, II, III, and so forth). In the virulent (disease-causing) forms of a strain, each bacterium is surrounded by a polysaccharide coat, which makes the bacterial colony appear smooth (S) when grown on an agar plate. Griffith found that these virulent forms occasionally mutated to nonvirulent forms, which lack a polysaccharide coat and produce a rough-appearing colony (R).

Griffith observed that small amounts of living type IIIS bacteria injected into mice caused the mice to develop pneumonia and die; when he examined the dead mice, he

TABLE 10.1	Base composition and ratios of bases in DNA from different sources						
Base Composition (percentage*)					**Ratio**		
Source of DNA	A	T	G	C	A/T	G/C	(A + G)/(T + C)
E. coli	26.0	23.9	24.9	25.2	1.09	0.99	1.04
Yeast	31.3	32.9	18.7	17.1	0.95	1.09	1.00
Sea urchin	32.8	32.1	17.7	18.4	1.02	0.96	1.00
Rat	28.6	28.4	21.4	21.5	1.01	1.00	1.00
Human	30.3	30.3	19.5	19.9	1.00	0.98	0.99

*Percentage in moles of nitrogenous constituents per 100 g-atoms of phosphate in hydrolysate corrected for 100% recovery.
Source: E. Chargaff and J. Davidson (eds.), *The Nucleic Acids*, Vol 1 (New York: Academic Press, 1955).

found large amounts of type IIIS bacteria in their blood (**Figure 10.3a**). When Griffith injected type IIR bacteria into mice, the mice lived, and no bacteria were recovered from their blood (**Figure 10.3b**). Griffith knew that boiling killed all bacteria and destroyed their virulence; when he injected large amounts of heat-killed type IIIS bacteria into mice, the mice lived, and no type IIIS bacteria were recovered from their blood (**Figure 10.3c**).

The results of these experiments were not unusual. However, Griffith got a surprise when he injected his mice with a small amount of living type IIR bacteria along with a large amount of heat-killed type IIIS bacteria. Because both the type IIR bacteria and the heat-killed type IIIS bacteria were nonvirulent, he expected these mice to live. Surprisingly, 5 days after the injections, the mice developed pneumonia and died (**Figure 10.3d**). When Griffith examined blood from the hearts of these mice, he observed live type IIIS

bacteria. Furthermore, these bacteria retained their type IIIS characteristics through several generations: their virulence was heritable.

Griffith considered all of the possible interpretations of his results. First, there was a possibility that he had not sufficiently sterilized the type IIIS bacteria and thus a few live bacteria had remained in the culture. Any live bacteria injected into the mice would have multiplied and caused pneumonia. Griffith knew that this possibility was unlikely because he had used only heat-killed type IIIS bacteria in the control experiment, and they never produced pneumonia in the mice.

A second interpretation was that the live type IIR bacteria had mutated to the virulent S form. Such a mutation would cause pneumonia in the mice, but it would produce type IIS bacteria, not the type IIIS that Griffith found in the dead mice. Because type II and type III bacteria differ in a number of traits, many mutations would be required for type II bacteria to mutate to type III bacteria, and the chance of all the mutations occurring simultaneously was impossibly low.

Griffith concluded that the type IIR bacteria had somehow been transformed, acquiring the genetic virulence of the dead type IIIS bacteria, and that this transformation had produced a permanent, genetic change in the bacteria. Although Griffith didn't understand the nature of this transformation, he theorized that some substance in the polysaccharide coat of the dead bacteria might be responsible. He called this substance the **transforming principle**.

> TRY PROBLEM 19

IDENTIFICATION OF THE TRANSFORMING PRINCIPLE

At the time of Griffith's report, Oswald Avery (see Figure 10.1) was a microbiologist at the Rockefeller Institute. At first, Avery was skeptical of Griffith's results, but after other microbiologists successfully repeated Griffith's experiments with other bacteria, Avery set out to understand the nature of the transforming substance.

After 10 years of research, Avery, Colin MacLeod, and Maclyn McCarty succeeded in isolating and partially purifying the transforming substance. They showed that it had a chemical composition closely matching that of DNA and quite different from that of proteins. Enzymes such as trypsin and chymotrypsin, known to break down proteins, had no effect on the transforming substance. Ribonuclease, an enzyme that destroys RNA, also had no effect. Enzymes capable of destroying DNA, however, eliminated the biological activity of the transforming substance (**Figure 10.4**). Avery, MacLeod, and McCarty showed that the transforming substance precipitated at about the same rate as purified DNA and that it absorbed ultraviolet light at the same wavelengths as DNA. These results, published in 1944, provided compelling evidence that the transforming principle—and therefore genetic information—resides in

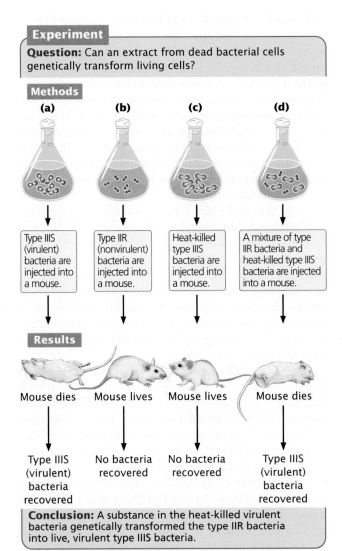

Experiment

Question: Can an extract from dead bacterial cells genetically transform living cells?

Methods

(a)	(b)	(c)	(d)
Type IIIS (virulent) bacteria are injected into a mouse.	Type IIR (nonvirulent) bacteria are injected into a mouse.	Heat-killed type IIIS bacteria are injected into a mouse.	A mixture of type IIR bacteria and heat-killed type IIIS bacteria are injected into a mouse.

Results

Mouse dies	Mouse lives	Mouse lives	Mouse dies
Type IIIS (virulent) bacteria recovered	No bacteria recovered	No bacteria recovered	Type IIIS (virulent) bacteria recovered

Conclusion: A substance in the heat-killed virulent bacteria genetically transformed the type IIR bacteria into live, virulent type IIIS bacteria.

10.3 Griffith's experiments demonstrated transformation in bacteria.

Experiment

Question: What is the chemical nature of the transforming substance?

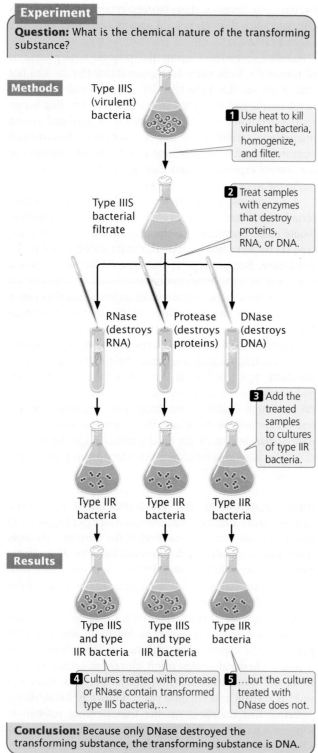

Methods

Type IIIS (virulent) bacteria

1 Use heat to kill virulent bacteria, homogenize, and filter.

Type IIIS bacterial filtrate

2 Treat samples with enzymes that destroy proteins, RNA, or DNA.

RNase (destroys RNA)

Protease (destroys proteins)

DNase (destroys DNA)

3 Add the treated samples to cultures of type IIR bacteria.

Type IIR bacteria

Type IIR bacteria

Type IIR bacteria

Results

Type IIIS and type IIR bacteria

Type IIIS and type IIR bacteria

Type IIR bacteria

4 Cultures treated with protease or RNase contain transformed type IIIS bacteria,...

5 ...but the culture treated with DNase does not.

Conclusion: Because only DNase destroyed the transforming substance, the transforming substance is DNA.

10.4 Avery, MacLeod, and McCarty's experiment revealed the chemical nature of the transforming principle.

DNA. However, new theories in science are rarely accepted on the basis of a single experiment, and many biologists continued to prefer the hypothesis that the genetic material is protein.

CONCEPTS

The process of transformation indicates that some substance—the transforming principle—is capable of genetically altering bacteria. Avery, MacLeod, and McCarty demonstrated that the transforming principle is DNA, providing the first evidence that DNA is the genetic material.

✔ CONCEPT CHECK 3

If Avery, MacLeod, and McCarty had found that samples of heat-killed bacteria treated with RNase and DNase transformed bacteria, but that samples treated with protease did not, what conclusion would they have drawn?

a. Protease carries out transformation.
b. RNA and DNA are the genetic materials.
c. Protein is the genetic material.
d. RNase and DNase are necessary for transformation.

THE HERSHEY–CHASE EXPERIMENT A second piece of evidence that indicated DNA was the genetic material resulted from a study of the T2 bacteriophage conducted by Alfred Hershey and Martha Chase. The T2 bacteriophage is a virus that infects the bacterium *Escherichia coli* (**Figure 10.5a**). As we saw in Section 9.4, a phage reproduces by attaching to the outer wall of a bacterial cell and injecting its DNA into the cell, where it replicates and directs the cell to synthesize phage proteins. The phage DNA becomes encapsulated within the phage proteins, producing progeny phages that lyse (break open) the cell and escape (**Figure 10.5b**).

At the time of the Hershey–Chase study (their paper was published in 1952), biologists did not understand exactly how phages reproduce. What they did know was that the T2 phage is approximately 50% protein and 50% DNA, that a phage infects a bacterial cell by first attaching to the cell wall, and that progeny phages are ultimately produced within the cell. Because the progeny carry the same traits as the infecting phage, genetic material from the infecting phage must be transmitted to the progeny, but how this genetic transmission takes place was unknown.

Hershey and Chase designed a series of experiments to determine whether the phage *protein* or the phage *DNA* is transmitted in phage reproduction. To follow the fates of protein and DNA, they used radioactive forms, or **isotopes**, of phosphorus and sulfur. A radioactive isotope can be used as a tracer to identify the location of a specific molecule because any molecule containing the isotope will be radioactive and therefore easily detected. DNA contains phosphorus, but not sulfur, so Hershey and Chase used a radioactive isotope of phosphorus (^{32}P) to follow phage DNA during reproduction. Protein contains sulfur, but not phosphorus, so they used a radioactive isotope of sulfur (^{35}S) to follow the protein.

Hershey and Chase grew one batch of *E. coli* in a medium containing ^{32}P and infected the bacteria with the T2 phage

(a)

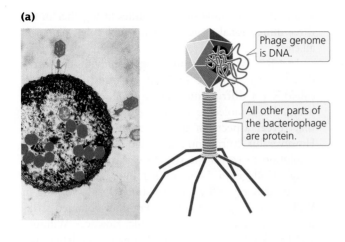

Phage genome is DNA.

All other parts of the bacteriophage are protein.

(b)

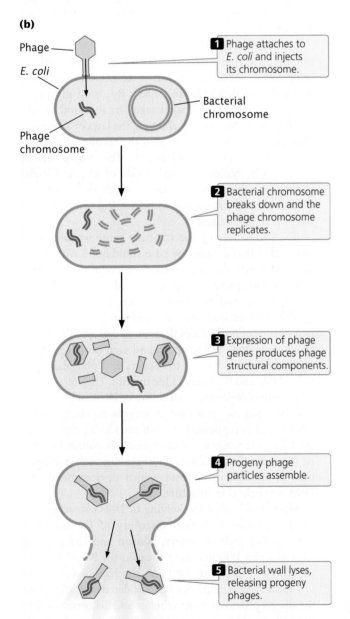

Phage

E. coli

Phage chromosome

Bacterial chromosome

1 Phage attaches to E. coli and injects its chromosome.

2 Bacterial chromosome breaks down and the phage chromosome replicates.

3 Expression of phage genes produces phage structural components.

4 Progeny phage particles assemble.

5 Bacterial wall lyses, releasing progeny phages.

10.5 T2 is a bacteriophage that infects E. coli. (a) T2 phage. (b) Its life cycle. [Micrograph: © Lee D. Simon/Science Source.]

so that all the progeny phages would have DNA labeled with ^{32}P (**Figure 10.6**). They grew a second batch of E. coli in a medium containing ^{35}S and infected these bacteria with T2 phage so that all the progeny phages would have proteins labeled with ^{35}S. Hershey and Chase then infected separate batches of unlabeled E. coli with the ^{35}S- and ^{32}P-labeled progeny phages. After allowing time for the phages to infect the E. coli cells, they placed the cells in a blender and sheared off the now-empty phage protein coats from the cell walls. They separated out the protein coats and cultured the infected bacterial cells.

In the case of the bacteria infected by phages labeled with ^{35}S, most of the radioactivity was detected in the phage protein coats, and little was detected in the cells. Furthermore, when new phages emerged from the cells, they contained almost no ^{35}S (see Figure 10.6). This result indicated that the protein component of a phage does not enter the cell and is not transmitted to progeny phages.

In contrast, when Hershey and Chase infected bacteria with ^{32}P-labeled phages and then removed the phage protein coats, the bacteria were radioactive. Most significantly, after the bacterial cells were lysed and new progeny phages emerged, many of those phages emitted radioactivity, demonstrating that DNA from the infecting phages had been passed on to the progeny phages (see Figure 10.6). These results confirmed that DNA, not protein, is the genetic material of phages. ▸ **TRY PROBLEM 24**

CONCEPTS

Using radioactive isotopes, Hershey and Chase traced the movement of DNA and protein during phage infection of bacteria. They demonstrated that DNA, not protein, enters the bacterial cell during phage reproduction and that only DNA is passed on to progeny phages.

✔ **CONCEPT CHECK 4**

Could Hershey and Chase have used a radioactive isotope of carbon instead of ^{32}P? Why or why not?

Watson and Crick's Discovery of the Three-Dimensional Structure of DNA

These experiments on the nature of the genetic material set the stage for one of the most important advances in the history of biology: the discovery of the three-dimensional structure of DNA by James Watson and Francis Crick in 1953.

Before Watson and Crick's breakthrough, much of the basic chemistry of DNA had already been determined by Miescher, Kossel, Levene, Chargaff, and others, who had established that DNA consists of nucleotides and that each nucleotide contains a sugar, a nitrogenous base, and a phosphate

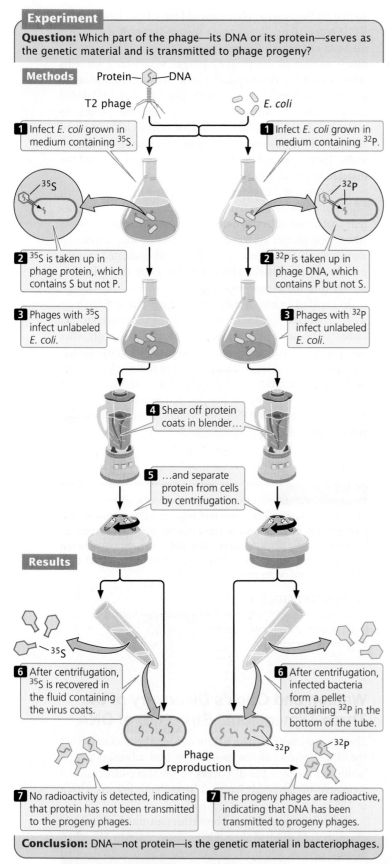

Experiment

Question: Which part of the phage—its DNA or its protein—serves as the genetic material and is transmitted to phage progeny?

Methods

Protein —— DNA

T2 phage   E. coli

1 Infect E. coli grown in medium containing ^{35}S.

1 Infect E. coli grown in medium containing ^{32}P.

^{35}S

^{32}P

2 ^{35}S is taken up in phage protein, which contains S but not P.

2 ^{32}P is taken up in phage DNA, which contains P but not S.

3 Phages with ^{35}S infect unlabeled E. coli.

3 Phages with ^{32}P infect unlabeled E. coli.

4 Shear off protein coats in blender…

5 …and separate protein from cells by centrifugation.

Results

^{35}S

6 After centrifugation, ^{35}S is recovered in the fluid containing the virus coats.

6 After centrifugation, infected bacteria form a pellet containing ^{32}P in the bottom of the tube.

^{32}P

^{32}P

Phage reproduction

7 No radioactivity is detected, indicating that protein has not been transmitted to the progeny phages.

7 The progeny phages are radioactive, indicating that DNA has been transmitted to progeny phages.

Conclusion: DNA—not protein—is the genetic material in bacteriophages.

10.6 Hershey and Chase demonstrated that DNA carries the genetic information in bacteriophages.

group. However, how the nucleotides fit together in the three-dimensional structure of the molecule was not at all clear.

In 1947, William Astbury began studying the three-dimensional structure of DNA by using a technique called **X-ray diffraction** (**Figure 10.7**), in which X-rays beamed at a molecule are reflected in specific patterns that reveal aspects of the structure of the molecule. However, his diffraction images did not provide enough resolution to reveal the structure. A research group at King's College in London, led by Maurice Wilkins, also used X-ray diffraction to study DNA. Working in Wilkins's laboratory, Rosalind Franklin obtained strikingly better images of the molecule. However, Wilkins's and Franklin's progress in developing a complete structure of the molecule was impeded by personal discord between them.

Watson and Crick investigated the structure of DNA not by collecting new data, but by using all available information about the chemistry of DNA to construct molecular models (**Figure 10.8a**). By using the excellent X-ray diffraction images taken by Rosalind Franklin (**Figure 10.8b**) and by applying the laws of structural chemistry, they were able to limit the number of possible structures that DNA could assume. They tested various structures by building models made of wire and metal plates. With their models, they were able to see whether a structure was compatible with chemical principles and with the X-ray images.

The key to solving the structure came when Watson recognized that an adenine base could bond with a thymine base and that a guanine base could bond with a cytosine base; these pairings accounted for the base ratios that Chargaff had discovered earlier. The model developed by Watson and Crick showed that DNA consists of two strands of nucleotides that run in opposite directions (are antiparallel) and wind around each other to form a right-handed helix, with the sugars and phosphates on the outside and the bases in the interior. They recognized that the double-stranded structure of DNA, with its specific base pairing, provided an elegant means by which DNA could be replicated. Watson and Crick published an electrifying description of their model in *Nature* in 1953. At the same time, Wilkins and Franklin each published their X-ray diffraction data, which demonstrated that DNA was helical in structure.

Many have called the solving of DNA's structure the most important biological discovery of the twentieth century. For their discovery, Watson and Crick, along with Maurice Wilkins, were awarded the Nobel Prize in chemistry in 1962. Rosalind Franklin had died of cancer in 1958 and thus could not be considered a candidate for the shared prize, but many scholars and historians believe that she should receive equal credit for solving the structure of DNA.

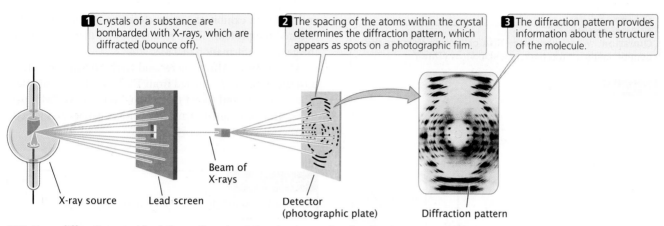

1 Crystals of a substance are bombarded with X-rays, which are diffracted (bounce off).

2 The spacing of the atoms within the crystal determines the diffraction pattern, which appears as spots on a photographic film.

3 The diffraction pattern provides information about the structure of the molecule.

Beam of X-rays

X-ray source   Lead screen   Detector (photographic plate)   Diffraction pattern

10.7 X-ray diffraction provides information about the structures of molecules. [Science Source.]

Following the discovery of DNA's structure, much research was focused on how genetic information is encoded within the base sequence and how this information is copied and expressed. Even today, the details of DNA structure and function continue to be the subject of active research.

THINK-PAIR-SHARE Questions 2 and 3

CONCEPTS

By collecting existing information about the chemistry of DNA and building molecular models, Watson and Crick were able to discover the three-dimensional structure of the DNA molecule.

✔ CONCEPT CHECK 5

What did Watson and Crick use to help solve the structure of DNA?

a. X-ray diffraction images   c. Models of DNA
b. Laws of structural chemistry   d. All the above

RNA As Genetic Material

In most organisms, DNA carries the genetic information. A few viruses, however, use RNA, not DNA, as their genetic material. This was demonstrated in 1956 by Heinz Fraenkel-Conrat and Bea Singer, who worked with the tobacco mosaic virus (TMV), which infects and causes disease in tobacco plants (**Figure 10.9**). The tobacco mosaic virus possesses a single molecule of RNA surrounded by a helically arranged cylinder of protein molecules. Fraenkel-Conrat found that, after separating the RNA and the protein of TMV, he could remix the RNA and protein of different strains of TMV and obtain intact, infectious viral particles.

With Singer, Fraenkel-Conrat then created hybrid viruses by mixing RNA and protein from different strains of TMV. When these hybrid viruses infected tobacco leaves, new viral particles were produced. The new viral progeny were identical with the strain from which the RNA had been isolated

(a)

(b)

10.8 James Watson and Francis Crick (a) developed a three-dimensional model of the structure of DNA based in part on X-ray diffraction images taken by Rosalind Franklin (b).

[Part a: A. Barrington/Science Photo Library/Science Source. Part b: Science Source.]

Question: What substance—RNA or protein—carries the genetic material in tobacco mosiac virus (TMV)?

Methods

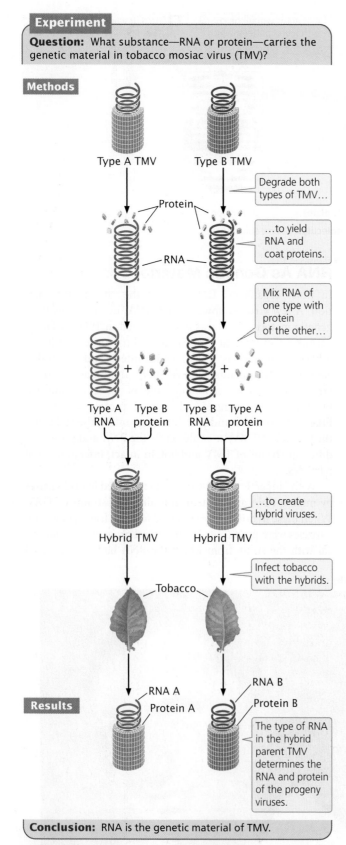

Type A TMV   Type B TMV

Degrade both types of TMV…

Protein

…to yield RNA and coat proteins.

RNA

Mix RNA of one type with protein of the other…

Type A   Type B   Type B   Type A
RNA   protein   RNA   protein

…to create hybrid viruses.

Hybrid TMV   Hybrid TMV

Infect tobacco with the hybrids.

Tobacco

RNA A   RNA B
Protein A   Protein B

Results

The type of RNA in the hybrid parent TMV determines the RNA and protein of the progeny viruses.

Conclusion: RNA is the genetic material of TMV.

10.9 Fraenkel-Conrat and Singer's experiment demonstrated that RNA carries the genetic information in tobacco mosaic virus.

and did not exhibit the characteristics of the strain that donated the protein. These results showed that RNA carries the genetic information in TMV.

Also in 1956, Alfred Gierer and Gerhard Schramm demonstrated that RNA isolated from TMV is sufficient to infect tobacco plants and direct the production of new TMV particles. This finding confirmed that RNA carries the genetic instructions in this virus. **> TRY PROBLEM 18**

CONCEPTS

RNA serves as the genetic material in some viruses.

10.3 DNA Consists of Two Complementary and Antiparallel Nucleotide Strands That Form a Double Helix

DNA, though relatively simple in structure, has an elegance and beauty unsurpassed by other large molecules. It is useful to consider the structure of DNA at three levels of increasing complexity, known as the primary, secondary, and tertiary structures of DNA. The primary structure of DNA refers to its nucleotide structure and how the nucleotides are joined together. The secondary structure refers to DNA's stable three-dimensional configuration, the helical structure worked out by Watson and Crick. In Chapter 11, we will consider DNA's tertiary structures, which are the complex packing arrangements of double-stranded DNA in chromosomes.

The Primary Structure of DNA

The primary structure of DNA consists of a string of nucleotides joined together by phosphodiester linkages.

NUCLEOTIDES DNA is typically a very long molecule and is therefore termed a macromolecule. For example, within each human chromosome is a single DNA molecule that, if stretched out straight, would be several centimeters in length, thousands of times longer than the cell itself. In spite of its large size, DNA has quite a simple structure: it is a polymer—that is, a chain made up of many repeating units linked together. The repeating units of DNA are nucleotides, each comprising three parts: (1) a sugar, (2) a phosphate group, and (3) a nitrogen-containing base.

The sugars of nucleic acids—called pentose sugars—have five carbon atoms, numbered 1′, 2′, 3′, 4′, and 5′ (**Figure 10.10**). The sugars of DNA and RNA are slightly different in structure. RNA's sugar, called **ribose**, has a hydroxyl group (OH) attached to the 2′-carbon atom, whereas DNA's sugar, or **deoxyribose**, has a hydrogen atom (H) at this position and therefore contains one oxygen atom fewer overall.

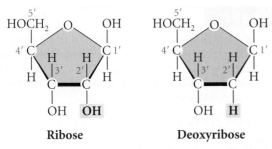

10.10 A nucleotide contains either a ribose sugar (in RNA) or a deoxyribose sugar (in DNA). The carbon atoms of the sugars are assigned primed numbers.

This difference gives rise to the names ribonucleic acid (RNA) and *deoxy*ribonucleic acid (DNA). This minor chemical difference is recognized by most of the cellular enzymes that interact with DNA or RNA, thus providing specific functions for each nucleic acid. Furthermore, the additional oxygen atom in the RNA nucleotide makes it more reactive and less chemically stable than DNA. For this reason, DNA is better suited to serve as the long-term carrier of genetic information.

The second component of a nucleotide is its **nitrogenous base**, which may be either of two types—a **purine** or a **pyrimidine** (**Figure 10.11**). Each purine consists of a six-member ring attached to a five-member ring, whereas each pyrimidine consists of a six-member ring only. Both DNA and RNA contain two purines, **adenine** and **guanine** (A and G), which differ in the positions of their double bonds and in

the groups attached to the six-member ring. Three pyrimidines are common in nucleic acids: **cytosine** (C), **thymine** (T), and **uracil** (U). Cytosine is present in both DNA and RNA; however, thymine is restricted to DNA, and uracil is found only in RNA. The three pyrimidines differ in the groups or atoms attached to the carbon atoms of the ring and in the number of double bonds in the ring. In a nucleotide, the nitrogenous base always forms a covalent bond with the 1′-carbon atom of the sugar (see Figure 10.10). A deoxyribose or a ribose sugar and a base together are referred to as a **nucleoside**.

The third component of a nucleotide is its **phosphate group**, which consists of a phosphorus atom bonded to four oxygen atoms (**Figure 10.12**). Phosphate groups are found in every nucleotide and frequently carry a negative charge, which makes DNA acidic. The phosphate group is always bonded to the 5′-carbon atom of the sugar (see Figure 10.10) in a nucleotide.

The DNA nucleotides are properly known as **deoxyribonucleotides**, or deoxyribonucleoside 5′-monophosphates. Because there are four types of bases, there are four different kinds of DNA nucleotides (**Figure 10.13**). The equivalent RNA nucleotides are termed **ribonucleotides**, or ribonucleoside 5′-monophosphates. RNA molecules sometimes contain additional rare bases, which are modified forms of the four common bases. These modified bases will be discussed in more detail when we examine the function of RNA molecules in Chapter 14. The names for DNA bases, nucleotides, and nucleosides are shown in **Table 10.2**.

❯ **TRY PROBLEM 26**

Purine
(basic structure)

Pyrimidine
(basic structure)

Adenine (A)

Guanine (G)

Cytosine (C)

Thymine (T)
(present in DNA)

Uracil (U)
(present in RNA)

10.11 A nucleotide contains either a purine or a pyrimidine base. The atoms of the rings in the bases are assigned unprimed numbers.

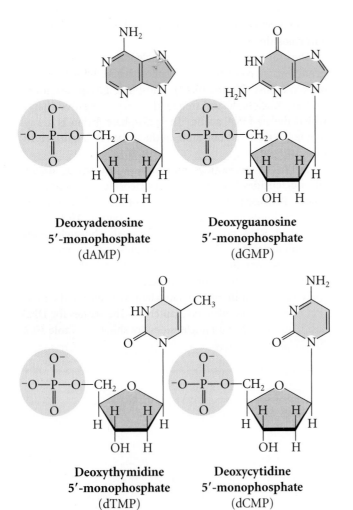

Phosphate

10.12 A nucleotide contains a phosphate group.

Deoxyadenosine
5′-monophosphate
(dAMP)

Deoxyguanosine
5′-monophosphate
(dGMP)

Deoxythymidine
5′-monophosphate
(dTMP)

Deoxycytidine
5′-monophosphate
(dCMP)

10.13 There are four types of DNA nucleotides. These nucleotides are deoxyribonucleoside 5′-monophosphates.

CONCEPTS

The primary structure of DNA consists of a string of nucleotides. Each nucleotide consists of a five-carbon sugar, a phosphate group, and a nitrogenous base. There are two types of DNA bases: purines (adenine and guanine) and pyrimidines (thymine and cytosine).

✔ CONCEPT CHECK 6

How do the sugars of RNA and DNA differ?

a. RNA has a six-carbon sugar; DNA has a five-carbon sugar.
b. The sugar of RNA has a hydroxyl group that is not found in the sugar of DNA.
c. RNA contains uracil; DNA contains thymine.
d. DNA's sugar has a phosphorus atom; RNA's sugar does not.

POLYNUCLEOTIDE STRANDS DNA is made up of many nucleotides connected by covalent bonds, which join the 5′-phosphate group of one nucleotide to the 3′-hydroxyl group of the next nucleotide (**Figure 10.14**). (Note that the structures shown in Figure 10.14 are flattened into two dimensions, although the molecule itself is three-dimensional, as shown in Figure 10.15a below.) These bonds, called **phosphodiester linkages**, are strong covalent bonds; a series of nucleotides linked in this way constitutes a **polynucleotide strand**. The backbone of the polynucleotide strand is composed of alternating sugars and phosphate groups; the bases project away from the long axis of the strand. The negative charges of the phosphate groups are frequently neutralized by their association with positive charges on proteins, metals, or other molecules.

An important characteristic of the polynucleotide strand is its direction, or polarity. At one end of the strand, a free phosphate group (unattached on one side) is attached to the 5′-carbon atom of the sugar in the nucleotide. This end of the strand is therefore referred to as the **5′ end**. The other end of the strand, referred to as the **3′ end**, has a free hydroxyl group attached to the 3′-carbon atom of the sugar. RNA nucleotides are also connected by phosphodiester linkages to form similar polynucleotide strands (see Figure 10.14).

TABLE 10.2	Names of DNA bases, nucleotides, and nucleosides			
	Adenine	**Guanine**	**Thymine**	**Cytosine**
Base symbol	A	G	T	C
Nucleotide	Deoxyadenosine 5′ monophosphate	Deoxyguanosine 5′ monophosphate	Deoxythymidine 5′ monophosphate	Deoxycytidine 5′ monophosphate
Nucleotide symbol	dAMP	dGMP	dTMP	dCMP
Nucleoside	Deoxyadenosine	Deoxyguanosine	Deoxythymidine	Deoxycytidine
Nucleoside symbol	dA	dG	dT	dC

CONCEPTS

The nucleotides of DNA are joined into polynucleotide strands by phosphodiester bonds that connect the 3'-carbon atom of one nucleotide to the 5'-phosphate group of the next. Each polynucleotide strand has polarity, with a 5' end and a 3' end.

Secondary Structures of DNA

The secondary structure of DNA refers to its three-dimensional configuration—its fundamental helical structure. DNA's secondary structure can assume a variety of configurations, depending on its base sequence and the conditions in which it is placed.

THE DOUBLE HELIX A fundamental characteristic of DNA's secondary structure is that it consists of two polynucleotide strands wound around each other—it's a *double helix*. The sugar–phosphate linkages are on the outside of the helix, and the bases are stacked in the interior of the molecule (see Figure 10.14). The two polynucleotide strands run in opposite directions—they are **antiparallel**, which means that the 5' end of one strand is opposite the 3' end of the other strand.

The two strands are held together by two types of molecular forces. Hydrogen bonds link the bases on opposite strands (see Figure 10.14). These bonds are relatively weak compared with the covalent phosphodiester bonds that connect the sugars and phosphate groups of adjoining nucleotides on the same strand. As we will see, several important functions of DNA require the separation of its

DNA polynucleotide strands   **RNA polynucleotide strand**

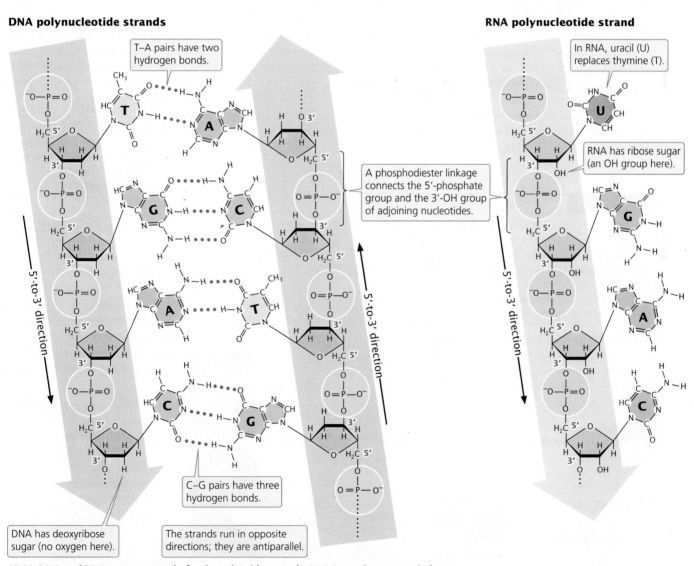

10.14 DNA and RNA are composed of polynucleotide strands. DNA is usually composed of two polynucleotide strands, although single-stranded DNA is found in some viruses.

two nucleotide strands, and this separation can be readily accomplished because of the relative ease of breaking and reestablishing the hydrogen bonds.

The nature of the hydrogen bond imposes a limitation on the types of bases that can pair. Adenine normally pairs only with thymine through two hydrogen bonds, and cytosine normally pairs only with guanine through three hydrogen bonds (see Figure 10.14). Because three hydrogen bonds form between C and G and only two hydrogen bonds form between A and T, C–G pairing is stronger than A–T pairing. The specificity of the base pairing means that wherever there is an A on one strand, there must be a T in the corresponding position on the other strand, and wherever there is a G on one strand, a C must be on the other. The two polynucleotide strands of a DNA molecule are therefore not identical, but rather **complementary DNA strands**. The complementary nature of the two nucleotide strands provides for efficient and accurate DNA replication, as we will see in Chapter 12.

The second force that holds the two DNA strands together is the interaction between the stacked base pairs in the interior of the molecule. Stacking means that adjacent bases are aligned so that their rings are parallel and stack on top of one another. The stacking interactions stabilize the DNA molecule but do not require that any particular base follow another. Thus, the base sequence of the DNA molecule is free to vary, allowing DNA to carry genetic information.

> **TRY PROBLEMS 32 AND 37**

CONCEPTS

DNA consists of two polynucleotide strands. The sugars and phosphate groups of each polynucleotide strand are on the outside of the molecule, and the bases are in the interior. Hydrogen bonding joins the bases of the two strands: guanine pairs with cytosine, and adenine pairs with thymine. The two polynucleotide strands of a DNA molecule are complementary and antiparallel.

✔ CONCEPT CHECK 7

The antiparallel nature of DNA refers to
a. its charged phosphate groups.
b. the pairing of bases on one strand with bases on the other strand.
c. the formation of hydrogen bonds between bases from opposite strands.
d. the opposite direction of the two strands of nucleotides.

DIFFERENT SECONDARY STRUCTURES As we have seen, DNA normally consists of two polynucleotide strands that are antiparallel and complementary (exceptions are the single-stranded DNA molecules found in a few viruses). The precise three-dimensional shape of the molecule can vary, however, depending on the conditions in which the DNA is placed and, in some cases, on the base sequence itself.

The three-dimensional structure of DNA described by Watson and Crick is termed the **B-DNA** structure (**Figure 10.15**). This structure exists when plenty of water surrounds the molecule and there are no unusual base sequences in the DNA—conditions that are likely to be present in cells. The B-DNA structure is the most stable configuration for a random sequence of nucleotides under physiological conditions, and most evidence suggests that it is the predominant structure in the cell.

B-DNA is a right-handed helix, meaning that it has a clockwise spiral. There are approximately 10 base pairs (bp) per 360-degree rotation of the helix, so each base pair is twisted 36 degrees relative to the adjacent bases (see Figure 10.15b). The base pairs are 0.34 nanometers (nm) apart, so each complete rotation of the molecule encompasses 3.4 nm. The diameter of the helix is 2 nm, and the bases are perpendicular to the long axis of the DNA molecule. A space-filling model shows that B-DNA has a slim and elongated structure (see Figure 10.15a). The spiraling of the nucleotide strands creates major and minor grooves in the helix, features that are important for the binding of some proteins that regulate the expression of genetic information (see Chapter 16).

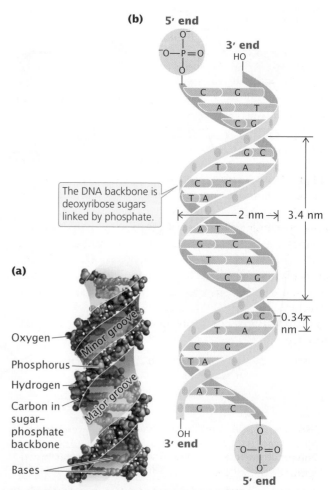

10.15 B-DNA consists of a right-handed helix with approximately 10 bases per turn. (a) Space-filling model of B-DNA showing major and minor grooves. (b) Diagrammatic representation.

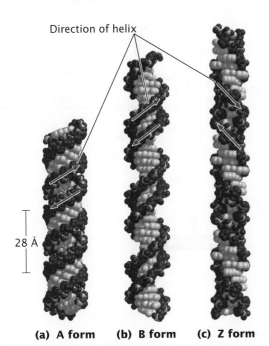

Direction of helix

28 Å

(a) **A form**   (b) **B form**   (c) **Z form**

10.16 DNA can assume several different secondary structures.
[After J. M. Berg, J. L. Tymoczko, and L. Stryer, *Biochemistry*, 6th ed.
(New York: W. H. Freeman and Company, 2002), pp. 785 and 787.]

Another secondary structure that DNA can assume is the **A-DNA** structure, which exists if less water is present. Like B-DNA, A-DNA is a right-handed helix (**Figure 10.16a**), but it is shorter and wider than B-DNA (**Figure 10.16b**), and its bases are tilted away from the main axis of the molecule. A-DNA has been detected in some DNA–protein complexes and in spores of some bacteria.

A radically different secondary structure, called **Z-DNA** (**Figure 10.16c**), forms a left-handed helix. In this structure, the sugar–phosphate backbone zigzags back and forth, giving rise to its name. A Z-DNA structure can result if the molecule contains particular base sequences, such as stretches of alternating C and G nucleotides. Researchers have found that Z-DNA-specific antibodies bind to regions of the DNA that are being transcribed into RNA, suggesting that Z-DNA may play some role in gene expression. Additional secondary structures of DNA (C-DNA, D-DNA, etc.) can form under specialized laboratory conditions or in DNA with specific base sequences.

THINK-PAIR-SHARE Question 4

CONCEPTS

DNA can assume different secondary structures, depending on the conditions in which it is placed and on its base sequence. B-DNA is thought to be the most common configuration in the cell.

✔ CONCEPT CHECK 8
How does Z-DNA differ from B-DNA?

CONNECTING CONCEPTS

Genetic Implications of DNA Structure

Watson and Crick's great contribution was their elucidation of the genotype's chemical structure, which made it possible for geneticists to begin to examine genes directly, instead of looking only at the phenotypic consequences of gene action. The determination of the structure of DNA led to the birth of molecular genetics—the study of the chemical and molecular nature of genetic information.

Watson and Crick's structure did more than create the potential for molecular genetic studies, however; it was an immediate source of insight into key genetic processes. At the beginning of this chapter, four fundamental requirements for genetic material were identified. Watson and Crick's structure showed how DNA met those requirements.

1. First, genetic material must be capable of carrying large amounts of information. Watson and Crick's model suggested that genetic instructions are encoded in the base sequence of DNA, the only variable part of the molecule.

2. A second necessary property of genetic material is its ability to replicate faithfully. The complementary polynucleotide strands of DNA make this replication possible. Watson and Crick proposed that in replication, the two polynucleotide strands unzip, breaking the weak hydrogen bonds between them, and each strand serves as a template on which a new strand is synthesized. The specificity of the base pairing means that only one possible sequence of bases—the complementary sequence—can be synthesized from each template strand. Newly replicated double-stranded DNA molecules are therefore identical with the original double-stranded DNA molecule (see Chapter 12 on DNA replication).

3. A third essential property of genetic material is the ability to be expressed as a phenotype. DNA expresses its genetic instructions by first transferring its information to an RNA molecule in a process termed **transcription** (see Chapter 13). The term *transcription* is appropriate because, although the information is transferred from DNA to RNA, the information remains in the language of nucleic acids. In some cases, the RNA molecule then transfers the genetic information to a protein by specifying its amino acid sequence. This process is termed **translation** (see Chapter 15) because the information must be *translated* from the language of nucleotides into the language of amino acids.

4. A fourth essential property of genetic material is the capacity to vary. The variation in DNA, as we have seen, consists of differences in the sequence of bases found among different individuals.

We can now identify three major pathways of information flow in the cell (**Figure 10.17a**): in **replication**, information passes from one DNA molecule to other DNA molecules; in transcription, information passes from DNA to RNA; and in translation, information passes from RNA to protein. This concept of information flow was formalized by Francis Crick in a concept that he called the **central dogma** of molecular biology. The central dogma states that genetic information passes from DNA to protein in a one-way information pathway. We now realize, however, that the central dogma is an oversimplification.

(a) Major information pathways

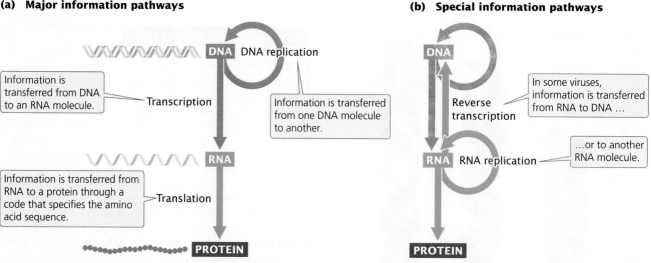

(b) Special information pathways

10.17 Pathways of information transfer within the cell.

In addition to the three general information pathways of replication, transcription, and translation, other transfers may take place in certain organisms or under special circumstances. Retroviruses (see Chapter 9) and some transposable elements (see Chapter 18) transfer information from RNA to DNA (in **reverse transcription**), and some RNA viruses transfer information from RNA to RNA (in **RNA replication**); (**Figure 10.17b**).

THINK-PAIR-SHARE Question 5

10.4 Special Structures Can Form in DNA and RNA

Sequences *within* a single strand of nucleotides may be complementary to each other and able to pair by forming hydrogen bonds, producing double-stranded regions (**Figure 10.18**). This internal base pairing imparts a secondary structure to a single-stranded molecule. One common type of secondary structure found in single strands of nucleotides is a **hairpin**, which forms when sequences of nucleotides on the same strand are inverted complements (see Figure 10.18a). A hairpin consists of a region of paired bases (the stem) and intervening unpaired bases (a loop). When the complementary sequences are contiguous, a stem is formed with no loop (see Figure 10.18b). RNA molecules may contain numerous hairpins, which allow them to fold up into complex structures (see Figure 10.18c). Secondary structures play important roles in the functions of many RNA molecules, as we will see in Chapters 14 and 15.

DNA sequences can also sometimes form three-stranded (triplex) structures, called **H-DNA**, when some of the DNA unwinds and a single polynucleotide strand

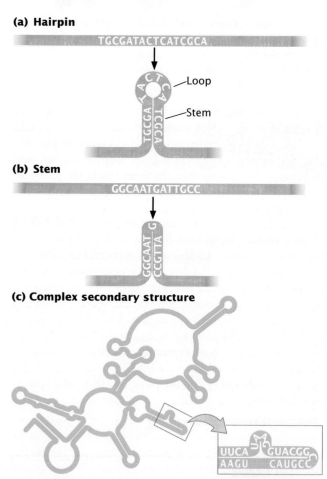

10.18 Both DNA and RNA can form special secondary structures.
(a) A hairpin consisting of a region of paired bases (which form the stem) and a region of unpaired bases between the complementary sequences (which form a loop at the end of the stem). (b) A stem with no loop. (c) Secondary structure of the RNA component of RNase P of *E. coli*. RNA molecules often have complex secondary structures.

from one part of the molecule pairs with double-stranded DNA from another part of the molecule (**Figure 10.19**). This is possible because, under certain conditions, one base can simultaneously pair with two other bases. H-DNA often occurs in long sequences containing only purine bases or only pyrimidine bases. Some triplex structures consist of one strand of purines paired with two strands of pyrimidines; other triplex structures consist of one strand of pyrimidines paired with two strands of purines. Sequences capable of adopting an H-DNA conformation are common in mammalian genomes, and evidence suggests that H-DNA occurs under natural conditions. Recent research has demonstrated that H-DNA breaks more readily than double-stranded DNA, which then leads to higher rates of mutation in regions where H-DNA structures occur. Quadruplex structures involving four strands of DNA can also occur under certain conditions. > TRY PROBLEM 38

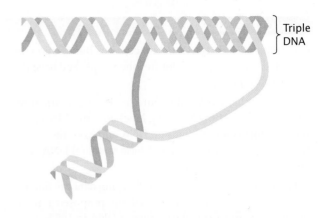

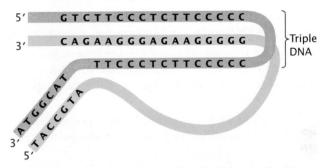

10.19 H-DNA arises when three polynucleotide strands pair.

CONCEPTS

In DNA and RNA, base pairing between nucleotides on the same strand produces special secondary structures such as hairpins. Triple-stranded DNA structures can arise when a single strand of DNA pairs with double-stranded DNA.

✔ CONCEPT CHECK 9

Hairpins are formed in DNA as a result of
a. sequences on the same strand that are inverted and complementary.
b. sequences on the opposite strand that are complements.
c. sequences on the same strand that are identical.
d. sequences on the opposite strand that are identical.

The extent of cytosine methylation varies among eukaryotic organisms: in most animal cells, about 5% of the cytosine bases are methylated, but there is no methylation of cytosine in yeast, and more than 50% of the cytosine bases in some plants are methylated. Why eukaryotic organisms differ so widely in their degree of methylation is not clear.

CONCEPTS

Methyl groups may be added to certain bases in DNA. Both prokaryotic and eukaryotic DNA can be methylated. In eukaryotes, cytosine bases are most often methylated to form 5-methylcytosine, and methylation is often related to gene expression.

The primary structure of DNA can also be modified in various ways. One such modification is **DNA methylation**, a process in which methyl groups (CH_3) are added (by specific enzymes) to certain positions on the nitrogenous bases. Bacterial DNA is frequently methylated to distinguish it from foreign, unmethylated DNA that may be introduced by viruses; bacteria use proteins called restriction enzymes to cut up any unmethylated viral DNA (see Chapter 19). In eukaryotic cells, methylation is often related to gene expression. Sequences that are methylated typically show low levels of transcription while sequences lacking methylation are actively being transcribed (see Chapter 17). Methylation can also affect the three-dimensional structure of the DNA molecule and is responsible for some epigenetic effects (see Chapter 21).

Adenine and cytosine are commonly methylated in bacteria. In eukaryotic DNA, cytosine bases are sometimes methylated to form **5-methylcytosine** (**Figure 10.20**).

5-Methylcytosine

10.20 In eukaryotic DNA, cytosine bases are often methylated to form 5-methylcytosine.

CONCEPTS SUMMARY

■ Genetic material must contain complex information, be replicated accurately, code for the phenotype, and have the capacity to vary.

■ Evidence that DNA is the source of genetic information came from the finding by Avery, MacLeod, and McCarty that transformation depends on DNA and from the demonstration by Hershey and Chase that viral DNA is passed on to progeny phages.

■ James Watson and Francis Crick, using data provided by Rosalind Franklin and Maurice Wilkins, proposed a model for the three-dimensional structure of DNA in 1953.

■ The results of experiments with tobacco mosaic virus showed that RNA carries genetic information in some viruses.

■ A DNA nucleotide consists of a deoxyribose sugar, a phosphate group, and a nitrogenous base. An RNA nucleotide consists of a ribose sugar, a phosphate group, and a nitrogenous base.

■ The bases of a DNA nucleotide are of two types: purines (adenine and guanine) and pyrimidines (cytosine and thymine). RNA contains the pyrimidine uracil instead of thymine.

■ Nucleotides are joined together by phosphodiester linkages to form a polynucleotide strand. Each polynucleotide strand has a free phosphate group at its 5′ end and a free hydroxyl group at its 3′ end.

■ DNA consists of two nucleotide strands that wind around each other to form a double helix. The sugars and phosphates lie on the outside of the helix, and the bases are stacked in the interior. The two strands are joined together by hydrogen bonding between bases in each strand. The two strands are antiparallel and complementary.

■ DNA molecules can form a number of different secondary structures, depending on the conditions in which the DNA is placed and on its base sequence.

■ The structure of DNA has several important genetic implications. Genetic information resides in the base sequence of DNA, which ultimately specifies the amino acid sequence of proteins. Complementarity of the bases on DNA's two strands allows genetic information to be replicated.

■ The central dogma of molecular biology proposes that information flows in a one-way direction, from DNA to RNA to protein. Exceptions to the central dogma are now known.

■ Pairing between bases on the same nucleotide strand can lead to hairpins and other special secondary structures.

■ DNA may be modified by the addition of methyl groups to the nitrogenous bases.

IMPORTANT TERMS

nucleotide (p. 290)
Chargaff's rules (p. 290)
transforming principle (p. 291)
isotope (p. 292)
X-ray diffraction (p. 294)
ribose (p. 296)
deoxyribose (p. 296)
nitrogenous base (p. 297)
purine (p. 297)
pyrimidine (p. 297)
adenine (A) (p. 297)
guanine (G) (p. 297)
cytosine (C) (p. 297)

thymine (T) (p. 297)
uracil (U) (p. 297)
nucleoside (p. 297)
phosphate group (p. 297)
deoxyribonucleotide (p. 297)
ribonucleotide (p. 297)
phosphodiester linkage (p. 298)
polynucleotide strand (p. 298)
5′ end (p. 298)
3′ end (p. 298)
antiparallel (p. 299)
complementary DNA
    strands (p. 300)

B-DNA (p. 300)
A-DNA (p. 301)
Z-DNA (p. 301)
transcription (p. 301)
translation (p. 301)
replication (p. 301)
central dogma (p. 301)
reverse transcription (p. 302)
RNA replication (p. 302)
hairpin (p. 302)
H-DNA (p. 302)
DNA methylation (p. 303)
5-methylcytosine (p. 303)

ANSWERS TO CONCEPT CHECKS

1. Without knowledge of the structure of DNA, an understanding of how genetic information was encoded or expressed was impossible.

2. c

3. c

4. No, because carbon is found in both protein and nucleic acid.

5. d

6. b

7. d

8. Z-DNA has a left-handed helix; B-DNA has a right-handed helix. The sugar–phosphate backbone of Z-DNA zigzags back and forth, whereas the sugar–phosphate backbone of B-DNA forms a smooth, continuous ribbon.

9. a

WORKED PROBLEMS

Problem 1

The percentage of cytosine in a double-stranded DNA molecule is 40%. What is the percentage of thymine?

≫ Solution Strategy

What information is required in your answer to the problem?
The percentage of thymine in the DNA molecule.

What information is provided to solve the problem?

- The DNA molecule is double stranded.
- The percentage of cytosine is 40%.

For help with this problem, review:
The Primary Structure of DNA and Secondary Structures of DNA in Section 10.3.

≫ Solution Steps

If C = 40%, then G also must be 40%. The total percentage of C + G is therefore 40% + 40% = 80%. All the remaining bases must be either A or T, so the total percentage of A + T = 100% − 80% = 20%; because the percentage of A equals the percentage of T, the percentage of T is 20%/2 = 10%.

> **Recall:** In double-stranded DNA, A pairs with T, whereas G pairs with C; so the percentage of A equals the percentage of T, and the percentage of G equals the percentage of C.

Problem 2

Which of the following relations will be true for the percentage of bases in double-stranded DNA?

a. $C + T = A + G$ **b.** $\dfrac{C}{A} = \dfrac{T}{G}$

≫ Solution Strategy

What information is required in your answer to the problem?

Whether $C + T = A + G$ and $\dfrac{C}{A} = \dfrac{T}{G}$ are true.

What information is provided to solve the problem?

- The DNA is double stranded.
- Ratios of different groups of bases.

For help with this problem, review:
The Primary Structure of DNA and Secondary Structures of DNA in Section 10.3.

≫ Solution Steps

An easy way to determine whether the relations are true is to arbitrarily assign percentages to the bases, remembering that, in double-stranded DNA, A = T and G = C. For example, if the percentages of A and T are each 30%, then the percentages of G and C are each 20%. We can substitute these values into the equations to see if the relations are true.

a. 20 + 30 = 30 + 20. This relation is true.

b. $^{20}/_{30} \neq {}^{30}/_{20}$. This relation is not true.

COMPREHENSION QUESTIONS

Section 10.1

1. What four general characteristics must the genetic material possess?

Section 10.2

2. Briefly outline the history of our knowledge of the structure of DNA until the time of Watson and Crick. Which do you think were the principal contributions and developments?

3. What experiments demonstrated that DNA is the genetic material?

4. What is transformation? How did Avery and his colleagues demonstrate that the transforming principle is DNA?

5. How did Hershey and Chase show that DNA is passed to new phages in phage reproduction?

6. Why was the discovery of DNA structure so important?

Section 10.3

7. Draw and identify the three parts of a DNA nucleotide.

8. How does an RNA nucleotide differ from a DNA nucleotide?

9. How does a purine differ from a pyrimidine? What purines and pyrimidines are found in DNA and RNA?

10. Draw a short segment of a single polynucleotide strand, including at least three nucleotides. Indicate the polarity of the strand by identifying the 5′ end and the 3′ end.

11. Which bases are capable of forming hydrogen bonds with each other?

12. What different types of chemical bonds are found in DNA and where are they found?

13. What are some of the important genetic implications of the DNA structure?

14. What are the three major pathways of information flow within the cell?

Section 10.4

15. What are hairpins and how do they form?

16. What is DNA methylation?

APPLICATION QUESTIONS AND PROBLEMS

Introduction

17. The introduction to this chapter, which describes the sequencing of 4000-year-old DNA, emphasizes DNA's extreme stability. What aspects of DNA's structure contribute to the stability of the molecule? Why is RNA less stable than DNA?

Section 10.2

*18. Match the researchers (a–j) with the discoveries listed.

a. Kossel

b. Fraenkel-Conrat

c. Watson and Crick

d. Levene

e. Miescher

f. Hershey and Chase

g. Avery, MacLeod, and McCarty

h. Griffith

i. Franklin and Wilkins

j. Chargaff

____ Took X-ray diffraction pictures used in constructing the structure of DNA.

____ Determined that DNA contains nitrogenous bases.

____ Identified DNA as the genetic material in bacteriophages.

____ Discovered regularity in the ratios of different bases in DNA.

____ Determined that DNA is responsible for transformation in bacteria.

____ Worked out the helical structure of DNA by building models.

____ Discovered that DNA consists of repeating nucleotides.

____ Determined that DNA is acidic and high in phosphorus.

____ Conducted experiments showing that RNA can serve as the genetic material in some viruses.

____ Demonstrated that heat-killed material from bacteria can genetically transform live bacteria.

*19. A student mixes some heat-killed type IIS *Streptococcus pneumoniae* bacteria with live type IIR bacteria and injects the mixture into a mouse. The mouse develops pneumonia and dies. The student recovers some type IIS bacteria from the dead mouse. If this is the only experiment conducted by the student, has the student demonstrated that transformation has taken place? What other explanations might explain the presence of the type IIS bacteria in the dead mouse?

20. Predict what would happen if Griffith had mixed some heat-killed type IIIS bacteria and some heat-killed type IIR bacteria and injected these into a mouse. Would the mouse have contracted pneumonia and died? Explain why or why not.

21. Explain how heat-killed type IIIS bacteria in Griffith's experiment genetically altered the live type IIR bacteria. (Hint: See the discussion of transformation in Chapter 9.)

22. What results would you expect if the bacteriophage that Hershey and Chase used in their experiment had contained RNA instead of DNA?

23. Which of the processes of information transfer illustrated in **Figure 10.17** are required for the T2 phage reproduction illustrated in **Figure 10.5**?

*24. Imagine that you are a student in Alfred Hershey and Martha Chase's lab in the late 1940s. You are given five test tubes containing *E. coli* bacteria infected with T2 bacteriophages that have been labeled with either ^{32}P or ^{35}S. Unfortunately, you forget to mark the tubes and are now uncertain about which were labeled with ^{32}P and which with ^{35}S. You place the contents of each tube in a blender and turn it on for a few seconds to shear off the phage protein coats. You then centrifuge the contents to separate the protein coats and the cells. You check for the presence of radioactivity and obtain the following results. Which tubes contained *E. coli* infected with ^{32}P-labeled phage? Explain your answer.

Tube number	Radioactivity present in
1	Cells
2	Protein coats
3	Protein coats
4	Cells
5	Cells

25. **Figure 10.9** illustrates Fraenkel-Conrat and Singer's experiment on the genetic material of TMV. What results would you expect in this experiment if protein carried the genetic information of TMV instead of RNA?

Section 10.3

*26. DNA molecules of different sizes are often separated with the use of a technique called electrophoresis (see Chapter 19). With this technique, DNA molecules are placed in a gel, an electrical current is applied to the gel, and the DNA molecules migrate toward the positive (+) pole of the current. What aspect of its structure causes a DNA molecule to migrate toward the positive pole?

*27. Each nucleotide pair of a DNA double helix weighs about 1×10^{-21} g. The human body contains approximately 0.5 g of DNA. How many nucleotide pairs of DNA are in the human body? If you assume that all the DNA in human cells is in the B-DNA form, how far would the DNA reach if stretched end to end?

28. One nucleotide strand of a DNA molecule has the base sequence illustrated below.

5′–ATTGCTACGG–3′

Give the base sequence and label the 5′ and 3′ ends of the complementary DNA nucleotide strand.

*29. Erwin Chargaff collected data on the proportions of nitrogenous bases from the DNA of a variety of different organisms and tissues (E. Chargaff, in *The Nucleic Acids: Chemistry and Biology*, vol. 1, E. Chargaff and J. N. Davidson, Eds. New York: Academic Press, 1955). The following data are from the DNA of several organisms analyzed by Chargaff.

Erwin Chargaff. [Horst Tappe/Getty Images.]

Organism and tissue	Percentage			
	A	G	C	T
Sheep thymus	29.3	21.4	21.0	28.3
Pig liver	29.4	20.5	20.5	29.7
Human thymus	30.9	19.9	19.8	29.4
Rat bone marrow	28.6	21.4	20.4	28.4
Hen erythrocytes	28.8	20.5	21.5	29.2
Yeast	31.7	18.3	17.4	32.6
E. coli	26.0	24.9	25.2	23.9
Human sperm	30.9	19.1	18.4	31.6
Salmon sperm	29.7	20.8	20.4	29.1
Herring sperm	27.8	22.1	20.7	27.5

a. For each organism, compute the ratio of (A + G)/(T + C) and the ratio of (A + T)/(C + G).

b. Are these ratios constant or do they vary among the organisms? Explain why.

c. Is the (A + G)/(T + C) ratio different for the sperm samples? Would you expect it to be? Why or why not?

30. Boris Magasanik collected data on the amounts of the bases of RNA isolated from a number of sources, expressed relative to a value of 10 for adenine (B. Magasanik, in *The Nucleic Acids: Chemistry and Biology*, vol. 1, E. Chargaff and J. N. Davidson, Eds. New York: Academic Press, 1955).

Organism and tissue	Amount			
	A	G	C	U
Rat liver nuclei	10	14.8	14.3	12.9
Rabbit liver nuclei	10	13.6	13.1	14.0
Cat brain	10	14.7	12.0	9.5
Carp muscle	10	21.0	19.0	11.0
Yeast	10	12.0	8.0	9.8

a. For each organism, compute the ratio of (A + G)/(U + C).

b. How do these ratios compare with the (A + G)/(T + C) ratio found in DNA (see Problem 29)? Explain.

31. Which of the following relations or ratios would be true for a double-stranded DNA molecule?

a. $A + T = G + C$

b. $A + T = T + C$

c. $A + C = G + T$

d. $\dfrac{A + T}{C + G} = 1.0$

e. $\dfrac{A + G}{C + T} = 1.0$

f. $\dfrac{A}{C} = \dfrac{G}{T}$

g. $\dfrac{A}{G} = \dfrac{C}{T}$

h. $\dfrac{A}{T} = \dfrac{G}{C}$

*32. If a double-stranded DNA molecule is 15% thymine, what are the percentages of all the other bases?

33. Suppose that each of the bases in DNA were capable of pairing with any other base. What effect would this capability have on DNA's capacity to serve as the source of genetic information?

34. Heinz Shuster collected the following data on the base composition of ribgrass mosaic virus (H. Shuster, in *The Nucleic Acids: Chemistry and Biology*, vol. 3, E. Chargaff and J. N. Davidson, Eds. New York: Academic Press, 1955). On the basis of this information, is the hereditary information of the ribgrass mosaic virus RNA or DNA? Is it likely to be single stranded or double stranded?

Ribgrass mosaic virus. [Leibniz Insitute for Age Research, Fritz Lipmann-Institute.]

	Percentage				
	A	G	C	T	U
Ribgrass mosaic virus	29.3	25.8	18.0	0.0	27.0

*35. The relative amounts of each nitrogenous base are tabulated here for four different viruses. For each virus listed in the following table, indicate whether its genetic material is DNA or RNA and whether it is single stranded or double stranded. Explain your reasoning.

Virus	T	C	U	G	A
I	0	12	9	12	9
II	23	16	0	16	23
III	34	42	0	18	39
IV	0	24	35	27	17

*36. A B-DNA molecule has 1 million nucleotide pairs. How many complete turns of the helix are there in this molecule?

*37. For entertainment on a Friday night, a genetics professor proposed that his children diagram a polynucleotide strand of DNA. Having learned about DNA in preschool, his 5-year-old daughter was able to draw a polynucleotide strand, but she made a few mistakes. The daughter's diagram (represented here) contained at least 10 mistakes.

a. Make a list of all the mistakes in the structure of this DNA polynucleotide strand.

b. Draw the correct structure for the polynucleotide strand.

Section 10.4

*38. Write a sequence of bases in an RNA molecule that will produce a hairpin structure.

CHALLENGE QUESTIONS

Section 10.1

*39. Suppose that an automated, unmanned probe is sent into deep space to search for extraterrestrial life. After wandering for many light-years among the far reaches of the universe, this probe arrives on a distant planet and detects life. The chemical composition of life on this planet is completely different from that of life on Earth, and its genetic material is not composed of nucleic acids. What predictions can you make about the chemical properties of the genetic material on this planet?

Section 10.2

40. How might ^{32}P and ^{35}S be used to demonstrate that the transforming principle is DNA? Briefly outline an experiment that would show that DNA, rather than protein, is the transforming principle.

Section 10.3

41. Researchers have proposed that early life on Earth used RNA as its source of genetic information and that DNA eventually replaced RNA as the source of genetic information. What aspects of DNA structure might make it better suited than RNA to be the genetic material?

42. Scientists have reportedly isolated short fragments of DNA from fossilized dinosaur bones hundreds of millions of years old. The technique used to isolate this DNA is the polymerase chain reaction, which is capable of amplifying very small amounts of DNA a millionfold (see Chapter 19). Critics have claimed that the DNA isolated from dinosaur bones is not purely of ancient origin, but instead has been contaminated by DNA from present-day organisms such as bacteria, mold, or humans. What precautions, analyses, and control experiments could be carried out to ensure that DNA recovered from fossils is truly of ancient origin?

THINK-PAIR-SHARE QUESTIONS

Section 10.1

1. Suppose that proteins, instead of nucleic acids, had evolved as the carriers of genetic information. How well would proteins satisfy the four requirements for the genetic material listed in Section 10.1?

Section 10.2

2. Isaac Newton said, "If I have seen further, it is by standing on the shoulders of giants." How does this statement apply to Watson and Crick?

3. Compare and contrast Gregor Mendel's scientific method and approach to science (see Chapter 3) with that of Watson and Crick. How did they differ? Were there any similarities?

Section 10.3

4. How does its structure enable DNA to function effectively as the genetic material? Give some specific examples.

5. Chapter 1 considered the theory of the inheritance of acquired characteristics and noted that this theory is no longer accepted. Is the central dogma consistent with the theory of the inheritance of acquired characteristics? Why or why not?

Chromosome Structure and Organelle DNA

Child from a Romanian orphanage. Research demonstrated that children who lived in state-run orphanages had shorter telomeres than children from foster homes. [Jenny Matthews/Alamy.]

Telomeres and Childhood Adversity

Within each of our cells are 46 chromosomes, exquisitely complex structures of DNA and protein that carry the coding instructions for all of our traits. These chromosomes are passed down from our parents and constitute the basis of heredity, the passage of traits from one generation to the next. But chromosomes don't just carry a record of our genetic legacy. They also carry a record—in the lengths of their telomeres—of the stresses we encounter.

Telomeres are special protective structures found at the ends of each of our chromosomes. Like the small plastic tips that keep the ends of a shoelace from unraveling, telomeres prevent chromosomes from being degraded at their ends. In spite of the protection of the telomeres, chromosomes of most cells shorten progressively with each cell division. Due to a quirk of DNA replication, most cells are unable to copy the very end of each linear chromosome (see Chapter 12 for a full discussion of the end-replication problem). Hence, with each round of replication, a chromosome becomes shorter, until it is so reduced that the cell stops dividing, becomes inactive, and eventually dies. For most cells, this shortening of telomeres limits the number of divisions possible. Exceptions occur in the germ-line cells that produce future generations, certain stem cells, and—unfortunately—many cancer cells that have escaped normal constraints on cell division.

Because telomeres become shorter with each cell division, much research has focused on determining if telomere length is indicative of biological aging. Although the relation between telomere length and aging is complex and not fully understood, considerable evidence suggests that telomeres do shorten with age, and that processes that lead to premature telomere shortening are associated with features of aging. In 2011, geneticists studying this phenomenon observed that hardships encountered early in life can play a part in shortening our telomeres.

To study the effects of early life experience on telomere length, geneticists studied 100 children living in state-run orphanages in Romania. At an early age, some of these children were placed in foster homes; others remained in the orphanages. Previous studies had demonstrated that children in such orphanages receive less individual attention and care compared with children growing up with natural or foster parents, and institutional care is assumed to be more stressful than foster care.

When the children were 6 to 10 years old, the researchers collected samples of their DNA and measured the length of their telomeres. The results were striking: children who remained in the orphanages had significantly shorter telomeres than those who spent time

E. coli bacterium

Bacterial chromosome

11.1 The DNA in *E. coli* is about 1000 times as long as the cell itself.

in foster care. The researchers concluded that telomere length is affected by childhood adversity: children reared in stressful environments are more likely to have shorter telomeres than those raised in less stressful environments. Several other studies have found a similar association between telomere length in adults and early childhood stresses, such as abuse and chronic illness. How stress affects telomeres and results in their shortening is not known, but the research documents that chromosomes are more than just a repository of our genetic information: their structure is also affected by our environment.

THINK-PAIR-SHARE

■ Most cells are unable to copy the ends of chromosomes, and therefore chromosomes shorten with each cell division. This limits the number of times a cell can divide. In germ cells and stem cells, however, an enzyme called telomerase lengthens the telomeres and prevents chromosome shortening. Thus, these cells are not limited in the number of times they can divide. All cells have the gene for telomerase, but most somatic cells don't express it, and they produce no telomerase. Why don't somatic cells express telomerase and have unlimited division?

■ The introduction to this chapter discussed recent research showing that children who experience early childhood stresses have shorter telomeres. How might this information be used in a practical sense?

In this chapter, we examine the molecular structure of chromosomes and of the DNA found in cytoplasmic organelles. The first part of the chapter focuses on a storage problem: how to cram tremendous amounts of DNA into the limited confines of a cell. Even in those organisms that have the smallest amounts of DNA, the length of the genetic material far exceeds the length of the cell. Thus, cellular DNA must be highly folded and tightly packed. But this packing itself creates problems: it renders the DNA inaccessible, unable to be copied or read. Functional DNA must be capable of partly unfolding and expanding so that individual genes can undergo replication and transcription. The flexible, dynamic nature of DNA packing that allows it to meet these challenges is a major theme of this chapter. We first consider supercoiling, an important tertiary structure of DNA found in both prokaryotic and eukaryotic cells. After a brief look at the bacterial chromosome, we examine the structure of eukaryotic chromosomes. We pay special attention to the working parts of a chromosome—specifically, centromeres and telomeres. We also consider the types of DNA sequences present in many eukaryotic chromosomes.

The second part of this chapter focuses on the organization of DNA sequences found in mitochondria and chloroplasts. The uniparental pattern of inheritance exhibited by genes found in these organelles was discussed in Chapter 5; here, we examine molecular aspects of organelle DNA. We briefly consider the structures of mitochondria and chloroplasts, the inheritance of traits encoded by their genes, and the evolutionary origin of these organelles. We then examine the general characteristics of **mitochondrial DNA** (mtDNA), followed by a discussion of the organization and function of different types of mitochondrial genomes. Finally, we turn to

chloroplast DNA (cpDNA) and examine its characteristics, organization, and function.

11.1 Large Amounts of DNA Are Packed into a Cell

The packaging of tremendous amounts of genetic information into the small space within a cell has been called the ultimate storage problem. Consider the chromosome of the bacterium *E. coli*, a single molecule of DNA with approximately 4.6 million base pairs. Stretched out straight, this DNA would be about 1000 times as long as the cell within which it resides (**Figure 11.1**). Human cells contain more than 6 billion base pairs of DNA, which would measure over 2 m (over 6 feet) stretched end to end. Even the DNA in the smallest human chromosome would stretch 14,000 times the length of the cell nucleus. Clearly, DNA molecules must be tightly packed to fit into such small spaces.

The structure of DNA can be considered at three hierarchical levels: the primary structure of DNA is its nucleotide sequence; the secondary structure is the double-stranded helix; and the tertiary structure is the higher-order folding that allows DNA to be packed into the confined space of a cell. The primary and secondary structures of DNA were described in Chapter 10; in this section, our focus is its tertiary structure.

CONCEPTS

Chromosomal DNA exists in the form of very long molecules that are tightly packed to fit into the small confines of a cell.

Supercoiling

One type of DNA tertiary structure is **supercoiling**, which takes place when the DNA helix is subjected to strain by being overwound or underwound. B-DNA (see Chapter 10) is in its lowest-energy state when it has approximately 10 bp per turn of its helix. In this **relaxed state**, a stretch of 100 bp of DNA would assume about 10 complete turns (**Figure 11.2a**). If energy is used to add or remove any turns, strain is placed on the molecule, causing the helix to supercoil, or twist on itself. Molecules that are overrotated exhibit **positive supercoiling** (**Figure 11.2b**). Underrotated molecules exhibit **negative supercoiling** (**Figure 11.2c**). Supercoiling is a partial solution to the cell's DNA packing problem because supercoiled DNA occupies less space than relaxed DNA.

Supercoiling takes place when the strain of overrotating or underrotating cannot be compensated by the turning of the ends of the double helix, which is the case if the DNA is circular—that is, there are no free ends. If the ends *can* turn freely, they will simply turn as extra rotations are added or removed, and the molecule will spontaneously revert to the relaxed state. Both bacterial and eukaryotic DNA usually fold into loops stabilized by proteins (which prevent free rotation of the ends; see Figure 11.3 below), and supercoiling takes place within the loops.

Supercoiling relies on **topoisomerases**, enzymes that add or remove rotations from the DNA helix by temporarily breaking the nucleotide strands, rotating the ends around each other, and then rejoining the broken ends. Thus, topoisomerases can both induce and relieve supercoiling, although not all topoisomerases do both.

Most DNA found in cells is negatively supercoiled. This state has two advantages over relaxed DNA. First, negative supercoiling makes the separation of the two strands of DNA easier during replication and transcription. Negatively supercoiled DNA is underrotated, so separation of the two strands during replication and transcription is more rapid and requires less energy. Second, the supercoiled DNA can be packed into a smaller space than can relaxed DNA.

CONCEPTS

Overrotation or underrotation of a DNA double helix places strain on the molecule, causing it to supercoil. Supercoiling is controlled by topoisomerase enzymes. Most cellular DNA is negatively supercoiled, which eases the separation of nucleotide strands during replication and transcription and allows the DNA to be packed into small spaces.

✔ CONCEPT CHECK 1

A DNA molecule 300 bp long has 20 complete rotations. This DNA molecule is

a. positively supercoiled.
b. negatively supercoiled.
c. relaxed.

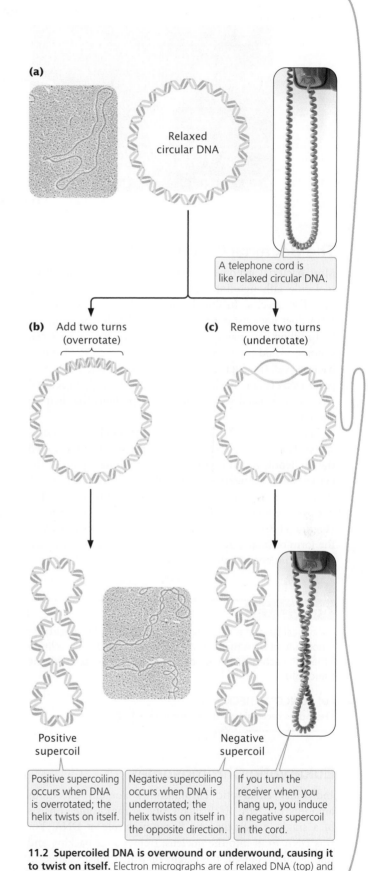

(a)

Relaxed circular DNA

A telephone cord is like relaxed circular DNA.

(b) Add two turns (overrotate)

(c) Remove two turns (underrotate)

Positive supercoil

Negative supercoil

| Positive supercoiling occurs when DNA is overrotated; the helix twists on itself. | Negative supercoiling occurs when DNA is underrotated; the helix twists on itself in the opposite direction. | If you turn the receiver when you hang up, you induce a negative supercoil in the cord. |

11.2 Supercoiled DNA is overwound or underwound, causing it to twist on itself. Electron micrographs are of relaxed DNA (top) and supercoiled DNA (bottom). [Photographs: Dr. Gopal Murti/Phototake.]

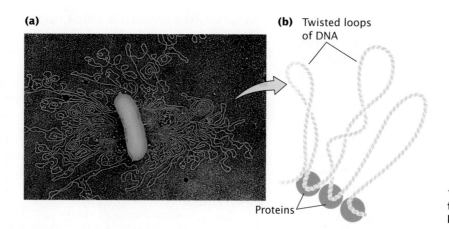

(a)

(b) Twisted loops of DNA

Proteins

11.3 Bacterial DNA is highly folded into a series of twisted loops. [Part a: G. Murti/Science Source.]

The Bacterial Chromosome

Most bacterial genomes consist of a single circular DNA molecule, although linear DNA molecules have been found in a few species. In circular bacterial chromosomes, the DNA does not exist in an open, relaxed circle; the 3 million to 4 million base pairs of DNA found in a typical bacterial genome would be much too large in this state to fit into a bacterial cell (see Figure 11.1). Unlike eukaryotic DNA, bacterial DNA is not attached to histone proteins, but bacterial DNA is associated with a number of proteins that help to compact it.

When a bacterial cell is viewed with an electron microscope, its DNA frequently appears as a distinct clump, called the **nucleoid**, which is confined to a definite region of the cytoplasm. If a bacterial cell is broken open gently, its DNA spills out in a series of twisted loops (**Figure 11.3a**). The ends of the loops are most likely held in place by proteins (**Figure 11.3b**). Many bacteria contain additional DNA in the form of small circular molecules called plasmids, which replicate independently of the chromosome (see Chapter 9).

> ### CONCEPTS
>
> A typical bacterial chromosome consists of a large, circular molecule of DNA that forms a series of twisted loops. Within the cell, bacterial DNA appears as a distinct clump, called the nucleoid.
>
> ✔ **CONCEPT CHECK 2**
> How does bacterial DNA differ from eukaryotic DNA?

Eukaryotic Chromosomes

Individual eukaryotic chromosomes contain enormous amounts of DNA. Each eukaryotic chromosome consists of a single, extremely long linear molecule of DNA. For this DNA to fit into the nucleus, tremendous packing and folding are required, the extent of which must change in the course of the cell cycle. The chromosomes are in an elongated, relatively uncondensed state during interphase

(see p. 25 in Chapter 2), but the term *relatively* is important here. Although the DNA of interphase chromosomes is less tightly packed than the DNA of mitotic chromosomes, it is still highly condensed; it's just *less* condensed. In the course of the cell cycle, the level of DNA packing changes: chromosomes progress from a highly packed state to a state of extreme condensation, which is necessary for chromosome movement in mitosis and meiosis. DNA packing also changes locally during replication and transcription, when the two nucleotide strands must unwind so that particular base sequences are exposed. Thus, the packing of eukaryotic DNA (its tertiary chromosomal structure) is not static, but changes regularly in response to cellular processes.

CHROMATIN Eukaryotic DNA in the cell is closely associated with proteins. This complex of DNA and proteins is called *chromatin*. The two basic types of chromatin are **euchromatin**, which undergoes the normal process of condensation and decondensation in the cell cycle, and **heterochromatin**, which remains in a highly condensed state throughout the cell cycle, even during interphase. Euchromatin constitutes the majority of the chromosomal material and is where most transcription takes place. All chromosomes have permanent heterochromatin (called *constitutive heterochromatin*) at the centromeres and telomeres; the Y chromosome also consists largely of constitutive heterochromatin. Heterochromatin may also occur during certain developmental stages; this material is referred to as *facultative heterochromatin*. For example, facultative heterochromatin occurs along one entire X chromosome in female mammals when that X becomes inactivated (see pp. 98–99 in Chapter 4). In addition to remaining condensed throughout the cell cycle, heterochromatin is characterized by a general lack of transcription, the absence of crossing over, and replication late in the S phase. Differences between euchromatin and heterochromatin are summarized in **Table 11.1**.

THINK-PAIR-SHARE Question 1 👥

The most abundant proteins in chromatin are the *histones*, which are small, positively charged proteins of five major types: H1, H2A, H2B, H3, and H4. All histones have a high

TABLE 11.1	Characteristics of euchromatin and heterochromatin	
Characteristic	**Euchromatin**	**Heterochromatin**
Chromatin condensation	Less condensed	More condensed
Location	On chromosome arms	At centromeres, telomeres, and other specific places
Type of sequences	Unique sequences	Repeated sequences*
Presence of genes	Many genes	Few genes*
When replicated	Throughout S phase	Late S phase
Transcription	Often	Infrequent
Crossing over	Common	Uncommon

*Applies only to constitutive heterochromatin.

percentage of arginine and lysine, positively charged amino acids that give the histones a net positive charge. These positive charges attract the negative charges on the phosphates of DNA; this attraction holds the DNA in contact with the histones. A heterogeneous assortment of **nonhistone chromosomal proteins** is also found in eukaryotic chromosomes. At times, variant histones, with somewhat different amino acid sequences, are incorporated into chromatin in place of one of the major histone types. These variants alter chromatin structure and influence its function. For example, some specific variant histones are associated with actively transcribed DNA; these variants are assumed to make the chromatin more open and accessible to the enzymes and proteins that carry out transcription. ▶ **TRY PROBLEM 18**

CONCEPTS

Chromatin, which consists of DNA complexed with proteins, is the material that makes up eukaryotic chromosomes. The most abundant of these proteins are the five types of positively charged histone proteins: H1, H2A, H2B, H3, and H4. Variant histones may at times be incorporated into chromatin in place of the normal histone types.

✔ CONCEPT CHECK 3

Neutralizing their positive charges would have which effect on the histone proteins?

a. They would bind the DNA tighter.
b. They would bind less tightly to the DNA.
c. They would no longer be attracted to each other.
d. They would cause supercoiling of the DNA.

THE NUCLEOSOME Chromatin has a highly complex structure with several levels of organization (**Figure 11.4**).

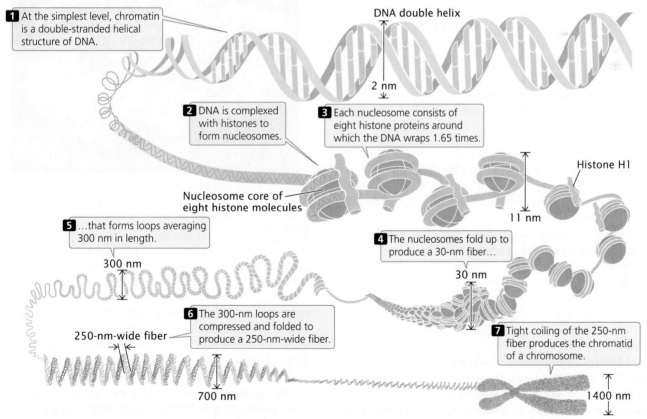

1 At the simplest level, chromatin is a double-stranded helical structure of DNA.

DNA double helix

2 nm

2 DNA is complexed with histones to form nucleosomes.

3 Each nucleosome consists of eight histone proteins around which the DNA wraps 1.65 times.

Nucleosome core of eight histone molecules

Histone H1

11 nm

5 ...that forms loops averaging 300 nm in length.

300 nm

4 The nucleosomes fold up to produce a 30-nm fiber...

30 nm

6 The 300-nm loops are compressed and folded to produce a 250-nm-wide fiber.

250-nm-wide fiber

7 Tight coiling of the 250-nm fiber produces the chromatid of a chromosome.

700 nm

1400 nm

11.4 Chromatin has a highly complex structure with several levels of organization.

The simplest level is the double-helical structure of DNA discussed in Chapter 10. At a more complex level, the DNA molecule is closely associated with proteins and is highly folded to produce a chromosome.

When chromatin is isolated from the nucleus of a cell and viewed with an electron microscope, it frequently looks like beads on a string (**Figure 11.5a**). If a small amount of nuclease is added to this structure, the enzyme cleaves the "string" between the "beads," leaving individual beads attached to about 200 bp of DNA (**Figure 11.5b**). If more nuclease is added, the enzyme chews up all the DNA between the beads and leaves a core of proteins attached to a fragment of DNA (**Figure 11.5c**). Such experiments demonstrated that chromatin is not a random association of proteins and DNA, but has a fundamental repeating structure.

The core of protein and DNA produced by digestion with nuclease enzymes is the simplest level of chromatin structure, the **nucleosome** (see Figure 11.4). The nucleosome is a core particle consisting of DNA wrapped about two times around an octamer of eight histone proteins (two copies each of H2A, H2B, H3, and H4), much like thread wound around a spool (**Figure 11.5d**). The DNA in direct contact with the histone octamer is between 145 and 147 bp in length.

Each of the histone proteins that make up the nucleosome core particle has a flexible "tail," containing from 11 to 37 amino acids, which extends out from the nucleosome. Positively charged amino acids in the tails of the histones interact with the negative charges of the phosphates on the DNA, keeping the DNA and histones tightly associated. The tails of one nucleosome may also interact with neighboring nucleosomes, which facilitates compaction of the nucleosomes themselves. Chemical modifications of the histone tails bring about changes in chromatin structure (discussed in the next section) that are necessary for gene expression.

The fifth type of histone, H1, is not a part of the nucleosome core particle, but plays an important role in nucleosome structure. H1 binds to 20–22 bp of DNA where the DNA joins and leaves the histone octamer (see Figure 11.4) and helps to lock the DNA into place, acting as a clamp around the nucleosome.

Each nucleosome encompasses about 167 bp of DNA. Nucleosomes are located at regular intervals along the DNA molecule and are separated from one another by **linker DNA**, which varies in size among cell types; in most cells, linker DNA comprises from about 30 to 40 bp. Nonhistone chromosomal proteins may be associated with this linker DNA, and a few also appear to bind directly to the core particle.

Although histone proteins are not present in bacteria, archaea have two types of histones that are similar to the H3 and H4 histones found in eukaryotes. These archaeal histones associate in groups of four, forming a tetramer instead of the octamers seen in eukaryotes. The DNA of archaea wraps around the tetramers of histones to form structures similar to the nucleosomes of eukaryotes, each encompassing about 60 bp of DNA. The nucleosomes of archaea occupy specific positions along the DNA, much as they do in eukaryotes, which suggests that they may play a role in gene expression (see Chapter 17). **▶ TRY PROBLEMS 19 AND 21**

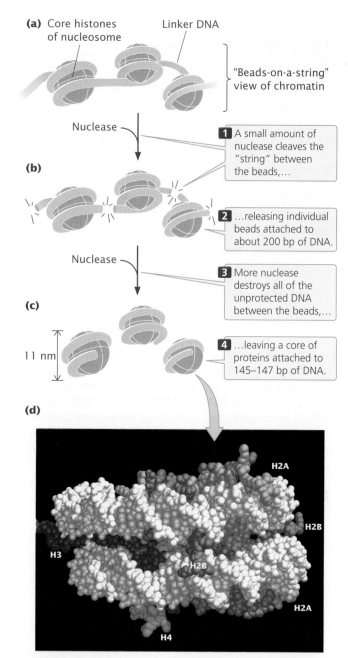

(a) Core histones of nucleosome · Linker DNA · "Beads-on-a-string" view of chromatin

Nuclease

(b)

1 A small amount of nuclease cleaves the "string" between the beads,…

2 …releasing individual beads attached to about 200 bp of DNA.

Nuclease

(c)

3 More nuclease destroys all of the unprotected DNA between the beads,…

11 nm

4 …leaving a core of proteins attached to 145–147 bp of DNA.

(d)

H2A · H2B · H3 · H2B · H2A · H4

11.5 The nucleosome is the fundamental repeating unit of chromatin. Part *d* shows a space-filling model of the core particle, which consists of two copies each of H2A, H2B, H3, and H4, around which DNA (white) coils. [Part d: Reprinted by permission from Macmillan Publishers Ltd. From K. Luger et al., "Crystal structure of the nucleosome core particle at 2.8 Å resolution," *Nature* 389:251. © 1997. Courtesy of T. H. Richmond, permission conveyed through Copyright Clearance Center, Inc.]

HIGHER-ORDER CHROMATIN STRUCTURE When chromatin is in a condensed form, nucleosomes fold on themselves to form a dense, tightly packed structure (see Figure 11.4) that makes up a fiber with a diameter of about 30 nm. Some recent research suggests that the 30-nm fiber consists of two interwound stacks of nucleosomes with the linker DNA between successive nucleosomes crisscrossing back and forth in the interior of the fiber. However, other structures of the 30-nm fiber have also been proposed, and indeed, different structures may exist under different conditions within the nucleus.

The precise structure of chromatin above the level of the 30-nm fiber is poorly understood. The next level of chromatin structure is a series of loops of the 30-nm fiber (see Figure 11.4), each anchored at its base by proteins. On average, each loop encompasses some 20,000 to 100,000 bp of DNA and is about 300 nm in length, but the individual loops vary considerably. The 300-nm loops are packed and folded to produce a 250-nm-wide fiber. Tight helical coiling of the 250-nm fiber, in turn, produces the structure that appears in metaphase: an individual chromatid approximately 700 nm in width. You can view the different levels of chromatin structure in **Animation 11.1**.

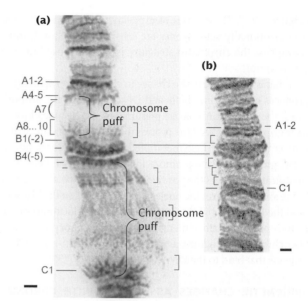

11.6 Chromosome puffs are regions of relaxed chromatin where active transcription is taking place. (a) Chromosome puffs on giant polytene chromosomes isolated from the salivary glands of larval *Drosophila*. (b) The corresponding region without chromosome puffs. The density of the stain correlates with the density of the chromatin. [Courtesy of Dmitri Novikov.]

CONCEPTS

The nucleosome consists of a core particle of eight histone proteins and the DNA that wraps around them. A single H1 histone associates with each core particle. Nucleosomes are separated by linker DNA. Nucleosomes fold to form a 30-nm chromatin fiber, which appears as a series of loops that pack to create a 250-nm fiber. Helical coiling of the 250-nm fiber produces a chromatid.

✔ CONCEPT CHECK 4

How many copies of the H2B histone would be found in chromatin containing 50 nucleosomes?

a. 5   c. 50
b. 10   d. 100

Changes in Chromatin Structure

Although eukaryotic DNA must be tightly packed to fit into the cell nucleus, it must also periodically unwind to undergo transcription and replication. Polytene chromosomes and DNase I sensitivity, discussed in the following sections, are visible evidence of the dynamic nature of chromatin structure.

POLYTENE CHROMOSOMES Giant chromosomes called **polytene chromosomes** are found in certain tissues of *Drosophila* larvae and some other organisms (**Figure 11.6**). Polytene chromosomes have provided researchers with evidence of the changing nature of chromatin structure.

These large, unusual chromosomes arise when repeated rounds of DNA replication take place without accompanying cell divisions, producing thousands of copies of DNA that lie side by side. When polytene chromosomes are stained with dyes, numerous bands are revealed. Under certain conditions, the bands may exhibit **chromosome puffs**—localized swellings of the chromosome. Each puff is a region of the chromatin that has a relaxed structure and, consequently, a more open state. Research indicates that chromosome puffs are regions of active transcription. This correlation between the occurrence of transcription and the relaxation of chromatin at a puff site indicates that chromatin structure undergoes dynamic change associated with gene activity.

THINK-PAIR-SHARE Question 2

DNase I SENSITIVITY A second piece of evidence indicating that chromatin structure changes with gene activity is sensitivity to DNase I, an enzyme that digests DNA. The ability of this enzyme to digest DNA depends on chromatin structure: when DNA is tightly bound to histone proteins, it is less sensitive to DNase I, whereas unbound DNA is more sensitive to DNase I. The results of experiments that examined the effect of DNase I on specific globin genes in chick embryos showed that DNase sensitivity is correlated with gene activity. Globin genes encode several types of hemoglobin, which are expressed in the erythroblasts (precursors of red blood cells) of chickens at different stages of development

(**Figure 11.7**). These types of experiments demonstrate that transcriptionally active genes are sensitive to DNase I, indicating that the chromatin structure is more exposed during transcription.

What is the nature of the change in chromatin structure that produces chromosome puffs and DNase I sensitivity? In both cases, the chromatin relaxes; presumably, the histones loosen their grip on the DNA. One process that alters chromatin structure is acetylation. Enzymes called acetyltransferases attach acetyl groups to lysine amino acids on the histone tails. This modification reduces the positive charges that normally exist on lysine and destabilizes the nucleosome structure, so the histones hold the DNA less tightly. Other chemical modifications of the histone proteins, such as methylation and phosphorylation, also alter chromatin structure, as do special chromatin-remodeling proteins that bind to the DNA. ▶ TRY PROBLEM 20

EPIGENETIC CHANGES ASSOCIATED WITH CHROMATIN MODIFICATIONS We have now seen how chromatin structure can be altered by chemical modification of the histone proteins. A number of other changes can also affect

chromatin structure, including the methylation of DNA (see Chapter 10), the use of variant histone proteins in the nucleosome, and the binding of nonhistone proteins to DNA and chromatin. Although these changes do not alter the DNA sequence, they often have major effects on the expression of genes, which will be discussed in more detail in Chapter 17.

Some changes in chromatin structure are retained through cell division, so that they are passed on to future generations of cells and even occasionally to future generations of organisms. Stable alterations of chromatin structure that may be passed on to descendant cells or individuals are frequently referred to as **epigenetic changes** or simply as *epigenetics* (see Chapter 5). For example, the *agouti* locus helps determine coat color in mice: parents that have identical DNA sequences but have different degrees of methylation on their DNA may give rise to offspring with different coat colors (**Figure 11.8**). Such epigenetic changes have been observed in a number of organisms and are responsible for a variety of phenotypic effects. Unlike mutations, epigenetic changes do not alter the DNA sequence, are capable of being reversed, and are often influenced by environmental factors (see Chapter 21 for more detail).

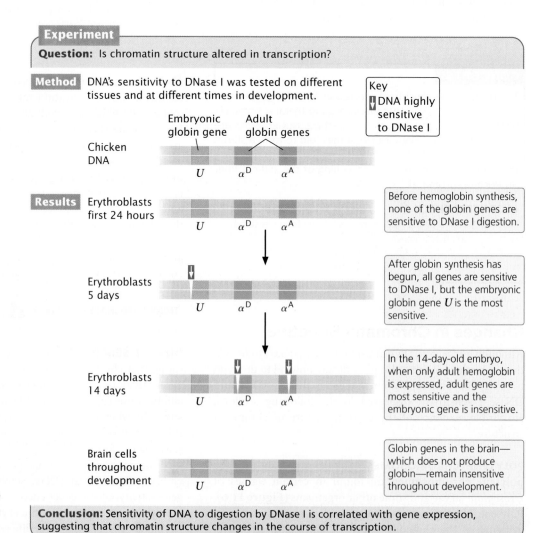

11.7 DNase I sensitivity is correlated with the transcription of globin genes in chick embryos. The *U* gene encodes embryonic hemoglobin; the α^D and α^A genes encode adult hemoglobin. The red arrows indicate sites of DNase I digestion.

Experiment

Question: Is chromatin structure altered in transcription?

Method DNA's sensitivity to DNase I was tested on different tissues and at different times in development.

Key
⇩ DNA highly sensitive to DNase I

Chicken DNA

Embryonic globin gene — Adult globin genes

U α^D α^A

Results

Erythroblasts first 24 hours

U α^D α^A

Before hemoglobin synthesis, none of the globin genes are sensitive to DNase I digestion.

Erythroblasts 5 days

U α^D α^A

After globin synthesis has begun, all genes are sensitive to DNase I, but the embryonic globin gene U is the most sensitive.

Erythroblasts 14 days

U α^D α^A

In the 14-day-old embryo, when only adult hemoglobin is expressed, adult genes are most sensitive and the embryonic gene is insensitive.

Brain cells throughout development

U α^D α^A

Globin genes in the brain—which does not produce globin—remain insensitive throughout development.

Conclusion: Sensitivity of DNA to digestion by DNase I is correlated with gene expression, suggesting that chromatin structure changes in the course of transcription.

11.8 Variation in DNA methylation at the *agouti* locus produces different coat colors in mice. [Cropley et al. "Germ-line epigenetic modification of the murine Avy allele by nutritional supplementation," PNAS November 14, 2006 vol. 103 no. 46 17308–17312, © 2006 National Academy of Sciences, USA.]

CONCEPTS

Epigenetic changes are alterations of chromatin or DNA structure that do not include changes in the base sequence but are stable and passed on to descendant cells or organisms. Some epigenetic changes result from alterations of chromatin structure.

11.2 Eukaryotic Chromosomes Possess Centromeres and Telomeres

Chromosomes, as we have seen, segregate in mitosis and meiosis and remain stable over many cell divisions. These properties of chromosomes arise in part from special structural features, including centromeres and telomeres.

Centromere Structure

The centromere, a constricted region of the chromosome, is the attachment site for the kinetochore and for spindle microtubules; this chromosome structure is necessary for proper chromosome movement in mitosis and meiosis (see Figure 2.7). The essential role of the centromere in chromosome movement was recognized by early geneticists, who observed the consequences of chromosome breakage. When a chromosome break produces two fragments, one with a centromere and one without, the chromosome fragment containing the centromere attaches to a spindle microtubule and moves to the spindle pole. The fragment lacking a centromere fails to connect to a spindle microtubule and is usually lost because it fails to move into the nucleus of a daughter cell during mitosis (**Figure 11.9**).

What are the key features of the centromere? In *Drosophila*, *Arabidopsis*, and humans, centromeres span hundreds of thousands of base pairs. Most of the centromere is made up of heterochromatin. Surprisingly, there are no specific sequences that are found in all centromeres, which raises the question of what exactly determines where the centromere is. Research suggests that most centromeres are not defined by DNA sequence, but rather by epigenetic changes in chromatin structure. Nucleosomes in

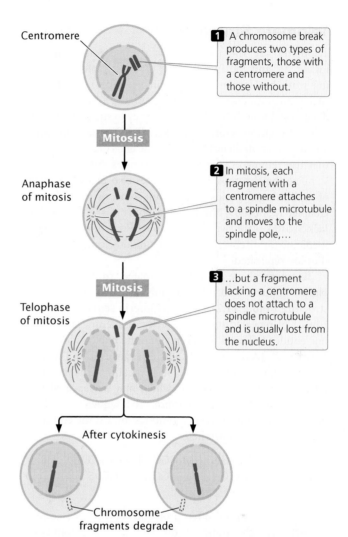

1 A chromosome break produces two types of fragments, those with a centromere and those without.

2 In mitosis, each fragment with a centromere attaches to a spindle microtubule and moves to the spindle pole,…

3 …but a fragment lacking a centromere does not attach to a spindle microtubule and is usually lost from the nucleus.

Centromere

Anaphase of mitosis

Mitosis

Telophase of mitosis

Mitosis

After cytokinesis

Chromosome fragments degrade

11.9 Chromosome fragments that lack centromeres are lost in mitosis.

the centromeres of most eukaryotes have a variant histone protein called CenH3, which takes the place of the usual H3 histone. The CenH3 variant histone is required for the assembly of proteins associated with the kinetochore; the presence of the CenH3 histone probably alters the nucleosome and chromatin structure, allowing kinetochore proteins to bind and spindle microtubules to attach.

CONCEPTS

The centromere is a region of the chromosome to which spindle microtubules attach. Centromeres display considerable variation in their DNA sequences and are distinguished by epigenetic alterations to chromatin structure, including the use of a variant H3 histone in the nucleosome.

✔ **CONCEPT CHECK 5**

What happens to a chromosome that loses its centromere?

Telomere Structure

Telomeres are the natural ends of a chromosome (see Figure 2.7 and the introduction to this chapter). Pioneering work by Hermann Muller (in fruit flies) and Barbara McClintock (in corn) showed that chromosome breaks produce unstable ends that have a tendency to stick together and enable the chromosome to be degraded. Because attachment and degradation do not happen to the ends of a chromosome that has telomeres, the telomeres must serve as caps that stabilize the chromosome. Telomeres also provide a means of replicating the ends of a chromosome, as we will see in Chapter 12. In 2009, Elizabeth Blackburn, Carol Greider, and Jack Szostak were awarded the Nobel Prize in physiology or medicine for discovering the structure of telomeres and how they are replicated.

Telomeres have now been isolated from protozoans, plants, humans, and other organisms; most are similar in structure (**Table 11.2**). These **telomeric sequences** usually consist of repeated units of a series of adenine or thymine nucleotides followed by several guanine nucleotides, taking the form $5'-(A\ or\ T)_mG_n-3'$, where m ranges from 1 to 4 and n is 2 or more. For example, the repeating unit in human telomeres is 5'–TTAGGG–3', which may be repeated from hundreds to thousands of times. The sequence is always oriented with the string of Gs and Cs toward the end of the chromosome, as shown here:

toward 5'–TTAGGGTTAGGGTTAGGG–3' end of
centro- ← 3'–AATCCCAATCCCAATCCC–5' → chromo-
mere   some

The G-rich strand often protrudes beyond the complementary C-rich strand at the end of the chromosome (**Figure 11.10a**), in which case it is called the G-rich 3' overhang. The G-rich 3' overhang in the telomeres of

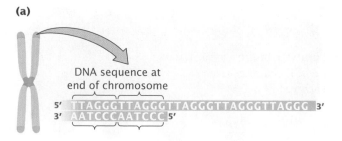

(a)

DNA sequence at end of chromosome

5' TTAGGGTTAGGGTTAGGGTTAGGGTTAGGG 3'
3' AATCCCAATCCC 5'

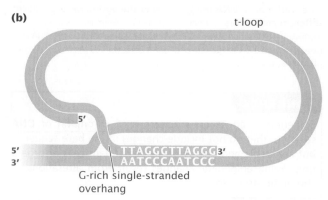

(b)   t-loop

5'
5'
3'   TTAGGGTTAGGG 3'
            AATCCCAATCCC
G-rich single-stranded
overhang

11.10 DNA at the ends of eukaryotic chromosomes consists of telomeric sequences. (a) The G-rich strand at the telomere is longer than the C-rich strand. (b) In some cells, the G-rich strand folds over and pairs with a short stretch of DNA to form a t-loop.

mammals is from 50 to 500 nucleotides long. Special proteins bind to this G-rich single-stranded sequence, protecting the telomere from degradation and preventing the ends of chromosomes from sticking together. A multiprotein complex called **shelterin** binds to telomeres and protects the ends of the DNA from being inadvertently repaired as a double-stranded break in the DNA. In some cells, the G-rich 3' overhang may fold over and pair with a short stretch of DNA to form a structure called a t-loop, which also functions in protecting the end of the telomere from degradation (**Figure 11.10b**).

TABLE 11.2	DNA sequences typically found in telomeres of various organisms
Organism	Sequence
Tetrahymena (protozoan)	5'–TTGGGG–3'
	3'–AACCCC–5'
Saccharomyces (yeast)	$5'-T_{1-6}GTG_{2-3}-3'$
	$3'-A_{1-6}CAC_{2-3}-5'$
Caenorhabditis (nematode)	5'–TTAGGC–3'
	3'–AATCCG–5'
Vertebrate	5'–TTAGGG–3'
	3'–AATCCC–5'
Arabidopsis (plant)	5'–TTTAGGG–3'
	3'–AAATCCC–5'

Source: V. A. Zakian, *Science* 270:1602, 1995.

CONCEPTS

A telomere is the stabilizing end of a chromosome. At the end of each telomere are many short telomeric sequences.

✔ CONCEPT CHECK 6

Which of the following is a characteristic of DNA sequences at the telomeres?

a. One strand consists of guanine and adenine (or thymine) nucleotides.
b. They consist of repeated sequences.
c. One strand protrudes beyond the other, creating some single-stranded DNA at the end.
d. All of the above

11.3 Eukaryotic DNA Contains Several Classes of Sequence Variation

Eukaryotic organisms vary dramatically in the amount of DNA per cell, a quantity termed an organism's **C-value** (Table 11.3). Each cell of a fruit fly, for example, contains 35 times the amount of DNA found in a cell of the bacterium *E. coli*. In general, eukaryotic cells contain more DNA than prokaryotic cells do, but variation among eukaryotes in their C-values is huge. Human cells contain more than 10 times the amount of DNA found in *Drosophila* cells, whereas some salamander cells contain 20 times as much DNA as human cells. Clearly, these differences in C-value cannot be explained simply by differences in organismal complexity. So what is all the extra DNA in eukaryotic cells doing? This question has been termed the **C-value paradox**. We do not yet have a complete answer to the C-value paradox, but analysis of eukaryotic DNA sequences has revealed a complexity that is absent from prokaryotic DNA.

The Denaturation and Renaturation of DNA

The first clue that eukaryotic DNA contains several types of sequences not present in prokaryotic DNA came from studies in which double-stranded DNA was separated and then allowed to reassociate. When double-stranded DNA in solution is heated, the hydrogen bonds that hold the two nucleotide strands together are weakened, and with enough heat, the two strands separate completely, a process called **denaturation** or *melting*. The temperature at which DNA denatures, called the **melting temperature** (T_m), depends on the base sequence of the particular sample of DNA: G–C base pairs have three hydrogen bonds, whereas A–T base pairs have only two, so the separation of G–C pairs requires more heat (energy) than does the separation of A–T pairs.

The denaturation of DNA by heating is reversible: if single-stranded DNA is slowly cooled, single strands will collide and hydrogen bonds will form again between complementary base pairs, producing double-stranded DNA. This reaction is called **renaturation** or *reannealing*.

TABLE 11.3	Genome sizes of various organisms
Organism	**Approximate Genome Size (bp)**
λ (bacteriophage)	50,000
Escherichia coli (bacterium)	4,640,000
Saccharomyces cerevisiae (yeast)	12,000,000
Arabidopsis thaliana (plant)	125,000,000
Drosophila melanogaster (insect)	170,000,000
Homo sapiens (human)	3,200,000,000
Zea mays (corn)	4,500,000,000
Amphiuma (salamander)	765,000,000,000

Two single-stranded molecules of DNA from different sources, such as different organisms, will anneal if they are complementary; this process is termed **hybridization**. For hybridization to take place, the two strands from the different sources do not have to be complementary at all their bases—just at enough bases to hold the two strands together. The extent of hybridization between DNA from two species can be used to measure the similarity of their nucleic acid sequences and to assess their evolutionary relationship. The rate at which hybridization takes place also provides information about the sequence complexity of DNA.

> **TRY PROBLEM 25**

Types of DNA Sequences in Eukaryotes

Eukaryotic DNA consists of at least three classes of sequences: unique-sequence DNA, moderately repetitive DNA, and highly repetitive DNA. **Unique-sequence DNA** consists of sequences that are present only once or, at most, a few times in the genome. This DNA includes sequences that encode proteins, as well as a great deal of DNA whose function is unknown. Genes that are present in a single copy constitute roughly 25% to 50% of the protein-encoding genes in most multicellular eukaryotes. Other genes within unique-sequence DNA are present in several similar, but not identical, copies and are collectively referred to as a **gene family**. Most gene families arose through duplication of an existing gene and include just a few member genes, but some, such as those that encode immunoglobulin proteins in vertebrates, contain hundreds of members. The genes that encode β-like globins are another example of a gene family. In humans, there are six β-globin genes, clustered together on chromosome 11. The polypeptides encoded by these genes join with α-globin polypeptides to form hemoglobin molecules, which transport oxygen in the blood.

Other sequences, called **repetitive DNA**, exist in many copies. Some eukaryotic organisms have large amounts of repetitive DNA; for example, almost half of the human genome consists of repetitive DNA. A major class of repetitive DNA is **moderately repetitive DNA**, which typically consists of sequences from 150 to 300 bp in length (although they may be longer) that are repeated many thousands of times. Some of these sequences perform important functions for the cell; for example, multiple copies of the genes for ribosomal RNAs (rRNAs) and transfer RNAs (tRNAs) make up a part of the moderately repetitive DNA. However, the function of much moderately repetitive DNA is unknown, and indeed, it may have no function.

Moderately repetitive DNA itself can be divided into two types of repeats. **Tandem repeats** appear one after another and tend to be clustered at particular locations on the chromosomes. **Interspersed repeats** are scattered throughout the genome. An example of an interspersed repeat is the *Alu* sequence, an approximately 300-bp sequence that is present more than a million times and constitutes 11% of the human genome, although it has no obvious cellular function.

Short repeats such as the *Alu* sequence are called **SINEs** (**short interspersed elements**). Longer interspersed repeats consisting of several thousand base pairs are called **LINEs** (**long interspersed elements**). One class of LINEs, called LINE1, constitutes about 17% of the human genome. Most interspersed repeats are the remnants of transposable elements, sequences that can multiply and move (see Chapter 18).

The other major class of repetitive DNA is **highly repetitive DNA**. These short sequences, often less than 10 bp in length, are present in hundreds of thousands to millions of copies that are repeated in tandem and clustered in certain regions of the chromosome, especially in centromeres and telomeres. Highly repetitive DNA is sometimes called *satellite DNA* because its proportions of the four bases differ from those of other DNA sequences and, therefore, it separates as a satellite fraction when centrifuged at high speeds in a density gradient (see pp. 341–342 in Chapter 12). Highly repetitive DNA is rarely transcribed into RNA. Although it may contribute to centromere and telomere function, most highly repetitive DNA has no known function.

Organization of Genetic Information in Eukaryotes

DNA-hybridization reactions and, more recently, direct sequencing of eukaryotic genomes have not only revealed the types of sequences we have just described, but have also told us a lot about how genetic information is organized within chromosomes. We now know that the density of genes varies greatly among and within chromosomes. For example, human chromosome 19 has a high density of genes, with about 26 genes per million base pairs. Chromosome 13, on the other hand, has only about 6.5 genes per million base pairs. Gene density can also vary among different regions of the same chromosome: some parts of the long arm of chromosome 13 have only 3 genes per million base pairs, whereas other parts have almost 30 genes per million base pairs. And the short arm of chromosome 13 contains almost no genes, consisting entirely of heterochromatin.

CONCEPTS

Eukaryotic DNA comprises three major classes: unique-sequence DNA, moderately repetitive DNA, and highly repetitive DNA. Unique-sequence DNA consists of sequences that exist in one or a few copies; moderately repetitive DNA consists of sequences that may be several hundred base pairs in length and are present in thousands to hundreds of thousands of copies. Highly repetitive DNA consists of very short sequences repeated in tandem and is present in hundreds of thousands to millions of copies. The density of genes varies greatly among and even within chromosomes.

✔ CONCEPT CHECK 7

Most of the genes that encode proteins are found in
a. unique-sequence DNA.   c. highly repetitive DNA.
b. moderately repetitive DNA. d. All of the above

11.4 Organelle DNA Has Unique Characteristics

As we have seen, eukaryotic chromosomes reside within the nucleus and have a complex structure consisting of DNA and associated histone proteins. However, some DNA found in eukaryotic cells occurs outside the nucleus, has a very different organization, and exhibits a different pattern of inheritance from nuclear DNA. This DNA occurs in mitochondria and chloroplasts, which are membrane-bounded organelles located in the cytoplasm of eukaryotic cells (**Figure 11.11**).

Mitochondrion and Chloroplast Structure

Mitochondria are present in almost all eukaryotic cells, whereas chloroplasts are found in plants, algae, and some protists. Both organelles generate ATP, the universal energy carrier of cells.

Mitochondria are tubular structures that are from 0.5 to 1.0 micrometer (μm) in diameter, about the size of a typical

Mitochondrion   **Chloroplast**

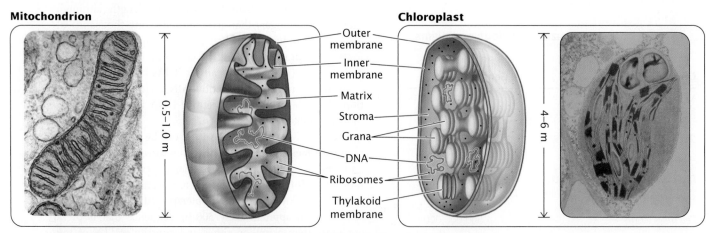

Outer membrane
Inner membrane
Matrix
Stroma
Grana
DNA
Ribosomes
Thylakoid membrane

0.5–1.0 m   4–6 m

11.11 Comparison of the structures of mitochondria and chloroplasts. [Left: Don W. Fawcett/Science Source. Right: Biophoto Associates/Science Source.]

bacterium, whereas chloroplasts are typically from about 4 to 6 μm in diameter. Both are surrounded by two membranes enclosing a region (called the matrix in mitochondria and the stroma in chloroplasts) that contains enzymes, ribosomes, RNA, and DNA. In mitochondria, the inner membrane is highly folded; embedded within it are the enzymes that catalyze electron transport and oxidative phosphorylation. Chloroplasts have a thylakoid membrane, which is highly folded and stacked to form aggregates called grana. This membrane bears the pigments and enzymes required for photophosphorylation. New mitochondria and chloroplasts arise by the division of existing organelles; these divisions take place throughout the cell cycle and are independent of mitosis and meiosis.

Mitochondria and chloroplasts possess DNA that encodes some polypeptides used by the organelles, as well as rRNAs found in the ribosomes and the tRNAs needed for the translation of these proteins. The genes for most of the 900 or so structural proteins and enzymes found in mitochondria, however, are actually encoded by nuclear DNA; the mitochondrial genome typically encodes only a few proteins and a few rRNA and tRNA molecules needed for mitochondrial protein synthesis.

The Endosymbiotic Theory

Chloroplasts and mitochondria are similar to bacteria in many ways. This resemblance is not superficial; indeed, there is compelling evidence that these organelles evolved from bacteria. The **endosymbiotic theory** (**Figure 11.12**) proposes that mitochondria and chloroplasts were once free-living bacteria that became internal inhabitants (endosymbionts) of early eukaryotic cells. It is assumed that over evolutionary time, many of the endosymbiont's original genes were subsequently lost (because nuclear genes existed that provided the same function) or were transferred to the nucleus.

A great deal of evidence supports the idea that mitochondria and chloroplasts originated as bacterial cells. Many modern single-celled eukaryotes (protists) are hosts to endosymbiotic bacteria. Mitochondria and chloroplasts are similar in size to present-day bacteria and possess their own DNA, which shares many characteristics with bacterial DNA. Mitochondria and chloroplasts possess ribosomes, some of which are similar in size and structure to bacterial ribosomes. In addition, antibiotics that inhibit protein synthesis in bacteria but do not affect protein synthesis in eukaryotic cells do inhibit protein synthesis in these organelles.

The strongest evidence for the endosymbiotic theory comes from studies of DNA sequences, which demonstrate that sequences in mtDNA and cpDNA are more closely related to sequences in the genes of bacteria than they are to those found in the eukaryotic nucleus. All of this evidence indicates that mitochondria and chloroplasts are more closely related to bacterial cells than they are to the eukaryotic cells in which they are now found.

> **CONCEPTS**
>
> Mitochondria and chloroplasts are membrane-bounded organelles of eukaryotic cells that generally possess their own DNA. The well-supported endosymbiotic theory proposes that these organelles began as free-living bacteria that developed stable endosymbiotic relations with early eukaryotic cells.
>
> ✔ **CONCEPT CHECK 8**
> What evidence supports the endosymbiotic theory?

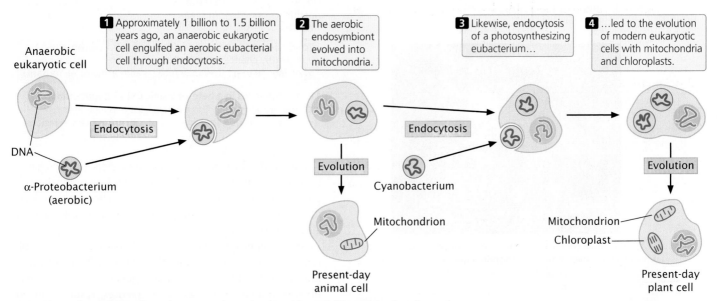

11.12 The endosymbiotic theory proposes that mitochondria and chloroplasts in eukaryotic cells arose from bacteria.

Uniparental Inheritance of Organelle-Encoded Traits

Mitochondria and chloroplasts are present in the cytoplasm and are usually inherited from a single parent. Thus, traits encoded by mtDNA and cpDNA exhibit uniparental inheritance (see Chapter 5). In animals, mtDNA is inherited almost exclusively from the female parent, although occasional male transmission of mtDNA has been documented. Maternal inheritance of animal mtDNA may be partly a function of gamete size: sperm are much smaller than eggs and hold fewer mitochondria. However, recent research has found that in some eukaryotes, paternal mitochondria are selectively eliminated by autophagy, a process in which mitochondria are digested by the cell. In these cases, paternal mitochondria are targeted for destruction, whereas maternal mitochondria are not; the mechanism that produces this difference is not known. Paternal inheritance of organelles is common in gymnosperms (conifers) and in a few angiosperms (flowering plants). Some plants even exhibit biparental inheritance of mtDNA and cpDNA.

REPLICATIVE SEGREGATION Individual cells may contain from dozens to hundreds of mitochondria and chloroplasts, each with numerous copies of the organelle genome, so each cell typically possesses from hundreds to thousands of copies of mitochondrial and chloroplast genomes (**Figure 11.13**). A mutation arising within one DNA molecule within one organelle generates a mixture of organelles within the cell, some with a mutant DNA sequence and others with a wild-type DNA sequence. The occurrence of two distinct varieties of DNA within the cytoplasm of a single cell is termed **heteroplasmy**. When a heteroplasmic cell divides, the organelles segregate randomly into the two progeny cells in a process called **replicative segregation** (**Figure 11.14**), and chance determines the proportion of mutant organelles in each cell. Although most progeny cells inherit a mixture of

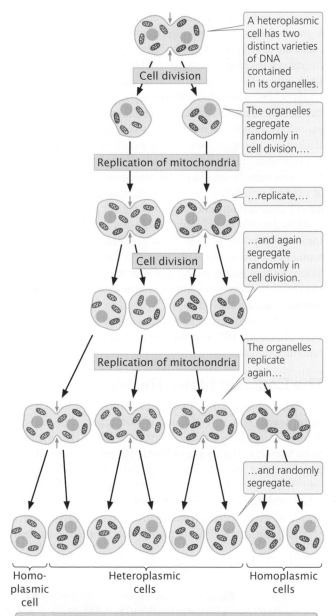

A heteroplasmic cell has two distinct varieties of DNA contained in its organelles.

Cell division

The organelles segregate randomly in cell division,…

Replication of mitochondria

…replicate,…

…and again segregate randomly in cell division.

Cell division

Replication of mitochondria

The organelles replicate again…

…and randomly segregate.

| Homo-plasmic cell | Heteroplasmic cells | Homoplasmic cells |

Conclusion: Most of the resulting cells are heteroplasmic, but, just by chance, some cells may receive only one type of organelle (e.g., they may receive all normal or all mutant).

11.14 Organelles in a heteroplasmic cell segregate randomly into the progeny cells. This diagram illustrates replicative segregation in mitosis; the same process also takes place in meiosis.

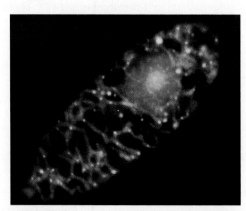

11.13 Individual cells may contain many mitochondria, each with several copies of the mitochondrial genome. Shown here is a cell of *Euglena gracilis*, a protist, stained so that the nucleus appears red, mitochondria green, and mtDNA yellow. [From Hayashi, Y. and K. Ueda, "The shape of mitochondria and the number of mitochondrial nucleoids during the cell cycle of *Euglena gracilis*," *Journal of Cell Science* (1989) 93, p. 565. © Company of Biologists. Permission conveyed through Copyright Clearance Center, Inc.]

mutant and wild-type organelles, some cells, just by chance, may receive organelles with only mutant or only wild-type sequences; the result, in which all organelles are genetically identical, is known as **homoplasmy**. Fusion of mitochondria also takes place frequently.

When replicative segregation takes place in somatic cells, it may create phenotypic variation within a single organism: different cells of the organism may possess different proportions of mutant and wild-type sequences, resulting in different degrees of phenotypic expression in different tissues. When replicative segregation takes place in the germ cells of a heteroplasmic cytoplasmic donor, there may be different phenotypes among the offspring.

The disease known as myoclonic epilepsy and ragged-red fiber (MERRF) syndrome is caused by a mutation in an mtDNA gene. In one case, a 20-year-old person who carried this mutation in 85% of his mtDNA sequences displayed a normal phenotype, whereas a cousin who had the mutation in 96% of his mtDNA sequences was severely affected. In diseases caused by mutations in mtDNA, the severity of the disease is frequently related to the proportion of mutant mtDNA sequences inherited at birth. **▶ TRY PROBLEM 27**

TRAITS ENCODED BY mtDNA A number of traits affected by organelle DNA have been studied. One of the first to be examined in detail was the phenotype produced by *petite* mutations in yeast (**Figure 11.15**). In the late 1940s, Boris Ephrussi and his colleagues noticed that when they grew yeast on solid medium, some colonies were much smaller than normal. Examination of these *petite* colonies revealed that the growth rates of the cells within the colonies were greatly reduced. The results of biochemical studies demonstrated that the *petite* mutants were unable to carry out aerobic respiration; they obtained all their energy from anaerobic metabolism (glycolysis and fermentation), which is much less efficient than aerobic respiration and results in a smaller colony size.

Some *petite* mutations are defects in nuclear DNA, but most *petite* mutations occur in mtDNA. Mitochondrial *petite* mutants often have large deletions in mtDNA or, in some cases, are missing mtDNA entirely. Much of the mtDNA sequence encodes enzymes that catalyze aerobic respiration; therefore, the *petite* mutants are unable to carry out aerobic respiration and cannot produce normal quantities of ATP, which inhibits their growth.

Another known mtDNA mutation occurs in *Neurospora* (see p. 430 in Chapter 15). Isolated by Mary Mitchell in 1952,

poky mutants grow slowly, display cytoplasmic inheritance, and have abnormal amounts of cytochromes. Cytochromes are protein components of the electron-transport chain of the mitochondria and play an integral role in the production of ATP. Most organisms have three primary types of cytochromes: cytochrome *a*, cytochrome *b*, and cytochrome *c*. *Poky* mutants have cytochrome *c*, but no cytochrome *a* or *b*. Like *petite* mutants, *poky* mutants are defective in ATP synthesis and therefore grow more slowly than do normal, wild-type cells. **▶ TRY PROBLEM 31**

In recent years, a number of genetic diseases that result from mutations in mtDNA (in addition to MERRF syndrome, mentioned earlier) have been identified in humans. Leber hereditary optic neuropathy (LHON) results from mutations in the mtDNA genes that encode electron-transport proteins. LHON typically leads to sudden loss of vision in middle age. Another disease caused by mtDNA mutations is neurogenic muscle weakness, ataxia, and retinitis pigmentosa (NARP), which is characterized by seizures, dementia, and developmental delay. Other mitochondrial diseases include Kearns–Sayre syndrome (KSS) and chronic external ophthalmoplegia (CEOP), both of which result in paralysis of the eye muscles, droopy eyelids, and, in severe cases, vision loss, deafness, and dementia. All of these diseases exhibit cytoplasmic inheritance and variable expression (see Chapter 5).

A trait in plants that is produced by mutations in mitochondrial genes is cytoplasmic male sterility, a mutant phenotype found in more than 140 different plant species and inherited only from the maternal parent. These mutations inhibit pollen development but do not affect female fertility.

A number of cpDNA mutants have also been discovered. One of the first to be recognized was the mutation responsible for leaf variegation in the four o'clock plant, *Mirabilis jalapa*, which was studied by Carl Correns in 1909 (see pp. 129–130 in Chapter 5). In the green alga *Chlamydomonas*, streptomycin-resistant mutations occur in cpDNA, and in higher plants, a number of mutants exhibiting altered pigmentation and growth have been traced to defects in cpDNA.

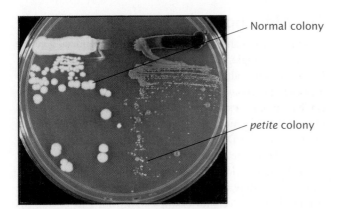

11.15 *Petite* mutations are deletions in mtDNA. Colonies of normal yeast cells and colonies of *petite* mutants are shown here. The *petite* mutants have large deletions in their mtDNA and are unable to carry out oxidative phosphorylation. [From Xin Jie Chen and G. Desmond Clark-Walker, "The mitochondrial genome integrity gene, MGII, of *Kluyveromyces lactis* encodes the p-subunit of F1-ATPase," *Genetics* 144: 1445–1454, Fig 1, 1996. © Genetics Society of America. Courtesy of Xin Jie Chen, Department of Biochemistry and Molecular Biology, SUNY Upstate Medical University, permission conveyed through Copyright Clearance Center, Inc.]

> **CONCEPTS**
>
> In most organisms, genes encoded by mtDNA and cpDNA are inherited from a single parent. A cell may contain more than one distinct type of mtDNA or cpDNA; in these cases, replicative segregation of the organelle DNA may produce phenotypic variation within a single organism, or it may produce different degrees of phenotypic expression among progeny.
>
> **✔ CONCEPT CHECK 9**
>
> In a few organisms, traits encoded by mtDNA can be inherited from either parent. This observation indicates that in these organisms,
>
> a. mitochondria do not exhibit replicative segregation.
> b. heteroplasmy is present.
> c. both sperm and eggs contribute cytoplasm to the zygote.
> d. there are multiple copies of mtDNA in each cell.

WORKED PROBLEM

A physician examines a young man who has a progressive muscle disorder and visual abnormalities. A number of the patient's relatives have the same condition, as shown in the pedigree below. The degree of expression of the trait is highly variable among members of the family: some are only slightly affected, whereas others developed severe symptoms at an early age. The physician concludes that this disorder is due to a mutation in the mitochondrial genome. Do you agree with the physician's conclusion? Why or why not? Could the disorder be due to a mutation in a nuclear gene? Explain your reasoning.

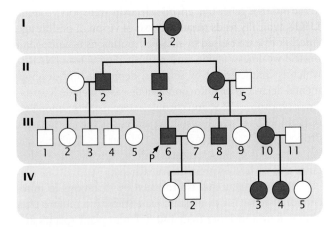

Solution Strategy

What information is required in your answer to the problem?

An explanation of whether and how a mutation in the mitochondrial genome could have caused the patient's disorder, as well as an explanation of whether and how the disorder could have been caused by a mutation in a nuclear gene.

What information is provided to solve the problem?

- The patient has a progressive muscle disorder and visual abnormalities.
- A pedigree for the patient's family.
- The trait exhibits variable expression among members of the family.

Solution Steps

The conclusion that the disorder is caused by a mutation in the mitochondrial genome is supported by the pedigree and by the observation of variable expression in affected members of the same family. The disorder is passed only from affected mothers to both their male and their female offspring; when fathers are affected, none of their children have the trait (as seen in the children of II-2 and III-6). This outcome is expected of traits determined by mutations in mtDNA because mitochondria are located in the cytoplasm and are usually inherited from a single parent (in humans,

the mother). The trait cannot be X-linked recessive because in that case, a cross between a female with the trait ($X^a X^a$) and a male without the trait (X^+Y) would *not* produce daughters with the trait ($X^a X^a$), which we see in III-10, IV-3, and IV-4. It cannot be X-linked dominant because in that case, II-2 and III-6 would have passed it to their daughters, who are unaffected (unless the trait exhibited incomplete penetrance).

The facts that some offspring of affected mothers do not show the trait (III-9 and IV-5) and that expression varies from one person to another suggest that the affected family members are heteroplasmic, with both mutant and wild-type mitochondria. Replicative segregation of mitochondria in meiosis may produce gametes having different proportions of mutant and wild-type mtDNA sequences, resulting in different degrees of phenotypic expression of the disorder among the offspring. Most likely, symptoms of the disorder develop when some minimum proportion of the mitochondria are mutant. Just by chance, some of the gametes produced by an affected mother contain few mutant mitochondria and result in offspring who lack the disorder.

Another possible explanation for the disorder is that it results from an autosomal dominant allele. When an affected (heterozygous) person mates with an unaffected (homozygous) person, about half of the offspring are expected to have the trait, but just by chance, some affected parents will have no affected offspring. Affected individuals II-2 and III-6 in the pedigree could have just happened to be male, and their sex could be unrelated to the mode of transmission. The variable expression could be explained by variable expressivity (see Chapter 5).

> » For more experience with the inheritance of organelle-encoded traits, try working **Problem 28** at the end of the chapter.

The Mitochondrial Genome

In most animals and fungi, the entire mitochondrial genome exists on a single, double-stranded, highly coiled, circular DNA molecule, although there may be many copies of this genome in each cell. This circular mitochondrial DNA molecule is similar in structure to a bacterial chromosome. Plant mitochondrial genomes often exist as a complex collection of multiple circular DNA molecules. In some species, the mitochondrial genome consists of a single, linear DNA molecule.

Each mitochondrion contains multiple copies of the mitochondrial genome, and a cell may contain many mitochondria. A typical rat liver cell, for example, has from 5 to 10 mtDNA molecules in each of about 1000 mitochondria, so each cell possesses from 5000 to 10,000 copies of the mitochondrial genome. Mitochondrial DNA constitutes about 1% of the total cellular DNA in a rat liver cell. Like bacterial chromosomes, mtDNA lacks the histone proteins normally associated with eukaryotic nuclear DNA, although it is complexed with other proteins that have some histone-like properties. The guanine–cytosine

(G–C) content of mtDNA is often sufficiently different from that of nuclear DNA that mtDNA can be separated from nuclear DNA by density-gradient centrifugation.

Mitochondrial genomes are small compared with nuclear genomes and vary greatly in size among different organisms (**Table 11.4**). The sizes of mitochondrial genomes of most species range from 15,000 bp to 65,000 bp, but those of a few species are much smaller (e.g., the genome of *Plasmodium falciparum*, the parasite that causes malaria, is only 6000 bp), while those of some plants are several million base pairs. There is no correlation, however, between genome size and number of genes. The number of genes is more constant than genome size: most species have only 40–50 genes. These genes encode five basic functions: respiration and oxidative phosphorylation, translation, transcription, RNA processing, and the import of proteins into the cell. Most of the variation in the size of mitochondrial genomes is due to differences in noncoding DNA sequences. As mentioned earlier, genes for most of the proteins and enzymes found in mitochondria are actually encoded by nuclear DNA.

HUMAN mtDNA Human mtDNA is a circular molecule encompassing 16,569 bp that encode 2 rRNAs, 22 tRNAs, and 13 proteins. The two nucleotide strands of the molecule differ in their base composition: the heavy (H) strand has more guanine nucleotides, whereas the light (L) strand has more cytosine nucleotides. The H strand is the template for both rRNAs, 14 of the 22 tRNAs, and 12 of the 13 proteins, whereas the L strand serves as the template for only 8 of the tRNAs and 1 protein. The **D loop** (**Figure 11.16**) is a region of the mtDNA that contains sites where replication and transcription of the mtDNA is initiated. Human mtDNA is highly economical in

TABLE 11.4	Sizes of mitochondrial genomes in selected organisms
Organism	**Size of mtDNA (bp)**
Pichia canadensis (fungus)	27,694
Podospora anserina (fungus)	100,314
Saccharomyces cerevisiae (fungus)	85,779*
Drosophila melanogaster (fruit fly)	19,517
Lumbricus terrestris (earthworm)	14,998
Xenopus laevis (frog)	17,553
Mus musculus (house mouse)	16,295
Homo sapiens (human)	16,569
Chlamydomonas reinhardtii (green alga)	15,758
Plasmodium falciparum (protist)	5,966
Paramecium aurelia (protist)	40,469
Arabidopsis thaliana (plant)	166,924
Cucumis melo (plant)	2,400,000

*Size varies among strains.

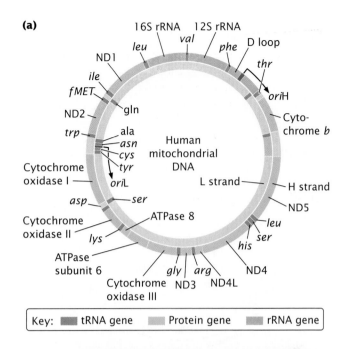

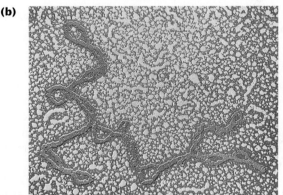

11.16 The human mitochondrial genome, consisting of 16,569 bp, is highly economical in its organization, with few sequences that do not code for RNA or protein. (a) The outer circle represents the heavy (H) strand, and the inner circle represents the light (L) strand. The origins of replication for the H and L strands are *ori*H and *ori*L, respectively. ND identifies genes that encode subunits of NADH dehydrogenase. (b) Electron micrograph of isolated mtDNA. [Part b: CNRI/Science Source.]

its organization: there are few noncoding nucleotides between the genes, and almost all the mRNA transcribed from it encodes proteins. Human mtDNA also contains very little repetitive DNA. The one region of human mtDNA that does contain some noncoding nucleotides is the D loop.

YEAST mtDNA The organization of yeast (*Saccharomyces cerevisiae*) mtDNA is quite different from that of human mtDNA. Although the yeast mitochondrial genome, with its 78,000 bp, is nearly five times as large, it encodes only six additional genes, for a total of 2 rRNAs, 25 tRNAs, and 16 proteins (**Figure 11.17**). Most of the extra DNA in the yeast mitochondrial genome consists of noncoding sequences found within and between genes.

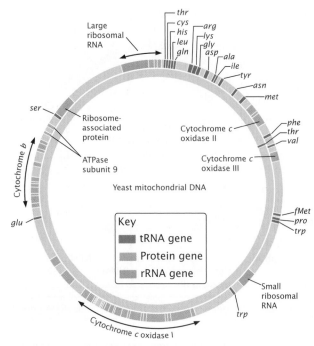

11.17 The yeast mitochondrial genome, consisting of 78,000 bp, contains much noncoding DNA.

FLOWERING-PLANT mtDNA Flowering plants (angiosperms) have the largest and most complex mitochondrial genomes known: their mitochondrial genomes range in size from 186,000 bp in white mustard to 2,400,000 bp in muskmelon. Even closely related plant species may differ greatly in the sizes of their mtDNA.

Part of the extensive size variation in the mtDNA of flowering plants can be explained by the presence of long sequences that are direct repeats. Crossing over between these repeats can generate multiple circular chromosomes of different sizes. The mitochondrial genome in turnips, for example, consists of a "master circle" consisting of 218,000 bp that has direct repeats. Homologous recombination between the repeats can generate two smaller circles of 135,000 bp and 83,000 bp (**Figure 11.18**). Other species contain several

direct repeats, providing possibilities for complex crossing-over events that may increase or decrease the number and sizes of the circles.

CONCEPTS

The mitochondrial genome consists of circular DNA with no associated histone proteins, although it is complexed with other proteins that have some histone-like properties. The sizes and structures of mtDNA differ greatly among organisms. Human mtDNA exhibits extreme economy, but mtDNAs found in yeast and flowering plants contain many noncoding nucleotides and repetitive sequences. In most flowering plants, mitochondrial DNA is large and typically has one or more large direct repeats that can recombine to generate smaller or larger molecules.

The Evolution of Mitochondrial DNA

As already mentioned, comparisons of mitochondrial DNA sequences with DNA sequences in bacteria strongly support a common bacterial origin for all mtDNA. Nevertheless, patterns of evolution seen in mtDNA vary greatly among different groups of organisms.

The sequences of vertebrate mtDNA exhibit an accelerated rate of evolution: the sequences in mammalian mtDNA, for example, typically change from 5 to 10 times faster than those in mammalian nuclear DNA. The accelerated rate of evolution seen in vertebrate mtDNA is due to its high mutation rate, which allows sequences to change more quickly. In spite of this high rate of sequence evolution, the numbers of genes present and the organization of vertebrate mitochondrial genomes are relatively constant. In contrast, the sequences of plant mtDNA evolve slowly, at a rate only $1/10$ that of the nuclear genome, but their gene content and organization change rapidly. The reason for these basic differences in rates of evolution is not yet known.

Mitochondrial DNA has been studied extensively to reconstruct patterns of evolution in humans and many other organisms. Some of the advantages of mtDNA for studying

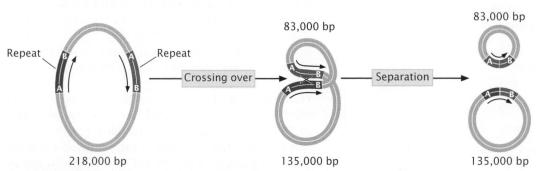

11.18 Size variation in plant mtDNA can be generated through recombination between direct repeats. In turnips, the mitochondrial genome consists of a "master circle" of 218,000 bp; crossing over between the direct repeats produces two smaller circles of 135,000 bp and 83,000 bp.

evolution include (1) the small size of mtDNA and its abundance in the cell; (2) the rapid evolution of mtDNA sequences in some organisms, which facilitates study of closely related groups; and (3) the maternal inheritance of mtDNA and lack of recombination, which makes it possible to trace female lines of descent. Samples of human mtDNA from thousands of people belonging to hundreds of different ethnic groups throughout the world have been analyzed. These mtDNA samples are helping to unravel many aspects of human evolution and history. For example, initial studies on mtDNA sequences led to the proposal that small groups of humans migrated out of Africa about 85,000 years ago and populated the rest of the world. This proposal, called the Out of Africa hypothesis or the African Replacement hypothesis, has now gained wide acceptance. The Out of Africa hypothesis is supported by additional studies of DNA sequences from the Y chromosome and nuclear genes. The use of mtDNA in evolutionary studies will be described in more detail in Chapter 26.

At conception, a mammalian zygote inherits approximately 100,000 copies of mtDNA from the egg. Because of the large number of mtDNA molecules in each cell and the high rate of mutation in mtDNA, most cells would be expected to contain a mixture of wild-type and mutant mtDNA molecules (heteroplasmy). However, heteroplasmy is rarely present: the copies of mtDNA in most individuals are genetically identical (homoplasmy). To account for the uniformity of mtDNA within individual mammals, geneticists hypothesize that, at some point in early development or gamete formation, mtDNA goes through some type of bottleneck, during which the mtDNAs within a cell are reduced to just a few copies, which then replicate and give rise to all subsequent copies of mtDNA. Through this process, genetic variation in mtDNA within a cell is eliminated. Recent studies have provided evidence that a bottleneck does exist, but there is contradictory evidence concerning where in development it arises.

THINK-PAIR-SHARE Question 3

CONCEPTS

All mtDNA appears to have evolved from a common bacterial ancestor, but the patterns of evolution seen in different mitochondrial genomes vary greatly. Vertebrate mtDNA exhibits rapid change in sequence but little change in gene content and organization, whereas the mtDNA of plants exhibits little change in sequence but much variation in gene content and organization. Mitochondrial DNA sequences are frequently used to study patterns of evolution.

Damage to Mitochondrial DNA Associated with Aging

The symptoms of many human genetic diseases caused by defects in mtDNA first appear in middle age or later and increase in severity as people age. One hypothesis to explain this pattern is related to the decline in oxidative phosphorylation with aging.

Oxidative phosphorylation is the process that generates ATP, the primary carrier of energy in the cell. This process takes place on the inner membrane of the mitochondrion and requires a number of different proteins, some encoded by mtDNA and others encoded by nuclear genes. Oxidative phosphorylation normally declines with age, and if it falls below a critical threshold, tissues do not make enough ATP to sustain vital functions and disease symptoms appear. Most people start life with an excess capacity for oxidative phosphorylation; this capacity decreases with age, but most people reach old age or die before the critical threshold is passed. People who are born with mitochondrial diseases carry mutations in their mtDNA that lower their capacity for oxidative phosphorylation. At birth, their capacity may be sufficient to support their ATP needs, but as their oxidative phosphorylation capacity declines with age, they cross the critical threshold and begin to experience disease symptoms.

Why does oxidative phosphorylation capacity decline with age? A possible explanation is that damage to mtDNA accumulates over time: deletions and base substitutions in mtDNA increase with age. For example, a common 5000-bp deletion in mtDNA is absent in normal heart muscle cells before the age of 40, but afterward, this deletion is present with increasing frequency. The same deletion is found at a low frequency in normal brain tissue before age 75, but is found in 11% to 12% of mtDNAs in the basal ganglia by age 80. People with mitochondrial genetic diseases may age prematurely because they begin life with damaged mtDNA.

The mechanism of age-related increases in mtDNA damage is not yet known. Oxygen radicals—highly reactive compounds that are natural by-products of oxidative phosphorylation—are known to damage DNA (see p. 531 in Chapter 18). Because mtDNA is physically close to the enzymes taking part in oxidative phosphorylation, mtDNA may be more prone to oxidative damage than is nuclear DNA. When mtDNA has been damaged, the cell's capacity to produce ATP drops.

Mitochondrial Replacement Therapy

A number of severe genetic disorders result from mutations in mtDNA, and 1 out of every 5000 babies is born with a mitochondrial disease. Geneticists have recently developed methods that allow a woman carrying an mtDNA mutation to give birth to a healthy child. Called **mitochondrial replacement therapy** (MRT), these methods combine the nuclear DNA of a female who carries a mitochondrial mutation with that of a sperm and the egg cytoplasm of a healthy donor, creating a "three-parent baby."

One type of MRT, termed pronuclear transfer, starts with an egg from a woman with an mtDNA mutation. The egg is fertilized in vitro (in the laboratory). After penetrating the egg, the sperm releases its nucleus into the egg, resulting in two nuclei (called pronuclei), one from the father and one from the mother. Using a very fine pipette, a technician removes the pronuclei and transfers them to an egg cell from a healthy donor from which the nucleus has been previously removed. Inside the donor egg, the pronuclei fuse, and the resulting embryo begins development. The developing embryo is then implanted into the uterus of the affected female.

In a related technique, called maternal spindle transfer, the nuclear material is removed from the unfertilized oocyte of an affected female and transferred to an enucleated donor egg. The resulting oocyte, containing nuclear genetic material from the affected mother and cytoplasm from the healthy donor, is fertilized in vitro. The resulting embryo is then implanted into the affected female.

Mitochondrial replacement therapy is now legally permissible for preventing mitochondrial disorders in the United Kingdom, but its use in the United States is still under consideration. Because it combines genetic material from three parents in the conception of a child, MRT has raised a number of ethical issues. There have also been some concerns about the possible developmental consequences of combining nuclear DNA and mtDNA from different people.

THINK-PAIR-SHARE Question 4

The Chloroplast Genome

Among plants, the chloroplast genome ranges in size from 80,000 to 600,000 bp, but most chloroplast genomes range from 120,000 to 160,000 bp (**Table 11.5**). Chloroplast DNA is usually a single, double-stranded DNA molecule that is circular, highly coiled, and lacks associated histone proteins. As in mtDNA, multiple copies of the chloroplast genome are found in each chloroplast, and there are multiple organelles per cell, so there are several hundred to several thousand copies of cpDNA in a typical plant cell.

The chloroplast genomes of a number of plant and algal species have been sequenced, and cpDNA is now recognized to be basically bacterial in its organization: the order of some groups of genes is the same as that observed in *E. coli*, and many chloroplast genes are organized into clusters similar to those found in bacteria. Many of the gene sequences in cpDNA are quite similar to those found in homologous bacterial genes.

Among vascular plants, chloroplast genomes are similar in gene content and gene order. A typical chloroplast genome encodes 4 rRNA genes, from 30 to 35 tRNA genes, a number of ribosomal proteins, many proteins engaged in photosynthesis, and several proteins with roles in nonphotosynthetic processes (**Figure 11.19**). A key protein encoded by cpDNA is ribulose-1,5-bisphosphate carboxylase-oxygenase (abbreviated RuBisCO), which participates in the fixation of carbon in photosynthesis. RuBisCO makes up about 50% of the protein found in green plants and is therefore considered the most abundant protein on Earth. It is a complex protein consisting of eight identical large subunits and eight identical small subunits. The large subunit is encoded by chloroplast DNA, whereas the small subunit is encoded by nuclear DNA. Much of cpDNA consists of noncoding sequences.

THE EVOLUTION OF CHLOROPLAST DNA The DNA sequences of chloroplasts are very similar to those found in cyanobacteria (a group of photosynthetic bacteria), so chloroplast genomes clearly have a bacterial ancestry. Overall, cpDNA sequences evolve slowly compared with sequences in nuclear DNA and some mtDNA. For most chloroplast genomes, size and gene organization are similar, although there are some notable exceptions. Because it evolves slowly and, like mtDNA, is inherited from only one parent, cpDNA is often useful for determining the evolutionary relationships among different plant species.

TABLE 11.5	Sizes of chloroplast genomes in selected organisms
Organism	**Size of cpDNA (bp)**
Euglena gracilis (protist)	143,172
Porphyra purpurea (red alga)	191,028
Chlorella vulgaris (green alga)	150,613
Marchantia polymorpha (liverwort)	121,024
Nicotiana tabacum (tobacco)	155,939
Zea mays (corn)	140,387
Pinus thunbergii (black pine)	119,707

CONCEPTS

Most chloroplast genomes consist of a single circular DNA molecule not complexed with histone proteins. Although there is considerable size variation among species, the chloroplast genomes found in most plants range from 120,000 to 160,000 bp. Chloroplast DNA sequences are most similar to DNA sequences in cyanobacteria, which supports the endosymbiotic theory of chloroplast origin.

✔ **CONCEPT CHECK 10**

In its organization, chloroplast DNA is most similar to

a. bacteria.

b. archaea.

c. nuclear DNA of plants.

d. nuclear DNA of primitive eukaryotes.

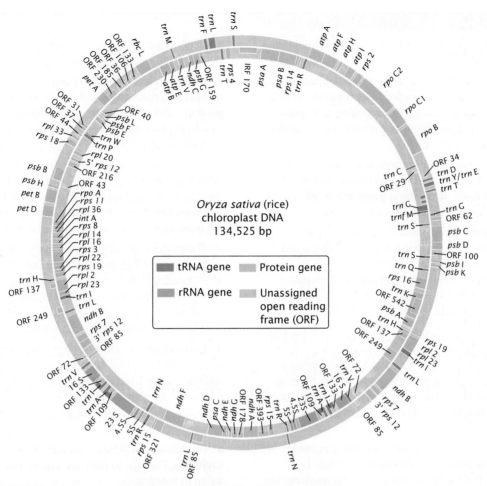

11.19 Chloroplast DNA of rice. An unassigned open reading frame (ORF) refers to an apparent protein-encoding gene—a sequence with a translation start and stop signal in the same reading frame—for which the protein has not yet been identified.

Movement of Genetic Information Between Nuclear, Mitochondrial, and Chloroplast Genomes

As we have seen, many of the proteins found in modern mitochondria and chloroplasts are encoded by nuclear genes, which suggests that much of the genetic material in the original endosymbionts has been transferred over evolutionary time to the nucleus. This assumption is also supported by the observation that some DNA sequences normally found in mtDNA have been detected in the nuclear DNA of some strains of yeast and corn. Likewise, chloroplast sequences have been found in the nuclear DNA of spinach. Furthermore, the sequences of those nuclear genes that encode organelle proteins are most similar to their bacterial counterparts.

There is also evidence that during evolution, genetic material has moved from chloroplasts to mitochondria. For example, DNA fragments from some rRNA genes that are normally found in cpDNA have been found in the mtDNA of corn. Sequences from the gene that encodes the large

subunit of RuBisCO, which is normally encoded by cpDNA, are duplicated in corn mtDNA. And there is even evidence that some nuclear genes have moved into mitochondrial genomes. The exchange of genetic material between the nuclear, mitochondrial, and chloroplast genomes over evolutionary time has given rise to the term "promiscuous DNA" to describe this phenomenon. The mechanism by which this exchange takes place is not entirely clear.

Some plants have acquired mtDNA from other plants through a process known as horizontal gene transfer (see pp. 266–267 in Chapter 9). One remarkable plant, *Amborella trichopoda*—a large shrub found in the rain forests of New Caledonia—has an enormous mitochondrial genome consisting of 3,866,039 bp of DNA. Its mitochondria contain the equivalent of six foreign mitochondrial genomes, acquired from green algae, mosses, and other flowering plants. Researchers propose that *A. trichopoda* captured mtDNA from other plants when it was covered by other plants and subsequently wounded, which led to the transfer of foreign mitochondria and then fusion of mitochondria within an *A. trichopoda* cell.

CONCEPTS SUMMARY

■ Chromosomes contain very long DNA molecules that are tightly packed.

■ Supercoiling results from strain produced when rotations are added to a relaxed DNA molecule or removed from it. Overrotation produces positive supercoiling; underrotation produces negative supercoiling. Supercoiling is controlled by topoisomerase enzymes.

■ A bacterial chromosome consists of a single, circular DNA molecule that is bound to proteins and exists as a series of large loops. It usually appears in the cell as a distinct clump known as the nucleoid.

■ Each eukaryotic chromosome contains a single, long linear DNA molecule that is bound to histone and nonhistone chromosomal proteins. Euchromatin undergoes the normal cycle of decondensation and condensation in the cell cycle. Heterochromatin remains highly condensed throughout the cell cycle.

■ The nucleosome is a core particle of eight histone proteins and the DNA that wraps around them. Nucleosomes are folded into a 30-nm fiber that forms a series of 300-nm-long loops; these loops are anchored at their bases by proteins. The 300-nm loops are condensed to form a fiber that is itself tightly coiled to produce a chromatid.

■ Chromosome regions that are undergoing active transcription are sensitive to digestion by DNase I, indicating that DNA is more exposed during transcription.

■ Epigenetic changes are stable alterations of gene expression that do not require changes in DNA sequences. Epigenetic changes can take place through alterations of chromatin structure.

■ Centromeres are chromosomal regions where spindle microtubules attach; chromosomes without centromeres are usually lost in the course of cell division. Most centromeres are defined by epigenetic changes to chromatin structure. Telomeres stabilize the ends of chromosomes.

■ Eukaryotic DNA comprises three classes of sequences. Unique-sequence DNA exists in very few copies. Moderately repetitive DNA consists of moderately long sequences that are repeated from hundreds to thousands of times. Highly repetitive DNA consists of very short sequences that are repeated in tandem from many thousands to millions of times.

■ Mitochondria and chloroplasts are eukaryotic organelles that possess their own DNA. The endosymbiotic theory proposes that mitochondria and chloroplasts originated as free-living bacteria that entered into a beneficial association with eukaryotic cells.

■ Traits encoded by mtDNA and cpDNA are usually inherited from a single parent, most often the mother. Replicative segregation of organelles in cell division may produce phenotypic variation among cells within an individual organism and among the offspring of a single female.

■ The mitochondrial genome usually consists of a single circular DNA molecule that lacks histone proteins. Mitochondrial DNA varies in size among different groups of organisms. Human mtDNA is highly economical, with few noncoding nucleotides. Fungal and plant mtDNAs contain much noncoding DNA between genes.

■ Comparisons of mtDNA sequences suggest that mitochondria evolved from a bacterial ancestor. Vertebrate mtDNA exhibits rapid change in sequence but little change in gene content and organization. Plant mtDNA exhibits little change in sequence but much variation in gene content and organization.

■ Mitochondrial DNA sequences are widely used to study evolution. Damage to mtDNA has been associated with aging in humans.

■ Mitochondrial replacement therapy can be used to transfer nuclear DNA from a woman with a mitochondrial disorder and from a sperm into an egg cell from a healthy donor, giving rise to a baby with genetic material from three parents. This procedure has raised ethical issues and safety concerns.

■ Chloroplast genomes consist of a single circular DNA molecule that lacks histone proteins and varies little in size. Each plant cell contains multiple copies of cpDNA. Chloroplast DNA sequences are most similar to those in cyanobacteria and tend to evolve slowly.

■ Through evolutionary time, many mitochondrial and chloroplast genes have moved to nuclear chromosomes. In some plants, there is evidence that copies of chloroplast genes have moved to the mitochondrial genome.

IMPORTANT TERMS

mitochondrial DNA (mtDNA) (p. 312)
chloroplast DNA (cpDNA) (p. 312)
supercoiling (p. 313)
relaxed state (p. 313)
positive supercoiling (p. 313)

negative supercoiling (p. 313)
topoisomerase (p. 313)
nucleoid (p. 314)
euchromatin (p. 314)
heterochromatin (p. 314)
nonhistone chromosomal protein (p. 315)

nucleosome (p. 316)
linker DNA (p. 316)
polytene chromosome (p. 317)
chromosome puff (p. 317)
epigenetic change (p. 318)
telomeric sequence (p. 320)

shelterin (p. 320)
C-value (p. 321)
C-value paradox (p. 321)
denaturation (p. 321)
melting temperature (T_m) (p. 321)
renaturation (p. 321)

hybridization (p. 321)
unique-sequence DNA
    (p. 321)
gene family (p. 321)
repetitive DNA (p. 321)
moderately repetitive DNA
    (p. 321)

tandem repeat (p. 321)
interspersed repeat
    (p. 321)
short interspersed element
    (SINE) (p. 322)
long interspersed element
    (LINE) (p. 322)

highly repetitive DNA
    (p. 322)
endosymbiotic theory
    (p. 323)
heteroplasmy (p. 324)
replicative segregation
    (p. 324)

homoplasmy (p. 324)
D loop (p. 327)
mitochondrial replacement
    therapy (p. 329)

ANSWERS TO CONCEPT CHECKS

1. b

2. Bacterial DNA is not complexed with histone proteins and is circular.

3. b

4. d

5. A chromosome that loses its centromere will not segregate into the nucleus in mitosis and is usually lost.

6. d

7. a

8. Many modern protists are hosts to endosymbiotic bacteria. Mitochondria and chloroplasts are similar in size to bacteria and have their own DNA, as well as ribosomes that are similar in size and shape to bacterial ribosomes. Antibiotics that inhibit protein synthesis in bacteria also inhibit protein synthesis in mitochondria and chloroplasts. Gene sequences in mtDNA and cpDNA are most similar to bacterial DNA sequences.

9. c

10. a

WORKED PROBLEMS

Problem 1

A diploid plant cell contains 2 billion base pairs of DNA.

a. How many nucleosomes are present in the cell?

b. Give the number of molecules of each type of histone protein associated with the genomic DNA.

≫ Solution Strategy

What information is required in your answer to the problem?
The number of nucleosomes per cell and the numbers of each type of histone protein associated with the DNA.

What information is provided to solve the problem?
The cell contains 2 billion base pairs of DNA.

For help with this problem, review:
The Nucleosome in Section 11.1.

≫ Solution Steps

> **Recall:** The repeating unit of the chromosome is a nucleosome, which consists of DNA complexed with histone proteins.

Each nucleosome encompasses about 200 bp of DNA: 145–147 bp of DNA wrapped around the histone core, 20–22 bp of DNA associated with the H1 protein, and another 30–40 bp of linker DNA.

a. To determine how many nucleosomes are present in the cell, we simply divide the total number of base pairs of DNA (2×10^9 bp) by the number of base pairs per nucleosome:

$$\frac{2 \times 10^9 \text{ base pairs}}{2 \times 10^2 \text{ base pairs/nucleosome}}$$

$$= 1 \times 10^7 \text{ nucleosomes}$$

Thus, there are approximately 10 million nucleosomes in the cell.

b. Each nucleosome includes two molecules each of H2A, H2B, H3, and H4 histones. Therefore, there are 2×10^7 molecules each of H2A, H2B, H3, and H4 histones. Each nucleosome has one copy of the H1 histone associated with it, so there are 1×10^7 molecules of H1.

Problem 2

Suppose that a new organelle is discovered in an obscure group of protists. This organelle contains a small DNA genome, and some scientists are arguing that, like chloroplasts and mitochondria, this organelle originated as a free-living eubacterium that entered into an endosymbiotic relation with the protist. Outline a research plan to determine whether the new organelle evolved from a free-living eubacterium. What kinds of data would you collect, and what predictions would you make if the theory were correct?

>> **Solution Strategy**

What information is required in your answer to the problem?
A research plan with the types of data you would collect as well as predictions.

What information is provided to solve the problem?
- The newly discovered organelle contains a small DNA genome.
- Other organelles have probably evolved through an endosymbiotic relationship.

For help with this problem, review:
The Endosymbiotic Theory, The Mitochondrial Genome, and The Chloroplast Genome in Section 11.4.

>> **Solution Steps**

We should examine the structure, organization, and sequences of the organelle genome. If the organelle shows only characteristics of eukaryotic DNA, then it most likely has a eukaryotic origin, but if it displays some characteristics of bacterial DNA, those observations support the theory of a bacterial origin.

We could start by examining the overall characteristics of the organelle DNA. If it has a bacterial origin, we might expect that the organelle genome will consist of a circular molecule and will lack histone proteins. We could compare the DNA sequences found in the organelle genome with homologous sequences from bacterial and eukaryotic genomes. If the theory of an endosymbiotic origin is correct, then the organelle sequences should be most similar to homologous sequences found in bacteria.

Recall: The endosymbiotic theory proposes that organelles evolved from bacteria.

COMPREHENSION QUESTIONS

Section 11.1

1. How does supercoiling arise? What is the difference between positive and negative supercoiling?

2. What functions does supercoiling serve for the cell?

3. Describe the composition and structure of the nucleosome.

4. Describe in steps how the double helix of DNA, which is 2 nm in width, gives rise to a chromosome that is 700 nm in width.

5. What are polytene chromosomes and chromosome puffs?

6. What are epigenetic changes and how are they brought about?

Section 11.2

7. Describe the function of the centromere. How are centromeres different from other regions of the chromosome?

8. Describe the function and molecular structure of a telomere.

9. What is the difference between euchromatin and heterochromatin?

Section 11.3

10. What is the C-value of an organism?

11. Describe the different classes of DNA sequence variation that exist in eukaryotes.

Section 11.4

12. Explain why many traits encoded by mtDNA and cpDNA exhibit considerable variation in their expression, even among members of the same family.

13. What is the endosymbiotic theory? How does it help to explain some of the characteristics of mitochondria and chloroplasts?

14. What evidence supports the endosymbiotic theory?

15. Briefly describe the organization of genes on the chloroplast genome.

16. What is meant by the term "promiscuous DNA"?

APPLICATION QUESTIONS AND PROBLEMS

Introduction

17. The introduction to this chapter discussed a study of telomere length in Romanian children. The study demonstrated that children raised in orphanages had shorter telomeres than children raised in foster homes. What effect, if any, do you think having shorter telomeres in childhood might have on adult life?

Section 11.1

*18. Compare and contrast prokaryotic and eukaryotic chromosomes. How are they alike and how do they differ?

*19. In a typical eukaryotic cell, would you expect to find more molecules of the H1 histone or more molecules of the H2A histone? Would you expect to find more molecules of H2A or more molecules of H3? Explain your reasoning.

*20. Based on the sensitivity of DNA to DNase I, as illustrated in **Figure 11.7**, which type of chicken hemoglobin (embryonic or adult) is likely to be produced in the highest quantity in the following tissues and developmental stages?

a. Erythroblasts during the first 24 hours

b. Erythroblasts at day 5

c. Erythroblasts at day 14

d. Brain cells throughout development

*21. A diploid human cell contains approximately 6.4 billion base pairs of DNA.

a. How many nucleosomes are present in such a cell? (Assume that the linker DNA encompasses 40 bp.)

b. How many histone proteins are complexed with this DNA?

*22. Would you expect to see more or less acetylation in regions of DNA that are sensitive to digestion by DNase I? Why?

23. Gunter Korge examined several proteins that are secreted from the salivary glands of *Drosophila melanogaster* during larval development (G. Korge. 1975. *Proceedings of the National Academy of Sciences of the United States of America* 72:4550–4554). One protein, called protein fraction 4, was encoded by a gene found by deletion mapping to be located on the X chromosome at position 3C. Korge observed that, about 5 hours before the first synthesis of protein fraction 4, an expanded and puffed-out region formed on the X chromosome at position 3C. This chromosome puff disappeared before the end of the third larval instar stage, when the synthesis of protein fraction 4 ceased. He observed that there was no puff at position 3C in a special strain of flies that lacked secretion of protein fraction 4. Explain these results. What is the chromosome puff at region 3, and why does its appearance and disappearance roughly coincide with the secretion of protein fraction 4?

24. Suppose a chemist develops a new drug that neutralizes the positive charges on the tails of histone proteins. What would be the most likely effect of this new drug on chromatin structure? Would this drug have any effect on gene expression? Explain your answers.

Section 11.3

*25. Which of the following two molecules of DNA has the lower melting temperature? Why?

AGTTACTAAAGCAATACATC
TCAATGATTTCGTTATGTAG

AGGCGGGTAGGCACCCTTA
TCCGCCCATCCGTGGGAAT

26. In a DNA-hybridization study, DNA was isolated from wheat, labeled with ^{32}P, and sheared into small fragments (S. K. Dutta et al. 1967. *Genetics* 57:719–727). Hybridizations between these labeled fragments and denatured DNA from different plant species were then compared. The following table gives the percentages of labeled wheat DNA that hybridized to DNA molecules of wheat, corn, radish, and cabbage.

Species	Percentage of wheat DNA hybridized relative to wheat
Wheat	100
Cabbage	23
Corn	63
Radish	30

What do these results indicate about the evolutionary differences among these plants?

Section 11.4

*27. A wheat plant that is light green in color is found growing in a field. Biochemical analysis reveals that chloroplasts in this plant produce only 50% of the chlorophyll normally found in wheat chloroplasts. Propose a set of crosses to determine whether the light-green phenotype is caused by a mutation in a nuclear gene or in a chloroplast gene.

*28. The following pedigree illustrates the inheritance of a rare neurological disease. What is the most likely mode of inheritance for this disorder? Explain your reasoning.

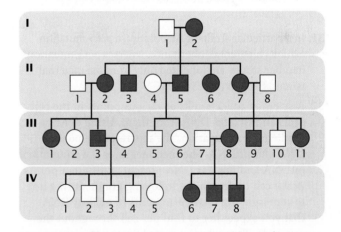

29. Assume that the muscle disorder whose inheritance is illustrated in the pedigree in the Worked Problem on p. 326 is a rare disease that results from a defect in mitochondrial DNA. If individual III-8 has a daughter, what is the probability that the daughter will inherit the muscle disorder from her affected parent?

30. Fredrick Wilson and his colleagues studied members of a large family who had low levels of magnesium in their blood (see the pedigree below). They argued that this

disorder (and associated high blood pressure and high cholesterol) is caused by a mutation in mtDNA (F. H. Wilson et al. 2004. *Science* 306:1190–1194).

a. What evidence suggests that a mutation in mtDNA is causing this disorder?

b. Could this disorder be caused by an autosomal dominant gene? Why or why not?

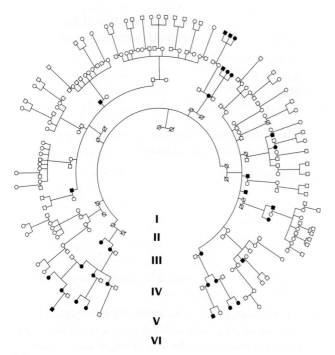

I
II
III
IV
V
VI

[After F. H. Wilson et al., 2004, *Science* 306:1190–1194.]

***31.** In a particular strain of *Neurospora*, a *poky* mutation exhibits biparental inheritance, whereas *poky* mutations in other strains are inherited only from the maternal parent. Explain these results.

***32.** A scientist collects cells at various points in the cell cycle and isolates DNA from them. Using density-gradient centrifugation, she separates the nuclear DNA and mtDNA. She then measures the amounts of mtDNA and nuclear DNA present at different points in the cell cycle. On the following graph, draw a line to represent the relative amounts of nuclear DNA that you expect her to find per cell throughout the cell cycle. Then, draw a dotted line on the same graph

to indicate the relative amounts of mtDNA that you would expect to see at different points throughout the cell cycle.

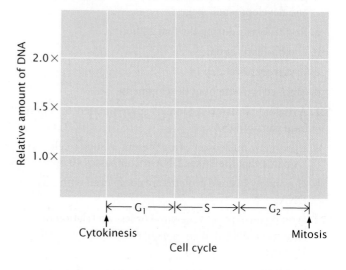

33. In 1979, bones found outside Ekaterinburg, Russia, were shown to be those of Tsar Nicholas and his family, who were executed in 1918 by a Bolshevik firing squad in the Russian Revolution (see the introduction to Chapter 14). To prove that the skeletons were those of the royal family, mtDNA was extracted from the bone samples, amplified by PCR, and compared with mtDNA from living relatives of the tsar's family.

a. Why was DNA from the mitochondria analyzed instead of nuclear DNA? What are some of the advantages of using mtDNA for this type of study?

b. Mitochondrial DNA from which living relatives would provide useful information for verifying that the skeletons were those of the royal family?

34. Antibiotics such as chloramphenicol, tetracycline, and erythromycin inhibit protein synthesis in bacteria, but have no effect on the synthesis of proteins encoded by eukaryotic nuclear genes. Cycloheximide inhibits the synthesis of proteins encoded by nuclear genes, but has no effect on bacterial protein synthesis. How might these compounds be used to determine which proteins are encoded by mitochondrial and chloroplast genomes?

CHALLENGE QUESTIONS

Section 11.1

35. An explorer discovers a strange new species of plant and sends some of the plant tissue to a geneticist to study. The geneticist isolates chromatin from the plant and examines it with an electron microscope. She observes

what appear to be beads on a string. She then adds a small amount of nuclease, which cleaves the string into individual beads that each contain 280 bp of DNA. After digestion with more nuclease, a 120-bp fragment of DNA remains attached to a core of histone proteins.

Analysis of the histone core reveals histones in the following proportions:

H1	12.5%
H2A	25%
H2B	25%
H3	0%
H4	25%
H7 (a new histone)	12.5%

On the basis of these observations, what conclusions could the geneticist make about the probable structure of the nucleosome in the chromatin of this plant?

Section 11.3

36. In DNA-hybridization experiments on six species of plants in the genus *Vicia*, DNA was isolated from each of the six species, denatured by heating, and sheared into small fragments (W. Y. Chooi. 1971. *Genetics* 68:213–230). In one experiment, DNA from each species and from *E. coli* was allowed to renature. The graph shows the results of this renaturation experiment.

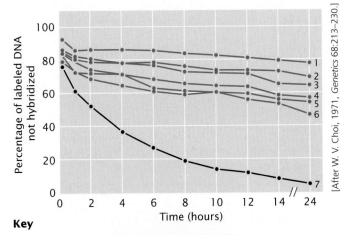

Key

1 = *V. melanops*, 2 = *V. sativa*, 3 = *V. benghalensis*,
4 = *V. atropurpurea*, 5 = *V. faba*, 6 = *V. narbonensis*, 7 = *E. coli*

[After W. V. Choi, 1971, *Genetics* 68:213–230.]

Fodder vetch (*Vicia sativa*).
[Bob Gibbons/Alamy.]

a. Can you explain why the *E. coli* DNA renatures at a much faster rate than does DNA from any of the *Vicia* species?

b. Notice that, for the *Vicia* species, the rate of renaturation is much faster in the first hour and then slows down. What might cause this initial rapid renaturation and the subsequent slowdown?

Section 11.4

37. Steven Frank and Laurence Hurst argued that a cytoplasmically inherited mutation in humans that has severe effects in males but no effect in females will not be eliminated from a population by natural selection because only females pass on mtDNA (S. A. Frank and L. D. Hurst. 1996. *Nature* 383:224). Using this argument, explain why males with Leber hereditary optic neuropathy are more severely affected than females.

38. In a study of a muscle disorder, several affected families exhibited vision problems, muscle weakness, and deafness (M. Zeviani et al. 1990. *American Journal of Human Genetics* 47:904–914). Analysis of the mtDNA from affected members of these families revealed that large numbers of their mtDNA molecules possessed deletions of varying lengths. Different members of the same family and even different mitochondria from the same person possessed deletions of different sizes, so the underlying defect appeared to be a tendency for the mtDNA of affected persons to have deletions. A pedigree of one of the families studied is shown below. The researchers concluded that this disorder is inherited as an autosomal dominant trait, and they mapped the disease-causing gene to a position on chromosome 10 in the nucleus.

a. What characteristics of the pedigree rule out inheritance of a trait encoded by a gene in the mtDNA as the cause of this disorder?

b. Explain how a mutation in a nuclear gene might lead to deletions in mtDNA.

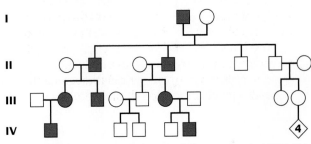

[After M. Zeviani et al., 1990, *American Journal of Human Genetics* 47:904–914.]

39. Mitochondrial DNA sequences have been detected in the nuclear genomes of many organisms, and cpDNA sequences are sometimes found in the mitochondrial genome. Propose a mechanism for how such "promiscuous DNA" might move between nuclear, mitochondrial, and chloroplast genomes.

THINK-PAIR-SHARE QUESTIONS

Section 11.1

1. The Y chromosome contains large amounts of constitutive heterochromatin. Why might there be more constitutive heterochromatin in the Y chromosome than in other chromosomes? (Hint: See Evolution of the Y Chromosome in Chapter 4.)

2. Polytene chromosomes are giant chromosomes found in cells of the salivary glands of larval fruit flies and some other organisms. Polytene chromosomes arise when repeated rounds of replication occur without cell division, producing thousands of copies of DNA that lie side by side. Why do polytene chromosomes occur in the cells of salivary glands of larval fruit flies, but not in other tissues? Propose some possible reasons. (If you are unfamiliar with the life cycle of fruit flies, see pp. A2–A3 in the Reference Guide to Model Genetic Organisms at end of this book.)

Section 11.4

3. The Out of Africa hypothesis (also called the African Replacement hypothesis) proposes that *Homo sapiens* arose in Africa and evolved there for several hundred thousand years. Then, some 85,000 years ago, a small band of *Homo sapiens* migrated out of Africa and populated the remainder of the world. Based on this hypothesis, what predictions would you make about worldwide human variation in mtDNA?

4. What are some possible social and ethical issues associated with mitochondrial replacement therapy (MRT)? Present arguments for and against approving MRT for prevention of the births of children with mitochondrial disorders.

DNA Replication and Recombination

The happy tree, *Camptotheca acuminata*, contains camptothecin, a substance used to treat cancer. Camptothecin inhibits cancer by blocking an important component of the replication machinery. [Johnny Pan/Getty Images.]

Topoisomerase, Replication, and Cancer

In 1966, Monroe Wall and Mansukh Wani found a potential cure for cancer in the bark of the happy tree (*Camptotheca acuminata*), a rare plant native to China. Wall and Wani were in the process of screening a large number of natural substances for anticancer activity, hoping to find chemicals that might prove effective in cancer treatment. They discovered that an extract from the happy tree was effective in treating leukemia in mice. Through chemical analysis, they were able to isolate the active compound, which was dubbed camptothecin.

In the 1970s, physicians administered camptothecin to patients with incurable cancers. Although the drug showed some anticancer activity, it had toxic side effects. Eventually, chemists synthesized several analogs of camptothecin that were less toxic and more effective in cancer treatment. Two of these analogs, topotecan and irinotecan, are used today for the treatment of ovarian cancer, small-cell lung cancer, and colon cancer.

For many years, the mechanism by which camptothecin compounds inhibited cancer was unknown. In 1985, almost 20 years after its discovery, scientists at Johns Hopkins University and Smith Kline and French Laboratories (now GlaxoSmithKline) showed that camptothecin worked by inhibiting an important component of the DNA-synthesizing machinery in humans, an enzyme called topoisomerase I.

Cancer chemotherapy is a delicate task because the target cells are the patient's own and the drugs must kill the cancer cells without killing the patient. One of the hallmarks of cancer is proliferation: the division of cancer cells is unregulated, and many cancer cells divide at a rapid rate, giving rise to tumors with the ability to grow and spread. As we learned in Chapter 2, before a cell can divide, it must successfully replicate its DNA so that each daughter cell receives an exact copy of the genetic material. Checkpoints in the cell cycle ensure that cell division does not proceed if DNA replication is inhibited or faulty, and many cancer treatments focus on interfering with the process of DNA replication.

DNA replication is a complex process that requires a large number of components, the actions of which must be intricately coordinated to ensure that DNA is accurately copied. An essential component of replication is topoisomerase. As the DNA unwinds in the course of replication, strain builds up ahead of the separating strands and the two strands writhe around each other, much as a rope knots up as you pull apart two of its strands. This writhing of the DNA is called supercoiling (see Section 11.1). If the supercoils are not removed, they eventually prevent strand separation, and replication comes to a halt. Topoisomerase enzymes remove the supercoils by clamping tightly to the DNA and breaking one or both of its strands. The strands then revolve around each other, removing the supercoiling and the strain. After the DNA has relaxed, the topoisomerase reseals the broken ends of the DNA strands.

Camptothecin works by interfering with topoisomerase I. The drug inserts itself into the gap created by the break in the DNA strand, blocking the topoisomerase from resealing the broken ends. Researchers originally assumed that camptothecin trapped the topoisomerase and blocked the action of other enzymes needed for synthesizing DNA. However, research now indicates that camptothecin poisons the topoisomerase so that it is unable to remove supercoils ahead of replication. Accumulating supercoils halt the replication machinery and prevent the proliferation of cancer cells. But like many other anticancer drugs, camptothecin also inhibits the replication of normal, noncancerous cells, which is why chemotherapy makes many patients sick.

THINK-PAIR-SHARE
- Chemotherapeutic agents such as camptothecin often have unwanted side effects. What are some common negative side effects of chemotherapy, and why do they arise?

This chapter focuses on DNA replication, the process by which a cell doubles its DNA before division. We begin with the basic mechanism of replication that emerged from the DNA structure discovered by Watson and Crick. We then examine several different modes of replication, the requirements of replication, and the universal direction of DNA synthesis. We also examine the enzymes and proteins that participate in the process. Finally, we consider the molecular details of recombination, which is closely related to replication and is essential for the segregation of homologous chromosomes, the production of genetic variation, and DNA repair.

12.1 Genetic Information Must Be Accurately Copied Every Time a Cell Divides

In a schoolyard game, a verbal message, such as "John's brown dog ran away from home," is whispered to a child, who runs to a second child and repeats the message. The message is relayed from child to child around the schoolyard until it returns to the original sender. Inevitably, the last child returns with an amazingly transformed message, such as "Joe Brown has a pig living under his porch." The larger the number of children playing the game, the more garbled the message becomes. This game illustrates an important principle: errors arise whenever information is copied, and the more times it is copied, the greater the potential number of errors.

A complex multicellular organism faces a problem analogous to that of the children in the schoolyard game: how to faithfully transmit genetic instructions each time its cells divide. The solution to this problem is central to replication. A single-celled human zygote contains 6.4 billion base pairs of DNA; even a low rate of error during copying, such as once per million base pairs, would result in 6400 mistakes made every time a cell divided—errors that would be compounded at each of the millions of cell divisions that take place in human development.

Not only must the copying of DNA be astoundingly accurate, it must also take place at breakneck speed. The single circular chromosome of *E. coli* contains about 4.6 million base pairs. At a rate of 1000 nucleotides per minute, replication of the entire chromosome would require over 3 days. Yet these

bacteria are capable of dividing every 20 minutes. *Escherichia coli* actually replicates its DNA at a rate of 1000 nucleotides per second, with less than one error in a billion nucleotides. How is this extraordinarily accurate and rapid process accomplished?

THINK-PAIR-SHARE Question 1

12.2 All DNA Replication Takes Place in a Semiconservative Manner

When Watson and Crick solved the three-dimensional structure of DNA in 1953 (see Figure 10.8), several important genetic implications were immediately apparent. The complementary nature of the two nucleotide strands in a DNA molecule suggested that, during replication, each strand can serve as a template for the synthesis of a new strand. The specificity of base pairing (adenine with thymine; guanine with cytosine) implied that only one sequence of bases can be specified by each template strand, and so the two DNA molecules built on the pair of templates will be identical with the original. This process is called **semiconservative replication** because each of the original nucleotide strands remains intact (conserved), despite their no longer being combined in the same molecule; thus, the original DNA molecule is half (semi) conserved during replication.

Initially, three models were proposed for DNA replication. In conservative replication (**Figure 12.1a**), the entire double-stranded DNA molecule serves as a template for a whole new molecule of DNA, and the original DNA molecule is *fully* conserved during replication. In dispersive replication (**Figure 12.1b**), both nucleotide strands break down (disperse) into fragments, which serve as templates for the synthesis of new DNA fragments, and then somehow reassemble into two complete DNA molecules. In this model, each resulting DNA molecule contains interspersed fragments of old and new DNA; none of the original molecule is conserved. Semiconservative replication (**Figure 12.1c**) is intermediate between these two models; the two nucleotide strands unwind, and each serves as a template for a new DNA molecule.

These three models allow different predictions to be made about the distribution of original DNA and newly synthesized

(a) Conservative replication   **(b) Dispersive replication**   **(c) Semiconservative replication**

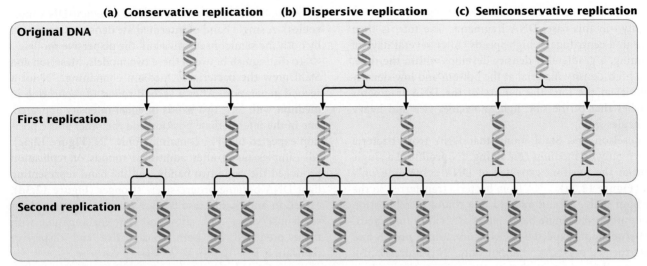

12.1 Three proposed models of DNA replication: conservative replication, dispersive replication, and semiconservative replication.

DNA after replication. With conservative replication, after one round of replication, 50% of the molecules would consist entirely of the original DNA and 50% would consist entirely of new DNA. After a second round of replication, 25% of the molecules would consist entirely of the original DNA and 75% would consist entirely of new DNA. With each additional round of replication, the proportion of molecules with new DNA would increase, although the number of molecules with the original DNA would remain constant. Dispersive replication would always produce hybrid molecules, containing some original and some new DNA, but the proportion of new DNA within the molecules would increase with each replication event. In contrast, with semiconservative replication, one round of replication would produce two hybrid molecules, each consisting of half original DNA and half new DNA. After a second round of replication, half the molecules would be hybrid and the other half would consist of new DNA only. Additional rounds of replication would produce more and more molecules consisting entirely of new DNA, but a few hybrid molecules would persist.

Meselson and Stahl's Experiment

To determine which of the three models of replication applied to *E. coli* cells, Matthew Meselson and Franklin Stahl needed a way to distinguish old and new DNA. They accomplished this by using two isotopes of nitrogen, ^{14}N (the common form) and ^{15}N (a rare, heavy form). Meselson and Stahl grew a culture of *E. coli* in a medium that contained ^{15}N as the sole nitrogen source; after many generations, all the *E. coli* cells had ^{15}N incorporated into all the purine and pyrimidine bases of their DNA (see Figure 10.11). Meselson and Stahl took a sample of these bacteria, switched the rest of the bacteria to a medium that contained only ^{14}N, and then took additional samples of bacteria over the next few cellular generations. In each sample, the bacterial DNA that was synthesized before the change in medium contained ^{15}N and

was relatively heavy, whereas any DNA synthesized after the switch contained ^{14}N and was relatively light.

Meselson and Stahl distinguished between the heavy ^{15}N-laden DNA and the light ^{14}N-containing DNA with the use of **equilibrium density gradient centrifugation** (**Figure 12.2**). In this technique, a centrifuge tube is filled

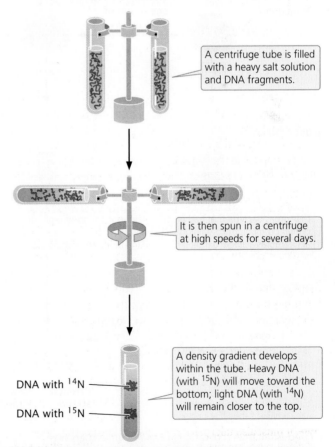

A centrifuge tube is filled with a heavy salt solution and DNA fragments.

It is then spun in a centrifuge at high speeds for several days.

DNA with ^{14}N

DNA with ^{15}N

A density gradient develops within the tube. Heavy DNA (with ^{15}N) will move toward the bottom; light DNA (with ^{14}N) will remain closer to the top.

12.2 Meselson and Stahl used equilibrium density gradient centrifugation to distinguish between heavy, ^{15}N-laden DNA and lighter, ^{14}N-laden DNA.

with a heavy salt solution and a substance of unknown density—in this case, DNA fragments. The tube is then spun in a centrifuge at high speeds. After several days of spinning, a gradient of density develops within the tube, with high-density material at the bottom and low-density material at the top. The density of the DNA fragments matches that of the salt: light molecules rise and heavy molecules sink.

Meselson and Stahl found that DNA from bacteria grown only on medium containing ^{15}N produced a single band at the position expected of DNA containing only ^{15}N (**Figure 12.3a**). DNA from bacteria transferred to the medium with ^{14}N and allowed one round of replication also produced a single band, but at a position intermediate between that expected of DNA containing only ^{15}N and that expected of DNA containing only ^{14}N (**Figure 12.3b**). This result is inconsistent with the conservative replication model, which predicts one heavy band (the original

DNA molecules) and one light band (the new molecules). A single band of intermediate density is predicted by both the semiconservative and the dispersive models.

To distinguish between these two models, Meselson and Stahl grew the bacteria in medium containing ^{14}N for a second generation. After a second round of replication in medium with ^{14}N, two bands of equal intensity appeared, one in the intermediate position and the other at the position expected of DNA containing only ^{14}N (**Figure 12.3c**). All samples taken after additional rounds of replication produced the same two bands, and the band representing light DNA became progressively stronger (**Figure 12.3d**). Meselson and Stahl's results were exactly as expected for semiconservative replication and were incompatible with those predicted for both conservative and dispersive replication. ▶ **TRY PROBLEM 22**

THINK-PAIR-SHARE Question 2

Experiment

Question: Which model of DNA replication—conservative, dispersive or semiconservative—applies to *E. coli*?

(a)

^{15}N medium

Transfer to ^{14}N; one round of replication

(b)

Second round of replication in ^{14}N medium

(c)

Additional rounds of replication in ^{14}N medium

(d)

Spin | Spin | Spin | Spin

Light (^{14}N)

Heavy (^{15}N)

DNA from bacteria that had been grown on medium containing ^{15}N appeared as a single band.

After one round of replication, the DNA appeared as a single band at intermediate weight.

After a second round of replication, DNA appeared as two bands, one light and the other intermediate in weight.

Samples taken after additional rounds of replication appeared as two bands, as in part c.

Original DNA

Parental strand   New strand

Conclusion: DNA replication in *E.coli* is semiconservative.

12.3 Meselson and Stahl demonstrated that DNA replication is semiconservative.

Meselson and Stahl convincingly demonstrated that replication in *E. coli* is semiconservative: each DNA strand serves as a template for the synthesis of a new DNA molecule.

✔ **CONCEPT CHECK 1**

How many bands of DNA would be expected in Meselson and Stahl's experiment after two rounds of *conservative* replication?

Modes of Replication

After Meselson and Stahl's work was published, investigators confirmed that other organisms also use semiconservative replication. No evidence was found for conservative or dispersive replication. There are, however, several different ways in which semiconservative replication can take place, differing principally in the nature of the template DNA—that is, whether it is linear or circular.

A segment of DNA that undergoes replication is called a **replicon**, and each replicon contains an **origin of replication**. Replication starts at the origin and continues until the entire replicon has been replicated. Bacterial chromosomes have a single origin of replication, whereas eukaryotic chromosomes contain many.

THETA REPLICATION A common mode of replication that takes place in circular DNA, such as that found in *E. coli* and other bacteria, is called **theta replication** (**Figure 12.4a**) because it generates an intermediate structure that resembles the Greek letter theta (θ). In Figure 12.4a and all subsequent figures in this chapter, the original (template) strand of DNA is shown in gray and the newly synthesized strand of DNA is shown in red.

In theta replication, double-stranded DNA begins to unwind at the origin of replication, producing single nucleotide strands that then serve as templates on which new DNA can be synthesized. The unwinding of the double helix generates a loop, termed a **replication bubble**. Unwinding may occur at one or both ends of the bubble, making it progressively larger. DNA replication on both of the template strands is simultaneous with unwinding. The point of unwinding, where the two strands separate from the double-stranded DNA helix, is called a **replication fork**.

If there are two replication forks, one at each end of the replication bubble, the forks proceed outward in both

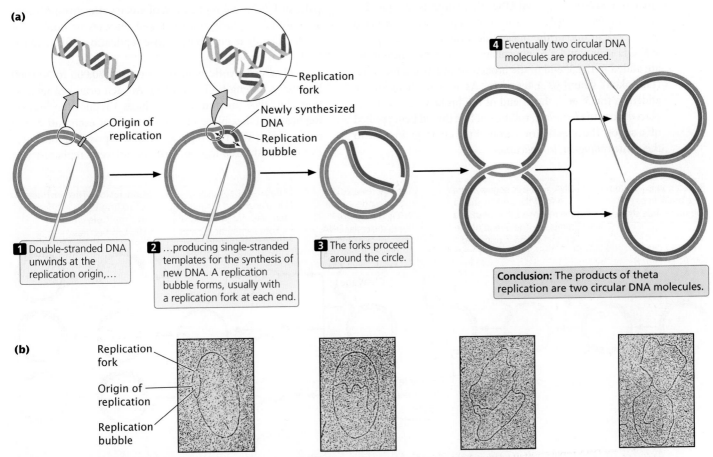

(a)

① Double-stranded DNA unwinds at the replication origin,…

Origin of replication

② …producing single-stranded templates for the synthesis of new DNA. A replication bubble forms, usually with a replication fork at each end.

Replication fork
Newly synthesized DNA
Replication bubble

③ The forks proceed around the circle.

④ Eventually two circular DNA molecules are produced.

Conclusion: The products of theta replication are two circular DNA molecules.

(b)

Replication fork
Origin of replication
Replication bubble

12.4 Theta replication is a type of replication common in *E. coli* and other organisms possessing circular DNA. [Part b: Bernhard Hirt, L'Institut Suisse de Recherche Expérimentale sur le Cancer.]

directions in a process called **bidirectional replication**, simultaneously unwinding and replicating the DNA until they eventually meet. If unidirectional replication with a single replication fork is present, it proceeds around the entire circle. Both bidirectional and unidirectional replication produce two complete circular DNA molecules, each consisting of one old and one new nucleotide strand.

John Cairns provided the first visible evidence of theta replication in 1963 by growing bacteria in the presence of radioactive nucleotides. After replication, each DNA molecule consisted of one "hot" (radioactive) strand and one "cold" (nonradioactive) strand. Cairns isolated DNA from the bacteria after replication, placed it on an electron-microscope grid, and then covered it with a photographic emulsion. Radioactivity present in the sample exposed the emulsion and produced a picture of the molecule (called an autoradiograph), in a manner similar to the light that exposes a photographic film. Because the newly synthesized DNA contained radioactive nucleotides, Cairns was able to produce electron micrographs of the replication process similar to those shown in **Figure 12.4b**.

ROLLING-CIRCLE REPLICATION Another form of replication, called **rolling-circle replication** (**Figure 12.5**), takes place in some viruses and in the F factor of *E. coli* (a small circle of extrachromosomal DNA that controls mating, discussed in Section 9.3). This form of replication is initiated by a break in one of the nucleotide strands, which exposes a 3′-OH group and a 5′-phosphate group. New nucleotides are added to the 3′ end of the broken strand, using the inner (unbroken) strand as a template. As new nucleotides are added to the 3′ end, the 5′ end of the broken strand is displaced from the template, rolling out like thread being pulled off a spool. The 3′ end grows around the circle, giving rise to the name *rolling-circle replication*.

The replication fork may continue around the circle a number of times, producing several linked copies of the same sequence. With each revolution around the circle, the growing 3′ end displaces the nucleotide strand synthesized in the preceding revolution. Eventually, the linear DNA molecule is cleaved from the circle, resulting in a double-stranded circular DNA molecule and a single-stranded linear DNA molecule. The linear molecule circularizes either before or after serving as a template for the synthesis of a complementary strand.

LINEAR EUKARYOTIC REPLICATION Circular DNA molecules that undergo theta or rolling-circle replication have a single origin of replication. Because of the limited size of these DNA molecules, replication starting from one origin can traverse the entire chromosome in a reasonable amount of time. The large linear chromosomes in eukaryotic cells, however, contain far too much DNA to be replicated speedily from a single origin. Eukaryotic replication proceeds at a rate ranging from 500 to 5000 nucleotides per minute at each replication fork (considerably more slowly than bacterial replication). Even at 5000 nucleotides per minute at each fork, DNA synthesis starting from a single origin would require 7 days to replicate a typical human chromosome consisting of 100 million base pairs of DNA. The replication of eukaryotic chromosomes actually takes place in a matter of minutes or hours, not days. This rate is possible because replication is initiated at thousands of origins.

Typical eukaryotic replicons are from 20,000 to 300,000 base pairs in length (**Table 12.1**). At each origin of replication, the DNA unwinds and produces a replication bubble. Replication takes place on both strands at each end of the bubble, with the two replication forks spreading outward. Eventually, the replication forks of adjacent replicons run

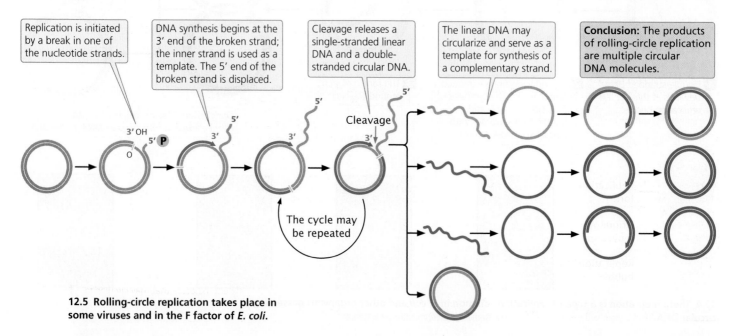

Replication is initiated by a break in one of the nucleotide strands.

DNA synthesis begins at the 3′ end of the broken strand; the inner strand is used as a template. The 5′ end of the broken strand is displaced.

Cleavage releases a single-stranded linear DNA and a double-stranded circular DNA.

The linear DNA may circularize and serve as a template for synthesis of a complementary strand.

Conclusion: The products of rolling-circle replication are multiple circular DNA molecules.

The cycle may be repeated

12.5 Rolling-circle replication takes place in some viruses and in the F factor of *E. coli*.

TABLE 12.1	Number and length of replicons	
Organism	Number of Replication Origins	Average Length of Replicon (bp)
Escherichia coli (bacterium)	1	4,600,000
Saccharomyces cerevisiae (yeast)	500	40,000
Drosophila melanogaster (fruit fly)	3,500	40,000
Xenopus laevis (frog)	15,000	200,000
Mus musculus (mouse)	25,000	150,000

Source: Data from B. L. Lewin, *Genes V* (Oxford: Oxford University Press, 1994), p. 536.

into each other, and the replicons fuse to form long stretches of newly synthesized DNA (**Figure 12.6**). Replication and fusion of all the replicons leads to two identical DNA molecules. Important features of theta replication, rolling-circle replication, and linear eukaryotic replication are summarized in **Table 12.2**. ▶ **TRY PROBLEM 23**

CONCEPTS

Theta replication, rolling-circle replication, and linear eukaryotic replication differ with respect to the initiation and progression of replication, but all produce new DNA molecules by semiconservative replication.

✔ CONCEPT CHECK 2

Which type of replication requires a break in the nucleotide strand to get started?
a. Theta replication
b. Rolling-circle replication
c. Linear eukaryotic replication
d. All of the above

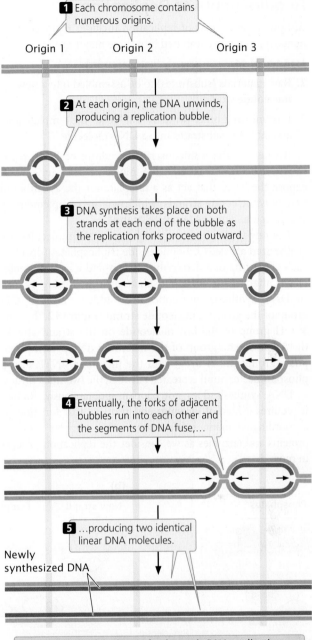

1 Each chromosome contains numerous origins.

Origin 1   Origin 2   Origin 3

2 At each origin, the DNA unwinds, producing a replication bubble.

3 DNA synthesis takes place on both strands at each end of the bubble as the replication forks proceed outward.

4 Eventually, the forks of adjacent bubbles run into each other and the segments of DNA fuse,…

5 …producing two identical linear DNA molecules.

Newly synthesized DNA

Conclusion: The products of eukaryotic DNA replication are two linear DNA molecules.

12.6 Linear DNA replication takes place in eukaryotic chromosomes.

TABLE 12.2	Characteristics of theta, rolling-circle, and linear eukaryotic replication				
Replication Model	DNA Template	Breakage of Nucleotide Strand	Number of Replicons	Unidirectional or Bidirectional	Products
Theta	Circular	No	1	Unidirectional or bidirectional	Two circular molecules
Rolling-circle	Circular	Yes	1	Unidirectional	One circular molecule and one linear molecule that may circularize
Linear eukaryotic	Linear	No	Many	Bidirectional	Two linear molecules

Requirements of Replication

Although the process of replication includes many components, they can be combined into three major groups:

1. A template consisting of single-stranded DNA

2. Raw materials (substrates) to be assembled into a new nucleotide strand

3. Enzymes and other proteins that "read" the template and assemble the substrates into a DNA molecule

Because of the semiconservative nature of DNA replication, a double-stranded DNA molecule must unwind to expose the bases that act as a template for the assembly of new polynucleotide strands, which will be complementary and antiparallel to the template strands.

The raw materials from which new DNA molecules are synthesized are deoxyribonucleoside triphosphates (dNTPs), each consisting of a deoxyribose sugar and a base (a nucleoside) attached to three phosphate groups (**Figure 12.7a**). In DNA synthesis, nucleotides are added to the 3′-OH group of the growing nucleotide strand (**Figure 12.7b**). The 3′-OH group of the last nucleotide on the strand attacks the 5′-phosphate group of the incoming dNTP. Two phosphate groups are cleaved from the incoming dNTP, and a phosphodiester bond is created between the two nucleotides.

DNA synthesis does not happen spontaneously. Rather, it requires a host of enzymes and proteins that function in a coordinated manner. We examine this complex array of proteins and enzymes as we consider the replication process in more detail.

CONCEPTS

DNA synthesis requires a single-stranded DNA template, deoxyribonucleoside triphosphates, a growing nucleotide strand, and a group of enzymes and proteins.

Direction of Replication

In DNA synthesis, new nucleotides are joined one at a time to the 3′ end of the newly synthesized strand. **DNA polymerases**, the enzymes that synthesize DNA, can add nucleotides only to the 3′ end of the growing strand (not the 5′ end), so new DNA strands always elongate in the same 5′-to-3′ direction (5′→3′). Because the two single-stranded DNA templates are antiparallel and strand elongation is always 5′→3′, if synthesis on one template proceeds from, say, right to left, then synthesis on the other template must proceed in the opposite direction, from left to right (**Figure 12.8**). As DNA unwinds during replication, the antiparallel nature of the two DNA strands means that one template is exposed in the 5′→3′ direction and the other template is exposed in the 3′→5′ direction. So how can synthesis take place simultaneously on both strands at the replication fork?

CONTINUOUS AND DISCONTINUOUS REPLICATION As the DNA unwinds, the template strand that is exposed in the 3′→5′ direction (the lower strand in Figures 12.8 and 12.9) allows the new strand to be synthesized continuously,

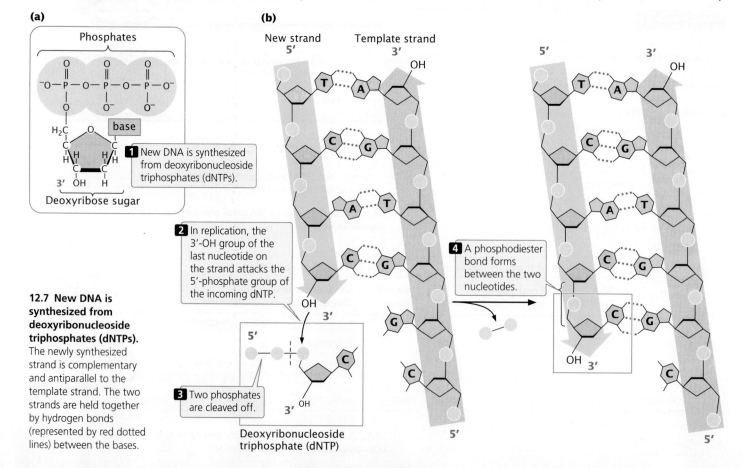

(a)

12.7 New DNA is synthesized from deoxyribonucleoside triphosphates (dNTPs). The newly synthesized strand is complementary and antiparallel to the template strand. The two strands are held together by hydrogen bonds (represented by red dotted lines) between the bases.

Phosphates

base

Deoxyribose sugar

1 New DNA is synthesized from deoxyribonucleoside triphosphates (dNTPs).

3 Two phosphates are cleaved off.

Deoxyribonucleoside triphosphate (dNTP)

(b)

New strand   Template strand

2 In replication, the 3′-OH group of the last nucleotide on the strand attacks the 5′-phosphate group of the incoming dNTP.

4 A phosphodiester bond forms between the two nucleotides.

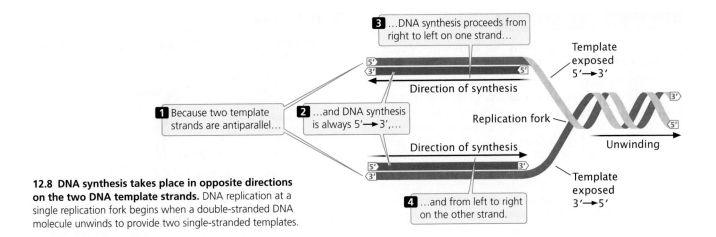

3 ...DNA synthesis proceeds from right to left on one strand...

Template exposed 5′→3′

Direction of synthesis

Replication fork

1 Because two template strands are antiparallel...

2 ...and DNA synthesis is always 5′→3′,...

Unwinding

Direction of synthesis

12.8 DNA synthesis takes place in opposite directions on the two DNA template strands. DNA replication at a single replication fork begins when a double-stranded DNA molecule unwinds to provide two single-stranded templates.

4 ...and from left to right on the other strand.

Template exposed 3′→5′

in the 5′→3′ direction. This new strand, which undergoes **continuous replication**, is called the **leading strand**.

The other template strand is exposed in the 5′→3′ direction (the upper strand in Figures 12.8 and 12.9). After a short length of the DNA has been unwound, synthesis must proceed 5′→3′; that is, in the direction *opposite* that of unwinding (**Figure 12.9**). Because only a short length of DNA needs to be unwound before synthesis on this strand gets started, the replication machinery soon runs out of template. By that time, more DNA has unwound, providing new template at the 5′ end of the new strand. DNA synthesis must start anew at the replication fork and proceed in the direction opposite that of the movement of the fork until it runs into the previously replicated segment of DNA. This process is repeated again and again, so synthesis of this strand is in short, discontinuous bursts. The newly made strand that undergoes **discontinuous replication** is called the **lagging strand**.

OKAZAKI FRAGMENTS The short lengths of DNA produced by discontinuous replication of the lagging strand are called **Okazaki fragments**, after Reiji Okazaki, who discovered them. In bacterial cells, each Okazaki fragment ranges from about 1000 to 2000 nucleotides in length; in eukaryotic cells, they are about 100 to 200 nucleotides long. Okazaki fragments on the lagging strand are linked together to create a continuous new DNA molecule. To see how replication occurs continuously on one strand and discontinuously on the other, view **Animation 12.1**.

Ⓐ

CONCEPTS

All DNA synthesis is 5′→3′, meaning that new nucleotides are always added to the 3′ end of the growing nucleotide strand. At each replication fork, synthesis of the leading strand proceeds continuously and that of the lagging strand proceeds discontinuously.

✔ CONCEPT CHECK 3

Discontinuous replication is a result of which property of DNA?

a. Complementary bases
b. Charged phosphate group
c. Antiparallel nucleotide strands
d. Five-carbon sugar

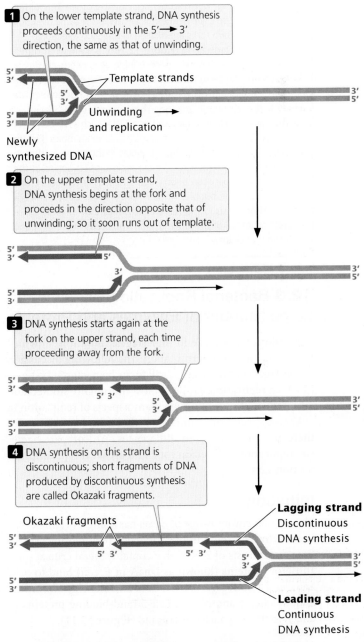

1 On the lower template strand, DNA synthesis proceeds continuously in the 5′→3′ direction, the same as that of unwinding.

Template strands

Unwinding and replication

Newly synthesized DNA

2 On the upper template strand, DNA synthesis begins at the fork and proceeds in the direction opposite that of unwinding; so it soon runs out of template.

3 DNA synthesis starts again at the fork on the upper strand, each time proceeding away from the fork.

4 DNA synthesis on this strand is discontinuous; short fragments of DNA produced by discontinuous synthesis are called Okazaki fragments.

Okazaki fragments

Lagging strand
Discontinuous DNA synthesis

Leading strand
Continuous DNA synthesis

12.9 DNA synthesis is continuous on one template strand of DNA and discontinuous on the other.

The Direction of Synthesis in Different Modes of Replication

Let's relate the direction of DNA synthesis to the modes of replication examined earlier. In theta replication (**Figure 12.10a**), the DNA unwinds at one particular location, the origin, and a replication bubble is formed. If the bubble has two replication forks, one at each end, synthesis takes place simultaneously at both forks (bidirectional replication). At each fork, synthesis on one of the template strands proceeds in the same direction as that of unwinding; this newly replicated strand is the leading strand and is replicated continuously. On the other template strand, synthesis proceeds in the direction opposite that of unwinding; this newly synthesized strand is the lagging strand and is replicated discontinuously. Focus on just one of the template strands within the bubble. Notice that synthesis on this template strand is continuous at one fork but discontinuous at the other. This difference arises because DNA synthesis is always in the same direction ($5' \rightarrow 3'$), but the two forks are moving in opposite directions.

Rolling-circle replication (**Figure 12.10b**) is somewhat different because there is no replication bubble. Replication begins at the 3' end of the broken nucleotide strand. Continuous replication takes place on the circular template as new nucleotides are added to this 3' end.

The replication of linear molecules of DNA, such as those found in eukaryotic cells, produces a series of replication bubbles (**Figure 12.10c**). DNA synthesis in these bubbles is the same as that in the single replication bubble of theta replication; it begins at the center of each replication bubble and proceeds at two forks, one at each end of the bubble. At both forks, synthesis of the leading strand proceeds in the same direction as that of unwinding, whereas synthesis of the lagging strand proceeds in the direction opposite that of unwinding. ▶ **TRY PROBLEM 25a–c**

12.3 Bacterial Replication Requires a Large Number of Enzymes and Proteins

Replication takes place in four stages: initiation, unwinding, elongation, and termination. The following discussion of the process of replication will focus on bacterial systems, in which replication has been most thoroughly studied and is best understood. Although many aspects of replication in eukaryotic cells are similar to those in bacterial cells, there are some important differences. We compare bacterial replication with eukaryotic and archaeal replication in Section 12.4.

Initiation

The circular chromosome of *E. coli* has a single origin of replication (*oriC*). The minimal sequence required for *oriC* to function consists of 245 bp that contain several critical sites. **Initiator proteins** (known as DnaA in *E. coli*) bind to *oriC* and cause a short section of DNA to unwind. This unwinding allows helicase and other single-strand-binding proteins to attach to the polynucleotide strand (**Figure 12.11**).

(a) Theta model

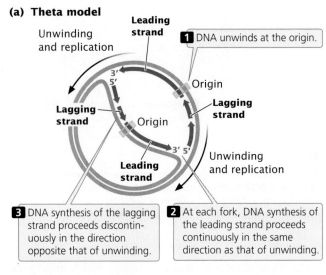

1 DNA unwinds at the origin.

2 At each fork, DNA synthesis of the leading strand proceeds continuously in the same direction as that of unwinding.

3 DNA synthesis of the lagging strand proceeds discontinuously in the direction opposite that of unwinding.

(b) Rolling-circle model

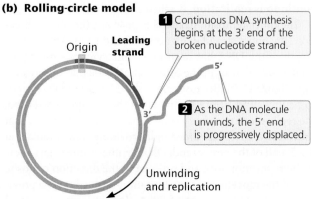

1 Continuous DNA synthesis begins at the 3' end of the broken nucleotide strand.

2 As the DNA molecule unwinds, the 5' end is progressively displaced.

(c) Linear eukaryotic replication

1 At each fork, the leading strand is synthesized continuously in the same direction as that of unwinding.

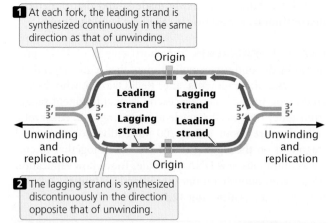

2 The lagging strand is synthesized discontinuously in the direction opposite that of unwinding.

12.10 The process of replication differs in theta replication, rolling-circle replication, and linear replication.

Unwinding

DNA synthesis requires a single-stranded template, and therefore double-stranded DNA must be unwound before DNA synthesis can take place. The cell relies on several proteins and enzymes to accomplish the unwinding.

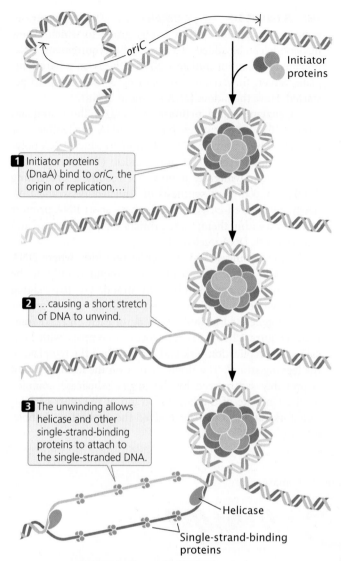

1 Initiator proteins (DnaA) bind to *oriC*, the origin of replication,...

2 ...causing a short stretch of DNA to unwind.

3 The unwinding allows helicase and other single-strand-binding proteins to attach to the single-stranded DNA.

Helicase

Single-strand-binding proteins

12.11 *E. coli* DNA replication begins when initiator proteins bind to *oriC*, the origin of replication.

DNA HELICASE A **DNA helicase** breaks the hydrogen bonds that exist between the bases of the two nucleotide strands of a DNA molecule. Helicase cannot *initiate* the unwinding of double-stranded DNA; the initiator protein first separates DNA strands at the origin, providing a short stretch of single-stranded DNA to which a helicase binds. Helicase binds to the lagging-strand template at each replication fork and moves in the 5′→3′ direction along this strand, thus also moving the replication fork (**Figure 12.12**).

SINGLE-STRAND-BINDING PROTEINS After DNA has been unwound by helicase, **single-strand-binding proteins** (**SSBs**) attach tightly to the exposed single-stranded DNA (see Figure 12.12). These proteins protect the single-stranded nucleotide chains and prevent the formation of secondary structures such as hairpins (see Figure 10.18) that would interfere with replication. Unlike many DNA-binding proteins, SSBs are indifferent to base sequence: they will bind to any single-stranded DNA. Single-strand-binding proteins form tetramers (groups of four); each tetramer covers from 35 to 65 nucleotides.

DNA GYRASE Another protein essential for the unwinding process is the enzyme **DNA gyrase**, a topoisomerase. As discussed in Chapter 11 and the introduction to this chapter, topoisomerases control the supercoiling of DNA. They come in two major types: type I topoisomerases alter supercoiling by making single-strand breaks in DNA, while type II topoisomerases create double-strand breaks. DNA gyrase is a type II topoisomerase. In replication, it reduces the torsional strain (torque) that builds up ahead of the replication fork as a result of unwinding (see Figure 12.12). It reduces torque by making a double-strand break in one segment of the DNA helix, passing another segment of the helix through the break, and then resealing the broken ends of the DNA. This action, which requires ATP, removes a twist in the DNA and reduces the supercoiling.

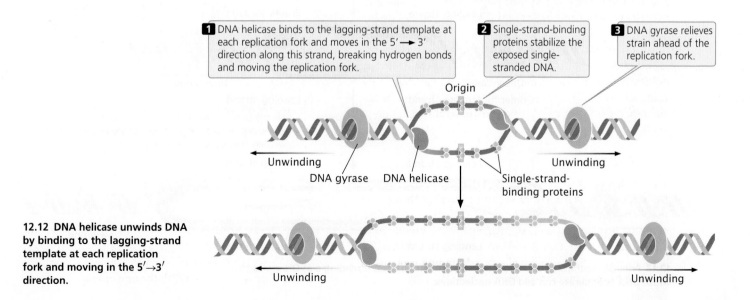

1 DNA helicase binds to the lagging-strand template at each replication fork and moves in the 5′ → 3′ direction along this strand, breaking hydrogen bonds and moving the replication fork.

2 Single-strand-binding proteins stabilize the exposed single-stranded DNA.

3 DNA gyrase relieves strain ahead of the replication fork.

Origin

Unwinding

Unwinding

DNA gyrase   DNA helicase

Single-strand-binding proteins

Unwinding

Unwinding

12.12 DNA helicase unwinds DNA by binding to the lagging-strand template at each replication fork and moving in the 5′→3′ direction.

A group of antibiotics called 4-quinolones kill bacteria by binding to DNA gyrase and inhibiting its action. The inhibition of DNA gyrase results in the cessation of DNA synthesis and bacterial growth. An example of a 4-quinolone is nalidixic acid, which was first introduced in the 1960s and is commonly used to treat urinary infections. Many bacteria have acquired resistance to quinolones through mutations in the gene for DNA gyrase.

CONCEPTS

Replication is initiated at an origin of replication, where an initiator protein binds and causes a short stretch of DNA to unwind. DNA helicase breaks hydrogen bonds at a replication fork, and single-strand-binding proteins stabilize the separated strands. DNA gyrase reduces the torsional strain that develops as the two strands of double-helical DNA unwind.

✔ **CONCEPT CHECK 4**

Place the following components in the order in which they are first used in the course of replication: helicase, single-strand-binding protein, DNA gyrase, initiator protein.

Elongation

In the elongation stage of replication, single-stranded DNA is used as a template for the synthesis of DNA. This process requires a series of enzymes.

THE SYNTHESIS OF PRIMERS All DNA polymerases require a nucleotide with a 3'-OH group to which a new nucleotide can be added. Because of this requirement, DNA polymerases cannot initiate DNA synthesis on a bare template; rather, they require an existing 3'-OH group to get started. How, then, does DNA synthesis begin?

An enzyme called **primase** synthesizes short stretches (about 10–12 nucleotides long) of RNA nucleotides, or **primers**, which provide a 3'-OH group to which DNA polymerases can attach DNA nucleotides. (Because primase is an RNA polymerase, it does not require a preexisting 3'-OH group to start synthesis of a nucleotide strand.) All newly synthesized DNA molecules have short RNA primers embedded within them; these primers are later removed and replaced with DNA nucleotides.

On the leading strand at a replication fork, where DNA synthesis is continuous, a primer is required only at the 5' end of the newly synthesized strand. On the lagging strand, where replication is discontinuous, a new primer must be generated at the beginning of each Okazaki fragment (**Figure 12.13**). Primase forms a complex with helicase at the replication fork and moves along the template of the lagging strand. The single primer on the leading strand is probably synthesized by the primase–helicase complex on the template of the lagging strand of the *other* replication fork, at the opposite end of the replication bubble.

▶ **TRY PROBLEM 30**

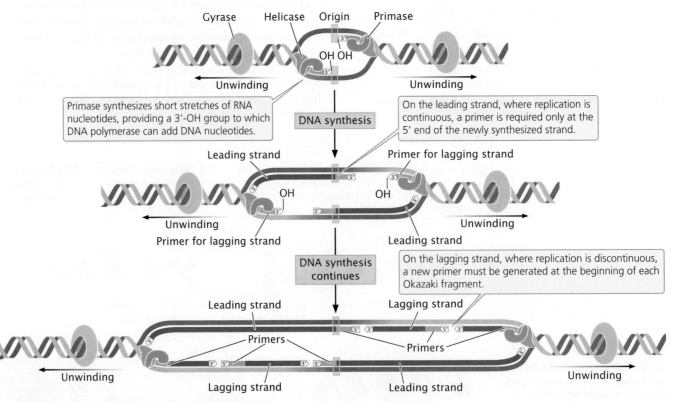

12.13 Primase synthesizes short stretches of RNA nucleotides, providing a 3'-OH group to which DNA polymerase can add DNA nucleotides.

Primase synthesizes a short stretch of RNA nucleotides (a primer), which provides a 3′-OH group for the attachment of DNA nucleotides to start DNA synthesis.

✔ **CONCEPT CHECK 5**

Primers are synthesized where on the lagging strand?
a. Only at the 5′ end of the newly synthesized strand
b. Only at the 3′ end of the newly synthesized strand
c. At the beginning of every Okazaki fragment
d. At multiple places within an Okazaki fragment

DNA SYNTHESIS BY DNA POLYMERASES After DNA has unwound and a primer has been added, DNA polymerases elongate the new polynucleotide strand by catalyzing DNA polymerization. The best-studied polymerases are those of *E. coli*, which has at least five different DNA polymerases. Two of them, DNA polymerase I and DNA polymerase III, carry out DNA synthesis in replication (**Table 12.3**); the other three have specialized functions in DNA repair.

DNA polymerase III is a large multiprotein complex that acts as the main workhorse of replication. DNA polymerase III synthesizes DNA by adding new nucleotides to the 3′ end of a growing DNA strand. This enzyme has two enzymatic activities (see Table 12.3). Its 5′→3′ polymerase activity allows it to add new nucleotides in the 5′→3′ direction. Its 3′→5′ exonuclease activity allows it to remove nucleotides in the 3′→5′ direction, enabling it to correct errors. If a nucleotide with an incorrect base is inserted into the growing DNA strand, DNA polymerase III uses its 3′→5′ exonuclease activity to back up and remove the incorrect nucleotide. It then resumes its 5′→3′ polymerase activity. These two functions together allow DNA polymerase III to efficiently and accurately synthesize new DNA molecules. DNA polymerase III has high *processivity*, which means that it is capable of adding many nucleotides to the growing DNA strand without releasing the template: it normally holds on to the template and continues synthesizing DNA until the template has been completely replicated. The high processivity of DNA polymerase III is ensured by one of the polypeptides that constitutes the enzyme. This ring-shaped polypeptide, termed the **β sliding clamp**, encircles the DNA and allows the polymerase to slide easily along the DNA template strand during replication. DNA polymerase III adds DNA nucleotides to the primer, synthesizing the DNA of both the leading and the lagging strands.

The first *E. coli* polymerase to be discovered, **DNA polymerase I**, also has 5′→3′ polymerase and 3′→5′ exonuclease activities (see Table 12.3), which allow the enzyme to synthesize DNA and to correct errors. Unlike DNA polymerase III, however, DNA polymerase I also possesses 5′→3′ exonuclease activity, which is used to remove the primers laid down by primase and replace them with DNA nucleotides by synthesizing in a 5′→3′ direction (see Figure 12.14 below). After DNA polymerase III has initiated synthesis at a primer and moved downstream, DNA polymerase I removes the RNA nucleotides of the primer, replacing them with DNA nucleotides. DNA polymerase I has lower processivity than DNA polymerase III. The removal and replacement of primers appears to constitute the main function of DNA polymerase I. DNA polymerases II, IV, and V function in DNA repair.

Despite their differences, all of *E. coli*'s DNA polymerases

1. synthesize any sequence specified by the template strand.

2. synthesize in the 5′→3′ direction by adding nucleotides to a 3′-OH group.

3. use dNTPs to synthesize new DNA.

4. require a 3′-OH group to initiate synthesis.

5. catalyze the formation of a phosphodiester bond by joining the 5′-phosphate group of the incoming nucleotide to the 3′-OH group of the preceding nucleotide on the growing strand, cleaving off two phosphates in the process.

6. produce newly synthesized strands that are complementary and antiparallel to the template strands.

7. are associated with a number of other proteins.

⟩ **TRY PROBLEM 27**

DNA polymerases synthesize DNA in the 5′→3′ direction by adding new nucleotides to the 3′ end of a growing nucleotide strand.

TABLE 12.3	Characteristics of DNA polymerases in *E. coli*			
DNA Polymerase	**5′→3′ Polymerase Activity**	**3′→5′ Exonuclease Activity**	**5′→3′ Exonuclease Activity**	**Function**
I	Yes	Yes	Yes	Removes and replaces primers
II	Yes	Yes	No	DNA repair; restarts replication after damaged DNA halts synthesis
III	Yes	Yes	No	Elongates DNA
IV	Yes	No	No	DNA repair
V	Yes	No	No	DNA repair; translesion DNA synthesis

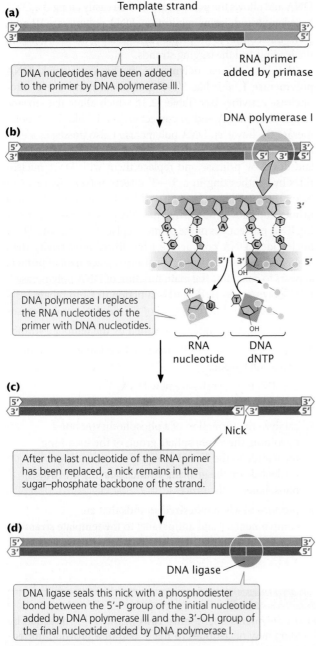

(a)

Template strand

DNA nucleotides have been added to the primer by DNA polymerase III.

RNA primer added by primase

(b)

DNA polymerase I

DNA polymerase I replaces the RNA nucleotides of the primer with DNA nucleotides.

RNA nucleotide   DNA dNTP

(c)

Nick

After the last nucleotide of the RNA primer has been replaced, a nick remains in the sugar–phosphate backbone of the strand.

(d)

DNA ligase

DNA ligase seals this nick with a phosphodiester bond between the 5′-P group of the initial nucleotide added by DNA polymerase III and the 3′-OH group of the final nucleotide added by DNA polymerase I.

12.14 DNA ligase seals the break left by DNA polymerase I in the sugar–phosphate backbone.

PRIMER REPLACEMENT AND DNA LIGASE After DNA polymerase III attaches a DNA nucleotide to the 3′-OH group on the last nucleotide of the RNA primer, each new DNA nucleotide then provides the 3′-OH group needed for the next DNA nucleotide to be added. This process continues as long as a template is available (**Figure 12.14a**). DNA polymerase I follows DNA polymerase III and, using its 5′→3′ exonuclease activity, removes the RNA primer. It then uses its 5′→3′ polymerase activity to replace the RNA nucleotides with DNA nucleotides. DNA polymerase I attaches the first nucleotide to the OH group at the 3′ end of the preceding

Okazaki fragment and then continues, in the 5′→3′ direction along the nucleotide strand, removing and replacing, one at a time, the RNA nucleotides of the primer (**Figure 12.14b**).

After polymerase I has replaced the last nucleotide of the RNA primer with a DNA nucleotide, a break remains in the sugar–phosphate backbone of the new DNA strand because the 3′-OH group of the last nucleotide to have been added by DNA polymerase I is not attached to the 5′-phosphate group of the first nucleotide added by DNA polymerase III (**Figure 12.14c**). This break is sealed by the enzyme **DNA ligase**, which catalyzes the formation of a phosphodiester bond without adding another nucleotide to the strand (**Figure 12.14d**). Some of the major enzymes and proteins required for bacterial DNA replication are summarized in **Table 12.4**.

CONCEPTS

After primers have been removed and replaced, the break in the sugar–phosphate backbone of the new DNA strand is sealed by DNA ligase.

✔ CONCEPT CHECK 6

Which bacterial enzyme removes the primers?

a. Primase
b. DNA polymerase I
c. DNA polymerase III
d. Ligase

TABLE 12.4	Components required for replication in bacterial cells
Component	**Function**
Initiator protein	Binds to origin and separates strands of DNA to initiate replication
DNA helicase	Unwinds DNA at replication fork
Single-strand-binding proteins	Attach to single-stranded DNA and prevent secondary structures from forming
DNA gyrase	Moves ahead of the replication fork, making and resealing breaks in the double-helical DNA to release the torque that builds up as a result of unwinding at the replication fork
DNA primase	Synthesizes a short RNA primer to provide a 3′-OH group for the attachment of DNA nucleotides
DNA polymerase III	Elongates a new nucleotide strand from the 3′-OH group provided by the primer
DNA polymerase I	Removes RNA primers and replaces them with DNA
DNA ligase	Joins Okazaki fragments by sealing breaks in the sugar–phosphate backbone of newly synthesized DNA

ELONGATION AT THE REPLICATION FORK Now that the major enzymatic components of elongation—DNA polymerases, helicase, primase, and ligase—have been introduced, let's consider how these components interact at the replication fork. Because the synthesis of the leading and lagging strands takes place simultaneously, two units of DNA polymerase III must be present at the replication fork, one for each strand. These two units of DNA polymerase III are connected (**Figure 12.15**); the lagging-strand template loops around so that it is in position for 5′→3′ replication. In this way, the DNA polymerase III complex is able to carry out 5′→3′ replication simultaneously on both templates, even though they run in opposite directions. After about 1000 bp of new DNA has been synthesized, DNA polymerase III releases the lagging-strand template, and a new loop forms (see Figure 12.15). Primase synthesizes a new primer on the lagging strand, and DNA polymerase III then synthesizes a new Okazaki fragment. See how replication takes place on both strands simultaneously by viewing **Animation 12.2**.

In summary, each active replication fork requires five basic components:

1. Helicase to unwind the DNA
2. Single-strand-binding proteins to protect the single nucleotide strands and prevent secondary structures
3. DNA gyrase to remove strain ahead of the replication fork
4. Primase to synthesize primers with a 3′-OH group at the beginning of each DNA fragment
5. DNA polymerase to synthesize the leading and lagging nucleotide strands

You can see how the different components of the replication process work together by viewing **Animation 12.3** and **Animation 12.4**.

Termination

In some DNA molecules, replication is terminated whenever two replication forks meet. In others, specific termination sequences (called *Ter* sites) block further replication. A termination protein, called Tus in *E. coli*, binds to these sequences, creating a Tus-*Ter* complex that blocks the movement of helicase, thus stalling the replication fork and preventing further DNA replication. Each Tus-*Ter* complex blocks a replication fork moving in one direction, but not the other.

The Fidelity of DNA Replication

Overall, the error rate in replication is less than one mistake per billion nucleotides. How is this incredible accuracy achieved?

DNA polymerases are very particular in pairing nucleotides with their complements on the template strand. Errors in nucleotide selection by DNA polymerase arise only about once per 100,000 nucleotides. Most of the errors that do arise in nucleotide selection are corrected in a second process called **proofreading**. When a DNA polymerase inserts an incorrect nucleotide into the growing strand, the 3′-OH group of the mispaired nucleotide is not correctly positioned

in the active site of the DNA polymerase for accepting the next nucleotide. The incorrect positioning stalls the polymerization reaction, and the 3′→5′ exonuclease activity of

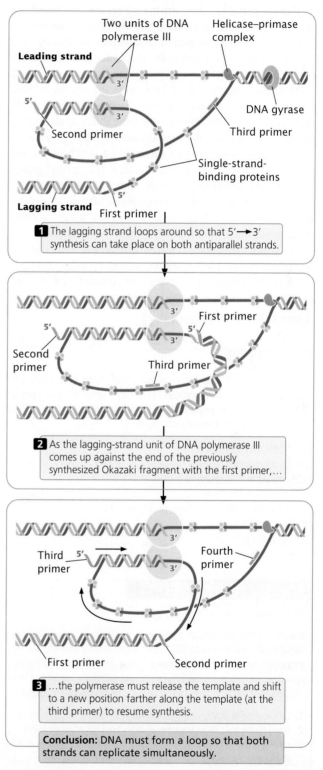

1 The lagging strand loops around so that 5′→3′ synthesis can take place on both antiparallel strands.

2 As the lagging-strand unit of DNA polymerase III comes up against the end of the previously synthesized Okazaki fragment with the first primer,…

3 …the polymerase must release the template and shift to a new position farther along the template (at the third primer) to resume synthesis.

Conclusion: DNA must form a loop so that both strands can replicate simultaneously.

12.15 During DNA replication in *E. coli*, the two units of DNA polymerase III are connected. The lagging-strand template forms a loop so that replication can take place on the two antiparallel DNA strands at the same time. Components of the replication machinery at the replication fork are shown at the top.

DNA polymerase removes the incorrectly paired nucleotide. DNA polymerase then inserts the correct nucleotide. Together, proofreading and nucleotide selection result in an error rate of only one in 10 million nucleotides.

A third process, called **mismatch repair** (discussed further in Chapter 18), corrects errors after replication is complete. Any incorrectly paired nucleotides remaining after replication produce a deformity in the secondary structure of the DNA; that deformity is recognized by enzymes, which excise the incorrectly paired nucleotide and use the original nucleotide strand as a template to replace the incorrect nucleotide. Mismatch repair thus requires the ability to distinguish between the old and the new strands of DNA because the enzymes need some way of determining which of the two incorrectly paired bases to remove. In *E. coli*, methyl groups (—CH₃) are added to particular nucleotide sequences, but only *after* replication. Thus, immediately after DNA synthesis, only the old DNA strand is methylated. It can therefore be distinguished from the newly synthesized strand, and mismatch repair takes place preferentially on the unmethylated nucleotide strand. No single process could produce this level of accuracy; a series of processes are required, each process catching errors missed by the preceding ones.

THINK-PAIR-SHARE Question 3

CONCEPTS

DNA replication is extremely accurate, with less than one error per billion nucleotides. The high level of accuracy in DNA replication is produced by precise nucleotide selection, proofreading, and mismatch repair.

✔ CONCEPT CHECK 7

Which mechanism requires the ability to distinguish between newly synthesized and template strands of DNA?

a. Nucleotide selection
b. DNA proofreading
c. Mismatch repair
d. All of the above

CONNECTING CONCEPTS

The Basic Rules of Replication

Bacterial DNA replication requires a number of enzymes (see Table 12.4), proteins, and DNA sequences that function together to synthesize a new DNA molecule. These components are important, but we must not become so immersed in the details of the process that we lose sight of the general principles of replication.

1. Replication is always semiconservative.
2. Replication begins at sequences called origins of replication.
3. DNA synthesis begins with the synthesis of short segments of RNA called primers.
4. The elongation of DNA strands is always in the 5′→3′ direction.
5. New DNA is synthesized from dNTPs; in the polymerization of DNA, two phosphate groups are cleaved from a dNTP, and

the resulting nucleotide is added to the 3′-OH group of the growing nucleotide strand.
6. Replication is continuous on the leading strand and discontinuous on the lagging strand.
7. New nucleotide strands are complementary and antiparallel to their template strands.
8. Replication takes place at very high rates and is astonishingly accurate, thanks to precise nucleotide selection, proofreading, and mismatch repair.

12.4 Eukaryotic DNA Replication Is Similar to Bacterial Replication but Differs in Several Aspects

Although eukaryotic replication resembles bacterial replication in many respects, replication in eukaryotic cells presents several additional challenges. First, the much greater size of eukaryotic genomes requires that replication be initiated at multiple origins. Second, eukaryotic chromosomes are linear, whereas prokaryotic chromosomes are circular. Third, the DNA template is associated with histone proteins in the form of nucleosomes, and nucleosome assembly must immediately follow DNA replication.

Eukaryotic Origins of Replication

Researchers first isolated eukaryotic origins of replication from yeast cells by demonstrating that certain DNA sequences confer the ability to replicate when transferred from a yeast chromosome to small circular pieces of DNA (plasmids). These **autonomously replicating sequences** (**ARSs**) enabled any DNA to which they were attached to replicate. They were subsequently shown to be the origins of replication in yeast chromosomes. The origins of replication of different eukaryotic organisms vary greatly in sequence, although they usually contain a number of A–T base pairs. A multiprotein complex, the **origin-recognition complex** (**ORC**), binds to origins and unwinds the DNA in those regions.

CONCEPTS

Eukaryotic DNA contains many origins of replication. At each origin, a multiprotein origin-recognition complex binds to initiate the unwinding of the DNA.

✔ CONCEPT CHECK 8

In comparison with prokaryotes, what are some differences in the genome structure of eukaryotic cells that affect how replication takes place?

DNA Synthesis and the Cell Cycle

In rapidly dividing bacteria, DNA replication is continuous. In eukaryotic cells, however, replication is coordinated with the cell cycle. Passage through the cell cycle, including the onset of replication, is controlled by cell cycle checkpoints. The important G_1/S checkpoint (see Section 2.2) holds the cell in G_1 until the DNA is ready to be replicated. After the G_1/S checkpoint is passed, the cell enters S phase and the DNA is replicated. A replication licensing system ensures that the DNA is not replicated again until after the cell has passed through mitosis.

The Licensing of DNA Replication

The use of thousands of origins of replication allows the entire eukaryotic genome to be replicated in a timely manner. The use of multiple origins, however, creates a special problem in the timing of replication: the entire genome must be precisely replicated once, and only once, in each cell cycle, so that no genes are left unreplicated and no genes are replicated more than once. How does a cell ensure that replication is initiated at thousands of origins only once per cell cycle?

The precise replication of DNA is accomplished by the separation of the initiation of replication into two distinct steps. In the first step, the origins are *licensed*—approved for replication. This step takes place early in the cell cycle, when **replication licensing factors** attach to each origin. In the second step, the replication machinery initiates replication at each licensed origin. The key is that the replication machinery functions only at licensed origins and that licensing occurs early in the cell cycle.

Licensing occurs in G_1 of interphase when the origin-recognition complex binds to an origin. ORC, with the help of two additional licensing factors, allows a complex called MCM2-7 (for minichromosome maintenance) to bind to an origin. Then, in S phase, the MCM2-7 complex associates with several cofactors and forms an active helicase that unwinds double-stranded DNA for replication. After replication has begun, several mechanisms prevent MCM2-7 from binding to DNA and reinitiating replication at origins until after mitosis has been completed.

Unwinding

Helicases that separate double-stranded DNA have been isolated from eukaryotic cells, as have single-strand-binding proteins and topoisomerases (which have a function equivalent to the DNA gyrase in bacterial cells). These enzymes and proteins are assumed to function in unwinding eukaryotic DNA in much the same way as their bacterial counterparts do.

Eukaryotic DNA Polymerases

Some significant differences between the processes of bacterial and eukaryotic replication are in the number and functions of DNA polymerases. Eukaryotic cells contain many more different DNA polymerases than bacteria do, which function in replication, recombination, and DNA repair.

Three DNA polymerases carry out most of nuclear DNA synthesis during replication: DNA polymerase α, DNA polymerase δ, and DNA polymerase ε (**Table 12.5**).

TABLE 12.5	DNA polymerases in eukaryotic cells		
DNA Polymerase	**$5' \rightarrow 3'$ Polymerase Activity**	**$3' \rightarrow 5'$ Exonuclease Activity**	**Cellular Function**
α (alpha)	Yes	No	Initiation of nuclear DNA synthesis and DNA repair; has primase activity
δ (delta)	Yes	Yes	Lagging-strand synthesis of nuclear DNA, DNA repair, and translesion DNA synthesis
ε (epsilon)	Yes	Yes	Leading-strand synthesis
γ (gamma)	Yes	Yes	Replication and repair of mitochondrial DNA
ξ (zeta)	Yes	No	Translesion DNA synthesis
η (eta)	Yes	No	Translesion DNA synthesis
θ (theta)	Yes	No	DNA repair
ι (iota)	Yes	No	Translesion DNA synthesis
κ (kappa)	Yes	No	Translesion DNA synthesis
λ (lambda)	Yes	No	DNA repair
μ (mu)	Yes	No	DNA repair
σ (sigma)	Yes	No	Nuclear DNA replication (possibly), DNA repair, and sister-chromatid cohesion
ϕ (phi)	Yes	No	Translesion DNA synthesis
Rev1	Yes	No	DNA repair

Note: The three polymerases listed at the top of the table are those that carry out nuclear DNA replication.

DNA polymerase α has primase activity and initiates nuclear DNA synthesis by synthesizing an RNA primer, followed by a short string of DNA nucleotides. After DNA polymerase α has laid down from 30 to 40 nucleotides, **DNA polymerase δ** completes replication on the lagging strand. Similar in structure and function to DNA polymerase δ, **DNA polymerase ε** replicates the leading strand. Other DNA polymerases take part in repair and recombination or catalyze the replication of organelle DNA.

Some DNA polymerases, such as DNA polymerase δ and DNA polymerase ε, are capable of replicating DNA at high speed and with high fidelity (few mistakes) because they have active sites that snugly and exclusively accommodate the four normal DNA nucleotides, adenosine, guanosine, cytidine, and thymidine monophosphates. As a result of this specificity, distorted DNA templates and abnormal bases are not readily accommodated within the active site of the enzyme. When these errors are encountered in the DNA template, the high-fidelity DNA polymerases stall and are unable to bypass the lesion.

Other DNA polymerases have lower fidelity, but are able to bypass distortions in the DNA template. These specialized **translesion DNA polymerases** generally have a more open active site and are able to accommodate and copy templates with abnormal bases, distorted structures, and bulky lesions. Thus, these specialized enzymes can bypass such errors, but because their active sites are more open and accommodating, they tend to make more errors. In replication, high-speed, high-fidelity enzymes are generally used until they encounter a replication block. At that point, one or more of the translesion DNA polymerases takes over, bypasses the lesion, and continues replicating a short section of DNA. Then the translesion polymerases detach from the replication fork, and high-fidelity polymerases resume replication with high speed and accuracy. DNA-repair enzymes often repair the errors produced by the translesion polymerases, although some of these errors may escape detection and lead to mutations.

THINK-PAIR-SHARE Question 4

Nucleosome Assembly

As we have seen, eukaryotic DNA is complexed with histone proteins to form nucleosomes, structures that contribute to the stability and packing of DNA (see Figure 11.4). In replication, chromatin structure is disrupted by the replication fork, but nucleosomes are quickly reassembled on the two new DNA molecules. Electron micrographs of eukaryotic DNA, such as that in **Figure 12.16**, show that recently replicated DNA is already covered with nucleosomes.

The creation of new nucleosomes requires three steps: (1) the disruption of the original nucleosomes on the parental DNA molecule ahead of the replication fork; (2) the redistribution of preexisting histones on the new DNA molecules; and (3) the addition of newly synthesized histones to complete the formation of new nucleosomes. Before replication, a single DNA molecule is associated with histone proteins. After replication and nucleosome assembly, two DNA molecules are associated with histone proteins. Do the original histones of a nucleosome remain together, attached to one of the new DNA molecules, or do they disassemble and mix with new histones on both DNA molecules?

Techniques similar to those employed by Meselson and Stahl to determine the mode of DNA replication were used to

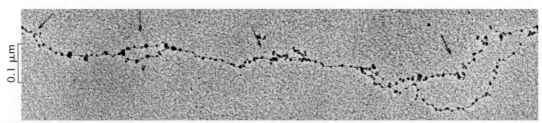

12.16 Nucleosomes are quickly reassembled on newly synthesized DNA. This electron micrograph of eukaryotic DNA in the process of replication clearly shows that newly replicated DNA is already covered with nucleosomes (dark circles). [Victoria Foe.]

address this question. Cells were cultivated for several generations in a medium containing amino acids labeled with a heavy isotope. The histone proteins incorporated these heavy amino acids and were dense (**Figure 12.17**). The cells were then transferred to a culture medium that contained amino acids labeled with a light isotope. Histones assembled after the transfer possessed the new, light amino acids and were less dense.

After replication, when the histone octamers were isolated and centrifuged in a density gradient, they formed a continuous band between the positions expected of high-density (old) and low-density (new) octamers. This finding indicates that newly assembled octamers consist of a mixture of old and new histones. Further evidence indicates that reconstituted nucleosomes appear on the new DNA molecules quickly after the new DNA emerges from the replication machinery.

The reassembly of nucleosomes during replication is facilitated by proteins called histone chaperones, which are associated with the helicase enzyme that unwinds the DNA. The histone chaperones accept old histones from the original DNA molecule and deposit them, along with newly synthesized histones, on the two new DNA molecules. Current evidence suggests that the original histone octamer is broken down into two H2A–H2B dimers (each dimer consisting of one H2A and one H2B) and a single H3–H4 tetramer (each tetramer consisting of two H3 histones and two H4 histones). The old H3–H4 tetramer is then transferred randomly to one of the new DNA molecules and serves as a foundation onto which either new or old copies of H2A–H2B dimers are added. Newly synthesized H3–H4 tetramers and H2A–H2B dimers are also added to each new DNA molecule to complete the formation of new nucleosomes. The assembly of the new nucleosomes is facilitated by a protein called chromatin-assembly factor 1 (CAF-1). ▶ **TRY PROBLEM 32**

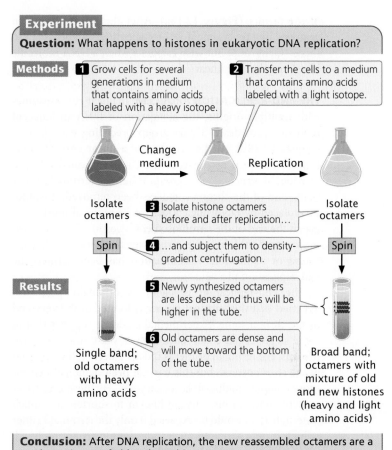

Experiment

Question: What happens to histones in eukaryotic DNA replication?

Methods
1 Grow cells for several generations in medium that contains amino acids labeled with a heavy isotope.

2 Transfer the cells to a medium that contains amino acids labeled with a light isotope.

Change medium

Replication

Isolate octamers

3 Isolate histone octamers before and after replication…

Spin

4 …and subject them to density-gradient centrifugation.

Isolate octamers

Spin

Results

5 Newly synthesized octamers are less dense and thus will be higher in the tube.

6 Old octamers are dense and will move toward the bottom of the tube.

Single band; old octamers with heavy amino acids

Broad band; octamers with mixture of old and new histones (heavy and light amino acids)

Conclusion: After DNA replication, the new reassembled octamers are a random mixture of old and new histones.

12.17 Experimental procedure for studying how nucleosomes dissociate and reassociate in the course of replication.

CONCEPTS

After DNA replication, new nucleosomes quickly reassemble on the two new molecules of DNA. Nucleosomes break down in the course of replication and reassemble from a mixture of old and new histones. The reassembly of nucleosomes during replication is facilitated by histone chaperones and chromatin-assembly factors.

The Location of Replication Within the Nucleus

The DNA polymerases that carry out replication are frequently depicted as moving down the DNA template, much as a locomotive travels along a train track. Evidence suggests that this view is incorrect. A more accurate view is that the polymerase is fixed in location and the template DNA is threaded through it, with newly synthesized DNA molecules emerging from the other end.

Techniques of fluorescence microscopy, which are able to reveal active sites of DNA synthesis, show that most replication in the nucleus of a eukaryotic cell takes place at a limited number of fixed sites, often referred to as replication factories. Time-lapse micrographs reveal that newly duplicated DNA is extruded from these particular sites. Similar results have been obtained for bacterial cells.

Replication at the Ends of Chromosomes

A fundamental difference between eukaryotic and bacterial replication arises because eukaryotic chromosomes are linear and thus have ends. As already stated, the 3′-OH group needed by DNA polymerases is provided at the initiation of replication by RNA primers that are synthesized by primase. This solution is temporary, however, because eventually, the primers must be removed and replaced by DNA nucleotides. In a circular DNA molecule, elongation around the circle eventually provides a 3′-OH group immediately in front of the primer (**Figure 12.18a**). After the primer has been removed, the replacement DNA nucleotides can be added to this 3'-OH group. But what happens when a DNA molecule is not circular, but linear?

THE END-REPLICATION PROBLEM In linear chromosomes with multiple origins, the elongation of DNA in adjacent replicons provides a 3'-OH group preceding each primer (**Figure 12.18b**). At the very end of a linear chromosome, however, there is no adjacent stretch of replicated DNA to provide this crucial 3'-OH group. When the terminal primer at the end of the chromosome has been removed, it cannot be replaced by DNA nucleotides, so its removal produces a gap at the end of the chromosome, suggesting that the chromosome should become progressively shorter with each round of replication. This situation has been termed the end-replication problem.

The end-replication problem, as originally proposed, assumed that the terminal primer is located at the very end of the chromosome. Experimental evidence suggests that in some single-celled eukaryotes, such as yeast and some protozoans, the terminal primer is indeed placed at the very end of the chromosome, but this has not been demonstrated for more complex multicellular eukaryotes. Furthermore, chromosome ends in humans are known to shorten at a much faster rate than would be expected if only the terminal primer (which is only about 10 nucleotides long) was not replaced. Research has now demonstrated that in replication of human chromosomes, the terminal primer is positioned not at the end of the chromosome, but rather some 70 to 100 nucleotides from the end (see Figure 12.18b). This means that 70 to 100 nucleotides of DNA at the end of the chromosome are not replicated during the division of somatic cells, and the chromosome shortens by this amount each time the cell divides.

TELOMERES AND TELOMERASE The end-replication problem suggests that chromosomes in eukaryotic cells should shorten with each cell division and eventually self-destruct. In single-celled eukaryotes, germ cells, and early embryonic cells, however, chromosomes do not shorten. So how are the ends of linear chromosomes in these cells replicated?

The ends of eukaryotic chromosomes—the telomeres—possess several unique features, one of which is the presence of many copies of a short repeated sequence. In humans, this telomeric repeat is TTAGGG (see Section 11.2). The strand containing this G-rich repeat typically protrudes beyond the complementary C-rich strand (**Figure 12.19a**; see also the section Telomere Structure in Chapter 11):

toward   5'–TTAGGGTTAGGGTTAGGG–3'   end of
centromere ← 3'–AATCCC–5'   → chromosome

The single-stranded protruding end of the telomere, known as the **G-rich 3' overhang**, can be extended by **telomerase**, an enzyme that has both a protein and an RNA component (also known as a ribonucleoprotein). The RNA

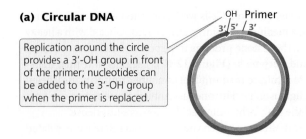

(a) Circular DNA

Replication around the circle provides a 3'-OH group in front of the primer; nucleotides can be added to the 3'-OH group when the primer is replaced.

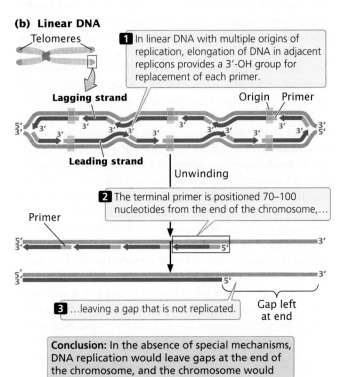

(b) Linear DNA

1 In linear DNA with multiple origins of replication, elongation of DNA in adjacent replicons provides a 3'-OH group for replacement of each primer.

2 The terminal primer is positioned 70–100 nucleotides from the end of the chromosome,...

3 ...leaving a gap that is not replicated.

Gap left at end

Conclusion: In the absence of special mechanisms, DNA replication would leave gaps at the end of the chromosome, and the chromosome would shorten each time the cell divides.

12.18 DNA synthesis at the ends of circular and linear chromosomes must differ.

component of the enzyme contains from 15 to 22 nucleotides that are complementary to the sequence on the G-rich strand. This RNA sequence pairs with the G-rich 3' overhang (**Figure 12.19b**) and provides a template for the synthesis of additional DNA copies of the repeats. DNA nucleotides are added to the 3' end of the G-rich strand one at a time (**Figure 12.19c**); after several nucleotides have been added, the RNA template moves down the DNA, and more nucleotides are added to the 3' end of the G-rich strand (**Figure 12.19d**). In this way, the telomerase can extend the 3' end of the chromosome without the use of a complementary DNA template (**Figure 12.19e**). How the complementary C-rich strand is synthesized (**Figure 12.19f**) is not clear. It may be synthesized by conventional replication, with DNA polymerase α synthesizing an RNA primer on the 5' end of the extended (G-rich) template. The removal of this primer once again leaves a gap at the 5' end of the chromosome, but this gap does not matter because the end of the chromosome is extended at each replication by telomerase, so the chromosome does not become shorter overall.

Telomerase is present in single-celled eukaryotes, germ cells, early embryonic cells, and certain proliferative somatic cells (such as bone-marrow cells and cells lining the intestine), all of which must undergo continuous cell division. Most somatic cells have little or no telomerase activity, and chromosomes in these cells progressively shorten with each cell division. These cells are capable of only a limited number of divisions; when the telomeres have shortened beyond a critical point, the chromosomes become unstable, have a tendency to undergo rearrangements, and are degraded. These events lead to cell death.

THINK-PAIR-SHARE Question 5

CONCEPTS

The ends of eukaryotic chromosomes are replicated by a ribonucleoprotein called telomerase. This enzyme adds extra nucleotides to the G-rich DNA strand of the telomere.

✔ **CONCEPT CHECK 10**

What would be the result if an organism's telomerase were mutated and nonfunctional?

a. No DNA replication would take place.

b. The DNA polymerase enzyme would stall at the telomere.

c. Chromosomes would shorten with each new generation.

d. RNA primers could not be removed.

TELOMERASE, AGING, AND DISEASE The shortening of telomeres may contribute to the process of aging. The telomeres of genetically engineered mice that lack a functional telomerase gene (and therefore do not express telomerase in either somatic or germ cells) undergo progressive shortening in successive generations. After several generations, these mice show some signs of premature aging, such as graying, hair loss, and delayed wound healing. Through genetic engineering, it is also possible to create somatic cells that express telomerase. In these cells, telomeres do not shorten, cell aging is inhibited, and the cells divide indefinitely.

Some of the strongest evidence that telomere length is related to aging comes from studies of telomeres in birds. In 2012, scientists in the United Kingdom measured telomere length in red blood cells taken from 99 zebra finches at various times during their lives. The scientists found a strong correlation between telomere length and longevity: birds with longer telomeres lived longer than birds with short telomeres. The strongest predictor of life span was telomere length early in life, at 25 days, which is roughly equivalent to human adolescence. Although these observations suggest that telomere length is associated with aging in some animals, the precise role of telomeres in *human* aging remains uncertain.

Some diseases are associated with abnormalities of telomere replication. People who have Werner syndrome, an autosomal recessive disease, show signs of premature aging that begin in adolescence or early adulthood, including wrinkled

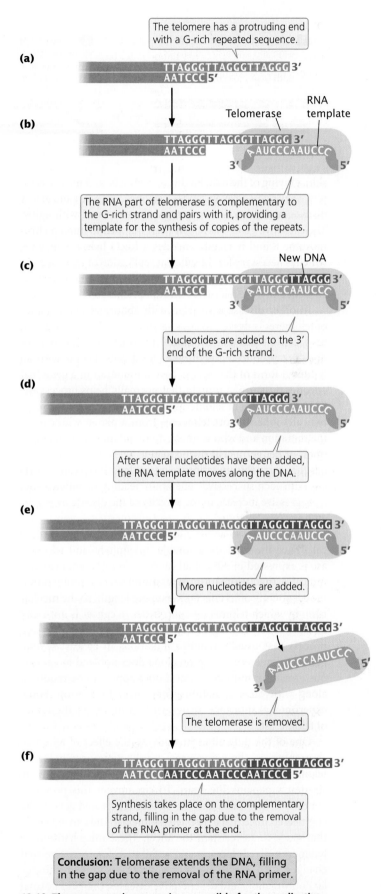

12.19 The enzyme telomerase is responsible for the replication of chromosome ends.

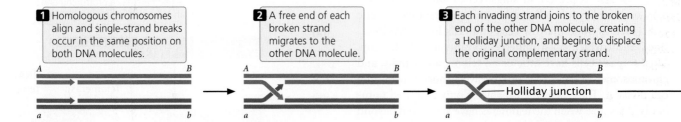

1 Homologous chromosomes align and single-strand breaks occur in the same position on both DNA molecules.

2 A free end of each broken strand migrates to the other DNA molecule.

3 Each invading strand joins to the broken end of the other DNA molecule, creating a Holliday junction, and begins to displace the original complementary strand.

Holliday junction

skin, graying of the hair, baldness, cataracts, and muscle atrophy. They often develop cancer, osteoporosis, heart and artery disease, and other ailments typically associated with aging. The causative gene, *WRN*, has been mapped to human chromosome 8 and normally encodes a RecQ helicase enzyme, which is necessary for the efficient replication of telomeres. In people who have Werner syndrome, this enzyme is defective; consequently, the telomeres shorten prematurely.

Another disease associated with abnormal maintenance of telomeres is dyskeratosis congenita (DKC), which leads to progressive bone-marrow failure, in which the bone marrow does not produce enough new blood cells. People with an X-linked form of the disease have a mutation in a gene that encodes dyskerin, a protein that normally helps process the RNA component of telomerase. People who have the disease typically inherit short telomeres from a parent who carries the mutation and who is unable to maintain telomere length in his or her germ cells owing to defective dyskerin. In families that carry this mutation, telomere length typically shortens with each successive generation, leading to anticipation, a progressive increase in the severity of the disease over generations (see Section 5.4).

Telomerase also appears to play a role in cancer. Cancer cells have the capacity to divide indefinitely, and telomerase is expressed in 90% of all cancers. Some recent evidence indicates that telomerase may stimulate cell proliferation independently of its effect on telomere length, so the mechanism by which telomerase contributes to cancer is not clear. As we will see in Chapter 23, cancer is a complex, multistep process that usually requires mutations in at least several genes. Telomerase activation alone does not lead to cancerous growth in most cells, but it does appear to be required, along with other mutations, for cancer to develop. Some experimental anticancer drugs work by inhibiting the action of telomerase.

One of the difficulties in studying the effect of telomere shortening on the aging process is that the expression of telomerase in somatic cells also promotes cancer, which may shorten a person's life span. To circumvent this problem, Antonia Tomas-Loba and her colleagues created genetically engineered mice that expressed telomerase and carried genes that made them resistant to cancer. These mice had longer telomeres, lived longer, and exhibited fewer age-related changes, such as skin alterations, decreases in neuromuscular coordination, and degenerative diseases, than did normal

mice. These results support the idea that telomere shortening contributes to aging. **TRY PROBLEM 34**

THINK-PAIR-SHARE Question 6

Replication in Archaea

The process of replication in archaea has a number of features in common with replication in eukaryotic cells. Many of the proteins taking part are more similar to those in eukaryotic cells than to those in eubacteria. Like eubacteria, some archaea have a single origin of replication, but many archaea have multiple origins, as eukaryotes do (although archaea have far fewer origins than are found in most eukaryotic chromosomes). The origins of archaea do not contain the typical sequences recognized by bacterial initiator proteins; instead, they have sequences that are similar to those found in some eukaryotic origins. The initiator proteins of archaea are also more similar to those of eukaryotes than to those of eubacteria. These similarities in replication between archaeal and eukaryotic cells reinforce the conclusion that the archaea are more closely related to eukaryotic cells than to the prokaryotic eubacteria.

12.5 Recombination Takes Place Through the Alignment, Breakage, and Repair of DNA Strands

Recombination is the exchange of genetic information between DNA molecules; when the exchange is between homologous DNA molecules, it is called **homologous recombination**. This process takes place in crossing over, in which homologous regions of chromosomes are exchanged and alleles are shuffled into new combinations (see Figure 7.5). Recombination is an extremely important genetic process because it increases genetic variation. Rates of recombination provide important information about linkage relations among genes, which is used to create genetic maps (see Figures 7.13 and 7.14). Recombination is also essential for some types of DNA repair (as we will see in Chapter 18).

Homologous recombination is a remarkable process: a nucleotide strand of one chromosome aligns precisely with a nucleotide strand of the homologous chromosome, breaks arise in corresponding regions of the two DNA molecules, parts of the molecules precisely change place, and then the pieces are correctly joined. In this complicated series of

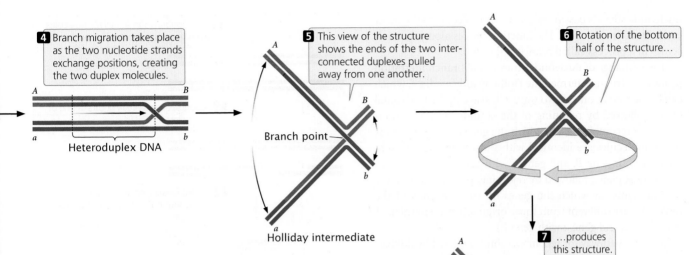

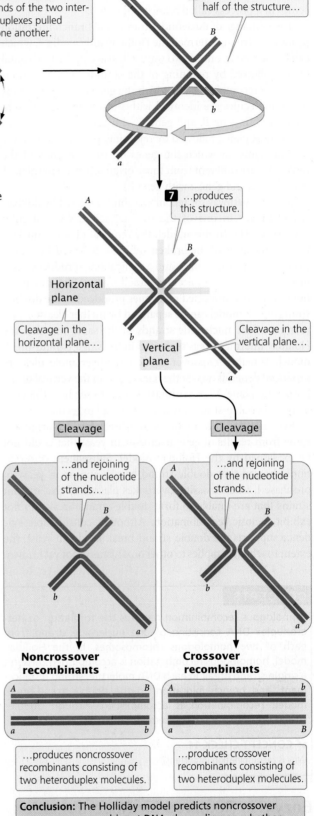

events, no genetic information is lost or gained. Although the precise molecular mechanism of homologous recombination is still not completely understood, the exchange is probably accomplished through the pairing of complementary bases. A single nucleotide strand of one chromosome pairs with the complementary strand of another, forming **heteroduplex DNA**, which is DNA consisting of nucleotide strands from different sources (see next section and Figure 12.20).

In meiosis, homologous recombination (crossing over) could theoretically take place before, during, or after DNA synthesis. Cytological, biochemical, and genetic evidence indicates that it takes place in prophase I of meiosis, whereas DNA replication takes place earlier, in interphase. Thus, crossing over must entail the breaking and rejoining of chromatids when homologous chromosomes are at the four-strand stage (see Figure 7.5). This section explores some theories about how the process of recombination takes place.

Models of Recombination

Homologous recombination takes place through several different pathways. One pathway is initiated by a single-strand break in each of two DNA molecules and includes the formation of a special structure called the **Holliday junction** (**Figure 12.20**). In this model, the double-stranded DNA molecules of two homologous chromosomes align precisely. A single-strand break in each of the DNA molecules provides a free end that invades and joins the free end of the other DNA molecule. Thus, strand invasion and joining take place on both DNA molecules, creating two heteroduplex DNAs, each consisting of one original strand plus one new strand from the other DNA molecule. The point at which the

12.20 The Holliday model of homologous recombination. In this model, recombination takes place through single-strand breaks, strand displacement, branch migration, and resolution of a single Holliday junction.

nucleotide strands pass from one DNA molecule to the other is the Holliday junction. The junction moves along the molecules in a process called branch migration.

The exchange of nucleotide strands and branch migration produce a structure termed the Holliday intermediate, which can be cleaved in one of two ways. Cleavage in the horizontal plane, followed by rejoining of the strands, produces noncrossover recombinants, in which the genes on the two ends of the molecules are identical with those originally present (gene *A* with gene *B*, and gene *a* with gene *b*). Cleavage in the vertical plane, followed by rejoining, produces crossover recombinants, in which the genes on the two ends of the molecules are different from those originally present (gene *A* with gene *b*, and gene *a* with gene *B*).

Another pathway for recombination is initiated by double-strand breaks in one of the two aligned DNA molecules (**Figure 12.21**). In this model, the removal of some nucleotides at the ends of the broken strands—followed by strand invasion, displacement, and replication—produces two heteroduplex DNA molecules joined by two Holliday junctions. The interconnected molecules produced in the double-strand-break model can be separated by further cleavage and reunion of the nucleotide strands in the same way that the Holliday intermediate is separated in the single-strand-break model. Whether crossover or noncrossover molecules are produced depends on whether cleavage is in the vertical or the horizontal plane. View **Animation 12.5** to see how the Holliday and double-strand break models lead to recombination.

Evidence for the double-strand-break model originally came from results of genetic crosses in yeast that could not be explained by the Holliday model. Subsequent observations showed that double-strand breaks appear in yeast in prophase I, when crossing over takes place, and that mutant strains that are unable to form double-strand breaks do not exhibit meiotic recombination. Although considerable evidence supports the double-strand-break model in yeast, the extent to which it applies to other organisms is not yet known.

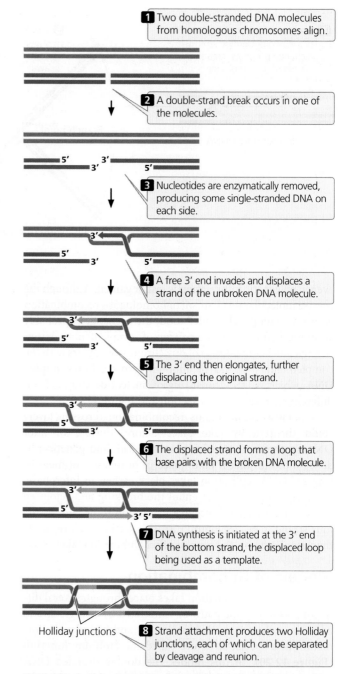

1. Two double-stranded DNA molecules from homologous chromosomes align.
2. A double-strand break occurs in one of the molecules.
3. Nucleotides are enzymatically removed, producing some single-stranded DNA on each side.
4. A free 3′ end invades and displaces a strand of the unbroken DNA molecule.
5. The 3′ end then elongates, further displacing the original strand.
6. The displaced strand forms a loop that base pairs with the broken DNA molecule.
7. DNA synthesis is initiated at the 3′ end of the bottom strand, the displaced loop being used as a template.
8. Strand attachment produces two Holliday junctions, each of which can be separated by cleavage and reunion.

Holliday junctions

12.21 The double-strand-break model of recombination. In this model, recombination takes place through a double-strand break in one DNA duplex, strand displacement, DNA synthesis, and the resolution of two Holliday junctions.

CONCEPTS

Homologous recombination requires the formation of heteroduplex DNA consisting of one nucleotide strand from each of two homologous chromosomes. In the Holliday model, homologous recombination is accomplished through a single-strand break in each DNA molecule, strand displacement, and branch migration. In the double-strand-break model, recombination is accomplished through double-strand breaks, strand displacement, and branch migration.

✔ **CONCEPT CHECK 11**

Why is recombination important?

Enzymes Required for Recombination

Recombination between DNA molecules requires the unwinding of DNA helices, the cleavage of nucleotide strands, strand invasion, and branch migration, followed by

further strand cleavage and union to remove Holliday junctions. Much of what we know about these processes arises from studies of gene exchange in *E. coli*. Although bacteria do not undergo meiosis, they do have a type of sexual reproduction (conjugation), in which one bacterium donates its chromosome to another (discussed more fully in Section 9.3). Subsequent to conjugation, the recipient bacterium has two chromosomes, which may undergo homologous recombination. Geneticists have isolated mutant strains of *E. coli* that are deficient in recombination; the study of these strains has resulted in the identification of genes and proteins that take part in bacterial recombination, revealing several different pathways by which it can take place.

Three genes that play pivotal roles in *E. coli* recombination are *recB*, *recC*, and *recD*, which encode three polypeptides that together form the RecBCD protein. This protein unwinds double-stranded DNA and is capable of cleaving nucleotide strands. The *recA* gene encodes the RecA protein; this protein allows invasion of a DNA helix by a single strand and the subsequent displacement of one of the original strands. In eukaryotes, the formation and branch migration of Holliday structures is facilitated by the enzyme Rad51.

In *E. coli*, *ruvA* and *ruvB* genes encode proteins that catalyze branch migration, and the *ruvC* gene produces a protein, called resolvase, that cleaves Holliday structures. Cleavage and resolution of Holliday structures in eukaryotes is carried out by an analogous enzyme called GEN1. Single-strand-binding proteins, DNA ligase, DNA polymerases, and DNA gyrase also play roles in various types of recombination, in addition to their functions in DNA replication.

> **CONCEPTS**
>
> A number of proteins have roles in recombination, including RecA, RecBCD, RuvA, RuvB, resolvase, single-strand-binding proteins, DNA ligase, DNA polymerases, and DNA gyrase.
>
> ✔ **CONCEPT CHECK 12**
>
> What is the function of resolvase in recombination?
> a. It unwinds double-stranded DNA.
> b. It allows a single DNA strand to invade a DNA helix.
> c. It displaces one of the original DNA strands during branch migration.
> d. It cleaves the Holliday structure.

Gene Conversion

As we have seen, homologous recombination is the mechanism that produces crossing over. It is also responsible for a related phenomenon known as **gene conversion**, a process of nonreciprocal genetic exchange that can produce abnormal ratios of gametes following meiosis. For example, an individual organism with genotype *Aa* is expected to produce ½ *A* gametes and ½ *a* gametes. Sometimes, however, meiosis in an *Aa* individual produces ¾ *A* and ¼ *a* or ¼ *A* and ¾ *a*. Gene conversion arises from heteroduplex formation that takes place in recombination. During heteroduplex formation, a single nucleotide strand of one chromosome pairs with a single strand of another chromosome. If the two strands in a heteroduplex come from chromosomes with different alleles, there will be a mismatch of bases in the heteroduplex DNA (**Figure 12.22**). Such mismatches are often repaired by the cell. Repair mechanisms frequently excise nucleotides on one of the strands and replace them with new DNA by using

the complementary strand as a template. In the process, one copy of an allele may be converted into the other allele, leading to a gene-conversion event (see Figure 12.22), depending on which strand serves as a template.

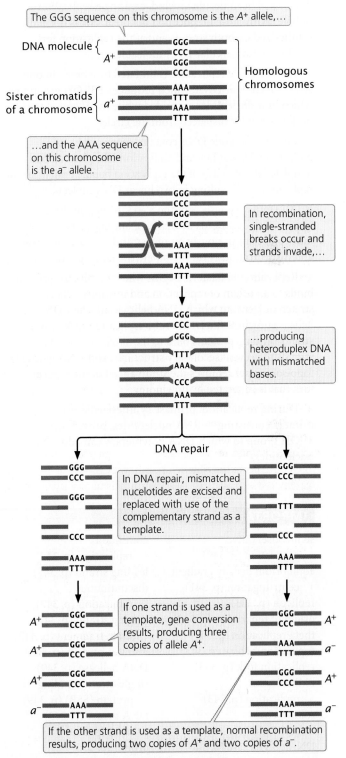

12.22 Gene conversion takes place through the repair of mismatched bases in heteroduplex DNA. For ease of illustration, only three nucleotide pairs are shown for each strand.

CONCEPTS SUMMARY

■ DNA replication is semiconservative: DNA's two nucleotide strands separate, and each serves as a template on which a new strand is synthesized.

■ In theta replication, the two nucleotide strands of a circular DNA molecule unwind, creating a replication bubble. Within each replication bubble, DNA is normally synthesized on both strands and at the replication fork, producing two circular DNA molecules.

■ Rolling-circle replication is initiated by a break in one strand of circular DNA, which produces a 3′-OH group to which new nucleotides are added while the 5′ end of the broken strand is displaced from the circle.

■ Linear eukaryotic DNA contains many origins of replication. Unwinding and replication take place on both templates at both ends of the replication bubble until adjacent replicons meet, resulting in two linear DNA molecules.

■ All DNA synthesis is in the 5′→3′ direction. Because the two nucleotide strands of DNA are antiparallel, replication takes place continuously on one strand (the leading strand) and discontinuously on the other (the lagging strand).

■ Replication in bacteria begins when an initiator protein binds to an origin of replication and unwinds a short stretch of DNA, to which DNA helicase attaches. DNA helicase unwinds the DNA at the replication fork, single-strand-binding proteins bind to the single nucleotide strands to prevent secondary structures, and DNA gyrase (a topoisomerase) removes the strain ahead of the replication fork that is generated by unwinding.

■ During replication, primase synthesizes short primers consisting of RNA nucleotides, providing a 3′-OH group to which DNA polymerase can add DNA nucleotides.

■ DNA polymerase adds new nucleotides to the 3′ end of a growing polynucleotide strand. Bacteria have two DNA polymerases that have primary roles in replication: DNA polymerase III, which synthesizes new DNA on the leading and lagging strands, and DNA polymerase I, which removes and replaces primers.

■ DNA ligase seals the breaks that remain in the sugar–phosphate backbone when the RNA primers are replaced by DNA nucleotides.

■ Several mechanisms ensure the high rate of accuracy in replication, including precise nucleotide selection, proofreading, and mismatch repair.

■ Precise replication at multiple origins in eukaryotes is ensured by licensing factors that must attach to an origin before replication can begin.

■ Eukaryotic nucleosomes are quickly assembled on new molecules of DNA; newly assembled nucleosomes consist of a random mixture of old and new histone proteins.

■ The ends of linear eukaryotic DNA molecules are replicated by the enzyme telomerase. The shortening of telomeres in somatic cells may contribute to aging.

■ The process of replication in archaea has some features in common with replication in eukaryotes.

■ Homologous recombination takes place through alignment of homologous DNA segments, breaks in nucleotide strands, and rejoining of the strands. Homologous recombination requires a number of enzymes and proteins.

■ Gene conversion is nonreciprocal genetic exchange and produces abnormal ratios of gametes.

IMPORTANT TERMS

semiconservative
   replication (p. 340)
equilibrium density gradient
   centrifugation (p. 341)
replicon (p. 343)
origin of replication (p. 343)
theta replication (p. 343)
replication bubble (p. 343)
replication fork (p. 343)
bidirectional
   replication (p. 343)
rolling-circle
   replication (p. 344)
DNA polymerase (p. 346)

continuous
   replication (p. 347)
leading strand (p. 347)
discontinuous
   replication (p. 347)
lagging strand (p. 347)
Okazaki fragment (p. 347)
initiator protein (p. 348)
DNA helicase (p. 349)
single-strand-binding
   protein (SSB) (p. 349)
DNA gyrase (p. 349)
primase (p. 350)
primer (p. 350)

DNA polymerase III (p. 351)
β sliding clamp (p. 351)
DNA polymerase I (p. 351)
DNA ligase (p. 352)
proofreading (p. 353)
mismatch repair (p. 354)
autonomously
   replicating sequence
   (ARS) (p. 354)
origin-recognition complex
   (ORC) (p. 354)
replication licensing
   factor (p. 355)
DNA polymerase α (p. 356)

DNA polymerase δ (p. 356)
DNA polymerase ε (p. 356)
translesion DNA
   polymerase (p. 356)
G-rich 3′ overhang (p. 358)
telomerase (p. 358)
homologous
   recombination (p. 360)
heteroduplex
   DNA (p. 361)
Holliday junction (p. 361)
gene conversion (p. 363)

ANSWERS TO CONCEPT CHECKS

1. Two bands

2. b

3. c

4. Initiator protein, helicase, single-strand-binding protein, DNA gyrase.

5. c

6. b

7. c

8. The size of eukaryotic genomes, the linear structure of eukaryotic chromosomes, and the association of DNA with histone proteins.

9. Error-prone DNA polymerases can bypass lesions in the DNA helix that stall accurate, high-speed DNA polymerases.

10. c

11. Recombination is important for genetic variation and for some types of DNA repair.

12. d

WORKED PROBLEMS

Problem 1

The following diagram represents the template strands of a replication bubble in a DNA molecule. Draw in the newly synthesized strands and identify the leading and lagging strands.

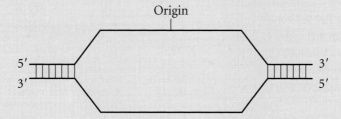

❯❯ Solution Strategy

What information is required in your answer to the problem?
The diagram above with the newly synthesized strands drawn in and the leading and lagging strands labeled.

What information is provided to solve the problem?
A diagram of the template DNA with 5′ and 3′ ends labeled.

For help with this problem, review:
Direction of Replication in Section 12.2 and Figure 12.10c.

❯❯ Solution Steps

> **Recall:** The two strands of DNA are antiparallel, so the newly synthesized strand should have a polarity opposite to the template strand.

To determine the leading and lagging strands, first note which end of each template strand is 5′ and which end is 3′. With a pencil, draw in the strands being synthesized on these templates, and identify their 5′ and 3′ ends.

Next, determine the direction of replication for each new strand, which must be 5′→3′. You might draw arrows on the new strands to indicate the direction of replication. After you have established the direction of replication for each strand, look at each replication fork and determine whether the direction of replication for a strand is the same as the direction of unwinding. The strand on which replication is in the same direction as that of unwinding is the leading strand. The strand on which replication is in the direction opposite that of unwinding is the lagging strand.

> **Recall:** DNA synthesis is always 5′ to 3′.

> **Hint:** Each replication fork should have one leading and one lagging strand.

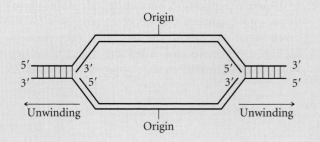

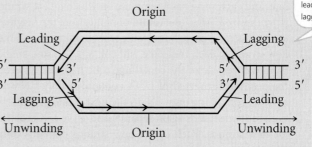

Problem 2

Consider the experiment conducted by Meselson and Stahl in which they used ^{14}N and ^{15}N in cultures of *E. coli* and equilibrium density gradient centrifugation. Draw pictures to represent the bands produced by bacterial DNA in the centrifuge tube before the switch to medium containing ^{14}N and after one, two, and three rounds of replication in that medium. Use separate sets of drawings to show the bands that would appear if replication were (a) semiconservative; (b) conservative; (c) dispersive.

> **Hint:** Review the distribution of new and old DNA in semiconservative, conservative, and dispersive replication in Figure 12.1.

≫ Solution Strategy

What information is required in your answer to the problem?

Drawings that represent the bands produced by bacterial DNA in centrifuge tubes before the switch to medium containing ^{14}N and after one, two, and three rounds of replication in that medium; thus, you should have drawings of four tubes for each model of replication. You will need separate sets of drawings for semiconservative, conservative, and dispersive replication.

What information is provided to solve the problem?

- The bacterial DNA was originally labeled with ^{15}N, and then the bacteria were switched to a medium with ^{14}N (see discussion of experiment on pp. 341–342).
- Original DNA will contain ^{15}N. Newly synthesized DNA will contain ^{14}N.
- Equilibrium density gradient centrifugation was performed before the bacteria were switched to ^{14}N and after one, two, and three rounds of replication following the switch.

For help with this problem, review:

Meselson and Stahl's Experiment in Section 12.2.

≫ Solution Steps

DNA labeled with ^{15}N will be denser than DNA labeled with ^{14}N; therefore, ^{15}N-labeled DNA will sink lower in the centrifuge tube. Before the switch to medium containing ^{14}N, all DNA in the bacteria will contain ^{15}N and will produce a single band in the lower end of the tube.

a. With semiconservative replication, the two strands separate, and each serves as a template on which a new strand is synthesized. After one round of replication, the original template strand of each molecule will contain ^{15}N and the new strand of each molecule will contain ^{14}N, so a single band will appear in the centrifuge tube halfway between the positions expected of DNA containing only ^{15}N and of DNA containing only ^{14}N. In the next round of replication, the two strands again separate and serve as templates for new strands. Each of the new strands contains only ^{14}N; thus, some DNA molecules will contain one strand

with the original ^{15}N and one strand with new ^{14}N, whereas the other molecules will contain two strands with ^{14}N. This labeling will produce two bands, one at the intermediate position and one at a higher position in the tube. Additional rounds of replication should produce increasing amounts of DNA that contains only ^{14}N; so the higher band will get darker.

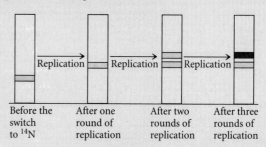

Before the switch to ^{14}N | After one round of replication | After two rounds of replication | After three rounds of replication

b. With conservative replication, the entire molecule serves as a template. After one round of replication, some molecules would consist entirely of ^{15}N, and others would consist entirely of ^{14}N; therefore, two bands would be present. Subsequent rounds of replication would increase the fraction of DNA consisting entirely of new ^{14}N; thus the upper band would get darker. However, the original DNA with ^{15}N would remain, and so two bands would still be present.

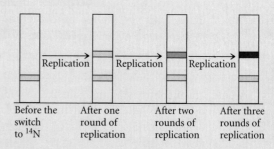

Before the switch to ^{14}N | After one round of replication | After two rounds of replication | After three rounds of replication

c. In dispersive replication, both nucleotide strands break down into fragments that serve as templates for the synthesis of new DNA. The fragments then reassemble into DNA molecules. After one round of replication, all DNA would contain approximately half ^{15}N and half ^{14}N, producing a single band halfway between the positions expected of DNA labeled with ^{15}N and of DNA labeled with ^{14}N. With further rounds of replication, the proportion of ^{14}N in each molecule would increase; therefore, a single hybrid band would remain, but its position in the tube would move upward. The band would also get darker as the total amount of DNA increased.

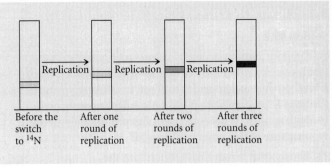

Before the switch to ^{14}N   After one round of replication   After two rounds of replication   After three rounds of replication

COMPREHENSION QUESTIONS

Section 12.2

1. What is semiconservative replication?

2. How did Meselson and Stahl demonstrate that replication in *E. coli* takes place in a semiconservative manner?

3. Draw a molecule of DNA undergoing theta replication. On your drawing, identify (a) origin of replication, (b) polarity (5′ and 3′ ends) of all template strands and newly synthesized strands, (c) leading and lagging strands, (d) Okazaki fragments, and (e) locations of primers.

4. Draw a molecule of DNA undergoing rolling-circle replication. On your drawing, identify (a) origin of replication, (b) polarity (5′ and 3′ ends) of all template and newly synthesized strands, (c) leading and lagging strands, (d) Okazaki fragments, and (e) locations of primers.

5. Draw a molecule of DNA undergoing eukaryotic linear replication. On your drawing, identify (a) origin of replication, (b) polarity (5′ and 3′ ends) of all template and newly synthesized strands, (c) leading and lagging strands, (d) Okazaki fragments, and (e) locations of primers.

6. What are the three major requirements of replication?

7. What substrates are used in DNA synthesis?

Section 12.3

8. List the different proteins and enzymes taking part in bacterial replication. Give the function of each in the replication process.

9. Why is DNA gyrase necessary for replication?

10. What similarities and differences exist in the enzymatic activities of DNA polymerases I and III? What is the function of each DNA polymerase in bacterial cells?

11. Why is primase required for replication?

12. What three mechanisms ensure the accuracy of replication in bacteria?

Section 12.4

13. How does replication licensing ensure that DNA is replicated only once at each origin per eukaryotic cell cycle?

14. In what ways is eukaryotic replication similar to bacterial replication, and in what ways is it different?

15. What is the end-replication problem? Why, in the absence of telomerase, do the ends of linear chromosomes get progressively shorter each time the DNA is replicated?

16. Outline in words and pictures how telomeres at the ends of eukaryotic chromosomes are replicated.

Section 12.5

17. Explain how the type of cleavage of the Holliday intermediate leads to noncrossover recombinants and crossover recombinants.

18. What are some of the enzymes taking part in recombination in *E. coli* and what roles do they play?

19. What is gene conversion? How does it arise?

APPLICATION QUESTIONS AND PROBLEMS

Section 12.2

20. Suppose a future scientist explores a distant planet and discovers a novel form of double-stranded nucleic acid. When this nucleic acid is exposed to DNA polymerases from *E. coli*, replication takes place continuously on both strands. What conclusion can you draw about the structure of this novel nucleic acid?

21. Phosphorus is required to synthesize the deoxyribonucleoside triphosphates used in DNA replication. A geneticist grows some *E. coli* in a medium containing nonradioactive phosphorus for many generations. A sample of the bacteria is then transferred to a medium that contains a radioactive isotope of phosphorus (^{32}P). Samples of the bacteria are removed immediately after the transfer and after one and two rounds of replication. Assume that newly synthesized DNA contains ^{32}P and the original DNA contains nonradioactive phosphorus. What will be the distribution of radioactivity in the DNA of the bacteria in each sample? Will radioactivity be detected in neither, one, or both strands of the DNA?

*22. A line of mouse cells is grown for many generations in a medium with ^{15}N. Cells in G_1 are then switched to a new medium that contains ^{14}N. Draw a pair of homologous chromosomes from these cells at the following stages, showing the two strands of DNA molecules found in the chromosomes. Use different colors to represent strands with ^{14}N and ^{15}N.

   a. Cells in G_1, before switching to medium with ^{14}N

   b. Cells in G_2, after switching to medium with ^{14}N

   c. Cells in anaphase of mitosis, after switching to medium with ^{14}N

   d. Cells in metaphase I of meiosis, after switching to medium with ^{14}N

   e. Cells in anaphase II of meiosis, after switching to medium with ^{14}N

*23. A circular molecule of DNA contains 1 million base pairs. If the rate of DNA synthesis at a replication fork is 100,000 nucleotides per minute, how much time will theta replication require to completely replicate the molecule, assuming that theta replication is bidirectional? How long will replication of this circular chromosome by rolling-circle replication take? Ignore replication of the displaced strand in rolling-circle replication.

24. A bacterium synthesizes DNA at each replication fork at a rate of 1000 nucleotides per second. If this bacterium completely replicates its circular chromosome by theta replication in 30 minutes, how many base pairs of DNA does its chromosome contain?

Section 12.3

*25. The following diagram represents a DNA molecule that is undergoing replication. Draw in the strands of newly synthesized DNA and identify (a) the polarity of the newly synthesized strands, (b) the leading and lagging strands, (c) Okazaki fragments, and (d) RNA primers.

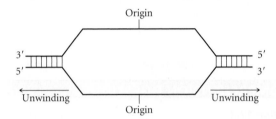

26. In **Figure 12.8**, which is the leading strand and which is the lagging strand?

*27. What would be the effect on DNA replication of mutations that destroyed each of the following activities of DNA polymerase I?

   a. $3' \rightarrow 5'$ exonuclease activity

   b. $5' \rightarrow 3'$ exonuclease activity

   c. $5' \rightarrow 3'$ polymerase activity

28. Which of the DNA polymerases shown in **Table 12.3** have the ability to proofread?

29. How would DNA replication be affected in a bacterial cell that is lacking DNA gyrase?

*30. If the gene for primase were mutated so that no functional primase was produced, what would be the effect on theta replication? On rolling-circle replication?

Section 12.4

31. Eukaryotic licensing factors prevent DNA replication from being initiated at origins more than once in the cell cycle. After replication has begun at an origin, a protein called Geminin inhibits licensing factors that are required for MCM2-7 to bind to an origin and initiate replication. Thus, when Geminin is present, MCM2-7 will not bind to an origin. At the end of mitosis, Geminin is degraded, allowing MCM2-7 to bind once again to DNA and relicense the origin. Marina Melixetian and her colleagues suppressed the expression of Geminin protein in human cells by treating the cells with small interfering RNAs (siRNAs) complementary to Geminin messenger RNA (M. Melixetian et al. 2004. *Journal of Cell Biology* 165:473–482). (Small interfering RNAs form a complex with proteins and pair with

complementary sequences on mRNAs; the complex then cleaves the mRNA, so there is no translation of the mRNA; see pp. 418–419 in Chapter 14). Forty-eight hours after treatment with siRNA, the Geminin-depleted cells were enlarged and contained a single giant nucleus. Analysis of DNA content showed that many of these Geminin-depleted cells were $4n$ or greater. Explain these results.

*32. What results would be expected in the experiment outlined in **Figure 12.17** if, during replication, all the original histone proteins remained on one strand of the DNA and new histones attached to the other strand?

33. A number of scientists who study cancer treatment have become interested in telomerase. Why? How might anticancer therapies that target telomerase work?

*34. The enzyme telomerase is part protein and part RNA. What would be the most likely effect of a large deletion in the gene that encodes the RNA component of telomerase? How would the function of telomerase be affected?

35. Dyskeratosis congenita (DKC) is a rare genetic disorder characterized by abnormal fingernails and skin pigmentation, the formation of white patches on the tongue and cheek, and progressive failure of the bone marrow. An autosomal dominant form of DKC results from mutations in the gene that encodes the RNA component of telomerase. Tom Vulliamy and his colleagues examined a series of families with autosomal dominant DKC (T. Vulliamy et al. 2004. *Nature Genetics* 36:447–449). They observed that the median age of onset of DKC in parents was 37 years, whereas the median age of onset in the children of affected parents was 14.5 years. Thus, DKC in these families arose at progressively younger ages in successive generations, a phenomenon known as anticipation (see p. 133 in Chapter 5). The researchers measured the telomere lengths of members of these families; the measurements are given in the accompanying table. Telomeres normally shorten with age, so telomere length was adjusted for age; the values given in the table are the differences between the actual length and the expected length based on age. Note that the values of all members of these families are negative, indicating that their telomeres are shorter than normal for their age; the more negative the number, the shorter the telomere.

Parent telomere length	Child telomere length
−4.7	−6.1
	−6.6
	−6.0
−3.9	−0.6
−1.4	−2.2
−5.2	−5.4
−2.2	−3.6
−4.4	−2.0
−4.3	−6.8
−5.0	−3.8
−5.3	−6.4
−0.6	−2.5
−1.3	−5.1
	−3.9
−4.2	−5.9

Note: The telomere lengths given in the table are the differences between the actual length and the expected length based on age. Negative values indicate shorter than expected telomeres.

a. How does the telomere length of the parents compare with the telomere length of the children? (Hint: Calculate the average telomere length of all parents and the average telomere length of all children.)

b. Explain why the telomeres of people with DKC are shorter than normal.

c. Explain why DKC arises at an earlier age in subsequent generations.

36. An individual is heterozygous at two loci (*Ee Ff*), and the two genes are in repulsion (see p. 181 in Chapter 7). Assume that single-strand breaks and branch migration occur at the positions shown below. Using different colors to represent the two homologous chromosomes, draw the noncrossover recombinant and crossover recombinant DNA molecules that will result from homologous recombination. (Hint: See **Figure 12.20**.)

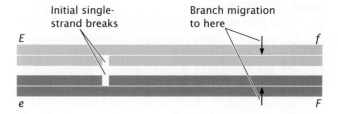

CHALLENGE QUESTIONS

Section 12.3

37. A conditional mutation expresses its mutant phenotype only under certain conditions (the restrictive conditions) and expresses the normal phenotype under other conditions (the permissive conditions). One type of conditional mutation is a temperature-sensitive mutation, which expresses the mutant phenotype only at certain temperatures.

   Strains of *E. coli* have been isolated that contain temperature-sensitive mutations in genes encoding different components of the replication machinery. In each of these strains, the protein produced by the mutated gene is nonfunctional under the restrictive conditions. You grow these strains under the permissive conditions and then abruptly switch them to the restrictive conditions. After one round of replication under the restrictive conditions, you isolate DNA from each strain and analyze it. What characteristics would you expect to see in the DNA isolated from a strain with a temperature-sensitive mutation in the gene that encodes each of the following proteins?

   a. DNA ligase

   b. DNA polymerase I

   c. DNA polymerase III

   d. Primase

   e. Initiator protein

Section 12.4

38. DNA topoisomerases play important roles in DNA replication and in supercoiling (see Chapter 11). These enzymes are also the targets for certain anticancer drugs (see the introduction to this chapter). Eric Nelson and his colleagues studied m-AMSA, one of the anticancer

compounds that acts on topoisomerase (E. M. Nelson, K. M. Tewey, and L. F. Liu. 1984. *Proceedings of the National Academy of Sciences of the United States of America* 81:1361–1365). They found that m-AMSA stabilizes an intermediate produced in the course of topoisomerase action. The intermediate consists of topoisomerase bound to the broken ends of the DNA. Breaks in DNA that are produced by anticancer compounds such as m-AMSA inhibit the replication of the cellular DNA and thus stop cancer cells from proliferating. Explain how m-AMSA and other anticancer agents that target topoisomerase enzymes taking part in replication might lead to DNA breaks and chromosome rearrangements.

*39. The regulation of replication is essential to genomic stability. Normally, the DNA is replicated just once in every eukaryotic cell cycle (in the S phase). Normal cells produce protein A, which increases in concentration in the S phase. In cells that have a mutated copy of the gene encoding protein A, the protein is not functional, and replication takes place continuously throughout the cell cycle, with the result that cells may have 50 times the normal amount of DNA. Protein B is normally present in G_1, but disappears from the cell nucleus during the S phase. In cells with a mutated copy of the gene encoding protein A, the levels of protein B fail to drop in the S phase and, instead, remain high throughout the cell cycle. When the gene for protein B is mutated, no replication takes place.

   Propose a mechanism for how protein A and protein B might normally regulate replication so that each cell gets the proper amount of DNA. Explain how mutation of these genes produces the effects just described.

THINK-PAIR-SHARE QUESTIONS

Section 12.1

1. In the 1996 movie *Multiplicity*, Doug (played by Michael Keaton) is a construction worker who wants to spend more time with his family. He meets a friendly scientist who has developed a method for cloning humans. Doug decides to make a clone of himself who can take over his work while he spends quality time with his family. The clone, named "Two," seems great at first, but later problems surface in his functioning. Two decides to make a clone of himself, so he won't have to work, and creates Three. A fourth clone is eventually made. Each successive clone seems to have more problems. Ignore, for the present, the technical difficulties and ethical problems with making an instantaneous adult clone of a

human. Is there any genetic validity to the premise that making a clone of a clone might create problems? If so, what might those problems be?

Section 12.2

2. In their experiment, could Meselson and Stahl have used two different isotopes of carbon, instead of ^{14}N and ^{15}N? Why or why not? What about two different isotopes of sulfur?

Section 12.3

3. DNA polymerases cannot act as primers for replication, yet primase and other RNA polymerases can. Some geneticists have speculated that the inability of DNA

polymerase to prime replication is a result of its proofreading function. This hypothesis argues that proofreading is essential for the faithful transmission of genetic information and that because DNA polymerases have evolved the ability to proofread, they cannot prime DNA synthesis. Explain why proofreading and priming functions in the same enzyme might be incompatible.

4. Okazaki fragments in bacterial cells are 1000–2000 nucleotides in length. Those in eukaryotic cells are much shorter, usually only 100–200 nucleotides in length. Propose some possible explanations for why Okazaki fragments in bacteria are longer than those in eukaryotes. What difference between the two cell types could lead to the difference in Okazaki fragment length?

Section 12.4

5. HeLa cells are a line of cells grown in laboratory culture that has been used extensively in research. This cell line was originally derived from malignant cervical cancer calls that were removed from a woman named Henrietta Lacks in 1951. They were subsequently grown in culture and shipped to research laboratories around the world, where they have been used in many important experiments. Like HeLa cells, many other cell lines were originally taken from cancerous tissue. Why are cancer cells often used for developing cell lines?

6. For centuries, people have searched for the fabled Fountain of Youth, said to confer the ability to forestall old age and remain young forever. To gain insight into how tackling aging by targeting telomerase might work, researchers looked at mice that lack a telomerase gene. These mice have shorter telomeres than normal mice and age prematurely. But when these mice are engineered to express telomerase in their somatic cells, their telomeres lengthen, and the effects of aging are reversed. This observation suggests that a drug that stimulates the expression of telomerase in somatic cells could prevent telomere shortening and stop the aging process. Do you think such a drug would work well in humans? What might be some potential side effects of such a drug?

Transcription

The death cap mushroom, *Amanita phalloides*, causes death by inhibiting the process of transcription. [© MAP/Jean-Yves Grospas/Age FotoStock America, Inc.]

Death Cap Poisoning

On November 8, 2009, 31-year-old Tomasa was hiking the Lodi Lake nature trail east of San Francisco with her husband and cousin when they came across some large white mushrooms that looked very much like the edible mushrooms that they enjoyed in their native Mexico. They picked the mushrooms and took them home, cooking and consuming them for dinner. Within hours, Tomasa and her family were sick and went to the hospital. They were later transferred to the critical care unit at California Pacific Medical Center in San Francisco, where Tomasa died of liver failure three weeks later. Her husband eventually recovered after a lengthy hospitalization; her cousin required a liver transplant to survive.

The mushrooms consumed by Tomasa and her family were *Amanita phalloides*, commonly known as death caps. A single death cap contains enough toxin to kill an adult human. The death rate among those who consume death caps is 22%; among children under the age of 10, it's more than 50%. Death cap mushrooms appear to be spreading in California, leading to a surge in the number of mushroom poisonings.

Death cap poisoning is insidious. Gastrointestinal symptoms—abdominal pain, cramping, vomiting, diarrhea—begin within 6 to 12 hours of consuming the mushrooms, but these symptoms usually subside within a few hours, and the patient seems to recover. Because of this initial remission, the poisoning is often not taken seriously until it's too late to pump the stomach and remove the toxin from the body. After a day or two, serious symptoms begin. Cells in the liver die, often causing permanent liver damage and, sometimes, death within a few days. There is no effective treatment, other than a liver transplant to replace the damaged organ.

How do death caps kill? Their deadly toxin, contained within the fruiting bodies that produce reproductive spores, is α-amanitin, which consists of a short peptide of eight amino acids that forms a circular loop. α-Amanitin is a potent inhibitor of RNA polymerase II, the enzyme that transcribes protein-encoding genes in eukaryotes. RNA polymerase II binds to genes and synthesizes RNA molecules that are complementary to the DNA template. In the process of transcription, the RNA polymerase moves down the DNA template, adding one nucleotide at a time to the growing RNA chain. α-Amanitin binds to RNA polymerase and jams the moving parts of the enzyme, interfering with its ability to move along the DNA template. In the presence of α-amanitin, RNA synthesis slows from its normal rate of several thousand nucleotides per minute to just a few nucleotides per minute. The results are catastrophic. Without transcription, protein synthesis—required for cellular function—ceases, and cells die. The liver, where the toxin accumulates, is irreparably damaged and stops functioning. In severe cases, the patient dies.

THINK-PAIR-SHARE
- Why would mushrooms produce a substance like α-amanitin? What function might this peptide have in mushrooms?

- Some cancer researchers have proposed using α-amanitin in cancer therapy. How might α-amanitin be used in the treatment of cancer?

Death cap poisoning illustrates the extreme importance of transcription and the central role that RNA polymerase plays in the process. This chapter is about transcription—the first step in the central dogma, the pathway of information transfer from DNA (genotype) to protein (phenotype). Transcription is a complex process that requires precursors to RNA nucleotides, a DNA template, and a number of protein components. As we examine the stages of transcription, try to keep all the details in perspective and focus on understanding how they relate to the overall purpose of transcription: the selective synthesis of an RNA molecule.

This chapter begins with a brief review of RNA structure and a discussion of the different classes of RNA. We then consider the major components required for transcription. Finally, we explore the process of transcription. At several points in the text, we pause to consider some general principles that emerge.

THINK-PAIR-SHARE Question 1

13.1 RNA, Consisting of a Single Strand of Ribonucleotides, Participates in a Variety of Cellular Functions

Before we begin our study of transcription, let's consider the past and present importance of RNA, review the structure of RNA, and examine some of the different types of RNA molecules.

An Early RNA World

Life requires two basic functions. First, living organisms must be able to store and faithfully transmit genetic information during reproduction. Second, they must have the ability to catalyze the chemical transformations that drive life processes. A long-held belief was that the functions of information storage and chemical transformation are handled by two entirely different types of molecules: genetic information is stored in nucleic acids, whereas chemical transformations are catalyzed by protein enzymes. This biochemical dichotomy created a dilemma: which came first, proteins or nucleic acids? If nucleic acids carry the coding instructions for proteins, how could proteins be generated without them? Nucleic acids are unable to copy themselves, so how could they be generated without proteins? If DNA and proteins each require the other, how could life begin?

This apparent paradox was answered in 1981 when Thomas Cech and his colleagues discovered that RNA can serve as a biological catalyst. They found that some RNA molecules from the protozoan *Tetrahymena thermophila* can excise 400 nucleotides from its RNA in the absence of any protein. Other examples of catalytic RNAs have now been discovered in different types of cells. Called **ribozymes**, these catalytic RNA molecules can cut out parts of their own sequences, connect some RNA molecules together, replicate others, and even catalyze the formation of peptide bonds between amino acids. The discovery of ribozymes complements other evidence suggesting that the original genetic material was RNA.

Self-replicating ribozymes probably first arose between 3.5 billion and 4 billion years ago and may have begun the evolution of life on Earth. Early life was probably an RNA world, where RNA molecules served both as carriers of genetic information and as catalysts that drove the chemical reactions needed to sustain and perpetuate life. These catalytic RNAs may have acquired the ability to synthesize protein-based enzymes, which are more efficient catalysts. With enzymes taking over more and more of the catalytic functions, RNA probably became relegated to the role of information storage and transfer. DNA, with its chemical stability and faithful replication, eventually replaced RNA as the primary carrier of genetic information. Nevertheless, RNA is either produced by or plays a vital role in many biological processes, including transcription, replication, RNA processing, and translation. Research in the past 20 years has also determined that newly discovered small RNA molecules play a fundamental role in many basic biological processes, demonstrating that life today is still very much an RNA world. These small RNA molecules will be discussed in more detail in Chapter 14.

CONCEPTS

Early in the history of life, RNA probably served both as the original genetic material and as biological catalysts.

The Structure of RNA

RNA, like DNA, is a polymer of nucleotides, each consisting of a sugar, a phosphate group, and a nitrogenous base, joined together by phosphodiester bonds (see Chapter 10). However, there are several important differences in the structures of DNA and RNA. Whereas DNA nucleotides contain deoxyribose sugars, RNA nucleotides have ribose sugars (**Figure 13.1a**). With a free hydroxyl group on the 2′-carbon atom of the ribose sugar, RNA is degraded rapidly under alkaline conditions. The deoxyribose sugar of DNA lacks this free hydroxyl group, so DNA is a more stable molecule. Another important difference is that the pyrimidine uracil is

present in RNA instead of thymine, one of the two pyrimidines found in DNA.

A final difference in the structures of DNA and RNA is that RNA usually consists of a single polynucleotide strand, whereas DNA normally consists of two polynucleotide strands joined by hydrogen bonding between complementary bases (although some viruses contain double-stranded RNA genomes, as discussed in Chapter 9). Although RNA is usually single stranded, short complementary regions within a nucleotide strand can pair and form secondary structures (**Figure 13.1b**). These RNA secondary structures are often called hairpins (or hairpin-loop or stem-loop structures). When two regions within a single RNA molecule pair up, the strands in those regions must be antiparallel and with pairing between cytosine and guanine and between adenine and uracil (although occasionally guanine pairs with uracil).

The formation of secondary structures plays an important role in RNA function. Secondary structure is determined by the base sequence of the nucleotide strand, so different RNA molecules can assume different structures. Because their structure determines their function, RNA molecules have the potential for tremendous variation in function. With its two complementary strands forming a helix, DNA is much more restricted in the range of secondary structures that it can assume and so has fewer functional roles in the cell. Similarities and differences in DNA and RNA structures are summarized in **Table 13.1**. ▶ **TRY PROBLEM 14**

Classes of RNA

RNA molecules perform a variety of functions in the cell. **Ribosomal RNA (rRNA)** and ribosomal protein subunits make up the ribosome, the site of protein assembly. We'll take a more detailed look at the ribosome in Chapter 14. **Messenger RNA (mRNA)** carries the coding instructions for a polypeptide chain from DNA to a ribosome. After attaching to the ribosome, an mRNA molecule specifies the sequence of the amino acids in a polypeptide chain and provides a template for the joining of those amino acids. Large precursor molecules, which are termed **pre-messenger RNAs**

(a)

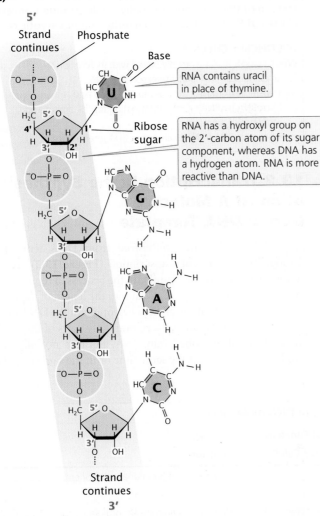

(b) Primary structure

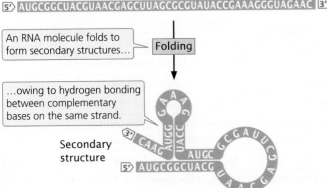

13.1 RNA has a primary and a secondary structure.

Characteristic	DNA	RNA
Composed of nucleotides	Yes	Yes
Type of sugar	Deoxyribose	Ribose
Presence of 2′-OH group	No	Yes
Bases	A, G, C, T	A, G, C, U
Nucleotides joined by phosphodiester bonds	Yes	Yes
Double or single stranded	Usually double	Usually single
Secondary structure	Double helix	Many types
Stability	Stable	Easily degraded

TABLE 13.1 — The structures of DNA and RNA compared

(**pre-mRNAs**), are the immediate products of transcription in eukaryotic cells. Pre-mRNAs are modified extensively before becoming mRNA and exiting the nucleus for translation into protein. Bacterial cells do not possess pre-mRNA; in these cells, transcription takes place concurrently with translation.

Transfer RNA (tRNA) serves as the link between the coding sequence of nucleotides in an mRNA molecule and the amino acid sequence of a polypeptide chain. Each tRNA attaches to one particular type of amino acid and helps to incorporate that amino acid into a polypeptide chain (as described in Chapter 15).

Additional classes of RNA molecules are found in the nuclei of eukaryotic cells. **Small nuclear RNAs (snRNAs)** combine with small protein subunits to form **small nuclear ribonucleoproteins (snRNPs**, affectionately known as "snurps"). Some snRNAs participate in the processing of RNA, converting pre-mRNA into mRNA. **Small nucleolar RNAs (snoRNAs)** take part in the processing of rRNA.

Two types of very small and abundant RNA molecules found in the cytoplasm of eukaryotic cells, termed **microRNAs (miRNAs)** and **small interfering RNAs (siRNAs)**, carry out RNA interference (RNAi), a process in which these small RNA molecules help trigger the degradation of mRNA or inhibit its translation into protein. More will be said about RNA interference in Chapter 14. Another class of small RNA molecules are **Piwi-interacting RNAs (piRNAs**; named after Piwi proteins, with which they interact). Found in mammalian testes, these RNA molecules are similar to miRNAs and siRNAs; they have a role in suppressing the expression of transposable elements (see Chapter 18) in reproductive cells. An RNA interference-like system has been discovered in prokaryotes, in which small **CRISPR RNAs (crRNAs)** assist in the destruction of foreign DNA molecules. **Long noncoding RNAs (lncRNAs)** are relatively long RNA molecules found in eukaryotes that do not code for proteins. They provide a variety of functions, including regulation of gene expression. Some of the different classes of RNA molecules are summarized in **Table 13.2**.

CONCEPTS

RNA differs from DNA in that RNA possesses a hydroxyl group on the 2'-carbon atom of its sugar, contains uracil instead of thymine, and is usually single stranded. Several classes of RNA exist within bacterial and eukaryotic cells.

✔ **CONCEPT CHECK 1**

Which class of RNA is correctly paired with its function?
a. Small nuclear RNA (snRNA): processes rRNA
b. Transfer RNA (tRNA): attaches to an amino acid
c. MicroRNA (miRNA): carries information for the amino acid sequence of a protein
d. Ribosomal RNA (rRNA): carries out RNA interference

13.2 Transcription Is the Synthesis of an RNA Molecule from a DNA Template

All cellular RNAs are synthesized from DNA templates through the process of transcription (**Figure 13.2**). Transcription is in many ways similar to the process of replication, but a fundamental difference relates to the length of the template used. In replication, all the nucleotides in the DNA molecule are copied, but in transcription, only parts of the DNA molecule are transcribed into RNA. Because not all gene products are needed at the same time or in the same cell, the

Class of RNA	Cell Type	Location of Function in Eukaryotic Cells*	Function
Ribosomal RNA (rRNA)	Prokaryotic and eukaryotic	Cytoplasm	Structural and functional components of the ribosome
Messenger RNA (mRNA)	Prokaryotic and eukaryotic	Nucleus and cytoplasm	Carries genetic code for proteins
Transfer RNA (tRNA)	Prokaryotic and eukaryotic	Cytoplasm	Helps incorporate amino acids into polypeptide chain
Small nuclear RNA (snRNA)	Eukaryotic	Nucleus	Processing of pre-mRNA
Small nucleolar RNA (snoRNA)	Eukaryotic	Nucleus	Processing and assembly of rRNA
MicroRNA (miRNA)	Eukaryotic	Nucleus and cytoplasm	Inhibits translation of mRNA
Small interfering RNA (siRNA)	Eukaryotic	Nucleus and cytoplasm	Triggers degradation of other RNA molecules
Piwi-interacting RNA (piRNA)	Eukaryotic	Nucleus and cytoplasm	Suppresses the transcription of transposable elements in reproductive cells
CRISPR RNA (crRNA)	Prokaryotic	—	Assists destruction of foreign DNA
Long noncoding RNA (lncRNA)	Eukaryotic	Nucleus and cytoplasm	Variety of functions

TABLE 13.2 Locations and functions of different classes of RNA molecules

*All eukaryotic RNAs are synthesized in the nucleus.

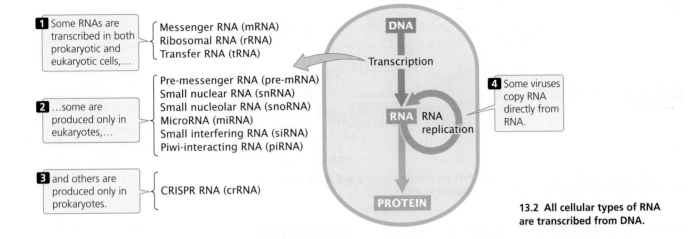

1 Some RNAs are transcribed in both prokaryotic and eukaryotic cells,...
- Messenger RNA (mRNA)
- Ribosomal RNA (rRNA)
- Transfer RNA (tRNA)

2 ...some are produced only in eukaryotes,...
- Pre-messenger RNA (pre-mRNA)
- Small nuclear RNA (snRNA)
- Small nucleolar RNA (snoRNA)
- MicroRNA (miRNA)
- Small interfering RNA (siRNA)
- Piwi-interacting RNA (piRNA)

3 and others are produced only in prokaryotes.
- CRISPR RNA (crRNA)

4 Some viruses copy RNA directly from RNA.

DNA → Transcription → RNA (RNA replication) → PROTEIN

13.2 All cellular types of RNA are transcribed from DNA.

constant transcription of all of a cell's genes would be highly inefficient. Furthermore, much of the DNA does not encode any functional product, and transcription of such sequences would be pointless. Transcription is, in fact, a highly selective process: individual genes are transcribed only as their products are needed. However, this selectivity imposes a fundamental problem on the cell: how to recognize individual genes and transcribe them at the proper time and place.

Like replication, transcription requires three major components:

1. A DNA template

2. The raw materials (ribonucleotide triphosphates) needed to build a new RNA molecule

3. The transcription apparatus, consisting of the proteins necessary for catalyzing the synthesis of RNA

The Template

In 1970, Oscar Miller, Jr., Barbara Hamkalo, and Charles Thomas used electron microscopy to demonstrate that RNA is transcribed from a DNA template. They broke open cells, extracted chromatin, and spread the chromatin onto a fine mesh grid. Under the electron microscope, they observed Christmas-tree-like structures, each consisting of a thin central fiber (the trunk of the tree) to which were attached strings (the branches) bearing granules (**Figure 13.3a**). The addition of deoxyribonuclease (an enzyme that degrades DNA) caused the central fibers to disappear, indicating that the "tree trunks" were DNA molecules. Ribonuclease (an enzyme that degrades RNA) removed the granular strings, indicating that the branches were RNA. Their conclusion was that each "Christmas tree" represents a gene undergoing transcription (**Figure 13.3b**). The transcription of each gene begins at the top of the tree; there, little of the DNA has been transcribed, and the RNA branches are short. As the transcription apparatus moves down the tree, transcribing more of the template, the RNA molecules lengthen, producing the long branches at the bottom.

(a)

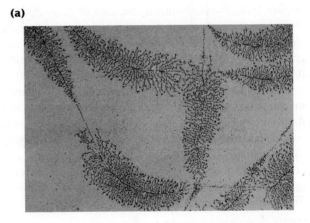

(b)

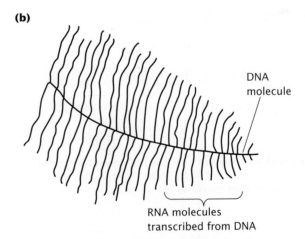

DNA molecule

RNA molecules transcribed from DNA

13.3 Under the electron microscope, DNA molecules undergoing transcription exhibit Christmas-tree-like structures. (a) Electron micrograph of transcription units. (b) The trunk of each "Christmas tree" (a transcription unit) represents a portion of a DNA molecule; the tree branches are RNA molecules that have been transcribed from the DNA. As the transcription apparatus moves down the DNA, transcribing more of the template, the RNA molecules become longer and longer. [Part a: © Phototake, Inc./Phototake.]

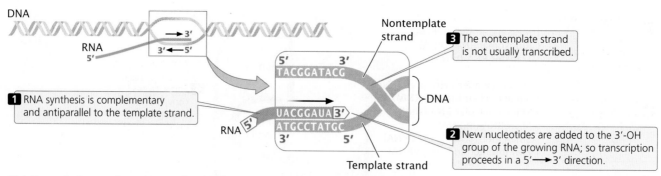

13.4 Transcription produces RNA molecules that are complementary and antiparallel to one of the two nucleotide strands of DNA, the template strand.

THE TRANSCRIBED STRAND The template for RNA synthesis, as for DNA synthesis, is a single strand of the DNA double helix. Unlike replication, however, the transcription of a gene takes place on only one of the two nucleotide strands of DNA (**Figure 13.4**). The nucleotide strand used for transcription is termed the **template strand**. The other strand, called the **nontemplate strand**, is not ordinarily transcribed. Thus, within a gene, only one of the nucleotide strands is normally transcribed into RNA (although there are some exceptions to this rule).

During transcription, an RNA molecule that is complementary and antiparallel to the DNA template strand is synthesized (see Figure 13.4). The RNA transcript has the same polarity and base sequence as the nontemplate strand, except that it contains U rather than T. In most organisms, each gene is transcribed from a single DNA strand, but different genes may be transcribed from different strands, as shown in **Figure 13.5**. ▶ **TRY PROBLEM 15**

THINK-PAIR-SHARE Question 2 👥

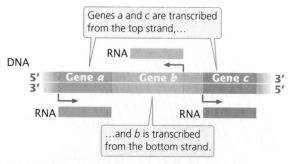

13.5 RNA is transcribed from one DNA strand. In most organisms, each gene is transcribed from a single DNA strand, but different genes may be transcribed from either DNA strand.

CONCEPTS

Within a single gene, only one of the two DNA strands, the template strand, is usually transcribed into RNA.

✔ **CONCEPT CHECK 2**

What is the difference between the template strand and the nontemplate strand?

THE TRANSCRIPTION UNIT A **transcription unit** is a stretch of DNA that encodes an RNA molecule and the sequences necessary for its transcription. How does the complex of enzymes and proteins that performs transcription—the transcription apparatus—recognize a transcription unit? How does it know which DNA strand to read and where to start and stop? This information is encoded by the DNA sequence.

Included within a transcription unit are three critical regions: a promoter, an RNA-coding region, and a terminator (**Figure 13.6**). The **promoter** is a DNA sequence that the transcription apparatus recognizes and binds. The promoter indicates which of the two DNA strands is to be read as the template and the direction of transcription. It also determines the transcription start site, the first nucleotide that will be transcribed into RNA. In many transcription units, the promoter is located next to the transcription start site but is not itself transcribed.

The second critical region of the transcription unit is the **RNA-coding region**, a sequence of DNA nucleotides that is copied into an RNA molecule. The third component of the transcription unit is the **terminator**, a sequence of nucleotides that signals where transcription is to end. Terminators are usually part of the RNA-coding sequence;

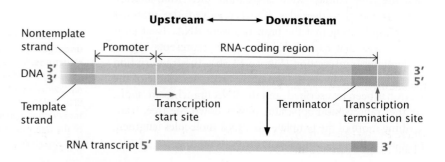

13.6 A transcription unit includes a promoter, an RNA-coding region, and a terminator.

transcription stops only after the terminator has been copied into RNA.

Molecular biologists often use the terms *upstream* and *downstream* to refer to the direction of transcription and the locations of nucleotide sequences surrounding the RNA-coding region. The transcription apparatus is said to move downstream during transcription: it binds to the promoter (which is usually upstream of the transcription start site) and moves toward the terminator (which is downstream of the start site).

When DNA sequences are written out, often the sequence of only one of the two strands is listed. Molecular biologists typically write the sequence of the nontemplate strand because it will be the same as the sequence of the RNA transcribed from the template strand (with the exception that U in RNA replaces T in DNA). By convention, the sequence of the nontemplate strand is written with the 5′ end on the left and the 3′ end on the right. The first nucleotide transcribed (the transcription start site) is numbered +1; nucleotides downstream of the start site are assigned positive numbers, and nucleotides upstream of the start site are assigned negative numbers. So, nucleotide +34 would be 34 nucleotides downstream of the start site, whereas nucleotide −75 would be 75 nucleotides upstream of the start site. There is no nucleotide numbered 0.

CONCEPTS

A transcription unit is a stretch of DNA that encodes an RNA molecule and the sequences necessary for its proper transcription. Each transcription unit includes a promoter, an RNA-coding region, and a terminator.

✔ CONCEPT CHECK 3

Which of the following phrases does *not* describe a function of the promoter?

a. Serves as sequence to which transcription apparatus binds
b. Determines the first nucleotide that is transcribed into RNA
c. Determines which DNA strand is template
d. Signals where transcription ends

The Substrate for Transcription

RNA is synthesized from **ribonucleoside triphosphates** (**rNTPs**), each consisting of a ribose sugar and a base (a nucleoside) attached to three phosphate groups (**Figure 13.7**). In RNA synthesis, nucleotides are added one at a time to the 3′-OH group of the growing RNA molecule. Two phosphate groups are cleaved from the incoming ribonucleoside triphosphate; the remaining phosphate group participates in a phosphodiester bond that connects the nucleotide to the growing RNA molecule. The overall chemical reaction for the addition of each nucleotide is

$$RNA_n + rNTP \rightarrow RNA_{n+1} + PP_i$$

where PP_i represents pyrophosphate. Nucleotides are always added to the 3′ end of the RNA molecule, and the direction

Triphosphate

13.7 Ribonucleoside triphosphates are substrates used in RNA synthesis.

of transcription is therefore 5′→3′ (**Figure 13.8**), the same as the direction of DNA synthesis during replication. Thus, the newly synthesized RNA is complementary and antiparallel to the template strand. Unlike DNA synthesis, RNA synthesis does not require a primer.

CONCEPTS

RNA is synthesized from ribonucleoside triphosphates. Transcription is 5′→3′: each new nucleotide is joined to the 3′-OH group of the last nucleotide added to the growing RNA molecule.

The Transcription Apparatus

As we have seen, DNA replication requires a number of different enzymes and proteins. Although transcription might initially appear to be quite different because a single enzyme—**RNA polymerase**—carries out all the required steps, the two processes on closer inspection, are actually similar. The action of RNA polymerase is enhanced by a number of accessory

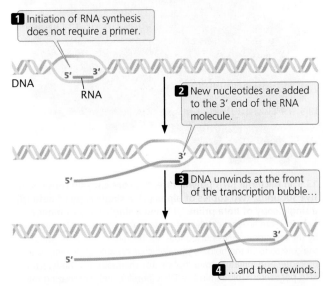

1 Initiation of RNA synthesis does not require a primer.

2 New nucleotides are added to the 3′ end of the RNA molecule.

3 DNA unwinds at the front of the transcription bubble...

4 ...and then rewinds.

13.8 In transcription, nucleotides are always added to the 3′ end of the RNA molecule.

proteins that join and leave the polymerase at different stages of the process. Each accessory protein is responsible for providing or regulating a special function. Thus, transcription, like replication, requires an array of proteins.

BACTERIAL RNA POLYMERASE Bacterial cells typically possess only one type of RNA polymerase, which catalyzes the synthesis of all classes of bacterial RNA: mRNA, tRNA, and rRNA. Bacterial RNA polymerase is a large multimeric enzyme (meaning that it consists of several polypeptide chains).

At the heart of most bacterial RNA polymerases are five subunits (individual polypeptide chains) that make up the **core enzyme**: two copies of a subunit called alpha (α) and single copies of subunits beta (β), beta prime (β'), and omega (ω) (**Figure 13.9**). The ω subunit is not essential for

transcription, but it helps stabilize the enzyme. The core enzyme catalyzes the elongation of the RNA molecule by the addition of RNA nucleotides. Other functional subunits join and leave the core enzyme at particular stages of the transcription process. The **sigma (σ) factor** controls the binding of RNA polymerase to the promoter. Without sigma, RNA polymerase initiates transcription at a random point along the DNA. After sigma has associated with the core enzyme (forming a **holoenzyme**), RNA polymerase binds stably only to the promoter and initiates transcription at the proper start site. Sigma is required only for promoter binding and initiation; after a few RNA nucleotides have been joined together, sigma usually detaches from the core enzyme. Many bacteria have multiple types of sigma factors; each type initiates the binding of RNA polymerase to a particular set of promoters.

Rifamycins are a group of antibiotics that kill bacterial cells by inhibiting RNA polymerase. These antibiotics are widely used to treat tuberculosis, a disease that kills almost 2 million people worldwide each year. The structures of bacterial and eukaryotic RNA polymerases are sufficiently different that rifamycins can inhibit bacterial RNA polymerases without interfering with eukaryotic RNA polymerases. Recent research has demonstrated that several rifamycins work by binding to the part of the bacterial RNA polymerase that clamps onto DNA and jamming it, thus preventing the RNA polymerase from interacting with the promoter on the DNA.

EUKARYOTIC RNA POLYMERASES Most eukaryotic cells possess three distinct types of RNA polymerase, each of which is responsible for transcribing a different class of RNA: **RNA polymerase I** transcribes rRNA; **RNA polymerase II** transcribes pre-mRNAs, snoRNAs, some miRNAs, and some snRNAs; and **RNA polymerase III** transcribes other small RNA molecules—specifically tRNAs, small rRNAs, some miRNAs, and some snRNAs (**Table 13.3**). RNA polymerases I, II, and III are found in all eukaryotes. Two additional RNA

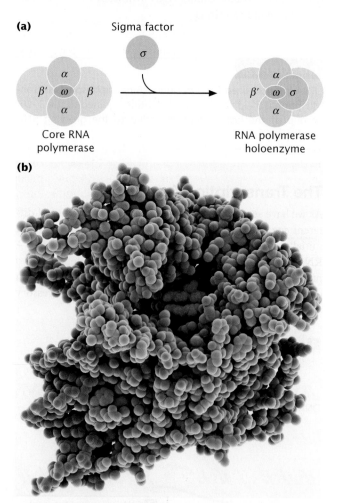

13.9 In bacterial RNA polymerase, the core enzyme consists of five subunits: two copies of alpha (α), a single copy of beta (β), a single copy of beta prime (β'), and a single copy of omega (ω). The core enzyme catalyzes the elongation of the RNA molecule by the addition of RNA nucleotides. (a) The sigma factor (σ) joins the core to form the holoenzyme, which is capable of binding to a promoter and initiating transcription. (b) The molecular model shows RNA polymerase (blue), binding DNA (purple), and synthesizing mRNA (red). [Part b: Laguna Design/Science Source.]

TABLE 13.3	Eukaryotic RNA polymerases	
Type	**Present in**	**Transcribes**
RNA polymerase I	All eukaryotes	Large rRNAs
RNA polymerase II	All eukaryotes	Pre-mRNA, some snRNAs, snoRNAs, some miRNAs
RNA polymerase III	All eukaryotes	tRNAs, small rRNAs, some snRNAs, some miRNAs
RNA polymerase IV	Plants	Some siRNAs
RNA polymerase V	Plants	RNA molecules taking part in heterochromatin formation

polymerases, **RNA polymerase IV** and **RNA polymerase V**, have been found in plants. RNA polymerases IV and V transcribe RNAs that play a role in DNA methylation and chromatin structure.

All eukaryotic polymerases are large multimeric enzymes, typically consisting of more than a dozen subunits. Some subunits are common to all RNA polymerases, whereas others are limited to one of the polymerases. As in bacterial cells, a number of accessory proteins bind to the core enzyme and affect its function.

> ### CONCEPTS
>
> Bacterial cells possess a single type of RNA polymerase, consisting of a core enzyme and other subunits that participate in various stages of transcription. Eukaryotic cells possess several distinct types of RNA polymerase that transcribe different kinds of RNA molecules.
>
> ### ✔ CONCEPT CHECK 4
> What is the function of the sigma factor?

13.3 Bacterial Transcription Consists of Initiation, Elongation, and Termination

Now that we've considered some of the major components of transcription, we're ready to take a detailed look at the process. Transcription can be conveniently divided into three stages:

1. Initiation, in which the transcription apparatus assembles on the promoter and begins the synthesis of RNA

2. Elongation, in which DNA is threaded through RNA polymerase and the polymerase unwinds the DNA and adds new nucleotides, one at a time, to the 3′ end of the growing RNA strand

3. Termination, the recognition of the end of the transcription unit and the separation of the RNA molecule from the DNA template

We first examine each of these steps in bacterial cells, in which the process is best understood; then we consider eukaryotic and archaeal transcription.

Initiation

Initiation comprises all the steps necessary to begin RNA synthesis, including (1) promoter recognition, (2) formation of a transcription bubble, (3) creation of the first bonds between rNTPs, and (4) escape of the transcription apparatus from the promoter.

Transcription initiation requires that the transcription apparatus recognize and bind to the promoter. At this step, the selectivity of transcription is enforced: the binding of

RNA polymerase to the promoter determines which parts of the DNA template are to be transcribed, and how often. Different genes are transcribed with different frequencies, and promoter binding is important in determining the frequency of transcription for a particular gene. Promoters also have different affinities for RNA polymerase. Even within a single promoter, affinity for RNA polymerase can vary with the passage of time, depending on the promoter's interaction with RNA polymerase and a number of other factors.

BACTERIAL PROMOTERS Essential information for the transcription apparatus—where it will start transcribing, which strand is to be read, and in what direction the RNA polymerase will move—is embedded in the nucleotide sequence of the promoter. In bacterial cells, promoters are usually adjacent to the RNA-coding region.

An examination of many promoters in *E. coli* and other bacteria reveals a general feature: although most of the nucleotides at most sites vary among these promoters, short stretches of nucleotides are common to many. Furthermore, the locations of these nucleotides relative to the transcription start site are similar in most promoters. These short stretches of common nucleotides are called consensus sequences. A **consensus sequence** is the set of the most commonly encountered nucleotides among sequences that possess considerable similarity, or *consensus* (**Figure 13.10**). The presence of consensus in a set of nucleotides usually implies that the sequence is associated with an important function.

▶ **TRY PROBLEM 21**

The most commonly encountered consensus sequence, found in almost all bacterial promoters, is centered about

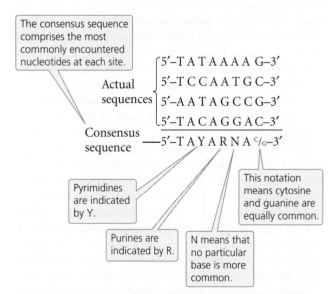

13.10 A consensus sequence consists of the most commonly encountered nucleotides at each site in a group of related sequences.

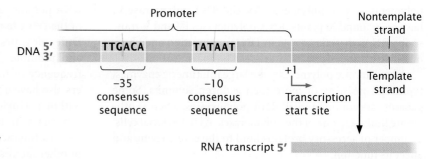

13.11 In bacterial promoters, consensus sequences are found upstream of the start site, approximately at positions −10 and −35.

10 bp upstream of the start site. Called the **−10 consensus sequence**, or sometimes the Pribnow box, this consensus sequence,

$$5'-TATAAT-3'$$
$$3'-ATATTA-5'$$

is often written simply as TATAAT (**Figure 13.11**). Remember that TATAAT is just the *consensus* sequence— representing the most commonly encountered nucleotides at each of these sites (see Figure 13.10). In most prokaryotic promoters, the actual sequence is not TATAAT.

Another consensus sequence common to most bacterial promoters is TTGACA, which lies approximately 35 nucleotides upstream of the start site and is termed the **−35 consensus sequence** (see Figure 13.11). The nucleotides on either side of the −10 and −35 consensus sequences and those between them vary greatly from promoter to promoter, suggesting that these nucleotides are not very important in promoter recognition.

The function of these consensus sequences in bacterial promoters has been studied by inducing mutations at various positions within the consensus sequences and observing the effect of the changes on transcription. These studies reveal that most base substitutions within the −10 and −35 consensus sequences reduce the rate of transcription; these substitutions are termed *down mutations* because they slow down the rate of transcription. Occasionally, a particular change in a consensus sequence increases the rate of transcription; such a change is called an *up mutation*.

The sigma factor, mentioned earlier, associates with the core RNA polymerase enzyme (**Figure 13.12a**) to form a holoenzyme, which binds to the −35 and −10 consensus sequences in the DNA promoter (**Figure 13.12b**). Although it binds only the nucleotides of the consensus sequences, the enzyme extends from −50 to +20 when bound to the promoter. The holoenzyme initially binds weakly to the promoter, but then undergoes a change in structure that allows it to bind more tightly and unwind the double-stranded DNA (**Figure 13.12c**). Unwinding begins within the −10 consensus sequence and extends downstream for about 14 nucleotides, including the start site (from nucleotides −12 to +2).

Some bacterial promoters contain a third consensus sequence that also takes part in the initiation of transcription. Called the **upstream element**, this sequence contains a number of A–T pairs and is found at about −40 to −60. A number of proteins may bind to sequences in and near the promoter; some stimulate the rate of transcription and others repress it. We will consider these proteins, which regulate gene expression, in Chapter 16. ⟩ **TRY PROBLEM 24**

CONCEPTS

A promoter is a DNA sequence that is adjacent to a gene and required for transcription. Promoters contain short consensus sequences that are important in the initiation of transcription.

✔ **CONCEPT CHECK 5**

What binds to the −10 consensus sequence found in most bacterial promoters?

a. The holoenzyme (core enzyme + sigma factor)
b. The sigma factor alone
c. The core enzyme alone
d. mRNA

INITIAL RNA SYNTHESIS Once the holoenzyme has bound to the promoter, RNA polymerase is positioned over the transcription start site (at position +1) and has unwound the DNA to produce a single-stranded template. The orientation and spacing of the consensus sequences on a DNA strand determine which strand will be the template for transcription and thereby determine the direction of transcription.

The position of the start site is determined not by the sequences located there, but by the locations of the consensus sequences, which position RNA polymerase so that the enzyme's active site is aligned for the initiation of transcription at +1. If the consensus sequences are artificially moved upstream or downstream, the location of the starting point of transcription correspondingly changes.

To begin the synthesis of an RNA molecule, RNA polymerase pairs the base at the start site on the DNA template strand with its complementary base on an rNTP (**Figure 13.12d**). No primer is required to initiate the synthesis of the 5′ end of the RNA molecule. Two of the three phosphate groups are cleaved from each rNTP as the nucleotide is added to the 3′ end of a growing RNA molecule. However, because the 5′ end of the first rNTP does not take

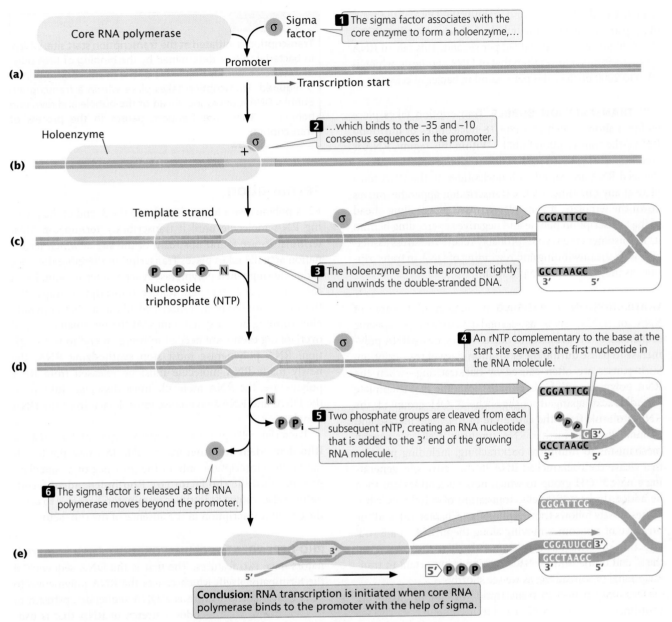

13.12 Transcription in bacteria is carried out by RNA polymerase, which must bind to the sigma factor to initiate transcription.

part in the formation of a phosphodiester bond, all three of its phosphate groups remain. An RNA molecule therefore possesses, at least initially, three phosphate groups at its 5′ end (**Figure 13.12e**).

Often, in the course of initiation, RNA polymerase repeatedly generates and releases short transcripts, from 2 to 6 nucleotides in length, while still bound to the promoter. This process, termed **abortive initiation**, occurs in both prokaryotes and eukaryotes. After several abortive initiation attempts, the polymerase synthesizes an RNA molecule from 9 to 12 nucleotides in length, which allows the polymerase to transition to the elongation stage.

Elongation

At the end of initiation, RNA polymerase undergoes a change in its conformation (shape) and thereafter is no longer able to bind to the consensus sequences in the promoter. This change allows the polymerase to escape from the promoter and begin transcribing downstream. The sigma factor is usually released after initiation, although some RNA polymerases may retain sigma throughout elongation.

As it moves downstream along the template, RNA polymerase progressively unwinds the DNA at the leading (downstream) edge of the transcription bubble, joining nucleotides to the growing RNA molecule according to the sequence of the template, and rewinds the DNA at the trailing (upstream) edge of the bubble. In bacterial cells at 37°C, about 40 nucleotides are added per second. This rate of RNA synthesis is much lower than that of DNA synthesis, which is 1000 to 2000 nucleotides per second in bacterial cells.

THE TRANSCRIPTION BUBBLE Transcription takes place within a short stretch of about 18 nucleotides of unwound DNA—the transcription bubble. Within this region, RNA is continuously synthesized. About 8 nucleotides of newly synthesized RNA are paired with nucleotides on the DNA template at any one time. As the transcription apparatus moves down the template, it generates positive supercoiling ahead of the transcription bubble and negative supercoiling behind it. Topoisomerase enzymes probably relieve the stress associated with the unwinding and rewinding of DNA in transcription, as they do in DNA replication.

TRANSCRIPTIONAL PAUSING A number of features of RNA or DNA, such as secondary structures, specific sequences, or the presence of nucleosomes, cause RNA polymerase to pause during the elongation stage of transcription. Such pauses are often caused by backtracking—when the RNA polymerase slides backward along the DNA template strand. Backtracking disengages the 3'-OH group of the RNA molecule from the active site of RNA polymerase and temporarily halts further RNA synthesis. Cells use several mechanisms to minimize backtracking, including proteins that cleave the backtracked RNA in the active site, generating a new 3'-OH group to which new nucleotides can then be added. In bacterial cells, translation of mRNA by ribosomes closely follows transcription (see Chapter 15), and the presence of ribosomes moving along the mRNA in a 5'→3' direction prevents backtracking of the RNA polymerase at the 3' end of the mRNA. Backtracking is important in transcriptional proofreading, as we see shortly.

Transitory pauses in transcription are important in the coordination of transcription and translation in bacteria (see the discussion of attenuation in Chapter 16), as well as in the coordination of RNA processing in eukaryotes. Pausing also affects the rate of RNA synthesis. Sometimes a pause may be stabilized by sequences in the DNA that ultimately lead to the termination of transcription.

ACCURACY OF TRANSCRIPTION Although RNA polymerase is quite accurate in incorporating nucleotides into the growing RNA chain, errors do occasionally arise. Research has demonstrated that RNA polymerase is capable of a type of proofreading in the course of transcription. When RNA polymerase incorporates a nucleotide that does not match the DNA template, it backtracks and cleaves the last two nucleotides (including the misincorporated nucleotide) from the growing RNA chain. RNA polymerase then proceeds forward, transcribing the DNA template again.

> ## CONCEPTS
>
> Transcription is initiated at the transcription start site, which, in bacterial cells, is determined by the binding of RNA polymerase to consensus sequences of the promoter. No primer is required. Transcription takes place within a transcription bubble. DNA is unwound ahead of the bubble and rewound behind it. There are frequent pauses in the process of transcription.

Termination

RNA polymerase adds nucleotides to the 3' end of the growing RNA molecule until it transcribes a terminator. Most terminators are found upstream of the site at which termination actually takes place. Transcription therefore does not suddenly stop when polymerase reaches a terminator, like a car stopping at a stop sign. Rather, transcription stops after the terminator has been transcribed, like a car that stops only after running over a speed bump. At the terminator, several overlapping events are needed to bring an end to transcription: RNA polymerase must stop synthesizing RNA, the newly made RNA molecule must be released from RNA polymerase, the RNA molecule must dissociate fully from the DNA, and RNA polymerase must detach from the DNA template.

Bacterial cells possess two major types of terminators. **Rho-dependent terminators** are able to cause the termination of transcription only in the presence of an ancillary protein called the **rho factor** (ρ). **Rho-independent terminators** (also known as intrinsic terminators) are able to cause the end of transcription in the absence of the rho factor.

RHO-DEPENDENT TERMINATORS Rho-dependent terminators have two features. The first is the DNA sequence of the terminator itself, which causes the RNA polymerase to pause. The second feature is a DNA sequence upstream of the terminator that encodes a stretch of RNA that is usually rich in cytosine nucleotides and devoid of any secondary structures. This RNA sequence, called the rho utilization (rut) site, serves as a binding site for the rho factor. Once rho binds to the RNA, it moves toward its 3' end, following the RNA polymerase (**Figure 13.13**). When RNA polymerase encounters the terminator, it pauses, allowing rho to catch up. The rho factor has helicase activity, which it uses to unwind the RNA–DNA hybrid in the transcription bubble, bringing transcription to an end.

RHO-INDEPENDENT TERMINATORS Rho-independent terminators, which make up about 50% of all terminators in prokaryotes, have two common features. First, they contain inverted repeats, which are sequences of nucleotides on the same strand that are inverted and complementary. When these inverted repeats are transcribed into RNA and bind to

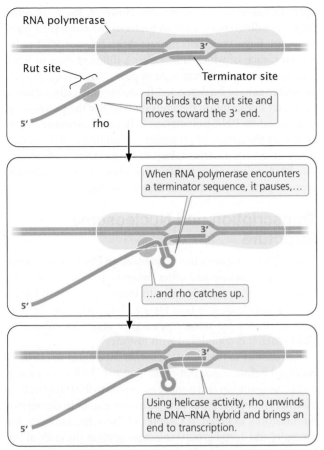

13.13 **The termination of transcription in some bacterial genes requires the presence of the rho factor.**

Within the figure (img_2):
- RNA polymerase
- Rut site
- 3′
- Terminator site
- 5′
- rho
- Rho binds to the rut site and moves toward the 3′ end.
- When RNA polymerase encounters a terminator sequence, it pauses,…
- 3′
- …and rho catches up.
- 5′
- 3′
- Using helicase activity, rho unwinds the DNA–RNA hybrid and brings an end to transcription.
- 5′

---

each other, a hairpin forms (**Figure 13.14**). Second, in rho-independent terminators, a string of seven to nine adenine nucleotides follows the inverted repeat in the template DNA. The transcription of these adenines produces a string of uracil nucleotides after the hairpin in the transcribed RNA.

The string of uracils in the RNA molecule causes the RNA polymerase to pause, allowing time for the hairpin structure to form. Evidence suggests that the formation of the hairpin destabilizes the DNA–RNA pairing, causing the RNA molecule to separate from its DNA template. Separation may be facilitated by the adenine–uracil base pairings, which are relatively weak compared with other types of base pairings. When the RNA transcript has separated from the template, RNA synthesis can no longer continue (see Figure 13.14).

> TRY PROBLEM 29

POLYCISTRONIC mRNA In bacteria, a group of genes is often transcribed into a single RNA molecule; such a molecule is termed **polycistronic mRNA**. Thus, polycistronic mRNA is produced when a single terminator is present at the end of a group of several genes that are transcribed together, instead of each gene having its own terminator.

Polycistronic mRNA does occur in some eukaryotes, such as *Caenorhabditis elegans*, but it is uncommon in eukaryotes.

You can view the process of transcription, including initiation, elongation, and termination, in **Animation 13.1**. The animation shows how the different parts of the transcription unit interact to bring about the complete synthesis of an RNA molecule.

THINK-PAIR-SHARE Question 3

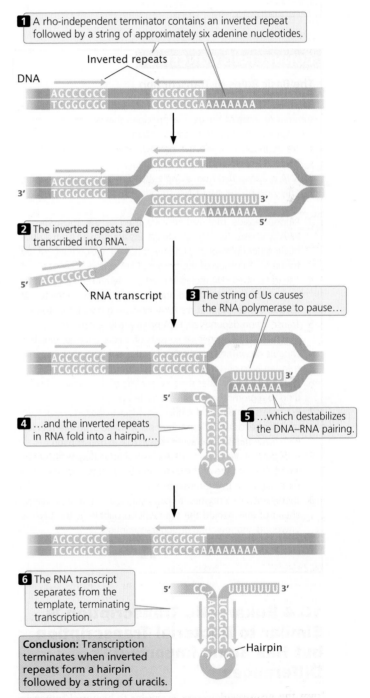

1 A rho-independent terminator contains an inverted repeat followed by a string of approximately six adenine nucleotides.

Inverted repeats

DNA

AGCCCGCC   GGCGGGCT
TCGGGCGG   CCGCCCGAAAAAAAA

2 The inverted repeats are transcribed into RNA.

RNA transcript

3 The string of Us causes the RNA polymerase to pause…

4 …and the inverted repeats in RNA fold into a hairpin,…

5 …which destabilizes the DNA–RNA pairing.

6 The RNA transcript separates from the template, terminating transcription.

Hairpin

Conclusion: Transcription terminates when inverted repeats form a hairpin followed by a string of uracils.

13.14 **Rho-independent termination in bacteria is a multistep process.**

CONNECTING CONCEPTS

The Basic Rules of Transcription

Before we examine the process of eukaryotic transcription, let's summarize some of the general principles that we've observed in our exploration of bacterial transcription.

1. Transcription is a selective process; only certain parts of the DNA are transcribed at any one time.

2. RNA is transcribed from a single strand of DNA. Within a gene, only one of the two DNA strands—the template strand—is usually copied into RNA.

3. Ribonucleoside triphosphates are used as the substrates in RNA synthesis. Two phosphate groups are cleaved from a ribonucleoside triphosphate, and the resulting nucleotide is joined to the 3'-OH group of the growing RNA strand.

4. The transcribed RNA molecule is antiparallel and complementary to the DNA template strand. Transcription is always in the 5'→3' direction, meaning that the RNA molecule grows at the 3' end.

5. Transcription depends on RNA polymerase, a large multimeric enzyme. RNA polymerase consists of a core enzyme, which is capable of synthesizing RNA, and other subunits that may join transiently to perform additional functions.

6. A sigma factor enables the core enzyme of RNA polymerase to bind to a promoter and initiate transcription.

7. Promoters contain short sequences crucial to the binding of RNA polymerase to DNA; these consensus sequences are interspersed with nucleotides that play no known role in transcription.

8. RNA polymerase binds to DNA at a promoter, begins transcribing at the start site of the gene, and ends transcription after a terminator has been transcribed.

9. Topoisomerase enzymes remove supercoiling that develops ahead of and behind the transcription bubble as the DNA is unwound and rewound during transcription.

THINK-PAIR-SHARE Question 4

13.4 Eukaryotic Transcription Is Similar to Bacterial Transcription but Has Some Important Differences

Transcription in eukaryotes is similar to bacterial transcription in that it includes initiation, elongation, and termination, and the basic principles of transcription already outlined apply to eukaryotic transcription. However, there are some important differences between bacterial and eukaryotic transcription. Eukaryotic cells possess three different RNA polymerases, each of which transcribes a different class of RNA and recognizes a different type of promoter. Thus, a generic promoter cannot be described for eukaryotic cells; rather, a promoter's description depends on whether the promoter is recognized by RNA polymerase I, II, or III. Another difference is in the nature of promoter recognition and initiation. Many accessory proteins take part in the binding of eukaryotic RNA polymerases to DNA templates, and the different types of promoters require different proteins.

Transcription and Nucleosome Structure

Transcription requires that sequences on DNA be accessible to RNA polymerase and other proteins. In eukaryotic cells, however, DNA is complexed with histone proteins in highly compressed chromatin (see Figure 11.4). How can the proteins necessary for transcription gain access to eukaryotic DNA when it is complexed with histones?

The answer to this question is that chromatin structure is modified before transcription so that the DNA is in a more open configuration and is more accessible to the transcription machinery. Several types of proteins have roles in chromatin modification, as noted in Chapter 11. Acetyltransferases, for example, add acetyl groups to amino acids at the ends of the histone proteins, which destabilize nucleosome structure and make the DNA more accessible. Other types of histone modification can also affect chromatin packing. In addition, chromatin-remodeling proteins may bind to the chromatin and displace nucleosomes from promoters and other regions important for transcription. We will take a closer look at the changes in chromatin structure associated with gene expression in Chapter 17.

Promoters

A significant difference between bacterial and eukaryotic transcription is the existence of three different eukaryotic RNA polymerases, which recognize different types of promoters. In bacterial cells, the holoenzyme (the RNA polymerase core enzyme plus the sigma factor) recognizes and binds directly to sequences in the promoter. In eukaryotic cells, promoter recognition is carried out by accessory proteins that bind to the promoter and then recruit a specific RNA polymerase (I, II, or III).

One class of accessory proteins comprises **general transcription factors**, which, along with RNA polymerase, form the **basal transcription apparatus**—a group of proteins that assembles near the transcription start site and is sufficient to initiate minimal levels of transcription. Another class of accessory proteins consists of **transcriptional activator proteins**, which bind to specific DNA sequences and bring about higher levels of transcription by stimulating the assembly of the basal transcription apparatus at the start site.

Here, we focus our attention on promoters recognized by RNA polymerase II, which transcribes the genes that encode proteins. A promoter for a gene transcribed by RNA polymerase II typically consists of two primary parts: the core promoter and the regulatory promoter.

CORE PROMOTER The **core promoter** is located immediately upstream of the gene (**Figure 13.15**) and is the site to which the basal transcription apparatus binds. The core promoter typically includes one or more consensus sequences. One of the most common of these sequences is the **TATA box**, which has the consensus sequence TATAAA and is located −25 to −30 bp upstream of the start site. Additional consensus sequences that may be found in the core promoters of genes transcribed by RNA polymerase II are shown in Figure 13.15. These consensus sequences are recognized by certain transcription factors, which bind to them and serve as a platform for the assembly of the basal transcription apparatus.

REGULATORY PROMOTER The **regulatory promoter** is located immediately upstream of the core promoter (see Figure 13.15). A variety of different consensus sequences can be found in regulatory promoters, and they can be mixed and matched in different combinations. Transcriptional activator proteins bind to these sequences and, either directly or indirectly, make contact with the basal transcription apparatus and affect the rate at which transcription is initiated. Transcriptional activator proteins may also regulate transcription by binding to more distant sequences called **enhancers**. The DNA between an enhancer and the promoter loops out so that transcriptional activator proteins bound to the enhancer can interact with the basal transcription apparatus at the core promoter. Enhancers will be discussed in more detail in Chapter 17.

POLYMERASE I AND III PROMOTERS RNA polymerase I and RNA polymerase III each recognize promoters that are distinct from those recognized by RNA polymerase II. For example, promoters for small rRNA and tRNA genes, transcribed by RNA polymerase III, contain **internal promoters** that are downstream of the start site and are transcribed into the RNA.

CONCEPTS

In eukaryotes, general transcription factors and RNA polymerase assemble on promoters to form the basal transcription apparatus, which binds to DNA near the transcription start site and is necessary for transcription to take place at minimal levels. Transcriptional activator proteins bind to other consensus sequences in promoters and enhancers and affect the rate of transcription.

✔ CONCEPT CHECK 7

What is the difference between the core promoter and the regulatory promoter?

a. Only the core promoter has consensus sequences.
b. The regulatory promoter is farther upstream of the gene.
c. General transcription factors bind to the core promoter; transcriptional activator proteins bind to the regulatory promoter.
d. Both b and c.

Initiation

Transcription in eukaryotes is initiated through the assembly of the transcription machinery on the promoter. This machinery consists of RNA polymerase II and a series of transcription factors that form a giant complex consisting of 50 or more polypeptides. Assembly of the transcription machinery begins when regulatory proteins bind DNA near the promoter and modify the chromatin structure so that transcription can take place. These and other regulatory proteins then recruit the basal transcription apparatus to the core promoter.

The basal transcription apparatus consists of RNA polymerase II, a series of general transcription factors, and a complex of proteins known as the mediator (**Figure 13.16**). The general transcription factors include TFIIA, TFIIB, TFIID, TFIIE, TFIIF, and TFIIH, in which TFII stands for

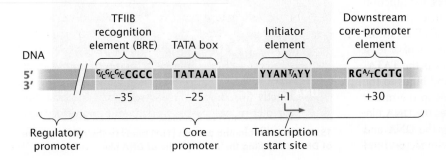

13.15 The promoters of genes transcribed by RNA polymerase II consist of two primary parts: a core promoter and a regulatory promoter. Both parts typically contain consensus sequences, but not all the consensus sequences shown here are found in all promoters.

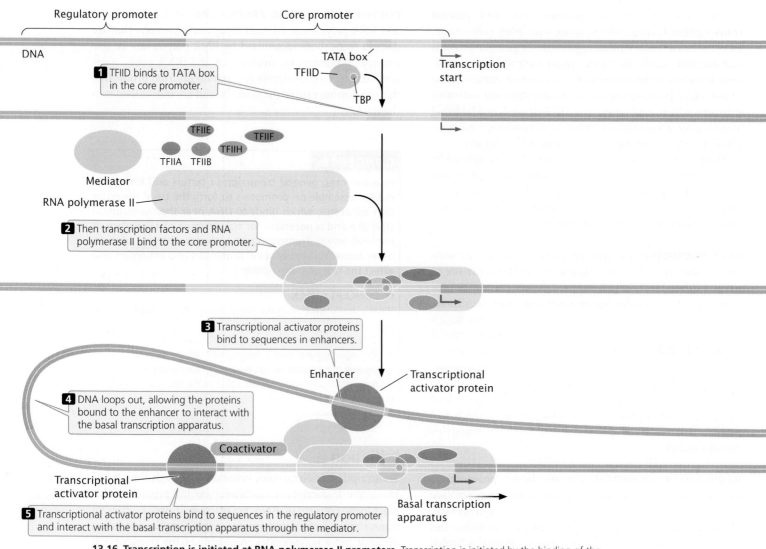

Regulatory promoter

Core promoter

DNA

1 TFIID binds to TATA box in the core promoter.

TATA box

TFIID

TBP

Transcription start

TFIIE

TFIIF

TFIIH

TFIIA   TFIIB

Mediator

RNA polymerase II

2 Then transcription factors and RNA polymerase II bind to the core promoter.

3 Transcriptional activator proteins bind to sequences in enhancers.

Enhancer

Transcriptional activator protein

4 DNA loops out, allowing the proteins bound to the enhancer to interact with the basal transcription apparatus.

Coactivator

Transcriptional activator protein

5 Transcriptional activator proteins bind to sequences in the regulatory promoter and interact with the basal transcription apparatus through the mediator.

Basal transcription apparatus

13.16 Transcription is initiated at RNA polymerase II promoters. Transcription is initiated by the binding of the TFIID transcription factor to the TATA box, followed by the binding of a preassembled holoenzyme containing general transcription factors, RNA polymerase II, and the mediator. TBP stands for TATA-binding protein.

transcription factor for RNA polymerase *II* and the final letter designates the individual factor.

RNA polymerase II and the general transcription factors assemble at the core promoter, forming a pre-initiation complex. Recall that in bacteria, the sigma factor recognizes and binds to the promoter sequence. In eukaryotes, the function of sigma is replaced by that of the general transcription factors. A first step in initiation is the binding of TFIID to the TATA box on the core promoter. TFIID consists of at least nine polypeptides. One of those polypeptides is the **TATA-binding protein** (**TBP**), which recognizes and binds to the TATA box. The TATA-binding protein binds to the minor groove of the DNA double helix and straddles the DNA like a molecular saddle (**Figure 13.17**), bending the DNA and partly unwinding it. Other general transcription factors bind

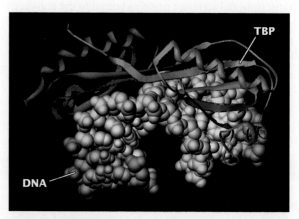

TBP

DNA

13.17 The TATA-binding protein (TBP) binds to the minor groove of DNA, straddling the double helix of DNA like a saddle.

to additional consensus sequences in the core promoter and to RNA polymerase II and position the polymerase over the transcription start site.

After the RNA polymerase and general transcription factors have assembled on the core promoter, conformational changes take place in both the DNA and the polymerase. These changes cause 11 to 15 bp of DNA surrounding the transcription start site to unwind, producing the single-stranded DNA that will serve as a template for transcription.

The DNA template strand is positioned within the active site of the RNA polymerase, creating a structure called the open complex. After the open complex has formed, the synthesis of RNA begins as phosphate groups are cleaved off rNTPs and nucleotides are joined together to form an RNA molecule. As in bacterial transcription, the RNA polymerase may generate and release several short RNA molecules in abortive initiation attempts before it initiates the synthesis of a full-length RNA molecule. ▶ TRY PROBLEM 34

CONCEPTS

Transcription in eukaryotes is initiated when the basal transcription apparatus, consisting of RNA polymerase, general transcription factors, and a mediator, assembles on the core promoter and becomes an open complex.

✔ **CONCEPT CHECK 8**

What is the role of TFIID in transcription initiation?

Elongation

After about 30 bp of RNA have been synthesized, the RNA polymerase leaves the promoter and begins the elongation stage of transcription. Many of the general transcription factors are left behind at the promoter, where they can serve to quickly reinitiate transcription with another RNA polymerase enzyme.

The molecular structure of eukaryotic RNA polymerase II and how it functions during elongation were revealed through the work of Roger Kornberg and his colleagues, for which Kornberg was awarded a Nobel Prize in chemistry in 2006. The RNA polymerase maintains a transcription bubble during elongation, in which about eight nucleotides of the growing RNA molecule remain paired with the DNA template strand. The DNA double helix enters a cleft in the polymerase and is gripped by jawlike extensions of the enzyme (**Figure 13.18**). The two strands of the DNA are unwound, and nucleotides that are complementary to the template strand are added to the growing 3' end of the RNA molecule. As it funnels through the polymerase, the DNA–RNA hybrid hits a wall of amino acids and bends at almost a right angle; this bend positions the end of the DNA–RNA hybrid at the active site of the polymerase, where new nucleotides

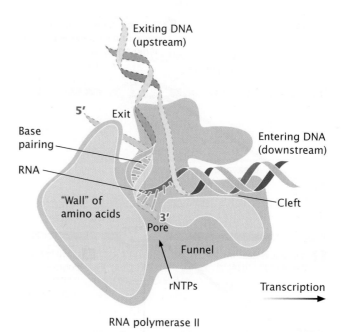

13.18 The structure of RNA polymerase II is a source of insight into its function. The DNA double helix enters RNA polymerase II through a cleft in the enzyme and unwinds. The DNA–RNA duplex is bent at a right angle, which positions the 3' end of the RNA at the active site of the enzyme. At the active site, new nucleotides are added to the 3' end of the growing RNA molecule.

are added to the 3' end of the growing RNA molecule. The newly synthesized RNA is separated from the DNA and runs through another cleft in the enzyme before exiting from the polymerase.

Termination

The three eukaryotic RNA polymerases use different mechanisms for termination. RNA polymerase I requires a termination factor similar to the rho factor used in terminating the transcription of some bacterial genes. Unlike rho, which binds to the newly transcribed RNA molecule, the termination factor for RNA polymerase I binds to a DNA sequence downstream of the terminator.

RNA polymerase III ends transcription after transcribing a terminator sequence that produces a string of uracil nucleotides in the RNA molecule. Recent research demonstrates that secondary structures, such as hairpins and stems, often occur upstream of the string of uracils and are necessary for termination. Thus, RNA polymerase III uses a termination mechanism similar to rho-independent termination in bacteria.

The termination of transcription by RNA polymerase II does not occur at specific sequences. Instead, RNA polymerase II often continues to synthesize RNA hundreds or even thousands of nucleotides past the coding sequence necessary to produce the mRNA. As we will see in Chapter 14, the pre-mRNA is cleaved at a specific site, designated by a

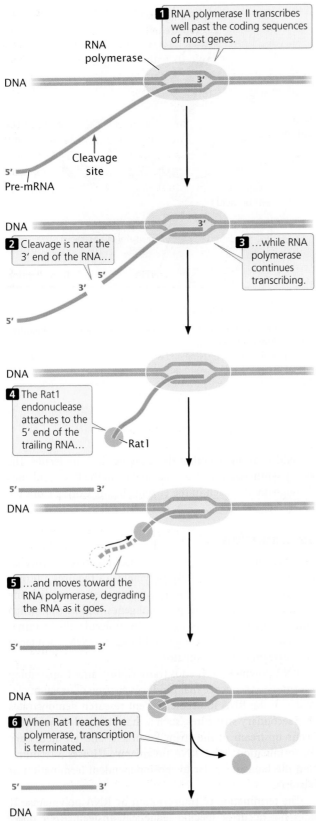

1 RNA polymerase II transcribes well past the coding sequences of most genes.

RNA polymerase

DNA

3'

Cleavage site

5'

Pre-mRNA

DNA

3'

2 Cleavage is near the 3' end of the RNA…

5'

3'

5'

3 …while RNA polymerase continues transcribing.

DNA

4 The Rat1 endonuclease attaches to the 5' end of the trailing RNA…

Rat1

5' ▬▬▬ 3'

DNA

5 …and moves toward the RNA polymerase, degrading the RNA as it goes.

5' ▬▬▬ 3'

DNA

6 When Rat1 reaches the polymerase, transcription is terminated.

5' ▬▬▬ 3'

DNA

13.19 Termination of transcription by RNA polymerase II requires the Rat1 exonuclease. Cleavage of the pre-mRNA produces a 5' end to which Rat1 attaches. Rat1 degrades the RNA molecule in the 5'→3' direction. When Rat1 reaches the polymerase, transcription is halted.

consensus sequence, while transcription is still taking place at the 3' end of the molecule. Cleavage cuts the pre-mRNA into two pieces: the mRNA that will eventually encode the protein and another piece of RNA that has its 5' end trailing out of the RNA polymerase (**Figure 13.19**). An enzyme (called Rat1 in yeast) attaches to the 5' end of this RNA and moves toward the 3' end, where the RNA polymerase continues the transcription of RNA. Rat1 is a 5'→3' exonuclease—an enzyme capable of degrading RNA in the 5'→3' direction. Like a guided torpedo, Rat1 homes in on the polymerase, chewing up the RNA as it moves. When Rat1 reaches the transcription machinery, transcription terminates. Note that this mechanism is similar to that of rho-dependent termination in bacteria (see Figure 13.13), except that rho does not degrade the RNA molecule.

THINK-PAIR-SHARE Question 5

CONCEPTS

The three RNA polymerases found in all eukaryotes use different mechanisms of termination. Transcription by RNA polymerase II is terminated when an exonuclease enzyme attaches to the cleaved 5' end of the RNA, moves down the RNA, and reaches the polymerase enzyme.

✔ CONCEPT CHECK 9

How are the processes of RNA polymerase II termination in eukaryotes and rho-dependent termination in bacteria similar, and how are they different?

13.5 Transcription in Archaea Is More Similar to Transcription in Eukaryotes Than to Transcription in Bacteria

Some 2 billion to 3 billion years ago, life diverged into three lines of evolutionary descent: the bacteria (also called eubacteria), the archaea, and the eukaryotes (see Chapter 2). Although bacteria and archaea are superficially similar—both are unicellular and lack a nucleus—the results of studies of their DNA sequences and other biochemical properties indicate that they are as distantly related to each other as they are to eukaryotes. The evolutionary distinctions between archaea, bacteria, and eukaryotes are clear. But did eukaryotes first diverge from an ancestral prokaryote, with a later separation of prokaryotes into bacteria and archaea, or did the archaea and the bacteria split first, with the eukaryotes later evolving from one of these groups?

Studies of transcription in bacteria, archaea, and eukaryotes have yielded important clues to the evolutionary relationships between these organisms. Archaea, like bacteria, have a single RNA polymerase, but this enzyme is most similar to the RNA polymerases of eukaryotes. As discussed earlier, bacterial RNA polymerase consists of 5 subunits, whereas eukaryotic RNA polymerases are much more complex; RNA polymerase II, for example, is composed of 12 subunits. Archaeal RNA polymerase is similarly complex, with 11 or more subunits. Furthermore, the amino acid sequence of archaeal RNA polymerase is similar to that of eukaryotic RNA polymerase II.

Archaeal promoters contain a consensus sequence similar to the TATA box found in eukaryotic promoters. The archaeal TATA box is found approximately 27 bp upstream of the transcription start site and helps to determine the location of the transcription start site, as it does in eukaryotes. Archaea possess a TATA-binding protein (TBP), which is a critical transcription factor for all three of the eukaryotic polymerases, but not for bacterial RNA polymerase. TBP binds the TATA box in archaea with the help of another transcription factor, TFIIB, which is also found in eukaryotes, but not in bacteria. However, some other regulators of transcription found in archaea are more similar to those found in bacteria, so transcription in archaea is not entirely eukaryotic in nature. As prokaryotes, archaea lack a nuclear membrane, but many species do produce histone proteins, which help compact the DNA and form nucleosome-like structures.

Thus, transcription, one of the most basic of life processes, has strong similarities in eukaryotes and archaea, suggesting that these two groups are more closely related to each other than either is to bacteria. This conclusion is supported by other data, including those obtained from a comparison of gene sequences.

> **CONCEPTS**
>
> The process of transcription in archaea has many similarities to transcription in eukaryotes.

CONCEPTS SUMMARY

- Early life used RNA both as the carrier of genetic information and as a biological catalyst.

- RNA is a polymer consisting of nucleotides joined together by phosphodiester bonds. Each RNA nucleotide consists of a ribose sugar, a phosphate group, and a nitrogenous base. In contrast to DNA, RNA contains the base uracil and is usually single stranded, which allows it to form secondary structures.

- Cells possess a number of different classes of RNA. Ribosomal RNA is a component of the ribosome, messenger RNA carries coding instructions for proteins, and transfer RNA helps incorporate amino acids into a polypeptide chain.

- The template for RNA synthesis is single-stranded DNA. In transcription, RNA synthesis is complementary and antiparallel to the DNA template strand. A transcription unit consists of a promoter, an RNA-coding region, and a terminator.

- The substrates for RNA synthesis are ribonucleoside triphosphates.

- RNA polymerase in bacterial cells consists of a core enzyme, which catalyzes the addition of nucleotides to an RNA molecule, and other subunits. The sigma factor controls the binding of the core enzyme to the promoter.

- Eukaryotic cells contain multiple types of RNA polymerases.

- The process of transcription consists of three stages: initiation, elongation, and termination.

- Transcription begins at the start site, which is determined by consensus sequences in a promoter. A short stretch of DNA is unwound near the start site, RNA is synthesized from a single template strand of DNA, and the DNA is rewound at the lagging end of the transcription bubble. RNA polymerases are capable of proofreading.

- RNA synthesis ceases after a terminator sequence has been transcribed. Bacterial cells have two types of terminators: rho-independent terminators and rho-dependent terminators.

- The initiation of transcription in eukaryotes requires the modification of chromatin structure. Different types of RNA polymerases in eukaryotes recognize different types of promoters.

- For genes transcribed by RNA polymerase II, general transcription factors, which bind to the core promoter, are part of the basal transcription apparatus. Transcriptional activator proteins bind to sequences in regulatory promoters and enhancers and interact with the basal transcription apparatus.

- The three RNA polymerases found in all eukaryotes use different mechanisms of termination.

- Transcription in archaea has many similarities to transcription in eukaryotes.

IMPORTANT TERMS

ribozyme (p. 374)
ribosomal RNA
 (rRNA) (p. 375)
messenger RNA
 (mRNA) (p. 375)
pre-messenger RNA
 (pre-mRNA) (p. 375)
transfer RNA
 (tRNA) (p. 376)
small nuclear RNA
 (snRNA) (p. 376)
small nuclear
 ribonucleoprotein
 (snRNP) (p. 376)
small nucleolar RNA
 (snoRNA) (p. 376)
microRNA
 (miRNA) (p. 376)

small interfering RNA
 (siRNA) (p. 376)
Piwi-interacting RNA
 (piRNA) (p. 376)
CRISPR RNA
 (crRNA) (p. 376)
long noncoding RNA
 (lncRNA) (p. 376)
template strand (p. 378)
nontemplate
 strand (p. 378)
transcription unit (p. 378)
promoter (p. 378)
RNA-coding region (p. 378)
terminator (p. 378)
ribonucleoside triphosphate
 (rNTP) (p. 379)
RNA polymerase (p. 379)

core enzyme (p. 380)
sigma (σ) factor (p. 380)
holoenzyme (p. 380)
RNA polymerase I (p. 380)
RNA polymerase II (p. 380)
RNA polymerase III (p. 380)
RNA polymerase IV (p. 381)
RNA polymerase V (p. 381)
consensus sequence (p. 381)
−10 consensus sequence
 (Pribnow box) (p. 382)
−35 consensus
 sequence (p. 382)
upstream element (p. 382)
abortive initiation (p. 383)
rho-dependent
 terminator (p. 384)
rho factor (ρ) (p. 384)

rho-independent
 terminator (p. 384)
polycistronic
 mRNA (p. 385)
general transcription
 factor (p. 387)
basal transcription
 apparatus (p. 387)
transcriptional activator
 protein (p. 387)
core promoter (p. 387)
TATA box (p. 387)
regulatory
 promoter (p. 387)
enhancer (p. 387)
internal promoter (p. 387)
TATA-binding protein
 (TBP) (p. 388)

ANSWERS TO CONCEPT CHECKS

1. b

2. The template strand is the DNA strand that is copied into an RNA molecule, whereas the nontemplate strand is not copied.

3. d

4. The sigma factor controls the binding of RNA polymerase to the promoter.

5. a

6. Inverted repeats followed by a string of adenine nucleotides.

7. d

8. TFIID binds to the TATA box and helps to center the RNA polymerase over the transcription start site.

9. Both processes use a protein that binds to the RNA molecule and moves down the RNA toward the RNA polymerase. They differ in that rho does not degrade the RNA, whereas Rat1 does so.

WORKED PROBLEMS

Problem 1

The accompanying diagram represents a sequence of nucleotides surrounding an RNA-coding sequence.

5′–CATGTT ... TTGATGT– [RNA-coding sequence] –GACGA ... TTTATA ... GGCGCGC–3′
3′–GTACAA ... AACTACA– [RNA-coding sequence] –CTGCT ... AAATAT ... CCGCGCG–5′

a. Is the RNA-coding sequence likely to be from a bacterial cell or from a eukaryotic cell? How can you tell?

b. Which DNA strand will serve as the template strand during the transcription of the RNA-coding sequence?

›› Solution Strategy

What information is required in your answer to the problem?

a. Whether the sequence is likely to be from a bacterial or eukaryotic cell, and why.

b. Which strand is the template strand.

What information is provided to solve the problem?

■ The nucleotide sequences of both strands of DNA.

■ The 5′ and 3′ ends of the strands.

For help with this problem, review:
The Template in Section 13.2, Bacterial Promoters in Section 13.3, and Promoters in Section 13.4.

›› Solution Steps

a. Bacterial and eukaryotic cells use the same DNA bases (A, T, G, and C), so the bases themselves provide no clue to the origin of the sequence. The RNA-coding sequence must be accompanied by a promoter, and bacterial and eukaryotic cells do differ in the consensus sequences found in their promoters; so we should examine the sequence for the presence of familiar consensus sequences. On the bottom strand to the right of the RNA-coding sequence, we find AAATAT, which, written in the conventional manner (5′ on the left), is 5′–TATAAA–3′. This sequence is the TATA box, which is found in most eukaryotic promoters. However, the sequence is also quite similar to the −10 consensus sequence (5′–TATAAT–3′) found in bacterial promoters.

Farther to the right on the bottom strand, we also see 5′–GCGCGCC–3′, which is the TFIIB recognition

> **Hint:** Review the consensus sequences found in bacterial and eukaryotic promoters in Figures 13.11 and 13.15.

element (BRE, see **Figure 13.15**) in eukaryotic RNA polymerase II promoters. No similar consensus sequence is found in bacterial promoters, so we can be fairly certain that this sequence is a eukaryotic promoter and an RNA-coding sequence.

b. The TATA box and BRE of RNA polymerase II promoters are upstream of the RNA-coding sequence, so RNA polymerase must bind to these sequences and then proceed downstream, transcribing the RNA-coding sequence. Thus, RNA polymerase must proceed from right (upstream) to left (downstream). The RNA molecule is always synthesized in the 5′→3′ direction and is antiparallel to the DNA template strand, so the template strand must be read 3′→5′. If the enzyme proceeds from right to left and reads the template in the 3′→5′ direction, the upper strand must be the template, as shown in the accompanying diagram.

> **Recall:** During transcription, the template strand is read 3′→5′.

Template strand	RNA-coding sequence	RNA polymerase
5′–CATGTT ... TTGATGT–		–GACGA ... TTTATA ... GGCGCGC–3′
3′–GTACAA ... AACTACA–		–CTGCT ... AAATAT ... CCGCGCG–5′

←————— **Direction of transcription**

Problem 2

Suppose that a consensus sequence in the regulatory promoter of a eukaryotic gene that encodes enzyme A were deleted. Which of the following effects would result from this deletion? Explain your reasoning.

a. Enzyme A would have a different amino acid sequence.

b. The mRNA for enzyme A would be abnormally short.

c. Enzyme A would be missing some amino acids.

d. The mRNA for enzyme A would be transcribed but not translated.

e. The amount of mRNA transcribed would be affected.

›› Solution Strategy

What information is required in your answer to the problem?
The result (*a*, *b*, *c*, *d*, or *e*) that would occur if a consensus sequence in the regulatory promoter were deleted.

What information is provided to solve the problem?
The deleted consensus sequence is in the regulatory promoter.

For help with this problem, review:
Section 13.4.

›› Solution Steps

The correct answer is part *e*. The regulatory promoter contains binding sites for transcriptional activator proteins. These sequences are not part of the RNA-coding sequence for enzyme A, so the mutation would have no effect on the length or the amino acid sequence of the enzyme, eliminating answers *a*, *b*, and *c*. The binding of transcriptional activator proteins to the regulatory promoter affects the amount of transcription that takes place through interactions with the basal transcription apparatus at the core promoter.

Section 13.1

1. Draw an RNA nucleotide and a DNA nucleotide, highlighting the differences. How is the structure of RNA similar to that of DNA? How is it different?

2. What are the major classes of cellular RNA?

3. Why is DNA more stable than RNA?

Section 13.2

4. What parts of DNA make up a transcription unit? Draw a typical bacterial transcription unit and identify its parts.

5. What is the substrate for RNA synthesis? How is this substrate modified and joined together to produce an RNA molecule?

6. Describe the structure of the bacterial RNA polymerase holoenzyme.

7. Give the names of the RNA polymerases found in eukaryotic cells and the types of RNA that they transcribe.

Section 13.3

8. What are the three basic stages of transcription? Describe what happens at each stage.

9. Draw a typical bacterial promoter and identify any common consensus sequences.

10. What are the two basic types of terminators found in bacterial cells? Describe the structure of each type.

Section 13.4

11. Compare the roles of general transcription factors and transcriptional activator proteins.

12. How are transcription and replication similar, and how are they different?

13. How is transcription different in bacteria and eukaryotes? How is it similar?

Section 13.1

*14. An RNA molecule has the following percentages of bases: A = 23%, U = 42%, C = 21%, and G = 14%.

   a. Is this RNA single stranded or double stranded? How can you tell?

   b. What would be the percentages of bases in the template strand of the DNA that contains the gene for this RNA?

Section 13.2

*15. The following diagram represents DNA that is part of the RNA-coding sequence of a transcription unit. The bottom strand is the template strand. Give the sequence found on the RNA molecule transcribed from this DNA and identify the 5′ and 3′ ends of the RNA.

   5′–ATAGGCGATGCCA–3′
   3′–TATCCGCTACGGT–5′ ← Template strand

16. For the RNA molecule shown in **Figure 13.1a**, write out the sequence of bases on the template and nontemplate strands of DNA from which this RNA is transcribed. Label the 5′ and 3′ ends of each strand.

17. The following sequence of nucleotides is found in a single-stranded DNA template:

   ATTGCCAGATCATCCCAATAGAT

   Assume that RNA polymerase proceeds along this template from left to right.

   a. Which end of the DNA template is 5′ and which end is 3′?

   b. Give the sequence and identify the 5′ and 3′ ends of the RNA transcribed from this template.

18. RNA polymerases carry out transcription much more slowly than DNA polymerases carry out replication. Why is speed more important in replication than in transcription?

19. Assume that a mutation occurs in the gene that encodes each of the following RNA polymerases. Match the mutation with its possible effects by placing the correct letter or letters in the blanks below. There may be more than one effect for each mutated polymerase.

A mutation in the gene that codes for	Effects
RNA polymerase I	_____
RNA polymerase II	_____
RNA polymerase III	_____

Possible effects

   a. tRNA is not synthesized

   b. Some ribosomal RNA is not synthesized

   c. Ribosomal RNA is not processed

   d. pre-mRNA is not processed

   e. Some mRNA molecules are not degraded

   f. pre-mRNA is not synthesized

Section 13.3

20. Provide the consensus sequence for the *first three* actual sequences shown in **Figure 13.10**.

***21.** Write the consensus sequence for the following set of nucleotide sequences.

AGGAGTT

AGCTATT

TGCAATA

ACGAAAA

TCCTAAT

TGCAATT

22. List at least five properties that DNA polymerases and RNA polymerases have in common. List at least three differences.

23. Most RNA molecules have *three* phosphate groups at the 5′ end, but DNA molecules never do. Explain this difference.

***24.** Write a hypothetical sequence of bases that might be found in the first 20 nucleotides of a promoter of a bacterial gene. Include both strands of DNA and identify the 5′ and 3′ ends of both strands. Be sure to include the transcription start site and any consensus sequences found in the promoter.

***25.** What would be the most likely effect of a mutation at the following locations in an *E. coli* gene?

a. −8   **c.** −20

b. −35   **d.** Start site

26. A strain of bacteria possesses a temperature-sensitive mutation in the gene that encodes the sigma factor. The mutant bacteria produce a sigma factor that is unable to bind to RNA polymerase at elevated temperatures. What effect will this mutation have on the process of transcription when the bacteria are raised at elevated temperatures?

27. On **Figure 13.5**, indicate the locations of the promoters and terminators for genes *a*, *b*, and *c*.

28. The following diagram represents a transcription unit on a DNA molecule.

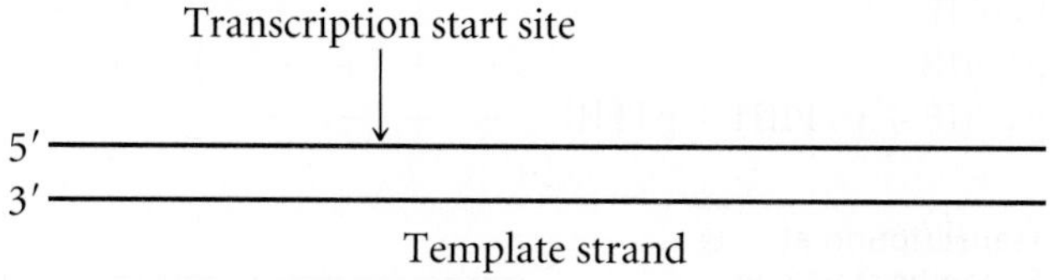

Transcription start site

5′ ————————————————

3′ ————————————————

Template strand

a. Assume that this DNA molecule is from a bacterial cell. Draw the approximate locations of the promoter and terminator for this transcription unit.

b. Assume that this DNA molecule is from a eukaryotic cell. Draw the approximate location of an RNA polymerase II promoter.

***29.** The following DNA nucleotides are found near the end of a bacterial transcription unit.

3′–AGCATACAGCAGACCGTTGGTCTGAAAAAAGCATACA–5′

a. Mark the point at which transcription will terminate.

b. Is this terminator rho independent or rho dependent?

c. Draw a diagram of the RNA that will be transcribed from this DNA, including its nucleotide sequence and any secondary structures that form.

30. A strain of bacteria possesses a temperature-sensitive mutation in the gene that encodes the rho subunit. At high temperatures, rho is not functional. When these bacteria are raised at elevated temperatures, which of the following effects would you expect to see? Explain your reasoning for accepting or rejecting each of these five options.

a. Transcription does not take place.

b. All RNA molecules are shorter than normal.

c. All RNA molecules are longer than normal.

d. Some RNA molecules are longer than normal.

e. RNA is copied from both DNA strands.

31. Suppose that the string of A nucleotides following the inverted repeat in a rho-independent terminator was deleted, but that the inverted repeat was left intact. How would this deletion affect termination? What would happen when RNA polymerase reached this region?

Section 13.4

32. The following diagram represents a transcription unit in a hypothetical DNA molecule.

5′ … TTGACA … TATAAT … 3′

3′ … AACTGT … ATATTA … 5′

a. On the basis of the information given, is this DNA from a bacterium or from a eukaryotic organism?

b. If this DNA molecule is transcribed, which strand will be the template strand and which will be the nontemplate strand?

c. Where, approximately, will the transcription start site be?

33. Computer programmers, working with molecular geneticists, have developed programs that can identify genes within long stretches of DNA sequence. Imagine that you are working with a programmer on such a project. On the basis of what you know about the process of transcription, what sequences should the program use to identify the beginning and end of a gene?

***34.** Through genetic engineering, a geneticist mutates the gene that encodes TBP in cultured human cells.

This mutation destroys the ability of TBP to bind to the TATA box. Predict the effect of this mutation on cells that possess it.

35. Elaborate repair mechanisms that prevent permanent mutations in DNA are associated with replication, yet no similar repair process is associated with transcription. Can you think of a reason for this difference between replication and transcription? (Hint: Think about the relative effects of a permanent mutation in a DNA molecule and one in an RNA molecule.)

CHALLENGE QUESTIONS

Section 13.3

36. Many genes in both bacteria and eukaryotes contain numerous sequences that can cause pauses in or premature termination of transcription. Nevertheless, the transcription of these genes within a cell normally produces multiple RNA molecules thousands of nucleotides long without pausing or premature termination. However, when a single round of transcription of these genes takes place in a test tube, RNA synthesis is frequently interrupted by pauses and premature terminations, which reduce the rate at which transcription takes place and frequently shorten the lengths of the mRNA molecules produced. Most pauses and premature terminations occur when RNA polymerase temporarily backtracks (i.e., backs up) for one or two nucleotides along the DNA. Experimental findings have demonstrated that most pauses and premature terminations disappear if several RNA polymerases are simultaneously transcribing the DNA molecule. Propose an explanation for this observation of faster transcription and longer mRNAs when the template DNA is being transcribed by multiple RNA polymerases.

Section 13.4

37. Enhancers are sequences that affect the initiation of the transcription of genes that are hundreds or thousands of nucleotides away. Transcriptional activator proteins that bind to enhancers usually interact directly with transcription factors at promoters by causing the intervening DNA to loop out. An enhancer of bacteriophage T4 does not function by looping of the DNA (D. R. Herendeen et al. 1992. *Science* 256:1298–1303). Propose some mechanisms other than DNA looping by which this enhancer might affect transcription at a gene thousands of nucleotides away.

*38. The locations of the TATA box in two species of yeast, *Saccharomyces pombe* and *Saccharomyces cerevisiae*, differ dramatically. The TATA box of *S. pombe* is about 30 nucleotides upstream of the transcription start site, similar to the location in most other eukaryotic cells. However, the TATA box of *S. cerevisiae* is 40 to 120 nucleotides upstream of the start site.

To better understand what sets the start site in these organisms, researchers at Stanford University conducted a series of experiments to determine which components of the transcription apparatus of these two species could be interchanged (Y. Li et al. 1994. *Science* 263:805–807). In these experiments, different general transcription factors and RNA polymerases were switched in *S. pombe* and *S. cerevisiae*, and the effects of each switch on the level of RNA synthesis and on the starting point of transcription were observed. The results from one set of experiments are shown in the table below. Components cTFIIB, cTFIIE, cTFIIF, cTFIIH are general transcription factors from *S. cerevisiae*. Components pTFIIB, pTFIIE, pTFIIF, pTFIIH are general transcription factors from *S. pombe*. Components cPol II and pPol II are RNA polymerase II from *S. cerevisiae* and *S. pombe*, respectively. The table indicates whether the component was present (+) or missing (−) in each experiment. In the accompanying gel, the presence of a band indicates that RNA was produced, and the position of the band indicates whether it was the length predicted when transcription begins 30 bp downstream of the TATA box or when it begins 40 to 120 bp downstream of the TATA box.

	Experiment						
Components	1	2	3	4	5	6	7
cTFIIE	+	−	+	+	+	+	−
cTFIIH + cTFIIF	+	−	+	+	+	−	+
cTFIIB + cPol II	+	−	−	−	−	−	−
pPol II	−	+	+	+	−	+	+
pTFIIB	−	+	+	−	+	+	+
pTFIIE + pTFIIH + pTFIIF	−	+	−	−	−	−	−

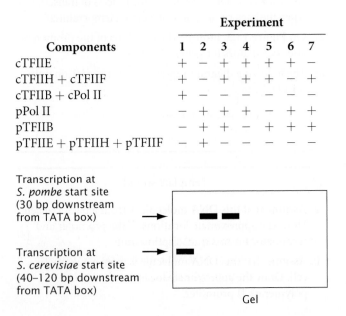

Transcription at *S. pombe* start site (30 bp downstream from TATA box) →

Transcription at *S. cerevisiae* start site (40–120 bp downstream from TATA box) →

Gel

a. What conclusion can you draw from these data about what components determine the start site for transcription?

b. What conclusions can you draw about the interactions of the different components of the transcription apparatus?

c. Propose a mechanism that might have caused the transcription start site in *S. pombe* to be about 30 bp downstream of the TATA box, while the start site for transcription in *S. cerevisiae* is 40 to 120 bp downstream of the TATA box.

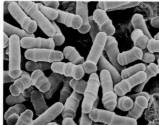

S. pombe. [Steve Gschmeissner/ Science Source.]

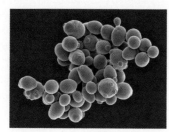

S. cerevisiae. [Steve Gschmeissner/ Science Source.]

39. Glenn Croston and his colleagues studied the relation between chromatin structure and transcription activity. In one set of experiments, they measured the level of in vitro transcription of a *Drosophila* gene by RNA polymerase II in the presence of DNA and various combinations of histone proteins (G. E. Croston et al. 1991. *Science* 251:643–649).

First, they measured the level of transcription of naked DNA, with no associated histone proteins. Then they measured the level of transcription after nucleosome octamers (without H1) were added to the DNA. The addition of the octamers caused the level of transcription to drop by 50%. When both nucleosome octamers and H1 proteins were added to the DNA, transcription was greatly repressed, dropping to less than 1% of that obtained with naked DNA, as shown in the table below.

GAL4-VP16 is a protein that binds to the DNA of certain eukaryotic genes. When GAL4-VP16 is added to DNA, the level of transcription by RNA polymerase II is greatly elevated.

Treatment	Relative amount of transcription
Naked DNA	100
DNA + octamers	50
DNA + octamers + H1	<1
DNA + GAL4-VP16	1000
DNA + octamers + GAL4-VP16	1000
DNA + octamers + H1 + GAL4-VP16	1000

Even in the presence of the H1 protein, GAL4-VP16 stimulates high levels of transcription.

Propose a mechanism by which the H1 protein represses transcription and by which GAL4-VP16 overcomes this repression. Explain how your proposed mechanism would produce the results obtained in these experiments.

THINK-PAIR-SHARE QUESTIONS

Introduction

1. RNA polymerase I is insensitive to α-amanitin, RNA polymerase II is highly sensitive, and RNA polymerase III is moderately sensitive. These three RNA polymerases have different functions (see **Table 13.3**) How might geneticists use α-amanitin as a tool in research? Propose a specific question that could be addressed using α-amanitin.

Section 13.2

2. Usually, within a gene, only one of the DNA strands is transcribed into RNA. Why haven't both strands in a DNA molecule evolved to carry genetic information? In other words, why might we not expect both strands to carry genetic information and be transcribed into RNA?

Section 13.3

3. The following diagram represents one of the Christmas-tree-like structures shown in **Figure 13.3**. On the diagram, identify parts *a* through *i*.

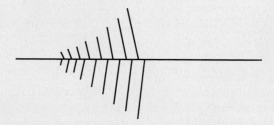

a. DNA molecule

b. 5′ and 3′ ends of the template strand of DNA

c. At least one RNA molecule

d. 5′ and 3′ ends of at least one RNA molecule

e. Direction of movement of the transcription apparatus on the DNA molecule

f. Approximate location of the promoter

g. Possible location of a terminator

h. Upstream and downstream directions

i. Molecules of RNA polymerase (use dots to represent these molecules)

4. Make a table listing similarities and differences between transcription and replication. List as many as you can think of.

Section 13.4

5. List some of the ways in which eukaryotic transcription is more complex than transcription in bacteria. Propose some possible reasons for the greater complexity of transcription in eukaryotic cells.

RNA Molecules and RNA Processing

Family of Tsar Nicholas Romanov II of Russia. The boy in front is the tsar's son Alexei, who suffered from hemophilia. [Mondadori Portfolio/Getty Images.]

A Royal Disease

On August 12, 1904, Tsar Nicholas Romanov II of Russia wrote in his diary: "A great never-to-be forgotten day when the mercy of God has visited us so clearly." That day, Alexei, Nicholas's first son and heir to the Russian throne, had been born.

At birth, Alexei was a large and vigorous baby with yellow curls and blue eyes, but at 6 weeks of age, he began spontaneously hemorrhaging from the navel. The bleeding persisted for several days and caused great alarm. As he grew and began to walk, Alexei often stumbled and fell, as all children do. Even his small scrapes bled profusely, and minor bruises led to significant internal bleeding. It soon became clear that Alexei had hemophilia.

Marked by slow clotting and excessive bleeding, hemophilia is caused by a mutation in one of several genes that encode proteins involved in the process of blood clotting. In people with hemophilia, minor injuries can result in life-threatening blood loss, and spontaneous bleeding into joints erodes the bone, with crippling consequences.

Alexei suffered from classic hemophilia, which is an X-linked genetic disorder. He inherited the hemophilia mutation from his mother, Alexandra, who was a carrier. The mutation appears to have originated with Queen Victoria of England (see Figure 6.8). In all, ten of Queen Victoria's male descendants suffered from hemophilia. Six female descendants, including her granddaughter Alexandra (Alexei's mother), were carriers.

During his childhood, Alexei experienced a number of severe bleeding episodes. The royal physicians were often helpless during these crises—they had no treatment that would stop the bleeding. At this time, the Russian Revolution broke out. Bolsheviks captured the tsar and his family and held them captive in the city of Yekaterinburg. On the night of July 16, 1918, a firing squad executed the royal family and their attendants, including Alexei and his four sisters. For many years, the bodies of the tsar's family were lost, but their skeletons were eventually recovered from two graves outside Yekaterinburg. Comparisons of mitochondrial DNA and nuclear DNA from the bones and from descendants of relatives of the family verified that the bones were indeed those of the royal family. Although this analysis determined the identity of the remains, the molecular nature of the royal hemophilia long remained unknown.

In 2009, geneticists analyzed DNA from the bones of the tsar's family to determine the genetic nature of the royal hemophilia. In people with X-linked hemophilia, a mutation commonly occurs in one of two genes on the X chromosome, either the gene for blood coagulation factor VIII or the gene for factor IX.

In eukaryotic cells, genes are often interrupted by noncoding sequences. Those parts of a gene that encode the amino acids of a protein are called exons, while the noncoding sequences are called introns. Initially, both exons and introns are transcribed into pre-mRNA, but, in a process called RNA splicing, the introns are later cut out and the exons pasted together. Examination of the sequences of the two clotting-factor genes in Alexandra's DNA revealed no mutations in the nucleotides that encode amino acids, but a mutation was present in one of the introns of her gene for factor IX. This mutation altered the splicing of the exons, creating a new stop signal in the coding sequence of the mRNA and producing a truncated, defective clotting factor IX. Both normal and mutant alleles were detected in Alexandra's DNA, as is expected of a heterozygous carrier. DNA from Alexei carried only the mutant allele, which caused his hemophilia.

THINK-PAIR-SHARE

- Classic hemophilia is inherited as an X-linked recessive condition. When a female is heterozygous for an X-linked recessive mutation such as hemophilia, what proportion of her male offspring are expected to have hemophilia? Examine the pedigree of the royal family in Figure 6.8. Calculate the proportion of affected male offspring born to females heterozygous for hemophilia. Carry out a chi-square goodness-of-fit test comparing the numbers of observed and expected male progeny of these females affected by hemophilia. What conclusion can you draw from the chi-square test?

- Hemophilia in the royal family was caused by the presence of a mutation in the gene for factor IX that altered the splicing of the exons and created a new stop signal in the mRNA for factor IX. Explain how a splicing mutation could produce a premature stop signal.

As illustrated by Alexei's hemophilia, RNA processing is critically important for the proper synthesis of proteins. In eukaryotic cells, RNA molecules are often extensively modified after transcription: for genes that encode proteins, a modified nucleotide called the 5′ cap is added to the 5′ end, a tail of adenine nucleotides is added to the 3′ end, and introns are cut out of the middle. In both prokaryotes and eukaryotes, rRNAs and tRNAs are also modified after transcription.

In Chapter 13, we focused on transcription—the process by which RNA molecules are synthesized. In this chapter, we examine the function and processing of RNA. We begin by taking a careful look at the nature of the gene. We then examine messenger RNA (mRNA), its structure, and how it is modified in eukaryotes after transcription. Then we turn to transfer RNA (tRNA), the adapter molecule that forms the interface between amino acids and mRNA in protein synthesis. We examine ribosomal RNA (rRNA), the structure and organization of rRNA genes, and how rRNAs are processed. Finally, we consider a newly discovered group of very small RNAs that play important roles in numerous biological functions.

As we explore the world of RNA and its role in gene function, we will see evidence of two important characteristics of this nucleic acid. First, RNA is extremely versatile, both structurally and biochemically. It can assume a number of different secondary structures, which provide the basis for its functional diversity. Second, RNA processing and function frequently include interactions between two or more RNA molecules.

14.1 Many Genes Have Complex Structures

What is a gene? As noted in Chapter 3, the definition of a gene often changes as we explore different aspects of heredity. A gene was defined in Chapter 3 as an inherited factor that determines a characteristic. This definition may have seemed vague because it says only what a gene does, not what a gene is. Nevertheless, this definition was appropriate at the time because our focus was on how genes influence the inheritance of traits. We did not have to consider the physical nature of the gene in learning the rules of inheritance.

Knowing something about the chemical structure of DNA and the process of transcription now enables us to be more precise about what a gene is. Chapter 10 described how genetic information is encoded in the base sequence of DNA: a gene consists of a set of DNA nucleotides. But how many nucleotides constitute a gene, and how is the information in those nucleotides organized? In 1902, Archibald Garrod correctly suggested that genes encode proteins. Proteins are made of amino acids, so a gene contains nucleotides that specify amino acids. Therefore, for many years, the working definition of a gene was a set of nucleotides that specifies the amino acid sequence of a protein. As geneticists learned more about the structure of genes, however, it became clear that this concept of a gene was an oversimplification.

Gene Organization

Early work on gene structure was carried out largely through the examination of mutations in bacteria and viruses. This research led Francis Crick to propose in 1958 that genes and proteins are **colinear**—that there is a direct correspondence between the nucleotide sequence of DNA and the amino acid sequence of a protein (**Figure 14.1**). The concept of colinearity suggests that the number of nucleotides in a gene should be proportional to the number of amino acids in the protein encoded by that gene. In a general sense, this concept is true for genes found in

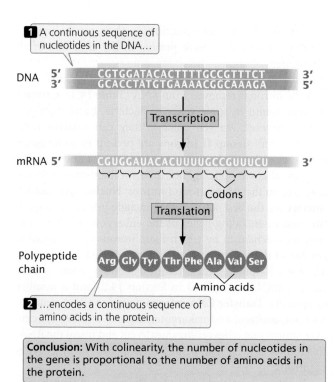

1 A continuous sequence of nucleotides in the DNA...

DNA 5′   CGTGGATACACTTTTGCCGTTTCT   3′
    3′   GCACCTATGTGAAAACGGCAAAGA   5′

Transcription

mRNA 5′   CGUGGAUACACUUUUGCCGUUUCU   3′

Codons

Translation

Polypeptide chain   Arg Gly Tyr Thr Phe Ala Val Ser

Amino acids

2 ...encodes a continuous sequence of amino acids in the protein.

Conclusion: With colinearity, the number of nucleotides in the gene is proportional to the number of amino acids in the protein.

14.1 The concept of colinearity suggests that a continuous sequence of nucleotides in DNA encodes a continuous sequence of amino acids in a protein.

bacterial cells and in many viruses, although these genes are slightly longer than would be expected if colinearity were strictly applied because the mRNAs encoded by the genes contain sequences at their ends that do not specify amino acids. At first, eukaryotic genes and proteins were also assumed to be colinear, but there were hints that eukaryotic gene structure is fundamentally different. Eukaryotic cells were found to contain far more DNA than is required to encode proteins (see Chapter 11). Furthermore, many large RNA molecules observed in the nucleus were absent from the cytoplasm, suggesting that nuclear RNAs undergo some type of change before they are exported to the cytoplasm.

Most geneticists were nevertheless surprised by the announcement in the 1970s that not all genes are continuous. Researchers observed four coding sequences in a gene from a eukaryote-infecting virus that were interrupted by nucleotides that did not specify amino acids. This discovery was made when the viral DNA was hybridized with the mRNA transcribed from it and the hybridized structure was examined using an electron microscope (**Figure 14.2**). The DNA was clearly much longer than the mRNA because regions of DNA looped out from the hybridized molecules. These regions of DNA contained nucleotide sequences that were absent from the coding nucleotides in the mRNA. Many other examples of interrupted genes were subsequently discovered; it quickly became apparent that most eukaryotic genes consist of stretches of coding and noncoding nucleotides.

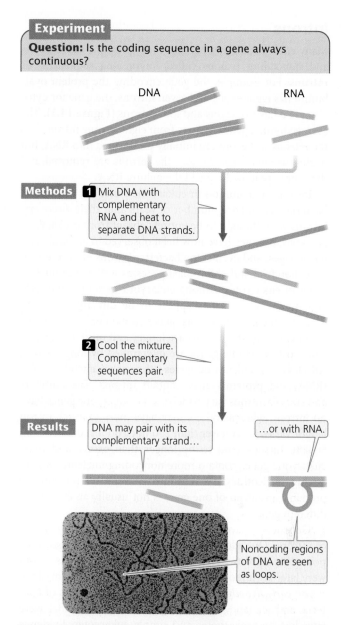

Experiment

Question: Is the coding sequence in a gene always continuous?

DNA   RNA

Methods **1** Mix DNA with complementary RNA and heat to separate DNA strands.

2 Cool the mixture. Complementary sequences pair.

Results   DNA may pair with its complementary strand...   ...or with RNA.

Noncoding regions of DNA are seen as loops.

Conclusion: Coding sequences in a gene may be interrupted by noncoding sequences.

14.2 The noncolinearity of eukaryotic genes was discovered by hybridizing DNA with the mRNA transcribed from it.
[From Susan M. Berget, Claire Moore, and Phillip A. Sharp, *Proc. Nati. Acad. Sci. USA* Vol. 74, No. 8, pp. 3171–3175, Fig. 4g August 1977 Biochemistry.]

CONCEPTS

When a continuous sequence of nucleotides in DNA encodes a continuous sequence of amino acids in a protein, the two are said to be colinear. In eukaryotes, not all genes are colinear with the proteins that they encode.

✔ CONCEPT CHECK 1

What evidence indicated that eukaryotic genes are not colinear with the proteins they encode?

Introns

Many eukaryotic genes contain coding regions called **exons** and noncoding regions called intervening sequences, or **introns**. For example, the gene encoding the protein ovalbumin has eight exons and seven introns; the gene for cytochrome *b* has five exons and four introns (**Figure 14.3**). The average human gene contains from eight to nine introns. All the introns and exons are initially transcribed into RNA, but during or after transcription, the introns are removed and the exons are joined to yield the mature RNA.

Introns are common in eukaryotic genes but are rare in bacterial genes. For a number of years after their discovery, introns were thought to be entirely absent from prokaryotic genomes, but they have now been observed in archaea, bacteriophages, and even some bacteria. Introns are present in mitochondrial and chloroplast genes as well as the nuclear genes of eukaryotes. Among eukaryotic genomes, the sizes and numbers of introns appear to be directly related to organismal complexity: yeast genes contain only a few short introns, *Drosophila* introns are longer and more numerous, and most vertebrate genes are interrupted by long introns. All classes of eukaryotic genes—those that encode rRNA, tRNA, and proteins—may contain introns. The numbers and sizes of introns vary widely: some eukaryotic genes have no introns, whereas others may have more than 60; intron length varies from fewer than 200 nucleotides to more than 50,000. Introns tend to be longer than exons, and most eukaryotic genes contain more noncoding nucleotides than coding nucleotides. Finally, most introns do not encode proteins: an intron of one gene is not usually an exon for a different gene.

Geneticists have long debated the evolutionary origin of introns. One idea, called the *intron late hypothesis*, proposes that introns were absent from ancient organisms but were later acquired by eukaryotes. Another idea, termed the *intron early hypothesis*, suggests that early ancestors of bacteria, archaea, and eukaryotes possessed introns that were later lost by prokaryotes and simple eukaryotes. Evidence suggests that introns have been lost *and* gained over evolutionary time. Many researchers now assume that the earliest eukaryotes possessed introns because divergent eukaryotes have introns in the same positions in their genes, suggesting that these introns were present in the ancestors of all eukaryotes.

There are four major types of introns, differentiated by how the intron is removed from RNA (**Table 14.1**). **Group I introns**, found in some genes of bacteria, bacteriophages, and eukaryotes, are self-splicing: they can catalyze their own removal. **Group II introns** are present in some genes of mitochondria, chloroplasts, archaea, and a few bacteria; they are also self-splicing, but their mechanism of splicing differs from that of the group I introns. **Nuclear pre-mRNA introns** are the best studied; they include introns located in the protein-encoding genes of the eukaryotic nucleus. The splicing mechanism by which these introns are removed is similar to that of the group II introns, but nuclear introns are not self-splicing; their removal requires small nuclear RNAs (snRNAs, discussed in Section 14.2) and a number of proteins. **Transfer RNA introns**, found in tRNA genes of bacteria, archaea, and eukaryotes, use yet another splicing mechanism that relies on enzymes to cut and reseal the RNA. In addition to these major types, there are several other types of introns.

We'll take a detailed look at the chemistry and mechanics of RNA splicing in Section 14.2. For now, we should keep

TABLE 14.1	Major types of introns	
Type of Intron	Location	Splicing Mechanism
Group I	Genes of bacteria, bacteriophages, and eukaryotes	Self-splicing
Group II	Genes of bacteria, archaea, and eukaryotic organelles	Self-splicing
Nuclear pre-mRNA	Protein-encoding genes in the nucleus of eukaryotes	Spliceosomal
tRNA	tRNA genes of bacteria, archaea, and eukaryotes	Enzymatic

Note: There are also several types of minor introns, including group III introns, twintrons, and archaeal introns.

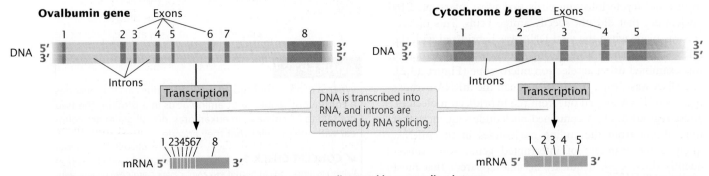

14.3 The coding sequences of many eukaryotic genes are disrupted by noncoding introns.

in mind two general characteristics of the splicing process: (1) the splicing of all pre-mRNA introns takes place in the nucleus; and (2) the order of exons in DNA is usually maintained in the spliced RNA; that is, the coding sequences of a gene may be split up, but they are not usually jumbled up.

> TRY PROBLEM 21

THINK-PAIR-SHARE Question 1

CONCEPTS

Many eukaryotic genes contain exons and introns. Both are transcribed into RNA, but introns are later removed by RNA processing. The numbers and sizes of introns vary from gene to gene. Introns are common in eukaryotic genes but uncommon in bacterial genes.

✔ **CONCEPT CHECK 2**

What are the four major types of introns?

The Concept of the Gene Revisited

How does the presence of introns affect our concept of a gene? To define a gene as a sequence of nucleotides that encodes amino acids in a protein no longer seems appropriate because this definition excludes introns, which do not specify amino acids. This definition also excludes the nucleotides that encode the 5′ and 3′ ends of an mRNA molecule, which, as we will see, are required for translation but do not encode amino acids. And defining a gene in these terms also excludes sequences that specify rRNA, tRNA, and other RNAs that do not encode proteins. Given our current understanding of DNA structure and function, we need a more precise definition of a gene.

Many geneticists have broadened the concept of a gene to include all sequences in DNA that are transcribed into a single RNA molecule. Defined this way, a gene includes all exons, introns, and those sequences at the beginning and end of the RNA that are not translated into protein. This definition also includes DNA sequences that encode rRNAs, tRNAs, and other types of non-messenger RNA. Some geneticists have expanded the definition of a gene even further to include the entire transcription unit—the promoter, the RNA-coding region, and the terminator. However, new evidence now calls into question even this definition. Recent research suggests that much of the genome is transcribed into RNA, although it is unclear what, if anything, much of this RNA does. What is certain is that the process of transcription is more complex than formerly thought, and defining a gene as a sequence that is transcribed into an RNA molecule is not as straightforward as formerly thought. The more we learn about the nature of genetic information, the more elusive the definition of a gene seems to become.

THINK-PAIR-SHARE Question 2

CONCEPTS

The discovery of introns forced a reevaluation of the definition of a gene. Today, a gene is often defined as a DNA sequence that encodes an RNA molecule or the entire DNA sequence required to transcribe and encode an RNA molecule.

14.2 Messenger RNAs, which Encode Proteins, Are Modified after Transcription in Eukaryotes

As soon as DNA was identified as the source of genetic information, it became clear that DNA cannot directly encode proteins. In eukaryotic cells, DNA resides in the nucleus, yet protein synthesis takes place in the cytoplasm. Geneticists recognized that an additional molecule must take part in transferring genetic information.

The results of studies of bacteriophage infection conducted in the late 1950s and early 1960s pointed to RNA as a likely candidate for this transport function. Bacteriophages inject their DNA into bacterial cells, where that DNA is replicated (see Chapter 9) and large amounts of phage protein are produced on the bacterial ribosomes. As early as 1953, Alfred Hershey discovered a type of RNA that was synthesized rapidly after bacteriophage infection. Findings from later studies showed that this short-lived RNA had a nucleotide composition similar to that of the phage DNA but quite different from that of the bacterial RNA. These observations were consistent with the idea that RNA was copied from DNA and that this RNA then directed the synthesis of proteins.

At the time, ribosomes were known to be *somehow* involved in protein synthesis, and much of the RNA in a cell was known to be in the form of ribosomes. Were ribosomes the agents by which genetic information was moved to the cytoplasm for the production of protein? Using isotopes of nitrogen and carbon and equilibrium density gradient centrifugation (see Figure 12.2), Sydney Brenner, François Jacob, and Matthew Meselson demonstrated in 1961 that they were not. They showed that new ribosomes are *not* produced during the burst of protein synthesis that accompanies bacteriophage infection (**Figure 14.4**). Thus, ribosomes could not be the carriers of the genetic information needed to produce new phage proteins.

In a related experiment, François Gros and his colleagues infected *E. coli* cells with bacteriophages while adding radioactively labeled uracil, which would become incorporated into newly produced phage RNA, to the medium in which they grew. Gros and his colleagues found that the newly produced phage RNA was short-lived, lasting only a few minutes, and was associated with ribosomes, but was distinct from them. They concluded that this short-lived RNA carries the genetic information for proteins to the ribosome, where translation occurs. The term *messenger RNA* was coined for this carrier.

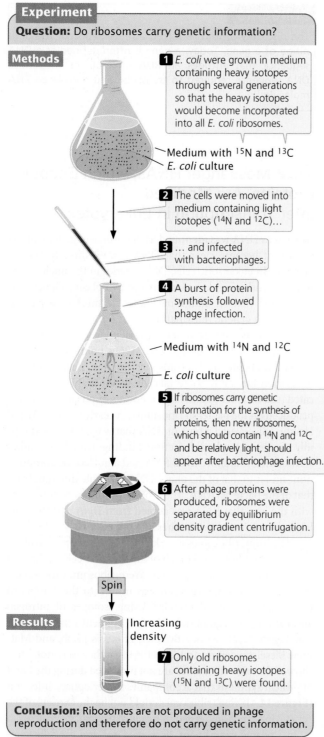

Question: Do ribosomes carry genetic information?

Methods

1 *E. coli* were grown in medium containing heavy isotopes through several generations so that the heavy isotopes would become incorporated into all *E. coli* ribosomes.

Medium with ^{15}N and ^{13}C
E. coli culture

2 The cells were moved into medium containing light isotopes (^{14}N and ^{12}C)...

3 ... and infected with bacteriophages.

4 A burst of protein synthesis followed phage infection.

Medium with ^{14}N and ^{12}C

E. coli culture

5 If ribosomes carry genetic information for the synthesis of proteins, then new ribosomes, which should contain ^{14}N and ^{12}C and be relatively light, should appear after bacteriophage infection.

6 After phage proteins were produced, ribosomes were separated by equilibrium density gradient centrifugation.

Spin

Results

Increasing density

7 Only old ribosomes containing heavy isotopes (^{15}N and ^{13}C) were found.

Conclusion: Ribosomes are not produced in phage reproduction and therefore do not carry genetic information.

14.4 Brenner, Jacob, and Meselson demonstrated that ribosomes do not carry genetic information.

The Structure of Messenger RNA

Messenger RNA functions as the template for protein synthesis. It carries genetic information from DNA to a ribosome, where it helps to assemble amino acids in their correct order. In bacteria, mRNA is transcribed directly from DNA, but in

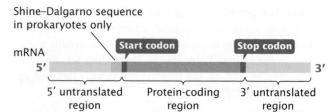

Shine–Dalgarno sequence
in prokaryotes only

mRNA

5' untranslated region | Protein-coding region | 3' untranslated region

14.5 Three primary regions of mature mRNA are the 5' untranslated region, the protein-coding region, and the 3' untranslated region.

eukaryotes, a pre-mRNA (also called the *primary transcript*) is first transcribed from DNA and then processed to yield the mature mRNA. We will reserve the term *mRNA* for RNA molecules that have been completely processed and are ready to undergo translation.

In an mRNA molecule, each amino acid of a protein is specified by a set of three nucleotides, called a **codon**. Both prokaryotic and eukaryotic mRNAs contain three primary regions (**Figure 14.5**). The first, the **5′ untranslated region** (5′ UTR; sometimes called the *leader*) is a sequence of nucleotides at the 5′ end of the mRNA that does not encode any of the amino acids of a protein. In bacterial mRNA, this region contains a consensus sequence (UAAGGAGGU) called the **Shine–Dalgarno sequence**, which serves as the ribosome-binding site during translation (see Chapter 15); it is found approximately seven nucleotides upstream of the first codon translated into an amino acid (called the start codon). The Shine–Dalgarno sequence is complementary to a sequence found in one of the RNA molecules that make up the ribosome, and it pairs with that sequence during translation. Eukaryotic mRNA has no equivalent RNA-binding consensus sequence in its 5′ untranslated region. In eukaryotic cells, ribosomes bind to a modified 5′ end of mRNA, as discussed later in this section.

The second primary region of mRNA is the **protein-coding region**, which comprises the codons that specify the amino acid sequence of the protein. The protein-coding region begins with a start codon and ends with a stop codon.

The third region of mRNA is the **3′ untranslated region** (3′ UTR; sometimes called the *trailer*), a sequence of nucleotides at the 3′ end of the mRNA that is not translated into amino acids. The 3′ UTR affects the stability of mRNA and helps regulate the translation of the mRNA protein-coding sequence. View **Animation 14.1** to see how mutations in different regions of a gene affect the flow of information from genotype to phenotype.

CONCEPTS

Messenger RNA molecules contain three primary regions: a 5′ untranslated region, a protein-coding region, and a 3′ untranslated region. The 5′ and 3′ untranslated regions do not encode any amino acids of a protein, but contain information that is important for RNA stability and the regulation of translation.

Pre-mRNA Processing

In bacterial cells, transcription and translation take place simultaneously: while the 3′ end of an mRNA is undergoing transcription, ribosomes attach to the Shine–Dalgarno sequence near the 5′ end and begin translation. Because transcription and translation are coupled, bacterial mRNA has little opportunity to be modified before protein synthesis. In contrast, transcription and translation are separated in both time and space in eukaryotic cells. Transcription takes place in the nucleus, whereas translation takes place in the cytoplasm; this separation provides an opportunity for eukaryotic RNA to be modified before it is translated. Indeed, eukaryotic mRNA is extensively altered after transcription. Changes are made to the 5′ end, the 3′ end, and the protein-coding region of the RNA molecule (**Table 14.2**).

ADDITION OF THE 5′ CAP One type of modification of eukaryotic pre-mRNA is the addition of a structure called a **5′ cap**. The cap is formed by the addition of an extra guanine nucleotide to the 5′ end of the mRNA and the addition of methyl groups (CH_3) to the base in that guanine and to the 2′-OH group of the sugar of one or more nucleotides at the 5′ end (**Figure 14.6**). The addition of the cap takes place rapidly after the initiation of transcription. The cap functions in the initiation of translation, as we'll see Chapter 15. Cap-binding proteins recognize the cap and attach to it; a ribosome then binds to these proteins and moves downstream along the mRNA until the start codon is reached and translation begins. The presence of a 5′ cap also increases the stability of mRNA and influences the removal of introns.

As noted in the discussion of transcription in Chapter 13, three phosphate groups are present at the 5′ end of all RNA molecules because phosphate groups are not cleaved from the first ribonucleoside triphosphate in the transcription reaction. The 5′ end of pre-mRNA can be represented as 5′–pppNpNpN ... , in which the letter N represents a ribonucleotide and p represents a phosphate. Shortly after the initiation of transcription, one of the three 5′ phosphate

groups is removed and a guanine nucleotide is added (see Figure 14.6). This guanine nucleotide is attached to the pre-mRNA by a unique 5′–5′ bond, which is quite different from the usual 5′–3′ phosphodiester bond that joins all the other nucleotides in RNA. One or more methyl groups are then added to the 5′ end; the first of these methyl groups is added to position 7 of the base of the terminal guanine nucleotide, making the base 7-methylguanine. Next, a methyl group may be added to the 2′ position of the sugar in the second and

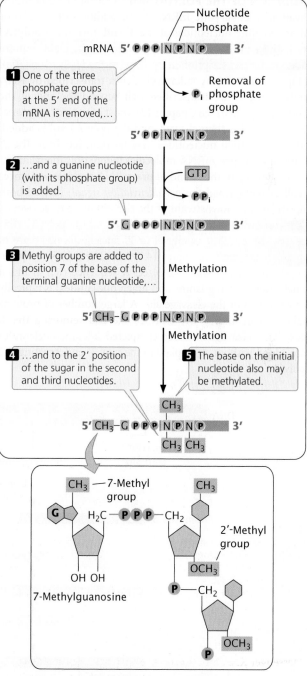

14.6 Most eukaryotic mRNAs have a 5′ cap. The cap consists of 7-methylguanosine attached to the pre-mRNA by a unique 5′–5′ bond (shown in detail in the bottom box).

TABLE 14.2	Posttranscriptional modifications to eukaryotic pre-mRNA
Modification	**Function**
Addition of 5′ cap	Facilitates binding of ribosome to 5′ end of mRNA, increases mRNA stability, enhances RNA splicing
3′ cleavage and addition of poly(A) tail	Increases stability of mRNA, facilitates binding of ribosome to mRNA
RNA splicing	Removes noncoding introns from pre-mRNA, facilitates export of mRNA to cytoplasm, allows for multiple proteins to be produced through alternative splicing
RNA editing	Alters nucleotide sequence of mRNA

third nucleotides (see Figure 14.6). Rarely, additional methyl groups may be attached to the bases of the second and third nucleotides of the pre-mRNA.

Several different enzymes take part in the addition of the 5′ cap. The initial step is carried out by an enzyme that associates with RNA polymerase II. Because neither RNA polymerase I nor RNA polymerase III have this associated enzyme, RNA molecules transcribed by these polymerases (rRNAs, tRNAs, and some snRNAs) are not capped.

ADDITION OF THE POLY(A) TAIL A second type of modification to eukaryotic mRNA is the addition of 50–250 or more adenine nucleotides at the 3′ end, forming a **poly(A) tail**. These nucleotides are not encoded in the DNA, but are added after transcription in a process termed *polyadenylation* (**Figure 14.7**). Many eukaryotic genes transcribed by RNA polymerase II are transcribed well beyond the end of the coding sequence (see Chapter 13); most of the extra material at the 3′ end is then cleaved, and the poly(A) tail is added. More than 1000 nucleotides may be removed from the 3′ ends of some pre-mRNA molecules before polyadenylation.

Processing of the 3′ end of pre-mRNA requires sequences (collectively termed the *polyadenylation signal*) located both upstream and downstream of the site where cleavage occurs. The consensus sequence AAUAAA (called the poly(A) consensus sequence) is usually 11 to 30 nucleotides upstream of the cleavage site (see Figure 14.7) and determines the point at which cleavage will take place. A sequence rich in uracil nucleotides (or in guanine and uracil nucleotides) is typically downstream of the cleavage site. A large number of proteins take part in finding the cleavage site and removing the 3′ end. After cleavage has been completed, adenine nucleotides are added without a template to the new 3′ end, creating the poly(A) tail.

The amount of protein produced by a cell depends not only on the amount of mRNA transcribed, but also on the stability of the mRNA that is transcribed. Thus, one way in which eukaryotic cells regulate the expression of genes is by controlling degradation of mRNA. The poly(A) tail and the 5′ cap are important in this process, conferring stability on those mRNAs that have them by protecting the mRNA from degradation by 5′ and 3′ exonucleases, thus increasing the time during which the mRNA remains intact and available for translation before it is degraded by these enzymes. The stability conferred by the poly(A) tail depends on proteins that attach to the tail and on its length. The poly(A) tail also facilitates attachment of the ribosome to the mRNA and plays a role in export of the mRNA to the cytoplasm.

Poly(U) tails are added to the 3′ ends of some mRNAs, microRNAs, and snRNAs. Although the function of poly(U) tails is still under investigation, evidence suggests that poly(U) tails on some mRNAs may facilitate their degradation. **▶ TRY PROBLEM 31**

CONCEPTS

Eukaryotic pre-mRNAs are processed at their 5′ and 3′ ends. A cap, consisting of a modified guanine nucleotide and several methyl groups, is added to the 5′ end. The 5′ cap facilitates the binding of a ribosome, increases the stability of the mRNA, and may affect the removal of introns. Processing at the 3′ end includes cleavage downstream of an AAUAAA consensus sequence and the addition of a poly(A) tail. The poly(A) tail confers stability on the mRNA, plays a role in its export to the cytoplasm, and facilitates attachment of the ribosome.

✔ CONCEPT CHECK 3

Why is it that pre-mRNAs are capped, but tRNAs and rRNAs are not?

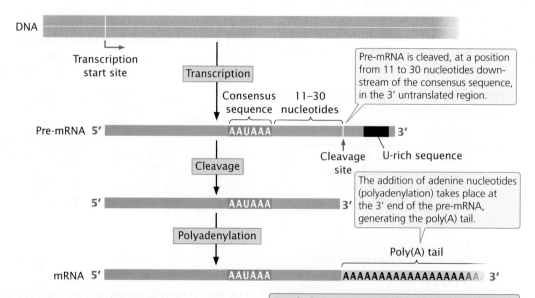

14.7 Most eukaryotic mRNAs have a 3′ poly(A) tail.

RNA Splicing

The other major type of modification of eukaryotic pre-mRNA is the removal of introns by **RNA splicing**. This modification takes place in the nucleus, before the RNA moves to the cytoplasm.

CONSENSUS SEQUENCES AND THE SPLICEOSOME
Splicing requires the presence of three sequences in the intron. One end of the intron is referred to as the **5′ splice site**, and the other end is the **3′ splice site** (**Figure 14.8**); these splice sites possess short consensus sequences. Most introns in pre-mRNAs begin with GU and end with AG, indicating that these sequences play a crucial role in splicing. Indeed, changing a single nucleotide at either of these sites prevents splicing.

The third sequence important for splicing is the **branch point**, which is an adenine nucleotide that lies 18–40 nucleotides upstream of the 3′ splice site (see Figure 14.8). The sequence surrounding the branch point does not have a strong consensus. Deletion or mutation of the adenine nucleotide at the branch point prevents splicing. These sequences—the 5′ splice site, the 3′ splice site, and the branch point—collectively termed the *splicing code*—determine where splicing will occur. Additional sequences in introns and exons may also affect the process of splicing and which splice sites are used (see Alternative Processing Pathways later in this section).

Splicing takes place within a large structure called the **spliceosome**, which is one of the largest and most complex of all molecular structures. The spliceosome consists of five RNA molecules and almost 300 proteins. The RNA components are small nuclear RNAs (snRNAs) ranging in length from 107 to 210 nucleotides; these snRNAs associate with proteins to form small nuclear ribonucleoprotein particles (snRNPs). Each snRNP contains a single snRNA molecule and multiple proteins. The spliceosome is composed of five snRNPs (U1, U2, U4, U5, and U6) and other proteins not associated with an snRNA.

CONCEPTS

Nuclear pre-mRNA introns contain three consensus sequences critical to splicing: a 5′ splice site, a 3′ splice site, and a branch point. The splicing of pre-mRNA takes place within a large complex called the spliceosome, which consists of snRNAs and proteins.

✔ **CONCEPT CHECK 4**
If a splice site were mutated so that splicing did not take place, what would be the effect on the mRNA?
a. It would be shorter than normal.
b. It would be longer than normal.
c. It would be the same length, but would encode a different protein.

THE PROCESS OF SPLICING Before splicing takes place, an intron lies between an upstream exon (exon 1) and a downstream exon (exon 2), as shown in **Figure 14.9**. Pre-mRNA is spliced in two distinct steps. In the first step of splicing, the pre-mRNA is cut at the 5′ splice site. This cut frees exon 1 from the intron, and the 5′ end of the intron attaches to the branch point; the intron folds back on itself, forming a structure called a **lariat**. In this reaction, the guanine nucleotide in the consensus sequence at the 5′ splice site bonds with the adenine nucleotide at the branch point through a transesterification reaction. In this reaction, both 5′ cleavage and lariat formation occur in a single step. As a result, the 5′ phosphate group of the guanine nucleotide becomes attached to the 2′-OH group of the adenine nucleotide at the branch point (see Figure 14.9).

In the second step of RNA splicing, a cut is made at the 3′ splice site, and simultaneously, the 3′ end of exon 1 becomes covalently attached (spliced) to the 5′ end of exon 2. The intron is released as a lariat. Eventually, a lariat debranching enzyme breaks the bond at the branch point, producing a linear intron that is rapidly degraded by nuclear enzymes. The mature mRNA, consisting of the exons spliced together, is exported to the cytoplasm, where it is translated.

These splicing reactions take place within the spliceosome, which assembles on the pre-mRNA in a step-by-step fashion (**Figure 14.10**). A crucial feature of the assembly process is a series of interactions between the snRNAs of the spliceosome and between the snRNAs and the mRNA. These interactions depend on complementary base pairing between the different RNA molecules and bring the essential components of the pre-mRNA and the spliceosome close together. Key catalytic steps in the splicing process are carried out by the snRNAs of the spliceosome.

First, snRNP U1 attaches to the 5′ splice site, and then U2 attaches to the branch point. A complex consisting of U4, U5, and U6 (which form a single snRNP) joins the spliceosome. This addition causes a conformational change in the spliceosome: the intron loops over, and the

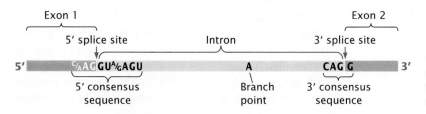

14.8 Splicing of pre-mRNA requires consensus sequences. Critical consensus sequences are present at the 5′ splice site and the 3′ splice site. A weak consensus sequence (not shown) exists at the branch point.

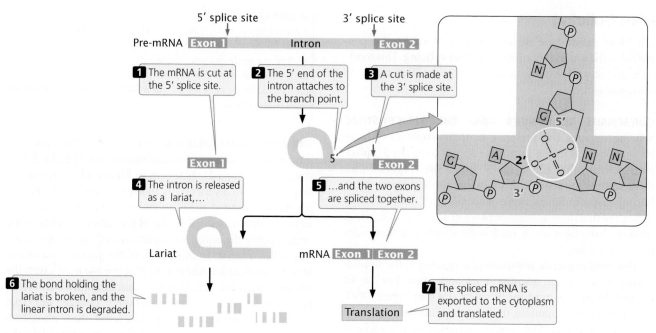

1 The mRNA is cut at the 5' splice site.

2 The 5' end of the intron attaches to the branch point.

3 A cut is made at the 3' splice site.

4 The intron is released as a lariat,...

5 ...and the two exons are spliced together.

6 The bond holding the lariat is broken, and the linear intron is degraded.

7 The spliced mRNA is exported to the cytoplasm and translated.

Lariat

mRNA Exon 1 Exon 2

Translation

14.9 The splicing of nuclear introns requires a two-step process.

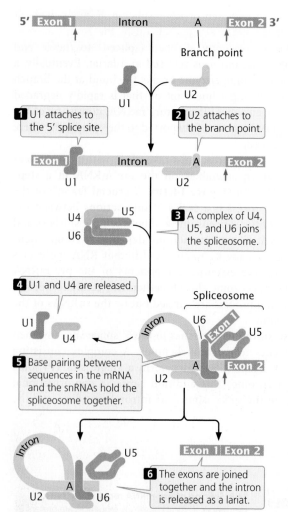

1 U1 attaches to the 5' splice site.

2 U2 attaches to the branch point.

3 A complex of U4, U5, and U6 joins the spliceosome.

4 U1 and U4 are released.

Spliceosome

5 Base pairing between sequences in the mRNA and the snRNAs hold the spliceosome together.

6 The exons are joined together and the intron is released as a lariat.

14.10 RNA splicing takes place within the spliceosome. The spliceosome assembles sequentially.

5' splice site is brought close to the branch point. Particles U1 and U4 dissociate from the spliceosome, and base pairing occurs between U6 and U2 and between U6 and the 5' splice site. The 5' splice site, 3' splice site, and branch point are now in close proximity, held together by the spliceosome. The two transesterification reactions take place, joining the two exons together and releasing the intron as a lariat. The U6 snRNP catalyzes both transesterification reactions.

THINK-PAIR-SHARE Question 3

Most mRNAs are produced from a single pre-mRNA molecule whose exons are spliced together. However, in a few organisms (principally nematodes and trypanosomes), mRNAs may be produced by splicing together sequences from two or more different RNA molecules; this process is called **trans-splicing**. In another variation of splicing, some long introns are removed in multiple steps through a process called **recursive splicing**.

Many human genetic diseases arise from mutations that affect pre-mRNA splicing; indeed, about 15% of single-base substitutions that result in human genetic diseases alter pre-mRNA splicing. Some of these mutations interfere with recognition of the normal 5' and 3' splice sites. Others create new splice sites, as was the case with the mutation that caused the royal hemophilia discussed in the introduction to this chapter. **TRY PROBLEM 26**

RNA splicing, which takes place in the nucleus, must be done before the RNA can move into the cytoplasm. Incompletely spliced RNAs remain in the nucleus until splicing is complete or until they are degraded. Immediately after splicing, a group of proteins called the exon-junction complex (EJC) is deposited approximately 20 nucleotides upstream of each exon–exon junction on the mRNA. The EJC promotes the export of the mRNA from the nucleus to the cytoplasm.

> **CONCEPTS**
>
> Splicing of nuclear pre-mRNA introns is a two-step process: (1) the 5′ end of the intron is cleaved from the upstream exon and attached to the branch point to form a lariat, and (2) the 3′ end of the intron is cleaved from the downstream exon and the ends of the two exons are spliced together. In the process, the intervening intron is removed. These reactions take place within the spliceosome.

MINOR SPLICING Some introns in the pre-mRNAs of multicellular eukaryotes are removed by a different process known as minor splicing. Introns that undergo minor splicing have different consensus sequences at the 5′ splice site and branch point and use a minor spliceosome, which consists of a somewhat different set of snRNAs. Some 700–800 genes in the human genome contain introns that undergo minor splicing.

SELF-SPLICING INTRONS Some introns are self-splicing—they possess the ability to remove themselves from an RNA molecule without the aid of enzymes or other proteins. These self-splicing introns fall into two major categories. Group I introns are found in a variety of genes, including some rRNA genes in protists, some mitochondrial genes in fungi, and even some bacterial and bacteriophage genes. Although the lengths of group I introns vary, all of them fold into a common secondary structure with nine hairpins (stem-loop structures; **Figure 14.11a**), which are necessary for splicing.

Group II introns, present in genes of bacteria, archaea, and eukaryotic organelles, also have the ability to self-splice. All group II introns also fold into secondary structures (**Figure 14.11b**). The splicing of group II introns is accomplished by a mechanism that has some similarities to the spliceosome-mediated splicing of nuclear pre-mRNA introns, and splicing generates a lariat structure. Because of these similarities, group II introns and nuclear pre-mRNA introns have been suggested to be evolutionarily related; perhaps the nuclear introns evolved from self-splicing group II introns and later adopted the proteins and snRNAs of the spliceosome to carry out the splicing reaction.

> **CONCEPTS**
>
> Some introns are removed by minor splicing. There are two types of self-splicing introns: group I introns and group II introns. These introns have complex secondary structures that enable them to catalyze their own excision from RNA molecules without the aid of enzymes or other proteins.

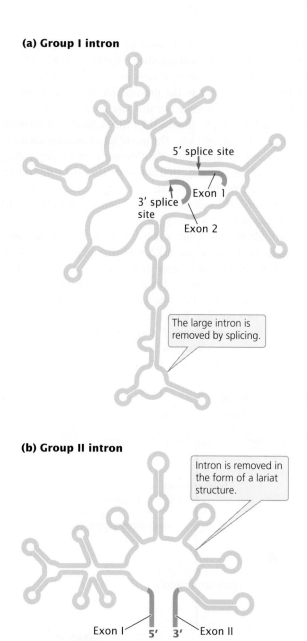

(a) Group I intron

5′ splice site

Exon 1

3′ splice site

Exon 2

The large intron is removed by splicing.

(b) Group II intron

Intron is removed in the form of a lariat structure.

Exon I   5′   3′   Exon II

14.11 Group I and group II introns fold into characteristic secondary structures.

Alternative Processing Pathways

A finding that complicates the view of a gene as a sequence of nucleotides that specifies the amino acid sequence of a protein (see The Concept of the Gene Revisited in Section 14.1) is the existence of **alternative processing pathways**. In these pathways, a single pre-mRNA can be processed in different ways to produce alternative types of mRNA, resulting in the production of different proteins from the same DNA sequence.

One type of alternative processing is **alternative splicing**, in which the same pre-mRNA can be spliced in more than one way to yield different mRNAs that are translated into different amino acid sequences and thus different proteins

(**Figure 14.12a**). Another type of alternative processing requires **multiple 3′ cleavage sites** (**Figure 14.12b**); in these cases, two or more potential sites for cleavage and polyadenylation are present in the pre-mRNA. In the example in Figure 14.12b, cleavage at the first site produces a relatively short mRNA compared with the mRNA produced through cleavage at the second site. The use of an alternative cleavage site may or may not produce a different protein, depending on whether the site is located before or after the stop codon. Alternative cleavage sites can also change the 3′ untranslated region of the mRNA, which influences the stability and translation of the mRNA.

Both alternative splicing and multiple 3′ cleavage sites can exist in the same pre-mRNA transcript. An example is seen in the mammalian gene that encodes calcitonin; this gene contains six exons and five introns (**Figure 14.13a**). The entire gene is transcribed into pre-mRNA (**Figure 14.13b**). There are two possible 3′ cleavage sites in the pre-mRNA. In cells of the thyroid gland, 3′ cleavage and polyadenylation take place after exon 4 to produce a mature mRNA consisting

of exons 1, 2, 3, and 4 (**Figure 14.13c**). This mRNA is translated into the hormone calcitonin, which is produced by the thyroid gland and regulates calcium concentrations in the blood. In brain cells, the *identical* pre-mRNA is transcribed from DNA, but 3′ cleavage and polyadenylation take place after exon 6, yielding an initial transcript that includes all six exons. During splicing, exon 4 is removed, so only exons 1, 2, 3, 5, and 6 are present in the mature mRNA (**Figure 14.13d**). When translated, this mRNA produces a protein called calcitonin-gene-related peptide (CGRP), which has an amino acid sequence quite different from that of calcitonin. CGRP causes dilation of blood vessels and can function in transmission of pain (some research suggests that CGRP is involved in the development of migraine headaches). Alternative splicing may produce different combinations of exons in the mRNA, but the order of the exons is not usually changed.

Alternative splicing is controlled by proteins and ribonucleoprotein particles that bind to sites within introns and exons and determine which splice and cleavage sites are used. Alternative processing of pre-mRNAs is common in

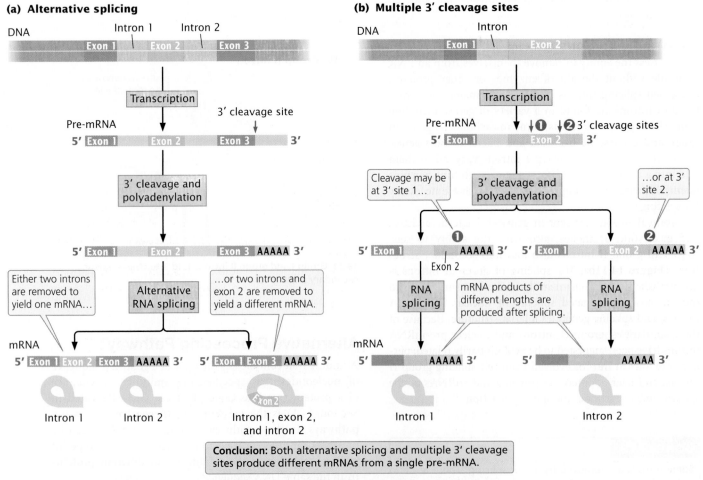

(a) Alternative splicing

(b) Multiple 3′ cleavage sites

Conclusion: Both alternative splicing and multiple 3′ cleavage sites produce different mRNAs from a single pre-mRNA.

14.12 Eukaryotic cells have alternative pathways for processing pre-mRNA. (a) With alternative splicing, pre-mRNA can be spliced in different ways to produce different mRNAs. (b) With multiple 3′ cleavage sites, there are two or more potential sites for cleavage and polyadenylation; use of the different sites produces mRNAs of different lengths.

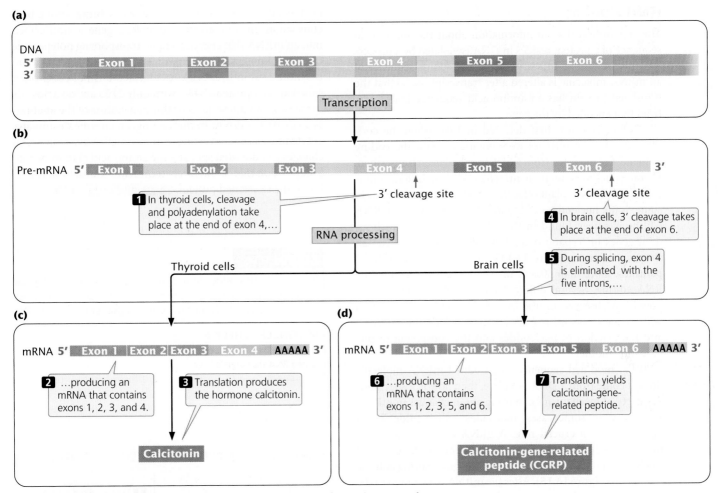

14.13 Pre-mRNA encoded by the gene for calcitonin undergoes alternative processing.

multicellular eukaryotes. For example, researchers estimate that the products of more than 95% of all human genes with multiple exons undergo alternative splicing and that 50% of human genes have multiple 3′ cleavage sites. Often the form of splicing differs between human tissues: human brain and liver tissues have more alternatively spliced mRNAs than other tissues. Sometimes splicing even varies from one person to another. Alternative processing pathways also contribute to gene regulation, as will be discussed in Chapter 17.

Alternative splicing may play a role in organismal complexity. The complete sequencing of the genomes of numerous organisms (see Chapter 20) has led to the conclusion that an organism's number of genes is not correlated with the organism's complexity. For example, fruit flies have only about 14,000 genes, whereas anatomically simpler nematode worms have 20,500 genes. The plant *Arabidopsis thaliana* has about 25,700 genes, more than humans have. If anatomically simple organisms have as many genes as complex organisms have, how is developmental complexity encoded in the genome? A possible answer is alternative processing, which can produce multiple proteins from a single gene and is an important source of protein diversity in vertebrates.

Recent research demonstrates that even closely related species often differ in how their pre-mRNAs are spliced. Alternative splicing appears to evolve relatively quickly and may be responsible for physical and behavioral differences between species, such as humans and chimpanzees, that are phenotypically different but genetically similar. Alternative splicing may therefore have played an important role in speciation (see Chapter 26).

CONCEPTS

Alternative splicing enables exons to be spliced together in different combinations to yield mRNAs that encode different proteins. Multiple 3′ cleavage sites allow pre-mRNA to be cleaved and polyadenylated at different sites.

✔ CONCEPT CHECK 5

Multiple 3′ cleavage sites result in
a. multiple genes of different lengths.
b. multiple pre-mRNAs of different lengths.
c. multiple mRNAs of different lengths.
d. All of the above

RNA Editing

The assumption that all information about the amino acid sequence of a protein resides in DNA is violated by a process called RNA editing. In **RNA editing**, the coding sequence of an mRNA molecule is altered after transcription, so that the translated protein has an amino acid sequence that differs from that encoded by the gene.

RNA editing was first detected in 1986 when the coding sequences of mRNAs were compared with the coding sequences of the DNA from which they had been transcribed. In some nuclear genes in mammalian cells and in some mitochondrial genes in plant cells, there had been substitutions in some of the nucleotides of the mRNA. More extensive RNA editing has been found in the mRNA for some mitochondrial genes in trypanosomes (one of which causes African sleeping sickness; **Figure 14.14**). In some mRNAs of these parasites, more than 60% of the sequence is determined by RNA editing. Several different types of RNA editing have now been observed in mRNAs, tRNAs, and rRNAs from a wide range of organisms; these types include the insertion and the deletion of nucleotides and the conversion of one base into another.

If the modified sequence in an edited mRNA molecule doesn't come from a DNA template, then how is it specified? A variety of mechanisms can bring about changes in mRNA sequences. In some cases, molecules called **guide RNAs** (**gRNAs**) play a crucial role. A gRNA contains sequences that are partly complementary to segments of the unedited mRNA, and the two molecules undergo base pairing at these sequences (**Figure 14.15**). After the mRNA is anchored to the gRNA, the mRNA undergoes cleavage and nucleotides are added, deleted, or altered according to the template provided by the gRNA. In other cases, enzymes bring about base conversion. In humans, for example, a gene is transcribed into an mRNA that encodes a lipid-transporting polypeptide called apolipoprotein-B100, which has 4563 amino acids and is synthesized in liver cells. A truncated form of the protein called apolipoprotein-B48—with only 2153 amino acids—is synthesized in intestinal cells through editing of the apolipoprotein-B100 mRNA. In this editing, an enzyme deaminates a cytosine base, converting it into uracil. This conversion changes a codon that specifies the amino acid glutamine into a stop codon that prematurely terminates translation, resulting in the shortened protein. ▶ **TRY PROBLEM 36**

CONCEPTS

Individual nucleotides in mRNA may be changed, added, or deleted by RNA editing. The amino acid sequence produced by the edited mRNA is not the same as that encoded by DNA.

✔ **CONCEPT CHECK 6**

What specifies the modified sequence of nucleotides found in an edited RNA molecule?

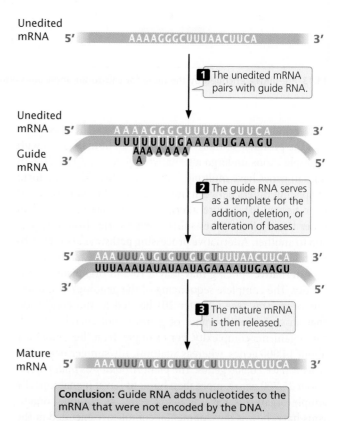

14.15 RNA editing is carried out by guide RNAs. The gRNA has sequences that are partly complementary to those of the unedited mRNA and pairs with it. After this pairing, the mRNA undergoes cleavage and new nucleotides are added, with sequences in the gRNA serving as a template. The ends of the mRNA are then rejoined.

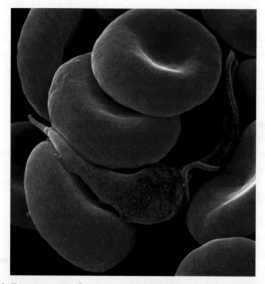

14.14 *Trypanosoma brucei* causes African sleeping sickness. Messenger RNA produced from mitochondrial genes of this parasite (in purple) undergoes extensive RNA editing. [David Spears/Last Refuge Ltd./Phototake.]

CONNECTING CONCEPTS

Eukaryotic Gene Structure and Pre-mRNA Processing

Chapters 13 and 14 have introduced a number of different components of genes and RNA molecules, including promoters, 5′ and 3′ untranslated regions, coding sequences (exons), introns, 5′ caps, and poly(A) tails. Let's see how some of these components are combined to create a typical eukaryotic gene and how a mature mRNA is produced from them.

The promoter, which typically lies upstream of the transcription start site, is necessary for transcription to take place, but is itself not usually transcribed when protein-encoding genes are transcribed by RNA polymerase II (**Figure 14.16a**). Farther upstream or downstream of the start site, there may be enhancers—DNA sequences involved in regulating transcription.

In transcription, all the nucleotides between the transcription start site and the terminator are transcribed into pre-mRNA, including exons, introns, and a long 3′ end that is later cleaved from the transcript (**Figure 14.16b**). Notice that the 5′ end of the first exon contains the 5′ UTR, and that the 3′ end of the last exon contains the 3′ UTR.

The pre-mRNA is then processed to yield a mature mRNA. The first step in this processing is the addition of a cap to the 5′ end of the pre-mRNA (**Figure 14.16c**). Next, the 3′ end is cleaved at a site downstream of the AAUAAA consensus sequence in the last exon (**Figure 14.16d**). Immediately after cleavage, a poly(A) tail is added to the 3′ end (**Figure 14.16e**). Finally, the introns are removed to yield the mature mRNA (**Figure 14.16f**). The mRNA now contains 5′ and 3′ untranslated regions, which are not translated into amino acids, and the nucleotides that carry the protein-coding sequences. You can explore the consequences of failed RNA processing by viewing and interacting with **Animation 14.2**.

The nucleotide sequence of a small gene (the human interleukin 2 gene), with these components identified, is presented in **Figure 14.17**.

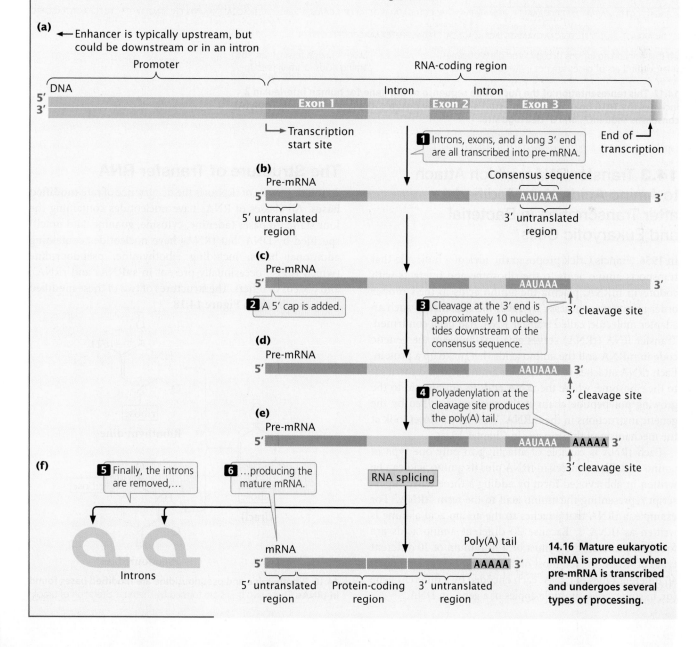

14.16 Mature eukaryotic mRNA is produced when pre-mRNA is transcribed and undergoes several types of processing.

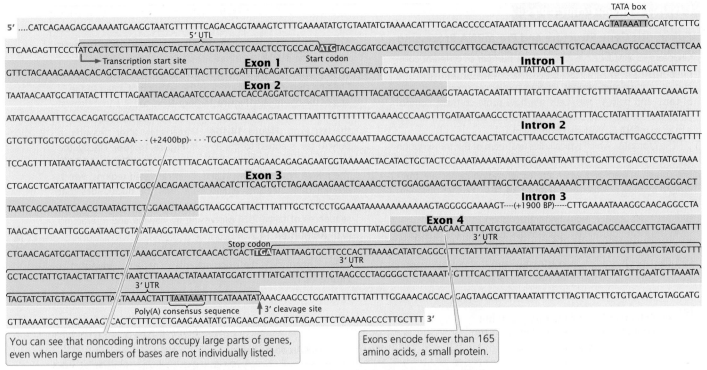

14.17 This representation of the nucleotide sequence of the gene for human interleukin 2 includes the TATA box, transcription start site, start and stop codons, introns, exons, poly(A) consensus sequence, and 3' cleavage site.

14.3 Transfer RNAs, which Attach to Amino Acids, Are Modified after Transcription in Bacterial and Eukaryotic Cells

In 1956, Francis Crick proposed the idea of a molecule that transports amino acids to the ribosome and interacts with codons in mRNA, placing the amino acids in their proper order in protein synthesis. By 1963, the existence of such an adapter molecule, called transfer RNA, had been confirmed. Transfer RNA (tRNA) serves as a link between the genetic code in mRNA and the amino acids that make up a protein. Each tRNA attaches to a particular amino acid and carries it to the ribosome, where the tRNA adds its amino acid to the growing polypeptide chain at the position specified by the genetic instructions in the mRNA. We'll take a closer look at the mechanism of this process in Chapter 15.

Each tRNA is capable of attaching to only one type of amino acid. The complex of tRNA plus its amino acid can be written in abbreviated form by adding a three-letter superscript representing the amino acid to the term "tRNA." For example, a tRNA that attaches to the amino acid alanine is written as tRNA[Ala]. Because 20 different amino acids are found in proteins, there must be a minimum of 20 different types of tRNA. In fact, most organisms possess at least 30 to 40 different types of tRNA, each encoded by a different gene (or, in some cases, multiple copies of a gene) in DNA.

The Structure of Transfer RNA

A unique feature of tRNAs is the occurrence of rare **modified bases**. All classes of RNAs have nucleotides containing the four standard bases (adenine, cytosine, guanine, and uracil) specified by DNA, but tRNAs have nucleotides containing additional bases, including ribothymine, pseudouridine (which is also occasionally present in snRNAs and rRNA), and dozens of others. The structures of two of these modified bases are shown in **Figure 14.18**.

14.18 Ribothymine and pseudouridine are modified bases found in tRNAs. These two bases are formed by chemical alteration of uracil.

If there are only four bases in DNA and all RNA molecules are transcribed from DNA, how do tRNAs acquire these additional bases? Modified bases arise from chemical changes made to the four standard bases after transcription. These changes are carried out by special **tRNA-modifying enzymes**. For example, the addition of a methyl group to uracil creates the modified base ribothymidine. The presence of modified bases affects the structure of the tRNA and how it interacts with the ribosome and with mRNA.

All tRNAs are similar in their secondary structure, a feature that is critical to tRNA function. Most tRNAs are short molecules containing between 74 and 95 nucleotides, some of which are complementary to each other and form intramolecular hydrogen bonds. As a result, each tRNA has a **cloverleaf** structure with four major arms (**Figure 14.19**). If we start at the top and proceed clockwise around the tRNA shown at the right in Figure 14.19, the four major arms are the acceptor arm, the TΨC arm, the anticodon arm, and the DHU arm. Three of the arms (the TΨC, anticodon, and DHU arms) consist of a hairpin formed by the pairing of complementary nucleotides. A loop lies at the terminus of the hairpin, where there is no nucleotide pairing.

Instead of a hairpin, the acceptor arm forms a stem that includes the 5′ and 3′ ends of the tRNA molecule. All tRNAs have the same sequence (CCA) at the 3′ end, where the amino acid attaches to the tRNA, so clearly this sequence is not responsible for specifying which amino acid will attach to the tRNA.

The TΨC arm is named for the bases of three nucleotides in the loop of this arm: thymine (T), pseudouridine (Ψ), and cytosine (C). The anticodon arm lies at the bottom of the tRNA. Three nucleotides at the end of this arm make up

the **anticodon**, which pairs with the corresponding codon on mRNA to ensure that the amino acids link in the correct order. The DHU arm is so named because it often contains the modified base dihydrouridine.

Although each tRNA molecule folds into a cloverleaf owing to the complementary pairing of bases, the cloverleaf is not the three-dimensional (tertiary) structure of tRNAs found in the cell. The results of X-ray crystallographic studies have shown that the cloverleaf folds on itself to form an L-shaped structure, as illustrated by the space-filling and ribbon models in Figure 14.19. Notice that the acceptor stem is at one end of the tertiary structure and the anticodon is at the other end.

Transfer RNA Gene Structure and Processing

The genes that produce tRNAs may be in clusters or scattered about the genome. In *E. coli*, the genes for some tRNAs are present in a single copy, whereas the genes for other tRNAs are present in several copies. Eukaryotic cells usually have many copies of each tRNA gene. All tRNA molecules in both bacterial and eukaryotic cells undergo processing after transcription.

In *E. coli*, several tRNAs are usually transcribed together as one large precursor tRNA, which is then cut up into pieces, each containing a single tRNA. Additional nucleotides may then be removed one at a time from the 5′ and 3′ ends of the tRNA in a process known as trimming. Base-modifying enzymes may then change some of the standard bases into modified bases (**Figure 14.20**). In some prokaryotes, the CCA sequences found at the 3′ ends of tRNAs are encoded

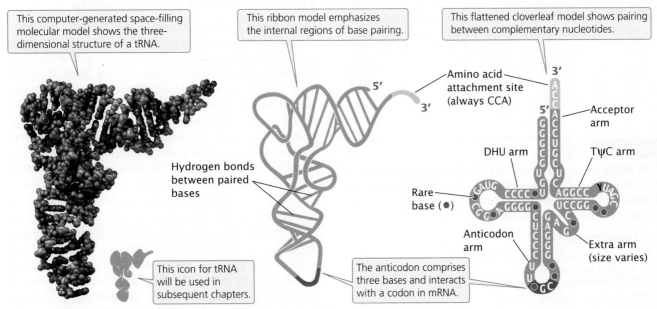

14.19 All tRNAs possess a common secondary structure, the cloverleaf. The base sequence in the flattened model is for tRNA^Ala.

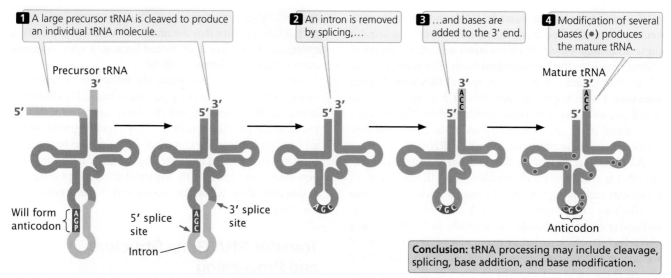

1 A large precursor tRNA is cleaved to produce an individual tRNA molecule.

2 An intron is removed by splicing,...

3 ...and bases are added to the 3' end.

4 Modification of several bases (●) produces the mature tRNA.

Precursor tRNA

Mature tRNA

Will form anticodon

5' splice site

3' splice site

Intron

Anticodon

Conclusion: tRNA processing may include cleavage, splicing, base addition, and base modification.

14.20 Transfer RNAs are processed in both bacterial and eukaryotic cells. Different tRNAs are modified in different ways. One example is shown here.

in the tRNA gene and are transcribed into the tRNA; in other prokaryotes and in eukaryotes, these sequences are added by a special enzyme that adds the nucleotides without the use of any template.

There is no generic processing pathway for all tRNAs: different tRNAs are processed in different ways. Eukaryotic tRNAs are processed in a manner similar to bacterial tRNAs: most are transcribed as part of larger precursors that are then cleaved, trimmed, and modified to produce mature tRNAs.

Some eukaryotic and archaeal tRNA genes possess introns of variable length that must be removed in processing. For example, about 40 of the 400 tRNA genes in yeast contain a single intron that is always found adjacent to the 3' side of the anticodon. These tRNA introns are shorter than those found in pre-mRNA and do not have the consensus sequences found at the intron–exon junctions of pre-mRNAs. The splicing process for tRNA genes is quite different from the spliceosome-mediated reactions that remove introns from protein-encoding genes.

CONCEPTS

All tRNAs are similar in size and have a common secondary structure known as the cloverleaf. Transfer RNAs contain modified bases and are extensively processed after transcription in both bacterial and eukaryotic cells.

✔ CONCEPT CHECK 7

How are rare bases incorporated into tRNAs?

a. Encoded by guide RNAs
b. By chemical changes to one of the standard bases
c. Encoded by rare bases in DNA
d. Encoded by sequences in introns

14.4 Ribosomal RNA, a Component of the Ribosome, Is Also Processed after Transcription

Within ribosomes, the genetic instructions contained in mRNA are translated into the amino acid sequences of polypeptides. Thus, ribosomes play an integral part in the transfer of genetic information from genotype to phenotype. We will examine the role of ribosomes in the process of translation in Chapter 15. Here, we consider ribosome structure and examine how ribosomes are processed before becoming functional.

The Structure of the Ribosome

The ribosome is one of the most abundant molecular complexes in the cell: a single bacterial cell may contain as many as 20,000 ribosomes, and eukaryotic cells possess even more. Ribosomes typically contain about 80% of the total cellular RNA. They are complex structures, each consisting of more than 50 different proteins and RNA molecules (**Table 14.3**). A functional ribosome consists of two subunits, a **large ribosomal subunit** and a **small ribosomal subunit**, each of which consists of one or more RNA molecules and a number of proteins. The sizes of the ribosomes and their RNA components are given in Svedberg (S) units (a measure of how rapidly an object sediments in a centrifugal field). (It is important to note that S units are not additive: combining a 10S structure and a 20S structure does not necessarily produce a 30S structure because the sedimentation rate is affected by the three-dimensional structure of an object as well as by its mass.) The three-dimensional structure of the bacterial ribosome has been elucidated in great detail through X-ray crystallography. More will be said about the ribosome's structure in Chapter 15.

TABLE 14.3	Composition of ribosomes in bacterial and eukaryotic cells			
Cell Type	Ribosome Size	Subunits	rRNA Component	Proteins
Bacterial	70S	Large (50S)	23S (2900 nucleotides), 5S (120 nucleotides)	31
		Small (30S)	16S (1500 nucleotides)	21
Eukaryotic	80S	Large (60S)	28S (4700 nucleotides), 5.8S (160 nucleotides), 5S (120 nucleotides)	49
		Small (40S)	18S (1900 nucleotides)	33

Note: The letter "S" stands for "Svedberg unit."

Ribosomal RNA Gene Structure and Processing

The genes for rRNA, like those for tRNA, may be present in multiple copies, and the numbers of copies vary among species (**Table 14.4**); all copies of an rRNA gene in a species are identical or nearly identical. In bacteria, copies of rRNA genes are dispersed, but in eukaryotic cells, they are clustered, with the genes arrayed in tandem, one after another.

Eukaryotic cells possess two types of rRNA genes: a large gene that encodes 18S rRNA, 28S rRNA, and 5.8S rRNA, and a small gene that encodes 5S rRNA. All three bacterial rRNAs (23S rRNA, 16S rRNA, and 5S rRNA) are encoded by a single type of gene.

Ribosomal RNA is processed after transcription in both bacterial and eukaryotic cells. In *E. coli*, each rRNA gene is transcribed into a 30S rRNA precursor (**Figure 14.21a**). This 30S precursor is methylated in several places, then cleaved and trimmed to produce 16S rRNA, 23S rRNA, and 5S

TABLE 14.4	Numbers of rRNA genes in different organisms
Species	Number of Copies of rRNA Genes per Genome
Escherichia coli	7
Yeast	100–200
Human	280
Frog	450

rRNA, along with one or more tRNAs. A series of enzymes bring about cleavage, methylation, and trimming.

Eukaryotic rRNAs undergo similar processing (**Figure 14.21b**). Small nucleolar RNAs (snoRNAs) help to cleave and modify eukaryotic rRNAs and assemble them into mature ribosomes. Like the snRNAs taking part in pre-mRNA splicing, snoRNAs associate with proteins

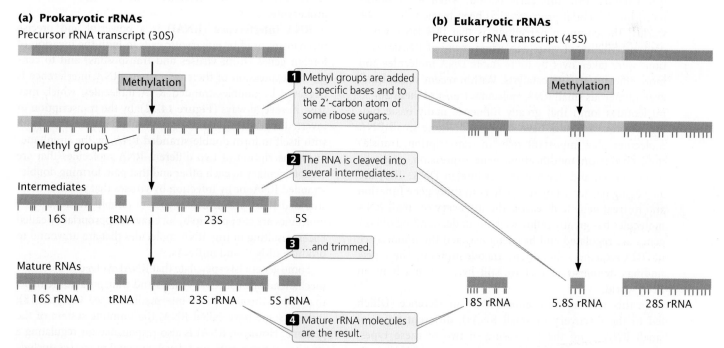

(a) Prokaryotic rRNAs
Precursor rRNA transcript (30S)

Methylation

Methyl groups

Intermediates

16S   tRNA   23S   5S

Mature RNAs

16S rRNA   tRNA   23S rRNA   5S rRNA

1 Methyl groups are added to specific bases and to the 2'-carbon atom of some ribose sugars.

2 The RNA is cleaved into several intermediates...

3 ...and trimmed.

4 Mature rRNA molecules are the result.

(b) Eukaryotic rRNAs
Precursor rRNA transcript (45S)

Methylation

18S rRNA   5.8S rRNA   28S rRNA

14.21 Ribosomal RNA is processed after transcription. Prokaryotic rRNA (a) and eukaryotic rRNA (b) are produced from precursor RNA transcripts that are methylated, cleaved, and processed to produce mature rRNAs. Eukaryotic 5S rRNA is transcribed separately from a different gene.

to form ribonucleoprotein particles (snoRNPs). The snoRNAs have extensive complementarity to the rRNA sequences in which modification takes place. Interestingly, some snoRNAs are encoded by sequences in the introns of protein-encoding genes. The processing of rRNA and ribosome assembly in eukaryotes, which takes place in the nucleolus, requires over 200 proteins, called assembly factors.

CONCEPTS

A ribosome is a complex structure consisting of several rRNA molecules and many proteins. Each functional ribosome consists of a large and a small subunit. Ribosomal RNAs in both bacterial and eukaryotic cells are modified after transcription. In eukaryotes, rRNA processing is carried out by small nucleolar RNAs.

✔ **CONCEPT CHECK 8**

What types of changes take place in rRNA processing?
a. Methylation of bases
b. Cleavage of a larger precursor
c. Trimming of nucleotides from the ends of rRNAs
d. All of the above

14.5 Small RNA Molecules Participate in a Variety of Functions

Much evidence suggests that the first genetic material was RNA and that early life was dominated by RNA molecules (see Chapter 13). This early period, when RNA dominated life's essential processes, has been termed the "RNA world." The common perception is that this RNA world died out billions of years ago, when many of RNA's functions were taken over by more stable DNA molecules and more efficient protein catalysts. Within recent years, however, numerous small RNA molecules (most of them 20–30 nucleotides long) that greatly influence many basic biological processes have been discovered. These small RNA molecules play important roles in transcription, translation, chromatin modification, gene expression, development, cancer, and defense against foreign DNA. They are also being harnessed by researchers to study gene function and to treat genetic diseases. The discovery of small RNA molecules has greatly influenced our understanding of how genes are regulated and has demonstrated the importance of DNA sequences that do not encode proteins. These new findings demonstrate that we still live very much in an RNA world.

In this section, we examine RNA interference (which led to the discovery of small RNAs), different types of small RNAs, and the processing of two of these types, microRNAs (miRNAs) and small interfering RNAs (siR-NAs). In Chapter 17, we will look further at the role of small RNAs in controlling gene expression; in Chapter 19, we will see how small RNAs are being used as tools in biotechnology.

RNA Interference

In 1998, Andrew Fire, Craig Mello, and their colleagues observed a strange phenomenon. They were inhibiting the expression of genes in the nematode *Caenorhabditis elegans* by injecting single-stranded RNA molecules that were complementary to a gene's DNA sequence. Called antisense RNA, such molecules were known to inhibit gene expression by binding to mRNA sequences and inhibiting their translation. However, Fire, Mello, and colleagues found that even more potent gene silencing was triggered when double-stranded RNA was injected into the nematodes. This finding was puzzling, because no mechanism by which double-stranded RNA could inhibit translation was known. Several other, previously described types of gene silencing were also found to be triggered by double-stranded RNA. These initial studies led to the discovery of small RNA molecules that are important in gene silencing. For their discovery of RNA interference, Fire and Mello were awarded the Nobel Prize in physiology or medicine in 2006.

Subsequent research revealed an astonishing array of small RNA molecules with important cellular functions in eukaryotes, which now include at least three major classes: small interfering RNAs (siRNAs), microRNAs (miRNAs), and Piwi-interacting RNAs (piRNAs). These small RNAs are found in many eukaryotes and are responsible for a variety of different functions, including the regulation of gene expression, defense against viruses, suppression of transposons, and modification of chromatin structure. An analogous group of small RNAs with silencing functions—called CRISPR RNAs (crRNAs)—have been detected in prokaryotes.

RNA interference (RNAi) is a powerful and precise mechanism used by eukaryotic cells to limit the invasion of foreign genes (from viruses and transposons) and to censor the expression of their own genes. RNA interference is triggered by double-stranded RNA molecules, which may arise in several ways (**Figure 14.22**): by the transcription of inverted repeats into an RNA molecule that then base pairs with itself to form double-stranded RNA; by the simultaneous transcription of two different RNA molecules that are complementary to each other and that pair, forming double-stranded RNA; or by infection by viruses that make double-stranded RNA. All of these types of double-stranded RNA molecules are chopped up by an enzyme appropriately called Dicer, resulting in tiny RNA molecules that are unwound to produce siRNAs and miRNAs.

Some geneticists speculate that RNAi evolved as a defense mechanism against RNA viruses and transposable elements that move through RNA intermediates (see Chapter 18); indeed, some have called RNAi the immune system of the genome. However, RNAi is also responsible for regulating a number of key genetic and developmental processes, including changes in chromatin structure, translation, cell fate and proliferation, and cell death. Geneticists also use the RNAi machinery as an effective tool for blocking the expression of specific genes (see Chapter 19).

Small Interfering RNAs and MicroRNAs

Two abundant classes of RNA molecules that function in RNA interference in eukaryotes are small interfering RNAs and microRNAs. Although these two types of RNA differ in how they originate (**Table 14.5**; see also Figure 14.22), they have a number of features in common, and their functions overlap considerably. Both are about 22 nucleotides long. Small interfering RNAs arise from the cleavage of mRNAs, RNA transposons, and RNA viruses. Some miRNAs are cleaved from RNA molecules transcribed from sequences that encode miRNA only, but others are encoded in the introns and exons of mRNAs. Each miRNA is cleaved from a single-stranded RNA precursor that forms small hairpins, whereas multiple siRNAs are produced from the cleavage of an RNA duplex consisting of two different RNA molecules.

Usually, siRNAs have exact complementarity with their target mRNA or DNA sequences, whereas miRNAs often have limited complementarity with their target mRNAs. Small interfering RNAs suppress gene expression by degrading mRNA or inhibiting transcription, while miRNAs often suppress gene expression by inhibiting translation. Finally, miRNAs usually silence genes that are distinct from those from which the miRNAs were transcribed, whereas siRNAs typically silence the genes from which the siRNAs were transcribed. Note, however, that these differences between siRNAs and miRNAs are not hard and fast, and scientists are increasingly finding small RNAs that exhibit characteristics of both. For example, some miRNAs (such as those in plants) have exact complementarity with mRNA sequences and cleave those sequences—characteristics that are usually associated with siRNAs.

Both siRNA and miRNA molecules combine with proteins to form an **RNA-induced silencing complex (RISC)** (see Figure 14.22). Key to the functioning of RISCs is a protein called Argonaute. The RISC pairs with an mRNA molecule that possesses a sequence complementary to its siRNA or miRNA component, and the RISC then either cleaves the mRNA, leading to its degradation, or represses translation of the mRNA. Some siRNAs also serve as guides for the methylation of complementary sequences in DNA, and others alter chromatin structure, both of which affect transcription. To see how small interfering RNAs and microRNAs affect gene expression, see **Animation 14.3**.

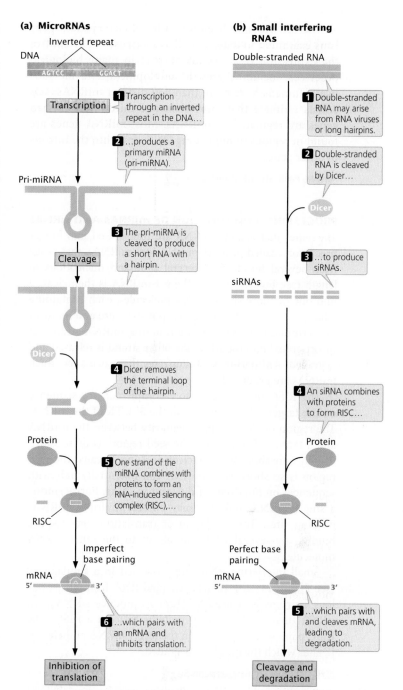

14.22 Small interfering RNAs and microRNAs are produced from double-stranded RNAs. Although microRNAs (a) and small interfering RNAs (b) differ in how they originate, they share a number of features, and their functions overlap.

TABLE 14.5	Differences between siRNAs and miRNAs	
Feature	**siRNA**	**miRNA**
Origin	mRNA, transposon, or virus	RNA transcribed from distinct gene
Cleavage of	RNA duplex or single-stranded RNA that forms long hairpins	Single-stranded RNA that forms short hairpins of double-stranded RNA
Size	21–25 nucleotides	21–25 nucleotides
Action	Degradation of mRNA, inhibition of transcription, chromatin modification	Degradation of mRNA, inhibition of translation, chromatin modification
Target	Genes from which they were transcribed	Genes other than those from which they were transcribed

MicroRNAs have been found in all eukaryotic organisms examined to date, as well as in viruses: they control the expression of genes taking part in many biological processes, including growth, development, and metabolism. Humans have more than 450 distinct miRNAs; scientists estimate that more than one-third of all human genes are regulated by miRNAs. Most miRNA genes are found in regions of noncoding DNA or within the introns of other genes.

THINK-PAIR-SHARE Question 4

PROCESSING AND FUNCTION OF miRNAs AND siRNAs

The genes that encode miRNA are transcribed into longer precursors, called primary miRNA (pri-miRNA), that range from several hundred to several thousand nucleotides in length (see Figure 14.22a). The pri-miRNA is then cleaved into one or more smaller RNA molecules, each containing a hairpin. Dicer binds to the hairpin structure and removes the terminal loop. One of the resulting miRNA strands is incorporated into the RISC; the other strand is released and degraded. **Animation 14.3** illustrates the process by which miRNAs are produced.

The RISC then attaches to a complementary sequence on the target mRNA, usually in the 3′ UTR of the mRNA. The region of close complementarity between the miRNA and a target mRNA, called the seed region, is quite short, usually only about seven nucleotides long. Because the seed region is so short, each miRNA can potentially pair with sequences on hundreds of different mRNAs. Furthermore, a single mRNA molecule may possess multiple miRNA-binding sites. The inhibition of translation may require binding by several RISC complexes to the same mRNA molecule.

Small interfering RNAs are processed in a similar way (see Figure 14.22b): double-stranded RNA from viruses or long hairpins is cleaved by Dicer to produce siRNAs. An siRNA then combines with proteins to form a RISC. The RISC pairs with sequences on the target mRNA and cleaves it, after which the mRNA is degraded.

THINK-PAIR-SHARE Question 5

CONCEPTS

Small interfering RNAs and microRNAs are tiny RNAs produced when larger, double-stranded RNA molecules are cleaved by the enzyme Dicer. Small interfering RNAs and microRNAs participate in a variety of processes, including mRNA degradation, the inhibition of translation, the methylation of DNA, and chromatin remodeling.

✔ **CONCEPT CHECK 9**

How do siRNAs and miRNAs target specific mRNAs for degradation or for the repression of translation?

Piwi-Interacting RNAs

Piwi-interacting RNAs (piRNAs), which were discovered in 2001, differ from siRNAs and miRNAs in several ways. They are somewhat longer than siRNAs or miRNAs, consisting of 24 to 31 nucleotides, and are derived from long, single-stranded RNA transcripts. Furthermore, Dicer is not involved in the production of piRNAs.

Piwi-interacting RNAs combine with Piwi proteins, which are related to Argonaute, to suppress the expression and movement of transposons in the germ cells of animals. The term Piwi stands for *P Element Induced Wimpy testis*, which refers to infertility in *Drosophila* associated with mutations in genes that encode Piwi proteins. Although the mechanism of transposon silencing by piRNAs is not fully understood, we know that it includes the degradation of mRNA transcribed from transposons, changes in chromatin structure that inhibit the transcription of transposons, and inhibition of the translation of proteins encoded by transposons. Piwi-interacting RNAs are also expressed in somatic tissues, where their function is less well understood; evidence suggests that they play a role in transposon silencing, stem-cell maintenance, tissue regeneration, and epigenetic processes.

CRISPR RNA

After the discovery of small RNAs in eukaryotes, similar small RNAs, called CRISPR RNAs (crRNAs), were discovered in prokaryotes. CRISPR RNAs are encoded by DNA sequences found in bacterial and archaeal genomes termed *clustered regularly interspaced short palindromic repeats*. Palindromic sequences are sequences that read the same forward and backward on two complementary DNA strands (see Chapter 4). A CRISPR array consists of a series of such palindromic sequences, separated by spacers that are derived from the DNA of foreign bacteriophage or plasmid genomes. For example, in the bacterium *Pseudomonas aeruginosa*, CRISPR palindromic repeats are 28 bp in length, separated by 32-bp spacers.

CRISPR RNAs combine with Cas (CRISPR-associated) proteins to provide defense against the invasion of specific foreign DNA molecules, such as DNA originating from bacteriophages and plasmids (see Chapter 9). Because they target specific DNA molecules, CRISPR-Cas systems have been compared to the immune systems of vertebrates. CRISPR-Cas systems are widespread in prokaryotes, occurring in 50% of bacterial species and 90% of archaea.

The action of a CRISPR-Cas system takes place in three stages: acquisition, expression, and interference. In acquisition, which is still incompletely understood, foreign DNA from a bacteriophage or plasmid that enters the cell is identified, processed, and inserted into the CRISPR array as a new spacer between the palindromic repeats (**Figure 14.23a**). The spacer DNA then serves as a memory of this foreign DNA.

In the second stage—expression—the CRISPR array is transcribed into a long CRISPR precursor RNA (**Figure 14.23b**), which is then cleaved by Cas proteins and processed into crRNAs, each of which contains one spacer

(a) Acquisition

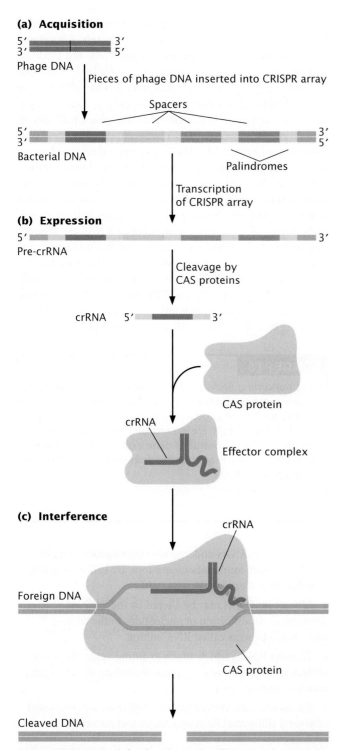

(b) Expression

(c) Interference

14.23 CRISPR RNAs defend prokaryotic cells against invasion by foreign DNA, such as DNA from bacteriophages and plasmids.

sequence that is homologous to the foreign DNA. The crRNA then combines with a Cas protein to form an effector complex.

The final stage (interference) occurs when foreign DNA from the same bacteriophage or plasmid enters the cell again. The effector complex binds to the foreign DNA through base paring between the crRNA and the foreign DNA sequence.

The Cas protein then cleaves the foreign DNA, rendering it nonfunctional (**Figure 14.23c**). In this way, crRNAs serve as an adaptive RNA defense system against foreign invaders.

CONCEPTS

Piwi-interacting RNAs are found in the germ cells of animals and inhibit transposons. CRISPR RNAs are found in prokaryotes, where they function in defense against foreign DNA.

14.6 Long Noncoding RNAs Regulate Gene Expression

For many years, our knowledge of RNA was limited to those molecules that play a central role in the synthesis of proteins: mRNAs, tRNAs, and rRNAs. Later, small nuclear RNAs that participate in the posttranscriptional processing of RNA (snRNAs and snoRNAs) were added to the list. Starting in the late 1990s, geneticists began to recognize that numerous small RNAs (siRNAs, miRNAs, piRNAs, and crRNAs) were also abundant and fundamentally important to cell function. More recently, it has become apparent that much of the eukaryotic genome is transcribed—although only about 1% of the human genome, for example, directly encodes proteins, over 80% of it is transcribed, producing many long RNA molecules that do not encode proteins. Called **long noncoding RNAs (lncRNAs)**, these RNAs are typically over 200 nucleotides in length and can be as many as 100,000 nucleotides long. They lack an open reading frame—a sequence with a start and a stop codon that is translated into a protein. Thousands of lncRNAs have been discovered in the last five years. It is estimated that the human genome generates more than 10,000 different lncRNAs. The DNA sequences that encode them, along with other DNA of unknown function, have been called "the dark matter of the genome."

Although the function of many lncRNAs is still unclear, there is increasing evidence that at least some play a role in controlling gene expression. Some lncRNAs interact with proteins that regulate transcription. For example, a lncRNA called lincRNA-p21 interacts with a protein called p53, a transcription factor that activates numerous genes, including genes involved in control of the cell cycle, cell death, and other cellular functions. By repressing p53, lincRNA-p21 affects the transcription of hundreds of genes. Other lncRNAs modify chromatin structure, which also regulates transcription (see Chapter 17).

Some lncRNAs have binding sites for proteins or miRNAs that interact with mRNAs; in these cases, the lncRNAs serve as decoys for their attachment, thereby limiting the numbers of miRNAs or proteins that are available to bind to mRNAs. For example, a lncRNA called NORAD (for *no*ncoding *R*NA *a*ctivated by *D*NA *d*amage), which is present in hundreds of copies in most mammalian cells, is important in maintaining chromosome stability. NORAD has multiple binding sites for

PUMILIO proteins, which bind to and repress the stability and translation of mRNAs involved in mitosis, DNA repair, and DNA replication. The NORAD lncRNAs normally sequester PUMILIO proteins so that they are not available to bind to mRNA and repress these processes. Researchers have found that inactivation of NORAD lncRNA leads to an increase in PUMILIO proteins, causing chromosome instability and aneuploidy. Thus, the NORAD lncRNA plays an essential role in maintaining chromosome stability in mammalian cells by serving as a decoy for PUMILIO proteins.

One of the best-studied lncRNAs is *Xist* RNA, which plays a central role in dosage compensation in mammalian cells (described in Chapter 4). To balance the expression of X-linked genes in males (with one X chromosome) and females (with two X chromosomes), one of the X chromosomes in each mammalian female cell is inactivated. Which X chromosome is inactivated is random and is determined early in development; once inactivated, this chromosome remains inactive through multiple rounds of cell division. *Xist* RNA is transcribed only from the X chromosome destined to become inactive; the *Xist* RNA coats that chromosome and recruits proteins that methylate histones in the chromatin. Methylation of the chromatin inhibits the transcription of genes on the inactive X chromosome. At least two additional lncRNAs act to regulate the expression of *Xist* RNA.

Still other lncRNAs are complementary to mRNA sequences and function by base pairing with those mRNAs and preventing their translation or splicing. Evidence suggests that still other lncRNAs bring about genomic imprinting (see Chapters 4 and 21). Imprinting occurs when a gene is expressed differently depending on whether it is inherited from the male or the female parent. Many clusters of imprinted genes contain sequences that encode lncRNAs, and evidence suggests that some imprinted genes are controlled by lncRNAs.

Noncoding RNAs called **enhancer RNAs** (**eRNAs**) are transcribed from enhancer sequences and play a role in regulating the expression of protein-encoding genes, probably by facilitating the looping of DNA between enhancers and promoters (see Chapter 13). Another recently discovered group of noncoding RNAs consists of circular RNAs. Large numbers of these RNAs accumulate in cells, probably because their circular nature makes them resistant to the action of enzymes that normally degrade linear RNA molecules at their 5′ or 3′ ends. Many of these circular noncoding RNAs are thought to arise through errors in splicing in which, for example, the 5′ end of an exon is erroneously joined to the 3′ end of the same exon. Some circular RNAs contain numerous binding sites for miRNAs. Like the lncRNAs mentioned earlier, these RNAs can serve as decoys for miRNAs, preventing them from attaching to and inhibiting the translation of mRNAs.

CONCEPTS

Long noncoding RNAs are long RNA molecules that do not encode proteins. Evidence increasingly suggests that many of these molecules function in the control of gene expression. Enhancer RNAs are transcribed from enhancers and also play a role in control of gene expression. Circular noncoding RNAs, which are numerous in some cells, may serve as decoys for miRNAs.

CONCEPTS SUMMARY

■ A gene is often defined as a sequence of DNA nucleotides that is transcribed into a single RNA molecule, although the widespread occurrence and complexity of transcription makes this definition problematic.

■ Introns—noncoding sequences that interrupt the coding sequences (exons) of genes—are common in eukaryotic cells but rare in bacterial cells.

■ An mRNA molecule has three primary regions: a 5′ untranslated region, a protein-coding region, and a 3′ untranslated region.

■ Bacterial mRNA is translated immediately after transcription and undergoes little processing. The pre-mRNA of a eukaryotic protein-encoding gene is extensively processed: a modified guanine nucleotide and methyl groups, collectively termed the cap, are added to the 5′ end of pre-mRNA; the 3′ end is cleaved and a poly(A) tail is added; and introns are removed.

■ Introns are removed from pre-mRNA within a structure called the spliceosome, which is composed of several small nuclear RNAs and proteins.

■ Some introns found in rRNA genes and mitochondrial genes are self-splicing.

■ Some pre-mRNAs undergo alternative processing, in which different combinations of exons are spliced together or different 3′ cleavage sites are used.

■ Messenger RNAs may be altered by the addition, deletion, or modification of nucleotides in the coding sequence, a process called RNA editing.

■ Transfer RNAs, which attach to amino acids, are short molecules that assume a common secondary structure and contain modified bases.

■ Ribosomes, the sites of protein synthesis, are composed of several ribosomal RNA molecules and numerous proteins.

■ Small interfering RNAs, microRNAs, Piwi-interacting RNAs, and CRISPR RNAs play important roles in gene silencing and in a number of other biological processes.

■ Long noncoding RNAs are RNA molecules that do not encode proteins. Evidence increasingly suggests that many of these molecules function in the control of gene expression.

■ Circular noncoding RNAs may serve as decoys for miRNAs.

IMPORTANT TERMS

colinearity (p. 400)
exon (p. 402)
intron (p. 402)
group I intron (p. 402)
group II intron (p. 402)
nuclear pre-mRNA
  intron (p. 402)
transfer RNA intron (p. 402)
codon (p. 404)
5′ untranslated region
  (5′ UTR) (p. 404)
Shine–Dalgarno
  sequence (p. 404)

protein-coding
  region (p. 404)
3′ untranslated region
  (3′ UTR) (p. 404)
5′ cap (p. 405)
poly(A) tail (p. 406)
RNA splicing (p. 407)
5′ splice site (p. 407)
3′ splice site (p. 407)
branch point (p. 407)
spliceosome (p. 407)
lariat (p. 407)
trans-splicing (p. 408)

recursive splicing (p. 408)
alternative processing
  pathway (p. 409)
alternative splicing (p. 409)
multiple 3′ cleavage
  sites (p. 410)
RNA editing (p. 412)
guide RNA (gRNA) (p. 412)
modified base (p. 414)
tRNA-modifying
  enzyme (p. 415)
cloverleaf (p. 415)
anticodon (p. 415)

large ribosomal
  subunit (p. 416)
small ribosomal
  subunit (p. 416)
RNA interference
  (RNAi) (p. 418)
RNA-induced
  silencing complex
  (RISC) (p. 419)
long noncoding RNA
  (lncRNA) (p. 421)
enhancer RNA
  (eRNA) (p. 422)

ANSWERS TO CONCEPT CHECKS

1. When DNA was hybridized to the mRNA transcribed from it, regions of DNA that did not correspond to RNA looped out.

2. Group I introns, group II introns, nuclear pre-mRNA introns, and transfer RNA introns.

3. A protein that adds the 5′ cap is associated with RNA polymerase II, which transcribes pre-mRNAs, but is absent from RNA polymerases I and III, which transcribe rRNAs and tRNAs, respectively.

4. b

5. c

6. Guide RNA

7. b

8. d

9. An siRNA or miRNA combines with proteins to form a RISC, which then pairs with mRNA through complementary pairing between bases on the siRNA or miRNA and bases on the mRNA.

WORKED PROBLEMS

Problem 1

DNA from a eukaryotic gene was isolated, denatured, and hybridized to the mRNA transcribed from the gene; the hybridized structure was then observed with an electron microscope. The adjoining diagram shows the structure that was observed.

a. How many introns and exons are there in this gene? Explain your answer.

b. Identify and label the exons and introns in this hybridized structure.

≫ Solution Strategy

What information is required in your answer to the problem?

a. The number of introns and exons and how you arrived at your answer.

b. The location of the introns and exons labeled on the figure.

What information is provided to solve the problem?

- The DNA and mRNA are from a eukaryote.
- The DNA was denatured and hybridized to the mRNA.
- A diagram of the hybridized structure.

For help with this problem, review:
Introns in Section 14.1 and Figure 14.2.

›› Solution Steps

> **Recall:** Introns are noncoding sequences found within eukaryotic genes.

a. Each of the loops represents a region in which sequences in the DNA do not have corresponding sequences in the RNA; these regions are introns. There are five loops in the hybridized structure; so there must be five introns in the DNA and six exons.

b.

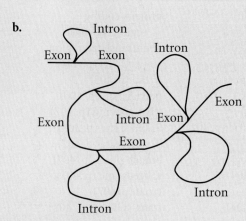

> **Hint:** The number of introns will be one less than the number of exons.

Problem 2

Draw a typical bacterial mRNA and the gene from which it was transcribed. Identify the 5′ and 3′ ends of the RNA and DNA molecules, as well as the following regions or sequences:

a. Promoter

b. 5′ untranslated region

c. 3′ untranslated region

d. Protein-coding sequence

e. Transcription start site

f. Terminator

g. Shine–Dalgarno sequence

h. Start and stop codons

> **Hint:** Review the structure of a transcription unit in Figure 13.6 and the structure of mRNA in Figure 14.5.

›› Solution Strategy

What information is required in your answer to the problem?

A drawing of an mRNA molecule and the gene from which it is transcribed. The 5′ and 3′ ends of the mRNA and DNA molecules. Locations of the listed structures on the drawing.

What information is provided to solve the problem?

- The gene is from a bacterium.
- Different parts of the DNA and RNA that are to be labeled.

For help with this problem, review:

The Template in Section 13.2 and The Structure of Messenger RNA in Section 14.2.

›› Solution Steps

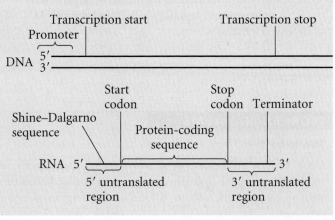

COMPREHENSION QUESTIONS

Section 14.1

1. What is the concept of colinearity? In what way is this concept fulfilled in bacterial and eukaryotic cells?

2. What are some characteristics of introns?

3. What are the four basic types of introns? In which organisms are they found?

Section 14.2

4. What are the three primary regions of mRNA sequences in bacterial cells?

5. What is the function of the Shine–Dalgarno consensus sequence?

6. What is the 5′ cap? How is the 5′ cap added to eukaryotic pre-mRNA? What is the function of the 5′ cap?

7. How is the poly(A) tail added to pre-mRNA? What is the purpose of the poly(A) tail?

8. What makes up the spliceosome? What is the function of the spliceosome?

9. Explain the process of pre-mRNA splicing in nuclear genes.

10. Describe two types of alternative processing pathways. How do these pathways lead to the production of multiple proteins from a single gene?

11. What is RNA editing? Explain the role of guide RNAs in RNA editing.

12. Summarize the different types of processing that can take place in pre-mRNA.

Section 14.3

13. What are some of the modifications in tRNA that take place through processing?

Section 14.4

14. Describe the basic structure of ribosomes in bacterial and in eukaryotic cells.

15. Explain how rRNA is processed.

Section 14.5

16. What is the origin of small interfering RNAs, microRNAs, and Piwi-interacting RNAs? What do these RNA molecules do in the cell?

17. What are some similarities and differences between siRNAs and miRNAs?

18. What role do CRISPR-Cas systems naturally play in bacteria?

19. Outline the three stages of CRISPR-Cas action.

20. Explain how some lncRNAs serve as molecular decoys for RNA-binding proteins and miRNAs.

APPLICATION QUESTIONS AND PROBLEMS

Section 14.1

*21. Duchenne muscular dystrophy is caused by a mutation in a gene that comprises 2.5 million base pairs and specifies a protein called dystrophin. However, less than 1% of the gene actually encodes the amino acids in the dystrophin protein. On the basis of what you now know about gene structure and RNA processing in eukaryotic cells, provide a possible explanation for the large size of the dystrophin gene.

22. What would happen in the experiment illustrated in **Figure 14.2** if the DNA and RNA that are mixed together came from very different organisms, for example a worm and a pig?

23. For the ovalbumin gene shown in **Figure 14.3**, where would the 5′ untranslated region and 3′ untranslated regions be located in the DNA and in the RNA?

Section 14.2

24. How do the mRNAs of bacterial cells and the pre-mRNAs of eukaryotic cells differ? How do the mature mRNAs of bacterial and eukaryotic cells differ?

25. Are the 5′ untranslated regions (5′ UTR) of eukaryotic mRNAs encoded by sequences in the promoter, exon, or intron of the gene? Explain your answer.

*26. Draw a typical eukaryotic gene and the pre-mRNA and mRNA derived from it. Assume that the gene contains three exons. Identify the following items and, for each item, give a brief description of its function:

a. 5′ untranslated region
b. Promoter
c. AAUAAA consensus sequence
d. Transcription start site
e. 3′ untranslated region
f. Introns
g. Exons
h. Poly(A) tail
i. 5′ cap

27. How would the deletion of the Shine–Dalgarno sequence affect a bacterial mRNA?

28. What would be the most likely effect of moving the AAUAAA consensus sequence shown in **Figure 14.7** 10 nucleotides upstream?

29. How would the deletion of the following sequences or features most likely affect a eukaryotic pre-mRNA?

a. AAUAAA consensus sequence
b. 5′ cap
c. Poly(A) tail

30. Suppose that a mutation occurs in the middle of a large intron of a gene encoding a protein. What would be the most likely effect of the mutation on the amino acid sequence of that protein? Explain your answer.

*31. A geneticist induces a mutation in a cell line growing in the laboratory. The mutation occurs in a gene that encodes a protein that participates in the cleavage and polyadenylation of eukaryotic mRNA. What will be the immediate effect of this mutation on RNA molecules in the cultured cells?

32. A geneticist induces a mutation in a cell line growing in the laboratory. The mutation occurs in a gene that encodes a protein that binds to the poly(A) tail of eukaryotic mRNA. What will be the immediate effect of this mutation in the cultured cells?

33. A geneticist isolates a gene that contains eight exons. He then isolates the mature mRNA produced by this gene. After making the DNA single stranded, he mixes the single-stranded DNA with the mRNA. Some of the single-stranded DNA hybridizes (pairs) with the complementary mRNA. Draw a picture of what the DNA–RNA hybrids will look like under the electron microscope.

34. A geneticist discovers that two different proteins are encoded by the same gene. One protein has 56 amino acids, and the other has 82 amino acids. Provide a possible explanation for how the same gene can encode both of these proteins.

35. Suppose that a 20-bp deletion occurs in the middle of exon 2 of the gene depicted in **Figure 14.12a**. What will be the likely effect of this deletion in the proteins produced by alternative splicing?

*36. Explain how each of the following processes complicates the concept of colinearity.

 a. Trans-splicing

 b. Alternative splicing

 c. RNA editing

Section 14.5

37. RNA interference may be triggered when inverted repeats are transcribed into an RNA molecule that then folds to form double-stranded RNA. Write out a sequence of inverted repeats within an RNA molecule. Using a diagram, show how the RNA with the inverted repeats can fold to form double-stranded RNA.

38. In the early 1990s, Carolyn Napoli and her colleagues were working on petunias, attempting to genetically engineer a variety with dark purple petals by introducing numerous copies of a gene that encodes purple pigment in the flower petals (C. Napoli, C. Lemieux, and R. Jorgensen. 1990. *Plant Cell* 2:279–289). Their thinking was that extra copies of the gene would cause more purple pigment to be produced and would result in a petunia with an even darker hue of purple. However, much to their surprise, many of the plants carrying extra copies of the purple gene were completely white or had only patches of color. Molecular analysis revealed that the amount of mRNA produced by the purple gene was reduced 50-fold in the engineered plants compared with wild-type plants. Somehow, the introduction of extra copies of the purple gene silenced both the introduced copies and the plant's own purple genes. Provide a possible explanation for how the introduction of numerous copies of the purple gene silenced all copies of the purple gene.

White petunia. [roger ashford/Alamy.]

CHALLENGE QUESTIONS

Section 14.2

39. Alternative splicing takes place in more than 95% of the human protein-encoding genes with multiple exons. Researchers have found that how a pre-mRNA is spliced is affected by the pre-mRNA's promoter sequence (D. Auboeuf et al. 2002. *Science* 298:416–419). In addition, factors that affect the rate of elongation by the RNA polymerase during transcription affect the type of splicing that takes place. These findings suggest that the process of transcription affects splicing. Propose one or more mechanisms that would explain how transcription might affect alternative splicing.

40. Duchenne muscular dystrophy (DMD) is an X-linked recessive genetic disease caused by mutations in the gene that encodes dystrophin, a large protein that plays an important role in the development of normal muscle fibers. The dystrophin gene is immense, spanning 2.5 million base pairs, and includes 79 exons and 78 introns. Many of the mutations that cause DMD produce premature stop codons, which bring protein synthesis to a halt, resulting in a greatly shortened and nonfunctional form of dystrophin. Some geneticists have proposed treating DMD patients by causing the spliceosome to skip the exon containing the stop codon. Exon skipping would produce a protein that is somewhat shortened (because an exon is skipped and some amino acids are missing), but might still result in a protein that had some function (A. Goyenvalle et al. 2004. *Science* 306:1796–1799). Propose a possible mechanism to bring about exon skipping for the treatment of DMD.

41. In eukaryotic cells, a poly(A) tail is normally added to pre-mRNA, but not to rRNA or tRNA. With the use of recombinant DNA techniques, a protein-encoding gene (which is normally transcribed by RNA polymerase II) can be connected to a promoter for RNA polymerase I. This hybrid gene is subsequently transcribed by RNA polymerase I, and the appropriate pre-mRNA is produced, but this pre-mRNA is not cleaved at the 3′ end, and a poly(A) tail is not added. Propose a mechanism to explain how the type of promoter found at the 5′ end of a gene can affect whether a poly(A) tail is added to the 3′ end.

42. SR proteins are essential to proper spliceosome assembly and are known to take part in the regulation of alternative splicing. Surprisingly, the role of SR proteins in splice-site selection and alternative splicing is affected by the promoter used for the transcription of the pre-mRNA. For example, through genetic engineering, RNA polymerase II promoters that have somewhat different sequences can be created. When pre-mRNAs with exactly the same sequences are transcribed from two different RNA polymerase II promoters that differ slightly in sequence, which promoter is used can affect how the pre-mRNA is spliced. Propose a mechanism by which the DNA sequence of an RNA polymerase II promoter could affect alternative splicing of pre-mRNA.

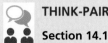

THINK-PAIR-SHARE QUESTIONS

Section 14.1

1. Eukaryotic genes are often interrupted by noncoding introns. What might be some possible reasons that organisms have evolved introns? And why might other organisms lose introns?

2. Suppose that you are at a party on Friday night, relaxing after your big genetics exam. Someone comes up to you and, hearing that you just finished your genetics exam, says, "What exactly is a gene?" How would you respond? What are the strengths and weaknesses of your definition of a gene?

Section 14.2

3. Many human genetic diseases are caused by mutations that occur at splice sites. Propose some ways that mutations at the 5′ splice site, 3′ splice site, and branch point might disrupt splicing and alter the phenotype.

Section 14.5

4. Small RNA molecules are involved in numerous genetic processes, including replication, translation, mRNA processing and degradation, inhibition of translation, chromatin modification, and protection against viruses and transposable elements. However, small DNA molecules have little or no role in these functions. Why has RNA and not DNA evolved to carry out these functions?

5. The number of microRNAs encoded by the genome varies widely among organisms: some species have many miRNA genes and other species have relatively few. Researchers have determined the number of miRNA genes possessed by different species and have made the following observations:

a. The number of miRNA genes found on a chromosome is not correlated with chromosome length. In other words, longer chromosomes do not necessarily have more miRNA genes.

b. Most species show a strong positive correlation between the number of miRNA genes on a chromosome and the number of non-protein-encoding genes on that chromosome. In other words, chromosomes with more non-protein-encoding genes have more miRNA genes.

c. Many species display a strong positive correlation between the number of miRNA genes on a chromosome and the number of protein-encoding genes on that chromosome. In other words, chromosomes with more protein-encoding genes have more miRNA genes.

Propose possible explanations for these observations.

The Genetic Code and Translation

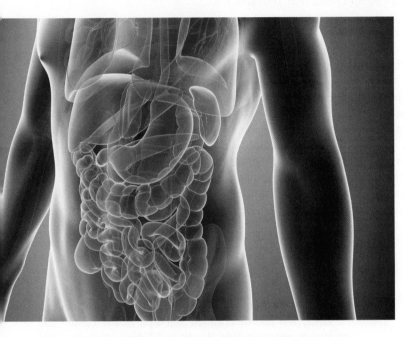

The spleen, an organ found in the upper abdomen, plays an important role in defense against infection. Isolated congenital asplenia is an autosomal dominant condition in which children are born without a spleen. [Sebastian Kaulitzki/Shutterstock.]

A Child Without a Spleen

The spleen is an often underappreciated organ. Brownish in color and weighing about a third of a pound, it sits in the left upper part of your abdomen, storing blood and filtering out bacteria and old blood cells. The spleen is underappreciated because it's widely believed that you can live without a spleen. Indeed, many people who lose their spleen to automobile accidents and other trauma do survive, although they are at increased risk of infection. But a young child without a spleen is in serious trouble. A small group of children are born without spleens; these kids are highly susceptible to life-threatening bacterial infections, and many die in childhood. This rare disorder, known as isolated congenital asplenia (ICA), is inherited as an autosomal dominant trait.

Except for the absence of a spleen, children with ICA are unaffected. But their immune function is severely compromised. When infected with bacteria that the immune system normally eliminates, these children develop raging infections that quickly spread throughout the body. Even when treated with modern antibiotics, they often die.

In 2013, an international team led by scientists from Rockefeller University discovered the genetic cause of ICA. Using the power of DNA sequencing, they examined all the coding DNA of 23 individuals with ICA and compared their DNA sequences with those of 508 individuals with normal spleens. Statistical analysis pointed to differences in one particular gene that was associated with ICA, a gene encoding ribosomal protein SA (RPSA). The RPSA protein is one of the 33 proteins that make up the small subunit of the ribosome, the organelle responsible for protein synthesis. How a defect in the *RPSA* gene results in the absence of a spleen is not known. Diseases such as ICA, which result from defective ribosomes, are referred to as ribosomopathies.

Many, but not all, individuals with ICA have mutations in *RPSA*, indicating that other genes may also be involved in the disorder. The researchers found several different types of mutations in *RPSA* associated with ICA: some caused premature stop codons, halting translation before a functional protein could be made; one was a frameshift mutation, a change that alters the way the mRNA sequence is read during translation; others changed the amino acid sequence of the RPSA protein.

One interesting but unanswered question is why a defect in *RPSA* affects only the spleen. Inherited mutations in *RPSA* occur in every cell of the body, and protein synthesis—carried out by ribosomes—is essential for numerous life processes, yet these mutations affect only the development of the spleen. Why aren't other organs altered? Why aren't numerous physiological functions affected? Scientists are still studying these important questions.

solated congenital asplenia illustrates the extreme importance of translation, the process of protein synthesis, which is the focus of this chapter. We begin by examining the molecular relation between genotype and phenotype. Next, we study the genetic code—the instructions that specify the amino acid sequence of a protein—and then examine the mechanism of translation. Our primary focus is protein synthesis in bacterial cells, but we also examine some features of this process in eukaryotic cells. At the end of the chapter, we look at some additional aspects of protein synthesis.

15.1 Many Genes Encode Proteins

The first person to suggest the existence of a relation between genotype and proteins was English physician Archibald Garrod. In 1908, Garrod correctly proposed that genes encode enzymes, but unfortunately, his theory made little impression on his contemporaries. Not until the 1940s, when George Beadle and Edward Tatum examined the genetic basis of biochemical pathways in the bread mold *Neurospora*, did the relation between genes and proteins become widely accepted. Beadle and Tatum's work helped define the relation between genotype and phenotype by leading to the one gene, one enzyme hypothesis, the idea that each gene encodes a separate enzyme.

THINK-PAIR-SHARE Question 1

The One Gene, One Enzyme Hypothesis

Beadle and Tatum used *Neurospora* to study the biochemical results of mutations. *Neurospora* is easy to cultivate in the laboratory, and the main vegetative part of the fungus is haploid, which allows the effects of otherwise recessive mutations to be easily observed (**Figure 15.1**).

Wild-type *Neurospora* grows on minimal medium, which contains only inorganic salts, nitrogen, a carbon source such as sucrose, and the vitamin biotin. The fungus can synthesize all the biological molecules that it needs from these basic compounds. However, mutations may arise that disrupt fungal growth by destroying the fungus's ability to synthesize one or more essential biological molecules.

These nutritionally deficient mutants, termed auxotrophs (see Chapter 9), cannot grow on minimal medium, but they can grow on medium that contains the substance that they are no longer able to synthesize.

Beadle and Tatum first irradiated spores of *Neurospora* to induce mutations (**Figure 15.2**). Then they placed the spores in different culture tubes with complete medium (medium containing all the biological substances needed for growth). These spores grew into fungi and produced spores by mitosis. Next, they transferred spores from each culture to tubes containing minimal medium. Fungi with auxotrophic mutations did not grow on the minimal medium, which allowed Beadle and Tatum to identify cultures that possessed mutations.

Once they had determined that a particular culture had an auxotrophic mutation, Beadle and Tatum set out to determine the specific effect of the mutation. They transferred spores of each mutant strain from complete medium to a series of tubes (see Figure 15.2), each of which contained minimal medium plus one of a variety of essential biological molecules, such as an amino acid. If the spores in a tube grew, Beadle and Tatum were able to identify the added substance as the biological molecule whose synthesis had been affected

15.1 Beadle and Tatum used the fungus *Neurospora*, which has a complex life cycle, to work out the relation of genes to proteins. [Namboori B. Raju, Stanford University.]

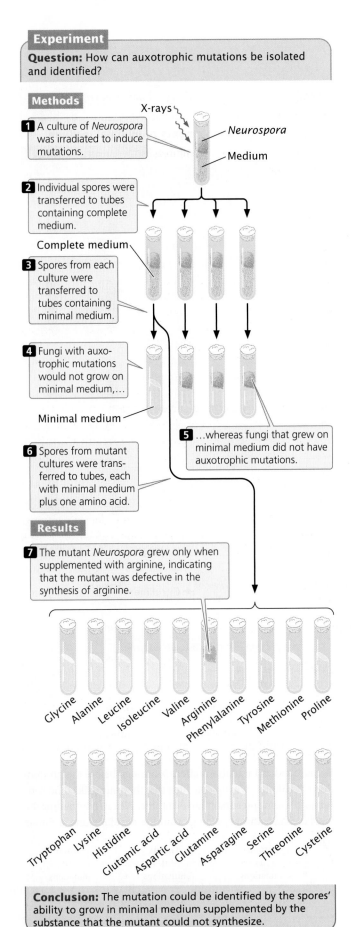

Experiment

Question: How can auxotrophic mutations be isolated and identified?

Methods

X-rays

Neurospora

Medium

1 A culture of *Neurospora* was irradiated to induce mutations.

2 Individual spores were transferred to tubes containing complete medium.

Complete medium

3 Spores from each culture were transferred to tubes containing minimal medium.

4 Fungi with auxotrophic mutations would not grow on minimal medium,...

Minimal medium

5 ...whereas fungi that grew on minimal medium did not have auxotrophic mutations.

6 Spores from mutant cultures were transferred to tubes, each with minimal medium plus one amino acid.

Results

7 The mutant *Neurospora* grew only when supplemented with arginine, indicating that the mutant was defective in the synthesis of arginine.

Glycine Alanine Leucine Isoleucine Valine Arginine Phenylalanine Tyrosine Methionine Proline

Tryptophan Lysine Histidine Glutamic acid Aspartic acid Glutamine Asparagine Serine Threonine Cysteine

Conclusion: The mutation could be identified by the spores' ability to grow in minimal medium supplemented by the substance that the mutant could not synthesize.

15.2 Beadle and Tatum developed a method for isolating auxotrophic mutants in *Neurospora*.

by the mutation. For example, an auxotrophic mutant that would grow only on minimal medium to which arginine had been added must have possessed a mutation that disrupts the synthesis of arginine.

Adrian Srb and Norman H. Horowitz patiently applied this procedure to genetically dissect the multistep biochemical pathway of arginine synthesis (**Figure 15.3**). They first isolated a series of auxotrophic mutants whose growth required arginine. Then they tested these mutants for their ability to grow on minimal medium supplemented with three compounds: ornithine, citrulline, and arginine. From the results, they were able to place the mutants into three groups on the basis of which of the substances allowed growth (**Table 15.1**).

Based on these results, Srb and Horowitz proposed that the biochemical pathway leading to the amino acid arginine has at least three steps:

$$\text{precursor} \xrightarrow{\overset{\text{Step}}{1}} \text{ornithine} \xrightarrow{\overset{\text{Step}}{2}} \text{citrulline} \xrightarrow{\overset{\text{Step}}{3}} \text{arginine}$$

They concluded that the mutations in group I affect step 1 of this pathway, mutations in group II affect step 2, and mutations in group III affect step 3. But how did they know that the order of the compounds in the biochemical pathway was correct?

Notice that if step 1 is blocked by a mutation, then the addition of either ornithine or citrulline allows growth because these compounds can still be converted into arginine (see Figure 15.3). Similarly, if step 2 is blocked, the addition of citrulline allows growth, but the addition of ornithine has no effect. If step 3 is blocked, the spores will grow only if arginine is added to the medium. The underlying principle is that an auxotrophic mutant cannot synthesize any compound that comes after the step blocked by a mutation.

Using this reasoning with the information in Table 15.1, we can see that the addition of arginine to the medium allows all three groups of mutants to grow. Therefore, biochemical steps affected by all the mutants precede the step that results in arginine. The addition of citrulline allows group I and

TABLE 15.1	Growth of arginine auxotrophic mutants on minimal medium with various supplements		
Mutant Strain	**Ornithine**	**Citrulline**	**Arginine**
Group I	+	+	+
Group II	−	+	+
Group III	−	−	+

Note: A plus sign (+) indicates growth; a minus sign (−) indicates no growth.

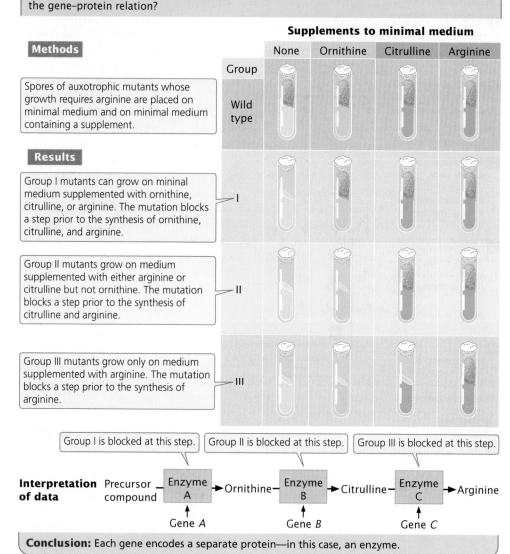

15.3 Method used to determine the relation between genes and enzymes in *Neurospora*. The biochemical pathway shown here leads to the synthesis of arginine in *Neurospora*. Steps in the pathway are catalyzed by enzymes affected by mutations.

group II mutants to grow, but not group III mutants; therefore, group III mutations must affect a biochemical step that takes place after the production of citrulline but before the production of arginine:

$$\text{citrulline} \xrightarrow[\text{mutations}]{\substack{\text{Group} \\ \text{III}}} \text{arginine}$$

The addition of ornithine allows the growth of group I mutants, but not group II or group III mutants; thus, mutations in groups II and III affect steps that come after the production of ornithine. We've already established that group II mutations affect a step before the production of citrulline; so group II mutations must block the conversion of ornithine into citrulline:

$$\text{ornithine} \xrightarrow[\text{mutations}]{\substack{\text{Group} \\ \text{II}}} \text{citrulline} \xrightarrow[\text{mutations}]{\substack{\text{Group} \\ \text{III}}} \text{arginine}$$

Because group I mutations affect some step before the production of ornithine, we can conclude that they must affect the conversion of some precursor into ornithine. We can now outline the biochemical pathway yielding ornithine, citrulline, and arginine:

$$\text{precursor} \xrightarrow[\text{mutations}]{\substack{\text{Group} \\ \text{I}}} \text{ornithine} \xrightarrow[\text{mutations}]{\substack{\text{Group} \\ \text{II}}} \text{citrulline} \xrightarrow[\text{mutations}]{\substack{\text{Group} \\ \text{III}}} \text{arginine}$$

Importantly, this procedure does not necessarily detect all steps in a pathway; rather, it detects only the steps that produce the compounds tested.

Using mutations and this type of reasoning, Beadle, Tatum, and others were able to identify the genes that control several biosynthetic pathways in *Neurospora*. They established that each step in a pathway is controlled by a different enzyme, as shown in Figure 15.3 for the arginine pathway. In addition, by conducting genetic crosses and mapping experiments (see Chapter 7), they were able to demonstrate that mutations affecting any one step in a pathway always occurred at the same chromosomal location. Beadle and Tatum reasoned that mutations affecting a particular biochemical step occurred at a single locus that encoded a particular enzyme. This idea became known as the **one gene, one enzyme hypothesis**: genes function by encoding enzymes, and each gene encodes a separate enzyme. Although the genes Beadle and Tatum examined encoded enzymes, many genes encode proteins that are not enzymes, so more generally, their idea was that each gene encodes a protein. When research findings showed that some proteins are composed of more than one polypeptide chain and that different polypeptide chains are encoded by separate genes, this model was modified to become the **one gene, one polypeptide hypothesis**. ▶ TRY PROBLEM 16

THINK-PAIR-SHARE Question 2

CONCEPTS

Beadle and Tatum's studies of biochemical pathways in the fungus *Neurospora* helped define the relation between genotype and phenotype by establishing the one gene, one enzyme hypothesis, the idea that each gene encodes a separate enzyme. This idea was later modified to the one gene, one polypeptide hypothesis.

✔ **CONCEPT CHECK 1**

Auxotrophic mutation 103 grows on minimal medium supplemented with A, B, or C; mutation 106 grows on medium supplemented with A or C, but not B; and mutation 102 grows only on medium supplemented with C. What is the order of A, B, and C in a biochemical pathway?

The Structure and Function of Proteins

Proteins are central to all living processes (**Figure 15.4**). Many proteins are enzymes, the biological catalysts that drive the chemical reactions of the cell; others are structural components, providing scaffolding and support for membranes, filaments, bone, and hair. Some proteins help transport substances; others have a regulatory, communication, or defense function.

AMINO ACIDS All proteins are polymers composed of **amino acids**, linked end to end. Twenty common amino acids are found in proteins; these amino acids are shown in **Figure 15.5** with both their three-letter and one-letter abbreviations (other amino acids sometimes found in proteins are modified forms of these common amino acids). All of the common amino acids are similar in structure: each consists of a central carbon atom bonded to an amino group, a hydrogen atom, a carboxyl group, and an R (radical) group that differs for each amino acid.

The amino acids in proteins are joined together by **peptide bonds** (**Figure 15.6**) to form **polypeptide** chains; a protein consists of one or more polypeptide chains. Like nucleic acids, polypeptides have polarity under physiological conditions: one end (often called the *amino end*) has a free amino group (NH_3^+) and the other end (the *carboxyl end*) has a free carboxyl group (COO^-). Proteins consist of 50 or more amino acids; some have as many as several thousand.

PROTEIN STRUCTURE Like that of nucleic acids, the molecular structure of proteins has several levels of organization. The *primary structure* of a protein is its sequence of amino acids (**Figure 15.7a**). Through interactions between neighboring amino acids, a polypeptide chain folds and twists into a *secondary structure* (**Figure 15.7b**). Two common secondary structures found in proteins are the beta (β) pleated sheet and the alpha (α) helix.

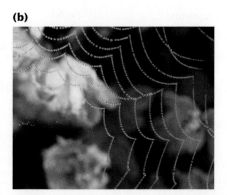

15.4 Proteins serve a number of biological functions. (a) The light produced by fireflies is the result of a light-producing reaction between luciferin and ATP catalyzed by the enzyme luciferase. (b) The protein fibroin is the major structural component of spider webs. (c) Castor beans contain a highly toxic protein called ricin. [Part a: Darwin Dale/Science Source. Part b: Rosemary Calvert/Imagestate/Media Bakery. Part c: Paroli Galperti/© Cuboimages/Photoshot.]

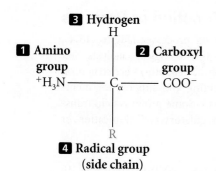

3 Hydrogen

1 Amino group

2 Carboxyl group

^+H_3N———C_α——— COO^-

4 Radical group (side chain)

15.5 The 20 common amino acids that make up proteins have similar structures. Each amino acid consists of a central carbon atom ($C\alpha$) attached to (1) an amino group (NH_3^+), (2) a carboxyl group (COO^-), (3) a hydrogen atom (H), and (4) a radical group, designated R. In the structures shown here, the parts in black are common to all amino acids and the parts in red are the R groups.

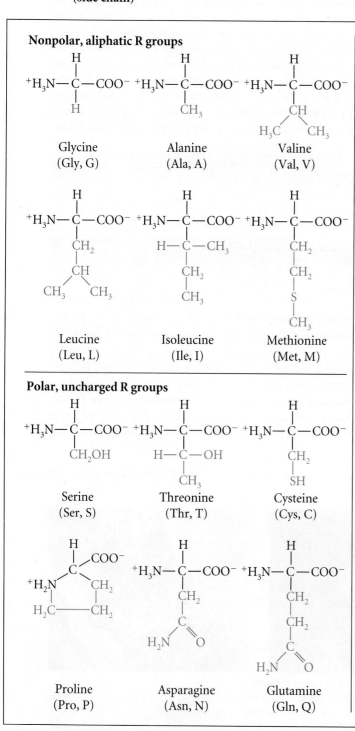

Nonpolar, aliphatic R groups

Glycine (Gly, G)

Alanine (Ala, A)

Valine (Val, V)

Leucine (Leu, L)

Isoleucine (Ile, I)

Methionine (Met, M)

Polar, uncharged R groups

Serine (Ser, S)

Threonine (Thr, T)

Cysteine (Cys, C)

Proline (Pro, P)

Asparagine (Asn, N)

Glutamine (Gln, Q)

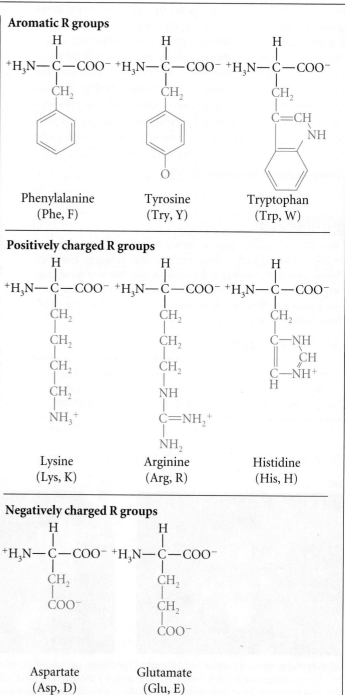

Aromatic R groups

Phenylalanine (Phe, F)

Tyrosine (Try, Y)

Tryptophan (Trp, W)

Positively charged R groups

Lysine (Lys, K)

Arginine (Arg, R)

Histidine (His, H)

Negatively charged R groups

Aspartate (Asp, D)

Glutamate (Glu, E)

$$^+H_3N-CH-C-O^-\quad ^+H-N-CH-COO^-$$

with R¹ and R², with O double bond, and H on nitrogen

$$\downarrow H_2O$$

$$^+H_3N-CH-C-N-CH-COO^-$$

Peptide bond

15.6 Amino acids are joined together by peptide bonds. In a peptide bond (pink shading), the carboxyl group of one amino acid is covalently attached to the amino group of another amino acid.

Secondary structures interact and fold further to form a *tertiary structure* (**Figure 15.7c**), which is the overall, three-dimensional shape of the protein. The secondary and tertiary structures of a protein are largely determined by the primary structure—the amino acid sequence—of the protein. Finally, some proteins consist of two or more polypeptide chains that associate to produce a *quaternary structure* (**Figure 15.7d**). Many proteins have an additional level of organization defined by domains. A *domain* is a group of amino acids that forms a discrete functional unit within the protein. For example, there are several different types of protein domains that function in DNA binding (see Chapter 16). Gene sequences that encode protein domains are discussed in Chapter 20.

THINK-PAIR-SHARE Question 3

CONCEPTS

The products of many genes are proteins whose functions produce the traits specified by these genes. Proteins are polymers consisting of amino acids linked by peptide bonds. The amino acid sequence of a protein is its primary structure. This structure folds to create the secondary and tertiary structures; two or more polypeptide chains may associate to create a quaternary structure.

✔ **CONCEPT CHECK 2**

What primarily determines the secondary and tertiary structures of a protein?

15.2 The Genetic Code Determines How the Nucleotide Sequence Specifies the Amino Acid Sequence of a Protein

In 1953, James Watson and Francis Crick solved the structure of DNA and identified its base sequence as the carrier of genetic information (see Chapter 10). However, the way in which the base sequence of DNA specifies the amino acid sequences of proteins (the genetic code) remained elusive for another 10 years.

One of the first questions about the genetic code to be addressed was how many nucleotides are necessary to specify a single amino acid. The set of nucleotides that encode a single amino acid—the basic unit of the genetic code—is called a *codon* (see Chapter 14). Many early investigators recognized that codons must contain a minimum of three nucleotides. Each nucleotide position in mRNA can be occupied by one of four bases: A, G, C, or U. If a codon consisted

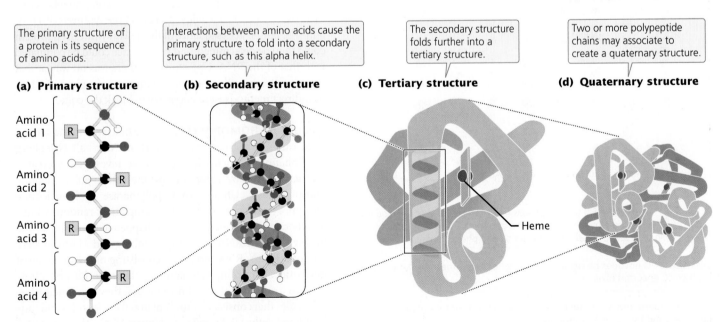

The primary structure of a protein is its sequence of amino acids.

Interactions between amino acids cause the primary structure to fold into a secondary structure, such as this alpha helix.

The secondary structure folds further into a tertiary structure.

Two or more polypeptide chains may associate to create a quaternary structure.

(a) Primary structure   **(b) Secondary structure**   **(c) Tertiary structure**   **(d) Quaternary structure**

Amino acid 1
Amino acid 2
Amino acid 3
Amino acid 4

Heme

15.7 Proteins have several levels of structural organization.

Experiment

Question: What amino acids are specified by codons composed of only one type of base?

Methods

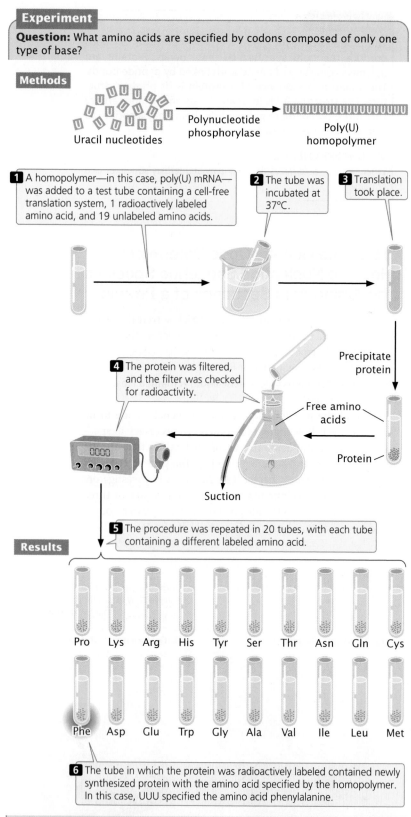

1 A homopolymer—in this case, poly(U) mRNA—was added to a test tube containing a cell-free translation system, 1 radioactively labeled amino acid, and 19 unlabeled amino acids.

2 The tube was incubated at 37°C.

3 Translation took place.

Uracil nucleotides → Polynucleotide phosphorylase → Poly(U) homopolymer

Precipitate protein

4 The protein was filtered, and the filter was checked for radioactivity.

Free amino acids

Suction

Protein

5 The procedure was repeated in 20 tubes, with each tube containing a different labeled amino acid.

Results

Pro Lys Arg His Tyr Ser Thr Asn Gln Cys

Phe Asp Glu Trp Gly Ala Val Ile Leu Met

6 The tube in which the protein was radioactively labeled contained newly synthesized protein with the amino acid specified by the homopolymer. In this case, UUU specified the amino acid phenylalanine.

Conclusion: UUU encodes phenylalanine; in other experiments, AAA encoded lysine, and CCC encoded proline.

15.8 Nirenberg and Matthaei developed a method for identifying the amino acid specified by an RNA homopolymer.

---

of a single nucleotide, only four different codons (A, G, C, and U) would be possible, which is not enough to encode the 20 different amino acids commonly found in proteins. If codons were made up of two nucleotides each (i.e., GU, AC, etc.), there would be $4 \times 4 = 16$ possible codons—still not enough to encode all 20 amino acids. With three nucleotides per codon, there are $4 \times 4 \times 4 = 64$ possible codons, which is more than enough to specify 20 different amino acids. Therefore, a *triplet code* requiring three nucleotides per codon would be the most efficient way to encode all 20 amino acids. Using mutations in bacteriophages, Francis Crick and his colleagues confirmed in 1961 that the genetic code is indeed a triplet code. **▶ TRY PROBLEMS 20 AND 21**

THINK-PAIR-SHARE Question 4

CONCEPTS

The genetic code is a triplet code, in which three nucleotides encode each amino acid in a protein.

✔ **CONCEPT CHECK 3**

A codon is
a. one of three nucleotides that encode an amino acid.
b. three nucleotides that encode an amino acid.
c. three amino acids that encode a nucleotide.
d. one of four bases in DNA.

Breaking the Genetic Code

Once it had been firmly established that the genetic code consists of codons that are three nucleotides in length, the next step was to determine which groups of three nucleotides specify which amino acids. Logically, the easiest way to break the code would have been to determine the base sequence of a piece of RNA, add it to a test tube containing all the components necessary for translation, and allow it to direct the synthesis of a protein. The amino acid sequence of the newly synthesized protein could then be determined, and its sequence could be compared with that of the RNA. Unfortunately, there was no way at that time to determine the nucleotide sequence of a piece of RNA, so indirect methods were necessary to break the code.

THE USE OF HOMOPOLYMERS The first clues to the genetic code came in 1961, from the work of Marshall Nirenberg and Johann Heinrich Matthaei. These investigators created synthetic RNAs by using an enzyme called polynucleotide phosphorylase. Unlike RNA polymerase, polynucleotide phosphorylase does not require a template; it randomly links together any RNA nucleotides that happen to be available. The first synthetic mRNAs used by Nirenberg and Matthaei were homopolymers, RNA molecules consisting of a single type of nucleotide. For example, by adding polynucleotide phosphorylase to a solution of uracil nucleotides, they generated RNA molecules that consisted entirely of uracil nucleotides and thus contained only UUU codons (**Figure 15.8**). These poly(U)

RNAs were then added to 20 test tubes, each containing the components necessary for translation and all 20 amino acids. A different amino acid was radioactively labeled in each of the 20 tubes. Radioactive protein appeared in only one of the tubes—the one containing labeled phenylalanine (see Figure 15.8). This result showed that the codon UUU specifies the amino acid phenylalanine. The results of similar experiments using poly(C) and poly(A) RNA demonstrated that CCC encodes proline and AAA encodes lysine; for technical reasons, the results from poly(G) were uninterpretable.

THE USE OF RANDOM COPOLYMERS To gain information about additional codons, Nirenberg and his colleagues created synthetic RNAs containing two or three different bases. Because polynucleotide phosphorylase incorporates nucleotides randomly, these RNAs contain random mixtures of the bases and are thus called random copolymers. For example, when adenine and cytosine nucleotides were mixed with polynucleotide phosphorylase, the RNA molecules produced had eight different codons: AAA, AAC, ACC, ACA, CAA, CCA, CAC, and CCC. These poly(AC) RNAs produced proteins containing six different amino acids: asparagine, glutamine, histidine, lysine, proline, and threonine.

The proportions of the different amino acids in the proteins produced depended on the ratio of the two nucleotides used in creating the random copolymers, and the theoretical probability of finding a particular codon could be calculated from the ratios of the bases. If a 4 : 1 ratio of C to A were used in making the RNA, then the probability of C being in any given position in a codon would be $^4/_5$ and the probability of A being in it would be $^1/_5$. With random incorporation of bases, the probability of any one of the codons with two Cs and one A (CCA, CAC, or ACC) would be $^4/_5 \times ^4/_5 \times ^1/_5 = ^{16}/_{125} = 0.13$, or 13%, and the probability of any codon with two As and one C (AAC, ACA, or CAA) would be $^1/_5 \times ^1/_5 \times ^4/_5 = ^4/_{125} = 0.032$, or about 3%. Therefore, an amino acid encoded by two Cs and one A should be more common than an amino acid encoded by two As and one C.

By comparing the percentages of amino acids in proteins produced by random copolymers with the theoretical frequencies expected for the codons, Nirenberg and his colleagues could derive information about the base *composition* of the codons. These experiments revealed nothing, however, about the codon base *sequence*; histidine was clearly encoded by a codon with two Cs and one A, but whether that codon was ACC, CAC, or CCA was unknown. There were other problems with this method: the theoretical calculations depended on the random incorporation of bases, which did not always occur; furthermore, because the genetic code is redundant, sometimes several different codons specify the same amino acid.

THE USE OF RIBOSOME-BOUND tRNAs To overcome the limitations of random copolymers, Nirenberg and Philip Leder developed another technique in 1964 that used ribosome-bound tRNAs. They found that a very short

sequence of mRNA—even one consisting of a single codon—would bind to a ribosome. The codon on the short mRNA would then base pair with the matching anticodon on a tRNA that carried the amino acid specified by the codon (**Figure 15.9**). When short mRNAs that were bound to ribosomes were mixed with tRNAs and amino acids, and this

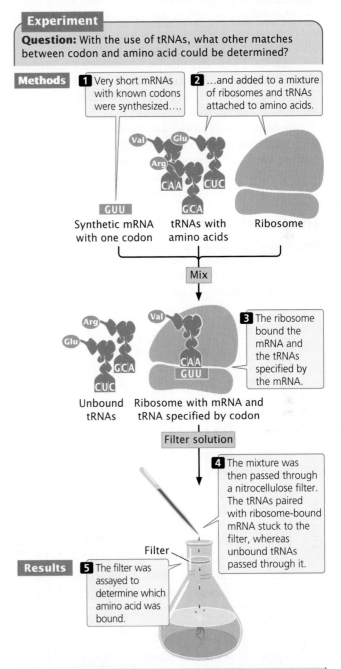

Experiment

Question: With the use of tRNAs, what other matches between codon and amino acid could be determined?

Methods

1 Very short mRNAs with known codons were synthesized....

2 ...and added to a mixture of ribosomes and tRNAs attached to amino acids.

GUU
Synthetic mRNA with one codon

tRNAs with amino acids

Ribosome

Mix

3 The ribosome bound the mRNA and the tRNAs specified by the mRNA.

Unbound tRNAs

Ribosome with mRNA and tRNA specified by codon

Filter solution

4 The mixture was then passed through a nitrocellulose filter. The tRNAs paired with ribosome-bound mRNA stuck to the filter, whereas unbound tRNAs passed through it.

Filter

Results **5** The filter was assayed to determine which amino acid was bound.

Conclusion: When an mRNA with GUU was added, the tRNAs on the filter were bound to valine; therefore, the codon GUU specifies valine. Many other codons were determined by using this method.

15.9 Nirenberg and Leder used ribosome-bound tRNAs to provide additional information about the genetic code.

mixture was passed through a nitrocellulose filter, the tRNAs that were paired with the ribosome-bound mRNA stuck to the filter, whereas unbound tRNAs passed through it. The advantage of this system was that it could be used with very short synthetic mRNA molecules that could be synthesized with a known sequence. Nirenberg and Leder synthesized more than 50 short mRNAs with known codons and added them individually to a mixture of ribosomes and tRNAs with amino acids. They then isolated the tRNAs that were bound to the mRNAs and ribosomes and determined which amino acids were present on the bound tRNAs. For example, synthetic mRNA with the codon GUU retained a tRNA to which valine was attached, whereas mRNAs with the codons UGU and UUG did not. Using this method, Nirenberg and his colleagues were able to determine the amino acids encoded by more than 50 codons.

Other experiments provided additional information about the genetic code, and it was fully deciphered by 1968. Let's examine some of the features of the code, which is so important to modern biology that Francis Crick compared its place to that of the periodic table of the elements in chemistry.

The Degeneracy of the Code

One amino acid is encoded by three consecutive nucleotides in mRNA, and each nucleotide can have one of four possible bases (A, G, C, or U), so there are $4^3 = 64$ possible codons (**Figure 15.10**). Three of these codons are stop codons, which specify the end of translation, as we'll see shortly. Thus, 61 codons, called **sense codons**, encode amino acids. Because there are 61 sense codons and only 20 different amino acids commonly found in proteins, the code contains more information than is needed to specify the amino acids and thus is said to be **degenerate**. This expression does not mean that the genetic code is depraved; *degenerate* is a term that Francis Crick borrowed from quantum physics, where it describes multiple physical states that have equivalent meaning. The degeneracy of the genetic code means that the code is redundant: amino acids may be specified by more than one codon. Only tryptophan and methionine are encoded by a single codon (see Figure 15.10). Other amino acids are specified by two or more codons, and some, such as leucine, are specified by six different codons. Codons that specify the same amino acid are said to be **synonymous codons**, just as synonymous words are different words that have the same meaning.

As we learned in Chapter 14, tRNAs serve as adapter molecules that bind particular amino acids and deliver them to a ribosome, where the amino acids are then assembled into polypeptide chains. Each type of tRNA attaches to a single type of amino acid. The cells of most organisms possess about 30 to 50 different tRNAs, yet there are only 20 different amino acids commonly found in proteins. Thus, some amino acids are carried by more than one tRNA. Different tRNAs that accept the same amino acid but have different anticodons are called **isoaccepting tRNAs**.

Even though some amino acids can pair with multiple (isoaccepting) tRNAs, there are still more codons than anticodons. One anticodon can pair with different codons through flexibility in base pairing at the third position of the codon. Examination of Figure 15.10 reveals that many synonymous codons differ only in the third position. For example, serine is encoded by the codons UCU, UCC, UCA, and UCG, all of which begin with UC. When the codon of the mRNA and the anticodon of the tRNA join (**Figure 15.11**), the first (5′) base of the codon forms hydrogen bonds with

Second base

		U	C	A	G	
First base	**U**	UUU Phe UUC UUA Leu UUG	UCU UCC Ser UCA UCG	UAU Tyr UAC UAA Stop UAG Stop	UGU Cys UGC UGA Stop UGG Trp	U C A G
	C	CUU CUC Leu CUA CUG	CCU CCC Pro CCA CCG	CAU His CAC CAA Gln CAG	CGU CGC Arg CGA CGG	U C A G
	A	AUU AUC Ile AUA AUG Met	ACU ACC Thr ACA ACG	AAU Asn AAC AAA Lys AAG	AGU Ser AGC AGA Arg AGG	U C A G
	G	GUU GUC Val GUA GUG	GCU GCC Ala GCA GCG	GAU Asp GAC GAA Glu GAG	GGU GGC Gly GGA GGG	U C A G

(Third base indicated on right side of table)

15.10 The genetic code consists of 64 codons. The amino acids specified by each codon are given in their three-letter abbreviation. The codons are written 5′→3′, as they appear in the mRNA. AUG is an initiation (start) codon as well as the codon for methionine; UAA, UAG, and UGA are termination (stop) codons.

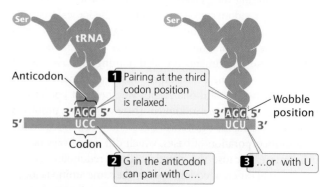

15.11 Wobble may exist in the pairing of a codon and anticodon. The mRNA and tRNA pair in an antiparallel fashion. Pairing at the first and second codon positions is in accord with the Watson-and-Crick rules (A with U, G with C); however, pairing rules are relaxed at the third position of the codon, and G on the anticodon can pair with either U or C on the codon in this example.

the third (3′) base of the anticodon, strictly according to the Watson-and-Crick base-pairing rules: A with U; C with G. Next, the middle bases of codon and anticodon pair, also strictly following the Watson-and-Crick rules. After these pairs have bonded, the third bases pair weakly, and there may be flexibility, or **wobble**, in their pairing.

In 1966, Francis Crick developed the wobble hypothesis, which proposed that some nonstandard pairings of bases could take place at the third position of a codon. For example, a G in the anticodon may pair with either a C or a U in the third position of the codon (**Table 15.2**; see p. 415 in Chapter 14). The important thing to remember about wobble is that it allows some tRNAs to pair with more than one mRNA codon; thus, from 30 to 50 tRNAs can pair with 61 sense codons. Some codons are synonymous through wobble. **> TRY PROBLEM 26**

THINK-PAIR-SHARE Question 5

TABLE 15.2	The wobble rules, indicating which bases in the third position (3′ end) of an mRNA codon can pair with which bases at the first position (5′ end) of a tRNA anticodon

First Position of Anticodon	Third Position of Codon	Pairing
C	G	Anticodon 3′–X—Y—C–5′ | | | 5′–Y—X—G–3′ Codon
G	U or C	Anticodon 3′–X—Y—G–5′ | | | 5′–Y—X—U–3′ C Codon
A	U	Anticodon 3′–X—Y—A–5′ | | | 5′–Y—X—U–3′ Codon
U	A or G	Anticodon 3′–X—Y—U–5′ | | | 5′–Y—X—A–3′ G Codon

I (inosine)*	A, U, or C	Anticodon
		3′–X—Y—I–5′
		| | |
		5′–Y—X—A–3′
		        U
		        C
		Codon

*Inosine is one of the modified bases found in tRNAs.

CONCEPTS

The genetic code includes 61 sense codons that specify the 20 common amino acids. The code is degenerate, meaning that some amino acids are encoded by more than one codon. Isoaccepting tRNAs are tRNAs with different anticodons that specify the same amino acid. Wobble at the third position of the codon allows different codons to specify the same amino acid.

✔ CONCEPT CHECK 4

Through wobble, a single _____ can pair with more than one _____.

a. codon, anticodon
b. group of three nucleotides in DNA, codon in mRNA
c. tRNA, amino acid
d. anticodon, codon

The Reading Frame and Initiation Codons

Findings from early studies of the genetic code indicated that the code is generally **nonoverlapping**. An overlapping code would be one in which a single nucleotide might be included in more than one codon, as follows:

Nucleotide sequence   A U A C G A G U C

Nonoverlapping code   A U A C G A G U C
                        Ile   Arg   Val

Overlapping code   A U A C G A G U
            Ile
            U A C
            Tyr
              A C G
              Thr

Usually, however, each nucleotide is part of a single codon. A few overlapping genes are found in viruses, but codons within the same gene do not overlap, and the genetic code is generally considered to be nonoverlapping.

For any sequence of nucleotides, there are three potential sets of codons—three ways in which the sequence can be read in groups of three. Each different way of reading the sequence is called a **reading frame**, and any sequence of nucleotides has three potential reading frames. The three reading frames have completely different sets of codons and therefore specify proteins with entirely different amino acid sequences. Thus, it is essential for the translation machinery to use the correct reading frame. How is the correct reading frame established? The reading frame is set by the **initiation codon** (or **start codon**), which is the first codon of the mRNA to specify an amino acid. After the initiation codon, the other codons are read as successive groups of three nucleotides. No bases are skipped between the codons, so there are no punctuation marks to separate the codons.

The initiation codon is usually AUG, although GUG and UUG are used on rare occasions. The initiation codon is not just a sequence that marks the beginning of translation; it also specifies an amino acid. In bacterial cells, the first AUG encodes a modified type of methionine, *N*-formylmethionine; thus, all proteins in bacteria initially begin with this amino acid, but its formyl group (or, in some cases, the entire amino acid) may be removed after the protein has been synthesized. When the codon AUG is at an internal position in a gene, it encodes unformylated methionine. In archaeal and eukaryotic cells, AUG specifies unformylated methionine, both at the initiation position and at internal positions. In both bacteria and eukaryotes, there are different tRNAs for the initiator methionine (designated $tRNA_i^{fMet}$ in bacteria and $tRNA_i^{Met}$ in eukaryotes) and internal methionine (designated $tRNA^{Met}$).

Termination Codons

Three codons—UAA, UAG, and UGA—do not encode amino acids. These codons, which signal the end of translation in both bacterial and eukaryotic cells, are called **stop codons**, **termination codons**, or **nonsense codons**. No tRNAs have anticodons that pair with termination codons.

THINK-PAIR-SHARE Question 6

TABLE 15.3	Some exceptions to the universal genetic code		
Genome	**Codon**	**Universal Code**	**Altered Code**
Bacterial DNA			
Mycoplasma capricolum	UGA	Stop	Trp
Mitochondrial DNA			
Human	UGA	Stop	Trp
Human	AUA	Ile	Met
Human	AGA, AGG	Arg	Stop
Yeast	UGA	Stop	Trp
Trypanosomes	UGA	Stop	Trp
Plants	CGG	Arg	Trp
Nuclear DNA			
Tetrahymena	UAA	Stop	Gln
Paramecium	UAG	Stop	Gln

The Universality of the Code

For many years, the genetic code was assumed to be **universal**, meaning that each codon specifies the same amino acid in all organisms. We now know that the genetic code is mostly, but not completely, universal; some exceptions have been found. Most of these exceptions are termination codons, but there are a few cases in which one sense codon substitutes for another. Many of these exceptions are found in mitochondrial genes; some nonuniversal codons have also been detected in the nuclear genes of protozoans and in bacterial DNA (**Table 15.3**). One study of bacteria and bacteriophages isolated from 1776 environmental samples found nonuniversal codons in a substantial fraction, suggesting that nonuniversal codons may be more common than previously thought. ▶ **TRY PROBLEM 22**

THINK-PAIR-SHARE Question 7

CONCEPTS

Each sequence of nucleotides possesses three potential reading frames. The correct reading frame is set by the initiation codon. The end of a protein-coding sequence is marked by a termination codon. With a few exceptions, all organisms use the same genetic code.

✔ **CONCEPT CHECK 5**

Do the initiation and termination codons specify amino acids? If so, which ones?

CONNECTING CONCEPTS

Characteristics of the Genetic Code

We have now considered a number of characteristics of the genetic code. Let's take a moment to review these characteristics.

1. The genetic code consists of a sequence of nucleotides in DNA or RNA. There are four letters in the code, corresponding to the four bases—A, G, C, and U (T in DNA).
2. The genetic code is a triplet code. Each amino acid is encoded by a sequence of three consecutive nucleotides, called a codon.
3. The genetic code is degenerate: of 64 codons, 61 codons encode only 20 amino acids in proteins (3 codons are termination codons). Some codons are synonymous, specifying the same amino acid.
4. Isoaccepting tRNAs are tRNAs with different anticodons that accept the same amino acid. Wobble allows the anticodon on one type of tRNA to pair with more than one type of codon on mRNA.
5. The genetic code is generally nonoverlapping; each nucleotide in an mRNA sequence belongs to a single reading frame.
6. The reading frame is set by an initiation codon, which is usually AUG.
7. When a reading frame has been set, codons are read as successive groups of three nucleotides.
8. Any one of three termination codons (UAA, UAG, or UGA) can signal the end of translation; no amino acids are encoded by the termination codons.
9. The genetic code is almost universal.

15.3 Amino Acids Are Assembled into a Protein Through Translation

Now that we are familiar with the genetic code, we can begin to study how amino acids are assembled into proteins. Because more is known about translation in bacteria than in eukaryotes, we will focus primarily on bacterial translation. In most respects, eukaryotic translation is similar, although some significant differences will be noted.

Remember that only mRNAs are translated into proteins. Translation takes place on ribosomes; indeed, ribosomes can be thought of as moving protein-synthesizing machines. Through a variety of techniques, a detailed view of the structure of the ribosome has been produced in recent years, which has greatly improved our understanding of translation. A ribosome attaches near the 5′ end of an mRNA strand and moves toward the 3′ end, translating the codons as it goes (**Figure 15.12**). Synthesis begins at the amino end of the protein, and the protein is elongated by the addition of new amino acids to the carboxyl end. Protein synthesis includes a series of RNA–RNA interactions: interactions between the mRNA and the rRNA that hold the mRNA in the ribosome, between the codon on the mRNA and the anticodon on the tRNA, and between the tRNA and the rRNAs of the ribosome.

Protein synthesis can be conveniently divided into four stages: (1) tRNA charging, in which tRNAs bind to amino acids; (2) initiation, in which the components necessary for translation are assembled at the ribosome; (3) elongation, in which amino acids are joined, one at a time, to the growing polypeptide chain; and (4) termination, in which protein synthesis halts at the termination codon and the translation components are released from the ribosome.

The Binding of Amino Acids to Transfer RNAs

The first stage of translation is the binding of tRNA molecules to their appropriate amino acids. As we have seen, each tRNA is specific for a particular amino acid. All tRNAs have the sequence CCA at the 3′ end, and the carboxyl group (COO^-) of the amino acid is attached to the adenine nucleotide at the 3′ end of the tRNA (**Figure 15.13**). If each tRNA is specific for a particular amino acid, but all amino acids are attached to the same nucleotide (A) at the 3′ end of a tRNA, how does a tRNA link up with its appropriate amino acid?

The key to specificity between an amino acid and its tRNA is a set of enzymes called **aminoacyl-tRNA synthetases**. A cell has 20 different aminoacyl-tRNA synthetases, one for each of the 20 amino acids. Each synthetase recognizes a particular amino acid as well as all the tRNAs that accept that amino acid. Its recognition of the appropriate amino acid is based on the different sizes, charges, and R groups of the amino acids. Its recognition of the appropriate tRNAs depends on the nucleotide sequences of the tRNAs. Researchers have identified which nucleotides are important in recognition by synthetases by altering different nucleotides in a particular tRNA and determining whether the altered tRNA is still recognized by its synthetase (**Figure 15.14**).

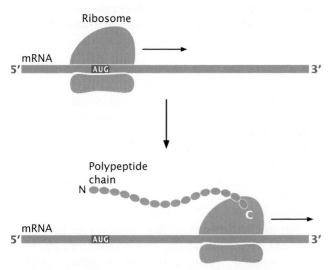

15.12 The translation of an mRNA molecule takes place on a ribosome. The letter N represents the amino end of the protein; C represents the carboxyl end.

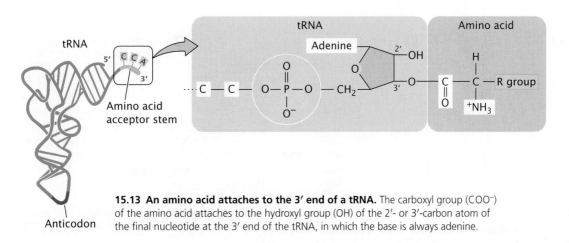

15.13 An amino acid attaches to the 3′ end of a tRNA. The carboxyl group (COO^-) of the amino acid attaches to the hydroxyl group (OH) of the 2′- or 3′-carbon atom of the final nucleotide at the 3′ end of the tRNA, in which the base is always adenine.

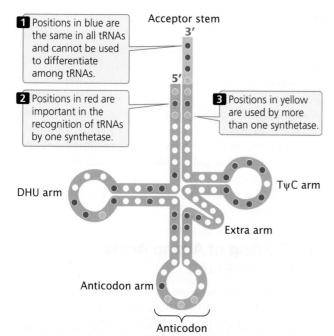

1 Positions in blue are the same in all tRNAs and cannot be used to differentiate among tRNAs.

2 Positions in red are important in the recognition of tRNAs by one synthetase.

3 Positions in yellow are used by more than one synthetase.

Acceptor stem

3'

5'

DHU arm

TψC arm

Extra arm

Anticodon arm

Anticodon

15.14 Certain positions on tRNA molecules are recognized by the appropriate aminoacyl-tRNA synthetase.

The attachment of a tRNA to its appropriate amino acid, termed **tRNA charging**, requires energy, which is supplied by adenosine triphosphate (ATP):

amino acid + tRNA + ATP → aminoacyl-tRNA + AMP + PP$_i$

This reaction takes place in two steps (**Figure 15.15**). To identify the resulting aminoacylated (charged) tRNA, we write the three-letter abbreviation for the amino acid in front of the tRNA; for example, the amino acid alanine (Ala) attaches to its tRNA (tRNA^{Ala}), giving rise to its aminoacyl-tRNA (Ala-tRNA^{Ala}).

Errors in tRNA charging are rare: they occur in only about 1 in 10,000 to 1 in 100,000 reactions. This fidelity is due in part to the presence of editing (proofreading) activity in many of the aminoacyl-tRNA synthetases. Editing activity detects incorrectly paired amino acids and removes them from the tRNAs. Some antifungal chemical agents work by trapping tRNAs in the editing site of the synthetases, preventing their release and thus inhibiting the process of translation in the fungi.

CONCEPTS

Amino acids are attached to their specific tRNAs by aminoacyl-tRNA synthetases in a two-step reaction that requires ATP.

✔ CONCEPT CHECK 6

Amino acids bind to which part of the tRNA?
a. Anticodon
b. DHU arm
c. 3' end
d. 5' end

The Initiation of Translation

The second stage in the process of protein synthesis is initiation. At this stage, all the components necessary for protein synthesis assemble: (1) mRNA; (2) the small and large subunits of the ribosome; (3) a set of three proteins called initiation factors; (4) initiator tRNA with *N*-formylmethionine attached (fMet-tRNA$_i^{fMet}$); and (5) guanosine triphosphate (GTP). Initiation comprises three major steps. First, mRNA binds to the small subunit of the ribosome. Second, initiator tRNA binds to the mRNA through base pairing between the initiation codon and the anticodon. Third, the large ribosomal subunit joins the initiation complex. Let's look at each of these steps more closely.

INITIATION IN BACTERIA The functional ribosome of bacteria exists as two subunits, the small 30S subunit and the large 50S subunit (**Figure 15.16a**). An mRNA molecule can bind to the small ribosomal subunit only when the subunits are separate. **Initiation factor 3** (**IF-3**) binds to the small ribosomal subunit and prevents the large subunit from binding during initiation (**Figure 15.16b**). Another factor, **initiation factor 1** (**IF-1**), enhances the disassociation of the large and small ribosomal subunits.

Where on the mRNA does the ribosome bind during initiation of translation? Key sequences on the mRNA required for ribosome binding have been identified by techniques designed to allow a ribosome to bind to mRNA, but not to proceed with protein synthesis; the ribosome is thereby stalled at the initiation site where binding occurs. A ribonuclease is added, which degrades all the mRNA

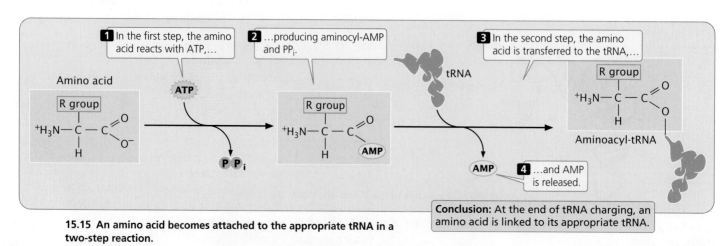

1 In the first step, the amino acid reacts with ATP,...

2 ...producing aminocyl-AMP and PP$_i$.

3 In the second step, the amino acid is transferred to the tRNA,...

Amino acid

R group

$^+H_3N-C-C\overset{O}{\underset{O^-}{=}}$

H

ATP

PP$_i$

R group

$^+H_3N-C-C\overset{O}{=}$

H

AMP

tRNA

R group

$^+H_3N-C-C\overset{O}{\underset{O}{=}}$

H

Aminoacyl-tRNA

AMP

4 ...and AMP is released.

Conclusion: At the end of tRNA charging, an amino acid is linked to its appropriate tRNA.

15.15 An amino acid becomes attached to the appropriate tRNA in a two-step reaction.

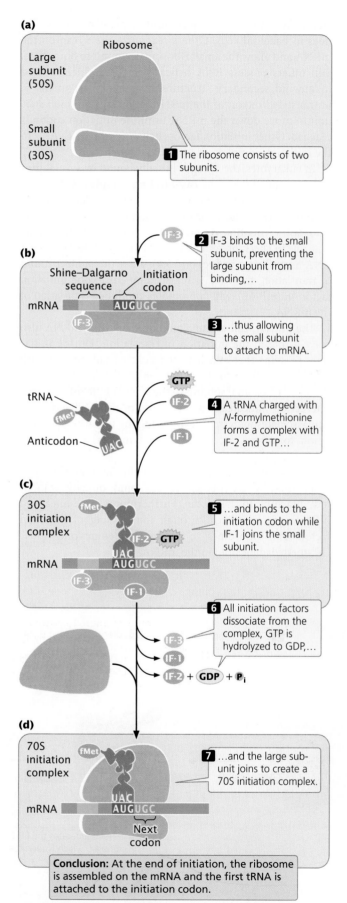

(a)

Ribosome

Large subunit (50S)

Small subunit (30S)

1 The ribosome consists of two subunits.

(b)

Shine–Dalgarno sequence   Initiation codon

mRNA   AUGUGC

2 IF-3 binds to the small subunit, preventing the large subunit from binding,...

3 ...thus allowing the small subunit to attach to mRNA.

GTP

IF-2

tRNA

fMet

Anticodon

UAC

IF-1

4 A tRNA charged with *N*-formylmethionine forms a complex with IF-2 and GTP...

(c)

30S initiation complex

fMet

IF-2 – GTP

UAC

mRNA   AUGUGC

IF-3

IF-1

5 ...and binds to the initiation codon while IF-1 joins the small subunit.

6 All initiation factors dissociate from the complex, GTP is hydrolyzed to GDP,...

IF-3

IF-1

IF-2 + **GDP** + **P**ᵢ

(d)

70S initiation complex

fMet

UAC

mRNA   AUGUGC

Next codon

7 ...and the large subunit joins to create a 70S initiation complex.

Conclusion: At the end of initiation, the ribosome is assembled on the mRNA and the first tRNA is attached to the initiation codon.

15.16 The initiation of translation requires several initiation factors and GTP.

Shine–Dalgarno sequence

mRNA **5′** AUGUAC **UAAGGAGGU** UGUA **AUG** GAACAAGACG **3′**

AUUCCUCCA

Initiation codon

16S rRNA **3′**   **5′**

15.17 The Shine–Dalgarno sequence in mRNA is required for the attachment of the small ribosomal subunit.

except the region covered by the ribosome. The intact mRNA can then be separated from the ribosome and studied. The sequence covered by the ribosome during initiation is 30 to 40 nucleotides long and includes the AUG initiation codon. Within the ribosome-binding site is the Shine–Dalgarno sequence (**Figure 15.17**; see also Chapter 14), a consensus sequence that is complementary to a sequence of nucleotides at the 3′ end of 16S rRNA (part of the small ribosomal subunit). During initiation, the nucleotides in the Shine–Dalgarno sequence pair with their complementary nucleotides in the 16S rRNA, allowing the small ribosomal subunit to attach to the mRNA and positioning the ribosome directly over the initiation codon. These ribosome-binding sequences are within the 5′ untranslated region of the mRNA.

The initiator tRNA, fMet-tRNAᵢ^fMet, attaches to the initiation codon (**Figure 15.16c**). This attachment requires **initiation factor 2** (**IF-2**), which forms a complex with GTP.

At this point, the initiation complex consists of (1) the small ribosomal subunit; (2) the mRNA; (3) the initiator tRNA with its amino acid (fMet-tRNAᵢ^fMet); (4) one molecule of GTP; and (5) several initiation factors. These components are collectively known as the **30S initiation complex** (see Figure 15.16c). In the final step of initiation, IF-3 dissociates from the small subunit, allowing the large ribosomal subunit to join the initiation complex. The molecule of GTP (provided by IF-2) is hydrolyzed to guanosine diphosphate (GDP), and the initiation factors dissociate from the complex (**Figure 15.16d**). When the large subunit has joined the initiation complex, the complex is called the **70S initiation complex**.

INITIATION IN EUKARYOTES Similar events take place in the initiation of translation in eukaryotic cells, but there are some important differences. In bacterial cells, sequences in 16S rRNA of the small ribosomal subunit bind to the Shine–Dalgarno sequence in mRNA. No analogous consensus sequence exists in eukaryotic mRNA. Instead, the cap at the 5′ end of eukaryotic mRNA plays a critical role in the initiation of translation. In a series of steps, the small subunit of the eukaryotic ribosome, initiation factors, and the initiator tRNA with its amino acid (Met-tRNAᵢ^Met) form an initiation complex that recognizes the cap and binds there. The initiation complex then moves along (scans) the mRNA until it locates the first AUG codon.

The identification of the start codon is facilitated by the presence of a consensus sequence (called the Kozak sequence) that surrounds the start codon:

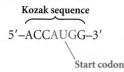

Another important difference is that eukaryotic initiation requires at least 12 initiation factors. Some of these factors keep the ribosomal subunits separated, just as IF-3 does in bacterial cells. Others recognize the 5′ cap on the mRNA and allow the small ribosomal subunit to bind there. Still others possess RNA helicase activity, which is used to unwind secondary structures that may exist in the 5′ untranslated region of the mRNA, allowing the small subunit to move down the mRNA until the initiation codon is reached. Other initiation factors help bring Met-tRNA$_i$^{Met} to the initiation complex.

In eukaryotes, the 5′ cap is initially bound by several proteins, one of which is the **cap-binding complex** (**CBC**). The CBC aids in exporting the mRNA from the nucleus and then promotes the "pioneer," or initial, round of translation in the cytoplasm. This first round of translation plays an important role in checking for errors in the mRNA (see Messenger RNA Surveillance in Section 15.4). After the pioneer round of translation, the CBC is replaced by eukaryotic initiation factor 4E (eIF-4E), which promotes continued translation of the mRNA.

The poly(A) tail at the 3′ end of eukaryotic mRNA also plays a role in the initiation of translation. During initiation, proteins that attach to the poly(A) tail interact with proteins that bind to the 5′ cap, enhancing the binding of the small ribosomal subunit to the 5′ end of the mRNA. This interaction indicates that the 3′ end of mRNA bends over and associates with the 5′ cap during the initiation of translation, forming a circular structure known as a closed loop (**Figure 15.18**).

A few eukaryotic mRNAs contain internal ribosome entry sites, where ribosomes can bind directly without first attaching to the 5′ cap. Furthermore, some uncapped mRNAs are translated through the binding of initiation factors and ribosomes to modified adenine nucleotides (N^6-methyladenosine) in the mRNA.

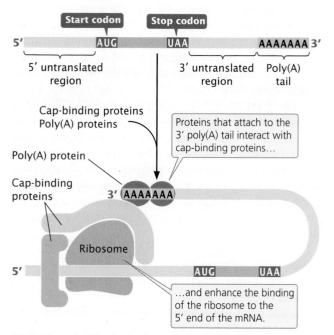

15.18 The poly(A) tail of eukaryotic mRNA plays a role in the initiation of translation.

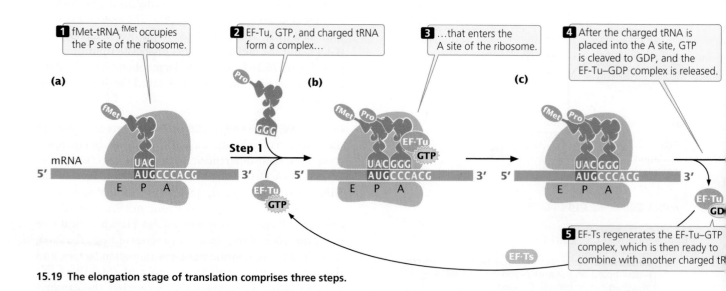

15.19 The elongation stage of translation comprises three steps.

Elongation

The next stage in protein synthesis is elongation, in which amino acids are joined to create a polypeptide chain. Elongation requires (1) the 70S initiation complex just described; (2) tRNAs charged with their amino acids; (3) several elongation factors; and (4) GTP.

A ribosome has three sites that can be occupied by tRNAs: the **aminoacyl (A) site**, the **peptidyl (P) site**, and the **exit (E), site** (**Figure 15.19a**). The initiator tRNA immediately occupies the P site (the only site to which the fMet-tRNA$_i^{fMet}$ is able to bind), but all other tRNAs first enter the A site. At the end of initiation, the ribosome is attached to the mRNA, and fMet-tRNA$_i^{fMet}$ is positioned over the AUG start codon in the P site; the adjacent A site is unoccupied (see Figure 15.19a).

Elongation takes place in three steps. In the first step (**Figure 15.19b**), a charged tRNA binds to the A site. This binding takes place when **elongation factor Tu (EF-Tu)** joins with GTP and then with a charged tRNA to form a three-part complex. This complex enters the A site of the ribosome, where the anticodon on the tRNA pairs with the codon on the mRNA. Once the charged tRNA is in the A site, GTP is cleaved to form GDP, and the EF-Tu–GDP complex is released (**Figure 15.19c**). **Elongation factor Ts (EF-Ts)** regenerates EF-Tu–GDP to EF-Tu–GTP. In eukaryotic cells, a similar set of reactions delivers a charged tRNA to the A site.

The second step of elongation is the formation of a peptide bond between the amino acids that are attached to tRNAs in the P and A sites (**Figure 15.19d**). The formation of this peptide bond releases the amino acid in the P site from its tRNA. Peptide-bond formation occurs within the large ribosomal subunit. The catalytic activity that creates the peptide bond is a property of one of the rRNA components of the large subunit (the 23S rRNA in bacteria, the 28S RNA in eukaryotes); this rRNA acts as a ribozyme (see p. 374 in Chapter 13).

The third step in elongation is **translocation** (**Figure 15.19e**), the movement of the ribosome down the mRNA in the 5′→3′ direction. This step, which positions the ribosome over the next codon, requires **elongation factor G (EF-G)** and the hydrolysis of GTP to GDP. Because the tRNAs in the P and A sites are still attached to the mRNA by codon–anticodon pairing, they do not move with the ribosome as it translocates. Consequently, the ribosome shifts so that the tRNA that previously occupied the P site now occupies the E site, from which it then moves into the cytoplasm, where it can be recharged with another amino acid. Translocation also causes the tRNA that occupied the A site (which is attached to the growing polypeptide chain) to occupy the P site, leaving the A site open. Thus, the progress of each tRNA through the ribosome in the course of elongation can be summarized

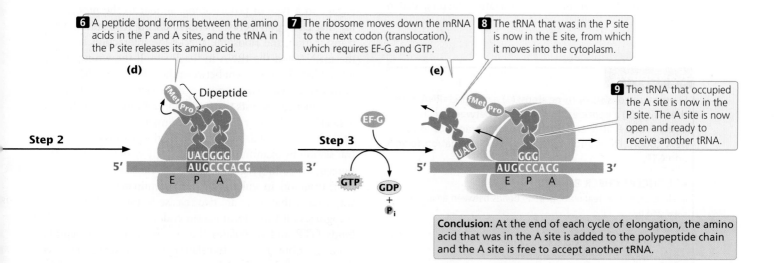

6 A peptide bond forms between the amino acids in the P and A sites, and the tRNA in the P site releases its amino acid.

7 The ribosome moves down the mRNA to the next codon (translocation), which requires EF-G and GTP.

8 The tRNA that was in the P site is now in the E site, from which it moves into the cytoplasm.

9 The tRNA that occupied the A site is now in the P site. The A site is now open and ready to receive another tRNA.

(d)

Dipeptide

Step 2

UACGGG
AUGCCCACG
5′   3′
E   P   A

Step 3

EF-G

GTP → GDP + P$_i$

(e)

fMet Pro

UAC
GGG
AUGCCCACG
5′   3′
E   P   A

Conclusion: At the end of each cycle of elongation, the amino acid that was in the A site is added to the polypeptide chain and the A site is free to accept another tRNA.

as follows: cytoplasm → A site → P site → E site → cytoplasm. As stated earlier, the initiator tRNA is an exception: it attaches directly to the P site and never occupies the A site.

After translocation, the A site of the ribosome is empty and ready to receive the tRNA specified by the next codon. The elongation cycle (see Figure 15.19b–e) repeats itself: a charged tRNA and its amino acid occupy the A site, a peptide bond is formed between the amino acids in the A and P sites, and the ribosome translocates to the next codon. Throughout the cycle, the polypeptide chain remains attached to the tRNA in the P site. Another protein, called elongation factor P (EF-P), enhances the translation of proteins that contain consecutive copies of the amino acid proline. If EF-P is absent, ribosomes often stall during the translation of such polyproline-containing proteins.

Messenger RNAs, although single stranded, often contain secondary structures formed by the pairing of complementary bases on different parts of the mRNA molecule (see Figure 13.1b). As the ribosome moves along the mRNA, these secondary structures are unwound by helicase activity located in the small ribosomal subunit.

Researchers have developed methods for following a single ribosome as it translates individual codons of an mRNA molecule. These studies have revealed that translation does not take place in a smooth, continuous fashion. Each translocation of the ribosome typically requires less than a tenth of a second, but sometimes there are distinct pauses, often lasting a few seconds, between translocations. Thus, translation takes place in a series of quick translocations interrupted by brief pauses. In addition to the short pauses between translocation events, translation may be interrupted by longer pauses—lasting from 1 to 2 minutes—that may play a role in regulating the process of translation.

Elongation in eukaryotic cells takes place in a manner similar to that in bacteria. Eukaryotes possess at least three elongation factors, one of which also acts in initiation and termination. Another of these elongation factors, called eukaryotic elongation factor 2 (eEF-2), is the target of a toxin produced by the bacterium that causes diphtheria, a disease that until recently was a leading killer of children. The diphtheria toxin inhibits eEF-2, preventing the translocation of the ribosome along the mRNA, and protein synthesis ceases.

Termination

Protein synthesis ends when the ribosome translocates to a termination codon. Because there are no tRNAs with anticodons complementary to the termination codons, no tRNA enters the A site of the ribosome (**Figure 15.20a**). Instead, proteins called **release factors** bind to the ribosome (**Figure 15.20b**). *Escherichia coli* has three release factors: **RF-1**, **RF-2**, and **RF-3**. Release factor 1 binds to the termination codons UAA and UAG, and RF-2 binds to UGA and UAA. The binding of RF-1 or RF-2 to the A site of the ribosome promotes the cleavage of the tRNA in the P site from the polypeptide chain and the release of the polypeptide. Release factor 3 forms a complex with GTP and binds to the ribosome. This complex brings about a conformational change in the ribosome, which releases RF-1 or RF-2 from the A site and causes the tRNA in the P site to move to the E site; in the process, GTP is hydrolyzed to GDP. Additional factors help bring about the release of the tRNA from the P site, the release of the mRNA from the ribosome, and the dissociation of the ribosome (**Figure 15.20c**).

It is important to note that the termination codon is not located at the 3′ end of the mRNA; rather, the termination codon is followed by a number of nucleotides that constitute the 3′ untranslated region (UTR) of the mRNA. The 3′ UTR often contains sequences that affect the stability of the mRNA and influence whether translation takes place (see Chapter 17).

THINK-PAIR-SHARE Question 8

Recent research shows that some bacterial ribosomes engage in a type of proofreading, similar to the proofreading that DNA polymerases perform during replication. After translocation, the ribosome checks the interaction between the mRNA and the tRNA in the P site. If the wrong tRNA was added, the alignment between the mRNA and tRNA will be incorrect, which triggers premature termination of translation. Evidence suggests that an important function of RF-3 is to help bring about termination of translation when the wrong tRNA has been used. You can explore the process of bacterial translation by examining the consequences of various mutations in the coding region of a gene in **Animation 15.1**.

Translation in eukaryotic cells terminates in a similar way, except that there are two release factors: eRF-1, which recognizes all three termination codons, and eRF-2, which binds GTP and stimulates the release of the polypeptide from the ribosome. No translational proofreading, such as that stimulated by RF-3 in bacteria, has been observed in eukaryotic cells. **▶ TRY PROBLEMS 30 AND 34**

CONCEPTS

Elongation consists of three steps: (1) a charged tRNA enters the A site, (2) a peptide bond is created between amino acids in the A and P sites, and (3) the ribosome translocates to the next codon. Elongation requires several elongation factors and GTP.

✔ CONCEPT CHECK 8

In elongation, the creation of peptide bonds between amino acids is catalyzed by

a. rRNA.
b. a protein in the small ribosomal subunit.
c. a protein in the large ribosomal subunit.
d. tRNA.

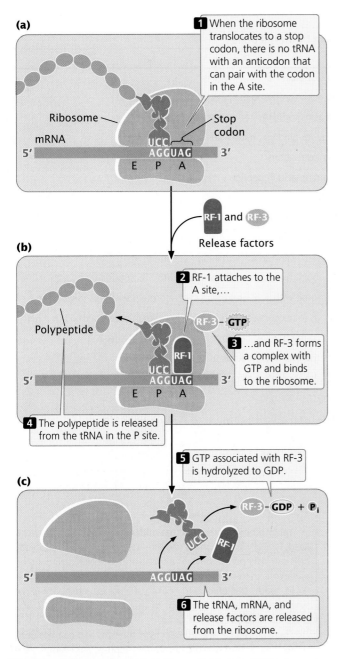

(a)

1 When the ribosome translocates to a stop codon, there is no tRNA with an anticodon that can pair with the codon in the A site.

Ribosome
mRNA
5′ UCC AGGUAG 3′
E P A
Stop codon

RF-1 and RF-3
Release factors

(b)

2 RF-1 attaches to the A site,…

Polypeptide

RF-3 — GTP

3 …and RF-3 forms a complex with GTP and binds to the ribosome.

RF-1
5′ UCC AGGUAG 3′
E P A

4 The polypeptide is released from the tRNA in the P site.

5 GTP associated with RF-3 is hydrolyzed to GDP.

(c)

RF-3 — GDP + P_i

UCC RF-1
5′ AGGUAG 3′

6 The tRNA, mRNA, and release factors are released from the ribosome.

15.20 Translation ends when a stop codon is encountered. Because UAG is the termination codon in this illustration, the release factor is RF-1.

CONCEPTS

Translation ends when the ribosome reaches a termination codon. Release factors bind to the termination codon, causing the release of the polypeptide from the last tRNA, of the tRNA from the ribosome, and of the mRNA from the ribosome.

The overall process of protein synthesis, including tRNA charging, initiation, elongation, and termination, is summarized in **Figure 15.21**. The components taking part in this process are listed in **Table 15.4**.

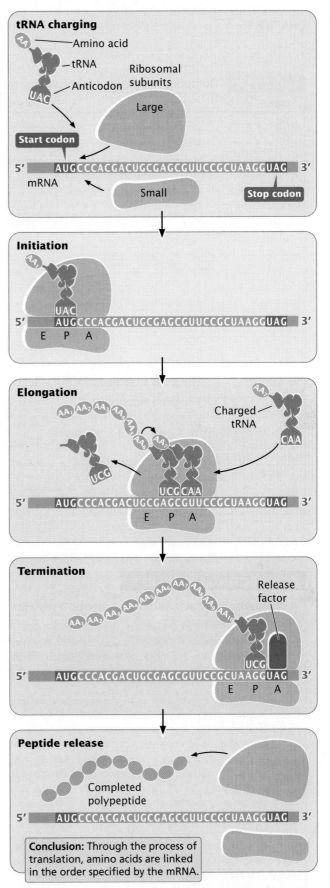

15.21 Translation requires tRNA charging, initiation, elongation, and termination. In the process of translation, amino acids are linked together in the order specified by mRNA to create a polypeptide chain. A number of initiation, elongation, and release factors take part in the process, and energy is supplied by ATP and GTP.

TABLE 15.4	Components required for protein synthesis in bacterial cells	
Stage	**Component**	**Function**
tRNA charging	Amino acids	Building blocks of proteins
	tRNAs	Deliver amino acids to ribosomes
	Aminoacyl-tRNA synthetases	Attach amino acids to tRNAs
	ATP	Provides energy for binding amino acids to tRNAs
Initiation	mRNA	Carries coding instructions
	fMet-tRNA$_i^{fMet}$	Provides first amino acid in peptide
	30S ribosomal subunit	Attaches to mRNA
	50S ribosomal subunit	Stabilizes tRNAs and amino acids
	Initiation factor 1	Enhances dissociation of large and small subunits of ribosome
	Initiation factor 2	Binds GTP; delivers fMet-tRNA$_i^{fMet}$ to initiation codon
	Initiation factor 3	Binds to 30S subunit and prevents association with 50S subunit
Elongation	70S initiation complex	Functional ribosome with A, P, and E sites where protein synthesis takes place
	Charged tRNAs	Bring amino acids to ribosome and help assemble them in order specified by mRNA
	Elongation factor Tu	Binds GTP and charged tRNA; delivers charged tRNA to A site
	Elongation factor Ts	Regenerates active elongation factor Tu
	Elongation factor G	Stimulates translocation of ribosome to next codon
	GTP	Provides energy
	23S rRNA in large ribosomal subunit	Creates peptide bond between amino acids in A site and P site
Termination	Release factors 1, 2, and 3	Bind to ribosome when stop codon is reached and terminate translation

CONNECTING CONCEPTS

A Comparison of Bacterial and Eukaryotic Translation

We have now considered the process of translation in bacterial cells and noted some distinctive differences that exist in eukaryotic cells. Let's reflect on some of the important similarities and differences between protein synthesis in bacterial and in eukaryotic cells.

First, we should emphasize that the genetic code of bacterial and eukaryotic cells is virtually identical; the only difference is in the amino acid specified by the initiation codon. In bacterial cells, the initiation codon AUG encodes a modified type of methionine, N-formylmethionine, whereas in eukaryotic cells, AUG encodes unformylated methionine. One consequence of the fact that bacteria and eukaryotes use the same genetic code is that eukaryotic genes can be translated in bacterial systems, and vice versa; this feature makes genetic engineering possible, as we will see in Chapter 19.

Another difference is that transcription and translation take place simultaneously in bacterial cells, but the nuclear envelope separates these processes in eukaryotic cells. The physical separation of transcription and translation has important implications for the control of gene expression in eukaryotes, which we will consider in Chapter 17, and it allows for extensive modification of eukaryotic mRNAs, as discussed in Chapter 14.

Yet another difference is that mRNA in bacterial cells is short-lived, typically lasting only a few minutes, but mRNA in eukaryotic cells can last hours or days. The 5′ cap and 3′ poly(A) tail found on eukaryotic mRNAs add to their stability (see Chapter 14).

In both bacterial and eukaryotic cells, aminoacyl-tRNA synthetases attach amino acids to their appropriate tRNAs, and the chemical process is the same.

There are significant differences in the sizes and compositions of bacterial and eukaryotic ribosomal subunits. For example, the large subunit of the eukaryotic ribosome contains three rRNAs, whereas that of the bacterial ribosome contains only two (see Table 14.3). These differences allow antibiotics and other substances to inhibit bacterial translation while having no effect on the translation of eukaryotic nuclear genes, as we will see in Section 15.4.

Other fundamental differences lie in the process of initiation. In bacterial cells, the small ribosomal subunit attaches directly to the region surrounding the start codon through base pairing between the Shine–Dalgarno sequence in the 5′ UTR of the mRNA and a sequence at the 3′ end of the 16S rRNA. In contrast, the small subunit of a eukaryotic ribosome first binds to proteins attached to the 5′ cap on mRNA and then migrates down the mRNA, scanning the sequence until it encounters the first initiation codon. Additionally, a larger number of initiation factors take part in eukaryotic initiation than in bacterial initiation.

Elongation and termination are similar in bacterial and eukaryotic cells, although different elongation and termination factors are used. In both types of organisms, mRNAs are translated multiple times and are simultaneously attached to several ribosomes, forming polyribosomes, as discussed in Section 15.4.

Less is known about the process of translation in archaea, but they appear to possess a mixture of bacterial and eukaryotic features. Because archaea lack nuclear membranes, transcription and translation take place simultaneously, just as they do in bacterial cells. Archaea use unformylated methionine as the initiator amino acid, a characteristic of eukaryotic translation. Some of the initiation and release factors in archaea are similar to those found in bacteria, whereas others are similar to those found in eukaryotes. Finally, some of the antibiotics that inhibit translation in bacteria have no effect on translation in archaea, providing further evidence of the fundamental differences between bacteria and archaea.

15.4 Additional Properties of RNA and Ribosomes Affect Protein Synthesis

Now that we have considered the process of translation in some detail, let's examine some additional aspects of protein synthesis and the translation machinery.

The Three-Dimensional Structure of the Ribosome

The central role of the ribosome in protein synthesis was recognized in the 1950s, and numerous aspects of its structure have been studied since then. Nevertheless, many aspects remained a mystery until detailed, three-dimensional reconstructions were completed. In 2009, the Nobel Prize in chemistry was awarded to Venkatraman Ramakrishnan, Thomas Steitz, and Ada Yonath for their research on the molecular structure of the ribosome.

Figure 15.22a shows a model of the *E. coli* ribosome at low resolution. A high-resolution image of the bacterial ribosome structure as determined by X-ray crystallography is represented in the model depicted in **Figure 15.22b**. An mRNA molecule is bound to the small ribosomal subunit, and tRNAs are located in the A, P, and E sites (**Figure 15.22c**) that bridge the small and large subunits (see Figure 15.22a). Initiation factors 1 and 3 bind to sites on the outside of the small subunit. EF-Tu, EF-G, and other factors complexed with GTP interact with a *factor-binding center*. High-resolution crystallographic images provide information indicating that a *decoding center* resides in the small subunit (the decoding center cannot be seen in Figure 15.22b). This center senses the fit between the codon on the mRNA and the anticodon on the incoming charged tRNA. Only tRNAs with the correct anticodon are bound tightly by the ribosome. These structural analyses indicate that peptide-bond formation occurs within the large ribosomal subunit and is carried out by the 23S rRNA molecule. These analyses also reveal that a tunnel connects the site of peptide-bond formation with the back of the ribosome; the growing polypeptide chain passes through this tunnel to the outside of the ribosome. The tunnel can accommodate about 35 amino acids of the growing polypeptide chain.

Polyribosomes

In both prokaryotic and eukaryotic cells, mRNA molecules are translated simultaneously by multiple ribosomes (**Figure 15.23**). The resulting structure—an mRNA with several ribosomes attached—is called a **polyribosome** (or often just a *polysome*). Each ribosome successively attaches to the ribosome-binding site at the 5′ end of the mRNA and moves toward the 3′ end; the polypeptide associated with each ribosome becomes progressively longer as the ribosome moves along the mRNA. In prokaryotic cells, transcription and translation are simultaneous; multiple ribosomes may be attached to the 5′ end of the mRNA while transcription is still taking place at the 3′ end, as shown in Figure 15.23. In eukaryotes, transcription and translation are separated in time and space; transcription takes place in the nucleus and translation takes place in the cytoplasm.

CONCEPTS

In both prokaryotic and eukaryotic cells, multiple ribosomes may translate a single mRNA molecule simultaneously, generating a structure called a polyribosome.

✔ CONCEPT CHECK 9

In a polyribosome, the polypeptides associated with which ribosomes will be the longest?

a. Those at the 5′ end of mRNA
b. Those at the 3′ end of mRNA
c. Those in the middle of mRNA
d. All polypeptides will be the same length.

Messenger RNA Surveillance

The accurate transfer of genetic information from one generation to the next and from genotype to phenotype is critical for the proper development and functioning of an organism. Consequently, cells have evolved a number of quality-control mechanisms to ensure the accuracy of information transfer. Protein synthesis is no exception: several mechanisms, collectively termed **mRNA surveillance**, exist to detect and deal with errors in mRNAs that may create problems in the course of translation. These mechanisms keep the cell from wasting resources translating aberrant mRNAs and prevent the production of truncated proteins, which may be toxic to the cell.

NONSENSE-MEDIATED mRNA DECAY A common type of mutation is the alteration of a codon that specifies an

(a)

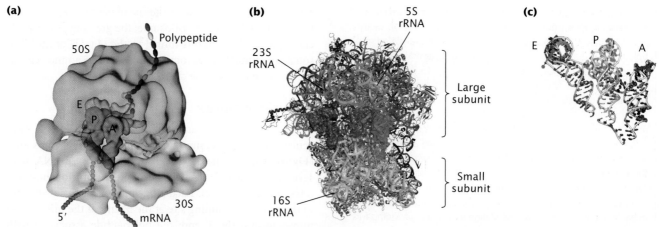

(b)

(c)

15.22 Structure of the ribosome. (a) Low-resolution model of the ribosome, showing the A, P, and E sites where tRNAs, the mRNA, and the growing polypeptide chain reside. (b) High-resolution model of the ribosome. The tRNA in the E site is reddish brown, the tRNA in the P site is red, and the tRNA in the A site is green. The 16S rRNA is shown in yellow, the 23S rRNA in cyan, and the 5S rRNA in green. Small ribosomal subunit proteins are shown in orange, whereas large ribosomal subunit proteins are shown in purple. (c) Positions of tRNAs in E, P, and A sites of the ribosome shown in part *b*. [Part b: MRC Lab of Molecular Biology, Wellcome Images.]

amino acid to become a termination codon (called a nonsense mutation, see Chapter 18). A nonsense mutation does not affect transcription, but translation ends prematurely when the termination codon is encountered. The resulting protein is truncated and often nonfunctional. Nonsense mutations often arise in eukaryotic transcription when one or more exons are skipped or improperly spliced. Improper splicing leads to the deletion or addition of nucleotides in the mRNA, which alters the reading frame and often introduces premature termination codons.

To prevent the synthesis of aberrant proteins resulting from nonsense mutations, eukaryotic cells have evolved a mechanism called **nonsense-mediated mRNA decay (NMD)**, which results in the rapid elimination of mRNA containing premature termination codons. The mechanism responsible for NMD is still poorly understood. In mammals, it appears to entail proteins that bind to exon–exon junctions. These exon-junction proteins may interact with enzymes that degrade the mRNA. One possibility is that the first ribosome to translate an mRNA (in the pioneer round of translation) removes the exon-junction proteins, thus protecting the mRNA from degradation. However, when the ribosome encounters a premature termination codon, the ribosome does not traverse the entire mRNA, and some of the exon-junction proteins are not removed, resulting in NMD. Nonsense-mediated mRNA decay does not occur in bacteria. ▶ **TRY PROBLEM 38**

STALLED RIBOSOMES AND NONSTOP mRNAs A problem that occasionally arises in translation is the stalling of a ribosome on mRNA before translation is terminated. This situation can arise when a mutation in DNA changes a termination codon into a codon that specifies an amino acid. It can

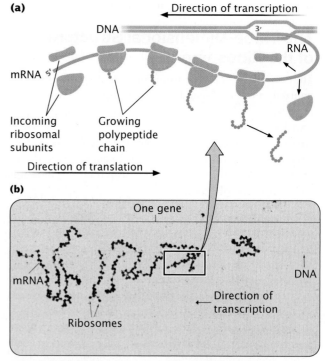

15.23 An mRNA molecule may be translated simultaneously by several ribosomes. (a) Four ribosomes are translating an mRNA molecule; the ribosomes move from the 5′ end to the 3′ end of the mRNA. (b) In this electron micrograph, the long horizontal filament is DNA, the dark-staining spheres are polyribosomes, and the thin filaments connecting the ribosomes are mRNAs. Transcription of the DNA is proceeding from right to left; the mRNAs on the right are shorter than those left. Each mRNA is being translated by multiple ribosomes. [Part b: From O. L. Miller, Jr., B. A. Hamkalo, and C. A. Thomas, Jr. "Visualization of Bacterial Genes in Action," *Science* 169(1970): 392. Reprinted with permission from AAAS, permission conveyed through Copyright Clearance Center, Inc.]

also arise when transcription terminates prematurely, producing a truncated mRNA lacking a termination codon. In these cases, the ribosome reaches the end of the mRNA without encountering a termination codon, whereupon it stalls, still attached to the mRNA. Attachment to the mRNA prevents the ribosome from being recycled for use on other mRNAs; if such occurrences are frequent, the result is a shortage of ribosomes that diminishes overall levels of protein synthesis.

Bacteria have evolved a kind of molecular tow truck called **transfer–messenger RNA** (**tmRNA**) that removes stalled ribosomes. This RNA molecule has properties of both tRNA and mRNA; its tRNA component is normally charged with the amino acid alanine. When a ribosome becomes stalled on an mRNA, EF-Tu delivers the tmRNA to the ribosome's A site, where it acts as a surrogate tRNA (**Figure 15.24**). A peptide bond is created between the amino acid in the P site of the ribosome and the alanine (attached to tmRNA) now in the A site, transferring the polypeptide chain to the tmRNA and releasing the tRNA in the P site.

The ribosome then resumes translation, switching from the original, aberrant mRNA to the mRNA part of tmRNA. It adds 10 amino acids encoded by the tmRNA, and then reaches a termination codon at the 3′ end of the tmRNA, which terminates translation and releases the ribosome. The added amino acids act as a tag that targets the incomplete polypeptide chain for degradation. Some evidence suggests that the tmRNA also targets the aberrant mRNA for degradation. How stalled ribosomes are recognized by the tmRNA is not clear, but this method is efficient at recycling stalled ribosomes and eliminating abnormal proteins that result from truncated transcription.

Eukaryotes have evolved a different mechanism to deal with mRNAs that are missing termination codons. Instead of restarting the stalled ribosome and degrading the abnormal protein that results, eukaryotic cells use a mechanism called **nonstop mRNA decay**, which results in the rapid degradation of abnormal mRNA. In this mechanism, the codon-free A site of the stalled ribosome is recognized by a special protein, which binds to the A site and recruits other proteins, which then degrade the mRNA from its 3′ end.

NO-GO DECAY Another mRNA surveillance system found in eukaryotes is **no-go decay** (**NGD**), which helps remove stalled ribosomes resulting from secondary structures in the mRNA, chemical damage to the mRNA, premature stop codons, and ribosomal defects. A series of proteins bring about termination, recycling of the ribosomes, and degradation of the mRNA.

CONCEPTS

Cells possess mRNA surveillance mechanisms to detect and eliminate mRNA molecules containing errors that create problems in the course of translation.

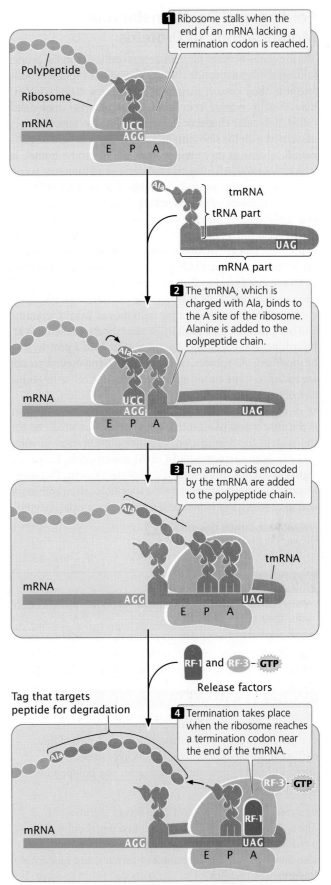

15.24 In bacteria, tmRNA allows stalled ribosomes to resume translation.

1 Ribosome stalls when the end of an mRNA lacking a termination codon is reached.

2 The tmRNA, which is charged with Ala, binds to the A site of the ribosome. Alanine is added to the polypeptide chain.

3 Ten amino acids encoded by the tmRNA are added to the polypeptide chain.

Tag that targets peptide for degradation

4 Termination takes place when the ribosome reaches a termination codon near the end of the tmRNA.

RF-1 and RF-3 – GTP
Release factors

Folding and Posttranslational Modifications of Proteins

The functions of many proteins depend critically on the proper folding of the polypeptide chain. Some proteins spontaneously fold into their correct shapes, but for others, correct folding may initially require the participation of other molecules, called **molecular chaperones**. Some molecular chaperones are associated with the ribosome and fold newly synthesized polypeptide chains as they emerge from the ribosome tunnel, in which case protein folding takes place as translation is ongoing.

Many proteins must be modified after translation to become functional. Proteins in both prokaryotic and eukaryotic cells often undergo alterations following translation, which are termed posttranslational modifications. A number of different types of modifications are possible. Some proteins are synthesized as part of larger precursor proteins that must be cleaved and trimmed by enzymes before the proteins can become functional. As mentioned earlier, the formyl group, or the entire methionine, may be removed from the amino end of a protein. Some proteins require the attachment of carbohydrates for activation. Amino acids within a protein may be modified: phosphates, carboxyl groups, and methyl groups are added to some amino acids. In eukaryotic cells, the amino end of a protein is often acetylated after translation.

A common posttranslational modification in eukaryotes is the attachment of a protein called ubiquitin, which targets the protein for degradation. Another modification of some proteins is the removal of 15 to 30 amino acids, called the **signal sequence**, at the amino end of the protein. The signal sequence helps direct a protein to a specific location within the cell, after which the sequence is removed by special enzymes.

THINK-PAIR-SHARE Question 9

> **CONCEPTS**
>
> Many proteins undergo posttranslational modifications after their synthesis.

Translation and Antibiotics

Antibiotics are drugs that kill bacteria. To make an effective antibiotic, not just any poison will do: the trick is to kill the bacteria without harming the patient.

Translation is frequently the target of antibiotics because it is essential to all living organisms and differs significantly between bacterial and eukaryotic cells. A number of antibiotics bind selectively to bacterial ribosomes and inhibit specific steps in translation, but do not affect eukaryotic ribosomes. Tetracyclines, for instance, are a class of antibiotics that bind to the A site of a bacterial ribosome and block the entry of charged tRNAs, yet they have no effect on eukaryotic ribosomes. Neomycin binds to the ribosome near the A site and induces translational errors, probably by causing mistakes in the binding of charged tRNAs to the A site. Chloramphenicol binds to the large ribosomal subunit and blocks peptidebond formation. Streptomycin binds to the small ribosomal subunit and inhibits initiation, and erythromycin blocks translocation. Although chloramphenicol and streptomycin are potent inhibitors of translation in bacteria, they do not inhibit translation in archaea.

The three-dimensional structure of puromycin resembles that of the 3′ end of a charged tRNA, which permits puromycin to enter the A site of a ribosome efficiently and inhibit the entry of tRNAs. A peptide bond can form between the puromycin molecule in the A site and an amino acid on the tRNA in the P site of the ribosome, but puromycin cannot bind to the P site, and translocation does not take place, blocking further elongation of the protein. Because tRNA structure is similar in all organisms, puromycin inhibits translation in both bacterial and eukaryotic cells; consequently, puromycin kills eukaryotic cells along with bacteria and is sometimes used in cancer therapy to destroy tumor cells.

Although these antibiotics act by blocking specific steps in translation, it is important to recognize that antibiotics may also have other effects that contribute to the killing of bacteria. Because different antibiotics block different steps in protein synthesis, antibiotics are frequently used to study protein synthesis.

CONCEPTS SUMMARY

- George Beadle and Edward Tatum developed the one gene, one enzyme hypothesis, which proposed that each gene specifies one enzyme; this hypothesis was later modified to become the one gene, one polypeptide hypothesis.

- Proteins are composed of 20 different amino acids. The amino acids in a protein are linked together by peptide bonds. The resulting polypeptide chains fold and associate to produce the secondary, tertiary, and quaternary structures of proteins.

- Solving the genetic code required several different approaches, including the use of synthetic mRNAs with random sequences and short mRNAs that bind ribosomes and tRNAs.

- The genetic code is a triplet code: three nucleotides specify a single amino acid. It is also degenerate (meaning that more than one codon may specify an amino acid), nonoverlapping, and universal (almost).

- Different tRNAs (isoaccepting tRNAs) may accept the same amino acid. Different codons may pair with the same anticodon through wobble, which can exist at the third position of the codon and allows some nonstandard pairing of bases in this position.

■ The reading frame is set by the initiation codon. The end of the protein-coding sequence of an mRNA is marked by one of three termination codons.

■ Protein synthesis comprises four steps: (1) the binding of amino acids to the appropriate tRNAs, (2) initiation, (3) elongation, and (4) termination.

■ The binding of an amino acid to the appropriate tRNA requires the presence of a specific aminoacyl-tRNA synthetase and ATP. The amino acid is attached by its carboxyl end to the 3′ end of the tRNA.

■ In bacterial translation initiation, the small ribosomal subunit attaches to the mRNA and is positioned over the initiation codon. It is joined by the initiator tRNA and its associated amino acid (*N*-formylmethionine in bacterial cells) and, later, by the large ribosomal subunit. Initiation requires several initiation factors and GTP.

■ In elongation, a charged tRNA enters the A site of a ribosome, a peptide bond is formed between the amino acids in the A and P sites, and the ribosome moves (translocates) along the mRNA to the next codon. Elongation requires several elongation factors as well as GTP.

■ Translation is terminated when the ribosome encounters one of the three termination codons. Release factors and GTP are required to bring about termination.

■ Each mRNA molecule may be simultaneously translated by several ribosomes, producing a structure called a polyribosome.

■ Cells possess mRNA surveillance mechanisms that eliminate mRNAs with errors that may create problems in translation.

■ Some proteins undergo posttranslational modification.

■ Many antibiotics work by interfering with translation because many aspects of translation differ in bacteria and eukaryotes.

IMPORTANT TERMS

one gene, one enzyme
  hypothesis (p. 433)
one gene, one polypeptide
  hypothesis (p. 433)
amino acid (p. 433)
peptide bond (p. 433)
polypeptide (p. 433)
sense codon (p. 438)
degenerate genetic
  code (p. 438)
synonymous
  codon (p. 438)
isoaccepting
  tRNA (p. 438)
wobble (p. 439)
nonoverlapping
  genetic code (p. 439)

reading frame (p. 439)
initiation (start)
  codon (p. 440)
stop (termination or
  nonsense)
  codon (p. 440)
universal genetic
  code (p. 440)
aminoacyl-tRNA
  synthetase (p. 441)
tRNA charging (p. 441)
initiation factors (IF-1,
  IF-2, IF-3) (pp. 442–443)
30S initiation
  complex (p. 443)
70S initiation
  complex (p. 443)

cap-binding complex
  (CBC) (p. 444)
aminoacyl (A)
  site (p. 445)
peptidyl (P)
  site (p. 445)
exit (E) site (p. 445)
elongation factor Tu
  (EF-Tu) (p. 445)
elongation factor Ts
  (EF-Ts) (p. 445)
translocation (p. 445)
elongation factor G
  (EF-G) (p. 445)
release factors (RF-1,
  RF-2, RF-3) (p. 446)
polyribosome (p. 449)

mRNA surveillance
  (p. 449)
nonsense-mediated
  mRNA decay
  (NMD) (p. 450)
transfer–messenger RNA
  (tmRNA) (p. 451)
nonstop mRNA
  decay (p. 451)
no-go decay
  (NGD) (p. 451)
molecular
  chaperone (p. 452)
signal sequence (p. 452)

ANSWERS TO CONCEPT CHECKS

1. B → A → C

2. The amino acid sequence (primary structure) of the protein

3. b

4. d

5. The initiation codon in bacteria encodes *N*-formylmethionine; in eukaryotes, it encodes methionine. Termination codons do not specify amino acids.

6. c

7. The Shine–Dalgarno sequence

8. a

9. b

WORKED PROBLEMS

Problem 1

A series of auxotrophic mutants were isolated in *Neurospora*. Examination of fungi containing these mutations revealed that they grew on minimal medium to which various compounds (A, B, C, D) were added; growth responses to each of the four compounds are presented in the following table. Give the order of compounds A, B, C, and D in a biochemical pathway. Outline a biochemical pathway that includes these four compounds and indicate which step in the pathway is affected by each of the mutations.

Mutation number	Compound			
	A	B	C	D
134	+	+	−	+
276	+	+	+	+
987	−	−	−	+
773	+	+	+	+
772	−	−	−	+
146	+	+	−	+
333	+	+	−	+
123	−	+	−	+

›› Solution Strategy

What information is required in your answer to the problem?

The order of compounds A–D in a biochemical pathway; for each mutation, which step in the pathway is blocked by the mutation.

What information is provided to solve the problem?

For each mutant, whether it grew on minimal medium to which compounds A, B, C, or D were added.

For help with this problem, review:

The One Gene, One Enzyme Hypothesis in Section 15.1.

Hint: Group the mutations by which compounds allow them to grow.

›› Solution Steps

Mutation		Compound			
Group	Number	A	B	C	D
I	276	+	+	+	+
	773	+	+	+	+
II	134	+	+	−	+
	146	+	+	−	+
	333	+	+	−	+
III	123	−	+	−	+
IV	987	−	−	−	+
	772	−	−	−	+

Hint: If a compound is added after the block, it will allow the mutant to grow; if a compound is added before the block, it will have no effect.

Mutants in group I will grow if compound A, B, C, or D is added to the medium, so these mutations must block a step before the production of all four compounds:

Group II mutants will grow if compound A, B, or D is added, but not if compound C is added. Thus, compound C comes before A, B, and D, and group II mutations block the conversion of compound C into one of the other compounds:

Group III mutants will grow if compound B or D is added, but not if compound A or C is added. Thus, group III mutations block steps that follow the production of A and C; we have already determined that compound C precedes A in the pathway, so A must be the next compound in the pathway:

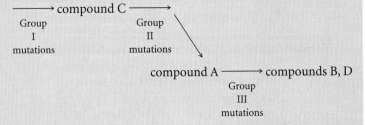

Finally, mutants in group IV will grow if compound D is added, but not if compound A, B, or C is added. Thus, compound D is the fourth compound in the pathway, and mutations in group IV block the conversion of B into D:

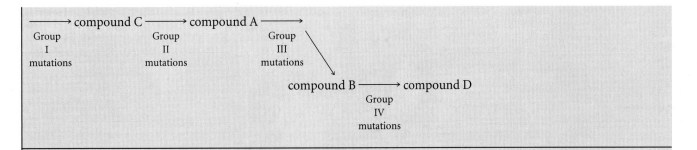

Problem 2

A template strand in bacterial DNA has the following base sequence:

$$5'-AGGTTTAACGTGCAT-3'$$

What amino acids are encoded by this sequence?

›› Solution Strategy

What information is required in your answer to the problem?
The list of amino acids encoded by the given sequence.

What information is provided to solve the problem?

- The DNA sequence of the template strand.
- The 5′ and 3′ ends of the template sequence.
- The amino acids encoded by different codons (**Figure 15.10**).

For help with this problem, review:
The Degeneracy of the Code in Section 15.2.

›› Solution Steps

To answer this question, we must first work out the mRNA sequence that will be transcribed from this DNA sequence:

DNA template strand: 5′–AGGTTTAACGTGCAT–3′
mRNA copied from DNA: 3′–UCCAAAUUGCACGUA–5′

An mRNA is translated 5′→3′, so it will be helpful if we turn the RNA molecule around with the 5′ end on the left:

  mRNA copied from DNA: 5′–AUGCACGUUAAACCU–3′

The codons consist of groups of three nucleotides that are read successively after the first AUG codon; using **Figure 15.10**, we can determine that the amino acids are

$$5'-AUG-CAC-GUU-AAA-CCU-3'$$
$$\downarrow \quad \downarrow \quad \downarrow \quad \downarrow \quad \downarrow$$
$$fMet-His-Val-Lys-Pro$$

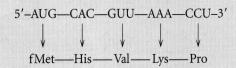

Recall: The mRNA is antiparallel and complementary to the DNA template strand.

Problem 3

The following triplets constitute anticodons found on a series of tRNAs. Name the amino acid carried by each of these tRNAs.

a. 5′–UUU–3′

b. 5′–GAC–3′

c. 5′–UUG–3′

d. 5′–CAG–3′

›› Solution Strategy

What information is required in your answer to the problem?
The amino acid carried by each tRNA.

What information is provided to solve the problem?

- The sequence on the anticodon of each tRNA.
- The amino acids encoded by different codons (**Figure 15.10**).

For help with this problem, review:
The Degeneracy of the Code in Section 15.2.

›› Solution Steps

To solve this problem, we first determine the codons with which these anticodons pair and then look up the amino acid specified by each codon in **Figure 15.10**. For part *a*, the anticodon is 5′–UUU–3′. According to the

Recall: Codons are antiparallel and complementary to the anticodons.

wobble rules, U in the first position of the anticodon can pair with either A or G in the third position of the codon, so there are two codons that can pair with this anticodon:

Anticodon:	5′–UUU–3′
Codon:	3′–AAA–5′
Codon:	3′–GAA–5′

Listing these codons in the conventional manner, with the 5′ end on the left, we have

Codon:	5′–AAA–3′
Codon:	5′–AAG–3′

According to **Figure 15.10**, both codons specify the amino acid lysine (Lys). Recall that the wobble in the third position allows more than one codon to specify the same amino acid,

so any wobble that exists should produce the same amino acid as the standard base pairings would, and we do not need to figure the wobble to answer this question. The answers for parts *b*, *c*, and *d* are

b.
Anticodon:	5′–GAC–3′
Anticodon:	3′–CAG–5′
Codon:	5′–GUC–3′ encodes Val

c.
Anticodon:	5′–UUG–3′
Anticodon:	3′–GUU–5′
Codon:	5′–CAA–3′ encodes Gln

d.
Anticodon:	5′–CAG–3′
Anticodon:	3′–GUC–5′
Codon:	5′–CUG–3′ encodes Leu

COMPREHENSION QUESTIONS

Section 15.1

1. What is the one gene, one enzyme hypothesis? Why was this hypothesis an important advance in our understanding of genetics?

Section 15.2

2. What different methods were used to help break the genetic code? What did each method reveal and what were the advantages and disadvantages of each one?

3. What are isoaccepting tRNAs?

4. What is the significance of the fact that many synonymous codons differ only in the third nucleotide position?

5. Define the following terms as they apply to the genetic code:

   a. Reading frame
   b. Overlapping code
   c. Nonoverlapping code
   d. Initiation codon
   e. Termination codon
   f. Sense codon
   g. Nonsense codon
   h. Universal code
   i. Nonuniversal codons

6. How is the reading frame of a nucleotide sequence set?

Section 15.3

7. How are tRNAs linked to their corresponding amino acids?

8. What role do the initiation factors play in protein synthesis?

9. How does the process of initiation differ in bacterial and eukaryotic cells?

10. Give the elongation factors used in bacterial translation and explain the role played by each factor in translation.

11. What events bring about the termination of translation?

12. Compare and contrast the process of protein synthesis in bacterial and eukaryotic cells, giving similarities and differences in the process of translation in these two types of cells.

Section 15.4

13. How do prokaryotic cells overcome the problem of a stalled ribosome on an mRNA that has no termination codon? How do eukaryotic cells solve this problem?

14. What are some types of posttranslational modification of proteins?

15. Explain how some antibiotics work by affecting the process of protein synthesis.

APPLICATION QUESTIONS AND PROBLEMS

Section 15.1

*16. Sydney Brenner isolated *Salmonella typhimurium* mutants that were implicated in the biosynthesis of tryptophan and would not grow on minimal medium. When these bacterial mutants were tested on minimal medium to which one of four compounds (indole glycerol phosphate, indole, anthranilic acid, and tryptophan) had been added, the growth responses shown in the following table were obtained.

Mutant	Minimal medium	Anthranilic acid	Indole glycerol phosphate	Indole	Tryptophan
trp-1	−	−	−	−	+
trp-2	−	−	+	+	+
trp-3	−	−	−	+	+
trp-4	−	−	+	+	+
trp-6	−	−	−	−	+
trp-7	−	−	−	−	+
trp-8	−	+	+	+	+
trp-9	−	−	−	−	+
trp-10	−	−	−	−	+
trp-11	−	−	−	−	+

Give the order of indole glycerol phosphate, indole, anthranilic acid, and tryptophan in a biochemical pathway leading to the synthesis of tryptophan. Indicate which step in the pathway is affected by each of the mutations.

17. Compounds I, II, and III are in the following biochemical pathway:

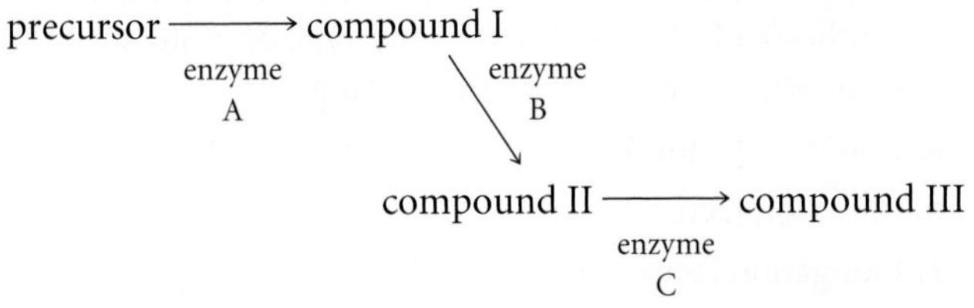

precursor ⟶ compound I
            enzyme
              A   ⟶ enzyme
                             B
        compound II ⟶ compound III
                   enzyme
                     C

Mutation *a* inactivates enzyme A, mutation *b* inactivates enzyme B, and mutation *c* inactivates enzyme C. Mutants, each having one of these defects, were tested on minimal medium to which compound I, II, or III was added. Fill in the results expected of these

tests by placing a plus sign (+) for growth or a minus sign (−) for no growth in the table below.

Strain with mutation	Minimal medium to which is added		
	Compound I	Compound II	Compound III
a			
b			
c			

Section 15.2

18. A geneticist conducts the experiment outlined in **Figure 15.8**, but this time she combines guanine nucleotides (instead of uracil) with polynucleotide phosphorylase. Radioactively labeled protein should appear in which tube?

19. For the experiment outlined in **Figure 15.8**, could Nirenberg and Matthaei have substituted RNA polymerase instead of polynucleotide phosphorylase without otherwise modifying the experiment? Why or why not?

*20. Assume that the number of different types of bases in RNA is four. What would be the minimum codon size (number of nucleotides) required to specify all amino acids if the number of different types of amino acids in proteins were (a) 2, (b) 8, (c) 17, (d) 45, (e) 75?

*21. How many codons would be possible in a triplet code if only three bases (A, C, and U) were used?

*22. Referring to the genetic code presented in **Figure 15.10**, give the amino acids specified by the following bacterial mRNA sequences.

a. 5′–AUGUUUAAAUUUAAAUUUUGA–3′

b. 5′–AGGGAAAUCAGAUGUAUAUAUAUAUAUGA–3′

c. 5′–UUUGGAUUGAGUGAAACGAUGGAUGAAAG AUUUCUCGCUUGA–3′

d. 5′–GUACUAAGGAGGUUGUAUGGGUUAGGGG ACAUCAUUUUGA–3′

23. A nontemplate strand of bacterial DNA has the following base sequence. What amino acid sequence will be encoded by this sequence?

5′–ATGATACTAAGGCCC–3′

24. The following amino acid sequence is found in a tripeptide: Met-Trp-His. Give all possible nucleotide sequences on the mRNA, on the template strand of

DNA, and on the nontemplate strand of DNA that can encode this tripeptide.

25. How many different mRNA sequences can encode a polypeptide chain with the amino acid sequence Met-Leu-Arg? (Be sure to include the stop codon.)

*26. A series of tRNAs have the following anticodons. Consider the wobble rules listed in **Table 15.2** and give all possible codons with which each tRNA can pair.

   a. 5′–GGC–3′

   b. 5′–AAG–3′

   c. 5′–IAA–3′

   d. 5′–UGG–3′

   e. 5′–CAG–3′

27. A researcher creates random copolymers of three nucleotides each by mixing polynucleotide phosphorylase with guanine and adenine nucleotides in a ratio of 5 guanine nucleotides to 1 adenine. Give the different copolymers produced and their theoretical proportions.

28. Assume that the nucleotide at the 5′ end of the first tRNA's anticodon (the tRNA on the left) in **Figure 15.11** was mutated from G to U. Give all codons with which the new, mutated anticodon could pair.

29. Which of the following amino acid changes could result from a mutation that changed a single base? For each change that could result from the alteration of a single base, determine which position of the codon (first, second, or third nucleotide) in the mRNA must be altered for the change to result.

   a. Leu → Gln

   b. Phe → Ser

   c. Phe → Ile

   d. Pro → Ala

   e. Asn → Lys

   f. Ile → Asn

Section 15.3

*30. Arrange the following components of translation in the approximate order in which they would appear or be used in prokaryotic protein synthesis:

   70S initiation complex

   30S initiation complex

   release factor 1

   elongation factor G

   initiation factor 3

   elongation factor Tu

   fMet-tRNA$_i^{fMet}$

31. Examine the tRNA in **Figure 15.14**. What do you think would be the potential effect of a mutation in the part of

the tRNA gene that encodes (a) the acceptor stem; (b) the anticodon; (c) one of the red nucleotide positions?

32. The following diagram illustrates a step in the process of translation. Identify the following elements on the diagram.

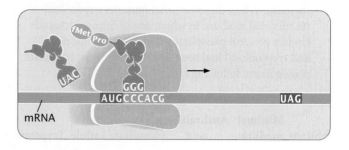

   a. 5′ and 3′ ends of the mRNA

   b. A, P, and E sites

   c. Start codon

   d. Stop codon

   e. Amino and carboxyl ends of the newly synthesized polypeptide chain

   f. Approximate location of the next peptide bond that will be formed

   g. Place on the ribosome where release factor 1 will bind

*33. Refer to the diagram in Problem 32 to answer the following questions.

   a. What will be the anticodon of the next tRNA added to the A site of the ribosome?

   b. What will be the next amino acid added to the growing polypeptide chain?

*34. A synthetic mRNA added to a cell-free protein-synthesizing system produces a peptide with the following amino acid sequence: Met-Pro-Ile-Ser-Ala. What would be the effect on translation if the following components were omitted from the cell-free protein-synthesizing system? What, if any, type of protein would be produced? Explain your reasoning.

   a. Initiation factor 3

   b. Initiation factor 2

   c. Elongation factor Tu

   d. Elongation factor G

   e. Release factors RF-1, RF-2, and RF-3

   f. ATP

   g. GTP

35. For each of the following sequences, place a check mark in the appropriate space to indicate the process *most immediately* affected by deleting the sequence. Choose only one process for each sequence (i.e., one check mark per sequence).

	Process most immediately affected by deletion			
Sequence deleted	**Replication**	**Transcription**	**RNA processing**	**Translation**
a. *ori* site	_____	_____	_____	_____
b. 3′ splice site consensus	_____	_____	_____	_____
c. Poly(A) tail	_____	_____	_____	_____
d. Terminator	_____	_____	_____	_____
e. Start codon	_____	_____	_____	_____
f. −10 consensus	_____	_____	_____	_____
g. Shine–Dalgarno	_____	_____	_____	_____

36. MicroRNAs are small RNA molecules that bind to the 3′ end of mRNA and suppress translation (see Section 14.5). Some eukaryotic mRNAs have internal ribosome-binding sites downstream of the 5′ cap, where ribosomes normally bind. In one investigation, miRNAs did not suppress translation by ribosomes that had attached to internal ribosome-binding sites (R. S. Pillai et al. 2005. *Science* 309:1573–1576). What does this finding suggest about how miRNAs suppress translation?

37. Give the amino acid sequence of the protein encoded by the mRNA in **Figure 15.21**.

Section 15.4

*38. Mutations that introduce stop codons cause a number of genetic diseases. For example, from 2% to 5% of the people who have cystic fibrosis possess a mutation that causes a premature stop codon in the gene encoding the cystic fibrosis transmembrane conductance regulator (CFTR). This premature stop codon produces a truncated form of CFTR that is nonfunctional and results in the symptoms of cystic fibrosis (see Section 5.1). One possible way to treat people with genetic diseases caused by these types of mutations is to trick the ribosome into reading through the stop codon, inserting an amino acid in its place. Although the protein produced may have one altered amino acid, it is more likely to be at least partly functional than is the truncated protein produced when the ribosome stalls at the stop codon. Indeed, geneticists have conducted clinical trials of a drug called PTC124 on people with cystic fibrosis. This drug interferes with the ribosome's ability to correctly read stop codons (C. Ainsworth. 2005. *Nature* 438:726–728). On the basis of what you know about the mechanism of nonsense-mediated mRNA decay (NMD), would you expect NMD to pose a problem for this type of treatment? Why or why not?

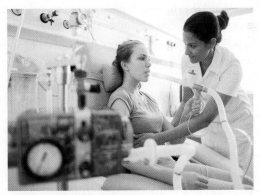

Patient with cystic fibrosis undergoing respiratory therapy. [SCIENCE PHOTO LIBRARY/Getty Images.]

CHALLENGE QUESTIONS

Section 15.2

39. The redundancy of the genetic code means that some amino acids are specified by more than one codon. For example, the amino acid leucine is specified by six different codons. Within a genome, synonymous codons are not present in equal numbers; some synonymous codons appear much more frequently than others, and the preferred codons differ among species. For example, in one species, the codon UUA might be used most often to specify leucine, whereas in another species, the codon CUU might be used most often. Speculate on a reason for this bias in codon usage and why the preferred codons are not the same in all organisms.

Section 15.3

40. In what ways are spliceosomes and ribosomes similar? In what ways are they different? Can you suggest some possible reasons for their similarities?

*41. Several experiments were conducted to obtain information about how the eukaryotic ribosome recognizes the AUG start codon. In one experiment, the gene that encodes methionine initiator tRNA (tRNA$_i^{Met}$) was located and changed; specifically, the nucleotides that specify the anticodon on tRNA$_i^{Met}$ were mutated so that the anticodon in the tRNA was 5′–CCA–3′ instead of 5′–CAU–3′. When this mutated gene was placed in a eukaryotic cell, protein synthesis took place, but the proteins produced were abnormal. Some of these proteins contained extra amino acids, and others contained fewer amino acids than normal.

a. What do these results indicate about how the ribosome recognizes the starting point for translation in eukaryotic cells? Explain your reasoning.

b. If the same experiment had been conducted on bacterial cells, what results would you expect?

c. Explain why some of the proteins produced contained extra amino acids while others contained fewer amino acids than normal.

THINK-PAIR-SHARE QUESTIONS

Section 15.1

1. Archibald Garrod was an English physician who first proposed that genes encode enzymes. Like the work of Gregor Mendel (see Chapter 3), his discovery had little impact on his contemporaries and was not widely accepted until many years later. Why are important discoveries in science sometimes not accepted immediately? Why does it often take years before they are generally accepted by other scientists?

2. Do you think the one gene, one polypeptide hypothesis is correct today? Why or why not?

3. Compare and contrast the structure and function of proteins and nucleic acids. How are they similar and how are they different? How do their different structures contribute to their different functions?

Section 15.2

4. A triplet code with three nucleotides per codon is the most efficient way to encode the 20 different amino acids. Why would cells be expected to use the most efficient code? In other words, what might be the advantages of using an efficient code?

5. Some evidence suggests that synonymous codons are not really synonymous. Sometimes mutations that change a codon to a synonymous codon produce an effect on the phenotype, even though the amino acid

sequence of the protein is the same. Propose some ways in which synonymous codons might have different phenotypic effects.

6. There are three termination codons (UAA, UAG, UGA) but usually only one initiation codon (AUG) is used. Propose some possible reasons for why there are more termination codons than initiation codons.

7. Exceptions to the universality of the genetic code were once thought to be rare, but more and more exceptions are being found. As mentioned in the text, one study found that a significant proportion of bacteria in the environment use nonuniversal codons. Can you think of any advantages for an organism using nonuniversal codons?

Section 15.3

8. Structural analysis of bacterial release factor 1 (RF-1) and release factor 2 (RF-2) reveals that these proteins are similar in size and shape to a tRNA molecule. This similarity has sometimes been called molecular mimicry. Why might RF-1 and RF-2 have evolved to mimic tRNAs?

Section 15.4

9. In humans, there may be three times as many proteins as genes. If each gene encodes a protein, how can there be more proteins than genes?

Control of Gene Expression in Bacteria

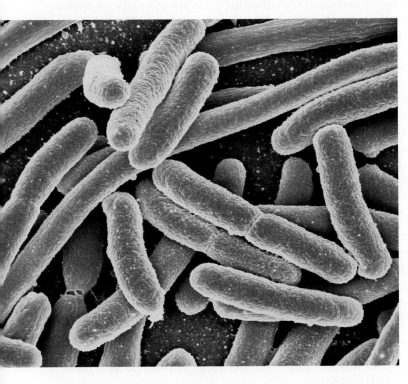

The expression of genes in bacteria is often regulated through operons, groups of genes that are transcribed as units. Shown here is *Escherichia coli*, a common bacterium found in the intestinal tracts of mammals. [Pasieka/Science Source.]

Operons and the Noisy Cell

In 2011, geneticists from around the world celebrated the 50th anniversary of the operon. An operon is a group of genes that share a common promoter and are transcribed as a unit, producing a single mRNA molecule that encodes several proteins. Typically, these proteins interact in some way; for example, the *trp* operon in the bacterium *Escherichia coli* encodes components of three enzymes that work together to synthesize the amino acid tryptophan. Operons control the expression of genes such as these by regulating their transcription. Genes in bacteria and archaea are often organized into operons, but operons are much less common in eukaryotes.

The operon was discovered through the elegant research of François Jacob and Jacques Monod, who worked at opposite ends of the attic floor of the Pasteur Institute in Paris. Jacob was studying bacteriophage λ, a virus that infects *E. coli*; Monod was analyzing the properties of β-galactosidase, an enzyme *E. coli* uses to metabolize the sugar lactose. One summer evening in 1958, Jacob had a flash of inspiration—he saw a connection between the research taking place at the two ends of the attic. Jacob recognized that the genes that induce phage reproduction were controlled in the same way as the genes that control production of β-galactosidase in *E. coli*. This realization led to an important collaboration between Jacob and Monod; together, they eventually uncovered the structure and function of the *lac* operon. In 1965, they were awarded the Nobel Prize in physiology or medicine, along with their collaborator Andre Lwoff.

Following Jacob and Monod's discovery of the *lac* operon, other operons were discovered in bacteria, and a great deal of research was focused on the mechanism of operon function. Despite extensive research on *how* operons work, much less was known about *why* operons exist: Why do prokaryotes have them while most eukaryotes don't? Why are some genes included in operons while others are not? These questions intrigued Oleg Igoshin, at Rice University, and Christian Ray, at the University of Texas MD Anderson Cancer Center. Igoshin and Ray are computational biologists, a new breed of scientists who use complex mathematics to study fundamental problems of biology. They took what might seem like an unlikely approach to the question of why operons exist. Instead of growing bacteria, inducing mutations, and examining DNA, they developed a series of mathematical models of gene networks that could be run on a computer. Using these models, they looked at how genes functioned when grouped into operons and when regulated separately.

Igoshin and Ray knew that random fluctuations in the levels of transcription and translation occur naturally. Because of these fluctuations—noise in the system—the amounts of different proteins can vary widely; thus, the amount of a protein produced might be more or less than is optimal for cell growth and survival. Igoshin and Ray hypothesized that coordinating the transcription of several genes through an operon structure might reduce noise in the system and permit more finely tuned control over gene expression.

To test their hypothesis, Igoshin and Ray ran computer models for six different types of interactions between the products of genes of the types that are potentially found in operons. Their models showed that for some types of protein interactions, grouping genes together in an operon decreases the biochemical noise. For other types of protein interactions, grouping the genes in operons actually increases the noise. Thus, the operon structure has the potential to increase or decrease noise depending on which genes are grouped together.

Igoshin and Ray then examined genes that are actually found in operons in *E. coli*, and they discovered that operons containing genes whose interactions decrease noise were more common than expected on a random basis. Conversely, operons whose gene interactions increase noise were less common than expected. They concluded that operons have evolved as a way for the cell to couple transcription of multiple genes so as to reduce biochemical noise in the cell, allowing the cell to more finely tune the relative proportions of the proteins encoded by the operon. Igoshin and Ray speculated that operons are less common in eukaryotes because the larger volume of eukaryotic cells reduces the effect of random fluctuations and, perhaps, because eukaryotes have other mechanisms (such as changes in chromatin structure) to couple the transcription of multiple genes.

THINK-PAIR-SHARE

- François Jacob and Jacques Monod began their successful collaboration, in part, because they worked on the same floor of the Pasteur Institute. In today's technology-filled world, where people can easily communicate by email and videoconferencing, is physical proximity still of any value in collaborating in research? Why or why not?

- What other mechanisms, in addition to operons, might allow groups of genes to be regulated together?

This chapter is the first of two chapters about **gene regulation**, the mechanisms and systems that control the expression of genes. In this chapter, we consider systems of gene regulation in bacteria. We begin by considering the necessity for gene regulation, the levels at which gene expression is controlled, and the difference between genes and regulatory elements. We then examine the structure and function of operons, and we consider gene regulation in some specific examples of operons, including the *lac* operon studied by Jacob and Monod. Finally, we discuss several types of bacterial gene regulation that are facilitated by RNA molecules. In Chapter 17, we will discuss mechanisms of gene regulation in eukaryotic genomes.

THINK-PAIR-SHARE Question 1

16.1 The Regulation of Gene Expression Is Critical for All Organisms

A major theme of molecular genetics is the central dogma, which states that genetic information flows from DNA to RNA to proteins (see Figure 10.17). Although the central dogma

provided a molecular basis for the connection between genotype and phenotype, it failed to address a critical question: How is the flow of information along this molecular pathway *regulated*?

Consider *E. coli*, a bacterium that resides in your large intestine. Your eating habits completely determine the nutrients available to this bacterium: it can neither seek out nourishment when nutrients are scarce nor move away when confronted with an unfavorable environment. The bacterium makes up for its inability to alter the external environment by being internally flexible. For example, if glucose is present, *E. coli* uses it to generate ATP; if there's no glucose, it uses lactose, arabinose, maltose, xylose, or any of a number of other sugars. When amino acids are available, *E. coli* uses them to synthesize proteins; if a particular amino acid is absent, *E. coli* produces the enzymes needed to synthesize that amino acid. Thus, *E. coli* responds to environmental changes by rapidly altering its biochemistry. This biochemical flexibility, however, has a high price. Constantly producing all the enzymes necessary for every environmental condition would be energetically expensive. So how does *E. coli* maintain biochemical flexibility while optimizing energy efficiency?

The answer is through gene regulation. Bacteria carry the genetic information for synthesizing many proteins, but only a subset of that genetic information is expressed at any time. When the environment changes, new genes are expressed, and proteins appropriate for the new environment are synthesized. For example, if a carbon source appears in the environment, genes encoding enzymes that take up and metabolize that carbon source are quickly transcribed and translated. When that carbon source disappears, the genes that encode those enzymes are shut off.

Multicellular eukaryotic organisms face a further challenge. Individual cells in a multicellular organism are specialized for particular tasks. The proteins produced by a nerve cell, for example, are quite different from those produced by a white blood cell. But although they differ in shape and function, a nerve cell and a white blood cell still carry the same genetic instructions.

A multicellular organism's challenge is to bring about the differentiation of cells that have a common set of genetic instructions (the process of development). This challenge is met through gene regulation: all of an organism's cells carry the same genetic information, but only a subset of genes are expressed in each cell type; genes needed for other cell types are not expressed. Gene regulation is therefore the key to both unicellular flexibility and multicellular specialization, and it is critical to the success of all living organisms.

> ### CONCEPTS
>
> In bacteria, gene regulation maintains internal flexibility, turning genes on and off in response to environmental changes. In multicellular eukaryotic organisms, gene regulation also brings about cell differentiation.

The mechanisms of gene regulation were first investigated in bacterial cells, in which the availability of mutants and the ease of laboratory manipulation made it possible to unravel these mechanisms. When the study of these mechanisms in eukaryotic cells began, eukaryotic gene regulation seemed very different from bacterial gene regulation. However, as more and more information has accumulated about gene regulation, a number of common themes have emerged. Today, many aspects of gene regulation in bacterial and eukaryotic cells are recognized to be similar. Before examining specific elements of bacterial gene regulation (in this chapter) and eukaryotic gene regulation (in Chapter 17), we briefly consider some themes of gene regulation common to all organisms.

Genes and Regulatory Elements

In considering gene regulation in both bacteria and eukaryotes, we must distinguish between the DNA sequences that are transcribed and the DNA sequences that regulate the expression of other sequences. **Structural genes** encode proteins that are used in metabolism or biosynthesis or that play a structural role in the cell. On the other hand, the products of **regulatory genes**—either RNA or proteins—interact with other DNA sequences and affect the transcription or translation of those sequences. In many cases, the products of regulatory genes are DNA-binding proteins or RNA molecules that affect gene expression.

Bacteria and eukaryotes use regulatory genes to control the expression of many of their structural genes. However, a few structural genes, particularly those that encode essential cellular functions (often called housekeeping genes) are expressed continually and are therefore said to be **constitutive**. Constitutive genes are not regulated.

We will also encounter DNA sequences that are not transcribed at all, but still play a role in regulating genes and other DNA sequences. These **regulatory elements** affect the expression of DNA sequences to which they are physically linked. Regulatory elements are common in both bacterial and eukaryotic cells, and much of gene regulation in both types of organisms takes place through the action of proteins produced by regulatory genes that recognize and bind to regulatory elements.

The regulation of gene expression can occur through processes that stimulate gene expression, termed *positive control*, or through processes that inhibit gene expression, termed *negative control*. Bacteria and eukaryotes use both positive and negative control mechanisms to regulate their genes.

THINK-PAIR-SHARE Question 2

> ### CONCEPTS
>
> Structural genes encode proteins, while regulatory elements are DNA sequences that are not transcribed, but affect the expression of genes. Positive control mechanisms stimulate gene expression, whereas negative control mechanisms inhibit gene expression.
>
> ✔ **CONCEPT CHECK 1**
> What is a constitutive gene?

Levels of Gene Regulation

In both bacteria and eukaryotes, genes can be regulated at a number of levels along the pathway of information flow from genotype to phenotype (**Figure 16.1**).

First, genes can be regulated through the alteration of DNA or chromatin structure; this type of gene regulation takes place primarily in eukaryotes. Modifications to DNA or its packaging can help to determine which sequences are available for transcription or the rate at which sequences are transcribed. DNA methylation and changes in chromatin structure are two processes that play a pivotal role in gene regulation.

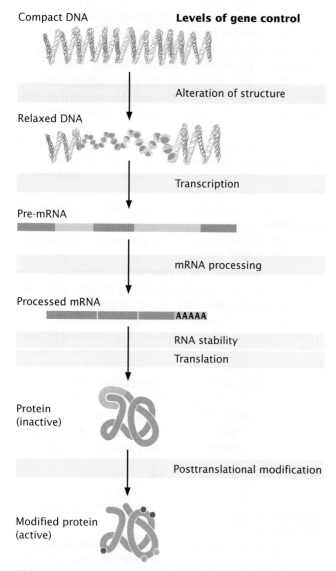

Levels of gene control

Compact DNA

Alteration of structure

Relaxed DNA

Transcription

Pre-mRNA

mRNA processing

Processed mRNA

RNA stability

Translation

Protein
(inactive)

Posttranslational modification

Modified protein
(active)

16.1 Gene expression can be controlled at multiple levels.

A second point at which a gene can be regulated is at the level of transcription. For the sake of cellular economy, limiting the production of a protein early in the process makes sense, and transcription is an important point of gene regulation in both bacterial and eukaryotic cells.

A third potential point of gene regulation is mRNA processing. Eukaryotic mRNA is extensively modified before it is translated: a 5′ cap is added, the 3′ end is cleaved and polyadenylated, and introns are removed (see Chapter 14). These modifications determine the movement of the mRNA into the cytoplasm, whether the mRNA can be translated, the rate of translation, and the amino acid sequence of the protein produced, as well as mRNA stability. There is growing evidence that a number of regulatory mechanisms in eukaryotic cells operate at the level of mRNA processing. In prokaryotes, however, in which these types of mRNA processing are largely absent, control of gene expression occurs primarily at other points.

A fourth point for the control of gene expression is the regulation of mRNA stability. The amount of protein produced depends not only on the amount of mRNA synthesized, but also on the rate at which the mRNA is degraded.

A fifth point of gene regulation is at the level of translation, a complex process requiring a large number of enzymes, protein factors, and RNA molecules (see Chapter 15). All of these factors, as well as the availability of amino acids, affect the rate at which proteins are produced and therefore provide points at which gene expression can be controlled. Translation can also be affected by sequences in mRNA, such as those in the 5′ and 3′ untranslated regions.

Finally, many proteins are modified after translation (see Chapter 15), and these modifications affect whether the proteins become active; therefore, genes can be regulated through processes that affect posttranslational modification. Gene expression can be affected by regulatory activities at any or all of these points.

CONCEPTS

Gene expression can be controlled at any of a number of levels along the molecular pathway from DNA to protein, including DNA or chromatin structure, transcription, mRNA processing, mRNA stability, translation, and posttranslational modification.

✔ **CONCEPT CHECK 2**

Why is transcription a particularly important level of gene regulation in both bacteria and eukaryotes?

DNA-Binding Proteins

Much of gene regulation in bacteria and eukaryotes is accomplished by proteins that bind to DNA sequences and affect their expression. These regulatory proteins generally have discrete functional parts—called **domains**, typically consisting of 60 to 90 amino acids—that are responsible for binding to DNA. Within a domain, only a few amino acids actually make contact with the DNA. These amino acids (most commonly asparagine, glutamine, glycine, lysine, and arginine) often form hydrogen bonds with the bases or interact with the sugar–phosphate backbone of the DNA. Many DNA-binding proteins have additional domains that can bind other molecules, such as other regulatory proteins. By physically attaching to DNA, these proteins can affect the expression of a gene. Most DNA-binding proteins bind dynamically, which means that they transiently bind and unbind DNA and other regulatory proteins. Thus, although they may spend most of their time bound to DNA, they are never permanently attached. This dynamic nature means that other molecules can compete with DNA-binding proteins for regulatory sites on the DNA.

DNA-binding proteins can be grouped into several distinct types on the basis of characteristic structures, called *motifs*, found within the binding domain. Motifs are simple

(a) Helix-turn-helix

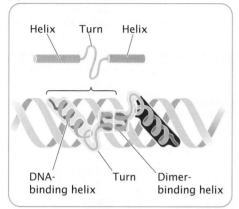

Helix   Turn   Helix

DNA-binding helix   Turn   Dimer-binding helix

(b) Zinc fingers

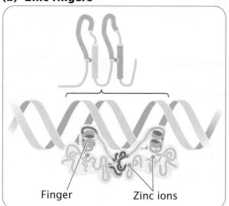

Finger   Zinc ions

(c) Leucine zipper

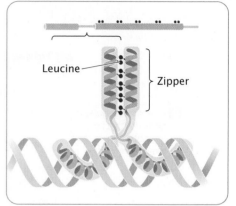

Leucine   Zipper

16.2 DNA-binding proteins can be grouped into several types on the basis of characteristic structures, or motifs.

structures, such as alpha helices, that can fit into the major groove of the DNA double helix. For example, the helix-turn-helix motif (**Figure 16.2a**), consisting of two alpha helices connected by a turn, is common in bacterial regulatory proteins. The zinc-finger motif (**Figure 16.2b**), shared by many eukaryotic regulatory proteins, consists of a loop of amino acids containing a zinc ion. The leucine zipper (**Figure 16.2c**) is another motif found in a variety of eukaryotic DNA-binding proteins. These common DNA-binding motifs and others are summarized in **Table 16.1**.

CONCEPTS

Regulatory proteins that bind DNA have common motifs that interact with the double-helical structure of DNA.

✔ CONCEPT CHECK 3

How do amino acids in DNA-binding proteins interact with DNA?
a. By forming covalent bonds with DNA bases
b. By forming hydrogen bonds with DNA bases
c. By forming covalent bonds with DNA sugars

16.2 Operons Control Transcription in Bacterial Cells

A significant difference between bacterial and eukaryotic gene control lies in the organization of functionally related genes. As we saw in the introduction to this chapter, many bacterial genes that have related functions are clustered together and under the control of a single promoter. These genes are often transcribed together into a single mRNA molecule. A group of bacterial structural genes that are transcribed together, along with their promoter and additional sequences that control their transcription, is called an **operon**. The operon regulates the expression of the structural genes by controlling transcription, which, in bacteria, is usually the most important level of gene regulation.

Operon Structure

The organization of a typical operon is illustrated in **Figure 16.3**. At one end of the operon is a set of structural genes, shown in Figure 16.3 as gene *a*, gene *b*, and gene *c*. These structural genes are transcribed into a single mRNA,

---

TABLE 16.1	Common DNA-binding motifs		
Motif	**Location**	**Characteristics**	**Binding Site in DNA**
Helix-turn-helix	Bacterial regulatory proteins; related motifs in eukaryotic proteins	Two alpha helices	Major groove
Zinc finger	Eukaryotic regulatory and other proteins	Loop of amino acids with zinc at base	Major groove
Steroid receptor	Eukaryotic proteins	Two perpendicular alpha helices with zinc surrounded by four cysteines	Major groove and DNA backbone
Leucine zipper	Eukaryotic transcription factors	Helix of leucine and a basic arm; two leucines interdigitate	Two adjacent major grooves
Helix-loop-helix	Eukaryotic proteins	Two alpha helices separated by a loop of amino acids	Major groove
Homeodomain	Eukaryotic regulatory proteins	Three alpha helices	Major groove

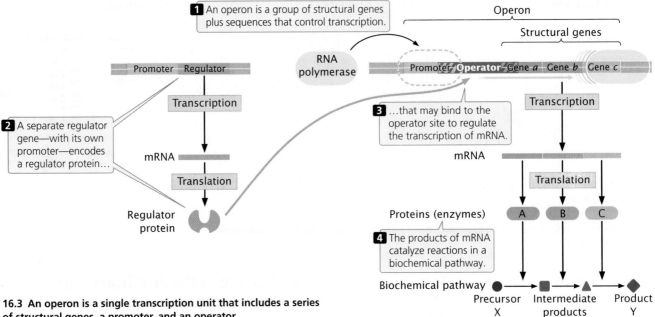

1 An operon is a group of structural genes plus sequences that control transcription.

2 A separate regulator gene—with its own promoter—encodes a regulator protein...

3 ...that may bind to the operator site to regulate the transcription of mRNA.

4 The products of mRNA catalyze reactions in a biochemical pathway.

16.3 An operon is a single transcription unit that includes a series of structural genes, a promoter, and an operator.

which is translated to produce enzymes A, B, and C. These enzymes carry out a series of biochemical reactions that convert precursor molecule X into product Y. The transcription of structural genes *a*, *b*, and *c* is under the control of a single promoter, which lies upstream of the first structural gene. RNA polymerase binds to the promoter and then moves downstream, transcribing the structural genes.

A **regulator gene** helps to control the expression of the structural genes of the operon by increasing or decreasing their transcription. Although it affects operon function, the regulator gene is not considered part of the operon. The regulator gene has its own promoter and is transcribed into a short mRNA, which is translated into a small protein. This **regulator protein** can bind to a region of the operon called the **operator** and affect whether transcription can take place. The operator usually overlaps the 3′ end of the promoter and sometimes the 5′ end of the first structural gene (see Figure 16.3).

> **CONCEPTS**
>
> Functionally related genes in bacterial cells are frequently clustered together in a single transcription unit termed an operon. A typical operon includes several structural genes, a promoter for those structural genes, and an operator to which the product of a regulator gene binds.
>
> **✔ CONCEPT CHECK 4**
>
> What is the difference between a structural gene and a regulator gene?
>
> a. Structural genes are transcribed into mRNA, but regulator genes aren't.
> b. Structural genes have complex structures; regulator genes have simple structures.
> c. Structural genes encode proteins that function in the structure of the cell; regulator genes carry out metabolic reactions.
> d. Structural genes encode proteins; regulator genes control the transcription of structural genes.

Negative and Positive Control: Inducible and Repressible Operons

There are two types of transcriptional control: **negative control**, in which a regulatory protein is a repressor, binding to DNA and inhibiting transcription; and **positive control**, in which a regulatory protein is an activator, stimulating transcription. Operons can also be either inducible or repressible. **Inducible operons** are those in which transcription is normally off (not taking place); something must happen to induce transcription, or turn it on. **Repressible operons** are those in which transcription is normally on (taking place); something must happen to repress transcription, or turn it off. In the next few sections, we will consider several varieties of these basic control mechanisms.

NEGATIVE INDUCIBLE OPERONS The regulator gene for a negative inducible operon encodes an active *repressor* protein that readily binds to the operator (**Figure 16.4a**). Because the operator site overlaps the promoter site, the binding of this protein to the operator physically blocks the binding of RNA polymerase to the promoter and prevents transcription. For transcription to take place, something must happen to prevent the binding of the repressor to the operator. This type of system is said to be *inducible* because transcription is normally off (inhibited) and must be turned on (induced).

Transcription of a negative inducible operon is turned on when a small molecule called an **inducer** is present and binds to the repressor (**Figure 16.4b**). Regulatory proteins frequently have two binding sites: one that binds to DNA and another that binds to a small molecule such as an inducer. The binding of the inducer (precursor V in Figure 16.4b) alters the shape of the repressor, preventing it from binding to DNA. Proteins such as this repressor, which change shape upon binding to another molecule, are called **allosteric proteins**.

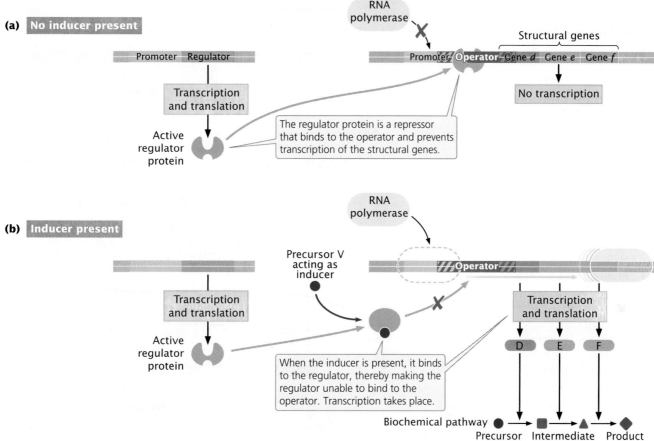

16.4 Some operons are inducible.

When the inducer is absent, the repressor binds to the operator, the structural genes are not transcribed, and enzymes D, E, and F (which metabolize precursor V) are not synthesized (see Figure 16.4a). This mechanism is an adaptive one: because no precursor V is present, synthesis of the enzymes would be wasteful because they would have no substrate to metabolize. As soon as precursor V becomes available, some of it binds to the repressor, rendering the repressor inactive and unable to bind to the operator. RNA polymerase can now bind to the promoter and transcribe the structural genes. The resulting mRNA is then translated into enzymes D, E, and F, which convert substrate V into product W (see Figure 16.4b). So, an operon with negative inducible control regulates the synthesis of the enzymes economically: the enzymes are synthesized only when their substrate (V) is available.

Inducible operons usually control proteins that carry out degradative processes—proteins that break down molecules. For these types of proteins, inducible control makes sense because the proteins are not needed unless the substrate (which is broken down by the proteins) is present.

NEGATIVE REPRESSIBLE OPERONS Some operons with negative control are *repressible*, meaning that transcription

normally takes place and must be turned off, or repressed. The regulator protein that acts on this type of operon is also a repressor, but it is synthesized in an *inactive* form that cannot by itself bind to the operator. Because no repressor is bound to the operator, RNA polymerase readily binds to the promoter, and transcription of the structural genes takes place (**Figure 16.5a**).

To turn transcription off, something must happen to make the repressor active. When a small molecule called a **corepressor** is present, it binds to the repressor and makes it capable of binding to the operator. In the example illustrated (see Figure 16.5a), the product (U) of the metabolic reaction controlled by the operon is the corepressor. As long as the level of product U is high, it is available to bind to the repressor and activate it, preventing transcription (**Figure 16.5b**). With the operon repressed, enzymes G, H, and I are not synthesized, and no more U is produced from precursor T. However, when all of product U is used up, the repressor is no longer activated by product U and cannot bind to the operator. The inactivation of the repressor allows the transcription of the structural genes and the synthesis of enzymes G, H, and I, resulting in the conversion of precursor T into product U. Like inducible operons, repressible operons are economical: the proteins they encode are synthesized only as needed.

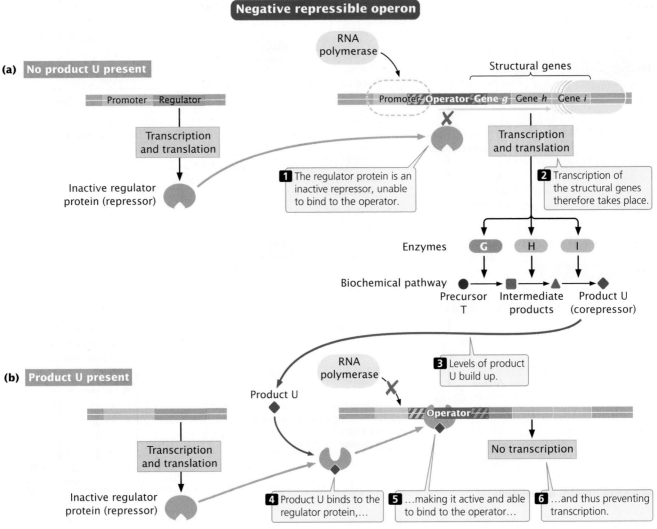

Negative repressible operon

(a) No product U present

1 The regulator protein is an inactive repressor, unable to bind to the operator.

2 Transcription of the structural genes therefore takes place.

Inactive regulator protein (repressor)

Enzymes   G   H   I

Biochemical pathway   Precursor T → Intermediate products → Product U (corepressor)

(b) Product U present

3 Levels of product U build up.

Inactive regulator protein (repressor)

4 Product U binds to the regulator protein,...

5 ...making it active and able to bind to the operator...

6 ...and thus preventing transcription.

16.5 Some operons are repressible.

Repressible operons usually control proteins that carry out the biosynthesis of molecules needed in the cell, such as amino acids. For these types of operons, repressible control makes sense because the product produced by the proteins is always needed by the cell. Thus, these operons are normally on and are turned off only when there are adequate amounts of the product already present.

Note that both the inducible and the repressible operon systems that we have just considered are forms of negative control, in which the regulator protein is a repressor.

▷ TRY PROBLEM 11

POSITIVE CONTROL In positive transcriptional control, the regulatory protein involved is an activator: it binds to DNA (usually at a site other than the operator) and stimulates transcription. For example, in the case of the *lac* operon, the catabolite activator protein (CAP) binds to the promoter and increases the efficiency with which RNA polymerase can bind the promoter and transcribe the structural genes (as we'll see later in this section).

THINK-PAIR-SHARE Question 3

CONCEPTS

There are two basic types of transcriptional control: negative and positive. In negative control, when a regulator protein (repressor) binds to DNA, transcription is inhibited; in positive control, when a regulator protein (activator) binds to DNA, transcription is stimulated. Some operons are inducible: their transcription is normally off and must be turned on. Other operons are repressible: their transcription is normally on and must be turned off.

✔ CONCEPT CHECK 5

The regulator protein that acts on a negative repressible operon is synthesized as

a. an active activator.   c. an active repressor.

b. an inactive activator.   d. an inactive repressor.

The *lac* Operon of *E. coli*

François Jacob and Jacques Monod first described the "operon model" for the genetic control of lactose metabolism in *E. coli* in 1961. Their work, and subsequent research on the genetics of lactose metabolism, established the operon as the basic unit of transcriptional control in bacteria. Despite the fact that, at the time, no methods were available for determining nucleotide sequences, Jacob and Monod deduced the structure of the operon *genetically* by analyzing the interactions of mutations that interfered with the normal regulation of lactose metabolism. We examine the effects of some of these mutations after seeing how the *lac* operon regulates lactose metabolism.

LACTOSE METABOLISM Lactose is a major carbohydrate found in milk; it can be metabolized by *E. coli* bacteria that reside in the mammalian gut. Lactose does not easily diffuse across the *E. coli* cell membrane and must be actively transported into the cell by the protein lactose permease (**Figure 16.6**). To use lactose as an energy source, *E. coli* must then break it into glucose and galactose, a reaction catalyzed by the enzyme β-galactosidase. This enzyme can also convert lactose into allolactose, a compound that plays an important role in regulating lactose metabolism. A third enzyme, thiogalactoside transacetylase, is also produced by the *lac* operon, but its function in lactose metabolism is not yet clear. One possible function is detoxification, preventing the accumulation of thiogalactosides that are transported into the cell along with lactose by permease.

REGULATION OF THE *lac* OPERON The *lac* operon of *E. coli* is an example of a negative inducible operon. The enzymes β-galactosidase, permease, and transacetylase are encoded by adjacent structural genes in the *lac* operon (**Figure 16.7a**) and have a common promoter (*lacP* in Figure 16.7a). β-Galactosidase is encoded by the *lacZ* gene, permease by the *lacY* gene, and transacetylase by the *lacA* gene. When lactose is absent from the medium in which *E. coli* grows, few molecules of each protein are produced. If lactose is added to the medium and glucose is absent, the rate of synthesis of all three proteins simultaneously increases about a thousandfold within 2 to 3 minutes. This boost in protein synthesis, which results from the simultaneous transcription of *lacZ*, *lacY*, and *lacA*, exemplifies **coordinate induction**, the simultaneous synthesis of several proteins stimulated by a specific molecule, the inducer (**Figure 16.7b**).

Although lactose might appear to be the inducer in this case, allolactose is actually responsible for induction. Upstream of *lacP* is a regulator gene, *lacI*, which has its own promoter (P_I). The *lacI* gene is transcribed into a short mRNA that is translated into a repressor. The repressor consists of four identical polypeptides and has two types of binding sites; one type of site binds to allolactose and the other binds to DNA. In the absence of lactose (and, therefore, allolactose), the repressor binds to the *lac* operator *lacO* (see Figure 16.7a). Jacob and Monod mapped the operator to a position adjacent to the *lacZ* gene; nucleotide sequencing has demonstrated that the operator actually overlaps the 3′ end of the promoter and the 5′ end of *lacZ* (**Figure 16.8**).

When the repressor is bound to the operator, the binding of RNA polymerase is blocked, and transcription is prevented. When lactose is present, some of it is converted into allolactose, which binds to the repressor and causes the repressor to be released from the operator. RNA polymerase then binds to the promoter and moves down the DNA molecule, transcribing the structural genes. In the presence of lactose, then, the repressor is inactivated, the binding of RNA polymerase is no longer blocked, the transcription of *lacZ*, *lacY*, and *lacA* takes place, and the *lac* proteins are produced.

Have you spotted the flaw in the explanation just given for the induction of *lac* protein synthesis? You might recall that permease is required to transport lactose into the cell. If the *lac* operon is repressed and no permease is being produced, how does lactose get into the cell to inactivate the repressor and turn on transcription? Furthermore, the inducer is actually allolactose, which must be produced from lactose by β-galactosidase. If β-galactosidase production is repressed, how can lactose metabolism be induced?

The answer is that repression never *completely* shuts down transcription of the *lac* operon. Even with active repressor bound to the operator, there is a low level of transcription, and a few molecules of β-galactosidase, permease, and transacetylase are synthesized. When lactose appears in the medium, the permease that is present transports a small amount of lactose into the cell. There, the few molecules of β-galactosidase that are present convert some of the lactose into allolactose, which then induces transcription.

Several compounds related to allolactose can also bind to the *lac* repressor and induce transcription of the *lac* operon. One such inducer is isopropylthiogalactoside (IPTG). Although IPTG inactivates the repressor and allows the transcription of *lacZ*, *lacY*, and *lacA*, this inducer is not metabolized by β-galactosidase; for this reason, IPTG is often used in research to examine the effects of induction independent of metabolism.

THINK-PAIR-SHARE Question 4

CONCEPTS

The *lac* operon of *E. coli* controls the transcription of three genes needed in lactose metabolism: the *lacZ* gene, which encodes β-galactosidase; the *lacY* gene, which encodes lactose permease; and the *lacA* gene, which encodes thiogalactoside transacetylase. The *lac* operon is a negative inducible operon: a regulator gene produces a repressor that binds to the operator and prevents the transcription of the structural genes. The presence of allolactose inactivates the repressor and allows the transcription of the *lac* operon.

✔ CONCEPT CHECK 6

In the presence of allolactose, the *lac* repressor
a. binds to the operator.
b. binds to the promoter.
c. cannot bind to the operator.
d. binds to the regulator gene.

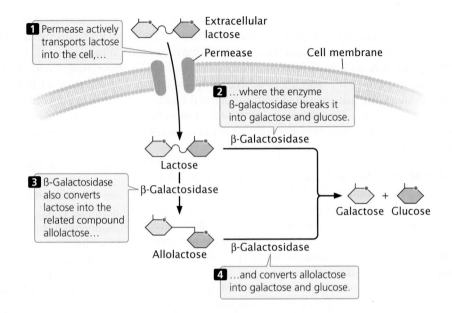

1 Permease actively transports lactose into the cell,…

Extracellular lactose

Permease

Cell membrane

2 …where the enzyme β-galactosidase breaks it into galactose and glucose.

β-Galactosidase

Lactose

3 β-Galactosidase also converts lactose into the related compound allolactose…

β-Galactosidase

Allolactose

β-Galactosidase

Galactose + Glucose

4 …and converts allolactose into galactose and glucose.

16.6 Lactose, a major carbohydrate found in milk, consists of two six-carbon sugars linked together.

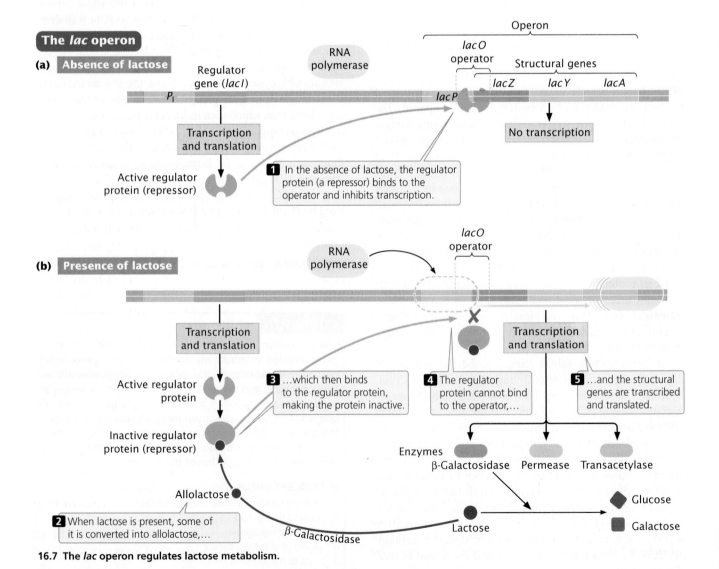

The *lac* operon

(a) **Absence of lactose**

RNA polymerase

Regulator gene (*lacI*)

Operon

lacO operator

Structural genes

lacZ   *lacY*   *lacA*

P_I

lacP

Transcription and translation

Active regulator protein (repressor)

No transcription

1 In the absence of lactose, the regulator protein (a repressor) binds to the operator and inhibits transcription.

(b) **Presence of lactose**

RNA polymerase

lacO operator

Transcription and translation

Active regulator protein

Transcription and translation

3 …which then binds to the regulator protein, making the protein inactive.

4 The regulator protein cannot bind to the operator,…

5 …and the structural genes are transcribed and translated.

Inactive regulator protein (repressor)

Enzymes

β-Galactosidase   Permease   Transacetylase

Allolactose

Glucose

2 When lactose is present, some of it is converted into allolactose,…

β-Galactosidase

Lactose

Galactose

16.7 The *lac* operon regulates lactose metabolism.

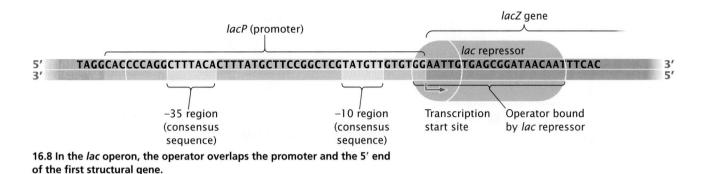

16.8 In the *lac* operon, the operator overlaps the promoter and the 5′ end of the first structural gene.

lac Mutations

Jacob and Monod worked out the structure and function of the *lac* operon by analyzing mutations that affected lactose metabolism. To help define the roles of the different components of the operon, they used **partial-diploid** strains of *E. coli*. The cells of these strains possessed two different DNA molecules: the full bacterial chromosome and an extra piece of DNA. Jacob and Monod created these strains by allowing conjugation to take place between two bacteria. In conjugation, a small circular piece of DNA (the F plasmid) is transferred from one bacterium to another (see Chapter 9). The F plasmid used by Jacob and Monod contained the *lac* operon, so the recipient bacterium became partly diploid, possessing two copies of the *lac* operon. By using different combinations of mutations on the bacterial and plasmid DNA, Jacob and Monod determined that some parts of the *lac* operon are cis acting (able to control the expression of genes only on the same piece of DNA), whereas other parts are trans acting (able to control the expression of genes on other DNA molecules).

STRUCTURAL-GENE MUTATIONS Jacob and Monod first discovered some mutant strains that had lost the ability to synthesize either β-galactosidase or permease (they did not study in detail the effects of mutations on the transacetylase enzyme, so transacetylase will not be considered here). The mutations in those mutant strains mapped to the *lacZ* or *lacY* structural genes and altered the amino acid sequences of the proteins encoded by the genes. These mutations clearly affected the *structure* of the proteins, but not the regulation of their synthesis.

Through the use of partial diploids, Jacob and Monod were able to establish that mutations of the *lacZ* and *lacY* genes were independent and usually affected only the product of the gene in which the mutation occurred. Partial diploids with *lacZ⁺ lacY⁻* on the bacterial chromosome and *lacZ⁻ lacY⁺* on the plasmid functioned normally, producing β-galactosidase and permease in the presence of lactose. (The genotype of a partial diploid is written by separating the genes on each DNA molecule with a slash: *lacZ⁺ lacY⁻/lacZ⁻ lacY⁺*.) In these partial diploids, a single functional β-galactosidase gene (*lacZ⁺*) was sufficient to produce β-galactosidase; whether the functional β-galactosidase gene was coupled to a functional (*lacY⁺*) or a defective (*lacY⁻*) permease gene made no difference. The same was true of the *lacY⁺* gene.

REGULATOR-GENE MUTATIONS Jacob and Monod also isolated mutations that affected the *regulation* of protein production. Mutations in the *lacI* gene affect the production of both β-galactosidase and permease because the genes for both proteins are in the same operon and are regulated coordinately by the *lac* repressor protein.

Some of these mutations were constitutive, causing the *lac* proteins to be produced all the time, whether lactose was present or not. Such mutations in the regulator gene were designated *lacI⁻*. The construction of partial diploids demonstrated that a *lacI⁺* gene is dominant over a *lacI⁻* gene; a single copy of *lacI⁺* (genotype *lacI⁺/lacI⁻*) was sufficient to bring about normal regulation of protein production. Furthermore, *lacI⁺* restored normal control to an operon even if the operon was located on a different DNA molecule, showing that *lacI⁺* can be trans acting. A partial diploid with genotype *lacI⁺ lacZ⁻/lacI⁻ lacZ⁺* functioned normally, synthesizing β-galactosidase only when lactose was present (**Figure 16.9**). In this strain, the *lacI⁺* gene on the bacterial chromosome was functional, but the *lacZ⁻* gene was defective; on the plasmid, the *lacI⁻* gene was defective, but the *lacZ⁺* gene was functional. The fact that a *lacI⁺* gene could regulate a *lacZ⁺* gene located on a different DNA molecule indicated to Jacob and Monod that the *lacI⁺* gene product was able to operate on either the plasmid or the chromosome.

Some *lacI* mutations isolated by Jacob and Monod prevented transcription from taking place even in the presence of lactose. These mutations were referred to as superrepressors (*lacIˢ*) because they produced defective repressor proteins that could not be inactivated by an inducer. A *lacIˢ* mutation produced a repressor with an

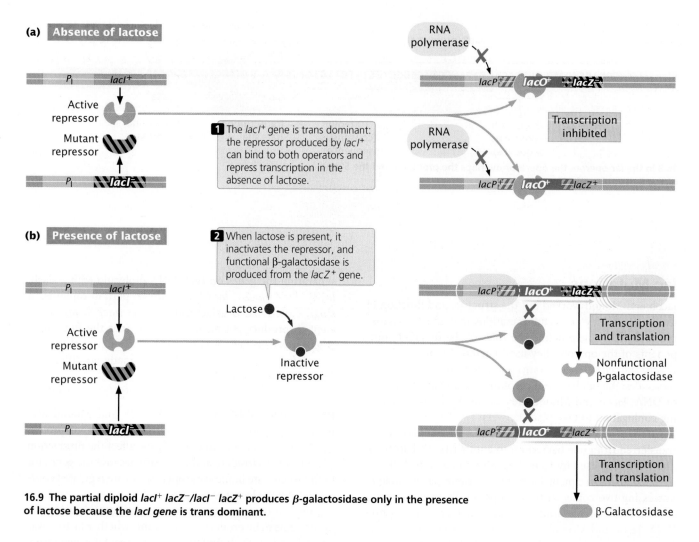

16.9 The partial diploid *lacI⁺ lacZ⁻/lacI⁻ lacZ⁺* **produces β-galactosidase only in the presence of lactose because the *lacI* gene is trans dominant.**

altered inducer-binding site; consequently, the inducer was unable to bind to the repressor, and the repressor was always able to attach to the operator and prevent transcription of the *lac* genes. Superrepressor mutations were dominant over *lacI⁺*; partial diploids with genotype *lacIˢ lacZ⁺ lacY⁺/lacI⁺ lacZ⁺ lacY⁺* were unable to synthesize either β-galactosidase or permease, whether or not lactose was present (**Figure 16.10**).

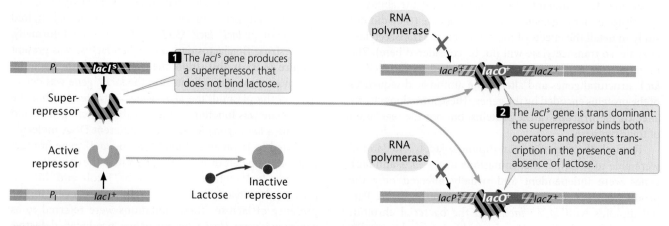

16.10 The partial diploid *lacIˢ lacZ⁺/lacI⁺ lacZ⁺* fails to produce β-galactosidase in the presence and absence of lactose because the *lacIˢ* gene encodes a superrepressor.

OPERATOR MUTATIONS Jacob and Monod mapped another class of constitutive mutations to a site adjacent to *lacZ*. These mutations occurred at the operator and were referred to as *lacO*ᶜ (*O* stands for operator and "c" for constitutive). The *lacO*ᶜ mutations altered the sequence of DNA at the operator so that the repressor protein was no longer able to bind to it. A partial diploid with genotype *lacI⁺ lacO*ᶜ *lacZ⁺/lacI⁺ lacO⁺ lacZ⁺* exhibited constitutive synthesis of β-galactosidase, indicating that *lacO*ᶜ is dominant over *lacO⁺*.

Analysis of other partial diploids showed that the *lacO* gene is cis acting, affecting only genes on the same DNA molecule. For example, a partial diploid with genotype *lacI⁺ lacO⁺ lacZ⁻/lacI⁺ lacO*ᶜ *lacZ⁺* was constitutive, producing β-galactosidase in the presence or absence of lactose (**Figure 16.11a**), but a partial diploid with genotype *lacI⁺ lacO⁺ lacZ⁺/lacI⁺ lacO*ᶜ *lacZ⁻* produced β-galactosidase only in the presence of lactose (**Figure 16.11b**). In the constitutive partial diploid (*lacI⁺ lacO⁺ lacZ⁻/lacI⁺ lacO*ᶜ *lacZ⁺*; see Figure 16.11a), the

(a) Partial diploid *lacI⁺ lacO⁺ lacZ⁻/lacI⁺ lacO*ᶜ *lacZ⁺*

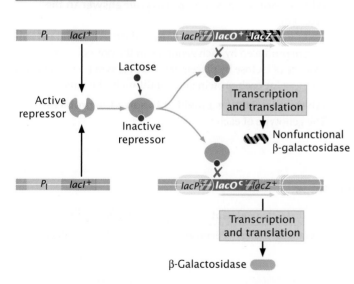

(b) Partial diploid *lacI⁺ lacO⁺ lacZ⁺/lacI⁺ lacO*ᶜ *lacZ⁻*

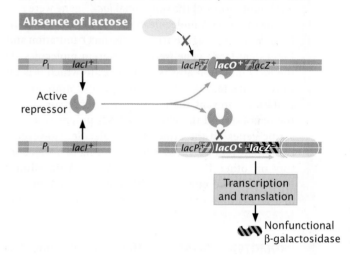

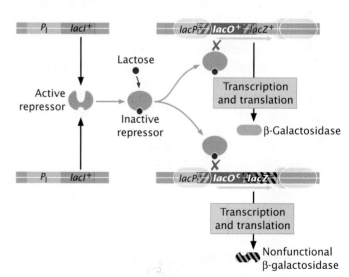

16.11 Mutations in *lacO* are constitutive and cis acting. (a) The partial diploid *lacI⁺ lacO⁺ lacZ⁻/lacI⁺ lacO*ᶜ *lacZ⁺* is constitutive, producing β-galactosidase in the presence and absence of lactose. (b) The partial diploid *lacI⁺ lacO⁺ lacZ⁺/lacI⁺ lacO*ᶜ *lacZ⁻* is inducible (produces β-galactosidase only when lactose is present), demonstrating that the *lacO* gene is cis acting.

TABLE 16.2	Characteristics of *lac* operon mutations			
Type	**Location**	**Cis/Trans**	**Effect**	
Structural-gene mutations	*lacZ, lacY*	Affect only *lacZ* or *lacY*	Alter amino acid sequence of protein encoded by gene in which mutation occurs	
Regulator-gene mutations	*lacI*	Trans	Affect transcription of structural genes	
Operator mutations	*lacO*	Cis	Affect transcription of structural genes	
Promoter mutations	*lacP*	Cis	Affect transcription of structural genes	

lacO^c mutation and the functional *lacZ*⁺ gene were present on the same DNA molecule, but in *lacI*⁺ *lacO*⁺ *lacZ*⁺/*lacI*⁺ *lacO*^c *lacZ*[−] (see Figure 16.11b), the *lacO*^c mutation and the functional *lacZ*⁺ gene were on different molecules. Thus, the *lacO* mutation affects only genes to which it is physically connected, as is true of all operator mutations. Such mutations prevent the binding of a repressor protein to the operator and thereby allow RNA polymerase to transcribe genes on the same DNA molecule. However, they cannot prevent a repressor from binding to normal operators on other DNA molecules. Watch **Animation 16.1** to observe the effects of different combinations of *lacI* and *lacO* mutations on the expression of the *lac* operon.

> **TRY PROBLEM 21**

PROMOTER MUTATIONS Mutations affecting lactose metabolism have also been isolated at the promoter; these mutations, designated *lacP*[−], interfere with the binding of RNA polymerase to the promoter. Because this binding is essential for the transcription of the structural genes, *E. coli* strains with *lacP*[−] mutations don't produce *lac* proteins either in the presence or in the absence of lactose. Like operator mutations, *lacP*[−] mutations are cis acting and thus affect only genes on the same DNA molecule. The partial diploid *lacI*⁺ *lacP*⁺ *lacZ*⁺/*lacI*⁺ *lacP*[−] *lacZ*⁺ exhibits normal synthesis of β-galactosidase, whereas *lacI*⁺ *lacP*[−] *lacZ*⁺/*lacI*⁺ *lacP*⁺ *lacZ*[−] fails to produce β-galactosidase

whether or not lactose is present. The different types of mutations that occur in the *lac* operon are summarized in **Table 16.2**.

WORKED PROBLEM

Make a table and, for the *E. coli* strains with the following *lac* genotypes, use a plus sign (+) to indicate the synthesis of β-galactosidase and permease and a minus sign (−) to indicate no synthesis of those proteins when lactose is absent and when it is present.

a. *lacI*⁺ *lacP*⁺ *lacO*⁺ *lacZ*⁺ *lacY*⁺
b. *lacI*⁺ *lacP*⁺ *lacO*^c *lacZ*[−] *lacY*⁺
c. *lacI*⁺ *lacP*[−] *lacO*⁺ *lacZ*⁺ *lacY*[−]
d. *lacI*⁺ *lacP*⁺ *lacO*⁺ *lacZ*[−] *lacY*[−]/
     lacI[−] *lacP*⁺ *lacO*⁺ *lacZ*⁺ *lacY*⁺

Solution Strategy

What information is required in your answer to the problem?
An indication of whether or not β-galactosidase and permease are produced by each genotype in the presence and in the absence of lactose by placement of a plus sign (+) or a minus sign (−) for each protein and condition in the table.

What information is provided to solve the problem?
The genotype of each strain.

Solution Steps

	Lactose absent		Lactose present	
Genotype of strain	**β-Galactosidase**	**Permease**	**β-Galactosidase**	**Permease**
a. *lacI*⁺ *lacP*⁺ *lacO*⁺ *lacZ*⁺ *lacY*⁺	−	−	+	+
b. *lacI*⁺ *lacP*⁺ *lacO*^c *lacZ*[−] *lacY*⁺	−	+	−	+
c. *lacI*⁺ *lacP*[−] *lacO*⁺ *lacZ*⁺ *lacY*[−]	−	−	−	−
d. *lacI*⁺ *lacP*⁺ *lacO*⁺ *lacZ*[−] *lacY*[−]/ *lacI*[−] *lacP*⁺ *lacO*⁺ *lacZ*⁺ *lacY*⁺	−	−	+	+

a. All the genes possess normal sequences, so the *lac* operon functions normally: when lactose is absent, the repressor protein binds to the operator and inhibits the transcription of the structural genes, so β-galactosidase and permease are not produced. When lactose is present, some of it is converted into allolactose, which binds to the repressor and makes it inactive; thus, the repressor does not bind to the operator, so the structural genes are transcribed and β-galactosidase and permease are produced.

b. The structural *lacZ* gene is mutated, so β-galactosidase is not produced under any conditions. The *lacO* gene has a constitutive mutation, which means that the repressor is unable to bind to *lacO*, so transcription takes place at all times. Therefore, permease is produced in both the presence and the absence of lactose.

c. In this strain, the promoter is mutated, so RNA polymerase is unable to bind to it, and transcription does not take place. Therefore, β-galactosidase and permease are not produced under any conditions.

d. This strain is a partial diploid, which possesses two copies of the *lac* operon—one on the bacterial chromosome and the other on a plasmid. The copy of the *lac* operon represented in the upper part of the genotype (before the slash) has mutations in both the *lacZ* and the *lacY* genes, so it is not capable of encoding β-galactosidase or permease under any conditions. The copy represented in the lower part of the genotype (after the slash) has a defective regulator gene, but the normal regulator gene in the upper copy produces a repressor protein that is capable of diffusing to other molecules (trans acting), so in the absence of lactose, it binds to the lower copy and inhibits transcription. Therefore, no β-galactosidase or permease is produced when lactose is absent. In the presence of lactose, the repressor cannot bind to the operator, so the structural genes in the lower copy of the *lac* operon are transcribed, and β-galactosidase and permease are produced.

> » Now try your own hand at predicting the outcome of different *lac* mutations by working **Problem 19** at the end of the chapter.

Positive Control and Catabolite Repression

As we've seen, *E. coli* and many other bacteria metabolize glucose preferentially, even in the presence of lactose and other sugars. They do so because glucose enters glycolysis without further modification and therefore requires less energy to metabolize than do other sugars. When glucose is available, genes that participate in the metabolism of other sugars are turned off through a process known as **catabolite repression**. Efficient transcription of the *lac* operon, for example, takes place only if lactose is present and glucose

is absent. But how is the expression of the *lac* operon influenced by glucose? What brings about catabolite repression?

In spite of being termed "repression," which might suggest negative control, catabolite repression actually results from positive control in response to glucose. (This regulation is in addition to the negative control brought about by repressor binding to the operator of the *lac* operon when lactose is absent.) This positive control is accomplished through the binding of a protein called the **catabolite activator protein** (**CAP**) to a site that is about 22 nucleotides long and is located within or slightly upstream of the promoter of the *lac* genes (**Figure 16.12**). RNA polymerase does not bind efficiently to many promoters unless CAP is first bound to the DNA. Before CAP can bind, however, it must form a complex with a modified nucleotide called **adenosine-3′,5′-cyclic monophosphate** (cyclic AMP, or **cAMP**), which is important in cellular signaling processes in both bacterial and eukaryotic cells. In *E. coli*, the concentration of cAMP is regulated so that it is inversely proportional to the level of available glucose. High concentrations of glucose within the cell lower the amount of cAMP, so little cAMP–CAP complex is available to bind to the DNA. Consequently, RNA polymerase has poor affinity for the *lac* promoter, and little transcription of the *lac* structural genes takes place. Low concentrations of glucose stimulate high levels of cAMP, resulting in increased cAMP–CAP binding to DNA. This increase enhances the binding of RNA polymerase to the *lac* promoter and increases transcription of the *lac* genes approximately 50-fold.

The catabolite activator protein exerts positive control on more than 20 operons of *E. coli*. The response to CAP varies among these operons; some are activated by low levels of CAP, whereas others require higher levels. CAP contains a helix-turn-helix DNA-binding motif (see Figure 16.2a), and when it binds at the CAP site on DNA, it causes the DNA helix to bend (**Figure 16.13**). The bent helix enables CAP to facilitate the binding of RNA polymerase at the promoter and the initiation of transcription. ▶ **TRY PROBLEM 14**

THINK-PAIR-SHARE Question 5

CONCEPTS

In spite of its name, catabolite repression is a type of positive control. The catabolite activator protein (CAP), complexed with cAMP, binds to a site at or near the *lac* promoter and stimulates the binding of RNA polymerase. Cellular levels of cAMP are controlled by glucose; a low glucose level increases the abundance of cAMP and enhances the transcription of the *lac* structural genes.

✔ CONCEPT CHECK 7

What is the effect of high levels of glucose on the *lac* operon?

a. Transcription is stimulated.
b. Little transcription takes place.
c. Transcription is not affected.
d. Transcription may be stimulated or inhibited, depending on the levels of lactose.

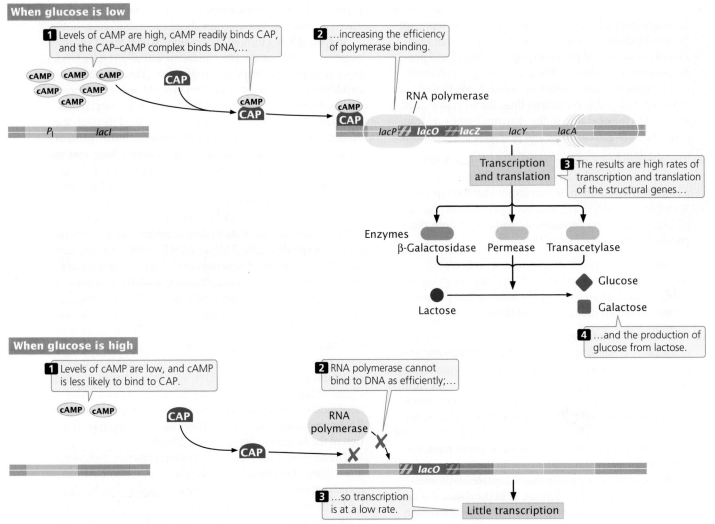

16.12 The catabolite activator protein (CAP) binds to the promoter of the *lac* operon and stimulates transcription. CAP must form a complex with adenosine-3′,5′-cyclic monophosphate (cAMP) before binding to the promoter. The binding of cAMP–CAP to the promoter activates transcription by facilitating the binding of RNA polymerase. Levels of cAMP are inversely related to glucose levels: low concentrations of glucose stimulate high concentrations of cAMP, and high concentrations of glucose stimulate low concentrations of cAMP.

The *trp* Operon of *E. coli*

The *lac* operon just discussed is an inducible operon, one in which transcription does not normally take place and must be turned on. Other operons are repressible: transcription in these operons is normally turned on and must be repressed when the products of its structural genes are not needed. The tryptophan (*trp*) operon in *E. coli*, which controls the biosynthesis of the amino acid tryptophan, is an example of a negative repressible operon.

The *trp* operon contains five structural genes (*trpE*, *trpD*, *trpC*, *trpB*, and *trpA*) that produce the components of three enzymes (two of the enzymes consist of two polypeptide chains). These enzymes convert a precursor called chorismate into tryptophan (**Figure 16.14**). The first structural gene, *trpE*, contains a long 5′ untranslated region (5′ UTR)

that is transcribed, but does not encode any of these polypeptides. Instead, this 5′ UTR plays an important role in another regulatory mechanism, which we'll describe in Section 16.3. Upstream of the 5′ UTR is the *trp* promoter. When tryptophan levels are low, RNA polymerase binds to the promoter and transcribes the five structural genes into a single mRNA, which is then translated into the enzymes that convert chorismate into tryptophan.

Some distance from the *trp* operon is a regulator gene, *trpR*, which encodes a repressor protein that is normally inactive (see Figure 16.14). Like the *lac* repressor, the tryptophan repressor has two binding sites, one that binds to the operator and another that binds to tryptophan (the corepressor). Binding with tryptophan causes a conformational change in the repressor that makes it capable of binding to the operator,

(a)

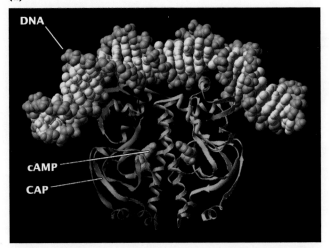

(b)

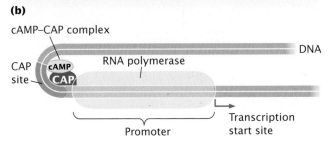

16.13 The binding of the cAMP–CAP complex to DNA produces a sharp bend in DNA that activates transcription.

which overlaps the promoter. When the operator is occupied by the repressor, RNA polymerase cannot bind to the promoter, and the structural genes cannot be transcribed. Thus, when cellular levels of tryptophan are low, transcription of the *trp* operon takes place and more tryptophan is synthesized; when cellular levels of tryptophan are high, transcription of the *trp* operon is inhibited and the synthesis of more tryptophan does not take place.

CONCEPTS

The *trp* operon is a negative repressible operon that controls the biosynthesis of tryptophan. In a repressible operon, transcription is normally turned on and must be repressed: in the case of the *trp* operon, this is accomplished through the binding of tryptophan to the repressor, which renders the repressor active. The active repressor then binds to the operator and prevents RNA polymerase from transcribing the structural genes.

✔ CONCEPT CHECK 8

In the *trp* operon, what happens to the *trp* repressor in the absence of tryptophan?

a. It binds to the operator and represses transcription.

b. It cannot bind to the operator, and transcription takes place.

c. It binds to the regulator gene and represses transcription.

d. It cannot bind to the regulator gene, and transcription takes place.

16.3 Some Operons Regulate Transcription Through Attenuation, the Premature Termination of Transcription

We've now seen several different ways in which a cell regulates the initiation of transcription in an operon. Some operons have an additional level of control that affects the *continuation* of transcription rather than its initiation. In **attenuation**, transcription begins at the transcription start site, but termination takes place prematurely, before the RNA polymerase even reaches the structural genes. Attenuation takes place in a number of operons that encode enzymes participating in the biosynthesis of amino acids.

Attenuation in the *trp* Operon of *E. coli*

We can understand the process of attenuation most easily by looking at one of the best-studied examples, which is found in the *trp* operon of *E. coli*. The *trp* operon is unusual in that it is regulated both by repression and by attenuation. Most operons are regulated by one of these mechanisms, but not by both of them.

Attenuation first came to light when Charles Yanofsky and his colleagues made several observations in the early 1970s that indicated that repression at the operator is not the only method of regulation of the *trp* operon. They isolated a series of mutant *E. coli* strains that exhibited high levels of transcription, but in which control at the operator was unaffected, suggesting that some mechanism other than repression at the operator was controlling transcription. Furthermore, they observed that two mRNAs of different sizes were transcribed from the *trp* operon: a long mRNA containing sequences encoded by the structural genes and a much shorter mRNA of only 140 nucleotides. These observations led Yanofsky to propose that a mechanism that causes premature termination of transcription also regulates transcription in the *trp* operon.

Close examination of the *trp* operon reveals a region of 162 nucleotides that corresponds to the long 5′ UTR of the mRNA (mentioned in Section 16.2) transcribed from the *trp* operon (**Figure 16.15a**). The 5′ UTR (also called a leader) contains four regions with the following complementarities: region 1 is complementary to region 2, region 2 is complementary to region 3, and region 3 is complementary to region 4 (**Figure 16.15b**). These complementarities allow the 5′ UTR to fold into two different secondary structures (**Figure 16.15c**). Only one of these secondary structures causes attenuation.

One of the secondary structures that can form in the 5′ UTR contains one hairpin produced by the base pairing of regions 1 and 2 and another hairpin produced by the base pairing of regions 3 and 4 (see Figure 16.15c, left). Notice that a string of uracil nucleotides follows the 3+4 hairpin. Not coincidentally, the structure of a bacterial intrinsic (rho-independent) terminator (see Figure 13.14) includes a hairpin followed by a string of uracil nucleotides; this secondary structure in the 5′ UTR of the *trp* operon is indeed a terminator of transcription and is called an **attenuator**. The attenuator forms when cellular levels of tryptophan are high,

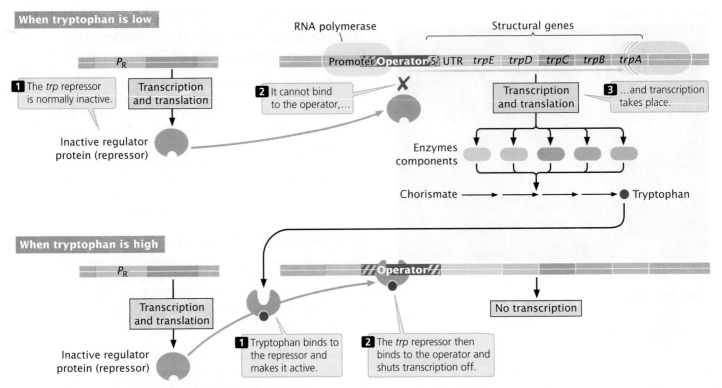

16.14 The *trp* operon controls the biosynthesis of the amino acid tryptophan in *E. coli*.

causing transcription to be terminated before the *trp* structural genes can be transcribed.

When cellular levels of tryptophan are low, however, the alternative secondary structure of the 5′ UTR is produced by the base pairing of regions 2 and 3 (see Figure 16.15c, right). This base pairing also produces a hairpin, but this hairpin is not followed immediately by a string of uracil nucleotides, so this structure does *not* function as a terminator. RNA polymerase continues past the 5′ UTR into the coding regions of the structural genes, and the enzymes that synthesize tryptophan are produced. Because it prevents the termination of transcription, the 2+3 structure is called an **antiterminator**.

To summarize, the 5′ UTR of the *trp* operon can fold into one of two secondary structures. When the tryptophan level is high, the 3+4 structure forms, transcription is terminated within the 5′ UTR, and no additional tryptophan is synthesized. When the tryptophan level is low, the 2+3 structure forms, transcription continues through the structural genes, and tryptophan is synthesized. The critical question, then, is this: Why does the 3+4 structure arise when the level of tryptophan in the cell is high, whereas the 2+3 structure arises when the level is low?

To answer this question, we must take a closer look at the nucleotide sequence of the 5′ UTR. At the 5′ end, upstream of region 1, is a ribosome-binding site (see Figure 16.15a). Region 1 encodes a small protein. Within the coding sequence for this protein are two UGG codons, which specify the amino acid tryptophan, so tryptophan is required for the translation of this 5′ UTR sequence. The small protein encoded by this part of the 5′ UTR is presumed to be

unstable; its only apparent function is to control attenuation. Although it was stated in Chapter 14 that 5′ UTR are not translated into proteins, the 5′ UTR of operons subject to attenuation are exceptions to this rule. The precise timing and interaction of transcription and translation in the 5′ UTR determine whether attenuation takes place.

TRANSCRIPTION WHEN TRYPTOPHAN LEVELS ARE LOW
Let's first consider what happens when intracellular levels of tryptophan are low. Recall that in prokaryotic cells, transcription and translation are coupled: while transcription is taking place at the 3′ end of the mRNA, translation is initiated at the 5′ end. RNA polymerase begins transcribing the DNA, producing region 1 of the 5′ UTR (**Figure 16.16a**). Closely following RNA polymerase, a ribosome binds to the 5′ UTR and begins to translate the protein-coding region. Meanwhile, RNA polymerase is transcribing region 2 (**Figure 16.16b**). Region 2 is complementary to region 1, but because the ribosome is translating region 1, the nucleotides in regions 1 and 2 cannot base pair.

RNA polymerase begins to transcribe region 3, and the ribosome reaches the Trp codons (UGG) in region 1. When it reaches the Trp codons, the ribosome stalls (**Figure 16.16c**) because the level of tryptophan is low and tRNAs charged with tryptophan are scarce or unavailable. The ribosome sits at the Trp codons, awaiting the arrival of a tRNA charged with tryptophan. Stalling of the ribosome does not, however, hinder transcription; RNA polymerase continues to move along the DNA, and transcription gets ahead of translation.

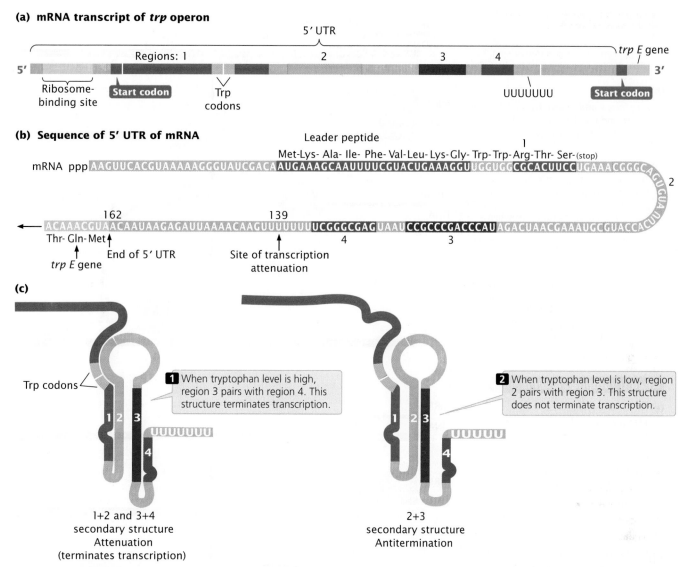

16.15 Two different secondary structures can be formed by the 5′ UTR of the mRNA transcript of the *trp* operon.

Because the ribosome is stalled at the Trp codons in region 1, region 2 is free to base pair with region 3, forming the 2+3 hairpin (**Figure 16.16d**). This hairpin does not cause termination, so transcription continues. Because region 3 is already paired with region 2, the 3+4 hairpin (the attenuator) never forms, so attenuation does not take place, and transcription continues. RNA polymerase continues along the DNA, past the 5′ UTR, transcribing all the structural genes into mRNA, which is translated into the enzymes encoded by the *trp* operon. These enzymes then synthesize tryptophan.

TRANSCRIPTION WHEN TRYPTOPHAN LEVELS ARE HIGH Now let's see what happens when intracellular levels of tryptophan are high. Once again, RNA polymerase begins transcribing the DNA, producing region 1 of the 5′ UTR (**Figure 16.16e**). Closely following RNA polymerase, a ribosome binds to the 5′ UTR and begins to translate the protein-coding region (**Figure 16.16f**). When the ribosome reaches the two Trp codons, it doesn't slow or stall because tryptophan is abundant and tRNAs charged with tryptophan are readily available (**Figure 16.16g**). This point is critical to note: because tryptophan is abundant, translation can keep up with transcription.

As it moves past region 1, the ribosome partly covers region 2 (**Figure 16.16h**); meanwhile, RNA polymerase completes the transcription of region 3. Although regions 2 and 3 are complementary, the ribosome physically blocks their pairing.

RNA polymerase continues to move along the DNA, eventually transcribing region 4 of the 5′ UTR. Region 4 is complementary to region 3, and because region 3 cannot base pair with region 2, it pairs with region 4. The pairing of regions 3 and 4 (see Figure 16.16h) produces the attenuator, and transcription terminates just beyond region 4. The structural genes are not transcribed, no tryptophan-producing enzymes are translated, and no additional tryptophan is synthesized.

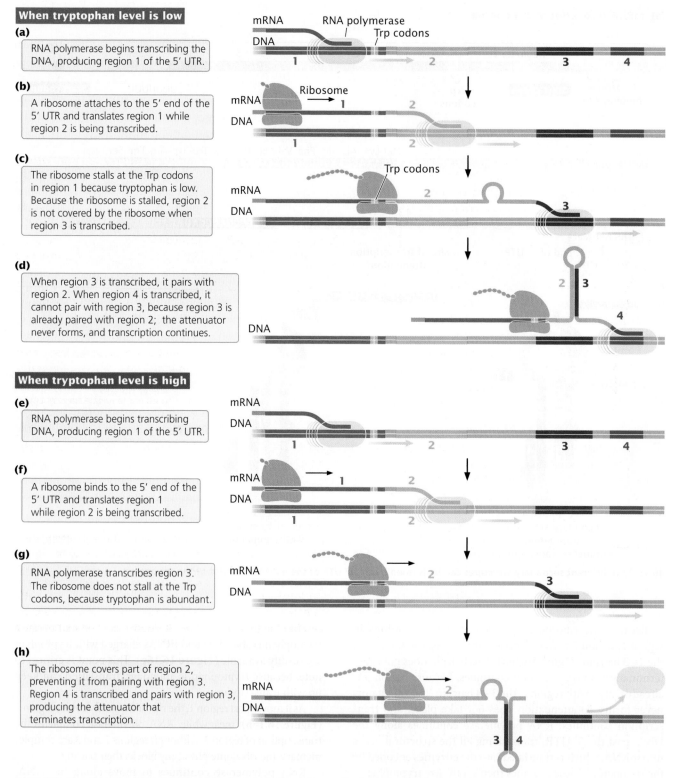

When tryptophan level is low

(a)

RNA polymerase begins transcribing the DNA, producing region 1 of the 5' UTR.

(b)

A ribosome attaches to the 5' end of the 5' UTR and translates region 1 while region 2 is being transcribed.

(c)

The ribosome stalls at the Trp codons in region 1 because tryptophan is low. Because the ribosome is stalled, region 2 is not covered by the ribosome when region 3 is transcribed.

(d)

When region 3 is transcribed, it pairs with region 2. When region 4 is transcribed, it cannot pair with region 3, because region 3 is already paired with region 2; the attenuator never forms, and transcription continues.

When tryptophan level is high

(e)

RNA polymerase begins transcribing DNA, producing region 1 of the 5' UTR.

(f)

A ribosome binds to the 5' end of the 5' UTR and translates region 1 while region 2 is being transcribed.

(g)

RNA polymerase transcribes region 3. The ribosome does not stall at the Trp codons, because tryptophan is abundant.

(h)

The ribosome covers part of region 2, preventing it from pairing with region 3. Region 4 is transcribed and pairs with region 3, producing the attenuator that terminates transcription.

16.16 Whether the premature termination of transcription (attenuation) takes place in the _trp_ operon depends on the cellular level of tryptophan.

TABLE 16.3	Events in the process of attenuation				
Intracellular Level of Tryptophan	Ribosome Stalls at Trp Codons	Position of Ribosome When Region 3 Is Transcribed	Secondary Structure of 5′ UTR	Termination of Transcription of *trp* Operon
High	No	Covers region 2	3+4 hairpin	Yes
Low	Yes	Covers region 1	2+3 hairpin	No

Important events in the process of attenuation are summarized in **Table 16.3**. Try pausing the ribosome at the Trp codons for different lengths of time in **Animation 16.2** and see what effect the pause has on transcription.

A key factor controlling attenuation is the number of tRNA molecules charged with tryptophan because their availability is what determines whether the ribosome stalls at the Trp codons. A second factor is the synchronization of transcription and translation, which is critical to attenuation. Synchronization is achieved through a site located in region 1 of the 5′ UTR where RNA polymerase pauses. When this site is transcribed, the RNA folds into a secondary structure that inhibits further transcription. Thus, the RNA polymerase stops temporarily at the pause site, allowing time for a ribosome to bind to the 5′ end of the mRNA. As the ribosome approaches the secondary structure in the RNA, it disrupts that structure and allows transcription to continue. Translation then closely follows transcription. It is important to point out that ribosomes do not traverse the convoluted hairpins of the 5′ UTR to translate the structural genes. Ribosomes that attach to the 5′ end of region 1 of the mRNA encounter a stop codon at the end of region 1. New ribosomes translating the structural genes attach to a different ribosome-binding site located near the beginning of the *trpE* gene (see Figure 16.15a). **▶ TRY PROBLEM 27**

THINK-PAIR-SHARE Questions 6 and 7

Why Does Attenuation Take Place in the *trp* Operon?

Why do bacteria need attenuation in the *trp* operon? Shouldn't repression at the operator prevent transcription from taking place when tryptophan levels in the cell are high? Why does the cell have two types of control? Part of the answer is that repression is never complete. Some transcription is initiated even when the *trp* repressor is active; repression reduces transcription only as much as 70-fold. Attenuation can further reduce transcription another 8- to 10-fold, so together, the two processes are capable of reducing transcription of the *trp* operon more than 600-fold. Both mechanisms provide *E. coli* with a much finer degree of control over tryptophan synthesis than either could achieve alone.

Another reason for this dual control is that attenuation and repression respond to different signals: repression responds to cellular levels of tryptophan, whereas attenuation responds to the availability of tRNAs charged with tryptophan. There might be circumstances in which a cell's ability to respond to these different signals is advantageous. Finally, the *trp* repressor affects several operons in addition to the *trp* operon. At an early stage in the evolution of *E. coli*, the *trp* operon may have been controlled only by attenuation. The *trp* repressor may have evolved primarily to control the other operons and only incidentally affects the *trp* operon.

Attenuation is a difficult process to grasp because you must simultaneously visualize two dynamic processes—transcription and translation—and it's easy to confuse them. Remember that attenuation refers to the early termination of *transcription*, not translation (although events in translation bring about the termination of transcription). Attenuation often causes confusion because we know that transcription must precede translation. We're comfortable with the idea that transcription might affect translation, but it's harder to imagine that the effects of translation could influence transcription, as they do in attenuation. The reality is that transcription and translation are closely coupled in prokaryotic cells, and events in one process can easily affect the other.

CONCEPTS

The *trp* operon regulates transcription through attenuation as well as through repression. In attenuation, transcription is initiated but is terminated prematurely. When tryptophan levels are low, the ribosome stalls at Trp codons, and transcription continues. When tryptophan levels are high, the ribosome does not stall at Trp codons, and the 5′ UTR adopts a secondary structure that terminates transcription before the structural genes can be transcribed.

✔ CONCEPT CHECK 9

Attenuation results when which regions of the 5′ UTR pair?
a. 1 and 3
b. 2 and 3
c. 2 and 4
d. 3 and 4

16.4 Other Sequences Control the Expression of Some Bacterial Genes

Several types of sequences outside of operons affect the expression of some genes in bacteria. These sequences include bacterial enhancers and several types of RNA regulators.

Bacterial Enhancers

An enhancer is a DNA element that affects transcription but, in contrast to promoters, is typically found some distance from the gene whose transcription it affects (see Chapter 17). Enhancers were originally described in eukaryotes, but research now indicates that some also occur in bacteria and archaea.

Like enhancers in eukaryotes, bacterial enhancers contain binding sites for proteins that increase the rate of transcription at promoters that are distant from the enhancer. They do this by causing the DNA between the promoter and the enhancer to loop out, so that the transcription activator bound to the enhancer directly interacts with RNA polymerase at the promoter. Enhancers are position independent, meaning that they can be moved without affecting their ability to enhance transcription. Most bacterial enhancers are found upstream of genes that use a special type of sigma factor (see Chapter 13) known as sigma 54 (σ^{54}). Enhancers will be discussed in more detail in Chapter 17.

> **CONCEPTS**
>
> Bacterial enhancers increase the rate of transcription at genes that are distant from the enhancer.

Antisense RNA

Some RNA molecules are complementary to particular sequences on mRNAs. These molecules, called **antisense RNA**, control gene expression by binding to sequences on mRNA and inhibiting its translation. Translational control by antisense RNA is seen in the regulation of the *ompF* gene of *E. coli* (**Figure 16.17a**). This gene encodes an outer-membrane protein that functions as a channel for the passive diffusion of small polar molecules, such as water and ions, across the membrane. Under most conditions, the *ompF* gene is transcribed

and translated and the OmpF protein is synthesized. However, when the osmolarity of the medium increases, the cell depresses the production of OmpF protein to help maintain cellular osmolarity. A regulatory gene named *micF*—for *m*RNA-*i*nterfering *c*omplementary RNA—is activated, and *micF* RNA is produced (**Figure 16.17b**). The *micF* RNA, an antisense RNA, binds to a complementary sequence in the 5′ UTR of the *ompF* mRNA and inhibits the binding of the ribosome. This inhibition reduces the amount of translation (see Figure 16.17b), which results in fewer OmpF proteins in the outer membrane and thus reduces the movement of substances across the membrane owing to the changes in osmolarity. Research has detected many antisense RNA molecules in a wide range of bacterial species. **› TRY PROBLEM 30**

THINK-PAIR-SHARE Question 8

Riboswitches

We have seen that operons of bacteria contain DNA sequences (promoters and operator sites) where the binding of small molecules induces or represses transcription. Some mRNA molecules contain regulatory sequences called **riboswitches**, where regulatory molecules can bind and affect gene expression by influencing the formation of secondary structures in the mRNA (**Figure 16.18**). Riboswitches were first discovered in 2002 and now appear to be common in bacteria, regulating about 4% of all bacterial genes. They are also present in archaea, fungi, and plants.

Riboswitches, which are typically found in the 5′ UTR of the mRNA, can fold into compact RNA secondary structures with a base stem and several branching hairpins. In some cases, a small regulatory molecule binds to the riboswitch and stabilizes a terminator, which causes premature termination of transcription. In other cases, the binding of a regulatory molecule stabilizes a secondary structure that masks the

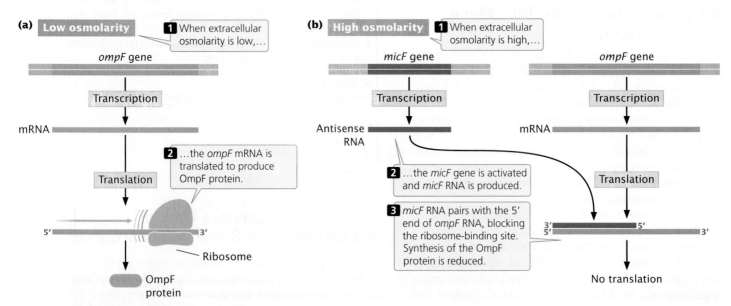

16.17 Antisense RNA can regulate translation.

(a) Regulatory protein present

Regulatory protein

Riboswitch

mRNA

Ribosome-binding site

Gene

5'   3'

A regulatory protein binds to the riboswitch and stabilizes a secondary structure...

...that masks a ribosome-binding site.

No translation takes place.

No translation

(b) Regulatory protein absent

Without the regulatory protein, the riboswitch assumes an alternative secondary structure...

...that makes the ribosome-binding site available.

5'

Ribosome-binding site

Gene

3'

Translation takes place.

Translation

16.18 Riboswitches are RNA sequences in mRNA that affect gene expression.

ribosome-binding site, preventing the initiation of translation. When not bound by the regulatory molecule, the riboswitch assumes an alternative structure that eliminates the premature terminator or makes the ribosome-binding site available.

A RIBOSWITCH CONTROLS THE SYNTHESIS OF VITAMIN B12 An example of a riboswitch is seen in the bacterial genes that encode enzymes that play roles in the synthesis of vitamin B12. The genes for these enzymes are transcribed into an mRNA molecule that contains a riboswitch. When the activated form of vitamin B12—called coenzyme B12—is present, it binds to the riboswitch, and the mRNA folds into a secondary structure that obstructs the ribosome-binding site. Consequently, no translation of the mRNA takes place. In the absence of coenzyme B12, the mRNA assumes a different secondary structure. This secondary structure does not obstruct the ribosome-binding site, so translation is initiated, the enzymes are synthesized, and vitamin B12 is produced. For some riboswitches, the regulatory molecule acts as a repressor (as just described) by inhibiting transcription or translation; for others, the regulatory molecule acts as an inducer by causing the formation of a secondary structure that allows transcription or translation to take place.

RNA-Mediated Repression Through Ribozymes

Another type of gene control is carried out by mRNA molecules called ribozymes, which possess catalytic activity (see Chapter 14). Termed RNA-mediated repression, this type of control has been demonstrated in the *glmS* gene of the bacterium *Bacillus subtilis*. Transcription of this gene produces an mRNA molecule that encodes the enzyme glutamine:fructose-6-phosphate amidotransferase (**Figure 16.19**), which helps synthesize a small sugar called

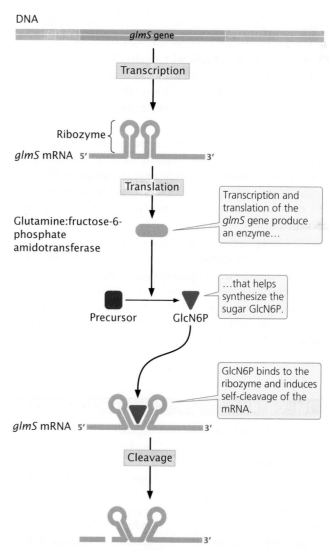

DNA

glmS gene

Transcription

Ribozyme

glmS mRNA   5'   3'

Translation

Glutamine:fructose-6-phosphate amidotransferase

Transcription and translation of the *glmS* gene produce an enzyme...

Precursor   GlcN6P

...that helps synthesize the sugar GlcN6P.

glmS mRNA   5'   3'

GlcN6P binds to the ribozyme and induces self-cleavage of the mRNA.

Cleavage

3'

16.19 Ribozymes, when bound by small regulatory molecules, can induce the cleavage and degradation of mRNA.

glucosamine-6-phosphate (GlcN6P). Within the 5′ UTR of the *glmS* mRNA are about 75 nucleotides that act as a ribozyme. When GlcN6P is absent from the cell, the *glmS* gene is transcribed and translated to produce the enzyme, which synthesizes GlcN6P. However, when sufficient GlcN6P is present, the sugar binds to the ribozyme sequence of the mRNA, which then induces self-cleavage of the mRNA—the ribozyme breaks the sugar–phosphate backbone of the RNA. This cleavage prevents translation of the mRNA.

CONCEPTS

In bacterial cells, antisense RNA can inhibit translation by binding to complementary sequences in the 5′ UTR of mRNA and preventing the attachment of a ribosome. Riboswitches are sequences in mRNA molecules that bind regulatory molecules and induce changes in the secondary structure of the mRNA that affect its translation. In RNA-mediated repression, a ribozyme sequence on the mRNA induces self-cleavage and degradation of the mRNA when bound by a regulatory molecule.

CONCEPTS SUMMARY

■ Gene expression can be controlled at different levels along the molecular pathway from DNA to protein, including the alteration of DNA or chromatin structure, transcription, mRNA processing, mRNA stability, translation, and posttranslational modification. Much of gene regulation takes place through the action of regulatory proteins that recognize and bind to regulatory elements in DNA.

■ Genes in bacterial cells are typically clustered into operons—groups of functionally related structural genes and the sequences that control their transcription. Structural genes in an operon are transcribed together as a single mRNA molecule.

■ In negative control, a repressor protein binds to DNA and inhibits transcription. In positive control, an activator protein binds to DNA and stimulates transcription. In inducible operons, transcription is normally off and must be turned on; in repressible operons, transcription is normally on and must be turned off.

■ The *lac* operon of *E. coli* is a negative inducible operon. In the absence of lactose, a repressor binds to the operator and prevents the transcription of genes that encode β-galactosidase, permease, and transacetylase. When lactose is present, some of it is converted into allolactose, which binds to and inactivates the repressor, allowing the structural genes to be transcribed.

■ Positive control of the *lac* operon and other operons occurs through catabolite repression. When complexed with cyclic AMP, the catabolite activator protein binds to a site at or near the promoter and stimulates the transcription of the structural genes. Levels of cAMP are inversely correlated with glucose, so low levels of glucose stimulate transcription and high levels inhibit transcription.

■ The *trp* operon of *E. coli* is a negative repressible operon that controls the biosynthesis of tryptophan.

■ Attenuation causes premature termination of transcription. It takes place through the close coupling of transcription and translation and depends on the secondary structure of the 5′ UTR sequence.

■ Bacterial enhancers increase the rate of transcription at genes that are distant from the enhancer.

■ Antisense RNAs are complementary to sequences in mRNA and may inhibit translation by binding to these sequences, thereby preventing the attachment or progress of the ribosome.

■ When bound by a regulatory molecule, riboswitches in an mRNA molecule induce changes in the secondary structure of the mRNA that affect its translation. Some mRNAs possess ribozyme sequences that induce self-cleavage and degradation when bound by a regulatory molecule.

IMPORTANT TERMS

gene regulation (p. 462)
structural gene (p. 463)
regulatory gene (p. 463)
constitutive (p. 463)
regulatory element (p. 463)
domain (p. 464)
operon (p. 465)
regulator gene (p. 466)

regulator protein (p. 466)
operator (p. 466)
negative control (p. 466)
positive control (p. 466)
inducible operon (p. 466)
repressible operon (p. 466)
inducer (p. 466)
allosteric protein (p. 466)

corepressor (p. 467)
coordinate
  induction (p. 469)
partial diploid (p. 471)
catabolite
  repression (p. 475)
catabolite activator protein
  (CAP) (p. 475)

adenosine-3′,5′-cyclic
  monophosphate
  (cAMP) (p. 475)
attenuation (p. 477)
attenuator (p. 477)
antiterminator (p. 478)
antisense RNA (p. 482)
riboswitch (p. 482)

ANSWERS TO CONCEPT CHECKS

1. A constitutive gene is not regulated and is expressed continually.

2. Transcription is the first step in the process of information transfer from DNA to protein. For cellular efficiency, gene expression is often regulated early in the process of protein production.

3. b

4. d

5. d

6. c

7. b

8. b

9. d

WORKED PROBLEMS

Problem 1

The *fox* operon, which has sequences *A*, *B*, *C*, and *D* (which may represent either structural genes or regulatory sequences), encodes enzymes 1 and 2. Mutations in sequences *A*, *B*, *C*, and *D* have the following effects, where a plus sign (+) indicates that the enzyme is synthesized and a minus sign (−) indicates that the enzyme is not synthesized.

Mutation in sequence	Fox absent		Fox present	
	Enzyme 1	Enzyme 2	Enzyme 1	Enzyme 2
No mutation	−	−	+	+
A	−	−	−	+
B	−	−	−	−
C	−	−	+	−
D	+	+	+	+

a. Is the *fox* operon inducible or repressible?

b. Indicate which sequence (*A*, *B*, *C*, or *D*) is part of the following components of the operon. Each sequence should be used only once.

Regulator gene _____   Structural gene for enzyme 1 _____

Promoter _____   Structural gene for enzyme 2 _____

⟩⟩ Solution Strategy

What information is required in your answer to the problem?

a. Whether the *fox* operon is inducible or repressible.

b. Which sequence represents each part of the operon.

What information is provided to solve the problem?

For each mutation, whether enzyme 1 and enzyme 2 are produced in the presence and absence of Fox.

For help with this problem, review:

Section 16.2.

⟩⟩ Solution Steps

a. When no mutations are present, enzymes 1 and 2 are produced in the presence of Fox but not in its absence,

indicating that the operon is inducible and that Fox is the inducer.

b. The mutation in *A* allows the production of enzyme 2 in the presence of Fox, but enzyme 1 is not produced in the presence or absence of Fox, and so *A* must have a mutation in the structural gene for enzyme 1. With the mutation in *B*, neither enzyme is produced under any conditions, so this mutation most likely occurs in the promoter and prevents RNA polymerase from binding. The mutation in *C* affects only enzyme 2, which is not produced in the presence or absence of Fox; enzyme 1 is produced normally (only in the presence of Fox), so the mutation in *C* most likely occurs in the structural gene for enzyme 2. The mutation in *D* is constitutive, allowing the production of enzymes 1 and 2 whether or not Fox is present. This mutation most likely occurs in the regulator gene, producing a defective repressor that is unable to bind to the operator under any conditions.

Regulator gene _D_

Promoter   _B_

Structural gene for enzyme 1 _A_

Structural gene for enzyme 2 _C_

Problem 2

A mutation occurs in the 5′ UTR of the *trp* operon that reduces the ability of region 2 to pair with region 3. What will the effect of this mutation be when the tryptophan level is high? When the tryptophan level is low?

›› Solution Strategy

What information is required in your answer to the problem?

The effect of the mutation when tryptophan is high and when it is low.

What information is provided to solve the problem?

- The mutation occurs in the 5′ UTR of the *trp* operon.
- The mutation reduces the ability of region 2 to pair with region 3.

For help with this problem, review:

Section 16.3.

›› Solution Steps

When the tryptophan level is high, regions 2 and 3 do not normally pair, and therefore the mutation will have no effect. When the tryptophan level is low, however, the ribosome normally stalls at the Trp codons in region 1 and does not cover region 2, so regions 2 and 3 are free to pair, which prevents regions 3 and 4 from pairing and forming a terminator, which would end transcription. If regions 2 and 3 cannot pair, then regions 3 and 4 will pair even when tryptophan is low, and attenuation will always take place. Therefore, no tryptophan will be synthesized even in the absence of tryptophan.

Hint: Review Figure 16.16 for a summary of attenuation.

COMPREHENSION QUESTIONS

Section 16.1

1. Why is gene regulation important for bacterial cells?

2. Name six different levels at which gene expression might be controlled.

Section 16.2

3. Draw a picture illustrating the general structure of an operon and identify its parts.

4. What is the difference between positive and negative control? What is the difference between inducible and repressible operons?

5. Briefly describe the *lac* operon and how it controls the metabolism of lactose.

6. What is catabolite repression? How does it allow a bacterial cell to use glucose in preference to other sugars?

Section 16.3

7. What is attenuation? What are the mechanisms by which the attenuator forms when tryptophan levels are high and the antiterminator forms when tryptophan levels are low?

Section 16.4

8. What is antisense RNA? How does it control gene expression?

9. What are riboswitches? How do they control gene expression? How do riboswitches differ from RNA-mediated repression?

APPLICATION QUESTIONS AND PROBLEMS

Section 16.1

10. Examine **Figure 16.2b**. Why do you think the motif of the DNA-binding protein shown is called a zinc-finger motif?

Section 16.2

*11. For each of the following types of transcriptional control, indicate whether the protein produced by the regulator gene will be synthesized initially as an active repressor or as an inactive repressor.

a. Negative control in a repressible operon

b. Negative control in an inducible operon

12. A mutation at the operator prevents the regulator protein from binding. What effect will this mutation have in the following types of operons?

a. Regulator protein is a repressor of a repressible operon.

b. Regulator protein is a repressor of an inducible operon.

13. The *blob* operon produces enzymes that convert compound A into compound B. The operon is controlled by regulator gene *S*. Normally, the enzymes are synthesized only in the absence of compound B. If gene *S* is mutated, the enzymes are synthesized in the presence *and* in the absence of compound B. Does gene *S* produce a regulator protein that exhibits positive or negative control? Is this operon inducible or repressible?

*****14.** A mutation prevents the catabolite activator protein (CAP) from binding to the promoter in the *lac* operon. What will the effect of this mutation be on the transcription of the operon?

15. Transformation is a process in which bacteria take up new DNA released by dead cells and integrate it into their own genomes (see p. 265 in Chapter 9). In *Streptococcus pneumoniae* (which causes many cases of pneumonia, inner-ear infections, and meningitis), the ability to carry out transformation requires from 105 to 124 genes, collectively termed the *com* regulon. The *com* regulon is activated in response to a protein called competence-stimulating peptide (CSP), which is produced by the bacteria and exported into the surrounding medium. When enough CSP accumulates, it attaches to a receptor on the bacterial cell membrane, which then activates a regulator protein that stimulates the transcription of genes within the *com* regulon and sets in motion a series of reactions that ultimately result in transformation. Does the *com* regulon in *Streptococcus pneumoniae* exhibit positive or negative control? Explain your answer.

16. Under which of the following conditions would a *lac* operon produce the greatest amount of β-galactosidase? The least? Explain your reasoning.

	Lactose present	Glucose present
Condition 1	Yes	No
Condition 2	No	Yes
Condition 3	Yes	Yes
Condition 4	No	No

17. A mutant strain of *E. coli* produces β-galactosidase in both the presence *and* the absence of lactose. Where in the operon might the mutation in this strain be located?

18. Examine **Figure 16.7**. What would be the effect of a drug that altered the structure of allolactose so that it was unable to bind to the regulator protein?

*****19.** For *E. coli* strains with the *lac* genotypes shown below, use a plus sign (+) to indicate the synthesis of β-galactosidase and permease and a minus sign (−) to indicate no synthesis of the proteins.

Genotype of strain	Lactose absent		Lactose present	
	β-Galactosidase	Permease	β-Galactosidase	Permease
lacI⁺ lacP⁺ lacO⁺ lacZ⁺ lacY⁺	_____	_____	_____	_____
lacI⁻ lacP⁺ lacO⁺ lacZ⁺ lacY⁺	_____	_____	_____	_____
lacI⁺ lacP⁺ lacOᶜ lacZ⁺ lacY⁺	_____	_____	_____	_____
lacI⁻ lacP⁺ lacO⁺ lacZ⁺ lacY⁻	_____	_____	_____	_____
lacI⁻ lacP⁻ lacO⁺ lacZ⁺ lacY⁺	_____	_____	_____	_____
lacI⁺ lacP⁺ lacO⁺ lacZ⁻ lacY⁺/ lacI⁻ lacP⁺ lacO⁺ lacZ⁺ lacY⁻	_____	_____	_____	_____
lacI⁻ lacP⁺ lacOᶜ lacZ⁺ lacY⁺/ lacI⁺ lacP⁺ lacO⁺ lacZ⁻ lacY⁻	_____	_____	_____	_____
lacI⁻ lacP⁺ lacO⁺ lacZ⁺ lacY⁻/ lacI⁺ lacP⁻ lacO⁺ lacZ⁻ lacY⁺	_____	_____	_____	_____
lacI⁺ lacP⁻ lacOᶜ lacZ⁻ lacY⁺/ lacI⁻ lacP⁺ lacO⁺ lacZ⁺ lacY⁻	_____	_____	_____	_____
lacI⁺ lacP⁺ lacO⁺ lacZ⁺ lacY⁺/ lacI⁺ lacP⁺ lacO⁺ lacZ⁺ lacY⁺	_____	_____	_____	_____
lacIˢ lacP⁺ lacO⁺ lacZ⁺ lacY⁻/ lacI⁺ lacP⁺ lacO⁺ lacZ⁻ lacY⁺	_____	_____	_____	_____
lacIˢ lacP⁻ lacO⁺ lacZ⁻ lacY⁺/ lacI⁺ lacP⁺ lacO⁺ lacZ⁺ lacY⁺	_____	_____	_____	_____

20. Give all possible genotypes of a *lac* operon that produces, or fails to produce, β-galactosidase and permease under the following conditions. Do not give partial-diploid genotypes.

	Lactose absent		Lactose present	
	β-Galactosidase	Permease	β-Galactosidase	Permease
a.	−	−	+	+
b.	−	−	−	+
c.	−	−	+	−
d.	+	+	+	+
e.	−	−	−	−
f.	+	−	+	−
g.	−	+	−	+

*21. Explain why mutations in the *lacI* gene are trans in their effects, but mutations in the *lacO* gene are cis in their effects.

22. Which strand of DNA (upper or lower) in **Figure 16.8** is the template strand? Explain your reasoning.

23. The *mmm* operon, which has sequences A, B, C, and D (which may be structural genes or regulatory sequences), encodes enzymes 1 and 2. Mutations in sequences A, B, C, and D have the following effects, where a plus sign (+) indicates that the enzyme is synthesized and a minus sign (−) indicates that the enzyme is not synthesized.

Mutation in sequence	mmm absent		mmm present	
	Enzyme 1	Enzyme 2	Enzyme 1	Enzyme 2
No mutation	+	+	−	−
A	−	+	−	−
B	+	+	+	+
C	+	−	−	−
D	−	−	−	−

a. Is the *mmm* operon inducible or repressible?

b. Indicate which sequence (A, B, C, or D) is part of the following components of the operon:

Regulator gene _____

Promoter _____

Structural gene for enzyme 1 _____

Structural gene for enzyme 2 _____

24. Ellis Engelsberg and his colleagues examined the regulation of genes taking part in the metabolism of arabinose, a sugar (E. Engelsberg et al. 1965. *Journal of Bacteriology* 90:946–957). Four structural genes encode enzymes that help metabolize arabinose (genes A, B, D, and E). An additional sequence C is linked to genes A, B, and D. These are in the order D-A-B-C. Gene E is distant from the other genes. Engelsberg and his colleagues isolated mutations at the C sequence that affected the expression of structural genes A, B, D, and E. In one set of experiments, they created various genotypes at A and C and determined whether arabinose isomerase (the enzyme encoded by gene A) was produced in the presence or absence of arabinose (the substrate of arabinose isomerase) by cells with these genotypes. Results from this experiment are shown in the following table, where a plus sign (+) indicates that the arabinose isomerase was synthesized and a minus sign (−) indicates that the enzyme was not synthesized.

Genotype	Arabinose absent	Arabinose present
1. $C^+ A^+$	−	+
2. $C^- A^+$	−	−
3. $C^- A^+/C^+ A^-$	−	+
4. $C^c A^-/C^- A^+$	+	+

a. On the basis of these results, is the C sequence an operator or a regulator gene? Explain your reasoning.

b. Do these experiments suggest that the arabinose operon is negatively or positively controlled? Explain your reasoning.

c. What type of mutation is C^c?

25. In *E. coli*, three structural genes (A, D, and E) encode enzymes A, D, and E, respectively. Sequence O is an operator. The sequences are in the order O-A-D-E on the chromosome. The three enzymes encoded by A, D, and E catalyze the biosynthesis of valine. Mutations at the A, D, E, and O were isolated to study the production of enzymes A, D, and E when cellular levels of valine were low (T. Ramakrishnan and E. A. Adelberg. 1965. *Journal of Bacteriology* 89:654–660). Amounts of the enzymes produced by partial-diploid *E. coli* with various combinations of mutations are shown in the following table.

	Genotype	Amount of enzyme produced		
		E	D	A
1.	$E^+ D^+ A^+ O^+/$ $E^+ D^+ A^+ O^+$	2.40	2.00	3.50
2.	$E^+ D^+ A^+ O^-/$ $E^+ D^+ A^+ O^+$	35.80	38.60	46.80
3.	$E^+ D^- A^+ O^-/$ $E^+ D^+ A^- O^+$	1.80	1.00	47.00
4.	$E^+ D^+ A^- O^-/$ $E^+ D^- A^+ O^+$	35.30	38.00	1.70
5.	$E^- D^+ A^+ O^-/$ $E^+ D^- A^+ O^+$	2.38	38.00	46.70

a. Is the regulator protein that binds to the operator of this operon a repressor (negative control) or an activator (positive control)? Explain your reasoning.

b. Are genes *A*, *D*, and *E* all under the control of operator *O*? Explain your reasoning.

c. Propose an explanation for the small amount of enzyme E produced by genotype 3.

Section 16.3

26. At which level of gene regulation shown in **Figure 16.1** does attenuation occur?

*27. Listed in parts *a* through *g* are some mutations that were found in the 5′ UTR of the *trp* operon of *E. coli*. What will the most likely effect of each of these mutations be on the transcription of the *trp* structural genes?

a. A mutation that prevents the binding of the ribosome to the 5′ end of the mRNA 5′ UTR

b. A mutation that changes the Trp codons in region 1 of the mRNA 5′ UTR into codons for alanine

c. A mutation that creates a stop codon early in region 1 of the mRNA 5′ UTR

d. Deletions in region 2 of the mRNA 5′ UTR

e. Deletions in region 3 of the mRNA 5′ UTR

f. Deletions in region 4 of the mRNA 5′ UTR

g. Deletion of the string of adenine nucleotides that follows region 4 in the 5′ UTR

28. Some mutations in the *trp* 5′ UTR increase termination by the attenuator. Where might these mutations occur, and how might they affect the attenuator?

29. Some of the mutations of the type mentioned in Problem 28 have an interesting property: they prevent the formation of the antiterminator that normally takes place when the tryptophan level is low. In one of these mutations, the AUG start codon for translation of the 5′ UTR has been deleted. How might this mutation prevent antitermination from taking place?

Section 16.4

*30. Several examples of antisense RNA regulating translation in bacterial cells have been discovered. Molecular geneticists have also used antisense RNA to artificially control transcription in both bacterial and eukaryotic genes. If you wanted to inhibit the transcription of a bacterial gene with antisense RNA, what sequences might the antisense RNA contain?

CHALLENGE QUESTION

Section 16.3

31. Would you expect to see attenuation in the *lac* operon and other operons that control the metabolism of sugars? Why or why not?

 THINK-PAIR-SHARE QUESTIONS

Introduction

1. What are some advantages and disadvantages of using mathematical models to study biological processes such as the function of operons?

Section 16.1

2. What types of genes would you expect to be constitutive?

Section 16.2

3. Suppose that an operon that exhibited positive control was inducible. Would the regulator gene in this case produce an active or an inactive activator protein? What would most likely turn transcription on? What about an operon with positive control that was repressible? In this case, what would turn transcription off? Explain your reasoning.

4. When during the development of most mammals is lactose available for use by *E. coli* in the gut? How does the availability of lactose to *E. coli* in humans differ from that seen in other mammals?

5. A strain of *E. coli* has the genotypes shown below at the *lac* operon, where *I* = regulator gene, *P* = promoter, *O* = operator, *Z* = β-galactosidase gene, and *Y* = permease gene. The superscript $^+$ indicates a wild-type allele, ^{c} indicates a constitutive mutation, and $^-$ indicates a defective mutation. For each genotype, indicate whether the enzyme will be synthesized or not synthesized when lactose is present or absent by placing a + for *synthesis occurring* and a − for *synthesis not occurring* in the appropriate blank. Explain your reasoning for each answer.

		Inducer (lactose) absent		Inducer (lactose) present	
	Genotype	β-Galactosidase	Permease	β-Galactosidase	Permease
a.	*lacI$^+$ lacP$^+$ lacO$^+$ lacZ$^+$ lacY$^+$*	_____	_____	_____	_____
b.	*lacI$^+$ lacP$^+$ lacO^c lacZ$^+$ lacY$^+$*	_____	_____	_____	_____
c.	*lacI$^-$ lacP$^+$ lacO$^+$ lacZ$^-$ lacY$^+$*	_____	_____	_____	_____
d.	*lacI$^-$ lacP$^+$ lacO$^+$ lacZ$^+$ lacY$^+$/* *lacI$^+$ lacP$^-$ lacO$^+$ lacZ$^+$ lacY$^-$*	_____	_____	_____	_____
e.	*lacI$^-$ lacP$^+$ lacO^c lacZ$^-$ lacY$^+$/* *lacI$^+$ lacP$^+$ lacO$^+$ lacZ$^+$ lacY$^+$*	_____	_____	_____	_____
f.	*lacI$^+$ lacP$^-$ lacO^c lacZ$^+$ lacY$^-$/* *lacI$^-$ lacP$^+$ lacO$^+$ lacZ$^-$ lacY$^+$*	_____	_____	_____	_____

Section 16.3

6. Indicate whether each of the following events occurs when tryptophan is high or when tryptophan is low by placing a check in each of the appropriate blanks. Share your reasoning.

Event	Tryptophan high	Tryptophan low
Ribosome does not stall at Trp codons	_____	_____
Region 2 of the leader pairs with region 3	_____	_____
Ribosome covers part of region 2 of leader	_____	_____
Transcription is terminated before structural genes are transcribed	_____	_____

7. What would be the effect on tryptophan production of mutations in region 1 of the leader of the *trp* operon that changed the Trp codons to stop codons? Explain your reasoning.

Section 16.4

8. How might antisense RNAs be used as antibiotics?

Control of Gene Expression in Eukaryotes

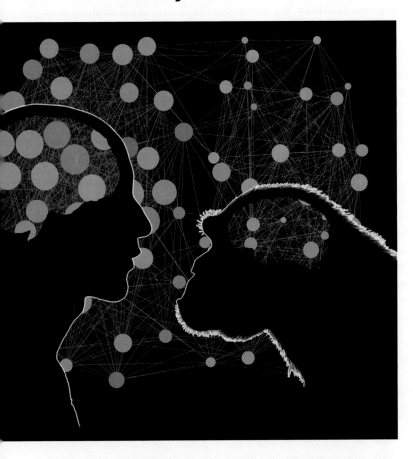

Changes in a relatively small number of regulatory sequences help produce the large phenotypic differences between humans and chimpanzees. Illustrated here is a network of interacting genes encoding transcription factors that are differentially expressed in the brains of humans and chimpanzees and control the expression of other genes. Red circles represent transcription factors that are more highly expressed in the human brain; green circles represent transcription factors that are more highly expressed in the chimpanzee brain. [Edwin Hadley, University of Illinois.]

Genetic Differences That Make Us Human

Over 140 years ago, Charles Darwin proposed that humans share a recent common ancestor with the African great apes. Today, there is significant evidence that our closest living relative is the chimpanzee (*Pan troglodytes*). Fossil evidence indicates that humans and chimpanzees diverged genetically only 5 million to 7 million years ago—a mere blink of the eye in evolutionary time. Yet humans and chimpanzees differ in a large number of anatomical, physiological, behavioral, and cognitive traits. For example, there are numerous differences in the structure of the backbone, pelvis, skull, jaw, teeth, hands, and feet of humans and chimpanzees. The size of the human brain is more than twice that of the chimpanzee brain, and humans exhibit complex language and cultural characteristics not seen in chimpanzees. Indeed, the degree of phenotypic difference between chimpanzees and humans is so large that scientists in the past placed them in entirely different primate families (humans in the family Hominidae and chimpanzees in the family Pongidae), although today biologists often put them together in the family Hominidae.

In spite of the large phenotypic gulf between humans and chimpanzees, sequencing of their genomes reveals that their DNA is remarkably similar. There is only a 1% difference in individual base pairs between the two species, along with a 3% difference in insertions and deletions. Thus, 96% of the DNA of humans and chimpanzees is identical. But clearly humans are not chimpanzees. How, then, did humans and chimpanzees come to be so different? Where are the genes that make us human?

A possible answer to this paradox was proposed by geneticists Mary-Claire King and A. C. Wilson in 1975. Using the limited techniques available at the time (comparison of amino acids in proteins and DNA hybridization studies), King and Wilson concluded that humans and chimpanzees differed at only about 1% of their DNA sequences. To explain how these very small genetic changes could account for the extensive physical and behavioral differences between humans and chimpanzees, King and Wilson suggested that the genetic variations that make us human are concentrated in regulatory sequences—those parts of the genome that control the expression of other genes. In this way, small genetic changes might influence the expression of numerous other genes and affect the phenotypes of many traits simultaneously. Unfortunately, there were no techniques available at the time to examine regulatory sequences to test their hypothesis.

Fast-forward to 2009. Using cutting-edge techniques of genomic research and bioinformatics, Katja Nowick and her colleagues at the University of Illinois and the Norwegian University of Science and Technology identified a group of 90 transcription factors whose expression differed significantly between humans and chimpanzees. As discussed in Chapter 13, transcription factors are proteins that bind to DNA and facilitate or repress the synthesis of RNA, the first step in the process of information transfer from genotype to phenotype. Each transcription factor may affect the expression of multiple genes, so a small genetic change affecting the expression of a single transcription factor can influence many additional genes. The differences that Nowick and her colleagues found in the expression of transcription factors were particularly pronounced in brain tissue, where they may account for the large differences in neural and cognitive function between humans and chimpanzees.

Many of the transcription factors that Nowick and her colleagues identified were Krüppel-associated box domain zinc finger proteins (KRAB-ZFPs), transcription factors that bind to specific DNA sequences and bring about changes in chromatin structure. As we'll see in this chapter, changes in chromatin structure are often involved in regulation of gene expression in eukaryotes. Other studies have demonstrated that KRAB-ZFPs have evolved rapidly in humans, probably because they were favored by natural selection. The transcription factors identified by Nowick and colleagues clustered into two distinct but interconnected regulatory networks, which control energy metabolism, transcription, vesicular transport, and protein modification.

These results support the idea first proposed by King and Wilson: that changes in a relatively small number of regulatory sequences affect the expression of numerous genes in humans and chimpanzees and help produce the large differences we see in their anatomy, brain size, cognition, and behavior.

THINK-PAIR-SHARE

■ Based on their anatomical and behavioral differences, biologists previously placed humans and chimpanzees in different families, but now DNA sequence data suggest that they are closely related and should be placed in the same family. What are some advantages of using DNA sequences to decide how to classify organisms? What might be some disadvantages?

■ If 4% of human and chimpanzee DNA differs, how many base pairs differ between the two species? (Hint: The human genome contains 3.2 billion base pairs of DNA.) Is this a large or a small difference? Explain your reasoning.

This chapter is about gene regulation in eukaryotic cells—the very changes that help to make humans unique. Gene regulation in eukaryotes typically takes place at multiple levels. We begin by considering how gene expression is influenced by changes in chromatin structure, which can be altered by several different mechanisms. We next consider the initiation of transcription in eukaryotes, which is controlled by the complex interactions of several types of proteins and the DNA regulatory elements to which they bind. We examine several ways in which gene expression is controlled through the processing, degradation, and translation of mRNA. We end by revisiting some of the similarities and differences in gene regulation in bacteria and eukaryotes.

17.1 Eukaryotic Cells and Bacteria Share Many Features of Gene Regulation but Differ in Several Important Ways

As discussed in Chapter 16, many features of gene regulation are shared by bacterial and eukaryotic cells. For example, in both types of cells, DNA-binding proteins influence the ability of RNA polymerase to initiate transcription. However, there are also some differences. First, many bacterial and archaeal genes are organized into operons and are transcribed into a single RNA molecule. Although some operon-like gene clusters have been found in worms and even in some primitive chordates, most eukaryotic genes have their own promoters and are transcribed separately. Second, chromatin structure affects gene expression in eukaryotic cells; DNA must unwind from the histone proteins before transcription can take place. Third, the presence of the nuclear membrane in eukaryotic cells separates transcription and translation in time and space. Therefore, the regulation of gene expression in eukaryotic cells is characterized by a greater diversity of mechanisms that act at different points in the transfer of information from DNA to protein.

Eukaryotic gene regulation is less well understood than bacterial gene regulation, partly owing to the larger genomes of eukaryotes, their greater sequence complexity, and the difficulty of isolating and manipulating mutations that can be used in the study of gene regulation. Nevertheless, great advances in our understanding of the regulation of eukaryotic genes have been made in recent years.

17.2 Changes in Chromatin Structure Affect the Expression of Eukaryotic Genes

One type of gene control in eukaryotic cells is accomplished through the modification of chromatin structure. In the nucleus, histone proteins associate to form octamers, around which helical DNA tightly coils to create chromatin (see Figure 11.4). In a general sense, this chromatin structure itself represses gene expression. For a gene to be transcribed, proteins called transcription factors (see Section 17.3) must bind to the DNA. Other regulator proteins and RNA polymerase must also bind to the DNA for transcription to take place. How can these proteins bind with DNA wrapped tightly around histone proteins? The answer is that, before transcription, chromatin structure changes so that the DNA becomes more accessible to the transcription machinery.

DNase I Hypersensitivity

As genes become transcriptionally active, regions around the genes become highly sensitive to the action of DNase I (see pp. 317–318 in Chapter 11). These regions, called **DNase I hypersensitive sites**, frequently develop about 1000 nucleotides upstream of the transcription start site, suggesting that the chromatin in these regions adopts a more open configuration during transcription. This relaxation of the chromatin structure allows regulatory proteins access to binding sites on the DNA. Indeed, many DNase I hypersensitive sites correspond to known binding sites for regulatory proteins. At least three different processes affect gene regulation by altering chromatin structure: (1) chromatin remodeling; (2) the modification of histone proteins; and (3) DNA methylation. Each of these mechanisms will be discussed in the sections that follow.

THINK-PAIR-SHARE Question 1

> **CONCEPTS**
>
> Sensitivity to DNase I in regions around transcriptionally active genes indicates that the chromatin in these regions assumes an open configuration before transcription.

Chromatin Remodeling

Some transcription factors and other regulatory proteins alter chromatin structure without altering the chemical structure of the histones directly. These proteins are called **chromatin-remodeling complexes**. They bind directly to particular sites on DNA and reposition the nucleosomes, allowing other transcription factors and RNA polymerase to bind to promoters and initiate transcription (**Figure 17.1**).

One of the best-studied examples of a chromatin-remodeling complex is SWI–SNF, which is found in yeast, humans, *Drosophila*, and other eukaryotes. This complex uses energy derived from the hydrolysis of ATP to reposition

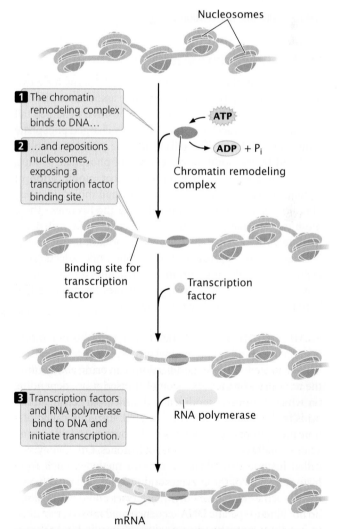

17.1 Chromatin-remodeling complexes reposition the nucleosomes, allowing transcription factors and RNA polymerase to bind to promoters and initiate transcription.

nucleosomes, exposing promoters in the DNA to the action of other regulatory proteins and RNA polymerase.

Evidence suggests at least two mechanisms by which chromatin-remodeling complexes reposition nucleosomes. First, some complexes cause the nucleosome to slide along the DNA, allowing DNA that was wrapped around the nucleosome to occupy a position in between nucleosomes, where it is more accessible to proteins affecting gene expression (see Figure 17.1). Second, some complexes cause conformational changes in the DNA, in nucleosomes, or in both so that DNA that is bound to the nucleosome assumes a more exposed configuration.

Chromatin-remodeling complexes are targeted to specific DNA sequences by transcriptional activators or repressors that attach to a complex and then bind to the promoters of specific genes. There is also evidence that chromatin-remodeling complexes work together with enzymes that alter histones, such as acetyltransferase enzymes (which we'll describe shortly), to change chromatin structure and expose DNA for transcription.

THINK-PAIR-SHARE Question 2

Histone Modification

The histones in the octamer core of a nucleosome have two domains: (1) a globular domain that associates with the other histones and the DNA, and (2) a positively charged tail domain that interacts with the negatively charged phosphate groups on the DNA. The tails of histone proteins are often modified by the addition or removal of phosphate groups, methyl groups, or acetyl groups. Another modification of histones is ubiquitination, in which a small molecule called ubiquitin is added to or removed from the histones. These modifications have sometimes been collectively called the **histone code** because they encode information that affects how genes are expressed. The histone code affects gene expression by altering chromatin structure directly or, in some cases, by providing recognition sites for proteins that bind to DNA and regulate transcription.

METHYLATION OF HISTONES One type of histone modification is the addition of methyl groups (CH_3) to the tails of histone proteins. These modifications can bring about either the activation or the repression of transcription, depending on which histone is modified and which particular amino acids in the histone tail are methylated. Enzymes called histone methyltransferases add methyl groups to specific amino acids (usually lysine or arginine) of histones. Other enzymes, called histone demethylases, remove methyl groups from histones. Many of the enzymes and proteins that modify histones, such as histone methyltransferases and demethylases, do not bind to specific DNA sequences and must be recruited to specific chromatin sites. Sequence-specific binding proteins, preexisting histone modifications, and RNA molecules serve to recruit histone-modifying enzymes to specific sites.

A common modification is the addition of three methyl groups to lysine 4 in the tail of the H3 histone protein, abbreviated H3K4me3 (K is the abbreviation for lysine). Histones containing the H3K4me3 modification are frequently found near promoters of transcriptionally active genes. Studies have identified proteins that recognize and bind to H3K4me3, including nucleosome remodeling factor (NURF). NURF and other proteins that recognize H3K4me3 have a common protein-binding domain that binds to the H3 histone tail and then alters chromatin packing, allowing transcription to take place. Research has also demonstrated that some transcription factors that are necessary for the initiation of transcription (see Chapter 13 and Section 17.3) bind directly to H3K4me3.

ACETYLATION OF HISTONES Another type of histone modification that affects chromatin structure is acetylation, the addition of acetyl groups (CH_3CO) to histones. The acetylation of histones usually stimulates transcription. For example, the addition of a single acetyl group to lysine 16 in the tail of the H4 histone prevents the formation of the 30-nm chromatin fiber (see Figure 11.4), causing the chromatin to be in an open configuration that makes the DNA available for transcription. In general, acetyl groups destabilize chromatin structure, allowing transcription to take place (**Figure 17.2**). Acetyl groups are added to histone proteins by acetyltransferase enzymes; other enzymes called deacetylases strip acetyl groups from histones and restore chromatin structure, which represses transcription. Certain transcription factors (see Chapter 13) and other proteins that regulate transcription either have acetyltransferase activity themselves or attract acetyltransferases to DNA.

ACETYLATION OF HISTONES AND FLOWERING IN *ARABIDOPSIS* The importance of histone acetylation in gene regulation is demonstrated by the control of flowering in *Arabidopsis thaliana*, a plant with a number of

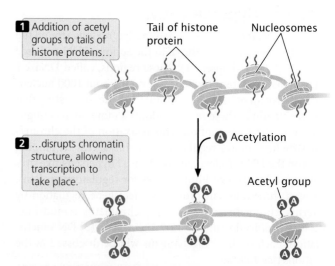

1 Addition of acetyl groups to tails of histone proteins…

Tail of histone protein   Nucleosomes

A Acetylation

2 …disrupts chromatin structure, allowing transcription to take place.

Acetyl group

17.2 The acetylation of histone proteins alters chromatin structure and permits some transcription factors to bind to DNA.

characteristics that make it an excellent genetic model for plant systems (see the Reference Guide to Model Genetic Organisms at the end of this book). The time at which flowering takes place is critical to the life of a plant: if flowering is initiated at the wrong time of year, pollinators may not be available to fertilize the flowers, or environmental conditions may be unsuitable for the survival and germination of the seeds. Consequently, flowering time in most plants is carefully regulated in response to multiple internal and external cues, such as plant size, photoperiod, and temperature.

Among the many genes that control flowering in *Arabidopsis* is *flowering locus C* (*FLC*), which plays an important role in suppressing flowering until after an extended period of cold (a process called vernalization). The *FLC* gene encodes a regulatory protein that represses the activity of other genes that affect flowering (**Figure 17.3**). As long as *FLC* is transcriptionally active, flowering remains suppressed.

The activity of *FLC* is controlled by another locus called *flowering locus D* (*FLD*), the key role of which is to stimulate flowering by repressing the action of *FLC*. In essence, flowering is stimulated because *FLD* represses the repressor.

How does *FLD* repress *FLC*? *FLD* encodes a deacetylase enzyme, which removes acetyl groups from histone proteins in the chromatin surrounding *FLC* (see Figure 17.3). The removal of these acetyl groups alters the chromatin structure and inhibits transcription. The inhibition of transcription prevents *FLC* from being transcribed and removes its repression on flowering. In short, *FLD* stimulates flowering in *Arabidopsis* by deacetylating the chromatin that surrounds *FLC*, thereby removing its inhibitory effect on flowering. ❯ TRY PROBLEM 18

CHROMATIN IMMUNOPRECIPITATION Our understanding of how changes in chromatin structure are associated with gene expression, and of how DNA-binding proteins affect transcription, has been greatly advanced by the use of a technique called chromatin immunoprecipitation (ChIP). This technique allows researchers to determine the locations within the genome where a specific protein interacts with DNA. ChIP has been used to determine the locations of histones that have undergone modifications as well as where transcription factors and other proteins bind to promoters and enhancers.

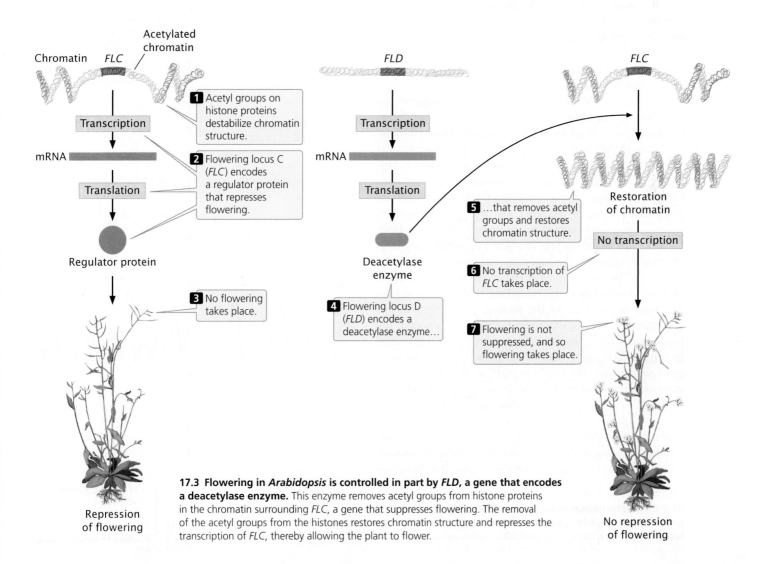

17.3 Flowering in *Arabidopsis* is controlled in part by *FLD*, a gene that encodes a deacetylase enzyme. This enzyme removes acetyl groups from histone proteins in the chromatin surrounding *FLC*, a gene that suppresses flowering. The removal of the acetyl groups from the histones restores chromatin structure and represses the transcription of *FLC*, thereby allowing the plant to flower.

The basic idea of ChIP is that a particular protein and the DNA to which it is bound are isolated, the protein and DNA are then separated, and the DNA sequence to which the protein was formerly bound is identified. The technique has provided a powerful means of determining the genome-wide locations of modified histones and the binding sites for transcription factors and other proteins that affect transcription.

One version of ChIP, called crosslinked ChIP (XChIP), is used for identifying the binding sites of transcription factors and other proteins that bind to DNA. In this procedure (**Figure 17.4**), the protein and associated DNA are temporarily crosslinked, which means that they are treated with formaldehyde or UV light to create covalent bonds between the DNA and protein. The crosslinking holds the DNA and protein together so that the DNA to which the protein is bound can be isolated along with the protein. After crosslinking, the cell is lysed, and the chromatin is broken into pieces by digestion with an enzyme or by mechanical shearing. Antibodies specific for a particular protein—such as a specific transcription factor—are then applied. The antibodies attach to the protein–DNA complexes as well as to a solid substrate (usually small beads). These beads can then be used to precipitate the protein–DNA complex. After the protein–DNA complex is precipitated, the crosslinking is reversed, separating the DNA and the protein. The protein is removed by an enzyme that digests protein but not DNA, leaving fragments of the DNA to which the protein was bound.

The location of the resulting DNA fragments within the genome can then be determined by several different methods. One method, termed ChIP-Seq, determines the base sequences of the fragments using next-generation sequencing technologies (see Chapter 19). When the chromatin is broken into pieces and subjected to immunoprecipitation (see Figure 17.4), many short overlapping fragments of DNA are precipitated. Some of the sequences within each fragment are those that were covered by the protein; others are DNA on either side of the binding site. When the overlapping fragments are sequenced, the base pairs that were covered by the protein will be present in more copies and will be sequenced more often (resulting in more sequence reads) than base pairs that were not covered by the protein, producing peaks of sequence reads at those sites where the protein was bound. The locations of these peaks within the genome can be determined by comparing the sequences within the peaks with a reference genome sequence. The results provide information about the genomic locations of binding sites for the specific protein.

Another version of ChIP, called native ChIP (nChIP), does not use crosslinking. It is often used for finding the locations of modified histone proteins. In this case, crosslinking is not required because the DNA and the histones are naturally linked by the nucleosome structure. The chromatin

17.4 Chromatin immunoprecipitation (ChIP) can be used to identify the DNA binding sites of a specific protein and the locations of modified histone proteins. Shown here is the method for crosslinked chromatin immunoprecipitation (XChIP).

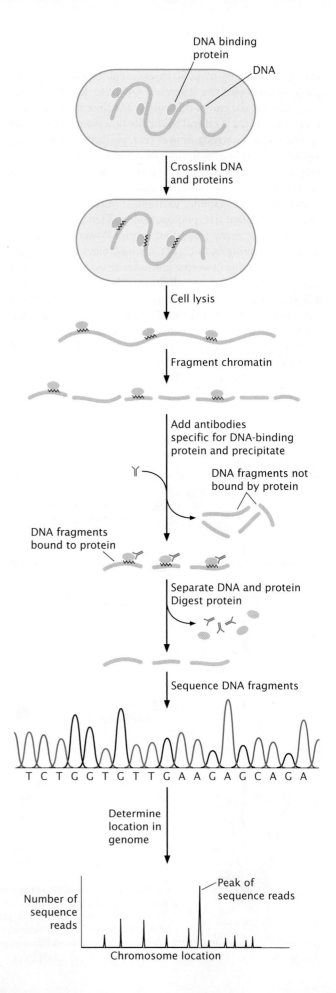

is isolated from the cell and fragmented, and antibodies to a particular protein—usually a specific modified histone—are used to precipitate the protein–DNA complexes. The histones and DNA are separated, the histones are digested, and the DNA fragments to which the modified histones were attached are sequenced or otherwise identified.

ChIP analysis has been used to determine the locations of modified histones that activate or repress transcription. As mentioned earlier, the H3K4me3 histone modification is associated with promoters of active genes. ChIP analysis has successfully identified locations of this modified histone in the human genome, helping to identify active promoters in different tissues.

CONCEPTS

The tails of histone proteins are often modified by the addition or removal of phosphate groups, methyl groups, or acetyl groups. These modifications alter chromatin structure and affect the transcription of genes. Chromatin immunoprecipitation can be used to identify the locations within the genome of modified histones and binding sites for regulatory proteins.

DNA Methylation

Another change in chromatin structure associated with transcription is the methylation of cytosine bases, which yields 5-methylcytosine (see Figure 10.20). The methylation of cytosine in DNA is distinct from the methylation of histone proteins mentioned earlier. Heavily methylated DNA is associated with the repression of transcription in vertebrates and plants, whereas transcriptionally active DNA is usually unmethylated in these organisms. Abnormal patterns of methylation are also associated with some types of cancer.

DNA methylation is most common on cytosine bases adjacent to guanine nucleotides (CpG, where p represents the phosphate group in the DNA backbone), so two methylated cytosines sit diagonally across from each other on opposite strands:

$$5'-\overset{m}{C}\,G-3'$$
$$3'-G\,\underset{m}{C}-5'$$

DNA regions with many CpG sequences are called **CpG islands** and are commonly found near transcription start sites. While genes are not being transcribed, these CpG islands are often methylated, but the methyl groups are removed before the initiation of transcription. CpG methylation is also associated with long-term gene repression, such as that of the inactivated X chromosome of female mammals (see Chapter 4).

Evidence indicates that an association exists between DNA methylation and the deacetylation of histones, both of which repress transcription. Certain proteins that bind tightly to methylated CpG sequences form complexes with other proteins that act as histone deacetylases. In other words, methylation appears to attract deacetylases, which remove acetyl groups from the histone tails, stabilizing the nucleosome structure and repressing transcription.

Demethylation of DNA allows acetyltransferases to add acetyl groups, disrupting nucleosome structure and permitting transcription.

Changes in chromatin structure that affect gene expression, such as the mechanisms just described (chromatin remodeling, histone modification, and DNA methylation), are examples of the phenomenon of epigenetics: alterations to DNA and chromatin structure that affect traits and are passed on to other cells or generations but are not caused by changes in the DNA base sequence. Epigenetic changes and their effects on gene regulation will be discussed in more detail in Chapter 21. ▶ **TRY PROBLEM 19**

CONCEPTS

Chromatin structure can be altered by methylation of DNA. In eukaryotes, DNA methylation often results in 5-methylctyosine at CpG dinucleotides. DNA methylation is usually associated with repression of transcription.

✔ **CONCEPT CHECK 1**

What are some of the processes that affect gene regulation by altering chromatin structure?

17.3 The Initiation of Transcription Is Regulated by Transcription Factors and Transcriptional Regulator Proteins

We have just considered one level at which gene expression is controlled: the alteration of chromatin and DNA structure. We now turn to another important level of control: control through the binding of proteins to DNA sequences that affect transcription. Transcription is an important level of control in eukaryotic cells, and its control requires a number of different types of proteins and regulatory elements.

The initiation of eukaryotic transcription was discussed in detail in Chapter 13. Recall that general transcription factors are part of the *basal transcription apparatus*, the complex of RNA polymerase, transcription factors, and other proteins that assembles to carry out transcription. The basal transcription apparatus binds to a *core promoter* located immediately upstream of a gene and is capable of minimal levels of transcription, but *transcriptional regulator proteins* are required to bring about normal levels of transcription. These proteins bind to a *regulatory promoter*, which is located upstream of the core promoter (**Figure 17.5**), and to *enhancers*, which may be located some distance from the gene. Some transcriptional regulator proteins are activators, stimulating transcription; others are repressors, inhibiting transcription.

Transcriptional Activators and Coactivators

Transcriptional activator proteins stimulate and stabilize the basal transcription apparatus at the core promoter. The activators may interact directly with the basal transcription apparatus or act on it indirectly through coactivator proteins.

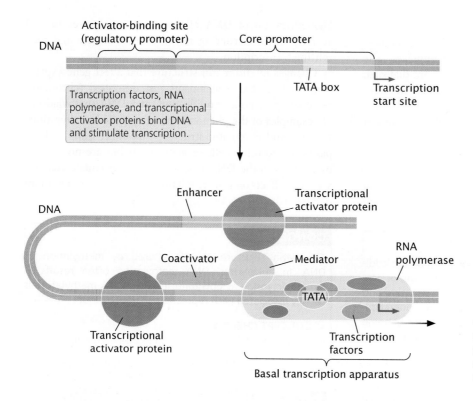

17.5 **Transcriptional activator proteins bind to sites on DNA and stimulate transcription.** Most act by stimulating or stabilizing the assembly of the basal transcription apparatus.

Some activators and coactivators, as well as general transcription factors, also have acetyltransferase activity and so further stimulate transcription by altering chromatin structure.

Transcriptional activator proteins have two distinct functions (see Figure 17.5). First, they are capable of binding DNA at a specific base sequence, usually a consensus sequence in a regulatory promoter or enhancer; for this function, most transcriptional activator proteins contain one or more DNA-binding motifs, such as the helix-turn-helix, zinc finger, or leucine zipper (see Figure 16.2). Their second function is to interact with other components of the basal transcription apparatus and influence the rate of transcription.

Regulatory promoters typically contain several different consensus sequences to which different transcriptional activators can bind. On different promoters, activator binding sites are mixed and matched in different combinations (**Figure 17.6**), so each promoter is regulated by a unique combination of transcriptional activator proteins.

The binding of transcriptional activator proteins to the consensus sequences in the regulatory promoter affects the assembly or stability of the basal transcription apparatus at the core promoter. One of the components of the basal transcription apparatus is a complex of proteins called the **mediator** (see Figure 17.5), which interacts with RNA polymerase. Transcriptional regulator proteins that bind to sequences in the regulatory promoter (or enhancer) make contact with the mediator and affect the rate at which transcription is initiated. Some regulatory promoters contain sequences that are bound by transcriptional repressor proteins, which lower the rate of transcription through inhibitory interactions with the mediator.

THINK-PAIR-SHARE Question 3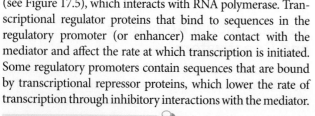

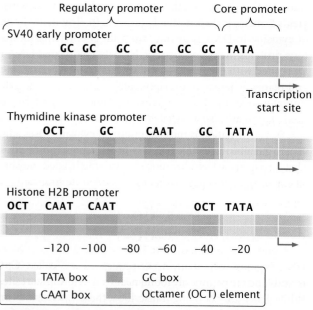

17.6 **The consensus sequences in the promoters of three eukaryotic genes illustrate the principle that these sequences are mixed and matched in different combinations in different promoters.** A different transcriptional activator protein binds to each consensus sequence, so each promoter responds to a unique combination of activator proteins.

REGULATION OF GALACTOSE METABOLISM THROUGH GAL4 An example of a transcriptional activator protein is GAL4, which regulates the transcription of several yeast genes whose products metabolize galactose. Like the genes in the *lac* operon, the genes that control galactose metabolism are

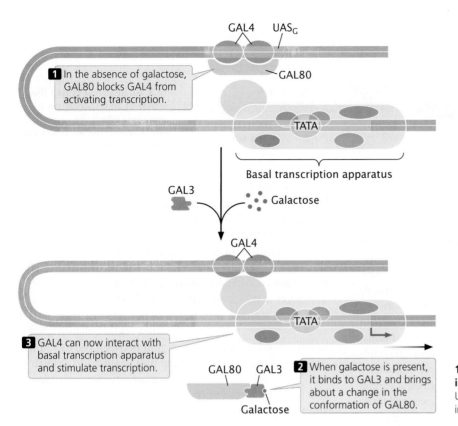

1 In the absence of galactose, GAL80 blocks GAL4 from activating transcription.

Basal transcription apparatus

GAL3

Galactose

GAL4

3 GAL4 can now interact with basal transcription apparatus and stimulate transcription.

GAL80   GAL3

Galactose

2 When galactose is present, it binds to GAL3 and brings about a change in the conformation of GAL80.

17.7 Transcription is activated by GAL4 in response to galactose. GAL4 binds to the UAS$_G$ site and controls the transcription of genes in galactose metabolism.

inducible: when galactose is absent, these genes are not transcribed and the proteins that break down galactose are not produced; when galactose is present, the genes are transcribed and the enzymes are synthesized. GAL4 contains several zinc fingers (DNA-binding motifs) and binds to a DNA sequence called UAS$_G$ (upstream activating sequence for GAL4). UAS$_G$ exhibits the properties of an enhancer—a regulatory sequence that lies some distance from the regulated gene. When bound to UAS$_G$, GAL4 activates the transcription of yeast genes needed for metabolizing galactose. GAL4 and a number of other transcriptional activator proteins contain multiple amino acids with negative charges that form an *acidic activation domain*. These acidic activators stimulate transcription by enhancing the assembly of the basal transcription apparatus.

A particular region of GAL4 binds another protein called GAL80, which regulates the activity of GAL4 in the presence of galactose. When galactose is absent, GAL80 binds to GAL4, preventing GAL4 from activating transcription (**Figure 17.7**). When galactose is present, however, the sugar binds to another protein called GAL3, which interacts with GAL80, causing a conformational change in GAL80 so that it can no longer bind GAL4. The GAL4 protein is then free to bind to UAS$_G$ and activate the transcription of the genes whose products metabolize galactose.

Transcriptional Repressors

Some transcriptional regulator proteins in eukaryotic cells act as repressors, inhibiting transcription. These repressors bind to sequences in the regulatory promoter or to distant

sequences called *silencers*, which, like enhancers, are position and orientation independent. Unlike repressors in bacteria, most eukaryotic repressors do not directly block RNA polymerase. These repressors may compete with transcriptional activators for binding sites on the DNA: when a site is occupied by an activator, transcription is activated, but if a repressor occupies that site, there is no activation. Alternatively, a repressor may bind to sites near an activator binding site and prevent the activator from contacting the basal transcription apparatus. A third possible mechanism of repressor action is direct interference with the assembly of the basal transcription apparatus, thereby blocking the initiation of transcription.

THINK-PAIR-SHARE Question 4

CONCEPTS

Transcriptional regulator proteins in eukaryotic cells can influence the initiation of transcription by affecting the stability or assembly of the basal transcription apparatus. Some transcriptional regulator proteins are activators that stimulate transcription; others are repressors that inhibit the initiation of transcription.

✔ CONCEPT CHECK 2

Most transcriptional activator proteins affect transcription by interacting with

a. introns.
b. the basal transcription apparatus.
c. DNA polymerase.
d. the terminator.

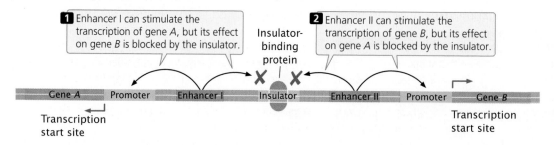

1 Enhancer I can stimulate the transcription of gene *A*, but its effect on gene *B* is blocked by the insulator.

2 Enhancer II can stimulate the transcription of gene *B*, but its effect on gene *A* is blocked by the insulator.

17.8 An insulator blocks the action of an enhancer on a promoter when the insulator lies between the enhancer and the promoter.

Enhancers and Insulators

Enhancers are regulatory elements that affect the transcription of distant genes. For example, an enhancer that regulates the gene encoding the alpha chain of the T-cell receptor is located 69,000 bp downstream of the gene's promoter; some vertebrate enhancers act over distances of millions of base pairs. The exact position and orientation of an enhancer relative to the promoter it regulates are not critical to its function (see Chapter 13). Enhancers are often a few hundred base pairs in length and contain multiple binding sites for transcriptional activator proteins; a typical enhancer contains about 10 binding sites for proteins that regulate transcription. Enhancers often regulate genes in a cell type–specific manner, meaning that they turn on different sets of genes in different cell types and help to establish the traits that characterize cells of different tissues. Some enhancer-like elements are found in prokaryotes.

How can an enhancer affect the initiation of transcription at a promoter that is tens of thousands of base pairs away? In many cases, the binding of transcriptional regulator proteins to the enhancer causes the DNA between the enhancer and the promoter to loop out, bringing the promoter and enhancer close to each other, so that the transcriptional regulator proteins are able to interact directly with the basal transcription apparatus at the core promoter (see Figure 17.5). Some enhancers may be attracted to promoters by proteins that bind to sequences in the regulatory promoter and "tether" the enhancer close to the core promoter. Enhancers may also affect transcription by undergoing modifications that alter chromatin structure.

Many enhancers are themselves transcribed into short RNA molecules called enhancer RNAs (eRNAs). Enhancer RNA is often produced bidirectionally, which means that RNA polymerase transcribes eRNA in both directions, copying a different DNA strand in each direction. These RNA molecules do not encode proteins, but transcription of eRNAs may play a role in enhancer function. Evidence suggests that transcription of enhancers is often associated with transcription at the promoters that the enhancers affect. How transcription at an enhancer might affect transcription occurring at a distant promoter is not clear. The enhancer might recruit RNA polymerase, which might then be transferred to the promoter when the enhancer interacts with the promoter. Alternatively, transcription of the enhancer might allow the chromatin to adopt a more open configuration, which would then facilitate transcription at nearby promoters. The existence of eRNAs

might also represent nothing more than transcriptional noise: because the enhancer is physically close to a promoter, where the concentration of RNA polymerase is high, the enhancer might get transcribed by accident.

Most enhancers are capable of stimulating any promoter in their vicinity. Their effects are limited, however, by **insulators** (also called *boundary elements*), which are DNA sequences that block the effects of enhancers in a position-dependent manner. If an insulator lies between an enhancer and a promoter, it blocks the action of the enhancer, but if an insulator lies outside the region between the two, it has no effect (**Figure 17.8**). Specific proteins bind to insulators and play a role in their blocking activity. Some insulators also limit the spread of changes in chromatin structure that affect transcription. **▶ TRY PROBLEM 23**

THINK-PAIR-SHARE Question 5

CONCEPTS

Some transcriptional regulator proteins bind to enhancers, which are regulatory elements that are distant from the gene whose transcription they stimulate. Insulators are DNA sequences that block the action of enhancers.

✔ CONCEPT CHECK 3

How does the binding of transcriptional regulator proteins to enhancers affect transcription at genes that are thousands of base pairs away?

Regulation of Transcriptional Stalling and Elongation

As we've seen, transcription in eukaryotes is often regulated through factors that affect the initiation of transcription, including changes in chromatin structure, transcription factors, and transcriptional regulator proteins. Research indicates that transcription may also be controlled through factors that affect stalling and elongation by RNA polymerase after transcription has been initiated.

The basal transcription apparatus—consisting of RNA polymerase, transcription factors, and other proteins—assembles at the core promoter. When the initiation of transcription has taken place, RNA polymerase moves downstream, transcribing the structural gene and producing an RNA product. At some genes, RNA polymerase initiates transcription and transcribes 24 to 50 nucleotides of RNA, but then pauses or stalls. For example, stalling is observed at genes that encode **heat-shock proteins** in *Drosophila*—proteins that help to prevent damage by stressors such as extreme heat. Heat-shock proteins are produced by a large number of different genes. During times of environmental stress, transcription of all the heat-shock genes is greatly elevated. In the absence of stress, RNA polymerase initiates transcription at heat-shock genes in *Drosophila*, but stalls downstream of the transcription start site. Stalled polymerases are released when stress is encountered, allowing rapid transcription of the genes and the production of heat-shock proteins that facilitate adaptation to the stressful environment.

Stalling was formerly thought to take place at only a small number of genes, but research now indicates that stalling is widespread throughout eukaryotic genomes, occurring at 30%–50% of genes. Several factors that promote stalling have been identified. One of these factors is a protein called negative elongation factor (NELF), which binds to RNA polymerase and causes it to stall after initiation. Another protein, called positive transcription elongation factor b (P-TEFb), relieves stalling and promotes elongation by phosphorylating NELF and RNA polymerase, perhaps by causing NELF to dissociate from the polymerase.

> ### CONCEPTS
>
> At some genes, RNA polymerase may pause or stall downstream of the promoter. Regulatory factors affect stalling and the elongation of transcription.

Coordinated Gene Regulation

Although most eukaryotic cells do not possess operons, several eukaryotic genes may be activated by the same stimulus. Groups of bacterial genes are often coordinately expressed (turned on and off together) because they are physically clustered in an operon and share a promoter, but coordinately expressed genes in eukaryotic cells are not clustered. How, then, is the transcription of eukaryotic genes coordinated?

Genes that are coordinately expressed in eukaryotic cells are able to respond to the same stimulus because they share short regulatory sequences in their promoters or enhancers. For example, different eukaryotic heat-shock genes possess a common regulatory sequence upstream of their transcription start sites. Such regulatory sequences are called **response elements**; they are short stretches of DNA that typically contain the same consensus sequences (**Table 17.1**) at varying distances from the genes being regulated. The response elements are binding sites for transcriptional activator proteins, which bind to the response elements and elevate transcription. If the same response element is present at multiple genes, it allows all of those genes to be activated by the same stimulus.

A single eukaryotic gene may be regulated by several different response elements. For example, the metallothionein gene encodes a protein that protects cells from the toxicity of heavy metals by binding to those metals and removing them from cells. Under normal conditions, the basal transcription apparatus assembles around the TATA box just upstream of the transcription start site for the metallothionein gene, but the apparatus alone is capable of only low rates of transcription. Other response elements found upstream of the metallothionein gene contribute to increasing its rate of transcription. For example, several copies of a metal response element (MRE) lie upstream of the metallothionein gene (**Figure 17.9**). The presence of heavy metals stimulates the binding of transcriptional activator proteins to the MRE, which elevates the rate of transcription of the metallothionein gene. Because there are multiple copies of the MRE, high rates of transcription are induced by metals. Two enhancers are also located in the region upstream of the metallothionein gene. One of these enhancers contains a response element known as TRE, which stimulates transcription in the presence of a protein called AP1. A third response element, called GRE, is located approximately 250 nucleotides upstream of the metallothionein gene and stimulates transcription in response to certain hormones.

This example illustrates a common feature of eukaryotic transcriptional control: a single gene may be activated by several different response elements found in both promoters and enhancers. Multiple response elements allow the same gene to be activated by different stimuli. At the same time, the presence of the same response element in different genes allows a single stimulus to activate multiple genes. In this way, response elements allow complex biochemical responses in eukaryotic cells.

TABLE 17.1	Some response elements found in eukaryotic cells	
Response Element	**Responds to**	**Consensus Sequence***
Heat-shock element	Heat and other stresses	CNNGAANNTCCNNG
Glucocorticoid response element	Glucocorticoids	TGGTACAAATGTTCT
Phorbol ester response element	Phorbol esters	TGACTCA
Serum response element	Serum	CCATATTAGG

Source: After B. Lewin, *Genes IV* (Oxford University Press, 1994), p. 880.
*N represents any nucleotide.

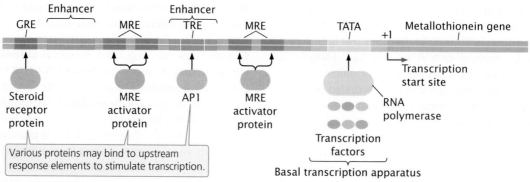

17.9 Multiple response elements are found in the region upstream of the metallothionein gene. The basal transcription apparatus assembles near the TATA box in the gene's promoter. In response to the presence of heavy metals, transcriptional activator proteins bind to several metal response elements (MREs) and stimulate transcription. Another response element, TRE, is the binding site for transcription factor AP1, which is stimulated by phorbol esters. In response to glucocorticoid hormones, steroid receptor proteins bind to a third response element, GRE, located approximately 250 nucleotides upstream of the metallothionein gene and stimulate transcription.

17.4 Some Eukaryotic Genes Are Regulated by RNA Processing and Degradation

In bacteria, transcription and translation take place simultaneously. In eukaryotes, transcription takes place in the nucleus, and the pre-mRNAs then undergo processing before being moved to the cytoplasm for translation, which allows opportunities for gene control after transcription. Consequently, posttranscriptional gene regulation assumes an important role in eukaryotic cells. RNA processing and degradation is a common level of gene regulation in eukaryotes.

Gene Regulation Through RNA Splicing

Alternative splicing allows pre-mRNA to be spliced in multiple ways, generating different proteins in different tissues or at different times in development (see Chapter 14). Many eukaryotic genes undergo alternative splicing; for example, it is estimated that 95% of all human genes with multiple exons are alternatively spliced. Thus, the regulation of splicing is an important means of controlling gene expression in eukaryotic cells.

As we saw in Section 14.2, proper splicing is dependent on the presence of consensus sequences at the 5′ splice site, the 3′ splice site, and the branch point. These consensus sequences determine the precise locations of introns and exons. Additional sequences called *exonic/intronic splicing enhancers* and *splicing silencers* help to promote or repress the use of particular splice sites during the process of RNA splicing, resulting in alternative splicing outcomes. Proteins and ribonucleoprotein particles bind to these sequences and promote or repress splice-site selection. One group of proteins involved in splice-site selection consists of the **SR** (serine- and arginine-rich) **proteins**, which have two protein domains: one domain is an RNA-binding region and the other contains alternating

serine and arginine amino acids. The precise mechanism by which SR proteins influence the choice of splice sites is poorly understood. One model suggests that SR proteins bind to splicing enhancers on the pre-mRNA and stimulate the attachment of small nuclear ribonucleoproteins (snRNPs), which then commit the site to splicing.

ALTERNATIVE SPLICING IN THE T-ANTIGEN GENE The T-antigen gene of the mammalian virus SV40 is a well-studied example of alternative splicing. This gene is capable of encoding two different proteins, the large T and small t antigens, which interact with different proteins in the host cell to ensure viral propagation. Which of the two proteins is produced depends on which of two alternative 5′ splice sites is used in RNA splicing (**Figure 17.10**). The use of one 5′ splice site produces an mRNA that encodes the large T antigen, whereas the use of the other 5′ splice site (which is farther downstream) produces an mRNA that encodes the small t antigen. An SR protein called splicing factor 2 (SF2) enhances the production of mRNA encoding the small t antigen (see Figure 17.10). Splicing factor 2 stimulates the binding of snRNPs to the 5′ splice site, one of the earliest steps in RNA splicing (see Chapter 14).

ALTERNATIVE SPLICING IN *DROSOPHILA* SEXUAL DEVELOPMENT Another example of regulation of gene expression by alternative mRNA splicing is sex determination in fruit flies. Sex differentiation in *Drosophila* arises from a cascade of gene regulation. When two X chromosomes are present, a female-specific promoter is activated early in development and stimulates the transcription of the *sex-lethal* (*Sxl*) gene (**Figure 17.11**). The protein encoded by *Sxl* regulates the splicing of the pre-mRNA transcribed from another gene called *transformer* (*tra*). The splicing pattern of *tra* pre-mRNA results in the production of the Tra protein (see Figure 17.11). Together with another protein (Tra-2), Tra stimulates the female-specific splicing of pre-mRNA from

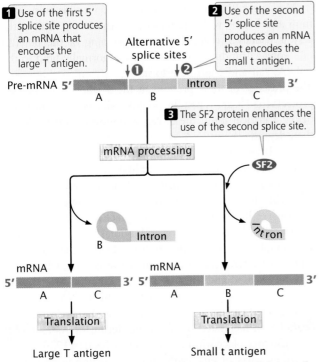

1 Use of the first 5′ splice site produces an mRNA that encodes the large T antigen.

2 Use of the second 5′ splice site produces an mRNA that encodes the small t antigen.

Alternative 5′ splice sites

Pre-mRNA 5′ — A | B | Intron | C — 3′

3 The SF2 protein enhances the use of the second splice site.

mRNA processing

SF2

Intron B

Intron

mRNA 5′ — A | C — 3′   mRNA 5′ — A | B | C — 3′

Translation

Translation

Large T antigen

Small t antigen

17.10 Alternative splicing leads to the production of the small t antigen or the large T antigen in the mammalian virus SV40.

yet another gene called *doublesex* (*dsx*). This event produces a female-specific Dsx protein, which causes the embryo to develop female characteristics.

In male embryos, which have a single X chromosome (see Figure 17.11), the promoter of the *Sxl* gene is inactive, so the

gene is not transcribed, and no Sxl protein is produced. In the absence of Sxl protein, *tra* pre-mRNA is spliced at a different 3′ splice site to produce a nonfunctional form of Tra protein (**Figure 17.12**). In turn, the presence of this nonfunctional Tra in males causes *dsx* pre-mRNAs to be spliced differently from those in females, and a male-specific Dsx protein is produced (see Figure 17.11). This event causes the development of male-specific traits.

In summary, the Tra, Tra-2, and Sxl proteins regulate alternative splicing patterns that produce male and female phenotypes in *Drosophila*. Exactly how these proteins regulate alternative splicing is not yet known, but the Sxl protein (produced only in females) might possibly block the upstream splice site on the *tra* pre-mRNA. This blockage would force the spliceosome to use the downstream 3′ splice site, which causes the production of Tra protein and eventually results in female traits (see Figure 17.12). **▶ TRY PROBLEM 24**

THINK-PAIR-SHARE Question 6

> **CONCEPTS**
>
> Eukaryotic genes can be regulated through the control of mRNA processing. The selection of alternative splice sites leads to the production of different proteins.

The Degradation of RNA

The amount of a protein that is synthesized depends on the amount of the corresponding mRNA that is available for translation. The amount of available mRNA, in turn, depends on both the rate of mRNA synthesis and the rate of

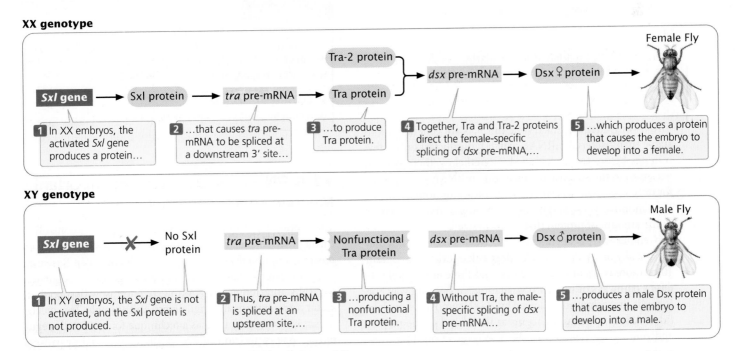

XX genotype

Female Fly

Sxl gene → Sxl protein → *tra* pre-mRNA → Tra protein

Tra-2 protein

dsx pre-mRNA → Dsx ♀ protein →

1 In XX embryos, the activated *Sxl* gene produces a protein...

2 ...that causes *tra* pre-mRNA to be spliced at a downstream 3′ site...

3 ...to produce Tra protein.

4 Together, Tra and Tra-2 proteins direct the female-specific splicing of *dsx* pre-mRNA,...

5 ...which produces a protein that causes the embryo to develop into a female.

XY genotype

Male Fly

Sxl gene → ✗ No Sxl protein   *tra* pre-mRNA → Nonfunctional Tra protein   *dsx* pre-mRNA → Dsx ♂ protein →

1 In XY embryos, the *Sxl* gene is not activated, and the Sxl protein is not produced.

2 Thus, *tra* pre-mRNA is spliced at an upstream site,...

3 ...producing a nonfunctional Tra protein.

4 Without Tra, the male-specific splicing of *dsx* pre-mRNA...

5 ...produces a male Dsx protein that causes the embryo to develop into a male.

17.11 Alternative splicing controls sex determination in *Drosophila*.

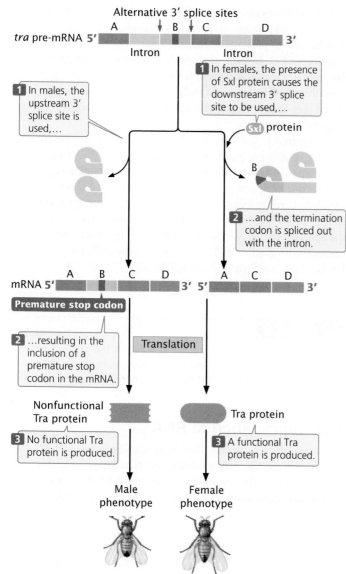

17.12 Alternative splicing of *tra* pre-mRNA. Two alternative 3′ splice sites are present.

mRNA degradation. Eukaryotic mRNAs are generally more stable than bacterial mRNAs, which typically last only a few minutes before being degraded. Nonetheless, there is great variation in the stability of eukaryotic mRNAs: some persist for only a few minutes, whereas others last for hours, days, or even months. These variations can produce large differences in the amount of protein that is synthesized.

Cellular RNA is degraded by ribonucleases, enzymes that specifically break down RNA. Most eukaryotic cells contain 10 or more types of ribonucleases, and there are several different pathways of mRNA degradation. In one pathway, the 5′ cap is first removed, followed by 5′→3′ removal of nucleotides. A second pathway begins at the 3′ end of the mRNA and removes nucleotides in the 3′→5′ direction. In a third pathway, the mRNA is cleaved at internal sites.

Messenger RNA degradation from the 5′ end is most common and begins with the removal of the 5′ cap. This pathway is usually preceded by the shortening of the poly(A) tail. Poly(A)-binding proteins (PABPs) normally bind to the poly(A) tail and contribute to its stability-enhancing effect. The presence of these proteins at the 3′ end of the mRNA protects the 5′ cap. When the poly(A) tail has been shortened below a critical limit, the 5′ cap is removed, and nucleases then degrade the mRNA by removing nucleotides from the 5′ end. These observations suggest that the 5′ cap and the 3′ poly(A) tail of eukaryotic mRNA physically interact with each other, most likely by the poly(A) tail bending around so that the PABPs make contact with the 5′ cap (see Figure 15.18).

Much of RNA degradation takes place in specialized complexes called P bodies. However, P bodies appear to be more than simply destruction sites for RNA. Evidence suggests that P bodies can temporarily store mRNA molecules, which may later be released and translated. Thus, P bodies help control the expression of genes by regulating which RNA molecules are degraded and which are sequestered for later release. RNA degradation facilitated by small interfering RNAs (siRNAs) may also take place within P bodies (see Section 17.5).

Other parts of eukaryotic mRNA, including sequences in the 5′ untranslated region (5′ UTR), the coding region, and the 3′ UTR, also affect mRNA stability. Some short-lived eukaryotic mRNAs have one or more copies of the consensus sequence 5′–AUUUAUAA–3′, referred to as the AU-rich element, in the 3′ UTR. Messenger RNAs containing AU-rich elements are degraded by a mechanism in which microRNAs take part (see Section 17.5). ▶ **TRY PROBLEM 26**

> **CONCEPTS**
>
> The stability of mRNA influences gene expression by affecting the amount of mRNA available to be translated. The stability of mRNA is affected by the 5′ cap, the poly(A) tail, the 5′ UTR, the coding region, and sequences in the 3′ UTR.
>
> ✔ **CONCEPT CHECK 4**
> How does the poly(A) tail affect mRNA stability?

17.5 RNA Interference Is an Important Mechanism of Gene Regulation

The expression of a number of eukaryotic genes is controlled through RNA interference, also known as *RNA silencing* or *posttranscriptional gene silencing* (see Chapter 14). Research suggests that as many as 30% of human genes are regulated by RNA interference. RNA interference is widespread in eukaryotes, existing in fungi, plants, and animals. This mechanism is also widely used as a technique for artificially regulating gene expression in genetically engineered organisms (see Chapter 19).

(a)

Double-stranded RNA

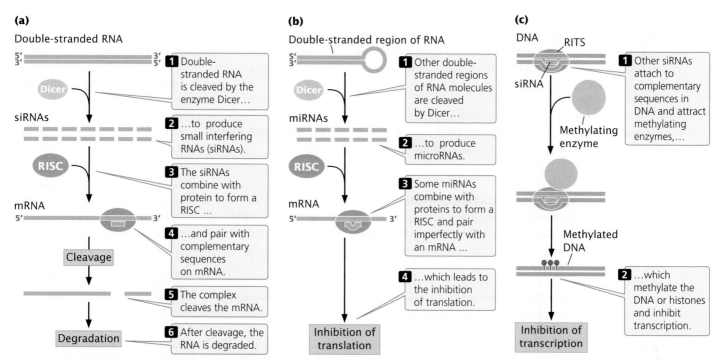

1 Double-stranded RNA is cleaved by the enzyme Dicer...

2 ...to produce small interfering RNAs (siRNAs).

3 The siRNAs combine with protein to form a RISC ...

4 ...and pair with complementary sequences on mRNA.

5 The complex cleaves the mRNA.

6 After cleavage, the RNA is degraded.

(b)

Double-stranded region of RNA

1 Other double-stranded regions of RNA molecules are cleaved by Dicer...

2 ...to produce microRNAs.

3 Some miRNAs combine with proteins to form a RISC and pair imperfectly with an mRNA ...

4 ...which leads to the inhibition of translation.

(c)

1 Other siRNAs attach to complementary sequences in DNA and attract methylating enzymes,...

2 ...which methylate the DNA or histones and inhibit transcription.

17.13 RNA silencing leads to the degradation of mRNA or to the inhibition of translation or transcription. (a) Small interfering RNAs (siRNAs) degrade mRNA by cleavage. (b) MicroRNAs (miRNAs) lead to the inhibition of translation. (c) Some siRNAs bring about methylation of histone proteins or DNA, inhibiting transcription.

Small Interfering RNAs and MicroRNAs

RNA interference is triggered by very small RNA molecules known as microRNAs (miRNAs) and small interfering RNAs (siRNAs), depending on their origin and mode of action (see Chapter 14). An enzyme called Dicer cleaves and processes double-stranded RNA to produce single-stranded siRNAs or miRNAs 21–25 nucleotides long (**Figure 17.13**), which combine with proteins to form an RNA-induced silencing complex (RISC). The RNA component of RISC then pairs with complementary base sequences in specific mRNA molecules, most often with sequences in the 3′ UTR of the mRNA. Small interfering RNAs tend to base pair perfectly with mRNAs, whereas microRNAs often form less-than-perfect pairings.

Mechanisms of Gene Regulation by RNA Interference

Small interfering RNAs and microRNAs regulate gene expression through at least four distinct mechanisms: (1) cleavage of mRNA, (2) inhibition of translation, (3) transcriptional silencing, and (4) degradation of mRNA.

RNA CLEAVAGE RISCs that contain an siRNA (and some that contain an miRNA) pair with an mRNA molecule and cleave the mRNA near the middle of the bound siRNA

(see Figure 17.13a). This cleavage is carried out by a protein that is sometimes referred to as Slicer. After cleavage, the mRNA is further degraded. Thus, the presence of siRNAs and miRNAs increases the rate at which mRNAs are broken down and decreases the amount of protein produced.

INHIBITION OF TRANSLATION Some miRNAs regulate genes by inhibiting the translation of complementary mRNAs (see Figure 17.13b). For example, an important gene in flower development in *Arabidopsis thaliana* is *APETALA2*. The expression of this gene is regulated by an miRNA that base pairs with nucleotides in the coding region of *APETALA2* mRNA and inhibits its translation.

The exact mechanism by which miRNAs repress translation is still poorly understood, but some research suggests that they can inhibit the initiation step of translation as well as steps after initiation, such as ribosome stalling or premature termination. Many mRNAs have multiple miRNA-binding sites, and translation is most efficiently inhibited when multiple miRNAs are bound to the mRNA.

TRANSCRIPTIONAL SILENCING Some siRNAs silence transcription by altering chromatin structure. These siRNAs combine with proteins to form a complex called RITS (for *R*NA-*i*nduced *t*ranscriptional *s*ilencing; see Figure 17.13c), which is analogous to RISC. The siRNA component of RITS then binds to a complementary sequence in DNA or in an

RNA molecule in the process of being transcribed, where it represses transcription by attracting enzymes that methylate the tails of histone proteins. The addition of methyl groups to the histones causes them to bind DNA more tightly, restricting the access of the proteins and enzymes necessary to carry out transcription (see Histone Modification in Section 17.2). Some miRNAs bind to complementary sequences in DNA and attract enzymes that methylate the DNA directly, which also leads to the suppression of transcription (see DNA Methylation in Section 17.2).

SLICER-INDEPENDENT DEGRADATION OF mRNA A final mechanism by which miRNAs regulate gene expression is by triggering the decay of mRNA through a process that does not require Slicer activity. This mechanism plays a role in the degradation of short-lived mRNAs that contain an AU-rich element in their 3′ UTR. AU-rich elements are found, for example, in mRNAs that encode cytokines, proteins that play an important role in the body's response to infection and injury. Overproduction of cytokines, or their production at the wrong time, can lead to shock and autoimmune disease, so precise regulation of cytokine synthesis is critical. Researchers have identified an miRNA with a sequence that is complementary to the AU-rich element. This miRNA binds to the AU-rich element and, in a way that is not yet fully understood, brings about the degradation of the mRNA in a process that requires Dicer and RISC.

The Control of Development by RNA Interference

Much of development in multicellular eukaryotes is controlled through gene regulation: different genes are turned on and off at specific times (see Chapter 22). In fact, when miRNAs were first discovered, researchers noticed that a mutation in an miRNA in *C. elegans* caused a developmental defect. Recent research has demonstrated that miRNA molecules are key factors in controlling development in animals, including humans, and plants. For example, the vertebrate heart develops through the programmed differentiation and proliferation of cardiac muscle cells, which are controlled by a specific miRNA termed miR-1-1.

Recent studies demonstrate that, through their effects on gene expression, miRNAs play important roles in many diseases and disorders. For example, a genetic form of hearing loss has been associated with a mutation in the gene that encodes an miRNA. Other miRNAs are associated with heart disease. One miRNA, called miR-1-2, is highly expressed in heart muscle. Mice genetically engineered to express only 50% of the normal amount of miR-1-2 frequently have holes in the wall that separates the left and right ventricles, a common congenital heart defect seen in newborn humans. Overexpression of another miRNA, called miR-1, in the hearts of adult mice causes cardiac arrhythmia—irregular electrical activity of the heart that can be life-threatening in humans. And finally, numerous studies have demonstrated that abnormal expression of miRNAs plays a role in many cancers; miRNAs are now being used for cancer prognosis and experimental treatment.

RNA Crosstalk

Recent research has identified an additional layer of gene regulation—called RNA crosstalk—that arises from interactions and competition among miRNAs, mRNAs, and other RNA molecules. These interactions occur because different RNA molecules often share binding sites for the same miRNA. If the miRNA is in limited supply, then the RNAs with shared miRNA binding sites compete with one another for the limited copies of the miRNA.

The RNA molecules that compete for miRNAs include some noncoding RNAs. In Chapter 14, we discussed long noncoding RNAs (lncRNAs), which are RNA molecules over 200 nucleotides long that do not encode a protein. Some lncRNAs have multiple binding sites for miRNAs that are shared with mRNAs. The lncRNAs serve as molecular decoys, attracting the miRNAs and tying them up so that they are not available to bind to mRNA. Some recently discovered circular RNAs (see Chapter 14) also have multiple binding sites for miRNAs and similarly serve as decoys for miRNAs.

The same type of competition can occur among RNA molecules that share binding sites for proteins that function in RNA splicing, stability, and translation. RNA molecules that engage in this type of crosstalk are collectively referred to as competing endogenous RNAs (ceRNAs). Numerous examples of ceRNAs have been discovered, and RNA crosstalk may be an important component of gene regulation.

CONCEPTS

RNA interference is initiated by double-stranded RNA molecules that are cleaved and processed. The resulting siRNAs or miRNAs combine with proteins to form complexes that bind to complementary sequences in mRNA or DNA. The siRNAs and miRNAs affect gene expression by cleaving mRNA, inhibiting translation, altering chromatin structure, or triggering RNA degradation. Different RNA molecules that share binding sites for miRNAs may compete among themselves for available miRNAs, creating an additional layer of gene regulation known as RNA crosstalk.

✔ CONCEPT CHECK 5

In RNA silencing, siRNAs and miRNAs usually bind to which part of the mRNA molecules that they control?

a. 5′ UTR
b. Coding region
c. 3′ poly(A) tail
d. 3′ UTR

17.6 The Expression of Some Genes Is Regulated by Processes That Affect Translation or by Modifications of Proteins

Ribosomes, charged tRNAs, initiation factors, and elongation factors are all required for the translation of mRNA molecules (see Chapter 15). The availability of these components affects the rate of translation and therefore influences gene expression. For example, the activation of T lymphocytes (T cells) is critical to the development of immune responses to viruses (see Chapter 22). T cells are normally in the G_0 stage of the cell cycle and not actively dividing. Upon exposure to viral antigens, however, specific T cells become activated and undergo rapid proliferation (**Figure 17.14**). Activation begins with a 7- to 10-fold increase in protein synthesis that causes the cells to enter the cell cycle and proliferate. This global burst of protein synthesis does not require an increase in mRNA synthesis. Instead, it is due to the increased availability of initiation factors, which allow ribosomes to bind to mRNA and begin translation. This increase in initiation factors leads to more translation of the existing mRNA molecules, increasing the overall amount of protein synthesized. In a similar way, insulin stimulates protein synthesis by increasing the availability of initiation factors. Initiation factors exist in inactive forms and, in response to various cell signals, can be activated by chemical modifications of their structure, such as phosphorylation.

In addition to increasing overall protein synthesis, translational control can also target the synthesis of specific proteins. For example, research has demonstrated that eukaryotic initiation factor 3 (eIF3), which functions generally to initiate eukaryotic translation, can also bind to secondary structures in the 5′ untranslated regions of specific mRNAs and either stimulate or repress their translation, depending on the particular sequence to which it binds. The initiation of translation in some mRNAs is regulated by proteins that bind to the 5′ UTR of certain mRNAs and inhibit the binding of ribosomes, in a manner similar to the binding of repressor proteins to operators that prevents the transcription of structural genes in prokaryotes. The translation of other mRNAs is affected by the binding of proteins to sequences in the 3′ UTR.

As we saw in Section 17.3, the presence of extreme heat or other stressors induces the synthesis of heat-shock proteins, which reduce cellular damage resulting from those stressors. Transcription of heat-shock genes is rapidly increased by the release of stalled RNA polymerases when stress is encountered. Translational control also plays a role in the induction of heat-shock proteins. In response to heat shock, adenosine nucleotides in the 5′ UTRs of mRNAs transcribed from heat-shock genes are preferentially methylated to form N^6-methyladenosine. The presence of N^6-methyladenosine then promotes cap-independent initiation of translation (see p. 444 in Chapter 15), leading to the synthesis of heat-shock proteins.

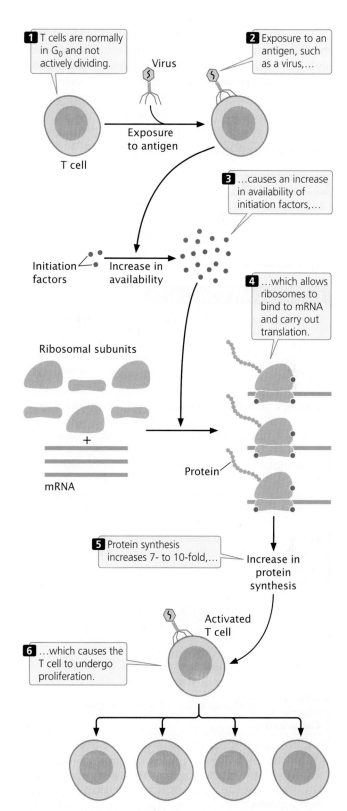

17.14 The expression of some eukaryotic genes is regulated by the availability of components required for translation. In this example, exposure to an antigen stimulates increased availability of initiation factors and a subsequent increase in protein synthesis, leading to T-cell proliferation.

Many eukaryotic proteins are extensively modified after translation by the selective cleavage and trimming of amino acids from the ends, by acetylation, or by the addition of phosphate groups, carboxyl groups, methyl groups, carbohydrates, or ubiquitin (a small protein) (see Chapter 15). Control of these modifications, which affect the transport, function, stability, and activity of the proteins, may also play a role in gene expression.

THINK-PAIR-SHARE Questions 7 and 8

CONCEPTS

The availability of ribosomes, charged tRNAs, and initiation and elongation factors may affect the rate of translation. Translation of some mRNAs is regulated by proteins that bind to the 5′ and 3′ untranslated regions of the mRNA.

CONNECTING CONCEPTS

A Comparison of Bacterial and Eukaryotic Gene Control

Now that we have considered the major types of gene regulation in bacteria (in Chapter 16) and eukaryotes (in this chapter), let's consider some of the similarities and differences in bacterial and eukaryotic gene control.

1. Much of gene regulation in bacterial cells is at the level of transcription (although it does exist at other levels). Gene regulation in eukaryotic cells takes place at multiple levels, including chromatin structure, transcription, mRNA processing, translation, RNA stability, and posttranslational modification of proteins.

2. Complex biochemical and developmental events in bacterial and eukaryotic cells may require a cascade of gene regulation, in which the activation of one set of genes stimulates the activation of another set.

3. Much of gene regulation in both bacterial and eukaryotic cells is accomplished through proteins that bind to specific sequences in DNA. Regulatory proteins come in a variety of types, but most can be characterized according to a small set of DNA-binding motifs.

4. Chromatin structure plays a role in eukaryotic (but not bacterial) gene regulation. In general, condensed chromatin represses gene expression; chromatin structure must therefore be altered before transcription can take place. Chromatin structure is altered by chromatin-remodeling complexes, modification of histone proteins, and DNA methylation.

5. Modifications to chromatin structure in eukaryotes may lead to epigenetic changes, which are changes that affect gene expression and are passed on to other cells or future generations.

6. In bacterial cells, many genes are clustered in operons and are coordinately expressed by transcription into a single mRNA molecule. In contrast, many eukaryotic genes have their own promoters and are transcribed independently of other genes. Coordinated gene regulation in eukaryotic cells often takes place through common response elements located in the promoters and enhancers of genes. Different genes that share the same response element are influenced by the same regulatory protein.

7. Regulator proteins that affect transcription exhibit two basic types of control: *repressors* inhibit transcription (negative control) and *activators* stimulate transcription (positive control). Both negative control and positive control are found in bacterial and eukaryotic cells.

8. The initiation of transcription is a relatively simple process in bacterial cells, and regulator proteins function by blocking or stimulating the binding of RNA polymerase to DNA. In contrast, eukaryotic transcription requires complex machinery that includes RNA polymerase, general transcription factors, and transcriptional activators and repressors, allowing transcription to be influenced by multiple factors.

9. Some eukaryotic transcriptional regulator proteins function at a distance from the gene by binding to enhancers, causing the formation of a loop in the DNA, which brings the promoter and enhancer into close proximity. Some sequences analogous to enhancers that act on distant genes have been described in bacterial cells, but they appear to be less common.

10. The greater time lag between transcription and translation in eukaryotic cells than in bacterial cells allows mRNA stability and mRNA processing to play larger roles in eukaryotic gene regulation.

11. Complementary RNA molecules (antisense RNA) may act as regulators of gene expression in bacteria. Regulation by siRNAs and miRNAs, which is extensive in eukaryotes, is absent from bacterial cells.

These similarities and differences in gene regulation in bacteria and eukaryotes are summarized in **Table 17.2**.

TABLE 17.2	Comparison of gene control in bacteria and eukaryotes	
Characteristic	**Bacterial Gene Control**	**Eukaryotic Gene Control**
Levels of regulation	Primarily transcription	Many levels
Cascades of gene regulation	Present	Present
DNA-binding proteins	Important	Important
Role of chromatin structure	Absent	Important
Presence of operons	Common	Uncommon
Negative and positive control	Present	Present
Initiation of transcription	Relatively simple	Relatively complex
Enhancers	Less common	More common
Transcription and translation	Occur simultaneously	Occur separately
Regulation by small RNAs	Rare	Common

CONCEPTS SUMMARY

- Eukaryotic cells differ from bacteria in several ways that affect gene regulation, including, in eukaryotes, the absence of operons, the presence of chromatin, and the presence of a nuclear membrane.

- In eukaryotic cells, chromatin structure represses gene expression. In transcription, chromatin structure may be altered by chromatin-remodeling complexes that reposition nucleosomes and by modifications of histone proteins, including acetylation, phosphorylation, and methylation. The methylation of DNA also affects transcription.

- The initiation of eukaryotic transcription is controlled by general transcription factors that are part of the basal transcription apparatus and by transcriptional regulator proteins that stimulate or repress normal levels of transcription by binding to regulatory promoters and enhancers.

- Enhancers affect the transcription of distant genes. Transcriptional regulator proteins bind to enhancers and interact with the basal transcription apparatus by causing the intervening DNA to loop out.

- DNA sequences called insulators limit the action of enhancers by blocking their action in a position-dependent manner.

- Some regulatory factors cause RNA polymerase to stall downstream of the promoter.

- Coordinately controlled genes in eukaryotic cells respond to the same factors because they share response elements that are stimulated by the same transcriptional activator.

- Gene expression in eukaryotic cells can be influenced by RNA processing.

- Gene expression can be regulated by changes in RNA stability. The 5′ cap, the poly(A) tail, the 5′ UTR, the coding region, and sequences in the 3′ UTR are important in controlling the stability of eukaryotic mRNAs.

- Proteins that bind to the 5′ and 3′ ends of eukaryotic mRNA can affect its translation.

- RNA interference plays an important role in eukaryotic gene regulation. Small RNA molecules (siRNAs and miRNAs) cleaved from double-stranded RNA combine with proteins and bind to complementary sequences on mRNA or DNA. These complexes cleave RNA, inhibit translation, silence transcription, or affect RNA degradation. RNA molecules with shared miRNA binding sites may compete for available copies of miRNAs, creating an added level of gene regulation.

- Control of the posttranslational modification of proteins may play a role in gene expression.

IMPORTANT TERMS

DNase I hypersensitive site (p. 493)
chromatin-remodeling complex (p. 493)
histone code (p. 494)
CpG island (p. 497)
mediator (p. 498)
insulator (p. 500)
heat-shock protein (p. 501)
response element (p. 501)
SR protein (p. 502)

ANSWERS TO CONCEPT CHECKS

1. Three of these processes are chromatin remodeling, the modification of histone proteins (e.g., methylation and acetylation of histones), and DNA methylation.

2. b

3. The DNA between the enhancer and the promoter loops out, so that regulatory proteins bound to the enhancer are able to interact directly with the basal transcription apparatus.

4. The poly(A) tail stabilizes the 5′ cap, which must be removed before the mRNA molecule can be degraded from the 5′ end.

5. d

WORKED PROBLEM

What would be the effect of a mutation that causes a poly(A)-binding protein to be nonfunctional?

›› Solution Strategy

What information is required in your answer to the problem?

The effect of a mutation that eliminates the function of a poly(A)-binding protein.

What information is provided to solve the problem?

- A mutation occurs in a gene that encodes a poly(A)-binding protein (PABP).
- The mutation causes the PABP to be nonfunctional.

For help with this problem, review:

The Degradation of RNA in Section 17.4.
Addition of the Poly(A) Tail in Section 14.2.

›› Solution Steps

Messenger RNA can be degraded from the 5′ end, from the 3′ end, or through internal cleavage. Degradation from the 5′ end requires the removal of the 5′ cap and is usually preceded by the shortening of the poly(A) tail. Poly(A)-binding proteins bind to the poly(A) tail and prevent it from being shortened. Thus, the presence of these proteins on the poly(A) tail protects the 5′ cap, which prevents RNA degradation. If the gene for a poly(A)-binding protein were mutated in such a way that nonfunctional PABP was produced, the protein would not bind to the poly(A) tail of the mRNAs with which it interacts. The tail would be shortened prematurely, the 5′ cap removed, and the mRNA degraded more easily. The end result would be less mRNA and thus less protein synthesis.

> **Recall:** The poly(A) tail affects the stability of mRNA.

COMPREHENSION QUESTIONS

Introduction

1. How similar are the genomes of humans and chimpanzees? What genetic changes might be responsible for the large differences in the anatomy, physiology, and behavior of humans and chimpanzees?

Section 17.1

2. List some important differences between bacterial and eukaryotic cells that affect the way in which genes are regulated.

Section 17.2

3. Where are DNase I hypersensitivity sites found, and what do they indicate about the nature of chromatin?

4. What changes take place in chromatin structure, and what role do these changes play in eukaryotic gene regulation?

5. What is the histone code?

6. How is chromatin immunoprecipitation used to determine the locations of histone modifications in the genome?

Section 17.3

7. Briefly explain how transcriptional activator and repressor proteins affect the level of transcription of eukaryotic genes.

8. What is an enhancer? How does it affect the transcription of distant genes?

9. What is an insulator?

10. What is a response element? How do response elements bring about the coordinated expression of eukaryotic genes?

Section 17.4

11. Outline the role of alternative splicing in the control of sex differentiation in *Drosophila*.

12. What role does RNA stability play in gene regulation? What controls RNA stability in eukaryotic cells?

Section 17.5

13. Briefly list some of the ways in which siRNAs and miRNAs regulate genes.

Section 17.6

14. How does bacterial gene regulation differ from eukaryotic gene regulation? How are they similar?

APPLICATION QUESTIONS AND PROBLEMS

Section 17.2

15. Malaria, one of the most pervasive and destructive of all infectious diseases, is caused by protozoan parasites of the genus *Plasmodium*, which are transmitted from person to person by mosquitoes. *Plasmodium* parasites are able to evade the host immune system by constantly altering the expression of their *var* genes, which encode *Plasmodium* surface antigens (L. H. Freitas-Junior et al. 2005. *Cell* 121:25–36). Individual *var* genes are expressed when chromatin structure is disrupted by chemical changes in histone proteins. What type of chemical changes in the histone proteins might be responsible for these changes in gene expression?

16. A geneticist is trying to determine how many genes are found in a 300,000-bp region of DNA. Analysis shows that four different areas within the 300,000-bp region have H3K4me3 modifications. What might their presence suggest about the number of genes located there?

17. In a line of human cells grown in culture, a geneticist isolates a temperature-sensitive mutation at a locus that encodes an acetyltransferase enzyme; at temperatures above 38°C, the mutant cells produce a nonfunctional form of the enzyme. What would be the most likely effect of this mutation if the cells were grown at 40°C?

*__18.__ What would be the most likely effect of deleting *flowering locus D* (*FLD*) in *Arabidopsis thaliana*?

*__19.__ X31b is an experimental compound that is taken up by rapidly dividing cells. Research has shown that X31b stimulates the methylation of DNA. Some cancer researchers are interested in testing X31b as a possible drug for treating prostate cancer. Offer a possible explanation for why X31b might be an effective anticancer drug.

Section 17.3

20. How do repressors that bind to silencers in eukaryotes differ from repressors that bind to operators in bacteria?

21. Examine **Figure 17.7**. What would be the effect on transcription if a mutation occurred in the gene that encodes GAL3, so that no functional GAL3 was produced?

22. What would be the effect of moving the insulator shown in **Figure 17.8** to a position between enhancer II and the promoter for gene B?

*__23.__ An enhancer is surrounded by four genes (*A*, *B*, *C*, and *D*), as shown in the accompanying diagram. An insulator lies between gene *C* and gene *D*. On the basis of the positions of the genes, the enhancer, and the insulator, the transcription of which genes is most likely to be stimulated by the enhancer? Explain your reasoning.

Section 17.4

*__24.__ What will be the effect on sexual development in newly fertilized *Drosophila* embryos if the following genes are deleted?

a. *sex-lethal*

b. *transformer*

c. *doublesex*

25. Examine **Figure 17.12**. What would be the effect of a mutation that eliminated the downstream 3′ splice site at the end of exon B in the *tra* pre-mRNA?

*__26.__ Some eukaryotic mRNAs have an AU-rich element in the 3′ untranslated region. What would be the effect on gene expression if this element were mutated or deleted?

27. A strain of *Arabidopsis thaliana* possesses a mutation in the *APETALA2* gene. As a result of this mutation, much of the 3′ UTR of the mRNA transcribed from the gene is deleted. What is the most likely effect of this mutation on the expression of the *APETALA2* gene?

Section 17.5

28. Suppose a geneticist introduced a small interfering RNA (siRNA) that was complementary to the *FLC* mRNA in **Figure 17.3**. What would be the effect on flowering of *Arabidopsis*? Explain your answer.

CHALLENGE QUESTIONS

Section 17.3

29. The yeast gene *SER3*, whose product has a role in serine biosynthesis, is repressed during growth in nutrient-rich medium, so little transcription takes place, and little SER3 enzyme is produced, under these conditions. In an investigation of the repression of the

SER3 gene, a region of DNA upstream of *SER3* was found to be heavily transcribed when *SER3* is repressed (J. A. Martens, L. Laprade, and F. Winston. 2004. *Nature* 429:571–574). Within this upstream region is a promoter that stimulates the transcription of an RNA molecule called *SRG1* RNA (for *SER3* regulatory gene 1). This RNA molecule has none of the sequences necessary for translation. Mutations in the promoter for *SRG1* result in the disappearance of *SRG1* RNA, and these mutations remove the repression of *SER3*. When RNA polymerase binds to the *SRG1* promoter, the polymerase travels downstream, transcribing the *SGR1* RNA, and passes through and transcribes the promoter for *SER3*. This activity leads to the repression of *SER3*. Propose a possible explanation for how the transcription of *SGR1* might repress the transcription of *SER3*. (Hint: Remember that the *SGR1* RNA does not encode a protein.)

Section 17.5

30. A common feature of many eukaryotic mRNAs is the presence of a rather long 3′ UTR, which often contains consensus sequences. Creatine kinase B (CK-B) is an important enzyme in cellular metabolism. Certain cells—termed U937D cells—have lots of CK-B mRNA, but no CK-B enzyme is present. In these cells, the 5′ end of the CK-B mRNA is bound to ribosomes, but the mRNA is apparently not translated. Something inhibits the translation of the CK-B mRNA in these cells.

Researchers introduced numerous short segments of RNA containing only 3′ UTR sequences into U937D cells. As a result, the U937D cells began to synthesize the CK-B enzyme, but the total amount of CK-B mRNA did not increase. The introduction of short segments of other RNA sequences did not stimulate the synthesis of CK-B; only the 3′ UTR sequences turned on the translation of the enzyme.

On the basis of these results, propose a mechanism for the inhibition of CK-B translation in the U937D cells. Explain how the introduction of short segments of RNA containing the 3′ UTR sequences might remove that inhibition.

THINK-PAIR-SHARE QUESTIONS

Section 17.2

1. Recent research has shown that activation of a topoisomerase enzyme (see Chapter 11) leads to greater expression of some genes in neurons. These genes remain silent until topoisomerase is activated; they then are rapidly transcribed. Propose a mechanism for how topoisomerase might stimulate gene expression.

2. Mutations in genes encoding chromatin-remodeling complexes have been identified in high frequencies in some human cancers. For example, mutations in components of the human SWI–SNF remodeling complex were found in 19% of tumors in a group of cancers. How might a mutation in a chromatin-remodeling complex contribute to cancer?

Section 17.3

3. What do you think would be the overall effect of reducing the amount of mediator present in a cell?

4. Some DNA nucleotides are located within the coding regions of genes, where they specify the amino acid sequence of a protein. Other nucleotides are found in regulatory elements, where they serve as binding sites for regulatory proteins, such as general transcription factors, transcriptional activators, and transcriptional repressors. It has long been assumed that sequences within coding regions and sequences within regulatory elements are independent, but recent research has determined that about 15% of human codons—dubbed duons—serve both to encode amino acids and as binding sites for regulatory proteins. How might the presence of duons affect which synonymous codons are used in the genetic code?

5. Mammals are anatomically and physiologically more complex than roundworms, yet both organisms have approximately the same number of genes, about 20,000. Some biologists have argued that mammals and other vertebrates have evolved increased complexity by means of pleiotropy—in which each gene encodes multiple characteristics—and that pleiotropy was made possible by the evolution of additional enhancers. Propose an explanation for how additional enhancers might produce increased pleiotropy.

Section 17.4

6. Research has demonstrated that the composition of a gene's promoter can affect alternative splicing of its RNA transcript. For example, promoters that are activated by certain transcriptional activator proteins cause one form of alternative splicing; when these activators are not present, a different form of splicing takes place. Propose some possible mechanisms by which the promoter could affect alternative splicing.

Section 17.6

7. As discussed in Chapter 15, some eukaryotic mRNAs have internal ribosome binding sites, where ribosomes can bind without first attaching to the 5′ cap. Yet all eukaryotic mRNAs are capped, and cap-dependent translation is highly efficient. What might be the advantage of having internal ribosome binding sites if the cap permits the initiation of translation and all eukaryotic mRNAs are capped?

8. Some research suggests that translation of proteins is regulated by ribosome stalling: ribosomes attach to mRNA and then stall; elongation commences only in response to a specific stimulus. What might be an advantage to gene regulation by ribosome stalling?

Gene Mutations and DNA Repair

Lou Gehrig and Expanding Nucleotide Repeats

Lou Gehrig at bat. Gehrig, who played baseball for the New York Yankees from 1923 to 1939, was diagnosed with amyotrophic lateral sclerosis, a disease that in some people is caused by an expanding nucleotide repeat mutation. [Transcendental Graphics/Getty Images.]

Lou Gehrig was the finest first baseman ever to play major league baseball. A left-handed power hitter who grew up in New York City, Gehrig played for the New York Yankees from 1923 to 1939. Throughout his career, he lived in the shadow of his teammates Babe Ruth and Joe DiMaggio, but Gehrig was a great hitter in his own right: he compiled a lifetime batting average of .340 and drove in more than 100 runs every season for 13 years. During his career, he batted in 1991 runs and hit a total of 23 grand slams (home runs with bases loaded). But Gehrig's greatest baseball record, which stood for more than 50 years and has been broken only once—by Cal Ripken, Jr., in 1995—is his record of playing 2130 consecutive games.

In the 1938 baseball season, Gehrig fell into a strange slump. For the first time since his rookie year, his batting average dropped below .300, and in the World Series that year, he managed only four hits—all singles. Nevertheless, he finished the season convinced that he was undergoing a temporary slump that he would overcome in the next season. He returned to training camp in 1939 with high spirits. When the season began, however, it was clear to everyone that something was terribly wrong. Gehrig had no power in his swing; he was awkward and clumsy at first base. His condition worsened, and on May 2, he voluntarily removed himself from the lineup. The Yankees sent Gehrig to the Mayo Clinic for diagnosis. On June 20, his medical report was made public: Lou Gehrig was suffering from a rare, progressive disease known as amyotrophic lateral sclerosis (ALS). Within two years, he was dead. Since then, ALS has commonly been known as Lou Gehrig disease.

Gehrig experienced symptoms typical of ALS: progressive weakness and wasting of skeletal muscles due to degeneration of the motor neurons. Most cases of ALS are sporadic, appearing in people with no family history of the disease. However, about 10% of cases run in families, and in these cases the disease is inherited as an autosomal dominant trait. ALS shares a number of features in common with another neurological disease called frontotemporal dementia (FTD); in fact, FTD occurs alongside ALS in some families, suggesting a common genetic basis underlying these two disorders.

Mutations in several genes can cause familial cases of ALS and FTD, the most common of which occur in a gene on chromosome 9 called *chromosome 9 open reading frame 72* (*C9orf72*). The alterations of *C9orf72* that are associated with ALS and FTD belong to an unusual group of mutations called expanding nucleotide repeats, in which the number of copies of a set of nucleotides is increased. Most people have somewhere between 2 and 23 repeats of the nucleotide sequence GGGGCC in their *C9orf72* gene, but this number is massively expanded in some people with ALS, who typically possess 700 to 1600 repeats of the sequence.

How the expansion of the GGGGCC repeat in *C9orf72* leads to symptoms of ALS and FTD is unknown, but recent research demonstrates that the repeats are translated into one or more proteins that are toxic to nerve cells. The repeats are translated in an unusual and intriguing way: the GGGGCC sequences on both the template and nontemplate strands of the gene are transcribed into RNAs that are translated without a start codon. Because there is no start codon to set the reading frame, all three reading frames on both mRNAs are translated into proteins, resulting in five proteins, each with a different series of repeating dipeptides: glycine-alanine, glycine-proline, proline-alanine, glycine-arginine, and proline-arginine.

To determine how the repeats might produce the disease, geneticists engineered a series of premature stop codons into the template and nontemplate strands of the *C9orf72* gene, so that RNA would be transcribed from the repeats but, because of the engineered stop codons, would not be translated into a protein. They inserted both the original repeat sequence and the engineered repeat sequence into fruit flies. The unaltered repeats caused neurodegeneration and early death in the fruit flies, but the engineered repeats had no effect. These results suggest that the toxicity of the repeat sequence was the result of the protein it encoded and not simply a product of the RNA alone (although the RNA may also be somewhat toxic). Further research suggested that proteins with the glycine-arginine and proline-arginine dipeptides are responsible for the neurodegeneration that occurs in ALS and FTD. The toxicity of these proteins may result from the fact that they mimic RNA-binding proteins and interfere with splicing of pre-mRNA and the processing of rRNA. This research provides important insight into the pathology of these diseases and suggests possible future targets for treatment.

THINK-PAIR-SHARE

- Propose some ways that the new information provided by research on the role of the GGGGCC repeat in ALS might be used to design potential treatments for the disease.

- Using the genetic code illustrated in Figure 15.10, show how translation of the GGGGCC repeat without a start codon results in the production of five proteins with different dipeptide repeats. (Hint: Consider all reading frames of the two RNAs copied from this sequence.)

The story of ALS and expanding nucleotide repeats illustrates the central importance of studying mutations: the analysis of mutants is often a source of key insights into diseases and important biological processes. This chapter focuses on gene mutations—on how these errors in genetic instructions arise and how they are studied. We begin with a brief examination of the different types of mutations, including their phenotypic effects, how they can be suppressed, and their rates of occurrence. The next section explores how mutations can arise spontaneously during and after the course of DNA replication, as well as how chemicals and radiation can induce them. After discussing the analysis of mutations, we turn to transposable elements, DNA sequences that are capable of moving within the genome and that often produce mutations when they do so. Finally, we take a look at DNA repair and some of the diseases that arise when DNA repair is defective.

18.1 Mutations Are Inherited Alterations in the DNA Sequence

DNA is a highly stable molecule that is replicated with amazing accuracy (as we saw in Chapters 10 and 12), but changes in DNA structure and errors of replication do take place. A **mutation** is defined as an inherited change in genetic information; the descendants that inherit the change may be cells or organisms.

The Importance of Mutations

Mutations are both the sustainer of life and the cause of great suffering. On the one hand, mutation is the source of all genetic variation, the raw material of evolution. The ability of organisms to adapt to environmental change depends on the presence of genetic variation in natural populations, and genetic variation is produced by mutation. On the other hand, many mutations have detrimental effects, and mutation is the source of many diseases and disorders.

Much of the study of genetics focuses on how genetic variants produced by mutation are inherited; genetic crosses are meaningless if all individual members of a species are identically homozygous for the same alleles. Much of Gregor Mendel's success in unraveling the principles of inheritance can be traced to his use of carefully selected variants of the garden pea. Similarly, Thomas Hunt Morgan and his students discovered many basic principles of genetics by analyzing mutant fruit flies.

Mutations are also useful for examining fundamental biological processes. Finding or creating mutations that affect different components of a biological system and studying their effects can often lead to a better understanding of the system. This method, referred to as genetic dissection, is analogous to figuring out how an automobile works by breaking different parts of a car and observing the effects; for example, smash the radiator and the engine overheats, revealing that the radiator cools the engine. The use of mutations to disrupt function can likewise be a source of insight into biological processes. For example, geneticists have begun to unravel the molecular details of development by studying mutations that interrupt various embryonic stages in *Drosophila* (see Chapter 22). Scientists have also used analysis of mutations to reveal the different parts of the *lac* operon (discussed in Chapter 16) and how they function in gene regulation. Although breaking "parts" to determine their function might seem like a crude approach to understanding a system, it is actually a very powerful one and has been used extensively in biochemistry, developmental biology, physiology, and behavioral science. But this method is *not* recommended for learning how your car works!

THINK-PAIR-SHARE Question 1

Categories of Mutations

In multicellular organisms, we can distinguish between two broad categories of mutations: somatic mutations and germ-line mutations. **Somatic mutations** arise in somatic tissues, which do not produce gametes (**Figure 18.1**). When a somatic cell with a mutation divides (by mitosis), the mutation is passed on to the daughter cells, leading to a population of genetically identical cells (a clone). The earlier in development that a somatic mutation takes place, the larger the clone of cells that contain the mutation will be.

Because of the huge number of cells present in a typical eukaryotic organism, somatic mutations are numerous. For example, there are about 10^{14} cells in the human body. Typically, a mutation arises once in every million cell divisions, so hundreds of millions of somatic mutations must arise in each person. Many somatic mutations have no obvious effect on the phenotype of the organism because the function of the mutant cell is taken over by a normal cell, or the mutant cell dies and is replaced by normal cells. However, cells with a somatic mutation that stimulates cell division can increase in number and spread; this type of mutation can give rise to cells with a selective advantage and is the basis for cancer (see Chapter 23). Somatic mutations are also associated with some other diseases, including hemimegalencephaly, in which just one hemisphere of the brain is enlarged, usually resulting in epilepsy. And somatic mutations can lead to mosaicism, in which different tissues within the body have different genetic information (see Chapter 6).

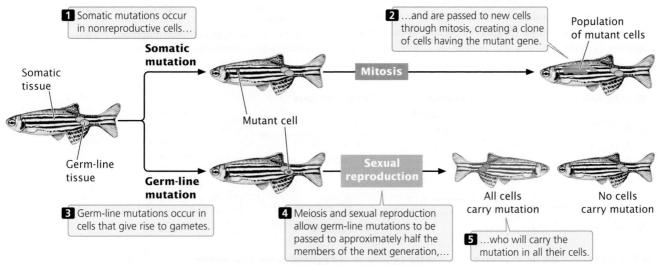

1 Somatic mutations occur in nonreproductive cells…

2 …and are passed to new cells through mitosis, creating a clone of cells having the mutant gene.

Population of mutant cells

Somatic tissue

Somatic mutation

Mutant cell

Mitosis

Germ-line tissue

Germ-line mutation

3 Germ-line mutations occur in cells that give rise to gametes.

Sexual reproduction

4 Meiosis and sexual reproduction allow germ-line mutations to be passed to approximately half the members of the next generation,…

All cells carry mutation

No cells carry mutation

5 …who will carry the mutation in all their cells.

18.1 The two basic classes of mutations are somatic mutations and germ-line mutations.

Germ-line mutations arise in cells that ultimately produce gametes. A germ-line mutation can be passed to future generations, producing offspring that carry the mutation in all their somatic and germ-line cells (see Figure 18.1). When we speak of mutations in multicellular organisms, we're usually talking about germ-line mutations.

Historically, mutations have been partitioned into those that affect a single gene, called *gene mutations*, and those that affect the number or structure of chromosomes, called *chromosome mutations*. This distinction arose because chromosome mutations could be observed directly, by looking at chromosomes with a microscope, whereas gene mutations could be detected only by observing their phenotypic effects. Now, DNA sequencing allows direct observation of gene mutations, and chromosome mutations are distinguished from gene mutations somewhat arbitrarily on the basis of the size of the DNA lesion. Nevertheless, it is practical to use *chromosome mutation* for a large-scale genetic alteration that affects chromosome structure or the number of chromosomes and to use **gene mutation** for a relatively small DNA lesion that affects a single gene. This chapter focuses on gene mutations; chromosome mutations were discussed in Chapter 8.

Types of Gene Mutations

There are a number of ways to classify gene mutations. Some classification schemes are based on the nature of the phenotypic effect, others are based on the causative agent of the mutation, and still others focus on the molecular nature of the defect. Here, we will categorize mutations primarily on the basis of their molecular nature, but we will also encounter some terms that relate the causes and the phenotypic effects of mutations.

BASE SUBSTITUTIONS The simplest type of gene mutation is a **base substitution**, the alteration of a single nucleotide in the DNA (**Figure 18.2a**). There are two types of base substitutions. In a **transition**, a purine is replaced by a different purine or, alternatively, a pyrimidine is replaced by a different pyrimidine (**Figure 18.3**). In a **transversion**, a purine is replaced by a pyrimidine or a pyrimidine is

replaced by a purine. The number of possible transversions (see Figure 18.3) is twice the number of possible transitions, but transitions arise more frequently because transforming a purine into a different purine or a pyrimidine into a different pyrimidine is easier than transforming a purine into a pyrimidine, or vice versa. **▶ TRY PROBLEM 18**

INSERTIONS AND DELETIONS Another class of gene mutations contains **insertions** and **deletions** (collectively called *indels*): the addition or removal, respectively, of one or more nucleotide pairs (**Figure 18.2b, c**). Although base substitutions are often assumed to be the most common type of mutation, molecular analysis has revealed that insertions and deletions are often more frequent. Insertions and deletions within sequences that encode proteins may lead to **frameshift mutations**: changes in the reading frame (see pp. 439–440 in Chapter 15) of the gene. Frameshift mutations usually alter all amino acids encoded by the nucleotides following the mutation, so they generally have drastic effects on the phenotype. Some frameshifts also introduce

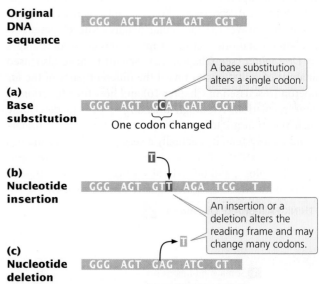

18.2 Three basic types of gene mutations are base substitutions, insertions, and deletions.

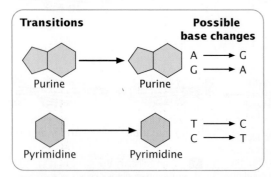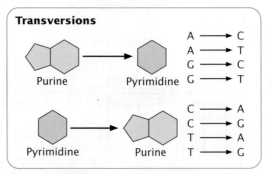

18.3 A transition is the substitution of a purine for a purine or of a pyrimidine for a pyrimidine; a transversion is the substitution of a pyrimidine for a purine or of a purine for a pyrimidine.

premature stop codons, terminating protein synthesis early and resulting in a shortened (truncated) protein. Not all insertions and deletions lead to frameshifts, however; insertions and deletions consisting of any multiple of three nucleotides leave the reading frame intact, although the addition or removal of one or more amino acids may still affect the phenotype. Indels that do not affect the reading frame are called **in-frame insertions** and **in-frame deletions**.

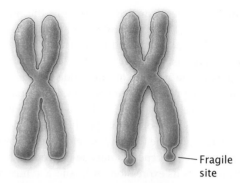

18.4 Fragile-X syndrome is associated with a characteristic constriction (fragile site) on the long arm of the X chromosome.

CONCEPTS

Gene mutations are changes in a single gene. They can be base substitutions (in which a single pair of nucleotides is altered) or insertions or deletions (in which nucleotides are added or removed). A base substitution can be a transition (substitution of like bases) or a transversion (substitution of unlike bases). Insertions and deletions often lead to a change in the reading frame of a gene.

✔ CONCEPT CHECK 1

Which of the following changes is a transition base substitution?
a. Adenine is replaced by thymine.
b. Cytosine is replaced by adenine.
c. Guanine is replaced by adenine.
d. Three nucleotide pairs are inserted into DNA.

EXPANDING NUCLEOTIDE REPEATS Mutations in which the number of copies of a set of nucleotides increases are called **expanding nucleotide repeats**. This type of mutation was first observed in 1991 in a gene called *FMR-1*, which causes fragile-X syndrome, the most common hereditary cause of intellectual disability. The disorder is so named because, when specially treated cells from people with the condition are examined under a microscope, the tip of each long arm of the X chromosome is attached by a

slender-appearing part of the chromosome (**Figure 18.4**). The normal *FMR-1* allele (not containing the mutation) has 60 or fewer copies of the sequence CGG, but in people with fragile-X syndrome, the allele may harbor hundreds or even thousands of copies.

Expanding nucleotide repeats have been found in almost 30 human diseases, several of which are listed in **Table 18.1**. Most of these diseases are caused by the expansion of a set of three nucleotides (called a trinucleotide), most often CNG, where N can be any nucleotide. However, some diseases are caused by repeats of four, five, and even twelve nucleotides. The number of copies of the repeat often correlates with the severity or age of onset of the disease. The number of copies of the repeat also correlates with its instability: when more repeats are present, the probability of expansion to even more repeats increases. This association between the number of copies of nucleotide repeats, the severity of the resulting disease, and the probability of expansion leads to a phenomenon known as anticipation (see Chapter 5), in which diseases caused by expanding nucleotide repeats become more

---

TABLE 18.1	Examples of human genetic diseases caused by expanding nucleotide repeats		
		Number of Copies of Repeat	
Disease	**Repeated Sequence**	**Normal Range**	**Disease Range**
Spinal and bulbar muscular atrophy	CAG	11–33	40–62
Fragile-X syndrome	CGG	6–54	50–1500
Jacobsen syndrome	CGG	11	100–1000
Spinocerebellar ataxia (several types)	CAG	4–44	21–130
Autosomal dominant cerebellar ataxia	CAG	7–19	37–220
Myotonic dystrophy	CTG	5–37	44–3000
Huntington disease	CAG	9–37	37–121
Friedreich ataxia	GAA	6–29	200–900
Dentatorubral-pallidoluysian atrophy	CAG	7–25	49–75
Myoclonus epilepsy of the Unverricht–Lundborg type	CCCCGCCCCGCG	2–3	12–13
Amyotrophic lateral sclerosis	GGGGCC	2–23	700–1600

severe in each generation. Less commonly, the number of nucleotide repeats may decrease within a family. Expanding nucleotide repeats have also been observed in some microbes and plants.

Increases in the number of nucleotide repeats can produce disease symptoms in different ways. In several diseases (e.g., Huntington disease), the nucleotide expansion occurs within the coding part of a gene, producing a toxic protein that has extra glutamine (the amino acid encoded by CAG). In other diseases, the repeat is outside the coding region of a gene and affects its expression. In fragile-X syndrome, the additional copies of the nucleotide repeat cause the DNA to become methylated, which turns off the transcription of an essential gene.

Some evidence suggests that expansion of nucleotide repeats occurs in the course of DNA replication and appears to be related to the formation of hairpins and other special secondary structures that form in single-stranded DNA consisting of nucleotide repeats. Such structures may interfere with normal replication by causing strand slippage, misalignment of the sequences, or stalling of replication. One model of how hairpin formation in the repeats might result in their expansion is shown in **Figure 18.5**. Watching **Animation 18.1** will help you understand how copies of nucleotide repeats increase in number. Other models of repeat expansion that occur through transcription and DNA repair have also been proposed. Many aspects of this phenomenon are not well understood, including why repeat expansion occurs in some people and not in others.

> ### CONCEPTS
>
> Expanding nucleotide repeats are mutations in which the number of copies of a set of repeated nucleotides increases over time. These mutations are associated with several human genetic diseases.

Functional Effects of Mutations

Another way that mutations are classified is by their functional effects. At the most general level, we can distinguish a mutation on the basis of its phenotype compared with the wild-type phenotype. A mutation that alters the wild-type phenotype is called a **forward mutation**, whereas a **reverse mutation** (a *reversion*) changes a mutant phenotype back into the wild type.

Geneticists use other terms to describe the effects of mutations on protein structure. A base substitution that results in a different amino acid in the protein is referred to as a **missense mutation** (**Figure 18.6a**). A **nonsense mutation** changes a sense codon (one that specifies an amino acid) into a nonsense codon (one that terminates translation), as shown in **Figure 18.6b**. If a nonsense mutation occurs early

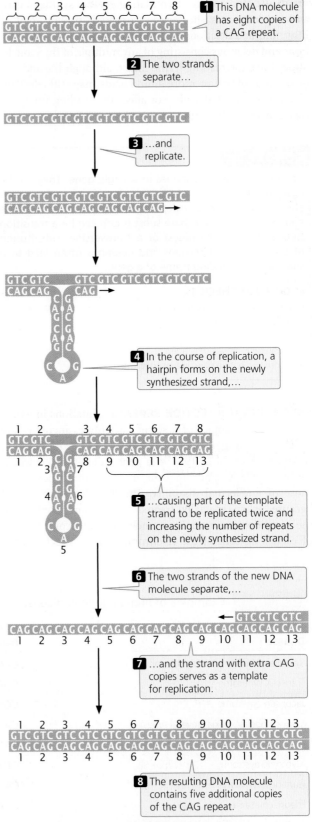

18.5 A model of how the number of copies of a nucleotide repeat may increase in replication.

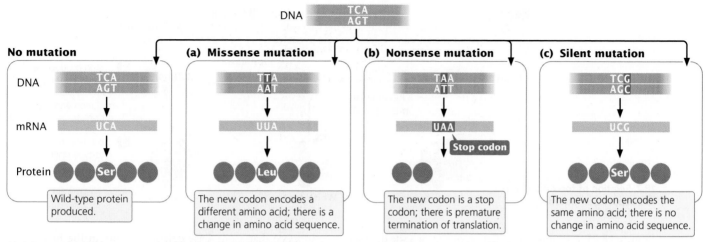

18.6 Base substitutions can cause (a) missense, (b) nonsense, or (c) silent mutations.

in the mRNA sequence, the protein will be truncated and usually nonfunctional.

Because of the redundancy of the genetic code, some different codons specify the same amino acid. A **silent mutation** changes a codon to a synonymous codon that specifies the same amino acid (**Figure 18.6c**), altering the DNA sequence without changing the amino acid sequence of the protein. Not all silent mutations, however, are truly silent: some do have phenotypic effects. They may have phenotypic effects, for example, when different tRNAs (called isoaccepting tRNAs; see Chapter 15) bind to different synonymous codons. Because some isoaccepting tRNAs are more abundant than others, which synonymous codon is used may affect the rate of protein synthesis. The rate of protein synthesis, in turn, can influence the phenotype by affecting the amount of protein present in the cell and, in a few cases, the folding of the protein. Other silent mutations may alter nucleotides that serve as binding sites for regulatory proteins or alter sequences near exon–intron junctions that affect mRNA splicing (see Chapter 14). Still other silent mutations can influence the binding of miRNAs to complementary sequences in the mRNA, which determines whether the mRNA is translated (see Chapter 14).

A **neutral mutation** is a missense mutation that alters the amino acid sequence of a protein but does not significantly change its function. Neutral mutations occur when one amino acid is replaced by another that is chemically similar, or when the affected amino acid has little influence on protein function. For example, some neutral mutations occur in the genes that encode hemoglobin; although these mutations alter the amino acid sequence of hemoglobin, they do not affect its ability to transport oxygen.

Loss-of-function mutations cause the complete or partial absence of normal protein function. A loss-of-function mutation is one that so alters the structure of the protein that the protein no longer works correctly, or one that occurs in regulatory regions that affect the transcription, translation, or splicing of the protein. Loss-of-function mutations are frequently recessive, in which case an individual diploid organism must be homozygous for the mutation before the effects of the loss of the functional protein can be exhibited. The mutations that cause cystic fibrosis are loss-of-function mutations: these mutations produce a nonfunctional form of the CFTR protein, which normally regulates the movement of chloride ions into and out of the cell (see Chapter 5).

In contrast, a **gain-of-function mutation** causes the cell to produce a protein or gene product whose function is not normally present. The result could be an entirely new gene product or one produced in an inappropriate tissue or at an inappropriate time in development. For example, a mutation in a gene that encodes a receptor for a growth factor might cause the mutated receptor to stimulate growth all the time, even in the absence of the growth factor. Gain-of-function mutations are frequently dominant in their expression because a single copy of the mutation leads to the presence of a new gene product.

Other mutations are **conditional mutations**, which are expressed only under certain conditions. For example, some conditional mutations affect the phenotype only at elevated temperatures. Still others are **lethal mutations**, which cause premature death (see Chapter 5). **❯ TRY PROBLEM 22**

THINK-PAIR-SHARE Question 2

Suppressor Mutations

A **suppressor mutation** is a genetic change that hides or suppresses the effect of another mutation. This type of mutation is different from a reverse mutation, in which the mutated site is changed back to the original wild-type sequence

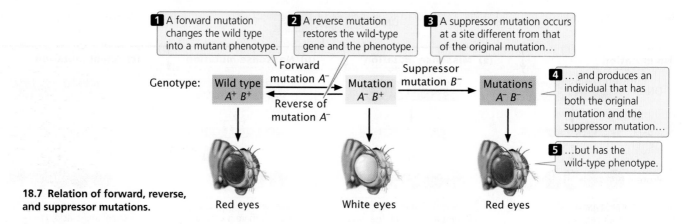

18.7 Relation of forward, reverse, and suppressor mutations.

Red eyes   White eyes   Red eyes

(**Figure 18.7**). A suppressor mutation occurs at a site distinct from the site of the original mutation; thus, an individual with a suppressor mutation is a double mutant, possessing both the original mutation and the suppressor mutation but exhibiting the phenotype of the nonmutated wild type. Geneticists distinguish between two classes of suppressor mutations: intragenic and intergenic.

INTRAGENIC SUPPRESSOR MUTATIONS An **intragenic suppressor mutation** takes place in the same gene that contains the mutation being suppressed. It may work in any of several ways. The suppressor may change a second nucleotide in the same codon altered by the original mutation, producing a codon that specifies the same amino acid that was specified by the original, nonmutated codon (**Figure 18.8**).

Intragenic suppressors may also work by suppressing a frameshift mutation. If the original mutation, for example, is a one-base deletion, then the addition of a single base elsewhere in the gene will restore the former reading frame. Consider the following nucleotide sequence on the template strand of DNA and the amino acids that it encodes:

DNA   3′–AAA   TCA   CTT   GGC   GTA   CAA–5′

mRNA   5′–UUU   AGU   GAA   CCG   CAU   GUU–3′

Amino acids   Phe   Ser   Glu   Pro   His   Val

Suppose that a one-base deletion occurs in the first nucleotide of the second codon. This deletion shifts the reading frame by one nucleotide and alters all the amino acids that follow the mutation:

One-nucleotide deletion

DNA   3′–AAA   TͨCAC   TTG   GCG   TAC   AA–5′

mRNA   5′–UUU   GUG   AAC   CGC   AUG   UU–3′

Amino acids   Phe   Val   Asn   Arg   Met

If a single nucleotide is added to the third codon (the suppressor mutation), the reading frame is restored, although two of the amino acids differ from those specified by the original sequence:

One-nucleotide insertion

DNA   3′–AAA   CAC   TTT   GGC   GTA   CAA–5′

mRNA   5′–UUU   GUG   AAA   CCG   CAU   GUU–3′

Amino acids   Phe   Val   Lys   Pro   His   Val

Similarly, a mutation due to an insertion may be suppressed by a subsequent deletion in the same gene.

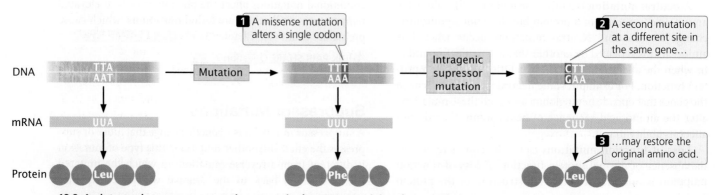

18.8 An intragenic suppressor mutation occurs in the gene containing the mutation being suppressed.

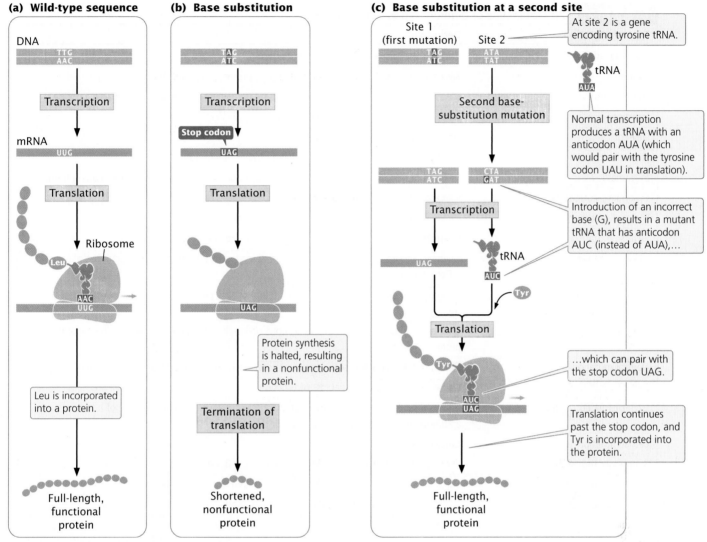

(a) Wild-type sequence

DNA

| TTG |
| AAC |

Transcription

mRNA

| UUG |

Translation

Ribosome

Leu

| AAC |
| UUG |

Leu is incorporated into a protein.

Full-length, functional protein

(b) Base substitution

| TAG |
| ATC |

Transcription

Stop codon

| UAG |

Translation

| UAG |

Protein synthesis is halted, resulting in a nonfunctional protein.

Termination of translation

Shortened, nonfunctional protein

(c) Base substitution at a second site

Site 1 (first mutation)   Site 2

| TAG | | ATA |
| ATC | | TAT |

At site 2 is a gene encoding tyrosine tRNA.

tRNA

| AUA |

Second base-substitution mutation

| TAG | | CTA |
| ATC | | GAT |

Normal transcription produces a tRNA with an anticodon AUA (which would pair with the tyrosine codon UAU in translation).

Transcription

| UAG | tRNA

| AUC |

Tyr

Introduction of an incorrect base (G), results in a mutant tRNA that has anticodon AUC (instead of AUA),…

Translation

Tyr

…which can pair with the stop codon UAG.

| AUC |
| UAG |

Translation continues past the stop codon, and Tyr is incorporated into the protein.

Full-length, functional protein

18.9 An intergenic suppressor mutation occurs in a gene other than the one bearing the original mutation that it suppresses. (a) A wild-type sequence produces a full-length, functional protein. (b) A base substitution at a site in that gene produces a premature stop codon, resulting in a truncated, nonfunctional protein. (c) A base substitution at a site in another gene, which in this case encodes tRNA, alters the anticodon of tRNA^{Tyr}; tRNA^{Tyr} can then pair with the stop codon produced by the original mutation, allowing tyrosine to be incorporated into the protein and translation to continue.

A third way in which an intragenic suppressor mutation may work is by making compensatory changes in the protein. A first missense mutation can alter the folding of a polypeptide chain by changing the way in which amino acids in the protein interact with one another. A second missense mutation at a different site (the suppressor mutation) can re-create the original folding pattern by restoring the interactions between the amino acids.

INTERGENIC SUPPRESSOR MUTATIONS An **intergenic suppressor mutation**, in contrast, occurs in a gene other than the one bearing the original mutation that it suppresses. These mutations, also known as *extragenic suppressor*

mutations, sometimes work by changing the way the mRNA is translated. In the example illustrated in **Figure 18.9a**, the original DNA sequence is AAC (UUG in the mRNA) and specifies leucine. This sequence mutates to A<u>T</u>C (U<u>A</u>G in mRNA), a stop codon (**Figure 18.9b**). The ATC nonsense mutation could be suppressed by a second mutation in a different gene that encodes a tRNA; this second mutation would result in a tRNA anticodon capable of pairing with the UAG stop codon (**Figure 18.9c**). For example, the gene that encodes the tRNA for tyrosine (tRNA^{Tyr}), which has the anticodon AUA, might be mutated to have the anticodon AU<u>C</u>, which would then pair with the UAG stop codon. Instead of translation terminating at the UAG codon, tyrosine would be

TABLE 18.2	Characteristics of different types of mutations
Type of Mutation	**Definition**
Base substitution	Changes the base of a single DNA nucleotide
Transition	Base substitution in which a purine replaces a purine or a pyrimidine replaces a pyrimidine
Transversion	Base substitution in which a purine replaces a pyrimidine or a pyrimidine replaces a purine
Insertion	Addition of one or more nucleotides
Deletion	Deletion of one or more nucleotides
Frameshift mutation	Insertion or deletion that alters the reading frame of a gene
In-frame deletion or insertion	Deletion or insertion of a multiple of three nucleotides that does not alter the reading frame
Expanding nucleotide repeats	Increases the number of copies of a set of nucleotides
Forward mutation	Changes the wild-type phenotype to a mutant phenotype
Reverse mutation	Changes a mutant phenotype back to the wild-type phenotype
Missense mutation	Changes a sense codon into a different sense codon, resulting in the incorporation of a different amino acid in the protein
Nonsense mutation	Changes a sense codon into a nonsense (stop) codon, causing premature termination of translation
Silent mutation	Changes a sense codon into a synonymous codon, leaving the amino acid sequence of the protein unchanged
Neutral mutation	Changes the amino acid sequence of a protein without altering its ability to function
Loss-of-function mutation	Causes a complete or partial loss of function
Gain-of-function mutation	Causes the appearance of a new trait or function or causes the appearance of a trait in inappropriate tissue or at an inappropriate time
Lethal mutation	Causes premature death
Suppressor mutation	Suppresses the effect of an earlier mutation at a different site
Intragenic suppressor mutation	Suppresses the effect of an earlier mutation within the same gene
Intergenic suppressor mutation	Suppresses the effect of an earlier mutation in another gene

inserted into the protein, and a full-length protein would be produced, although tyrosine would now substitute for leucine. The effect of this change would depend on the role of this amino acid in the overall structure of the protein, but the effect of the suppressor mutation would probably be less detrimental than the effect of the nonsense mutation, which would halt translation prematurely.

Because cells in many organisms have multiple copies of tRNA genes, other nonmutated copies of tRNA^{Tyr} would remain available to recognize tyrosine codons in mRNA transcripts. We might expect that the tRNAs encoded by the gene with the suppressor mutation just described would suppress the normal stop codons at the ends of other coding sequences as well as the one in the transcript of the original mutant gene, resulting in the production of longer-than-normal proteins, but this event does not usually take place.

Intergenic suppressor mutations can also work through gene interactions (see Chapter 5). For example, polypeptide chains that are produced by two different genes may interact to produce a functional protein. A mutation in one gene

may alter the encoded polypeptide such that the interaction between the two polypeptides is destroyed, in which case a functional protein is not produced. A suppressor mutation in the second gene may produce a compensatory change in its polypeptide, therefore restoring the original interaction. Characteristics of some of the different types of mutations are summarized in **Table 18.2**.

CONCEPTS

A suppressor mutation overrides the effect of an earlier mutation at a different site. An intragenic suppressor mutation occurs within the *same* gene that contains the original mutation; an intergenic suppressor mutation occurs in a *different* gene.

✔ **CONCEPT CHECK 2**

How is a suppressor mutation different from a reverse mutation?

WORKED PROBLEM

A gene encodes a protein with the following amino acid sequence:

Met-Arg-Cys-Ile-Lys-Arg

A mutation of a single nucleotide alters the amino acid sequence to

Met-Asp-Ala-Tyr-Lys-Gly-Glu-Ala-Pro-Val

A second single-nucleotide mutation occurs in the same gene and suppresses the effects of the first mutation (an intragenic suppressor). With the original mutation and the intragenic suppressor present, the protein has the following amino acid sequence:

Met-Asp-Gly-Ile-Lys-Arg

What is the nature and location of the first mutation and of the intragenic suppressor mutation?

Solution Strategy

What information is required in your answer to the problem?

The type and location of the first mutation and the intragenic suppressor.

What information is provided to solve the problem?

■ The amino acid sequence of the protein encoded by the original nonmutated gene.

■ The amino acid sequence of the protein encoded by the mutated gene.

■ The amino acid sequence of the protein encoded by the mutated gene and the intragenic suppressor.

Solution Steps

The first mutation alters the reading frame, because all amino acids after Met are changed, including the stop codon (which results in a longer protein). Insertions and deletions affect the reading frame; the original mutation consists of a single-nucleotide insertion or deletion in the second codon. The intragenic suppressor restores the reading frame; the intragenic suppressor also is most likely a single-nucleotide insertion or deletion. If the first mutation is an insertion, the suppressor must be a deletion; if the first mutation is a deletion, then the suppressor must be an insertion. Notice that the protein produced by the suppressor still differs from the original protein at the second and third amino acids, but the second amino acid produced by the suppressor is the same as that in the protein produced by the original mutation. Thus, the suppressor mutation must have occurred in the third codon, because the suppressor does not alter the second amino acid.

» For more practice with analyzing mutations, try working **Problem 23** at the end of the chapter.

Mutation Rates

The frequency with which a wild-type allele at a locus changes into a mutant allele is referred to as the **mutation rate**. The mutation rate is generally expressed as the number of mutations per biological unit, which may be mutations per cell division, per gamete, or per round of replication. For example, achondroplasia is a type of hereditary dwarfism in humans that results from a dominant mutation. On average, about four achondroplasia mutations arise in every 100,000 gametes, and so the mutation rate is $^4/_{100,000}$, or 0.00004 mutations per gamete. The mutation rate provides information about how often a mutation arises.

FACTORS AFFECTING MUTATION RATES Calculations of mutation rates are affected by three factors. First, they depend on the frequency with which changes in DNA take place. Mutations can arise as spontaneous molecular changes in DNA, or they can be induced by chemical, biological, or physical agents in the environment.

The second factor influencing the mutation rate is the probability that when an alteration in DNA takes place, it will be repaired. Most cells possess a number of mechanisms for repairing altered DNA, so most alterations are corrected before they are replicated. If these repair systems are effective, mutation rates will be low; if they are faulty, mutation rates will be elevated. Some mutations increase the overall rate of mutation at other genes; these mutations usually occur in genes that encode components of the replication machinery or DNA-repair enzymes.

The third factor is the probability that a mutation will be detected. When DNA is sequenced, all mutations are potentially detectable. In practice, however, mutations are usually detected by their phenotypic effects. Some mutations may appear to arise at a higher rate simply because they are easier to detect.

VARIATION IN MUTATION RATES We can draw several general conclusions about mutation rates, though they vary among genes and among species (**Table 18.3**). First, spontaneous mutation rates are low for all organisms studied. Typical mutation rates for bacterial genes range from about 1 to 100 mutations per 10 billion cells (from 1×10^{-8} to 1×10^{-10}). The mutation rates for most eukaryotic genes are a bit higher, from about 1 to 10 mutations per million gametes (from 1×10^{-5} to 1×10^{-6}). These higher values in eukaryotes may be due to the fact that the rates are calculated *per gamete*, and that several cell divisions are required to produce a gamete, whereas mutation rates in prokaryotic cells are calculated *per cell division*.

The differences in mutation rates among species may be due to differing abilities to repair mutations, unequal exposures to mutagens, or biological differences in rates of spontaneous mutations. Even within a single species, spontaneous mutation rates vary among genes. The reason for this variation is not entirely understood, but some regions of DNA are known hotspots for mutations.

TABLE 18.3   Mutation rates of different genes in different organisms

Organism	Mutation	Rate	Unit
Bacteriophage T2	Lysis inhibition	1×10^{-8}	Per replication
	Host range	3×10^{-9}	
Escherichia coli	Lactose fermentation	2×10^{-7}	Per cell division
	Histidine requirement	2×10^{-8}	
Neurospora crassa	Inositol requirement	8×10^{-8}	Per asexual spore
	Adenine requirement	4×10^{-8}	
Corn	Kernel color	2.2×10^{-6}	Per gamete
Drosophila	Eye color	4×10^{-5}	Per gamete
	Allozymes	5.14×10^{-6}	
Mouse	Albino coat color	4.5×10^{-5}	Per gamete
	Dilution coat color	3×10^{-5}	
Human	Huntington disease	1×10^{-6}	Per gamete
	Achondroplasia	1×10^{-5}	
	Neurofibromatosis (Michigan)	1×10^{-4}	
	Hemophilia A (Finland)	3.2×10^{-5}	
	Duchenne muscular dystrophy (Wisconsin)	9.2×10^{-5}	

Recent research suggests that fewer mutations occur in DNA sequences that are associated with nucleosomes (see Chapter 11). Reduced mutation rates may occur in these sequences because DNA associated with nucleosomes is less exposed to mutagens, but they could also be explained by the effect of nucleosomes on DNA repair, recombination, or replication, all of which influence the rate of mutation.

Several recent studies have measured mutation rates directly by sequencing genes of organisms over a number of generations. These new studies suggest that mutation rates are often higher than those previously measured on the basis of changes in phenotype. In one study, geneticists sequenced randomly chosen stretches of DNA in the nematode *Caenorhabditis elegans* and found about 2.1 mutations per genome per generation, which was 10 times higher than previous estimates based on phenotypic changes. The researchers found that about half of the mutations were insertions and deletions.

Recent genome sequencing has also provided more accurate information about mutation rates in humans. Several sequencing studies suggest that the overall rate of base substitutions in humans is about 1×10^{-8} mutations per base pair per generation. Other research suggests that each person carries approximately 100 new loss-of-function germ-line mutations.

THINK-PAIR-SHARE Question 3

ADAPTIVE MUTATION As will be discussed in Chapters 24 through 26, genetic variation is critical for evolutionary change that brings about adaptation to new environments.

New genetic variants arise primarily through mutation. For many years, genetic variation was assumed to arise randomly and at rates that are independent of the need for adaptation. However, some evidence suggests that stressful environments—where adaptation may be necessary to survive—can induce more mutations in bacteria, a process that has been termed **adaptive mutation**. The idea of adaptive mutation has been intensely debated; critics counter that most mutations are expected to be deleterious, and therefore increased mutagenesis would probably be harmful most of the time.

CONCEPTS

The mutation rate is the frequency with which a specific mutation arises. Rates of mutations are generally low and are affected by environmental and genetic factors.

✔ CONCEPT CHECK 3
What three factors affect mutation rates?

18.2 Mutations May Be Caused by a Number of Different Factors

Mutations result from both internal and external factors. Those that occur under normal conditions are termed **spontaneous mutations**, whereas those that result from changes caused by environmental chemicals or radiation are **induced mutations**.

Spontaneous Replication Errors

Replication is amazingly accurate: less than one error in a billion nucleotides arises in the course of DNA synthesis (see Chapter 12). However, spontaneous replication errors do occasionally occur.

THINK-PAIR-SHARE Question 4 👥

TAUTOMERIC SHIFTS The primary cause of spontaneous replication errors was at one time thought to be tautomeric shifts, in which the positions of protons (hydrogen atoms) in the DNA bases change. Each of the four DNA bases exists in different chemical forms, called *tautomers*. The two tautomeric forms of each base are in dynamic equilibrium, although one form is much more common than the other. The standard Watson-and-Crick base pairings—adenine with thymine, and cytosine with guanine—occur between the common forms of the bases, but if the bases are in their rare tautomeric forms, other base pairings are possible. For example, the common form of cytosine pairs with guanine, but the rare tautomer of cytosine pairs with adenine.

Watson and Crick proposed that tautomeric shifts might produce mutations, and for many years their proposal was the accepted model for spontaneous replication errors. However, there has never been convincing evidence that the rare tautomers are the cause of spontaneous mutations.

MISPAIRING DUE TO OTHER STRUCTURES Mispairings often arise through wobble (see Chapter 15), in which normal, protonated, and other forms of the bases are able to pair because of flexibility in the DNA helical structure (**Figure 18.10**). These structures have been detected in DNA molecules and are now thought to be responsible for many of the mispairings in replication.

INCORPORATED ERRORS AND REPLICATED ERRORS When a mispaired base has been incorporated into a newly synthesized nucleotide chain, an **incorporated error** is said to have occurred. Suppose that, in replication, thymine (which normally pairs with adenine) mispairs with guanine through wobble (**Figure 18.11**). In the next round of replication, the two mismatched bases separate, and each serves as template for the synthesis of a new nucleotide strand. This time, thymine pairs with adenine, producing another copy of the original DNA

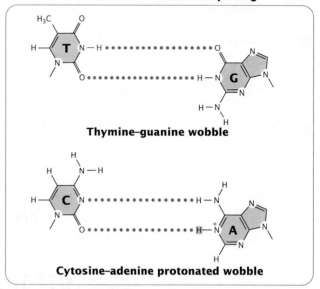

Non-Watson-and-Crick base pairing

Thymine–guanine wobble

Cytosine–adenine protonated wobble

18.10 Nonstandard base pairings can occur as a result of the flexibility in DNA structure. Thymine and guanine in their normal forms can pair through wobble. Cytosine and adenine can pair through wobble when adenine is protonated (has an extra hydrogen atom).

sequence. On the other strand, however, the incorrectly incorporated guanine serves as the template and pairs with cytosine, producing a new DNA molecule that has an error: a C • G pair in place of the original T • A pair (a T • A → C • G base substitution). The original incorporated error (the T–G mispairing) leads to a **replicated error** (the C • G base pair instead of the original T • A base pair), which creates a permanent mutation because all the base pairings are correct and there is no way for repair systems to detect the error.

CAUSES OF DELETIONS AND INSERTIONS Small insertions and deletions can arise spontaneously in replication and crossing over. **Strand slippage** can occur when one nucleotide strand forms a small loop (**Figure 18.12**). If the looped-out nucleotides are on the newly synthesized strand, an insertion results. At the next round of replication, the insertion will be replicated, and both strands will contain the insertion. If the looped-out nucleotides are on the template strand, then the newly replicated strand will have a deletion, and this deletion will be perpetuated in subsequent rounds of replication.

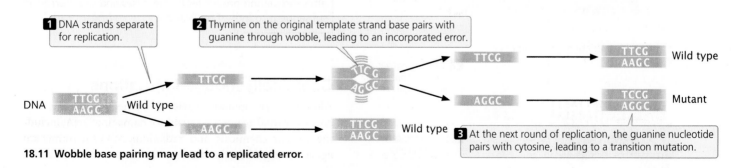

1 DNA strands separate for replication.

2 Thymine on the original template strand base pairs with guanine through wobble, leading to an incorporated error.

DNA — TTCG / AAGC — Wild type

TTCG

AGGC

TTCG — TTCG / AAGC — Wild type

TTCG — TTCG / AAGC — Wild type

AGGC — TCCG / AGGC — Mutant

3 At the next round of replication, the guanine nucleotide pairs with cytosine, leading to a transition mutation.

18.11 Wobble base pairing may lead to a replicated error.

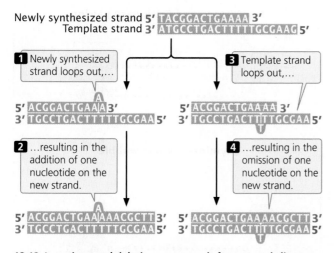

18.12 Insertions and deletions may result from strand slippage.

Another process that produces insertions and deletions is unequal crossing over (see Section 8.2). In normal crossing over, the homologous sequences of the two DNA molecules align, and crossing over produces no net change in the number of nucleotides in either molecule. Misaligned pairing, however, can cause unequal crossing over, which results in one DNA molecule with an insertion and the other with a deletion (**Figure 18.13**).

CONCEPTS

Spontaneous replication errors arise from altered base structures and from wobble. Small insertions and deletions can occur through strand slippage during replication and through unequal crossing over.

Spontaneous Chemical Changes

In addition to spontaneous mutations that arise in replication, mutations also result from spontaneous chemical changes in DNA. One such change is **depurination**, the loss of a purine

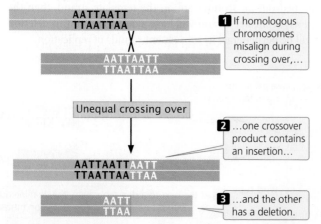

18.13 Unequal crossing over produces insertions and deletions.

base from a nucleotide. Depurination results when the covalent bond connecting the purine to the 1'-carbon atom of the deoxyribose sugar breaks (**Figure 18.14a**), producing an apurinic site, a nucleotide that lacks its purine base. An apurinic site cannot act as a template for a complementary base in replication. In the absence of base-pairing constraints, an incorrect nucleotide (most often adenine) is incorporated into the newly synthesized DNA strand opposite the apurinic site (**Figure 18.14b**), frequently leading to an incorporated error. The incorporated error is then transformed into a replicated error at the next round of replication. Depurination is a common cause of spontaneous mutation; a mammalian cell in culture loses approximately 10,000 purines every day. Loss of pyrimidine bases also occurs, but at a much lower rate than depurination.

Another spontaneously occurring chemical change that takes place in DNA is **deamination**, the loss of an amino group (NH_2) from a base. Deamination may be spontaneous or may be induced by mutagenic chemicals. Deamination can alter the pairing properties of a base: the deamination of cytosine, for example, produces uracil (**Figure 18.15a**), which pairs with adenine in replication. After another round of replication, the adenine will pair with thymine, creating a T•A pair in place of the original C•G pair (C•G → U•A → T•A); this chemical change is a transition mutation. This type of mutation is usually prevented by enzymes that remove uracil whenever it is found in DNA. The ability of these enzymes to recognize the product of cytosine deamination may explain why thymine, not uracil, is found in DNA. In mammals, including humans, some cytosine bases in DNA are naturally methylated and exist in the form of 5-methylcytosine (5mC) (see Figure 10.20). When deaminated, 5mC becomes thymine (**Figure 18.15b**). Because thymine pairs with adenine in replication, the deamination of 5-methylcytosine changes an original C•G pair to T•A (C•G → 5mC•G → T•G → T•A). Consequently, C•G → T•A transitions are frequent in mammalian cells, and 5mC sites are mutation hotspots in humans. **▸ TRY PROBLEM 27**

CONCEPTS

Some mutations arise from spontaneous alterations in DNA structure, such as depurination and deamination, which can alter the pairing properties of the bases and cause errors in subsequent rounds of replication.

Chemically Induced Mutations

Although many mutations arise spontaneously, a number of environmental agents are capable of damaging DNA, including certain chemicals and radiation. Any environmental agent that significantly increases the rate of mutation above the spontaneous rate is called a **mutagen**.

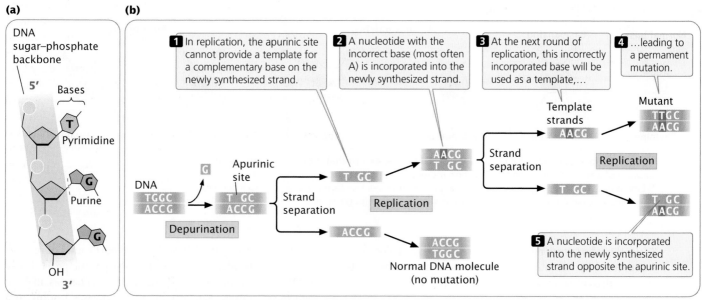

18.14 Depurination (the loss of a purine base from a nucleotide) may lead to a base substitution. (a) Depurination occurs when the covalent bond connecting a purine to the 1′ carbon of the sugar is broken (indicated by dotted red line). (b) Replication of a template strand with an apurinic site may lead to an incorporated error.

The first discovery of a chemical mutagen was made by Charlotte Auerbach, who started her career in Berlin researching the development of mutants in *Drosophila*. Faced with increasing anti-Semitism in Nazi Germany, Auerbach emigrated to Britain in 1933. There she continued her research on *Drosophila* and, in 1940, began a collaboration with pharmacologist John Robson at the University of Edinburgh on the mutagenic effects of mustard gas, which had been used as a chemical weapon in World War I. The experimental conditions were crude: they heated liquid mustard gas over a Bunsen burner on the roof of the pharmacology building and exposed the flies to the gas in a large chamber. After developing serious burns on her hands from the gas, Auerbach let others carry out the exposures, and she analyzed the flies. Auerbach and Robson showed that mustard gas is indeed a powerful mutagen, reducing the viability of gametes and increasing the numbers of mutations seen in the offspring of exposed flies. Because the research was part of the secret war effort, publication of their findings was delayed until 1947.

BASE ANALOGS One class of chemical mutagens consists of **base analogs**, chemicals with structures similar to those of any of the four standard nitrogenous bases of DNA. DNA polymerases cannot distinguish these analogs from the standard bases, so if base analogs are present during replication, they may be incorporated into newly synthesized DNA molecules. For example, 5-bromouracil (5BU) is an analog of thymine; it has the same structure as thymine except that it has a bromine (Br) atom on the 5-carbon atom instead of a methyl group (**Figure 18.16a**). Normally, 5BU pairs with adenine just as thymine does, but it occasionally mispairs with guanine (**Figure 18.16b**), leading to a transition (T•A → 5BU•A → 5BU•G → C•G), as shown in **Figure 18.17**. Through mispairing, 5BU can also be incorporated into a newly synthesized DNA strand opposite guanine. In the next round of replication, 5BU pairs with adenine, leading to another transition (G•C → G•5BU → A•5BU → A•T).

Another mutagenic chemical is 2-aminopurine (2AP), which is a base analog of adenine. Normally, 2AP base pairs with thymine, but it may mispair with cytosine, causing a transition (T•A → T•2AP → C•2AP → C•G). Alternatively, 2AP may be incorporated through mispairing into the newly synthesized DNA opposite cytosine and then later pair with thymine, leading to a C•G → C•2AP → T•2AP → T•A transition.

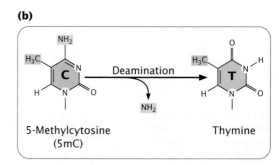

18.15 Deamination alters DNA bases.

(a)

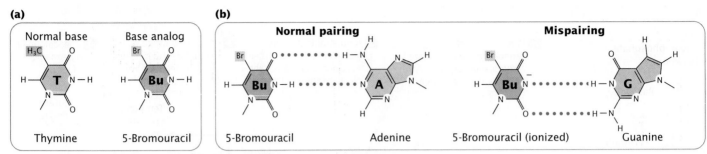

(b)

18.16 5-Bromouracil (a base analog) resembles thymine, except that it has a bromine atom in place of a methyl group on the 5-carbon atom. Because of the similarity in their structures, 5-bromouracil may be incorporated into DNA in place of thymine. Like thymine, 5-bromouracil normally pairs with adenine, but when ionized, it may pair with guanine through wobble.

In the laboratory, mutations caused by base analogs can be reversed by treatment with the same analog or by treatment with a different analog.

ALKYLATING AGENTS Alkylating agents are chemicals that donate alkyl groups, such as methyl (CH_3) and ethyl (CH_3—CH_2) groups, to nucleotide bases. For example, ethylmethylsulfonate (EMS) adds an ethyl group to guanine, producing O^6-ethylguanine, which pairs with thymine (**Figure 18.18a**). Thus, EMS produces $C \bullet G \rightarrow T \bullet A$ transitions. EMS is also capable of adding an ethyl group to thymine, producing 4-ethylthymine, which then pairs with guanine, leading to a $T \bullet A \rightarrow C \bullet G$ transition. Because EMS produces both $C \bullet G \rightarrow T \bullet A$ and $T \bullet A \rightarrow C \bullet G$ transitions, mutations produced by EMS can be reversed by additional treatment with EMS. Mustard gas is another alkylating agent.

DEAMINATING CHEMICALS In addition to its spontaneous occurrence (see Figure 18.15), deamination can be induced by some chemicals. For instance, nitrous acid deaminates

cytosine, creating uracil, which in the next round of replication pairs with adenine (**Figure 18.18b**), producing a $C \bullet G \rightarrow T \bullet A$ transition mutation. Nitrous acid also changes adenine into hypoxanthine, which pairs with cytosine, leading to a $T \bullet A \rightarrow C \bullet G$ transition. And nitrous acid also deaminates guanine, producing xanthine, which pairs with cytosine just as guanine does; however, xanthine can also pair with thymine, leading to a $C \bullet G \rightarrow T \bullet A$ transition. Nitrous acid produces exclusively transition mutations, and because both $C \bullet G \rightarrow T \bullet A$ and $T \bullet A \rightarrow C \bullet G$ transitions are produced, these mutations can be reversed with nitrous acid.

HYDROXYLAMINE Hydroxylamine is a very specific base-modifying mutagen that adds a hydroxyl group to cytosine, converting it into hydroxylaminocytosine (**Figure 18.18c**). This conversion increases the frequency of a rare tautomer that pairs with adenine instead of guanine and leads to $C \bullet G \rightarrow T \bullet A$ transitions. Because hydroxylamine acts only on cytosine, it does not generate $T \bullet A \rightarrow C \bullet G$ transitions; thus, hydroxylamine will not reverse the mutations that it produces. **▶ TRY PROBLEM 25**

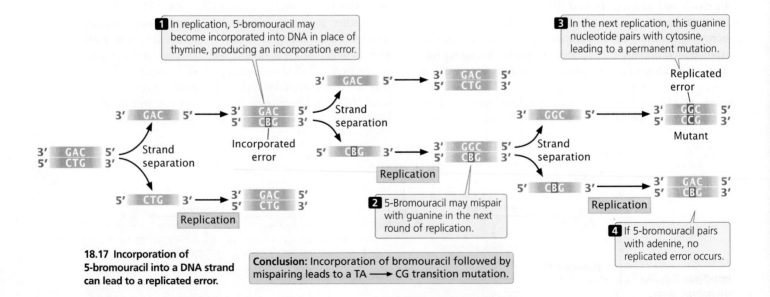

18.17 Incorporation of 5-bromouracil into a DNA strand can lead to a replicated error.

1 In replication, 5-bromouracil may become incorporated into DNA in place of thymine, producing an incorporation error.

2 5-Bromouracil may mispair with guanine in the next round of replication.

3 In the next replication, this guanine nucleotide pairs with cytosine, leading to a permanent mutation.

4 If 5-bromouracil pairs with adenine, no replicated error occurs.

Conclusion: Incorporation of bromouracil followed by mispairing leads to a TA ⟶ CG transition mutation.

	Original base	Mutagen	Modified base	Pairing partner	Type of mutation
(a)	Guanine	EMS Alkylation	O^6-Ethylguanine	Thymine	CG ⟶ TA TA ⟶ CG
(b)	Cytosine	Nitrous acid (HNO_2) Deamination	Uracil	Adenine	CG ⟶ TA TA ⟶ CG
(c)	Cytosine	Hydroxylamine (NH_2OH) Hydroxylation	Hydroxylamino-cytosine	Adenine	CG ⟶ TA

18.18 Chemicals may alter DNA bases. Shown here are some examples of mutations produced by chemical agents.

OXIDATIVE RADICALS Reactive forms of oxygen (including superoxide radicals, hydrogen peroxide, and hydroxyl radicals) are produced in the course of normal aerobic metabolism as well as by radiation, ozone, peroxides, and certain drugs. These reactive forms of oxygen damage DNA and induce mutations by bringing about chemical changes in DNA. For example, oxidation converts guanine into 8-oxy-7,8-dihydrodeoxyguanine (**Figure 18.19**), which frequently mispairs with adenine instead of cytosine, causing a G • C → T • A transversion.

INTERCALATING AGENTS Proflavin, acridine orange, ethidium bromide, and dioxin are **intercalating agents** (**Figure 18.20a**), which produce mutations by sandwiching themselves (intercalating) between adjacent bases in DNA, distorting the three-dimensional structure of the helix and causing single-nucleotide insertions and deletions in replication (**Figure 18.20b**). These insertions and deletions frequently produce frameshift mutations, so the mutagenic effects of intercalating agents are often severe. Because intercalating agents generate both insertions and deletions, they can reverse mutations they produce.

Guanine   8-Oxy-7,8-dihydrodeoxyguanine (may mispair with adenine)

Oxidative radicals

18.19 Oxidative radicals convert guanine into 8-oxy-7,8-dihydrodeoxyguanine.

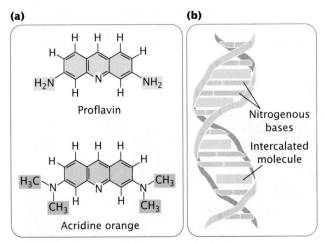

Proflavin

Acridine orange

Nitrogenous bases

Intercalated molecule

18.20 Intercalating agents are mutagens. Intercalating agents, such as proflavin and acridine orange (a), insert themselves between adjacent bases in DNA, distorting the three-dimensional structure of the DNA double helix (b).

Mutagenic chemicals can produce mutations by a number of mechanisms. Base analogs are incorporated into DNA and frequently pair with the wrong base. Alkylating agents, deaminating chemicals, hydroxylamine, and oxidative radicals change the structure of DNA bases, thereby altering their pairing properties. Intercalating agents wedge between the bases and cause single-base insertions and deletions in replication.

✔ CONCEPT CHECK 4

Base analogs are mutagenic because of which characteristic?

a. They produce changes in DNA polymerase that cause it to malfunction.
b. They distort the structure of DNA.
c. They are similar in structure to the normal bases.
d. They chemically modify the normal bases.

Radiation

In 1927, Hermann Muller demonstrated that mutations in fruit flies could be induced by X-rays. The results of subsequent studies showed that X-rays greatly increase mutation rates in all organisms. Because of their high energies, X-rays, gamma rays, and cosmic rays are all capable of penetrating tissues and damaging DNA. These forms of radiation, called ionizing radiation, dislodge electrons from the atoms they encounter, changing stable molecules into free radicals and reactive ions, which then alter the structures of bases and break phosphodiester bonds in DNA. Ionizing radiation also frequently results in double-strand breaks in DNA. Attempts to repair these breaks can produce chromosome mutations (discussed in Chapter 8).

Ultraviolet (UV) light has less energy than ionizing radiation and does not dislodge electrons, but is nevertheless highly mutagenic. Pyrimidine bases readily absorb UV light, which causes chemical bonds to form between adjacent pyrimidine molecules on the same strand of DNA, creating **pyrimidine dimers** (**Figure 18.21a**). Pyrimidine dimers consisting of two thymine bases (called thymine dimers) are

most frequent, but cytosine dimers and thymine–cytosine dimers can also form. These dimers are bulky lesions that distort the configuration of DNA (**Figure 18.21b**) and often block replication. Most pyrimidine dimers are immediately repaired by mechanisms discussed in Section 18.5, but some escape repair and inhibit replication and transcription.

When pyrimidine dimers block replication, cell division is inhibited and the cell usually dies; for this reason, UV light kills bacteria and is an effective sterilizing agent. For a mutation—a hereditary error in the genetic instructions—to occur, the replication block must be overcome. Bacteria can sometimes circumvent replication blocks produced by pyrimidine dimers and other types of DNA damage by means of the **SOS system**. This system can overcome these blocks and allow replication to proceed, but in the process, it makes numerous mistakes and greatly increases the rate of mutation. Indeed, the very reason that replication can proceed in the presence of a block is that the enzymes in the SOS system do not strictly adhere to the base-pairing rules. The trade-off is that replication continues and the cell survives, but only by sacrificing the normal accuracy of DNA synthesis.

Ionizing radiation such as X-rays and gamma rays damages DNA by dislodging electrons from atoms; these electrons then break phosphodiester bonds and alter the structure of bases. Ultraviolet light causes mutations primarily by producing pyrimidine dimers that disrupt replication and transcription. The SOS system enables bacteria to overcome replication blocks, but introduces mistakes in replication.

18.3 Mutations Are the Focus of Intense Study by Geneticists

Because mutations often have detrimental effects, they are frequently studied by geneticists. These studies have included the development of tests to determine the mutagenic properties of chemical compounds and the investigation of human populations tragically exposed to high levels of radiation.

Detecting Mutagens with the Ames Test

People in industrial societies are surrounded by a multitude of artificially produced chemicals: more than 50,000 different chemicals are in commercial and industrial use today, and from 500 to 1000 new chemicals are introduced each year. Some of these chemicals are potential **carcinogens** (substances capable of causing cancer). Many natural products are also potentially carcinogenic. One method for testing the cancer-causing potential of substances is to administer them to laboratory animals (rats or mice) and compare the incidence of cancer in the treated animals with that in control

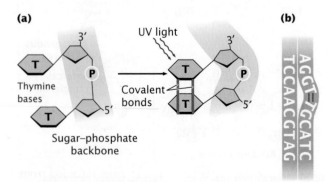

18.21 Pyrimidine dimers result from ultraviolet light.
(a) Formation of a thymine dimer. (b) A thymine dimer distorts the DNA molecule.

animals. Unfortunately, these tests are time-consuming and expensive. Furthermore, the ability of a substance to cause cancer in rodents is not always indicative of its effect on humans. After all, we aren't rats!

In 1974, Bruce Ames developed a simple test for evaluating the potential of chemicals to cause cancer. The **Ames test** is based on the principle that both cancer and mutations result from damage to DNA, and the results of experiments have demonstrated that 90% of known carcinogens are also mutagens. Ames proposed that mutagenesis in bacteria could serve as an indicator of carcinogenesis in humans.

The Ames test uses auxotrophic strains of the bacterium *Salmonella typhimurium* with traits that make them particularly susceptible to mutagens. These strains have defects in the lipopolysaccharide coat, which normally protects the bacteria from chemicals in the environment, and their DNA-repair systems have been inactivated. A recent version of the test (called Ames II) uses several auxotrophic strains that detect different types of base substitutions. Other strains detect different types of frameshift mutations.

Some compounds are not active carcinogens but can be converted into cancer-causing compounds in the body. To make the Ames test sensitive to such *potential* carcinogens, a compound to be tested is first incubated in mammalian liver extract that contains metabolic enzymes.

Each strain of bacteria used in the test carries a his^- mutation, which renders it unable to synthesize the amino acid histidine. When the bacteria are plated onto medium that lacks histidine (**Figure 18.22**), only bacteria that have undergone a reverse mutation of the histidine gene ($his^- \rightarrow his^+$) are able to synthesize histidine and grow on the medium, which makes these mutations easy to detect. Different dilutions of a chemical to be tested are added to the bacteria and plates inoculated with the bacteria. The number of mutated bacterial colonies that appear on each plate is compared with the number that appear on control plates with no chemical (i.e., that arose through spontaneous mutation). Any chemical that significantly increases the number of colonies appearing on a treated plate is mutagenic and probably also carcinogenic.

The Ames test has been applied to thousands of chemicals and commercial products. An early demonstration of its usefulness was the discovery, in 1975, that many hair dyes sold in the United States contained compounds that were mutagenic to bacteria. These compounds were then removed from most hair dyes.

CONCEPTS

The Ames test uses his^- strains of bacteria to test chemicals for their ability to produce $his^- \rightarrow his^+$ mutations. Because mutagenic activity and carcinogenic potential are closely correlated, the Ames test is widely used to screen chemicals for their cancer-causing potential.

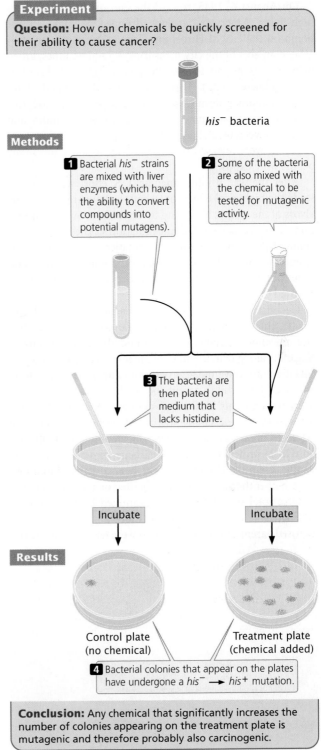

Experiment

Question: How can chemicals be quickly screened for their ability to cause cancer?

his^- bacteria

Methods

1 Bacterial his^- strains are mixed with liver enzymes (which have the ability to convert compounds into potential mutagens).

2 Some of the bacteria are also mixed with the chemical to be tested for mutagenic activity.

3 The bacteria are then plated on medium that lacks histidine.

Incubate   Incubate

Results

Control plate (no chemical)   Treatment plate (chemical added)

4 Bacterial colonies that appear on the plates have undergone a $his^- \rightarrow his^+$ mutation.

Conclusion: Any chemical that significantly increases the number of colonies appearing on the treatment plate is mutagenic and therefore probably also carcinogenic.

18.22 The Ames test is used to identify chemical mutagens.

Effects of Radiation Exposure in Humans

People are routinely exposed to low levels of radiation from cosmic, medical, and environmental sources, but there have also been tragic events that produced exposures of much higher degree.

On August 6, 1945, toward the end of World War II, a high-flying American airplane dropped a single atomic bomb on the city of Hiroshima, Japan. The explosion devastated an area of the city measuring 4.5 square miles, killed between 90,000 and 140,000 people, and injured almost as many (**Figure 18.23**). Three days later, the United States dropped another atomic bomb on the city of Nagasaki, this time destroying an area measuring 1.5 square miles and killing between 60,000 and 80,000 people. Huge amounts of radiation were released during these explosions, and many people were exposed.

After the war, a joint Japanese–U.S. effort was made to study the biological effects of radiation exposure on the survivors of the atomic blasts and their children. Somatic mutations were assessed by studying radiation sickness and cancer among the survivors; germ-line mutations were assessed by looking at birth defects, chromosome abnormalities, and gene mutations in children born to people that had been exposed to radiation.

Geneticist James Neel and his colleagues examined almost 19,000 children of people who were within 2000 meters (1.2 miles) of the center of the atomic blast at Hiroshima or Nagasaki, along with a similar number of children whose parents did not receive radiation exposure. Radiation doses were estimated for a child's parents on the basis of careful assessment of the parents' location, posture, and position at the time of the blast. A blood sample was collected from each child, and gel electrophoresis was used to investigate amino acid substitutions in 28 proteins. When rare variants were detected, blood samples from the child's parents were also analyzed to establish whether the variant was an inherited or a new mutation.

Of a total of 289,868 genes examined by Neel and his colleagues, only one mutation was found in the children of exposed parents; no gene mutations were found in the control group. From these findings, a mutation rate of 3.4×10^{-6} was estimated for the children whose parents were exposed to the blasts, which is within the range of spontaneous mutation rates observed for other eukaryotes. Neel and his colleagues also examined the frequency of chromosome mutations, the frequency of aneuploidy, and the sex ratios of children born to exposed parents. There was no evidence in any of these assays for increased mutations among the children of the people who were exposed to radiation from the atomic explosions, suggesting that germ-line mutations were not elevated.

Animal studies clearly show that radiation causes germ-line mutations, so why was there no apparent increase in germ-line mutations among the inhabitants of Hiroshima and Nagasaki? The exposed parents did exhibit an increased incidence of leukemia and other types of cancers, so somatic mutations were clearly induced. The answer to the question is not known, but the lack of germ-line mutations may be due to the fact that those people who received the largest radiation doses died soon after the blasts. Additional insights into the genetic effects of radiation have come from studies of people exposed to radiation in the Chernobyl nuclear accident in 1986 and other nuclear accidents, as well as to radiation used in medicine and industry.

THINK-PAIR-SHARE Question 5

18.4 Transposable Elements Can Cause Mutations

Transposable elements—DNA sequences that can move about in the genome—are often a cause of mutations. These mobile DNA elements have been given a variety of names, including transposons, transposable genetic elements, movable genes, controlling elements, and jumping genes. They are found in the genomes of all organisms and are abundant in many, including our own. Most transposable elements are able to insert themselves at many different locations in the genome, relying on mechanisms that are distinct from homologous recombination. Through their movement (transposition), transposable elements often cause mutations, either by inserting into a gene and disrupting it or by promoting chromosome rearrangements such as deletions, duplications, and inversions (see Chapter 8).

General Characteristics of Transposable Elements

There are many different types of transposable elements: some have simple structures, encompassing only those sequences necessary for their own transposition, whereas others have complex structures and encode a number of functions not directly related to transposition. Despite this variation, many transposable elements have certain features in common.

18.23 Hiroshima was destroyed by an atomic bomb on August 6, 1945. The atomic explosion produced many somatic mutations among the survivors. [© Bettmann/Corbis.]

Short **flanking direct repeats** from 3 to 12 bp long are present on both sides of most transposable elements. The sequences of these repeats vary, but their length is constant for each type of transposable element. These repeats are not a part of the transposable element and do not travel with it. Rather, they are generated in the process of transposition at the point of insertion. Flanking direct repeats are created when staggered cuts are made in the target DNA, as shown in **Figure 18.24**. These staggered cuts leave short single-stranded pieces of DNA on either side of the transposable element. Replication of these single-stranded pieces creates the flanking direct repeats.

At the ends of many, but not all, transposable elements are **terminal inverted repeats**—sequences from 9 to 40 bp in length that are inverted complements of each other. For example, the following sequences are inverted repeats:

$$5'–ACAGTTCAG \ldots CTGAACTGT–3'$$
$$3'–TGTCAAGTC \ldots GACTTGACA–5'$$

On the same strand, the two sequences are not simple inversions, as their name might imply; rather, they are both inverted and complementary. (Notice that the sequence from left to right in the top strand is the same as the sequence from right to left in the bottom strand.) Terminal inverted repeats are recognized by enzymes that catalyze transposition, and they are required for transposition to take place. **Figure 18.25** summarizes the general characteristics of transposable elements. **▶ TRY PROBLEM 33**

1 Staggered cuts are made in the target DNA.

CGTCGATAG
GCAGCTATC

CGTCGAT
GC
AG
AGCTATC

Transposable element

2 A transposable element inserts itself into the DNA.

CGTCGAT
GC
AG
AGCTATC

Gaps filled in by DNA polymerase

3 The staggered cuts leave short, single-stranded pieces of DNA.

CGTCGAT   TCGATAG
GCAGCTA   AGCTATC

Flanking direct repeats

4 Replication of this single-stranded DNA creates the flanking direct repeats.

18.24 Flanking direct repeats are generated when a transposable element inserts into DNA.

CONCEPTS

Transposable elements are mobile DNA sequences that often cause mutations. There are many different types of transposable elements; most generate short flanking direct repeats at the target sites as they insert. Many transposable elements also possess short terminal inverted repeats.

✔ CONCEPT CHECK 5

How are flanking direct repeats created in transposition?

The Process of Transposition

As mentioned above, **transposition** is the movement of a transposable element from one location to another. Several different mechanisms are used for transposition in both prokaryotic and eukaryotic cells. Nevertheless, all types of transposition share several steps: (1) staggered breaks are made in the target DNA (see Figure 18.24); (2) the transposable element is joined to single-stranded ends of the target DNA; and (3) DNA is replicated at the single-strand gaps. A **transposase** enzyme, often encoded by the transposable element, is used to make the staggered breaks in DNA and to integrate the transposable element into a new site.

Some transposable elements transpose as DNA and are referred to as **DNA transposons** (or *Class II transposable*

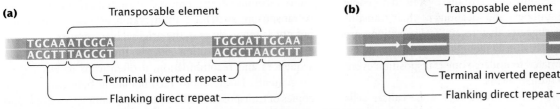

(a) Transposable element

TGCAAATCGCA   TGCGATTGCAA
ACGTTTAGCGT   ACGCTAACGTT

Terminal inverted repeat

Flanking direct repeat

(b) Transposable element

Terminal inverted repeat

Flanking direct repeat

18.25 Many transposable elements have common characteristics. (a) Most transposable elements generate flanking direct repeats on each side of the point of insertion into target DNA. Many transposable elements also possess terminal inverted repeats. (b) These representations of direct and inverted repeats are used in illustrations throughout this chapter.

elements). Other transposable elements, called **retrotransposons** or *Class I transposons*, transpose through an RNA intermediate. In this case, RNA is transcribed from the transposable element (DNA) and is then copied back into DNA by a special enzyme called reverse transcriptase. Transposable elements found in bacteria are DNA transposons. Both DNA transposons and retrotransposons are found in eukaryotes, although retrotransposons are more common.

Among DNA transposons, transposition may be replicative or nonreplicative. In **replicative transposition** (also called *copy-and-paste transposition*), a new copy of the transposable element is introduced at a new site while the old copy remains behind at the original site, so the number of copies of the transposable element increases as a result of transposition. In **nonreplicative transposition** (*cut-and-paste transposition*), the transposable element is excised from the old site and inserted at a new site without any increase in the number of its copies. Nonreplicative transposition requires the replication of only the few nucleotides that constitute the flanking direct repeats. Retrotransposons use replicative transposition only.

CONTROL OF TRANSPOSITION Many organisms limit transposition by methylating the DNA in regions where transposons are common. DNA methylation usually suppresses transcription (see Chapter 17), preventing the production of the transposase enzyme necessary for transposition. Alterations of chromatin structure are also used to prevent the transcription of transposons. In other cases, translation of the transposase mRNA is controlled. Some animals use small RNA molecules called Piwi-interacting RNAs (piRNAs, see Chapter 14) to silence transposons; piRNAs combine with Piwi proteins and inhibit the expression of transposon sequences.

TRANSPOSITION IN HUMANS About 45% of the human genome comprises sequences that are related to transposable elements, mostly retrotransposons (some research suggests that almost two-thirds of the human genome consists of transposable elements). Researchers previously assumed that most of these transposable elements are inactive and that little transposition occurs today, although it clearly took place extensively during our past evolution. More recently, however, researchers have begun to map copies of transposons across the human genome and have discovered that people often differ in the numbers and locations of their transposons. This finding suggests that recent transpositions are more common than previously thought. The L1 transposon, for example, is estimated to undergo one transposition event in about every 100 human births.

Research has also demonstrated that some cancer cells have elevated levels of transposition, probably because patterns of DNA methylation that normally inhibit transposition are disrupted in these cells.

CONCEPTS

Transposition may take place through DNA or through an RNA intermediate. In replicative transposition, a new copy of the transposable element inserts in a new location and the old copy stays behind; in nonreplicative transposition, the old copy excises from the old site and moves to a new site. Transposition through an RNA intermediate requires reverse transcription to integrate into the target site. Many cells regulate transposition by a variety of mechanisms.

The Mutagenic Effects of Transposition

Because transposable elements can insert into genes and disrupt their function, transposition is generally mutagenic. In fact, more than half of all spontaneously occurring mutations in *Drosophila* result from the insertion of a transposable element in or near a functional gene.

A number of cases of human genetic disease have been traced to the insertion of a transposable element into a vital gene. For example, insertion of the L1 transposable element into the gene for blood-clotting factor VIII has caused hemophilia. Although most mutations resulting from transposition are detrimental, transposition may occasionally activate a gene or change the phenotype of the cell in a beneficial way. For instance, bacterial transposable elements sometimes carry genes that encode antibiotic resistance, and several transposable elements have created mutations that confer insecticide resistance in insects.

A dramatic example of the mutagenic effect of transposable elements is seen in the color of grapes, which come in black, red, and white varieties (**Figure 18.26**). Black and red grapes result from the production of red pigments, called anthocyanins, in the skin, which are lacking in white grapes. White grapes resulted from a mutation in black grapes that turned off the production of anthocyanins. This mutation consisted of the insertion of a 10,422-bp retrotransposon called *Gret1* near a gene that promotes the production of anthocyanins. The *Gret1* retrotransposon apparently disrupted sequences that regulate the gene, effectively shutting down anthocyanin production and producing a white grape with no anthocyanins. Interestingly, red grapes resulted from a second mutation that occurred in white grapes (see Figure 18.26). This mutation (probably resulting from faulty recombination) removed most, but not all, of the retrotransposon, switching anthocyanin production back on, but not as intensely as in the original black grapes.

Because transposition entails the exchange of DNA sequences and recombination, it often leads to DNA rearrangements. Homologous recombination between multiple copies of transposons can lead to duplications, deletions, and inversions, as shown in **Figure 18.27**. The *Bar* mutation in *Drosophila* (see Figure 8.6) is a tandem duplication thought to have arisen through homologous recombination between

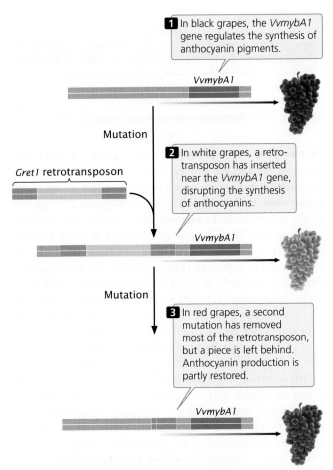

18.26 White and red color in grapes resulted from insertion and deletion of a retrotransposon, respectively.

two copies of a transposable element present in different locations on the X chromosome. Similarly, recombination between copies of the transposable element *Rider* caused a duplication that results in elongate fruit in tomatoes.

DNA rearrangements can also be caused by the excision of transposable elements in nonreplicative transposition. If the broken DNA is not repaired properly, a DNA rearrangement can be generated.

Because most transposable elements insert randomly into DNA sequences, they provide researchers with a powerful tool for inducing mutations throughout the genome, allowing them to determine the functions of genes, study genetic phenomena, and map genes. Furthermore, because the transposable element being used has a known sequence, it can serve as a "tag" for locating a gene in which a mutation has occurred. For example, researchers engineered a transposable element named *Sleeping Beauty* to induce mutations in mice and used it to search for genes that cause cancer. *Sleeping Beauty* was introduced into a strain of mice that produce the transposase needed for transposition. The transposable element inserted randomly into different locations in the genome; occasionally, it inserted into a gene that protects against cancer and destroyed its function. By looking for the

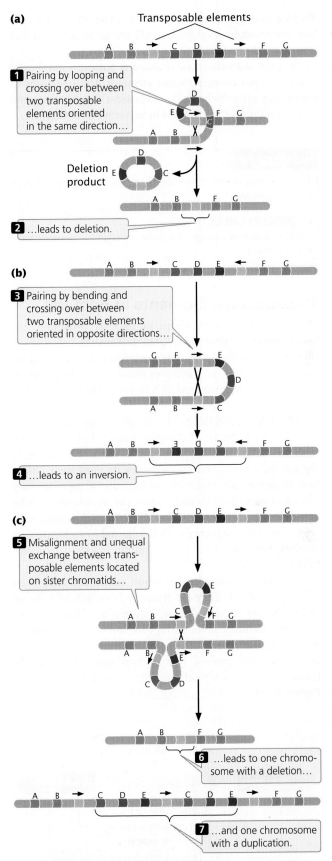

18.27 Many chromosomal rearrangements are generated by transposition.

Sleeping Beauty sequence in the DNA from the tumor cells that subsequently developed, geneticists identified a number of genes that protect against cancer.

Bacteria and eukaryotic organisms possess a number of different types of transposable elements, the structures of which vary extensively. Next, we consider the structures and types of transposable elements in bacteria and eukaryotes.

> **CONCEPTS**
>
> Transposable elements frequently cause mutations and DNA rearrangements.
>
> ✔ **CONCEPT CHECK 6**
>
> Briefly explain how transposition causes mutations and chromosome rearrangements.

Transposable Elements in Bacteria

The DNA transposons found in bacteria constitute two major groups: (1) simple transposable elements, called insertion sequences, that carry only the information required for movement, and (2) more complex transposable elements that contain DNA sequences not directly related to transposition.

INSERTION SEQUENCES The simplest type of transposable element in bacterial chromosomes and plasmids is an **insertion sequence (IS)**. This type of element carries only the genetic information necessary for its movement. Insertion sequences are common constituents of bacteria; they can also infect plasmids and viruses and, in this way, can be passed from one cell to another. Geneticists designate each type of insertion sequence with *IS* followed by an identifying number. For example, *IS1* is a common insertion sequence found in *E. coli*.

A number of different insertion sequences have been found in bacteria. They are typically from 800 to 2000 bp in length and possess the two hallmarks of transposable elements: terminal inverted repeats and the generation of flanking direct repeats at the site of insertion. Most insertion sequences contain one or two genes that encode transposase. *IS1*, a typical insertion sequence, is shown in **Figure 18.28**. ▶ TRY PROBLEM 36

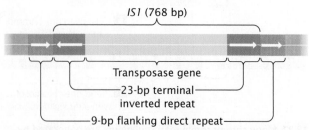

18.28 Insertion sequences such as *IS1* are simple transposable elements found in bacteria.

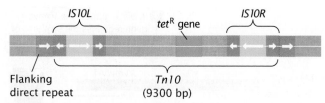

18.29 *Tn*10 is a composite transposon found in bacteria.

COMPOSITE TRANSPOSONS Any segment of DNA that becomes flanked by two copies of an insertion sequence may itself transpose, in which case it is called a **composite transposon**. Each type of composite transposon is designated by the abbreviation *Tn* followed by a number. The composite transposon *Tn10*, for example, consists of about 9300 bp that carry a gene for tetracycline resistance between two *IS10* insertion sequences (**Figure 18.29**). The two insertion sequences each have terminal inverted repeats, so the composite transposon also ends in inverted repeats. Composite transposons also generate flanking direct repeats at their sites of insertion (see Figure 18.29). The insertion sequences at the ends of a composite transposon may be in the same orientation, or they may be inverted relative to each other (as in *Tn10*).

The insertion sequences at the ends of a composite transposon are responsible for its transposition. The DNA between the insertion sequences is not required for movement and may carry additional information (such as a gene for antibiotic resistance, as in *Tn10*). Presumably, composite transposons evolve when one insertion sequence transposes to a location close to another of the same type. The transposase produced by one of the insertion sequences then catalyzes the transposition of both, allowing them to move together and carry along the DNA that lies between them. In some composite transposons (such as *Tn10*), one of the insertion sequences may be defective, so its movement depends on the transposase produced by the other.

NONCOMPOSITE TRANSPOSONS Some transposable elements in bacteria lack insertion sequences and are referred to as **noncomposite transposons**. Noncomposite transposons possess a gene for transposase and have terminal inverted repeats. For instance, the noncomposite transposon *Tn3* carries genes for transposase and resolvase (an enzyme that functions in recombination), plus a gene that encodes the enzyme β-lactamase, which provides resistance to the antibiotic ampicillin.

TRANSPOSING BACTERIOPHAGES A few bacteriophage genomes reproduce by transposition and use transposition to insert themselves into a bacterial chromosome in their lysogenic cycle. The best-studied transposing bacteriophage is Mu (**Figure 18.30**). Although Mu does not possess terminal inverted repeats, it does generate short (5-bp) flanking direct repeats when it inserts randomly into bacterial DNA. Mu replicates through transposition and causes mutations at the site of insertion, properties characteristic of transposable elements.

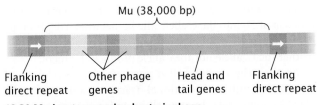

Mu (38,000 bp)

Flanking direct repeat | Other phage genes | Head and tail genes | Flanking direct repeat

18.30 Mu is a transposing bacteriophage.

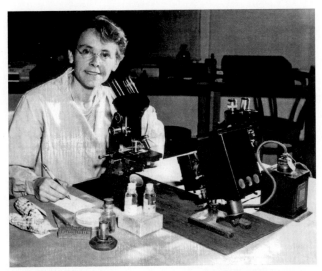

18.31 Barbara McClintock was the first to discover transposable elements. [Topham/The Image Works.]

CONCEPTS

Insertion sequences are prokaryotic transposable elements that carry only the information needed for transposition. A composite transposon consists of two insertion sequences plus intervening DNA. Noncomposite transposons in bacteria lack insertion sequences but have terminal inverted repeats and carry information not related to transposition. All of these transposable elements generate flanking direct repeats at their points of insertion.

✔ **CONCEPT CHECK 7**

Which type of transposable element possesses terminal inverted repeats?

a. Insertion sequence
b. Composite transposon
c. Noncomposite transposon
d. All of the above

Transposable Elements in Eukaryotes

Eukaryotic transposable elements can be divided into two groups. One group is structurally similar to transposable elements in bacteria, typically ending in short inverted repeats and transposing as DNA; examples include the *P* elements in *Drosophila* and the *Ac* and *Ds* elements in corn (maize). The other group comprises retrotransposons: these elements use RNA intermediates, and many are similar in structure and movement to retroviruses (see Section 9.4). The structure, function, and genomic sequences of some retrotransposons show that they are evolutionarily related to retroviruses. Although their mechanism of movement is fundamentally different from that of other transposable elements, retrotransposons also generate flanking direct repeats at the point of insertion. Retrotransposons include the *Ty* elements in yeast, the *copia* elements in *Drosophila*, and the *Alu* sequences in humans.

Ac AND Ds ELEMENTS IN CORN Transposable elements were first identified in corn more than 50 years ago by Barbara McClintock (**Figure 18.31**). McClintock spent much of her long career studying their properties, and her work stands among the landmark discoveries of genetics. Her results, however, were misunderstood and ignored for many years. Not until molecular techniques were developed in the late 1960s and 1970s was the importance of transposable

elements widely accepted. The significance of McClintock's early discoveries was finally recognized in 1983, when she was awarded the Nobel Prize in physiology or medicine.

McClintock's discovery of transposable elements had its genesis in the early work of Rollins A. Emerson on the corn genes that caused variegated (multicolored) kernels. Most corn kernels are either wholly pigmented or colorless (yellow), but Emerson noted that some yellow kernels had spots or streaks of color (**Figure 18.32**). He proposed that these kernels resulted from an unstable mutation: a mutation in the wild-type gene for pigment produced a colorless kernel, but in some cells, the mutation reverted to the wild type, causing a spot of pigment. However, Emerson didn't know why this mutation was unstable.

McClintock discovered that the cause of the unstable mutation was a gene that moved. She noticed that chromosome breakage in corn often occurred at a gene that she called *Dissociation (Ds)*, but only if another gene, the *Activator (Ac)*,

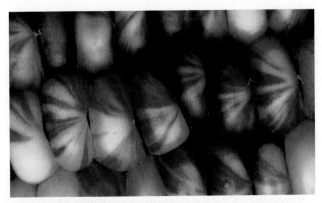

18.32 Variegated (multicolored) kernels in corn are caused by mobile genes. The study of variegated corn kernels led Barbara McClintock to discover transposable elements. [Matt Meadows/Getty Images.]

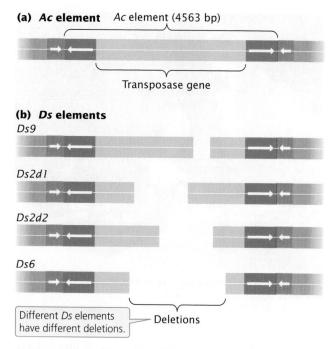

(a) *Ac* element   *Ac* element (4563 bp)

Transposase gene

(b) *Ds* elements

Ds9

Ds2d1

Ds2d2

Ds6

Different *Ds* elements have different deletions. — Deletions

18.33 *Ac* and *Ds* are transposable elements in corn.

was also present. Occasionally, the genes moved together to a different chromosomal location. McClintock called these moving genes controlling elements because they controlled the expression of other genes.

Since the significance of McClintock's work was recognized, *Ac* and *Ds* elements in corn have been examined in detail. They are DNA transposons that possess terminal inverted repeats and generate flanking direct repeats at the points of insertion. Each *Ac* element contains a single gene that encodes a transposase enzyme (**Figure 18.33a**). Thus, *Ac* elements are *autonomous*—able to transpose. *Ds* elements are *Ac* elements containing one or more deletions that have inactivated the transposase gene (**Figure 18.33b**). Unable to transpose on their own (*nonautonomous*), *Ds* elements can transpose in the presence of *Ac* elements because they still possess terminal inverted repeats recognized by Ac transposase.

Each kernel in an ear of corn is an individual offspring, originating as an ovule fertilized by a pollen grain. A kernel's pigment pattern is determined by several loci. A pigment-encoding allele at one of these loci can be designated *C*, and an allele at the same locus that does not produce pigment can be designated *c*. A kernel with genotype *cc* will be colorless—that is, yellow or white (**Figure 18.34a**); a kernel with genotype *CC* or *Cc* will produce pigment and be purple (**Figure 18.34b**).

In some kernels, a *Ds* element, transposing under the influence of a nearby *Ac* element, may insert into the *C* allele, destroying its ability to produce pigment (**Figure 18.34c**). An allele inactivated by a transposable element is designated by a subscript t; in this case, the allele would be C_t. If a kernel is

initially heterozygous with genotype *Cc*, then after the transposition of *Ds* into the *C* allele, the kernel will have genotype $C_t c$. This kernel will be colorless (white or yellow) because neither the C_t allele nor the *c* allele produces pigment.

As the original one-celled corn embryo develops and divides by mitosis, additional transpositions may take place in some cells. In any cell in which the transposable element excises from the C_t allele and moves to a new location, the *C* allele may be rendered functional again: all cells derived from those in which this event has taken place will have the genotype *Cc* and be purple. The presence of these pigmented cells, surrounded by the colorless ($C_t c$) cells, produces a purple spot or streak (called a sector) in the otherwise yellow kernel (**Figure 18.34d**). The size of the sector varies depending on when the excision of the transposable element from the C_t allele takes place. If excision is early in development, then many cells will contain the functional *C* allele, and the pigmented sector will be large; if excision is late in development, few cells will have the functional *C* allele, and the pigmented sector will be small.

❯ **TRY PROBLEM 42**

TRANSPOSABLE ELEMENTS IN *DROSOPHILA* A number of different transposable elements are found in *Drosophila*. One family of *Drosophila* transposable elements comprises the *P* elements. Most functional *P* elements are about 2900 bp long, although shorter *P* elements containing deletions also exist. Each *P* element possesses terminal inverted repeats and generates flanking direct repeats at the site of insertion. Like transposable elements in bacteria, *P* elements are DNA transposons.

Each *P* element encodes both a transposase and a repressor of transposition. The role of the repressor in controlling transposition is demonstrated dramatically in **hybrid dysgenesis**, which is the sudden appearance of numerous mutations, chromosome aberrations, and sterility in the offspring of a cross between a P^+ male fly (*with P elements*) and a P^- female fly (*without them*). The reciprocal cross between a P^+ female and a P^- male produces normal offspring.

Hybrid dysgenesis arises from a burst of transposition when *P* elements are introduced into a cell that does not possess them. In a cell that contains *P* elements, a repressor protein in the cytoplasm inhibits transposition. When a P^+ female produces eggs, the repressor protein is incorporated into the egg cytoplasm, which prevents further transposition in the embryo and thus prevents mutations from arising. The resulting offspring are fertile as adults (**Figure 18.35a**). However, a P^- female does not produce the repressor protein, so none is stored in the cytoplasm of her eggs. Sperm contain little or no cytoplasm, so a P^+ male does not contribute the repressor protein to his offspring. When eggs from a P^- female are fertilized by sperm from a P^+ male, the absence of repression allows the *P* elements contributed by the sperm to undergo rapid

(a) Genotype *cc*: no transposition

1 Cells with genotype *cc* produce no pigment,...

2 ...resulting in a colorless (yellow or white) kernel.

Ac   Ds   c

Phenotype

Yellow kernel

c

(b) Genotype *Cc*: no transposition

3 Cells with genotype *Cc* produce pigment,...

4 ...resulting in a pigmented (purple) kernel.

Ac   Ds   C

Purple kernel

c

(c) Genotype *Cc* ⟶ *C_tc*: transposition

5 An *Ac* element produces transposase,...

6 ...which stimulates transposition of a *Ds* element into the *C* allele...

Ac   Ds   C

c

C_t

Ac

Yellow kernel

c

7 ...and disrupts its pigment-producing function.

8 The resulting cells have genotype $C_t c$ and are colorless.

(d) Genotype *C_tc* ⟶ *C_tc/Cc*: mosaic (transposition during development)

9 An *Ac* element produces transposase,...

10 ...which stimulates further transposition of the *Ds* element in some cells.

Ac   Ds

c

Ac   C

Early transposition

Late transposition

Variegated kernel

c

11 As *Ds* transposes, it leaves the *C* allele, restoring the allele's function.

12 A cell in which *Ds* has transposed out of the *C* allele will produce pigment, generating spots of color in an otherwise colorless kernel.

Conclusion: Variegated corn kernels result from the excision of *Ds* elements from genes controlling pigment production during development.

18.34 Transposition results in variegated corn kernels.

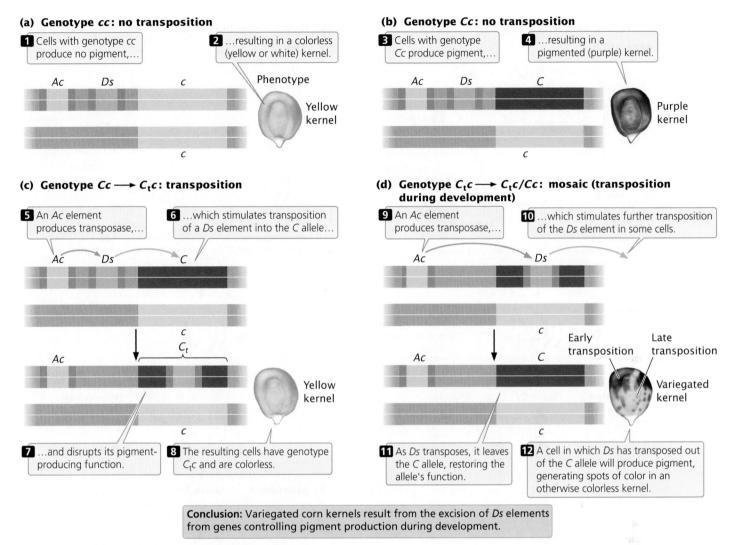

transposition in the embryo, causing hybrid dysgenesis (**Figure 18.35b**).

Hybrid dysgenesis and *P* elements have attracted geneticists' attention because *P* elements appear to have arisen within populations of *Drosophila melanogaster* within the past 50 years and may play a role in the species' evolution. Other species of *Drosophila* lack *P* elements, which are also completely absent from laboratory strains of *D. melanogaster* originally collected from the wild prior to the 1960s. Today, however, most wild populations of *D. melanogaster* have *P* elements. Laboratory strains collected during the 1970s are mixed: some have *P* elements and some do not. These observations suggest that *P* elements arose sometime during the past century and spread quickly throughout all wild populations of *D. melanogaster*. Because crosses between males with *P* elements and females without them cause sterility, *P* elements and similar transposable elements have the potential to serve as reproductive isolating mechanisms between populations and may play a role in bringing about

speciation (see Chapter 26). Thus, these observations support the idea that transposable elements play important roles in evolution. **▶ TRY PROBLEM 38**

TRANSPOSABLE ELEMENTS IN HUMANS One of the most common transposable elements in the human genome is *Alu*. Every human cell contains more than 1 million related, but not identical, copies of *Alu* in its chromosomes. *Alu* sequences are similar to that of the gene that encodes the 7S RNA molecule, which transports newly synthesized proteins across the endoplasmic reticulum. *Alu* sequences create short flanking direct repeats when they insert into DNA and have characteristics that suggest that they have transposed through an RNA intermediate.

Alu sequences belong to a class of repetitive sequences found frequently in mammalian and some other genomes. These sequences, collectively referred to as short interspersed elements (SINEs), constitute about 11% of the human genome. Most SINEs are copies of transposable elements

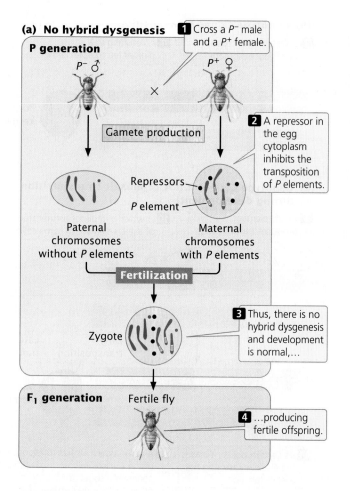

18.35 Hybrid dysgenesis in *Drosophila* is caused by the transposition of *P* elements. [After W. Y. Chooi, *Genetics* 68:213–230, 1971.]

that have been shortened at the 5′ end, probably because the reverse-transcription process used in their transposition terminated before the entire sequence was copied. SINEs have been identified as the cause of mutations in more than 20 cases of human genetic disease.

The human genome also has many transposons classified as long interspersed elements (LINEs), which are somewhat more similar in structure to retroviruses. Like SINEs, most LINEs in the human genome have been shortened at the 5′ end. The longest LINEs are usually about 6000 bp, but because most copies are shortened, the average LINE is only about 900 bp. There are approximately 900,000 copies of LINEs in the human genome, collectively constituting 21% of the total human DNA.

CONCEPTS

A great variety of transposable elements exist in eukaryotes. Some resemble transposable elements in prokaryotes, having terminal inverted repeats, and transpose as DNA. Others are retrotransposons that transpose through an RNA intermediate.

✔ CONCEPT CHECK 8

Hybrid dysgenesis results when

a. a male fly with *P* elements (P^+) mates with a female fly that lacks *P* elements (P^-).

b. a P^- male mates with a P^+ female.

c. a P^+ male mates with a P^+ female.

d. a P^- male mates with a P^- female.

Types of Transposable Elements

Now that we have looked at some examples of transposable elements, let's review their major types (**Table 18.4**).

Transposable elements can be divided into two major classes on the basis of their structure and movement. Class I comprises the retrotransposons, which possess terminal direct repeats and transpose through RNA intermediates. They generate flanking direct repeats at their points of insertion when they transpose into DNA. Retrotransposons do not encode transposase, but some types are similar in structure to retroviruses and carry sequences that produce reverse transcriptase. Transposition takes place when transcription produces an RNA intermediate, which is then transcribed into DNA by reverse transcriptase and inserted into the target site. Examples of retrotransposons include *Ty* elements in yeast and *Alu* sequences in humans. Retrotransposons are not found in prokaryotes.

Class II consists of DNA transposons that possess terminal inverted repeats and transpose as DNA. Like Class I transposons, they generate flanking direct repeats at their points of insertion into DNA. Unlike Class I transposons, all active forms of Class II transposable elements encode transposase, which is required for their movement. Some also encode resolvase, repressors, and other proteins. Their transposition may be replicative or nonreplicative, but they never use RNA intermediates. Examples of transposable elements in this class include insertion sequences and all complex transposons in bacteria, *Ac* and *Ds* elements in corn, and *P* elements in *Drosophila*.

Transposable Elements in Genome Evolution

Transposable elements have clearly played an important role in shaping the genomes of many organisms. About 50% of all spontaneous mutations in *Drosophila*, for example, are due to transposition, and much of the tremendous variation in genome size found among eukaryotic organisms is due to differences in numbers of copies of transposable elements. Homologous recombination between copies of transposable elements has been an important force in producing gene duplications and other chromosome rearrangements. Furthermore, some transposable elements may carry extra DNA with them when they transpose to a new site, providing the potential to move DNA sequences that regulate genes to new sites, where they may alter the expression of genes.

TRANSPOSABLE ELEMENTS AS GENOMIC PARASITES As we have seen, transposable elements that undergo replicative transposition leave a copy behind when they transpose to a new location and therefore increase in number within a genome with the passage of time. This ability to replicate and spread suggests that many transposable elements serve no purpose for the cell; they exist simply because they are capable of replicating and spreading. The insertion of transposable elements into a gene often destroys its function, with harmful consequences for the cell. Furthermore, expenditure of the time and energy required to replicate large numbers of transposable elements is likely to place a metabolic burden on the cell. Thus, transposable elements can be thought of as genomic parasites that provide no benefit to the cell and may even be harmful.

DOMESTICATION OF TRANSPOSABLE ELEMENTS Although many transposable elements may be genomic parasites, some have clearly evolved to serve useful purposes for their host cells. These transposons are sometimes referred to as domesticated, implying that their parasitic tendencies have been replaced by properties useful to the cell. For example, the mechanism that generates antibody diversity in the immune systems of vertebrates (see Chapter 22) probably evolved from a transposable element. Immune-system cells called lymphocytes have the ability to unite several DNA segments that encode antigen-recognition proteins. This mechanism may have arisen from a transposable element that inserted into the germ line of a vertebrate ancestor some 450 million years ago.

Transposable elements have also been played an important role in the evolution of corn. Modern corn (maize) was domesticated from teosinte in Central America more than 8000 years ago. One of the important genetic differences between teosinte and corn involves a gene called *tb1*, which encodes a transcriptional regulator that represses the growth of side branches. In corn, transcription of *tb1* is elevated compared with that in teosinte, with the result that modern corn is more upright and less branched than teosinte. Recent research has demonstrated that the elevated transcription

TABLE 18.4	Characteristics of two major classes of transposable elements			
	Structure	**Genes Encoded**	**Transposition**	**Examples**
Class I (retrotransposons)	Long terminal direct repeats; short flanking direct repeats at target site	Reverse-transcriptase gene (and sometimes others)	By RNA intermediate	*Ty* (yeast) *copia* (*Drosophila*) *Alu* (human)
Class II	Short terminal inverted repeats; short flanking direct repeats at target site	Transposase gene (and sometimes others)	Through DNA (replicative or nonreplicative)	*IS1* (*E. coli*) *Tn3* (*E. coli*) *Ac, Ds* (corn) *P* elements (*Drosophila*)

of *tb1* in corn is the result of a transposable element called *Hopscotch* that inserted into regulatory sequences that control *tb1* transcription. This insertion was probably present in teosinte as a variant and was selected by humans during the domestication of corn because it produced a more desirable plant shape. Transposable elements may also play a role in speciation, the process by which new species arise (see the earlier discussion of hybrid dysgenesis).

THINK-PAIR-SHARE Question 6

> ### CONCEPTS
>
> Many transposable elements appear to be genomic parasites, existing in large numbers because of their ability to increase in copy number. Such increases in copy numbers of transposable elements have contributed to the large size of many eukaryotic genomes. In several cases, transposable elements have become adapted for specific cellular functions.

18.5 A Number of Pathways Can Repair DNA

The integrity of DNA is under constant assault from radiation, chemical mutagens, and spontaneously arising changes. In spite of these damaging agents, the rate of mutation remains remarkably low, thanks to the efficiency with which DNA is repaired.

There are a number of complex pathways for repairing DNA, but several general statements can be made about DNA repair. First, most DNA-repair mechanisms require two nucleotide strands of DNA because most replace whole nucleotides, and a template strand is needed to specify the base sequence.

A second general feature of DNA repair is redundancy, meaning that many types of DNA damage can be corrected by more than one repair system. This redundancy illustrates the extreme importance of DNA repair to the survival of the cell: if a mistake escapes one repair system, it's likely to be repaired by another system, ensuring that almost all mistakes are corrected.

Much research has been conducted on mechanisms of DNA repair, and the 2015 Nobel Prize in chemistry was awarded to Tomas Lindahl, Paul Modrich, and Aziz Sancar for their pioneering work on DNA repair. Here, we consider five general mechanisms of DNA repair: mismatch repair, direct repair, base-excision repair, nucleotide-excision repair, and repair of double-strand breaks.

Mismatch Repair

Replication is extremely accurate: each new copy of DNA has less than one error per billion nucleotides. However, in the process of replication, mismatched bases are incorporated into the new DNA with a frequency of about 10^{-4} to 10^{-5}; this discrepancy shows that most of the errors that initially arise are

corrected and never become permanent mutations. Some of these corrections are made in proofreading (see pp. 353–354 in Chapter 12) by the DNA polymerases.

Many incorrectly inserted nucleotides that escape detection by proofreading are corrected by *mismatch repair*. Incorrectly paired bases are detected and corrected by mismatch-repair enzymes. In addition, the mismatch-repair system corrects small unpaired loops in the DNA, such as those caused by strand slippage in replication (see Figure 18.12). However, some nucleotide repeats may form secondary structures on the unpaired strand (see Figure 18.5), allowing them to escape detection by the mismatch-repair system.

After an incorporated error has been recognized, the mismatch-repair enzymes cut out a section of the newly synthesized strand and fill the gap with new nucleotides by using the original DNA strand as a template. For this strategy to work, mismatch repair must have some way of distinguishing between the old and the new strands of the DNA so that the incorporated error, but not part of the original strand, is removed.

The proteins that carry out mismatch repair in *E. coli* differentiate between old and new strands by the presence of methyl groups on certain sequences of the old strand. After replication, adenine nucleotides in the sequence GATC are methylated. The process of methylation is delayed, however, so immediately after replication, the old strand is methylated and the new strand is not (**Figure 18.36a**). The mismatch-repair complex brings an unmethylated GATC sequence close to the mismatched bases. It nicks the unmethylated strand at the GATC site (**Figure 18.36b**) and degrades the strand between the nick and the mismatched bases (**Figure 18.36c**). DNA polymerase and DNA ligase fill in the gap on the unmethylated strand with correctly paired nucleotides (**Figure 18.36d**).

Mismatch repair in eukaryotic cells is similar to that in *E. coli*, but how the old and new strands are recognized in eukaryotic cells is not known. In some eukaryotes, such as yeast and fruit flies, there is no detectable methylation of DNA, and yet mismatch repair still takes place. Humans who possess mutations in mismatch-repair genes often exhibit elevated somatic mutation rates and are frequently susceptible to colon cancer.

> ### CONCEPTS
>
> Mismatched bases and other DNA lesions are corrected by mismatch repair. Enzymes cut out a section of the newly synthesized strand of DNA and replace it with new nucleotides.
>
> ✔ **CONCEPT CHECK 9**
>
> Mismatch repair in *E. coli* distinguishes between old and new strands of DNA on the basis of
>
> a. differences in the base composition of the two strands.
> b. modification of histone proteins.
> c. base analogs on the new strand.
> d. methyl groups on the old strand.

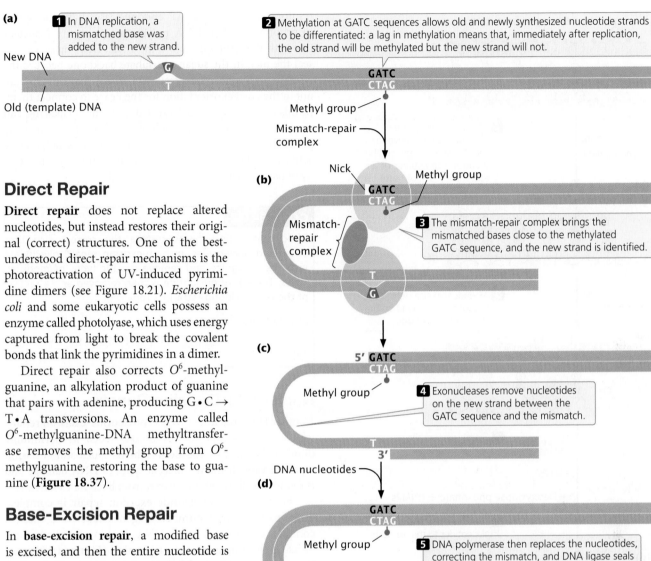

(a)

1 In DNA replication, a mismatched base was added to the new strand.

2 Methylation at GATC sequences allows old and newly synthesized nucleotide strands to be differentiated: a lag in methylation means that, immediately after replication, the old strand will be methylated but the new strand will not.

New DNA

Old (template) DNA

GATC
CTAG

Methyl group

Mismatch-repair complex

(b)

Nick

Methyl group

GATC
CTAG

Mismatch-repair complex

3 The mismatch-repair complex brings the mismatched bases close to the methylated GATC sequence, and the new strand is identified.

(c)

5' GATC
CTAG

Methyl group

4 Exonucleases remove nucleotides on the new strand between the GATC sequence and the mismatch.

3'

DNA nucleotides

(d)

GATC
CTAG

Methyl group

5 DNA polymerase then replaces the nucleotides, correcting the mismatch, and DNA ligase seals the nick in the sugar–phosphate backbone.

T
A

18.36 Many incorrectly inserted nucleotides that escape proofreading are corrected by mismatch repair.

Direct Repair

Direct repair does not replace altered nucleotides, but instead restores their original (correct) structures. One of the best-understood direct-repair mechanisms is the photoreactivation of UV-induced pyrimidine dimers (see Figure 18.21). *Escherichia coli* and some eukaryotic cells possess an enzyme called photolyase, which uses energy captured from light to break the covalent bonds that link the pyrimidines in a dimer.

Direct repair also corrects O^6-methylguanine, an alkylation product of guanine that pairs with adenine, producing $G \cdot C \rightarrow T \cdot A$ transversions. An enzyme called O^6-methylguanine-DNA methyltransferase removes the methyl group from O^6-methylguanine, restoring the base to guanine (**Figure 18.37**).

Base-Excision Repair

In **base-excision repair**, a modified base is excised, and then the entire nucleotide is replaced. The excision of modified bases is catalyzed by a set of enzymes called DNA glycosylases, each of which recognizes and removes a specific type of modified base by cleaving the bond that links that base to the 1'-carbon atom of a deoxyribose sugar (**Figure 18.38a**). Uracil glycosylase, for example, recognizes and removes uracil produced by the deamination of cytosine. Other glycosylases recognize hypoxanthine, 3-methyladenine, 7-methylguanine, and other modified bases.

After the modified base has been removed from a nucleotide, an enzyme called AP (apurinic or apyrimidinic) endonuclease cuts the phosphodiester bond, and other enzymes remove the deoxyribose sugar (**Figure 18.38b**). DNA polymerase then adds one or more new nucleotides to the exposed 3'-OH group (**Figure 18.38c**), replacing a section of nucleotides on the damaged strand. The nick in the sugar–phosphate backbone is sealed by DNA ligase (**Figure 18.38d**), and the original intact sequence is restored (**Figure 18.38e**).

Bacteria use DNA polymerase I to replace excised nucleotides, but eukaryotes use DNA polymerase β, which has no proofreading ability and tends to make mistakes. On average, DNA polymerase β makes one mistake per 4000 nucleotides inserted. About 20,000 to 40,000 base modifications per day are repaired by base excision, so DNA polymerase β may introduce as many as 10 errors per day into the human genome. How are these errors corrected? Recent research

O^6-Methylguanine

Methyltransferase

Guanine

18.37 Direct repair restores the original structures of nucleotides.

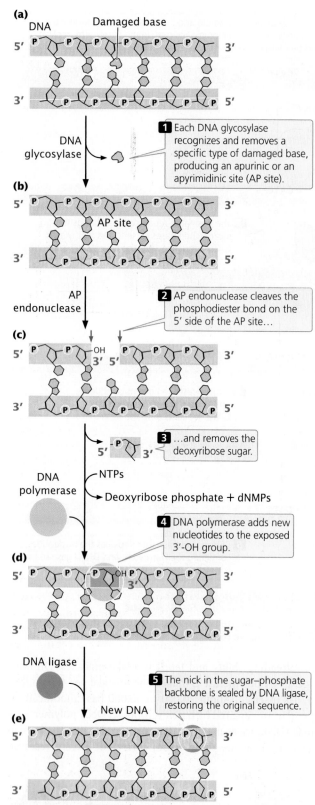

(a)

DNA   Damaged base

5′ ... 3′

3′ ... 5′

DNA glycosylase

1 Each DNA glycosylase recognizes and removes a specific type of damaged base, producing an apurinic or an apyrimidinic site (AP site).

(b)

5′ ... 3′

AP site

3′ ... 5′

AP endonuclease

2 AP endonuclease cleaves the phosphodiester bond on the 5′ side of the AP site...

(c)

5′ ...OH 3′ 5′ ... 3′

3′ ... 5′

3 ...and removes the deoxyribose sugar.

5′ -P 3′

DNA polymerase

NTPs

Deoxyribose phosphate + dNMPs

4 DNA polymerase adds new nucleotides to the exposed 3′-OH group.

(d)

5′ ...OH P 3′

3′ ... 5′

DNA ligase

5 The nick in the sugar–phosphate backbone is sealed by DNA ligase, restoring the original sequence.

New DNA

(e)

5′ ... 3′

3′ ... 5′

18.38 Base-excision repair excises modified bases and then replaces one or more nucleotides.

results show that some AP endonucleases have the ability to proofread. When DNA polymerase β inserts a nucleotide with the wrong base into the DNA, DNA ligase cannot seal the nick in the sugar–phosphate backbone because the 3′-OH and 5′-phosphate groups of adjacent nucleotides are not in the correct orientation for the ligase to connect them. In this case, AP endonuclease 1 detects the mispairing and uses its 3′→5′ exonuclease activity to excise the incorrectly paired base. DNA polymerase β then uses its polymerase activity to fill in the missing nucleotide. In this way, the fidelity of base-excision repair is maintained.

CONCEPTS

Direct-repair mechanisms restore the correct structures of altered nucleotides. In base-excision repair, glycosylase enzymes recognize and remove specific types of modified bases. The entire nucleotide is then removed and a section of the polynucleotide strand is replaced.

✔ CONCEPT CHECK 10

How does direct repair differ from mismatch repair and base-excision repair?

Nucleotide-Excision Repair

Another DNA-repair pathway is **nucleotide-excision repair**, which removes bulky DNA lesions (such as pyrimidine dimers) that distort the double helix. Nucleotide-excision repair can repair many different types of DNA damage and is found in cells of all organisms, from bacteria to humans.

The process of nucleotide-excision repair is complex; in humans, a large number of genes take part. First, a complex of enzymes scans DNA, looking for distortions of its three-dimensional configuration (**Figure 18.39a, b**). When a distortion is detected, additional enzymes separate the two nucleotide strands at the damaged region, and single-strand-binding proteins stabilize the separated strands (**Figure 18.39c**). Next, the sugar–phosphate backbone of the damaged strand is cleaved on both sides of the damage (**Figure 18.39d**). Part of the damaged strand is peeled away by helicase enzymes (**Figure 18.39e**), and the gap is filled in by DNA polymerase and sealed by DNA ligase (**Figure 18.39f**).

CONCEPTS

Nucleotide-excision repair removes and replaces many types of damaged DNA that distort the DNA structure. The two strands of DNA are separated, a section of the DNA containing the distortion is removed, DNA polymerase fills in the gap, and DNA ligase seals the filled-in gap.

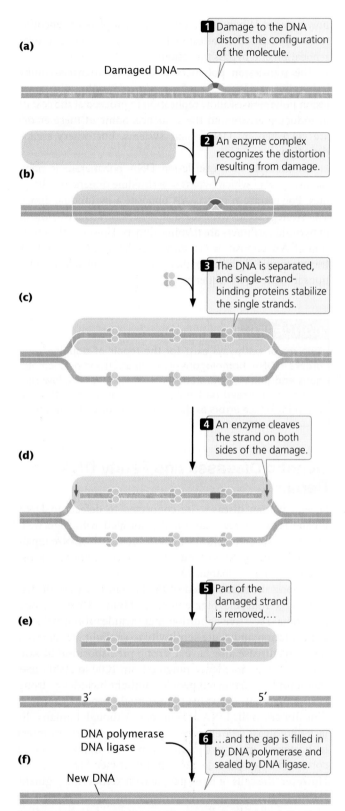

(a)

Damaged DNA

1 Damage to the DNA distorts the configuration of the molecule.

(b)

2 An enzyme complex recognizes the distortion resulting from damage.

(c)

3 The DNA is separated, and single-strand-binding proteins stabilize the single strands.

(d)

4 An enzyme cleaves the strand on both sides of the damage.

(e)

5 Part of the damaged strand is removed,…

3′   5′

DNA polymerase
DNA ligase

(f)

New DNA

6 …and the gap is filled in by DNA polymerase and sealed by DNA ligase.

18.39 Nucleotide-excision repair removes bulky DNA lesions that distort the double helix.

CONNECTING CONCEPTS

The Basic Pathway of DNA Repair

We have now examined several different mechanisms of DNA repair. What do these mechanisms have in common? How are they different? Most mechanisms of DNA repair depend on the presence of two strands because nucleotides in the damaged area are removed and replaced. Nucleotides are replaced in mismatch repair, base-excision repair, and nucleotide-excision repair, but are not replaced by direct-repair mechanisms.

Repair mechanisms that include nucleotide removal use a common four-step pathway:

1. Detection: The damaged section of the DNA is recognized.
2. Excision: Endonucleases nick the sugar–phosphate backbone on one or both sides of the DNA damage, and one or more nucleotides are removed.
3. Polymerization: DNA polymerase adds nucleotides to the newly exposed 3′-OH group by using the other strand as a template and replacing damaged nucleotides (and frequently some undamaged ones as well).
4. Ligation: DNA ligase seals the nicks in the sugar–phosphate backbone.

The primary differences in the mechanisms of mismatch, base-excision, and nucleotide-excision repair are in the details of detection and excision. In base-excision and mismatch repair, a single nick is made in the sugar–phosphate backbone on one side of the damage; in nucleotide-excision repair, nicks are made on both sides of the DNA lesion. In base-excision repair, DNA polymerase displaces the old nucleotides as it adds new nucleotides to the 3′ end of the nick; in mismatch repair, the old nucleotides are degraded; and in nucleotide-excision repair, the old nucleotides are displaced by helicase enzymes. All three mechanisms use DNA polymerase and ligase to fill in the gap produced by the excision and removal of damaged nucleotides.

Repair of Double-Strand Breaks

A common type of DNA damage is a double-strand break, in which both strands of the DNA helix are broken. Double-strand breaks are caused by ionizing radiation, oxidative radicals, and other DNA-damaging agents. These types of breaks are particularly detrimental to the cell because they stall DNA replication and may lead to chromosome rearrangements, such as deletions, duplications, inversions, and translocations. There are two major pathways for repairing double-strand breaks: homologous recombination and nonhomologous end joining.

HOMOLOGOUS RECOMBINATION Homologous recombination repairs a broken DNA molecule by using the identical or nearly identical genetic information contained in another DNA molecule, usually a sister chromatid, the same mechanism employed in the process of homologous recombination that is responsible for crossing over (see Chapter 12). Homologous recombination begins with the removal of some nucleotides at the broken ends, followed by strand invasion, displacement, and replication (see Figure 12.21). Many of the same enzymes that carry out crossing over are used in the repair of double-strand breaks by homologous recombination: two such enzymes are BRCA1 and BRCA2. The genes that encode these proteins are frequently mutated in breast-cancer cells.

NONHOMOLOGOUS END JOINING Nonhomologous end joining repairs double-strand breaks without using a homologous template. This pathway is often used when the cell is in G_1 and a sister chromatid is not available for repair through homologous recombination. Nonhomologous end joining uses proteins that recognize the broken ends of DNA, bind to the ends, and then join them together. Nonhomologous end joining is more error prone than homologous recombination and often leads to deletions, insertions, and translocations. A poorly understood alternative type of end joining is microhomology-mediated end joining (MMEJ), which relies on very small (5–25 bp) homologies near the ends of broken DNA molecules to recognize the ends and rejoin them. Because the regions of homology are so short, this mechanism is also more error prone than homologous recombination. The most common DNA-repair mechanisms are summarized in **Table 18.5**.

Translesion DNA Polymerases

As discussed in Chapter 12, the high-fidelity DNA polymerases that normally carry out replication operate at high speed and, like a high-speed train, require a smooth track—an undistorted template. Some mutations, such as pyrimidine dimers, produce distortions in the three-dimensional structure of the DNA helix, blocking replication by the high-speed

TABLE 18.5	Summary of common DNA-repair mechanisms
Repair System	**Type of Damage Repaired**
Mismatch	Replication errors, including mispaired bases and strand slippage
Direct	Pyrimidine dimers; other specific types of alterations
Base excision	Abnormal bases, modified bases, and pyrimidine dimers
Nucleotide excision	DNA damage that distorts the double helix, including abnormal bases, modified bases, and pyrimidine dimers
Homologous recombination	Double-strand breaks
Nonhomologous end joining	Double-strand breaks

polymerases. When distortions of the template are encountered, specialized translesion DNA polymerases take over replication and bypass the lesions.

The translesion polymerases are able to bypass bulky lesions but, in the process, often make errors. Thus, the translesion polymerases allow replication to proceed at the cost of introducing errors into the sequence. Some of these errors are corrected by DNA-repair systems, but others escape detection.

An example of a translesion DNA polymerase is polymerase η (eta), which bypasses pyrimidine dimers in eukaryotes. Polymerase η inserts AA opposite a pyrimidine dimer. This strategy seems to be reasonable because about two-thirds of pyrimidine dimers are thymine dimers. However, the insertion of AA opposite a CT dimer results in a $C \cdot G \rightarrow T \cdot A$ transversion. Polymerase η therefore tends to introduce mutations into the DNA sequence.

CONCEPTS

Two major pathways exist for the repair of double-strand breaks in DNA: homologous recombination and nonhomologous end joining. Translesion DNA polymerases allow replication to proceed past bulky distortions in the DNA, but often introduce errors as they bypass the distorted region.

Genetic Diseases and Faulty DNA Repair

Several human diseases are connected to defects in DNA repair. These diseases are often associated with high incidences of specific cancers because defects in DNA repair lead to increased rates of mutation. This phenomenon is discussed further in Chapter 23.

Among the best studied of the human DNA-repair diseases is xeroderma pigmentosum (**Figure 18.40**), a rare autosomal recessive condition that includes abnormal skin pigmentation and acute sensitivity to sunlight. People who have this disease also have a strong predisposition to skin cancer, with an incidence ranging from 1000 to 2000 times that found in unaffected people. Sunlight includes a strong UV component, so exposure to sunlight produces pyrimidine dimers in the DNA of skin cells. Although human cells lack photolyase (the enzyme that repairs pyrimidine dimers in bacteria), most pyrimidine dimers in humans can be corrected by nucleotide-excision repair (see Figure 18.39). However, the cells of most people with xeroderma pigmentosum have defective nucleotide-excision repair, so many of their pyrimidine dimers go uncorrected and may lead to cancer.

Xeroderma pigmentosum can result from defects in several different genes. Some individuals with xeroderma pigmentosum have mutations in a gene encoding the protein that recognizes and binds to damaged DNA; others have

TABLE 18.6	Genetic diseases associated with defects in DNA-repair systems	
Disease	**Symptoms**	**Genetic Defect**
Xeroderma Pigmentosum	Freckle-like spots on skin, sensitivity to sunlight, predisposition to skin cancer	Defects in nucleotide-excision repair
Cockayne syndrome	Dwarfism, sensitivity to sunlight, premature aging, deafness, intellectual disability	Defects in nucleotide-excision repair
Trichothiodystrophy	Brittle hair, skin abnormalities, short stature, immature sexual development, characteristic facial features	Defects in nucleotide-excision repair
Hereditary nonpolyposis colon cancer	Predisposition to colon cancer	Defects in mismatch repair
Fanconi anemia	Increased skin pigmentation, abnormalities of skeleton, heart, and kidneys, predisposition to leukemia	Possibly defects in the repair of inter-strand crosslinks
Li-Fraumeni syndrome	Predisposition to cancer in many different tissues	Defects in DNA damage response
Werner syndrome	Premature aging, predisposition to cancer	Defect in homologous recombination

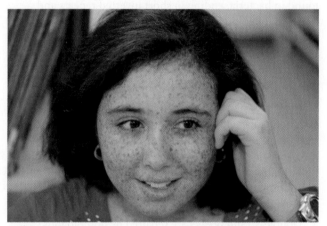

18.40 Xeroderma pigmentosum results from defects in DNA repair. The disease is characterized by freckle-like spots on the skin (shown here) and a predisposition to skin cancer. [© Stephane AUDRAS/ REA/Redux.]

mutations in a gene encoding helicase. Still others have defects in the genes whose products play a role in cutting the damaged strand on the 5′ or 3′ sides of the pyrimidine dimer. Some people have a slightly different form of the disease (xeroderma pigmentosum variant) owing to mutations

in the gene encoding polymerase η, the translesion DNA polymerase that bypasses pyrimidine dimers.

Another genetic disease caused by faulty DNA repair is an inherited form of colon cancer called hereditary nonpolyposis colon cancer (HNPCC). It is one of the most common hereditary cancers, accounting for about 15% of colon cancers. Research findings indicate that HNPCC arises from mutations in the proteins that carry out mismatch repair (see Figure 18.36). Some genetic diseases associated with defective DNA repair are listed in **Table 18.6**.

▶ TRY PROBLEM 44

THINK-PAIR-SHARE Question 7

CONCEPTS

Defects in DNA repair are the underlying cause of several genetic diseases. Many of these diseases are characterized by a predisposition to cancer.

✔ CONCEPT CHECK 11

Why are defects in DNA repair often associated with increases in cancer?

CONCEPTS SUMMARY

- Mutations are heritable changes in genetic information. Somatic mutations occur in somatic cells; germ-line mutations occur in cells that give rise to gametes.

- The simplest type of mutation is a base substitution, a change in a single base pair of DNA. Transitions are base substitutions in which purines are replaced by purines or pyrimidines are replaced by pyrimidines. Transversions are base substitutions in which a purine replaces a pyrimidine or a pyrimidine replaces a purine.

- Insertions are additions of nucleotides, and deletions are removals of nucleotides. These mutations often change the reading frame of the gene.

- Expanding nucleotide repeats are mutations in which the number of copies of a set of nucleotides increases with the passage of time; they are responsible for several human genetic diseases.

- A missense mutation alters the coding sequence so that one amino acid is substituted for another. A nonsense mutation changes a codon that specifies an amino acid into a stop codon. A silent mutation produces a synonymous codon that specifies the same amino acid as the original sequence, whereas a neutral mutation alters the amino acid sequence but does not change the functioning of the protein.

■ A suppressor mutation reverses the effect of a previous mutation at a different site and may be intragenic (within the same gene as the original mutation) or intergenic (within a different gene).

■ The mutation rate is the frequency with which a particular mutation arises. Mutation rates are influenced by both genetic and environmental factors.

■ Some mutations occur spontaneously. These mutations include the mispairing of bases in replication and spontaneous depurination and deamination.

■ Insertions and deletions can arise from strand slippage in replication or from unequal crossing over.

■ Base analogs can become incorporated into DNA in the course of replication and pair with the wrong base in subsequent replication events. Alkylating agents, deaminating chemicals, and hydroxylamine lead to mutations by modifying the chemical structure of bases. Intercalating agents insert into the DNA molecule and cause single-nucleotide insertions and deletions. Oxidative reactions alter the chemical structures of bases.

■ Ionizing radiation is mutagenic, altering base structures and breaking phosphodiester bonds. Ultraviolet light produces pyrimidine dimers, which block replication. Bacteria use the SOS system to overcome replication blocks produced by pyrimidine dimers and other lesions in DNA.

■ The Ames test uses bacteria to assess the mutagenic potential of chemical substances.

■ Transposable elements are mobile DNA sequences that insert into many locations within a genome and often cause mutations and DNA rearrangements.

■ Most transposable elements have two common characteristics: terminal inverted repeats and short flanking direct repeats generated at the point of insertion.

■ Transposable elements may transpose as DNA or through the production of an RNA intermediate that is then reverse transcribed into DNA. In replicative transposition, the transposable element is copied and the copy moved to a new site; in nonreplicative transposition, the transposable element excises from the old site and moves to a new site.

■ Insertion sequences are small bacterial transposable elements that carry only the information needed for their own movement. Composite transposons in bacteria are more complex elements that consist of DNA between two insertion sequences. Some complex transposable elements in bacteria do not contain insertion sequences.

■ DNA transposons in eukaryotic cells are similar to those found in bacteria, ending in short inverted repeats and producing flanking direct repeats at the point of insertion. Other eukaryotic transposons are retrotransposons, similar in structure to retroviruses, that transpose through RNA intermediates.

■ Transposons have played an important role in genome evolution.

■ Most damage to DNA is corrected by DNA-repair mechanisms. These mechanisms include mismatch repair, direct repair, base-excision repair, and nucleotide-excision repair. Most repair pathways require two strands of DNA. The various pathways exhibit some redundancy in the types of damage repaired.

■ Double-strand breaks are repaired by homologous recombination and nonhomologous end joining. Translesion DNA polymerases allow replication to proceed past bulky distortions in the DNA.

■ Defects in DNA repair are the underlying cause of several genetic diseases.

IMPORTANT TERMS

mutation (p. 516)
somatic mutation (p. 517)
germ-line mutation (p. 518)
gene mutation (p. 518)
base substitution (p. 518)
transition (p. 518)
transversion (p. 518)
insertion (p. 518)
deletion (p. 518)
frameshift mutation (p. 518)
in-frame insertion (p. 519)
in-frame deletion (p. 519)
expanding nucleotide repeat (p. 519)
forward mutation (p. 520)
reverse mutation (reversion) (p. 520)
missense mutation (p. 520)
nonsense mutation (p. 520)
silent mutation (p. 521)
neutral mutation (p. 521)
loss-of-function mutation (p. 521)
gain-of-function mutation (p. 521)
conditional mutation (p. 521)
lethal mutation (p. 521)
suppressor mutation (p. 521)
intragenic suppressor mutation (p. 522)
intergenic suppressor mutation (p. 523)
mutation rate (p. 525)
adaptive mutation (p. 526)
spontaneous mutation (p. 526)
induced mutation (p. 526)
incorporated error (p. 527)
replicated error (p. 527)
strand slippage (p. 527)
depurination (p. 528)
deamination (p. 528)
mutagen (p. 528)
base analog (p. 529)
intercalating agent (p. 531)
pyrimidine dimer (p. 532)
SOS system (p. 532)
carcinogen (p. 532)
Ames test (p. 533)
transposable element (p. 534)
flanking direct repeat (p. 535)

terminal inverted	replicative	composite	direct repair (p. 545)
repeat (p. 535)	transposition (p. 536)	transposon (p. 538)	base-excision
transposition (p. 535)	nonreplicative	noncomposite	repair (p. 545)
transposase (p. 535)	transposition (p. 536)	transposon (p. 538)	nucleotide-excision
DNA transposon (p. 535)	insertion	hybrid dysgenesis (p. 540)	repair (p. 546)
retrotransposon (p. 536)	sequence (IS) (p. 538)		

ANSWERS TO CONCEPT CHECKS

1. c

2. A reverse mutation restores the original phenotype by changing the DNA sequence back to the wild-type sequence. A suppressor mutation restores the phenotype by causing an additional change in the DNA at a site that is different from that of the original mutation.

3. The frequency with which changes arise in DNA, how often these changes are repaired by DNA-repair mechanisms, and our ability to detect the mutation.

4. c

5. In transposition, staggered cuts are made in DNA at the point of insertion, leaving short single-stranded pieces of DNA on either side of the transposable element. Later, replication of these pieces creates flanking direct repeats on either side of the inserted transposable element.

6. Transposition often results in mutations because the transposable element inserts into a gene, destroying its function. Chromosome rearrangements arise because transposition includes the breaking and exchange of DNA sequences. Additionally, multiple copies of a transposable element may undergo homologous recombination, producing chromosome rearrangements.

7. d

8. a

9. d

10. Direct-repair mechanisms return an altered base to its correct structure without removing and replacing nucleotides. Mismatch repair and base-excision repair remove and replace nucleotides.

11. Changes in DNA structure may not undergo repair in people with defects in DNA-repair mechanisms. Consequently, increased numbers of mutations occur at all their genes, including mutations that predispose them to cancer. This observation indicates that cancer arises from mutations in DNA.

WORKED PROBLEMS

Problem 1

The mutations produced by the following numbered compounds can be reversed by the substances shown. What conclusions can you draw about the nature of the mutations originally produced by these compounds?

	Mutations produced by compound	Reversed by			
		5-Bromouracil	EMS	Hydroxyl amine	Acridine orange
a.	1	Yes	Yes	No	No
b.	2	Yes	Yes	Some	No
c.	3	No	No	No	Yes
d.	4	Yes	Yes	Yes	Yes

» Solution Strategy

What information is required in your answer to the problem?
The types of mutations produced by each compound.

What information is provided to solve the problem?
Which substances reverse the mutations produced by each compound.

For help with this problem, review:
Chemically Induced Mutations in Section 18.2.

» Solution Steps

a. Mutations produced by compound 1 are reversed by 5-bromouracil, which produces both A•T → G•C and G•C → A•T transitions, which tells us that compound 1 produces single-base substitutions that may include the generation of either A•T or G•C base pairs. The mutations produced by compound 1 are also reversed by EMS, which, like 5-bromouracil, produces both A•T → G•C and G•C → A•T transitions; so no additional information is provided here. Hydroxylamine does not reverse the mutations produced by compound 1. Because hydroxylamine produces only C•G → T•A transitions, we know that compound 1 does not generate C•G base pairs. Acridine orange, an intercalating agent that produces frameshift mutations, also does not reverse the mutations, revealing that compound 1 produces only single-base substitutions, not insertions or deletions. In summary, compound 1 appears to cause single-base substitutions that generate T•A but not G•C base pairs.

b. Compound 2 generates mutations that are reversed by 5-bromouracil and EMS, indicating that it may produce G•C or A•T base pairs. Some of these mutations are reversed by hydroxylamine, which produces only C•G → T•A transitions, indicating that some of the mutations produced by compound 2 generate C•G base pairs. None of the mutations are reversed by acridine orange, so compound 2 does not induce insertions or deletions. In summary, compound 2 produces single-base substitutions that generate both G•C and A•T base pairs.

c. Compound 3 produces mutations that are reversed only by acridine orange, so compound 3 appears to produce only insertions and deletions.

d. Compound 4 produces mutations that are reversed by 5-bromouracil, EMS, hydroxylamine, and acridine orange, indicating that this compound produces single-base substitutions, which include the generation of G•C and A•T base pairs, as well as insertions and deletions.

Problem 2

Certain repeated sequences in eukaryotes are flanked by short direct repeats, suggesting that they originated as transposable elements. These same sequences lack introns and possess a string of thymine nucleotides at one end. Have these elements transposed as DNA or through RNA sequences? Explain your reasoning.

›› Solution Strategy

What information is required in your answer to the problem?

Whether the transposable elements transposed as DNA or through an RNA intermediate, and why you drew this conclusion.

What information is provided to solve the problem?

- The elements are flanked by short direct repeats.
- The elements lack introns, and each has a string of Ts at one end.

For help with this problem, review:

Transposition in Section 18.4.

›› Solution Steps

The absence of introns and the string of thymine nucleotides (which would be complementary to adenine nucleotides in RNA) at one end are characteristics of processed RNA. These similarities to RNA suggest that the elements were originally transcribed into mRNA, processed to remove the introns and add a poly(A) tail, and then reverse transcribed into complementary DNA that was inserted into the chromosome.

> **Recall:**
> Pre-mRNAs in eukaryotes are processed: a 5′ cap is added, introns are removed, and a 3′ poly(A) tail is added.

COMPREHENSION QUESTIONS

Section 18.1

1. What is the difference between a transition and a transversion? Which type of base substitution is more common?

2. Briefly describe expanding nucleotide repeats. How do they account for the phenomenon of anticipation?

3. What is the difference between a missense mutation and a nonsense mutation? Between a silent mutation and a neutral mutation?

4. Briefly describe two different ways in which intragenic suppressors can reverse the effects of mutations.

Section 18.2

5. How do insertions and deletions arise?

6. How do base analogs lead to mutations?

7. How do alkylating agents, nitrous acid, and hydroxylamine produce mutations?

Section 18.3

8. What is the purpose of the Ames test? How are *his⁻* bacteria used in this test?

Section 18.4

9. What general characteristics are found in many transposable elements?

10. How does a retrotransposon move?

11. Draw the structure of a typical insertion sequence and identify its parts.

12. Explain how *Ac* and *Ds* elements produce variegated corn kernels.

13. What are some differences between class I and class II transposable elements?

14. Why are transposable elements often called genomic parasites?

Section 18.5

15. List at least three different types of DNA repair and briefly explain how each is carried out.

16. What are the two major mechanisms for the repair of double-strand breaks? How do they differ?

APPLICATION QUESTIONS AND PROBLEMS

Section 18.1

17. A codon that specifies the amino acid Gly undergoes a single-base substitution to become a nonsense mutation. In accord with the genetic code given in **Figure 15.10**, is this mutation a transition or a transversion? At which position of the codon does the mutation occur?

*18. Refer to the genetic code in **Figure 15.10** to answer the following questions:

   a. If a single transition occurs in a codon that specifies Phe, what amino acids can be specified by the mutated sequence?

   b. If a single transversion occurs in a codon that specifies Phe, what amino acids can be specified by the mutated sequence?

   c. If a single transition occurs in a codon that specifies Leu, what amino acids can be specified by the mutated sequence?

   d. If a single transversion occurs in a codon that specifies Leu, what amino acids can be specified by the mutated sequence?

19. Hemoglobin is a complex protein that contains four polypeptide chains. The normal hemoglobin found in adults—called adult hemoglobin—consists of two alpha and two beta polypeptide chains, which are encoded by different loci. Sickle-cell hemoglobin, which causes sickle-cell anemia, arises from a mutation in the beta chain of adult hemoglobin. Adult hemoglobin and sickle-cell hemoglobin differ in a single amino acid: the sixth amino acid from one end in adult hemoglobin is glutamic acid, whereas sickle-cell hemoglobin has valine at this position. After consulting the genetic code provided in **Figure 15.10**, indicate the type and location of the mutation that gave rise to sickle-cell anemia.

20. The following nucleotide sequence is found on the template strand of DNA. First, determine the amino acids of the protein encoded by this sequence by using the genetic code provided in **Figure 15.10**. Then give the altered amino acid sequence of the protein that will be found in each of the following mutations:

Sequence
of DNA
template
↳   3′–TAC TGG CCG TTA GTT GAT ATA ACT–5′
↱   1   24
Nucleotide
number

   a. Mutant 1:   A transition at nucleotide 11

   b. Mutant 2:   A transition at nucleotide 13

   c. Mutant 3:   A one-nucleotide deletion at nucleotide 7

   d. Mutant 4:   A T → A transversion at nucleotide 15

   e. Mutant 5:   An addition of TGG after nucleotide 6

   f. Mutant 6:   A transition at nucleotide 9

21. Draw a hairpin structure like that shown in **Figure 18.5** for the repeated sequence found in fragile-X syndrome (see **Table 18.1**).

*22. A polypeptide has the following amino acid sequence:

Met-Ser-Pro-Arg-Leu-Glu-Gly

The amino acid sequence of this polypeptide was determined in a series of mutants listed in parts *a* through *e*. For each mutant, indicate the type of mutation that occurred in the DNA (single-base substitution, insertion, deletion) and the phenotypic effect of the mutation (nonsense mutation, missense mutation, frameshift, etc.).

   a. Mutant 1:   Met-Ser-Ser-Arg-Leu-Glu-Gly

   b. Mutant 2:   Met-Ser-Pro

   c. Mutant 3:   Met-Ser-Pro-Asp-Trp-Arg-Asp-Lys

   d. Mutant 4:   Met-Ser-Pro-Glu-Gly

   e. Mutant 5:   Met-Ser-Pro-Arg-Leu-Leu-Glu-Gly

*23. A gene encodes a protein with the following amino acid sequence:

Met-Trp-His-Arg-Ala-Ser-Phe

A mutation occurs in the gene. The mutant protein has the following amino acid sequence:

Met-Trp-His-Ser-Ala-Ser-Phe

An intragenic suppressor mutation restores the amino acid sequence to that of the original protein:

Met-Trp-His-Arg-Ala-Ser-Phe

Give at least one example of base changes that could produce the original mutation and the intragenic suppressor. (Consult the genetic code in **Figure 15.10**.)

24. A gene encodes a protein with the following amino acid sequence:

Met-Lys-Ser-Pro-Ala-Thr-Pro

A nonsense mutation caused by a single-base-pair substitution occurs in this gene, resulting in a protein with the amino acid sequence Met-Lys. An intergenic suppressor mutation allows the gene to produce the full-length protein. With the original mutation and the intergenic suppressor present, the gene now produces a protein with the following amino acid sequence:

Met-Lys-Cys-Pro-Ala-Thr-Pro

Give the location and nature of the original mutation and of the intergenic suppressor.

Section 18.2

*25. Can nonsense mutations be reversed by hydroxylamine? Why or why not?

26. The following nucleotide sequence is found in a short stretch of DNA:

5′–ATGT–3′
3′–TACA–5′

If this sequence is treated with hydroxylamine, what sequences will result after replication?

*27. The following nucleotide sequence is found in a short stretch of DNA:

5′–AG–3′
3′–TC–5′

a. Give all the mutant sequences that can result from spontaneous depurination in this stretch of DNA.

b. Give all the mutant sequences that can result from spontaneous deamination in this stretch of DNA.

28. In many eukaryotic organisms, a significant proportion of cytosine bases are naturally methylated to 5-methylcytosine. Through evolutionary time, the proportion of AT base pairs in the DNA of these organisms increases. Can you suggest a possible mechanism for this increase?

Section 18.3

*29. A chemist synthesizes four new chemical compounds in the laboratory and names them PFI1, PFI2, PFI3, and PFI4. He gives the PFI compounds to a geneticist friend and asks her to determine their mutagenic potential. The geneticist finds that all four are highly mutagenic. She also tests the capacity of mutations produced by the PFI compounds to be reversed by other known mutagens and obtains the following results. What conclusions can you make about the nature of the mutations produced by these compounds?

Mutations produced by	Reversed by			
	2-Amino-purine	Nitrous acid	Hydroxyl-amine	Acridine orange
PFI1	Yes	Yes	Some	No
PFI2	No	No	No	No
PFI3	Yes	Yes	No	No
PFI4	No	No	No	Yes

30. Mary Alexander studied the effects of radiation on mutation rates in the sperm of *Drosophila melanogaster*. She exposed *Drosophila* larvae to either 3000 roentgens (r) or 3975 r of radiation, collected the adult males that developed from irradiated larvae, and mated them with nonirradiated females that were homozygous for recessive alleles at eight loci. She then counted the number of F_1 flies that carried a new mutation at each locus. All mutant flies that appeared were used in subsequent crosses to determine if their mutant phenotypes were heritable. For the *roughoid* locus, she obtained the following results (M. L. Alexander. 1954. *Genetics* 39:409–428):

Group	Number of offspring	Offspring with a mutation at the *roughoid* locus
Control (0 r)	45,504	0
Irradiated (3000 r)	49,512	5
Irradiated (3975 r)	50,159	16

a. Calculate the mutation rates at the *roughoid* locus for the control group and the two groups of irradiated flies.

b. On the basis of these data, do you think radiation has any effect on mutation? Explain your answer.

31. What conclusion would you draw if the numbers of bacterial colonies in **Figure 18.22** were the same on the control plate and the treatment plate? Explain your reasoning.

32. A genetics instructor designs a laboratory experiment to study the effects of UV radiation on mutation in bacteria. In the experiment, the students spread bacteria on petri plates, expose the plates to UV light for different lengths of time, place the plates in an incubator for 48 hours, and then count the number of colonies that appear on each plate. The bacteria that have received more UV radiation should have more pyrimidine dimers, which block replication; thus, fewer colonies should appear on the plates exposed to UV light for longer periods. Before the students carry out the experiment, the instructor warns them that while the bacteria are in the incubator, the students must not open the incubator door unless the room is darkened. Why should the bacteria not be exposed to light?

Section 18.4

*33. A particular transposable element generates flanking direct repeats that are 4 bp long. Give the sequence that will be found on both sides of the transposable element if this transposable element inserts at the position indicated on each of the following sequences.

a. Transposable
element

5′–ATTCGAACTGACCGATCA–3′

b. Transposable
element

5′–ATTCGAACTGACCGATCA–3′

34. White eyes in *Drosophila melanogaster* result from an X-linked recessive mutation. Occasionally, white-eyed mutants give rise to offspring that possess white eyes with small red spots. The number, distribution, and size of the red spots are variable. Explain how a transposable element could be responsible for this spotting phenomenon.

35. What factor might potentially determine the length of the flanking direct repeats that are produced in transposition?

*36. Which of the following pairs of sequences might be found at the ends of an insertion sequence?

a. 5′–GGGCCAATT–3′   and   5′–CCCGGTTAA–3′

b. 5′–AAACCCTTT–3′   and   5′–AAAGGGTTT–3′

c. 5′–TTTCGAC–3′   and   5′–CAGCTTT–3′

d. 5′–ACGTACG–3′   and   5′–CGTACGT–3′

e. 5′–GCCCCAT–3′   and   5′–GCCCAT–3′

37. Explain why the corn kernel in **Figure 18.34d** is variegated, with some areas colored and some areas lacking pigment.

*38. Two different strains of *Drosophila melanogaster* are mated in reciprocal crosses. When strain A males are crossed with strain B females, the progeny are normal. However, when strain A females are crossed with strain B males, there are many mutations and chromosome rearrangements in the gametes of the F_1 progeny, and the F_1 generation is effectively sterile. Explain these results.

39. An insertion sequence contains a large deletion in its transposase gene. Under what circumstances would this insertion sequence be able to transpose?

40. Zidovudine (AZT) is a drug used to treat patients with AIDS. AZT works by blocking the reverse-transcriptase enzyme used by the human immunodeficiency virus (HIV), the causative agent of AIDS. Do you expect that AZT would have any effect on transposable elements? If so, what type of transposable elements would be affected, and what would be the most likely effect?

41. A transposable element is found to encode a reverse transcriptase enzyme. On the basis of this information, what conclusions can you draw about the likely structure and method of transposition of this element?

*42. A geneticist examines an ear of corn in which most kernels are yellow, but he finds a few kernels with purple spots, as shown here. Give a possible explanation for the appearance of the purple spots in these otherwise yellow kernels, accounting for the different sizes of the spots. (Hint: See the section on *Ac* and *Ds* elements in corn.)

Section 18.5

43. Which DNA–repair mechanism would most likely correct the incorporated error labeled by balloon 2 in **Figure 18.11**?

*44. A plant breeder wants to isolate mutants in tomatoes that are defective in DNA repair. However, this breeder does not have the expertise or equipment to study enzymes in DNA-repair systems. How can the breeder identify tomato plants that are deficient in DNA repair? What are the traits to look for?

CHALLENGE QUESTIONS

Section 18.1

45. Robert Bost and Richard Cribbs studied a strain of E. coli (*araB14*) that possessed a nonsense mutation in the structural gene that encodes L-ribulokinase, an enzyme that allows the bacteria to metabolize the sugar arabinose (R. Bost and R. Cribbs. 1969. *Genetics* 62:1–8). From the *araB14* strain, they isolated some bacteria that possessed mutations that caused them to revert back to the wild type. Genetic analysis of these revertants showed that they possessed two different suppressor mutations. One suppressor mutation (*R1*) was linked to the original mutation in L-ribulokinase and probably occurred at the same locus. By itself, this mutation allowed the production of L-ribulokinase, but the enzyme produced was not as effective in metabolizing arabinose as the enzyme encoded by the wild-type allele. The second suppressor mutation (*Su*ᴮ) was not linked to the original mutation. In conjunction with the *R1* mutation, *Su*ᴮ allowed the production of L-ribulokinase, but *Su*ᴮ by itself was not able to suppress the original mutation.

   a. On the basis of this information, are the *R1* and *Su*ᴮ mutations intragenic suppressors or intergenic suppressors? Explain your reasoning.

   b. Propose an explanation for how *R1* and *Su*ᴮ restore the ability of *araB14* to metabolize arabinose and why *Su*ᴮ is able to more fully restore this ability.

46. Achondroplasia is an autosomal dominant disorder characterized by disproportionate short stature: the legs and arms of people with achondroplasia are short compared with the head and trunk. The disorder is due to a base substitution in the gene, located on the short arm of chromosome 4, that encodes fibroblast growth factor receptor 3 (FGFR3).

Although achondroplasia is clearly inherited as an autosomal dominant trait, more than 80% of the people who have achondroplasia are born to parents with normal stature. This high percentage indicates that most cases are caused by newly arising mutations; these cases (not inherited from an affected parent) are referred to as *sporadic*. Studies have demonstrated that sporadic cases of achondroplasia are almost always caused by mutations inherited from the father (paternal mutations). In addition,

A family of three who have achondroplasia. [AP Photo/ Gail Burton.]

the occurrence of achondroplasia is higher among the children of older fathers; approximately 50% of children with achondroplasia are born to fathers older than 35 years of age. There is no association with maternal age. The mutation rate for achondroplasia (about 4×10^{-5} mutations per gamete) is high compared with those for other genetic disorders. Explain why most spontaneous mutations for achondroplasia are paternal in origin and why the occurrence of achondroplasia is higher among older fathers.

47. Tay–Sachs disease is a severe autosomal recessive genetic disease that produces deafness, blindness, seizures, and, eventually, death at 2 to 3 years of age. The disease results from a defect in the *HEXA* gene, which encodes hexosaminidase A. This enzyme normally degrades G_{M2} gangliosides. In the absence of hexosaminidase A, G_{M2} gangliosides accumulate in the brain. The results of molecular studies showed that the most common mutation causing Tay–Sachs disease is a 4-bp insertion that produces a downstream premature stop codon. Results of further studies have revealed that the transcription of the *HEXA* gene is normal in people who have Tay–Sachs disease, but the *HEXA* mRNA is unstable. Propose a mechanism to account for how a premature stop codon could cause mRNA instability.

48. *Ochre* and *amber* are two distinct nonsense mutations. Before the genetic code was worked out, Sydney Brenner, Anthony O. Stretton, and Samuel Kaplan applied different types of mutagens to bacteriophages in an attempt to determine the bases present in the codons responsible for *amber* and *ochre* mutations. They knew that the *ochre* and *amber* mutations were suppressed by different types of suppressor mutations, which demonstrated that each is a different stop codon. They obtained the following results:

   (1) A single-base substitution could convert an *ochre* mutation into an *amber* mutation.
   (2) Hydroxylamine induced both *ochre* and *amber* mutations in wild-type phages.
   (3) 2-Aminopurine caused *ochre* to mutate to *amber*.
   (4) Hydroxylamine did not cause *ochre* to mutate to *amber*.

   These data do not allow the complete nucleotide sequence of the *amber* and *ochre* codons to be worked out, but they do provide some information about the bases found in the nonsense mutations.

   a. What conclusions about the bases found in the codons of *amber* and *ochre* mutations can be made from these observations?

   b. Of the three nonsense codons (UAA, UAG, UGA), which represents the *ochre* mutation?

Section 18.4

49. Marilyn Houck and Margaret Kidwell proposed that P elements were carried from *Drosophila willistoni* to *Drosophila melanogaster* by mites that feed on fruit flies (M. A. Houck et al. 1991. *Science* 253:1125–1129). What evidence do you think would be required to demonstrate that *D. melanogaster* acquired *P* elements in this way? Propose a series of experiments to provide such evidence.

Section 18.5

50. Trichothiodystrophy is a human inherited disorder characterized by premature aging, including osteoporosis, osteosclerosis, early graying, infertility, and reduced life span. The results of studies showed that the mutation that causes this disorder occurs in a gene that encodes a DNA helicase. Propose a mechanism for how a mutation in a DNA helicase might cause premature aging. Be sure to relate the symptoms of the disorder to possible functions of the helicase enzyme.

THINK-PAIR-SHARE QUESTIONS

Section 18.1

1. Are mutations good or bad? Explain your response to this question.

2. Explain why loss-of-function mutations are frequently recessive, whereas gain-of-function mutations are frequently dominant.

3. As discussed, the overall rate of mutations in humans is estimated to be about 1×10^{-8} mutations per base pair per generation. How many new mutations would you expect each person to carry, on average, based on this mutation rate? Other studies have estimated that each person carries about 100 new loss-of-function mutations. How does this number compare with your estimate of the number of mutations based on the mutation rate? What might account for any differences?

Section 18.2

4. What accounts for the amazingly accurate replication of DNA, which keeps the mutation rate low?

Section 18.3

5. To determine whether radiation associated with the atomic bombings of Hiroshima and Nagasaki produced recessive germ-line mutations, scientists examined the sex ratio of the children of the survivors of the blasts. Can you explain why an increase in germ-line mutations might be expected to alter the sex ratio?

Section 18.4

6. Discuss the differences between transposable elements that are genomic parasites and transposable elements that have become domesticated. Do you think these terms (genomic parasite and domesticated transposable elements) are good choices for these types of elements? Why or why not?

Section 18.5

7. Research has shown that more mutations accumulate in regions of a chromosome that consist of compact chromatin, such as heterochromatin. Offer an explanation for why mutation rates would be higher where chromatin is more compact.

Molecular Genetic Analysis and Biotechnology

For over 40 years, entertainer Jerry Lewis hosted the Jerry Lewis MDA Labor Day Telethon, raising almost $2.5 billion to support research, education, and medical services for neuromuscular diseases, including muscular dystrophy. CRISPR-Cas9 genome editing techniques are being developed for possible use in treating Duchenne muscular dystrophy. [Chris Farina/Getty Images.]

Editing the Genome with CRISPR-Cas9

One of the most remarkable of all DNA sequences is the human gene for dystrophin. This gene, located on the X chromosome, is enormous, encompassing 2.2 million base pairs, including 79 exons and 78 introns. Its product, dystrophin, connects the cytoskeleton of a muscle cell to its extracellular matrix and is critical for muscle contraction.

Mutations in the dystrophin gene are the cause of Duchenne muscular dystrophy, a fatal muscle disorder that strikes nearly 1 in 3500 boys. At birth, these boys are healthy, but muscle weakness begins to appear between 3 and 5 years of age. As the disorder progresses, the child stumbles frequently, has difficulty climbing stairs, and is unable to rise from a sitting position. In time, the arm and leg muscles become progressively weaker. By age 11, most boys with the disorder must use a wheelchair, and by age 20, many have died. At present, there is no cure for the disease.

Duchenne muscular dystrophy was first recognized in 1852, and the disease was fully described in 1861 by Benjamin A. Duchenne, a Paris physician. Even before Mendel's laws were discovered, physicians noticed its X-linked pattern of inheritance, remarking that the disease developed almost exclusively in males and seemed to be inherited through unaffected mothers. In spite of this early recognition of its hereditary basis, the molecular basis of Duchenne muscular dystrophy remained a mystery for many years. In 1985, Louis Kunkel and his colleagues at Harvard Medical School observed that the X chromosome of a boy with Duchenne muscular dystrophy had a visible deletion. Reasoning that this boy's disease was caused by the absence of a gene within that deletion, they used the deletion to pinpoint the location on the X chromosome of a gene that, when mutated or absent, caused Duchenne muscular dystrophy. Kunkel and his colleagues eventually located and cloned the piece of DNA responsible for the disease and discovered the dystrophin gene.

Ever since the discovery of the dystrophin gene, researchers and physicians have dreamed of treating Duchenne muscular dystrophy with gene therapy—restoring dystrophin production by inserting a corrected copy of the dystrophin gene into the patient's cells—but the size of the dystrophin gene has made this approach challenging. Recently, however, geneticists have developed a novel and interesting approach to gene therapy that makes use of a new technique for editing the genome.

Many of the mutations that cause Duchenne muscular dystrophy consist of duplications or deletions that cause frameshift mutations, leading to premature stop codons that produce a truncated, nonfunctional dystrophin. The huge dystrophin protein consists of several domains. Some of these domains are not essential for muscle function. Researchers proposed that if the exons where mutations occurred could be removed, the premature stop codons would be eliminated. The resulting protein would be missing some amino acids, but a dystrophin that provided some muscle function might still be produced.

The trick to making this strategy work is the ability to selectively edit DNA sequences that control the splicing of pre-mRNA. Until recently, that wasn't possible. But within the past few years, a powerful new tool, called CRISPR-Cas9, has been developed that allows precision editing of the genome. In recent experiments, CRISPR-Cas9 was used to engineer changes in the genomes of mice that bring about exon skipping in the dystrophin gene, demonstrating the potential feasibility of this approach.

The geneticists used a mouse model of Duchenne muscular dystrophy called *mdx* mice. These mice possess a premature stop codon in exon 23 of the dystrophin gene and exhibit many of the symptoms of the human disorder. The CRISPR-Cas9 system consists of a protein called Cas9 that cuts DNA, along with two RNA molecules that can be genetically engineered to guide Cas9 to specific sequences—in this case, to the ends of exon 23. The geneticists loaded viruses with genetic material encoding these CRISPR-Cas9 components, then injected *mdx* mice with the viruses.

The results were remarkable. Delivered by the virus, CRISPR-Cas9 entered muscle cells and produced double-stranded cuts on either side of exon 23. Exon 23, along with the disease-causing premature stop codon it contained, was removed from the genomes of some cells. The result was synthesis of a partially functional dystrophin protein that enhanced muscle function in the mice. The mice weren't completely cured, and much work needs to be done before these methods can be applied to humans, but the results are encouraging and suggest that CRISPR-Cas9 might be used in a similar way to help restore muscle function to boys with Duchenne muscular dystrophy.

THINK-PAIR-SHARE

■ Why does Duchenne muscular dystrophy affect only boys? Is it possible for it to occur in girls? If so, how might this take place?

■ Becker muscular dystrophy (BMD) is another type of muscular dystrophy that results from mutations in the dystrophin gene, but in which the symptoms are less severe than those of Duchenne muscular dystrophy (DMD). About 60% of the mutations that cause both DMD and BMD are deletions. In BMD, but not DMD, some multiple of three base pairs is usually deleted. Propose an explanation for why the symptoms of BMD are usually less severe than those seen in DMD and how this difference might be related to the types of deletions.

The possibility of using CRISPR-Cas9 to genetically engineer a treatment for muscular dystrophy illustrates the power of molecular techniques for manipulating DNA sequences. This chapter introduces some of the techniques used in molecular genetic analysis. We begin by considering the challenges of working at the molecular level. We then examine a number of methods used to analyze and alter DNA, placing emphasis on several transformative techniques that are used to cut, clone, amplify, and sequence DNA. Finally, we explore some of the applications of molecular genetic analysis.

19.1 Genetics Has Been Transformed by the Development of Molecular Techniques

A vast array of molecular methods is now available for probing the nature of hereditary information and revealing the details of genetic processes. These molecular techniques have drastically altered the way in which genes are studied. Previously, information about the structure and organization of genes was gained by examining their phenotypic effects, but molecular genetic analysis now allows the nucleotide sequences themselves to be read. This analysis has provided new information about the structure and function of genes and has altered many fundamental concepts of genetics. Our detailed understanding of genetic processes such as replication, transcription, translation, RNA processing, and gene regulation has been obtained through the use of molecular genetic techniques. These techniques are used in many other fields as well, including biochemistry, microbiology, developmental biology, neurobiology, evolution, and ecology. Molecular genetic techniques are also being used to create a number of commercial products, including drugs, hormones, enzymes, and crops (**Figure 19.1**). The **biotechnology** industry has grown up around the use of these techniques to develop new products. In medicine, molecular genetic analysis is being used to probe the nature of cancer, diagnose genetic and infectious diseases, produce drugs, and treat hereditary disorders.

Key Innovations in Molecular Genetics

The field of molecular genetics had its inception with the discovery of the three-dimensional structure of DNA by Watson, Crick, Franklin, and Wilkins in 1953. Since that discovery, the field has witnessed at least four major transformations, each driven by the development of powerful new techniques that provided important new insights into basic genetic processes. These seminal transformations include (1) the development of recombinant DNA technology, which allowed DNA from different sources to be combined; (2) the invention of the polymerase chain reaction, which allowed very small quantities of specific DNA fragments to be quickly amplified; (3) the development of quick and accurate methods of determining DNA sequences; and (4) the engineering of CRISPR-Cas systems for accurate and efficient editing of genome sequences. In this chapter, we discuss each of these methods, along with supporting techniques and some of their applications.

THINK-PAIR-SHARE Question 1

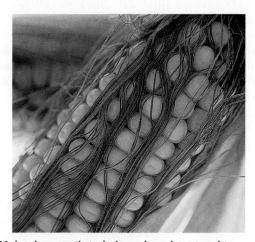

19.1 Molecular genetic techniques have been used to create genetically modified crops. Genetically engineered corn now constitutes 92% of all corn grown in the United States. [Chris Knapton/ Science Source.]

CONCEPTS

Techniques of molecular genetics are used to locate, analyze, alter, sequence, study, and recombine DNA sequences. These techniques are used to probe the structure and function of genes, address questions in many areas of biology, create commercial products, and diagnose and treat diseases. Four key innovations that have revolutionized genetics include recombinant DNA technology, the polymerase chain reaction, DNA sequencing technology, and genome editing methods.

Working at the Molecular Level

The manipulation of genes at the molecular level presents a serious challenge, often requiring strategies that may not, at first, seem obvious. The basic problem is that genes are minute and every cell contains thousands of them. Individual nucleotides cannot be seen, and no physical features mark the beginning or the end of a gene.

Let's consider a typical situation faced by a molecular geneticist. Suppose we want to use bacteria to produce large quantities of a human protein. The first and most formidable problem is to find the gene that encodes the desired protein. A haploid human genome consists of 3.2 billion base pairs of DNA. Let's assume that the gene that we want to isolate is 3000 bp long. Our target gene occupies only one-millionth of the genome, so searching for our gene in the huge expanse of genomic DNA is more difficult than looking for the proverbial needle in a haystack. But even if we are able to locate the gene, how do we separate it from the rest of the DNA?

If we succeed in locating and isolating the desired gene, we will next need to insert it into a bacterial cell. Linear fragments of DNA are quickly degraded by bacteria, so the gene must be inserted in a stable form. It must also be able to replicate successfully or it will not be passed on when the cell divides. If we succeed in transferring our gene to bacteria in a stable form, we must still ensure that the gene is properly transcribed and translated. Furthermore, the methods used to isolate and transfer genes are inefficient: of a million bacterial cells that are subjected to these procedures, only *one* cell might successfully take up and express the human gene. So we must search through many cells to find the one containing the recombinant DNA. We are back to the problem of the needle in a haystack.

Although these problems might seem insurmountable, molecular techniques have been developed to overcome them. Human genes are now routinely transferred to and expressed in bacterial cells.

CONCEPTS

Molecular genetic analyses require special techniques because individual genes make up a tiny fraction of the cellular DNA and cannot be seen.

19.2 Molecular Techniques Are Used to Cut and Visualize DNA Sequences

Often a first step in the molecular analysis of a DNA segment or gene is to isolate it from the remainder of the DNA so that further analyses can be carried out. In the sections that follow, we examine three groups of molecular techniques that are used to cut DNA segments: restriction enzymes, engineered nucleases, and CRISPR-Cas systems.

Recombinant DNA Technology

Techniques for accurately and efficiently cleaving DNA helped to usher in the first major revolution in molecular genetics: the development of recombinant DNA technology. In 1973, a group of scientists produced the first organisms with recombinant DNA molecules. The group, led by Stanley Cohen, at Stanford University, and Herbert Boyer, at the University of California, San Francisco School of Medicine, inserted a piece of DNA from one plasmid into another, creating an entirely new, recombinant DNA molecule. They then introduced the recombinant plasmid into *E. coli* cells. These experiments ushered in one of the most momentous revolutions in the history of science.

Recombinant DNA technology is a set of molecular techniques for locating, isolating, altering, and studying DNA segments. The term *recombinant* is used because frequently the goal is to combine DNA from two distinct sources. Genes from two different bacteria might be joined, for example, or a human gene might be inserted into a viral chromosome. Commonly called **genetic engineering**, recombinant DNA technology now encompasses many molecular techniques that can be used to analyze, alter, and recombine virtually any DNA sequences from any number of sources.

Restriction Enzymes

A key event in the development of recombinant DNA technology was the discovery in the late 1960s of **restriction enzymes** (also called **restriction endonucleases**), which recognize specific nucleotide sequences in DNA and make double-stranded cuts at those sequences (called *restriction sites*). Restriction enzymes are produced naturally by bacteria and are used in defense against viruses. A bacterium protects its own DNA from a restriction enzyme by modifying the recognition sequence, usually by adding methyl groups to its DNA.

Several different types of restriction enzymes have been isolated from bacteria. Type II restriction enzymes recognize specific sequences and cut the DNA at defined sites within or near the recognition sequence. Virtually all molecular genetics work is done with type II restriction enzymes; discussions of restriction enzymes throughout this book refer to type II enzymes.

More than 800 different restriction enzymes that recognize and cut DNA at more than 100 different sequences have been isolated from bacteria. Many of these enzymes are commercially available; examples of some commonly used restriction enzymes are given in **Table 19.1**. The name of each restriction enzyme begins with an abbreviation that signifies its bacterial origin.

The sequences recognized by restriction enzymes are usually from 4 to 8 bp long; most of these enzymes recognize a sequence of 4 or 6 bp. Most recognition sequences are palindromic—sequences that read the same (5′ to 3′) on the two complementary DNA strands.

Some restriction enzymes make staggered cuts in the DNA. For example, *Hin*dIII recognizes the following sequence:

$$\downarrow$$
$$5'-AAGCTT-3'$$
$$3'-TTCGAA-5'$$
$$\uparrow$$

*Hin*dIII cuts the sugar–phosphate backbone of each strand at the point indicated by the arrow, generating fragments with short, single-stranded overhanging ends:

$$5'-A \qquad AGCTT-3'$$
$$3'-TTCGA \qquad A-5'$$

Such ends are called **cohesive ends**, or *sticky ends*, because they are complementary to each other and can spontaneously pair to connect the fragments. Thus, DNA fragments with sticky ends can be "glued" together: any two such fragments cleaved by the same enzyme will have complementary ends and will pair (**Figure 19.2**). When their cohesive ends have paired, the two DNA fragments can be joined together permanently by DNA ligase, which seals nicks between the sugar–phosphate groups of the fragments.

Not all restriction enzymes produce staggered cuts and sticky ends (see Figure 19.2a). *Pvu*II cuts in the middle of its recognition sequence, and the cuts on the two strands are directly opposite each other, producing blunt-ended fragments that must be joined together in other ways:

$$\downarrow$$
$$5'-CAGCTG-3'$$
$$3'-GTCGAC-5'$$
$$\uparrow$$

$$\downarrow$$

$$5'-CAG \qquad CTG-3'$$
$$3'-GTC \qquad GAC-5'$$

TABLE 19.1	Characteristics of some common type II restriction enzymes used in recombinant DNA technology		
Enzyme	**Microorganism from Which Enzyme Is Produced**	**Recognition Sequence**	**Type of Fragment End Produced**
*Bam*HI	*Bacillus amyloliquefaciens*	↓ 5′–GGATCC–3′ 3′–CCTAGG–3′ ↑	Cohesive
*Cof*I	*Clostridium formicoaceticum*	↓ 5′–GCGC–3′ 3′–CGCG–5′ ↑	Cohesive
*Eco*RI	*Escherichia coli*	↓ 5′–GAATTC–3′ 3′–CTTAAG–5′ ↑	Cohesive
*Eco*RII	*Escherichia coli*	↓ 5′–CCAGG–3′ 3′–GGTCC–5′ ↑	Cohesive
*Hae*III	*Haemophilus aegyptius*	↓ 5′–GGCC–3′ 3′–CCGG–5′ ↑	Blunt
*Hind*III	*Haemophilus influenzae*	↓ 5′–AAGCTT–3′ 3′–TTCGAA–5′ ↑	Cohesive
*Pvu*II	*Proteus vulgaris*	↓ 5′–CAGCTG–3′ 3′–GTCGAC–5′ ↑	Blunt

Note: The first three letters of the abbreviation for each restriction enzyme refer to the bacterial species from which the enzyme was isolated (e.g., *Eco* refers to *E. coli*). A fourth letter may refer to the strain of bacteria from which the enzyme was isolated (the "R" in *Eco*RI indicates that this enzyme was isolated from the RY13 strain of *E. coli*). Roman numerals that follow the letters identify different enzymes from the same species.

The sequences recognized by a restriction enzyme are located randomly within the genome. Accordingly, there is a relation between the length of the recognition sequence and the number of times it is present in a genome: there will be fewer longer recognition sequences than shorter ones because the probability of the occurrence of a particular sequence consisting of, say, six specific bases is less than the probability of the occurrence of a particular sequence of four specific bases. Therefore, restriction enzymes that recognize longer sequences will cut a given piece of DNA into fewer and longer fragments than will restriction enzymes that recognize shorter sequences.

Restriction enzymes are used whenever DNA fragments must be cut or joined. In a typical restriction reaction, a concentrated solution of purified DNA is placed in a small tube with a buffer solution and a small amount of restriction enzyme. The reaction mixture is then heated at the optimal temperature for

the enzyme, usually 37°C. Often within an hour, the enzyme cuts the appropriate restriction sites in all the DNA molecules, producing a set of DNA fragments. **› TRY PROBLEM 29**

CONCEPTS

Type II restriction enzymes cut DNA at specific base sequences that are palindromic. Some restriction enzymes make staggered cuts, producing DNA fragments with cohesive ends; others cut both strands straight across, producing blunt-ended fragments. There are fewer long recognition sequences in DNA than short recognition sequences.

✔ **CONCEPT CHECK 1**

Where do restriction enzymes come from?

(a)

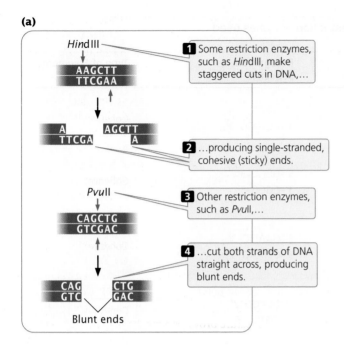

1. Some restriction enzymes, such as *Hind*III, make staggered cuts in DNA,...

2. ...producing single-stranded, cohesive (sticky) ends.

3. Other restriction enzymes, such as *Pvu*II,...

4. ...cut both strands of DNA straight across, producing blunt ends.

Blunt ends

(b)

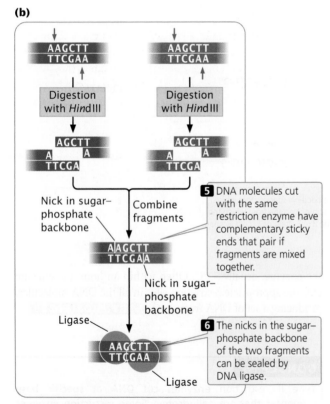

Digestion with *Hind*III

Digestion with *Hind*III

Nick in sugar–phosphate backbone

Combine fragments

5. DNA molecules cut with the same restriction enzyme have complementary sticky ends that pair if fragments are mixed together.

Nick in sugar–phosphate backbone

Ligase

6. The nicks in the sugar–phosphate backbone of the two fragments can be sealed by DNA ligase.

Ligase

19.2 Restriction enzymes make double-stranded cuts in DNA, producing cohesive, or sticky, ends.

Engineered Nucleases

One limitation of restriction enzymes is that because their recognition sequences are short, typically 4–8 bp in length,

they occur at random, many times within a genome. For example, consider the 6-bp recognition sequence for *Bam*HI:

$$5'-GGATCC-3'$$
$$3'-CCTAGG-5'$$

There are over 47,000 *Bam*HI restriction sites in the human genome. Cutting human DNA with *Bam*HI or another restriction enzyme results in thousands of fragments. Thus, it is impossible to precisely cut genomic DNA at a single location with restriction enzymes.

To overcome this problem of numerous cuts, geneticists have sought nucleases that recognize longer DNA sequences. In recent years, geneticists have designed complex enzymes, termed **engineered nucleases**, that are capable of making unique double-stranded cuts in DNA at predetermined sequences. Engineered nucleases consist of the part of a restriction enzyme that cleaves DNA nonspecifically, coupled with another protein that recognizes and binds to a specific DNA sequence; the particular sequence to which the protein binds is determined by the protein's amino acid sequence. By altering the amino acid sequence of the binding protein, geneticists can custom-design a nuclease to bind to and cut any particular DNA sequence.

Engineered nucleases include **zinc-finger nucleases (ZFNs)**, which use a DNA-binding domain called a zinc finger (see Section 16.1) attached to a restriction enzyme, most often *Fok*I. In these nucleases, several zinc-finger domains are combined in an array; the most commonly used ZFNs have three of these domains, which together recognize a 9-bp DNA sequence. To make double-stranded cuts in DNA, ZFNs are used in pairs, which increases their specificity to 18 or more base pairs—enough to ensure that most genomes are cut only once. Another type of engineered nuclease is a **transcription activator–like effector nuclease (TALEN)**, in which a protein that normally binds to sequences in promoters is attached to the *Fok*I restriction enzyme. The binding protein consists of a series of repeats whose amino acid sequence determines the DNA base sequence that it recognizes.

The key to both ZFNs and TALENs is the ability to alter the amino acid sequence of the binding proteins in these nucleases in such a way that they bind to a specific target DNA sequence. For ZFNs, there is no simple correspondence between the amino acid sequence of the binding protein and the DNA sequence to which it binds, so each ZFN must be custom-designed for a particular sequence. For TALENs, a more straightforward relation exists between the amino acid sequence of the binding protein and the base sequence of the target DNA, but making specific proteins for individual DNA sequences is laborious and costly.

CRISPR-Cas Genome Editing

Another molecular tool for precisely cutting DNA is the **CRISPR-Cas system** that has been developed in recent years. This technique has revolutionized the field of genetics, providing a powerful way of editing the genome that has now been applied to DNA sequences in bacteria, yeast, nematodes, plants, fruit flies, zebrafish, mice, rats, monkeys, humans, and many more organisms.

CRISPR-Cas IMMUNITY IN BACTERIA AND ARCHAEA

CRISPR-Cas systems occur naturally in bacteria and archaea and are used to protect these organisms against bacteriophages, plasmids, and other invading DNA elements (see Chapters 9 and 14). CRISPR RNAs (crRNAs) are encoded by DNA sequences called Clustered Regularly Interspaced Short Palindromic Repeats (CRISPR). A CRISPR array consists of a series of such palindromic sequences (sequences that read the same forward and backward on two complementary DNA strands) separated by unique spacers, which consist of sequences derived from bacteriophages or foreign plasmids (see Figure 14.23a). When bacteriophage or plasmid DNA enters a prokaryotic cell, proteins cut up the foreign DNA and insert bits of it into a CRISPR array, which then serves as a memory of the invader. The DNA sequences in bacteriophage or plasmid DNA that match the spacer elements in the CRISPR array are referred to as *protospacers*.

The CRISPR array is transcribed into a long precursor CRISPR RNA (pre-crRNA), which is cleaved into short crRNAs. The crRNAs combine with proteins called CRISPR-associated (Cas) proteins to form effector complexes. The Cas proteins have nuclease activity—the ability to cut DNA. If the same foreign DNA (the protospacer) enters the cell in the future, a CRISPR-Cas complex recognizes and attaches to it. The crRNA in the complex binds to its complementary sequence in the foreign DNA, and the Cas protein cleaves the foreign DNA, rendering it nonfunctional. In this way, CRISPR-Cas serves as an adaptive RNA defense system that remembers and destroys foreign invaders.

GENOME EDITING WITH CRISPR-Cas

A variety of CRISPR-Cas systems have been found in different bacterial and archaeal species. Geneticists have engineered some of these systems to serve as molecular editing tools. The most widely used system is CRISPR-Cas9, derived from the bacterium *Streptococcus pyogenes*. This system naturally requires two RNA molecules, crRNA and another RNA molecule termed tracrRNA. These two RNAs pair and then combine with Cas9 to form an effector complex. To facilitate the use of this system in genome editing, researchers have engineered the crRNA and the tracrRNA into a single guide RNA (sgRNA) (**Figure 19.3**). A 20-nucleotide region of the sgRNA pairs with DNA, although the nucleotides within an 8–12-nucleotide "seed" sequence are most important in pairing. By altering the sequence of the sgRNA, it is possible to direct the action of the effector complex to any specific DNA sequence desired. This relatively long recognition sequence makes CRISPR-Cas9 much more specific than restriction enzymes, meaning that it can be directed to unique sites within the genome.

An important feature of the CRISPR-Cas9 system is the required presence of a sequence in the target DNA (the DNA to be cleaved) called a protospacer-adjacent motif (PAM; see Figure 19.3). These sequences are short and have a weak consensus (the PAM for the *S. pyogenes* system is 5′–NGG–3′, where N represents any base and G represents guanine), so they occur at numerous random places throughout most genomes. The CRISPR-Cas9 effector complex first associates with a PAM, and then Cas9 unwinds the DNA nearby. This unwinding allows the sgRNA to pair with its complementary sequence in the DNA. Once the target DNA and sgRNA are paired, Cas9 makes double-stranded cuts in the DNA. Because PAMs are short, they occur frequently within most genomes. For example, the CRISPR-Cas9 PAM is estimated to occur, on average, about once every 8 bp in the human genome.

REPAIR OF BREAKS PRODUCED BY CRISPR-Cas9 Once the DNA of a cell has been cleaved by CRISPR-Cas9, the cell immediately uses its DNA-repair mechanism to try to repair the break. This feature provides a mechanism for editing the target sequence. There are two major pathways by which double-strand breaks are repaired within cells (see Section 18.5). One, called nonhomologous end joining, joins together the two ends of DNA without using any template. This process tends to introduce small insertions and deletions when the two ends are joined, a side effect that allows geneticists to disable a gene. The CRISPR-Cas9 system can be

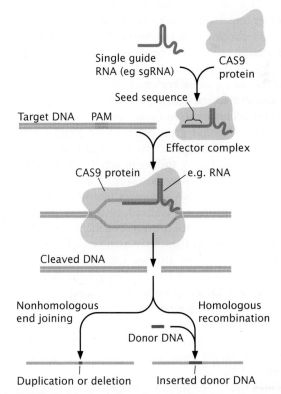

19.3 CRISPR-Cas9 is a technique for precisely editing the genome.

targeted to a specific gene; when that gene is cleaved by Cas9 and then repaired by nonhomologous end joining, the introduction of insertions or deletions at the break site often produces frameshift mutations that disrupt the coding sequence and disable the gene (see Figure 19.3).

The other mechanism used by cells to repair double-strand breaks is homologous recombination (see Section 18.5). This mechanism functions when a DNA template is provided for repairing the break. A researcher can provide a donor piece of double-stranded DNA that has ends complementary to the sequences at the ends of the break made by Cas9; homologous recombination may insert the donor DNA sequence into the break. In this way, researchers can selectively insert a desired sequence into a genome (see Figure 19.3). Unfortunately, homologous recombination is not highly efficient, and often the two ends are connected without insertion of the donor DNA.

THINK-PAIR-SHARE Question 2

ADVANTAGES OF CRISPR-Cas9 An advantage of CRISPR-Cas9 over restriction enzymes is the relatively long nucleotide sequence recognized by the sgRNA, which allows researchers to produce unique cuts within genomic DNA. By changing the sequence of the sgRNA, precise edits can be made almost anywhere in the genome. Modifying the sgRNA to match a particular sequence in the genome is much simpler than trying to alter a DNA-binding protein, such as the proteins in zinc-finger nucleases and TALENS. Another important feature of CRISPR-Cas9 is the ability to use this technology in intact cells; mRNA for the Cas9 protein and specific sgRNAs can be introduced into many types of cells, where they are expressed and carry out their editing function. In a process known as multiplexing, it is possible to introduce several sgRNAs into a cell simultaneously and carry out multiple cuts in a single step.

CRISPR-Cas has great potential for genetic engineering and biotechnology. It can be applied to many different species, including species in which other methods of DNA manipulation have not worked well. It can be used to introduce new DNA sequences into whole animals and humans. For example, it is already being used to induce specific mutations in mice to create genetic models of human diseases, which can then serve as powerful research tools for the study of those diseases. CRISPR-Cas is also being developed as a tool for correcting genetic defects (gene therapy) and has the potential for treating infectious diseases by eliminating viral DNA from human cells. It will enable genetic modification of crops and domestic animals, in which it can be used to create very specific genetic alterations that benefit yield and produce characteristics that improve cultivation.

Researchers are already developing modifications of CRISPR-Cas systems to provide additional functions. For example, the Cas9 protein can be modified so that it makes single-stranded cuts in DNA, which are more likely to be repaired by homologous recombination and other precise DNA-repair mechanisms. Nucleases of other CRISPR systems are being used for different types of cutting. For example, a Cas protein called Cpf1, which is used in some CRISPR systems, makes staggered cuts that produce complementary sticky ends like those produced by some restriction enzymes. The production of complementary sticky ends makes repair more accurate and enables the insertion of desired sequences with complementary sticky ends.

When a PAM for Cas9 is not available near a site where cleavage is desired, Cas proteins from different species that recognize different PAMs can be used. There is the potential to use a modified CRISPR-Cas system (with its cleavage function inactivated) as a general RNA-guided device for other functions, such as targeted transcriptional activation and transcriptional silencing. Transcriptional activator proteins or chromatin-remodeling proteins can be tethered to Cas to bring about transcription at specific sites in the genome.

THINK-PAIR-SHARE Question 3

LIMITATIONS AND CHALLENGES OF CRISPR-Cas9 EDITING A limitation in the use of CRISPR-Cas9 for genome editing is the potential for off-target cleavage. Some mismatches between the sgRNA and the complementary DNA sequence are tolerated by the Cas9 protein, so the system may cleave DNA at sites other than the desired target sequence. Unfortunately, off-target cleavage depends on many factors and varies among cell types, so predicting where and when cleavage will occur has been difficult. Much work has been concentrated on engineering the sgRNA and Cas9 protein to be more selective in pairing and cutting so as to reduce undesirable off-target cuts. For example, geneticists have produced a high-fidelity Cas9 by altering its amino acid sequence so that nonspecific interaction between the Cas protein and the DNA backbone is weakened, thereby forcing the nuclease to depend more on base pairing between the sgRNA and the DNA and increasing its specificity. These changes eliminated nearly all off-target cleavage by the high-fidelity Cas9.

If CRISPR-Cas is applied to multicellular embryos or whole organisms, there is also the potential to create genetic mosaics (see Section 6.2). Genome editing with CRISPR-Cas is not 100% efficient, meaning that DNA in some cells is edited and DNA in other cells is not. It can therefore produce tissues that are mosaics, in which different cells have different DNA sequences. The effects of this mosaicism could vary tremendously, depending on what genes are edited, what tissues are involved, and how efficient the editing is.

Another potential difficulty is getting the CRISPR-Cas components into a cell. This can often be done with cells in culture by transfection, in which cells are bathed in a solution containing DNA or mRNA encoding the Cas protein and the sgRNA and treated to make their membranes more permeable to foreign DNA. While transfection works in cell culture, getting the CRISPR-Cas components into intact organisms is more challenging. Sometimes the sgRNA and Cas protein are directly injected into fertilized egg cells, but most cells are too small for direct injection. Cas9 is a relatively large protein that does not easily pass across the cell membrane. Some other CRISPR-associated nucleases are smaller and easier to get into cells; these smaller nucleases are being exploited for use in delivery.

CONCERNS WITH THE USE OF CRISPR-Cas9 EDITING
One concern about CRISPR-Cas9 technology is that its ease of use and potential to alter almost any DNA sequence might mean that it could be used to genetically modify humans in ethically questionable ways. The CRISPR-Cas system has already been used to carry out genome editing in human embryos, although these embryos were not capable of completing development and producing live-born humans. Nevertheless, it is clear that the technology can be used for genetically modifying humans, and this observation has raised considerable debate. Most ethical concern focuses on germ-line editing (in which genes in reproductive cells are altered) because the edited genes will be passed on to future generations, affecting the future gene pool of the species.

There are also safety concerns about applying CRISPR-Cas to humans. Off-target cuts could induce mutations that might lead to cancer or other medical problems. Some researchers have pointed to the potential dangers of using CRISPR-Cas to edit animals and plants that are released into the wild.

In early 2015, a group of leading scientists and ethicists met in Napa, California, to discuss the scientific, legal, and ethical implications of genome editing using CRISPR-Cas technology. This group strongly discouraged all attempts at human germ-line modification until the implications of CRISPR-Cas editing could be discussed more widely by scientific and governmental organizations. The group encouraged transparent research, educational forums on the new technology, and the convening of representative developers and users to discuss its social, ethical, and scientific implications. Other groups have echoed many of these same recommendations.

THINK-PAIR-SHARE Question 4

> **CONCEPTS**
>
> The development of CRISPR-Cas9 technology provides a powerful means of cutting and editing the genome. CRISPR-Cas9 combines a single guide RNA with a nuclease, which together attach to specific DNA sequences and make double-stranded cuts at specific locations. Repair of these cuts by nonhomologous end joining or homologous recombination provides the means to introduce alterations to the genome. This technology has been used in many organisms and cell types, and it has the potential for additional applications, but its use has raised a number of ethical concerns.

Separating and Viewing DNA Fragments

We have now discussed several techniques for precisely cutting a DNA molecule. After cutting, it is often necessary to separate the resulting fragments and to verify that the DNA was altered in the expected fashion. Gel electrophoresis provides a way to separate and visualize DNA fragments.

Electrophoresis is a standard technique for separating molecules on the basis of their size and electrical charge. There are a number of different types of electrophoresis; to separate DNA molecules, **gel electrophoresis** is used. The gel most often used is a porous gel made from agarose (a polysaccharide isolated from seaweed), which is melted in a buffer solution and poured into a plastic mold. As it cools, the agarose solidifies, making a gel that looks something like stiff gelatin.

Small wells are made at one end of the gel, solutions of DNA fragments are placed in the wells (**Figure 19.4a**), and an electrical current is passed through the gel. Because the phosphate group on each DNA nucleotide carries a negative charge, the DNA fragments migrate toward the positive end of the gel. During this migration, the porous gel acts as a sieve, separating the DNA fragments by size. Small DNA fragments migrate more rapidly than do large ones, so, over 1–2 hours, the fragments separate on the basis of their size. Typically, DNA fragments of known length (size standards) are placed in one of the wells. By comparing the migration distance of the unknown fragments with the distance traveled by the size standards, a researcher can determine the approximate size of the unknown fragments (**Figure 19.4b**).

The DNA fragments are still too small to see, so the problem of visualizing them must be addressed. Visualization can be accomplished in several ways. The simplest procedure is to stain the gel with a dye specific for nucleic acids, such as ethidium bromide, which wedges itself tightly (intercalates) between the bases of DNA and fluoresces when exposed to UV light, producing brilliant bands on the gel (**Figure 19.4c**). Alternatively, DNA fragments can be visualized by adding a label to the DNA before it is placed in the gel. For example, chemical labels can be detected by adding antibodies or other substances that carry a dye and will attach to the relevant DNA, which can then be visualized directly.

> **CONCEPTS**
>
> DNA fragments can be separated, and their sizes determined, with the use of gel electrophoresis. The fragments can be viewed by using a dye that is specific for nucleic acids or by labeling the fragments with a chemical tag.
>
> ✔ **CONCEPT CHECK 2**
>
> DNA fragments that are 500 bp, 1000 bp, and 2000 bp in length are separated by gel electrophoresis. Which fragment will migrate farthest in the gel?
> a. The 2000-bp fragment
> b. The 1000-bp fragment
> c. The 500-bp fragment
> d. All will migrate equal distances.

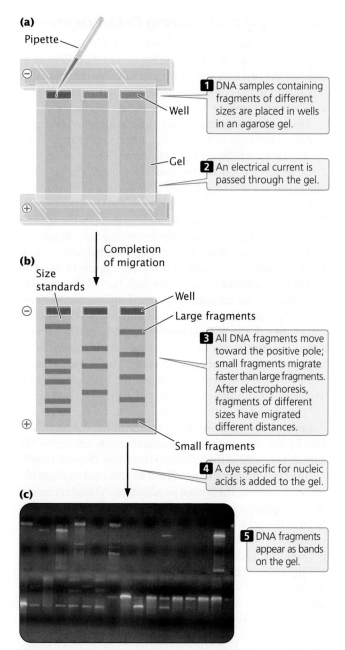

(a)
Pipette

Well

Gel

1 DNA samples containing fragments of different sizes are placed in wells in an agarose gel.

2 An electrical current is passed through the gel.

Completion of migration

(b)
Size standards

Well

Large fragments

3 All DNA fragments move toward the positive pole; small fragments migrate faster than large fragments. After electrophoresis, fragments of different sizes have migrated different distances.

Small fragments

4 A dye specific for nucleic acids is added to the gel.

(c)

5 DNA fragments appear as bands on the gel.

19.4 Gel electrophoresis can be used to separate DNA molecules on the basis of their size and electrical charge. [Photograph: Klaus Guldbrandsen/Science Source.]

Locating DNA Fragments with Probes

If a small piece of DNA, such as a plasmid, is cut by a restriction enzyme, the few fragments produced can be seen as distinct bands on an electrophoretic gel. In contrast, if genomic DNA from a cell is cut by a restriction enzyme, thousands of fragments of different sizes are produced. A restriction enzyme that recognizes a four-base sequence would theoretically cut about once every 256 bp. The human genome, with 3.2 billion base pairs, would generate more than 12 million fragments when cut by this restriction enzyme. When separated by electrophoresis and stained, this large set of fragments would appear as a continuous smear on the gel because

of the presence of so many fragments of differing sizes. Usually, researchers are interested in only a few of these fragments, perhaps those carrying a specific gene. How can they locate the desired fragments in such a large pool of DNA?

One approach is to use a **probe**, which is often a DNA or RNA molecule with a base sequence complementary to a sequence in the gene of interest. The bases on the probe will pair only with the bases on a complementary sequence, so, if suitably labeled, the probe can be used to locate a specific gene or other DNA sequence. To use a probe, a researcher first cuts the DNA into fragments by using one or more restriction enzymes and then separates the fragments with gel electrophoresis. Next, the separated fragments must be denatured and transferred to a permanent solid medium (such as a nitrocellulose or nylon membrane). **Southern blotting** (named after Edwin M. Southern) is one technique used to transfer the denatured (single-stranded) fragments from a gel to a permanent solid medium.

After the single-stranded DNA fragments have been transferred, the membrane is placed in a hybridization solution containing a labeled probe. The probe binds to (hybridizes with) any DNA fragments on the membrane that bear complementary sequences. Often, a probe binds to only a part of a DNA fragment, so the DNA fragment may contain sequences not found in the probe. The membrane is then washed to remove any unbound probe; a biochemical method reveals the presence of the bound probe.

RNA can be transferred from a gel to a solid support by a related procedure called **Northern blotting** (not named after anyone, but capitalized to match Southern blotting). Hybridization with a probe can reveal the size of a particular mRNA molecule, its relative abundance, or the tissues in which it is transcribed. **Western blotting** is the transfer of protein from a gel to a membrane. Here, the probe is usually an antibody, used to determine the size of a particular protein and the pattern of the protein's expression.

CONCEPTS

Labeled probes, which are sequences of RNA or DNA that are complementary to the sequence of interest, can be used to locate individual genes or sequences among DNA fragments separated by electrophoresis.

✔ **CONCEPT CHECK 3**

How do Northern and Western blotting differ from Southern blotting?

19.3 Specific DNA Fragments Can Be Amplified

Many of the methods used to manipulate and analyze DNA sequences cannot be carried out on single molecules, requiring instead numerous copies of a specific DNA fragment. A major problem in working at the molecular level is that each gene is a tiny fraction of the total cellular DNA. Because each gene is so rare, it must be isolated and amplified before it can be studied. There are two basic approaches to amplifying a specific DNA fragment: replicating the DNA within cells (in vivo), or replicating the DNA enzymatically outside of cells (in vitro).

In the in vivo approach, a DNA fragment is inserted into a bacterial cell and the cell is allowed to replicate the DNA. Each time the cell divides, one or more copies of the DNA fragment are passed on to each daughter cell. Most bacterial cells divide rapidly, so within a short time (usually a few days), a large number of genetically identical cells are produced, each carrying one or more copies of the DNA fragment. The cells are then lysed to release their DNA, and the desired fragment is isolated from the rest of the bacterial DNA. This procedure is termed **gene cloning** because identical copies (clones) of the original piece of DNA are replicated within bacterial cells.

For many years, all amplification of DNA was done by gene cloning. A major disadvantage of gene cloning is the time required: the process of inserting the DNA into bacteria, selecting and growing the bacterial cells that have incorporated it, and isolating the amplified DNA usually requires several days. Gene cloning is also relatively labor-intensive, requiring a number of steps that are difficult to automate. An advantage is that, because it uses the cell's high-fidelity replication machinery, gene cloning typically copies DNA with great accuracy.

The second, in vitro, approach is to amplify DNA enzymatically in a test tube outside of cells. This amplification is done with the **polymerase chain reaction** (**PCR**), a technique first developed in 1983 by Kary Mullis. The basis of PCR is DNA replication catalyzed by a DNA polymerase. Because a DNA molecule consists of two nucleotide strands, each of which can serve as a template, the amount of DNA

doubles with each replication event. PCR allows DNA fragments to be amplified a billionfold within just a few hours, and it can be used with extremely small amounts of original DNA, even a single molecule. The polymerase chain reaction revolutionized molecular biology and is now one of the most widely used molecular techniques.

We first discuss PCR because it is the most widely used technique for amplifying DNA fragments today. We then return to gene cloning, which, in addition to its use in amplifying DNA, is often employed to manipulate gene sequences.

The Polymerase Chain Reaction

Replication by PCR requires a DNA template (the fragment of interest, or target DNA) from which a new DNA strand can be copied, and a pair of single-stranded primers, each with a 3'-OH group to which new nucleotides can be added. The primers used in PCR are short fragments of DNA, typically 17–25 nucleotides long, that are complementary to known sequences on the template. To carry out PCR, researchers begin with a solution that includes the target DNA, DNA polymerase, all four deoxyribonucleoside triphosphates (dNTPs—the substrates for DNA polymerase), primers, and magnesium ions and other salts that are necessary for the reaction to proceed.

A typical polymerase chain reaction includes three steps (**Figure 19.5**). In step 1, a starting solution of DNA is heated to 90°–100°C to break the hydrogen bonds between the strands and thus produce the necessary single-stranded templates. The reaction mixture is held at this temperature for only a minute or two. In step 2, the DNA solution is cooled quickly to 30°–65°C and held at this temperature for a minute or less. During this short interval, the DNA strands

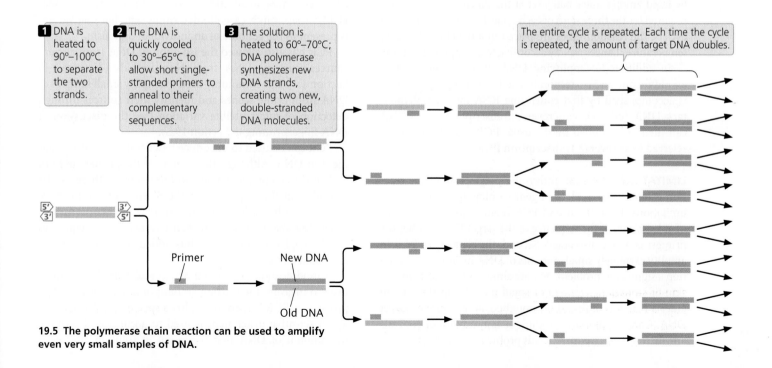

1 DNA is heated to 90°–100°C to separate the two strands.

2 The DNA is quickly cooled to 30°–65°C to allow short single-stranded primers to anneal to their complementary sequences.

3 The solution is heated to 60°–70°C; DNA polymerase synthesizes new DNA strands, creating two new, double-stranded DNA molecules.

The entire cycle is repeated. Each time the cycle is repeated, the amount of target DNA doubles.

Primer

New DNA

Old DNA

19.5 The polymerase chain reaction can be used to amplify even very small samples of DNA.

do not have a chance to reanneal, but the primers are able to attach to the template strands. In step 3, the solution is heated for a minute or less to 72°C, the temperature at which DNA polymerase can synthesize new DNA strands. At the end of the cycle, two new double-stranded DNA molecules are produced for each original molecule of DNA.

The whole cycle is then repeated. With each cycle, the amount of target DNA doubles, so the amount of target DNA increases geometrically. One molecule of DNA increases to more than 1000 molecules in 10 PCR cycles, to more than 1 million molecules in 20 cycles, and to more than 1 billion molecules in 30 cycles. Each cycle is completed within a few minutes, so a large amplification of DNA can be achieved within a few hours. To see how the polymerase chain reaction quickly increases the number of copies of a DNA fragment, view **Animation 19.1**.

Two key innovations facilitated the use of PCR in the laboratory. The first was the discovery of a DNA polymerase that is stable at the high temperatures used in step 1. The DNA polymerase from *E. coli* that was originally used in PCR denatures at 90°C, so fresh enzyme had to be added to the reaction mixture in *each* cycle, which slowed the process considerably. This obstacle was overcome when DNA polymerase was isolated from the bacterium *Thermus aquaticus*, which lives in the boiling springs of Yellowstone National Park. This enzyme, dubbed ***Taq* polymerase**, is remarkably stable at high temperatures and is not denatured in the strand-separation step; it can be added to the reaction mixture at the beginning of the PCR process and will continue to function through many cycles.

The second key innovation was the development of automated thermal cyclers: machines that bring about the rapid temperature changes necessary for the different steps of PCR. Originally, tubes containing reaction mixtures were moved by hand among water baths set at the different temperatures required for the three steps of each cycle. In automated thermal cyclers, the reaction tubes are placed in a metal block whose temperature changes rapidly according to a computer program.

In addition to amplifying DNA, PCR can be used to amplify sequences corresponding to RNA. This amplification is accomplished by first converting RNA into complementary DNA (cDNA) with reverse transcriptase. The cDNA can then be amplified by the usual PCR. This technique is referred to as **reverse-transcription PCR**.

LIMITATIONS OF PCR Today, the polymerase chain reaction is often used in place of gene cloning, but it has several limitations. First, the use of PCR requires prior knowledge of at least part of the sequence of the target DNA so that the primers can be constructed. Second, the capacity of PCR to amplify extremely small amounts of DNA makes contamination a significant problem. Minute amounts of DNA from the skin of laboratory workers or small particles in the air can enter a reaction tube and be amplified along with the target DNA. Careful laboratory technique and the use of controls are necessary to circumvent this problem.

A third limitation of PCR is accuracy. Unlike other DNA polymerases, *Taq* polymerase does not have the capacity to proofread (see pp. 353–354 in Chapter 12), and under standard PCR conditions, it incorporates an incorrect nucleotide about once every 20,000 bp. New heat-stable DNA polymerases with proofreading capacity have been isolated that give more accurate PCR results.

A fourth limitation of PCR is that the size of the fragments that can be amplified by standard *Taq* polymerase is usually less than 2000 bp. By using a combination of *Taq* polymerase and a DNA polymerase with proofreading capacity, and by modifying the reaction conditions, investigators have been successful in extending PCR amplification to larger fragments, but even these larger fragments are limited in length to 50,000 bp.

APPLICATIONS OF PCR In spite of its limitations, PCR has become one of the most widely used tools within molecular biology. Because the primers used in PCR are specific for known DNA sequences, PCR can be used to detect the presence of a particular DNA sequence in a sample. For example, PCR is often used to detect the presence of viruses in blood samples by adding primers complementary to known viral DNA sequences to the reaction mixture. If viral DNA is present, the primers will attach to it, and a DNA fragment of a known length will be amplified. The presence of a DNA fragment of that length on a gel indicates the presence of viral DNA in the blood sample. Modern diagnostic tests for infection with HIV, the causative agent of AIDS (see Section 9.4), use this type of PCR amplification of HIV sequences.

Another common application of PCR is to identify genetic variation in natural populations. PCR can be used to amplify specific segments of DNA, which are then analyzed with other molecular tools that identify differences in nucleotide sequences. In addition, primers specific to a particular DNA variant can be used to determine whether that variant is present in the genome of an individual organism.

PCR has also allowed the isolation of DNA from ancient sources, such as DNA from Neanderthals (discussed in Section 1.1). It is commonly used to amplify small samples of DNA from crime scenes and identify their sources through detection of microsatellite variation (see the discussion of DNA fingerprinting in Section 19.5).

PCR can be used to introduce new sequences into a fragment of DNA. Although the 3′ ends of the primers used in PCR must be complementary to the template, there can be flexibility in the sequences at the 5′ ends of the primers. Primers can be designed to contain new restriction sites or other desirable sequences at their 5′ ends. These sequences will be copied by PCR and thus added to the ends of the amplified DNA.

A modification of PCR, known as **real-time PCR**, can be used to measure the starting amount of nucleic acid. In this procedure, PCR is used to amplify a specific DNA fragment, and a sensitive instrument is used to accurately determine the amount of DNA that is present in solution after each

PCR cycle. A fluorescent probe that is specific to the DNA sequence of interest is frequently used in the reaction, so that only the DNA of interest is measured. The technique is called real-time PCR because the amount of DNA amplified is measured as the reaction proceeds.

Often, real-time PCR is combined with reverse-transcription PCR to measure the amount of mRNA in a sample, allowing biologists to determine the level of gene expression in different cells and under different conditions. For example, researchers interested in how gene expression changes in response to the administration of a drug often use real-time PCR to measure the amount of mRNA produced by specific genes in cells exposed to the drug and compare it with the amount of mRNA produced by the same genes in control cells with no drug exposure.

CONCEPTS

The polymerase chain reaction is an enzymatic, in vitro method for rapidly amplifying DNA. In this process, DNA is heated to separate the two strands, short primers attach to the target DNA, and DNA polymerase synthesizes new DNA strands from the primers. Each cycle of PCR doubles the amount of DNA. PCR has a number of important applications in molecular biology.

✔ CONCEPT CHECK 4

Why is the use of a heat-stable DNA polymerase important to the success of PCR?

Gene Cloning

Despite the widespread use of PCR, some DNA sequences are still amplified by gene cloning. In addition, gene cloning provides a powerful means of altering and manipulating DNA sequences. It is often used to alter cells so that they have desired properties or produce substances of commercial value. Here, we outline some of the methods used to manipulate DNA sequences and introduce them into bacterial cells.

Many cloning techniques require a **cloning vector**, a stable, replicating DNA molecule to which a foreign DNA fragment can be attached for introduction into a cell. An effective cloning vector has three important characteristics (**Figure 19.6**): (1) an origin of replication, which ensures that the vector is replicated within the cell; (2) selectable markers, which enable any cells containing the vector to be selected or identified; and (3) one or more unique restriction sites into which a DNA fragment can be inserted. The restriction sites used for cloning must be unique; if a vector is cut at multiple recognition sites, several pieces of DNA are generated, and getting those pieces back together in the correct order is possible, but extremely difficult.

PLASMID VECTORS Plasmids, circular DNA molecules that exist naturally in bacteria (see Chapter 9), are commonly used as vectors for cloning DNA fragments in bacteria.

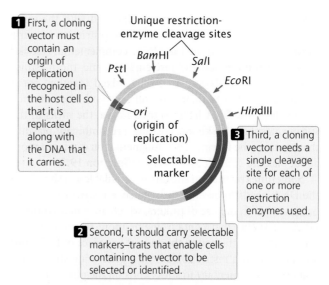

1 First, a cloning vector must contain an origin of replication recognized in the host cell so that it is replicated along with the DNA that it carries.

Unique restriction-enzyme cleavage sites

*Bam*HI   *Sal*I
*Pst*I   *Eco*RI
*Hind*III

ori (origin of replication)

Selectable marker

3 Third, a cloning vector needs a single cleavage site for each of one or more restriction enzymes used.

2 Second, it should carry selectable markers–traits that enable cells containing the vector to be selected or identified.

19.6 An ideal cloning vector has an origin of replication, one or more selectable markers, and recognition sites for one or more restriction enzymes.

They contain origins of replication and are therefore able to replicate independently of the bacterial chromosome. The plasmids typically used in cloning have been artificially constructed from larger, naturally occurring bacterial plasmids and have multiple unique restriction sites and selectable markers as well as an origin of replication (**Figure 19.7**).

The easiest method for inserting a DNA sequence into a plasmid vector is to cut the foreign DNA (containing the

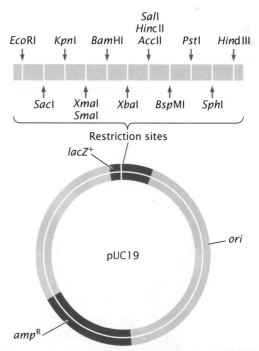

19.7 The pUC19 plasmid is a typical cloning vector. This artificially constructed plasmid contains a cluster of unique restriction sites, an origin of replication, and two selectable markers: an ampicillin-resistance gene and a *lacZ* gene.

DNA fragment of interest) and the plasmid with the same restriction enzyme (**Figure 19.8**). If the restriction enzyme makes staggered cuts in the DNA, complementary sticky ends are produced on the foreign DNA and the plasmid. When DNA and plasmids are then mixed together, some of the foreign DNA fragments will pair with the cut ends of the plasmids. DNA ligase is used to seal the nicks in the sugar–phosphate backbone, creating a recombinant plasmid that contains the foreign DNA fragment. You can learn more about plasmid cloning by viewing **Animation 19.2**.

Ⓐ

Sometimes restriction sites are not available at a site where the DNA needs to be cut. In that case, a restriction site can be created with the use of **linkers**, which are small synthetic DNA fragments that contain one or more restriction sites. Linkers can be attached to the ends of any piece of DNA and are then cut by a restriction enzyme, generating sticky ends that are complementary to sticky ends on the plasmid.

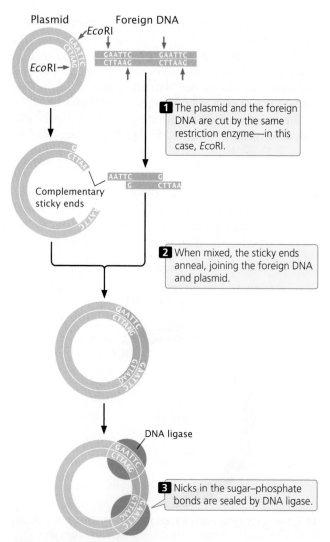

1 The plasmid and the foreign DNA are cut by the same restriction enzyme—in this case, *Eco*RI.

2 When mixed, the sticky ends anneal, joining the foreign DNA and plasmid.

3 Nicks in the sugar–phosphate bonds are sealed by DNA ligase.

19.8 A foreign DNA fragment can be inserted into a plasmid with the use of restriction enzymes.

TRANSFORMATION Once a DNA fragment of interest has been placed inside a plasmid, the plasmid must be introduced into bacterial cells. This task is usually accomplished by *transformation*, the mechanism by which bacterial cells take up DNA from the external environment (see Section 9.3). Some types of cells undergo transformation naturally; others must be treated chemically or physically before they will undergo transformation. Inside the cells, the plasmids replicate and multiply as the cells themselves multiply.

SCREENING CELLS FOR RECOMBINANT PLASMIDS Cells bearing recombinant plasmids can be detected with the use of the selectable markers on the plasmid. Genes that confer resistance to an antibiotic are commonly used as selectable markers: any cell that contains such a gene will be able to live in the presence of the antibiotic, which normally kills bacterial cells.

One way to screen cells for the presence of a recombinant plasmid is to use a plasmid that contains a fragment of the *lacZ* gene—a small part of the front end of the gene (**Figure 19.9**). This partial *lacZ* gene contains a series of unique restriction sites into which a piece of DNA can be inserted. The bacteria that are to be transformed by the plasmid also have special features that make the presence of the recombinant plasmid evident: they are *lacZ⁻*, missing the front end of the *lacZ* gene but containing its back end. Without the plasmid, these bacteria are unable to synthesize β-galactosidase. If no foreign DNA has been inserted into the partial *lacZ* gene of a plasmid taken up by a cell, the front end of the gene (provided by the plasmid) and the back end of the gene (provided by the bacterium) work together within the cell to produce β-galactosidase (see Section 16.2). If foreign DNA has been successfully inserted into the restriction site, it disrupts the front end of the *lacZ* gene, and β-galactosidase is *not* produced. The plasmid also usually contains a selectable marker, which may be a gene that confers resistance to an antibiotic such as ampicillin.

Bacteria that are *lacZ⁻* are exposed to the plasmids and then plated on medium that contains ampicillin. Only cells that have been successfully transformed and thus contain a plasmid with the ampicillin-resistance gene will survive and grow. Some of these cells contain an intact plasmid, whereas others possess a recombinant plasmid. The medium also contains the chemical X-gal, which produces a blue substance when cleaved. Bacterial cells with an intact original plasmid—without an inserted DNA fragment—have a functional *lacZ* gene (the front end of which is provided by the plasmid and the back end by the bacterium). These bacteria can synthesize β-galactosidase, which cleaves X-gal and turns the bacteria blue. Bacterial cells with a recombinant plasmid, however, have had the front end of the *lacZ* gene disrupted by the inserted DNA; they do not synthesize β-galactosidase and remain white. (Recall that the cells' own *lacZ* genes have been inactivated, so only bacteria with the plasmid turn blue.) Thus, the color of a bacterial colony allows quick determination of whether a recombinant or intact plasmid is

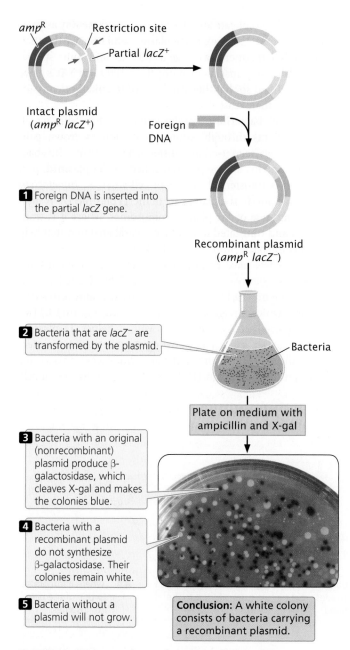

1 Foreign DNA is inserted into the partial *lacZ* gene.

Recombinant plasmid
(*amp*^R *lacZ*⁻)

2 Bacteria that are *lacZ*⁻ are transformed by the plasmid.

Bacteria

Plate on medium with ampicillin and X-gal

3 Bacteria with an original (nonrecombinant) plasmid produce β-galactosidase, which cleaves X-gal and makes the colonies blue.

4 Bacteria with a recombinant plasmid do not synthesize β-galactosidase. Their colonies remain white.

5 Bacteria without a plasmid will not grow.

Conclusion: A white colony consists of bacteria carrying a recombinant plasmid.

19.9 The *lacZ* gene can be used to screen for bacteria containing recombinant plasmids. An artificially constructed plasmid carries a fragment of the *lacZ* gene and an ampicillin-resistance gene. [Photograph courtesy of Edvotek.]

present in the cells. After cells with the recombinant plasmid have been identified, they can be grown in large numbers to replicate the inserted fragment of DNA.

Plasmids make ideal cloning vectors, but they can hold only DNA fragments of less than about 15 kb in size. When large DNA fragments are inserted into a plasmid vector, the plasmid becomes unstable and tends to lose DNA spontaneously.

OTHER CLONING VECTORS A number of other vectors have been developed for cloning larger pieces of DNA in bacteria. For example, bacteriophage λ, which infects *E. coli*, can be used to clone up to 23,000 bp of foreign DNA; it transfers DNA into bacterial cells with high efficiency. **Cosmids** are plasmids that are packaged in empty viral protein coats and transferred to bacteria by viral infection. They can carry more than twice as much foreign DNA as can a phage vector. **Bacterial artificial chromosomes** (**BACs**) are vectors originally constructed from the F plasmid (which controls mating and the transfer of genetic material in some bacteria; see Section 9.3) and can hold very large fragments of DNA (as long as 300,000 bp). **Table 19.2** compares the properties of plasmids, phage λ, cosmids, and BACs.

Sometimes the goal in gene cloning is not just to replicate the gene, but also to produce the protein it encodes. To ensure transcription and translation, a foreign gene is usually inserted into an **expression vector**, which, in addition to the usual origin of replication, restriction sites, and selectable markers, contains sequences required for transcription and translation in bacterial cells (**Figure 19.10**).

Although manipulating genes in bacteria is simple and efficient, the goal may be to transfer a gene into eukaryotic cells. For example, it might be desirable to transfer a gene conferring herbicide resistance into a crop plant, or to transfer a gene for clotting factor into a person with hemophilia. Many eukaryotic proteins are modified after translation (e.g., carbohydrate groups may be added). Such modifications are essential for proper function, but bacteria do not have the capacity to carry them out; thus, a functional protein can be produced only in a eukaryotic cell.

A number of cloning vectors have been developed that allow the insertion of genes into eukaryotic cells. Special plasmids have been developed for cloning in yeast, and

TABLE 19.2	Comparison of plasmids, phage λ, cosmids, and bacterial artificial chromosomes as cloning vectors		
Cloning Vector	**Size of DNA That Can Be Cloned**	**Method of Propagation**	**Introduction to Bacteria**
Plasmid	As large as 15 kb	Plasmid replication	Transformation
Phage λ	As large as 23 kb	Phage reproduction	Phage infection
Cosmid	As large as 44 kb	Plasmid reproduction	Phage infection
Bacterial artificial chromosome	As large as 300 kb	Plasmid reproduction	Electroporation

Note: 1 kb = 1000 bp. Electroporation is the use of electrical pulses to increase the permeability of a membrane.

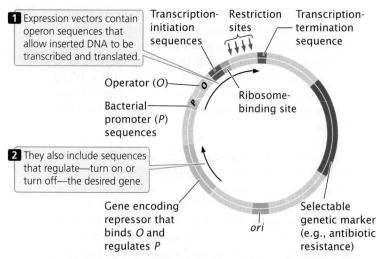

1 Expression vectors contain operon sequences that allow inserted DNA to be transcribed and translated.

Transcription-initiation sequences

Restriction sites

Transcription-termination sequence

Operator (*O*)

Bacterial promoter (*P*) sequences

Ribosome-binding site

2 They also include sequences that regulate—turn on or turn off—the desired gene.

Gene encoding repressor that binds *O* and regulates *P*

ori

Selectable genetic marker (e.g., antibiotic resistance)

19.10 To ensure transcription and translation, a foreign gene may be inserted into an expression vector, in this example, an *E. coli* expression vector.

retroviral vectors have been developed for cloning in mammals. A **yeast artificial chromosome** (**YAC**) is a DNA molecule that has a yeast origin of replication, a pair of telomeres, and a centromere. These features ensure that YACs are stable and that they replicate and segregate in the same way as yeast chromosomes. YACs are particularly useful because they can carry DNA fragments as large as 600 kb, and some special YACs can carry inserts of more than 1000 kb. YACs have been modified so that they can be used in eukaryotic organisms other than yeast.

The soil bacterium *Agrobacterium tumefaciens*, which invades plants through wounds and induces crown galls (tumors), has been used to transfer genes to plants. This bacterium contains a large plasmid called the **Ti plasmid**, part of which is transferred to a plant cell when *A. tumefaciens* infects a plant. In the plant, part of the Ti plasmid DNA integrates into one of the plant chromosomes, where it is transcribed and translated to produce several enzymes that help support the bacterium (**Figure 19.11a**).

Geneticists have engineered a vector that contains the flanking sequences required to transfer DNA (called TL and TR, see Figure 19.11a), a selectable marker, and restriction sites into which foreign DNA can be inserted (**Figure 19.11b**). When placed in *A. tumefaciens* with a helper Ti plasmid (which carries sequences allowing the bacterium to infect the plant cell, which the artificial vector lacks), this vector will transfer the foreign DNA that it carries into a plant cell,

(a)

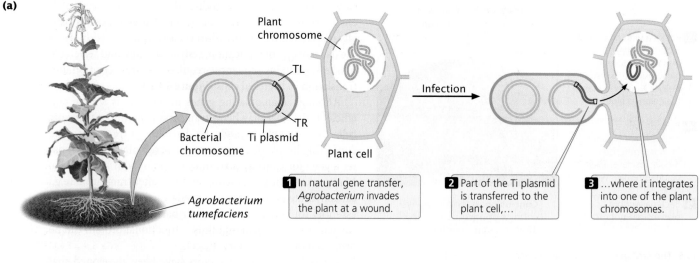

Plant chromosome

TL

TR

Bacterial chromosome

Ti plasmid

Plant cell

Infection

Agrobacterium tumefaciens

1 In natural gene transfer, *Agrobacterium* invades the plant at a wound.

2 Part of the Ti plasmid is transferred to the plant cell,...

3 ...where it integrates into one of the plant chromosomes.

(b)

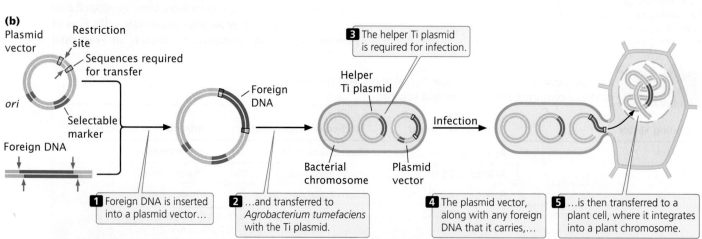

Plasmid vector

Restriction site

Sequences required for transfer

ori

Selectable marker

Foreign DNA

Foreign DNA

3 The helper Ti plasmid is required for infection.

Helper Ti plasmid

Infection

Bacterial chromosome

Plasmid vector

1 Foreign DNA is inserted into a plasmid vector...

2 ...and transferred to *Agrobacterium tumefaciens* with the Ti plasmid.

4 The plasmid vector, along with any foreign DNA that it carries,...

5 ...is then transferred to a plant cell, where it integrates into a plant chromosome.

19.11 The Ti plasmid can be used to transfer genes into plants. Flanking sequences TL and TR are required for the transfer of the DNA segment from bacteria to the plant cell.

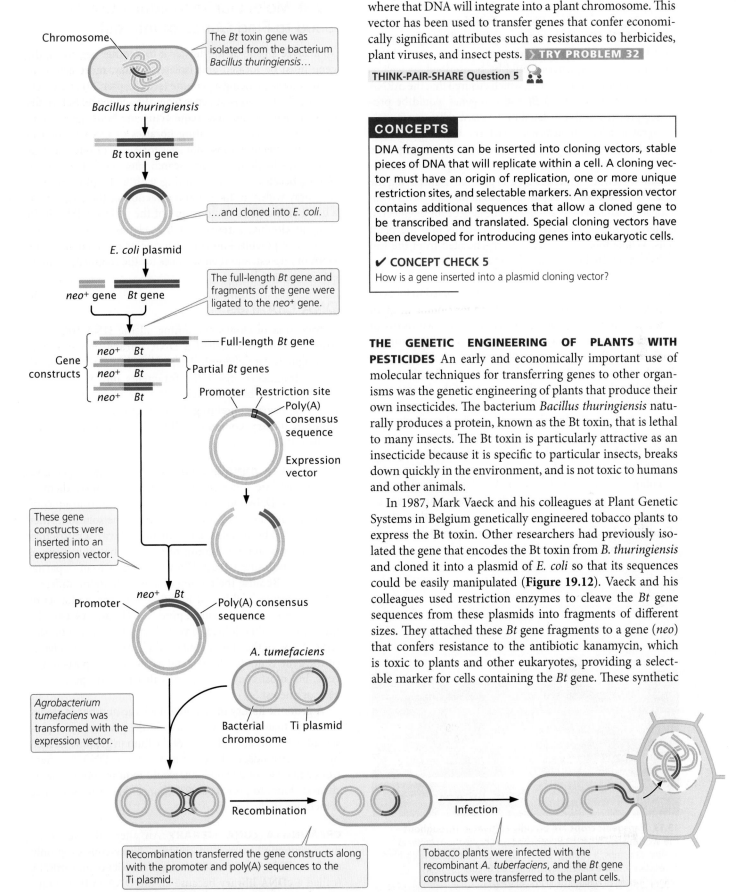

where that DNA will integrate into a plant chromosome. This vector has been used to transfer genes that confer economically significant attributes such as resistances to herbicides, plant viruses, and insect pests. **▶ TRY PROBLEM 32**

THINK-PAIR-SHARE Question 5 👥

> ### CONCEPTS
>
> DNA fragments can be inserted into cloning vectors, stable pieces of DNA that will replicate within a cell. A cloning vector must have an origin of replication, one or more unique restriction sites, and selectable markers. An expression vector contains additional sequences that allow a cloned gene to be transcribed and translated. Special cloning vectors have been developed for introducing genes into eukaryotic cells.
>
> #### ✔ CONCEPT CHECK 5
> How is a gene inserted into a plasmid cloning vector?

THE GENETIC ENGINEERING OF PLANTS WITH PESTICIDES An early and economically important use of molecular techniques for transferring genes to other organisms was the genetic engineering of plants that produce their own insecticides. The bacterium *Bacillus thuringiensis* naturally produces a protein, known as the Bt toxin, that is lethal to many insects. The Bt toxin is particularly attractive as an insecticide because it is specific to particular insects, breaks down quickly in the environment, and is not toxic to humans and other animals.

In 1987, Mark Vaeck and his colleagues at Plant Genetic Systems in Belgium genetically engineered tobacco plants to express the Bt toxin. Other researchers had previously isolated the gene that encodes the Bt toxin from *B. thuringiensis* and cloned it into a plasmid of *E. coli* so that its sequences could be easily manipulated (**Figure 19.12**). Vaeck and his colleagues used restriction enzymes to cleave the *Bt* gene sequences from these plasmids into fragments of different sizes. They attached these *Bt* gene fragments to a gene (*neo*) that confers resistance to the antibiotic kanamycin, which is toxic to plants and other eukaryotes, providing a selectable marker for cells containing the *Bt* gene. These synthetic

19.12 The *Bt* toxin gene, which encodes an insecticide, was isolated from bacteria and transferred to tobacco plants.

sequences, called genetic constructs, contained fragments of the *Bt* gene of different lengths linked to a *neo* gene (see Figure 19.12). The constructs were inserted into an expression vector that contained a promoter, which ensured that the introduced sequences would be transcribed, as well as poly(A) consensus sequences, which ensured that the mRNA produced from the fused *Bt* and *neo* genes would be processed properly and translated in the plant cells.

Agrobacterium tumefaciens bacteria were then transformed with the expression vectors. After the vectors were inside the bacteria, sequences on the expression vector recombined with sequences on a Ti plasmid, transferring the gene constructs to the Ti plasmid. Small discs of leaves from a tobacco plant were infected with the genetically engineered bacteria, which transferred the Ti plasmids into the plant cells. Whole tobacco plants were regenerated from the leaf disks and selected for resistance to kanamycin.

The resulting transgenic plants were then tested for resistance to insect pests. Leaves of the plants were fed to tobacco hornworms, whose mortality rates were monitored. High hornworm mortality was observed on about two-thirds of the plants containing fragments of the *Bt* gene. Interestingly, none of the plants containing the full-length *Bt* gene produced any insect-killing toxins; apparently, the plant cells were better able to translate the fragments of the *Bt* gene than to translate the entire gene. The transgenic plants producing Bt toxin grew normally. They were interbred to produce F$_1$ plants, which also exhibited insect and kanamycin resistance, indicating that the introduced genes were stably integrated into the plant chromosomes.

Following the research done by Vaeck and his colleagues, other researchers used similar methods to introduce the gene for Bt toxin into cotton, tomatoes, corn, and other plants (**Figure 19.13**). Today, transgenic plants expressing the Bt toxin are used broadly throughout the world.

19.13 Transgenic crops are broadly cultivated throughout the world. Shown here is Bt corn and unmodified corn growing in alternate strips, a strategy used to try to prevent the development of resistance to the genetically engineered Bt toxin by pests.
[John L. Obermeyer, Purdue Extension Entomology.]

19.4 Molecular Techniques Can Be Used to Find Genes of Interest

To analyze a gene or to transfer it to another organism, the gene must be located and isolated. Today, most genes are located by sequencing a genome (see Chapter 20) and determining the locations of genes from the sequence. Before the development of low-cost sequencing methods, genes were often located by first creating libraries of DNA sequences and then screening those libraries for genes of interest. This approach—to clone first and search later—is called *shotgun cloning* because it is like hunting with a shotgun: the pellets spray widely in the general direction of the target, with a good chance that one or more of the pellets will hit it. In shotgun cloning, a researcher first clones a large number of DNA fragments, knowing that one or more contains the DNA of interest, and then searches for the fragment of interest among the clones.

DNA Libraries

A collection of clones containing all the DNA fragments from one source is called a **DNA library**. For example, we might isolate genomic DNA from human cells, break it into fragments, insert the fragments into vectors, and clone them in bacterial cells. The set of bacterial colonies or phages containing these fragments is a human **genomic library**, containing all the DNA sequences found in the human genome.

CREATING A GENOMIC LIBRARY To create a genomic library, cells are collected and lysed, which causes them to release their DNA and other cellular contents into an aqueous solution, and the DNA is extracted from the solution. After the DNA has been isolated, it is incubated with a restriction enzyme for a limited amount of time so that only *some* of the restriction sites in each DNA molecule are cut (a partial digestion). Because the cutting of sites is random, different DNA molecules will be cut in different places, and a set of overlapping fragments will be produced (**Figure 19.14**). The fragments are then joined to vectors and transferred to bacteria. A few of the clones contain the entire gene of interest (if the gene is not too large), and a few contain parts of the gene, but most contain fragments that have no part of the gene of interest.

A genomic library must contain a large number of clones to ensure that all DNA sequences in the genome are represented in the library. A library of the human genome formed by using cosmids, each carrying a random DNA fragment from 35,000 to 44,000 bp long, would require about 350,000 cosmid clones to provide a 99% chance that every sequence is included in the library.

CREATING A cDNA LIBRARY An alternative to creating a genomic library is to create a library consisting only of those DNA sequences that are transcribed into mRNA (called a **cDNA library** because all the DNA in this library

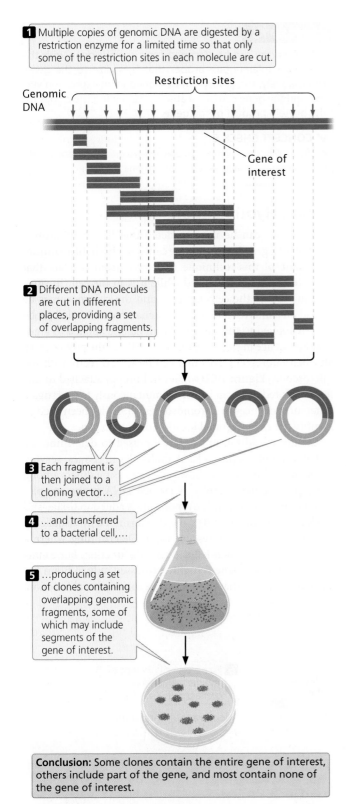

1 Multiple copies of genomic DNA are digested by a restriction enzyme for a limited time so that only some of the restriction sites in each molecule are cut.

Restriction sites

Genomic DNA

Gene of interest

2 Different DNA molecules are cut in different places, providing a set of overlapping fragments.

3 Each fragment is then joined to a cloning vector…

4 …and transferred to a bacterial cell,…

5 …producing a set of clones containing overlapping genomic fragments, some of which may include segments of the gene of interest.

Conclusion: Some clones contain the entire gene of interest, others include part of the gene, and most contain none of the gene of interest.

19.14 A genomic library contains all the DNA sequences found in an organism's genome.

is *complementary* to mRNA). Much of eukaryotic DNA consists of sequences that are not transcribed into mRNA, and these sequences are not represented in a cDNA library. A cDNA library has two additional advantages. First, it is enriched with fragments from actively transcribed genes. Second, introns do not interrupt the cloned sequences; introns would pose a problem when the goal is to produce a eukaryotic protein in bacteria because most bacteria have no means of removing the introns.

The disadvantage of a cDNA library is that it contains only sequences that are present in mature mRNA. Sometimes, researchers are interested in sequences that are not transcribed, such as those in promoters and enhancers, and these sequences will not be present in a cDNA library. Furthermore, a cDNA library contains only those genes expressed in the tissue from which the RNA was isolated, and the frequency of a particular DNA sequence in a cDNA library depends on the abundance of the corresponding mRNA in that tissue. So, if a particular gene is not expressed, or is expressed at low frequency, in a particular tissue, it may be absent from a cDNA library prepared from that tissue. In contrast, almost all genes are present at the same frequency in a genomic library.

To create a cDNA library, messenger RNA is first extracted from cells and separated from other types of cellular RNA (tRNA, rRNA, snRNA, etc.). The mRNA molecules are then copied into cDNA using reverse transcriptase, an enzyme isolated from retroviruses (see Section 9.4) that synthesizes single-stranded complementary DNA using RNA as a template. The resulting RNA–DNA hybrid molecule is finally converted into a double-stranded cDNA molecule by DNA polymerase. **▶ TRY PROBLEM 36**

CONCEPTS

One method of finding a gene is to create and screen a DNA library. A genomic library is created by cutting genomic DNA into overlapping fragments and cloning each fragment in a separate bacterial cell. A cDNA library is created from mRNA that is converted into cDNA and cloned in bacteria.

SCREENING DNA LIBRARIES Once a DNA library is created, it can be screened to find a gene or sequence of interest. The screening procedure used depends on what is known about the gene.

The first step in screening is to plate the clones of the library. If a plasmid or cosmid vector was used to construct the library, the bacterial cells are diluted and plated so that each bacterium grows into a distinct colony. If a phage vector was used, the phages are allowed to infect a lawn of bacteria on a petri plate. Each resulting plaque or bacterial colony contains a single cloned DNA fragment that must be screened for the gene of interest.

A common way to screen DNA libraries is with probes (**Figure 19.15**). But how is a probe obtained when the gene of interest has not yet been isolated? One option is to use a similar gene from another organism as the probe.

For example, if we wanted to screen a human genomic library for the growth-hormone gene and a growth-hormone gene had already been isolated from rats, we could use a purified rat-gene sequence as a probe to find the human gene for growth hormone. Successful hybridization does not require perfect complementarity between the probe and the target sequence, so a related sequence can often be used as a probe.

Alternatively, synthetic probes can be created if the protein produced by the gene of interest has been isolated and its amino acid sequence has been determined. With the use of the genetic code, possible nucleotide sequences of a small region of the gene can be deduced from the amino acid sequence of the protein. Although only one nucleotide sequence in the gene encodes a particular protein, the existence of synonymous codons means that the same protein could be produced by several different sequences, and it is impossible to know which one is correct. To overcome this problem, a mixture of all the possible nucleotide sequences is used as a probe. To minimize the number of sequences required in the mixture, a region of the protein with relatively little degeneracy in its codons is selected. When part of the DNA sequence of the gene has been determined, a set of DNA probes can be synthesized chemically by using an automated machine known as an oligonucleotide synthesizer.

Yet another method of screening a DNA library is to look for the protein product of a gene. This method requires that the DNA library be cloned in an expression vector. The clones can be tested for the presence of the protein by using an antibody that recognizes the protein or by using a chemical test for the protein. This method depends on the existence of an antibody or test for the protein produced by the gene of interest. DNA libraries can also be screened using PCR or by sequencing, as mentioned earlier in this chapter.

CONCEPTS

A DNA library can be screened for a specific gene with the use of complementary probes that hybridize to the gene. Alternatively, the library can be cloned in an expression vector, and the gene can be located by examining the clones for the protein product of the gene.

✔ **CONCEPT CHECK 6**

Briefly explain how synthetic probes are created to screen a DNA library when the protein encoded by the gene is known.

In Situ Hybridization

DNA probes can be used to determine the chromosomal location of a gene in a technique called **in situ hybridization**. The name of this technique is derived from the fact that DNA (or RNA) is visualized while it is in the cell (in situ). It requires that the cells be fixed and the chromosomes be spread on a microscope slide and denatured. A labeled probe is then applied to the slide, just as it can be applied to a gel. In fluorescence in situ hybridization (FISH), the probes carry attached fluorescent dyes that can be seen directly with the microscope (**Figure 19.16a**). Several probes attached to different colored dyes can be used simultaneously to investigate different sequences or chromosomes. FISH has been used to identify the chromosomal location of human genes.

In situ hybridization can also be used to determine the distribution of specific mRNA molecules in tissues, serving as a source of insight into how gene expression differs among cell types (**Figure 19.16b**). A labeled DNA probe complementary to a specific mRNA molecule is added to tissue, and the location of the bound probe is determined by visualizing the radioactive or fluorescent label. Determining where a gene is expressed often helps define its function. For example, finding that a gene is highly expressed only in brain tissue might suggest that the gene has a role in neural function.

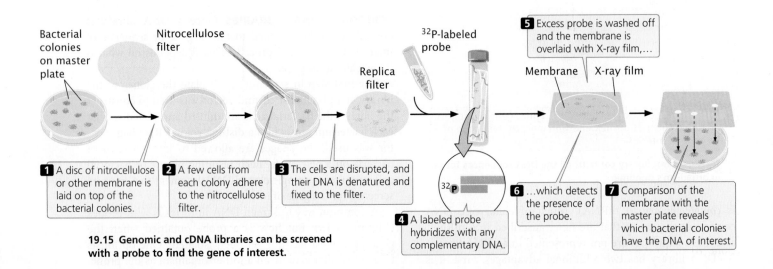

19.15 Genomic and cDNA libraries can be screened with a probe to find the gene of interest.

Bacterial colonies on master plate — Nitrocellulose filter — Replica filter — ³²P-labeled probe — Membrane | X-ray film

1 A disc of nitrocellulose or other membrane is laid on top of the bacterial colonies.

2 A few cells from each colony adhere to the nitrocellulose filter.

3 The cells are disrupted, and their DNA is denatured and fixed to the filter.

4 A labeled probe hybridizes with any complementary DNA.

5 Excess probe is washed off and the membrane is overlaid with X-ray film,...

6 ...which detects the presence of the probe.

7 Comparison of the membrane with the master plate reveals which bacterial colonies have the DNA of interest.

(a)

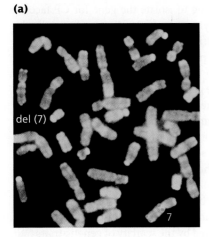

(b)

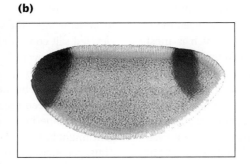

19.16 With in situ hybridization, DNA probes are used to determine the chromosomal or cellular location of a gene or its product. (a) A probe attached to a green fluorescent dye is specific to a region of chromosome 7, revealing a deletion on one copy of chromosome 7. (b) In situ hybridization is used to detect the presence of mRNA from the *tailless* gene in a *Drosophila* embryo. [Part a: Addenbrookes Hospital/Science Source. Part b: Courtesy of L. Tsuda.]

Positional Cloning

For many genes with important functions, no associated protein product has yet been identified. The biochemical bases of many human genetic diseases, for example, are still unknown. How can these genes be isolated? One approach is to first determine the general location of the gene on a chromosome by using recombination frequencies derived from crosses or pedigrees (see Chapter 7). After the chromosomal region where the gene is found has been pinpointed, genes in that region can be cloned and identified. Then other techniques can be used to determine which of the "candidate" genes might be the one that causes the disease. This approach—isolation of genes on the basis of their position on a gene map—is called **positional cloning**.

In the first step of positional cloning, geneticists use mapping studies (see Chapter 7) to establish linkage between molecular markers and a phenotype of interest, such as a human disease or a desirable physical trait in a plant or animal. Demonstration of linkage between the phenotype and one or more molecular markers tells us which chromosome carries the locus that codes for the phenotype and its general location on that chromosome.

The next step is to narrow down the location of the gene by using additional molecular markers clustered in the chromosomal region where the gene resides. Clones that cover that region can then be isolated from a genomic library. With the use of a technique called **chromosome walking** (**Figure 19.17**), it is possible to progress from clones of molecular markers to linked clones, one of which might contain the gene of interest. The basis of chromosome walking is the fact that a genomic library consists of a set of *overlapping* DNA fragments (see Figure 19.14). We start with a cloned gene marker that is close to the new gene of interest so that the "walk" will be as short as possible. One end of the clone of a neighboring marker (clone A in Figure 19.17) is used to make a complementary probe. This probe is used to screen the genomic library to find a second clone (clone B) that overlaps with the first and extends in the

direction of the gene of interest. This second clone is isolated and purified, and a probe is prepared from its end. The second probe is used to screen the library for a third clone (clone C) that overlaps with the second. In this way, one can systematically "walk" toward the gene of interest, one clone at a time. A related technique, called chromosome jumping, allows one to move from more distantly linked markers to clones that contain a sequence of interest.

After clones that cover the delineated chromosome region have been obtained by chromosome walking or jumping, all genes located within the region are identified. Genes can be distinguished from other sequences by the presence of characteristic features, such as consensus sequences in the

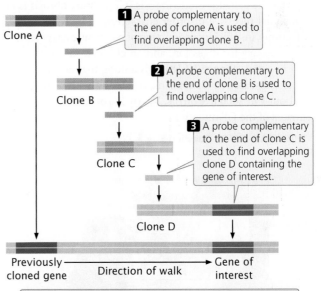

Conclusion: By making probes complementary to areas of overlap between cloned fragments in a genomic library, we can connect a gene of interest to a previously mapped, linked gene.

19.17 In chromosome walking, neighboring genes are used to locate a gene of interest.

promoter, and a start codon and a stop codon within the same reading frame. After candidate genes have been identified, they can be evaluated to determine which is most likely to be the gene of interest. The expression pattern of the gene—where and when it is transcribed—can often provide clues about its function; for example, genes for neurological disease would probably be expressed in the brain. Geneticists often look in the coding regions of candidate genes for mutations among individuals with the phenotype of interest. More will be said about determining the function of genes in the sections that follow and in Chapter 20.

CONCEPTS

Positional cloning allows researchers to isolate a gene without having knowledge of the biochemical basis of its phenotype. Linkage studies are used to map the locus producing a phenotype of interest to a particular chromosome region. Chromosome walking and jumping can be used to progress from molecular markers to clones containing sequences that cover the chromosome region. Candidate genes within the region are then evaluated to determine whether they encode the phenotype of interest.

✔ CONCEPT CHECK 7

How are candidate genes that are identified by positional cloning evaluated to determine whether they encode the phenotype of interest?

ISOLATING THE GENE FOR CYSTIC FIBROSIS BY POSITIONAL CLONING The first gene responsible for a human genetic disease that was isolated entirely by positional cloning was the gene for cystic fibrosis (CF). Cystic fibrosis is an autosomal recessive disorder characterized by chronic lung infections, insufficient production of pancreatic enzymes that are necessary for digestion, and an elevated salt concentration in sweat (**Figure 19.18**). It is among the most common genetic diseases in Caucasians, occurring with a frequency of about 1 in 2000 live births. Nearly 5% of all Caucasians are carriers of the CF mutation.

THINK-PAIR-SHARE Question 6

19.18 Cystic fibrosis was the first genetic disease for which the causative gene was isolated entirely by positional cloning. Annual medical review of a 22-year-old woman with cystic fibrosis. Here, she is breathing into a spirometer. [Burger/Phanie/Science Source.]

Geneticists attempting to isolate the gene for CF faced a formidable task. The symptoms of the disease, especially the elevated salt concentration in sweat, suggested that the gene for CF somehow takes part in the movement of ions into and out of the cell, but no information was available about the protein encoded by the gene. At the time, the human genome had not yet been sequenced. Analyses of pedigrees showed that CF is inherited as an autosomal recessive trait, and so the gene might be located on any one of the 22 pairs of autosomes. Thus, geneticists were seeking an unknown gene—probably encompassing a few thousand or tens of thousands of base pairs—among the 3.2 billion base pairs of the human genome.

Researchers began by looking for associations between the inheritance of CF and that of other traits (**Figure 19.19**). Early studies were limited by the scarcity of genetic traits that varied and could be used for gene-mapping studies, but in the 1980s, advances in molecular biology provided a large number of molecular markers that could be used for linkage analysis (see p. 199 in Chapter 7). Geneticists collected pedigrees of families in which several members had CF. They compared the inheritance of CF with that of molecular markers among the members of these families, looking for evidence of linkage. The gene for CF was found to be closely linked to two markers, MET and D7S8, located on the long arm of chromosome 7. MET and D7S8 are separated by about 1.5 map units (see Chapter 7). In the human genome, each map unit roughly corresponds to 1 million base pairs, so the gene for CF had to be located somewhere within a stretch of 1.5 million base pairs of DNA—a huge expanse of sequence.

Further linkage studies with additional markers were carried out to more precisely delineate where in the 1.5 million-base-pair region the CF gene lies. Researchers selected additional molecular markers from the region surrounding MET and D7S8 and performed studies of linkage between these new markers and CF (see Figure 19.19). These studies identified two additional markers, D7S122 and D7S340, that are closely linked to CF. Furthermore, they showed that the order of the four markers is MET-D7S340-D7S122-D7S8 and that the CF gene lies very close to D7S122 and D7S340. This finding narrowed the region in which the gene for CF lies to about 500,000 bp.

At this stage, geneticists began isolating clones of sequences from the delineated region. Starting from the molecular markers, they used a combination of chromosome walking and chromosome jumping to identify clones from human genomic libraries that completely covered the region of interest (see Figure 19.19). An examination of sequences within these clones revealed the presence of four genes in the region encompassed by the linked markers (see Figure 19.19). Additional studies were then carried out to better characterize these candidate genes. Three of the candidate genes were eventually eliminated, either because linkage studies suggested that they were not closely linked with the inheritance of CF or because analysis of their sequences or their expression patterns suggested they were not involved in CF.

Hybridization studies were carried out with the one remaining gene to determine where it was expressed.

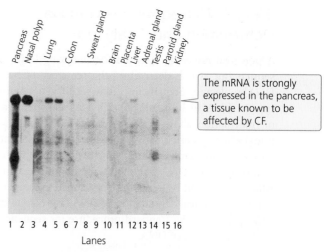

The mRNA is strongly expressed in the pancreas, a tissue known to be affected by CF.

19.20 A candidate for the cystic fibrosis gene is expressed in pancreatic, respiratory, and sweat-gland tissues—tissues that are affected by the disease. Shown is a Northern blot of mRNA produced by the candidate gene in different tissues. These data provided evidence that the candidate gene is in fact the gene that causes cystic fibrosis. [From Riordan, J. et al. "Identification of the cystic fibrosis gene: cloning and characterization of complementary DNA," *Science* 245: 1066–1073, © 1989. Reprinted with permission from AAAS, permission conveyed through Copyright Clearance Center, Inc.]

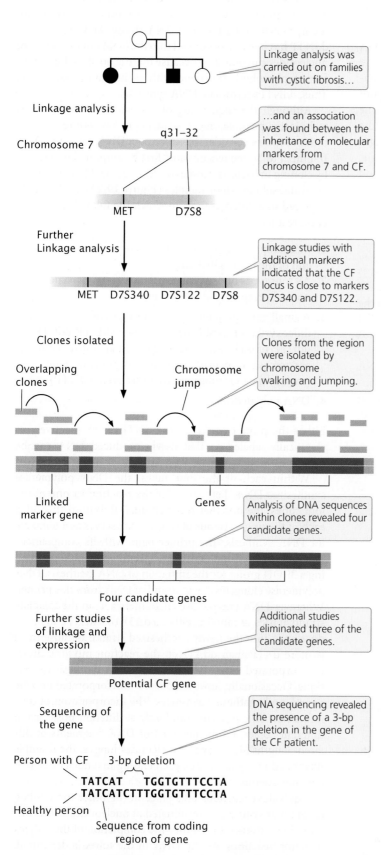

Linkage analysis was carried out on families with cystic fibrosis...

...and an association was found between the inheritance of molecular markers from chromosome 7 and CF.

Linkage studies with additional markers indicated that the CF locus is close to markers D7S340 and D7S122.

Clones from the region were isolated by chromosome walking and jumping.

Analysis of DNA sequences within clones revealed four candidate genes.

Additional studies eliminated three of the candidate genes.

DNA sequencing revealed the presence of a 3-bp deletion in the gene of the CF patient.

Person with CF   3-bp deletion

TATCAT TGGTGTTTCCTA
TATCATCTTTGGTGTTTCCTA

Healthy person

Sequence from coding region of gene

19.19 The gene for cystic fibrosis was located by positional cloning.

Messenger RNA was isolated from different tissues and probed with sequences from the candidate gene. The gene showed high levels of expression in the pancreas, lungs, and sweat glands (**Figure 19.20**), tissues known to be affected by CF.

Copies of the candidate gene from a person with CF and from an unaffected person were then sequenced, and the sequence data were examined for differences that might represent a mutation causing CF. The findings revealed that the person with CF had a 3-bp deletion in the coding region of the gene, while the unaffected person did not have this deletion. The deletion resulted in the absence of a phenylalanine from the protein encoded by the candidate gene. Then, for a large sample of patients with CF, geneticists used PCR to amplify the region of the gene where the deletion was found; 68% of the CF patients had this deletion. Subsequent studies demonstrated that the remaining CF patients possessed mutations at other locations within the candidate gene, thus proving that the candidate gene was indeed the locus that caused CF.

Researchers eventually demonstrated that the gene for CF encodes a membrane protein that controls the movement of chloride into and out of cells; it is known today as the *cystic fibrosis transmembrane conductance regulator* (*CFTR*) gene. Patients with CF have two mutated copies of *CFTR*, which cause the chloride channels to remain closed. Chloride ions build up in the cell, leading to the formation of thick mucus and the symptoms of the disease.

19.5 DNA Sequences Can Be Determined and Analyzed

A powerful molecular method for analyzing DNA is **DNA sequencing**, which determines the sequence of bases in a DNA molecule. Sequencing allows the genetic information in DNA to be read, providing an enormous amount of information about gene structure and function. In the mid-1970s, Frederick Sanger and his colleagues created the dideoxy-sequencing method, which is based on the elongation of DNA by DNA polymerase; at about the same time, Allan Maxam and Walter Gilbert developed a second method based on chemical degradation of DNA. The Sanger method quickly became the standard procedure for sequencing any purified fragment of DNA.

Dideoxy Sequencing

The dideoxy (or Sanger) method of DNA sequencing is based on replication. The fragment to be sequenced is used as a template to make a series of new DNA molecules. In the process, replication is sometimes, but not always, terminated when a specific base is encountered, producing DNA strands of different lengths, each of which ends in a known base.

The method relies on the use of a special substrate for DNA synthesis. Normally, DNA is synthesized from deoxyribonucleoside triphosphates (dNTPs), which have an OH group on the 3'-carbon atom (**Figure 19.21a**). In the Sanger method, special nucleotides, called **dideoxyribonucleoside triphosphates** (**ddNTPs; Figure 19.21b**), are also used as substrates.

The ddNTPs are identical with dNTPs, except that they lack a 3'-OH group. In the course of DNA synthesis, ddNTPs are incorporated into a growing DNA strand. However, after a ddNTP has been incorporated into the DNA strand, no more nucleotides can be added because there is no 3'-OH group to form a phosphodiester bond with an incoming nucleotide. Thus, ddNTPs terminate DNA synthesis.

Although the sequencing of a single DNA molecule is technically possible, most sequencing procedures in use today require a considerable amount of DNA; thus, any DNA fragment to be sequenced must first be amplified by PCR or by cloning in bacteria. Copies of the target DNA are then isolated and split into four samples (**Figure 19.22**). Each sample is placed in a different tube, to which the following ingredients are added:

1. Many copies of a primer that is complementary to one end of the target DNA strand

2. All four types of deoxyribonucleoside triphosphates, the normal precursors of DNA synthesis

3. A small amount of one of the four types of dideoxyribonucleoside triphosphates (ddATP, ddCTP, ddGTP, or ddTTP), which will terminate DNA synthesis as soon as it is incorporated into any growing chain (each of the four tubes receives a different ddNTP)

4. DNA polymerase

Either the primer or one of the dNTPs is radioactively or chemically labeled so that newly produced DNA can be detected.

Within each of the four tubes, the DNA polymerase synthesizes DNA. Let's consider the reaction in one of the four tubes: the one that received ddATP. Within this tube, each of the single strands of target DNA serves as a template for DNA synthesis. The primer pairs with its complementary sequence at one end of each template strand, providing a 3'-OH group for the initiation of DNA synthesis. DNA polymerase elongates a new strand of DNA from this primer. Wherever DNA polymerase encounters a T on the template strand, it uses, at random, either a dATP or a ddATP to introduce an A into the newly synthesized strand. Because there is more dATP than ddATP in the reaction mixture, dATP is incorporated most often, allowing DNA synthesis to continue. Occasionally, however, ddATP is incorporated into the strand, and synthesis terminates. The incorporation of ddA into a new strand occurs randomly at different positions in different copies, producing a set of DNA fragments of different lengths (12, 7, and 2 nucleotides long in the example illustrated in Figure 19.22), each ending in a nucleotide that contains adenine.

Equivalent reactions take place in the other three tubes, except that synthesis is terminated at nucleotides with a different base in each tube. After the completion of the polymerization reactions, all the DNA in the tubes is denatured, and the single-stranded products of each reaction are separated by gel electrophoresis.

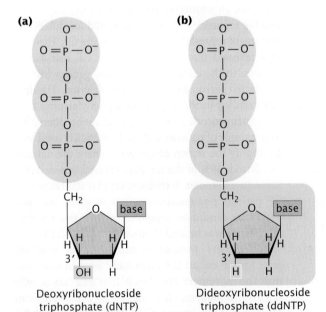

(a)

(b)

Deoxyribonucleoside triphosphate (dNTP)

Dideoxyribonucleoside triphosphate (ddNTP)

19.21 The dideoxy-sequencing reaction requires a special substrate for DNA synthesis. (a) Structure of deoxyribonucleoside triphosphate, the normal substrate for DNA synthesis. (b) Structure of dideoxyribonucleoside triphosphate, which lacks an OH group on the 3'-carbon atom.

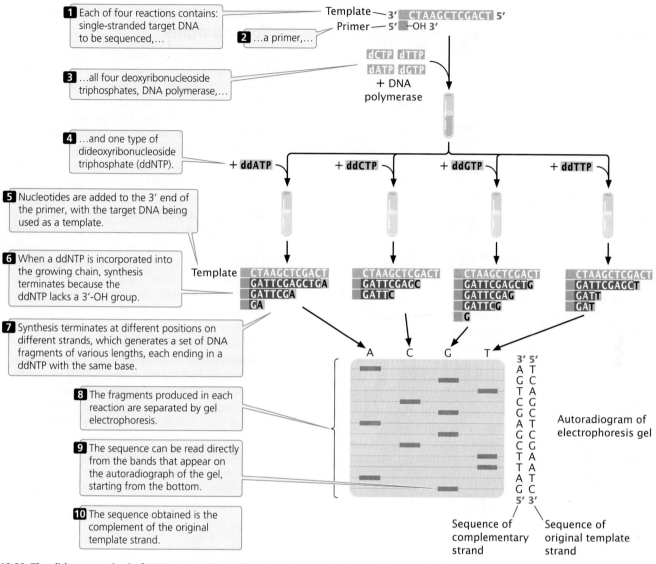

1 Each of four reactions contains: single-stranded target DNA to be sequenced,...

2 ...a primer,...

Template — 3′ CTAAGCTCGACT 5′
Primer — 5′ —OH 3′

3 ...all four deoxyribonucleoside triphosphates, DNA polymerase,...

dCTP dTTP
dATP dGTP
+ DNA polymerase

4 ...and one type of dideoxyribonucleoside triphosphate (ddNTP).

+ **ddATP**   + **ddCTP**   + **ddGTP**   + **ddTTP**

5 Nucleotides are added to the 3′ end of the primer, with the target DNA being used as a template.

6 When a ddNTP is incorporated into the growing chain, synthesis terminates because the ddNTP lacks a 3′-OH group.

Template

CTAAGCTCGACT   CTAAGCTCGACT   CTAAGCTCGACT   CTAAGCTCGACT
GATTCGAGCTGA   GATTCGAGC   GATTCGAGCTG   GATTCGAGCT
GATTCGA   GATTC   GATTCGA   GATT
GA   G   GAT

7 Synthesis terminates at different positions on different strands, which generates a set of DNA fragments of various lengths, each ending in a ddNTP with the same base.

A   C   G   T

8 The fragments produced in each reaction are separated by gel electrophoresis.

9 The sequence can be read directly from the bands that appear on the autoradiograph of the gel, starting from the bottom.

3′ 5′
A T
G C
T A
C G
G C
A T
G C
C G
T A
T A
A T
G C
5′ 3′

Autoradiogram of electrophoresis gel

10 The sequence obtained is the complement of the original template strand.

Sequence of complementary strand   Sequence of original template strand

19.22 The dideoxy method of DNA sequencing is based on the termination of DNA synthesis.

The contents of the four tubes are separated side by side on an acrylamide gel (which allows finer separation than agarose) so that DNA strands differing in length by only a single nucleotide can be distinguished. After electrophoresis, the locations, and therefore the sizes, of the DNA strands in the gel are revealed by the radioactive or chemical labels.

Reading the DNA sequence is the simplest and shortest part of the procedure. In Figure 19.22, you can see that the band closest to the bottom of the gel is from the tube that contained the ddGTP reaction, which means that the first nucleotide added was guanine (G). The next band up is from the tube that contained ddATP; so the next nucleotide in the sequence is adenine (A), and so forth. In this way, the sequence is read from the bottom to the top of the gel, with the nucleotides near the bottom corresponding to the 5′ end of the newly synthesized DNA strand and those near the top corresponding to the 3′ end. Keep in mind that the sequence obtained is not that of the target DNA but that of

its *complement*. To see dideoxy sequencing in action, view **Animation 19.3.** ▶ TRY PROBLEM 39

For many years, DNA sequencing was done largely by hand and was laborious and expensive. Today, sequencing is usually carried out by automated machines that use fluorescent dyes and laser scanners to sequence thousands of base pairs in a few hours (**Figure 19.23**). The dideoxy-sequencing reaction is also used here, but the ddNTPs used in the reaction are labeled with fluorescent dyes, and a different colored dye is used for each type of ddNTP. In this case, the four sequencing reactions can take place in the same test tube and can be placed in the same well for electrophoresis. The sequencing devices carry out electrophoresis in gel-containing capillary tubes. The different-sized fragments produced by the sequencing reaction separate within a tube and migrate past a laser beam and detector. As the fragments pass the laser, their fluorescent dyes are activated. Each colored dye emits fluorescence of a characteristic wavelength, which is read by an optical scanner. The information is fed

1 A single-stranded DNA fragment whose base sequence is to be determined (the template) is isolated.

2 Each of the four ddNTPs is tagged with a different fluorescent dye, and the Sanger sequencing reaction is carried out.

3 The fragments that end in the same base have the same colored dye attached.

4 The products are denatured, and the DNA fragments produced by the reaction are loaded into a single well on an electrophoresis gel. The fragments migrate through the gel according to size,...

5 ...and the fluorescent dye on the DNA is detected by a laser beam.

6 Each fragment appears as a peak on the computer printout; the color of the peak indicates which base is present.

7 The sequence information is read directly into the computer.

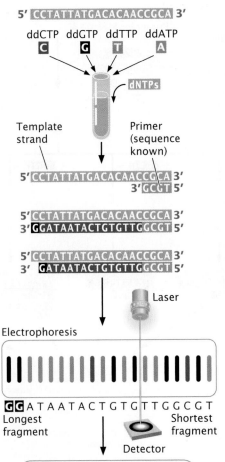

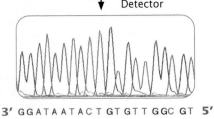

19.23 The dideoxy-sequencing method can be automated.

into a computer for interpretation, and the results are printed out as a set of peaks on a graph (see Figure 19.23). Automated sequencing machines may contain 96 or more capillary tubes, allowing from 50,000 to 60,000 bp of sequence to be read in a few hours.

CONCEPTS

The dideoxy-sequencing method uses ddNTPs, which terminate DNA synthesis at specific bases.

✔ CONCEPT CHECK 8

In the dideoxy-sequencing reaction, what terminates DNA synthesis at a particular base?

a. The absence of a base on the ddNTP halts the DNA polymerase.

b. The ddNTP causes a break in the sugar–phosphate backbone.

c. DNA polymerase will not incorporate a ddNTP into the growing DNA strand.

d. The absence of a 3′-OH group on the ddNTP prevents the addition of another nucleotide.

Next-Generation Sequencing Technologies

Newer methods, called **next-generation sequencing technologies**, have made sequencing hundreds of times faster and less expensive than the traditional Sanger sequencing method. Most next-generation sequencing technologies do sequencing in parallel, which means that hundreds of thousands or even millions of DNA fragments can be sequenced simultaneously, allowing, for example, a human genome to be sequenced in days instead of years.

ILLUMINA SEQUENCING Illumina sequencing employs a technology that is similar in principle to that used in dideoxy sequencing. Special nucleotides are used that have a fluorescent tag attached, with a different colored tag for each type of nucleotide. Each nucleotide also has a chemical group (a terminator) that, once incorporated into the growing DNA chain, prevents the incorporation of any additional nucleotides. This is similar to termination caused by dideoxyribonucleotides in Sanger sequencing. In this case, however, the terminator is reversible—it can be chemically removed.

To carry out sequencing, the target DNA is first cleaved into millions of short overlapping fragments. The fragments are attached to a slide and then amplified, creating clusters of up to 1000 copies of each fragment in close proximity on the slide. The fragments are then denatured, and a solution of primers, DNA polymerase, and the special nucleotides is added. The primer attaches to each DNA template, and the first nucleotide is incorporated into the newly synthesized strand. The solution is washed away, and the tag on the incorporated nucleotide is excited with a laser, which causes it to fluoresce. As mentioned above, each type of nucleotide (A, T, G, or C) has a different colored fluorescent tag, so the color of the light produced reveals the type of the nucleotide just added. The terminator and the fluorescent tag are then chemically removed, and the process is repeated. As the nucleotides are added one at a time, the sequence is read as a series of flashes of colored light from each cluster of DNA. Hundreds of thousands of DNA clusters, each consisting of copies of a different DNA fragment, are sequenced simultaneously, allowing large amounts of DNA to be sequenced in a short time.

PYROSEQUENCING Several other forms of next-generation sequencing have been developed. One type, called pyrosequencing, is based on DNA synthesis: nucleotides are added one at a time in the order specified by template DNA and the addition of a particular nucleotide is detected with a flash of light, which is generated as the nucleotide is added.

To carry out pyrosequencing, DNA to be sequenced is first fragmented (**Figure 19.24a**). An adaptor, consisting of

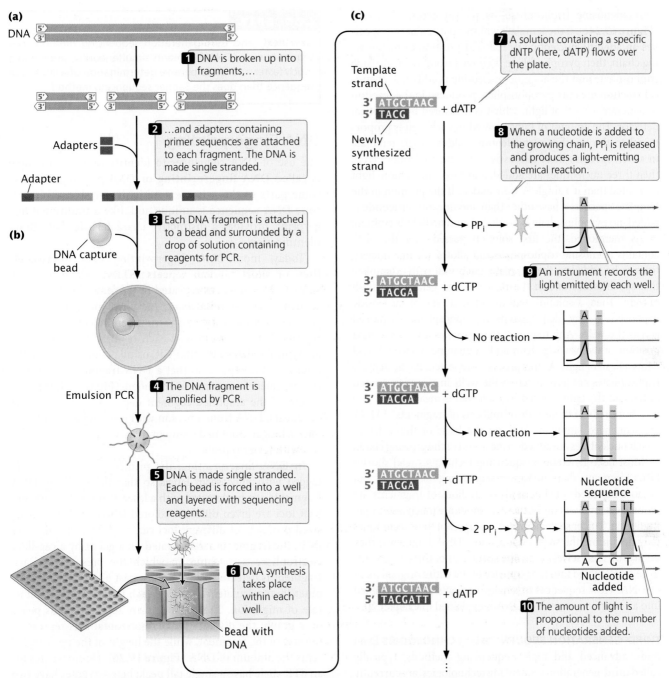

19.24 Next-generation sequencing methods are able to simultaneously determine the sequences of hundreds of thousands or millions of DNA fragments. Pyrosequencing is illustrated here.

a short string of nucleotides, is added to each fragment. The adaptor provides a known sequence to prime a PCR reaction. The DNA fragments are then made single stranded. In one version of pyrosequencing, each fragment is then attached to a separate bead and surrounded by a droplet of solution containing the reagents for PCR (**Figure 19.24b**). The bead is used to hold the DNA and, later, to deposit it on a plate for the sequencing reaction, as we'll see shortly. Within the droplet, the fragment is amplified by PCR, and the copies of the fragment remain attached to the bead. After amplification, the beads are mixed with DNA polymerase and are deposited

on a plate containing more than a million wells (microscopic holes). Each bead is deposited into a different well.

The sequencing reaction that takes place in each well is based on DNA synthesis. Recall from Chapter 12 that deoxyribonucleoside triphosphate—the substrate for DNA synthesis—consists of a deoxyribose sugar attached to a base and three phosphate groups. In the process of DNA synthesis, two phosphate groups (forming pyrophosphate, or PP_i) are cleaved off, and the resulting nucleotide is attached to the 3' end of the growing DNA chain. A solution containing one particular type of deoxynucleoside triphosphate—say, deoxyadenosine triphosphate—is passed across the wells (**Figure 19.24c**). If the template within a particular well specifies an adenine nucleotide in the next position of the growing chain, then pyrophosphate is cleaved from the nucleoside triphosphate and the adenine nucleotide is added. A chemical reaction uses the pyrophosphate produced in the reaction to generate a flash of light, which is measured by an optical detector. The amount of light emitted in each well is proportional to the number of nucleotides added: if the template in a well specifies three successive adenine nucleotides (As), then three nucleotides are added, and three times more light is emitted than if a single A were added. If the position in the template specifies a base other than adenine, no nucleotide is added, no pyrophosphate is produced, and no light is emitted.

As mentioned, the first solution passed over the plate contains adenosine triphosphate and allows adenine nucleotides to be added to the template. Each well with a template that specifies adenine in the next position generates a flash of light. Then, a solution with a different type of nucleoside triphosphate—say, deoxyguanosine triphosphate—is passed across the wells. Any fragment that specifies a G in the next position of its growing chain adds a guanine nucleotide and emits a flash of light. As this process is repeated, the nucleotide triphosphates are passed across the wells in a predetermined order, and the light emitted by each well is measured. In this way, hundreds of thousands or millions of fragments of DNA are sequenced simultaneously on the basis of the order in which nucleotides are added to the 3′ end of the growing chain.

Most next-generation sequencing techniques read shorter DNA fragments than the Sanger sequencing method can, but because hundreds of thousands or millions of fragments are sequenced simultaneously, these methods are much faster than traditional Sanger sequencing technology. But these techniques only determine the sequences of short DNA fragments; they do not, by themselves, allow the sequences of these fragments to be reassembled into the sequence of the entire original piece of DNA. How sequences of small fragments are reassembled into a continuous stretch of DNA is explained in Chapter 20.

THIRD-GENERATION SEQUENCING TECHNOLOGIES Even more advanced and rapid sequencing methods, typically called third-generation sequencing technologies, are currently under development. Nanopore sequencing, for example, can determine the sequence of individual molecules of DNA. In this method, a single strand of DNA is passed through a tiny hole—a nanopore—in a membrane. As the molecule passes, one nucleotide at a time, through the nanopore, it disrupts an electrical current in the membrane, and the nature of the disruption is affected by the shape of the nucleotide passing through the nanopore. Each of the four bases of DNA causes a characteristic disruption, so the sequence of DNA can be read by analyzing the membrane current as the strand passes, one nucleotide at a time, through the nanopore. Hundreds of thousands of nanopores can be created on a single chip, so that many DNA fragments can be read simultaneously.

CONCEPTS

New next- and third-generation sequencing methods can sequence many DNA fragments simultaneously, providing a much faster and less expensive determination of a DNA base sequence than does the Sanger sequencing method.

DNA Fingerprinting

The use of DNA sequences to identify individual people is called **DNA fingerprinting** or *DNA profiling*. Because some parts of the genome are highly variable, each person's DNA sequence is unique and, like a traditional fingerprint, provides a distinctive characteristic that allows identification.

Today, most DNA fingerprinting uses **microsatellites**, or **short tandem repeats** (**STRs**), which are very short DNA sequences repeated in tandem. These repeated sequences are found at many loci throughout the human genome. People vary in the number of copies of these repeats that they possess at each of these loci. The STRs are typically detected with PCR, using primers flanking the microsatellite repeats so that a DNA fragment containing the repeated sequences is amplified (**Figure 19.25**). The length of the amplified segment depends on the number of repeats; DNA from a person with more repeats will produce a longer amplified segment than will DNA from a person with fewer repeats.

In DNA fingerprinting, the primers used in PCR are tagged with fluorescent labels so that the resulting DNA fragments can be detected with a laser. Primers for different STR loci are given different colored labels so that similar-sized products of different loci can be differentiated. After PCR, the fragments are separated on a gel or by a capillary electrophoresis machine, in which the presence of each fragment is detected as it migrates past a laser, and a computer then calculates the size of each fragment based on its rate of migration. The fragments are represented as peaks on a graph; the distance on the horizontal axis represents the size of the fragment, while the height of the peak represents the amount of DNA (**Figure 19.26**). Homozygotes for an STR allele have a single tall peak; heterozygotes have two shorter peaks. When several different microsatellite loci are examined, the probability of two people having the same set of patterns becomes vanishingly small, unless they are identical twins.

The Federal Bureau of Investigation has developed a system using 13 STR loci (**Table 19.3**) that are commonly used for identifying people and solving crimes. These loci make up the Combined DNA Index System (CODIS). Each STR locus in CODIS has a large number of alleles and is located on a different human chromosome, so variation at each locus assorts independently. When all 13 CODIS loci are used together, the probability of two randomly selected people having the same DNA profile is less than 1 in 10 billion.

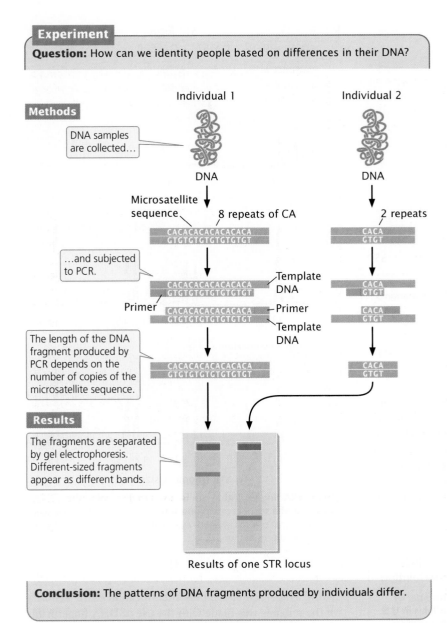

Question: How can we identity people based on differences in their DNA?

Methods

Individual 1   Individual 2

DNA samples are collected...

DNA   DNA

Microsatellite sequence   8 repeats of CA   2 repeats

CACACACACACACACA
GTGTGTGTGTGTGTGT

CACA
GTGT

...and subjected to PCR.

CACACACACACACACA
GTGTGTGTGTGTGTGT   Template DNA

CACA
GTGT

Primer

CACACACACACACACA
GTGTGTGTGTGTGTGT   Primer

CACA
GTGT

Template DNA

The length of the DNA fragment produced by PCR depends on the number of copies of the microsatellite sequence.

CACACACACACACACA
GTGTGTGTGTGTGTGT

CACA
GTGT

Results

The fragments are separated by gel electrophoresis. Different-sized fragments appear as different bands.

Results of one STR locus

Conclusion: The patterns of DNA fragments produced by individuals differ.

19.25 DNA fingerprinting can be used to identify people.

In a typical application, DNA fingerprinting might be used to confirm that a suspect was present at the scene of a crime. A sample of DNA from blood, semen, hair, or other body tissue is collected from the crime scene. If the sample is very small, PCR can be used to amplify it so that enough DNA is available for testing. Additional DNA samples are collected from one or more suspects. The pattern of DNA fragments produced by DNA fingerprinting of the sample is then compared with the patterns produced by DNA fingerprinting of the DNA from the suspect. A match between the sample and a suspect can provide evidence that the suspect was present at the scene of the crime (**Figure 19.27**).

Since its introduction in the 1980s, DNA fingerprinting has helped convict a number of suspects in murder and rape

cases. Suspects in other cases have been proved innocent when their DNA failed to match that from the crime scenes. Initially, calculations of the odds of a match (the probability that two people could have the same pattern) were controversial, and there were concerns about quality control (such as the accidental contamination of samples and the reproducibility of results) in laboratories where DNA analysis is done. Today, DNA fingerprinting has become widely accepted as an important tool in forensic investigations. In addition to its application in solving crimes, DNA fingerprinting is used to assess paternity, study genetic relationships among individual organisms in natural populations, identify specific strains of pathogenic bacteria, and identify human remains.

CONCEPTS

DNA fingerprinting detects genetic differences among individuals by analyzing highly variable regions of chromosomes.

✔ **CONCEPT CHECK 9**

How are microsatellites detected?

IDENTIFYING PEOPLE WHO DIED IN THE COLLAPSE OF THE WORLD TRADE CENTER WITH DNA FINGERPRINTING On the morning of September 11, 2001, terrorists hijacked two passenger planes and flew them into the World Trade Center towers in New York City. The catastrophic damage and ensuing fire led, within a few hours, to the complete collapse of all 110 floors of both towers, killing almost 3000 building occupants and rescue personnel. The tremendous destructive force generated by the collapse of the towers, with their 425,000 cubic yards of concrete and 200,000 tons of steel, pulverized many of the bodies beyond recognition.

In the days immediately following the World Trade Center collapse, forensic scientists began the task of identifying the remains of those who perished. The goal was to provide evidence for the ongoing criminal investigation of the attack and to identify the remains for families and friends of the victims. This task was unprecedented in scope and difficulty. There was no complete list of victims (such as a passenger list in an airline crash) with which investigators could match the remains. In all, almost 20,000 individual remains were found, varying from whole bodies to tiny fragments of charred bone. The remains had been subjected

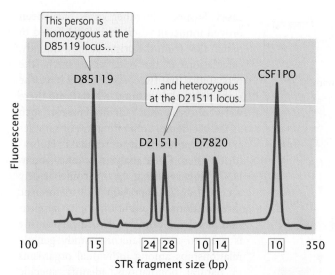

19.26 A DNA profile represents the pattern of DNA fragments produced by performing PCR on the STR loci. This profile shows the results from four STR loci (D8S1179, D21S11, D7S820, and CSF1P0). The number below each peak represents the number of STRs in that DNA fragment.

to fires with temperatures exceeding 1000°C that burned for more than 3 months. The collapse of the buildings intermixed many victims' remains, and many body fragments were not recovered for months, during which time they were exposed to dust, water, bacteria, and decay. The usual means of victim identification—personal items, fingerprints, dental records—were of little use for most of the

TABLE 19.3	Characteristics of the 13 STR loci used in CODIS for DNA fingerprinting		
Locus Name	Chromosome	Number of Repeats	Number of Alleles in U.S. Population
CSF1PO	5	5–17	10
FGA	4	12–51	23
TH01	11	3–14	8
TPOX	2	4–16	8
VWA	12	10–25	10
D3S1358	3	6–26	10
D5S818	5	4–29	9
D7S820	7	5–16	11
D8S1179	8	6–20	11
D13S317	13	5–17	8
D16S539	16	4–17	7
D18S51	18	5–40	19
D21S11	21	12–43	22

Source: J. M. Butler and C. R. Hill, *Forensic Science Review* 24:15–26, 2012.

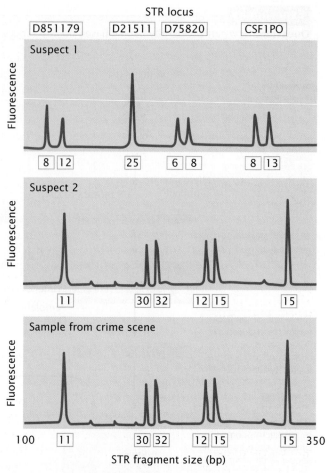

19.27 DNA fingerprinting can be used to determine the presence of a suspect at a crime scene. The DNA profile of suspect 2 matches that of DNA evidence collected at the crime scene. Shown here are results from four STR loci.

World Trade Center remains. Identification of the majority of the remains was made with the use of DNA fingerprinting (**Figure 19.28**).

DNA was first extracted from the tissue samples using sterile techniques to prevent cross-contamination between samples. After the DNA had been extracted, PCR was used to amplify STR loci in the CODIS system (see previous section). The DNA fingerprint generated from each sample was compared with that of DNA extracted from reference samples, such as victims' toothbrushes and blood samples, provided by families and friends. If the DNA from a tissue sample had the same alleles at all 13 loci as the reference sample, then a positive identification was made. When no reference sample was available, investigators collected DNA from victims' family members and tried to match the DNA profiles of remains to those of relatives; in this case, some, but not all, STR alleles would match.

Unfortunately, many of the remains were so badly degraded that little DNA remained and one or more of the STR loci could not be amplified. For these remains, DNA

19.28 DNA fingerprinting was used to identify the remains of people who died in the collapse of the World Trade Center. The New York City Office of Chief Medical Examiner used automated DNA fingerprinting to match DNA from human remains recovered from the World Trade Center with DNA extracted from reference samples, such as blood samples and toothbrushes, of possible victims. [Scott Gries/ Getty Images.]

fingerprinting was also carried out on mitochondrial DNA (see Chapter 11). Because there are many mitochondria per cell and each mitochondrion contains multiple DNA molecules, there are many more copies of mitochondrial DNA per cell than nuclear DNA; mitochondrial DNA has been successfully extracted and analyzed from ancient remains. Alone, fingerprinting from mitochondrial DNA was insufficient to provide identification with a high degree of confidence (because there are not as many sequences that vary among people as in the CODIS loci), but when it was used in conjunction with data from at least some STR loci, a positive identification could often be made.

Used in combination, these techniques allowed the remains of many victims to be positively identified. However, despite the heroic efforts of hundreds of molecular geneticists, forensic anthropologists, and medical examiners, no positively identified remains were recovered for almost half of the people who are thought to have died in the disaster.

19.6 Molecular Techniques Are Increasingly Used to Analyze Gene Function

In the preceding sections, we learned about the powerful molecular techniques that are available for amplifying, isolating, recombining, and sequencing DNA. Although these techniques provide a great deal of information about the organization and nature of gene sequences, the ultimate goal of many molecular studies is to better understand the function of these sequences. In this section, we explore some advanced molecular techniques that are frequently used to determine the functions of genes and to better understand the genetic processes that these sequences undergo.

Forward and Reverse Genetics

The traditional approach to the study of gene function begins with the identification of mutant organisms. For example, suppose a geneticist is interested in genes that affect cardiac function in mammals. A first step would be to find individuals—perhaps mice—that have hereditary defects in heart function. The mutations causing the cardiac problems in the mice could then be mapped, and the implicated genes could be isolated and sequenced. The proteins produced by the genes could then be predicted from the gene sequences and isolated. Finally, the biochemistry of the proteins could be studied and their role in heart function discerned. This approach, which begins with a phenotype (a mutant individual) and proceeds to a gene that encodes the phenotype, is called **forward genetics**.

An alternative approach is to begin with a genotype—a DNA sequence—and proceed to the phenotype by altering the sequence or inhibiting its expression. A geneticist might begin with a gene of unknown function, induce mutations in it, and then observe the effect of these mutations on the phenotype of the organism. This approach is called **reverse genetics**. Today, both forward and reverse genetic approaches are widely used in analyses of gene function.

Creating Random Mutations

Forward genetics depends on the identification and isolation of random mutations that affect a phenotype of interest. Early in the study of genetics, geneticists were forced to rely on naturally occurring mutations, which are usually rare and can be detected only if large numbers of organisms are examined. The discovery of mutagens—environmental factors that increase the rate of mutation (see Section 18.2)—provided a means of increasing the number of mutants in experimental populations of organisms. One of the first examples of experimentally created mutations was Hermann Muller's use of X-rays in 1927 to induce X-linked mutations in *Drosophila melanogaster*.

Today, radiation, chemical mutagens, and transposable elements are all used to create mutations for genetic analysis. To determine all genes that might affect a phenotype, it is desirable to create mutations in as many genes as possible—that is, to saturate the genome with mutations. This procedure, known as a mutagenesis screen, will be described in Chapter 20.

Targeted Mutagenesis

Reverse genetics depends on the ability to create mutations, not at random, but in particular DNA sequences, and then to study the effects of these mutations on the organism. Mutations are induced at specific locations through a process called **targeted mutagenesis**.

A number of different strategies have been developed for targeted mutagenesis. The CRISPR-Cas9 system (see Section 19.2) can be used to create mutations at specific sites. Single guide RNAs are designed to be complementary to a target sequence. The sgRNA and Cas9 are inserted into a cell, where Cas9 is guided to the target site by the sgRNA and makes double-stranded cuts at the site. Short duplications and deletions are often introduced when the cuts are repaired by nonhomologous end joining. Alternatively, donor DNA containing a specific desired mutation can be provided along with CRISPR-Cas9; the donor DNA will be inserted at the target site by homologous recombination.

Another strategy, called **site-directed mutagenesis**, that is used in bacteria is to cut out a short sequence of nucleotides with restriction enzymes and replace it with a synthetic oligonucleotide, a short DNA fragment that contains the desired mutated sequence. The success of this method depends on the availability of restriction sites flanking the sequence to be altered.

If appropriate restriction sites are not available, **oligonucleotide-directed mutagenesis** can be used (**Figure 19.29**). In this method, a single-stranded oligonucleotide is produced that differs from the target sequence by one or a few bases. Because they differ in only a few bases, the target DNA and the oligonucleotide will pair. When successfully paired with the target DNA, the oligonucleotide can act as a primer to initiate DNA synthesis, which produces a double-stranded molecule with a mismatch in the primer region. When this DNA is transferred to bacterial cells, the mismatched bases will be repaired by bacterial enzymes. About half of the time the normal bases will be changed into mutant bases, and about half of the time the mutant bases will be changed into normal bases. The bacteria are then screened for the presence of the mutant gene.

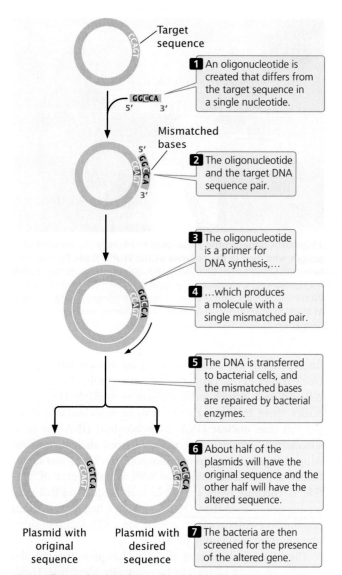

19.29 Oligonucleotide-directed mutagenesis is used to study gene function when appropriate restriction sites are not available.

CONCEPTS

Forward genetics begins with a phenotype and detects and analyzes the genotype that causes that phenotype. Reverse genetics begins with a gene sequence and determines the phenotype it encodes. Particular mutations can be introduced at specific sites within a gene by means of CRISPR-Cas editing and site-directed and oligonucleotide-directed mutagenesis.

✔ **CONCEPT CHECK 10**

A geneticist interested in immune function induces random mutations in a number of specific genes in mice and then determines which of the resulting mutant mice have impaired immune function. This procedure is an example of

a. forward genetics.
b. reverse genetics.
c. both forward and reverse genetics.
d. neither forward nor reverse genetics.

Transgenic Animals

Another way in which gene function can be analyzed is by adding DNA sequences of interest to the genome of an organism that normally lacks such sequences and then observing the effect of the introduced sequences on the organism's phenotype. This method is a form of reverse genetics. An organism that has been permanently altered by the addition of a DNA sequence to its genome is said to be *transgenic*, and the foreign DNA that it carries is called a **transgene** (**Figure 19.30**). Here, we consider techniques for the creation of transgenic mice, which are often used in the study of the function of human genes because they can be genetically manipulated in ways that are impossible with humans and because, as mammals, they are more similar to humans than are fruit flies, fish, and other model genetic organisms.

The oocytes of mice and other mammals are large enough that DNA can be injected into them directly. Immediately after penetration by a sperm, a fertilized mouse egg contains two pronuclei, one from the sperm and one from the egg; these pronuclei later fuse to form the nucleus of the embryo. Mechanical devices can manipulate extremely fine, hollow glass needles to inject DNA directly into one of

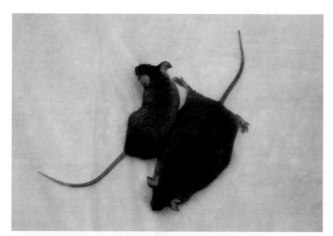

19.30 The genome of a transgenic organism has been permanently altered by genetic engineering. The mouse on the right is transgenic for a mutation that causes obesity. [Dr. Liangyou Rui, Professor, Molecular & Integrative Physiology, University of Michigan.]

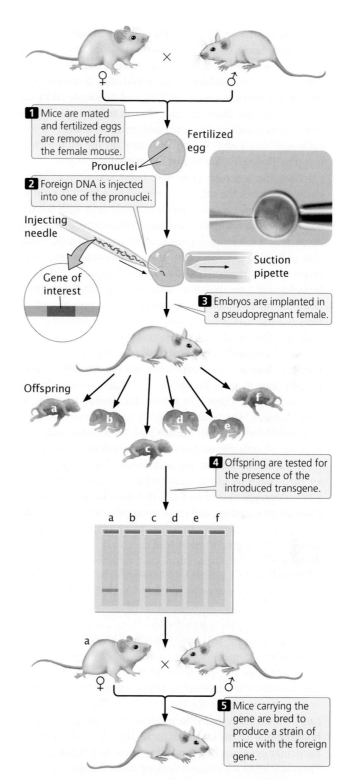

19.31 Transgenic animals have genomes that have been permanently altered through recombinant DNA technology. In the photograph, a mouse embryo is being injected with DNA. [Photograph: Chad Baker/Thomas Northcut/Getty Images.]

the pronuclei of a fertilized egg (**Figure 19.31**). Typically, a few hundred copies of cloned, linear DNA are injected into a pronucleus. In a few of the injected eggs, copies of the cloned DNA integrate randomly into one of the chromosomes through a process called nonhomologous recombination. After injection, the embryos are implanted in a pseudopregnant female: a surrogate mother that has been physiologically prepared for pregnancy by mating with a vasectomized male.

Only about 10% to 30% of the embryos survive, and of those that do survive, only a few carry a copy of the cloned DNA stably integrated into a chromosome. Nevertheless, if several hundred embryos are injected and implanted, there is a good chance that one or more mice whose chromosomes contain the foreign DNA will be born. Moreover, because the DNA was injected at the one-cell stage of the embryo, these mice usually carry the cloned DNA in every cell of their bodies, including their reproductive cells, and will therefore pass the foreign DNA on to their progeny. Through interbreeding, a strain of mice that carry the foreign gene can be created.

Transgenic mice have proved useful in the study of gene function. For example, proof that the *SRY* gene (see Chapter 4) is the male-determining gene in mice was obtained by injecting a copy of the *SRY* gene into XX embryos and observing that these mice developed as males. In addition, researchers have created a number of transgenic mouse strains that serve as experimental models for human genetic diseases.

Knockout Mice

A useful variant of the transgenic approach is to produce mice in which a normal gene has been not just mutated, but fully disabled. These animals, called **knockout mice**, are particularly helpful in determining the function of a gene: the phenotype of the knockout mouse often gives a good indication of the function of the missing gene.

The creation of knockout mice begins when a normal gene is cloned in bacteria and then "knocked out," or disabled. There are a number of ways to disable a gene, but a common method is to insert a gene called *neo*, which confers resistance to the antibiotic G418, into the middle of the target gene (**Figure 19.32**). The insertion of *neo* both disrupts the target gene and provides a convenient marker for finding copies of the disabled gene. In addition, a second gene, usually the herpes simplex viral thymidine kinase (*tk*) gene, is linked to the disabled gene. The disabled gene is then transferred to cultured mouse embryonic cells, where it may exchange places with the normal copy on the mouse chromosome through homologous recombination.

After the disabled gene has been transferred to the embryonic cells, the cells are screened by adding the antibiotic G418 to their growth medium. Only cells with the disabled gene containing the *neo* insert will survive. Because the frequency of nonhomologous recombination is higher than that of homologous recombination, and because the intact target gene is replaced by the disabled copy only through homologous recombination, a means to select for the rarer homologous recombinants is required. The presence of the viral *tk* gene makes the cells sensitive to the antibiotic gancyclovir. Thus, transfected cells that grow on medium containing G418 and gancyclovir contain the *neo* gene (and thus the disabled target gene), but not the adjacent *tk* gene, because the *tk* gene has been eliminated in the double-crossover event. These cells contain the desired homologous recombinants. The nonhomologous recombinants (random insertions) contain both the *neo* and the *tk* genes, and these transfected cells die on the selection medium owing to the presence of gancyclovir. The surviving cells are injected into an early-stage mouse embryo, which is then implanted in a pseudopregnant mouse. Cells in the embryo carrying the disabled gene and normal embryonic cells carrying the wild-type gene will develop together, producing a chimera—a mouse that is a genetic mixture of the two cell types. The production of chimeric mice is not itself desirable, but replacing all the cells of the embryo with injected cells is impossible.

Chimeric mice can be easily identified if the injected embryonic cells came from a black mouse and the embryos into which they are injected came from a white mouse; the resulting chimeras will have variegated black and white fur. Some of the chimeras may have the knockout gene in their germ-line cells and can transmit it to the next generation. The chimeras are crossed to white mice; any black progeny are heterozygous for the knockout. The black progeny can then be intercrossed to produce some progeny that are homozygous for the knockout gene. The effects of disabling a particular gene can be observed in these homozygous mice. Three scientists who helped develop the techniques for creating knockout mice—Mario Capecchi, Oliver Smithies, and Martin Evans—were awarded the Nobel Prize in physiology or medicine in 2007 for their work.

Additional techniques have been developed for knocking out genes only in certain tissues or at certain times; the

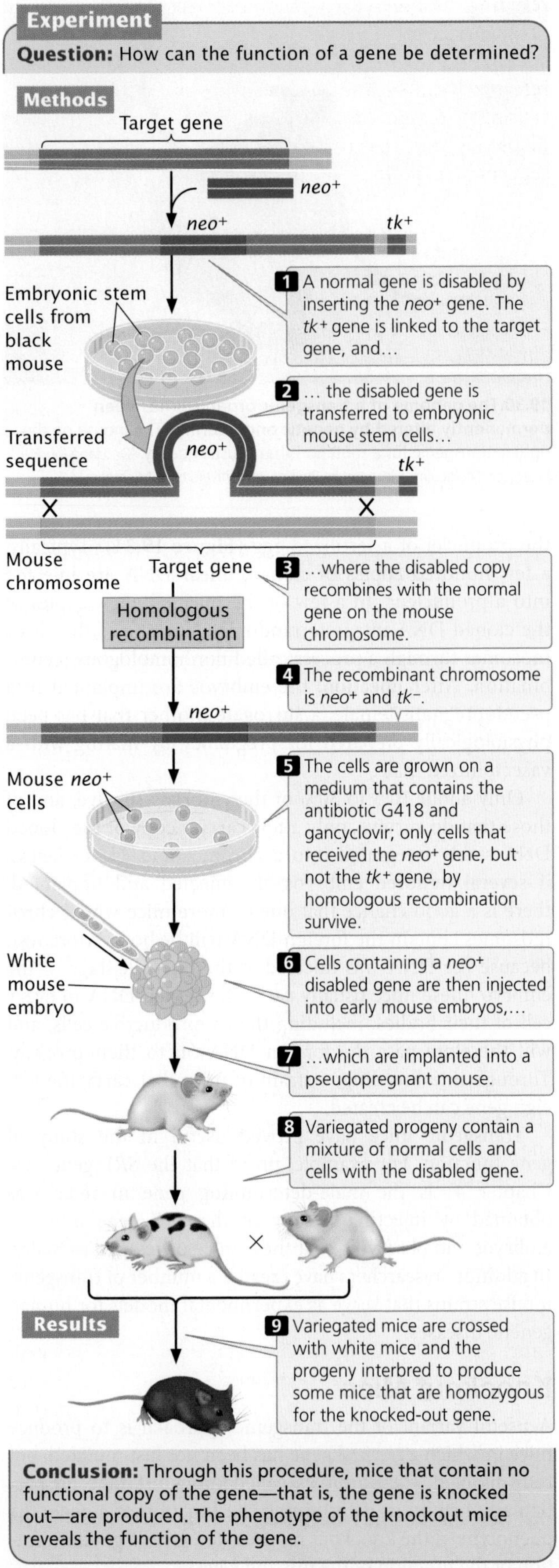

19.32 Knockout mice possess a genome in which a gene has been disabled.

Experiment

Question: How can the function of a gene be determined?

Methods

Target gene

neo⁺

neo⁺   *tk*⁺

Embryonic stem cells from black mouse

1 A normal gene is disabled by inserting the *neo*⁺ gene. The *tk*⁺ gene is linked to the target gene, and…

2 …the disabled gene is transferred to embryonic mouse stem cells…

Transferred sequence   *neo*⁺

tk⁺

Mouse chromosome   Target gene

3 …where the disabled copy recombines with the normal gene on the mouse chromosome.

Homologous recombination

4 The recombinant chromosome is *neo*⁺ and *tk*⁻.

neo⁺

Mouse *neo*⁺ cells

5 The cells are grown on a medium that contains the antibiotic G418 and gancyclovir; only cells that received the *neo*⁺ gene, but not the *tk*⁺ gene, by homologous recombination survive.

White mouse embryo

6 Cells containing a *neo*⁺ disabled gene are then injected into early mouse embryos,…

7 …which are implanted into a pseudopregnant mouse.

8 Variegated progeny contain a mixture of normal cells and cells with the disabled gene.

Results

9 Variegated mice are crossed with white mice and the progeny interbred to produce some mice that are homozygous for the knocked-out gene.

Conclusion: Through this procedure, mice that contain no functional copy of the gene—that is, the gene is knocked out—are produced. The phenotype of the knockout mice reveals the function of the gene.

resulting mice are called conditional knockouts. A variant of the knockout procedure is to insert a particular DNA sequence at a known chromosome location. For example, researchers might insert the sequence of a human disease-causing allele into the same locus in mice, creating a precise mouse model of the human disease. Mice that carry inserted sequences at specific locations are called **knock-in mice**.

CONCEPTS

A transgenic mouse is produced by injecting cloned DNA into the pronucleus of a fertilized egg and then implanting the egg in a female mouse. In knockout mice, the injected DNA contains a mutation that disables a gene. Inside the mouse embryo, the disabled copy of the gene can exchange with the normal copy of the gene through homologous recombination.

✔ **CONCEPT CHECK 11**

What is the advantage of using the *neo* gene to disrupt the function of a gene in knockout mice?

a. The *neo* gene produces an antibiotic that kills unwanted cells.

b. The *neo* gene is the right size for disabling other genes.

c. The *neo* gene provides a selectable marker for finding cells that contain the disabled gene.

d. The *neo* gene produces a toxin that inhibits transcription of the target gene.

Silencing Genes with RNAi

In the preceding sections, we have seen that introducing mutations or new DNA sequences into the genome and observing the resulting phenotypes can provide information about the function of the altered or introduced DNA. We can also analyze gene function by temporarily turning a gene off and observing the effect of the absence of the gene product on the phenotype. For many years, there was no such method for selectively affecting gene expression. However, the discovery of siRNAs (small interfering RNAs) and miRNAs (microRNAs; see Chapters 14 and 17) has provided powerful tools for controlling the expression of individual genes.

Recall that siRNAs are small RNA molecules that combine with proteins to form an RNA-induced silencing complex (RISC). In a process called RNA interference, or RNAi, the RISC pairs with complementary sequences on mRNA and either cleaves the mRNA or prevents it from being translated. Molecular geneticists have exploited this natural machinery for turning off the expression of specific genes. The effect of silencing a gene with siRNAs can often be a source of insight into the gene's function.

The first step in using RNAi technology is to design siRNAs that will be recognized and cleaved by Dicer (the protein that processes siRNAs; see Section 14.5). The complementary sequence must be unique to the target mRNA and must not be found on other mRNAs so that the siRNA will not inhibit nontarget mRNAs. Computer programs are often used to design optimal siRNAs.

After the siRNA sequence has been designed, it must be synthesized. One way to create an siRNA is to use an oligonucleotide synthesizer to synthesize a DNA fragment corresponding to the siRNA sequence. The synthesized oligonucleotide can be cloned into a plasmid expression vector between two strong promoters (**Figure 19.33**). *Escherichia coli* are then transformed with the plasmid. Within the bacteria, transcription from the two promoters will proceed in both directions, producing two complementary RNA molecules that will pair to form a double-stranded RNA molecule recognized by Dicer. Alternatively, double-stranded RNA sequences can be synthesized directly.

The next task is to deliver the double-stranded siRNA to the target cells. Delivery can be accomplished in a variety of ways, depending on the cell type. The model genetic organism *Caenorhabditis elegans* (see p. A6) can be fed *E. coli* (their natural food) containing the expression vector. Transcription within the bacteria produces double-stranded RNA, which the worms ingest and incorporate into their cells. Alternatively, double-stranded siRNA can be injected directly into the cells or the body cavity. Yet another approach is to synthesize a short sequence of DNA that has internal complementarity so that, when transcribed, its RNA product folds up into a short hairpin RNA (shRNA) with a double-stranded section. Within a cell, the shRNAs are processed by Dicer to produce siRNAs that bring about gene silencing. DNA sequences containing siRNAs can be introduced into a vector using standard cloning techniques, and the vector can be used to deliver the DNA to a cell. An advantage of this approach is that, with the addition of a DNA sequence, the RNAi sequence has the potential to become a permanent part of the cell's genome and be passed on to progeny.

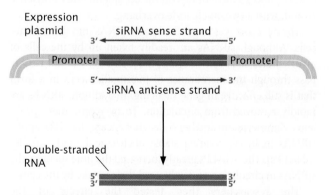

19.33 Small interfering RNAs can be produced by cloning DNA sequences corresponding to the siRNAs between two strong promoters. When cloned into an expression vector, both DNA strands will be transcribed and the complementary RNA molecules will anneal to form double-stranded RNA that will be processed into siRNA by Dicer.

One of the major advantages of siRNA for controlling gene expression is that it acts in trans—that is, a single copy of an siRNA gene will shut down expression of both copies of the target gene. Another advantage is that the target gene remains intact, and therefore the silencing effects are reversible. ▸ **TRY PROBLEM 42**

Using RNAi to Treat Human Disease

In addition to its value in determining gene function, RNAi holds potential as a therapeutic agent for the treatment of human diseases.

TREATMENT OF HIGH CHOLESTEROL WITH RNAi Research has examined the potential of RNAi for the treatment of cholesterol metabolism disorders. Although cholesterol is essential for life, too much cholesterol is unhealthy: high blood cholesterol is a major contributor to heart disease, the leading cause of death in the United States. Cholesterol is normally transported throughout the body in the form of small particles called lipoproteins, which consist of a core of lipids surrounded by a shell of phospholipids and proteins (see Figure 6.6). The ApoB protein is an essential part of lipoproteins. Some people possess genetic mutations that cause elevated levels of ApoB, which predispose them to coronary artery disease. Findings from studies suggest that lowering the amount of ApoB can reduce the number of lipoproteins and lower blood cholesterol in these people, as well as in people who have elevated cholesterol for other reasons.

In 2006, Tracy Zimmermann and her colleagues at Alnylam Pharmaceuticals and Protiva Biotherapeutics demonstrated that RNAi could be used to reduce the levels of ApoB and blood cholesterol in nonhuman primates. The investigators first created siRNAs that targeted *apoB* gene expression (apoB-siRNAs) on the basis of the known sequence of the gene. The apoB-siRNAs were synthesized in the laboratory and consisted of 21 nucleotides on the sense strand (the strand that was complementary to the *apoB* mRNA) and 23 nucleotides on the complementary antisense strand, with a two-nucleotide overhang.

The next task was to get the apoB-siRNAs into the animals' cells. Although siRNAs are readily taken up by the cells of invertebrates such as *C. elegans*, most siRNAs will not readily pass through the membranes of mammalian cells in a form that is still effective in gene silencing. In addition, siRNAs are rapidly removed from circulation. To overcome these problems, Zimmermann and her colleagues encapsulated the apoB-siRNAs in lipids, creating stable nucleic-acid–lipid particles (SNALPs). The SNALPs greatly increased the time spent by the siRNAs in circulation and enhanced their uptake by the cells.

The researchers then tested the effects of the apoB-siRNAs on the synthesis of the ApoB protein and on cholesterol levels. They injected cynomolgus monkeys (**Figure 19.34a**) with SNALPs containing apoB-siRNAs at two doses: 1 mg/kg and 2.5 mg/kg. In addition, they injected a third group of monkeys with saline as a control. The apoB-siRNAs clearly silenced the *apoB* gene: 48 hours after treatment, *apoB* mRNA in the liver was reduced by 68% for monkeys receiving the 1-mg dose and 90% for monkeys receiving the 2.5-mg dose.

When the researchers examined serum cholesterol levels in the monkeys, they found that monkeys receiving the apoB-siRNAs had significant reductions in blood-cholesterol levels (**Figure 19.34b**). Importantly, they observed no negative effects of the siRNA treatment. Although preliminary, this study suggests that siRNAs have potential for future treatment of human diseases.

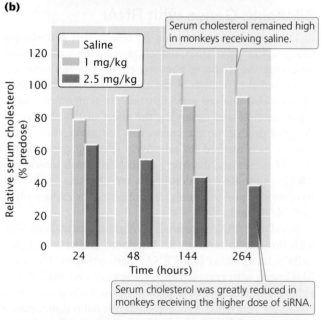

(a)

(b)

19.34 Transfer of siRNAs for the ApoB protein into cynomolgus monkeys significantly lowered blood-cholesterol levels. (a) Cynomolgus monkey. (b) Different groups of monkeys were given saline (control), a 1-mg/kg dose, or a 2.5-mg/kg dose of siRNAs. [Part a: Tony Camacho/Science Source. Part b: After T. S. Zimmerman, *Nature* 441:112, 2006, Figure 3b.]

19.7 Biotechnology Harnesses the Power of Molecular Genetics

In addition to providing valuable new information about the nature and function of genes, molecular genetic techniques have many practical applications, including the production of pharmaceutical products and other chemicals, specialized bacteria, agriculturally important plants, and genetically engineered farm animals. These techniques are also used extensively in medical testing and are even being used to correct human genetic defects. Hundreds of companies now specialize in developing products through genetic engineering, and many large multinational corporations have invested enormous sums of money in molecular genetic research. As discussed earlier, DNA analysis is also used in criminal investigations and for the identification of human remains.

Pharmaceutical Products

The first commercial products to be developed with the use of genetic engineering were pharmaceutical products used in the treatment of human diseases and disorders. In 1979, Eli Lilly and Company began selling human insulin produced with the use of recombinant DNA technology. The gene for human insulin was inserted into plasmids and transferred to bacteria, which then produced human insulin. Previously, diabetes was treated with insulin isolated from pig and cow pancreases; a few patients developed allergic reactions to this foreign protein. Insulin produced by recombinant bacteria has the advantage of being the same as that produced in the human body. Other pharmaceutical products produced through recombinant DNA technology include human growth hormone (for children with growth deficiencies) and tissue plasminogen activator (used to dissolve blood clots in heart-attack patients).

Specialized Bacteria

Bacteria play important roles in many industrial processes, including the production of ethanol from plant material, the leaching of minerals from ore, and the treatment of sewage and other wastes. The bacteria used in these processes have been modified by genetic engineering so that they work more efficiently. New strains of bacteria are being developed that will break down toxic chemicals and pollutants, enhance oil recovery, increase nitrogen uptake by plants, and inhibit the growth of pathogenic bacteria and fungi.

Agricultural Products

Recombinant DNA technology has had a major effect on agriculture, where it is now used to create crop plants and domestic animals with valuable traits. For example, plant pathologists had recognized for many years that plants infected with mild strains of viruses are resistant to infection by virulent strains. Using this knowledge, geneticists created plants with resistance to viruses by transferring genes for viral proteins to the plant cells. A genetically engineered squash, called Freedom II, carries genes from the watermelon mosaic virus 2 and the zucchini yellow mosaic virus, which protect the squash against viral infections.

Another objective has been to genetically engineer pest resistance into plants to reduce dependence on chemical pesticides. As discussed in Section 19.3, a gene from the bacterium *Bacillus thuringiensis* that produces an insecticidal toxin has been transferred into corn, tomatoes, potatoes, cotton, and other plants. These Bt crops are now grown worldwide. Other genes that confer resistance to viruses and herbicides have been introduced into a number of crop plants; resistance to herbicides allows growers to use herbicides for weed control without harming the crops. During 2015, over 17 million farmers worldwide planted 180 million hectares of genetically engineered crops. In the United States, 92% of all corn, 94% of all cotton, and 94% of all soybeans grown in 2015 were genetically engineered.

Recombinant DNA technology is also being applied to domestic animals. The first animal to be genetically modified for food production in the United States—a rapidly growing salmon—was approved for commercial use in 2015. These salmon carry a foreign growth-hormone gene and promoter; the transgenic fish grow year-round instead of just during warm months, reaching market size more quickly and with less feed than wild salmon.

Transgenic animals are also being developed to carry genes that encode pharmaceutical products. Some eukaryotic proteins must be modified after translation, and only other eukaryotes (not bacteria) are capable of carrying out the modifications. For example, a gene for human clotting factor VIII was attached to the regulatory region of the sheep gene for β-lactoglobulin, a milk protein. The fused gene was injected into sheep embryos, creating transgenic sheep that produced the human clotting factor, which is used to treat hemophilia, in their milk. And through genetic engineering, scientists have created transgenic chickens that express a small RNA that blocks infection by avian influenza virus.

In the future, CRISPR-Cas systems will increasingly be used to precisely edit the genomes of domestic plants and animals. For example, CRISPR-Cas9 has already been used to create transgenic wheat that is resistant to powdery mildew, a fungus that infects wheat. Because of its ease of use and low cost, CRISPR-Cas could be applied to less widely used, specialty crops and animals, potentially widening the use of genetically modified organisms. The ability to make very precise genetic changes with CRISPR-Cas, however, makes these alterations difficult to detect and track once they are made, raising questions about the labeling of products as genetically modified.

The genetic engineering of agricultural products is controversial. One area of concern focuses on the potential effects of releasing novel organisms produced by genetic engineering into the environment. There are many examples in which nonnative organisms released into a new environment have caused ecological disruption because they are free of their predators and other natural control mechanisms. Genetic engineering normally transfers only short sequences of DNA, which are small relative to the genetic differences that often exist between species, but even small genetic differences may alter ecologically important traits that might affect the ecosystem.

Another area of concern is the effect of genetically engineered crops on biodiversity. In the largest field test of genetically engineered plants ever conducted, scientists cultivated beets, corn, and rapeseed that were genetically engineered to resist herbicide along with traditional crops on 200 test plots throughout the United Kingdom. They then measured the biodiversity of native plants and animals in the plots. They found that the genetically engineered plants were highly successful in suppressing weeds; however, plots with genetically engineered beets and rapeseed had significantly fewer insects that feed on weeds. For example, plots with genetically engineered rapeseed had 24% fewer butterflies than did plots with traditional crops.

There is also concern that transgenic organisms may hybridize with native organisms and transfer their genetically engineered traits. For example, herbicide resistance engineered into crop plants might be transferred to weeds, which would then be resistant to the herbicides that are now used for their control. Some studies have detected hybridization between genetically engineered crops and wild populations of plants. For example, evidence suggests that transgenic rapeseed (*Brassica napus*) has hybridized with the weed *Brassica rapa* in Canada. Other concerns focus on health-safety matters associated with the presence of engineered products in natural foods. Some critics have advocated required labeling of all genetically engineered foods that contain transgenic DNA or protein. Such labeling is required in countries of the European Union, but not in the United States.

On the other hand, the use of genetically engineered crops and domestic animals has potential benefits. Genetically engineered crops that are pest resistant have the potential to reduce the use of environmentally harmful chemicals, and research findings indicate that lower amounts of pesticides are being used in the United States as a result of the adoption of transgenic plants. Studies conducted in China show that when Bt crops are used, farmers spray less chemical insecticide, allowing more predatory insects to survive and provide natural pest control. Transgenic crops also increase yields, providing more food per acre, which reduces the amount of land that must be used for agriculture. Genetically engineered plants offer the potential for the greater yields that may be necessary to feed the world's future population.

THINK-PAIR-SHARE Question 7

Genetic Testing

The identification and cloning of many important disease-causing human genes have allowed the development of probes for detecting disease-causing mutations. Prenatal testing is already available for many genetic disorders (see Section 6.4). Additionally, presymptomatic genetic tests for adults and children are available for an increasing number of disorders. Some genetic tests are now being offered directly to consumers, without requiring the participation of a health-care provider. These direct-to-consumer genetic tests are potentially available for many genetic conditions. The U.S. Food and Drug Administration has limited the sale of direct-to-consumer genetic tests for medical conditions, and today, only a few such medical tests are available in the United States.

The growing availability of genetic tests raises a number of ethical and social questions. For example, is it ethical to test for genetic diseases for which there is no cure or treatment? Other ethical and legal questions concern the confidentiality of test results. Who should have access to the results of genetic testing? Should relatives who also might be at risk be informed of the results of genetic testing?

Another set of concerns is related to the accuracy of genetic tests. For many genetic diseases, the only predictive tests available are those that identify a *predisposing* mutation in DNA, but many genetic diseases can be caused by dozens or hundreds of different mutations. Probes that detect common mutations can be developed, but they will fail to detect rare mutations and may give a false negative result. Short of sequencing the entire gene—which is expensive and time consuming—there is no way to identify all predisposed people. These questions and concerns are currently the focus of intense debate by ethicists, physicians, scientists, and patients.

Gene Therapy

Perhaps the ultimate application of recombinant DNA technology is **gene therapy**, the direct transfer of genes into human patients to treat disease. Today, thousands of patients have received gene therapy, and many clinical trials are under way. Gene therapy has been used as an experimental treatment for genetic diseases, cancer, heart disease, and even some infectious diseases such as AIDS. A number of different methods for transferring genes into human cells

TABLE 19.4	Vectors used in gene therapy	
Vector	**Advantages**	**Disadvantages**
Retrovirus	Efficient transfer	Transfers DNA only to dividing cells, inserts randomly; risk of producing wild-type viruses
Adenovirus	Transfers to nondividing cells	Causes immune reaction
Adeno-associated virus	Does not cause immune reaction	Holds small amount of DNA; hard to produce
Herpesvirus	Can insert into cells of nervous system; does not cause immune reaction	Hard to produce in large quantities
Lentivirus	Can accommodate large genes	Safety concerns
Liposomes and other lipid-coated vectors	No replication; does not stimulate immune reaction	Low efficiency
Direct injection	No replication; directed toward specific tissues	Low efficiency; does not work well within some tissues
Pressure treatment	Safe, because tissues are treated outside the body and then transplanted into the patient	Most efficient with small DNA molecules
Gene gun (small DNA-coated gold particles are shot into tissue)	No vector required	Low efficiency

Source: After E. Marshall, Gene therapy's growing pains, *Science* 269:1050–1055, 1995.

are currently under development. Commonly used vectors include genetically modified retroviruses, adenoviruses, and adeno-associated viruses (**Table 19.4**). Researchers are exploring the use of CRISPR-Cas systems in gene therapy.

In spite of the growing number of clinical trials for gene therapy, significant problems remain in transferring foreign genes into human cells, getting them expressed, and limiting immune responses to the gene products and the vectors used to transfer the genes. There are also concerns about the safety of gene therapy. In 1999, a patient participating in a gene-therapy trial had a fatal immune reaction after he was injected with a viral vector carrying a gene to treat his metabolic disorder. In addition, five children who underwent gene therapy for severe combined immunodeficiency disease developed leukemia that appeared to be directly related to the insertion of the retroviral vector into cancer-causing genes. Despite these setbacks, gene-therapy research has moved ahead. Unequivocal results demonstrating positive benefits from gene therapy for several different diseases have now been published.

Gene therapy conducted to date has targeted only non-reproductive, or somatic, cells. Correcting genetic defects in these cells (termed *somatic gene therapy*) may provide positive benefits to patients, but will not affect the genes of future generations. Gene therapy that alters reproductive, or germ-line, cells (termed *germ-line gene therapy*) is technically possible, but raises a number of significant ethical issues because it has the capacity to alter the gene pool of future generations.

THINK-PAIR-SHARE Question 8

CONCEPTS

Gene therapy is the direct transfer of genes into humans to treat disease. Gene therapy is now being used to treat genetic diseases, cancer, and infectious diseases.

✔ **CONCEPT CHECK 13**

What is the difference between somatic gene therapy and germ-line gene therapy?

CONCEPTS SUMMARY

■ Techniques of molecular genetics are used to locate, analyze, alter, sequence, study, and recombine DNA sequences. Four key innovations that have revolutionized genetics are recombinant DNA technology, the polymerase chain reaction, DNA sequencing technology, and genome editing methods.

■ Recombinant DNA technology is a set of molecular techniques for locating, isolating, altering, and studying DNA segments.

■ Restriction endonucleases are enzymes that make double-stranded cuts in DNA at specific base sequences.

- Engineered nucleases are proteins that are capable of making double-stranded cuts at specific DNA sequences. They include zinc-finger nucleases (ZFNs) and transcription activator–like effector nucleases (TALENs).

- The CRISPR-Cas9 system can be used to cut and edit the genome. It combines a single guide RNA with a nuclease, which together attach to DNA sequences and make double-stranded cuts at specific locations.

- DNA fragments can be separated by gel electrophoresis and visualized by staining the gel with a dye that is specific for nucleic acids or by labeling the fragments with a radioactive or chemical tag.

- The polymerase chain reaction is a method for amplifying DNA enzymatically without cloning. A solution containing DNA is heated, so that the two DNA strands separate, and is then quickly cooled, allowing primers to attach to the template DNA. The solution is then heated again, and DNA polymerase synthesizes new strands from the primers. Each time the cycle is repeated, the amount of DNA doubles.

- In gene cloning, a gene or a DNA fragment is placed into a bacterial cell, where it will be multiplied as the cell divides.

- Plasmids, small circular pieces of DNA, are often used as vectors to ensure that a cloned gene is stable and is replicated within the recipient cells. Expression vectors contain sequences necessary for foreign DNA to be transcribed and translated.

- Genes can be isolated by creating a DNA library: a set of bacterial colonies or viral plaques that each contain a different cloned fragment of DNA. A genomic library contains the entire genome of an organism; a cDNA library contains DNA fragments complementary to the mRNAs expressed in a cell.

- In situ hybridization can be used to determine the chromosomal location of a gene or the distribution of the mRNA produced by a gene.

- Positional cloning uses linkage relations to determine the location of genes without any knowledge of their products.

- The dideoxy (Sanger) method of DNA sequencing uses special substrates for DNA synthesis (dideoxyribonucleoside triphosphates, ddNTPs) that terminate synthesis after they are incorporated into the newly synthesized DNA. Next-generation and third-generation sequencing methods sequence many DNA fragments simultaneously, providing a much faster and less expensive determination of a DNA sequence.

- Short tandem repeats (STRs), or microsatellites, are used to identify people by DNA fingerprinting.

- Forward genetics begins with a phenotype and conducts analyses to locate the responsible genes. Reverse genetics starts with a DNA sequence and conducts analyses to determine its phenotypic effect.

- Targeted mutagenesis can be used to produce mutations at specific sites in DNA, allowing genes to be tailored for a particular purpose.

- Transgenic animals, produced by injecting DNA into fertilized eggs, contain foreign DNA that is integrated into a chromosome. Knockout mice are mice in which a normal gene is disabled.

- RNA interference is used to silence the expression of specific genes.

- Techniques of molecular genetics are being used to create products of commercial importance, to develop diagnostic tests, and to treat diseases.

- In gene therapy, diseases are being treated by altering the genes of human cells.

IMPORTANT TERMS

biotechnology (p. 561)
recombinant DNA technology (p. 562)
genetic engineering (p. 562)
restriction enzyme (p. 562)
restriction endonuclease (p. 562)
cohesive end (p. 562)
engineered nuclease (p. 564)
zinc-finger nuclease (ZFN) (p. 564)
transcription activator–like effector nuclease (TALEN) (p. 564)
CRISPR-Cas system (p. 564)
gel electrophoresis (p. 567)

probe (p. 568)
Southern blotting (p. 568)
Northern blotting (p. 568)
Western blotting (p. 568)
gene cloning (p. 569)
polymerase chain reaction (PCR) (p. 569)
Taq polymerase (p. 570)
reverse-transcription PCR (p. 570)
real-time PCR (p. 570)
cloning vector (p. 571)
linker (p. 572)
cosmid (p. 573)
bacterial artificial chromosome (BAC) (p. 573)

expression vector (p. 573)
yeast artificial chromosome (YAC) (p. 574)
Ti plasmid (p. 574)
DNA library (p. 576)
genomic library (p. 576)
cDNA library (p. 576)
in situ hybridization (p. 578)
positional cloning (p. 579)
chromosome walking (p. 579)
DNA sequencing (p. 582)
dideoxyribonucleoside triphosphate (ddNTP) (p. 582)
next-generation sequencing technology (p. 584)
DNA fingerprinting (p. 586)

microsatellite (p. 586)
short tandem repeat (STR) (p. 586)
forward genetics (p. 589)
reverse genetics (p. 589)
targeted mutagenesis (p. 589)
site-directed mutagenesis (p. 590)
oligonucleotide-directed mutagenesis (p. 590)
transgene (p. 590)
knockout mouse (p. 591)
knock-in mouse (p. 593)
gene therapy (p. 596)

ANSWERS TO CONCEPT CHECKS

1. Restriction enzymes exist naturally in bacteria, which use them in defense against viruses.

2. c

3. Southern blotting is used to transfer DNA from a gel to a solid medium. Northern blotting is used to transfer RNA from a gel to a solid medium, and Western blotting is used to transfer protein from a gel to a solid medium.

4. A heat-stable DNA polymerase enzyme is important to the success of PCR because the first step of the reaction requires that the solution be heated to 90°–100°C to separate the two DNA strands. Most enzymes are denatured at this temperature. With the use of a heat-stable polymerase, the enzyme can be added at the beginning of the reaction and will function throughout multiple cycles.

5. The gene and plasmid are cut with the same restriction enzyme and mixed together. DNA ligase is used to seal nicks in the sugar–phosphate backbone.

6. With the use of the genetic code and the amino acid sequence of the protein, possible nucleotide sequences that cover a small region of the gene can be deduced. A mixture of all the possible nucleotide sequences that might encode the protein, taking into consideration synonymous codons, is used to probe the library.

To minimize the number of sequences required, a region of the protein that has relatively little degeneracy in its codons is selected.

7. The expression pattern of the gene can be examined, and the coding region of copies of the gene from individuals with the mutant phenotype can be compared with the coding region of wild-type individuals.

8. d

9. By using PCR with primers that flank the region containing tandem repeats.

10. b

11. c

12. Possible concerns include (a) ecological damage caused by introducing novel organisms into the environment; (b) negative effects of transgenic organisms on biodiversity; (c) possible spread of transgenes to native organisms by hybridization; and (d) health effects of eating genetically modified foods.

13. Somatic gene therapy modifies genes only in somatic tissue, and these modifications cannot be inherited. Germ-line gene therapy alters genes in germ-line cells; these alterations have the potential to be passed on to future generations.

WORKED PROBLEMS

Problem 1

A molecule of double-stranded DNA that is 5 million base pairs long has a base composition that is 62% G + C. How many times, on average, are restriction sites for the following restriction enzymes likely to be present in this DNA molecule?

a. *Bam*HI (recognition sequence is GGATCC)

b. *Hin*dIII (recognition sequence is AAGCTT)

c. *Hpa*II (recognition sequence is CCGG)

›› Solution Strategy

What information is required in your answer to the problem?

The number of restriction sites likely to be present in the DNA molecule for each of the specified restriction enzymes.

What information is provided to solve the problem?

- The size of the DNA molecule.
- The G + C base composition of the DNA molecule.
- The recognition sequence for each restriction enzyme.

For help with this problem, review:

Restriction Enzymes in Section 19.2.

›› Solution Steps

The percentages of G and C are equal in double-stranded DNA; so, if G + C = 62%, then %G = %C = 62%/2 = 31%. The percentage of A + T = (100% − G − C) = 38%, and %A = %T = 38%/2 = 19%. To determine the probability of finding a particular base sequence, we use the multiplication rule (see Section 3.2), multiplying together the probability of finding each base at a particular site.

a. The probability of finding the sequence GGATCC = 0.31 × 0.31 × 0.19 × 0.19 × 0.31 × 0.31 = 0.0003333. To determine the average number of GGATCC recognition

Hint: If you know the percentage of any base in the DNA, you can determine the percentages of all the other bases because G = C and A = T.

Recall: The multiplication rule states that the probability of two or more independent events is calculated by multiplying their independent probabilities.

sequences in a 5 million-base-pair piece of DNA, we multiply 5,000,000 bp × 0.00033 = 1666.5 recognition sequences.

b. The number of AAGCTT recognition sequences is 0.19 × 0.19 × 0.31 × 0.31 × 0.19 × 0.19 × 5,000,000 = 626 recognition sequences.

c. The number of CCGG recognition sequences is 0.31 × 0.31 × 0.31 × 0.31 × 5,000,000 = 46,176 recognition sequences.

Problem 2

You are given the following DNA fragment to sequence: 5′–GCTTAGCATC–3′. You first clone the fragment in bacterial cells to produce sufficient DNA for sequencing. You isolate the DNA from the bacterial cells and carry out the dideoxy-sequencing method. You then separate the products of the polymerization reactions by gel electrophoresis. Draw the bands that should appear on the gel from the four sequencing reactions.

≫ Solution Strategy

What information is required in your answer to the problem?
The positions of the bands on the sequencing gel.

What information is provided to solve the problem?
The base sequence of the DNA fragment to be sequenced.

For help with this problem, review:
Dideoxy Sequencing in Section 19.5.

≫ Solution Steps

Recall: In dideoxy sequencing, a new DNA strand is synthesized, and that strand is what is sequenced. Thus, the bands that appear on the gel represent the complement of the original sequence.

The first task is to write out the sequence of the newly synthesized fragment, which will be complementary and antiparallel to the original fragment. The original sequence is 5′–GCTTAGCATC–3′, so the newly synthesized sequence will be:

Original (template) sequence: 5′–GCTTAGCATC–3′

Newly synthesized sequence: 3′–CGAATCGTAG–5′

Thus, the sequence of the newly synthesized strand, written 5′→3′ is 5′–GATGCTAAGC–3′. Bands representing this sequence will appear on the gel, with the bands representing nucleotides near the 5′ end of the molecule at the bottom of the gel.

Hint: The smallest fragments, those nearest the 5′ end of the newly synthesized strand, will migrate most rapidly and will appear closest to the bottom (negative end) of the gel.

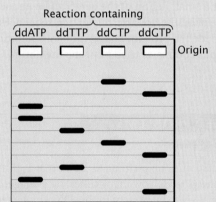

Reaction containing

COMPREHENSION QUESTIONS

Section 19.1

1. List some of the effects and practical applications of molecular genetic analyses.

Section 19.2

2. What feature is commonly seen in the sequences recognized by type II restriction enzymes?

3. What normal role do restriction enzymes play in bacteria? How do bacteria protect their own DNA from the action of restriction enzymes?

4. How is CRISPR-Cas used to introduce specific changes into DNA sequences?

5. Explain how gel electrophoresis is used to separate DNA fragments of different lengths.

6. After DNA fragments have been separated by gel electrophoresis, how can they be visualized?

Section 19.3

7. Briefly explain how the polymerase chain reaction is used to amplify a specific DNA sequence. What are some of the limitations of PCR?

8. What is real-time PCR?

9. Give three important characteristics of cloning vectors.

10. Briefly describe two different methods for inserting foreign DNA into plasmids, giving the strengths and weaknesses of each method.

11. Briefly explain how an antibiotic-resistance gene and the *lacZ* gene can be used to determine which cells contain a particular plasmid.

Section 19.4

12. How does a genomic library differ from a cDNA library?

13. How are probes used to screen DNA libraries? Explain how a synthetic probe can be prepared when the protein product of a gene is known.

14. Briefly explain in situ hybridization, giving some applications of this technique.

15. Briefly explain how a gene can be isolated through positional cloning.

16. Explain how chromosome walking can be used to find a gene.

Section 19.5

17. What is the purpose of the dideoxyribonucleoside triphosphates in the dideoxy-sequencing reaction?

18. What is DNA fingerprinting? What types of sequences are examined in DNA fingerprinting?

Section 19.6

19. How does a reverse genetic approach differ from a forward genetic approach?

20. Briefly explain how site-directed mutagenesis is carried out.

21. What are knockout mice, how are they produced, and for what are they used?

22. How is RNA interference used in the analysis of gene function?

Section 19.7

23. What is gene therapy?

APPLICATION QUESTIONS AND PROBLEMS

Section 19.2

24. CRISPR-Cas9 was first developed as a molecular tool in 2012; during the next few years, its use in molecular biology exploded, as scientists around the world began applying it to many different research problems, and hundreds of research papers describing its application were published. Explain why CRISPR-Cas is such a powerful tool in molecular genetics.

*25. Suppose that a geneticist discovers a new restriction enzyme in the bacterium *Aeromonas ranidae*. This restriction enzyme is the first to be isolated from this bacterial species. Using the standard convention for abbreviating restriction enzymes, give this new restriction enzyme a name (for help, see footnote to **Table 19.1**).

26. How often, on average, would you expect a type II restriction endonuclease to cut a DNA molecule if the recognition sequence for the enzyme had 5 bp? (Assume that the four types of bases are equally likely to be found in the DNA and that the bases in a recognition sequence are independent.) How often would the endonuclease cut the DNA if the recognition sequence had 8 bp?

*27. A microbiologist discovers a new type II restriction endonuclease. When DNA is digested by this enzyme, fragments that average 1,048,500 bp in length are produced. What is the most likely number of base pairs in the recognition sequence of this enzyme?

28. Will restriction sites for an enzyme that has 4 bp in its recognition sequence be closer together, farther apart, or similarly spaced, on average, compared with those of an enzyme that has 6 bp in its recognition sequence? Explain your reasoning.

*29. About 60% of the base pairs in the human genome are AT. If the human genome has 3.2 billion base pairs

of DNA, about how many times will the following restriction sites be present?

a. *Bam*HI (recognition sequence is 5′–GGATCC–3′)

b. *Eco*RI (recognition sequence is 5′–GAATTC–3′)

c. *Hae*III (recognition sequence is 5′–GGCC–3′)

*30. A linear piece of DNA has the following *Eco*RI restriction sites:

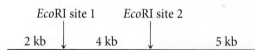

a. This piece of DNA is cut by *Eco*RI, the resulting fragments are separated by gel electrophoresis, and the gel is stained with ethidium bromide. Draw a picture of the bands that will appear on the gel.

b. If a mutation that alters *Eco*RI site 1 occurs in this piece of DNA, how will the banding pattern on the gel differ from the one that you drew in part *a*?

c. If mutations that alter *Eco*RI sites 1 and 2 occur in this piece of DNA, how will the banding pattern on the gel differ from the one that you drew in part *a*?

d. If 1000 bp of DNA were inserted between the two restriction sites, how would the banding pattern on the gel differ from the one that you drew in part *a*?

e. If 500 bp of DNA between the two restriction sites were deleted, how would the banding pattern on the gel differ from the one that you drew in part *a*?

Section 19.3

31. Compare and contrast the use of PCR and gene cloning for amplifying DNA fragments. What are the advantages and disadvantages of each method?

*32. Which vectors (plasmid, phage λ, cosmid, bacterial artificial chromosome) can be used to clone a continuous fragment of DNA with the following lengths?

   a. 4 kb

   b. 20 kb

   c. 35 kb

   d. 100 kb

33. A geneticist uses a plasmid for cloning that has the *lacZ* gene and a gene that confers resistance to penicillin. The geneticist inserts a piece of foreign DNA into a restriction site that is located within the *lacZ* gene and uses the plasmid to transform bacteria. Explain how the geneticist can identify bacteria that contain a copy of a plasmid with the foreign DNA.

34. In **Figure 19.12**, what is the purpose of the *neo*⁺ gene that is attached to the *Bt* gene?

Section 19.4

35. Suppose that you have just graduated from college and have started working at a biotechnology firm. Your first job assignment is to clone the pig gene for the hormone prolactin. Assume that the pig gene for prolactin has not yet been isolated, sequenced, or mapped; however, the mouse gene for prolactin has been cloned, and the amino acid sequence of mouse prolactin is known. Briefly explain two different strategies that you might use to find and clone the pig gene for prolactin.

*36. A molecular biologist wants to isolate a gene from a scorpion that encodes the deadly toxin found in its stinger, with the ultimate purpose of transferring this

[Sahara Nature/Fotolia LLC.]

gene to bacteria and producing the toxin for use as a commercial pesticide. Isolating the gene requires a DNA library. Should the molecular biologist create a genomic library or a cDNA library? Explain your reasoning.

*37. A protein has the following amino acid sequence:

Met-Tyr-Asn-Val-Arg-Val-Tyr-Lys-Ala-Lys-
                    Trp-Leu-Ile-His-Thr-Pro

You wish to make a set of probes to screen a cDNA library for the sequence that encodes this protein. Your probes should be at least 18 nucleotides in length.

   a. Which amino acids in the protein should be used to construct the probes so that the least degeneracy results? (Consult the genetic code in **Figure 15.10**.)

   b. How many different probes must be synthesized to be certain that you will find the cDNA sequence that specifies the protein?

Section 19.5

38. Suppose that you want to sequence the following DNA fragment:

5′–TCCCGGGAAA-primer site–3′

You first use PCR to amplify the fragment, so that there is sufficient DNA for sequencing. You carry out dideoxy sequencing and then separate the products of the sequencing reactions by gel electrophoresis. Draw the bands that should appear on the gel from the four sequencing reactions.

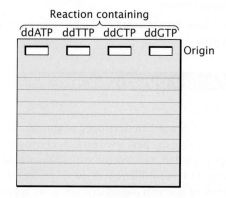

*39. Suppose that you are given a short fragment of DNA to sequence. You amplify the fragment with PCR and set up a series of four dideoxy reactions. You then separate the products of the reactions by gel electrophoresis and obtain the following banding pattern:

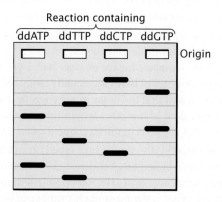

Write out the base sequence of the original fragment that you were given.

Original sequence: 5′– _____ –3′

40. The accompanying photo shows a sequencing gel from the original study that first sequenced the cystic fibrosis

gene (J. R. Riordan et al. 1989. *Science* 245:1066–1073). From the photo, determine the sequence of the normal copy of the gene and the sequence of the mutated copy of the gene. Identify the location of the mutation that causes cystic fibrosis. (Hint: The CF mutation is a 3-bp deletion.)

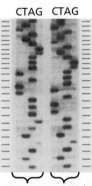

CTAG   CTAG

DNA from a healthy person   DNA from a person with CF

[From Riordan, J. et al. "Identification of the cystic fibrosis gene: cloning and characterization of complementary DNA," *Science* 245: 1066–1073, © 1989. Reprinted with permission from AAAS, permission conveyed through Copyright Clearance Center, Inc.]

Section 19.6

41. You have discovered a gene in mice that is similar to a gene in yeast. How might you determine whether this gene is essential for development in mice?

*__**42.**__ Andrew Fire, Craig Mello, and their colleagues were among the first to examine the effects of double-stranded RNA (dsRNA) on gene expression (A. Fire

et al. 1998. *Nature* 391:806–811). In one experiment, they used a transgenic strain of *C. elegans* into which a gene (*gfp*) for a green fluorescent protein had been introduced. They injected some worms with double-stranded RNA complementary to coding sequences of the *gfp* gene and injected other worms with double-stranded RNA complementary to the coding region of a different gene (*unc22A*) that encodes a muscle protein. The accompanying photographs show larval and adult progeny of the injected worms. Green fluorescent protein appears as bright spots in the photographs.

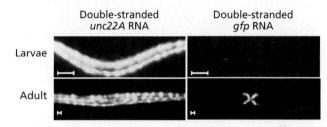

[Reprinted by permission of Macmillan Publishers Ltd. Andrew Fire, Craig Mello et al. "Potent and specific genetic interference by double-stranded RNA in *Caenorhabditis elegans*," *Nature* 391: 806–811. © 1998, permission conveyed through Copyright Clearance Center, Inc.]

a. Explain these results.

b. Fire and Mello conducted another experiment in which they injected double-stranded RNA complementary to the introns and promoter sequences of the *gfp* gene. What results would you expect with this experiment? Explain your answer.

CHALLENGE QUESTION

Section 19.6

43. Suppose that you are hired by a biotechnology firm to produce a strain of giant fruit flies by using recombinant DNA technology so that genetics students will not be forced to strain their eyes when looking at tiny flies. You go to the library and learn that growth in fruit flies is normally inhibited by a hormone called shorty substance P (SSP). You decide that you can produce giant fruit flies if you can somehow turn off the production of SSP. Shorty substance P is synthesized from a compound called XSP in a single-step reaction catalyzed by the enzyme runtase:

$$XSP \xrightarrow[\text{Runtase}]{} SSP$$

A researcher has already isolated cDNA for runtase and has sequenced it, but the location of the runtase gene in the *Drosophila* genome is unknown. In attempting to devise a strategy for turning off the production of SSP using standard recombinant DNA techniques, you discover that deleting, inactivating, or otherwise mutating this DNA sequence in *Drosophila* turns out to be extremely difficult. Therefore, you must restrict your genetic engineering to gene augmentation (adding new genes to cells). Describe the methods that you will use to turn off SSP and produce giant flies.

THINK-PAIR-SHARE QUESTIONS

Section 19.1

1. What were some key innovations prior to Watson, Crick, Franklin, and Wilkins's discovery of the three-dimensional structure of DNA that were important in shaping the study of genetics?

Section 19.2

2. Now that you understand how the CRISPR-Cas9 system works, think back to the experiments discussed in the introduction to this chapter, in which researchers used CRISPR-Cas9 genome editing to treat mice with Duchenne muscular dystrophy. Why did the researchers choose to cut out the entire exon 23 in the mice with the disorder? Why not replace the specific mutation using a donor piece of DNA and homologous recombination? Propose some possible explanations.

3. Propose some specific uses of a modified CRISPR-Cas system as a general RNA-guided device for altering cellular functions. What might these functions be, and how could CRISPR-Cas be used to study them?

4. Some people have argued that editing the genomes of human embryos is ethically defensible as long as the embryos are not allowed to develop past an early stage and thus will never result in the birth of genetically modified humans. Others have argued that no genome editing should be carried out on human embryos. Present arguments for and against using genome editing on human embryos.

Section 19.3

5. Make a table comparing the advantages and disadvantages of PCR and gene cloning for amplifying DNA fragments.

Section 19.4

6. As discussed in the text, almost 5% of Caucasians carry a recessive gene for cystic fibrosis. Until recently, most people with cystic fibrosis died in childhood or early adulthood. Why is cystic fibrosis so frequent in spite of the fact that until recently it was usually lethal?

Section 19.7

7. Much of the controversy over genetically engineered foods has centered on whether special labeling should be required on all products made from genetically modified crops. Some people have advocated labeling that identifies the product as having been made from genetically modified plants. Others have argued that food labeling should be required to identify only the ingredients, not the process by which they were produced. Choose a side in this issue and justify your stand.

8. Why is it that somatic gene therapy is allowed in many countries and yet germ-line gene therapy is often restricted? What are some arguments for and against germ-line gene therapy?

SaplingPlus   Self-study tools that will help you practice what you've learned and reinforce this chapter's concepts are available online. Go to www.macmillanlearning.com/PierceGenetics6e.

Genomics and Proteomics

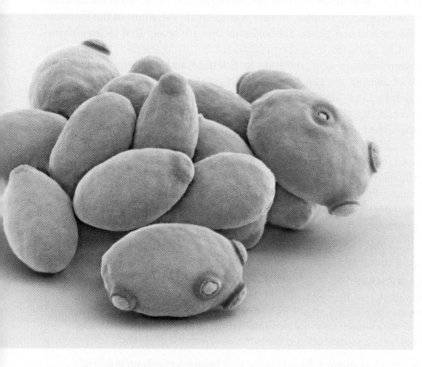

Researchers and students synthesized an artificial chromosome for baker's yeast, *Saccharomyces cerevisiae*, shown here. [Micrograph: Thomas Deerinck, NCMIR/Science Source.]

Building a Chromosome for Class

In the spring of 2014, an international group of scientists reported the first synthesis of a completely artificial eukaryotic chromosome. When substituted for chromosome III in *Saccharomyces cerevisiae*, the synthetic chromosome functioned well. Remarkably, much of the work in constructing this artificial chromosome was carried out by undergraduate students at Johns Hopkins University, who completed the task as a part of their Build a Genome class.

Saccharomyces cerevisiae, commonly known as baker's yeast, has been used for thousands of years for making bread and brewing beer, and today it plays an important role in the food and chemical industries. It's also an important model genetic organism (see the Reference Guide to Model Genetic Organisms at the end of this book). Baker's yeast, a single-celled eukaryote, is easy to cross and grow in the laboratory. It was the first eukaryote to have its genome completely sequenced, and a number of molecular techniques have been developed to manipulate its DNA. Scientists are now working to create a completely synthetic genome for *S. cerevisiae*, a flexible genome that can be altered at will to endow yeast with selected genes and special properties.

The first step in creating a synthetic genome for yeast was synthesis of an artificial chromosome III. Scientists began by designing the new chromosome on the computer, using the basic sequence of natural chromosome III but deleting nonessential DNA, including introns, repeated sequences near the telomeres, transposable elements, and transfer RNA genes. They designed the artificial chromosome with special sequences that allow easy deletion of genes and rearrangement of the genome. They also changed all the TAG stop codons to TAA, freeing up the TAG codon for future use in encoding an additional, artificial amino acid.

Once the sequence of the chromosome was digitally designed, actual construction began. As a first step, students in the Build a Genome class stitched together small pieces of DNA into 750-bp building blocks. Each student started with a series of overlapping 60–79-bp DNA fragments called oligonucleotides, which were synthesized by machine and provided to the class by a commercial firm. The students assembled these fragments into building blocks using methods based on the polymerase chain reaction. After verifying that the correct sequences were joined together, the students cloned the building blocks in bacteria to make many copies. Scientists then assembled the building blocks into 127 larger, overlapping fragments of approximately 7000 bp, called minichunks. Finally, they systematically replaced the native sequences of yeast chromosome III with the minichunks of artificial DNA until a completely synthetic chromosome was achieved.

The resulting artificial chromosome, dubbed synIII, consisted of 272,871 bp of DNA, replacing the original 316,617 bp of the natural chromosome. When inserted into yeast cells in place of the natural chromosome III, synIII functioned normally, replicating and expressing its genetic instructions. Cells with synIII had the same morphology as natural yeast cells and grew well. There appeared to be no fitness differences between synIII cells and natural yeast, and the genomes of the synIII cells were stable over 125 mitotic divisions.

In the design of synIII, geneticists inserted special sequences, called loxPsym sites, on either side of many of its genes, providing an easy way to create variations of the chromosome by deleting and rearranging genes. The loxPsym sites are recognized by an enzyme called Cre, which, when activated by a chemical added by the researcher, brings about recombination between loxPsym sites, resulting in deletions and rearrangements of the genes. This method of altering the genome is called Synthetic Chromosome Rearrangement and Modification by LoxP-Mediated Evolution, or SCRaMbLE for short. One reason scientists wanted to create an artificial yeast genome is so that they could systematically take out groups of genes to determine the minimal genome necessary for a cell to function. The SCRaMbLE system greatly facilitates this effort.

Perhaps the greatest impact of the project was on the students, who helped to advance the field of genetics while taking an undergraduate course. In a now-expanded effort called the Synthetic Yeast Genome Project, additional students and scientists from schools around the world are synthesizing the other 15 chromosomes of yeast, with the goal of creating an entirely synthetic genome.

THINK-PAIR-SHARE

- What are some possible research questions and practical applications that could be addressed by creating organisms with artificial chromosomes and synthetic genomes? What might be some potential safety, environmental, social, and ethical concerns about creating organisms with synthetic genomes?

- What do you think about taking a course that focuses entirely on building an artificial chromosome? Would this be an effective way to learn genetics? Propose some arguments for and against learning genetics in this way.

Genomics is the field of genetics that attempts to understand the content, organization, function, and evolution of the genetic information contained in whole genomes. The field of genomics is at the cutting edge of modern biology; information resulting from research in this field has made significant contributions to human health, agriculture, and numerous other areas. It has provided gene sequences necessary for producing medically important proteins through recombinant DNA technology, and comparisons of genome sequences from different organisms are leading to a better understanding of evolution and the history of life. It provides the potential for manipulating genomes, as illustrated by the story of building an artificial chromosome for yeast.

We begin this chapter by examining genetic and physical maps and methods for sequencing entire genomes. Next, we explore functional genomics—how genes are identified in genome sequences and how their functions are defined. The sequence of a genome, by itself, is of limited use, and now that the sequencing of genomes has become routine, much of genomics is currently focused on deciphering the functions of the sequences that are obtained. We also discuss how genome sequence and organization vary among organisms and how genomes evolve. In the final section of the chapter, we consider proteomics, the study of the complete set of proteins found in a cell.

20.1 Structural Genomics Determines the DNA Sequences and Organization of Entire Genomes

Structural genomics is the study of the organization and sequence of the genetic information contained within a genome, providing the basic DNA sequence information that is used in functional and evolutionary studies. An early step in characterizing a genome may be to prepare genetic and physical maps of its chromosomes. These maps provide information about the relative locations of genes, molecular markers, and chromosome segments, which is often essential for positioning chromosome segments and aligning stretches of sequenced DNA into a whole-genome sequence. These maps are also the foundation for positional cloning (see Section 19.4), in which genes that influence specific traits are located and sequenced.

Genetic Maps

Everyone has used a map at one time or another. Maps are indispensable for finding a new friend's house, the way to an unfamiliar city in your state, or the location of a country. Each of these examples requires a map with a different scale. To find a friend's house, you would probably use a city street map; to find your way to an unknown city, you might pick up a state highway map; to find a country such as Kazakhstan, you would need a world atlas. Similarly, navigating a genome requires maps of different types and scales.

CONSTRUCTION OF GENETIC MAPS Genetic maps (also called *linkage maps*) provide a rough approximation of the locations of genes relative to the locations of other known genes (**Figure 20.1**). These maps are based on the genetic process of recombination (hence the name *genetic map*). The basic principles of constructing genetic maps are discussed in detail in Chapter 7. In short, individual organisms of known genotype are crossed, and the frequency of recombination between loci is determined by examining the progeny. If the recombination frequency between two loci is 50%, then the loci are found on different chromosomes or are far apart on the same chromosome. If the recombination frequency is less than 50%, the loci are found close together on the same chromosome (they belong to the same linkage group). For linked genes, the rate of recombination is proportional to the physical distance between the loci. Distances on genetic maps are measured in recombination frequencies (centiMorgans [cM] or map units [m.u.]). Data from multiple two-point or three-point crosses can be integrated into linkage maps of whole chromosomes.

For many years, genes could be detected only by observing their influence on a trait (the phenotype), and the construction of genetic maps was limited by the availability of single-locus traits that could be examined for evidence of recombination. Eventually, this limitation was overcome by the development of molecular techniques, such as the analysis of restriction sites, the polymerase chain reaction, and DNA sequencing (see Chapter 19), which are able to provide molecular markers that can be used to construct and refine genetic maps.

LIMITATIONS OF GENETIC MAPS Genetic maps have several limitations, the first of which is resolution, or detail. The human genome includes 3.2 billion base pairs of DNA and has a total genetic distance of about 4000 cM, an average of 800,000 bp/cM. Even if a marker were present every centiMorgan (which is unrealistic), their resolution in regard to the physical structure of the DNA would still be quite low. In other words, the detail of a genetic map is very limited. A second problem with genetic maps is that they do not always accurately correspond to physical distances between genes. Genetic maps are based on rates of crossing over, which vary somewhat from one part of a chromosome to another; so the distances on a genetic map are approximations of

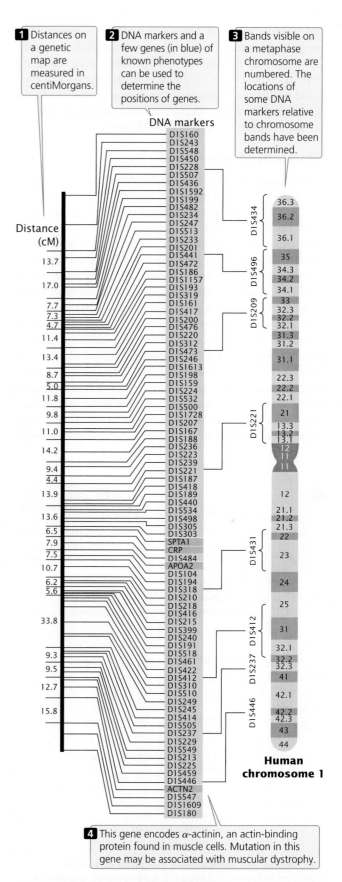

1 Distances on a genetic map are measured in centiMorgans.

2 DNA markers and a few genes (in blue) of known phenotypes can be used to determine the positions of genes.

3 Bands visible on a metaphase chromosome are numbered. The locations of some DNA markers relative to chromosome bands have been determined.

4 This gene encodes α-actinin, an actin-binding protein found in muscle cells. Mutation in this gene may be associated with muscular dystrophy.

Human chromosome 1

20.1 Genetic maps are based on rates of recombination. Shown here is a genetic map of human chromosome 1.

real physical distances along a chromosome. **Figure 20.2** compares the genetic map of chromosome III of yeast with a physical map determined by DNA sequencing. There are some discrepancies in the distances between genes and even in the positions of some genes. In spite of these limitations, genetic maps have been critical to the development of physical maps and the sequencing of whole genomes.

Physical Maps

Physical maps are based on the direct analysis of DNA, and they place genes in relation to distances measured in number of base pairs, kilobases, or megabases. A common type of physical map is one that connects isolated pieces of genomic DNA that have been cloned in bacteria or yeast (**Figure 20.3**). Physical maps generally have higher resolution and are more accurate than genetic maps. A physical map is analogous to a neighborhood map that shows the location of every house along a street, whereas a genetic map is analogous to a highway map that shows the general locations of major towns and cities.

One of the techniques that has been used for creating physical maps is restriction mapping, which determines the positions of restriction sites in DNA. When a piece of DNA is cut (digested) with a restriction enzyme and the fragments are separated by gel electrophoresis, the number of restriction sites in the DNA and the distances between them can be determined by the number and positions of bands on the gel

(see p. 568 in Chapter 19). However, this information does not tell us the order or the precise locations of the restriction sites. To map restriction sites, a sample of the DNA is cut with one restriction enzyme, and another sample of the same DNA is cut with a different restriction enzyme. A third sample is cut with both restriction enzymes together (a double digest). The DNA fragments produced by these restriction digests are then separated by gel electrophoresis, and their sizes are compared. Overlap in the sizes of the fragments produced by the digests can be used to position the restriction sites on the original DNA molecule (see the Worked Problem at the end of this chapter). Most restriction mapping is done with several restriction enzymes, used alone and in various combinations, producing many restriction fragments. With long pieces of DNA (greater than 30 kb), computer programs are used to determine the restriction maps, and restriction mapping may be facilitated by labeling one end of a large DNA fragment with a radioactive tag or by identifying the end with the use of a probe.

Physical maps, such as restriction maps of DNA fragments or even whole chromosomes, are often created for genomic analysis. These lengthy maps are created by combining maps of shorter, overlapping DNA fragments. A number of additional techniques exist for creating physical maps: determining the positions of short unique sequences of DNA on a chromosome; in situ hybridization, by which markers can be visually mapped to locations on chromosomes; and DNA sequencing.

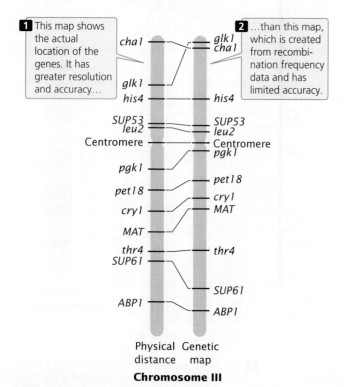

1 This map shows the actual location of the genes. It has greater resolution and accuracy...

2 ...than this map, which is created from recombination frequency data and has limited accuracy.

Physical distance | Genetic map

Chromosome III

20.2 Genetic and physical maps may differ in relative distances between, and even in the positions of, genes on a chromosome. Genetic and physical maps of yeast chromosome III reveal such differences.

CONCEPTS

Both genetic and physical maps provide information about the relative positions of and distances between genes, molecular markers, and chromosome segments. Genetic maps are based on rates of recombination and are measured in recombination frequencies (centiMorgans or map units). Physical maps are based on physical distances and are measured in base pairs.

✔ **CONCEPT CHECK 1**

What are some of the limitations of genetic maps?

Sequencing an Entire Genome

The first genomes to be sequenced were small virus genomes. The genome of bacteriophage λ, consisting of 49,000 bp, was completed in 1982. In 1995, the first genome of a free-living organism (*Haemophilus influenzae*) was sequenced by Craig Venter and Claire Fraser of The Institute for Genomic Research (TIGR) and Hamilton Smith of Johns Hopkins University. This bacterium has a small genome of 1.8 million base pairs (**Figure 20.4**). By 1996, the genome of the first eukaryotic organism (yeast) had been determined, followed by the genomes of *Escherichia coli* (1997), *Caenorhabditis elegans*

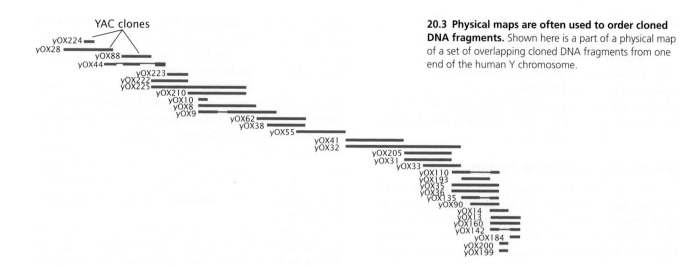

20.3 Physical maps are often used to order cloned DNA fragments. Shown here is a part of a physical map of a set of overlapping cloned DNA fragments from one end of the human Y chromosome.

(1998), *Drosophila melanogaster* (2000), and *Arabidopsis thaliana* (2000). Sequencing of the human genome was completed in the spring of 2003.

The ultimate goal of structural genomics is to determine the ordered nucleotide sequences of entire genomes of organisms, providing the sequence information that is used to answer other questions. In Chapter 19, we considered some of the methods used to sequence small fragments of DNA. The main obstacle to sequencing a whole genome is the immense size of most genomes. Bacterial genomes are usually at least several million base pairs long; many eukaryotic genomes are billions of base pairs long and are distributed among dozens of chromosomes. Furthermore, for technical reasons, sequencing cannot begin at one end of a chromosome and continue straight through to the other end; only small fragments of DNA—usually no more than 500 to 700 nucleotides—are sequenced at one time. Therefore, determining the sequence for an entire genome requires that the DNA be broken into

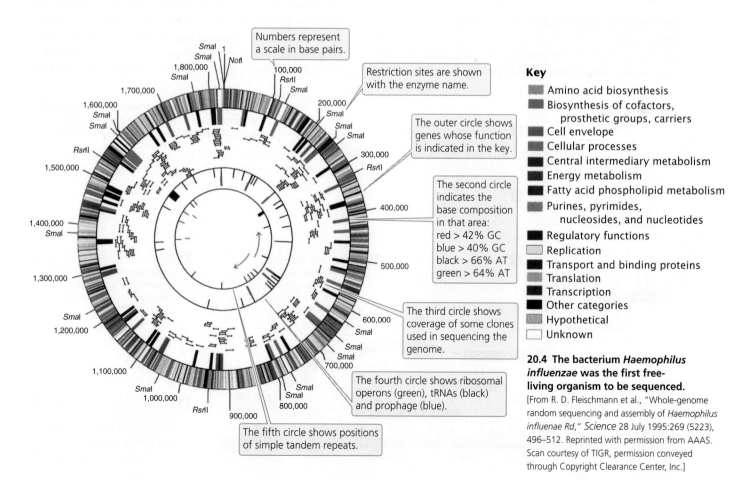

20.4 The bacterium *Haemophilus influenzae* was the first free-living organism to be sequenced.

[From R. D. Fleischmann et al., "Whole-genome random sequencing and assembly of *Haemophilus influenae* Rd," *Science* 28 July 1995:269 (5223), 496–512. Reprinted with permission from AAAS. Scan courtesy of TIGR, permission conveyed through Copyright Clearance Center, Inc.]

thousands or millions of small overlapping fragments that can then be sequenced. The difficulty lies in putting these short sequences back together in the correct order. Two different approaches have been used to assemble the sequenced fragments into a complete genome: map-based sequencing and whole-genome shotgun sequencing. Here, we consider these two approaches in the context of the Human Genome Project, in which both approaches were used.

The Human Genome Project

By 1980, methods for mapping and sequencing DNA fragments had been sufficiently developed that geneticists began seriously proposing that the entire human genome could be sequenced. An international collaboration was planned to undertake the Human Genome Project; initial estimates suggested that 15 years and $3 billion would be required to accomplish the task.

The Human Genome Project officially began in October 1990. Initial efforts focused on developing new and automated methods for cloning and sequencing DNA and on generating detailed physical and genetic maps of the human genome. The effort was a public project consisting of the international collaboration among the 20 research groups and hundreds of individual researchers who formed the International Human Genome Sequencing Consortium. This group used a map-based strategy for sequencing the human genome.

MAP-BASED SEQUENCING In **map-based sequencing**, short sequenced fragments are assembled into a whole-genome sequence by first creating detailed genetic and physical maps of the genome, which provide known locations of genetic markers (restriction sites, genes, or known DNA sequences) at regularly spaced intervals along each chromosome. These markers are later used to help align the short sequenced fragments in their correct order.

Once the genetic and physical maps are available, chromosomes or large pieces of chromosomes are separated by pulsed-field gel electrophoresis (PFGE) or by flow cytometry. Standard gel electrophoresis cannot separate very large pieces of DNA, such as whole chromosomes, but PFGE can separate large molecules of DNA or whole chromosomes in a gel by periodically alternating the orientation of an electrical current. In flow cytometry, chromosomes are sorted optically by size.

Each chromosome (or sometimes the entire genome) is then cut up by partial digestion with restriction enzymes (**Figure 20.5**). Partial digestion means that the restriction enzymes are allowed to act for only a limited time so that not all restriction sites in every DNA molecule are cut. Thus, partial digestion produces a large set of overlapping DNA fragments, which are then cloned with the use of cosmids, yeast artificial chromosomes (YACs), or bacterial artificial chromosomes (BACs; see Section 19.3).

Next, these large-insert clones are put together in their correct order on the chromosome (see Figure 20.5). This assembly can be done in several ways. One method relies on the availability of a high-density map of genetic markers (variable sequences that can be detected). A complementary DNA probe is made for each genetic marker, and a library of the large-insert clones is screened with the probe, which hybridizes to any colony containing a clone with the marker. The library is then screened for neighboring markers.

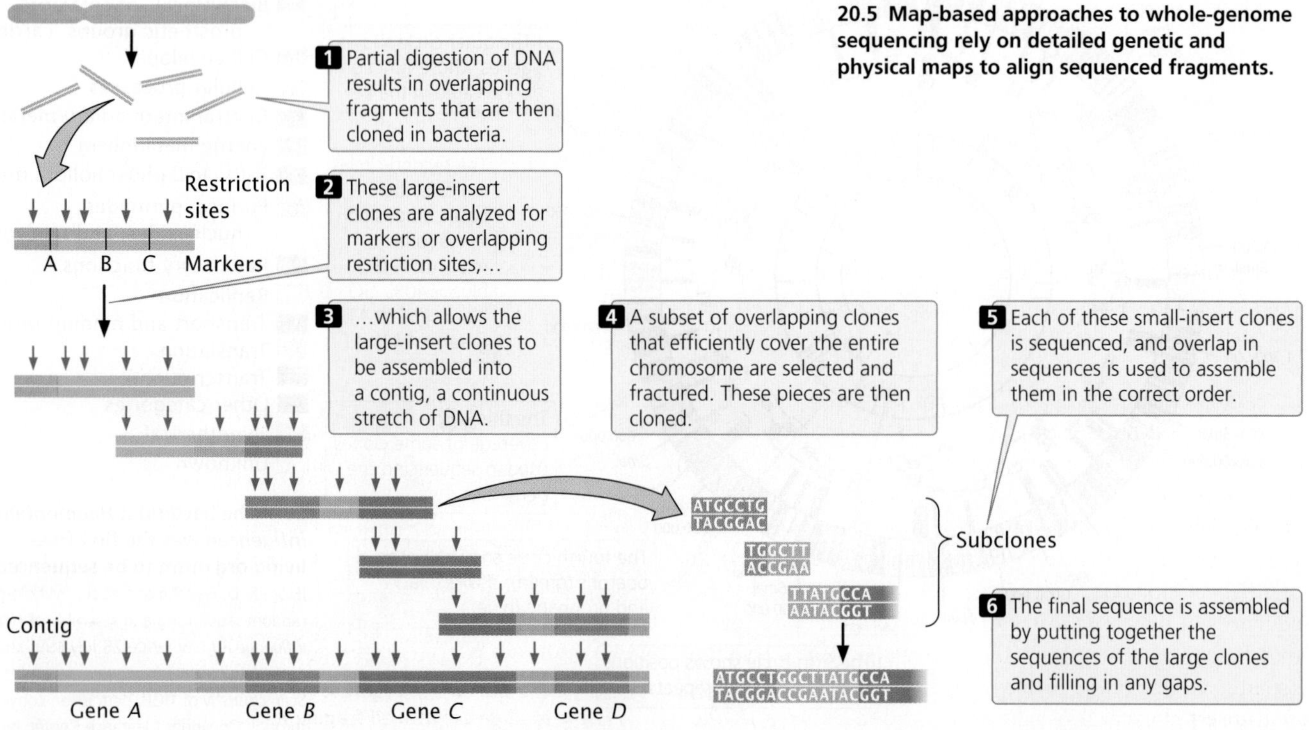

20.5 Map-based approaches to whole-genome sequencing rely on detailed genetic and physical maps to align sequenced fragments.

1 Partial digestion of DNA results in overlapping fragments that are then cloned in bacteria.

2 These large-insert clones are analyzed for markers or overlapping restriction sites,...

3 ...which allows the large-insert clones to be assembled into a contig, a continuous stretch of DNA.

4 A subset of overlapping clones that efficiently cover the entire chromosome are selected and fractured. These pieces are then cloned.

5 Each of these small-insert clones is sequenced, and overlap in sequences is used to assemble them in the correct order.

Restriction sites

A B C Markers

Contig

Gene A   Gene B   Gene C   Gene D

ATGCCTG
TACGGAC

TGGCTT
ACCGAA

TTATGCCA
AATACGGT

Subclones

6 The final sequence is assembled by putting together the sequences of the large clones and filling in any gaps.

ATGCCTGGCTTATGCCA
TACGGACCGAATACGGT

Because the clones are much larger than the markers used as probes, some clones will have more than one marker. For example, clone A might have markers M1 and M2, clone B markers M2, M3, and M4, and clone C markers M4 and M5. Such a result would indicate that these clones contain areas of overlap, as shown here:

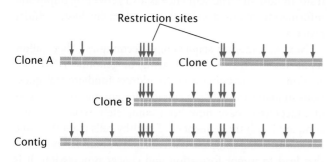

A set of two or more overlapping DNA fragments that form a contiguous stretch of DNA is called a **contig**. This approach was used in 1993 to create a contig consisting of 196 overlapping YAC clones (see Figure 20.3) of the human Y chromosome.

The order of clones can also be determined without the use of preexisting genetic maps. For example, each clone can be cut with a series of restriction enzymes and the resulting fragments then separated by gel electrophoresis. This method generates a unique set of restriction fragments, called a fingerprint, for each clone. The restriction patterns for the clones are stored in a database. A computer program is then used to examine the restriction patterns of all the clones and look for areas of overlap. The overlap is then used to arrange the clones in order, as shown here:

Restriction sites

Clone A   Clone C

Clone B

Contig

Other genetic markers can be used to help position contigs along the chromosome. **▶ TRY PROBLEM 27**

When the large-insert clones have been assembled in the correct order on the chromosome, a subset of overlapping clones that efficiently cover the entire chromosome can be chosen for sequencing; the goal is to select the minimum number of clones that is necessary to represent the chromosome. Each of the selected large-insert clones is fractured into smaller overlapping fragments, which are themselves cloned (see Figure 20.5). These smaller clones (called small-insert clones or subclones) are then sequenced. The sequences of the small-insert clones are examined for overlap, which allows them to be correctly assembled to give the sequence of the large-insert clones. Enough overlapping small-insert clones are usually sequenced to ensure that the

entire genome is sequenced several times. Finally, the whole genome is assembled by putting together the sequences of all overlapping contigs (see Figure 20.5). Often, gaps in the genome sequence still exist and must be filled in by using other methods.

The International Human Genome Sequencing Consortium used a map-based approach to sequencing the human genome. Many copies of the human genome were cut up into fragments of about 150,000 bp each, which were inserted into bacterial artificial chromosomes. Yeast artificial chromosomes and cosmids had been used in early stages of the project but did not prove to be as stable as the BAC clones, although YAC clones were instrumental in putting together some of the larger contigs. Restriction fingerprints and other genetic markers were used to assemble the BAC clones into contigs, which were positioned on the chromosomes with the use of genetic markers and probes. The individual BAC clones were then sheared into smaller overlapping fragments and sequenced, and the whole genome was assembled by putting together the sequence of the BAC clones.

WHOLE-GENOME SHOTGUN SEQUENCING In 1998, Craig Venter announced that he would lead a company called Celera Genomics in a private effort to sequence the human genome. He proposed using a shotgun sequencing approach, which he suggested would be quicker than the map-based approach employed by the International Human Genome Sequencing Consortium. In **whole-genome shotgun sequencing** (**Figure 20.6**), small-insert clones

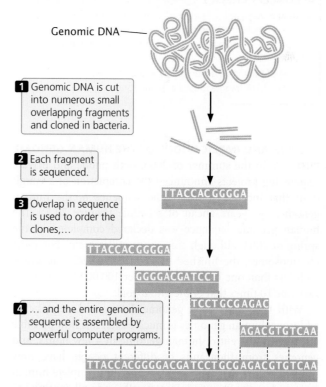

Genomic DNA

1 Genomic DNA is cut into numerous small overlapping fragments and cloned in bacteria.

2 Each fragment is sequenced.

3 Overlap in sequence is used to order the clones,...

TTACCACGGGGA

4 ... and the entire genomic sequence is assembled by powerful computer programs.

TTACCACGGGGA
GGGGACGATCCT
TCCTGCGAGAC
AGACGTGTCAA

TTACCACGGGGACGATCCTGCGAGACGTGTCAA

20.6 Whole-genome shotgun sequencing uses sequence overlap to align sequenced fragments.

are prepared directly from genomic DNA and sequenced. Powerful computer programs then assemble the entire genome by examining overlap among the small-insert clones. One advantage of shotgun sequencing is that the small-insert clones can be placed in plasmids, which are simple and easy to manipulate. The requirement for overlap means that most of the genome must be sequenced many times. The average number of times a nucleotide in the genome is sequenced is called the sequencing coverage. For example, 10× coverage means that an average nucleotide in the genome has been sequenced 10 times.

Shotgun sequencing was initially used for assembling small genomes such as those of bacteria. When Venter proposed the use of this approach for sequencing the human genome, it was not at all clear that the approach could successfully assemble a complex genome consisting of billions of base pairs. Today, virtually all genomes are sequenced using the whole-genome shotgun approach.

20.7 Craig Venter (left), President of Celera Genomics, and Francis Collins (right), Director of the National Human Genome Research Institute, NIH, announce the completion of a rough draft of the human genome at a press conference in Washington, D.C., on June 26, 2000. [Alex Wong/Newsmakers/Getty Images.]

CONCEPTS

Sequencing a genome requires breaking it up into small overlapping fragments whose DNA sequences can be determined in a sequencing reaction. In map-based sequencing, sequenced fragments are ordered into the final genome sequence with the use of genetic and physical maps. In whole-genome shotgun sequencing, the genome is assembled by means of overlap in the sequences of small fragments.

✔ **CONCEPT CHECK 2**

A contig is
a. a set of molecular markers used in gene mapping.
b. a set of overlapping fragments that form a continuous stretch of DNA.
c. a set of fragments generated by a restriction enzyme.
d. a small DNA fragment used in sequencing.

RESULTS AND IMPLICATIONS OF THE HUMAN GENOME PROJECT In the summer of 2000, both public and private sequencing projects announced the completion of a rough draft that included most of the sequence of the human genome, five years ahead of schedule (**Figure 20.7**). The human genome sequence was declared completed in the spring of 2003, although some gaps still remain. For most chromosomes, the finished sequence is 99.999% accurate, with less than one base-pair error per 100,000 bp, an accuracy rate 10 times that of the initial goal.

With the first human genome determined, sequencing additional human genomes is much easier. It is now possible to sequence an entire human genome in a few hours, and genomes from thousands of different people have now been sequenced. The cost of sequencing a complete human genome has also dropped dramatically and will continue to fall as sequencing technology improves.

The availability of the complete sequence of the human genome is proving to be of great benefit. The sequence has provided tools for detecting and mapping genetic variants across the human genome, greatly facilitating gene mapping in humans. For example, as we will see shortly, several million sites at which people differ in a single nucleotide (called single-nucleotide polymorphisms) have now been identified, and these sites are being used in genome-wide association studies to locate genes that affect diseases and traits in humans. The sequence is also providing important information about development and many basic cellular processes.

Next-generation sequencing techniques that allow rapid and inexpensive sequencing of genomic DNA (see Section 19.5) are being used to address fundamental questions in many areas. For example, the genomes of a number of cancer cells have now been completely sequenced and compared with the sequences of healthy cells from the same person, allowing complete determination of all the mutations that lead to tumor formation and cancer progression. It is now possible to sequence the genomes of single cells, providing information about somatic mutations and how, for example, cells within a tumor change over time. The complete genome of an unborn baby has been sequenced from fetal DNA isolated from its mother's blood.

Researchers have reported the analysis of genomes from over 2600 Icelanders, providing a spectacular picture of the genetic composition of a nation. The 1000 Genomes Project, completed in 2015, sequenced and compared the genomes of 2504 individuals from 26 populations in Africa, East Asia, Europe, South Asia, and the Americas, with the goal of detecting most of the common variations that exist in the human species. Sequencing of the complete genomes of parents and their children has allowed a direct estimate of mutation rates.

Genomes have been sequenced from DNA extracted from the fossils of long-extinct animals, such as a 700,000-year-old Pleistocene horse. DNA has been isolated and genomes sequenced from the bones of ancient humans, including Neanderthals and Denisovans (a little-known group of humans that appear to be closely related to Neanderthals). Comparisons of the modern human genome with the genomes of these and other species are adding to our understanding of human evolution. For example, research now suggests that the genomes of Europeans and Asians consist of 1% to 4% Neanderthal DNA, which was transferred to modern humans when interbreeding took place between the two species.

Along with the many potential benefits of having complete human sequence information, there are concerns about it being misused. With the knowledge gained from genome sequencing, many more genes for diseases, disorders, and behavioral and physical traits have been identified, increasing the number of genetic tests that can be performed to make predictions about the future phenotype and health of a person. There is concern that information from genetic testing might be used to discriminate against people who are carriers of disease-causing genes or who might be at risk for some future disease. This matter has been addressed to some extent in the United States with the passage of the Genetic Information Nondiscrimination Act (see Section 6.4), which prohibits health insurers and employers from using genetic information to make decisions about health-insurance coverage and employment (although it does not apply to life, disability, or long-term care insurance). Questions also arise about who should have access to a person's genome sequence. What about relatives, who have similar genomes and might also be at risk for some of the same diseases? There are also questions about the use of this information to select for specific traits in future offspring. All of these concerns are legitimate and must be addressed if we are to use the information from genome sequencing responsibly.

THINK-PAIR-SHARE Question 1

CONCEPTS

The Human Genome Project was an effort to sequence the entire human genome. Rough drafts of the sequence were completed by two competing teams in 2000. The entire sequence was published in 2003. The ability to rapidly sequence human genomes raises a number of ethical questions.

What Exactly Is the Human Genome?

As we have seen, the human genome sequence was determined first as rough drafts in 2000 and then in "final" form in 2003. This sequence has been of immense importance in research, medicine, and biotechnology. However,

geneticist Ken Weiss has argued that what people refer to as "the human genome" does not actually exist: there is not a single genome that represents the human species. What was determined by the Human Genome Project is a *reference* genome, which was based on DNA donated from not one, but several different individuals, whose identities are kept confidential for ethical reasons. In reality, there are billions of human genomes; every person's genome is likely to be unique, differing at least a little from every other person's genome. At millions of nucleotide positions across the genome, people vary in the particular base that occurs, with some people, for example, possessing a T and others possessing a G. Some people even have extensive segments of DNA (called structural variants) that are missing, duplicated, or rearranged. Furthermore, we know that there is not even a single genome for a person; genomic research has demonstrated that individual cells within the same person vary in their DNA sequences due to somatic mutations.

It's important to keep in mind that the official human reference genome is not the ancestral sequence for all humans, nor even the most common genome sequence among humans. Rather, it an arbitrary sequence, based on the genetic idiosyncrasies of those individuals whose DNA was sequenced in the Human Genome Project. And even the reference genome keeps changing as corrections and better sequences for problematic regions are obtained.

THINK-PAIR-SHARE Question 2

Single-Nucleotide Polymorphisms

Genome sequencing has identified a number of sites where humans differ in their genome sequences. Imagine that you are riding in an elevator with a random stranger. How much of your genome do you have in common with this person? Studies of variation in the human genome indicate that you and the stranger will be identical at about 99.9% of your DNA sequences. The difference between you and the stranger is very small in *relative* terms, but because the human genome is so large (3.2 billion base pairs), you and the stranger will be different at more than 3 million base pairs of your genomic DNA. These differences are what make each of us unique, and they greatly affect our physical features, our health, and possibly even our intelligence and personality.

A site in the genome where individual members of a species differ in a single base pair is called a **single-nucleotide polymorphism** (SNP, pronounced "snip"). Arising through mutation, SNPs are inherited as allelic variants in the same way as alleles that produce phenotypic differences (such as blood types), although SNPs do not usually produce phenotypic differences. Single-nucleotide polymorphisms are numerous and are present throughout genomes. In a comparison of the same chromosome from two different people, a SNP can be found approximately every 1000 bp.

THINK-PAIR-SHARE Question 3

HAPLOTYPES AND LINKAGE DISEQUILIBRIUM Most SNPs present within a population arose once from a single mutation that occurred on a particular chromosome in a particular individual and subsequently spread throughout the population. Thus, each SNP is initially associated with other SNPs (as well as with other types of genetic variants or alleles) that were present on the particular chromosome on which the mutation arose. The specific set of SNPs and other genetic variants observed on a single chromosome or part of a chromosome is called a **haplotype** (**Figure 20.8**). SNPs within a haplotype are physically linked and therefore tend to be inherited together. New haplotypes can arise through mutation or crossing over, which breaks up the particular set of SNPs in a haplotype. Because the rate of crossing over is proportional to the physical distance between genes, SNPs and other genetic variants that are located close together on a chromosome will be strongly associated as haplotypes. The nonrandom association between genetic variants within a haplotype is called *linkage disequilibrium* (see Section 7.3). Because the SNPs in a haplotype are inherited together, a haplotype consisting of thousands of SNPs can be identified with the use of only a few SNPs. The few SNPs used to identify a haplotype are called **tag-SNPs**. There are about 10 million SNPs in the human population, but because of linkage disequilibrium, these SNPs can be grouped into a much smaller number of haplotypes. Therefore, a relatively small number of SNPs—perhaps only 100,000—can be used to identify most of the haplotypes in humans.

THE USE OF SINGLE-NUCLEOTIDE POLYMORPHISMS Because of their variability and widespread occurrence throughout the genome, SNPs are valuable as markers in linkage studies. When a SNP is physically close to a disease-causing locus, it will tend to be inherited along with the disease-causing allele. Therefore, people with the disease will tend to have different SNPs than healthy people. A comparison between SNP haplotypes in people with a disease and in healthy people can reveal the presence of genes that affect the disease; because the disease gene and the SNP are closely linked, the location of the disease-causing gene can be determined from the locations of associated SNPs.

An international effort, called the International HapMap Project, cataloged and mapped SNPs and other genetic variants that could be used to identify common haplotypes in human populations for use in linkage and family studies. The project cataloged a total of 4.6 million SNPs from human populations. These SNPs are spread over all 23 human chromosomes. The study found alleles from most of the common SNPs in four ethnic groups (African, Japanese, Chinese, and European), although the frequencies of the alleles vary considerably among human populations. The greatest genetic diversity of SNPs is found within Africans, which is consistent with many other studies that suggest that humans first evolved in Africa.

Data from the HapMap Project have been used to provide important information about the function and evolution of the human genome. For example, studies of linkage disequilibrium with the use of SNPs have determined that recombination does not take place randomly across the chromosomes: there are numerous recombination hotspots, where more recombination takes place than expected merely by chance. The distribution of SNPs is also providing information about regions of the human genome that have evolved quickly in the recent past, providing clues to how humans have responded to natural selection.

GENOME-WIDE ASSOCIATION STUDIES Many common diseases are caused by complex interactions among multiple genes. The availability of SNPs has greatly facilitated the search for these genes. **Genome-wide association studies** use numerous SNPs scattered across the genome to find genes of interest. Soon after the completion of the first phase of the HapMap Project, researchers used its SNP data to conduct a genome-wide association study of age-related macular degeneration, one of the leading causes of blindness among the elderly. In this study, researchers genotyped 96 people with macular degeneration and 50 unaffected people for more than 100,000 SNPs that blanketed the genome. The study revealed a strong association between the disease and a gene on chromosome 1 that encodes complement factor H, which has a role in immune function. This finding suggests that macular degeneration might be treatable with drugs that affect complement proteins. Another study uncovered a major gene associated with Crohn disease, a common inflammatory disorder of the gastrointestinal tract. Other studies using genome-wide scans of SNPs have identified genes that contribute to heart disease, bone density, prostate cancer, diabetes, and many more medical conditions.

20.8 A haplotype is a specific set of SNPs and other genetic variants observed on a single chromosome or part of a chromosome. Chromosomes 1a, 1b, 1c, and 1d represent different copies of a chromosome that might be found in a population.

In one successful application of SNPs for finding disease associations, researchers genotyped 17,000 people in the United Kingdom for 500,000 SNPs in 2007. They detected strong associations between 24 genes or chromosome segments and the incidence of seven common diseases, including coronary artery disease, Crohn disease, rheumatoid arthritis, bipolar disorder, hypertension, and two types of diabetes. The importance of this study is its demonstration that genome-wide association studies using SNPs could successfully locate genes that contribute to complex diseases caused by multiple genetic and environmental factors.

Within the past few years, SNPs have been used in genome-wide association studies to successfully locate genes that influence many additional traits, such as height, body mass index, the age of puberty and menopause in women, variation in facial features, skin pigmentation, eye color, glaucoma, and even susceptibility to infectious diseases such as meningococcal disease and tuberculosis. Unfortunately, the genes identified often explain only a modest proportion of the genetic influence on the trait. For example, one huge genome-wide association study combined data from over 100,000 people in an attempt to locate genes encoding blood lipids that are involved in cardiovascular disease. Although the study identified 95 different loci associated with lipid traits, these genes corresponded to only 25%–30% of the total genetic variation in these traits.

The missing DNA sequences that encode the majority of the genetic variation in such traits—sometimes called the "dark matter of the genome"—have, thus far, remained largely undetected. The low percentage of variation explained by most current genome-wide association studies means that the genes identified are not, by themselves, useful predictors of the risk of inheriting the disease or trait. Nevertheless, the identification of specific genes that influence a disease or trait can lead to a better understanding of the biological processes that produce the phenotype.

Copy-Number Variations

A diploid person normally possesses two copies of every gene, one inherited from the mother and one inherited from the father. Nevertheless, studies of the human genome have revealed differences among people in the number of copies of large DNA sequences (greater than 1000 bp); these variations are called **copy-number variations** (**CNVs**). Copy-number variations may include deletions, causing some people to have only a single copy of a sequence, or duplications, causing some people to have more than two copies. A study of CNVs in 270 people from four populations identified a surprising number of these types of variants: more than 1447 genomic regions varied in copy number, encompassing 12% of the human genome. But only a small part of the human genome was surveyed in this study, so the CNVs detected are only a small subset of all that exist. Many of the CNVs encompassed large regions of DNA, often several hundred thousand base pairs in length. Thus, people differ not only at millions of individual SNPs, but also in the number of copies of many larger segments of the genome.

Most CNVs contain multiple genes and potentially affect the phenotype by altering gene dosage and by changing the position of sequences, which may affect the regulation of nearby genes. Indeed, several studies have found associations between CNVs and disease, and even between CNVs and normal phenotypic variation in human populations. For example, variations in the number of copies of the *UGT2B17* gene contribute to differences in testosterone metabolism among individuals and affect the risk of prostate cancer. Copy-number variations have been associated with Crohn disease, rheumatoid arthritis, psoriasis, schizophrenia, autism, diabetes, and intellectual disability.

> **CONCEPTS**
>
> In addition to collecting sequence data, genome projects are collecting databases of nucleotides that vary among individual organisms (single-nucleotide polymorphisms, SNPs) and variations in the number of copies of sequences (copy-number variations).
>
> ✔ **CONCEPT CHECK 3**
> What did the HapMap Project accomplish?

Bioinformatics

Complete genome sequences have now been determined for numerous organisms, and many additional projects are under way. These studies are producing tremendous quantities of sequence data. Cataloging, storing, retrieving, and analyzing this huge data set are major challenges of modern genetics. **Bioinformatics** is an interdisciplinary field that combines molecular biology and computer science; it centers on the development of databases, computer-search algorithms, gene-prediction software, and other analytical tools that are used to make sense of DNA-, RNA-, and protein-sequence data. Bioinformatics develops and applies these tools to "mine the data"; that is, to extract the useful information from sequencing projects. The development and use of algorithms and computer software for analyzing sequence data have helped to make molecular biology a more quantitative field. The storage of sequence data in publicly available online databases enables scientists and students throughout the world to access this tremendous resource.

A number of databases have been established for the collection and analysis of DNA- and protein-sequence information. Primary databases contain the sequence information, along with information that describes the source of the sequence and its determination. Secondary databases contain the results of analyses carried out on the primary sequence data, such as information about particular sequence patterns, variations, mutations, and evolutionary relationships.

After a genome has been sequenced, one of the first tasks is to identify potential genes within the sequence.

Unfortunately, there are no universal characteristics that mark the beginning and end of a gene. The enormous amount of DNA in a genome and the complexities of gene structure make finding genes within a sequence a difficult task. Computer programs have been developed to look for specific sequences in DNA that are associated with certain genes. There are two general approaches to finding genes. The *ab initio* approach scans the sequence looking for features that are usually found within a gene. For example, protein-encoding genes are characterized by an open reading frame, which includes a start codon and a stop codon in the same reading frame. Specific sequences mark the splice sites at the beginnings and ends of introns; other specific sequences are present in promoters immediately upstream of start codons. The comparative approach looks for similarity between a new sequence and the sequences of all known genes. If a match is found, then the new sequence is assumed to be a similar gene. Some of these computer programs are capable of examining databases of protein sequences to see if there is evidence that a potential gene is expressed.

It is important to note that the programs that have been developed to identify genes on the basis of DNA sequence are not perfect. Therefore, the numbers of genes reported in most genome-sequencing projects are estimates. The presence of multiple introns, alternative splicing, multiple copies of some genes, and much noncoding DNA between genes makes accurate identification and counting of genes difficult.

After a gene has been identified, it must be **annotated**, which means linking its sequence information to other information about its function and expression, the protein it encodes, and similar genes in other species. There are a number of methods of probing a gene's function, which will be discussed in Section 20.2. Computer programs are available for determining whether similar sequences have already been found, either in the same species or in different species. The most widely used of these programs is the Basic Local Alignment Search Tool (BLAST). To conduct a BLAST search, a researcher submits a query sequence, and the program searches the database for any other sequences that have regions of high similarity to the query sequence. The program returns all sequences in the database that are similar, along with information about the degree of similarity and the significance of the match (how likely the similarity would be to occur by chance alone).

Additional databases contain information on sequence diversity: how and where a genome varies among individual organisms. The human genome has now been mapped for millions of SNPs, and information about the frequency of these variants in different populations is being collected. Additional databases have been developed to catalog all known mutations causing particular diseases; the Human Variome Project is an effort to collect and make available all genetic variations that affect human health. Important information about the expression patterns of the thousands of genes found in genomes is also being compiled. **▶ TRY PROBLEM 29**

CONCEPTS

Bioinformatics is an interdisciplinary field that combines molecular biology and computer science. It develops databases of DNA, RNA, and protein sequences and tools for analyzing those sequences.

✔ CONCEPT CHECK 4

The *ab initio* approach finds genes by looking for

a. common sequences found in most genes.

b. similarity in sequence with known genes.

c. mRNA with the use of in situ hybridization.

d. mutant phenotypes.

Metagenomics

Advances in sequencing technology, which have made sequencing faster and less expensive, now provide the possibility of sequencing not just the genomes of individual species, but the genomes of entire communities of organisms. **Metagenomics** is an emerging field in which the genome sequences of an entire group of organisms that inhabit a common environment are sampled and determined.

Thus far, metagenomics has been applied largely to microbial communities. It provides the ability to address two important issues: (1) the identification and study of microbes that cannot be cultured in the laboratory and (2) the study of the community structure of microorganisms.

Traditionally, bacteria have been studied by growing and analyzing them in the laboratory. However, many bacteria cannot be cultured with the use of laboratory techniques. Metagenomics analyzes microbial communities by extracting DNA from the environment, determining its sequences, and reconstructing community composition and function on the basis of those sequences. This technique allows the identification and genetic analysis of species that cannot be grown in the laboratory and have never been studied by traditional microbiological methods. The entire genomes of some dominant species have been reconstructed from environmental samples, providing scientists with a great deal of information on the biology of these microbes.

An early metagenomic study analyzed the microbial community found in acid drainage from a mine and determined that this community consisted of only a few dominant bacterial species. Another study, called the Global Ocean Sampling Expedition, followed the route of Darwin's voyage on H.M.S. *Beagle* in the 1800s. Scientists collected ocean samples and used metagenomic methods to determine the microbial communities they contained. In this study, scientists cataloged sequences for more than 6 million proteins, including more than 1700 new protein families. Some important results have already emerged from metagenomic studies. Analyses of bacterial genomes found in ocean samples led to the discovery of proteorhodopsin proteins, which are light-driven proton pumps. Subsequent research demonstrated that these proteins are found in diverse microbial groups and are a major source of energy flux in the world's oceans.

Other metagenomic studies have examined the genes of bacteria that inhabit the human intestinal tract. These bacteria, along with those that inhabit the skin and other parts of the human body, are termed the human **microbiome**. The microbiome of a typical person includes over 100 trillion cells—more than 10 times the number of human cells—and contains 100 times as many genes as the human genome. The National Microbiome Initiative, unveiled in 2016, is a project to advance scientific knowledge and understanding of microbiomes.

Research is demonstrating that the human microbiome plays an important role in human health. One study examined the gut microflora of obese and lean people. Two groups of bacteria are common in the human gut: Bacteroidetes and Firmicutes. Researchers discovered that obese people have relatively more Firmicutes than do lean people and that the proportion of Firmicutes decreases in obese people who lose weight on a low-calorie diet. These same results were observed in obese and lean mice. In an elegant experiment, researchers transferred bacteria from obese to lean mice. Lean mice that received bacteria from obese mice extracted more calories from their food and stored more fat, suggesting that gut microflora might play some role in obesity.

Ecological communities have traditionally been characterized on the basis of the species they contain. Metagenomics affords the possibility of characterizing communities on the basis of their component genes, an approach that has been termed gene-centric. A gene-centric approach leads to new questions: Are certain types of genes more common in some communities than others? Are some genes essential for energy flow and nutrient recycling within a community? Because of the larger sizes of their genomes, eukaryotic communities have not yet been the focus of these approaches, but many researchers predict that they will be in the future.

CONCEPTS

Metagenomic studies examine the genomes of communities of organisms that inhabit a common environment. This approach has been applied to microbial communities and allows the composition and genetic makeup of the community to be determined without cultivating and isolating individual species. Metagenomic studies are a source of important new insights into microbial communities.

Synthetic Biology

Our ability to sequence and study whole genomes, coupled with our increased understanding of the genetic information required for basic biological processes, now provides the possibility of creating—entirely from scratch—novel organisms that have never before existed. **Synthetic biology** is a new field that seeks to design organisms that might provide useful functions, such as microbes that provide clean energy or break down toxic wastes.

Synthetic biologists have already mixed and matched parts from different organisms to synthesize microbes. In 2002, geneticists re-created the poliovirus by joining together pieces of DNA that were synthesized in the laboratory. Even more impressively, in 2010, Daniel Gibson and his colleagues synthesized from scratch the complete 1.08 million-base-pair genome of the bacterium *Mycoplasma mycoides*. They started with a thousand pieces of DNA that were synthesized in the laboratory and joined them together in successively larger pieces until they had assembled a complete copy of the genome. Within their synthetic genome they included a set of DNA sequences that spelled out—in code—an email address, the names of the researchers who participated in the project, and several well-known quotations. Finally, the researchers transplanted the artificial genome into a cell of a different bacterial species, *M. capricolum*, whose original genome had been removed. The new cell then began expressing the traits specified by the synthetic genome. Synthetic biology has also been extended to eukaryotic cells, as we saw in the introduction to this chapter. In 2014, geneticists created a synthetic chromosome that successfully replaced a natural chromosome in yeast cells, and efforts are currently under way to engineer a completely synthetic genome in yeast.

These types of experiments have raised a number of concerns. The ability to tailor-make novel genomes and to mix and match parts from different organisms creates the potential for the synthesis of dangerous microbes that might create ecological havoc if they escaped from the laboratory or that could be used in biological warfare or bioterrorism. Ongoing discussions among geneticists, ethicists, security experts, and politicians are addressing these concerns and whether synthetic genomes can be safely made and used.

THINK-PAIR-SHARE Question 4

20.2 Functional Genomics Determines the Functions of Genes by Using Genomic Approaches

A genome sequence is, by itself, of limited use. Merely knowing the sequence of a genome would be like having a huge set of encyclopedias without being able to read: you could recognize the different letters, but the text would be meaningless. **Functional genomics** characterizes what sequences do—their function. The goals of functional genomics include the identification of all the RNA molecules transcribed from a genome, called the **transcriptome** of that genome, and all the proteins encoded by the genome, called the **proteome**. Functional genomics uses both bioinformatics and laboratory-based experimental approaches in its effort to define the functions of DNA sequences.

In Chapter 19, we considered several methods for identifying genes and assessing their functions, including in situ hybridization, experimental mutagenesis, and the use of transgenic animals and knockouts. Those methods can be applied to individual genes and can provide important information about the locations and functions of genetic information. In this section, we focus primarily on methods that rely on knowing the sequences of other genes and on methods that can be applied to large numbers of genes simultaneously.

Predicting Function from Sequence

The nucleotide sequence of a gene can be used to predict the amino acid sequence of the protein it encodes. The protein can then be synthesized or isolated and its properties studied to determine its function. However, this biochemical approach to understanding gene function is both time-consuming and expensive. A major goal of functional genomics has been to develop computational methods that allow gene function to be identified from DNA sequence alone, bypassing the laborious process of isolating and characterizing individual proteins.

HOMOLOGY SEARCHES One computational method (often the first employed) for determining gene function is to conduct a homology search, which relies on comparisons of DNA and protein sequences from the same species and from different species. Genes that are evolutionarily related, referred to as **homologous** genes, are likely to have similar sequences. Homologous genes found in different species that evolved from the same gene in a common ancestor are called **orthologs** (**Figure 20.9**). For example, both mouse and human genomes contain a gene that encodes the alpha subunit of hemoglobin; the mouse and human alpha-hemoglobin genes are said to be orthologs because both genes evolved from an alpha-hemoglobin gene in a common mammalian ancestor. Homologous genes in the same species (arising by duplication of a single gene in the evolutionary past) are called **paralogs** (see Figure 20.9). Within the human genome is a gene that encodes the alpha subunit of hemoglobin and another, homologous gene that encodes the beta subunit of hemoglobin. These paralogs arose because an ancestral gene underwent duplication and the resulting two genes diverged through evolutionary time, giving rise to the alpha- and beta-subunit genes (see Figure 26.16). Homologous genes (both orthologs and paralogs) often have the same or related functions, so, after a function has been assigned to a particular gene, it can provide a clue to the function of a homologous gene.

Databases containing genes and proteins found in a wide array of organisms are available for homology searches. Powerful computer programs, such as BLAST, have been developed for scanning these databases to look for particular sequences. Suppose a geneticist sequences a genome and locates a gene that encodes a protein of unknown function. A homology search conducted on databases containing the DNA or protein sequences of other organisms may identify one or more orthologous sequences. If a function is known for a protein encoded by one of those sequences, that could provide information about the function of the newly

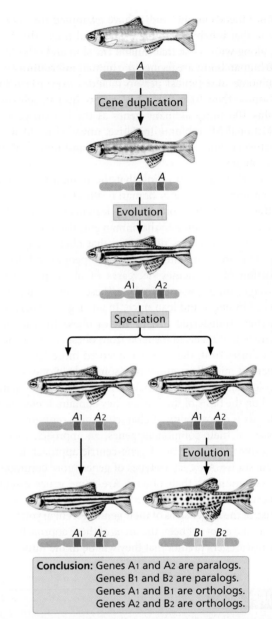

Conclusion: Genes A1 and A2 are paralogs.
Genes B1 and B2 are paralogs.
Genes A1 and B1 are orthologs.
Genes A2 and B2 are orthologs.

20.9 Homologous genes are evolutionarily related. Orthologs are homologous genes found in different species; paralogs are homologous genes in the same species.

discovered protein. Sometimes searches do not identify homologous genes; in that case, researchers may look for parts of a gene that are similar to parts of other genes with known function. **▶ TRY PROBLEM 33**

PROTEIN DOMAINS Complex proteins often contain regions, called **protein domains**, that have specific shapes or functions. For example, certain DNA-binding proteins attach to DNA in the same way; these proteins share a domain that provides the DNA-binding function. Some DNA-binding proteins have a domain called a zinc finger, consisting of a loop of amino acids containing a zinc ion (see Chapter 16). Each protein domain has an arrangement of amino acids common to that domain (see Figure 16.2). Protein domains have been relatively stable over evolutionary time; for example, protein domains found in humans are also found in other multicellular eukaryotes. There is probably a limited, though large, number of protein domains, which have mixed and matched through evolutionary time to yield the protein diversity seen in present-day organisms.

Many protein domains have been characterized and had their molecular functions determined. The sequence from a newly identified gene can be scanned against a database of known domains. If the gene sequence encodes one or more domains whose functions have been previously determined, the functions of those domains can provide important information about possible functions of the new gene.

CONCEPTS

The function of an unknown gene can sometimes be determined by finding genes with a similar sequence whose function is known. A gene's function may also be determined by identifying functional domains in the protein it encodes.

✔ CONCEPT CHECK 5

What is the difference between orthologs and paralogs?

a. Orthologs are homologous sequences; paralogs are analogous sequences.

b. Orthologs are more similar than paralogs.

c. Orthologs are in the same species; paralogs are in different species.

d. Orthologs are in different species; paralogs are in the same species.

Gene Expression

Often researchers are interested not just in a DNA sequence, but in how that sequence is transcribed or expressed. The expression of single genes can be studied by isolating individual RNA molecules with the use of probes or by in situ hybridization (see Section 19.4), but those methods are slow and labor-intensive when the expression profiles of many genes are required. Genomic methods provide the ability to examine the expression of thousands of genes. This approach—examining the expression of the genome—is termed **transcriptomics**.

MICROARRAYS One approach to transcriptomics uses **microarrays**, which rely on nucleic acid hybridization (see Chapter 19). In this procedure, known DNA fragments are used as probes to find complementary sequences (**Figure 20.10**). Numerous known DNA fragments are fixed to a solid support in an orderly pattern, or array, usually as a series of dots. These DNA fragments (the probes) usually correspond to known genes from a particular organism. An array containing tens of thousands of probes can be applied to a glass slide or silicon chip just a few square centimeters in size.

After the microarray has been constructed, mRNA, DNA, or cDNA isolated from experimental cells is labeled with fluorescent nucleotides and applied to the array. Any of the DNA or RNA molecules that are complementary to probes on the array will hybridize with them and emit fluorescence, which can be detected by an automated scanner.

THE USE OF MICROARRAYS Used with cDNA, microarrays can provide information about the expression of thousands of genes, enabling scientists to determine which genes are active in particular tissues. They can also be used to investigate how gene expression changes in the course of biological processes such as development or disease progression. For example, breast cancer affects 1 in 8 women in the United States, and half of those women die from it. Current treatment depends on a number of factors, including a woman's age, the size of the tumor, the characteristics of the tumor cells, and whether the cancer has already spread to nearby lymph nodes. Many women whose cancer has not spread are treated by removal of the tumor and radiation therapy, yet the cancer later reappears in some of the women thus treated. These women might benefit from more aggressive treatment when the cancer is first detected.

Using microarrays, researchers examined the expression patterns of 25,000 genes from primary tumors of 78 young women who had breast cancer (see Figure 20.10). Messenger RNA from the cancer cells and from noncancer cells was converted into cDNA and labeled with red fluorescent nucleotides and green fluorescent nucleotides, respectively. The labeled cDNAs were mixed and hybridized to a DNA chip, which contained DNA probes from numerous genes. Hybridization of the red (cancer) and green (noncancer) cDNAs was proportional to the relative amounts of mRNA in the samples. The fluorescence of each spot was assessed with microscopic scanning and appeared as a single color. Red indicated the overexpression of a gene in the cancer cells relative to its expression in the noncancer cells (more red-labeled cDNA hybridized), whereas green indicated the underexpression of a gene in the cancer cells relative to its expression in the noncancer cells (more green-labeled cDNA hybridized). Yellow indicated equal expression in both types of cells (equal hybridization of red- and green-labeled cDNAs), and no color indicated no expression in either type of cell.

In 34 of the 78 patients, the cancer later spread to other sites; the other 44 patients remained free of breast cancer for five years after their initial diagnoses. The researchers identified a subset of 70 genes whose expression patterns in the initial tumors accurately predicted whether the cancer would later spread (see Figure 20.10). This degree of prediction was much higher than that of traditional predictive measures, which are based on the size and appearance of the tumor.

Researchers have also used microarrays to examine the expression of microRNAs (miRNAs) in human cancers. Recent research indicates that miRNAs are frequently expressed abnormally in cancerous tissue and may contribute

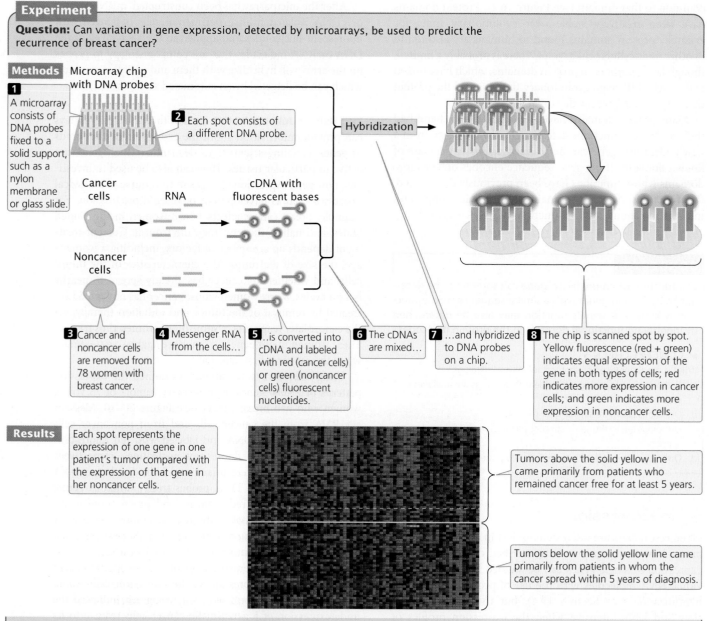

Experiment

Question: Can variation in gene expression, detected by microarrays, be used to predict the recurrence of breast cancer?

Methods

Microarray chip with DNA probes

1 A microarray consists of DNA probes fixed to a solid support, such as a nylon membrane or glass slide.

2 Each spot consists of a different DNA probe.

Hybridization →

Cancer cells

RNA

cDNA with fluorescent bases

Noncancer cells

3 Cancer and noncancer cells are removed from 78 women with breast cancer.

4 Messenger RNA from the cells...

5 ...is converted into cDNA and labeled with red (cancer cells) or green (noncancer cells) fluorescent nucleotides.

6 The cDNAs are mixed...

7 ...and hybridized to DNA probes on a chip.

8 The chip is scanned spot by spot. Yellow fluorescence (red + green) indicates equal expression of the gene in both types of cells; red indicates more expression in cancer cells; and green indicates more expression in noncancer cells.

Results

Each spot represents the expression of one gene in one patient's tumor compared with the expression of that gene in her noncancer cells.

Tumors above the solid yellow line came primarily from patients who remained cancer free for at least 5 years.

Tumors below the solid yellow line came primarily from patients in whom the cancer spread within 5 years of diagnosis.

Conclusion: Seventy genes were identified whose expression patterns accurately predicted the recurrence of breast cancer within 5 years of treatment.

20.10 Microarrays can be used to examine gene expression associated with disease progression. Each row in the microarray represents a tumor from one patient. [Expression data matrix reprinted by permission from Macmillan Publishers Ltd. Van'T Veer, Laura J., et al., "Gene expression profiling predicts clinical outcome of breast cancer," *Nature* 415:532. © 2002, permission conveyed through Copyright Clearance Center, Inc.]

to the progression of cancer (see Section 23.2). For example, one study using microarrays found that several miRNAs were overexpressed in cancerous cervical tissue compared with normal cervical tissue, while other miRNAs were underexpressed (**Figure 20.11**). Other studies using microarrays have demonstrated that miRNA expression is associated with resistance of tumors to chemotherapy and radiation, and that miRNA expression can be used to predict the responses of some tumors to cancer treatment. Results such as these

suggest that gene-expression data obtained from microarrays can be a powerful tool in cancer research and treatment. The products of the genes that show differences in expression are being examined as possible targets for drug therapy.

❯ TRY PROBLEM 34

Microarrays that allow the detection of specific alleles, SNPs, and even particular proteins have also been created. Importantly, not all DNA molecules bind equally to microarrays, and so microarrays may sometimes overestimate or

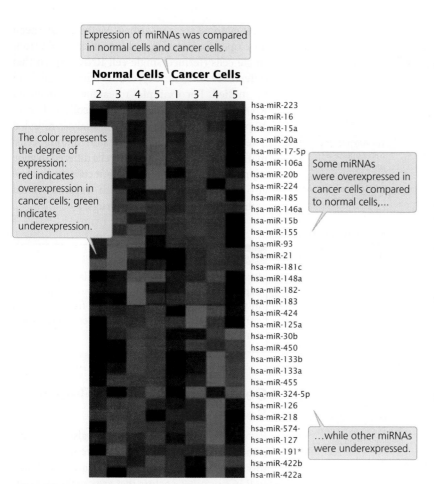

Expression of miRNAs was compared in normal cells and cancer cells.

The color represents the degree of expression: red indicates overexpression in cancer cells; green indicates underexpression.

Some miRNAs were overexpressed in cancer cells compared to normal cells,...

...while other miRNAs were underexpressed.

20.11 Microarrays have been used to compare the expression of miRNAs in cancerous cervical cells with that in normal cervical cells. [Data from Wang X., Tang S., Le S.-Y., Lu R., Rader J. S., et al. (2008) Aberrant Expression of Oncogenic and Tumor-Suppressive MicroRNAs in Cervical Cancer Is Required for Cancer Cell Growth. *PLoSONE* 3(7): e2557. doi:10.1371/journal.pone.0002557.]

underestimate the expression of specific genes. Thus, verification of the results of microarrays through other methods is desirable.

CONCEPTS

Microarrays, consisting of DNA probes attached to a solid support, can be used to determine which RNA and DNA sequences are present in a mixture of nucleic acids. They are capable of determining which RNA molecules are being synthesized and can thus be used to examine patterns of gene expression.

RNA SEQUENCING Microarrays, while providing a powerful means of simultaneously examining the expression of thousands of genes, have some significant limitations. First, one must have prior knowledge of gene sequences in order to construct the array, and sometimes this information does

not exist. Another limitation is that similar sequences will sometimes hybridize to the same probe on a microarray, creating cross-hybridization artifacts. A third problem is that microarrays provide a limited ability to quantify the degree of gene expression.

Another approach to the study of gene expression has been made possible by the development of rapid, low-cost next-generation sequencing methods. This approach determines the presence of RNA molecules in a cell by sequencing cDNAs copied from cellular RNA molecules. Termed **RNA sequencing**, or RNA-Seq, this approach provides detailed information about gene expression, including the types and number of RNA molecules produced by transcription, the presence of alternatively processed RNA molecules, differential expression of the two alleles in a diploid individual, and different RNA molecules generated by bidirectional or overlapping transcription of DNA sequences.

RNA sequencing can be carried out in several ways but usually involves at least five steps (**Figure 20.12**): (1) isolation of the RNA molecules of interest from cells; (2) conversion of the RNA to complementary DNA (cDNA) sequences; (3) fragmentation and preparation of the cDNAs for sequencing; (4) sequencing of the cDNA; and (5) assembly of the sequence reads into RNA transcripts.

First, the cells are lysed and the RNA chemically extracted (Figure 20.12a). This process yields the total RNA of the cell, which includes mRNA, pre-mRNA, tRNAs, rRNA, a variety of small RNAs such as microRNAs and siRNAs, and long noncoding RNAs. In most cells, the vast majority of RNA consists of rRNA, which often constitutes over 90% of the total cellular RNA. Usually, rRNA is not the focus of study; if not removed, rRNA will be the majority of the RNA molecules that are sequenced and will reduce the detection of other RNA molecules that are of more interest. The specific RNA of interest is then separated out for sequencing (Figure 12.12b). Once isolated, the RNA must be converted to cDNA (Figure 20.12c). This is done with the enzyme reverse transcriptase, which synthesizes a complementary DNA sequence from an RNA template.

After the RNA molecules are converted to cDNA, the cDNA molecules are prepared for sequencing (Figure 20.12d). They are broken into overlapping fragments of a size suitable for DNA sequencing, usually about 200 bp in length (sometimes the RNA is fragmented before conversion to cDNA). Adaptors, short DNA sequences necessary for amplifying and sequencing the DNA, are then added to the ends of the fragments. The cDNA is amplified using

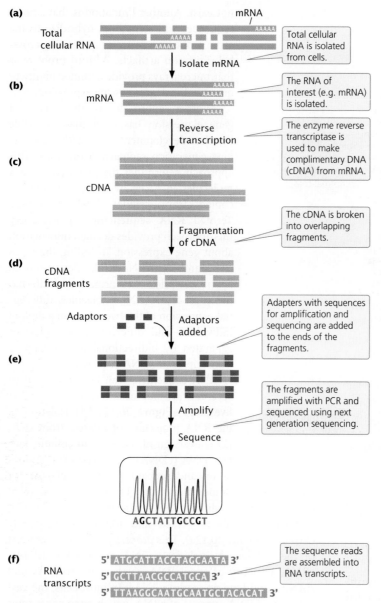

(a)

Total cellular RNA

mRNA

> Total cellular RNA is isolated from cells.

Isolate mRNA

(b)

mRNA

> The RNA of interest (e.g. mRNA) is isolated.

Reverse transcription

> The enzyme reverse transcriptase is used to make complimentary DNA (cDNA) from mRNA.

(c)

cDNA

Fragmentation of cDNA

> The cDNA is broken into overlapping fragments.

(d)

cDNA fragments

Adaptors

Adaptors added

> Adapters with sequences for amplification and sequencing are added to the ends of the fragments.

(e)

Amplify

Sequence

> The fragments are amplified with PCR and sequenced using next generation sequencing.

AGCTATTGCCGT

(f)

RNA transcripts

5' ATGCATTACCTAGCAATA 3'

5' GCTTAACGCCATGCA 3'

5' TTAAGGCAATGCAATGCTACACAT 3'

> The sequence reads are assembled into RNA transcripts.

20.12 RNA sequencing can be used to determine the expression of genes. Cellular RNA is isolated, converted to cDNA, and sequenced, providing information on the RNA transcripts present in a cell.

the polymerase chain reaction and sequenced using a next-generation sequencing platform (Figure 20.12e) such as an Illumina sequencer.

The final step is converting the readouts from the sequencing machine into full-length RNA transcripts (Figure 20.12f), which is accomplished by comparing overlap in the sequenced fragments. This process is often facilitated by mapping the sequence reads onto a reference genome sequence obtained by DNA sequencing.

RNA sequencing is now widely used to examine patterns of gene expression. An early application of this method was used to create a transcriptome map of yeast and revealed, surprisingly, that 75% of the nonrepetitive sequences in the

yeast genome are transcribed. Methods have been developed for isolating and sequencing RNA from single cells (termed single-cell RNA-Seq) so that differences in gene expression among cells can be studied. For example, researchers sequenced RNA from 4645 individual cells collected from 19 patients with melanoma. The analysis revealed heterogeneity in gene expression among cells: malignant and nonmalignant cells differed in gene expression. Within the malignant cells, there were also differences, some of which were associated with resistance to anticancer drugs.

CONCEPTS

RNA sequencing is used to determine the expression of genes throughout the genome. In this approach, RNA is isolated from cells and converted to cDNA, and the resulting cDNA fragments are sequenced.

Gene Expression and Reporter Sequences

Patterns of gene expression can also be determined visually by using a reporter sequence. In this approach, genomic fragments are first cloned in BACs or other vectors that are capable of holding the coding region of a gene plus its regulatory sequences. The coding region of a gene whose expression is to be studied is then replaced with a reporter sequence, which encodes an easily observed product. For example, a commonly used reporter sequence encodes a green fluorescent protein (GFP) from jellyfish. The cloned fragment is then inserted into an embryo, creating a transgenic organism. The regulatory sequences of the cloned gene ensure that it is expressed at the appropriate time and in the appropriate tissue within the transgenic organism, but the product of its expression is a green fluorescent pigment, which is easily observed (**Figure 20.13**).

This technique is being used to study the expression patterns of genes that affect brain function. In the Gene Expression Nervous System Atlas (GENSAT) project, scientists are systematically replacing the coding regions of hundreds of genes with the GFP reporter sequence and observing their patterns of expression in transgenic mice. The goal is to produce a comprehensive atlas of gene expression in the mouse brain. This project has already shed light on where in the brain several genes that play roles in inherited neurological disorders are expressed. Another functional genomics project, the Allen Brain Atlas, has compiled information on the expression patterns of 20,000 genes in the mouse brain.

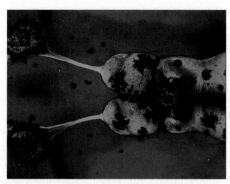

20.13 A reporter sequence can be used to examine expression of a gene. Expression of the neural-specific β-tubulin gene in the brain of a tadpole is revealed by green fluorescent protein. [Photo by Miranda Gomperts, Courtesy Enrique Amaya, The Amaya Lab.]

Genome-Wide Mutagenesis

As discussed in Chapter 18, one of the best methods for determining the function of a gene is to examine the phenotypes of individual organisms that possess a mutation in the gene. Traditionally, genes encoding naturally occurring variations in a phenotype were mapped, the causative genes were isolated, and their products were studied. But this procedure was limited by the number of naturally occurring mutations and the difficulty of mapping genes with a limited number of chromosomal markers. The number of naturally occurring mutations can be increased by exposing organisms to mutagens, and the accuracy of mapping is increased dramatically by the availability of mapped molecular markers, such as SNPs. These two methods—random inducement of mutations on a genome-wide basis and mapping with molecular markers—are coupled and automated in a **mutagenesis screen**.

Genome-wide mutagenesis screens can be used to search for all genes affecting a particular function or trait. For example, mutagenesis screens of mice are being used to identify genes that play roles in cardiovascular function. When such genes are located in mice, homology searches are carried out to determine whether similar genes exist in humans. Those genes can then be studied to better understand cardiac disease in humans.

To conduct a mutagenesis screen, random mutations are induced in a population of organisms, creating new phenotypes. Mutations can be induced by exposing the organisms to radiation, chemical mutagens, or transposable elements (DNA sequences that insert randomly into the DNA) (see Chapter 18). The procedure for a typical mutagenesis screen is illustrated in **Figure 20.14**. Here, male zebrafish are treated with ethylmethylsulfonate (EMS), a chemical that induces germ-line mutations in their sperm. The treated males are then mated with wild-type female fish. A few of the offspring will be heterozygous for mutations induced by EMS. The offspring are screened for any mutant phenotypes that might be the products of dominant mutations expressed in these heterozygous fish. Recessive mutations are not expressed in

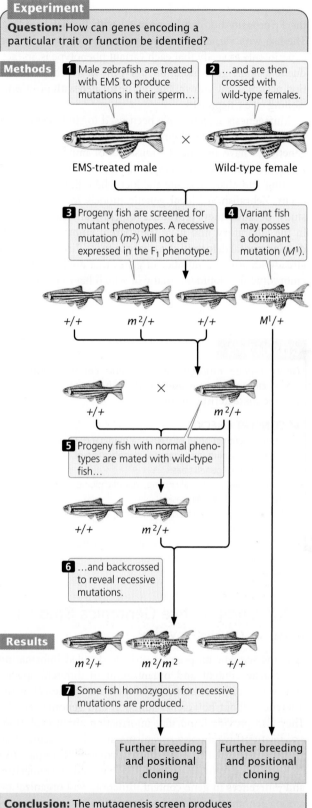

Experiment

Question: How can genes encoding a particular trait or function be identified?

Methods

1 Male zebrafish are treated with EMS to produce mutations in their sperm...

2 ...and are then crossed with wild-type females.

EMS-treated male   ×   Wild-type female

3 Progeny fish are screened for mutant phenotypes. A recessive mutation (m^2) will not be expressed in the F_1 phenotype.

4 Variant fish may posses a dominant mutation (M^1).

$+/+$ $m^2/+$ $+/+$ $M^1/+$

$+/+$ × $m^2/+$

5 Progeny fish with normal phenotypes are mated with wild-type fish...

$+/+$ $m^2/+$

6 ...and backcrossed to reveal recessive mutations.

Results

$m^2/+$ m^2/m^2 $+/+$

7 Some fish homozygous for recessive mutations are produced.

Further breeding and positional cloning

Further breeding and positional cloning

Conclusion: The mutagenesis screen produces fish with a mutation affecting the trait. These fish are further analyzed to identify the mutation.

20.14 Genes affecting a particular characteristic or function can be identified by a genome-wide mutagenesis screen. In this illustration, M^1 represents a dominant mutation and m^2 represents a recessive mutation.

the F_1 progeny but can be revealed with further breeding. The fish with variant phenotypes undergo further breeding experiments to verify that each variant phenotype is, in fact, due to a single-gene mutation. The gene in which the mutation occurs is not known, but can be located with positional cloning (see Chapter 19).

Mutagenesis screens have been used to study genes that control vertebrate development. A team of developmental geneticists have produced thousands of mutations in the zebrafish that affect development and are systematically locating and characterizing the loci where these mutations occur. Zebrafish are ideal genetic models for this type of study because they reproduce quickly, are easily reared in the laboratory, and have transparent embryos in which developmental deformities are easy to spot. This research has already identified a number of genes that are important in embryonic development, many of which have counterparts in humans.

CONCEPTS

Genome-wide mutagenesis screening coupled with positional cloning can be used to identify genes that affect a specific characteristic or function.

✔ **CONCEPT CHECK 6**

Which is the correct order of steps in a mutagenesis screen?
a. Positional cloning, mutagenesis, identification of mutants, verification of genetic basis.
b. Mutagenesis, positional cloning, identification of mutants, verification of genetic basis.
c. Mutagenesis, identification of mutants, verification of genetic basis, positional cloning.
d. Identification of mutants, positional cloning, mutagenesis, verification of genetic basis.

20.3 Comparative Genomics Studies How Genomes Evolve

Genome-sequencing projects provide detailed information about gene content and organization in different species and even in different members of the same species, allowing inferences about how genes function and genomes evolve. They also provide important information about evolutionary relationships among organisms and about factors that influence the speed and direction of evolution. **Comparative genomics** is the field of genomics that studies similarities and differences in gene content, function, and organization among genomes of different organisms.

Prokaryotic Genomes

Thousands of prokaryotic genomes have now been sequenced. Most prokaryotic genomes consist of a single circular chromosome. However, there are exceptions, such as *Vibrio cholerae*, the bacterium that causes cholera, which has two circular chromosomes, and *Borrelia burgdorferi*, which has one large linear chromosome and 21 smaller chromosomes.

GENOME SIZE AND NUMBER OF GENES The total amount of DNA in prokaryotic genomes ranges from 490,885 bp in *Nanoarchaeum equitans*, an archaean that lives entirely within another archaean, to 9,105,828 bp in *Bradyrhizobium japonicum*, a soil bacterium (**Table 20.1**). Although this range in genome size might seem extensive, it is much less than the enormous range of genome sizes seen in eukaryotes, which can vary from a few million base pairs to hundreds of billions of base pairs. *Escherichia coli*, the bacterium most widely used for genetic studies, has a fairly typical genome size for a prokaryote at 4.6 million base pairs. Archaea and bacteria are similar in their ranges of genome size. Surprisingly, genome size also varies extensively within some species; for example, different strains of *E. coli* vary in genome size by more than 1 million base pairs.

Among prokaryotes, the number of genes typically varies from 1000 to 2000, but some species have as many as 8300 and others as few as 480. Interestingly, the density of genes is rather constant across all species, with an average of about one gene per 1000 bp. Thus, prokaryotes with larger genomes have more genes, in contrast to eukaryotes, in which there is little association between genome size and number of genes, as we'll see shortly. The evolutionary factors that determine the sizes of prokaryotic genomes (as well as eukaryotic genomes) are still largely unknown. However, the time required for cell division is frequently longer in organisms with larger genomes because of the time required to replicate the DNA before division. Thus, selection may favor smaller genomes in organisms that occupy environments where rapid reproduction is advantageous.

Prokaryotes with the smallest genomes tend to be species that occupy restricted habitats, such as bacteria that live inside other organisms. The constant environment and the metabolic functions supplied by the host organism may allow these bacteria to survive with fewer genes. Bacteria with large genomes tend to occupy highly complex and variable environments, such as the soil or root nodules of plants. In these environments, genes that are needed only occasionally may be useful. In complex environments where resources are abundant, there may be little need for rapid division (which favors smaller genomes).

An example of a large-genome bacterium occupying a complex environment is *Streptomyces coeliocolor*, a filamentous bacterium with a genome size of 8,677,507 bp. This species has been called the "Boy Scout" bacterium because it has a diverse set of genes and therefore follows the Scout motto: Be prepared. In addition to the usual housekeeping genes for genetic functions such as replication, transcription, and translation, its genome contains a number of genes for the breakdown of complex carbohydrates, allowing it

TABLE 20.1	Characteristics of representative prokaryotic genomes that have been completely sequenced	
Species	Size (Millions of Base Pairs)	Number of Predicted Genes
Archaea		
Archaeoglobus fulgidus	2.18	2407
Methanobacterium thermoautotrophicum	1.75	1869
Nanoarchaeum equitans	0.490	536
Eubacteria		
Bacillus subtilis	4.21	4100
Bradyrhizobium japonicum	9.11	8317
Escherichia coli	4.64	4289
Haemophilus influenzae	1.83	1709
Mycobacterium tuberculosis	4.41	3918
Mycoplasma genitalium	0.58	480
Staphylococcus aureus	2.88	2697
Vibrio cholerae	4.03	3828

Source: Data from the Genome Atlas of the Center for Biological Sequence Analysis, www.cbs.dtu.dk/services/GenomeAtlas.

to consume decaying matter from plants, animals, insects, fungi, and other bacteria. It has genes for a large number of proteins that produce secondary metabolites (breakdown products) that can function as antibiotics and protect against desiccation and low temperatures. A large genome with lots of genes appears to be beneficial in the complex environment that this species inhabits.

HORIZONTAL GENE TRANSFER Bacteria possess a number of mechanisms by which they can gain and lose DNA. DNA can be lost through simple deletion and can be gained by gene duplication and through the insertion of transposable elements. Another mechanism for gaining new genetic information is horizontal gene transfer, a process by which both closely and distantly related bacterial species periodically exchange genetic information over evolutionary time. Such exchanges may take place through the bacterial uptake of DNA from the environment (transformation), through the exchange of plasmids, and through viral vectors (transduction). Horizontal gene transfer has been recognized for some time, but analyses of many microbial genomes now indicate that it is more extensive than was formerly recognized. The widespread occurrence of horizontal gene transfer has caused some biologists to question whether distinct species even exist in bacteria (see Section 9.3 for more discussion of this matter).

Bacteria may even transfer genetic information to human genomes. When the human genome was first sequenced, researchers reported evidence of bacterial DNA in its sequence, but that conclusion was later questioned. More recently, researchers have reexamined this issue and have found that the human genome contains as many as 145 genes

from bacteria, archaea, viruses, fungi, and other organisms. The amount of bacterial DNA is much higher in some types of cancer cells, suggesting either that cancer cells are more susceptible to horizontal gene transfer or that the insertion of bacterial DNA contributes to the progression of cancer.

CONCEPTS

Comparative genomics compares the content and organization of whole genomes of different organisms. Prokaryotic genomes are small, usually ranging from 1 million to 4 million base pairs of DNA and containing 1000 to 2000 genes. Prokaryotes with the smallest genomes tend to occupy restricted habitats, whereas those with the largest genomes are usually found in complex environments. Horizontal gene transfer has played a major role in bacterial genome evolution.

✔ CONCEPT CHECK 7

What is the relation between genome size and gene number in prokaryotes?

Eukaryotic Genomes

The genomes of many eukaryotic organisms have been completely sequenced, including those of numerous fungi and protozoans, cucumbers, papayas, corn, rice, sorghum, grapevines, strawberries, silkworms, a number of fruit-fly species, aphids, mosquitoes, anemones, tunicates, ctenophores, mice, rats, dogs, pigs, sheep, cows, horses, minke whales, over 200 species of birds, orangutans, gorillas, chimpanzees, humans, and many more species. Even the genomes of some extinct organisms have now been sequenced, including those of woolly mammoths and Neanderthals. It is important to note, however, that even though the genomes of these organisms have been "completely sequenced," many of the final assembled sequences contain gaps, and regions of heterochromatin may not have been sequenced at all. Thus, the sizes of eukaryotic genomes are often estimates, and the number of base pairs given as the genome size for a particular species may vary. Predicting the number of genes that are present in a eukaryotic genome is also difficult, and estimates may vary depending on the assumptions made and the particular gene-finding software used.

GENOME SIZE AND NUMBER OF GENES The genomes of eukaryotic organisms (**Table 20.2**) are larger than those of prokaryotes, and in general, multicellular eukaryotes have more DNA than do simple, single-celled eukaryotes such as yeast (see p. 321 in Chapter 11). However, there is no close relation between genome size and complexity among the multicellular eukaryotes. For example, the mosquito *Anopheles gambiae* and the fruit fly *Drosophila melanogaster* are both insects with similar structural complexity, yet the mosquito has 60% more DNA than the fruit fly.

In general, eukaryotic genomes also contain more genes than do prokaryotic genomes (although some large bacteria have more genes than single-celled yeast), and the genomes of multicellular eukaryotes have more genes than do the genomes of single-celled eukaryotes. In contrast to prokaryotes, there is no correlation between genome size and number of genes

in eukaryotes. Nor is the number of genes among multicellular eukaryotes obviously related to phenotypic complexity: humans have more genes than do invertebrates, but only twice as many as fruit flies, and fewer than the plant *A. thaliana*. The nematode *C. elegans* has more genes than *D. melanogaster*, but is less complex. The pufferfish has only about one-tenth the amount of DNA present in humans and mice, but about as many genes. Eukaryotic genomes contain multiple copies of many genes, indicating that gene duplication has been an important process in genome evolution.

Most of the DNA in multicellular eukaryotes is noncoding, and many genes are interrupted by introns. In the more complex eukaryotes, both the number and the length of the introns are greater.

SEGMENTAL DUPLICATIONS AND MULTIGENE FAMILIES Many eukaryotic genomes, especially those of multicellular organisms, are filled with **segmental duplications**, duplicated regions greater than 1000 bp that are almost identical in sequence. For example, about 4% of the human genome consists of segmental duplications. In most segmental duplications, the two copies are found on the same chromosome (intrachromosomal duplications), but in others, the two copies are found on different chromosomes (interchromosomal duplications). In the human genome, the average size of segmental duplications is 15,000 bp.

Segmental duplications arise from processes that generate chromosome duplications, such as unequal crossing over (see Section 8.2). After a segmental duplication arises,

TABLE 20.2	Characteristics of representative eukaryotic genomes that have been completely sequenced	
Species	**Genome Size (Millions of Base Pairs)**	**Number of Predicted Genes**
Saccharomyces cerevisiae (yeast)	12	6,144
Physcomitrella patens (moss)	480	38,354
Arabidopsis thaliana (thale-cress plant)	125	25,706
Zea mays (corn)	2,400	32,000
Hordeum vulgare (barley)	5,100	26,159
Caenorhabditis elegans (nematode)	103	20,598
Drosophila melanogaster (fruit fly)	170	13,525
Anopheles gambiae (mosquito)	278	14,707
Danio rerio (zebrafish)	1,465	22,409
Takifugu rubripes (tiger pufferfish)	329	22,089
Xenopus tropicalis (clawed frog)	1,510	18,429
Anolis carolinensis (anole lizard)	1,780	17,792
Mus musculus (mouse)	2,627	26,762
Pan troglodytes (chimpanzee)	2,733	22,524
Homo sapiens (human)	3,223	~20,000

Source: Data from the Ensembl Web site: http://useast.ensembl.org/index.html and plants.ensembl.org/index.html.

it promotes further duplication by causing misalignment among the duplicated regions. Segmental duplication plays an important role in evolution by giving rise to new genes. After a segmental duplication arises, the original copy of the segment can continue its function while the new copy undergoes mutation. These changes may eventually lead to a new function. The importance of gene duplication in genome evolution is demonstrated by the large number of multigene families that exist in many eukaryotic genomes. A **multigene family** is a group of evolutionarily related genes that arose through repeated duplication and evolution of an ancestral gene. For example, the globin gene family in humans consists of 13 genes that encode globinlike molecules, most of which produce proteins that carry oxygen. An even more spectacular example is the human olfactory multigene family, which consists of about 1000 genes that encode olfactory receptor molecules used in our sense of smell.

NONCODING DNA Most eukaryotic organisms contain vast amounts of DNA that does not encode proteins. For example, only about 1.5% of the human genome consists of DNA that directly specifies the amino acids of proteins. The function of the remaining DNA sequences, called noncoding DNA, has long been in question. Some research has suggested that much of this DNA has no function. For example, Marcelo Nóbrega and his colleagues genetically engineered mice that were missing large chromosomal regions with no protein-encoding genes (called **gene deserts**). In one experiment, they created mice that were missing a 1,500,000-bp gene desert from mouse chromosome 3; in another, they created mice missing an 845,000-bp gene desert from chromosome 19. Remarkably, these mice appeared healthy and were indistinguishable from control mice. The researchers concluded that large regions of the mammalian genome can be deleted without major phenotypic effects and may, in fact, be superfluous.

Other research, however, has suggested that gene deserts may contain sequences that have a functional role. For example, genome-wide association studies demonstrated that DNA sequences contained within a gene desert on human chromosome 9 are associated with coronary artery disease, and subsequent studies have demonstrated the presence of 33 enhancers in this gene desert.

In 2002, the Encyclopedia of DNA Elements (ENCODE) project was undertaken to determine whether noncoding DNA has any function. Researchers cataloged all nucleotides within the genome that provide some function, including sequences that encode proteins and RNA molecules and those that serve as control sites for gene expression. This 10-year project was carried out by a team of over 400 scientists from around the world. In a series of papers published in 2012, the ENCODE project concluded that at least 80% of the human genome is involved in some type of gene function. Many of the functional sequences consisted of sites where proteins bind and influence the expression of genes. However, the ENCODE conclusion that 80% of the genome

TABLE 20.3	Percentages of eukaryotic genomes consisting of interspersed repeats derived from transposable elements
Organism	**Percentage of Genome**
Arabidopsis thaliana (thale-cress plant)	10.5
Zea mays (corn)	85.0
Caenorhabditis elegans (nematode)	6.5
Drosophila melanogaster (fruit fly)	3.1
Takifugu rubripes (tiger pufferfish)	2.7
Homo sapiens (human)	44.4

is functional has been challenged by other researchers, and some of ENCODE's findings remain controversial.

TRANSPOSABLE ELEMENTS A substantial part of the genomes of most multicellular organisms consists of moderately and highly repetitive sequences (see Section 11.3), and the percentage of repetitive sequences is usually higher in those species with larger genomes (**Table 20.3**). Most of these repetitive sequences appear to have arisen through transposition. In the human genome, 45% of the DNA is derived from transposable elements, many of which are defective and no longer able to move. In corn, 85% of the genome is derived from transposable elements.

PROTEIN DIVERSITY In spite of only a modest increase in gene number, vertebrates have considerably more protein diversity than do invertebrates. One way to measure protein diversity is by counting the number of protein domains, which are characteristic parts of proteins that are often associated with a function (see Section 20.2). Vertebrate genomes do not encode significantly more protein domains than do invertebrate genomes; for example, there are 1262 domains in humans compared with 1035 in fruit flies (**Table 20.4**). However, the existing domains in humans are assembled into more combinations, leading to many more types of proteins.

TABLE 20.4	Number of estimated protein domains encoded by some eukaryotic genomes
Species	**Number of Predicted Protein Domains**
Saccharomyces cerevisiae (yeast)	851
Arabidopsis thaliana (thale-cress plant)	1012
Caenorhabditis elegans (nematode)	1014
Drosophila melanogaster (fruit fly)	1035
Homo sapiens (human)	1262

Source: Number of genes and protein-domain families from the International Human Genome Sequencing Consortium, Initial sequencing and analysis of the human genome, *Nature* 409:860–921, Table 23, 2001.

HOMOLOGOUS GENES An obvious and remarkable trend seen in eukaryotic genomes is the degree of homology among genes found in even distantly related species. For example, mice and humans share about 99% of their genes. About 50% of the genes in fruit flies are homologous to genes in humans, and even in plants, about 18% of the genes are homologous to those found in humans.

COLLINEARITY BETWEEN RELATED GENOMES One of the features of genome evolution revealed by comparing the genome sequences of different organisms is that many genes are present in the same order in related genomes, a phenomenon that is sometimes termed collinearity. The reason for collinearity among genomes is that they are descended from a common ancestral genome, and evolutionary forces have maintained the same gene order in the genomes of descendants. Genomic studies of grasses—plants in the family Poaceae—illustrate the principle of collinearity.

Grasses comprise more than 10,000 species, including economically important crops such as rice, wheat, barley, corn, millet, oats, and sorghum. Taken together, grasses make up about 60% of the world's food production. The genomes of these species vary greatly in chromosome number and size. For example, chromosome number in grasses ranges from 4 to 266; the rice genome consists of only about 460 million base pairs, whereas the genome of wheat contains 17 billion base pairs. In spite of these large differences in chromosome number and genome size, the positions and order of many genes within the genomes are remarkably conserved. Over evolutionary time, regions of DNA between the genes (intergenic regions) have increased, decreased, and undergone rearrangements, whereas the genes themselves have stayed relatively constant in order and content. An example of the collinear relationships among genes in rice, sorghum, and *Brachypodium* (wild grass) is shown in **Figure 20.15.** ❯ **TRY PROBLEM 38**

THINK-PAIR-SHARE Question 5

CONCEPTS

Genome size varies greatly among eukaryotic species. For multicellular eukaryotic organisms, there is no clear relation between organismal complexity and amount of DNA or number of genes. A substantial part of the genome in eukaryotic organisms consists of repetitive DNA, much of which is derived from transposable elements. Many eukaryotic genomes share homologous genes, and genes are often in the same order in the genomes of related organisms.

✔ **CONCEPT CHECK 8**

Segmental duplications play an important role in evolution by
a. giving rise to new genes and multigene families.
b. keeping the number of genes in a genome constant.
c. eliminating repetitive sequences produced by transposition.
d. controlling the base content of the genome.

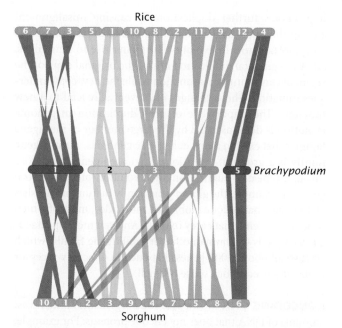

20.15 Collinear relationships among blocks of genes found in rice, sorghum, and *Brachypodium* (wild grass). Each colored band represents a block of genes that is collinear between chromosomes of the three species.

The Human Genome

The human genome, which is fairly typical of mammalian genomes, has been extensively studied and analyzed because of its importance to human health and evolution. It is 3.2 billion base pairs in length, but only about 1.5% of it encodes proteins. Active genes are often separated by vast regions of noncoding DNA, much of which consists of repeated sequences derived from transposable elements.

The average gene in the human genome is approximately 27,000 bp in length, with about 9 exons (**Table 20.5**). (One exceptional gene has 234 exons.) The introns of human genes are much longer, and there are more of them, than in other genomes (**Figure 20.16**). The human genome does not encode substantially more protein domains (see Table 20.4), but its domains are combined in more ways to produce a

TABLE 20.5	Average characteristics of genes in the human genome
Characteristic	**Average**
Number of exons	8.8
Size of internal exon	145 bp
Size of intron	3,365 bp
Size of 5′ untranslated region	300 bp
Size of 3′ untranslated region	770 bp
Size of coding region	1,340 bp
Total length of gene	27,000 bp

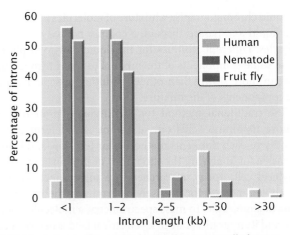

20.16 The introns of genes in humans are generally longer than the introns of genes in nematodes and fruit flies.

relatively diverse proteome. Gene functions encoded by the human genome are presented in **Figure 20.17**. A single gene often encodes multiple proteins through alternative splicing; each gene encodes, on average, two or three different mRNAs.

Gene density varies among human chromosomes; chromosomes 17, 19, and 22 have the highest densities and chromosomes X, Y, 4, 13, and 18 have the lowest densities. Some proteins encoded by the human genome that are not found in other animals include those affecting immune function; neural development, structure, and function; intercellular and intracellular signaling pathways in development; hemostasis; and apoptosis.

Transposable elements are much more common in the human genome than in nematode, plant, and fruit-fly genomes (see Table 20.3). The human genome contains a variety of types of transposable elements, including LINEs, SINEs, retrotransposons, and DNA transposons (see Section 18.4). Most appear to be evolutionarily old and are defective, containing mutations and deletions making them no longer capable of transposition.

20.4 Proteomics Analyzes the Complete Set of Proteins Found in a Cell

DNA sequence data offer tremendous insight into the biology of an organism, but they are not the whole story. Many genes encode proteins, and proteins carry out the vast majority of the biochemical reactions that shape the phenotype of an organism. Although proteins are encoded by DNA sequences, many proteins undergo modifications after translation, and in more complex eukaryotes there are many more proteins than genes. Thus, in recent years, molecular biologists have turned their attention to analysis of the protein content of cells. Their ultimate goal is to determine the proteome: the complete set of proteins found in a given cell. The study of the proteome is termed **proteomics**.

Plans are under way to identify and characterize all proteins in the human body, an effort that has been called the **Human Proteome Project**. The project would catalog the proteins present in each cell type, the location of each protein within the cell, and which other proteins each interacts with. Many researchers feel that this information will be of immense benefit in identifying drug targets, understanding the biological basis of disease, and understanding the molecular basis of many biological processes.

Determination of Cellular Proteins

The basic procedure for characterizing the proteome is first to separate the proteins found in a cell and then to identify and quantify the individual proteins. One method for separating proteins is **two-dimensional polyacrylamide gel electrophoresis (2D-PAGE)**, in which the proteins are

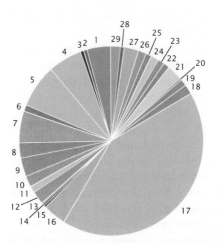

1. Miscellaneous
2. Viral protein
3. Transfer or carrier protein
4. Transcription factor
5. Nucleic acid enzyme
6. Signaling molecule
7. Receptor
8. Kinase
9. Select regulatory molecule
10. Transferase

11. Synthase and synthetase
12. Oxidoreductase
13. Lyase
14. Ligase
15. Isomerase
16. Hydrolase
17. Molecular function unknown
18. Transporter
19. Intracellular transporter
20. Select calcium-binding protein

21. Proto-oncogene
22. Structural protein of muscle
23. Motor
24. Ion channel
25. Immunoglobulin
26. Extracellular matrix
27. Cytoskeletal structural protein
28. Chaperone
29. Cell adhesion

20.17 Functions for many human genes have yet to be determined. Proportion of the circle occupied by each color represents the proportion of genes affecting various known and unknown functions.

separated in one dimension by charge, separated in a second dimension by mass, and then stained (**Figure 20.18a**). This procedure separates the different proteins into spots, with the size of each spot proportional to the amount of protein present. A typical 2D-PAGE gel may contain several hundred to several thousand spots (**Figure 20.18b**).

Because 2D-PAGE does not detect some proteins in low abundance and is difficult to automate, researchers have turned to liquid chromatography for separating proteins.

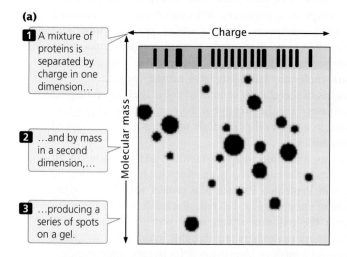

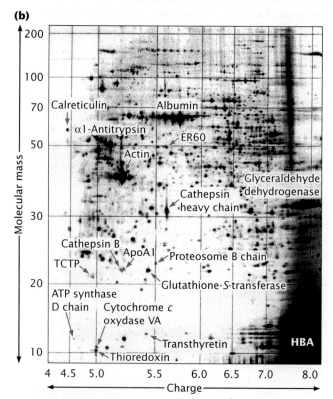

20.18 Two-dimensional acrylamide gel electrophoresis (2D-PAGE) can be used to separate cellular proteins.
[After G. Gibson and S. Muse. 2004. *A Primer of Genome Science*, 2e. Sinauer Associates, Inc. p. 274, Fig. 5.4.]

In liquid chromatography, a mixture of molecules is dissolved in a liquid and passed through a column packed with solid particles. Different affinities for the liquid and solid phases cause some components of the mixture to travel through the column more slowly than others, resulting in separation of the components of the mixture.

The traditional method for identifying a protein is to remove its amino acids one at a time and determine the identity of each one. This method is far too slow and labor-intensive for analyzing the thousands of proteins present in a typical cell. Today, researchers use **mass spectrometry**, which is a method for precisely determining the molecular mass of a molecule. In mass spectrometry, a molecule is ionized and its migration rate in an electrical field is determined. Because small molecules migrate more rapidly than larger molecules, the migration rate can accurately determine the mass of the molecule.

To analyze a protein with mass spectrometry, researchers first digest it with the enzyme trypsin, which cleaves the protein into smaller peptide fragments, each containing several amino acids (**Figure 20.19a**). Mass spectrometry is then used to separate the peptides on the basis of their mass-to-charge ratio (**Figure 20.19b**). This separation produces a profile of peaks, in which each peak corresponds to the mass-to-charge ratio of one peptide (**Figure 20.19c**). A computer program then searches through a database of proteins to find a match between the profile generated and the profile of a known protein (**Figure 20.19d**), allowing the protein in the sample to be identified. Using bioinformatics, the computer creates "virtual digests" and predicts the profiles of all the proteins found in a genome, given the DNA sequences of the protein-encoding genes.

Mass spectrometric methods can also be used to measure the amount of each protein identified. With recent advances, researchers can now carry out "shotgun" proteomics, which eliminates most of the initial protein-separation stage. In this procedure, a complex mixture of proteins (such as those from a tissue sample) is digested and analyzed with mass spectrometry. The computer program then sorts out the proteins present in the sample by analyzing the peptide profiles.

Mary Lipton and her colleagues used this approach to study the proteome of *Deinococcus radiodurans*, an exceptional bacterium that is able to withstand high doses of ionizing radiation that are lethal to all other organisms. The genome of *D. radiodurans* had already been sequenced. Lipton and her colleagues extracted proteins from the bacteria, digested them with trypsin into small peptide fragments, separated the fragments with liquid chromatography, and then used mass spectrometry to determine the proteins from the peptide fragments. They were able to identify 1910 proteins, which represented more than 60% of the proteins predicted on the basis of the genome sequence.

Deciphering the proteome of even a single cell is a challenging task. Every cell contains a complete sequence of genes, but different cells express vastly different proteins. Each gene may produce a number of different proteins through alternative processing (see Chapter 14) and post-translational protein processing (see Chapter 15). A typical human cell contains as many as 100,000 different proteins that vary greatly in abundance, and no technique such as PCR can be used to easily amplify proteins.

Affinity Capture

Proteomics concerns not just the identification of all proteins in a cell, but also an understanding of how these proteins interact and how their expression varies with the passage of time. Researchers have developed a number of techniques for identifying proteins that interact within the cell. In **affinity capture**, an antibody (see Figure 22.22) to a specific protein is used to capture one protein from a complex mixture of proteins. The protein captured will "pull down" with it any proteins with which it physically interacts. The pulled-down mixture of proteins can then be analyzed by mass spectrometry to identify the proteins. Various modifications of affinity capture and other techniques can be used to determine the complete set of protein interactions in a cell, termed the **interactome**.

Protein Microarrays

Protein–protein interactions can also be analyzed with **protein microarrays (Figure 20.20)**, which are similar to the microarrays used for examining gene expression. With this technique, a large number of different proteins are applied

(a)

A protein is treated with the enzyme trypsin,...

Trypsin

...which breaks it into short peptides.

(b)

Accelerator

The peptides are analyzed with a mass spectrometer, which determines their mass-to-charge ratio.

Mass spectrometer

Detector

(c)

Counts

Mass (*m/z*)

A profile of peaks is produced.

(d)

A computer program compares the profile with those of known and predicted proteins.

GIVPPTKVWYRAITNDEKTSLAFR

A match identifies the protein.

20.19 Mass spectrometry is used to identify proteins.

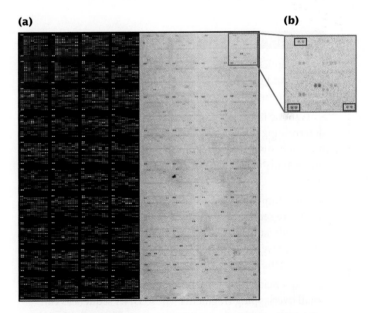

20.20 Protein microarrays can be used to examine interactions among proteins. (a) A microarray containing 4400 proteins found in yeast. (b) The array was probed with an enzyme that phosphorylates proteins to determine which proteins serve as substrate for the enzyme. Dark spots represent proteins that were phosphorylated by the enzyme. Proteins that phosphorylate themselves (autophosphorylate) are included in each block of the microarray (shown in blue boxes) to serve as reference points. [Republished with permission of Elsevier, from D. Hall, J. Ptacek, and M. Snyder, "Protein microarray technology," *Mechanisms of Ageing and Development*, 128 (2007) 161–167. Permission conveyed through Copyright Clearance Center, Inc.]

to a solid support in an orderly array of spots, with each spot containing a different protein. In one application, each spot contains an antibody for a different protein, labeled with a tag that fluoresces when bound. An extract of tissue is applied to the microarray. A spot of fluorescence appears when a protein in the extract binds to an antibody, indicating the presence of that particular protein in the tissue.

Structural Proteomics

The high-resolution three-dimensional structure of a protein provides a great deal of useful information. It is often a source of insight into the function of an unknown protein; it may also suggest the locations of active sites and provide information about other molecules that interact with the protein. Furthermore, knowledge of a protein's structure often suggests targets for potential drugs that might interact with the protein. Therefore, one goal of proteomics is to determine the structure of every protein found in a cell.

Two procedures are currently used to solve the structures of complex proteins: (1) X-ray crystallography, in which crystals of the protein are bombarded with X-rays and the diffraction patterns of the X-rays are used to determine the structure (see Section 10.2), and (2) nuclear magnetic resonance (NMR), which provides information on the positions of specific atoms within a molecule by using the magnetic properties of their nuclei.

Both X-ray crystallography and NMR require human intervention at many stages and are too slow for determining the structures of the thousands of proteins that may exist within a cell. Because the structures of hundreds of thousands of proteins are required for studies of the proteome, researchers ultimately hope to be able to predict the structure of a protein from its amino acid sequence. Such predictions are not possible at the present time, but the hope is that, if enough high-resolution structures are solved, it will be possible in the future to model a protein's structure from the amino acid sequence alone. As scientists work on automated methods that will speed the structural determination of proteins, bioinformaticians are developing better computer programs for predicting protein structure from sequence.

CONCEPTS

The proteome is the complete set of proteins found in a cell. Techniques of protein separation and mass spectrometry are used to identify the proteins present within a cell. Affinity capture and microarrays are used to determine sets of interacting proteins. Structural proteomics attempts to determine the structures of all proteins.

✔ **CONCEPT CHECK 9**

Why is knowledge of a protein's structure important?

CONCEPTS SUMMARY

■ Genomics is the field of genetics that attempts to understand the content, organization, and function of the genetic information contained in whole genomes.

■ Genetic maps position genes relative to other genes by determining rates of recombination and are measured in recombination frequencies. Physical maps are based on the physical distances between genes and are measured in base pairs.

■ The Human Genome Project was an effort to determine the entire sequence of the human genome. The project began officially in 1990; rough drafts of the human genome sequence were completed in 2000. The final draft of the human genome sequence was completed in 2003.

■ Sequencing a whole genome requires breaking it into small overlapping fragments whose DNA sequences can be determined in sequencing reactions. The individual sequences can be ordered into a whole-genome sequence with a map-based approach, in which fragments are assembled by using previously created genetic and physical maps, or with a whole-genome shotgun approach, in which overlap between fragments is used to assemble them into a whole-genome sequence. Today, almost all genomes are sequenced using whole-genome shotgun sequencing.

■ Single-nucleotide polymorphisms are single-base differences in DNA between individual organisms and are

valuable as markers in linkage studies. Individual organisms may also have differences in their numbers of copies of DNA sequences, called copy-number variations.

■ Bioinformatics is a synthesis of molecular biology and computer science that develops tools to store, retrieve, and analyze DNA-, RNA-, and protein-sequence data.

■ Metagenomics studies the genomes of entire groups of organisms. Synthetic biology is developing techniques for creating genomes and organisms.

■ Homologous genes are evolutionarily related. Orthologs are homologous sequences found in different species, whereas paralogs are homologous sequences found in the same species. Gene function may be determined by looking for homologous sequences (both orthologs and paralogs) whose function has been previously determined.

■ A microarray consists of known DNA fragments, fixed in an orderly pattern to a solid support, that act as probes. When a solution containing a mixture of DNA or RNA is applied to the array, any nucleic acid that is complementary to one of the probes will bind to that probe. Microarrays can be used to monitor the expression of thousands of genes simultaneously.

■ RNA sequencing analyzes the expression of genes throughout the genome. In this approach, RNA is isolated from cells and converted to cDNA, and the resulting cDNA fragments are sequenced.

■ By linking a reporter sequence to the regulatory sequences of a gene, the expression pattern of the gene can be observed by looking for the product of the reporter sequence. Genes affecting a particular function can also be identified through whole-genome mutagenesis screens.

■ Most prokaryotic species have between 1 million and 4 million base pairs of DNA and from 1000 to 2000 genes. The density of genes in prokaryotic genomes is relatively uniform, with about one gene per 1000 bp. Horizontal gene transfer (the movement of genes between different species) has been an important evolutionary process in prokaryotes.

■ Eukaryotic genomes are larger and more variable in size than prokaryotic genomes. There is no clear relation between organismal complexity and the amount of DNA or number of genes among multicellular organisms. Much of the genomes of eukaryotic organisms consist of repetitive DNA. Transposable elements are very common in most eukaryotic genomes.

■ Proteomics determines the protein content of a cell and the functions of those proteins. Proteins within a cell can be separated and identified with the use of mass spectrometry. Structural proteomics attempts to determine the three-dimensional shapes of proteins.

IMPORTANT TERMS

genomics (p. 606)
structural genomics (p. 606)
genetic (linkage) map (p. 607)
physical map (p. 608)
map-based sequencing (p. 610)
contig (p. 611)
whole-genome shotgun sequencing (p. 611)
single-nucleotide polymorphism (SNP) (p. 613)

haplotype (p. 614)
tag-SNP (p. 614)
genome-wide association study (p. 614)
copy-number variation (CNV) (p. 615)
bioinformatics (p. 615)
annotation (p. 616)
metagenomics (p. 616)
microbiome (p. 617)
synthetic biology (p. 617)
functional genomics (p. 617)
transcriptome (p. 617)

proteome (p. 617)
homologous gene (p. 618)
orthologous gene (p. 618)
paralogous gene (p. 618)
protein domain (p. 618)
transcriptomics (p. 619)
microarray (p. 619)
RNA sequencing (p. 621)
mutagenesis screen (p. 623)
comparative genomics (p. 624)
segmental duplication (p. 626)

multigene family (p. 627)
gene desert (p. 627)
proteomics (p. 629)
Human Proteome Project (p. 629)
two-dimensional polyacrylamide gel electrophoresis (2D-PAGE) (p. 629)
mass spectrometry (p. 630)
affinity capture (p. 631)
interactome (p. 631)
protein microarray (p. 631)

ANSWERS TO CONCEPT CHECKS

1. Accuracy and resolution.

2. b

3. It cataloged and mapped SNPs and other human genetic variants that could be used to identify common haplotypes in human populations.

4. a

5. d

6. c

7. Prokaryotic species with larger genomes generally have more genes than species with smaller genomes, so gene density is quite constant.

8. a

9. Structure often provides important information about how a protein functions and the types of proteins with which it is likely to interact.

WORKED PROBLEM

Problem

A linear piece of DNA that is 30 kb long is first cut with *Bam*HI, then with *Hpa*II, and, finally, with both *Bam*HI and *Hpa*II together. Fragments of the following sizes are obtained from these reactions:

*Bam*HI: 20-kb, 6-kb, and 4-kb fragments

*Hpa*II: 21-kb and 9-kb fragments

*Bam*HI and *Hpa*II: 20-kb, 5-kb, 4-kb, and 1-kb fragments

Draw a restriction map of the 30-kb piece of DNA, indicating the locations of the *Bam*HI and *Hpa*II restriction sites.

≫ Solution Strategy

What information is required in your answer to the problem?

A map that includes the number and relative locations of restriction sites for *Bam*HI and *Hpa*II and the distances in base pairs between the sites.

What information is provided to solve the problem?

- The piece of DNA is 30 kb long.
- The sizes of the fragments produced when the DNA is cut with *Bam*HI, with *Hpa*II, and with both enzymes together.

For help with this problem, review:

Physical Maps in Section 20.1.

≫ Solution Steps

Note: This problem can be solved correctly through a variety of approaches; this solution applies one possible approach.

Hint: For linear DNA, the number of restriction sites is one less than the number of fragments produced.

Hint: Look for fragments in the double digest that sum to the length of a fragment present in the single digest.

When cut by *Bam*HI alone, the linear piece of DNA is cleaved into three fragments; so there must be two *Bam*HI restriction sites. When cut with *Hpa*II alone, a clone of the same piece of DNA is cleaved into only two fragments; so there is a single *Hpa*II site.

Let's begin to determine the locations of these sites by examining the *Hpa*II fragments. Notice that the 21-kb fragment produced when the DNA is cut by *Hpa*II is not present in the fragments produced when the DNA is cut by *Bam*HI and *Hpa*II together (the double digest); this result indicates that the 21-kb *Hpa*II fragment has within it a *Bam*HI site. If we examine the fragments produced by the double digest, we see that the 20-kb and 1-kb fragments sum to 21 kb; so a *Bam*HI site must be 20 kb from one end of the fragment and 1 kb from the other end.

*Bam*HI site

20 kb   1 kb

Similarly, we see that the 9-kb *Hpa*II fragment does not appear in the double digest and that the 5-kb and 4-kb fragments in the double digest add up to 9 kb; so another *Bam*HI site must be 5 kb from one end of this fragment and 4 kb from the other end.

*Bam*HI site

5 kb   4 kb

Now, let's examine the fragments produced when the DNA is cut by *Bam*HI alone. The 20-kb and 4-kb fragments are also present in the double digest; so neither of these fragments contains an *Hpa*II site. The 6-kb fragment, however, is not present in the double digest, and the 5-kb and 1-kb fragments in the double digest sum to 6 kb; so this fragment contains an *Hpa*II site that is 5 kb from one end and 1 kb from the other end.

*Hpa*II site

5 kb   1 kb

We have accounted for all the restriction sites, but we must still determine the order of the sites on the original 30-kb fragment. Notice that the 5-kb fragment must be adjacent to both the 1-kb and the 4-kb fragments; so it must be in between those two fragments.

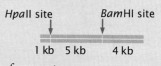

*Hpa*II site   *Bam*HI site

1 kb   5 kb   4 kb

We have also established that the 1-kb and 20-kb fragments are adjacent; because the 5-kb fragment is on one side, the 20-kb fragment must be on the other, completing the restriction map:

Hint: Two fragments in the double digest that were produced by cutting a fragment in the single digest must be adjacent to one another.

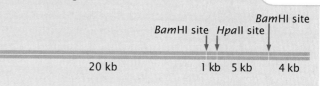

*Bam*HI site   *Hpa*II site   *Bam*HI site

20 kb   1 kb   5 kb   4 kb

COMPREHENSION QUESTIONS

Section 20.1

1. What is the difference between a genetic map and a physical map? Which generally has higher resolution and accuracy and why?

2. What is the difference between a map-based approach to sequencing a whole genome and a whole-genome shotgun approach?

3. What are some of the ethical concerns arising out of the information produced by the Human Genome Project?

4. What is a single-nucleotide polymorphism (SNP)? How are SNPs used in genomic studies?

5. What is a haplotype? How do different haplotypes arise?

6. What is linkage disequilibrium? How does it result in haplotypes?

7. How is a genome-wide association study carried out?

8. What is copy-number variation? How does it arise?

9. How are genes recognized within genome sequences?

Section 20.2

10. What are homologous sequences? What is the difference between orthologs and paralogs?

11. Describe several different methods for inferring the function of a gene by examining its DNA sequence.

12. What is a microarray? How can it be used to obtain information about gene function?

13. How are RNA molecules sequenced?

14. Explain how a reporter sequence can be used to provide information about the expression pattern of a gene.

15. Briefly outline how a mutagenesis screen is carried out.

Section 20.3

16. What is the relation between genome size and gene number in prokaryotes?

17. What is horizontal gene transfer? How might it take place between different species of bacteria?

18. Genome size varies considerably among multicellular organisms. Is this variation closely related to the number of genes and the complexity of the organism? If not, what accounts for some of this variation?

19. More than half of the genome of *Arabidopsis thaliana* consists of duplicated sequences. What mechanisms are thought to have been responsible for these extensive duplications?

20. What is a segmental duplication?

21. The human genome does not encode substantially more protein domains than do invertebrate genomes, yet it encodes many more proteins. How are more proteins encoded when the number of domains does not differ substantially?

22. What is genomics, and how does structural genomics differ from functional genomics? What is comparative genomics?

Section 20.4

23. How does proteomics differ from genomics?

24. How is mass spectrometry used to identify proteins in a cell?

APPLICATION QUESTIONS AND PROBLEMS

Section 20.1

*25. A 22-kb piece of DNA has the following restriction sites:

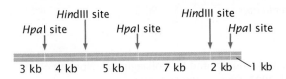

A batch of this DNA is first fully digested by *Hpa*I alone, then another batch is fully digested by *Hin*dIII alone, and finally, a third batch is fully digested by both *Hpa*I and *Hin*dIII together. The fragments resulting from each of the three digestions are placed in separate wells of an agarose gel, separated by gel electrophoresis, and stained by ethidium bromide. Draw the bands as they would appear on the gel.

*26. A linear piece of DNA that is 14 kb long is cut first by *Eco*RI alone, then by *Sma*I alone, and finally, by both *Eco*RI and *Sma*I together. The following results are obtained:

Digestion by *Eco*RI alone	Digestion by *Sma*I alone	Digestion by both *Eco*RI and *Sma*I
3-kb fragment	7-kb fragment	2-kb fragment
5-kb fragment	7-kb fragment	3-kb fragment
6-kb fragment		4-kb fragment
		5-kb fragment

Draw a map of the *Eco*RI and *Sma*I restriction sites on this 14-kb piece of DNA, indicating the relative positions of the restriction sites and the distances between them.

*27. The presence (+) or absence (−) of six sequences in each of five bacterial artificial chromosome (BAC) clones (A–E) is indicated in the following table. Using these markers, put the BAC clones in their correct order and indicate the locations of the numbered sequences within them.

	Sequences					
BAC clone	**1**	**2**	**3**	**4**	**5**	**6**
A	+	−	−	−	+	−
B	−	−	−	+	−	+
C	−	+	+	−	−	−
D	−	−	+	−	+	−
E	+	−	−	+	−	−

28. A linear piece of DNA was broken into random, overlapping fragments and each fragment was sequenced. The sequence of each fragment is shown below.

Fragment 1: 5′–TAGTTAAAAC–3′

Fragment 2: 5′–ACCGCAATACCCTAGTTAAA–3′

Fragment 3: 5′–CCCTAGTTAAAAC–3′

Fragment 4: 5′–ACCGCAATACCCTAGTT–3′

Fragment 5: 5′–ACCGCAATACCCTAGTTAAA–3′

Fragment 6: 5′–ATTTACCGCAAT–3′

On the basis of overlap in sequence, create a contig sequence of the original piece of DNA.

*29. How does the density of genes found on chromosome 22 compare with the density of genes found on chromosome 21, two similar-sized chromosomes? How does the number of genes on chromosome 22 compare with the number found on the Y chromosome?

To answer these questions, go to www.ensembl.org. Under the heading *Species*, select *Human*. On the next page, click on *View Karyotype*. Pictures of the human chromosomes will appear. Click on chromosome 22 and select *Chromosome Summary*. You will be shown a picture of this chromosome and histograms of known genes (colored bars). The total numbers of coding (protein-encoding) genes, along with the chromosome length in base pairs, are given in the table at the bottom of the diagram. Write down the total length of the chromosome and the number of coding genes.

Now go to chromosome 21 by selecting it from the Change Chromosome drop-down. Examine the total length and total number of protein-encoding genes for chromosome 21. Now do the same for the Y chromosome. Calculate the gene density (number of genes/length) for chromosomes 22, 21, and Y.

a. Which chromosome has the highest density and greatest number of genes? Which has the fewest?

b. Examine in more detail the genes at the tip of the short arm of the Y chromosome by clicking on the top bar in the histogram of genes. Jump to location view. A more detailed view will be shown. What known genes are found in this region? How many protein-encoding genes are there in this region?

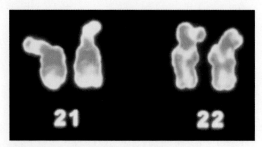

Human chromosomes 21 and 22.
[Leonard Lessin/Science Source.]

30. In recent years, honeybee colonies throughout North America have been decimated by colony collapse disorder (CCD), which results in the rapid deaths of worker bees. First noticed by beekeepers in 2004, the disorder has been responsible for the loss of 50% to 90% of beekeeping operations in the United States. Evidence suggests that CCD is caused by a pathogen. Diana Cox-Foster and her colleagues (D. Cox-Foster et al. 2007. *Science* 318:283–287) used a metagenomic approach to try to identify the causative agent of CCD by isolating DNA from normal honeybee hives and from hives that had experienced CCD. A number of different bacteria, fungi, and viruses were identified in the metagenomic analysis. The following table gives the percentage of CCD hives and non-CCD hives that tested positive for four potential pathogens identified in the metagenomic analysis. On the basis of these data, which potential pathogen appears most likely to be responsible for CCD? Explain your reasoning. Do these data prove that this pathogen is the cause of CCD? Explain.

Potential pathogen	CCD hives infected ($n = 30$)	Non-CCD hives infected ($n = 21$)
Israeli acute paralysis virus	83.3%	4.8%
Kashmir bee virus	100%	76.2%
Nosema apis	90%	47.6%
Nosema cernae	100%	80.8%

31. James Noonan and his colleagues (J. Noonan et al. 2005. *Science* 309:597–599) set out to study the genome sequence of an extinct species of cave bear. They extracted DNA from 40,000-year-old bones from a cave bear and used a metagenomic approach to isolate, identify, and sequence the cave-bear DNA. Why did they use a metagenomic approach when their objective was to sequence the genome of one species (the cave bear)?

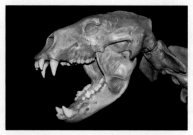

[Larry Miller/Science Source.]

Section 20.2

32. Explain why genes A2 and B2 in **Figure 20.9** are orthologs and not paralogs.

*33. Examine **Figure 26.16**. Are the epsilon (ε) and beta (β) genes on chromosome 11 orthologs or paralogs? Explain your answer.

*34. Microarrays can be used to determine relative levels of gene expression. In one type of microarray, hybridization of red (experimental) and green (control) labeled cDNAs is proportional to the relative amounts of mRNA in the samples. Red indicates the overexpression of a gene and green indicates the underexpression of a gene in the experimental cells relative to the control cells, yellow indicates equal expression in experimental and control cells, and no color indicates no expression in either experimental or control cells.

In one experiment, mRNA from a strain of antibiotic-resistant bacteria (experimental cells) is converted into cDNA and labeled with red fluorescent nucleotides; mRNA from a nonresistant strain of the same bacteria (control cells) is converted into cDNA and labeled with green fluorescent nucleotides. The cDNAs from the resistant and nonresistant cells are mixed and hybridized to a chip containing spots of DNA from genes 1 through 25. The results are shown in the illustration above. What conclusions can you draw about which genes might be implicated in antibiotic resistance in these bacteria? How might this information be used to design new antibiotics that are less vulnerable to resistance?

35. Of the genes in the microarray shown in the lower part of **Figure 20.10**, are most overexpressed or underexpressed in tumors from patients that remained cancer free for at least five years? Explain your reasoning.

36. What does the photograph in **Figure 20.13** reveal about the expression of β-tubulin?

Section 20.3

37. *Dictyostelium discoideum* is a soil-dwelling social amoeba: much of the time, the organism consists of single, solitary cells, but during times of starvation, amoebae come together to form aggregates that have many characteristics of multicellular organisms. Biologists have long debated whether *D. discoideum* is a unicellular or a multicellular organism. In 2005, its genome was completely sequenced. The accompanying table lists some genomic characteristics of *D. discoideum* and other eukaryotes (L. Eichinger et al. 2005. *Nature* 435:43–57).

a. On the basis of the organisms other than *D. discoideum* listed in the table, what are some differences in genome

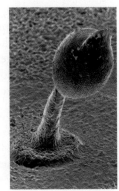

Dictyostelium discoideum.
[David Scharf/Science Source.]

Table for Problem 37 Genomic characteristics of *D. discoideum* and other eukaryotes							
Feature	*D. discoideum*	*P. falciparum*	*S. cerevisiae*	*A. thaliana*	*D. melanogaster*	*C. elegans*	*H. sapiens*
Organism	Amoeba	Protozoan	Yeast	Plant	Fruit fly	Nematode	Human
Cellularity	?	Uni	Uni	Multi	Multi	Multi	Multi
Genome size (millions of base pairs)	34	23	13	125	180	103	2,851
Number of genes	12,500	5,268	5,538	25,498	13,676	19,893	22,287
Average gene length (bp)	1,756	2,534	1,428	2,036	1,997	2,991	27,000
Genes with introns (%)	69	54	5	79	38	5	85
Mean number of introns	1.9	2.6	1.0	5.4	4.0	5.0	8.1
Mean intron size (bp)	146	179	nd*	170	nd*	270	3,365
Mean G + C (exons)	27%	24%	28%	28%	55%	42%	45%

*nd = Not determined.

characteristics between unicellular and multicellular organisms?

b. On the basis of these data, do you think the genome of *D. discoideum* is more like those of unicellular eukaryotes or more like those of multicellular eukaryotes? Explain your answer.

*38. How do the following genomic features of prokaryotic organisms compare with those of eukaryotic organisms? How do they compare among eukaryotes?

a. Genome size

b. Number of genes

c. Gene density (bp/gene)

d. Number of exons

39. A group of scientists sequenced the genomes of 12 species of *Drosophila* (*Drosophila* 12 Genomes Consortium. 2007. *Nature* 450:203–218). Data on genome sizes and numbers of protein-encoding genes from this study are given in the accompanying table. Plot the number of protein-encoding genes as a function of genome size for the 12 species of *Drosophila*. Is there a relation between genome size and number of genes in fruit flies? How does this compare with the relation between genome size and number of genes across all eukaryotes?

Characteristics of 12 *Drosophila* species genomes

Species	Genome size (millions of base pairs)	Number of protein-encoding genes
D. melanogaster	200	13,733
D. simulans	162	15,983
D. sechellia	171	16,884
D. yakuba	190	16,423
D. erecta	135	15,324
D. ananassae	217	15,276
D. pseudoobscura	193	16,363
D. persimilis	193	17,325
D. willistoni	222	15,816
D. virilis	364	14,680
D. mojavensis	130	14,849
D. grimshawi	231	15,270

Section 20.4

40. A scientist determines the complete genomes and proteomes of a liver cell and a muscle cell from the same person. Would you expect bigger differences in the genomes or in the proteomes of these two cell types? Explain your answer.

CHALLENGE QUESTIONS

Section 20.1

41. The genome of *Drosophila melanogaster*, a fruit fly, was sequenced in 2000. However, this "completed" sequence did not include most heterochromatin regions. The heterochromatin was not sequenced until 2007 (R. A. Hoskins et al. 2007. *Science* 316:1625–1628). Most completed genome sequences do not include heterochromatin. Why is heterochromatin usually not sequenced in genome-sequencing projects? (Hint: See Chapter 11 for a more detailed discussion of heterochromatin.)

42. In metagenomic studies, a comparison of ribosomal RNA sequences is often used to determine the number of different species present. What are some characteristics of ribosomal sequences that make them useful for determining what species are present?

*43. Some synthetic biologists have proposed creating an entirely new, free-living organism with a minimal genome, the smallest set of genes that allows for replication of the organism in a particular environment. This genome could be used to design and create, from "scratch," novel organisms that might perform specific tasks, such as the breakdown of toxic materials in the environment.

a. How might the minimal genome required for life be determined?

b. What, if any, social and ethical concerns might be associated with the construction of an entirely new organism with a minimal genome?

THINK-PAIR-SHARE QUESTIONS

Section 20.1

1. Increasingly, whole-genome sequencing of individuals is being done to help identify and treat medical conditions. Genome sequencing invariably identifies a number of variations, some common and some rare, that might be clinically relevant. For example, suppose a person had their genome sequenced to help determine their risk for cardiovascular disease and, just by chance, the sequence revealed that they carry one or more variants that predispose them to cancer or Alzheimer disease. Does the sequencing laboratory or physician have an obligation to report this finding, which was not the purpose of the sequencing and which the patient did not request? What about reporting variants for which no or limited information can be provided about their clinical significance? Does the answer to this question differ for sequencing done on children?

2. If "the human genome sequence" does not really exist, can you think of better ways in which we might represent the human genome? Propose some possibilities.

3. As pointed out in the text, you and a complete stranger are 99.9% identical in DNA sequence. But you also differ at more than 3 million base pairs. Is this a large or a small difference? What are some of the consequences of these similarities and differences?

4. Researchers systematically replaced 414 essential genes in yeast with similar genes from humans. Almost half of these transplants (47%) were successful: cells with the humanized gene were able to function and grow. What does this observation tell us about differences between yeast and humans? How might this information be used?

Section 20.3

5. The Japanese canopy plant (*Paris japonica*) has one of the largest of all eukaryotic genomes, with approximately 150 billion base pairs, about 50 times the size of the human genome. In contrast, the bladderwort *Utricularia gibba* has one of the smallest plant genomes, with only 82 million base pairs. What predictions can you make about the genomes of these two species?

Epigenetics

The Dutch Hunger Winter, a severe famine in the Netherlands at the end of World War II, had long-term effects on people who were conceived and developed prenatally during the famine. These effects are thought to be due to epigenetic alterations. Shown here is a young girl eating after the famine ended. [Keystone-France/Getty Images.]

Epigenetics and the Dutch Hunger Winter

In 1944, near the close of World War II, Allied armies marched north into the Netherlands to liberate the country from German occupation. The Allied advance, however, came to a halt at the Waal and Rhine rivers. To help the Allied cause, the Dutch people carried out a railway strike; in retaliation, the German army blocked all food supplies to cities in the north of the country, and the Dutch people starved. From October through April, no food reached the area, and food supplies quickly ran out. People subsisted on bread and potatoes. In November 1944, food rations provided less than 1000 calories per person per day; by April 1945, many people were getting less than 500 calories per day. Thousands died. The famine ended in May 1945, when Allied armies liberated the entire country.

Known as the Dutch Hunger Winter, this terrible episode of intense starvation provides a natural experiment for examining the consequences of famine for humans. Researchers have been particularly interested in the long-term effects of starvation on people who were conceived and developed prenatally during the famine. Numerous studies have examined these individuals, who are now adults. What the researchers found was remarkable: the caloric restriction that these people experienced while in the womb had life-long consequences. For example, men who were conceived and went through early development during the famine were twice as likely to be obese by age 18. Both men and women who underwent early development during the famine were more likely to be obese by age 50 and more likely to die from stroke and cardiovascular disease. They also experienced more cognitive decline in later life.

Scientists believe that these long-term effects are due to epigenetic alterations: modifications of chromatin and DNA that alter gene expression but do not change the base sequence of the DNA. One common epigenetic change is DNA methylation, the addition of methyl groups to cytosine bases in the DNA. Geneticists examined DNA from individuals who were conceived during the Dutch Hunger Winter and compared it with DNA from their same-sex siblings conceived at other times. They examined 15 loci that are involved in growth and metabolic disease. Of these loci, six showed significant differences in methylation patterns. At some loci, methylation was higher among the famine-exposed individuals, and at other loci, methylation was lower.

These results show that an early life event—caloric restriction during prenatal development—can bring about alterations in chromatin structure that have long-term consequences. Studies on rodents show that the phenotypic effects of nutrition can carry over for several generations. As this chapter explains, epigenetic alterations can be passed from cell to cell and sometimes from generation to generation. Epigenetic inheritance was not envisioned by Mendel nor, until recently, by most modern geneticists, but epigenetic processes appear to play an important role in the inheritance of many phenotypes. Today, the study of epigenetics is the focus of intensive research.

A question that this study raises is why limited calories during prenatal development would *increase* adult obesity and risk of dying from cardiovascular disease. One might expect just the opposite, that nutritional stress during fetal development would decrease adult body weight. Evolutionary biologists have proposed an explanation for this relation, which has also been observed in other studies. Termed the *thrifty phenotype hypothesis*, it proposes that when environmental conditions are poor during fetal development, they are usually likely to persist and to be poor in adult life as well. The fetus responds by developing a metabolically thrifty phenotype—eating as much as possible when food is available, minimizing energy expenditure, and lowering metabolic rate—because the prenatal environment predicts that there will be little food available in the future. This strategy was probably advantageous in the distant past, before agriculture, but it often backfires in modern society. Eating all you can, minimizing energy expenditure, and hoarding calories when food is plentiful often leads to obesity, heart disease, and diabetes.

THINK-PAIR-SHARE
- Among people who were prenatally exposed to famine in the Dutch Hunger Winter, obesity and increased risk of cardiovascular disease are seen only in those who were in early gestational stages during the famine; those individuals who were in the last trimester of prenatal development do not show these effects. Propose an explanation for why those in early stages of development were affected, but not those in later stages.

- What are some practical implications of the finding that caloric restriction has long-term negative effects on body weight, cardiovascular disease, and late-life cognitive decline?

This chapter is about epigenetics, the explanation for the effect of famine on the future health of the people who were conceived during the Dutch Hunger Winter. We begin by discussing the origin of the term *epigenetics* and what the term encompasses today. We then review the types of changes in chromatin that can occur and the major processes that alter chromatin structure. We also take a look at how changes in chromatin structure might be passed on to future cells and future generations. We then look at a number of epigenetic effects, including paramutation, behavioral effects, effects of chemicals, metabolic effects, effects on monozygotic twins, X inactivation, cell differentiation, and genomic imprinting. We end the chapter by discussing efforts to map the genome-wide pattern of epigenetic marks—the epigenome.

21.1 What Is Epigenetics?

The term *epigenetics* was first used by Conrad Waddington (**Figure 21.1**) in 1942 to describe how, through the process of development, a genotype produces a phenotype. In coining the term, Waddington combined the words *epigenesis*, the development of an embryo, with *genetics*, the study of genes and heredity. Waddington's goal was to encourage the merging of genetics and development. However, his use of the term preceded our modern understanding of DNA and chromosome structure, and today, *epigenetics* has taken on a narrower meaning.

21.1 Conrad Waddington first used the word *epigenetics* to refer to the development of a phenotype from a genotype.
[Godfrey Argent Studio/The Royal Society.]

The Greek root *epi* means "over" or "above"; the term *epigenetics* has come to represent the inheritance of variation above and beyond differences in DNA sequence. Today, *epigenetics* usually refers to the phenotypes and processes that are transmitted to other cells and sometimes to future generations, but are not the result of differences in the DNA base sequence. Many epigenetic effects are caused by changes in gene expression that result from alterations in chromatin structure or other aspects of DNA structure, such as DNA methylation. One definition of an epigenetic trait is a stably inherited phenotype resulting from changes in chromatin without alterations in the DNA sequence. Some have broadened the definition of epigenetics to refer to any alteration of chromatin or DNA structure that affects gene expression. Here, we will use *epigenetics* to refer to changes in gene expression or phenotype that are potentially heritable without alteration of the underlying DNA base sequence.

Many epigenetic changes are stable, persisting across cell divisions or even generations. However, epigenetic alterations can also be influenced by environmental factors. For example, environmental stress has been shown to alter methylation of the rat *Bdnf* gene, which encodes a growth factor that plays an important role in brain development. DNA methylation has been tied to the expression of genes and the phenotypes they produce. As we will see, altered DNA methylation is capable of being replicated across cell division, resulting in progeny with the new phenotype, although there is no corresponding difference in the DNA base sequence of individuals that "inherit" the new phenotype. The fact that epigenetic traits may be induced by environmental effects and transmitted to future generations has been interpreted by some to mean that through epigenetics, genes have memory—that environmental factors acting on individuals can have effects that are transmitted to future generations, as was seen with the effect of diet on obesity and cardiovascular disease in the introduction to this chapter. Epigenetics has been called "inheritance, but not as we know it."

Epigenetics is providing an explanation for how changes outside of the DNA sequence can influence the phenotype and how those changes can be heritable. It is also making important contributions to the study of behavior, environmental science, cancer, neurobiology, and pharmacology.

▶ TRY PROBLEM 2

THINK-PAIR-SHARE Question 1

CONCEPTS

Epigenetic effects are phenotypes that are passed to other cells and sometimes to future generations, but are not the result of differences in the DNA base sequence. The study of epigenetics is making important contributions to many areas of biology.

21.2 Several Molecular Processes Lead to Epigenetic Changes

Epigenetics alters the expression of genes; these alterations are stable enough to be transmitted through mitosis (and sometimes meiosis), but can also be changed. Most evidence suggests that epigenetic effects are brought about by physical changes in chromatin structure. In Chapter 11, we considered a number of chemical changes in DNA and histone proteins that affect chromatin structure, including DNA methylation, modification of histone proteins, and repositioning of nucleosomes. In Chapter 17, we discussed the role that these alterations have on the expression of genes. These chromatin changes are thought to play a role in epigenetic traits. Chapter 14 discussed small RNA molecules, some of which play an important role in bringing about epigenetic changes.

In this section, we consider three types of molecular mechanisms that alter chromatin structure and underlie many epigenetic phenotypes: (1) changes in patterns of DNA methylation; (2) chemical modifications of histone proteins; and (3) RNA molecules that affect chromatin structure and gene expression.

CONCEPTS

Many epigenetic phenotypes are the result of alterations in chromatin structure, mediated through three major processes: DNA methylation, histone modification, and RNA molecules.

DNA Methylation

The best-understood mechanism of epigenetic change is methylation of DNA. DNA methylation refers to the addition of methyl groups to the nucleotide bases. In eukaryotes, the predominant type of DNA methylation is the methylation of cytosine to produce 5-methylcytosine (**Figure 21.2a**). As discussed in Section 17.2, DNA methylation is often associated with repression of transcription.

DNA methylation often occurs on cytosine nucleotides that are immediately adjacent to guanine nucleotides, together referred to as CpG dinucleotides (where p represents the phosphate group that connects the C and G nucleotides). In CpG dinucleotides, cytosine nucleotides on the two DNA strands are diagonally across from each other. Usually both cytosine bases are methylated, so that methyl groups occur on both DNA strands, as shown below and in **Figure 21.2b**.

$$5'-\overset{m}{C}\,G-3'$$
$$3'-G\,\underset{m}{C}-5'$$

(a)

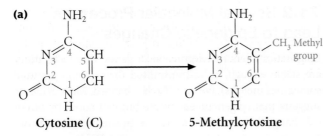

Cytosine (C) → 5-Methylcytosine

(b)

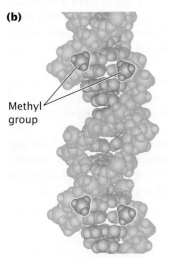

Methyl group

21.2 DNA methylation is a common epigenetic modification of chromatin. (a) Cytosine bases are often modified to form 5-methylcytosine. (b) Three-dimensional structure of DNA showing methylation of CpG dinucleotides.

In plants, DNA methylation also occurs at CpNpG trinucleotides, where N represents a nucleotide with any base.

Some DNA regions have many CpG dinucleotides and are referred to as CpG islands. In mammalian cells, CpG islands are often located in or near the promoters of genes. These CpG islands are usually not methylated when genes are being actively transcribed. However, methylation of CpG islands near a gene leads to repression of transcription.

Cells repress and activate genes by methylating and demethylating cytosine bases. Enzymes called DNA methyltransferases methylate DNA by adding methyl groups to cytosine bases to create 5-methylcytosine. Other enzymes, called demethylases, remove methyl groups, converting 5-methylcytosine back to cytosine (see Section 17.2).

MAINTENANCE OF METHYLATION The fact that epigenetic changes are passed to other cells and sometimes to future generations means that the changes in chromatin structure associated with epigenetic phenotypes must be faithfully maintained when chromosomes replicate. How are epigenetic changes retained and replicated through the process of cell division?

Methylation of CpG dinucleotides means that two methylated cytosine bases sit diagonally across from each other on opposite strands. Before replication, cytosine bases on both strands are methylated (**Figure 21.3**). Immediately after semiconservative replication, the cytosine base on the template strand is methylated, but the cytosine base on the newly replicated strand is unmethylated. Special methyltransferase enzymes recognize the hemimethylated state of CpG dinucleotides and add methyl groups to the unmethylated cytosine bases, resulting in two new DNA molecules that are fully methylated. In this way, the methylation pattern of DNA is maintained across cell division.

▶ TRY PROBLEM 25

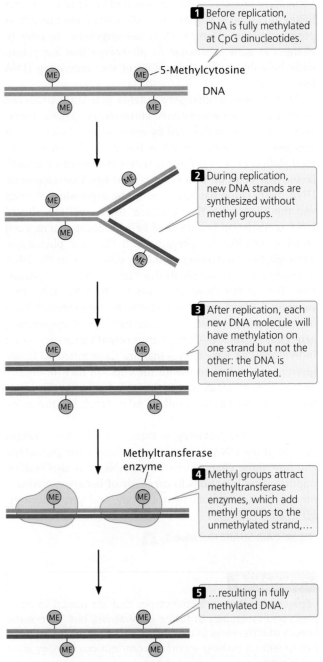

1 Before replication, DNA is fully methylated at CpG dinucleotides.

5-Methylcytosine

DNA

2 During replication, new DNA strands are synthesized without methyl groups.

3 After replication, each new DNA molecule will have methylation on one strand but not the other: the DNA is hemimethylated.

Methyltransferase enzyme

4 Methyl groups attract methyltransferase enzymes, which add methyl groups to the unmethylated strand,...

5 ...resulting in fully methylated DNA.

21.3 DNA methylation is stably maintained through DNA replication.

DNA METHYLATION IN HONEYBEES A remarkable example of epigenetics is seen in honeybees. Queen bees and worker bees are both female, but the resemblance ends there. The queen is large and develops functional ovaries, whereas workers are small and sterile (**Figure 21.4**). The queen goes on a mating flight and then spends the rest of her life reproducing, whereas workers spend all of their time collecting nectar and pollen, tending the queen, and raising her offspring. In spite of these differences in their anatomy, physiology, and behavior, queens and workers are genetically the same; both develop from ordinary eggs. How they differ is in diet: worker bees produce and feed a few female larvae a special substance called royal jelly, which causes these larvae to develop as queens. Other larvae are fed ordinary bee food, and they develop as workers. This simple difference in diet greatly affects gene expression, causing different genes to be activated in queens and in workers and resulting in very different sets of phenotypic traits.

How royal jelly affects gene expression has long been a mystery, but research now suggests that it changes an epigenetic mark. In 2008, Ryszard Kucharski and his colleagues demonstrated that royal jelly silences the expression of a key gene called *Dnmt3*, whose product normally adds methyl groups to DNA (**Figure 21.5**). With *Dnmt3* shut down, bee DNA is less methylated, and many genes that are normally silenced in workers are expressed, leading to the development of queen characteristics. Kucharski and his colleagues demonstrated the importance of DNA methylation in queen development by injecting into bee larvae small interfering RNAs (siRNAs; see Section 14.5) that specifically inhibited the expression of *Dnmt3*. These larvae had lower levels of DNA methylation, and many developed as queens with fully functional ovaries. This experiment demonstrated that royal jelly brings about epigenetic changes (less DNA methylation), which are transmitted through cell division and modify developmental pathways, eventually leading to a queen bee. **▶ TRY PROBLEM 24**

THINK-PAIR-SHARE Question 2

21.4 Epigenetic changes are responsible for differences in the phenotypes of honeybee (*Apis mellifera*) queens (left) and workers (right). [WILDLIFE GmbH/Alamy.]

REPRESSION OF TRANSCRIPTION BY DNA METHYLATION How does DNA methylation suppress gene expression? The methyl group of 5-methylcytosine sits within the major groove of the DNA molecule, which is recognized by many DNA-binding proteins. The presence of the methyl group in the major groove inhibits the binding of transcription factors and other proteins required for transcription to occur. The 5-methylcytosine also attracts certain proteins that directly repress transcription. In addition, DNA methylation attracts histone deacetylase enzymes, which remove acetyl groups from the tails of histone proteins, altering chromatin structure in a way that represses transcription (see Section 17.2).

CONCEPTS

Cytosine bases are often methylated to form 5-methyl-cytosine, which is associated with repression of transcription. DNA methylation is stably maintained through replication by methyltransferase enzymes that recognize the hemimethyl-ated state of CpG dinucleotides and add methyl groups to the unmethylated cytosine bases.

✔ CONCEPT CHECK 1

Which of the following is true of CpG islands?

a. They are methylated near promoters of actively transcribed genes.

b. They are unmethylated near promoters of actively transcribed genes.

c. Acetylation of CpG islands leads to repression of transcription.

d. CpG islands code for RNA molecules that activate transcription.

Histone Modifications

Epigenetic changes can also occur through modification of histone proteins. In eukaryotic cells, DNA is complexed to histone proteins in the form of nucleosomes, which are the basic repeating units of chromatin structure (see Chapter 11). Modifications of histones include the addition of phosphates, methyl groups, acetyl groups, and ubiquitin. Many of these modifications take place in the positively charged tails of the histone proteins, which interact with the DNA and affect chromatin structure. These modifications can occur at different amino acids on different histones and create more than 100 unique potential changes in the histones. Many of these modifications alter chromatin structure and affect transcription of genes (see Chapter 17). They may also serve as binding sites for proteins such as transcription factors that are required for transcription.

The addition of acetyl groups to amino acids in the histone tails (histone acetylation) generally destabilizes chromatin structure, causing it to assume a more open configuration, and is associated with increased transcription (see Figure 17.2). The addition of methyl groups to histones (histone methylation) also alters chromatin structure, but the effect varies depending on the specific amino acid that is methylated; some types of histone methylation are associated with increased transcription and other types are associated with decreased transcription.

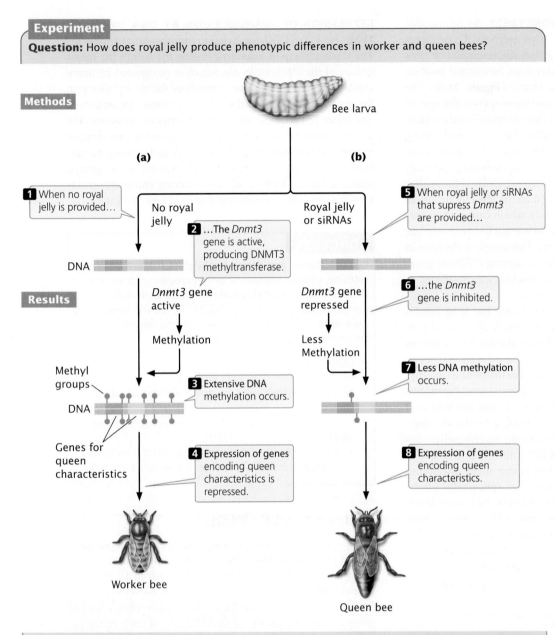

Question: How does royal jelly produce phenotypic differences in worker and queen bees?

Methods

Bee larva

(a)   **(b)**

1 When no royal jelly is provided…

No royal jelly

2 …The *Dnmt3* gene is active, producing DNMT3 methyltransferase.

DNA

Results

Dnmt3 gene active

Methylation

Methyl groups

DNA

3 Extensive DNA methylation occurs.

Genes for queen characteristics

4 Expression of genes encoding queen characteristics is repressed.

Worker bee

5 When royal jelly or siRNAs that supress *Dnmt3* are provided…

Royal jelly or siRNAs

6 …the *Dnmt3* gene is inhibited.

Dnmt3 gene repressed

Less Methylation

7 Less DNA methylation occurs.

8 Expression of genes encoding queen characteristics.

Queen bee

Conclusion: Differences in worker and queen bees result from epigenetic differences in DNA methylation.

21.5 Phenotypic differences in queen and worker honeybees result from epigenetic differences in DNA methylation. Royal jelly suppresses Dnm3t, which normally methylates DNA, leading to expression of genes that encode characteristics of the queen.

For example, the addition of three methyl groups to lysine 4 in the H3 histone (H3K4me3; K stands for lysine) is often found near transcriptionally active genes. Methylation of lysine 36 in the H3 histone (H3K36me3) is also associated with increased transcription. On the other hand, the addition of three methyl groups to lysine 9 in H3 (H3K9me3) or to lysine 20 in histone 4 (H4K20me3) is associated with repression of transcription. Many additional histone modifications, as well as changes to DNA that do not involve the base sequence, have been shown to be associated with the level of transcription. These types of modifications have been called **epigenetic marks**.

Histone modifications are added and removed by special proteins. The polycomb group (PcG) is a large group of proteins that repress transcription by modifying histones. These modifications alter chromatin structure so that the DNA is not accessible to transcription factors, RNA polymerase, and other proteins required for transcription. For example, polycomb repressive complex 2 (PRC2) adds two or three methyl groups to lysine 27 of histone H3, creating the H3K27me3 epigenetic mark, which represses transcription.

Many of the enzymes and proteins that produce epigenetic marks cannot bind to specific DNA sequences by themselves. Thus, they must be recruited to specific targets on the chromosome. Sequence-specific transcription factors, preexisting chromatin marks, and noncoding RNA molecules serve to recruit histone-modifying enzymes to specific sites.

Research indicates that single histone modifications, such as those mentioned here, do not individually determine the transcriptional activity of a gene. Rather, it is the combined presence of multiple epigenetic marks that determines the activity level. There is also considerable "crosstalk" between epigenetic marks: one histone mark may affect whether additional marks occur nearby and how they function. Crosstalk occurs because histone modifications attract enzymes and proteins that modify other histones.

Histone modifications not only affect transcription, but can also influence other molecular processes such as DNA repair and cell cycle checkpoint signaling (see pp. 698–702 in Chapter 23). For example, ubiquitination of histone H2B is required for repair of double-strand breaks in DNA. This modification leads to other histone modifications, such as methylation of lysine 79 of H3 (H3K79me); these modifications alter chromatin structure and allow access by proteins that repair double-strand breaks.

MAINTENANCE OF HISTONE MODIFICATIONS The process by which histone modifications are maintained across cell division is not as well understood as that of DNA methylation. There is no universal mechanism for maintaining histone modifications; different types of modifications are undoubtedly maintained by different mechanisms.

Several models have been proposed to explain how histone modifications are faithfully transmitted to daughter cells. During the process of DNA replication, nucleosomes are disrupted, and the original histone proteins are distributed randomly between the two new DNA molecules. Newly synthesized histones are then added to complete the formation of new nucleosomes (see Section 12.4). Some models assume that after replication, the epigenetic marks remain on the original histones, and that these marks recruit enzymes that make similar changes in the new histones. For example, PRC2, which adds the H3K27me3 epigenetic mark to histones, preferentially targets histones in chromatin that already contains an H3K27me3 mark, ensuring that any new nucleosomes that are added after replication also become methylated. In this way, the histone modifications can be maintained across cell division. An alternative model, supported by experimental evidence in *Drosophila* embryos, proposes that the epigenetic marks are lost during replication, but that the enzymes that bring about histone modifications remain attached to the original histones during the replication process and reestablish the marks on the original and new histones after replication is completed.

> **CONCEPTS**
>
> Modifications of histone proteins, including the addition of methyl groups, acetyl groups, phosphates, and ubiquitin, alter chromatin structure. Some of these modifications are passed to daughter cells during cell division and to future generations.

Epigenetic Effects of RNA Molecules

Evidence increasingly demonstrates that RNA molecules play an important role in bringing about epigenetic effects. The first discovered and still best understood example of RNA mediation of epigenetic change is X inactivation (mentioned in Section 4.3), in which a long noncoding RNA called *Xist* suppresses transcription on one of the X chromosomes in female mammals. Another example involves paramutation in corn, in which an epigenetically altered allele, by means of siRNAs, induces a change in another allele that then gets transmitted to future generations. Both of these examples will be discussed in Section 21.3.

Various mechanisms are involved in epigenetic changes produced by RNA molecules. In the case of X inactivation, the *Xist* RNA coats one X chromosome and then attracts PRC2, which deposits methyl groups on lysine 27 of histone H3, creating the H3K27me3 epigenetic mark, which alters chromatin structure and represses transcription.

Other examples of RNA-associated epigenetic phenotypes are produced by siRNA molecules that silence genes and transposable elements (see Section 18.4) by directing DNA methylation or histone modifications to specific DNA sequences. In addition, research has demonstrated that epigenetic processes such as methylation and histone modification influence the expression of microRNAs (see Section 14.5), which, in turn, play an important role in regulating other genes. MicroRNAs also control the expression of genes that produce epigenetic effects, such as those encoding enzymes that methylate DNA and modify histone proteins. How RNA-based epigenetic changes are maintained across cell divisions is less clear, although some apparently involve small RNAs that are transmitted through the cytoplasm.

> **CONCEPTS**
>
> RNA molecules bring about modifications of chromatin by a variety of processes.
>
> ✔ **CONCEPT CHECK 2**
>
> Which of the following is *not* a major mechanism of epigenetic change?
> a. DNA methylation
> b. Alteration of a DNA base sequence in a promoter
> c. Histone acetylation
> d. Action of RNA molecules

21.3 Epigenetic Processes Produce a Diverse Set of Effects

Initially, epigenetic mechanisms were thought to play a role in a small number of unusual phenotypes. However, research has now revealed that epigenetics underlies an impressive and ever-increasing array of biological phenomena. In this section, we look at some examples of these epigenetic effects.

Paramutation

One of the first examples of epigenetics was a curious phenotype that Alexander Brink described in corn in the 1950s. Brink was studying the *r1* locus, which helps to determine pigmentation in the seeds of corn. The R^r allele at this locus normally produces purple kernels, while the R^{st} allele codes for spotted kernels. Brink observed that when R^{st} was present in a genotype with the R^r allele, the R^{st} allele permanently altered the expression of the R^r allele, so that it, too, produced spotted seeds. This diminished effect of the altered R^r allele on pigmentation persisted for several generations, even in the absence of the R^{st} allele. Brink called this phenomenon paramutation.

Today, **paramutation** is defined as an interaction between two alleles that leads to a heritable change in the expression of one of the alleles. Surprisingly, paramutation produces these changes in phenotype without any alteration in the DNA base sequence of the converted allele. The phenomenon of paramutation has several important features. First, the newly established expression pattern of the converted allele is transmitted to future generations, even when the allele that brought about the alteration is no longer present with it. Second, the altered allele is now able to convert other alleles to the new phenotype. And third, there are no associated DNA sequence changes in the altered alleles. A number of examples of paramutation have now been discovered in different organisms, and geneticists have begun to unravel the molecular mechanism of this curious phenomenon.

PARAMUTATION IN CORN A few years after Brink reported paramutation at the *r1* locus in corn, another related example was discovered by Ed Coe, Jr. This case involved interaction between the alleles at the *b1* locus in corn, which also aids in determining pigmentation. Paramutation at the *b1* locus is more straightforward than at the *r1* locus, so we will use *b1* to examine the process and mechanism of paramutation.

The *b1* locus helps to determine the amount of purple anthocyanin that a corn plant produces. The locus actually encodes a transcription factor that regulates genes involved in anthocyanin production. Plants homozygous for the *B-I* allele (*B-I B-I*) show high expression of the *b1* locus and are dark purple (**Figure 21.6**). Plants homozygous for the *B'* allele (*B' B'*) show a lower expression of the *b1* locus and are lightly pigmented. However, the DNA sequences of the *B-I* and *B'* alleles are identical. Genetically identical alleles such as these, which produce heritable differences in phenotypes through epigenetic processes, are referred to as **epialleles**.

THINK-PAIR-SHARE Question 3

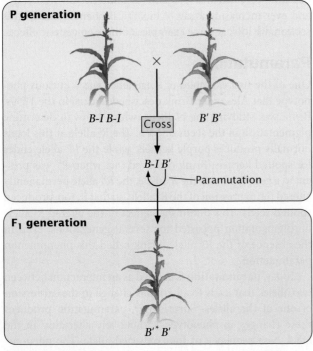

21.6 In paramutation at the *b1* locus in corn, a copy of the *B'* allele converts the *B-I* allele to *B'**, which has the same phenotype as *B'*. The *B-I B-I* genotype produces a pigmented plant, while *B' B'* and *B' B'** genotypes are lightly pigmented.

In plants that are heterozygous *B-I B'*, the *B-I* allele is converted to *B'*, with the result that the heterozygous plants are lightly pigmented (see Figure 21.6), just like the *B' B'* homozygotes. The newly converted allele is usually designated *B'**. Importantly, there is no functional difference between *B'* and *B'**; the *B'** allele is now fully capable of converting other *B-I* alleles into *B'** alleles in subsequent generations.

Research has demonstrated that one of the features required for paramutation at the *b1* locus is the presence of a series of seven tandem repeats of an 853-bp sequence located approximately 100,000 base pairs upstream of the coding sequence for the *b1* locus (**Figure 21.7**). The repeats do not encode any protein. Both the *B-I* and *B'* alleles have these tandem repeats, but the chromatin structure of the two alleles differs: the *B-I* allele has more open chromatin. The tandem repeats are required for high expression of the *B-I* allele and high pigment production. It has been suggested that the repeats act like an enhancer (see Chapter 17), stimulating transcription at the *b1* locus, but only when the chromatin surrounding the repeats is in an open configuration, as it is in the *B-I* allele. The more closed configuration in the *B'* allele may prevent the repeats from interacting with the promoter of *b1* and stimulating transcription. How the repeats might interact with the *B'* allele is not known.

The different chromatin states of *B-I* and *B'* may explain their different levels of expression, but how does the *B'* allele convert the *B-I* allele to *B'**? Although the mechanism is not completely understood, recent research demonstrates that the communication between *B'* and *B-I* probably occurs through the action of small RNA molecules.

The tandem repeats that are required for paramutation encode 25-nucleotide-long siRNAs. Some siRNAs are known to modify chromatin structure by directing DNA methylation to specific DNA sequences. Geneticists have isolated several genes in corn that are required for paramutation to take place; inactivating these genes eliminates paramutation. One of these genes is *mop1*, which encodes an RNA-directed RNA polymerase (an enzyme that synthesizes RNA from an RNA template). This gene is required to generate the siRNAs encoded by the tandem repeats, although it does not appear to be the enzyme that actually transcribes the DNA copies of the tandem repeats. Another gene required for paramutation, called *rmr1*, encodes a chromatin-remodeling protein. Thus, the current evidence suggests that siRNA molecules convert *B-I* to *B'** and that this conversion involves a change in the chromatin states of the alleles. Research also shows that transcription of the tandem repeats and generation of siRNAs from them are necessary, but not sufficient, for paramutation, so additional factors must be involved. It is also not clear how the production of the siRNAs is transmitted across generations.

PARAMUTATION IN MICE Several examples of paramutation have also been observed in mice. One involves the *Kit* locus, which encodes a tyrosine kinase receptor and functions in pigment production, germ cell development, and production of blood cells. Geneticists had earlier genetically engineered a mutant *Kit* allele (designated here as *Kit*^t), which carries a 3000-bp portion of the *lacZ* gene (see Chapter 16) inserted into the *Kit* locus. Mice that are homozygous for the wild-type allele (*Kit*^+*Kit*^+) have normal pigment. Mice that are homozygous for mutant *Kit* alleles (*Kit*^t*Kit*^t) die shortly after birth. Mice heterozygous for wild-type and mutant alleles (*Kit*^+*Kit*^t) have white tail tips and white feet (**Figure 21.8**). When a heterozygous mouse is crossed with a homozygous wild-type mouse, half of the progeny are homozygous (*Kit*^+*Kit*^+) and half are heterozygous (*Kit*^+*Kit*^t), as expected. However, many of the *Kit*^+*Kit*^+ mice develop white tails and feet, the phenotype expected of the heterozygotes. In the presence of the *Kit*^t allele in the heterozygote, the *Kit*^+ allele is altered so that it has the same phenotype as *Kit*^t. Mice with these altered alleles are designated as *Kit*^*. The altered *Kit*^* allele is stably transmitted to future generations, where it continues to produce white tails and feet. Some humans with a white spot in the forehead hair and areas of reduced pigment (called the piebald trait) have mutations in the *Kit* locus; other mutations in *Kit* result in a predisposition to some cancers.

Researchers have demonstrated that this example of paramutation is also mediated by RNA molecules, although the mechanism is likely to be different from that seen in paramutation in corn. In mice, the white tails and feet of individuals with the *Kit*^t allele appear to be caused by microRNAs (miRNAs) that degrade the *Kit* mRNA, and these miRNAs appear to be transmitted to future generations through the gametes. Researchers observed a twofold decrease in *Kit* mRNA in both the heterozygous mice and the *Kit*^* mice, which suggested that the white tails and feet of the heterozygotes are due to a reduction in the amount of *Kit* mRNA. To determine whether RNA is responsible for paramutation, they injected some wild-type embryos with RNA from *Kit*^+*Kit*^+ homozygotes and injected other wild-type embryos with RNA from *Kit*^+*Kit*^t heterozygotes. Among the mice that completed development, they observed white tails and feet more frequently in those injected with RNA from heterozygotes, which suggested that RNA from heterozygotes is capable of altering the *Kit*^+ allele of the wild-type mice (**Table 21.1**). The researchers then injected some wild-type embryos with miRNAs that degrade *Kit* mRNA. This treatment produced more mice with white tails and feet than the control treatment, in which they injected nonspecific miRNAs into wild-type embryos (see Table 21.1). The ability to produce the white tails and feet characteristic of the *Kit*^+*Kit*^t genotype by injecting miRNA into embryos suggests that this case of paramutation is associated with miRNA molecules that are transferred to the embryo via egg and sperm. However, many aspects of paramutation at the *Kit* locus are still poorly understood.

THINK-PAIR-SHARE Question 4 👥

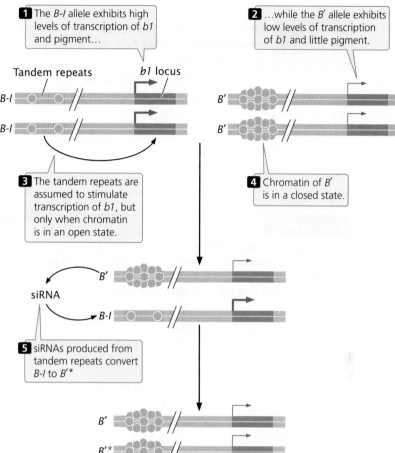

1 The *B-I* allele exhibits high levels of transcription of *b1* and pigment…

2 …while the *B'* allele exhibits low levels of transcription of *b1* and little pigment.

Tandem repeats   *b1* locus

3 The tandem repeats are assumed to stimulate transcription of *b1*, but only when chromatin is in an open state.

4 Chromatin of *B'* is in a closed state.

siRNA

5 siRNAs produced from tandem repeats convert *B-I* to *B'**

21.7 Paramutation at the *b1* locus in corn requires the presence of seven tandem repeats upstream of the *b1* locus.

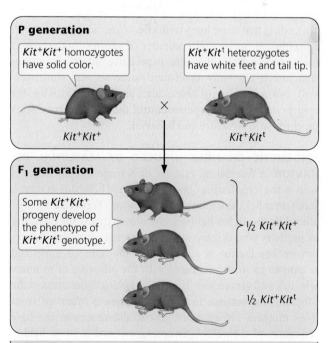

P generation

Kit^+*Kit*^+ homozygotes have solid color.

Kit^+*Kit*^t heterozygotes have white feet and tail tip.

Kit^+*Kit*^+   ×   *Kit*^+*Kit*^t

F₁ generation

Some *Kit*^+*Kit*^+ progeny develop the phenotype of *Kit*^+*Kit*^t genotype.

½ *Kit*^+*Kit*^+

½ *Kit*^+*Kit*^t

Conclusion: A cross between *Kit*^+*Kit*^+ and *Kit*^+*Kit*^t produces ½ *Kit*^+*Kit*^+ and ½ *Kit*^+*Kit*^t progeny, but some *Kit*^+*Kit*^+ develop the phenotype of heterozygotes.

21.8 Paramutation at the *Kit* locus in mice.

TABLE 21.1	Effects of injection of different types of RNA into wild-type (Kit^+Kit^+) mouse embryos	
Type of RNA Injected	**Presence of White Tail Tips and Feet**	
Kit^+Kit^+ mRNA	Uncommon	
Kit^+Kit^t mRNA	More common	
miRNA to Kit mRNA	More common	
Nonspecific miRNA	Uncommon	

CONCEPTS

Paramutation occurs when one allele creates a heritable alteration in another allele without any change in DNA sequence. Research suggests that paramutation in corn and mice is mediated through small RNA molecules.

✔ CONCEPT CHECK 3

Which of the following is a characteristic of paramutation?

a. One allele is able to alter another allele when both are present in a heterozygote.
b. Altered alleles must be passed on to future generations.
c. Altered alleles must be capable of altering other alleles in future generations.
d. All of the above.

Behavioral Epigenetics

Research has shown that life experiences, especially those early in life, can have long-lasting effects on behavior, in some cases into future generations. Increasingly, researchers are finding that these long-term effects are mediated through epigenetic processes. The number of studies that convincingly demonstrate that life experience alters chromatin structure is currently small (and some are still controversial), but a number of researchers are actively looking for epigenetic effects of experience and their long-term effects on chromatin structure and behavior.

EPIGENETIC CHANGES INDUCED BY MATERNAL BEHAVIOR A fascinating example of behavioral epigenetics is seen in the long-lasting effects of maternal behavior in rats. A mother rat licks and grooms her offspring (**Figure 21.9**), usually while she arches her back and nurses them. The offspring of mothers who display more licking and grooming behavior are less fearful as adults and show reduced hormonal responses to stress compared with the offspring of mothers who lick and groom less. These long-lasting differences in the offspring are not due to genetic differences inherited from their mothers—at least not genetic differences in the base sequences of their DNA. Offspring exposed to more licking and grooming develop a different pattern of DNA methylation than offspring exposed to less licking and grooming. These differences in DNA methylation affect the acetylation

21.9 Young rats exposed to more licking and grooming from their mothers develop different patterns of DNA methylation, which alters the expression of stress-response genes and makes them less fearful as adults compared with the offspring of mothers who lick and groom less. [Eric Isselee/Shutterstock.]

pattern of histone proteins. The altered acetylation pattern persists into adulthood and alters the expression of the glucocorticoid receptor gene, which plays a role in hormonal responses to stress. The expression of other stress-response genes is also affected.

To demonstrate the effect of altered chromatin structure on the stress responses of the offspring, researchers infused the brains of young rats with a deacetylase inhibitor, which prevents the removal of acetyl groups from histone proteins. After infusion of the deacetylase inhibitor, differences in DNA methylation and histone acetylation associated with maternal grooming behavior disappeared, as did the difference in responses to fear and stress in the young rats when they reached adulthood. This study demonstrates that the mother rat's licking and grooming behavior brings about epigenetic changes in the offspring's chromatin, which causes long-lasting differences in their behavior.

▶ **TRY PROBLEM 27**

EPIGENETIC EFFECTS OF EARLY STRESS Numerous studies have demonstrated that stress during childhood and adolescence produces a number of adverse effects that persist into adult life. For example, childhood abuse increases the probability that the child will experience depression, anxiety, and suicide as an adult. In one study, researchers examined the brains of 24 people who had committed suicide, half of whom had experienced childhood abuse. They found that those who had experienced childhood abuse had a greater degree of methylation of the glucocorticoid receptor gene, a gene involved in the stress response, than those who had not. Although the number of brains studied was small, the study suggests that early childhood stress can indeed cause epigenetic modifications to chromatin structure in humans.

Other studies have demonstrated that gene expression is affected by early life experience. For example, researchers found that children growing up in a lower socioeconomic environment before the age of 5 showed altered expression of over 100 genes related to immune function as adults. The introduction to Chapter 11 discusses the observation that early childhood stress—in the form of growing up in an orphanage—alters telomere length, a type of epigenetic change.

Research in mice suggests that epigenetic effects of stress may be mediated by small RNA molecules. In these studies, male mice were subjected to chronic stresses such as the odor of a predator, restraint, or noise. After they were exposed to the stress, the males were bred to females who had experienced no stress. Their offspring displayed a blunted hormonal response when stressed by restraint, a response called reduced hypothalamic–pituitary–adrenal (HPA) stress axis reactivity. Furthermore, these offspring exhibited altered expression of genes involved in the stress response.

The researchers found increased levels of nine miRNAs in the stressed fathers' sperm, which they believed conveyed the blunted hormonal response to the offspring. To test this hypothesis, they injected copies of these specific miRNAs into mouse embryos whose parents had not been exposed to stressful conditions and implanted the embryos into surrogate mothers. When these mice grew up, they exhibited the same blunted hormonal response as the offspring of stressed fathers, even though their parents had never been exposed to stressful conditions. Appropriate controls were included to ensure that the observed effect was not simply due to the injection procedure. The results demonstrated that miRNAs are involved in the epigenetic transmission of the altered hormonal response across generations. Using RNA sequencing to examine levels of transcription (see Section 20.2), the researchers also showed that the miRNAs brought about changes in gene expression associated with the altered stress response. Other studies have found that fragments of tRNAs passed through sperm similarly bring about epigenetic effects of diet that are passed from male mice to their offspring.

EPIGENETICS IN COGNITION A number of research studies have shown that abnormalities in DNA methylation are associated with disorders of development and intellectual ability in humans. These findings prompted researchers to look for effects of chromatin structure on learning, memory, and cognitive ability in mice and rats. One study found that training mice to avoid an aversive stimulus at a specific location reduced DNA methylation of the *Bdnf* gene, which encodes a growth factor that stimulates the growth of connections between neurons. When demethylated, the *Bdnf* gene was more active. When researchers injected a drug that inhibits demethylation into the brains of the trained mice, activity of the *Bdnf* gene was decreased, and the mice's memory of where the adverse stimulus occurred also decreased.

Another study found that a drug that promotes the acetylation of histone proteins improved learning and memory in mice that have a disorder similar to Alzheimer disease. Recall that acetylation of histones alters chromatin structure by loosening the association of DNA with histone proteins and stimulates transcription of many genes. Other studies have found that histone acetylation decreases with age in

mice, resulting in diminished expression of genes related to learning and memory. When researchers injected mice with a drug that is an inhibitor of deacetylase activity, acetylation of histones increased, transcription of genes involved in memory increased, and the memory of the mice improved. These studies suggest that changes in chromatin structure may be involved in memory and learning.

THINK-PAIR-SHARE Question 5

> ## CONCEPTS
>
> Studies are providing evidence that early life experiences can produce epigenetic changes that have long-lasting effects on behavior.

Epigenetic Effects of Environmental Chemicals

Because some chemicals are capable of modifying chromatin structure, researchers have looked for long-term effects of environmental toxicants on chromatin structure and epigenetic traits.

There has been much recent interest in chemicals called endocrine disruptors, which mimic or interfere with natural hormones. Endocrine disruptors are capable of interfering with processes regulated by natural hormones, such as sexual development and reproduction. One of these endocrine disruptors is vinclozolin, a common fungicide used to control fungal diseases in vegetables and fruits—particularly wine grapes—and to treat turf on golf courses. Vinclozolin acts as an antagonist at the androgen receptor; that is, vinclozolin and its metabolites mimic testosterone and bind to the androgen receptor, preventing testosterone from binding. But vinclozolin and its metabolites do not properly activate the receptor, and in this way, vinclozolin inhibits the action of androgens and prevents sperm production.

In one study, researchers found that exposure of embryonic male rats to vinclozolin led to reduced sperm production not only in the treated animals (when they reached puberty), but also in several subsequent generations. Increased DNA methylation was seen in the sperm of the males that were exposed to vinclozolin, and these patterns of methylation were inherited. This study and others have raised concerns that, through epigenetic changes, environmental exposure to some chemicals might have effects on the health of future generations. **❯ TRY PROBLEM 28**

THINK-PAIR-SHARE Question 6

> ## CONCEPTS
>
> Through epigenetic changes, environmental chemicals may have influences that extend to later generations.

Epigenetic Effects on Metabolism

In the introduction to this chapter, we saw that nutrition during prenatal development can have effects on health in later life. These types of epidemiological studies on humans are supported by laboratory studies of mice and rats. In one study, researchers fed inbred male mice either a normal (control) diet or a diet low in protein. They then bred mice in both groups to control females fed a normal diet. The males were then separated from the females and never had any contact with their offspring; their only contribution to the offspring was a set of paternal genes transferred through the sperm.

The offspring were raised and their lipid and cholesterol levels examined. The offspring of males fed a low-protein diet exhibited increased expression of genes involved in lipid and cholesterol metabolism, and a corresponding decrease in levels of cholesterol, compared with the offspring of males fed a normal diet. The researchers also observed numerous differences in DNA methylation in the offspring of the two groups of fathers, although no differences could be found in the methylation patterns of the sperm of the two groups of fathers. These results suggest that epigenetic changes altered the cholesterol metabolism of the offspring, although how the differences in methylation were transmitted from father to offspring was unclear.

In another study, researchers fed male rats a high-fat diet and, not surprisingly, the rats gained weight. The researchers then bred these males to females that had been fed a normal diet. The offspring were also fed a normal diet. The daughters of the male rats on the high-fat diet had normal weight, but as adults they developed a diabetes-like condition of impaired glucose tolerance and insulin secretion. The researchers observed that in the insulin-secreting pancreatic islet cells of the daughters, the expression of 642 genes involved in insulin secretion and glucose tolerance was altered, demonstrating that a father's diet affected gene expression in his daughters.

Epigenetic Effects in Monozygotic Twins

Monozygotic (identical) twins develop from a single egg fertilized by a single sperm that divides and gives rise to two zygotes (see Section 6.3). Thus, monozygotic twins are genetically identical, in the sense that they possess identical DNA sequences, but they often differ somewhat in appearance, health, and behavior. The nature of these differences in the phenotypes of identical twins is not well understood, but recent evidence suggests that at least some of these differences may be due to epigenetic changes. In one study, Mario Fraga, at the Spanish National Cancer Center, and his colleagues examined 80 pairs of identical twins and compared the degree and location of their DNA methylation and histone acetylation. They found that DNA methylation and histone acetylation in identical twin pairs were similar early in life, but that older twin pairs had remarkable differences in their overall content and distribution of DNA methylation and histone acetylation. Furthermore, these differences

affected gene expression in the twins. This research suggests that identical twins do differ epigenetically and that phenotypic differences between them may be caused by differential gene expression.

THINK-PAIR-SHARE Question 7

CONCEPTS

Phenotypic differences between genetically identical monozygotic twins may result from epigenetic effects.

✔ CONCEPT CHECK 4

What degree of differences would you expect to see in the DNA base sequences and epigenetic marks of monozygotic twins?

a. Similar differences in DNA base sequence and epigenetic marks
b. Greater differences in DNA base sequence than epigenetic marks
c. Greater differences in epigenetic marks than DNA base sequence
d. No differences in either DNA base sequence or epigenetic marks

X Inactivation

Early in the development of female mammals, one X chromosome in each cell is randomly inactivated to provide equal expression of X-linked genes in males and females (dosage compensation; see Section 4.3). Through this process, termed X inactivation, many genes on the inactivated X chromosome are permanently silenced and are not transcribed. Once a particular X chromosome is inactivated in a cell, that same X chromosome remains inactivated when the DNA is replicated, and the inactivation mark is passed on to daughter cells through mitosis. This phenomenon is responsible for the patchy distribution of black and orange pigment seen in tortoiseshell cats (see Figure 4.19). X inactivation is a type of epigenetic effect because it results in a stable change in gene expression that is passed on to other cells.

A great deal of research has demonstrated that which X chromosome is inactivated within a cell is controlled by a particular segment of the X chromosome called the **X-inactivation center**, which is 100,000 to 500,000 bp in length. Inactivation is initiated at the X-inactivation center and then spreads to the remainder of the inactivated X chromosome. Examination of the X-inactivation center led to the discovery of several genes that play a role in inactivating all but one X chromosome in each female cell and keeping the other X chromosome active (**Figure 21.10**).

The key player in X inactivation is a gene called *Xist* (for *X-inactive specific transcript*), which encodes a long noncoding RNA (lncRNA) that is 17,000 bp in length (**Figure 21.11**). As its name implies, this RNA molecule does not encode a protein. Instead, *Xist* lncRNA coats the X chromosome from which it was transcribed. *Xist* lncRNA then attracts polycomb repressive complex 2 (PRC2) and, eventually, polycomb repressive complex 1 (PRC1). These proteins produce epigenetic marks, such as H3K27me3, and other

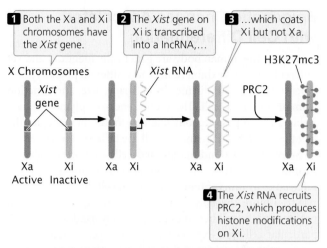

1 Both the Xa and Xi chromosomes have the *Xist* gene.

2 The *Xist* gene on Xi is transcribed into a lncRNA,…

3 …which coats Xi but not Xa.

4 The *Xist* RNA recruits PRC2, which produces histone modifications on Xi.

21.10 In X inactivation, the *Xist* gene on the inactive X produces a long noncoding RNA that coats the inactive X chromosome and suppresses transcription.

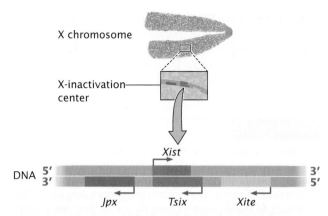

21.11 Several genes within the X-inactivation center interact to bring about inactivation of one X chromosome while keeping the other X chromosome active. Shown are the genes and X chromosome for the mouse.

histone modifications that repress transcription. Eventually, many CpG dinucleotides are methylated, leading to permanent silencing of the inactivated X chromosome.

In mice, there are two separate inactivation events. Soon after fertilization, when the embryo reaches the eight-cell stage, the X chromosome from the male parent is inactivated, while the maternal X chromosome remains active. This event is called imprinted X-chromosome inactivation. In the developing embryo, the paternal X chromosome is then reactivated during blastocyst maturation. Inactivation occurs again in early development, but now which X is inactivated is random: the X from the male parent and the X from the female parent are equally likely to be inactivated. From this point on, whichever X is inactivated remains silenced through subsequent cell divisions. However, some genes on the inactivated X chromosome escape inactivation and continue to be transcribed. How these genes escape X inactivation is not known. Interestingly, in marsupial mammals, the paternal X chromosome is the copy that remains permanently silenced in all cells.

As mentioned, X inactivation is brought about by the transcription of the *Xist* gene on the inactive X chromosome to produce *Xist* lncRNA, which coats the inactive X chromosome and leads to changes in chromatin structure that silence transcription. But what happens on the active X chromosome? Why isn't it coated by *Xist* RNA and silenced? Although all details of this process are not yet understood, recent research has demonstrated that there are several additional genes in the X-inactivation center that encode other lncRNAs. These lncRNAs help bring about inactivation of the inactive X while not silencing the active X (see Figure 21.10). One of these genes is the *Tsix* gene, which is transcribed on the active X chromosome. *Tsix* is antisense to *Xist*, which means that it overlaps with the *Xist* gene and is transcribed from the opposite strand (see Figure 21.11), producing a *Tsix* lncRNA that is complementary to *Xist*

lncRNA. Through several mechanisms, *Tsix* represses the expression of *Xist* on the active X chromosome. Another major player is a gene called *Jpx*, which encodes a lncRNA that stimulates transcription of *Xist* on the inactive X chromosome. Thus, *Xist* is controlled by two parallel switches with opposite effects: (1) *Jpx* stimulates *Xist* expression on the inactive X chromosome, causing *Xist* to be transcribed and leading to X inactivation; and (2) *Tsix* represses *Xist* on the active X chromosome, causing *Xist* not to be transcribed on that chromosome and preventing inactivation. Several other genes are also involved. A gene called *Xite* encodes a lncRNA that sustains *Tsix* expression on the active X chromosome. The major genes involved in the process of X inactivation are summarized in **Table 21.2**.

This complex process ensures that in each female cell, one X chromosome is inactivated and one remains active. Scientists have long recognized that X inactivation also involves some type of mechanism that is capable of counting X chromosomes, because all but one X chromosome in each cell

TABLE 21.2		Major genes involved in X inactivation
Gene	**Encodes**	**Action of Gene**
Xist	lncRNA	Coats inactive X chromosome and leads to silencing of transcription of many genes on the inactive X
Tsix	lncRNA	Inhibits transcription of *Xist* on active X chromosome
Jpx	lncRNA	Stimulates transcription of *Xist* on inactive X chromosome
Xite	lncRNA	Sustains *Tsix* expression on active X, which inhibits *Xist* and maintains transcription of genes on active X chromosome

is inactivated. Thus, the single X in the cells of an XY male remains active (no X inactivation occurs), and two X chromosomes are inactivated in XXX females (see discussion of Barr bodies in Section 4.3). The nature of this counting mechanism is not yet well understood. ❯ **TRY PROBLEM 30**

CONCEPTS

Epigenetic changes underlie X inactivation, in which one X chromosome in female cells is permanently silenced. X inactivation occurs through the action of several genes in the X-inactivation center that encode long noncoding RNAs. The products of these genes interact to ensure that one X chromosome is inactive and one remains active in each female cell.

✔ CONCEPT CHECK 5

What would be the effect of introducing into a female cell siRNAs that degrade *Xist* RNA?

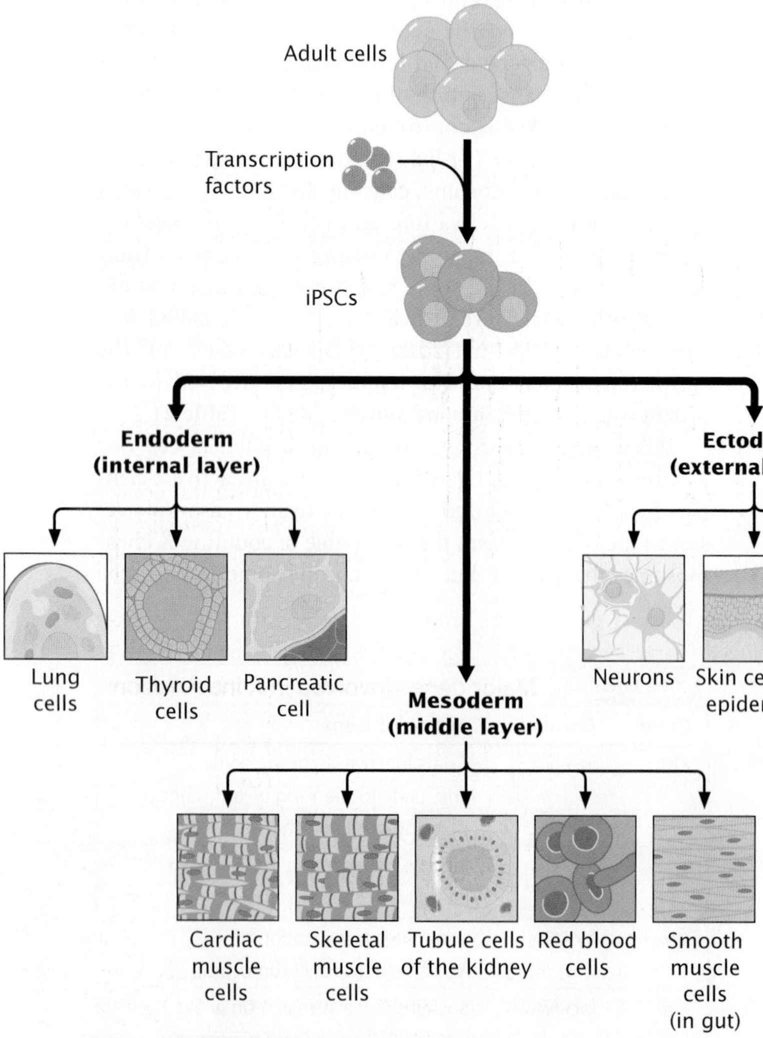

Epigenetic Changes Associated with Cell Differentiation

All cells in the human body are genetically identical, and yet different cell types exhibit remarkably different phenotypes—a nerve cell is quite distinct in its shape, size, and function from an intestinal cell. These differences in phenotypes are stable and are passed from one cell to another, despite the fact that the DNA sequences of all the cells are the same.

Stem cells are undifferentiated cells that are capable of forming every type of cell in an organism, a property referred to as **pluripotency**. As a stem cell divides and gives rise to a more specialized type of cell, the gene-expression program of the cell becomes progressively fixed, so that each particular cell type expresses only those genes necessary to carry out the functions of that cell type. Though the control of these cell-specific expression programs is not well understood, changes in DNA methylation and chromatin structure clearly play important roles in silencing some genes and activating others.

Stem cells provide a potential source of cells for regeneration of tissues, medical treatment, and research. In the past, embryos were the only source of stem cells with the capacity to differentiate into adult tissues, but because of ethical concerns about creating and using human embryos for harvesting stem cells, researchers have long sought the ability to induce adult somatic cells to dedifferentiate and revert to stem cells. Such cells are called **induced pluripotent stem cells** (**iPSCs**). Researchers have now successfully created iPSCs by treating fibroblasts (fully differentiated human connective-tissue cells) in culture with a cocktail of transcription factors (**Figure 21.12**), although less than 1% of the cells that are so treated actually revert to iPSCs. Transcription factors that induce pluripotency cause extensive epigenetic reprogramming, altering the patterns of DNA methylation and histone modifications that accumulate with cell differentiation. Recent research has shown, however, that iPSCs retain a memory of their past and are not completely equivalent to embryonic stem cells (those derived from embryos). One study found that although the DNA methylation patterns of iPSCs differ greatly from those of differentiated somatic cells, the iPSCs retained some methylation marks of the somatic cells, and the methylation of iPSCs was not identical with that of embryonic stem cells. Another study compared histone modifications of fibroblasts, iPSCs, and embryonic stem

21.12 Differentiated adult cells can be reprogrammed to form induced pluripotent stem cells (iPSCs), which are capable of differentiating into many different types of cells.

cells. The iPSCs and embryonic stem cells had many fewer H3K27me3 and H3K9me3 marks than did the fibroblasts, but researchers also found significantly more of these marks on the iPSCs than on the embryonic stem cells.

Genomic Imprinting

Diploid organisms usually possess two alleles at each autosomal locus, one allele inherited from the mother and one allele inherited from the father. For most genes, both alleles are expressed, and the effect of a particular allele on the phenotype is independent of which parent transmitted the allele to the offspring. However, for a few genes, the sex of the parent that contributed the allele influences how that allele is expressed—alleles inherited from the mother and from the father are not equivalent (**Figure 21.13**). This phenomenon, in which the sex of the parent that transmits the allele determines its expression, is termed genomic imprinting. For some imprinted genes, when the allele is inherited from the male parent it is expressed, but when it is inherited from the female it is silent; for other genes, when the allele is inherited from the female parent it is expressed, but when it is inherited from the male it is silent. As discussed in Section 5.3, genomic imprinting is thought to be due to different degrees of methylation of the alleles inherited from the two parents.

An interesting example of genomic imprinting involves crosses between a horse and a donkey. A cross between a male donkey and a female horse produces a mule, but a cross between a female donkey and a male horse produces a hinny. Mules and hinnies differ in appearance, physiology, and behavior; for example, hinnies are smaller than mules, have shorter ears, and have stronger legs. Because both mules and hinnies have the same genetic makeup (half donkey and half horse) and differ only in which sex conveys the horse and donkey genes, their differences are thought to be due to genomic imprinting. Studies of RNA from mules and hinnies demonstrate that they do indeed exhibit differential expression of some genes.

Previous research had suggested that the number of imprinted genes was limited, but more recent research suggests that the number is much higher. A study conducted by Christopher Gregg, at Harvard University, and his colleagues found that over 1300 genes in the mouse brain exhibited evidence of genomic imprinting. Many of these imprinted genes were not completely silenced; instead, there was biased expression, with one sex transmitting an allele that was more highly expressed than the allele transmitted by the other sex. Gregg and his colleagues also found that imprinting was highly variable; some genes were imprinted only in certain tissues or at certain times of development.

Genomic imprinting has a number of interesting parallels to X inactivation. Most imprinted genes are located in clusters of 3–12 genes that occur in a discrete region of a particular chromosome. Each cluster contains genes that encode proteins as well as genes that produce noncoding RNA. In each of the

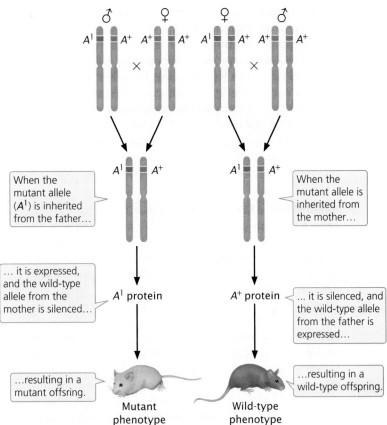

21.13 In genomic imprinting, the expression of an allele depends on whether it is inherited from the male or the female parent. In this case, the A^1 allele is expressed only when inherited from the male parent, but in other cases, an allele is expressed only when inherited from the female parent.

well-studied examples, there is an imprinting control region that determines imprinting; deletion of this region destroys the ability to imprint. In addition, the imprinting control region exhibits differences in chromatin modifications between alleles inherited from the male and female parents. Each imprinting cluster contains genes for one or more lncRNAs, which play an important role in imprinting and are themselves imprinted. For example, the gene for insulin-like growth factor 2 (*Igf2*) in humans exhibits genomic imprinting; the *Igf2* allele transmitted from the male parent is expressed, while the *Igf2* allele transmitted from the female is silenced (see Figure 5.18). Several lncRNAs produced by other genes in the imprinting control region are required for silencing of the female *Igf2* allele in the fetus, although how they bring about repression of transcription is not clear.

Many of the well-studied clusters of imprinted genes are associated with disorders that result from faulty imprinting. Beckwith-Wiedemann syndrome is one such disorder. Children with Beckwith-Wiedemann syndrome exhibit excessive growth during fetal development and early childhood. They also have unusual embryonic malignant tumors. Beckwith-Wiedemann syndrome is associated with imprinting in a cluster of genes on chromosome 11, including the *Igf2* gene. Individuals with Beckwith-Wiedemann syndrome often have small deletions on chromosome 11 that interfere with the normal process of imprinting. For example, *Igf2* is normally expressed only when inherited from the father, but in some children with Beckwith-Wiedemann syndrome, deletions within the imprinting center lead to expression of alleles from both parents. The result is that too much *Igf2* is produced, leading to excessive growth and cancer. Prader-Willi syndrome and Angelman syndrome are disorders that are due to imprinting defects on chromosome 15 (see Section 5.3).

IMPRINTING AND GENETIC CONFLICT Many imprinted genes affect early embryonic and fetal growth. One possible explanation for genomic imprinting is the **genetic conflict hypothesis**, which suggests that there are different and conflicting evolutionary pressures acting on maternal and paternal alleles for genes (such as *Igf2*) that affect fetal growth. From an evolutionary standpoint, paternal alleles that maximize the size of the offspring are favored because birth weight is strongly associated with infant survival and adult health. Thus, it is to the advantage of the male parent to pass on alleles that promote maximum fetal growth in his offspring. In contrast, maternal alleles that cause more limited fetal growth are favored: committing too many nutrients to any one fetus may limit a mother's ability to reproduce in the future, and giving birth to very large babies is also difficult and risky. The genetic conflict hypothesis predicts that genomic imprinting will evolve: paternal copies of genes that increase fetal growth should be maximally expressed, whereas maternal copies of the same genes should be less actively expressed or even silent. Indeed, *Igf2* follows this pattern: the paternal allele is active and promotes growth; the maternal allele is silent and does not contribute to growth. Recent findings demonstrate that the paternal copy of *Igf2* promotes fetal growth by directing maternal nutrients to the fetus through the placenta.

CONCEPTS

Genomic imprinting is caused by epigenetic differences in the alleles inherited from male and female parents. The genetic conflict hypothesis suggests that imprinting evolves because of conflicting evolutionary pressures acting on maternal and paternal alleles.

✔ **CONCEPT CHECK 6**

Which of the following is true of genomic imprinting?

a. The sex of the parent that transmits an allele affects the expression of the allele in the offspring.

b. The sex of the offspring affects the expression of an allele inherited from one of the parents.

c. The sex of the parent affects how an allele is transmitted to the offspring.

d. The sex of the offspring affects which allele is inherited from the parent.

21.4 The Epigenome

In 2003, researchers declared that essentially the entire human genome had been sequenced. This monumental achievement provided a wealth of information about how genetic information is encoded within the genome. Yet the DNA base sequence is only a partial record of heritable information. As we have discussed, additional epigenetic information is contained within chromatin structure—information that is heritable and affects how the DNA base sequence is expressed. The overall pattern of chromatin modifications in a genome has been termed the **epigenome**. Over the past few years, techniques have become available for detecting and describing epigenetic modifications across the genome.

DETECTING DNA METHYLATION A number of techniques have been developed for examining levels of DNA methylation. Some of these techniques rely on restriction endonucleases, enzymes that make double-stranded cuts in the DNA at specific base sequences (see Section 19.2). Some restriction enzymes are sensitive to methylation and will not cut a sequence that contains 5-methylcytosine, whereas other restriction enzymes are insensitive to methylation. By cutting DNA with enzymes that are sensitive to methylation and with enzymes that are not, and then analyzing the resulting fragments, overall patterns of methylation can be determined.

A more precise and widely used technique for analyzing DNA methylation is bisulfite sequencing (**Figure 21.14**). In this technique, genomic DNA is first treated with sodium bisulfite, which chemically converts unmethylated cytosine to uracil. Uracil is then detected as thymine during sequencing. However, 5-methylcytosine is not chemically altered by treatment with bisulfite, and it is detected as cytosine during sequencing (see Section 19.5 for a discussion of DNA sequencing). By sequencing genomic DNA with and without bisulfite treatment, researchers are able to determine the locations of all copies of 5-methylcytosine in the DNA.

DETECTING HISTONE MODIFICATIONS Histone modifications can be detected by breaking the chromatin into fragments and applying an antibody specific to a particular histone modification, a process called chromatin immunoprecipitation (abbreviated ChIP; see Section 17.2). The antibody mediates the precipitation of chromatin fragments with the histone modification, causing them to separate from chromatin fragments without the modification. The histones are then removed by digestion with an enzyme that degrades protein, but not DNA. The genomic location of the precipitated DNA fragment with which the histone was associated is then determined. In a technique called ChIP-Seq, the fragments are sequenced and their genomic locations determined by comparison of the sequence with a reference genome. This technique provides information about where in the genome histone modifications occur.

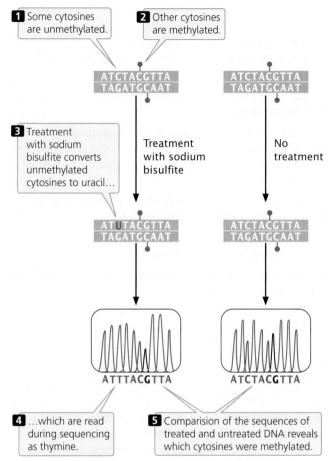

1 Some cytosines are unmethylated.

2 Other cytosines are methylated.

ATCTACGTTA
TAGATGCAAT

ATCTACGTTA
TAGATGCAAT

3 Treatment with sodium bisulfite converts unmethylated cytosines to uracil…

Treatment with sodium bisulfite

No treatment

ATUTACGTTA
TAGATGCAAT

ATCTACGTTA
TAGATGCAAT

ATTTACGTTA

ATCTACGTTA

4 …which are read during sequencing as thymine.

5 Comparison of the sequences of treated and untreated DNA reveals which cytosines were methylated.

21.14 Bisulfite sequencing can be used to determine the locations of 5-methylcytosines.

GENOME-WIDE EPIGENETIC MARKS Using the techniques we have just described, geneticists have compared the epigenomes of different types of cells. For example, researchers determined the distribution of 5-methylcytosine across the entire genome in two cell types: (1) an undifferentiated human stem cell and (2) a fibroblast. The researchers found widespread differences in the methylation patterns of these cells. The finding that patterns of DNA methylation vary among cell types supports the idea that cytosine methylation provides the means by which cell types stably maintain their differences during development. Other researchers have compared the epigenomes of cancer cells and normal cells and have observed distinct epigenetic marks associated with cancer.

Similarly, researchers have mapped the genomic locations of histone modifications in different cell types. These studies have detected specific histone modifications associated with promoters and enhancers of active genes. In one study, researchers mapped nine different epigenetic marks in nine different types of human cells. They were able to determine how the marks varied across cell types and to compare the marks associated with active and repressed genes. Because specific epigenetic marks are often associated with regulatory elements such as promoters and enhancers, researchers have used the presence of these marks to map the locations of these regulatory elements throughout the genome.

ANCIENT EPIGENOMES Neanderthals and Denisovans are two species of hominids that went extinct some 30,000 to 40,000 years ago. In recent years, DNA has been isolated from their bones and sequenced, providing a detailed view of their genomes. Research now provides complementary information about their epigenomes.

Traditional methods of detecting DNA methylation, such as bisulfite sequencing, do not work well with ancient DNA because it is often present in small amounts and is degraded. However, other methods, computational in nature, make it possible to determine regions of the DNA that were originally methylated. This is possible because, although DNA naturally decays after death, methylated and unmethylated cytosine bases do not decay in the same way. Unmethylated cytosine undergoes deamination to uracil (see Section 18.2), but methylated cytosine decays to thymine. These differences make it possible to determine the methylation status of the DNA prior to death. Before sequencing, uracil can be eliminated chemically, or a sequencing method that does not detect uracil can be used. Thus, those sites at which unmethylated cytosine has decayed to uracil are not read. Thymine, however, is read, so sites with methylated cytosine that decayed to thymine are read as thymine. This means that the ratio of thymine to cytosine bases that are read during sequencing will be higher in those regions of the DNA that were originally methylated than in regions that were not methylated.

By using the ratio of cytosine to thymine bases read during sequencing of Neanderthal and Denisovan DNA, scientists have determined the DNA methylation patterns of their genomes. The methylation patterns of over 99% of the genomes of these ancient hominids were similar to those observed in modern humans; this is not unexpected, given the close evolutionary relationships of Neanderthals, Denisovans, and *Homo sapiens*. However, there were a number of regions for which patterns of methylation differed among the species. One such region was in a group of five genes known as the *HOXD* cluster, which is an important regulator of limb development in mammals. These results suggest that this cluster of genes may have played an important role in the limb development of modern humans. Other differences in methylation were seen in enhancers and in transcription-factor binding sites. In addition, differential methylation in modern humans was seen more often in genes related to diseases, particularly neurological and psychiatric disorders such as autism and schizophrenia. This observation suggests that the epigenetic changes in gene expression that have evolved in modern humans might have increased the risk of some human diseases.

CONCEPTS

The epigenome is the complete set of chromatin modifications in a genome. Epigenomes can be examined by a number of different techniques, revealing that the epigenomes vary among cell types. Epigenomes of ancient hominids have been determined.

CONCEPTS SUMMARY

■ The term *epigenetics*, first coined by Conrad Waddington, today refers to the mechanisms by which changes in phenotype are maintained in a cell, or passed to other cells or future generations, without a change in the base sequence of DNA.

■ Many epigenetic phenotypes result from changes in chromatin structure.

■ Epigenetic changes are stable but can be affected by environmental factors.

■ Three molecular mechanisms underlie many epigenetic phenotypes: (1) changes in patterns of DNA methylation; (2) chemical modifications of histone proteins; and (3) RNA molecules that affect chromatin structure and gene expression.

■ Some epigenetic effects result from DNA methylation, in which cytosine bases are methylated to form 5-methylcytosine. Methylation often occurs at CpG dinucleotides. The presence of 5-methylcytosine is associated with repression of transcription.

■ DNA regions with many CpG dinucleotides are referred to as CpG islands. Methylation of CpG islands near a gene often leads to repression of transcription.

■ DNA methylation is maintained across cell divisions by methyltransferase enzymes that recognize methylation of CpG dinucleotides on one strand of DNA and add methyl groups to the unmethylated cytosine bases on the other strand.

■ DNA methylation inhibits transcription by inhibiting the binding of transcription factors and other proteins required for transcription to occur. DNA methylation also attracts proteins that repress transcription and histone deacetylase enzymes that alter chromatin structure.

■ Modifications of histone proteins alter chromatin structure. Histone modifications may be maintained across cell divisions.

■ RNA molecules bring about modifications of chromatin by a variety of processes.

■ Paramutation is a heritable alteration of one allele by another allele without any change in DNA sequence. Paramutation in corn and mice is mediated through small RNA molecules.

■ Early life experiences can produce epigenetic changes that have long-lasting effects on behavior.

■ Environmental chemicals may produce epigenetic effects that are passed to later generations.

■ Epigenetic modifications have effects on metabolism that extend across generations.

■ Phenotypic differences between genetically identical monozygotic twins may result from epigenetic effects.

■ X inactivation occurs when one X chromosome in female cells is permanently silenced. Epigenetic changes bring about X inactivation and require the action of several genes that encode long noncoding RNAs.

■ Genomic imprinting occurs when the expression of a gene depends on which parent transmitted the gene. It is caused by epigenetic changes in chromatin structure that are passed to offspring. The genetic conflict hypothesis suggests that imprinting evolves because of conflicting evolutionary pressures acting on maternal and paternal alleles.

■ The epigenome is the complete set of chromatin modifications in a genome.

IMPORTANT TERMS

epigenetic mark (p. 646)
paramutation (p. 647)
epiallele (p. 648)

X-inactivation
   center (p. 652)
stem cell (p. 654)

pluripotency (p. 654)
induced pluripotent stem
   cell (iPSC) (p. 654)

genetic conflict
   hypothesis (p. 656)
epigenome (p. 656)

ANSWERS TO CONCEPT CHECKS

1. b

2. b

3. d

4. c

5. No *Xist* RNA would be present to coat the X chromosome, and X inactivation would not occur. Both X chromosomes would remain active.

6. a

WORKED PROBLEM

The *b1* allele encodes a transcription factor that stimulates production of anthocyanin, a purple pigment in plants. What would be the effect of deleting the seven tandem repeats that are located 100,000 bp upstream of the *b1* locus in corn?

>> Solution Strategy

What information is required in your answer to the problem?
The effects of deleting the repeats.

What information is provided to solve the problem?

- The *b1* locus encodes a transcription factor that stimulates anthocyanin production in plants.
- The seven tandem repeats are located upstream of the *b1* locus.

For help with this problem, review:
Paramutation in Corn in Section 21.3.

>> Solution Steps

The information provided in the text indicates that the *B-I B-I* plant normally has high expression of the *b1* locus, produces anthocyanin, and is purple in color. The tandem repeats are required for high transcription of the *b1* locus, which encodes a transcription factor that stimulates production of anthocyanin, a purple pigment. The tandem repeats act like an enhancer, stimulating transcription of the *b1* locus in the *B-I B-I* genotype. Without the enhancer-like action of the repeats, *b1* will be transcribed at minimal levels and little anthocyanin will be produced. This will result in lightly pigmented *B-I B-I* plants, the same phenotype as is usually seen in plants with genotype *B′ B′*.

The tandem repeats also encode 25-nucleotide-long siRNAs, which are required for paramutation (conversion of *B-I* alleles into *B′** alleles). Deletion of the tandem repeats would thus eliminate the ability of *B′* alleles to carry out paramutation.

> **Recall:** Enhancers stimulate transcription at genes that may be distant from the enhancer (see Chapter 17).

COMPREHENSION QUESTIONS

Introduction

1. What is the thrifty phenotype hypothesis? How does it help to explain the long-term effects of famine on people conceived during the Dutch Hunger Winter?

Section 21.1

*2. What are the important characteristics of an epigenetic trait?

Section 21.2

3. What three molecular mechanisms alter chromatin structure and are responsible for many epigenetic phenotypes?

4. What is the major form of DNA methylation that is seen in eukaryotes? At what type of DNA sequence is DNA methylation usually found?

5. How does DNA methylation repress transcription?

6. Briefly explain how patterns of DNA methylation are transmitted across cell divisions.

7. What types of histone modifications are responsible for epigenetic phenotypes?

Section 21.3

8. What is paramutation? What are the key features of this phenomenon?

9. Briefly describe paramutation at the *Kit* locus in mice. What evidence suggests that small RNA molecules play a role in this phenomenon?

10. What evidence suggests that cognition in mice is influenced by epigenetic changes?

11. Explain how vinclozolin acts as an endocrine disrupter.

12. Give an example of an epigenetic effect of diet on metabolism.

13. What evidence suggests that differences in monozygotic twins may be caused by epigenetic effects?

14. What makes X inactivation an epigenetic phenotype?

15. Briefly describe the molecular processes that cause one X chromosome in each female cell to be active and the other X chromosome to be inactivated.

16. What are induced pluripotent stem cells? How are they derived from adult somatic cells?

17. Define genomic imprinting.

18. What is the genomic conflict hypothesis for the origin of genomic imprinting?

Section 21.4

19. What is the epigenome?

APPLICATION QUESTIONS AND PROBLEMS

Introduction

20. The introduction to this chapter describes the long-term effects of famine on people conceived during the Dutch Hunger Winter.

    a. What evidence suggests that these are epigenetic effects?

    b. What additional evidence would help to demonstrate that these changes are due to epigenetic changes?

Section 21.1

21. How do epigenetic traits differ from traditional genetic traits, such as the differences in the color and shape of peas that Mendel studied?

Section 21.2

22. What would be required to prove that a phenotype is caused by an epigenetic change?

23. Which honeybee in **Figure 21.4** (the worker or the queen) will have more copies of 5-methylcytosine in its DNA? Explain your answer.

*24. What would be the effect of deleting the *Dnmt3* gene in honeybees?

*25. Much of DNA methylation in eukaryotes occurs at CpG dinucleotides, but some individual cytosine nucleotides are also methylated to form 5-methylcytosine. Considering what you know about the process by which DNA methylation at CpG dinucleotides is maintained across cell division, do you think that methylation at individual C nucleotides would also be maintained by the same process? Explain your reasoning.

Section 21.3

26. A cross between the F_1 individual in **Figure 21.6** and a plant with genotype *B-I B-I* will produce progeny with what phenotype?

*27. A scientist does an experiment in which she removes the offspring of rats from their mother at birth and has her genetics students feed and rear the offspring. Assuming that the students do not lick and groom the baby rats as the mother rats normally do, what long-term behavioral and epigenetic effects would you expect to see in the rats when they grow up?

*28. Pregnant female rats were exposed to a daily dose of 100 or 200 mg/kg of vinclozolin, a fungicide commonly used in the wine industry (M. D. Anway et al. 2005. *Science* 308:1466–1469). The F_1 offspring of the exposed female rats were interbred, producing F_2, F_3, and F_4 rats. None of the F_2, F_3, or F_4 rats were exposed to vinclozolin. Testes from the F_1–F_4 male rats were examined and compared with those of control rats descended from females that had not been exposed to vinclozolin. The effects detailed in the accompanying graphs were

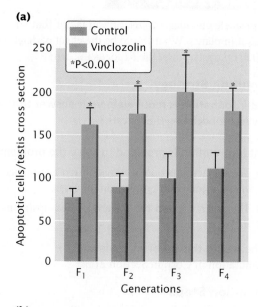

(a)

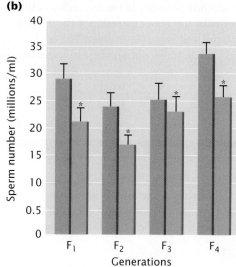

(b)

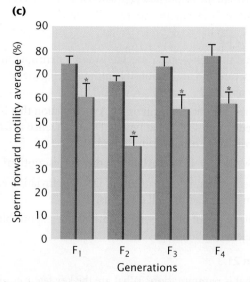

(c)

[Data from M. D. Anway et al. 2005. *Science* 308:1466–1469.]

seen in more than 90% of the F_1–F_4 male offspring. Furthermore, 8% of the F_1–F_4 males descended from vinclozolin-exposed females developed complete infertility, compared with 0% of the F_1–F_4 males descended from control females. Molecular analysis of the testes demonstrated that DNA methylation patterns differed between offspring of vinclozolin-exposed females and offspring of control females. Provide an explanation for the transgenerational effects of vinclozolin on male fertility.

29. Based on the information from studies of the long-term effects of diet on metabolism in mice, what might the epigenetic effects be on the children and grandchildren of people who were exposed to famine as children? Include in your answer the types of epigenetic changes in chromatin that you might expect to see and the phenotypic effects on lipid and cholesterol metabolism.

*30. What would be the effect on X inactivation of adding siRNAs that eliminated the products of each of the following genes?

a. *Xist*

b. *Jpx*

Section 21.4

31. A DNA fragment with the following base sequence has some cytosine bases that are methylated (indicated by C^*) and others that are unmethylated. To determine the locations of methylated and unmethylated cytosines, researchers sequenced this fragment both with and without treatment with sodium bisulfite. Give the sequence of bases that will be read with and without bisulfite treatment.

—ATCGC*GTTAC*GTTGC*GTCA—

32. A geneticist is interested in determining the locations of methylated cytosines within a fragment of DNA. She treats some copies of the fragment with sodium bisulfite and leaves some copies untreated. She then sequences the treated and untreated copies of the fragment and obtains the following results. Give the original sequence of the DNA fragment and indicate the locations of methylated cytosines.

Sequence without treatment: —AATTGCCCGATCGATTAAGCCA—

Sequence with treatment: —AATTGTTTGATCGATTAAGCTA—

CHALLENGE QUESTIONS

Section 21.3

33. Would the genomic conflict hypothesis be likely to explain genomic imprinting for genes involved in adult memory? Why or why not?

34. In recent years, techniques have been developed to clone mammals through a process called nuclear transfer, in which the nucleus of a somatic cell is transferred to an egg cell from which the nuclear material has been removed (see Chapter 22). Research has demonstrated that when a nucleus from a differentiated somatic cell is transferred to an egg cell, only a small percentage of the resulting embryos complete development, and many of those that do die shortly after birth. In contrast, when a nucleus from an undifferentiated embryonic stem cell is transferred to an egg cell, the percentage of embryos that complete development is significantly higher (W. M. Rideout, K. Eggan, and R. Jaenisch. 2001. *Science* 293:1095–1098). Propose a possible reason for why a higher percentage of cloned embryos develop successfully when the nucleus transferred comes from an undifferentiated embryonic stem cell.

Section 21.4

35. The use of embryonic stem cells has been proposed for replacing cells that are destroyed by disease or injury. Because of ethical concerns about creating and destroying embryos to produce embryonic stem cells, researchers have attempted to create induced pluripotent stem cells (iPSCs). In this chapter, we discussed studies showing that iPSCs retain some epigenetic marks from the differentiated adult cells from which they were derived. What implications might this research have for attempts to use iPSCs to regrow cells and tissues?

THINK-PAIR-SHARE QUESTIONS

Section 21.1

1. Epigenetics has been described as "inheritance, but not as we know it." Do you think this is a good definition? Why or why not?

Section 21.2

2. One reason that worker bees forgo their own reproduction to help their sister (the queen) reproduce is that female bees are more closely related to their sisters than they are to their own offspring. This quirk of genetics results from the fact that bees have haplodiploid sex determination, in which females are diploid, with a mother and a father, but males are haploid, developing from unfertilized eggs. Because males are haploid, they produce sperm by mitosis. Explain why haplodiploid sex determination causes females to be more closely related to their sisters than to their offspring.

Section 21.3

3. How are epialleles different from genetic alleles, such as those that encode differences in the shape of peas or blood types?

4. How is paramutation similar to normal gene mutation? How does it differ? Make a list of similarities and differences.

5. In the introduction to this chapter, we discussed the role of famine during early prenatal development and how, through epigenetic effects, it influenced the health of individuals as adults. One effect that has been observed in people who were conceived during the Dutch Hunger Winter is that they suffer more cognitive decline later in life. If these individuals were autopsied after death, what differences might you expect to see in the chromatin of their brain cells? What do the mouse studies of cognition and epigenetic effects suggest?

6. Some people have said that epigenetics provides genes with a memory. Explain how epigenetics allows genes to remember.

7. People often say that monozygotic twins are genetically identical. Do you think that is a correct statement? Present arguments for and against this statement.

Developmental Genetics and Immunogenetics

Alterations of key regulatory sequences can bring about major developmental changes. Mutations in an enhancer of the *Pitx1* gene have caused the loss of pelvic spines in freshwater populations of threespine sticklebacks (shown here). [Barrett & MacKay/All Canada Photos/Corbis.]

The Origin of Spineless Sticklebacks

Threespine sticklebacks (*Gasterosteus aculeatus*) are curious little fish. In spite of their small size—they reach only about 2 inches (5 cm) in length—they are heavily armored, with protective plates on the back, sides, and belly, and possess three spines on the dorsal surface, thus giving rise to their name. Each fish also possesses two impressive pelvic spines, which are anchored to the pelvic girdle and project out from the sides (**Figure 22.1**). The dorsal and pelvic spines make sticklebacks difficult to swallow, allowing them to survive in an environment where an unprotected 2-inch fish is an easily caught meal for numerous predators.

Most sticklebacks are marine, living in the ocean, but a few isolated populations can be found inland, in freshwater lakes. In North America, these freshwater populations originated 10,000 to 20,000 years ago, at the end of the last ice age, when marine sticklebacks invaded the lakes. Many lake populations of threespine sticklebacks have lost their armor and spines, probably because fish predators that might eat them are absent from the lakes, there is little calcium available to develop the plates and spines, and invertebrate predators found in the lakes catch the fish by grabbing onto their spines. Biologists have long been interested in how sticklebacks made the evolutionary transition from the marine to the freshwater environment: How did a heavily armored fish become spineless? Research conducted by developmental geneticists has begun to provide an answer to this question.

In 1998, geneticist David Kingsley, from Stanford University, began a collaboration with Dolph Schluter, an evolutionary biologist from the University of British Columbia. Their goal was to understand how threespine sticklebacks lost their pelvic spines during the evolutionary transition from marine to freshwater environments. The scientists crossed a female marine stickleback that possessed spines with a male from Paxton Lake, British Columbia, that lacked pelvic spines. All of the F_1 fish from this cross possessed pelvic spines. They then crossed two of the F_1 fish, producing a total of 375 F_2 progeny. These F_2 progeny showed a wide range of variation in their pelvic spines: some had fully developed spines, some lacked spines altogether, and others had varying degrees of spine reduction.

Kingsley and his colleagues then examined the association of pelvic spines in the F_2 progeny with their inheritance of genetic markers across the genome. They found that most of the variation in pelvic spines was associated with genetic markers from a particular region on chromosome 7. Interestingly, this same region contains *Pitx1*, a gene that is often expressed in the hind limbs of developing vertebrates. Mice with a mutation in the *Pitx1*

663

(a) Marine sticklebacks

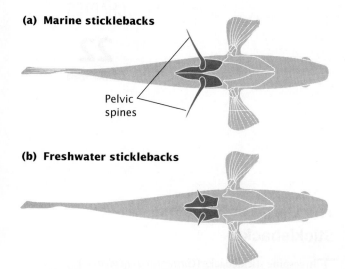

Pelvic spines

(b) Freshwater sticklebacks

22.1 Marine populations of threespine sticklebacks possess pelvic spines (a), but the spines are reduced or missing in sticklebacks from many freshwater lakes (b). [Information by permission from Macmillan Publishers Ltd: Neil H. Shubin and Randall D. Dahn, Evolutionary Biology: Lost and Found. *Nature* 428, 703–704 (15 April 2004), copyright 2004.]

gene often have reduced hind limbs as well as other developmental abnormalities. This observation suggested that mutations within the *Pitx1* gene might be responsible for the absence of pelvic spines in freshwater populations of sticklebacks. But when the researchers examined the DNA sequences of the *Pitx1* gene in marine and freshwater threespine sticklebacks, they found no differences that would alter the amino acid sequence of the protein encoded by the gene. What they did find was that the expression of the *Pitx1* gene differed in marine and freshwater fish. Sticklebacks with pelvic spines expressed *Pitx1* in the pelvic region; sticklebacks without pelvic spines expressed *Pitx1* in other tissues, but the gene was completely inactive in the pelvic region. This finding suggested that although the *Pitx1* gene itself is not mutated in spineless fish, mutations in regulatory elements that affect the expression of *Pitx1* might be the source of variation in the presence or absence of pelvic spines.

In 2010, Kingsley and his team located the mutation that causes the absence of spines in freshwater populations of threespine sticklebacks. They found an enhancer 500 bp upstream of the *Pitx1* gene that controls the tissue-specific expression of *Pitx1*. Enhancers are DNA sequences that promote the transcription of distant genes, often in a tissue-specific manner (see Chapter 17). The researchers determined that spineless fish from Paxton Lake possessed a deletion that removed this enhancer, thus preventing expression of *Pitx1* and ultimately resulting in the loss of pelvic spines. Spineless fish from other lakes in Canada, Alaska, and even Iceland also possessed deletions of the same enhancer, but fish from different lakes possessed different deletions. This observation suggests that during the course of evolution, pelvic spines have been lost multiple times through natural selection acting on different mutations that have the same phenotypic effect.

 THINK-PAIR-SHARE
- Some uninformed people might argue that studying the development of an obscure little fish like the threespine stickleback is a waste of time and money. Why would biologists study the genes that control development in stickleback fish? Give some reasons.

- Threespine sticklebacks that invaded freshwater lakes have lost their pelvic spines. What other adaptations might you expect to see in fish that have moved from a marine environment to a freshwater habitat?

The story of how the stickleback lost its spines illustrates that major anatomical alterations can occur through small genetic changes in key regulatory sequences that affect development, a theme throughout this chapter on the genetic control of development. The chapter begins with a consideration of how cell differentiation occurs, not through loss of genes, but rather through alteration of gene expression. We then discuss the genetic control of early development of *Drosophila* embryos, one of the best-understood developmental systems. We next consider the genetic control of floral structure in plants, another model system that has been well studied, and then take a more detailed look at programmed cell death and the use of development for understanding evolution. At the end of the chapter, we turn to the development of immunity and its genetic control.

22.1 Development Takes Place Through Cell Determination

Every multicellular organism begins life as a unicellular fertilized egg. This single-celled zygote undergoes repeated cell divisions, eventually producing millions or trillions of cells that constitute a complete adult organism. Initially, each cell in the early embryo is **totipotent**: it has the potential to develop into any cell type. Many cells in plants and fungi remain totipotent, but animal cells usually become committed to developing into specific types of cells after just a few early embryonic divisions. This commitment often comes well before a cell begins to exhibit any characteristics of a particular cell type; when the cell becomes committed, it does not normally reverse its fate and develop into a different cell type. A cell becomes committed by a process called **determination**.

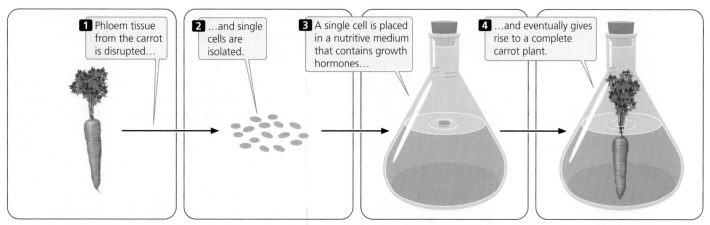

1 Phloem tissue from the carrot is disrupted…

2 …and single cells are isolated.

3 A single cell is placed in a nutritive medium that contains growth hormones…

4 …and eventually gives rise to a complete carrot plant.

22.2 Many kinds of plants can be cloned from isolated single cells. This type of experiment demonstrates that none of the original genetic material is lost during plant development.

For many years, the work of developmental biologists was limited to describing the changes that take place in the course of development because techniques for probing the intracellular processes behind these changes were not available. But in recent years, powerful genetic and molecular techniques have had a tremendous influence on the study of development; for example, DNA sequencing has provided a great deal of information about the nature and organization of DNA sequences that control developmental processes. In some model systems, such as *Drosophila* and *Arabidopsis*, the molecular mechanisms underlying developmental change are now beginning to be understood.

If all cells in a multicellular organism are derived from the same original cell, how do different cell types arise? Before the 1950s, two hypotheses were considered. One possibility was that throughout development, genes might be selectively lost or altered, causing different cell types to have different genomes. Alternatively, each cell might contain the same genetic information, but different genes might be expressed in each cell type. The results of early cloning experiments helped settle this issue.

Cloning Experiments on Plants

In the 1950s, Frederick Steward developed methods for cloning plants. He disrupted phloem tissue from the root of a carrot by separating and isolating individual cells, and then placed individual cells in a sterile medium that contained nutrients and other substances required for growth. Steward was successful in getting the cells to grow and divide, and he eventually obtained whole edible carrots from single cells (**Figure 22.2**). Because all parts of the plant were regenerated from a specialized phloem cell, he concluded that each phloem cell contained the genetic potential for a whole plant: none of the original genetic material was lost during determination.

Cloning Experiments on Animals

The results of other studies demonstrated that most animal cells also retain a complete set of genetic information during development. In 1952, Robert Briggs and Thomas King removed the nuclei from unfertilized oocytes of the frog

Lithobates (Rana) pipiens. They then isolated nuclei from frog blastulas (an early embryonic stage), injected these nuclei individually into the oocytes, and pricked the eggs with a needle to stimulate them to divide. Although most were damaged in the process, a few eggs developed into complete tadpoles that eventually metamorphosed into frogs.

In the late 1960s, John Gurdon used these methods to successfully clone a few frogs with nuclei isolated from the intestinal cells of tadpoles. This result suggested that the differentiated intestinal cells carried the genetic information necessary to encode traits found in all other cells. However, Gurdon's successful clonings may have resulted from the presence of a few undifferentiated stem cells in the intestinal tissue, which were inadvertently used as nucleus donors.

In 1997, researchers at the Roslin Institute of Scotland announced that they had successfully cloned a sheep using the genetic material from a differentiated cell of an adult animal. To perform this experiment, they fused an udder cell from a white-faced Finn Dorset ewe with an enucleated egg cell and stimulated the egg electrically to initiate development. After growing the embryo in the laboratory for a week, they implanted it into a Scottish black-faced surrogate mother. Dolly, the first mammal cloned from an adult cell, was born on July 5, 1996 (**Figure 22.3**). Since then, a number

22.3 In 1996, researchers at the Roslin Institute of Scotland successfully cloned a sheep named Dolly. They used the genetic material from a differentiated cell of an adult animal. [Paul Clements/AP.]

of other animals, including sheep, goats, mice, rabbits, cows, pigs, horses, mules, dogs, and cats, have been cloned from differentiated adult cells. Importantly, although Dolly and other cloned mammals contain the same *nuclear* genetic material as their donor-cell parent, they are not identical for *cytoplasmic* genes, such as those on the mitochondrial chromosome, because the cytoplasm is donated by both the donor cell and the enucleated egg cell.

These cloning experiments demonstrated that genetic material is not lost or permanently altered during development; therefore, development must require the selective expression of genes. But how do cells regulate their gene expression in a coordinated manner to give rise to a complex multicellular organism? Research has now begun to provide some answers to this important question.

THINK-PAIR-SHARE Questions 1 and 2

CONCEPTS

The successful cloning of plants and animals from single differentiated cells demonstrates that genes are not lost or permanently altered during development.

✔ CONCEPT CHECK 1

Scientists have cloned some animals by injecting a nucleus from an early embryo into an enucleated egg cell. Does this outcome demonstrate that genetic material is not lost during development? Why or why not?

22.2 Pattern Formation in *Drosophila* Serves as a Model for the Genetic Control of Development

Pattern formation consists of the developmental processes that lead to the shape and structure of complex multicellular organisms. One of the best-studied systems for the genetic control of pattern formation is the early embryonic development of *Drosophila melanogaster*. Geneticists have isolated a large number of mutations in fruit flies that influence all aspects of their development, and molecular analysis of these mutations has provided a great deal of information about how genes control early development in *Drosophila*.

The Development of the Fruit Fly

An adult fruit fly possesses three basic body parts: head, thorax, and abdomen (**Figure 22.4**). The thorax consists of three segments: the first thoracic segment carries a pair of legs; the second thoracic segment carries a pair of legs and a pair of wings; and the third thoracic segment carries a pair of legs and the halteres (rudiments of the second pair of wings found in most other insects). The abdomen consists of eight segments.

When a *Drosophila* egg has been fertilized, its diploid nucleus (**Figure 22.5a**) immediately divides nine times without division of the cytoplasm, creating a single, multinucleate cell (**Figure 22.5b**). These nuclei are scattered throughout the cytoplasm, but later migrate toward the periphery of the

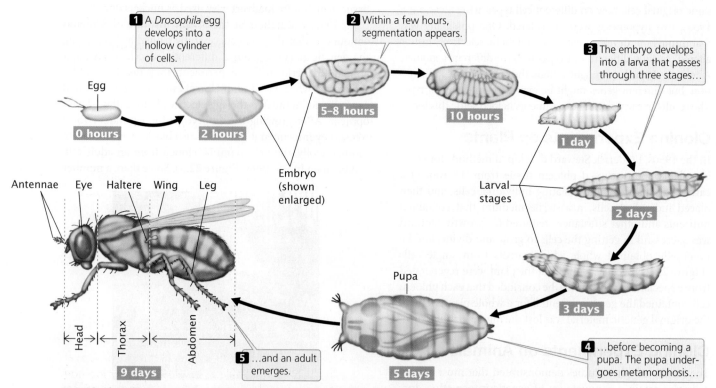

22.4 The fruit fly *Drosophila melanogaster* passes through three larval stages and a pupal stage before developing into an adult fly. The three major body parts of the adult are head, thorax, and abdomen.

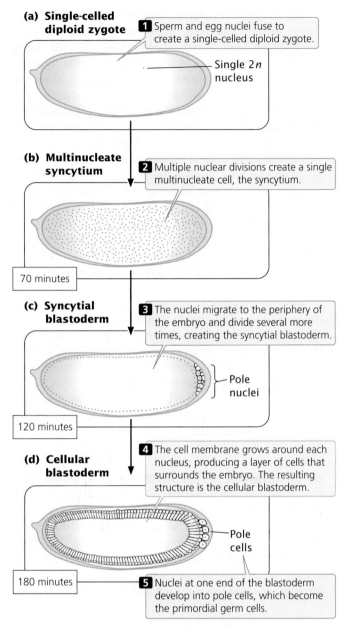

(a) Single-celled diploid zygote

1 Sperm and egg nuclei fuse to create a single-celled diploid zygote.

Single 2n nucleus

(b) Multinucleate syncytium

2 Multiple nuclear divisions create a single multinucleate cell, the syncytium.

70 minutes

(c) Syncytial blastoderm

3 The nuclei migrate to the periphery of the embryo and divide several more times, creating the syncytial blastoderm.

Pole nuclei

120 minutes

(d) Cellular blastoderm

4 The cell membrane grows around each nucleus, producing a layer of cells that surrounds the embryo. The resulting structure is the cellular blastoderm.

Pole cells

180 minutes

5 Nuclei at one end of the blastoderm develop into pole cells, which become the primordial germ cells.

22.5 Early development of a *Drosophila* embryo.

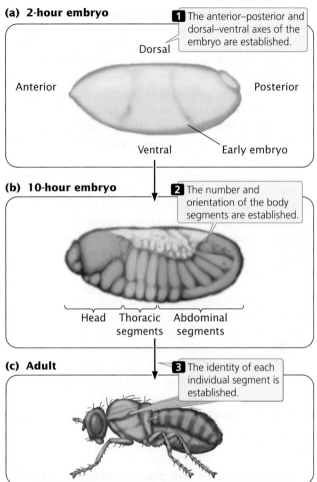

(a) 2-hour embryo

1 The anterior–posterior and dorsal–ventral axes of the embryo are established.

Dorsal

Anterior

Posterior

Ventral

Early embryo

(b) 10-hour embryo

2 The number and orientation of the body segments are established.

Head | Thoracic segments | Abdominal segments

(c) Adult

3 The identity of each individual segment is established.

22.6 In an early *Drosophila* embryo, the major body axes are established, the number and orientation of the body segments are determined, and the identity of each individual segment is established. Different sets of genes control each of these three stages.

TABLE 22.1	Stages in the early development of fruit flies and the genes that control each stage
Developmental Stage	**Genes**
Establishment of main body axes	Egg-polarity genes
Determination of number and polarity of body segments	Segmentation genes
Establishment of identity of each segment	Homeotic genes

embryo and divide several more times (**Figure 22.5c**). Next, the cell membrane grows inward and around each nucleus, creating a layer of approximately 6000 cells at the outer surface of the embryo (**Figure 22.5d**). Nuclei at one end of the embryo develop into pole cells, which eventually give rise to germ cells.

The early embryo then undergoes further development in three distinct stages: (1) the anterior–posterior axis and the dorsal–ventral axis of the embryo are established (**Figure 22.6a**); (2) the number and orientation of the body segments are determined (**Figure 22.6b**); and (3) the identity of each individual segment is established (**Figure 22.6c**). Different sets of genes control each of these three stages (**Table 22.1**).

Egg-Polarity Genes

The **egg-polarity genes** (genes that determine polarity, or direction) play a crucial role in establishing the two main axes of development in fruit flies. You can think of these axes as the longitude and latitude of development: any location in the *Drosophila* embryo can be defined in relation to these two axes.

The egg-polarity genes are transcribed into mRNAs in the course of egg formation (oogenesis) in the maternal parent, and these mRNAs become incorporated into the cytoplasm of the egg. The mRNAs are translated into proteins that play an important role in determining the anterior–posterior and dorsal–ventral axes of the embryo. Because the mRNAs of the polarity genes are produced by the female parent and influence the phenotype of the offspring, the traits encoded by them are examples of genetic maternal effect (see Chapter 5).

There are two sets of egg-polarity genes: one set determines the anterior–posterior axis, and the other determines the dorsal–ventral axis. These genes work by setting up concentration gradients of morphogens within the developing embryo. A **morphogen** is a protein that varies in concentration and elicits different developmental responses at different concentrations. Egg-polarity genes function by producing proteins that become asymmetrically distributed in the cytoplasm, giving the egg polarity. This asymmetrical distribution may take place in a couple of ways. An mRNA may be localized to particular regions of the egg cell, leading to an abundance of the protein in those regions when the mRNA is translated. Alternatively, the mRNA may be randomly distributed, but the protein that it encodes may become asymmetrically distributed by a transport system that delivers it to particular regions of the cell, by regulation of its translation, or by its removal from particular regions by selective degradation.

DETERMINATION OF THE DORSAL–VENTRAL AXIS The dorsal–ventral axis defines the back (dorsum) and belly (ventrum) of a fly (see Figure 22.6). At least 12 different genes determine this axis, one of the most important being a gene called *dorsal*. The *dorsal* gene is transcribed and translated in the maternal ovary, and the resulting mRNA and protein are transferred to the egg during oogenesis. In a newly laid egg, mRNA and protein encoded by the *dorsal* gene are uniformly distributed throughout the cytoplasm, but after the nuclei have migrated to the periphery of the embryo (see Figure 22.5c), Dorsal protein becomes redistributed. Along one side of the embryo, Dorsal protein remains in the cytoplasm; this side will become the dorsal surface. Along the other side, Dorsal protein is taken up into the nuclei; this side will become the ventral surface. At this point, there is a smooth gradient of increasing nuclear Dorsal concentration from the dorsal to the ventral side (**Figure 22.7**).

The nuclear uptake of Dorsal protein is thought to be governed by a protein called Cactus, which binds to Dorsal protein and traps it in the cytoplasm. The presence of yet another protein, called Toll, leads to the phosphorylation of Cactus, causing it to be degraded. Where Cactus is degraded, Dorsal is released and can move into the nucleus. Together, Cactus and Toll regulate the nuclear distribution of Dorsal protein, which in turn determines the dorsal–ventral axis of the embryo.

Inside the nucleus, Dorsal protein acts as a transcription factor, binding to regulatory sites on the DNA and activating or repressing the expression of other genes (**Table 22.2**). High nuclear concentrations of Dorsal protein (as in cells on the ventral side of the embryo) activate a gene called *twist*, which causes ventral tissues to develop. Low nuclear concentrations of Dorsal protein (as in cells on the dorsal side of the embryo) activate a gene called *decapentaplegic*, which specifies dorsal structures. In this way, the ventral and dorsal sides of the embryo are determined.

DETERMINATION OF THE ANTERIOR–POSTERIOR AXIS One of the most important early developmental events is the determination of the anterior (head) and posterior (butt) ends of an animal. Here, we consider several key genes that establish the anterior–posterior axis in the *Drosophila* embryo (**Table 22.3**). An important gene in this regard is *bicoid*, which is first transcribed in the maternal ovary during oogenesis. The *bicoid* mRNA and Bicoid protein become

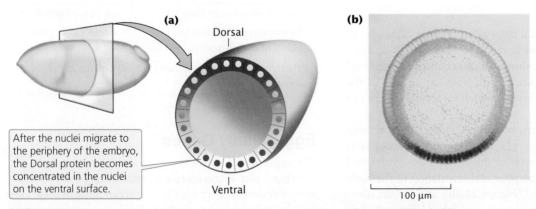

(a)

Dorsal

After the nuclei migrate to the periphery of the embryo, the Dorsal protein becomes concentrated in the nuclei on the ventral surface.

Ventral

(b)

100 µm

22.7 Dorsal protein in the nuclei helps to determine the dorsal–ventral axis of the *Drosophila* embryo. (a) Relative concentrations of Dorsal protein in the cytoplasm and nuclei of cells in the early *Drosophila* embryo. (b) Micrograph of a cross section of an embryo showing the Dorsal protein, darkly stained, in the nuclei along the ventral surface. [Part b: Max Planck Institute for Developmental Biology.]

TABLE 22.2	Key genes that control the development of the dorsal–ventral axis in fruit flies and their action	
Gene	**Where Expressed**	**Action of Gene Product**
dorsal	Ovary	Affects the expression of genes such as *twist* and *decapentaplegic*
cactus	Ovary	Traps Dorsal protein in the cytoplasm
toll	Ovary	Leads to the phosphorylation of Cactus, which is then degraded, releasing Dorsal to move into the nuclei of ventral cells
twist	Embryo	Takes part in the development of mesodermal tissues*
decapentaplegic	Embryo	Takes part in the development of gut structures

*One of the three primary tissue layers in the early embryo.

incorporated into the anterior end of the cytoplasm of the egg. After fertilization, *bicoid* mRNA and protein form a concentration gradient along the anterior–posterior axis of the embryo, with high concentrations at the anterior end and lower concentrations at the posterior end (**Figure 22.8a**). The high concentration of Bicoid protein at the anterior end induces the development of anterior structures, such as the head of the fruit fly. It stimulates the development of anterior structures by binding to regulatory sequences in the DNA and influencing the expression of other genes. One of the most important genes stimulated by Bicoid protein is *hunchback*, which is required for the development of the head and thoracic structures of the fruit fly.

TABLE 22.3	Some key genes that determine the anterior–posterior axis in fruit flies	
Gene	**Where Expressed**	**Action**
bicoid	Ovary	Regulates expression of genes responsible for anterior structures; stimulates *hunchback*
nanos	Ovary	Regulates expression of genes responsible for posterior structures; inhibits translation of *hunchback* mRNA
hunchback	Embryo	Regulates transcription of genes responsible for anterior structures

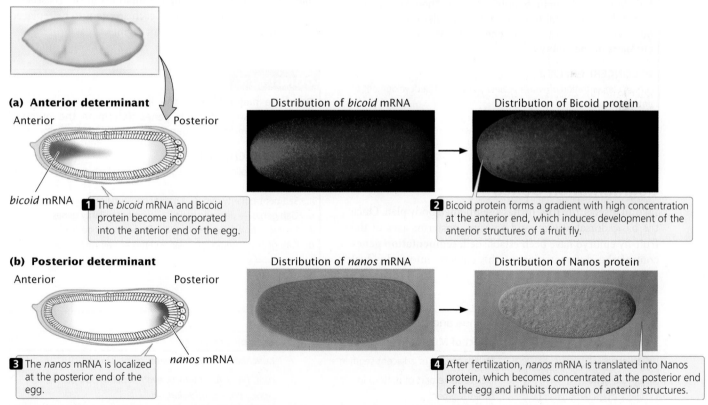

(a) Anterior determinant

Anterior   Posterior

bicoid mRNA

1 The *bicoid* mRNA and Bicoid protein become incorporated into the anterior end of the egg.

Distribution of *bicoid* mRNA

Distribution of Bicoid protein

2 Bicoid protein forms a gradient with high concentration at the anterior end, which induces development of the anterior structures of a fruit fly.

(b) Posterior determinant

Anterior   Posterior

3 The *nanos* mRNA is localized at the posterior end of the egg.

nanos mRNA

Distribution of *nanos* mRNA

Distribution of Nanos protein

4 After fertilization, *nanos* mRNA is translated into Nanos protein, which becomes concentrated at the posterior end of the egg and inhibits formation of anterior structures.

22.8 The anterior–posterior axis in a *Drosophila* embryo is determined by concentrations of Bicoid and Nanos proteins. [Part a (center, right): Ali-Murthy, Z. and T. B. Kornberg (2016) Bicoid gradient formation and function in the *Drosophila* pre-syncytial blastoderm. *eLIFE*: DOI: dx.doi.org/10.7554/eLife.13222; Published and distributed under the terms of the Creative Commons Attribution License; https://creativecommons.org/licenses/by/4.0/. Part b (center, right): Courtesy of E. R. Gavis, L. K. Dickenson, and R. Lehman, Massachusetts Institute of Technology.]

The development of the anterior–posterior axis is also greatly influenced by a gene called *nanos*, an egg-polarity gene that acts at the posterior end of the embryo. The *nanos* gene is transcribed in the ovary of the adult female, and the resulting mRNA becomes localized at the posterior end of the egg (**Figure 22.8b**). After fertilization, *nanos* mRNA is translated into Nanos protein, which diffuses slowly toward the anterior end. The Nanos protein concentration gradient is opposite that of the Bicoid protein: Nanos is most concentrated at the posterior end of the embryo and is least concentrated at the anterior end. Nanos protein inhibits the formation of anterior structures by repressing the translation of *hunchback* mRNA. The synthesis of Hunchback protein is therefore stimulated at the anterior end of the embryo by Bicoid protein and repressed at the posterior end by Nanos protein. This combined stimulation and repression results in a Hunchback protein concentration gradient along the anterior–posterior axis that, in turn, affects the expression of other genes and helps determine the anterior and posterior structures.

> ### CONCEPTS
>
> The major axes of development in early fruit-fly embryos are established as a result of initial differences in the distribution of specific mRNAs and proteins encoded by genes in the female parent (genetic maternal effect). These differences in distribution establish concentration gradients of morphogens, which cause different genes to be activated in different parts of the embryo.
>
> ✔ **CONCEPT CHECK 2**
>
> A high concentration of which protein stimulates the development of anterior structures?
>
> a. Dorsal   c. Bicoid
> b. Toll   d. Nanos

Segmentation Genes

Like all insects, the fruit fly has a segmented body plan. Once the basic dorsal–ventral and anterior–posterior axes of the fruit-fly embryo have been established, **segmentation genes** control the differentiation of the embryo into individual segments. These genes affect the number and organization of the segments, and mutations in them usually disrupt whole sets of segments. The approximately 25 segmentation genes in *Drosophila* are transcribed after fertilization, so they don't exhibit genetic maternal effect, and their expression is regulated by the Bicoid and Nanos protein gradients.

The segmentation genes fall into three classes, as shown in **Table 22.4** and **Figure 22.9**. The three classes act sequentially, affecting progressively smaller regions of the embryo. First, the products of the egg-polarity genes activate or repress **gap genes**, which divide the embryo into broad regions. The gap genes, in turn, regulate **pair-rule genes**, which affect the development of pairs of segments. Finally, the pair-rule genes influence **segment-polarity genes**, which guide the development of individual segments.

Gap genes define large sections of the embryo; mutations in these genes eliminate whole groups of adjacent segments. Mutations in the *Krüppel* gene, for example, cause the absence of several adjacent segments. Pair-rule genes define regional sections of the embryo and affect alternate segments. Mutations in the *even-skipped* gene cause the deletion of even-numbered segments, whereas mutations in the *fushi tarazu* gene cause the absence of odd-numbered segments. Segment-polarity genes affect the orientation of segments. Mutations in these genes cause part of each segment to be deleted and replaced by a mirror image of part or all of an adjacent segment. For example, mutations in the *gooseberry* gene cause the posterior half of each segment to be replaced by the anterior half of an adjacent segment.

> ### CONCEPTS
>
> Once the major axes of the fruit-fly embryo have been established, segmentation genes determine the number, orientation, and basic organization of the body segments.
>
> ✔ **CONCEPT CHECK 3**
>
> Which of the following is the sequence in which the segmentation genes act?
>
> a. Segment-polarity genes → gap genes → pair-rule genes
> b. Gap genes → pair-rule genes → segment-polarity genes
> c. Segment-polarity genes → pair-rule genes → gap genes
> d. Gap genes → segment-polarity genes → pair-rule genes

TABLE 22.4	Segmentation genes and the effects of mutations in them	
Class of Gene	**Effect of Mutations**	**Examples of Genes**
Gap genes	Delete groups of adjacent segments	*hunchback, Krüppel, knirps, giant, tailless*
Pair-rule genes	Delete same part of pattern in every other segment	*runt, hairy, fushi tarazu, even-skipped, odd-paired, sloppy paired, odd-skipped*
Segment-polarity genes	Affect polarity of segment; part of segment replaced by mirror image of part of another segment	*engrailed, wingless, gooseberry, cubitus interruptus, patched, hedgehog, disheveled, costal-2, fused*

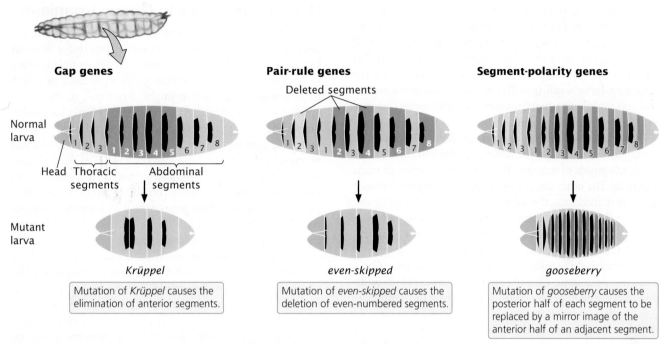

Gap genes

Pair-rule genes

Deleted segments

Segment-polarity genes

Normal larva

Head Thoracic segments Abdominal segments

Mutant larva

Krüppel

Mutation of *Krüppel* causes the elimination of anterior segments.

even-skipped

Mutation of *even-skipped* causes the deletion of even-numbered segments.

gooseberry

Mutation of *gooseberry* causes the posterior half of each segment to be replaced by a mirror image of the anterior half of an adjacent segment.

22.9 Segmentation genes control the differentiation of the *Drosophila* embryo into individual segments. Gap genes affect large sections of the embryo. Pair-rule genes affect alternate segments. Segment-polarity genes affect the orientation of segments.

Homeotic Genes in *Drosophila*

After the segmentation genes have established the number and orientation of the segments, **homeotic genes** become active and determine the *identity* of individual segments. Eyes normally arise only on the head segment, whereas legs develop only on the thoracic segments. The products of homeotic genes activate other genes that encode these segment-specific characteristics. Mutations in the homeotic genes cause body parts to appear in the wrong segments.

In the late 1940s, Edward Lewis began to study homeotic mutations in *Drosophila*—mutations that cause bizarre rearrangements of body parts. Mutations in the *Antennapedia* gene, for example, cause legs to develop on the head of a fly

in place of the antennae (**Figure 22.10**). Homeotic genes create addresses for the cells of particular segments, telling the cells where they are within the regions defined by the segmentation genes. When a homeotic gene is mutated, the address is wrong, and cells in the segment develop as though they were somewhere else in the embryo.

Homeotic genes in *Drosophila* are expressed after fertilization and are activated by specific concentrations of the proteins produced by the gap, pair-rule, and segment-polarity genes. The homeotic gene *Ultrabithorax* (*Ubx*), for example, is activated when the concentration of Hunchback protein (a product of a gap gene) is within certain values. These concentrations exist only in the middle region of the embryo; so *Ubx* is expressed only in the middle segments.

(a)

(b)

22.10 The homeotic mutation *Antennapedia* substitutes legs for the antennae of a fruit fly. (a) Normal, wild-type antennae. (b) *Antennapedia* mutant. [F. Rudolf Turner, Ph.D., Indiana University.]

The homeotic genes in animals encode regulatory proteins that bind to DNA; each of these genes contains a subset of nucleotides, called a **homeobox**, that is similar in all homeotic genes. The homeobox encodes 60 amino acids that serve as a DNA-binding domain; this domain is related to the helix-turn-helix motif (see Figure 16.2a). Homeoboxes are also present in segmentation genes and other genes that play a role in spatial development.

There are two major clusters of homeotic genes in *Drosophila*. One cluster, the **Antennapedia complex**, affects the development of the adult fly's head and anterior thoracic segments. The other cluster, the **bithorax complex**, includes genes that influence the adult fly's posterior thoracic and abdominal segments. Together, the *bithorax* and *Antennapedia* gene complexes are termed the **homeotic complex** (**HOM-C**). In *Drosophila*, the *bithorax* complex comprises three genes and the *Antennapedia* complex has five; all are located on the same chromosome (**Figure 22.11**). In addition to these eight genes, HOM-C contains many sequences that regulate them. Remarkably, the order of the genes in HOM-C is the same as the order in which the genes are expressed along the anterior–posterior axis of the body. The genes that are expressed in the more anterior segments are found at one end of the complex, whereas those expressed in the more posterior end of the embryo are found at the other end of the complex (see Figure 22.11).

CONCEPTS

Homeotic genes help determine the identity of individual segments in *Drosophila* embryos by producing regulatory proteins that bind to DNA and activate other genes. Each homeotic gene contains a consensus sequence called a homeobox, which encodes the DNA-binding domain.

✔ **CONCEPT CHECK 4**

Mutations in homeotic genes often cause
a. the deletion of segments.
b. the absence of structures.
c. too many segments.
d. structures to appear in the wrong place.

Homeobox Genes in Other Organisms

After homeotic genes in *Drosophila* had been isolated and cloned, molecular geneticists set out to determine whether similar genes exist in other animals. Probes complementary to the homeobox of *Drosophila* genes were used to search for homologous genes that might play a role in the development of other animals. The search was hugely successful: homeobox-containing genes have been found in all animals, including nematodes, beetles, sea urchins, frogs, birds, and mammals. Genes with homeoboxes have even been discovered in fungi and plants, indicating that the homeobox arose early in the evolution of eukaryotes. One group of homeobox genes comprises the **Hox genes**, which include the homeotic complex of *Drosophila* that we have just described. *Hox* genes are found in all animals except sponges.

In vertebrates, there are usually four clusters of *Hox* genes, each of which contains from 9 to 11 genes. Mammalian *Hox* genes, like those in *Drosophila*, encode transcription factors that help determine the identity of body regions along an anterior–posterior axis. The *Hox* genes of other organisms often exhibit the same relation between order on the chromosome and order of their expression along the anterior–posterior axis of the embryo as that of *Drosophila* (**Figure 22.12**), but this pattern is not universal. For example, the tunicate *Oikopleura dioica* (a primitive relative of vertebrates) has 9 *Hox* genes, but the genes are scattered throughout the genome, in contrast to the clustered arrangement seen in most animals. Despite a lack of physical clustering, the *Hox* genes in *O. dioica* are expressed in the same anterior–posterior order as that seen in vertebrates.

The *Hox* genes of vertebrates also exhibit a relation between their order on the chromosome and the timing of their expression: genes at one end of the complex (those expressed at the anterior end) are expressed early in development, whereas genes at the other end (those expressed at the posterior end) are expressed later. If a *Hox* gene is experimentally moved to a new location within the *Hox*-gene complex, it is expressed in the appropriate tissue, but the timing of its expression is altered, suggesting that the timing of gene expression is controlled by the physical location of a gene within the complex. Although the mechanism

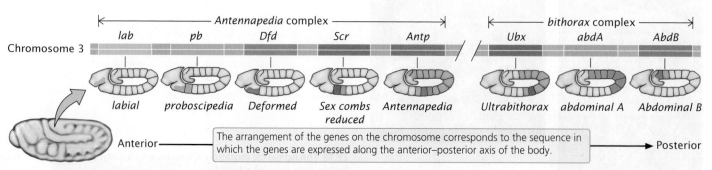

22.11 Homeotic genes, which determine the identity of individual segments in *Drosophila*, are present in two complexes. The *Antennapedia* complex has five genes, and the *bithorax* complex has three genes.

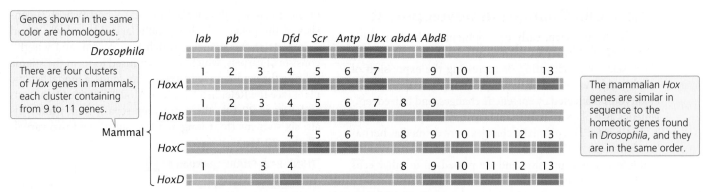

Genes shown in the same color are homologous.

There are four clusters of *Hox* genes in mammals, each cluster containing from 9 to 11 genes.

The mammalian *Hox* genes are similar in sequence to the homeotic genes found in *Drosophila*, and they are in the same order.

22.12 *Hox* genes in mammals are similar to those found in *Drosophila*. Here, the complexes are arranged so that genes with similar sequences lie in the same column. See Figure 22.11 for the full names of the *Drosophila* genes.

of this sequential control is not well understood, evidence suggests that, in mice, it involves a progressive change in the methylation patterns of histone proteins, an epigenetic change that alters chromatin structure and affects transcription (see Chapter 21). Recent studies have also identified microRNA (miRNA) genes (see Section 14.5) within the *Hox*-gene clusters, and evidence suggests that miRNAs play a role in controlling the expression of some *Hox* genes.

THINK-PAIR-SHARE Question 3

Hox genes and their expression are often correlated with anatomical differences among animals, and *Hox* genes are believed to play an important role in the evolution of animals. For example, the lancet *Branchiostoma* (another primitive relative of vertebrates) has a simple body form and possesses only 14 *Hox* genes in a single cluster, whereas some fishes, with much more complex body architecture, have as many as 48 *Hox* genes in seven clusters.

CONNECTING CONCEPTS

The Control of Development

Development is a complex process consisting of numerous events that must take place in a highly specific sequence. Studies of fruit flies and other organisms reveal that this process is regulated by a large number of genes. In *Drosophila*, the dorsal–ventral axis and the anterior–posterior axis are established by maternal genes; these genes encode mRNAs and proteins that are localized to specific regions within the egg and cause specific genes to be expressed in different regions of the embryo. The protein products of these genes then stimulate other genes, which in turn stimulate yet other genes in a cascade of control. As might be expected, most of the gene products in the cascade are regulatory proteins, which bind to DNA and activate other genes.

In the course of development, successively smaller regions of the embryo are determined (**Figure 22.13**). In *Drosophila*, first, the major axes and regions of the embryo are established by egg-polarity genes. Next, patterns within each region are determined by the action of segmentation genes: the gap genes define large sections of the embryo, the pair-rule genes define regional

sections and affect alternate segments, and the segment-polarity genes affect individual segments. Finally, the homeotic genes provide each segment with a unique identity. Initial gradients in proteins and mRNA stimulate localized gene expression, which produces more finely located gradients that stimulate even more localized gene expression. Developmental regulation thus becomes more and more narrowly defined.

The processes by which limbs, organs, and tissues form (called morphogenesis) are less well understood, although this pattern of generalized-to-localized gene expression is encountered frequently. ▶ **TRY PROBLEM 22**

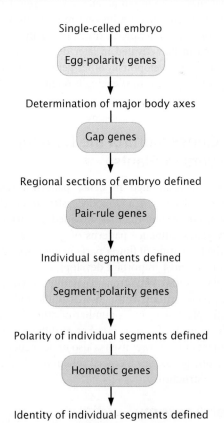

22.13 A cascade of gene regulation establishes the polarity and identity of individual segments of *Drosophila*. In development, successively smaller regions of the embryo are determined.

Epigenetic Changes in Development

As we have just seen, early development in the fruit fly is controlled in large part by the products of certain key genes selectively activating or repressing the expression of other genes. As discussed in Chapter 21, gene expression in eukaryotes is affected by epigenetic changes, and indeed, epigenetics also plays an important role in development.

In Chapter 21, epigenetic changes were defined as heritable alterations to DNA and chromatin structure—alterations that affect gene expression and are passed on to other cells but are not changes to the DNA base sequence. In the course of development, a single-celled zygote divides and gives rise to many cells, which differentiate and acquire the characteristics of specific organs and tissues. Each type of cell eventually expresses a different subset of genes, producing the proteins needed for that cell type, and this program of gene expression is passed on when the differentiated cell divides.

The gene-expression program of cells that make up a particular organ or tissue type is often defined by epigenetic marks. As development and differentiation proceed, cells acquire epigenetic changes that turn specific sets of genes on and off. In Chapter 21, we considered several types of epigenetic marks, including DNA methylation, the modification of histone proteins, and changes due to small RNA molecules. These epigenetic changes help determine which genes are expressed by a cell. In early stages of development, genes that may be required at later stages are often held in a transiently silent state by histone modifications. Histone modifications are generally flexible and easily reversed, so these genes can be activated in later developmental stages. In the course of development, other genes are permanently silenced; this longer-term silencing is often accomplished by DNA methylation.

22.3 Genes Control the Development of Flowers in Plants

We have now examined in detail pattern formation in *Drosophila*, which serves as a model system of development. Another model system that has provided important insight into how genes influence patterns of growth and development is the formation of flower parts in angiosperms.

One of the most important developmental events in the life of a plant is the switch from vegetative growth to flowering. The precise timing of this switch is affected by season, day length, plant size, and a number of other factors and is under the control of a large number of different genes. The development of the flower itself is also under genetic control, and homeotic genes play a crucial role in the determination of the floral structures.

Flower Anatomy

A flower is made up of four concentric rings of modified leaves, called whorls. The outermost whorl (whorl 1) consists of the green, leaflike sepals. The next whorl (whorl 2) consists of the petals, which typically lack chlorophyll. Whorl 3 consists of the stamens, which bear pollen, and whorl 4 consists of carpels, which are often fused to form the stigma bearing the ovules. In wild-type *Arabidopsis*, a model genetic plant (see the Reference Guide to Model Genetic Organisms and **Figure 22.14**), there are four sepals, four white petals, six stamens (four long and two short), and two carpels (**Figure 22.15a**).

THINK-PAIR-SHARE Question 4

Genetic Control of Flower Development

Elliot Meyerowitz and his colleagues conducted a series of experiments in the late 1980s and early 1990s to examine the genetic basis of flower development in *Arabidopsis*. They began by isolating and analyzing homeotic mutants in *Arabidopsis*. Homeotic mutations were actually first identified in plants in 1894, when William Bateson noticed that the floral parts of plants occasionally appeared in the wrong place: he found, for example, flowers in which stamens grew in the place where petals normally grow.

Meyerowitz and his coworkers used these types of mutants to reveal the genes that control flower development. They were able to place the homeotic mutations that they isolated into three groups on the basis of their effect on floral structure. Class A mutants had carpels instead of sepals in the first whorl and stamens instead of petals in the second whorl (**Figure 22.15b**). The third whorl consisted of stamens, and the fourth whorl consisted of carpels, the normal pattern.

22.14 The flower produced by *Arabidopsis thaliana* has four sepals, four white petals, six stamens, and two carpels.

[Darwin Dale/Science Source.]

Experiment

Question: How do genes control the development of flower structures?

Methods Isolate and analyze homeotic mutants that affect flower development.

Results

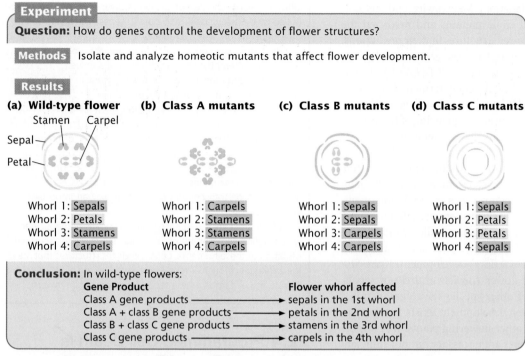

(a) Wild-type flower

Whorl 1: Sepals
Whorl 2: Petals
Whorl 3: Stamens
Whorl 4: Carpels

(b) Class A mutants

Whorl 1: Carpels
Whorl 2: Stamens
Whorl 3: Stamens
Whorl 4: Carpels

(c) Class B mutants

Whorl 1: Sepals
Whorl 2: Sepals
Whorl 3: Carpels
Whorl 4: Carpels

(d) Class C mutants

Whorl 1: Sepals
Whorl 2: Petals
Whorl 3: Petals
Whorl 4: Sepals

Conclusion: In wild-type flowers:

Gene Product	Flower whorl affected
Class A gene products ⟶	sepals in the 1st whorl
Class A + class B gene products ⟶	petals in the 2nd whorl
Class B + class C gene products ⟶	stamens in the 3rd whorl
Class C gene products ⟶	carpels in the 4th whorl

22.15 Analysis of homeotic mutants in *Arabidopsis thaliana* led to an understanding of the genes that determine floral structures in plants.

Class B mutants had sepals in the first and second whorls and carpels in the third and fourth whorls (**Figure 22.15c**). The final group, class C mutants, had sepals and petals in the first and second whorls, respectively, as is normal, but had petals in the third whorl and sepals in the fourth whorl (**Figure 22.15d**).

Meyerowitz and his colleagues concluded that each class of mutants was missing the product of a gene, or the products of a set of genes, critical to proper flower development: class A mutants were missing gene A activity, class B mutants were missing gene B activity, and class C mutants were missing gene C activity. They hypothesized that the class A genes are active in the first and second whorls. Class A gene products alone cause the first whorl to differentiate into sepals, and together with class B gene products, they cause the second whorl to develop into petals. Class C gene products, together with class B gene products, induce the third whorl to develop into stamens. Class C genes alone cause the fourth whorl to become carpels. The products of the different gene classes and their effects are summarized in the conclusion of Figure 22.15.

To explain their results, they also proposed that the genes of some classes affect the activities of others. Where class A is active, class C is repressed, and where class C is active, class A is repressed. Additionally, if a mutation inactivates class A, then class C becomes active, and vice versa. Class A genes are normally expressed in whorls 1 and 2, class B genes are expressed in whorls 2 and 3, and class C genes are expressed in whorls 3 and 4 (**Figure 22.16**).

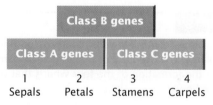

	Class B genes		
Class A genes		Class C genes	
1	2	3	4
Sepals	Petals	Stamens	Carpels

22.16 Expression of class A, B, and C genes varies among the structures of a flower.

The interaction of these three classes of genes explains the different classes of mutants in Figure 22.15. For example, class A mutants are lacking class A gene products, and therefore class C genes are active in all tissues because when A is inactivated, C becomes active. Therefore, whorl 1, with only class C gene products, will consist of carpels; whorl 2, with class C and class B gene products, will produce stamens; whorl 3, with class B and class C gene products, will produce stamens; and whorl 4, with only class C gene activity, will produce carpels (see Figure 22.15b):

Class C (in the absence of class A) gene products	→ carpels (1st whorl)
Class B + class C (in the absence of class A) gene products	→ stamens (2nd whorl)
Class B + class C gene products	→ stamens (3rd whorl)
Class C gene products	→ carpels (4th whorl)

To confirm this explanation, Meyerowitz and his colleagues bred double and triple mutants and predicted the outcome. The resulting flower structures fit their predictions. In subsequent studies, they isolated the genes of each class. There are two class A genes, termed *APETALA1* (*AP1*) and *APETALA2* (*AP2*); two class B genes, termed *APETALA3* (*AP3*) and *PISTILLATA* (*PI*); and one class C gene, termed *AGAMOUS* (*AG*). The cloning and sequencing of these genes revealed that all are MADS-box genes—genes whose products function as transcription factors and affect the expression of other genes. MADS-box genes in plants play a role similar to that of homeobox genes in animals, although MADS-box genes and homeobox genes are not homologous.

The results of other studies have demonstrated the presence of an additional group of genes, called *SEPALLATA* (*SEP*), that are expressed in whorls 2, 3, and 4, and they, too, are required for normal floral development. If the *SEP* genes are defective, the flower consists entirely of sepals. Findings from studies of other species have demonstrated that this system of flower development exists not only in *Arabidopsis*, but also in other flowering plants. It is important to note that these genes are necessary but not sufficient for proper flower development; other genes also take part in determining the identity of the different parts of flowers.

▶ TRY PROBLEM 25

CONCEPTS

Homeotic genes control the development of floral structures in plants. The products of homeotic genes interact to determine the formation of the four whorls that constitute a complete flower.

✔ CONCEPT CHECK 5

What types of flower structures would you expect to see in whorls 1 through 4 of a mutant plant that failed to produce both class A and class B gene products?

a. Carpels, stamens, stamens, carpels
b. Sepals, sepals, carpels, carpels
c. Sepals, sepals, sepals, sepals
d. Carpels, carpels, carpels, carpels

CONNECTING CONCEPTS

Comparison of Development in *Drosophila* and Flowers

We have now considered two very different model systems of development: the development of body form and pattern in fruit flies and the development of flower structures in angiosperms. In spite of their differences, these two systems exhibit similarities in how genes control development.

First, both pattern formation in *Drosophila* and flower development in plants are controlled by numerous genes that interact in complex ways. For example, we saw in *Drosophila* how a large complex of genes successively defines smaller and smaller regions of the fruit-fly embryo and how genes at one level stimulate and inhibit genes at other levels. Similarly, flower development is controlled by class A, class B, and class C genes. The action of each class depends on which products of other classes are present, and the individual genes of each class interact in complex ways to control the differentiation of each whorl of a flower.

Another common feature of development in fruit flies and in flowers is that many of the genes involved in these processes function by influencing the expression of other genes. In both fruit flies and flowers, there is a cascade of development in which gene products stimulate other genes, which in turn stimulate yet other genes. Many of the gene products are regulatory proteins that bind to DNA and affect the transcription of other genes. For example, *Hox* genes in *Drosophila* and MADS-box genes in flowers encode transcription factors that play an important role in development.

A final similarity is that each system contains homeotic genes, which define the identity of particular structures or segments. Mutation of these homeotic genes produces structures that are in the wrong place, such as legs where antennas are normally found in fruit flies or carpels where sepals usually occur in flowers.

22.4 Programmed Cell Death Is an Integral Part of Development

An important aspect of development is the death of cells. Cell death shapes many body parts in the course of development: it is responsible for the disappearance of a tadpole's tail during metamorphosis and causes the removal of tissue between the digits to produce the human hand. Cell death is also used to eliminate dangerous cells that have escaped normal controls (see the section on mutations in cell-cycle control genes and cancer in Chapter 23).

APOPTOSIS Cell death in animals is often initiated by the cell itself in a process called **apoptosis**, or programmed cell death. In this process, a cell's DNA is degraded, its nucleus and cytoplasm shrink, and blebbing occurs: the cell membrane bulges outward and breaks off into vesicles, taking some cytoplasm with it. These vesicles, and the cell itself, undergo phagocytosis by other cells without any leakage of cellular contents or damage to nearby cells (**Figure 22.17a**). Cells that are injured, on the other hand, die by a relatively uncontrolled process called *necrosis*. In this process, a cell swells and bursts, spilling its contents over neighboring cells and eliciting an inflammatory response (**Figure 22.17b**). Apoptosis is essential to embryogenesis; most multicellular animals cannot complete development if the process is inhibited.

REGULATION OF APOPTOSIS Surprisingly, most cells are programmed to undergo apoptosis and will survive only if the internal death program is continuously held in check. Apoptosis is highly regulated and depends on numerous signals from inside and outside the cell. Geneticists have identified a number of genes that have roles in various stages of the regulation of apoptosis. Some of these genes encode enzymes called **caspases**, which cleave other proteins at specific sites

(a) Apoptosis   **(b) Necrosis**

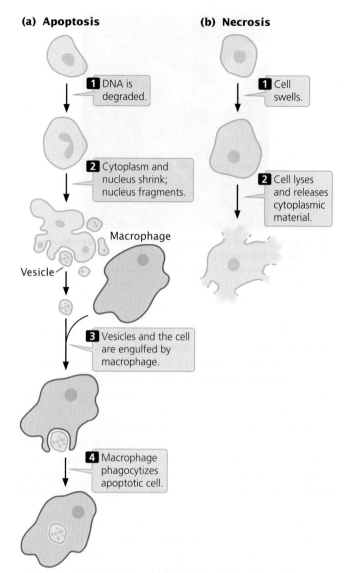

1 DNA is degraded.

1 Cell swells.

2 Cytoplasm and nucleus shrink; nucleus fragments.

2 Cell lyses and releases cytoplasmic material.

Macrophage

Vesicle

3 Vesicles and the cell are engulfed by macrophage.

4 Macrophage phagocytizes apoptotic cell.

22.17 Programmed cell death by apoptosis is distinct from uncontrolled cell death through necrosis.

(after aspartic acid). Each caspase is synthesized as a large, inactive precursor (a procaspase) that is activated by cleavage, often by another caspase. When one caspase is activated, it cleaves other procaspases, which trigger even more caspase activity. The resulting cascade of caspase activity eventually cleaves proteins essential to cell function, such as those supporting the nuclear membrane and cytoskeleton. Caspases also cleave a protein that normally keeps an enzyme that degrades DNA (DNase) in an inactive form. Cleavage of this protein activates DNase and leads to the breakdown of cellular DNA, which eventually leads to cell death.

Procaspases and other proteins required for cell death are continuously produced by healthy cells, so the potential for cell suicide is always present. A number of different signals can trigger apoptosis; for instance, infection by a virus can activate immune-system cells to secrete substances onto an infected cell, causing that cell to undergo apoptosis. This process is believed to be a defense mechanism designed to prevent the reproduction and spread of viruses. Similarly, DNA damage can induce apoptosis and thus prevent the replication of mutated sequences. Damage to mitochondria or the accumulation of a misfolded protein in the endoplasmic reticulum also stimulate programmed cell death.

APOPTOSIS IN DEVELOPMENT Apoptosis plays a critical role in development. As animals develop, excess cells are often produced and then later culled by apoptosis to produce the proper number of cells required for an organ or a tissue. In some cases, whole structures are created that are later removed by apoptosis. For example, early mammalian embryos develop both male and female reproductive ducts, but the Wolffian ducts degenerate in females and the Mullerian ducts degenerate in males. Apoptosis also plays an important role in the development of immunity (see Section 22.6), in which lymphocytes that recognize the body's own cells normally undergo apoptosis so that they do not attack self-tissues.

During embryonic development in *Drosophila*, large numbers of cells die. Three genes in *Drosophila* activate caspases that are essential for apoptosis: *reaper* (*rpr*), *grim*, and *head involution defective* (*hid*). Embryos possessing a deletion that removes all three genes exhibit no apoptosis and die in the course of embryogenesis with an excess of cells. Numerous other genes also affect the process of apoptosis.

Apoptosis is also crucial in metamorphosis, the process by which larval structures are transformed into adult structures. For example, the large salivary glands of larval fruit flies regress during metamorphosis. The hormone ecdysone stimulates metamorphosis, including the onset of apoptosis. Ecdysone induces the expression of *rpr* and *hid* and inhibits the expression of other genes, which then leads to apoptosis of salivary gland cells.

THINK-PAIR-SHARE Question 5

APOPTOSIS IN DISEASE The symptoms of many diseases and disorders are caused by apoptosis or, in some cases, its absence. In neurodegenerative disorders such as Parkinson disease and Alzheimer disease, symptoms are caused by a loss of neurons through apoptosis. In heart attacks and stroke, some cells die through necrosis, but many others undergo apoptosis. Cancer is often stimulated by mutations in genes that regulate apoptosis, leading to a failure of apoptosis that would normally eliminate cancer cells (see Chapter 23).

CONCEPTS

Cells are capable of apoptosis (programmed cell death), a highly regulated process that depends on enzymes called caspases. Apoptosis plays an important role in animal development and is associated with a number of diseases.

✔ **CONCEPT CHECK 6**

How does cell death from apoptosis differ from cell death from necrosis?

22.5 The Study of Development Reveals Patterns and Processes of Evolution

"Ontogeny recapitulates phylogeny" is a familiar phrase that was coined in the 1860s by German zoologist Ernst Haeckel to describe his belief—now considered an oversimplification—that during their development (ontogeny) organisms repeat their evolutionary history (phylogeny). According to Haeckel's belief, a human embryo passes through fish, amphibian, reptilian, and mammalian stages before developing human traits. Scientists have long recognized that organisms do not pass through the adult stages of their ancestors during their development, but the embryos of these related organisms often display similarities.

COMMON GENES IN DEVELOPMENTAL PATHWAYS

Although ontogeny does not precisely recapitulate phylogeny, many evolutionary biologists today are turning to the study of development for a better understanding of the processes and patterns of evolution. Sometimes called "evo-devo," the study of evolution through the analysis of development is revealing that the same genes often shape developmental pathways in distantly related organisms. Biologists once thought that segmentation in vertebrates and invertebrates was only superficially similar, but we now know that in both *Drosophila* and the primitive chordate *Branchiostoma*, the *engrailed* gene divides the embryo into specific segments. A gene called *Distal-less*, which creates the legs of a fruit fly, plays a role in the development of crustacean branched appendages. This same gene stimulates body outgrowths in many other organisms, from polychaete worms to starfish. Another example is *Pitx1*, whose role in controlling the presence and absence of pelvic spines in sticklebacks was discussed in the introduction to this chapter. This same gene is found in mice, where it also affects hind limb development, and in humans, where mutations of the gene have been associated with the development of club foot.

An amazing example of how the same genes in distantly related organisms can shape similar developmental pathways is seen in the development of eyes in fruit flies, mice, and humans. Walter Gehring and his collaborators examined the effect of the *eyeless* gene in *Drosophila*, which is required for proper development of the fruit-fly eye. Gehring and his coworkers genetically engineered cells that expressed *eyeless* in parts of the fly where the gene is not normally expressed. When these flies hatched, they had eyes on their wings, antennae, and legs (**Figure 22.18**). These structures were not just tissue that resembled eyes, but complete eyes with a cornea, cone cells, and photoreceptors that responded to light, although the flies could not use these eyes to see because they lacked a connection to the nervous system.

The *eyeless* gene has counterparts in mice and humans that affect the development of mammalian eyes. There is a striking similarity between the *eyeless* gene of *Drosophila* and

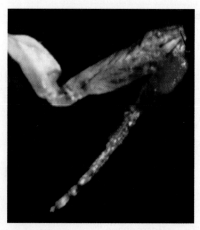

22.18 Expression of the *eyeless* gene causes the development of an eye on the leg of a fruit fly. Genes similar to *eyeless* also control eye development in mice and humans. [U. Kloter & G. Halder/VIB Center for the Biology of Disease, KU Leuven.]

the *Small eye* gene that exists in mice. In mice, a mutation in one copy of *Small eye* causes small eyes; a mouse that is homozygous for the *Small eye* mutation has no eyes. There is also a similarity between the *eyeless* gene in *Drosophila* and the *Aniridia* gene in humans; a mutation in *Aniridia* produces a severely malformed human eye. Similarities in the sequences of *eyeless*, *Small eye*, and *Aniridia* suggest that all three genes evolved from a common ancestral sequence and that a common pathway underlies eye development in flies, mice, and humans. This possibility is surprising because the eyes of insects and mammals were thought to have evolved independently.

Similar genes may be part of a developmental pathway common to two different species but have quite different effects. For example, a *Hox* gene called *Abdominal B* (*AbdB*) helps define the posterior end of a *Drosophila* embryo; a similar group of genes in birds divides the wing into three segments. In another example, the *short gastrulation* (*sog*) gene in fruit flies stimulates cells to assume a ventral orientation in the embryo, but the expression of a similar gene called *chordin* in vertebrates causes cells to assume a dorsal orientation, exactly the opposite of the situation in fruit flies. In vertebrates, *toll* genes encode proteins called Toll-like receptors, which bind to molecules on pathogens and stimulate the immune system. In fruit flies, the *toll* gene similarly functions in immunity, but it also encodes a protein that helps determine the dorsal–ventral axis, as mentioned in Section 22.2. The theme emerging from these studies is that a small, shared set of genes may underlie many basic developmental processes in many different organisms.

EVOLUTION THROUGH CHANGE IN GENE EXPRESSION

Another principle revealed by studies in evo-devo is that many major evolutionary adaptations are accomplished not through changes in the types of proteins produced, but through changes in the expression of genes that encode proteins that regulate development. This principle is illustrated in the introduction to this chapter, where we saw that deletion of an enhancer that stimulates the *Pitx1* gene in sticklebacks is responsible for the evolution of pelvic spine reduction in freshwater populations of the fish.

This principle is also seen in the evolution of blind cavefish. The Mexican tetra, *Astyanax mexicanus*, normally occurs in surface waters of streams and rivers in Texas and northern Mexico. Some 10,000 years ago, a few tetras migrated into caves. In the total darkness of their cave environment, vision was not needed, and with the passage of time, these tetras lost their eyes. Today, some 30 distinct populations of Mexican tetras are totally blind and eyeless (**Figure 22.19**), whereas surface-dwelling populations of the same species have retained normal eye development.

How did Mexican tetras lose their eyes? Mexican tetra zygotes begin to develop eyes, just like their surface-dwelling cousins, but about 24 hours after fertilization, eye development is aborted, and the cells that were destined to become the lens spontaneously die. The absence of the lens prevents other components of the eye, such as the cornea and iris, from developing. The optic cup and retina form, but their growth is retarded, and photoreceptor cells never differentiate. The degenerate eye sinks into the orbit and is eventually covered by a flap of skin.

Blind Mexican tetras have the same genes as surface-dwelling Mexican tetras; how they differ is in gene expression. Two genes, named *sonic hedgehog* (*shh*) and *tiggy-winkle hedgehog* (*twhh*), are more widely expressed in the eye primordium of blind cavefish than in surface fish. (The original *hedgehog* gene was named for a mutant phenotype in *Drosophila*, in which the mutant embryo is covered with denticles like a hedgehog. The *sonic hedgehog* gene is named after the video game character, and the *tiggy-winkle hedgehog* gene is named for Mrs. Tiggy-Winkle, a character from Beatrix Potter's books.) The expanded expression of *shh* and *twhh* activates the transcription of other genes, which cause lens cells to undergo apoptosis, and the lens degenerates.

When geneticists injected *twhh* or *shh* mRNA into the embryos of surface fish, the development of the lens was aborted, and adults that developed from these embryos were missing eyes. When drugs were used to inhibit the expression of *twhh* and *shh* in cavefish embryos, eye development in these fish was partly restored. These results demonstrate that eye development in Mexican tetras is regulated by the precise expression of *shh* and *twhh*. Overexpression of one or both of these genes in the cavefish induces the death of lens cells and aborts normal eye development. A small increase in the transcription of either gene during embryonic development results in a major anatomical change in the adult fish, a change that has allowed these fish to adapt to the darkness of the cave environment. ▶ **TRY PROBLEM 27**

Another example of differences in gene expression bringing about evolutionary change is seen in Darwin's finches, a group of closely related bird species found on the Galápagos Islands (see Figure 26.6). The species differ primarily in the size and shape of their beaks: ground finches have deep and wide beaks, cactus finches have long and pointed beaks, and warbler finches have sharp and thin beaks. These differences are associated with diet, and evolutionary changes in beak shape and size have taken place in the past when climate changes brought about shifts in the abundance of food items.

To investigate the underlying genetic basis of these evolutionary changes, Arkhat Abzhanov and his colleagues used microarrays (see Section 20.2) to examine differences in the transcription levels of several thousand genes in five species of Darwin's finches. They found differences in the expression of a gene that encodes a protein called calmodulin (CaM); the gene that encodes CaM was more highly expressed in the long and pointed beak of cactus-finch embryos than in the beaks of the other species. CaM takes part in a process called calcium signaling, which is known to affect many aspects of development. When Abzhanov and his coworkers activated calcium signaling in developing chicken embryos, the chickens had longer beaks, resembling those of the cactus finch. Thus, these researchers were able to reproduce, at least in part, the evolutionary difference that distinguishes cactus finches. This experiment shows that changes in the expression of a single gene in the course of development can produce significant anatomical differences in adults.

In these studies and others, the combined efforts of developmental biologists, geneticists, and evolutionary biologists are providing important insights into how evolution takes place. Although Haeckel's euphonious phrase "ontogeny recapitulates phylogeny" was incorrect, evo-devo is proving that development can reveal much about the process of evolution.

22.6 The Development of Immunity Occurs Through Genetic Rearrangement

As we have seen in our consideration of animal and plant development, a basic principle of developmental biology is that every somatic cell carries an identical set of genetic information and no genes are lost in development. Although this principle holds for most cells, there are some important exceptions, one of which concerns genes that encode

22.19 Mexican tetras that live in caves have lost their eyes through a developmental change in gene expression.
[Mark Smith/Science Source.]

immune function in vertebrates. In the development of immunity, individual segments of certain genes are rearranged into different combinations, producing immune-system cells that contain different genetic information and are each adapted to attack one particular foreign substance. This rearrangement and loss of genetic material is key to the power of our immune systems to protect us against almost any conceivable foreign substance.

The immune system provides protection against infection by bacteria, viruses, fungi, and parasites. The focus of an immune response is an **antigen**, defined as any molecule (usually a protein) that elicits an immune reaction. The immune system is remarkable in its ability to recognize an almost unlimited number of potential antigens. The body is full of proteins, so it is essential that the immune system be able to distinguish between self-antigens and foreign antigens. Occasionally, the ability to make this distinction breaks down, and the body produces an immune reaction to its own antigens, resulting in an **autoimmune disease (Table 22.5)**.

The Organization of the Immune System

The immune system contains a number of different components and uses several mechanisms to provide protection against pathogens, but most immune responses can be grouped into two major classes: humoral immunity and cellular immunity. Although it is convenient to think of these classes as separate systems, they interact with and influence each other significantly.

HUMORAL IMMUNITY Immune function is carried out by specialized blood cells called lymphocytes, which are a type of white blood cell. **Humoral immunity** centers on the

TABLE 22.5	Examples of autoimmune diseases
Disease	**Tissues Attacked**
Graves disease, Hashimoto thyroiditis	Thyroid gland
Rheumatic fever	Heart muscle
Systemic lupus erythematosus	Joints, skin, and other organs
Rheumatoid arthritis	Joints
Insulin-dependent diabetes mellitus	Insulin-producing cells in pancreas
Multiple sclerosis	Myelin sheath around nerve cells

production of antibodies by specialized lymphocytes called **B cells (Figure 22.20)**, which mature in the bone marrow. **Antibodies** are proteins that circulate in the blood and other body fluids, where they bind to specific antigens and mark them for destruction by phagocytic cells. Antibodies also activate a set of proteins called *complement* that help to lyse cells and attract macrophages.

CELLULAR IMMUNITY T cells (see Figure 22.20) are specialized lymphocytes that mature in the thymus and respond only to antigens found on the surfaces of the body's own cells. These lymphocytes are responsible for the second type of immune response, **cellular immunity**.

After a pathogen such as a virus has infected a host cell, some viral antigens appear on the cell's surface. Proteins called **T-cell receptors** on the surfaces of T cells bind to these antigens and mark the infected cell for destruction. T-cell receptors must simultaneously bind a foreign antigen and

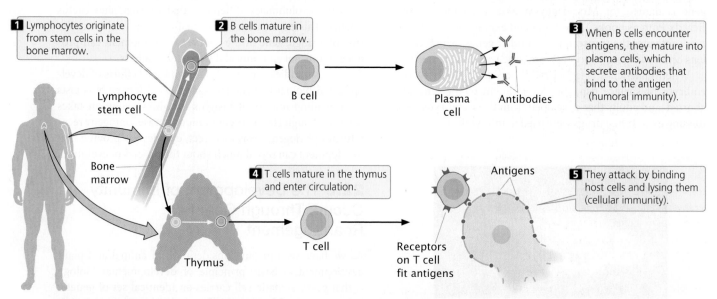

1 Lymphocytes originate from stem cells in the bone marrow.

2 B cells mature in the bone marrow.

3 When B cells encounter antigens, they mature into plasma cells, which secrete antibodies that bind to the antigen (humoral immunity).

Lymphocyte stem cell

Bone marrow

B cell

Plasma cell

Antibodies

4 T cells mature in the thymus and enter circulation.

5 They attack by binding host cells and lysing them (cellular immunity).

Antigens

T cell

Receptors on T cell fit antigens

Thymus

22.20 Immune responses are classified as humoral immunity, in which antibodies are produced by B cells, and cellular immunity, which is produced by T cells.

a self-antigen called a **major histocompatibility complex (MHC) antigen** on the host-cell surface (discussed later in this section). Not all T cells attack cells having foreign antigens; some help regulate immune responses, providing communication among different components of the immune system.

CLONAL SELECTION How can the immune system recognize an almost unlimited number of foreign antigens? Remarkably, each mature lymphocyte is genetically programmed to attack one and only one specific antigen: each mature B cell produces antibodies against a single antigen, and each T cell is capable of attaching to only one type of foreign antigen.

If each lymphocyte is specific for only one type of antigen, how does an immune response develop? The **theory of clonal selection** states that, initially, there is a large pool of millions of different lymphocytes, each capable of binding to only one antigen (**Figure 22.21**), so that millions of different foreign antigens can be detected. To illustrate clonal selection, let's imagine that a foreign protein enters the body. Only a few lymphocytes in the pool will be specific for this particular foreign antigen. When one of these lymphocytes encounters the foreign antigen and binds to it, that lymphocyte is stimulated to divide. The lymphocyte proliferates rapidly, producing a large population of genetically identical cells—a clone—each of which is specific for that particular antigen.

This initial proliferation of antigen-specific B and T cells is known as a **primary immune response** (see Figure 22.21); in most cases, the primary response destroys the foreign antigen. Following the primary immune response, most of the lymphocytes in the clone die, but a few continue to circulate in the body. These **memory cells** may remain in circulation for years, or even for the rest of a person's life. Should the same antigen reappear at some time in the future, memory cells specific to that antigen become activated and quickly give rise to another clone of cells capable of binding to the antigen. The rise of this second clone is termed a **secondary immune response** (see Figure 22.21). This ability to quickly produce a second clone of antigen-specific cells permits the long-lasting immunity that often follows recovery from a disease. For example, people who have chicken pox usually have lifelong immunity to the disease. The secondary immune response is also the basis for vaccination, which stimulates a primary immune response to an antigen and results in memory cells that can quickly produce a secondary response if that same antigen appears in the future.

Three sets of proteins are used in immune responses: antibodies, T-cell receptors, and the major histocompatibility antigens. The next section explores how the enormous diversity in these proteins is generated.

THINK-PAIR-SHARE Question 6

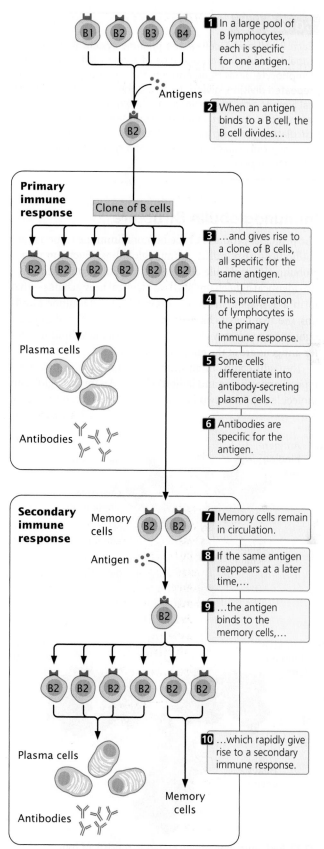

22.21 An immune response to a specific antigen is produced through clonal selection.

Immunoglobulin Structure

The principal products of the humoral immune response are antibodies—also called immunoglobulins. Each immunoglobulin (Ig) molecule consists of four polypeptide chains—two identical light chains and two identical heavy chains—that form a Y-shaped structure (**Figure 22.22**). Disulfide bonds link the two heavy chains in the stem of the Y and attach a light chain to a heavy chain in each arm of the Y. Binding sites for antigens are at the ends of the two arms.

The light chains of an immunoglobulin are of two basic types: kappa chains and lambda chains. An immunoglobulin molecule can have two kappa chains or two lambda chains,

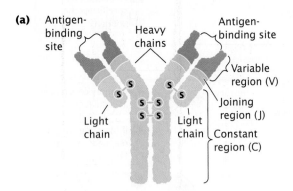

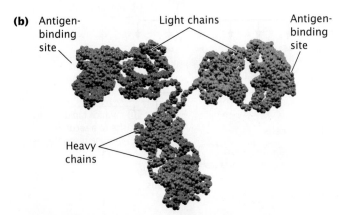

22.22 Each immunoglobulin molecule consists of four polypeptide chains—two light chains and two heavy chains—that combine to form a Y-shaped structure. (a) Structure of an immunoglobulin. (b) Folded, space-filling model.

but it cannot have one of each type. Both the light and the heavy chains have a variable region at one end and a constant region at the other end; the variable regions of different immunoglobulin molecules vary in their amino acid sequence, whereas the constant regions of different immunoglobulins are similar in sequence. The variable regions of both light and heavy chains make up the antigen-binding regions and specify the type of antigen to which the antibody can bind. The constant regions of the heavy chains at the base of the Y-shaped antibody help to determine the function of the antibody and its ability to communicate with other parts of the immune system. For example, these constant regions are important in the uptake and destruction of antigen-bound antibodies by phagocytic cells.

The Generation of Antibody Diversity

The human immune system is capable of making antibodies against virtually any antigen that might be encountered in a person's lifetime: each person is capable of making at least 10^{11} different antibody molecules. Antibodies are proteins, so the amino acid sequences of all 10^{11} potential antibodies must be encoded in the human genome. However, there are fewer than 1×10^5 genes in the human genome and, in fact, only about 3×10^9 total base pairs—so how can this huge diversity of antibodies be encoded?

The answer lies in the fact that antibody genes are composed of several distinct types of segments. There are a number of copies of each type of segment, each differing slightly from the others. In the maturation of a lymphocyte, segments of each type are joined together to create an immunoglobulin gene. The particular copy of each segment used is random, and because there are multiple copies of each type, there are many possible combinations of the segments. A limited number of segments can therefore encode a huge diversity of antibodies.

To illustrate this process of antibody assembly, let's consider the immunoglobulin light chains. Kappa and lambda chains are encoded by separate genes on different chromosomes. Each of these genes is composed of three types of segments: *V*, for variable; *J*, for joining; and *C*, for constant. The *V* segments encode most of the variable region of a light chain, the *C* segment encodes the constant region of the chain, and the *J* segments encode a short set of nucleotides that join the *V* and *C* segments together. The number of *V*, *J*, and *C* segments differs among species. For the human kappa gene, there are 30 to 35 different functional *V* gene segments, 5 different *J* gene segments, and a single *C* gene segment, all of which are present in the germ-line DNA (**Figure 22.23a**). The *V* gene segments, which are about 400 bp in length, are all located on the same chromosome and are separated from one another by about 7000 bp. The *J* gene segments are each about 30 bp in length and together encompass about 1400 bp.

Initially, an immature lymphocyte inherits all of the *V* gene segments and all of the *J* gene segments present in the germ line. In the maturation of the lymphocyte, somatic recombination within a single chromosome moves one of the *V* gene segments to a position next to one of the *J* gene segments. In **Figure 22.23b**, V_2 (the second of approximately 35 different *V* gene segments) undergoes somatic recombination, which places it next to J_3 (the third of 5 *J* gene segments); the intervening segments are lost.

After somatic recombination has taken place, the light-chain gene is transcribed into pre-mRNA that contains one *V* segment and several *J* segments, along with the *C* segment (**Figure 22.23c**). The resulting pre-mRNA is processed (**Figure 22.23d**) to produce a mature mRNA that contains transcripts of only one *V*, one *J*, and one *C* segment; this mRNA is translated into a functional light chain (**Figure 22.23e**). In this way, each mature human B cell produces a unique type of kappa chain, and different B cells produce slightly different kappa chains, depending on the combination of *V* and *J* segments that are joined together.

The gene that encodes the lambda light chain is organized in a similar way, but differs from the kappa-chain gene in the number of copies of the different segments. Somatic recombination takes place among the segments in the same way as that in the kappa gene, generating many possible combinations of lambda chains. The gene that encodes the immunoglobulin

heavy chain is also arranged in *V*, *J*, and *C* segments, but this gene possesses *D* (for diversity) segments as well. Thus, many different types of light and heavy chains are possible.

Somatic recombination is brought about by RAG1 and RAG2 proteins, which generate double-strand breaks at specific nucleotide sequences, called recombination signal sequences, that flank the *V*, *D*, *J*, and *C* gene segments. DNA-repair proteins then process the ends of particular segments and join them together.

Other mechanisms in addition to somatic recombination add to antibody diversity. First, each type of light chain can potentially combine with each type of heavy chain to make a functional immunoglobulin molecule, increasing the amount of possible variation in antibodies. Second, the recombination process that joins *V*, *J*, *D*, and *C* gene segments in the developing B cell is imprecise, and a few random nucleotides are frequently lost or gained at the junctions of the recombining segments. This **junctional diversity** greatly enhances variation among antibodies. A third mechanism that adds to antibody diversity is **somatic hypermutation**, a process that leads to a high mutation rate in the antibody genes. This process is initiated when cytosine bases are deaminated and converted into uracil. The uracil bases are detected and replaced

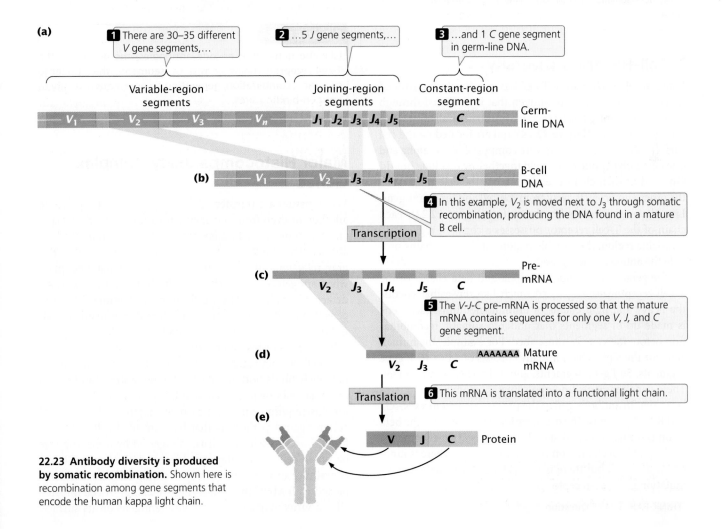

(a)

1. There are 30–35 different *V* gene segments,…
2. …5 *J* gene segments,…
3. …and 1 *C* gene segment in germ-line DNA.

Variable-region segments Joining-region segments Constant-region segment

V_1 V_2 V_3 V_n J_1 J_2 J_3 J_4 J_5 *C* Germ-line DNA

(b) V_1 V_2 J_3 J_4 J_5 *C* B-cell DNA

4. In this example, V_2 is moved next to J_3 through somatic recombination, producing the DNA found in a mature B cell.

Transcription

(c) V_2 J_3 J_4 J_5 *C* Pre-mRNA

5. The *V-J-C* pre-mRNA is processed so that the mature mRNA contains sequences for only one *V*, *J*, and *C* gene segment.

(d) V_2 J_3 *C* AAAAAAA Mature mRNA

Translation 6. This mRNA is translated into a functional light chain.

(e) V J C Protein

22.23 Antibody diversity is produced by somatic recombination. Shown here is recombination among gene segments that encode the human kappa light chain.

by DNA-repair mechanisms (see Section 18.5) that are error prone and often replace the original cytosine with a different base, leading to a mutation. These processes—somatic recombination, combining of heavy and light chains, junctional diversity, and somatic hypermutation—provide the possibility of producing at least 10^{11} different antibody specificities in humans. Through these processes, each lymphocyte comes to possess a unique set of genetic information (different from that in other lymphocytes) that encodes an antibody specific to a particular antigen. ▸ TRY PROBLEM 28

CONCEPTS

The genes encoding the chains that make up antibodies are organized in segments, and germ-line DNA contains multiple versions of each segment. The many possible combinations of *V*, *J*, and *D* segments permit an immense variety of different antibodies to be generated. This diversity is augmented by different combinations of light and heavy chains, the random addition and deletion of nucleotides at the junctions of the segments, and the high mutation rates in immunoglobulin genes.

✔ CONCEPT CHECK 7

How does somatic recombination differ from alternative splicing of RNA?

T-Cell-Receptor Diversity

Like a B cell, each mature T cell has genetically determined specificity for one type of antigen that is mediated through the cell's receptors. T-cell receptors are structurally similar to immunoglobulins (**Figure 22.24**) and are located on the cell surface. Most T-cell receptors are composed of one alpha and one beta polypeptide chain held together by disulfide bonds. One end of each chain is embedded in the cell membrane; the other end projects away from the cell and binds antigens. Like the immunoglobulin chains (see Figure 22.22), each chain of the T-cell receptor possesses a constant region and a variable region; the variable regions of the two chains provide the antigen-binding site.

The genes that encode the alpha and beta chains of the T-cell receptor are organized much like those that encode the heavy and light chains of immunoglobulins: each gene is made up of segments that undergo somatic recombination before the gene is transcribed. For example, the human gene for the alpha chain initially consists of 44 to 46 *V* gene segments, 50 *J* gene segments, and a single *C* gene segment. These gene segments undergo somatic recombination similar to that in antibody genes; the process also requires RAG1 and RAG2 proteins. The organization of the gene for the beta chain is similar, except that it also contains *D* gene segments. Alpha and beta chains combine randomly, and there is junctional diversity, but there is no evidence for somatic hypermutation in T-cell-receptor genes.

THINK-PAIR-SHARE Question 7

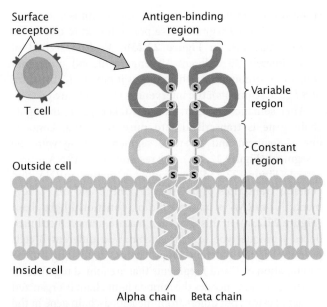

22.24 A T-cell receptor is composed of two polypeptide chains, each having a variable and a constant region. Most T-cell receptors are composed of alpha and beta polypeptide chains held together by disulfide bonds. One end of each chain traverses the cell membrane; the other end projects away from the cell and binds antigens.

CONCEPTS

Like the genes that encode antibodies, the genes for the T-cell-receptor chains consist of segments that undergo somatic recombination, generating an enormous diversity of antigen-binding sites.

Major Histocompatibility Complex Genes

When tissues are transferred from one vertebrate species to another, or even from one member to another within a species, the transplanted tissues are usually rejected by the host animal. The results of early studies demonstrated that this graft rejection is due to an immune response that takes place when antigens on the surface of cells of the grafted tissue are detected and attacked by T cells in the host animal. The antigens that elicit graft rejection are referred to as histocompatibility antigens, and they are encoded by a cluster of genes called the major histocompatibility complex (MHC).

T cells are activated only when the T-cell receptor simultaneously binds both its specific foreign antigen and one of the host cell's own histocompatibility antigens. The reason for this requirement is not clear; it may reserve T cells for action against pathogens that have invaded cells. When a foreign body, such as a virus, is ingested by a macrophage or other cell, partly digested pieces of the foreign body, containing antigens, are displayed on the cell's surface complexed with MHC molecules (**Figure 22.25**). A cell infected with a virus may also express viral antigens on its surface.

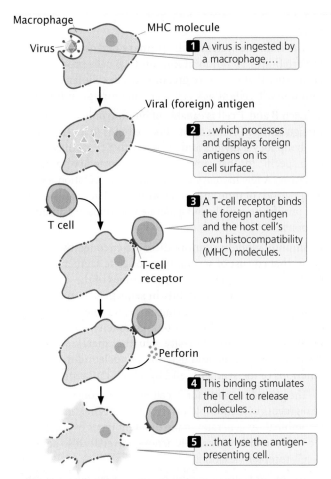

Macrophage

Virus

MHC molecule

1 A virus is ingested by a macrophage,…

Viral (foreign) antigen

2 …which processes and displays foreign antigens on its cell surface.

T cell

3 A T-cell receptor binds the foreign antigen and the host cell's own histocompatibility (MHC) molecules.

T-cell receptor

Perforin

4 This binding stimulates the T cell to release molecules…

5 …that lyse the antigen-presenting cell.

22.25 T cells are activated by binding both to a foreign antigen and to a histocompatibility antigen on the surface of a self-cell.

Through their T-cell receptors, T cells bind to both the histocompatibility protein and the foreign antigen and secrete substances that either destroy the antigen-containing cell, activate other B and T cells, or do both.

The MHC genes are among the most variable genes known: there are more than 100 different alleles for some MHC loci. Because each person possesses five or more MHC loci and because many alleles are possible at each locus, no two people (with the exception of identical twins) produce the same set of histocompatibility antigens. The variation in histocompatibility antigens provides each of us with a unique identity for our own cells, which allows our immune systems to distinguish self from nonself. This variation is also the cause of rejection in organ transplants.

CONCEPTS

The MHC genes encode proteins that provide identity to the cells of each individual. To bring about an immune response, a T-cell receptor must simultaneously bind both a histocompatibility (self) antigen and its specific foreign antigen.

Genes and Organ Transplants

For a person with a seriously impaired organ, a transplant operation may offer the only hope of survival. Successful transplantation requires more than the skills of a surgeon; it also requires a genetic match between the patient and the person donating the organ. The fate of transplanted tissue depends largely on the type of antigens present on the surface of its cells. Because foreign tissues are usually rejected by the host, the successful transplantation of tissues between different people is very difficult. Tissue rejection can be partly inhibited by drugs that interfere with cellular immunity. Unfortunately, this treatment can create serious problems for transplant patients because they may have difficulty fighting off common pathogens and may thus die of infection. The only other option for controlling the immune reaction is to carefully match the donor and the recipient, maximizing their genetic similarities.

The tissue antigens that elicit the strongest immune reaction are the very ones used by the immune system to mark its own cells: those encoded by the major histocompatibility complex. In humans, the MHC spans a region of more than 3 million base pairs on chromosome 6 and has many alleles, providing different MHC antigens on the cells of each person and allowing the immune system to recognize foreign cells. The severity of an immune rejection of a transplanted organ depends on the number of MHC antigens on the cells of the transplanted tissue that fail to match those of the recipient. The ABO red-blood-cell antigens are also important because they elicit a strong immune reaction. The ideal donor is the patient's own identical twin, who will have exactly the same MHC and ABO antigens. Unfortunately, most patients don't have an identical twin. The next-best donor is a sibling with the same MHC and ABO antigens. If a sibling is not available, donors from the general population are considered. An attempt is made to match as many of the MHC antigens of the donor and recipient as possible, and immunosuppressive drugs are used to control rejection due to the mismatches. The long-term success of transplants depends on the closeness of the match. Survival rates after kidney transplants (the most successful of the major organ transplants) increase from 63% with zero or one MHC match to 90% with four matches.

Scientists have now been successful in inducing adult cells to lose their specialized characteristics and return to an undifferentiated state (induced pluripotent stem cells; see Section 21.3), in which they are capable of developing into many different cell types. In the future, it may be possible to create pluripotent stem cells from a person's adult cells, then grow those cells into tissues or organs that could be transplanted back into the same person. Such cells would have the same MHC antigens as the original cell, avoiding the immune rejection that occurs with transplants between different people.

THINK-PAIR-SHARE Question 8

CONCEPTS SUMMARY

■ Each multicellular organism begins as a single cell that has the potential to develop into any cell type. As development proceeds, cells become committed to particular fates. The results of cloning experiments demonstrated that this process arises from differential gene expression.

■ In the early *Drosophila* embryo, pattern formation is brought about through a cascade of genetic control.

■ The dorsal–ventral and anterior–posterior axes of the *Drosophila* embryo are established by egg-polarity genes, which are expressed in the female parent and produce mRNAs and proteins that are deposited in the egg cytoplasm. Initial differences in the distribution of these molecules regulate gene expression in various parts of the embryo. The dorsal–ventral axis is defined by a concentration gradient of the Dorsal protein, and the anterior–posterior axis is defined by concentration gradients of the Bicoid and Nanos proteins.

■ Three types of segmentation genes act sequentially to determine the number and organization of the embryonic segments in *Drosophila*. The gap genes establish large sections of the embryo, the pair-rule genes affect alternate segments, and the segment-polarity genes affect the organization of individual segments. Homeotic genes then define the identity of individual *Drosophila* segments.

■ Homeotic genes control the development of flower structure. Groups of genes interact to determine the identity of the four whorls found in a complete flower.

■ Apoptosis, or programmed cell death, is a highly regulated process that depends on caspases—proteins that cleave other proteins. Apoptosis plays an important role in the development of many animals.

■ The immune system is the primary defense network in vertebrates. In humoral immunity, B cells produce antibodies that bind foreign antigens; in cellular immunity, T cells attack cells carrying foreign antigens.

■ Each B and T cell is capable of binding only one type of foreign antigen. When a lymphocyte binds to an antigen, the lymphocyte divides and gives rise to a clone of cells, each specific for that same antigen—the primary immune response. A few memory cells remain in circulation for long periods of time and, on exposure to that same antigen, can proliferate rapidly and generate a secondary immune response.

■ Immunoglobulins (antibodies) are encoded by genes that consist of several types of gene segments; germ-line DNA contains multiple copies of these gene segments, which differ slightly in sequence. Somatic recombination randomly brings together one version of each segment to produce a single complete gene, allowing many combinations. Diversity is further increased by the random addition and deletion of nucleotides at the junctions of the segments and by a high mutation rate.

■ The germ-line genes for T-cell receptors consist of segments with multiple varying copies. Somatic recombination generates many different types of T-cell receptors in different cells. Junctional diversity also adds to T-cell-receptor variation.

■ The major histocompatibility complex encodes a number of antigens. The MHC antigens allow the immune system to distinguish self from nonself. Each locus for the MHC contains many alleles.

IMPORTANT TERMS

totipotency (p. 664)
determination (p. 664)
egg-polarity gene (p. 667)
morphogen (p. 668)
segmentation gene (p. 670)
gap gene (p. 670)
pair-rule gene (p. 670)
segment-polarity
  gene (p. 670)
homeotic gene (p. 671)
homeobox (p. 672)

Antennapedia
  complex (p. 672)
bithorax complex (p. 672)
homeotic complex
  (HOM-C) (p. 672)
Hox gene (p. 672)
apoptosis (p. 676)
caspase (p. 676)
antigen (p. 680)
autoimmune
  disease (p. 680)

humoral immunity (p. 680)
B cell (p. 680)
antibody (p. 680)
T cell (p. 680)
cellular immunity (p. 680)
T-cell receptor (p. 680)
major histocompatibility
  complex (MHC)
  antigen (p. 681)
theory of clonal
  selection (p. 681)

primary immune
  response (p. 681)
memory cell (p. 681)
secondary immune
  response (p. 681)
somatic recombination
  (p. 683)
junctional diversity (p. 683)
somatic hypermutation
  (p. 683)

ANSWERS TO CONCEPT CHECKS

1. No, it does *not* prove that genetic material is not lost during development, because differentiation has not yet taken place in an early embryo. The early embryo would still be likely to contain all its genes, and its nucleus could therefore give rise to a complete animal. The use of specialized cells, such as a cell from an udder, does prove that genes are not lost during development because if they were lost, there would be no cloned animal.

2. c

3. b

4. d

5. d

6. In cell death from necrosis, the cell swells and bursts, causing an inflammatory response. In cell death through apoptosis, the cell's DNA is degraded, its nucleus and cytoplasm shrink, and the cell is phagocytized, without leakage of cellular contents.

7. Somatic recombination takes place through the rearrangement of DNA segments, so each lymphocyte has a different sequence of nucleotides in its DNA. Alternative splicing (see Section 14.2) takes place through the rearrangement of RNA sequences in pre-mRNA; there is no change in the DNA that encodes the pre-mRNA. The generation of antibody diversity requires both somatic recombination of DNA sequences and alternative splicing of pre-mRNA sequences.

WORKED PROBLEMS

Problem 1

If a fertilized *Drosophila* egg is punctured at the anterior end and a small amount of cytoplasm is allowed to leak out, what will be the most likely effect on the development of the fly embryo?

›› Solution Strategy

What information is required in your answer to the problem?
The likely effects on development of removing cytoplasm from the anterior end of a fertilized fly egg.

What information is provided to solve the problem?
Cytoplasm is removed from the anterior end.

For help with this problem, review:
Egg-Polarity Genes in Section 22.2.

›› Solution Steps

The egg-polarity genes determine the major axes of development in the *Drosophila* embryo. One of these genes is *bicoid*, which is transcribed in the maternal ovary. After fertilization Bicoid protein forms a concentration gradient along the anterior–posterior axis of the embryo. The high concentration of Bicoid protein at the anterior end induces the development of anterior structures, such as the head of the fruit fly. If the anterior end of the egg is punctured, cytoplasm containing high concentrations of *bicoid* mRNA and Bicoid protein will leak out, reducing the concentration of Bicoid protein at the anterior end. The result will be that the embryo fails to develop head and thoracic structures at the anterior end.

Recall: The *bicoid* gene is an egg-polarity gene that helps determine the anterior–posterior axis of the developing embryo.

Problem 2

The immunoglobulin molecules of a particular mammalian species have kappa and lambda light chains and heavy chains. The kappa gene consists of 250 *V* and 8 *J* segments. The lambda gene contains 200 *V* and 4 *J* segments. The gene for the heavy chain consists of 300 *V*, 8 *J*, and 4 *D* segments. If just somatic recombination and random combinations of light and heavy chains are taken into consideration, how many different types of antibodies can be produced by this species?

›› Solution Strategy

What information is required in your answer to the problem?
The number of different types of antibodies that can be produced if somatic recombination and random combinations of light and heavy chains are considered.

What information is provided to solve the problem?
- The kappa gene has 250 *V* and 8 *J* segments.
- The lambda gene has 200 *V* and 4 *J* segments.
- The heavy chain has 300 *V*, 8 *J*, and 4 *D* segments.

For help with this problem, review:
The Generation of Antibody Diversity in Section 22.6.

>> **Solution Steps**

Hint: The number of each type of light chain consists of the number of *V* segments times the number of *J* segments.

For the kappa light chain, there are $250 \times 8 = 2000$ combinations; for the lambda light chain, there are $200 \times 4 = 800$ combinations; so a total of 2800 different types of light chains are possible. For the heavy chains, there are $300 \times 8 \times 4 = 9600$ possible types. Any of the 2800 light chains can combine with any of the 9600 heavy chains; so there are $2800 \times 9600 = 26{,}880{,}000$ different types of antibodies possible from somatic recombination and random chain combination alone. Junctional diversity and somatic hypermutation would greatly increase this diversity.

Hint: To determine the number of different types of antibodies, multiply the number of possible light chains by the number of possible heavy chains.

COMPREHENSION QUESTIONS

Section 22.1

1. What experiments suggested that genes are not lost or permanently altered in development?

Section 22.2

2. Briefly explain how the Dorsal protein is redistributed in the formation of the *Drosophila* embryo and how this redistribution helps to establish the dorsal–ventral axis of the fruit fly.

3. Briefly describe how the *bicoid* and *nanos* genes help to determine the anterior–posterior axis of the fruit fly.

4. List the three major classes of segmentation genes and outline the function of each.

5. What role do homeotic genes play in the development of fruit flies?

Section 22.3

6. How do class A, B, and C genes in plants work together to determine the structures of the flower?

Section 22.4

7. What is apoptosis and how is it regulated?

Section 22.6

8. Explain how each of the following processes contributes to antibody diversity.
   a. Somatic recombination
   b. Junctional diversity
   c. Hypermutation

9. What is the function of the MHC antigens? Why are the genes that encode these antigens so variable?

APPLICATION QUESTIONS AND PROBLEMS

Section 22.1

10. If telomeres are normally shortened after each round of replication in somatic cells (see Chapter 12), what prediction would you make about the length of telomeres in Dolly, the first cloned sheep?

Section 22.2

11. A drug causes the degradation of Cactus protein. What would be the effect of administering this drug to developing *Drosophila* embryos?

12. What would be the effect of deleting the *toll* gene in *Drosophila* embryos?

13. Why is it that mutations in *bicoid* and *nanos* exhibit genetic maternal effect in *Drosophila* (a mutation in the maternal parent produces a phenotype that shows up in the offspring; see Section 5.3), but mutations in *runt* and *gooseberry* do not? (Hint: See **Tables 22.3** and **22.4**.)

*14. Give examples of genes that affect development in fruit flies by regulating gene expression at the level of (a) transcription and (b) translation.

15. Using **Figure 22.6**, indicate the stage at which segmentation genes, homeotic genes, and egg-polarity genes would have an effect on development.

16. What would be the most likely effect on development of puncturing the posterior end of a *Drosophila* egg, allowing a small amount of cytoplasm to leak out, and then injecting that cytoplasm into the anterior end of another egg?

17. Christiane Nüsslein-Volhard and her colleagues carried out several experiments in an attempt to understand what determines the anterior and posterior ends of a *Drosophila* larva (reviewed in C. Nüsslein-Volhard, H. G. Frohnhofer, and R. Lehmann. 1987. *Science* 238:1675–1681). They isolated fruit flies with mutations in the *bicoid* gene (*bcd⁻*). These flies produced embryos that lacked a head and thorax. When they transplanted cytoplasm from the anterior end of an egg from a wild-type female into the anterior end of an egg from a mutant *bicoid* female, normal head and thorax development took place in the embryo. However, transplanting cytoplasm from the posterior end of an

egg from a wild-type female into the anterior end of an egg from a *bicoid* female had no effect. Explain these results in regard to what you know about proteins that control the determination of the anterior–posterior axis.

18. What would be the most likely result of injecting *bicoid* mRNA into the posterior end of a *Drosophila* embryo and inhibiting the translation of *nanos* mRNA?

19. What would be the most likely effect of inhibiting the translation of *hunchback* mRNA throughout a *Drosophila* embryo?

*20. Molecular geneticists have performed experiments in which they altered the number of copies of the *bicoid* gene in flies, affecting the amount of Bicoid protein produced.

   a. What would be the effect on development of an increased number of copies of the *bicoid* gene?

   b. What would be the effect of a decreased number of copies of *bicoid*? Justify your answers.

21. What would be the most likely effect on fruit-fly development of a deletion in the *nanos* gene?

*22. Give an example of a gene found in each of the categories of genes (egg-polarity, gap, pair-rule, and so forth) listed in **Figure 22.13**.

23. In Chapter 1, we considered preformationism, the early idea about heredity that suggested that inside the egg or sperm is a tiny adult called a homunculus, with all the features of an adult human in miniature. According to this idea, the homunculus simply enlarges during development. What types of evidence presented in this chapter prove that preformationism is false?

Section 22.3

24. Explain how (a) the absence of class B gene expression produces the flower structures seen in class B mutants (see **Figure 22.15c**) and (b) the absence of class C gene expression produces the structures seen in class C mutants (see **Figure 22.15d**).

*25. What would you expect a flower to look like in a plant that lacked both class A and class B genes? In a plant that lacked both class B and class C genes?

26. What will be the flower structure of a plant in which expression of the following genes is inhibited in the specified whorls?

   a. Expression of class B genes is inhibited in the second whorl, but not in the third whorl.

   b. Expression of class C genes is inhibited in the third whorl, but not in the fourth whorl.

   c. Expression of class A genes is inhibited in the first whorl, but not in the second whorl.

   d. Expression of class A genes is inhibited in the second whorl, but not in the first whorl.

Section 22.5

*27. William Jeffrey and his colleagues crossed surface-dwelling Mexican tetras that had fully developed eyes with cave-dwelling blind Mexican tetras. The progeny from this cross had uniformly small eyes compared with those of surface fish (Y. Yamamoto, D. W. Stock, and W. R. Jeffrey. 2004. *Nature* 431:844–847). What prediction can you make about the expression of *shh* in these progeny at the embryonic stage relative to its expression in embryonic surface fish?

Section 22.6

*28. In a particular species, the gene for the kappa light chain has 200 *V* segments and 4 *J* segments. In the gene for the lambda light chain, this species has 300 *V* segments and 6 *J* segments. If only the antibody diversity arising from somatic recombination is taken into consideration, how many different types of light chains are possible?

29. Based on the information provided in **Figure 22.21**, what would be the likely effect of a mutation that prevented the formation of memory cells?

CHALLENGE QUESTIONS

Section 22.2

30. As we have learned in this chapter, the Nanos protein inhibits the translation of *hunchback* mRNA, lowering the concentration of Hunchback protein at the posterior end of a fruit-fly embryo and stimulating the differentiation of posterior characteristics. The results of experiments have demonstrated that the action of Nanos on *hunchback* mRNA depends on the presence of an 11-base sequence that is located in the 3′ untranslated region (3′ UTR) of *hunchback* mRNA. This sequence has been termed the Nanos response element (NRE). There are two copies of NRE in the 3′ UTR of *hunchback* mRNA. If a copy of NRE is added to the 3′ UTR of another mRNA produced by a different gene,

that mRNA is repressed by Nanos. The repression is greater if several NREs are added. On the basis of these observations, propose a mechanism for how Nanos inhibits Hunchback translation.

31. Given the distribution of *Hox* genes among animals, what would you predict about the number and type of *Hox* genes in the common ancestor of all animals?

Section 22.6

32. Ataxia-telangiectasis (ATM) is a rare genetic neurodegenerative disease. About 20% of people with ATM develop acute lymphocytic leukemia or lymphoma, cancers of the immune-system cells. Cells in many of these cancers exhibit chromosome

rearrangements, with chromosome breaks occurring at antibody and T-cell-receptor genes (A. L. Bredemeyer et al. 2006. *Nature* 442:466–470). Many people with ATM also have a weakened immune system, which makes them susceptible to respiratory infections. Research has shown that the locus that causes ATM has a role in the repair of double-strand breaks. Explain why people who have a genetic defect in the repair of double-strand breaks might have a high incidence of chromosome rearrangements in their immune-system cells and why their immune systems might be weakened.

THINK-PAIR-SHARE QUESTIONS

Section 22.1

1. Although a number of different animals have been successfully cloned, the process of creating cloned animals is very inefficient. Typically, only about 0.1% to 3% of attempts result in a live-born animal. What might be some of the reasons for this low rate of success?

2. Research has now demonstrated that techniques that have been developed for cloning animals could be used to clone human embryos, although no living human clones have been produced. There is widespread consensus that cloning a live human would be unethical. However, some people have suggested creating cloned embryos from which stem or pluripotent cells, which have the ability to develop into any tissue or organ, could be extracted. The stem and pluripotent cells might then be used to create genetically identical tissues and organs that could then be transplanted back into the donor to treat failed organs and diseases. What might be some reasons for or against this type of therapeutic cloning?

Section 22.2

3. Propose one or more explanations for why *Hox* genes exhibit a relation between their order on the chromosome and the timing of their expression, with *Hox* genes at one end of the complex expressed early and those at the other end expressed later.

Section 22.3

4. Flowers are complex and beautiful structures that are highly variable among plant species, differing in color and size and in the number and arrangement of their basic parts: the sepals, petals, stamens, and carpels. Why do plants go to the trouble of producing these complex structures, and why do flowers vary so much among different plant species?

Section 22.4

5. The frog *Xenopus laevis* has often served as a model system for the study of apoptosis. Can you think of some reasons that frogs are particularly good models for the study of apoptosis?

Section 22.6

6. Childhood vaccinations are required by law in all U.S. states and in many countries to prevent the outbreak and spread of infectious diseases. Some people object to required vaccinations and feel that whether to vaccinate or not should be a personal choice. What are some arguments for and against requiring everyone to be vaccinated?

7. Severe combined immunodeficiency disease is a life-threatening genetic disease in which children are born with defective B-cell and T-cell function. These children are susceptible to severe infections, such as pneumonia, meningitis, or blood-borne infections. They can also be infected and sickened by live viruses that are present in some vaccines. Children with some forms of severe combined immunodeficiency disease have mutations in their *RAG1* or *RAG2* genes. Provide a possible explanation for how mutations in these genes might lead to the symptoms of this disorder.

8. In the novel *Chromosome 6*, by Robin Cook, a biotechnology company genetically engineers individual bonobos (a type of chimpanzee) to serve as future organ donors for clients. The genes of the bonobos are altered so that no tissue rejection takes place when their organs are transplanted into a client. What genes would need to be altered for this scenario to work? Explain your answer.

Cancer Genetics

Villa designed by Renaissance architect Andrea Palladio, for whom the
***palladin* gene is named.** Palladin encodes an essential component of a cell's
cytoskeleton; when mutated, **palladin** contributes to the spread of pancreatic
cancer. [Gianni Dagli Orti/The Art Archive at Art Resource, NY.]

Palladin and the Spread of Cancer

Pancreatic cancer is among the most serious of
all cancers. With about 53,000 new cases each
year in the United States, it is only the twelfth most
common form of the disease, but it is the third
leading cause of death due to cancer, killing more
than 41,000 people each year. Most people with
pancreatic cancer survive less than 6 months after
they are diagnosed; only about 7% survive more
than 5 years. A primary reason for pancreatic
cancer's lethality is its propensity to spread rapidly to
the lymph nodes and other organs. Most symptoms
don't appear until the cancer is advanced and has
invaded other organs. So what makes pancreatic
cancer so likely to spread?

In 2006, researchers identified a key gene that
contributes to the development of pancreatic
cancer, which proved to be an important source
of insight into the disease's aggressive nature.
Geneticists at the University of Washington
in Seattle found a unique family in which
nine members over three generations had been diagnosed with pancreatic cancer
(**Figure 23.1**). Nine additional family members had precancerous growths that were
likely to develop into pancreatic cancer. In this family, pancreatic cancer was inherited
as an autosomal dominant trait.

By using gene-mapping techniques, the geneticists determined that the gene
causing pancreatic cancer in the family was located within a region on the long arm of
chromosome 4. Unfortunately, this region encompasses 16 million base pairs and includes
250 genes.

To determine which of the 250 genes in the delineated region might be responsible
for cancer in the family, researchers designed a unique microarray (see Chapter 20)
that contained sequences from the region. They used this microarray to examine gene
expression in pancreatic tumors and precancerous growths in family members, as well
as in sporadic pancreatic tumors in other people and in normal pancreatic tissue from
unaffected people. The researchers reasoned that the cancer gene might be overexpressed
or underexpressed in the tumors relative to normal tissue. Data from the microarray
revealed that the most overexpressed gene in the pancreatic tumors and precancerous
growths was a gene encoding a critical component of the cytoskeleton—a gene called
palladin. Sequencing demonstrated that all members of the family with pancreatic cancer
had an identical mutation in exon 2 of the *palladin* gene.

The *palladin* gene is named for Renaissance architect Andrea Palladio because it plays
a central role in the architecture of the cell. The Palladin protein functions as a scaffold
for the binding of the other cytoskeletal proteins that are necessary for maintaining cell
shape, movement, and differentiation. The ability of cancer cells to spread is directly
related to their cytoskeleton: cells that spread typically have poor cytoskeletal architecture,

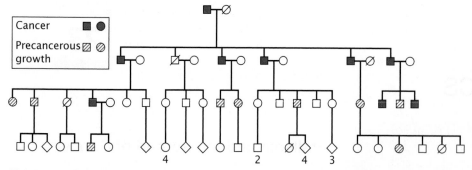

23.1 Pancreatic cancer is inherited as an autosomal dominant trait in a family that possesses a mutated *palladin* gene. [Data from K. L. Pogue et al., *PLoS Medicine* 3:2216–2228, 2006.]

which enables them to detach easily from a primary tumor mass and migrate through other tissues. To determine whether mutations in the *palladin* gene affect cell mobility, researchers genetically engineered cells with a mutated copy of the *palladin* gene and tested the ability of these cells to migrate. The cells with a mutated *palladin* gene were 33% more efficient at migrating than were cells with the normal *palladin* gene, demonstrating that the *palladin* gene contributes to the ability of pancreatic cancer cells to spread.

> **THINK-PAIR-SHARE**
> - Pancreatic cancer is clearly inherited as an autosomal dominant trait in the family illustrated in Figure 23.1. Yet most cases of pancreatic cancer are sporadic, appearing as isolated cases in families with no obvious inheritance. How can a trait be strongly inherited in one family and not inherited in another?
>
> - Is it correct to say that the *palladin* gene causes cancer? Everyone has a *palladin* gene, but not everyone gets pancreatic cancer. What might be a more accurate way to talk about the link between the *palladin* gene and cancer?

The discovery of *palladin*'s link to pancreatic cancer illustrates the power of modern molecular genetics to unravel the biological nature of cancer. In this chapter, we examine the genetic nature of cancer, a disease that is fundamentally genetic but is often not inherited. We begin by considering the multiple genetic alterations that are required to transform a normal cell into a cancerous one. We then consider some of the types of genes that contribute to cancer, including oncogenes and tumor-suppressor genes, genes that control the cell cycle, genes encoding DNA-repair systems and telomerase, and genes such as *palladin* that contribute to the spread of cancer. Next, we take a look at epigenetic changes associated with cancer, and as an example, we examine how specific genes contribute to the progression of colon cancer. Finally, we discuss chromosome mutations associated with cancer and the role of viruses in some cancers.

THINK-PAIR-SHARE Question 1

23.1 Cancer Is a Group of Diseases Characterized by Cell Proliferation

About one of every five women and one of every four men in the United States will die from cancer, and cancer treatments cost billions of dollars per year. Cancer is not a single disease; rather, it is a heterogeneous group of disorders characterized by the presence of cells that do not respond to the normal controls on division. Cancer cells divide rapidly and continuously, creating tumors that crowd out normal cells and eventually rob healthy tissues of nutrients (**Figure 23.2**).

The cells of an advanced tumor can separate from the tumor and travel to distant sites in the body, where they may take up

(a)

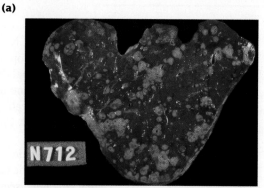

(b)

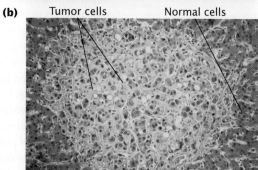

Tumor cells   Normal cells

23.2 Abnormal proliferation of cancer cells produces a tumor that crowds out normal cells. (a) Metastatic breast cancer masses (white protrusions) growing in a human liver. (b) A light micrograph of a liver section with tumors. The cancer cells are the light, pale-stained cells; the darker cells are healthy liver cells. [CNRI/Science Source.]

residence and develop into new tumors. The most common cancers in the United States are those of the prostate gland, breast, lung, colon and rectum, and blood (**Table 23.1**).

THINK-PAIR-SHARE Question 2

Tumor Formation

Normal cells grow, divide, mature, and die in response to a complex set of internal and external signals. A normal cell receives both stimulatory and inhibitory signals, and its growth and division are regulated by a delicate balance between these opposing forces. In a cancer cell, one or more of these signals has been disrupted, which causes the cell to proliferate at an abnormally high rate. As they lose their response to the normal controls, cancer cells gradually lose their regular shape and boundaries, eventually forming a distinct mass of abnormal cells—a tumor. If the tumor cells remain localized, the tumor is said to be benign; if the cells invade other tissues, the tumor is said to be **malignant**. Cells that travel to other sites in the body, where they establish secondary tumors, have undergone **metastasis**.

Cancer As a Genetic Disease

Cancer arises as a result of fundamental defects in the regulation of cell division, and its study therefore has significance not only for public health, but also for our basic understanding of cell biology. Through the years, many ideas have been put forth to explain cancer, but we now recognize that most, if not all, cancers arise from defects in DNA.

EVIDENCE FOR THE GENETIC THEORY OF CANCER Early observations suggested that cancer might result from genetic damage. First, many agents that cause mutations, such as ionizing radiation and chemicals, also cause cancer (are carcinogens; see Section 18.3). Second, some cancers are consistently associated with particular chromosome abnormalities. About 90% of people with chronic myeloid leukemia, for example, have a reciprocal translocation between chromosome 22 and chromosome 9. Third, some specific types of cancers tend to run in families. Retinoblastoma, a rare childhood cancer of the retina, appears with high frequency in a few families, in which it is inherited as an autosomal dominant trait, suggesting that a single gene is responsible for these cases of the disease.

Although these observations hinted that genes play some role in cancer, the theory of cancer as a genetic disease had several significant problems. If cancer is inherited, then every cell in the body should receive the cancer-causing gene, and therefore every cell should become cancerous. In the types of cancer that run in families, however, tumors typically appear only in certain tissues and often only when the person reaches an advanced age. Finally, many cancers do not run in families at all, and even in those cancers that generally do, isolated cases crop up in families with no history of the disease.

KNUDSON'S MULTISTEP MODEL OF CANCER In 1971, Alfred Knudson proposed a model to explain the genetic basis of cancer. Knudson was studying retinoblastoma, which usually develops in only one eye but occasionally appears in both. Knudson found that when retinoblastoma appears in both eyes, onset is at an early age, and that many children with bilateral retinoblastoma have close relatives who also have the disease.

Knudson proposed that retinoblastoma results from two separate genetic defects, both of which are necessary for cancer to develop (**Figure 23.3**). He suggested that in the cases in which the disease affects just one eye, a single cell in one eye undergoes two successive mutations. Because the chance of these two mutations occurring in the same cell is remote, retinoblastoma is rare and typically develops in only one eye. Knudson proposed that children with bilateral retinoblastoma inherit one of the two mutations required for the cancer, and so every cell contains this initial mutation.

TABLE 23.1	Estimated incidences of various cancers and cancer mortality in the United States in 2016	
Type of Cancer	**New Cases per Year**	**Deaths per Year**
Breast	249,260	40,890
Lung and bronchus	224,390	158,080
Prostate	180,890	26,120
Colon and rectum	134,490	49,190
Lymphoma	81,080	21,270
Bladder	76,960	16,390
Melanoma	76,380	10,130
Thyroid	64,300	1,980
Kidney	62,700	14,240
Leukemia	60,140	24,406
Uterus	60,050	10,470
Pancreas	53,070	41,780
Oral cavity and pharynx	48,330	9,570
Liver	39,230	27,170
Myeloma	30,330	12,650
Stomach	26,370	10,730
Brain and nervous system	23,770	16,050
Ovary	22,280	14,240
Esophagus	16,910	15,690
Larynx	13,430	3,620
Uterine cervix	12,990	4,120
Cancers of soft tissues including heart	12,310	4,990
All cancers	1,685,210	595,690

Source: American Cancer Society, *Cancer Facts and Figures, 2016* (Atlanta: American Cancer Society, 2016), p. 4.

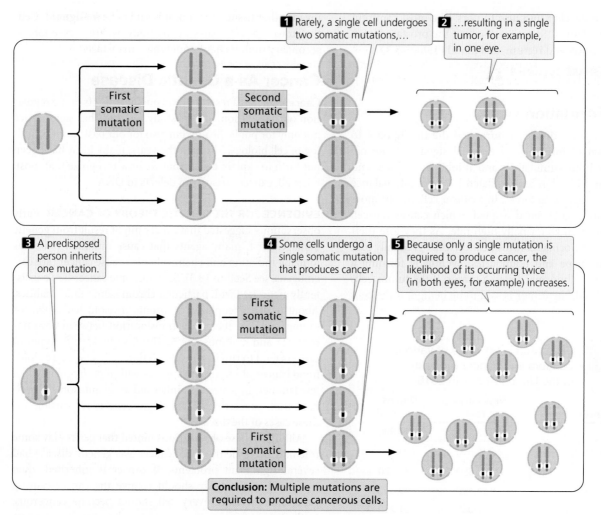

1 Rarely, a single cell undergoes two somatic mutations,…

2 …resulting in a single tumor, for example, in one eye.

First somatic mutation

Second somatic mutation

3 A predisposed person inherits one mutation.

4 Some cells undergo a single somatic mutation that produces cancer.

5 Because only a single mutation is required to produce cancer, the likelihood of its occurring twice (in both eyes, for example) increases.

First somatic mutation

First somatic mutation

Conclusion: Multiple mutations are required to produce cancerous cells.

23.3 Alfred Knudson proposed that retinoblastoma results from two separate genetic defects, both of which are necessary for cancer to develop.

In these cases, all that is required for cancer to develop is for one eye cell to undergo the second mutation. Because each eye possesses millions of cells, the probability that the second mutation will occur in at least one cell of each eye is high, so tumors may occur in both eyes at an early age.

Knudson's proposal suggested that cancer is the result of a multistep process that requires several mutations. If one or more of the required mutations are inherited, fewer additional mutations are required to produce cancer, and the cancer tends to run in families. Knudson's idea has been called the "two-hit hypothesis" because, in retinoblastoma, only two mutations are necessary to cause a tumor. In most cancers, however, more than two mutations are involved in the transformation of normal cells into cancer cells. In the case of retinoblastoma, the two required mutations occur at the same locus (both alleles become mutated), but mutations at different loci are required for the development of many other cancers. The idea that cancer results from multiple mutations turns out to be correct for most cancers.

Knudson's genetic theory of cancer has been confirmed by the identification of genes that, when mutated, cause cancer. Today, we recognize that cancer is fundamentally a genetic disease, although few cancers are actually inherited. Most tumors arise from somatic mutations that accumulate over a person's life span, either through spontaneous mutation or in response to environmental mutagens.

THINK-PAIR-SHARE Question 3

THE CLONAL EVOLUTION OF TUMORS Cancer begins when a single somatic cell undergoes a mutation that causes the cell to divide at an abnormally rapid rate. The cell proliferates, giving rise to a clone of cells, each of which carries the same mutation. Because the cells of the clone divide more rapidly than normal, they soon outgrow other cells. An additional somatic mutation that arises in some of the clone's cells may further enhance the ability of those cells to proliferate, and cells carrying both mutations soon become the most common cells in the clone. Eventually, they may be overtaken by cells that contain yet more mutations that enhance proliferation. In this process, called **clonal evolution**, the tumor cells accumulate somatic mutations that allow them to become increasingly more aggressive in their proliferative properties (**Figure 23.4**).

The rate of clonal evolution depends on the frequency with which new mutations arise. Any genetic defect that allows more mutations to arise will accelerate cancer progression. Genes that regulate DNA repair are often found to have been mutated in the cells of advanced cancers, and inherited disorders of DNA repair are usually characterized by increased incidences of cancer. Because DNA-repair mechanisms normally eliminate many of the mutations that arise, cells with defective DNA-repair systems are more likely to retain mutations than are normal cells, including

mutations in genes that regulate cell division. Xeroderma pigmentosum, for example, is a rare disorder caused by a defect in DNA repair (see Section 18.5). People with this condition have elevated rates of skin cancer when exposed to sunlight (which induces mutation). Similarly, breast cancer can be caused by mutations in *BRCA1* and *BRCA2*, two genes that function in DNA repair.

Mutations in genes that affect chromosome segregation may also contribute to the clonal evolution of tumors. Many cancer cells are aneuploid (contain extra or missing copies of individual chromosomes; see Chapter 8), and clearly, chromosome mutations can contribute to cancer progression by duplicating some genes (those on extra chromosomes) and eliminating others (those on deleted chromosomes). Cellular defects that interfere with chromosome separation increase aneuploidy and may therefore accelerate cancer progression.

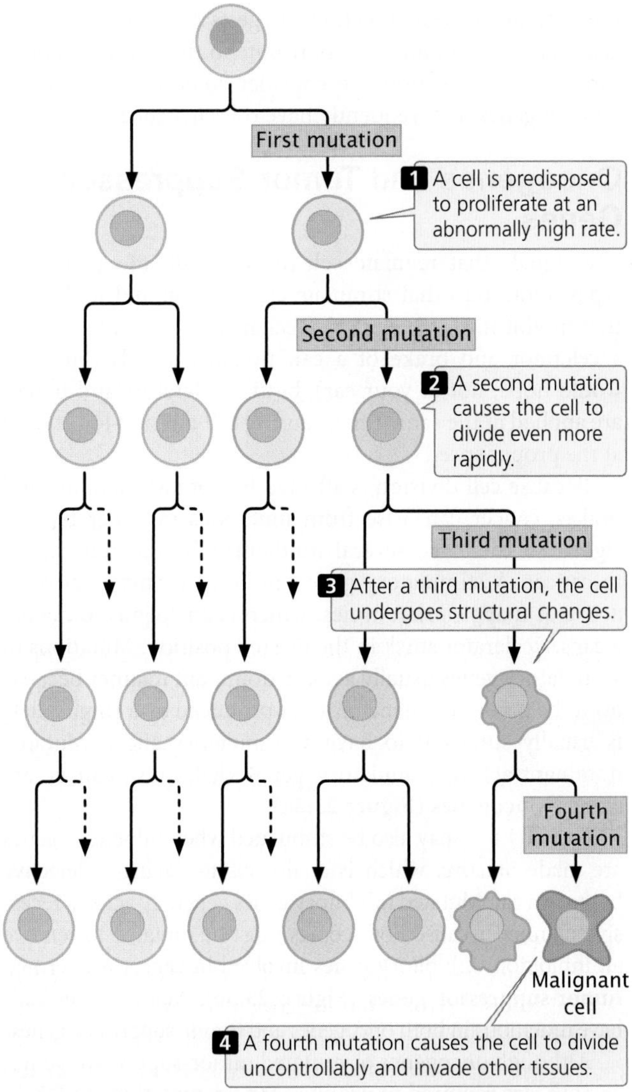

23.4 Through clonal evolution, tumor cells acquire multiple mutations that allow them to become increasingly more aggressive and proliferative. To conserve space, a dashed arrow is used to represent a second cell of the same type in each case.

First mutation

1 A cell is predisposed to proliferate at an abnormally high rate.

Second mutation

2 A second mutation causes the cell to divide even more rapidly.

Third mutation

3 After a third mutation, the cell undergoes structural changes.

Fourth mutation

Malignant cell

4 A fourth mutation causes the cell to divide uncontrollably and invade other tissues.

CONCEPTS

Cancer is fundamentally a genetic disease. Mutations in several genes are usually required to produce cancer. If one of those mutations is inherited, fewer somatic mutations are necessary for cancer to develop, and the person may have a predisposition to cancer. Clonal evolution is the accumulation of mutations in a clone of cells.

✔ CONCEPT CHECK 1

How does the multistep model of cancer explain the observation that sporadic cases of retinoblastoma usually appear in only one eye, whereas inherited forms of the cancer appear in both eyes?

The Role of Environmental Factors in Cancer

Although cancer is a genetic disease, most cancers are not inherited, and many are influenced by environmental factors. The role of environmental factors in cancer is suggested by differences in the incidence of specific cancers throughout the world (**Table 23.2**). The results of studies show that migrant populations typically take on the cancer incidence of their host country. For example, the overall rates of colon cancer are considerably lower in Japan than in Hawaii. However, within a single generation after migration to Hawaii, Japanese people develop colon cancer at rates similar to or exceeding those of Caucasian Hawaiians. The increased cancer rate among the migrants is due to the fact that they are exposed to the same environmental factors as the natives are.

A number of environmental factors contribute to cancer, but those that have the greatest effects include tobacco use, diet, obesity, alcohol, and UV radiation (**Table 23.3**). Other environmental factors that induce cancer are certain types of chemicals, such as benzene (used as an industrial solvent), benzo[a]pyrene (found in cigarette smoke),

TABLE 23.2	Examples of geographic variation in the incidence of cancer	
Type of Cancer	**Location**	**Incidence Rate***
Lip	Canada (Newfoundland)	15.1
	Brazil (Fortaleza)	1.2
Nasopharynx	Hong Kong	30.0
	United States (Utah)	0.5
Colon	United States (Iowa)	30.1
	India (Mumbai)	3.4
Lung	United States (New Orleans, African Americans)	110.0
	Costa Rica	17.8
Prostate	United States (Utah)	70.2
	China (Shanghai)	1.8
Bladder	United States (Connecticut, Whites)	25.2
	Philippines (Rizal)	2.8
All cancer	Switzerland (Basel)	383.3
	Kuwait	76.3

Source: C. Muir et al., *Cancer Incidence in Five Continents*, vol. 5 (Lyon: International Agency for Research on Cancer, 1987), Table 12-2.
*The incidence rate is the age-standardized rate in males per 100,000 population.

TABLE 23.3	Percentage of cancer cases in the United Kingdom attributed to environmental factors
Factor	**Percentage of Cancer Cases**
Tobacco	19.4
Diet	9.2
Overweight and obesity	5.5
Alcohol	4.0
Occupation	3.7
Radiation (UV)	3.5
Infections	3.1
Radiation (ionizing)	1.8
All environmental factors	42.7

Source: Data from Parkin, D. M., L. Boyd, and L. C. Walker. 2011. Fraction of cancer attributable to lifestyle and environmental factors in the UK in 2010. *British Journal of Cancer* 105:S77–S81.

and polychlorinated biphenyls (PCBs; used in industrial transformers and capacitors). Most environmental factors associated with cancer cause somatic mutations that stimulate cell division or otherwise affect the process of cancer progression.

THINK-PAIR-SHARE Question 4

Environmental factors may interact with genetic predispositions to cancer. Lung cancer, for example, is clearly associated with smoking, an environmental factor. Genome-wide association studies (see Section 20.1) revealed that variation at several genes predisposes some people to smoking-induced lung cancer. Variants at some of these genes cause people to be more likely to become addicted to smoking. Other predisposing genes encode receptors that bind potential carcinogens in cigarette smoke. **> TRY PROBLEM 23**

23.2 Mutations in Several Types of Genes Contribute to Cancer

As we have learned, cancer is a disease caused by alterations in DNA. However, there are many different types of genetic alterations that may contribute to the development of cancer. More than 350 different human genes have been identified that contribute to cancer; the actual number is probably much higher. Research on mice suggests that more than 2000 genes can, when mutated, contribute to the development of cancer. In this section, we consider some of the different types of genes that frequently have roles in cancer.

Oncogenes and Tumor-Suppressor Genes

The signals that regulate cell division fall into two basic types: molecules that stimulate cell division and molecules that inhibit it. These control mechanisms are similar to the accelerator and brake of a car. In normal cells (but, one would hope, not in your car), both accelerators and brakes are applied at the same time, causing cell division to proceed at the proper speed.

Because cell division is affected by both accelerators and brakes, cancer can arise from mutations in either type of signal, so there are several fundamentally different routes to cancer. A stimulatory gene can be made hyperactive or active at inappropriate times, which is analogous to having a car's accelerator stuck in the floored position. Mutations in stimulatory genes usually act in a dominant manner because even the amount of gene product produced by a single allele is usually sufficient to have a stimulatory effect. Mutated dominant-acting stimulatory genes that cause cancer are termed **oncogenes** (**Figure 23.5a**).

Cell division may also be stimulated when inhibitory genes are made *inactive*, which is analogous to having a defective brake in a car. Mutated inhibitory genes generally act in a recessive manner because both copies must be mutated to remove all inhibition. Inhibitory genes involved in cancer are termed **tumor-suppressor genes** (**Figure 23.5b**). Many cancer cells have mutations in both oncogenes and tumor-suppressor genes.

Although oncogenes or mutated tumor-suppressor genes, or both, are required to produce cancer, mutations in DNA-repair genes can increase the likelihood of acquiring mutations in those genes. Having mutated DNA-repair genes is analogous to having a lousy car mechanic who does not make the necessary repairs on a broken accelerator or brake.

(a) Oncogenes

Dominant-acting mutation

Homozygous wild type (+/+)

Heterozygous (+/−)

Mutation in either allele

Normal growth-stimulating factors

Hyperactive stimulatory factor

Normal stimulatory factor

Normal cell division

Excessive cell proliferation

1 Proto-oncogenes normally produce factors that stimulate cell division.

2 Mutant alleles (oncogenes) tend to be dominant: one copy of the mutant allele is sufficient to induce excessive cell proliferation.

(b) Tumor-suppressor genes

Recessive-acting mutation

Homozygous wild type (+/+)

Homozygous (−/−)

Mutation in both alleles (or mutation in one and deletion in one)

Normal growth-limiting factors

No inhibitory factor

No inhibitory factor

Normal cell division

Excessive cell proliferation

3 Tumor-suppressor genes normally produce factors that inhibit cell division.

4 Mutant alleles are recessive (both alleles must be mutated to produce excessive cell proliferation).

23.5 Both oncogenes and tumor-suppressor genes contribute to cancer, but they differ in their modes of action and dominance.

ONCOGENES Oncogenes were the first cancer-causing genes to be identified. In 1909, a farmer brought physician Peyton Rous a hen with a large connective-tissue tumor (sarcoma) growing on its breast. When Rous injected pieces of this tumor into other hens, they also developed sarcomas. Rous conducted experiments that demonstrated that the tumors were being transmitted by a retrovirus, which became known as the Rous sarcoma virus. A number of other cancer-causing viruses were subsequently isolated from various animal tissues. These viruses were generally assumed to carry a cancer-causing gene that was transferred to the host cell. The first oncogene, called *src*, was isolated from the Rous sarcoma virus in 1970.

In 1975, Michael Bishop, Harold Varmus, and their colleagues began to use probes for viral oncogenes to search for related sequences in normal cells. They discovered that the genomes of all normal cells carry DNA sequences that are closely related to oncogenes. These normal cellular genes are called **proto-oncogenes**. They are responsible for basic cellular functions in normal cells, but when mutated, they become oncogenes that contribute to the development of cancer. When a virus infects a cell, a proto-oncogene may become incorporated into the viral genome through recombination. Within the viral genome, the proto-oncogene may mutate to an oncogene that, when inserted back into a host cell, causes rapid cell division and cancer. While viruses are capable of converting proto-oncogenes into oncogenes, most proto-oncogenes are mutated to form oncogenes without the involvement of a virus.

Many oncogenes have been identified by experiments in which selected fragments of DNA are added to cells in culture. Some of the cells take up the DNA, and if these cells become cancerous, then the DNA fragment that was added to the culture must contain an oncogene. The fragments can then be sequenced and the oncogene identified. A large number of oncogenes have now been discovered (**Table 23.4**). About 90% of all cancer genes are thought to be dominant oncogenes.

TUMOR-SUPPRESSOR GENES Tumor-suppressor genes are more difficult to identify than oncogenes because they *inhibit* cancer and are recessive; both alleles must be mutated before the inhibition of cell division is removed. Because the *failure* of their function promotes cell proliferation, tumor-suppressor genes cannot be identified by adding them to cells and looking for cancer. About 10% of cancer-causing genes

TABLE 23.4	Some oncogenes and functions of their corresponding proto-oncogenes	
Gene	Normal Function	Cancer in Which Gene Is Mutated
erbB	Part of growth factor receptor	Many types of cancer
fos	Transcription factor	Osteosarcoma and endometrial carcinoma
jun	Transcription factor, cell cycle control	Lung cancer, breast cancer
myc	Transcription factor	Lymphomas, leukemias, neuroblastoma
ras	GTP binding and GTPase	Many types of cancer
sis	Growth factor	Glioblastomas and other cancers
src	Protein tyrosine kinase	Many types of cancer

are thought to be tumor-suppressor genes, but their impact is substantial: most cancers contain one or more mutated tumor-suppressor genes.

Defects in both copies of a tumor-suppressor gene are usually required to cause cancer. An organism can inherit one defective copy of a tumor-suppressor gene (be heterozygous for a cancer-causing mutation) and not have cancer because the remaining normal allele produces the tumor-suppressing product. However, these heterozygotes are often predisposed to cancer because the inactivation or loss of the one remaining normal allele is all that is required to completely eliminate the tumor-suppressor gene product. Inactivation of the remaining wild-type allele of a heterozygote is referred to as the **loss of heterozygosity**. A common mechanism for the loss of heterozygosity is a deletion on the chromosome that carried the normal copy of the tumor-suppressor gene (**Figure 23.6**). Sometimes, however, mutations in a tumor-suppressor gene can act in a dominant fashion; this may occur when the mutation alters the sequence or expression of the tumor-suppressor protein in such a way that it gains a new function that contributes to cancer.

Among the first tumor-suppressor genes to be identified was the gene that causes retinoblastoma. In 1985, Raymond White and Webster Cavenne showed that large segments of chromosome 13 were missing in cells from retinoblastoma tumors, and later, the *RB* tumor-suppressor gene was isolated from those segments. Another example of a tumor-suppressor gene is *BRCA1*, mutations of which are associated with increased risk of breast and ovarian cancer. *BRCA1* produces a protein that normally helps in the repair of double-strand breaks in DNA by homologous recombination (see Section 18.5). The protein also acts as a transcription factor and interacts with histone deacetylase enzymes, which affect transcription. A number of tumor-suppressor genes have now been discovered (**Table 23.5**).

TABLE 23.5	Some tumor-suppressor genes and their normal functions	
Gene	Normal Function	Cancer in Which Gene Is Mutated
APC	Scaffold protein, interacts with microtubules	Colorectal
BRCA1	DNA repair, transcription factor	Breast and ovarian
CDKN2A	Regulates cell division	Melanoma
NF1	GTPase activator	Neurofibromatosis
p53	Regulates cell division, apoptosis, DNA repair, and other functions	Many types of cancer
RB	Regulates cell division	Retinoblastoma and many other cancers

Sometimes the mutation or loss of a single allele of a recessive tumor-suppressor gene is sufficient to cause cancer. This effect—the appearance of a normally recessive trait in an individual cell or organism that is heterozygous for that trait—is called **haploinsufficiency**. This phenomenon is thought to be due to dosage effects: the heterozygote, with only one normal allele, produces only half as much of the product encoded by the tumor-suppressing gene as the homozygote does. Normally, this amount is sufficient for the cellular processes that prevent tumor formation, but it is less than the optimal amount, and other factors may sometimes combine with the lowered amount of tumor-suppressor product to cause cancer. **▶ TRY PROBLEM 24**

CONCEPTS

Proto-oncogenes are genes that control normal cellular functions; when mutated, they become oncogenes that stimulate cell proliferation. They tend to be dominant in their action. Tumor-suppressor genes normally inhibit cell proliferation; when mutated, they allow cells to proliferate. Tumor-suppressor genes tend to be recessive in their action. Individual organisms that are heterozygous for tumor-suppressor genes are often predisposed to cancer.

✔ CONCEPT CHECK 2

Why are oncogenes usually dominant in their action, whereas tumor-suppressor genes are recessive?

Genes That Control the Cell Cycle

The cell cycle is the normal process by which cells undergo growth and division. Normally, progression through the cell cycle is tightly regulated so that cells divide only when additional cells are needed, when all the components necessary for division are present, and when the DNA has been replicated without damage. Sometimes, however, errors arise in

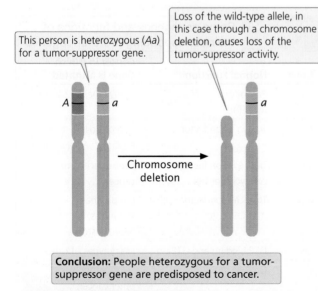

This person is heterozygous (*Aa*) for a tumor-suppressor gene.

Loss of the wild-type allele, in this case through a chromosome deletion, causes loss of the tumor-suppressor activity.

A   *a*   *a*

Chromosome deletion

Conclusion: People heterozygous for a tumor-suppressor gene are predisposed to cancer.

23.6 The loss of heterozygosity often leads to cancer in a person heterozygous for a tumor-suppressor gene.

one or more of the components that regulate the cell cycle. These errors often cause cells to divide at inappropriate times or rates, leading to cancer. Indeed, many proto-oncogenes and tumor-suppressor genes function normally by helping to control the cell cycle. Before considering how errors in this system contribute to cancer, we must first understand how the cell cycle is usually regulated.

CONTROL OF THE CELL CYCLE As discussed in Chapter 2, the cell cycle is the series of stages through which a cell passes from one cell division to the next. Cells that are actively dividing pass through the G_1, S, and G_2 phases of interphase and then move directly into the M phase, in which cell division takes place. Nondividing cells pass from G_1 into the G_0 stage, in which they are functional but not actively growing or dividing. Progression from one stage of the cell cycle to another is influenced by a number of internal and external signals and is regulated at key points in the cycle, called checkpoints.

Key events of the cell cycle are controlled by **cyclin-dependent kinases (CDKs)**. Kinases are enzymes that phosphorylate (add phosphate groups to) other proteins. In some cases, phosphorylation activates the other protein; in others, it inactivates the other protein. As their name implies, CDKs are functional only when associated with another type of protein, called a **cyclin**. The levels of cyclins oscillate over the course of the cell cycle; when bound to a CDK, a cyclin specifies which proteins the CDK will phosphorylate. Each cyclin appears at a specific point in the cell cycle, usually because its synthesis and destruction are regulated by another cyclin. Cyclins and CDKs are called by different names in different organisms; here, we use the terms applied to these molecules in mammals.

G_1-TO-S TRANSITION Let's begin by looking at the G_1-to-S transition. As mentioned earlier, cell cycle checkpoints ensure that all cellular components are present and in good working order before the cell proceeds to the next stage of the cycle. The G_1/S checkpoint is at the end of G_1, just before the cell enters the S phase and replicates its DNA. The cell is prevented from passing through the G_1/S checkpoint by the retinoblastoma (RB) protein (**Figure 23.7**), which binds to a transcription factor called E2F and keeps it inactive. During G_1, cyclin D and cyclin E continuously increase in concentration and combine with their associated CDKs. Cyclin-D–CDK and cyclin-E–CDK both phosphorylate molecules of RB. By late in G_1, the phosphorylation of RB is completed, which inactivates RB. Without the inhibitory effects of RB, E2F is released. The E2F transcription factor stimulates the transcription of genes that produce enzymes necessary for the replication of DNA, and the cell moves into the S phase of the cell cycle.

G_2-TO-M TRANSITION Regulation of the G_2-to-M transition is similar to that of the G_1-to-S transition. In the G_2-to-M transition, cyclin B combines with a CDK to form an inactive complex called *mitosis-promoting factor* (MPF).

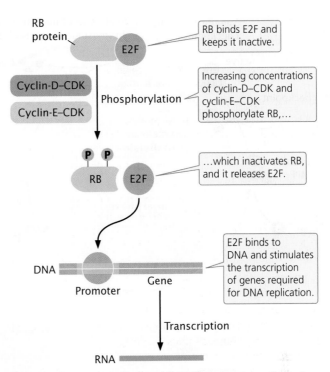

23.7 The RB protein helps control a cell's progression through the G_1/S checkpoint by binding transcription factor E2F.

After MPF has been formed, it must be activated by the removal of a phosphate group (**Figure 23.8a**). During G_1, cyclin B levels are low, so the amount of MPF is also low. As more cyclin B is produced, it combines with CDK to form increasing amounts of MPF. Near the end of G_2, the amount of active MPF reaches a critical level, which commits the cell to divide. The MPF concentration continues to increase, reaching a peak in mitosis.

The active form of MPF phosphorylates other proteins, which then bring about many of the events associated with mitosis, such as nuclear-membrane breakdown, mitotic spindle formation, and chromosome condensation. At the end of metaphase, cyclin B is abruptly degraded, which lowers the amount of MPF and, by initiating anaphase, sets in motion a chain of events that ultimately brings mitosis to a close (**Figure 23.8b**). In brief, high levels of active MPF stimulate mitosis, and low levels of MPF bring a return to interphase conditions.

The G_2/M checkpoint is at the end of G_2, before the cell enters mitosis. A number of factors stimulate the synthesis of cyclin B and the activation of MPF, whereas other factors inhibit MPF. Together, these factors ensure that mitosis is not initiated until conditions are appropriate for cell division. For example, DNA damage inhibits the activation of MPF; consequently, the cell is arrested in G_2 and does not undergo division.

SPINDLE-ASSEMBLY CHECKPOINT Yet another checkpoint, called the spindle-assembly checkpoint, functions in metaphase. This checkpoint delays the onset of anaphase until all chromosomes are aligned on the metaphase plate

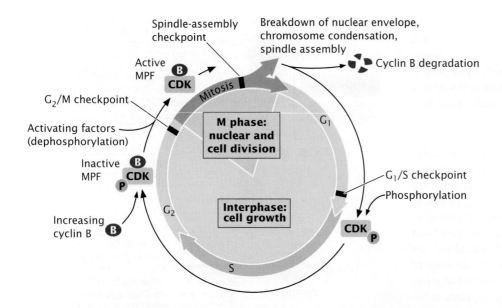

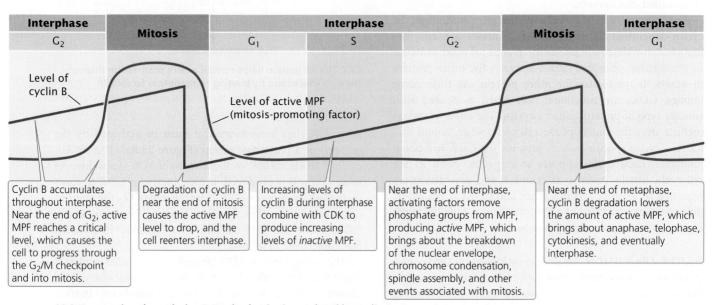

23.8 Progression through the G₂/M checkpoint is regulated by cyclin B.

and sister kinetochores are attached to spindle microtubules from opposite poles. If all chromosomes are not properly aligned, the checkpoint blocks the destruction of cyclin B. The persistence of cyclin B keeps MPF active and maintains the cell in a mitotic state. An additional checkpoint controls the cell's exit from mitosis.

MUTATIONS IN CELL CYCLE CONTROL AND CANCER
Many cancers are caused by defects in the cell cycle's regulatory machinery. For example, mutations in the gene that encodes the RB protein—which normally holds the cell in G₁ until the DNA is ready to be replicated—are associated with many cancers, including retinoblastoma. When the *RB* gene is mutated, cells pass through the G₁/S checkpoint without the normal controls that prevent cell proliferation. The gene

that encodes cyclin D (which stimulates the passage of cells through the G₁/S checkpoint) is overexpressed in about 50% of all breast cancers, as well as in some cases of esophageal and skin cancer. Likewise, the tumor-suppressor gene *p53*, which is mutated in about 75% of all colon cancers, regulates a potent inhibitor of CDK activity.

Some proto-oncogenes and tumor-suppressor genes have roles in apoptosis, a process of programmed cell death in which the cell's DNA is degraded, its nucleus and cytoplasm shrink, and the cell undergoes phagocytosis by other cells without the leakage of its contents. Cells have the ability to assess themselves, and if they are abnormal or damaged, they normally undergo apoptosis (see Section 22.4). Cancer cells frequently have chromosome mutations, DNA damage, or other cellular anomalies that would normally stimulate apoptosis and prevent their proliferation. Many of these cells have mutations in genes that regulate apoptosis and therefore do not undergo programmed cell death. The ability of a cell to initiate apoptosis in response to DNA damage, for example, depends on *p53*, which is inactive in many human cancers.

CONCEPTS

Progression through the cell cycle is controlled at checkpoints, which are regulated by interactions between cyclins and cyclin-dependent kinases. Genes that control the cell cycle are frequently mutated in cancer cells.

✔ CONCEPT CHECK 3

What would be the most likely effect of a mutation that causes cyclin B to be unable to bind to CDK?

a. Cells pass through the G_2/M checkpoint and enter mitosis even when DNA has not been replicated.

b. Cells never pass through the G_1/S checkpoint.

c. Cells pass through mitosis more quickly than nonmutated cells.

d. Cells fail to pass through the G_2/M checkpoint and do not enter mitosis.

SIGNAL-TRANSDUCTION PATHWAYS Whether cells pass through the cell cycle and continue to divide is influenced by a large number of internal and external signals. External signals are initiated by hormones and growth factors. These molecules are often unable to pass through the cell membrane because of their size or charge; they exert their effects by binding to receptors on the cell surface. This binding triggers a series of intracellular reactions that then carry the message to the nucleus or another site within the cell. This type of system, in which an external signal triggers a cascade of intracellular reactions that ultimately produce a specific response, is called a **signal-transduction pathway**. Defects in signal-transduction pathways are often associated with cancer.

A signal-transduction pathway begins with the binding of an external signaling molecule to a specific receptor that is embedded in the cell membrane. Receptors in signal-transduction pathways usually have three parts: (1) an extracellular domain that protrudes from the cell and binds the signaling molecule; (2) a transmembrane domain that passes across the membrane and conducts the signal to the interior of the cell; and (3) an intracellular domain that extends into the cytoplasm and, upon the binding of the signaling molecule, undergoes a chemical or conformational change that is transmitted to molecules of the signal-transduction pathway in the cytoplasm. The binding of a signaling molecule to the membrane-bound receptor activates a protein in the pathway. On activation, this protein activates the next molecule in the pathway, often by adding or removing phosphate groups or causing changes in the conformation of the protein. The newly activated protein activates the next molecule in the pathway, and in this way, the signal is passed along through a cascade of reactions and ultimately produces the response, such as stimulating or inhibiting the cell cycle.

In the past decade, much research has been conducted to determine the pathways by which various signals influence the cell cycle. To illustrate signal transduction, let's consider the Ras signal-transduction pathway, which plays an important role in control of the cell cycle. Each Ras protein cycles between an active form and an inactive form. In its inactive form, a Ras protein is bound to guanosine diphosphate (GDP); in its active form, it is bound to guanosine triphosphate (GTP).

The Ras signal-transduction pathway is activated when a growth factor, such as epidermal growth factor (EGF), binds to a receptor on the cell membrane (**Figure 23.9**). The binding of EGF causes a conformational change in the receptor and the addition of phosphate groups to it. The addition of the phosphate groups allows adaptor molecules to bind to the receptor. These adaptor molecules link the receptor with an inactive molecule of Ras protein. The adaptor molecules activate Ras by stimulating it to release GDP and bind GTP. The newly activated Ras protein then binds to an inactive form of another protein, called Raf, and activates it. After activating Raf, the Ras protein hydrolyzes GTP to GDP, returning to its inactive form.

Activated Raf then sets in motion a cascade of reactions that ends in the activation of a protein called MAP kinase. Activated MAP kinase moves into the nucleus and phosphorylates a number of transcription factors, which then stimulate the transcription of genes taking part in the cell cycle. In this way, the original external signal promotes cell division. A number of other signal-transduction pathways that affect the cell cycle and cell proliferation have been identified.

Because signal-transduction pathways help control the cell cycle, defects in their components often contribute to cancer. For example, genes that encode Ras proteins are frequently oncogenes, and mutations in these genes are often found in cancer cells: 95% of tumors of the pancreas and 45% of colorectal tumors have mutations in *ras* genes. Mutations in these genes produce mutant Ras proteins that are permanently activated and continuously stimulate cell division.

CONCEPTS

Molecules outside the cell often bring about intracellular responses by binding to a transmembrane receptor and stimulating a cascade of intracellular reactions known as a signal-transduction pathway. Many molecules in the pathway are proteins that alternate between active and inactive forms. Defects in signal-transduction pathways are often associated with cancer.

✔ CONCEPT CHECK 4

Ras proteins are activated when they

a. bind GTP.

b. release GTP.

c. bind GDP.

d. undergo acetylation.

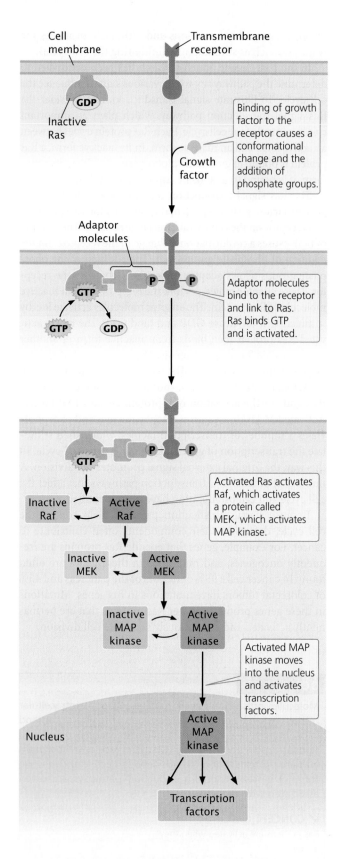

23.9 The Ras signal-transduction pathway conducts signals from growth factors and hormones to the nucleus and stimulates the cell cycle. Mutations in this pathway often contribute to cancer.

DNA-Repair Genes

As we have seen, cancer arises from the accumulation of multiple mutations in a single cell. Some cancer cells have normal rates of mutation, and multiple mutations accumulate because each mutation gives the cell a replicative advantage over cells without the mutations. Other cancer cells have higher-than-normal rates of mutation in all of their genes, which leads to more frequent mutation of oncogenes and tumor-suppressor genes. What might be the source of these high rates of mutation in some cancer cells?

Two processes control the rate at which mutations arise within a cell: (1) the rate at which errors arise during and after the course of DNA replication and (2) the efficiency with which these errors are corrected. The error rate in replication is controlled by the fidelity of DNA polymerases and other proteins involved in the replication process (see Chapter 12). However, defects in genes encoding replication proteins have not been strongly linked to cancer.

The mutation rate is also strongly affected by whether errors are corrected by DNA-repair systems (see Section 18.5). Defects in genes that encode components of these repair systems have been consistently associated with a number of cancers. People with xeroderma pigmentosum, for example, are defective in nucleotide-excision repair, an important DNA-repair system that normally corrects DNA damage caused by a number of mutagens, including ultraviolet light. Likewise, about 13% of colorectal, endometrial, and stomach cancers have cells that are defective in mismatch repair, another major DNA-repair system in the cell.

A particular type of colon cancer, called hereditary nonpolyposis colon cancer, is inherited as an autosomal dominant trait. In families with this condition, a person can inherit one mutated and one normal allele of a gene that controls mismatch repair. The normal allele provides sufficient levels of the protein product for mismatch repair to function, but it is highly likely that this normal allele will become mutated or lost in at least a few cells. If it does so, there is no mismatch repair, and these cells undergo higher-than-normal rates of mutation, leading to defects in oncogenes and tumor-suppressor genes that cause the cells to proliferate.

Defects in DNA-repair systems may also contribute to the generation of chromosome rearrangements and genomic instability. Many DNA-repair systems make single- and double-strand breaks in the DNA as a part of the repair process. If these breaks are not repaired properly, then chromosome rearrangements often result.

Genes That Regulate Telomerase

Another factor that may contribute to the progression of cancer is the inappropriate activation of the enzyme telomerase. Telomeres are special sequences at the ends of eukaryotic chromosomes. Recall from Chapter 12 that in most somatic cells, the ends of chromosomes cannot be replicated, and the telomeres become shorter with each cell division. This shortening eventually leads to the destruction of the chromosomes and cell death, so somatic cells are capable of only a limited number of cell divisions.

In germ cells and stem cells, telomerase replicates the chromosome ends (see pp. 357–360 in Chapter 12), thereby maintaining the telomeres, but this enzyme is not normally expressed in somatic cells. In many tumor cells, however, sequences that regulate the expression of the telomerase gene are mutated, allowing the enzyme to be expressed, and the cell is capable of unlimited cell division. This mutation allows cancer cells to divide indefinitely.

Neuroblastoma, for example, is a tumor of the sympathetic nervous systems that occurs in children. Many of these tumors are low risk, regressing spontaneously or responding to treatment. On the other hand, about half are high risk, do not respond to treatment, and often are lethal. In an attempt to differentiate between low-risk and high-risk neuroblastoma tumors, researchers sequenced the entire genome of a series of these tumors. They found that many of the high-risk tumors had chromosome rearrangements or other mutations that increased the expression of telomerase. Thus, expression of telomerase appears to be a characteristic of many high-risk neuroblastomas.

Genes That Promote Vascularization and the Spread of Tumors

Another important group of factors that contribute to the progression of cancer includes genes that affect the growth and spread of tumors. Oxygen and nutrients, which are essential to the survival and growth of tumors, are supplied by blood vessels, and the growth of new blood vessels (angiogenesis) is important to tumor progression. Angiogenesis is stimulated by growth factors and others proteins encoded by genes whose expression is carefully regulated in normal cells. In tumor cells, genes encoding these proteins are often overexpressed compared with normal cells, and inhibitors of angiogenesis-promoting factors may be inactivated or underexpressed. At least one inherited cancer—von Hippel–Lindau disease, in which people develop multiple types of tumors—is caused by the mutation of a gene that affects angiogenesis.

In the development of many cancers, the primary tumor gives rise to cells that spread to distant sites, producing secondary tumors. This process of metastasis, which is the cause of death in 90% of human cancer deaths, is influenced by cellular changes induced by somatic mutation. As discussed in the introduction to this chapter, the *palladin* gene, when mutated, contributes to the metastasis of pancreatic tumors. By using microarrays to measure levels of gene expression in tumors, researchers have identified other genes that are transcribed at a significantly higher rate in metastatic cells than in nonmetastatic cells. For example, one study detected a set of 95 genes that were overexpressed or underexpressed in a population of metastatic breast-cancer cells that were strongly metastatic to the lung, compared with a population of cells that were only weakly metastatic to the lung.

Genes that contribute to metastasis often encode components of the extracellular matrix and the cytoskeleton. Others encode adhesion proteins, which help hold cells together.

Advances in sequencing technology have now made it possible to study how metastasized tumor cells differ from normal cells as well as from those of the primary tumor from which they were derived. In one experiment, researchers sequenced the entire genome of cells from a metastasized breast-cancer tumor and compared it with the genome of normal cells from the same person and with the genome of the primary tumor, which had been removed from the patient nine years earlier. The researchers found 32 different somatic mutations in the coding regions of genes from the metastasized tumor cells, 19 of which were not detected in the primary tumor. This finding suggests that the metastasized tumor underwent considerable genetic changes in its nine-year evolution from the primary tumor. In contrast, another study of a breast-cancer metastasis found only two mutations that were not present in the primary tumor, but in this case, the metastasis had evolved in only one year.

A long-standing view has been that metastasized tumors originate from a single cell that breaks off from the primary tumor and travels to a new site. A recent study that carried out genome sequencing of metastatic prostate tumors has revealed that metastases are sometimes composed of cells from more than one subclone found within the primary tumor. This finding shows that the metastases were seeded by multiple cells from the primary tumor, a process termed polyclonal seeding. There was also evidence that some metastatic tumors received cells from the primary tumor as well as from metastatic tumors at other sites in the body, a process called cross-seeding.

CONCEPTS

Mutations in genes that encode components of DNA-repair systems are often associated with cancer; these mutations increase the rate at which mutations are retained and result in an increased number of mutations in proto-oncogenes, tumor-suppressor genes, and other genes that contribute to cell proliferation. Mutations that allow telomerase to be expressed in somatic cells and those that affect vascularization and metastasis can also contribute to cancer progression.

✔ **CONCEPT CHECK 5**

Which type of mutation in telomerase is associated with cancer cells?

a. Mutations that produce an inactive form of telomerase
b. Mutations that decrease the expression of telomerase
c. Mutations that increase the expression of telomerase
d. All of the above

MicroRNAs and Cancer

MicroRNAs (miRNAs) are a class of small RNA molecules that pair with complementary sequences on mRNA and degrade the mRNA or inhibit its translation (see Section 14.5).

Given the fact that miRNAs are important in controlling gene expression and development, it is not surprising that they are also associated with tumor development. Many tumor cells exhibit widespread reduction in the expression of many miRNAs. Researchers have genetically engineered mouse tumor cells that lacked the machinery to generate miRNAs and found that these cells showed enhanced tumor progression when implanted into mice. Interestingly, this effect was seen only in cells that had already initiated tumor development, suggesting that miRNAs play a role in later stages of tumor progression.

Lowered levels of miRNAs may contribute to cancer by allowing oncogenes that are normally controlled by the miRNAs to be expressed at high levels. For example, let-7 miRNA normally controls the expression of the *ras* oncogene, probably by binding to complementary sequences in the 3′ untranslated region of *ras* mRNA and inhibiting translation. In lung-cancer cells, levels of let-7 miRNA are often low, allowing the Ras protein to be highly expressed, which then leads to the development of lung cancer.

A transcription factor called c-MYC is often expressed at high levels in cancer cells. Evidence suggests that c-MYC helps to drive cell proliferation and the development of cancer. Among other effects, c-MYC binds to the promoters of miRNA genes and decreases their transcription, decreasing the abundance of the miRNAs. Some of these miRNAs are known to suppress tumor development. Research has shown that if, through genetic manipulation, the miRNAs are expressed at high levels, the development of tumors decreases. All of these findings suggest that altered expression of miRNAs plays an important role in cancer.

In other cases, overexpression of miRNAs has been associated with cancer. For example, several miRNAs have been implicated in the process of metastasis. A particular miRNA called miR-10b has been associated with the formation of metastatic breast tumors. In one experiment, investigators manipulated a line of breast-cancer cells so that miR-10b was overexpressed. When the manipulated cells were injected into mice, many of the mice developed metastatic tumors, demonstrating that high levels of this miRNA appear to promote the spread of cancer cells. Further study revealed that in humans, levels of miR-10b are elevated in metastatic tumors compared with breast tumors in metastasis-free patients. Research shows that miR-10b regulates the expression of a number of other genes, including some that are known to suppress the spread of tumor cells. Other miRNAs are known to inhibit metastasis.

Cancer Genome Projects

Formed in 2008, the International Cancer Genome Consortium coordinates efforts to determine the genome sequences of tumors. A goal of the consortium is to completely sequence 500 tumors from each of 50 different types of cancer, along with the genomes of normal tissues from the same patients. This effort is producing important results,

revealing the numbers and types of mutations that are associated with particular cancers. The hope is that new cancer-causing genes will be identified, which will lead to a better understanding of the nature of cancer and suggest new targets for cancer treatment. Another research project, called The Cancer Genome Atlas (TCGA), began in the United States in 2005. This project seeks to provide a comprehensive genomic analysis of over 100 different types of cancer cells, including sequencing of all exons across the genome as well as characterization of mRNA expression, DNA methylation, copy number variations, and microRNAs of the tumor cells.

The genomes of a number of tumors have been sequenced, and many more are currently being sequenced. For example, the entire genome of a small-cell lung carcinoma (a type of lung cancer) was sequenced in 2010 and compared with the genome of normal cells from the same person. More than 22,000 base-pair mutations were identified in the tumor, of which 134 were within protein-encoding genes. The tumor also possessed 58 chromosome rearrangements and 334 copy-number variations (see Chapter 20). In another study that was part of TCGA, researchers examined mRNA expression, microRNAs, DNA methylation, and copy number variations in 489 ovarian adenocarcinomas and sequenced the DNA of exons from 316 of the tumors. Almost all of the tumors contained mutations in *p53*, a tumor-suppressor gene involved in DNA repair and cell cycle control. Mutations in *BRCA1* and *BRCA2*, two tumor-suppressor genes that are also mutated in breast cancer, occurred in 22% of the tumors. Mutations in seven other genes occurred statistically more often than in normal cells. The results suggested several new approaches for drug treatment of ovarian cancer.

Another series of studies sequenced a number of genes in samples of malignant gliomas, an incurable and deadly form of brain cancer. The researchers examined DNA sequences, copy number variations, DNA methylation, and RNA expression in these tumors. The analyses revealed mutations in several genes that appear to be important in the development of glioma tumors. All of these genomic studies are providing new insight in the genetic basis of cancer.

The mutations found in cancer genomes can be divided into two types: drivers and passengers. **Drivers** are mutations that drive the cancer process: they directly contribute to the development of cancer. Drivers include mutations in oncogenes, tumor-suppressor genes, DNA-repair genes, and the other types of genes discussed in this chapter that contribute to the progression of cancer. **Passengers** are mutations that arise randomly in the process of tumor development and do not contribute to cancer progression. Many passengers are located in introns (regions between genes) and in other DNA that is not transcribed and translated, but they can also arise within protein-encoding genes. A major challenge is to determine which of the numerous mutations found in tumors are drivers that actually contribute to the development of cancer and which are passengers, with no effect.

THINK-PAIR-SHARE Question 5

23.3 Epigenetic Changes Are Often Associated with Cancer

Epigenetic changes—alterations to chromatin structure that affect gene expression (see Chapter 21)—are seen in many cancer cells. Two broad lines of evidence suggest that epigenetic changes play an important role in cancer progression. First, genes encoding proteins that are important regulators of epigenetic changes are often mutated in some types of cancer. For example, almost 90% of cases of follicular lymphoma exhibit mutations in the *MLL2* gene, which encodes a histone methyltransferase; this enzyme adds methyl groups to DNA, a type of epigenetic modification that alters chromatin structure and affects transcription. Similarly, the *UTX* gene, which encodes a histone demethylase (an enzyme that removes methyl groups from histone proteins), is mutated in a number of different types of cancer.

A second line of evidence suggesting that epigenetic alterations are important in cancer comes from recent genomic studies that have compared the chromatin structure of cancer cells with that of normal cells from the same individual. These studies often find that the cancer cells have significant alterations to their DNA methylation patterns and histone structure. One type of epigenetic alteration often observed in cancer cells is an overall lower level of DNA methylation (hypomethylation). As discussed in Chapter 17, DNA methylation is often associated with repression of transcription. It is assumed that hypomethylation leads to transcription of oncogenes, which then stimulate cancer. Some evidence also suggests that hypomethylation causes chromosome instability, a hallmark of many tumors. Tumor cells from mice that have been genetically engineered to have reduced DNA methylation show increased gains and losses of chromosomes, but how hypomethylation might cause chromosome instability is unclear.

A number of studies have observed that although the overall level of DNA methylation is often lower in cancer cells, some specific CpG islands (see Chapter 17) have extra methylation (are hypermethylated). For example, one study found that 5% to 10% of normally unmethylated CpG islands located at promoters become abnormally methylated in cancer cells. This excess methylation may inhibit transcription of tumor-suppressor genes, thus stimulating the development of cancer. Methylation of the promoter of the *Apaf-1* gene is seen in many malignant melanoma cells. *Apaf-1* helps bring about apoptosis in cells with damaged DNA; methylation of its promoter reduces the expression of *Apaf-1*, interrupting the process of apoptosis and allowing abnormal cancer cells to survive.

Research has also demonstrated that the histone proteins in nucleosomes, the fundamental units of chromatin, are often abnormally modified in cancer cells. Modification of histone proteins, including methylation and acetylation, alters chromatin structure and affects whether transcription occurs. Genome-wide patterns of histone acetylation are often altered in cancer cells. Epigenetic processes are receiving increasing attention from cancer researchers because they may be amenable to drug therapy.

CONCEPTS

Epigenetic changes, including DNA methylation and histone modification, are often associated with cancer.

✔ **CONCEPT CHECK 6**

Hypermethylation is thought to contribute to cancer by
a. inhibiting DNA replication.
b. inhibiting the expression of tumor-suppressor genes.
c. stimulating the translation of oncogenes.
d. stimulating telomerase.

23.4 Colorectal Cancer Arises Through the Sequential Mutation of a Number of Genes

Most cancers arise from mutations in several genes, often a combination of oncogenes and tumor-suppressor genes. Colorectal cancer is an excellent example of how the accumulation of successive genetic defects can lead to cancer.

Colorectal cancer arises in the cells lining the colon and rectum. More than 134,000 new cases of colorectal cancer are diagnosed each year in the United States, where this cancer is responsible for almost 50,000 deaths annually. If detected early, colorectal cancer can be treated successfully; consequently, there has been much interest in identifying the molecular events responsible for the initial stages of this cancer.

Colorectal cancer is thought to originate as benign tumors called adenomatous polyps (**Figure 23.10**). Initially, these polyps are microscopic, but in time they enlarge, and their cells acquire the abnormal characteristics of cancer cells. In the later stages of the disease, the tumor may invade the muscle layer surrounding the gut and metastasize. The progression of the disease is slow: it may take 10 to 35 years for a benign tumor to develop into a malignant tumor.

Most cases of colorectal cancer are sporadic, developing in people with no family history of the disease, but a few families display a clear genetic predisposition to it. In one form of hereditary colon cancer, known as familial adenomatous polyposis coli, hundreds or thousands of polyps develop in the colon and rectum; if these polyps are not removed, one or more almost invariably become malignant.

Because polyps and tumors of the colon and rectum can be easily observed and removed with a colonoscope (a fiber-optic instrument used to view the interior of the rectum and colon), much is known about the progression of colorectal cancer, and some of the genes responsible for its clonal evolution have been identified. Mutations in these genes are responsible for the successive steps of colorectal-cancer

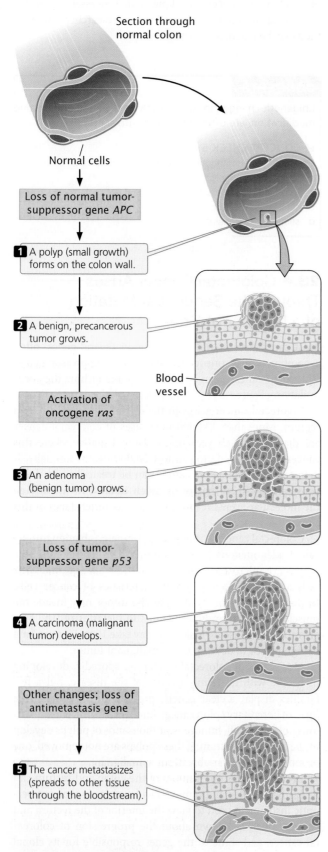

Section through normal colon

Normal cells

↓

Loss of normal tumor-suppressor gene *APC*

↓

1 A polyp (small growth) forms on the colon wall.

↓

2 A benign, precancerous tumor grows.

Blood vessel

↓

Activation of oncogene *ras*

↓

3 An adenoma (benign tumor) grows.

↓

Loss of tumor-suppressor gene *p53*

↓

4 A carcinoma (malignant tumor) develops.

↓

Other changes; loss of antimetastasis gene

↓

5 The cancer metastasizes (spreads to other tissue through the bloodstream).

23.10 Mutations in multiple genes contribute to the progression of colorectal cancer.

progression. Among the earliest steps is a mutation that inactivates the *APC* gene; this mutation increases the rate of cell division, leading to polyp formation (see Figure 23.10). A person with familial adenomatous polyposis coli inherits one defective copy of the *APC* gene, and defects in this gene are associated with the numerous polyps that appear in those who have this disorder. Mutations in *APC* are also found in polyps that develop in people who do not have familial adenomatous polyposis coli.

Mutations of the *ras* oncogene usually occur later, in larger polyps consisting of cells that have acquired some genetic mutations. As discussed earlier in this chapter, the normal *ras* proto-oncogene is a key player in a signal-transduction pathway that relays a signal from growth factors to the nucleus, where the signal stimulates cell division. When *ras* is mutated, the protein it produces continually relays a stimulatory signal for cell division even when growth factors are absent.

Mutations in tumor-suppressor gene *p53* and other genes appear still later in tumor progression; these mutations are rare in polyps but are common in malignant cells. Many colorectal cancers have mutations in *p53*. Because *p53* prevents the replication of cells with genetic damage and affects proper chromosome segregation, mutations in *p53* can allow a cell to rapidly acquire further gene and chromosome mutations, which then contribute to further proliferation and invasion of surrounding tissues.

Recent research also shows that inflammation plays a role in the development of colorectal cancer. For example, people with inflammatory bowel disease are at increased risk for developing colorectal cancer. (Inflammatory bowel disease is chronic inflammation of the digestive tract, often accompanied by severe diarrhea, pain, fatigue, and weight loss.) Although the precise mechanism by which inflammation contributes to cancer is unknown, immune-system cells and the proteins they secrete (see Section 22.6) are likely to affect many of the steps in the progression of cancer.

The sequence of steps just outlined is not the only route to colorectal cancer, and the mutations need not occur in the order presented here. However, this sequence is a common pathway by which colon and rectal cells become cancerous.

23.5 Changes in Chromosome Number and Structure Are Often Associated with Cancer

Most tumors contain cells with chromosome mutations. For many years, geneticists argued about whether these chromosome mutations were the cause or the result of cancer. Some types of tumors are consistently associated with *specific* chromosome mutations; for example, most cases of chronic myelogenous leukemia are associated with a specific reciprocal translocation between chromosomes 22 and 9. These types of associations suggest that chromosome mutations contribute to the cause of the cancer. Yet many cancers are not associated with specific types of chromosome abnormalities, and individual *gene* mutations are now known to contribute to many types of cancer. Nevertheless, as we have noted, chromosome instability is a general feature of cancer cells, causing them to accumulate chromosome mutations, which then affect individual genes that may contribute to the cancer process. Thus, chromosome mutations appear to both *cause* cancer and *result* from it.

At least three types of chromosome rearrangements—deletions, inversions, and translocations—are associated with certain types of cancer. Deletions can result in the loss of one or more tumor-suppressor genes. Inversions and translocations contribute to cancer in several ways. First, the chromosome breaks that accompany these mutations can lie within tumor-suppressor genes, disrupting their function and leading to cell proliferation.

Second, translocations and inversions can bring together sequences from two different genes, generating a fusion protein that stimulates some aspect of the cancer process. Fusion proteins are seen in most cases of chronic myelogenous leukemia, which affects bone-marrow cells. Most people with chronic myelogenous leukemia have a reciprocal translocation between the long arm of chromosome 22 and the tip of the long arm of chromosome 9 (**Figure 23.11**). This translocation produces a shortened chromosome 22, called the Philadelphia chromosome because it was first discovered in Philadelphia. At the end of a normal chromosome 9 is a potential cancer-causing gene called c-*ABL*. As a result of the translocation, part of the c-*ABL* gene is fused with the *BCR* gene from chromosome 22. The protein produced by this *BCR*–c-*ABL* fusion gene is much more active than the protein produced by the normal c-*ABL* gene; the fusion protein stimulates increased, unregulated cell division and eventually leads to leukemia.

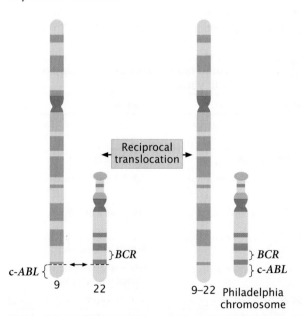

23.11 A reciprocal translocation between chromosomes 9 and 22 causes chronic myelogenous leukemia.

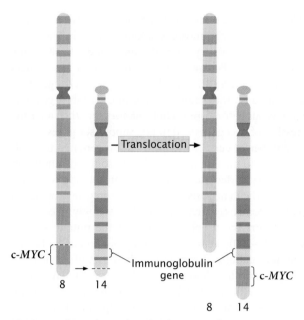

23.12 A reciprocal translocation between chromosomes 8 and 14 causes Burkitt lymphoma.

A third mechanism by which chromosome rearrangements can produce cancer is the transfer of a potential cancer-causing gene to a new location, where it is activated by different regulatory sequences. Burkitt lymphoma is a cancer of the B cells, the lymphocytes that produce antibodies (see Section 22.6). Many people with Burkitt lymphoma possess a reciprocal translocation between chromosome 8 and chromosome 2, 14, or 22 (**Figure 23.12**). This translocation relocates a gene called c-*MYC* from the tip of chromosome 8 to a position on chromosome 2, 14, or 22 that is next to a gene that encodes an immunoglobulin protein. At this new location, c-*MYC*, a cancer-causing gene, comes under the control of regulatory sequences that normally activate the production of immunoglobulins, and c-*MYC* is therefore expressed in B cells. The c-MYC protein stimulates the division of the B cells and leads to Burkitt lymphoma.

CONCEPTS

Many tumors contain a variety of chromosome mutations. Some types of tumors are associated with specific deletions, inversions, and translocations. Deletions can eliminate or inactivate genes that control the cell cycle; inversions and translocations can cause breaks in genes that suppress tumors, fuse genes to produce cancer-causing proteins, or move genes to new locations where they are under the influence of different regulatory sequences.

✔ CONCEPT CHECK 7

Chronic myelogenous leukemia is usually associated with which type of chromosome rearrangement?

a. Duplication   c. Inversion

b. Deletion   d. Translocation

Most advanced tumors contain cells that exhibit a dramatic variety of chromosome anomalies, including extra chromosomes, missing chromosomes, and chromosome rearrangements (**Figure 23.13**). Some cancer researchers believe that cancer is initiated when genetic changes take place that cause the genome to become unstable, generating numerous chromosome abnormalities that then alter the expression of oncogenes and tumor-suppressor genes.

A number of genes that contribute to genomic instability and lead to missing or extra chromosomes (aneuploidy) have now been identified. Aneuploidy in somatic cells usually arises when chromosomes do not segregate properly in mitosis. Normal cells have a spindle-assembly checkpoint that monitors the proper assembly of the mitotic spindle (see pp. 699–700); if chromosomes are not properly attached to the microtubules at metaphase, the onset of anaphase is blocked. Some aneuploid cancer cells contain mutant alleles of genes that encode proteins with roles in this checkpoint; these cells enter anaphase despite the improper assembly of the spindle, or lack of it, and chromosome abnormalities result. For example, mutations in *RB* increase aneuploidy by increasing the expression of a protein called Mad2, which is a critical component of the spindle-assembly checkpoint.

Mutations in genes that encode parts of the mitotic spindle apparatus may also contribute to abnormal segregation and lead to chromosome abnormalities. The *APC* gene, as we have seen, is a tumor-suppressor gene that is often mutated in colon-cancer cells (see Section 23.4). The APC protein has several functions, one of which is to interact with the ends of the microtubules that associate with the kinetochore. Dividing mouse cells that have defective copies of the *APC* gene give rise to cells with many chromosome anomalies.

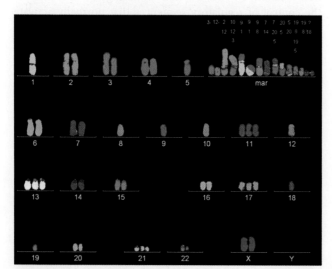

23.13 Cancer cells often possess chromosome abnormalities, including extra chromosomes, missing chromosomes, and chromosome rearrangements. Shown here are chromosomes from a colon-cancer cell, which has numerous chromosome abnormalities. For comparison, see the normal karyotype in Figure 2.6. [Courtesy Dr. Peter Duesberg, UC Berkeley.]

The *p53* gene, in addition to its many other functions, plays a role in the duplication of the centrosome, which is required for proper formation of the spindle and for chromosome segregation. Normally, the centrosome duplicates once per cell cycle. If *p53* is mutated or missing, however, the centrosome may undergo extra duplications, resulting in the unequal segregation of chromosomes. In this way, mutation of the *p53* gene may generate chromosome mutations that contribute to cancer. The *p53* gene is also a tumor-suppressor gene that prevents cell division when the DNA is damaged.

› TRY PROBLEM 31

23.6 Viruses Are Associated with Some Cancers

As mentioned earlier in this chapter, viruses are responsible for a number of cancers in animals, and there is evidence that viruses contribute to at least a few cancers in humans (**Table 23.6**). For example, about 95% of all women with cervical cancer are infected with **human papillomavirus (HPV)**. Similarly, infection with the virus that causes hepatitis B increases the risk of liver cancer in some people. The Epstein–Barr virus, which is responsible for mononucleosis, has been linked to several types of cancer that are prevalent in parts of Africa, including Burkitt lymphoma.

Retroviruses and Cancer

Many of the viruses that cause cancer in animals are retroviruses. In Section 23.2, we saw how studies of the Rous sarcoma retrovirus in chickens led to the identification of oncogenes in humans. Retroviruses sometimes cause cancer by mutating and rearranging host genes, converting proto-oncogenes into oncogenes (**Figure 23.14a**). Another way in which viruses can contribute to cancer is by altering the expression of host genes (**Figure 23.14b**). Retroviruses often contain strong promoters to ensure that their own genetic

TABLE 23.6	Some human cancers associated with viruses
Virus	**Cancer**
Human papillomavirus (HPV)	Cervical, penile, and vulvar cancers
Hepatitis B virus	Liver cancer
Human T-cell leukemia virus 1 (HTLV-1)	Adult T-cell leukemia
Human T-cell leukemia virus 2 (HTLV-2)	Hairy-cell leukemia
Epstein–Barr virus	Burkitt lymphoma, nasopharyngeal cancer, Hodgkin lymphoma
Human herpesvirus	Kaposi sarcoma
Merkel cell polyomavirus	Merkel cell carcinoma

Note: Some of these associations between cancer and viruses exist only in certain populations and geographic areas.

(a)

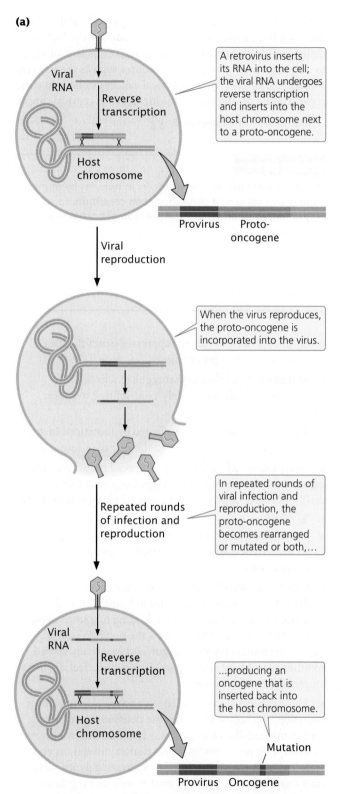

A retrovirus inserts its RNA into the cell; the viral RNA undergoes reverse transcription and inserts into the host chromosome next to a proto-oncogene.

When the virus reproduces, the proto-oncogene is incorporated into the virus.

In repeated rounds of viral infection and reproduction, the proto-oncogene becomes rearranged or mutated or both,…

…producing an oncogene that is inserted back into the host chromosome.

23.14 Retroviruses can cause cancer by (a) mutating or rearranging proto-oncogenes or (b) inserting strong promoters near proto-oncogenes.

(b)

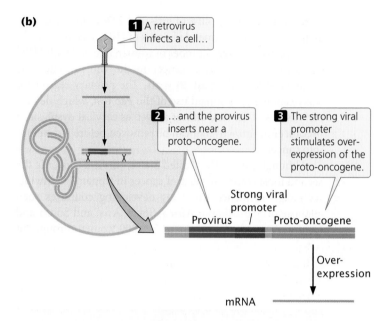

1 A retrovirus infects a cell…

2 …and the provirus inserts near a proto-oncogene.

3 The strong viral promoter stimulates over-expression of the proto-oncogene.

material is transcribed by the host cell. If the provirus inserts near a proto-oncogene, viral promoters can stimulate high levels of expression of the proto-oncogene, leading to cell proliferation.

There are only a few retroviruses that cause cancer in humans. HTLV-1, the first human retrovirus discovered, is associated with human adult T-cell leukemia. Other human cancers are associated with DNA viruses, which, like retroviruses, integrate into the host chromosome but, unlike retroviruses, do not use reverse transcription.

Human Papillomavirus and Cervical Cancer

Human papillomavirus is a DNA virus that causes warts and other types of benign tumors of epithelial cells. More than 100 different types of HPV are known, about 30 of which are sexually transmitted and cause genital warts. A few of them are associated with cervical cancer. In the United States, 70% of the cases of cervical cancer are caused by HPV-16 and HPV-18. These viruses cause cancer by producing proteins that attach to and inactivate *RB* and *p53*. As we have seen, the proteins encoded by these two genes play key roles in the regulation of the cell cycle. When these proteins are inactivated, cells are stimulated to progress through the cell cycle and divide without the normal controls that prevent cell proliferation.

Most sexually active women and men in the United States are infected with HPV, but only a small number of women with HPV will ever develop cervical cancer. The risk of infection by HPV can be reduced by limiting the number of sexual partners and by a vaccine that was approved by the U.S. Food and Drug Administration for women in 2006 and for men in 2009. This vaccine is highly effective against four of the most common HPVs associated with cervical cancer. The U.S. Centers for Disease Control recommends that all boys and girls aged 11 or 12 years be vaccinated. A national survey conducted in the United States in 2014 reported that 60% of girls aged 13–17 years had received at least one dose of the vaccine, but only 42% of boys aged 13–17 years received at least one dose. About 40% of girls and 22% of boys received all three of the recommended doses of HPV vaccine.

In spite of the large number of women infected with HPV, the incidence of cervical cancer in the United States has declined 75% in the past 40 years. The primary reason for this decline is widespread use of the Pap test, which detects early stages of cervical cancer as well as cervical dysplasia, a precancerous growth that can be removed before it develops into cancer.

Although rare in the United States, cervical cancer is the second most common cause of cancer in women worldwide, with high incidences in many developing countries such as those of Sub-Saharan Africa, South Asia, and South and Central America. An estimated 510,000 women throughout the world develop cervical cancer each year, and 288,000 die from the disease. The primary reason for the high incidence and death rate in developing regions is lack of access to cervical-cancer screening procedures such as the Pap test. Making the vaccine against cervical cancer available in these regions could greatly reduce the incidence of cervical cancer.

THINK-PAIR-SHARE Question 6

CONCEPTS

Viruses contribute to a few cancers in humans by mutating and rearranging host genes that then contribute to cell proliferation, or by altering the expression of host genes.

CONCEPTS SUMMARY

■ Cancer is fundamentally a genetic disease, arising from somatic mutations in multiple genes that affect cell division and proliferation. If one or more mutations are inherited, then fewer additional mutations are required for cancer to develop.

■ A mutation that allows a cell to divide rapidly provides the cell with a growth advantage; that cell gives rise to a clone of cells having the same mutation. Within this clone, other mutations may occur that provide additional growth advantages; cells with these additional mutations become dominant in the clone. In this way, the clone evolves.

■ Environmental factors play an important role in the development of many cancers by increasing the rate of somatic mutations.

■ Oncogenes are dominant-acting mutated copies of normal genes (proto-oncogenes) that stimulate cell division. Tumor-suppressor genes normally inhibit cell division; recessive mutations in these genes may contribute to cancer. Sometimes, the mutation of a single copy of a tumor-suppressor gene is sufficient to cause cancer, a phenomenon known as haploinsufficiency.

■ The cell cycle is controlled by cyclins and cyclin-dependent kinases. Mutations in genes that control progression through the cell cycle are often associated with cancer.

■ Signal-transduction pathways conduct external signals that bring about intracellular responses. These pathways consist of a series of proteins that are activated and inactivated in a cascade of reactions. Mutations in the components of signal-transduction pathways can disrupt the cell cycle and may contribute to cancer.

■ Defects in DNA-repair genes often increase the overall mutation rate of other genes, leading to defects in proto-oncogenes and tumor-suppressor genes that can contribute to cancer progression.

■ Mutations in sequences that regulate telomerase allow cells to divide indefinitely, contributing to cancer progression.

■ Tumor progression is also affected by mutations in genes that promote vascularization and the spread of tumors.

■ Many tumor cells exhibit a widespread reduction in the expression of many miRNAs, suggesting that a reduction in miRNA control of gene expression may play a role in tumor progression.

■ Epigenetic changes in chromatin structure, including DNA methylation and histone modification, are often associated with cancer.

■ Colorectal cancer offers a model system for understanding tumor progression in humans. Initial mutations stimulate cell division, leading to a small benign polyp. Additional mutations allow the polyp to enlarge, invade the muscle layer of the gut, and eventually spread to other sites. Mutations in particular genes affect different stages of this progression.

■ Some cancers are associated with specific chromosome mutations, including chromosome deletions, inversions, and translocations. Deletions may cause cancer by removing or disrupting genes that suppress tumors; inversions and translocations may break tumor-suppressing genes, or they may move genes to positions next to different regulatory sequences, which alter their expression.

■ Mutations in some genes cause or allow chromosomes to segregate improperly, leading to aneuploidy that can contribute to cancer.

■ Viruses are associated with some cancers; they contribute to cell proliferation by mutating and rearranging host genes or by altering the expression of host genes.

IMPORTANT TERMS

malignant (p. 693)
metastasis (p. 693)
clonal evolution
    (p. 695)
oncogene (p. 696)

tumor-suppressor
    gene (p. 696)
proto-oncogene (p. 697)
loss of heterozygosity (p. 698)
haploinsufficiency (p. 698)

cyclin-dependent kinase
    (CDK) (p. 699)
cyclin (p. 699)
signal-transduction
    pathway (p. 701)

driver (p. 704)
passenger (p. 704)
human papillomavirus
    (HPV) (p. 708)

ANSWERS TO CONCEPT CHECKS

1. Retinoblastoma results from at least two separate genetic defects, both of which are necessary for cancer to develop. In sporadic cases, two successive mutations must occur in a single cell, which is unlikely, and therefore the cancer typically affects only one eye. In people who have inherited one of the two required mutations, every cell contains this mutation, and so a single additional mutation is all that is required for cancer to develop. Given the millions of cells in each eye, there is a high probability that the second mutation will occur in at least one cell of each eye, producing tumors in both eyes and the inheritance of this type of retinoblastoma.

2. Oncogenes have a stimulatory effect on cell proliferation. Mutations in oncogenes are usually dominant because a

mutation in a single copy of the gene is usually sufficient to produce a stimulatory effect. Tumor-suppressor genes inhibit cell proliferation. Mutations in tumor-suppressor genes are generally recessive because both copies must be mutated to remove all inhibition.

3. d

4. a

5. c

6. b

7. d

WORKED PROBLEM

In some cancer cells, a specific gene has been duplicated many times. Is this gene likely to be an oncogene or a tumor-suppressor gene? Explain your reasoning.

›› Solution Strategy

What information is required in your answer to the problem?
Whether the duplicated gene is likely to be an oncogene or a tumor-suppressor gene, and why.

What information is provided to solve the problem?
In cancer cells, the gene has been duplicated many times.

For help with this problem, review:
Oncogenes and Tumor-Suppressor Genes in Section 23.2.

›› Solution Steps

The gene is likely to be an oncogene. Oncogenes stimulate cell proliferation and act in a dominant manner. Therefore, extra copies of an oncogene will result in cell proliferation and cancer. Tumor-suppressor genes, on the other hand, suppress cell proliferation and act in a recessive manner; a single copy of a tumor-suppressor gene is sufficient to prevent cell proliferation. Therefore, extra copies of the tumor-suppressor gene will not lead to cancer.

> **Recall:** An oncogene is an accelerator of cell division, while a tumor-suppressor gene is a brake.

COMPREHENSION QUESTIONS

Section 23.1

1. What types of evidence indicate that cancer arises from genetic changes?

2. How is cancer different from most other types of genetic diseases?

3. Outline Knudson's two-hit hypothesis of retinoblastoma and describe how it helps to explain unilateral and bilateral cases of retinoblastoma.

4. Briefly explain how cancer progresses through clonal evolution.

Section 23.2

5. What is the difference between an oncogene and a tumor-suppressor gene? Give some examples of the functions of proto-oncogenes and tumor-suppressor genes in normal cells.

6. What is haploinsufficiency? How might it affect cancer risk?

7. How do cyclins and CDKs differ? How do they interact in controlling the cell cycle?

8. Briefly outline the events that control the progression of cells through the G_1/S checkpoint in the cell cycle.

9. Briefly outline the events that control the progression of cells through the G_2/M checkpoint in the cell cycle.

10. What is a signal-transduction pathway? Why are mutations in components of signal-transduction pathways often associated with cancer?

11. How is the Ras protein activated and inactivated?

12. Why do mutations in genes that encode DNA-repair enzymes often produce a predisposition to cancer?

13. What role do telomeres and telomerase play in cancer progression?

Section 23.3

14. How is an epigenetic change different from a mutation?

15. How is DNA methylation related to cancer?

Section 23.4

16. Briefly outline some of the genetic changes commonly associated with the progression of colorectal cancer.

Section 23.5

*17. Explain how chromosome deletions, inversions, and translocations can cause cancer.

18. Briefly outline how the Philadelphia chromosome leads to chronic myelogenous leukemia.

19. What is genomic instability? Give some ways in which genomic instability may arise.

Section 23.6

*20. How do viruses contribute to cancer?

APPLICATION QUESTIONS AND PROBLEMS

Introduction

21. What characteristics of the pedigree shown in **Figure 23.1** suggest that pancreatic cancer in this family is inherited as an autosomal dominant trait?

Section 23.1

22. If cancer is fundamentally a genetic disease, how might an environmental factor such as smoking cause cancer?

*23. Both genes and environmental factors contribute to cancer. Prostate cancer is 39 times more common among people from Utah than among people from Shanghai (see **Table 23.2**). Briefly outline how you might determine whether these differences in the incidence of prostate cancer are due to differences in the genetic makeup of the two populations or to differences in their environments.

Section 23.2

*24. The *palladin* gene, which plays a role in pancreatic cancer (see the introduction to this chapter), is said to be an oncogene. Which of its characteristics suggest that it is an oncogene rather than a tumor-suppressor gene?

25. Mutations in the *RB* gene are often associated with cancer. Explain how a mutation that results in a nonfunctional RB protein contributes to cancer.

26. Cells in a tumor contain mutated copies of a particular gene that promotes tumor growth. Gene therapy can be used to introduce a normal copy of this gene into the tumor cells. Would you expect this therapy to be effective if the mutated gene were an oncogene? A tumor-suppressor gene? Explain your reasoning.

27. What would be the effect on the cell cycle of a drug that inhibited each of the following?

   a. MPF

   b. Cyclin-E–CDK

   c. Cyclin-D–CDK

28. What would be the effect of a drug that inhibited the breakdown of cyclin B?

Section 23.3

29. David Seligson and his colleagues examined levels of histone protein modification in prostate tumors and their association with clinical outcomes (D. B. Seligson et al. 2005. *Nature* 435:1262–1266). They used antibodies to stain for acetylation at three different sites and for methylation at two different sites on histone proteins. They found that the degree of histone acetylation and methylation helped predict whether prostate cancer would return within 10 years in the patients who had a prostate tumor removed. Explain how acetylation and methylation might be associated with tumor recurrence in prostate cancer. (Hint: See Chapter 17.)

30. Some cancers have been treated with drugs that demethylate DNA. Explain how these drugs might work. Do you think the cancer-causing genes that responded to the demethylation are likely to have been oncogenes or tumor-suppressor genes? Explain your reasoning.

Section 23.5

*31. Some cancers are consistently associated with the deletion of a particular part of a chromosome. Does the deleted region contain an oncogene or a tumor-suppressor gene? Explain.

Section 23.6

32. Assume that the provirus in **Figure 23.14** inserts just upstream of a tumor-suppressor gene. Would this insertion be likely to cause cancer? Why or why not?

CHALLENGE QUESTIONS

Section 23.2

33. Many cancer cells are immortal (will divide indefinitely) because they have mutations that allow telomerase to be expressed. How might this knowledge be used to design anticancer drugs?

34. Bloom syndrome is an autosomal recessive disease that exhibits haploinsufficiency. A recent survey showed that people heterozygous for mutations at the *BLM* locus are at increased risk of colon cancer. Suppose you are a genetic counselor. A young woman is referred to you whose mother has Bloom syndrome; the young woman's father has no family history of Bloom syndrome. The young woman asks whether she is likely to experience any other health problems associated with her family history of Bloom syndrome. What advice would you give her?

35. Imagine that you discover a large family in which bladder cancer is inherited as an autosomal dominant trait. Briefly outline a series of studies that you might conduct to identify the gene that causes bladder cancer in this family.

THINK-PAIR-SHARE QUESTIONS

Introduction

1. The mutation associated with pancreatic cancer in the family in **Figure 23.1** was located in an exon of the *palladin* gene. In general, would you expect to find more cancer-causing mutations in exons or in introns? Explain your answer.

Section 23.1

2. The chapter points out that about one of every five women and one of every four men in the United States will die from cancer. Why are rates of death from cancer different in men and women? Provide some possible explanations.

3. A couple has one child with bilateral retinoblastoma. The mother is free from cancer, but the father has unilateral retinoblastoma and he has a brother who has bilateral retinoblastoma.

a. If the couple has another child, what is the probability that this next child will have retinoblastoma?

b. If the next child has retinoblastoma, is it likely to be bilateral or unilateral?

c. Explain why the father's case of retinoblastoma is unilateral, whereas his son's and brother's cases are bilateral.

4. **Table 23.3** lists occupation as an environmental factor associated with cancer. What occupations do you think might be associated with higher rates of cancer? What about these occupations might create higher cancer rates?

Section 23.2

5. Drivers are mutations that drive the cancer process; passengers are mutations that arise randomly in the process of tumor development and do not contribute to the progression of cancer. How might researchers go about distinguishing between drivers and passengers? Propose some different approaches.

Section 23.6

6. In 2007, then–Texas governor Rick Perry became the first governor in the United States to mandate by executive order the vaccination of all Texas girls aged 11–12 for HPV, arguing that the vaccine prevented cancer. Some conservatives criticized Perry, arguing that the vaccine would encourage sexual promiscuity among children and young adults. The Texas legislature overturned the executive order, and Perry later reversed himself, saying the order was a mistake. Do you think states should require boys and girls to be vaccinated for HPV? Discuss reasons for and against such a requirement.

SaplingPlus   Self-study tools that will help you practice what you've learned and reinforce this chapter's concepts are available online. Go to www.macmillanlearning.com/PierceGenetics6e.

Quantitative Genetics

Methods of quantitative genetics coupled with molecular techniques have been used to identify a gene that determines oil content in corn.
[Jim Craigmyle/Corbis.]

Corn Oil and Quantitative Genetics

In 2016, the world's population was 7.4 billion. The United Nations projects that by 2050, it will increase by another 2.3 billion. Feeding those billions of additional people will be a major challenge for agriculture in the next few decades. Crop plants will have to provide most of the calories and nutrients required for the world's future population. Furthermore, because of dwindling petroleum supplies and concerns about global warming, plants are also increasingly being used as sources of biofuels, placing additional demands on crop production.

To help meet the need for increased crop yields, plant breeders are using the latest genetic techniques in their quest to develop higher-yielding, more efficient crop plants. The power of this approach is demonstrated by research aimed at increasing the oil content of corn. The oil content of corn is an inherited characteristic, but its inheritance is more complex than that of the characteristics we have studied so far; it is not a simple single-gene characteristic like seed shape in peas. Numerous genes and environmental factors contribute to the oil content of corn. For characteristics such as oil content, several loci frequently interact, and their expression is affected by environmental factors.

Can the inheritance of a complex characteristic such as oil content be studied? Is it possible to predict the oil content of a plant on the basis of its breeding? The answer to both questions is yes—at least in part—but these questions cannot be addressed with the methods we have used for simple genetic characteristics. Instead, we must use statistical procedures that have been developed for analyzing complex characteristics. The genetic analysis of complex characteristics such as the oil content of corn is known as **quantitative genetics**.

In 2008, geneticists used a combination of quantitative genetic and molecular techniques to identify a gene that plays a key role in controlling the oil content of corn. First, they conducted crosses between high-oil and low-oil corn plants to identify chromosomal regions that play a role in determining oil production. Chromosomal regions containing genes that influence a quantitative trait are termed quantitative trait loci (QTLs). Through these crosses, the geneticists located several QTLs that affected oil content; one of them was on corn chromosome 6.

Fine-scale gene mapping further narrowed the QTL down to a small region of 4.2 map units (centiMorgans) on chromosome 6. Researchers sequenced DNA from the region and found that it contained five genes, one of which was *DGAT1-2*, a gene known to encode an enzyme that catalyzes the final step in a pathway for triacylglycerol biosynthesis. DNA from the *DGAT1-2* gene in a high-oil-producing strain of corn contained an insertion of a codon that added phenylalanine to the enzyme—an insertion that was missing from a low-oil-producing strain. The researchers confirmed the effect of the additional phenylalanine codon on oil production by producing transgenic corn that contained the extra codon; these transgenic strains produced more oil than transgenic strains without the additional codon. Interestingly, the additional phenylalanine codon is present in wild relatives of corn, suggesting that the codon was lost in the process of domestication or subsequent breeding of modern varieties.

This research suggests that the oil content of corn and other plants might be increased by genetically modifying their *DGAT1-2* genes to contain the additional codon for phenylalanine. Other studies that similarly combine quantitative and molecular analyses have led to the identification of genes that increase the vitamin A content of rice and increase sugar production in tomatoes.

THINK-PAIR-SHARE

- What are some other problems associated with the increasing human population, and how is genetics being used to address them?

- Researchers produced transgenic corn that contained an extra codon for phenylalanine and had a high oil content. Suppose an existing commercial variety of corn was modified in the same way and sold to farmers. Would you consider this variety to be a genetically modified organism (GMO; see Chapter 19)? Should the corn produced by this variety be labeled as a genetically modified food product?

This chapter is about the genetic analysis of complex characteristics such as the oil content of corn. We begin by considering the differences between quantitative and qualitative genetic characteristics and why the expression of some characteristics varies continuously. We'll see that quantitative characteristics are often influenced by many genes, each of which has a small effect on the phenotype. Next, we'll examine statistical procedures for describing and analyzing quantitative characteristics. We'll consider how much of phenotypic variation can be attributed to genetic and environmental influences. Finally, we'll look at the effects of selection on quantitative characteristics.

24.1 Quantitative Characteristics Are Influenced by Alleles at Multiple Loci

Qualitative, or discontinuous, characteristics possess only a few distinct phenotypes (**Figure 24.1a**); these characteristics are the types studied by Mendel (e.g., round and wrinkled peas) and have been the focus of our attention thus far. Many characteristics, however, vary continuously along a scale of measurement, exhibiting many overlapping phenotypes (**Figure 24.1b**). They are referred to as *continuous characteristics*, and they are also called *quantitative characteristics* because any individual's phenotype must be described by a quantitative measurement. Examples of quantitative characteristics include height, weight, and blood pressure in humans, growth rate in mice, seed weight in plants, and milk production in cattle.

Quantitative characteristics arise from two phenomena. First, many are polygenic: they are influenced by genes at many loci. If many loci take part, many genotypes are possible, resulting in many different phenotypes. Second, quantitative characteristics often arise when environmental factors affect the phenotype because environmental variation results in a single genotype producing a range of phenotypes. Most continuously varying characteristics are *both* polygenic *and* influenced by environmental factors, and these characteristics are said to be multifactorial.

The Relation Between Genotype and Phenotype

For some discontinuous characteristics, the relation between genotype and phenotype is straightforward: each genotype produces a single phenotype, and most phenotypes are encoded by a single genotype. Dominance and epistasis may allow different genotypes to produce the same phenotype, but the relation remains simple. This simple relation between genotype and phenotype allowed Mendel to decipher the basic rules of inheritance from his crosses with pea plants; it also permits us to predict the outcome of genetic crosses and to assign genotypes to individuals.

For quantitative characteristics, the relation between genotype and phenotype is usually more complex. If the characteristic is polygenic, many different genotypes are possible, several of which may produce the same phenotype. For instance, consider a plant whose height is determined

(a) Discontinuous characteristic

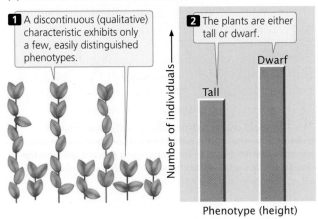

1 A discontinuous (qualitative) characteristic exhibits only a few, easily distinguished phenotypes.

2 The plants are either tall or dwarf.

Number of individuals →

Tall   Dwarf

Phenotype (height)

(b) Continuous characteristic

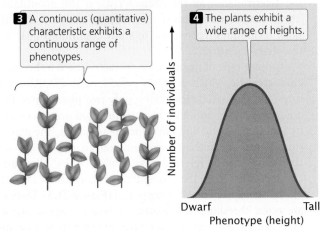

3 A continuous (quantitative) characteristic exhibits a continuous range of phenotypes.

4 The plants exhibit a wide range of heights.

Number of individuals →

Dwarf   Tall

Phenotype (height)

24.1 Discontinuous and continuous characteristics differ in the number of phenotypes exhibited.

TABLE 24.1	Hypothetical example of plant height determined by pairs of alleles at each of three loci	
Plant Genotype	**Doses of Hormone**	**Height (cm)**
A^-A^- B^-B^- C^-C^-	0	10
A^+A^- B^-B^- C^-C^-	1	11
A^-A^- B^+B^- C^-C^-		
A^-A^- B^-B^- C^-C^+		
A^+A^+ B^-B^- C^-C^-	2	12
A^-A^- B^+B^+ C^-C^-		
A^-A^- B^-B^- C^+C^+		
A^+A^- B^+B^- C^-C^-		
A^+A^- B^-B^- C^+C^-		
A^-A^- B^+B^- C^+C^-		
A^+A^+ B^+B^- C^-C^-	3	13
A^+A^+ B^-B^- C^+C^-		
A^+A^- B^+B^+ C^-C^-		
A^-A^- B^+B^+ C^+C^-		
A^+A^- B^-B^- C^+C^+		
A^-A^- B^+B^- C^+C^+		
A^+A^- B^+B^- C^+C^-		
A^+A^+ B^+B^+ C^-C^-	4	14
A^+A^+ B^+B^- C^+C^-		
A^+A^- B^+B^+ C^+C^-		
A^-A^- B^+B^+ C^+C^+		
A^+A^+ B^-B^- C^+C^+		
A^+A^- B^+B^- C^+C^+		
A^+A^+ B^+B^+ C^+C^-	5	15
A^+A^- B^+B^+ C^+C^+		
A^+A^+ B^+B^- C^+C^+		
A^+A^+ B^+B^+ C^+C^+	6	16

Note: Each + allele contributes 1 cm in height above a baseline of 10 cm.

by three loci (*A*, *B*, and *C*), each of which has two alleles. Assume that one allele at each locus (A^+, B^+, and C^+) encodes a plant hormone that causes the plant to grow 1 cm above its baseline height of 10 cm. The other allele at each locus (A^-, B^-, and C^-) does not encode a plant hormone and thus does not contribute to additional height. If we consider only the two alleles at a single locus, 3 genotypes are possible (A^+A^+, A^+A^-, and A^-A^-). If all three loci are taken into account, there are a total of $3^3 = 27$ possible multilocus genotypes (A^+A^+ B^+B^+ C^+C^+, A^+A^- B^+B^+ C^+C^+, etc.). Although there are 27 genotypes, they produce only seven phenotypes (10 cm, 11 cm, 12 cm, 13 cm, 14 cm, 15 cm, and 16 cm in height) because some of the genotypes produce the same phenotype (**Table 24.1**). For example, genotypes A^+A^- B^-B^- C^-C^-, A^-A^- B^+B^- C^-C^-, and A^-A^- B^-B^- C^+C^- each have one allele that encodes a plant hormone. Each of these genotypes produces one dose of the hormone and results in a plant that is 11 cm tall. Even in this simple example with only three loci, the relation between genotype and phenotype is quite complex. The more loci encoding a characteristic, the greater the complexity. As the number of loci encoding a characteristic

increases, the number of potential phenotypes increases, and differences between individual phenotypes become more difficult to distinguish.

The influence of environment on a characteristic can also complicate the relation between genotype and phenotype. When environmental factors affect the phenotype, a single genotype can produce a range of phenotypes. Furthermore, the phenotypic ranges of different genotypes can overlap, making it difficult to know whether individuals differ in phenotype because of genetic or environmental differences (**Figure 24.2**).

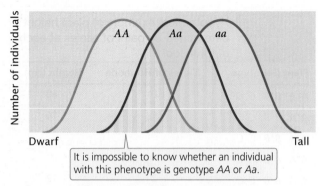

24.2 For a quantitative characteristic influenced by environmental factors, each genotype may produce a range of possible phenotypes. In this hypothetical example, the phenotypes produced by genotypes *AA*, *Aa*, and *aa* overlap.

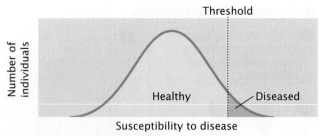

24.3 Threshold characteristics display only two possible phenotypes—the trait is either present or absent—but they are quantitative because the underlying susceptibility to the characteristic varies continuously. When the susceptibility exceeds a threshold value, the characteristic is expressed.

In summary, the simple relation between genotype and phenotype that exists for many qualitative (discontinuous) characteristics is absent for quantitative characteristics, and it is impossible to assign a genotype to an individual on the basis of its phenotype alone. The methods used for analyzing qualitative characteristics (examining the phenotypic ratios of progeny from a genetic cross) will not work for quantitative characteristics. Our goal remains the same: we wish to make predictions about the phenotypes of offspring produced in a genetic cross. We may also want to know how much of the variation in a characteristic results from genetic differences and how much results from environmental differences. To answer these questions, we must turn to statistical methods that allow us to make predictions about the inheritance of phenotypes in the absence of information about the underlying genotypes. ▶ **TRY PROBLEM 16**

Types of Quantitative Characteristics

Before we look more closely at polygenic characteristics and relevant statistical methods, we need to more clearly define what is meant by a quantitative characteristic. Thus far, we have considered only quantitative characteristics that vary continuously in a population. A *continuous characteristic* can theoretically assume any value between two extremes; the number of phenotypes is limited only by our ability to precisely measure them. Human height is a continuous characteristic because, within certain limits, people can theoretically have any height. Although the number of phenotypes possible with a continuous characteristic is infinite, we often group similar phenotypes together for convenience; we may say that two people are both 5 feet 11 inches tall, although careful measurement may show that one is slightly taller than the other.

Some characteristics are not continuous but are nevertheless considered quantitative because they are determined by multiple genetic and environmental factors. **Meristic characteristics**, for instance, are measured in whole numbers. An example is litter size: a female mouse can have 4, 5, or 6 pups, but not 4.13 pups. A meristic characteristic has

a limited number of distinct phenotypes, but the underlying determination of the characteristic is still quantitative. These characteristics must therefore be analyzed with the same techniques that we use to study continuous quantitative characteristics.

Another type of quantitative characteristic is a **threshold characteristic**, which is simply present or absent. Although threshold characteristics exhibit only two phenotypes, they are considered quantitative because they, too, are determined by multiple genetic and environmental factors. The expression of the characteristic depends on an underlying susceptibility (usually referred to as liability or risk) that varies continuously. When the susceptibility is larger than a threshold value, a specific trait is expressed (**Figure 24.3**). Diseases are often threshold characteristics because many factors, both genetic and environmental, contribute to disease susceptibility. If enough of the susceptibility factors are present, the disease develops; otherwise, it is absent. Although we focus on the genetics of continuous characteristics in this chapter, the same principles apply to many meristic and threshold characteristics.

Just because a characteristic can be measured on a continuous scale does not mean that it exhibits quantitative variation. One of the characteristics studied by Mendel was the height of pea plants, which can be described by measuring the length of a plant's stem. However, Mendel's particular plants exhibited only two distinct phenotypes (some were tall and others short), and these differences were determined by alleles at a single locus. The differences that Mendel studied were therefore discontinuous in nature.

CONCEPTS

Characteristics for which the phenotypes vary continuously are quantitative characteristics. For most quantitative characteristics, the relation between genotype and phenotype is complex. Some characteristics for which the phenotypes do not vary continuously are also considered quantitative because they are influenced by multiple genes and environmental factors.

Polygenic Inheritance

After the rediscovery of Mendel's work in 1900, questions soon arose about the inheritance of continuously varying characteristics. These characteristics had already been the focus of a group of biologists and statisticians, led by Francis Galton, who used statistical procedures to examine the inheritance of characteristics such as human height and intelligence. The results of these studies showed that quantitative characteristics were at least partly inherited, although the mechanism of inheritance was not yet known. Some biometricians argued that the inheritance of quantitative characteristics could not be explained by Mendelian principles, whereas others believed that Mendel's principles acting on numerous genes (polygenes) could adequately account for the inheritance of quantitative characteristics.

This conflict began to be resolved by the work of Wilhelm Johannsen, who showed that continuous variation in the weight of beans was influenced by both genetic and environmental factors. George Udny Yule, a mathematician, proposed in 1906 that several genes acting together could produce continuous characteristics. This hypothesis was later confirmed by Herman Nilsson-Ehle, working on wheat and tobacco, and by Edward East, working on corn. The argument was finally laid to rest in 1918, when Ronald Fisher demonstrated that the inheritance of quantitative characteristics could indeed be explained by the cumulative effects of many genes, each following Mendel's rules.

Kernel Color in Wheat

To illustrate how multiple genes acting on a characteristic can produce a continuous range of phenotypes, let's examine one of the first demonstrations of polygenic inheritance. Nilsson-Ehle studied kernel color in wheat and found that the intensity of red pigmentation was determined by three unlinked loci, each of which had two alleles.

NILSSON-EHLE'S CROSS Nilsson-Ehle obtained several homozygous varieties of wheat that differed in color. Like Mendel, he performed crosses between these homozygous varieties and studied the ratios of phenotypes in the progeny. In one experiment, he crossed a variety of wheat that possessed white kernels with a variety that possessed purple (very dark red) kernels and obtained the following results:

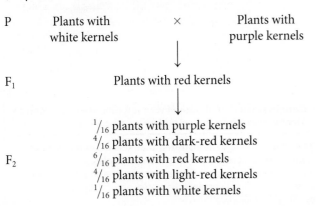

INTERPRETATION OF THE CROSS Nilsson-Ehle interpreted this phenotypic ratio as the result of the segregation of alleles at two loci (although he found alleles at three loci that affect kernel color, the two varieties used in this cross differed at only two of the loci). He proposed that there were two alleles at each locus: one that produced red pigment and another that produced no pigment. We'll designate the alleles that encoded pigment A^+ and B^+ and the alleles that encoded no pigment A^- and B^-. Nilsson-Ehle recognized that the effects of the genes were additive. Each gene seemed to contribute equally to color, so the overall phenotype could be determined by adding the effects of all the alleles, as shown in the following table:

Genotype	Doses of pigment	Phenotype
$A^+A^+\ B^+B^+$	4	Purple
$A^+A^+\ B^+B^-$ $A^+A^-\ B^+B^+$	3	Dark red
$A^+A^+\ B^-B^-$ $A^-A^-\ B^+B^+$ $A^+A^-\ B^+B^-$	2	Red
$A^+A^-\ B^-B^-$ $A^-A^-\ B^+B^-$	1	Light red
$A^-A^-\ B^-B^-$	0	White

Notice that the purple and white phenotypes are each encoded by a single genotype, but the other phenotypes may result from several different genotypes.

From these results, we see that five phenotypes are possible when alleles at two loci influence the phenotype and the effects of the genes are additive. If alleles at more than two loci influenced the phenotype, more phenotypes would be possible, and the color would appear to vary continuously between white and purple. If environmental factors influenced the characteristic, individuals of the same genotype would vary somewhat in color, making it even more difficult to distinguish between discrete phenotypic classes. Luckily, environment played little role in determining kernel color in Nilsson-Ehle's crosses, and only a few loci encoded color, so Nilsson-Ehle was able to distinguish among the different phenotypic classes. This ability allowed him to see the Mendelian nature of the characteristic.

Let's now see how Mendel's principles explain the ratio obtained by Nilsson-Ehle in his F_2 progeny. Remember that Nilsson-Ehle crossed the homozygous purple variety ($A^+A^+\ B^+B^+$) with the homozygous white variety ($A^-A^-\ B^-B^-$), producing F_1 progeny that were heterozygous at both loci ($A^+A^-\ B^+B^-$). This is a dihybrid cross, like those that we worked in Chapter 3, except that both loci encode the same trait. Each of the F_1 plants possessed two pigment-producing alleles, so each plant received two doses of pigment, which produced red kernels. The types and proportions of progeny expected in the F_2 can be found by applying Mendel's principles of segregation and independent assortment.

Let's first examine the effects of each locus separately. At the first locus, two heterozygous F_1s are crossed ($A^+A^- \times A^+A^-$). As we learned in Chapter 3, when two heterozygotes are crossed, we expect progeny in the proportions $\frac{1}{4} A^+A^+$, $\frac{1}{2} A^+A^-$, and $\frac{1}{4} A^-A^-$. At the second locus, two heterozygotes are also crossed, and again, we expect progeny in the proportions $\frac{1}{4} B^+B^+$, $\frac{1}{2} B^+B^-$, and $\frac{1}{4} B^-B^-$.

To obtain the probability of combinations of genes at both loci, we must use the multiplication rule (see Chapter 3), the use of which assumes Mendel's principle of independent assortment. The expected proportion of F_2 progeny with genotype $A^+A^+ B^+B^+$ is the product of the probability of obtaining genotype A^+A^+ ($\frac{1}{4}$) and the probability of obtaining genotype B^+B^+ ($\frac{1}{4}$), or $\frac{1}{4} \times \frac{1}{4} = \frac{1}{16}$ (**Figure 24.4**). The probabilities of each of the phenotypes can then be obtained by adding the probabilities of all the genotypes that produce that phenotype. For example, the red phenotype is produced by three genotypes:

Genotype	Phenotype	Probability
$A^+A^+ \, B^-B^-$	**Red**	$\frac{1}{16}$
$A^-A^- \, B^+B^+$	**Red**	$\frac{1}{16}$
$A^+A^- \, B^+B^-$	**Red**	$\frac{1}{4}$

Thus, the overall probability of obtaining red kernels in the F_2 progeny is $\frac{1}{16} + \frac{1}{16} + \frac{1}{4} = \frac{6}{16}$. Figure 24.4 shows that the phenotypic ratio expected in the F_2 is $\frac{1}{16}$ purple, $\frac{4}{16}$ dark red, $\frac{6}{16}$ red, $\frac{4}{16}$ light red, and $\frac{1}{16}$ white. This phenotypic ratio is precisely what Nilsson-Ehle observed in his F_2 progeny, demonstrating that the inheritance of a continuously varying characteristic such as kernel color is indeed explained by Mendel's basic principles.

CONCLUSIONS AND IMPLICATIONS Nilsson-Ehle's crosses demonstrated that the difference between the inheritance of genes influencing quantitative characteristics and the inheritance of genes influencing discontinuous characteristics is in the *number* of loci that determine the characteristic. When multiple loci affect a characteristic, more genotypes are possible, so the relation between the genotype and the phenotype is less obvious. As the number of loci affecting a characteristic increases, the number of phenotypic classes in the F_2 increases (**Figure 24.5**).

Several conditions of Nilsson-Ehle's crosses greatly simplified the polygenic inheritance of kernel color and made it possible for him to recognize the Mendelian nature of the characteristic. First, genes affecting color segregated at only two or three loci. If genes at many loci had been segregating, he would have had difficulty in distinguishing the phenotypic classes. Second, the genes affecting kernel color had strictly additive effects, making the relation between genotype and phenotype simple. Third, environment played almost no role in the phenotype; had environmental factors modified the phenotypes, distinguishing between the five phenotypic classes would have been difficult. Finally, the loci that

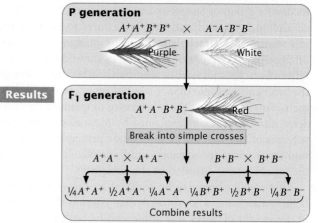

Experiment

Question: How is a continous trait, such as kernel color in wheat, inherited?

Methods Cross wheat with white kernels and wheat with purple kernels. Intercross the F_1 to produce F_2.

P generation

$A^+A^+B^+B^+$ $\times$ $A^-A^-B^-B^-$
Purple — White

Results

F_1 generation

$A^+A^- B^+B^-$ Red

Break into simple crosses

$A^+A^- \times A^+A^-$ $\quad$ $B^+B^- \times B^+B^-$

$\frac{1}{4}A^+A^+$ $\frac{1}{2}A^+A^-$ $\frac{1}{4}A^-A^-$ $\quad$ $\frac{1}{4}B^+B^+$ $\frac{1}{2}B^+B^-$ $\frac{1}{4}B^-B^-$

Combine results

F_2 generation — **Number of pigment genes** — **Phenotype**

$\frac{1}{4}A^+A^+$
- $\frac{1}{4}B^+B^+$ → $\frac{1}{4}\times\frac{1}{4}=\frac{1}{16}$ $A^+A^+B^+B^+$ — 4 — Purple
- $\frac{1}{2}B^+B^-$ → $\frac{1}{4}\times\frac{1}{2}=\frac{2}{16}$ $A^+A^+B^+B^-$ — 3 — Dark red
- $\frac{1}{4}B^-B^-$ → $\frac{1}{4}\times\frac{1}{4}=\frac{1}{16}$ $A^+A^+B^-B^-$ — 2 — Red

$\frac{1}{2}A^+A^-$
- $\frac{1}{4}B^+B^+$ → $\frac{1}{2}\times\frac{1}{4}=\frac{2}{16}$ $A^+A^-B^+B^+$ — 3 — Dark red
- $\frac{1}{2}B^+B^-$ → $\frac{1}{2}\times\frac{1}{2}=\frac{4}{16}$ $A^+A^-B^+B^-$ — 2 — Red
- $\frac{1}{4}B^-B^-$ → $\frac{1}{2}\times\frac{1}{4}=\frac{2}{16}$ $A^+A^-B^-B^-$ — 1 — Light red

$\frac{1}{4}A^-A^-$
- $\frac{1}{4}B^+B^+$ → $\frac{1}{4}\times\frac{1}{4}=\frac{1}{16}$ $A^-A^-B^+B^+$ — 2 — Red
- $\frac{1}{2}B^+B^-$ → $\frac{1}{4}\times\frac{1}{2}=\frac{2}{16}$ $A^-A^-B^+B^-$ — 1 — Light red
- $\frac{1}{4}B^-B^-$ → $\frac{1}{4}\times\frac{1}{4}=\frac{1}{16}$ $A^-A^-B^-B^-$ — 0 — White

Combine common phenotypes

F_2 ratio

Frequency	Number of pigment genes	Phenotype
$\frac{1}{16}$	4	Purple
$\frac{4}{16}$	3	Dark red
$\frac{6}{16}$	2	Red
$\frac{4}{16}$	1	Light red
$\frac{1}{16}$	0	White

Conclusion: Kernel color in wheat is inherited according to Mendel's principles acting on alleles at two loci.

24.4 Nilsson-Ehle demonstrated that kernel color in wheat is inherited according to Mendelian principles. The ratio of phenotypes in the F_2 can be determined by breaking the dihybrid cross into two simple single-locus crosses and combining the results by using the multiplication rule.

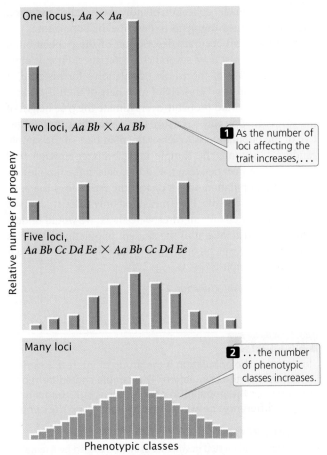

One locus, $Aa \times Aa$

Two loci, $Aa\,Bb \times Aa\,Bb$

1 As the number of loci affecting the trait increases,…

Five loci,
$Aa\,Bb\,Cc\,Dd\,Ee \times Aa\,Bb\,Cc\,Dd\,Ee$

Many loci

2 …the number of phenotypic classes increases.

Relative number of progeny

Phenotypic classes

24.5 Crossing heterozygotes for different numbers of loci affecting a characteristic results in different numbers of phenotypic classes.

that affects the characteristic. This equation provides us with a possible means of determining the number of loci influencing a quantitative characteristic.

To illustrate the use of this equation, assume that we cross two different homozygous varieties of pea plants that differ in height by 16 cm, interbreed the F_1, and find that approximately $\frac{1}{256}$ of the F_2 are similar to one of the original homozygous parental varieties. This outcome would suggest that four loci with segregating pairs of alleles ($\frac{1}{256} = (\frac{1}{4})^4$) are responsible for the height difference between the two varieties. Because the two homozygous strains differ in height by 16 cm, and there are four loci, each of which has two alleles (eight alleles in all), each of the alleles contributes 16 cm/8 = 2 cm in height.

This method of determining the number of loci affecting phenotypic differences requires the use of homozygous strains, which may be difficult to obtain in some organisms. It also assumes that all the genes influencing the characteristic have equal effects, that their effects are additive, and that the loci are unlinked. For many polygenic characteristics, these assumptions are not valid, and so this method of determining the number of genes affecting a characteristic has limited application. **▶ TRY PROBLEM 19**

CONCEPTS

The same principles determine the inheritance of quantitative and discontinuous characteristics, but more genes take part in the determination of quantitative characteristics.

✔ CONCEPT CHECK 1

Briefly explain how the number of genes influencing a polygenic trait can be determined.

Nilsson-Ehle studied were not linked, so the genes assorted independently. Nilsson-Ehle was fortunate: for many polygenic characteristics, these simplifying conditions are not present, and the Mendelian inheritance of these characteristics is not obvious. **▶ TRY PROBLEM 17**

THINK-PAIR-SHARE Question 1

Determining Gene Number for a Polygenic Characteristic

The proportion of F_2 individuals that resemble one of the original parents can be used to estimate the number of genes affecting a polygenic trait. When two individuals homozygous for different alleles at a single locus are crossed ($A^1A^1 \times A^2A^2$) and the resulting F_1 are interbred ($A^1A^2 \times A^1A^2$), one-fourth of the F_2 should be homozygous like each of the original parents. If the original parents are homozygous for different alleles at two loci, as are those in Nilsson-Ehle's crosses, then $\frac{1}{4} \times \frac{1}{4} = \frac{1}{16}$ of the F_2 should resemble one of the original homozygous parents. Generally, $(\frac{1}{4})^n$ will be the proportion of individuals among the F_2 progeny that should resemble each of the original homozygous parents, where n equals the number of loci with a segregating pair of alleles

24.2 Statistical Methods Are Required for Analyzing Quantitative Characteristics

Because quantitative characteristics are described by a measurement and are influenced by multiple factors, their inheritance must be analyzed statistically. This section explains the basic concepts of statistics that are used to analyze quantitative characteristics.

Distributions

An understanding of the genetic basis of any characteristic begins with a description of the numbers and kinds of phenotypes present in a group of individuals. Phenotypic variation in a group can be conveniently represented by a **frequency distribution**, which is a graph of the frequencies (numbers or proportions) of the different phenotypes (**Figure 24.6**). In a typical frequency distribution, phenotypic classes are plotted on the horizontal (x) axis, and the numbers (or proportions) of individuals in each class are plotted on the vertical (y) axis. A frequency distribution is a concise method of summarizing all phenotypes of a quantitative characteristic.

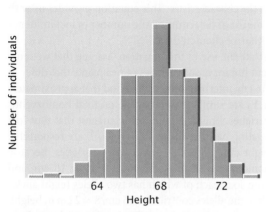

24.6 A frequency distribution is a graph that displays the numbers or proportions of different phenotypes present in a group of individuals. Phenotypic values (here, height in inches) are plotted on the horizontal axis, and the numbers (or proportions) of individuals with each phenotype are plotted on the vertical axis.

Connecting the points of a frequency distribution with a line creates a curve that is characteristic of the distribution. Many quantitative characteristics exhibit a symmetrical (bell-shaped) curve called a **normal distribution** (**Figure 24.7a**). Normal distributions arise when a large number of independent factors contribute to a measurement, as is often the case in quantitative characteristics. Two other common types of distributions (skewed and bimodal) are illustrated in **Figure 24.7b** and **c**.

Samples and Populations

Biologists frequently need to describe the distribution of phenotypes exhibited by a group of individuals. We might want to describe the height of students at the University of Texas (UT), but UT has more than 40,000 students, and measuring all of them is not practical. Scientists are constantly confronted with this problem: the group of interest, called the **population**, is too large for a complete census. One solution is to measure a smaller collection of individuals, called a **sample**, and use those measurements to describe the population.

To provide an accurate description of the population, a good sample must have several characteristics. First, it must be representative of the whole population—for instance, if our sample consisted entirely of members of the UT basketball team, we would probably overestimate the true height of the students. One way to ensure that a sample is representative of the population is to select the members of the sample randomly. Second, the sample must be large enough that chance differences between the individuals in the sample and the overall population do not distort the estimate of the population measurements. If we measured only three students at UT and just by chance all three were short, we would underestimate the true height of the student population. Statistics can tell us how much confidence to have in estimates based on random samples.

THINK-PAIR-SHARE Question 2

CONCEPTS

In statistics, the population is the group of interest; a sample is a subset of the population. The sample should be representative of the population and large enough to minimize chance differences between the population and the sample.

✔ CONCEPT CHECK 2

A geneticist is interested in whether asthma is caused by a mutation in the *DS112* gene. The geneticist collects DNA from 120 people with asthma and 100 healthy people, and she sequences the DNA. She finds that 35 of the people with asthma and none of the healthy people have a mutation in the *DS112* gene. What is the population in this study?

a. The 120 people with asthma
b. The 100 healthy people
c. The 35 people with a mutation in the *DS112* gene
d. All people with asthma

(a) Sugar beet percentage of sucrose

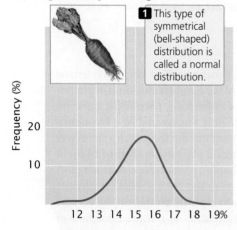

1 This type of symmetrical (bell-shaped) distribution is called a normal distribution.

Frequency (%)

20

10

12 13 14 15 16 17 18 19%

(b) Squash fruit length

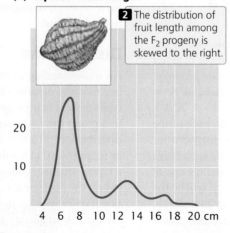

2 The distribution of fruit length among the F$_2$ progeny is skewed to the right.

20

10

4 6 8 10 12 14 16 18 20 cm

(c) Earwig forceps length

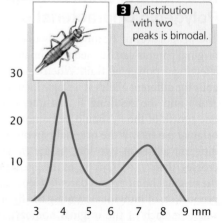

3 A distribution with two peaks is bimodal.

30

20

10

3   4   5   6   7   8   9 mm

24.7 Distributions of phenotypes can assume several different shapes.

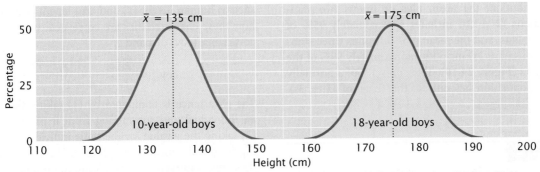

24.8 The mean provides information about the center of a distribution. The distributions of the heights of both 10-year-old and 18-year-old boys are normal, but they have different locations along a continuum of height, which makes their means different.

The Mean

The **mean**, also called the average, is a statistic that provides information about the center of a distribution. If we measured the heights of 10-year-old and 18-year-old boys and plotted a frequency distribution for each group, we would find that both distributions are normal, but the two distributions would be centered at different heights, and this difference would be indicated by their different means (**Figure 24.8**).

Suppose that we have five measurements of height in centimeters: 160, 161, 167, 164, and 165. If we represent a group of measurements as x_1, x_2, x_3, and so forth, then the mean ($\bar{x}$) is calculated by adding all the individual measurements and dividing by the total number of measurements in the sample (n):

$$\bar{x} = \frac{x_1 + x_2 + x_3 + \ldots + x_n}{n} \qquad (24.1)$$

In our example, $x_1 = 160$, $x_2 = 161$, $x_3 = 167$, and so forth. The mean height ($\bar{x}$) equals

$$\bar{x} = \frac{160 + 161 + 167 + 164 + 165}{5} = \frac{817}{5} = 163.4$$

A shorthand way to represent this formula is

$$\bar{x} = \frac{\sum x_i}{n} \qquad (24.2)$$

or

$$\bar{x} = \frac{1}{n}\sum x_i \qquad (24.3)$$

where the symbol Σ means "the summation of" and x_i represents the individual x values.

The Variance and Standard Deviation

A statistic that provides key information about a distribution is its **variance**, which indicates the variability of a group of measurements, or how spread out the distribution is. Distributions can have the same mean but different variances (**Figure 24.9**). The larger the variance, the greater the spread of measurements in a distribution about its mean.

The variance (s^2) is defined as the average squared deviation from the mean:

$$s^2 = \frac{\sum (x_i - \bar{x})^2}{n - 1} \qquad (24.4)$$

To calculate the variance, we (1) subtract the mean from each measurement and square the value obtained, (2) add all of these squared deviations together, and (3) divide that sum by the number of original measurements minus 1.

Another statistic that is closely related to the variance is the **standard deviation** (s), which is defined as the square root of the variance:

$$s = \sqrt{s^2} \qquad (24.5)$$

Whereas the variance is expressed in units squared, the standard deviation is in the same units as the original measurements; so the standard deviation is often preferred for describing the variability of a measurement.

Because a normal distribution is symmetrical, its mean and standard deviation are sufficient to describe its shape. The mean plus or minus one standard deviation ($\bar{x} \pm s$) includes approximately 66% of the measurements in a

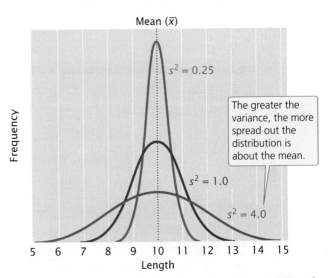

24.9 The variance provides information about the variability of a group of phenotypes. Shown here are three distributions with the same mean but different variances.

normal distribution; the mean plus or minus two standard deviations ($\bar{x} \pm 2s$) includes approximately 95% of the measurements, and the mean plus or minus three standard deviations ($\bar{x} \pm 3s$) includes approximately 99% of the measurements (**Figure 24.10**). Thus, only 1% of a normally distributed population lies outside the range of ($\bar{x} \pm 3s$).

▶ TRY PROBLEM 22

CONCEPTS

The mean and the variance are statistics that describe a distribution of measurements: the mean provides information about the location of the center of a distribution, and the variance provides information about its variability.

✔ CONCEPT CHECK 3

The measurements of a distribution with a higher _____ will be more spread out.

a. mean
b. variance
c. standard deviation
d. variance and standard deviation

Correlation

The mean and the variance can be used to describe an individual characteristic, but geneticists are frequently interested in more than one characteristic. Often, two or more characteristics vary together. For instance, both the number and the weight of eggs produced by hens are important to the poultry industry. These two characteristics are not independent of each other. There is an inverse relation between egg number and weight: hens that lay more eggs produce smaller eggs. This kind of relation between two characteristics is called a **correlation**. When two characteristics are correlated, a change in one characteristic is likely to be associated with a change in the other.

Correlations between characteristics are measured by a **correlation coefficient** (designated r), which measures the strength of their association. Consider two characteristics,

such as human height (x) and arm length (y). To determine how these characteristics are correlated, we first obtain the covariance (cov) of x and y:

$$\text{cov}_{xy} = \frac{\sum (x_i - \bar{x})(y_i - \bar{y})}{n - 1} \tag{24.6}$$

The covariance is computed by (1) taking the x value for an individual and subtracting from it the mean of x ($\bar{x}$); (2) taking the y value for the same individual and subtracting from it the mean of y ($\bar{y}$); (3) multiplying the results of these two subtractions; (4) adding the results for all the xy pairs; and (5) dividing this sum by $n - 1$ (where n equals the number of xy pairs).

The correlation coefficient (r) is obtained by dividing the covariance of x and y by the product of the standard deviations of x and y:

$$r = \frac{\text{cov}_{xy}}{s_x s_y} \tag{24.7}$$

A correlation coefficient can theoretically range from -1 to $+1$. A positive value indicates that there is a direct association between the variables (**Figure 24.11a**): as one variable increases, the other variable also tends to increase. A positive correlation exists for human height and weight: tall people tend to weigh more. A negative correlation coefficient indicates that there is an inverse relation between the two variables (**Figure 24.11b**): as one variable increases, the other tends to decrease (as is the case for egg number and egg weight in chickens).

The absolute value of the correlation coefficient (the size of the coefficient, ignoring its sign) provides information about the strength of association between the variables. A coefficient of -1 or $+1$ indicates a perfect correlation between the variables, meaning that a change in x is always accompanied by a proportional change in y. Correlation coefficients close to -1 or close to $+1$ indicate a strong association between the variables: a change in x is almost always associated with a proportional change in y, as seen in **Figure 24.11c**. On the other hand, a correlation coefficient closer to 0 indicates a weak correlation: a change in x is associated with a change in y, but not always (**Figure 24.11d**). A correlation of 0 indicates that there is no association between the variables (**Figure 24.11e**).

A correlation coefficient can be computed for two variables measured for the same individual, such as height (x) and weight (y). A correlation coefficient can also be computed for a single variable measured for pairs of individuals. For example, we can calculate for a fish species the correlation between the number of vertebrae of a parent (x) and the number of vertebrae of its offspring (y), as shown in **Figure 24.12**. This approach is often used in quantitative genetics.

A correlation between two variables indicates only that the variables are associated; it does not imply a cause-and-effect relation. Nor does correlation mean that the values of two variables are the same; it means only that a change in one variable is associated with a proportional change in the other variable. For example, the x and y variables in the following

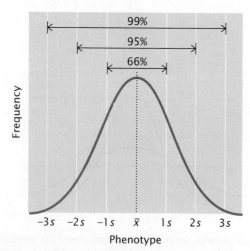

24.10 The proportions of a normal distribution occupied by plus or minus one, two, and three standard deviations from the mean.

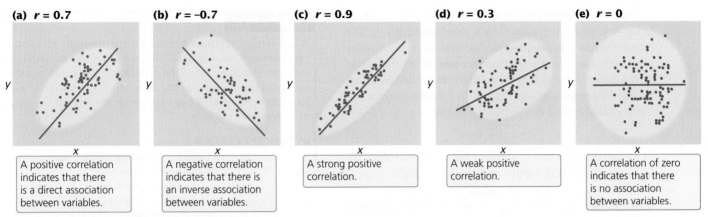

(a) *r* = 0.7

y

x

A positive correlation indicates that there is a direct association between variables.

(b) *r* = −0.7

y

x

A negative correlation indicates that there is an inverse association between variables.

(c) *r* = 0.9

y

x

A strong positive correlation.

(d) *r* = 0.3

y

x

A weak positive correlation.

(e) *r* = 0

y

x

A correlation of zero indicates that there is no association between variables.

24.11 The correlation coefficient describes the relation between two or more variables.

list are almost perfectly correlated, with a correlation coefficient of 0.99.

x value	*y* value
12	123
14	140
10	110
6	61
3	32
Average: 9	90

A strong correlation is found between these *x* and *y* variables; larger values of *x* are always associated with larger values of *y*. Note that the *y* values are about 10 times as large as the corresponding *x* values, so, although *x* and *y* are correlated, they are not identical. The distinction between correlation and identity becomes important when we consider the effects of heredity and environment on the correlation of characteristics. **>TRY PROBLEM 24**

THINK-PAIR-SHARE Question 3

Regression

Correlation provides information only about the strength and direction of association between variables. However, we often want to know more than whether two variables are associated: we want to be able to predict the value of one variable given a value of the other.

A positive correlation exists between the body weight of parents and the body weight of their offspring; this correlation exists in part because genes influence body weight, and parents and offspring have genes in common. Because of this association between parental and offspring phenotypes, we can predict the weight of an individual on the basis of the weights of its parents. This type of statistical prediction is called **regression**. This technique plays an important role in quantitative genetics because it allows us to predict the characteristics of offspring of a given mating, even without knowledge of the genotypes that encode those characteristics.

Regression can be understood by plotting a series of *x* and *y* values. **Figure 24.13** illustrates the relation between

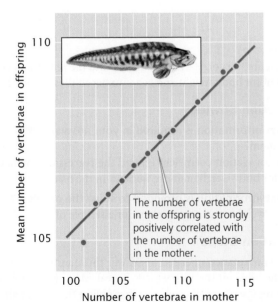

The number of vertebrae in the offspring is strongly positively correlated with the number of vertebrae in the mother.

Mean number of vertebrae in offspring

110

105

100   105   110   115

Number of vertebrae in mother

24.12 A correlation coefficient can be computed for a single variable measured for pairs of individuals. Here, the numbers of vertebrae in mothers and offspring of the fish *Zoarces viviparus* are compared.

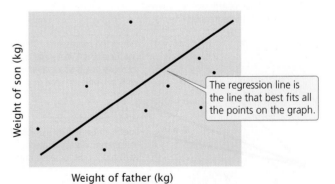

Weight of son (kg)

Weight of father (kg)

The regression line is the line that best fits all the points on the graph.

24.13 A regression line defines the relation between two variables. Illustrated here is a regression of the weights of fathers against the weights of sons. Each father–son pair is represented by a point on the graph: the *x* value of a point is the father's weight and the *y* value of the point is the son's weight.

the weight of a father (x) and the weight of his son (y). Each father–son pair is represented by a point on the graph. The overall relation between these two variables is depicted by the regression line, which is the line that best fits all the points on the graph (deviations of the points from the line are minimized). The regression line defines the relation between the x and y variables and can be represented by

$$y = a + bx \qquad (24.8)$$

In Equation 24.8, x and y represent the x and y variables (in this case, the father's weight and the son's weight, respectively). The variable a is the y intercept of the regression line, which is the expected value of y when x is 0. The variable b is the slope of the regression line, also called the **regression coefficient**.

Positioning a regression line by eye is not only very difficult, but unlikely to be accurate when there are many points scattered over a wide area. Fortunately, the regression coefficient and the y intercept can be obtained mathematically. The regression coefficient (b) can be computed from the covariance of x and y (cov_{xy}) and the variance of x (s_x^2) by

$$b = \frac{cov_{xy}}{s_x^2} \qquad (24.9)$$

The regression coefficient indicates how much y increases, on average, per increase in x. Several regression lines with different regression coefficients are illustrated in **Figure 24.14**. Notice that as the regression coefficient increases, the slope of the regression line increases.

After the regression coefficient has been calculated, the y intercept can be calculated by substituting the regression coefficient and the mean values of x and y into the following equation:

$$a = \bar{y} - b\bar{x} \qquad (24.10)$$

The regression equation ($y = a + bx$; Equation 24.8) can then be used to predict the value of any y given the value of x.

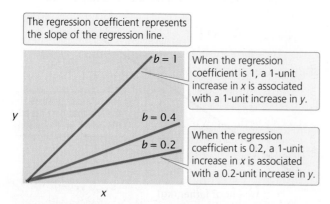

24.14 The regression coefficient, *b*, represents the change in *y* per unit change in *x*. Shown here are regression lines with different regression coefficients.

WORKED PROBLEM

The body weights of 11 female fish and the numbers of eggs they produce are given in the following table:

Weight (mg)	Eggs (thousands)
x	y
14	61
17	37
24	65
25	69
27	54
33	93
34	87
37	89
40	100
41	90
42	97

What are the correlation coefficient and the regression coefficient for body weight and egg number in these 11 fish?

Solution Strategy

What information is required in your answer to the problem?
The correlation coefficient (r) and the regression coefficient (b) for body weight and egg number in the fish.

What information is provided to solve the problem?
Body weights and egg numbers for a sample of 11 fish.

Solution Steps

The computations needed to answer this question are given in the table on the facing page. To calculate the correlation and regression coefficients, we first obtain the sum of all the x_i values (Σx_i) and the sum of all the y_i values (Σy_i); these sums are shown in the last row of the table. We can calculate

A Weight (mg)	B	C	D Eggs (thousands)	E	F	G
x	$x_i - \bar{x}$	$(x_i - \bar{x})^2$	y	$y_i - \bar{y}$	$(y_i - \bar{y})^2$	$(x_i - \bar{x})(y_i - \bar{y})$
14	−16.36	267.65	61	−15.55	241.80	254.40
17	−13.36	178.49	37	−39.55	1564.20	528.39
24	−6.36	40.45	65	−11.55	133.40	73.46
25	−5.36	28.73	69	−7.55	57.00	40.47
27	−3.36	11.29	54	−22.55	508.50	75.77
33	2.64	6.97	93	16.45	270.60	43.43
34	3.64	13.25	87	10.45	109.20	38.04
37	6.64	44.09	89	12.45	155.00	82.67
40	9.64	92.93	100	23.45	549.90	226.06
41	10.64	113.21	90	13.45	180.90	143.11
42	11.64	135.49	97	20.45	418.20	238.04
$\Sigma x_i = 334$		$\Sigma(x_i - \bar{x})^2 = 932.55$	$\Sigma y_i = 842$		$\Sigma(y_i - \bar{y})^2 = 4188.70$	$\Sigma(x_i - \bar{x})(y_i - \bar{y}) = 1743.84$

Source: R. R. Sokal and F. J. Rohlf, *Biometry*, 2d ed. (San Francisco: W. H. Freeman and Company, 1981).

the means of the two variables by dividing the sums by the number of measurements, which is 11:

$$\bar{x} = \frac{\Sigma x_i}{n} = \frac{334}{11} = 30.36$$

$$\bar{y} = \frac{\Sigma y_i}{n} = \frac{842}{11} = 76.55$$

After the means have been calculated, the deviations of each value from the mean are computed; these deviations are shown in columns B and E of the table. The deviations are then squared (columns C and F) and summed (last row of columns C and F). Next, the products of the deviation of the *x* values and the deviation of the *y* values $((x_i - \bar{x})(y_i - \bar{y}))$ are calculated; these products are shown in column G, and their sum is shown in the last row of column G.

To calculate the covariance, we use Equation 24.6:

$$\text{cov}_{xy} = \frac{\Sigma(x_i - \bar{x})(y_i - \bar{y})}{n - 1} = \frac{1743.84}{10} = 174.38$$

To calculate the correlation and the regression requires the variances and standard deviations of *x* and *y*:

$$s_x^2 = \frac{\Sigma(x_i - \bar{x})^2}{n - 1} = \frac{932.55}{10} = 93.26$$

$$s_x = \sqrt{s_x^2} = \sqrt{93.26} = 9.66$$

$$s_y^2 = \frac{\Sigma(y_i - \bar{y})^2}{n - 1} = \frac{4188.70}{10} = 418.87$$

$$s_y = \sqrt{s_y^2} = \sqrt{418.87} = 20.47$$

We can now compute the correlation and regression coefficients as shown here.

Correlation coefficient:

$$r = \frac{\text{cov}_{xy}}{s_x s_y} = \frac{174.38}{9.66 \times 20.47} = 0.88$$

Regression coefficient:

$$b = \frac{\text{cov}_{xy}}{s_x^2} = \frac{174.38}{93.26} = 1.87$$

>> Practice your understanding of correlation and regression by working **Problem 25** at the end of the chapter.

Applying Statistics to the Study of a Polygenic Characteristic

Edward East carried out an early statistical study of polygenic inheritance on the length of flowers in tobacco (*Nicotiana longiflora*). He obtained two varieties of tobacco that differed in flower length: one variety had a mean flower length of 40.5 mm, and the other had a mean flower length of 93.3 mm (**Figure 24.15**). Each of these two varieties had been inbred for many generations and was homozygous at all loci contributing to flower length. Thus, there was no genetic variation in the original parental strains; the small variations in flower length within each strain were due to environmental effects on flower length.

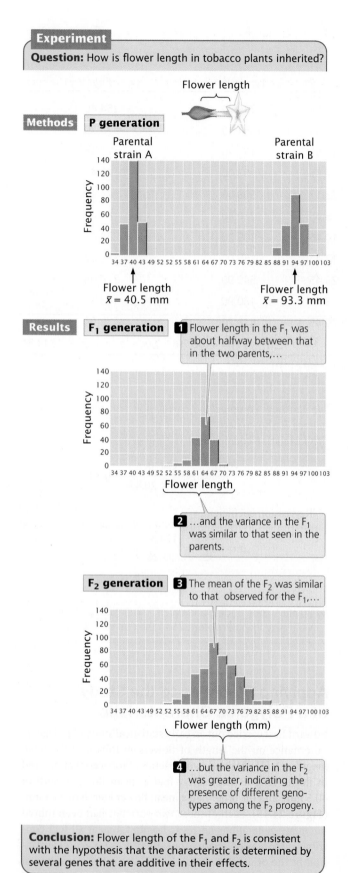

Experiment

Question: How is flower length in tobacco plants inherited?

Methods

P generation

Flower length

Parental strain A

Parental strain B

Flower length
$\bar{x} = 40.5$ mm

Flower length
$\bar{x} = 93.3$ mm

Results

F₁ generation

1 Flower length in the F₁ was about halfway between that in the two parents,…

2 …and the variance in the F₁ was similar to that seen in the parents.

F₂ generation

3 The mean of the F₂ was similar to that observed for the F₁,…

Flower length (mm)

4 …but the variance in the F₂ was greater, indicating the presence of different genotypes among the F₂ progeny.

Conclusion: Flower length of the F₁ and F₂ is consistent with the hypothesis that the characteristic is determined by several genes that are additive in their effects.

24.15 Edward East conducted an early statistical study of the inheritance of flower length in tobacco.

When East crossed the two strains, he found that the mean flower length of the F₁ was about halfway between the mean flower lengths of the two parents (see Figure 24.15), as would be expected if the genes determining the difference between the two strains were additive in their effects. The variance of flower length in the F₁ was similar to that seen in the parental strains, because all the F₁ had the same genotype, as did each parental strain (the F₁ were all heterozygous at the genes that differed between the two parental varieties).

East then interbred the F₁ to produce F₂ progeny. The mean flower length of the F₂ was similar to that of the F₁, but the variance of the F₂ was much greater (see Figure 24.15). This greater variation indicates that not all of the F₂ progeny had the same genotype.

East selected some F₂ plants and interbred them to produce F₃ progeny. He found that flower length in the F₃ depended on flower length in the plants selected as their parents. This finding demonstrated that flower-length differences in the F₂ were partly genetic and were therefore passed to the next generation. None of the 444 F₂ plants raised by East exhibited flower lengths similar to those of the two parental strains. This result suggested that more than four loci with pairs of alleles affected flower length in his varieties, because four allelic pairs are expected to produce 1 of 256 progeny $[(^1/_4)^4 = ^1/_{256}]$ having one or the other of the original parental phenotypes.

24.3 Heritability Is Used to Estimate the Proportion of Variation in a Trait That Is Genetic

In addition to being polygenic, quantitative characteristics are frequently influenced by environmental factors. Knowing how much of the variation in a quantitative characteristic is due to genetic differences and how much is due to environmental differences is often useful. The proportion of the total phenotypic variation that is due to genetic differences is known as **heritability**.

Consider a dairy farmer who owns several hundred milk cows. The farmer notices that some cows consistently produce more milk than others. The nature of these differences is important to the profitability of his dairy operation. If the differences in milk production are largely genetic in origin, then the farmer may be able to boost milk production by selectively breeding the cows that produce the most milk. On the other hand, if the differences are largely environmental in origin, selective breeding will have little effect, and the farmer might better boost milk production by adjusting the environmental factors associated with higher milk production. To determine the extent of genetic and environmental influences on variation in a characteristic, phenotypic variation in that characteristic must be partitioned into components attributable to different factors.

Phenotypic Variance

To determine how much of the phenotypic variation in a population is due to genetic factors and how much is due to environmental factors, we must first have some quantitative measure of the phenotype under consideration. Consider a population of wild plants that differ in size. We could collect a representative sample of plants from the population, weigh each plant in the sample, and calculate the mean and variance of plant weight. This **phenotypic variance** is represented by V_P.

COMPONENTS OF PHENOTYPIC VARIANCE Phenotypic variance can be divided into several components. First, some of the phenotypic variance may be due to differences in genotypes among individual members of the population. These differences are termed the **genetic variance** and are represented by V_G.

Second, some of the differences in phenotype may be due to environmental differences among the plants; these differences are termed the **environmental variance**, V_E. Environmental variance includes differences that result from environmental factors such as the amount of light or water that the plant receives; it also includes random differences in development that cannot be attributed to any specific factor. Any variation in phenotype that is not inherited is, by definition, a part of the environmental variance.

Third, **genetic–environmental interaction variance** (V_{GE}) arises when the effect of a gene depends on the specific environment in which it is found. An example is shown in **Figure 24.16**. In a dry environment, genotype AA produces a plant that averages 12 g in weight, and genotype aa produces a smaller plant that averages 10 g. In a wet environment, genotype aa produces the larger plant, averaging 24 g in weight, whereas genotype AA produces a plant that averages 20 g. In this example, there are clearly differences in the two

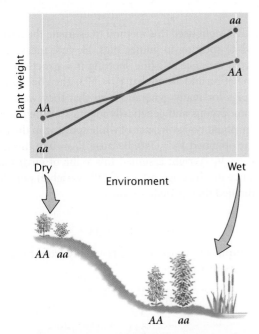

24.16 Genetic–environmental interaction variance arises when the effect of a gene depends on the specific environment in which the organism is found. In this theoretical example, the genotype affects plant weight, but the environmental conditions determine which genotype produces the heavier plant.

environments: both genotypes produce heavier plants in the wet environment. There are also differences in the weights of the two genotypes, but the relative performances of the genotypes depend on whether the plants grow in a wet or a dry environment. In this case, the influences on phenotype cannot be neatly allocated to genetic and environmental components because the expression of the genotype depends on the environment in which the plant grows. The phenotypic variance must therefore include a component that accounts for the way in which genetic and environmental factors interact.

In summary, the total phenotypic variance can be apportioned into three components:

$$V_P = V_G + V_E + V_{GE} \qquad (24.11)$$

COMPONENTS OF GENETIC VARIANCE Genetic variance can be further subdivided into components consisting of different types of genetic effects. First, **additive genetic variance** (V_A) comprises the additive effects of genes on the phenotype, which can be summed to determine the overall effect on the phenotype (see Section 24.1). For example, suppose that in a plant, allele A^1 contributes 2 g in weight and allele A^2 contributes 4 g. If the effects of the alleles are strictly additive, then the genotypes would have the following weights:

$$A^1A^1 = 2 + 2 = 4\text{ g}$$
$$A^1A^2 = 2 + 4 = 6\text{ g}$$
$$A^2A^2 = 4 + 4 = 8\text{ g}$$

The genes studied by Nilsson-Ehle that affect kernel color in wheat were additive in this way. It is the additive genetic variance that primarily determines the resemblance between parents and offspring. For example, if all of the phenotypic variance is due to additive genetic variance, then the average phenotype of the offspring will be exactly intermediate between the phenotypes of the parents.

Second, there is **dominance genetic variance** (V_D) when some genes have a dominance component. In this case, the alleles at a locus are not additive; rather, the effect of an allele depends on the identity of the other allele at that locus. For example, with a dominant allele (T), genotypes TT and Tt have the same phenotype. Here, we cannot simply add the effects of the alleles together because the effect of the t allele is masked by the presence of the T allele. Instead, we must add a component (V_D) to the genetic variance to account for the way in which alleles interact.

Third, genes at different loci may interact in the same way that alleles at the same locus interact. When this gene interaction takes place, the effects of the genes are not additive. Coat color in Labrador retrievers, for example, exhibits gene interaction, as described in Section 5.2: genotypes $BB\ ee$ and $bb\ ee$ both produce yellow dogs because the effects of alleles at the B locus are masked when two e alleles are present at the E locus. With gene interaction, we must add a third component, called **gene interaction variance** (V_I), to the genetic variance:

$$V_G = V_A + V_D + V_I \qquad (24.12)$$

SUMMARY EQUATION We can now integrate these components into one equation to represent all the potential contributions to the phenotypic variance:

$$V_P = V_A + V_D + V_I + V_E + V_{GE} \qquad (24.13)$$

This equation provides us with a model that describes the potential causes of differences we observe among individual phenotypes. It's important to note that this model deals strictly with observable *differences* (variance) in phenotypes among individual members of a population; it says nothing about the absolute values of characteristics or about the underlying genotypes that produce these differences.

Types of Heritability

The model of phenotypic variance that we've just developed can be used to determine how much of the phenotypic variance in a characteristic is due to genetic differences. **Broad-sense heritability** (H^2) represents the proportion of phenotypic variance that is due to genetic variance. It is calculated by dividing the genetic variance by the phenotypic variance:

$$\text{broad-sense heritability} = H^2 = \frac{V_G}{V_P} \qquad (24.14)$$

The symbol H^2 represents broad-sense heritability because it is a measure of variance, which is given in units squared.

Broad-sense heritability can potentially range from 0 to 1. A value of 0 indicates that none of the phenotypic variance results from differences in genotype and all of the differences in phenotype result from environmental variation. A value of 1 indicates that all of the phenotypic variance results from differences in genotype. A heritability value between 0 and 1 indicates that both genetic and environmental factors influence the phenotypic variance.

Often, geneticists are more interested in the proportion of the phenotypic variance that results from the additive genetic variance because, as mentioned earlier, the additive genetic variance primarily determines the resemblance between parents and offspring. **Narrow-sense heritability** (h^2) is equal to the additive genetic variance divided by the phenotypic variance:

$$\text{narrow-sense heritability} = h^2 = \frac{V_A}{V_P} \qquad (24.15)$$

> **TRY PROBLEM 26**

Calculating Heritability

Now that we have considered the components that contribute to phenotypic variance and developed a general concept of heritability, we can ask how to go about estimating these different components and calculating heritability. There are several ways to measure the heritability of a characteristic. They include eliminating one or more variance components, comparing the resemblance of parents and offspring, comparing the phenotypic variances of individuals with different degrees of relatedness, and measuring the response to selection (this last method will be discussed in Section 24.4). The mathematical theory that underlies these calculations of heritability is complex, so we focus here on developing a general understanding of how heritability is measured.

HERITABILITY BY ELIMINATION OF VARIANCE COMPONENTS One way of calculating broad-sense heritability is to eliminate one of the components of phenotypic variance. We have seen that $V_P = V_G + V_E + V_{GE}$. If we eliminate all environmental variance ($V_E = 0$), then $V_{GE} = 0$ (because if either V_G or V_E equals 0, no genetic–environmental interaction can take place), and $V_P = V_G$. In theory, we might make V_E equal to 0 by ensuring that all individuals are raised in exactly the same environment, but in practice, it is virtually impossible. Instead, we could make V_G equal to 0 by raising genetically identical individuals, causing V_P to be equal to V_E. In a typical experiment, we might raise cloned or highly inbred, identically homozygous individuals in a defined environment and measure their phenotypic variance to estimate V_E. We might then raise a group of genetically variable individuals and measure their phenotypic variance (V_P). Using the V_E calculated for the genetically identical individuals, we could obtain the genetic variance of the variable individuals by subtraction:

$$V_{G\,(\text{of genetically varying individuals})}$$
$$= V_{P\,(\text{of genetically varying individuals})} - V_{E\,(\text{of genetically identical individuals})} \qquad (24.16)$$

The broad-sense heritability of the genetically variable individuals would then be calculated as follows:

$$H^2 = \frac{V_{G(\text{of genetically varying individuals})}}{V_{P(\text{of genetically varying individuals})}} \qquad (24.17)$$

Sewall Wright used this method to estimate the heritability of white spotting in guinea pigs. He first measured the phenotypic variance of white spotting in a genetically variable population and found that $V_P = 573$. Then he inbred the guinea pigs for many generations so that they were essentially homozygous and genetically identical. When he measured the phenotypic variance of white spotting in the inbred group, he obtained $V_P = 340$. Because $V_G = 0$ in this group, their $V_P = V_E$. Wright assumed this value of environmental variance for the original (genetically variable) population and estimated their genetic variance:

$$V_P - V_E = V_G$$
$$573 - 340 = 233$$

He then estimated broad-sense heritability from the genetic and phenotypic variance:

$$H^2 = \frac{V_G}{V_P}$$
$$H^2 = \frac{233}{573} = 0.41$$

This value implies that 41% of the variation in white spotting in guinea pigs in Wright's population was due to differences in genotype.

Estimating heritability with this method assumes that the environmental variance of genetically identical individuals is the same as the environmental variance of genetically variable individuals, which may not be true. Additionally, this approach can be applied only to organisms for which it is possible to create genetically identical individuals.

⟩ TRY PROBLEM 31

HERITABILITY BY PARENT–OFFSPRING REGRESSION Another method of estimating heritability is to compare the phenotypes of parents with those of their offspring. When genetic differences are responsible for phenotypic variance, offspring should resemble their parents more than they resemble unrelated individuals because offspring and parents share some genes that help determine their phenotype. Correlation and regression can be used to analyze the association of phenotypes in different individuals.

To calculate the narrow-sense heritability of a characteristic in this way, we first measure the characteristic in a series of parents and offspring. The data are arranged into families, and the mean parental phenotype is plotted against the mean offspring phenotype (**Figure 24.17**). Each data point in the graph represents one family; the value on the x (horizontal) axis is the mean phenotypic value of the parents in a family, and the value on the y (vertical) axis is the mean phenotypic value of the offspring in the family.

Let's begin by assuming that there is no narrow-sense heritability for the characteristic ($h^2 = 0$), meaning that genetic differences do not contribute to the phenotypic differences among individuals. In this case, offspring will be no more similar to their parents than they are to unrelated individuals, and the data points will be scattered randomly, generating a regression coefficient of 0 (see Figure 24.17a). Next, let's assume that all of the phenotypic differences are due to additive genetic differences ($h^2 = 1$). In this case, the mean phenotype of the offspring will be equal to the mean phenotype of the parents, and the regression coefficient will be 1 (see Figure 24.17b). Finally, if genes and environment both contribute to the phenotypic differences, both heritability and the regression coefficient will lie between 0 and 1 (see Figure 24.17c). The regression coefficient therefore provides information about the magnitude of heritability.

A complex mathematical proof (which we will not go into here) demonstrates that, in a regression of the mean phenotype of the offspring against the mean phenotype of the parents, narrow-sense heritability (h^2) equals the regression coefficient (b):

$$h^2 = b \text{ (regression of offspring mean against mean of both parents)} \quad (24.18)$$

An example of calculating heritability by regression of the phenotypes of parents and offspring is illustrated in **Figure 24.18**.

Sometimes, only the phenotype of one parent is known. In a regression of the mean offspring phenotype against the phenotype of only one parent, the narrow-sense heritability equals twice the regression coefficient:

$$h^2 = 2b \text{ (regression of offspring mean against mean of one parent)} \quad (24.19)$$

With only one parent, the heritability is twice the regression coefficient because only half the genes of the offspring come from that one parent; thus, we must double the regression coefficient to obtain the full heritability.

HERITABILITY AND DEGREES OF RELATEDNESS A third method for calculating heritability is to compare the phenotypes of individuals with different degrees of relatedness. This method is based on the concept that the more closely related two individuals are, the more genes they share.

Monozygotic (identical) twins share 100% of their genes, whereas dizygotic (nonidentical) twins share, on average, 50% of their genes (see Section 6.3). If genes are important in determining variation in a characteristic, then monozygotic twins should be more similar in that characteristic than dizygotic twins. By using correlation to compare the phenotypes

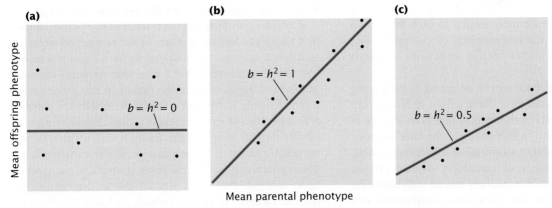

24.17 Narrow-sense heritability, h^2, equals the regression coefficient, b, in a regression of the mean phenotype of offspring against the mean phenotype of their parents. (a) There is no relation between the mean parental phenotype and the mean offspring phenotype. (b) The mean offspring phenotype is the same as the mean parental phenotype. (c) Both genes and environment contribute to phenotypic differences.

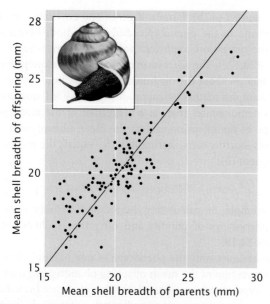

24.18 The heritability of shell breadth in snails can be determined by regression of the mean phenotype of offspring against the mean phenotype of their parents. The regression coefficient, which equals the heritability, is 0.70. [From L. M. Cook, *Evolution* 19:86–94, 1965.]

of monozygotic twins and of dizygotic twins, we can estimate broad-sense heritability. A rough estimate of broad-sense heritability can be obtained by taking twice the difference between the correlation coefficients for a quantitative characteristic in monozygotic and dizygotic twins:

$$H^2 = 2(r_{MZ} - r_{DZ}) \qquad (24.20)$$

where r_{MZ} equals the correlation coefficient among monozygotic twins and r_{DZ} equals the correlation coefficient among dizygotic twins. For example, suppose we found the correlation of height among the two members of monozygotic twin pairs (r_{MZ}) to be 0.9 and the correlation of height among the two members of dizygotic twin pairs (r_{DZ}) to be 0.5. The broad-sense heritability for height would be $H^2 = 2(0.9 - 0.5) = 2(0.4) = 0.8$. This calculation assumes that the two members of a monozygotic twin pair experience environments that are no more similar to each other than those experienced by the two members of a dizygotic twin pair. This assumption is often not met when twins have been reared together.

Narrow-sense heritability can be estimated by comparing the phenotypic variances for a characteristic in full siblings (who have both parents in common as well as an average of 50% of their genes) and in half siblings (who have only one parent in common and thus an average of 25% of their genes).

All of these estimates of heritability depend on the assumption that the environments of related individuals are not more similar than those of unrelated individuals. This assumption is difficult to meet in human studies because related people are usually reared together. Therefore, heritability estimates for humans should always be viewed with caution. **▶ TRY PROBLEM 35**

The Limitations of Heritability

Knowledge of heritability has great practical value because it allows us to statistically predict the phenotypes of offspring on the basis of their parents' phenotypes. It also provides useful information about how characteristics will respond to selection (see Section 24.4). In spite of its importance, heritability is frequently misunderstood. It does not provide information about an individual's genes or the environmental factors that control the development of a characteristic, and it says nothing about the nature of differences between groups. This section outlines some limitations and common misconceptions of broad- and narrow-sense heritability.

HERITABILITY DOES NOT INDICATE THE DEGREE TO WHICH A CHARACTERISTIC IS GENETICALLY DETERMINED Heritability is the proportion of the phenotypic variance that is due to genetic variance; it says nothing about the degree to which genes determine a characteristic. Heritability indicates only the degree to which genes determine *variation* in a characteristic. The determination of a characteristic and the determination of variation in a characteristic are two very different things.

Consider polydactyly (the presence of extra digits) in rabbits, which can be caused either by environmental factors or by a dominant gene. Suppose that we have a group of rabbits that are all homozygous for a gene that produces the usual numbers of digits. None of the rabbits in this group carries a gene for polydactyly, but a few of the rabbits are polydactylous because of environmental factors. Broad-sense heritability for polydactyly in this group is zero because there is no genetic variation for polydactyly; all the variation is due to environmental factors. However, it would be incorrect for us to conclude that genes play no role in determining the number of digits in rabbits. Indeed, we know that there are specific alleles that can produce extra digits (although these alleles are not present in the group of rabbits under consideration). Heritability indicates nothing about whether genes control the development of a characteristic; it provides information only about causes of the variation in a characteristic within a defined group.

AN INDIVIDUAL DOES NOT HAVE HERITABILITY Broad- and narrow-sense heritabilities are statistical values based on the genetic and phenotypic variances found in a *group* of individuals. Heritability cannot be calculated for an individual, and heritability has no meaning for a specific individual. Suppose that we calculate the narrow-sense heritability of adult body weight for the students in a biology class and obtain a value of 0.6. We could conclude that 60% of the variation in adult body weight among the students in this class is determined by additive genetic variance. We should not, however, conclude that 60% of any particular student's body weight is due to additive genes.

THERE IS NO UNIVERSAL HERITABILITY FOR A CHARACTERISTIC The heritability value for a characteristic is specific to a given population in a given environment. Recall that broad-sense heritability is genetic variance divided by phenotypic variance. Genetic variance depends on which alleles are present, which often differs between populations. In the example of polydactyly in rabbits, there were no alleles for polydactyly in the group of rabbits we considered, so the heritability of the characteristic was zero. A different group of rabbits might contain many alleles for polydactyly, and the heritability of the characteristic might be high in that group.

Environmental differences can also affect heritability because V_P is composed of both genetic and environmental variance. When the environmental factors that affect a characteristic differ between two groups, the heritabilities for the two groups often differ as well.

Because heritability is specific to a defined population in a given environment, it is important not to extrapolate heritabilities from one population to another. For example, human height is determined by environmental factors (such as nutrition and health) and by genes. If we measured the heritability of height in a developed country, we might obtain a value of 0.8, indicating that the variation in height in this population is largely genetic. Height has a high heritability in this population because most people have adequate nutrition and health care (V_E is low), so most of the phenotypic variation in height is genetically determined. It would be incorrect for us to assume, however, that height has a high heritability in all human populations. In developing countries, there may be more variation in a range of environmental factors; some people may enjoy good nutrition and health, whereas others may have a diet deficient in protein and suffer from diseases that affect stature. If we measured the heritability of height in such a country, we would undoubtedly obtain a lower value than we observed in the developed country because there would be more environmental variation, so the genetic variance in height would constitute a smaller proportion of the phenotypic variation, making the heritability lower. The important point to remember is that heritability must be calculated separately for each population and each environment.

EVEN WHEN HERITABILITY IS HIGH, ENVIRONMENTAL FACTORS CAN INFLUENCE A CHARACTERISTIC A high heritability value does not mean that environmental factors cannot influence the expression of a characteristic. High heritability indicates only that the environmental variation to which the population is *currently* exposed is not responsible for variation in the characteristic. Let's look again at human height. In most developed countries, the heritability of human height is high, indicating that genetic differences are responsible for most of the variation in height. It would be wrong for us to conclude, however, that human height cannot be changed by alterations in the environment. Indeed, height decreased in several European cities during World War II owing to hunger and disease, and height can be increased dramatically by the administration of growth hormone to children. The absence of environmental variation in a characteristic does not mean that the characteristic will not respond to environmental change.

HERITABILITIES INDICATE NOTHING ABOUT DIFFERENCES AMONG POPULATIONS A common misconception about heritability is that it provides information about the nature of population differences in a characteristic. Heritability is specific to a given population in a given environment, so it cannot be used to draw conclusions about why populations differ in a characteristic.

Suppose that we measure the heritability of human height in two groups. One group is from a small town in a developed country, where everyone consumes a high-protein diet. Because there is little variation in the environmental factors that affect human height in this group, and because there is some genetic variation, the heritability of height in this group is high. The second group comprises the inhabitants of a single village in a developing country. These people consume only 25% as much protein as those in the first group, so their average adult height is several centimeters less than that of the group from the developed country. Again, there is little variation in the environmental factors that determine height in this group because everyone in the village eats the same types of food and is exposed to the same diseases. Because there is little environmental variation and there is some genetic variation, the heritability of height in this group is also high.

Thus, the heritability of height in both groups is high, and the average height differs considerably between the two groups. We might be tempted to conclude that the difference in height between the two groups is genetically based—that the people in the developed country are genetically taller than the people in the developing country. This conclusion is obviously wrong, however, because these differences in height are due largely to diet—an environmental factor. Heritability provides no information about the causes of differences between populations.

These limitations of heritability have often been ignored, particularly in arguments about the possible social implications of genetic differences among humans. Soon after Mendel's principles of heredity were rediscovered, some geneticists began to claim that many human behavioral characteristics are determined entirely by genes. This claim led to debates about whether characteristics such as human intelligence are determined by genes or environment. Many of the early claims of genetically based human behavior were based on poor research; unfortunately, the results of these studies were often accepted at face value and led to a number of eugenic laws that discriminated against certain groups of people (see Section 25.4). Today, geneticists recognize that many behavioral characteristics are influenced by a complex interaction of genes and environment, and that separating genetic effects from those of the environment is very difficult.

The results of a number of modern studies indicate that human intelligence, as measured by IQ and other intelligence tests, has a moderately high heritability (usually from 0.4 to 0.8). On the basis of this observation, some people have argued that intelligence is innate and that enhanced educational opportunities cannot boost intelligence. This argument is based on the misconception that, when heritability is high, changing the environment will not alter the characteristic. In addition, because heritability values for intelligence range from 0.4 to 0.8, a considerable amount of the variance in intelligence originates from environmental differences.

Another argument based on a misconception about heritability is that ethnic differences in measures of intelligence are genetically based. Because the results of some genetic studies show that IQ has moderately high heritability, and because other studies find differences in average IQ among ethnic groups, some people have suggested that ethnic differences in IQ are genetically based. As in the example of the effects of diet on human height, heritability provides no information about causes of differences among groups; it indicates only the degree to which phenotypic variance within a single group is genetically based. High heritability for a characteristic does not mean that phenotypic differences in that characteristic among ethnic groups are genetic. We should also remember that because separating genetic and environmental effects in humans is so difficult, heritability estimates themselves may be unreliable. The limitations of heritability are summarized in **Table 24.2**. ▶ **TRY PROBLEM 34**

THINK-PAIR-SHARE Questions 4 and 5

> ### CONCEPTS
>
> Heritability provides information only about the degree to which *variation* in a characteristic is genetically determined. There is no universal heritability for a characteristic; heritability is specific to a given population in a specific environment. Environmental factors can potentially affect characteristics with high heritability, and heritability says nothing about the nature of differences among populations in a characteristic.
>
> ### ✔ CONCEPT CHECK 6
> Suppose that you just learned that the narrow-sense heritability of blood pressure measured among a group of African Americans in Detroit, Michigan, is 0.4. What does this heritability tell us about genetic and environmental contributions to blood pressure?

TABLE 24.2	Limitations of heritability
1.	Heritability does not indicate the degree to which a characteristic is genetically determined.
2.	An individual does not have heritability.
3.	There is no universal heritability for a characteristic.
4.	Even when heritability is high, environmental factors can influence a characteristic.
5.	Heritabilities indicate nothing about the nature of population differences in a characteristic.

Locating Genes That Affect Quantitative Characteristics

The statistical methods we have just described can be used both to make predictions about the average phenotype expected in offspring and to estimate the overall contribution of genes to variation in a characteristic. These methods do not, however, allow us to identify and determine the influence of individual genes that affect quantitative characteristics. As discussed in the introduction to this chapter, chromosomal regions containing genes that control polygenic characteristics are referred to as **quantitative trait loci** (**QTLs**). Although quantitative genetics has made important contributions to basic biology and to plant and animal breeding, the past inability to identify QTLs and measure their individual effects severely limited the application of quantitative genetic methods.

MAPPING QTLs In recent years, numerous genetic markers have been identified and mapped with the use of molecular techniques, making it possible to identify QTLs by linkage analysis. The underlying idea is simple: if the inheritance of a genetic marker is associated consistently with the inheritance of a particular phenotype, such as increased height, then that marker must be linked to a QTL that affects height. The key is to have enough genetic markers so that QTLs can be detected throughout the genome. With the introduction of microsatellites and single-nucleotide polymorphisms (see Sections 19.5 and 20.1), variable markers are now available for mapping QTLs in a number of different organisms (**Figure 24.19**).

24.19 QTL mapping is used to identify genes that influence many important quantitative traits, including muscle mass in pigs. [USDA.]

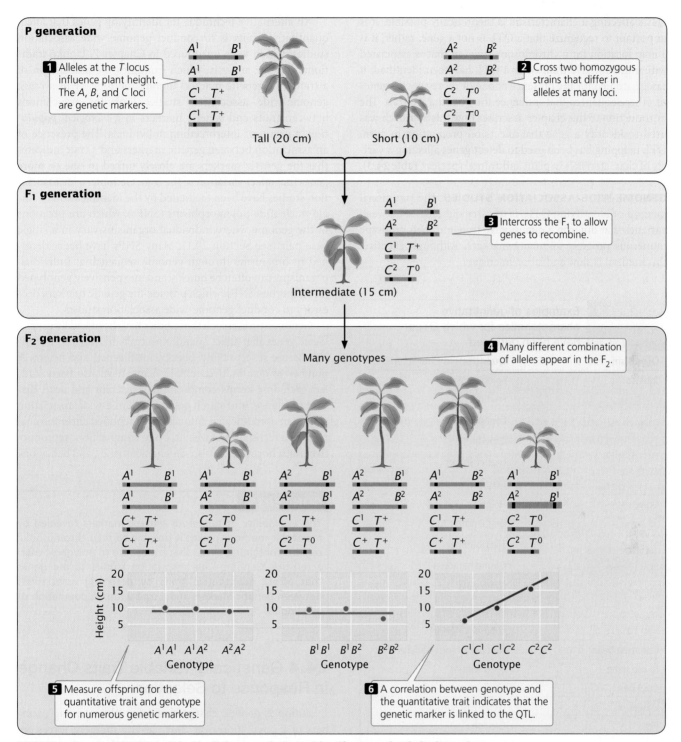

24.20 Mapping quantitative trait loci by linkage analysis can identify genes that help determine differences in quantitative traits. Genotypes at the *C* locus are associated with the inheritance of differences in plant height, indicating that a QTL for height (the *T* locus) is closely linked to the *C* locus.

A common procedure for mapping QTLs is to cross two homozygous strains that differ in alleles at many loci (**Figure 24.20**). The resulting F_1 progeny are then intercrossed or backcrossed to allow the genes to recombine through independent assortment and crossing over. Genes on different chromosomes and genes that are far apart on the same chromosome will recombine freely; genes that are closely linked will be inherited together. The F_2 progeny are measured for one or more quantitative characteristics; at the same time, they are genotyped for numerous genetic markers that span the genome. Any correlation between the inheritance of a particular marker allele and that of a quantitative phenotype suggests that a QTL is physically linked to that marker. If enough markers are used, the detection of all the

QTLs affecting a characteristic is theoretically possible. It is important to recognize that a QTL is not a gene; rather, it is a map location for a chromosomal region that is associated with a quantitative trait. After a QTL has been identified, it can be studied for the presence of one or more specific genes or other sequences that influence the quantitative trait. The introduction to this chapter describes how this approach was used to identify a gene that affects oil production in corn. QTL mapping has been used to detect genes affecting a variety of characteristics in plant and animal species (**Table 24.3**).

GENOME-WIDE ASSOCIATION STUDIES The traditional method of identifying QTLs is to carry out crosses between varieties that differ in a quantitative trait and then genotype numerous progeny for many markers. Although effective, this method is slow and labor-intensive.

TABLE 24.3	Examples of quantitative characteristics for which QTLs have been detected
Organism	**Quantitative Characteristic**
Tomato	Soluble solids
	Fruit mass
	Fruit pH
	Growth
	Leaflet shape
	Height
Corn	Height
	Leaf length
	Tiller number
	Glume hardness
	Grain yield
	Number of ears
	Thermotolerance
Common bean	Number of root nodules
Mung bean	Seed weight
Cow pea	Seed weight
Wheat	Preharvest sprout
Pig	Growth
	Length of small intestine
	Average back fat
	Abdominal fat
Mouse	Epilepsy
Rat	Hypertension
Dog	Body size

Source: After S. D. Tanksley, Mapping polygenes, *Annual Review of Genetics* 27:218, 1993. QTLs for body size in dogs from A. R. Boyko et al., A simple genetic architecture underlies morphological variation in dogs, *PLoS Biology* 8: e1000451, 2010.

An alternative technique for identifying genes that affect quantitative traits is to conduct genome-wide association studies, which were introduced in Chapter 7. Unlike traditional linkage analysis, which examines the association of a trait with genetic markers among the *progeny of a cross*, genome-wide association studies look for associations between traits and genetic markers in a *biological population*, a group of interbreeding individuals. The presence of an association between genetic markers and a trait indicates that the genetic markers are closely linked to one or more genes that affect variation in the trait. Genome-wide association studies have been facilitated by the identification of single-nucleotide polymorphisms (SNPs), which are positions in the genome where individual organisms vary in a single base pair (see Section 20.1). Many SNPs have been identified in organisms through genome sequencing. Individual organisms can often be quickly and inexpensively genotyped for numerous SNPs, which provide the genetic markers necessary to conduct genome-wide association studies.

Genome-wide association studies have been widely used to locate genes that affect quantitative traits in humans, including disease susceptibility, obesity, intelligence, and height. A number of quantitative traits in plants have also been studied, including kernel composition, size, color and taste, disease resistance, and starch quality. Genome-wide association studies in domesticated animals have identified chromosomal regions affecting body weight, body composition, reproductive traits, hormone levels, hair characteristics, and behaviors.

> **CONCEPTS**
>
> The availability of numerous genetic markers revealed by molecular methods makes it possible to map chromosomal regions containing genes that contribute to polygenic characteristics. Genome-wide association studies locate genes that affect quantitative traits by detecting associations between genetic markers and a trait within a population of individuals.

24.4 Genetically Variable Traits Change in Response to Selection

Evolution is genetic change that takes place among members of a population over time. Several different forces are potentially capable of bringing about evolution, and we will explore these forces and the process of evolution more fully in Chapter 25. Here, we consider how one of these forces—natural selection—can bring about genetic change in a quantitative characteristic.

Charles Darwin proposed the idea of natural selection in his book *On the Origin of Species* in 1859. **Natural selection** arises through the differential reproduction of individuals with different genotypes. Because of the genes they possess, some individuals produce more offspring than others. The more successful reproducers give rise to more offspring, which inherit the genes that confer a reproductive advantage. Thus, the frequencies of the genes that confer a reproductive advantage increase with the passage of time, and the population evolves. Natural selection is among the most important of the forces that bring about evolutionary change. Through natural selection, organisms become genetically suited to their environments; as environments change, groups of organisms change in ways that make them better able to survive and reproduce.

For thousands of years, humans have practiced a form of selection by promoting the reproduction of organisms with traits they perceive as desirable. This form of selection, called **artificial selection**, has produced the domesticated plants and animals that make modern agriculture possible. The power of artificial selection—the first application of genetic principles by humans—is illustrated by the tremendous diversity of shapes, colors, and behaviors of modern domestic dogs (**Figure 24.21**).

Predicting the Response to Selection

When a quantitative characteristic is subjected to natural or artificial selection, it frequently changes with the passage of time, provided that there is genetic variation for that characteristic in the population. Let's return to the dairy farmer who wants to increase milk production among the cows in his herd. Variation at several loci potentially affects milk production in cows; some alleles at these loci confer high milk production, whereas others confer low milk production. The dairy farmer breeds only those cows in his herd that have the highest milk production. If there is genetic variation for milk production in his herd (i.e., there are different alleles at the loci that control milk production), the mean milk production in the offspring of the selected cows should be higher than the mean milk production of the original herd. This increased production is due to the fact that the selected cows possess more alleles for high milk production than does the average cow, and these alleles are passed on to the offspring. Thus, the offspring of the selected cows possess a higher proportion of alleles for high milk production, and therefore produce more milk, than the average cow in the original herd.

The extent to which a characteristic subjected to selection changes in one generation is termed the **response to selection**. Suppose that the average cow in a dairy herd produces 80 liters of milk per week. A farmer selects for increased milk production by breeding the highest milk producers, and the female progeny of those selected cows produce 100 liters of milk per week on average. The response to selection is calculated by subtracting the mean phenotype of the original population (80 liters) from the mean phenotype of the offspring (100 liters); in our example, the response to selection is 100 − 80 = 20 liters of milk per week.

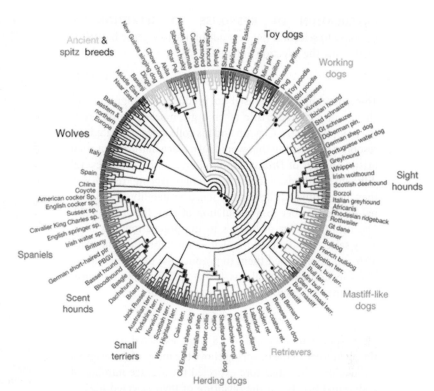

24.21 Artificial selection has produced the tremendous diversity of shape, size, color, and behavior seen today among breeds of domestic dogs. This diagram depicts the evolutionary relationships among wolves and different breeds of dogs as determined from analyses of DNA sequences. [Reprinted by permission from Macmillan Publishers Ltd.: B. M. von Holdt et al., *Nature*, 464 (7290), 898–902. © 2010. Permission conveyed through Copyright Clearance Center, Inc.]

FACTORS INFLUENCING RESPONSE TO SELECTION The response to selection is determined primarily by two factors. First, it is affected by narrow-sense heritability, which largely determines the degree of resemblance between parents and offspring. When narrow-sense heritability is high, offspring will tend to resemble their parents; conversely, when narrow-sense heritability is low, there will be little resemblance between parents and offspring.

The second factor that determines the response to selection is how much selection there is. If the farmer is very stringent in the choice of parents and breeds only the highest milk producers in the herd (say, the top two cows), then all the offspring will receive genes for high milk production. If the farmer is less selective and breeds the top twenty milk producers in the herd, then the offspring will not carry as many genes for high milk production, and on average, they will not produce as much milk as the offspring of the top two producers. The response to selection depends on the phenotypic difference of the individuals that are selected as parents; this phenotypic difference is measured by the **selection differential**, defined as the difference between the mean phenotype of the selected parents and the mean phenotype of the original population. If the average milk production of the original herd is 80 liters and the farmer breeds cows with an average milk production of 120 liters, then the selection differential is 120 − 80 = 40 liters.

CALCULATION OF RESPONSE TO SELECTION The response to selection (R) depends on narrow-sense heritability (h^2) and the selection differential (S):

$$R = h^2 \times S \qquad (24.21)$$

This equation can be used to predict the magnitude of change in a characteristic when a given selection differential is applied. G. A. Clayton and his colleagues estimated the response to selection that would take place in the abdominal bristle number of *Drosophila melanogaster*. By using several different methods, including parent–offspring regression, they first estimated the narrow-sense heritability of abdominal bristle number in a population of fruit flies to be 0.52. The mean number of bristles in the population was 35.3. They selected individual flies with a mean bristle number of 40.6 and intercrossed them to produce the next generation. The selection differential was $40.6 - 35.3 = 5.3$; so they predicted that the response to selection would be

$$R = 0.52 \times 5.3 = 2.8$$

The response to selection of 2.8 is the expected increase in the phenotype of the offspring above the mean of the original population. They therefore expected the average number of abdominal bristles in the offspring of their selected flies to be $35.3 + 2.8 = 38.1$. Indeed, they found an average bristle number of 37.9 in these flies.

ESTIMATING HERITABILITY FROM RESPONSE TO SELECTION Rearranging Equation 24.21 provides another way to calculate narrow-sense heritability:

$$h^2 = \frac{R}{S} \qquad (24.22)$$

In this way, h^2 can be calculated by conducting a response-to-selection experiment. First, the selection differential is obtained by subtracting the population mean from the mean of the selected parents. The selected parents are then interbred, and the mean phenotype of their offspring is measured. The difference between the mean of the offspring and that of the original population is the response to selection, which can be used with the selection differential to estimate heritability. Heritability determined by a response-to-selection experiment is usually termed **realized heritability**. If certain assumptions are met, realized heritability is identical with narrow-sense heritability.

One of the longest-running selection experiments is a study of oil and protein content in corn kernels (**Figure 24.22**). This experiment began at the University of Illinois on 163 ears of corn with an oil content ranging from 4% to 6%. Corn plants with high oil content and corn plants with low oil content were selected and interbred. Response to selection for increased oil content (the upper line in Figure 24.22) reached about 20%, whereas response to selection for decreased oil content reached a lower limit near zero. Genetic analyses of the high- and low-oil-content strains revealed that

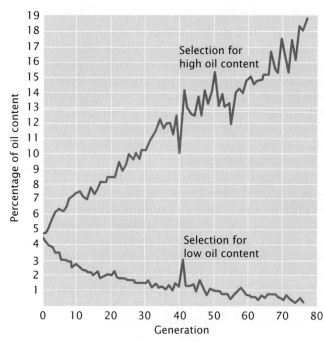

24.22 In a long-term response-to-selection experiment, selection for oil content in corn increased oil content in one line to about 20%, whereas it almost eliminated it in another line.

at least 20 loci take part in determining oil content, one of which we explored in the introduction to this chapter.

▶ **TRY PROBLEM 39**

THINK-PAIR-SHARE Question 6

CONCEPTS

The response to selection is influenced by narrow-sense heritability and the selection differential.

✔ **CONCEPT CHECK 7**

The narrow-sense heritability for a trait is 0.4 and the selection differential is 0.5. What is the predicted response to selection?

Limits to the Response to Selection

When a characteristic has been selected for many generations, the response may eventually level off, and the characteristic may no longer respond to selection (**Figure 24.23**). A potential reason for this leveling off is that the genetic variation in the population is exhausted; at some point, all individuals in the population may have become homozygous for the alleles that encode the selected trait. When there is no more additive genetic variance, heritability equals zero, and no further response to selection can take place.

Sometimes, the response to selection levels off even while some genetic variation remains in the population. This leveling off takes place because natural selection opposes further change in the characteristic. The response to selection for small body size in mice, for example, eventually levels off because

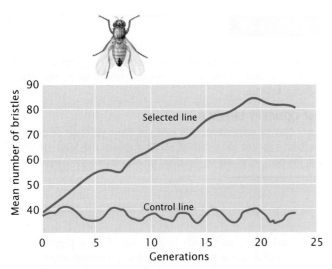

24.23 The response of a population to selection often levels off at some point. In a response-to-selection experiment in which fruit flies were selected for increased bristle number in females, the number of bristles increased steadily for about 20 generations and then leveled off.

the smallest animals are sterile and cannot pass on their genes for small body size. In this case, artificial selection for small size is opposed by natural selection for fertility, and the population can no longer respond to the artificial selection.

THINK-PAIR-SHARE Question 7

Correlated Responses to Selection

Often, when a specific trait is selected, other traits change at the same time. This type of associated response is due to the fact that the traits are encoded by the same genes.

Two or more characteristics are often correlated. For example, in many plants, plant size and number of seeds produced exhibit a positive correlation: larger plants, on average, produce more seeds than smaller plants. This type of correlation is called a **phenotypic correlation** because the association is between two phenotypes of the same individual. Phenotypic correlations may be due to environmental or genetic correlations. Environmental correlations refer to two or more characteristics that are influenced by the same environmental factor. Moisture availability, for example, may affect both the size of a plant and the number of seeds it produces. Plants growing in environments with lots of water are large and produce many seeds, whereas plants growing in environments with limited water are small and have few seeds.

Alternatively, a phenotypic correlation may result from a **genetic correlation**, which means that the genes affecting two characteristics are associated. The primary genetic cause of phenotypic correlations is pleiotropy: the effect of one gene on two or more characteristics (see Section 5.5). In humans, for example, many body structures respond to growth hormone, and there are genes that affect the amount of growth hormone secreted by the pituitary gland. People with certain genotypes produce high levels of growth hormone, which

increases both height and hand size. Those with other genotypes produce lower levels of growth hormone, which leads to both short stature and small hands. Height and hand size are therefore phenotypically correlated in humans, and this correlation is due to a genetic correlation—the fact that both characteristics are affected by the same genes that control the amount of growth hormone. Genetically speaking, height and hand size are the same characteristic because they are the phenotypic manifestation of a single set of genes. When two characteristics are influenced by the same genes, they are genetically correlated.

Genetic correlations are quite common (**Table 24.4**) and may be positive or negative. A positive genetic correlation between two characteristics means that genes that cause an increase in one characteristic also produce an increase in the other characteristic. Thorax length and wing length in *Drosophila* are positively correlated because the genes that increase thorax length also increase wing length. A negative genetic correlation means that genes that cause an increase in one characteristic produce a decrease in the other characteristic. Milk yield and percentage of butterfat are negatively correlated in cattle: genes that cause higher milk production result in milk with a lower percentage of butterfat.

Genetic correlations are important in animal and plant breeding because they produce a correlated response to selection, which means that when one characteristic is selected, genetically correlated characteristics also change. Correlated responses to selection are due to the fact that both characteristics are influenced by the same genes; selection for one characteristic causes a change in the genes affecting that characteristic, and because those genes also affect the second characteristic, they cause it to change at the same time. Correlated responses may well be undesirable and may limit our ability to alter a characteristic by selection.

TABLE 24.4	Genetic correlations in various organisms	
Organism	**Characteristics**	**Genetic Correlation**
Cattle	Milk yield and percentage of butterfat	−0.38
Pig	Weight gain and back-fat thickness	0.13
	Weight gain and efficiency	0.69
Chicken	Body weight and egg weight	0.42
	Body weight and egg production	−0.17
	Egg weight and egg production	−0.31
Mouse	Body weight and tail length	0.29
Fruit fly	Abdominal bristle number and sternopleural bristle number	0.41

Source: After D. S. Falconer and T. F. C. Mackay, *Introduction to Quantitative Genetics* (New York: Pearson, 1996), p. 314.

From 1944 to 1964, for example, domesticated turkeys were subjected to intense selection for growth rate and body size. At the same time, fertility, egg production, and egg hatchability all declined. These correlated responses were due to negative genetic correlations between body size and fertility; eventually, these genetic correlations limited the extent to which the growth rate of turkeys could respond to selection. Genetic correlations may also limit the ability of natural populations to respond to selection in the wild and adapt to their environments. ▶ TRY PROBLEM 44

CONCEPTS

Genetic correlations result from pleiotropy. When two characteristics are genetically correlated, selection for one characteristic will produce a correlated response in the other characteristic.

✔ **CONCEPT CHECK 8**

In a herd of dairy cattle, milk yield and the percentage of butterfat exhibit a genetic correlation of −0.38. If greater milk yield is selected in this herd, what will be the effect on the percentage of butterfat?

CONCEPTS SUMMARY

■ Quantitative genetics focuses on the inheritance of complex characteristics whose phenotypes vary continuously. For many quantitative characteristics, the relation between genotype and phenotype is complex because many genes, as well as environmental factors, influence a characteristic.

■ The individual genes that influence a polygenic characteristic follow the same Mendelian principles that govern discontinuous characteristics, but because many genes participate, the expected ratios of phenotypes are obscured.

■ A frequency distribution, in which phenotypes are represented on one axis and the numbers or proportions of individuals possessing each phenotype are represented on the other axis, is a convenient means of summarizing the phenotypes found in a group of individuals.

■ A population is the group of interest, and a sample is a subset of the population used to describe it.

■ The mean and variance provide key information about a distribution: the mean gives the central location of the distribution, and the variance provides information about how the phenotype varies within a group.

■ The correlation coefficient measures the direction and strength of association between two variables. Regression can be used to predict the value of one variable on the basis of the value of a correlated variable.

■ Phenotypic variance in a characteristic can be divided into components that are due to additive genetic variance, dominance genetic variance, gene interaction variance, environmental variance, and genetic–environmental interaction variance.

■ Broad-sense heritability is the proportion of phenotypic variance that is due to genetic variance. Narrow-sense heritability is the proportion of phenotypic variance that is due to additive genetic variance.

■ Heritability provides information only about the degree to which variation in a characteristic results from genetic differences. Heritability is based on the variation present within a group of individuals; an individual does not have heritability. The heritability of a characteristic varies among populations and among environments. Even if the heritability for a characteristic is high, the characteristic may still be altered by changes in the environment. Heritabilities provide no information about the nature of differences among populations in a characteristic.

■ Quantitative trait loci (QTL) are chromosomal regions containing genes that influence polygenic characteristics. QTLs can be mapped by examining the association between the inheritance of a quantitative characteristic and the inheritance of genetic markers. Genes influencing quantitative traits can also be located with the use of genome-wide association studies.

■ The amount of change in a quantitative characteristic in a single generation when subjected to selection (the response to selection) is directly related to the selection differential and narrow-sense heritability.

■ A genetic correlation may be present when the same gene affects two or more characteristics (pleiotropy). Genetic correlations produce correlated responses to selection.

IMPORTANT TERMS

quantitative genetics (p. 715)
meristic characteristic (p. 718)
threshold characteristic (p. 718)
frequency distribution (p. 721)
normal distribution (p. 722)
population (p. 722)
sample (p. 722)
mean (p. 723)
variance (p. 723)
standard deviation (p. 723)
correlation (p. 724)
correlation coefficient (p. 724)
regression (p. 725)
regression coefficient (p. 726)
heritability (p. 728)
phenotypic variance (p. 729)
genetic variance (p. 729)
environmental variance (p. 729)
genetic–environmental interaction variance (p. 729)
additive genetic variance (p. 729)

dominance genetic	narrow-sense	response to	phenotypic
variance (p. 729)	heritability (p. 730)	selection (p. 737)	correlation (p. 739)
gene interaction	quantitative trait locus	selection	genetic
variance (p. 729)	(QTL) (p. 734)	differential (p. 737)	correlation (p. 739)
broad-sense	natural selection (p. 736)	realized	
heritability (p. 730)	artificial selection (p. 737)	heritability (p. 738)	

ANSWERS TO CONCEPT CHECKS

1. Cross two individuals that are each homozygous for different genes affecting the traits and then intercross the resulting F_1 progeny to produce the F_2. Determine what proportion of the F_2 progeny resemble one of the original homozygotes in the P generation. This proportion should be $(1/4)^n$, where n equals the number of loci with segregating pairs of alleles that affect the characteristic.

2. d

3. d

4. b

5. a

6. It indicates that about 40% of the differences in blood pressure among African Americans in Detroit are due to additive genetic differences. It neither provides information about the heritability of blood pressure in other groups of people nor indicates anything about the nature of differences in blood pressure between African Americans in Detroit and people in other groups.

7. 0.2

8. The percentage of butterfat will decrease.

WORKED PROBLEMS

Problem 1

Seed weight in a particular plant species is determined by pairs of alleles at two loci (a^+a^- and b^+b^-) that are additive and equal in their effects. Plants with genotype $a^-a^-\ b^-b^-$ have seeds that average 1 g in weight, whereas plants with genotype $a^+a^+\ b^+b^+$ have seeds that average 3.4 g in weight. A plant with genotype $a^-a^-\ b^-b^-$ is crossed with a plant of genotype $a^+a^+\ b^+b^+$.

a. What is the predicted weight of seeds from the F_1 progeny of this cross?

b. If the F_1 plants are intercrossed, what are the expected seed weights and proportions of the F_2 plants?

›› Solution Strategy

What information is required in your answer to the problem?

- The predicted weight of seeds from the F_1 progeny.
- The expected seed weights and their proportions among the F_2 progeny.

What information is provided to solve the problem?

- Seed weight is determined by pairs of alleles at two loci (a^+a^- and b^+b^-).
- The alleles are additive and equal in their effects.
- Plants with genotype $a^-a^-\ b^-b^-$ have seeds that average 1 g in weight.
- Plants with genotype $a^+a^+\ b^+b^+$ have seeds that average 3.4 g in weight.
- A plant with genotype $a^-a^-\ b^-b^-$ is crossed with a plant of genotype $a^+a^+\ b^+b^+$.

For help with this problem, review:
Polygenic Inheritance in Section 24.1.

›› Solution Steps

The difference in average seed weight between the two parental genotypes is $3.4\ \text{g} - 1\ \text{g} = 2.4\ \text{g}$. These two genotypes differ in four genes, so each gene difference contributes an additional $2.4\ \text{g}/4 = 0.6\ \text{g}$ of weight to the 1-g seed weight of a plant ($a^-a^-\ b^-b^-$) that has none of the contributing genes.

> **Hint:** Because the alleles are equal and additive, each allele contributes the same amount to seed weight.

The cross between the two homozygous genotypes produces the F_1 and F_2 progeny shown on the next page.

a. The F_1 are heterozygous at both loci ($a^+a^-\ b^+b^-$) and possess two genes that contribute an additional 0.6 g each to the 1-g seed weight of a plant that has no contributing genes. Therefore, the seeds of the F_1 should average $1\ \text{g} + 2(0.6\ \text{g}) = 2.2\ \text{g}$.

b. The F_2 have the following phenotypes and proportions: $1/16$ 1 g; $4/16$ 1.6 g; $6/16$ 2.2 g; $4/16$ 2.8 g; and $1/16$ 3.4 g.

P $\quad a^- a^- b^- b^- \quad \times \quad a^+ a^+ b^+ b^+$

$\qquad$ 1 g $\qquad\qquad$ 3.4 g

F$_1$ $\quad a^+ a^- b^+ b^-$

$\qquad$ 2.2 g

> **Recall:** The probability of each two-locus genotype can be determined by multiplying the probability of the single-locus genotypes.

	Genotype	Probability	Number of contributing genes	Average seed weight
	$a^- a^- b^- b^-$	$\frac{1}{4} \times \frac{1}{4} = \frac{1}{16}$	0	$1\,g + (0 \times 0.6\,g) = 1\,g$
	$a^+ a^- b^- b^-$	$\frac{1}{2} \times \frac{1}{4} = \frac{1}{8}$ $\Big\} \frac{2}{8} = \frac{4}{16}$	1	$1\,g + (1 \times 0.6\,g) = 1.6\,g$
	$a^- a^- b^+ b^-$	$\frac{1}{4} \times \frac{1}{2} = \frac{1}{8}$		
F$_2$	$a^+ a^+ b^- b^-$	$\frac{1}{4} \times \frac{1}{4} = \frac{1}{16}$		
	$a^- a^- b^+ b^+$	$\frac{1}{4} \times \frac{1}{4} = \frac{1}{16}$ $\Big\} \frac{2}{16} + \frac{1}{4} + = \frac{6}{16}$	2	$1\,g + (2 \times 0.6\,g) = 2.2\,g$
	$a^+ a^- b^+ b^-$	$\frac{1}{2} \times \frac{1}{2} = \frac{1}{4}$		
	$a^+ a^+ b^+ b^-$	$\frac{1}{4} \times \frac{1}{2} = \frac{1}{8}$ $\Big\} \frac{2}{8} = \frac{4}{16}$	3	$1\,g + (3 \times 0.6\,g) = 2.8\,g$
	$a^+ a^- b^+ b^+$	$\frac{1}{4} \times \frac{1}{2} = \frac{1}{8}$		
	$a^+ a^+ b^+ b^+$	$\frac{1}{4} \times \frac{1}{4} = \frac{1}{16}$	4	$1\,g + (4 \times 0.6\,g) = 3.4\,g$

Problem 2

A farmer is raising rabbits. The average body weight in his population of rabbits is 3 kg. The farmer selects the 10 largest rabbits in his population, whose average body weight is 4 kg, and interbreeds them. If the narrow-sense heritability of body weight in the rabbit population is 0.7, what is the expected body weight among the offspring of the selected rabbits?

›› Solution Strategy

What information is required in your answer to the problem?

The expected body weight of the offspring of the selected rabbits.

What information is provided to solve the problem?

- The average body weight in the population is 3 kg.
- The average body weight of selected rabbits is 4 kg.
- The narrow-sense heritability of body weight is 0.7.

For help with this problem, review:

Predicting the Response to Selection in Section 24.4.

›› Solution Steps

The farmer has carried out a response-to-selection experiment. The selection differential equals the difference in average weight between the selected rabbits and the entire population: $4\,kg - 3\,kg = 1\,kg$. The narrow-sense heritability is given as 0.7, so the expected response to selection is $R = h^2 \times S = 0.7 \times 1\,kg = 0.7\,kg$. This value is the increase in weight that is expected in the offspring of the selected parents, so the average weight of the offspring is expected to be $3\,kg + 0.7\,kg = 3.7\,kg$.

> **Recall:** The response to selection equals the selection differential multiplied by narrow-sense heritability.

COMPREHENSION QUESTIONS

Section 24.1

1. How does a quantitative characteristic differ from a discontinuous characteristic?

2. Briefly explain why the relation between genotype and phenotype is frequently complex for quantitative characteristics.

3. Why do polygenic characteristics have many phenotypes?

Section 24.2

4. Explain the relation between a population and a sample. What characteristics should a sample have to be representative of the population?

5. What information do the mean and variance provide about a distribution?

6. How is the standard deviation related to the variance?

7. What information does the correlation coefficient provide about the association between two variables?

8. What is regression? How is it used?

Section 24.3

9. List all the components that contribute to the phenotypic variance and define each component.

10. How do broad-sense and narrow-sense heritabilities differ?

11. Briefly outline some of the ways in which heritability can be calculated.

12. Briefly describe common misunderstandings or misapplications of the concept of heritability.

13. Briefly explain how genes affecting a polygenic characteristic are located with the use of QTL mapping.

Section 24.4

14. How is the response to selection related to narrow-sense heritability and the selection differential? What information does the response to selection provide?

15. Why does the response to selection often level off after many generations of selection?

APPLICATION QUESTIONS AND PROBLEMS

Section 24.1

*16. For each of the following characteristics, indicate whether it would be considered a discontinuous characteristic or a quantitative characteristic. Briefly justify your answer.

a. Kernel color in a strain of wheat, in which two codominant alleles segregating at a single locus determine the color. Thus, there are three phenotypes present in this strain: white, light red, and medium red.

b. Body weight in a family of Labrador retrievers. An autosomal recessive allele that causes dwarfism is present in this family. Two phenotypes are recognized: dwarf (less than 13 kg) and normal (greater than 23 kg).

c. Presence or absence of leprosy. Susceptibility to leprosy is determined by multiple genes and numerous environmental factors.

d. Number of toes in guinea pigs, which is influenced by genes at many loci.

e. Number of fingers in humans. Extra (more than five) fingers are caused by the presence of an autosomal dominant allele.

*17. Assume that plant weight is determined by a pair of alleles at each of two independently assorting loci (A and a, B and b) that are additive in their effects. Further assume that each allele represented by an uppercase letter contributes 4 g to weight and that each allele represented by a lowercase letter contributes 1 g to weight.

a. If a plant with genotype $AA\ BB$ is crossed with a plant with genotype $aa\ bb$, what weights are expected in the F_1 progeny?

b. What is the distribution of weight expected in the F_2 progeny?

18. Assume that three loci, each with two alleles (A and a, B and b, C and c), determine the difference in height between two homozygous strains of a plant. These genes are additive and equal in their effects on plant height. One strain ($aa\ bb\ cc$) is 10 cm in height. The other strain ($AA\ BB\ CC$) is 22 cm in height. The two strains are crossed, and the resulting F_1 are interbred to produce F_2 progeny. Give the phenotypes and the expected proportions of the F_2 progeny.

*19. A farmer has two homozygous varieties of tomatoes. One variety, called Little Pete, has fruits that average only 2 cm in diameter. The other variety, Big Boy, has fruits that average a whopping 14 cm in diameter. The farmer crosses Little Pete and Big Boy; he then intercrosses the F_1 to produce F_2 progeny. He grows 2000 F_2 tomato plants and doesn't find any F_2 offspring that produce fruits as small as Little Pete or as large as Big Boy. If we assume that the difference between these varieties in fruit size is produced by genes with equal and additive effects, what can we conclude about the minimum number of loci with pairs of alleles determining the difference in fruit size between the two varieties?

20. Seed size in a plant is a polygenic characteristic. A grower crosses two pure-breeding varieties of the plant and measures seed size in the F_1 progeny. She then backcrosses the F_1 plants to one of the parental varieties and measures seed size in the backcross progeny. The grower finds that seed size in the backcross progeny has a higher variance than does seed size in the F_1 progeny. Explain why the backcross progeny are more variable.

Section 24.2

21. The following data are the numbers of digits per foot in 25 guinea pigs. Construct a frequency distribution for these data.

4, 4, 4, 5, 3, 4, 3, 4, 4, 5, 4, 4, 3, 2, 4, 4, 5, 6, 4, 4, 3, 4, 4, 4, 5

*22. Ten male Harvard students were weighed in 1916. Their weights are given here in kilograms. Calculate the mean, variance, and standard deviation for these weights.

51, 69, 69, 57, 61, 57, 75, 105, 69, 63

23. Among a population of tadpoles, the correlation coefficient for size at metamorphosis and time required for metamorphosis is −0.74. On the basis of this correlation, what conclusions can you draw about the relative sizes of tadpoles that metamorphose quickly and those that metamorphose more slowly?

*24. Body weight and length were measured on six mosquito fish; these measurements are given in the following table. Calculate the correlation coefficient for weight and length in these fish.

[A. Hartl/Age Fotostock America, Inc.]

Wet weight (g)	Length (mm)
115	18
130	19
210	22
110	17
140	20
185	21

25. The heights of mothers and daughters are given in the following table:

Height of mother (in)	Height of daughter (in)
64	66
65	66
66	68
64	65
63	65
63	62
59	62
62	64
61	63
60	62

   a. Calculate the correlation coefficient for the heights of the mothers and daughters.

   b. Using regression, predict the expected height of a daughter whose mother is 67 inches tall.

Section 24.3

*26. Phenotypic variation in the tail length of mice has the following components:

Additive genetic variance (V_A)	= 0.5
Dominance genetic variance (V_D)	= 0.3
Gene interaction variance (V_I)	= 0.1
Environmental variance (V_E)	= 0.4
Genetic–environmental interaction variance (V_{GE})	= 0.0

   a. What is the narrow-sense heritability of tail length?

   b. What is the broad-sense heritability of tail length?

27. The narrow-sense heritability of ear length in Reno rabbits is 0.4. The phenotypic variance (V_P) is 0.8, and the environmental variance (V_E) is 0.2. What is the additive genetic variance (V_A) for ear length in these rabbits?

28. Assume that human ear length is influenced by multiple genetic and environmental factors. Suppose you measure ear length in three groups of people, in which group A consists of five unrelated people, group B consists of five siblings, and group C consists of five first cousins.

   a. With the assumption that the environments of all three groups are similar, which group should have the highest phenotypic variance? Explain why.

   b. Is it realistic to assume that the environmental variance for each group is similar? Explain your answer.

29. A characteristic has a narrow-sense heritability of 0.6.

   a. If the dominance variance (V_D) increases and all other variance components remain the same, what will happen to narrow-sense heritability? Will it increase, decrease, or remain the same? Explain.

   b. What will happen to broad-sense heritability? Explain.

   c. If the environmental variance (V_E) increases and all other variance components remain the same, what will happen to narrow-sense heritability? Explain.

   d. What will happen to broad-sense heritability? Explain.

30. Flower color in the varieties of pea plants studied by Mendel is controlled by alleles at a single locus. A group of peas homozygous for purple flowers is grown. Careful study of the plants reveals that all their flowers are purple, but there is some variation in the intensity of the purple color. What would the estimated heritability be for this variation in flower color? Explain your answer.

*31. A graduate student is studying a population of bluebonnets along a roadside. The plants in this population are genetically variable. She counts the seeds produced by each of 100 plants and measures the mean and variance of seed number. The variance is 20. Selecting one plant, the student takes cuttings from it and cultivates them in a greenhouse, eventually producing many genetically identical clones of the same plant. She then transplants these clones into the roadside population, allows them to grow for one year, and then counts the seeds produced by each of the cloned plants. The student finds that the variance in seed number among these cloned plants is 5. From the phenotypic variances of the genetically variable and the genetically identical plants, she calculates the broad-sense heritability.

[Purestock/Getty Images.]

   a. What is the broad-sense heritability of seed number for the roadside population of bluebonnets?

   b. What might cause this estimate of heritability to be inaccurate?

32. Many researchers have estimated the heritability of human traits by comparing the correlation coefficients of monozygotic and dizygotic twins (see pp. 731–732). One of the assumptions made in using this method is that monozygotic twin pairs experience environments that are no more similar to each other than those experienced by dizygotic twin pairs. How might this assumption be violated? Give some specific examples of how the environments of two monozygotic twins might be more similar than the environments of two dizygotic twins.

33. What conclusion can you draw from **Figure 24.18** about the proportion of phenotypic variation in shell breadth that is due to genetic differences? Explain your reasoning.

*34. A genetics researcher determines that the broad-sense heritability of height among Southwestern University undergraduate students is 0.90. Which of the following conclusions would be reasonable? Explain your answer.

a. Sally is a Southwestern University undergraduate student, so 10% of her height is determined by nongenetic factors.

b. Ninety percent of variation in height among all undergraduate students in the United States is due to genetic differences.

c. Ninety percent of the height of Southwestern University undergraduate students is determined by genes.

d. Ten percent of the variation in height among Southwestern University undergraduate students is determined by variation in nongenetic factors.

e. Because the heritability of height among Southwestern University students is so high, any change in the students' environment will have minimal effect on their height.

*35. The length of the middle joint of the right index finger was measured in 10 sets of parents and their adult offspring. The mean parental lengths and the mean offspring lengths for each family are listed in the accompanying table. Calculate the regression coefficient for regression of mean offspring length against mean parental length and estimate the narrow-sense heritability for this characteristic.

Mean parental length (mm)	Mean offspring length (mm)
30	31
35	36
28	31
33	35
26	27
32	30
31	34
29	28
40	38
33	34

36. Assume that in **Figure 24.14**, x equals the mean phenotype of the parents and y equals the mean phenotype of the offspring. Which line represents the highest heritability? Explain your answer.

37. *Drosophila buzzatii* is a fruit fly that feeds on the rotting fruits of cacti in Australia. Timothy Prout and Stuart Barker calculated the heritabilities of body size, as measured by thorax length, for a natural population of *D. buzzatii* raised in the wild and for a population of *D. buzzatii* collected in the wild but raised in the laboratory (T. Prout and J. S. F. Barker. 1989. *Genetics* 123:803–813). They found the following heritabilities:

Population	Heritability of body size ($\pm$ standard error)
Wild population	0.0595 ± 0.0123
Laboratory-reared population	0.3770 ± 0.0203

Why do you think that the heritability measured in the laboratory-reared population is higher than that measured in the natural population raised in the wild?

38. Mr. Jones is a pig farmer. For many years, he has fed his pigs the food left over from the local university cafeteria, which is known to be low in protein, deficient in vitamins, and downright untasty. However, the food is free, and his pigs don't complain. One day a salesman from a feed company visits Mr. Jones. The salesman claims that his company sells a new, high-protein, vitamin-enriched feed that enhances weight gain in pigs. Although the feed is expensive, the salesman claims that the increased weight gain of the pigs will more than pay for the cost of the feed, increasing Mr. Jones's profit. Mr. Jones responds that he took a genetics class at the university and that he has conducted some genetic experiments on his pigs; specifically, he has calculated the narrow-sense heritability of weight gain for his pigs and found it to be 0.98. Mr. Jones says that this heritability value indicates that 98% of the variance in weight gain among his pigs is determined by genetic differences, and therefore the new pig feed can have little effect on the growth of his pigs. He concludes that the feed would be a waste of his money. The salesman doesn't dispute Mr. Jones's heritability estimate, but he still claims that the new feed can significantly increase weight gain in Mr. Jones's pigs. Who is correct and why?

Section 24.4

*39. Joe is breeding cockroaches in his dorm room. He finds that the average wing length in his population of cockroaches is 4 cm. He chooses the six cockroaches that have the largest wings; the average wing length among these selected cockroaches is 10 cm. Joe interbreeds these selected cockroaches. From earlier studies, he knows that the narrow-sense heritability for wing length in his population of cockroaches is 0.6.

a. Calculate the selection differential and expected response to selection for wing length in these cockroaches.

b. What should be the average wing length of the progeny of the selected cockroaches?

40. Three characteristics in beef cattle—body weight, fat content, and tenderness—are measured, and the following variance components are estimated:

	Body weight	Fat content	Tenderness
V_A	22	45	12
V_D	10	25	5
V_I	3	8	2
V_E	42	64	8
V_{GE}	0	0	1

In this population, which characteristic would respond best to selection? Explain your reasoning.

41. A rancher determines that the average amount of wool produced by a sheep in her flock is 22 kg per year. In an attempt to increase the wool production of her flock, the rancher picks the five male and five female sheep that produce the most wool; the average amount of wool produced per sheep by those selected sheep is 30 kg. She interbreeds these selected sheep and finds that the average wool production among their progeny is 28 kg. What is the narrow-sense heritability for wool production among the sheep in the rancher's flock?

42. A strawberry farmer determines that the average weight of individual strawberries produced by plants in his garden is 2 g. He selects the 10 plants that produce the largest strawberries; the average weight of strawberries produced by these selected plants is 6 g. He interbreeds these selected plants. The progeny of these selected plants produce strawberries that weigh an average of 5 g. If the farmer were to select plants that produce strawberries with an average weight of 4 g, what would be the predicted weight of strawberries produced by the progeny of those selected plants?

43. Has the response to selection leveled off in the strain of corn selected for high oil content shown in **Figure 24.22**? What does this observation suggest about genetic variation in the strain selected for high oil content?

***44.** The narrow-sense heritability of wing length in a population of *Drosophila melanogaster* is 0.8. The narrow-sense heritability of head width in the same population is 0.9. The genetic correlation between wing length and head width is −0.86. If a geneticist selects for increased wing length in these flies, what will happen to head width?

45. Pigs have been domesticated from wild boars. Would you expect to find higher heritability for weight among domesticated pigs or wild boars? Explain your answer.

CHALLENGE QUESTIONS

Section 24.1

46. Bipolar disorder is a psychiatric illness with a strong hereditary basis, but the exact mode of its inheritance is not known. Research has shown that siblings of patients with bipolar disorder are more likely to develop the disorder than are siblings of unaffected people. Findings from one study demonstrated that the ratio of bipolar brothers to bipolar sisters is higher when the patient is male than when the patient is female. In other words, relatively more brothers of patients with bipolar disorder also have the disease when the patient is male than when the patient is female. What does this observation suggest about the inheritance of bipolar disorder?

Section 24.3

47. We have explored some of the difficulties in separating the genetic and environmental components of human behavioral characteristics. Considering these difficulties and what you know about calculating heritability, propose an experimental design for accurately measuring the heritability of musical ability.

Section 24.4

48. Eugene Eisen selected for increased 12-day litter weight (total weight of a litter of offspring 12 days after birth) in a population of mice (E. J. Eisen. 1972. *Genetics* 72:129–142). The 12-day litter weight of the population steadily increased, but then leveled off after about 17 generations. At generation 17, Eisen took one family of mice from the selected population and reversed the selection procedure: in this group, he selected for *decreased* 12-day litter weight. This group immediately responded to the reversed selection: the 12-day litter weight dropped 4.8 g within 1 generation and dropped 7.3 g after 5 generations. On the basis of the results of the reverse selection, what is the most likely explanation for the leveling off of 12-day litter weight in the original population?

[J & C Sohns/Age Fotostock America, Inc.]

THINK-PAIR-SHARE QUESTIONS

Section 24.1

1. To locate genes associated with quantitative traits, geneticists often use QTL mapping (see the introduction to this chapter). To carry out QTL mapping, a first step is crossing two strains that differ in a quantitative trait, such as a strain of corn with high oil content and a strain with low oil content. The F_1 progeny of this cross are then interbred or backcrossed to produce an F_2 generation. Researchers then look for statistical associations between genetic markers and the value of the quantitative trait (e.g., oil content) in the F_2 generation. Why do the geneticists look for statistical associations in the F_2 plants? Why not use the F_1 progeny?

Section 24.2

2. What is the value of a random sample? Will random sampling always ensure that a sample is representative of a population? Explain your answer.

3. A researcher studying alcohol consumption in North American cities finds a significant correlation coefficient of 0.85 between the number of preachers in a city and total alcohol consumption for the city. Is it reasonable for the researcher to conclude that the preachers are consuming most of the alcohol? Why or why not? If not, how would you account for the significant correlation?

Section 24.3

4. A student who has just learned about quantitative genetics says, "Heritability estimates are worthless! They don't tell you anything about the genes that affect a characteristic. They don't provide any information about the types of offspring to expect from a cross. Heritability estimates measured in one population can't be used for other populations, so they don't even give you any general information about how much of a characteristic is genetically determined. Heritabilities don't do anything but make undergraduate students sweat during tests." How would you respond to this statement? Is the student correct? What good are heritabilities, and why do geneticists bother to calculate them?

5. Do you think genetics influences a person's knowledge of the game of football? Explain your answer.

Section 24.4

6. We have learned that the response to selection is equal to the selection differential times the narrow-sense heritability, and that the narrow-sense heritability includes only the additive genetic variance. Why aren't the dominance genetic variance and the gene interaction variance included? Why don't they contribute to the genetic variation that is acted on by selection?

7. A geneticist selects for increased body weight in a population of fruit flies that she is raising in her laboratory. She measures body weight in her population and selects the five heaviest males and the five heaviest females and uses them as the parents for the next generation. From the progeny produced by these parents, she selects the five heaviest males and five heaviest females and mates them. She repeats this procedure each generation. The average body weight of flies in the original population was 1.1 mg. The flies respond to selection, and their body weight steadily increases. After 20 generations of selection, the average body weight is 2.3 mg. However, after about 20 generations, the response to selection in subsequent generations levels off, and the average body weight of the flies no longer increases. At this point, the geneticist takes a long vacation; while she is gone, the fruit flies in her population interbreed randomly. When she returns from vacation, she finds that the average body weight of the flies in the population has decreased to 2.0 mg.

   Provide an explanation for why the response to selection leveled off after 20 generations. Why did the average size of the fruit flies decrease when selection was no longer applied during the geneticist's vacation?

Population Genetics

Wolves on Isle Royale have been studied for over 50 years. Inbreeding and low genetic variation has led to the population's decline. [PA Images/Alamy.]

The Wolves of Isle Royale

Rugged, isolated, and pristine, Isle Royale is surrounded by the deep, cold waters of northern Lake Superior. The island is a United States national park and wilderness area, covering only a little over 500 square kilometers of boreal forests and wetlands. Though small and isolated, Isle Royale is famous for its wolves and is the home of the longest-running study of predator–prey dynamics.

Moose first made it to Isle Royale in the early 1900s, most likely by swimming the 24 km of water that separates the island from the mainland. Because these moose had no natural predators, their numbers on Isle Royale mushroomed; the island's forest subsequently declined, as moose decimated the birch and aspen trees by browsing. But this situation changed in 1949, when a single breeding pair of wolves trekked across the frozen lake in winter and set up residence on the island. The wolves preyed on the moose and kept their population in check. With fewer moose, the island's trees rebounded. When biologists initiated their study of moose–wolf interaction in 1959, there were 20 wolves and about 500 moose on the island.

Over the next 50 years, the numbers of wolves and moose on Isle Royale waxed and waned. In 1980, the wolf population reached an all-time high of 50 animals. But the population crashed shortly afterward due to an epidemic of canine parvovirus introduced by a visiting domestic dog. After the epidemic, wolf numbers climbed from 14 to about 20, but remained low throughout the 1980s and early 1990s. The wolf population suffered from inbreeding and low genetic variation. As we will see in this chapter, inbreeding and the effects of chance factors operating in small populations (called genetic drift) lead to a loss of genetic variation, which often results in low fertility, birth defects, and lowered resistance to disease.

Then, in 1997, a remarkable event occurred. A lone male wolf, later dubbed Old Gray Guy for his light-colored fur, migrated across the winter ice to the island. He joined the Isle Royale wolves and introduced fresh genes that invigorated the population. Old Gray Guy was a prolific breeder. He eventually produced 34 offspring, and his genes came to dominate the population—by 2008, 59.4% of the wolf genes on the island were his. The population climbed to over 30 wolves, and moose numbers declined. Old Gray died in 2006. Unfortunately, he had introduced a limited number of new genes. No more new wolves joined the population, and levels of inbreeding reached a new high (**Figure 25.1**). There was no reproduction in 2012, and by early 2013, only 8 wolves were left on the island. Their number dropped further in the following years—by 2016, only 2 wolves remained, a male and female pair. Genetic studies revealed that this pair was highly inbred: they were half siblings (born to the same mother) as well as father and daughter. Freed from wolf predation, the moose population boomed again.

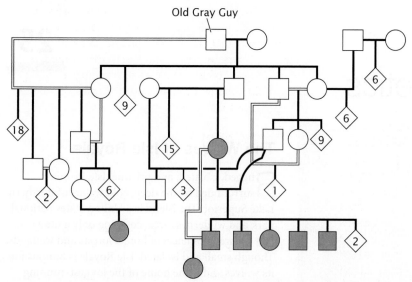

25.1 Pedigree of some of the wolves of Isle Royale. The shaded squares and circles represent wolves that were alive in 2012. Double lines represent matings between closely related individuals.

A factor complicating the future of wolves on Isle Royale is climate warming. Genetic studies suggest that Old Gray Guy was not the only wolf to migrate to the island since the population was established in 1949. Several wolves probably joined the population during its history. Migration is possible only during cold winters, when the lake waters freeze for several days to weeks at a time and wolves can walk from the mainland to the island. During the 1960s, ice bridges formed every few years, but more recently they have formed much less frequently; climatologists project that Lake Superior will be largely ice free by 2040, cutting off all future wolf migration.

As the number of wolves on Isle Royale became precariously low, some biologists proposed that additional wolves be transported to the island to counteract the effects of inbreeding. The introduction of new genetic variation into an inbred population, called **genetic rescue**, has been carried out for other inbred populations; it often dramatically improves the health of such populations and can better ensure their long-term survival. Others felt that nature should be left to take its course, even if it led to the extinction of the Isle Royale wolf population. In 2014, the National Park Service, which traditionally avoids human intervention in wilderness areas, decided against introducing new wolves to rescue the population. As of 2016, the National Park Service is debating the long-term fate of wolves on Isle Royale.

THINK-PAIR-SHARE

- Imagine that you are asked to manage a population of wolves that have taken up residence on an island off the coast of Alaska. Because prey resources are limited, the island will support only about 50 wolves at any one time. What steps would you take to prevent inbreeding and genetic drift in the population in the future?

- Do you think the National Park Service should have brought in new wolves to Isle Royale in an attempt to rescue the population, or do you think that it was correct in allowing natural processes to proceed? Provide some possible reasons for and against attempting genetic rescue.

The wolves of Isle Royale illustrate several important principles of genetics: Small populations lose genetic variation over time through inbreeding and genetic drift, often with catastrophic consequences for survival and reproduction. Migration, whether occurring naturally or through genetic rescue, introduces new genetic variation that counteracts the effects of genetic drift and inbreeding. These principles have important implications not only for wildlife management, but also for the evolution of organisms in the natural world.

This chapter introduces *population genetics*, the branch of genetics that studies the genetic makeup of groups of individuals and how a group's genetic composition changes over time. Population geneticists usually focus their attention on a **Mendelian population**, which is a group of interbreeding, sexually reproducing individuals that have a common set of genes—the **gene pool**. A population evolves through changes in its gene pool; therefore, population genetics is also the study of evolution. Population geneticists study the variation in alleles within and between groups and the evolutionary forces responsible for shaping the patterns of genetic variation found in nature. In this chapter, we learn how the gene pool of a population is described and what factors are responsible for shaping it.

25.1 Genotypic and Allelic Frequencies Are Used To Describe the Gene Pool of a Population

An obvious and pervasive feature of life is variation. Students in a typical college class vary in eye color, hair color, skin pigmentation, height, weight, facial features, blood type, and susceptibility to numerous diseases and disorders. No two students in the class are likely to be the same in appearance.

Humans are not unique in their extensive variation (**Figure 25.2a**); almost all organisms exhibit phenotypic variation. For instance, lady beetles are highly variable in their spotting patterns (**Figure 25.2b**), mice vary in color, snails have different numbers of stripes on their shells, and plants vary in their susceptibility to pests. Much of this phenotypic variation is hereditary. Recognition of the extent of phenotypic variation led Charles Darwin to the idea of evolution through natural selection. Genetic variation is the basis of all evolution, and the extent of genetic variation within a population affects its potential to adapt to environmental change.

In fact, even more genetic variation exists in populations than is visible in the phenotype. A great deal of variation exists at the molecular level owing, in part, to the redundancy of the genetic code, which allows different codons to specify the same amino acid. Thus, two members of a population can produce the same protein even if their DNA sequences are different. There is also variation in DNA sequences between genes and introns within genes, which do not encode proteins; some of this variation is thought to have little effect on the phenotype. Although this variation may not affect the phenotype, it is often useful for determining evolutionary relationships among organisms and understanding the evolutionary forces that have shaped a species.

Mathematical Models for Understanding Genetic Variation

An important but frequently misunderstood tool used for investigating the nature of genetic variation in populations is the mathematical model. Let's take a moment to consider what a model is and how it can be used. A mathematical model usually describes a process as an equation. Factors that may influence the process are represented by variables in the equation; the equation defines the way in which the variables influence the process. Most models are simplified representations of a process because simultaneous consideration of all the factors involved is impossible; some factors must be ignored in order to examine the effects of others. At first, a model might consider only one or a few factors, but after their effects are understood, the model can be improved by the addition of more details. Even a simple model, however, can be a source of valuable insight into how a process is influenced by key variables.

Before we can explore the evolutionary forces that shape genetic variation, we must be able to describe the genetic structure of a population. The usual way of describing this structure is to enumerate the types and frequencies of genotypes and alleles in a population. A frequency is simply a proportion or a percentage, usually expressed as a decimal fraction. For example, if 20% of the alleles at a particular locus in a population are *A*, we would say that the frequency of the *A* allele in the population is 0.20. For large populations, for which a determination of the genotypes of all individual members is impractical, a sample of the population is usually taken (see Section 24.2) and the genotypic and allelic frequencies are calculated for this sample. The genotypic and allelic frequencies of the sample are then used to represent the gene pool of the population.

Calculating Genotypic Frequencies

To calculate a **genotypic frequency**, we simply add up the number of individuals possessing a genotype and divide by the total number of individuals in the sample (*N*). For a locus with three genotypes, *AA*, *Aa*, and *aa*, the frequency (*f*) of each genotype is

(a)

(b)

25.2 All organisms exhibit genetic variation. (a) Extensive variation among humans. (b) Variation in the spotting patterns of Asian lady beetles. [Part a: Michael Dwyer/Alamy.]

$$f(AA) = \frac{\text{number of } AA \text{ individuals}}{N}$$

$$f(Aa) = \frac{\text{number of } Aa \text{ individuals}}{N} \qquad (25.1)$$

$$f(aa) = \frac{\text{number of } aa \text{ individuals}}{N}$$

The sum of all the genotypic frequencies always equals 1.

Calculating Allelic Frequencies

The gene pool of a population can also be described in terms of **allelic frequencies**. There are always fewer alleles than genotypes, so the gene pool of a population can be described in fewer terms when allelic frequencies are used. In a sexually reproducing population, the genotypes are only temporary assemblages of the alleles. As described by Mendel's principle of segregation, individual alleles, not genotypes, are passed from generation to generation through the gametes, and genotypes re-form from alleles in each generation. Thus, the types and numbers of alleles, rather than genotypes, have real continuity from one generation to the next and make up the gene pool of a population.

Allelic frequencies can be calculated from (1) the numbers or (2) the frequencies of the genotypes. To calculate the allelic frequency from the numbers of the genotypes, we count the number of copies of a particular allele present at a locus in a sample and divide by the total number of all alleles in the sample:

$$\text{frequency of an allele} = \frac{\text{number of copies of the allele}}{\text{number of copies of all alleles at the locus}} \quad (25.2)$$

For a locus with only two alleles (A and a), the frequencies of the alleles are usually represented by the symbols p and q. The frequencies can be calculated as follows:

$$p = f(A) = \frac{2n_{AA} + n_{Aa}}{2N}$$

$$q = f(a) = \frac{2n_{aa} + n_{Aa}}{2N} \quad (25.3)$$

where n_{AA}, n_{Aa}, and n_{aa} represent the numbers of AA, Aa, and aa individuals, and N represents the total number of individuals in the sample. To obtain the number of copies of the allele in the numerator of the equation, we add twice the number of homozygotes (because each has two copies of the allele for which the frequency is being calculated) to the number of heterozygotes (because each has a single copy of the allele). We divide by $2N$ because each diploid individual has two alleles at a locus. The sum of the allelic frequencies always equals 1 ($p + q = 1$); so, after p has been obtained, q can be determined by subtraction: $q = 1 - p$.

Alternatively, allelic frequencies can be calculated from the genotypic frequencies. This method is useful if the genotypic frequencies have already been calculated and the numbers of the different genotypes are not available. To calculate an allelic frequency from genotypic frequencies, we add the frequency of the homozygote for each allele to half the frequency of the heterozygote (because half of the heterozygote's alleles are of each type):

$$p = f(A) = f(AA) + \frac{1}{2}f(Aa)$$

$$q = f(a) = f(aa) + \frac{1}{2}f(Aa) \quad (25.4)$$

We obtain the same values of p and q whether we calculate the allelic frequencies from the numbers of the genotypes (Equation 25.3) or from the genotypic frequencies (Equation 25.4). A sample calculation of allelic frequencies is provided in the next Worked Problem. **▶ TRY PROBLEM 16**

LOCI WITH MULTIPLE ALLELES We can use the same principles to determine the frequencies of alleles for loci with more than two alleles. To calculate the allelic frequencies from the numbers of genotypes, we count up the number of copies of an allele by adding twice the number of homozygotes to the number of heterozygotes that possess the allele and divide this sum by twice the number of individuals in the sample. For a locus with three alleles (A^1, A^2, and A^3) and six genotypes (A^1A^1, A^1A^2, A^2A^2, A^1A^3, A^2A^3, and A^3A^3), the frequencies (p, q, and r) of the alleles are

$$p = f(A^1) = \frac{2n_{A^1A^1} + n_{A^1A^2} + n_{A^1A^3}}{2N}$$

$$q = f(A^2) = \frac{2n_{A^2A^2} + n_{A^1A^2} + n_{A^2A^3}}{2N} \quad (25.5)$$

$$r = f(A^3) = \frac{2n_{A^3A^3} + n_{A^1A^3} + n_{A^2A^3}}{2N}$$

Alternatively, we can calculate the frequencies of multiple alleles from the genotypic frequencies by extending Equation 25.4. Once again, we add the frequency of the homozygote to half the frequency of each heterozygous genotype that possesses the allele:

$$p = f(A^1A^1) + \frac{1}{2}f(A^1A^2) + \frac{1}{2}f(A^1A^3)$$

$$q = f(A^2A^2) + \frac{1}{2}f(A^1A^2) + \frac{1}{2}f(A^2A^3) \quad (25.6)$$

$$r = f(A^3A^3) + \frac{1}{2}f(A^1A^3) + \frac{1}{2}f(A^2A^3)$$

X-LINKED LOCI To calculate allelic frequencies for genes at X-linked loci, we apply these same principles. However, we must remember that a female possesses two X chromosomes and therefore has two X-linked alleles, whereas a male has only a single X chromosome and one X-linked allele.

Suppose there are two alleles at an X-linked locus, X^A and X^a. Females may be either homozygous (X^AX^A or X^aX^a) or heterozygous (X^AX^a). All males are hemizygous (X^AY or X^aY). To determine the frequency of the X^A allele (p), we first count the number of copies of X^A: we multiply the number of X^AX^A females by two and add the number of X^AX^a females and the number of X^AY males. We then divide the sum by the total number of alleles at the locus, which is twice the total number of females plus the number of males:

$$p = f(X^A) = \frac{2n_{X^AX^A} + n_{X^AX^a} + n_{X^AY}}{2n_{\text{females}} + n_{\text{males}}} \quad (25.7a)$$

Similarly, the frequency of the X^a allele is

$$q = f(X^a) = \frac{2n_{X^aX^a} + n_{X^AX^a} + n_{X^aY}}{2n_{\text{females}} + n_{\text{males}}} \quad (25.7b)$$

The frequencies of X-linked alleles can also be calculated from genotypic frequencies by adding the frequency of the females that are homozygous for the allele, half the frequency of the females that are heterozygous for the allele, and the frequency of males that are hemizygous for the allele:

$$p = f(X^A) = f(X^A X^A) + \tfrac{1}{2} f(X^A X^a) + f(X^A Y)$$
$$q = f(X^a) = f(X^a X^a) + \tfrac{1}{2} f(X^A X^a) + f(X^a Y)$$
(25.8)

If you remember the logic behind these calculations, you can determine allelic frequencies for any set of genotypes, and it will not be necessary to memorize all of the formulas.

> **TRY PROBLEM 18**

CONCEPTS

Population genetics concerns the genetic composition of a population and how it changes over time. The gene pool of a population can be described by the frequencies of genotypes and alleles in the population.

✔ CONCEPT CHECK 1

What are some advantages of using allelic frequencies to describe the gene pool of a population instead of using genotypic frequencies?

WORKED PROBLEM

The human MN blood-type antigens are determined by two codominant alleles, L^M and L^N (see p. 112 in Chapter 5). The MN blood types and corresponding genotypes of 398 Finns from Karjala are tabulated here.

Phenotype	Genotype	Number
M	$L^M L^M$	182
MN	$L^M L^N$	172
N	$L^N L^N$	44

Source: Data from W. C. Boyd, *Genetics and the Races of Man* (Boston: Little, Brown, 1950).

Calculate the genotypic and allelic frequencies at the MN locus for the Karjala population.

Solution Strategy

What information is required in your answer to the problem?
The genotypic and allelic frequencies of the population.

What information is provided to solve the problem?
The numbers of the different MN genotypes in the sample.

Solution Steps

The genotypic frequencies for the population are calculated with the following formula:

genotypic frequency

$$= \frac{\text{number of individuals with genotype}}{\text{total number of individuals in sample } (N)}$$

$$f(L^M L^M) = \frac{\text{number of } L^M L^M \text{ individuals}}{N} = \frac{182}{398} = 0.457$$

$$f(L^M L^N) = \frac{\text{number of } L^M L^M \text{ individuals}}{N} = \frac{172}{398} = 0.432$$

$$f(L^N L^N) = \frac{\text{number of } L^N L^N \text{ individuals}}{N} = \frac{44}{398} = 0.111$$

The genotypic frequencies sum to 1.0. The allelic frequencies can be calculated from either the numbers or the frequencies of the genotypes. To calculate allelic frequencies from numbers of genotypes, we count the number of copies of each allele and divide by the number of copies of all alleles at that locus:

$$\text{frequency of an allele} = \frac{\text{number of copies of the allele}}{\text{number of copies of all alleles}}$$

$$p = f(L^M) = \frac{(2n_{L^M L^M}) + (n_{L^M L^N})}{2N}$$

$$= \frac{2(182) + 172}{2(398)} = \frac{536}{796} = 0.673$$

$$q = f(L^N) = \frac{(2n_{L^N L^N}) + (n_{L^M L^N})}{2N}$$

$$= \frac{2(44) + 172}{2(398)} = \frac{260}{796} = 0.327$$

To calculate the allelic frequencies from genotypic frequencies, we add the frequency of the homozygote for that genotype to half the frequency of each heterozygote possessing that allele:

$$p = f(L^M) = f(L^M L^M) + \tfrac{1}{2} f(L^M L^N)$$
$$= 0.457 + \tfrac{1}{2}(0.432) = 0.673$$

$$q = f(L^N) = f(L^N L^N) + \tfrac{1}{2} f(L^M L^N)$$
$$= 0.111 + \tfrac{1}{2}(0.432) = 0.327$$

>> Now try your hand at calculating genotypic and allelic frequencies by working **Problem 17** at the end of the chapter.

25.2 The Hardy–Weinberg Law Describes the Effect of Reproduction on Genotypic and Allelic Frequencies

The primary goal of population genetics is to understand the processes that shape a population's gene pool. First, we must ask what effects reproduction and Mendelian principles have on the genotypic and allelic frequencies: How do the segregation of alleles in gamete formation and the combining of alleles in fertilization influence the gene pool? The answer to this question lies in the **Hardy–Weinberg law**, among the most important principles of population genetics.

The Hardy–Weinberg law was formulated independently by G. H. Hardy and Wilhelm Weinberg in 1908 (similar conclusions were reached by several other geneticists at about the same time). The law is actually a mathematical model that evaluates the effect of reproduction on the genotypic and allelic frequencies of a population. It makes several simplifying assumptions about the population and provides two key predictions if these assumptions are met. For an autosomal locus with two alleles, the Hardy–Weinberg law can be stated as follows:

> **Assumptions** If a population is large, randomly mating, and not affected by mutation, migration, or natural selection, then
>
> **Prediction 1** the allelic frequencies of a population do not change; and
>
> **Prediction 2** the genotypic frequencies stabilize (will not change) after one generation in the proportions p^2 (the frequency of AA), $2pq$ (the frequency of Aa), and q^2 (the frequency of aa), where p equals the frequency of allele A and q equals the frequency of allele a.

The Hardy–Weinberg law indicates that, when its assumptions are met, reproduction alone does not alter allelic or genotypic frequencies and the allelic frequencies determine the frequencies of genotypes.

The statement that the genotypic frequencies stabilize after one generation means that they may change after the first generation, because one generation of random mating is required to produce Hardy–Weinberg proportions of the genotypes. Afterward, the genotypic frequencies, like the allelic frequencies, do not change as long as the population continues to meet the assumptions of the Hardy–Weinberg law. When genotypes are in the expected proportions of p^2, $2pq$, and q^2, the population is said to be in **Hardy–Weinberg equilibrium**.

CONCEPTS

The Hardy–Weinberg law describes how reproduction and Mendelian principles affect the allelic and genotypic frequencies of a population.

✔ CONCEPT CHECK 2

Which of the following statements is not an assumption of the Hardy–Weinberg law?

a. The allelic frequencies (p and q) are equal.
b. The population is randomly mating.
c. The population is large.
d. Natural selection has no effect.

Genotypic Frequencies at Hardy–Weinberg Equilibrium

How do the assumptions of the Hardy–Weinberg law lead to genotypic proportions of p^2, $2pq$, and q^2? Mendel's principle of segregation says that each individual organism possesses two alleles at a locus and that each of those two alleles has an equal probability of passing into a gamete. Thus, the frequencies of alleles in gametes will be the same as the frequencies of alleles in the parents. Suppose that we have a Mendelian population in which the frequencies of alleles A and a are p and q, respectively. These frequencies will also be those in the gametes. If mating is random (one of the assumptions of the Hardy–Weinberg law), the gametes will come together in random combinations, which can be represented by a Punnett square (**Figure 25.3**).

The multiplication rule (see Section 3.2) can be used to determine the probability of various gametes pairing. For example, the probability of a sperm containing allele A is p and the probability of an egg containing allele A is p. Applying the multiplication rule, we find that the probability that these two gametes will combine to produce an AA homozygote is $p \times p = p^2$. Similarly, the probability of a sperm containing allele a combining with an egg containing allele a to produce an aa homozygote is $q \times q = q^2$. An Aa heterozygote can be produced in one of two ways: (1) a sperm containing allele A may combine with an egg containing allele a ($p \times q$) or (2) an egg containing allele A may combine with a sperm containing allele a ($p \times q$). Thus, the probability of alleles A and a combining to produce an Aa heterozygote is $2pq$. In summary, whenever the frequencies of alleles in a randomly mating population are p and q, the frequencies of the genotypes in the next generation will be p^2, $2pq$, and q^2. Figure 25.3 demonstrates that only a single generation of random mating is required to produce the Hardy–Weinberg genotypic proportions.

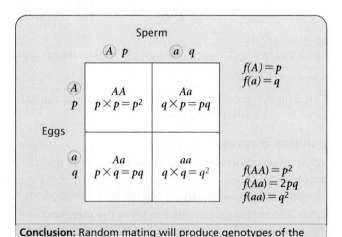

Conclusion: Random mating will produce genotypes of the next generation in proportions $p^2(AA)$, $2pq(Aa)$, and $q^2(aa)$.

25.3 Random mating produces genotypes in the proportions p^2, $2pq$, and q^2. Note that this square represents mating in a population, not an individual cross.

Closer Examination of the Hardy–Weinberg Law

Before we consider the implications of the Hardy–Weinberg law, we need to take a closer look at the three assumptions that it makes about a population. First, it assumes that the population is large. How big is "large"? Theoretically, the Hardy–Weinberg law requires that a population be infinitely large in size, but this requirement is obviously unrealistic. In practice, many large populations have genotypes in the predicted Hardy–Weinberg proportions, and significant deviations arise only when population size is rather small. In Section 25.4, we will examine the effects of small population size on allelic frequencies.

The second assumption of the Hardy–Weinberg law is that members of the population mate randomly with respect to genotype, which means that each genotype mates relative to its frequency. For example, suppose that three genotypes are present in a population in the following proportions: $f(AA) = 0.6, f(Aa) = 0.3$, and $f(aa) = 0.1$. With random mating, the frequency of mating between two AA homozygotes ($AA \times AA$) will be equal to the product of their frequencies: $0.6 \times 0.6 = 0.36$; whereas the frequency of mating between two aa homozygotes ($aa \times aa$) will be only $0.1 \times 0.1 = 0.01$.

The third assumption of the Hardy–Weinberg law is that the allelic frequencies of the population are not affected by natural selection, migration, or mutation. Although mutation occurs in every population, its rate is so low that it has little short-term effect on the predictions of the Hardy–Weinberg law (although it may shape allelic frequencies over long periods when no other forces are acting). Although natural selection and migration are significant factors in real populations, we must remember that the purpose of the Hardy–Weinberg law is to examine only the effect of reproduction on the gene pool. When this effect is known, the effects of other factors (such as migration and natural selection) can be examined.

A final point is that the assumptions of the Hardy–Weinberg law apply to a single locus. No real population mates randomly for all traits, and no population is completely free of natural selection for all traits. The Hardy–Weinberg law, however, does not require random mating and the absence of selection, migration, and mutation for all traits; it requires these conditions only for the locus under consideration. A population may be in Hardy–Weinberg equilibrium for one locus but not for others.

Implications of the Hardy–Weinberg Law

The Hardy–Weinberg law has several important implications for the genetic structure of a population. One implication is that a population cannot evolve if it meets the Hardy–Weinberg assumptions, because evolution consists of change in the allelic frequencies of a population. Therefore, the Hardy–Weinberg law tells us that reproduction alone will not bring about evolution. Other processes—such as mutation, migration, and natural selection—or chance events are required for populations to evolve.

A second important implication is that when a population is in Hardy–Weinberg equilibrium, the genotypic frequencies are determined by the allelic frequencies. When a population is not in Hardy–Weinberg equilibrium, we have no basis for predicting the genotypic frequencies. Although we can always determine the allelic frequencies from the genotypic frequencies (see Equation 25.3), the reverse (determining the genotypic frequencies from the allelic frequencies) is possible only when the population is in Hardy–Weinberg equilibrium.

For a locus with two alleles, the frequency of the heterozygote is greatest when allelic frequencies are between 0.33 and 0.66 and is at a maximum when allelic frequencies are each 0.5 (**Figure 25.4**). The heterozygote frequency never exceeds 0.5 when the population is in Hardy–Weinberg equilibrium. Furthermore, when the frequency of one allele is low, homozygotes for that allele will be rare, and most of the copies of a rare allele will be present in heterozygotes. As you can see from Figure 25.4, when the frequency of allele a is 0.2, the frequency of aa homozygotes is only 0.04 (q^2), but the frequency of Aa heterozygotes is 0.32 ($2pq$); 80% of the a alleles are in heterozygotes. Use **Animation 25.1** to examine the effect of allelic frequencies ⋯⋯ Ⓐ on genotypic frequencies when a population is in Hardy–Weinberg equilibrium.

A third implication of the Hardy–Weinberg law is that a single generation of random mating produces the equilibrium frequencies of p^2, $2pq$, and q^2. The fact that genotypes are in Hardy–Weinberg proportions does not prove that the population is free from natural selection, mutation, and migration. It means only that these forces have not acted since the last time random mating took place.

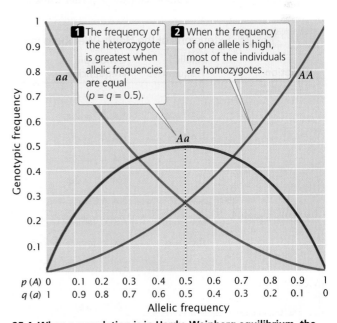

25.4 When a population is in Hardy–Weinberg equilibrium, the proportions of genotypes are determined by the frequencies of alleles.

A final implication is that when a population is not in Hardy–Weinberg equilibrium, one of the assumptions of the law has not been met, although without further investigation it will not be apparent which assumption has been violated. Finding that a population is not in equilibrium often leads to other studies to determine what evolutionary forces are acting on the population.

Extensions of the Hardy–Weinberg Law

The Hardy–Weinberg expected proportions can also be applied to multiple alleles and X-linked alleles (**Table 25.1**). With multiple alleles, the genotypic frequencies expected at equilibrium are the square of the allelic frequencies. For an autosomal locus with three alleles, the equilibrium genotypic frequencies are $(p + q + r)^2 = p^2 + 2pq + q^2 + 2pr + 2qr + r^2$. For an X-linked locus with two alleles, X^A and X^a, the equilibrium frequencies of the female genotypes are $(p + q)^2 = p^2 + 2pq + q^2$, where p^2 is the frequency of X^AX^A, $2pq$ is the frequency of X^AX^a, and q^2 is the frequency of X^aX^a. Males have only a single X-linked allele, so the frequencies of the male genotypes are p (frequency of X^AY) and q (frequency of X^aY). These proportions are those of the genotypes among males and among females, rather than the proportions among the entire population. Thus, p^2 is the expected proportion of females with the genotype X^AX^A; if females make up 50% of the population, then the expected proportion of this genotype in the entire population is $0.5 \times p^2$.

Notice that the frequency of an X-linked recessive trait among males is q, whereas the frequency among females is q^2. When an X-linked allele is uncommon, the trait will therefore be much more frequent in males than in females. Consider hemophilia A, a blood-clotting disorder caused by an X-linked recessive allele with a frequency (q) of approximately 1 in 10,000, or 0.0001. At Hardy–Weinberg equilibrium, this frequency will also be the frequency of the disease among males. The frequency of the disease among females, however, will be $q^2 = (0.0001)^2 = 0.00000001$, which is only 1 in 100 million. Hemophilia is 10,000 times more frequent in males than in females.

Testing for Hardy–Weinberg Proportions

To determine whether a population's genotypes are in Hardy–Weinberg equilibrium, the genotypic frequencies expected under the Hardy–Weinberg law must be compared with the observed genotypic frequencies. To do so, we first calculate the allelic frequencies, then find the expected genotypic frequencies by using the square of the allelic frequencies, and finally, compare the observed and expected genotypic frequencies by using a chi-square goodness-of-fit test.

WORKED PROBLEM

Jeffrey Mitton and his colleagues found three genotypes (R^2R^2, R^2R^3, and R^3R^3) at a locus encoding the enzyme peroxidase in ponderosa pine trees growing at Glacier Lake, Colorado. The observed numbers of these genotypes are given in the following table.

Genotypes	Number observed
R^2R^2	135
R^2R^3	44
R^3R^3	11

Are the ponderosa pine trees at Glacier Lake in Hardy–Weinberg equilibrium at the peroxidase locus?

Solution Strategy

What information is required in your answer to the problem?
The results of a chi-square test to determine whether the population is in Hardy–Weinberg equilibrium.

What information is provided to solve the problem?
The numbers of the different genotypes in a sample of the population.

Solution Steps

If the frequency of the R^2 allele equals p and the frequency of the R^3 allele equals q, the frequency of the R^2 allele is

$$p = f(R^2) = \frac{(2n_{R^2R^2}) + (n_{R^2R^3})}{2N} = \frac{2(135) + 44}{2(190)} = 0.826$$

The frequency of the R^3 allele is obtained by subtraction:

$$q = f(R^3) = 1 - p = 0.174$$

TABLE 25.1	Extensions of the Hardy–Weinberg law	
Situation	**Allelic Frequencies**	**Genotypic Frequencies**
Three alleles	$f(A^1) = p$	$f(A^1A^1) = p^2$
	$f(A^2) = q$	$f(A^1A^2) = 2pq$
	$f(A^3) = r$	$f(A^2A^2) = q^2$
		$f(A^1A^3) = 2pr$
		$f(A^2A^3) = 2qr$
		$f(A^3A^3) = r^2$
X-linked alleles	$f(X^1) = p$	$f(X^1X^1 \text{ female}) = p^2$
	$f(X^2) = q$	$f(X^1X^2 \text{ female}) = 2pq$
		$f(X^2X^2 \text{ female}) = q^2$
		$f(X^1Y \text{ male}) = p$
		$f(X^2Y \text{ male}) = q$

For X-linked female genotypes, the frequencies are the proportions among all females; for X-linked male genotypes, the frequencies are the proportions among all males.

The frequencies of the genotypes expected under Hardy–Weinberg equilibrium are then calculated by using p^2, $2pq$, and q^2:

$$R^2R^2 = p^2 = (0.826)^2 = 0.683$$
$$R^2R^3 = 2pq = 2(0.826)(0.174) = 0.287$$
$$R^3R^3 = q^2 = (0.174)^2 = 0.03$$

Multiplying each of these expected genotypic frequencies by the total number of observed genotypes in the sample (190), we obtain the numbers expected for each genotype:

$$R^2R^2 = 0.683 \times 190 = 129.8$$
$$R^2R^3 = 0.287 \times 190 = 54.5$$
$$R^3R^3 = 0.03 \times 190 = 5.7$$

By comparing these expected numbers with the observed numbers of each genotype, we see that there are more R^2R^2 and R^3R^3 homozygotes and fewer R^2R^3 heterozygotes in the population than we expect at equilibrium.

A chi-square goodness-of-fit test (see Section 3.4) is used to determine whether the differences between the observed and the expected numbers of each genotype are due to chance:

$$\chi^2 = \sum \frac{(\text{observed} - \text{expected})^2}{\text{expected}}$$

$$= \frac{(135 - 129.8)^2}{129.8} + \frac{(44 - 54.5)^2}{54.5} + \frac{(11 - 5.7)^2}{5.7}$$

$$= 0.21 + 2.02 + 4.93 = 7.16$$

The calculated chi-square value is 7.16. To obtain the probability associated with this chi-square value, we determine the appropriate degrees of freedom.

Up to this point, the chi-square test for assessing Hardy–Weinberg equilibrium has been identical with the chi-square tests that we used in Chapter 3 to assess progeny ratios in a genetic cross, in which the degrees of freedom were $n - 1$, and n equaled the number of expected genotypes. For the Hardy–Weinberg test, however, we must subtract an additional degree of freedom because the expected numbers are based on the observed allelic frequencies; therefore, the observed numbers are not completely free to vary. In general, the degrees of freedom for a chi-square test of Hardy–Weinberg equilibrium equal the number of expected genotypic classes minus the number of associated alleles. For this particular Hardy–Weinberg test, the degrees of freedom are $3 - 2 = 1$.

After we have calculated both the chi-square value and the degrees of freedom, the probability associated with this value can be sought in a chi-square table (see Table 3.7). With 1 degree of freedom, a chi-square value of 7.16 has a probability between 0.01 and 0.001. It is very unlikely that the peroxidase genotypes observed at Glacier Lake are in Hardy–Weinberg equilibrium.

> >> For additional practice, determine whether the genotypic frequencies in **Problem 22** at the end of the chapter are in Hardy–Weinberg equilibrium.

Estimating Allelic Frequencies with the Hardy–Weinberg Law

A practical use of the Hardy–Weinberg law is that it allows us to calculate allelic frequencies when dominance is present. For example, cystic fibrosis is a life-threatening autosomal recessive disorder characterized by frequent and severe respiratory infections, incomplete digestion, and abnormal sweating (see Section 5.1). Among North American Caucasians, the incidence of the disease is approximately 1 person in 2000. The formula for calculating allelic frequencies (see Equation 25.3) requires that we know the numbers of homozygotes and heterozygotes, but cystic fibrosis is a recessive disease, so we cannot easily distinguish between homozygous unaffected persons and heterozygous carriers. Although molecular tests are available for identifying heterozygous carriers of the cystic fibrosis gene, the low frequency of the disease makes widespread screening impractical. In such situations, the Hardy–Weinberg law can be used to estimate the allelic frequencies.

If we assume that a population is in Hardy–Weinberg equilibrium with regard to this locus, then the frequency of the recessive genotype (aa) is q^2, and the frequency of the recessive allele is the square root of that genotypic frequency:

$$q = \sqrt{f(aa)} \qquad (25.9)$$

If the frequency of cystic fibrosis in North American Caucasians is approximately 1 in 2000, or 0.0005, then $q = \sqrt{0.0005} = 0.02$. Thus, about 2% of the alleles in the Caucasian population encode the defective protein that causes cystic fibrosis. We can calculate the frequency of the normal allele by subtracting: $p = 1 - q = 1 - 0.02 = 0.98$. After we have calculated p and q, we can use the Hardy–Weinberg law to determine the frequencies of homozygous unaffected people and heterozygous carriers of the cystic fibrosis allele:

$$f(AA) = p^2 = (0.98)^2 = 0.960$$
$$f(Aa) = 2pq = 2(0.02)(0.98) = 0.0392$$

Thus, about 4% (1 of 25) of Caucasians are heterozygous carriers of the allele that causes cystic fibrosis. >> **TRY PROBLEM 25**

THINK-PAIR-SHARE Question 1

25.3 Nonrandom Mating Affects the Genotypic Frequencies of a Population

An assumption of the Hardy–Weinberg law is that mating is random with respect to genotype. Although it does not alter the frequencies of alleles, nonrandom mating affects the way in which alleles combine to form genotypes and alters the genotypic frequencies of a population.

We can distinguish between two types of nonrandom mating. **Positive assortative mating** refers to a tendency for like individuals to mate. For example, humans exhibit positive assortative mating for height: tall people mate preferentially with other tall people; short people mate preferentially with other short people. **Negative assortative mating** refers to a tendency for unlike individuals to mate. If people engaged in negative assortative mating for height, tall and short people would preferentially mate. Assortative mating is usually for a particular trait and affects only those genes that encode the trait (and genes closely linked to them).

THINK-PAIR-SHARE Question 2

One form of nonrandom mating is **inbreeding**, which is preferential mating between related individuals. Inbreeding is actually positive assortative mating for relatedness, but it differs from other types of assortative mating because it affects all genes, not just those that determine the trait for which the mating preference exists. Inbreeding causes a departure from the Hardy–Weinberg equilibrium frequencies of p^2, $2pq$, and q^2. More specifically, it leads to an increase in the proportion of homozygotes and a decrease in the proportion of heterozygotes in a population. **Outcrossing** is preferential mating between unrelated individuals.

In a diploid organism, a homozygous individual has two copies of the same allele. These two copies may be the same in *state*, which means that the two alleles are alike in structure and function but do not have a common origin. Alternatively, the two alleles in a homozygous individual may be the same because they are identical by *descent*; that is, the copies are descended from a single allele that was present in an ancestor (**Figure 25.5**). If we go back far enough in time, many alleles are likely to be identical by descent, but for calculating the effects of inbreeding, we consider identity by descent going back only a few generations.

Inbreeding is usually measured by the **inbreeding coefficient**, designated F, which is a measure of the probability that two alleles are identical by descent. Inbreeding coefficients can range from 0 to 1. A value of 0 indicates that mating is occurring randomly in a large population; a value of 1 indicates that all alleles are identical by descent. Inbreeding coefficients can be calculated from analyses of pedigrees, or they can be determined from the reduction in

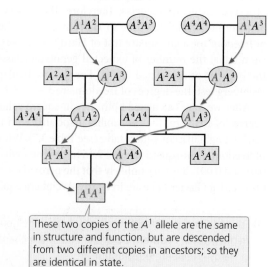

25.5 Individuals may be homozygous by descent or by state. Inbreeding is a measure of the probability that two alleles are identical by descent.

(a) Alleles identical by descent

These two copies of the A^1 allele are descended from the same copy in a common ancestor; so they are identical by descent.

(b) Alleles identical by state

These two copies of the A^1 allele are the same in structure and function, but are descended from two different copies in ancestors; so they are identical in state.

the heterozygosity of a population. Although we will not go into the details of how F is calculated, an understanding of how inbreeding affects genotypic frequencies is important.

When inbreeding takes place, the proportion of heterozygotes decreases by $2Fpq$, and half of this value (Fpq) is *added* to the proportion of each homozygote each generation. The frequencies of the genotypes will then be

$$f(AA) = p^2 + Fpq$$

$$f(Aa) = 2pq - 2Fpq \qquad (25.10)$$

$$f(aa) = q^2 + Fpq$$

Consider a population that reproduces by self-fertilization ($F = 1$). We will assume that this population begins with genotypic frequencies in Hardy–Weinberg proportions (p^2, $2pq$, and q^2). With selfing, each homozygote produces only progeny of the same homozygous genotype ($AA \times AA$ produces all AA; $aa \times aa$ produces all aa), whereas only half the progeny of a heterozygote have the same genotype as the parent ($Aa \times Aa$ produces $\frac{1}{4}\,AA$, $\frac{1}{2}\,Aa$, and $\frac{1}{4}\,aa$). Selfing therefore reduces the proportion of heterozygotes in the population by half with each generation, until all genotypes in the population are homozygous (**Table 25.2** and **Figure 25.6**). Although all individuals in the population are homozygous at this point, both types of homozygotes (AA and aa) are still present, provided neither homozygote has an advantage over the other (there are no selective differences). ▶ **TRY PROBLEM 28**

For most outcrossing species, close inbreeding is harmful because it increases the proportion of homozygotes and thereby boosts the probability that deleterious and lethal recessive alleles will combine to produce homozygotes with a harmful trait. Assume that a recessive allele (a) that causes a genetic disease has a frequency (q) of 0.01. If the population mates randomly ($F = 0$), the frequency of individuals affected by the disease (aa) will be $q^2 = 0.01^2 = 0.0001$; so only 1 in 10,000 individuals will have the disease. However, if $F = 0.25$ (the equivalent of brother–sister

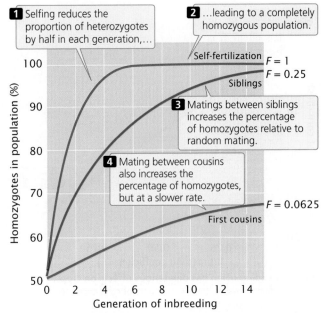

1 Selfing reduces the proportion of heterozygotes by half in each generation,...

2 ...leading to a completely homozygous population.

3 Matings between siblings increases the percentage of homozygotes relative to random mating.

4 Mating between cousins also increases the percentage of homozygotes, but at a slower rate.

Self-fertilization $F = 1$

$F = 0.25$

Siblings

First cousins $F = 0.0625$

25.6 Inbreeding increases the percentage of homozygous individuals in a population.

mating), then the expected frequency of the aa homozygote is $q^2 + Fpq = (0.01)^2 + (0.25)(0.99)(0.01) = 0.0026$; thus, the genetic disease is 26 times as frequent at this level of inbreeding. This increased appearance of lethal and deleterious traits with inbreeding is termed **inbreeding depression**; the more intense the inbreeding, the more severe the inbreeding depression. To see the effects of inbreeding on genotypic frequencies, view the Mini-Tutorial in **Animation 25.1**.

The harmful effects of inbreeding have been recognized for thousands of years and may be the basis of cultural taboos against mating between close relatives. William Schull and James Neel found that for each 10% increase in F, the mean IQ of Japanese children dropped six points. Child mortality also increases with close inbreeding (**Table 25.3**); children of first cousins have a 40% increase in mortality over that seen among the children of unrelated people. Similarly, studies conducted on over 11,000 babies born between 2007 and 2011 in Bradford, Great Britain, found that the risk of congenital anomalies was twice as high among children whose

TABLE 25.2	Generational increase in frequency of homozygotes in a self-fertilizing population starting with $p = q = 0.5$		
	Genotypic Frequencies		
Generation	AA	Aa	aa
1	$\frac{1}{4}$	$\frac{1}{2}$	$\frac{1}{4}$
2	$\frac{1}{4} + \frac{1}{8} = \frac{3}{8}$	$\frac{1}{4}$	$\frac{1}{4} + \frac{1}{8} = \frac{3}{8}$
3	$\frac{3}{8} + \frac{1}{16} = \frac{7}{16}$	$\frac{1}{8}$	$\frac{3}{8} + \frac{1}{16} = \frac{7}{16}$
4	$\frac{7}{16} + \frac{1}{32} = \frac{15}{32}$	$\frac{1}{16}$	$\frac{7}{16} + \frac{1}{32} = \frac{15}{32}$
n	$\dfrac{1 - (\frac{1}{2})^n}{2}$	$(\frac{1}{2})^n$	$\dfrac{1 - (\frac{1}{2})^n}{2}$
∞	$\frac{1}{2}$	0	$\frac{1}{2}$

TABLE 25.3	Effects of inbreeding on Japanese children	
Genetic Relationship of Parents	**F**	**Mortality of Children (through 12 years of age)**
Unrelated	0	0.082
Second cousins	0.016 ($\frac{1}{64}$)	0.108
First cousins	0.0625 ($\frac{1}{16}$)	0.114

Source: Data from D. L. Hartl and A. G. Clark, *Principles of Population Genetics*, 2d ed. (Sunderland, Mass.: Sinauer Associates, 1989), Table 2. Original data from W. J. Schull and J. V. Neel, *The Effects of Inbreeding on Japanese Children* (New York: Harper & Row, 1965).

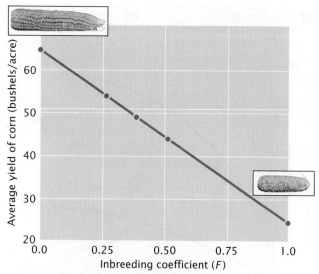

25.7 Inbreeding often has deleterious effects on crops.
As inbreeding increases, the average yield of corn decreases.

parents were first cousins than among those whose parents were unrelated. Inbreeding also has deleterious effects on crops (**Figure 25.7**) and domestic animals.

Inbreeding depression is most often studied in humans, as well as in plants and animals reared in captivity, but the negative effects of inbreeding may be more severe in natural populations. Julie Jimenez and her colleagues collected wild mice from a natural population in Illinois and bred them in the laboratory for three to four generations. Laboratory matings were chosen so that some mice had no inbreeding, whereas others had an inbreeding coefficient of 0.25. When both types of mice were released back into the wild, the weekly survival rate of the inbred mice was only 56% of that of the non-inbred mice. Inbred male mice also continuously lost body weight after release into the wild, whereas non-inbred male mice initially lost weight, but then regained it within a few days after release.

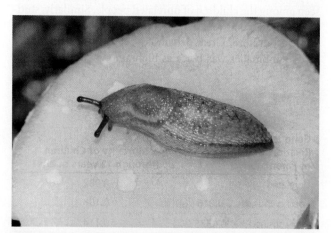

25.8 Although inbreeding is generally harmful, a number of inbreeding organisms are successful. Shown here is the terrestrial slug *Arion circumscriptos*, an inbreeding species that causes damage in greenhouses and flower gardens. [Huetter, C/Age Fotostock America, Inc.]

In spite of the fact that inbreeding is generally harmful for outcrossing species, a number of plants and animals regularly inbreed and are successful (**Figure 25.8**). As stated earlier, inbreeding increases homozygosity, and eventually all individuals in the population become homozygous. If a species undergoes inbreeding for a number of generations, many deleterious recessive alleles are weeded out by natural or artificial selection so that the population becomes homozygous for beneficial alleles. In this way, the harmful effects of inbreeding may eventually be eliminated, leaving a population that is homozygous for beneficial traits. In addition, inbreeding helps to preserve groups of genes (called co-adapted gene complexes) that exhibit gene interaction and work well together in a specific environment. Outcrossing, on the other hand, breaks up these co-adapted gene complexes and leads to the recombination of genes that might not work well together. In this way, inbreeding may be favored.

THINK-PAIR-SHARE Question 3

CONCEPTS

Nonrandom mating alters the frequencies of genotypes but not the frequencies of alleles. Inbreeding is preferential mating between related individuals. With inbreeding, the frequency of homozygotes increases, whereas the frequency of heterozygotes decreases.

✔ CONCEPT CHECK 5

What is the effect of outcrossing on a population?
a. Allelic frequencies change.
b. There will be more heterozygotes than predicted by the Hardy–Weinberg law.
c. There will be fewer heterozygotes than predicted by the Hardy–Weinberg law.
d. Genotypic frequencies will equal those predicted by the Hardy–Weinberg law.

25.4 Several Evolutionary Forces Can Change Allelic Frequencies

The Hardy–Weinberg law indicates that allelic frequencies do not change as a result of reproduction. The processes that bring about change in allelic frequencies include mutation, migration, genetic drift (random effects due to small population size), and natural selection.

Mutation

Before evolution can take place, genetic variation must exist within a population; consequently, all evolution depends on processes that generate genetic variation. Although new *combinations* of existing genes may arise through recombination in meiosis, all genetic variants ultimately arise through mutation.

THE EFFECT OF MUTATION ON ALLELIC FREQUENCIES

Mutation can influence the rate at which one genetic variant increases at the expense of another. Consider a single locus in a population of 25 diploid individuals. Each individual possesses two alleles at the locus under consideration, so the gene pool of the population consists of 50 allele copies. Let's assume that there are two different alleles, designated G^1 and G^2, with frequencies p and q, respectively. If there are 45 copies of G^1 and 5 copies of G^2 in the population, $p = 0.90$ and $q = 0.10$. Now suppose that a mutation changes a G^1 allele into a G^2 allele. After this mutation, there are 44 copies of G^1 and 6 copies of G^2, and the frequency of G^2 has increased from 0.10 to 0.12. Mutation has changed the allelic frequencies.

If copies of G^1 continue to mutate to G^2, the frequency of G^2 will increase and the frequency of G^1 will decrease (**Figure 25.9**). The amount by which G^2 will change (Δq) as a result of mutation depends on (1) the rate of G^1-to-G^2

mutation (μ) and (2) p, the frequency of G^1 in the population. When p is large, there are many copies of G^1 available to mutate to G^2, and the amount of change will be relatively large. As more mutations occur and p decreases, there will be fewer copies of G^1 available to mutate to G^2. The change in G^2 as a result of mutation equals the mutation rate times the allelic frequency:

$$\Delta q = \mu p \tag{25.11}$$

As the frequency of p decreases as a result of mutation, the change in frequency due to mutation will be less and less.

So far, we have considered only the effects of $G^1 \rightarrow G^2$ forward mutations. Reverse $G^2 \rightarrow G^1$ mutations also occur at rate v, which will probably be different from the forward mutation rate, μ. Whenever a reverse mutation occurs, the frequency of G^2 decreases and the frequency of G^1 increases (see Figure 25.9). The rate of change due to reverse mutations equals the reverse mutation rate times the allelic frequency of G^2 ($\Delta q = vq$). The overall change in allelic frequencies is a balance between the opposing forces of forward mutation and reverse mutation:

$$\Delta q = \mu p - vq \tag{25.12}$$

REACHING EQUILIBRIUM OF ALLELIC FREQUENCIES

Consider a population that begins with a high frequency of G^1 and a low frequency of G^2. In this population, many copies of G^1 are initially available to mutate to G^2, and the increase in G^2 due to forward mutation will be relatively large. However, as the frequency of G^2 increases as a result of forward mutations, fewer copies of G^1 are available to mutate, so the number of forward mutations decreases. On the other hand, few copies of G^2 are initially available to undergo reverse mutation to G^1, but as the frequency of G^2 increases, the number of copies of G^2 available to undergo reverse mutation to G^1 increases, so the number of genes undergoing reverse mutation will increase. Eventually, the number of genes undergoing forward mutation will be counterbalanced by the number of genes undergoing reverse mutation. At this point, the increase in q due to forward mutation will be equal to the decrease in q due to reverse mutation, and there will be no net change in allelic frequency ($\Delta q = 0$), in spite of the fact that forward and reverse mutations continue to occur. A point at which there is no change in the allelic frequencies of a population is referred to as an **equilibrium** (see Figure 25.9). At mutational equilibrium, the frequency of G^2 ($\hat{q}$) will be

$$(\hat{q}) = \frac{\mu}{\mu + v} \tag{25.13}$$

This final equation tells us that the allelic frequency at equilibrium is determined solely by the forward (μ) and reverse (v) mutation rates. **TRY PROBLEM 30**

25.9 Recurrent mutation changes allelic frequencies. Forward and reverse mutations eventually lead to equilibrium.

SUMMARY OF EFFECTS OF MUTATION When the only evolutionary force acting on a population is mutation, allelic frequencies change over time because some alleles mutate into others. Eventually, these allelic frequencies reach equilibrium and are determined only by the forward and reverse mutation rates. The Hardy–Weinberg law tells us that when the allelic frequencies reach equilibrium, the genotypic frequencies will also remain the same.

The mutation rates for most genes are low, so change in allelic frequencies due to mutation in one generation is very small, and long periods are required for a population to reach mutational equilibrium. For example, if the forward and reverse mutation rates for alleles at a locus are 1×10^{-5} and 0.3×10^{-5} per generation, respectively (rates that have actually been measured at several loci in mice), and the allelic frequencies are $p = 0.9$ and $q = 0.1$, then the net change in allelic frequency per generation due to mutation is

$$\Delta q = \mu p - \nu q$$
$$= (1 \times 10^{-5})(0.9) - (0.3 \times 10^{-5})(0.1)$$
$$= 8.7 \times 10^{-6}$$
$$= 0.0000087$$

Therefore, change due to mutation in a single generation is extremely small, and as the frequency of p drops as a result of mutation, the amount of change will become even smaller (**Figure 25.10**). The effect of typical mutation rates on Hardy–Weinberg equilibrium is negligible, and many generations are required for a population to reach mutational equilibrium. Nevertheless, if mutation is the only force acting on a population for long periods, mutation rates will determine allelic frequencies.

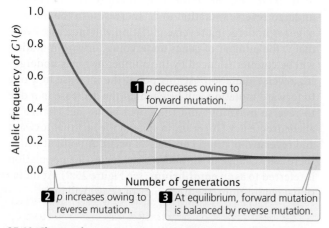

25.10 Change due to recurrent mutation slows as the frequency of *p* drops. Allelic frequencies approach mutational equilibrium at typical low mutation rates. The allelic frequency of G^1 decreases as a result of forward ($G^1 \rightarrow G^2$) mutation at rate of 0.0001 and increases as a result of reverse ($G^2 \rightarrow G^1$) mutation at rate of 0.00001. Owing to the low rate of mutations, eventual equilibrium takes many generations to be reached.

CONCEPTS

Recurrent mutation causes changes in the frequencies of alleles. At equilibrium, allelic frequencies are determined by forward and reverse mutation rates. Because mutation rates are low, the effect of mutation on allelic frequencies per generation is very small.

✔ CONCEPT CHECK 6

When a population is in mutational equilibrium, which of the following is true?

a. The number of forward mutations is greater than the number of reverse mutations.

b. No forward or reverse mutations occur.

c. The number of forward mutations is equal to the number of reverse mutations.

d. The population is in Hardy–Weinberg equilibrium.

Migration

Another process that may bring about change in a population's allelic frequencies is the influx of genes from other populations, commonly called **migration** or **gene flow**. One of the assumptions of the Hardy–Weinberg law is that migration does not take place, but many natural populations do experience migration from other populations. The overall effect of migration is twofold: (1) it prevents populations from becoming genetically different from one another, and (2) it increases genetic variation within populations.

THE EFFECT OF MIGRATION ON ALLELIC FREQUENCIES

Let's consider the effects of migration by looking at a simple, unidirectional model of migration between two populations that differ in the frequency of an allele *a*. Suppose that the frequency of this allele in population I is q_I and in population II is q_{II} (**Figure 25.11**). In each generation, a representative sample of the individuals in population I migrates to population II and reproduces, adding its alleles to population II's gene pool. Migration is only from population I to population II (unidirectional), and all the assumptions of the Hardy–Weinberg law (large population size, random mating, etc.) apply, except the absence of migration.

After migration, population II consists of two types of individuals. Some are migrants; they make up proportion *m* of population II, and they carry alleles from population I, so the frequency of allele *a* in the migrants is q_I. The other individuals in population II are the original residents. If the migrants make up proportion *m* of population II, then the residents make up proportion $1 - m$; because the residents originated in population II, the frequency of allele *a* in this group is q_{II}. After migration, the frequency of allele *a* in the merged population II (q'_{II}) is

$$q'_{II} = q_I(m) + q_{II}(1 - m) \qquad (25.14)$$

Population I
$f(a) = q_I$

Population II
$f(a) = q_{II}$

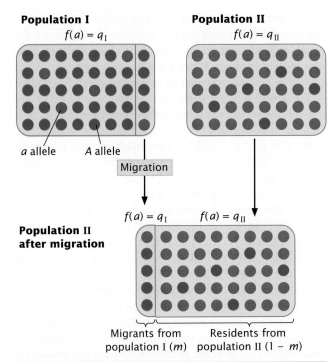

a allele   *A* allele

Migration

$f(a) = q_I$ $f(a) = q_{II}$

Population II after migration

Migrants from population I (*m*)   Residents from population II (1 − *m*)

Conclusion: The frequency of allele *a* in population II after migration is $q'_{II} = q_I m + q_{II}(1 - m)$.

25.11 The amount of change in allelic frequencies due to migration between populations depends on the difference between the populations in their allelic frequencies and on the extent of migration. Shown here is a model of the effect of unidirectional migration on allelic frequencies. The frequency of allele *a* in the source population (population I) is q_I. The frequency of this allele in the recipient population (population II) is q_{II}.

where $q_I(m)$ is the contribution to q made by the copies of allele *a* in the migrants and $q_{II}(1 - m)$ is the contribution to q made by the copies of allele *a* in the residents. The change in the allelic frequency due to migration (Δq) will be

$$\Delta q = m(q_I - q_{II}) \qquad (25.15)$$

Equation 25.15 summarizes the factors that determine the amount of change in allelic frequency due to migration. The amount of change in q is directly proportional to the amount of migration (m); as the amount of migration increases, the change in allelic frequency increases. The magnitude of change is also affected by the difference in allelic frequencies between the two populations ($q_I - q_{II}$); when the difference is large, the change in allelic frequency will be large.

With each generation of migration, the allelic frequencies of the two populations become more and more similar until, eventually, the frequency of population II equals that of population I. When $q_I - q_{II} = 0$, there will be no further change in the allelic frequency of population II, in spite of the fact that migration continues. If migration between two populations takes place for a number of generations with no other evolutionary forces present, an equilibrium is reached at which the allelic frequencies of the recipient population equal those of the source population.

The simple model of unidirectional migration between two populations just outlined can be expanded to accommodate multidirectional migration between several populations.

THINK-PAIR-SHARE Question 4

THE OVERALL EFFECT OF MIGRATION Migration has two major effects. First, it causes the gene pools of different populations to become more similar. Later in this section, we will see how genetic drift and natural selection lead to genetic differences between populations; migration counteracts these evolutionary forces and tends to keep populations homogeneous in their allelic frequencies. Second, migration adds genetic variation to populations. Different alleles may arise in different populations owing to rare mutational events, and these alleles can be spread to new populations by migration, increasing the genetic variation within the recipient population. **> TRY PROBLEM 34**

THINK-PAIR-SHARE Question 5

CONCEPTS

Migration causes changes in the allelic frequencies of a population by introducing alleles from other populations. The amount of change due to migration depends on both the amount of migration and the difference in allelic frequencies between the source and the recipient populations. Migration decreases genetic differences between populations and increases genetic variation within populations.

✔ CONCEPT CHECK 7

In each generation, 10 random individuals migrate from population A to population B. What will happen to allelic frequency q as a result of migration when q is equal in populations A and B?

a. q in A will decrease.

b. q in B will increase.

c. q will not change in either A or B.

d. q in B will become q^2.

Genetic Drift

The Hardy–Weinberg law assumes random mating in an infinitely large population; only when population size is infinite will the gametes carry genes that perfectly represent the parental gene pool. But no real population is infinitely large, and when population size is limited, the gametes that unite to form individuals of the next generation carry a sample of the alleles present in the parental gene pool. Just by chance, the composition of this sample often deviates from that of the parental gene pool, and this deviation may cause allelic frequencies to change. The smaller the gametic sample, the greater the chance that its composition will deviate from that of the parental gene pool.

The role of chance in altering allelic frequencies is analogous to the flip of a coin. Each time we flip a coin, we have a

50% chance of getting a head and a 50% chance of getting a tail. If we flip a coin 1000 times, the observed ratio of heads to tails will be very close to the expected 50 : 50 ratio. If, however, we flip a coin only 10 times, there is a good chance that we will obtain not exactly five heads and five tails, but maybe seven heads and three tails, or eight tails and two heads. This kind of deviation from an expected ratio due to limited sample size is referred to as **sampling error**.

Sampling error arises when gametes unite to produce progeny. Many organisms produce a large number of gametes, but when population size is small, a limited number of gametes unite to produce the individuals of the next generation, and chance influences which alleles are present in this limited sample. In this way, sampling error may lead to **genetic drift**, or changes in allelic frequencies. Because the deviations from the expected ratios are random, the direction of change is unpredictable. We can nevertheless predict the magnitude of the changes.

THE MAGNITUDE OF GENETIC DRIFT The effects of genetic drift over time can be viewed in two ways. First, we can see how genetic drift changes the allelic frequencies of a single population. Second, we can see how it affects differences that accumulate among a series of populations. Imagine that we have 10 small populations, all beginning with the same allelic frequencies of $p = 0.5$ and $q = 0.5$. When genetic drift occurs in a population, allelic frequencies within the population will change, but because genetic drift is random, the ways in which allelic frequencies change in each population will not be the same. In some populations, p may increase as a result of chance. In other populations, p may decrease as a result of chance. In time, the allelic frequencies in the 10 populations will become different: the populations will diverge genetically over time. The change in allelic frequencies within each population and the genetic divergence among populations are due to the same force: the random change in allelic frequencies. Thus, the magnitude of genetic drift can be assessed either by examining the change in allelic frequencies within a single population or by examining the magnitude of genetic differences that accumulate among populations.

The amount of genetic drift can be estimated from the variance in allelic frequency. Variance (s^2) is a statistical measure that describes the variability of a trait (see p. 723 in Chapter 24). Suppose that we observe a large number of separate populations, each with N individuals and allelic frequencies of p and q. After one generation of random mating, genetic drift, expressed in terms of the variance in allelic frequency among the populations (s_p^2) will be

$$s_p^2 = \frac{pq}{2N} \tag{25.16}$$

The amount of change resulting from genetic drift (the variance in allelic frequency) is determined by two parameters: the allelic frequencies (p and q) and the population size (N). Genetic drift is maximal when p and q are equal (each 0.5). For example, assume that a population consists of 50 individuals. When the allelic frequencies are equal ($p = q = 0.5$), the variance in allelic frequency (s_p^2) will be $(0.5 \times 0.5)/(2 \times 50) = 0.0025$. In contrast, when $p = 0.9$ and $q = 0.1$, the variance in allelic frequency will be only 0.0009. Genetic drift will also be greater when the population size is small. If $p = q = 0.5$, but the population size is only 10 instead of 50, then the variance in allelic frequency becomes $(0.5 \times 0.5)/(2 \times 10) = 0.0125$, which is five times greater than when the population size is 50.

This divergence of populations through genetic drift is strikingly illustrated in the results of an experiment carried out by Peter Buri on fruit flies (**Figure 25.12**). Buri examined the frequencies of two alleles (bw^{75} and bw) that affect eye color in fruit flies. He set up 107 replicate populations, each consisting of eight males and eight females. He began each population with a frequency of bw^{75} equal to 0.5. He allowed the flies within each replicate population to mate randomly, and each generation, he randomly selected eight male and eight female flies to be the parents of the next generation. He followed the changes in the frequencies of the two alleles over 19 generations. In one population, for example, the average frequency of bw^{75} (p) over the 19 generations was 0.53125. We can use Equation 25.16 to calculate the expected variance in allelic frequency due to genetic drift. The frequency of the bw allele (q) is $1 - p = 1 - 0.53125 = 0.46875$. The population size ($N$) equals 16. The expected variance in allelic frequency is therefore

$$\frac{pq}{2N} = \frac{0.53125 \times 0.46875}{2 \times 16} = 0.0156$$

which was very close to the actual observed variance of 0.0151.

The effect of population size on genetic drift is illustrated by a study conducted by Luca Cavalli-Sforza and his colleagues. They studied variation in blood types among villagers in the Parma Valley of Italy, where migration between villages was limited. They found that variation in allelic frequency was greatest between small, isolated villages in the upper valley, but was less between larger villages and towns farther down the valley. This result is exactly what we expect with genetic drift: there should be more genetic drift, and thus more variation, among populations when population sizes are small.

For ecological and demographic studies, population size is usually defined as the number of individuals in a group. However, the evolution of a gene pool depends only on those individuals who contribute genes to the next generation. Population geneticists usually define population size as the equivalent number of breeding adults, referred to as the **effective population size** (N_e). Several factors determine the equivalent number of breeding adults, including

Question: What effect does genetic drift have on the genetic composition of populations?

Methods Buri examined the frequencies of two alleles (bw^{75} and bw) that affect *Drosophila* eye color in 107 replicate small populations over 19 generations.

Results

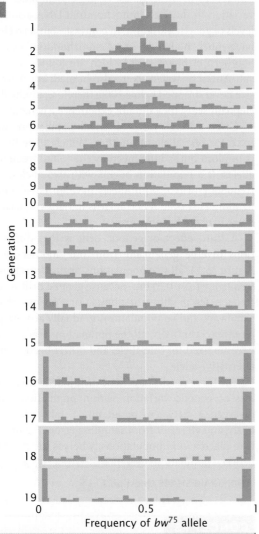

Conclusion: As a result of genetic drift, allelic frequencies in the different populations diverged and often became fixed for one allele or the other.

25.12 Populations diverge in allelic frequency and become fixed for one allele as a result of genetic drift. In Buri's study of two eye-color alleles (bw^{75} and bw) in *Drosophila*, each population consisted of eight males and eight females and began with the frequency of bw^{75} equal to 0.5.

the sex ratio, the variation between individuals in number of offspring produced, fluctuations in population size, the age structure of the population, and whether mating is random.

CONCEPTS

Genetic drift is change in allelic frequencies due to chance. The amount of change in allelic frequencies due to genetic drift is inversely related to the effective population size (the equivalent number of breeding adults in a population).

✔ CONCEPT CHECK 8

Which of the following statements describes an example of genetic drift?
a. Allele *g* for fat production increases in a small population because birds with more body fat have higher survivorship in a harsh winter.
b. Random mutation increases the frequency of allele *A* in one population but not in another.
c. Allele *R* reaches a frequency of 1.0 because individuals with genotype *rr* are sterile.
d. Allele *m* is lost when a virus kills all but a few individuals and just by chance, none of the survivors possess allele *m*.

CAUSES OF GENETIC DRIFT All genetic drift arises from sampling error, but there are several different ways in which sampling error can arise. First, a population may be reduced in size for a number of generations because of limitations in space, food, or some other critical resource. Genetic drift in a small population over multiple generations can significantly affect the composition of a population's gene pool.

A second way that sampling error can arise is through the **founder effect**, which results from the establishment of a population by a small number of individuals; the population of wolves on Isle Royale, discussed in the introduction to this chapter, underwent a founder effect. Although the population may increase and become quite large, the genes carried by all its members are derived from the few genes originally present in the founders (assuming no migration or mutation). Chance events affecting which genes were present in the founders have an important influence on the makeup of the entire population.

A third way in which genetic drift arises is through a **genetic bottleneck**, which develops when a population undergoes a drastic reduction in size. An example is seen in northern elephant seals (**Figure 25.13**). Before 1800, thousands of northern elephant seals were found along the California coast, but hunting between 1820 and 1880 devastated the population. By 1884, as few as 20 seals survived on a remote beach on Isla de Guadalupe, west of Baja California, Mexico. Restrictions on hunting enacted by the United States and Mexico allowed the seals to recover, and there are now more than 100,000 seals in the population. All the seals in the population today are genetically similar, however, because they have only those genes that were carried by the few survivors of the population bottleneck.

THINK-PAIR-SHARE Question 6

THE EFFECTS OF GENETIC DRIFT Genetic drift has several important effects on the genetic composition of a population. First, it produces change in allelic frequencies within

25.13 Northern elephant seals underwent a severe genetic bottleneck between 1820 and 1880. Today, these seals have low levels of genetic variation. [PhotoDisc/Getty Images.]

a population. Because genetic drift is random, the frequency of any allele is just as likely to increase as it is to decrease and will wander with the passage of time (hence the name *genetic drift*). **Figure 25.14** illustrates a computer simulation of genetic drift in five populations over 30 generations, starting with $q = 0.5$ and maintaining a constant population size of 10 males and 10 females. The allelic frequencies in these populations change randomly from generation to generation.

A second effect of genetic drift is to reduce genetic variation within populations. Through random change, an allele may eventually reach a frequency of either 1 or 0, at which point all individuals in the population are homozygous for one allele. When an allele has reached a frequency of 1, we

say that it has reached **fixation**. Other alleles are lost (reach a frequency of 0) and can be restored only by migration from another population or by mutation. Fixation, then, leads to a loss of genetic variation within a population. Such losses can be seen in the northern elephant seals just described. Today, these seals have low levels of genetic variation; a study of 24 protein-encoding genes found no individual or population differences in these genes. A subsequent study of sequence variation in the seals' mitochondrial DNA also revealed low levels of genetic variation. In contrast, southern elephant seals have much higher levels of mitochondrial DNA variation. Southern elephant seals, which are found in Antarctic and sub-Antarctic waters, were also hunted, but their population size never dropped below 1000; therefore, unlike the northern elephant seals, they did not experience a genetic bottleneck.

Given enough time, all small populations will become fixed for one allele or another. Which allele becomes fixed is random but is influenced by the initial frequencies of the alleles. If a population begins with two alleles, each with a frequency of 0.5, both alleles have an equal probability of fixation. However, if one allele is initially more common, it is more likely to become fixed.

A third effect of genetic drift is that different populations diverge genetically from one another over time. In Figure 25.14, all five populations begin with the same allelic frequency ($q = 0.5$) but, because genetic drift is random, the frequencies in different populations do not change in the same way, and so populations gradually acquire genetic differences. Eventually, all the populations reach fixation; some become fixed for one allele, and others become fixed for the alternative allele.

The three results of genetic drift (allelic frequency change, loss of genetic variation within populations, and genetic divergence between populations) take place simultaneously, and all result from sampling error. The first two results take place *within* populations, whereas the third takes place *between* populations. ❯ **TRY PROBLEM 35**

THINK-PAIR-SHARE Question 7

> ### CONCEPTS
>
> Genetic drift results from continuous small population size, the founder effect (in a population established by a few founders), or a genetic bottleneck (in a population whose size has been drastically reduced). Genetic drift causes change in allelic frequencies within a population, a loss of genetic variation through the fixation of alleles, and genetic divergence between populations.

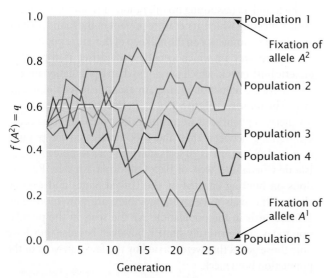

25.14 Genetic drift changes allelic frequencies within populations, leading to a reduction in genetic variation through fixation and genetic divergence among populations. Shown here is a computer simulation of changes in the frequency of allele A^2 (q) in five different populations due to random genetic drift. Each population consists of 10 males and 10 females and begins with $q = 0.5$.

Natural Selection

A final process that brings about changes in allelic frequencies is natural selection, the differential reproduction of genotypes (see Section 24.4). Natural selection takes place when individuals with adaptive traits produce a greater number

25.15 Natural selection leads to adaptations, such as those seen in the polar bears that inhabit the extreme Arctic environment. These bears blend into the snowy background, which helps them in hunting seals. The hairs of their fur stay erect even when wet, and thick layers of blubber provide insulation, which protects against subzero temperatures. Their digestive tracts are adapted to a seal-based carnivorous diet. [Digital Vision/Getty Images.]

of offspring than do individuals not carrying such traits. If the adaptive traits have a genetic basis, they are inherited by the offspring and appear with greater frequency in the next generation. A trait that provides a reproductive advantage thereby increases over time, enabling populations to become better suited to their environments—better adapted. Natural selection is unique among evolutionary forces in that it promotes adaptation (**Figure 25.15**).

FITNESS AND THE SELECTION COEFFICIENT The effect of natural selection on the gene pool of a population depends on the fitness values of the genotypes in the population. **Fitness** is defined as the relative reproductive success of a genotype. Here, the term *relative* is critical: fitness is the reproductive success of one genotype compared with the reproductive successes of other genotypes in the population.

Fitness values (W) range from 0 to 1. Suppose the average number of viable offspring produced by three genotypes is

Genotypes:	A^1A^1	A^1A^2	A^2A^2
Mean number of offspring produced:	10	5	2

To calculate fitness for each genotype, we take the mean number of offspring produced by a genotype and divide it by the mean number of offspring produced by the most prolific genotype:

$$
\begin{array}{ccc}
A^1A^1 & A^1A^2 & A^2A^2 \\
W_{11} = \dfrac{10}{10} = 1.0 & W_{12} = \dfrac{5}{10} = 0.5 & W_{22} = \dfrac{2}{10} = 0.2 \quad (25.17)
\end{array}
$$

The fitness of genotype A^1A^1 is designated W_{11}, that of A^1A^2 is W_{12}, and that of A^2A^2 is W_{22}.

A related variable is the **selection coefficient** (s), which is the relative intensity of selection against a genotype. We usually speak of selection for a particular genotype, but keep in mind that, when selection is *for* one genotype, selection is automatically *against* at least one other genotype. The selection coefficient is equal to $1 - W$, so the selection coefficients for the preceding three genotypes are

$$
\begin{array}{cccc}
 & A^1A^1 & A^1A^2 & A^2A^2 \\
\text{Selection coefficient } (1 - W): & s_{11} = 0 & s_{12} = 0.5 & s_{22} = 0.8
\end{array}
$$

CONCEPTS

Natural selection is the differential reproduction of genotypes. It is measured as fitness, which is the reproductive success of a genotype compared with that of other genotypes in a population.

✔ CONCEPT CHECK 9

The average numbers of offspring produced by three genotypes are $GG = 6$; $Gg = 3$, $gg = 2$. What is the fitness of Gg?

a. 3
b. 0.5
c. 0.3
d. 0.27

THE GENERAL SELECTION MODEL With selection, differential fitness among genotypes leads to changes in the frequencies of the genotypes over time, which, in turn, lead to changes in the frequencies of the alleles that make up the genotypes. We can predict the effect of natural selection on allelic frequencies by using a general selection model, which is outlined in **Table 25.4**. Use of this model requires

TABLE 25.4	Method for determining changes in allelic frequency due to selection		
	A^1A^1	A^1A^2	A^2A^2
Initial genotypic frequencies	p^2	$2pq$	q^2
Fitnesses	W_{11}	W_{12}	W_{22}
Proportional contribution of genotypes to population	p^2W_{11}	$2pqW_{12}$	q^2W_{22}
Relative genotypic frequency after selection	$\dfrac{p^2W_{11}}{\overline{w}}$	$\dfrac{2pqW_{12}}{\overline{w}}$	$\dfrac{q^2W_{22}}{\overline{w}}$

Note: $\overline{w} = p^2W_{11} + 2pqW_{12} + q^2W_{22}$
Allelic frequencies after selection: $p' = f(A^1) = f(A^1A^1) + \frac{1}{2}f(A^1A^2)$; $q' = 1 - p'$.

TABLE 25.5   Formulas for calculating change in allelic frequencies with different types of selection

Type of selection	Fitness Values			Change in q
	A^1A^1	A^1A^2	A^2A^2	
Selection against a recessive trait	1	1	$1-s$	$\dfrac{-spq^2}{1-sq^2}$
Selection against a dominant trait	1	$1-s$	$1-s$	$\dfrac{-spq^2}{1-s+sq^2}$
Selection against a trait with no dominance	1	$1-\frac{1}{2}s$	$1-s$	$\dfrac{-\frac{1}{2}spq}{1-sq}$
Selection against both homozygotes (overdominance)	$1-s_{11}$	1	$1-s_{22}$	$\dfrac{pq(s_{11}p-s_{22}q)}{1-s_{11}p^2-s_{22}q^2}$

knowledge of both the initial allelic frequencies and the fitness values of the genotypes. It assumes that mating is random and that the only force acting on a population is natural selection. The general selection model can be used to calculate the allelic frequencies after any type of selection. It is also possible to work out formulas for determining the change in allelic frequencies when selection is against recessive, dominant, or codominant traits, as well as traits in which the heterozygote has highest fitness (**Table 25.5**).

CONCEPTS

The change in allelic frequency due to selection can be determined for any type of genetic trait by using the general selection model.

WORKED PROBLEM

Let's apply the general selection model in Table 25.4 to a real example. Alcohol is a common substance in rotting fruit, where fruit-fly larvae grow and develop; larvae use the enzyme alcohol dehydrogenase (ADH) to detoxify the alcohol. In some fruit-fly populations, two alleles are present at the locus that encodes ADH: Adh^F, which encodes a form of the enzyme that migrates rapidly (fast) on an electrophoretic gel; and Adh^S, which encodes a form of the enzyme that migrates slowly on an electrophoretic gel. Female fruit flies with different Adh genotypes produce the following numbers of offspring when alcohol is present:

Genotype	Mean number of offspring
Adh^F/Adh^F	120
Adh^F/Adh^S	60
Adh^S/Adh^S	30

a. Calculate the fitnesses of females having these genotypes.

b. If a population of fruit flies has an initial frequency of Adh^F equal to 0.2, what will its frequency be in the next generation when alcohol is present?

Solution Strategy

What information is required in your answer to the problem?

a. The fitnesses of females of each genotype.

b. The frequency of Adh^F in the next generation when alcohol is present.

What information is provided to solve the problem?

- The mean numbers of offspring produced by each genotype.

- The population has an initial allelic frequency of 0.2.

Solution Steps

a. First, we must calculate the fitnesses of the three genotypes. Fitness is the relative reproductive output of a genotype and is calculated by dividing the mean number of offspring produced by that genotype by the mean number of offspring produced by the most prolific genotype. The fitnesses of the three Adh genotypes are therefore

Genotype	Mean number of offspring	Fitness
Adh^F/Adh^F	120	$W_{FF}=\dfrac{120}{120}=1$
Adh^F/Adh^S	60	$W_{FS}=\dfrac{60}{120}=0.5$
Adh^S/Adh^S	30	$W_{SS}=\dfrac{30}{120}=0.25$

	Adh^F/Adh^F	Adh^F/Adh^S	Adh^S/Adh^S
Initial genotypic frequencies:	$p^2 = (0.2)^2 = 0.04$	$2pq = 2(0.2)(0.8) = 0.32$	$q^2 = (0.8)^2 = 0.64$
Fitnesses:	$W_{FF} = 1$	$W_{FS} = 0.5$	$W_{SS} = 0.25$
Proportional contribution of genotypes to population:	$p^2 W_{FF} = (0.04)(1) = 0.04$	$2pq W_{FS} = (0.32)(0.5) = 0.16$	$q^2 W_{SS} = (0.64)(0.25) = 0.16$

b. To calculate the frequency of the Adh^F allele after selection, we can apply the table method. In the first row of the table above, we record the initial genotypic frequencies before selection has acted. If mating has been random (an assumption of the model), the genotypes will have the Hardy–Weinberg equilibrium frequencies of p^2, $2pq$, and q^2. In the second row of the table above, we put the fitness values of the corresponding genotypes. The proportion of the population represented by each genotype after selection is obtained by multiplying the initial genotypic frequency times its fitness (third row of Table 25.4). Now the genotypes are no longer in Hardy–Weinberg equilibrium.

The mean fitness ($\bar{w}$) of the population is the sum of the proportional contributions of the three genotypes: $(\bar{w}) = p^2 W_{11} + 2pq W_{12} + q^2 W_{22} = 0.04 + 0.16 + 0.16 = 0.36$ The mean fitness ($\bar{w}$) is the average fitness of all individuals in the population and allows the frequencies of the genotypes after selection to be obtained.

The frequency of a genotype after selection will be equal to its proportional contribution divided by the mean fitness of the population ($p^2 W_{11}/\bar{w}$ for genotype A^1A^1, $2pq W_{12}/\bar{w}$ for genotype A^1A^2, and $q^2 W_{22}/\bar{w}$ for genotype A^2A^2), as shown in the fourth line of Table 25.4. We can now add these values to our table as shown below:

	Adh^F/Adh^F	Adh^F/Adh^S	Adh^S/Adh^S
Initial genotypic frequencies:	$p^2 = (0.2)^2 = 0.04$	$2pq = 2(0.2)(0.8) = 0.32$	$q^2 = (0.8)^2 = 0.64$
Fitnesses:	$W_{FF} = 1$	$W_{FS} = 0.5$	$W_{SS} = 0.25$
Proportional contribution of genotypes to population:	$p^2 W_{FF} = (0.04)(1) = 0.04$	$2pq W_{FS} = (0.32)(0.5) = 0.16$	$q^2 W_{SS} = (0.64)(0.25) = 0.16$
Relative genotypic frequency after selection:	$\dfrac{p^2 w_{FF}}{\bar{w}} = \dfrac{0.04}{0.36} = 0.11$	$\dfrac{2pq w_{FS}}{\bar{w}} = \dfrac{0.16}{0.36} = 0.44$	$\dfrac{q^2 w_{SS}}{\bar{w}} = \dfrac{0.16}{0.36} = 0.44$

After the new genotypic frequencies have been calculated, the new allelic frequency of Adh^F (p') can be determined by using the now-familiar formula of Equation 25.4:

$$p' = f(Adh^F) = f(Adh^F/Adh^F) + \tfrac{1}{2} f(Adh^F/Adh^S)$$
$$= 0.11 + \tfrac{1}{2}(0.44) = 0.33$$

and that of q' can be obtained by subtraction:

$$q' = 1 - p'$$
$$= 1 - 0.33 = 0.67$$

We predict that the frequency of Adh^F will increase from 0.2 to 0.33.

> » For more practice with the selection model, try **Problem 37** at the end of this chapter.

THE RESULTS OF SELECTION The results of selection depend on the fitnesses of the genotypes in a population. In a population with three genotypes (A^1A^1, A^1A^2, and A^2A^2)

with fitnesses W_{11}, W_{12}, and W_{22}, we can identify six different types of natural selection (**Table 25.6**).

In type 1 selection, a dominant allele A^1 confers a fitness advantage; in this case, the fitnesses of genotypes A^1A^1 and A^1A^2 are equal and higher than the fitness of A^2A^2 ($W_{11} = W_{12} > W_{22}$). Because both the heterozygote and the A^1A^1 homozygote have copies of the A^1 allele and produce more offspring than the A^2A^2 homozygote does, the frequency of the A^1 allele will increase over time, and the frequency of the A^2 allele will decrease. This form of selection, in which one allele or trait is favored over another, is termed **directional selection**.

Type 2 selection is directional selection against a dominant allele A^1 ($W_{11} = W_{12} < W_{22}$). In this case, the A^2 allele increases and the A^1 allele decreases. Type 3 and type 4 selection are also directional selection, but in these cases, there is incomplete dominance, and the heterozygote has a fitness that is intermediate between the two homozygotes ($W_{11} > W_{12} > W_{22}$ for type 3; $W_{11} < W_{12} < W_{22}$ for type 4). When A^1A^1 has the highest fitness (type 3), the A^1 allele increases and the A^2 allele decreases over time. When A^2A^2 has the highest fitness (type 4), the A^2 allele increases and

TABLE 25.6 | **Types of natural selection**

Type	Fitness Relation	Form of Selection	Result
1	$W_{11} = W_{12} > W_{22}$	Directional selection against recessive allele A^2	A^1 increases, A^2 decreases
2	$W_{11} = W_{12} < W_{22}$	Directional selection against dominant allele A^1	A^2 increases, A^1 decreases
3	$W_{11} > W_{12} > W_{22}$	Directional selection against incompletely dominant allele A^2	A^1 increases, A^2 decreases
4	$W_{11} < W_{12} < W_{22}$	Directional selection against incompletely dominant allele A^1	A^2 increases, A^1 decreases
5	$W_{11} < W_{12} > W_{22}$	Overdominance	Stable equilibrium, both alleles maintained
6	$W_{11} > W_{12} < W_{22}$	Underdominance	Unstable equilibrium

Note: W_{11}, W_{12}, and W_{22} represent the fitnesses of genotypes A^1A^1, A^1A^2, and A^2A^2, respectively.

the A^1 allele decreases over time. Eventually, all four types of directional selection lead to fixation of the favored allele and elimination of the other allele, as long as no other evolutionary forces act on the population.

The last two types of selection (types 5 and 6) occur in special situations and lead to equilibrium, at which there is no further change in allelic frequency. Type 5 selection is referred to as **overdominance** or *heterozygote advantage*. Here, the heterozygote has higher fitness than either homozygote ($W_{11} < W_{12} > W_{22}$). With overdominance, both alleles are favored in the heterozygote, and neither allele is eliminated from the population. Initially, the allelic frequencies may change because one homozygote has higher fitness than the other; the direction of change depends on the fitness values of the two homozygotes. The allelic frequencies change with overdominant selection until a stable equilibrium is reached, at which point there is no further change. The allelic frequency at equilibrium ($\hat{q}$) depends on the fitnesses (usually expressed as selection coefficients) of the two homozygotes:

$$\hat{q} = f(A^2) = \frac{s_{11}}{s_{11} + s_{22}} \qquad (25.18)$$

where s_{11} represents the selection coefficient of the A^1A^1 homozygote and s_{22} represents the selection coefficient of the A^2A^2 homozygote.

An example of overdominance is sickle-cell anemia in humans, a disease that results from a mutation in one of the genes that encodes hemoglobin. People who are homozygous for the sickle-cell mutation produce only sickle-cell hemoglobin, have severe anemia, and often incur tissue damage. People who are heterozygous—with one normal copy and one mutated copy of the gene—produce both normal and sickle-cell hemoglobin, but their red blood cells contain enough normal hemoglobin to prevent sickle-cell anemia. However, heterozygotes are resistant to malaria and thus have higher fitness than do homozygotes for normal hemoglobin or homozygotes for sickle-cell hemoglobin.

The last type of selection (type 6) is **underdominance**, in which the heterozygote has lower fitness than either homozygote ($W_{11} > W_{12} < W_{22}$). Underdominance leads to an unstable equilibrium; here, allelic frequencies do not change as long as they are at equilibrium, but if they are disturbed from the equilibrium point by some other evolutionary force, they will move away from equilibrium until one allele eventually becomes fixed. To see the effects of natural selection on allelic and genotypic frequencies, view the Mini-Tutorial in **Animation 25.1.** ▶ **TRY PROBLEM 38**

> **CONCEPTS**
>
> Natural selection changes allelic frequencies; the direction and magnitude of change depend on the intensity of selection, the dominance relations of the alleles, and the allelic frequencies. Directional selection favors one allele over another and eventually leads to fixation of the favored allele. Overdominance leads to a stable equilibrium with maintenance of both alleles in the population. Underdominance produces an unstable equilibrium because the heterozygote has lower fitness than either homozygote.

CHANGE IN THE FREQUENCY OF A RECESSIVE ALLELE DUE TO NATURAL SELECTION The rate at which selection changes allelic frequency depends on the allelic frequency itself. If allele A^2 is lethal and recessive, $W_{11} = W_{12} = 1$, whereas $W_{22} = 0$. The frequency of the A^2 allele decreases over time (because A^2A^2 homozygotes produce no offspring), and the rate of decrease is proportional to the frequency of the allele. When the frequency of the allele is high, the change in each generation is relatively large, but as the frequency of the allele drops, a higher proportion of A^2 alleles are in heterozygous genotypes, where they are immune to the action of natural selection (because the heterozygotes have the same phenotype as the favored homozygote). Thus, selection against a rare recessive allele is very inefficient, and its removal from the population is slow.

The relation between the frequency of a recessive allele and its rate of change under natural selection has an important implication. Some people believe that the survival and reproduction of patients with rare recessive genetic diseases will cause the disease gene to increase, eventually leading to degeneration of the human gene pool. This mistaken belief was the basis of eugenic laws that were passed in the early part of the twentieth century prohibiting the marriage of persons with certain genetic conditions and allowing the involuntary sterilization of others. However, most copies of rare recessive alleles are present in heterozygotes, and selection against homozygotes has little effect on the frequency of a recessive allele. There are many ethical issues associated with eugenics, but regardless, the reproduction of homozygotes for a recessive trait has little effect on the frequency of the disorder.

THINK-PAIR-SHARE Question 8

MUTATION AND NATURAL SELECTION Recurrent mutation and natural selection act as opposing forces on detrimental alleles: mutation increases their frequency and natural selection decreases their frequency. Eventually, these two forces reach an equilibrium in which the number of alleles added by mutation is balanced by the number of alleles removed by selection.

The frequency of a deleterious *recessive* allele at equilibrium ($\hat{q}$) is equal to the square root of the mutation rate divided by the selection coefficient:

$$\hat{q} = \sqrt{\frac{\mu}{s}} \qquad (25.19)$$

The frequency of a deleterious *dominant* allele at equilibrium can be shown to be

$$\hat{q} = \frac{\mu}{s} \qquad (25.20)$$

Achondroplasia is a common type of human dwarfism that results from a dominant allele. People with this condition are fertile, although they produce only about 74% as many children as are produced by people without achondroplasia.

The fitness of people with achondroplasia therefore averages 0.74, and the selection coefficient (s) is $1 - W$, or 0.26. If we assume that the mutation rate for achondroplasia is about 3×10^{-5} (a typical mutation rate in humans), then we can predict that the equilibrium frequency for the achondroplasia allele will be

$$\hat{q} = \frac{0.00003}{0.26} = 0.0001153$$

This frequency is close to the actual frequency of the condition. ▶ **TRY PROBLEM 41**

CONCEPTS

Mutation and natural selection act as opposing forces on detrimental alleles: mutation tends to increase their frequency and natural selection tends to decrease their frequency, eventually producing an equilibrium.

CONNECTING CONCEPTS

The General Effects of Forces That Change Allelic Frequencies

You now know that four evolutionary forces can bring about change in the allelic frequencies of a population: mutation, migration, genetic drift, and natural selection. Their short- and long-term effects on allelic frequencies are summarized in **Table 25.7**. In some cases, the change continues until one allele is eliminated and the other becomes fixed in the population. Genetic drift and directional selection eventually result in fixation, provided that these forces are the only ones acting on a population. With the other evolutionary forces, allelic frequencies change until an equilibrium point is reached, after which there is no further change in allelic frequencies. Mutation, migration, and some forms of natural selection can lead to stable equilibria.

These evolutionary forces affect both genetic variation within populations and genetic divergence between populations. Evolutionary forces that maintain or increase genetic variation within populations are listed in the upper-left quadrant of **Figure 25.16** These forces include some types of natural selection, such as overdominance, in which both alleles are favored. Mutation and migration also increase genetic variation within populations

---

TABLE 25.7	Effects of different evolutionary forces on allelic frequencies within populations	
Force	**Short-Term Effect**	**Long-Term Effect**
Mutation	Change in allelic frequencies	Equilibrium reached between forward and reverse mutations
Migration	Change in allelic frequencies	Equilibrium reached when allelic frequencies of source and recipient population are equal
Genetic drift	Change in allelic frequencies	Fixation of one allele
Natural selection	Change in allelic frequencies	Directional selection: fixation of one allele
		Overdominant selection: equilibrium reached
		Underdominant selection: unstable equilibrium

	Within populations	Between populations
Increase genetic variation	**Mutation Migration Some types of natural selection**	**Mutation Genetic drift Some types of natural selection**
Decrease genetic variation	**Genetic drift Some types of natural selection**	**Migration Some types of natural selection**

25.16 Mutation, migration, genetic drift, and natural selection have different effects on genetic variation within populations and on genetic divergence between populations.

because they introduce new alleles into the population. Evolutionary forces that decrease genetic variation within populations are listed in the lower-left quadrant of Figure 25.16. These forces include genetic drift, which decreases variation through the fixation of alleles, and some forms of natural selection, such as directional selection.

These same evolutionary forces also affect genetic divergence between populations. Natural selection increases divergence between populations if different alleles are favored in different

populations, but it can also decrease divergence between populations by favoring the same allele in different populations. Mutation almost always increases divergence between populations because different mutations arise in each population. Genetic drift also increases divergence between populations because changes in allelic frequencies due to genetic drift are random and are likely to proceed in different directions in separate populations. Migration, on the other hand, decreases divergence between populations because it makes populations more similar in their genetic composition.

Migration and genetic drift act in opposite directions: migration increases genetic variation within populations and decreases divergence between populations, whereas genetic drift decreases genetic variation within populations and increases divergence between populations. Mutation increases both variation within populations and divergence between populations. Natural selection can either increase or decrease variation within populations, and it can increase or decrease divergence between populations.

An important point to keep in mind is that real populations are simultaneously affected by many evolutionary forces. In this chapter, we examined the effects of mutation, migration, genetic drift, and natural selection in isolation so that the influence of each process would be clear. In the real world, however, populations are commonly affected by several evolutionary forces at the same time, and evolution results from the complex interplay of numerous processes.

CONCEPTS SUMMARY

■ Population genetics examines the genetic composition of groups of individuals and how their composition changes with time.

■ A Mendelian population is a group of interbreeding, sexually reproducing individuals, whose set of genes constitutes the population's gene pool. Evolution takes place through changes in this gene pool.

■ A population's genetic composition can be described by its genotypic and allelic frequencies.

■ The Hardy–Weinberg law describes the effects of reproduction and Mendelian principles on the allelic and genotypic frequencies of a population. It assumes that a population is large, randomly mating, and free from the effects of mutation, migration, and natural selection. When these conditions are met, the allelic frequencies do not change, and the genotypic frequencies stabilize after one generation in the Hardy–Weinberg equilibrium proportions p^2, $2pq$, and q^2, where p and q equal the frequencies of the alleles.

■ Nonrandom mating affects the frequencies of genotypes, but not those of alleles.

■ Inbreeding, a type of positive assortative mating, increases the frequency of homozygotes while decreasing the frequency of heterozygotes. Inbreeding is frequently detrimental because it increases the appearance of lethal and deleterious recessive traits.

■ Mutation, migration, genetic drift, and natural selection can change allelic frequencies.

■ Recurrent mutation eventually leads to an equilibrium, at which allelic frequencies are determined by the relative rates of forward and reverse mutation. Change due to mutation in a single generation is usually very small because mutation rates are low.

■ Migration, the movement of genes between populations, increases the amount of genetic variation within populations and decreases the difference in allelic frequencies between populations.

■ Genetic drift is change in allelic frequencies due to chance. Genetic drift arises when a population consists of a small number of individuals, is established by a small number of founders, or undergoes a major reduction in size. Genetic drift changes allelic frequencies, reduces genetic variation within populations, and causes genetic divergence among populations.

■ Natural selection is the differential reproduction of genotypes; it is measured by the relative reproductive successes (fitnesses) of genotypes. The effects of natural selection on allelic frequency can be determined by applying the general selection model. Directional selection leads to the fixation of one allele. The rate of change in allelic frequency due to selection depends on the intensity of selection, the dominance relations, and the initial frequencies of the alleles.

■ Mutation and natural selection can produce an equilibrium in which the number of new alleles introduced by mutation is balanced by the elimination of alleles through natural selection.

IMPORTANT TERMS

genetic rescue (p. 750)
Mendelian
    population (p. 750)
gene pool (p. 750)
genotypic
    frequency (p. 751)
allelic frequency (p. 752)
Hardy–Weinberg
    law (p. 754)

Hardy–Weinberg
    equilibrium (p. 754)
positive assortative
    mating (p. 758)
negative assortative
    mating (p. 758)
inbreeding (p. 758)
outcrossing (p. 758)
inbreeding coefficient (p. 758)

inbreeding
    depression (p. 759)
equilibrium (p. 761)
migration (gene flow)
    (p. 762)
sampling error (p. 764)
genetic drift (p. 764)
effective population
    size (p. 764)

founder effect (p. 765)
genetic bottleneck (p. 765)
fixation (p. 766)
fitness (p. 767)
selection coefficient (p. 767)
directional
    selection (p. 769)
overdominance (p. 770)
underdominance (p. 770)

ANSWERS TO CONCEPT CHECKS

1. There are fewer alleles than genotypes, so the gene pool can be described by fewer parameters when allelic frequencies are used. Additionally, the genotypes are temporary assemblages of alleles that break down each generation; the alleles are passed from generation to generation in sexually reproducing organisms.

2. a

3. c

4. 0.10

5. b

6. c

7. c

8. d

9. b

WORKED PROBLEM

A recessive allele for red hair (r) has a frequency of 0.2 in population I and a frequency of 0.01 in population II. A famine in population I causes a number of people in population I to migrate to population II, where they reproduce randomly with the members of population II. Geneticists estimate that, after migration, 15% of the people in population II consist of people who migrated from population I. What will be the frequency of red hair in population II after the migration?

≫ Solution Strategy

What information is required in your answer to the problem?

The frequency of red hair in population II after the migration.

What information is provided to solve the problem?

- The allele for red hair (r) is recessive.
- The initial frequency of r in population I is 0.2.
- The initial frequency of r in population II is 0.01.
- After migration, 15% of the people in population II consist of people who migrated from population I.

For help with this problem, review:

Migration in Section 25.4.

≫ Solution Steps

From Equation 25.14, the allelic frequency in a population after migration (q'_{II}) is

$$q'_{II} = q_I(m) + q_{II}(1 - m)$$

where q_I and q_{II} are the allelic frequencies in population I (migrants) and population II (residents), respectively, and m is the proportion of population II that consists of migrants. In this problem, the frequency of red hair is 0.2 in population I and 0.01 in population II. Because 15% of population II consists of migrants, $m = 0.15$. Substituting these values into Equation 25.14, we obtain

$$q'_{II} = 0.2(0.15) + (0.01)(1 - 0.15) = 0.03 + 0.0085 = 0.0385$$

which is the expected frequency of the allele for red hair in population II after the migration. Red hair is a recessive trait; if mating is random for hair color, the frequency of red hair in population II after migration will be

$$f(rr) = q^2 = (0.0385)^2 = 0.0015$$

Recall: With random mating, the expected frequencies of the genotypes are p^2 (RR), $2pq$ (Rr), and q^2 (rr).

COMPREHENSION QUESTIONS

Section 25.1

1. What is a Mendelian population? How is the gene pool of a Mendelian population usually described?

Section 25.2

2. What are the predictions given by the Hardy–Weinberg law?

3. What assumptions must be met for a population to be in Hardy–Weinberg equilibrium?

4. What is random mating?

5. Give the genotypic frequencies expected under the Hardy–Weinberg law for (a) an autosomal locus with three alleles and (b) an X-linked locus with two alleles.

Section 25.3

6. Define inbreeding and briefly describe its effects on a population.

Section 25.4

7. What determines the allelic frequencies at mutational equilibrium?

8. What factors affect the magnitude of change in allelic frequencies due to migration?

9. Define genetic drift and give three ways in which it can arise. What effect does genetic drift have on a population?

10. What is effective population size? How does it affect the amount of genetic drift?

11. Define natural selection and fitness.

12. Briefly describe the differences between directional selection, overdominance, and underdominance. Describe the effect of each type of selection on the allelic frequencies of a population.

13. What factors affect the rate of change in allelic frequency due to natural selection?

14. Compare and contrast the effects of mutation, migration, genetic drift, and natural selection on genetic variation within populations and on genetic divergence between populations.

APPLICATION QUESTIONS AND PROBLEMS

Section 25.1

15. How would you respond to someone who said that models are useless in studying population genetics because they represent oversimplifications of the real world?

*16. Voles (*Microtus ochrogaster*) were trapped in fields in southern Indiana and genotyped for a locus that encodes transferrin (a blood protein). The following numbers of genotypes were recorded, where T^E and T^F represent different alleles.

[Tom McHugh/Science Source.]

$T^E T^E$	$T^E T^F$	$T^F T^F$
407	170	17

Calculate the genotypic and allelic frequencies of the transferrin locus for this population.

17. Jean Manning, Charles Kerfoot, and Edward Berger studied genotypic frequencies at the phosphoglucose isomerase (GPI) locus in the cladoceran *Bosmina longirostris* (a small crustacean known as a water flea). They collected 176 of the animals from a single location in Union Bay in Seattle, Washington, and determined their GPI genotypes by using electrophoresis (J. Manning, W. C. Kerfoot, and E. M. Berger. 1978. *Evolution* 32:365–374).

Genotype	Number
$S^1 S^1$	4
$S^1 S^2$	38
$S^2 S^2$	134

Determine the genotypic and allelic frequencies for this population.

*18. Orange coat color in cats is due to an X-linked allele (X^O) that is codominant with the allele for black (X^+). When genotypes at the orange locus were determined for a sample of cats in Minneapolis and St. Paul, Minnesota, the following data were obtained:

$X^O X^O$ females	11
$X^O X^+$ females	70
$X^+ X^+$ females	94
$X^O Y$ males	36
$X^+ Y$ males	112

Calculate the frequencies of the X^O and X^+ alleles for this population.

Section 25.2

19. Use the graph shown in **Figure 25.4** to determine which genotype is most frequent when the frequency of the *A* allele is

a. 0.2

b. 0.5

c. 0.8

20. A total of 6129 North American Caucasians were blood typed for the MN locus, which is determined by two codominant alleles, L^M and L^N. The following data were obtained:

Blood type	Number
M	1787
MN	3039
N	1303

Use a chi-square test to determine whether this population is in Hardy–Weinberg equilibrium at the MN locus.

21. Assume that the phenotypes of the lady beetles shown in **Figure 25.2b** are encoded by the following genotypes:

Phenotype	Genotype
All black	*BB*
Some black spots	*Bb*
No black spots	*bb*

a. For the lady beetles shown in the figure, calculate the frequencies of the genotypes and the frequencies of the alleles.

b. Use a chi-square test to determine whether the lady beetles shown are in Hardy–Weinberg equilibrium.

*22. Most black bears (*Ursus americanus*) are black or brown in color. However, occasional white bears of this species appear in some populations along the coast of British Columbia. Kermit Ritland and his colleagues determined that white coat color in these bears results from a recessive mutation (*G*) caused by a single nucleotide replacement in which guanine substitutes for adenine at the melanocortin-1 receptor locus (*mc1r*), the same locus responsible for red hair in humans (K. Ritland, C. Newton, and H. D. Marshall. 2001. *Current Biology* 11:1468–1472). The wild-type allele at this locus (*A*) encodes black or brown color. Ritland and his colleagues collected samples from bears on three islands and determined their genotypes at the *mc1r* locus:

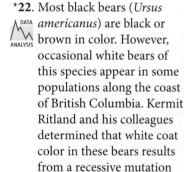

[Wendy Shattil/Alamy.]

Genotype	Number
AA	42
AG	24
GG	21

a. What are the frequencies of the *A* and *G* alleles in these bears?

b. Give the genotypic frequencies expected if the population is in Hardy–Weinberg equilibrium.

c. Use a chi-square test to compare the number of observed genotypes with the number expected under Hardy–Weinberg equilibrium. Is this population in Hardy–Weinberg equilibrium? Explain your reasoning.

23. Genotypes of leopard frogs from a population in central Kansas were determined for a locus (*M*) that encodes the enzyme malate dehydrogenase. The following numbers of genotypes were observed:

Genotype	Number
M^1M^1	20
M^1M^2	45
M^2M^2	42
M^1M^3	4
M^2M^3	8
M^3M^3	6
Total	125

a. Calculate the genotypic and allelic frequencies for this population.

b. What would the expected numbers of genotypes be if the population were in Hardy–Weinberg equilibrium?

24. Full color (*D*) in domestic cats is dominant over dilute color (*d*). Of 325 cats observed, 194 have full color and 131 have dilute color.

a. If this population of cats is in Hardy–Weinberg equilibrium for the dilution locus, what is the frequency of the dilute (*d*) allele?

b. How many of the 194 cats with full color are likely to be heterozygous?

*25. Tay–Sachs disease is an autosomal recessive disorder. Among Ashkenazi Jews, the frequency of Tay–Sachs disease is 1 in 3600. If the Ashkenazi population is mating randomly with respect to the Tay–Sachs gene, what proportion of the population consists of heterozygous carriers of the Tay–Sachs allele?

26. In the plant *Lotus corniculatus*, cyanogenic glycoside protects against insect pests and even grazing by cattle. The presence of this glycoside in an individual plant is due to a simple dominant allele. A population of *L. corniculatus* consists of 77 plants that possess cyanogenic glycoside and 56 that lack the compound. What is the frequency of the dominant allele responsible for the presence of cyanogenic glycoside in this population?

27. Color blindness in humans is an X-linked recessive trait. Approximately 10% of the men in a particular population are color blind.

   a. If mating is random with respect to the color-blindness locus, what is the frequency of the color-blindness allele in this population?

   b. What proportion of the women in this population are expected to be color blind?

   c. What proportion of the women in this population are expected to be heterozygous carriers of the color-blindness allele?

Section 25.3

*28. The human MN blood type is determined by two codominant alleles, L^M and L^N. The frequency of L^M in Eskimos on a small Arctic island is 0.80.

   a. If random mating takes place in this population, what are the expected frequencies of the M, MN, and N blood types on the island?

   b. If the inbreeding coefficient for this population is 0.05, what are the expected frequencies of the M, MN, and N blood types on the island?

29. Demonstrate mathematically that full-sib mating ($F = 1/4$) reduces heterozygosity by $1/4$ with each generation.

Section 25.4

*30. The forward mutation rate for piebald spotting in guinea pigs is 8×10^{-5}; the reverse mutation rate is 2×10^{-6}. If no other evolutionary forces are assumed to be acting, what is the expected frequency of the allele for piebald spotting in a population that is in mutational equilibrium?

31. For three years, Gunther Schlager and Margaret Dickie estimated the forward and reverse mutation rates for five loci in mice that encode various aspects of coat color by examining more than 5 million mice for spontaneous mutations (G. Schlager and M. M. Dickie. 1966. *Science* 151:205–206). They detected the following numbers of mutations at the *dilute* locus:

	Number of gametes examined	Number of mutations detected
Forward mutations	260,675	5
Reverse mutations	583,360	2

Calculate the forward and reverse mutation rates at this locus. If these mutations rates are representative of rates in natural populations of mice, what would the expected equilibrium frequency of *dilute* mutations be?

32. In **Figure 25.11**, each blue dot represents one copy of the *A* allele and each red dot represents one copy of

the *a* allele. Calculate the frequencies of the *A* allele in population II before and after migration. Explain why the frequency of *A* in population II changed after migration.

33. In German cockroaches, curved wing (*cv*) is recessive to normal wing (*cv$^+$*). Bill, who is raising cockroaches in his dorm room, finds that the frequency of the gene for curved wings in his cockroach population is 0.6. In his friend Joe's apartment, the frequency of the gene for curved wings is 0.2. One day Joe visits Bill in his dorm room, and several cockroaches jump out of Joe's hair and join the population in Bill's room. Bill estimates that, now, 10% of the cockroaches in his dorm room are individual roaches that jumped out of Joe's hair. What is the new frequency of curved wings among cockroaches in Bill's room?

*34. A population of water snakes is found on an island in Lake Erie. Some of the snakes are banded and some are unbanded; banding is caused by an autosomal allele that is recessive to an allele for no bands. The frequency of banded snakes on the island is 0.4, whereas the frequency of banded snakes on the mainland is 0.81. One summer, a large number of snakes migrate from the mainland to the island. After this migration, 20% of the island population consists of snakes that came from the mainland.

   a. If both the mainland population and the island population are assumed to be in Hardy–Weinberg equilibrium for the alleles that affect banding, what is the frequency of the allele for bands on the island and on the mainland before migration?

   b. After migration has taken place, what is the frequency of the allele for the banded phenotype on the island?

*35. Pikas are small mammals that live at high elevations on the talus slopes of mountains. Most populations located on mountaintops in Colorado and Montana in North America are isolated from one another: the pikas don't occupy the low-elevation habitats that separate the mountaintops and don't venture far from the talus slopes. Thus, there is little gene flow between populations. Furthermore, each population is small in size and was founded by a small number of pikas.

   A group of population geneticists proposes to study the amount of genetic variation in a series of pika populations and to compare the allelic frequencies in different populations. On the basis of the biology and distribution of pikas, predict what the population geneticists will find concerning the within- and between-population genetic variation.

36. What proportion of the populations shown in **Figure 25.14** reached fixation by generations 10, 25,

and 30? How does the proportion of populations that reach fixation due to genetic drift change over time?

*37. In a large, randomly mating population, the frequency of the allele (*s*) for sickle-cell hemoglobin is 0.028. The results of studies have shown that people with the following genotypes at the beta-chain locus produce the following average numbers of offspring:

Genotype	Average number of offspring produced
SS	5
Ss	6
ss	0

a. What will the frequency of the sickle-cell allele (*s*) be in the next generation?

b. What will the frequency of the sickle-cell allele be at equilibrium?

*38. Two chromosome inversions are commonly found in populations of *Drosophila pseudoobscura*: Standard (*ST*) and Arrowhead (*AR*). When the flies are treated with the insecticide DDT, the genotypes for these inversions exhibit overdominance, with the following fitnesses:

Genotype	Fitness
ST/ST	0.47
ST/AR	1
AR/AR	0.62

What will the frequencies of *ST* and *AR* be after equilibrium has been reached?

39. In a large, randomly mating population, the frequency of an autosomal recessive lethal allele is 0.20. What will the frequency of this allele be in the next generation if all homozygotes die before reproducing?

40. The larvae of the fruit fly *Drosophila melanogaster* normally feed on rotting fruit, which may ferment and produce high concentrations of alcohol. Douglas Cavener and Michael Clegg studied allelic frequencies at the locus encoding alcohol dehydrogenase (*Adh*) in experimental populations of *D. melanogaster* (D. R. Cavener and M. T. Clegg. 1981. *Evolution* 35:1–10). The experimental populations were established from wild-caught flies and were raised in cages in the laboratory. Two control populations (C1 and C2) were raised on a standard cornmeal–molasses–agar diet. Two ethanol populations (E1 and E2) were raised on a cornmeal–molasses–agar diet to which was added 10% ethanol. The four populations were periodically sampled to determine the frequencies of two alleles at the alcohol dehydrogenase locus, *Adh*^S and *Adh*^F. The frequencies

of these alleles in the four populations are shown in the accompanying graph.

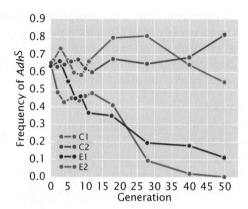

a. On the basis of these data, what conclusion might you draw about the evolutionary forces that are affecting the *Adh* alleles in these populations?

b. Cavener and Clegg measured the viability of the different *Adh* genotypes in the ethanol environment and obtained the following values:

Genotype	Relative viability
Adh^F/*Adh*^F	0.932
Adh^F/*Adh*^S	1.288
Adh^S/*Adh*^S	0.596

Using these relative viabilities, calculate fitnesses for the three genotypes. If a population has an initial frequency of $p = f(Adh^F) = 0.5$, what is the expected frequency of *Adh*^F in the next generation in the ethanol environment on the basis of these fitness values?

*41. A certain form of congenital glaucoma is caused by an autosomal recessive allele. Assume that the mutation rate is 10^{-5} and that people with this condition produce, on the average, only about 80% of the offspring produced by people who do not have glaucoma.

a. At equilibrium between mutation and selection, what will the frequency of the gene for congenital glaucoma be?

b. What will the frequency of the disease be in a randomly mating population that is at equilibrium?

42. Examine **Figure 25.16**. Which evolutionary forces

a. cause an increase in genetic variation both within and between populations?

b. cause a decrease in genetic variation both within and between populations?

c. cause an increase in genetic variation within populations but cause a decrease in genetic variation between populations?

CHALLENGE QUESTIONS

Section 25.2

43. The frequency of allele *A* in a population is 0.8 and the frequency of allele *a* is 0.2. If the population mates randomly with respect to this locus, give all the possible matings among the genotypes at this locus and the expected proportion of each type of mating.

Section 25.4

44. The Barton Springs salamander is an endangered species found only in three adjacent springs in the city of Austin, Texas. There is growing concern that a chemical spill on a nearby freeway could pollute the spring and wipe out the species. To provide a source of salamanders to repopulate the spring in the event of such a catastrophe, a proposal has been made to establish a captive breeding population of the salamander in a local zoo. You are asked to provide a plan for the establishment of this captive breeding population, with the goal of maintaining as much of the genetic variation of the species as possible. What factors might cause loss of genetic variation in the establishment of the captive population? How could loss of such variation be prevented? With the assumption that only a limited number of salamanders can be maintained in captivity, what procedures should be instituted to ensure the long-term maintenance of as much variation as possible?

 THINK-PAIR-SHARE QUESTIONS

Section 25.2

1. Miguel says that the Hardy–Weinberg law is only theoretical, of no practical value, and totally worthless because populations will never be large, randomly mating for all traits, and free from all evolutionary forces. Thus, he asserts that populations will never be in Hardy–Weinberg equilibrium. Barbara says she knows the Hardy–Weinberg law is important. Who is correct and why?

Section 25.3

2. Name some traits for which you think humans exhibit nonrandom mating. Is mating for these traits likely to be positive or negative assortative mating? Name some traits for which you think people mate randomly. How could you test whether mating for these traits is random?

3. Assume that the frequency of an allele that causes an autosomal recessive disease is 0.03. What is the probability of a person having the disease when his or her parents are unrelated (i.e., mating is random)? What is the probability of having the disease when the parents are first cousins? What about when the parents are second cousins? (Hint: See **Table 25.3**.)

Section 25.4

4. **Figure 25.11** presents a simple mathematical model of migration. What are the assumptions of this model? Write down as many as you can think of. How reasonable are these assumptions?

5. Is migration good or bad for populations? Defend your answer.

6. What historical, social, religious, cultural, and economic factors promote genetic drift in humans? Can you think of some specific human groups in which genetic drift is likely to have occurred?

7. Examine **Figure 25.12**, which illustrates Buri's experiment on genetic drift in populations of *Drosophila*. List evidence that you see for the effects of genetic drift in the results of this experiment.

8. After the rediscovery of Mendel's principles of inheritance in the early twentieth century, eugenics became popular in many countries, including the United States. The eugenic movement proposed to improve the human species through the application of genetics. A number of eugenic laws were passed, prohibiting the marriage of people with certain conditions and allowing for the involuntary sterilization of others. Eventually this practice fell into disfavor because of ethical concerns and because many of the laws were based on bad science. Some people think that the increasing use of modern genetic techniques to test for genetic diseases and traits and to manipulate reproductive outcomes (see Chapters 6 and 19) represents the resurgence of eugenics. What are some ethical concerns associated with encouraging the birth of people with desirable traits and discouraging the birth of those with undesirable traits?

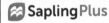 Self-study tools that will help you practice what you've learned and reinforce this chapter's concepts are available online. Go to www.macmillanlearning.com/PierceGenetics6e.

Evolutionary Genetics

Some chimpanzees, like humans, have the ability to taste phenylthiocarbamide (PTC), whereas others do not. Recent research indicates that the PTC taster polymorphism evolved independently in humans and chimpanzees. [FLPA/Alamy.]

Taster Genes in Spitting Apes

Almost every student of biology knows about the taster test. The teacher passes out small pieces of paper impregnated with a compound called phenylthiocarbamide (PTC), and the students, following the teacher's instructions, put the paper in their mouths. The reaction is always the same: a number of the students immediately spit the paper out, repelled by the bitter taste of PTC. A few students, however, can't taste the PTC and continue to suck on the paper, wondering what all the spitting is about. Variation among individuals in a trait such as the ability to taste PTC is termed a polymorphism.

The ability to taste PTC is inherited as an autosomal dominant trait in humans. The frequencies of taster and nontaster alleles have been estimated in hundreds of human populations worldwide. Almost all populations have both tasters and nontasters, but the frequencies of the two alleles vary widely.

PTC is not found in nature, but the ability to taste it is strongly correlated with the ability to taste other naturally occurring bitter compounds, some of which are toxic. The ability to taste PTC has also been linked to dietary preferences and may be associated with susceptibility to certain diseases, such as thyroid deficiency. These observations suggest that natural selection has played a role in the evolution of the taster trait.

Some understanding of the evolution of the taster trait was gained when well-known population geneticist Ronald A. Fisher and his colleagues took a trip to the zoo in 1939. Fisher wondered whether other primates also might have the ability to taste PTC. To answer this question, he prepared some drinks with different concentrations of PTC and set off for the zoo with his friends, fellow biologists Edmund (E. B.) Ford and Julian Huxley. At the zoo, the PTC-laced drinks were offered to eight chimpanzees and one orangutan. Fisher and his friends were initially concerned that they might not be able to tell whether the apes could taste the PTC. That concern disappeared, however, when the first one sampled the drink and immediately spat on Fisher. Of the eight chimpanzees tested, six were tasters and two were nontasters.

The observation that chimpanzees and humans both have the PTC taster polymorphism led Fisher and his friends to assume that the polymorphism arose in a common ancestor of humans and chimpanzees, which passed it to both species. However, they had no way to test their hypothesis. Sixty-five years later, geneticists armed with the latest molecular genetic techniques were able to determine the origin of the PTC taster polymorphism and test the hypothesis of Fisher and his friends.

Molecular studies revealed that our ability to taste PTC is controlled by alleles at the *TAS2R38* locus, a 1000-bp gene found on chromosome 7. This locus encodes receptors for bitter compounds and is expressed in the cells of our taste buds. One common allele encodes a receptor that confers the ability to taste PTC; an alternative allele encodes a receptor that does not respond to PTC.

Recent research has demonstrated that PTC taste sensitivity in chimpanzees is also controlled by alleles at the *TAS2R38* locus. However, much to the investigators' surprise, the taster alleles in humans and in chimpanzees are not the same at the molecular level. In the human taster and nontaster alleles, nucleotide differences at three positions affect which amino acids are present in the taste-receptor protein. In chimpanzees, none of these nucleotide differences are present. Instead, a mutation in the initiation codon produces the nontaster allele. This substitution eliminates the normal initiation codon, and the ribosome initiates translation at an alternative downstream initiation codon, resulting in the production of a shortened receptor protein that fails to respond to PTC.

What these findings mean is that Fisher and his friends were correct that humans and chimpanzees both have the PTC taster polymorphism, but were incorrect in their hypothesis about its origin: humans and chimpanzees evolved the PTC taster polymorphism independently.

THINK-PAIR-SHARE

- Explain how natural selection might be responsible for the PTC taster polymorphism. Why might some populations have a higher frequency of the taster allele than others?

- What does the finding that the PTC taster polymorphism evolved independently in humans and chimpanzees suggest about the evolutionary forces responsible for the PTC taster polymorphism? For example, does it suggest that natural selection might be involved? Explain your answer.

This chapter is about the genetic basis of evolution. As illustrated by the PTC taster polymorphism, evolutionary genetics has a long history but has been transformed in recent years by the application of powerful molecular genetic techniques. In Chapter 25, we considered the evolutionary forces that bring about change in the allelic frequencies of a population: mutation, migration, genetic drift, and selection. In this chapter, we examine some specific ways in which these forces shape the genetic makeup of populations and bring about long-term evolution. We begin by looking at how the process of evolution depends on genetic variation and how genetic variation in natural populations is studied. We then turn to the evolutionary changes that result in the appearance of new species, and we see how evolutionary histories (phylogenies) can be constructed. We end the chapter by taking a look at patterns of evolutionary change at the molecular level.

26.1 Evolution Occurs Through Genetic Change within Populations

The concept of evolution is one of the foundational principles of all of biology. Theodosius Dobzhansky, an important early leader in the field of evolutionary genetics, once remarked, "Nothing in biology makes sense except in the light of evolution." Indeed, evolution is an all-encompassing theory that helps to make sense of much of the natural world, from the sequences of DNA found in our cells to the types of organisms that surround us. The evidence for evolution is overwhelming. Evolution has been directly observed numerous times; for example, hundreds of different insect species evolved resistance to common pesticides that were introduced after World War II. The theory of evolution is supported by the fossil record, comparative anatomy, embryology, the distribution of plants and animals (biogeography), and molecular genetics.

THINK-PAIR-SHARE Question 1

Biological Evolution

In spite of its vast importance to all fields of biology, evolution is often misunderstood and misinterpreted. In our society, the term *evolution* frequently refers to any type of change. However, **evolution**, in the biological sense, refers only to a specific type of change: genetic change taking place in a group of organisms. Two aspects of this definition should be emphasized. First, biological evolution includes genetic change only. Many nongenetic changes take place in living organisms, such as the development of a complex, intelligent person from a single-celled zygote. Although remarkable, this change isn't evolution because it does not include genetic changes. Second, biological evolution takes place in *groups* of organisms. An individual organism does not evolve; what evolves is the gene pool common to a group of organisms.

Evolution as a Two-Step Process

Evolution can be thought of as a two-step process. In the first step, genetic variation arises. Genetic variation has its origin in the processes of mutation, which produces new alleles, and recombination, which shuffles alleles into new combinations. Both of these processes are random and produce genetic variation continually, regardless of evolution's requirement for it. The second step in the process of evolution is change in the frequencies of genetic variants. The various evolutionary forces discussed in Chapter 25 cause some alleles in the gene pool to increase in frequency and other alleles to decrease in frequency. This shift in the composition of the gene pool common to a group of organisms constitutes evolutionary change.

We can differentiate between two types of evolution that take place within a group of organisms connected by reproduction (**Figure 26.1**). **Anagenesis** refers to evolution taking place in a single lineage (a group of organisms connected by ancestry) over time. Another type of evolution is **cladogenesis**, the splitting of one lineage into two. When a lineage splits, the two branches no longer have a common gene pool and evolve independently of each other. New species arise through cladogenesis. ▶ TRY PROBLEM 20

THINK-PAIR-SHARE Question 2

CONCEPTS

Biological evolution is genetic change that takes place within a group of organisms. Anagenesis is evolution that takes place within a single lineage; cladogenesis is the splitting of one lineage into two.

✔ CONCEPT CHECK 1

Briefly describe the two steps by which the process of evolution takes place.

Evolution in Bighorn Sheep

Rocky Mountain bighorn sheep (*Ovis canadensis*) are among North America's most spectacular animals, characterized by the male's magnificent horns that curve gracefully back over the ears, spiraling down and back up beside the face (**Figure 26.2**). Two hundred years ago, bighorn sheep were numerous throughout western North America, ranging from Mexico to southern Alberta and from Colorado to California. Meriwether Lewis and William Clark reported numerous

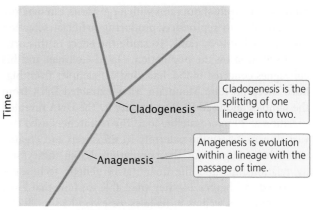

26.1 Anagenesis and cladogenesis are two different types of evolutionary change. Anagenesis is change within an evolutionary lineage; cladogenesis is the splitting of lineages.

sightings of these animals in their expedition across the western United States from 1804 to 1806. Before 1900, there were an estimated 2 million bighorn sheep in North America. But today, fewer than 70,000 of these beautiful animals remain, their numbers decimated by loss of habitat, competition from livestock, diseases carried by domesticated sheep, and hunting.

Bighorn sheep have not just decreased in number as a result of hunting; they have also evolved. Researchers have been studying the evolution of bighorn sheep in response to trophy hunting at Ram Mountain, Canada, since 1973. This population was intensively hunted until 1996, when hunting regulations increased the minimum size of the horns of rams (males) that could be shot, after which the number of rams killed by hunters each year decreased dramatically. In 2011, hunting at Ram Mountain ceased entirely. Trophy

26.2 Microsatellite variation has been used to study the response of bighorn sheep to selective pressure on horn size due to trophy hunting. [Stockbyte/Getty Images.]

hunters selectively shoot rams with large horns, often before they are able to reproduce, producing artificial selection for smaller horns in rams. To study the effect of this artificial selection on the population, David Coltman and his colleagues collected blood, hair, and ear samples from bighorn sheep on Ram Mountain. They extracted DNA from the samples and used PCR to amplify it. The DNA revealed variation at 20 microsatellite loci. The researchers used this variation to determine paternity in the sheep and created a detailed pedigree for the population. Using these family relationships and the quantitative genetic techniques described in Chapter 24, they were able to show that horn size in rams (as well as in females, or ewes) had moderate heritability.

The researchers observed that, in response to selective pressure imposed by the hunters, the horn length of rams decreased by almost 30% between 1973 and 1996. After hunting pressure decreased in 1996, horn length rebounded, increasing by about 13%. Through their genetic analysis, the researchers demonstrated that the decrease in horn length during the period of intensive hunting was an evolutionary response. Although a similar conclusion from an earlier study was criticized by some geneticists, who suggested that factors other than selection might account for the decrease in horn length, a 2016 analysis of the data confirmed that the evolutionary changes observed were due to selection. Ironically, the killing of trophy rams with large horns on Ram Mountain led to a decrease in the very traits prized by the hunters.

26.2 Many Natural Populations Contain High Levels of Genetic Variation

Because genetic variation must be present for evolution to take place, evolutionary biologists have long been interested in the amounts of genetic variation in natural populations and the forces that control the amount and nature of that variation. For many years, they could not examine genes directly and were limited to studying the phenotypes of organisms. Although genetic variation could not be quantified directly, studies of phenotypes suggested that many organisms within the same species harbor considerable genetic variation. Organisms in nature exhibit tremendous phenotypic variation. For example, butterflies of the species *Panaxia dominula* (**Figure 26.3**) differ in spotting patterns, frogs within the species *Lithobates* (*Rana*) *pipiens* display variation in color and spotting patterns, field mice (*Peromyscus maniculatus*) differ in the darkness of their fur, and humans vary in blood types, to mention just a few. Crosses revealed that some of these traits were inherited as simple genetic traits, but for most traits, the precise genetic basis was complex and unknown, preventing early evolutionary geneticists from quantifying the genetic variation in natural populations.

As we saw in Section 24.4, a population's response to selection depends on narrow-sense heritability, which is a measure of the additive genetic variation of a trait within a population. Many organisms respond to artificial selection carried out by humans, suggesting that the populations

Normal homozygotes

Heterozygotes

Recessive bimacula phenotype

26.3 To study genetic variation, early evolutionary geneticists were forced to rely on phenotypic traits that had a simple genetic basis. Variation in the spotting patterns of the butterfly *Panaxia dominula* is an example.

of these organisms contain a great deal of additive genetic variation. For example, humans have used artificial selection to produce numerous dog breeds, which vary tremendously in size, shape, color, and behavior (see Figure 24.21). Early studies of chromosome variations in *Drosophila* and plants also suggested that genetic variation in natural populations is plentiful and widespread.

Molecular Variation

In recent years, advances in molecular genetics have made it possible to investigate evolutionary change directly by analyzing protein and nucleic acid sequences. Techniques such as protein electrophoresis, analysis of microsatellite variation, and DNA sequencing have revolutionized population and evolutionary genetic studies.

ADVANTAGES OF MOLECULAR DATA Molecular data offer a number of advantages for studying the process and patterns of evolution:

Molecular data are genetic. Evolution results from genetic change over time. Many anatomical, behavioral, and physiological traits have a genetic basis, but the relation between the underlying genes and the trait may be complex. Variation in protein and nucleic acid sequences has a clear genetic basis that is often easy to interpret.

Molecular methods can be used with all organisms. Early studies of population genetics relied on simple genetic traits such as human blood types, banding patterns in snails, or spotting patterns in butterflies (see Figure 26.3), which are restricted to a small collection of organisms. However, all living organisms have proteins and nucleic acids, so molecular data can be collected from any organism.

Molecular methods can be applied to a huge amount of genetic variation. An enormous amount of data can be accessed by molecular methods. The human genome, for example, contains 3.2 billion base pairs of DNA, which constitutes a large pool of information about our evolution.

All organisms can be compared with the use of molecular data. Trying to assess the evolutionary history of distantly related organisms is often difficult because they have few characteristics in common. The evolutionary relationships between angiosperms were traditionally assessed by comparing floral anatomy, whereas the evolutionary relationships of bacteria were determined by their nutritional and staining properties. Because plants and bacteria have so few structural characteristics in common, evaluating how they are related to each other was difficult in the past. All organisms have certain molecular traits in common, such as ribosomal RNA sequences and some fundamental proteins. These molecules offer a valid basis for comparisons among all organisms.

Molecular data are quantifiable. Protein- and nucleic acid–sequence data are precise, accurate, and quantifiable, which facilitates the objective assessment of evolutionary relationships.

Molecular data often provide information about the process of evolution. Molecular data can reveal important clues about the process of evolution. For example, the results of a study of DNA sequences revealed that one type of insecticide resistance in mosquitoes probably arose from a single mutation that subsequently spread throughout the world.

The amount of molecular information is large and growing. Today, the very large collection of DNA and protein sequences that are available in publically accessible databases can be used for making evolutionary comparisons and inferring mechanisms of evolution.

DNA SEQUENCE DATA The rapid and inexpensive methods for sequencing DNA that are now available (see Chapter 19) allow the detection of genetic variation in almost any population. DNA sequence data often reveal processes that influence evolution, and they are invaluable for determining the evolutionary relationships of different organisms. The use of PCR means that data can be obtained from a very small initial sample of DNA, which often facilitates studies of wild populations. For example, genetic variation in wolf populations (see the introduction to Chapter 25) has been measured and analyzed from DNA isolated from samples of wolf scat (feces).

In one example of how molecular variation has been used in evolutionary studies, Alfred Roca and his colleagues used DNA sequencing to reassess the genetic relationships among African elephants. They obtained tissue samples from 195 elephants by shooting them with needlelike darts that fell to the ground after hitting an elephant, but retained a small plug of skin. From the skin samples, the scientists sequenced 1732 base pairs of DNA from four nuclear genes. Their analysis revealed large genetic differences between forest elephants and savannah elephants, suggesting that there is limited gene flow (migration) between these two groups of elephants. On the basis of these results, the scientists proposed that two different species of elephants exist in Africa.

POPULATION VARIATION Numerous studies have demonstrated that most populations of organisms possess large amounts of variation in their protein and DNA sequences. Individuals within populations are not identical in their DNA sequences, but instead differ at many nucleotide sites.

One explanation for the extensive molecular variation that is observed in natural populations is the **neutral-mutation hypothesis**, which proposes that much molecular variation is adaptively neutral; that is, that individuals with different molecular variants have equal fitness. This hypothesis, first developed by Motoo Kimura, suggests that most variations in DNA and protein sequences are functionally equivalent. Because they are functionally equivalent, natural selection does not differentiate between them, and their evolution is shaped largely by the random processes of genetic drift and mutation. The neutral-mutation hypothesis does not preclude the importance of natural selection; rather, it suggests that when selection occurs, it is largely directional, favoring the "best" allele while eliminating others. Thus, natural selection is viewed as an evolutionary force that largely limits variation.

Research has shown, however, that some genetic variation is maintained by natural selection. In these cases, genetic variants are not functionally equivalent; instead, they result in phenotypic effects that cause differences in reproduction (fitness differences). Selection that maintains variation is called **balancing selection**. Overdominance, in which the heterozygote has higher fitness than either homozygote (see Section 25.4), is one type of balancing selection. With overdominance, both alleles are maintained in a population because heterozygous individuals have the highest fitness and possess both alleles, so neither allele becomes fixed in the population. An example of overdominance is variation in the β-globin gene of humans that results in sickle-cell anemia: individuals heterozygous for an allele that causes sickle-cell anemia are resistant to malaria and thus have higher fitness than either homozygotes for the normal allele or homozygotes for the sickle-cell allele (see p. 770 in Chapter 25). Thus, both sickle-cell and normal alleles have been maintained in populations exposed to malaria. Because balancing selection maintains genetic variation within natural populations, when it is present, there will be much variation present in the population.

CONCEPTS

Molecular techniques and data offer a number of advantages for evolutionary studies. The neutral-mutation hypothesis proposes that much molecular variation is neutral with regard to natural selection and is shaped largely by mutation and genetic drift. Variation within populations is also maintained by balancing selection.

✔ CONCEPT CHECK 2

Which statement about the neutral-mutation hypothesis is true?
a. All proteins are functionless.
b. Natural selection plays no role in evolution.
c. Many molecular variants are functionally equivalent.
d. All of the above

26.3 New Species Arise Through the Evolution of Reproductive Isolation

In Section 26.2, we discussed two types of evolution: anagenesis (change within a lineage) and cladogenesis (the splitting of lineages). Cladogenesis occurs through speciation, the process by which one population separates into two distinct evolutionary groups.

The term *species* literally means "kind" or "appearance"; **species** are different kinds or types of living organisms. In many cases, species differences are easy to recognize: a horse is clearly a different species than a chicken. Sometimes, however, species differences are not so clear. Some species of *Plethodon* salamanders are so similar in appearance that they can be distinguished only by looking at their proteins or genes.

The concept of a species has two primary uses in biology. First, a species is a particular type of organism to which a unique name has been given. For effective communication, biologists must use a standard set of names for the organisms that they study, and species names serve that purpose. When a geneticist talks about conducting crosses with *Drosophila melanogaster*, other biologists immediately understand which organism was used. The second use of the term *species* is in an evolutionary context: a species is considered an evolutionarily independent group of organisms.

The Biological Species Concept

What kinds of differences are required to consider two organisms different species? A widely used definition of a species is the **biological species concept**, first fully developed by evolutionary biologist Ernst Mayr in 1942. Mayr was primarily interested in the biological characteristics that are responsible for separating organisms into independently evolving units. He defined a species as a group of organisms whose members are capable of interbreeding with one another but are reproductively isolated from the members of other species. In other words, members of the same species have the biological potential to exchange genes, and members of different species cannot exchange genes. Because different species do not exchange genes, each species evolves independently.

Not all biologists adhere to the biological species concept, and there are several problems associated with it. For example, reproductive isolation, on which the biological species concept is based, cannot be determined from fossils, and in practice, it is often difficult to determine whether even living species are biologically capable of exchanging genes. There are many examples of organisms that are accepted as different species that sometimes exchange genes, or *hybridize*. Furthermore, the biological species concept cannot be applied to asexually reproducing organisms, such as bacteria. In practice, many species are distinguished on the basis of phenotypic (usually anatomical) differences. Biologists often assume that phenotypic differences represent underlying genetic differences; if the phenotypes of two organisms are quite different, then they probably cannot and do not interbreed in nature.

Because of these problems, some biologists have proposed alternative definitions for a species. For example, the *morphospecies concept* defines a species based entirely on phenotypic (morphological) similarities and differences. The *phylogenetic species concept* defines a species as the smallest recognizable group that has a unique evolutionary history. In this chapter, we use the biological species concept because it is widely used and is based on reproductive differences.

Reproductive Isolating Mechanisms

The key to species differences under the biological species concept is reproductive isolation: the existence of biological characteristics that prevent genes from being exchanged between different species. Any biological factor or mechanism that prevents gene exchange is termed a **reproductive isolating mechanism**.

PREZYGOTIC REPRODUCTIVE ISOLATING MECHANISMS Some species are separated by **prezygotic reproductive isolating mechanisms**, which prevent gametes from two different species from fusing and forming a hybrid zygote. In **ecological isolation**, members of two species do not encounter one another, and therefore do not reproduce with one another, because they have different ecological niches, living in different habitats and interacting with the environment in different ways. For example, some species of forest-dwelling birds feed and nest in the forest canopy, whereas others confine their activities to the forest floor. Because they never come into contact, these birds are reproductively isolated from one another. Other species are separated by **behavioral isolation**, differences in behavior that prevent interbreeding. Many male frogs attract females of the same species by using a unique, species-specific call. Two closely related frog species may use the same pond, but never interbreed, because females are attracted only to the call of their own species.

Another type of prezygotic reproductive isolation is **temporal isolation**, in which reproduction in different species takes place at different times of the year. Some species of plants do not exchange genes because they flower at different times of the year. **Mechanical isolation** results from anatomical differences that prevent successful copulation. This type of isolation is seen in many insects, in which closely related species differ in their male and female genitalia, so that copulation between them is physically impossible. Finally, some species are separated by **gametic isolation**, in which mating between individuals of different species may take place, but the gametes do not form zygotes. Male gametes may not survive in the female reproductive tract, or may not be attracted to the female gametes. In other cases, male and female gametes meet, but are too incompatible to fuse to form a zygote. Gametic isolation is seen in many plants, in which pollen from one species cannot fertilize the ovules of another species.

POSTZYGOTIC REPRODUCTIVE ISOLATING MECHANISMS Other species are separated by **postzygotic reproductive isolating mechanisms**, in which gametes of two species may fuse and form a zygote, but there is no gene flow between the two species, either because the resulting hybrids are inviable or sterile or because reproduction breaks down in subsequent generations.

If prezygotic reproductive isolating mechanisms fail or have not yet evolved, mating between two organisms of different species may take place, and a hybrid zygote containing genes from two different species may be formed. In many cases, such species are still separated by **hybrid inviability**, in which incompatibility between the genomes of the two species prevents the hybrid zygote from developing. Hybrid inviability is seen in some groups of frogs, in which mating between different species and fertilization take place, but the resulting embryos never complete development.

Other species are separated by **hybrid sterility**, in which hybrid embryos complete development, but are sterile, so that genes are not passed between the two parental species. Donkeys and horses frequently mate and produce viable offspring—mules—but most mules are sterile, so there is no gene flow between donkeys and horses (but see Problem 47 at the end of Chapter 8).

Finally, some closely related species are capable of mating and producing viable and fertile F_1 progeny. However, genes do not flow between the two species because of **hybrid breakdown**, in which further crossing of the hybrids produces inviable or sterile offspring. For example, crosses among species of African cichlid fishes produce F_2 progeny that have lower viability than F_1 progeny or nonhybrid progeny. Hybrid breakdown is thought to be due the breakup of gene interaction as alleles from the different species mix in different combination in the F_2 and later generations.

Many species are separated by multiple prezygotic and postzygotic reproductive isolating mechanisms. The different types of reproductive isolating mechanisms are summarized in **Table 26.1**.

THINK-PAIR-SHARE Questions 3 and 4

CONCEPTS

The biological species concept defines a species as a group of potentially interbreeding organisms that are reproductively isolated from the members of other species. Under this definition, species are separated by prezygotic or postzygotic reproductive isolating mechanisms.

✔ **CONCEPT CHECK 3**

Which of the following is an example of postzygotic reproductive isolation?

a. Sperm of species A dies in the oviduct of species B before fertilization can take place.

b. Hybrid zygotes between species A and B are spontaneously aborted early in development.

c. The mating seasons of species A and B do not overlap.

d. Males of species A are not attracted to the pheromones produced by the females of species B.

TABLE 26.1	Types of reproductive isolating mechanisms
Type	**Characteristics**
Prezygotic	**Acts before a zygote has formed**
Ecological	Differences in habitat; individuals do not meet
Temporal	Reproduction takes place at different times
Mechanical	Anatomical differences prevent copulation
Behavioral	Differences in mating behavior prevent mating
Gametic	Gametes are incompatible or not attracted to each other
Postzygotic	**Acts after a zygote has formed**
Hybrid inviability	Hybrid zygote does not survive to reproduction
Hybrid sterility	Hybrid is sterile
Hybrid breakdown	F_1 hybrids are viable and fertile, but F_2 are inviable or sterile

Modes of Speciation

Speciation is the process by which new species arise. In terms of the biological species concept, speciation comes about through the evolution of reproductive isolating mechanisms.

There are two principal ways in which new species arise. **Allopatric speciation** occurs when a geographic barrier splits a population into two groups and blocks the exchange of genes between them. The interruption of gene flow then leads to the evolution of genetic differences that result in reproductive isolation. **Sympatric speciation** is speciation that arises in the absence of any external barrier to gene flow; reproductive isolating mechanisms evolve within a single population.

ALLOPATRIC SPECIATION Allopatric speciation is initiated when a geographic barrier splits a population into two or more groups and prevents gene flow between those groups (**Figure 26.4a**). Geographic barriers can take a number of forms. Uplifting of a mountain range may split a population of lowland plants into separate groups on each side of the mountains. Oceans serve as effective barriers for many types of terrestrial organisms, separating individuals on different islands from one another and from those on the mainland. Rivers often separate populations of fish located in separate drainages. The erosion of mountains may leave populations of alpine plants isolated on separate mountain peaks.

After two populations have been separated by a geographic barrier that prevents gene flow between them, they evolve independently (**Figure 26.4b**). Their genetic isolation allows each population to accumulate genetic differences from the other population; these genetic differences arise through natural selection, unique mutations, and genetic drift (if the populations are small). This genetic differentiation may eventually lead to prezygotic and postzygotic reproductive isolation. It is important to note that both types of reproductive isolation arise simply as a consequence of genetic divergence.

If the geographic barrier that once separated the two populations disappears, or if individuals are able to disperse over it, the populations come into secondary contact (**Figure 26.4c**). At this point, several outcomes are possible. If limited genetic differentiation has taken place during the separation of the populations, reproductive isolating mechanisms may not have evolved or may be incomplete. Genes will flow between the two populations, eliminating any genetic differences that did arise, and the populations will remain a single species.

A second possible outcome is that genetic differentiation during separation has led to prezygotic reproductive isolating mechanisms; in this case, the two populations have become different species. A third possible outcome is that during their time apart, some genetic differentiation has taken place between the populations, leading to incompatibility in their genomes and postzygotic reproductive

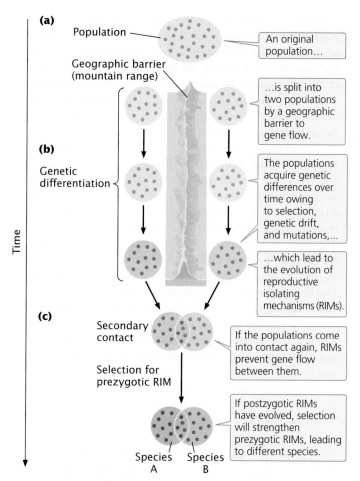

26.4 Allopatric speciation is initiated by a geographic barrier to gene flow between two populations.

isolation. If postzygotic isolating mechanisms have evolved, any mating between individuals from the different populations will produce hybrid offspring that are inviable or sterile. Individuals that mate only with members of their own population will therefore have higher fitness than individuals that mate with members of the other population, so natural selection will increase the frequency of any trait that prevents interbreeding between members of the two populations. Over time, prezygotic reproductive isolating mechanisms will evolve. In short, if some postzygotic reproductive isolation exists, natural selection will favor the evolution of prezygotic reproductive isolating mechanisms to prevent wasted reproductive effort by individuals mating with members of the other population. This process of postzygotic reproductive isolation leading to the evolution of prezygotic isolating mechanisms is termed *reinforcement*. Mathematical models of evolution support the idea that reinforcement can increase reproductive isolation. Although reinforcement is difficult to prove in natural populations, some field studies have provided evidence that reinforcement does occur in nature.

A number of variations in this general model of allopatric speciation are possible. Many new species probably arise when a small group of individuals becomes geographically isolated from the main population; for example, a few individuals of a mainland population might migrate to a geographically isolated island. In this situation, founder effect and genetic drift play a larger role in the evolution of genetic differences between the populations.

SPECIATION IN DARWIN'S FINCHES An excellent example of allopatric speciation can be found in Darwin's finches, a group of birds that Charles Darwin discovered on the Galápagos Islands during his voyage aboard the *Beagle*. The Galápagos Islands form an archipelago located about 900 km off the coast of South America (**Figure 26.5**). Consisting of more than a dozen large islands and many smaller ones, the Galápagos formed from volcanoes that erupted over a geological hot spot that has remained stationary while the geological plate over it moved eastward over the past 3 million years. The movement of the plate pulled newly formed islands eastward, so the islands to the east (San Cristóbal and Española) are older than those to the west (Isabela and Fernandina). Over time, the number of islands in the archipelago increased as new volcanoes arose.

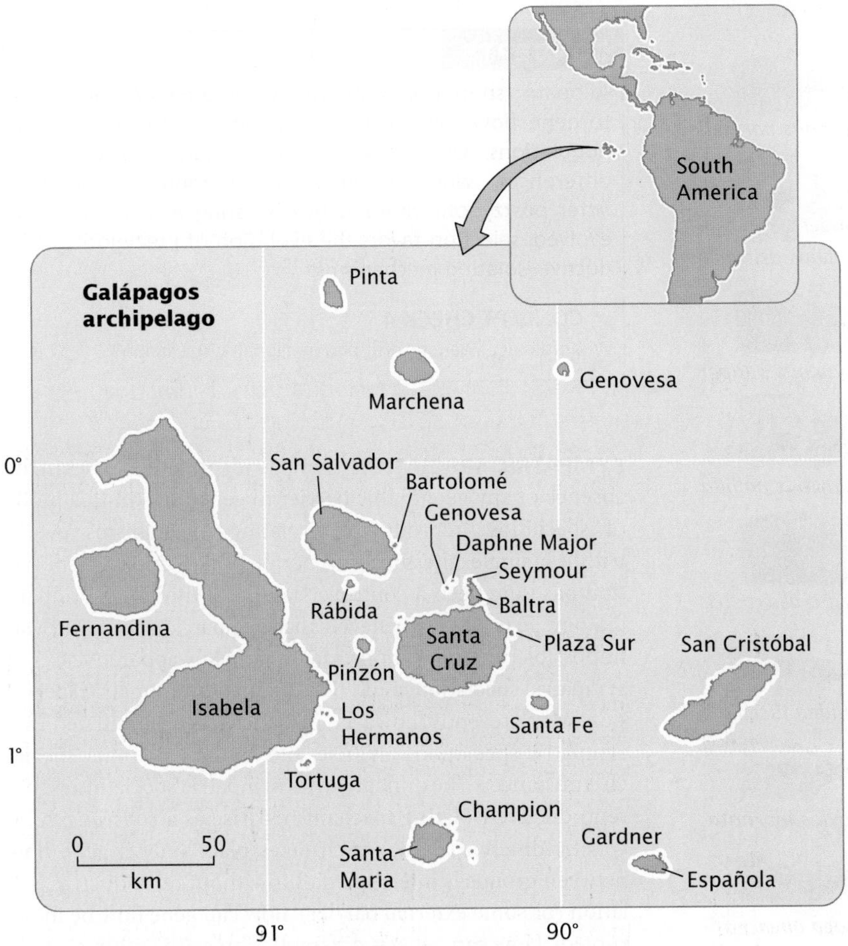

26.5 The Galápagos Islands are geologically young and are volcanic in origin. The oldest islands are to the east. [Information from *Philosophical Transactions of the Royal Society of London, Series B* 351:756–772, 1996.]

Darwin's finches consist of 14 species that are found on various islands in the Galápagos archipelago (**Figure 26.6**). An additional species is found on Cocos Island, which is some 780 km to the north of the Galápagos. The birds vary in the shapes and sizes of their beaks, which are adapted for eating different types of food. Recent studies of the development of finch embryos have helped to reveal some of the molecular details of how differences in beak shapes have evolved (see Section 22.5). Genetic studies have demonstrated that all the finches are closely related and that they evolved from a single ancestral species that migrated to the islands from the coast of South America some 2 million to 3 million years ago. The evolutionary relationships among the 14 Galápagos species, based on studies of DNA microsatellite data, are depicted in Figure 26.6. Most of the species are separated by a behavioral isolating mechanism (song in particular), but some of the species can and do hybridize in nature.

The first finches to arrive in the Galápagos probably colonized one of the larger eastern islands. A breeding population became established and increased over time. At some point, a few birds dispersed to another island, where they were effectively isolated from the original population, and established a new population. The new population underwent genetic differentiation owing to genetic drift and adaptation to the local conditions of the island, and it eventually became reproductively isolated from the original population. Individual birds from the new population then dispersed to other islands and gave rise to additional species. This process was repeated many times. Occasionally, newly evolved species dispersed to an island where another species was already present, giving rise to secondary contact between the species. Today, many of the islands have more than one resident finch species.

Researchers have now sequenced the entire genomes of all 14 Galápagos species and the Cocos Island species, providing a rich data set with which to interpret the history of the group. The evolutionary relationships based on these new data agreed, for the most part, with phylogenies based on earlier data (see Figure 26.6), but the genome sequences revealed some additional complexities. For example, the birds traditionally classified as *Geospiza difficilis* (the sharp-beaked ground finch), which are found on different islands, appear to comprise at least three different species. The researchers also found genetic evidence of extensive interbreeding between some species, which gave rise to species with a mix of ancestry.

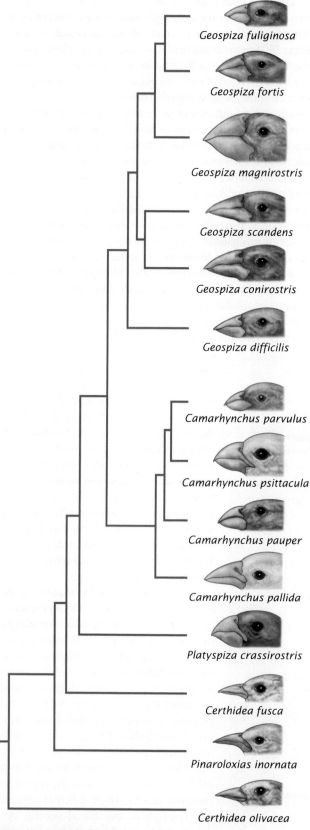

Geospiza fuliginosa

Geospiza fortis

Geospiza magnirostris

Geospiza scandens

Geospiza conirostris

Geospiza difficilis

Camarhynchus parvulus

Camarhynchus psittacula

Camarhynchus pauper

Camarhynchus pallida

Platyspiza crassirostris

Certhidea fusca

Pinaroloxias inornata

Certhidea olivacea

26.6 The Galápagos Islands are home to 14 species of Darwin's finches, which evolved from a single ancestral species that migrated to the islands and underwent repeated allopatric speciation. This phylogeny is based on DNA microsatellite variation. [After B. R. Grant and P. R. Grant, *Bioscience* 53:965–975, 2003.]

The researchers also examined the DNA sequences associated with differences in beak shape and found variations at 15 chromosomal regions that were associated with differences between blunt and pointed beaks. Six of these regions contained genes that had been previously associated with cranial or facial development in birds or mammals, including the calmodulin gene discussed in Section 22.5. The strongest association was with the *Aristaless-like homeobox 1 (ALX1)* gene, which encodes a transcription factor that plays a role in cranial and facial development in mice and humans. Almost all birds with blunt beaks were homozygous for one haplotype at this locus, whereas most birds with sharp beaks were homozygous for a different haplotype.

The ages of the 14 finch species have been estimated with data from mitochondrial DNA. As **Figure 26.7** shows, there is a strong correspondence between the number of species present at various times in the past and the number of islands in the archipelago. This correspondence is one of the most compelling pieces of evidence for the theory that the different species of finches arose through allopatric speciation.

THINK-PAIR-SHARE Question 5

> **CONCEPTS**
>
> Allopatric speciation is initiated when a geographic barrier to gene flow splits a single population into two or more populations. Over time, the populations evolve genetic differences, which bring about reproductive isolation. After postzygotic reproductive isolating mechanisms have evolved, selection favors the evolution of prezygotic reproductive isolating mechanisms.
>
> ✔ **CONCEPT CHECK 4**
> What role does genetic drift play in allopatric speciation?

SYMPATRIC SPECIATION Sympatric speciation arises in the absence of any geographic barrier to gene flow; in this mode of speciation, reproductive isolating mechanisms evolve within a single interbreeding population. Sympatric speciation has long been a controversial topic within evolutionary biology. Ernst Mayr believed that sympatric speciation was impossible, and he demonstrated that many apparent cases of sympatric speciation could be explained by allopatric speciation. More recently, however, evidence has accumulated that sympatric speciation can arise, and has arisen, under special circumstances. The difficulty with sympatric speciation is that reproductive isolating mechanisms arise as a *consequence* of genetic differentiation, which takes place only if gene flow between groups is interrupted. But without reproductive isolation (or some external barrier), how can gene flow be interrupted? How can genetic differentiation arise within a single group that is freely exchanging genes?

Most models of sympatric speciation assume that genetic differentiation is initiated by selection favoring different phenotypes taking place within a single population. For example,

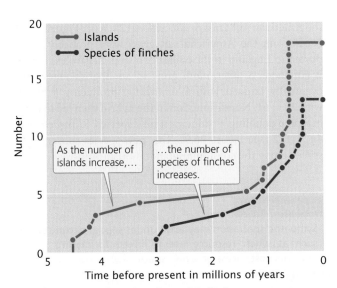

26.7 The number of species of Darwin's finches present at various times in the past corresponds with the number of islands in the Galápagos archipelago. [Data from P. R. Grant, B. R. Grant, and J. C. Deutsch, Speciation and hybridization in island birds, *Philosophical Transactions of the Royal Society of London, Series B* 351:765–772, 1996.]

imagine that one homozygote (A^1A^1) is strongly favored on one resource (perhaps the plant species that is host to an insect) and the other homozygote (A^2A^2) is favored on a different resource (perhaps a different host plant). Heterozygotes (A^1A^2) have lower fitness than either homozygote on both resources. In this situation, natural selection will favor genotypes at other loci that cause positive assortative mating (matings between like individuals, see Section 25.3), so that matings between A^1A^1 and A^2A^2, which would produce A^1A^2 offspring with low fitness, will be avoided.

Now imagine that alleles at a second locus affect mating behavior, such that C^1C^1 individuals prefer mating only with other C^1C^1 individuals, and C^2C^2 individuals prefer mating with other C^2C^2 individuals. If alleles at the A locus are nonrandomly associated with alleles at the C locus so that only A^1A^1 C^1C^1 individuals and A^2A^2 C^2C^2 individuals exist, then gene flow between individuals using the different resources will be restricted, allowing the two groups to evolve further genetic differences that might lead to reproductive isolation and sympatric speciation.

The difficulty with this model is that recombination quickly breaks up the nonrandom associations between genotypes at the two loci, producing individuals such as A^1A^1 C^2C^2, which would prefer to mate with A^2A^2 C^2C^2 individuals. This mating would produce all A^1A^2 offspring, which do poorly on both resources. Thus, even limited recombination will prevent the evolution of the mating-preference genes.

CONCEPTS

Sympatric speciation arises within a single interbreeding population without any geographic barrier to gene flow.

REPRODUCTIVE ISOLATION IN APPLE MAGGOT FLIES

Sympatric speciation is more probable if the genes that affect resource utilization also affect mating preferences. Genes that affect both resource utilization and mating preference are indeed present in some host races—populations of specialized insect species that feed on different host plants. Researchers have studied what appear to be initial stages of speciation in host races of the apple maggot fly (*Rhagoletis pomonella*, **Figure 26.8**). Flies of this species feed on the fruits of a specific host tree. Mating takes place on and near the fruits, and the flies lay their eggs on the ripened fruits, where their larvae grow and develop. *Rhagoletis pomonella* originally existed only on the fruits of hawthorn trees, which are native to North America; 150 years ago, *R. pomonella* was first observed on cultivated apples, which are related to hawthorns but a different species. Infestations of apples by this new apple host race of *R. pomonella* quickly spread, and today, many apple trees throughout North America are infested with the flies.

The apple host race of *R. pomonella* probably originated when a few flies acquired a mutation that allowed them to feed on apples instead of hawthorn fruits. Because mating takes place on and near the fruits, flies that use apples are more likely to mate with other flies that use apples, leading to genetic isolation between flies using hawthorns and those using apples. Indeed, researchers found that some genetic differentiation has already taken place between the two host races. Flies lay their eggs on ripening fruit, so there has been strong selection on the flies to synchronize their reproduction with the period when the fruit of their host species is ripening. Apples ripen several weeks earlier than hawthorns. Correspondingly, the peak mating period of the apple host race is 3 weeks earlier than that of the hawthorn host race. This difference in the timing of reproduction between apple and hawthorn races has further reduced gene flow—to about 4%—between the two host races. Analyses of protein variation and DNA microsatellites have demonstrated that the apple and hawthorn races now differ significantly at a number of loci. These differences have evolved in the past 150 years, and evolution appears to be ongoing.

26.8 Host races of the apple maggot fly (*Rhagoletis pomonella*) have evolved some reproductive isolation without any geographic barrier to gene flow. [Joseph Berger, Bugwood.org.]

In an effort to better understand the genetic changes that took place as *R. pomonella* shifted from hawthorns to apples, researchers carried out a selection experiment. They collected larvae from the hawthorn race of *R. pomonella* and exposed them as pupae to either 7 warm days (their normal conditions) or 32 warm days (conditions experienced by flies that breed on earlier-fruiting apples) prior to winter temperatures. As a control, they also collected larvae from the apple race and exposed them to the 7-day treatment that represents normal conditions for the hawthorn race. DNA samples from the surviving flies were isolated and their genomes sequenced.

Over 32,000 nucleotide sites were identified where flies differed in which DNA base was present (single-nucleotide polymorphisms, or SNPs). Of these, a total of 2245 SNPs showed significant differences in frequencies of the bases present between the short (7-day) and long (32-day) pre-winter treatments; the locations of these differences were scattered across the genome. The genetic changes that were observed in the selection experiment mirrored genetic differences that are seen today in natural populations of the hawthorn and apple races. Apparently, natural selection in the experiment increased the frequency of genetic variants that occur naturally within the hawthorn race to produce the differences that are now found between the apple and hawthorn races. These observations suggest that the selection experiment re-created, in a single generation, the early evolutionary changes that occurred in *R. pomonella* as it shifted from hawthorns to apples 150 years ago. The results also suggest that numerous genetic differences scattered across the genome accompanied the transition of the flies from hawthorns to apples.

Although genetic differentiation has taken place between apple and hawthorn host races of *R. pomonella* and some degree of reproductive isolation has evolved between them, reproductive isolation is not yet complete, and speciation has not fully taken place. **▶ TRY PROBLEM 21**

SPECIATION THROUGH POLYPLOIDY A special type of sympatric speciation takes place through polyploidy. Polyploid organisms have more than two sets of chromosomes (3n, 4n, 5n, etc.). As discussed in Section 8.4, allopolyploidy often arises when two diploid species hybridize, producing 2n hybrid offspring, and nondisjunction in one of the hybrid offspring then produces a 4n tetraploid. Because this tetraploid contains exactly two copies of each chromosome, it is usually fertile and is reproductively isolated from the two parental species by differences in chromosome number (see Figure 8.27).

Numerous species of flowering plants are allopolyploids. Speciation through polyploidy was observed when it led to a new species of salt-marsh grass that arose along the coast of England around 1870. This polyploid contains genomes of the European salt grass *Spartina maritima* (2n = 60) and

the American salt grass *S. alterniflora* (2n = 62; **Figure 26.9**). Seeds from the American salt grass were probably transported to England in the ballast of a ship. Regardless of how it got there, *S. alterniflora* grew in an English marsh and eventually crossed with *S. maritima*, producing a hybrid with 2n = 61. Nondisjunction in the hybrid then led to chromosome doubling, producing a new species, *S. anglica*, with 4n = 122 (see Figure 26.9). This new species subsequently spread along the coast of England.

CONCEPTS

Sympatric speciation may arise under special circumstances, such as when resource use is linked to mating preference (in host races) or when species hybridization leads to allopolyploidy.

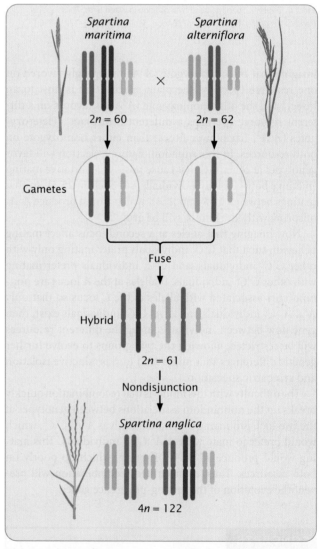

26.9 *Spartina anglica* **arose sympatrically through allopolyploidy.**

Genetic Differentiation Associated with Speciation

As we have seen, genetic differentiation leads to the evolution of reproductive isolating mechanisms, which restrict gene flow between populations and lead to speciation. How much genetic differentiation is required for reproductive isolation to take place? This question has received considerable study by evolutionary geneticists, but unfortunately, there is no universal answer. Some newly formed species differ in many genes, whereas others appear to have undergone divergence in just a few genes.

One group of organisms that has been extensively studied for genetic differences associated with speciation is the genus *Drosophila*. The *Drosophila willistoni* species complex consists of at least 12 species found in Central and South America in various stages of the process of speciation. Using analysis of genetic differences in proteins, Francisco Ayala and his colleagues genotyped flies from different geographic populations (populations that had limited genetic differences), subspecies (populations with considerable genetic differences), sibling species (newly arisen species), and nonsibling species (older species). The loci examined were used to estimate the overall degree of genetic differentiation and were not assumed to be directly involved in reproductive isolation. For each group, they computed a measure of genetic similarity, which ranges from 1 to 0 and represents the overall level of genetic differentiation (**Table 26.2**). They found that there was a general decrease in genetic similarity as flies evolve from geographic populations to subspecies to sibling species to nonsibling species. These data suggest that considerable genetic differentiation at many loci is required for speciation to arise. Another study of *D. simulans* and *D. melanogaster*, two species that produce inviable hybrids when crossed, suggested that at least 200 genes contribute to the inviability of hybrids between the two species.

However, other studies suggest that speciation can arise through changes in just a few genes. For example, *D. heteroneura* and *D. silvestris* are two species of Hawaiian fruit flies that exhibit behavioral reproductive isolation. The isolation is determined largely by differences in head shape: *D. heteroneura* has a hammer-shaped head with widely separated eyes that is recognized by females of the same species but rejected by *D. silvestris*

females. Genetic studies indicate that only a few loci (about 10) determine the differences in head shape.

In another study, researchers examined genetic differences between two closely related species of birds, the collared flycatcher (*Ficedula albicollis*) and pied flycatcher (*F. hypoleuca*), both of which occur in Europe. These two species are believed to have diverged from a common ancestor less than 2 million years ago, probably by allopatric speciation associated with glaciation. In areas where both species occur together, collared flycatchers and pied flycatchers sometimes mate, but the two species are separated by both pre- and postzygotic reproductive isolating mechanisms. For example, females prefer males with the plumage and song of their own species, and hybrids do not survive as well as offspring of parents that are both from the same species.

Researchers at Uppsala University in Sweden completely sequenced the genomes of 10 male birds of each species. They found that large parts of the genomes of the two species were the same: for example, only about 5 out of 1000 nucleotides differed between species, compared with 4 out of 1000 nucleotides within species. The differences that did exist between the species were not uniform across the genome; rather, the differences tended to be concentrated in a few "divergence islands," which had sequence differences up to 50 times greater than the average for the genome. A few divergence islands were found on each chromosome, and they occurred primarily near centromeres and telomeres, suggesting that differences in chromosome structure play an important role in speciation. The divergence islands tended to show few differences within species, indicating that natural selection had favored different genes in each region.

CONCEPTS

Some newly arising species have a considerable number of genetic differences; others have few genetic differences.

26.4 The Evolutionary History of a Group of Organisms Can Be Reconstructed by Studying Changes in Homologous Characteristics

The evolutionary relationships among a group of organisms are termed a **phylogeny**. Because most evolution takes place over long periods and is not amenable to direct observation, biologists must reconstruct phylogenies by inferring the evolutionary relationships among present-day organisms. The discovery of fossils of ancestral organisms can aid in the reconstruction of phylogenies, but the fossil record is often too poor to be of much help. Thus, biologists are

TABLE 26.2	Genetic similarity in groups of the *Drosophila willistoni* complex
Group	**Mean Genetic Similarity**
Geographic populations	0.970
Subspecies	0.795
Sibling species	0.517
Nonsibling species	0.352

often restricted to analyses of characteristics in present-day organisms to determine their evolutionary relationships. In the past, phylogenies were reconstructed on the basis of phenotypic characteristics—often, anatomical traits. Today, molecular data, including protein and DNA sequences, are frequently used for this purpose.

Phylogenies are reconstructed by inferring changes that have taken place in homologous characteristics: those that have evolved from the same character in a common ancestor. For example, although the front leg of a mouse and the wing of a bat look different and have different functions, close examination of their structure and development reveals that they are indeed homologous; both evolved from the forelimb of an early mammal that was an ancestor to both mouse and bat. And, because mouse and bat have these homologous features and others in common, we know that they are both mammals. Similarly, DNA sequences are homologous if two present-day sequences evolved from a single sequence found in an ancestor. For example, all eukaryotic organisms have a gene for cytochrome c, an enzyme that helps carry out oxidative respiration. This gene is assumed to have arisen in a single organism in the distant past and to have been passed down to descendants of that early ancestor. Today, all copies of the cytochrome c gene are homologous because they all evolved from the same original copy in the distant ancestor of all organisms that possess this gene.

A graphical representation of a phylogeny is called a **phylogenetic tree**. As shown in **Figure 26.10**, a phylogenetic tree depicts the evolutionary relationships among different organisms, in a manner similar to a pedigree that represents the genealogical relationships among family members. A phylogenetic tree consists of branches and nodes. The **branches** are the evolutionary connections between organisms. In some phylogenetic trees, the lengths of the branches represent the amount of evolutionary divergence that has taken place. The **nodes** are the points where the branches split; they represent common ancestors that existed before divergence took place. In most cases, the nodes represent past ancestors that are inferred from the analysis. When one internal node represents a common ancestor to all other nodes on the tree, the tree is said to be **rooted**. Trees are often rooted by including in the analysis one or more organisms that are distantly related to all the others; such a distantly related organism is referred to as an *outgroup*.

Phylogenetic trees are created to depict the evolutionary relationships among organisms; they are also created to depict the evolutionary relationships among DNA sequences. The latter type of phylogenetic tree is termed a **gene tree** (**Figure 26.11**). **> TRY PROBLEM 26**

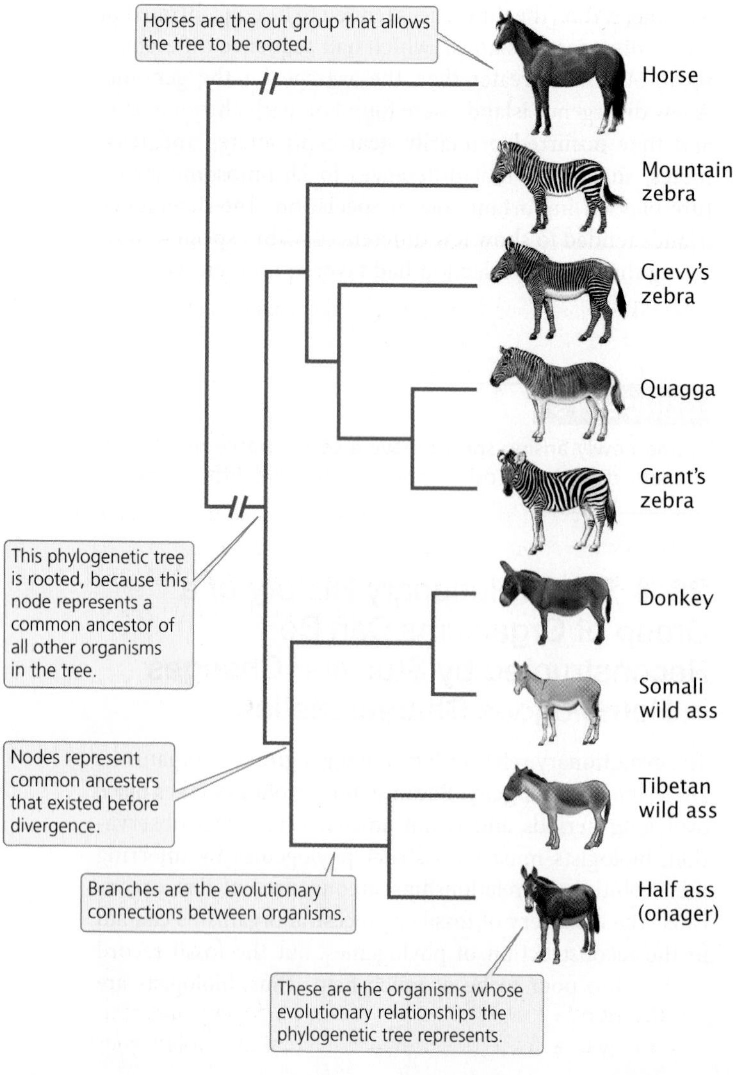

26.10 A phylogenetic tree is a graphical representation of the evolutionary relationships among a group of organisms. This phylogeny of zebras and asses is based on DNA sequences from 20,374 protein-encoding genes obtained by complete sequencing of the genomes of the species. Domesticated horses are used as an outgroup to root the tree. [Data from Jonsson, H. 2014. Speciation with gene flow in equids despite extensive chromosomal plasticity. *Proceedings of the National Academy of Science,* 111:18655-18660.]

CONCEPTS

A phylogeny represents the evolutionary relationships among a group of organisms. It is often depicted graphically by a phylogenetic tree, which consists of nodes representing the organisms and branches representing their evolutionary connections.

✔ **CONCEPT CHECK 5**

Which of the following features is found in a rooted tree but not in an unrooted tree?
a. Terminal nodes
b. Internal nodes
c. A common ancestor to all other nodes
d. Branch lengths that represent the amount of evolutionary divergence between nodes

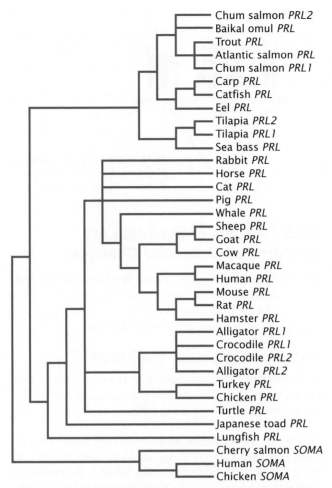

26.11 A gene tree can be used to represent the evolutionary relationships among a group of genes. This gene tree is a rooted tree. *PRL* represents a prolactin gene; *PRL1* and *PRL2* are two different prolactin genes found in the same organism; and *SOMA* represents a somatropin gene, which is related to prolactin genes. [Data from M. P. Simmons and J. V. Freudestein, Uninode coding vs. gene tree parsimony for phylogenetic reconstruction using duplicate genes, *Molecular Phylogenetics and Evolution* 23:488, 2002.]

The Alignment of Homologous Sequences

Today, phylogenetic trees are often constructed from DNA sequence data. This construction requires that homologous sequences be compared. Thus, a first step is to identify homologous genes and properly align their nucleotide bases. Consider the following short sequences that might be found in two different organisms (most actual sequences used would be much longer):

Nucleotide position	1 2 3 4 5 6 7 8
Gene *X* from species A	5′–A T T G C G A A–3′
Gene *X* from species B	5′–A T G C C A A C–3′

These two sequences can be aligned in several possible ways. We might assume that there have been base substitutions at positions 3, 4, 6, and 8:

Nucleotide position	1 2 3 4 5 6 7 8
Gene *X* from species A	5′–A T T G C G A A–3′
Gene *X* from species B	5′–A T G C C A A C–3′

Alternatively, we might assume that a nucleotide at position 3 has been inserted or deleted, generating a gap in the sequence of species B, and that there has been a single nucleotide substitution at position 6:

Nucleotide position	1 2 3 4 5 6 7 8
Gene *X* from species A	5′–A T T G C G A A–3′
Gene *X* from species B	5′–A T – G C C A A C–3′

The second alignment requires fewer evolutionary steps (a deletion or insertion plus one base substitution) than does the first alignment (four base substitutions).

Sequence alignments are usually made by computer programs that include assumptions about which types of change are more likely to take place. If two sequences have undergone a great deal of divergence, generating alignments can be difficult.

The Construction of Phylogenetic Trees

Consider a simple phylogeny that depicts the evolutionary relationships among three organisms: humans, chimpanzees, and gorillas. Charles Darwin originally proposed that chimpanzees and gorillas were closely related to humans, and modern research supports a close relationship between these three species. There are three possible phylogenetic trees for humans, chimpanzees, and gorillas (**Figure 26.12**). The goal of the evolutionary biologist is to determine which of the trees is correct. Molecular data applied to this question strongly suggest a close relationship between humans and chimpanzees.

To understand the difficulty in constructing phylogenetic trees, let's consider for a moment the number of possible trees that might exist for a group of organisms. The number of possible rooted trees for a group of organisms is

$$\text{number of rooted trees} = \frac{(2N - 3)!}{2^{N-2}(N - 2)!}$$

where N equals the number of organisms included in the phylogeny, and the ! symbol stands for factorial, the product of all the integers from N to 1. Substituting values of N into this equation, we find the following numbers of possible rooted trees:

Number of organisms included in phylogeny (N)	Number of rooted trees
2	1
3	3
4	15
5	105
10	34,459,425
20	8.2×10^{21}

As the number of organisms in the phylogeny increases beyond just a few, the number of possible rooted trees

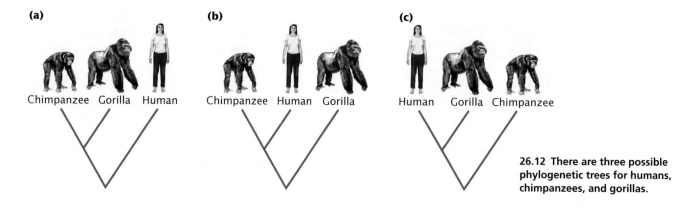

26.12 There are three possible phylogenetic trees for humans, chimpanzees, and gorillas.

becomes astronomically large. Clearly, choosing the best tree by directly comparing all the possibilities is impossible.

There are several different approaches to inferring evolutionary relationships and constructing phylogenetic trees. In one approach, termed the *distance approach*, evolutionary relationships are inferred on the basis of the overall degree of similarity between organisms. Typically, a number of different phenotypic characteristics or gene sequences are examined, and the organisms are grouped on the basis of their overall similarity, taking into consideration all the examined characteristics and sequences. A second approach, called the *maximum parsimony approach*, infers phylogenetic relationships on the basis of the fewest number of evolutionary changes that must have taken place since the organisms last had an ancestor in common. A third approach, called *maximum likelihood* and *Bayesian methods*, infers phylogenetic relationships on the basis of which phylogeny maximizes the probability of obtaining the set of characteristics exhibited by the organisms. In this approach, a phylogeny with a higher probability of producing the observed characteristics in the organisms studied is preferred over a phylogeny with a lower probability. Maximum likelihood and Bayesian methods incorporate models of how evolutionary change takes place. For example, they incorporate the probability that one particular DNA base will be substituted for a different base.

With all three approaches to constructing phylogenies, several different numerical methods are available for the construction of phylogenetic trees. All include certain assumptions that help limit the number of different trees that must be considered; most rely on computer programs that compare phenotypic characteristics or sequence data to sequentially group organisms in the construction of the tree.

CONCEPTS

Molecular data can be used to infer phylogenies (evolutionary histories) of groups of living organisms. The construction of phylogenies requires the proper alignment of homologous DNA sequences. Several different approaches are used to reconstruct phylogenies, including distance methods, maximum parsimony methods, and maximum likelihood and Bayesian methods.

26.5 Patterns of Evolution Are Revealed by Molecular Changes

Our ability to analyze genetic variation at the molecular level has revealed a number of evolutionary processes and features that were formerly unsuspected. This section considers several aspects of evolution at the molecular level.

Rates of Molecular Evolution

Findings from molecular studies of numerous genes have demonstrated that different genes, and even different parts of the same gene, may evolve at different rates.

RATES OF NUCLEOTIDE SUBSTITUTION Rates of evolutionary change in nucleotide sequences are usually measured as the rate of nucleotide substitution, which is the number of substitutions taking place per nucleotide site per year within a population. To calculate the rate of nucleotide substitution, we begin by looking at homologous sequences from different organisms. We first align the homologous sequences, and then compare the sequences and determine the number of nucleotides that differ between the two sequences. We might compare the growth-hormone sequences for mice and rats, which diverged from a common ancestor some 15 million years ago. From the number of nucleotide differences between their growth-hormone genes, we compute the number of nucleotide substitutions that must have taken place since they diverged. Because the same site may have mutated more than once, the number of nucleotide substitutions is larger than the number of nucleotide differences between two sequences; special mathematical methods have been developed for inferring the actual number of substitutions likely to have taken place.

When we have the number of nucleotide substitutions per nucleotide site, we can use the total amount of evolutionary time that separates the two organisms (usually obtained from the fossil record) to obtain an overall rate of nucleotide substitution. For the mouse and rat growth-hormone genes, the overall rate of nucleotide substitution is approximately 8×10^{-9} substitutions per site per year.

RATES OF NONSYNONYMOUS AND SYNONYMOUS SUBSTITUTION Nucleotide changes in a gene that alter the amino acid sequence of a protein are referred to as nonsynonymous substitutions. Nucleotide changes, particularly those at the third position of a codon, that do not alter the amino acid sequence of a protein are called synonymous substitutions. The rate of nonsynonymous substitution varies widely among mammalian genes. The rate for the α-actin protein is only 0.01×10^{-9} substitutions per site per year, whereas the rate for interferon γ is 2.79×10^{-9}, almost 300 times higher. The rate of synonymous substitution also varies among genes, but not as much as the nonsynonymous rate. For most protein-encoding genes, the rate of synonymous substitution is considerably higher than the nonsynonymous rate (**Table 26.3**) because synonymous mutations have little or no effect on fitness—that is, they are selectively neutral. Nonsynonymous mutations, on the other hand, alter the amino acid sequence of the protein and, in many cases, are detrimental to the fitness of the organism; most of these mutations are eliminated by natural selection.

THINK-PAIR-SHARE Question 6

SUBSTITUTION RATES FOR DIFFERENT PARTS OF A GENE Different parts of a gene also evolve at different rates. The highest rates of substitution occur in those regions of the gene that have the least effect on function, such as the

third position of a codon, flanking regions, and introns (**Figure 26.13**). The 5′ and 3′ flanking regions of genes are not transcribed into RNA; therefore, substitutions in these regions do not alter the amino acid sequence of the protein, although they may affect gene expression (see Chapters 16 and 17). Rates of substitution in introns are nearly as high as those in flanking regions. Although these nucleotides do not encode amino acids, introns must be spliced out of the pre-mRNA for a functional protein to be produced, and particular sequences are required at the 5′ splice site, 3′ splice site, and branch point for correct splicing (see Chapter 14).

Substitution rates are somewhat lower in the 5′ and 3′ untranslated regions of a gene. These regions are transcribed into RNA but do not encode amino acids. The 5′ untranslated region contains the ribosome-binding site, which is essential for translation, and the 3′ untranslated region

TABLE 26.3	Rates of nonsynonymous and synonymous substitutions in mammalian genes based on human–rodent comparisons	
Gene	**Nonsynonymous Rate (per site per 10^9 years)**	**Synonymous Rate (per site per 10^9 years)**
α-Actin	0.01	3.68
β-Actin	0.03	3.13
Albumin	0.91	6.63
Aldolase A	0.07	3.59
Apoprotein E	0.98	4.04
Creatine kinase	0.15	3.08
Erythropoietin	0.72	4.34
α-Globin	0.55	5.14
β-Globin	0.80	3.05
Growth hormone	1.23	4.95
Histone 3	0.00	6.38
Immunoglobulin heavy chain (variable region)	1.07	5.66
Insulin	0.13	4.02
Interferon α1	1.41	3.53
Interferon γ	2.79	8.59
Luteinizing hormone	1.02	3.29
Somatostatin-28	0.00	3.97

Source: After W. Li and D. Graur, *Fundamentals of Molecular Evolution* (Sunderland, Mass.: Sinauer Associates, 1991), p. 69, Table 1.

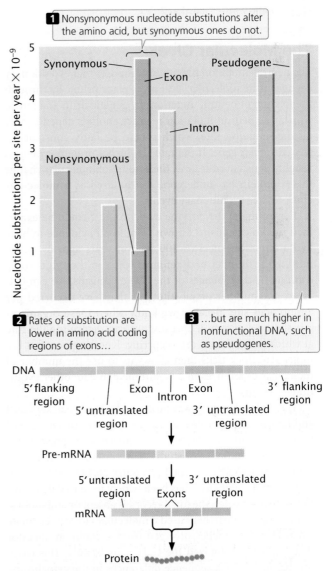

26.13 Different parts of genes evolve at different rates. The highest rates of nucleotide substitution are in sequences that have the least effect on protein function.

contains sequences that may function in regulating mRNA stability and translation, so substitutions in these regions may have deleterious effects on organismal fitness and may not be tolerated.

The lowest rates of substitution are seen for nonsynonymous changes in exons because these substitutions always alter the amino acid sequence of the protein and are often deleterious. High rates of substitution occur in *pseudogenes*, most of which are duplicate copies of genes that have been rendered nonfunctional by mutations. Such genes no longer produce a functional product, so mutations in pseudogenes have little effect on the fitness of the organism.

In summary, there is a relation between the function of a DNA sequence and its rate of evolution: the highest rates of change are found where changes have the least effect on function. This observation fits with the neutral-mutation hypothesis, which predicts that molecular variation is not affected by natural selection. **▶ TRY PROBLEM 29**

THINK-PAIR-SHARE Question 7

The Molecular Clock

The neutral-mutation hypothesis proposes that evolutionary change at the molecular level takes place primarily through the fixation of neutral mutations by genetic drift. The rate at which one neutral mutation replaces another depends only on the mutation rate, which should be fairly constant for any particular gene. If the rate at which a protein evolves is roughly constant over time, the amount of molecular change that a protein has undergone can be used as a **molecular clock** to date evolutionary events.

For example, the enzyme cytochrome *c* could be examined in two organisms known from fossil evidence to have had a common ancestor 400 million years ago. By determining the number of differences between the cytochrome *c* amino acid sequences of the two organisms, we can calculate the number of substitutions that have occurred per amino acid site. Because we know when these organisms last shared a common ancestor, we can determine the rate at which substitutions are occurring. Knowing how fast the molecular clock ticks then allows us to use the number of molecular differences in cytochrome *c* to date other evolutionary events.

The molecular clock was proposed by Emile Zuckerkandl and Linus Pauling in 1965 as a possible means of dating evolutionary events on the basis of molecules in present-day organisms. A number of studies have examined the rate of evolutionary change in proteins (**Figure 26.14**) and in genes, and the molecular clock has been widely used to date evolutionary events when the fossil record is absent or ambiguous. For example, researchers used a molecular clock to estimate when Darwin's finches diverged from a common ancestor that originally colonized the Galápagos Islands. This clock was based on DNA sequence differences in the cytochrome *b* gene. The researchers concluded that the ancestor of

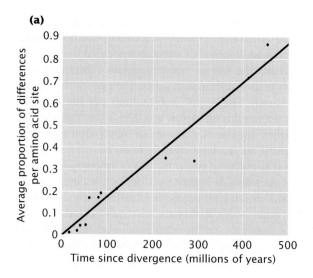

(a)

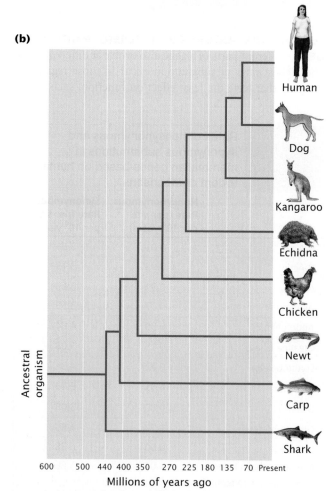

(b)

26.14 The molecular clock is based on the assumption of a constant rate of change in protein or DNA sequences. (a) Relation between the rate of amino acid substitution and time since divergence, based in part on the amino acid sequences of α-globin from the eight species shown in part *b*. The rate of evolution in protein and DNA sequences has been used as a molecular clock to date past evolutionary events. (b) Phylogeny of eight of the species that are plotted in part *a* and their approximate times of divergence based on the fossil record.

Darwin's finches arrived in the Galápagos and began diverging some 2 million to 3 million years ago. The results of several studies have shown, however, that the molecular clock does not always tick at a constant rate, particularly over shorter time periods, and this method remains controversial.

CONCEPTS

Different genes and different parts of the same gene evolve at different rates. Those parts of genes that have the least effect on function tend to evolve at the highest rates. The idea that individual proteins and genes evolve at a constant rate and that the differences in the sequences of present-day organisms can be used to date past evolutionary events is referred to as the molecular clock.

✔ **CONCEPT CHECK 6**

In general, which of the following is expected to exhibit the lowest rate of evolutionary change?

a. Synonymous changes in amino acid–coding regions of exons
b. Nonsynonymous changes in amino acid–coding regions of exons
c. Introns
d. Pseudogenes

Evolution Through Changes in Gene Regulation

One of the challenges of evolutionary biology is understanding the genetic basis of adaptation. Many evolutionary changes occur with relatively little genetic change. For example, humans and chimpanzees differ greatly in anatomy, physiology, and behavior, yet they differ at only about 4% of their DNA sequences (see the introduction to Chapter 17). Evolutionary biologists have long assumed that many anatomical differences result not from the evolution of new genes, but rather from relatively small DNA differences that alter the expression of existing genes. Recent research in evolutionary genetics has focused on how evolution occurs through alteration of gene expression.

An example of adaptation that has occurred through changes in regulatory sequences is seen in the evolution of pigmentation in *Drosophila melanogaster* fruit flies in Africa. Most fruit flies are light tan in color, but flies in some African populations have much darker abdomens. These darker flies usually occur in mountainous regions at high elevations. Indeed, 59% of pigmentation variation among populations within Sub-Saharan Africa can be explained by differences in elevation (**Figure 26.15**). Researchers have demonstrated that these differences are genetically determined and that natural selection has favored darker pigmentation at high elevations. High-elevation populations are exposed to lower temperatures, and the darker pigmentation is assumed to help flies absorb more solar radiation and better regulate their body temperature in these environments.

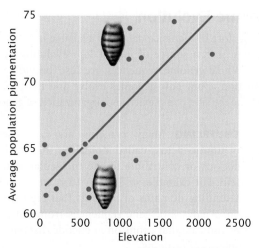

26.15 Sub-Saharan African populations of the fruit fly *Drosophila melanogaster* exhibit a positive association between pigmentation and elevation.

How did flies at high elevations evolve darker color? Genetic studies indicate that the dark abdominal pigmentation seen in flies from these populations results from variation at or near a locus called *ebony*. The *ebony* locus encodes a multifunctional enzyme that produces a yellow exoskeleton; the absence of this enzyme produces a dark phenotype. Sequencing of the *ebony* locus of flies from light and dark populations found no differences in the coding region of the *ebony* gene. However, molecular analysis revealed a marked reduction in the amount of *ebony* mRNA in darker flies, suggesting that the difference in pigmentation is due not to mutations in the *ebony* gene itself, but rather to changes in its expression. Further investigation detected genetic differences within an enhancer that is about 3600 bp upstream of the *ebony* gene. Dark and light flies differed in over 120 nucleotides scattered over 2400 bp of the enhancer. By experimentally creating enhancers with different combinations of these mutations, researchers determined that five of the mutations are responsible for the majority of the difference in pigmentation.

The results of these studies suggest that over time, high-elevation populations accumulated multiple mutations in the enhancer, which reduced the expression of the *ebony* locus and caused darker pigmentation. Further analysis suggested that these mutations were added sequentially. Some of the mutations are widespread throughout Africa; it is assumed that these existing mutations were favored by natural selection in high-elevation populations and increased in frequency because they helped the flies thermoregulate in colder environments. Other mutations that are seen only in the high-elevation populations probably arose as new mutations within those populations and were quickly favored by natural selection.

THINK-PAIR-SHARE Question 8

Genome Evolution

The vast store of sequence data now available in DNA databases has been a source of insight into evolutionary processes. Whole-genome sequences are also providing new information about how genomes evolve and the processes that shape the size, complexity, and organization of genomes.

EXON SHUFFLING Many proteins are composed of groups of amino acids, called domains, that have discrete functions or contribute to the molecular structure of a protein. For example, in Section 16.1, we considered the DNA-binding domains of proteins that regulate gene expression. Analyses of gene sequences from eukaryotic organisms indicate that exons often encode discrete functional domains of proteins.

Some genes elongated and evolved new functions when one or more exons duplicated and underwent divergence. For example, the human serum-albumin gene is made up of three copies of a sequence that encodes a protein domain consisting of 195 amino acids. Additionally, the genes that encode human immunoglobulins have undergone repeated tandem duplications, creating many similar *V*, *J*, *D*, and *C* segments (see Section 22.6) that enable the immune system to respond to almost any foreign substance that enters the body.

A comparison of DNA sequences from different genes reveals that new genes have repeatedly evolved through a process called **exon shuffling**, in which exons of different genes are exchanged, creating genes that are mosaics of other genes. For example, tissue plasminogen activator (TPA) is an enzyme that contains four domains of three different types, called kringle, growth factor, and finger. Each domain is encoded by a different exon. The gene for TPA is believed to have acquired its exons from other genes that encode different proteins: the kringle exon came from the plasminogen gene, the growth-factor exon came from the epidermal growth factor gene, and the finger exon came from the fibronectin gene. The mechanism by which exon shuffling takes place is poorly understood, but new proteins with different combinations of functions encoded by other genes apparently have repeatedly evolved by this mechanism.

THINK-PAIR-SHARE Question 9

GENE DUPLICATION New genes have also evolved through the duplication of whole genes and their subsequent divergence. This process creates **multigene families**, sets of genes that are similar in sequence but encode different products. For example, humans possess 13 different genes found on chromosomes 11 and 16 that encode globinlike molecules, which take part in oxygen transport (**Figure 26.16**). All of these genes have a similar structure, with three exons separated by two introns. They are assumed to have evolved through repeated duplication and divergence from a single globin gene in a distant ancestor. This ancestral gene is thought to have been most similar to the present-day myoglobin gene, and it was probably first duplicated to produce an α/β-globin precursor gene and the myoglobin gene. The α/β-globin gene then underwent another duplication to give rise to a primordial α-globin gene and a primordial β-globin gene. Subsequent duplications led to multiple α-globin and β-globin genes. Similarly, vertebrates contain four clusters of *Hox* genes, each comprising from 9 to 11 genes. *Hox* genes play an important role in development (see pp. 672–673 in Chapter 22).

Some gene families include genes that are arrayed in tandem on the same chromosome; others are dispersed among different chromosomes. Gene duplication is a common occurrence in eukaryotic genomes; for example, about 5% of the human genome consists of duplicated segments.

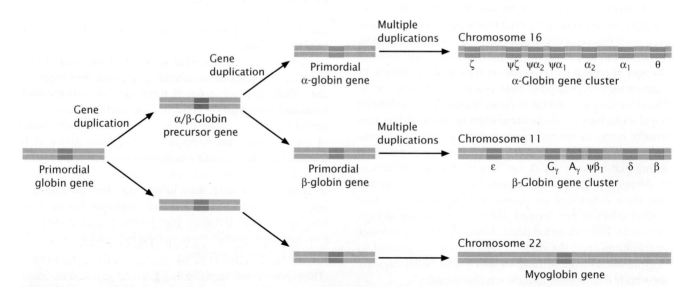

26.16 The human globin gene family has evolved through successive gene duplications.

Gene duplication provides a mechanism for the addition of new genes with novel functions. Once a gene has been duplicated, there are two copies of that gene, one of which is then free to change and potentially take on a new function. The extra copy of the gene may, for example, become active at a different time in development, or be expressed in a different tissue, or even diverge and encode a protein containing different amino acids. The most common outcome of gene duplication, however, is that one copy acquires a mutation that renders it nonfunctional and becomes a pseudogene. Pseudogenes are common in the genomes of complex eukaryotes; the human genome is estimated to contain as many as 20,000 pseudogenes.

WHOLE-GENOME DUPLICATION In addition to the duplication of individual genes, whole genomes of some organisms have been duplicated in the past. For example, a comparison of the genome of the yeast *Saccharomyces cerevisiae* with the genomes of other fungi reveals that *S. cerevisiae*, or one of its immediate ancestors, underwent a whole-genome duplication that generated two copies of every gene. Many of the copies subsequently acquired new functions; others acquired mutations that destroyed their original function and then diverged into random DNA sequences.

Whole-genome duplication can take place through polyploidy, as we saw in Section 26.3. During their evolution, plants have undergone a number of whole-genome duplications in this way. While polyploidy is less frequent in animals, genetic evidence suggests that several whole-genome duplication events have occurred during animal evolution. In 1970, Susumu Ohno proposed that early vertebrates underwent two rounds of genome duplication. Called the 2R hypothesis, this idea has been controversial, but recent data from genome sequencing have provided support for it. Among vertebrates, polyploidy has been particularly common in amphibians.

HORIZONTAL GENE TRANSFER Traditionally, scientists assumed that organisms acquire their genomes through vertical transmission—transfer through the reproduction of genetic information from parents to offspring—and most phylogenetic trees assume vertical transmission of genetic information. Findings from DNA sequence studies reveal that DNA sequences are sometimes transmitted by horizontal gene transfer, in which DNA is transferred between individuals of different species (see Section 9.3). This process is especially common among bacteria, and there are a number of documented cases in which genes have been transferred from bacteria to eukaryotes. The extent of horizontal gene transfer among eukaryotic organisms is controversial, as there are few well-documented cases. Horizontal gene transfer can obscure phylogenetic relationships and make the reconstruction of phylogenetic trees difficult.

One apparent case of horizontal gene transfer among eukaryotes is the presence in some aphids of genes for enzymes that synthesize carotenoids. Carotenoids are colored compounds produced by bacteria, archaea, fungi, and plants. Many animals also possess carotenoids, but they lack the enzymes necessary to make the compounds themselves; in almost all cases, animals obtain carotenoids from their food.

Aphids—small insects that feed on plants—possess carotenoids, which are responsible for color differences between and within species. Some aphids are green and contain α-, β-, and γ-carotene, which are all yellow carotenoids. Other aphids are red or brown and contain lycopene or torulene, carotenoids that are red. One species, the pea aphid (*Acyrthosiphon pisum*), contains both green and red individuals, and these color differences are genetically inherited. Many researchers previously assumed that the color differences were due to carotenoids that were acquired in the aphids' food.

The sequencing of the entire genome of *A. pisum* provided researchers with the opportunity to determine whether the aphids possess their own genes for carotenoid synthesis. Examination of the genome sequence revealed that pea aphids have several genes that code for carotenoid-synthesizing enzymes. Interestingly, these genes are closely related to carotenoid-synthesizing genes found in some fungi. The evidence suggests that in the distant past, an aphid acquired its carotenoid genes from a fungus through horizontal gene transfer and then passed the genes to other aphids through vertical transmission.

> **CONCEPTS**
>
> New genes may evolve through the duplication of exons, shuffling of exons, duplication of genes, and duplication of whole genomes. Genes can be passed among different species through horizontal gene transfer.

CONCEPTS SUMMARY

- Evolution is genetic change that takes place within a group of organisms. It is a two-step process: (1) genetic variation arises, and (2) genetic variants change in frequency.

- Anagenesis refers to change within a single lineage; cladogenesis is the splitting of one lineage into two.

- Molecular methods offer a number of advantages for the study of evolution.

- Most natural populations contain large amounts of genetic variation. The neutral-mutation hypothesis proposes that most molecular variants are selectively neutral and that their evolution is shaped largely by mutation and genetic drift. Some molecular variation is maintained by balancing selection.

■ A species can be defined as a group of organisms that are capable of interbreeding with one another and are reproductively isolated from the members of other species.

■ Species are prevented from exchanging genes by prezygotic or postzygotic reproductive isolating mechanisms.

■ Allopatric speciation arises when a geographic barrier prevents gene flow between two populations. With the passage of time, the two populations acquire genetic differences that may lead to reproductive isolation.

■ Sympatric speciation arises when reproductive isolation exists in the absence of any geographic barrier. It may arise under special circumstances.

■ Some species arise only after populations have undergone considerable genetic differentiation; others arise after changes have taken place in only a few genes.

■ A phylogeny can be represented by a phylogenetic tree, consisting of nodes that represent organisms and branches that represent their evolutionary connections.

■ Approaches to constructing phylogenetic trees include the distance approach, the maximum parsimony approach, and the maximum likelihood and Bayesian methods.

■ Different parts of a gene show different amounts of genetic variation. In general, those parts that have the least effect on function evolve at the highest rates.

■ The molecular-clock hypothesis proposes a constant rate of molecular change, providing a means of dating evolutionary events by looking at sequence differences between present-day organisms.

■ Genome evolution takes place through the duplication and shuffling of exons, the duplication of genes to form multigene families, whole-genome duplication, and horizontal transfer of genes.

IMPORTANT TERMS

evolution (p. 780)
anagenesis (p. 781)
cladogenesis (p. 781)
neutral-mutation hypothesis (p. 783)
balancing selection (p. 783)
species (p. 784)
biological species concept (p. 784)
reproductive isolating mechanism (p. 784)

prezygotic reproductive isolating mechanism (p. 784)
ecological isolation (p. 784)
behavioral isolation (p. 785)
temporal isolation (p. 785)
mechanical isolation (p. 785)
gametic isolation (p. 785)

postzygotic reproductive isolating mechanism (p. 785)
hybrid inviability (p. 785)
hybrid sterility (p. 785)
hybrid breakdown (p. 785)
speciation (p. 786)
allopatric speciation (p. 786)
sympatric speciation (p. 786)

phylogeny (p. 791)
phylogenetic tree (p. 792)
branch (p. 792)
node (p. 792)
rooted tree (p. 792)
gene tree (p. 792)
molecular clock (p. 796)
exon shuffling (p. 798)
multigene family (p. 798)

ANSWERS TO CONCEPT CHECKS

1. First, genetic variation arises. Then various evolutionary forces cause changes in the frequency of genetic variants.

2. c

3. b

4. Genetic drift can bring about changes in the allelic frequencies of a population and lead to genetic differences among populations. Genetic differentiation is the cause of postzygotic and prezygotic reproductive isolation between populations that leads to speciation.

5. c

6. b

COMPREHENSION QUESTIONS

Section 26.1

1. How is biological evolution defined?

2. What are the two steps in the process of evolution?

3. How is anagenesis different from cladogenesis?

Section 26.2

4. What are some of the advantages of using molecular data in evolutionary studies?

5. What is the key difference between the neutral-mutation hypothesis and balancing selection?

6. Describe some of the methods that have been used to study variation in DNA.

Section 26.3

7. What is the biological species concept?

8. What is the difference between prezygotic and postzygotic reproductive isolating mechanisms? List some different types of each.

9. What is the basic difference between the allopatric and sympatric modes of speciation?

10. Briefly outline the process of allopatric speciation.

11. What are some of the difficulties with sympatric speciation?

12. Briefly explain how switching from hawthorn fruits to apples has led to genetic differentiation and partial reproductive isolation in *Rhagoletis pomonella*.

Section 26.4

13. Draw a simple phylogenetic tree and identify a node, a branch, and an outgroup.

14. Briefly describe differences among the distance approach, the maximum parsimony approach, and the maximum likelihood approach to the reconstruction of phylogenetic trees.

Section 26.5

15. Outline the different rates of evolution that are typically seen in different parts of a protein-encoding gene. What might account for these differences?

16. What is the molecular clock?

17. What is exon shuffling? How can it lead to the evolution of new genes?

18. What is a multigene family? What processes produce multigene families?

19. Define horizontal gene transfer. What problems does it cause for evolutionary biologists?

APPLICATION QUESTIONS AND PROBLEMS

Section 26.1

*20. The following illustrations represent two different patterns of evolution. Briefly discuss the differences in these two patterns with regard to how evolutionary change (on the *x* axis) occurs with respect to time (on the *y* axis).

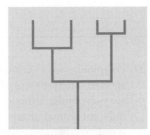

Evolutionary change   Evolutionary change

Section 26.3

*21. Which of the isolating mechanisms listed in **Table 26.1** have partly evolved between apple and hawthorn host races of *Rhagoletis pomonella*, the apple maggot fly?

22. We considered the sympatric evolution of reproductive isolating mechanisms in host races of *Rhagoletis pomonella*, the apple maggot fly. The wasp *Diachasma alloeum* parasitizes apple maggot flies, laying its eggs on the larvae of the flies. Immature wasps hatch from the eggs and feed on the fly larvae. Research by Andrew Forbes and his colleagues (Forbes et al. 2009. *Science* 323:776–779) demonstrated that wasps that parasitize the apple race of *R. pomonella* are genetically differentiated from those that parasitize the hawthorn race. They also found that wasps that prey on the apple race of the flies are attracted to odors from apples, whereas wasps that prey on the hawthorn race are attracted to odors from hawthorn fruits. Propose an explanation for how genetic differences might have evolved between the wasps that parasitize the two host races of *R. pomonella*. How might these differences lead to speciation in the wasps?

23. Which of the salt-marsh grasses in **Figure 26.9** is a polyploid?

   a. *Spartina maritima*

   b. *Spartina alterniflora*

   c. Hybrid between *Spartina maritima* and *Spartina alterniflora*

   d. *Spartina anglica*

   e. Both c and d

Section 26.4

24. How many rooted trees are theoretically possible for a group of seven organisms? How many for twelve organisms?

25. Align the sequences below so as to maximize their similarity. What is the minimum number of evolutionary steps that separate these two sequences?

   TTGCAAAC
   TGAAACTG

*26. Michael Bunce and his colleagues in England, Canada, and the United States extracted and sequenced mitochondrial DNA from fossils of Haast's eagle, a gigantic eagle that was driven to extinction 700 years ago when humans first arrived in New Zealand (M. Bunce et al. 2005. *PLOS Biology* 3:44–46). Using mitochondrial DNA sequences from living eagle species and those from Haast's-eagle fossils, they created the accompanying phylogenetic tree. On this phylogenetic tree, identify (a) all nodes; (b) one example of a branch; and (c) the outgroup.

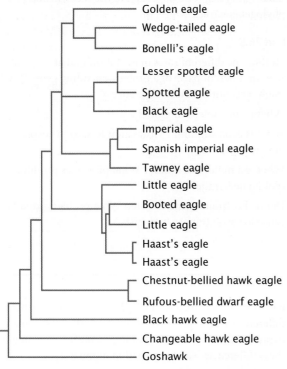

Golden eagle
Wedge-tailed eagle
Bonelli's eagle
Lesser spotted eagle
Spotted eagle
Black eagle
Imperial eagle
Spanish imperial eagle
Tawney eagle
Little eagle
Booted eagle
Little eagle
Haast's eagle
Haast's eagle
Chestnut-bellied hawk eagle
Rufous-bellied dwarf eagle
Black hawk eagle
Changeable hawk eagle
Goshawk

[After M. Bunce et al., *PLOS Biology* 3:44–46, 2005.]

27. On the basis of the phylogeny of Darwin's finches shown in **Figure 26.6**, predict which two species in each of the following groups will be the most similar genetically.

   a. *Camarhynchus parvulus, Camarhynchus psittacula, Camarhynchus pallida*

   b. *Camarhynchus parvulus, Camarhynchus pallida, Platyspiza crassirostris*

   c. *Geospiza difficilis, Geospiza conirostris, Geospiza scandens*

   d. *Camarhynchus parvulus, Certhidea fusca, Pinaroloxias inornata*

Section 26.5

28. If a gene had similar rates of nonsynonymous and synonymous substitutions, what might that suggest about the evolution of that gene?

*29. Based on the information provided in **Figure 26.13**, do introns or 3′ untranslated regions of a gene have higher rates of nucleotide substitution? Explain why.

CHALLENGE QUESTIONS

Section 26.3

30. Explain why natural selection may cause prezygotic reproductive isolating mechanisms to evolve if postzygotic reproductive isolating mechanisms are already present, but can never cause the evolution of postzygotic reproductive isolating mechanisms.

31. Polyploidy is very common in flowering plants: approximately 40% of all flowering plant species are polyploids. Although polyploidy exists in many different animal groups, it is much less common. Why is polyploidy more common in plants than in animals? Give one or more possible reasons.

THINK-PAIR-SHARE QUESTIONS

Section 26.1

1. Consider Theodosius Dobzhansky's remark, "Nothing in biology makes sense except in the light of evolution." Why is evolution so important to the field of biology? Give some specific examples of how the theory of evolution helps us to make sense of biology.

2. Evolution is often misunderstood and misinterpreted. What are some common misconceptions about evolution? How are these misconceptions wrong?

Section 26.3

3. Examine **Table 26.1**, which lists different types of reproductive isolating mechanisms. Try to come up with at least one example of an organism that is reproductively isolated from other species by each of the mechanisms listed in the table. Are some groups of organisms more likely to exhibit one type of mechanism than another? Give some examples.

4. The biological species concept is based on the assumption that species are reproductively isolated and do not share genes. And yet a number of organisms that are considered different species hybridize (mate and exchange genes). Hybridization between different species is more common in plants than in animals. Propose some possible reasons for this difference.

5. One of Darwin's finches, the medium ground finch (*Geospiza fortis*), is found on the small island of Daphne Major. These finches are seed-eating birds. A major drought occurred on the island in 1977. Following the drought, the average beak size of medium ground finches had increased about 3%–4%. Why might a drought lead to an evolutionary change in beak size? Propose a hypothesis and explain how you could go about testing it.

Section 26.5

6. In most cases, the rate of synonymous substitution for a gene is higher than the rate of nonsynonymous substitution. Sometimes, however, the rate of nonsynonymous substitution is higher. When would you expect to see this? What might bring it about?

7. Do the data in **Figure 26.13** support the predictions of the neutral-mutation hypothesis, or do they suggest that balancing selection is occurring? Explain your reasoning.

8. What changes, if any, would you predict would occur in the pigmentation of *Drosophila melanogaster* with increased global warming? What type of genetic changes would you expect to see? Be as specific as you can.

9. Propose an evolutionary explanation for introns. Why might they have evolved?

[Dr. Jeremy Burgess/Science Source.]
[© Leszczynski, Zigmund/Animals Animals.]

[Sinclair Stammers/Science Source.]

[Eye of Science/Science Source.]

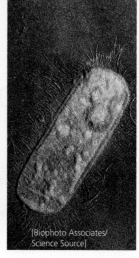
[Steve Gschmeissner/Science Photo Library/Alamy.]

[Biophoto Associates/Science Source]
[© Gelia I Dreamstime.com]

Model genetic organisms possess characteristics that make them useful for genetic studies. Shown are several organisms commonly used in genetic studies.

Reference Guide to Model Genetic Organisms

What do Lou Gehrig, the finest first baseman in major league history, and Stephen Hawking, the world's most famous theoretical physicist, have in common? They both suffered or suffer from amyotrophic lateral sclerosis (ALS, also known as Lou Gehrig disease), a degenerative neurological disease that leads to progressive weakness and wasting of skeletal muscles.

Most cases of ALS are sporadic, appearing in people who have no family history of the disease such as in Lou Gehrig, discussed in the introduction to Chapter 18. However, about 10% of the cases run in families and are inherited. In 2004, geneticists discovered a large Brazilian family with multiple cases of ALS. Genetic analysis revealed that ALS in this family is due to a mutation in a gene called *VABP*, which encodes a vesicle-associated membrane protein. This gene is distinct from the *C9orf72* gene that causes ALS in other families (see Chapter 18) illustrating the principle that genetic diseases can be caused by mutations in several different genes.

To better understand how mutations in *VABP* lead to the symptoms of ALS, geneticists turned to an unlikely subject—the fruit fly *Drosophila melanogaster*. Fruit flies don't have ALS, but they do possess a gene very similar to *VABP*. Using a wide array of techniques that have been developed for genetically manipulating fruit flies, geneticists created transgenic flies with the mutant sequence of the *VABP* gene that causes ALS in humans. These flies are a disease model for ALS and are being used to better understand what the gene *VABP* does normally and how its disruption can lead to ALS.

The field of genetics has been greatly influenced and shaped by a few key organisms—called model genetic organisms—whose characteristics make them particularly amenable to genetic studies. The use of *Drosophila* for studying ALS in humans illustrates the power of this approach. Because features of genetic systems are common to many organisms, research conducted on one species can often be a source of insight into the genetic systems of other species. This commonality of genetic function means that geneticists can focus their efforts on model organisms that are easy to work with and likely to yield results.

Model genetic organisms possess life cycles and genomic features that make them well suited to genetic study and analysis. Some key features possessed by many model genetic organisms include:

- a short generation time, and so several generations of genetic crosses can be examined in reasonable time;
- the production of numerous progeny, which allows genetic ratios to be easily observed;
- the ability to carry out and control genetic crosses in the organism;
- the ability to be reared in a laboratory environment, requiring little space and few resources to maintain;
- the availability of numerous genetic variants; and
- an accumulated body of knowledge about their genetic systems.

In recent years, the genomes of many model genetic organisms have been completely sequenced, greatly facilitating their use in genetic research.

Not all model organisms possess all of these characteristics. However, each model genetic organism has one or more features that make it useful for genetic analysis. For example, corn cannot be easily grown in the laboratory (and usually isn't) and it has a relatively long generation time, but it produces numerous progeny and there are many genetic variants of corn available for study.

This reference guide highlights six model genetic organisms with important roles in the development of genetics: the fruit fly *(Drosophila melanogaster)*, bacterium *(Escherichia coli)*, roundworm *(Caenorhabditis elegans)*, thale cress plant *(Arabidopsis thaliana)*, house mouse *(Mus musculus)*, and yeast *(Saccharomyces cerevisiae)*. These six organisms have been widely used in genetic research and instruction. A number of other organisms also are used as model systems in genetics, including corn *(Zea mays)*, zebrafish *(Danio rerio)*, clawed frog *(Xenopus lavis)*, bread mold *(Neurospora crassa)*, rat *(Rattus norvegicus)*, and Rhesus macaque *(Macaca mulatta)*, just to mention a few.

The Fruit Fly *Drosophila melanogaster*

*D*rosophila melanogaster, a fruit fly, was among the first organisms used for genetic analysis and, today, it is one of the most widely used and best known genetically of all organisms. It has played an important role in studies of linkage, epistasis, chromosome genetics, development, behavior, and evolution. Because all organisms use a common genetic system, understanding a process such as replication or transcription in fruit flies helps us to understand these same processes in humans and other eukaryotes.

Drosophila is a genus of more than 1000 described species of small flies (about 1 to 2 mm in length) that frequently feed and reproduce on fruit, although they rarely cause damage and are not considered economic pests. The best known and most widely studied of the fruit flies is *D. melanogaster*, but genetic studies have been extended to many other species of the genus as well. *D. melanogaster* first began to appear in biological laboratories about 1900. After first taking up breeding experiments with mice and rats, Thomas Hunt Morgan began using fruit flies in experimental studies of heredity at Columbia University. Morgan's laboratory, located on the top floor of Schermerhorn Hall, became known as the Fly Room (see Figure 4.11b). To say that the Fly Room was unimpressive is an understatement. The cramped room, only about 16 by 23 feet, was filled with eight desks, each occupied by a student and his experiments. The primitive laboratory equipment consisted of little more than milk bottles for rearing the flies and hand-held lenses for observing their traits. Later, microscopes replaced the hand-held lenses, and crude incubators were added to maintain the fly cultures, but even these additions did little to increase the physical sophistication of the laboratory. Morgan and his students were not tidy: cockroaches were abundant (living off spilled *Drosophila* food), dirty milk bottles filled the sink, ripe bananas—food for the flies—hung from the ceiling, and escaped fruit flies hovered everywhere. In spite of its physical limitations, the Fly Room was the source of some of the most important research in the history of biology. There was daily excitement among the students, some of whom initially came to the laboratory as undergraduates. The close quarters facilitated informality and the free flow of ideas. Morgan and the Fly Room illustrate the tremendous importance of "atmosphere" in producing good science. Morgan and his students eventually used *Drosophila* to elucidate many basic principles of heredity, including sex-linked inheritance, epistasis, multiple alleles, and gene mapping.

ADVANTAGES OF *D. MELANOGASTER* AS A MODEL GENETIC ORGANISM *Drosophila*'s widespread use in genetic studies is no accident. The fruit fly has a number of characteristics that make it an ideal subject for genetic investigations. Compared with other organisms, it has a relatively short generation time; fruit flies will complete an entire generation in about 10 days at room temperature, and so several generations can be studied within a few weeks. Although *D. melanogaster* has a short generation time, it possesses a complex life cycle, passing through several different developmental stages, including egg, larva, pupa, and adult. A female fruit fly is capable of mating within 8 hours of emergence and typically begins to lay eggs after about 2 days. Fruit flies also produce a large number of offspring, laying as many as 400 to 500 eggs in a 10-day period. Thus, large numbers of progeny can be obtained from a single genetic cross.

Another advantage is that fruit flies are easy to culture in the laboratory. They are usually raised in small glass vials or bottles and are fed easily prepared, pastelike food consisting of bananas or corn meal and molasses. Males and females are readily distinguished and virgin females are easily isolated, facilitating genetic crosses. The flies are small, requiring little space—several hundred can be raised in a half-pint bottle—but they are large enough for many mutations to be easily observed with a hand lens or a dissecting microscope.

Finally, *D. melanogaster* is the organism of choice for many geneticists because it has a relatively small genome consisting of 175 million base pairs of DNA, which is only about 5% of the size of the human genome. It has four pairs of chromosomes: three pairs of autosomes and one pair of sex chromosomes. The X chromosome (designated chromosome 1) is large and acrocentric, whereas the Y chromosome is large and submetacentric, although it contains very little genetic information. Chromosomes 2 and 3 are large and metacentric; chromosome 4 is a very small acrocentric chromosome. In the salivary glands, the chromosomes are very large (see p. 317 in Chapter 11), making *Drosophila* an excellent subject for chromosome studies. In 2000, the complete genome of *D. melanogaster* was sequenced, followed by the sequencing of the genome of *D. pseudoobscura* in 2005 and the genomes of 10 additional *Drosophila* genomes in 2007. *Drosophila* continues today to be one of the most versatile and powerful of all genetic model organisms. ∎

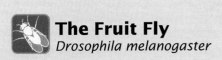

The Fruit Fly
Drosophila melanogaster

ADVANTAGES

- Small size
- Short generation time of 10 days at room temperature
- Each female lays 400–500 eggs
- Easy to culture in laboratory
- Small genome
- Large chromosomes
- Many mutations available

STATS

Taxonomy:	Insect
Size:	2–3 mm in length
Anatomy:	3 body divisions, 6 legs, 1 pair of wings
Habitat:	Feeds and reproduces on fruit

Life Cycle

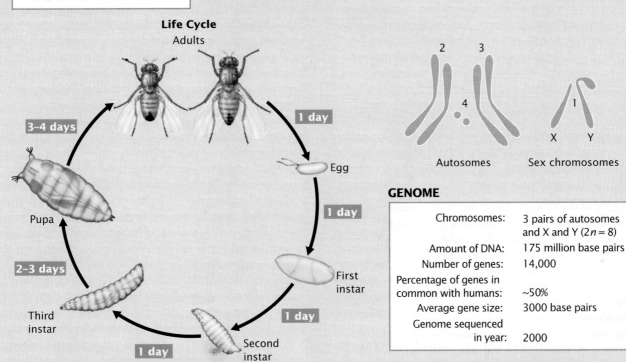

Adults

1 day

Egg

1 day

First instar

1 day

Second instar

1 day

Third instar

2–3 days

Pupa

3–4 days

Autosomes   Sex chromosomes

2   3   4   1   X   Y

GENOME

Chromosomes:	3 pairs of autosomes and X and Y ($2n = 8$)
Amount of DNA:	175 million base pairs
Number of genes:	14,000
Percentage of genes in common with humans:	~50%
Average gene size:	3000 base pairs
Genome sequenced in year:	2000

CONTRIBUTIONS TO GENETICS

- Basic principles of heredity including sex-linked inheritance, multiple alleles, epistatsis, gene mapping, etc.
- Mutation research
- Chromosome variation and behavior
- Population genetics
- Genetic control of pattern formation
- Behavioral genetics

The Bacterium *Escherichia coli*

The most widely studied prokaryotic organism and one of the best genetically characterized of all species is the bacterium *Escherichia coli*. Although some strains of *E. coli* are toxic and cause disease, most are benign and reside naturally in the intestinal tracts of humans and other warm-blooded animals. *E. coli* was first described by Theodore Escherich in 1885 but, for many years, the assumption was that all bacteria reproduced only asexually and that genetic crosses were impossible. In 1946, Joshua Lederberg and Edward Tatum demonstrated that *E. coli* undergoes a type of sexual reproduction; their finding initiated the use of *E. coli* as a model genetic organism. A year later, Lederberg published the first genetic map of *E. coli* based on recombination frequencies and, in 1952, William Hays showed that mating between bacteria is asymmetrical, with one bacterium serving as genetic donor and the other as genetic recipient.

ADVANTAGES OF *E. COLI* AS A MODEL GENETIC ORGANISM *Escherichia coli* is one of the true workhorses of genetics; its twofold advantage is rapid reproduction and small size. Under optimal conditions, this organism can reproduce every 20 minutes and, in a mere 7 hours, a single bacterial cell can give rise to more than 2 million descendants. One of the values of rapid reproduction is that enormous numbers of cells can be grown quickly, and so even very rare mutations will appear in a short period. Consequently, numerous mutations in *E. coli*, affecting everything from colony appearance to drug resistance, have been isolated and characterized.

Escherichia coli is easy to culture in the laboratory in liquid medium (see Figure 9.1a) or on solid medium within petri plates (see Figure 9.1b). In liquid culture, *E. coli* cells will grow to a concentration of a billion cells per milliliter, and trillions of bacterial cells can be easily grown in a single test tube. When *E. coli* cells are diluted and spread onto the solid medium of a petri dish, individual bacteria reproduce asexually, giving rise to a concentrated clump of 10 million to 100 million genetically identical cells, called a colony. This colony formation makes it easy to isolate genetically pure strains of the bacteria.

THE *E. COLI* GENOME The *E. coli* genome is on a single chromosome and—compared with those of humans, mice, plants, and other multicellular organisms—is relatively small, consisting of 4,638,858 base pairs. If stretched out straight, the DNA molecule in the single *E. coli* chromosome would be 1.6 mm long, almost a thousand times as long as the *E. coli* cell within which it resides (see Figure 11.1). To accommodate this huge amount of DNA within the confines of a single cell, the *E. coli* chromosome is highly coiled and condensed. The information within the *E. coli* chromosome also is compact, having little noncoding DNA between and within the genes and having few sequences for which there is more than one copy. The *E. coli* genome contains an estimated 4300 genes, many of which have no known function. These "orphan genes" may play important roles in adapting to unusual environments, coordinating metabolic pathways, organizing the chromosome, or communicating with other bacterial cells. The haploid genome of *E. coli* makes it easy to isolate mutations because there are no dominant genes at the same loci to suppress and mask recessive mutations.

LIFE CYCLE OF *E. COLI* Wild-type *E. coli* is prototrophic and can grow on minimal medium that contains only glucose and some inorganic salts. Under most conditions, *E. coli* divides about once an hour, although in a richer medium containing sugars and amino acids, it will divide every 20 minutes. It normally reproduces through simple binary fission, in which the single chromosome of a bacterium replicates and migrates to opposite sides of the cell, followed by cell division, giving rise to two identical daughter cells (see Figure 2.5). Mating between bacteria, called conjugation, is controlled by fertility genes normally located on the F plasmid (see p. 256). In conjugation, one bacterium donates genetic material to another bacterium, followed by genetic recombination that integrates new alleles into the bacterial chromosome. Genetic material can also be exchanged between strains of *E. coli* through transformation and transduction (see Figure 9.7).

GENETIC TECHNIQUES WITH *E. COLI* *Escherichia coli* is used in a number of experimental systems in which fundamental genetic processes are studied in detail. For example, in vitro translation systems contain within a test tube all the components necessary to translate the genetic information of a messenger RNA molecule into a polypeptide chain. Similarly, in vitro systems containing components from *E. coli* cells allow transcription, replication, gene expression, and many other important genetic functions to be studied and analyzed under controlled laboratory conditions.

Escherichia coli is also widely used in genetic engineering (recombinant DNA; see Chapter 19). Plasmids have been isolated from *E. coli* and genetically modified to create effective vectors for transferring genes into bacteria and eukaryotic cells. Often, new genetic constructs (DNA sequences created in the laboratory) are assembled and cloned in *E. coli* before transfer to other organisms. Methods have been developed to introduce specific mutations within *E. coli* genes, and so genetic analysis no longer depends on the isolation of randomly occurring mutations. New DNA sequences produced by recombinant DNA can be introduced by transformation into special strains of *E. coli* that are particularly efficient (competent) at taking up DNA.

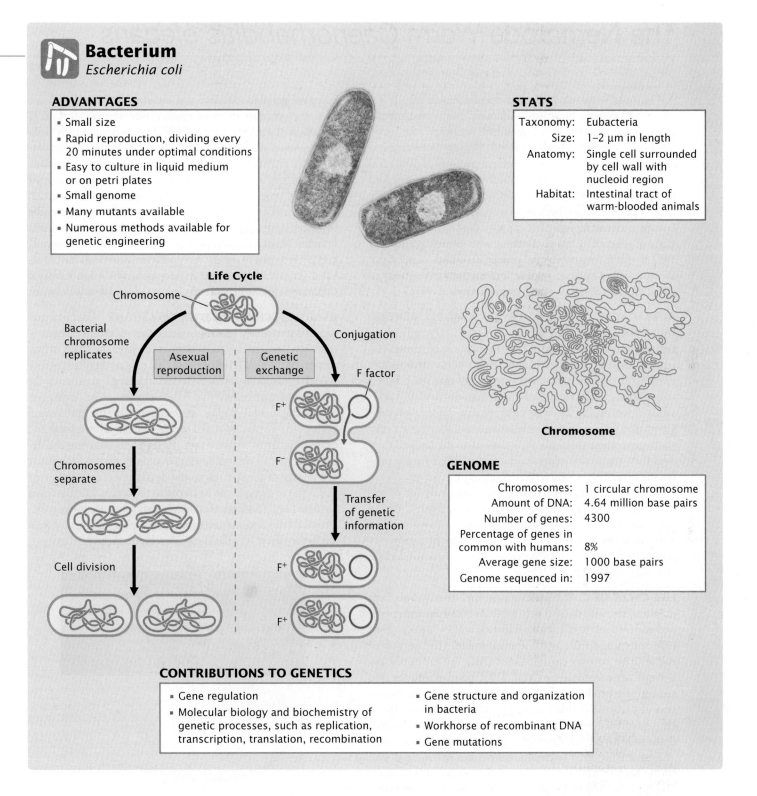

Bacterium
Escherichia coli

ADVANTAGES

- Small size
- Rapid reproduction, dividing every 20 minutes under optimal conditions
- Easy to culture in liquid medium or on petri plates
- Small genome
- Many mutants available
- Numerous methods available for genetic engineering

STATS

Taxonomy:	Eubacteria
Size:	1–2 μm in length
Anatomy:	Single cell surrounded by cell wall with nucleoid region
Habitat:	Intestinal tract of warm-blooded animals

Life Cycle

Chromosome

Bacterial chromosome replicates

Asexual reproduction

Genetic exchange

Conjugation

F factor

F⁺

F⁻

Chromosomes separate

Transfer of genetic information

Cell division

F⁺

F⁺

Chromosome

GENOME

Chromosomes:	1 circular chromosome
Amount of DNA:	4.64 million base pairs
Number of genes:	4300
Percentage of genes in common with humans:	8%
Average gene size:	1000 base pairs
Genome sequenced in:	1997

CONTRIBUTIONS TO GENETICS

- Gene regulation
- Molecular biology and biochemistry of genetic processes, such as replication, transcription, translation, recombination
- Gene structure and organization in bacteria
- Workhorse of recombinant DNA
- Gene mutations

Because of its powerful advantages as a model genetic organism, *E. coli* has played a leading role in many fundamental discoveries in genetics, including elucidation of the genetic code, probing the nature of replication, and working out the basic mechanisms of gene regulation. ■

The Nematode Worm *Caenorhabditis elegans*

You may be asking, What is a nematode, and why is it a model genetic organism? Although rarely seen, nematodes are among the most abundant organisms on Earth, inhabiting soils throughout the world. Most are free living and cause no harm, but a few are important parasites of plants and animals, including humans. Although *Caenorhabditis elegans* has no economic or medical importance, it has become widely used in genetic studies because of its simple body plan, ease of culture, and high reproductive capacity. First introduced to the study of genetics by Sydney Brenner, who formulated plans in 1962 to use *C. elegans* for the genetic dissection of behavior, this species has made important contributions to the study of development, cell death, aging, and behavior.

ADVANTAGES OF *C. ELEGANS* AS A MODEL GENETIC ORGANISM An ideal genetic organism, *C. elegans* is small, easy to culture, and produces large numbers of offspring. The adult *C. elegans* is about 1 mm in length. Most investigators grow *C. elegans* on agar-filled petri plates that are covered with a lawn of bacteria, which the nematodes devour. Thousands of worms can be easily cultured in a single laboratory. Compared with most multicellular animals, they have a very short generation time, about 3 days at room temperature. And they are prolific reproducers, with a single female producing from 250 to 1000 fertilized eggs in 3 to 4 days.

Another advantage of *C. elegans*, particularly for developmental studies, is that the worm is transparent, allowing easy observation of internal development at all stages. It has a simple body structure, with a small, invariant number of somatic cells: 959 cells in a mature hermaphroditic female and 1031 cells in a mature male.

LIFE CYCLE OF *C. ELEGANS* Most mature adults are hermaphrodites, with the ability to produce both eggs and sperm and undergo self-fertilization. A few are male, which produce only sperm and mate with hermaphrodites. The hermaphrodites have two sex chromosomes (XX); the males possess a single sex chromosome (XO). Thus, hermaphrodites that self-fertilize produce only hermaphrodites (with the exception of a few males that result from nondisjunction of the X chromosomes). When hermaphrodites mate with males, half of the progeny are XX hermaphrodites and half are XO males.

Eggs are fertilized internally, either from sperm produced by the hermaphrodite or from sperm contributed by a male. The eggs are then laid, and development is completed externally. Approximately 14 hours after fertilization, a larva hatches from the egg and goes through four larval stages—termed L1, L2, L3, and L4—that are separated by molts. The L4 larva undergoes a final molt to produce the adult worm. Under normal laboratory conditions, worms will live for 2 to 3 weeks.

THE *C. ELEGANS* GENOME Geneticists began developing plans in 1989 to sequence the genome of *C. elegans*, and the complete genome sequence was obtained in 1998. Compared with the genomes of most multicellular animals, that of *C. elegans*, at 103 million base pairs of DNA, is small, which facilitates genomic analysis. The availability of the complete genome sequence provides a great deal of information about gene structure, function, and organization in this species. For example, the process of programmed cell death (apoptosis, see Chapter 22) plays an important role in development and in the suppression of cancer. Apoptosis in *C. elegans* is remarkably similar to that in humans. Having the complete genome sequence of *C. elegans*, and given its ease of genetic manipulation, geneticists have identified genes that participate in apoptosis, which has increased our understanding of apoptosis in humans and its role in cancer.

GENETIC TECHNIQUES WITH *C. ELEGANS* Chemical mutagens are routinely used to generate mutations in *C. elegans*—mutations that are easy to identify and isolate. The ability of hermaphrodites to self-fertilize means that progeny homozygous for recessive mutations can be obtained in a single generation; the existence of males means that genetic crosses can be carried out.

Developmental studies are facilitated by the transparent body of the worms. As stated earlier, *C. elegans* has a small and exact number of somatic cells. Researchers studying the development of *C. elegans* have meticulously mapped the entire cell lineage of the species, so the developmental fate of every cell in the adult body can be traced to the original single-celled fertilized egg. Developmental biologists often use lasers to destroy (ablate) specific cells in a developing worm and then study the effects on physiology, development, and behavior.

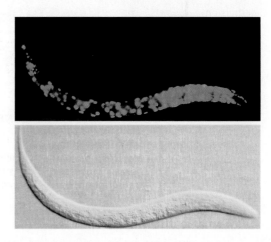

Figure 1 A sequence for the green fluorescent protein (GFP) has been used to visually determine the expression of genes inserted into *C. elegans* (lower photograph). The gene for GFP is injected into the ovary of a worm and becomes incorporated into the worm genome. The expression of this transgene produces GFP, which fluoresces green (upper photograph). [Huaqi Jiang, Rong Guo, and Jo Anne Powell-Coffman. The *Caenorhabditis elegans* hif-1 gene encodes a bHLH-PAS protein that is required for adaptation to hypoxia. *PNAS* 98: 7916-7921, 2001. ©2001 National Academy of Sciences, U.S.A.]

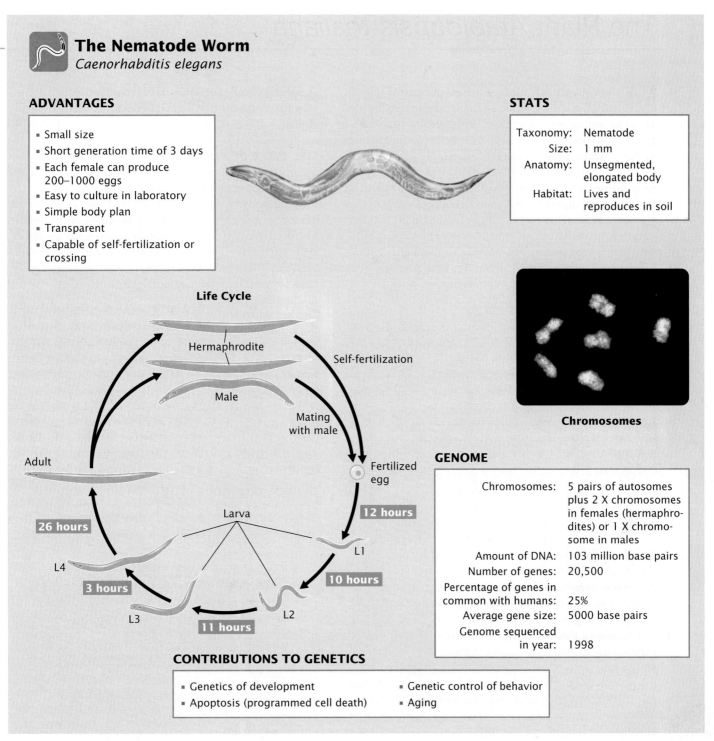

The Nematode Worm
Caenorhabditis elegans

ADVANTAGES

- Small size
- Short generation time of 3 days
- Each female can produce 200–1000 eggs
- Easy to culture in laboratory
- Simple body plan
- Transparent
- Capable of self-fertilization or crossing

STATS

Taxonomy:	Nematode
Size:	1 mm
Anatomy:	Unsegmented, elongated body
Habitat:	Lives and reproduces in soil

Life Cycle

Hermaphrodite

Self-fertilization

Male

Mating with male

Fertilized egg

12 hours

Adult

26 hours

Larva

L1

10 hours

L4

3 hours

L3

11 hours

L2

Chromosomes

GENOME

Chromosomes:	5 pairs of autosomes plus 2 X chromosomes in females (hermaphrodites) or 1 X chromosome in males
Amount of DNA:	103 million base pairs
Number of genes:	20,500
Percentage of genes in common with humans:	25%
Average gene size:	5000 base pairs
Genome sequenced in year:	1998

CONTRIBUTIONS TO GENETICS

- Genetics of development
- Apoptosis (programmed cell death)
- Genetic control of behavior
- Aging

[Photograph courtesy of William Goodyer and Monique Zetka.]

RNA interference has proved to be an effective tool for turning off genes in *C. elegans*. Geneticists inject double-stranded copies of RNA that is complementary to specific genes; the double-stranded RNA then silences the expression of these genes through the RNAi process. The worms can even be fed bacteria that have been genetically engineered to express the double-stranded RNA, thus avoiding the difficulties of microinjection.

Transgenic worms can be produced by injecting DNA into the ovary, where the DNA becomes incorporated into the oocytes. Geneticists have created a special reporter gene that produces the jellyfish green fluorescent protein (GFP). When this reporter gene is injected into the ovary and becomes inserted into the worm genome, its expression produces GFP, which fluoresces green, allowing the expression of the gene to be easily observed (**Figure 1**). ■

The Plant *Arabidopsis thaliana*

Much of the early work in genetics was carried out on plants, including Mendel's seminal discoveries in pea plants as well as important aspects of heredity, gene mapping, chromosome genetics, and quantitative inheritance in corn, wheat, beans, and other plants. However, by the mid-twentieth century, many geneticists had turned to bacteria, viruses, yeast, *Drosophila*, and mouse genetic models. Because a good genetic plant model did not exist, plants were relatively neglected, particularly for the study of molecular genetic processes.

This changed in the last part of the twentieth century with the widespread introduction of a new genetic model organism, the plant *Arabidopsis thaliana*. *A. thaliana* was identified in the sixteenth century, and the first mutant was reported in 1873; but this species was not commonly studied until the first detailed genetic maps appeared in the early 1980s. Today, *Arabidopsis* figures prominently in the study of genome structure, gene regulation, development, and evolution in plants, and it provides important basic information about plant genetics that is applied to economically important plant species.

ADVANTAGES OF *ARABIDOPSIS* AS A MODEL GENETIC ORGANISM   The thale cress *Arabidopsis thaliana* is a member of the Brassicaceae family and grows as a weed in many parts of the world. Except in its role as a model genetic organism, *Arabidopsis* has no economic importance, but it has a number of characteristics that make it well suited to the study of genetics. As an angiosperm, it has features in common with other flowering plants, some of which play critical roles in the ecosystem or are important sources of food, fiber, building materials, and pharmaceutical agents. *Arabidopsis*'s chief advantages are its small size (maximum height of 10–20 cm), prolific reproduction, and small genome.

Arabidopsis thaliana completes development—from seed germination to seed production—in about 6 weeks. Its small size and ability to grow under low illumination make it ideal for laboratory culture. Each plant is capable of producing from 10,000 to 40,000 seeds, and the seeds typically have a high rate of germination; so large numbers of progeny can be obtained from single genetic crosses.

THE *ARABIDOPSIS* GENOME   A key advantage for molecular studies is *Arabidopsis*'s small genome, which consists of only 125 million base pairs of DNA on five pairs of chromosomes, compared with 2.5 billion base pairs of DNA in the maize genome and 16 billion base pairs in the wheat genome. The genome of *A. thaliana* was completely sequenced in 2000, providing detailed information about gene structure and organization in this species. A number of variants of *A. thaliana*—called ecotypes—that vary in shape, size, physiological characteristics, and DNA sequence are available for study.

LIFE CYCLE OF *ARABIDOPSIS*   The *Arabidopsis* life cycle is fairly typical of most flowering plants (see Figure 2.22). The main, vegetative part of the plant is diploid; haploid gametes are produced in the pollen and ovaries. When a pollen grain lands on the stigma of a flower, a pollen tube grows into the pistil and ovary. Two haploid sperm nuclei contained in each pollen grain travel down the pollen tube and enter the embryo sac. There, one of the haploid sperm cells fertilizes the haploid egg cell to produce a diploid zygote. The other haploid sperm cell fuses with two haploid nuclei to form the $3n$ endosperm, which provides tissue that will nourish the growing embryonic plant. The zygotes develop within the seeds, which are produced in a long pod.

Under appropriate conditions, the embryo germinates and begins to grow into a plant. The shoot grows upward and the roots downward, a compact rosette of leaves is produced, and, under the right conditions, the shoot enlarges and differentiates into flower structures. At maturity, *A. thaliana* is a low-growing plant with roots, a main shoot with branches that bear mature leaves, and small white flowers at the tips of the branches.

GENETIC TECHNIQUES WITH *ARABIDOPSIS*   A number of traditional and modern molecular techniques are commonly used with *Arabidopsis* and provide it with special advantages for genetic studies. *Arabidopsis* can self-fertilize, which means that any recessive mutation appearing in the germ line can be recovered in the immediate progeny. Cross-fertilization also is possible by removing the anther from one plant and dusting pollen on the stigma of another plant—essentially the same technique used by Gregor Mendel with pea plants (see Figure 3.5).

As already mentioned, many naturally occurring variants of *Arabidopsis* are available for study, and new mutations can be produced by exposing its seeds to chemical mutagens, radiation, or transposable elements that randomly insert into genes. The large number of offspring produced by *Arabidopsis* facilitates screening for rare mutations.

Genes from other organisms can be transferred to *Arabidopsis* by the Ti plasmid from the bacterium *Agrobacterium tumefaciens*, which naturally infects plants and transfers the Ti plasmid to plant cells (see Chapter 19). After transfer to a plant cell, the Ti plasmid randomly inserts into the DNA of the plant that it infects, thereby generating mutations in the plant DNA in a process called *insertional mutagenesis*. Geneticists have modified the Ti plasmid to carry a *GUS* gene, which has no promoter of its own. The *GUS* gene encodes

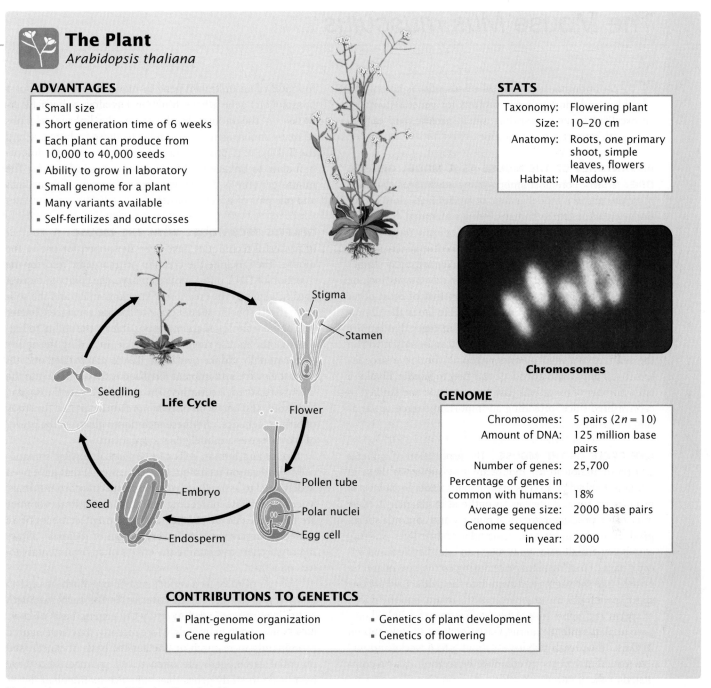

The Plant
Arabidopsis thaliana

ADVANTAGES

- Small size
- Short generation time of 6 weeks
- Each plant can produce from 10,000 to 40,000 seeds
- Ability to grow in laboratory
- Small genome for a plant
- Many variants available
- Self-fertilizes and outcrosses

STATS

Taxonomy:	Flowering plant
Size:	10–20 cm
Anatomy:	Roots, one primary shoot, simple leaves, flowers
Habitat:	Meadows

Chromosomes

GENOME

Chromosomes:	5 pairs ($2n = 10$)
Amount of DNA:	125 million base pairs
Number of genes:	25,700
Percentage of genes in common with humans:	18%
Average gene size:	2000 base pairs
Genome sequenced in year:	2000

Life Cycle

Seedling

Stigma

Stamen

Flower

Pollen tube

Polar nuclei

Egg cell

Embryo

Seed

Endosperm

CONTRIBUTIONS TO GENETICS

- Plant-genome organization
- Gene regulation
- Genetics of plant development
- Genetics of flowering

[Photograph courtesy of Anand P Tyagi and Luca Comai.]

an enzyme that converts a colorless compound (X-Glu) into a blue dye. Because the *GUS* gene has no promoter, it is expressed only when inserted into the coding sequence of a plant gene. When that happens, the enzyme encoded by *GUS* is synthesized and converts X-Gluc into a blue dye that stains the cell. This dye provides a means to visually determine the expression pattern of a gene that has been interrupted by Ti DNA, producing information about the expression of genes that are mutated by insertional mutagenesis. ■

The Mouse *Mus musculus*

The common house mouse, *Mus musculus*, is among the oldest and most valuable subjects for genetic study. It's an excellent genetic organism—small, prolific, and easy to keep, with a short generation time.

ADVANTAGES OF THE MOUSE AS A MODEL GENETIC ORGANISM Foremost among many advantages that *Mus musculus* has as a model genetic organism is its close evolutionary relationship to humans. Being a mammal, the mouse is genetically, behaviorally, and physiologically more similar to humans than are other organisms used in genetics studies, making the mouse the model of choice for many studies of human and medical genetics. Other advantages include a short generation time compared with that of most other mammals. *Mus musculus* is well adapted to life in the laboratory and can be easily raised and bred in cages that require little space; thus several thousand mice can be raised within the confines of a small laboratory room. Mice have large litters (8–10 pups), and are docile and easy to handle. Finally, a large number of mutations have been isolated and studied in captive-bred mice, providing an important source of variation for genetic analysis.

LIFE CYCLE OF THE MOUSE The production of gametes and reproduction in the mouse are very similar to those in humans. Diploid germ cells in the gonads undergo meiosis to produce sperm and oocytes, as outlined in Chapter 2. Male mice begin producing sperm at puberty and continue sperm production throughout the remainder of their lives. Starting at puberty, female mice go through an estrus cycle about every 4 days. If mating takes place during estrus, sperm are deposited into the vagina and swim into the oviduct, where one sperm penetrates the outer layer of the ovum, and the nuclei of sperm and ovum fuse. After fertilization, the diploid embryo implants into the uterus. Gestation typically takes about 21 days. Mice reach puberty in about 5 to 6 weeks and will live for about 2 years. A complete generation can be completed in about 8 weeks.

THE MOUSE GENOME The mouse genome contains about 2.6 billion base pairs of DNA, which is similar in size to the human genome. Mice and humans also have similar numbers of genes. For most human genes, there are homologous genes in the mouse. An important tool for determining the function of an unknown gene in humans is to search for a homologous gene whose function has already been determined in the mouse. Furthermore, the linkage relations of many mouse genes are similar to those in humans, and the linkage relations of genes in mice often provide important clues to linkage relations among genes in humans. The mouse genome is distributed across 19 pairs of autosomes and one pair of sex chromosomes.

GENETIC TECHNIQUES WITH THE MOUSE A number of powerful techniques have been developed for use in the mouse. They include the creation of transgenic mice by the injection of DNA into a mouse embryo, the ability to disrupt specific genes by the creation of knockout mice, and the ability to insert specific sequences into specific loci (see Chapter 19). These techniques are made possible by the ability to manipulate the mouse reproductive cycle, including the ability to hormonally induce ovulation, isolate unfertilized oocytes from the ovary, and implant fertilized embryos back into the uterus of a surrogate mother. The ability to create transgenic, knockout, and knock-in mice has greatly facilitated the study of human genetics, and these techniques illustrate the power of the mouse as a model genetic organism.

Mouse and human cells can be fused, allowing somatic-cell hybridization techniques (see Chapter 7) that have been widely used to assign human genes to specific chromosomes. Mice also tolerate inbreeding well, and inbred strains of mice are easily created by brother–sister mating. Members of an inbred strain are genetically very similar or identical, allowing researchers to examine the effects of environmental factors on a trait.

The use of mice as a model genetic organism has led to many important genetic discoveries. In the early twentieth century, mice were used to study the genetic basis of coat-color variation in mammals. More recently, they have figured prominently in research on the genetic basis of cancer, and potential carcinogens are often tested in mice. Mice have been used to study genes that influence mammalian development, including mutations that produce birth defects in humans. A large number of mouse models of specific human diseases have been created—in some cases, by isolating and inbreeding mice with naturally occurring mutations and, in other cases, by using knockout and knock-in techniques to disable and modify specific genes. ■

The Mouse
Mus musculus

ADVANTAGES

- Closely related to humans
- Small size
- Rapid reproduction
- Easy to rear in the laboratory
- Tolerates inbreeding

STATS

Taxonomy:	Mammal
Size:	2–3 inches 20 grams
Anatomy:	Typical rodent body plan
Habitat:	Fields, houses, and other human structures

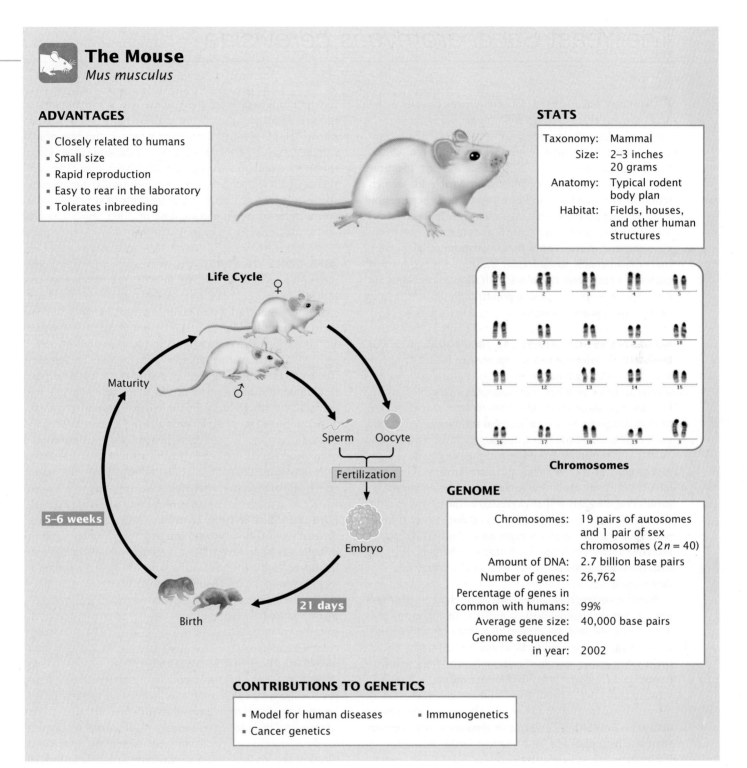

Life Cycle

♀

Maturity

♂

Sperm   Oocyte

Fertilization

Embryo

5–6 weeks

21 days

Birth

Chromosomes

GENOME

Chromosomes:	19 pairs of autosomes and 1 pair of sex chromosomes ($2n = 40$)
Amount of DNA:	2.7 billion base pairs
Number of genes:	26,762
Percentage of genes in common with humans:	99%
Average gene size:	40,000 base pairs
Genome sequenced in year:	2002

CONTRIBUTIONS TO GENETICS

- Model for human diseases
- Cancer genetics
- Immunogenetics

[Photograph courtesy of Ellen C. Akeson and Muriel T. Davisson, The Jackson Laboratory, Bar Harbor, Maine.]

The Yeast *Saccharomyces cerevisiae*

Common baker's yeast *(Saccharomyces cerevisiae)* has been widely adopted as a simple model system for the study of eukaryotic genetics. Long used for baking bread and making beer, yeast has more recently been utilized for the production of biofuels. Louis Pasteur identified *S. cerevisiae* as the microorganism responsible for fermentation in 1857, and its use in genetic analysis began about 1935. Having been the subject of extensive studies in classical genetics for many years, yeast genes are well known and characterized. At the same time, yeast's unicellular nature makes it amenable to molecular techniques developed for bacteria. Thus, yeast combines both classical genetics and molecular biology to provide a powerful model for the study of eukaryotic genetic systems.

ADVANTAGES OF YEAST AS A MODEL GENETIC ORGANISM The great advantage of yeast is that it not only is a eukaryotic organism, with genetic and cellular systems similar to those of other, more complex eukaryotes such as humans, but also is unicellular, with many of the advantages of manipulation found with bacterial systems. Like bacteria, yeast cells require little space, and large numbers of cells can be grown easily and inexpensively in the laboratory.

Yeast exists in both diploid and haploid forms. When haploid, the cells possess only a single allele at each locus, which means that the allele will be expressed in the phenotype; unlike the situation in diploids, there is no dominance through which some alleles mask the expression of others. Therefore, recessive alleles can be easily identified in haploid cells, and then the interactions between alleles can be examined in the diploid cells.

Another feature that makes yeast a powerful genetic model system is that, subsequent to meiosis, all of the products of a meiotic division are present in a single structure called an ascus (see the next subsection) and remain separate from the products of other meiotic divisions. The four cells produced by a single meiotic division are termed a tetrad. In most organisms, the products of different meiotic divisions mix, and so identification of the results of a single meiotic division is impossible. For example, if we were to isolate four sperm cells from the testes of a mouse, it is extremely unlikely that all four would have been produced by the same meiotic division. Having tetrads separate in yeast allows us to directly observe the effects of individual meiotic divisions on the types of gametes produced and to more easily identify crossover events. The genetic analysis of a tetrad is termed tetrad analysis.

Yeast has been subjected to extensive genetic analysis, and thousands of mutants have been identified. In addition, many powerful molecular techniques developed for manipulating genetic sequences in bacteria have been adapted for use in yeast. Yet, in spite of a unicellular structure and ease of manipulation, yeast cells possess many of the same genes found in humans and other complex multicellular eukaryotes, and many of these genes have identical or similar functions in these eukaryotes. Thus, the genetic study of yeast cells often contributes to our understanding of other, more complex eukaryotic organisms, including humans.

LIFE CYCLE OF YEAST As stated earlier, *Saccharomyces cerevisiae* can exist as either haploid or diploid cells. Haploid cells usually exist when yeast is starved for nutrients and reproduce mitotically, producing identical, haploid daughter cells through budding. Yeast cells can also undergo sexual reproduction. There are two mating types, a and α; haploid cells of different mating types fuse and then undergo nuclear fusion to create a diploid cell. The diploid cell is capable of budding mitotically to produce genetically identical diploid cells. Starvation induces the diploid cells to undergo meiosis, resulting in four haploid nuclei, which become separated into different cells, producing haploid spores. Because the four products of meiosis (a tetrad) are enclosed in a common structure, the ascus, all the products of a single meiosis can be isolated (tetrad analysis).

THE YEAST GENOME *Saccharomyces cerevisiae* has 16 pairs of typical eukaryotic chromosomes. The rate of recombination is high, giving yeast a relatively long genetic map compared with those of other organisms. The genome of *S. cerevisiae* contains 12 million base pairs, plus the 2 million to 3 million base pairs of rRNA genes. In 1996, *S. cerevisiae* was the first eukaryotic organism whose genome was completely sequenced.

GENETIC TECHNIQUES WITH YEAST One advantage of yeast to researchers is the use of plasmids to transfer genes or DNA sequences of interest into cells. Yeast cells naturally possesses a circular plasmid, named 2μ, that is 6300 bp long and is transmitted to daughter cells in mitosis and meiosis. This plasmid has an origin of replication recognized by the yeast replication system, and so it replicates autonomously in the cell. The 2μ plasmid has been engineered to provide an efficient vector for transferring genes into yeast. In other cases, bacterial plasmids have been adapted for use in yeast. Some of them undergo homologous recombination with the yeast chromosome, transferring their sequences to the yeast chromosome. Shuttle vectors, which can be propagated in both bacteria and yeast, are particularly effective. Such vectors make it possible to

Yeast
Saccharomyces cerevisiae

ADVANTAGES

- Unicellular eukaryote
- Short cell cycle of 90 minutes
- Exists in haploid and diploid forms
- All products of meiosis are in a single structure

STATS

Taxonomy:	Fungus
Size:	2 μm
Anatomy:	Single cell
Habitat:	Surfaces of plants?

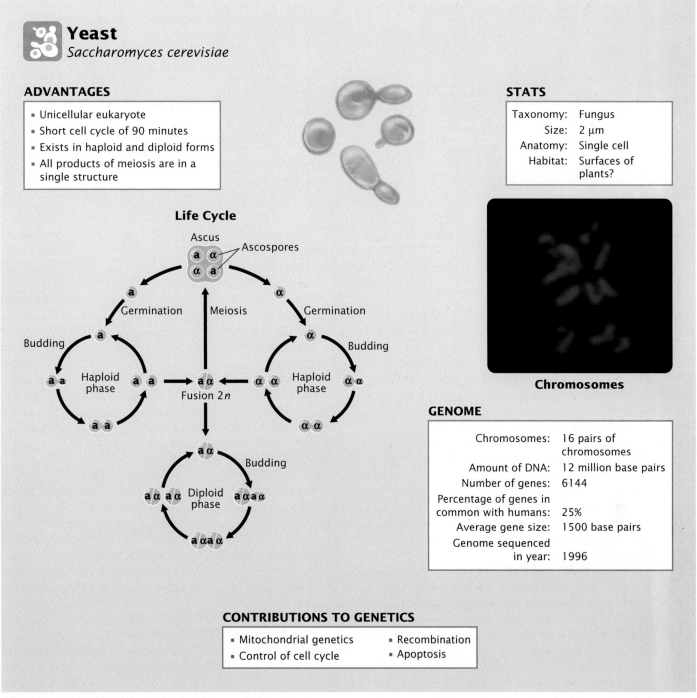

Chromosomes

Life Cycle

GENOME

Chromosomes:	16 pairs of chromosomes
Amount of DNA:	12 million base pairs
Number of genes:	6144
Percentage of genes in common with humans:	25%
Average gene size:	1500 base pairs
Genome sequenced in year:	1996

CONTRIBUTIONS TO GENETICS

- Mitochondrial genetics
- Control of cell cycle
- Recombination
- Apoptosis

[Photograph courtesy of Mara Stewart and Dean Dawson, Department of Microbiology and Molecular Biology, Sackler School of Biomedical Sciences, Tufts University.]

construct and manipulate gene sequences in bacteria, where often more powerful techniques are available for genetic manipulation and selection, and then transfer the gene sequences into yeast cells, where their function can be tested.

Plasmids are limited in the size of DNA fragments that they can carry (see pp. 571–573 in Chapter 19). Yeast artificial chromosomes (YACs)—engineered DNA fragments that contain centromeric and telomeric sequences and segregate like chromosomes in meiosis and mitosis—overcome this limitation; YACs can hold DNA fragments as large as several hundred thousand base pairs. ■

Working with Fractions: A Review

Fractions are commonly used in working genetics problems because we are often concerned with probabilities and with the proportions of progeny resulting from genetic crosses. At some point in the past, you undoubtedly learned how to manipulate fractions, but perhaps that was a long time ago. The following brief review is designed to help you recall some basic operations with fractions.

Characteristics of Fractions

What is a fraction? You probably remember that a fraction is used to represent a proportion or part of a whole. It is a number presented in the form $\frac{a}{b}$, sometimes written as a/b. The top part of the fraction (the part above the line, a in this case) is called the **numerator**. The bottom part of the fraction (the part below the line, b in this case) is called the **denominator**.

We can think of the fraction $\frac{a}{b}$ as a out of b, as in 3 out of 4 $\left(\frac{3}{4}\right)$ of the progeny of a cross have yellow seeds. Another way to think of the fraction $\frac{a}{b}$ is that it means a divided by b. For example, $\frac{3}{4} = 3 \div 4 = 0.75$. We can always convert a fraction into a decimal number by simply dividing the numerator by the denominator.

In simple fractions, which are the kind most often used in genetics problems, a and b are integers—positive or negative whole numbers. A simple fraction might be $\frac{3}{4}$ or $\frac{7}{125}$, but not $\frac{0.7}{1.5}$. Also, remember that the denominator of a fraction cannot be zero, because the fraction $\frac{a}{b}$ means that a is divided by b, and a number cannot be divided by zero. You will probably also remember that any number divided by itself is 1, so any fraction with the same number as the numerator and the denominator is equal to 1:

$$\frac{2}{2} = 1.0$$

$$\frac{7}{7} = 1.0$$

$$\frac{16}{16} = 1.0$$

The size of a fraction depends on both the numerator and the denominator. If the numerators of two fractions are the same but the denominators differ, the fraction with the larger denominator is smaller. Thus, $\frac{3}{8}$ ($= 0.375$) is smaller than $\frac{3}{4}$ ($= 0.750$). The opposite is true of the numerator. If the denominators of two fractions are the same but the numerators differ, the fraction with the larger numerator is larger. For example, $\frac{14}{16}$ ($= 0.875$) is larger than $\frac{3}{16}$ ($= 0.188$).

Equivalent Fractions

Now that we've discussed some of the features of fractions, let's review equivalent fractions. The same number can be represented by many different fractions. For example, the fractions $\frac{1}{2}$, $\frac{2}{4}$, and $\frac{4}{8}$ all equal 0.5. These fractions, which have different numbers in the numerator and the denominator but are numerically equivalent, are called equivalent fractions. The proportions represented by equivalent fractions are the same:

$$\frac{1}{2} = 0.5$$

$$\frac{2}{4} = 0.5$$

$$\frac{4}{8} = 0.5$$

We can produce equivalent fractions by multiplying both the numerator and the denominator by the same number, which is the same as multiplying by one:

$$\frac{1}{2} = \frac{1 \times 2}{2 \times 2} = \frac{2}{4} = 0.5$$

$$\frac{2}{4} = \frac{2 \times 2}{4 \times 2} = \frac{4}{8} = 0.5$$

In this way, we can create fractions that have the same denominator. This ability will be important when we need to add and subtract fractions (see Adding and Subtracting Fractions below). For example, suppose we want to convert $\frac{1}{3}$ and $\frac{1}{4}$ into equivalent fractions, both with the same denominator. We could multiply $\frac{1}{3}$ by $\frac{4}{4}$ and $\frac{1}{4}$ by $\frac{3}{3}$, so that both proportions now have 12 in the denominator:

$$\frac{1}{3} = \frac{1 \times 4}{3 \times 4} = \frac{4}{12}$$

$$\frac{1}{4} = \frac{1 \times 3}{4 \times 3} = \frac{3}{12}$$

Our new equivalent fraction $\left(\frac{4}{12}\right)$ is equal in value to the original fraction $\left(\frac{1}{3}\right)$, but has 12 in the denominator. Likewise, $\frac{3}{12}$ is equal in value to the original fraction $\left(\frac{1}{4}\right)$, but also has 12 in the denominator.

Simplifying Fractions

We often want to put fractions into their simplest equivalent forms. Simplifying fractions, for example, may help us to see that two fractions are equivalent.

To simplify a fraction, the numerator and denominator must first be factored. A factor of a number is a number that the original number can be divided by evenly with no remainder. For example, 4 is a factor of 12 because 4 will divide into 12 exactly 3 times $\left(\frac{12}{4} = 3\right)$. However, 5 is not a factor of 12 because 5 cannot divide into 12 evenly $\left(\frac{12}{5} = 2.4\right)$. Thus, 4 can be factored as $4 = 2 \times 2$. Similarly, 8 can be factored as $8 = 2 \times 2 \times 2$, and 15 can be factored as $15 = 3 \times 5$.

Fractions can be simplified when the numerator and denominator have a common factor:

$$\frac{4}{16} = \frac{4 \times 1}{4 \times 4}$$

Because $\frac{4}{4} = 1$, the 4s in both the numerator and denominator cancel out:

$$\frac{4}{16} = \frac{\cancel{4} \times 1}{\cancel{4} \times 4} = \frac{1}{4}$$

Thus, $\frac{4}{16}$ can be simplified to $\frac{1}{4}$.

Now that we've discussed some features of fractions, let's review how we carry out simple mathematical operations with fractions, such as addition, subtraction, multiplication, and division. These are the primary operations you will need to use as you work genetics problems.

Adding and Subtracting Fractions

Adding and subtracting fractions is easy, but first they must have the same denominator. To add two or more fractions with the same denominator, add the numerators and place the sum over the common denominator:

$$\frac{1}{4} + \frac{1}{4} = \frac{2}{4}$$

$$\frac{3}{8} + \frac{2}{8} + \frac{1}{8} = \frac{6}{8}$$

Similarly, to subtract two or more fractions with the same denominator, subtract the numerators and place the result over the common denominator:

$$\frac{6}{8} - \frac{1}{8} = \frac{5}{8}$$

$$\frac{13}{16} - \frac{3}{16} - \frac{1}{16} = \frac{9}{16}$$

When the fractions you want to add or subtract don't have the same denominator, then you must convert them into equivalent fractions that have the same denominator (see Equivalent Fractions above). For example, to add $\frac{1}{2}$ and $\frac{1}{3}$, convert both fractions into equivalent fractions with the common denominator 6:

$$\frac{1}{2} = \frac{1 \times 3}{2 \times 3} = \frac{3}{6}$$

$$\frac{1}{3} = \frac{1 \times 2}{3 \times 2} = \frac{2}{6}$$

Once the fractions have the same denominator, they can be added by summing the numerators and putting the result over the common denominator:

$$\frac{3}{6} + \frac{2}{6} = \frac{5}{6}$$

The same is true when subtracting fractions—they must have the same denominator. For example, to subtract $\frac{1}{3}$ from $\frac{7}{8}$, first convert the two fractions into equivalent fractions that have a common denominator, such as 24:

$$\frac{7}{8} = \frac{7 \times 3}{8 \times 3} = \frac{21}{24}$$

$$\frac{1}{3} = \frac{1 \times 8}{3 \times 8} = \frac{8}{24}$$

Then subtract the numerators and put the result over the common denominator.

$$\frac{21}{24} - \frac{8}{24} = \frac{13}{24}$$

Multiplying and Dividing Fractions

Multiplying two or more fractions is simple. You multiply the numerators of the fractions and then multiply the denominators of the fractions:

$$\frac{1}{2} \times \frac{3}{4} = \frac{1 \times 3}{2 \times 4} = \frac{3}{8}$$

$$\frac{3}{4} \times \frac{3}{4} = \frac{3 \times 3}{4 \times 4} = \frac{9}{16}$$

$$\frac{1}{2} \times \frac{3}{4} \times \frac{3}{4} = \frac{1 \times 3 \times 3}{2 \times 4 \times 4} = \frac{9}{32}$$

Dividing fractions is a bit more complicated. Suppose we want to divide one fraction by another fraction, such as $\frac{1}{4}$ divided by $\frac{3}{4}$. First take the reciprocal of the divisor (the second fraction, or $\frac{3}{4}$). Taking the reciprocal means that you reverse the numbers in the numerator and the denominator. For example, the reciprocal of $\frac{3}{4}$ is $\frac{4}{3}$. Once you have the reciprocal, multiply the dividend (the first fraction, or $\frac{1}{4}$) by the reciprocal of the divisor:

$$\frac{1}{4} \times \frac{4}{3} = \frac{4}{12}$$

Another way to think about dividing fractions is to represent the original division ($\frac{1}{4}$ divided by $\frac{3}{4}$) as a fraction itself, with $\frac{1}{4}$ as the numerator and $\frac{3}{4}$ as the denominator:

$$\frac{\dfrac{1}{4}}{\dfrac{3}{4}}$$

To carry out the division, you multiply both the numerator and denominator by the reciprocal of the denominator ($\frac{4}{3}$), which is the equivalent of multiplying by 1:

$$\frac{\dfrac{1}{4} \times \dfrac{4}{3}}{\dfrac{3}{4} \times \dfrac{4}{3}} = \frac{\dfrac{4}{12}}{\dfrac{12}{12}}$$

The denominator now equals $\frac{12}{12}$, which is 1, so the equation simplifies to

$$\frac{\dfrac{4}{12}}{\dfrac{12}{12}} = \frac{\dfrac{4}{12}}{1} = \frac{4}{12}$$

This is the same result we obtained when we multiplied the dividend by the reciprocal of the divisor.

Practice Problems

1. What is the denominator in the following fraction: $\frac{6}{16}$? What is the numerator?

2. What is the decimal value of the following fraction: $\frac{8}{32}$?

3. Which of the following fractions represents a larger number, $\frac{7}{16}$ or $\frac{9}{16}$?

4. Which of the following fractions represents a larger number, $\frac{10}{32}$ or $\frac{10}{64}$?

5. Convert the following fractions into equivalent fractions that have the same denominator:

a. $\dfrac{1}{2}$ and $\dfrac{2}{3}$

b. $\dfrac{4}{7}$ and $\dfrac{5}{9}$

c. $\dfrac{3}{4}$ and $\dfrac{7}{12}$

6. Simplify the following fractions:

a. $\dfrac{6}{9}$

b. $\dfrac{10}{25}$

c. $\dfrac{4}{64}$

d. $\dfrac{12}{32}$

7. Add or subtract the following fractions:

a. $\dfrac{5}{8} + \dfrac{2}{8}$

b. $\dfrac{15}{16} - \dfrac{9}{16}$

c. $\dfrac{2}{3} + \dfrac{1}{16}$

d. $\dfrac{53}{64} - \dfrac{5}{16}$

8. Multiply the following fractions:

a. $\dfrac{3}{4} \times \dfrac{1}{8}$

b. $\dfrac{3}{16} \times \dfrac{2}{3}$

c. $\dfrac{3}{8} \times \dfrac{13}{32}$

9. Divide the following fractions:

a. $\dfrac{1}{4}$ by $\dfrac{1}{2}$

b. $\dfrac{3}{4}$ by $\dfrac{3}{8}$

c. $\dfrac{1}{16}$ by $\dfrac{7}{8}$

Answers to Practice Problems

1. Denominator $= 16$; numerator $= 6$

2. 0.25

3. $\dfrac{9}{16}$

4. $\dfrac{10}{32}$

5.

a. $\dfrac{1}{2} = \dfrac{1 \times 3}{2 \times 3} = \dfrac{3}{6}; \dfrac{2}{3} = \dfrac{2 \times 2}{3 \times 2} = \dfrac{4}{6}$

b. $\dfrac{4}{7} = \dfrac{4 \times 9}{7 \times 9} = \dfrac{36}{63}; \dfrac{5}{9} = \dfrac{5 \times 7}{9 \times 7} = \dfrac{35}{63}$

c. $\dfrac{3}{4} = \dfrac{3 \times 3}{4 \times 3} = \dfrac{9}{12}; \dfrac{7}{12}$

6.

a. $\dfrac{6}{9} = \dfrac{3 \times 2}{3 \times 3} = \dfrac{2}{3}$

b. $\dfrac{10}{25} = \dfrac{5 \times 2}{5 \times 5} = \dfrac{2}{5}$

c. $\dfrac{4}{64} = \dfrac{4}{4 \times 16} = \dfrac{1}{16}$

d. $\dfrac{12}{32} = \dfrac{4 \times 3}{4 \times 8} = \dfrac{3}{8}$

7.

a. $\dfrac{7}{8}$

b. $\dfrac{6}{16}$, which simplifies to $\dfrac{3}{8}$

c. $\dfrac{2}{3} = \dfrac{2 \times 16}{3 \times 16} = \dfrac{32}{48}; \dfrac{1}{16} = \dfrac{1 \times 3}{16 \times 3} = \dfrac{3}{48};$

$\dfrac{32}{48} + \dfrac{3}{48} = \dfrac{35}{48}$

d. $\dfrac{5}{16} = \dfrac{5 \times 4}{16 \times 4} = \dfrac{20}{64}; \dfrac{53}{64} - \dfrac{20}{64} = \dfrac{33}{64}$

8.

a. $\dfrac{3}{4} \times \dfrac{1}{8} = \dfrac{3}{32}$

b. $\dfrac{3}{16} \times \dfrac{2}{3} = \dfrac{6}{48}$, which simplifies to $\dfrac{1}{8}$

c. $\dfrac{3}{8} \times \dfrac{13}{32} = \dfrac{39}{256}$

9.

a. $\dfrac{1}{4} \times \dfrac{2}{1} = \dfrac{2}{4}$, which simplifies to $\dfrac{1}{2}$

b. $\dfrac{3}{4} \times \dfrac{8}{3} = \dfrac{24}{12}$, which simplifes to 2

c. $\dfrac{1}{16} \times \dfrac{8}{7} = \dfrac{8}{112}$, which simplifes to $\dfrac{1}{14}$

Glossary

abortive initiation Process during initiation of transcription in which RNA polymerase repeatedly generates and releases short transcripts, from 2 to 6 nucleotides in length, while still bound to the promoter. Occurs in both prokaryotes and eukaryotes.

acentric chromatid Chromatid that lacks a centromere; produced when crossing over takes place within a paracentric inversion. The acentric chromatid does not attach to a spindle microtubule and does not segregate in meiosis or mitosis; so it is usually lost after one or more rounds of cell division.

acrocentric chromosome Chromosome in which the centromere is near one end, producing a long arm at one end and a knob, or satellite, at the other end.

activator *See* **transcriptional activator protein.**

adaptive mutation Process by which a specific environment induces mutations that enable organisms to adapt to that environment.

addition rule Rule stating that the probability of any of two or more mutually exclusive events occurring is calculated by adding the probabilities of the individual events.

additive genetic variance Component of genetic variance that can be attributed to the additive effects of different genotypes.

adenine (A) Purine base in DNA and RNA.

adenosine-3′,5′-cyclic monophosphate (cAMP) Modified nucleotide that functions in catabolite repression. Low levels of glucose stimulate high levels of cAMP; cAMP then attaches to CAP, which binds to the promoters of certain operons and stimulates transcription.

adjacent-1 segregation Type of segregation that takes place in a heterozygote for a translocation. If the original, nontranslocated chromosomes are N_1 and N_2 and the chromosomes containing the translocated segments are T_1 and T_2, then adjacent-1 segregation takes place when N_1 and T_2 move toward one pole and T_1 and N_2 move toward the opposite pole.

adjacent-2 segregation Type of segregation that takes place in a heterozygote for a translocation. If the original, nontranslocated chromosomes are N_1 and N_2 and the chromosomes containing the translocated segments are T_1 and T_2, then adjacent-2 segregation takes place when N_1 and T_1 move toward one pole and T_2 and N_2 move toward the opposite pole.

A-DNA Right-handed helical structure of DNA that exists when little water is present.

affinity capture Use of an antibody to capture one protein from a complex mixture of proteins. The captured protein will "pull down" with it any proteins with which it interacts, which can then be analyzed by mass spectrometry to identify the proteins.

allele One of two or more alternative forms of a gene.

allelic frequency Proportion of a particular allele within a population.

allopatric speciation Speciation that arises when a geographic barrier first splits a population into two groups and blocks the exchange of genes between them.

allopolyploidy Condition in which the sets of chromosomes of a polyploid individual are derived from two or more species.

allosteric protein Protein that changes its conformation upon binding with another molecule.

alternate segregation Type of segregation that takes place in a heterozygote for a translocation. If the original, nontranslocated chromosomes are N_1 and N_2 and the chromosomes containing the translocated segments are T_1 and T_2, then alternate segregation takes place when N_1 and N_2 move toward one pole and T_1 and T_2 move toward the opposite pole.

alternation of generations Complex life cycle in plants that alternates between the diploid sporophyte stage and the haploid gametophyte stage.

alternative processing pathway One of several pathways by which a single pre-mRNA can be processed in different ways to produce alternative types of mRNA.

alternative splicing Process by which a single pre-mRNA can be spliced in more than one way to produce different types of mRNA.

Ames test Test in which special strains of bacteria are used to evaluate the potential of chemicals to cause cancer.

amino acid Repeating unit of proteins; consists of an amino group, a carboxyl group, a hydrogen atom, and a variable R group.

aminoacyl (A) site One of the three sites in a ribosome occupied by a tRNA during translation. All charged tRNAs (with the exception of the initiator tRNA) first enter the A site.

aminoacyl-tRNA synthetase Enzyme that attaches an amino acid to a tRNA. Each aminoacyl-tRNA synthetase is specific for a particular amino acid.

amniocentesis Procedure used for prenatal genetic testing to obtain a sample of amniotic fluid from a pregnant woman. A long sterile needle is inserted through the abdominal wall into the amniotic sac to obtain the fluid.

amphidiploidy Type of allopolyploidy in which two different diploid genomes are combined such that every chromosome has one and only one homologous partner and the genome is functionally diploid.

anagenesis Evolutionary change within a single lineage.

anaphase Stage of mitosis in which chromatids separate and move toward the spindle poles.

anaphase I Stage of meiosis I in which homologous chromosomes separate and move toward the spindle poles.

anaphase II Stage of meiosis II in which chromatids separate and move toward the spindle poles.

aneuploidy Change from the wild type in the number of individual chromosomes; most often an increase or decrease of one or two chromosomes.

annotation Linking of the sequence information of a gene that has been identified to other information about the gene's function and expression, about the protein encoded by the gene, and about similar genes in other species.

***Antennapedia* complex** Cluster of five homeotic genes in fruit flies that affects the development of the adult fly's head and anterior thoracic segments.

antibody Protein produced by a B cell that circulates in the blood and other body fluids. An antibody binds to a specific antigen and marks the antigen for destruction by a phagocytic cell.

anticipation Increasing severity or earlier age of onset of a genetic trait in succeeding generations. For example, symptoms of a genetic disease may become more severe as the trait is passed from generation to generation.

anticodon Sequence of three nucleotides in tRNA that pairs with the corresponding codon in mRNA in translation.

antigen A molecule that is recognized by the immune system and elicits an immune response.

antigenic drift The appearance of new strains of a rapidly evolving virus because of mutations.

antigenic shift Major change in a viral genome through the reassortment of genetic material from two different strains of the virus.

antiparallel The orientation of the two polynucleotide strands of the DNA double helix in opposite directions.

antisense RNA Small RNA molecule that base pairs with a complementary DNA or RNA sequence and affects its functioning.

antiterminator Protein or DNA sequence that inhibits the termination of transcription.

apoptosis Programmed cell death, in which a cell degrades its own DNA, the nucleus and cytoplasm shrink, and the cell undergoes phagocytosis by other cells without leakage of its contents.

archaea One of the three primary divisions of life, consisting of unicellular organisms with prokaryotic cell structure.

artificial selection Selection practiced by humans.

attenuation Type of gene regulation in some bacterial operons in which transcription is initiated but terminates prematurely before the transcription of the structural genes.

attenuator Secondary structure that forms in the 5′ untranslated region of some operons and causes the premature termination of transcription.

autoimmune disease Disease characterized by an abnormal immune response to a person's own (self) antigen.

autonomous element Transposable element that is fully functional and able to transpose on its own.

autonomously replicating sequence (ARS) DNA sequence that confers the ability to replicate; contains an origin of replication.

autopolyploidy Condition in which all the sets of chromosomes of a polyploid individual are derived from a single species.

autosome Chromosome that is the same in males and females; a nonsex chromosome.

auxotrophic Possessing a nutritional mutation that disrupts the ability to synthesize an essential biological molecule; an auxotrophic bacterium cannot grow on minimal medium but can grow on minimal medium to which has been added the biological molecule that it cannot synthesize.

backcross Cross between an F_1 individual and one of the parental (P) genotypes.

bacteria One of the three primary divisions of life, consisting of prokaryotic unicellular organisms; also called **eubacteria**.

bacterial artificial chromosome (BAC) Cloning vector used in bacteria that is capable of carrying DNA fragments as large as 500 kb.

bacterial colony Clump of genetically identical bacteria derived from a single bacterial cell that undergoes repeated rounds of division.

bacteriophage Virus that infects bacterial cells.

balancing selection Natural selection that maintains genetic variation.

Barr body Inactivated X chromosome that appears as a condensed, darkly staining structure in most cells of female placental mammals.

basal transcription apparatus Complex of transcription factors, RNA polymerase, and other proteins that assemble on the promoter and are capable of initiating minimal levels of transcription.

base *See* **nitrogenous base**.

base analog Chemical substance that has a structure similar to that of one of the four standard bases of DNA and may be incorporated into newly synthesized DNA molecules in replication.

base-excision repair DNA repair that first excises modified bases and then replaces the entire nucleotide.

base substitution Mutation in which a single pair of bases in DNA is altered.

B cell Type of lymphocyte that produces humoral immunity; matures in the bone marrow and produces antibodies.

B-DNA Right-handed helical structure of DNA that exists when water is abundant; the secondary structure described by Watson and Crick and probably the most common DNA structure in cells.

behavioral isolation Reproductive isolation due to differences in behavior that prevent interbreeding.

β sliding clamp A ring-shaped polypeptide component of DNA polymerase III that encircles DNA during replication and allows the polymerase to slide along the DNA template strand.

bidirectional replication Replication at both ends of a replication bubble.

bioinformatics Interdisciplinary field that combines molecular biology and computer science; develops databases and computational tools to store, retrieve, and analyze nucleic acid– and protein-sequence data.

biological species concept Definition of a species as a group of organisms whose members are capable of interbreeding with one another but are reproductively isolated from the members of other species. Because different species do not exchange genes, each species evolves independently. Not all biologists adhere to this concept.

biotechnology Use of biological processes, particularly molecular genetics and recombinant DNA technology, to produce products of commercial value.

bithorax complex Cluster of three homeotic genes in fruit flies that influences the adult fly's posterior thoracic and abdominal segments.

bivalent A synapsed pair of homologous chromosomes consisting of four chromatids; also called a **tetrad**.

blending inheritance Early concept of heredity proposing that offspring possess a mixture of the traits from both parents.

branch A connection between nodes in a phylogenetic tree representing an evolutionary connection between organisms.

branch migration Movement of a cross bridge along two DNA molecules.

branch point Adenine nucleotide in nuclear pre-mRNA introns that lies 18 to 40 nucleotides upstream of the 3′ splice site.

broad-sense heritability Proportion of phenotypic variance that can be attributed to genetic variance.

cap-binding complex A group of proteins in eukaryotes that binds to the 5′ cap and initiates translation. Aids in exporting mRNA from the nucleus to the cytoplasm and promotes the initial (pioneer) round of translation.

carcinogen A substance capable of causing cancer.

caspase Enzyme that cleaves other proteins and regulates apoptosis. Each caspase is synthesized as a large, inactive precursor (a procaspase) that is activated by cleavage, often by another caspase.

catabolite activator protein (CAP) Protein that functions in catabolite repression. When bound with cAMP, CAP binds to the promoters of certain operons and stimulates transcription.

catabolite repression System of gene control in some bacterial operons in which glucose is used preferentially and the metabolism of other sugars is repressed in the presence of glucose.

cDNA (complementary DNA) library Collection of bacterial colonies or phages containing DNA fragments that have been produced by reverse transcription of cellular mRNA.

cell cycle Stages through which a cell passes from one cell division to the next.

cell line Genetically identical cells that divide indefinitely and can be cultured in the laboratory.

cell theory Theory stating that all life is composed of cells, that cells arise only from other cells, and that the cell is the fundamental unit of structure and function in living organisms.

cellular immunity Type of immunity resulting from T cells, which recognize antigens found on the surfaces of self cells.

centiMorgan (cM) *See* **map unit**.

central dogma Concept that genetic information passes from DNA to RNA to protein in a one-way information pathway.

centriole Cytoplasmic organelle consisting of microtubules; present at each pole of the spindle apparatus in animal cells.

centromere Constricted region on a chromosome that stains less strongly than the rest of the chromosome; serves as the attachment point for spindle microtubules.

centromeric sequence DNA sequence found in functional centromeres.

centrosome Structure from which the spindle apparatus develops: contains the centriole.

Chargaff's rules Rules developed by Erwin Chargaff and his colleagues concerning the ratios of bases in DNA.

checkpoint A key transition point at which progression to the next stage in the cell cycle is regulated.

chiasma (pl., chiasmata) Point of attachment between homologous chromosomes at which crossing over took place.

chi-square goodness of fit test Statistical test used to evaluate how well a set of observed values fit the expected values. The probability associated with a calculated chi-square value is the probability that the differences between the observed and the expected values are due to chance.

chloroplast DNA (cpDNA) DNA in chloroplasts; has many characteristics in common with eubacterial DNA and typically consists of a circular molecule that lacks histone proteins and encodes some of the rRNAs, tRNAs, and proteins found in chloroplasts.

chorionic villus sampling (CVS) Procedure used for prenatal genetic testing in which a small piece of the chorion (the outer layer of the placenta) is removed from a pregnant woman. A catheter is inserted through the vagina and cervix into the uterus; suction is then applied to remove the sample.

chromatin Material found in the eukaryotic nucleus; consists of DNA and proteins.

chromatin-remodeling complex Complex of proteins that alters chromatin structure without acetylating histone proteins.

chromatin-remodeling protein Protein that binds to a DNA sequence and disrupts chromatin structure, causing the DNA to become more accessible to RNA polymerase and other proteins.

chromosome Structure consisting of DNA and associated proteins that carries and transmits genetic information.

chromosome deletion Loss of a chromosome segment.

chromosome duplication Mutation that doubles a segment of a chromosome.

chromosome inversion Rearrangement in which a segment of a chromosome has been inverted 180 degrees.

chromosome jumping Method of locating a gene by moving from a gene on a cloned fragment to sequences on distantly linked fragments.

chromosome mutation Difference from the wild type in the number or structure of one or more chromosomes; often affects many genes and has large phenotypic effects.

chromosome puff Localized swelling of a polytene chromosome; a region of chromatin in which DNA has unwound and is undergoing transcription.

chromosome rearrangement Change from the wild type in the structure of one or more chromosomes.

chromosome theory of heredity Theory stating that genes are located on chromosomes.

chromosome walking Method of locating a gene by using partly overlapping genomic clones to move in steps from a previously cloned, linked gene to the gene of interest.

cis configuration *See* **coupling configuration**.

cladogenesis Evolution in which one lineage is split into two.

clonal evolution Process by which mutations that enhance the ability of cells to proliferate predominate in a clone of cells, allowing the clone to become increasingly rapid in growth and increasingly aggressive in proliferation properties.

cloning vector Stable, replicating DNA molecule to which a foreign DNA fragment can be attached for transfer to a host cell.

cloverleaf Secondary structure common to all tRNAs.

coactivator Protein that cooperates with an activator of transcription. In eukaryotic transcriptional control, coactivators often physically interact with transcriptional activators and the basal transcription apparatus.

codominance Type of allelic interaction in which the heterozygote simultaneously expresses traits of both homozygotes.

codon Sequence of three nucleotides that encodes one amino acid in a protein.

coefficient of coincidence Ratio of observed double crossovers to expected double crossovers.

cohesin Molecule that holds the two sister chromatids of a chromosome together. The breakdown of cohesin at the centromeres enables the chromatids to separate in anaphase of mitosis and anaphase II of meiosis.

cohesive end Short, single-stranded overhanging end on a DNA molecule produced when the DNA is cut by certain restriction enzymes; also called a sticky end. Cohesive ends are complementary and can spontaneously pair to rejoin DNA fragments that have been cut with the same restriction enzyme.

colinearity Concept of direct correspondence between the nucleotide sequence of a gene and the continuous sequence of amino acids in a protein.

colony *See* **bacterial colony**.

comparative genomics Comparative studies of the genomes of different organisms.

competence The ability to take up DNA from the environment (to be transformed).

complementary DNA strands Nucleotide strands of DNA in which each purine on one strand pairs with a specific pyrimidine on the opposite strand (A pairs with T, and G pairs with C).

complementation Manifestation of two different mutations in the heterozygous condition as the wild-type phenotype; indicates that the mutations are at different loci.

complementation test Test designed to determine whether two different mutations are at the same locus (are allelic) or at different loci (are nonallelic). Two individuals that are homozygous for two independently derived mutations are crossed, producing F_1 progeny that are heterozygous for the mutations. If the mutations are at the same locus, the F_1 will have a mutant phenotype. If the mutations are at different loci, the F_1 will have a wild-type phenotype.

complete dominance Type of dominance in which the same phenotype is expressed in homozygotes (*AA*) and in heterozygotes (*Aa*); only the dominant allele is expressed in a heterozygote.

complete linkage Linkage between genes that are located close together on the same chromosome with no crossing over between them.

complete medium Medium used to culture bacteria or other microorganisms that contains all the nutrients required for growth and reproduction, including those normally synthesized by the organism. Auxotrophic mutants can grow on complete medium.

composite transposon Type of transposable element in bacteria that consists of two insertion sequences flanking a segment of DNA.

compound heterozygote An individual with two different recessive alleles at a locus that result in a recessive phenotype.

concept of dominance Principle of heredity discovered by Mendel stating that when two different alleles are present in a genotype, only one allele may be expressed in the phenotype. The dominant allele is the allele that is expressed, and the recessive allele is the allele that is not expressed.

concordance Percentage of twin pairs in which both twins have a particular trait.

concordant Refers to a pair of twins both of whom have a particular trait.

condensins A group of proteins that bind to chromosomes as a cell enters prophase, causing the chromosomes to become more compact and visible under a light microscope.

conditional mutation Mutation that is expressed only under certain conditions.

conditional probability Probability that is modified by additional information that another event has occurred.

conjugation Mechanism by which genetic material can be exchanged between bacterial cells. In conjugation, two bacteria lie close together, and a cytoplasmic connection forms between them. A plasmid, or sometimes a part of the bacterial chromosome, passes through this connection from one cell to the other.

consanguinity Mating between related individuals.

consensus sequence Sequence that comprises the most commonly encountered nucleotides found at a specific location in DNA or RNA.

−10 consensus sequence (Pribnow box) Consensus sequence (TATAAT) found in most bacterial promoters approximately 10 bp upstream of the transcription start site.

−35 consensus sequence Consensus sequence (TTGACA) found in many bacterial promoters approximately 35 bp upstream of the transcription start site.

constitutive gene A gene that is expressed continually without regulation.

constitutive mutation A mutation that causes the continuous transcription of one or more structural genes.

contig Set of overlapping DNA fragments that have been assembled in the correct order to form a continuous stretch of DNA sequence.

continuous characteristic Characteristic that displays a large number of possible phenotypes that are not easily distinguished, such as human height.

continuous replication Replication of the leading strand of DNA in the same direction as that of unwinding, allowing new nucleotides to be added continuously to the $3'$ end of the new strand as the template is exposed.

coordinate induction The simultaneous synthesis of several enzymes stimulated by a single environmental factor.

copy-number variation (CNV) Difference among individual organisms in the number of copies of any large DNA sequence (larger than 1000 bp).

core enzyme Set of five subunits at the heart of most bacterial RNA polymerases that, during transcription, catalyzes the elongation of the RNA molecule by the addition of RNA nucleotides; consists of two copies of a subunit called alpha (α) and single copies of subunits beta (β), beta prime (β'), and omega (ω).

corepressor Substance that inhibits transcription in a repressible system of gene regulation; usually a small molecule that binds to a repressor protein and alters it so that the repressor is able to bind to DNA and inhibit transcription.

core promoter DNA sequence located immediately upstream of a eukaryotic gene, to which the basal transcription apparatus binds.

correlation Degree of association between two or more variables.

correlation coefficient Statistic that measures the degree of association between two or more variables. A correlation coefficient can range from −1 to +1. A positive value indicates a direct relation between the variables; a negative correlation indicates an inverse relation. The absolute value of the correlation coefficient provides information about the strength of association between the variables.

cosmid Cloning vector consisting of a plasmid packaged in a viral protein coat that can be transferred to bacteria by viral infection; can carry large pieces of DNA into bacteria.

cotransduction Process in which two or more genes are transferred together from one bacterial cell to another by a virus. Only genes located close together on a bacterial chromosome will be cotransduced.

cotransformation Process in which two or more genes are taken up together during cell transformation.

coupling (cis) configuration Arrangement of linked genes in which the wild-type alleles of two or more genes are on one chromosome and their mutant alleles are on the homologous chromosome.

CpG island DNA region that contains many copies of a cytosine base followed by a guanine base; often found near transcription start sites in eukaryotic DNA. The cytosine bases in CpG islands are commonly methylated when genes are inactive but are demethylated before the initiation of transcription.

CRISPR-Cas system A molecular tool used for precise editing of DNA that relies on the action of CRISPR RNAs and Cas proteins.

CRISPR RNAs (crRNAs) Small RNA molecules found in prokaryotes that assist in the destruction of foreign DNA.

cross bridge In a heteroduplex DNA molecule, the point at which each nucleotide strand passes from one DNA molecule to the other.

crossing over Exchange of genetic material between homologous but nonsister chromatids.

cruciform Structure formed by the pairing of inverted repeats on both strands of double-stranded DNA.

C-value Haploid amount of DNA found in a cell of an organism.

C-value paradox The absence of a relation between the C-values (genome sizes) of eukaryotes and organismal complexity.

cyclin A key protein in the control of the cell cycle; combines with a cyclin-dependent kinase (CDK). The levels of cyclin rise and fall in the course of the cell cycle.

cyclin-dependent kinase (CDK) A key protein in the control of the cell cycle; combines with cyclin.

cytokinesis Process by which the cytoplasm of a cell divides.

cytoplasmic inheritance Inheritance of characteristics encoded by genes located in the cytoplasm. Because the cytoplasm is usually contributed entirely by one parent, most cytoplasmically inherited characteristics are inherited from only one parent.

cytosine (C) Pyrimidine base in DNA and RNA.

deamination Loss of an amino group (NH_2) from a base.

degenerate genetic code Refers to the fact that the genetic code contains more codons than are needed to specify all 20 common amino acids.

deletion Mutation in which one or more nucleotides are deleted from a DNA sequence.

deletion mapping Technique for determining the chromosomal location of a gene by studying the association of its phenotype or product with particular chromosome deletions.

denaturation (melting) Process that separates the strands of double-stranded DNA when DNA is heated.

deoxyribonucleotide Basic building block of DNA, consisting of deoxyribose, a phosphate group, and a nitrogenous base.

deoxyribose Five-carbon sugar in DNA; lacks a hydroxyl group on the 2′-carbon atom.

depurination Break in the covalent bond connecting a purine base to the 1′-carbon atom of deoxyribose, resulting in the loss of the purine base.

determination Process by which a cell becomes committed to developing into a particular cell type.

diakinesis Fifth substage of prophase I in meiosis. In diakinesis, chromosomes contract, the nuclear membrane breaks down, and the spindle forms.

dicentric bridge Structure produced when the two centromeres of a dicentric chromatid are pulled toward opposite poles, stretching the dicentric chromosome across the center of the nucleus. Eventually, the dicentric bridge breaks as the two centromeres are pulled apart.

dicentric chromatid Chromatid that has two centromeres; produced when crossing over takes place within a paracentric inversion. The two centromeres of the dicentric chromatid are frequently pulled toward opposite poles in mitosis or meiosis, breaking the chromosome.

dideoxyribonucleoside triphosphate (ddNTP) Special substrate for DNA synthesis used in the Sanger dideoxy sequencing method;

identical with dNTP (the usual substrate for DNA synthesis) except that it lacks a 3′-OH group. The incorporation of a ddNTP into DNA terminates DNA synthesis.

dihybrid cross Cross between two individuals that differ in two characteristics—more specifically, a cross between individuals that are homozygous for different alleles at two loci (*AA BB* × *aa bb*); also refers to a cross between two individuals that are both heterozygous at two loci (*Aa Bb* × *Aa Bb*).

dioecious Belonging to a species whose individual members have either male or female reproductive structures.

diploid Possessing two sets of chromosomes (two genomes).

diplotene Fourth substage of prophase I in meiosis. In diplotene, centromeres of homologous chromosomes move apart, but the homologs remain attached at chiasmata.

directional selection Selection in which one trait or allele is favored over another.

direct repair DNA repair in which modified bases are changed back into their original structures.

direct-to-consumer genetic test Test for a genetic condition that can be purchased directly by a consumer without the involvement of a physician or other health-care provider.

discontinuous characteristic Characteristic that exhibits only a few, easily distinguished phenotypes. An example is seed shape in which seeds are either round or wrinkled.

discontinuous replication Replication of the lagging strand of DNA in the direction opposite that of unwinding, which means that DNA must be synthesized in short stretches (Okazaki fragments).

discordant Refers to a pair of twins of whom one twin has a particular trait and the other does not.

displaced duplication Chromosome rearrangement in which the duplicated segment is some distance from the original segment, either on the same chromosome or on a different one.

dizygotic twins Nonidentical twins that arise when two different eggs are fertilized by two different sperm.

D loop Region of mitochondrial DNA that contains an origin of replication and promoters; it is displaced during the initiation of replication, leading to the name displacement, or D, loop.

DNA fingerprinting Technique used to identify individuals by examining their DNA sequences.

DNA gyrase Topoisomerase enzyme in *E. coli* that relieves the torsional strain that builds up ahead of the replication fork.

DNA helicase Enzyme that unwinds double-stranded DNA by breaking hydrogen bonds.

DNA library Collection of clones containing all the DNA fragments from one source.

DNA ligase Enzyme that catalyzes the formation of a phosphodiester bond between adjacent 3′-OH and 5′-phosphate groups in a DNA molecule.

DNA methylation Modification of DNA by the addition of methyl groups to certain positions on the bases.

DNA polymerase Enzyme that synthesizes DNA.

DNA polymerase I Bacterial DNA polymerase that removes RNA primers and replaces them with DNA nucleotides.

DNA polymerase II Bacterial DNA polymerase that takes part in DNA repair; restarts replication after synthesis has halted because of DNA damage.

DNA polymerase III Bacterial DNA polymerase that synthesizes new nucleotide strands by adding new nucleotides to the 3′-OH group provided by the primer.

DNA polymerase IV Bacterial DNA polymerase that probably takes part in DNA repair.

DNA polymerase V Bacterial DNA polymerase that probably takes part in DNA repair.

DNA polymerase α Eukaryotic DNA polymerase that initiates replication.

DNA polymerase β Eukaryotic DNA polymerase that participates in DNA repair.

DNA polymerase δ Eukaryotic DNA polymerase that replicates the lagging strand during DNA synthesis; also carries out DNA repair and translesion DNA synthesis.

DNA polymerase ε Eukaryotic DNA polymerase that replicates the leading strand during DNA synthesis.

DNA polymerase γ Eukaryotic DNA polymerase that replicates mitochondrial DNA. A γ-like DNA polymerase replicates chloroplast DNA.

DNase I hypersensitive site Chromatin region that becomes sensitive to digestion by the enzyme DNase I.

DNA sequencing Process of determining the sequence of bases along a DNA molecule.

DNA transposon Transposable element that transposes as DNA.

domain Functional part of a protein.

dominance genetic variance Component of genetic variance that can be attributed to dominance (interaction between genes at the same locus).

dominant Refers to an allele or a phenotype that is expressed in homozygotes (*AA*) and in heterozygotes (*Aa*); only the dominant allele is expressed in a heterozygote phenotype.

dosage compensation Equalization in males and females of the amount of protein produced by X-linked genes. In placental mammals, dosage compensation is accomplished by the random inactivation of one X chromosome in the cells of females.

double fertilization Fertilization in plants; includes the fusion of a sperm cell with an egg cell to form a zygote and the fusion of a second sperm cell with the polar nuclei to form an endosperm.

double-strand-break model Model of homologous recombination in which a DNA molecule undergoes double-strand breaks.

Down syndrome (trisomy 21) Human condition characterized by variable degrees of intellectual disability, characteristic facial features, some retardation of growth and development, and an increased incidence of heart defects, leukemia, and other abnormalities; caused by the duplication of all or part of chromosome 21.

driver Mutation found in a cancer cell that contributes to the process of cancer development.

ecological isolation Reproductive isolation in which different species live in different habitats and interact with the environment in different ways, so that their members do not encounter one another and do not reproduce with one another.

Edward syndrome (trisomy 18) Human condition characterized by severe intellectual disability, low-set ears, a short neck, deformed feet, clenched fingers, heart problems, and other disabilities; results from the presence of three copies of chromosome 18.

effective population size Effective number of breeding adults in a population; influenced by the number of individuals contributing genes to the next generation, their sex ratio, variation between individuals in reproductive success, fluctuations in population size, the age structure of the population, and whether mating is random.

egg Female gamete.

egg-polarity genes Set of genes that determine the major axes of development in an early fruit-fly embryo. One set of egg-polarity genes determines the anterior–posterior axis and another determines the dorsal–ventral axis.

electrophoresis *See* **gel electrophoresis.**

elongation factor G (EF-G) Protein that combines with GTP and is required for movement of the ribosome along the mRNA during translation.

elongation factor Ts (EF-Ts) Protein that regenerates elongation factor Tu in the elongation stage of translation.

elongation factor Tu (EF-Tu) Protein taking part in the elongation stage of translation; forms a complex with GTP and a charged tRNA and then delivers the charged tRNA to the ribosome.

endosymbiotic theory Theory stating that some membrane-bounded organelles, such as mitochondria and chloroplasts, originated as free-living eubacterial cells that entered into an endosymbiotic relation with a eukaryotic host cell and evolved into the present-day organelles; supported by a number of similarities in structure and sequence between organelle and eubacterial DNA.

engineered nuclease Protein consisting of part of a restriction enzyme, which cleaves DNA, combined with another protein that recognizes and binds to a specific DNA sequence; capable of making double-stranded cuts to the DNA at a predetermined DNA sequence. Engineered nucleases can be custom designed to bind to and cut any particular DNA sequence.

enhancer Sequence that stimulates maximal transcription of distant genes; affects only genes on the same DNA molecule (is cis acting), contains short consensus sequences, is not fixed in relation to the transcription start site, can stimulate almost any promoter in its vicinity, and may be upstream or downstream of the gene. The function of an enhancer is independent of sequence orientation.

enhancer RNA (eRNA) Type of noncoding RNA transcribed from enhancer sequences that may play a role in regulating the expression of protein-encoding genes.

environmental variance Component of phenotypic variance that is due to environmental differences among individual members of a population.

epialleles Alleles that do not differ in their base sequence but have epigenetic differences that produce heritable variations in phenotypes.

epigenetic change A stable alteration of chromatin structure that may be passed on to descendant cells or individuals. *See also* **epigenetics.**

epigenetic mark Heritable change in DNA or chromatin structure that does not involve alteration of the base sequence and that brings about changes in gene expression.

epigenetics Phenomena due to alterations in DNA that do not include changes in the base sequence; often affect the way in which DNA sequences are expressed. Such alterations are often stable and heritable in the sense that they are passed to descendant cells or individuals.

epigenome All epigenetic modifications within the genome of an individual organism.

episome Plasmid capable of integrating into a bacterial chromosome.

epistasis Type of gene interaction in which a gene at one locus masks or suppresses the effects of a gene at a different locus.

epistatic gene Gene that masks or suppresses the effect of a gene at a different locus.

equilibrium Situation in which no further change takes place; in population genetics, refers to a population in which allelic frequencies do not change.

equilibrium density gradient centrifugation Method used to separate molecules or organelles of different density by centrifugation.

eubacteria *See* **bacteria.**

euchromatin Chromatin that undergoes condensation and decondensation in the course of the cell cycle.

eukaryotes One of the three primary divisions of life, consisting of organisms whose cells have a complex structure including a nuclear envelope and membrane-bounded organelles. Eukaryotes include unicellular and multicellular forms.

evolution Genetic change that takes place in a group of organisms.

exit (E) site One of the three sites in a ribosome occupied by a tRNA during translation. In the elongation stage, the tRNA moves from the peptidyl (P) site to the E site, from which it then exits the ribosome.

exon Coding region of a gene that is interrupted by introns; after transcription and posttranscriptional processing, the exons remain in mRNA.

exon shuffling Process by which exons of different genes are exchanged and mixed into new combinations, creating new genes that are mosaics of other preexisting genes; has been important in the evolution of eukaryotic genes.

expanding nucleotide repeat Type of mutation in which the number of copies of a set of nucleotides (most often three nucleotides) increases in succeeding generations.

expression vector Cloning vector containing DNA sequences such as a promoter, a ribosome-binding site, and transcription initiation and termination sites that allow DNA fragments inserted into the vector to be transcribed and translated.

expressivity Degree to which a trait is expressed.

familial Down syndrome Human condition caused by a Robertsonian translocation in which the long arm of chromosome 21 is translocated to another chromosome; tends to run in families.

fertilization Fusion of gametes (sex cells) to form a zygote.

fetal cell sorting Separation of fetal cells from maternal blood. Genetic testing on the fetal cells can provide information about genetic diseases and disorders in the fetus.

F (fertility) factor Episome of *E. coli* that controls conjugation and gene exchange between cells. The F factor contains an origin of replication and genes that enable the bacterium to undergo conjugation.

F$_1$ (first filial) generation Offspring of the initial parents (P) in a genetic cross.

F$_2$ (second filial) generation Offspring of the F$_1$ generation in a genetic cross; the third generation of a genetic cross.

first polar body One of the products of meiosis I in oogenesis; contains half the chromosomes but little of the cytoplasm.

fitness Reproductive success of a genotype relative to that of other genotypes in a population.

5-methylcytosine Modified form of the base cytosine, containing a methyl group (CH_3) on the 5′ carbon.

5′ cap Modified 5′ end of eukaryotic mRNA, consisting of an extra (methylated) nucleotide and methyl groups at the 2′ position of the ribose sugar in one or more subsequent nucleotides; plays a role in the binding of the ribosome to mRNA and affects mRNA stability and the removal of introns.

5′ end End of a polynucleotide chain at which a phosphate group is attached to the 5′-carbon atom of the sugar in the nucleotide.

5′ splice site The 5′ end of an intron where cleavage takes place in RNA splicing.

5′ untranslated region (5′ UTR) Sequence of nucleotides at the 5′ end of mRNA; does not encode the amino acids of a protein.

fixation Point at which one allele reaches a frequency of 1 in a population. At this point, all members of the population are homozygous for that allele.

flanking direct repeat Short, directly repeated sequence produced on either side of a transposable element when the element inserts into DNA.

forward genetics Traditional approach to the study of gene function that begins with a mutant phenotype and proceeds to a gene that encodes the phenotype.

forward mutation Mutation that alters a wild-type phenotype.

founder effect Sampling error that arises when a population is established by a small number of individuals; leads to genetic drift.

fragile site Constriction or gap that appears at a particular location on a chromosome when cells are cultured under special conditions.

fragile-X syndrome A form of X-linked intellectual disability that appears primarily in males; associated with a fragile site that results from an expanding trinucleotide repeat.

frameshift mutation Mutation that alters the reading frame of a gene.

frequency distribution Graphical way of representing values. In genetics, the values found in a group of individuals are usually displayed as a frequency distribution. Typically, the phenotypic values are plotted on the horizontal (x) axis and the numbers (or proportions) of individuals with each value are plotted on the vertical (y) axis.

functional genomics Area of genomics that studies the functions of the genetic information contained within genomes.

G$_0$ (gap 0) Nondividing stage of the cell cycle.

G$_1$ (gap 1) Stage of interphase in the cell cycle in which the cell grows and develops.

G$_2$ (gap 2) Stage of interphase in the cell cycle that follows DNA replication. In G$_2$, the cell prepares for division.

gain-of-function mutation Mutation that produces a new trait or causes a trait to appear in inappropriate tissues or at inappropriate times in development.

gametic isolation Reproductive isolation due to the incompatibility of gametes. Mating between members of different species may take place, but the gametes do not form zygotes. Seen in many plants in which pollen from one species cannot fertilize the ovules of another species.

gametophyte Haploid phase of the life cycle in plants.

gap genes In fruit flies, a set of segmentation genes that define large sections of the embryo. Mutations in gap genes usually eliminate whole groups of adjacent segments.

gel electrophoresis Technique for separating charged molecules (such as proteins or nucleic acids) on the basis of molecular size or charge, or both.

gene Inherited factor that helps determine a trait; often defined at the molecular level as a DNA sequence that is transcribed into an RNA molecule.

gene cloning Insertion of DNA fragments into bacteria in such a way that the fragments will be stable and will be copied by the bacteria.

gene conversion Process of nonreciprocal genetic exchange that can produce abnormal ratios of gametes following meiosis.

gene desert A region of a genome that is gene poor—that is, a long stretch of DNA, possibly consisting of hundreds of thousands to millions of base pairs, that is completely devoid of any known genes or other functional sequences.

gene family *See* **multigene family**.

gene flow *See* **migration**.

gene interaction Interaction between genes at different loci that affect the same characteristic.

gene interaction variance Component of genetic variance that can be attributed to gene interaction (interaction between genes at different loci).

gene mutation Mutation that affects a single gene or locus.

gene pool Total of all genes in a population.

generalized transduction Transduction in which any gene can be transferred from one bacterial cell to another by a virus.

general transcription factor Protein that binds to a eukaryotic promoter near the transcription start site and is a part of the basal transcription apparatus that initiates transcription.

gene regulation Mechanisms and processes that control the phenotypic expression of genes.

gene therapy Use of recombinant DNA to treat a disease or disorder by altering the genetic makeup of the patient's cells.

genetic bottleneck Sampling error that arises when a population undergoes a drastic reduction in size; leads to genetic drift.

genetic conflict hypothesis Suggestion that genomic imprinting evolved because different and conflicting pressures act on maternal and paternal alleles of genes that affect fetal growth. For example, paternally derived alleles often favor maximum fetal growth, whereas maternally derived alleles favor less than maximum fetal growth because of the high cost of fetal growth to the mother.

genetic correlation Phenotypic correlation due to an effect of the same gene or genes on two or more characteristics.

genetic counseling Educational process that attempts to help patients and family members deal with all aspects of a genetic condition.

genetic drift Change in allelic frequencies due to sampling error.

genetic engineering *See* **recombinant DNA technology**.

genetic–environmental interaction variance Component of phenotypic variance that results from an interaction between genotype and environment that causes genotypes to be expressed differently in different environments.

Genetic Information Nondiscrimination Act (GINA) U.S. law prohibiting health insurers from using genetic information to make decisions about health-insurance coverage and rates; prevents employers from using genetic information in employment decisions; also prevents health insurers and employers from asking or requiring a person to take a genetic test.

genetic (linkage) map Map of the relative distances between genetic loci, markers, or other chromosome regions determined by rates of recombination; measured in recombination frequencies or map units.

genetic marker Any gene or DNA sequence used to identify a location on a genetic or physical map.

genetic maternal effect Determination of the phenotype of an offspring not by its own genotype, but by the nuclear genotype of its mother.

genetic mosaicism Condition in which regions of tissue within a single individual have different chromosome constitutions.

genetic rescue Introduction of new genetic variation into an inbred population in an effort to improve the health of the population and increase its chances of long-term survival.

genetic variance Component of phenotypic variance that is due to genetic differences among individual members of a population.

gene tree Phylogenetic tree representing the evolutionary relationships among a set of genes.

genic sex determination Sex determination in which the sexual phenotype is specified by genes at one or more loci, but there are no obvious differences in the chromosomes of males and females.

genome Complete set of genetic instructions for an organism.

genome-wide association study A study that looks for nonrandom associations between the presence of a trait and alleles at many different loci scattered across a genome—that is, for associations between traits and particular suites of alleles in a population.

genomic imprinting Differential expression of a gene that depends on the sex of the parent that transmitted the gene.

genomic library Collection of bacterial colonies or phages containing DNA fragments that constitute the entire genome of an organism.

genomics Study of the content, organization, and function of genetic information in whole genomes.

genotype The set of alleles possessed by an individual organism.

genotypic frequency Proportion of a particular genotype within a population.

germ-line mutation Mutation in a germ-line cell (one that gives rise to gametes).

germ-plasm theory Theory stating that cells in the reproductive organs carry a complete set of genetic information.

G_2/M checkpoint Checkpoint in the cell cycle near the end of G_2; after this checkpoint has been passed, the cell undergoes mitosis.

G-rich 3′ overhang A guanine-rich sequence of nucleotides that protrudes beyond the complementary C-rich strand at the end of a chromosome.

group I introns A class of introns in some ribosomal RNA genes that are capable of self-splicing.

group II introns A class of introns in some protein-encoding genes that are capable of self-splicing and are found in mitochondria, chloroplasts, and a few eubacteria.

G_1/S checkpoint Checkpoint in the cell cycle near the end of G_1; after this checkpoint has been passed, DNA replicates, and the cell is committed to dividing.

guanine (G) Purine base in DNA and RNA.

guide RNA (gRNA) RNA molecule that serves as a template for an alteration made in mRNA during RNA editing.

gynandromorph Individual organism that is a genetic mosaic for the sex chromosomes, possessing tissues with different sex-chromosome constitutions.

gyrase *See* **DNA gyrase**.

hairpin Secondary structure formed when sequences of nucleotides on the same polynucleotide strand are complementary and pair with each other.

haploid Possessing a single set of chromosomes (one genome).

haploinsufficiency Appearance of a mutant phenotype in an individual cell or organism that is heterozygous for a normally recessive trait.

haploinsufficient gene Gene that must be present in two copies for normal function. If one copy of the gene is missing, a mutant phenotype is produced.

haplotype A specific set of linked genetic variants or alleles on a single chromosome or on part of a chromosome.

Hardy–Weinberg equilibrium Frequencies of genotypes when the conditions of the Hardy–Weinberg law are met.

Hardy–Weinberg law Principle of population genetics stating that in a large, randomly mating population not affected by mutation, migration, or natural selection, allelic frequencies will not change and genotypic frequencies will stabilize after one generation in the proportions p^2 (the frequency of *AA*), $2pq$ (the frequency of *Aa*), and q^2 (the frequency of *aa*), where p equals the frequency of allele *A* and q equals the frequency of allele *a*.

H-DNA DNA structure consisting of three nucleotide strands (triplex DNA); can occur when a single nucleotide strand from one part of a DNA molecule pairs with double-stranded DNA from another part of the molecule.

heat-shock protein Protein produced by many cells in response to extreme heat and other stresses that helps the cells prevent damage from such stressing agents.

helicase *See* **DNA helicase**.

hemizygosity Possession of a single allele at a locus. Males of organisms with XX-XY sex determination are hemizygous for X-linked loci because their cells possess a single X chromosome.

heritability Proportion of phenotypic variation that is due to genetic differences. *See also* **broad-sense heritability**, **narrow-sense heritability**.

hermaphroditism Condition in which an individual organism possesses both male and female reproductive structures. True hermaphrodites produce both male and female gametes.

heterochromatin Chromatin that remains in a highly condensed state throughout the cell cycle; found at the centromeres and telomeres of most chromosomes.

heteroduplex DNA DNA consisting of two strands, each of which is from a different chromosome.

heterogametic sex The sex (male or female) that produces two types of gametes with respect to sex chromosomes. For example, in the XX-XY sex-determining system, the male produces both X-bearing and Y-bearing gametes.

heterokaryon Cell possessing two nuclei derived from different cells through cell fusion.

heteroplasmy Presence of two or more distinct variants of DNA within the cytoplasm of a single cell.

heterozygote advantage *See* **overdominance**.

heterozygote screening Testing of members of a population to identify heterozygous carriers of a disease-causing allele who are healthy but have the potential to produce children who have the disease.

heterozygous Having two different alleles at a locus.

highly repetitive DNA DNA that consists of short sequences that are present in hundreds of thousands to millions of copies and are clustered in certain regions of chromosomes.

histone Low-molecular-weight protein found in eukaryotes that complexes with DNA to form chromosomes.

histone code Modifications of histone proteins, such as the addition or removal of phosphate groups, methyl groups, or acetyl groups, that encode information affecting how genes are expressed.

Holliday intermediate Structure that forms in homologous recombination; consists of two duplex molecules connected by a cross bridge.

Holliday junction Special structure resulting from homologous recombination that is initiated by single-strand breaks in a DNA molecule.

holoenzyme Complex of an enzyme and other protein factors necessary for its complete function.

homeobox Conserved subset of nucleotides found in homeotic genes.

homeotic complex (HOM-C) Major cluster of homeotic genes in fruit flies; consists of the *Antennapedia* complex, which affects the development of the adult fly's head and anterior segments, and the *bithorax* complex, which affects the adult fly's posterior thoracic and abdominal segments.

homeotic genes Genes that determine the identity of individual segments or parts in an early embryo. Mutations in homeotic genes cause body parts to appear in the wrong places.

homogametic sex The sex (male or female) that produces gametes that are all alike with regard to sex chromosomes. For example, in the XX-XY sex-determining system, the female produces only X-bearing gametes.

homologous genes Evolutionarily related genes descended from a gene in a common ancestor.

homologous pair A pair of chromosomes that are alike in structure and size and that carry genetic information for the same set of hereditary characteristics. One chromosome of a homologous pair is inherited from the male parent and the other is inherited from the female parent.

homologous recombination Exchange of genetic information between homologous DNA molecules.

homoplasmy Presence of only one version of DNA within the cytoplasm of a single cell.

homozygous Having two identical alleles at a locus.

horizontal gene transfer Transfer of genes from one organism to another by a mechanism other than reproduction.

Hox **gene** Gene that contains a homeobox.

human papillomavirus (HPV) Virus associated with cervical cancer.

Human Proteome Project Project with the goal of identifying and characterizing all proteins in the human body.

humoral immunity Type of immunity resulting from antibodies produced by B cells.

hybrid breakdown Reproductive isolating mechanism in which closely related species are capable of mating and producing viable and fertile F_1 progeny, but genes do not flow between the two species because further crossing of the hybrids produces inviable or sterile offspring.

hybrid dysgenesis Sudden appearance of numerous mutations, chromosome aberrations, and sterility in the offspring of a cross between a male fruit fly that possesses *P* elements and a female fly that lacks them.

hybrid inviability Reproductive isolating mechanism in which mating between two organisms of different species takes place and hybrid offspring are produced, but are not viable.

hybridization Pairing of two partly or fully complementary single-stranded polynucleotide chains.

hybrid sterility Reproductive isolating mechanism in which hybrid embryos complete development, but are sterile; exemplified by mating between donkeys and horses to produce a mule, a viable but usually sterile offspring.

hypostatic gene Gene that is masked or suppressed by the action of a gene at a different locus.

identical twins *See* **monozygotic twins**.

inbreeding Mating between related individuals that takes place more frequently than expected by chance.

inbreeding coefficient Measure of inbreeding; the probability (ranging from 0 to 1) that two alleles are identical by descent.

inbreeding depression Decreased fitness arising from inbreeding; often due to the increased expression of lethal or deleterious recessive traits.

incomplete dominance Type of dominance in which the phenotype of the heterozygote is intermediate between the phenotypes of the two homozygotes.

incomplete linkage Linkage between genes that exhibit some crossing over; intermediate in its effects between independent assortment and complete linkage.

incomplete penetrance A case in which some individuals possess the genotype for a trait but do not express the expected phenotype.

incorporated error Incorporation of a damaged nucleotide or mismatched base pair into a DNA molecule.

independent assortment *See* **principle of independent assortment**.

induced mutation Mutation that results from environmental agents, such as chemicals or radiation.

induced pluripotent stem cells (iPSCs) Adult cells that have been artificially induced to dedifferentiate and revert to pluripotent stem cells capable of becoming many types of cells.

inducer Substance that stimulates transcription in an inducible system of gene regulation; usually a small molecule that binds to a repressor protein and alters that repressor so that it can no longer bind to DNA and inhibit transcription.

inducible operon Operon in which transcription is normally turned off, so that something must take place for transcription to be induced, or turned on.

in-frame deletion Deletion of some multiple of three nucleotides, which does not alter the reading frame of the gene.

in-frame insertion Insertion of some multiple of three nucleotides, which does not alter the reading frame of the gene.

inheritance of acquired characteristics Early notion of inheritance proposing that acquired traits are passed to descendants.

initiation (start) codon The codon in mRNA that specifies the first amino acid (fMet in bacterial cells; Met in eukaryotic cells) of a protein; most commonly AUG.

initiation factor 1 (IF-1) Protein required for the initiation of translation in bacterial cells; enhances the dissociation of the large and small subunits of the ribosome.

initiation factor 2 (IF-2) Protein required for the initiation of translation in bacterial cells; forms a complex with GTP and the charged initiator tRNA and then delivers the charged tRNA to the initiation complex.

initiation factor 3 (IF-3) Protein required for the initiation of translation in bacterial cells; binds to the small subunit of the ribosome and prevents the large subunit from binding during initiation.

initiator protein Protein that binds to an origin of replication and unwinds a short stretch of DNA, allowing helicase and other single-strand-binding proteins to bind and initiate replication.

insertion Mutation in which nucleotides are added to a DNA sequence.

insertion sequence (IS) Simple type of transposable element found in bacteria and their plasmids that contains only the information necessary for its own movement.

in situ hybridization Method used to determine the chromosomal location of a gene or other specific DNA fragment or the tissue distribution of an mRNA by using a labeled probe that is complementary to the sequence of interest.

insulator DNA sequence that blocks, or insulates, the effect of an enhancer; must be located between the enhancer and the promoter to have blocking activity; may also limit the spread of changes in chromatin structure.

integrase Enzyme that inserts prophage, or proviral, DNA into a chromosome.

interactome Complete set of protein interactions in a cell.

intercalating agent Molecule that is about the same size as a nucleotide and may become sandwiched between adjacent bases in DNA, distorting the three-dimensional structure of the DNA helix and causing single-nucleotide insertions and deletions in replication.

interference Degree to which one crossover interferes with additional crossovers.

intergenic suppressor mutation Suppressor mutation that occurs in a gene (locus) that is different from the gene containing the mutation it suppresses.

interkinesis Period between meiosis I and meiosis II.

internal promoter Promoter located within the sequences of DNA that are transcribed into RNA.

interphase Major phase of the cell cycle between cell divisions. In interphase, the cell grows, develops, and prepares for cell division.

interspersed repeats Repeated sequences that are found at multiple locations throughout the genome.

intragenic suppressor mutation Suppressor mutation that occurs in the same gene (locus) as the mutation that it suppresses.

intron Noncoding sequence between coding regions in a eukaryotic gene; removed from the RNA after transcription.

inverted repeats Sequences on the same strand that are inverted and complementary.

isoaccepting tRNAs Different tRNAs with different anticodons that specify the same amino acid.

isotopes Different forms of an element that have the same number of protons and electrons but differ in the number of neutrons in the nucleus.

junctional diversity Addition or deletion of nucleotides at the junctions of gene segments brought together in the somatic recombination of genes that encode antibodies and T-cell receptors.

karyotype The complete set of chromosomes possessed by an organism; usually presented as a picture of a complete set of its metaphase chromosomes.

kinetochore Set of proteins that assemble on the centromere, providing the point of attachment for spindle microtubules.

Klinefelter syndrome Human condition in which cells contain one or more Y chromosomes along with multiple X chromosomes (most commonly XXY but may also be XXXY, XXXXY, or XXYY). People with Klinefelter syndrome are male in appearance but frequently possess small testes, some breast enlargement, and reduced facial and pubic hair; often taller than normal and sterile, most have normal intelligence.

knock-in mouse Mouse that carries a foreign DNA sequence inserted at a specific chromosome location.

knockout mouse Mouse in which a normal gene has been disabled ("knocked out").

lagging strand DNA strand that is replicated discontinuously.

large ribosomal subunit The larger of the two subunits of a functional ribosome.

lariat Loop-like structure created in the splicing of nuclear pre-mRNA when the 5′ end of an intron is attached to the branch point.

leading strand DNA strand that is replicated continuously.

leptotene First substage of prophase I in meiosis. In leptotene, chromosomes contract and become visible.

lethal allele Allele that causes the death of an individual organism, often early in development, so that the organism does not appear in the progeny of a genetic cross. A recessive lethal allele kills individuals that are homozygous for the allele; a dominant lethal allele kills both heterozygotes and homozygotes.

lethal mutation Mutation that causes premature death.

LINE *See* **long interspersed element**.

linkage analysis Gene mapping based on the detection of physical linkage between genes, as measured by the rate of recombination in the progeny of a cross.

linkage disequilibrium Nonrandom association between genetic variants within a haplotype.

linkage group A group of linked genes.

linkage map *See* **genetic map**.

linked genes Genes located close together on the same chromosome.

linker Small synthetic DNA fragment that contains one or more restriction sites; can be attached to the ends of any piece of DNA and used to insert it into a plasmid vector.

linker DNA Stretch of DNA separating two nucleosomes.

locus (pl., loci) Position on a chromosome where a specific gene is located.

lod (logarithm of odds) score Logarithm of the ratio of the probability of obtaining a set of observations, assuming a specified degree of linkage, to the probability of obtaining the same set of observations with independent assortment; used to assess the likelihood of linkage between genes from pedigree data.

long interspersed element (LINE) Long DNA sequence repeated many times and interspersed throughout the genome.

long noncoding RNA (lncRNA) A class of relatively long RNA molecules found in eukaryotes that do not code for proteins but provide a variety of other functions, including regulation of gene expression.

loss-of-function mutation Mutation that causes the complete or partial absence of normal function.

loss of heterozygosity Inactivation or loss of the wild-type allele in a heterozygote.

Lyon hypothesis Proposal by Mary Lyon in 1961 that one X chromosome in each female cell becomes inactivated (a Barr body) and that which of the X chromosomes is inactivated is random and varies from cell to cell.

lysogenic cycle Life cycle of a bacteriophage in which phage genes first integrate into the bacterial chromosome and are not immediately transcribed and translated.

lytic cycle Life cycle of a bacteriophage in which phage genes are transcribed and translated, new phage particles are produced, and the host cell is lysed.

major histocompatibility complex (MHC) antigens A large and diverse group of antigens found on the surfaces of cells that mark those cells as self; encoded by a large cluster of genes known as the major histocompatibility complex. T cells simultaneously bind to foreign and MHC antigens.

malignant tumor Tumor consisting of cells that are capable of invading other tissues.

map-based sequencing Method of sequencing a genome in which sequenced fragments are assembled into the correct sequence in contigs with the use of genetic or physical maps.

mapping function Mathematical function that relates recombination frequencies to actual physical distances between genes.

map unit (m.u.) Unit of measure for distances on a genetic map; also called a **centiMorgan**. 1 map unit equals a recombination frequency of 1%.

mass spectrometry Method for precisely determining the mass of a molecule by using the migration rate of an ionized molecule in an electrical field.

maternal blood screening test Method of screening for genetic conditions in a fetus by examining levels of certain substances in the blood of the mother. For example, the level of α-fetoprotein in maternal blood provides information about the probability that a fetus has a neural-tube defect.

mean Statistic that describes the center of a frequency distribution of measurements; calculated by dividing the sum of all measurements by the number of measurements; also called the average.

mechanical isolation Reproductive isolation resulting from anatomical differences that prevent successful copulation.

mediator Complex of proteins that is one of the components of the basal transcription apparatus.

megaspore One of the four products of meiosis in plants.

megasporocyte Diploid reproductive cell in the ovary of a plant that undergoes meiosis to produce haploid macrospores.

meiosis Process by which the chromosomes of a eukaryotic cell divide to give rise to haploid reproductive cells. Consists of two divisions: meiosis I and meiosis II.

meiosis I First phase of meiosis. In meiosis I, chromosome number is reduced by half.

meiosis II Second phase of meiosis. Events in meiosis II are similar to those in mitosis.

melting *See* **denaturation**.

melting temperature Midpoint of the melting range of DNA.

memory cell Long-lived lymphocyte that is among the clone of cells generated when a foreign antigen is encountered. If the same antigen is encountered again, the memory cells quickly divide and give rise to another clone of cells specific for that particular antigen.

Mendelian population A group of interbreeding, sexually reproducing individuals.

meristic characteristic Characteristic whose phenotype varies in whole numbers, such as number of vertebrae, but may be underlain by continuous genetic variation.

merozygote *See* **partial diploid**.

messenger RNA (mRNA) RNA molecule that carries genetic information for the amino acid sequence of a protein.

metacentric chromosome Chromosome in which the two chromosome arms are approximately the same length.

metagenomics An emerging field of genetics in which the genome sequences of a group of organisms inhabiting a common environment are sampled and determined.

metaphase Stage of mitosis in which chromosomes align in the center of the cell.

metaphase I Stage of meiosis I in which homologous pairs of chromosomes align in the center of the cell.

metaphase II Stage of meiosis II in which individual chromosomes align in the center of the cell.

metaphase plate Plane in a cell between two spindle poles. In metaphase, chromosomes align on the metaphase plate.

metastasis The movement of cells that separate from malignant tumors to other sites, where they establish secondary tumors.

microarray Ordered array of DNA fragments, fixed to a solid support, which serve as probes to detect the presence of complementary sequences; often used to assess the expression of genes in various tissues and under different conditions.

microbiome Complete set of all bacteria found in a particular environment; for example, the human microbiome consists of all bacteria found in and on the human body.

microRNA (miRNA) Small RNA molecules, typically 21 or 22 bp in length, produced by cleavage of double-stranded RNA arising from small hairpins within RNA that is mostly single stranded. The miRNAs combine with proteins to form a complex that binds (imperfectly) to mRNA molecules and inhibits their translation.

microsatellite Very short DNA sequence repeated in tandem; also called a **short tandem repeat**.

microspore Haploid product of meiosis in plants.

microsporocyte Diploid reproductive cell in the stamen of a plant; undergoes meiosis to produce four haploid microspores.

microtubule Long fiber composed of the protein tubulin; plays an important role in the movement of chromosomes in mitosis and meiosis.

migration Movement of genes from one population to another; also called **gene flow**.

minimal medium Medium used to culture bacteria or other microorganisms that contains only the nutrients required by prototrophic (wild-type) cells—typically, a carbon source, essential elements such as nitrogen and phosphorus, certain vitamins, and other required ions and nutrients.

mismatch repair Process that corrects mismatched nucleotides in DNA after replication has been completed. Enzymes excise incorrectly paired nucleotides from the newly synthesized strand and use the original nucleotide strand as a template for replacing them.

missense mutation Mutation that alters a codon in mRNA, resulting in a different amino acid in the protein encoded.

mitochondrial DNA (mtDNA) DNA in mitochondria; has some characteristics in common with eubacterial DNA and typically consists of a circular molecule that lacks histone proteins and encodes some of the rRNAs, tRNAs, and proteins found in mitochondria.

mitochondrial replacement therapy Methods that combine the nuclear DNA of a woman carrying a mtDNA mutation with DNA of a sperm and the egg cytoplasm of a healthy donor, allowing her to give birth to a child free from mitochondrial disease.

mitosis Process by which the nucleus of a eukaryotic cell divides.

mitosis-promoting factor (MPF) Protein functioning in the control of the cell cycle; consists of a cyclin combined with cyclin-dependent kinase (CDK). Active MPF stimulates mitosis.

mitotic spindle Array of microtubules that radiate from two poles; moves chromosomes in mitosis and meiosis.

model genetic organism An organism that is widely used in genetic studies because it has characteristics, such as short generation time and large numbers of progeny, that make it particularly useful for genetic analysis.

moderately repetitive DNA DNA consisting of sequences 150 to 300 bp in length that are repeated thousands of times.

modified base Any of several rare bases found in some RNA molecules. Such bases are modified forms of the standard bases (adenine, guanine, cytosine, and uracil).

molecular chaperone Molecule that assists in the proper folding of another molecule.

molecular clock Use of molecular differences to estimate the time of evolutionary divergence between organisms; assumes a roughly constant rate at which one neutral mutation replaces another.

molecular genetics Study of the chemical nature of genetic information and how it is encoded, replicated, and expressed.

molecular motor Specialized protein that moves cellular components.

monoecious Refers to an individual organism that has both male and female reproductive structures.

monohybrid cross Cross between two individuals that differ in a single characteristic—more specifically, a cross between individuals that are homozygous for different alleles at the same locus ($AA \times aa$); also refers to a cross between two individuals that are heterozygous for two alleles at a single locus ($Aa \times Aa$).

monosomy Absence of one of the chromosomes of a homologous pair.

monozygotic twins Twins that arise when a single egg fertilized by a single sperm splits into two separate embryos; also called identical twins.

morphogen Molecule whose concentration gradient affects the developmental fate of surrounding cells.

M (mitotic) phase The major phase of the cell cycle that encompasses active cell division; includes mitosis (nuclear division) and cytokinesis (cytoplasmic division).

mRNA surveillance Mechanisms for the detection and elimination of mRNAs that contain errors that may create problems in the course of translation.

multifactorial characteristic Characteristic determined by multiple genes and environmental factors.

multigene family Set of genes similar in sequence that arose through repeated duplication events and often encode different protein products.

multiple alleles Presence of more than two alleles at a locus in a group of diploid individuals; however, each individual member of the group has only two of the possible alleles.

multiple 3′ cleavage sites The presence of more than one 3′ cleavage site on a single pre-mRNA, which allows cleavage and polyadenylation to take place at different sites, producing mRNAs of different lengths.

multiplication rule Rule stating that the probability of two or more independent events occurring together is calculated by multiplying the probabilities of each of the individual events.

mutagen Any environmental agent that significantly increases the rate of mutation above the spontaneous rate.

mutagenesis screen Method for identifying genes that influence a specific phenotype. Random mutations are induced in a population of organisms, and individual organisms with mutant phenotypes are identified. These individual organisms are crossed to determine the genetic basis of the phenotype and to map the location of mutations that cause the phenotype.

mutation Heritable change in genetic information.

mutation rate Frequency with which a gene changes from the wild-type allele to a mutant allele; generally expressed as the number of mutations per biological unit (that is, mutations per cell division, per gamete, or per round of replication).

narrow-sense heritability Proportion of phenotypic variance that can be attributed to additive genetic variance.

natural selection Differential reproduction of individuals with different genotypes.

negative assortative mating Mating between unlike individuals that is more frequent than would be expected by chance.

negative control Gene regulation in which the binding of a regulatory protein to DNA inhibits transcription (the regulatory protein is a repressor).

negative supercoiling Tertiary structure that forms when strain is placed on a DNA helix by underwinding.

neutral mutation Mutation that changes the amino acid sequence of a protein but does not alter the function of the protein.

neutral-mutation hypothesis Proposal that much of the molecular variation seen in natural populations is adaptively neutral and unaffected by natural selection; that is, that individuals with different molecular variants have equal fitnesses.

newborn screening Testing of newborn infants for certain genetic disorders.

next-generation sequencing technologies Sequencing methods, such as pyrosequencing, that are capable of simultaneously determining the sequences of many DNA fragments; these technologies are much faster and less expensive than the Sanger dideoxy sequencing method.

nitrogenous base Nitrogen-containing base that is one of the three parts of a nucleotide.

node Point in a phylogenetic tree where branches split; represents a common ancestor.

no-go decay (NGD) An mRNA surveillance system in eukaryotes that helps remove stalled ribosomes resulting from secondary structures in mRNA, chemical damage to mRNA, premature stop codons, and ribosomal defects.

nonautonomous element Transposable element that cannot transpose on its own but can transpose in the presence of an autonomous element of the same family.

noncomposite transposon Type of transposable element in bacteria that lacks insertion sequences, possesses a gene for transposase, and has terminal inverted repeats.

nondisjunction Failure of homologous chromosomes or sister chromatids to separate in meiosis or mitosis.

nonhistone chromosomal protein One of a heterogeneous assortment of nonhistone proteins in chromatin.

noninvasive prenatal genetic diagnosis Genetic test performed on a fetus without taking a tissue sample from the fetus; usually performed by testing fetal DNA found within the maternal blood.

nonoverlapping genetic code Refers to the fact that, generally, each nucleotide is a part of only one codon and encodes only one amino acid in a protein.

nonreciprocal translocation Movement of a chromosome segment to a nonhomologous chromosome or chromosomal region without any (or with unequal) reciprocal exchange of segments.

nonrecombinant (parental) gamete Gamete that contains only the original combinations of alleles present in the parents.

nonrecombinant (parental) progeny Progeny that possess only the original combinations of traits possessed by the parents.

nonreplicative transposition Type of transposition in which a transposable element excises from an old site and moves to a new site, resulting in no net increase in the number of copies of the transposable element.

nonsense codon *See* **stop codon.**

nonsense-mediated mRNA decay (NMD) Process that brings about the rapid elimination of mRNA that has a premature stop codon.

nonsense mutation Mutation that changes a sense codon (one that specifies an amino acid) into a stop codon.

nonstop mRNA decay Mechanism in eukaryotic cells for dealing with ribosomes stalled at the 3′ end of an mRNA that lacks a termination codon, in which a protein binds to the A site of the stalled ribosome and recruits other proteins that degrade the mRNA from the 3′ end.

nontemplate strand The DNA strand that is complementary to the template strand; not ordinarily used as a template during transcription.

normal distribution Common type of frequency distribution that exhibits a symmetrical, bell-shaped curve; usually arises when a large number of independent factors contribute to the measured value.

Northern blotting Process by which RNA is transferred from a gel to a solid support such as a nitrocellulose or nylon filter.

nuclear envelope Membrane that surrounds the genetic material in eukaryotic cells to form a nucleus; segregates the DNA from other cellular contents.

nuclear matrix Network of protein fibers in the nucleus; holds the nuclear contents in place.

nuclear pre-mRNA introns A class of introns in nuclear protein-encoding genes that are removed by spliceosome-mediated splicing.

nucleoid Bacterial DNA confined to a definite region of the cytoplasm.

nucleoside Ribose or deoxyribose bonded to a nitrogenous base.

nucleosome Basic repeating unit of chromatin, consisting of a core of eight histone proteins (two each of H2A, H2B, H3, and H4) and about 146 bp of DNA that wraps around the core about two times.

nucleotide Repeating unit of DNA or RNA made up of a sugar, a phosphate group, and a nitrogenous base.

nucleotide-excision repair DNA repair that removes bulky DNA lesions and other types of DNA damage.

nucleus Compartment in eukaryotic cells that is enclosed by the nuclear envelope and contains the chromosomes.

nullisomy Absence of both chromosomes of a homologous pair $(2n - 2)$.

Okazaki fragment Short stretch of newly synthesized DNA produced by discontinuous replication on the lagging strand; these fragments are eventually joined together.

oligonucleotide-directed mutagenesis Method of site-directed mutagenesis that uses an oligonucleotide to introduce a mutant sequence into a DNA molecule.

oncogene Dominant-acting gene that stimulates cell division, leading to the formation of tumors and contributing to cancer; arises from a mutated copy of a normal cellular gene (proto-oncogene).

one gene, one enzyme hypothesis Proposal by Beadle and Tatum that each gene encodes a separate enzyme.

one gene, one polypeptide hypothesis Modification of the one gene, one enzyme hypothesis; proposes that each gene encodes a separate polypeptide chain.

oogenesis Egg production in animals.

oogonium Diploid cell in the ovary; capable of undergoing meiosis to produce an egg cell.

open reading frame (ORF) Continuous sequence of DNA nucleotides that contains a start codon and a stop codon in the same reading frame; is assumed to be a gene that encodes a protein but, in many cases, the protein has not yet been identified.

operator DNA sequence in an operon of a bacterial cell to which a regulator protein binds; this binding affects the rate of transcription of the structural genes.

operon Set of structural genes in a bacterial cell, along with their common promoter and other sequences (such as an operator) that control their transcription.

origin of replication Site where DNA synthesis is initiated.

origin-recognition complex (ORC) Multiprotein complex that binds to an origin of replication and unwinds the DNA around it to initiate DNA replication.

orthologs Homologous genes found in different species that evolved from the same gene in a common ancestor.

outcrossing Mating between unrelated individuals that is more frequent than would be expected by chance.

overdominance Selection in which the heterozygote has higher fitness than either homozygote; also called **heterozygote advantage**.

ovum Final product of oogenesis.

pachytene Third substage of prophase I in meiosis. The synaptonemal complex forms during pachytene.

pair-rule genes Set of segmentation genes in fruit flies that define regional sections of the embryo and affect alternate segments. Mutations in these genes often cause the deletion of every other segment.

palindrome Sequence of nucleotides that reads the same on complementary strands; inverted repeats.

pangenesis Early concept of heredity proposing that particles carry genetic information from different parts of the body to the reproductive organs.

paracentric inversion Chromosome inversion that does not include the centromere in the inverted region.

paralogs Homologous genes in the same species that arose through the duplication of a single ancestral gene.

paramutation Epigenetic change in which one allele of a genotype alters the expression of another allele; the altered expression persists for several generations, even after the altering allele is no longer present.

parental gamete *See* **nonrecombinant (parental) gamete**.

parental progeny *See* **nonrecombinant (parental) progeny**.

partial diploid Bacterial cell that possesses two copies of some genes, one on the bacterial chromosome and the other on an extra piece of DNA (usually a plasmid); also called a **merozygote**.

passenger Mutation found in a cancer cell that does not contribute to the development of cancer.

Patau syndrome (trisomy 13) Human condition characterized by severe intellectual disability, a small head, sloping forehead, small eyes, cleft lip and palate, extra fingers and toes, and other disabilities; results from the presence of three copies of chromosome 13.

pedigree Pictorial representation of a family history outlining the inheritance of one or more traits or diseases.

penetrance Percentage of individuals with a particular genotype that express the phenotype expected of that genotype.

peptide bond Chemical bond that connects amino acids in a protein.

peptidyl (P) site One of the three sites in a ribosome occupied by a tRNA during translation. In the elongation stage, tRNAs move from the aminoacyl (A) site into the P site.

pericentric inversion Chromosome inversion that includes the centromere in the inverted region.

P (parental) generation First set of parents in a genetic cross.

phage *See* **bacteriophage**.

phenocopy Phenotype produced by environmental effects that is the same as the phenotype produced by a genotype.

phenotype Appearance or manifestation of a characteristic.

phenotypic correlation Correlation between two or more phenotypes in the same individual.

phenotypic variance Measure of the degree of phenotypic difference among a group of individuals; composed of genetic, environmental, and genetic–environmental interaction variances.

phosphate group A phosphorus atom attached to four oxygen atoms; one of the three components of a nucleotide.

phosphodiester linkage A strong covalent bond that joins the $5'$-phosphate group of one nucleotide to the $3'$-hydroxyl group of the next nucleotide in a polynucleotide strand.

phylogenetic tree Graphical representation of a phylogeny in the form of a branching diagram.

phylogeny Evolutionary relationships among a group of organisms or genes.

physical map Map of physical distances between loci, genetic markers, or other chromosome segments; measured in base pairs.

pilus (pl., pili) Extension of the surface of some bacteria that allows conjugation to take place. When a pilus on one cell makes contact with a receptor on another cell, the pilus contracts and pulls the two cells together.

Piwi-interacting RNA (piRNA) Small RNA molecule belonging to a class named after Piwi proteins, with which these molecules interact; similar to microRNA and small interfering RNA and thought to have a role in suppressing the expression of transposable elements in reproductive cells.

plaque Clear patch of lysed cells on a continuous layer of bacteria on the agar surface of a petri plate; each plaque represents a single original phage that multiplied and lysed many cells.

plasmid Small, circular DNA molecule found in bacterial cells that is capable of replicating independently from the bacterial chromosome.

pleiotropy Ability of a single gene to influence multiple phenotypes.

pluripotency The property of being undifferentiated, with the capacity to form every type of cell in an organism.

poly(A)-binding protein (PABP) Protein that binds to the poly(A) tail of eukaryotic mRNA and makes the mRNA more stable. There are several types of PABP.

poly(A) tail String of adenine nucleotides added to the 3′ end of a eukaryotic mRNA after transcription.

polycistronic mRNA Single RNA molecule transcribed from a group of several genes; uncommon in eukaryotes.

polygenic characteristic Characteristic encoded by genes at many loci.

polymerase chain reaction (PCR) Method of enzymatically amplifying DNA fragments.

polynucleotide strand Series of nucleotides linked together by phosphodiester bonds.

polypeptide Chain of amino acids linked by peptide bonds; also called a protein.

polyploidy Possession of more than two sets of chromosomes.

polyribosome Messenger RNA molecule with several ribosomes attached to it.

polytene chromosome Giant chromosome in the salivary glands of *Drosophila melanogaster*. Each polytene chromosome consists of a number of DNA molecules lying side by side.

population (1) In statistics, the group of interest; often represented by a subset called a sample. (2) A group of members of the same species.

population genetics Study of the genetic composition of populations and how their gene pools change over time.

positional cloning Method that allows for the isolation and identification of a gene by examining the cosegregation of a phenotype with previously mapped genetic markers.

position effect Dependence of the expression of a gene on the gene's location in the genome.

positive assortative mating Mating between like individuals that is more frequent than would be expected by chance.

positive control Gene regulation in which the binding of a regulatory protein to DNA stimulates transcription (the regulatory protein is an activator).

positive supercoiling Tertiary structure that forms when strain is placed on a DNA helix by overwinding.

posttranslational modification Alteration of a protein after translation; may include cleavage from a larger precursor protein, the removal of amino acids, and the attachment of other molecules to the protein.

postzygotic reproductive isolating mechanism Reproductive isolating mechanism that operates after gametes from two different species have fused to form a zygote, either because the resulting hybrids are inviable or sterile or because reproduction breaks down in subsequent generations.

preformationism Early concept of inheritance proposing that a miniature adult (homunculus) resides in either the egg or the sperm and increases in size in development, and that all traits are inherited from the parent that contributes the homunculus.

preimplantation genetic diagnosis (PGD) Genetic testing on an embryo produced by in vitro fertilization before implantation of the embryo in the uterus.

pre-messenger RNA (pre-mRNA) Eukaryotic RNA molecule that is modified after transcription to become mRNA.

presymptomatic genetic testing Testing to determine whether a person has inherited a disease-causing gene before the symptoms of the disease have appeared.

prezygotic reproductive isolating mechanism Reproductive isolating mechanism that prevents gametes from two different species from fusing and forming a hybrid zygote.

primary Down syndrome Human condition caused by the presence of three copies of chromosome 21.

primary immune response Initial clone of cells specific for a particular antigen and generated when the antigen is first encountered by the immune system.

primary oocyte Oogonium that has entered prophase I.

primary spermatocyte Spermatogonium that has entered prophase I.

primary structure of a protein The amino acid sequence of a protein.

primase Enzyme that synthesizes a short stretch of RNA on a DNA template; functions in replication to provide a 3′-OH group for the attachment of a DNA nucleotide.

primer Short stretch of RNA on a DNA template; provides a 3′-OH group for the attachment of a DNA nucleotide at the initiation of replication.

principle of independent assortment (Mendel's second law) Principle of heredity discovered by Mendel that states that genes encoding different characteristics (genes at different loci) separate independently; applies only to genes located on different chromosomes or to genes far apart on the same chromosome.

principle of segregation (Mendel's first law) Principle of heredity discovered by Mendel that states that each diploid individual possesses two alleles at a locus and that these two alleles separate when gametes are formed, one allele going into each gamete.

probability Likelihood of the occurrence of a particular event; more formally, the number of times that a particular event occurs divided by the number of all possible outcomes. Probability values range from 0 to 1.

proband A person having a trait or disease for whom a pedigree is constructed.

probe Known sequence of DNA or RNA that is complementary to a sequence of interest and will pair with it; used to find specific DNA sequences.

prokaryote Unicellular organism with a simple cell structure. Prokaryotes include bacteria (eubacteria) and archaea.

prometaphase Stage of mitosis in which the nuclear membrane breaks down and the spindle microtubules attach to the chromosomes.

promoter DNA sequence to which the transcription apparatus binds so as to initiate transcription; indicates the direction of transcription, which of the two DNA strands is to be read as the template, and the starting point of transcription.

proofreading Process by which DNA polymerases remove and replace incorrectly paired nucleotides in the course of replication.

prophage Phage genome that is integrated into a bacterial chromosome.

prophase Stage of mitosis in which the chromosomes contract and become visible, the cytoskeleton breaks down, and the mitotic spindle begins to form.

prophase I Stage of meiosis I in which chromosomes condense and pair, crossing over takes place, the nuclear membrane breaks down, and the mitotic spindle forms.

prophase II Stage of meiosis after interkinesis in which chromosomes condense, the nuclear membrane breaks down, and the spindle forms. Some cells skip this stage.

protein-coding region The part of mRNA consisting of the nucleotides that specify the amino acid sequence of a protein.

protein domain Region of a protein that has a specific shape or function.

protein kinase Enzyme that adds phosphate groups to other proteins.

protein microarray Large number of different proteins applied to a solid support as a series of spots, each containing a different protein; used to analyze protein–protein interactions.

proteome Set of all proteins found in a cell.

proteomics Study of the proteome.

proto-oncogene Normal cellular gene that controls cell division. When mutated, it may become an oncogene and contribute to cancer progression.

prototrophic Capable of using a carbon source, essential elements such as nitrogen and phosphorus, certain vitamins, and other required ions and nutrients to synthesize all the compounds needed for growth and reproduction; a prototrophic bacterium can grow on minimal medium.

provirus DNA copy of viral DNA or RNA that is integrated into the host chromosome and replicated along with the host chromosome.

pseudoautosomal region Small region of the X and Y chromosomes that contains homologous gene sequences.

pseudodominance Expression of a normally recessive allele owing to a deletion on the homologous chromosome.

Punnett square Shorthand method of determining the outcome of a genetic cross. On a grid, the gametes of one parent are written along the upper edge and the gametes of the other parent are written along the left-hand edge. Within the cells of the grid, the alleles in the gametes are combined to form the genotypes of the offspring.

purine Type of nitrogenous base in DNA and RNA. Adenine and guanine are purines.

pyrimidine Type of nitrogenous base in DNA and RNA. Cytosine, thymine, and uracil are pyrimidines.

pyrimidine dimer Structure in which a bond forms between two adjacent pyrimidine molecules on the same strand of DNA; disrupts normal hydrogen bonding between complementary bases and distorts the normal configuration of the DNA molecule.

quantitative characteristic Continuous characteristic; displays a large number of possible phenotypes or is encoded by multiple genetic factors.

quantitative genetics Genetic analysis of complex characteristics or characteristics influenced by multiple genetic factors.

quantitative trait locus (QTL) A gene or chromosomal region that contributes to the expression of quantitative characteristics.

quaternary structure of a protein Interaction of two or more polypeptides to form a functional protein.

reading frame Particular way in which a nucleotide sequence is read in groups of three nucleotides (codons) in translation; begins with a start codon and ends with a stop codon.

realized heritability Narrow-sense heritability measured from a response-to-selection experiment.

real-time PCR Modification of the polymerase chain reaction that quantitatively determines the amount of starting nucleic acid; the amount of DNA amplified is measured as the reaction proceeds.

reannealing *See* **renaturation**.

recessive Refers to an allele or phenotype that is expressed only in homozygotes (*aa*); the recessive allele is not expressed in a heterozygote (*Aa*) phenotype.

reciprocal crosses Pair of crosses in which the phenotypes of the male and female parents are reversed. For example, in one cross, a tall male is crossed with a short female, and in the other cross, a short male is crossed with a tall female.

reciprocal translocation Reciprocal exchange of segments between two nonhomologous chromosomes.

recombinant DNA technology Set of molecular techniques for locating, isolating, altering, combining, and studying DNA segments; also commonly called **genetic engineering**.

recombinant gamete Gamete that possesses new combinations of alleles.

recombinant progeny Progeny formed from recombinant gametes that possess new combinations of traits.

recombination Process that produces new combinations of alleles on a chromatid.

recombination frequency Proportion of recombinant progeny produced in a cross.

recursive splicing A variation of splicing in which some long introns are removed in multiple steps.

regression Analysis of how one variable changes in response to another variable.

regression coefficient Statistic that measures how much one variable changes, on average, with a unit change in another variable.

regulator gene Gene associated with an operon in bacterial cells that encodes a protein or RNA molecule that functions in controlling the transcription of one or more structural genes.

regulator protein Protein produced by a regulator gene that binds to another DNA sequence and controls the transcription of one or more structural genes.

regulatory element DNA sequence that affects the transcription of other DNA sequences to which it is physically linked.

regulatory gene DNA sequence that encodes a protein or RNA molecule that interacts with DNA sequences and affects their transcription or translation or both.

regulatory promoter DNA sequence located immediately upstream of the eukaryotic core promoter; contains consensus sequences to which transcriptional regulator proteins bind.

relaxed state of DNA Energy state of a DNA molecule when there is no structural strain on the molecule.

release factor Protein required for the termination of translation; binds to a ribosome when a stop codon is reached and stimulates the release of the polypeptide chain, the tRNA, and the mRNA from the ribosome. Eukaryotic cells require two release factors (eRF-1 and eRF-2), whereas *E. coli* requires three (RF-1, RF-2, and RF-3).

renaturation Process by which two complementary single-stranded DNA molecules pair; also called **reannealing**.

repetitive DNA DNA sequences that exist in multiple copies in a genome.

replicated error An incorporated error that is replicated, leading to a permanent mutation.

replication Process by which DNA is synthesized from a single-stranded nucleotide template.

replication bubble Segment of a DNA molecule that is unwinding and undergoing replication.

replication fork Point at which a double-stranded DNA molecule separates into two single strands that serve as templates for replication.

replication licensing factor Protein that ensures that replication takes place only once at each origin of replication; required at the origin before replication can be initiated and removed after the DNA has been replicated.

replication origin *See* **origin of replication**.

replication terminus Point at which replication stops.

replicative segregation Random segregation of organelles into progeny cells in cell division. If two or more versions of an organelle are present in the parent cell, chance determines the proportion of each type that will segregate into each progeny cell.

replicative transposition Type of transposition in which a copy of a transposable element moves to a new site while the original copy remains at the old site; increases the number of copies of the transposable element.

replicon Unit of replication consisting of DNA from the origin of replication to the point at which replication on either side of the origin ends.

repressible operon Operon in which transcription is normally turned on, so that something must take place for transcription to be repressed, or turned off.

repressor Regulatory protein that binds to a DNA sequence and inhibits transcription.

reproductive isolating mechanism Any biological factor or mechanism that prevents gene exchange.

repulsion (trans) configuration Arrangement of two linked genes in which each of a homologous pair of chromosomes contains one wild-type (dominant) allele and one mutant (recessive) allele.

response element DNA sequence shared by the promoters or enhancers of several eukaryotic genes to which a regulatory protein can bind to stimulate the coordinate transcription of those genes.

response to selection The amount of change in a characteristic in one generation owing to selection; equals the selection differential times the narrow-sense heritability.

restriction endonuclease *See* **restriction enzyme**.

restriction enzyme Enzyme that recognizes particular base sequences in DNA and makes double-stranded cuts nearby; also called a **restriction endonuclease**.

restriction mapping Method of determining the locations of sites cut by restriction enzymes in a piece of DNA.

retrotransposon Type of transposable element in eukaryotic cells that possesses some characteristics of retroviruses and transposes through an RNA intermediate.

retrovirus Virus that injects its RNA genome into a host cell, where reverse transcription produces a complementary, double-stranded DNA molecule from the RNA template; the DNA copy then integrates into the host chromosome to form a provirus.

reverse duplication Duplication of a chromosome segment in which the sequence of the duplicated segment is inverted relative to the sequence of the original segment.

reverse genetics A molecular approach to the study of gene function that begins with a genotype (a DNA sequence) and proceeds to the phenotype by altering the sequence or by inhibiting its expression.

reverse mutation (reversion) Mutation that changes a mutant phenotype back into the wild type.

reverse transcriptase Enzyme capable of synthesizing complementary DNA from an RNA template.

reverse transcription Synthesis of DNA from an RNA template.

reverse-transcription PCR Technique that amplifies sequences corresponding to RNA; reverse transcriptase is used to convert RNA into complementary DNA, which can then be amplified by the usual polymerase chain reaction.

reversion *See* **reverse mutation**.

rho-dependent terminator Sequence in bacterial DNA that requires the presence of the rho factor to terminate transcription.

rho factor (ρ) A protein that binds to bacterial RNA polymerase and facilitates the termination of transcription of some genes.

rho-independent terminator Sequence in bacterial DNA that does not require the presence of the rho factor to terminate transcription.

ribonucleoside triphosphate (rNTP) Substrate of RNA synthesis; consists of ribose, a nitrogenous base, and three phosphate groups linked to the 5′-carbon atom of the ribose. In transcription, two of the phosphates are cleaved, producing an RNA nucleotide.

ribonucleotide Basic building block of RNA, consisting of ribose, a phosphate group, and a nitrogenous base.

ribose Five-carbon sugar in RNA; has a hydroxyl group attached to the 2′-carbon atom.

ribosomal RNA (rRNA) RNA molecule that is a structural component of the ribosome.

riboswitch Regulatory sequence in an RNA molecule. When an inducer molecule binds to the riboswitch, the binding changes the configuration of the RNA molecule and alters the expression of the RNA, usually by affecting the termination of transcription or by affecting translation.

ribozyme RNA molecule that can act as a biological catalyst.

RNA-coding region Sequence of DNA nucleotides that encodes an RNA molecule.

RNA editing Process in which the protein-coding sequence of an mRNA is altered after transcription, so that the amino acids specified by the altered mRNA are different from those encoded by the gene.

RNA-induced silencing complex (RISC) Complex of a small interfering RNA (siRNA) or microRNA (miRNA) with proteins that can cleave mRNA, leading to degradation of the mRNA or repressing its translation.

RNA interference (RNAi) Process in which cleavage of double-stranded RNA produces small RNAs (siRNAs or miRNAs) that bind to mRNAs containing complementary sequences and bring about their cleavage and degradation.

RNA polymerase Enzyme that synthesizes RNA from a DNA template during transcription.

RNA polymerase I Eukaryotic RNA polymerase that transcribes large ribosomal RNA molecules (18S rRNA and 28S rRNA).

RNA polymerase II Eukaryotic RNA polymerase that transcribes pre-mRNA, some small nuclear RNAs, and some microRNAs.

RNA polymerase III Eukaryotic RNA polymerase that transcribes tRNA, small ribosomal RNAs (5S rRNA), some small nuclear RNAs, and some microRNAs.

RNA polymerase IV RNA polymerase that transcribes small interfering RNAs in plants.

RNA polymerase V RNA polymerase that transcribes RNAs that play a role in heterochromatin formation in plants.

RNA replication Process in some viruses by which RNA is synthesized from an RNA template.

RNA sequencing A method in which cDNA molecules are copied from the RNA molecules present in a cell and sequenced.

RNA silencing *See* **RNA interference**.

RNA splicing Process by which introns are removed from RNA and exons are joined together.

Robertsonian translocation Translocation in which the long arms of two acrocentric chromosomes become joined to a common centromere, resulting in a chromosome with two long arms and usually another chromosome with two short arms.

rolling-circle replication Replication of circular DNA that is initiated by a break in one of the nucleotide strands, producing a double-stranded circular DNA molecule and a single-stranded linear DNA molecule, the latter of which may circularize and serve as a template for the synthesis of a complementary strand.

rooted tree Phylogenetic tree in which one node represents the common ancestor of all other organisms (nodes) on the tree. In a rooted tree, all the organisms depicted have a common ancestor.

R plasmid Plasmid possessing genes that confer antibiotic resistance on any cell that contains the plasmid.

sample Subset used to describe a population.

sampling error Deviations from expected ratios due to chance occurrences when the sample size is small.

secondary immune response Generation of a clone of cells when a memory cell encounters an antigen; provides long-lasting immunity.

secondary oocyte One of the products of meiosis I in oogenesis; receives most of the cytoplasm.

secondary spermatocyte Product of meiosis I in male animals.

secondary structure of a protein Regular folding arrangement of amino acids in a protein. Common secondary structures found in proteins include the alpha helix and the beta pleated sheet.

second polar body One of the products of meiosis II in oogenesis; contains a set of chromosomes but little of the cytoplasm.

segmental duplications Duplicated chromosome segments larger than 1000 bp.

segmentation genes Set of about 25 genes in fruit flies that control the differentiation of the embryo into individual segments, affecting the number and organization of the segments. Mutations in segmentation genes usually disrupt whole sets of segments.

segment-polarity genes Set of segmentation genes in fruit flies that affect the organization of segments. Mutations in segment-polarity genes cause part of each segment to be deleted and replaced by a mirror image of part or all of an adjacent segment.

segregation *See* **principle of segregation**.

selection coefficient Measure of the relative intensity of selection against a genotype; equals 1 minus fitness.

selection differential Difference in phenotype between selected individuals and the average of the entire population.

semiconservative replication Replication in which the two nucleotide strands of DNA separate and each serves as a template for the synthesis of a new strand. All DNA replication is semiconservative.

sense codon Codon that specifies an amino acid in a protein.

separase Molecule that cleaves cohesin molecules, which hold the sister chromatids together.

sequential hermaphroditism Phenomenon in which the sex of an individual organism changes in the course of its lifetime; the organism is male at one age or developmental stage and female at a different age or stage.

70S initiation complex Final complex formed in the initiation of translation in bacterial cells; consists of the small and large subunits of the ribosome, mRNA, and initiator tRNA charged with fMet.

sex Sexual phenotype: male or female.

sex chromosomes Chromosomes that differ in number or morphology in males and females.

sex determination Specification of sex (male or female). Sex-determining mechanisms include chromosomal, genic, and environmental sex-determining systems.

sex-determining region Y (*SRY*) gene Gene on the Y chromosome that triggers male development.

sex-influenced characteristic Characteristic encoded by autosomal genes that are more readily expressed in one sex. For example, an autosomal dominant gene may have higher penetrance in males than in females, or an autosomal gene may be dominant in males but recessive in females.

sex-limited characteristic Characteristic encoded by autosomal genes and expressed in only one sex. Both males and females carry genes for sex-limited characteristics, but the characteristics appear in only one of the sexes.

sex-linked characteristic Characteristic determined by a gene or genes on sex chromosomes.

shelterin Multiprotein complex that binds to mammalian telomeres and protects the ends of the DNA from being inadvertently repaired as a double-strand break in the DNA.

Shine–Dalgarno sequence Consensus sequence found in the bacterial 5′ untranslated region of mRNA; contains the ribosome-binding site.

short interspersed element (SINE) Short DNA sequence repeated many times and interspersed throughout the genome.

short tandem repeat (STR) *See* **microsatellite**.

sigma (σ) factor Subunit of bacterial RNA polymerase that allows the RNA polymerase to recognize a promoter and initiate transcription.

signal sequence Sequence of 15–30 amino acids found at the amino end of some eukaryotic proteins that directs the protein to a specific location in the cell; usually cleaved from the protein.

signal-transduction pathway System in which an external signal (initiated by a hormone or growth factor) triggers a cascade of intracellular reactions that ultimately produce a specific response.

silencer Sequence that has many of the properties possessed by an enhancer but represses transcription.

silent mutation Change in the nucleotide sequence of DNA that does not alter the amino acid sequence of a protein.

single-nucleotide polymorphism (SNP) A single-base-pair difference in DNA sequence between individual members of a species.

single-strand-binding (SSB) protein Protein that binds to single-stranded DNA during replication and prevents it from annealing with a complementary strand and forming secondary structures.

sister chromatids Two copies of a chromosome that are held together at the centromere. Each chromatid consists of a single DNA molecule.

site-directed mutagenesis Targeted mutagenesis technique used in bacteria in which a short sequence of nucleotides is cut out with restriction enzymes and replaced with a synthetic oligonucleotide.

small interfering RNA (siRNA) Single-stranded RNA molecule (usually 21 to 25 nucleotides in length) produced by the cleavage and processing of double-stranded RNA that binds to complementary sequences in mRNA and brings about the cleavage and degradation of the mRNA. Some siRNAs bind to complementary sequences in DNA and bring about their methylation.

small nuclear ribonucleoprotein (snRNP) Structure found in the nuclei of eukaryotic cells that consists of small nuclear RNA (snRNA) and protein; functions in the processing of pre-mRNA.

small nuclear RNA (snRNA) Small RNA molecule found in the nuclei of eukaryotic cells; functions in the processing of pre-mRNA.

small nucleolar RNA (snoRNA) Small RNA molecule found in the nuclei of eukaryotic cells; functions in the processing of rRNA and in the assembly of ribosomes.

small ribosomal subunit The smaller of the two subunits of a functional ribosome.

somatic-cell hybridization Fusion of somatic cells of different types.

somatic hypermutation High rate of somatic mutation such as that in genes encoding antibodies.

somatic mutation Mutation in a cell that does not give rise to gametes.

somatic recombination Recombination in somatic cells, such as maturing lymphocytes, among segments of genes that encode antibodies and T-cell receptors.

SOS system System of proteins and enzymes that allows a cell to replicate its DNA in the presence of a distortion in DNA structure; makes numerous mistakes in replication and increases the rate of mutation.

Southern blotting Process by which DNA is transferred from a gel to a solid support such as a nitrocellulose or nylon filter.

specialized transduction Transduction in which genes near special sites on the bacterial chromosome are transferred from one bacterium to another; requires lysogenic bacteriophages.

speciation Process by which new species arise. *See also* **allopatric speciation, sympatric speciation**.

species Term applied to different kinds or types of living organisms. *See also* **biological species concept**.

spermatid Immediate product of meiosis II in spermatogenesis; matures to sperm.

spermatogenesis Sperm production in animals.

spermatogonium Diploid cell in the testis; capable of undergoing meiosis to produce a sperm.

S (synthesis) phase Stage of interphase in the cell cycle. In S phase, DNA replicates.

spindle-assembly checkpoint Checkpoint in the cell cycle near the end of metaphase; after this checkpoint has been passed, the cell enters anaphase.

spindle microtubule Microtubule that moves chromosomes in mitosis and meiosis.

spindle pole Point from which spindle microtubules radiate.

spliceosome Large complex consisting of several RNAs and many proteins that splices protein-encoding pre-mRNA; contains five small nuclear ribonucleoprotein particles (U1, U2, U4, U5, and U6).

spontaneous mutation Mutation that arises from natural changes in DNA structure or from errors in replication.

sporophyte Diploid phase of the life cycle in plants.

SR proteins A group of serine- and arginine-rich proteins that regulate alternative splicing of pre-mRNA.

standard deviation Statistic that describes the variability of a group of measurements; the square root of the variance.

start codon *See* **initiation codon**.

stem cell Undifferentiated cell that is capable of forming every type of cell in an organism.

sticky end *See* **cohesive end**.

stop (termination or nonsense) codon Codon in mRNA that signals the end of translation. The three common stop codons are UAA, UAG, and UGA.

strand slippage Slipping of the template and newly synthesized strands in replication in which one of the strands loops out from the other and nucleotides are inserted or deleted on the newly synthesized strand.

structural gene DNA sequence that encodes a protein that functions in metabolism or biosynthesis or that has a structural role in the cell.

structural genomics Area of genomics that studies the organization and sequence of information contained within genomes; sometimes used by protein chemists to refer to the determination of the three-dimensional structure of proteins.

structural variants Collective term for chromosome rearrangements and copy-number variations.

submetacentric chromosome Chromosome in which the centromere is displaced toward one end, producing a short arm and a long arm.

supercoiling Tertiary structure that forms when strain is placed on a DNA helix by overwinding or underwinding. *See also* **positive supercoiling, negative supercoiling**.

suppressor mutation Mutation that hides or suppresses the effect of another mutation at a nucleotide site that is distinct from the site of the original mutation.

sympatric speciation Speciation arising in the absence of any geographic barrier to gene flow, in which reproductive isolating mechanisms evolve within a single interbreeding population.

synapsis Close pairing of homologous chromosomes.

synaptonemal complex Three-part structure that develops between synapsed homologous chromosomes.

synonymous codons Different codons that specify the same amino acid.

synthetic biology A field that seeks to design organisms that might provide functions useful to humanity.

tag single-nucleotide polymorphism (tag-SNP) Single-nucleotide polymorphism used to identify a haplotype.

tandem duplication Chromosome rearrangement in which a duplicated chromosome segment is adjacent to the original segment.

tandem repeats Type of moderately repetitive DNA in which sequences are repeated one after another; tend to be clustered at specific locations on a chromosome.

***Taq* polymerase** DNA polymerase commonly used in PCR reactions. Isolated from the bacterium *Thermus aquaticus*, the enzyme is stable at high temperatures, so it is not denatured during the strand-separation step of the cycle.

targeted mutagenesis Induction of mutations in particular DNA sequences to study their effects.

TATA-binding protein (TBP) Polypeptide chain found in several different transcription factors that recognizes and binds to sequences in eukaryotic promoters.

TATA box Consensus sequence (TATAAAA) commonly found in eukaryotic RNA polymerase II promoters; usually located 25 to 30 bp upstream of the transcription start site. The TATA box determines the start point for transcription.

T cell Type of lymphocyte that produces cellular immunity; originates in the bone marrow and matures in the thymus.

T-cell receptor Receptor found on the surface of a T cell that simultaneously binds a foreign antigen and a self-antigen.

telocentric chromosome Chromosome in which the centromere is at or very near one end.

telomerase Ribonucleoprotein enzyme that replicates the ends (telomeres) of eukaryotic chromosomes. The RNA part of the enzyme has a template that is complementary to repeated sequences in the telomere and pairs with them, providing a template for the synthesis of additional copies of the repeats.

telomere Stable end of a eukaryotic chromosome.

telomeric sequence Sequence found at the ends of a chromosome; consists of many copies of short, simple sequences repeated one after the other.

telophase Stage of mitosis in which the chromosomes arrive at the spindle poles, the nuclear membrane re-forms, and the chromosomes relax and lengthen.

telophase I Stage of meiosis I in which chromosomes arrive at the spindle poles.

telophase II Stage of meiosis II in which chromosomes arrive at the spindle poles.

temperate phage Bacteriophage that can undergo the lysogenic cycle, in which the phage DNA integrates into the bacterial chromosome and remains in an inactive state.

temperature-sensitive allele Allele that is expressed only at certain temperatures.

template strand The strand of DNA that is used as a template during transcription. The RNA synthesized during transcription is complementary and antiparallel to the template strand.

temporal isolation Reproductive isolation in which the reproduction of different groups takes place at different times of the year, so that there is no gene flow between groups; exemplified by species of plants that flower at different times of the year and thus do not exchange genes.

terminal inverted repeats Sequences found at both ends of a transposable element that are inverted complements of one another.

termination codon *See* **stop codon.**

terminator Sequence of DNA nucleotides that causes the termination of transcription.

tertiary structure of a protein Higher-order folding of amino acids in a protein to form the overall three-dimensional shape of the molecule.

testcross Cross between an individual with an unknown genotype and an individual with the homozygous recessive genotype.

tetrad *See* **bivalent.**

tetrasomy Presence of two extra copies of a chromosome ($2n + 2$).

TFIIB recognition element (BRE) Consensus sequence [(G or C)(G or C)(G or C)CGCC] found in some RNA polymerase II

core promoters; usually located from 32 to 38 bp upstream of the transcription start site.

theory of clonal selection Theory that explains the generation of primary and secondary immune responses; proposes that the binding of a B cell to an antigen stimulates the cell to divide, giving rise to a clone of genetically identical cells, all of which are specific for the antigen.

theta replication Replication of circular DNA that is initiated by the unwinding of the two nucleotide strands, producing a replication bubble. Unwinding continues at one or both ends of the bubble, making it progressively larger. DNA replication on both of the template strands is simultaneous with unwinding until the two replication forks meet.

30S initiation complex Initial complex formed in the initiation of translation in bacterial cells; consists of the small subunit of the ribosome, mRNA, initiator tRNA charged with fMet, GTP, and initiation factors 1, 2, and 3.

three-point testcross Cross between an individual heterozygous at three loci and an individual homozygous for recessive alleles at those loci.

3′ end End of a polynucleotide chain at which an OH group is attached to the 3′-carbon atom of the sugar in the nucleotide.

3′ splice site The 3′ end of an intron where cleavage takes place in RNA splicing.

3′ untranslated region (3′ UTR) Sequence of nucleotides at the 3′ end of mRNA; does not encode the amino acids of a protein, but affects both the stability of the mRNA and its translation.

threshold characteristic Characteristic that has only two phenotypes (presence and absence) but whose expression depends on an underlying susceptibility that varies continuously.

thymine (T) Pyrimidine base in DNA, but not in RNA.

Ti plasmid Large plasmid isolated from the bacterium *Agrobacterium tumefaciens* and used to transfer genes to plant cells.

topoisomerase Enzyme that adds or removes rotations in a DNA helix by temporarily breaking nucleotide strands; controls the degree of DNA supercoiling.

totipotency The potential of a cell to develop into any other cell type.

trans configuration *See* **repulsion configuration.**

transcription Process by which RNA is synthesized from a DNA template.

transcription activator–like effector nuclease (TALEN) An engineered nuclease in which a protein of a type that normally binds to promoters is attached to a restriction enzyme.

transcriptional activator protein Protein in eukaryotic cells that binds to consensus sequences in regulatory promoters or enhancers and initiates transcription by stimulating the assembly of the basal transcription apparatus.

transcription bubble Region of a DNA molecule that has unwound to expose a single-stranded template, which is being transcribed into RNA.

transcription factor Protein that binds to DNA sequences in eukaryotic cells and affects transcription.

transcription start site The first DNA nucleotide that is transcribed into an RNA molecule.

transcription unit Sequence of nucleotides in DNA that encodes a single RNA molecule and the sequences necessary for its transcription; normally contains a promoter, an RNA-coding sequence, and a terminator.

transcriptome Set of all RNA molecules transcribed from a genome.

transcriptomics Study of the expression of the genome.

transducing phage Phage that contains a piece of the chromosome of its bacterial host inside the phage coat.

transductant Bacterial cell that has received genes from another bacterium through transduction.

transduction Type of gene exchange that takes place when a virus carries genes from one bacterium to another. After it is inside the cell, the newly introduced DNA may undergo recombination with the bacterial chromosome.

transesterification Chemical reaction in some RNA-splicing reactions.

transfer–messenger RNA (tmRNA) An RNA molecule that has properties of both mRNA and tRNA; functions in rescuing ribosomes that are stalled at the end of an mRNA molecule.

transfer RNA (tRNA) RNA molecule that carries an amino acid to the ribosome and transfers it to a growing polypeptide chain in translation.

transfer RNA introns A class of introns in tRNA genes whose splicing relies on enzymes.

transformant Cell that has received genetic material through transformation.

transformation Mechanism by which DNA found in the environment is taken up by a cell. After transformation, recombination may take place between the introduced genes and the cellular chromosome.

transforming principle Substance responsible for transformation. DNA is the transforming principle.

transgene Foreign gene or other DNA fragment carried in germ-line DNA.

transgenic mouse Mouse whose genome contains a foreign gene or genes added by employing recombinant DNA methods.

transition Base substitution in which a purine is replaced by a different purine or a pyrimidine is replaced by a different pyrimidine.

translation Process by which a protein is assembled from information contained in mRNA.

translesion DNA polymerase Specialized DNA polymerase that is able to replicate DNA through distorted structures and bulky lesions that halt other DNA polymerases; often makes more errors during DNA synthesis than other DNA polymerases.

translocation (1) Movement of a chromosome segment to a nonhomologous chromosome or to a region within the same chromosome. (2) Movement of a ribosome along mRNA in the course of translation.

translocation carrier Individual organism heterozygous for a chromosome translocation.

transmission genetics Field of genetics that encompasses the basic principles of genetics and how traits are inherited.

transposable element DNA sequence capable of moving from one site to another within the genome through a mechanism that differs from that of homologous recombination.

transposase Enzyme encoded by many types of transposable elements that is required for their transposition. The enzyme makes single-strand breaks at each end of the transposable element and on either side of the target sequence where the element inserts.

transposition Movement of a transposable element from one site to another. *See also* **replicative transposition, nonreplicative transposition**.

trans-splicing Process of splicing together exons from two or more pre-mRNAs.

transversion Base substitution in which a purine is replaced by a pyrimidine or a pyrimidine is replaced by a purine.

trihybrid cross A cross between two individuals that differ in three characteristics (*AA BB CC* × *aa bb cc*); also refers to a cross between two individuals that are both heterozygous at three loci (*Aa Bb Cc* × *Aa Bb Cc*).

triplet code Refers to the fact that three nucleotides encode each amino acid in a protein.

triple-X syndrome Human condition in which cells contain three X chromosomes. A person with triple-X syndrome has a female phenotype without distinctive features other than a tendency to be tall and thin; a few such women are sterile, but many menstruate regularly and are fertile.

trisomy Presence of an extra copy of a chromosome ($2n + 1$).

trisomy 8 Presence of three copies of chromosome 8; in humans, results in intellectual disability, contracted fingers and toes, low-set malformed ears, and a prominent forehead.

trisomy 13 *See* **Patau syndrome**.

trisomy 18 *See* **Edward syndrome**.

trisomy 21 *See* **Down syndrome**.

tRNA charging Chemical reaction in which an aminoacyl-tRNA synthetase attaches an amino acid to its corresponding tRNA.

tRNA-modifying enzyme Enzyme that creates a modified base in tRNA by catalyzing a chemical change in the standard base.

tubulin Protein found in microtubules.

tumor-suppressor gene Gene that normally inhibits cell division. Recessive mutations in such genes often contribute to cancer.

Turner syndrome Human condition in which cells contain a single X chromosome and no Y chromosome (XO). People with Turner syndrome are female in appearance but do not undergo puberty and have poorly developed female secondary sex characteristics; most are sterile but have normal intelligence.

two-dimensional polyacrylamide gel electrophoresis (2D-PAGE) Method for separating proteins in which the proteins are separated in one dimension by charge, separated in a second dimension by mass, and then stained. Each of the resulting spots is proportional to the amount of protein present.

two-point testcross Cross between an individual heterozygous at two loci and an individual homozygous for recessive alleles at those loci.

ultrasonography Procedure for visualizing a fetus in which high-frequency sound is beamed into the uterus; sound waves that encounter dense tissue bounce back and are transformed into a picture of the fetus.

unbalanced gamete Gamete that has a variable number of chromosomes; some chromosomes may be missing and others may be present in more than one copy.

underdominance Selection in which the heterozygote has lower fitness than either homozygote.

unequal crossing over Misalignment of the two DNA molecules during crossing over, resulting in one DNA molecule with an insertion and the other with a deletion.

uniparental disomy Inheritance of both chromosomes of a homologous pair from a single parent.

unique-sequence DNA DNA sequence that is present only once or a few times in a genome.

universal genetic code Refers to the fact that particular codons specify the same amino acids in almost all organisms.

upstream element Consensus sequence found in some bacterial promoters that contains a number of A–T pairs and is located about 40 to 60 bp upstream of the transcription start site.

uracil (U) Pyrimidine base in RNA, but not normally in DNA.

variance Statistic that describes the variability of a group of measurements.

virulent phage Bacteriophage that reproduces only through the lytic cycle and kills its host cell.

virus Noncellular replicating structure consisting of nucleic acid surrounded by a protein coat; can replicate only within a host cell.

Western blotting Process by which protein is transferred from a gel to a solid support such as a nitrocellulose or nylon filter.

whole-genome shotgun sequencing Method of sequencing a genome in which sequenced fragments are assembled into the correct sequence in contigs by using only the overlaps in sequence.

wild type The trait or allele that is most commonly found in natural (wild) populations.

wobble Base pairing between codon and anticodon in which there is nonstandard pairing, usually at the third (3′) position of the codon; allows more than one codon to pair with the same anticodon.

X-inactivation center Segment of the X chromosome at which inactivation of one X chromosome is initiated in female cells.

X-linked characteristic Characteristic determined by a gene or genes on the X chromosome.

X-ray diffraction Method for analyzing the three-dimensional shape and structure of a chemical substance in which crystals of the substance are bombarded with X-rays, which hit the crystals, bounce off, and produce a pattern of spots on a detector. The pattern of the spots produced on the detector provides information about the molecular structure.

yeast artificial chromosome (YAC) Cloning vector consisting of a DNA molecule with a yeast origin of replication, a pair of telomeres, and a centromere. YACs can carry very large pieces of DNA (as large as several hundred thousand base pairs) and replicate and segregate as yeast chromosomes do.

Y-linked characteristic Characteristic determined by a gene or genes on the Y chromosome.

Z-DNA Secondary structure of DNA characterized by 12 bases per turn, a left-handed helix, and a sugar–phosphate backbone that zigzags back and forth.

zinc-finger nuclease (ZFN) An engineered nuclease consisting of an array of zinc-finger domains attached to a restriction enzyme.

zygotene Second substage of prophase I in meiosis. In zygotene, chromosomes enter into synapsis.

Answers to Selected Problems

Chapter 1

1. In the Hopi culture, people with albinism were considered special and given special status. Because extensive exposure to sunlight could be damaging or deadly, Hopi males with albinism did no agricultural work. Males with albinism had more children, thus increasing the frequency of the albino mutation. Finally, the small size of the Hopi population may have helped increase the frequency of the albino mutation owing to chance.

17. Evolution is genetic change over time. For evolution to occur, genetic variation must first arise, and then evolutionary forces change the proportions of genetic variants over time. Genetic variation is therefore the basis of all evolutionary change.

18. (a) Transmission genetics; (b) population genetics; (c) population genetics; (d) molecular genetics; (e) molecular genetics; (f) transmission genetics.

22. Genetics is old in the sense that humans have been aware of hereditary principles for thousands of years and have applied them since the beginning of agriculture and the domestication of plants and animals. It is very young in the sense that the fundamental principles were not uncovered until Mendel's time, and the discovery of the structure of DNA and recombinant DNA techniques have occurred only within the last 60 years.

23. (a) Germ-plasm theory; (b) preformationism; (c) inheritance of acquired characteristics; (d) pangenesis.

25. (a) Both cell types have genomes encoded in nucleic acids, and the two cell types have similar machinery for the replication, transcription, and translation of those genomes. Eukaryotic cells have a nucleus containing chromosomal DNA and possess internal membrane-bounded organelles; prokaryotic cells have neither of these features.
(b) A gene is a basic unit of hereditary information. Alleles are alternative forms of a gene.
(c) The genotype is the set of genes or alleles inherited by an organism from its parent(s). The expression of the genes of a particular genotype, through interaction with environmental factors, produces the phenotype, an observable trait.
(d) Both are nucleic acid polymers that encode genetic information. RNA contains a ribose sugar, whereas DNA contains a deoxyribose sugar. RNA contains uracil as one of the four bases, whereas DNA contains thymine. The other three bases are common to both DNA and RNA. Finally, DNA usually consists of two complementary nucleotide strands, whereas RNA is single stranded.
(e) Chromosomes are structures consisting of DNA and associated proteins. The DNA contains the genetic information.

26.

Type of albinism	Phenotype	Gene mutated
OCA2	Pigment reduced in skin, hair, and eyes, but small amount of pigment acquired with age; visual problems	OCA2
OCA1B	General absence of pigment in hair, skin, and eyes, but may be small amount of pigment; does not vary with age; visual problems	Tyrosinase
OCA1A	Complete absence of pigment; visual problems	Tyrosinase
OCA3	Some pigment present, but sun sensitivity and visual problems	Tyrosinase-related protein 1
OASD	Lack of pigment in the eyes and deafness later in life	Unknown
OA1	Lack of pigment in the eyes but normal elsewhere	GPR143
ROCA	Bright copper red coloration in skin and hair of Africans; dilution of color in iris	Tyrosinase-related protein 1
OCA4	Reduced pigmentation	MATP

28. All genomes must have the ability to store complex information and to vary. The genetic material of any organism must be stable, be replicated precisely, and be transmitted faithfully to the progeny, but must be capable of mutating.

Chapter 2

20. (a) The two chromatids of a chromosome
(b) The two chromosomes of a homologous pair
(c) Cohesin
(d) The enzyme separase
(e) The hands of the two blind men
(f) If one man failed to grasp his sock, use of the knife to cut the string holding them together would be difficult. The two socks of a pair would not be separated, and both would end up in one man's bag. Similarly, if each chromatid is not attached to a spindle microtubule, and if the two chromatids are not pulled in opposite directions, they will not separate, and both will migrate to the same cell. This cell will have two copies of one chromosome.

24.

Stage	Number of cells at each stage	Proportion of cells at each stage	Average duration (hours)
Interphase	160	0.80	19.2
Prophase	20	0.10	2.4
Prometaphase	6	0.03	0.72
Metaphase	2	0.01	0.24
Anaphase	7	0.035	0.84
Telophase	5	0.025	0.6
Totals	200	1.0	24

The average duration of M phase can be determined by adding up the hours spent in each stage of mitosis. In these cells, M phase lasts 4.8 hours. Metaphase requires 0.24 hours, or 14.4 minutes.

28. (a) 12 chromosomes and 24 DNA molecules; (b) 12 chromosomes and 24 DNA molecules; (c) 12 chromosomes and 24 DNA molecules; (d) 12 chromosomes and 24 DNA molecules; (e) 12 chromosomes and 12 DNA molecules; (f) 6 chromosomes and 12 DNA molecules; (g) 12 chromosomes and 12 DNA molecules; (h) 6 chromosomes and 6 DNA molecules.

30. (a) The diploid number of chromosomes is 6.
(b) The left-hand cell is in anaphase I of meiosis; the middle cell is in anaphase of mitosis; the right-hand cell is in anaphase II of meiosis.
(c) Anaphase I, 6 chromosomes and 12 DNA molecules; anaphase of mitosis, 12 chromosomes and 12 DNA molecules; anaphase II of meiosis, 6 chromosomes and 6 DNA molecules.

31. (a) 7.3 pg; **(b)** 14.6 pg; **(c)** 14.6 pg; **(d)** 3.7 pg; **(e)** 14.6 pg; **(f)** 7.3 pg.

32. (a) If cohesin fails to form early in mitosis, the sister chromatids could separate prior to anaphase, resulting in improper segregation of chromosomes to daughter cells.
(b) If shugoshin is absent during meiosis, the cohesin at the centromere could be broken, allowing for the separation of sister chromatids during anaphase I.
(c) If shugoshin is not broken down, the cohesin at the centromere will remain protected from degradation. The intact cohesin will prevent the sister chromatids from separating during anaphase II of meiosis.
(d) If separase is defective, homologous chromosomes and sister chromatids will not separate in meiosis and mitosis, resulting in some cells that have too few chromosomes and some cells that have too many chromosomes.

33. Prophase of mitosis: 24 chromosomes; prophase I of meiosis: 24 chromosomes.

35. The house fly. The number of different combinations of chromosomes that are possible in the gametes is $2n$, where n is equal to the number of homologous pairs of chromosomes. For the fruit fly, $2^4 = 16$. For the house fly, $2^6 = 64$.

36. (a) Metaphase I

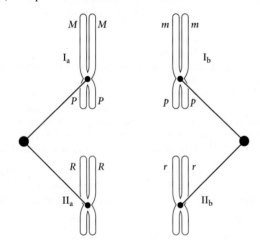

(b) Gametes

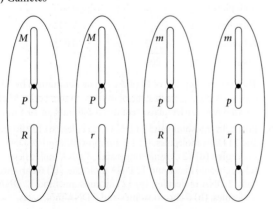

38.

Cell type	Number of chromosomes	Number of DNA molecules
(a) Spermatogonium	64	64
(b) First polar body	32	64
(c) Primary oocyte	64	128
(d) Secondary spermatocyte	32	64

40. (a) No. The first polar body and the secondary oocyte are the result of meiosis I, which produces two nonidentical cells. The first polar body and the secondary oocyte will contain only one member of each original chromosome pair. Additionally, crossing over that took place in prophase I will have generated new and different arrangements of genetic material on those chromatids that participated in crossing over.
(b) No. The second polar body and the ovum will contain copies of the same members of the homologous pairs of chromosomes that separated in meiosis. However, because of crossing over, the sister chromatids that separated in anaphase II and gave rise to the ovum and second polar body are no longer identical.

43. Because meiosis takes place only in diploid cells, haploid male bees do not undergo meiosis. Male bees produce sperm only through mitosis. Haploid cells that divide mitotically produce haploid cells.

Chapter 3

14. (a) The parents are *RR* (orange fruit) and *rr* (cream fruit). All the F_1 are *Rr* (orange). The F_2 are 1 *RR* : 2 *Rr* : 1 *rr* and have a phenotypic ratio of 3 orange to 1 cream.
(b) Half of the progeny are homozygous and have orange fruit (*RR*), and half are heterozygous and have orange fruit (*Rr*).
(c) Half of the progeny are heterozygous and have orange fruit (*Rr*), and half are homozygous and have cream fruit (*rr*).

17. (a) Although the white female gave birth to the offspring, her eggs were produced by the ovary from the black female guinea pig. The transplanted ovary produced only eggs containing the allele for black coat color. Like most mammals, guinea pig females produce primary oocytes early in development, and thus the transplanted ovary already contained primary oocytes produced by the black female guinea pig.
(b) *Ww*.
(c) The production of black offspring suggests that the allele for black coat color was passed to the offspring from the transplanted ovary, in agreement with the germ-plasm theory. If pangenesis were correct, the offspring should have been white. The white-coat alleles would have traveled to the transplanted ovary and then into the white female's gametes. The absence of any white offspring indicates that pangenesis did not occur.

18. (a) Female parent is $i^B i^B$; male parent is $I^A i^B$.
(b) Both parents are $i^B i^B$.
(c) Male parent is $i^B i^B$; female parent is most likely $I^A I^A$ or possibly $I^A i$, but a heterozygous female in this mating is unlikely to have produced eight blood-type-A kittens owing to chance alone.
(d) Both parents are $I^A i^B$.
(e) Either both parents are $I^A I^A$ or one parent is $I^A I^A$ and the other parent is $I^A i^B$. The blood type of the offspring does not allow a determination of the precise genotype of either parent.
(f) Female parent is $i^B i^B$; male parent is $I^A i^B$.

21. Yes. The ram and the lamb must be homozygous for the normal allele (*l*) because both have the normal fleece phenotype. Because the lamb receives only a single allele (*l*) from the ram, the ewe must have contributed the other recessive *l* allele. Therefore, the ewe must be heterozygous for lustrous fleece. In summary:

Lustrous fleece ewe × Normal fleece ram
(*Ll*)   (*ll*)
↓
Normal fleece lamb
(*ll*)

22. **(a)** Sally is *Aa*, Sally's mother is *Aa*, Sally's father is *aa*, and Sally's brother is *aa*; **(b)** $\frac{1}{2}$; **(c)** $\frac{1}{2}$.

25. **(a)** $\frac{1}{6}$; **(b)** $\frac{1}{3}$; **(c)** $\frac{1}{2}$; **(d)** $\frac{5}{6}$.

26. **(a)** $\frac{1}{18}$; **(b)** $\frac{1}{36}$; **(c)** $\frac{11}{36}$; **(d)** $\frac{1}{6}$; **(e)** $\frac{1}{4}$; **(f)** $\frac{3}{4}$.

27. **(a)** $\frac{1}{128}$; **(b)** $\frac{1}{64}$; **(c)** $\frac{7}{128}$; **(d)** $\frac{35}{128}$; **(e)** $\frac{35}{128}$.

29. Parents:

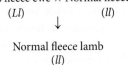

F$_1$ generation:

F$_2$ generation:

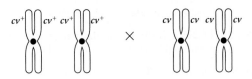

30. **(a)** G$_1$, one copy; G$_2$, two copies; metaphase of mitosis, two copies; metaphase I of meiosis, two copies; after cytokinesis of meiosis, one copy but only in half of the cells produced by meiosis. (The remaining half will not contain the *B* allele.)
(b) G$_1$, two copies; G$_2$, four copies; metaphase of mitosis, four copies; metaphase I of meiosis, four copies; metaphase II, two copies; after cytokinesis of meiosis, one copy.

33. **(a)** $\frac{9}{16}$ black and curled, $\frac{3}{16}$ black and normal, $\frac{3}{16}$ gray and curled, and $\frac{1}{16}$ gray and normal.
(b) $\frac{1}{4}$ black and curled, $\frac{1}{4}$ black and normal, $\frac{1}{4}$ gray and curled, $\frac{1}{4}$ gray and normal.

34. **(a)** $\frac{1}{2}$ (*Aa*) × $\frac{1}{2}$ (*Bb*) × $\frac{1}{2}$ (*Cc*) × $\frac{1}{2}$ (*Dd*) × $\frac{1}{2}$ (*Ee*) = $\frac{1}{32}$.
(b) $\frac{1}{2}$ (*Aa*) × $\frac{1}{2}$ (*bb*) × $\frac{1}{2}$ (*Cc*) × $\frac{1}{2}$ (*dd*) × $\frac{1}{4}$ (*ee*) = $\frac{1}{64}$.
(c) $\frac{1}{4}$ (*aa*) × $\frac{1}{2}$ (*bb*) × $\frac{1}{4}$ (*cc*) × $\frac{1}{2}$ (*dd*) × $\frac{1}{4}$ (*ee*) = $\frac{1}{256}$.
(d) No offspring have this genotype.

37. **(a)** Gametes from *Aa Bb* individual:

Gametes from *aa bb* individual:

(b) Progeny at G$_1$:

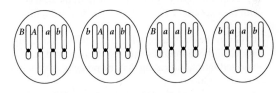

Progeny at G$_2$:

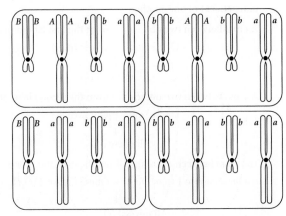

Progeny at prophase of mitosis:

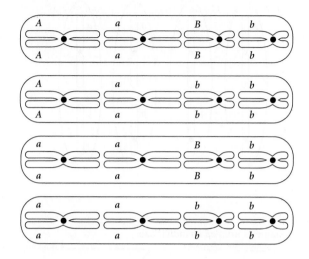

The order of chromosomes on the metaphase plate can vary.

38. **(a)** The results are consistent with the burnsi phenotype being recessive to the pipiens phenotype.
(b) Let B represent the burnsi allele and B^+ represent the pipiens allele.

$$\text{burnsi } (BB^+) \times \text{burnsi } (BB^+)$$
$$\text{burnsi } (BB^+) \times \text{pipiens } (B^+B^+)$$
$$\text{burnsi } (BB^+) \times \text{pipiens } (B^+B^+)$$

(c) burnsi × burnsi ($BB^+ \times BB^+$) cross: $\chi^2 = 3.26$, df $= 1$, $0.1 > P > 0.05$.
burnsi × pipiens ($BB^+ \times B^+B^+$) cross: $\chi^2 = 1.78$, df $= 1$, $P > 0.05$.
Second burnsi × pipiens ($BB^+ \times B^+B^+$) cross: $\chi^2 = 0.46$, df $= 1$, $P > 0.05$.

41. **(a)** A chi-square test comparing the fit of the observed data with the expected 1 : 1 : 1 : 1 ratio yields $\chi^2 = 35$, df $= 3$, $P < 0.005$.
(b) Given the chi-square value, it is unlikely that chance produced the differences between the observed and the expected ratios, indicating that the progeny are not in a 1 : 1 : 1 : 1 ratio.
(c) The number of plants with the $cc\,ff$ genotype is fewer than expected. Some of the plants with the $cc\,ff$ genotype may have died before the progeny were counted.

43. The alleles for obesity from both laboratories are recessive. However, they are located at different gene loci that are independent of one another.

A cross between two obese mice from the first laboratory is

$$\text{Obese } (o_1o_1) \times \text{Obese } (o_1o_1)$$
$$\downarrow$$
$$\text{All obese } (o_1o_1)$$

A cross between two obese mice from the second laboratory is

$$\text{Obese } (o_2o_2) \times \text{Obese } (o_2o_2)$$
$$\downarrow$$
$$\text{All obese } (o_2o_2)$$

The cross between obese mice from the two laboratories is

$$\text{Obese mouse 1 } (o_1o_1\ O_2O_2) \times \text{Obese mouse 2 } (O_1O_1\ o_2o_2)$$
$$\downarrow$$
$$\text{All normal } (O_1o_1\ O_2o_2)$$

Chapter 4

4. XX-XY system: males are heterogametic and produce gametes with either an X chromosome or a Y chromosome. ZZ-ZW system: females are heterogametic and produce gametes with either a Z or a W chromosome.

15. **(a)** Female; **(b)** male; **(c)** male, sterile; **(d)** female; **(e)** male; **(f)** female; **(g)** metafemale; **(h)** male; **(i)** intersex; **(j)** female; **(k)** metamale, sterile; **(l)** metamale; **(m)** intersex.

17. **(a)** Female; **(b)** male; **(c)** male; **(d)** female; **(e)** male.

21. Her father's mother and her father's father, but not her mother's mother or her mother's father.

22. Bridges's exceptional white-eyed females were X^wX^wY, and the red-eyed males were X^+Y. The results of crossing such a white-eyed female with a red-eyed male are shown in **Figure 4.13**. Meiosis in this female generates 45% X^wY, 45% X^w, 5% X^wX^w, and 5% Y gametes. Meiosis in the red-eyed male generates 50% X^+ and 50% Y gametes.

	0.5 X^+	**0.5 Y**
0.45 X^wY	0.225 X^wX^+Y red-eyed females	0.225 X^wYY white-eyed males
0.45 X^w	0.225 X^wX^+ red-eyed females	0.225 X^wY white-eyed males
0.45 X^wX^w	0.025 $X^wX^wX^+$ metafemale, dies	0.025 X^wX^wY white-eyed females
0.45 Y	0.025 X^+Y red-eyed males	0.025 YY dies

23. **(a)** Yes; **(b)** yes; **(c)** no; **(d)** no.

24. **(a)** F_1: $\frac{1}{2}\,X^+Y$ (gray males), $\frac{1}{2}\,X^+X^y$ (gray females); F_2: $\frac{1}{4}\,X^+Y$ (gray males), $\frac{1}{4}\,X^yY$ (yellow males), $\frac{1}{4}\,X^+X^y$ (gray females), $\frac{1}{4}\,X^+X^+$ (gray females).
(b) F_1: $\frac{1}{2}\,X^yY$ (yellow males), $\frac{1}{2}\,X^+X^y$ (gray females); F_2: $\frac{1}{4}\,X^+Y$ (gray males), $\frac{1}{4}\,X^yY$ (yellow males), $\frac{1}{4}\,X^+X^y$ (gray females), $\frac{1}{4}\,X^yX^y$ (yellow females).
(c) F_2: $\frac{1}{4}\,X^+Y$ (gray males), $\frac{1}{4}\,X^yY$ (yellow males), $\frac{1}{4}\,X^+X^+$ (gray females), $\frac{1}{4}\,X^+X^y$ (gray females).
(d) F_3: $\frac{1}{8}$ gray males, $\frac{3}{8}$ yellow males, $\frac{5}{16}$ gray females, and $\frac{3}{16}$ yellow females.

26. John has grounds for suspicion. The color-blind daughter must be X^cX^c and must have inherited a color-blindness allele from each parent. Because John has normal color vision, he must be X^+Y and therefore could not pass on a color-blindness allele to his daughter.

A remote alternative possibility is that the daughter is XO, having inherited a recessive color-blindness allele from her mother and no sex chromosome from her father. In that case, the daughter would have Turner syndrome and color blindness. Another unlikely possibility is that a new mutation in one of John's gametes produced a new X-linked color-blindness allele that he passed on to the daughter.

If Cathy had given birth to a color-blind son, then John would have no grounds for suspicion. The son would have inherited John's Y chromosome and the color-blindness X chromosome from Cathy.

28. Because Bob must have inherited his Y chromosome from his father, and his father has normal color vision, a nondisjunction event in the paternal lineage cannot account for Bob's genotype. Bob's mother must be heterozygous X^+X^c

because she has normal color vision, and she must have inherited an X^c chromosome from her color-blind father. For Bob to inherit two X^c chromosomes from his mother, the egg must have arisen from a nondisjunction in meiosis II. In meiosis I, the homologous X chromosomes separate, so one cell has the X^+ chromosome and the other has X^c. The failure of sister chromatids to separate in meiosis II would then result in an egg with two copies of X^c.

33. F_1: $^1/_2$ Z^bZ^+ (normal males), $^1/_2$ Z^bW (bald females).
F_2: $^1/_4$ Z^+Z^b (normal males), $^1/_4$ Z^+W (normal females), $^1/_4$ Z^bZ^b (bald males), $^1/_4$ Z^bW (bald females).

38. **(a)** F_1: all males have miniature wings and red eyes ($X^mY\ s^+s$), and all females have long wings and red eyes ($X^+X^m\ s^+s$). F_2: $^3/_{16}$ male, normal, red; $^1/_{16}$ male, normal, sepia; $^3/_{16}$ male, miniature, red; $^1/_{16}$ male, miniature, sepia; $^3/_{16}$ female, normal, red; $^1/_{16}$ female, normal, sepia; $^3/_{16}$ female, miniature, red; $^1/_{16}$ female, miniature, sepia.
(b) F_1: all females have long wings and red eyes ($X^{m+}X^m\ s^+s$), and all males have long wings and red eyes ($X^{m+}Y\ s^+s$). F_2: $^3/_{16}$ males, long wings, red eyes; $^1/_{16}$ males, long wings, sepia eyes; $^3/_{16}$ males, miniature wings, red eyes; $^1/_{16}$ males, miniature wings, sepia eyes; $^6/_{16}$ females, long wings, red eyes; $^2/_{16}$ females, long wings, sepia eyes.

40. Assuming that this form of color blindness is an X-linked trait, the mother must be X^cX^c, and the father must be X^+Y. Normally, all the sons should be color blind, and all the daughters should have normal color vision. The most likely way to have a daughter who is color blind would be for her not to have inherited an X^+ from her father. The observation that the color-blind daughter is short in stature and has failed to undergo puberty is consistent with Turner syndrome (XO). The color-blind daughter would then be X^cO.

41. **(a)** 1; **(b)** 0; **(c)** 0; **(d)** 1; **(e)** 1; **(f)** 2; **(g)** 0; **(h)** 2; **(i)** 3.

44. **(a)** X inactivation occurs randomly in each of the cells of the early embryo, then is maintained in the mitotic progeny cells. The irregular patches of skin lacking sweat glands arose from skin precursor cells that inactivated the X chromosome with the normal allele.
(b) The X-inactivation event occurs randomly in each of the cells of the early embryo. Even in identical twins, different skin precursor cells inactivate different X chromosomes, resulting in different distributions of patches lacking sweat glands.

45. If enlarged testes is an autosomal dominant trait, it appears only in males because only males have testes. An autosomal dominant trait would be passed from a heterozygous male to approximately half of his sons and could be passed through females. A Y-linked trait is passed from a father to all of his sons and cannot be passed through females. To determine whether the trait is Y linked or autosomal dominant, cross the male with enlarged testes with a female that has no history of enlarged testes in her family (so that it is unlikely she is carrying a gene for enlarged testes). Then cross the F_1 females with normal males. If enlarged testes appear in some of the offspring, the trait cannot be Y-linked.

Chapter 5

13. **(a)** The results of the crosses indicate that cremello and chestnut are pure-breeding traits (homozygous). Palomino is a heterozygous trait that produces a 1 : 2 : 1 ratio when palominos are crossed with each other. The simplest hypothesis consistent with these results is incomplete dominance, with palomino as the phenotype of the heterozygotes resulting from chestnuts crossed with cremellos.

(b) Let C^B = chestnut, C^W = cremello. The parents and offspring of these crosses have the following genotypes: chestnut = C^BC^B; cremello = C^WC^W; palomino = C^BC^W.

Cross	Offspring
$C^BC^W \times C^BC^W$	13 C^BC^W, 6 C^BC^B, 5 C^WC^W
$C^BC^B \times C^BC^B$	16 C^BC^B
$C^WC^W \times C^WC^W$	13 C^WC^W
$C^BC^W \times C^BC^B$	8 C^BC^W, 9 C^BC^B
$C^BC^W \times C^WC^W$	11 C^BC^W, 11 C^WC^W
$C^BC^B \times C^WC^W$	23 C^BC^W

15. 0.15, or 15%.

17. **(a)** The 2 : 1 ratio in the progeny of the two spotted hamsters suggests lethality, and the 1 : 1 ratio in the progeny of a spotted hamster and a hamster without spots indicates that spotting is a heterozygous phenotype.
(b) Because spotting is a heterozygous phenotype, obtaining Chinese hamsters that breed true for spotting is impossible.

19. **(a)** $^3/_4$ red, $^1/_4$ purple; **(b)** $^1/_2$ red, $^1/_4$ purple, $^1/_4$ green; **(c)** $^3/_4$ red, $^1/_4$ purple; **(d)** $^1/_2$ purple, $^1/_2$ green; **(e)** $^1/_2$ red, $^1/_2$ purple.

24. The child's genotype has an allele for blood-type B and an allele for blood-type N that could not have come from the mother and must have come from the father. Therefore, the child's father must have an allele for type B and an allele for type N. George, Claude, and Henry are eliminated as possible fathers because they lack an allele for either type B or type N.

26. **(a)** Brown backcross fish are $Bb\ Rr$; blue fish are $Bb\ rr$; red fish are $bb\ Rr$; and white fish are $bb\ rr$.
(b) $\chi^2 = 0.5$, df = 3; $0.9 < P < 0.975$.
We cannot reject the hypothesis and therefore assume that the backcross progeny appear in a 1 : 1 : 1 : 1 ratio.
(c) All progeny would be $bb\ Rr$, or red fish.
(d) F_1 will be all brown: $Bb\ Rr$.
Backcross progeny will be $^1/_2$ brown and $^1/_2$ red.

29. **(a)** Labrador retrievers vary in two loci, B and E. Black dogs have dominant alleles at both loci ($B\_\ E\_$), brown dogs have $bb\ E\_$, and yellow dogs have $B\_\ ee$ or $bb\ ee$. Because all the puppies were black, all of them must have inherited a dominant B allele from the yellow parent and a dominant E allele from the brown parent. The brown female parent must have been $bb\ EE$, and the yellow male must have been $BB\ ee$. All the black puppies were $Bb\ Ee$.
(b) Mating two yellow Labradors will produce yellow puppies only. Mating two brown Labradors will produce either brown puppies only, if at least one of the parents is homozygous EE, or $^3/_4$ brown and $^1/_4$ yellow puppies if both parents are heterozygous Ee.

31. Let A and B represent the two loci. The F_1 heterozygotes are $Aa\ Bb$. The F_2 are $A\_\ B\_$ disc-shaped, $A\_\ bb$ spherical, $aa\ B\_$ spherical, $aa\ bb$ long.

33. **(a)** According to the information in **Table 5.3**, Irish setters are $BB\ ee\ SS$ and A^s or a^t. The B permits expression of black pigment, but the ee genotype prevents black color on the body coat, resulting in a reddish color except on the nose and in the eyes. The S prevents spotting, resulting in a uniform coat color.
(b) Poodles are SS. Because the dominant S allele prevents spotting, no puppies from matings with poodles will have spotting.

(c) St. Bernards are $a^y a^y\ BB$, and Dobermans are $a^t a^t\ EE\ SS$. The offspring will be genotype $a^y a^t\ B\_\ E\_\ S\_$. Because a^y, specifying yellow, is dominant over a^t, and the E allele allows expression of the A genotype throughout, the offspring will have yellow coats.

(d) Rottweilers are $a^t a^t\ BB\ EE\ SS$, and Labrador retrievers are $A^s A^s\ SS$. The offspring will be $A^s a^t\ B\_\ E\_\ SS$. The combination of the dominant A^s and E alleles should create solid coats.

35. Among females, $^3/_4$ hornless, $^1/_4$ horned.

39. **(a)** False. Genetic maternal effect means that the phenotype of the individual is determined solely by the genotype of the individual's mother. So we know Martha's mother must have been ss because Martha is sinistral. If Martha was produced as a result of self-fertilization, then Martha must indeed be ss. But if Martha was produced by cross-fertilization, then we cannot know Martha's genotype without more information.
(b) True. Martha's mother is ss, so Martha must be either s^+s or ss.
(c) False. Because we do not know Martha's genotype, we cannot yet predict the phenotype of her offspring.
(d) False. If Martha is s^+s, then all her offspring will be dextral. If Martha is ss, then all her offspring will be sinistral.
(e) False. Martha's mother's phenotype is determined by the genotype of her mother (Martha's maternal grandmother). We know that Martha's mother's genotype must have been ss, so her mother's mother had at least one s allele. But we cannot know if she was a heterozygote or homozygous ss.
(f) True. Because Martha's mother must have been ss, all her progeny must be sinistral.

45. Accounting for the phenocopies, we have 50% (subtracting the $^1/_3$ that are phenocopies from the 75%) of the puppies with the autosomal dominant genotype for long ears and 50% with the recessive genotype. Therefore, one parent is homozygous recessive, and the other parent is a heterozygote.

Chapter 6

10. **(1)** A person may be aware of a genetic disease or risk factor in his or her family. **(2)** An older woman may be pregnant or contemplating pregnancy and may need information about risks and options for prenatal genetic testing. **(3)** A person may have tested positive for a genetic disease or risk factor and may need help interpreting the test results. **(4)** A person or couple may have a child with a genetic disease, or may be caregivers for a person with a genetic disease, and require counseling on treatment options and management of the disease.

12. Amniocentesis is relatively safe, but results are not available until the 17th or 18th week of pregnancy. Chorionic villus sampling has a slightly higher risk of complications, including fetal injury, but results are available several weeks earlier.

15. Direct-to-consumer genetic tests are tests purchased by consumers without an order from a medical professional. They may test for genetic conditions or predispositions, paternity, or ancestry. There are concerns that the tests may not be accurate, and that the people who buy them may not be equipped to properly interpret or understand the information they provide. There is also a concern about the privacy or confidentiality of the individual ordering the test.

16. GINA prohibits discrimination by employers or health-insurance companies based on information obtained by genetic testing. Requiring an individual to undergo genetic testing for employment or health insurance is also prohibited.

18. Study of human genetics is necessary to understand and overcome human genetic diseases. Because of humans' long life span, relatively large body size, and unique lifestyle and behaviors, animal models are nonexistent or insufficient for many genetic disorders. The careful preservation of marriage, birth, death, and health records in many societies provides a wealth of data for genetic analysis. The completion of the Human Genome Project now facilitates mapping and identification of human genes. Humans have a strong sense of identity and worth as individuals, and they wish to understand how an individual's genetic profile contributes to that person's health, behavior, abilities and disabilities, and individual future prospects. The study of human genetics can also reveal the historical origins and anthropological relationships of individuals and populations.

19. **(a)**

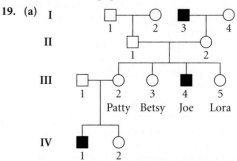

(b) X-linked recessive; **(c)** zero; **(d)** $^1/_4$; **(e)** $^1/_4$.

22. **(a)** Only males have the disease, it skips generations, and unaffected female carriers have both affected and unaffected sons. These observations are consistent with a recessive X-linked trait. Y-linked traits are transmitted directly from father to son and do not skip generations.
(b) We will use X^+ to denote the normal X allele and X^d to denote the Dent allele.
I: 1 is X^+Y; 2 is X^+X^d.
II: 1 and 5 are X^dY, 7 and 8 are X^+X^d, and the rest do not have the disease allele.
III: 2 and 3 are carriers with X^+X^d, 4–7 are X^dY, and the rest do not have the disease allele.
IV: 2 is X^dY; 1 is X^+Y.

24. **(a)** Autosomal dominant. Affected males pass the trait to both sons and daughters. It does not skip generations, all affected individuals have affected parents, and it is extremely unlikely that multiple unrelated individuals mating into the pedigree are carriers of a rare trait.
(b) X-linked dominant. Affected females pass the trait to sons and daughters, but affected males pass the trait to all daughters and no sons.
(c) Y-linked. Affects only males and is passed from father to all sons.
(d) X-linked recessive or sex-limited autosomal dominant. We can eliminate Y linkage because affected males do not pass the trait to their sons. X-linked recessive inheritance is consistent with the pattern of unaffected female carriers producing both affected and unaffected sons and affected males producing unaffected female carriers, but no affected sons. Sex-limited autosomal dominant inheritance is also consistent with unaffected heterozygous females producing affected heterozygous sons, unaffected homozygous recessive sons, and unaffected heterozygous or homozygous recessive daughters. We could distinguish between these two possibilities if we had enough data to determine whether affected males have both affected and unaffected sons, as expected from

autosomal dominant inheritance, or whether affected males have only unaffected sons, as expected from X-linked recessive inheritance. Unfortunately, this pedigree shows only two sons of affected males. In both cases, the sons are unaffected, consistent with X-linked recessive inheritance, but two male progeny are not enough to conclude that affected males cannot produce affected sons.

(e) Autosomal recessive. Unaffected parents produce affected progeny. The affected daughter must have inherited recessive alleles from both unaffected parents, so the trait must be autosomal. If it were X-linked, her father would show the trait.

30. Migraine headaches: genetic and environmental. Markedly greater concordance in monozygotic twins, who are 100% genetically identical, than in dizygotic twins, who are 50% genetically identical, is indicative of a genetic influence. However, only 60% concordance for monozygotic twins indicates that environmental factors also play a role.
Eye color: genetic. Concordance is greater in monozygotic twins than in dizygotic twins. Monozygotic twins have 100% concordance, indicating that environment has no detectable influence.
Measles: no detectable genetic influence. No difference in concordance between monozygotic and dizygotic twins. Some environmental influence can be detected because monozygotic twins show less than 100% concordance.
Club foot: genetic and environmental. Markedly greater concordance in monozygotic twins indicates genetic influence. A strong environmental influence is indicated by the high discordance in monozygotic twins.
High blood pressure: genetic and environmental. Markedly greater concordance in monozygotic twins indicates genetic influence. Less than 100% concordance in monozygotic twins indicates environmental influence.
Handedness: no genetic influence. The concordance is the same in monozygotic and dizygotic twins, indicating no genetic influence. Less than 100% concordance in monozygotic twins indicates environmental influence.
Tuberculosis: no genetic influence. Concordance is the same in monozygotic and dizygotic twins. The importance of environmental influence is indicated by the very low concordance in monozygotic twins.

33. The data suggest that schizophrenia has a strong genetic component. The biological parents of adoptees with schizophrenia are far more likely to have schizophrenia than are their adoptive parents, to whom they are genetically unrelated, despite the fact that the adoptees share their environment with their adoptive parents. Another possibility is that the high frequency of schizophrenia in the biological parents of the adoptees simply reflects a greater likelihood that parents with schizophrenia will give up their children for adoption. This possibility is ruled out by the observation that the biological parents of adoptees who do not have schizophrenia do not show a similar high frequency of schizophrenia compared with adoptive parents.

Chapter 7

13. **(a)** If pattern baldness were sex-influenced, we would see independent assortment between genetic markers on the X chromosome and pattern baldness. Sex-influenced traits are encoded by autosomal genes, which assort independently of X-linked genes because they are on different chromosomes.

(b) If pattern baldness were X-linked recessive, it would be encoded by a gene on the X chromosome. If this gene and the X-linked markers were close together on the X chromosome, they would not assort independently. However, if the gene for pattern baldness and the X-linked markers were located far apart, so that crossing over between them occurred every meiosis, they might assort independently.

15. **(a)** $1/4$ wild-type eyes, wild-type wings; $1/4$ red eyes, wild-type wings; $1/4$ wild-type eyes, white-banded wings; $1/4$ red eyes, white-banded wings.

(b) The F_1 heterozygotes inherited a chromosome with alleles for red eyes and white-banded wings ($re\ wb$) from one parent and a chromosome with alleles for wild-type eyes and wild-type wings ($re^+\ wb^+$) from the other parent. These are therefore the phenotypes of the nonrecombinant progeny, present in the highest numbers. The recombinants are the 19 with red eyes, wild-type wings and the 16 with wild-type eyes, white-banded wings.

$$\text{recombination frequency} = \frac{\text{number of recombinant progeny}}{\text{total number of progeny}} \times 100\%$$
$$= (19 + 16)/879 \times 100\%$$
$$= 4.0\%$$

The distance between the genes is 4 map units.

16. The genes are linked and have not assorted independently.

17.
Heart-shaped, numerous spines	33.7%
Normal, few spines	33.7%
Heart-shaped, few spines	16.3%
Normal, numerous spines	16.3%

21. The distances between the genes are indicated by the recombination rates. Because loci R and L_2 have the highest recombination rate, they must be the farthest apart, and W_2 is in the middle. The order of the genes is R, W_2, L_2.

27.
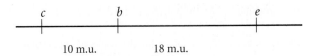

Gene f is unlinked to either of these groups; it is on a third linkage group.

29. **(a)** All the progeny receive $p \quad sh\text{-}1 \quad Hb^2$ from the male parent, shown as the lower chromosome in each progeny type. The upper chromosomes are the products of meiosis in the heterozygous parent and are identified in the table. The nonrecombinants have $p \quad sh\text{-}1 \quad Hb^2$ or $P \quad Sh\text{-}1 \quad Hb^1$; the double crossovers have $p \quad Sh\text{-}1 \quad Hb^2$ or $P \quad sh\text{-}1 \quad Hb^1$. The two classes differ in the $Sh\text{-}1$ locus; therefore, $Sh\text{-}1$ is the middle locus.

(b) P and $Sh\text{-}1$: Recombinants have $P \quad sh\text{-}1$ or $p \quad Sh\text{-}1$.

$$\text{Recombination frequency} = (57 + 45 + 1)/708$$
$$= 0.145 = 14.5 \text{ m.u.}$$

$Sh\text{-}1$ and Hb: Recombinants have $Sh\text{-}1 \quad Hb^2$ or $sh\text{-}1 \quad Hb^1$.

$$\text{Recombination frequency} = (6 + 5 + 1)/708$$
$$= 0.017 = 1.7 \text{ m.u.}$$

(c) Expected double crossovers = recombination frequency 1 × recombination frequency 2 × total progeny = 0.145 × 0.017 × 708 = 1.7
Coefficient of coincidence = observed double crossovers/expected double crossovers = 1/1.7 = 0.59
Interference = 1 − coefficient of coincidence = 0.41

36. This lod score indicates that the probability of observing this degree of association if the marker is linked to asthma is 100 times higher than if the marker has no linkage to asthma. A lod score of 3, for a ratio of 1000, is generally considered convincing evidence of linkage.

38. Enzyme 1 is located on chromosome 9. Chromosome 9 is the only chromosome that is present in the cell lines that produce enzyme 1 and absent in the cell lines that do not produce enzyme 1.

Enzyme 2 is located on chromosome 4. Chromosome 4 is the only chromosome that is present in cell lines that produce enzyme 2 (C and D) and absent in cell lines that do not produce enzyme 2 (A and B).

Enzyme 3 is located on the X chromosome. The X chromosome is the only chromosome present in the three cell lines that produce enzyme 3 and absent in the cell line that does not produce enzyme 3.

39.

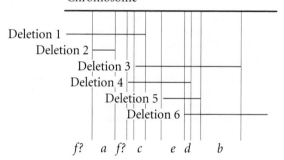

The location of *f* is ambiguous; it could be in either location shown in the deletion map.

Chapter 8

19. **(a)** Duplications; **(b)** polyploidy; **(c)** deletions; **(d)** inversions; **(e)** translocations.

20. **(a)** Tandem duplication of AB; **(b)** displaced duplication of AB; **(c)** paracentric inversion of DEF; **(d)** deletion of B; **(e)** deletion of FG; **(f)** paracentric inversion of CDE; **(g)** pericentric inversion of ABC; **(h)** duplication and inversion of DEF; **(i)** duplication of CDEF, inversion of EF.

23. **(a)** $\frac{1}{3}$ white-eyed Notch females, $\frac{1}{3}$ wild-type females, and $\frac{1}{3}$ wild-type males
(b) $\frac{1}{3}$ red-eyed Notch females, $\frac{1}{3}$ wild-type females, and $\frac{1}{3}$ white-eyed males
(c) $\frac{1}{3}$ white-eyed Notch females, $\frac{1}{3}$ white-eyed females, and $\frac{1}{3}$ white-eyed males

25. **(a)**

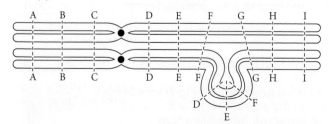

(b)

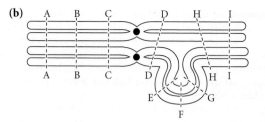

(c)

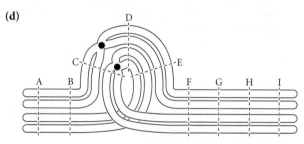

(d)

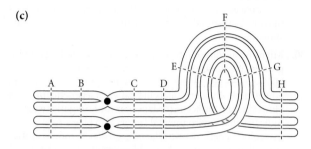

27. If crossing over occurs within the pericentric inversion at a rate of 26% of meioses, then 13% of the woman's oocytes will have duplication or deficient chromosome 8, and the probability of the couple having a child with a syndrome caused by the crossing over is 13%.

28. **(a)**

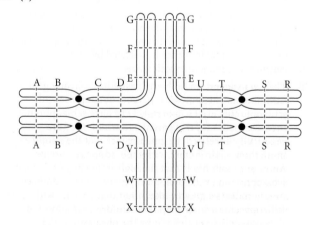

(b) Alternate:

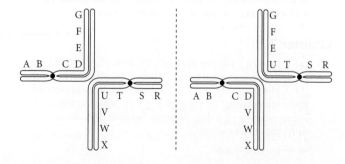

Adjacent-1:

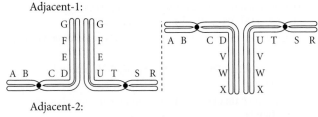

Adjacent-2:

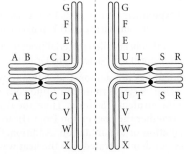

(c) Alternate: Gametes contain either both normal or both translocation chromosomes, and are all viable.

$$AB \cdot CDEFG + RS \cdot TUVWX$$

and

$$AB \cdot CDVWX + RS \cdot TUEFG$$

Adjacent-1: Gametes contain one normal and one translocation chromosome, resulting in the duplication of some genes and a deficiency of others.

$$AB \cdot CDEFG + RS \cdot TUEFG$$

and

$$AB \cdot CDVWX + RS \cdot TUVWX$$

Adjacent-2 (rare): Gametes contain one normal and one translocation chromosome, with the duplication of some genes and a deficiency of others.

$$AB \cdot CDEFG + AB \cdot CDVWX$$

and

$$RS \cdot TUVWX + RS \cdot TUEFG$$

29. The nondisjunction event took place during meiosis II of the egg.

32. The high incidence of Down syndrome in Bill's family and among Bill's relatives is consistent with familial Down syndrome, caused by a Robertsonian translocation of chromosome 21. Bill and his sister, who are unaffected, are unaffected carriers of the translocation and have 45 chromosomes. Their children and Bill's brother, who have Down syndrome, have 46 chromosomes. From the information given, there is no reason to suspect that Bill's wife Betty has any chromosome abnormalities. Therefore, statement *d* is most likely correct. All other statements are incorrect.
(a) Incorrect. Bill is an unaffected carrier of the translocation and has 45 chromosomes.
(b) Incorrect. There is no reason to suspect that Bill's wife Betty has any chromosome abnormalities.
(c) Incorrect. Bill and Betty's children have familial Down syndrome and have 46 chromosomes.
(d) Correct. Bill's sister is an unaffected carrier of the translocation and has 45 chromosomes.
(e) Incorrect. Bill is an unaffected carrier of the translocation and has 45 chromosomes.
(f) Incorrect. There is no reason to suspect that Bill's wife Betty has any chromosome abnormalities.
(g) Incorrect. Bill's brother has familial Down syndrome and has 46 chromosomes.

34. **(a)** Possible gamete types: (i) normal chromosome 13 and normal chromosome 22; (ii) translocated chromosome 13 + 22; (iii) translocated chromosome 13 + 22 and normal chromosome 22; (iv) normal chromosome 13; (v) normal chromosome 13 and translocated chromosome 13 + 22; (vi) normal chromosome 22.
(b) Zygote types: (i) 13, 13, 22, 22; normal; (ii) 13, 13 + 22, 22; translocation carrier; (iii) 3, 13 + 22, 22, 22; trisomy 22; (iv) 13, 13, 22; monosomy 22; (v) 13, 13, 13 + 22, 22; trisomy 13; (vi) 13, 22, 22; monosomy 13.
(c) 50%.

38. **(a)** Such allotriploids could have 1*n* from species I and 2*n* from species II for 3*n* = 18; alternatively, they could have 2*n* from species I and 1*n* from species II for 3*n* = 15.
(b) 4*n* = 28.
(c) 2*n* + 1 = 9.
(d) 2*n* − 1 = 13.
(e) 2*n* + 2 = 10.
(f) Allotetraploids must have chromosomes from both species, and total 4*n*. There are three possible combinations for such allotetraploids: 2*n* from each: 2(4) + 2(7) = 22; 1*n* from species I + 3*n* from species II: 1(4) + 3(7) = 25; 3*n* from species I + 1*n* from species II: 3(4) + 1(7) = 19.

Chapter 9

17. Isolate DNA from water samples taken from the two streams. Sequence the DNA and identify the bacterial species based on differences in their DNA.

22. Distances between genes are in minutes.

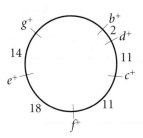

25. Only genes located near one another on the bacterial chromosome will be cotransformed. However, by performing transformation experiments and screening for different pairs of cotransformed genes, bacterial genes can be mapped. Gene pairs that are never cotransformed must be relatively far apart on the chromosome, while gene pairs that are cotransformed are more closely linked. From the data, we see that gene a^+ is cotransformed with both e^+ and d^+. However, genes d^+ and e^+ are not cotransformed, indicating that a^+ and e^+ are more closely linked than d^+ and e^+. Gene a^+ is not cotransformed with either b^+ or c^+, yet gene e^+ is. This observation indicates that gene e^+ is more closely linked to genes b^+ and c^+ than is gene a^+. The orientation of genes b^+ and c^+ relative to e^+ cannot be determined from the data provided.

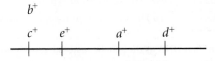

30. The original infecting phages were wild type ($a^+ \ b^+$) and doubly mutant ($a^- \ b^-$). The $a^+ \ b^-$ and $a^- \ b^+$ plaques are produced by recombinant phages.

The frequency of recombination is the total number of recombinant plaques divided by the total number of plaques: $677/4854 = 0.14$.

31. **(a)** Transductants were initially screened for the presence of $proC^+$. Thus, only $proC^+$ transductants were identified.
(b) The wild-type genotypes ($proC^+ proA^+ proB^+ proD^+$) represent single transductants of $proC^+$. Both the $proC^+ proA^- proB^+ proD^+$ and $proC^+ proA^+ proB^- proD^+$ genotypes represent cotransductants of $proC^+, proA^-$ and $proC^+, proB^-$.
(c) Both $proA$ and $proB$ were cotransduced with $proC$ at about the same rate, which suggests that they are similar in distance from $proC$. The $proD$ gene was never cotransduced with $proC$, suggesting that $proD$ is more distant from $proC$ than $proA$ and $proB$. However, there were a small number of all types of cotransductants, and the absence of cotransduction between $proD$ and $proC$ might be due to chance.

32.

Plaque phenotype	Expected number	
$c^+ m^+$	460	(nonrecombinant)
$c^- m^-$	460	(nonrecombinant)
$c^+ m^-$	40	(recombinant)
$c^- m^+$	40	(recombinant)
Total plaques	1000	

34. **(a)** The recombination frequency between r_2 and h is $^{13}/_{180} = 0.072$, or 7.2%. The recombination frequency between r_{13} and h is $^3/_{216} = 0.014$, or 1.4%.

(b)

```
        h    r₁₃              r₂
        +----+----------------+
       
        1.4 m.u.   5.8 m.u.
```

```
        r₂              h    r₁₃
        +---------------+----+
       
         7.2 m.u.   1.4 m.u.
```

35. **(a)**

```
        h⁺         st⁺          c⁺
        +----------+------------+
```

(b)

```
    h⁺           st⁺                        c⁺
    +------------+--------------------------+
   
     7.1 m.u.              24.1 m.u.
```

(c) Coefficient of coincidence $= \dfrac{(6+5)}{(0.071 \times 0.241 \times 942)}$

$= 0.68.$

Interference $= 1 - 0.68 = 0.32$.

Chapter 10

18. Took X-ray diffraction pictures used in constructing the structure of DNA: **(i)** Franklin and Wilkins

Determined that DNA contains nitrogenous bases: **(a)** Kossel

Identified DNA as the genetic material in bacteriophages: **(f)** Hershey and Chase

Discovered regularity in the ratios of different bases in DNA: **(j)** Chargaff

Determined that DNA is responsible for transformation in bacteria: **(g)** Avery, MacLeod, and McCarty

Worked out the helical structure of DNA by building models: **(c)** Watson and Crick

Discovered that DNA consists of repeating nucleotides: **(d)** Levene

Determined that DNA is acidic and high in phosphorus: **(e)** Miescher

Conducted experiments showing that RNA can serve as the genetic material in some viruses: **(b)** Fraenkel-Conrat

Demonstrated that heat-killed material from bacteria can genetically transform live bacteria: **(h)** Griffith

19. No, the student has not demonstrated that transformation has taken place. A single mutation could convert the IIR strain into the virulent IIS strain. Thus, the student cannot determine whether the conversion from IIR to IIS is due to transformation or to a mutation. Additionally, the student has not demonstrated that the heat was sufficient to kill all the IIS bacteria. A second useful control experiment would have been to inject the heat-killed IIS into mice and see if any of the IIS bacteria survived the heat treatment.

24. Tubes 1, 4, and 5. The DNA of the bacteriophage contains phosphorus and the protein contains sulfur. When the bacteriophages infect a cell, they inject their DNA into the cell, but the protein coats stay on the surface of the cell. The phage protein coats are sheared off in the blender, while the bacterial cells containing the phage DNA form a pellet at the bottom of the tube. Thus, tubes containing bacteria infected with ^{35}S-labeled bacteriophages will have radioactivity associated with the protein coats, whereas tubes containing bacteria infected with ^{32}P-labeled bacteriophages will have radioactivity associated with the cells.

26. The phosphate backbone of DNA molecules typically carries a negative charge, thus making the DNA molecules attractive to the positive pole of the current.

27. Approximately 5×10^{20} nucleotide pairs. Stretched end to end, the DNA would reach 1.7×10^8 km.

29. **(a)**

Organism and tissue	$(A + G)/(T + C)$	$(A + T)/(C + G)$
Sheep thymus	1.03	1.36
Pig liver	0.99	1.44
Human thymus	1.03	1.52
Rat bone marrow	1.02	1.36
Hen erythrocytes	0.97	1.38
Yeast	1.00	1.80
E. coli	1.04	1.00
Human sperm	1.00	1.67
Salmon sperm	1.02	1.43
Herring sperm	1.04	1.29

(b) The $(A + G)/(T + C)$ ratio of about 1.0 is constant for these organisms. Each of them has a double-stranded genome. The percentage of purines should equal the percentage of pyrimidines in double-stranded DNA, which means that $(A + G) = (T + C)$. The $(A + T)/(C + G)$ ratios are not constant. The number of A–T base pairs relative to the number of C–G base pairs is unique to each organism and can vary among the different organisms.

(c) The (A + G)/(T + C) ratio is about the same for the three sperm samples. Although the sperm is haploid, its DNA is still double stranded. As in part *b*, the percentage of purines should equal the percentage of pyrimidines.

32. Adenine = 15%; guanine = 35%; cytosine = 35%.

35. Virus I is a double-stranded RNA virus. Uracil is present, indicating an RNA genome. As expected for a double-stranded genome, the amounts of adenine and uracil are equal, as are the amounts of guanine and cytosine.

Virus II is a double-stranded DNA virus. The presence of thymine indicates that the viral genome is DNA. As expected for a double-stranded DNA molecule, the amounts of adenine and thymine are equal, as are the amounts of guanine and cytosine.

Virus III is a single-stranded DNA virus. The presence of thymine indicates a DNA genome. However, the amounts of thymine and adenine are unequal, as are the amounts of guanine and cytosine, which suggests a single-stranded DNA molecule.

Virus IV is a single-stranded RNA virus. The presence of uracil indicates an RNA genome. However, the amount of adenine does not equal that of uracil, and the amount of guanine does not equal that of cytosine, which suggests a single-stranded genome.

36. 100,000

37. **(a)**

   (1) Neither 5′-carbon of the two sugars is directly linked to phosphorus.

   (2) Neither 5′-carbon of the two sugars has a hydroxyl group attached.

   (3) Neither sugar molecule has oxygen in its ring structure between the 1′ and 4′ carbons.

   (4) In both sugars, the 2′-carbon has a hydroxyl group attached, which does not occur in deoxyribonucleotides.

   (5) At the 3′ position in both sugars, only hydrogen is attached, as opposed to a hydroxyl group.

   (6) The 1′-carbon of both sugars has a hydroxyl group, as opposed to just a hydrogen attached.

   (b)

38. Hairpins often consist of a stem and loop. The stem consists of inverted complementary RNA sequences and the loop consists of a region of noncomplementary sequence. The inverted complements form the stem structure, and the loop of the hairpin is formed by the noncomplementary sequences.

   5′–UGCAU–3′ … unpaired nucleotides … 5′–AUGCA–3′

39. The genetic material will contain complex information, replicate faithfully, and encode the phenotype. Even if the genetic material on the planet is not DNA, it must have these properties. Additionally, the genetic material will be stable.

Chapter 11

18. Prokaryotic chromosomes are usually circular, whereas eukaryotic chromosomes are linear. A single prokaryotic chromosome generally contains the entire genome, whereas the eukaryotic genome is divided into multiple chromosomes. Prokaryotic chromosomes are generally much smaller than eukaryotic chromosomes and have only a single origin of DNA replication, whereas eukaryotic chromosomes contain multiple origins of DNA replication. Eukaryotic chromosomes contain DNA packaged into nucleosomes; nucleosomes are absent from prokaryotic chromosomes (although both bacterial and archaeal DNA is complexed with proteins). The condensation state of eukaryotic chromosomes varies with the cell cycle.

19. Because each nucleosome contains two molecules of H2A and only one molecule of H1, eukaryotic cells will have more H2A than H1. Because each nucleosome contains two molecules of H2A and two molecules of H3, eukaryotic cells should have equal amounts of these two histones.

20. **(a)** None; **(b)** embryonic; **(c)** adult; **(d)** none.

21. **(a)** 3.2×10^7; **(b)** 2.9×10^8.

22. More acetylation. Regions of DNase I sensitivity are less condensed than DNA that is not sensitive to DNase I, the sensitive DNA is less tightly associated with nucleosomes, and it is in a more open state. Such a state is associated with the acetylation of lysine in the histone tails. Acetylation eliminates the positive charge of the lysine and reduces the affinity of the histone for the negatively charged phosphates of the DNA backbone.

25. The upper molecule, because it has a higher percentage of A–T base pairs, will have a lower melting temperature. A–T base pairs have two hydrogen bonds and are thus less stable than G–C base pairs, which have three hydrogen bonds.

27. If the mutation is located within a chloroplast gene, then it should be inherited only from the female parent. We would expect the following results, no matter which trait is dominant:

   Wild-type male × light-green female → offspring all light green

   Light-green male × wild-type female → offspring all wild type

   If the mutation is in a nuclear gene, then both parents can pass the light-green mutation to their offspring. If the wild type is dominant, then we would expect the following results:

Wild-type male × light-green female → offspring all wild type

Light-green male × wild-type female → offspring all wild type

If the light-green phenotype is dominant, then we would expect the following results:

Wild-type male × light-green female → offspring all light green

Light-green male × wild-type female → offspring all light green

28. The pedigree indicates that the neurological disorder is a cytoplasmically inherited trait. Only females pass the trait to their offspring. The trait does not appear to be sex-specific in that males as well as females can have the disorder.

31. Maternal inheritance indicates that a *poky* mutation has a mitochondrial origin because mitochondrial genomes are inherited maternally in *Neurospora*. Biparental inheritance suggests that the *poky* mutation is of nuclear origin. In *Neurospora*, nuclear genes exhibit biparental inheritance.

32. Amounts of nuclear DNA should increase during only the S phase before declining at cytokinesis. Mitochondrial DNA levels should increase throughout the cell cycle before declining at cytokinesis.

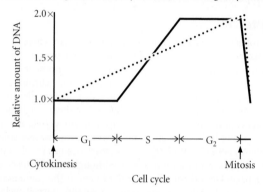

Chapter 12

22. **(a)** Cells in G_1, before switching to medium with ^{14}N

(b) Cells in G_2, after switching to medium with ^{14}N

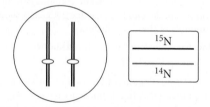

(c) Cells in anaphase of mitosis, after switching to medium with ^{14}N

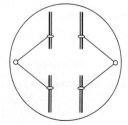

(d) Cells in metaphase I of meiosis, after switching to medium with ^{14}N

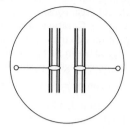

(e) Cells in anaphase II of meiosis, after switching to medium with ^{14}N

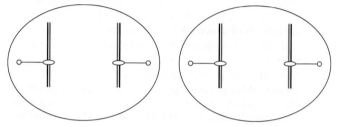

23. Theta replication, 5 minutes; rolling-circle replication, 10 minutes.

25.

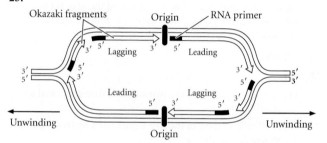

27. **(a)** More errors in replication; **(b)** primers would not be removed; **(c)** primers that had been removed would not be replaced.

30. Primase is required for replication initiation in theta replication. If primase is nonfunctional, then replication initiation would not take place, and no replication would occur. Rolling-circle replication does not require primase. A single-strand break within one strand provides a 3′-OH group to which nucleotides can be added, so rolling-circle replication could occur without a functional primase.

32. Two distinct bands. If the original histones remained on one strand, then we would expect to see the octamers with original histones nearer the bottom of the centrifuge tube in a distinct band. Octamers with newly synthesized histones would be lighter and would appear in a distinct band higher in the tube.

34. The RNA component of telomerase is needed to provide a template for synthesizing the telomeric sequences at the ends of the chromosomes. A large deletion would probably prevent telomere synthesis.

39. Protein B may be needed for the successful initiation of replication at origins of replication. Protein B is present at the beginning of S phase, but disappears by the end of it. Protein A may be responsible for removing or inactivating protein B. As levels of protein A increase, levels of protein B decrease, preventing extra initiation events. When protein A is mutated, it can no longer inactivate protein B; thus successive rounds of replication can begin, owing to the high levels of protein B. When protein B is mutated, it cannot assist initiation, and replication ceases.

Chapter 13

14. (a) Single stranded. If it was double stranded, we would expect nearly equal percentages of adenine and uracil, as well as equal percentages of guanine and cytosine. In this RNA molecule, the percentages of these bases are not equal.
(b) A = 42%, T = 23%, C = 14%, and G = 21%

15. 5′–AUAGGCGAUGCCA–3′

21. ^T_AGCAATT

24.

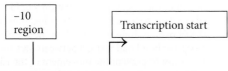

5′–GGACTA<u>TATGAT</u>GCGGCCCAT–3′
3′–CCTGATATACTACGCCGGGTA–5′

The −10 consensus sequence, or Pribnow box, has the consensus sequence TATAAT. However, few bacterial promoters actually contain the exact consensus sequence. The transcription start site is about 10 bp downstream of the −10 consensus sequence.

25. (a) Would probably affect the −10 consensus sequence, which would most likely decrease transcription.
(b) Would probably affect the −35 consensus sequence, reducing or inhibiting transcription.
(c) Unlikely to have any effect on transcription.
(d) Would have little effect on transcription.

29. (a)

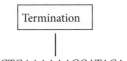

3′–AGCATACAGCAGACCGTTGGTCTGAAAAAAGCATACA–5′

(b) A rho-independent terminator.
(c)

```
              A
          C       A
          G C
          G C
          U A
          C G
          U A
      5′–UCGUAUGUCG CUUUU–3′
```

34. If TBP cannot bind to the TATA box, then genes with these promoters will be transcribed at very low levels or not at all.

38. (a) From experiment 3 and its corresponding lane in the gel, pTFIIB and the RNA polymerase from *S. pombe* (pPol II) appear to have been sufficient to determine the transcription start site. The *S. pombe* start site was used even though many of the other transcription factors were from *S. cerevisiae*.
(b) When TFIIE and TFIIH were individually exchanged, transcription was affected. However, the paired exchange of TFIIE and TFIIH allowed for transcription, suggesting that TFIIE and TFIIH undergo a species-specific interaction essential for transcription or that the absence of this interaction inhibits transcription. Similar results were observed in the paired exchange of TFIIB and RNA polymerase, which suggests that species-specific interaction between TFIIB and RNA polymerase is needed for transcription.
(c) TFIIB probably interacts both with RNA polymerase and with other general transcription factors necessary for initiation at the promoter. The data indicate that TFIIB and RNA polymerase II must be from the same species for transcription to take place. TFIIB potentially assembles downstream of the TATA box but in association with other transcription factors located at or near the TATA box. Possibly, these downstream interactions between TFIIB and the other transcription factors stimulate conformational changes in RNA polymerase and the DNA sequence, enabling transcription to begin at the appropriate start site. In experiment 3, both the RNA polymerase and TFIIB were from *S. pombe*, leading to the positioning of the RNA polymerase at the *S. pombe* transcription start site by pTFIIB.

Chapter 14

21. The large size of the dystrophin gene is probably due to the presence of many intervening sequences, or introns, within the coding region of the gene.

26.

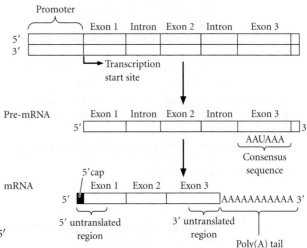

(a) Lies upstream of the translation start codon. In bacteria, the ribosome-binding site, or Shine–Dalgarno sequence, is within the 5′ untranslated region. However, eukaryotic mRNA does not have an equivalent sequence, and a eukaryotic ribosome binds at the 5′ cap of the mRNA molecule.
(b) A DNA sequence recognized and bound by the transcription apparatus to initiate transcription.
(c) Lies downstream of the coding region of the gene; determines the location of the 3′ cleavage site in the pre-mRNA molecule.
(d) First nucleotide transcribed into RNA; located 25–30 nucleotides downstream of the TATA box.
(e) A sequence of nucleotides located at the 3′ end of the mRNA that is not translated into proteins. However, it does affect the stability and translation of the mRNA.
(f) Noncoding DNA sequences that occur within coding regions of a gene; removed during processing of the pre-mRNA.
(g) Coding regions of a gene.
(h) A sequence of adenine nucleotides added to the 3′ end of the pre-mRNA; affects mRNA stability and the binding of the ribosome to the mRNA.
(i) Functions in binding of the ribosome to the mRNA and in mRNA stability.

31. The mutation will probably prevent 3′ cleavage and addition of the poly(A) tail, resulting in more rapid degradation of the mRNA. Less of the encoded protein will be produced.

36. (a) In trans-splicing, exons from different genes are spliced together during RNA processing. As a result, the amino acid sequence of the translated protein is encoded by two or more different genes. According to the concept of colinearity, we would have expected the DNA sequence of a single gene to correspond directly to the amino acid sequence of the protein it encodes. **(b)** Different mature mRNAs can be produced from a single gene by alternative splicing. Thus, different proteins can be encoded within the same gene, as opposed to one gene corresponding to one protein, as is predicted by the concept of colinearity. **(c)** In RNA editing, genetic information is added to the pre-mRNA after it is transcribed; the mature mRNA therefore contains information that was not part of the DNA from which it was transcribed. As a result, the nucleotide sequence of the gene does not correspond to the amino acid sequence of the protein—a clear violation of the concept of colinearity.

Chapter 15

16. The mutations can be assembled into four groups:

    Group 1 mutants (*trp-1, trp-10, trp-11, trp-9, trp-6,* and *trp-7*) can grow only on minimal medium supplemented with tryptophan.

    Group 2 mutants (*trp-3*) can grow on minimal medium supplemented with either tryptophan or indole.

    Group 3 mutants (*trp-2* and *trp-4*) can grow on minimal medium supplemented with tryptophan, indole, or indole glycerol phosphate.

    Group 4 mutants (*trp-8*) can grow on minimal medium supplemented with the addition of tryptophan, indole, indole glycerol phosphate, or anthranilic acid.

Group 4		Group 3		Group 2		Group 1
Precursor →	anthranilic acid	→	indole glycerol phosphate	→	indole	→ tryptophan

20. **(a)** 1; **(b)** 2; **(c)** 3; **(d)** 3; **(e)** 4.

21. 3^3, or 27.

22. **(a)** amino–fMet-Phe-Lys-Phe-Lys-Phe–carboxyl
    (b) amino–fMet-Tyr-Ile-Tyr-Ile–carboxyl
    (c) amino–fMet-Asp-Glu-Arg-Phe-Leu-Ala–carboxyl
    (d) amino–fMet-Gly–carboxyl (The stop codon UAG follows the codon for glycine.)

26. **(a)** 3′–CCG–5′ or 3′–UCG–5′
    (b) 3′–UUC–5′
    (c) 3′–AUU–5′ or 3′–UUU–5′ or 3′–CUU–5′
    (d) 3′–ACC–5′ or 3′–GCC–5′
    (e) 3′–GUC–5′

30. initiation factor 3
    fMet-tRNA$_i^{fMet}$
    30S initiation complex
    70S initiation complex
    elongation factor Tu
    elongation factor G
    release factor 1

33. **(a)** 3′–UGC–5′; **(b)** threonine.

34. **(a)** The lack of IF-3 would prevent protein synthesis. IF-3 separates the large and small ribosomal subunits, which is required for the initiation of translation. The absence of IF-3 would mean that translation would not be initiated and no proteins would be synthesized.

(b) No translation would take place. IF-2 is necessary for the initiation of translation. The lack of IF-2 would prevent fMet-tRNA$_i^{fMet}$ from being delivered to the small ribosomal subunit, thus blocking translation.
(c) Although translation would be initiated by the delivery of fMet-tRNA$_i^{fMet}$ to the ribosome–mRNA complex, no other amino acids would be delivered to the ribosome. EF-Tu, which binds to GTP and the charged tRNA, is necessary for elongation. This three-part complex enters the A site of the ribosome. If EF-Tu were not present, the charged tRNA would not enter the A site, and translation would stop.
(d) EF-G is necessary for the translocation (movement) of the ribosome along the mRNA in the 5′→3′ direction. Once a peptide bond had formed between Met and Pro, the lack of EF-G would prevent the movement of the ribosome along the mRNA, so no new codons would be read. The formation of the dipeptide Met-Pro does not require EF-G.
(e) The release factors RF-1 and RF-2 recognize the stop codons and bind to the ribosome at the A site. They then interact with RF-3 to promote cleavage of the peptide from the tRNA at the P site. The absence of the release factors would prevent the termination of translation at the stop codon.
(f) ATP is required for tRNAs to be charged with amino acids by aminoacyl-tRNA synthetases. Without ATP, the charging would not take place, and no amino acids would be available for protein synthesis.
(g) GTP is required for the initiation, elongation, and termination of translation. If GTP were absent, protein synthesis would not take place.

38. NMD should not be a problem. NMD is thought to be dependent on exon-junction proteins that are normally removed from the mRNA by the movement of the ribosomes during translation, but trigger NMD when they are not removed. If the first ribosome to read the mRNA inserts an amino acid for the stop codon due to the action of PTC124, then it should not stall at the stop codon and should remove any exon-junction proteins. The result is that the mRNA will be stabilized and protected from NMD, thus allowing translation to continue.

41. **(a)** These results suggest that the initiator tRNA plays a role in selection of the start codon. In eukaryotes, the ribosome attaches to the 5′ cap and then scans the mRNA until it locates the AUG start codon. When the anticodon on the tRNA$_i^{Met}$ was mutated, it appears that a different start codon was selected, because protein synthesis occurred, but the resulting protein had more or fewer amino acids than normal.
(b) The initiation of translation in bacteria occurs in a different way—it requires that the 16S RNA of the small ribosomal subunit interact with the Shine–Dalgarno sequence. This interaction lines up the ribosome over the initiation codon. If the anticodon of tRNA$_i^{fMet}$ were mutated, no initiator tRNA would pair with the initiation codon, and no protein synthesis would take place.
(c) When the anticodon of the tRNA$_i^{Met}$ is mutated to 5′–CCA–3′, initiation takes place at the first 5′–UGG–3′ codon (complementary to 5′–CCA–3′) encountered by the scanning ribosome. In some mRNAs, this codon will occur before the normal 5′–AUG–3′ start codon, so extra amino acids will be added to the protein; in other mRNAs, it will occur after the normal start codon, resulting in a protein with fewer amino acids.

Chapter 16

11. **(a)** Inactive repressor; **(b)** active repressor.

14. RNA polymerase will bind the *lac* promoter poorly, significantly decreasing the transcription of the *lac* structural genes.

19.

Genotype of strain	Lactose absent		Lactose present	
	β-Galactosidase	Permease	β-Galactosidase	Permease
lacI⁺ lacP⁺ lacO⁺ lacZ⁺ lacY⁺	−	−	+	+
lacI⁻ lacP⁺ lacO⁺ lacZ⁺ lacY⁺	+	+	+	+
lacI⁺ lacP⁺ lacOᶜ lacZ⁺ lacY⁺	+	+	+	+
lacI⁻ lacP⁺ lacO⁺ lacZ⁺ lacY⁻	+	−	+	−
lacI⁻ lacP⁻ lacO⁺ lacZ⁺ lacY⁺	−	−	−	−
lacI⁺ lacP⁺ lacO⁺ lacZ⁻ lacY⁺ / lacI⁻ lacP⁺ lacO⁺ lacZ⁺ lacY⁻	−	−	+	+
lacI⁻ lacP⁺ lacOᶜ lacZ⁺ lacY⁺ / lacI⁺ lacP⁺ lacO⁺ lacZ⁻ lacY⁻	+	+	+	+
lacI⁻ lacP⁺ lacO⁺ lacZ⁺ lacY⁻ / lacI⁺ lacP⁻ lacO⁺ lacZ⁻ lacY⁺	−	−	+	−
lacI⁺ lacP⁻ lacOᶜ lacZ⁻ lacY⁺ / lacI⁻ lacP⁺ lacO⁺ lacZ⁺ lacY⁻	−	−	+	−
lacI⁺ lacP⁺ lacO⁺ lacZ⁺ lacY⁺ / lacI⁺ lacP⁺ lacO⁺ lacZ⁺ lacY⁺	−	−	+	+
lacIˢ lacP⁺ lacO⁺ lacZ⁺ lacY⁻ / lacI⁺ lacP⁺ lacO⁺ lacZ⁻ lacY⁺	−	−	−	−
lacIˢ lacP⁻ lacO⁺ lacZ⁻ lacY⁺ / lacI⁺ lacP⁺ lacO⁺ lacZ⁺ lacY⁺	−	−	−	−

21. The *lacI* gene encodes the *lac* repressor protein, which can diffuse within the cell and attach to any operator. It can therefore affect the expression of genes on the same molecule or on a different molecule of DNA. The *lacO* gene encodes the operator. It affects the binding of RNA polymerase to DNA and therefore affects the expression of genes only on the same molecule of DNA.

27. **(a)** No gene expression; **(b)** transcription of the structural genes only when alanine levels are low; **(c)** no transcription; **(d)** no transcription; **(e)** transcription will proceed; **(f)** transcription will proceed; **(g)** transcription will proceed.

30. To block transcription, you need to disrupt the action of RNA polymerase either directly or indirectly. Antisense RNA containing sequences complementary to the gene's promoter should inhibit the binding of RNA polymerase. If transcription initiation requires the assistance of an activator protein, then antisense RNA complementary to the activator protein–binding site of the gene could also disrupt transcription.

Chapter 17

18. It is likely that flowering will not occur. The protein encoded by *FLD* is a deacetylase enzyme that removes acetyl groups from histones surrounding the flowering locus C (*FLC*). Once the acetyl groups are removed, the chromatin structure within this region is restored, which inhibits transcription from the *FLC* locus. *FLC* encodes a transcriptional activator whose expression activates other genes that suppress flowering. If *FLC* transcription is active, then flowering will not occur.

19. DNA methylation is associated with transcriptional repression. If X31b is taken up by rapidly dividing cancer cells, it could stimulate methylation of DNA sequences, leading to transcriptional repression of genes. The repression of transcription could slow the growth of the cancer cells.

23. The action of an enhancer on the promoter of a gene is blocked when the insulator is located between the enhancer and the promoter. It is likely that transcription of genes A, B, and C will be stimulated by the enhancer because there is no insulator between these genes and the enhancer. Transcription of gene D will not be stimulated because the insulator is located between gene D and the enhancer.

24. The fruit flies will develop **(a)** male characteristics; **(b)** male characteristics; **(c)** both male and female characteristics.

26. The presence of AU-rich elements is associated with rapid degradation of mRNA molecules that contain them through RNA interference. If the AU-rich element were deleted, then miRNA would not be able to bind to it, and mRNA degradation would not be initiated. It is likely that this mRNA molecule would be more stable, resulting in increased expression of the protein encoded by the mRNA.

Chapter 18

18. **(a)** Leucine, serine, or phenylalanine; **(b)** isoleucine, tyrosine, leucine, valine, or cysteine; **(c)** phenylalanine, proline, serine, or leucine; **(d)** methionine, phenylalanine, valine, arginine, tryptophan, leucine, isoleucine, tyrosine, histidine, or glutamine, or a stop codon could result as well.

22. **(a)** A single-base-pair substitution resulting in a missense mutation.
(b) A single-base-pair substitution resulting in a nonsense mutation.
(c) The deletion of a single nucleotide resulting in a frameshift mutation.
(d) A six-base-pair deletion resulting in the elimination of two amino acids (Arg and Leu) from the protein.
(e) The insertion of three nucleotides resulting in the addition of a Leu codon.

23. Four of the six Arg codons could be mutated by a single-base-pair substitution to produce a Ser codon. All of these codons could undergo a second mutation at a different site to produce the amino acid Arg.

Original Arg codon	Ser codon	Restored Arg codon
CGU	AGU	AGG or AGA
CGC	AGC	AGG or AGA
AGA	AGU	CGU
	AGC	CGC
AGG	AGC	CGC
	AGU	CGU

25. No, hydroxylamine cannot reverse nonsense mutations. Hydroxylamine modifies cytosine-containing nucleotides and can result only in C • G → T • A transition mutations. In a stop codon, the C • G → T • A transition will result only in a different stop codon.

27. **(a)** The strand contains two purines, adenine and guanine. Because repair of depurination typically results in adenine being substituted for the missing purine, only the loss of the guanine by depurination will result in a mutant sequence.

$$5'\text{–AG–}3' \qquad \text{to} \qquad 5'\text{–AA–}3'$$
$$3'\text{–TC–}5' \qquad\qquad\quad 3'\text{–TT–}5'$$

(b) Deamination of guanine, cytosine, and adenine can occur. However, the deamination of only cytosine and adenine are likely to result in mutant sequences because the deamination products can form improper base pairs. Deaminated guanine does not pair with thymine but can still form two hydrogen bonds with cytosine, thus no change will occur.

$$5'\text{–AG–}3' \quad \text{if A is deaminated, then} \quad 5'\text{–GG–}3'$$
$$3'\text{–TC–}5' \qquad\qquad\qquad\qquad\qquad 3'\text{–CC–}5'$$

$$5'\text{–AG–}3' \quad \text{if C is deaminated, then} \quad 5'\text{–AA–}3'$$
$$3'\text{–TC–}5' \qquad\qquad\qquad\qquad\qquad 3'\text{–TT–}5'$$

29. PFI1 causes transitions, PFI2 causes transversions or large deletions, PFI3 causes transitions, and PFI4 causes single-base insertions or deletions.

33. The flanking repeat is in boldface type.

(a) 5′–ATTCGAAC**TGAC**[transposable element]**TGAC**CGATCA–3′
(b) 5′–ATT**CGAA**[transposable element]**CGAA**CTGACCGATCA–3′

36. The pairs of sequences in **(b)** and **(d)** are inverted repeats because they are both reversed and complementary and might be found at the ends of insertion sequences.

38. These results could be explained by hybrid dysgenesis, with strain B harboring *P* elements and strain A having no *P* elements.

42. The appearance of purple spots of varying sizes in these few yellow corn kernels could be explained by transposition. The yellow kernels may be due to inactivation of a pigment gene by insertion of a *Ds* element in the plant bearing this ear. Because the *Ds* element cannot transpose on its own, the mutant allele is stable in the absence of *Ac*, and the plant produces yellow kernels when fertilized by pollen from the same strain (lacking *Ac*). However, a few kernels may have been fertilized by pollen from a different strain with an active *Ac* element. The *Ac* element could then mobilize transposition of the *Ds* element out of the pigment gene, restoring the pigment gene's function. Excision of the *Ds* element earlier in kernel development would produce larger clones of cells producing purple pigment. Excision later in kernel development would produce smaller clones of purple cells.

44. The breeder can look for plants that have increased levels of mutation in either their germ-line or somatic tissues. Potentially mutant plants may have been exposed to standard mutagens that damage DNA. If they are defective in DNA repair, they should have higher rates of mutation.

Chapter 19

25. *Ara*I

27. 10

29. **(a)** 460,800; **(b)** 1,036,800; **(c)** 5,120,000.

30.

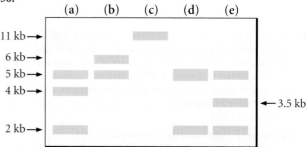

32. **(a)** Plasmid; **(b)** phage λ; **(c)** cosmid; **(d)** bacterial artificial chromosome.

36. A cDNA library, created from mRNA isolated from the venom gland. Bacteria cannot splice introns, so cDNA sequence that has been reverse transcribed from mRNA and therefore has no intron sequences is needed. The venom gland should be the source of the mRNA used for cDNA synthesis because that is where the toxin gene is likely to be transcribed and abundant mRNA for the toxin produced.

37. **(a)** Val-Tyr-Lys-Ala-Lys-Trp; **(b)** 128.

39. 5′–NGCATCAGTA–3′

42. **(a)** The figure shows that both larvae and adults whose parents were injected with dsRNA for *unc22A* express high levels of GFP protein. However, larvae and adults whose parents were injected with dsRNA for *gfp* do not express GFP. These results indicate that dsRNA specifically inhibits expression of the gene corresponding to the dsRNA, but not unrelated genes. **(b)** Injection of dsRNA corresponding to introns and promoter sequences would have little effect on *gfp* gene expression because RNA interference works by targeting mRNA. Introns and promoter sequences are not present in mRNA.

Chapter 20

25.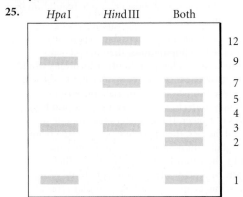

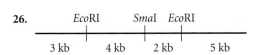

26.

	EcoRI		SmaI	EcoRI	
3 kb		4 kb		2 kb	5 kb

27.

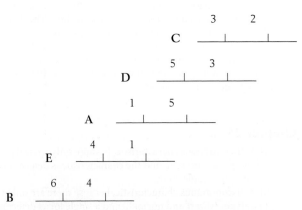

29. **(a)** Chromosome 22 has the highest density and greatest number of genes, whereas the Y chromosome has the lowest density and fewest genes. **(b)** The known genes found in this region (0–1,000,000 bp) are *PLCXD1*, *GTPBP6*, *PPP2R3B*, and *SHOX*.

33. Paralogs, because both evolved from duplication of a primordial β-globin gene.

34. Genes 2 and 24 are expressed at far higher levels in the antibiotic-resistant bacteria than in the nonresistant cells. Conversely, genes 4, 17, and 22 are downregulated. All of these genes may be involved in antibiotic resistance. The upregulated genes may be involved in metabolism of the antibiotic or may perform functions that are inhibited by the antibiotic. The downregulated genes may be involved in uptake of the antibiotic or represent a cellular mechanism that accentuates the potency of the antibiotic. Characterization of these genes might lead to information regarding the mechanism of antibiotic resistance, and thus to the design of new antibiotics that can circumvent this resistance mechanism.

38. **(a)** Genome size: 0.5 million to over 9 million base pairs in prokaryotes; 12 million (yeast) to hundreds of billions in eukaryotes. **(b)** Number of genes: From 500 to 8000 in prokaryotes; from 6000 to 32,000 in eukaryotes. **(c)** Gene density (bp/gene): Approximately 1000 bp/gene in prokaryotes; varies from 2000 bp/gene to greater than 100,000 bp/gene in eukaryotes. **(d)** Number of exons: With few exceptions, prokaryotic genes have zero or only one exon; multicellular eukaryotic genes typically have multiple exons.

43. **(a)** The minimal genome required might be determined by examining simple free-living organisms having small genomes to determine which genes they share. Mutations could then be systematically induced to determine which of the shared genes are essential for these organisms to survive. Genes could then be deleted one by one. Elimination of any of the essential genes would result in loss of viability. Alternatively, essential genes could be assembled through synthetic biology, creating entirely novel organisms with different sets of genes. The ability of these synthetic organisms to survive and reproduce could then be evaluated to determine which genes are essential. **(b)** Among the social and ethical concerns would be the ethics of creating entirely new organisms, and the question of whether such novel synthetic organisms could or would be used to develop pathogens for biological warfare or terrorism. There would also be uncertainty about the new organisms' effects on ecosystems if they were released or escaped.

Chapter 21

2. An epigenetic trait is one that is stable and that is passed on to descendant cells or offspring, but does not involve changes in the DNA base sequence. Many epigenetic traits are caused by change in gene expression resulting from modifications to chromatin. Although stable, many epigenetic traits can be influenced by environmental factors.

24. All female bees would develop characteristics of queens, regardless of whether they were fed royal jelly.

25. No. At CpG dinucleotides, two cytosine nucleotides sit diagonally across from each other on the two DNA strands, and both are methylated. The presence of 5-methylcytosine on both strands is required for maintaining methylation after replication. Following replication, one strand of each new DNA molecule is methylated and one is not. Special methyltransferase enzymes recognize the methyl group on one strand and methylate the cytosine on the other strand, perpetuating the methylated state of the DNA. Individual cytosine nucleotides will not have a nucleotide on the opposite strand that can be methylated following replication. Therefore, no new methyl groups will be added by the methyltransferases after replication.

27. We would expect to see differences in DNA methylation and histone acetylation that alter the expression of genes involved in responses to stress. We would also expect that as adults, the rats would show increased fear and heightened hormonal responses to stress.

28. Because only the original pregnant females were exposed to vinclozolin, the effects on the sperm of F_2–F_4 mice cannot be explained by direct effects of vinclozolin on male fertility. Because of the high frequency (90%) of the mice affected in F_2–F_4, it appears unlikely that the effects are due to mutations induced by vinclozolin. These transgenerational effects are most likely due to epigenetic changes. This conclusion is supported by the different DNA methylation patterns of the F_1–F_4 offspring of vinclozolin-exposed females. DNA methylation is known to affect chromatin structure and is responsible for some epigenetic effects.

30. (a) Both X chromosomes would be active.
    (b) Both X chromosomes would be active.

Chapter 22

14. (a) The products of *bicoid* and *dorsal* affect embryonic polarity by regulating the transcription of target genes.
    (b) The product of *nanos* regulates the translation of *hunchback* mRNA.

20. (a) Higher levels of maternal *bicoid* mRNA in the anterior cytoplasm of eggs. After fertilization, embryos would have higher levels of Bicoid protein and thus enlarged anterior and thoracic structures.
    (b) Lower levels of Bicoid protein and embryos with small head structures.

22. Egg-polarity (genetic maternal effect): *bicoid, nanos*. Gap: *hunchback, Krüppel*. Pair-rule: *even-skipped, fushi tarazu*. Segment-polarity: *gooseberry*. Homeotic: *labial, Antennapedia, Ultrabithorax, abdominal A*.

25. A plant that lacked class A and class B genes would express only class C genes in all four whorls, resulting in flowers with only carpels. A plant that lacked both class B and class C genes would express only class A genes in all four whorls, and result in flowers with only sepals.

27. Widespread expression of *shh* in the eye primordium of blind cavefish causes degeneration of lens cells. In these F_1 progeny of blind cavefish and surface fish, the expression of *shh* is intermediate between that in blind cavefish and that in surface fish, resulting in small eyes.

28. 2600

Chapter 23

17. Deletions can cause the loss of one or more tumor-suppressor genes. Inversions and translocations can inactivate tumor-suppressor genes if the chromosome breaks that accompany them lie within tumor-suppressor genes. A translocation can also place a proto-oncogene in a new location where it is activated by different regulatory sequences, causing the overexpression or unregulated expression of the proto-oncogene. Inversions and translocations can also bring parts of two different genes together, causing the synthesis of a novel fusion protein that is oncogenic.

20. After its integration into a host genome, a retrovirus promoter can drive the overexpression of a cellular proto-oncogene. Alternatively, the integration of a retrovirus can inactivate a tumor-suppressor gene. A few retroviruses carry oncogenes that are altered versions of host proto-oncogenes. Other viruses, such as human papillomavirus, express gene products (proteins or RNA molecules) that interfere with host cell cycle regulation by inactivating tumor-suppressor proteins.

23. If the differences in cancer rates are due to genetic differences in the two populations, then people who migrated from Utah or Shanghai to other locations would have rates of prostate cancer incidence similar to those of people who stayed in Utah or Shanghai. Moreover, different ethnic groups in Utah or Shanghai would have different rates of cancer. If the cancer rates are due to environmental factors, then people who migrated from Utah or Shanghai would have rates of prostate cancer determined by their new locations and not by their place of origin, and people from different ethnic groups living in the same location would have similar rates of cancer.

24. Because oncogenes promote cell proliferation, they act in a dominant manner. In contrast, mutations in tumor-suppressor genes cause loss of function and usually act in a recessive manner. When introduced into cells, the mutated *palladin* gene increases cell migration. Such a dominant effect suggests that *palladin* is an oncogene.

31. The deleted region contains a tumor-suppressor gene. Tumor suppressors act as inhibitors of cell proliferation. The deletion of tumor-suppressor genes therefore permits the uncontrolled cell proliferation that is characteristic of cancer. Oncogenes, on the other hand, function as stimulators of cell division. Deletion of oncogenes therefore prevents cell proliferation and usually cannot cause cancer.

Chapter 24

16. (a) A discontinuous characteristic, because only a few distinct phenotypes are present, and the characteristic is determined by alleles at a single locus.
    (b) A discontinuous characteristic, because there are only two phenotypes (dwarf and normal), and a single locus determines the characteristic.
    (c) A quantitative characteristic, because susceptibility is a continuous trait determined by multiple genes and environmental factors (an example of a quantitative phenotype with a threshold effect).
    (d) A quantitative characteristic, because it is determined by many loci (an example of a meristic characteristic).
    (e) A discontinuous characteristic, because only a few distinct phenotypes are determined by alleles at a single locus.

17. (a) All weigh 10 grams.
    (b) $\frac{1}{16}$ weigh 16 grams, $\frac{4}{16}$ weigh 13 grams, $\frac{6}{16}$ weigh 10 grams, $\frac{4}{16}$ weigh 7 grams, and $\frac{1}{16}$ weigh 4 grams.

19. That six or more loci take part.

22. The sum of the weights is 676; dividing by 10 students yields a mean of 67.6 kg.

    The variance is $s^2 = \dfrac{\sum (x_i - \bar{x})^2}{n - 1} = 2024.4/9 = 224.9$

    The standard deviation $= s = \sqrt{s^2} = 15$

24. The correlation coefficient r is calculated from the formula

    $r = \dfrac{\text{cov}_{xy}}{s_x s_y}$

    $\text{cov}_{xy} = 72$
    $s_x = 40.33$
    $s_y = 1.87$

    $r = \dfrac{72}{(40.33 \times 1.87)} = 0.95$

26. (a) 0.38; (b) 0.69.

31. (a) 0.75.
    (b) Its inaccuracy might be due to a difference between the environmental variance of the genetically identical population and that of the genetically diverse population.

34. The only reasonable conclusion is (d). Statement (a) is not justified because the heritability value applies not to absolute height, nor to an individual, but to the variance in height among Southwestern undergraduates. Statement (b) is not justified because heritability has been determined only for Southwestern University students; students at other universities, with different ethnic backgrounds and from different regions of the country, may have different heritabilities of height. Statement (c) is again not justified because heritability refers to the variance of height rather than to absolute height. Statement (e) is not justified because heritability has been determined for the range of variation in nongenetic factors experienced by the population under study; environmental variation outside this range (such as severe malnutrition) may have profound effects on height.

35. The regression coefficient and the narrow-sense heritability are each 0.8.

39. (a) Selection differential $= 6$ cm; response to selection $= 3.6$ cm.
    (b) The average wing length of the progeny should be 7.6 cm.

44. Head width will decrease. These two traits have high negative genetic correlation; therefore, selection for one trait will affect the other trait inversely.

Chapter 25

16. $f(T^ET^E) = 0.685$; $f(T^ET^F) = 0.286$; $f(T^FT^F) = 0.029$; $f(T^E) = 0.828$; $f(T^F) = 0.172$.

18. Add up all the X^O or X^+ alleles and divide by the total number of X^O and X^+ alleles.

The number of X^O alleles $= 2(X^OX^O) + (X^OX^+) + (X^OY) = 22 + 70 + 36 = 128$
The number of X^+ alleles $= 2(X^+X^+) + (X^OX^+) + (X^+Y) = 188 + 70 + 112 = 370$
$f(X^O) = 128/(128 + 370) = 128/498 = 0.26$
$f(X^+) = 370/498 = 0.74$

22. (a) $f(A) = 0.62$
       $f(G) = 0.38$
    (b) Expected genotypic frequencies:
       $f(AA) = 0.384$
       $f(AG) = 0.471$
       $f(GG) = 0.144$
    (c)

Genotype	Observed	Expected	$O - E$	$(O - E)^2$	$(O - E)^2/E$
AA	42	33	9	81	2.45
AG	24	41	17	289	7.05
GG	21	13	8	64	4.92

$\chi = \sum (O - E)^2/E = 14.42$

The degrees of freedom is $3 - 2 = 1$.

The p value is much less than 0.05; therefore, we reject the hypothesis that these genotypic frequencies may be expected under Hardy–Weinberg equilibrium.

25. The frequency of heterozygous carriers $= 2pq = 2(0.983)(0.017) = 0.033$; approximately 1 in 30 are carriers.

28. (a) 0.64 for M, 0.32 for MN, and 0.04 for N.
    (b) $f(L^ML^M) = 0.648$
       $f(L^ML^N) = 0.304$
       $f(L^NL^N) = 0.048$

30. The frequency at equilibrium is 0.98.

34. (a) On the island before the migration, $q^2 = 0.4$; $q = 0.63$. On the mainland, $q^2 = 0.81$; $q = 0.9$.
    (b) After the migration, $q_{new} = 0.68$.

35. The small population sizes and founder effects will cause strong genetic drift. The geneticists will find considerable variation between populations in allelic frequencies. Within populations, the same factors, coupled with inbreeding, will cause loss of genetic variation and a high degree of homozygosity.

37. (a) 0.032; (b) 0.145.

38. At equilibrium, the frequency of AR will be 0.58, and the frequency of ST will be 0.42.

41. (a) 0.0071; (b) 5×10^{-5}.

Chapter 26

20. The first illustration shows both anagenesis and cladogenesis occurring gradually over time. In the second illustration, little anagenesis (change within a lineage) occurs, and most evolution is associated with cladogenesis—that is, most evolution occurs quickly when one lineage splits into two.

21. Ecological (different host plants) and temporal (different times of mating).

26. **(a)** The nodes are all the points where lineages split.
(b) The branches are the horizontal lines connecting nodes; the thick blue line illustrates one example of a branch.
(c) The outgroup is the goshawk, represented by the bottom branch and node in the figure.

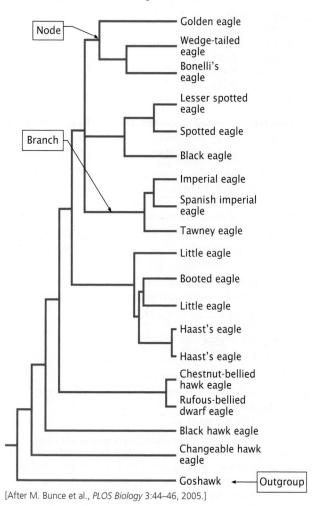

[After M. Bunce et al., *PLOS Biology* 3:44–46, 2005.]

29. Introns. Higher rates of substitution are typically observed in those gene regions that have the least function because natural selection limits variation in functional parts of genes. While the 3′ untranslated region of a gene does not encode amino acids, it does contain sequences that play a role in mRNA stability and translation. Within an intron, only sequences at the 5′ end, 3′ end, and branch point function in splicing.

Note: Page numbers followed by f indicate figures; those followed by t indicate tables. Page numbers preceded by A refer to the *Reference Guide to Model Genetic Organisms*, and those followed by B refer to *Working with Fractions: A Review*.

A site, in ribosome, 445–446, 445f, 450f
AbdB gene, 678
ABO blood group antigens, 115–116, 116f, 118–119
  Bombay phenotype and, 118–119
  nail-patella syndrome and, 198–199, 198f
  in organ transplantation, 685
Abortive initiation, in transcription, 383
Abzahanov, Arkaht, 679
Ac elements, 539–541, 540f
Acentric chromatids, 226
Acetylation, histone, 318, 494–495, 494f, 645–647, 651
  detection of, 656
Achondroplasia (dwarfism), 159t, 771
Acidic activation domains, 499
Acquired immunodeficiency syndrome, 274–275, 275f, 276f
Acquired traits, inheritance of, 51
Acridine orange, as mutagen, 531, 531f
Acrocentric chromosomes, 22, 22f, 218
Activator gene, 539–540
Adaptive mutations, 526
Addition, of fractions, B2
Addition rule, 57–58, 57f
Additive genetic variance, 729
Adenine, 297f, 298, 298t, 299f, 300.
    See also Base(s)
Adenosine monophosphate (AMP)
  in catabolite repression, 475
  in translation, 442, 442f
Adenosine triphosphate (ATP)
  oxidative phosphorylation and, 329
  in translation, 442, 442f
Adenosine-3′,5′ cyclic monophosphate (cAMP), in catabolite repression, 475
Adermatoglyphia, 145–146, 146f
Adjacent-1 segregation, 228f, 229
Adjacent-2 segregation, 228f, 229
A-DNA, 301, 301f
Adoption studies, 156–157
  of obesity, 156, 157f
Affinity capture, 631
African Replacement hypothesis, 329
African sleeping sickness, 412, 412f
Agar plates, 254, 254f
Age, maternal, aneuploidy and, 235, 235f
Aging
  apoptosis and, 676–677
  mitochondrial DNA and, 329
  premature, 359–360
  telomerase and, 359–360

telomere shortening and, 359–360
  in Werner syndrome, 359–360
agouti locus, coat color in mice and, 318, 319f
Agriculture. *See also* Breeding; Crop plants
  genetic engineering in, 574f, 575–576, 575f, 595–596
  genetics in, 3, 3f, 7–8, 8f, 9
    quantitative, 715–716
  recombinant DNA technology in, 217–218
Agrobacterium tumefaciens, in cloning, 574, 574f
AIDS, 274–275, 275f, 276f
Alanine, 434f
Albinism. *See also* Color/pigmentation
  gene interaction and, 120–121
  in Hopi, 1–2, 2f
  inheritance of, 58–59
  in snails, 120–121, 121f
Alkylating agents, as mutagens, 530
Alleles, 12, 21. *See also* Gene(s)
  in crossing-over, 29, 30f
  definition of, 50, 50t
  dominant, 53, 53f, 110–113
  lethal, 114
  letter notation for, 52, 60–61
  molecular nature of, 54
  multiple, 114–116, 116f
  recessive, 53
  segregation of, 55f, 62f, 65–67, 174–175
  temperature-sensitive, 133–134, 133f, 134f
  wild-type, 60
Allelic fixation, 766
Allelic frequency
  calculation of, 752–753
  at equilibrium, 754, 761, 761f
  estimation of, 757–758
  fixation and, 766
  genetic drift and, 763–766, 766f, 771–772, 771t
  Hardy-Weinberg law and, 757
  migration and, 762–763, 763f, 771–772, 771t
  for multiple alleles, 752
  mutations and, 760–762, 761f, 771–772, 771t
  natural selection and, 766–772, 767t, 768t, 770t, 771t, 772f
  nonrandom mating and, 758–760
  rate of change for, 770–771, 772f
  for X-linked alleles, 752–753
Allelic series, 114
Allen Brain Atlas, 622
Allolactose, 469, 470f
Allopatric speciation, 786–787, 786f
Allopolyploidy, 238–239, 241t
  speciation and, 790
Allosteric proteins, 466
Alpha chains, of T-cell receptor, 684, 684f
Alpha (α) helix, 300–301, 300f, 433, 435f
α-Amanitin, 373–374
α-fetoprotein, 160

Alternate segregation, 228f, 229
Alternation of generations, 38, 38f
Alternative processing pathways, 409–411, 411f
Alternative splicing, 409–411, 411f
  in gene regulation, 502–503
Alu sequences, 541
Alzheimer disease, 651
Amborella trichopoda, 331
Ames, Bruce, 533
Ames test, 532–533, 533f
Amino acids, 433
  assembly into proteins, 441–449.
    See also Translation
  definition of, 433
  in domains, 464
  in genetic code, 435–440, 438f
  peptide bonds of, 433, 435f, 445
  proportions of, 437
  sequence of, 435, 435f
  structure of, 433, 434f
  triplet code and, 436
  types of, 434f
Aminoacyl (A) site, in ribosome, 444f, 445–446, 450f
Aminoacyl-tRNA, synthesis of, 441, 442f
Aminoacyl-tRNA synthetases, 441, 442f
Amniocentesis, 158–159, 160f
AMP (adenosine monophosphate)
  cyclic, in catabolite repression, 475
  in translation, 442, 442f
Amphidiploidy, 238–239
Amyotrophic lateral sclerosis, 515–516
Anagenesis, 781
Anaphase
  in meiosis, 29, 30f, 31f, 32t
  in mitosis, 25f, 26, 26t, 32t
Androgen insensitivity syndrome, 89
Anemia
  Fanconi, 549t
  sickle-cell, 159t, 770, 783–784
Aneuploidy, 219–220, 219f, 230–235, 241t
  autosomal, 232
  cancer and, 235
  causes of, 230, 231f
  definition of, 230
  in Down syndrome, 232–235, 233f
  in Edward syndrome, 235
  effects of, 230–232
  in evolution, 241
  in humans, 232–235
  maternal age and, 235, 235f
  mosaicism and, 236, 236f
  in Patau syndrome, 235
  of sex chromosomes, 232
    in Klinefelter syndrome, 88, 88f, 99
    in Turner syndrome, 87–88, 87f, 99
  in trisomy 8, 235
  types of, 230
  uniparental disomy and, 235–236

Angelman syndrome, 131–132
Angiogenesis, in cancer, 703
Angiosperms. *See also Arabidopsis thaliana* (mustard plant); Plant(s)
 flower color in
  inheritance of, 110–112, 111f
  lethal alleles and, 114
 flower development in, 674–676, 674f, 675f
 flower length in, 727–728, 728f
 mitochondrial DNA in, 328
Animals
 breeding of, 8, 595
  artificial selection and, 737, 737f
  genetic correlations in, 739–740, 739t
  inbreeding in, 760
 cloning of, 635f, 665–666
 coat color in. *See* Coat color
 sexual reproduction in, 36–37, 37f
 transgenic, 590–593, 591f, 592f, 595
Aniridia gene, 678
Annotated genes, 616
Anopheles gambiae (mosquito), genome of, 626t
Antennapedia complex, 671f, 672
Antibiotics
 resistance to, gene transfer and, 264
 RNA polymerase and, 380
 translation and, 452
Antibodies (immunoglobulins), 680, 680f, 681f, 682–684
 diversity of, 682–684
 structure of, 680f, 682
Antibody diversity, 682–684, 683f, 684f
Anticipation, 133, 519–520
Anticodons, 415, 415f
Antigen(s), 680, 680f, 682f
 blood group, 115–116, 116f
  ABO, 115–116, 116f, 118–119, 198–199, 685
   Bombay phenotype and, 118–119
  MN, 112
  nail-patella syndrome and, 198–199, 198f
  in transplantation, 685
 major histocompatibility complex, 681
  in transplantation, 685
 T, alternative splicing in, 502
Antigenic drift, 276
Antigenic shift, 276
Antiparallel DNA strands, 299f
Antisense RNA, in gene regulation, 418, 482, 482f
Antiterminators, 478
Apaf-1 gene, in melanoma, 705
APC gene, in colon cancer, 706, 708
Aphids, carotenoid synthesis in, 799
Apis mellifera, DNA methylation in, 645, 645f, 646f
ApoB protein, in hypercholesterolemia, 594
Apolipoprotein-B100, 412
Apoptosis
 in cancer, 700–701
 in development, 676–677, 677f
Apple maggot flies, evolution of, 789–790, 789f
Arabidopsis thaliana (mustard plant)
 flower development in, 674–676, 674f, 675f

gene regulation in, 494–495
 genome of, 626, 626t, 627t
 as model genetic organism, 6, 6f, A8–A9
 translation in, inhibition of, 505
Archaea, 19, 253. *See also* Prokaryotes
 genome of, 625, 625t
 histones of, 316
 replication in, 360
 transcription in, 390–391
 translation in, 449
Arginine, 434f
 synthesis of, 431–432
Artificial chromosomes. *See also* Bacterial artificial chromosomes (BACs); Yeast artificial chromosomes (YACs)
 creation of, 605–606
 in genome sequencing, 610–611
 as vectors, 573, 573t
Artificial selection, 737, 737f. *See also* Breeding
Asexual reproduction, polyploidy and, 240
Asparagine, 434f
Aspartate, 434f
Asplenia, 429
Assembly factors, 418
Assortative mating, 758–760
 in sympatric speciation, 789
Astbury, William, 294
Asthma, twin studies of, 156
Atomic bomb, radiation exposure from, 534, 534f
ATP (adenosine triphosphate)
 oxidative phosphorylation and, 329
 in translation, 442, 442f
Attenuation, 477–481, 480f, 481t
 in *trp* operon, 477–481
Attenuator, 477–478, 480f
Auerbach, Charlotte, 529
AU-rich elements, 504, 506
Autoimmune diseases, 680, 680t
Autonomously replicating sequences, 354
Autopolyploidy, 237–238, 237f, 238f, 240, 241t
Autosomal aneuploidy, 232
Autosomal traits
 dominant, 149–150, 153t
 recessive, 148–149, 148f, 153t
Autosomes, 84. *See also* Sex chromosomes
Auxotrophs, 254, 254f, 430–431
Avery, Oswald, 289f, 291
Avian influenza, 276, 277f
Ayala, Francisco, 791

B cells, 680, 680f–682f
Bacillus subtilis, transcription in, 483–484, 483f
Bacillus thuringiensis, 575–576, 575f
 in agriculture, 595
Backcross, 56
Backtracking, by RNA polymerase, 384
Bacteria, 251–267. *See also Escherichia coli* (bacterium)
 antibiotic resistance in, 264
 auxotrophic, 254, 254f, 430–431
 cell reproduction in, 20, 21f. *See also* Replication

cell structure in, 18f, 19
 chloroplast origin from, 323, 323f
 conjugation in, 257, 257f–264f, 258–264, 262t, 417. *See also* Conjugation
 CRISPR-Cas systems in, 267, 561
 culture of, 254–255, 254f
 defense mechanisms of, 267
 diversity of, 253
 DNA in, 19, 19f, 312f, 314, 314f
 endosymbiotic theory and, 323, 323f
 essential functions of, 252
 eubacteria, 19, 253. *See also* Eubacteria
 gene mapping in, 262–264, 263f, 265–266, 266f, 269–273, 271f
 gene regulation in, 461–490
  vs. in eukaryotes, 482, 492–493, 508, 508t
 gene transfer in, 257–267
  by conjugation, 257, 257f–264f, 258–264
  horizontal, 266–267, 625, 799
  by transduction, 257, 257f, 269–271, 269f–271f
  by transformation, 257, 257f, 265–266, 265f, 266f
 genes of, 19, 19f, 255–256, 255f, 256f, 266
  number of, 624–625, 625t
 in genetic studies, 252t
 genetically modified, industrial uses of, 595
 gut, obesity and, 617
 insertion sequences in, 538, 538f
 mitochondrial origin from, 323, 323f
 mutant strains of, 254, 254f
 overview of, 252–253
 phages and, 267–273. *See also* Bacteriophage(s)
 population of, 253
 prokaryotic, 12, 18–20. *See also* Prokaryotes
 prototrophic, 254
 recombination in, 362–363
 replication in, 341–344, 343f, 344f, 345t, 348–354
 transcription in, 381–386, 390–391
 transduction in, 257f, 258, 269–271, 269f–271f. *See also* Transduction
 transformation in, 257–258, 257f, 265–266, 265f, 266f
 translation in, 441–449
 transposable elements in, 538–539, 538f
 viral invasion of, 267
 vs. eukaryotes, 253
 wild-type, 254
Bacterial artificial chromosomes (BACs)
 in genome sequencing, 610–611
 reporter sequences and, 622
 as vectors, 573, 573t
Bacterial chromosomes, 19, 255, 255f, 256f, 312f
 DNA packaging in, 312–314, 314f
Bacterial colonies, 254, 254f
Bacterial cultures, 254–255, 254f
Bacterial enhancers, 482
Bacterial genome, 18f, 19f, 255, 255f, 256f, 266, 624–625
 sequencing of, 625t
Bacterial plasmids, 255–256, 255f, 256f. *See also* Plasmid(s)

Bacteriophage(s), 267–273
  in bacterial gene mapping, 271f, 272–273
  culture of, 268–269, 269f
  DNA in, 292–293, 293f, 294f
  experimental advantages of, 267
  experimental techniques with, 268–269
  gene mapping in, 272–273
  in Hershey-Chase experiment, 292–293, 293f, 294f
  lambda, 573, 573t
  life cycle of, 267–268, 268f
  prophage, 268, 268f
  reproduction of, 292–293, 293f, 294f
  T2, 292–293, 293f, 294f
  temperate, 268, 268f
  transducing, 269–271
  transposition in, 538–539
  as vectors, 573, 573t
  virulent, 268, 268f
Balancing selection, 783–784
Baldness, male-pattern baldness, 173–174
Bananas, genome sequencing for, 217–218
Banding, chromosome, 219
Bar mutations, 221, 221f, 536–537
Barr bodies, 98, 98f, 99t
Barry, Joan, 115–116
Basal transcription apparatus, 379–381, 380f, 380t, 382f, 387, 388f, 488f, 497–499, 498f, 499f, 500
  mediator in, 498
Base(s), 290, 290t, 297, 297f, 298t, 299f, 300f.
    See also Nucleotides and specific bases
  Chargaff's rules for, 290, 290t
  methylation of, 303, 497. See also DNA methylation
  modified, 414–415, 415f
  purine, 297, 297f, 298t, 299f
  pyrimidine, 297, 297f, 298t, 299f
  ratios of, 290, 290t
  in RNA, 374–375, 375t
  sequence of, 301
  in Watson-Crick model, 294
Base analogs, as mutagens, 529–530, 530f
Base pairing
  in codons, 438–439, 438f, 439t
  deamination and, 528, 529f
  depurination and, 528, 529f
  in double helix, 299–302, 299f
  internal, 302
  mutations and, 527, 527f
    Ames test and, 532–533, 533f
  nonstandard, 527, 527f
  tautomeric shifts and, 527
  in transcription, 383–384
  in translation of genetic code, 438–439, 438f, 439t
  in trp operon, 477
  wobble in, 438f, 439, 439t, 527, 527f
Base substitution, 518, 520, 521f, 523f, 524t, 529f, 530f
Base-excision repair, 545–546, 548t
Basic Local Alignment Search Tool (BLAST), 616, 618
Bateson, William, 175, 674
Baur, Erwin, 114
Bayesian method, 794

B-DNA, 300, 300f, 301f
Beadle, George, 430–433
Beadle-Tatum one gene, one enzyme hypothesis, 430–433
Beads-on-a-string chromatin, 316, 316f
Beckwith-Wiedemann syndrome, 656
Bees, DNA methylation in, 645, 645f, 646f
Behavioral epigenetics, 650–651
Behavioral isolation, 785, 785t
Belling, John, 231
Beta chains, of T-cell receptor, 684, 684f
Beta (β) pleated sheet, 433, 435f
Beta (β) sliding clamp, 351
bicoid gene, 668–669, 669f, 669t
Bidirectional replication, 344
Bighorn sheep, 781–782, 781f
Binary fission, 19f, 20, 21f
Binomial expansion, 59
Biodiversity
  DNA analysis and, 5
  genetic engineering and, 595–596
Bioinformatics, 615–616
Biological species concept, 784
Biology, genetics in, 4
Biosphere, DNA in, 5
Biotechnology, 3–4, 561.
    See also Recombinant DNA technology
Bird flu, 276, 277f
Birds, telomeres in, 359
Bishop, Michael, 697
Bisulfite sequencing, for DNA methylation, 656, 657f
bithorax complex, 672, 672f
Bivalent, 29
Blackburn, Elizabeth, 320
Blakeslee, A. Francis, 230–231
BLAST searches, 616, 618
Blending inheritance, 8–9, 10t
Blind men's riddle, 17–18
Blond hair, inheritance of, 47–48
Blood group antigens
  ABO, 115–116, 116f, 118–119
    Bombay phenotype and, 118–119
    nail-patella syndrome and, 198–199, 198f
    in transplantation, 685
  MN, 112
Blood tests, maternal, 160–161
Blood transfusions, ABO antigens and, 115–116, 116f
Body weight, adoption studies of, 156, 157f
Bombay phenotype, 118–119
Bonds
  hydrogen, in DNA, 299–300, 299f
  peptide, 433, 435f, 444f, 445
  phosphodiester, 298, 299f
Boundary elements, 500, 500f
Boveri, Theodor, 54
Boy Scout bacterium, genome of, 624–625
Boycott, Arthur, 109–110
Bradyrhizobium japonicum, genome of, 624, 625t
Branch diagrams, 64–65, 64f, 65f
Branch point, 407
Branches, of phylogenetic trees, 792, 792f
BRCA gene, 698, 698t

Bread mold (Neurospora crassa), 6, 430–433, 430f
Breast cancer, 704
  BRCA gene in, 698, 698t
  microarrays and, 619–620
  miRNA in, 619–620, 620f
Breeding
  of animals, 8, 595
  artificial selection and, 737, 737f
  genetic correlations in, 739–740, 739t
  inbreeding and, 758–760, 759f, 760f
  of plants, 3, 3f, 8, 8f, 238–239, 239f, 760
  quantitative genetics in, 715–716
Brenner, Sydney, 403
Bridges, Calvin, 87, 90–91, 183, 230
Briggs, Robert, 665
Brink, Alexander, 647
Broad-sense heritability, 730
5-Bromouracil, as mutagen, 529, 530f
Brown, Robert, 9
Bt toxin, 575–576, 575f
  in agriculture, 595
Bulbar muscular atrophy, 519t
Buri, Peter, 764
Burkitt lymphoma, 707, 707f

C banding, 219, 219f
C segment, 682–684, 682f, 683f
C9orf72 gene, 515–516
cactus gene, 668, 669t
Caenorhabditis elegans (nematode)
  genome of, 626, 626t, 627t
  as model genetic organism, 6, 6f, A6–A7
  RNA interference in, 418
Cairns, John, 344
Calcitonin, 410
Calcium signaling, 679
Calmodulin, 679
cAMP, in catabolite repression, 475
Camptotheca acuminata (happy tree), 339
Camptothecin, 339–340
Cancer
  abnormal cell growth in, 677, 700–701
  Ames test and, 532–533, 533f
  aneuploidy in, 235
  angiogenesis in, 703
  apoptosis in, 700–701
  breast, 619–620, 620f, 698, 698t, 704
  Burkitt lymphoma, 707, 707f
  cervical, 708, 708t, 709–710
  chemotherapy for, 339–340
  chromosome abnormalities in, 698, 706–708, 707f
  clonal evolution of, 694–695, 695f
  colorectal, 549, 549t, 702, 705–706, 706f
  DNA methylation in, 303, 705
  DNA repair in, 702
  environmental factors in, 695–696, 696t
  epigenetic processes in, 705
  faulty DNA repair in, 549, 549t
  as genetic disease, 693–695
  genomic instability in, 708
  haploinsufficiency in, 698
  incidence of, 693t
  Knudson's multistep model of, 693–694, 694f
  loss of heterozygosity in, 698, 698f

metastasis in, 693, 703
microarrays and, 619–620
microRNA and, 506, 619–620, 620f, 621f, 703–704
mutations in, 532–534, 533f, 693–695, 702, 704, 706–708, 707f
oncogenes in, 697, 697f, 697t
pancreatic, 691–692, 692f
replication in, 339–340
retroviruses and, 273
signaling in, 701
skin, in xeroderma pigmentosum, 695, 702
spread of, 691–692
stimulatory genes in, 696
telomerase in, 360, 702–703
topoisomerase and, 339–340
tumor-suppressor genes in, 697–698, 697f, 698t
viruses and, 708–710, 708t, 709f
in Von Hippel-Lindau disease, 703
The Cancer Genome Atlas, 704
Cap-binding complex, 444, 444f
Capecchi, Mario, 592
Capsicum annuum (pepper), fruit color in, 117–118, 117f
Carcinogens
  Ames test for, 532–533, 533f
  environmental, 695–696, 696t
Cardiac disease, microRNA in, 506
Carotenoids, in aphids, 799
Carriers, translocation, 234, 234f
Caspases, 676–677
Catabolite activator protein, 475, 476f, 477f
Catabolite repression, 475, 476f
Cats, coat color in, 99, 99f
Cavalli-Sforza, Luca, 764
Cavefish, eye development in, 679
Cavenne, Webster, 698
cDNA libraries
  creation of, 576–577
  screening of, 577–578
Cech, Thomas, 374
Cell(s)
  competent, 265
  diploid, 21, 22f
  eukaryotic, 12, 18f, 19, 19f. *See also* Eukaryotes
  genetic material in, 18f, 19, 19f
  haploid, 21
  information pathways in, 301–302, 302f
  prokaryotic, 12, 18f, 19. *See also* Prokaryotes
  reproduction of, 20–23
  structure of, 18f, 19
  totipotent, 664
Cell cycle. *See also* Meiosis; Mitosis
  centromeres in, 319
  checkpoints in, 23, 23f, 699–700, 700f
  chromosome movement in, 26
  chromosome number in, 27–28, 27f
  definition of, 23
  DNA molecules in, counting of, 27–28, 27f
  DNA synthesis in, 24, 26, 27–28
  G_0 phase of, 23f, 24, 26t
  G_1 phase of, 23f, 24, 26t
  G_2 phase of, 23f, 24, 26t

genetic consequences of, 26–28
interphase in, 23f, 24, 25f, 26t
M phase of, 23, 23f, 24–26, 25f, 26t
molecular motors in, 26
overview of, 23–26, 23f, 26t
regulation of, 698–701
  in cancer, 700–701
  mutations in, 700–701
  signal transduction in, 701
replication in, 23f, 24, 26t, 354–355. *See also* Replication
S phase of, 23f, 24, 26, 26t
stages of, 23–26, 23f, 25f, 26t
Cell differentiation, epigenetics and, 654–655
Cell division
  cytokinesis, 23, 23f
  in eukaryotes, 20–26. *See also* Cell cycle
  in meiosis, 28–40
  in mitosis, 23–26, 23f, 26t, 35, 35f
  in prokaryotes, 19f, 20
Cell growth, in cancer, 694–695, 700–701
Cell lines, 201
Cell signaling
  calcium in, 679
  in cancer, 701
  in cell cycle regulation, 701
Cell theory, 10, 10t
Cellular immunity, 680–681, 680f
CentiMorgan (cM), 187
Central bearded dragon, sex determination in, 81–82, 85
Central dogma, 301
Centrifugation, equilibrium density gradient, 341–342, 341f
Centriole, 24
Centromeres, 22, 22f, 319
  in chromosome movement, 319, 319f
  definition of, 319
  structure of, 319f
Centrosomes, 24, 25f
Cerebellar ataxia, 519t
Cervical cancer, 708, 708t, 709–710
CFTR gene, 112–113, 116, 581
Chaperones
  histone, 357
  molecular, 452
Chaplin, Charlie, 115–116
Characteristics
  definition of, 50t
  vs. traits, 51. *See also* Traits
Chargaff, Erwin, 290, 293
Chargaff's rules, 290, 290t
Chase, Martha, 292–293
Checkpoints, in cell cycle, 23, 23f, 698–700, 700f
Chemicals, environmental, epigenetic effects of, 651
Chemotherapy, cancer, 339
Cheng, Keith, 6
Chernobyl nuclear accident, 534
Chiasma, 29
Chickens, feathering patterns in, 127, 128f
Childhood adversity
  epigenetic effects of, 650–651
  telomeres and, 311–312
Chimeras, 154. *See also* Genetic mosaicism

Chimeric mice, 529
Chimpanzees, genome in, vs. human genome, 491–492
ChIP (chromatin immunoprecipitation), 495–497, 657
ChIP-Seq, 656
χ^2 (chi-square) distribution, critical values for, 68t
Chi-square test
  for crosses, 67–69, 68t, 69f,
  goodness-of-fit, 67–69,
  for Hardy-Weinberg proportions, 756–757
  of independence, 185–186, 186f
Chloroplast(s)
  endosymbiotic theory and, 323, 323f
  structure of, 322–323, 322f
Chloroplast DNA (cpDNA), 129, 322–323
  evolution of, 330
  gene structure and organization in, 330–331
  inheritance of, 324
  mutations in, 129–130, 324–325
Chloroplast genome, 330–331, 330t, 331f
Chloroplast-encoded traits, inheritance of, 325
Cholesterol
  elevated serum, 594
  in familial hypercholesterolemia, 149–150, 150f
  epigenetics and, 652
chordin gene, 678
Chorionic villus sampling, 159–160, 160f
Chromatids
  acentric, 226
  dicentric, 226
  nonsister, in crossing over, 30f, 31–34
  sister, 22f, 23
    counting of, 27–28
    separation of, 25f, 26, 26t, 32, 34–35, 35
Chromatin, 19, 19f, 314–315, 316f, 317f
  epigenetic changes in, 318–319, 497, 643–644, 644f. *See also* DNA methylation
  nucleosome in, 315–317, 316f
  proteins in, 314–315, 315f
  structure of, 19, 19f, 314–315, 315f, 316f, 317
    changes in, 317–319, 386, 493–497
    DNAse I and, 317–318, 318f, 493
    gene expression and, 493–497
    levels of, 315f, 317
    transcriptional modification of, 317–318, 386
    epigenetic effects of, 318–319
  types of, 314–315
Chromatin immunoprecipitation (ChIP), 495–497, 656
Chromatin-assembly factor 1 (CAF-1), 357
Chromatin-remodeling complexes, 493–494
Chromatin-remodeling proteins, 386
Chromosomal disorders. *See also* Genetic diseases
  prenatal diagnosis of, 159t
Chromosomal proteins, nonhistone, 315
Chromosomal puffs, 317, 317f

Chromosome(s)
  acrocentric, 22, 22f, 218
  bacterial, 19, 19f, 20, 255, 255f, 256f, 312f, 314, 314f
    DNA packaging in, 312, 314, 314f
  bacterial artificial
    in genome sequencing, 610–611
    reporter sequences and, 622
    as vectors, 573, 573t
  chromatin in. See Chromatin
  condensation of, 29
  counting of, 27–28, 27f
  coupling configurations for, 181–182, 181f
  crossing over of. See Crossing over
  daughter
    bacterial, 253
    formation of, 24, 25f, 253
  in diploid organisms, 21, 22f
  DNA packaging in, 21–22
    in bacterial chromosome, 312, 314, 314f
    in eukaryotic chromosome, 312–313, 312f, 314–318, 315f–317f
  ends of, 22–23, 22f, 320
    replication of, 358–360, 358f
  eukaryotic, 19, 19f, 20–23, 322
    DNA packaging in, 312–313, 312f, 314–318, 315f–317f
  fragile sites on, 229, 229f
  gene density in, 629
  gene location on, 13, 13f, 183.
      See also Gene loci
  in haploid organisms, 21
  homologous pairs of, 21, 22f, 35
    random separation of, 33–34
  karyotypes and. See Karyotypes
  mapping of. See Gene mapping
  in meiosis, 28–40, 53–54, 53f
  metacentric, 22, 22f, 218
  in mitosis, 24–26, 25f, 26t
  morphology of, 218–219, 218f, 219f
  movement of, 25f, 26
    centromeres in, 319, 319f
  nondisjunction of. See Nondisjunction
  number of, 21, 22f
    abnormal, 230–241. See also Aneuploidy; Polyploidy
  origins of replication of, 23
  Philadelphia, 707
  polytene, 317, 317f
  prokaryotic, 19, 19f, 20
  proteins in, 314–318. See also Chromatin; Protein(s)
    histone, 314–315, 315t
    nonhistone, 315
  random distribution of, 33–34, 33f
  replication of, 22f, 23, 25f. See also Cell cycle
  in repulsion, 181–182, 181f
  segregation of. See Segregation
  sets of, 21, 22f
  sex. See Sex chromosomes
  shortening of, 311–312
  staining of, 218–219, 219f
  structure of, 21–23, 22f, 218–219, 218f, 219f, 314–318
  submetacentric, 22, 22f, 218
  synapsis of, 29

  telocentric, 22, 22f, 218
  telomeric sequences in, 22–23, 22f, 320, 320t
  Y, in sex determination, 83–85, 84f, 87–89
  yeast artificial
    in genome sequencing, 610–611
    as vectors, 574
Chromosome banding, 219, 219f
Chromosome jumping, 579–580
Chromosome maps. See Gene mapping
Chromosome mutations. See Mutations, chromosome
Chromosome rearrangements, 220–230, 223t, 241t. See also Mutations
  in cancer, 706–708, 707f
  copy-number variations, 154, 229–230, 615
  deletions, 222–224, 223f, 223t, 224f, 241t, 518–519, 518f, 524t, 527–528, 528f
  duplications, 219f, 220–222, 223t, 241t.
      See also Duplications
  in evolution, 241
  inversions, 224–227, 224f–226f, 241t
  phenotypic consequences of, 220–222, 222f, 223t
  translocations, 227–229, 227f, 228f, 241t.
      See also Translocation
  transposition and, 536–537
Chromosome theory of heredity, 54, 83–84, 174–175, 183
  nondisjunction and, 90–92, 92f
Chromosome walking, 579–580, 579f
Chronic myelogenous leukemia, 707, 707f
Chukchi, DNA of, 287–288
Circular chromosome, bacterial, 255
Circular DNA
  bacterial, 255, 255f
  replication in, 357–358
Cis configuration, 181–182, 181f
Cladogenesis, 781
Clark, William, 781
Classical genetics, 5, 5f
Cleavage, in RNA processing, 362, 405t, 406, 406f, 410, 505
Cleft lip and palate, 159t
Climate change, population genetics and, 750
Clonal evolution, 694–695, 695f
Clonal selection, 681, 681f
Cloning, 569, 571–575
  of animals, 635f, 665–666
  disadvantages of, 569
  of DNA library, 578
  gene, 571–575
  linkers in, 572
  of plants, 665
  positional, 579–580, 581f
  restriction sites in, 571, 571f
  selectable markers in, 571, 571f
  transformation in, 572
  vs. polymerase chain reaction, 569
Cloning vectors, 517–572
  bacterial artificial chromosome, 573
  bacteriophage, 573
  cosmid, 573, 573t
  for eukaryotes, 573–574
  expression, 573, 574f
  plasmid, 571–572, 571f, 572f, 573t
  selectable markers for, 571

  yeast artificial chromosome, 574
Cloverleaf structure, 415, 415f
c-MYC gene, in Burkitt lymphoma, 707
Coat color
  in cats, 99f
  in dogs, 118–119, 124–125, 125f
  epigenetic changes in, 318–319, 319f
  gene interaction and, 118–119, 124–125, 125f
  lethal alleles and, 114, 114f
  in mice, 114, 318–319, 319f
  in rabbits, 134, 134f
Cock feathering, 127, 128f
Cockayne syndrome, 549t
CODIS system, 586, 588–589, 588t
Codominance, 112, 112t
Codons, 404, 404f, 414f
  base composition of, 437
  base pairing in, 438–439, 438f, 439t
  base sequence in, 437
  in genetic code, 435–440, 438f, 439t, 440t
  nonsense, 440
  reading frames and, 439–440
  sense, 438
  start (initiation), 404
  stop (termination), 404, 414f, 440, 447f
  synonymous, 438
Coe, Ed, Jr., 648
Coefficient of coincidence, 193–195
Cognition, epigenetic effects in, 651
Cohesin, 34–35, 36f
Cohesive (sticky) ends, 562, 564f
Colds, rhinoviruses and, 276–277
Colinearity
  of genes and proteins, 400–401, 401f, 628
  genomic, 628
Collins, Francis, 612f
Colon cancer, 702, 705–706, 706f
  faulty DNA repair in, 549, 549t
Colonies, bacterial, 254, 254f
Color blindness, 92–94, 93f, 221, 221f
Color/pigmentation
  body, in D. melanogaster, 797, 797f
  coat. See Coat color
  eye
    in D. melanogaster, 89–91, 91f, 92f, 124, 190–195, 190f, 192f, 194f, 764, 765f
    as X-linked characteristic, 89–91, 91f, 92f
  feather, 94, 95f, 115, 115f
  flower
    inheritance of, 110–112, 111f
    lethal alleles and, 114
  fruit
    epistasis and, 119–120, 120f
    gene interaction for, 117–118, 119–120, 120f
    incomplete dominance and, 111–112, 111f
  gene interaction and, 117–125, 120f, 121f
  hair, 47–48
  inheritance of, 47–48, 58–59, 719–721, 720f
  leaf, cytoplasmic inheritance and, 129–130, 130f
  snail, 120–121, 121f
  temperature-dependent, 134, 134f
  wheat kernel, 719–721, 720f

Combined DNA Index System (CODIS), 586, 588–589, 588t

Common cold, rhinoviruses and, 276–277

Common slipper limpet, sex determination in, 85–86, 86f

Comparative genomics, 624–629
 eukaryotic genomes and, 625–629
 prokaryotic genomes and, 624–625

Competence, cellular, 265

Complementary DNA strands, 300

Complementation, 124

Complementation tests, 124

Complete dominance, 111, 111f, 112t

Complete linkage, 178f, 182

Complete medium, 254

Composite transposons, 538

Compound heterozygotes, 116

Concept of dominance, 53f, 54.
 See also Dominance

Concordance, in twin studies, 155, 155t

Condensation, chromosomal, 29

Conditional mutations, 521

Conditional probability, 58

Congenital asplenia, 429

Conjugation, 257, 257f–264f, 258–264, 262t, 471
 in Davis's U-tube experiment, 259, 259f
 definition of, 257
 F′ cells in, 261, 262t
 F⁺ cells in, 259–262, 260f, 261f, 262t
 F⁻ cells in, 259–262, 260f–262f, 262t
 F factor in, 256, 256f, 259–262, 260f–262f, 262t
 Hfr cells in, 260–262, 261f, 262f, 262t
 interrupted, in gene mapping, 262–264
 in Lederberg-Tatum experiment, 258–259

Consanguinity, 148–149, 148f

Consensus sequences
 in bacteria, 381–382
 in eukaryotes, 387, 389, 414f, 498, 498f
 poly(A). *See* Poly(A) tail
 Shine-Dalgarno, 404, 443, 443f, 448.
  See also Shine-Dalgarno sequences
 in splicing, 407, 407f

Conservative replication, 340–341, 341f.
 See also Replication

Constitutive genes, 463

Contigs, 611

Continuous characteristics. *See* Quantitative characteristics

Continuous replication, 346–347, 347f

Controlling elements, 540.
 See also Transposable elements

Coordinate induction, 469, 470f

Copy-number variations, 154, 229–230, 615.
 See also Deletion(s); Duplications

Core enzyme, 380, 380f

Core promoter, 387, 497

Corepressors, 467

Corn. *See* Maize

Correlation, 724–725, 725f

Correlation coefficient, 724–725, 725f

Correns, Carl, 49, 129, 325

Cosmid vectors, 573, 573t

Cotransduction, 270

Cotransformation, 265–266

Coupling (cis) configuration, 181–182, 181f

cpDNA. *See* Chloroplast DNA (cpDNA)

CpG dinucleotides, in DNA methylation, 643–644, 644f

CpG islands, 497, 644

Creighton, Harriet, 183, 183f

Crepidula fornicata (limpet), sex determination in, 85–86, 86f

C-rich strand, of telomere, 320, 320f

Crick, Francis, 10–11, 288, 293–294, 295f, 301, 400, 414, 438, 561

Cri-du-chat syndrome, 223t, 224

CRISPR RNA (crRNA), 267, 376, 376t, 377f, 418, 420–421, 421f, 565

CRISPR-Cas systems/CRISPR-Cas9
 in bacterial immunity, 565
 in gene therapy, 597
 in genetic engineering, 566
 in genome editing, 11–12, 267, 559–560, 561, 564–567, 565–567, 565f
  advantages and disadvantages of, 566–567
  ethical concerns for, 567

Criss-cross inheritance, 93f

Crop plants. *See also* Plant(s)
 breeding of, 3, 3f, 8, 8f, 238–239, 239f
  inbreeding in, 760
  quantitative genetics in, 715–716
 genetically modified, 3, 3f, 595–596
 herbicide-resistant, 595–596

Crosses, 31–34, 32f, 32t, 35f
 addition rule for, 57–58, 57f
 backcross, 56
 bacteriophage, 272–273
 branch diagrams for, 64–65, 64f, 65f
 chi-square test for, 67–69, 68t, 69f
 cis configuration in, 181–182, 181f
 dihybrid, 61–67, 62f–65f
 genotypic ratios in, 61, 61t
 with linked genes, 174–198.
  See also Linkage
  coupling in, 181–182, 181f
  notation for, 176
  predicting outcome of, 184
  recombination frequency for, 180, 184
  in repulsion, 181–182, 181f
  testcrosses for, 177, 178f
 meiosis and, 54–55
 monohybrid, 51–61
 multiple-loci, 61–67
 multiplication rule for, 57, 57f
 notation for, 54, 60–61, 176
 outcome prediction for, 56–60, 184
 phenotypic ratios for, 61, 61t
  observed vs. expected, 67–70, 69f
 probability rules for, 57, 57f, 64–65, 64f, 65f
 Punnett square for, 56, 56f, 58
 reciprocal, 52
 recombination frequencies for, 180, 184, 187–188, 193
 testcrosses, 60, 65, 65f, 177, 178f, 185–186, 186f. *See also* Testcrosses
 three-point, 189–195
 two-point, 188–189
  gene mapping with, 188–189
 with unlinked genes, 67–70, 68t, 69f

Crossing over, 29f, 30f, 31–34, 32t.
  See also Crosses; Crossovers
 among three genes, 189–195
 coupling configurations in, 181–182, 181f
 definition of, 29
 dominance and, 53f, 54
 genetic diversity and, 30–34
 homologous recombination in, 360–361, 361f
 with incompletely linked genes, 178–180, 182
 independent assortment and, 62–64, 62f, 174–175, 176f, 178–180, 178f, 182, 185–186, 186f
 within inversions, 225–226, 225f, 226f
 with linked genes, 174–198, 178f, 182.
  See also Linked genes
 nonindependent assortment and, 174–175, 175f
 recombination frequencies and, 180, 184, 187–188, 193
 recombination in, 176f, 180, 182, 184.
  See also Recombination
 repulsion and, 181–182, 181f
 segregation in, 53–54, 56f, 174–175.
  See also Segregation
 trans configuration in, 181–182, 181f
 transformation and, 265–266
 unequal, 221
  chromosome mutations and, 221, 221f, 528, 528f

Crosslinked ChIP, 496

Crossovers
 double, 187, 189f, 193–195
  coefficient of coincidence for, 193–195
  four-strand, 197–198, 197f
  within inversions, 226, 226f
  three-strand, 197–198, 197f
  two-strand, 197–198, 197f
 hotspots for, 203
 interference between, 193–195
 mapping of, 190f, 192–193, 192f
 multiple, 197–198, 197f
 predicted number of, 193–195
 three-gene, 189–195, 189f
 two-strand, 197–198, 197f

Cultures
 bacterial, 254f
 bacteriophage, 268–269, 269f

C-value, 321

C-value paradox, 321

Cyclic AMP, in catabolite repression, 475

Cyclin, 699

Cyclin-D-CDK, 699, 699f

Cyclin-dependent kinases, 699

Cyclin-E-CDK, 699, 699f

CYP2C9, warfarin and, 162–163

Cysteine, 434f

Cystic fibrosis
 inheritance of, 112–113, 116, 236
 prenatal diagnosis of, 159t, 757
  gene isolation in, 580–581
  positional cloning in, 580–581, 581f

Cystic fibrosis transmembrane conductance regulator (CFTR) gene, 112–113, 116, 581

Cytochrome *c*, in evolutionary studies, 796
Cytokines, 506
Cytokinesis, 23, 26t
Cytology, history of, 9–10
Cytoplasmic inheritance, 128–130, 129f, 130f, 132t
Cytosine, 297, 297f, 298t, 299f, 300, 528. *See also* Base(s)
   methylation of, 303, 303f, 497, 643–644, 644f. *See also* DNA methylation

D loop, 314f, 327
D segment, 682–684, 682f, 683f
dAMP (deoxyadenosine 5′ monophosphate), 298f
Danio rerio (zebrafish), 6–7, 7f
   genome of, 626t
Darwin, Charles, 10, 145, 491, 736, 751, 787, 793
Darwin's finches, 679, 796–797
   allopatric speciation in, 787–788, 788f
Databases, bioinformatic, 615–616, 618
Datura stramonium (Jimson weed), aneuploidy in, 230–231, 232f
Daughter chromosomes
   bacterial, 253
   formation of, 25f, 26, 253
Davis, Bernard, 259
Dawkins, Richard, 4
dCMP (deoxycytidine 5′ monophosphate), 298f
ddNTP (dideoxyribonucleoside triphosphate), in DNA sequencing, 582
De Vries, Hugo, 49
Deacetylases, 494, 495f
Deacetylation, histone, 494, 495f
   detection of, 656
Deamination
   induced, 530, 531f
   spontaneous, 528, 529f
Death cap poisoning, 373–374
decapentaplegic gene, 668, 669t
Decoding center, 449
Degenerate code, 438–439
Deinococcus radiodurans, 630
Deletion(s), 221, 222–224, 223f, 223t, 224f, 241t, 518–519, 518f, 524t, 527–528
   in cancer, 707
   in-frame, 518–519, 524t
Deletion mapping, 201, 202f
Dementia, frontotemporal, 515–516
Denisovans, genomes of, 657
Denominator, B1
Dentatorubral-pallidoluysian atrophy, 519t
Deoxyadenosine 5′ monophosphate (dAMP), 298f
Deoxycytidine 5′ monophosphate (dCMP), 298f
Deoxyguanosine 5′ monophosphate (dGMP), 298f
Deoxyribonucleoside triphosphates (dNTPs)
   in DNA sequencing, 582
   in replication, 346, 346f
Deoxyribonucleotides, 297, 298f, 298t
Deoxyribose, 296–297, 375

Deoxythymidine 5′ monophosphate (dTMP), 297, 298f, 298t
Depurination, mutations and, 528, 529f
Determination, 664
Development, 663–690
   in *Arabidopsis thaliana*, 674–676, 674f, 675f
   cloning experiments in, 635f, 665–666
   in *D. melanogaster*, 666–674, 667f–669f, 671f–673f. *See also Drosophila melanogaster* (fruit fly), pattern formation in
   determination in, 664
   evolution and, 678–679
   of eye, 678, 679
   gene rearrangement in, 679–685
   gene regulation in, 4, 4f, 667f–669f, 671f–673f
   ontogeny recapitulates phylogeny in, 678
   programmed cell death in, 676–677, 677f
   RNA interference in, 506
   sexual
     precocious puberty in, 127, 128f
     sequential hermaphroditism in, 86, 86f
   in spineless sticklebacks, 663–664, 678
   totipotent cells in, 664
dGMP (deoxyguanosine 5′ monophosphate), 298f
Diakinesis, 29, 29f
Diastrophic dysplasia, 3f
Dicentric bridge, 226
Dicentric chromatids, 226
Dicer, 505
Dideoxy sequencing, 582–584, 582f–584f
Dideoxyribonucleoside triphosphate (ddNTP), in DNA sequencing, 582
Diet, epigenetics and, 641–642, 652
Dihybrid crosses, 61–67, 62f–65f
   phenotypic ratios from, 122, 122t
Dillon, Robert T., 121
Dioecious organisms, 83
Diphtheria, 446
Diploid, partial, 471
Diploid cells, 21, 22f
Diplotene, 29, 29f
Direct repair, 545, 548t
Directional gene transfer, 264, 264f
Directional selection, 769–770
Direct-to-consumer genetic tests, 163
Discontinuous (qualitative) characteristics, 134, 716, 716f, 717f
Discontinuous replication, 346–347, 347f
Discordance, 155
Diseases
   autoimmune, 680, 680t
   genetic. *See* Genetic diseases
   neurodegenerative, apoptosis in, 677
Disomy, uniparental, 235–236
Dispersive replication, 340–341, 341f
Displaced duplications, 220
Dissociation gene, 539–540
Distal-less gene, 678
Distance approach, for evolutionary relationships, 794
Distributions, in statistical analysis, 721–722, 722f, 723f

Diver, C., 109–110
Division, of fractions, B2
Dizygotic twins, 154, 154f
DNA, 12, 287–309
   A form of, 301, 301f
   B form of, 300, 300f, 301f
   bacterial, 19, 19f, 312f, 314, 314f
   in bacteriophages, 292–293, 293f, 294f
   bases in, 290, 290t, 297, 297f, 298t, 300f, 302–303. *See also* Base(s)
   in biosphere, 5
   in cell cycle, 23f, 24, 26t, 27–28, 27f, 698–699
   cellular amounts of, 321t
   chloroplast, 129–130, 322–325, 330–331. *See also* Chloroplast DNA (cpDNA)
   circular
     bacterial, 255, 255f, 256f
     replication in, 357–358, 358f
   coiling of, 21–22
   damage to
     age-related, 329
     repair of. *See* DNA repair
   denaturation of, 321
   double helix of, 299–300, 300f
   early studies of, 288–293, 289f
   eukaryotic, 19, 19f, 314–318
   fetal, in prenatal diagnosis, 161
   as genetic material, 290–293
   H form of, 302–303, 303f
   hairpin in, 302
   heteroduplex, 361, 363
   highly repetitive, 321
   information transfer via, 301–302, 302f
   key characteristics of, 288
   linker, 316
   measurement of, 27–28, 27f
   melting temperature of, 321
   microsatellite, in DNA fingerprinting, 586
   mitochondrial, 129, 326–328. *See also* Mitochondrial DNA (mtDNA)
   mobile. *See* Transposable elements
   moderately repetitive, 321
   noncoding, 321, 627
   nucleotides of, 297, 297f–299f, 298t
   packaging of, 21–22, 312–318, 314f
     in bacteria, 314, 314f
     in eukaryotes, 314–318, 315f, 316f
   palindromes in, 96
   prokaryotic, 19, 19f
   promiscuous, 331
   quadruplex, 303
   in relaxed state, 313
   renaturation of, 321
   repetitive, 321
   replication of. *See* Replication
   secondary structures of, 299–301, 299f, 300f, 301f, 302–303, 302f
   single-stranded, 302
   strands of. *See* Polynucleotide strands
   structure of, 296–303
     discovery of, 293–295, 295f
     genetic implications of, 301–302
     hierarchical nature of, 312
     primary, 296–299, 297f–299f, 312

secondary, 293–295, 299–301, 299f, 300f, 301f, 302, 302f, 312
tertiary, 312, 313f
tetranucleotide theory of, 290
vs. RNA structure, 374–375, 376t
Watson-Crick model of, 293–295, 295f, 301–302
supercoiling of, 313, 313f, 339–340
synthesis of. See Replication
total amount of, 5
in transformation, 265
as transforming principle, 290–292, 292f
in transposition, 535–536.
See also Transposable elements; Transposons
triplex, 302–303, 303f
unique-sequence, 321
unwinding of
in recombination, 362, 363
in replication, 346–347, 347f, 348–350, 349f, 355, 362–363
X-ray diffraction studies of, 294, 295f
Z form of, 301, 301f
DNA amplification, 568–576
cloning in, 571–576, 571f–575f
polymerase chain reaction in, 569–571, 570f
DNA fingerprinting, 586–589, 587f–589f, 588t
DNA gyrase
in recombination, 363
in replication, 349–350, 349f, 353
DNA helicase, 349, 349f, 353, 353f
DNA hybridization, 321, 568
DNA libraries, 576–578
cDNA, 576–578, 578f
cloning of, 578
creation of, 576–577
genomic, 576, 577f
screening of, 577–578, 578f
DNA ligase
in recombination, 363
in replication, 352, 352f, 353f
DNA methylation, 303, 303f, 497, 643–645, 644f
in ancient epigenomes, 657
in cancer, 705
cognitive effects of, 651
detection of, 656, 657f
epigenetics and, 643–645, 644f–646f, 656.
See also Epigenetics
in gene regulation, 497
in genomic imprinting, 132
maintenance of, 644, 644f
RNA silencing and, 528
in transcription repression, 497, 645
DNA methyltransferases, 644
DNA mismatch repair, 354, 363, 544, 545f, 548t
DNA polymerase(s)
in base-excision repair, 545–546, 546f
in DNA sequencing, 582–584
high-fidelity, 356
low-fidelity, 356
in recombination, 363
in replication, 346, 351t, 352f, 355–356
in bacteria, 351–353, 351t
in eukaryotes, 355–356
translesion, 356, 548

DNA polymerase α, 355–356
DNA polymerase δ, 355–356
DNA polymerase ε, 355–356
DNA polymerase I, 351–353
DNA polymerase III, 351–353
DNA probes, 568
in DNA library screening, 577–578, 578f
in in situ hybridization, 203, 203f, 578, 579f
selection of, 577–578
DNA profiling, 586–589, 587f–589f, 588t
DNA proofreading, 353–354, 544
DNA repair, 544–549
base-excision, 545–546, 546f, 548t
cancer and, 702
in crossing over, 29, 30f
direct, 545, 548t
of double-strand breaks, 547–548, 565–566
in genetic diseases, 548–549, 549t
homologous recombination in, 547–548, 566
mismatch, 354, 363, 544, 545f, 548t
nonhomologous end joining in, 548
nucleotide-excision, 546, 547f, 548t
in genetic disease, 549t
SOS system in, 532
steps in, 547
DNA sequence alignments, in phylogenetic trees, 792–794
DNA sequence analysis, 561
DNA fingerprinting in, 586–589, 587f–589f
DNA sequencing in, 582–586
in gene mapping, 203
oligonucleotide-directed mutagenesis in, 590, 590f
site-directed mutagenesis in, 590
DNA sequence variation, 783
DNA sequences, types of, 321–323
DNA sequencing, 582–586
automated, 583–584
dideoxy (Sanger) method of, 582–584, 582f–584f
in evolutionary studies, 783
in gene mapping, 203, 582–586, 582f–584f
in Human Genome Project, 610–613
Illumina, 584
nanopore, 586
next-generation technologies for, 584–586, 585f
pyrosequencing in, 584–586, 585f
third-generation technology for, 586
DNA template
in replication, 340, 346–347, 347f
in transcription, 377–379, 378f
DNA transposons, 534, 538, 538f.
See also Transposable elements; Transposons
DNA-binding motifs, 464–465, 465f, 465t, 498
DNA-binding proteins, 349, 463, 464–465, 465f, 465t, 498
DNase I, chromatin structure and, 317–318, 318f, 493
DNase I hypersensitive sites, 493
DNAse I sensitivity, 317–318, 318f
dNTPs (deoxyribonucleoside triphosphates)
in DNA sequencing, 582
in replication, 346, 346f

Dobzhansky, Theodosius, 780
Dogs, coat color in, 118–119, 124–125, 125f, 126t
Dolly (cloned sheep), 635f, 665–666
Domains, 464
protein function and, 618–619
Dominance, 53f, 54, 110–113, 111f, 112t
characteristics of, 112
codominance and, 111f, 112, 112t
complete, 111, 111f, 112t
incomplete, 110–112, 111f, 112t
phenotype level and, 112–113
Dominance genetic variance, 729
Dominant epistasis, 119–120, 120f
Dominant traits
autosomal, 149–150, 149f, 153t
inheritance of, 54
X-linked, 151f, 153t
Donkeys, hinnies and, 655
dorsal gene, 668
Dosage compensation, 97–98, 231–232
Double crossovers. See Crossovers, double
Double fertilization, 39f, 40
Double helix, 299–300, 300f
Double-strand breaks, 362, 362f
repair of, 547–548, 565–566.
See also DNA repair
Down, John Langdon, 232–235
Down mutations, 382
Down syndrome, 232–235, 233f–235f
familial, 233–234, 233f, 234f
maternal age and, 235, 235f
prenatal diagnosis of, 161
primary, 233, 233f
Drivers, in cancer, 704
Drosophila heteroneura, 791
Drosophila melanogaster (fruit fly)
Bar mutations in, 221, 221f, 536–537
body color in, evolution of, 797, 797f
eye color in, 89–91, 91f, 92f, 124
gene mapping for, 190–195, 190f, 192f, 194f
genetic drift and, 764, 765f
eye development in, 678, 678f
eye size in, 221, 221f
genetic drift in, 764, 765f
genetic map for, 190–195, 190f, 192f, 194f
genome of, 626, 626t, 627t
gynandromorphic, 236
life cycle of, 666–667, 667f, A2–A3
as model genetic organism, 6, 6f, A2–A3
mosaicism in, 236, 236f
Notch mutation in, 224, 224f
pattern formation in, 666–674
anterior-posterior axis in, 668–669, 669t
dorsal-ventral axis in, 668, 668f, 669t
egg-polarity genes in, 667–670, 667t
gap genes in, 670, 670t, 671f
homeotic genes in, 671–672, 671f–673f
pair-rule genes in, 670, 670t, 671f
segmentation genes in, 670, 671f
segment-polarity genes in, 670, 670t, 671f
stages of, 666–667, 667f
vs. in flowers, 676

sex determination in, 87, 87t
  alternative splicing in, 502–503, 503f, 504f
  speciation and, 791
  temperature-dependent gene expression in, 133, 133f
  transposable elements in, 540–541, 542f
  X-linked characteristics in, 89–91, 91f, 92f
Drosophila silvestris, 791
Drosophila simulans, 791
Drosophila spp., speciation and, 791
Drosophila willistoni, 791, 791t
Drug development
  genetics in, 3–4
  recombinant DNA technology in, 595
Drug resistance, gene transfer and, 264
Drug therapy, genetic testing in, 162–163
Ds elements, 539–540, 540f
dTMP (deoxythymidine 5′ monophosphate), 298f
Duchenne, Benjamin A., 559
Duchenne muscular dystrophy, 202, 509–510
Ducks
  feather color in, 115f
  plumage patterns in, 115, 115f
Duplicate recessive epistasis, 120–121
Duplications, 219f, 220–222
  chromosome, 220–222, 223t, 241t
  displaced, 220
  in evolution, 222
  gene, multigene families and, 799–800
  interchromosomal, 220
  intrachromosomal, 220
  reverse, 220
  segmental, 220, 222
    in genome, 626–627
  tandem, 220
  whole-genome, 799
Dutch Hunger Winter, 641–642
Dwarfism (achondroplasia), 159t, 771
Dyskeratosis congenita, 360
Dyskerin, 360
Dystrophin, 559–560

E site, in ribosome, 445–446, 445f, 447f, 450f
East, Edward, 719, 727–728
Ecological isolation, 784–785, 785t
Edward syndrome, 235
Effective population size, 764
Eggplant, fruit color in, 111–112, 111f
Egg-polarity genes, 667–670, 667t
Electrophoresis, 567, 568f
  in proteomics, 629–630, 630f
  in restriction mapping, 608
Elephants, genetic variation in, 783
Elongation
  stalled, 500–501
  in transcription, 383–384, 389, 389f, 500–501
  in translation, 444f, 445–446, 447, 449
Elongation factor G, 445
Elongation factor Ts, 445
Elongation factor Tu, 445
Emerson, Rollins A., 539
Encyclopedia of DNA Elements (ENCODE), 627
Endocrine disruptors, 651

Endonucleases, restriction, 562–563, 563t, 564f
  in gene mapping, 608
Endosymbiotic theory, 323, 323f
Engineered nucleases, 564
engrailed gene, 678
Enhancer(s)
  bacterial, 482
  eukaryotic, 387, 500, 500f
  exonic/intronic splicing, 502
Enhancer RNA (eRNA), 422, 500
env gene, 273–275
Environmental factors
  in adoption studies, 156–157
  epigenetic, 651
  in gene expression, 133–135, 133f, 134f
  genotype-phenotype relationship and, 716–718. *See also* Genotype-phenotype relationship
  in heritability, 733
  in twin studies, 156
Environmental sex determination, 85–86, 86f, 86t
Environmental variance, 729
Enzyme(s). *See also specific enzymes*
  deficiencies of, 134. *See also* Genetic diseases
  gene expression and, 433
  in recombination, 362–363
  in replication, 351–353, 352t, 355–356. *See also* DNA polymerase(s)
  restriction, 562–563, 563t, 564f
    in gene mapping, 608
  in RNA editing, 412, 412f
    tRNA-modifying, 415
  in translation, 441
Ephrussi, Boris, 325
Epialleles, 648
Epigenetic changes, 318–319
Epigenetic marks, 646
  genome-wide, 657
Epigenetic traits, 643
Epigenetics, 11, 132–133, 641–662
  behavioral, 650–651
  cancer and, 705
  cell differentiation and, 654–655, 654f
  chromatin modification and, 318–319, 497, 643–644, 644f
  definition of, 318, 643
  development and, 674
  diet and, 641–642, 652
  DNA methylation and, 643–645, 644f, 645f, 646f, 656, 657f
  environmental chemicals and, 651
  genomic imprinting and, 132–133, 655–656, 655f
  histone modifications and, 645–647, 656
  overview of, 642–643
  paramutation and, 647–650
    in corn, 648, 648f
    in mice, 648–649, 648f, 649f, 650t
  RNA and, 647
  stem cells and, 654–655, 654f
  stress effects and, 650–651
  in twins, 652
  X inactivation and, 652–654, 653f, 653t
Epigenomes, 656–657, 657f
Epilepsy, in twins, 155, 155t

Episomes, 256
Epistasis, 118–121
  definition of, 117
  dominant, 119–120, 120f
  recessive, 118–119
    duplicate, 120–121
Epistatic gene, 117
Epstein-Barr virus, cancer and, 708, 708t
Equilibrium
  for allelic frequencies, 761, 761f
  Hardy-Weinberg, 754
  unstable, 770
Equilibrium density gradient centrifugation, 341–342, 341f
Equivalent fractions, B1
eRNA (enhancer RNA), 422, 500
Escherichia coli (bacterium). *See also* Bacteria
  DNA mismatch repair in, 544
  DNA polymerases in, 351–353, 351t, 352t
  gene regulation in, 462
  genome of, 624
    sequencing of, 625t
  in Hershey-Chase experiment, 292–293, 293f, 294f
  lac operon in, 468–475
  as model genetic organism, 6, 6f, A4–A5
  partial diploid strains of, 471
  recombination in, 362–363
  replication in, 341–344, 343f, 344f, 345t
  ribosome in, 448–449, 450f
  T2 bacteriophage in, 292–293, 293f, 294f
  transformation in, 256f, 257–258
  tRNA in, 415–416, 416f
Ethical issues
  in genetic testing, 164, 613
  in genome editing, 567
  in genome sequencing, 613
Ethylmethanesulfonate (EMS), as mutagen, 530, 531f
Eubacteria, 19, 253. *See also* Bacteria; Prokaryotes
  genome of, 625t
  mitochondrial origin from, 323f
  transcription in, 390–391
Euchromatin, 314
Eukaryotes, 12, 18–19, 18–23, 18f, 19f
  cell reproduction in, 20–23.
    See also Cell cycle
  cell structure in, 18f, 19
  chromosomes in, 19, 19f, 20–23, 322
    DNA packaging in, 312–313, 312f, 314–318, 315f–317f
  definition of, 18–19
  DNA in, 19, 19f, 314–318, 322
  gene regulation in, 491–513
    vs. in bacteria, 482, 492–493, 508, 508t
  genes of, 19, 625–629
  genome of, 19f, 625–629, 626t, 627t
  replication in, 344–345, 345f, 345t, 354–360
  sexual reproduction in, 81–85, 83–85
  vs. bacteria, 253
Eukaryotic elongation factor 2 (eEF-2), in diphtheria, 446
Evans, Martin, 592
even-skipped gene, 670, 670t, 671f
Evo-devo, 678

Evolution
    alternative splicing in, 411
    anagenesis in, 781
    catalytic RNA in, 374
    of chloroplast DNA, 330
    cladogenesis in, 781
    clonal, 694–695, 695f
    Darwinian theory and, 10
    definition of, 780
    development and, 678–679
    DNA sequence variation and, 783
    duplications in, 222
    gene expression in, 678–679
    gene regulation in, 796f, 797, 797f
    as genetic change, 13
    genetic variation and, 4, 543–544, 751,
        771–772. See also Population genetics
    of genome, 798–799, 798f
    of globin genes, 222
    horizontal gene transfer in, 625
    intron early hypothesis and, 402
    intron late hypothesis and, 402
    inversions in, 226
    of mitochondrial DNA, 328–329
    molecular clock and, 796–797, 797f
    molecular data for, 782–783
    molecular variation and, 782–783
    mutations in, 241
    natural selection in, 736–737, 766–772.
        See also Natural selection; Selection
    of operons, 462
    phylogenetics and, 791–794
    population variation and, 782–784
    protein variation and, 783–784
    rate of, 794–796
    reproductive isolation in, 784–785
    response to selection in, 737–740
    of sex chromosomes, 95–96
    speciation in, 784–791. See also Speciation
    transcription in, 390–391
    transposable elements in, 543
    as two-step process, 781
    of viruses, 274–275, 275f
Evolutionary relationships, phylogenetic trees
        for, 792–794, 792f–794f
Exit (E) site, in ribosomes, 445–446, 445f,
        447f, 450f
Exon shuffling, 798
Exonic/intronic splicing enhancers, 502
Exon-junction complex, 408–409
Exons, 402, 407–411, 407f–411f, 414f
Expanding trinucleotide repeats, 519–520,
        519t, 520f, 524t
    in amyotrophic lateral sclerosis, 515–516
    in frontotemporal dementia, 515–516
Expression vectors, 573, 574f
Expressivity, 113, 113f
Eye color
    in D. melanogaster, 89–91, 91f, 92f, 124
        gene mapping for, 190–195, 190f,
            192f, 194f
        genetic drift and, 764, 765f
        as X-linked characteristic, 89–91, 91f, 92f
Eye development
    in cavefish, 679
    in Drosophila melanogaster, 678

Eye size, in D. melanogaster,
        221, 221f
eyeless gene, 678, 678f

F' cells, 262t
F+ cells, 259–262, 260f, 261f, 262t
F− cells, 259–262, 260f–262f, 262t
F factor, 256, 256f, 259–262, 262t
    replication in, 344, 344f
F (filial) generations, in monohybrid crosses,
        51–52, 52f, 53f
F prime cells, 260
Factor-binding center, 449
Familial adenomatous polyposis coli, 705–706
Familial Down syndrome, 233–234, 233f,
        234f. See also Down syndrome
Familial hypercholesterolemia, 149–150, 150f
Familial vitamin D–resistant rickets, 152
Fanconi anemia, 549t
Feather color
    in ducks, 115, 115f
    in Indian blue peafowl, 94, 95f
Feathering, in cock vs. hen, 127, 128f
Fertility (F) factor, 256, 256f, 259–262,
        260f–262f, 262t
    replication in, 344, 344f
Fertilization
    in animals, 37, 37f
    definition of, 28
    double, 39f, 40
    in plants, 38–40, 39f
Fetal cell sorting, 161
Fetal DNA, in prenatal diagnosis, 161
Fetal ultrasonography, 158, 159f
Ficedula albicollis, 791
Ficedula hypoleuca, 791
Filial (F) generations, in monohybrid crosses,
        51–52, 52f, 53f
Finches, Darwin's, 679, 796–797
    allopatric speciation in, 787–788, 788f
Fingerprints, absence of, 145–146, 146f
Fire, Andrew, 418
Firmicutes, obesity and, 617
First polar body, 36, 37f
FISH (fluorescence in situ hybridization) in
        gene mapping, 203, 203f
Fisher, Ronald A., 10, 719, 779–780
Fitness, 767
5′ cap, 405–406, 405f, 405t
    in translation, 441, 441f, 442f, 444, 444f
5′ splice site, 407, 407f
5′ untranslated region, 404, 404f, 444, 444f
    in trp operon, 476, 479f
Fixation, allelic, 766
Flanking direct repeats, 535, 535f, 538f
    in composite transposons, 538, 538f
    in insertion sequences, 538, 538f
FLC (flowering locus C), 495
FLD (flowering locus D), 495, 495f
Flemming, Walther, 10
Flower color
    inheritance of, 110–112
    lethal alleles and, 114
Flower development, 674–676, 674f, 675f
Flower length, inheritance of, 727–728, 728f
flowering locus C, 495, 495f

flowering locus D, 495, 495f
Flowering plants. See Angiosperms; Plant(s)
Fluorescence in situ hybridization (FISH), in
        gene mapping, 203, 203f
Fly Room, 90f
Ford, Charles, 87
Ford, E. B., 779
Forensics, DNA fingerprinting in, 586–589,
        587f–589f, 588t
Forward genetics, 589
Forward mutations, 520, 524t
Fossil DNA, genome sequencing from, 613
Founder effect, 765
Four-o'clocks (Mirabilis jalapa), leaf
        variegation in, 129–130, 130f, 325
Fractions, B1–B3
    adding, B2
    dividing, B2
    multiplying, B2
    simplifying, B1–B2
    subtracting, B2
Fraenkel-Conrat, Heinz, 295
Fraenkel-Conrat-Singer experiment,
        295–296, 296f
Fragile sites, 229, 229f
Fragile-X syndrome, 229, 519, 519f, 519t, 520
Frameshift mutations, 518–519, 524t
    in Ames test, 533
Franklin, Rosalind, 11, 288, 294, 295f
Fraser, Claire, 608
Free radicals
    aging and, 329
    as mutagens, 531, 531f
Frequency
    allelic, 752
    definition of, 751
    genotypic, 751
Frequency distribution, 721–722, 722f
Friedreich's ataxia, 519t
Frontotemporal dementia, 515–516
Fruit color
    epistasis and, 119–120, 120f
    gene interaction for, 117–118, 120f
    incomplete dominance and, 111–112, 111f
Fruit fly. See Drosophila melanogaster
        (fruit fly)
Functional genetic analysis, 589–595
Functional genomics, 617–624
    homology searches in, 618
    microarrays in, 619–621, 620f, 621f
    mutagenesis screens in, 623–624, 623f
    reporter sequences in, 622, 623f
fushi tarazu gene, 670
Fusion proteins, in cancer, 707

G banding, 219, 219f
G_0 phase, 23f, 24, 26t
G_1 phase, 23f, 24, 26t
G_1/S checkpoint, 23f, 24, 355, 699, 700f
G_2 phase, 23f, 24, 26t
G_2/M checkpoint, 23f, 24, 699, 700f
gag gene, 273–275
Gain-of-function mutations, 521, 524t
GAL4, 498–499, 499f
Galactose metabolism, gene regulation in,
        498–499, 499f

β-Galactosidase, 469

Galápagos Islands, Darwin's finches of, 679, 787–788, 787f, 796–797

Gallo, Robert, 274

Galton, Francis, 145, 719

Gametes
 nonrecombinant (parental), 177, 179f
 recombinant, 177, 179f
 size of, 83, 83f
 unbalanced, 237

Gametic isolation, 785, 785t

Gametophyte, 38, 38f

Gap genes, 670, 670t, 671f

Garrod, Archibald, 1, 400, 430

Gasterosteus aculeatus, evolution of, 663–664

Gehring, Walter, 678

Gel electrophoresis, 567, 568f
 in proteomics, 629–630, 630f
 in restriction mapping, 608

Gender, vs. sex, 83

Gene(s). *See also* Genome(s); Protein(s)
 and specific genes
 allelic, 12, 50. *See also* Alleles
 annotated, 616
 bacterial, 19, 19f, 255, 255f, 256f, 266
  number of, 624–625, 625t
 Bt, 575–576, 575f
 cloning of, 571–575. *See also* Cloning
 coding vs. noncoding regions of, 401
 colinear, 400–401, 401f, 628
 constitutive, 463
 definition of, 50, 50t, 403
 during development, 665–666
 distance between, recombination
  frequencies and, 187–188, 193,
  197–198, 198f
 dosage of, 221–222
 duplication of, multigene families and,
  798–799
 egg-polarity, 667–670, 667t
 epistatic, 117
 eukaryotic, 19, 625–629
 evolution of, 798–799, 798f
 functionally related, 465. *See also* Operons
 functions of
  DNA sequence and, 617–619.
   See also Functional genomics
  in humans, 629f
  in prokaryotes, 625t
 as fundamental unit of heredity, 12
 gap, 670, 670t, 671f
 haploinsufficient, 224, 698
 homeotic (homeobox), 671–673
  in *D. melanogaster*, 671–672, 671f–673f
  in flower development, 674–676, 674f,
   675f
  in vertebrates, 672–673, 673f
 homologous, 618, 618f
  in eukaryotes vs. prokaryotes, 628
 hypostatic, 117
 identification of, functional genomic
  techniques for, 617–624. *See also*
  Functional genomics
 interrupted, 401
 isolation of, molecular techniques for,
  576–581

jumping. *See* Transposable elements
linked, 174–198. *See also* Linkage; Linked
 genes
location of. *See* Gene loci
major histocompatibility complex,
 684–685, 685f
movable. *See* Transposable elements
in multigene families, 627
nucleotide substitutions in, rate of,
 794, 795t
number of
 developmental complexity and, 411
 gene dosage and, 231
 for polygenic characteristic, 721
 in prokaryotes, 624–625, 625t
oncogenes, 273, 697
organization of, 400–401, 401f
orthologous, 618
pair-rule, 670, 671f
paralogous, 618
prokaryotic, 19
regulator, 466
regulatory, 463
 mutations in, 471–472, 474f
segmentation, 670, 671f
segment-polarity, 670, 670t, 671f
size of, in humans, 628–629, 629f
structural, 463. *See also* Operons
 mutations in, 471
structure of, 400–403
taster, 779–780
tumor-suppressor, 697–698, 697f, 698t
viral, 20, 20f
vs. traits, 12. *See also* Genotype-phenotype
 relationship

Gene cloning, 569, 571–575. *See also* Cloning

Gene conversion, 363

Gene density, 322, 629

Gene deserts, 627

Gene dosage
 gene number and, 231
 unbalanced, 221–222

Gene expression, 4, 4f
 chromatin structure and, 318–319,
  493–497
 environmental effects on, 133–135
 enzymes and, 433
 epistatic, 118–121
 in evolution, 679
 expressivity in, 113, 113f
 functional genomics and, 617–624.
  See also Functional genomics
 genomic imprinting and, 131–133, 132t
 long noncoding RNA in, 421–422
 microarrays and, 619–621, 620f, 621f
 penetrance in, 113, 113f
 phenotype and. *See* Genotype-phenotype
  relationship
 regulation of. *See* Gene regulation
 reporter sequences and, 622, 623f
 RNA sequencing and, 621–622, 622f
 transcriptomics and, 619

Gene Expression Nervous System Atlas
 (GENSAT) project, 622

Gene families, 321

Gene flow, 762–763, 763f

Gene interaction, 117–125
 albinism and, 120–121, 121f
 definition of, 117
 dog coat color and, 124–125, 125f, 126t
 epistasis and, 118–121
 novel phenotypes from, 117–118, 117f
 phenotypic ratios from, 122, 122t

Gene interaction variance, 729

Gene loci, 13, 50, 183
 definition of, 50t
 methods of finding, 576–581.
  See also Gene mapping
 quantitative trait, 715–716, 719–720,
  734–736, 736t
  mapping of, 734–736, 735f

Gene mapping, 174–203
 in bacteria, 262–264, 263f, 265–266,
  269–273, 271f
 chromosome walking and, 579–580, 579f
 coefficient of coincidence in, 193–195
 with cotransformation, 265–266, 266f
 crossover locations in, 190f, 192–193, 192f
 in *D. melanogaster*, 190–195, 190f,
  192f, 194f
 deletion, 201–202, 202f
 DNA sequencing in, 203, 582–586,
  582f–584f
 double crossovers and, 187, 189f, 193–195
  two-strand, 197–198, 197f
  three-strand, 197–198, 197f
  four-strand, 197–198, 197f
 in eukaryotes, 174–203
 gene order in, 191–192, 264
 genetic maps in, 187–199, 607–608,
  607f, 608f
 genetic markers in, 199
 genome sequencing and, 608–613
 for histone modifications, 656
 in humans, 198–199
 in situ hybridization in, 203, 203f
 interference in, 193–195
 with interrupted conjugation, 262–264,
  271–272
 lod scores and, 199
 map units for, 187, 607
 in phages, 272–273
 physical maps in, 187, 200–203, 608, 608f
 quantitative trait loci in, 734–736, 735f
 recombination frequencies in, 180, 184,
  187–188, 193, 197–198, 273, 607.
   See also Recombination frequencies
 restriction, 608, 608f
 single-nucleotide polymorphisms in,
  613–615, 614f
 somatic-cell hybridization in, 201, 201f,
  202f
 with three-point testcrosses, 189–195
 with transduction, 269–272, 271f
 with transformation, 265–266, 266f,
  271–272
 with two-point testcrosses, 188–189
 in viruses, 272–273

Gene microarrays, 619–621, 620f, 621f

Gene mutations. *See also* Mutations
 vs. chromosome mutations, 518

Gene pool, 750

Gene regulation. *See also* Gene expression
  alternative splicing in, 502–503, 503f, 504f
  antisense RNA in, 418, 482, 482f
  attenuation in, 477–481, 480f, 481t
  in bacteria, 461–490
    vs. in eukaryotes, 482, 492–493, 508, 508t
  boundary elements in, 500, 500f
  catabolite repression and, 475, 476f
  chromatin structure and, 318, 493–497
  coordinate induction in, 469, 470f
  coordinated, 501
  definition of, 462
  in development, 4, 4f, 666–674, 667f–669f, 671f–673f
  DNA methylation in, 303, 303f, 497
  DNA-binding proteins in, 349, 464–465, 465f, 465t, 498
  enhancers in, 500, 500f
  in eukaryotes, 491–513, 508
    vs. in bacteria, 482, 492–493, 508, 508t
  in evolution, 796f, 797, 797f
  in galactose metabolism, 498–499, 499f
  gene silencing in, 418–419
  histone acetylation in, 494–495, 494f
  inducers in, 466–467
  insulators in, 500, 500f
  levels of, 463–464, 464f
  mRNA processing in, 502–505, 502f, 503f, 505f
  negative, 463, 466–467, 467f, 468f
  operators in, 466, 466f
  operons in, 465–481
    lac, 468–475, 470f
  overview of, 462–465
  positive, 463, 468
  posttranslational, 507–508
  regulatory elements in, 463
  repressors in, 469, 470f, 499.
    See also Repressors
  response elements in, 501, 502f
  riboswitches in, 482–483, 483f
  ribozymes in, 483–484, 483f
  RNA crosstalk in, 506
  RNA degradation in, 503–504
  RNA silencing (interference) in, 499, 504–506, 593–594
  RNA-mediated repression in, 483–484, 483f
  transcriptional, 464, 464f
  transcriptional activator proteins in, 387
  translational, 507–508
Gene therapy, 4, 11, 596–597
  CRISPR-Cas systems in, 597
  germ-line, 597
  for hypercholesterolemia, 594
  RNA interference in, 594, 594f
  somatic, 597
Gene transfer
  antibiotic resistance and, 264
  bacterial, 257–267, 625, 799
  in biotechnology. *See* Recombinant DNA technology
  by conjugation, 257, 257f–264f, 258–264, 262t, 271–272
  directional, 264
  gene order in, 264
  horizontal, 266–267, 625, 799

  natural, 264
  by transduction, 257f, 258, 269–272, 271f
  by transformation, 257–258, 257f, 265–266, 265f, 266f, 271–272
Gene trees, 792, 793f
Gene-environment interactions, 133–135, 133f, 134f
General selection model, 767–768
General transcription factors, 387
Generalized transduction, 269–270, 270f, 271f
Genetic analysis, functional, 589–595
Genetic bottleneck, 765
Genetic code, 435–440
  amino acids in, 438f
  in bacteria vs. eukaryotes, 448–449
  breaking of, 436–438
  characteristics of, 440, 448
  codons in, 435–440, 438f, 439t, 440t.
    See also Codons
  degeneracy of, 438–439, 438f
  diagram of, 438f
  exceptions to, 439t, 440, 440t
  overlapping, 439
  reading frames for, 439–440
  triplet, 436
  universality of, 440
Genetic conflict hypothesis, 656
Genetic correlation, 739–740, 739t
Genetic counseling, 157–164, 157t
Genetic crosses. *See* Crosses
Genetic diagnosis. *See* Genetic testing
Genetic differentiation, speciation and, 784–791. *See also* Speciation
Genetic diseases
  achondroplasia (dwarfism), 159t, 771
  age-related mtDNA damage and, 329
  albinism, 1–2
  Angelman syndrome, 131–132
  Beckwith-Wiedemann syndrome, 656
  cancer as, 693–695. *See also* Cancer
  chromosomal, 159t
  cleft lip and palate, 159t
  cri-du-chat syndrome, 223t, 224
  cystic fibrosis, 112–113, 116, 159t, 236, 580–581, 580f, 581f, 757
  cytoplasmic inheritance of, 130
  diagnosis of, 158–164, 159t
    postnatal, 162–163, 162t
    prenatal, 158–161, 159t
  diastrophic dysplasia, 3f
  dyskeratosis congenita, 360
  environmental factors in, 134
  expanding trinucleotide repeats in, 519–520, 519t
  familial hypercholesterolemia, 149–150, 150f
  faulty DNA repair in, 548–549, 549t
  gene therapy for, 4, 11
  genetic counseling for, 157–158, 157t
  genetic testing for. *See* Genetic testing
  genome-wide association studies for, 614–615
  genomic imprinting and, 131–133, 132t, 656
  hemophilia, 8, 151, 151f, 159t, 399–400
  hypophosphatemia, 152

  isolated congenital asplenia, 429
  Leber hereditary optic neuropathy, 130, 325
  Lesch-Nyhan syndrome, 159t
  macular degeneration, 614
  microRNA in, 506
  mitochondrial, 130, 325, 329–330
  muscular dystrophy, 202, 509–510
  myoclonic epilepsy and ragged-red fiber disease, 325
  nail-patella syndrome, 198–199, 198f
  neural-tube defects, 159t
  neurofibromatosis, 227
  osteogenesis imperfecta, 159t
  pedigree analysis of, 147–154
  phenylketonuria, 134, 159t
  Prader-Willi syndrome, 131–132, 236
  prevention of, 329–330
  screening for. *See* Genetic testing
  sickle-cell anemia, 159t
  single-nucleotide polymorphisms in, 614–615
  Tay-Sachs disease, 149, 159t
  telomerase and, 359–360
  transposable elements in, 536
  Waardenburg syndrome, 147–148, 148f
  Werner syndrome, 359–360
  Williams-Beuren syndrome, 223t, 224
  Wolf-Hirschhorn syndrome, 223t, 224
  xeroderma pigmentosum, 548–549, 549f, 549t
Genetic dissection, 517
Genetic diversity. *See* Genetic variation
Genetic drift, 763–766, 765f, 766f, 771–772, 771t
  allelic frequencies and, 765–766, 766f, 771t
  causes of, 765
  definition of, 764
  effects of, 765–766, 771t
  magnitude of, 764
Genetic engineering, 562. *See also* Recombinant DNA technology
  with pesticides, 575–576
Genetic Information Nondiscrimination Act, 164, 613
Genetic maps, 187–199, 607–608, 607f, 608f. *See also* Gene mapping
Genetic markers
  in gene mapping, 199
  Y-linked, 96–97
Genetic material
  developmental conservation of, 666
  DNA as, 290–292
  early studies of, 288–293, 289f
  key characteristics of, 288, 301–302
  RNA as, 295–296, 296f
Genetic maternal effect, 110, 130–131, 131f, 132t
  egg-polarity genes and, 667–670
Genetic mosaicism, 154, 236, 236f
  CRISPR-Cas genome editing and, 566
  in trisomy 8, 235
Genetic mutations. *See* Mutations
Genetic recombination. *See* Recombination
Genetic rescue, 750

Genetic studies, 145–172
  bacteria in, 252t. *See also* Bacteria
  human
    adoption studies, 156–157, 157f
    difficulties in, 146–147
    pedigree analysis, 147–154
    twin studies, 155–156, 155t
  model organisms for, 6–7, 6f, 7f.
    See also Model genetic organisms
  viruses in, 252t. *See also* Viruses
Genetic testing, 158–164, 159t, 596
  direct-to-consumer, 163
  discrimination and, 164
  ethical issues in, 161, 613
  legal issues in, 164, 613
  pharmacogenetic, 162–163
  postnatal, 162–163
  prenatal, 158–161, 159f, 159t, 160f
  presymptomatic, 162
  privacy and, 164
  test interpretation in, 163
Genetic variance, 729–730. *See also* Variance
  additive, 729
  dominance, 729
Genetic variation, 4, 30–34, 543–544, 751
  allelic fixation and, 766
  chromosome distribution and, 33–34, 33f
  crossing over and, 31–34
  DNA sequence, 783
  evolution and, 4, 771–772, 782–784
  genetic drift and, 763–766
  measures of, 783
  migration and, 762–763, 763f
  mutations and, 526, 543, 760–762, 761f.
    See also Mutations
  neutral-mutation hypothesis for, 783
  random separation of homologous
    chromosomes and, 33–34
  recombination and, 360–361
  sexual reproduction and, 28–40
  theories of, 783–784
  universality of, 751
Genetically modified plants, 3, 3f, 595–596
Genetic-environmental interaction variance,
  729, 729f
Genetics
  in agriculture, 3, 3f, 7–8, 8f
  applications of, 3–4, 11–12
  bacterial, 253–267
  basic concepts of, 2–7, 4, 12–13
  in biology, 4
  commercial applications of, 3–4
  cutting edge, 11
  developmental, 663–690
  divisions of, 5, 5f
  in evolution, 4, 10, 13
  forward, 589
  future of, 11–12
  historical perspective on, 8–11
  importance of, 2–7
  in medicine, 4, 11–12
  model organisms in, 6–7, 6f, A–1–A–13.
    See also Model genetic organisms
  in modern era, 11–12
  molecular, 4f, 5, 5f
  notation in. *See* Notation

population, 5, 5f. *See also* Population genetics
quantitative, 715–747. *See also* Quantitative
  genetics
reverse, 589
transmission (classical), 5, 5f
universality of, 4
viral, 267–277
Genic sex determination, 85, 86t
Genome(s). *See also* Gene(s)
  of *Arabidopsis thaliana*, 626, 626t, 627t
  of bacteria, 18f, 19, 255, 257f, 266, 624–625
    sequencing of, 625t
    size of, 624–625, 625t
  of *Caenorhabditis elegans*, 626
  of chloroplasts, 330–331, 330t, 331f
  colinear, 628
  of *D. melanogaster*, 626, 626t, 627t
  dark matter of, 615
  definition of, 4
  duplication of, 799
  of *E. coli*, 624
  of eukaryotes, 18f, 19, 625–629, 626t, 627t
    segmental duplications in, 626–627
  evolution of, 798–799, 798f
  exchange of genetic information
    between, 331
  gene number in, 411
  of *Homo sapiens*, 628–629, 629f
  human
    as reference genome, 613
    vs. chimpanzee, 491–492
  mitochondrial, 326–328, 327f, 327t
  of Neanderthals and Denisovans, 657
  of plants, 627t
  of prokaryotes, 624–625, 625t
  sequencing of, 11, 11f, 608–613
    for communities of organisms, 616–617
    ethical issues in, 613
    gene-centric approach in, 617
    in Human Genome Project, 609f,
      610–613, 612f
    map-based, 610–611, 610f
    metagenomic, 616–617
    in model genetic organisms, 608–609
    single-nucleotide polymorphisms in,
      614–615, 614f
    whole-genome shotgun, 611–612, 611f
  size of, 321t
    developmental complexity and, 411
    in prokaryotes, 624–625, 625t
  of *Streptomyces coeliocolor*, 624–625
  synthetic, 605–606
  of viruses, 273
  of yeast, 327, 328f, 626t, 627t
Genome editing, ethical concerns about, 567
Genome editing, CRISPR-Cas systems/
  CRISPR-Cas9 in, 11–12, 267, 559–560,
  561, 564–567
Genome-wide association studies, 47,
  199–200, 200f, 614, 736
Genomic imprinting, 131–133, 132f, 132t,
  655–656, 655f
  genetic conflict hypothesis and, 656
  genetic diseases and, 131–133, 656
  long noncoding RNA in, 421–422
  X-inactivation and, 656

Genomic instability, in cancer, 708
Genomic libraries, 576
  creation of, 577f
  screening of, 577–578
Genomics
  applications of, 629–632
  comparative, 624–629.
    See also Comparative genomics
  definition of, 606
  functional, 617–624. *See also* Functional
    genomics
  structural, 606–617. *See also* Structural
    genomics
Genotype
  definition of, 12, 50t, 51
  expression of. *See* Gene expression
  inheritance of, 51. *See also* Inheritance
Genotype-phenotype relationship, 12, 51,
  430–433. *See also* Gene expression
  continuous characteristics and, 134–135
  cytoplasmic inheritance and, 128–130, 130f
  environmental influences on, 133–135,
    133f, 134f, 717
  expressivity and, 113, 113f
  gene interaction and, 117–125.
    See also Gene interaction
  genetic maternal effect and, 130–131, 131f
  genetic variation and, 751. *See also* Genetic
    variation
  heritability and, 728–736
  mutations and, 521
  one gene, one enzyme hypothesis and,
    430–433
  penetrance and, 113, 113f
  polygenic inheritance and, 719, 720f, 722f
  quantitative traits and, 716–718, 718f
  sex influences on, 126–133, 132t
Genotypic frequency
  calculation of, 751
  Hardy-Weinberg law and, 754
  nonrandom mating and, 758–760
Genotypic ratios, 61, 61t
  observed vs. expected, 67–70, 69f
Germ-line gene therapy, 597
Germ-line mutations, 517f, 518
Germ-plasm theory, 8f, 9f, 10
Gibson, Daniel, 617
Gierer, Alfred, 296
Gilbert, Walter, 11, 582
Global Ocean Sampling Expedition, 616
Global warming, population genetics
  and, 750
Globin genes, evolution of, 222,
  798–799, 798f
Glucocorticoid response element (GRE),
  501, 501t, 502f
Glucose metabolism, catabolite repression
  and, 475, 476f
Glutamate, 434f
Glutamine, 434f
Glycine, 434f
G_{M2} gangliosidase, in Tay-Sachs disease, 149
Goats, bearding of, 126–127, 127f
Goodness-of-fit chi-square test, 67–70, 68t,
  69f, 185. *See also* Chi-square test
gooseberry gene, 670, 670t, 671f

Graft rejection, 685
GRE (glucocorticoid response element), 501
Green Revolution, 3, 3f
Gregg, Christopher, 655
Greider, Carol, 320
Gret1 retrotransposon, 536
Grew, Nehemiah, 9
G-rich strand, of telomere, 320, 320f
G-rich 3′ overhang, 358
Griffith, Fred, 290–291
gRNA, 412
Gros, François, 403
Group I/II introns, 402, 402t
Guanine, 297, 297f, 298t, 299f, 300.
        See also Base(s)
Guanosine triphosphate (GTP), in
        translation, 442, 445
Guide RNA, 412
Gurdon, John, 665
Gut bacteria, obesity and, 617
Gynandromorphs, 236, 236f
Gyrase
    in recombination, 363
    in replication, 349–350, 349f, 353

H3K4me3, 494
Haeckel, Ernst, 678
Haemophilus influenzae, genome sequencing
        for, 608, 609f
Hair color, inheritance of, 47–48
Hairpins, 302, 302f
    in attenuation, 477–478, 479f, 481t
    in transcription termination, 385, 477–478,
        479f, 481t
    trinucleotide repeats and, 519–520, 520, 520f
    in *trp* operon, 477–478, 479f, 481t
Haldane, John B. S., 10
Hamkalo, Barbara, 377
Haploid cells, 21
Haploinsufficiency, 224
Haplotypes, 200, 614, 614f
HapMap project, 614
Happy tree, 339
Hardy, G. H., 754
Hardy-Weinberg equilibrium, 754
    genotypic frequencies at, 754
    statement of, 754
    testing for, 756–757
Hardy-Weinberg law, 753–758
    allelic frequencies and, 757
    extensions of, 756
    genotypic frequencies and, 754
    implications of, 755–756
H-DNA, 302–303
Heart disease, microRNA in, 506
Heat-shock proteins, 501, 501t, 507
Heavy chains, immunoglobulin, 682, 682f
    antibody diversity and, 682–684
Helicase, in replication, 349, 349f, 353, 353f
Helix
    alpha, 300f, 301–302, 433, 435f
    double, 299–300, 300f
Helix-loop-helix, 465t
Helix-turn-helix, 465, 465f, 465t, 498
Helper T cells, in HIV infection, 275, 276f
Hemings, Sally, 97

Hemizygote
    definition of, 90
    X chromosome inactivation and, 98–99
Hemophilia, 8, 151, 151f, 159t
    inheritance of, 399–400
Hen feathering, 127, 128f
Henking, Hermann, 84
Hereditary nonpolyposis colon cancer,
        549, 706
    faulty DNA repair in, 549t
Heredity. *See also* Inheritance
    chromosome theory of, 54–55, 83–84,
        174–175, 183
        nondisjunction and, 90–92, 92f
    gene as fundamental unit of, 12
    molecular basis of, 293–295.
        See also DNA; RNA
    principles of, 47–79
    sex influences on, 126–133, 132t
Heritability, 728–736
    broad-sense, 730
    calculation of, 730–732
        by degrees of relatedness, 731–732
        by elimination of variance components,
            730–731
        by parent-offspring regression, 731, 731f
    definition of, 728
    environmental factors in, 733
    individual vs. group, 733
    of intelligence, 734
    limitations of, 732–734, 734t
    misconceptions about, 732–734
    narrow-sense, 730
    phenotypic variance and, 728–730
    population differences and, 733
    realized, 738
    response to selection and, 737–740
    specificity of, 733
    summary equation for, 730
Hermaphroditism, 83
    sequential, 86, 86f
Hershey, Alfred, 272–273, 292–293, 403
Hershey-Chase experiment, 292–293,
        293f, 294f
Heterochromatin, 314
Heteroduplex DNA, 361, 363
Heterogametic sex, 84
Heterokaryons, 201, 201f
Heteroplasmy, 324, 329
Heterozygosity
    autosomal dominant traits and, 149–150,
        149f, 153t
    autosomal recessive traits and, 148–149,
        148f, 153t
    definition of, 50t, 51
    dominance and, 110–112, 111f
    inbreeding and, 758–760, 759f
    loss of, in cancer, 698, 698f
    nonrandom mating and, 758–760,
        759f, 760f
Heterozygote, compound, 116
Heterozygote advantage, 770
Heterozygote screening, 162
Hexosaminidase deficiency, in Tay-Sachs
        disease, 149
Hfr cells, 261f, 262f, 262t

Highly repetitive DNA, 321
Hillmer, Alex, 173–174
Hinnies, 655
Hiroshima, atomic bombing of, 534, 534f
Histidine, 434f
Histone(s), 19, 19f, 22, 314–315, 315t
    acetylation/deacetylation of, 318, 494–495,
        494f, 645–647, 651
    of archaea, 316
    methylation of, 494, 645–647, 651
    modification of
        detection of, 656
        epigenetic changes and, 645–647, 656
        maintenance of, 647
        mapping of, 656
    in nucleosome, 315f, 316, 356–357
    variant, 315
Histone chaperones, 357
Histone code, 494
HIV (human immunodeficiency virus
        infection), 274–275, 275f, 276f
Holandric (Y-linked) traits, 94–97
Holliday intermediate, 362
Holliday junction, 361–362, 362f
Holliday model, 361–362, 361f
Holoenzymes, 380, 386
Homeoboxes, 672, 672f
Homeodomains, 465t
Homeotic complex (HOM-C), 672, 672f
Homeotic (homeobox) genes
    in *D. melanogaster*, 671–672, 671f–673f
    in flower development, 674–676, 674f, 675f
    in vertebrates, 672–673, 673f
Homo sapiens. See also under Human
    genome of, 626t, 628–629, 629f
Homogametic sex, 84
Homologous genes, 618, 618f, 628
Homologous pairs, 21, 22f, 35
    random separation of, 33–34
Homologous recombination, 360–361, 361f
    in DNA repair, 547–548, 566
    gene conversion and, 363
Homologous traits, phylogenies and, 791–794
Homoplasmy, 324, 329
Homozygosity
    autosomal dominant traits and, 149f,
        150, 153t
    autosomal recessive traits and, 148–149,
        148f, 153t
    definition of, 50t, 51
    dominance and, 110–113, 111f
    inbreeding and, 758f, 759–760, 759f
    nonrandom mating and, 758–760, 759f
    by state vs. descent, 758, 758f
Homunculus, 8, 9f
Honeybees, DNA methylation in, 645,
        645f, 646f
Hooke, Robert, 8
Hopi, albinism in, 1–2, 2f
Hoppe-Seyler, Ernst Felix, 289
Horizontal gene transfer, 266–267, 331,
        625, 799
Hormones, in sex determination, 89
Horowitz, Norman H., 431
Horses, hinnies and, 655
Hotspots, recombination, 203

Hox genes, 672–673, 672f
Hrdlieka, Ales, 1
HTLV-1, 708t, 709
Human chorionic gonadotropin (HCG), in prenatal screening, 161
Human Genome Project, 610–613.
    See also Genome(s), sequencing of
Human immunodeficiency virus (HIV), 274–275, 275f, 276f
Human papillomavirus (HPV), cervical cancer and, 708, 708t, 709–710
Human Proteome project, 629
Human Variome Project, 616
Humoral immunity, 680, 680f
hunchback gene, 669t, 670
Huntington disease, 519t, 520
Huxley, Julian, 779
Hybrid breakdown, 785, 785t
Hybrid dysgenesis, 540–541
Hybrid inviability, 785, 785t
Hybrid sterility, 785, 785t
Hybridization. *See also* Breeding
    allopolyploidy and, 238–239, 239f
    DNA, 321
    in situ, 203, 203f
    of plants, 9
    somatic-cell, 201, 201f, 202f
Hydrogen bonds, in DNA, 299–300, 299f
Hydroxylamine, as mutagen, 530, 531f
Hypercholesterolemia, 594
    epigenetics and, 652
    familial, 149–150, 150f
Hypermethylation, in cancer, 705
Hypomethylation, in cancer, 705
Hypophosphatemia, 152
Hypostatic gene, 117

Igoshin, Oleg, 461–462
Illumina sequencing, 584
Immigration delay disease, 145–146, 146f
Immune response
    primary, 681
    secondary, 681
    T-cell activation in, 507
    in transplantation, 685
Immune system
    autoimmune disease and, 680, 680t
    organization of, 680–682, 680f, 681f
    spleen in, 429
Immunity
    antibody diversity and, 682–684, 683f
    cellular, 680–681, 680f
    clonal selection in, 681, 681f
    CRISPR-Cas, 565
    genetic rearrangement in, 679–685
    humoral, 680, 680f
    T-cell receptor diversity and, 684
Immunodeficiency states, 274–275, 275f, 276f
Immunoglobulins (antibodies), 680f, 681, 681f, 682–684
    diversity of, 682–684
    structure of, 680f, 682
Imprinting, genomic.
    See Genomic imprinting
In situ hybridization, 578, 579f
    in gene mapping, 203, 203f

In vitro fertilization, preimplantation genetic diagnosis and, 161
Inbreeding, 758–760, 758f–760f
    in Isle Royale wolves, 749–750
Inbreeding depression, 759
Incomplete dominance, 110–112, 111f, 112t, 113f
Incomplete linkage, 178–180, 182
Incomplete penetrance, 113, 113f
Incorporated errors, 527, 527f
Indels, 518–519, 518f
Independent assortment, 34, 61–67, 62f, 117, 174–175, 182
    chi-square test for, 185–186, 186f
    definition of, 62–64
    interchromosomal recombination and, 182.
        See also Recombination
    in meiosis, 63, 63f
    testcrosses for, 185–186, 186f
    vs. complete linkage, 177, 178f
    vs. nonindependent assortment, 174–175, 177
    vs. segregation, 53t
Indian blue peafowl, feather color in, 94, 95f
Induced mutations, 526, 528–532, 589.
        See also Mutations
    in mutagenesis screens, 623–624
Induced pluripotent stem cells, 654, 654f
Inducers, 466–467
Inducible operons, 466–467, 467f, 468
    definition of, 466
    negative, 466–467
    lac operon as, 469. *See also* *lac* operon
Induction, coordinate, 469, 470f
Influenza virus, 276, 276t, 277f
In-frame deletions, 519, 524t
In-frame insertions, 519, 524t
Inheritance, 47–79. *See also* Heredity
    of acquired characteristics, 8–9, 10t, 51
    anticipation in, 133, 519–520
    blending, 8–9, 10t
    chromosome theory of, 54–55, 83–84, 174–175, 183
        nondisjunction and, 90–92, 92f
    codominance in, 112, 112t
    of continuous characteristics, 134–135
    criss-cross, 93f
    cytoplasmic, 128–130, 129f, 130f, 132t
    of dominant traits, 53f, 54, 110–113, 111f, 112t, 148f, 149–150, 149f, 151–152, 152f, 153t
    early concepts of, 8–11
    gene interactions and, 117–125
    of genotype vs. phenotype, 51
    incomplete dominance in, 110–112, 111f
    incomplete penetrance in, 113, 113f
    of linked genes, 174–203. *See also* Linkage; Recombination
    Mendelian, 10, 10t, 49–51
    in monohybrid crosses, 51–61. *See also* Monohybrid crosses
    polygenic, 719
    of quantitative characteristics, 58–59, 719–721, 720f
    of recessive traits, 54, 148–149, 148f, 150–151, 151f, 153t

segregation in, 53–54, 56f, 174–175
    sex-linked, 82–83, 89–99.
        See also Sex-linked traits
    studies of. *See* Genetic studies
    uniparental, 235–236, 324–325
    of Y-linked traits, 152, 153t
Initiation codons, 440
Initiation factors, in translation, 442–443, 443f, 507
Initiator proteins, 348, 349f
Insertion(s), 518–519, 518f, 524t, 527–528, 528f
    in-frame, 519, 524t
Insertion sequences, 538, 538f
    in composite transposons, 538, 538f
Insulators, 500, 500f
Integrase, 273
Intelligence
    epigenetics and, 651
    heritability of, 734
    inbreeding and, 759
Interactome, 631
Intercalating agents, as mutagens, 531, 531f
Interchromosomal duplications, 220
Interchromosomal recombination, 182
Interference, 193–195
Intergenic suppressor mutations, 523–524, 523f, 524t
Interkinesis, 29, 32t
Interleukin 2 gene, 414f
Internal promoters, 387
International Cancer Genome Consortium, 704
International HapMap project, 614
International Human Genome Sequencing Consortium, 611
Interphase
    in meiosis, 29, 32t
    in mitosis, 23f, 24, 25f, 26t. *See also* Cell cycle
Interrupted conjugation, in gene mapping, 262–264
Interrupted genes, 401
Interspersed repeat sequences, 321–322
Intrachromosomal duplications, 220
Intrachromosomal recombination, 182
Intragenic suppressor mutations, 522–523, 524t
Intron(s), 402–403, 402f, 402t, 414f
    group I, 402, 402t, 409, 409f
    group II, 402, 402t, 409, 409f
    nuclear pre-mRNA, 402, 402t, 403, 414f
    self-splicing, 402, 402t, 409, 409f.
        See also Splicing
    size of, in humans, 628, 629f
    tRNA, 402, 402t, 416, 416f
Intron early hypothesis, 402
Intron late hypothesis, 402
Inversions, 224–227, 224f–226f, 241t
    in cancer, 707
    in evolution, 226
Inverted repeats, 535, 535f
    in composite transposons, 538
    in insertion sequences, 538f
    in noncomposite transposons, 538
    in rho-independent terminator, 384–385
Ionizing radiation, 532.
    See also Radiation exposure

IQ
  epigenetics and, 651
  heritability of, 734
  inbreeding and, 759
Irinotecan, 339
Isle Royale wolves, 749–750
Isoaccepting tRNA, 438
Isolated congenital asplenia, 429
Isoleucine, 434f
Isopropylthiogalactoside (IPTG), 469
Isotopes, 292

J segment, 682f, 683f
  antibody diversity and, 682–684
Jacob, François, 263, 403, 461, 468–469,
      470–474
Jacobsen syndrome, 519t
Japan, atomic bombing of, 534, 534f
Jefferson, Thomas, 97
Jimson weed, aneuploidy in, 230–231, 232f
Johannsen, Wilhelm, 50, 719
Jpx gene, 653, 653f, 653t
Jumping genes. *See* Transposable elements
Junctional diversity, 683

Kappa chains, 682
Karpechenko, George, 239
Karyotypes, 218–219
  definition of, 218
  in genetic testing, 159–160
  human, 218–219, 218f
  preparation of, 218–219
Kenny, Eimear, 47
Kimura, Motoo, 783
Kinases, cyclin-dependent, 699
Kinetochores, 22, 22f, 24
King, Thomas, 665
Kingsley, David, 663–664
Kit allele, in paramutation in mice,
      648–649, 649f
Klinefelter syndrome, 88, 88f, 99, 232
Knock-in mice, 593
Knockout mice, 591–593
Knudson, Alfred, 693
Knudson's multistep cancer model,
      693–694, 694f
Kossel, Albrecht, 289–290, 289f, 293
Kozak sequences, 444
Krüppel gene, 670, 670t, 671f
Krüppel-associated box domain zinc
      fingers, 492
Kucharski, Ryszard, 645
Kunkel, Louis, 559
Kuroda, Reiko, 110

lac enzymes, induction of, 469
lac mutations, 471–474, 474f, 475f
  operator, 473–474, 473f
  promoter, 474
  regulator-gene, 471–472
  structural-gene, 471
lac operon, 468–475, 470f
  catabolite repression and, 475, 476f
  discovery of, 461
  mutations in, 471–474, 474f, 475f
  regulation of, 469, 470f

lac promoter, 469
  mutations in, 474
lac repressors, 469, 470f
lacA gene, 469, 470f
lacI gene, 469
lacO gene, 469
lacP gene, 469, 470f
Lactose, 470f
  metabolism of, 469, 470f.
      See also lac operon
lacY gene, 469, 470f
lacZ gene, 469, 470f
  in cloning, 572–573
  mutations in, 471–474
Lagging strand, in replication, 347, 347f
Lambda chains, 682
Lambda phage (phage λ).
      See also Bacteriophage(s)
  as vector, 573, 573t
Large ribosomal subunit, 416
Lariat, 408f, 407l
Leading strand, in replication, 347, 347f
Leaf variegation, cytoplasmic inheritance and,
      129–130, 130f, 325
Leber hereditary optic neuropathy, 130, 325
Leder, Philip, 437–438
Lederberg, Joshua, 258–259, 269–270
Lederberg-Zinder experiment,
      269–271, 269f
Legal issues, in genetic testing, 164, 613
Leprosy, 251–252
Leptotene, 29, 29f
Lesch-Nyhan syndrome, 159t
Lethal alleles, 114
Lethal mutations, 521, 524t
Leucine, 434f
Leucine zipper, 465, 465f, 465t, 498
Leukemia, 707, 707f
Levene, Phoebus Aaron, 289f, 290, 293
Lewis, Edward, 671
Lewis, Meriwether, 781
Libraries. *See* DNA libraries
Li-Fraumeni syndrome, 549t
Ligase
  in recombination, 363
  in replication, 352, 352f, 353f
Light chains, immunoglobulin, 682, 682f
  antibody diversity and, 682–684
Linear eukaryotic replication, 344–345, 345f,
      345t, 348, 348f
LINEs (long interspersed elements), 322, 542
Linkage
  chi-square test for, 67–70, 68t, 69f, 185–186
  complete, 177, 178f, 182
  incomplete, 177, 178f, 182
  independent assortment and, 177, 178f
  lod scores and, 199
  testcross for, 177, 178f
  three-gene, 189–195
Linkage analysis, 199, 607f, 608f.
      See also Gene mapping
  single-nucleotide polymorphisms in, 200,
      613–615, 614f
Linkage disequilibrium, 200, 614
Linkage groups, 175
  in two-point crosses, 188–189

Linkage maps, 187–199, 607–608, 607f.
      See also Gene mapping
Linked genes, 174–198
  complete linkage of, 177, 178f, 182
  crosses with, 174–198. *See also* Crosses
      coupling in, 181–182, 181f
      notation for, 176
      predicting outcome of, 184
      recombination frequency for, 180, 184
      in repulsion, 181–182, 181f
      testcrosses for, 177, 178f
  crossing over with, 178–180, 179f
  definition of, 175
  incomplete linkage of, 178–180, 179f, 182
  recombination frequency for, 180, 184
Linker DNA, 316
Linkers, in cloning, 572
Lipoproteins
  elevated serum, 594
  in familial hypercholesterolemia,
      149–150, 150f
  epigenetics and, 652
Lipton, Mary, 630
lncRNA (long noncoding RNA), 375,
      421–422, 506
Loci, gene. *See* Gene loci
Lod scores, 199
Long interspersed elements (LINEs), 322, 542
Long noncoding RNA (lncRNA), 375,
      421–422, 506
Loss of heterozygosity, in cancer, 698, 698f
Loss-of-function mutations, 521, 524t
Lou Gehrig disease, 515–516
Lwoff, Andrew, 461
Lymnaea peregra, shell coiling in, 109–110,
      109f, 130–131, 131f
Lymphocytes, 680–681, 683f. *See also* T cell(s)
  B, 680, 680f–681f
  T, in HIV infection, 275, 276f
Lymphoma, Burkitt, 707, 707f
Lyon, Mary, 98, 98f
Lyon hypothesis, 98–99
Lysine, 434f
Lysogenic life cycle, viral, 267–268, 268f
Lytic life cycle, viral, 267–268, 268f

M phase, in cell cycle, 23, 23f, 24–26, 25f, 26t.
      See also Mitosis
MacLeod, Colin, 289f, 291
Macular degeneration, 614
Maize
  paramutation in, 648, 648f, 649f
  recombination in, 183
  transposable elements in, 539f, 540f, 541f
Major histocompatibility complex (MHC)
      antigens, 681
  in transplantation, 685
Major histocompatibility complex (MHC)
      genes, 684–685, 685f
Male-limited precocious puberty, 127, 128f
Male-pattern baldness, 173–174
Malignant tumors, 693. *See also* Cancer
Map unit (m.u.), 187, 607
Map-based sequencing, 610–611, 610f
Mapping functions, 197–198, 198f
Maps. *See* Gene mapping

Markers, Y-linked, 96–97

Mass spectrometry, in proteomics, 630–631, 631f

Maternal age, aneuploidy and, 235, 235f

Maternal blood testing, 160–161

Maternal effect, 110

Maternal spindle transfer, 330

Mathematical models, in population genetics, 751

Mating
  assortative, 759–760
    in sympatric speciation, 789
  nonrandom, 758–760

Matthaei, Johann Heinrich, 436–437

Maxam, Allan, 582

Maximum likelihood approach, 794

Maximum parsimony approach, 794

Mayr, Ernst, 788

McCarty, Maclyn, 289f, 291

McClintock, Barbara, 183, 183f, 320, 539–540, 539f

McClung, Clarence E., 84

MCM2-7, 355

Mean, 723, 723f

Mechanical isolation, 785, 785t

Media, culture, 254f

Mediator in transcription apparatus, 498

Medicine, genetics in, 4, 11–12

Megaspores, 38, 39f

Megasporocytes, 38, 39f

Meiosis, 13, 28–40, 28f, 29f–38f
  in animals, 36–38, 37f
  cell division in, 28–29, 28f
  crossing over in, 29, 29f, 30f, 31–34, 32t
  definition of, 28
  genetic consequences of, 33f
  genetic crosses and, 54–55
  genetic variation and, 30–34, 32f
  independent assortment in, 62f, 63
  inversions in, 225–226, 225f, 226f
  overview of, 28–29, 29f–31f
  in plants, 38–40, 38f, 39f
  regulation of, 699–700
  segregation in, 53–54, 56f, 174–175
  stages of, 28–29, 29f–31f, 32t
  translocation in, 227–229
  vs. mitosis, 28, 34, 34t, 35f

Mello, Craig, 418

Melting temperature, of DNA, 321

Memory cells, 681

Mendel, Gregor, 9, 10, 10f, 49–50, 49f, 517

Mendelian inheritance, 10, 10t, 49–51.
  See also Inheritance
  first law of, 53–54, 53f, 53t.
  See also Segregation
  polygenic, 719–721, 720f, 722f
  second law of, 53f, 53t, 61–67, 62f.
  See also Independent assortment

Mendelian population, 750

Meristic characteristics, 718

Merozygotes, 261

Meselson, Matthew, 341–342, 403

Meselson-Stahl experiment, 341–343, 341f, 342f

Messenger RNA. *See* mRNA (messenger RNA)

Metacentric chromosomes, 22, 22f, 218

Metagenomics, 616–617

Metal response element, 500

Metallothionein gene, regulation of, 501

Metaphase
  in meiosis, 29, 30f, 31f, 32t
  in mitosis, 24–26, 25f, 26t

Metaphase plate, 24

Metastasis, 693, 703

Methionine, 434f

Methylation
  DNA. *See* DNA methylation
  histone, 494, 645–647, 651

5-Methylcytosine, 303, 303f, 643–644, 644f

Mexican tetras, eye development in, 679

Meyerowitz, Elliot, 674–675

MHC antigens, 681
  in transplantation, 685

MHC genes, 684–685, 685f

Mice
  chimeric, 592
  genetic techniques with, 590–593
  knock-in, 593
  knockout, 591–593, 592f
  as model genetic organisms, 6, 6f, A10–A11
  paramutation in, 648–649, 649f, 650t
  transgenic, 590–593, 591f, 592f
  yellow, 114

Microarrays
  gene, 619–621, 620f, 621f
  protein, 631–632, 631f

Microbial communities, genome sequencing for, 616–617

MicroRNA. *See* miRNA (microRNA)

Microsatellites, in DNA fingerprinting, 586

Microspores, 38, 39f

Microsporocytes, 38, 39f

Microtubules, spindle, 22, 22f, 24, 25f, 26

Miescher, Johann Friedrich, 289, 289f, 293

Migration, 762–763
  allelic frequency and, 763f, 771–772, 771t

Miller, Oscar, Jr., 377

Minichromosome maintenance (MCM) complex, 355

Minimal media, 254

Minor splicing, 409

miRNA (microRNA), 376, 376t, 377f, 419–420, 419f, 419t. *See also* RNA
  in cancer, 506, 619–620, 620f, 621f, 703–704
  functions of, 419–420, 506
  in gene regulation, 505–506
  in genetic disorders, 506
  processing of, 419–420
  RNA crosstalk and, 506
  in RNA silencing, 505–506, 593–594
  vs. small interfering RNA, 419, 419t

Mismatch repair, 354, 544, 545f, 548t

Missense mutations, 520, 521f, 524t

Mitchell, Mary, 325

Mitochondria
  endosymbiotic theory and, 323, 323f
  mutations affecting, 324–325
  structure of, 322–323, 322f

Mitochondrial diseases, 130, 325
  aging and, 329
  prevention of, 329–330

Mitochondrial DNA (mtDNA), 129
  age-related changes in, 329
  cellular amounts of, 326
  evolution of, 328–329
  in humans, 327, 327f
    aging and, 329
  inheritance of, 324–325, 324f
  in plants, 327t, 328, 328f, 331
  replication of, 324–325
  structure and organization of, 326–328, 327f, 328f
  telomeres and, 311–312
  in yeast, 327, 327t, 328f

Mitochondrial genome, 326–328, 327f, 327t, 328f

Mitochondrial replacement therapy, 329–330

Mitosis, 13, 23–26, 23f, 26t
  as cell cycle phase, 23–26, 23f, 25f, 26t
  chromosome movement during, 25f, 26
  definition of, 28
  regulation of, 699–700
  stages of, 23f, 24–26, 25f, 26t
  unequal crossing over in, mutations and, 221, 221f, 528
  vs. meiosis, 28, 34, 34t, 35f

Mitosis promoting factor, 699

Mitotic spindle, 22, 22f, 24, 25f.
  See also under Spindle
  in cancer, 699
  centromeres and, 319

Mitton, Jeffrey, 756

MN blood group antigens, 112

Mobile DNA. *See* Transposable elements

Model genetic organisms, 6–7, 6f, A1–A13
  Arabidopsis thaliana, 6, 6f, A8–A9
  Caenorhabditis elegans, 6, 6f, A6–A7
  Drosophila melanogaster, 6, 6f, A2–A3
  Escherichia coli, 6, 6f, A4–A5
  genome of, 626, 626t, 627t
  genome sequencing for, 608–609
  Mus musculus, 6, 6f, A10–A11
  Saccharomyces cerevisiae, 6, 6f, A12–A13

Moderately repetitive DNA, 321

Modified bases, 414–415, 414f

Modified ratios, 122, 122t

Molecular chaperones, 452

Molecular clock, 796–797, 796f

Molecular evolution. *See also* Evolution
  DNA sequence variation and, 783
  molecular clock and, 796f
  rate of, 794–796

Molecular genetic analysis, 560–604.
  See also Recombinant DNA technology
  applications of, 561
  challenges in, 561
  key innovations in, 561
  recombinant DNA technology in, 562.
  See also Recombinant DNA technology
  techniques of, 561

Molecular genetics, 5, 5f

Molecular markers, in gene mapping, 199

Molecular phylogenies, 791–794, 792f–794f

Molecular variation, evolution and, 782–783

Mollusks, sex determination in, 85–86, 86f

Monod, Jacques, 461, 468–469, 470–474
Monoecious organisms, 83
Monohybrid crosses, 51–59.
    See also Crosses; Inheritance
  F_1 generation in, 51–52, 52f
  F_2 generation in, 52, 52f
  F_3 generation in, 53f, 54
  P generation in, 51, 52f
  reciprocal, 52
Monosomy, 230, 241t
Monozygotic twins, 154, 154f
  epigenetic effects in, 652
Morgan, Thomas Hunt, 10, 89–90, 90f, 174, 187, 517
Morphogens, 668
Mosaicism, 154, 236, 236f
  CRISPR-Cas genome editing and, 566
  in trisomy 8, 235
Motifs, DNA-binding, 464–465, 465f, 465t, 498
Mouse. *See* Mice
mRNA (messenger RNA), 375, 376t, 403–413.
    See also RNA
  cDNA libraries and, 577, 578f
  degradation of, in gene regulation, 503–504
  functions of, 404
  polyribosomal, 449, 450f
  protein-coding region of, 404
  ribosomes and, 403, 404f, 449, 450f
  RNA crosstalk and, 506
  stability of, in gene regulation, 503–504
  structure of, 404, 404f
  synthetic, in genetic-code experiments, 436–438, 436f, 437f
  transfer, 451, 451f
  in translation, 441–449.
    See also Translation
mRNA decay, nonstop, 451
mRNA processing, 405–413, 405f–413f.
    See also Pre-mRNA; RNA processing
  alternative pathways for, 409–411, 411f
  in gene regulation, 502–505, 502f–504f, 505f
  splicing in, 407–409
  steps in, 413, 413f
mRNA surveillance, 449–451
mtDNA. *See* Mitochondrial DNA (mtDNA)
Mu bacteriophage, transposition in, 538–539, 539f
Muller, Hermann, 320, 532
Mullerian-inhibiting substance, 88
Mullis, Kary, 11, 569
Multifactorial characteristics, 135
Multigene families, 627, 798, 798f
Multiple 3′ cleavage sites, 410, 414f
Multiple alleles, 114–116, 115f, 116f
Multiple crossovers, 197–198, 197f
Multiple-loci crosses, 61–67
Multiplication, of fractions, B2
Multiplication rule, 57, 57f, 754
Mus musculus. *See* Mice
Musa acuminata (banana), genome sequencing for, 217–218
Muscular dystrophy, 202, 509–510
Mushroom poisoning, 373–374
Mustard gas, as mutagen, 529, 530

Mustard plant. *See Arabidopsis thaliana* (mustard plant)
Mutagen(s), 528–532
  Ames test for, 532–533, 533f
  chemical, 528–532, 529f–531f
  definition of, 528
Mutagenesis
  genome-wide, 623–624
  oligonucleotide-directed, 590, 590f
  site-directed, 590
  targeted, 589–590, 590f
Mutagenesis screens, 623–624, 623f
Mutations, 13, 515–557.
    See also specific genes
  aging and, 329
  alkylating agents and, 530, 531f
  allelic, 124
  allelic frequencies and, 760–762, 761f, 771–772, 771t
  Ames test for, 532–533, 533f
  aneuploid, 230–235. *See also* Aneuploidy
  anticipation and, 133, 519–520
  auxotrophic, 430–433, 431f, 432f
  base analogs and, 529–530, 530f
  base mispairing and, 527, 527f
  base substitution, 518, 518f, 521f, 523f, 524t, 529–530, 529f, 530f
  Beadle-Tatum experiment with, 430–433, 431f, 432f
  in cancer, 532–534, 533f, 693–695, 702, 706–708, 706f
  causes of, 526–532
  in cell cycle regulation, 700–701
  chloroplast, 324–325
  chromosome, 217–249, 219–241, 241t, 518
    aneuploid, 219–220, 219f, 230–235, 241t.
      See also Aneuploidy
    in cancer, 700–701, 706f
    definition of, 218
    fragile-site, 229, 229f
    lethality of, 231, 232
    maternal age and, 235, 235f
    polyploid, 219f, 220, 230, 236–241.
      See also Polyploidy
    rate of, 231
    types of, 219–220, 219f, 241t
    unequal crossing over and, 221, 221f, 528, 528f
    vs. gene mutations, 518
  in cis configuration, 181–182, 181f
  classification of, 517–518
  clonal evolution and, 694–695, 695f
  complementation tests for, 124
  conditional, 521
  constitutive, 471
  in coupling, 181–182, 181f
  deamination and, 528, 529f, 530
  definition of, 516
  deletion, 518–519, 518f, 524t
  depurination and, 528, 529f
  DNA repair and, 544–549, 548t
  down, 382
  in evolution, 241
  expanding trinucleotide repeats and, 519–520, 519t, 520f, 524t

  experimental uses of, 517
  forward, 520, 522f, 524t
  frameshift, 518–519, 524t
  gain-of-function, 521, 524t
  gene, 518
  in genetic analysis, 517
  germ-line, 517f, 518
  hydroxylating agents and, 531, 531f
  importance of, 516–517
  incorporated errors and, 527, 527f, 528
  induced, 526, 528–532, 589
    chemical changes and, 528–532
    in mutagenesis screens, 623–624, 623f
  insertion, 518–519, 518f, 524t
  intergenic suppressor, 524t
  intragenic suppressor, 524t
  lac, 471–474, 474f, 475f
  lethal, 521, 524t
  location of, 124
  loss-of-function, 521, 524t
  missense, 520, 521f, 524t
  mitochondrial, 324–325, 325f
  natural selection and, 771
  neutral, 521, 524t
    genetic variation and, 783, 796
  nonsense, 449–450, 520, 521f, 524t
  one gene, one enzyme hypothesis and, 430–433, 431f, 432f
  organelle, 324–325, 325f
  oxidative, 531, 531f
  petite, 325, 325f
  phenotypic effects of, 521
  poky, 325
  radiation-induced, 532, 532f, 533–534
  rates of, 525–526
  replicated errors and, 527, 527f
  in repulsion, 181–182, 181f
  reverse, 520, 522f, 524t
  silent, 521, 521f, 524t
  single-nucleotide polymorphisms and, 613–615, 614f
  somatic, 517, 517f
  SOS system and, 532
  spontaneous, 527–528
    chemical changes and, 529f–531f
  strand slippage and, 527, 528f
  study of, 532–534, 533f
  suppressor, 521–524, 522f, 524t
  tautomeric shifts and, 527
  in trans configuration, 181–182, 181f
  transition, 518, 518f, 524t, 528
  translesion DNA polymerases and, 548
  transposable elements and, 538f
  transposition and, 536–538, 538f
  transversion, 518, 518f, 524t
  unequal crossing over and, 221, 221f, 528, 528f
  up, 382
Mycobacterium leprae, 251–252
Myles, Sean, 47
Myoclonic epilepsy and ragged-red fiber disease (MERRF), 324–325
Myoclonic epilepsy of Unverricht-Lundborg type, 519t
Myotonic dystrophy, 519t

Nagasaki, atomic bombing of, 534
Nail-patella syndrome, 198–199, 198f
Nanoarchaeum equitans, genome of, 624, 625t
Nanopore sequencing, 584–586
nanos gene, 669f, 669t, 670
Narrow-sense heritability, 730
Native Americans, albinism in, 1–2, 2f
Native ChIP, 497–498
Natural selection, 736–737, 766–772
    allelic frequency and, 766–772, 767t, 768t,
        770t, 771t, 772f
    balancing, 783–784
    directional, 769–770
    fitness and, 767
    general selection model and, 767–768
    mutation and, 771
    selection coefficient and, 767
Neanderthals, genomes of, 657
Neel, James, 534, 759
Negative assortative mating, 758
Negative control, transcriptional, 466–468,
    467f, 468f
Negative inducible operons, 466–467, 467f
    lac operon as, 469
Negative repressible operons, 467–468, 468f
    trp operon as, 476–481, 478f, 479f
Negative supercoiling, 313, 313f
Nematode. *See Caenorhabditis elegans*
    (nematode)
Neonatal screening, for genetic diseases, 162
Neural-tube defects, 159t, 160
Neurodegenerative diseases, apoptosis in, 677
Neurofibromatosis, 227
Neurospora crassa (bread mold), 6, 430–433,
    431f, 432f
Neutral mutations, 521, 524t
    genetic variation and, 783, 796
Neutral-mutation hypothesis, 783, 796
Newborn screening, for genetic diseases, 162
Next-generation sequencing technologies,
    584–586, 585f
Nilsson-Ehle, Herman, 719–721
Nirenberg, Marshall, 436–438
Nitrogenous bases. *See* Base(s)
Nitrous acid, as mutagen, 530, 531, 531f
Nóbrega, Marcelo, 627
Nodes, on phylogenetic tree, 792, 792f
No-go decay, 451
Noncoding DNA, 321–322, 627
Noncomposite transposons, 538
Nondisjunction
    aneuploidy and, 230
    chromosome theory of inheritance and,
        90–92, 92f
    definition of, 91
    Down syndrome and, 233
    maternal age and, 235
    mosaicism and, 235
    polyploidy and, 236–241, 237f
Nonhistone chromosomal proteins, 315
Nonhomologous end joining, 548
Nonindependent assortment, 175f
Nonoverlapping genetic code, 439
Nonrandom mating, 758–760
Nonreciprocal translocations, 227, 241t
Nonrecombinant gametes, 179–180, 179f

Nonrecombinant progeny, 177, 179–180,
    179f, 180f
Nonreplicative transposition, 536
Nonsense codons, 440
Nonsense mutations, 449–450, 520, 521f, 524t
Nonsense-mediated mRNA decay,
    449–450, 450
Nonstop mRNA decay, 451
Nonsynonymous substitutions, 794–795,
    795f, 795t
Nontemplate strand, 378
NORAD, 421–422
Normal distribution, 722, 722f
Northern blotting, 568
Notation
    for alleles, 52, 60–61
    for crosses, 54–55, 60–61
    for X-linked genes, 54–55, 60–61
Notch mutation, 224, 224f
Nousbeck, Janna, 145
Nowick, Katja, 492
Nuclear envelope, 18f, 19
Nuclear magnetic resonance, in structural
    proteomics, 632
Nuclear pre-mRNA introns, 402, 402t, 403
Nucleases, engineered, 564
Nucleic acids. *See also* DNA; RNA
    protein and, 374
Nuclein, 289
Nucleoids, 314
Nucleosides, 297, 297f, 298t, 299f
Nucleosome(s), 315–317, 315f, 316f
    assembly of, in replication, 356–357
Nucleosome remodeling factor, 494
Nucleotide(s). *See also* Base(s)
    in codons, 435–436
    deamination of, 528, 529f, 531f
    definition of, 290
    depurination of, mutations and, 528, 529f
    discovery of, 290
    DNA, 297, 297f–299f, 298t. *See also*
        Polynucleotide strands
    evolutionary rates for, 794–795, 795t
    in genetic code, 435–436, 438–439, 438f,
        439t, 440t. *See also* Codons
    reading frames for, 439–440, 518
    RNA, 295–296, 378–379
        addition of, in transcription, 379f
    sequence of, protein function and, 617–619
    structure of, 290, 290t
Nucleotide substitutions, rates of, 794–795, 795t
Nucleotide-excision repair, 546, 547f, 548t
    in genetic disease, 548, 549t
Nucleus, 18f, 19
Nullisomy, 230, 241t
Numerator, B1
Nutrition, epigenetics and, 641–642, 652

Obesity
    adoption studies of, 156–157, 157f
    genetic factors in, 156–157, 157f
    gut bacteria and, 617
Okazaki, Reiji, 347
Okazaki fragments, 347, 347f
Oligonucleotide-directed mutagenesis, 590, 590f
Oncogenes, 273, 697

One gene, one enzyme hypothesis, 430–433,
    431f, 432f
1000 Genomes Project, 612
"Ontogeny recapitulates phylogeny," 678
Oocytes, 36–37, 37f
Oogenesis, 36–37, 37f
Oogonia, 36, 37f
Operators, 466, 466f
Operons, 461–462, 465–481
    definition of, 465
    discovery of, 461
    evolution of, 462
    functions of, 461–462
    inducible
        definition of, 466
        negative, 466–468, 467f, 469
        lac, 468–475, 470f
            mutations in, 471–474
    promoters in, 466, 466f, 469
    regulator proteins in, 466
    regulatory genes in, 463, 466
        mutations in, 471–474, 474f
    repressible, 467–468, 478f
        negative, 467–468, 468f, 476–481, 479f
    structural genes in, 463, 465–466
        mutations in, 471
    structure of, 465–466
    trp, 476–481, 478f
Organ transplantation, immune
    response in, 685
Organelle(s). *See also* Chloroplast(s);
        Mitochondria
    endosymbiotic theory and, 323, 323f
    inheritance of, 324, 324f
    mutations affecting, 324–325, 325f
    origin of, 323, 323f
    structure of, 322–323, 322f
Organelle-encoded traits, 324–325
Origin of replication, 23
    in cloning vector, 571, 571f
Origin-recognition complex, 354
Orthologs, 618, 618f
Oryza sativa (rice), genome of, 331f
Osteogenesis imperfecta, 159t
Out of Africa hypothesis, 329
Outcrossing, 758
Ovary, in plants, 38
Overdominance, 770
Ovum, 37, 37f
Oxidation
    aging and, 329
    mutations and, 531, 531f
Oxidative phosphorylation, aging and, 329

P bodies, 504
P elements, 540–541
P (parental) generation, in monohybrid
    crosses, 51, 52f
p53, 700, 706, 709
Pachytene, 29, 29f
Pair-rule genes, 670, 670t, 671f
Palindromic sequences
    clustered regularly interspaced short.
        See under CRISPR
    on Y chromosome, 96

palladin gene, 681–692

Palladio, Andrea, 691

Pancreatic cancer, 661–662, 692f

Pangenesis, 8, 9f, 10t

Paracentric inversions, 224, 225f, 241t

Paralogs, 618, 618f

Paramutation, 647–650
  in corn, 648, 648f, 649f
  in mice, 648–649, 649f, 650t

Parental gametes, 179–180, 179f

Parental (P) generation, in monohybrid
  crosses, 51

Parental progeny, 177, 179–180, 179f, 180f

Parsimony approach, for evolutionary
  relationships, 794

Parthenogenesis, 240

Partial diploid, 471

Patau syndrome, 235

Pattern formation
  in *Arabidopsis*, 674–676, 674f, 675f
  in *D. melanogaster*, 666–674
    anterior-posterior axis in, 668–669, 669t
    dorsal-ventral axis in, 668, 668f, 669t
    egg-polarity genes in, 667–670, 667t
    gap genes in, 670, 670t, 671f
    homeobox genes in, 671–672, 672f, 673f
    pair-rule genes in, 670, 670t, 671f
    segmentation genes in, 670, 671f
    segment-polarity genes in, 670,
     670t, 671f
    stages of, 666–667, 667f
    vs. in flowers, 676

Pauling, Linus, 796

Pavo cristatus (Indian blue peafowl), feather
  color in, 94, 95f

Peas, Mendel's experiments with, 49–54

Pedigree, 147

Pedigree analysis, 147–154
  autosomal dominant traits in, 149–150,
    149f, 153t
  autosomal recessive traits in, 148–149,
    148f, 153t
  proband in, 147, 148f
  symbols in, 141, 147f
  X-linked traits in
    dominant, 151–152, 152f, 153t
    recessive, 150–151, 151f, 153t
  Y-linked traits in, 152, 153t

Penetrance
  definition of, 113
  incomplete, 113

Pentaploidy, 236. *See also* Polyploidy

Pentose sugars, 296–297, 297f

Pepper plant, fruit color in, 117–118, 117f

Peptide bonds, 433, 435f, 444f, 445

Peptidyl (P) site, in ribosome, 444f,
  445–446, 447f

Pericentric inversions, 224, 224f, 226f, 241t

Permease, 469

Pesticides, genetic engineering with, 575–576

Petal color. *See* Flower color

petite mutations, 325, 325f

Petri plates, 254, 254f

Phage(s). *See also* Bacteriophage(s)
  lambda, as vector, 573, 573t

Pharmacogenetic testing, 162–163

Pharmacology
  genetics and, 3–4
  recombinant DNA technology and, 595

Phenocopy, 134

Phenotype. *See also* Traits
  definition of, 12, 50t, 51
  expression of, 51, 117–118
    gene interaction and, 117–122.
      See also Gene interaction
  factors affecting, 51
  genotype and, 12, 51.
    See also Genotype-phenotype
    relationship
  mutations and, 521
  novel, from gene interactions, 117–118, 117f

Phenotypic correlation, 739

Phenotypic ratios, 61, 61t
  from gene interaction, 122, 122t
  observed vs. expected, 67–70, 69f

Phenotypic variance, 728–730, 729f.
    See also Heritability

Phenylalanine, 434f

Phenylketonuria (PKU), 134, 159t

Phenylthiocarbamide, taste sensitivity for,
  779–780

Philadelphia chromosome, 707

Phorbol ester response element, 501t

Phosphate groups, 297, 298f

Phosphodiester linkages, 298, 299f

Phylogenetic trees, 792–794, 792f–794f

Phylogeny, 791–794
  ontogeny and, 678

Physa heterostropha, albinism in, 121, 121f

Physical maps, 187, 200–203, 608, 608f.
    See also Gene mapping
  definition of, 187

Pigmentation. *See* Color/pigmentation

Pili, sex, 259, 260f

piRNA (Piwi-interacting RNA), 376, 420

Pisum sativum (pea), Mendel's experiments
  with, 49–54

Pitx1 gene, 663–664, 678, 679

Piwi-interacting RNA (piRNA), 376, 420

Plant(s)
  alternation of generations in, 38, 38f
  aneuploidy in, 230–231, 232f
  breeding of, 3, 3f, 8, 8f, 238–239, 239f
    genetic correlations in, 739–740, 739t
    inbreeding and, 760
    quantitative genetics in, 715–716
  chloroplast DNA in, 129–130, 322–323,
    322f, 323–324, 330–331.
      See also Chloroplast DNA (cpDNA)
  cloning of, 665
  cytoplasmic inheritance in, 129–130, 130f
  cytoplasmic male sterility in, 325
  flower color in
    inheritance of, 110–112, 111f
    lethal alleles and, 114
  flower development in, 674–676, 674f, 675f
  flower length in, inheritance of,
    727–728, 728f
  gene transfer in, Ti plasmid for, 574, 574f
  genetically engineered, 3, 3f, 595–596
  genome of, 626t, 627t
  herbicide-resistant, 595–596

life cycle of, 38, 38f

Mendelian inheritance in, 49–54, 50f,
  52f, 53f

mitochondrial DNA in, 327t, 328, 328f, 331

pattern formation in, 674–676, 674f, 675f

pest-resistant, 595–596

polyploidy in, 236, 240. *See also* Polyploidy

sexual reproduction in, 38–40, 39f

viruses in, 273

Plantains, 217

Plaque, recombinant, 268f, 269, 272–273, 272f

Plasmid(s), 19
  bacterial, 255–256, 257f
  R, antibiotic resistance and, 264
  Ti, as cloning vector, 574, 574f

Plasmid vectors, 571–572, 571f, 572f, 573t
  selectable markers for, 571–572

Plating, 254
  replica, 254–255

Pleiotropy, 135
  genetic correlations and, 739

Plumage patterns
  in chickens, 127, 128f
  in ducks, 115, 115f
  in Indian blue peafowl, 94, 95f

Pluripotency, epigenetics and, 654–655

Pogona vitticeps, sex determination in,
  81–82, 85

Poisoning, mushroom, 373–374

poky mutations, 325

pol gene, 273–275

Polar bodies, 36–37, 37f

Polyacrylamide gel electrophoresis,
  629–630, 630f

Polyadenylation, of pre-mRNA, 406

Poly(A)-binding proteins, 504

Polycistronic RNA, 385

Polydactyly, 113, 113f

Polygenic characteristics, 135, 719–721,
  720f, 722f. *See also* Quantitative
  (continuous) characteristics
  definition of, 716
  gene number determination for, 721
  inheritance of, 719–721
  statistical analysis of, 727–728

Polygeny, 135

Polymerase chain reaction (PCR), 561,
  569–571, 569f
  applications of, 570–571
  key innovations in, 570
  limitations of, 570
  real-time, 570–571
  reverse-transcription, 570, 571
  steps in, 569–570, 569f

Polynucleotide strands, 298
  antiparallel, 299, 299f
  cohesive (sticky) ends of, 562, 564f
  complementary, 300
  definition of, 298
  in double helix, 299, 300f
  5′ end of. *See* 5′ end (cap)
  lagging, 347, 347f
  leading, 347, 347f
  nontemplate, 378, 379
  quadruple, 303
  reannealing of, 321

slippage of, 527, 528f
template, 378, 378f
3′ end of. *See* 3′ end
transcribed, 378, 378f
triple, 302–303, 303f
unwinding of
in recombination, 363
in replication, 346–347, 347f, 348–350,
349f, 355, 362–363
Polypeptides, 433
Polyploidy, 219f, 220, 230, 236–241, 241t
allopolyploidy, 238–239, 239f
autopolyploidy, 236f, 237–238, 237f
definition of, 230
in evolution, 241
in humans, 240–241
in plants vs. animals, 236, 240
significance of, 240–241
in speciation, 790
Polyribosomes, 449, 450f
Poly(A) tail, 406f, 414f
in RNA processing, 405t, 406, 406f
in translation, 444
Polytene chromosomes, 317, 317f
Poly-X females, 88
Population(s), 722
genetic structure of, 752
Mendelian, 750
migration of, 762–763
Population genetics, 5, 5f
allelic frequency and, 752–753.
See also Allelic frequency
definition of, 750
DNA sequence variation and, 783
effective population size and, 764
evolution and, 780–782
founder effect and, 765
genetic bottleneck and, 765
genetic drift and, 763–766, 765f, 766f,
771–772, 771t
genetic variation and, 751.
See also Genetic variation
genome evolution and, 798–799
genotypic frequency and, 751
Hardy-Weinberg law and, 753–758
of Isle Royale wolves, 749–750
mathematical models in, 751
migration and, 762–763, 763f, 771–772, 771t
mutations and, 760–762, 761f,
771–772, 771t
natural selection and, 766–772, 771t
nonrandom mating and, 758–760
phylogenies and, 791–794
Population size
effective, 764
genetic drift and, 764–765
Population variation, evolution and, 782–784
Position effect, 225
Positional cloning, 579–580, 581f
in cystic fibrosis, 580–581, 581f
Positive assortative mating, 758
Positive control, transcriptional, 468
Positive supercoiling, 313, 313f
Posttranscriptional gene silencing, 418–419,
419f, 499, 504–506, 593–594
in gene therapy, 594, 594f

Posttranslational processing, 452
in gene regulation, 507–508
Postzygotic reproductive isolating
mechanisms, 785, 785t
Prader-Willi syndrome, 131–132, 223t, 236
Precocious puberty, male-limited, 127, 128f
Preformationism, 8, 9f, 10t
Pregnancy-associated plasma protein, 161
Preimplantation genetic diagnosis, 161
Pre-mRNA, 375–376, 377f.
See also mRNA; RNA
processing of, 405–406
addition of 5′ cap in, 405–406, 405f, 405t
addition of poly(A) tail in, 405t, 406, 406f
alternative pathways for, 409–411,
411f, 419f
in gene regulation, 502–505,
502f–504f, 504f
polyadenylation in, 405t, 406, 406f
RNA editing in, 412, 412f
splicing in, 407–409. *See also* Splicing
steps in, 413, 413f
Pre-mRNA introns, 402, 402t, 403
Prenatal caloric restriction, epigenetic
changes and, 641–642
Prenatal genetic testing, 158–161, 159f, 159t,
160f. *See also* Genetic testing
noninvasive, 161
preimplantation, 161
Prenatal mitochondrial replacement therapy,
329–330
Presymptomatic genetic testing, 162
Prezygotic reproductive isolating
mechanisms, 784–785, 785t
Pribnow box, 382, 382f
Primary Down syndrome, 233, 233f.
See also Down syndrome
Primary immune response, 681
Primary miRNA (pri-miRNA), 420
Primary oocytes, 36, 37f
Primary spermatocyte, 36, 37f
Primase, 350f, 351, 353
Primers
in DNA sequencing, 582
in replication, 350f, 351, 352, 352f
pri-miRNA (primary miRNA), 420
Principle of independent assortment.
See Independent assortment
Principle of segregation. *See* Segregation
Privacy, genetic testing and, 164
Probability
addition rule for, 57–58, 57f
binomial expansion and, 59
chi-square test and, 67–70, 68t, 69f
conditional, 58
definition of, 56–57
multiplication rule for, 57, 57f
Probability method
for dihybrid crosses, 64–65
for monohybrid crosses, 56–60, 57f
Proband, 147, 148f
Probes, 568
in chromosome walking, 579–580, 579f
in DNA library screening, 577–578, 578f
in in situ hybridization, 203, 203f
selection of, 577–578

Proflavin, as mutagen, 531, 531f
Programmed cell death
in cancer, 699–700
in development, 676–677, 677f
Prokaryotes, 12, 18–19, 18f, 19f.
See also Archaea; Bacteria; Eubacteria
cell reproduction in, 20, 21f
cell structure in, 18f, 19, 19f
chromosomes of, 20, 21f
definition of, 18
DNA in, 19, 19f
gene regulation in, 462–464
genes of, 19
genome of, 624–625
sequencing of, 625t
Proline, 434f
Prometaphase, 24, 25f, 26t
Promiscuous DNA, 331
Promoters, 378, 378f
bacterial, 381–382
consensus sequences in, 381–382, 382f
core, 387
definition of, 378
eukaryotic, 386–387
internal, 387
lac, 469
mutations in, 474
in operon, 465, 466f, 469
regulatory, 387
RNA polymerase I, 387
RNA polymerase II, 387–388
trp, 476–481, 478f, 479f
Pronuclear transfer, 329–330
Proofreading
in replication, 353–354, 544
in transcription, 384
Prophages, 268
Prophase
in meiosis, 29, 29f–31f, 32t
in mitosis, 24, 25f, 26t
Protein(s). *See also* Gene(s)
allosteric, 466
amino acids in, 435–440.
See also Amino acids
antiterminator, 478, 479f
catabolite activator, 475, 476f, 477f
chromatin-remodeling, 386
colinearity with genes, 400–401
diversity of, 627
DNA-binding, 349, 464–465, 465f,
465t, 498
evolution of, 798–799, 798f
folding of, 452
functions of, 433, 433f
prediction from sequence, 618–619.
See also Functional genomics
fusion, in cancer, 707
heat-shock, 501, 501t, 507
histone. *See* Histone(s)
identification of, 629–631
information transfer to, 301–302, 302f
initiator, 348, 349f
nonhistone chromosomal, 315
nucleic acids and, 374
posttranslational modifications of, 452,
507–508

in recombination, 362–363
regulator, 466
single-strand-binding, 349, 349f, 363
structure of, 416f, 433–435, 435f
    determination of, 629–631
synthesis of, 441–449. *See also* Translation
TATA-binding, 387, 388f, 391
transcriptional activator, 387
transcriptional antiterminator, 478
variation in, 783–784.
    See also Genetic variation
Protein domains, 464
Protein microarrays, 631–632, 631f
Protein-coding region, of mRNA, 404
Proteome, 617
    definition of, 629
Proteomics, 12, 629–632
Proto-oncogenes, 697, 697t
Protospacers, 565
Prototrophic bacteria, 254
Proviruses, 273, 274f
Pseudoautosomal regions, 85
Pseudodominance, 224
Pseudogenes, 796, 799
Pseudouridine, 414–415, 414f
PTC, taste sensitivity for, 779–780
Puberty, male-limited precocious,
    127, 128f
Puffs, chromosomal, 317, 317f
PUMILIO proteins, 422
Punnett, Reginald C., 56, 175
Punnett square, 56, 56f, 58
Purines, 297–298, 297f, 298t
Pyrimidines, in DNA, 297–298, 297f, 298t
Pyrosequencing, 584–586, 585f

Q banding, 219, 219f
Quad screen, 161
Quadruplex DNA, 303
Qualitative (discontinuous) characteristics,
    134, 716, 716f, 717f
Quantitative (continuous) characteristics,
    135, 716–728. *See also* Polygenic
    characteristics
analytic methods for, 721–728.
    See also Statistical analysis
genome-wide association studies for, 736
genotype-phenotype relationship and,
    716–718, 717t, 718f
heritability of, 728–736.
    See also Heritability
inheritance of, 58–59, 135, 719–721, 720f
meristic, 718
origin of, 715–716
statistical analysis of, 721–728
threshold, 718, 718f
types of, 718
vs. qualitative traits, 716, 717f
Quantitative genetics, 715–747
    definition of, 715
Quantitative trait loci (QTLs), 715–716,
    734–736, 736t
    definition of, 715, 734
    mapping of, 734–736, 735f
Quaternary structure, of proteins,
    435, 435f

R banding, 219, 219f
R plasmids, antibiotic resistance and, 264
Rabbits, coat color in, 134, 134f
Radiation
    ionizing, 532
    ultraviolet, 532
Radiation exposure
    in Japan, 534, 534f
    mutations and, 532, 532f, 533–534
    in Russia, 534
Ramakrishnan, Venkatraman, 449
ras oncogene, in colon cancer, 706
Ras signal-transduction pathways, 701, 702f
RAT-1 exonuclease, 390, 390f
Ratios, phenotypic/genotypic, 61, 61t
    from gene interaction, 122, 122t
    observed vs. expected, 67–70, 69f
Ray, Christian, 461–462
RB protein, 699, 699f
Reading frames, 439–440, 518
Realized heritability, 738
Real-time polymerase chain reaction,
    570–571
Reannealing, of DNA, 321
RecBCD protein, 363
Receptors, steroid, 465t
Recessive epistasis, 118–119
    duplicate, 120–121
Recessive traits
    inheritance of, 54, 148–149, 148f, 150–151,
        151f, 153t
    X-linked, 150–151, 151f, 153t
Reciprocal crosses, 52
Reciprocal translocations, 227, 228f, 241t
    in cancer, 707, 707f
Recombinant DNA technology, 3–4, 559–604.
        See also Molecular genetic analysis
    in agriculture, 217–218, 574f, 575–576,
        575f, 595–596
    challenges facing, 561
    chromosome walking in, 579–580, 579f
    cloning in, 571–575. *See also* Cloning
    concerns about, 595–596
    CRISPR-Cas systems in, 566
    definition of, 562
    DNA fingerprinting in, 586–589, 587f–589f
    DNA hybridization in, 568
    DNA libraries in, 576–578
    DNA sequencing in, 582–586
    in drug development, 595
    gel electrophoresis in, 567, 568f
    in gene identification, 576–581
    in gene mapping, 582–585f
    in genetic testing, 596
    in situ hybridization in, 578, 579f
    knockout mice in, 591–593, 592f
    mutagenesis in, 589–590
    Northern blotting in, 568
    polymerase chain reaction in, 569–571, 569f
    probes in, 568
    restriction enzymes in, 562–563, 563t, 564f
    Southern blotting in, 568
    transgenic animals in, 590–593
    Western blotting in, 568
Recombinant gametes, 179–180, 179f
Recombinant plaques, 268f, 269, 272–273

Recombinant progeny, 179–180, 179f
Recombination, 32, 174–203
    in bacteria, 362–363
    cleavage in, 362
    crossing over and, 29, 30f, 175f, 176f
    definition of, 174, 360
    double-strand break model of, 362, 362f
    enzymes in, 362–363
    Holliday model of, 361–362, 361f
    homologous, 360–361, 361f
    hotspots for, 203
    independent assortment and, 34, 62–64,
        62f, 117, 174–175, 175f
    interchromosomal, 182
    intrachromosomal, 182
    inversions and, 224–227, 225f
    nonindependent assortment and,
        174–175, 175f
    physical basis of, 183
    rate variation in, 203
    somatic, 683, 684
    three-gene, 189–195
    two-gene, 174–189
Recombination frequencies, 180, 184
    calculation of, 180, 193, 273
    definition of, 180
    distance between genes and, 187–188,
        197–198, 198f
    gene mapping with, 187–188, 193,
        197–198, 607
Recursive splicing, 408
Regression, 725–727, 725f, 726f
Regression coefficient, 726, 726f
Regulator genes, 466
Regulator proteins, 466
Regulatory domains, 464
Regulatory elements, 463
Regulatory genes, 463
    mutations in, 471–472, 474f
Regulatory promoter, 387, 387f
Reinforcement, 786
Relaxed-state DNA, 313, 313f
Release factors, 446, 447f
Renaturation, of DNA, 321
Repetitive DNA, 321
Replica plating, 254–255
Replicated errors, 527, 527f
Replication, 301–302, 302f, 339–360
    accuracy of, 340, 353–354, 527
    in archaea, 360
    autonomously replicating sequences in, 354
    in bacteria, 341–344, 343f, 344f, 345t,
        348–354
    base pairing in. *See* Base(s)
    basic rules of, 354
    bidirectional, 344
    in cancer, 339–340
    in cell cycle, 23, 24, 26t, 27–28, 27f,
        354–355, 699–700
    at chromosome ends, 358–360, 358f, 359f
    in circular vs. linear DNA, 358–360, 358f
    conservative, 340–341, 341f
    continuous, 346–347, 347f
    definition of, 301
    deoxyribonucleoside triphosphates in,
        346, 346f

direction of, 343–344, 346–348, 347f
discontinuous, 346–347, 347f
dispersive, 340–341, 341f
DNA gyrase in, 349–350, 349f
DNA helicase in, 349, 349f
DNA ligase in, 352, 352f
DNA polymerases in
  in bacteria, 346, 351–353, 351t
  in eukaryotes, 355–356, 355t
DNA template in, 340, 346–347, 347f
elongation in, 350–353
in eukaryotes, 20–26, 344–345, 345f, 345t, 354–360, 363
information transfer via, 301–302, 302f
initiation of, 348, 349f
lagging strand in, 347, 347f
leading strand in, 347, 347f
licensing of, 355
linear, 345f, 345t, 348f
linear eukaryotic, 344–345, 345f, 345t
mechanisms of, 348–360
Meselson-Stahl experiment and, 341–343, 341f, 342f
mismatch repair in, 354
of mitochondrial DNA, 324–325
modes of, 343–345
nucleosome assembly in, 356–357, 356f
nucleotide selection in, 353–354
Okazaki fragments in, 347, 347f
origin of, 23
origin-recognition complex in, 354
plasmid, 255–256, 256f
primers in, 350f, 351, 352, 352f
proofreading in, 353–354, 544
rate of, 340, 344
requirements of, 346
RNA, 301–302, 302f
rolling-circle, 344, 344f, 345t, 348
semiconservative, 340–348, 341f
site of, 357
spontaneous errors in, 527.
    See also Mutations
stages of, 348–354
telomerase in, 358–360, 359f
termination of, 353
theta, 343–344, 343f, 345t, 348, 348f
transcription apparatus in, 379–381
in transposition, 535f, 536
unwinding in, 346–347, 347f
  in bacteria, 348–350, 349f
  in eukaryotes, 355, 362–363
viral, 273, 274f
in viruses, 344, 345t
Replication blocks, 532, 532f
Replication bubble, 343, 343f, 344, 344f, 348
Replication errors, 353–354
  mutations and, 527f
Replication fork, 343–344, 343f, 344, 344f, 347f, 348, 353
Replication licensing factor, 355
Replication origin, 343, 344f, 345t, 348, 349f
Replicative segregation, 324–325, 324f
Replicative transposition, 535f, 536
Replicons, 343, 344–345, 345t
Reporter sequences, 622, 623f
Repressible operons, 466, 467–468

negative, 467–468, 468f
  trp operon as, 476–481, 476f, 479f
Repressors
  bacterial, 469, 470f
  eukaryotic, 499
  lac, 469, 470f
  trp, 470f
Reproduction
  asexual, polyploidy and, 240
  cellular, 20–26. *See also* Cell cycle; Cell division
  sexual, 28–40. *See also* Meiosis; Sexual reproduction
Reproductive isolating mechanisms, 784–785, 785t
Repulsion (trans) configuration, 181–182, 181f
Response elements, 501, 501t
Response to selection, 737–740
  calculation of, 738
  factors affecting, 737
  genetic correlation and, 739–740, 740t
  limits to, 738–739, 739t
  phenotypic correlation and, 739
Restriction enzymes (endonucleases), 267, 562–563, 563t, 564f
  in gene mapping, 608, 608f
Restriction fragment length polymorphisms (RFLPs), 199
Restriction mapping, 608
Restriction sites, in cloning, 571–572
Retinoblastoma, 693–694, 698
Retinoblastoma protein, 699, 699f
Retrotransposons, 536, 537f, 539, 543, 543t.
    See also Transposable elements
  in *D. melanogaster*, 540–541
  in humans, 541–542
  in yeast, 539
Retroviruses, 273, 274–275, 274f–276f
  cancer-associated, 708–709, 708t, 709f
  human immunodeficiency virus, 274–275, 275f, 276f
Reverse duplications, 220
Reverse genetics, 589
Reverse mutations (reversions), 520, 522f, 524t
Reverse transcriptase, 273
Reverse transcription, 273, 302, 302f
Reverse-transcription polymerase chain reaction, 570, 571
Reversions, 520, 522f, 524t
Rhagoletis pomenella, evolution of, 789–790, 789f
Rhinoviruses, 276–277
Rho factor, 384
Rho-dependent terminator, 384
Rho-independent terminator, 384–385
Ribonucleases, in mRNA degradation, 504
Ribonucleoproteins, small nuclear, 376, 376t, 377f, 418–419
Ribonucleoside triphosphates (rNTPs), 379, 379f
Ribonucleotides, 297, 298f, 298t
Ribose, 296–297
Ribosomal protein SA (RPSA), 429
Ribosomal RNA. *See* rRNA (ribosomal RNA)

Ribosomal subunits, 416, 442
Ribosome(s)
  bacterial, 417t
  decoding center in, 449
  eukaryotic, 417t
  factor-binding center in, 449
  mRNA and, 403, 404f, 507
  in polyribosomes, 449, 450f
  stalled, 450–451
  structure of, 416, 441, 441f, 449, 450f
  translation on, 441, 441f, 507–508.
      See also Translation
  tRNA binding sites on, 444, 444f, 445–446
Riboswitches, 482–483, 483f
Ribothymidine, 414f
Ribothymine, 414–415
Ribozymes, 374, 483–484, 483f
Rickets, vitamin D–resistant, 152
Rifamycin, 380
RNA, 12
  antisense, in gene regulation, 482, 482f
  as biological catalyst, 374
  classes of, 375–376
  CRISPR, 267, 376, 376t, 377f, 418, 420–421, 421f, 565. *See also* CRISPR-Cas systems/CRISPR-Cas9
  degradation of, in gene regulation, 503–504
  developmental conservation of, 666
  enhancer, 422, 500
  epigenetic changes and, 647
  functions of, 376t
  as genetic material, 295–296, 296f
  guide, 412
  information transfer via, 301–302, 302f
  location of, 376t
  long noncoding, 375, 421–422, 506
  messenger. *See* mRNA (messenger RNA)
  micro. *See* miRNA (microRNA)
  polycistronic, 385
  posttranscriptional processing of.
      See RNA processing
  primeval, 374
  replication of, 301–302, 302f
  ribosomal. *See* rRNA (ribosomal RNA)
  secondary structures in, 375
  small cytoplasmic, 377f
  small interfering. *See* siRNA (small interfering RNA)
  small nuclear. *See* snRNA (small nuclear RNA)
  small nucleolar. *See* snoRNA (small nucleolar RNA)
  splicing of, 402–403, 407–409.
      See also Splicing
  structure of, 374–375, 375f
  synthesis of. *See* Transcription
  synthetic, in genetic-code experiments, 436–438, 436f, 437f
  transfer. *See* tRNA (transfer RNA)
  in translation, 441–449. *See also* Translation
  Xist, 422, 647
RNA amplification, polymerase chain reaction in, 570
RNA cleavage, 362, 405t, 406, 406f, 410, 505
RNA crosstalk, 506

RNA editing, 412, 412f
RNA interference (RNAi), 418–419, 419f, 499, 504–506, 593–594
  in gene therapy, 594, 594f
RNA polymerase(s), 379–381
  antibiotics and, 380
  bacterial, 380, 380f, 384
  eukaryotic, 380–381, 380t, 387–388, 389f
  stalled, 500–501
  in transcription apparatus, 379–381, 380f, 387
RNA polymerase I, 380, 380t
RNA polymerase I promoters, 387
RNA polymerase II, 380, 380t
  structure of, 389, 389f
RNA polymerase II promoters, 386–387
RNA polymerase III, 380, 380t
RNA polymerase III promoters, 380t
RNA polymerase IV, 381
RNA polymerase V, 381
RNA probes. See Probes
RNA processing, 403–413. See also Pre-mRNA, processing of
  alternative pathways for, 409–411, 411f
  in gene regulation, 502–505, 502f–504f, 504f
  of mRNA, 405f–413f
  splicing in, 407–409, 409f. See also Splicing
  steps in, 413, 413f
  in tRNA, 415–416, 416f
RNA sequencing, 621–622, 622f
RNA silencing (RNA interference), 418–419, 419f, 499, 504–506, 528, 593–594
  in gene therapy, 594, 594f
RNA viruses, 273–277, 274f–277f
RNA world, 374, 418
RNA-coding region, 378, 378f
RNAi (RNA interference), 418–419, 419f, 499, 504–506, 593–594
  in gene therapy, 594, 594f
RNA-induced silencing complex (RISC), 419, 505
RNA-induced transcriptional silencing, 505–506
rNTPs (ribonucleoside triphosphates), 379, 379f
Robertsonian translocations, 227, 227f
  aneuploidy and, 230, 233–234
  in Down syndrome, 233–234
Robson, John, 529
Roca, Alfred, 783
Rocky Mountain bighorn sheep, 781–782, 781f
Roesch, Luiz, 253
Rolling-circle replication, 344, 344f, 345t, 348, 348f
Romanov dynasty, hemophilia in, 399–400
Rooted phylogenetic trees, 792
Rotman, Raquel, 272–273
Roundworms. See Caenorhabditis elegans (nematode)
Rous, Peyton, 697
Rous sarcoma virus, 697, 708
Royal jelly, 645
RPSA (ribosomal protein SA), 429

rRNA (ribosomal RNA), 375, 376t, 377f, 416–418. See also RNA
  bacterial, 417t
  eukaryotic, 417t
  genes produced by, 417–418
  processing of, 417–418
  structure of, 416, 417t
Russia, Chernobyl nuclear accident in, 534

S phase, of cell cycle, 23f, 24, 26t
Saccharomyces cerevisiae (yeast). See also Yeast
  genome of, 327, 328t, 626t, 627t
    duplication of, 241
  as model genetic organism, 6, 6f, A12–A13
  synthetic genome for, 605–606
Salmonella typhimurium
  in Ames test, 533
  transduction in, 269–270
Sample, in statistics, 722
Sampling errors, 764
Sanger, Frederick, 11, 582
Sanger's DNA sequencing method, 582–584, 582f–584f
Saqqaq, DNA of, 287–288
Saunders, Edith Rebecca, 175
Schizosaccharomyces pombe. See Yeast
Schleiden, Matthias Jacob, 10
Schramm, Gerhard, 296
Schull, William, 759
Schwann, Theodor, 10
Screening, for genetic diseases. See also Genetic testing
  postnatal, 162–163
  prenatal, 160–161
Screens, mutagenesis, 623–624, 623f
scRNA (small cytoplasmic RNA), 377f. See also RNA
Second polar body, 37, 37f
Secondary immune response, 681
Secondary oocyte, 36, 37f
Secondary spermatocyte, 36, 37f
Secondary structures
  in attenuation, 477–478, 479f, 481t
  in DNA, 299f, 300–301, 301f, 302–303, 302f, 376t
  in proteins, 433–435, 435f
  in RNA, 375, 376t
Segmental duplications, 220, 222
  in genome, 626–627
Segmentation genes, 670, 671f
Segment-polarity genes, 670, 670t, 671f
Segregation, 53–54, 53f, 56f, 62, 62f, 174–175
  adjacent-1, 228f, 229
  adjacent-2, 228f, 229
  alternate, 228f, 229
  chi-square test for, 67–70, 68t, 69f, 185–186
  independent assortment and, 34, 62–64, 62f, 174–175
  recombination and, 174–175, 176f. See also Recombination
  replicative, 324–325, 324f
  telomeric sequences in, 320f
  translocations and, 227–229, 228f
  vs. independent assortment, 53t

Selection
  artificial, 737, 737f. See also Breeding
  natural, 736–737, 766–772. See also Natural selection
  response to. See Response to selection
Selection coefficient, 767
Selection differential, 737–738
Self-splicing introns, 402, 402t, 409, 409f. See also Splicing
Semiconservative replication, 340–348, 341f. See also Replication
Sense codons, 438
Separase, 35
Sequence alignments, in phylogenetic trees, 792–794
Sequential hermaphroditism, 86, 86f
Serine, 416f
Serum response element, 501t
70S initiation complex, 443, 443f
Sex
  definition of, 83
  gamete size and, 79f, 83
  heredity and, 126–133
  heterogametic, 84
  homogametic, 84
  vs. gender, 83
Sex chromosomes
  aneuploidy of, 232
    in Klinefelter syndrome, 88, 88f, 99
    in Turner syndrome, 87–88, 87f, 99
  definition of, 84
  evolution of, 95–96
  X, 89–99
    abnormal number of, 97–99
    discovery of, 84
    inactivation of, 97–99, 98f, 99f, 231–232
    in Klinefelter syndrome, 88, 88f, 99
    in sex determination, 83–89, 84f. See also Sex determination
    structure of, 82f
    in triple-X syndrome, 88
    in Turner syndrome, 87–88, 87f, 99
  Y, 94–96
    characteristics of, 96
    discovery of, 84
    evolution of, 95–96
    function of, 94–96
    genetic markers on, 88–89, 96–97
    in Klinefelter syndrome, 88, 88f, 99
    as male-determining gene, 88–89
    palindromic sequences on, 96
    in sex determination, 83–89, 84f, 85f
    structure of, 82f
  Z, 85, 94, 95f
Sex determination, 81–108
  abnormalities in, 87–88, 87f, 88f
  in central bearded dragon, 81–82, 85
  chromosomal, 83–89
    XX-XO, 84, 86t
    XX-XY, 84–85, 84f, 86t, 97
    ZZ-ZW, 81–82, 85, 86t, 94
  in Crepidula fornicata, 85–86, 86f
  in D. melanogaster, 87, 87t
    alternative splicing in, 502–503, 503f, 504f
  definition of, 83

environmental, 85–86, 86f, 86t
genic, 85, 86t
hormonal, 89
in humans, 87–89
sequential hermaphroditism in, 86, 86f
X chromosome in, 83–89, 84f, 85f
Y chromosome in, 83–89, 84f, 85f
Y gene in, 89f
Sex pili, 259, 260f
Sex ratio, 84, 84f
Sex selection, 1:1 sex ratio in, 84, 84f
Sex-determining region Y gene, 89, 89f
Sex-influenced traits, 126–127, 127f, 132t
Sex-limited traits, 127, 128f, 132t
Sex-linked traits, 83, 89–99, 132t
    chromosome theory of inheritance and,
        90–91, 92f
    definition of, 89
    early studies of, 89–91
    inheritance of, 97
    nondisjunction and, 90–91, 92f
    recognition of, 97
    X-linked, 89–99
        chromosome inactivation and, 98f, 99f
        color blindness as, 92–93, 92–94, 93f
        criss-cross inheritance of, 93f
        in D. melanogaster, 89–91, 91f, 92f
        dominant, 151–152, 152f, 153t
        dosage compensation and, 97–99
        eye color as, 89–91, 91f, 92f
        inheritance of, 97, 150–151, 150f, 151f,
            152f, 153t
        nondisjunction and, 90–91, 92f
        notation for, 94
        recessive, 150–151, 151f, 153t
        recognition of, 97
        X chromosome inactivation and,
            97–99, 98f, 99f
        Y-linked, 89, 94–97, 152, 153t
            inheritance of, 97
            notation for, 94
            recognition of, 97
        Z-linked, 94, 95f
Sexual development
    precocious puberty in, 127, 128f
    sequential hermaphroditism in, 86, 86f
Sexual reproduction, 28–40
    in animals, 36–37, 37f
    in eukaryotes, 81–85, 84f
    fertilization in, 28
    genetic variation and, 30–34
    meiosis and, 28–40. See also Meiosis
    in plants, 38–40, 39f
Sheep, bighorn, 781–782, 781f
Shell coiling, in snails, 109–110, 109f
    genetic maternal effect and,
        130–131, 131f
Shelterin, 320
shh gene, 679
Shine-Dalgarno sequences, 404, 404f
    in translation, 443, 443f, 448
short gastrulation gene, 678
Short interspersed elements (SINEs), 322,
    541–542
Short tandem repeats (microsatellites), in
    DNA fingerprinting, 586

Shotgun sequencing, whole-genome,
    611–612, 611f
Shugosin, 35
Sickle-cell anemia, 159t, 770, 783–784
Sigma (σ) factor, 380, 380f
Signal sequences, 452
Signaling
    calcium, 679
    in cancer, 701
    in cell cycle regulation, 701
Signal-transduction pathways, 701, 702f
Silencers, 499
    splicing, 502
Silent mutations, 521, 521f, 524t
SINEs (short interspersed elements), 322,
    541–542
Singer, Bea, 295
Single-nucleotide polymorphisms (SNPs),
    146, 200, 614–615, 614f
    in genetic diseases, 614–615
    identification of, 736
    tag, 614
Single-strand–binding proteins
    in recombination, 363
    in replication, 349, 349f
siRNA (small interfering RNA), 376, 376t,
    377f, 419–420. See also RNA
    in paramutation, 648
    in RNA silencing, 505–506, 593–594
    vs. microRNA, 419, 419t
Sister chromatids, 22f, 23
    counting of, 27–28
    separation of, 25f, 26, 26t, 28, 29, 32–33,
        34–35
Site-directed mutagenesis, 590
SIV_cpz (simian immunodeficiency virus
    chimpanzee), 274–275, 275f
Skin cancer, in xeroderma pigmentosum,
    695, 702
Sleeping Beauty transposable element, 537
Slicer, 505, 506
Small cytoplasmic RNA (scRNA), 377f.
    See also RNA
Small eye gene, 678
Small interfering RNA. See siRNA
Small nuclear ribonucleoproteins (snRNPs),
    376, 376t, 377f, 418–420
Small nuclear RNA (snRNA), 376, 376t, 377f,
    407–408, 418–420. See also RNA
Small ribosomal subunit, 416
Smith, Hamilton, 608
Smithies, Oliver, 592
Snails
    albinism in, 120–121, 121f
    shell coiling in, 109–110, 109f,
        130–131, 131f
snoRNA (small nucleolar RNA), 376, 376t,
    377f, 417–418. See also RNA
snRNA (small nuclear RNA), 376, 376t, 377f,
    407–408, 418–420. See also RNA
snRNPs (small nuclear ribonucleoproteins),
    376, 376t, 377f, 418–420
sog gene, 678
Solomon Islanders, blond hair in, 47–48
Somatic gene therapy, 597
Somatic hypermutation, 683–684

Somatic mutations, 517, 517f
Somatic recombination, 683, 684
Somatic-cell hybridization, 201, 201f, 202f
sonic hedge gene, 679
SOS system, 532
Southern blotting, 568
Spartina anglica (salt grass), evolution of,
    70f, 790
Specialized transduction, 270–271
Speciation, 784–791
    allopatric, 786–787, 786f
    alternative splicing in, 411
    in Darwin's finches, 787–788, 787f
    definition of, 786
    genetic differentiation and, 791
    polyploidy in, 790
    sympatric, 786, 788–789, 790f
Species
    biological species concept and, 784
    definition of, 784
    reproductive isolation and, 784–785, 785t
Sperm
    in animals, 36, 37f
    in plants, 38–40
Spermatids, 36, 37f
Spermatocytes, 36, 37f
Spermatogenesis, 36, 37f
    vs. oogenesis, 36–37
Spermatogonia, 36, 37f
Spinal muscular atrophy, 519t
Spindle, mitotic, 22, 22f, 24, 25f
    in cancer, 699
    centromeres and, 319
Spindle microtubules, 22, 22f, 24, 25f, 26
Spindle-assembly checkpoint, 24–26,
    699–700
Spineless sticklebacks, development in,
    663–664, 678
Spinocerebellar ataxia, 519t
Spleen, absence of, 429
Spliceosome, 407–408, 408f
Splicing, 407–409
    alternative, 409–411, 411f
        in gene regulation, 502–503, 503f, 504f
    branch point in, 407
    consensus sequences in, 407, 407f
    minor, 409
    nonsense mutations and, 449–450
    recursive, 408
    self-splicing and, 402, 402t
    sites of, 407–408, 408f
    spliceosome in, 407–408, 408f
    steps in, 407–409, 407f–411f
    trans-splicing, 408
Splicing code, 407
Splicing enhancers, exonic/intronic, 502
Splicing silencers, 502
Spontaneous mutations, 527–528
Sporophytes, 39, 39f
SR proteins, in splicing, 502, 503f
Srb, Adrian, 431
SRY gene, 89, 89f
Stahl, Franklin, 341–342
Staining, chromosome, 218–219, 219f
Standard deviation, 723–724, 724f
Start codons, 404, 414f

Statistical analysis, 721–728
  correlation in, 724–725, 725f
  frequency distribution in, 721–722, 722f
  frequency in, 751–752
  mean in, 723, 723f
  normal distribution in, 722, 722f
  of polygenic traits, 727–728
  population in, 722
  of quantitative characteristics, 721–728
  regression, 725–727, 725f, 726f
  sample in, 722
  sampling errors in, 764
  standard deviation in, 723–724, 724f
  variance in, 723–724, 723f
Steitz, Thomas, 449
Stem, 302, 302f
Stem cells
  epigenetics and, 654–655
  induced pluripotent, 654–655, 654f
Steroid receptors, 465t
Stevens, Nettie, 84, 183
Steward, Frederick, 665
Sticky (cohesive) ends, 562, 564f
Stop codons, 404, 414f, 440, 447f
Strand slippage, 527, 528f
Streptococcus pneumoniae, transformation in, 290–291
Streptomyces coeliocolor, genome of, 624–625
Stress
  epigenetic effects of, 650–651
  telomeres and, 311–312
Structural genes, 463, 465–466.
    See also Operons
  mutations in, 471
Structural genomics, 606–617
  bioinformatics and, 615–616
  copy-number variations and, 154, 229–230, 516
  definition of, 606
  DNA sequencing and, 582–584
  genetic maps and, 607–608, 607f, 608f
  Human Genome Project and, 610–613
  physical maps and, 608
  single-nucleotide polymorphisms and, 613–615
Structural proteomics, 632
Submetacentric chromosomes, 22, 22f, 218
Subtraction, of fractions, B2
Sugars, nucleic acid, 296–297, 297f
Supercoiled DNA, 313, 313f, 339–340
Suppressor mutations, 521–524, 524t
  intergenic, 523–524, 523f
  intragenic, 522–523, 522f
Sutton, Walter, 10, 54, 175, 183
Swine flu, 276, 277f
SWI-SNF complex, 493–494
Symbols
  for alleles, 52, 60–61
  for crosses, 54–55, 60–61
  for X-linked genes, 54–55, 60–61
Sympatric speciation, 786, 788–789, 790f
Synapsis, 29
Synaptonemal complex, 29, 29f
Synonymous codons, 438
Synonymous substitutions, 794–795, 795f, 795t

Synthetic biology, 617
Synthetic genomes, 605–606
Szostak, Jack, 320

T antigen, alternative splicing in, 502, 503f
T cell(s), 680–681, 680f
  activation of, 507, 684–685, 685f
  in HIV infection, 275, 276f
T2 phage, 292–293, 293f, 294f
Tag single-nucleotide polymorphisms (SNPs), 614
Tandem duplications, 220
Tandem repeat sequences, 321
Taq polymerase, 570
Targeted mutagenesis, 589–590, 590f
Taster genes, 779–780
TATA box, 387, 391, 405f
TATA-binding protein, 388, 388f, 391
TATAAT consensus sequence, 382, 382f
Tatum, Edward, 258–259, 430–433
Tautomeric shifts, mutations and, 527
Tay-Sachs disease, 149, 159t, 162
T-cell receptor, 680–681
  antigen binding by, 684, 685f
  structure of, 684, 684f
T-cell receptor diversity, 684
Telocentric chromosomes, 22, 22f, 218
Telomerase, 358–360
  in aging, 359–360
  in cancer, 360, 702–703
  definition of, 358
  disease and, 359–360
  in replication, 358–360, 359f
Telomere(s), 22–23, 22f, 320
  aging and, 359–360
  childhood adversity and, 311–312
  in replication, 358–360, 359f
  structure of, 320
Telomeric repeats, 358
Telomeric sequences, 320, 320f, 320t
Telophase
  in meiosis, 29, 31f, 32t
  in mitosis, 25f, 26, 26t
Temperate phage, 268, 268f
Temperature-sensitive alleles, 133–134
Template strand, in transcription, 378
Temporal isolation, 785, 785t
−10 consensus sequence, 382, 382f
Terminal inverted repeats, 535, 535f
  in composite transposons, 538
  in insertion sequences, 538f
  in noncomposite transposons, 538
  in rho-independent terminator, 384–385
Termination codons, 404, 405f, 440, 447f
Terminators
  in bacteria, 378–379, 378f
    transcriptional, 384–385
  in eukaryotes, 389–390
  rho-dependent/independent, 384–385
  transcriptional, 477–481
Tertiary structure, of proteins, 435, 435f
Testcrosses, 60, 177
  dihybrid, 65, 65f
  for independent assortment, 185–186, 186f
  with linked genes, 177, 179f
  monohybrid, 60. *See also* Crosses

three-point, 189–195
two-point, gene mapping with, 188–189
Testis
  in androgen insensitivity syndrome, 89
  spermatogenesis in, 36, 37f
Testosterone, in sex determination, 89
Tetrad, 29
Tetranucleotide theory, 290
Tetraploidy, 236. *See also* Polyploidy
Tetras, eye development in, 679
Tetrasomy, 230, 241t
Theory of clonal selection, 681
Theta replication, 343–344, 343f, 345t, 348, 348f
Thiogalactoside transacetylase, 469
30S initiation complex, 443, 443f
−35 consensus sequence, 382, 382f
Thomas, Charles, 377
3′ cleavage, in RNA processing, 405t, 406, 406f
3′ cleavage site, 405f, 410, 411f
3′ end, 298
  in replication, 346–347, 347f, 358, 358f
  in RNA processing, 405t, 406, 406f
  in transcription, 382, 382f
  in translation, 441, 441f, 442f, 444, 444f
3′ splice site, 407, 407f
3′ untranslated region, 404, 404f, 444, 444f
Three-point testcross, 189–195
  steps in, 195
Threonine, 416f
Threshold characteristics, 718, 718f
Thrifty phenotype hypothesis, 642
Thymine, 297, 297f, 298t, 299f, 300, 528.
    See also Base(s)
Ti plasmid, as cloning vector, 574, 574f
tiggy-winkle hedgehog gene, 679
tmRNA (transfer-messenger RNA), 451, 451f.
    See also RNA
Tn10 composite transposon, 538
Tobacco mosaic virus, 295–296, 296f
toll gene, 668, 669t, 678
Tomas-Loba, Antonia, 360
Topoisomerases
  in crossover interference, 195
  in supercoiling, 313
Topotecan, 339
Totipotent cells, 664
Traits. *See also* Phenotype
  acquired, inheritance of, 8–9, 10t, 51
  autosomal, dominant, 149–150, 153t
  definition of, 50t, 51
  discontinuous, 134, 716, 716f, 717f
  dominant, 54
    autosomal, 149–150, 153t
    X-linked, 151–152, 152f, 153t
  epigenetic, 643
  heritability of, 728–736. *See also* Heritability
  homologous, phylogenies and, 791–794
  meristic, 718
  multifactorial, 135
  organelle-encoded, 324–325
  pleiotropic, 135
  polygenic, 135, 716, 720f, 722f, 727–728.
    See also Polygenic characteristics
  inheritance of, 719–721

qualitative, 134, 716, 717, 717f
quantitative, 134–135, 716–728.
      See also Quantitative (continuous)
      characteristics
recessive, 54
      autosomal, 148–149, 148f, 153t
      X-linked, 150–151, 151f, 153t
sex-influenced, 126–127, 127f, 132t
sex-limited, 127, 128f, 132t
sex-linked, 89–99. *See also* Sex-linked traits
vs. characteristics, 51
vs. genes, 12. *See also* Genotype-phenotype
      relationship
X-linked, 89–99, 150–151, 151f, 153t. *See
      also* X-linked traits
Y-linked, 89, 94–97, 152, 153t. *See also*
      Y-linked traits
Z-linked, 94, 95f
Trans configuration, 181–182, 181f
Transcription, 13, 301–302, 302f, 373–398
      accuracy of, 384
      activators in, 387
      in archaea, 390–391
      backtracking in, 384
      in bacteria, 381–386
      basic rules of, 386
      chromatin modification in, 317f, 318, 386,
            493–497
      consensus sequences in
            in bacteria, 381–382, 382f
            in eukaryotes, 387, 389
      coupled to translation, 405, 448, 449, 482
      direction of, 379
      DNA template in, 377–379, 378f
      DNAse I in, 317–318, 318f
      downstream, 379
      elongation in
            in bacteria, 383–384
            in eukaryotes, 389, 389f
      enhancers in, 387
      essential components of, 377
            in bacteria, 390–391
            in eukaryotes, 386–390
            vs. in archaea and eubacteria, 390–391
      evolutionary relationships and, 390–391
      holoenzymes in, 380, 386
      information transfer via, 301–302, 302f
      initiation of
            abortive, 383
            in bacteria, 381–383
            in eukaryotes, 387–389, 497–500,
                  498f–500f
            regulation of, 465–477
      in *lac* operon, 469
      nontemplate strand in, 378
      nucleotide addition in, 379, 379f
      numbering system for, 379
      pausing in, 384, 500–501
      promoters in, 378, 378f
            in bacteria, 381–382
            in eukaryotes, 386–387
      proofreading in, 384
      regulation of. *See also* Gene regulation
            in bacteria, 461–490
            in eukaryotes, 491–513
      repression by methylation, 497, 645

reverse, 273, 302, 302f
ribonucleoside triphosphates in, 379, 379f
RNA polymerases in, 379–381, 380f
      in bacteria, 380, 380f, 383, 384
      in eukaryotes, 386–387
sigma (σ) factor in, 380, 380f
stages of, 413, 413f
      in bacteria, 365–386
      in eukaryotes, 387–390
stalling in, 500–501
start site for, 382, 405f
substrate for, 379
template strand in, 378, 378f
termination of
      in bacteria, 384–386
      in eukaryotes, 389–390
      premature, 477–481
      regulation of, 477–481
transcribed strand in, 378, 378f
transcription apparatus in, 379–381,
      380f, 388f
upstream, 379
upstream element in, 382
Transcription activator–like nuclease, 564
Transcription apparatus, 379–381, 380f, 380t,
      382f, 387, 388f, 488f, 497–499, 498f,
      499f, 500
      mediator in, 498
Transcription bubble, 384
Transcription factors, 387, 389–390, 497–501
Transcription unit, 378–379, 378f
Transcriptional activator proteins, 387,
      497–499, 498f
Transcriptional coactivator proteins, 497–499
Transcriptional repressors. *See* Repressors
Transcriptomes, 617
Transcriptomics, 619. *See also* Gene
      expression
Transducing phages, 269–272
Transductants, 270
Transduction, 257f, 258, 269–272, 269f–271f
      cotransduction, 270
      in gene mapping, 269–272, 271f
      generalized, 269–272, 270f, 271f
      specialized, 270–271
Transfer RNA. *See* tRNA (transfer RNA)
Transfer-messenger RNA (tmRNA), 451,
      451f. *See also* RNA
Transformants, 265
Transformation
      in bacteria, 257–258, 257f, 265–266,
            265f, 266f
      in cloning, 572
      in gene mapping, 265–266, 266f
Transforming principle, 290–292, 292f
Transfusions, ABO antigens and, 115–116, 116f
Transgenes, 590
Transgenic animals, 590–593, 591f, 592f, 595
Transition mutations, 524t, 528
Transitions, 518, 518f, 519t
Translation, 13, 301, 302f, 441–449
      accuracy in, 449–451
      antibiotics and, 452
      in archaea, 449
      in bacteria vs. eukaryotes, 443–444,
            448–449

coupled to transcription, 405, 448, 449, 482
elongation in, 445–446, 445f, 449
in gene regulation, 507–508
information transfer via, 301–302, 302f
inhibition of, 505
initiation of, 442–443, 443f, 444f, 448,
      448t, 507
mRNA surveillance in, 449–451
pauses in, 446
polyribosomes in, 449, 450f
posttranslational protein modification and,
      452, 507–508
ribosome as site of, 441, 441f
stages of, 441–447, 448t
stalled ribosomes in, 450–451
termination of, 446–449, 447f, 448t, 449
translocation in, 445–446, 445f
tRNA charging in, 442, 442f
Translesion DNA polymerases, 356, 548
Translocation(s), 227–229, 227f, 228f, 290t,
      445–446, 445f
      in cancer, 707, 707f
      in Down syndrome, 232–235, 233f, 234f
      nonreciprocal, 227, 241t
      reciprocal, 227, 228f, 241t, 707, 707f
      Robertsonian, 227, 227f
            aneuploidy and, 230, 233–234
            in Down syndrome, 233–234
Translocation carriers, 234, 234f
Transmission genetics, 5f, 6
Transplantation, immune response in, 685
Transposable elements, 534–544, 535f, 627,
      627t, 629. *See also* Transposons
      in bacteria, 538–539, 538f
      in bacteriophages, 538, 539f
      characteristics of, 534–535
      Class I, 536, 543, 543t *See also*
            Retrotransposons
      Class II, 535, 543, 543t
      in *D. melanogaster*, 540–541, 542f
      domestication of, 543–544
      in eukaryotes, 539–543
      in evolution, 629
      flanking direct repeats and, 535, 535f
      functions of, 543–544
      general characteristics of, 535f
      in genetic diseases, 536
      in genome evolution, 543–544
      genomic content of, 627, 627t, 629
      as genomic parasites, 543
      in maize, 539–540, 540f, 541f
      movement of, 535. *See also* Transposition
      mutations and, 536–538, 538f
      terminal inverted repeats and, 535, 535f
      in yeast, 538
Transposase, 535, 538
      in composite transposons, 538, 538f
      in insertion sequences, 538, 538f
Transposition, 535–544.
      See also Transposable elements
      in bacteria, 538–539, 538f
      in bacteriophages, 538, 539f
      copy-and-paste, 536
      cut-and-paste, 536
      definition of, 535
      in humans, 536

hybrid dysgenesis in, 540–541
mechanisms of, 535–536
mutagenic effects of, 536–538, 538f, 543
nonreplicative, 536
regulation of, 536
replicative, 536
through RNA intermediate.
See Retrotransposons
Transposons, 535–536, 538f.
See also Transposable elements
composite, 538
noncomposite, 538
Trans-splicing, 408
Transversions, 518, 518f, 524t
Trichothiodystrophy, 549t
Trinucleotide repeats, expanding, 519–520, 519t, 520f, 524t
Triplet code, 436. *See also* Genetic code
Triplex DNA, 302–303, 303f
Triple-X syndrome, 88
Triploidy, 236. *See also* Polyploidy
Trisomy, 230
uniparental disomy and, 236
Trisomy 8, 235
Trisomy 13, 235
Trisomy 18, 235
Trisomy 21, 232–235, 233f.
See also Down syndrome
Triticum aestivum (wheat)
kernel color in, inheritance of, 719–721, 720f
polyploidy in, 240
tRNA (transfer RNA), 376, 376t, 377f, 414–416. *See also* RNA
aminoacylated, 442, 442f
definition of, 414
genes for, 415–416
in genetic-code experiments, 437–438, 437f
isoaccepting, 438
processing of, 415–416
ribosome bindings sites on, 444, 445f
structure of, 414–416, 415f
in translation, 441–442, 442f.
See also Translation
tRNA binding sites, 444, 445f
tRNA charging, 442, 442f
tRNA introns, 402, 402t
tRNA-modifying enzymes, 415
trp operon, 476–481, 478f, 479f
attenuation in, 477–481, 479f, 481t
trp promoter, 476–477, 478f, 479f
Trypanosoma brucei, 412, 412f
Tryptophan, 416f
Tryptophan operon. *See trp* operon
Tsix gene, 653, 653f, 653t
TTGACA consensus sequence, 382, 382f
Tubulin subunits, 24, 26f
Tumors. *See* Cancer
Tumor-suppressor genes, 697–698, 697f, 698t, 704
in colorectal cancer, 706
Turner syndrome, 87–88, 87f, 99, 232, 236
Tus-*Ter* complex, 353
twhh (tiggy-winkle hedgehog) gene, 679
Twin(s)
dizygotic, 154, 154f

monozygotic, 154, 154f
epigenetic effects in, 652
Twin studies, 155–156
of asthma, 156
concordance in, 155, 155t
of epilepsy, 155, 155t
twist gene, 668, 669f
Two-dimensional polyacrylamide gel electrophoresis (2D-PAGE), in proteomics, 629–630, 630f
Two-point testcrosses, gene mapping with, 188–189
Ty elements, 539
Tyrosinase-related protein 1 gene *(TYRP1)*, 47–48
Tyrosine, 416f

Ultrabithorax gene, 671
Ultrasonography, fetal, 158, 159f
Ultraviolet light, as mutagen, 532, 532f
Unbalanced gametes, 237
Underdominance, 770
Unequal crossing over, mutations and, 221, 221f, 527f, 528
Uniparental disomy, 235–236
Uniparental inheritance, 235–236, 324–325
Unique-sequence DNA, 321
Universal genetic code, 440
Up mutations, 382
Upstream element, 382
Uracil, 297, 297f
U-tube experiment, 259, 259f

V segment, 682f, 683f
antibody diversity and, 682–684
Vaeck, Mark, 575
Valine, 416f
Variable expressivity, 113, 113f
Variance, 723–724, 723f
definition of, 723
environmental, 729
gene interaction, 729
genetic, 729–730
additive, 729
dominance, 729
genetic-environmental interaction, 729, 729f
phenotypic, 728–730, 729, 729f
Variation. *See* Genetic variation
Varmus, Harold, 697
Vectors
cloning. *See* Cloning vectors
in gene therapy, 595–596, 596–597, 597t
Venter, Craig, 608, 612f
Vertebrates, genome of, duplication of, 241
Victim identification, DNA fingerprinting in, 587–589, 588t, 589f
Vinclozolin, as endocrine disruptor, 651
Virulent phage, 268, 268f
Viruses, 20, 20f, 267–277
animal, 273–277
bacterial, 267–273. *See also* Bacteriophage(s)
bacterial defense mechanisms against, 267
cancer-associated, 708–710, 708t, 709f
definition of, 267

diversity of, 267, 268f
enveloped, 267
essential functions of, 252
evolution of, 274–275, 275f
gene mapping in, 272–273
genes of, 20, 20f
genetic material in, 295–296
in genetic studies, 252t
genome of, 273
human immunodeficiency, 274–275, 275f, 276f
influenza, 276
overview of, 252
plant, 273
proviruses and, 273, 274f
retroviruses, 273, 274–275, 274f, 275f
rhinoviruses, 276–277
RNA, 273–277, 274f–277f
RNA in, 295–296
in transduction, 257f, 258, 269–272, 269f–271f
Vision, color blindness and, 92–93, 93f, 221, 221f
Vitamin B12, synthesis of, 483
Vitamin D–resistant rickets, 152
Von Hippel-Lindau disease, 703
Von Tschermak, Erich, 49

Waardenburg syndrome, 147–148, 148f
Waddington, Conrad, 642, 642f
Wall, Monroe, 339
Wani, Mansukh, 339
Warfarin, genetic testing for, 162–163
Watson, James, 10–11, 288, 293–294, 295f, 561
Watson-Crick model, 293–295, 295f, 301–302
Weight, adoption studies of, 156, 157f
Weinberg, Wilhelm, 754
Weismann, August, 10
Weiss, Kurt, 613
Werner syndrome, 359–360, 549t
Western blotting, 568
Wethington, Amy R., 121
Wheat
kernel color in, inheritance of, 719–721, 720f
polyploidy in, 240
White, Raymond, 698
Whole-genome duplication, 799
Whole-genome shotgun sequencing, 611–612, 611f
Wild-type alleles, 60
Wild-type bacteria, 254
Wilkins, Maurice, 11, 288, 294, 561
Williams-Beuren syndrome, 223t, 224
Wilson, A. C., 491–492
Wilson, Edmund, 84, 183
Wobble, 438f, 439, 439t, 527f
mutations and, 527, 527f
Wolf-Hirschhorn syndrome, 223t, 224
Wollman, Elie, 263
Wolves, Isle Royale, 749–750
Worm. *See Caenorhabditis elegans* (nematode)
Wright, Sewall, 10

X chromosome, 89–99
    abnormal number of, 97–99
    discovery of, 84
    inactivation of, 97–99, 98f, 99f, 231–232
        epigenetics and, 652–654, 653f, 653t
    in Klinefelter syndrome, 88, 88f, 99
    in sex determination, 83–89, 84f, 85f
    structure of, 82f
    in triple-X syndrome, 88
    in Turner syndrome, 87–88, 87f, 99
    in XXY males, 88
X:A ratio, 87, 87t
Xenopus laevis (clawed frog), 6
Xeroderma pigmentosum, 548–549, 549f,
        549t, 695, 702
X-inactivation center, 652, 653f
Xist gene, 99, 647, 652, 653f, 653t
Xist RNA, 422, 647
Xite gene, 653, 653f, 653t
X-linked genes
    dosage compensation for, 97–99
    notation for, 94
X-linked traits, 89–99
    color blindness as, 92–93, 93f
    in *D. melanogaster*, 89–91, 91f, 92f
    dominant, 151–152, 152f, 153t
    dosage compensation and, 97–99
    eye color as, 89–91, 91f, 92f
    inheritance of, 97, 150–151, 150f, 151f,
        152f, 153t
    nondisjunction and, 90–91, 92f

    notation for, 94
    recessive, 150–151, 151f, 153t
    recognition of, 97
    X chromosome inactivation and, 97–99,
        98f, 99f
X-ray(s). *See also* Radiation exposure
    mutations and, 532
X-ray crystallography, in structural
        proteomics, 632
X-ray diffraction, 294, 295f
XX-XO sex determination, 84, 86t
XX-XY sex determination, 84–85, 86t
XYY males, 88

Y chromosome, 88, 94–97
    characteristics of, 94–97, 96
    discovery of, 84
    evolution of, 95–96
    genetic markers on, 96–97
    in Klinefelter syndrome, 88, 88f, 99
    as male-determining gene, 88
    palindromic sequences on, 96
    in sex determination, 83–89, 84f, 85f
    SRY gene on, 89, 89f
    structure of, 82f
    in XXY males, 88
Yanofsky, Charles, 477
Yeast
    genome of, 327, 328f, 626t, 627t
    mitochondrial DNA in, 327, 327t, 328f
    as model genetic organism, 6, 6f

    petite mutations in, 325, 325f
    synthetic genome for, 605–606
    transposable elements in, 538
Yeast artificial chromosomes (YACs)
    creation of, 605–606
    in genome sequencing, 610–611
    as vectors, 574
Yellow mice, 114
Y-linked markers, 96–97
Y-linked traits, 89, 94–97, 152, 153t
    inheritance of, 97
    notation for, 94
    recognition of, 97
Yonath, Ada, 449
Yule, George Udny, 719

Z chromosome, 85, 94, 95f
Z-DNA, 301, 301f
Zea mays (corn), 6
Zebrafish, 6–7, 7f
    genome of, 626t
Zimmerman, Tracy, 594
Zinc fingers, 465, 465f, 465t, 498
    Krüppel-associated box domain, 492
Zinc-finger nucleases, 564
Zinder, Norton, 269–270
Z-linked traits, 94, 95f
Zuckerandl, Emile, 796
Zygotene, 29, 29f
ZZ-ZW sex determination, 81–82,
        85, 86t, 94